Infrared absorption bands useful for structural assignments[a]

X—H Stretching Deformation Bands:

Bond	cm^{-1}	(μm)
O—H	3650–3100	(2.74–3.22)
N—H	3550–3100	(2.82–3.22)
≡C—H	3330	(3.00)
=C—H	3085–3025	(3.24–3.31)
Ar—H	3030	(3.30)
$-\overset{\vert}{\underset{\vert}{C}}-H$	2960–2870	(3.38–3.49)
S—H	2500	(4.00)

X=Y Stretching Deformation Bands:

Bond	cm^{-1}	(μm)
C=O[b]	1850–1650	(5.40–6.05)
C=NR[b]	1690–1640	(5.92–6.10)
C=C[b]	1680–1600	(5.95–6.25)
N=N	1630–1575	(6.14–6.35)
N=O	1650–1500	(6.06–6.66)
⌬	1600–1450 (four bands)	(6.25–6.90)

X≡Y and X=Y=Z Stretching Bands:

Bond	cm^{-1}	(μm)
C≡N	2260–2240	(4.43–4.46)
RC≡CR′	2260–2190	(4.43–4.56)
RC≡CH	2140–2100	(4.68–4.76)
C=C=O	2170–2130	(4.61–4.70)
C=C=C	1975–1950	(5.07–5.13)

C—H Bending Deformation Bands in the Fingerprint Region:

Structure	cm^{-1} (μm)
(R)(H)C=C(H)(R)	~970 (10.30)
(R)(R)C=C(H)(H)	~900 (11.11)
(R)(R)C=C(H)(R)	~825 (12.11)
(R)(H)C=C(R)(H)	~670 (14.90)
(R)(H)C=C(H)(H)	~1000 and 900 (10.0 and 11.1)

[a] Data are given in wave numbers (cm^{-1}) and, in parentheses, micrometers (μm).
[b] Conjugation with a C=C or aryl group lowers the frequency about 30 cm^{-1}

C. David Gutsche
Department of Chemistry
Washington University, St. Louis

Daniel J. Pasto
Department of Chemistry
University of Notre Dame

Fundamentals of Organic Chemistry

Prentice-Hall, Inc., Englewood Cliffs, New Jersey

Gutsche, C. David
Fundamentals of organic chemistry.

1. Chemistry, Organic. I. Pasto, Daniel J., joint author. II. Title.
QD251.2.G87 547 74-9714
ISBN: 0-13-333443-0

Dedicated to the University

An institution rooted in antiquity; unbroken in lineage for seven centuries; survivor of the vicissitudes of time; accumulator, purveyor, and creator of mankind's knowledge; one of civilization's sturdiest bulwarks against an uncertain future.

Current printing: 10 9 8 7 6 5 4 3 2 1

Printed in the United States of America

PRENTICE-HALL INTERNATIONAL, INC., *London*
PRENTICE-HALL OF AUSTRALIA, PTY. LTD., *Sydney*
PRENTICE-HALL OF CANADA, LTD., *Toronto*
PRENTICE-HALL OF INDIA PRIVATE LIMITED, *New Delhi*
PRENTICE-HALL OF JAPAN, INC., *Tokyo*

Illustration Acknowledgments

The authors are indebted to the following sources for permission to reproduce illustrations:

Edgar Fahs Smith Memorial Collection in the History of Chemistry, University of Pennsylvania.

Dickerson, R. E. and Irving Geis, *The Structure and Action of Proteins*, Harper & Row, Publishers, N.Y., 1970.

Authors' Acknowledgments

When a textbook is written in a field that has reached maturity, such as the field of organic chemistry, it is inevitable that other textbooks and compendia provide invaluable sources of information in its preparation. Our first acknowledgment, therefore, is to the many chemists who have preceded us in this task of authorship and from whose efforts we have drawn innumerable facts, theories, and ideas.

During the several years that the manuscript for this text was in preparation a number of professional colleagues read completely through one or another of its various editions. Included in this group are Professors Joseph M. Baven (William Rainey Harper College), Robert A. Benkeser (Purdue University), Weston T. Borden (University of Washington), Peter Campbell (New York University), Charles P. Casey (University of Wisconsin), David L. Dalrymple (University of Delaware), James Duncan (University of Notre Dame), Walter J. Genseler (Boston University), Glen L. Keldser (Clarion State College), Robert L. Lyle (University of New Hampshire), James A. Marshall (Northwestern University), Kenneth L. Marsi (State University of California at Long Beach), and Audrey S. Monahan (University of Connecticut). To these readers we are indebted for their perceptive criticisms and suggestions, many of which were incorporated in the final edition of the manuscript.

Other professional colleagues read smaller segments of the text, segments dealing with areas of their particular expertise. Included in this group of critics are Professors Robert Barker (University of Iowa), Robert G. Bergman (California Institute of Technology), Oscar P. Chilson (Washington University), Ernest L. Eliel (University of North Carolina), Peter P. Gaspar (Washington University), Lindsay Helmholz (Washington University), Alfred G. Hortmann (Washington University), Joseph L. Kurz (Washington University), and David Lipkin (Washington University). To these individuals we are indebted for advice and corrections which were invaluable in improving the level of accuracy and clarity in various portions of the text.

To our undergraduate students who were subjected to the several editions of the embryonic text as it progressed through its gestation period, we owe a special debt of gratitude. Not only did the majority of these students enter into this experiment in the happy spirit of experimentation, but a number of them proved to be extraordinarily helpful in spotting errors that escaped the attention of most other readers. Particularly outstanding in this respect are Steven Paspak, James Sherman, and Jeffrey Skolnik whose sharp eyes and quick brains materially improved the accuracy of the final text.

To our graduate students and postdoctoral associates who were engaged as unpaid proofreaders of the manuscript at various intermediate stages of preparation and of the text at the galley proof stage we also owe our appreciation. Included in this group of willing helpers are Dr. Ernest Brody, Chi-wan Chen, George Gettys, Dr. T. C. Kung, Hon-pen Lau, Lee Melhado, and Gregory

Roberts, the last named individual performing this task with unusual skill that provided us with a number of critical corrections and valuable suggestions.

For our arbiters of style we relied on Lee Melhado, a gifted graduate student and a former English major, and Alice Gutsche, wife of one of the authors, whose fine sensitivity for the English language proved to be immensely helpful in improving the clarity of expression in many places. We are further indebted to Alice Gutsche who, in addition to her editorial function, also carried out the important and demanding task of indexing the entire text. Indeed, her substantial involvement as reader, critic, and indexer establish her as the "silent" partner in this venture.

Finally, acknowledgment is made to those who contributed in less direct ways. Between the initial scribblings of the authors and the final manuscript for the publisher came numerous intervening editions, each requiring long and careful secretarial effort. Among the several individuals who helped in this respect, special commendation goes to Renate Keith, whose rapid and accurate typing materially aided this project. Two other individuals who were also indispensable to the writing of the text are Dr. Frank Williamson and Norma Frick, whose collective willingness and ability to undertake administrative responsibility made it possible for one of the authors (CDG) to work on the book while at the same time acting, ostensibly, as chairman of a department.

To all of the many people who, in these various ways, have helped with the preparation of this textbook we express our deep gratitude. Because of their contributions it is a much better book than it might otherwise have been.

Contents

Preface

Over the course of the last 150 years organic chemistry has grown from a vulnerable infant to a mature and robust giant. To comprehend in detail all of its many facets now requires years of study. Many textbooks have been written to provide the starting point for this study, each of them with the intention of capturing the essence of the giant in a few thousand well-chosen words. The present book is no exception in this respect, for we hope to give the reader some understanding and appreciation of *all* of the facets of organic chemistry as well as a reasonably detailed comprehension of those particular areas that are of special interest and importance to students preparing to enter scientific professions in which a knowledge of organic chemistry is an essential ingredient. To achieve this, the subject matter has been organized in a spiral fashion in which there are several turns to the spiral.

Among the many problems attending the presentation of any large body of knowledge, the two that loom largest are those of where to start and how to arrange the material. To some extent the second of these, the problem of arrangement, is solved by the nature of the material itself. There is a considerable degree of linearity to the subject matter of organic chemistry, one idea building on another, so certain patterns of presentation automatically emerge. However, there is also a certain degree of circularity, topic C requiring a knowledge of topic B, topic B requiring a knowledge of topic A, but topic A requiring a knowledge of topic C. To minimize the difficulties arising from this situation, the present text surveys all of the basic principles of organic chemistry quickly and concisely and then, building from this foundation, reconsiders certain segments in greater detail. While the problem of circularity remains, it is minimized by reducing the size of the circle or, perhaps more accurately, by replacing one big circle by a series of smaller circles. With the spiral scheme for organizing the subject matter of organic chemistry, the reader can perceive the overall pattern into which fit the many facts of the field faster than he could with the strictly linear presentation.

Part I of the book presents a general survey of organic chemistry and deals with the structure, nomenclature, stereochemistry, physical properties, and chemical properties of the important classes of organic compounds. Although focusing principally on the structural aspects of compounds, it gives an overall view of the entire field of organic chemistry and provides the framework to which Part II adds detail, especially with respect to the mechanisms, scope, and limitations of organic reactions. To borrow from the language of psychology, the approach to the learning of organic chemistry that is used in this text might be designated as "gestalt" in character, the term "gestalt" referring to a system in which the phenomena are integrated in such a fashion as to constitute a functional unit with properties not derivable from its parts—more colloquially referred to as the whole being greater than the sum of its parts. Thus, Part II requires Part I for its foundation, but Part I becomes strengthened through a study of Part II, each part reinforcing the other.

Part III presents the organic chemistry of natural products, a topic of special interest to those who will eventually deal, in one form or another, with living systems. The emphasis in Part III is primarily on the structure of natural products, and the material in Part I alone provides an adequate preparation for understanding it. Some readers and teachers, therefore, may choose to change the sequence and consider some or all of Part III immediately after Part I, deferring Part II to a later point. Both sequences of presentation have been class-tested during the past several years, and the book appears to be equally well suited to either alternative.

Part IV treats several topics in a more advanced manner than Parts I, II, and III and provides a transition to the material that one encounters in higher-level courses in organic chemistry. It is directed primarily to those who plan to continue their study of organic chemistry. Except in the most rapidly paced elementary courses, most of this material would not be included in the two semester sequence.

The intention of this book, which is a teaching text rather than an encyclopaedic reference book, is to furnish a well-documented survey of modern organic chemistry and a secure foundation for more advanced study of organic chemistry. Encyclopaedic textbooks are available, and these, along with advanced treatises and original articles in the chemical literature, are recommended supplementary reading. Students planning to enter chemically oriented professions are urged to actively explore these additional sources of information, many of which are cited at the end of each chapter.

An often-voiced sentiment of students asserts that "organic chemistry is simply a memory course." That it does, indeed, exercise one's ability to remember cannot be denied. There are few, if any, intellectual endeavors that don't make considerable demands on this remarkable faculty of man. But the idea that it is *simply* a feat of memory is, in our experience, quickly dispelled by the spiral presentation of the subject matter. Since the spiral approach allows the student to quickly perceive the "grand plan" and, as a consequence, to be able quite soon in the study of organic chemistry to appreciate the logical connections between the facts that are presented, the apparent arbitrariness of having to remember the fact disappears. The process of committing the fact to memory becomes logical rather than rote and, concomitantly, less onerous. For many students it even becomes an enjoyable experience, and we hope that this will be true for the readers of this textbook. *Bon voyage!*—C. DAVID GUTSCHE, DANIEL J. PASTO.

PART I

A Survey of Organic Chemistry: The Structure of Organic Compounds

Part I, as indicated in the Preface, provides a general survey of organic chemistry and includes discussions of structure, nomenclature, stereochemistry, physical properties, and chemical properties of the important classes of organic compounds. Following Chapter 1, which presents a capsule survey and history of organic chemistry, Chapters 2–8 deal with the carbon frameworks of organic compounds. Chapters 9–17 then consider the chemistry of the various compounds that are formed when functional groups are attached to these carbon frameworks. Chapter 18 furnishes a summary of some of the topics covered in the first 17 chapters, including tables of physical properties, summaries of the rules of nomenclature, and guidelines for structure elucidation and synthesis. Part I is concluded by a detailed chapter on stereochemistry, the elementary principles of which are first introduced in Chapter 3.

The relative emphasis placed on the various topics covered in Part I will vary from one reader to another and from one instructor to another. Perhaps most sensitive in this respect is the material in Chapter 6 dealing with the spectral properties of organic compounds, the extent to which it is covered depending on the aims and goals of the particular student and the particular course. Premedical students and students in the Humanities, who are primarily interested in acquiring an overall appreciation of organic chemistry, may be less inclined to delve into the intricacies of spectra than chemistry majors, biochemistry majors, and others who will eventually use organic chemistry as a practical tool in their work. The book has been written in such a fashion that spectra can be emphasized, if that is the reader's and instructor's choice, or be dealt with rather cursorily, or even omitted entirely without substantially interfering with the subsequent presentation.

The suggestion that spectral properties can be deemphasized does not mean that they are trivial or incidental. On the contrary, spectral measurements are an absolutely essential ingredient of modern organic chemistry as it is conducted in the laboratory, and for persons who intend to use organic chemistry in a practical fashion it is necessary that a good working knowledge of spectroscopy be acquired. But, to be able to read and appreciate most of the material in this book one does not have to be conversant with spectra. Clearly, though, the more one knows about all the facets of a subject, the deeper is his overall understanding and the greater is his capacity for savoring the nuances of the subject.

Chapter 1

The scope of organic chemistry

Organic chemistry gives me the impression of a primeval tropical forest, full of the most remarkable things.

FRIEDRICH WÖHLER (1840)

1.1 A Capsule View of the Scope of Organic Chemistry

Several centuries ago, when alchemists still dominated the chemical scene, it was thought that certain compounds found in nature required a mysterious "vital force" for their synthesis. Compounds of this type came to be called *organic*, and the study of them came to be called *organic chemistry*. Although the notion of a "vital force" has long since been discarded and although most of the compounds that today we class as "organic" have been synthesized in the laboratory rather than in nature, the use of the term has persisted.

The "modern" era of organic chemistry began to take shape at the beginning of the nineteenth century. Even in its infancy organic chemistry attracted the attention of some of the greatest chemists of the day and inspired their awe. Friedrich Wöhler, who played a key role in guiding the discipline in its earliest days, commented that "organic chemistry is full of the most remarkable things, a monstrous and boundless thicket . . . ." In the century and a half that has elapsed since Wöhler's time many more remarkable things have been added to organic chemistry, and we might suppose that today the thicket is even more monstrous and boundless than ever. Fortunately, the reverse is true. Although the territory that organic chemistry now encompasses is indeed large, its boundaries are sufficiently well defined and its terrain sufficiently well perceived that an understanding and appreciation of its details is easily within the grasp of everyone; the boundless has become bounded and the thicket has been carefully pruned.

To gain a brief overall glimpse of the territory called organic chemistry, let us trace the development of our knowledge of a particular organic compound that has probably been known to man since the dawn of recorded history, the compound that we now call acetic acid. Acetic acid is the principal nonaqueous component of vinegar, which derives its name from a combination of the French words *vin* ("wine") and *aigre* ("sour"), *i.e.*, "sour wine," indicating that one of the natural sources of vinegar is wine that has turned sour as the result of the action of certain microorganisms. For centuries, vinegar has been used for flavoring and preserving foods. It was prescribed by Hippocrates in ancient Greece as a medicine and was used during the American Civil War in the treatment of scurvy. Today acetic acid is produced on an enormous scale, the United States alone manufacturing over 2 billion pounds in 1971. In addition to the continued culinary use of vinegar, acetic acid now finds application in the manufacture of a host of useful materials, including fibers, fabrics, plastics, and medicines. It is a typical organic compound whose chemistry we shall now quickly explore.

Sometime near the beginning of the present millenium, a medieval scholastic and early alchemist named Geber discovered that vinegar can be distilled and that the major fraction of the distillate is water. He also discovered that a smaller fraction is obtained, which is a stronger, more concentrated vinegar than the original material. For several hundred years, however, this early experiment went unheeded. Not until the seventeenth and eighteenth centuries was distillation equipment sufficiently improved to allow more highly concentrated vinegar solutions to be prepared. Finally, in 1798 Tobias Lowitz combined distillation with other means of purification and succeeded in preparing a water-free sample of vinegar, which he obtained as a crystalline solid melting at 16°C. This material acquired the German name *eisessig*, meaning "solid vinegar," and it is the compound that we refer to in English as acetic acid. With the preparation of a pure sample of acetic acid, the first phase in the study of the organic chemistry of this compound was complete, *viz*., the separation of vinegar into its component parts (about 94% water and 6% acetic acid) followed by the purification of each of these constituents.

With a pure sample of acetic acid available, it then became possible to carry out the next phase of the investigation, *viz*., the determination of the elemental composition of the compound. Building on Antoine Lavoisier's contributions to analytical chemistry at the end of the eighteenth century, Jöns Jacob Berzelius added improvements springing from his own fertile mind, and in 1814 he succeeded in establishing that acetic acid contains carbon, hydrogen, and oxygen atoms in the ratio 1:2:1; in other words, he determined that the *empirical formula* for acetic acid is CH_2O. Let us see how he did this.

Commencing his analysis by carefully weighing out a sample of pure acetic acid, Berzelius then burned the sample (combusted it), thereby converting it to carbon dioxide and water. By weighing the carbon dioxide and water that were formed and by knowing the weight of the original sample, he was able to determine the relative amounts of carbon, hydrogen, and oxygen present in acetic acid by means of the following procedures. Let us assume, for example, that the weight of the acetic acid sample was 3.26 g and that this produced 4.74 g of carbon dioxide and 1.92 g of water upon combustion.* From these data, the

* With the equipment available to Berzelius it was necessary to trap the carbon dioxide and water in weighing tubes filled with appropriate reagents. For example, magnesium perchlorate traps water, and sodium hydroxide traps carbon dioxide, as indicated in the following picture of a typical apparatus for carrying out an analysis of the type described. In a modern combustion analysis, milligram rather than gram amounts of sample would be used; *e.g*., a 3.26 mg sample of acetic acid would produce 4.74 mg of carbon dioxide and 1.82 mg of water.

percentage of carbon, hydrogen, and oxygen in acetic acid can be calculated as follows:

$$\text{wt. of C} = \text{wt. of CO}_2\left(\frac{\text{at. wt. of C}}{\text{mol. wt. of CO}_2}\right) = (1.74)\frac{12}{44} = 1.29$$

$$\left(\frac{\text{wt. of C}}{\text{wt. of sample}}\right)100 = \left(\frac{1.29}{3.26}\right)100 = 39.6\%\ \text{C}$$

$$\text{wt. of H} = \text{wt. of H}_2\text{O}\left(\frac{\text{at. wt. of H} \times 2}{\text{mol. wt. of H}_2\text{O}}\right) = (1.92)\frac{2}{18} = 0.213$$

$$\left(\frac{\text{wt. of H}}{\text{wt. of sample}}\right)100 = \left(\frac{0.213}{3.26}\right)100 = 6.53\%\ \text{H}$$

The sum of the percentages of carbon and hydrogen is only 46.13% of the weight of the sample, however, indicating that at least one other element must be present in acetic acid. Since most of the likely possibilities were eliminated by the various qualitative experiments that Berzelius carried out on acetic acid, in which he tested for the presence of halogens, nitrogen, and sulfur, the most probable "other element" is oxygen. Subtracting 46.13% from 100%, then, leaves 53.87% as the oxygen content. Division of each of these percentages by the atomic weight of the element, followed by normalization of these results to the least abundant element present in the compound (*i.e.*, the lowest percent abundance), yields the following figures:

$$39.6/12 = 3.30 \text{ for C} \qquad 3.30/3.30 = 1.00 \text{ for C}$$

$$6.53/1 = 6.53 \text{ for H} \qquad 6.53/3.30 = 1.98 \text{ for H}$$

$$53.87/16 = 3.36 \text{ for O} \qquad 3.36/3.30 = 1.02 \text{ for O}$$

The normalized value for carbon is 1.00, that for hydrogen is 1.98, which is almost 2.00, and that for oxygen is 1.02, which is almost 1.00; this suggests an empirical formula of CH_2O. If the quantitative analysis were absolutely ac-

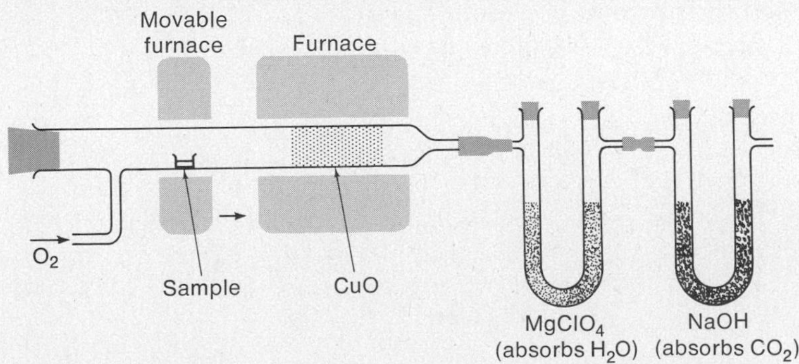

A recent modification of the combustion analysis dispenses with the weighing tube procedure and simply passes the combustion gases (containing water, carbon dioxide, and other oxides, if other elements are present) through a gas-phase chromatographic unit. The amount of water, carbon dioxide, and other oxides that are present can be read directly from the plot produced by the recorder attached to the instrument.

curate for a CH_2O formula, the normalized values for carbon, hydrogen, and oxygen would, of course, have been exactly 1.00, 2.00, and 1.00, respectively. Experimental error is a constant companion of chemical analysis, however, and the chemist must exercise a certain amount of judgment and intuition in interpreting his data. A slavish adherence to the data that Berzelius actually obtained in his analysis, for instance, could have led him to postulate the empirical formula $C_{50}H_{99}O_{51}$, which corresponds to a highly unlikely possibility.

The empirical formula provides information concerning the *relative numbers* of the various atoms comprising a compound but not the absolute numbers. To ascertain the latter it is necessary to determine the molecular weight of the compound and on the basis of this value decide whether the *molecular formula* is identical with the empirical formula or is some integral multiple of it. For this to be achieved with acetic acid a number of years had to elapse, because the concept of molecular weight was still clouded and confused in 1814. Eventually, however, clarification came, and various methods were devised for determining molecular weights; when these were applied to acetic acid, values in the vicinity of 60 were obtained. Since CH_2O has a formula weight of only 30, this means that the molecular formula of acetic acid must be twice the empirical formula; *viz.*, $C_2H_4O_2$.

We turn now to the third and most challenging phase of the investigation of the organic chemistry of acetic acid, a phase involving the determination of the structure of the compound. How are the two carbons, the four hydrogens, and the two oxygens of acetic acid connected to each other? What is the shape and geometry of the molecule? These are difficult questions to answer. In fact, the unraveling of the intricacies of the molecular arrays of chemical compounds, which is fundamental to the entire subject of chemistry, provides the focal point for the discussion in Part I of this text. At the moment we can do no more than sketch some of the considerations that come into play in establishing the structure of even so simple a molecule as acetic acid.

Let us begin our structure proof by recognizing that hydrogen is monovalent, oxygen is divalent, and carbon is tetravalent in the overwhelming majority of known compounds. Operating from this premise, we can then write a variety of combinations of two carbons, four hydrogens, and two oxygens in which these valencies are preserved; *i.e.*,

```
    H  H             H                H  H               H  O—H
    |  |             |                |  |               |  |
  H—C—C—H          H—C—O           H—C—C—O—H             C=C
    |  |             |  |              \ /               |  |
    O—O              O—C—H              O                H  O—H
                        |
    (a)                 H              (c)                (d)
                      (b)

    H  O          H  H           O       H              H  O
    |  ||         |  |           ||      |              |  ||
  H—C—C—H         C=C          H—C—O—C—H           H—C—C—O—H
    |             |  |                   |              |
    O             O  O                   H              H
    |             |  |
    H             H  H              (g)                (h)

   (e)           (f)
```

To choose among these eight possible representations, we must consider a variety of types of information concerning the chemical and physical characteristics of acetic acid, no single one of which provides a definitive structure proof by itself. Rather, it is the self-consistency of a whole network of facts that ultimately allows us to choose one structural representation in preference to all others. Without going into detail at this point, suffice it to say that the unusually high boiling point of acetic acid, the rather high acidity of the compound, the fact that it can be prepared by the oxidation of grain alcohol,

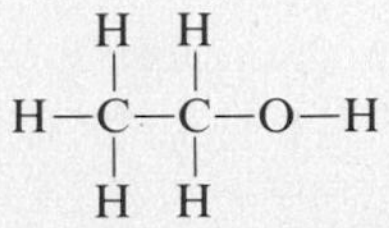

```
    H  H
    |  |
H—C—C—O—H
    |  |
    H  H
```

and various other chemical and physical attributes that it possesses are all more commensurate with structure (h) than with structures (a)–(g). Consequently, structure (h) is the one that is generally accepted as the proper representation of how the two carbons, four hydrogens, and two oxygens are connected in the particular molecule that we call acetic acid.

One of the tenets of structure determination states that the structure of a compound cannot be considered as unequivocally established until the compound has been constructed (*i.e.*, synthesized) from other compounds of known structure by chemical conversions of known type. We can see, in retrospect, that this was accomplished for acetic acid at a very early stage in the development of organic chemistry, for in 1845 a chemist named Hermann Kolbe prepared acetic acid starting with elemental carbon and chlorine (Fig. 1.1). At the time of the synthesis, however, it could hardly have been considered a structure proof, for in 1845 the whole concept of structure was still a hazy and ill-defined concept (see Section 1.3). Nevertheless, the sequence of reactions shown in Fig. 1.1 does constitute a synthesis of acetic acid, and it illustrates the fourth phase in the study of organic chemistry, which involves the conversion of one organic chemical species to another. This very important aspect of organic chemistry can be subdivided into several categories; one deals with organic reactions (*i.e.*, chemical

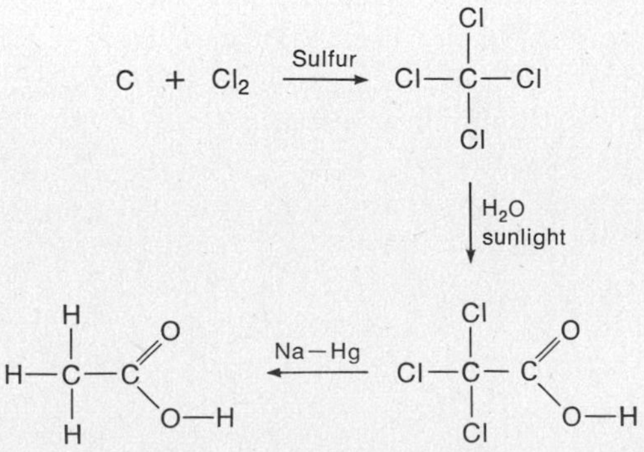

Fig. 1.1. Hermann Kolbe's synthesis of acetic acid from elemental carbon, carried out in 1845.

conversions) in which the focal point is the synthesis of compounds; another deals with organic reactions in which the focal point is their mechanism (*i.e.*, how they take place); and a third deals with organic reactions in which the focal point is a combination of their synthesis applications and mechanisms.

In each of the four phases in the study of organic chemistry discussed so far, the physical properties of acetic acid have played an implicit part. Let us now designate a fifth phase of the study of organic chemistry, which deals explicitly with the measurement of physical properties of organic molecules; let us also, for brevity, include physiological properties in this same category.

We can conclude from this brief survey that organic chemistry comprises (a) the isolation and purification of organic compounds, (b) the qualitative and quantitative analysis of the elemental composition of organic compounds, (c) the determination of the structure of organic compounds, (d) the study of organic reactions (their mechanisms, their scope and limitations, their application to organic synthesis), and (e) the study of the physical and physiological properties of organic compounds. It is a big field! But, if its broad outlines are appreciated at the outset, and if its several areas are then studied a step at a time, it is an entirely comprehensible field. The purpose of Part I, therefore, is to provide the reader as quickly as possible with a moderately detailed view of the entire field of organic chemistry. Then, with this background information to work from, a more detailed treatment in Part II is given to those areas that are passed over rather quickly the first time around.

1.2 The Origins of Organic Chemistry

Many, although not all, cosmologists hold the theory that our present universe started at a single point in space, its birth resulting from the explosion of some kind of densely packed matter—the so-called "Big Bang Theory." This event, which is placed on the human time scale as occurring 15 billion years ago, resulted in the formation of a number of entities, including neutrons, protons, and electrons. Coalescence of these electrons, protons, and neutrons then yielded hydrogen atoms, which are by far the most prevalent chemical component of today's universe. But further coalescence also occurred, producing other elements, compounds, and untold billions of more or less dense objects poetically referred to as the heavenly bodies. Thus, according to this theory, the universe has evolved from a single entity to a multitude of entities, some more organized than others. The stars, the hot bodies of the universe, represent one aspect or one stage in cosmic evolution; the planets, the cold bodies of the universe, represent another aspect or stage of this evolution. And, on at least one of these planets, still another aspect of evolution is in progress—an evolution that originally was chemical in character but which, subsequently, changed in some mysterious way to become biological. It has been said of this transition from chemical to biological evolution that "the advent of life on the planet Earth was the most improbable and most significant event in the history of the universe." From the standpoint of the readers of this book along with all the other inhabitants of the planet Earth, the significance of the event does, indeed, seem vouchsafed.

To those of past generations who thought about the problem of the origin of life on earth and who were not bound in their thinking by the proscriptions of religion, the event most certainly seemed to be a completely improbable one—a lucky accident defying all the rules of logic and chance. More recently, however, a new attitude has developed, which asserts that life should not be regarded as a fortuitous event but should be viewed as an integral part in the process of an orderly evolution of the universe. It is held that life on Earth is merely a particular manifestation in the evolution of compounds, particularly the compounds of carbon. Credence to parts of this hypothesis has actually been provided by laboratory experiments involving chemical synthesis under the conditions that are thought to have prevailed on the young Earth.

When hydrogen, the most prevalent element in the universe, undergoes fusion reactions, heavier elements are formed—primarily helium, but other elements as well, including carbon, nitrogen, and oxygen. Even under laboratory conditions, carbon can be made to combine with hydrogen to form methane, CH_4; thus, a widespread distribution in the universe of this and other carbon-containing compounds is not surprising. Almost all the heavenly bodies that have lent themselves to investigation in this regard have, in fact, revealed the presence of hydrocarbons; methane, for example, is present in cosmic dust clouds, comets, meteorites, and the atmospheres of Jupiter, Saturn, Uranus, and Neptune. In similar fashion, water (formed by combination of hydrogen with oxygen) and ammonia (formed by combination of hydrogen with nitrogen) can be viewed as natural constituents of what has been referred to as the "primeval broth" that existed on Earth 5 billion years ago. That these simple ingredients might be the building blocks for compounds which recent man has labeled "organic" has been demonstrated by laboratory experiments. In 1955, Harold C. Urey and Sidney Miller at the University of Chicago found that when an electric discharge (simulated lightning) was passed through a gaseous mixture of CO_2, CH_4, NH_3, H_2, and H_2O, a mixture of more complex compounds was formed, a mixture of "organic" compounds called α amino acids. Many subsequent investigations of syntheses under "prebiotic conditions" have been carried out, and convincing proof has been supplied that from these simple building blocks a host of "small" compounds (monomers) can be prepared—including the aforementioned amino acids, sugars, purines, nucleotides, and vitamins. Furthermore, it has been shown that "large" compounds (polymers) characteristic of living things, *e.g.*, polypeptides (proteins), polysaccharides, and polynucleotides (nucleic acids), can also be produced from smaller compounds under prebiotic conditions.

These experiments, carried out over the past decade and a half, suggest beyond reasonable doubt that the primeval broth probably contained myriad compounds, including complex monomers as well as high molecular weight polymers. Thus, the organization of electrons, protons, and neutrons to form atoms, the organization of atoms to form simple molecules, such as methane, ammonia, and water, and the subsequent organization of these simple compounds into more complex substances can be directly demonstrated under laboratory conditions. At some point in the evolutionary scheme a new degree of order entered into the system; random polymers became ordered polymers, and organized polymers became oriented into cell-like structures. These transitions have not yet been demonstrated in the laboratory, and how they took place remains a secret—truly, the mystery of life. The transitions occurred, however, and after many eons these early "living" systems evolved into *Homo sapiens*, an organism endowed with the remarkable ability of analyzing itself and its environment.

The analytical endeavor that *Homo sapiens* calls "organic chemistry" is an exceedingly recent arrival on the scene. It is approximately a century and a half old, and in anything resembling its matured state it is barely over a century old. At the beginning of the nineteenth century, not long after George Washington had fathered the United States, when Beethoven was creating his greatest symphonies and string quartets, organic chemistry was also being sired. Acquiring a shady reputation from the very start, however, there were two fathers—Friedrich Wöhler and Justus Liebig. These remarkable individuals, so opposite in temperament, were brought together by a common interest in a pair of compounds called fulminic acid and cyanic acid, which possess the same molecular formula. Prior to the investigations of Wöhler and Liebig, it had been accepted as axiomatic that the *composition* of a substance uniquely defined the *structure* of that substance. It seemed inconceivable to chemists of that day that two materials might have the same composition and yet possess different properties. The striking differences between fulminic and cyanic acid, however, left no doubt as to this possibility, and with this realization the concept of structure as somehow being related to the *arrangement* of the elements within a molecule came into being. With this, in 1830, organic chemistry was born, although many chemists, including Berzelius, the greatest figure of the day, were unwilling to acknowledge the birth for several years.

Friedrich Wöhler
(1800–1882)

Wöhler and Liebig next collaborated on an investigation of a series of substances derived from a material called benzoic acid, a compound to be discussed in detail in a later section of this book. Although they had no real knowledge of the structures of these substances, they were able to show that a certain unit, which they named the "benzoyl unit," was common to all members of the series. The possibility of a *group of atoms* acting as a *unit* in an organic compound was a revolutionary idea, and this investigation has been cited as one of the great achievements in the history of organic chemistry. It had an immediate and profound impact on the subsequent development of the science.

Justus Liebig
(1803–1873)

During this early period, when the concept of structural isomerism (*i.e.*, two compounds of identical molecular formula but different properties) was just dawning, when the possibility of organic groups (*e.g.*, units common to a series of compounds) was being proven, when the concepts of molecular weights were still hazy, and when a good structural theory was still a quarter of a century away, it is not surprising that the synthetic side of organic chemistry was practically nonexistent. Indeed, as has already been mentioned, it was supposed that organic compounds could not be synthesized from inorganic materials outside of a living system—*i.e.*, that organic compounds require a living cell or a "vital force" for their synthesis. In 1828, however, Wöhler achieved the conversion in the laboratory of ammonium cyanate, which is a recognized inorganic material, to urea, which is a product from certain living systems; and in 1845, Kolbe prepared acetic acid from elemental carbon, hydrogen, and oxygen via the series of reactions shown in Fig. 1.1. The requirement of a living cell and a vital force for the synthesis of organic compounds thus had to be abandoned, although the name "organic" survived its now discredited ancestry. Organic chemistry today is best characterized as the chemistry of compounds that contain carbon atoms, hydrogen atoms, and often other elements as well.

1.3 A Capsule History of Organic Chemistry

The earliest epoch of organic chemistry, described in Section 1.2, might be characterized as its "one-dimensional phase." At the time of the acetic acid synthesis, for instance, chemists had no concept of structure as we conceive it today. One can read completely through Kolbe's paper in the German journal known as the *Annalen* without once encountering the now familiar formula for acetic acid.

The theories of Kekulé and Couper regarding the *structure* of organic compounds were to change this situation and bring about another revolution in the science. This story is both exciting, and sad. In Kirkintilloch, Scotland, there is a plaque bearing the following inscription: "This plaque marks the birthplace of Archibald Scott Couper, born 1831, died 1892, whose brilliant pioneering contributions to chemical theory have won for him international renown, and whose genius, stifled by an early illness was denied the opportunity of consummation." It was Couper who, in 1858, first formulated in writing a theory of organic chemical structure, proposing the idea that carbon can enter into chemical combination with itself, *i.e.*, that a carbon–carbon bond can exist. Through the procrastination of his superior, the well-known chemist Adolphe Wurtz, however, Couper's manuscript was not communicated to the French academy until after August Kekulé, another young chemist of the day, had presented similar ideas in a publication appearing in the *Annalen*. Grieved by his failure to receive due recognition, Couper succumbed to a nervous breakdown, which forced him into retirement until his death many years later. This unfortunate turn of events, combined with Kekulé's fascinating and compelling personality and the fact that Kekulé had been able to couch his ideas in language more accessible to the chemists of his era, led to an almost complete neglect of Couper's contributions. Only in the twentieth century was he reinstated to his rightful place in history. Clearly, science is not always objective; its laborers in the vineyard suffer the slings and arrows of outrageous fortune along with the rest of humanity.

Archibald Scott Couper
(1831–1892)

Ten years after the first structural theories of Kekulé and Couper, Kekulé proposed a formula for the molecule known as benzene. The story is one of the most often quoted episodes of science. The following are Kekulé's own words:*

> I was sitting writing at my textbook but the work did not progress; my thoughts were elsewhere. I turned my chair to the fire and dozed. Again the atoms were gamboling before my eyes. This time the smaller groups kept modestly in the background. My mental eye, rendered more acute by repeated visions of the kind, could now distinguish larger structures of manifold conformation: long rows, sometimes more closely fitted together, all twining and twisting in snake-like motion. But look! What was that? One of the snakes had siezed hold of its own tail, and the form whirled mockingly before my eyes. As if by a flash of lightning I awoke; and this time also I spent the rest of the night in working out the consequences of this hypothesis.

Friedrich August Kekulé
(1829–1896)

* A. Kekulé, *Chemische Berichte*, **23**, 1302 (1890); transl. by O. T. Benfey in *Journal of Chemical Education*, **35**, 21 (1958).

Thus, did Kekulé hit on the idea that the six carbon atoms of benzene are arranged in a ring—an animal biting its own tail. It is perhaps difficult today to realize how these ideas changed men's concepts and how, probably more than any other single event in the history of organic chemistry, they served to guide the science along the path that it has taken in the century following.

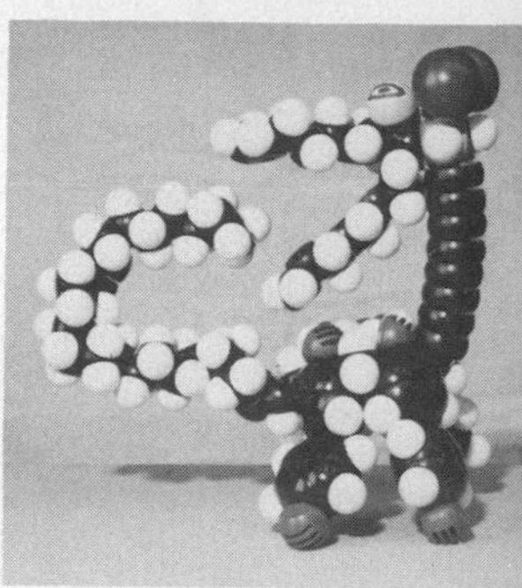

By 1860, then, organic chemistry had developed the concept of structural isomerism, of structural units, and of carbon atoms, hydrogen atoms, oxygen atoms, etc., bonded to each other in certain definite and distinguishable fashions. It had grown from a concept of no dimension prior to the nineteenth century, through concepts of one dimension in the early part of the nineteenth century, to a concept of two dimensions in 1860. Extension to the third dimension became the contribution of the chemists of the latter portion of that century. To us today this would seem to be a logical, even obvious, development, and it is surprising to read of eminent chemists of the day who opposed the theory of the tetrahedral carbon atom that was advanced by Jacobus van't Hoff and Joseph LeBel in 1874. Kolbe, the first chemist to synthesize acetic acid, greeted the theory with the following comments:*

> In a recently published paper I pointed out that one of the causes of the present day retrogression of chemical research in Germany is the lack of fundamental chemical knowledge. Under this lack no small number of our professors of chemistry are laboring, with great harm to the science. A consequence of this is the spread of the weed of the apparently scholarly and clever, but actually trivial and stupid natural philosophy, which was displaced fifty years ago by exact natural science, but which is now brought forth again out of the storeroom harboring the errors of the human mind, by pseudoscientists who try to smuggle it, like a fashionably dressed and freshly rouged prostitute, into good society where it does not belong. Anyone to whom this concern seems exaggerated may read the book by Messrs. van't Hoff and Herrmann on the Arrangement of Atoms in Space which has recently appeared and which overflows with fantastic foolishness . . . . A Dr. J. H. van't Hoff of the Veterinary School at Utrecht has no liking, it seems, for exact chemical investigations. He has considered it more convenient to mount Pegasus (apparently borrowed from the Veterinary School) and to proclaim in his book how the atoms appear to him to be arranged in space when he is on the chemical Mt. Parnassus which he has reached by bold flight. It is indicative of the present day, in which critics are few and hated, that two practically unknown chemists, one from a veterinary school and the other from an agricultural institute, judge with such assurance the most important problems of chemistry, which may well never be solved—in particular, the question of the spatial arrangement of atoms—and undertake their answer with such courage as to astonish the real scientists.

Hermann Kolbe
(1818–1887)

Kolbe, as we now realize, was wrong in his judgment, and the theory of the tetrahedral carbon has been accepted for a century. Even before van't Hoff and LeBel's theory, an Italian chemist by the name of Emanuelo Paterno made use

* H. Kolbe, *Journal für praktische Chemie* (Neue Folge), **15**, 473 (1877); transl. by G. W. Wheland, in *Advanced Organic Chemistry*, 3rd ed., John Wiley & Sons, Inc., New York, 1960, p. 197.

of a mechanical model in which the four groups attached to carbon were arranged tetrahedrally. And, organic chemists have been using mechanical models ever since with an ever expanding array from which to choose (see Fig. 1.2).

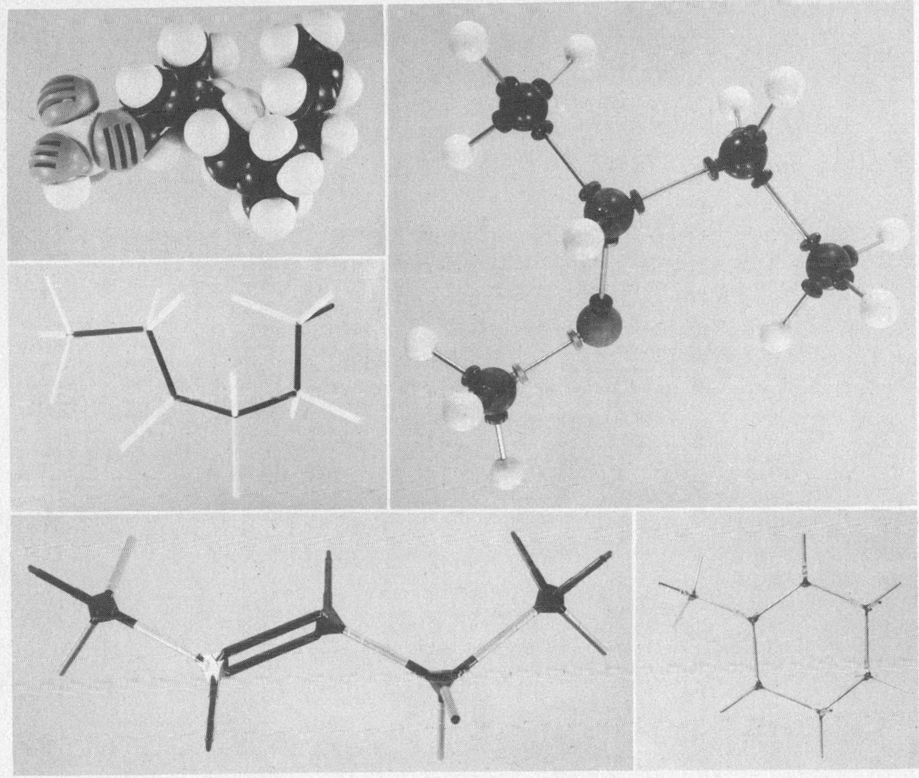

Fig. 1.2. Five examples of currently available molecular models.

Although in 1874 many subtleties of organic structure still remained to be discovered and interpreted, the formulation and eventual acceptance of the theory of the three-dimensional tetrahedral carbon atom constituted the finishing stroke for the broad outlines of the structural theory of organic compounds. The developments immediately thereafter, which still continue today, were, therefore, directed largely to synthesis possibilities. In 1860, the great French chemist Marcelin Berthelot, spurred by the theories of Kekulé and Couper, published a book concerned with organic synthesis. It was Berthelot's aim in this publication to direct the attention of chemists to the general principles and methods of synthetic organic chemistry and to stimulate investigation in the field of organic synthesis so as to prove beyond all question that compounds identical with those produced by plants and animals can be synthesized from inorganic or mineral matter. In this aim he was enormously successful, and activity on the synthesis scene proceeded at an ever-increasing pace for the remainder of the century.

Concurrent with Berthelot's attempt to focus the attention of organic chemists on the synthesis of natural products, another development was taking place that had a profound impact on the worlds of both chemistry and commerce. In 1855, A. W. Hoffman came to London from Germany to become director of the newly founded Royal College of Chemistry. The following year, a young man named William Henry Perkin enrolled in chemistry; quickly captured by Hoffman's enthusiasm for organic chemistry, he began experiments under his tutelage. So enterprising was Perkin that, when the appointment to an assistantship curtailed his research activities in Hoffman's laboratory, he outfitted a laboratory in his own home. It was here, during the Easter vacation of 1856, that Perkin undertook a synthesis of the natural product known as quinine. The outline for this synthesis, extraordinarily naïve by modern standards, included an oxidation of a compound known as toluidine as the final step in the sequence. When the oxidation was carried out, however, quinine was not produced, and instead "dirty, slimy precipitates formed." Perkin's first impulse was to "throw the rubbish away," but second thoughts prevailed and he went back and engaged in a more careful investigation of the product. The result was the isolation of the first dyestuff ever produced from a coal tar chemical—the now well-known aniline purple or mauve. Realizing the possibilities of his discovery, Perkin quit school, rounded up some capital, and set up a plant to manufacture aniline purple. Again, it is difficult for us to realize today what a tremendous undertaking this was. Nowhere in the world was there anything resembling a commercial synthetic organic installation, and everything for its operation had to be devised. Aniline had to be prepared from benzene, and benzene was available only in poor quality and at the high price of $1.25 a gallon. Nitric acid for the nitration of benzene had to be manufactured from Chile saltpeter, and special apparatus for the reduction of nitrobenzene to aniline had to be invented. But Perkin persisted and eventually marketed the dye, which started a business that made millionaires of some, forced bankruptcy on others, provided the basis for future organic chemical industries, and encouraged the liaison between the science and the commerce of chemistry. So enticing were the prospects of the dye industry, in fact, that a major emphasis of chemists in the period from 1860 to 1880 involved the discovery of reactions which might lead to new dyes.

William Henry Perkin
(1838–1907)

Perkin's interest in the dye industry eventually waned, and when he had acquired sufficient money for his simple wants, he sold his factory and again occupied himself with pure research. Fortune was destined to strike yet again, for in 1907, in the course of an investigation of the chemistry of salicylic acid and its derivatives, he hit on the synthesis of coumarin, and thus became the first person to prepare a naturally occurring perfume. This, like aniline purple a half-century earlier, eventually appeared as an item of commerce under the name of Jockey Club and Odor of New Mown Hay, somewhat more earthy appellations than the boudoir names of today's fragrances.

The two decades from 1860 to 1880 saw organic chemistry put on a rational, two-dimensional, structural basis at the beginning of this period and expanded to a three-dimensional concept of structure in 1874. During these years many reactions were added to the armamentum of the organic chemist, and a chemical industry developed which, among other things, provided the laboratory chemist with many previously unavailable starting materials. Thus the stage was set for the appearance of one of organic chemistry's greatest stars, Emil Fischer. Ap-

Emil Fischer
(1852–1919)

pointed to an assistantship in 1874 under the tutelage of Adolph von Baeyer, Fischer discovered just a year later the compound called phenylhydrazine and used this compound to unravel the mysteries of the structure of the sugars, a task that occupied him for the next two decades. While still a student, Fischer had pledged his life to the solution of the problems of the chemistry of the carbohydrates and the proteins, and he honored this pledge with commitment and brilliance. Directing his attention first to the carbohydrates and later to the proteins, Fischer brought to bear an originality of mind, a skill at devising apparatus, and a capacity for organization that has left, to this day, an imprint on the art and science of organic chemistry.

In retrospect, it can be seen that the advent of the First World War signaled the beginning of significant changes in the character of organic chemistry. Gradually, the nationality of the outstanding chemists came to include Englishmen and Americans; the center of chemical industry shifted from Germany to the United States; slowly, the intellectual attitude toward organic chemistry changed from the strictly empirical approach to include an appreciation for reaction mechanism. The seeds of mechanistic studies, planted early in the twentieth century by the English chemist Arthur Lapworth, sprouted and grew with the nourishment of other English chemists, notably Robert Robinson and Christopher Ingold. Taking cognizance of Lapworth's pioneering work and realizing the importance of the then recent ideas of Gilbert N. Lewis, Irving Langmuir, and Walter Kossel regarding the place of the electron in the explanation of atomic and molecular structure, Robinson and Ingold added their own vast knowledge of chemistry and their own intuition and constructed the framework of what is now known as the theory of organic reactions. Robinson, more devoted to structure elucidation and synthesis, outlined his ideas in the qualitative sort of language more appealing to the classical organic chemist. Ingold, more devoted to quantitative measurement and theoretical concepts, outlined his ideas in the sort of language more appealing to the physical chemist. It is the coming to maturity of theoretical organic chemistry and its ability to communicate with synthetic organic chemistry that has been a major achievement in the middle of the twentieth century, particularly during the period following the Second World War.

Organic chemistry, born at the beginning of the nineteenth century, progressed for the remainder of the century at the leisurely pace characteristic of that era. The increasing connections between chemistry and commerce, a liaison strengthened by two world conflicts in the twentieth century, has brought about a growing awareness of the importance of organic chemistry and hastened the pace by emphasizing the role of change, innovation, and invention. The number of persons engaged in the various areas of organic chemistry has increased enormously over the past several decades; the extent of industrial involvement has gone steadily upward; the advent of new instruments, such as those for measuring ultraviolet, visible, infrared, mass, and nuclear magnetic resonance spectra, those for effecting chromatographic separations, and those for exploring the three-dimensional structures of compounds via X-ray diffraction characteristics, have revolutionized the analysis of mixtures, the purification of compounds, and the determination of structures. Structural organic chemistry can truly be said now to be approaching a state of perfection, although new techniques yet to be invented will undoubtedly reveal even more subtle aspects of this facet of organic chemistry. Synthetic organic chemistry has also come along at a remarkable pace, and with the synthesis of the enzyme ribonuclease in 1968, a 77 unit polynucleotide gene segment in 1970, and a 126 unit polynucleotide gene segment in 1973, the construction in the laboratory of a molecule capable of self-duplication may not be far off. Theoretical organic chemistry, representing the newest dimension of the field, has also taken giant strides in recent years; although many of the obstacles to the theoretical analysis of complex chemical systems still seem insurmountable, the impact of theory on practice in a field where intuition has played such an important role cannot be minimized. The intermingling of structure determination, analysis, synthesis, and theory has elevated organic chemistry to a new plane of sophistication, which the rest of the chapters in the book will attempt to delineate in more detailed fashion.

1.4 A Case History of an Organic Reaction

If structures are the vocabulary of organic chemistry, organic reactions are its sentences and organic syntheses its essays. Before we can write sentences and essays, we must, of course, learn vocabulary, and this is no less true for organic chemistry than it is for a language. Our first attention, therefore, will be directed to the structure of organic compounds. As a preview of the interesting topic of organic reactions, though, let us quickly look into the case history of one particular reaction, *viz.*, the one involving phenol and formaldehyde.

Phenol is probably familiar to many readers as a substance that they have encountered while in the dentist's chair. It is often used as a dental antiseptic, and its characteristic taste and odor are instantly recognizable by anyone who has experienced its use in this respect. Formaldehyde also will be recognized by many readers, for its odor often permeates biology buildings, where it is used as a tissue fixative and embalming medium for biological specimens.

Adolph von Baeyer, one of the great chemists of the latter part of the nineteenth century, discovered that when phenol and formaldehyde are mixed in the presence of aqueous sodium hydroxide a reaction takes place. From his reaction mixtures, however, he was able to obtain nothing but a thick, resinous material whose structure he was unable to determine. It remained for a chemist named Manasse to discover in 1894 that if the conditions of the reaction are carefully controlled in a certain way it is possible to isolate a compound consisting simply of the chemical attachment of one phenol (C_6H_6O) to one formaldehyde (CH_2O) to produce a compound known as saligenin ($C_7H_8O_2$):

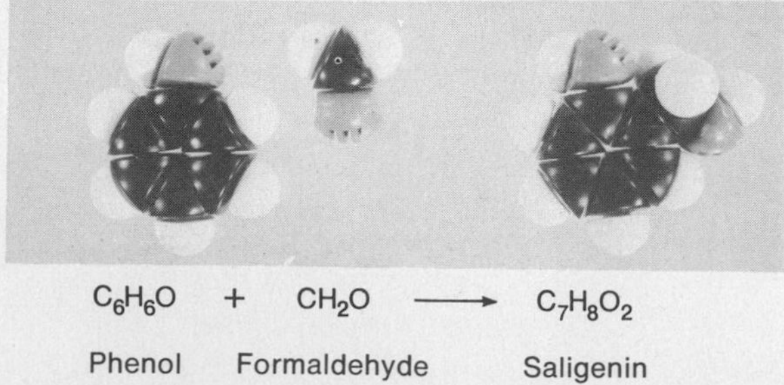

$$C_6H_6O + CH_2O \longrightarrow C_7H_8O_2$$

Phenol Formaldehyde Saligenin

It turns out that saligenin was not a new compound in 1894, for it had been prepared 30 years earlier by a different route involving the action of sodium amalgam on a compound called salicylaldehyde. In fact, saligenin is even older than this, for nature has been producing it for eons in the bark of the willow tree, where it is present in combination with the sugar glucose. The material, which we now call salicin, was known in the form of a crude extract by the ancients who used it as an antipyretic (fever-reducing agent). Purification and structure determination carried out early in the twentieth century established the structure of salicin and showed that it can be converted to saligenin and glucose by the action of water:

$$\underset{\textbf{Salicin}}{C_{13}H_{18}O_7} + H_2O \longrightarrow \underset{\textbf{Saligenin}}{C_7H_8O_2} + \underset{\textbf{Glucose}}{C_6H_{12}O_6}$$

More recent biochemical experiments have shown that salicin is formed in the willow tree from saligenin and glucose, and that the saligenin is synthesized by a process analogous to the condensation of phenol and formaldehyde. To be sure, the living cell uses milder conditions than the chemist in the laboratory, relying on nature's catalysts, the enzymes, for help. But, it is clear that the chemistry that takes place in nature often closely resembles that which takes place in the test tube.

Our story of the reaction of phenol and formaldehyde does not stop here, however. Some years after Adolph von Baeyer's experience with the reaction, a Belgian-born American named Leo Baekeland decided to take a closer look at the material that Baeyer had dismissed as intractable and, therefore, uninteresting. Although Baekeland was not entirely successful in establishing the structure of this material, he was able to show that it had a very high molecular weight and that it had possibilities for use as a commercial material. So, he patented his method for making this hard, resinous substance from phenol and formaldehyde and in 1909 initiated the industrial production of a type of plastic that came to be known as Bakelite. As everyone knows, plastics are today one of our major commodities. Enormous quantities are produced, and although many other plastics have come on the scene since Baekeland's product, Bakelite continues to be made in large volume. Thus, where one trail in the investigation of an organic reaction led to the unraveling of some of the mysteries of natural products, another trail led to the production of new and useful materials for man's consumption (and, incidentally, to great wealth for Baekeland and others).

With the industrial importance of the phenol-formaldehyde reaction having been established, the study of the details of this reaction became enticing not only to those interested in chemistry *per se* but also to those interested in improving the commercial characteristics of the product. As a result, a great deal of laboratory time was devoted to the study of the "mechanism" by which phenol and formaldehyde react to form the high molecular weight polymer. It was in the pursuit of such studies that still another trail of chemical endeavor accidentally opened up. When a compound called *t*-butylphenol is used instead of phenol in the reaction with formaldehyde it turns out that under certain conditions the reaction goes neither to the saligenin type of compound nor to the high molecular weight polymer. Instead, it goes to a compound derived from four molecules of the phenol and four molecules of formaldehyde; structural determinations show that this compound (called a cyclic tetramer) possesses a cup-like or basin-like shape:

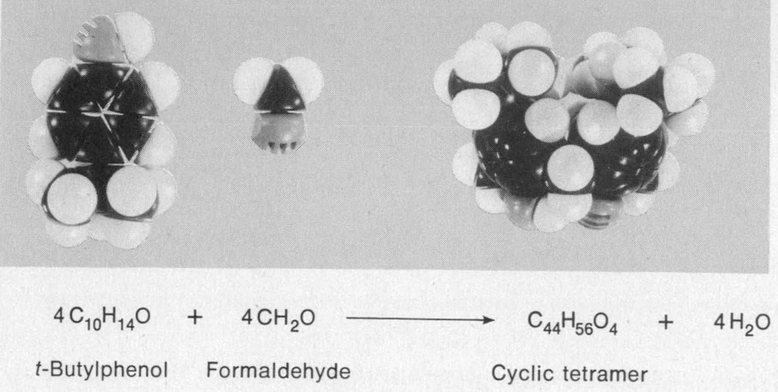

$$4\,C_{10}H_{14}O \quad + \quad 4\,CH_2O \longrightarrow C_{44}H_{56}O_4 \quad + \quad 4\,H_2O$$

t-Butylphenol Formaldehyde Cyclic tetramer

The particular interest that organic chemists have in these cyclic tetramers stems from the ability of the cyclic tetramers to capture other molecules in their cup-like interior, resembling enzymes in this respect. The cyclic tetramers from phenols and formaldehyde, therefore, are being investigated as "enzyme models" in the hope of providing chemists with greater insight into how the enzymes work as catalysts.

As this example of the phenol-formaldehyde reaction illustrates, the pursuit of organic reactions can go in many directions. It might lead to the synthesis of a natural product, it might result in a salable and profitable commodity, it might provide compounds that have inherently interesting chemical and catalytic properties, or it might lead in any of a number of other directions. Very often the chemist doesn't know along which pathway he will be taken when he starts the investigation of an organic reaction, and this is one of the great charms and challenges of organic chemistry. To be able to meet this challenge and to reap the full benefit of these charms, one must first know something about the structures of the compounds being dealt with. The use only of names and molecular formulas is hardly a sufficient basis for any real understanding of what is happening in an organic reaction. For the next several hundred pages, therefore, we shall devote our attention to the structural details of organic chemistry so as to build the foundation that will then allow us to discuss reactions such as the one between phenol and formaldehyde in considerably more detail than has been possible in this brief case history of an organic reaction.

REFERENCES

For more detailed information on the history of organic chemistry, consult the following books and articles.

BENFEY, O. T., *From Vital Force to Structural Formulas*, Houghton Mifflin Company, Boston, 1964. A well-written account of the emergence of the structural theory of organic chemistry with an emphasis on the developments during the period 1830–1860.

BENFEY, O. T., Kekulé–Couper Centennial, *Journal of Chemical Education*, **36**, 319 (1959).

BROWN, H. C., Foundations of the Structural Theory, *Journal of Chemical Education*, **36**, 104 (1959).

CAMPAIGNE, E., Wöhler and the Overthrow of Vitalism, *Journal of Chemical Education*, **32**, 403 (1955).

FARRAR, W. V., and K. R. FARRAR, Faith and Doubt: The Theory of Structure in Organic Chemistry, *Proceedings of the Chemical Society*, 285 (1959).

FINDLAY, A., *A Hundred Years of Chemistry*, The Macmillan Company, New York, 1937. A very readable, short account of the history of chemistry, with good sections on organic chemistry and industrial chemistry, among others.

HARTMAN, L., Wöhler and the Vital Force, *Journal of Chemical Education*, **34**, 141 (1957).

HIEBERT, E. N., The Experimental Basis of Kekulé's Valence Theory, *Journal of Chemical Education*, **36**, 320 (1959).

IHDE, A. J., *The Development of Modern Chemistry*, Harper & Row, Publishers, New York, 1964. A long (851 pages) and detailed account of the history of chemistry.

MACKLE, H., The Evolution of Valence Theory and Bond Symbolism, *Journal of Chemical Education*, **31**, 618 (1954).

PALMER, W. G., *A History of the Concept of Valency to 1930*, Cambridge University Press, New York, 1965. An interesting, biographically annotated history of the concept of valency, starting with John Dalton and continuing to the introduction of the wave mechanical theory.

PARTINGTON, J. R., *A Short History of Chemistry*, 3rd ed., Harper Brothers, New York, 1960. A relatively short (385 pages) history of chemistry from ancient Greece to the recent past.

WINDERLICH, R., History of the Chemical Sign Language, *Journal of Chemical Education*, **30**, 58 (1953).

For more detailed information on the evolution of life on Earth, consult the following books and articles.

CALVIN, M., *Chemical Evolution: Molecular Evolution Towards the Origin of Living Systems on the Earth and Elsewhere*, Oxford University Press, London, 1969. A discussion of chemical evolution by one of the pioneers in the field.

KEOSIAN, J., *The Origin of Life*, 2nd. ed., Reinhold Publishing Corp., New York, 1968. A review of the theories, including the abiotic synthesis of organic compounds.

LEHNINGER, A. L., *Biochemistry*, Worth Publishers, Inc., New York, 1970, pp. 769–92. A short, succinct, and excellent review of the topic.

OPARIN, A. I., *Life: Its Nature, Origin, and Development*, Academic Press, Inc., New York, 1962. A review of the field by one of its early experimenters.

PONNAMPERUMA, C., ed., *Exobiology*, Elsevier, New York, 1972. A series of articles by various leaders in the field.

THURBER, J., The Sea and the Shore, in *Further Fables for Our Time*, Simon and Schuster, Inc., New York, 1956, p. 1. A short history of how life on land started, as perceived by one of America's greatest humorists.

PROBLEMS

1. A gaseous compound boiling at $-23°$ gives the following analytical data. Combustion of a 4.85-mg sample yields 9.28 mg of carbon dioxide and 5.72 mg of water, and a 210-mg sample occupies 114 ml of volume at 22° and 742 torr. Calculate an empirical formula and a molecular formula for the compound, and suggest a reasonable structure for it.
2. The molecular weight determination to establish the molecular formula for alcohol is, in a sense, superfluous, for any integral multiple of the empirical formula C_2H_6O corresponds to an impossible compound. Explain why this is so, taking cognizance of the fact that carbon is tetravalent, oxygen is divalent, and hydrogen is monovalent.
3. The simplest hydrocarbon is methane, which has the formula CH_4. The next higher member of this series is the compound called ethane, which has the empirical formula CH_3 and the molecular formula C_2H_6. Explain why this empirical formula is unlikely to be the molecular formula for a stable compound. The three-carbon compound called propane has the molecular formula C_3H_8. What is its empirical formula? The four-carbon compound called butane has the molecular formula C_4H_{10}. What is its empirical formula? Can a constant structural increment between methane, ethane, propane, and butane be perceived? What is it? Predict the molecular formula for the hydrocarbon containing 10 carbon atoms.
4. Combustion of a 5.20-mg sample of a compound boiling at 69° produces 15.95 mg of carbon dioxide and 7.66 mg of water. Vaporization of a 0.256-g sample produces 86.5 ml of gas, measured at 80° and 750 torr. From these data, calculate an empirical formula and a molecular formula for the compound.
5. It might have been more appropriate to the modern scene if we had chosen marijuana rather than acetic acid as the compound around which to build the introduction in Section 1.1. Not only for legal and sociological reasons but also for pedagogical reasons as well, however, acetic acid is a better choice, for tetrahydrocannabinol, the active ingredient of marijuana, is a far more complex molecule than acetic acid. On the basis of the following analytical data, calculate the empirical and molecular formula for tetrahydrocannabinol. Combustion of a 4.385-mg sample yields 12.89 mg of carbon dioxide and 3.77 mg of water. The compound is not sufficiently volatile for a Dumas molecular weight determination, and other methods for molecular weight determination must be employed. By the osmometric technique, whereby the vapor pressure difference between a pure solvent (*i.e.*, benzene) and a solution of a known concentration (*i.e.*, 31 mg of tetrahydrocannabinol in 10 ml of benzene) is measured, the molecular weight is found to be 315 $\pm$ 5 daltons.
6. Thiamine (Vitamin B_1), one of the essential growth factors for man, is shown by qualitative analysis to contain chlorine, nitrogen, and sulfur (in addition to carbon, hydrogen, and possibly oxygen). When a 9.218-mg sample of the compound is dissolved in water and treated with silver nitrate, 4.385 mg of silver chloride precipitates. Fusion of a 12.025-mg sample with sodium followed by solution in water and treatment with silver nitrate yields 15.36 mg of a mixture of silver chloride and silver sulfide. Combustion of thiamine in a carbon and hydrogen apparatus produces 8.535 mg of carbon dioxide and 2.411 mg of water from a 4.851-mg sample. Combustion in a nitrogen apparatus produces 1.38 ml of molecular nitrogen (at standard temperature and pressure) from a 9.314-mg sample. On the basis of these data, calculate an empirical formula for thiamine. A molecular weight determination gives a value of 301 $\pm$ 5 daltons. What is the molecular formula?

COMMENTS ON PROBLEMS

Problems are to the readers of textbooks what grindstones are to axes; grindstones sharpen edges, and textbook problems sharpen wits. To most effectively sharpen your wits it is important that you work as many of the problems at the end of each chapter as possible. Some of these will be easy and may tax your memory but not your ingenuity; others will be more difficult and should tax both. Generally, the easiest problems come at the beginning of the set and the more difficult ones toward the end. Frequently, however, similar types of problems are grouped together, starting with the easier ones and concluding with the more difficult ones. In such instances, easy problems may occur at several places within the set, even at the very end.

Answers to all of the problems in the text are contained in the supplementary volume "Student Guide to Fundamentals of Organic Chemistry" by Gutsche and Pasto.

Chapter **2**

Atomic and molecular structure: The building blocks for carbon frameworks

> ...this is the day we celebrate Bohr
> Who gave us the complementarity law
> That gives the correspondence (as Bohr said before)
> That holds in the shell as well as the core
> That possesses the compound levels galore
> That makes up the spectrum
> That's due to the modes
> That belong to the drop
> That looks like the nucleus
> That sits in the atom
> That Bohr built.
>
> by R. E. PEIERLS, to celebrate
> Niels Bohr's seventieth birthday (1955)

2.1 Early Views of Atomic Structure and Chemical Bonding

In Section 1.2, we defined organic chemistry as the study of compounds in which carbon is, in most cases, combined with itself, almost always combined with hydrogen, and in many cases combined with other elements as well. The recognition of the possibility of the carbon-carbon bond laid the foundation for the structural theory of organic chemistry (see Section 1.3), and it is the existence of such bonds that makes possible the virtually limitless array of organic compounds. The first real insight into the several types of carbon-carbon bonds was provided by the theory of the tetrahedral carbon atom. Viewing the carbon atom as a regular tetrahedron, van't Hoff and LeBel suggested in 1874 that carbon might bond with itself in three different ways; *viz*., a point-to-point combination (corresponding to what is now recognized as a carbon-carbon single bond), an edge-to-edge combination (corresponding to what is now recognized as a carbon-carbon double bond), and a face-to-face combination (corresponding to what is now recognized as a carbon-carbon triple bond). On the basis of this simple, yet ingenious, geometric interpretation, most of the essential features of the three-dimensional characteristics of organic compounds can, in fact, be explained and predicted. Rather than pursue the topic of carbon-carbon bonding in the terms of the original van't Hoff-LeBel theory, however, we shall focus the following discussion on the ideas that have been generated in the twentieth century regarding atomic and molecular structure.

For a complete understanding of chemical structure, it would be necessary to start with a consideration of how the nucleus of an atom is constructed. The atomic and molecular phenomena that are of interest to the organic chemist, however, do not involve any nuclear changes but only changes in certain of the characteristics of the electrons surrounding the nuclei. Thus, although the presence of the nucleus is essential to the existence of chemical phenomena, the nucleus remains invariant during virtually all organic chemical changes, and its structure can therefore be neglected. The distribution of the electrons surrounding the nucleus, though, is by no means invariant during chemical processes, and it is with the "electronic structure" of atoms and molecules that the following discussion is concerned.

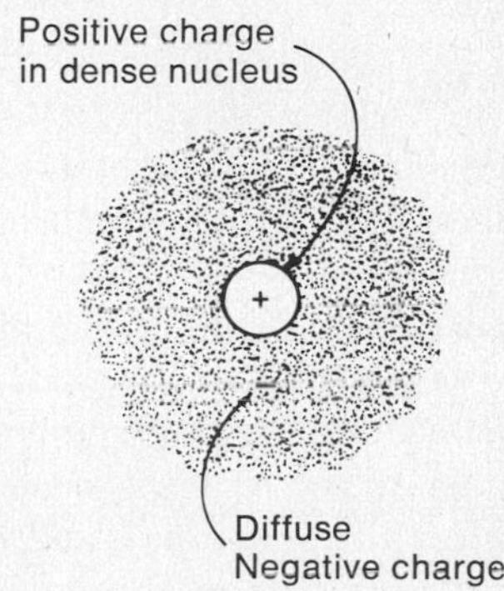

Rutherford Atom

With the discovery of the electron by J. J. Thomson in 1897, the first plank in the foundation of modern structural theory was put firmly in place, although earlier experiments had already pointed to the probable existence of such a particle. From this discovery and from the work and ideas of Ernest Rutherford, came a working model of an atom as an entity composed of a small, positively charged nucleus surrounded by a cloud of negatively charged electrons. This model suffered from certain theoretical difficulties, however, and in 1913 a new interpretation of atomic structure was announced by Niels Bohr. Basing his ideas on the quantum theory of energy, which had been formulated by Max Planck at the turn of the century, Bohr postulated that the electron could move about the nucleus only in a series of discrete energy states with each state of motion corresponding to a definite energy. The physical picture that emerged from these ideas imputed to the atom a structure analogous to that of a solar system with the nucleus acting as the sun and the electrons as the planets. But the "Bohr atom," too, encountered difficulties; although it provided a quantitatively satisfactory interpretation for the hydrogen atom, it failed for heavier atoms and was also inadequate for the interpretation of chemical bonding. After only 12 years of life it was replaced by still another theory called the "wave mechanical atom," which retained the quantum basis of the Bohr theory but added a new concept, as discussed in the next section. The wave mechanical atom has proved to be the hardiest of the lot and survives to this day with no apparent signs of undue wear and tear. Since it provides the currently accepted interpretation of atomic and molecular structure, let us consider it in some detail.

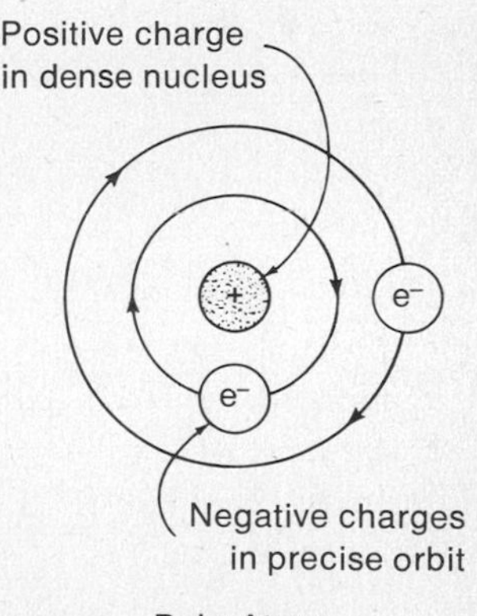

Bohr Atom

2.2 *The Wave Mechanical Atom (Wave-Particle Duality: Schrödinger Treatment)*

In 1924, Louis de Broglie, arguing from theory rather than from experiment and taking cognizance of Albert Einstein's ideas concerning the photon, advanced the postulate that moving particles have wave motions associated with them. What may have seemed at the time to be a bizarre notion was substantiated by experiment only three years later when the phenomenon of electron diffraction was demonstrated by Davisson and Germer in the United States and by J. J. Thomson and Reid in England. Thus electrons could be shown by one set of experiments to be ponderable entities with a measurable mass and by another set of experiments to behave like waves. This wave-particle duality, sometimes seen as a conflict, is now a generally accepted fact of the physical world and is the foundation for the wave mechanical interpretation of the atom.

On the basis of Einstein's postulate for photons, de Broglie suggested that the relation between the momentum of a particulate electron and the length of the wave associated with the electron is given by the equation $p = h/\lambda$, where p is the momentum of the electron, h is Planck's constant (6.624×10^{-27} erg sec), and λ is the length of the wave associated with the electron. On the assumption, then, that the electron can be treated as a *particle* having *wavelike* properties, the problem of constructing the appropriate formalism for its mathematical description became one of finding suitable mathematical expressions to describe a three-dimensional wave motion on which certain constraints are placed. Expressions had, of course, already been worked out years earlier for a variety of motions, such as those associated with a vibrating string, and many of these could be carried directly over to the mathematical description of the wave motion associated with an electron. Using these and modifying them in the appropriate fashion, scientists obtained an equation that could describe various properties of the electronic system. It turns out that in solving this expression, first devised by Erwin Schrödinger and known as the *Schrödinger equation*, three mathematical constants (constants of integration) automatically enter in. These are called quantum numbers, and they are categorized and interpreted as follows: (a) the principal quantum number, designated as n, has integral values 1, 2, 3, . . .(b) the angular momentum quantum number (azimuthal quantum number), l, has integral values from zero up to one less than the principal quantum number; and (c) the magnetic quantum number, m, has integral values ranging from $-l$, through zero, to $+l$ (see Table 2.1). The principal quantum number determines the total

Table 2.1 QUANTUM NUMBERS

Principal quantum number, n	1	2		3		
Angular momentum quantum number, l	0	0	1	0	1	2
Magnetic quantum number, m	0	0	−1, 0, +1	0	−1, 0, +1	−2, −1, 0, +1, +2

energy of the particular electron associated with the nucleus, and the three quantum numbers together characterize the shape and spatial orientation of the volume of space in which the electron is traveling. It eventually became apparent from the work of Uhlenbeck, Goudsmit, and Pauli that still another quantum number was necessary in addition to the three that come directly out of the Schrödinger treatment. Called the spin quantum number and designated as s, it defines an intrinsic magnetic property (*i.e.*, the spin angular momentum) of the electron and has the value of $+\frac{1}{2}$ or $-\frac{1}{2}$.

2.3 *The Application of Wave Mechanics to Atomic Systems*

Of particular interest in considering the details of atomic and molecular structure are the physical interpretations that can be attached to the mathematical solutions of the Schrödinger equation. To introduce this topic, let us first turn to the simple analogy of a vibrating string. For a string of a particular length, density, and tautness, a certain frequency of vibration can be induced by plucking the string near its midpoint. At any particular instant in time, the string will occupy a certain definite position in space, but at different times it will occupy different positions. If, on a single plot, we draw the positions of the string for, let us say, 50 observations, the picture would look something like Fig. 2.1(a); this is a time-averaged picture of the vibrating string, a standing wave representation. The standing wave shown in Fig. 2.1(a) corresponds to the fundamental vibrational frequency of the string; it contains no nodes* between its two ends, the fixed position of the string at the ends being defined as a "boundary condition."

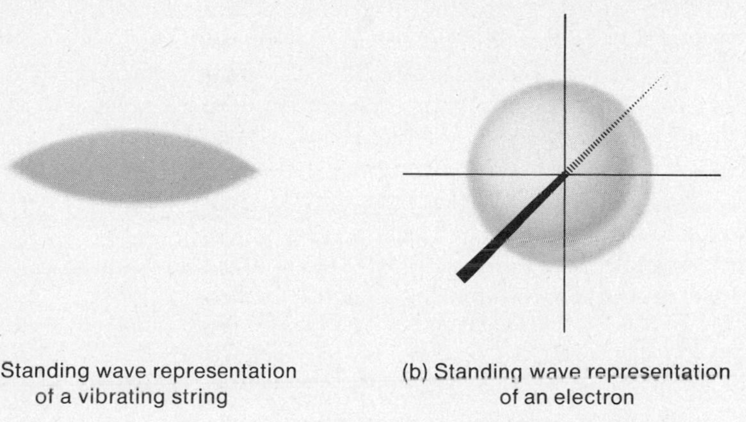

Fig. 2.1. Standing wave representations of the fundamental mode of a vibrating string and an electron associated with a nucleus.

The wave characteristics of an electron traveling around a nucleus can be translated into a physical picture in a fashion roughly analogous to that of the vibrating string. If we locate the positions of the electron at particular instants in time and indicate these positions by dots, a standing wave representation emerges. It turns out that the shape of the standing wave corresponding to the electron in its lowest energy state is a sphere with no nodes [Fig. 2.1(b)]. Standing wave representations such as these are called *atomic orbitals* (abbreviated AO), and they are characterized in the wave mechanical expressions that describe them by the three quantum numbers, n, l, and m. Atomic orbitals with spherical symmetry are called *s orbitals*, and the one with the lowest energy is designated as a 1*s orbital*; its quantum numbers are $n = 1$, $l = 0$, and $m = 0$.

According to the precepts of the Heisenberg uncertainty principle we cannot know the exact position of an electron at a precise moment in time. We are

---

* A node is defined as a region (point, plane, or surface) at which the mathematical sign of the wave function changes; *i.e.*, the region in which the amplitude of the wave function passes through zero. Thus, by this definition, the fixed positions of the string at its ends (*i.e.*, at the "boundary condition") are not characterized as nodes.

forced to deal in probabilities, and the square of the wave function, ψ_1^2 multiplied by a volume element, $d\tau$, is interpreted as expressing the probability of finding an electron in a particular volume of space at a particular time. For example, if $\psi^2\,d\tau$ is 0.1 in a volume element that is 0.4 Å from the nucleus, the probability of finding the electron in this volume element is one chance in ten; *i.e.*, 10% of the time the electron will be found in this region of space. If we show what is called the "radial probability" of finding the electron somewhere within a 1*s* oribtal by plotting the quantity $4\pi r^2\psi^2$ against the distance from the nucleus, *r*, we obtain the graph shown at the right. This diagram indicates that the probability of finding the electron at the nucleus approaches zero, that there is a maximum probability of finding it at a certain distance from the nucleus, and that the probability diminishes at greater distances than this optimum value and is zero at infinite distance. The zero probability at infinite distance from the nucleus corresponds to a "boundary condition" analogous to the fixed ends of the vibrating string. In drawing atomic orbital representations we are taking an arbitrarily chosen cut-off point along this probability distribution curve and showing the outer surface of the orbital at that point. If, for example, we take the cut-off point indicated by the dashed vertical line *a* in the plot shown above, the outer surface of the atomic orbital corresponds to the region of maximum electron density; if we take the cut-off point indicated by the dashed vertical line *b* the orbital is larger, and the region of maximum electron density is closer to the nucleus than the outer surface of the orbital.

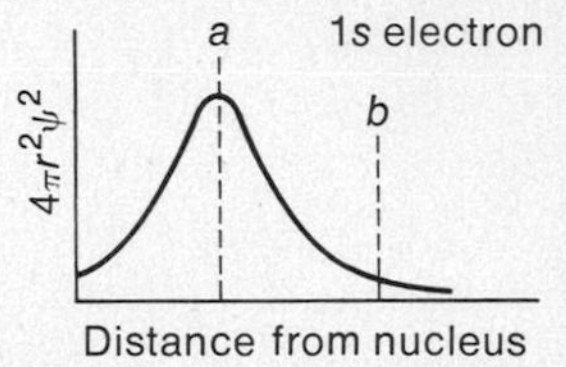

As players of stringed instruments know, if a vibrating string is lightly touched at its midpoint, its pitch goes up an octave. The string is said to be vibrating in its first overtone mode. If we plot the form and position of the vibrating string over a period of time, we obtain a standing wave representation in which there is a node midway between the extremities of the string [Fig. 2.2(a)]. At the node the string is motionless, and its amplitude is zero. Similarly, perhaps with the help of the fingers of Maxwell's demon,* the wave characteristics of the electron can be changed, although the situation is somewhat more complex because of the three-dimensional character of the electron wave. In fact, two different types of atomic orbitals are possible for the first overtone mode, each of these orbitals containing a nodal surface. One of these orbitals possesses spherical symmetry, is specified as a 2*s* orbital, and is characterized by the quantum numbers, $n = 2$, $l = 0$, and $m = 0$ [Fig. 2.2(b)]; the other possesses axial symmetry, is specified as a 2*p* orbital, and is characterized by the quantum numbers $n = 2$, $l = 1$, and $m = 1, 0,$ and -1 [Fig. 2.2(c)].

A plot of $4\pi r^2\psi^2$ against the distance from the nucleus for an electron in a 2*s* orbital produces the graph shown at the right. This diagram indicates that there is a significant probability of finding the electron at a certain distance away from the nucleus, that the probability then falls to zero as we move further away from the nucleus but subsequently rises to a second and even higher maximum, after which it again diminishes, going to zero at infinite distance. The first point at which it falls to zero corresponds to a node in the wave function and represents

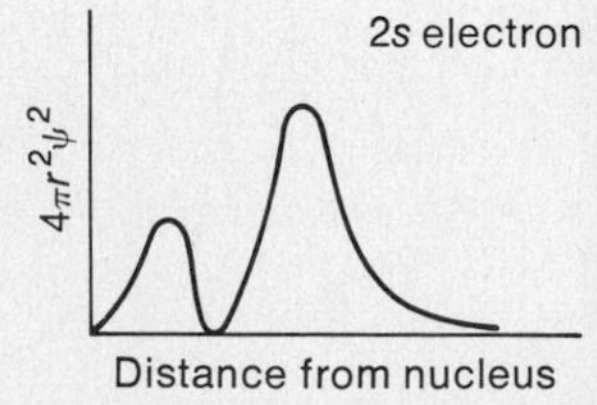

* A Maxwell's demon is defined as "a hypothetical being of intelligence and molecular proportions imagined by J. C. Maxwell to illustrate limitations of the second law of thermodynamics."

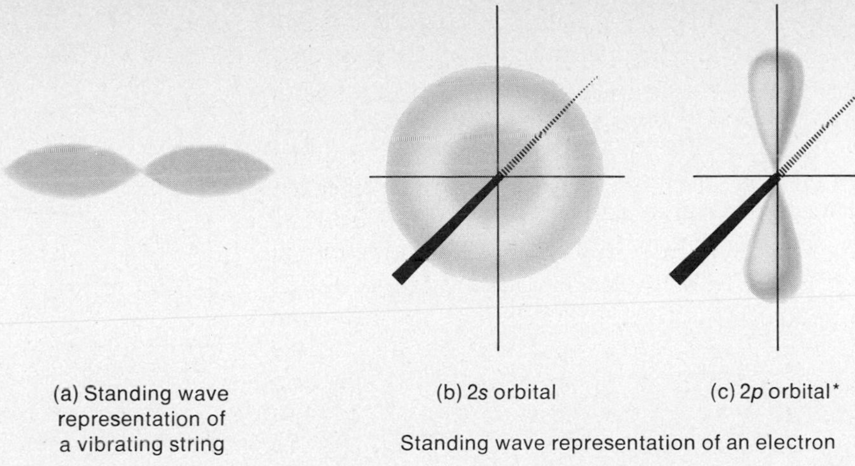

(a) Standing wave representation of a vibrating string

(b) 2*s* orbital

(c) 2*p* orbital*

Standing wave representation of an electron

Fig. 2.2. Standing wave representation of the first overtone mode of a vibrating string and standing wave representations of electrons in 2*s* and 2*p* orbitals.

a spherical surface around the nucleus on which there is zero probability of finding the electron. The second point at which it falls to zero, *i.e.*, at infinite distance, corresponds to the "boundary condition."

A plot of $4\pi r^2\psi^2$ against the distance from the nucleus for an electron in a 2*p* orbital produces the graph shown at the right. This diagram indicates that there is zero probability of finding the electron at the nucleus, that there is maximum probability of finding the electron at a certain distance from the nucleus, and that at greater distances than this optimum value the probability diminishes and falls to zero at infinite distance. Thus, the 2*p* orbital has a node (designated as a nodal plane) at the nucleus and the usual "boundary condition" at infinite distance from the nucleus.

2*p* electron

$4\pi r^2\psi^2$

Distance from nucleus

Orbitals such as those of the 1*s* and 2*s* type in which the angular momentum quantum number is zero are spherically symmetrical and can be oriented in space in only one way. Orbitals in which the angular quantum number is greater than zero, however, do not have spherical symmetry and can have distinguishable

* The 2*p* orbital is more correctly represented in the following manner:

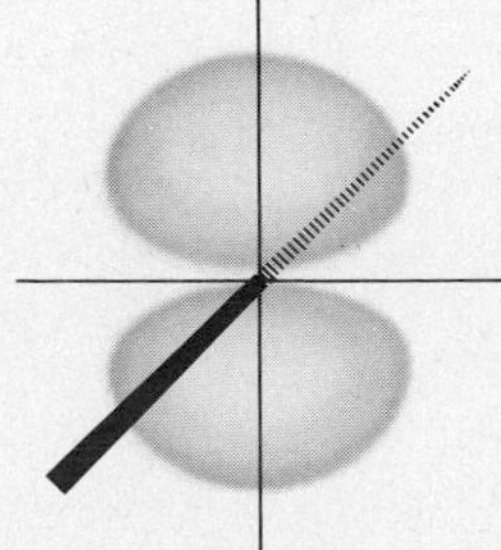

For convenience, however, the representation shown in Fig. 2.2(c) is more generally used, and this custom will be observed thoughout the text.

orientations in space. Orbitals of the $2p$ type, for example, have axial symmetry and, as specified by the three values for the magnetic momentum quantum number (*i.e.*, 1, 0, and -1), have three orientations in space, *viz.* along the three Cartesian coordinates (Fig. 2.3). To differentiate one orbital from another, the designations p_x, p_y, and p_z (*i.e.*, $2p_x$, $2p_y$, and $2p_z$ in the case of $2p$ orbitals) are used. Although differently oriented in space, all the p orbitals having the same principal quantum number n (*i.e.*, 2 for $2p$ orbitals) have the same energy.* This is greater than the energy of the corresponding $2s$ orbitals which, in turn, is greater than that of the $1s$ orbital.

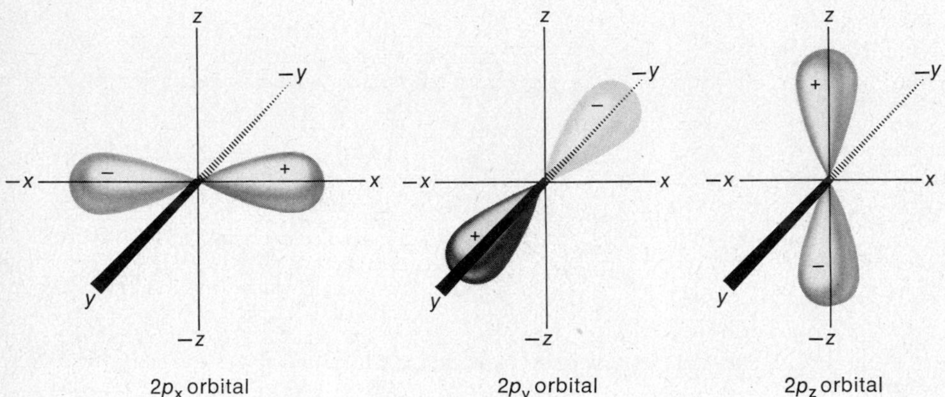

Fig. 2.3. The orientation in space of $2p$ orbitals.

In orbitals having one or more nodes, the mathematical "sign" of the wave function *changes at the node.* This can be illustrated in the case of p orbitals by placing the orbitals in a set of Cartesian coordinates in such a fashion that the nodal points (*i.e.*, the nucleus) are at the origin and the orbitals are oriented along the three axes. It is evident that one half of each of the orbitals (*i.e.*, one of its "lobes") extends in a direction designated by (+) (*i.e.*, $+x$, $+y$, or $+z$) and that the other half extends in a direction designated by (−) (*i.e.*, $-x$, $-y$, or $-z$), the change in sign occurring at the origin (*i.e.*, the nucleus). In the case of a $2s$ orbital the inner portion is generally designated as (−) and the outer portion as (+), the change in sign occurring at the surface between these two concentric spheres (*i.e.*, at the spherical node). A $1s$ orbital is designated as either entirely (+) or entirely (−). For reasons that will become apparent when molecular bonding is discussed, it is often convenient to indicate these signs in the pictorial representations of the orbitals. Remember, however, that the (+) and (−) designations in this context have *nothing whatsoever to do with electrical charge*; they are *mathematical* signs that arise as a consequence of the mathematics used to describe atomic and molecular systems.

If a vibrating string is touched a third of the way from one end it emits a tone that is an octave and five notes higher than the fundamental. It is said to be

* Systems which are identical in energy are often referred to as being *degenerate.* Thus, the $2p_x$, $2p_y$, and $2p_z$ orbitals constitute a degenerate set.

vibrating in its second overtone mode, and its standing wave has two nodes between the fixed ends of the string. Similarly, the wave form of an electron traveling around a nucleus can acquire additional nodes as the energy of the system is increased. For example, 3*s* orbitals, characterized by the quantum numbers, $n = 3$, $l = 0$, and $m = 0$, are spherical in shape and possess two nodes, both of which are spherical nodal surfaces on which the probability of finding the electron is zero (Fig. 2.4). Orbitals of the 3*p* type, characterized by the quantum numbers, $n = 3$, $l = 1$, and $m = 1, 0$, and -1, similarly have two nodes, one a nodal surface possessing a dumbbell shape and the second a nodal plane passing through the nucleus (Fig. 2.4). Orbitals of the 3*d* type, characterized by the quantum numbers, $n = 3$, $l = 2$, and $m = 2, 1, 0, -1$, and -2, also possess two nodes and have the shapes and orientations in space shown in Fig. 2.4.

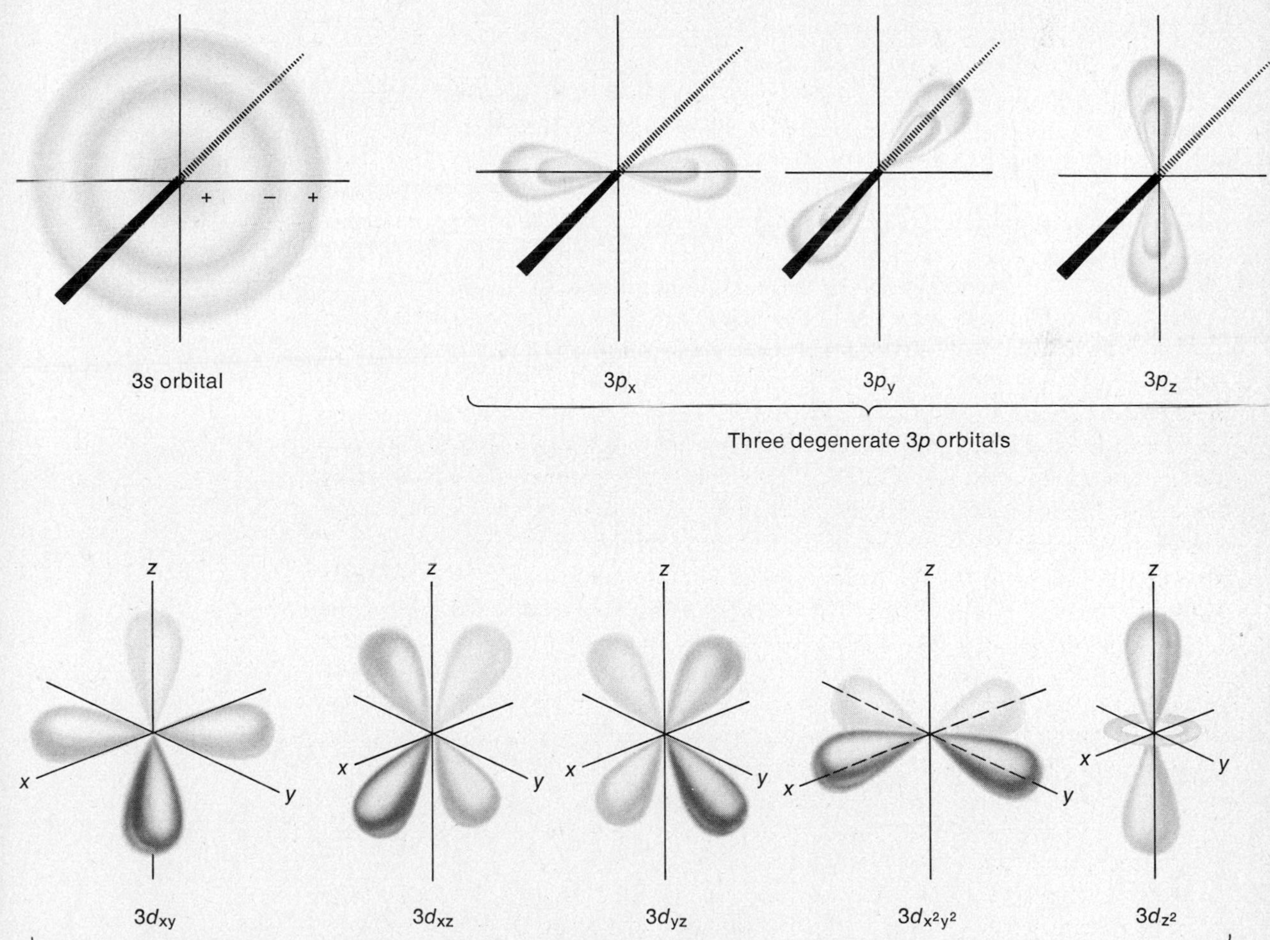

Fig. 2.4. Standing wave representations of electrons in 3*s*, 3*p*, and 3*d* orbitals.

Strings can vibrate in third overtone frequencies, fourth overtone frequencies, etc. Similarly, electrons associated with nuclei can have wave forms that are characterized by increasing numbers of nodes. Almost all atomic and molecular systems discussed in this book, however, involve electrons with wave forms possessing none or only one node; *i.e.*, 1*s*, 2*s*, and 2*p* orbitals. Only rarely will 3*s*, 3*p*, and 3*d* orbitals be considered.

To consider a further consequence of the wave mechanical treatment of the atom, let us again turn to the analogy of the vibrating string. As already mentioned, a vibrating string can be made to sound an octave higher (first overtone), an octave and five notes higher (second overtone), etc. However, notes between the fundamental tone and the first octave, or notes between the first octave and the octave and a fifth *cannot* be made to sound except by changing some characteristic of the string, such as its length, tautness, or density. For a string of specified length, tautness, and density only certain notes can be generated; only certain energy levels are available to the system. The *energy of the system is quantized.* No matter how much energy is added, the system cannot be made to sound any notes other than the fundamental and the various overtones. The energy of an electron traveling around a nucleus is similarly quantized. The energy levels associated with *s* orbitals, *p* orbitals, *d* orbitals, etc. are available to the system, but intermediate energy levels are not. Changes in energy of electrons, just as in vibrating strings, take place in jumps—in quanta.

If we assume that an unperturbed electronic system will seek its most stable situation, we can elaborate the electronic configuration for the atoms in the periodic chart, taking cognizance of the Pauli exclusion principle,* which states that a given atomic orbital cannot contain more than two electrons (Fig. 2.5). Beginning with carbon, sets of degenerate 2*p* orbitals (see footnote on p. 28) are available for electron occupancy, and another rule has been established to handle the situation. Friedrich Hund postulated (hence "Hund's rule") that if two or more degenerate levels are available, each level will become half-filled before any one of them becomes completely filled. Thus, carbon with a nuclear charge of six is pictured in its ground state† as being surrounded by six electrons: two in the 1*s* orbital, two in the 2*s* orbital, one in the $2p_x$ orbital, and one in the $2p_y$ orbital (*i.e.*, rather than both of the *p* electrons being in the same orbital).

PROBLEM 2-1

What are the four quantum numbers for each of the 16 electrons surrounding the sulfur nucleus in the ground state of the atom?

---

* The Pauli exclusion principle states that in a given atomic system no two electrons can have all four quantum numbers the same; thus, if *n*, *l*, and *m* are the same (*i.e.*, two electrons in the same orbital) the electrons must have different spin quantum numbers and are said to be spin paired.

† The lowest energy state of a system is referred to as its "ground state." As will be seen in later discussions, electrons can be elevated from the lowest energy orbital available to them to higher energy orbitals. The atom in the latter condition is said to be in an electronically excited state.

Period	Atom	1s	2s	$2p_x$	$2p_y$	$2p_z$	3s	$3p_x$	$3p_y$	$3p_z$
1	H	1								
	He	2								
2	Li	2	1							
	Be	2	2							
	B	2	2	1						
	C	2	2	1	1					
	N	2	2	1	1	1				
	O	2	2	2	1	1				
	F	2	2	2	2	1				
	Ne	2	2	2	2	2				
3	Na	2	2	2	2	2	1			
	Mg	2	2	2	2	2	2			
	Al	2	2	2	2	2	2	1		
	Si	2	2	2	2	2	2	1	1	
	P	2	2	2	2	2	2	1	1	1
	S	2	2	2	2	2	2	2	1	1
	Cl	2	2	2	2	2	2	2	2	1
	Ar	2	2	2	2	2	2	2	2	2

Fig. 2.5. Electronic configurations of ground states for atoms of periods 1, 2, and 3 of the periodic table.

2.4 *The Application of Wave Mechanics to Molecular Systems: The Hydrogen Molecule*

Before confronting the problem of describing bonding in complex organic molecules, we must first consider some of the concepts of chemical bonding in very simple molecules, such as hydrogen. The simplest statement of the wave mechanical theory of molecular structure says that chemical bonding results from the overlap of atomic orbitals to form molecular orbitals (abbreviated MO). Molecular orbitals are described by molecular wave functions comparable to those that are used for atomic orbitals. They must meet the same mathematical criteria that hold for atomic orbitals, and they are ascribed the same kinds of properties. Thus, they are defined by a form and an energy, and they can contain a maximum of two electrons. From the molecular wave function, Ψ, the probability of finding an electron in a particular region within the *molecule* can be calculated.

Not unexpectedly, the solution of the mathematical descriptions of molecular orbitals is more complicated than that of atomic orbitals, and a particularly difficult problem is to find mathematically suitable wave functions. One approach to this problem has been to retain as much of the mathematical apparatus of the atomic orbitals as possible by expressing the molecular orbital wave functions in terms of a combination of atomic orbital wave functions—the so-called *linear combination of atomic orbitals* (LCAO) method. Thus, the wave function for a diatomic molecule such as hydrogen becomes $\Psi_{\text{molecule}} = \psi_A \pm \psi_B$, where ψ_A and ψ_B refer to the atomic wave functions of atoms A and B between which the

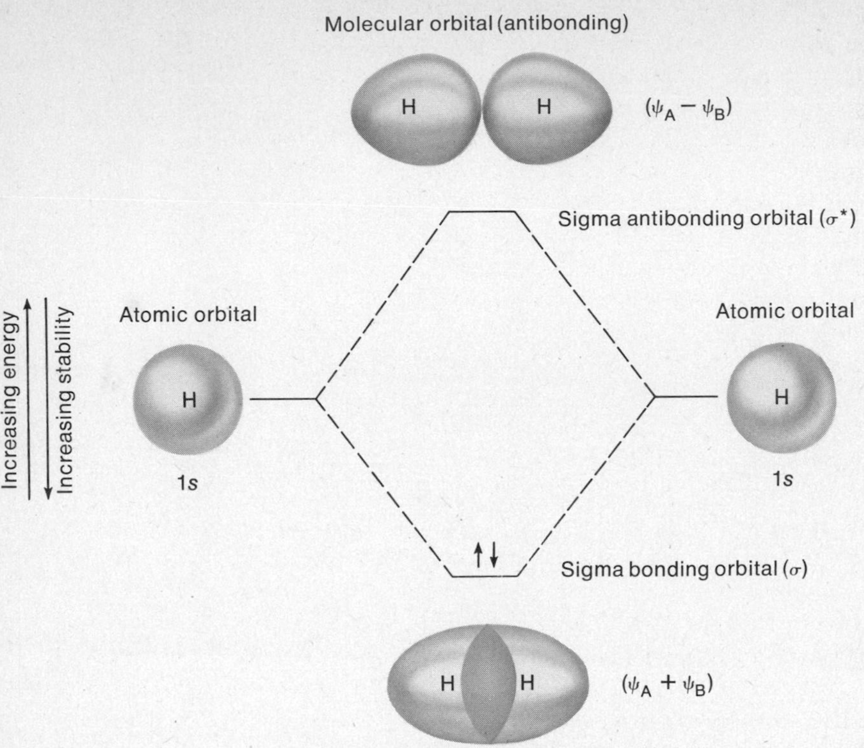

Fig. 2.6. An energy diagram illustrating the formation of bonding and antibonding molecular orbitals by the combination of atomic orbitals of hydrogen atoms. The electron occupancy of the neutral hydrogen molecule is indicated by a pair of arrows in the bonding orbital, the heads of the arrows pointing in opposite directions to indicate that the electrons are spin-paired.

bond is formed.* Solutions of this expression for the hydrogen molecule lead to two molecular orbitals, one corresponding to $\Psi_{MO} = \psi_A + \psi_B$ and the other to $\Psi_{MO} = \psi_A - \psi_B$; *i.e.*, a combination of atomic orbitals having the *same* sign and a combination of atomic orbitals having the *opposite* signs. The combination of orbitals with the same sign corresponds to the lower energy situation and is designated as a bonding molecular orbital. It is an orbital with an ellipsoid shape possessing no nodes (Fig. 2.6) and is designated as a *sigma* (symbolized by σ) orbital. The combination of atomic orbitals with opposite signs corresponds to the higher energy situation and is designated as an antibonding molecular orbital. It is a dumbbell shaped orbital possessing a nodal plane passing through the midpoint between the hydrogen nuclei (Fig. 2.6), and is designated as σ^*.

In constructing the molecular orbital wave functions, we have ignored the electrons that must occupy the molecular orbitals. After the molecular orbitals have been constructed, the electrons are then placed with paired spins in the

* More correctly, $\psi_{\text{molecule}} = c_A\psi_A + c_B\psi_B$ where c_A and c_B are coefficients that indicate how much of the atomic orbital is used in the mixing process.

lowest energy orbitals available. Thus, when one electron is added to the molecular orbitals of the hydrogen system (H_2), it enters the lower-energy, bonding orbital to form the molecular ion H_2^+. If a second electron is then added, it enters the same orbital and spin pairs with the first to form the neutral molecule H_2. The total bond energy of this system is the sum of the contributions of each of the electrons in the bonding orbital. If, now, a third electron is added to the system to give the molecular ion H_2^-, it must enter the higher-energy, antibonding orbital. As a result, the strength of the H—H bond arising from the electrons in the lower-energy, bonding orbital is approximately half-canceled by the electron in the higher-energy, antibonding orbital; thus, H_2^- has approximately the same bond strength as H_2^+. Addition of a fourth electron, of course, destroys the bonding completely, and the energy of $H_2^{=}$ is equivalent to a pair of separated hydrogen anions, 2 H^-.

The distribution of electron density in the region of the nuclei comprising a bond is expressed by the molecular orbitals in which the electrons reside. Thus, in the bonding orbital of hydrogen the electron density is greatest in the region *between* the two hydrogen nuclei, whereas in the antibonding orbital the electron density is at a minimum in this region.

2.5 Hybridization of the Atomic Orbitals on Carbon

The ground state electronic configuration of carbon is described as $1s^2 2s^2 2p_x^1 2p_y^1$. Since the 1*s* and 2*s* orbitals are doubly occupied, they have less "bonding capacity" than the $2p_x$ and $2p_y$ orbitals, which are singly occupied. This electronic configuration would suggest that carbon might be expected to behave as a divalent element. The evidence is overwhelming, however, that in virtually all its compounds carbon behaves as a tetravalent element. Some adjustment in the wave mechanical treatment, therefore, is necessary to accommodate the features of bound carbon.

One of the techniques by which this adjustment has been made is known as *hybridization of atomic orbitals*. Just as molecular orbitals can be constructed by a linear combination of atomic orbitals, new types of atomic orbitals can also be constructed by a linear combination of two or more atomic orbitals. For instance, a linear combination of 2*s* and 2*p* orbitals can be achieved by the appropriate mathematical mixing of wave functions to give hybrid orbitals equal in number to the number of atomic orbitals that were mixed. Thus, if one 2*s* orbital is mixed with one 2*p* orbital, a pair of equivalent (degenerate) hybrid orbitals (designed as *sp* orbitals) is obtained; if one 2*s* orbital is mixed with two 2*p* orbitals, three equivalent hybrid orbitals (designated as sp^2 orbitals) are obtained; if one 2*s* orbital is mixed with three 2*p* orbitals, four equivalent hybrid orbitals (designated as sp^3 orbitals) are obtained. It is also possible to mix 2*s* and 2*p* orbitals in other ways; for example, one 2*s* orbital and three 2*p* orbitals can be mixed to give a pair of $sp^{8/3}$ orbitals and a pair of $sp^{10/3}$ orbitals. For the present discussion, however, only hybrid orbitals of carbon in which integral amounts of *p* are present will be considered—*viz*., *sp*, sp^2, and sp^3. All these hybrid orbitals have the same general shape (Fig. 2.7), and resemble a very lopsided dumbbell; the dumbbell shape is a reflection of the *p* orbital component, and the much larger lobe of the dumbbell is a reflection of the *s* orbital component. Since the

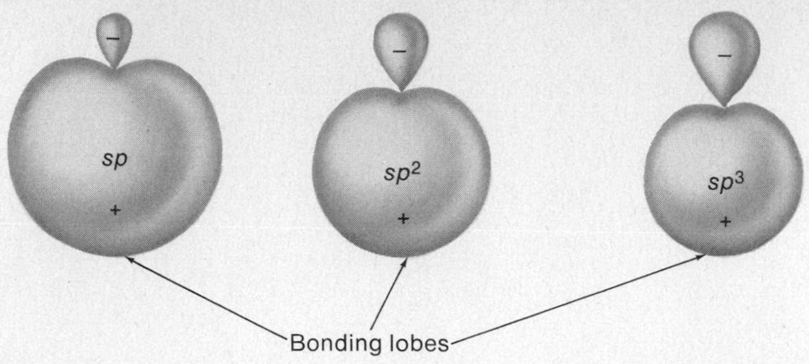

Fig. 2.7. Atomic orbitals hybridized in the *sp*, sp^2, and sp^3 fashion.

strength of a bond is directly related to the extent to which atomic orbitals can overlap, the larger lobe of sp^n hybrid orbitals is the one that engages in bonding. Accordingly, it is called the bonding lobe and is arbitrarily given a (+) sign. Carbon atoms hybridized in the *sp*, sp^2, and sp^3 fashion provide the building blocks for constructing the frameworks of organic compounds.

PROBLEM 2-2

Is it possible to have a carbon atom containing four sp^4-hybridized orbitals? Six sp^2-hybridized orbitals? Three sp^2-hybridized orbitals? Explain.

Let us now explore in more detail the characteristics of these building blocks. In the sp^3-hybridized carbon atom, the four electrons in the second level (*i.e.*, possessing the principal quantum number $n = 2$) occupy hybrid orbitals, and the two electrons in the first level are spin paired in the 1*s* orbital. It turns out to be a natural consequence of the mathematical description of hybridization that the four hybrid orbitals, equivalent in energy, are directed in space to the four corners of a regular tetrahedron. Thus, the nineteenth-century concept of van't Hoff and LeBel regarding the tetrahedral carbon atom finds a twentieth-century rationalization in the wave mechanical treatment of atomic structure. If a carbon atom hybridized in the sp^3 fashion engages in orbital overlap with four hydrogen atoms, the resulting molecule is called methane and has the tetrahedral structure shown in Fig. 2.8.* Using the approach described for the formation of the hydrogen molecule, we see that the linear combination of the four hybrid

---

* In this and many subsequent drawings of molecules, lightly shaded orbitals project *away* from the viewer (recede away from the plane of the page), darkly shaded orbitals project *toward* the viewer (project out of the plane of the page), and medium shaded orbitals are *in the plane* of the page.

orbitals of carbon with four 1*s* orbitals of the hydrogen atoms produces a total of eight molecular orbitals, the wave functions of which encompass all five atoms. The eight electrons of methane occupy the four lowest-energy, bonding orbitals, which, when summed by appropriate mathematical procedures, impart to methane its physical properties. Although a molecular orbital description of this sort (*i.e.*, a "delocalized" orbital description) provides the most accurate picture of the bonding of a molecule, for the organic chemist the simpler representation shown in Fig. 2.8 (*i.e.*, using "localized" orbitals) is generally sufficient. Only occasionally will it be necessary to consider molecular orbital descriptions in the complete, delocalized fashion.

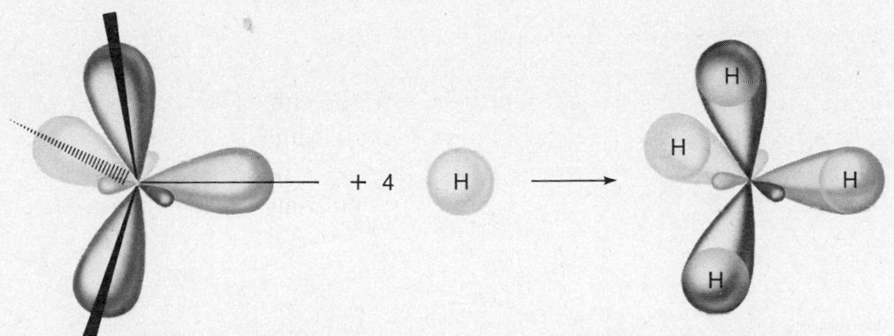

Fig. 2.8. Localized orbital representation of methane, formed from sp^3-hybridized carbon atom and four hydrogen atoms.

An sp^3 orbital of one carbon atom may overlap with an sp^3 orbital of another carbon atom to form a carbon-carbon single bond (C—C bond). If the remaining three sp^3 hybrid orbitals on each of these carbons then overlap with hydrogen orbitals, the molecule known as ethane is produced (Fig. 2.9). The process of C—C bond formation by the overlap of the orbitals from sp^3-hybridized carbon atoms can proceed further, three carbon atoms yielding a three-carbon chain, four carbon atoms yielding a four-carbon chain, etc. Ramifications of these possibilities are discussed in the subsequent section on alkanes. For the moment, it is sufficient simply to remember that in the sp^3-hybridized carbon atom the hybrid

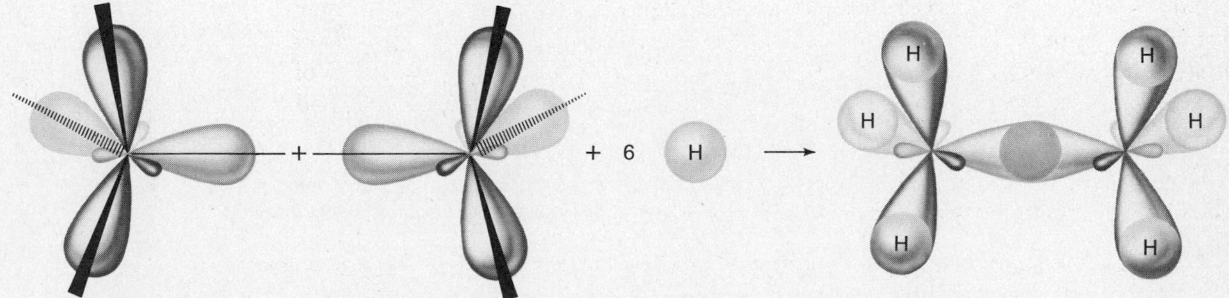

Fig. 2.9. Localized orbital representation of ethane, formed from two sp^3-hybridized carbons and six hydrogens.

orbitals are directed to the corners of a regular tetrahedron, the angle between the orbitals being 109°28′ (referred to as the interorbital angle).

In the sp^2-hybridized carbon atom, only three of the electrons in the valence level reside in the hybrid orbitals, and the geometry is different from that of the sp^3-hybridized carbon. The hybrid orbitals occupy the same plane and are separated from one another by an interorbital angle of 120°. The unhybridized $2p$ orbital is perpendicular to the plane defined by the three hybrid sp^2 orbitals. When two sp^2-hybridized carbon atoms are brought together with the proper orientation relative to one another, axial overlap of a hybrid orbital on each of the carbon atoms produces a *sigma* bond; lateral overlap of the remaining unhybridized $2p$ orbitals produces a second bond called a *pi* bond (π bond). The combination of a σ and a π bond is known as a *double bond*, and is often denoted simply as C=C. If the remaining two sp^2 hybrid orbitals on each of the carbon atoms form bonds with hydrogen atoms, the molecule known as ethylene is produced (Fig. 2.10). The various structural consequences of the presence of a double bond in carbon frameworks are discussed in detail in Chapter 4.

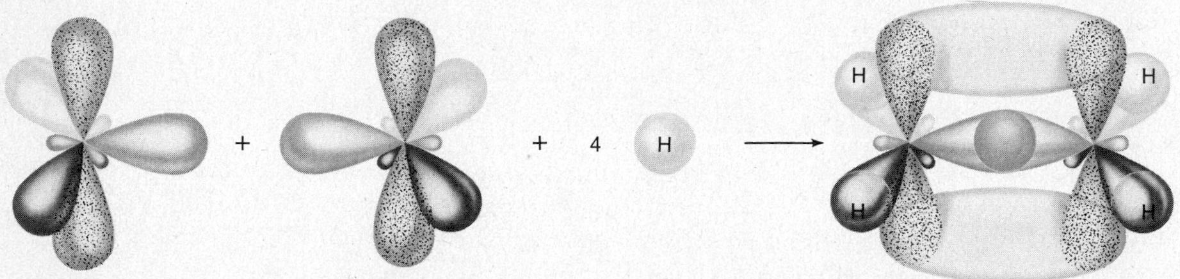

Fig. 2.10. Localized orbital representation of ethylene, formed from two sp^2-hybridized carbons and four hydrogens.

The molecular orbital wave functions and energy diagram for ethylene differ substantially from those of a saturated hydrocarbon* such as methane and merit further comment. As previously indicated, ethylene contains two different types of bonds, designated as σ (C—H and C—C bonds) and π bonds. All the σ bonds reside in the same plane. The π bond, formed by interaction of the p orbitals on the carbon atoms, is perpendicular to this plane and possesses a nodal plane in the plane of the σ molecular orbitals. The π bond, therefore, does not interact with the σ bonds and is said to be orthogonal to them. The interaction of a pair of adjacent $2p$ orbitals produces two π molecular orbitals, one of which does not contain a node between the carbons and one of which does. The first of these is the bonding orbital (designated as π), and the second is the antibonding orbital (designated as π^*) (Fig. 2.11). In the ground state of ethylene, the two

* Compounds containing only single bonds are said to be "saturated," indicating that they have no capability of adding other reagents to their bonds. In contrast, compounds containing double or triple bonds are said to be "unsaturated," indicating that they do have the capability of adding reagents across the double or triple bonds.

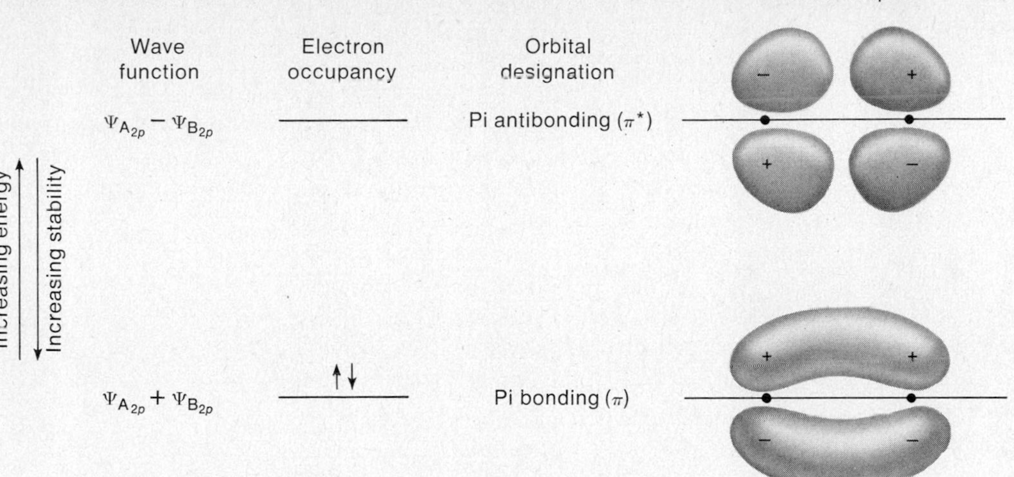

Fig. 2.11. An energy diagram illustrating the bonding and antibonding molecular orbitals for the π system of ethylene.

electrons of the π bond occupy the bonding π molecular orbital, as Fig. 2.11 illustrates.

In the *sp*-hybridized carbon atom, only two of the electrons in the second level reside in hybrid orbitals; the other two electrons are located in unhybridized 2*p* orbitals. The geometry of this system is different from that of either the sp^3- or sp^2-hybridized carbon atoms; the two hybrid orbitals are oriented along the same axis (separated from each other by 180°), and the 2*p* orbitals are oriented along the other pair of axes, mutually perpendicular to each other and to the hybrid orbitals. As a consequence of this arrangement, even greater bonding between a pair of carbon atoms is possible, for in addition to the σ bond derived by overlap of a pair of *sp* hybrid orbitals, both of the 2*p* orbitals on each carbon can engage in lateral overlap to give two π bonds. The result is a *triple bond* between the carbon atoms, often denoted as C≡C. If the remaining *sp* hybrid orbitals on each of the carbons overlap with 1*s* orbitals of hydrogen atoms, the molecule known as acetylene is produced (Fig. 2.12). The various structural

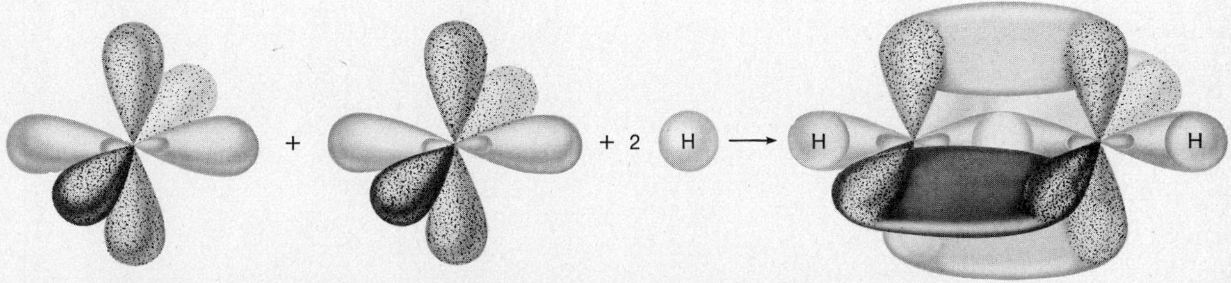

Fig. 2.12. Localized orbital representation of acetylene, formed from two *sp*-hybridized carbons and two hydrogens.

consequences of triple bonding in carbon frameworks are discussed in detail in Chapter 4.

2.6 *Bond Structure and the Representation of Carbon Frameworks*

The most condensed representation of a compound is its molecular formula, *e.g.*, $C_2H_4O_2$ for acetic acid. Except in a few simple molecules, such as methane (CH_4), molecular formulas convey little or no structural information. The simplest representation that, at least to the experienced organic chemist, does convey structural information is one in which the various groupings in the molecule are specified. Thus, the formula C_2H_6 for ethane conveys less information than the formula CH_3CH_3, the latter explicitly stating that a pair of carbons are singly bonded and that three hydrogens are attached to each of the carbons. Even more explicit are bond representations, in which the bonds between atoms are actually shown by lines or by pairs of dots, each line representing a pair of

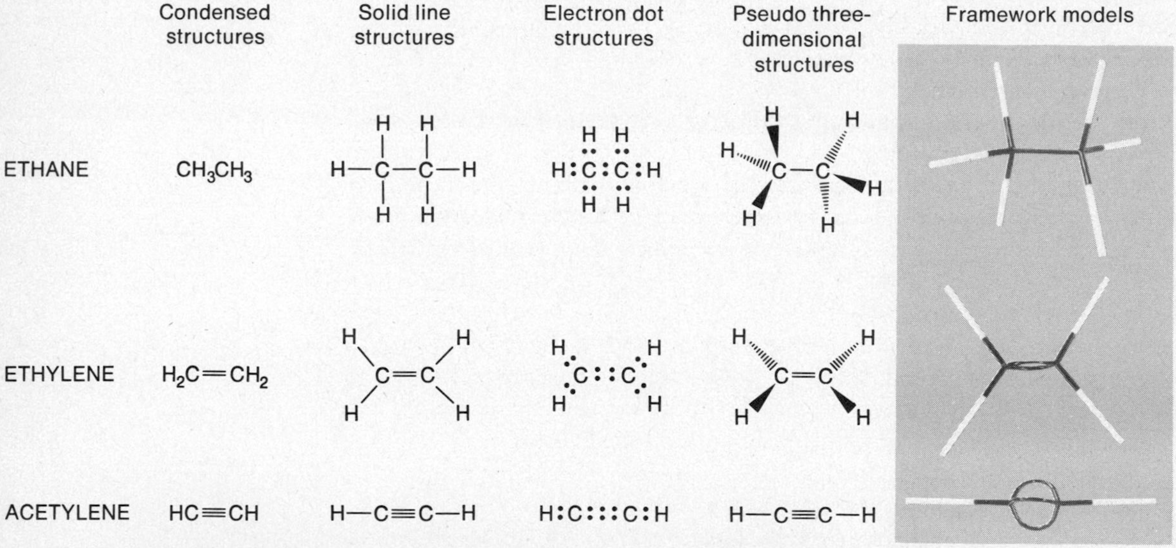

	Condensed structures	Solid line structures	Electron dot structures	Pseudo three-dimensional structures	Framework models
ETHANE	CH_3CH_3				
ETHYLENE	$H_2C{=}CH_2$				
ACETYLENE	HC≡CH	H—C≡C—H	H:C:::C:H	H—C≡C—H	

Fig. 2.13. Condensed structures, solid-line structures, electron-dot structures, pseudo three-dimensional structures, and molecular models of ethane, ethylene, and acetylene.

electrons and each dot representing a single electron (Fig. 2.13). These are two-dimensional representations, however, whereas molecules are actually three-dimensional objects. Even more informative, therefore, are *pseudo three-dimensional* representations, in which the orientations of the bonds in space are depicted by (a) solid uniform lines (—) if the bond is in the plane of the page, (b) solid "flying wedges" (◄) if the bond is projecting in front of the page, and (c) cross hatched "flying wedges" (⦙⦙⦙) if the bond is projecting behind the page (Fig. 2.13).

PROBLEM 2-3

Using molecular models, convince yourself that all the following pseudo three-dimensional representations are for the identical object and simply correspond to different orientations in space.

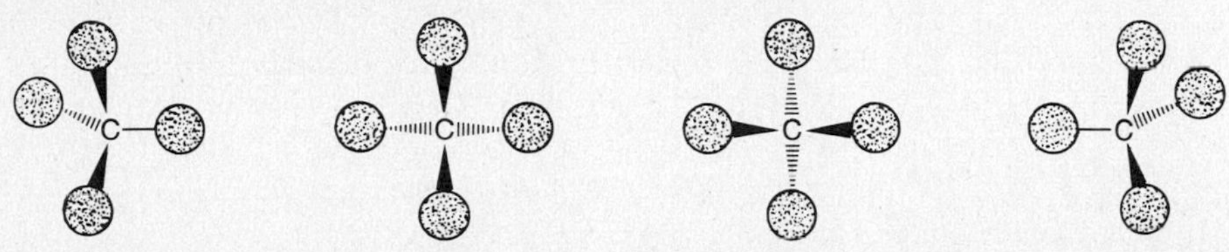

2.7 *Molecular Models*

Superior to any two-dimensional rendering of molecular structure is a three-dimensional model, and a variety of these have been invented and used by chemists over the years (*cf.*, Fig. 1.2). Molecular models generally fall into one of two classes: framework or space filling.

Framework models (sometimes referred to as stick and ball models) represent the *nuclear centers* either by the ends of the sticks that are used to represent the bonds or by a ball whose diameter is some small fraction of a bond length; the actual space occupied by the atom is not explicitly indicated. In the most useful framework models, the distances between nuclear centers (*i.e.*, bond lengths) are scaled so that reasonably accurate estimates of these distances can be obtained by simple measurement with a ruler. The particular advantages of framework models are (a) bond distances are easily measured if the models are properly scaled, (b) bond angles are easily discerned, and (c) the nuclear positions are easily located and the shape of the skeleton of the molecule is quickly discerned. Included among the currently available framework models are the following:

Dreiding models: obtained from Rinco Instrument Co., Greenville, Ill.
Cenco–Peterson models: obtained from Central Scientific Co., Chicago, Ill.
Prentice-Hall framework models: obtained from Prentice-Hall, Inc., Englewood Cliffs, N.J.
HGS Molecular models: obtained from W. A. Benjamin, Inc., Reading, Mass.

Although framework models provide an excellent view of the skeletal shape of a molecule, they fail to depict its outer surface, *i.e.*, its peripheral shape. Space-filling models have been devised in which the actual space (*i.e.*, the van der Waals radius) occupied by the atoms within a molecule is indicated to scale. These models show how close various portions within the molecule actually are, and they provide a more or less accurate indication of the extent to which steric interference exists within the molecule. Included among the currently available space-filling models are the following:

Stuart–Briegleb models: obtained from Arthur S. LaPine Co., Chicago, Ill.
Ealing models: obtained from Ealing Corp., Cambridge, Mass.
Corey–Pauling–Koltun models: obtained from Ealing Corp., Cambridge, Mass.

The framework and space-filling models complement each other, and both are exceedingly useful to the organic chemist. For the beginner, however, the framework models are preferable, for not only are they less expensive but they also are much more useful in depicting the backbone structures of molecules, which is our concern throughout the remainder of Part I. *The student is urged, therefore, to acquire a set of framework models as an invaluable aid for studying the three-dimensional characteristics of the many molecules that are about to be discussed.*

2.8 Polarization of Bonds

Implicit in the discussion of bonding between carbon and hydrogen or between carbon and carbon is the notion that the electron density is evenly distributed between the nuclei comprising the bond. Bonds in which this is true are characterized as *covalent*, and for most C—H and C—C bonds this is reasonably accurate. Even in some C—H bonds, however, and certainly in bonds between carbon and other elements or between hydrogen and other elements, the covalent character of the bond may be reduced, and the bond may possess a certain amount of dipolar character. For example, the compounds containing hydrogen attached

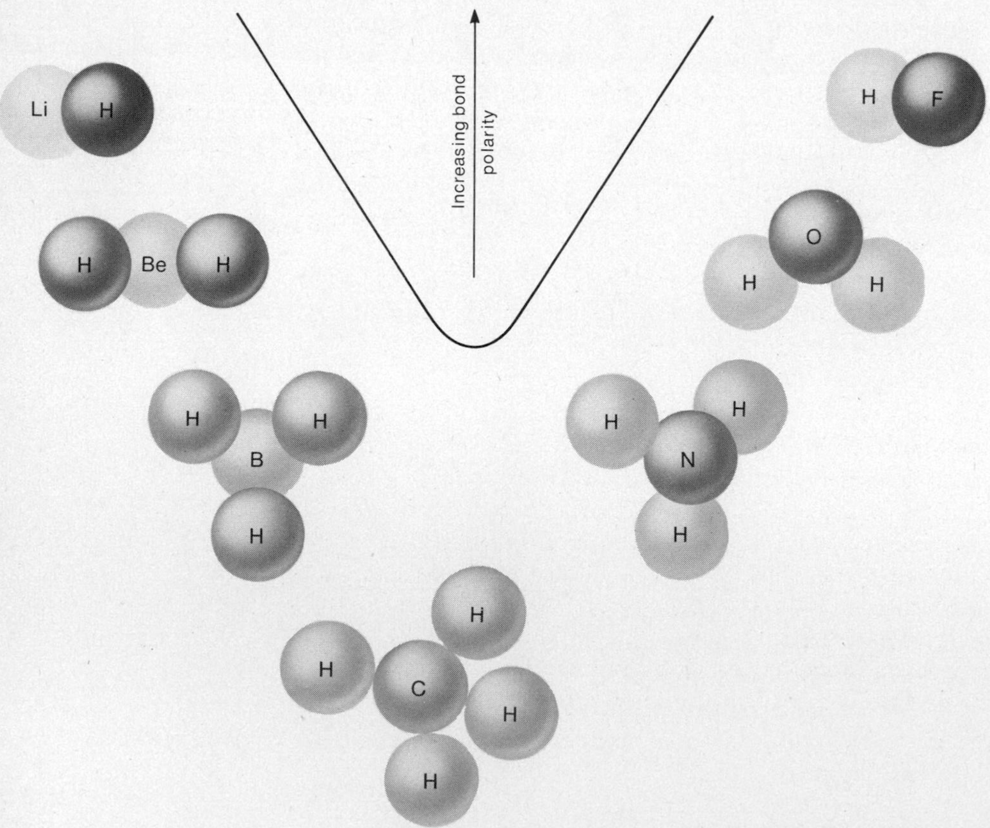

Fig. 2.14. Representations of the distribution of electron density in the hydrides of the first-row elements.

to the first-row elements of the periodic table show striking differences in dipolar character across the row from left to right. In LiH, the bond is strongly polarized in the direction of Li^+ and H^-. Charge separation then diminishes as the row is traversed through BeH_2 and BH_3 to CH_4, where the bond is essentially homopolar and almost completely covalent. Charge separation increases again, however, through NH_3 and OH_2 to FH, the polarity now being reversed; *i.e.*, FH is polarized in the direction of F^- and H^+. Electron density representations of the hydrides of the first row elements are shown in Fig. 2.14, in which the relative electron densities around the first-row elements and the hydrogen are inidcated by the intensities of the shading; the heavier the shading the greater is the electron density on the atom. The polarity of a bond between two atoms is directly related to the electronegativities of the atoms comprising the bond. Tables of electronegativities are available, the one most often used having been devised by Linus Pauling (Table 2.2). The numbers in this table have no fundamental significance, but the differences in the numbers for particular pairs of atoms provide an indication of the relative polarity of the bond connecting those atoms. For example, the C—H bond is predicted to be only slightly polar, carbon having a value of 2.5 on the Pauling scale and hydrogen a value of 2.1. The C—F bond, on the other hand, is predicted to be strongly dipolar, for the difference in Pauling electronegativity values for carbon and fluorine is 1.5.

Table 2.2 PAULING SCALE OF ELECTRONEGATIVITIES OF SELECTED ATOMS

H (2.1)							
	Li (1.0)	Be (1.5)	B (2.0)	C (2.5)	N (3.0)	O (3.5)	F (4.0)
				Si (1.8)	P (2.1)	S (2.5)	Cl (3.0)
							Br (2.8)

The use of the solid line for representing C—H and C—C bonds has already been mentioned, the implication of the solid line being that the bond is an essentially covalent one. As the polarity of a bond increases, however, the solid line representation becomes less and less satisfactory; the ionic representations Li^+H^- and H^+F^- are much closer to the truth. A compound of intermediate polarity, such as BeH_2, BH_3, NH_3, and OH_2, then, must be considered as some mixture of covalent bonding (for which the solid line is appropriate) and ionic bonding (for which the ion pair is appropriate). This mixture or composite (represented, for example, as $H—F \leftrightarrow H^+F^-$) is called a *resonance hybrid* and is discussed in greater detail in Chapter 4.

In similar fashion, although usually to a lesser degree, carbon may acquire cationic characteristics when attached to a strongly electron-withdrawing (electronegative) atom such as fluorine, or anionic characteristics when attached to a strongly electron-releasing (electropositive) atom such as lithium. Even in carbon-carbon bonds, a certain amount of dipolar character may be present if the carbons are differently substituted. In many instances, however, it is convenient to overlook these minor corrections to the covalent character of the bond and to consider bond polarity only in those instances where it provides insight into the chemical and physical behavior of the compound.

PROBLEM 2-4

Which of the following compounds do you think is least well represented by a covalent bond structure? Which is best represented by a covalent bond structure? Explain.

Br_2 LiBr

```
   H
   |
H—C—Br
   |
   H
```

2.9 Bond Strengths and Bond Lengths

In general, an atomic system represents a less stable, more energetic condition than the corresponding molecular system. Thus, when a pair of hydrogen atoms coalesce to form a hydrogen molecule, heat is evolved. This can be quantitatively measured and is found to be 104 kcal for every 2 g of molecular hydrogen formed. The value of 104 kcal, then, is a measure of the strength of the H—H bond; when the process is viewed in the opposite direction, it is a measure of the amount of energy that must be added to convert a mole of hydrogen molecules to hydrogen atoms. Similar measurements can be made for other molecules, and the resulting data give information concerning the comparative strengths of various types of bonds. These data are sometimes expressed as *bond energies* (or empirical bond energies) and sometimes as *bond dissociation energies*. For diatomic molecules there is no difference between these quantities, but there may be significant differences in molecules in which two or more bonds of the same type are involved. For instance, the heat required to dissociate 1 mole of methane (CH_4) to 1 g-atom of carbon and 4 g-atoms of hydrogen is 398 kcal. Since there are four C—H bonds in CH_4, the average bond energy is $\frac{398}{4} = 99.5$ kcal/mole. Implicit in this calculation is the thought that the ease of dissociation of $CH_4 \rightarrow \cdot CH_3 + H\cdot$, of $\cdot CH_3 \rightarrow \cdot\dot{C}H_2 + H\cdot$, of $\cdot\dot{C}H_2 \rightarrow \cdot\dot{\underset{\cdot}{C}}H + H\cdot$, and of $\cdot\dot{\underset{\cdot}{C}}H \rightarrow \cdot\dot{\underset{\cdot}{C}}\cdot + H\cdot$ are all equal; however, they are *not* equal. In actual fact, the energy required to effect the first dissociation, $CH_4 \rightarrow \cdot CH_3 + H\cdot$, is not 99.5 kcal/mole but is 102 kcal/mole, and this value is called the *bond dissociation energy*. Although in many instances it is a more useful value than the average bond energy, the average bond energies are often suitable for simple calculations of approximate energy differences. Some representative bond energy values are shown in Table 2.3.

The term "energy" is frequently used in discussions of chemical phenomena, sometimes in seemingly confusing and contradictory fashions. For example, we refer to the strongest bond in a molecule as being associated with the lowest energy orbital but at the same time assign it the greatest bond energy. Of course, there is really no contradiction if we realize that the energy level of an orbital is

Table 2.3 BOND ENERGIES IN KCAL/MOLE FOR VARIOUS TYPES OF SINGLE, DOUBLE, AND TRIPLE BONDS

	Homopolar Bond		Bond to Hydrogen		Bond to Carbon	
Hydrogen:	H—H	104	H—H	104	C—H	99
Halogen:	F—F	37	H—F	135	C—F	116
	Cl—Cl	58	H—Cl	103	C—Cl	81
	Br—Br	46	H—Br	87	C—Br	68
	I—I	36	H—I	71	C—I	51
Oxygen:	O—O	35	H—O	111	C—O	85
	O=O	119			C=O	179
					C≡O	256
Nitrogen:	N—N	39	H—N	93	C—N	73
	N=N	100			C=N	147
	N≡N	226			C≡N	213
Carbon:	C—C	83				
	C=C	146				
	C≡C	200				

Bond energies for specific types of C—H bonds

H_3C—H	102	—C(—)(—)—H	90
H_2C(—)—H	97	=C(\\)—H	106
		≡C—H	121
HC(—)(—)—H	94	C_6H_5—H	112

related to the amount of energy that is *released* when the bond forms, and the bond energy is a measure of the amount of energy that must be *added* to sever the bond; *i.e.*, the two statements bear a reciprocal relationship. Thus, the lower the energy content of a molecule the greater is the bond energy and the more stable is the molecule. As we shall see in Chapter 8, compounds such as methane can be cleaved to carbon and hydrogen ($CH_4 \rightarrow C + 4\,H$) or oxidized to carbon dioxide and water ($CH_4 + 2\,O_2 \rightarrow CO_2 + 2\,H_2O$). In the first process heat (energy) is required, and in the second process heat (energy) is released. Thus, the system $C + 4\,H$ is the most energetic (bond energy the lowest), the system $CO_2 + 2\,H_2O$ is the least energetic (bond energy the highest), and the system CH_4 is somewhere between these extremes (Fig. 2.15).

PROBLEM 2-5

Arrange the following systems in order of *decreasing* energy: (a) $4\,C + 10\,H$; (b) C_4H_{10}; (c) $2\,C + 5\,H$; (d) $2\,C_2H_5$. Explain the reasons for your arrangement.

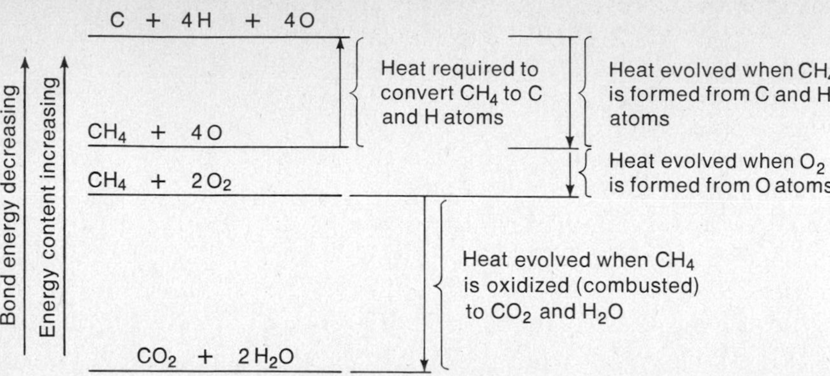

Fig. 2.15. Energy relationships in the system $C + 4H\cdot + 4O\cdot \rightarrow CO_2 + 2\,H_2O$.

The covalent bond has been described in the previous section in terms of the overlap of atomic orbitals to form molecular orbitals. Accepting this premise, one's intuitive feeling would be that the greater is the overlap of the atomic orbitals the stronger should be the bond; the mathematics of the wave mechanical treatment do, in fact, support this intuition. For *complete* overlap of the atomic orbitals of a pair of atoms, however, the interacting nuclei would have to come into coincidence with one another, which, of course, is impossible. Even a very close approach of nuclei is precluded by the enormous forces of repulsion that come into play at small internuclear distances. Thus, there is a competition between the forces of bonding that tend to pull the nuclei closer to each other and the forces of repulsion that tend to push the nuclei apart. This imparts a vibrational motion to the bond, the bond length being greater one moment and smaller the next. The *average value* between the greatest bond length (*i.e.*,

Table 2.4 BOND LENGTHS (INTERNUCLEAR DISTANCES) IN ANGSTROM UNITS

	Homopolar		Bond to Hydrogen		Bond to Carbon	
Single Bonds:	H—H	0.74	H—H	0.74	C—H	1.09
	F—F	1.42	H—F	0.92	C—F	1.38
	Cl—Cl	1.98	H—Cl	1.27	C—Cl	1.78
	Br—Br	2.28	H—Br	1.41	C—Br	1.90
	I—I	2.67	H—I	1.61	C—I	2.14
	O—O	1.48	H—O	0.96	C—O	1.41
	N—N	1.45	H—N	1.07	C—N	1.47
	C—C	1.54	H—C	1.09	C—C	1.54
Double Bonds:	O=O	1.21			C=O	1.20
	N=N		—C(—)—H (sp³ C—H)	1.094	C=N	1.28
	C=C	1.34			C=C	1.34
Triple Bonds:	N≡N	1.10			C≡N	1.16
	C≡C	1.20	=C—H	1.079		
			≡C—H	1.057		

nuclear centers farthest apart) and the smallest bond length (*i.e.*, nuclear centers closest to each other) is what is usually referred to as the *bond length.* Some representative values are shown in Table 2.4. The rate at which the bond length changes between these extremes can be measured; this will be discussed in Chapter 6 under the topic of infrared spectroscopy.

The three building blocks of carbon frameworks, the sp^3-, sp^2- and *sp*-hybridized carbon atoms, possess hybrid orbitals that are qualitatively similar but quantitatively different. The quantitative difference is manifested both by bond strengths and lengths. For example, the bond energy of a $H—C^{sp^3}$ bond is 90–102 kcal/mole, that for a $H—C^{sp^2}$ bond is 106 kcal/mole, and that for a $H—C^{sp}$ bond is 121 kcal/mole. Coincident with this increase in bond energy is a decrease in C—H bond length—*viz.*, 1.094 Å for $H—C^{sp^3}$, 1.079 Å for $H—C^{sp^2}$ and 1.057 Å for $H—C^{sp}$. Similarly, carbon-carbon bonds show a variation in length as a function of the type of hybrid that comprises the bond, as the data in Table 2.5 illustrate.

Table 2.5 CARBON–CARBON BOND LENGTHS AS A FUNCTION OF HYBRIDIZATION OF CARBON ATOMS

Type of Bond	Specific Example[a]	Bond Length, Å
sp^3-sp^3	$H_3C \overset{\downarrow}{-} CH_3$	1.530
sp^3-sp^2	$H_3C \overset{\downarrow}{-} CH{=}CH_2$	1.510
sp^2-sp^2	$H_2C{=}CH \overset{\downarrow}{-} CH{=}CH_2$	1.466
sp^3-sp	$H_3C \overset{\downarrow}{-} C{\equiv}CH$	1.456
sp^2-sp	$H_2C{=}CH \overset{\downarrow}{-} C{\equiv}CH$	1.432
sp-sp	$HC{\equiv}C \overset{\downarrow}{-} C{\equiv}CH$	1.374

[a] The arrows point to the bond whose length is specified.

2.10 Study Guide

To help the reader review what he or she has read in the text, we shall give brief outlines including certain key words and concepts from the discussion at the conclusion of each chapter. It is important to acquire an understanding of each new topic as it is presented, for organic chemistry is a cumulative subject; new facts and concepts build on previously presented facts and concepts. The study guides, therefore, should be used as check lists for determining which facts and concepts are well in hand and which need additional attention.

I. Basic principles of atomic structure
  A. Wave-particle duality
  B. Wave mechanical atom; quantum numbers
    1. Principal quantum number (*n*)
    2. Angular momentum quantum number (*l*)
    3. Magnetic quantum number (*m*)
    4. Spin quantum number (*s*)
  C. Atomic orbitals
    1. Significance of atomic wave function
    2. Types of atomic orbitals—1*s*, 2*p*, 3*d*, etc.

3. Atomic orbitals and electron distribution in atoms
4. Hybridization of atomic orbitals—sp^3, sp^2, and sp hybrids of the carbon atom

II. Basic principles of molecular structure
   A. Molecular orbitals
      1. The hydrogen molecule
      2. Construction from atomic orbitals (LCAO method)
      3. Significance of molecular orbital wave function
      4. Types of molecular orbitals—σ orbital, π orbital
      5. Bonding and antibonding orbitals
   B. Bonding in organic compounds
      1. Single bonds: C—C and C—H
      2. Double bonds: C=C
      3. Triple bonds: C≡C

III. Representation of chemical bonds
   A. Condensed representations of organic compounds
   B. Solid line representations
   C. Electron dot representations
   D. Pseudo-three-dimensional representations
   E. Molecular models: framework models and space-filling models

IV. Characteristics of chemical bonds
   A. Polarization of bonds
   B. Bond lengths
   C. Bond strengths; bond dissociation energies

REFERENCES

Numerous books dealing with the wave mechanical approach to atomic and molecular structure are available. All these are very much more advanced in their treatment of the subject than this chapter. The following are representative examples.

CARTMELL, E., and G. W. A. FOWLES, *Valency and Molecular Structure*, 3rd. ed., Butterworths & Co. Ltd., London, 1966. A relatively short (315 p.), nonmathematical, easily comprehended introduction to atomic and molecular structure.

COULSON, C. A. *Valence*, Oxford Press, London, 1961. A very readable, essentially nonmathematical treatment of atomic orbital, molecular orbital, and valence bond theory that is easily accessible to the beginning student.

DEWAR, M. J. S., *The Molecular Orbital Theory of Organic Chemistry*, McGraw-Hill Book Company, New York, 1969. A mathematically sophisticated discussion of quantum methanics and its applications to molecular systems.

KARPLUS, M., and R. N. PORTER, *Atoms and Molecules: An Introduction for Students of Physical Chemistry*, W. A. Benjamin, Inc., New York, 1970. A rather long (620 p.) discussion of atomic and molecular structure directed to the advanced undergraduate student with a good background in mathematics.

LIBERLES, A., *Introduction to Molecular Orbital Theory*, Holt, Rinehart & Winston, Inc., New York, 1966. A fairly short (198 p.) and simple introduction to molecular orbital theory, representing an expansion on the approach in the book by J. D. Roberts (see below).

RIGGS, N. V., *Quantum Chemistry; Elementary Principles and Methods*, The Macmillan Company, Collier-Macmillan Ltd., London, 1969. A moderately short (243 p.) treatment of the principles of quantum chemistry at a middle level of sophistication.

ROBERTS, J. D., *Notes on Molecular Orbital Calculations*, W. A. Benjamin, Inc., New York, 1961. A short (156 p.) presentation of the rudiments of molecular orbital theory and calculations discussed in a fashion readily comprehended by the beginning student.

SALEM, L., *The Molecular Orbital Theory of Conjugated Systems*, W. A. Benjamin, Inc., New York, 1966. A 576 page treatise for the reader who already has some background in molecular orbital theory.

STREITWIESER, A., *Molecular Orbital Theory for Organic Chemists*, John Wiley & Sons, Inc., New York, 1961. A 489 page book that first develops the background of molecular orbital theory in a moderately sophisticated mathematical fashion and then applies molecular orbital methods to the interpretation of the properties and reactions of molecules.

PROBLEMS

1. Using the tetrahedral representation for the carbon atom, *i.e.*, , draw the van't Hoff–LeBel structures resulting; from a point-to-point, an edge-to-edge, and a face-to-face combination of two tetrahedra representing ethane, ethylene, and acetylene.

2. The standard A to which orchestras are supposed to tune is 440 Hz. What are the frequencies for the first, second, and third overtones of A? Are these notes higher or lower than the standard A?

3. Show why H_2^- is more stable than the separated species H^- and H·.

4. Write the electronic configurations for the elements in the fourth row of the periodic table—*viz.*, K, Ca, Sc, Ti, etc., through Kr. The elements Sc through Zn are called transition metals. Why?

5. For the following equilibria, predict whether the equilibrium constant will be greater or less than 1 (*i.e.*, whether the equilibrium favors the species on the right or the left), and explain your prediction in terms of atomic and molecular orbital considerations.

(a) $H^+ + e^- \rightleftharpoons H\cdot$
(b) $He + e^- \rightleftharpoons He^-$
(c) $2\,H\cdot \rightleftharpoons H_2$
(d) $2\,H^- \rightleftharpoons H_2^{=}$
(e) $H_2^- + e^- \rightleftharpoons H_2^{=}$
(f) $H_2^+ + e^- \rightleftharpoons H_2$
(g) $2\,He \rightleftharpoons He_2$
(h) $\cdot\dot{\underset{\cdot}{C}}\cdot + 3\,H\cdot \rightleftharpoons \cdot CH_3$
(i) $\cdot CH_3 + H\cdot \rightleftharpoons CH_4$
(j) $2\ :CH_2 \rightleftharpoons CH_2{=}CH_2$
(k) $CH_3^- + CH_3^+ \rightleftharpoons CH_3CH_3$
(l) $CH_3^- + CH_3^- \rightleftharpoons C_2H_6^{=}$

6. By means of measurements with scale molecular models, estimate the distance in angstroms between the following pairs of hydrogen atoms.

(a) The hydrogen atoms in methane.
(b) The hydrogen atoms on adjacent carbon atoms of ethane, indicating a maximum and a minimum value.

(c) The hydrogen atoms on the same carbon atom and adjacent carbon atoms in ethylene.
(d) The hydrogen atoms in acetylene.

7. Using the table of bond energies (Table 2.3), calculate the total bond energies for the following molecules:
(a) Methane
(b) Ethane
(c) Ethylene
(d) Acetylene
(e) Ethanol (C_2H_5OH)
(f) Ammonia
(g) Water

8. On the basis of bond energy considerations, predict whether the following reactions will have equilibrium constants greater or less than unity:
(a) $H_2C{=}CH_2 + H_2O \rightleftharpoons CH_3CH_2OH$
(b) $CH_3CH_3 \rightleftharpoons H_2C{=}CH_2 + H_2$
(c) $HC{\equiv}CH + H_2 \rightleftharpoons H_2C{=}CH_2$
(d) $2\,C_2H_6 + 7\,O_2 \rightleftharpoons 4\,CO_2 + 6\,H_2O$

9. In general, it requires approximately 15 times more energy to twist a double bond around its axis than a single bond. Explain the reason for this.

Chapter 3

Carbon frameworks containing sp^3-hybridized carbon atoms: The alkanes

It is best to do things systematically, since we are only human, and disorder is our worst enemy.

HESIOD (8th century B.C.)

3.1 Classification of Organic Compounds

Over 2 million organic compounds have been characterized,* and many more are being added to this group every day. Just as it is convenient to identify the flora and fauna of the Earth by individual names or as belonging to certain phyla and classes, so it is also convenient, and indeed necessary, to identify organic compounds in various ways. Each organic compound, as we shall see, can be assigned a unique name; each compound can also be assigned to a specific class or family of compounds on the basis of structural characteristics. The assimilation of knowledge in any field is greatly facilitated by principles that are sufficiently broad and general to allow the viewing of each new fact not as an isolated event but as a segment to be fitted into an overall pattern. The classification of compounds into families, in which the members possess similar structures and similar chemical properties, provides one of the important generalizing principles of organic chemistry. The first of these families (classes) to be discussed is called the *alkanes* or *paraffin hydrocarbons*, the term paraffin having been chosen because of the general chemical inertness of these compounds (Latin, *parum affinis*, "slight affinity"). All the bonds in alkanes are single bonds (*i.e.*, C—C and C—H), and all the carbon atoms are sp^3-hybridized. Alkanes all conform to the general formula C_nH_{2n+2}; the simplest compound has the formula CH_4, the next member the formula C_2H_6, the next member C_3H_8, *etc.* Because

* The word "characterize" in this context means "to establish the structure of."

compounds of this type contain only single bonds they are often referred to as *saturated hydrocarbons*; the chemical basis for this statement will become apparent in later discussions.

Members of any given class possess certain features in common with each other. As we well know, however, the members of human families are by no means identical. Similarly, the members of any particular class of organic compounds do not show identical physical or chemical behavior, although, in general, the similarities are striking. One of the aims of the following discussion is to stress these similarities as a useful generalizing principle, while, at the same time, pointing out the divergent behavior shown by certain members of each class. In the minds of many novices, the divergencies are one of the exasperating aspects of organic chemistry; for the more experienced organic chemist, however, they are one of its charms.

Our discussion of the alkanes commences with the one-carbon compound *methane*, at which point the topics of *bond angles* and *equivalences of hydrogens* are introduced. We then progress to the two-carbon compound *ethane*, and discuss the concepts of *conformational isomerism*; to the three-carbon compound *propane*, where the idea of *homologous series* is introduced; to the four-carbon compound *butane* and the topic of *structural isomerism*; to the five-carbon compound *pentane* and the six-carbon compound *hexane*, where we deal with the principles of *nomenclature*; and to seven-, eight-, and nine-carbon compounds, in which the possibility of *mirror image isomerism* exists. Following this we discuss *cycloalkanes* particularly emphasizing again the concepts of structural isomerism, mirror image isomerism, and conformational isomerism.

3.2 The Structure of Methane: Bond Angles and the Equivalences of Hydrogens

Methane, the parent member of the alkanes, has the molecular formula CH_4 and is constructed from one sp^3-hybridized carbon atom and four hydrogen atoms (four single bonds), as illustrated in Fig. 2.8. It is the smallest stable organic molecule, and its size fades into insignificance when compared with some of its

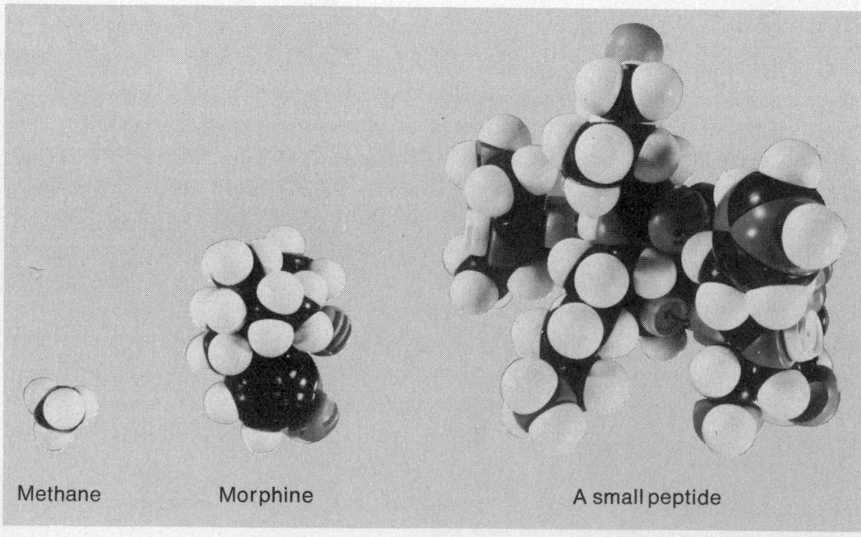

larger brethren. It is an historically important molecule, and as a probable constituent of the primeval broth it may have been a building block from which such molecules as morphine and peptides were constructed. On the basis of abundant experimental evidence, methane is now recognized to be a symmetrical molecule in which the four hydrogens are attached to the central carbon by tetrahedrally directed bonds of equal length. Employing the solid line representation we use structure (a) to depict methane.

$$\begin{array}{c} H_c \\ | \\ H_d{-}C{-}H_b \\ | \\ H_a \end{array}$$

(a)

It might appear from this representation that the distance between H_a and H_b is different from the distance between H_a and H_c. Of course, it isn't, for methane is a *three-dimensional molecule* with *tetrahedral symmetry*, and the distance between all the hydrogens is *exactly the same*. To prove this to yourself, inspect a framework model of methane; it should be immediately apparent that this last statement is true. If we look at the model of methane in one orientation in space (a) and then in another orientation in space (b), we are unable to perceive any difference between the two representations, except for the fact that we have arbitrarily labeled the hydrogens with subscripts a, b, c, and d. The hydrogens are said to be *equivalent*, for each one is related to the remainder of the molecule in precisely the same fashion.

H_d H_c C—H_b H_a **(a)** → Turn 120° around this axis → H_a H_d C—H_b H_c **(b)**

It is important to understand this concept of equivalence, for it will appear in numerous places throughout our discussion of molecular structure. It is discussed in detail in Section 19.10.

Because of the symmetry of the methane molecule, the concept of the sp^3-hybridized carbon atom provides an accurate description of its geometry and correctly predicts that all the H—C—H angles are 109°28′. The majority of molecules, however, lack the symmetry of the methane molecule and depart, to a greater or lesser extent, from the concept of the strictly sp^3-hybridized carbon atom. For example, if two of the hydrogens of methane are replaced by chlorine atoms, the resulting CH_2Cl_2 molecule has a Cl—C—Cl bond angle of 112° and an H—C—H bond angle of 106°. Two of the bonds in the molecule have slightly more p character than an sp^3 hybrid orbital provides, and two of the bonds have slightly less. Fortunately, the departure from strict tetrahedral geometry is usually no more than a few degrees in most compounds containing sp^3-type-hybridized carbon atoms. Consequently, the general statements concerning the structure and chemistry of these compounds need not be significantly modified by these small departures from ideality.

PROBLEM 3-1

If we replace one of the hydrogens of methane with a chlorine atom, the molecule called methyl chloride is obtained. It is represented by the solid line structure

```
     H
     |
  H—C—Cl
     |
     H
```

Assuming that methyl chloride retains a true tetrahedral configuration, are the distances between the chlorine and the three hydrogen atoms the same or different? Are the distances between the three hydrogen atoms the same or different? Explain.

PROBLEM 3-2

If we replace two of the hydrogens of methane with chlorine atoms, the molecule CH_2Cl_2 is obtained. Are the three structures shown for this molecule the same or different? Explain.

```
    Cl           Cl           H
    |            |            |
 H—C—Cl       H—C—H       Cl—C—Cl
    |            |            |
    H            Cl           H
```

3.3 The Structure of Ethane: Conformational Isomerism

The two-carbon alkane called ethane has the molecular formula C_2H_6 and contains two sp^3-hybridized carbons (see Fig. 2.9). If a framework model of ethane is inspected, it is apparent that there are an infinite number of orientations which the two CH_3 groups can take with respect to each other. Let us, for the moment, concentrate our attention on just one of the possibilities, specifically the one shown in Fig. 3.1, in which the hydrogens of the CH_3 groups are as close to each other as possible. This structure can be represented by a "three-quarters view" or by a "head-on view," in which the viewer looks directly down the axis of the carbon–carbon bond (often referred to as a *Newman projection*). Quite clearly, the hydrogens on the "front" CH_3 group are directly in front of the hydrogens on the "back" CH_3 group, eclipsing them. It was, of course, an arbitrary choice to put the CH_3 groups of ethane in this particular orientation. Let us, then, look at another orientation, in which the hydrogens of the CH_3 groups are as far from each other as possible, as represented by the "three-quarters view" and "head-on view" in Fig. 3.1. In this case, the hydrogens no longer eclipse each other and are said to be *staggered*.

Compounds possessing the same number and kinds of atoms and the same molecular weight (*i.e.*, the same molecular formula) but differing in structure and properties are called *isomers* (Greek, *isos*, "equal"; *meros*, "part"; *i.e.*, having equal parts). The inorganic substance ammonium cyanate and the organic substance urea, for example, are isomers, as Wöhler demonstrated on the occasion of the birth of organic chemistry in 1828. A more subtle kind of isomerism is

Eclipsed Conformation

"Three quarters view" "Head on view"

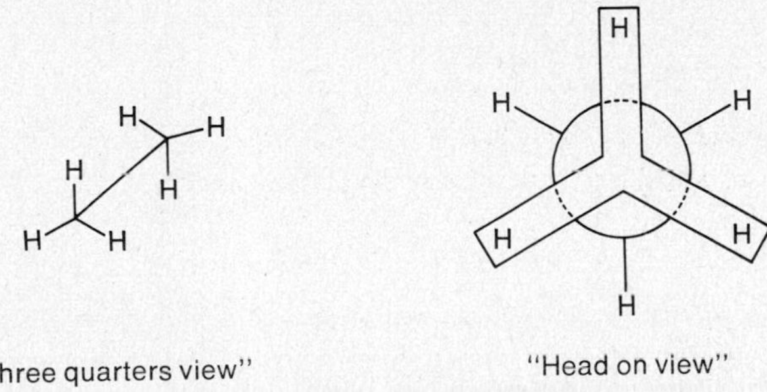

Fig. 3.1. "Three-quarters view" ("sawhorse view") and "head-on view" (Newman projection) representations of the eclipsed and staggered conformations of ethane.

illustrated by the eclipsed and staggered forms of ethane. Clearly, we are dealing with the same molecule in both of these forms but not with precisely identical structures. The eclipsed and staggered forms differ in their geometry as a result of twisting (*i.e.*, rotating) around the axis of a C—C bond. They are said to be *conformational isomers* (also called *conformers* or *rotamers*); any particular twist form is called a *conformation*. Thus, ethane is able to assume an eclipsed conformation and a staggered conformation. To induce the conversion of one to the other requires a certain amount of energy, and a plot of the energy content of an ethane molecule as a function of its conformation is shown in Fig. 3.2. The idea that the staggered conformation is at a lower energy level than the eclipsed conformation is intuitively appealing, for the greater distance between the hydrogens in the staggered conformation produces less interference between the hydrogens than in the eclipsed conformation. Intramolecular encounter of this sort is often called "nonbonded" interference, indicating that the interacting groups (*e.g.*, the hydrogens in the case of ethane) are not directly bonded: the strain arising from this encounter is called "nonbonded strain."*

* The reader is cautioned against too literal an interpretation of the term "nonbonded strain," because the exact nature of the barrier to rotation around a single bond (*i.e.*, the torsional barrier) is not yet clearly understood. It may, at least in part, be ascribed to steric interference between nonbonded atoms (*i.e.*, a van der Waals repulsion), but other factors are also important.

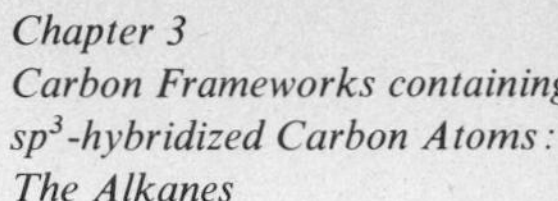

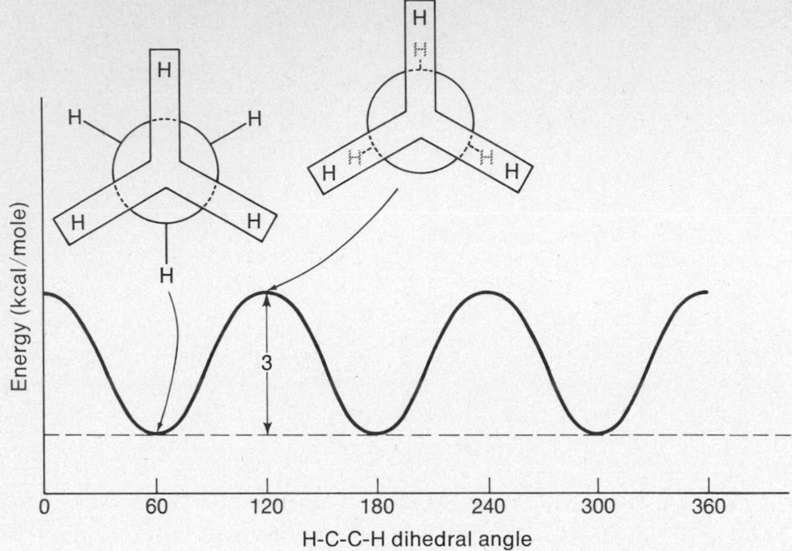

Fig. 3.2. The energy of the ethane molecule as a function of conformation.

The difference in energy between the eclipsed and staggered conformations is about 3 kcal/mole, and if this amount of energy is not available to a collection of ethane molecules, the majority will exist *only* in the staggered form. At temperatures approaching absolute zero this situation prevails, but at ordinary room temperature sufficient thermal energy is available to the system to induce easy and rapid interconversion of the conformations. The staggered conformation, lower in energy, will always be more heavily populated than the eclipsed conformation, but the lifetime of any particular molecule in any particular conformation will be short ($<10^{-6}$ sec). There is, in other words, rapid rotation around the C—C bond, and for most purposes ethane can be considered as a rapidly equilibrating mixture of molecules in the eclipsed and staggered conformations along with the infinite number of additional conformations falling between these extremes.

Although the number of possible conformations is infinite, it is convenient to focus attention on only a few of them—in particular, those that are at the top and bottom of the energy plots. In ethane there are two: the eclipsed conformation and the staggered conformation. The rate of interconversion of conformational isomers is so rapid under ordinary conditions that it is impossible to isolate and study individual conformers. For this reason, there has been an understandable tendency to overlook conformational differences and to ascribe "free rotation" to groups attached by single bonds. For most structural considerations this notion is justified. Remember, however, that the rotation is *not* completely free, that there are differences in energy between various conformations, and that for certain chemical considerations these differences are significant. Conformation, as we shall see in later sections, can play a decisive role in the chemical and physical behavior of certain compounds.

In all the conformations of ethane, the six hydrogens are equivalent to one another. For example, no difference between the representation of the staggered conformation (a) and the staggered conformation (b) can be discerned, except for the fact that the hydrogens have been arbitrarily labeled; the six hydrogens constitute an equivalent set.

Twist 120° around C—C bond

(a) (b)

3.4 The Structure of Propane: Homologous Series

The three-carbon member of the alkane family contains three sp^3-hybridized carbons, has the molecular formula C_3H_8, and is called propane. Looking back at its lower molecular weight relatives, methane and ethane, we notice that the molecular formula of propane differs from each of these by an integral number of CH_2 units; *i.e.*, one CH_2 unit in the case of ethane and two CH_2 units in the case of methane. Propane is said to be related to these other members of the family in a homologous sense, *i.e.*, propane is the immediate higher *homolog* of ethane and the double homolog of methane. Members of a group in which an integral number of CH_2 groups (called methylene groups) constitutes the only difference between them are said to belong to a *homologous series.* All alkanes with the formula C_nH_{2n+2} thus belong to a homologous series. Similar relationships exist in other families of organic compounds, as will be seen in the following discussions. In the more general sense, a series of compounds in which each member differs from its neighbor by *any* constant structural increment (*i.e.*, not necessarily a CH_2 group) is called a homologous series, although the most usual application of the term involves systems in which the structural increment is a CH_2 group.

Assuming that carbon is tetravalent and that hydrogen is monovalent, there is only one possible way in which three carbons and eight hydrogens can be assembled, as represented by the solid line and pseudo-three-dimensional representations in Fig. 3.3. This, of course, is true for ethane and methane as well,

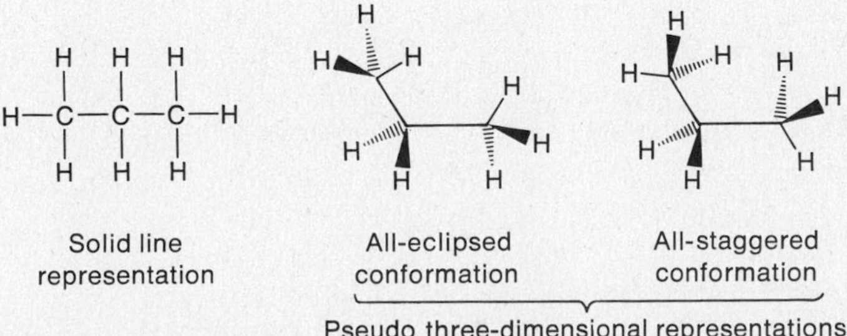

Fig. 3.3. Representations of the propane molecule.

the molecular formula uniquely defining the structure of each compound. The only type of isomerism that can be exhibited by these compounds is conformational, and the bond rotation is sufficiently rapid under most conditions that this subtlety of structure is not ordinarily given explicit recognition.

The six hydrogens on the two CH_3 portions of the propane molecule are *not* equivalent to one another in any single conformation. For example, if we orient a model of propane so that we are looking along the C—C bond that joins one of the CH_3 carbons and the adjacent CH_2 carbon, and then draw a Newman projection for some particular conformation [*e.g.*, conformation (a)], it is clear that the distances between the hydrogens on the "front" CH_3 carbon and the "back" CH_3 carbon are not all the same. Now, if we rotate the front CH_3 portion by 120° and draw the Newman projection [conformation (b)], and then rotate this CH_3 portion another 120° and draw another Newman projection [conformation (c)],

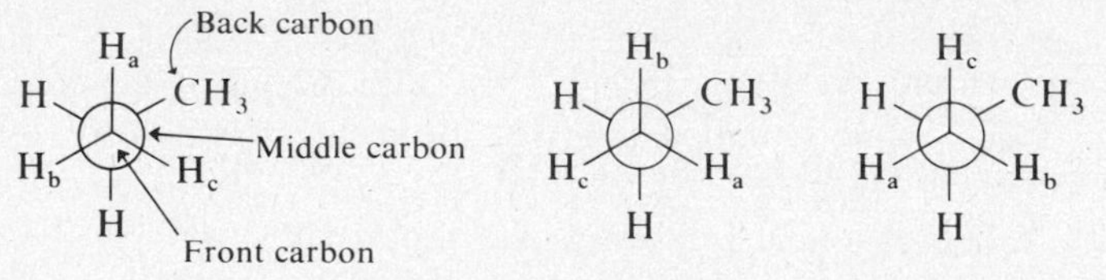

(a) (b) (c)

we find that in all three conformations two of the hydrogens on the "front" carbon are close to the "back" CH_3 group and one is farther away. It turns out, therefore, that the *averaged* relationship of H_a, H_b, and H_c to the "back" CH_3 group is the same. Thus, in conformation (a), H_a and H_c are closer than H_b to CH_3; in conformation (b), H_b and H_a are closer than H_c to CH_3; in conformation (c), H_c and H_b are closer than H_a to CH_3. Taking the average of these three conformations, two-thirds of the time H_a is close to CH_3, two-thirds of the time H_b is close to CH_3, and two-thirds of the time H_c is close to CH_3—*viz.*, the same for all three hydrogens. Because of rapid rotation around C—C bonds, all six hydrogens on the two CH_3 groups are equivalent on a time-averaged basis, although they are not inherently equivalent. By a similar argument, we can demonstrate that the two hydrogens on the CH_2 portion of propane are equivalent on a time-averaged basis. Under no circumstance, however, are the six CH_3 hydrogens equivalent to the two CH_2 hydrogens. Thus, the methyl and methylene hydrogens comprise structurally different sets. The consequences of this difference will become apparent in the discussion in Section 3.6.

PROBLEM 3-3

In which of the following structures do all the hydrogens constitute an equivalent set?

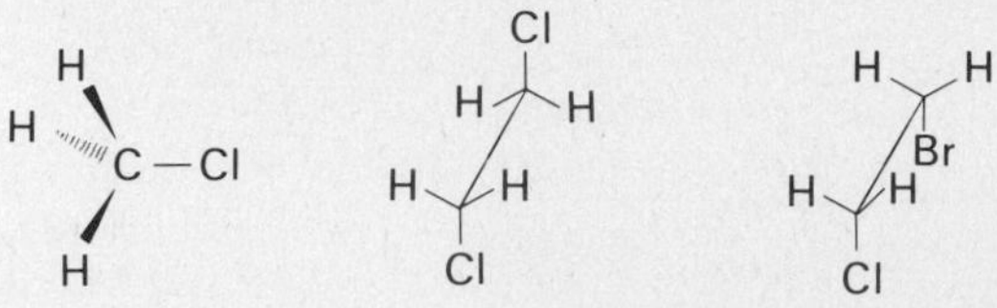

3.5 *The Structure of Butane: Structural Isomerism*

Although the molecular formulas CH_4, C_2H_6, and C_3H_8 accommodate only one structure in each case, the molecular formula C_4H_{10} permits the four sp^3-hybridized carbons and the ten hydrogens to be assembled in two different structures called *structural isomers* (Fig. 3.4).

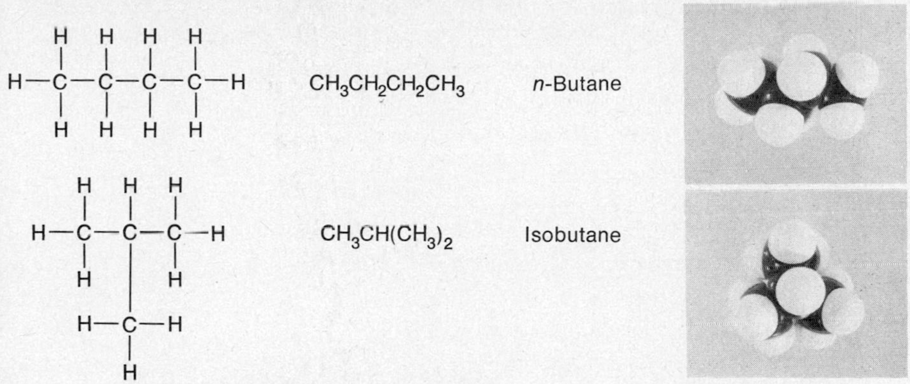

```
    H   H   H   H
    |   |   |   |
H — C — C — C — C — H        CH3CH2CH2CH3        n-Butane
    |   |   |   |
    H   H   H   H

      H   H   H
      |   |   |
  H — C — C — C — H          CH3CH(CH3)2         Isobutane
      |   |   |
      H   |   H
          |
      H — C — H
          |
          H
```

Fig. 3.4. Condensed and solid line representations of *n*-butane and isobutane, and space-filling models of these isomers.

The concept of structure was developed to provide a rationalization for variations in the properties of compounds. If the structures of a pair of compounds are different, one or more of the measurable properties are different. Thus, physically differentiable forms of the C_4H_{10} hydrocarbon called butane are predictable and do, indeed, exist. Boiling points (one form boils at $-12°$ and the other at $-0.5°$), freezing points (one form freezes at $-138°$ and the other at $-145°$), infrared spectral characteristics (see Section 7.2), and various other measurable properties of the two butanes show, beyond doubt, that the two materials are different. Since both of these compounds have the formula C_4H_{10} and, therefore, are butanes, some further nomenclatural distinction is clearly necessary to give each of them a unique designation. To do this, we affix the prefix *n*- (standing for "normal") to the isomer having the straight-chain framework, and the prefix *i*- (standing for "iso") or iso- to the isomer having the branched-chain framework.

Let us now look at the structure of *n*-butane in finer detail by focusing our attention on the C—C bond in the middle of the molecule. We note that when we hold one-half of the molecule fixed in space and twist the other half of the molecule around the C-2/C-3 bond, four conformations can be easily distinguished: (1) an eclipsed conformation, designated as *synperiplanar*, in which one methyl group eclipses the other; (2) a staggered conformation, designated as *synclinal*, in which the methyl groups are separated by a dihedral angle of 60°; (3) an eclipsed conformation, designated as *anticlinal*, in which a methyl group eclipses a hydrogen; and (4) a staggered conformation, designated as *antiperiplanar*, in which the methyl groups are separated by a dihedral angle of 180°.* A plot showing

* The synclinal conformation is also referred to as the "gauche" or "skew" conformation, and the antiperiplanar conformation as the "anti," "trans," or "transoid" conformation. The systematic names are preferred, however.

the energy of the *n*-butane molecule as a function of the conformation around the C-2/C-3 bond (Fig. 3.5) reveals that the energy of these four conformations increases in the following order: antiperiplanar < synclinal < anticlinal < synperiplanar. Although all conformations are populated to some extent at temperatures above absolute zero, the population of the antiperiplanar conformation is largest and the synperiplanar conformation is smallest. The difference in energy between the antiperiplanar and synclinal forms is small enough, however, that at room temperature both are appreciably populated—about 68% in the antiperiplanar form and 32% in the synclinal form. The rate of interconversion among all these conformations is very rapid at room temperature. So, although we can single out certain conformations and give them names, we should realize that under ordinary laboratory conditions, *n*-butane is a mixture of all these conformers; the concept of "free rotation" around the C—C bond can be accepted in defining the structure of butane for most purposes.

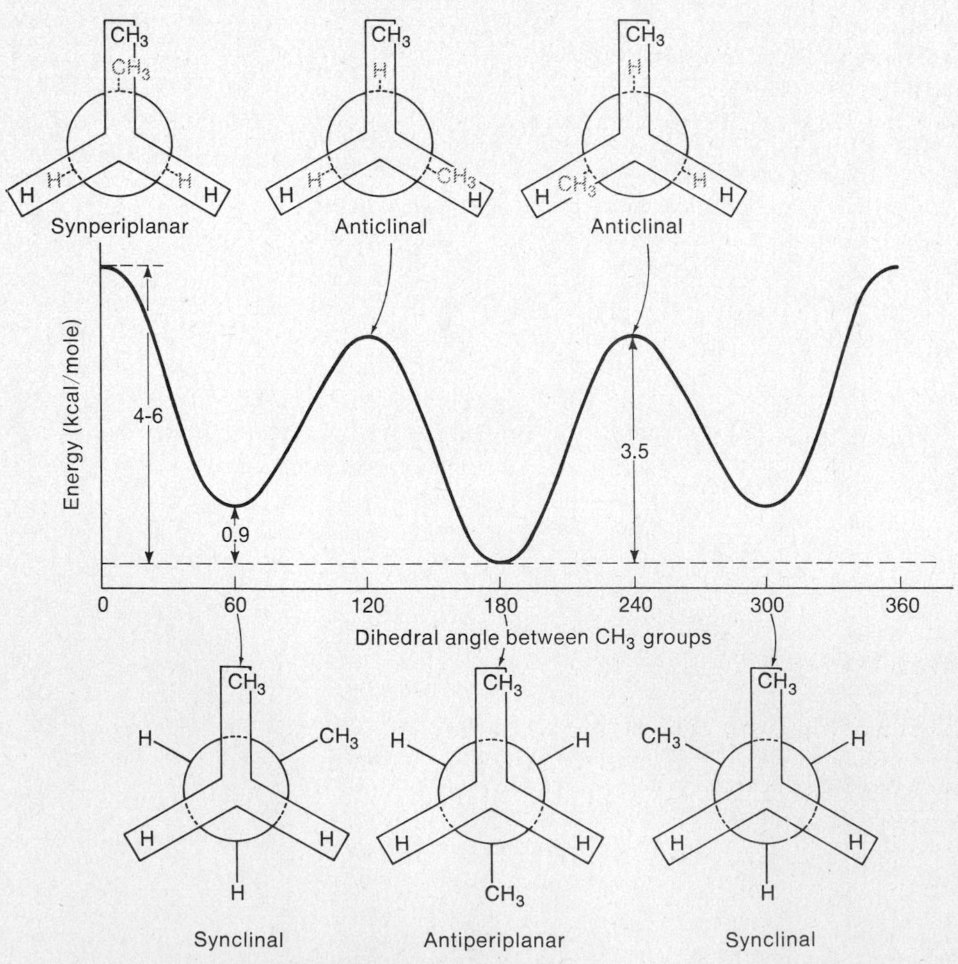

Fig. 3.5. Conformations of *n*-butane.

PROBLEM 3-4

At 4°K, would you expect to find a higher or lower concentration of the antiperiplanar conformation of butane than at room temperature? What would you predict to be the situation at 500° C? Explain.

Employing an argument similar to the one used in Section 3.4 to show that the six CH_3 hydrogens of propane belong to one equivalent set and the two CH_2 hydrogens of propane to a different equivalent set, we can show that the six CH_3 hydrogens of *n*-butane constitute an equivalent set and the four CH_2 hydrogens another equivalent set. Similarly, we can show that in isobutane, the nine CH_3 hydrogens constitute one equivalent set, and the CH hydrogen constitutes the lone member of a second equivalent set.

3.6 The Structure of Pentanes and Hexanes: Nomenclature of the Alkanes

As the number of carbon atoms in an alkane increases, the number of structural possibilities also increases. For the molecule pentane, composed of 5 sp^3-hybridized carbons and 12 hydrogens, three modes of combination are possible, and the compounds to which these correspond are called *n*-pentane, isopentane, and neopentane (Fig. 3.6).

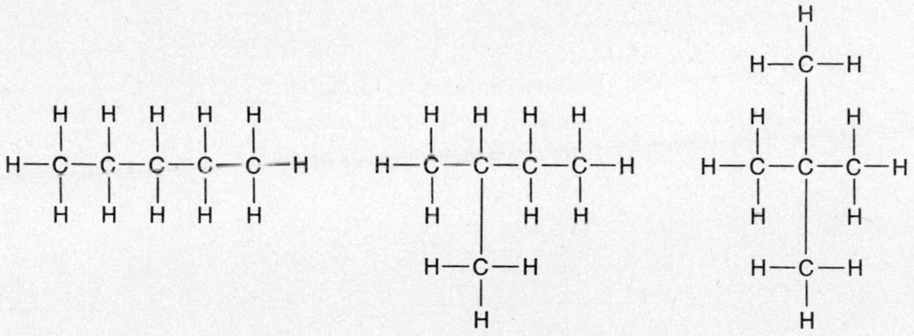

$CH_3CH_2CH_2CH_2CH_3$ — *n*-Pentane

$(CH_3)_2CHCH_2CH_3$ — Isopentane

$(CH_3)_4C$ — Neopentane

Fig. 3.6. Solid line and condensed representations of the pentanes.

For the molecule hexane, composed of 6 sp^3-hybridized carbons and 14 hydrogens, there are five ways in which these 20 atoms can be put together, as illustrated by the condensed representations in Fig. 3.7. Only three of the hexane structures, however, can be named on the basis of the prefixes that have been applied to the pentanes—*i.e.*, *n*-, iso-, and neo-. To name the other two isomers of hexane it is necessary either to invent more prefixes or to adopt some other system. The latter choice has been made, because the number of structural isomers escalates rapidly as more carbons are added to the system and the number of prefixes that would be necessary becomes astronomical; C_7H_{16} (heptane) has

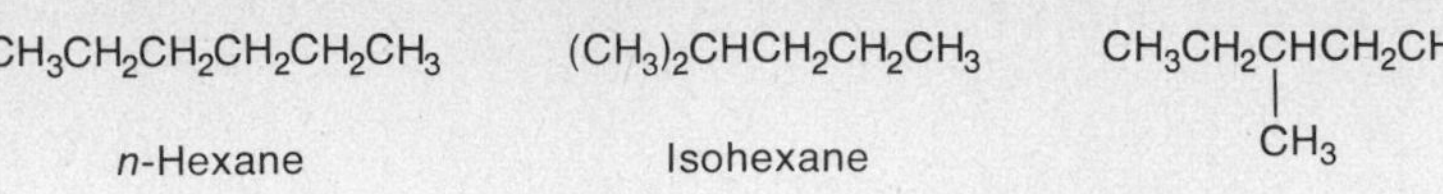

$(CH_3)_2CHCH(CH_3)_2$ $\quad$ $(CH_3)_3CCH_2CH_3$

Neohexane

Fig. 3.7. Condensed representations for the hexanes.

9 structural isomers; C_8H_{18} (octane) has 18; $C_{10}H_{22}$ has 75; $C_{20}H_{42}$ has over 350,000; $C_{40}H_{82}$ has over 10^{13}; etc. To accommodate to this situation, chemists have invented systematic nomenclature systems, two of which we shall discuss.

PROBLEM 3-5

Draw the structures of all the compounds having a C_7H_{16} molecular formula. How many of these can be named by means of the appropriate prefixes illustrated in Figs. 3.6 and 3.7?

Before considering the details of the nomenclature systems that have been applied to the alkanes, however, we must first discuss the concept of the "group." One of the landmarks in the structural theory of chemistry took place in the 1830's, when Wöhler and Liebig (see p. 10) realized that a certain unit, which they called a benzoyl unit, was common to a series of compounds derived from benzoic acid. Portions of a whole structure that recur as units of invariant structure are referred to as *groups*. In alkanes, these recurring units of invariant structure are called *alkyl groups*,* and they correspond to the entire alkane minus *one* hydrogen atom. Thus, the alkyl group derived from methane is $—CH_3$ and is called *methyl*; that from ethane is $—C_2H_5$ and is called *ethyl*; that from propane is $—C_3H_7$ and is called *propyl*; etc. As discussed in Section 3.4, however, there are two different "kinds" of hydrogen atoms in propane. Thus, removal of one of the terminal hydrogens yields $CH_3CH_2\overset{|}{C}H_2$, whereas removal of one of the internal hydrogens yields $CH_3\overset{|}{C}HCH_3$. The first of these alkyl groups is designated as the *n*-propyl group and the second as the isopropyl (or *i*-propyl) group. To illustrate this difference by anticipating the discussion in the next section, we see that if a chlorine atom is substituted for one of the terminal hydrogens of propane, the product ($CH_3CH_2CH_2Cl$) is called *n*-propyl chloride, whereas if it is substituted for one of the middle hydrogens the product

---

* Alkyl groups are also called alkyl radicals. To avoid confusion with the term "free radical" (see Section 7.9), the term "group" rather than "radical" is used throughout the text.

($(CH_3)_2CHCl$) is called isopropyl chloride. In similar fashion, four types of butyl groups can be obtained from butane, including $CH_3CH_2CH_2CH_2$— (*n*-butyl), $CH_3CH_2CH(—)CH_3$ (*sec*-butyl),

$$(CH_3)_2CHCH_2—$$

(isobutyl), and $(CH_3)_3C—$ (*t*-butyl). The prefixes *sec*- (standing for "secondary") and *t*- (standing for "tertiary") are derived from the following designations of types of carbon atoms: (a) if a carbon is attached to three hydrogens it is called a "primary carbon"; (b) if it is attached to two hydrogens it is called a "secondary carbon;" (c) if it is attached to only one hydrogen it is called a "tertiary carbon;" and (d) if it carries no hydrogen but is attached to four carbons it is called a "quaternary carbon." The corresponding alkyl group is obtained by the removal of a hydrogen at the position in question. Thus, CH_3CH_3 contains primary carbon atoms and gives $CH_3CH_2—$ which is a primary alkyl group.

$$H_3C—CH_2—CH(CH_3)—C(CH_3)_2—CH_3$$

Primary Secondary Tertiary Quaternary

The structures and designations of the commonly encountered alkyl groups are shown in Fig. 3.8. As originally defined, the *neo*- prefix is applicable only to the five-carbon alkyl group and the *iso*- prefix only to alkyl groups having six or fewer carbon atoms. These restrictions are frequently disregarded, however, particularly in the case of the *iso*- prefix, which is often affixed to alkyl groups having seven or more carbons and carrying a pair of methyl groups attached to the penultimate carbon; *i.e.*,

$$(CH_3)_2CH(CH_2)_n—$$

Similarly, the *neo*- prefix is sometimes affixed to alkyl groups having six or more carbons and carrying three methyl groups attached to the penultimate carbon; *i.e.*,

$$CH_3—C(CH_3)(CH_2)(CH_2)_n—$$

The names of the lower members of the alkanes, as well as their derived groups, have already been used without comment. The names of the first four members originated in various ways. Methane is related to methyl alcohol, called "wood alcohol", so named because it can be obtained by the distillation

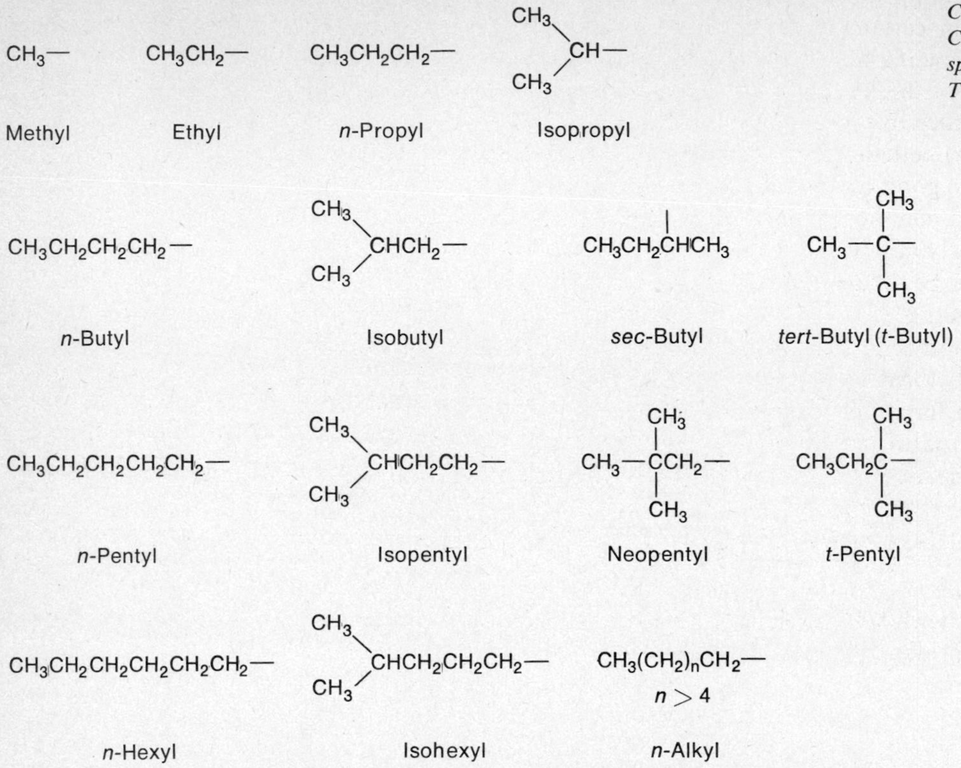

CH_3— Methyl

CH_3CH_2— Ethyl

$CH_3CH_2CH_2$— *n*-Propyl

$(CH_3)_2CH$— Isopropyl

$CH_3CH_2CH_2CH_2$— *n*-Butyl

$(CH_3)_2CHCH_2$— Isobutyl

$CH_3CH_2CH(—)CH_3$ *sec*-Butyl

$(CH_3)_3C$— *tert*-Butyl (*t*-Butyl)

$CH_3CH_2CH_2CH_2CH_2$— *n*-Pentyl

$(CH_3)_2CHCH_2CH_2$— Isopentyl

$(CH_3)_3CCH_2$— Neopentyl

$CH_3CH_2C(CH_3)_2$— *t*-Pentyl

$CH_3CH_2CH_2CH_2CH_2CH_2$— *n*-Hexyl

$(CH_3)_2CHCH_2CH_2CH_2$— Isohexyl

$CH_3(CH_2)_nCH_2$— $n > 4$ *n*-Alkyl

Fig. 3.8. Condensed representations of commonly encountered alkyl groups and their designations.

of wood (Greek, *methy*, "wine"; *hylé*, "wood"). Ethane is the two-carbon relative of ethyl ether, so named because of its flammable characteristics (Greek *aithein*, "to kindle or blaze"). Propane is related to the three-carbon compound propionic acid, so named because it is the first compound to exhibit "fatty acid properties" (Greek, *proto*, "first"; *piōn*, "fat"). Butane is related to butyric acid, so named because it is found in rancid butter (Latin, *butyrum*, "butter"). From the five-carbon compound on, however, the names are systematically based on the appropriate Greek or Latin number prefix, to which the suffix *-ane* is added (to designate the alkane) or *-yl* (to designate the group) (see Table 3.1; for a more complete list, see Table 18.2).

Table 3.1 NAMES OF ALKANES AND ALKYL GROUPS

Carbons	Alkane	Alkyl Group	Carbons	Alkane	Alkyl Group
1	Methane	Methyl	7	Heptane	Heptyl
2	Ethane	Ethyl	8	Octane	Octyl
3	Propane	Propyl	9	Nonane	Nonyl
4	Butane	Butyl	10	Decane	Decyl
5	Pentane	Pentyl	11	Undecane	Undecyl
6	Hexane	Hexyl	12	Dodecane	Dodecyl

One method for naming more complex alkanes, somewhat limited in its utility but still useful in certain instances, is designated the "methane-based system." In this system, compounds are considered as derivatives of methane in which one or more of the hydrogens of methane have been substituted by alkyl groups. For instance, propane can be called dimethylmethane, *n*-butane can be called ethylmethylmethane, neopentane can be called tetramethylmethane, etc., although there is no good reason for using these names in place of the more common ones. Where a common name does not apply, however, the methane-based system is frequently applicable. Thus, the two isomers of hexane that are unnamed in Fig. 3.7 can be designated as diethylmethylmethane and isopropyldimethylmethane, respectively.

A more systematic nomenclature that is applicable to *all* alkanes is based on rules that were originally formulated in 1892 at a meeting of chemists in Geneva, Switzerland, and which have since been modified at subsequent meetings of a group known as the International Union of Pure and Applied Chemistry (IUPAC). The approach that was devised at these meetings was originally called the Geneva System, but more recently has come to be called the IUPAC System. The IUPAC rules for naming alkanes are as follows:

1. The longest continuous chain in the alkane is selected as the basis for designating what is called the "parent compound," and the actual compound is then considered to be a derivative of the parent compound. For instance, isohexane,

$$\begin{array}{l} CH_3 \diagdown \\ \quad\quad CHCH_2CH_2CH_3 \\ CH_3 \diagup \end{array}$$

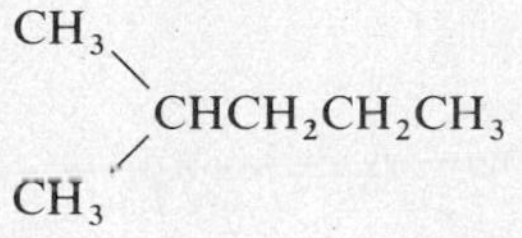

has as its longest chain a five-carbon sequence; it is, therefore, named as a methyl-substituted pentane.

2. The positions of the substituents are designated by numbers, the parent alkane being numbered from one of its two ends in accordance with rule 3. For instance, isohexane is designated as 2-methylpentane, and the other pentane-derived isomer of hexane is designated as 3-methylpentane; *i.e.*,

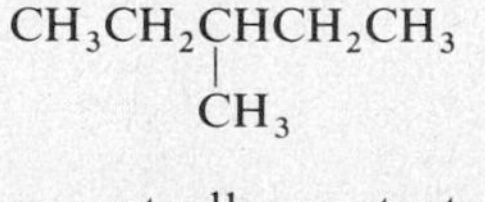

$$\begin{array}{c} CH_3CH_2\underset{\displaystyle CH_3}{\underset{|}{C}}HCH_2CH_3 \end{array}$$

3. The numbering of the parent alkane starts from the end that gives the substituent the lower number. For instance, isohexane is 2-methylpentane and not 4-methylpentane. If two or more substituents are present, the numbering of the parent alkane starts from the end that gives the lower number to the substituent that is encountered first. For example,

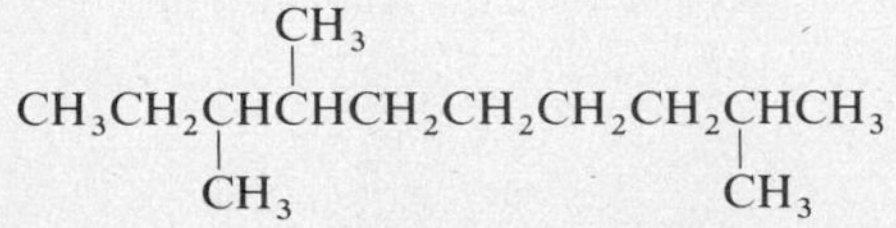

$$CH_3CH_2\underset{\displaystyle CH_3}{\underset{|}{C}}H\overset{\displaystyle CH_3}{\overset{|}{C}}HCH_2CH_2CH_2CH_2\underset{\displaystyle CH_3}{\underset{|}{C}}HCH_3$$

is 2,7,8-trimethyldecane and not 3,4,9-trimethyldecane (even though the sum of the numbers is lower in this name). If identical groups are encountered at identical distances from the two ends, groups further along the chain then determine the numbering. For example,

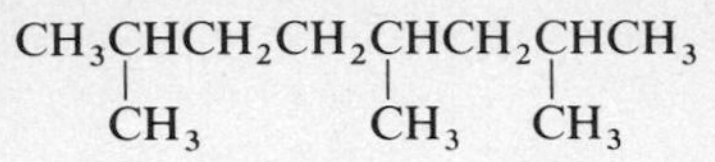

$$\begin{array}{l} CH_3\underset{|}{C}HCH_2CH_2\underset{|}{C}HCH_2\underset{|}{C}HCH_3 \\ \quad\; CH_3 \qquad\qquad CH_3 \;\; CH_3 \end{array}$$

is 2,4,7-trimethyloctane rather than 2,5,7-trimethyloctane.

4. The choice among parent chains of equal length is made in favor of the most highly substituted one. For example,

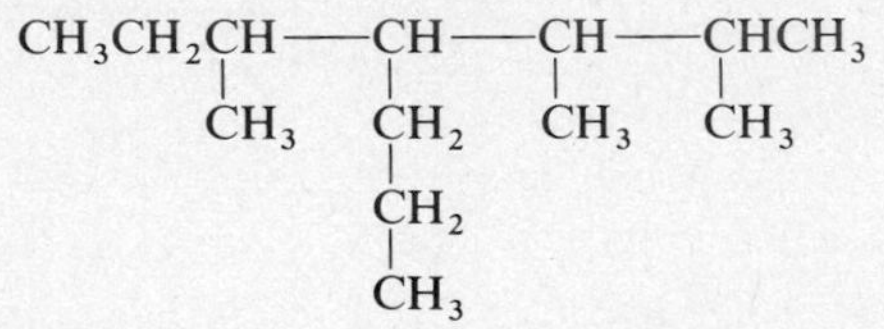

$$\begin{array}{l} CH_3CH_2\underset{|}{C}H—\underset{|}{C}H—\underset{|}{C}H—\underset{|}{C}HCH_3 \\ \qquad\quad CH_3 \;\; CH_2 \;\; CH_3 \;\; CH_3 \\ \qquad\qquad\quad\;\; | \\ \qquad\qquad\;\; CH_2 \\ \qquad\qquad\quad\;\; | \\ \qquad\qquad\;\; CH_3 \end{array}$$

is correctly named 2,3,5-trimethyl-4-propylheptane* and not 4-*sec*-butyl-2,3-dimethylheptane or 4-(1,2-dimethylpropyl)-3-methylheptane.

5. If the same alkyl group occurs two or more times as a substituent group, the prefix di, tri, tetra, etc., is used to designate the number present. For instance, neohexane

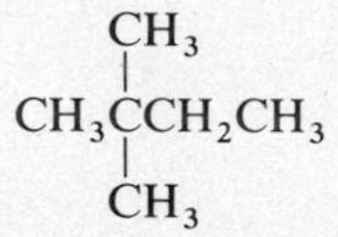

$$\begin{array}{c} CH_3 \\ | \\ CH_3CCH_2CH_3 \\ | \\ CH_3 \end{array}$$

can be named 2,2-dimethylbutane.

6. Substituents may be named by the use of the common alkyl group designations (see Fig. 3.8) when possible. Alternatively, however, substituents can be named according to the same rules just outlined. For instance,

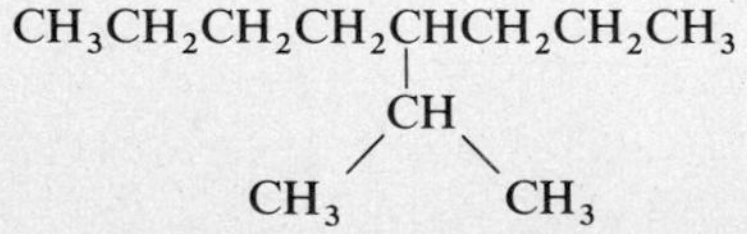

$$\begin{array}{l} CH_3CH_2CH_2CH_2\underset{|}{C}HCH_2CH_2CH_3 \\ \qquad\qquad\qquad\; CH \\ \qquad\qquad\quad\; / \quad \backslash \\ \qquad\qquad CH_3 \qquad CH_3 \end{array}$$

can be called 4-isopropyloctane or 4-(1-methylethyl)-octane. In the second of these names, the isopropyl group has been specified by the IUPAC rules, designating the parent portion of the substituent as the two-carbon chain (ethyl) and numbering the chain so that the *carbon directly attached to the main chain is 1.*

---

* Alkyl groups containing three or more carbon atoms are assumed in the IUPAC system to be unbranched (*i.e.*, *n*−) unless otherwise noted. Thus, it is sufficient to specify the substituent at the 4-position of this compound as "propyl" rather than "*n*-propyl."

7. If two or more different alkyl groups are attached to the parent compound, their sequence in naming follows alphabetical order. For example,

```
            CH3  CH2CH2CH3
             |    |
CH3CH2CHCHCHCH2CH2CH2CH3
             |
            CH2
             |
            CH3
```

is named 4-ethyl-3-methyl-5-propylnonane. The substituent groups are arranged alphabetically regardless of the number of each that are present; thus, dimethyl, trimethyl, etc., are indexed under "m" (for methyl) rather than under "d" (for di), "t" (for tri), etc. A compound group name, however, is treated as a unit; thus, the substituent group "1,2-dimethylpropyl" is indexed under "d" rather than "p". For example, the compound

```
                         CH3
                          |
                   CH3   CHCH3
                    |     |
            CH3    CH2   CHCH3
             |      |     |
CH3CH2C——CH—CH——CH2CH2CH2CH3
             |
            CH3
```

is named 5-(1,2-dimethylpropyl)-4-ethyl-3,3-dimethylnonane. Alkyl groups preceded by the suffixes *n*-, *sec*-, or *t*- are indexed according to the first letter in the name of the alkyl group. Isoalkyl and neoalkyl groups, on the other hand, are treated as a single word and are indexed under "i" and "n", respectively.

PROBLEM 3-6

Using the IUPAC system, name all the isomers of the heptanes for which you drew structures in Problem 3-5.

Nomenclature is an essential, although sometimes vexing, aspect of structural organic chemistry. Structural formulas do, of course, uniquely specify particular compounds and are themselves a kind of name. Many structural formulas, however, are so complex that to draw them out in detail as a nomenclature device is wasteful of time and space. Furthermore, they are pictures, not words, and are difficult to convey verbally. Although the systematic names may sometimes appear to be as space consuming as the structural formulas themselves (and in a few instances, actually are), this is generally not the case. Systematic names have the added virtue of allowing a strictly linear representation and the capability of being easily typed, printed, indexed, classified, scanned, and computerized. The nomenclature of organic compounds will, therefore, be stressed as one of the important facets of structural organic chemistry; by the time the reader has finished Part I, he should be reasonably adept at this often challenging game. A thorough grasp of the intricacies of nomenclature is an invaluable aid in acquiring a sound understanding of the structure of organic compounds.

3.7 *The Structures of 3-Methylhexane, 3,4-Dimethylhexane, and 3,4-Dimethylheptane: Mirror Image Isomerism*

Two types of isomerism have been considered to this point in the discussion—*viz*., conformational isomerism and structural isomerism. In one of the heptanes, still another kind of isomerism is possible as the result of the particular arrangement of groups around one of its carbon atoms. It is a kind of isomerism that was explicitly appreciated by the van't Hoff and LeBel theory of the tetrahedral carbon atom and intuitively understood, although expressed in terms somewhat foreign to the modern ear, some years earlier by Louis Pasteur. The particular heptane isomer designated as 3-methylhexane contains at C-3 a carbon atom that is substituted by *four different* groups, *i.e*., hydrogen, methyl, ethyl, and propyl. As a consequence of the tetrahedral orientation of these groups at the C-3 carbon, two *nonsuperimposable mirror image forms* of this compound exist (Fig. 3.9). Every real object, of course, has a mirror image. Sometimes the mirror

Nonsuperimposable mirror images — 3-Methylhexane

Superimposable mirror images — 3,3-Dimethylpentane

images are superimposable (*i.e*., identical), as, for instance, in the case of a baseball bat, a spoon, or a nail; but, in other instances they are nonsuperimposable, as, for instance, in the case of a golf club, a glove, or a propeller (the mirror image of a propeller with a right-handed helical twist is a propeller with a left-handed helical twist). It turns out that for alkanes containing one to six carbons, the mirror images of all the members are *superimposable* on the originals (disregarding conformational isomers—a point that will be discussed later in this section). This means that in these cases the *mirror images are identical* and cannot be differentiated in any conceivable way. For 3-methylhexane, however, the mirror images are *not* identical; they are *not* superimposable and can, in fact, be distinguished by appropriate physical probes.

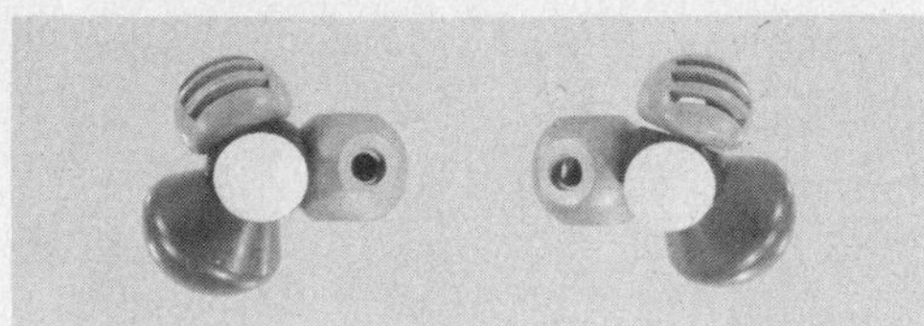

Nonsuperimposable mirror images

Fig. 3.9. Superimposable and nonsuperimposable mirror images of heptane isomers.

It may not be immediately apparent to the reader that the mirror image forms of 3-methylhexane are, indeed, *non*superimposable, or that the mirror image forms of its isomer 3,3-dimethylpentane are *identical* (*i.e.*, superimposable). To appreciate this fact, we must perceive the molecules as three-dimensional entities. Except for those rare individuals who are gifted with the ability to translate easily in their mind's eye the two-dimensional representations on the page to the three-dimensional reality, three-dimensional models of some sort are a necessary tool. The types of models available have already been mentioned (see Section 2.7), and the desirability of using such models is again emphasized. Without them most readers will have considerable difficulty in gaining a true understanding of the three-dimensional architecture of organic molecules.

PROBLEM 3-7

For each of the objects shown, draw the mirror image and indicate whether the mirror images are superimposable or nonsuperimposable.

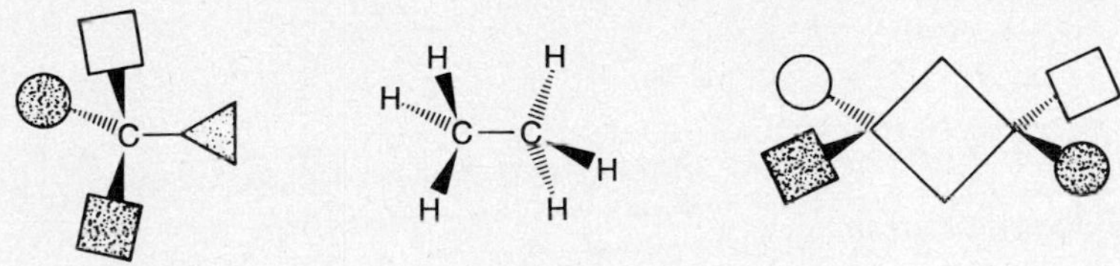

The physical probe that is most often used to detect a difference between a pair of compounds bearing a nonsuperimposable mirror image relationship to each other is a beam of plane-polarized light. When passed through a solution containing molecules of one of the mirror image forms, the plane of polarization is rotated. If the beam of polarized light is passed through a solution containing molecules of the other mirror image form, the plane of polarization is again rotated but in the opposite sense; *i.e.*, if one mirror image form rotates the plane of polarization in a clockwise direction, the other mirror image form rotates it to an equal extent in a counterclockwise direction. Because of this effect on plane-polarized light, compounds possessing isomerism of this type are sometimes referred to as optical isomers and are said to be *optically active.*

Any compound that can exist in nonsuperimposable mirror image forms has the potentiality for exhibiting optical activity. As we have just seen, this circumstance arises when a carbon atom in a molecule is attached to *four different* atoms or groups. It may also arise from other structural features. For example, the synclinal conformer of *n*-butane possesses a pair of mirror image forms (Fig. 3.10) that are nonsuperimposable, and a sample composed entirely of one or the other of these mirror images would show optical activity. This is impossible to demonstrate experimentally under ordinary conditions, however, because of the rapid rate of interconversion between the nonsuperimposable mirror image conformers. If it were possible by some laboratory legerdemain to prepare a sample containing only one of the mirror image forms of synclinal *n*-butane, it would be

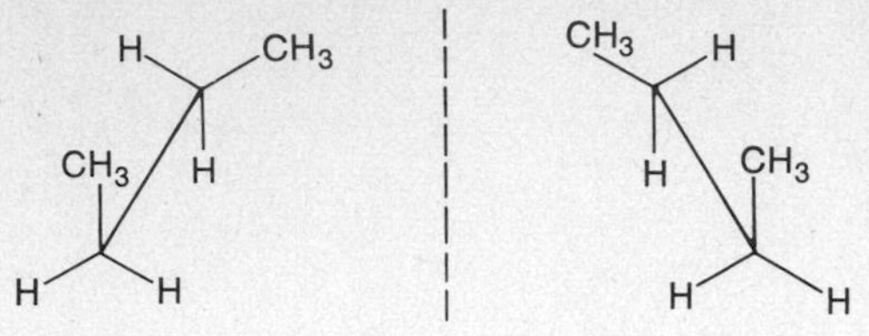

Fig. 3.10. Nonsuperimposable mirror forms of synclinal *n*-butane.

necessary to measure its optical activity almost instantaneously. Within a small fraction of a second at room temperature, an equilibrium mixture is established, containing *equal* amounts of the two synclinal conformations (along with other conformations, such as the antiperiplanar, anticlinal, synperiplanar, etc.); as a result, optical activity vanishes. Although mirror image isomerism arising from conformational isomerism is theoretically possible for many compounds, it is experimentally demonstrable only in those cases in which the rate of rotation around the C—C bond is slow enough to permit each of the nonsuperimposable mirror image forms to exist for an appreciable length of time (*i.e.*, at least long enough for a measurement of the effect on plane-polarized light). The rare circumstances in which this occurs will be discussed in detail in Chapter 19. In the majority of instances, the rate of C—C bond rotation is sufficiently rapid that mirror image isomerism arising from conformational isomerism can be neglected.

Compounds that are structurally identical (*i.e.*, all the atoms are bonded in the same fashion) but that differ in the way in which the atoms within the molecule are oriented in space are called *stereoisomers* (Greek, *stereos*, "solid"). Conformational isomerism and mirror image isomerism both arise as a consequence of the spatial arrangements of atoms within molecules and, therefore, are varieties of stereoisomerism. The term "stereoisomerism," however, is generally restricted to mirror image isomerism, *cis, trans* isomerism (see p. 72), and (*Z, E*) isomerism (see p. 96), a custom that we shall follow here. One of the features giving rise to mirror image isomerism, as we have seen, is the presence of a carbon atom to which four different atoms or groups are attached. Such a carbon is designated as a *chiral carbon* [Greek, *χειρ* (*cheir*), "handedness"], and optically active molecules are called "*chiral molecules.*" The presence of chiral carbons, however, is not necessarily sufficient to render the entire molecule chiral and optically active. For example, 3,4-dimethylhexane possesses *two* chiral carbons, each of which is substituted with the same set of four different atoms and groups [*i.e.*, H, CH_3, C_2H_5, and $CH(CH_3)C_2H_5$]. If the groups around the two chiral carbons are attached in the fashion shown in structure I in Fig. 3.11, a pair of nonsuperimposable mirror image forms [I(a) and I(b)] is possible and mirror image isomerism is present. If the groups are attached in the fashion shown in structure II in Fig. 3.11, however, the two mirror image forms are *identical* (*i.e.*, superimposable), and only a single entity exists; mirror image isomerism is absent. Thus, 3,4-dimethylhexane has three stereoisomers, two of which are optically active [*i.e.*, I(a) and I(b)] and one of which is optically inactive (*i.e.*, II). The mirror image isomers I(a) and I(b) are said to be *enantiomers* of each other; the optically inactive isomer II, which is *not* the mirror image of either I(a) or I(b), is said to be a

Nonsuperimposable mirror images — I (a), (b) — Superimposable mirror images — II (a), (b)

Fig. 3.11. Stereoisomers of 3,4-dimethylhexane.

diastereomer of I(a) and I(b). Similar relationships prevail in all sets of stereoisomeric compounds, and any particular member can be specified as being *enantiomeric* or *diastereomeric* in relation to the other members of the set.

3,4-Dimethylheptane, the next higher homolog of 3,4-dimethylhexane, resembles its lower homolog in having two chiral carbons. In this case, however, the chiral carbons are *not* identically substituted. Consequently, for each of the arrangements of the groups around the chiral carbons a pair of nonsuperimposable mirror image forms exists, giving rise to *four* optically active stereoisomers (Fig. 3.12). The stereoisomeric relationships among the members of this set are shown in the following tabulation:

	I(a)	I(b)	II(a)	II(b)
I(a)	—	Enantiomer	Diastereomer	Diastereomer
I(b)	Enantiomer	—	Diastereomer	Diastereomer
II(a)	Diastereomer	Diastereomer	—	Enantiomer
II(b)	Diastereomer	Diasteromer	Enantiomer	—

Clearly, any member of a stereoisomeric set can be enantiomeric with *only one* other member of the set and is diastereomeric with *all* the remaining members of the set.

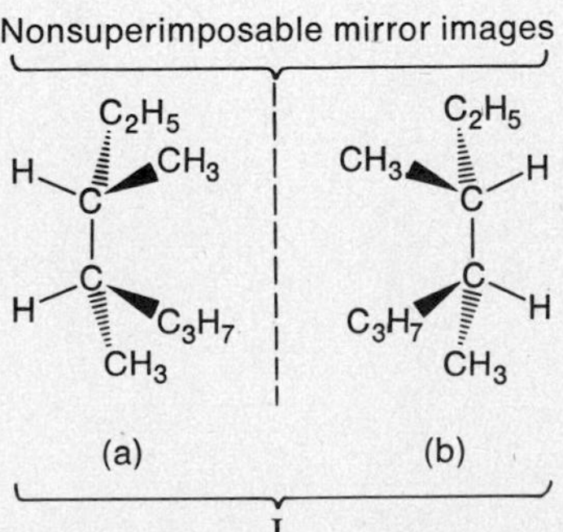

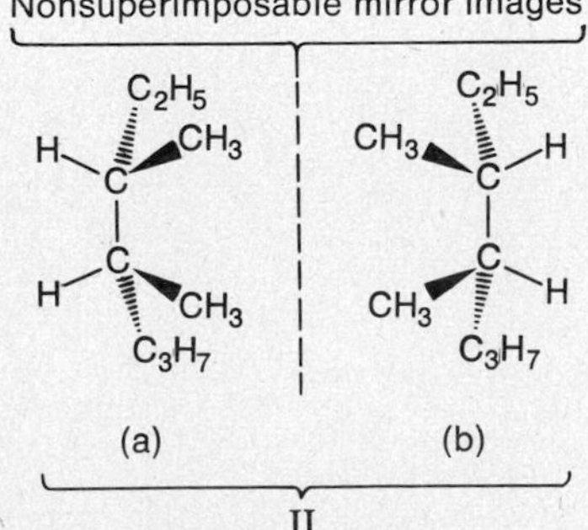

Fig. 3.12. Stereoisomers of 3,4-dimethylheptane.

PROBLEM 3-8

The smallest alkane in which mirror image stereoisomerism is possible, as we have seen, is one of the heptanes. If one of the hydrogens of an alkane is replaced by a chlorine atom (*i.e.*, a $C_nH_{2n-1}Cl$ compound), what is the simplest molecule in which nonsuperimposable mirror images are possible? If two of the hydrogens of an alkane are replaced by chlorine atoms (*i.e.*, $C_nH_{2n}Cl_2$ compound), what is the smallest molecule containing two chiral carbon atoms? How many stereoisomers are possible for this compound?

The two partners of an enantiomeric pair can be experimentally characterized on the basis of their effect on plane-polarized light. The enantiomer that rotates the plane to the right (*i.e.*, clockwise) is referred to as the *dextrorotatory* enantiomer and is specified as (+); the enantiomer that rotates the plane to the left (*i.e.*, counterclockwise) is referred to as the *levorotatory* enantiomer and is specified as (−). A mixture of equal numbers of molecules of (+)- and (−)-enantiomers is often referred to as a *racemate** and is specified as (±). Since the absolute magnitude of the dextrorotation and levorotation for a pair of enantiomers is identical, the *racemate is optically inactive*. The particular arrangement of the four substituents attached to a chiral atom is referred to as the *configuration* of the chiral center, the configuration of the (+)-enantiomer being the opposite (*i.e.*, the mirror image) of that of the (−)-enantiomer. The currently used nomenclature scheme for specifying the configuration of chiral centers is discussed in detail in Chapter 19.

The dextrorotatory and levorotatory forms for any enantiomeric pair have identical physical properties under ordinary conditions. For example, the melting point, boiling point, refractive index, spectral properties, solubility, etc., for (+)-3-methylhexane are *exactly the same* as those for (−)-3-methylhexane. Only if *chiral probes*, such as plane-polarized light, are used can any difference between enantiomers be detected. For this reason, it is difficult to separate enantiomers; the ordinary techniques of separation are to no avail, and the special methods discussed in Chapter 19 must be employed. Compounds related in a diastereomeric manner, on the other hand, *do* show differences in physical and chemical properties under ordinary conditions. For example, the melting point, boiling point, refractive index, spectral properties, solubility, etc., for the optically inactive isomer of 3,4-dimethylhexane (compound II in Fig. 3.11) are different from those of the optically active isomer [compounds I(a) and I(b) in Fig. 3.11]; compound II is diastereomeric with compounds I(a) and I(b) as well as the optically inactive racemate comprising I(a) and I(b). Similarly, the racemate [*i.e.*, the (±)-mixture of enantiomers] of 3,4-dimethylheptane represented by compound I in Fig. 3.12 is diastereomeric with the racemate represented by compound II in Fig. 3.12, and can be separated from it by ordinary techniques, such as distillation.

---

* The lowercase letters *d* and *l* are often used to designate dextrorotatory and levorotatory, respectively; thus, a racemate is designated as a *dl* compound with these prefixes.

But to separate the (+)-enantiomer from the (−)-enantiomer of either of the racemates requires special methods.

Stereochemistry is a fascinating and subtle facet of chemical structure, and it deserves a more detailed treatment than has been accorded by this brief discussion. Rather than delay the discussion of other general features of carbon frameworks by an extended consideration of stereochemistry at this point, however, we defer the more detailed treatment of this topic to Chapter 19.

3.8 Cyclic Alkanes Containing Three-Membered, Four-Membered, and Five-Membered Rings

All the alkanes discussed so far have been acyclic compounds (*i.e.*, containing *no* cyclic or ring structures). It is also possible, however, for sp^3-hybridized carbon* atoms to be arranged in a ring. Thus, three sp^3-hybridized carbons can be arranged in a three-membered ring, four sp^3-hybridized carbons in a four-membered ring, five sp^3-hybridized carbons in a five-membered ring, etc. [Fig. 3.13]. Compounds

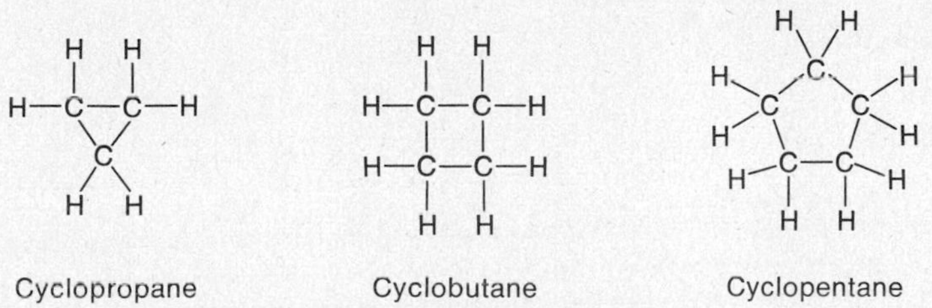

Cyclopropane　　Cyclobutane　　Cyclopentane

Fig. 3.13. Solid line representations for cyclic alkanes.

of this type are designated as carbocyclic† and are named by affixing the prefix "cyclo" to the name of the straight chain alkane containing the same number of carbons as the ring system; *i.e.*, C_3H_6 is cyclopropane, C_4H_8 is cyclobutane, and C_5H_{10} is cyclopentane, and the corresponding cycloalkyl groups $C_3H_5—$, $C_4H_7—$, and $C_5H_9—$ are cyclopropyl, cyclobutyl, and cyclopentyl, respectively.

Cyclopropane, of necessity, is a strictly planar molecule, but cyclobutane and cyclopentane are not so severely constrained. Data from thermodynamic and spectral studies indicate, in fact, that cyclobutane and cyclopentane are mixtures of rapidly oscillating, nonplanar conformers. Cyclobutane exists predominantly in a "butterfly" conformation (Fig. 3.14), in which the dihedral angle between the two wings is about 20°. Cyclopentane exists in bent ("envelope") and twisted ("twist") conformations, in which one or two carbons (Fig. 3.14) are out of the plane comprised by the remaining carbons of the ring; the identity of the "out-of-plane" carbon atoms constantly changes as the result of a twisting motion of the ring, a process known as "pseudorotation." In the planar forms of cyclobutane and cyclopentane, all the adjacent CH_2 groups are in eclipsed

---

* The designation sp^3-hybridized carbon is used, somewhat loosely, to designate carbons that are single bonded to four other atoms. In actual fact, the carbons in some cycloalkanes are not precisely sp^2-hybridized (see p. 51), and a better term in this context is "tetraliganded carbon," specifying that the carbon is attached (*i.e.*, liganded) to four other atoms.

† The term "carbocyclic" denotes a cyclic array of atoms in which all the atoms are carbon. Cyclic arrays in which atoms other than carbons are included are called "heterocyclic."

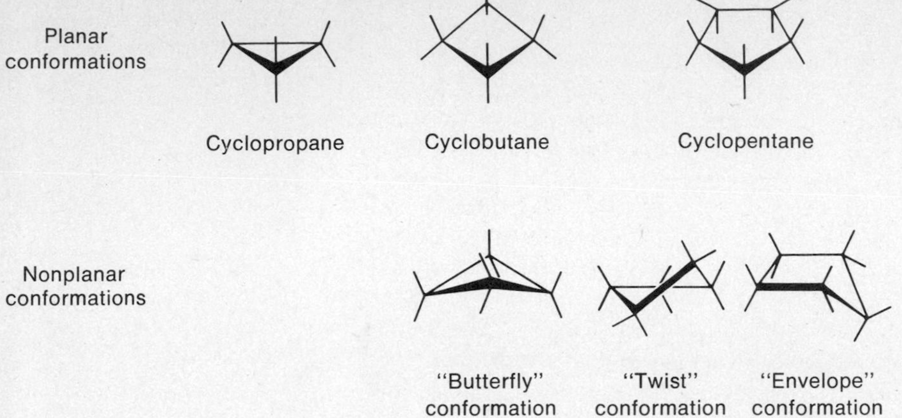

Fig. 3.14, Pseudo-three-dimensional representations for cyclopropane, cyclobutane, and cyclopentane.

conformations, and the nonbonded strain of this conformation is relieved to some extent when the molecule is twisted into a nonplanar form. Twisting, however, requires energy, because the C—C—C bond angles change. The nonplanar structure that is most stable, therefore, represents a compromise between the nonbonded strain (also called Pitzer strain) and twisting or torsional strain, generally called angle strain. On a time-averaged basis, cyclobutane and cyclopentane rings might be considered to be planar. Although this is a convenient assumption for certain structural discussions, remember that it is not really the case; at any given instant in time the rings are most likely to be nonplanar.

Cyclopropane, the smallest carbocyclic ring system, has C—C—C bond angles of 60°. A carbon that is hybridized in the sp^3 fashion, with an interorbital angle of 109°28′, cannot accommodate to the required bond angles of such a ring system, and several descriptions for the bonding in cyclopropane have been proposed to handle this problem. In all these descriptions the interorbital angle on the carbon atoms is shown to be greater than the internuclear angle of 60°, a situation that causes "bond-angle strain" (often designated as internal or I-strain).

Substitution of one of the hydrogens of cyclopropane by a methyl group gives methylcyclopropane, a compound for which only one structural isomer is possible. Substitution of two of the hydrogens of cyclopropane by methyl groups, however, allows two isomeric possibilities—*viz*., 1,1-dimethylcyclopropane and 1,2-dimethylcyclopropane. In the latter case, stereoisomerism also enters the picture; 1,2-dimethylcyclopropane contains two identically substituted chiral carbons, resembling 3,4-dimethylhexane (see p. 69) by existing in three stereoisomeric forms—two optically active forms and one optically inactive form (Fig. 3.15).

The prefixes *cis* and *trans* are frequently applied to disubstituted cycloalkanes to indicate the spatial relationship between the groups attached to the ring. The prefix *cis* (Latin, *cis*, "on this side") designates the compound in which the reference groups (*i.e.*, the methyl groups in 1,2-dimethylcyclopropane) are on the

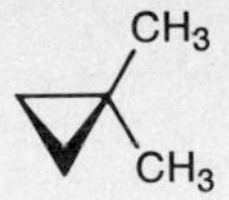

1,1-Dimethylcyclopropane

Nonsuperimposable mirror images
trans-1,2-Dimethylcyclopropane

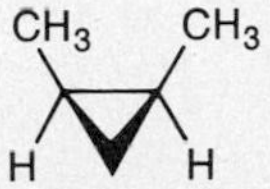

Optically inactive isomer
cis-1,2-Dimethylcyclopropane

Fig. 3.15. Structures of dimethylcyclopropanes.

same side of the ring; the prefix *trans* (Latin, *trans*, "across") designates the compounds in which the reference groups are on *opposite sides* of the ring.

Substitution of hydrogens on cyclobutane and cyclopentane by alkyl groups leads to structural possibilities similar to those for cyclopropane. Thus, there is only one methylcyclobutane and methylcyclopentane but several dimethylcyclobutanes and dimethylcyclopentanes. The dimethylcyclobutanes include 1,1-dimethylcyclobutane, *cis*- and *trans*-1,2-dimethylcyclobutane (the *trans* isomer has nonsuperimposable mirror image forms), and *cis*- and *trans*-1,3-dimethylcyclobutane; the dimethylcyclopentanes include 1,1-dimethylcyclopentane, *cis*- and *trans*-1,2-dimethylcyclopentane (the *trans* isomer has nonsuperimposable mirror image forms), and *cis*- and *trans*-1,3-dimethylcyclopentane (the *trans* isomer has nonsuperimposable mirror image forms). For practice, draw structures for each of these compounds.

PROBLEM 3-9

Draw the structures of all the monobromo cycloalkanes having the molecular formula C_4H_7Br. Which of these can have one, or more, chiral atoms? Draw the nonsuperimposable mirror image forms of these compounds.

3.9 Cyclic Alkanes Containing Six-Membered Rings

The history of chemists' insight into the structure of cyclohexane is an interesting example of how misinterpretations of theory and experiment can delay for many years the emergence of correct ideas. In the wake of the van't Hoff and LeBel theory of the tetrahedral carbon atom, Adolph von Baeyer in 1885 proposed planar structures for the cycloalkanes which, quite logically, were constructed from tetrahedral carbon atoms. He correctly ascribed angle strain to planar compounds such as cyclopropane and cyclobutane, in which the internuclear angle must be less than 109°28′. For cyclohexane, in extension of his hypothesis, he postulated that strain should also be present, because if the molecule is planar the internuclear angle must be 120°, *i.e.*, greater than the preferred tetrahedral angle. Hermann Sachse suggested five years later, however, that cyclohexane may *not* be a planar molecule but a puckered molecule, in which the tetrahedral angles are *retained.* He recognized the possibility of *two* puckered or nonplanar forms, structures to which present-day chemists have given the names of *chair form* and *boat form* (Fig. 3.16). Unfortunately, Sachse assumed that the chair

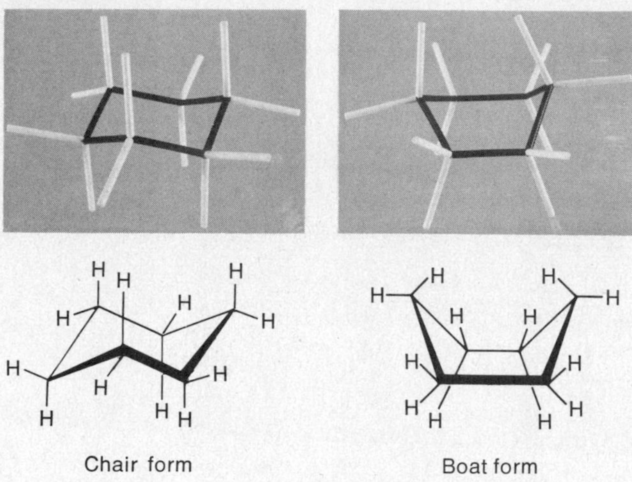

Fig. 3.16. Framework models and pseudo-three-dimensional representations of chair and boat conformations of cyclohexane.

form was a rigid, stable entity incapable of interconverting to other forms, and he postulated the existence of two stable, isolable isomers of *trans*-1,4-disubstituted cyclohexanes.

When subsequent experiments failed to produce such compounds, Sachse's theory fell into disrepute, and for several decades afterward, organic chemists accepted the theory of the average planar cyclohexane ring. As a result of the chemical experiments of Walter Hückel in 1926, and the various observations by physical chemists in the 1930's and 1940's the validity of the Sachse hypothesis was substantiated. Sachse's only error was in asserting that isomers of *trans*-1,4-disubstituted cyclohexanes should be stable and isolable. They do, indeed, exist, but they interconvert so rapidly (several hundred or more times a second at room temperature) that they cannot be isolated under ordinary conditions. In 1950, the English chemist Derek H. R. Barton gathered these pieces of information and presented them to the chemical world as the basis of what has come to be known as the conformational theory. For this contribution he was awarded a Nobel Prize in 1969, which he shared with Otto Hassel, who also contributed to our modern concepts of conformation by his X-ray crystallographic investigations of chlorocyclohexanes.

Recognizing that a molecule of cyclohexane is a nonplanar structure during most of its moments of existence, we must ask which of its nonplanar forms is most stable. We can answer this question without recourse to direct experiment if we simply consider the nature of the nonbonded interactions in the chair and boat

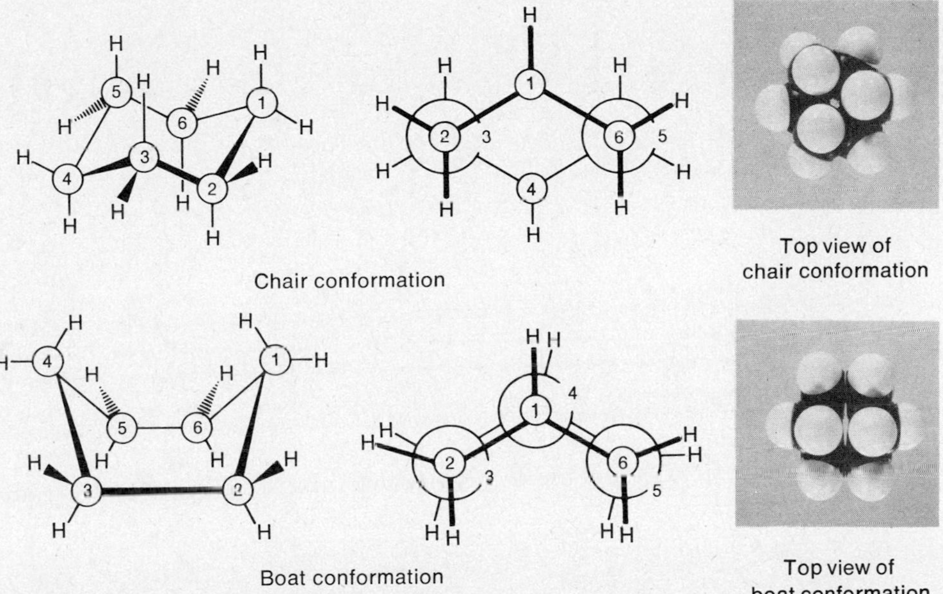

Fig. 3.17. Pseudo-three-dimensional projections and space-filling models of chair and boat conformations of cyclohexane.

forms (Fig. 3.17). If we focus attention on a four-carbon segment of the chair form of cyclohexane and consider its butane-like conformation,* we realize that it is synclinal. Similarly, each of the remaining five butane-like conformations of

Viewer

C-1
CH_2-
C-2
C-4 CH_2- C-3

Synclinal

the chair conformation is found to be synclinal. If we use as our point of reference a hypothetical cyclohexane molecule in which all the conformations are of the

* The reader is urged to resort to three-dimensional molecular models as an aid in appreciating the conformational interactions in cyclohexane.

antiperiplanar variety (a physical impossibility), the *additional* strain energy resulting from six synclinal conformations would be 6 × 0.9 kcal/mole [where 0.9 kcal/mole is the contribution from one synclinal interaction (see Fig. 3.5)] or 5.4 kcal/mole.

A similar consideration of the six butane-like segments for cyclohexane in the boat form shows that four have synclinal conformations but two have synperiplanar conformations:

Synperiplanar

Taking 4.5 kcal/mole as a minimum value for the difference in energy between a synperiplanar conformation and an antiperiplanar conformation, we see that the boat form differs from the hypothetical all-antiperiplanar cyclohexane by 12.6 kcal/mole (*i.e.*, 4 × 0.9 + 2 × 4.5). Clearly, the chair form is more stable than the boat form. The actual difference in energy, however, is less than the calculated amount of 7.2 kcal/mole (*i.e.*, 12.6 − 5.4), because the "true" boat form on which the calculations are based is not the boat conformation that the molecule actually assumes. Instead, the molecule assumes the "twist" boat form to minimize the unfavorable interactions present in the "true" boat form (Fig. 3.18).

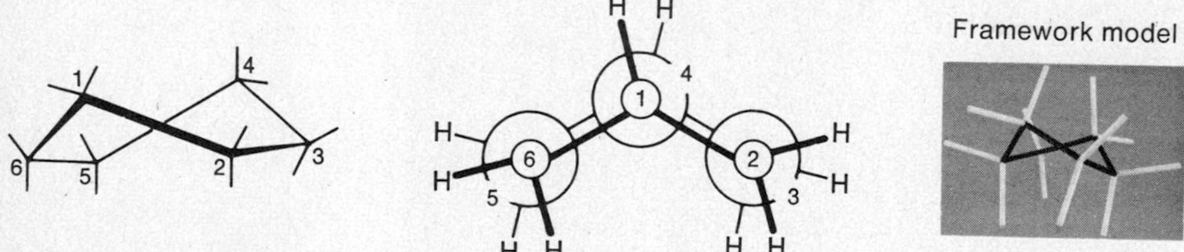

Fig. 3.18. Pseudo-three-dimensional and projection representations of twist boat conformation of cyclohexane.

Thus, the actual difference in energy between the chair form and the boat form is approximately 4.5 kcal/mole, which means that at room temperature only one molecule in about every thousand molecules is in the boat conformation; the other 999 are in the chair conformation.*

---

* Calculated from the expression $\Delta G = -2.303RT \log K$, where ΔG is the free energy difference between boat and chair conformations, R is the gas constant (in cal/deg mole), T is the absolute temperature, and K is the equilibrium constant: *e.g.*,

$$\log K = \frac{4{,}500}{2.303 \times 1.987 \times 300} \simeq 3$$

See Section 7.2.

Since the chair form is the conformation in which a cyclohexane molecule is most likely to be found at any given moment, a further consideration of the characteristics of this conformation is instructive. Inspection reveals that it contains hydrogens attached to the carbon framework in two different alignments; six are approximately perpendicular to the "average plane" of the ring, and the other six are roughly parallel with this "average plane." Those that are aligned perpendicular to the plane are called *axial hydrogens*, and we refer to the bonds as axial bonds. Those that are aligned in the plane are called *equatorial hydrogens*, and we refer to the bonds as equatorial bonds. The difference between axial and equatorial alignments assumes importance when substituents other than hydrogen are attached to the cyclohexane ring. For instance, methylcyclohexane can assume two different chair conformations (Fig. 3.19), and the

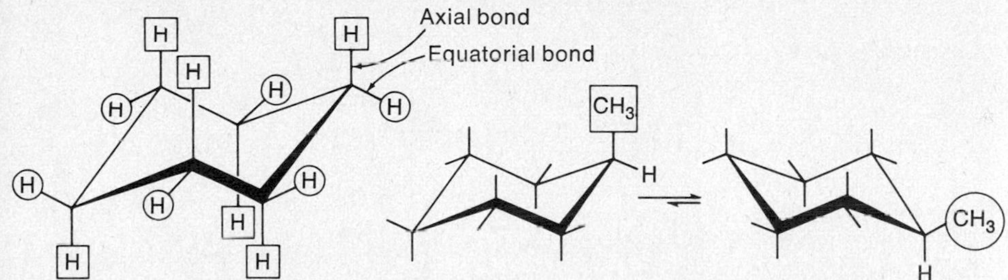

Fig. 3.19. Equatorial and axial bonds in cyclohexane and methylcyclohexane.

question is, "Which is the more stable?" We can again predict the answer without recourse to direct experiment. Considering the two additional butane-like segments that are introduced into the system by the presence of the methyl group, we find that if the methyl group is attached by an *equatorial* bond, two more antiperiplanar conformations are present, whereas if it is attached by an *axial* bond, two more synclinal conformations are present. From this we can calculate that the equatorial alignment is more stable than the axial alignment by two synclinal interactions (2×0.9 kcal/mole), which amounts to 1.8 kcal/mole. Thus, at room temperature, about 95% of the molecules have the methyl group in the equatorial alignment and only 5% in the axial alignment. The two forms are, however, readily interconvertible, and, as Sachse learned to his sorrow, it is not possible to isolate either form as a stable entity under ordinary conditions. Nevertheless, as is discussed in later chapters, the reaction characteristics of some molecules may be significantly influenced by their conformation. As might be anticipated, the magnitude of the equilibrium constant for the chair-boat interconversion of substituted cyclohexanes is dependent on the substituent. Thus, the equilibrium constant for the interconversion of methyl(axial)cyclohexane $\rightleftharpoons$ methyl(equatorial)cyclohexane is about 20; that for *t*-butyl(axial)cyclohexane $\rightleftharpoons$ *t*-butyl(equatorial)cyclohexane is 1000 or greater, whereas that for chloro(axial)cyclohexane $\rightleftharpoons$ chloro(equatorial)cyclohexane is only slightly greater than 2.

The conformation of methylcyclohexane in which the methyl group is axial not only has two additional synclinal interactions but also possesses a

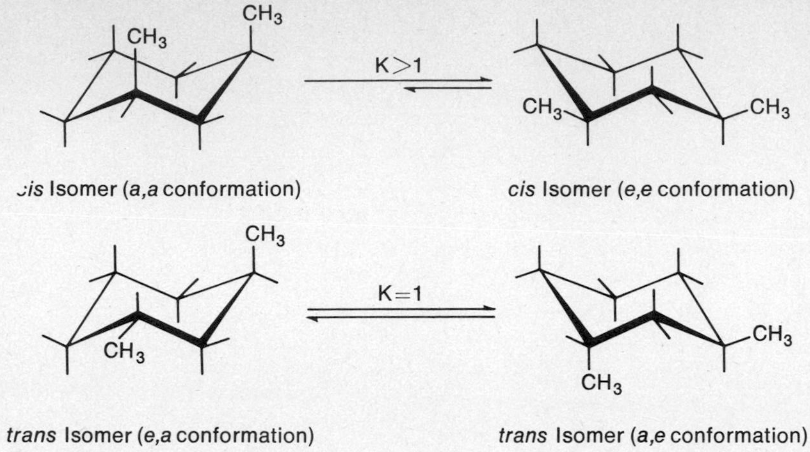

Fig. 3.20. Conformational isomers of *cis-* and *trans-*1, 3-dimethylcyclohexane (*e* stands for equatorial and *a* stands for axial).

certain amount of nonbonded strain arising from the interaction between the C-1 methyl group and the C-3 and C-5 hydrogens in the axial alignment. The magnitude of the strain resulting from 1,3-diaxial interactions increases, of course, as the interacting groups become larger. Thus, in *cis*-1,3-dimethylcyclohexane (Fig. 3.20), the preference of the diequatorial conformer over the diaxial conformer is somewhat enhanced by the 1,3-diaxial interaction between the methyl groups. In *trans*-1,3-dimethylcyclohexane, one of the methyl groups is in the axial alignment and the other is in the equatorial alignment. The rapidly interconverting chair conformers in this case are precisely equal in energy, the equilibrium constant for the process being 1.

PROBLEM 3-10

Using a set of framework molecular models, measure the distance (in millimeters) between the ends of the following pairs of bonds: (a) hydrogens axially aligned and attached to adjacent carbons; (b) hydrogens equatorially aligned and attached to adjacent carbons; (c) one hydrogen equatorially aligned and the other hydrogen axially aligned, attached to adjacent carbons; and (d) hydrogens axially aligned attached to C-1 and C-3.

3.10 *Cyclic Alkanes Containing Seven-Membered and Larger Rings*

Although the alkanes and cycloalkanes are among the least chemically reactive organic compounds, they undergo combustion to carbon dioxide and water in an atmosphere containing oxygen in a process familiar to anyone who has watched a candle burn, a gas stove operate, or an automobile run. The amount of heat evolved when a hydrocarbon is burned can be accurately measured and used to estimate the stability of the hydrocarbon (see Fig. 2.15). Thus, compounds that release a smaller amount of heat (*lower energy of combustion*) per CH_2 group are more stable than those that release a larger amount of heat (*higher energy of combustion*) per CH_2 group. The heats of combustion per CH_2 group for ethylene,

Table 3.2 Heats of combustion of cyclic alkanes

Cycloalkane	Heat of Combustion per CH_2 Group, kcal/mole
(Ethylene)	168.6
Cyclopropane	166.6
Cyclobutane	164.0
Cyclopentane	158.7
Cyclohexane	157.4
Cycloheptane	158.3
Cyclooctane	158.6
Cyclononane	158.8
Cyclodecane	158.6
Cyclopentadecane	157.5

cyclopropane, cyclobutane, cyclopentane, and cyclohexane (see Table 3.2) fall precisely in the order that would be predicted on the basis of the ideas set forth in Sections 3.8 and 3.9. Cyclohexane, for example, has no bond angle strain and less nonbonded strain than any of its smaller ring homologs, and, as a consequence, has the lowest heat of combustion per CH_2 group. Surprising, however, are the increased values noted for the higher homologs of cyclohexane. The compounds, cycloheptane through cyclotridecane (13 carbons in ring), belong to the "medium-size ring" group of cycloalkanes, and all contain more strain energy than cyclohexane. Although six sp^3-hybridized carbons fit perfectly into the chair form of a six-membered ring to give a minimum of nonbonded strain, seven to thirteen sp^3-hybridized carbons fit into rings somewhat less comfortably. Steric interactions between hydrogens *across the ring* are encountered, and the lowest energy conformation, as with the small ring compounds, represents a compromise between bond angle strain and nonbonded strain factors (Fig. 3.21). In compounds

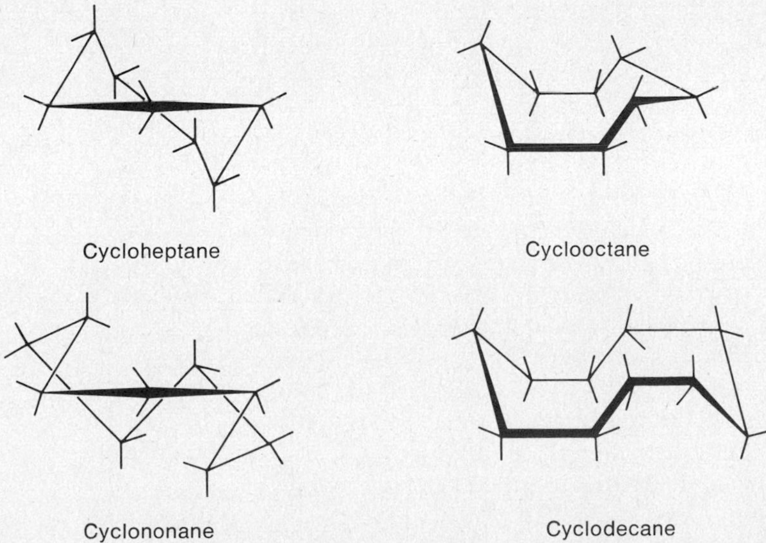

Fig. 3.21. Conformations of medium ring compounds.

containing rings larger than 13 carbons, the "hole" in the middle of the ring is large enough and the molecule flexible enough that *transannular* (Latin, *trans*, "across"; *annularis*, "ring") hydrogens do not interfere with each other; staggered conformations can be easily attained, and nonbonded strain is therefore minimized.

3.11 Polycyclic Alkanes

The reader may already have realized, perhaps as the result of experimenting with molecular framework models, that cyclic systems constructed from sp^3-hybridized carbons need not be limited to one ring. Polycyclic systems can also be constructed, and, indeed, a large number of compounds possessing this feature are known. Many important naturally occurring compounds contain polycyclic alkane rings.

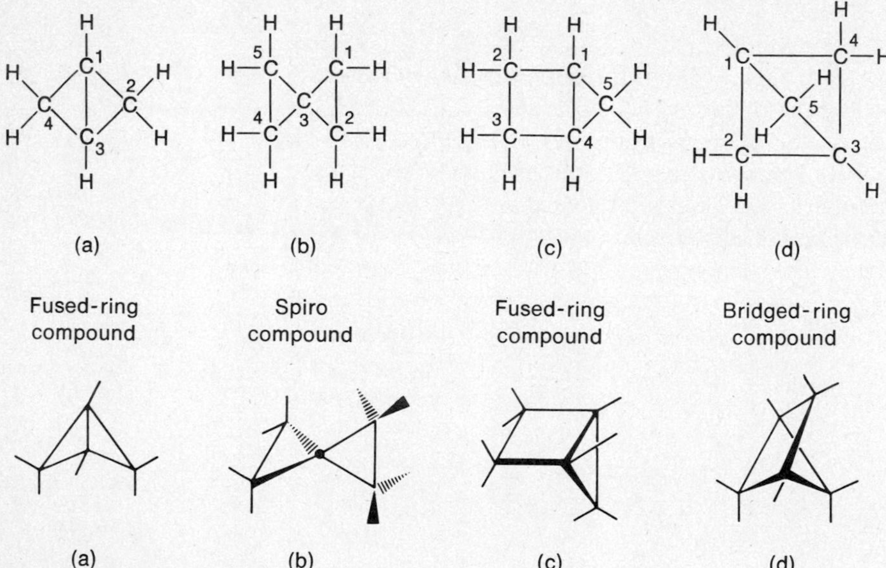

Fig. 3.22. Solid line and pseudo-three-dimensional representations of bicyclic compounds containing four and five carbon atoms.

The smallest possible bicyclic system consists of four sp^3-hybridized (tetraliganded*) carbon atoms joined in such a fashion that two cyclopropane rings are formed [Fig. 3.22(a)]. Although it possesses a large amount of strain energy and, as a consequence, is thermodynamically unstable relative to its less strained structural isomers, the compound has been prepared in the laboratory. Thermodynamic instability is usually reflected in chemical reactivity, and this compound is very readily converted to other substances. The next higher bicyclic homolog is a compound containing five sp^3-hybridized carbon atoms, for which three structural possibilities exist; *i.e.*, structures (b), (c), or (d) (Fig. 3.22). Bicyclic compounds in which both rings share *one common atom* [*e.g.*, C-3 in Fig. 3.22(b)],

* See footnote, p. 71.

are called "spiro" structures; those in which both rings share *two common atoms directly joined by a bond* are called "fused-ring" structures [*e.g.*, C-1 and C-3 in Fig. 3.22(a) and C-1 and C-4 in (c)]; and those in which both rings share *two common atoms not directly attached* [*e.g.*, C-1 and C-3 in Fig. 3.22(d)] are called "bridged-ring" structures.

Monocyclic alkanes are named simply by relating them to the corresponding acyclic alkanes having the same number of carbon atoms; *e.g.*, C_3H_6 is cyclopropane, C_6H_{12} is cyclohexane, and $C_{12}H_{24}$ is cyclododecane. This system of nomenclature cannot be carried over to the polycyclic series; therefore, the following scheme has been devised. The name for the bicyclic compound is assigned on the basis of the total number of *carbons* in the cyclic portion of the carbon framework, the prefix "spiro" or "bicyclo" is affixed, and the number of atoms in the chains between the carbons that are common to the two rings is specified in brackets placed between the prefix and suffix; the numbers are separated by periods and are arranged in *descending* order in the case of fused-ring and bridged-ring compounds and *ascending* order in the case of spiro compounds. In spiro compounds, only one carbon is common to both rings, and only two numbers are necessary to specify the lengths of the two bridges connected by the shared carbon. By convention, the smaller ring is numbered before the larger ring; the numbering of the system commences on the carbon atom of the smaller ring that is adjacent to the common carbon atom, and progresses around the smaller ring, through the common carbon atom, and around the larger ring (Fig. 3.23).

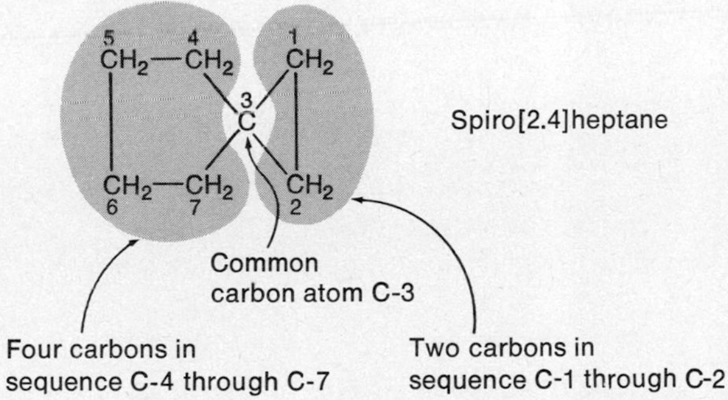

Fig. 3.23. Numbering system for spiro compounds.

In fused-ring and bridged-ring compounds, two carbons are common to the two rings, and three numbers are necessary to specify the system since there are three "pathways" (bridges) by means of which one can go from one of these carbon atoms to the other (designated as bridgehead carbons). In fused-ring compounds, one of these pathways acquires the designation "0" (*i.e.*, zero), because the bridgehead atoms are directly joined, but in bridged-ring compounds all three pathways involve one or more carbons. By convention, the numbering around the rings starts at one of the bridgehead carbons, progresses along the longest pathway first and then to the next longest pathway, and finally moves along the shortest pathway (Fig. 3.24). Compounds containing more than two rings can also be named in a comparable fashion. The scheme becomes more complicated in these instances, however, and we refer the reader to the summary of nomenclature in Chapter 18. The geometric arrangement of substituents in a fused- or bridged-ring system is specified by (a) the prefixes *cis* and *trans*, which denote the relative position in space between a pair of substituents, and (b) the prefixes *endo* and *exo*, which denote the orientation of the substituent relative to the longer of the two bridges *that do not contain the substituent.* In an *endo* (Greek, *endo*, "within") orientation, the substituent is closer to the longer bridge; in the *exo* (Greek, *exo*, "out of") orientation the substituent points away from the longer bridge.

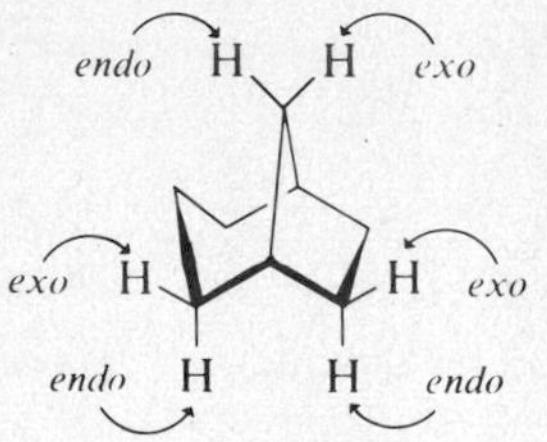

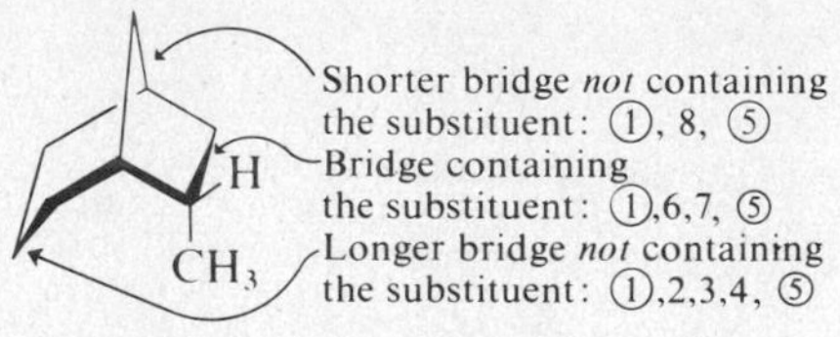

Shorter bridge *not* containing the substituent: ①, 8, ⑤
Bridge containing the substituent: ①,6,7, ⑤
Longer bridge *not* containing the substituent: ①,2,3,4, ⑤

***endo*-6-Methylbicyclo[3.2.1]octane**

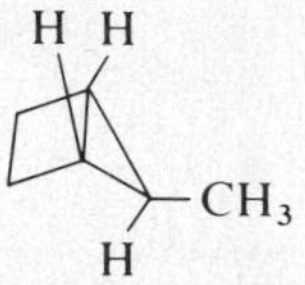

***exo*-5-Methylbicyclo[2.1.0]pentane**

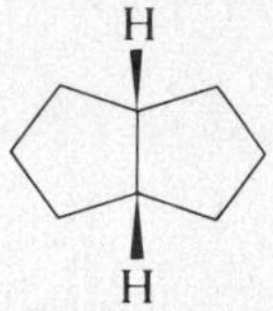

***cis*-Bicyclo[3.3.0]octane**

Two compounds that have played important roles in the study of conformational properties of molecules are the fused-ring compound bicyclo[4.4.0]decane (commonly called decalin) and the bridged-ring compound bicyclo[2.2.1]heptane (commonly called norbornane). As Hückel discovered in 1926, decalin can exist in two forms. In one of these, the second ring is attached to the first in the *trans* fashion by two equatorial bonds; in the other, the second ring is attached in the *cis* fashion by an equatorial and an axial bond.* Inspection of a framework model of the *cis* isomer shows that it is a flexible form and that the interconversion between the two equivalent chair forms is possible (Fig. 3.25). Inspection of the *trans* isomer, however, shows that it is incapable of being converted to the other chair-chair form, which would require a diaxial attachment of the second ring to the first. In this form of decalin, then, the conformation is fixed, and the molecule is rigid; *trans*-decalin, as a result, has provided a useful framework for studying the differences between substituents attached via equatorial and axial bonds. The bridged-ring compound norbornane is a similarly rigid molecule, in which the cyclohexane ring is forced to be in the boat conformation. These two molecules illustrate some of the constraints that may limit

---

* The geometric implications originally intended by the prefixes *cis* and *trans* tend to be obscured in flexible systems such as cyclohexane and decalin. The distance between the methyl groups in 1,2-dimethylcyclohexane, for example, is about the same in the *cis* compound and the *trans* diequatorial compound.

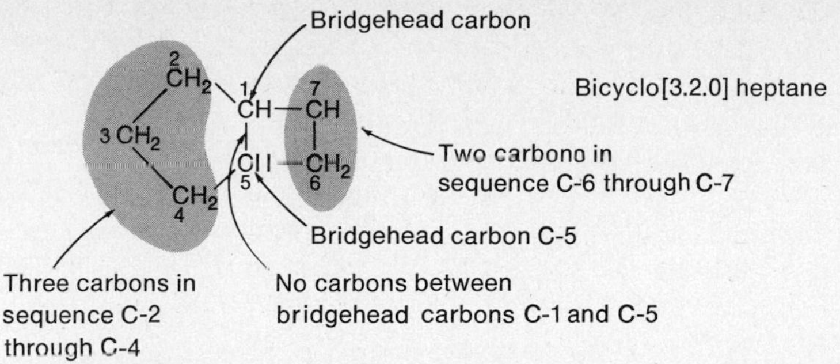

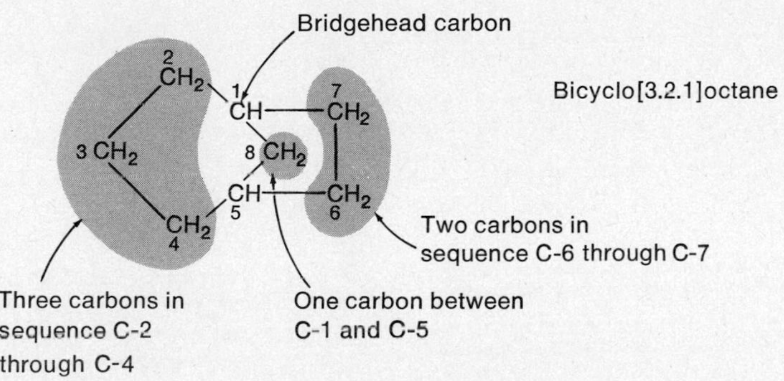

Fig. 3.24. Numbering system for fused-ring and bridged-ring bicyclic compounds.

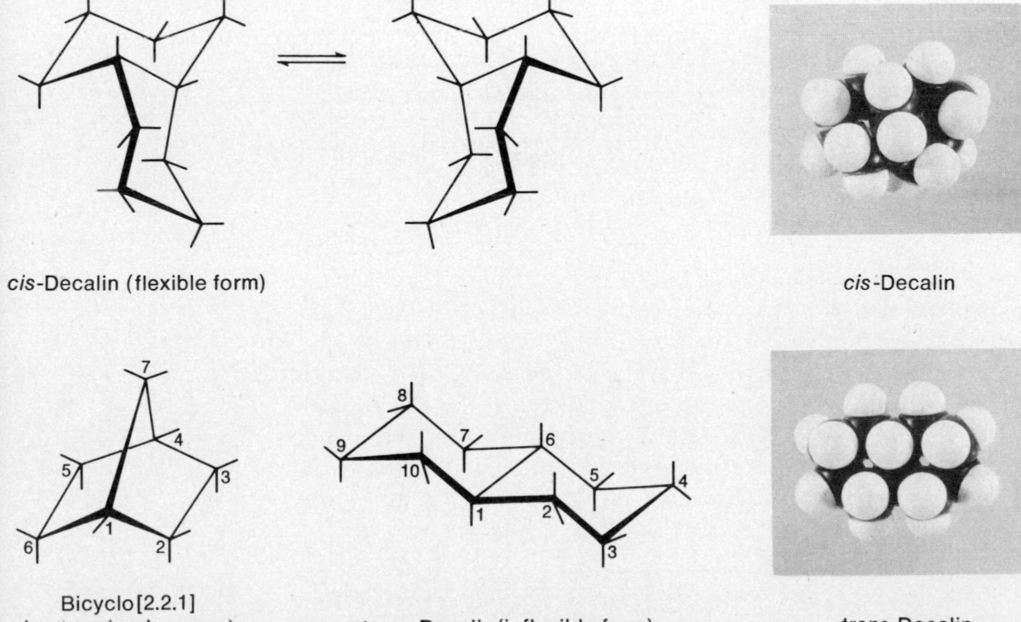

Fig. 3.25. Pseudo-three-dimensional representations of polycyclic alkanes.

the stereochemical and structural possibilities for a given system. Norbornane, for example, possesses fewer isomers than predicted, for it might be thought that a bicyclo[2.2.1]heptane structure should exist in which the bridging carbon is attached to the topside of the cyclohexane ring at C-1 and the bottomside at C-4. Such a molecule, however, is too strained to be an isolable entity, and only a single bicyclo[2.2.1]heptane is known. On the other hand, bicyclo[4.4.1]undecane, a system containing a methylene group that bridges a considerably larger and more flexible ring than cyclohexane, is known in two isomeric forms (Fig. 3.26).

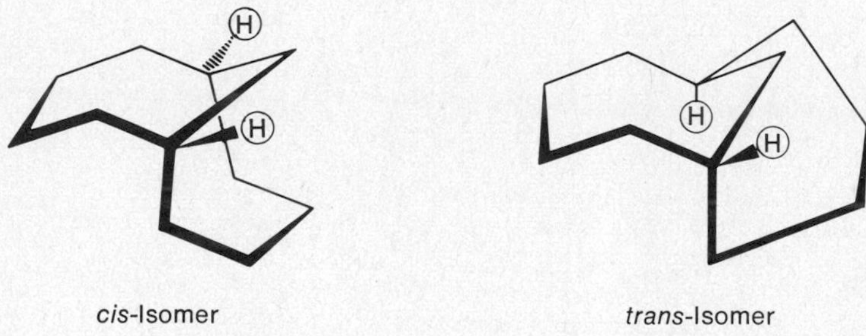

Fig. 3.26. *cis, trans*-Isomerism in bicyclo[4.4.1]undecane.

Recourse to molecular models illustrates this difference very clearly, and shows that the *trans* isomer of bicyclo[2.2.1]heptane is incapable of construction, whereas the *trans* isomer of bicyclo[4.4.1]undecane is easily built. However, whether a structure represents a compound that is sufficiently stable to be capable of isolation and characterization is not always predictable simply from a consideration of theory or from experiments with a set of mechanical models.

Although to this point in the discussion organic chemistry has been presented as an abstract game of building blocks, it should be realized that it is a game with rules inexorably established by nature. It is the chemist's task to discover these rules, which he does by laboratory experiments. Chemistry is an *experimental science*, and it is important to remember that *all* the facts and theories presented in this book are the result of the labors of thousands of men who have spent lifetimes in laboratories collecting data.

3.12 Hetero Atom Functional Groups

In addition to groups containing only carbon and hydrogen (*e.g.*, the methyl group), groups containing other atoms are also frequently encountered in organic compounds. Most common of these atoms, often referred to as hetero atoms, are nitrogen, oxygen, sulfur, and the halogens. As we shall see in later chapters, there are numerous ways in which hetero atoms appear in what are often referred to as functional groups, substituent groups, functions, or substituents. In this chapter we shall consider just a few representative examples of hetero atom groups—in particular, those that are single bonded to an sp^3-hybridized carbon atom.

Halogen atoms contain seven electrons in their valence shell, and, consequently, are monovalent. Halogens can be attached to an sp^3-hybridized carbon to give a class of compounds known as *alkyl halides* (Fig. 3.27). Oxygen contains

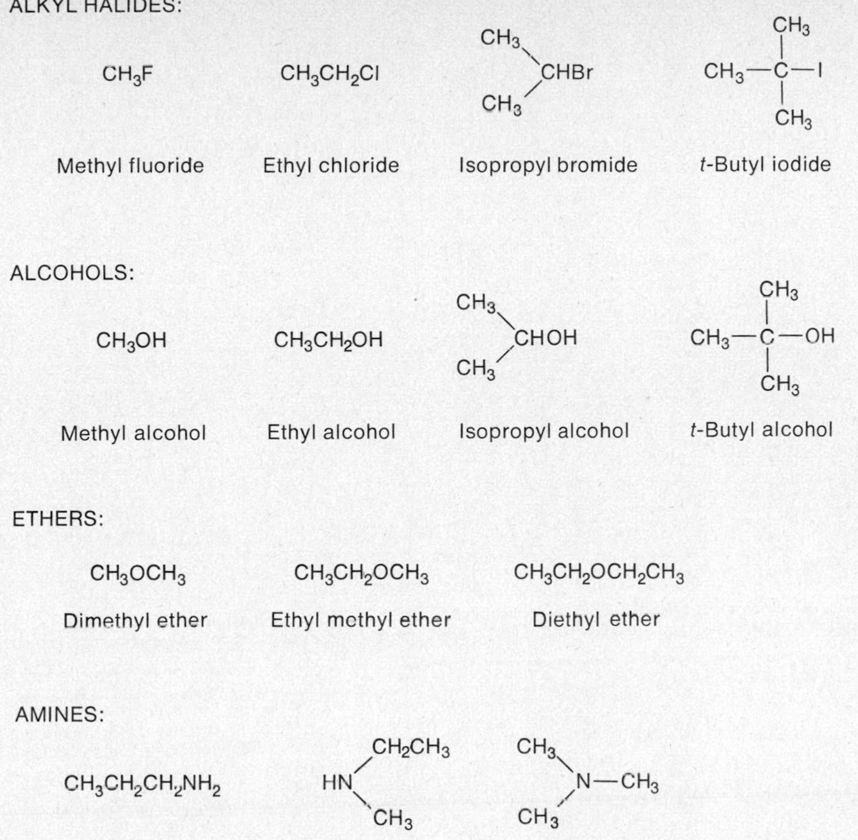

Fig. 3.27. Functional groups containing halogen, oxygen, or nitrogen singly bonded to sp^3-hybridized carbon atoms.

only six electrons in the outer valence shell and, as a divalent atom, can form two single bonds (*e.g.*, as in water) or, as we shall see in the next chapter, one double bond. When oxygen is attached to one hydrogen and one sp^3-hybridized carbon atom, a class of compounds known as *alcohols* is obtained. When oxygen is attached to two sp^3-hybridized carbon atoms, a class of compounds known as *ethers* is obtained (Fig. 3.27). Nitrogen has even greater bonding capacity than oxygen and, with five electrons in its outer valence shell, can form three single bonds, one single bond and a double bond, or one triple bond. When nitrogen is single bonded to one, two, or three sp^3-hybridized carbon atoms, a class of compounds known as *amines* is obtained (Fig. 3.27).

The unit consisting of one hydrogen and one oxygen (*i.e.*, —OH) is called a *hydroxyl group*, the unit consisting of one carbon framework and one oxygen (*i.e.*, —OR) is called an *alkoxy group*, and the unit consisting of two hydrogens and one nitrogen (*i.e.*, $—NH_2$) is called an *amino group*.

3.13 Study Guide

The following topics have been discussed in this chapter, and the reader should recognize all these terms and understand their importance and implications.

I. Basic principles of structure of alkanes and cycloalkanes
   A. sp^3-Hybridized (tetraliganded) carbon, the basic structural unit
   B. Modifications in sp^3 hybridization necessary for cyclopropane
   C. Structural isomerism
   D. Stereoisomerism
      1. *cis* and *trans* Isomerism in ring systems
      2. Mirror image or optical isomerism—nonsuperimposable mirror image forms
      3. Conformational isomers—synperiplanar, synclinal, anticlinal, and antiperiplanar conformations
         (a) Conformation of alkanes
         (b) Conformation of small-ring cycloalkanes
         (c) Conformation of cyclohexane—chair and boat forms, axial and equatorial bonds

II. Strain factors in molecules
   A. Types of strain
      1. Nonbonded, torsional strain
      2. Bond angle strain
   B. Dynamic properties of molecules
      1. "Free" rotation in alkanes
      2. Conformational mobility in cycloalkanes—concept of "average planar ring"

III. Structural representations
   A. Solid line representations
   B. Pseudo-three-dimensional representations
   C. Molecular models

IV. Nomenclature of alkanes and cycloalkanes
   A. Concept of equivalent hydrogens
      1. Dynamic equivalency (*i.e.*, equivalency of CH_3 hydrogens in $CH_3CH_2CH_3$)
      2. Static equivalency (*i.e.*, nonequivalency of CH_3 hydrogens with CH_2 hydrogens in $CH_3CH_2CH_3$)
   B. Concept of alkyl groups
      1. Structures of alkyl groups
      2. Names of alkyl groups
   C. Nomenclature of alkanes and cycloalkanes
      1. Common names
      2. IUPAC nomenclature
      3. "Methane-based" nomenclature
   D. Concept of the homologous series

V. Hetero atom functional groups
   A. Alkyl halides
   B. Alcohols
   C. Ethers
   D. Amines

REFERENCES

ALLINGER, N. L., and J. ALLINGER, *Structures of Organic Molecules*, in *Foundations of Modern Organic Chemistry Series*, Prentice-Hall, Inc., Englewood Cliffs, N. J., 1965. A short (128 p.) paperback volume written for the beginning student in organic chemistry.

BANKS, J. E., *Naming Organic Compounds*, W. B. Saunders Co., Philadelphia 1967. A slow, thorough, and self-contained introduction to the naming of monofunctional compounds but with little attention given to polyfunctional compounds.

CAHN, R. S., *An Introduction to Chemical Nomenclature*, 3rd. ed., Plenum Press, New York, 1968. A short but quite comprehensive survey of organic nomenclature that includes both monofunctional and polyfunctional types of compounds.

ELIEL, E. L., *Elements of Stereochemistry*, John Wiley & Sons, Inc., New York, 1969. A short paperback volume that introduces the rudiments of stereochemistry in an easily assimilable fashion.

HERZ, W., *The Shape of Carbon Compounds*, W. A. Benjamin, Inc., New York, 1964. A very brief and concise survey of organic chemistry written for the beginning student.

LAMBERT, J. B., The Shapes of Organic Molecules, *Scientific American*, Vol. 222, no. 1, p. 58 (1970). A short article written for the reader with no background in organic chemistry.

TRAYNHAM, J. G., *Organic Nomenclature: A Programmed Introduction*, in *Foundations of Modern Organic Chemistry Series*, Prentice-Hall, Inc., Englewood Cliffs, N.J., 1966. A workbook, with answers, to be used in conjunction with any elementary textbook of organic chemistry.

PROBLEMS

1. Draw the condensed representations and name all the possible isomers for the following molecular formulas containing only sp^3-hybridized carbon atoms: (a) C_5H_{12}; (b) C_5H_{10}; (c) C_5H_8; (d) C_6H_{10}.

2. Draw pseudo-three-dimensional representations of the following compounds:

(a)
$$\begin{array}{c} Cl \\ | \\ Cl-C-H \\ | \\ H \end{array}$$

(b)
$$\begin{array}{c} H \\ | \\ Cl-C-Cl \\ | \\ H \end{array}$$

(c)
$$\begin{array}{c} \;\;\;H\;\;H \\ \;\;\;|\;\;\;\;| \\ Cl-C-C-Cl \\ \;\;\;|\;\;\;\;| \\ \;\;\;H\;\;H \end{array}$$
(in the antiperiplanar conformation)

(d) $CH_3CH_2CH_2CH_2CH_3$
(in the all antiperiplanar conformation)

(e)
$$\begin{array}{c} Cl \\ | \\ H-C-CH_2-CH_3 \\ | \\ CH_3 \end{array}$$
(in nonsuperimposable mirror image forms)

(f) CH_3 CH_3 (1,3-dimethylcyclohexane ring drawing)
(*cis* and *trans* isomers)

3. Draw a structure for each of the following compounds.

(a) An alkane having two methyl groups.
(b) An alkane having two methyl groups and a *sec*-butyl group.
(c) An alkane having three methyl groups and an isopropyl group.
(d) An alkane having two *t*-butyl groups.
(e) A cycloalkane having one neopentyl group.
(f) An alkane with the molecular weight 86 possessing four methyl groups.
(g) An optically active compound containing one chlorine atom and having the molecular weight 106.
(h) An alkane with the molecular weight 100 and possessing primary, tertiary, and quaternary carbon atoms.

4. Draw the structures corresponding to the following names:

(a) Methane
(b) Propane
(c) Neopentane
(d) Isoheptane
(e) Cyclooctane
(f) Ethylcyclobutane
(g) Triethylmethane
(h) Tetramethylmethane
(i) *t*-Butylcyclohexane
(j) Cyclopropylcyclopentane
(k) 2-Methylbutane
(l) 2,3-Dimethylpentane
(m) 3-Isopropylpentane
(n) 4-Ethyl-2,3-dimethylhexane
(o) *cis*-3-Ethyl-1-methylcyclobutar
(p) 5-Cyclobutylmethyl-3,4-diethyl-3-(*cis*-methylcyclohexyl)-heptane

5. Draw the bicyclic structure corresponding to the following names:

(a) *cis*-Bicyclo[3.3.0]octane
(b) *trans*-4,4,8,8-Tetramethylbicyclo[5.1.0]octane
(c) 1,4-Dimethylbicyclo[2.2.2]octane.
(d) *endo*-6-Methylbicyclo[3.2.1]octane.
(e) 5-Methylspiro[3.4]octane.

6. Name each of the following structures according to the IUPAC system:

(a) $(CH_3)_2CHCH_2CH_3$ — drawn as two CH_3 groups bonded to $CHCH_2CH_3$

(b) $CH_3CH_2CH_2CHCH_2CH_2CH_3$ with CH_3 on the central CH

(c) $CH_3CH_2CHCH_2CH_3$ with CH_2CH_3 on the CH

(d) $CH_3CHCHCH_3$ with CH_3 above the second CH and CH_2—CH_3 below the first CH

(e) $CH_3CH_2CH—C—CH_2CH_3$ with CH_3 above the CH, and CH_3 above and CH_3 below the C

(f) cyclopentane—$CH(CH_3)_2$

(g) $CH_3CH_2CH_2CH$ bearing CH(CH_3)$_2$ above, bonded below to the C of $CH_3CH_2CCH_2CHCH_2CH_2CH_2CH_3$; the C also bears CH_3 below; the CH bears $CHCH_2CH_3$ below, which bears CH_3

(h) cyclopropane—$CH_2CH_2CH_3$

(i) 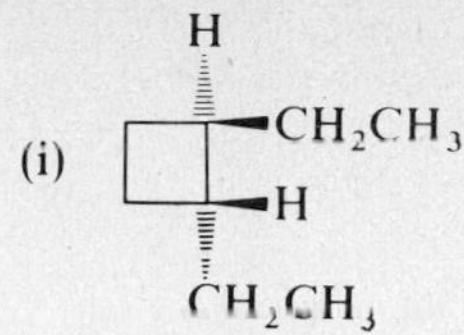

(j) 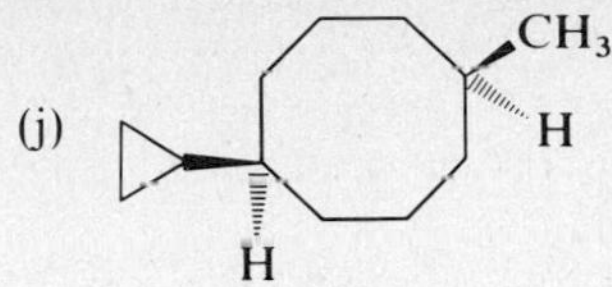

(k) 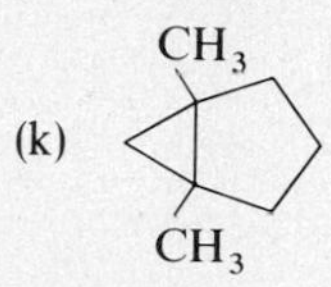

(l) 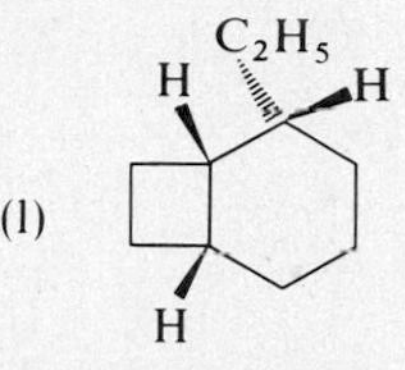

(m) 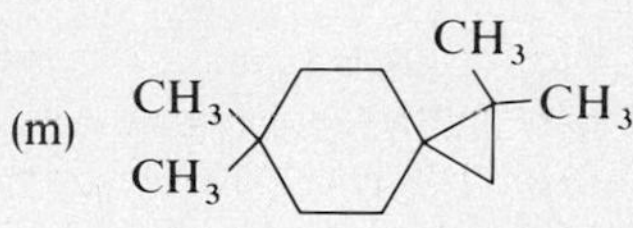

(n)

(o) 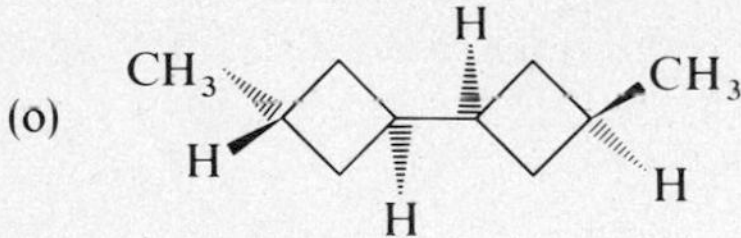

7. From the heats of combustion given in Table 3.2, calculate the total strain energies of cyclopropane, cyclobutane, and cyclopentane.

8. Using framework molecular models and your knowledge and intuition concerning organic structure gained thus far, predict which of the following compounds are likely to be stable enough to exist under ordinary conditions.

(a) 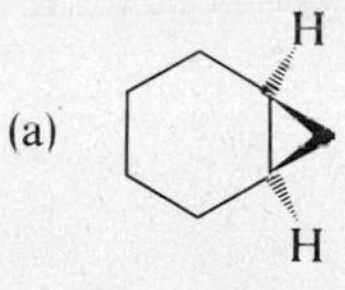

(b) 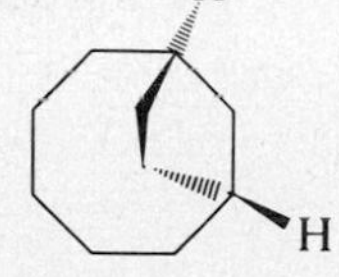

(c) 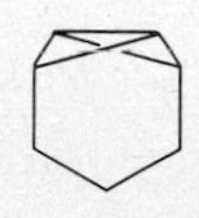

(d) 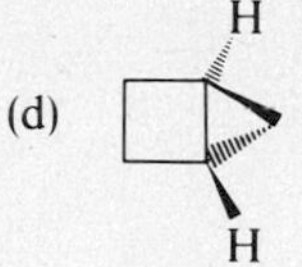

(e) 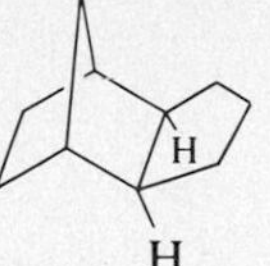

(f)

(g) 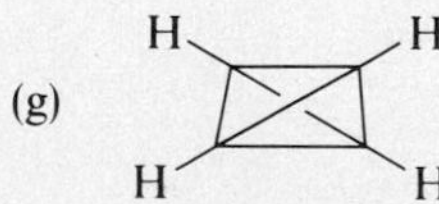

(h)

(i)

Cubane

(j)

Adamantane

9. Draw the three-dimensional representations for the mirror image forms for the following compounds, and indicate which are superimposable and which nonsuperimposable.
(a) The staggered conformation of ethane.

(b) The synclinal conformation of $ClCH_2CH_2Cl$.
(c) The antiperiplanar conformation of $ClCH_2CH_2Cl$.
(d) *trans*-1,2-Dimethylcyclopentane.
(e) *exo*-2-Methylbicyclo[2.2.1]heptane.

10. Explain why 1,2-dichloroethane cannot be isolated in an optically active form, even though its synclinal conformation can exist in a pair of nonsuperimposable mirror image forms (*i.e.* enantiomers)

11. 2-Bromo-3-chlorobutane can exist in the four stereoisomeric forms pictured.

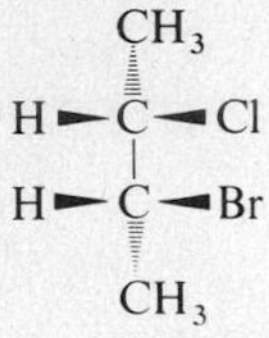

(i) 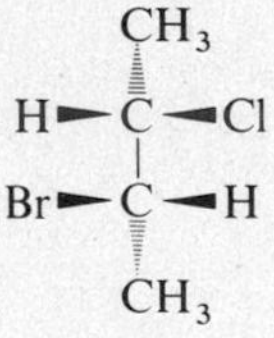 (ii) 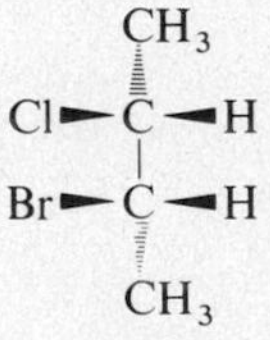 (iii) 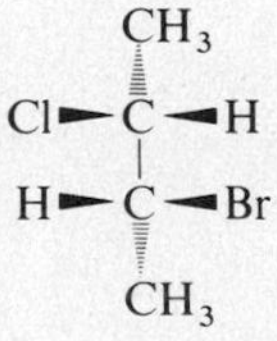 (iv)

(a) Which of these isomers bear an enantiomeric relationship?
(b) Which of these isomers bear a diastereomeric relationship?
(c) Which of these isomers have the same boiling point?
(d) Which of these isomers have different refractive indices?
(e) Will the water solubilities of isomers (ii) and (iv) be the same or different? Explain.
(f) Will isomers (i) and (ii) have the same or different effects on plane polarized light? Explain?
(g) Will isomers (i) and (iii) have the same or different effects on plane polarized light? Explain.

12. In each of the following pairs of compounds, indicate which possesses the higher heat of combustion.

(a) 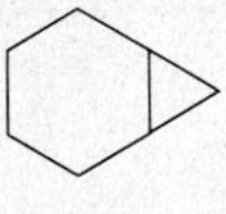 (i) or (ii)

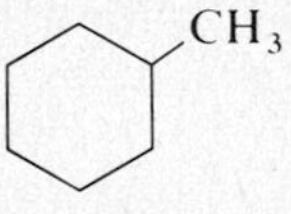

(b) 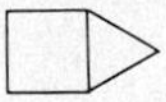 (i) or (ii)

(c) 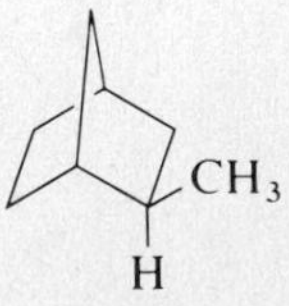 (i) or (ii)

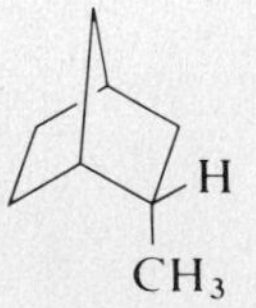

(d) 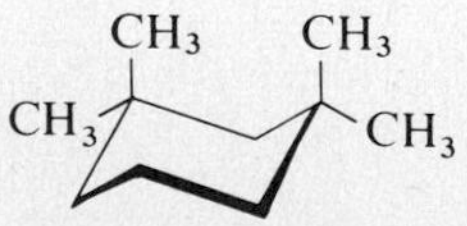 (i) or (ii)

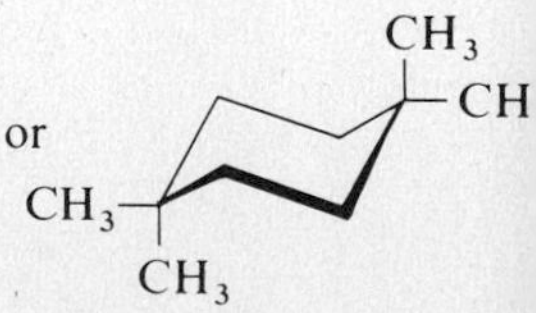

(e) 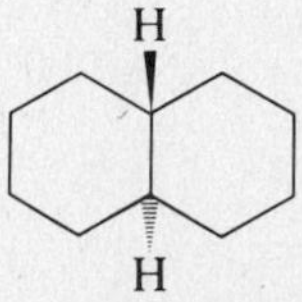 (i) or (ii)

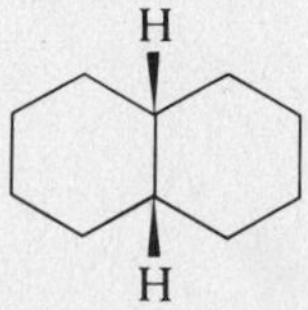

13. Draw the three-dimensional representations for the most stable conformations of the following compounds:

(a) $ClCH_2CH_2CH_3$

(b) $(CH_3)_2CHCH(CH_3)_2$

(c)

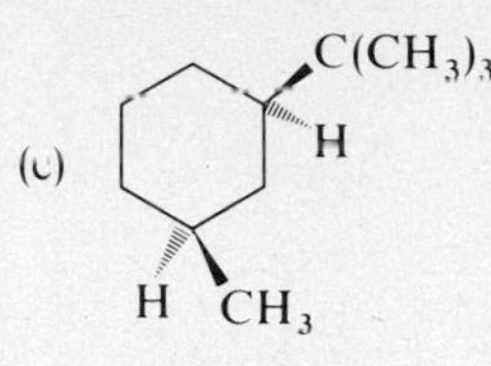

(d) $CH_3CH_2CHCH_3$ with $CH(CH_3)_2$ on the third carbon

(e)

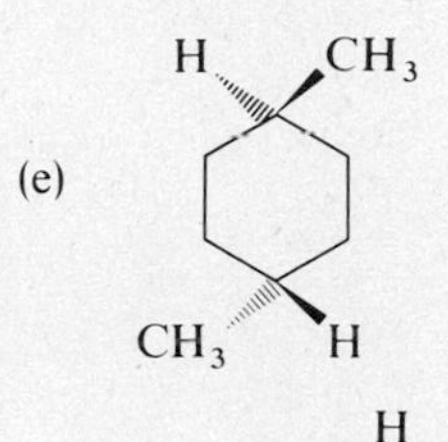

(f)

(g)

14. Draw structures for the following compounds:

(a) All the alcohols with the molecular formula $C_4H_{10}O$.
(b) All the ethers with the molecular formula $C_4H_{10}O$.
(c) All the amines with the molecular formula $C_4H_{11}N$.
(d) The nonsuperimposable mirror images of all the compounds in (a), (b), and (c) where this possibility exists.

15. Define each of the following terms as briefly and concisely as possible. Do not use a specific example in defining the term, but cite a typical example to illustrate the definition.

(a) Stereoisomer
(b) Paraffin hydrocarbon
(c) Nonbonded interference
(d) Newman projection
(e) Homologous series
(f) Synclinal conformation
(g) Enantiomer
(h) Diastereomer
(i) Optical activity
(j) Structural isomer

Chapter 4

Carbon frameworks containing sp^2- and sp-hybridized carbon atoms: alkenes, alkynes, and arenes

"There's no use trying," she said: "one *can't* believe impossible things." "I dare say you haven't had much practice," said the Queen. "When I was your age, I always did it for half an hour a day. Why, sometimes I've believed as many as six impossible things before breakfast."

LEWIS CARROLL
Alice Through the Looking Glass

4.1 Compounds Containing sp^2-Hybridized Carbon Atoms: Alkenes

Ethylene* possesses the molecular formula C_2H_4 and is the first member of a homologous series known as the alkenes, characterized by the general formula C_nH_{2n} and the presence of a C=C bond. The compounds in this class are also referred to as (a) *olefins*, which is a term derived from the name "olefiant gas" (Latin, *oleum*, "oil"; *facio*, "to make"), given to ethylene because it reacts with chlorine to form an "oil," and (b) *unsaturated hydrocarbons*, a term signifying that the compounds contain less hydrogen than the saturated alkane analogs; *e.g.*, ethylene reacts with hydrogen to produce ethane, $C_2H_4 + H_2 \rightarrow C_2H_6$.

The double bond between the carbon atoms in ethylene can, as we have seen (see Fig. 2.10), be interpreted in terms of a combination of a σ bond and a π bond. According to this construction, ethylene is predicted to have a H—C—H bond angle of 120°, the measured angle of 117.5° being reasonably close to this value.

---

* Although "ethene" is the correct systematic name for the parent member of this series, the designation "ethylene" will be used in the present discussion because it is in common parlance among chemists.

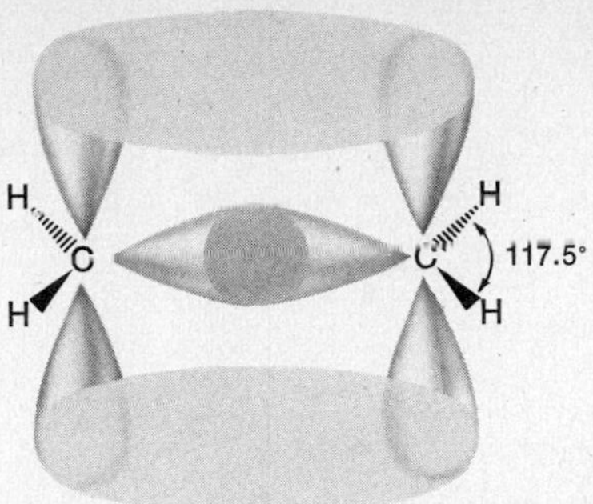

The next two higher homologs of ethylene are (a) propene, with the molecular formula C_3H_6 (isomeric with cyclopropane), for which only one structure is possible, and (b) butene, with the molecular formula C_4H_8 (isomeric with cyclobutane and methylcyclopropane), for which four isomeric structures are possible (Fig. 4.1). Three of the butenes are constructed from a linear carbon framework

$CH_3(H)C{=}CH_2$	$H_2C{=}CH(CH_2CH_3)$	$CH_3(H)C{=}C(H)CH_3$	$CH_3(H)C{=}C(CH_3)H$	$H_2C{=}C(CH_3)_2$
	(a)	(b)	(c)	(d)
Propene	Ethylethylene 1-Butene	*cis*-1,2-Dimethylethylene *cis*-2-Butene	*trans*-1,2-Dimethylethylene *trans*-2-Butene	1,1-Dimethylethylene 2-Methylpropene

Isomers of butene

Fig. 4.1. Structures of propene and the butenes.

[structures (a), (b), and (c)] and one from a branched carbon framework [structure (d)]. Structures (a) and (b) [or (c)] are structural isomers, often referred to as *positional isomers* in recognition of the difference in the location of the double bond. Structures (b) and (c), on the other hand, are stereoisomers, differing from each other only in the spatial arrangement of the atoms within the molecule. If we employ the prefixes referred to earlier (see Fig. 3.15), the *cis* compound is the one having the methyl groups on the same side of the double bond, and the *trans* compound is the one having the methyl groups on the opposite sides of the double bond.

We might view *cis*-2-butene and *trans*-2-butene as conformational isomers. Whereas C—C bonds generally rotate more or less freely at room temperature, thereby allowing the facile interconversion of conformers, C=C bonds require a large amount of energy to promote rotation and conformational interconversion. As a result, the *cis* and *trans* forms are stable, isolable entities that are interconverted only with considerable difficulty. They bear a diastereomeric relationship to each other, and this type of stereoisomerism can be designated as *double-bond diastereoisomerism.** An important property of diastereomers,

* Until recently, this was usually designated as "geometric isomerism." The term is not properly descriptive, however, for *all* stereoisomerism is geometric in the sense that it arises as a consequence of the geometry of the molecule. Alternative designations are "double-bond *cis,trans* isomerism" and "(*Z*), (*E*) isomerism" (see p. 96).

regardless of whether they are in an alkene system (*i.e.*, *cis*- and *trans*-2-butene), a cycloalkane system (*i.e.*, *cis*- and *trans*-1,2-dimethylcyclopropane), or an alkane system (*i.e.*, the 3,4-dimethylhexanes), is that they differ in *all* physical and chemical characteristics. For example, *cis*-2-butene boils at 4°C and *trans*-2-butene boils at 1°C, and can be separated from each other by ordinary fractional distillation.

The three straight-chain butenes also differ in stability (total energy). If equilibrium is established among the three isomers, the *trans*-2-butene is present in the greatest quantity (*i.e.*, it is the most stable), *cis*-2-butene is present in somewhat lesser quantity, and 1-butene is the least abundant (*i.e.*, it is the least stable). Why the *trans* isomer should be more stable than the *cis* isomer is readily understood if one considers the nonbonded strain energy present in the two molecules. In the *cis* isomer, the two methyl groups are eclipsed, causing greater steric strain than that arising from the eclipsing of the hydrogens with the methyl groups in the *trans* isomer. That the "internal" alkenes should be more stable than the "terminal" alkene is not so obvious, however, and the reasons for this difference are not yet clear. In general, the greater is the number of alkyl groups attached to the C=C moiety, the greater is the stability of the alkene.

The C=C bond differs from the C—C bond not only in its resistance to rotation but also in length and strength. Although the double bond is not quite twice as strong as a single bond, it is significantly stronger, nevertheless, and requires 146 kcal/mole to dissociate it (in contrast to the 83 kcal/mole required for a typical C—C bond). Also, the C=C bond is significantly shorter (1.33 Å) than the C—C bond (1.54 Å). The geometry around the sp^2-hybridized carbon atom is different from that around the sp^3-hybridized carbon atom. The H—C—H bond angle between the hydrogen atoms in methane is 109° 28′, and that between the hydrogens at a carbon atom of ethylene is 117° 30′. Clearly, substituents attached to an sp^3-hybridized carbon are closer to each other than those attached to an sp^2-hybridized carbon, and this difference sometimes has an important influence on the chemical characteristics of the compound.

Two systems of nomenclature are used for alkenes. One is based on ethylene as the parent compound, and the other is based on the IUPAC rules. In the "ethylene-based" method, alkenes are named as substituted ethylenes by prefix designations that indicate the structure of the substituents and the carbon to which they are attached (*i.e.*, C-1 or C-2 of ethylene); if necessary, the prefixes *cis* or *trans* are also affixed. Thus, propylene is called methylethylene and compounds (a), (b), (c), and (d) of Fig. 4.1 are called ethylethylene, *cis*-1,2-dimethylethylene, *trans*-1,2-dimethylethylene, and 1,1-dimethylethylene, respectively. This system, although useful in certain relatively simple cases, is of limited application. The IUPAC system, which is completely general in its applicability, is usually preferred. According to the IUPAC rules, alkenes are named as follows:

1. The longest continuous chain in the alkene *containing the double bond* is selected to designate the parent compound, and the actual compound is then considered to be a derivative of the parent compound. The parent

compound is named according to alkane nomenclature but with the suffix "ane" replaced by "ene." For instance, $H_2C{=}C(CH_3){-}CH_3$ [Fig. 4.1(d)] has a three-carbon chain as its longest unit containing the double bond; it is named as a methyl-substituted propene.

2. The position of the double bond in the parent compound is designated by counting from the terminus that allows the lower number to be assigned to the first carbon of the double bond encountered along the chain; thus, the parent compound in $H_2C{=}C(CH_3)CH_3$ is 1-propene rather than 2-propene, and $H_2C{=}CHCH_2CH_3$ [Fig. 4.1(a)] is 1-butene rather than 3-butene. In simple alkenes such as these, the number is placed before the "alkene" descriptor. In more complex alkenes, however, it is frequently placed between the prefix "alk" and the suffix "ene"; *e.g.*, but-1-ene.
3. After the numbering of the parent compound has been established by the position of the double bond, the substituents are specified by names, assigned numbers, and ordered by the same rules that apply for alkanes (see p. 63). If necessary, the stereochemical designations *cis* or *trans* (or (*Z*) or (*E*)—see p. 96) are also affixed. For instance, compounds (a)-(d) in Fig. 4.1 are named 1-butene, *cis*-2-butene, *trans*-2-butene, and 2-methyl-1-propene (although the "1" is actually redundant and does not have to appear in this case), respectively.
4. Sometimes a double bond is present in a group that must be treated as a substituent rather than as part of the parent compound. If the substituent contains two or three carbon atoms, the following names can be used:

$H_2C{=}CH{-}$	$H_2C{=}CHCH_2{-}$	$CH_3CH{=}CH{-}$	$H_2C{=}C(CH_3){-}$
Vinyl	**Allyl**	**Propenyl** (***cis* or *trans***)	**Isopropenyl**

If the substituent contains four or more carbons, it is named by the IUPAC rules, using the procedure described, except that the numbering is determined by the point of attachment of the substituent to the parent compound rather than by position of the double bond with respect to the closer terminus; for example $\overset{5}{C}H_3\overset{4}{C}H{=}\overset{3}{C}H\overset{2}{C}H_2\overset{1}{C}H_2{-}$ is designated as 3-pentenyl rather than 2-pentenyl, the 1 position designating the point of attachment of the pentenyl group to the parent compound.

5. Sometimes it is necessary to include a group as a double-bonded substituent to the parent compound. In this circumstance, the group is named by replacing the suffix "ane" in the name of the appropriate hydrocarbon by "ylidene." Thus, $=CH_2$ is methylidene (or methylene), $=CHCH_3$ is ethylidene, $=C(CH_3)_2$ is isopropylidene, etc. The compound (cyclohexane ring)$=CHCH_3$ is named ethylidenecyclohexane.

6. Stereochemical designations are applied in the following ways. In simple cases such as the 2-butenes, the *cis* and *trans* prefixes can be applied without difficulty. However, in more complex compounds, such as

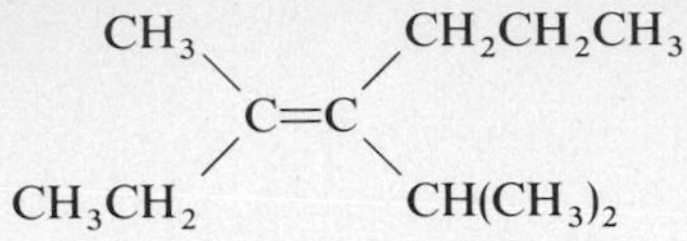

4-Isopropyl-3-methyl-3-heptene

problems arise with respect to which isomer is better labeled *cis* and which *trans*. To circumvent these problems, chemists have devised a system whereby atoms and groups are arranged in a sequence on the basis of their atomic number and atomic mass, *higher atomic number taking precedence over lower atomic number* and higher atomic mass taking precedence over lower atomic mass (in cases where the atomic number is the same). The groups attached to each of the carbon atoms of the double bond are then ranked on the basis of these "sequence rules," one of the groups at each of the carbons having a higher priority and the other group a lower priority. If the group of higher priority at C-1 of the double bond is on the same side as the group of higher priority at C-2, the prefix (*Z*) (German, *zusammen*, "together") is affixed; if the group of higher priority at C-1 of the double bond is on the opposite side from the group of higher priority at C-2 the prefix (*E*) (German, *entgegen*, "opposed to") is affixed:

(*Z*) configuration **(*E*) configuration**

Priority sequence: a > b and c > d

For example, in 2-butene the pair of groups attached to both alkene carbons are hydrogen and methyl. According to the sequence rule, carbon has higher priority than hydrogen; therefore, *cis*-2-butene is (*Z*)-2-butene and *trans*-2-butene is (*E*)-2-butene. In 4-isopropyl-3-methyl-3-heptene, however,

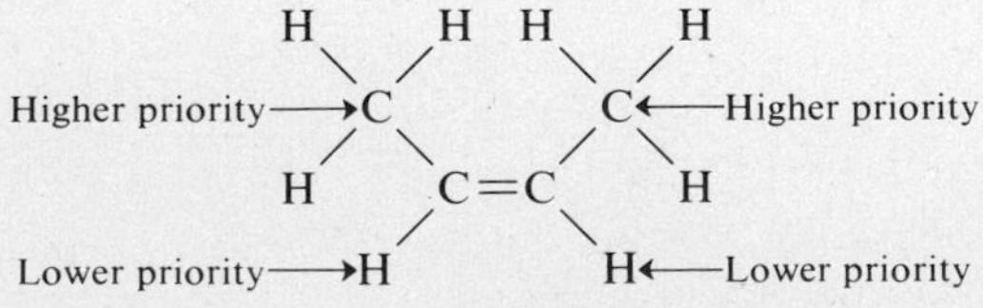

(*Z*)-2-butene

all four of the groups at the double bond are carbon moieties, and the priorities cannot be established on the basis of the atoms directly attached to the alkene carbons. In such cases, "outward exploration" along the framework of the attached groups is made until a point is reached where a decision on the basis of atomic number or mass number is possible. Thus, progressing outward in the ethyl group brings us to carbon at the second bond. Carbon takes precedence over hydrogen, so ethyl takes precedence over methyl. In the case of the *n*-propyl and

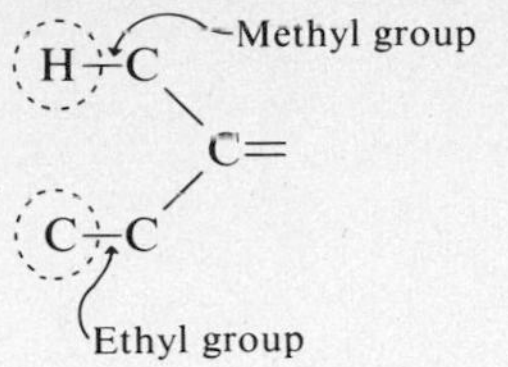

H—C ← Methyl group
C=
C—C ← Ethyl group

isopropyl groups, outward progression past the first carbon brings us to H, H, and C along the three pathways in the *n*-propyl group and to H, C, and C in the isopropyl group. Thus, two of the pathways are identical (*i.e.*, H and C) but the third is different (*i.e.*, C instead of H in the isopropyl group), and the isopropyl group takes precedence over the *n*-propyl group. By similar considerations, the order of precedence in the butyl series is

$$(CH_3)_3C— > CH_3CH_2\overset{|}{C}HCH_3 > (CH_3)_2CH\overset{|}{C}H_2 > CH_3CH_2CH_2\overset{|}{C}H_2.$$

The isomer of 4-isopropyl-3-methyl-3-heptene pictured on p. 96 is correctly named, therefore, as the (*Z*) isomer.

The stereochemical designations (*cis,trans* or (*Z*),(*E*)) are placed either at the beginning of the name or immediately before the number of the double bond to which they refer. In complicated names, the latter is preferred.

PROBLEM 4-1

Draw all the possible isomers of hexene, including the (*Z*)- and (*E*)-isomers. Using the IUPAC nomenclature, assign a name to each of these structures.

The sp^3- and sp^2-hybridized carbon atom building blocks can be admixed in a variety of ways, including incorporation into cyclic structures. The smallest cycloalkene is cyclopropene, constructed from a pair of sp^2-hybridized carbons, one sp^3-hybridized carbon, and four hydrogens (Fig. 4.2). Cyclopropene is a known compound, although it is more strained than cyclopropane, and, consequently, is less thermodynamically stable and more chemically reactive. The larger cycloalkenes (*e.g.*, cyclobutene, cyclohexene, cyclodecene, etc.) all are capable of existence, and many have been prepared in the laboratory. In smaller cycloalkenes, the double bond restricts the flexibility of the molecule. Thus, both cyclopropene and cyclobutene are planar, and cyclopentene and cyclohexene are closer to planarity than their corresponding cycloalkanes (Fig. 4.2). The bonds at C-3 and C-6 of the cyclohexene ring are altered from the equatorial and axial alignment present in cyclohexane and are sometimes referred to as *pseudo-equatorial* and *pseudoaxial* bonds.

Knowing that 2-butene can exist as both a *cis* and *trans* isomer [(*Z*) and (*E*) isomers], one might ask why the same possibility is not available to cyclobutene. A set of molecular framework models provides the answer. The strain imposed

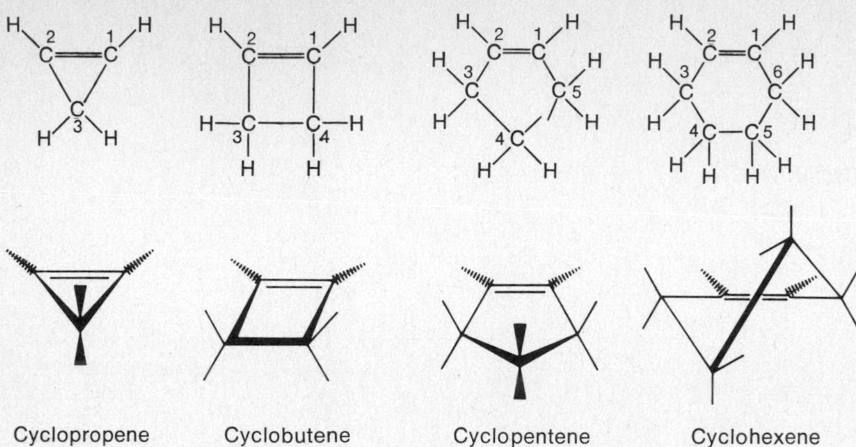

Fig. 4.2. Dashed-bond and pseudo-three-dimensional representations for cycloalkenes.

on the molecule by spanning the *trans* positions of the double bond by a two-carbon unit is so great that the molecule is incapable of existence. If, however, six or more carbon atoms span these positions a *trans*-cycloalkene is possible, such as, for example, *trans*-cyclooctene.* Another restriction of a similar type is encountered in certain bridged-ring bicyclic compounds containing a double bond at a bridgehead position. For example, bicyclo[2.2.1]hept-1(2)-ene [Fig. 4.3(d)] is a structure that is too strained to exist and represents a system that is said to follow Bredt's rule. Bredt's rule is an empirically derived hypothesis, which, in its original formulation, asserted that a stable bridged-ring structure cannot contain a double bond at a bridgehead position. In the light of more recent findings, however, Bredt's rule has been modified, for in the larger, more flexible homologs of fused-ring bicyclics, such as Fig. 4.3(e), the restriction no longer holds. In a modern formulation of the rule, the bicyclic system is represented in the fashion shown, in which the double bond is outside ring (a)-(b) (exocyclic) but inside rings (a)-(c) and (b)-(c) (endocyclic); if the ring in which the double bond has the *trans* configuration [*e.g.*, ring (a)-(c)] contains fewer than eight atoms, the system is too strained to be isolable. In Fig. 4.3(d), the *trans* double bond is contained in a six-membered ring; therefore, the compound is incapable of existence. In Fig. 4.3(e), however, the *trans* double bond is contained in a nine-membered ring, and the compound is capable of preparation and isolation.

PROBLEM 4-2

What is the smallest alkene in which nonsuperimposable mirror image isomers can exist? What is the smallest monochloro cycloalkene containing two chiral carbons? How many stereoisomers are possible for this compound?

* A *trans*-cycloheptene has been "trapped" as a reactive intermediate, but its lifetime is short and it is quickly converted to more stable compounds.

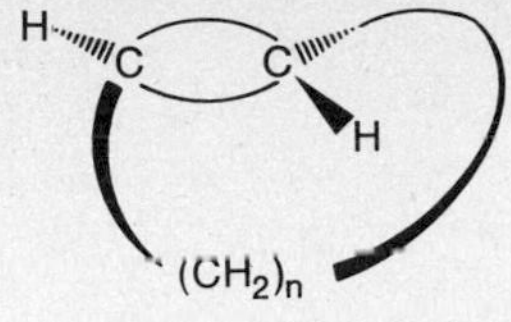

(a) *trans*-Cycloalkene

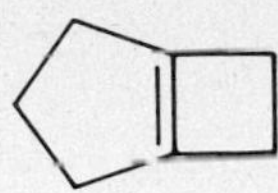

(b) Bicyclo[3.2.0]hept-1(5)-ene

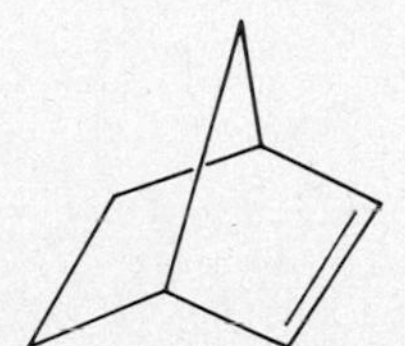

(c) Bicyclo[2.2.1]hept-2-ene

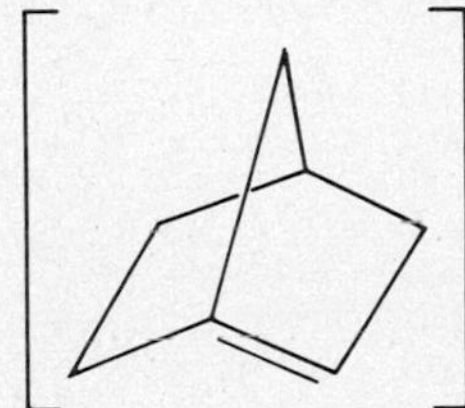

(d) Bicyclo[2.2.1]hept-1-ene

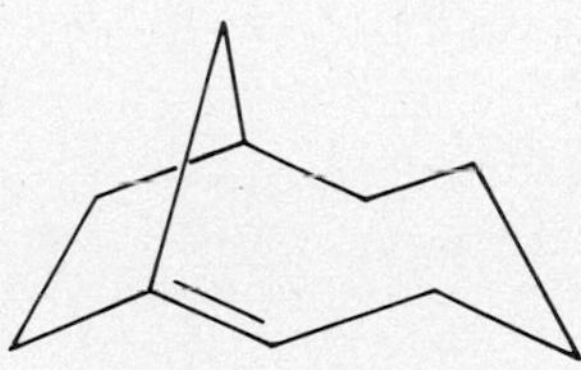

(e) Bicyclo[5.2.1]dec-1-ene

Fig. 4.3. Cycloalkenes.*

The nomenclature of substituted cycloalkenes warrants comment, because in this case there is no "end" to the carbon chain. The positions of attachment of substituents are designated by starting the numbering at the double bond, specifying the double-bond carbons as C-1 and C-2 and then continuing on around the ring. If the double bond is unsubstituted or disubstituted, C-1 is chosen so that the ring substituents are assigned the lower set of numbers. If the double bond is monosubstituted, the carbon bearing the substituent is specified as C-1.

3,5,5-Trimethylcyclohexene
(*not* 4,4,6-trimethylcyclohexene)

1,6-Dimethylcyclohexene
(*not* 1,2-dimethylcyclohexene or 2,3-dimethylcyclohexene)

4.2 *Compounds Containing sp-Hybridized Carbon Atoms: Alkynes and Cycloalkynes*

Acetylene, possessing the molecular formula C_2H_2, is the first member of a homologous series known as the alkynes which are characterized by the general formula C_2H_{2n-2} and the presence of a C≡C bond. The range of possibilities of structural isomers resulting from the attachment of alkyl groups to an acetylenic

* If one of the termini of a double bond in a bicyclic system is a bridgehead carbon, the other terminus is indicated by a number in parentheses to distinguish between the several possibilities. Thus,

is bicyclo[4.2.0]oct-1(2)-ene and is bicyclo[4.2.0]oct-1(8)-ene.

framework is similar to that of the alkenes, and compounds such as methylacetylene, dimethylacetylene, and ethylacetylene (Fig. 4.4) are well known. The lower members of the alkyne series possess fewer structural isomers than the corresponding alkenes, however, because the linear geometry of the acetylenic unit precludes (*Z*),(*E*)-type isomerism. The small ring cycloalkenes have no counterparts in the acetylene series, for the smallest stable cycloalkyne is cyclooctyne. There is experimental evidence that cycloheptyne and even cyclohexyne have a fleeting existence as transient intermediates, but they are too reactive to be isolated and studied.

$CH_3C{\equiv}CH$	$CH_3C{\equiv}CCH_3$	$CH_3CH_2C{\equiv}CH$
Methylacetylene Propyne	Dimethylacetylene 2-Butyne	Ethylacetylene 1-Butyne

Fig. 4.4. Solid line representations of alkynes.

The bond strength of 200 kcal/mole for the C≡C bond indicates that it is stronger than the double bond, although by a somewhat smaller increment (*i.e.*, 54 kcal/mole) than the increment between a double and single bond (*i.e.*, 63 kcal/mole). Similarly, the bond length of 1.20 Å is less than that of the C=C bond, although again by a smaller increment (*i.e.*, 0.13 Å) than that between the double and single bonds (*i.e.*, 0.21 Å). The bond length to attached substituents increases from alkynes to alkenes to alkanes (see Table 2.4 for C—H bonds). These differences have important consequences for chemical reactivity, although with some surprises, as we shall see in a later chapter.

The nomenclature of the alkynes presents nothing new. One system uses acetylene as the parent compound and follows the same rules that have been discussed for the methane based and ethylene based nomenclatures. Thus, the acyclic alkynes in Fig. 4.4 are named as methylacetylene, dimethylacetylene, and ethylacetylene, respectively. The other system follows the IUPAC rules in precisely the same fashion as for alkenes, except that the suffix "yne" replaces "ane" of the parent hydrocarbon. Thus, the acyclic alkynes in Fig. 4.4 are named propyne, 2-butyne, and 1-butyne, respectively. Cyclic alkynes are named by replacing the suffix "ane" of the corresponding cycloalkane by "yne" e.g., cyclooctyne, cyclododecyne, etc.

4.3 Compounds Containing Two or More Multiple Bonds

It is possible, of course, to have more than one multiply bonded group (often called an "unsaturated group") in the same molecule—*i.e.*, structures containing two or more C=C bonds, two or more C≡C bonds, or various combinations of C=C and C≡C bonds. Compounds containing two double bonds are called dienes, and three types of structures have been defined: compounds in which two double bonds share a common atom [Fig. 4.5(a)], designated as *cumulative dienes*; compounds in which two C=C bonds are directly attached via a single bond [Fig. 4.5(b)], designated as *conjugated dienes*; and compounds in which two C=C bonds are not directly connected but have one or more sp^3 carbon atoms between them [Fig. 4.5(c)], designated as *nonconjugated dienes* or isolated double-bond structures. In doubly unsaturated compounds containing a C≡C bond,

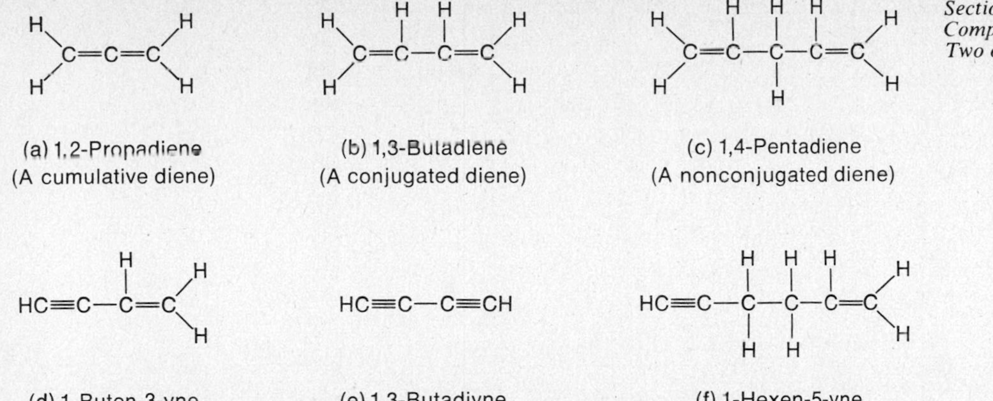

Fig. 4.5. Compounds containing two or more centers of unsaturation.

the cumulative structures are precluded, but conjugated structures [*e.g.*, Fig. 4.5(d) and (c)] and nonconjugated structures [*e.g.*, Fig. 4.5(f)] are still possible.

Multiply unsaturated compounds are usually best named by the IUPAC nomenclature, according to which the parent name is derived from the longest continuous chain that contains the *maximum number of unsaturated groups.* To the appropriate name, chosen by reference to alkane designations previously described, a suffix is added (*e.g.*, diene, triene, enyne, dienyne, diyne, etc.), which indicates the type(s) and number of unsaturated bonds that are present. Their location in the parent chain is specified by numbers placed either before the designation of the parent chain or between the designation of the parent chain and the suffix indicating the nature of the unsaturation. The numbering of the chain is determined by the multiple bond that is closer to one end of the molecule, despite whether it is C=C or C≡C. In assigning the suffix designations for the unsaturated bonds, however, we indicate the double bond before the triple bond (see Fig. 4.5). If the double and triple bonds are similarly situated in the chain (*e.g.*, compound (f) in Fig. 4.5), the double bond takes precedence over the triple bond in determining the numbering. We then name substituents, and designate their positions according to the rules that have been discussed for alkynes, alkenes, and alkanes. For example,

$CH_3CH{=}C{=}C(CH_3)_2$

2-Methyl-2,3-pentadiene or 2-Methylpenta-2,3-diene

$CH_3CH_2(H)C{=}C(H)\text{—}(H)C{=}C(H)CH_3$

***trans*-2,*cis*-4-Heptadiene or Hepta-*trans*-2,*cis*-4-diene; (*E*)-2,(*Z*)-4-Heptadiene or Hepta-(*E*)-2,(*Z*)-4-diene**

1-(4-Methyl-*trans*-2-pentenyl)-cyclohexene

$CH_3C{\equiv}CC(CH{=}CH_2){=}CHCH_2CH_3$

4-Vinylhept-4-en-2-yne

4.4 *Structure of Allene*

The parent compound in the cumulative series is known as allene (the IUPAC name is 1,2-propadiene). The building blocks for allene include a pair of sp^2-hybridized carbon atoms for the terminal carbons, an sp-hybridized carbon atom for the central carbon, and four hydrogen atoms. The π orbitals of the two double bonds in allene are perpendicular (orthogonal) to each other and do not interact; they are considered to be nonconjugated double bonds. The planes in which the terminal groups fall are also perpendicular to each other (see Fig. 4.6), one of the interesting consequences of which is that certain allenes, such as 1,3-dimethylallene, exist in nonsuperimposable mirror image forms. Chapter 19 gives a more detailed account of this kind of stereoisomerism.

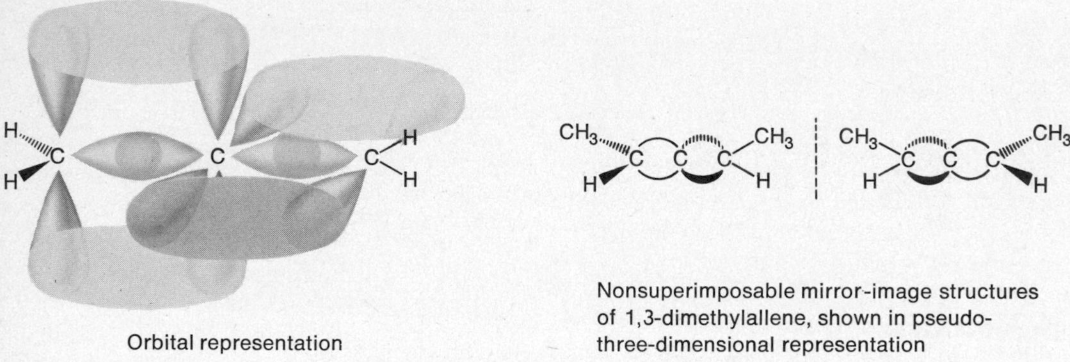

Fig. 4.6. Structure of allenes.

PROBLEM 4-3

Which compound has more stereoisomers: 2,3-pentadiene or 2,3-hexadiene? Explain.

4.5 *Structure of 1,3-Butadiene; π-Bond Delocalization*

1,3-Butadiene is the simplest of the conjugated polyenes and is constructed from four sp^2-hybridized carbons and six hydrogens. In what we shall refer to as its classical, localized bond structure it is represented as having six C—H bonds, two C=C bonds, and one C—C bond. Measurements of the bond length and bond strength of the C—C bond show it to be somewhat different from the C—C bond of ethane; it is shorter and stronger, and the question is, why? One explanation is that it is the result of hybridization differences; whereas the C—C bond of ethane is between a pair of sp^3-hybridized carbons, that in butadiene is between a pair of sp^2-hybridized carbons. An alternative explanation is that it arises from the overlap of the p orbitals at C-2 and C-3 to give partial double-bond character to the bond between C-2 and C-3, a phenomenon known as bond delocalization.

To introduce the concept of bond delocalization, let us look into the details of the molecular orbital approach to the representation of the π system of butadiene. If four sp^2-hybridized carbons and six hydrogens overlap to form C—C and C—H bonds, a four-carbon framework is obtained in which each carbon carries an unhybridized p orbital. For complete lateral overlap to take place between these p orbitals, it is necessary that the σ framework adopt either conformation (a) or (b) in Fig. 4.7, in which all the p orbitals are colinear with their neighbors.

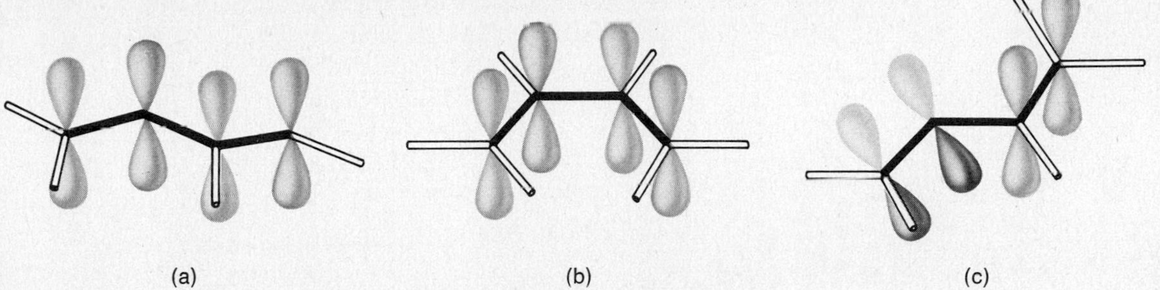

Fig. 4.7. Conformations of the σ framework of 1,3-butadiene.

Now, using the general rules for constructing molecular orbitals, let us look into the possibilities that arise when we combine the four atomic orbital wave functions for the $2p$ orbitals on each of the four carbon atoms in the butadiene system; *e.g.*,

$$\Psi_\pi = \psi_1 \pm \psi_2 \pm \psi_3 \pm \psi_4$$

where Ψ_π is the molecular orbital wave function for the π system, and ψ_1, ψ_2, ψ_3, and ψ_4 are the atomic orbital wave functions for the $2p$ orbitals on carbon atoms 1, 2, 3, and 4.* Mathematical manipulation of this expression by appropriate methods of approximation yields four molecular orbitals† for which standing wave representations are pictured in Fig. 4.8. The lowest energy orbital, Ψ_1, contains no nodes and is bonding between C-1 and C-2, C-2 and C-3, and C-3 and C-4 (note that the bonding between C-2 and C-3 is of the π type and exists in addition to the σ bond between C-2 and C-3). The next higher orbital, Ψ_2, contains one node; it is bonding between C-1 and C-2 and between C-3 and C-4,

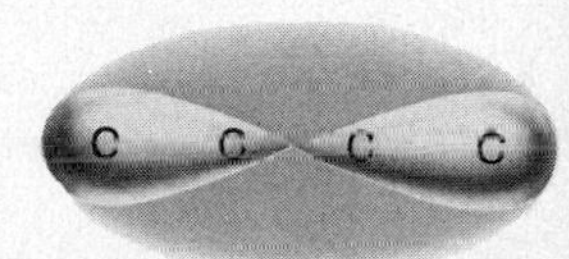

Composite of Ψ_1 and Ψ_2 for the π-system of 1,3-butadiene

---

* The molecular orbital wave function is more correctly given (see footnote on p. 32) as $\Psi_\pi = c_1\psi_1 \pm c_2\psi_2 \pm c_3\psi_3 \pm c_4\psi_4$ where c_1–c_4 are coefficients indicating how much of each atomic orbital is mixed into the molecular orbital.

† The number of orbitals generated by the linear combination of atomic orbitals is always equal to the number of atomic orbitals used. Molecular orbitals are characterized as *bonding* when the number of bonding interactions between adjacent atoms (*i.e.*, no change in the sign of the MO wave function) exceeds the number of antibonding interactions (*i.e.*, a change in sign in the MO wave function), as *nonbonding* when the number of bonding and antibonding interactions are equal, or as *antibonding* when the number of bonding interactions is less than the number of antibonding interactions. Nonbonding molecular orbitals also arise when the MO wave function has zero amplitude (*i.e.*, a node) right on the atom rather than between atoms as, for example, in the three-carbon π system shown.

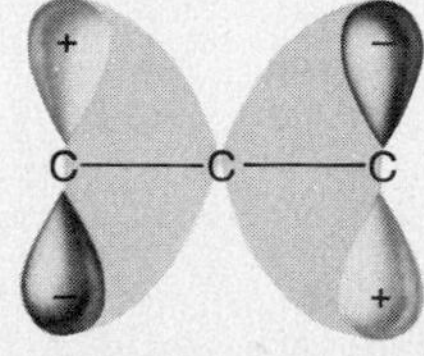

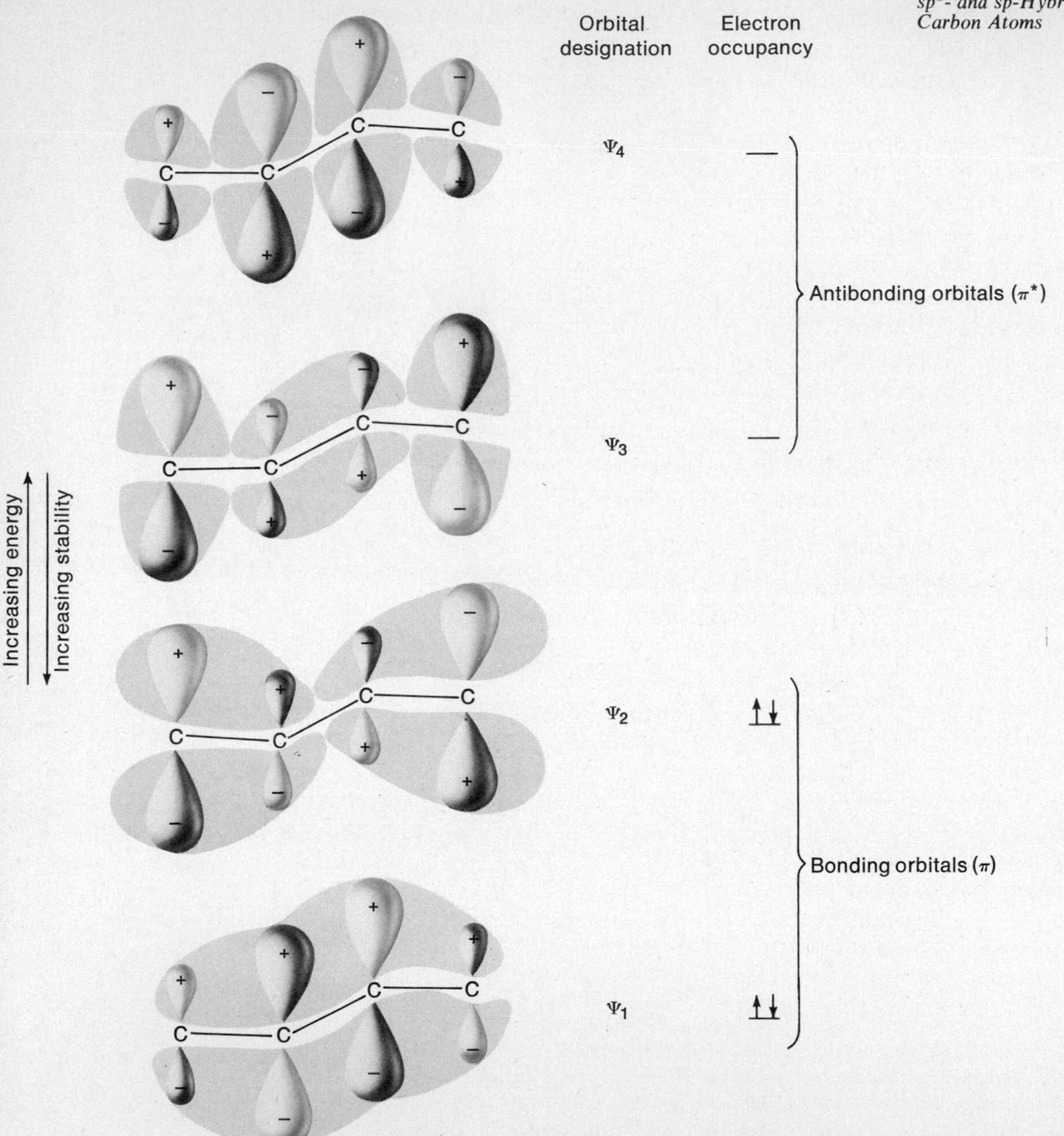

Fig. 4.8. An energy diagram illustrating the bonding and antibonding molecular orbitals for the π system of 1,3-butadiene.

but antibonding between C-2 and C-3. Orbitals Ψ_3 and Ψ_4 contain two and three nodes, respectively, and are antibonding.* Having constructed the molecular orbitals for the π system, we then add electrons to arrive at the neutral molecule butadiene. Four electrons are required, the first two going into Ψ_1 and the second two into Ψ_2 to give a molecule in which the π bonding is a composite of Ψ_1 and Ψ_2.

Thus, this treatment predicts the presence of some net π-bonding character between C-2 and C-3. To the extent that this is the case, the classical, localized bond structure is an inadequate representation, and a delocalized π system must be indicated in which electrons are *not* constrained to remain between only two nuclei but can be associated with a greater portion of the molecule. A certain amount of "double-bond character" is imputed to the bond between C-2 and C-3 of 1,3-butadiene, in accord with spectral measurements which indicate an increased barrier to rotation around this bond. Heat of combustion measurements, which show an increased bond strength, and X-ray diffraction measurements, which show a shortened bond (1.47 Å, as compared with 1.54 Å in ethane), are commensurate with this viewpoint.

4.6 Resonance Structures

The inadequacy of single structures to represent certain compounds properly was alluded to in Section 2.8 in connection with strongly polarized compounds, such as LiH and HF. A similar circumstance arises in 1,3-butadiene, where the classical, localized bond structure $H_2C{=}CH{-}CH{=}CH_2$ does not indicate that there is some "double-bond" character to the bond between C-2 and C-3. Although the molecular orbital interpretation of the butadiene structure accommodates this feature, its pictorial representation is cumbersome to draw and requires an artistic talent not possessed by most organic chemists. An easier pictorial representation comes from an alternative treatment of molecular bonding known as the *valence bond approach.*

In the valence bond treatment, in contrast to the molecular orbital treatment, explicit recourse is made to classical, localized bond structures. The wave function for the bonding in the molecule is constructed by a linear combination not of atomic orbitals, as in the molecular orbital method, but of *localized bond structures*; *i.e.*, $\Psi_{\text{molecule}} = \psi_{\text{structure-1}} \pm \psi_{\text{structure-2}}$. The real molecule is then considered to be a *hybrid* (usually referred to as a *resonance hybrid*) of the localized bond structures comprising the molecular orbital wave function. Although the valence bond method is mathematically more difficult than the molecular orbital method, it has the great virtue of retaining classical structures in the representation of delocalized bond structures, such as butadiene, and charge-separated structures, such as LiH and HF. Thus, according to the resonance concept,

---

* It might be argued that the antibonding character between the C-2 and C-3 positions in Ψ_2 just offsets the bonding character of Ψ_1 between these atoms, in which event the composite picture of the two orbitals would be precisely equivalent to that depicted by the classical structure. Calculations reveal, however, that the bonding interaction between C-2 and C-3 in orbital Ψ_1 is greater than the antibonding interaction between these carbons in orbital Ψ_2.

1,3-butadiene is a resonance hybrid of the following structures:

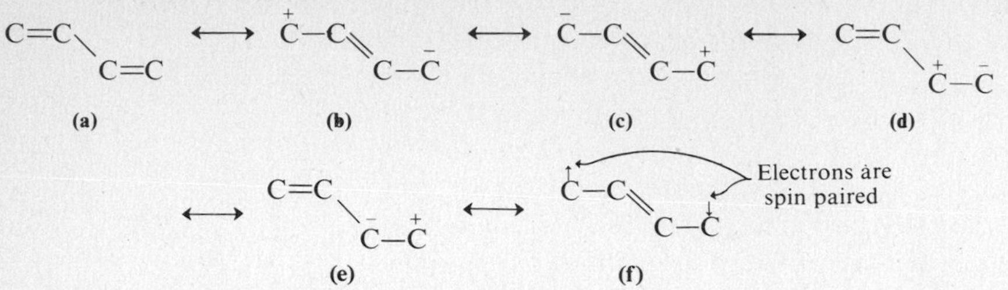

(a) ⟷ (b) ⟷ (c) ⟷ (d)

⟷ (e) ⟷ (f)

Electrons are spin paired

where the double headed arrow, ↔, carries the specific connotation that the structures in this collection are resonance structures.* No one of them, by itself, represents butadiene; the true structure is a hybrid of all these resonance structures, admixed in varying proportions. Furthermore, the hybrid is *not* to be construed as a system that is oscillating between structures (a), (b), (c), (d), (e), and (f). At all moments in time, butadiene is a single entity, which, according to the precepts of the valence bond approach, must be represented by a composite of resonance structures. The following two rules govern the drawing of resonance structures for various systems where localized bonding inadequately depicts the real structure.

1. The positions of *all* the atoms in the contributing structures must remain unchanged in space.† For example, cyclobutene is *not* a legitimate resonance structure of 1,3-butadiene, because the distance between two of the atoms (shown in bold face) is different in the two structures:

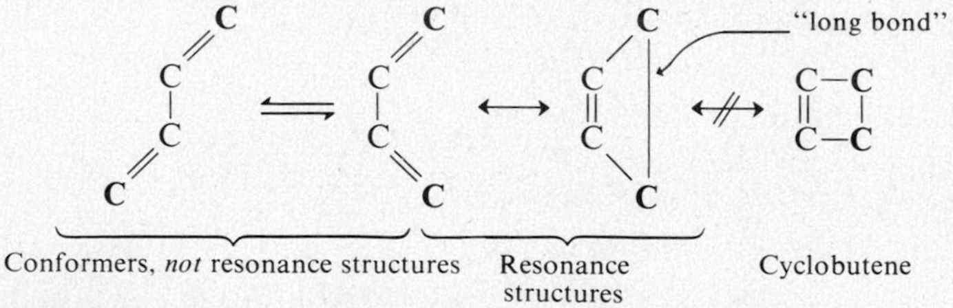

"long bond"

Conformers, *not* resonance structures — Resonance structures — Cyclobutene

2. The same number of unpaired electrons must be present in all resonance structures. In the case of 1,3-butadiene, the number of unpaired electrons in the resonance structures (a)–(d) is zero. A structure resembling (f) but containing two spin-unpaired electrons, for example, is not a legitimate resonance structure for this system.

---

* It is important to remember that the double-headed arrow should be reserved for this purpose. It should not, for instance, be used to indicate an equilibrium between species; for that, the symbol ⇌ should be used.

† There is a subtle difficulty in this requirement. In the resonance structures (a) and (b) for butadiene, for instance, there are differences in the bond character at any particular bond; in one structure, a single bond of 1.54-Å length changes to a double bond of 1.33-Å length in the other structure, necessitating what might appear to be slight changes in atomic positions. In the real molecule butadiene, the bond lengths are intermediate between C—C and C=C bond lengths; thus, in (*a*), the C-2/C-3 bond length is "compressed," whereas in (b), (c), and (d), the C-2/C-3 bond length is "elongated."

Electrons are spin paired

Electrons are spin unpaired (not a resonance structure of 1,3-butadiene)

A resonance hybrid is usually more stable than any of its contributing structures. The difference in energy between the resonance hybrid (delocalized bond structure) and the lowest energy, localized bond structure is called the *resonance stabilization energy* (or *resonance energy* or *delocalization energy*). For illustration, let us compare the heat that is evolved when 1,3-pentadiene and 1,4-pentadiene are burned to carbon dioxide and water; *i.e.*, their heats of combustion:

$$CH_3CH{=}CHCH{=}CH_2 \;+\; 7\,O_2 \longrightarrow 5\,CO_2 \;+\; 4\,H_2O \qquad \Delta H = -720.05 \text{ kcal/mole}$$

$$H_2C{=}CHCH_2CH{=}CH_2 \;+\; 7\,O_2 \longrightarrow 5\,CO_2 \;+\; 4\,H_2O \qquad \Delta H = -726.65 \text{ kcal/mole}$$

The extra 6.6 kcal/mole of heat that is evolved when 1,4-pentadiene is combusted is a measure of the greater total bond strength possessed by the conjugated diene. A comparison of the heat that is evolved when the dienes are hydrogenated to *n*-pentane shows a similar difference:

$$CH_3CH{=}CHCH{=}CH_2 \;+\; 2\,H_2 \longrightarrow CH_3CH_2CH_2CH_2CH_3 \qquad \Delta H = -54.1 \text{ kcal/mole}$$

$$H_2C{=}CHCH_2CH{=}CH_2 \;+\; 2\,H_2 \longrightarrow CH_3CH_2CH_2CH_2CH_3 \qquad \Delta H = -60.8 \text{ kcal/mole}$$

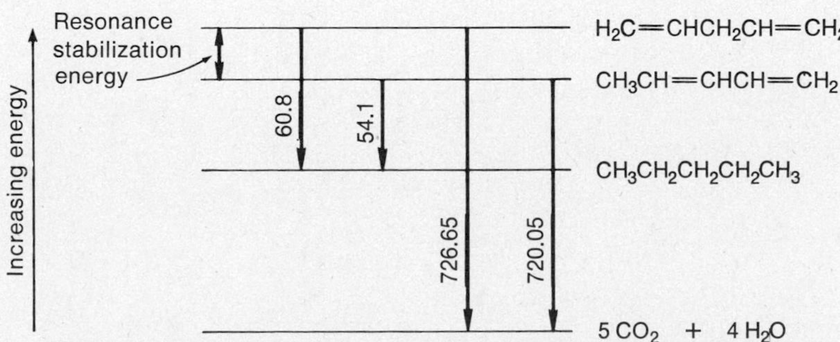

Fig. 4.9. Energy diagram for the combustion and hydrogenation of 1,3-pentadiene and 1,4-pentadiene. The numbers in the diagram indicate heats of reaction in kilocalories per mole.

The 6.7 kcal/mole additional heat that is evolved in the hydrogenation of 1,4-pentadiene indicates that the unconjugated diene is less stable than the conjugated diene by this amount, the value agreeing very closely with the one obtained from heat of combustion data (Fig. 4.9).*

Not all resonance structures contribute equally to the resonance hybrid, and it is useful to have some general rules-of-thumb as a basis for making predictions concerning the relative importance of the various contributing structures.

1. Structures containing complete octets of electrons are more stable than structures containing fewer than eight electrons in the valence shell. For example, resonance structure (a) for butadiene is more stable than resonance structures (b), (c), or (d), all of which contain one or more carbons in which fewer than eight electrons are present in the valence shell of carbon. (First-row and second-row elements are incapable of accommodating more than eight electrons in their valence shell, but third-row, and higher-row, elements permit an expansion of the octet by the use of *d* orbitals).
2. The greater is the number of covalent bonds, the greater is the contribution of the resonance structure to the resonance hybrid. The structure $H_2C{=}CHCH{=}CH_2$ of butadiene has more covalent bonds (*i.e.*, a total of 11, counting each double bond twice) than the structure $H_2\overset{+}{C}{-}CH{=}CH{-}\overset{-}{C}H_2$ (*i.e.*, a total of 10) and, accordingly, contributes more significantly.
3. Bonds between adjacent atoms contribute more significantly than bonds between nonadjacent atoms (sometimes called "long bonds"). Thus, the structure $H_2C{=}CHCH{=}CH_2$ of butadiene contributes more than the structure $H_2\overset{\uparrow}{C}{-}CH{=}CH{-}\overset{\downarrow}{C}H_2$, in which the lone electrons at C-1 and C-4 are spin paired and, therefore, bonded; *i.e.*,

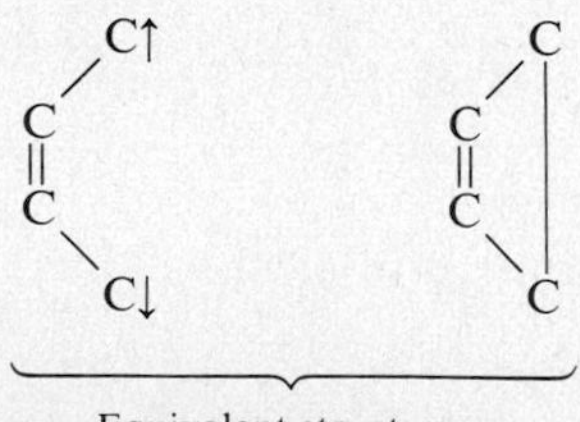

Equivalent structures

* Often, we cannot make a direct comparison between the conjugated diene and an appropriate nonconjugated analog. We then adopt another approach, as illustrated for 1,3-butadiene, where the heat of hydrogenation of 1-butene (multiplied by 2) is compared with the heat of hydrogenation of 1,3-butadiene:

$$2\,(CH_3CH_2CH{=}CH_2 \;+\; H_2) \longrightarrow 2\,(CH_3CH_2CH_2CH_3) \qquad \Delta H = (2 \times -30.3) = -60.6 \text{ kcal/mole}$$

$$H_2C{=}CHCH{=}CH_2 \;+\; 2\,H_2 \longrightarrow CH_3CH_2CH_2CH_3 \qquad \Delta H = -57.1 \text{ kcal/mole}$$

The difference of 3.5 kcal/mole between these heats of hydrogenation provides a measure of the resonance stabilization energy of 1,3-butadiene.

4. The greater is the number of charges in a structure, the lower is the contribution of that structure. Thus, $H_2\overset{+}{C}-CH=CH-\overset{-}{C}H_2$ contributes less than $H_2C=CHCH=CH_2$ but more than $H_2\overset{+}{C}-\overset{-}{C}H-\overset{+}{C}H-\overset{-}{C}H_2$.
5. In structures containing *unlike* charges, the closer the charges are the greater is the contribution of the structure. Conversely, in structures containing *like* charges, the farther apart the charges are the greater is the contribution of the structure. For example, $H_2C=CH-\overset{-}{C}H-\overset{+}{C}H_2$ contributes more than $H_2\overset{-}{C}-CH=CH-\overset{+}{C}H_2$.

Charge-separated resonance structures do not contribute very much to the resonance hybrid of butadiene; carbon–carbon bonds are essentially homopolar, so the molecule has a zero time-averaged dipole moment. In other types of bonds, however, charge-separated structures may contribute significantly to the resonance hybrid. For example, the series of hydrides of the first-row elements (see Fig. 2.14) are increasingly misrepresented by covalent bond structures as the position of the first-row element in the periodic table shifts to either the left or the right of carbon. Thus, although the contribution of charge-separated resonance structures in methane is extremely small, lithium hydride and hydrogen fluoride require a great deal of mixing in of the charge-separated structures in the resonance hybrid (Fig. 4.10).

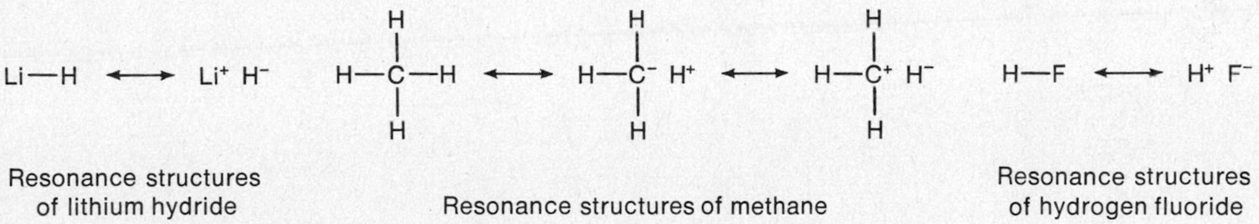

$$Li-H \longleftrightarrow Li^+\ H^-$$

$$CH_4\ (H-C-H \text{ with H above and below}) \longleftrightarrow H_3C^-\ H^+ \longleftrightarrow H_3C^+\ H^-$$

$$H-F \longleftrightarrow H^+\ F^-$$

Resonance structures of lithium hydride

Resonance structures of methane

Resonance structures of hydrogen fluoride

Fig. 4.10. Resonance structures of lithium hydride, methane, and hydrogen fluoride.

Resonance structures serve a useful purpose in describing the structures and interpreting the chemistry of many molecules, and it is essential that the chemist be thoroughly conversant with their use. Realize that molecular orbital theory and the valence bond theory (resonance theory) represent two different methods of approximation for solving the *same* problem, and that if carried through with sufficient refinement they yield identical answers. The particular virtue of the valence bond approach is its retention of localized bond representations. The reader is encouraged at this point to review Sections 4.5 and 4.6 before reading further to be sure that the concepts of molecular orbital and valence bond descriptions are well understood.

PROBLEM 4-4

The π system of 1,3,5-hexatriene is constructed from six *p* orbitals. How many molecular orbitals are obtained when these six *p* orbitals are combined by the LCAO method? Suggest the shapes that these orbitals have. How many of the orbitals are bonding and how many are antibonding? In the neutral molecule, how many electrons are in the π system of 1,3,5-hexatriene?

4.7 The Structure of Benzene

The extent of π bond delocalization in 1,3-butadiene is sufficiently small that we do not err greatly in representing this compound by its classical localized bond structure. Let us now look at a system that in no way can be adequately represented by a single classical, localized bond structure. That system is benzene, a molecule constructed from six sp^2-hybridized carbon atoms and six hydrogen atoms, in which the carbons are arranged in a cyclic hexagonal array. Since the angle between the hybrid orbitals of an sp^2-hybridized carbon is 120°, the hexagon is planar, and the *p* orbitals on each of the six carbon atoms are perfectly aligned to engage in lateral overlap (Fig. 4.11). Herein lies a problem, however; because the system is completely symmetrical, there is no more reason for effective overlap to take place with one neighbor than with the other. Therefore, the classical, localized bond representation (*i.e.*, 1,3,5-cyclohexatriene) is completely inadequate (Fig. 4.11).

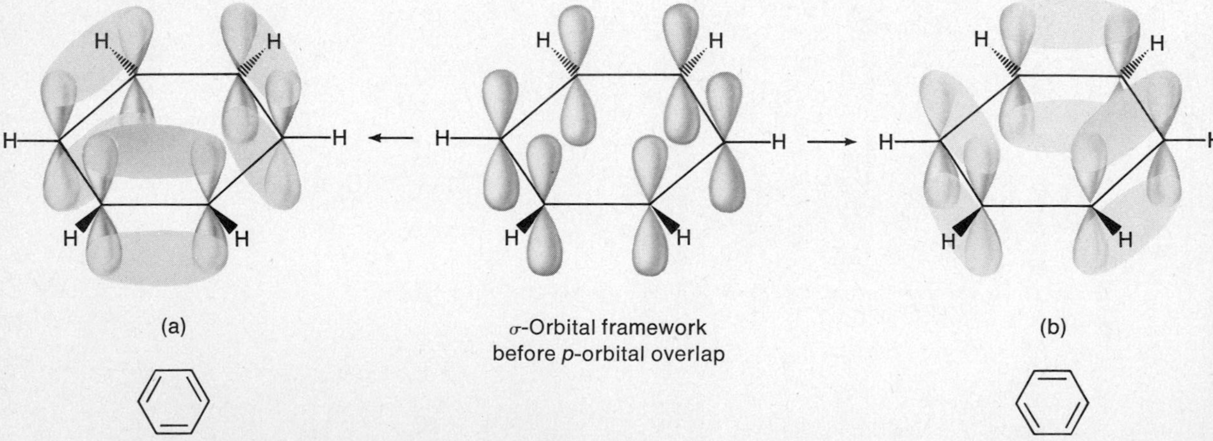

Fig. 4.11. Molecular orbital and abbreviated dashed-bond representations of localized π-bond structures of benzene.

The "benzene problem" is handled in the molecular orbital treatment in the same way that has been described for butadiene. Six molecular orbital wave functions are constructed from six 2*p* atomic orbital wave functions, one from each of the six carbon atoms. Because of the cyclic structure of this π system, it turns out that although there are six molecular orbitals there are only four energy levels, orbitals Ψ_2 and Ψ_3 having the same energy and orbitals Ψ_4 and Ψ_5 having the same energy (Fig. 4.12). Having constructed the orbitals for the π system of

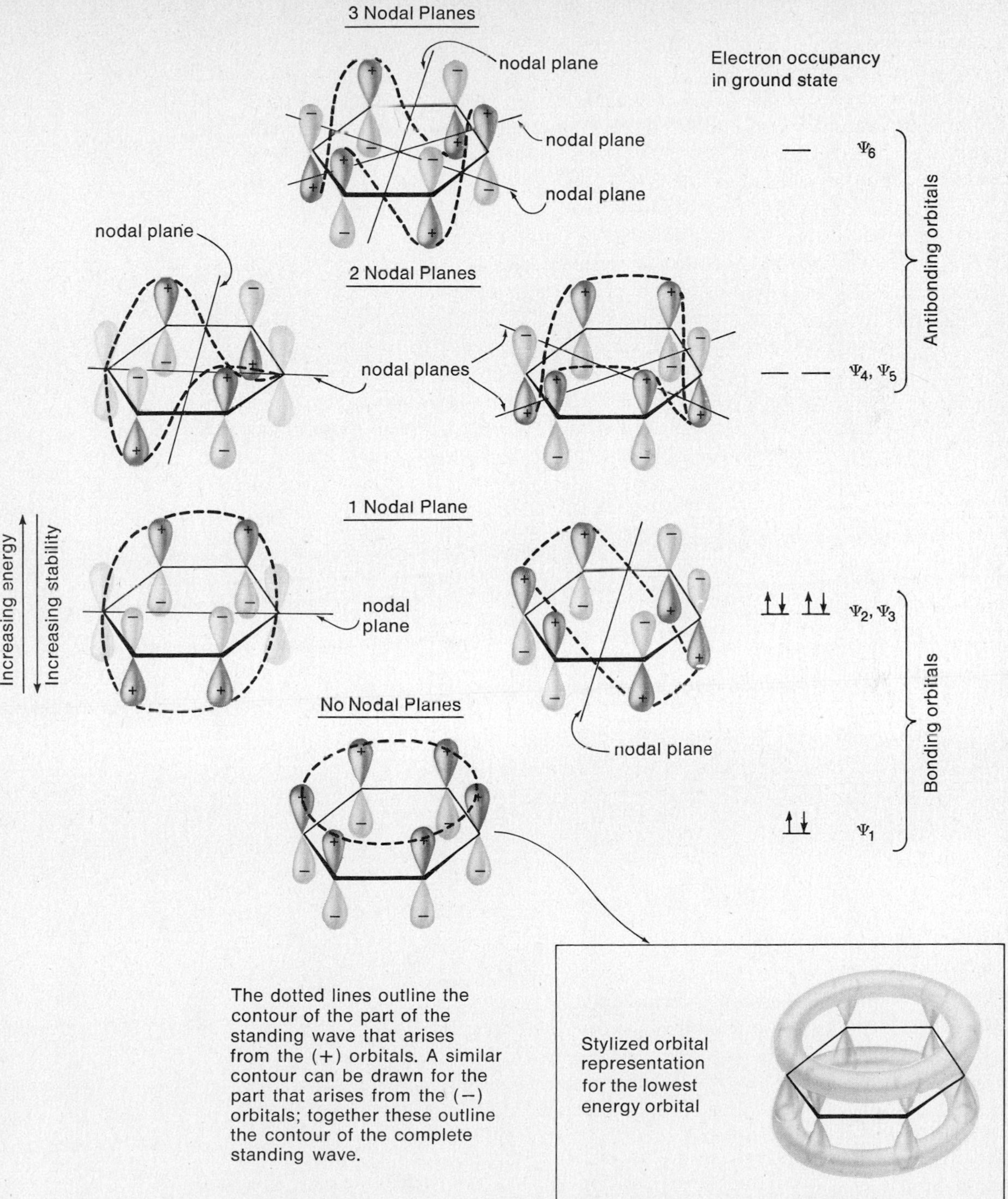

Fig. 4.12. An energy diagram illustrating the molecular orbitals in the π-system of benzene. The unshaded orbitals that contain neither a + or a − sign are orbitals which reside at the node of a molecular orbital.

benzene, we then add six electrons to give the neutral molecule. Two of these go into the lowest energy orbital Ψ_1, and the remaining four go into the equal energy orbitals (degenerate orbitals) Ψ_2 and Ψ_3. The structure that results has *identical* bonding between all six of the carbon atoms in the ring, bonding that is somewhere between a C—C and a C=C bond. X-ray diffraction measurements of benzene substantiate this picture and show *only one* carbon–carbon bond distance of 1.396 Å, a value that is closer to the C=C bond length of ethylene (1.34 Å) than the C—C bond length of ethane (1.54 Å). Thus, the electron system of benzene is *completely delocalized* and encompasses the entire ring.

The resonance hybrid structure just described for benzene is very different from the representation of benzene as cyclohexatriene. In fact, only by considering benzene to be a resonance hybrid comprising at least a pair of cyclohexatriene resonance structures (called *Kekulé structures*) can we come at all close to an accurate picture of the real molecule. For an even closer approach to the "truth," other resonance structures must also be added, such as the *Dewar structures* (containing "long bonds") and charge-separated structures, although these are much less important contributors than the Kekulé structures (Fig. 4.13).

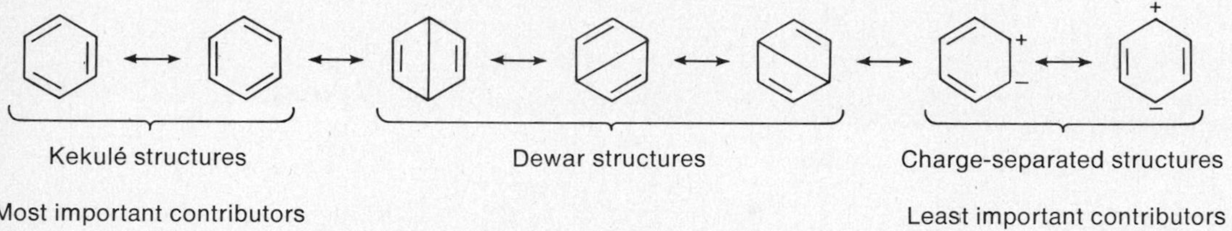

Fig. 4.13. Resonance structures of benzene.

As a shorthand device for representing the delocalized bond structure of benzene, it is customary to draw a hexagon in which a circle is inscribed.

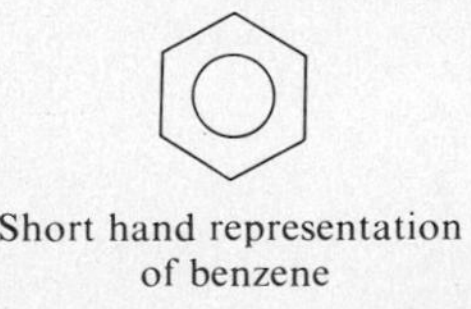

Short hand representation of benzene

We again stress the idea that benzene, represented as a resonance hybrid of two Kekulé structures (and other resonance structures), should *not* be imagined to be in one Kekulé form one moment and in the other Kekulé form the next, oscillating between the two. Instead, at all moments it is in a single form, which partakes of the characteristics of each of the Kekulé forms. For illustration, let us conduct an experiment with a can of blue paint, a can of yellow paint, and a pair of cylinders equipped so that they can be rotated rapidly on their axes. One of the cylinders we paint half-blue and half-yellow, the interface between the two colors being along the axis of the cylinder. We then mix the blue and yellow paints, and with the green paint that is produced we paint the other cylinder. Now, the cylinders are rotated rapidly, and to the ordinary eye they both appear to have approximately the same color, *viz.*, green. But one of them is green because of the rapid oscillation between blue and yellow (*i.e.*, the analog of rapidly oscillating Kekulé structures), whereas the other cylinder is green because it is inherently so (*i.e.*, it is not blue one moment and yellow the next but is green at all times, although having some of the attributes of blueness and some of the attributes of yellowness). The inherently green cylinder is the analog of the benzene molecule.

4.8 Aromaticity

Benzene is a compound that has a higher carbon-to-hydrogen ratio than any of the alkanes and most of the alkenes. It was originally obtained by decarboxylation of benzoic acid, which, in turn, had been isolated from a natural source called *gum bezoin.* Benzene is a volatile compound possessing a fragrant aroma; hence, it was called *aromatic*, a term that has since come to be applied to the large family of compounds bearing a structural resemblance to benzene. The physical and chemical characteristics associated with compounds of this family are called "aromatic properties" and include the following:

1. *Carbon-to-Hydrogen Ratio*: Aromatic compounds have a high ratio of carbon to hydrogen; *i.e.*, benzene is C_6H_6, naphthalene (see Section 4.9) is $C_{10}H_8$, phenanthrene (see Section 4.9) is $C_{14}H_{10}$, etc. The majority of nonaromatic compounds (generally referred to as aliphatic compounds) do not have such high C:H ratios. There are exceptions, however, such as the compounds $H_2C{=}C{=}C{=}CH_2$ and $HC{\equiv}C{-}C{\equiv}CH$, which have the molecular formulas C_4H_4 and C_4H_2, respectively.
2. *Bond Lengths*: The carbon–carbon bonds in the benzene ring are intermediate in length between double and single bonds and have a bond length of approximately 1.39 Å.
3. *Chemical Reactivity*: One of the characteristic chemical reactions of alkenes and alkynes is their tendency to undergo "addition reactions," whereas aromatic compounds, although "unsaturated" and possessing π bonds, tend instead to undergo "substitution reactions." These are discussed in Chapter 8.
4. *Spectral Characteristics*: Aromatic compounds have ultraviolet, infrared, and nuclear magnetic resonance absorption characteristics sufficiently different from those of aliphatic compounds to allow a clear distinction. These are discussed in Chapter 6.
5. *Resonance Stabilization Energy*: Molecular orbital calculations and measurements of heats of hydrogenation and heats of combustion reveal that benzene possesses approximately 36 kcal/mole of resonance stabilization energy (usually referred to simply as "resonance energy"). The resonance energy of benzene has been estimated by the indirect method (see p. 108) by comparing the heat of hydrogenation of 1 mole of benzene with that of 3 moles of cyclohexene. The heat of hydrogenation of benzene is found to be -49.8 kcal/mole and that for cyclohexene is found to be -28.6 kcal/mole. Assuming that the double bond of cyclohexene is an adequate model for a *localized* π system for benzene (i.e., cyclohexatriene), a value of 36.0 kcal/mole (*i.e.*, $3 \times 28.6 - 49.8$) of "extra stability" is calculated for benzene (see Fig. 4.14). A value that agrees very closely with

this has also been obtained from heat of combustion measurements of benzene and appropriate models for cyclohexatriene. The resonance stabilization energy (*i.e.*, 36 kcal/mole) for benzene is considerably greater than that for conjugated acyclic dienes and trienes (*i.e.*, 3–7 kcal/mole), and it imparts special physical and chemical properties to benzene.

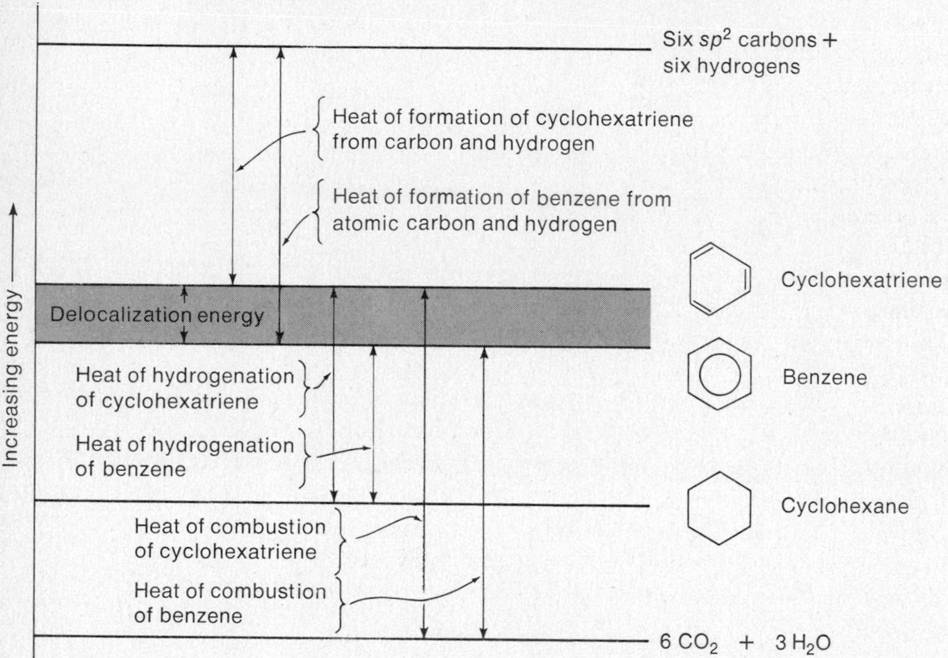

Fig. 4.14. Energy relationships between cyclohexatriene and benzene.

PROBLEM 4-5

Using the bond energy data in Table 2.3, calculate the amount of energy that is released when six carbon and six hydrogen atoms combine to form 1,3,5-cyclohexatriene. Knowing, however, that the actual molecule formed is benzene, and knowing that it has 36 kcal/mole of resonance energy, how would you correct this calculated value for the heat of reaction?

Knowing that benzene possesses "aromatic properties," we might predict that both the lower and higher vinylogs* of benzene should be similar in this respect. Neither cyclobutadiene nor cyclooctatetraene, however, is aromatic; both are very reactive compounds—cyclobutadiene so much so, in fact, that it cannot be isolated except at extremely low temperatures. The reasons for this

* A vinylog refers to a compound differing from another member of the series by one or more —CH=CH— units.

dramatic difference in properties between benzene on the one hand and cyclobutadiene and cyclooctatetraene on the other hand point up the special features that are required for aromaticity. An interpretation of this problem of aromaticity was put forth in 1931 by Eric Hückel, who approached it in terms of molecular orbital theory.

Molecular orbital energy diagrams for monocyclic, planar systems are always characterized by one, lowest-energy, bonding orbital. The higher-energy orbitals occur in sets of degenerate pairs until the highest-energy orbital is reached. If there is an even number of p orbitals contributing to the π system, there is a single highest-energy orbital; if there is an odd number of p orbitals contributing to the π system, there is a pair of degenerate highest-energy orbitals. Benzene, containing an even number of contributing p orbitals (*i.e.*, six), as we have seen, has one lowest-energy orbital, one highest-energy orbital, and two pairs of degenerate orbitals of intermediate energy. According to Hückel's interpretation of aromaticity, when any of the energy levels of such systems are completely filled (*i.e.*, two electrons in each of the orbitals of a given energy level), the system is especially stable. In the case of benzene, this is attained when the π system has acquired (a) 2 electrons (*i.e.*, the lowest-energy level filled), (b) 6 electrons (*i.e.*, first and second levels filled), (c) 10 electrons (*i.e.*, first, second, and third levels filled), etc. The first possibility corresponds to a benzene^{+4} cation, however, and the third possibility to a benzene^{-4} anion,* and systems containing such high charges are very unstable. Thus, the isolable species in this system is the neutral benzene molecule containing six electrons in the π system.

Let us now see how these ideas apply to cyclobutadiene, a molecule that has captured the attention of chemists for many years but has eluded synthesis until recently. Cyclobutadiene possesses an even number of p orbitals contributing to its cyclic, planar, π system and has one lowest-energy molecular orbital, a pair of degenerate orbitals, and one highest-energy orbital (Fig. 4.15).† According to the Hückel theory, an especially stable system (*i.e.*, an aromatic system) should be attained when two electrons are added, giving a cyclobutadiene^{+2} cation, or when six electrons are added, giving a cyclobutadiene^{-2} anion. The addition of four electrons, giving the neutral cyclobutadiene molecule, is predicted *not* to give an especially stabilized system, which is, indeed, the case. This is the reason that cyclobutadiene is so reactive and so difficult to synthesize.

Because there is only *one* lowest-energy orbital in a cyclic system but *pairs* of higher-energy orbitals, it is clear that *filled systems* occur when there are 2, 6, 10, 14,... electrons in the π system. This was recognized by Hückel, who stated that the necessary number of electrons which must be in a π system if it is to show aromaticity is $4n + 2$, where n is 0, 1, 2, 3,.... This generalization is now known as the *Hückel rule*. Systems containing $4n$ rather than $4n + 2$ electrons not only lack aromatic properties but in some instances may be even less stable than their acyclic analogs, in which case they are said to be *antiaromatic*.

---

* Realize that electrons can be added to unfilled orbitals even if they are high-energy, antibonding orbitals.

† The two degenerate orbitals have two bonding interactions and two antibonding interactions. Thus, the bonding and antibonding interactions cancel each other, and the orbital is nonbonding.

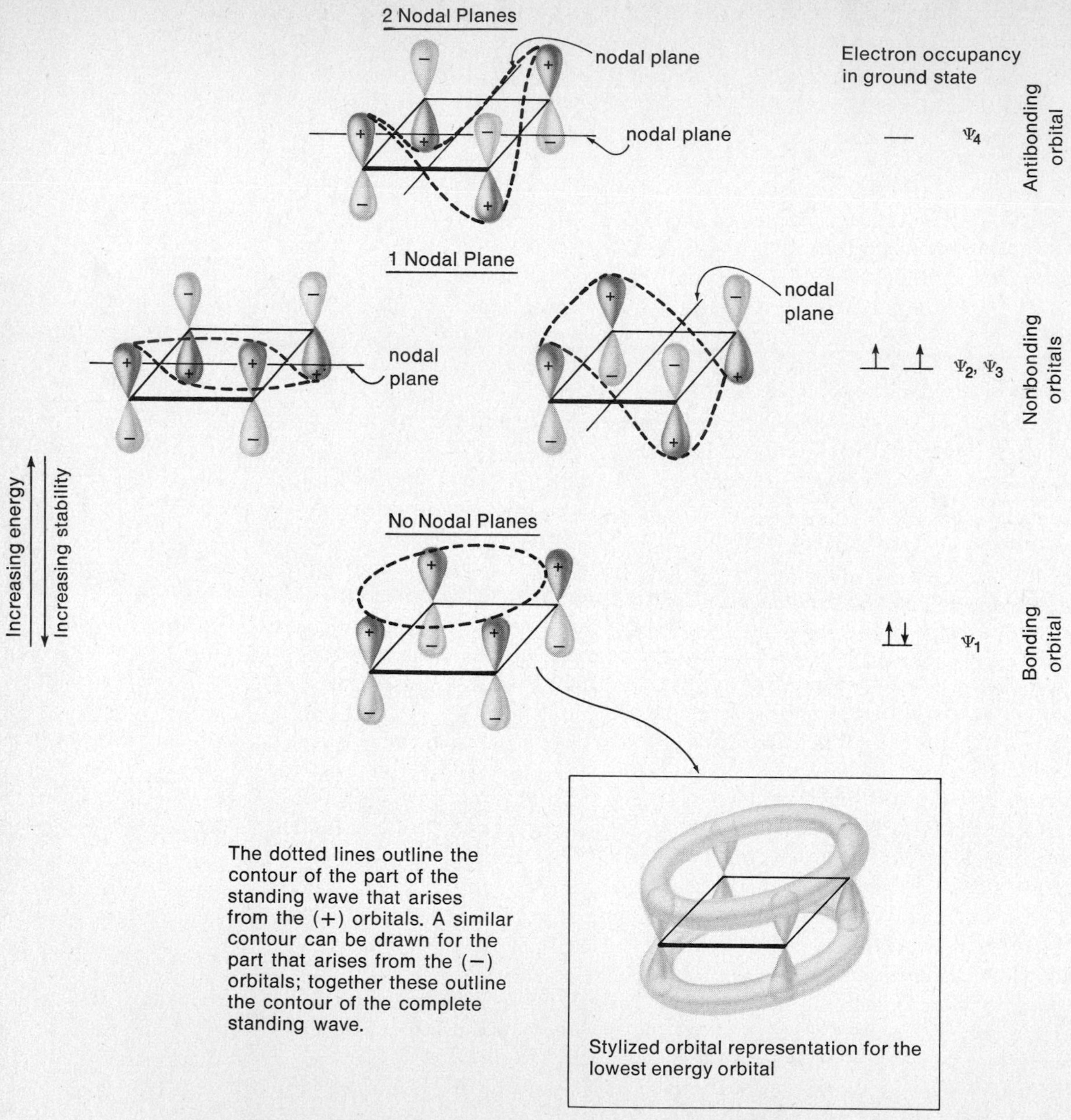

Fig. 4.15. An energy diagram illustrating the molecular orbitals of the π system of cyclobutadiene.

It is interesting to see how the Hückel rule applies to various other cyclic π systems. The molecular orbital energy diagram for a three-membered cyclic π system contains one lowest-energy bonding orbital and a pair of degenerate antibonding orbitals. According to the Hückel rule, this system should be aromatic if it contains two electrons. This corresponds not to a neutral molecule but to the cyclopropenyl cation (Fig. 4.16). Cyclopropenyl cations have been synthesized, and they do, indeed, possess at least some of the physical and chemical criteria for aromaticity.

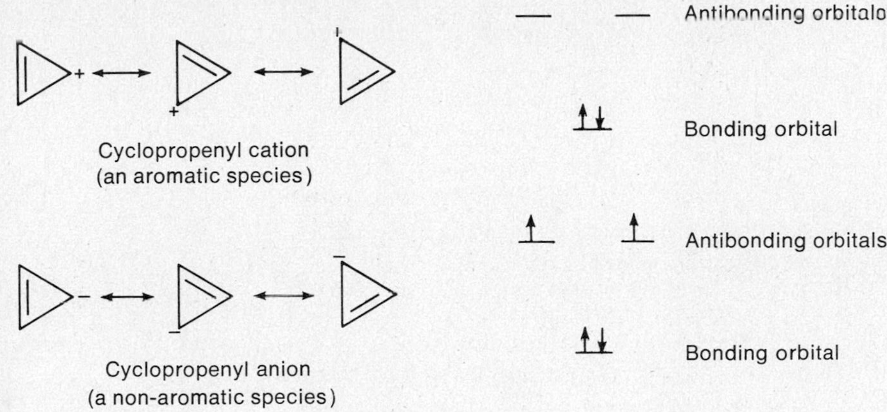

Fig. 4.16. Representation of the cyclopropenyl cation and anion.

A five-membered cyclic π system is predicted to possess aromatic character if it contains six electrons. This corresponds to the cyclopentadienyl anion, which is a well-known and very stable entity (Fig. 4.17). A seven-membered π system is also predicted to be aromatic if it contains six electrons. This corresponds to the cycloheptatrienyl cation (called a tropylium ion), which has been synthesized

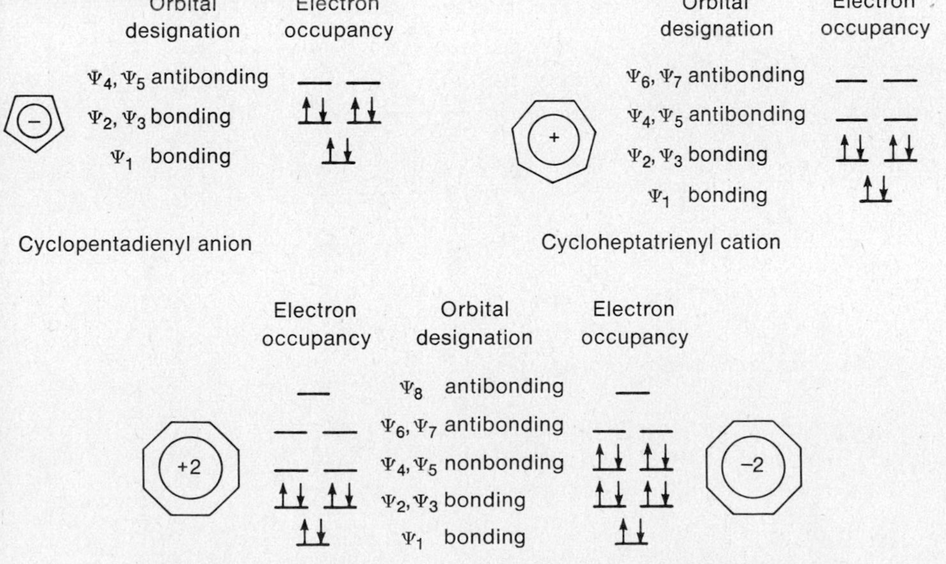

Fig. 4.17. Representations of the cyclic π systems possessing aromatic character.

and shown to be very stable (Fig. 4.17). Although cyclooctatetraene is nonaromatic, the dication (corresponding to a 6-electron system) obtained by removing 2 electrons and the dianion (corresponding to a 10-electron system) obtained by adding 2 electrons have both been prepared and been shown to possess certain aromatic properties (Fig. 4.17).

Systems such as cyclopropenyl cations, cyclopentadienyl anions, and cycloheptatrienyl cations, which possess aromatic character, have, like benzene, identical carbon–carbon bonds around the ring. Thus, the positive or negative charge in these systems is not localized at a particular carbon but is evenly distributed among all the carbons. We can indicate this by writing resonance structures, such as those shown in Fig. 4.16 for the cyclopropenyl cation. In similar fashion, five and seven resonance structures can be drawn for cyclopentadienyl anion and cycloheptatrienyl cation, respectively.

4.9 Aromatic Compounds

Benzene, the parent member of the "benzenoid aromatics," can have one or more of its hydrogen atoms replaced by the various types of carbon frameworks that have been discussed in previous sections. Thus, alkyl (and cycloalkyl) groups, alkenyl (and cycloalkenyl) groups, and alkynyl (and cycloalkynyl) groups can be attached to benzene to give compounds that are named as substituted benzenes (Fig. 4.18). Substituted benzenes are often represented in abbreviated form as C_6H_5-X, Ph-X, ϕ-X, and Ar-X; the first three designations (*i.e.*, C_6H_5, Ph, and ϕ) are specific for the benzene ring, whereas the last one indicates any aromatic ring.

Space-filling model of benzene

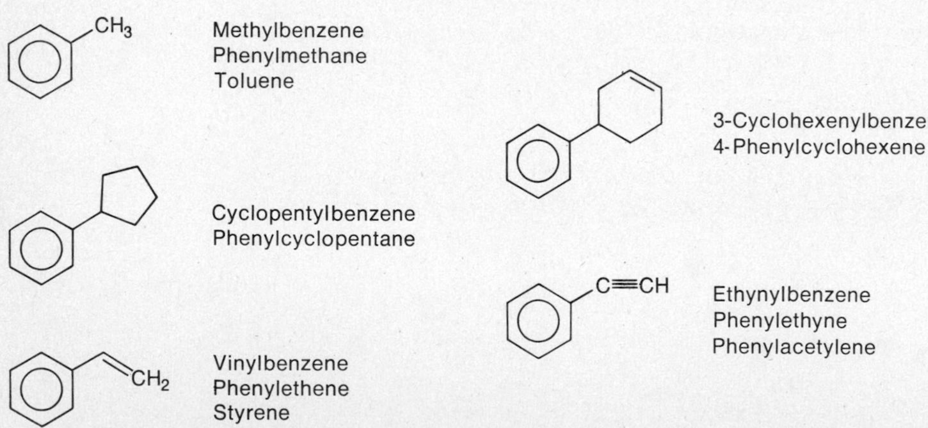

Fig. 4.18. Structures and names of substituted benzenes.

If two substituents are attached to a benzene ring, three structural isomers are possible, for the attachment can be at the 1,2 positions, the 1,3 positions, or the 1,4 positions. The positions are designated either by numbers or by the prefixes "ortho" (Greek, *orthos*, "straight"), "meta" (Greek, *meta*, "between"), and "para" (Greek, *para*, "beyond"), usually abbreviated as *o*-, *m*-, and *p*- (Fig. 4.19). Attachment of three identical substituents to benzene allows three structural isomers, *i.e.*, the 1,2,3, the 1,2,4, and the 1,3,5 isomers. If the three substituent groups are not identical, more structural isomers are possible, and the reader can

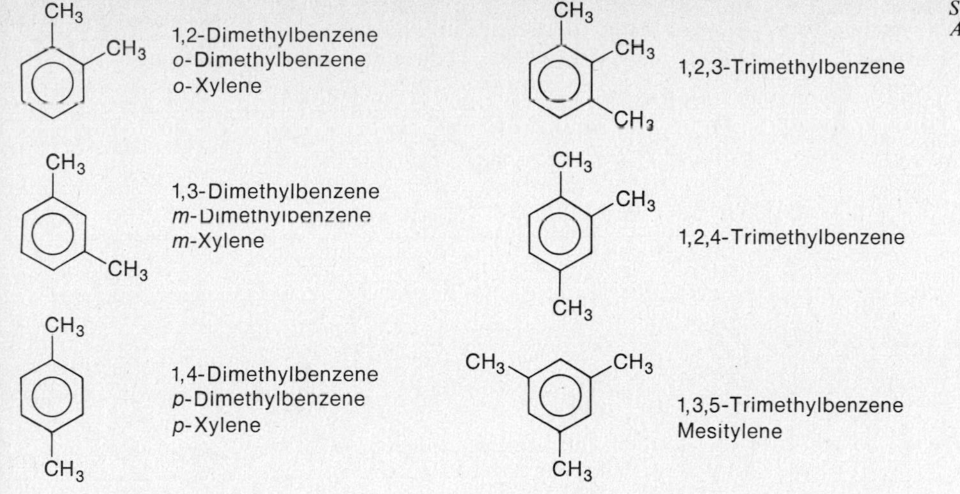

Fig. 4.19. Structures and names of dimethyl- and trimethylbenzenes.

test his understanding of structural isomerism in the benzene series by drawing these structural possibilities.

PROBLEM 4-6

Draw all the C_9H_{12} compounds containing a benzene ring.

Many compounds contain several aromatic rings attached to each other in various ways. Two or more benzene rings can be linked, for example, by single bonds to give compounds such as biphenyl, terphenyl, etc. (Fig. 4.20). The

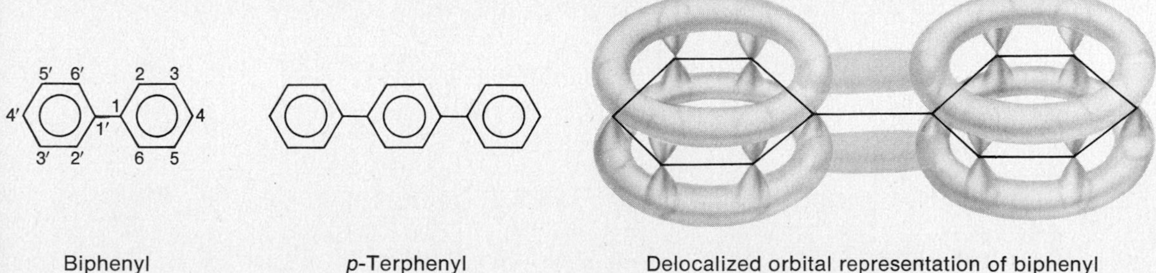

Fig. 4.20. Structures of biphenyl and terphenyl and an orbital representation of biphenyl.

benzene rings in these polyphenyls are conjugated with each other, and the bond between them is shortened and strengthened in a fashion comparable to the bond between C-2 and C-3 in butadiene. Alternatively, aromatic rings can be joined at a pair of adjacent carbons (*i.e.*, edgewise combination) to yield compounds such as naphthalene, anthracene, and chrysene (Fig. 4.21), in which 10, 14, 18 ... sp^2-hybridized carbons comprise planar frameworks consisting of 2, 3, 4, ... hexagonal benzenoid units.* These compounds contain π systems that are delocalized over the entire polycyclic ring system.

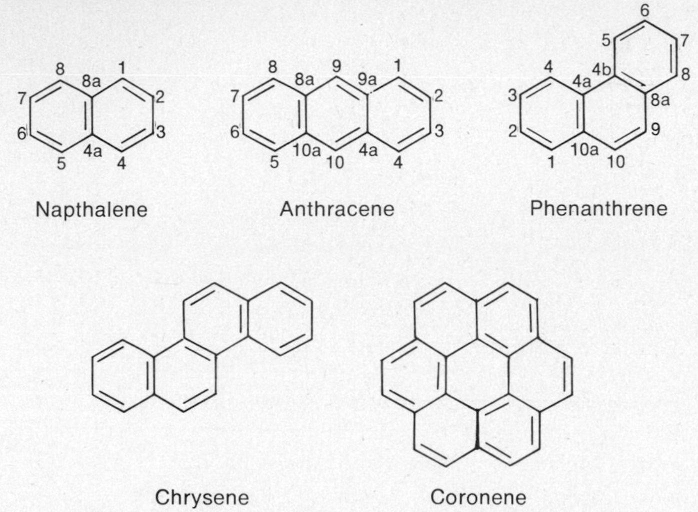

Napthalene

Anthracene

Phenanthrene

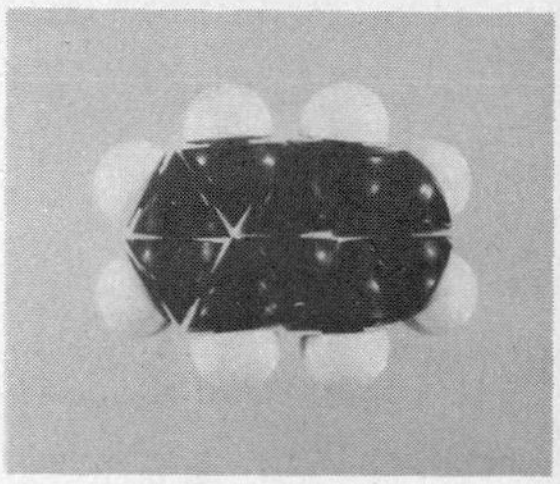

Space-filling model of napthalene

Chrysene

Coronene

Fig. 4.21. Structures of polycyclic aromatic hydrocarbons.

The proliferation of benzenoid rings can continue almost indefinitely. Graphite represents an extreme case, in which a huge number of sp^2-hybridized carbons are incorporated in a polyhexagon structure supporting a delocalized system. The planar, "two-dimensional" structure confers interesting properties on graphite, one of which is its efficacy as a lubricant.

PROBLEM 4-7

Draw all the $C_{18}H_{12}$ aromatic compounds containing four benzenoid rings.

Although it is not immediately obvious, there is another way in which 10 sp^2 carbon atoms can form a framework that is essentially planar and which, therefore, allows extensive π-bond delocalization. It is the fused-ring bicyclic framework containing a five-membered and a seven-membered ring. The compound, called azulene because of its deep blue color, possesses aromatic propeties.

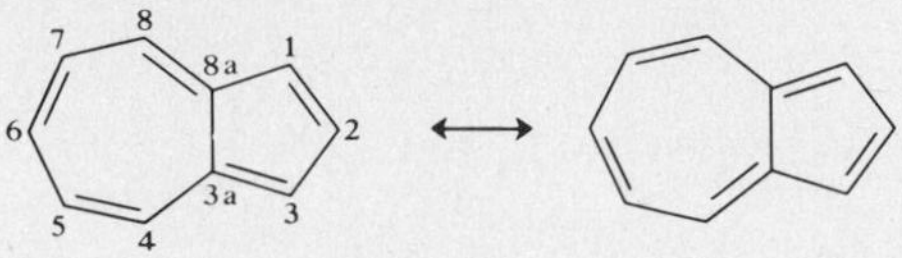

Azulene

---

* Polycyclic aromatic hydrocarbons, such as naphthalene, anthracene, and phenanthrene, are often called "fused-ring" aromatics.

Azulene, cyclopropenyl cation, cyclopentadienyl anion, cycloheptatrienyl cation, etc., are all examples of *nonbenzenoid aromatic compounds.*

Aromatic rings frequently provide the basis for assigning the parent name of the compound in which they are present. In some instances, however, the aromatic rings must be treated as substituent groups, in which case "yl" is added to a prefix made from the stem portion of the name; *e.g.*, naphthalene becomes naphthyl-, phenanthrene becomes phenanthryl-, and anthracene becomes anthryl-. Benzene, however, does *not* become benzyl- (which stands for the group C_6H_5—CH_2—) but becomes phenyl. As a generic group name, the term "arene" is often applied to the aromatic hydrocarbons; when present as substituents (*e.g.*, phenyl, naphthyl, etc.), they are referred to as *aryl groups.* As mentioned previously, the phenyl group is abbreviated by C_6H_5-, Ph-, or ϕ-; aryl groups are abbreviated by Ar.

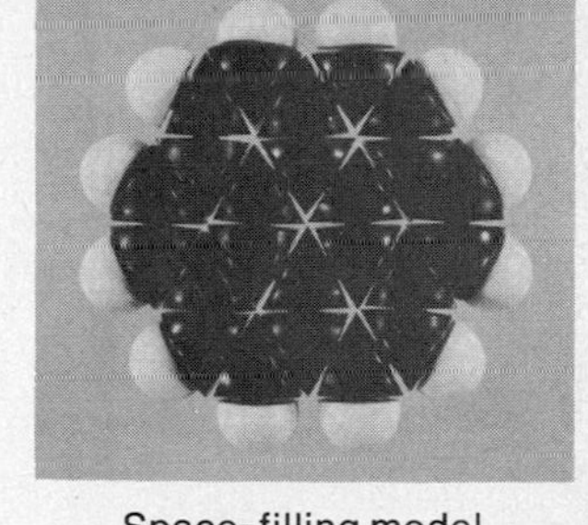

Space-filling model of coronene

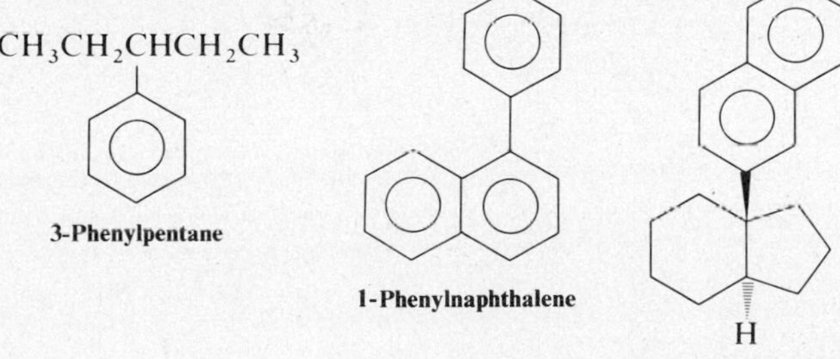

3-Phenylpentane

1-Phenylnaphthalene

trans-1-(2-Naphthyl)-bicyclo-[4.3.0]nonane

Aromatic rings represent the most highly unsaturated condition in which a stable cyclic system can exist. The addition of one or more equivalents of hydrogen to an aromatic ring system yields "hydroaromatic" compounds that are less unsaturated than the parent. Benzene, for example, can be reduced with one mole equivalent of hydrogen to 1,3-cyclohexadiene (also called 1,2-dihydrobenzene) or to 1,4-cyclohexadiene (also called 1,4-dihydrobenzene); naphthalene can be reduced with two mole equivalents of hydrogen to 1,2,3,4-tetrahydronaphthalene (also called tetralin) or with five mole equivalents of hydrogen to 1,2,3,4,4a,5,6,-7,8,8a-decahydronaphthalene (also called decalin).

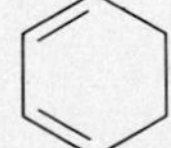

1,3-Cyclohexadiene
1,2-Dihydrobenzene

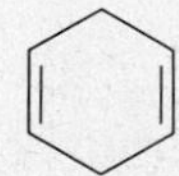

1,4-Cyclohexadiene
1,4-Dihydrobenzene

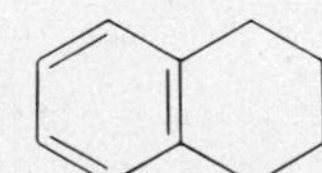

1,2,3,4-Tetrahydronaphthalene
Tetralin

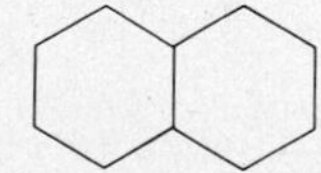

1,2,3,4,4a,5,6,7,8,8a-Decahydronaphthalene
Decalin (*cis* and *trans* isomers can exist)

4.10 *Hetero Atom Functional Groups*

Carbon atoms that are sp^2- and sp-hybridized can be attached to halogen, oxygen, and nitrogen atoms by single bonds (Fig. 4.22), to oxygen and nitrogen by double bonds (Fig. 4.23), and to nitrogen by triple bonds (Fig. 4.23).

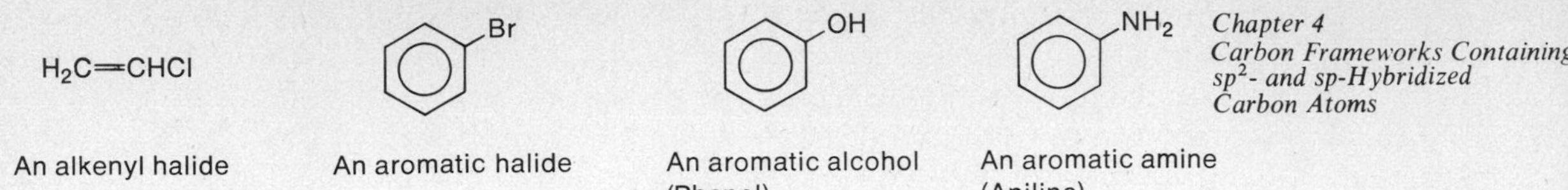

An alkenyl halide | An aromatic halide | An aromatic alcohol (Phenol) | An aromatic amine (Aniline)

Fig. 4.22. Functional groups single bonded to sp^2-hybridized carbon atoms.

When an sp^2-hybridized carbon atom is double bonded to an oxygen atom, a unit having the structure $>C{=}O$ is produced; it is called a *carbonyl group*. The carbonyl group occurs in several classes of compounds, including the following:

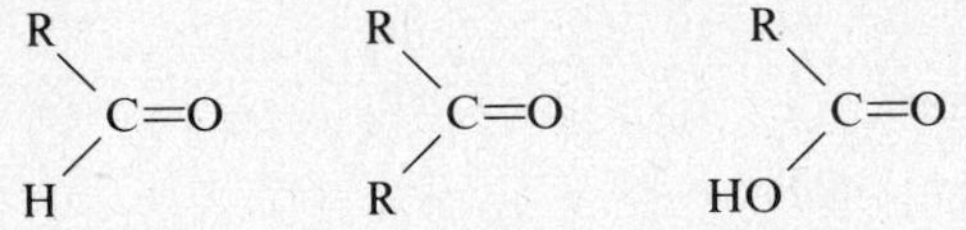

$$\underset{\textbf{Aldehydes}}{\text{R(H)C=O}} \qquad \underset{\textbf{Ketones}}{\text{R(R)C=O}} \qquad \underset{\textbf{Carboxylic acids}}{\text{R(HO)C=O}}$$

where R* can be any carbon framework (Fig. 4.23). When an *sp*-hybridized carbon atom is triply bonded to a nitrogen atom, a class of compounds (RCN) known as nitriles is obtained (Fig. 4.23). The $-C{\equiv}N$ entity is called a cyano group.

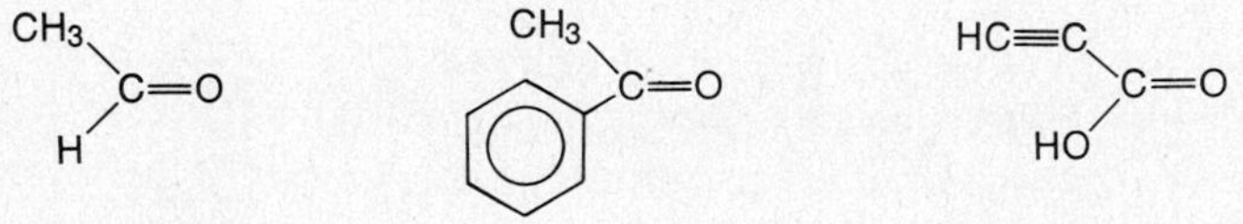

$CH_3(H)C{=}O$ — An aldehyde

$C_6H_5(CH_3)C{=}O$ — A ketone

$HC{\equiv}C(HO)C{=}O$ — A carboxylic acid

$CH_3CH_2CH_2C{\equiv}N$ — An alkyl nitrile

$CH_3{-}C_6H_4{-}C{\equiv}N$ — An aromatic nitrile

Fig. 4.23. Compounds containing carbonyl groups and cyano groups.

* The abbreviations "R" and "Ar" are often used to denote carbon frameworks. The abbreviation "Ar" is specific in its meaning, referring only to aromatic frameworks, such as phenyl, naphthyl, etc. The abbreviation "R," however, is less precise in its meaning, sometimes referring only to alkyl frameworks but sometimes referring to any of the four types of carbon frameworks. Since it is not feasible to make the meaning of "R" explicit in all instances where it is used in the text, it must be left to the reader to develop a sense of the intended meaning in any particular context.

4.11 Summary of Rules for Constructing Carbon Frameworks

Starting with carbon atoms hybridized in the sp^3, sp^2, and sp fashion, connecting these with one another to form C—C, C═C, and C≡C bonds, and then filling in the remaining valencies of the carbon atoms with the requisite number of hydrogen atoms, we can assemble an almost infinite variety of carbon frameworks. The details of these structures have been discussed, and the families into which these various frameworks can be classified have been indicated. It is useful at this point to summarize these ideas, because an understanding of the remainder of the book depends on a thorough knowledge of the structural details of these frameworks.

1. Although subject to modification to accommodate certain special cases, the carbon atoms in carbon frameworks can usually be considered to be hybridized in the sp^3, sp^2, or sp fashion.
2. The orbitals of sp^3-, sp^2-, and sp-hybridized carbon atoms are directed in space in a definite fashion, and the bonds resulting therefrom are, accordingly, oriented in a specified fashion: the hybrid orbitals on an sp^3-hybridized carbon are tetrahedrally oriented (angle = 109° 28′); those on an sp^2-hybridized carbon are coplanar (angle = 120°); and those on an sp-hybridized carbon are colinear (angle = 180°).
3. The hybrid orbitals of carbon have not only the capacity to overlap with orbitals from other elements, such as hydrogen, but also the special capacity of being able to overlap with orbitals on other carbon atoms. This makes possible the existence of multicarbon atom frameworks and endows the chemistry of carbon with enormous structural variation.
4. Although most of the virtually infinite number of possible combinations of carbon are capable of existence, a few are precluded because of the excessive strain that is present in the resulting framework. The following guidelines are useful in predicting whether such strain is likely to be present in any specific instance.
   (a) Isolable carbocyclic compounds containing seven or fewer carbons in the cycle cannot have C≡C or *trans*-substituted C═C bonds in the ring.
   (b) Isolable carbocyclic compounds containing seven or fewer carbons in the cycle cannot have a C═C═C moiety in the ring.
   (c) Isolable bicyclo[*a.b.c*]alkenes (for bicyclic nomenclature, see p. 81) cannot contain a double bond at a bridgehead carbon unless the ring in which the double bond has a *trans* configuration contains eight or more atoms.
   (d) Isolable bicyclo[*a.b.*1]alkanes cannot have a *trans* bridge between the bridgehead carbons unless both *a* and *b* are equal to or greater than 4.
   (e) Isolable bicyclo[*a*.1.0]alkanes cannot be *trans* fused unless *a* is equal to or greater than 5.

Except for these few constraints, almost all structures are possible, and even compounds that in an earlier day were thought to be incapable of existence because of inordinate strain have come to be known through synthesis. Included among the many "exotic" compounds that have appeared on the scene in recent years are those pictured in Fig. 4.24.

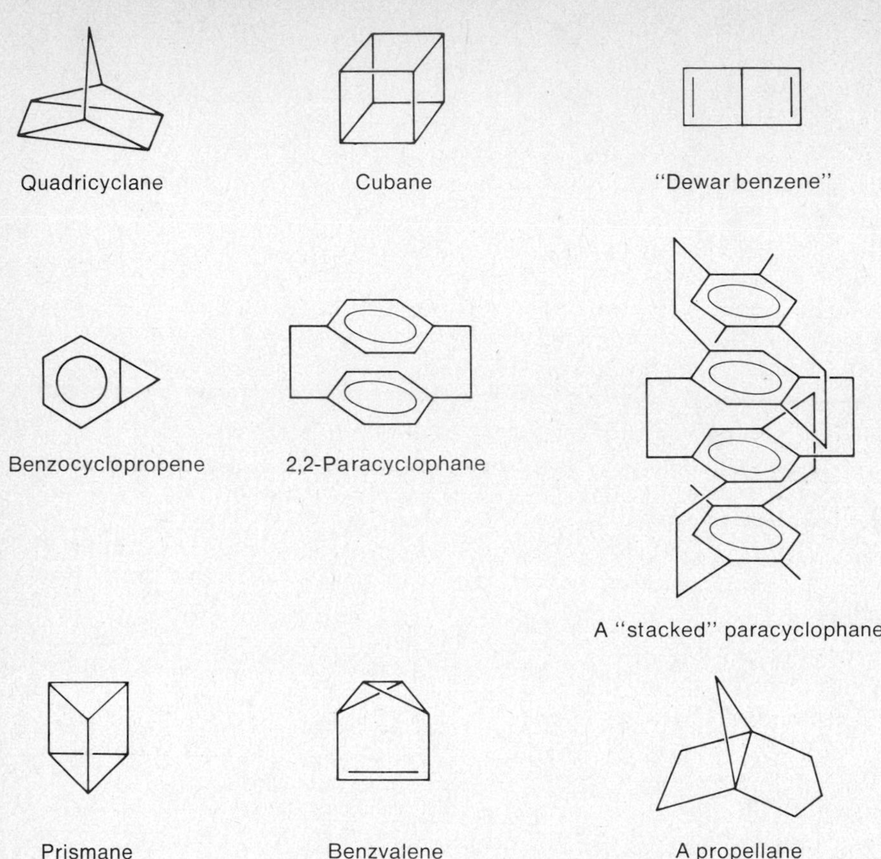

Fig. 4.24. Strained compounds that have been prepared by synthesis.

4.12 Study Guide

I. Basic principles of structures obtained from sp^2- and sp-hybridized carbons
  A. Structure of alkenes and cycloalkenes
    1. sp^2-Hybridized (triliganded) carbon, the basic structural unit
    2. Structural isomerism
    3. Double-bond diastereoisomerism
    4. Limitations in cycloalkenes
  B. Structure of alkynes and cycloalkynes
    1. sp-Hybridized (diliganded) carbon, the basic structural unit
    2. Structural isomerism
    3. Limitations in cycloalkynes
  C. Structure of polyunsaturated compounds
    1. Structural isomerism (cumulative, conjugated, and nonconjugated polyenes)
    2. Resonance structures for conjugated polyenes
    3. Limitations in cyclic compounds
  D. Structure of aromatic compounds
    1. Construction of delocalized molecular orbitals

2. Structural consequences of orbital delocalization
3. Resonance structures
4. Resonance stabilization energy
5. Criteria for aromaticity
6. Nonbenzenoid aromatic compounds (cyclopropenyl cation, cyclopentadienyl anion, cycloheptatrienyl cation, etc.)

II. Nomenclature
   A. Nomenclature of alkenes and cycloalkenes (IUPAC and "ethylene based")
   B. Nomenclature of double-bond diastereomers (*cis,trans* and (*Z*),(*E*))
   C. Nomenclature of alkynes and cycloalkynes (IUPAC and "acetylene based")
   D. Nomenclature of polyunsaturated compounds (IUPAC)
   E. Nomenclature of aromatic compounds (common names and IUPAC)

III. Properties of bonds
   A. Bond lengths of C=C, C≡C, and aromatic bonds
   B. Bond strengths of C=C, C≡C, and aromatic bonds

REFERENCES

The following references treat molecular orbital and valence bond (resonance) theory.

DEWAR, M. J. S., *The Molecular Orbital Theory of Organic Chemistry*, McGraw-Hill Book Company, New York, 1969. A mathematically sophisticated discussion of quantum mechanics and its applications to molecular systems.

LIBERLES, A., *Introduction to Molecular Orbital Theory*, Holt, Rinehart, & Winston, Inc., New York, 1966. A simple introduction to molecular orbital theory, representing an expansion on the approach in the book by J. D. Roberts (see below).

ORCHIN, M., and H. H. JAFFÉ, *The Importance of Antibonding Orbitals*, Houghton Mifflin Company, Boston, 1967. A very short (104 p.) treatment of atomic and molecular orbital theory in a simple, nonmathematical fashion.

ROBERTS, J. D., *Notes on Molecular Orbital Calculations*, W. A. Benjamin, Inc., New York, 1961. A short presentation of the rudiments of molecular orbital theory and calculations discussed in a fashion readily comprehended by the beginning student.

SALEM, L., *The Molecular Orbital Theory of Conjugated Systems*, W. A. Benjamin, Inc., New York, 1966. A 576 page treatise for the reader who already has some background in molecular orbital theory.

STREITWIESER, A., *Molecular Orbital Theory for Organic Chemists*, John Wiley & Sons, Inc., New York, 1961. A 489 page book that first develops the background of molecular orbital theory in a moderately sophisticated mathematical fashion and then applies molecular orbital methods to the interpretation of the properties and reactions of molecules.

WHELAND, G. W., *Resonance in Organic Chemistry*, John Wiley & Sons, Inc., New York, 1955. An 846 page classic that is the definitive presentation of the valence bond treatment of organic compounds.

The following reference sets forth the rules for the (*Z*), (*E*) nomenclature of alkenes.

IUPAC TENTATIVE RULES FOR THE NOMENCLATURE OF ORGANIC CHEMISTRY, SECTION E, Fundamental Stereochemistry, *Journal of Organic Chemistry*, **35**, 2849 (1970).

PROBLEMS

1. Draw three-dimensional representations for the following compounds:
(a) *trans*-2-Butene.
(b) $C_6H_5CH_2CH_3$.
(c) An eclipsed conformation of $HC{\equiv}CCH_2CH_3$.
(d) All the diastereomers of $CH_3CH{=}CH{-}CH{=}CHCH_3$.
(e) The most stable conformation of 3-methylcyclohexene.

2. Without designating the stereochemistry, name the following compounds by the IUPAC system in all cases and the ethylene-based or acetylene-based system where applicable.

(a) $CH_3CH_2CH{=}CH_2$
(b) $CH_3CH{=}CHCH_2CH_3$
(c) $(CH_3)_2C{=}CH_2$
(d) $(CH_3)_2CHCH{=}CHCH_3$
(e) $(CH_3)_3CCH{=}CHCH(CH_3)_2$
(f) $CH_3CH_2C{\equiv}CH$
(g) $(CH_3)_2CHC{\equiv}CCHCH(CH_3)_2$ (with CH_3 on the CH)
(h) $H_2C{=}CHCHCH{=}CH_2$ (with CH_3 on the central CH)
(i) $CH_3CH{=}CC{\equiv}CH$ (with CH_3 on the C)
(j) $HC{\equiv}CCH_2CHC{\equiv}CH$ (with CH_2CH_3 on the CH)

3. Using the IUPAC system, name the following compounds and include the appropriate stereochemical designation where necessary

(a) 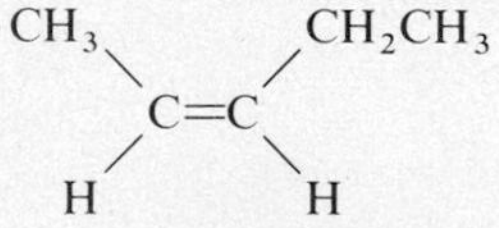 $CH_3(H)C{=}C(H)CH_2CH_3$ — CH_3 upper left, CH_2CH_3 upper right, H lower left, H lower right
(b) 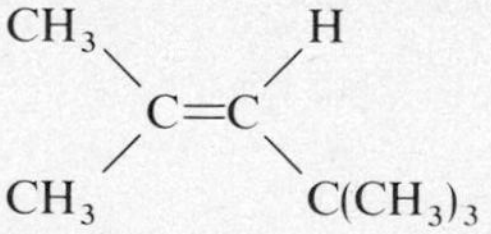 $(CH_3)_2C{=}C(H)C(CH_3)_3$ — CH_3 upper left, H upper right, CH_3 lower left, $C(CH_3)_3$ lower right
(c) $H_2C{=}CHCCH{=}CH_2$ (with two CH_3 on the central C) 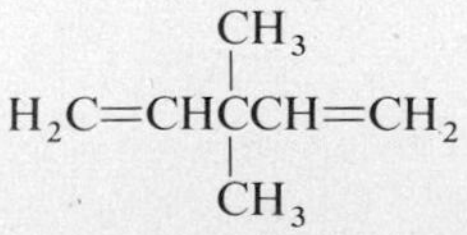
(d) 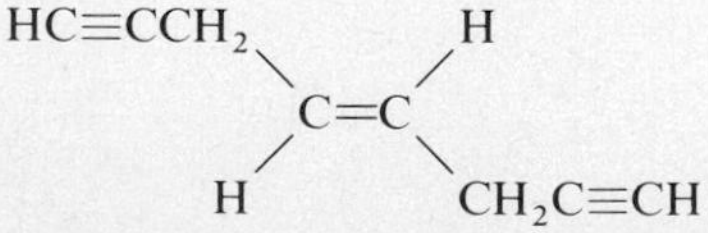 $HC{\equiv}CCH_2(H)C{=}C(H)CH_2C{\equiv}CH$ — $HC{\equiv}CCH_2$ upper left, H upper right, H lower left, $CH_2C{\equiv}CH$ lower right
(e) $H_2C{=}CHCH_2C{\equiv}CCH_2CH{=}CH_2$
(f) 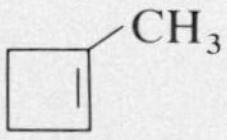 1-methylcyclobutene ring structure with CH_3 substituent

4. Draw the structures corresponding to the following names:

(a) 2,3-Dimethyl-l-butene
(b) 2-Methyl-2-butene
(c) Propyne
(d) 1,5-Hexadiyne
(e) 1-Methylcyclohexene
(f) Toluene
(g) *m*-Xylene
(h) Vinylbenzene
(i) Allylbenzene
(j) *o*-Propenyltoluene
(k) *p*-Diisopropenylbenzene
(l) 2-Neopentylnaphthalene
(m) *trans*-*t*-Butylmethylethylene
(n) 2,5-Dimethyl-*cis*-3-hexene
(o) 4-Ethyl-6-methyl-(Z)-5-octen-2-yne

(p) (Z)-1-Isopropyl-1,2-dimethylethylene
(q) Bicyclo[3.2.1]oct-2-ene
(r) Spiro[4.5]dec-2-ene
(s) 2-Isopropenylphenanthrene
(t) 3-Vinyl-1,3,5-hexatriene
(u) *endo*-6-Methyl-2-methylene-bicyclo[2.2.1]heptane
(v) 6-(*trans*-2-Butenyl)-*cis,trans*-2,8-undecadiene
(w) 2-Cyclopropylidenebutane

5. Using the IUPAC system, name the following compounds and include the appropriate stereochemical designation where necessary

(a) 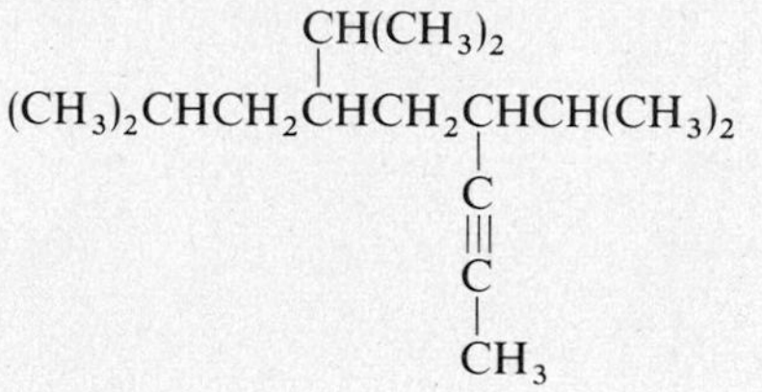 $(CH_3)_2CHCH_2CH(CH(CH_3)_2)CH_2CH(C{\equiv}CCH_3)CH(CH_3)_2$

(b) $(CH_3)_3CC{\equiv}CC{\equiv}CC(CH_3)_3$

(c) 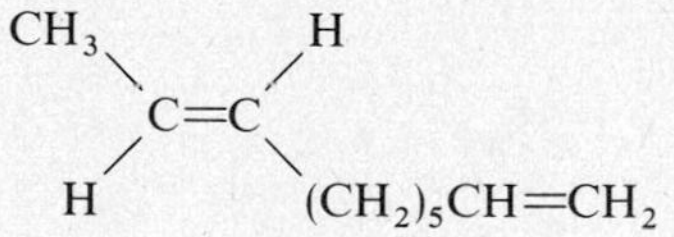 $CH_3(H)C{=}C(H)(CH_2)_5CH{=}CH_2$

(d) 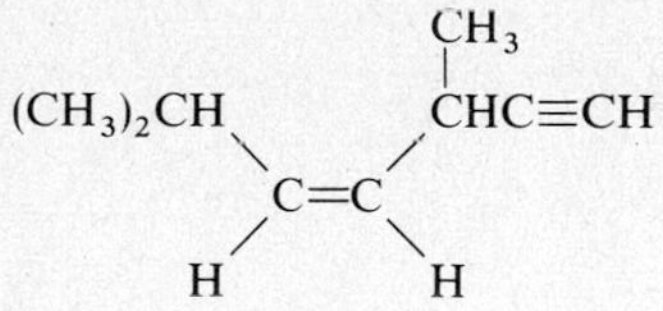 $(CH_3)_2CH(H)C{=}C(H)CH(CH_3)C{\equiv}CH$

(e) 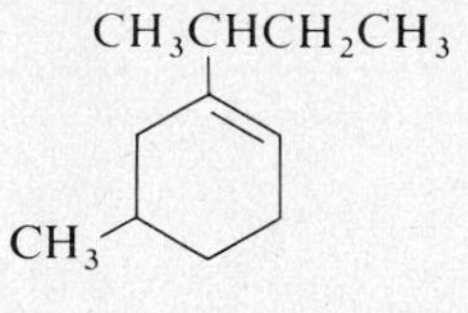 $CH_3CHCH_2CH_3$; CH_3

(f) 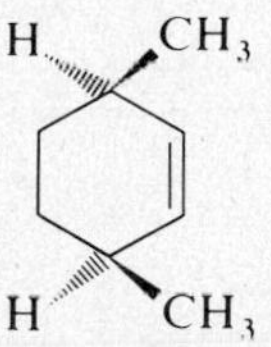 H, CH_3 ; H, CH_3

(g)

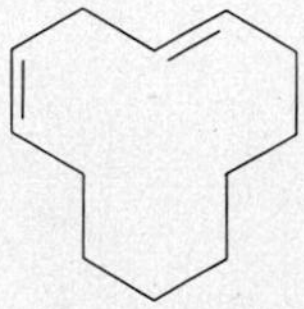

(h)

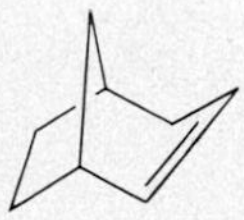

(i)

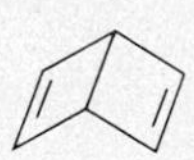

(j) 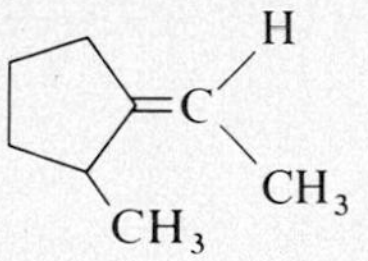 H, CH_3 ; CH_3

(k) 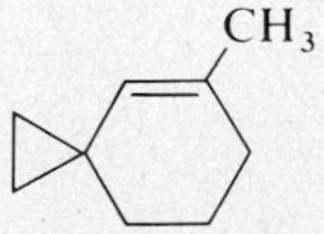 CH_3

6. Using whatever nomenclature you believe to be most appropriate, name the following aromatic compounds

(a) 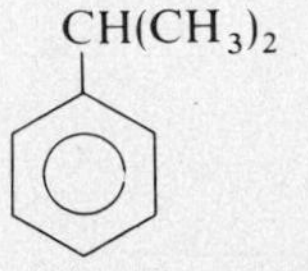 $CH(CH_3)_2$

(b) 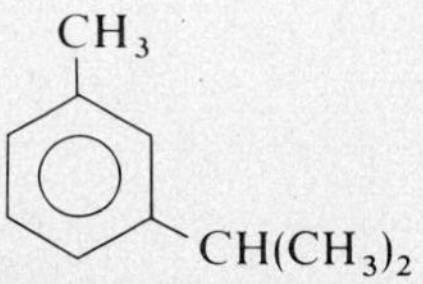 CH_3 ; $CH(CH_3)_2$

(c)

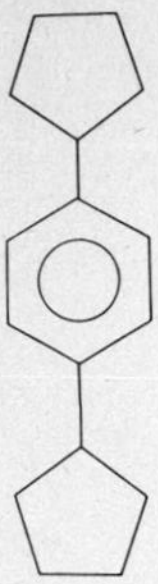

(d) 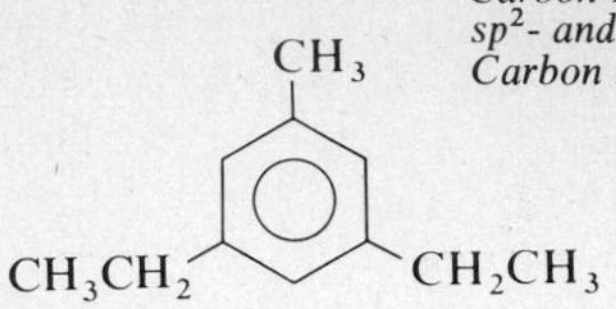 CH_3, CH_3CH_2, CH_2CH_3

(e) 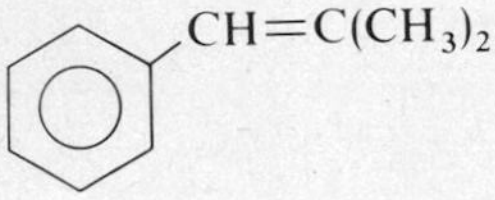 $CH{=}C(CH_3)_2$

(f) 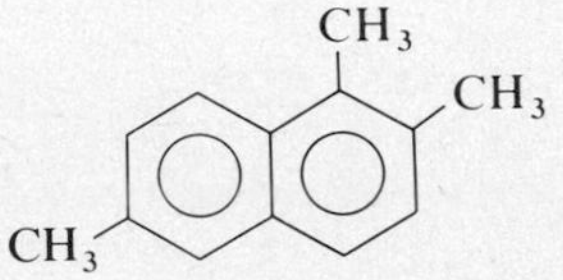 CH_3, CH_3, CH_3

(g) 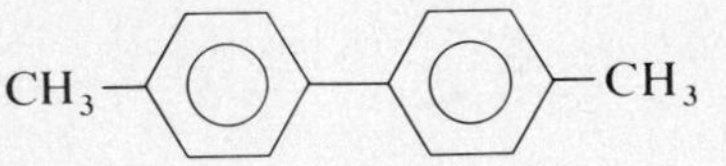 CH_3—, —CH_3

(h) CH_3, H, H

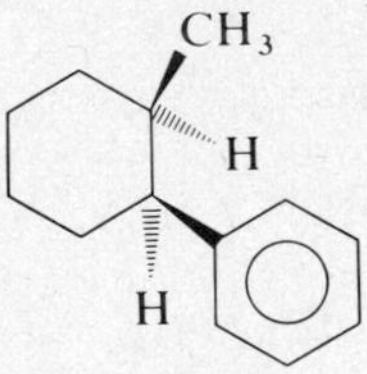

(i) 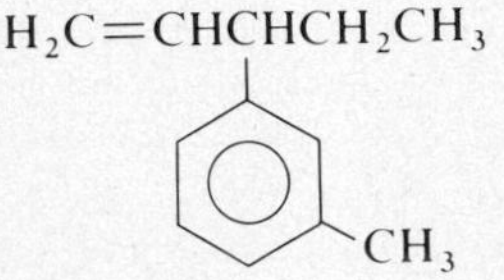 $H_2C{=}CHCHCH_2CH_3$, CH_3

(j) 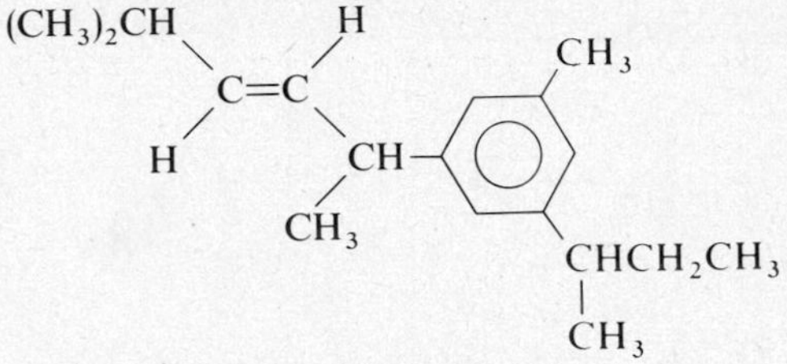 $(CH_3)_2CH$, H, C=C, H, CH—, CH_3, CH_3, $CHCH_2CH_3$, CH_3

(k) 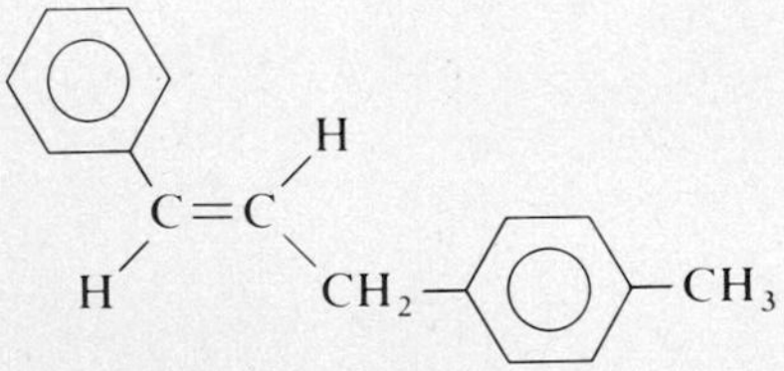 H, C=C, H, CH_2—, —CH_3

(l) 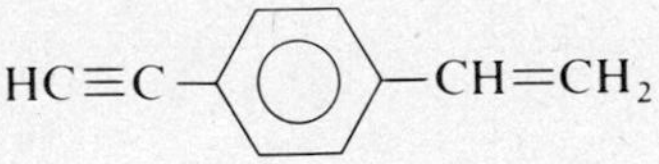 $HC{\equiv}C$—, —$CH{=}CH_2$

(m) 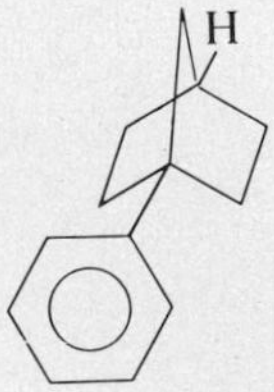 H

(n) 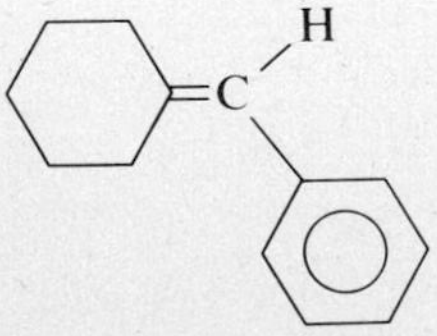 H, =C

7. Draw condensed representations for all the possible isomers corresponding to the following molecular formulas in which there are two or more sp^2- and/or sp-hybridized carbon atoms: (a) C_4H_6; (b) C_5H_{10}; (c) $C_{10}H_{12}$ containing one benzene ring and one C=C bond. Name each of the structures. Which of these would be most likely to be stable, isolable compounds, and which would be most likely to be unstable compounds?

8. Define each of the following in a concise and precise manner. Do not use a specific example for the definition, but cite an example along with the definition.

(a) π Bond
(b) Atomic orbital
(c) Molecular orbital
(d) sp^2-Hybridized carbon atom
(e) An alkene
(f) Diastereoisomerism
(g) Delocalization
(h) Aromaticity
(i) Resonance energy
(j) Conjugation
(k) Positional isomerism

9. The structures of five isomeric C_5H_8 compounds are shown:

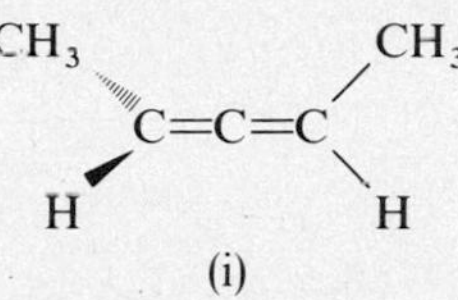

(i)

CH_3, CH_3, C=C=C, H, H
(ii)

$H_2C{=}CH$, CH_3, C=C, H, H
(iii)

$H_2C{=}CH$, H, C=C, H, CH_3
(iv)

$H_2C{=}CHCH_2CH{=}CH_2$
(v)

(a) Do (i) and (ii) have the same boiling point? Explain.
(b) Do (i) and (ii) have the same effect on plane polarized light? Explain.
(c) Do (iii) and (iv) have the same boiling point? Explain.
(d) Do (iii) and (iv) have the same effect on plane polarized light? Explain.
(e) How many stereoisomers for (v) can exist?
(f) Do (iv) and (v) have the same boiling point? Explain.

10. Predict which of the following structures might be expected to possess aromatic character, and explain the basis for your choice:

(a)
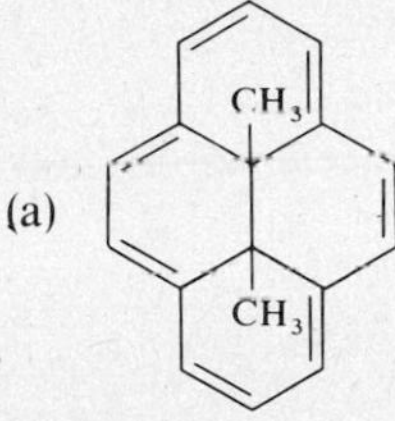

(b)
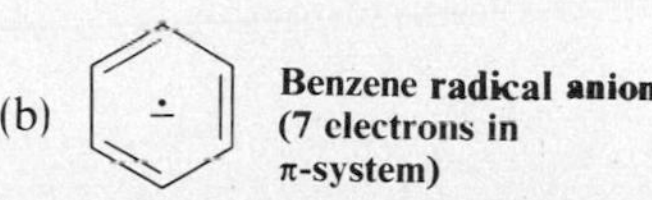
Benzene radical anion (7 electrons in π-system)

(c)
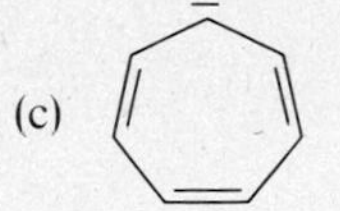

(d)

(e)
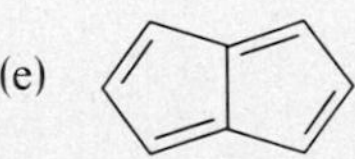

(f)
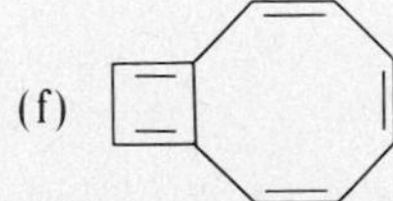

(g)
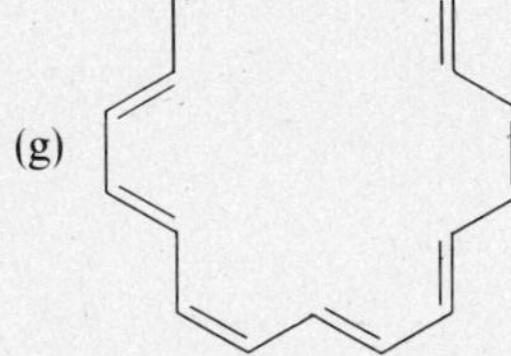

(h)

(i)

11. Predict the heats of hydrogenation (to the completely saturated cyclic systems) on the basis of the resonance energies (which are derived from heat of combustion data) for the following compounds:
(a) Naphthalene—resonance energy of 61.0 kcal/mole.
(b) Azulene—resonance energy of 33.0 kcal/mole.
(c) Anthracene—resonance energy of 83.5 kcal/mole.
(d) Phenanthrene—resonance energy of 91.3 kcal/mole.

12. The measured heats of combustion of 1,3-butadiene and 1,3-cyclopentadiene are 608.5 and 707.7 kcal/mole, respectively. The calculated heats of combustion of these molecules are 611.5 and 709.3 kcal/mole, respectively. On the basis of these data, infer which is the most stable conformation for butadiene (see Fig. 4.7).

13. Draw important resonance structures for each of the following compounds. Indicate which of the resonance structures you have drawn is the most and which the least important contributor to the resonance hybrid in each case.

(a) $H_2C{=}C{=}CHCH{=}CH_2$ (b) $HC{\equiv}CCH{=}CH_2$

(c) 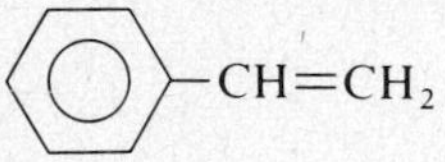 C_6H_5—$CH{=}CH_2$ (d) 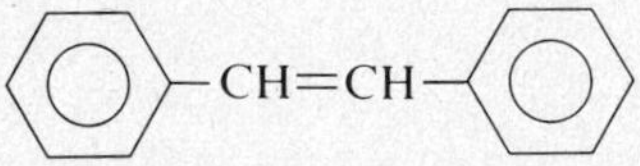 C_6H_5—$CH{=}CH$—C_6H_5

(e) 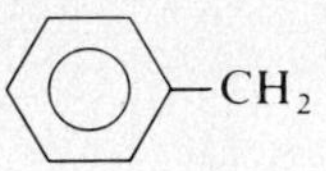 C_6H_5—$\overset{+}{C}H_2$ (f) C_6H_5—$CH{=}CH\overset{+}{C}H_2$

(g) (naphthalene) (h) (phenanthrene)

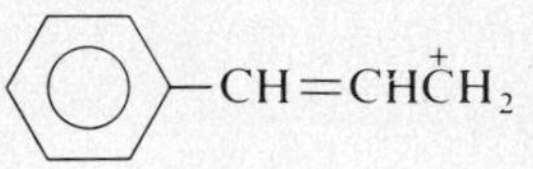

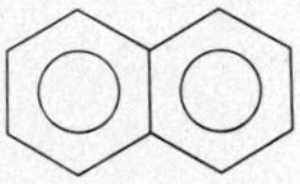

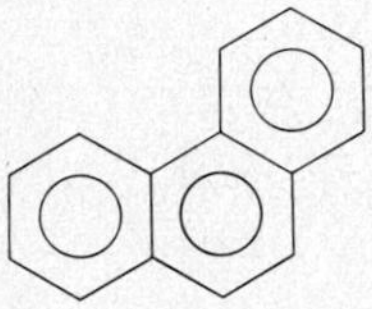

14. Draw the following structures:
(a) A 13-carbon ketone containing two benzene rings.
(b) All the nitriles having the molecular formula C_5H_9N.
(c) A pair of enantiomeric carboxylic acids having the molecular formula $C_5H_{10}O_2$.
(d) Three isomeric C_4H_6O aldehydes.

Chapter 5

The physical properties of organic compounds

Come, give us a taste of your quality.

WILLIAM SHAKESPEARE
Hamlet

A chemical compound is a real entity—a substance that occupies space—something that can be weighed, smelled, and tasted. The various chemical, physical, and physiological properties of compounds are the heart and soul of the study of chemistry in the laboratory. Lest we be lulled into thinking that chemistry exists simply as pictures on the pages of books (sometimes referred to as "paper chemistry"), we must remember that the representations with which we have dealt in the first four chapters are just a part of the language that chemists have invented; they are, so to speak, the vocabulary of organic chemistry. So far, we have dealt with the structures of organic compounds in a rather abstract fashion, considering the rules according to which the building blocks can be put together into various kinds of carbon frameworks. We have neglected, for the most part, the connection that must be made between an adequate structural representation and the physical and chemical properties of the substance being represented. In the present chapter, this connection will be established, and it will be continually reemphasized in subsequent chapters.

5.1 Intrinsic (Intramolecular) Properties of Compounds

Some of the intrinsic or intramolecular (Latin, *intra*, "within"; to be contrasted with Latin, *inter*, "between") properties of molecules have already been discussed, including bond strengths, bond lengths, and bond polarities (see Section 2.7). Bond polarities can be quantitatively expressed in terms of dipole moments (μ), which are the products of the charges on the atoms and the distance between the charges. When the charge is measured in electrostatic units and the distance in Ångstroms, the result is given in Debye units (D). Hydrogen chloride, for instance, possesses a permanent dipole moment of 1.03 Debyes, in which the hydrogen is the positive end and the chlorine is the negative end of the dipole. Chlorobenzene has a dipole moment of 1.70 D, the chlorine in this case also being the negative end of the dipole. If a second chlorine atom is attached to give *p*-dichlorobenzene, the measured dipole moment is zero (Fig. 5.1). Thus, what one actually measures

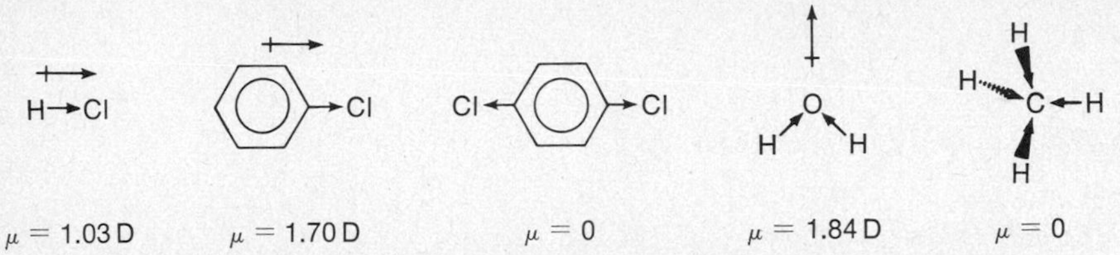

$\mu = 1.03$ D $\mu = 1.70$ D $\mu = 0$ $\mu = 1.84$ D $\mu = 0$

Fig. 5.1. Dipole moments of molecules. Arrows indicate the direction of the dipole, the head of the arrow being the negative end.

is a composite dipole moment that is the resultant of the dipole moments of all the individual bonds (usually referred to as "bond moments"). Only in the simple diatomic molecules does the measured dipole moment coincide with the bond dipole moment. By means of dipole moment measurements on appropriately chosen molecules, however, and by the use of various subtraction and addition techniques, the magnitude and sign of the dipole moments for a number of types of bonds have been assessed, as Table 5.1 lists.

Table 5.1 BOND DIPOLE MOMENTS[a]

Bond	Dipole Moment, D	Bond	Dipole Moment, D
H—H	0	H—C	0.3
O—O	0	C—C, C=C, C≡C	0
N—N	0	C—F	1.51
H—F	1.98	C—Cl	1.56
H—Cl	1.03	C—Br	1.48
H—Br	0.79	C—I	1.29
H—I	0.38	C—O	0.86
H—O	1.53	C=O	2.4
H—N	1.31		

[a] In each case the positively charged atom is shown at the left.

The bond moment is a vector quantity, so if the signs, magnitudes, and orientations of bonds within a molecule are known, the resultant dipole moment can be calculated. Or conversely, if the signs and magnitudes of the bond moments and the measured resultant dipole moment are known, the geometry of the molecule can be gleaned. For example, the bond moment for H—O is 1.53 D, and the measured dipole moment for water is 1.84 D; From these data, we can calculate that the H—O—H bond angle in water is about 104°. The measured dipole moment of methane is zero, which is commensurate with a completely symmetrical, tetrahedral structure (Fig. 5.1). Still another useful application of dipole moment measurements is in the assignment of structure to (*Z*) and (*E*) isomers. Thus, the (*Z*) isomer of 1,2-dichloroethylene has a finite dipole moment, whereas the (*E*) isomer has a zero dipole moment (Fig. 5.2).

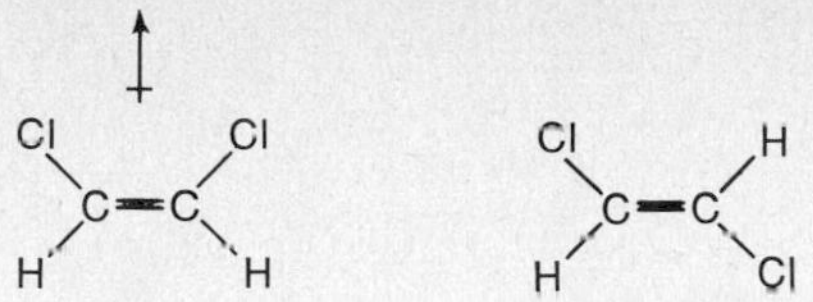

Cl Cl / C=C / H H — $\mu = 1.74$ D — (Z) isomer

Cl H / C=C / H Cl — $\mu = 0$ — (E) isomer

Fig. 5.2. Dipole moments of (*Z*)- and (*E*)-1,2-dichloroethylene.

PROBLEM 5-1

Arrange the following compounds in order of increasing dipole moment:

CH_3Br CH_3I CH_2Cl_2 CF_4

5.2 *Extrinsic (Intermolecular) Properties of Compounds*

In addition to the very strong forces that operate within molecules to hold nuclei in certain definite positions, there also exist forces that operate *between* molecules. These extrinsic or intermolecular forces are manifested in a variety of ways that can be measured as boiling points, melting points, density, and viscosity. The data obtained from such measurements supply useful information about molecules and provide critical yardsticks against which to measure the accuracy and utility of a structural theory.

5.2a. BOILING POINTS. The boiling point (abbreviated bp) of a substance can be defined as the temperature at which the thermal energy of the particles driving them apart from each other is equal to the intermolecular forces holding them together in the liquid state. The higher the temperature at which this occurs, the greater are the intermolecular forces. Knowing that unlike charges attract, one can correctly predict that compounds possessing permanent dipole moments will boil at higher temperatures than those without a significant dipole moment; water, for example, boils 261° higher than methane. Realizing that methane has a zero dipole moment, one might ask why it has a boiling point at all; why can it be condensed to a liquid? The explanation is found in the concept of the instantaneous or fluctuating dipole. Although the time-averaged dipole of a molecule may be zero, its value at any particular moment may be finite. Even a completely symmetrical molecule such as H_2 may, at a given instant in time, have an unsymmetrical distribution of electrons between the hydrogen atoms, H:H (*i.e.*, $\overset{\delta^-}{H}-\overset{\delta^+}{H}$), giving rise to a dipolar entity). A moment later, the electrons may be displaced in the opposite direction, H :H (*i.e.*, $\overset{\delta^+}{H}-\overset{\delta^-}{H}$), so that over a period of time one dipolar form will occur just as often as the other and the time-averaged dipole moment is zero. The "instantaneous" dipole moment in one hydrogen molecule, however, may "induce" a dipole moment in a nearby neighbor molecule, with the result that there is momentarily a dipole–dipole interaction (more properly referred to as an instantaneous dipole—induced dipole interaction). The intermolecular force resulting from the interaction is variously known as a dispersion force, London force, or van der Waals force, and it is operative in all molecular systems. Even highly symmetrical molecules, such as hydrogen and methane, can be condensed to liquids, although cooling to very low temperatures is required. Thus, every bond has a fluctuating dipole associated with it, and the greater the number of bonds within a molecule, the higher should be the boiling point. As the molecular weight increases in ascending a homologous series, therefore, the boiling points generally increase,

Table 5.2 Boiling points, melting points, and densities of representative alkanes, alkenes, alkynes, and arenes

Carbons	Alkane	BP	MP	Density	Alkene	BP	MP	Density
1	CH_4	−161°	−183°					
2	CH_3CH_3	−88°	−172°		$H_2C{=}CH_2$	−102°	−169°	
3	$CH_3CH_2CH_3$	−42°	−187°		$CH_3CH{=}CH_2$	−48°	−185°	
4	$CH_3CH_2CH_2CH_3$	0°	−138°	0.6012	$CH_3CH_2CH{=}CH_2$	−6.5°	−185°	0.5946
5	$CH_3(CH_2)_3CH_3$	36°	−130°	0·6262	$CH_3(CH_2)_2CH{=}CH_2$	29°	−138°	0.6411
6	$CH_3(CH_2)_4CH_3$	68°	−95°	0.6594	$CH_3(CH_2)_3CH{=}CH_2$	63°	−138°	0.6734
8	$CH_3(CH_2)_6CH_3$	125°	−56°	0.7025	$CH_3(CH_2)_5CH{=}CH_2$	121°	−101°	0.7149
10	$CH_3(CH_2)_8CH_3$	174°	−30°	0.7300	$CH_3(CH_2)_7CH{=}CH_2$	170°	−66°	0.7408
12	$CH_3(CH_2)_{10}CH_3$	216°	−10°	0.7487	$CH_3(CH_2)_9CH{=}CH_2$	213°	−35°	0.7584
14	$CH_3(CH_2)_{12}CH_3$	253°	6°	0.7627	$CH_3(CH_2)_{11}CH{=}CH_2$	246°	−12°	0.7852
18	$CH_3(CH_2)_{16}CH_3$	305°	28°	0.7768	$CH_3(CH_2)_{15}CH{=}CH_2$		17°	0.7891
20	$CH_3(CH_2)_{18}CH_3$	343°	37°	0.7886	$CH_3(CH_2)_{17}CH{=}CH_2$	341°	28°	0.7882
30	$CH_3(CH_2)_{28}CH_3$		65°	0.7750				

Carbons	Alkyne	BP	MP	Density	Arene	BP	MP	Density
2	$HC{\equiv}CH$	−75°	−82°					
3	$CH_3C{\equiv}CH$	−23°	−101°					
4	$CH_3CH_2C{\equiv}CH$	8.1°	−122°	0.6784				
5	$CH_3(CH_2)_2C{\equiv}CH$	39°	−90°	0.6909				
6	$CH_3(CH_2)_3C{\equiv}CH$	71°	−132°	0.7152	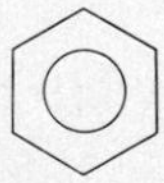	80°	5.5°	0.8787
8	$CH_3(CH_2)_5C{\equiv}CH$	125°	−80°	0.7457				
10	$CH_3(CH_2)_7C{\equiv}CH$	174°	−44°	0.7655	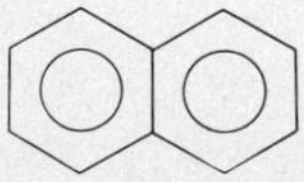	211°	80°	1.145
12	$CH_3(CH_2)_9C{\equiv}CH$	95°		0.7758				
14					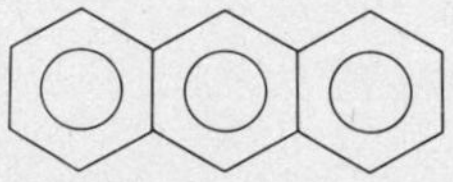	340°	216°	1.25
18	$CH_3(CH_2)_{15}C{\equiv}CH$	313°	22°	0.7955		Sublimes	335°	
20	$CH_3(CH_2)_{17}C{\equiv}CH$	314°		0.8181				

as shown by the data in Table 5.2. In such instances there is usually a consistent increase in boiling point if the character of the structure remains otherwise unchanged. If, however, comparisons are made between structures differing in shape, *e.g.*, in the nature of the chain branching, the boiling point increments may change capriciously. Thus, *n*-heptane boils at 98° and its eight-carbon straight chain homolog *n*-octane boils at 126°; the more highly branched eight-carbon homolog 2,2,3,3-tetramethylbutane, however, boils at 106°.

The magnitude of the van der Waals forces might also be expected to increase as the tightness with which the electrons are bound in the system diminishes. Since electrons in π bonds are more polarizable* than electrons in σ bonds, the presence of unsaturation in a molecule should increase the boiling point. Although this appears not to be the case when alkanes are compared with alkenes, alkynes do boil somewhat higher than the corresponding alkanes, even though the alkynes are four molecular weight units lighter. As the extent of conjugation increases, boiling points increase, as noted in the series 1-hexene (bp 63°), 1,3-hexadiene (bp 72°), 1,3,5-hexatriene (bp 79°), and benzene (bp 80°), despite the fact that the molecular weights decrease in the order given. Other examples of boiling point elevations arising from intermolecular attraction due to increased electron mobility are encountered in systems in which halogen, oxygen, and nitrogen atoms are also present. These will be considered in later chapters.

PROBLEM 5-2

Arrange the following compounds in the order of increasing boiling point:

$(CH_3)_2CHCH{=}CHCH_2CH(CH_3)_2$, $CH_3CH_2CH_2CH{=}CHCH_2CH_2CH_2CH_3$

```
          CH3     CH3
             \   /
    CH3        C        CH3
       \     /   \     /
         C ——————— C
       /             \
    CH3                CH3
```

5.2b. MELTING POINTS. The melting point (abbreviated mp) of a substance can be defined as the temperature at which the thermal energy of the particles driving them apart from each other is equal to the intermolecular forces holding them together in the solid state; it is the temperature at which the solid and liquid phase are in equilibrium. In similarity to the boiling point, the greater the intermolecular forces, the higher is the temperature at which this occurs. Correlations between structure and melting point are much less predictable than those between structure and boiling point, however. The forces that hold molecules in the crystalline lattice of the solid state are very short range and fall off by a power of about $1/d^7$, where d is the distance between molecules. As a

---

* Polarizability is a measure of the ease with which the distribution of electron density in an atomic or molecular system can be distorted.

result, the magnitude of this short range force is sensitive not only to dipole interactions and van der Waals forces *per se*, but also to the way in which molecules can "fit" next to each other; thus, steric characteristics are more important in determining a melting point than a boiling point. So, although molecular weight plays a significant role in determining melting points, as illustrated by the data in Table 5.2, molecular geometry also assumes importance. One of the most dramatic examples of this is the comparison of *n*-octane (which melts at 56°) with its much more symmetrical isomer 2,2,3,3-tetramethylbutane (which melts at 101°). As already noted, the latter compound boils at 106°, which means that it exists as a liquid only over a 5° range.

Although the correlation of structure with melting point is capricious, melting points have, nevertheless, been of inestimable use in the characterization of organic compounds. The significance and importance of melting points have been somewhat diminished in recent years by the advent of spectral methods, but for over a century they provided the most readily measurable and most specifically distinguishable physical attribute of an organic solid. Even today they have by no means fallen into disuse; in addition to being one of the several characterizing features of a compound, they afford an easy way for following the purification of a solid. In general, as a solid becomes increasingly pure, its melting range becomes higher and narrower. Pure compounds usually have a melting range (generally referred to as a melting point) of a degree or less. Implicit in this statement is the converse statement that the less pure is a compound, the lower and broader is its melting range. This has been taken advantage of for many years in the technique called the "mixed melting point" (or, perhaps more correctly, the "mixture melting point"). The basis for this procedure is simple. If two pure compounds are identical, the melting point of a mixture of them shows no difference from the melting point of the known compound; if two compounds are not identical, the mixed melting point shows a depression. If, then, authentic samples are available for comparison, the mixed melting point technique affords a simple way for establishing identity or nonidentity with the unknown compound.

5.2c. DENSITY AND VISCOSITY. The density of a compound is another consequence of intermolecular forces. Its magnitude can be considered to be an *intermolecular force per unit volume* of molecule. Thus, the density of hydrocarbons increases only slightly in going from the alkanes to the alkenes to the alkynes, and also increases only slightly as the molecular weight within any of these three homologous series increases.

Still another physical attribute that arises from intermolecular forces is the viscosity of liquids. Viscosity is partly a consequence of simple mechanical entanglements between molecules; the longer the molecule, the greater is the possibility for molecular entanglement and the higher is the viscosity. This has proved to be one of the useful means for assessing the molecular weights of various kinds of very large molecules known as polymers. That molecules do, indeed, entangle themselves has been demonstrated in an interesting way. When a cyclization reaction of a compound containing a 34-carbon chain is carried out in a solvent consisting of a 34-carbon cycloalkane, a small amount of the product consists of a C_{68} "compound," in which two interlocking rings are present (Fig. 5.3). This type of compound is called a catenane (Latin, *catena*, "chain") and further adds to the long list of structural possibilities noted in the previous chapters.

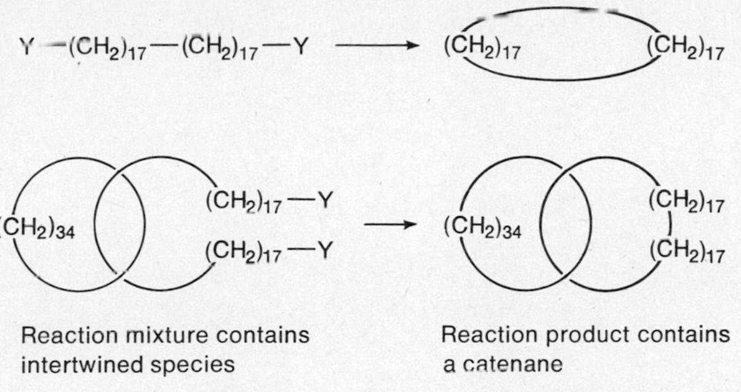

Fig. 5.3. Synthesis of a catenane.

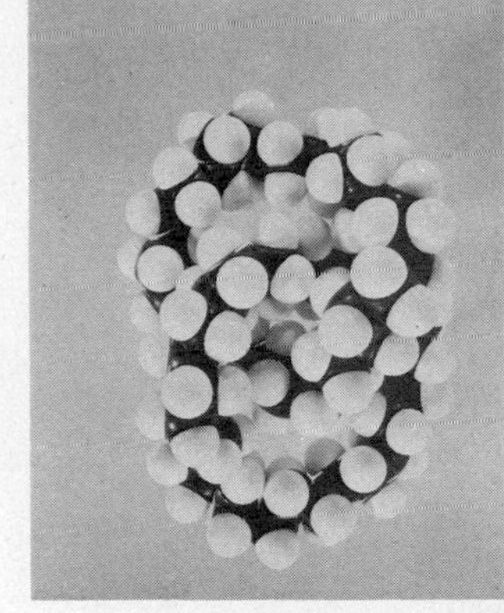

A catenane

5.2d. SOLUBILITY. A physical property that often assumes great importance when the chemistry of organic compounds is being studied is the solubility of one substance in another. In the majority of instances, reactions involving organic compounds are carried out in solution, *i.e.*, the organic reactants are added to some liquid in which they are mutually soluble. Therefore, it is necessary for practical reason to know something about solubilities; as we shall see in later chapters, such knowledge is also a useful guide in interpreting the mechanisms of reactions.

The rule of thumb that "like dissolves like" has served chemists for scores of years. The operation of this rule has undoubtedly been observed by many readers of this book who know that sodium chloride, a highly polar compound, is far more soluble in the polar solvent water than in the nonpolar solvent gasoline, and, conversely, that paraffin wax (a high molecular weight nonpolar alkane) is soluble in gasoline and insoluble in water. To explain these phenomena, we must explore the nature of the process of solution. When a liquid or a solid goes into solution, the intermolecular forces holding the molecules together in either the liquid or the solid state are disrupted by the solvent. If the process is to be thermodynamically feasible, however (*i.e.*, if it is to proceed to any extent), the solvent must provide the energy necessary to effect this disruption. It does so by making available to the material being dissolved an alternative set of intermolecular forces. Whereas in the original liquid or solid the intermolecular forces are between *identical* molecules, in the dissolved state the forces are between *nonidentical* molecules. Thus, when sodium chloride goes into aqueous solution, the very strong ion–ion interactions existing in the crystal are replaced by the ion–dipole interactions between sodium ions and water and between chloride ions and water. The dissolved ions are said to be solvated or, in this specific instance, hydrated, as illustrated in Fig. 5.4. Clearly, a strongly polar substance such as water is able to form an ion-dipole solvate with the ions, whereas a nonpolar substance such as gasoline, is not.

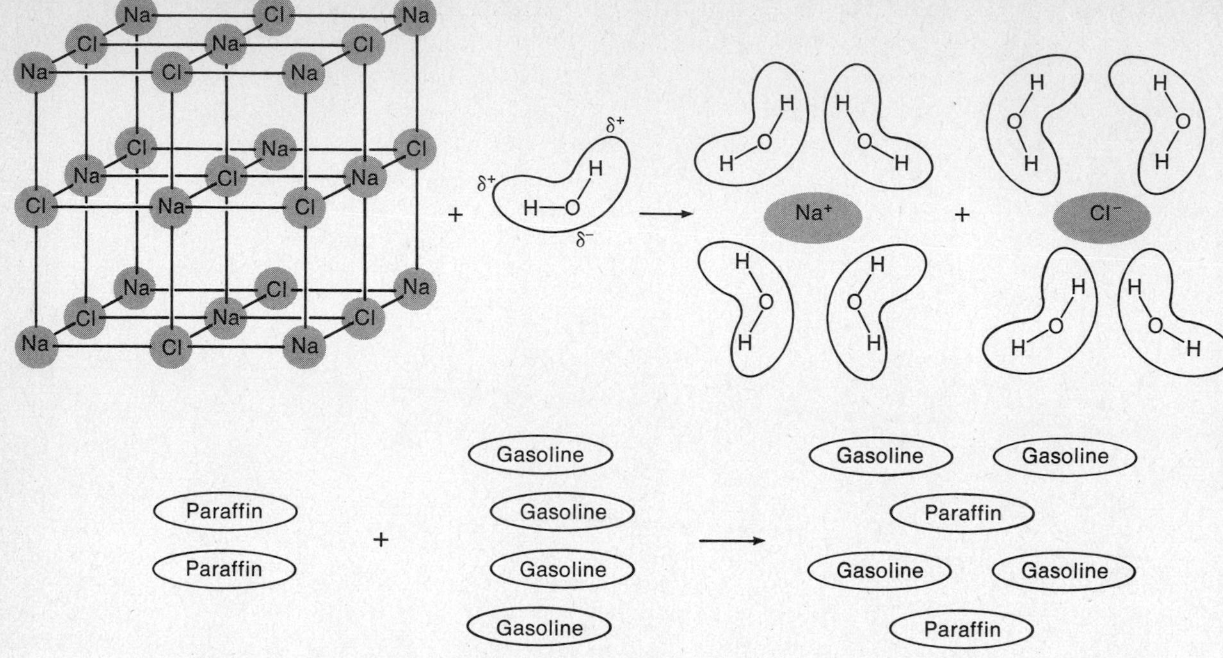

Fig. 5.4. Dissolution of NaCl in H_2O and paraffin wax in gasoline.

Water, in solvating sodium ions and chloride ions, must itself give up a certain amount of dipole–dipole interaction. It does so, however, because it gains an even stronger interaction in the solvation process. If, however, the entity requiring solvation is nonpolar, the loss of the original dipole–dipole interaction in water is *not* compensated by the solvation process. Consequently, water does not dissolve the nonionic material paraffin wax. On the other hand, the intermolecular forces holding gasoline molecules together are much weaker than those holding water together and, in fact, are roughly comparable to those involved in the solvation of paraffin molecules. Accordingly, paraffin dissolves in gasoline.

PROBLEM 5-3

What solvent would you use to separate the following pairs of substances: (a) table salt (NaCl) and butter; (b) vinegar (acetic acid) and castor oil; (c) sugar and sand; and (d) ethanol (C_2H_5OH) and eicosane ($C_{20}H_{42}$)?

5.2e. INDEX OF REFRACTION. A beam of light has an electrical field and a magnetic field associated with it. Atoms and molecules, containing electrons in motion, also have associated electric and magnetic fields; consequently, an interaction between a beam of light and a collection of molecules can occur in several ways. One possibility involves an interaction between the light and the electrons of the molecule to induce an electronic oscillation within the molecule.

The oscillating dipoles then reemit the absorbed energy almost instantaneously, although not necessarily in phase with the incident light. The recombination of the reemitted, phase-changed (*i.e.*, scattered) light with the incident beam results in the phenomenon known as refraction. The extent to which the light has been refracted can be measured and expressed as an *index of refraction.* Of the physical methods that have been discussed to this point, the index of refraction is probably the most sensitive measure of the polarizability of the electrons in the molecular system; the more polarizable are the electrons, the greater is their perturbation by the beam of light and the higher is the index of refraction, as illustrated by the data in Table 5.3.

Table 5.3 INDEX OF REFRACTION OF ALKANES, ALKENES, ALKYNES, AND ARENES

Compound	Refractive Index
$CH_3CH_2CH_2CH_2CH_2CH_3$	1.3748
$(CH_3)_3CCH_2CH_3$	1.3750
(*E*) $CH_3CH_2CH_2CH{=}CHCH_3$	1.3935
$H_2C{=}CHCH_2CH_2CH{=}CH_2$	1.4044
$CH_3CH_2CH_2C{\equiv}CCH_3$	1.4135
$CH_3CH_2CH_2CH{=}C{=}CH_2$	1.4298
$CH_3CH{=}CHCH{=}CHCH_3$	1.4516
(*Z*) $H_2C{=}CHCH{=}CHCH{=}CH_2$	1.4577
(*E*) $H_2C{=}CHCH{=}CHCH{=}CH_2$	1.5135
Benzene	1.5011

REFERENCES

For a general compilation of physical data, consult the following references.

Dictionary of Organic Compounds, 4th ed., Oxford University Press, New York, 1965. Five volumes with supplements.

Handbook of Chemistry and Physics, Chemical Rubber Co., Cleveland, new ed. annually.

GORDON, A. J. and R. A. FORD, *The Chemist's Companion—A Handbook of Practical Data, Techniques, and References*, John Wiley and Sons, New York, 1972.

Lange's Handbook of Chemistry, McGraw-Hill Book Company, New York, 11th ed., 1973.

The Merck Index; an Encyclopedia of Chemicals and Drugs, 9th ed., Merck and Co., Inc., Rahway, N.J., 1970.

YUKAWA, Y., ed., *Handbook of Organic Structural Analysis*, W. A. Benjamin, Inc., New York, 1965.

PROBLEMS

1. Van der Waals forces between molecules are estimated to be approximately 1 kcal/mole per CH_2 group. Knowing this, why can it be predicted that a hydrocarbon of the structure $CH_3(CH_2)_{150}CH_3$ would probably undergo carbon–carbon bond cleavage upon attempted distillation?

2. Predict which member of each of the following pairs of compounds would have the higher boiling point, and explain the reason for your choice:

(a) $CH_3(CH_2)_4CH_3$ and $CH_3(CH_2)_5CH_3$

(b) and

(c) $H_2C{=}CHCH_2CH{=}CHCH_2CH{=}CH_2$ and $CH_3CH{=}CHCH{=}CHCH{=}CHCH_3$

(d) $ClCH_2CH_2Cl$ and CH_3CHCl_2

(e) 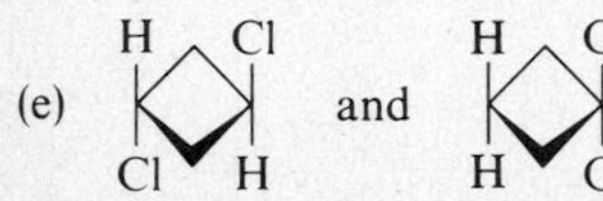

3. Given the bond angles and using the data in Table 5.1, calculate the dipole moments for the following compounds:

(a) $HC{\equiv}CCl$

(b) CH_3Cl (assume tetrahedral symmetry)

(c) $CHCl_3$ (assume tetrahedral symmetry)

(d) Cl Cl

4. Given the molecular dipole moment of 1.74 D and using the data in Table 5.1, calculate the Cl—C—C bond angle in *cis*-1,2-dichloroethylene.

5. 1,2,3,4,5,6-Hexachlorocyclohexane is more stable in the chair conformation than in the boat conformation. In the chair conformation, a number of isomers are possible with various equatorial and axial alignments of the attached chlorine atoms. Which of these would be predicted to have a zero dipole moment?

6. Define each of the following terms as briefly and concisely as possible. Do not use a specific example in defining the term, but cite a specific example to illustrate the definition.

(a) Intramolecular phenomenon

(b) Intermolecular phenomenon

(c) Dipole moment

(d) Boiling point

(e) Melting point

Chapter 6

Spectral properties of organic compounds

Nothing tends so much to the advancement of knowledge as the application of a new instrument.

SIR HUMPHREY DAVY
Collected Works

The electromagnetic spectrum (Fig. 6.1) covers an enormous range of frequencies and energies, from cosmic rays with frequencies of 10^{22} Hz* and energies of 10^9 kcal/mole† to radio waves with frequencies of 10^6 Hz and energies of 10^{-7} kcal/mole. One mode of interaction between a collection of molecules and electromagnetic radiation has already been discussed in connection with the refractive index (see Section 5.2e). Molecules also interact with electromagnetic radiation in other ways, giving rise to phenomena known as ultraviolet (uv) and visible light absorption, infrared (ir) absorption, and nuclear magnetic resonance (nmr) absorption. The nature of these phenomena and their utility to the organic chemist are discussed in this chapter, along with the related phenomena of X-ray crystallography and mass spectrometry.

6.1 Ultraviolet and Visible Absorption Spectroscopy

Electromagnetic radiation with frequencies in the ultraviolet and visible region can be absorbed by molecules, resulting in the excitation of an electron from a lower energy molecular orbital to a vacant, higher energy molecular orbital. A molecule that interacts with light in this fashion is converted from its electronic ground state, in which all the electrons occupy the lowest energy orbitals available to them, to an electronic excited state, in which one or more electrons occupy higher energy orbitals than in the ground state. These transitions can be defined in terms of where the electrons that are undergoing the excitation are situated in

* The preferred frequency unit is the Hertz (abbreviated Hz), which is synonymous with cycles per second (abbreviated cps); *i.e.*, 1 Hz = 1 cps.

† 6.02 × 10^{23} quanta of electromagnetic energy (*i.e.*, a "mole") is called an "Einstein."

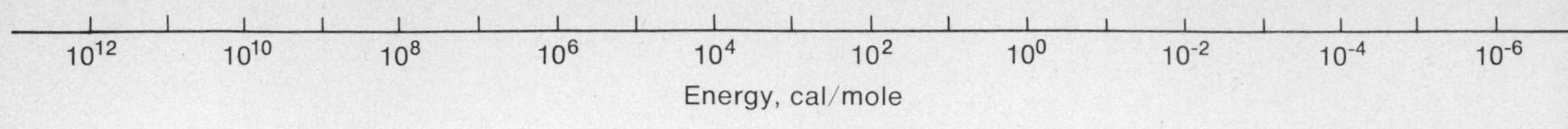

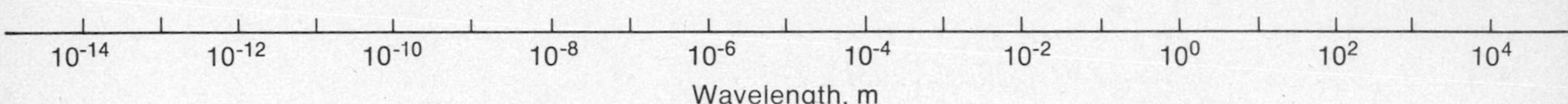

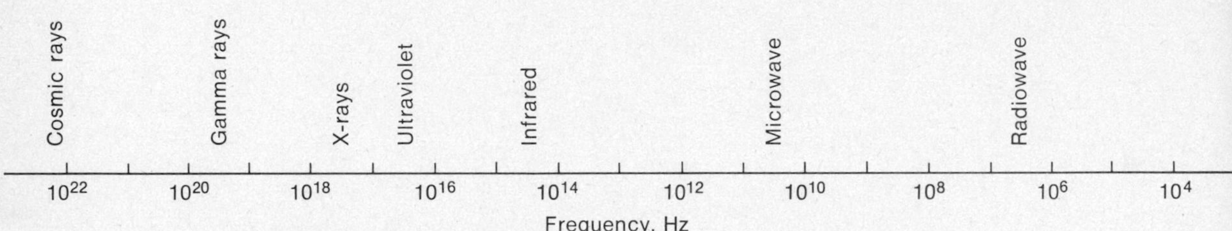

Fig. 6.1. The electromagnetic spectrum.

the ground state and in the excited state. For example, a transition of an electron from a bonding sigma orbital to the higher-energy antibonding sigma orbital is designated as $\sigma \rightarrow \sigma^*$; a transition of an electron from a bonding pi orbital to the higher-energy antibonding pi orbital is designated as $\pi \rightarrow \pi^*$; a transition of a nonbonding electron to an antibonding pi orbital is designated as $n \rightarrow \pi^*$.

Electronically excited molecules are unstable relative to the ground state, and they return more or less rapidly to the ground state. The excess energy of the excited state can be dissipated by the emission of light (fluorescence or phosphorescence), by the emission of heat, and/or by chemical reaction; all these phenomena are intriguing to the chemist. For the moment, however, we are interested only in the process in which the ground state is converted to the excited state by the absorption of energy; what happens to the molecule thereafter is of no concern if we simply want to measure the ultraviolet and visible absorption characteristics of the molecule. We can do this by means of instruments called spectrophotometers (spectrometers). We can determine at what wavelength(s) of light the molecule absorbs energy, and we can quantitatively measure how much light is absorbed at that wavelength. The wavelength of maximum absorption is symbolized by λ_{max} and is usually expressed in nanometers (nm); the magnitude of absorption is symbolized by ε_{max} and is usually expressed simply as a number referred to as the molecular extinction coefficient or the molar absorptivity. The molecular extinction coefficient (molar absorptivity) is equal to the ratio of the log of the intensity of the incident and emergent light (*i.e.*, log I_0/I, which is called the absorbence) at λ_{max} divided by the concentration (in moles per liter) and the thickness of the absorption cell (in centimeters);

$$\varepsilon_{max} = \frac{\log I_0/I}{(c)(l)}$$

PROBLEM 6-1

A 12-mg sample of a compound with molecular weight of 120 is dissolved in 1000 ml of alcohol. When the ultraviolet spectrum of this solution is measured in an absorption cell of 1.0-cm path length, an optical density (*i.e.*, $\log I_0/I$) of 1.26 is observed. What is the extinction coefficient (the molar absorptivity) of the absorption?

For practical reasons, it is very difficult to measure electronic absorptions below 200 nm; therefore, almost all the available data pertain to absorptions that occur above this wavelength. Electronic transitions associated with the electrons in single bonds and with those in "isolated" double bonds generally occur below 200 nm, and therefore are not readily measured. Systems having conjugated double bonds, however, undergo $\pi \rightarrow \pi^*$ transitions above 200 nm, and the longer the conjugated system the longer is the wavelength at which absorption takes place. For example, as we go up the series from ethylene to 1,3-butadiene to 1,3,5-hexatriene, the wavelength of absorption increases from 162 to 215 to 265 nm. In similar fashion, compounds containing carbon–carbon multiple bonds attached to carbon–oxygen double bonds (*e.g.*, C=C—C=O) absorb above 200 nm, and the wavelength of absorption increases as the length of the conjugated system increases (Table 6.1). Aromatic rings constitute a conjugated system. Benzene attached to a vinyl group or to a carbonyl group absorbs strongly in the region near 250 nm (Table 6.1). Thus, the absorption of ultraviolet and visible light by organic compounds finds its principal use in structural chemistry in discerning whether a conjugated system is present. For example, suppose that a compound having the molecular formula C_4H_6O has been isolated and we wish to ascertain its structure. Since the number of structural possibilities is limited to fewer than 30 stable compounds in this simple case, we might start by drawing them all.

STRUCTURES CONTAINING 4 CONTIGUOUS CARBONS:

No C=C *or* C≡C *bonds*:

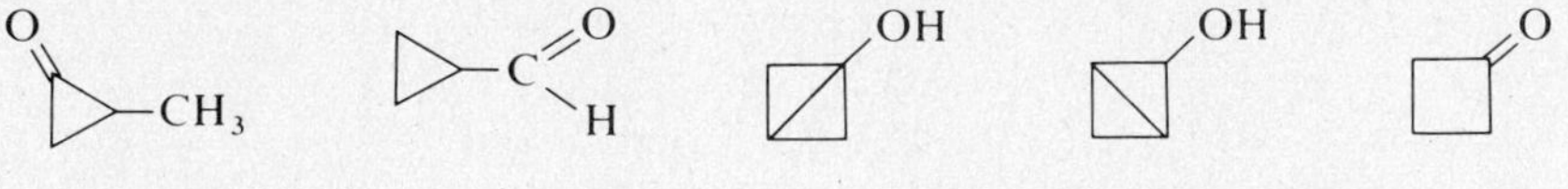

O CH$_3$ C=O H   OH   OH   O

One C=C *bond*:

$H_2C{=}CH\overset{O}{\overset{\|}{C}}CH_3$ $H_2C{=}CHCH_2CHO$ $CH_3CH{=}CHCHO$

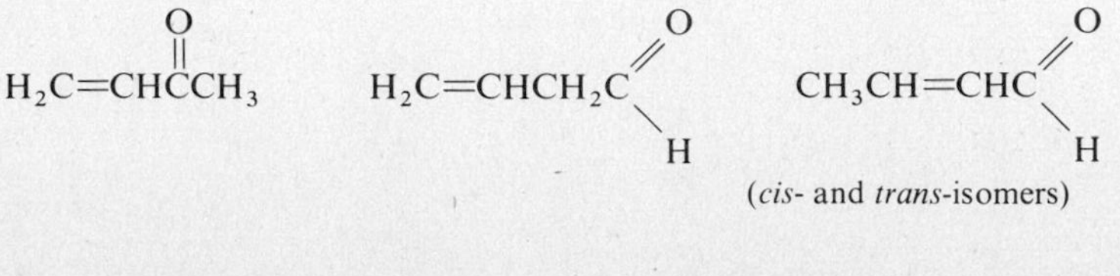

(*cis*- and *trans*-isomers)

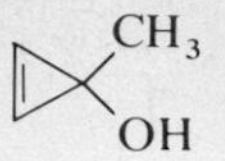

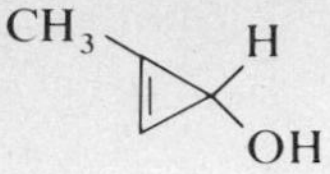

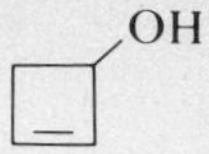

$CH_3CH_2CH{=}C{=}O$

Two C=C *bonds*:*

$H_2C{=}C{=}CHCH_2OH$

One C≡C *bond*:

$HC{\equiv}CCH(OH)CH_3$ $HC{\equiv}CCH_2CH_2OH$ $CH_3C{\equiv}CCH_2OH$

STRUCTURES CONTAINING AN OXYGEN IN A LINEAR FRAMEWORK:

$H_2C{=}CHOCH{=}CH_2$ $HC{\equiv}COCH_2CH_3$ $HC{\equiv}CCH_2OCH_3$ $CH_3C{\equiv}COCH_3$ $H_2C{=}C{=}CHOCH_3$

STRUCTURES CONTAINING OXYGEN IN A CYCLIC FRAMEWORK:

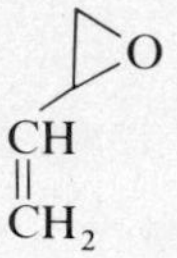

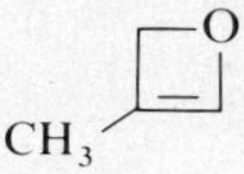

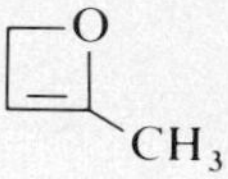

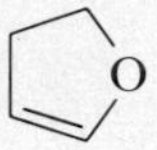

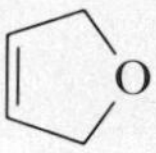

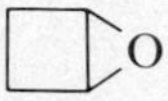

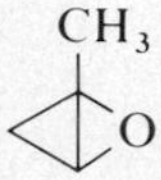

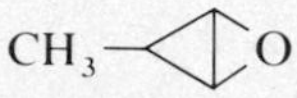

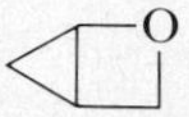

CH=CH₂ CH₃ CH₃ O O O O

CH₃

CH₃

(*exo-* and *endo-* isomers)

Among this group of more than two dozen possibilities, only two have a conjugated structure: *viz.*,

$H_2C{=}CHC(=O)CH_3$ and $CH_3CH{=}CHC(=O)H$

Consequently, if we obtain the ultraviolet-visible absorption spectrum of the C_4H_6O compound and find a strong absorption (*i.e.*, molar absorptivity of 5,000 or greater) in the 225-nm region, we are provided with good evidence that the compound is one of these two structures.† To decide between the conjugated ketone and the conjugated aldehyde, we need either additional

* Structures such as $H_2C{=}CHC(OH){=}CH_2$ and $CH_3CH_2C{\equiv}C{-}OH$, in which the hydroxyl group is attached to an sp^2- or sp-hybridized carbon, are unstable relative to the isomeric aldehyde or ketone (see Chapter 10).

† Exercise some caution, however, for a few other compounds in this group also absorb above 200 nm; compounds having C=O groups (conjugated or unconjugated) have weak absorptions in the 280 nm region; compounds containing a —O—C=C moiety absorb quite strongly in the 200 nm region.

Table 6.1 Ultraviolet and visible light absorption characteristics of unsaturated compounds

Compound	Position of Absorption, λ_{max}	Intensity of nm Absorption, ε
$H_2C{=}CH_2$	162	15,000
$RCH{=}CHCH{=}CHR$	215–225	10,000–25,000
$RCH{=}CHCH{=}CHCH{=}CHR$	265–275	30,000–55,000
$R(CH{=}CH)_4R$	305–315	50,000–75,000
$R(CH{=}CH)_6R$	Approx. 380	Approx. 150,000
$R(CH{=}CH)_8R$	Approx. 415	Approx. 200,000
	184	60,000
	203	7,400
	254	204
—$CH{=}CH_2$	248	14,000
	282	750
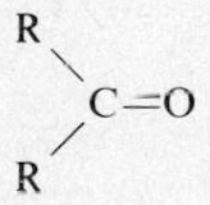 $R_2C{=}O$	270–295	Approx. 20
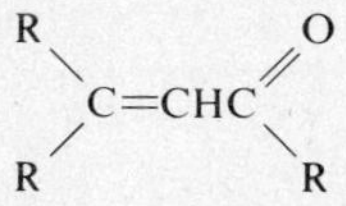 $R_2C{=}CHC({=}O)R$	215–240	4,000–13,000
	320	30
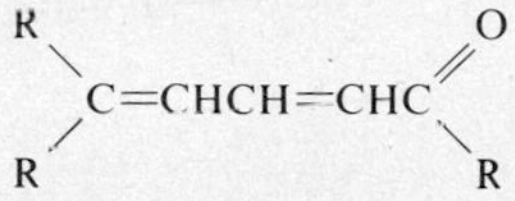 $R_2C{=}CHCH{=}CHC({=}O)R$	245–280	Approx. 20,000
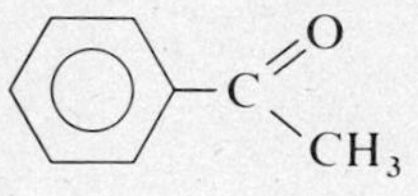 —$C({=}O)CH_3$	243	12,600

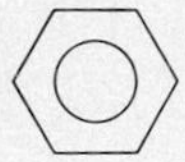

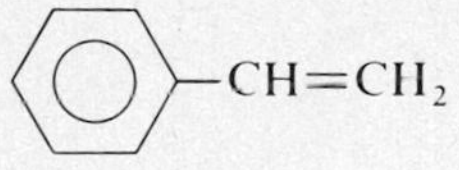

spectral data (*e.g.*, infrared and nuclear magnetic resonance spectra) or some chemical information.

Several examples of ultraviolet and visible spectra are illustrated in Fig. 6.2.

PROBLEM 6-2

Compounds A and B both have the molecular formula C_5H_6O and both absorb strongly in the ultraviolet region above 200 nm, with compound B absorbing at a longer wavelength than compound A. Suggest structures for these compounds.

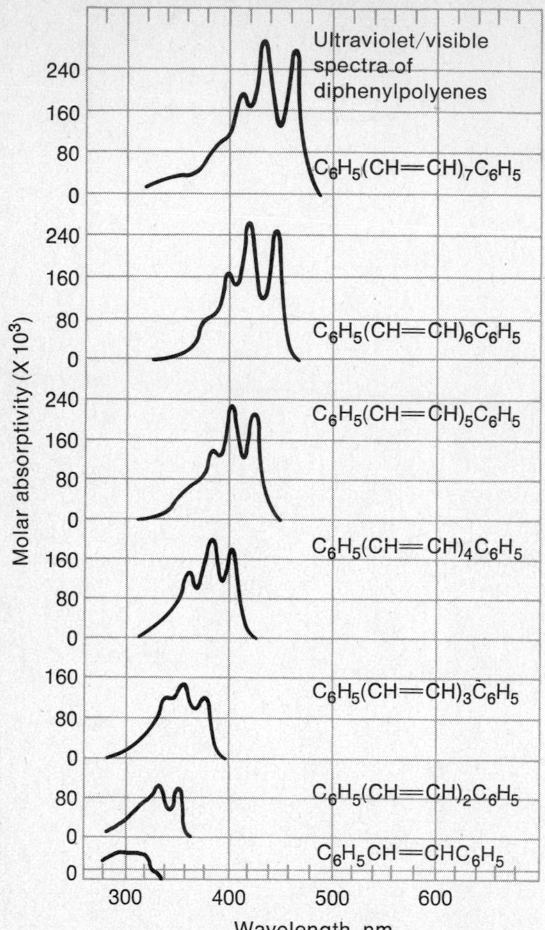

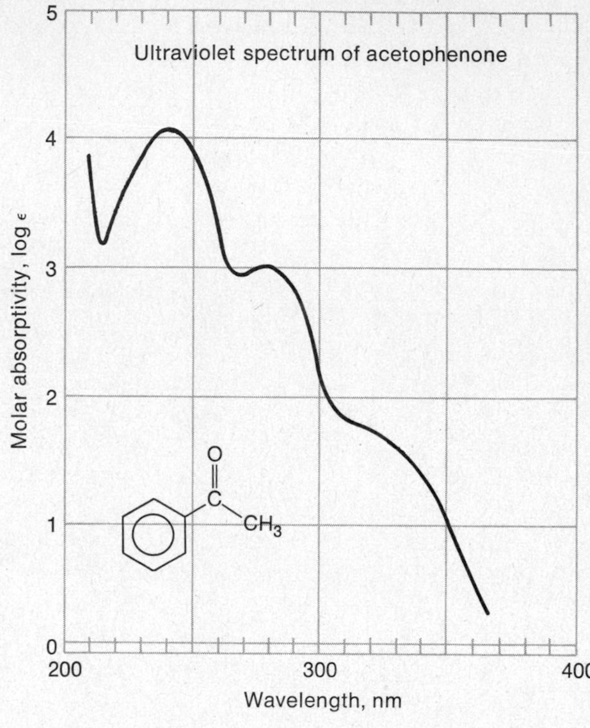

Fig. 6.2. Representative examples of ultraviolet and visible spectra.

6.2 *Infrared Spectroscopy*

Although bond lengths can be measured as the average distance of separation between adjacent nuclei, the actual distance at any moment is generally larger or smaller than the average value. The constant movement of adjacent atoms relative to each other constitutes another periodic, quantized motion within molecules that can absorb electromagnetic radiation. Interaction between organic molecules and light of frequency of about 10^{14} Hz (*i.e.*, infrared region) results in a transfer of energy from the incident beam to the molecule, leading to vibrationally excited molecules. The wavelengths at which this occurs can be measured (usually expressed in cm^{-1} or μm*) and recorded as the infrared spectrum of the compound.

---

* The designation cm^{-1} is called a wave number and is equal to the reciprocal of the wavelength expressed in centimeters. The designation μm is called a micrometer (10^{-6} meter). Wavelengths may also be indicated in μ (micron) equivalent to 1 μm.

The vibrational modes of bonds in a molecule can be divided into two major categories: (1) those involving a stretching and shrinking of the bond along its axis, called the stretching deformation; and (2) those involving a bending of the bond away from and toward the average bond axis, called the bending deformation. The frequencies associated with these kinds of deformations vary with the

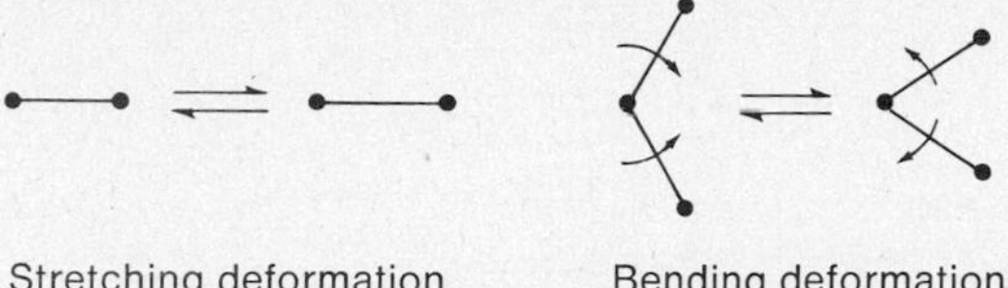

masses of the atoms comprising the bond and with the strength of the bond. As the weights of the atoms increase, the frequencies of the vibrations diminish; as the force constant* for the deformation increases, the frequency of absorption increases.

Even so simple a molecule as methane, containing only five atoms, has nine fundamental vibrational modes† available to it, each of which can be excited to a higher vibrational level by the absorption of energy characteristic for the particular vibration. Higher molecular weight compounds have many more vibrational modes, and their infrared spectra often contain several dozen absorption bands. Despite this complexity, certain portions of an infrared spectrum are often readily interpretable, allowing structural assignments to be made with considerable certainty. Figure 6.3 shows a plot of the portion of the infrared region that is ordinarily scanned with the modern instruments (*i.e.*, 4000–500

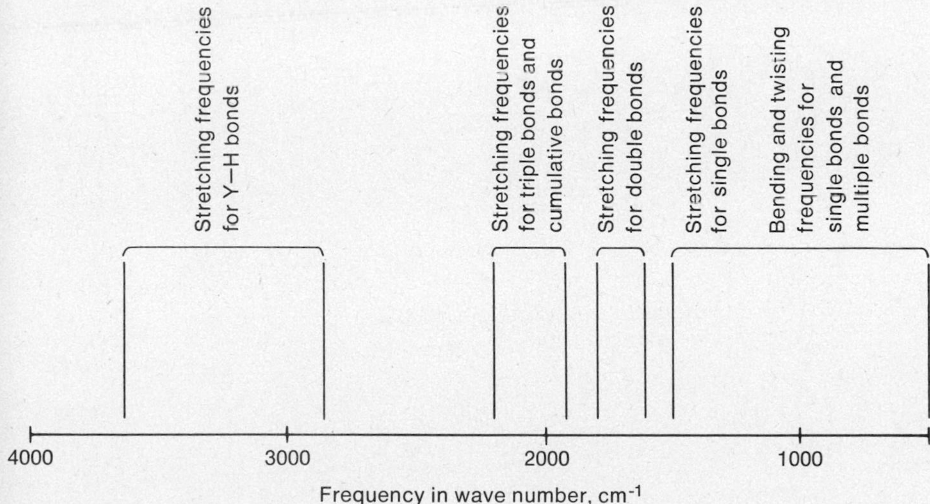

Fig. 6.3. Infrared absorption regions of organic compounds.

---

* The force constant is a measure of the energy required to distort the bond from its equilibrium position; it reflects the bond's resistance to stretching or bending.

† Calculated from the formula $3n - 6$, where n is the number of atoms in the molecule.

cm^{-1}; 2.5–20 μm), and it indicates the kinds of bonds that give rise to absorptions in the various regions of the spectrum.

Within each of the regions specified in Fig. 6.3, finer distinctions are possible. In the Y—H stretching region from 3600–2800 cm^{-1}, O—H, N—H, C—H, and S—H can be distinguished readily, and even various types of C—H bonds can be differentiated. As the hybridization on carbon changes from sp^3 to sp^2 to sp, the C—H bond strength increases (see Table 2.3), causing an increase in the C—H

Table 6.2 INFRARED ABSORPTION BANDS USEFUL FOR STRUCTURAL ASSIGNMENTS[a]

Group	Wave number (cm^{-1})	(μm)
X—H Stretching Deformation Bands:		
O—H	3650–3100	(2.74–3.22)
N—H	3550–3100	(2.82–3.22)
≡C—H	3330	(3.00)
=C—H	3085–3025	(3.24–3.31)
Ar—H	3030	(3.30)
—C(—)(—)—H (sp³ C—H)	2960–2870	(3.38–3.49)
S—H	2500	(4.00)
X=Y Stretching Deformation Bands:		
C=O[b]	1850–1650	(5.40–6.05)
C=NR[b]	1690–1640	(5.92–6.10)
C=C[b]	1680–1600	(5.95–6.25)
N=N	1630–1575	(6.14–6.35)
N=O	1650–1500	(6.06–6.66)
Benzene ring	1600–1450 (four bands)	(6.25–6.90)
X≡Y and X=Y=Z Stretching Bands:		
C≡N	2260–2240	(4.43–4.46)
RC≡CR′	2260–2190	(4.43–4.56)
RC≡CH	2140–2100	(4.68–4.76)
C=C=O	2170–2130	(4.61–4.70)
C=C=C	1975–1950	(5.07–5.13)
C—H Bending Deformation Bands in the Fingerprint Region:		
RHC=CHR (trans: R and H on C1; H and R on C2)	~970	(10.30)
$R_2C=CH_2$	~900	(11.11)
$R_2C=CHR$	~825	(12.11)
RHC=CHR (cis: R, R above; H, H below)	~670	(14.90)
$RHC=CH_2$	~1000 and 900	(10.0 and 11.1)

[a] Data are given in wave numbers (cm^{-1}) and, in parentheses, micrometers (μm).
[b] Conjugation with a C=C or aryl group lowers the frequency about 30 cm^{-1}.

stretching frequencies as the structure changes from $R—CH_3$ to $R_2C=CH_2$ to R—C≡CH (see Table 6.2). The carbon–carbon multiple bonds of allenes and acetylenes are readily detected by their absorption bands near 2200 and 2000 cm^{-1} respectively, whereas in the 1850- to 1500-cm^{-1} region, clear distinctions are generally possible between C=C, C=O, and C=N groups. Particularly useful are the finer differentiations possible with carbonyl compounds discussed in greater detail in Chapters 13–15. Aromatic compounds usually give rise to four characteristic absorption bands in the 1600- to 1450-cm^{-1} region. The 1400- to 1000-cm^{-1} region contains the stretching absorption bands of C—C, C—O, and C—N, as well as complex group deformations and C=C, C=O, and C=N bending deformation bands. Because of the large number of absorption bands that usually appear in this region, however, definitive assignments are difficult to make. A number of diagnostically useful absorption bands appear in the 1000- to 600-cm^{-1} region (often referred to as the "fingerprint" region), which arises from the bending deformations of vinyl and aromatic C—H bonds. For instance, the degree of substitution, as well as the stereochemistry of 1,2-disubstituted alkenes can be readily determined (see Table 6.2). As an example, the distinction between the structures

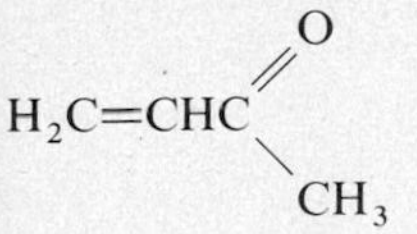

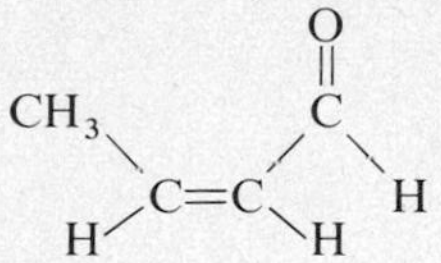

$$H_2C{=}CHC({=}O)CH_3 \qquad \text{(CH}_3\text{)(H)C{=}C(H)(CHO)} \quad \text{and} \quad \text{(CH}_3\text{)(H)C{=}C(H)(CHO)}$$

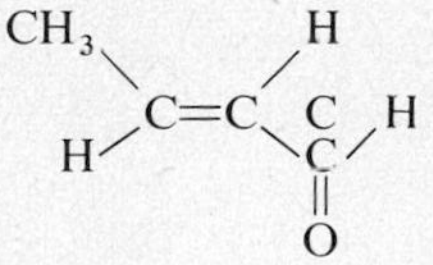

(see p. 144) can be made by means of the infrared spectral properties of the compound. The first compound, which possesses a $H_2C{=}C\!\!<^{H}$ moiety, has absorption bands near 1010 and 900 cm^{-1}, whereas the second compound possesses an absorption band near 670 cm^{-1} and the third compound has one near 970 cm^{-1}. Thus, by the combined use of ultraviolet and infrared spectroscopy, the compound possessing a molecular formula C_4H_6O and showing an intense absorption band near 225 nm in the ultraviolet and absorption bands near 1010 and 900 cm^{-1} in the infrared region, can be definitely assigned the structure

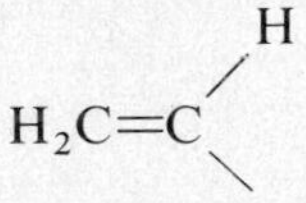

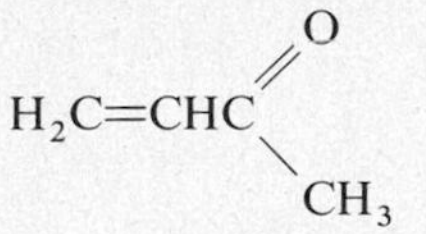

$$H_2C{=}CHC({=}O)CH_3$$

Several examples of infrared spectra are illustrated in Fig. 6.4.

PROBLEM 6-3

Compounds A, B, C, and D all have the molecular formula C_4H_6. Compound A has an infrared absorption near 2200 cm^{-1}, compound B has an absorption near 1950 cm^{-1}, compound C has an absorption near 1650 cm^{-1}, and compound D has no absorption in any of these regions. Suggest reasonable structures for compounds A–D that are in accord with these spectral data.

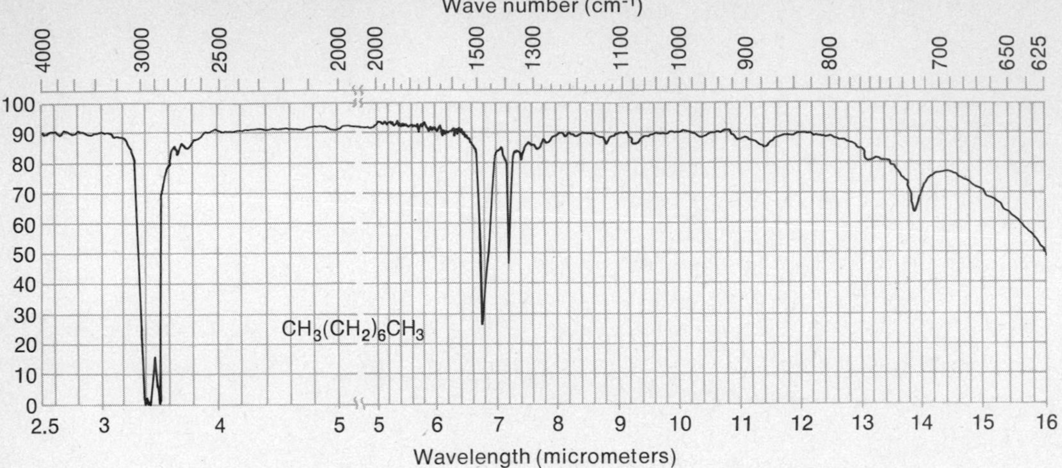

Octane: C—H stretch of CH_2 and CH_3 near 2900 cm^{-1} and C—H bend at 1470 and 1390 cm^{-1}; note the simplicity of the spectrum

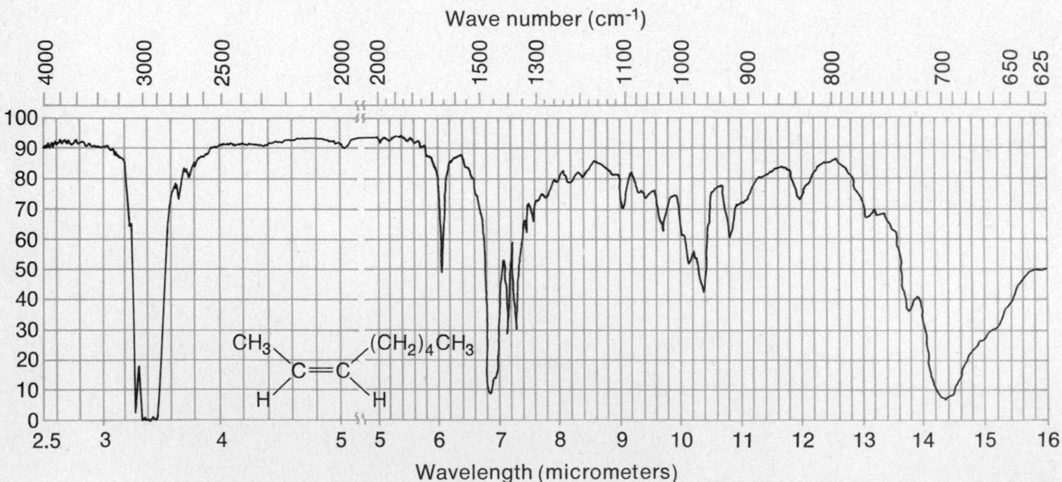

cis-2-Octene: =CH stretch on the high frequency edge of CH_2 and CH_3 stretch; note band near 700 cm^{-1} for *cis*-alkene

Fig. 6.4. Illustrative examples of infrared spectra.

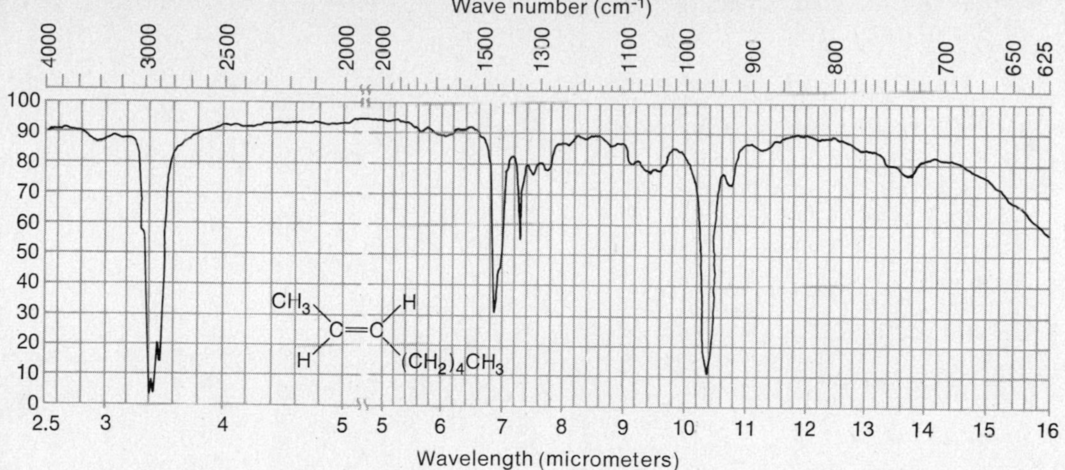

trans-2-Octene: Note the band at 965 cm^{-1} for *trans*-alkene

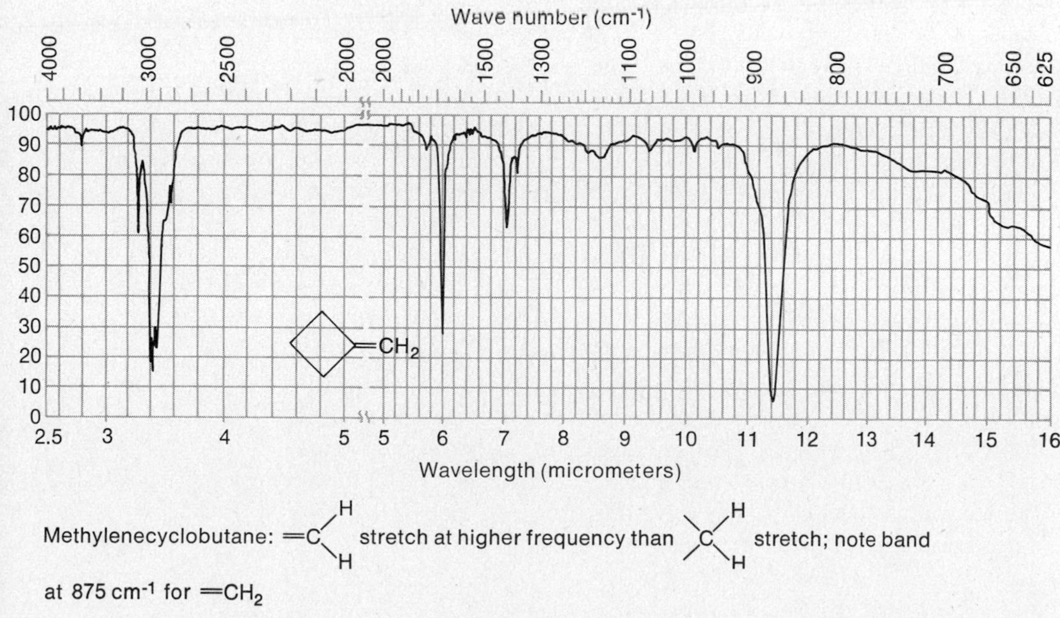

Methylenecyclobutane: $=CH_2$ stretch at higher frequency than $>CH_2$ stretch; note band at 875 cm^{-1} for $=CH_2$

Fig. 6.4. (continued)

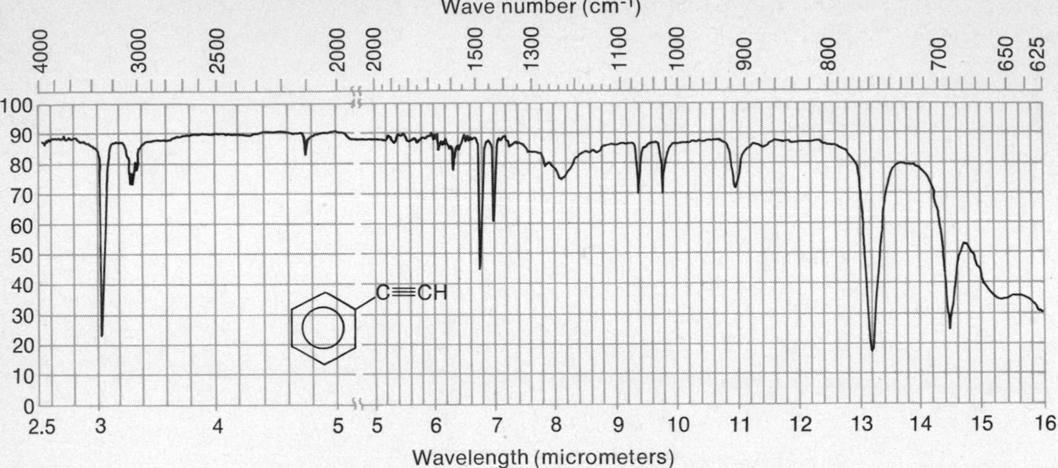

Phenylethyne: ≡C—H stretch at higher frequency than Ar—H stretch; note bands at 1600 (weak) and 1500 (strong) cm^{-1} for the aromatic ring and at 2110 cm^{-1} for the C≡C stretch

Fig. 6.4. (continued)

6.3 *Nuclear Magnetic Resonance* Spectroscopy*

6.3a. THE CHEMICAL SHIFT. Another type of transition occurring within some molecules but involving energies many orders of magnitude lower than those required to effect electronic transitions (*i.e.*, about 100 kcal/mole) and vibrational transitions (*i.e.*, about 10 kcal/mole) is associated with the nuclei themselves. Those nuclei which have either an *odd atomic number* or an *odd mass number* (or both) possess spin angular moments, and they show magnetic properties. In the simplest case of nuclei with spin number of $\frac{1}{2}$, the nucleus can be considered to have the properties of a bar magnet, capable of aligning itself in two ways in the presence of an external magnetic field—*i.e.*, *with the field*, so that its north pole is oriented toward the south pole of the external field, or *against the field*, so that its north pole is oriented toward the north pole of the external field. These alignments represent two different energy states (called *spin states*), the alignment with the field corresponding to the lower energy state and the alignment against the field corresponding to the higher energy state. If the precise amount of energy by which these two spin states differ is now added to the system, a transition from the orientation of the nucleus "with the field" to the orientation of the nucleus "against the field" can be induced. The phenomenon is called nuclear magnetic resonance and is analogous to electronic transitions (*i.e.*, ultraviolet-visible spectra) and vibrational transitions (*i.e.*, infrared spectra) in that it involves a quantized jump from a lower to a higher energy level. The precise region in the electromagnetic spectrum at which the nuclear spin transition takes place can be measured and recorded as a *nuclear magnetic resonance spectrum*.

* The term "resonance" in this context should not be confused with its use in the context of "resonance structures." In nuclear magnetic resonance, it refers to an actual physical phenomenon; in the valence bond theory, it refers to a mathematical concept.

Hydrogen, one of the atoms present in almost all organic compounds, has a nuclear spin of $\frac{1}{2}$ and, consequently, behaves in the manner just described. The most abundant carbon isotope, ^{12}C, on the other hand, has no nuclear magnetic properties and, therefore, cannot give rise to nuclear magnetic resonance. The nuclear magnetic resonance spectra of organic compounds that have been most useful, therefore, are the hydrogen (*i.e.*, ^{1}H) spectra. A small amount, about 1% of ^{13}C accompanies the major isotope of carbon; since its mass number is odd, it possesses nuclear magnetic properties and can undergo spin transitions. To an increasing extent as a result of improvements in instrumentation, ^{13}C magnetic resonance spectra also are coming to prominence, supplementing and extending the capabilities of ^{1}H magnetic resonance spectra.*

The measurement of an nmr spectrum is carried out by placing a glass tube containing a solution of the sample between the pole faces of a strong magnet. A beam of electromagnetic radiation in the radio frequency region is then passed through the sample, and the difference in intensity between the incident and emergent beams is measured (Fig. 6.5). By varying either the strength of the magnetic field or the frequency of the radiofrequency beam, a scan can be made

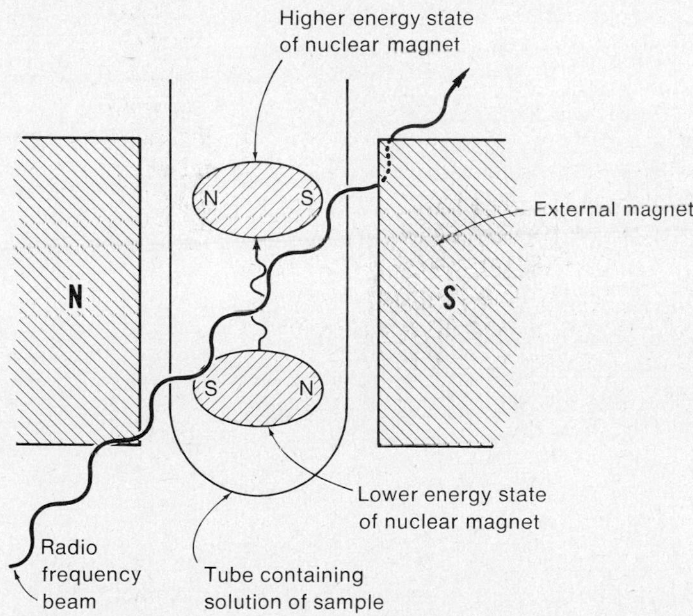

Fig. 6.5. The nuclear magnetic resonance phenomenon.

to see under what conditions energy is absorbed by the sample. When absorption occurs, nuclear magnetic resonance is said to have taken place. The results of an nmr measurement are usually plotted so that the position of the absorption (*i.e.*,

* ^{1}H nuclear magnetic resonance is frequently referred to as proton magnetic resonance (abbreviated pmr), and the hydrogens in organic compounds are frequently referred to as protons in nmr discussions although in the majority of instances they are covalently bonded and are not actually present at H^+. The term "hydrogen magnetic resonance" is a more accurate designation.

the particular combination of magnetic field strength and radio frequency) appears on the horizontal axis and the intensity of the absorption on the vertical axis. For example, when a measurement was performed with a sample of diethyl ether on one of the early nmr instruments, a plot of the sort shown in Fig. 6.6 was obtained. The appearance of only a single resonance might seem to suggest that all ten of the hydrogens in diethyl ether undergo nuclear spin transition at essentially the same place. Remarkable improvements have been made in nmr spectrometers, however, and it is now realized that the small perturbations in the magnetic field around any particular hydrogen atom caused by neighboring nuclei and electrons are sufficient to make hydrogens in different molecular environments come into resonance at *slightly different* frequencies. If the nmr spectrum of diethyl ether is measured in a more highly resolving instrument, therefore, two resonances are observed (Fig. 6.7). These arise because the magnetic field *at the nucleus* of the hydrogens of the CH_3 group is slightly different from the magnetic field *at the nucleus* of the hydrogens of the CH_2 group. This difference can be ascribed to differences in the electron density around these two types of hydrogens. Electrons themselves have magnetic properties, and in the presence of an external magnetic field they exert a "shielding effect," which makes the magnetic field at the nucleus somewhat lower than the applied field. Thus, the greater the electron density around a hydrogen, the greater is the shielding effect. In the compound $CH_3CH_2OCH_2CH_3$, the oxygen is electronegative (see Section 2.7) and tends to pull electrons toward it. This reduces the electron density around the CH_2 hydrogens (they are strongly "deshielded") and, to a lesser extent, the electron density around the CH_3 hydrogens (they are less strongly "deshielded"). As a result of this difference in deshielding, the magnetic field at the CH_2 and CH_3 hydrogens is different, and *two* resonances are produced.* This difference is called the *chemical shift*.

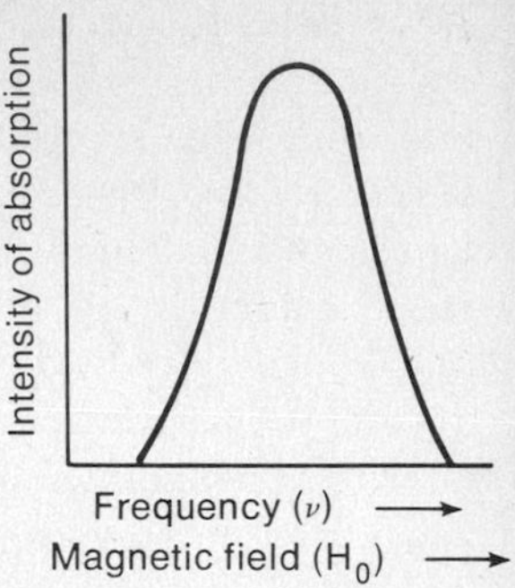

Fig. 6.6. Low resolution nm spectrum of $CH_3CH_2OCH_2CH$

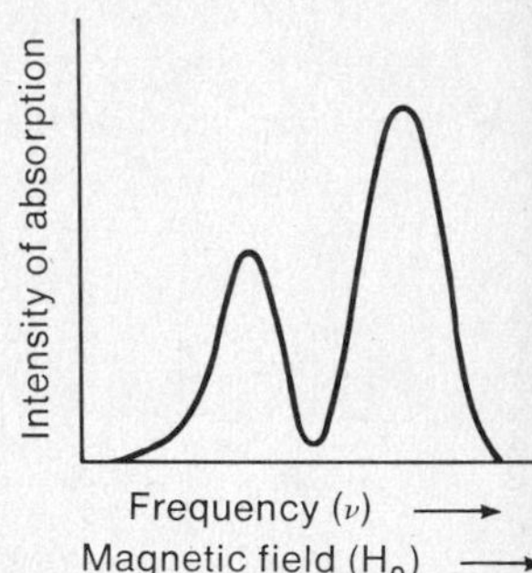

Fig. 6.7. Medium resolution nm spectrum of $CH_3CH_2OCH_2CH$

The magnitude of the chemical shift is usually expressed in terms of the difference between the resonance positions in the compound being measured and the resonance position of the protons in some reference compound. For the majority of organic compounds, the reference material chosen is tetramethylsilane, $(CH_3)_4Si$ (abbreviated TMS), which is the silicon analog of neopentane. The chemical shift is *field dependent*; the greater the magnetic field, the greater is the shift. For example, the downfield shifts of the CH_3 and CH_2 hydrogens in $CH_3CH_2OCH_2CH_3$ from the hydrogens in tetramethylsilane are 69 and 202 Hz, respectively, when the spectrum is measured at 60 MHz; they are 115 and 337 Hz, respectively, when the spectrum is measured at 100 MHz. To allow the chemical shift to be expressed in such a way that its value is independent of the conditions under which the measurement is made, the chemical shift parameter (the *δ value*) is defined as:

$$\delta = \left(\frac{\nu_{\text{compound}} - \nu_{\text{tetramethylsilane}}}{\nu_{\text{applied field}}}\right) \times 10^6$$

* Of course, the two CH_2 groups in ether are equivalent, and the two CH_3 groups are equivalent (see, *e.g.*, the discussion of equivalent groups in $CH_3CH_2CH_3$, on p. 56, and $CH_3CH_2CH_2CH_3$, on p. 59).

If we assign a ν value of zero to the reference compound tetramethylsilane, the values for the CH_3 and CH_2 hydrogens of ether are $69 \times 10^6/60 \times 10^6 = 1.15$ and $202 \times 10^6/60 \times 10^6 = 3.37$, respectively; the same values are obtained from the data at 100 MHz. Hydrogens in various structural environments in molecules have characteristic δ values (Table 6.3), and the nmr method provides the single most powerful tool that modern organic chemists have for routine structural determination. When applicable, X-ray diffraction methods (see Section 6.5) may be more revealing of structural details, but they cannot yet be called routine nor are they easily applied by the average chemist.

Table 6.3 NMR CHARACTERISTICS OF VARIOUS MOLECULAR GROUPINGS

Compound Type	δ Value	Compound Type	δ Value
$(CH_3)_4Si$	0.00	Halogen$-CH_3$	3.5 ± 1.2
CH_2—CH_2—CH_2 (cyclopropane)	0.22	$O-CH_3$	3.6 ± 0.3
		$RO-H$	0.5–5.5
CH_4	0.23	$=CH_2$ and $=CHR$	4.5–7.5
$-\overset{\vert}{\underset{\vert}{C}}-CH_3$	1.1 ± 0.1	$Ar-H$	7.4 ± 1.0
$-\overset{\vert}{\underset{\vert}{C}}-CH_2-\overset{\vert}{\underset{\vert}{C}}-$	1.3 ± 0.1	$ArO-H$	4.5 – 9.0
		$RCONH_2$	8.0 ± 0.1
$(C)_3C-H$	1.5 ± 0.1		
		$RCHO$	9.8 ± 0.3
		RCO_2H	11.6 ± 0.8
RNH_2	0.6–4.0		
		RSO_3H	11.9 ± 0.3
$=CCH_3$	1.75 ± 0.15		
$\equiv CCH_3$	1.80 ± 0.15		
$Ar-CH_3$	2.35 ± 0.15		
$\equiv CH$	1.8 ± 0.1		

Closer inspection of the spectrum of Fig. 6.7 reveals still another feature; *viz.*, the areas under the two resonance envelopes are unequal. A measurement shows that they are in the ratio of 3:2, corresponding exactly to the ratio of the number of hydrogens in the CH_3 and CH_2 groups. Thus, not only does the nmr spectrum show the chemical shifts for various types of hydrogens, but it also indicates the *number of hydrogens* in each of these various environments.

6.3b. SPIN–SPIN COUPLING. We can obtain additional information by using an even more highly resolving instrument. The nmr spectrum of $CH_3CH_2OCH_2CH_3$ in a modern high-resolution nmr spectrometer has the appearance shown in Fig. 6.8, in which each of the resonance envelopes associated with the CH_3 and CH_2 hydrogens, respectively, is split into a number of lines. The splitting patterns arise from the phenomenon known as *spin–spin coupling*, which is ascribed to the magnetic influence of neighboring hydrogens on the hydrogen that is undergoing the nmr transition. Just as the magnetic properties of electrons affect the magnetic field at the nucleus of nearby hydrogens, so neighboring hydrogens also influence this field. The magnitude of the influence, however, is a function of the spin state of the neighboring hydrogen; when the hydrogen is in a $+\frac{1}{2}$ spin state, this magnetic influence is different from when it is in a $-\frac{1}{2}$ spin state. The resonance patterns resulting from this phenomenon are called spin–spin splitting patterns, and the separation between the resonance lines in the pattern is called the *coupling constant*; it is symbolized by J and is designated in Hz units. Unlike the chemical shift, which is field dependent, the coupling constant is field *independent*; *i.e.*, it is the same regardless of the strength of the external magnetic field.

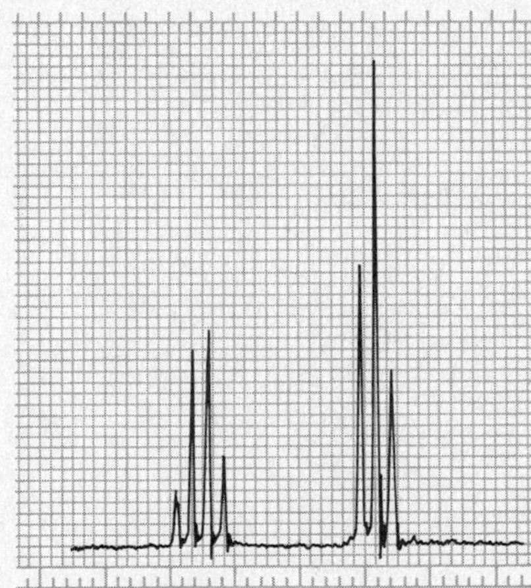

Fig. 6.8. High resolution nmr spectrum of $CH_3CH_2OCH_2CH_3$.

Before discussing spin–spin coupling in more detail, we should comment further about the ratio of the populations in the lower- and higher-energy spin states. Although the lower-energy spin state is preferred, its population is only very slightly higher than that of the higher-energy spin state. The phenomenon of nuclear magnetic resonance absorption depends, of course, on this difference in population, small though it may be, and it is one of the triumphs of instrumentation that methods for its detection have been devised. So far as spin–spin interactions are concerned, however, the difference is negligible and without any perceptible effect; we can assume the $+\frac{1}{2}$ and $-\frac{1}{2}$ spin states to be equally populated.

Before considering the splitting pattern in $CH_3CH_2OCH_2CH_3$, let us first interpret the phenomenon of spin–spin coupling in the compound shown below, which has the high resolution spectrum pictured in Fig. 6.9.

O
||
C
O S
C=C
H_a H_b

The separation between TMS and the centers of the two resonance envelopes is a measure of the chemical shift (δ) of H_a and H_b. The splitting of each of these resonances into two lines (called doublets) is the result of the magnetic interaction between the two neighboring hydrogens, whereby the spin states of one affect the transition energy of the other. For maintenance of the magnetic field at the nucleus of H_a at the level necessary for resonance, a different external field must be applied when H_b is in the $+\frac{1}{2}$ spin state than when it is in the $-\frac{1}{2}$ spin state. Since these two spin states are virtually equally probable, as mentioned, two equal intensity lines appear in the spectrum. Reciprocally, the transition energy of H_b is affected in a similar way by the spin states of H_a, and two equal intensity lines also appear for its resonance pattern.

We can now use the same considerations to explain the *triplet* and *quadruplet* patterns of $CH_3CH_2OCH_2CH_3$ (Fig. 6.8). In this case, *two* methylene hydrogens act as neighbors to the methyl group, and *three* methyl hydrogens act as neighbors to the methylene group, in contrast to the single-hydrogen neighbors in the compound shown in Fig. 6.9. The result of the increased number of hydrogens is

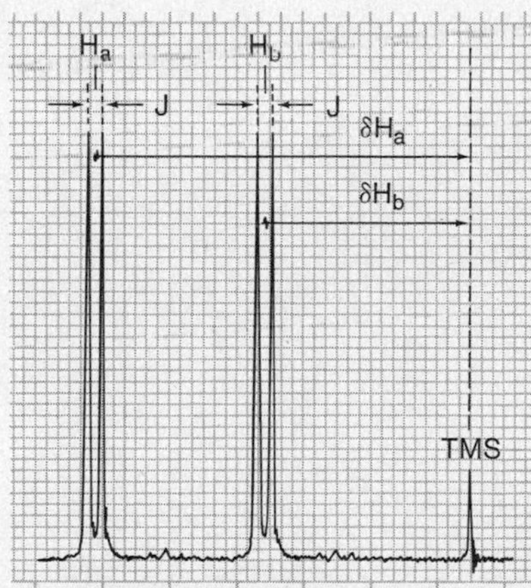

Fig. 6.9. High resolution nmr spectrum of

O
||
C
O S
C=C
H_a H_b

an increased number of possible spin states for each neighboring group. Thus, the methylene group, containing two hydrogens, can have both hydrogens in the $+\frac{1}{2}$ spin state (total spin of $+1$), one hydrogen in the $+\frac{1}{2}$ and the other in the $-\frac{1}{2}$ spin state (total spin of 0), or both hydrogens in the $-\frac{1}{2}$ spin state (total spin of -1)—*i.e.*, three different spin states. For the same reasons that spin–spin coupling of the hydrogens in the compound in Fig. 6.9 leads to doublets, the three spin states of the methylene group give rise to three resonance lines for the neighboring methyl group—*i.e.*, a triplet. The methyl group can combine the spin states of its three hydrogens in four different ways to give total spins of $+\frac{3}{2}$, $+\frac{1}{2}$, $-\frac{1}{2}$, and $-\frac{3}{2}$ (*i.e.*, four different spin states). Accordingly, four resonance lines (*i.e.*, a quartet) would be expected for the neighboring methylene group resonance, which is what is observed.

PROBLEM 6-4

Compounds A and B both have the molecular formula $C_2H_4Br_2$. Compound A has a singlet resonance in the nmr, and compound B has two resonance envelopes, one a doublet and the other a quartet. Suggest structures for compounds A and B.

Finally, we must consider the unequal peak heights in the triplet and quartet patterns of Fig. 6.8. The methyl and methylene resonances are split into three and four lines, respectively, but the ratio of the intensities within these resonance patterns is not 1:1 as was the case for the resonances in Fig. 6.9. This difference is a consequence of probabilities. Whereas there is only one way that a two-spin system can have a total spin of $+1$ (*i.e.*, both nuclei aligned with the field) or a total spin of -1 (*i.e.*, both nuclei aligned against the field), there are two ways that a two-spin system can have a total spin of 0 (*i.e.*, $+\frac{1}{2}$ and $-\frac{1}{2} = 0$; $-\frac{1}{2}$ and $+\frac{1}{2} = 0$). Consequently, it is twice as probable that the total spin will be 0 rather than either $+1$ or -1; this is reflected in the resonance pattern of the neighboring methyl group which appears as a triplet pattern with the relative intensities 1:2:1. Similar reasoning, depicted in Fig. 6.10, can account for the quartet pattern for the methylene resonance in which the ratios are 1:3:3:1.* Where equivalent sets of neighboring hydrogens are involved in the spin–spin coupling interaction, the number of lines in the resonance pattern is equal to $(n + 1)$, where n is the number of equivalent hydrogens. Only if the neighboring hydrogens constitute equivalent sets and only if the chemical shift between them is significantly larger than the spin coupling constant can the splitting patterns be interpreted in the simple manner just described. When the hydrogens are non-equivalent and when the chemical shifts are comparable in magnitude to the spin coupling constants, the patterns are more complex and less amenable to simple analysis (see, for example, the spectrum in Fig. 6.11).

---

* For example, if three pennies are tossed 8,000 times and the head-tail pattern noted each time, 1,000 times it will be all heads, 1,000 times it will be all tails, 3,000 times it will be two heads and one tail, and 3,000 times it will be one head and two tails.

Spin combination of protons In CH_2	Total spin number	Spin combination of protons In CH_3	Total spin number
↑↑	+1	↑↑↑	+3/2
↓↑	0	↑↑↓	+1/2
↑↓	0	↑↓↑	+1/2
↓↓	−1	↓↑↑	+1/2
		↓↓↑	−1/2
		↓↑↓	−1/2
		↑↓↓	−1/2
		↓↓↓	−3/2

Fig. 6.10. Total spin states for methylene and methyl groups.

PROBLEM 6-5

Draw the nmr spectra that you would predict for the two isomers of butane.

PROBLEM 6-6

In Problem 6-3 there are four possibilities for the C_4H_6 compound containing one double bond. Two of these show a two-hydrogen resonance in the vinyl region and no methyl resonance in the nmr; the third shows a two-hydrogen resonance in the vinyl region and a methyl resonance; the fourth shows a one-hydrogen resonance in the vinyl region and a methyl resonance. Suggest structures for these four compounds, and indicate what additional information might be obtained to provide definitive structure assignments.

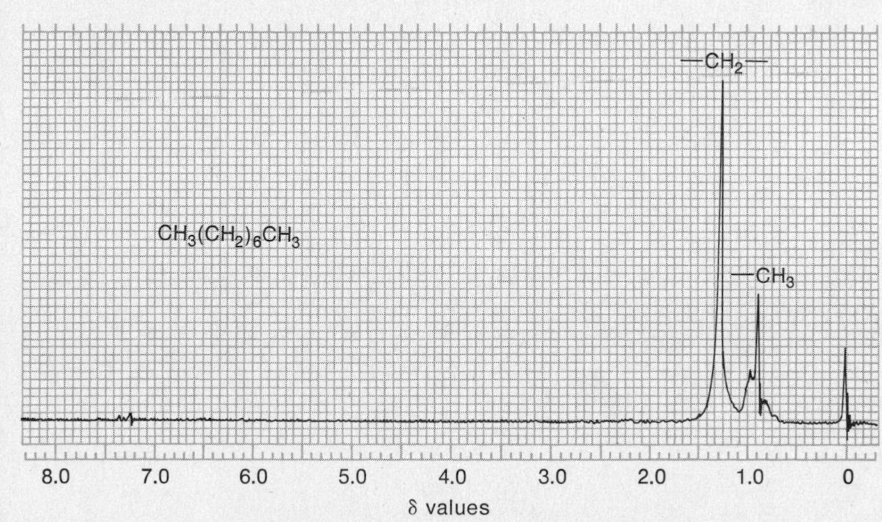

Fig. 6.11. Representative examples of nmr spectra.

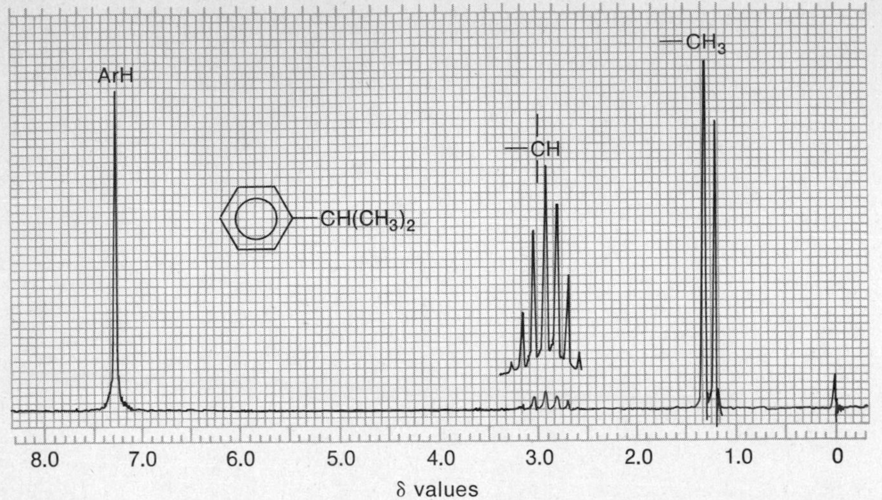

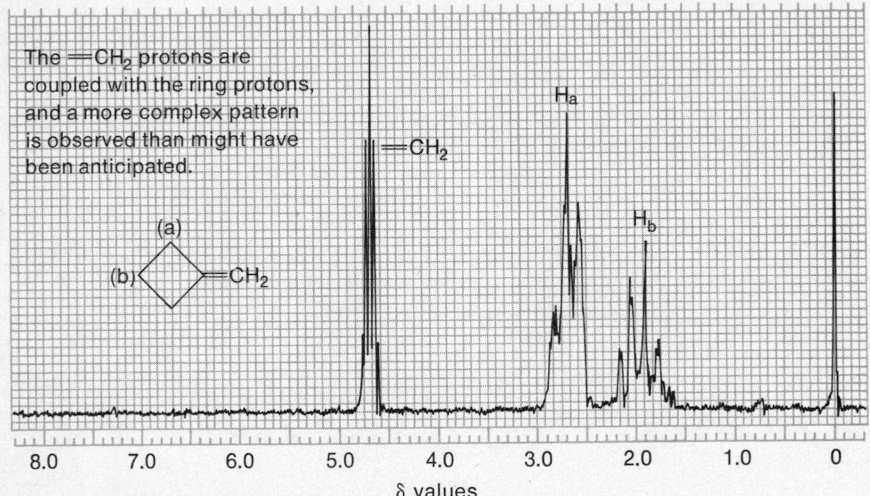

Fig. 6.11. (continued)

6.3c. GUIDELINES FOR THE INTERPRETATION OF NMR SPECTRA. The principal items of concern in the interpretation of nmr spectra are (1) the chemical shift which indicates the electron density at the nucleus undergoing the spin transition, (2) the spin–spin splitting patterns (*i.e.*, the number and relative intensities of the resonance lines) which indicate the number of adjacent hydrogens, and (3) the relative intensities of the resonance envelopes which indicate the number of hydrogens.

The following rules-of-thumb are useful in predicting the *magnitude of the chemical shift* for various types of structures:

1. The downfield shift (δ value) increases from $\mathbf{RCH_3}$ to $\mathbf{R_2CH_2}$ to $\mathbf{R_3CH}$.
2. The downfield shift (δ value) increases from alkyl to alkenyl to aryl. Alkynyl (*i.e.*, RC≡CH) has a much smaller downfield shift than might have been

expected because of a geometric dependence of the magnetic influence of neighboring groups, the phenomenon known as *diamagnetic anisotropy*.

3. The downfield shift (δ value) increases as the electronegativity of the neighboring group increases. Thus the downfield shift increases in the order: $CH_3Li < CH_3CH_3 < CH_3NH_2 < CH_3OH < CH_3F$.
4. The downfield shift (δ value) increases as the distance between the electronegative group and the hydrogen undergoing transition diminishes. Thus, the downfield shift increases from ROC—C—C—**H** to ROC—C—**H** to RO—C—**H**.

The *number of different resonances* (referred to as "envelopes" or "resonance envelopes") is equal to the number of sets of equivalent hydrogens in a molecule. For example, $(CH_3)_4C$ has only one set of equivalent hydrogens; *i.e.*, all 12 hydrogens bear an identical relation to the rest of the molecule, and only one resonance line is observed (coupling between identical nuclei is *not* observed). In the example discussed, $CH_3CH_2OCH_2CH_3$ has two sets of equivalent hydrogens and shows two resonance envelopes. The compound CH_3CH_2OH has three sets of equivalent hydrogens (three from CH_3, two from CH_2, and one from OH). The compound shown below has four sets of equivalent hydrogens, and four resonance envelopes are observed.

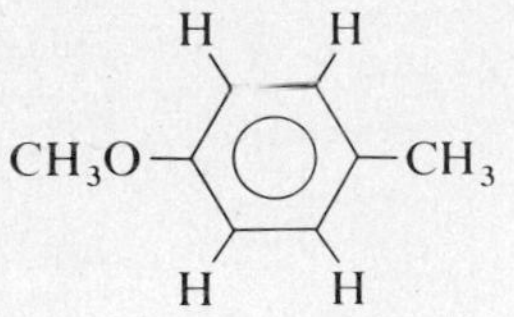

H H

CH_3O—⟨ring⟩—CH_3

H H

The spin–spin splitting patterns are determined by the number of neighboring nuclei that interact with the hydrogen (or equivalent set of hydrogens) undergoing transition, as illustrated in the following general cases:

1. *Simple* ($n + 1$) *cases.* In the simplest cases the number of lines is equal to the number of neighboring protons +1. Thus, the —CH_3 resonance in CH_3CHCl_2 is a doublet (*i.e.*, $1 + 1 = 2$) as the result of spin coupling with the neighboring —$CHCl_2$ hydrogen. The —$CHCl_2$ resonance in this compound appears as a quartet (*i.e.*, $3 + 1 = 4$) as the result of spin coupling with the three neighboring methyl hydrogens. In the compound $(CH_3)_2CHCl$, the six methyl hydrogens (an equivalent set) appear as a doublet, and the —CHCl hydrogen appears as a septet (*i.e.*, $6 + 1 = 7$). The ($n + 1$) rule is applicable to those cases in which *all* the hydrogens responsible for the spin–spin coupling are in the same equivalent set.
2. *Other cases.* If the hydrogens responsible for the spin–spin coupling are not all in the same equivalent set, the number of lines will be equal to $(n + 1)(n' + 1)(n'' + 1)$.* For example, in the compound $Cl_2CHCH_2CHBr_2$, the two hydrogens of the —CH_2— group constitute an equivalent set, but the

* An exception to this general rule arises when the values of the coupling constants between the equivalent sets of hydrogens and the hydrogens under observation are identical, in which case the number of resonance lines is equal to $n + n' + n'' + 1$.

two hydrogens in the $-CHCl_2$ and $-CHBr_2$ groups do not; they are in structurally different environments and have different chemical shifts. The $-CH_2-$ resonance envelope, therefore, will have $(1 + 1)(1 + 1)$ lines (*i.e.*, four lines). In the compound $ClCH_2CH_2CH_2Br$, the central $-CH_2-$ group will have a resonance envelope consisting of $(2 + 1)(2 + 1)$ lines (*i.e.*, nine lines). In 2-bromo-5-methyl-3-phenylcyclopenta-1,3-diene, the resonance of the circled hydrogen will appear as $(3 + 1)(1 + 1)(1 + 1)$ lines (*i.e.*, 16 lines).

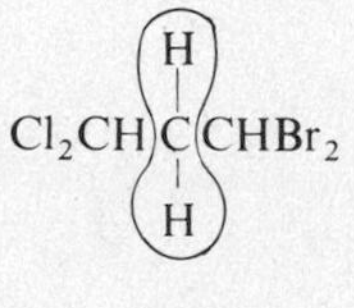

$Cl_2CH\overset{H}{\underset{H}{C}}CHBr_2$ **4 lines**

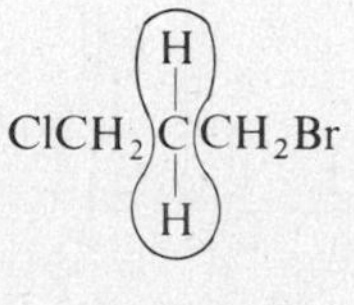

$ClCH_2\overset{H}{\underset{H}{C}}CH_2Br$ **9 lines**

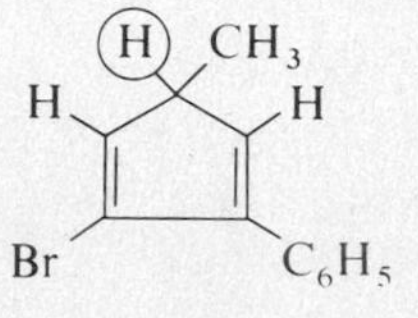

H, CH_3, H, H, Br, C_6H_5 **16 lines**

The *relative line intensities in splitting patterns* depend on the equivalences of the hydrogens responsible for the splitting, as illustrated in the following general cases:

1. In the $(n + 1)$ cases, the relative intensities are equal to the coefficients in the binomial expansion $(a + b)^n$, where n is the number of equivalent hydrogens. Thus, when $n = 1$, the ratio is 1:1; when $n = 2$, the ratio is 1:2:1; when $n = 3$, the ratio is 1:3:3:1; when $n = 4$, the ratio is 1:4:6:4:1; etc.
2. In $(1 + 1)(1 + 1)$ cases, the envelope is called a doublet of doublets, and all four lines are of equal intensity.
3. In $(2 + 1)(2 + 1)$ cases, the envelope is called a triplet of triplets, and in the idealized case consists of a 1:2:1:2:4:2:1:2:1 set of intensity ratios. In many cases, however, the lines of such a pattern may overlap, producing a less easily interpreted pattern.

The *magnitude of the spin–spin coupling constant* is a function of the distance and the type of bonds between the interacting nuclei.

1. The spin–spin interaction diminishes rapidly with distance, and its magnitude usually falls to less than 1 Hz if more than three single bonds intervene between the interacting nuclei. Thus, significant splitting occurs between the hydrogens in $Cl_2CHCHBr_2$ but not in $Cl_2CHOCHBr_2$.
2. Spin–spin interaction is generally transmitted more effectively through multiple bonds than through single bonds. Thus, the coupling constant between the allylic and vinyl hydrogens in the system $H-\overset{|}{\underset{|}{C}}-C{=}\underset{|}{C}-H$ is 1 – 1.5 Hz.
3. Spin–spin interaction often is not observed if a hydrogen is attached to a "labile" bond. For example, coupling between OH and CH_3 hydrogens in CH_3OH is generally not observed except under special conditions, because of an "exchange" process between the molecules; *i.e.*,

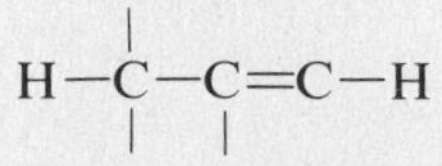

$$CH_3\mathbf{OH} + CH_3OH \rightleftharpoons CH_3OH + CH_3\mathbf{OH}$$

Under the conditions that ordinarily prevail, this exchange is rapid with respect to the time required for an nmr transition, which erases the spin–spin interaction. In scrupulously pure alcohols, however, and in certain solvents (such as dimethyl sulfoxide), the exchange is slowed to the point where spin–spin interaction is observed. Thus, under ordinary conditions, CH_3OH shows a single line resonance for the CH_3 group and a single line resonance for the OH group, but under the special conditions just noted, it shows a doublet for the CH_3 group and a 1:3:3:1 quartet for the OH hydrogen.

To give a simple illustration of the utility of the nmr method, let us again look at the structural possibilities for a C_4H_6O molecular formula (see pp. 143–4) and consider two specific examples. (1) One compound, which we shall specify as compound A, shows a three-hydrogen doublet (with slight splitting of each member of the doublet) at δ 2.03, one-hydrogen multiplets* at δ 6.13 and 6.87, and a one-hydrogen doublet at δ 9.48. (2) The other compound, which we shall specify as compound B, shows a three-hydrogen triplet at δ 1.20, a one-hydrogen singlet at δ 3.05, and a two-hydrogen quartet at δ 3.70. In looking over the structures shown on pp. 143–4, we see that only three have the CH_3CH moiety required to explain the three-hydrogen doublet that is present in compound A; *viz.*,

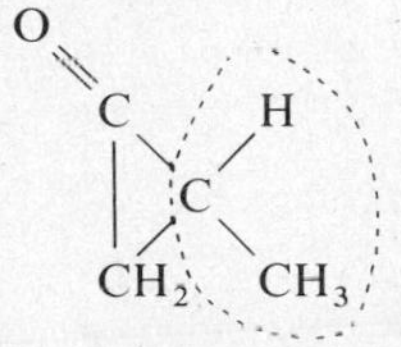

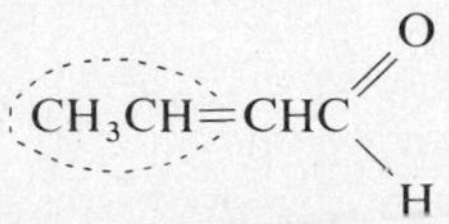

$CH_3CH{=}CHCHO$

2-Butenal

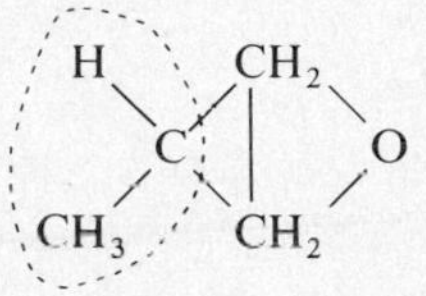

We readily choose in favor of 2-butenal (crotonaldehyde) on the basis of the resonances at δ 6.13 and 6.87 (vinyl hydrogens) and the resonance at δ 9.48 (aldehyde hydrogen). Similarly, from the group of structures shown on pp. 143–4,

---

* Resonance envelopes containing complex splitting patterns and/or lines that are poorly resolved are referred to as "multiplets." In the compound

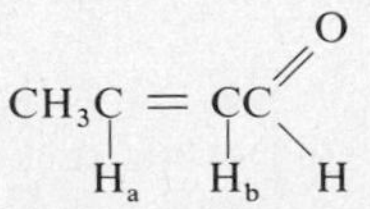

$CH_3C(H_a){=}C(H_b)CHO$

for example, the resonance from H_a is a double quartet (*i.e.*, quartet from splitting with the CH_3 hydrogens and doublet from splitting with H_b). The resonance from H_b is a double doublet (*i.e.*, doublet from splitting with H_a and doublet from splitting with the CHO hydrogen). The resonance from H_b may be further split by "long-range" coupling with the CH_3 group, the J values being about 1 Hz for this coupling. The chemical shifts of H_a and H_b are comparable, with the result that the resonance envelopes from these hydrogens fall in the same region of the spectrum. The overlay of the splitting patterns from the two "vinyl" hydrogens gives a complex pattern, difficult to analyze and simply referred to as a multiplet. Since the chemical shift is field dependent, the overlap of the resonances from H_a and H_b can be reduced by measuring the nmr spectrum at higher magnetic fields. Thus, a pattern that appears as a complex multiplet at 40 MHz may resolve into a clean set of resonance envelopes (*i.e.*, a double quartet and a double doublet) at 220 MHz.

only one has the CH_3CH_2 moiety required to explain the triplet-quartet pattern (at δ 1.20 and 3.70) present in compound B; *viz.*, $HC{\equiv}COCH_2CH_3$. Although the compound $CH_3CH_2CH{=}C{=}O$ also contains a CH_3CH_2 moiety, the CH_2 portion does not appear as a triplet because spin–spin splitting with the vinyl hydrogen occurs in addition to spin–spin splitting with the CH_3 hydrogens, to give either a quintet (if the coupling constants between the CH_3 and $CH{=}$ hydrogens are equal) or an octet.

Several examples of nmr spectra are illustrated in Fig. 6.11.

6.4 Mass Spectrometry

A molecular phenomenon to which the term "spectrum" is attached arises from the ability of molecules to form a variety of ions when bombarded with a beam of high-energy electrons. Cations, anions and neutral species are produced in this process, but in most instruments used today, only the cations are separated and measured. By directing these fast-moving cations, accelerated by negatively charged plates, through magnetic fields it is possible to separate them on the basis of their different mass-to-charge (m/e) ratios. In general, most of the cations that are formed are singly charged, so this process actually corresponds to a separation of fragments on the basis of the mass alone. A display of the masses and relative intensities of the fragments, therefore, is referred to as a *mass spectrum* of the compound. The initial step in the process is a collision of a high-energy electron with the molecule, resulting in the ejection of an electron and the formation of a radical cation (called the parent ion or the molecular ion) corresponding to the parent compound, *i.e.*, $M + e^- \rightarrow M^+ + 2e^-$. The mass number at which this cation appears in the collector, then, specifies the molecular weight of the molecule. This, in fact, is one of the extremely useful aspects of the mass spectral technique, for in many cases it provides the most accurate method known for determining molecular weights.

The mass spectrometer not only records the masses of the ions that reach the collector but also the relative abundances of ions of any particular mass. The carbon in organic compounds is actually only 98.92% ^{12}C, the remaining 1.08% being ^{13}C. Thus, if methane is introduced into the mass spectrometer, a parent ion at mass number 16 is observed along with another signal, about 1.1% as intense as that of the parent ion, at mass number 17. The smaller peak arises from the 1.08% of $^{13}CH_4$ present in ordinary methane. If ethane is placed in the mass spectrometer, a line corresponding to the parent ion at mass 30 will appear along with a less intense line at 31 corresponding to the parent ion +1 (the $M + 1$ ion). The relative intensity of the latter is about 2.2% that of the parent ion because, in contrast to methane, there is now twice the probability that any particular ethane molecule will have one ^{13}C atom in it. It is, of course, possible that any particular ethane molecule will have two ^{13}C atoms. The probability of this occurrence is much lower, however, and the $M + 2$ ion appears as only a very small signal. Propane gives a parent ion peak and an $M + 1$ peak of relative intensities of 100:3.3; butane a parent ion peak and an $M + 1$ peak of relative intensities of 100:4.4; etc. Thus, from a measurement of the relative intensities of the parent peak and the next higher mass peak, not only can we measure accurately the molecular weight of a compound but also we can estimate the number of carbon atoms in the molecule. Some mass spectrometers are so exceedingly accurate in their measurement of the mass number that the precise molecular weight that is measured can be uniquely associated with a specific molecular formula. For example, a molecular weight of 285.1365 corresponds uniquely to a molecular formula of $C_{17}H_{19}NO_3$ (*e.g.*, morphine — see p. 50).

In addition to the parent ion and the $M + 1$ ion in the mass spectrum, many other ions almost invariably appear. The parent ion possesses a great deal of excess energy when it is formed, which is dissipated by subsequent bond cleavage reactions into various "daughter ions." The daughter ions can, in turn, undergo further fragmentation. For example, *n*-butane produces the mass spectrum shown in Fig. 6.12. The peak for the parent ion is *not* the most abundant ion in this

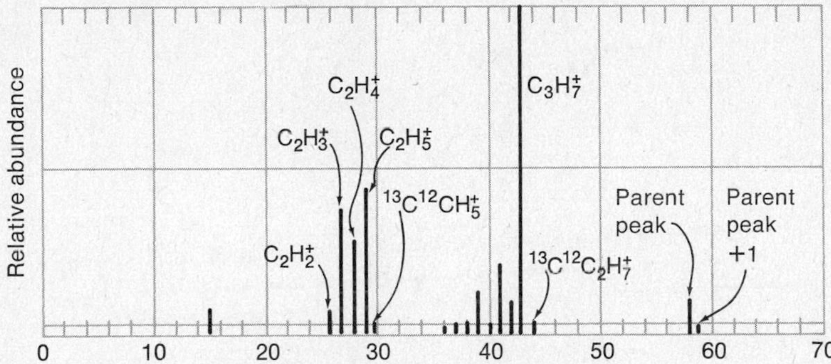

Fig. 6.12. Mass spectrum of *n*-butane.

spectrum. The major peak, referred to as the "base peak," has a mass number of 43 and corresponds to one of the cleavage products of butane, *viz.*, $C_3H_7^+$. Several other peaks also appear as major contributors to the spectrum, their intensities all exceeding that of the parent ion. This occurrence is not uncommon, and in some instances the parent peak may be missing entirely.* Usually, however, it is sufficiently strong that a molecular weight determination is possible, although an accurate assessment of the carbon content from the intensity ratio of $M/(M + 1)$ may be difficult. Analogously to the region in the infrared spectrum below 1000 cm^{-1}, it provides a "fingerprint" of the molecule and can be exceedingly useful for establishing the identity of a compound by comparison with a known sample. For those experienced in the art, the fragmentation patterns can also be used as a structural probe in a more absolute sense. Studies of the mass spectra of a large number of compounds have revealed certain characteristic features by means of which patterns can be related to structure. For additional details, the interested reader should consult the references cited at the end of the chapter.

* When this occurs, it is advisable to reduce the energy of the bombarding electrons, thereby reducing the violence of the collision and increasing the probability of observing the parent ion.

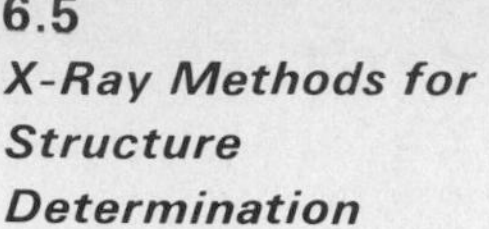

6.5 *X-Ray Methods for Structure Determination*

The ultraviolet/visible, infrared, and nuclear magnetic resonance spectra arise, as we have seen, as the result of the absorption of electromagnetic radiation by molecules. If the frequency of the radiation does not correspond to an absorption mode, however, the energy is only momentarily taken up by the molecule and is almost instantaneously reemitted, giving rise to the phenomenon of scattering (see Section 5.2e). In the special circumstance in which the centers that are responsible for the scattering are separated by a distance that is comparable to the wavelength of the incident light (remember, wavelength = speed of light/frequency), the phenomenon of diffraction can come into play. The scattered light from proximate points of scattering may suffer either reinforcement or cancellation depending on the phase of the light that is scattered from each of the centers. If the proximate points are arranged in a regular pattern, the result of these reinforcements and cancellations is the production of alternating regions

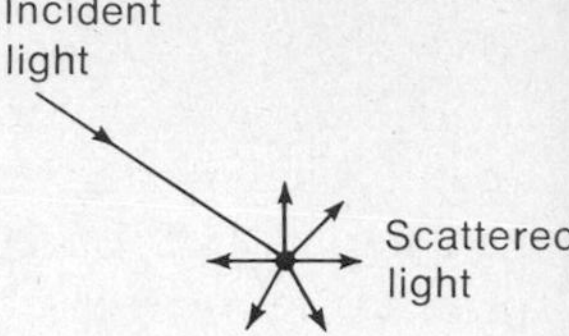

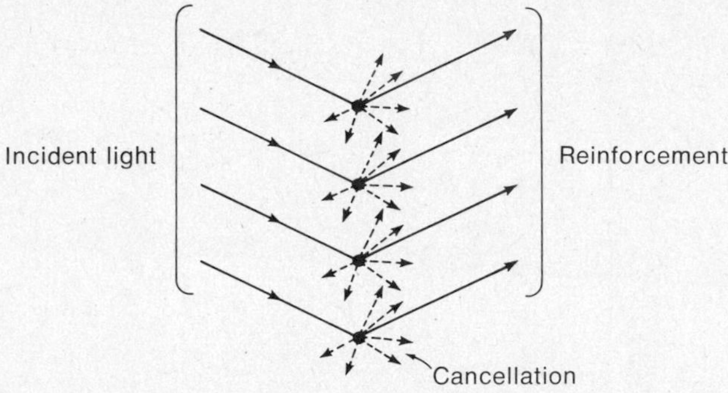

of high and low intensity of scattered radiation. For a crystalline compound, where the molecules comprising the crystal are arranged in some orderly, repeating, three-dimensional array, and where the atoms within each molecule are arranged in a definite fashion, the diffraction pattern resulting from the scattering has the appearance of a series of spots (Fig. 6.13).

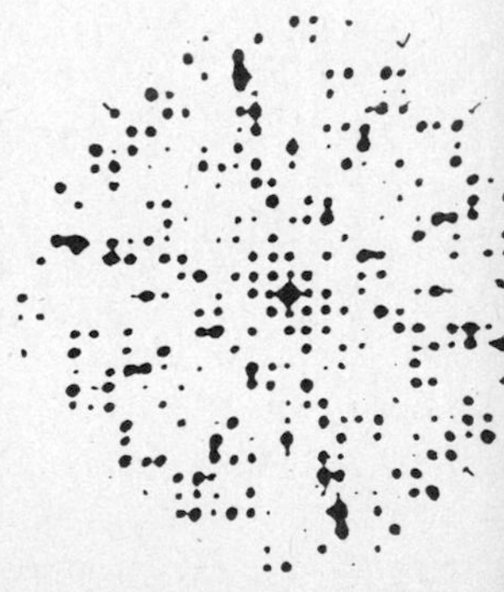

Fig. 6.13. X-ray diffractio pattern of a crystalline protein.

The requirement that the dimensions of the scattering centers be comparable to the wavelength of light if diffraction is to be observed is met in the case of molecules by using X rays of about 1 Å wavelength. Thus, when a beam of X rays is "reflected" from a crystalline protein, the diffraction pattern shown in Fig. 6.13 is obtained when the crystal and beam are oriented in a particular fashion. When the orientation of the crystal with respect to the beam is changed, another pattern is obtained, in which the spacing between the spots and the relative intensities of the spots are different. By obtaining patterns from a variety of orientations and then analyzing the spots as a function of their spacings and intensities in these orientations, it is possible to establish the three-dimensional structure of the molecule. Even for a relatively small molecule, however, it is necessary to analyze many hundreds or even thousands of spots, and this number escalates as the size and complexity of the molecule increases. With the most modern X-ray crystallographic instruments, the data on the spacings and intensities of the spots are fed directly into a computer that very swiftly performs some, although not all, of the analysis. As a consequence of the emergence of the computer-assisted X-ray spectrometer, the structure proof of molecules with molecular weights below 500 has become semiroutine; days or weeks may still be required, however, and in some cases a definitive answer cannot be obtained. For larger molecules, such as the biological polymers, the structural analysis is very much more demanding of time and ingenuity and may require weeks, months, or even years of work. It is time and effort that many scientists have been willing to invest, however, for X-ray analysis provides information unobtainable by any other known method. The impact of X-ray crystallography on chemistry in general and on biochemistry in particular has been enormous, and a technique that at one time was arcane in the minds of most chemists is rapidly taking its place alongside the spectral methods that have become part and parcel of every structure determination.

6.6 Determination of Structure from Spectral Data

One of the principal uses that organic chemists make of spectral data is in the determination of the structures of compounds. The advent of instruments for the measurement of ultraviolet/visible, infrared, and nmr spectra has revolutionized structural chemistry, and structure determinations that required weeks, months, or even years of work prior to the 1940's can now be completed in days or even hours. Although in all but the simplest cases, chemical as well as spectral information is usually necessary for a complete structure determination, spectra alone can often reveal a tremendous amount of information about the molecule, as we have seen in the example on pp. 143–4. To cite another illustrative example, let us explore the structure proof of a natural product called "aethusin," which has been isolated from *Aethusa cynapium L.* An elemental and quantitative analysis indicates the empirical formula to be $C_{13}H_{14}$, and the appearance of the parent ion in the mass spectrum at m/e 170 shows that this is also the molecular formula. The compound is a colorless oil, solidifying at 23° and possessing the following uv and ir spectra shown in the graphs on p. 168.

From the molecular formula alone, we can deduce the "degrees of unsaturation" in the molecule. If the compound were acyclic and completely saturated, its molecular formula would be $C_{13}H_{28}$ (*i.e.*, C_nH_{2n+2}). It falls short of this by 14 hydrogen atoms, and must, therefore, contain some combination of rings, double bonds, and/or triple bonds to account for this difference. Since each ring and each double bond will subtract two hydrogens from the C_nH_{2n+2} formula (a triple bond is equal to two double bonds*), the compound must contain $\frac{14}{2} = 7$ degrees of unsaturation. It might contain these as the combination of two rings, three double bonds, and a triple bond; or one ring, two double bonds, and two triple bonds; or any of several other possible combinations. The molecular formula sheds no light on the nature of the unsaturation but simply indicates how much is present.

The ultraviolet spectrum of aethusin indicates that it has an extended conjugated system, and the longest wavelength absorption at 335 nm suggests that it contains five unsaturated units in conjugation (*i.e.*, between $(C{=}C)_4$ and

* Note, for future reference, that a phenyl group has four degrees of unsaturation—*i.e.*, one ring and three double bonds.

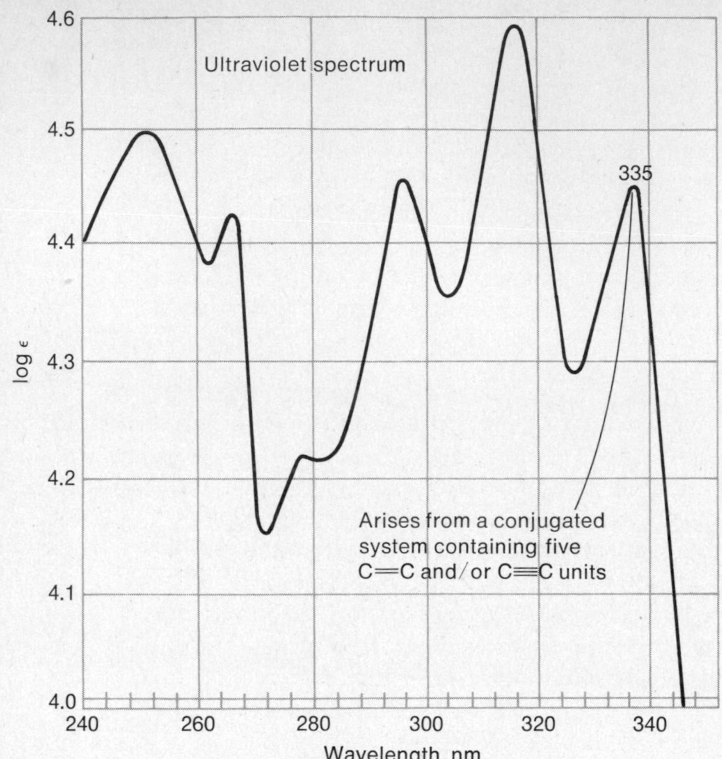
Ultraviolet spectrum
335
Arises from a conjugated system containing five C═C and/or C≡C units
log ε
4.6
4.5
4.4
4.3
4.2
4.1
4.0
240
260
280
300
320
340
Wavelength, nm

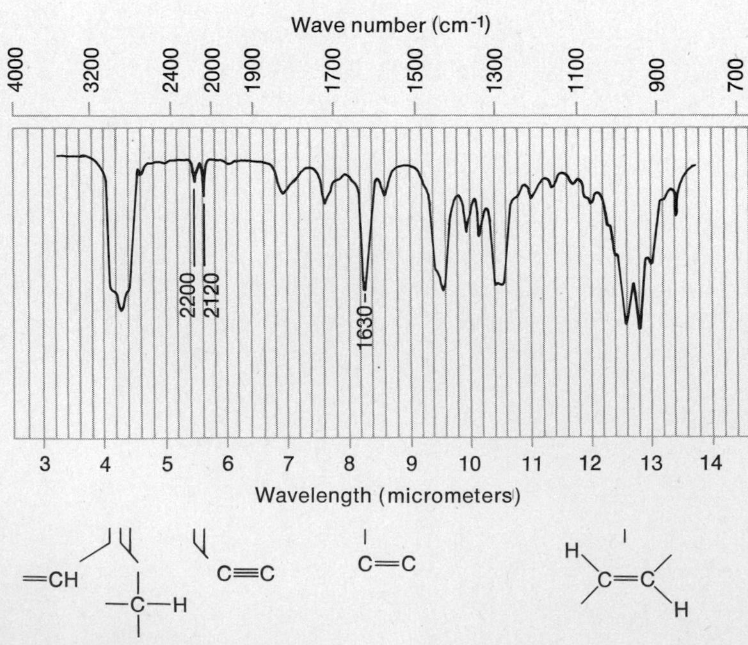
Wave number (cm-1)
4000
3200
2400
2000
1900
1700
1500
1300
1100
900
700
2200
2120
1630
3
4
5
6
7
8
9
10
11
12
13
14
Wavelength (micrometers)
═CH
C≡C
C═C
H
H

$(C{=}C)_6$ in Table 6.1). The infrared spectrum shows $=\overset{|}{C}-H$ and $-\overset{|}{\underset{|}{C}}-H$ stretching deformations at 3050 and 2950 cm^{-1}, $C{\equiv}C$ stretching deformations at 2200 and 2120 cm^{-1} (indicating two $C{\equiv}C$ groups), $C{=}C$ stretching deformation at 1630 cm^{-1} and the bending deformation of a *trans* disubstituted alkene at 985 cm^{-1} (Table 6.2). The absence of a C H stretching deformation band at 3300 cm^{-1} and the absence of a strong band at 900 cm^{-1} suggest that none of the $C{\equiv}C$ or $C{=}C$ groups is in a terminal position. On the basis of the information from the molecular formula, uv and ir, therefore, we can assert that at least 12 of the carbons in the molecule are in a chain in which there are three $C{=}C$ and two $C{\equiv}C$ groups—*viz.*,

$$C-C{\equiv}C-C{\equiv}C-C{\equiv}C-C{\equiv}C-C{\equiv}C-C$$

(two of the bonds must be $C{\equiv}C$). All that remains to complete the carbon framework is to place the last carbon, and the nmr spectrum provides the necessary clue for this. The triplet pattern at δ 1.12 establishes the presence of a $-CH_2CH_3$ moiety, possible only if the last carbon is attached at the end of the 12-carbon unit to give a completely straight chain compound. The presence of six vinyl hydrogens (complex resonance pattern at δ 6.0), another methyl group (doublet at δ 1.8), and a methylene group (quintet at δ 2.0) accounts for all 14 of

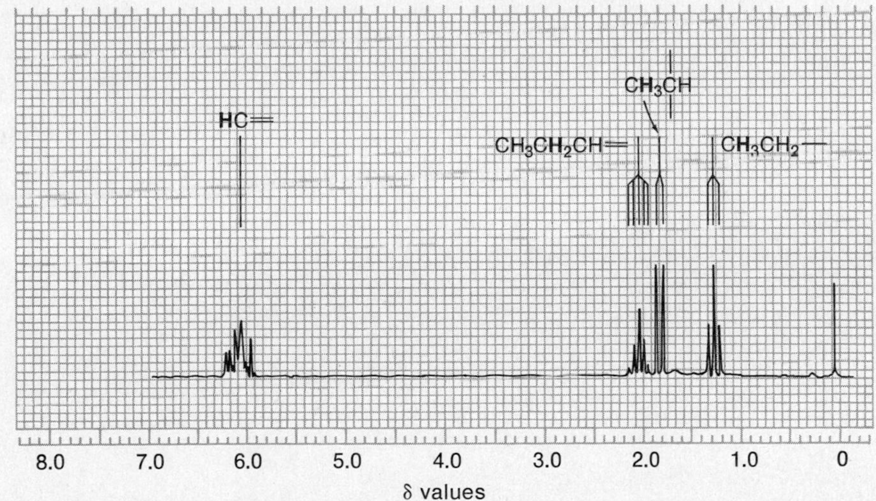

the hydrogens in the molecule. The doublet pattern for the methyl resonance at δ 1.8 indicates a $CH_3CH{=}C$ rather than a $CH_3C{\equiv}C$ moiety at one end of the molecule, and the quintet for the methylene resonance at δ 2.0 similarly indicates a $CH_3CH_2CH{=}C$ rather than a $CH_3CH_2C{\equiv}C$ moiety at the other end of the molecule. On the basis of these data, therefore, aethusin must be one of the following three structures:

$$CH_3CH{=}CHC{\equiv}CC{\equiv}CCH{=}CHCH{=}CHCH_2CH_3$$

$$CH_3CH{=}CHC{\equiv}CCH{=}CHC{\equiv}CCH{=}CHCH_2CH_3$$

$$CH_3CH{=}CHCH{=}CHC{\equiv}CC{\equiv}CCH{=}CHCH_2CH_3$$

The choice of the first of these three alternatives as the correct structure for aethusin requires additional data, and a definitive structural determination cannot be made on the basis of spectral data alone. Nevertheless, it is impressive to realize how, in the matter of the hour or so that would be required to carry out these determinations, the arrangement of 13 carbons and 14 hydrogens can be delimited to only three of the thousands of possible structures.

6.7 Study Guide

I. Electronic spectra (ultraviolet/visible spectra)
  A. Underlying phenomenon
    1. Electronic excitation
    2. Types of transitions
      (a) $\sigma \rightarrow \sigma^*$ (σ bonding $\rightarrow$ σ antibonding)
      (b) $\pi \rightarrow \pi^*$ (π bonding $\rightarrow$ π antibonding)
      (c) $n \rightarrow \pi^*$ (nonbonding $\rightarrow$ π antibonding)
    3. Relation between energy and wavelength (or frequency)
  B. Quantities measured
    1. Position of absorption (designated in nm); λ_{max}
    2. Intensity of absorption (designated as molar extinction coefficient); ε_{max}
  C. Major utility—for detection of conjugated systems
II. Vibrational spectra (infrared spectra)
  A. Underlying phenomenon
    1. Vibrational excitation
    2. Types of transitions
      (a) Stretching
      (b) Bending
    3. Relation between reduced mass of system, force constant of bond, and frequency of absorption
  B. Quantities measured—usually only position (designated in cm^{-1}, μm, or μ)
  C. Major utility—for the detection of certain functions
    1. X—H stretching frequencies at about 3600–2500 cm^{-1}
    2. X≡Y and X=Y=Z stretching frequencies at about 2250–2000 cm^{-1}
    3. X=Y stretching frequencies at about 1900–1600 cm^{-1}
    4. Other deformation modes in "fingerprint region" below 1000 cm^{-1}
III. Nuclear magnetic resonance spectra
  A. Underlying phenomenon
    1. Nuclear spin (in atoms with odd atomic number and/or odd mass number)
    2. Spin states in an external magnetic field
    3. Relation between frequency of radiation and strength of magnetic field
    4. Shielding by electrons; the chemical shift
    5. Interaction with neighboring hydrogens
      (a) Spin–spin coupling
      (b) The spin–spin coupling constant, J (designated in Hz)

(c) The $(n + 1)$ rule for equivalent hydrogens
(d) Splitting by nonequivalent hydrogens

B. Quantities measured
1. Chemical shift in δ
2. Splitting pattern (designated as singlet, doublet, triplet, quartet, double doublet, double triplet, etc.)
3. Coupling constant (symbolized by J and designated in Hz)
4. Relative areas under resonance envelopes

C. Major utility—see Table 6.3

IV. Mass spectra
A. Underlying phenomenon—electron bombardment to form ions
B. Quantities measured
1. m/e Ratios for all ions formed including the molecular ion
2. "Fragmentation" patterns
C. Major utility
1. Particularly useful for measuring mass of parent ion, providing the molecular weight of the compound
2. Measuring number of carbons present by comparing intensities of molecular ion peak and molecular ion + 1 peak
3. Fragmentation patterns often correlate with various structural features

V. X-ray crystallography
A. Underlying phenomenon—diffraction of light by atoms in crystals
B. Quantities measured—diffraction patterns
C. Major utility—assigning complete three-dimensional structures of compounds

REFERENCES

For more extensive and advanced treatment of spectra, consult the following references:

General References:

DYER, J. R., *Applications of Absorption Spectroscopy of Organic Compounds*, in Foundations of Modern Organic Chemistry Series, Prentice-Hall, Inc., Englewood Cliffs, N.J., 1965.

GORDON, A. J. and R. A. FORD, *The Chemist's Companion—A Handbook of Practical Data, Techniques, and References*, John Wiley and Sons, New York, 1972.

PASTO, D. J., and C. R. JOHNSON, *Organic Structure Determination*, Prentice-Hall, Inc., Englewood Cliffs, N.J., 1969.

SILVERSTEIN, R. M., and G. C. BASSLER, *Spectrometric Identification of Organic Compounds*, 2nd ed., John Wiley & Sons, Inc., New York, 1967.

WEST, W., ed., Chemical Applications of Spectroscopy, *Technique of Organic Chemistry*, vol. 9, Interscience Publishers, New York, 1956.

Ultraviolet and Visible Spectra:

GILLAM, A. E., E. S. STERN, and C. J. TIMMONS, *An Introduction to Electronic Absorption Spectroscopy in Organic Chemistry*, Edward Arnold, Publishers, Ltd., London, 1970.

JAFFÉ, H. H., and M. ORCHIN, *Theory and Applications of Ultraviolet Spectroscopy*, John Wiley & Sons, Inc., New York, 1962.

SCOTT, A. I., *Interpretation of the Ultraviolet Spectra of Natural Products*, The Macmillan Co., New York, 1964.

Infrared Spectra:

BELLAMY, L. J., *The Infrared Spectra of Complex Molecules*, 2nd ed., John Wiley & Sons, Inc., New York, 1958.

CONLEY, R. T., *Infrared Spectroscopy*, 2nd ed., Allyn and Bacon, Inc., Boston, 1972.

MELOAN, C. E., *Elementary Infrared Spectroscopy*, The Macmillan Company, New York, 1963.

NAKANISHI, K., *Infrared Absorption Spectroscopy*, Holden-Day, Inc., San Francisco, 1962.

RAO, C. N. R., *Chemical Applications of Infrared Spectroscopy*, Academic Press, Inc., New York, 1963.

Nuclear Magnetic Resonance Spectra:

BHACCA, N. S., and D. H. WILLIAMS, *Applications of NMR Spectroscopy in Organic Chemistry*, Holden-Day, Inc., San Francisco, 1964.

BIBLE, R. H., *Interpretation of NMR Spectra; an Empirical Approach*, Plenum Press, New York, 1965.

EMSLEY, J. W., J. FEENEY, and L. H. SUTCLIFFE, *High Resolution Nuclear Magnetic Resonance Spectroscopy*, Pergamon Press, New York, 1965.

JACKMAN, L. M., and S. STERNHELL, *Applications of Nuclear Magnetic Resonance Spectroscopy in Organic Chemistry*, 2nd ed., Pergamon Press, New York, 1969.

LEVY, G. S., and G. L. NELSON, *Carbon-13 Nuclear Magnetic Resonance for Organic Chemists*, Wiley–Interscience, Inc., New York, 1972.

MATHIESON, D. W., ed., *Nuclear Magnetic Resonance for Organic Chemists*, Academic Press, Inc., New York, 1967.

PAUDLER, W. W., *Nuclear Magnetic Resonance*, Allyn and Bacon, Inc., Boston, 1971.

POPLE, J. A., W. G. SCHNEIDER, and H. J. BERNSTEIN, *High-Resolution Nuclear Magnetic Resonance*, McGraw-Hill Book Company, New York, 1959.

ROBERTS, J. D., *Nuclear Magnetic Resonance; Applications to Organic Chemistry*, McGraw-Hill Book Company, New York, 1959.

STOTHERS, J. B., *Carbon-13 NMR Spectroscopy*, Academic Press, Inc., New York, 1972.

Mass Spectrometry:

BUDZIKIEWICZ, H., C. DJERASSI, and D. H. WILLIAMS, *Mass Spectrometry of Organic Compounds*, Holden-Day, Inc., San Francisco, 1967.

McLafferty, F. W., *Interpretation of Mass Spectra*, 2nd ed., W. A. Benjamin, Inc., Reading, Mass., 1973.

Shrader, S. R., *Introductory Mass Spectrometry*, Allyn and Bacon, Inc., Boston, 1971.

X-Ray Crystallography:

Bacon, G. E., *X-ray and Neutron Diffraction*, Pergamon Press, New York, 1966.

Stout, G. H., and L. H. Jensen, *X-Ray Structure Determination: A Practical Guide*, The Macmillan Company, New York, 1968.

Wilson, H. R., *Diffraction of X-Rays by Proteins, Nucleic Acids, and Viruses*, St. Martin's Press, New York, 1966.

PROBLEMS

1. The relationship between the frequency and the energy of electromagnetic radiation is given by the Einstein equation, $E = h\nu$ where E is the energy, h is Planck's constant (6.625×10^{-27} erg/sec), and ν is the frequency. Remembering that $\nu = c/\lambda$, where c is the speed of light (3×10^{10} cm/sec), and using the factor 1.439×10^{13} for converting ergs to kilocalories per mole, calculate the energy of the following transitions in kilocalories per mole.

(a) The uv transition at 253 nm.
(b) The uv transition at 350 nm.
(c) The visible transition at 550 nm.
(d) The ir transition at 3000 cm^{-1}.
(e) The ir transition at 1700 cm^{-1}
(f) The ir transition at 750 cm^{-1}.
(g) The nmr transition at 60 MHz.

2. For each of the molecular formulas listed, suggest a structure of a compound that (a) absorbs strongly in the 200–300-nm region of the ultraviolet, and (b) is essentially transparent in the 200–300-nm region of the ultraviolet:

(a) C_8H_{12}
(b) C_6H_8
(c) C_4H_6O
(d) $C_9H_{10}O$

3. Assuming the force constants for the C—H and C—D bonds to be the same, would you expect to find the C—D stretching deformation band of $C_6H_5CD_3$ at a higher or lower frequency than that in ordinary toluene? The bending deformation band for a C—H bond appears at a lower frequency than the stretching deformation band. Which deformation has the higher force constant?

4. Indicate how infrared spectra can be used to differentiate between the following isomeric compounds:

(a) $CH_3CH{=}CHCHO$ (CH₃CH=CHC(=O)H) and $CH_3C{\equiv}CCH_2OH$

(b) (C₆H₅)(H)C=C(H)(C₆H₅) [trans: H and C₆H₅ on the first carbon; C₆H₅ and H on the second] and (H)(C₆H₅)C=C(H)(C₆H₅) [cis]

(c) 1-cyclopentenyl methyl ketone (C(=O)CH₃ on the double-bond carbon) and 3-cyclopentenyl methyl ketone (C(=O)CH₃)

(d) dicyclohexylidenemethane (cyclohexylidene=C=cyclohexylidene) and 1-cyclohexenyl—CH_2—1-cyclohexenyl

(e) $(C_6H_5)_2C{=}C{=}C(CN)(CH_3)$ and $(C_6H_5)_2C{=}C{=}NCH{=}CH_2$

5. As discussed on p. 154, the chemical shift is field dependent. For example, the chemical shift for the methyl group of toluene is 132 Hz when the spectrum is measured in a 60-MHz spectrometer (relative to TMS) but 220 Hz when measured in a 100-MHz spectrometer.

(a) What is the chemical shift for a methyl group of toluene in a 40-MHz spectrometer?

(b) If the chemical shift of a particular resonance is 150 Hz in a 60-MHz spectrometer, what is its δ value?

(c) If the δ value of a particular group is 3.3, what is its chemical shift in Hz at 60, 100, and 220 MHz?

(d) What is the δ value of a resonance that occurs 63 Hz upfield from TMS in a 60-MHz spectrometer?

6. As discussed on p. 156 the spin–spin coupling constant is field independent, in contrast to the chemical shift. Knowing that the δ values for the methyl and methynyl protons in 2-bromopropane are δ 1.71 and 4.32, respectively, and that the spin–spin coupling constant, J, is 6 Hz, draw the spectrum of the compound at 60 and 100 MHz.

7. For each of the following compounds, indicate which hydrogens belong to equivalent sets, and draw the expected nmr spectrum.

(a) CH_3CH_3

(b) $CH_3CH_2CH_3$

(c) $CH_3CH_2CH_2CH_3$

(d) $(CH_3)_3C{-}C(CH_3)_3$

(e)

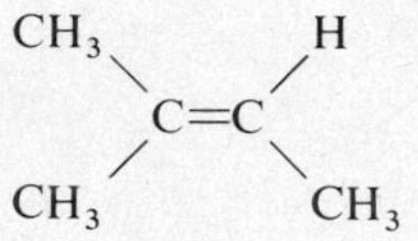

(f)

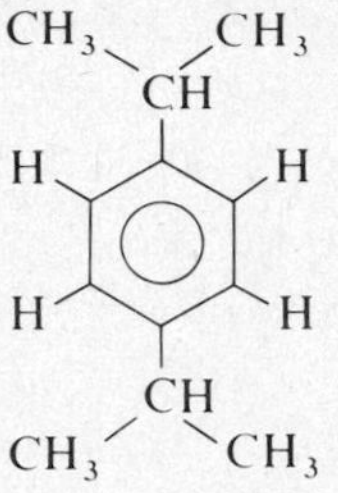

(g)

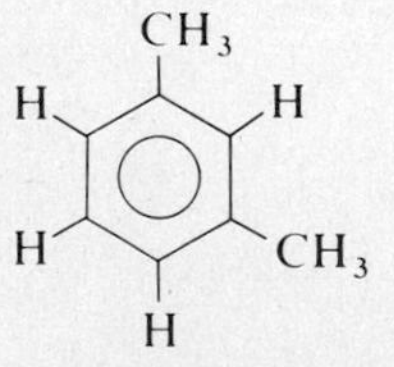

(h)

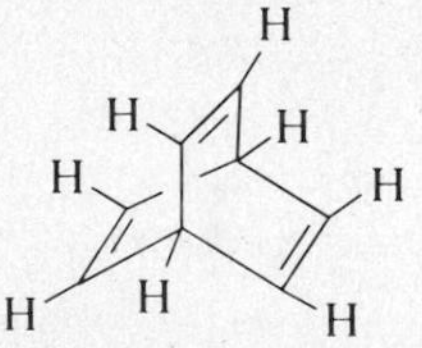

(i)

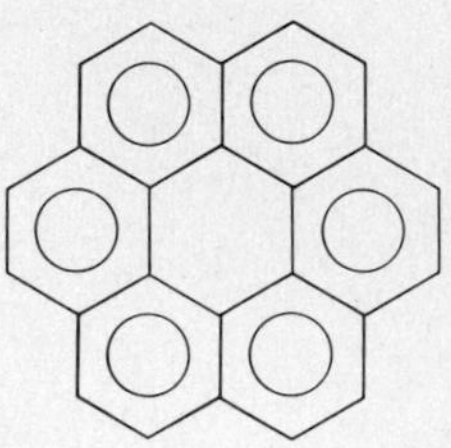

(j) $(CH_3)_3CC{\equiv}CCH_2CH_2{-}C_6H_5$

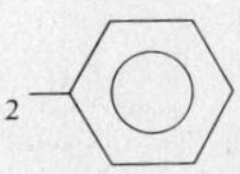

(k)

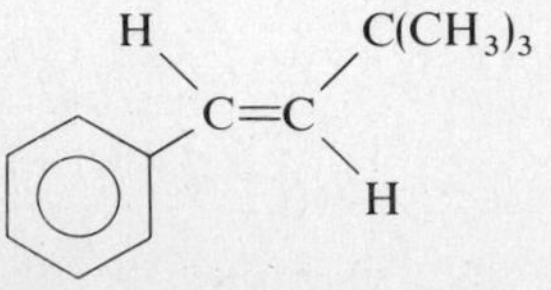

(l) $CH_3CH_2CH_2Cl$

(m)

8. The unexpectedly large downfield shift of the protons attached to the benzene ring is ascribed to the particular nature of the induced magnetic field associated with the benzene ring as a result of its aromaticity. It is said to possess a "ring current," the presence of which, as discerned by high δ values, provides a criterion for aromaticity. The δ values of the protons in cyclohexene, cyclooctatetraene, and the C_{18} annulene [see Problem 10(g), Chapter 4, for structure] are 5.57, 5.69, and 8.8, respectively. What conclusions can be drawn regarding their aromaticity?

9. Draw the structures of the compounds that most reasonably fit the following molecular formulas and nmr spectral data.

(a) Molecular formula C_4H_8

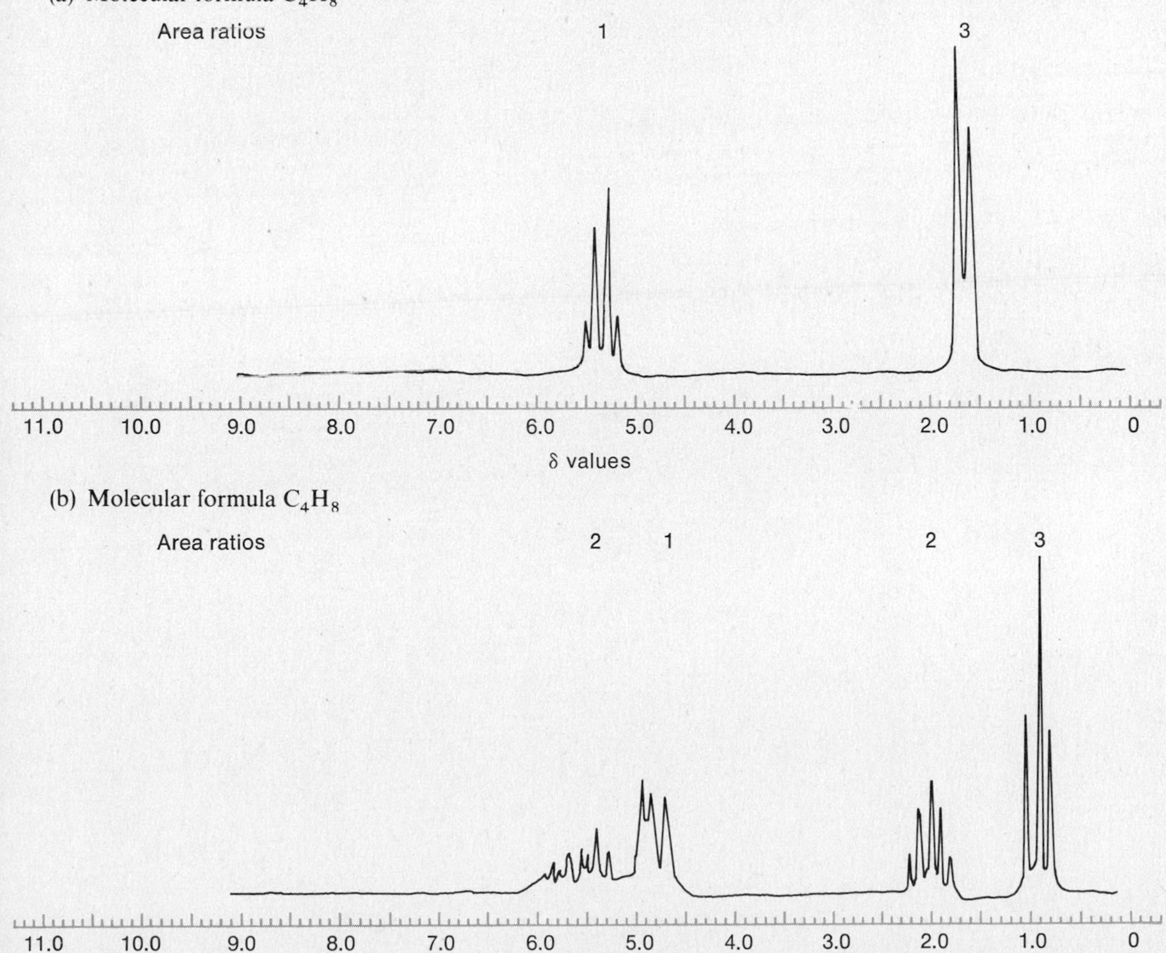

(b) Molecular formula C_4H_8

(c) Molecular formula $C_6H_4Cl_2$

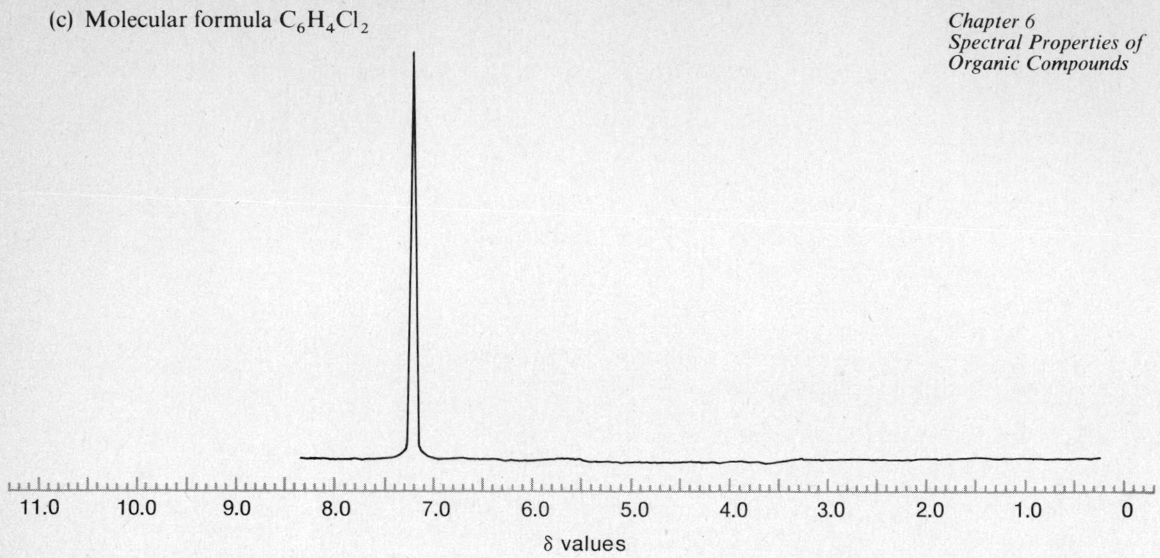

(d) Molecular formula $C_6H_4Cl_2$

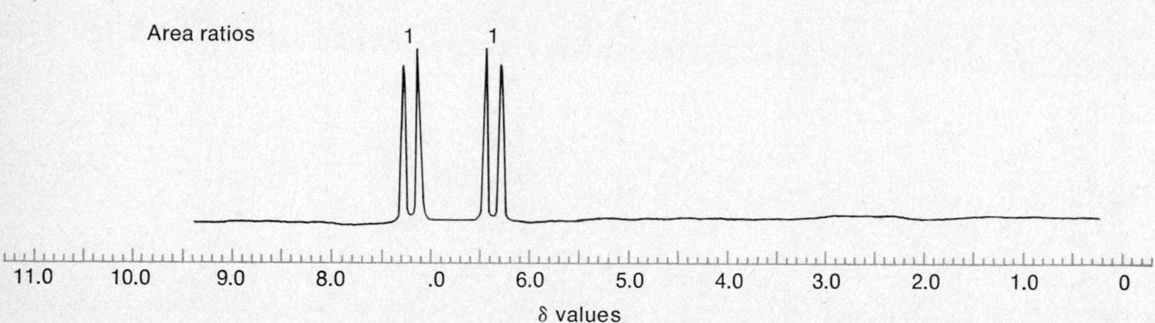

(e) Molecular formula $C_{10}H_{14}$

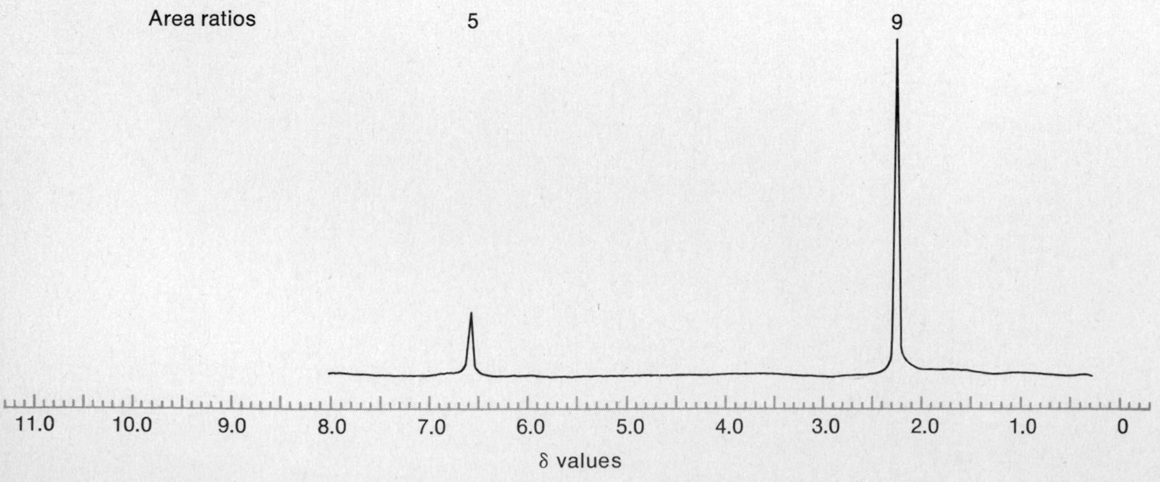

10. Indicate how nmr spectral determinations can be used to differentiate between the following isomeric pairs.

(a) $(CH_3)_2C{=}C(CH_3)_2$ and $(CH_3CH_2)_2C{=}CH_2$

(b) $C_6H_5{-}CH_2C(=O)H$ and $C_6H_5{-}C(=O)CH_3$

(c) $CH_3CH_2C(CH_3)_2CHO$ and $(CH_3)_3CCH_2CHO$

(d) 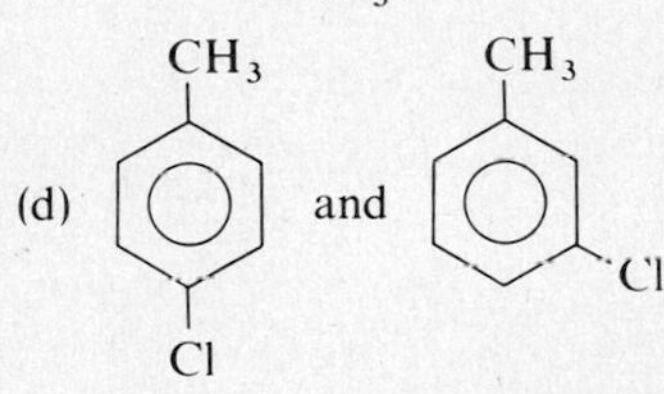 and $m{-}ClC_6H_4CH_3$

(e) $\triangleright{-}CH_3$ and $\square$

(f) $H_2C{=}CHCH{=}CH{-}C_6H_5$ and $CH_3C{\equiv}CCH_2{-}C_6H_5$

11. Using the indicated spectral determinations, indicate how the following pairs of isomeric compounds can be differentiated.

(a)	$C_6H_5{-}C(=O)CH_2CH_3$ and $C_6H_5{-}CH_2C(=O)CH_3$	Ultraviolet
(b)	$HC{\equiv}CCH_2{-}C_6H_5$ and $H_2C{=}C{=}CH{-}C_6H_5$	Infrared
(c)	$CH_3CH_2O{-}C_6H_5$ and $CH_3OCH_2{-}C_6H_5$	Nuclear magnetic resonance
(d)	$ClCH_2C{\equiv}CCH_3$ and $ClCH_2CH_2C{\equiv}CH$	Any method
(e)	$p{-}NC{-}C_6H_4{-}CH_2Cl$ and 3-Cl-5-CH_3-C_6H_3-CN	Any method

12. The natural product capillene has been isolated from *Artemisia capillaris.* It possesses the molecular formula $C_{12}H_{10}$, absorbs in the uv with λ_{max} 239 nm (ε_{max} = 5,000), shows ir bands at 2210 and 2160 cm^{-1}, and has the following nmr spectrum:

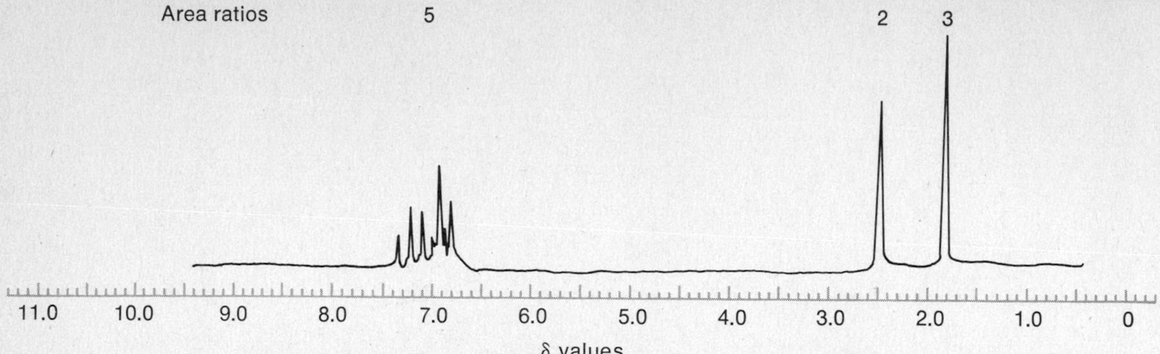

Suggest one or more structures commensurate with these data.

13. A natural product isolated from *Centaurea cyanus L.* is a solid, with a mp of 87°, that has the molecular formula $C_{13}H_8$. Its uv spectrum shows a series of strong absorptions, the one of longest wavelength appearing at 389 nm. The ir spectrum shows bands at 2190, 2120, 1000, and 925 cm^{-1}, and the nmr spectrum shows a 3-proton doublet at δ 1.8 ppm and a 5-proton multiplet at about δ 6.0. Suggest a structure commensurate with these data.

Chapter 7

Chemical reactions

All is change; all yields its place and goes.

EURIPIDES (422 B.C.)

It should be apparent from the discussions in the previous chapters that an almost infinite number of combinations of the building blocks of organic chemistry are possible. Although to date only 2 to 3 million of these combinations have actually been isolated from natural sources or synthesized in the laboratory, more are being added daily. One of the major preoccupations of the organic chemist, in fact, is the search for new ways in which to combine the building blocks, and a knowledge of the *reactions of organic compounds* is of primary importance in this endeavor.

A chemical reaction is a process in which one species is converted to another, a process in which a rearrangement in the positions of the nuclei and electrons in the system takes place. To understand a reaction in detail, it is necessary to have information concerning (a) the structures of the reactants and products, (b) the conditions under which the reaction occurs, and (c) an intimate description of the pathways by which the nuclei and electrons change positions as the reactants change to products. It is the last of these topics that is discussed in the present chapter, which deals with the general outlines of what are called the "mechanisms" of chemical reactions. However, before concerning ourselves with the actual pathways by which chemical conversions take place (*i.e.*, the dynamic aspects), we shall first consider the factors relating to the stabilities of reactants and products (*i.e.*, the static aspects).

7.1 Acid-Base Equilibria

The type of chemical equilibria that are probably most familiar to those readers who have had a course in general chemistry involve acids and bases. For example, when phosphoric acid is added to water, an interaction takes place in which a proton is transferred from the phosphoric acid to a water molecule, the latter behaving as a base; *i.e.*,

$$\underset{\textbf{An acid}}{H_3PO_4} + \underset{\textbf{A base}}{H_2O} \rightleftharpoons \underset{\textbf{An acid}}{H_3O^+} + \underset{\textbf{A base}}{H_2PO_4^-}$$

The products of this reaction (H_3O^+ and $H_2PO_4^-$) can themselves react with each other again to form H_3PO_4 and H_2O. Thus, a pair of acid-base reactions take place, one to form H_3O^+ and $H_2PO_4^-$ from H_3PO_4 and H_2O, and the other to form H_3PO_4 and H_2O from H_3O^+ and $H_2PO_4^-$. After the phosphoric acid and water have been mixed, a steady state is reached in which these two reactions proceed at the same rate; at this point, the concentrations of H_3PO_4, H_2O, H_3O^+, and $H_2PO_4^-$ remain invariant (at constant temperature). Under these conditions, the system is said to be in *dynamic equilibrium*, and it is of interest to the chemist to know the "position" of the equilibrium—*i.e.*, the concentrations of the various species present. This is best expressed in the form of an equilibrium constant, called a dissociation constant, which is simply the ratio of the concentrations of the products to the reactants, *i.e.*,

$$K'_a = \frac{[H_3O^+][H_2PO_4^-]}{[H_3PO_4][H_2O]}$$

where K'_a is commonly called the Brønsted acidity constant. In the case of phosphoric acid and water, the equilibrium concentrations when 1 mole of H_3PO_4 is diluted with water to a volume of 1000 ml are approximately $[H_3O^+] = 8.5 \times 10^{-2}$, $[H_2PO_4^-] = 8.5 \times 10^{-2}$, $[H_3PO_4] = 0.915$, and $[H_2O] = 55.5\ M$, from which a value of K'_a of 1.42×10^{-4} can be calculated. When the strengths of acids are compared by measuring the degree of dissociation in dilute aqueous solution, the water concentration is not included in the equation; *i.e.*, for phosphoric acid,

$$H_3PO_4 \rightleftharpoons H^+ + H_2PO_4^-$$

with

$$K_a = \frac{[H^+][H_2PO_4^-]}{[H_3PO_4]} = K'_a[H_2O]$$

in which case K_a (commonly called the *acid ionization constant*) has the value 7.88×10^{-3}. The ionization constants of acids are conveniently expressed as the $-\log K_a$ (or $\log 1/K_a$), which is called the pK_a. Thus, the pK_a for phosphoric acid is $\log 1/7.88 \times 10^{-3} = 2.12$.

PROBLEM 7-1

The equilibrium constant for the dissociation of *p*-nitrophenol in water is 3.3×10^{-8}. What is its pK_a? The pK_a of benzoic acid is 4.2. What is its dissociation constant? Is *p*-nitrophenol a stronger or weaker acid than benzoic acid? What is the ratio of their acidities?

All organic compounds containing one or more hydrogens share with phosphoric acid the ability to transfer, to a greater or lesser extent, a proton to a basic compound. In some cases this tendency is even greater than that of phosphoric acid; benzenesulfonic acid, for example, has a pK_a of -0.6. In others it is less; for example, benzoic acid has a pK_a of 4.2, phenol has a pK_a of 10.0, and methanol has a pK_a of 15.5. In some compounds, the acidities are so very much less than that of phosphoric acid (*e.g.*, ethylamine with a pK_a of 33 and methane with a pK_a of 40) that the quantitative differences become essentially qualitative. In Table 7.1, a list of pK_a values is given for a variety of reactions of the type

Table 7.1 pK_a VALUES FOR THE REACTION $AH + B \rightleftharpoons BH^+ + A^-$ (ACID DISSOCIATION CONSTANTS)

Reaction	pK_a
$FSO_3H + B \rightleftharpoons BH^+ + FSO_3^-$	−12
$HI + B \rightleftharpoons BH^+ + I^-$	−10
$HClO_4 + B \rightleftharpoons BH^+ + ClO_4^-$	−10
$H_2SO_4 + B \rightleftharpoons BH^+ + HSO_4^-$	− 9
$(H)(R)C{=}\overset{+}{O}H + B \rightleftharpoons BH^+ + (H)(R)C{=}O$	− 8
$HCl + B \rightleftharpoons BH^+ + Cl^-$	− 7
$(R)(R)C{=}\overset{+}{O}H + B \rightleftharpoons BH^+ + (R)(R)C{=}O$	− 7
$(R)(R)\overset{+}{O}{-}H + B \rightleftharpoons BH^+ + (R)(R)O$	− 3.5
$(H)(H)\overset{+}{O}{-}H + B \rightleftharpoons BH^+ + (H)(H)O$	− 1.7
$HNO_3 + B \rightleftharpoons BH^+ + NO_3^-$	− 1.3
$C_6H_5SO_3H + B \rightleftharpoons BH^+ + C_6H_5SO_3^-$	− 0.6
$CF_3CO_2H + B \rightleftharpoons BH^+ + CF_3CO_2^-$	0
$CCl_3CO_2H + B \rightleftharpoons BH^+ + CCl_3CO_2^-$	0.9
$Cl_2CHCO_2H + B \rightleftharpoons BH^+ + Cl_2CHCO_2^-$	1.3
$H_3PO_4 + B \rightleftharpoons BH^+ + H_2PO_4^-$	2.1
$ClCH_2CO_2H + B \rightleftharpoons BH^+ + ClCH_2CO_2^-$	2.8
$HF + B \rightleftharpoons BH^+ + F^-$	3.2
$HCO_2H + B \rightleftharpoons BH^+ + HCO_2^-$	3.7
$C_6H_5CO_2H + B \rightleftharpoons BH^+ + C_6H_5CO_2^-$	4.2
$C_6H_5\overset{+}{N}H_3 + B \rightleftharpoons BH^+ + C_6H_5NH_2$	4.6
$CH_3CO_2H + B \rightleftharpoons BH^+ + CH_3CO_2^-$	4.8
pyridinium ($C_5H_5\overset{+}{N}H$) $+ B \rightleftharpoons BH^+ +$ pyridine (C_5H_5N)	5.2
$C_6H_5SH + B \rightleftharpoons BH^+ + C_6H_5S^-$	7.8
$C_6H_5CO_3H + B \rightleftharpoons BH^+ + C_6H_5CO_3^-$	8.2
$CH_3COCH_2COCH_3 + B \rightleftharpoons BH^+ + CH_3CO\bar{C}HCOCH_3$	9.0
$HCN + B \rightleftharpoons BH^+ + CN^-$	9.1
$C_6H_5OH + B \rightleftharpoons BH^+ + C_6H_5O^-$	10.0
$HCO_3^- + B \rightleftharpoons BH^+ + CO_3^=$	10.2
$CH_3NH_3^+ + B \rightleftharpoons BH^+ + CH_3NH_2$	10.4
$CH_3CH_2SH + B \rightleftharpoons BH^+ + CH_3CH_2S^-$	10.5
$CH_3NO_2 + B \rightleftharpoons BH^+ + \bar{C}H_2NO_2$	11.0
$CH_3COCH_2CO_2C_2H_5 + B \rightleftharpoons BH^+ + CH_3CO\bar{C}HCO_2C_2H_5$	11.0
$CCl_3CH_2OH + B \rightleftharpoons BH^+ + CCl_3CH_2O^-$	12.2
$CH_2(CO_2C_2H_5)_2 + B \rightleftharpoons BH^+ + {}^-CH(CO_2C_2H_5)_2$	13.0
cyclopentadiene $+ B \rightleftharpoons BH^+ +$ cyclopentadienide anion	14.0
$CH_3CONH_2 + B \rightleftharpoons BH^+ + CH_3CONH^-$	15.0
$CH_3OH + B \rightleftharpoons BH^+ + CH_3O^-$	15.5
$H_2O + B \rightleftharpoons BH^+ + HO^-$	15.7
$CH_3CH_2OH + B \rightleftharpoons BH^+ + CH_3CH_2O^-$	17
$(CH_3)_3COH + B \rightleftharpoons BH^+ + (CH_3)_3CO^-$	19
$CH_3COCH_3 + B \rightleftharpoons BH^+ + CH_3COCH_2^-$	20
$C_6H_5C{\equiv}CH + B \rightleftharpoons BH^+ + C_6H_5C{\equiv}C^-$	21
$(C_6H_5)_2NH + B \rightleftharpoons BH^+ + (C_6H_5)_2N^-$	23
$HC{\equiv}CH + B \rightleftharpoons BH^+ + HC{\equiv}C^-$	26
$C_6H_5NH_2 + B \rightleftharpoons BH^+ + C_6H_5NH^-$	27
$(C_6H_5)_3CH + B \rightleftharpoons BH^+ + (C_6H_5)_3C^-$	31.5
$(C_6H_5)_2CH_2 + B \rightleftharpoons BH^+ + (C_6H_5)_2CH^-$	33
$CH_3CH_2NH_2 + B \rightleftharpoons BH^+ + CH_3CH_2NH^-$	33
$C_6H_5CH_3 + B \rightleftharpoons BH^+ + C_6H_5CH_2^-$	35
$NH_3 + B \rightleftharpoons BH^+ + NH_2^-$	36

Table 7.1 continued

Reaction	pK_a	Reaction	pK_a
$H_2CO_3 + B \rightleftharpoons BH^+ + HCO_3^-$	6.5	$H_2C{=}CH_2 + B \rightleftharpoons BH^+ + H_2C{=}CH^-$	36
$H_2S + B \rightleftharpoons BH^+ + HS^-$	7	$CH_4 + B \rightleftharpoons BH^+ + CH_3^-$	40
$O_2N{-}C_6H_4{-}OH + B \rightleftharpoons BH^+ + O_2N{-}C_6H_4{-}O^-$	7.2	$CH_3CH_3 + B \rightleftharpoons BH^+ + CH_3CH_2^-$	42
$C_6H_4(CO)_2NH + B \rightleftharpoons BH^+ + C_6H_4(CO)_2N^-$	7.4	$C_6H_6 + B \rightleftharpoons BH^+ + C_6H_5^-$	42
		$C_6H_{10}H_2 + B \rightleftharpoons BH^+ + C_6H_{10}^-{-}H$	45

$A{:}H + B{:} \rightleftharpoons B{:}H^+ + A{:}^-$, where $A{:}H$ is designated as the acid, $B{:}$ as the base, $A{:}^-$ as the "conjugate base" of $A{:}H$, and $B{:}H^+$ as the "conjugate acid" of $B{:}$.*

All the acids, HA, in Table 7.1 have the potentiality of transferring a proton to a base and are classed as *protic acids* (*i.e.*, proton-donating acids). A more general definition of acidity suggested by G. N. Lewis states that any substance, protic or aprotic, capable of gaining electrons is considered to be an acid. In protic acids, the electron-accepting entity is the proton; in aprotic acids (often called "Lewis acids"), it is some other electron-deficient element; *e.g.*, in aluminum chloride

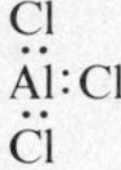

* The base species B: in the reactions listed in Table 7.1 is water for those cases in which the pK_a falls in the region 0–14. Acids with pK_a values less than 0 transfer the proton to water so completely that too little undissociated acid, HA, remains for accurate measurement: acids with pK_a values greater than 14 are too weak to transfer enough protons to water to allow an accurate measurement. For very strong and very weak acids, therefore, indirect methods are necessary whereby, through a series of measurements of equilibrium constants for acid-base pairs separated by 1–2 pK units, extrapolations beyond the 0–14 pK_a range can be made. For example, the equilibrium constant for the reaction $C_2H_5OH + HO^- \rightleftharpoons C_2H_5O^- + H_2O$ is measured in an aqueous system and found to be 0.05. Knowing the pK_a of water to be 15.7, the pK_a of C_2H_5OH can be calculated to be greater than this by the log of 1/0.05 (*i.e.*, 1.3 pK units) or 17.0. Then, the equilibrium constant for the reaction $(CH_3)_3COH + C_2H_5O^- \rightleftharpoons (CH_3)_3CO^- + C_2H_5OH$ is measured in an alcohol system (*i.e.*, nonaqueous) and found to be 0.01, from which the pK_a of *t*-butyl alcohol can be calculated to be 17.0 + log 1/0.01 or 19. And, in similar fashion, escalation up the pK_a scale to weaker and weaker acids, or down the scale beyond pK_a 0 to stronger and stronger acids is possible. The pK_a's calculated from actual concentrations of the species involved in the equilibrium are called "thermodynamic acidities."

The continuation of the approach thus outlined has a practical limit—*viz.*, the availability of solvent bases of sufficient basicity to remove the proton from the acidic species. For extremely weak acids, such as alkanes and alkenes, therefore, a different approach must be used in which the *rate* at which a proton is exchanged with tritium (a radioactive isotope of hydrogen) is measured. The pK_a's calculated in this fashion are called "kinetic acidities":

$$R{-}H + C_6H_{13}NHLi \xrightarrow{\text{Slow}} R^-Li^+ \xrightarrow[C_6H_{13}NH_2^*]{\text{Very fast}} R{-}H^* \quad (H^* \text{ is } {}^3H)$$

the electron-accepting entity is the aluminum atom, which has a valence shell containing only six electrons. Thus, a wide variety of substances in addition to protic acids (see Table 7.1) are classed as Lewis acids, including neutral compounds containing elements with unfilled valence shells (*i.e.*, $AlCl_3$, BF_3, $FeCl_3$, etc.), and cations (*i.e.*, Na^+, Ca^{+2}, Fe^{+3}, etc.). Conversely, a base is defined in the Lewis system as any substance capable of releasing electrons. This includes anions (*i.e.*, HO^-, CH_3O^-, Cl^-, etc.), neutral compounds containing nonbonded electrons (*i.e.*, $H_2\ddot{O}$, $(C_2H_5)_2\ddot{O}$, $(CH_3)_3N$:, etc), and compounds containing π electrons (*i.e.*, $R_2C{=}CR_2$, $RC{\equiv}CR$, ArH, etc.). For example, aluminum chloride reacts with $C_6H_5COC_6H_5$ (benzophenone) to form a solid that crystallizes in the form of long yellow needles melting at 130°. The reaction involves a partial transfer of electrons from the carbonyl group of benzophenone, acting as a Lewis base, to the unfilled outer valence shell of the aluminum in aluminum trichloride, acting as a Lewis acid; *i.e.*,

$$(C_6H_5)_2C{=}\ddot{O}: \; + \; AlCl_3 \longrightarrow (C_6H_5)_2C{=}\overset{+}{\ddot{O}}{-}\overset{-}{Al}Cl_3$$

Lewis base **Lewis acid** **Complex**

Lewis acids are also referred to as electrophiles (meaning "electron seeking") or as electrophilic reagents; Lewis bases are also referred to as nucleophiles (meaning "nucleus seeking") or nucleophilic reagents. Remember these terms, for they are often used interchangeably.

PROBLEM 7-2

Classify the following entities as Lewis acids or Lewis bases: benzene, $MgCl_2$, Li^+, $C_6H_5O^-$, H^+, $\overset{+}{N}H_3CH_2CO_2^-$.

7.2 General Features of Equilibrium Processes

Acid-base equilibria are examples of the fact that in a closed system *all* chemical reactions are equilibrium processes, although in many cases the equilibrium constants are either so small that the reactions barely proceed or so large that they go virtually to completion. Information concerning the magnitude of the equilibrium constant is useful to the chemist, for it provides an indication of the direction in which a reaction is expected to proceed. However, the equilibrium constant gives no indication of the *rate* at which the equilibrium can be established.

To illustrate some of the features of the equilibrium phenomenon, let us look at the reaction in which cyclobutene is isomerized to 1,3-butadiene. Although the reaction is too slow to proceed to any measurable extent at room temperature, at 200° it takes place moderately rapidly. Heat is evolved in the course of the reaction, and when the amount given off during the isomerization of *one mole* of cyclobutene is quantitatively measured, it is found to be about 4.4 kcal. This

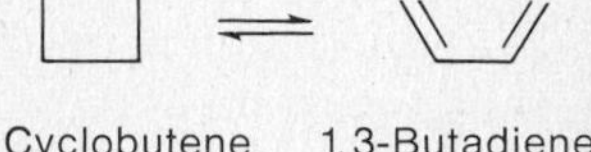

Cyclobutene 1,3-Butadiene

quantity is called the *enthalpy of reaction* (symbolized by ΔH), and it is a measure of the amount of heat that can be transferred between the reacting compound and its surroundings. The enthalpy, along with another quantity called the *entropy of reaction*, together determine the magnitude of the equilibrium constant for a reaction. Entropy (symbolized by ΔS), though, is more difficult to define. In a sense it is a measure of the difference in the flexibility, the organization, and the degrees of freedom between the reactant and the product. In the cyclobutene–butadiene reaction, it has a positive value (4.6 entropy units), commensurate with the idea that butadiene is more flexible and less rigid than cyclobutene. Multiplying the entropy value by the temperature (in degrees Kelvin) and subtracting this value from the enthalpy of reaction, we get a quantity called the *free energy of reaction* (symbolized by ΔG);

$$\Delta H - T\Delta S = \Delta G$$

The free energy and the equilibrium constant (K) for a reaction are related by the expression

$$\Delta G = -RT \ln K$$

where T is the absolute temperature and R is the universal gas constant (1.987 cal/mole-deg). Inspection of this expression shows that when the free energy of a reaction is zero, the equilibrium constant is 1; when the free energy is positive, the equilibrium constant is less than 1; and when the free energy is negative, the equilibrium constant is greater than 1. In the cyclobutene–butadiene reaction, the negative value for the enthalpy of reaction (ΔH is negative for an exothermic process) and the positive value for the entropy of reaction lead to a negative free energy of reaction and, consequently, an equilibrium constant appreciably greater than 1. In other words, the reaction proceeds, for all practical purposes, to completion once it is started. Why, though, if the equilibrium constant is so favorable for the conversion of cyclobutene to butadiene, does it require the input of a considerable amount of energy, by heating to 200°, to make it go? To answer this question we must explore the features that govern the *rate of a reaction*.

PROBLEM 7-3

Calculate the free energy of reaction for the conversion of cyclobutene to butadiene, given the enthalpy and entropy values cited. [*Note:* The enthalpy value is −4.4 kcal/mole (*i.e.*, −4400 calories/mole, the negative value indicating that heat is released in the process—*i.e.*, the reaction is exothermic). Is the sign of the free energy of reaction positive or negative for this reaction?

7.3 General Features of Rate Processes

The conversion of one type of molecule to a different type of molecule involves a reorganization of the nuclei and electrons of the system. To initiate this reorganization, we must add energy, even though the reorganized system may be lower in energy and more stable than the original system, leading to a net release of energy. Were energy not required to initiate the reorganization, no molecule would be stable for any appreciable length of time, and everything would be in a state of constant flux. Let us, then, look into the process whereby a molecule is "activated" to the point where it can undergo some kind of chemical conversion.

One of the simplest of chemical processes involves a diatomic molecule, which we shall symbolize as X—Y, reacting with an atom, which we shall symbolize as Z;

$$\text{X—Y} + \text{Z} \longrightarrow \text{Y—Z} + \text{X}$$

For the reactant molecule XY as well as the product molecule YZ, we can depict the energy of the system as a function of the distance between the two nuclei, *viz*., a bond length. Such a plot is called a Morse curve and has the appearance shown in Fig. 7.1. For a stable system, the forces of attraction between the nuclei

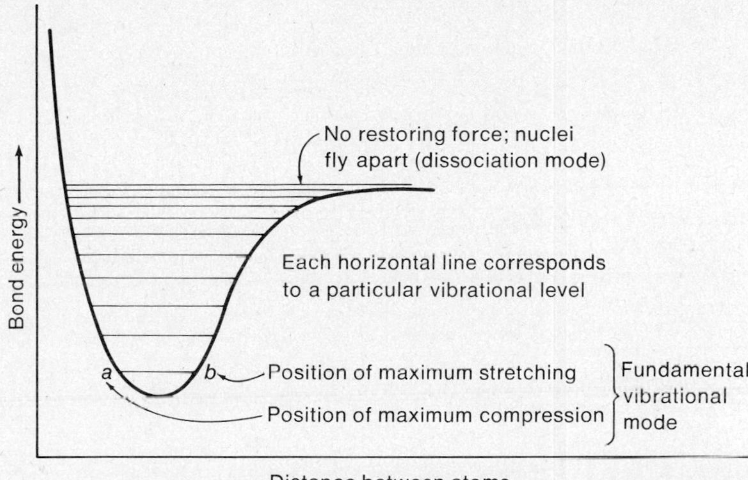

Fig. 7.1. A Morse curve, showing the change in potential energy as a function of the distance between a pair of adjacent atoms.

prevent the bond from stretching beyond a certain point (*e.g*., point *b* for the fundamental vibrational mode shown in Fig. 7.1), but if sufficient energy is added, the restoring force can no longer serve this purpose and the bond breaks. Thus, at a certain temperature the reactant molecule XY can be dissociated to atoms X and Y, and the product molecule YZ can be dissociated to atoms Y and Z. One conceivable pathway for transforming XY + Z to YZ + X, therefore, is to add enough energy to convert the system to the three atoms X, Y, and Z and then let the atoms recombine to form either the reactant XY or the product YZ. We can depict this course of events on a three-dimensional plot, showing the Morse curve for the XY molecule on one face and that for the YZ molecule on another face, as Fig. 7.2 illustrates.

Let us, though, consider other ways in which the transformation might occur. It might, for example, involve the interaction of Z with XY in such a fashion that the Y—Z bond starts to form *before* the X—Y bond is broken, leading to the intermediate structure X- -Y- -Z. If the energy of this structure is plotted on our

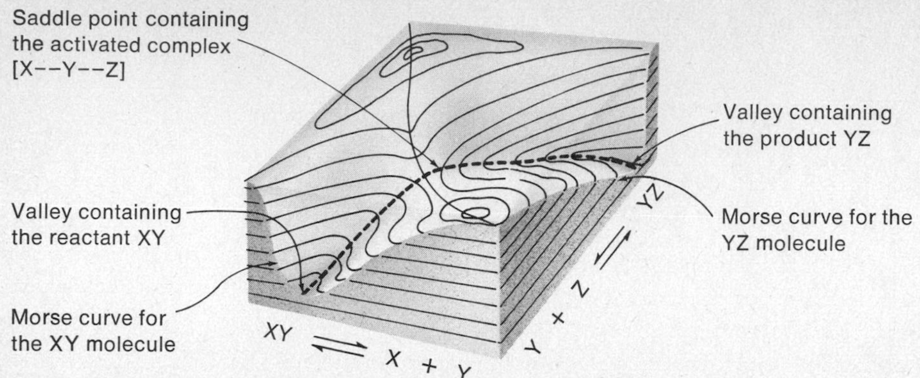

Fig. 7.2. Three-dimensional energy contour representation for the reaction XY + Z → YZ + X.

three-dimensional diagram (Fig. 7.2), it turns out that it is lower than that corresponding to the three-atom system X + Y + Z. If we plot the energy of the infinite number of possible arrangements of X--Y--Z (*i.e.*, different distances of separation of X, Y, and Z and different "bond" angles), one of the arrangements will have a lower energy than all the rest. This particular arrangement is called the "activated complex," and the system in this condition is said to be in the "transition state." Our three-dimensional plot now has the appearance of two valleys separated from each other by hills, the contour of the terrain between the valleys resembling that of a pass or saddle point in a range of mountains. The valleys represent the reactant and product, and the pass, which is the lowest point between the valleys, represents the activated complex. Just as Hannibal's army of men and elephants sought the lowest pass in crossing the Alps, a reacting system seeks the lowest energy pathway in progressing from reactant to product, as illustrated by the solid line in Fig. 7.2. Representations of reaction pathways such as that shown in Fig. 7.2 are difficult to draw, however, so what is usually done is to take a curving cross section along the reaction pathway in this three-dimensional figure and display it on a two-dimensional surface. The resulting projection, illustrated in Fig. 7.3, is called a *reaction coordinate diagram* or a *reaction energy profile.*

The interpretation of the reaction of XY with Z to form YZ and X can be extrapolated to other systems, such as the cyclobutene to butadiene conversion. The fact that in the cyclobutene–butadiene rearrangement we are dealing with a single reactant (a unimolecular reaction) rather than a pair of reactants (bimolecular reaction) does not alter the interpretation. We can depict the reactant cyclobutene and the product butadiene as being in the valleys in the plot shown in Fig. 7.2 and the activated complex, a structure somewhere between cyclobutene and butadiene, as being in the pass. A two-dimensional reaction coordinate diagram for this process looks precisely like that shown in Fig. 7.3, where the difference in energy between the cyclobutene and butadiene reflects the equilibrium constant for the process and where the difference in energy between cyclobutene and the activated complex is related to the rate at which this equilibrium is established.

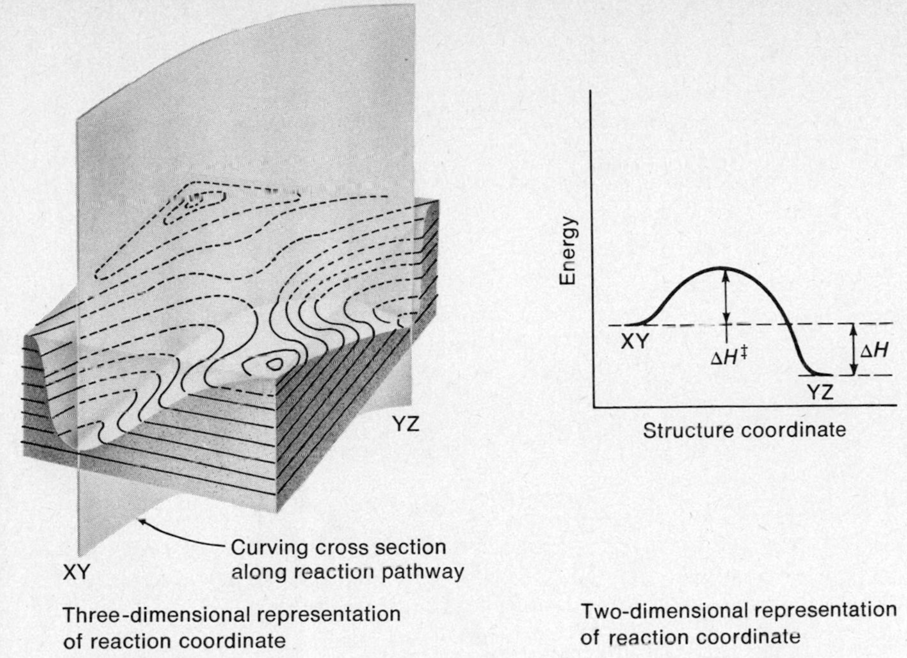

Fig. 7.3. Three-dimensional and two-dimensional representations of reaction coordinates.

Let us now look into the rate process in a more detailed fashion. As intuition suggests, the greater is the difference in energy between the reactant and the activated complex the more difficult it is to transform the reactant to the activated complex and the slower is the reaction. Borrowing the mathematical expressions as well as the terms from simple equilibrium processes, we refer to this difference in energy as the *enthalpy of activation*; it is symbolized by $\Delta H^{\ddagger}$ and is expressed in kilocalories per mole. Just as with simple equilibria, however, the enthalpy is not the sole determining factor. Another factor called the *entropy of activation*, symbolized by $\Delta S^{\ddagger}$, also enters in. The entropy of activation can be viewed as relating to the contour of the pass through which the reaction pathway leads. If the pass is narrow, the reaction is slower than if the pass is wide. Thus, we have the enthalpy of activation and the entropy of activation together determining the rate of a reaction, their combined influence defining the *free energy of activation*, symbolized by $\Delta G^{\ddagger}$. The activation parameters are related in precisely the same way that the simple equilibrium parameters are related (*i.e.*, $\Delta G^{\ddagger} = \Delta H^{\ddagger} - T\Delta S^{\ddagger}$), and the free energy of activation is related to the rate of reaction in a manner analogous to the relation between free energy and the equilibrium constant (*i.e.*, the greater is the positive value of the free energy of activation, the slower is the reaction). In depicting reaction coordinates it is more customary to plot the free energy on the vertical axis than the enthalpy as shown in Fig. 7.3, thus combining the contributions of the enthalpy and entropy to the activation process (*i.e.*, $\Delta G^{\ddagger}$) and the equilibrium process (*i.e.*, ΔG).

PROBLEM 7-4

If we were to consider the cyclobutene–butadiene conversion in the opposite direction (*i.e.*, the conversion of butadiene to cyclobutene), what would the reaction coordinate look like? Would the equilibrium constant for this process be greater or less than 1? Would the free energy of reaction be positive or negative?

PROBLEM 7-5

For the cyclobutene to butadiene conversion, the enthalpy of activation is $+32.7$ kcal/mole, and the entropy of activation is $+1.6$ entropy units. Calculate the value for the free energy of activation at 200°C. Does the positive entropy contribution make the reaction go faster or slower than if $\Delta S^{\ddagger}$ were zero? Considering the interconversion in the reverse direction (*i.e.*, see Problem 7-4), what is the value of the free energy of activation?

7.4 Reactive Intermediates of Organic Reactions

Many reactions are not adequately represented by diagrams such as Fig. 7.3 and require reaction coordinate diagrams showing two or more energy maxima corresponding to two or more activated complexes between reactant and product (Fig. 7.4). The activated complexes in a reaction involving one or more intermediates generally have different energies, indicated on the reaction coordinate diagrams by different heights. The valley between a pair of activated complexes corresponds to an intermediate structure that is formed along the way between reactant and product. Reaction intermediates, then, are species that are less stable than reactant or product but more stable than the activated complexes separating them from reactant and product. We can define their structural details in the same terms we use for stable organic molecules, although the experimental means for providing this definition may be quite difficult in the case of highly reactive, short-lived intermediates. In this section, we shall discuss the structures of the four types of intermediates most often encountered in organic reactions.

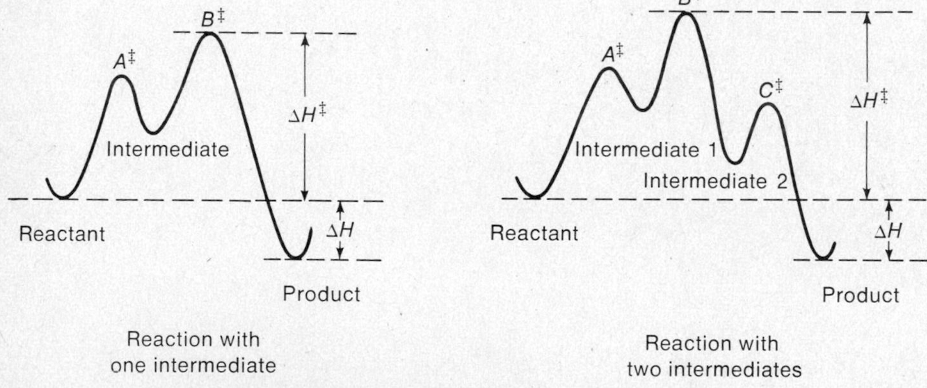

Fig. 7.4. Reaction coordinates for reactions involving intermediates.

PROBLEM 7-6

The reaction in which proteins are hydrolyzed (as part of the digestion process) under the influence of an enzyme known as chymotrypsin (found in the stomach) is postulated to involve no fewer than five steps, with the third step being the slowest (*i.e.*, the rate-determining step). Draw a reaction coordinate for this process.

The first of the reactive intermediates to be considered is known as a "carbonium ion" or "carbocation." It is defined as a triply liganded (see footnote, p. 71) carbon possessing only six electrons in the valence shell. In the majority of cases the carbon atom of a carbonium ion is sp^2-hybridized, and the stability of the carbonium ion is greatly diminished if the planar configuration required by the sp^2 carbon cannot be attained (Fig. 7.5). The stability of a carbonium ion also depends on the attached groups. Phenyl and vinyl groups, for example, have a considerable stabilizing effect, because interaction between the unoccupied *p* orbital of the cation and the π system of the benzene ring or the vinyl group allows the positive charge to be delocalized over the entire system. The delocalized system of the phenylcarbonium ion* is illustrated in Fig. 7.6 in orbital and resonance structure representation.

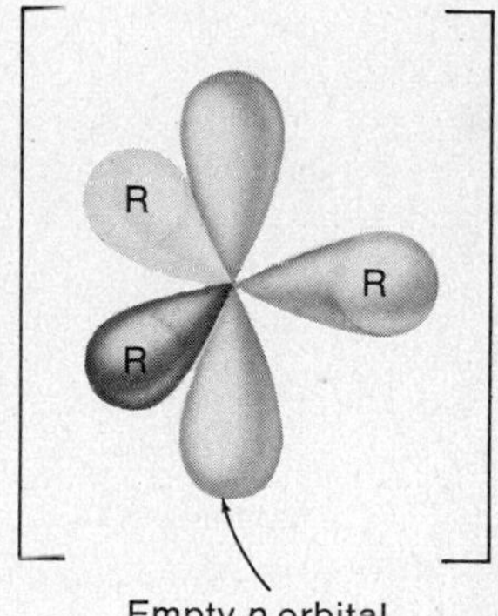

Fig. 7.5. Carbonium ion.

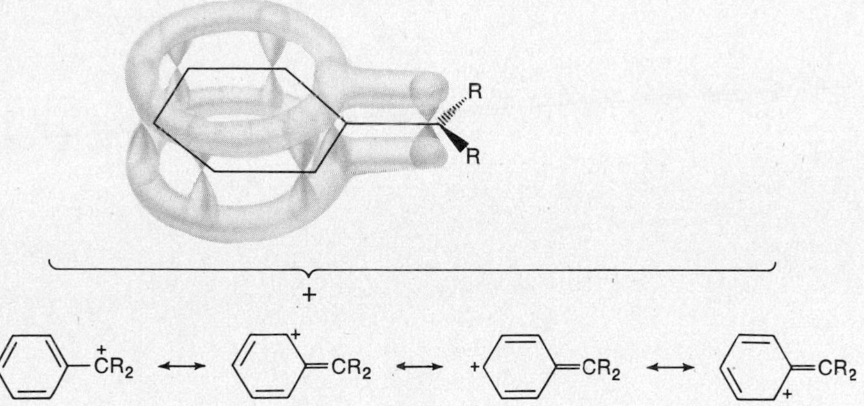

Fig. 7.6. Phenylcarbonium ion (benzyl cation).

In addition to the types of resonance structures we have already discussed (see Section 4.6), still another type is invoked to explain the stabilizing effect of methyl groups on carbonium ions. Numerous experimental data show that the order of stability of alkyl cations decreases from $(CH_3)_3C^+$ to $(CH_3)_2CH^+$ to $CH_3CH_2^+$ to CH_3^+. One explanation postulates that C—H bonds, like C=C

* The designation "carbonium ion" refers specifically to the species CH_3^+. Thus, the carbocation $C_6H_5CH_2^+$ is properly named phenylcarbonium ion or benzyl cation, but not benzylcarbonium ion.

bonds (including phenyl groups), can delocalize the positive charge. The resonance structures that are drawn to depict delocalization by C—H bonds are called *hyperconjugative resonance structures*, and the phenomenon is called hyperconjugation (Fig. 7.7). A more descriptive term applied to these structures is

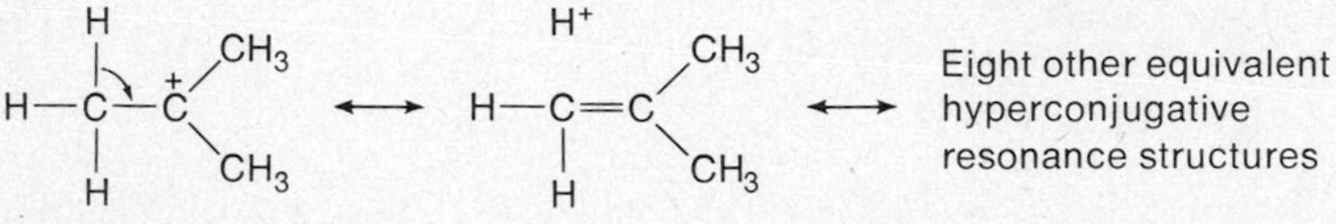

Fig. 7.7. Hyperconjugative resonance structures for trimethylcarbonium ion (*t*-butyl cation).

"no-bond resonance structures," indicating that to the extent that σ electrons have delocalized into a π-system, the bonding between the carbon and hydrogen has diminished. Obviously, considerable C—H bonding remains, for otherwise it would be pointless to consider a structure showing the hydrogen in the vicinity of the carbon. The bond strength is diminished, nevertheless, and the C—H bonds in *t*-butyl cation are, indeed, weaker than the C—H bonds in isobutane. Measurements of carbonium ion stabilities indicate that $C_6H_5CH_2^+$ and $(CH_3)_3C^+$ are about equal and that $C{=}C{-}CH_2^+$ and $(CH_3)_2CH^+$ are about equal, leading to the general stability sequence: $ArCH_2^+ \simeq R_3C^+ > C{=}C{-}CH_2^+ \simeq R_2CH^+ > RCH_2^+ > CH_3^+$.

Although much of the evidence that has been adduced in support of the existence of carbonium ions is beyond the scope of the present discussion* and is introduced at later points, we should mention one item. Compounds such as *t*-butyl bromide, $(CH_3)_3CBr$, undergo facile replacement of the bromine atom in reactions that are thought to involve trimethylcarbonium ion, $(CH_3)_3C^+$, as an intermediate. The related compound, 1-bromobicyclo[2.2.1]heptane (Fig. 7.8), however, is completely inert in these same reactions. A study of the framework models of these compounds indicates why. Whereas $(CH_3)_3C^+$ can, without difficulty, assume a planar configuration as required by the sp^2-hybridized central carbon, the bicyclic compound cannot; the constraints imposed by the bridged-ring system prevent the bridgehead carbon from becoming planar, except with an excessive amount of strain.

---

* Recent evidence supporting the existence of carbonium ions comes from nmr measurements of compounds such as *t*-butyl fluoride in a solvent composed of hydrogen fluoride and antimony pentafluoride. Whereas the methyl groups in *t*-butyl fluoride itself show a single-line resonance at δ 1.30, *t*-butyl fluoride dissolved in $HF{-}SbF_5$ shows this resonance at δ 4.35. This large shift is commensurate with the effect that is known to operate when an electron-withdrawing group is close to the hydrogens undergoing the nmr transition. Fluorine is, itself, a strongly electron-withdrawing group, but a positive charge is much stronger.

$$\underset{\delta\ 1.30\ \text{ppm}}{(CH_3)_3CF} \xrightarrow[HF]{SbF_5} \underset{\delta\ 4.35\ \text{ppm}}{(CH_3)_3C^+} + SbF_6^-$$

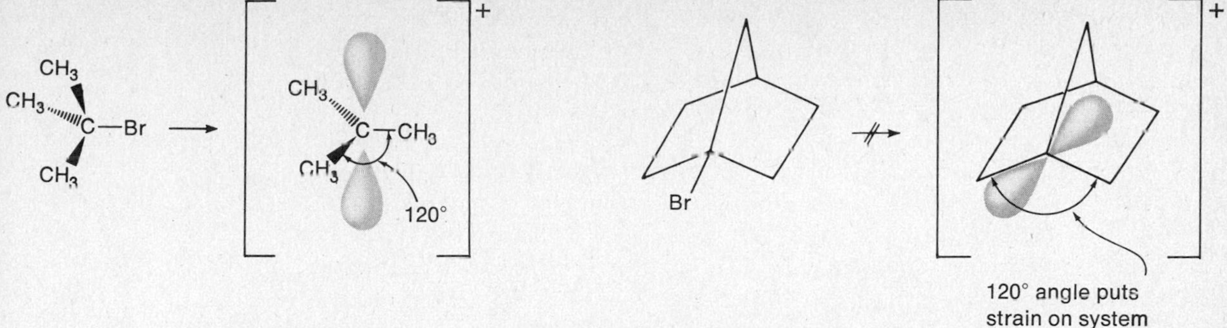

Fig. 7.8. Geometry of *t*-butyl cation and bicycloheptyl cation.

PROBLEM 7-7

Arrange the following carbocations in the order of decreasing stability:

The second reactive intermediate to be considered is called a "free radical." It is defined as a triply liganded carbon possessing seven electrons in the outer valence shell. Its structure differs from that of the carbonium ion (Fig. 7.5) in having one electron in the *p* orbital (Fig. 7.9). Thus, whereas the carbonium ion is a positively charged entity, the free radical is neutral. Both are electron-poor species, because they possess fewer than eight electrons in the valence shell. The same groups that stabilize carbonium ions also stabilize free radicals, the order of stability being $Ar\dot{C}H_2 > C{=}C{-}\dot{C}H_2 > R_3\dot{C} > R_2\dot{C}H > R\dot{C}H_2 > \dot{C}H_3$. Stabilization by phenyl groups (see Fig. 7.10 for an orbital and a resonance structure representation of the delocalized system in the benzyl radical) is so effective,

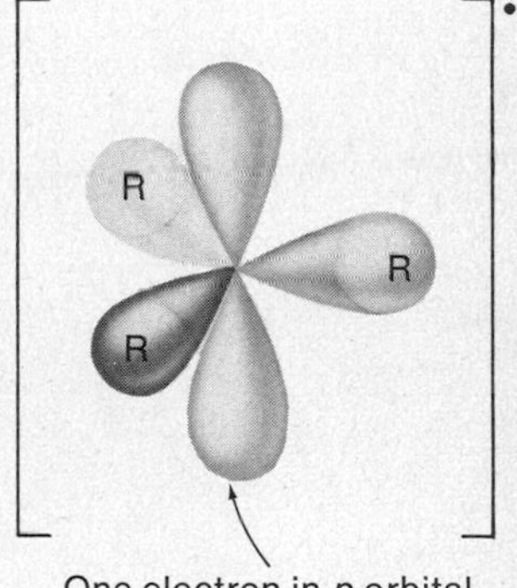

Fig. 7.9. Free radical.

Fig. 7.10. Resonance-stabilized benzyl radical.

in fact, that when three of them are attached to the radical center, the free radical is stable enough to be isolated and studied. The synthesis of such a compound was accomplished in 1900 by Moses Gomberg during his attempts to prepare hexaphenylethane ($(C_6H_5)_3C—C(C_6H_5)_3$). The properties of the compound that he isolated were so incommensurate with a hexaphenylethane structure that he was forced to postulate the existence of the triphenylmethyl radical, $(C_6H_5)_3C\cdot$. Attempts to make alkyl radicals had failed in the hands of numerous workers during the nineteenth century, so chemists were slow to accept the validity of Gomberg's hypothesis. But, eventually, his claim was substantiated, and triarylmethyl radicals are now represented by many examples. Alkyl free radicals, also, have been established as real entities. They are highly reactive and very shortlived, though, and the abortive efforts of earlier chemists to trap and study them are now readily understood.*

PROBLEM 7-8

Draw the resonance structures for the radical species

$$H_2C{=}CH\dot{C}H{-}C_6H_5$$

A third reactive intermediate to be considered is called a "carbanion," defined as a triply liganded carbon possessing eight electrons in the valence shell. It has two more electrons than the positively charged carbonium ion and one more electron than the neutral free radical, and is, therefore, negatively charged (Fig. 7.11). Whether it has sp^2 or sp^3 geometry depends on the attached substituents. If a phenyl group is attached, the sp^2 geometry is favored, for this allows the negative charge to be delocalized (Fig. 7.12). If groups of lesser ability to delocalize the negative charge are attached, the geometry is probably close to that of an sp^3-hybridized carbon, *i.e.*, the electron pair is in a hybrid orbital. Species of this type are configurationally mobile, however, and pass from one tetrahedral form to the other with great facility. The mobility has been demonstrated, for

---

* The most compelling evidence for the existence of free radicals comes from electron-spin resonance (esr) measurements [sometimes called electron paramagnetic resonance (epr) measurements]. The electron, like the hydrogen nucleus, has magnetic characteristics commensurate with a nuclear spin of $\frac{1}{2}$. In an external magnetic field, therefore, it has two spin states available to it, and when irradiated with the proper frequency of electromagnetic radiation it undergoes a transition from the lower-energy to the higher-energy state. The process is entirely comparable to the nmr phenomenon and differs only quantitatively because the electron is a considerably stronger magnet than the hydrogen nucleus. As a consequence, it comes into resonance at a different combination of external magnetic field strength and radio frequency. Despite the wide disparity in intensity of magnetic properties between electrons and hydrogen nuclei, they spin couple in the same fashion that hydrogen nuclei spin couple with each other (see Section 6.3). This provides a means, then, not only of discerning the presence of a free radical but of learning something about the distribution of the lone electron in the system. The esr spectrum of the triphenylmethyl radical, for instance, consists of 28 lines (six equivalent *o*-hydrogens → 7 lines; three equivalent *p*-hydrogens → 4 lines: $7 \times 4 = 28$).

example, by preparing an optically active compound such as the one shown in Fig. 7.13, converting it to a carbanion, and noting the loss in optical activity, which results from the configurational inversion and the eventual production of equal numbers of enantiomers (*i.e.*, racemization).

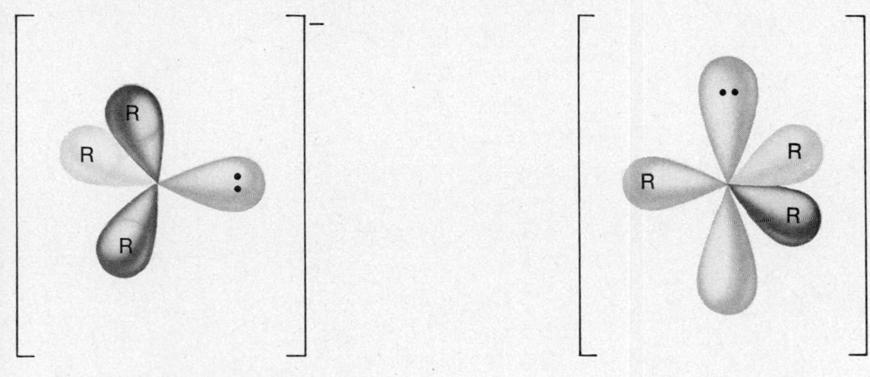

Fig. 7.11. Carbanion.

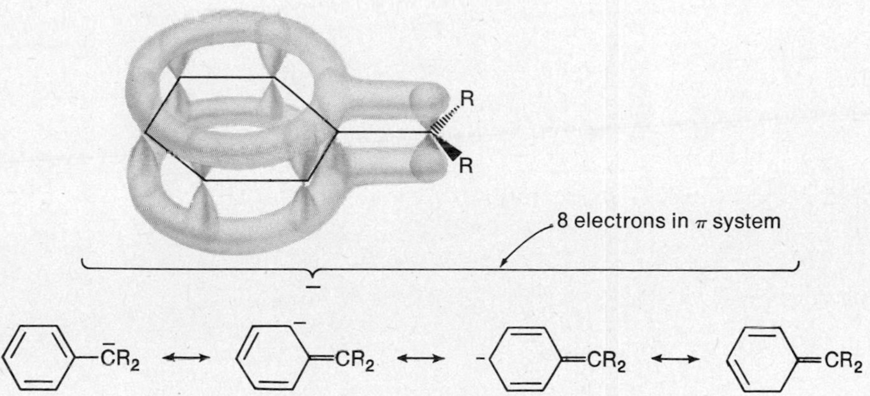

Fig. 7.12. Resonance-stabilized phenylcarbanion (benzyl anion).

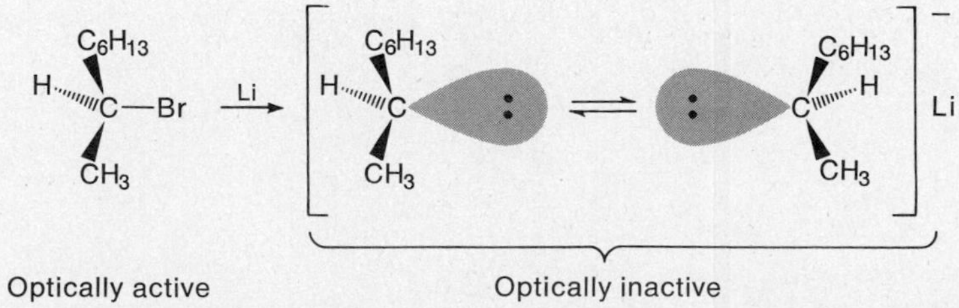

Optically active

Optically inactive

Fig. 7.13. Configurational instability of nondelocalized carbanion.

PROBLEM 7-9

One of the ways for generating carbanions involves treatment of halogen-containing compounds with magnesium, the reaction being described in the following terms: $RX + Mg \rightarrow R^{-}MgX^{+}$. When the magnesium-containing compound is treated with a protic material, the carbanion acquires a proton: $R^{-}MgX^{+} + H^{+} \rightarrow RH + MgX^{+}$. If an optically active halide is treated in this fashion, would you expect the product to be optically active or racemic? Explain.

Carbonium ions, free radicals, and carbanions all contain triply liganded carbons. We can consider them as being formed by the removal of one of the R groups in a R_4C compound with *both* electrons of the C—R bond (to form R_3C^+), with *one* of the electrons of the C—R bond (to form $R_3\dot{C}$), or with neither of the electrons of the C—R bond (to form R_3C^-). A fourth reactive intermediate, called a "carbene," can be formed by removing *two* of the R groups, one *with* the pair of electrons and one *without* the pair of electrons, to leave a doubly liganded carbon (R_2C:), which is a neutral entity and has six electrons in its valence shell (Fig. 7.14). A structural problem that does not arise in the triply-liganded species but is important in the carbenes is whether the two nonbonded

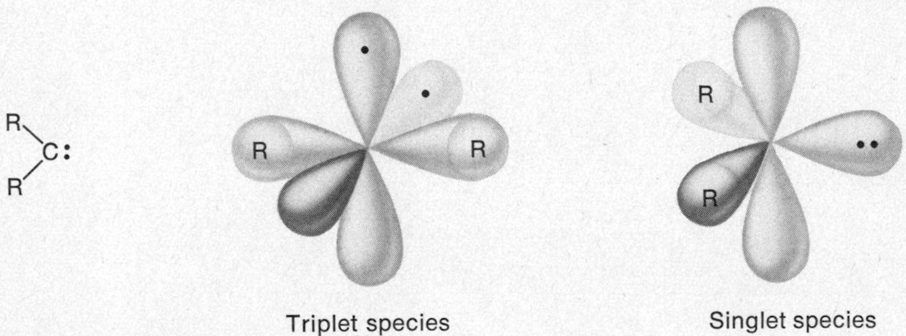

Fig. 7.14. Structures of carbenes.

electrons are in the same orbital or in different orbitals. The orbital occupancy is referred to as the *multiplicity of the electronic state* of the carbene. If both electrons are in the same orbital, they *must* be spin paired; the total spin number is zero, and the species is called a *singlet*. If the electrons are in different orbitals, they can have opposite spins (represented by ↑↓) or they can have the same spin (represented by ↑↑ or ↓↓), and the total spin number can be +1, 0, or −1; the species is called a *triplet*.*

* In a magnetic field, three orientations are available to the triplet carbene and it is possible to observe transitions between these states via esr techniques. The triplet form of diphenylcarbene, *e.g.*, has been detected and measured in this fashion. Singlet species, the reader should be reminded, have no esr spectra because *all* the electrons in the system are paired. Other techniques must be used, therefore, to study singlet carbenes, and consideration of these is beyond the scope of the present discussion.

7.5 *Catalysis of Reactions*

One of the methods for assessing the resonance energy of conjugated dienes, as we have seen (see p. 107), is to measure the heats of hydrogenation of compounds such as 1,3- and 1,4-pentadiene. It turns out, though, that if we simply mix either of these dienes with molecular hydrogen nothing happens; the system is almost indefinitely stable. Despite the favorable equilibrium constant (*i.e.*, reaction is exothermic), no reaction occurs because the free energy of activation is extremely high. But, if some finely divided platinum metal is added to the mixture of diene and hydrogen, a reaction takes place so rapidly that it may be difficult to control. If the platinum is inspected before and after the reaction, it is found to be unchanged, and if the same platinum is added to another mixture of diene and hydrogen the same exothermic reaction is again observed. Clearly, the platinum has survived the reaction unscathed and is acting as a catalyst, a catalyst being defined as a substance that alters the rate of a reaction without itself suffering any net change. What the catalyst does is to lower, in some fashion, the free energy of activation. In the case of the hydrogenation process, the platinum acts as a catalyst by absorbing the alkene and hydrogen molecules on its surface in a way that makes their combination proceed more readily, thereby reducing the free energy of activation. Thus, the reaction coordinate for the uncatalyzed reaction, represented by the solid line in Fig. 7.15(a), is replaced by the reaction coordinate for the catalyzed reaction, represented by the dashed line in Fig. 7.15(a). Although the *free energy of activation* is lowered by the catalyst, the *free energy* of the reaction remains unchanged. Catalysts alter the rate of a reaction but *not* the equilibrium constant.

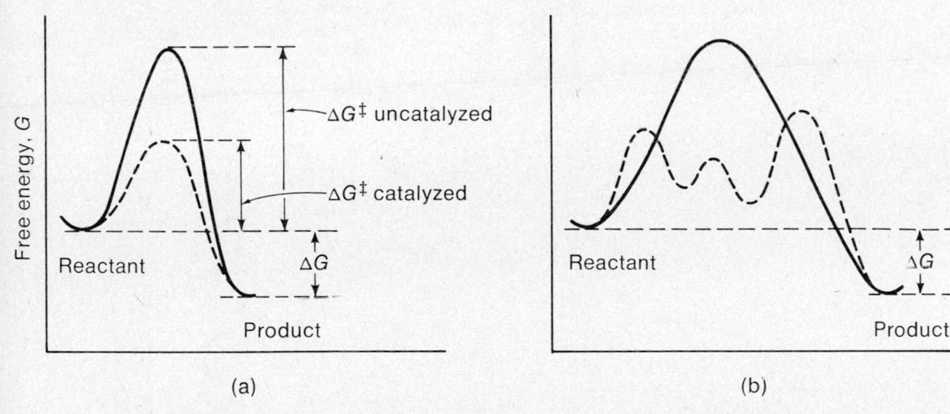

Fig. 7.15. Reaction coordinates for uncatalyzed and catalyzed reactions.

The lowering of the free energy of activation may involve the substitution of the high-energy step by just one lower energy step, as depicted in Fig. 7.15(a), or by a series of lower-energy steps, as depicted in Fig. 7.15(b). There is considerable evidence to indicate that many enzyme-catalyzed reactions use this latter device, changing a single-step reaction into a multistep process.

7.6 *The Mechanisms of Organic Reactions*

The art of organic chemistry is well over one hundred and fifty years old, but the science of organic chemistry, in some respects, is in its infancy. That organic chemistry is still very much an art will be appreciated by anyone who has carried out an organic synthesis in the laboratory and has found that it doesn't go exactly as had been described or anticipated. Not infrequently something goes awry, and it becomes necessary to resort to various methods of trial and error to circumvent the difficulty. In past generations, the rationale for any particular experiment was based principally on the intuition of the chemist. Although intuition is still an important ingredient in a good chemist's success, the rationale for an experiment today can be stated in much more exact terms than was possible even so short a time as two decades ago. This is the result of a very concerted effort on the part of organic chemists to study *reaction mechanisms*.

A reaction mechanism seeks to picture in as minute detail as possible the sequence of changes in the positions of nuclei and electrons that take place in going from reactant to product. The methods developed over the last century have provided the means for deducing, generally with great accuracy, the structures of reactants and products. However, they do not necessarily provide clear pictures of the entities that are involved *between* the reactant and product, and this is the central problem in the study of reaction mechanisms. In drawing deductions concerning a reaction mechanism, one needs to have at hand as much information as possible, and this will usually include knowledge of (a) the structures of all the reactants and all the products, (b) the stoichiometry of the reaction, (c) the kinetics of the reaction, (d) the stereochemistry of the reaction, (e) the correlation between reactant, structure, and reactivity, and (f) the correlation between solvent and reactivity. Part II of this book is devoted to a detailed consideration of these various topics as they relate to reactions and reaction mechanisms.

7.7 Study Guide

I. Thermodynamics
  A. Systems at equilibrium
    1. Equilibrium constants
      (a) For $A + B \rightleftharpoons AB$, $K = \dfrac{[AB]}{[A][B]}$
      (b) $pK = -\log K = \log 1/K$
    2. Acid-base equilibria
      (a) Equilibrium constants symbolized by K_a and pK_a
      (b) Concept of Lewis acids and Lewis bases
      (c) Definition of electrophile, electrophilic reagent, nucleophile, nucleophilic reagent
    3. Thermodynamic parameters
      (a) Enthalpy of reaction, ΔH
      (b) Entropy of reaction, ΔS
      (c) Free energy of reaction, ΔG
    4. Relationships between thermodynamic parameters
      (a) $\Delta G = \Delta H - T\Delta S$
      (b) $\Delta G = -RT \ln K$, where K is the equilibrium constant
  B. Systems in motion (chemical reactions)
    1. Concept of the activated complex (the transition state)

2. Parameters of activation
   (a) Enthalpy of activation, $\Delta H^{\ddagger}$
   (b) Entropy of activation, $\Delta S^{\ddagger}$
   (c) Free energy of activation, $\Delta G^{\ddagger}$
3. Reaction coordinates
   (a) Bond vibrations—Morse curve
   (b) Three-dimensional representation of reaction pathway
   (c) Two-dimensional representation of reaction pathway

II. Reactive intermediates
  A. Triliganded species
   1. Carbonium ions, R_3C^+
     (a) sp^2-Hybridized carbon
     (b) Planar geometry
     (c) Stability sequence: $\mathrm{Ar\overset{+}{C}H_2} \simeq \mathrm{R_3C^+} > \mathrm{C{=}C\overset{+}{C}H_2} \simeq \mathrm{R_2\overset{+}{C}H} > \mathrm{R\overset{+}{C}H_2} > \mathrm{\overset{+}{C}H_3}$
     (d) Concept of hyperconjugation
   2. Free radicals, $R_3C\cdot$
     (a) sp^2-Hybridized carbon (probably)
     (b) Planar geometry
     (c) Stability sequence: $\mathrm{Ar\dot{C}H_2} > \mathrm{C{=}C\dot{C}H_2} > \mathrm{R_3\dot{C}} > \mathrm{R_2\dot{C}H} > \mathrm{R\dot{C}H_2} > \mathrm{\dot{C}H_3}$
   3. Carbanions, R_3C^-
     (a) sp^3-Hybridized carbon in some cases; sp^2-hybridized carbon in others
     (b) Nonplanar geometry when sp^3-hybridized; planar when sp^2-hybridized
  B. Diliganded species, carbenes, R_2C:
   1. Hybridization—depends on spin state
   2. Spin state
     (a) Singlet—all electrons paired
     (b) Triplet—two unpaired electrons

III. Catalysis of reactions
  A. No necessary relation between rate of reaction and equilibrium constant of reaction
  B. Methods of catalysis
   1. Reduction of the activation energy of the slowest step (the rate-determining step)
   2. Replacement of the slow step by a series of faster steps

REFERENCES

For relatively elementary treatments of some of the topics discussed in this chapter, consult the following references:

KLOTZ, I. M., *Introduction to Chemical Thermodynamics*, W. A. Benjamin, Inc., New York, 1964.

MARGOLIS, E. J., *Chemical Principles in Calculations of Ionic Equilibria*, The Macmillan Company, New York, 1966.

WASER, J., *Basic Chemical Thermodynamics*, W. A. Benjamin, Inc., New York, 1966.

PROBLEMS

1. For each of the following reactions, specify the reactant that is the Lewis acid, the reactant that is the Lewis base, the product that is the conjugate acid, and the product that is the conjugate base.

(a) $HF + NaHCO_3 \longrightarrow NaF + H_2CO_3(\rightarrow H_2O + CO_2)$

(b) $CH_3\overset{O}{\overset{\|}{C}}CH_3 + (C_6H_5)_3CNa \longrightarrow CH_3\overset{O}{\overset{\|}{C}}CH_2Na + (C_6H_5)_3CH$

(c) $C_6H_6 + HF \longrightarrow \left[C_6H_6\text{—}H\right]^+ F^-$

(d) $C_5H_{10}O + AlCl_3 \longrightarrow C_5H_{10}\overset{+}{O}\text{—}\overset{-}{AlCl_3}$

(e) $(CH_3)_3N + BF_3 \longrightarrow (CH_3)_3\overset{+}{N}\text{—}\overset{-}{B}F_3$

2. Answer the following questions concerning equilibrium constants:

(a) The equilibrium constants for the second and third ionizations of H_3PO_4 are 7.5×10^{-8} and 4.8×10^{-13}, respectively. What are the pK_a values for these ionization constants?

(b) For the reaction $NH_4^+ + H_2O \rightleftharpoons NH_3 + H_3O^+$, the pK_a value is 9.2. What is the equilibrium constant?

(c) For the reaction $2\,H_2O \rightleftharpoons H_3O^+ + OH^-$, the equilibrium concentrations of H_3O^+ and OH^- are about 10^{-7}. Why is the pK_a value of water 15.7 rather than 14?

3. Using the pK_a values in Table 7.1, calculate the equilibrium constants for the following reactions:

(a) $NH_2^- + C_6H_5CH_3 \rightleftharpoons NH_3 + C_6H_5CH_2^-$
(b) $NH_2^- + CH_4 \rightleftharpoons NH_3 + CH_3^-$
(c) $NH_2^- + CH_3COCH_3 \rightleftharpoons NH_3 + CH_3COCH_2^-$

4. How many molecules of butadiene are present in 1 liter of gas at 20°C and 10^{-2} Torr (available with a good mechanical vacuum pump in the laboratory); at 10^{-5} Torr (available with a good mercury diffusion pump in the laboratory); at 10^{-12} Torr (about the maximum available with the best laboratory vacuum)?

5. Calculate the following thermodynamic values from the data given:

(a) ΔG for a reaction with $K = 127$ at 300°K
(b) K for a reaction in which $\Delta G = -27.6$ kcal/mole at 150°C
(c) $\Delta H^{\ddagger}$ for a reaction if $\Delta G^{\ddagger} = 35.2$ kcal/mole and $\Delta S^{\ddagger} = 10$ eu at 23°C

6. Draw reaction coordinates from the following information:

(a) Reactant more stable than product in a one-step reaction.

(b) Product more stable than reactant in a two-step reaction in which the rate-determining step is the conversion of intermediate to product.
(c) A three-step reaction, in which the formation of the second intermediate is the rate-determining step and in which the product is more stable than reactant.

7. Draw pseudo three-dimensional representations for the following reactive intermediates.

(a) $(CH_3)_3C^+$

(b) 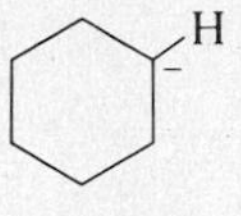 [cyclohexyl structure with H]

(c) 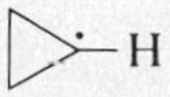 [cyclopropyl structure]–H

(d) $(C_6H_5)_3C\cdot$

8. Draw resonance structures for each of the following reaction intermediates:

(a) 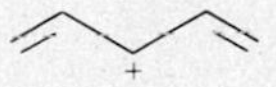 [pentadienyl cation structure]

(b) $HC{\equiv}C\bar{C}HCH_3$

(c) 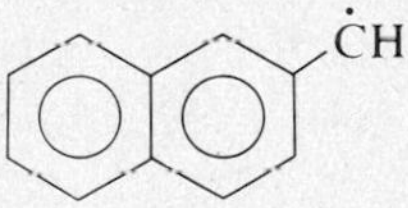 [naphthyl–$\dot{C}H_2$ structure]

(d) $H_2C{=}C{=}CH\ddot{C}CH_3$

(e) 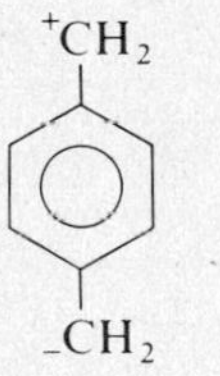 [benzene ring with $^+CH_2$ and $^-CH_2$ para]

(f) 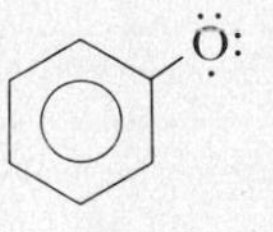 [phenyl–$\ddot{O}$: structure]

(g) $(CH_3CH_2)_3C^+$

(h) 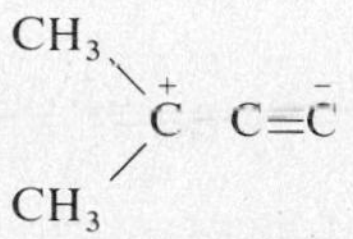 $(CH_3)_2\overset{+}{C}\;C{=}\bar{C}$

(i) $CH_3\overset{O}{\overset{\|}{C}}\bar{C}H\overset{O}{\overset{\|}{C}}CH_3$

(j) 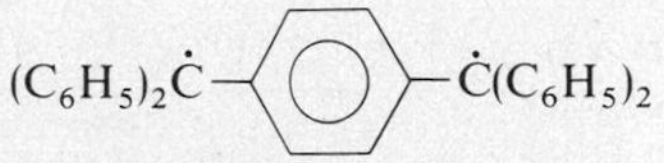 $(C_6H_5)_2\dot{C}$–[phenylene]–$\dot{C}(C_6H_5)_2$

9. At 100°, 1,3,5-cyclooctatriene and bicyclo[4.2.0]octa-2,4-diene are in thermal equilibrium, and spectral measurements indicate that about 85% of the mixture is the triene and 15% the bicyclic diene. What is the ΔG for the equilibrium?

10. The possibility of a biradical intermediate for the isomerization of cyclobutene to butadiene has been considered. Draw a reaction coordinate that would correspond to this situation, and draw resonance structures for the intermediate. It is estimated that a biradical of the sort that would be postulated for this reaction would have a lifetime of about 10^{-10} sec. What would be its lifetime compared with that of an activated complex?

11. Define each of the following terms as briefly and concisely as possible. Do not use a specific example in defining the term, but cite an example to illustrate the definition.

(a) Conjugate acid
(b) Lewis acid
(c) Protic acid
(d) Electrophile
(e) Nucleophile
(f) Free energy of reaction
(g) Free energy of activation
(h) Reaction coordinate
(i) Hyperconjugative resonance structure
(j) Carbonium ion
(k) Catalyst

Chapter 8

The reactions of carbon frameworks

> The chemists are a strange class of mortals impelled by an almost insane impulse to seek their pleasure among smoke and vapor, soot and flame, poison and poverty. Yet among these evils I seem to live so sweetly that may I die if I would change places with the Persian king.
>
> J. J. BECHER
> (*ca.* 1675)

Cyclobutene, as we have seen, can be isomerized to butadiene by heating to about 200°. At a higher temperature, butadiene decomposes to acetylene and hydrogen,

$$H_2C{=}CHCH{=}CH_2 \longrightarrow 2\,HC{\equiv}CH + H_2$$

and at a still higher temperature acetylene decomposes to atomic carbon and atomic hydrogen,

$$HC{\equiv}CH \longrightarrow 2\,C + 2\,H$$

The process in which substances are converted to their constituent atoms is called "cracking" and is common to all compounds, organic or inorganic. Also common to all organic compounds is the reaction with oxygen that yields the oxides of the constituent atoms in a process known as *combustion.* Cyclobutene, for example, reacts exothermally with oxygen to yield carbon dioxide and water,

$$2\,C_4H_6 + 11\,O_2 \longrightarrow 8\,CO_2 + 6\,H_2O$$

PROBLEM 8-1

Using the bond energy values of Table 2.3, calculate the heat of combustion of cyclobutene.

The importance of combustion reactions cannot be minimized; most of the heat and power (except that generated by falling water, wind, sun, compressed steam from within the earth, or nuclear sources) comes from the burning of the fossil fuels, oil and coal, and the energy that sustains life is the result of very complicated oxidative processes in which compounds of carbon are ultimately converted to carbon dioxide and water. From the standpoint of the chemist doing basic research, however, cracking and combustion reactions are usually not very interesting. Seldom does he carry them out intentionally, for his fancy is much more likely to be captured by reactions in which only a few bonds in the molecule are changed at a time. In general, it is reactions of this type that are useful to him in his efforts to synthesize new and interesting compounds.

This chapter, then, is concerned with *selective reactions*; *i.e.*, reactions in which relatively small changes occur in the transformation of reactants to products. Hundreds of such reactions are now known to the organic chemist, and many of these will be presented in the remaining chapters of the text. To introduce this large and important segment of organic chemistry, let us first focus our attention on some of the reactions that are characteristic of the various types of carbon frameworks that we have discussed in the previous chapters. To do this, we shall organize the discussion on the basis of four major types of reactions: *viz.*, *addition reactions*, *substitution reactions*, *elimination reactions*, and *oxidation reactions*.

8.1 Addition Reactions

8.1a. Addition reactions of alkenes. The most generally useful reactions of alkenes involve the addition of various reagents to the double bond, converting the double bond to a single bond in processes that we can represent as

$$\text{(a)}\quad \gt\text{C}=\text{C}\lt \;+\; \text{AB} \longrightarrow \text{A}-\overset{|}{\underset{|}{\text{C}}}-\overset{|}{\underset{|}{\text{C}}}-\text{B}, \quad \text{or} \quad \text{(b)}\quad \gt\text{C}=\text{C}\lt \;+\; \text{A} \longrightarrow \gt\text{C}\overset{\text{}}{-}\text{C}\lt \text{ (with A bridging both C)}$$

Illustrative examples of such reactions are shown in Fig. 8.1, with propene as the alkene undergoing the addition.

To be able to use reactions such as those shown in Fig. 8.1* in the design of organic syntheses, we must, of course, remember what kinds of reactants give what kinds of products. For example, we must remember that when bromine reacts with propene the product is 1,2-dibromopropane. Rather than learn each reaction as an isolated fact, however, it is essential that we discern general patterns by means of which we can predict with some certainty how a particular set of compounds will react even though we may not have encountered the specific compounds previously. For example, knowing how propene and bromine react, we can predict, correctly, that cyclohexene will react with chlorine to form

* Throughout the text we use the following conventions when writing organic reactions: (a) Reagents that are consumed in the reaction are placed above a horizontal arrow or to the right of a vertical arrow; (b) reagents that are not consumed in the reaction (*e.g.*, catalysts), solvents, and reaction conditions (*e.g.*, temperature) are placed below a horizontal arrow or to the left of a vertical arrow; (c) other information about the reaction (*e.g.*, its classification, the moieties that are lost or gained in the transformation of reactant to product) are bracketed and are placed on either side of the arrow.

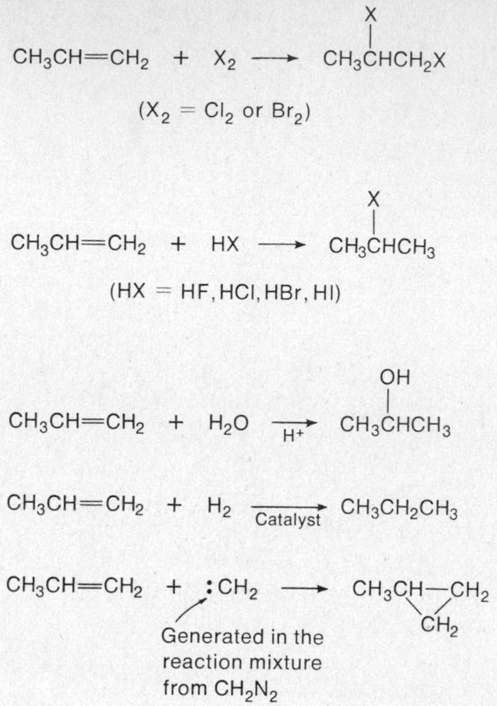

$$CH_3CH{=}CH_2 + X_2 \longrightarrow CH_3CH(X)CH_2X$$

$(X_2 = Cl_2 \text{ or } Br_2)$

$$CH_3CH{=}CH_2 + HX \longrightarrow CH_3CH(X)CH_3$$

$(HX = HF, HCl, HBr, HI)$

$$CH_3CH{=}CH_2 + H_2O \xrightarrow{H^+} CH_3CH(OH)CH_3$$

$$CH_3CH{=}CH_2 + H_2 \xrightarrow[\text{Catalyst}]{} CH_3CH_2CH_3$$

$$CH_3CH{=}CH_2 + {:}CH_2 \longrightarrow CH_3CH\text{—}CH_2 \text{ (ring closed by } CH_2\text{)}$$

(:CH_2) Generated in the reaction mixture from CH_2N_2

Fig. 8.1. Addition reactions of alkenes, as represented by propene.

1,2-dichlorocyclohexane. In fact, we can rely on this kind of reaction as a general diagnostic test for the presence of a carbon–carbon multiple bond in a molecule. Thus, if treatment of an unknown compound with a dilute solution of bromine in carbon tetrachloride (pale brown-red color) causes the solution to turn colorless, it is a reasonable inference that the compound contains a carbon–carbon multiple bond to which the bromine in the solution has added. To furnish substance to predictions and extrapolations of this sort, however, it is useful that we have some knowledge of *how the reactions take place—i.e.*, by what *mechanisms* they occur. The focus in this chapter, therefore, will be not only on the reactions that various carbon frameworks undergo but also, in certain selected instances, on the mechanisms of these reactions as well. The real emphasis on reaction mechanisms, however, will be deferred to Part II, and the few examples discussed in Chapters 8–18 are intended simply to adumbrate an aspect of organic chemistry that will be treated later in much fuller detail.

PROBLEM 8-2

Three pure compounds, known to be *n*-decane, 1-decene, and 1,4,7-decatriene, are contained in three bottles labelled A, B, and C. Suggest a simple chemical means for determining which compound is in which bottle.

Electrophilic Addition Reactions of Alkenes. For a molecule of hydrogen bromide to add to the double bond of propene to give bromopropane, it is necessary that the H—Br bond cleave, and there are three reasonable ways in which this might occur, *viz.*, $H{:}Br \rightleftharpoons H\cdot + Br\cdot$ (homolytic cleavage), $H{:}Br \rightleftharpoons H^+ + {:}Br^-$ (heterolytic cleavage), or $H{:}Br \rightleftharpoons H{:}^- + Br^+$ (heterolytic cleavage). Depending on the reaction conditions, the addition of HBr to propene can occur either by the homolytic cleavage route or by one of the heterolytic cleavage routes, as the following discussion will illustrate.

A considerable amount of experimental evidence indicates that when pure samples (in particular, samples containing no peroxide or other radical-forming compounds) of propene and hydrogen bromide are mixed, a reaction occurs which can be viewed as being initiated by the dissociation of hydrogen bromide molecules into protons and bromine anions. Attachment of the proton (an electrophile) to one of the carbons of the double bond (a Lewis base or nucleophile) to form a C—H bond initiates the addition reaction by yielding a carbonium ion which, in a second step, interacts with the bromide ion to form bromopropane (Fig. 8.2). Because the reaction is initiated by the attachment of an electrophile, the process is called an *electrophilic addition reaction.*

Step 1: Formation of electrophile

$$HBr \rightleftharpoons H^+ + Br^-$$

Step 2: Formation of carbonium ion intermediate

$$CH_3CH{=}CH_2 + H^+ \longrightarrow CH_3\overset{+}{C}HCH_3$$

Step 3: Formation of product

$$CH_3\overset{+}{C}HCH_3 + Br^- \longrightarrow CH_3\underset{}{\overset{\overset{Br}{|}}{C}}HCH_3$$

Fig. 8.2. Mechanism of the electrophilic addition of bromine to propene.

In the mechanism depicted in Fig. 8.2, we have shown the initial attachment of the proton to propene to occur at C-1, resulting in the formation of isopropyl bromide. Alternatively, we could have shown the initial attachment to occur at C-2, resulting in the formation of *n*-propyl bromide. We must ask, then, which of these possibilities is more likely to take place—*i.e.*, will isopropyl bromide or *n*-propyl bromide be the major product? It turns out that it is isopropyl bromide, and this can be explained in terms of relative carbonium ion stabilities. Secondary carbonium ions, formed by attachment of H^+ to C-1, are more stable than primary carbonium ions, formed by attachment of H^+ to C-2 (see p. 189), and this results in the preferential attachment of the bromide ion to C-2 leading to the formation of isopropyl bromide (Fig. 8.3).

The addition of hydrogen bromide involves the attachment of an anionic nucleophile in the final step. It is not necessary that the nucleophile be an anion, however, as illustrated by the acid-catalyzed addition of water to alkenes.

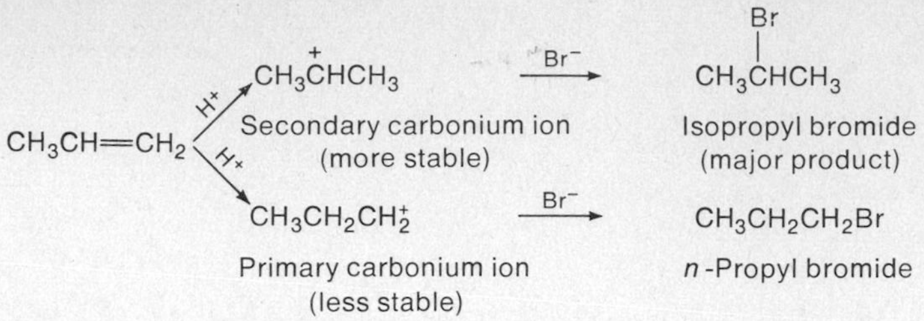

$$CH_3CH{=}CH_2 \xrightarrow{H^+} CH_3\overset{+}{C}HCH_3 \xrightarrow{Br^-} CH_3CH(Br)CH_3$$

Secondary carbonium ion (more stable) → Isopropyl bromide (major product)

$$CH_3CH{=}CH_2 \xrightarrow{H^+} CH_3CH_2CH_2^+ \xrightarrow{Br^-} CH_3CH_2CH_2Br$$

Primary carbonium ion (less stable) → *n*-Propyl bromide

Fig. 8.3. Preferential formation of isopropyl bromide in the addition of hydrogen bromide to propene.

Propene, for example, reacts with a dilute solution of sulfuric acid to form isopropyl alcohol in a process that involves (a) addition of a proton to propene to form the more stable secondary carbonium ion, (b) reaction of the carbonium ion with water to form a protonated form of the alcohol (called an oxonium ion), and (c) loss of a proton from the oxonium ion to yield isopropyl alcohol (Fig. 8.4). That the proton is simply serving a catalytic role in this process is apparent from the fact that although it adds to the propene in step (a) it is released in step (c).

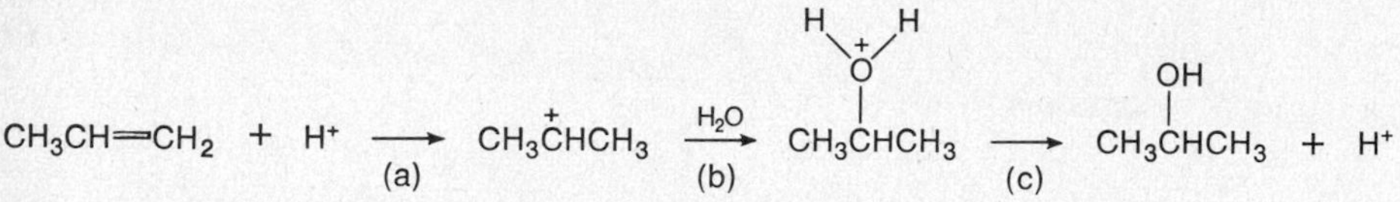

$$CH_3CH{=}CH_2 + H^+ \xrightarrow[(a)]{} CH_3\overset{+}{C}HCH_3 \xrightarrow[(b)]{H_2O} CH_3CH(\overset{+}{O}H_2)CH_3 \xrightarrow[(c)]{} CH_3CH(OH)CH_3 + H^+$$

Fig. 8.4. Mechanism of the acid-catalyzed addition of water to propene.

The fact that unsymmetrical reagents such as hydrogen bromide and water tend to add to alkenes to give the more highly branched rather than the less highly branched product was recognized many years ago by a Russian chemist Markovnikov, and the formation of isopropyl bromide and isopropyl alcohol from propene and hydrogen bromide and water, respectively, are examples of what has come to be known as "Markovnikov's rule." Based initially on empirical observations without recourse to mechanistic interpretation, the rule stated that the addition of a reagent HY to a double bond occurs in such a fashion that the H becomes attached to the carbon carrying the greater number of hydrogen atoms. Thus, in the case of propene, C-1 carries two hydrogen atoms and C-2 carries only one, leading to preferential attachment of the H of H—Br or H—OH to C-1 and the Y portion (*i.e.*, Br or OH) to C-2. With the advent of mechanistic interpretations of organic reactions, Markovnikov's rule can now be interpreted in terms of relative carbonium ion stabilities, as noted above.

Knowledge about the structural and mechanistic features of selective reactions often provides a way for deducing the structure of a product from the structures of the reactants. For example, knowing that hydrogen bromide adds to propene in Markovnikov fashion, we can predict with some certainty that hydrogen iodide should add to 2-methylpropene to yield *t*-butyl iodide:

$$(CH_3)_2C{=}CH_2 + HI \longrightarrow (CH_3)_3C{-}I$$

PROBLEM 8-3

Draw the structure of the major organic product from each of the following reactions: (a) cyclohexene $+HCl$; (b) 1-ethylcyclopentene $+H_2O$ containing a small amount of H_2SO_4; and (c) 2-phenylpropene $+Br_2$.

Free Radical Addition Reactions of Alkenes. Although the majority of addition processes involving alkenes occur via carbonium ion intermediates, a few follow a different pathway and involve radical intermediates. One of the most interesting of these is the addition of hydrogen bromide in the presence of peroxide reagents. In the absence of peroxides, hydrogen bromide adds to propene via the Markovnikov pathway to yield isopropyl bromide. In the presence of peroxides, however, the order of addition is reversed, and the product is *n*-propyl bromide; the addition in this case is said to be anti-Markovnikov (Fig. 8.5). This is interpreted in terms of initiation of the addition reaction by bromine atom, Br·, rather than by a proton, as is the case for electrophilic addition. The bromine atom, formed by the action of a free radical on hydrogen bromide (*e.g.*, RO· + HBr → ROH + Br·), adds to the double bond of propene at either C-1 to form a secondary radical

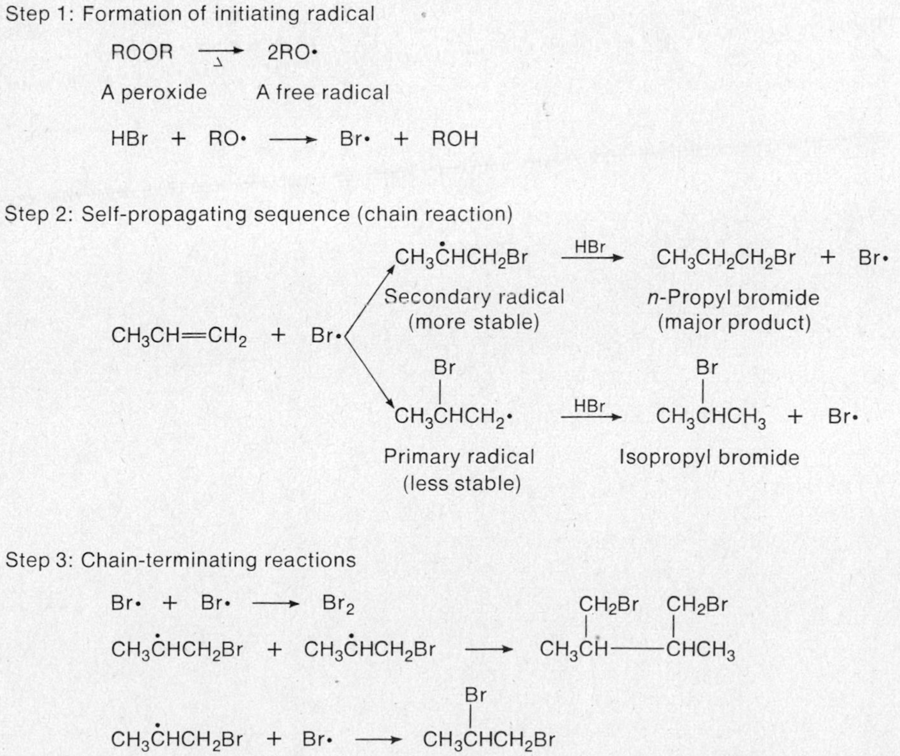

Step 1: Formation of initiating radical

$$\underset{\text{A peroxide}}{\text{ROOR}} \xrightarrow[\Delta]{} \underset{\text{A free radical}}{2\text{RO}\cdot}$$

$$\text{HBr} + \text{RO}\cdot \longrightarrow \text{Br}\cdot + \text{ROH}$$

Step 2: Self-propagating sequence (chain reaction)

$$CH_3CH{=}CH_2 + Br\cdot \longrightarrow \underset{\substack{\text{Secondary radical}\\ \text{(more stable)}}}{CH_3\dot{C}HCH_2Br} \xrightarrow{HBr} \underset{\substack{n\text{-Propyl bromide}\\ \text{(major product)}}}{CH_3CH_2CH_2Br} + Br\cdot$$

$$CH_3CH{=}CH_2 + Br\cdot \longrightarrow \underset{\substack{\text{Primary radical}\\ \text{(less stable)}}}{CH_3CH(Br)CH_2\cdot} \xrightarrow{HBr} \underset{\text{Isopropyl bromide}}{CH_3CH(Br)CH_3} + Br\cdot$$

Step 3: Chain-terminating reactions

$$Br\cdot + Br\cdot \longrightarrow Br_2$$

$$CH_3\dot{C}HCH_2Br + CH_3\dot{C}HCH_2Br \longrightarrow CH_3CH(CH_2Br)\text{—}CH(CH_2Br)CH_3$$

$$CH_3\dot{C}HCH_2Br + Br\cdot \longrightarrow CH_3CH(Br)CH_2Br$$

Fig. 8.5. Mechanism of the peroxide-induced (free radical) addition of hydrogen bromide to propene.

or C-2 to form a primary radical. Since the stabilities of radicals follow the same order as that of carbonium ions (see p. 191), the reaction occurs preferentially via the secondary radical, which subsequently collides with a molecule of hydrogen bromide to form *n*-propyl bromide and a bromine atom. Once this stage of the reaction sequence is reached, the process becomes self-propagating; every time a carbon radical abstracts a hydrogen from hydrogen bromide another bromine atom is formed, which adds to propene to form another carbon radical.*

PROBLEM 8-4

The initiator radical, RO·, reacts with hydrogen bromide to form ROH and Br· rather than ROBr and H·. Does this suggest that the bond energy of the O—Br bond is greater or less than 111 kcal/mole? Explain.

Such a sequence is called a *chain reaction*, and in theory it requires but a single initiating radical to convert all of the reactants to products. In actual practice, however, this does not occur, for various processes can break the chain. For example, the combination of two bromine atoms to form molecular bromine, the combination of two bromopropyl radicals to form a dibromohexane, or the combination of bromine atom and a bromopropyl radical to form 1,2-dibromopropane (Fig. 8.5) are all classed as *chain-terminating* processes. To the extent that they occur, new bromine atoms must be produced to keep the hydrogen bromide addition reaction in progress.

Addition of Hydrogen to Alkenes. Molecular hydrogen in the presence of a catalyst (see Section 7.5) such as finely divided platinum or palladium† adds to alkenes to yield the corresponding alkanes. Reactions of this sort, in which hydrogen is added to the system, are generally referred to as reduction reactions; propene (see Fig. 8.1) is said to be reduced to propane.

Addition of Carbenes to Alkenes. Carbenes (see Section 7.4), as neutral, diliganded carbon species containing only six electrons in the valence shell, can behave as Lewis acids, seeking to acquire a pair of electrons from the π

---

* Hydrogen bromide is unique among the halogen acids in possessing the ability to add to alkenes in anti-Markovnikov fashion via a free radical pathway. This can be explained in terms of the exothermic or endothermic character of the two steps involved in the chain sequence. Referring to Table 2.3, we can calculate that the addition step ($C{=}C + Br\cdot \rightarrow \dot{C}{-}C{-}Br$) is exothermic by 5 kcal/mole and that the subsequent step ($\dot{C}{-}C{-}Br + HBr \rightarrow H{-}C{-}C{-}Br + Br\cdot$) is exothermic by 12 kcal/mole. Thus, *both* steps are exothermic, so the reaction proceeds spontaneously. Similar calculations for hydrogen chloride show that although the addition step ($C{=}C + Cl\cdot \rightarrow \dot{C}{-}C{-}Cl$) is exothermic (by 18 kcal/mole), the subsequent step ($\dot{C}{-}C{-}Cl + HCl \rightarrow H{-}C{-}C{-}Cl + Cl\cdot$) is endothermic (by 4 kcal/mole). Thus, the second step in this case does not occur spontaneously but requires the input of energy, preventing a chain reaction from occurring.

† Metal catalysts, such as palladium and platinum, are often "supported" on various materials. Included among these "supports" are carbon (palladium on carbon is symbolized as Pd/C) and barium sulfate (palladium on barium sulfate is symbolized as $Pd/BaSO_4$).

system of an alkene. For example, propene reacts with carbene (generated *in situ**) by irradiation of diazomethane (CH_2N_2) to form methylcyclopropane (see Fig. 8.6).

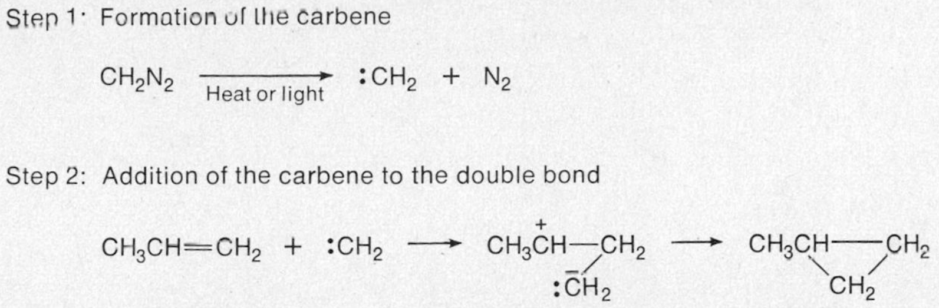

Step 1: Formation of the carbene

$$CH_2N_2 \xrightarrow[\text{Heat or light}]{} :CH_2 + N_2$$

Step 2: Addition of the carbene to the double bond

$$CH_3CH{=}CH_2 + :CH_2 \longrightarrow CH_3\overset{+}{C}H{-}CH_2(:\bar{C}H_2) \longrightarrow CH_3CH{-}CH_2{-}CH_2 \text{ (cyclopropane ring)}$$

Fig. 8.6. Formation of carbene and its addition to propene.

8.1b. Addition reactions of alkynes. The reactions of the C≡C bond generally resemble those of the C=C bond, although the second π bond in alkynes introduces new possibilities. The electrophilic addition of halogens and halogen acids to alkynes occurs in Markovnikov fashion to yield the corresponding vinyl halides as the initial products (Fig. 8.7). Since these compounds still contain a double bond, the possibility for the addition of a second equivalent of the

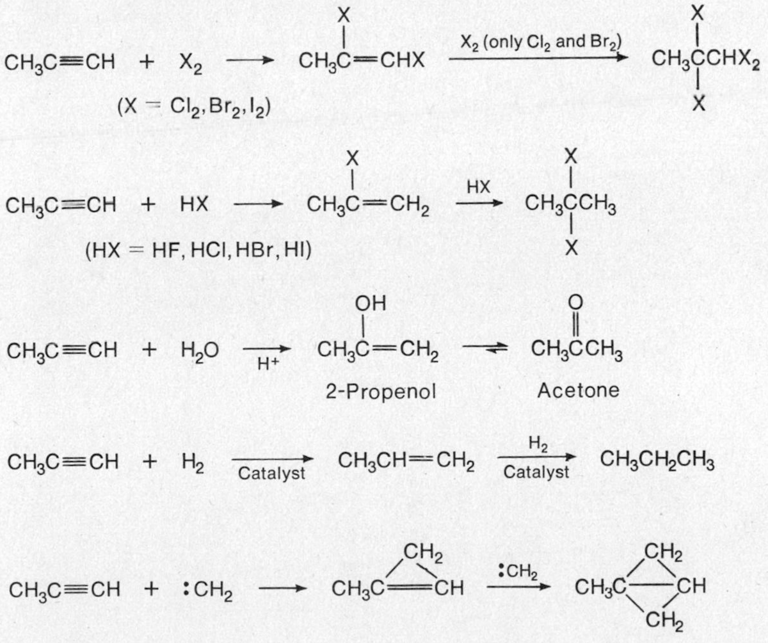

$$CH_3C{\equiv}CH + X_2 \longrightarrow CH_3CX{=}CHX \xrightarrow{X_2 \text{ (only } Cl_2 \text{ and } Br_2)} CH_3CX_2CHX_2$$

$(X = Cl_2, Br_2, I_2)$

$$CH_3C{\equiv}CH + HX \longrightarrow CH_3CX{=}CH_2 \xrightarrow{HX} CH_3CX_2CH_3$$

$(HX = HF, HCl, HBr, HI)$

$$CH_3C{\equiv}CH + H_2O \xrightarrow[H^+]{} \underset{\text{2-Propenol}}{CH_3C(OH){=}CH_2} \rightleftharpoons \underset{\text{Acetone}}{CH_3\overset{O}{\overset{\|}{C}}CH_3}$$

$$CH_3C{\equiv}CH + H_2 \xrightarrow[\text{Catalyst}]{} CH_3CH{=}CH_2 \xrightarrow[\text{Catalyst}]{H_2} CH_3CH_2CH_3$$

$$CH_3C{\equiv}CH + :CH_2 \longrightarrow CH_3C{=}CH \text{ (bridged by } CH_2) \xrightarrow{:CH_2} CH_3C{-}CH \text{ (bridged by two } CH_2)$$

Fig. 8.7. Addition reactions of alkynes, as represented by propyne.

* Compounds formed under conditions where they immediately react with something else in the mixture are said to be generated *in situ* (*Latin*, in its place).

addend exists. The second step is often enough slower than the first, however, that the reaction can be terminated at the alkene stage. The acid-catalyzed addition of water to alkynes deserves special comment, because the outcome is different from what would be expected by analogy with the comparable alkene reaction. Thus, treatment of propyne with water (in the presence of mercuric ion, which is a particularly effective catalyst) yields 2-propenol, which immediately rearranges to the more stable compound, acetone. In similar fashion, acetylene adds water to form the compound acetaldehyde ($HC{\equiv}CH + H_2O \rightarrow H_2C{=}CHOH \rightarrow CH_3CHO$), again the result of isomerization of the initially formed vinyl alcohol (considered in more detail in Chapter 10).

PROBLEM 8-5

Knowing that the C—Cl bond is polarized in the direction $\overset{\delta^+}{C}{-}\overset{\delta^-}{Cl}$, explain why the addition of a second equivalent of Cl_2 to $CH_3C{\equiv}CCH_3$ is slower than the addition of the first equivalent.

Like the alkenes, the alkynes (a) undergo peroxide-induced addition of hydrogen bromide in an anti-Markovnikov fashion, (b) add hydrogen in the presence of catalysts to form alkenes or alkanes, depending on how much hydrogen is supplied, and (c) react with carbene to form cyclopropenes or bicyclobutanes, depending on how much carbene is supplied (Fig. 8.7).

8.1c. ADDITION REACTIONS OF ARENES. Benzene and other aromatic compounds are far less prone to undergo addition reactions than are alkenes and alkynes, for their special stability arising from aromaticity is lost in the addition process. Nevertheless, there are some examples of aromatic addition reactions, the most useful one being the catalyzed addition of hydrogen. For example, benzene in the presence of molecular hydrogen and finely divided platinum yields cyclohexane, and naphthalene under comparable conditions yields tetrahydronaphthalene (tetralin) if two equivalents of hydrogen are supplied or decahydronaphthalene (decalin) if five equivalents of hydrogen are supplied (Fig. 8.8).

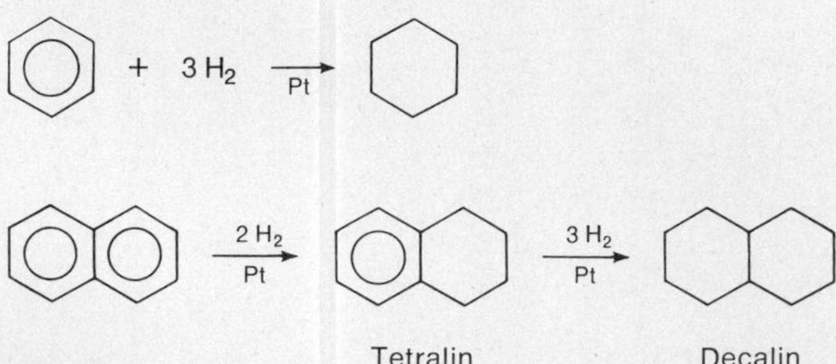

Fig. 8.8. Addition of hydrogen to aromatic compounds.

8.2a. SUBSTITUTION REACTIONS OF ALKANES. The most characteristic reaction of alkanes, which are much less reactive compounds than alkenes, alkynes, or arenes, is a substitution process in which a hydrogen is replaced by some other atom or group; $R-H + Y\cdot \rightarrow R-Y + H\cdot$. Among the several examples of reactions of this type, the generally most useful ones involve chlorine or bromine. For example, when a mixture of *n*-butane and chlorine is heated or irradiated, a reaction takes place to produce a mixture of 1-chlorobutane and 2-chlorobutane (Fig. 8.9). A variety of pieces of evidence indicate that this reaction commences

$$CH_3CH_2CH_2CH_3 + X_2 \longrightarrow CH_3CH_2CH_2CH_2X + (CH_3CH_2)(CH_3)CHX$$

n-Butane		28% (X = Cl)	72% (X = Cl)
		2% (X = Br)	98% (X = Br)

$$(CH_3)_2CHCH_3 + X_2 \longrightarrow (CH_3)_2CHCH_2X + (CH_3)_3CX$$

Isobutane		64% (X = Cl)	36% (X = Cl)
		1% (X = Br)	99% (X = Br)

Fig. 8.9. Halogenation of the butanes.

with the dissociation of molecular chlorine to chlorine atoms ($Cl_2 \rightleftharpoons Cl\cdot + Cl\cdot$), and that a chlorine atom then collides with a molecule of *n*-butane to abstract a hydrogen atom to form hydrogen chloride and a butyl radical. If the collision occurs at the C—H bond of the methyl group, a primary radical is produced ($CH_3CH_2CH_2\dot{C}H_2$); if it occurs at the C—H bond of the methylene carbon, a secondary radical is produced ($CH_3CH_2\dot{C}HCH_3$). Subsequent collisions of these radicals with molecular chlorine then yield 1-chlorobutane and 2-chlorobutane, respectively, along with chlorine atoms (Fig. 8.10). At this stage of the reaction sequence, we have the same situation that we encountered in the peroxide-induced addition of hydrogen bromide to alkenes, *viz.*, a self-propagating or chain reaction. As in that case, here also the chain does not continue indefinitely, for various chain-terminating reactions occur, requiring new chlorine atoms to be constantly supplied to keep the reaction going. However, as many as 10^4 cycles of the self-propagating steps may take place before the chain is broken.

The chlorination of *n*-butane produces 2-chlorobutane in greater amount than 1-chlorobutane, even though there are six methyl hydrogens and only four methylene hydrogens in the starting material. This can be interpreted in terms of either comparative C—H bond strengths (97 kcal/mole for the methyl C—H bonds *vs.* 94 kcal/mole for the methylene C—H bonds) or comparative radical stabilities (*i.e.*, $R_2CH\cdot$ is more stable than $RCH_2\cdot$; see p. 191). The ratio of secondary to primary alkyl halide in the halogenation of *n*-butane becomes even greater if we use bromine rather than chlorine as the halogenating agent. To understand the reasons for this, we should consider some of the thermodynamics of the process. In the first step of the self-propagating sequence, a C—H bond is broken and an H—X bond is formed. From the table of bond energies (see

Step 1: Formation of initiating radical

$$Cl_2 \xrightarrow[\text{Heat or light}]{} Cl\cdot + Cl\cdot$$

Step 2: Self-propagating sequence (chain reaction)

$$Cl\cdot + CH_3CH_2CH_2CH_3 \longrightarrow CH_3CH_2CH_2\dot{C}H_2 \text{ (Primary radical)} \xrightarrow{Cl_2} CH_3CH_2CH_2CH_2Cl + Cl\cdot$$

$$Cl\cdot + CH_3CH_2CH_2CH_3 \longrightarrow CH_3CH_2\dot{C}HCH_3 \text{ (Secondary radical)} \xrightarrow{Cl_2} CH_3CH_2CH(Cl)CH_3 + Cl\cdot$$

Step 3: Chain-terminating reactions

$$CH_3CH_2CH_2\dot{C}H_2 + CH_3CH_2CH_2\dot{C}H_2 \longrightarrow CH_3(CH_2)_6CH_3$$

$$Cl\cdot + Cl\cdot \longrightarrow Cl_2$$

$$CH_3CH_2CH_2\dot{C}H_2 + Cl\cdot \longrightarrow CH_3CH_2CH_2CH_2Cl$$

$$2\,CH_3CH_2CH_2\dot{C}H_2 \longrightarrow CH_3CH_2CH_2CH_3 + CH_3CH_2CH{=}CH_2$$

Fig. 8.10. Mechanism of the free radical halogenation of alkanes.

Table 2.3) we can calculate that this step is exothermic when chlorine atom is involved but endothermic when bromine atom is involved. In other words, the chlorine atom is a much more reactive species than bromine atom with respect to the abstraction of hydrogen from *n*-butane. It is often observed in chemical reactions that the greater the reactivity of a reagent (or an intermediate) the less is the discrimination that occurs between alternative sites for reaction. Thus, chlorine atom is much less discriminating between the methyl and methylene hydrogens than the bromine atom, and this is reflected in the product ratios of primary and secondary halides that are formed.

The halogenation of isobutane proceeds in a fashion similar to that of *n*-butane to give a mixture of isobutyl halide and *t*-butyl halide. Since $R_3C—H$ bonds are considerably weaker than $RCH_2—H$ bonds, we expect that attack by halogen will take place preferentially at the tertiary position. The production of twice as much isobutyl chloride as *t*-butyl chloride in the chlorination reaction, therefore, might appear to contradict this prediction. But, the contradiction is apparent rather than real; when a statistical correction is applied to take into account the fact that there are *nine* methyl hydrogens for every $R_3C—H$ hydrogen, the reaction is shown to favor substitution at the tertiary position by a factor of 5.06.* In the bromination of isobutane it is apparent, even without applying the statistical correction, that there is an overwhelming preference for bromination at the tertiary position.

Carbon–hydrogen bonds even weaker than those in *n*-butane and isobutane are encountered in compounds containing benzylic C—H (*i.e.*, $C_6H_5\overset{|}{\underset{|}{C}}—H$) and allylic C—H bonds (*i.e.*, $\gt C{=}\underset{|}{C}—\overset{|}{\underset{|}{C}}—H$), and compounds such as toluene and propene are particularly susceptible to free radical halogenation. A reagent that is very frequently used to carry out such reactions is N-bromosuccinimide (abbreviated NBS), which acts as a convenient source of bromine atoms (Fig. 8.11). N-Bromosuccinimide is quite specific in its ability to replace benzylic and allylic hydrogens by bromine in a process known as the *Wohl–Ziegler reaction.*

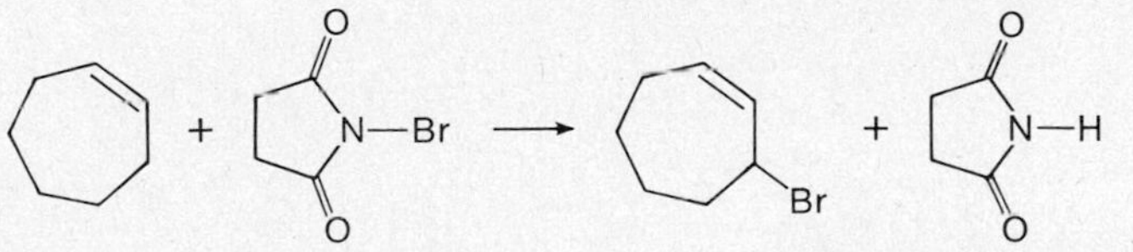

Fig. 8.11. Allylic bromination of cycloheptene with N-bromosuccinimide.

PROBLEM 8-6

A mixture containing 1 mole each of the following compounds was treated with 1 mole of bromine under conditions conducive to free radical halogenation. Arrange the compounds in the order of decreasing amount of monobromo product that you would expect to find in the bromination mixture.

$H_2C{=}CHCH_2CH_2CH_3$ $(CH_3)_2C{=}CHCH_3$ $\square$—CH_3 (methylcyclobutane) cyclopentane

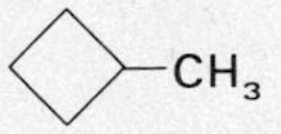

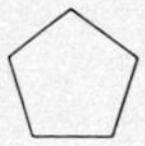

8.2b. SUBSTITUTION REACTIONS OF ALKENES. The carbon–carbon double bonds of alkenes are so much more vulnerable to attack than the carbon–hydrogen bonds that addition rather than substitution takes place in the vast majority of cases. Consequently, there are no substitution reactions occurring at sp^2-hybridized carbon atoms of alkenes that merit consideration at this point in the discussion.

8.2c. SUBSTITUTION REACTIONS OF ALKYNES. Extrapolating from the statement in Section 8.2b concerning addition *vs.* substitution reactions of alkenes, we might predict that alkynes would also fail to undergo substitution reactions. In fact, bond dissociation energies and bond lengths indicate that the C—H

* Multiplying the observed ratio of $\frac{64}{36}$ for primary *vs.* tertiary product by the statistical ratio of $\frac{1}{9}$ gives the statistically corrected product ratio of 0.198 or, reciprocally, 5.06.

bonds of 1-alkynes (RC≡CH) are shorter (about 1.20 Å) and stronger (about 121 kcal/mole) than the C—H bonds in alkenes (about 1.33 Å and 108–112 kcal/mole, respectively). Surprisingly, however, 1-alkynes are markedly more acidic (pK_a = 20–26) than alkenes (pK_a about 37) and alkanes (pK_a about 40–45) (see Table 7.1), and they undergo substitution reactions with metals such as sodium or with very strong bases such as potassium amide (KNH_2) to form metal acetylides;

$$RC{\equiv}CH + Na \longrightarrow RC{\equiv}CNa + H_2$$

$$RC{\equiv}CH + KNH_2 \longrightarrow RC{\equiv}CK + NH_3$$

The marked difference in acidity between 1-alkynes and alkenes or alkanes is ascribed to the different states of hybridization of their carbon atoms. The hybrid orbitals of the carbon of a triple bond possess more *s* character (50%) than the orbitals of carbons in alkenes (sp^2 has 33% *s*-character) and alkanes (sp^3 has 25% *s*-character). The greater the *s*-character of the hybrid orbital, the smaller is the orbital and the lower is energy, so electrons in an *sp*-hybrid orbital correspond to a more stable situation than electrons in sp^2- or sp^3-hybrid orbitals. This results in a decrease in the stability of carbanions in the following order: $HC{\equiv}\bar{C} > H_2C{=}\bar{C}H > H_3C\bar{C}H_2$.*

The carbanions produced by the removal of the acidic proton from 1-alkynes are highly reactive, triliganded species that seek to attach a fourth ligand. Thus, sodium and potassium acetylides react as nucleophilic reagents, acquiring protons from acids with pK_a values lower than that of acetylene ($HC{\equiv}CNa + H_2O \rightarrow HC{\equiv}CH + NaOH$) and engaging in other reactions typical of nucleophiles (which will be discussed in later chapters). An important example of the protonation reaction is the commercial process in which acetylene is produced by hydrolysis of calcium carbide, formed by the reaction of calcium oxide and carbon (Fig. 8.12). The conversion, in turn, of acetylene to a host of other organic compounds provides dramatic and explicit demonstration of man's ability to convert elemental carbon into organic materials.

The ≡C—H hydrogen of acetylenes can be replaced not only by metals above hydrogen in the emf series (*e.g.*, Na and K) but also by some metals below

$$3C + CaO \longrightarrow \underset{\text{Calcium carbide}}{CaC_2} + CO$$

$$CaC_2 + H_2O \longrightarrow \underset{\text{Acetylene}}{HC{\equiv}CH} + Ca(OH)_2$$

Fig. 8.12. Conversion of carbon to acetylene.

* The thermodynamic acidity of acetylene is based on the relative stabilities of the reactants and products and not on the rate at which the proton is removed and equilibrium attained. The very high C—H bond strength of the ≡C—H bond results in a rather slow attainment of equilibrium, as compared with acids of similar pK_a but lower X—H bond strength (*cf.* discussion of kinetic acidity in the footnote on p. 182).

hydrogen. Ammoniacal silver nitrate and ammoniacal cuprous chloride both react with compounds of the general structure $RC{\equiv}CH$ to form silver and copper acetylides, respectively. These substances are water-insoluble solids, so their formation is easily noted by the appearance of a precipitate. Thus, if the chemist is confronted with the problem of deciding, for example, whether structure $CH_3C{\equiv}CCH_3$ or $CH_3CH_2C{\equiv}CH$ correctly represents the compound in hand, the decision is readily made on the basis of the ammoniacal silver nitrate or cuprous chloride test: the second compound forms a water-insoluble metal acetylide, whereas the first compound does not. That is,

$$CH_3CH_2C{\equiv}CH + Ag(NH_3)_2NO_3 \longrightarrow CH_3CH_2C{\equiv}CAg + NH_4NO_3 + NH_3$$

$$CH_3CH_2C{\equiv}CH + Cu(NH_3)_2Cl \longrightarrow CH_3CH_2C{\equiv}CCu + NH_4Cl + NH_3$$

$$CH_3C{\equiv}CCH_3 + Ag(NH_3)_2NO_3 \text{ or } Cu(NH_3)_2Cl \longrightarrow \text{No reaction}$$

PROBLEM 8-7

Three pure compounds, known to be bicyclo[2.2.1]heptane, cycloheptene, and 1-heptyne, are contained in three bottles labeled A, B, and C. Suggest simple chemical tests for determining which compound is in which bottle.

8.2d. SUBSTITUTION REACTIONS OF ARENES. The most characteristic and generally most useful reactions of aromatic compounds are those in which an Ar—H bond is replaced by an Ar—Y bond in substitution processes such as those illustrated in Fig. 8.13. Each of these reactions follows the same mechanistic pathway, being initiated by the addition of an electrophilic species to one of the carbon atoms of the aromatic ring. In this respect the reactions resemble the electrophilic additions of alkenes and alkynes which are initiated in a similar fashion by the addition of an electrophile. In the subsequent step of the sequence, however, a difference occurs; whereas the cations obtained from alkenes and alkynes generally react with a nucleophile to form an addition product (see Figs. 8.2–8.4), the cations obtained by the addition of an electrophile to an aromatic system release a proton, thereby regenerating the aromatic system and gaining back the energy lost in forming the intermediate, nonaromatic carbonium ion (Fig. 8.14).

The particular kind of compound obtained in an electrophilic substitution reaction of benzene is dependent, of course, on the identity of the electrophile, E^+. For the five reactions illustrated in Fig. 8.13, the electrophiles are generated *in situ* in the following fashions:

Halogenation:

$$X_2 + FeX_3 \longrightarrow X^+ + FeX_4^-$$

Nitration:

$$HONO_2 + H_2SO_4 \longrightarrow HSO_4^- + H_2\overset{+}{O}{-}NO_2 \longrightarrow NO_2^+ + H_2O$$

Sulfonation:

$$HOSO_3H + H_2SO_4 \longrightarrow HSO_4^- + H_2\overset{+}{O}SO_3H \longrightarrow SO_3H^+ + H_2O$$

Friedel–Crafts alkylation:

$$RX + AlCl_3 \longrightarrow R^+ + AlCl_3X^-$$

Friedel–Crafts acylation:

$$RCOCl + AlCl_3 \longrightarrow RCO^+ + AlCl_4^-$$

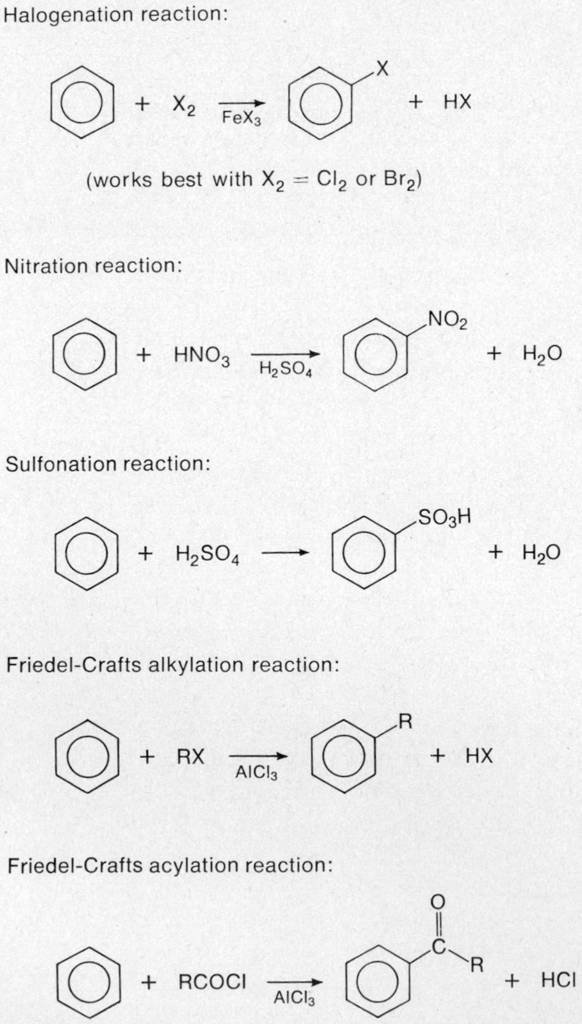

Halogenation reaction:

$$C_6H_6 + X_2 \xrightarrow[FeX_3]{} C_6H_5X + HX$$

(works best with $X_2 = Cl_2$ or Br_2)

Nitration reaction:

$$C_6H_6 + HNO_3 \xrightarrow[H_2SO_4]{} C_6H_5NO_2 + H_2O$$

Sulfonation reaction:

$$C_6H_6 + H_2SO_4 \longrightarrow C_6H_5SO_3H + H_2O$$

Friedel-Crafts alkylation reaction:

$$C_6H_6 + RX \xrightarrow[AlCl_3]{} C_6H_5R + HX$$

Friedel-Crafts acylation reaction:

$$C_6H_6 + RCOCl \xrightarrow[AlCl_3]{} C_6H_5C(=O)R + HCl$$

Fig. 8.13. Electrophilic substitution reactions of benzene.

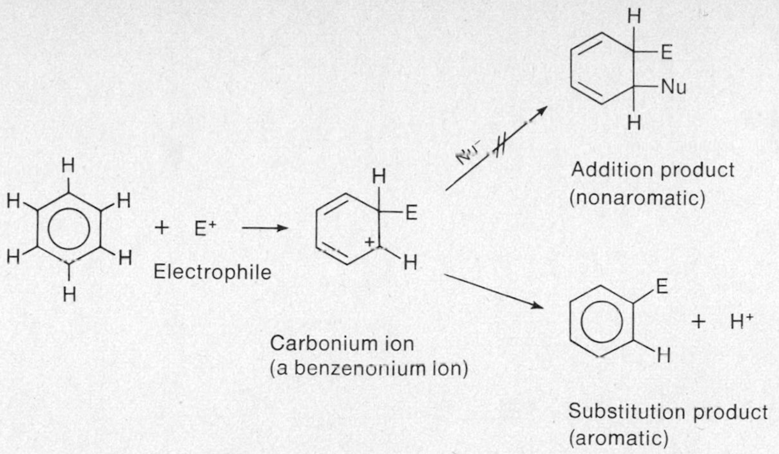

Fig. 8.14. Mechanism of electrophilic aromatic substitution reactions.

Thus, the key steps in the electrophilic substitution reactions of arenes are (a) formation of an electrophilic species, (b) interaction of the electrophile with the π-system of the arene to form a carbonium ion, and (c) expulsion of a proton to regenerate the aromatic ring. Reactions of this sort are exceedingly useful for the synthesis of a variety of substituted arenes and will be used frequently in later discussions.

PROBLEM 8-8

Benzene and cyclohexene both react with a dilute solution of bromine in carbon tetrachloride. Write balanced equations for both of these reactions in which bromine is consumed and the original solution loses its color. How can the reaction with bromine be used to distinguish between benzene and cyclohexene?

8.3 *Elimination Reactions*

Addition reactions, as we have seen in Section 8.1, proceed from a less saturated reactant to a more saturated product. Elimination reactions represent the reciprocal case and proceed from a more saturated reactant to a less saturated product. We have already alluded to one type of elimination reaction at the beginning of this chapter where we discussed the cracking reaction, in which carbon–hydrogen and carbon–carbon bonds are broken to yield, ultimately, atomic carbon and hydrogen. A comparable reaction more generally useful for laboratory synthesis involves the *partial* removal of hydrogen from six-membered alicyclic* compounds to yield aromatic compounds. For example, when cyclohexane is treated at an elevated temperature with platinum, palladium, sulfur, or selenium, hydrogen is removed and benzene is formed (Fig. 8.15). In similar fashion, tetralin and decalin can be converted to naphthalene.

* The term *alicyclic* is a contraction of *aliphatic* and *cyclic*. It refers to all carbocyclic (*i.e.*, containing only carbon in the ring) compounds that are not aromatic. Thus, cycloalkanes, cycloalkenes, cycloalkynes, cycloalkadienes, etc., are all included in the alicyclic category.

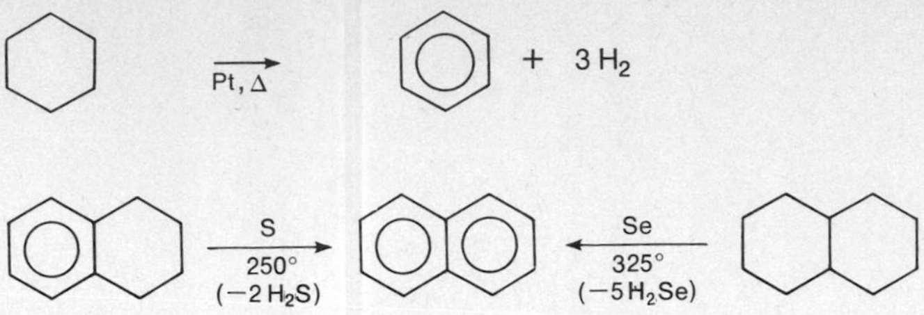

Pt, Δ
+ $3 H_2$
S 250° $(-2 H_2S)$
Se 325° $(-5 H_2Se)$

Fig. 8.15. Dehydrogenation of benzene, tetralin, and decalin.

The process of removing hydrogen from six-membered alicyclic compounds is called dehydrogenation and finds useful application in the determination of the structure of alicyclic compounds. Aromatic compounds are generally crystallizable and less stereochemically complicated than their alicyclic precursors; also, they are more likely to be known compounds or, if unknown, to be accessible by easy synthesis. From the structure of the dehydrogenated product, then, the structural details of the alicyclic compound can be inferred. Caution must be exercised in using such information, however, because rearrangements and loss of functional groups sometimes occur during dehydrogenation (see Fig. 34.10 for a typical example).

8.4 Oxidation Reactions

For the present discussion, we shall define oxidation reactions as processes in which C—H or C—C bonds are converted to C—O bonds or in which carbon–carbon multiple bonds are converted to C=O bonds.

8.4a. OXIDATION REACTIONS OF ALKANES. Alkanes are generally resistant to oxidation and can withstand the action even of strong oxidizing mixtures such as potassium permanganate in aqueous sodium hydroxide or sodium dichromate in sulfuric acid. If the alkane contains a tertiary C—H bond (*i.e.*, R_3C—H) or a benzylic or allylic C—H bond (*i.e.*, ArC—H or C=C—C—H), however, oxidation can proceed with some facility, and under appropriate conditions, in a

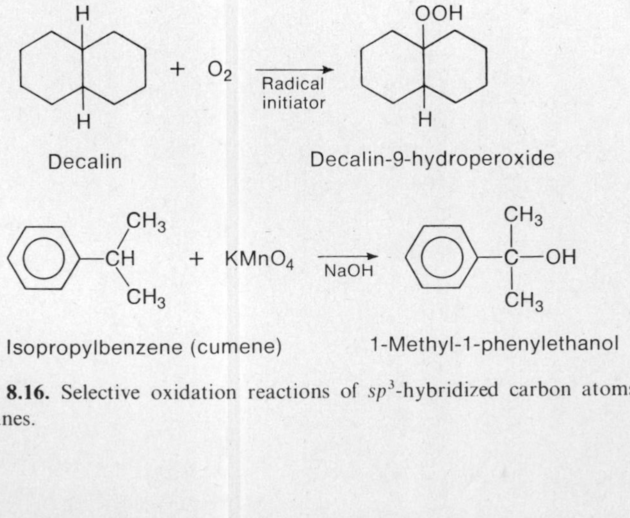

H, H + O_2 → (Radical initiator) OOH, H

Decalin — Decalin-9-hydroperoxide

CH_3, CH, CH_3 + $KMnO_4$ → (NaOH) CH_3, C—OH, CH_3

Isopropylbenzene (cumene) — 1-Methyl-1-phenylethanol

Fig. 8.16. Selective oxidation reactions of sp^3-hybridized carbon atoms in alkanes.

selective fashion. For example, decalin reacts with oxygen in the presence of free radical initiators to yield a compound in which oxygen has been introduced onto one of the carbons common to the two rings, and isopropylbenzene (cumene) reacts with potassium permanganate under very mild conditions to yield an alcohol (Fig. 8.16). Under more strenuous oxidizing conditions, isopropylbenzene is more extensively oxidized and yields benzoic acid, illustrating a type of reaction that is characteristic of all alkylbenzenes of the general structure $C_6H_5CH_2R$ and ArCHRR′ (Fig. 8.17). The resistance of the benzene ring to oxidation is a most useful tool in structure elucidation; the positions of the carboxylic acid groups (*i.e.*, the $-CO_2H$ groups) can be determined by methods discussed in Chapter 14, and this can reveal the points of attachment of the alkyl groups in the original molecule.

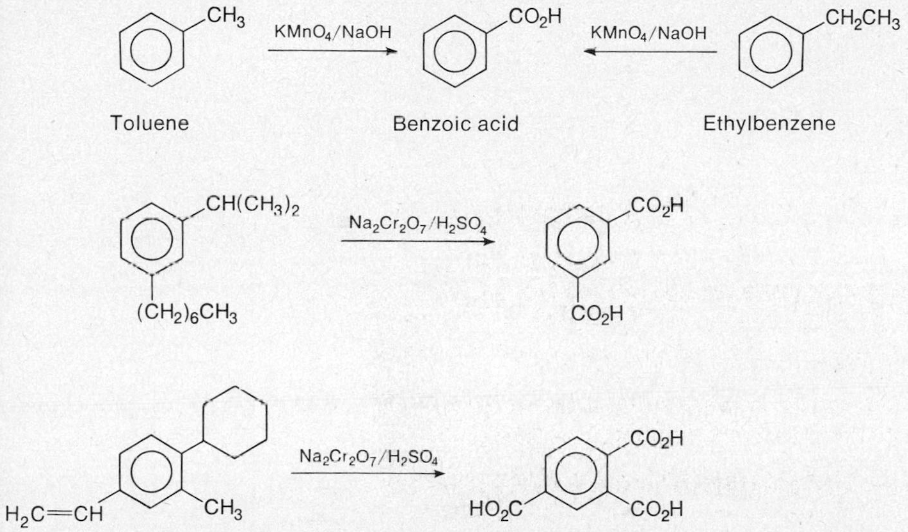

CH_3 — $KMnO_4/NaOH$ → CO_2H ← $KMnO_4/NaOH$ — CH_2CH_3

Toluene — Benzoic acid — Ethylbenzene

$CH(CH_3)_2$, $(CH_2)_6CH_3$ — $Na_2Cr_2O_7/H_2SO_4$ → CO_2H, CO_2H

$H_2C{=}CH$, CH_3 — $Na_2Cr_2O_7/H_2SO_4$ → CO_2H, HO_2C, CO_2H

Fig. 8.17. Oxidations of substituted benzenes.

8.4b. OXIDATION REACTIONS OF ALKENES AND ALKYNES. Alkenes and alkynes are far less resistant to oxidation than alkanes and undergo cleavage of the C=C or C≡C bond in the presence of a variety of oxidizing agents (Fig. 8.18). A particularly useful oxidizing agent is ozone, which converts alkenes of structure $RCH{=}CH_2$ or $RCH{=}CHR'$ to aldehydes, alkenes of structure $R_2C{=}CR'_2$ to ketones, and alkenes of structure $R_2C{=}CHR'$ to mixtures of an aldehyde and ketone. Alkynes also react with ozone, yielding carboxylic acids, but the reaction is slower than with alkenes, and may yield mixtures containing other compounds in addition to or instead of carboxylic acids. Ozonolysis provides a most useful way for locating the position of a C=C bond in a compound. For example, suppose that we have a sample of a material that we know to be either compound (a) or compound (b) in Fig. 8.19. If it is compound (a) both products of ozonolysis are ketones; if it is compound (b), one of the products of ozonolysis is a ketone and the other is an aldehyde. Spectral methods for distinguishing between aldehydes and ketones have already been discussed (see, *e.g.*, Table 6.3), and chemical methods for making this distinction are discussed in Chapter 13.

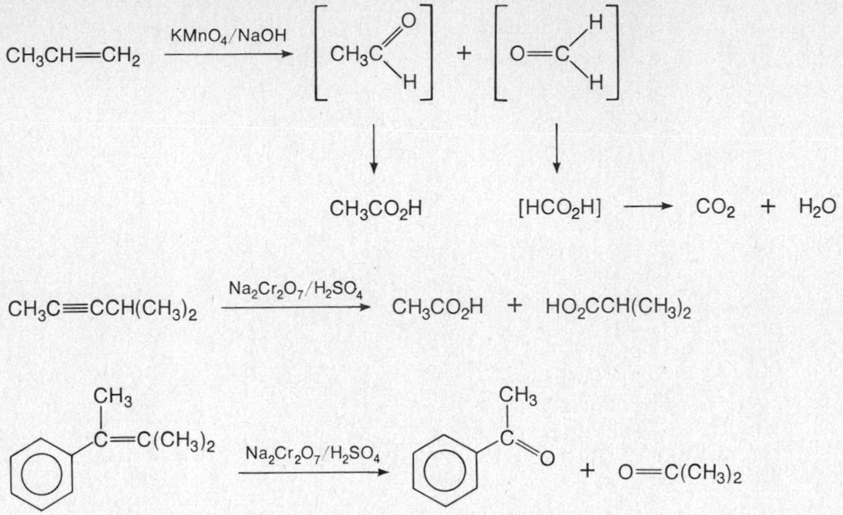

$$CH_3CH{=}CH_2 \xrightarrow{KMnO_4/NaOH} \left[CH_3C(=O)H\right] + \left[O{=}CH_2\right]$$

$$\left[CH_3C(=O)H\right] \longrightarrow CH_3CO_2H \qquad \left[O{=}CH_2\right] \longrightarrow [HCO_2H] \longrightarrow CO_2 + H_2O$$

$$CH_3C{\equiv}CCH(CH_3)_2 \xrightarrow{Na_2Cr_2O_7/H_2SO_4} CH_3CO_2H + HO_2CCH(CH_3)_2$$

$$C_6H_5C(CH_3){=}C(CH_3)_2 \xrightarrow{Na_2Cr_2O_7/H_2SO_4} C_6H_5C(CH_3){=}O + O{=}C(CH_3)_2$$

Fig. 8.18. Oxidation reactions of alkenes and alkynes.

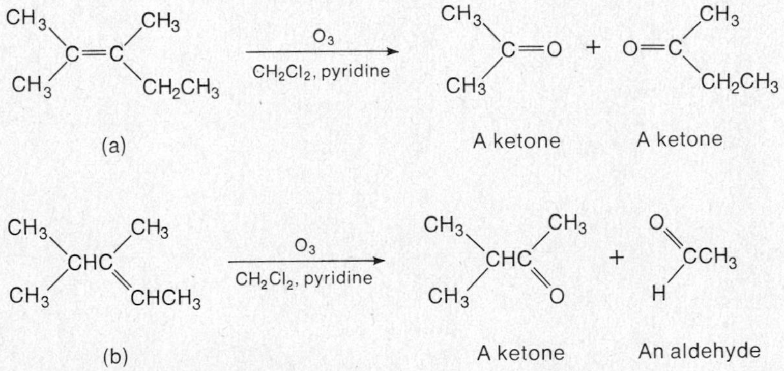

(a)

$$(CH_3)_2C{=}C(CH_3)CH_2CH_3 \xrightarrow[CH_2Cl_2,\ pyridine]{O_3} (CH_3)_2C{=}O + O{=}C(CH_3)CH_2CH_3$$

A ketone — A ketone

(b)

$$(CH_3)_2CHC(CH_3){=}CHCH_3 \xrightarrow[CH_2Cl_2,\ pyridine]{O_3} (CH_3)_2CHC(CH_3){=}O + O{=}C(H)CH_3$$

A ketone — An aldehyde

Fig. 8.19. Ozonolysis of isomeric dimethylpentenes.

PROBLEM 8-9

Compounds A, B, and C all have the molecular formula C_6H_8. Compound A yields three different products on ozonolysis, compound B yields two different products, and compound C yields a single product. Suggest structures for compounds A, B, and C on the basis of these data.

PROBLEM 8-10

Compounds A and B both have the molecular formula C_6H_6 and both undergo ozonolysis to yield a pair of products. Compound A forms a precipitate when treated with ammoniacal silver nitrate, and compound B does not. On the basis of this information, suggest structures for compounds A and B.

8.4c. OXIDATION REACTIONS OF ARENES. The conversion of alkylbenzenes to benzoic acids (see Fig. 8.17) suggests that the benzene ring is more resistant to oxidation than the alkyl side chains, which is indeed true when permanganate or chromate reagents are used as oxidants and when the compound contains at least one benzylic C—H bond. If the alkylbenzene contains no benzylic C—H bonds, the side chain is then far more resistant to oxidative cleavage, and benzoic acids are not obtained. For example, *t*-butylbenzene undergoes chromate oxidation to yield trimethylacetic acid as the initially formed product, which rapidly oxidizes further to yield acetic acid as the major product:

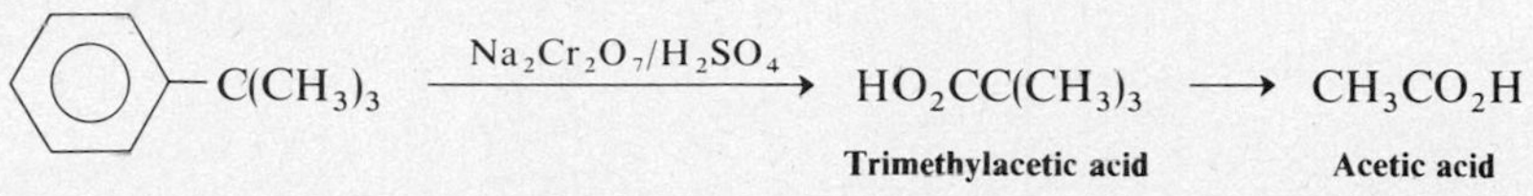

$$C_6H_5{-}C(CH_3)_3 \xrightarrow{Na_2Cr_2O_7/H_2SO_4} \underset{\textbf{Trimethylacetic acid}}{HO_2CC(CH_3)_3} \longrightarrow \underset{\textbf{Acetic acid}}{CH_3CO_2H}$$

If ozone rather than permanganate or chromate reagents is used as the oxidant, the relative reactivities of the benzene ring and the alkyl side chain are reversed, the benzene ring becoming the portion of the molecule more vulnerable to attack and yielding a carboxyl group (Fig. 8.20). The ozonolysis of a benzene ring requires a longer reaction time than is necessary for the oxidation of C=C and C≡C bonds, making it possible to cleave the latter with essentially no effect on the ring in alkyl-, alkenyl-, and alkynylbenzenes (Fig. 8.20).

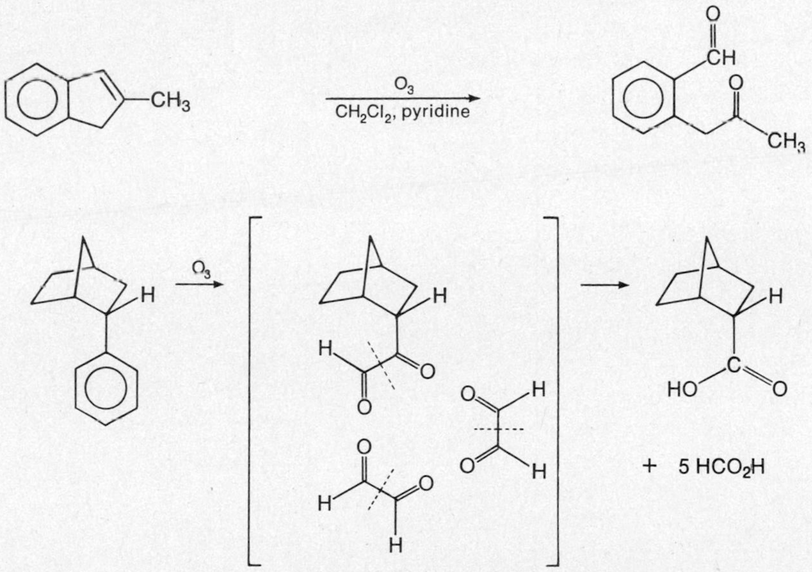

CH₃ — O_3 / CH_2Cl_2, pyridine → CH, O, CH₃

H — O_3 → [H, H, O, O, O, H, O, H, O, H, O, H] → H, HO, C, O + 5 HCO_2H

Fig. 8.20. Ozonolysis of alkenes and arenes.

Although benzene itself and alkyl-, alkenyl-, and alkynylbenzenes are resistant to ring oxidation, particularly in the presence of permanganate and chromate reagents, the presence of hydroxyl groups (—OH) and amino groups (—NH_2) on the ring greatly increases its susceptibility to oxidation. For example, *p*-hydroxytoluene, does not yield *p*-hydroxybenzoic acid when treated with strong oxidizing agents but yields mixtures of compounds in which the benzene ring has been destroyed. The initial point of oxidative attack in these cases is probably the O—H and N—H bonds, for *p*-methoxytoluene behaves in a normal fashion and yields *p*-methoxybenzoic acid on oxidation:

$$CH_3O-C_6H_4-CH_3 \xrightarrow{KMnO_4/NaOH} CH_3O-C_6H_4-CO_2H$$

p-Methoxytoluene　　　　*p*-Methoxybenzoic acid

PROBLEM 8-11

Suggest structures compatible with the following data:

(a) A C_8H_6 compound that forms a precipitate with ammoniacal cuprous chloride and yields benzoic acid when oxidized.

(b) A C_9H_{12} compound that undergoes electrophilic substitution with bromine to yield a $C_9H_{11}Br$ compound which has a two-hydrogen singlet, a three-hydrogen singlet, and a six-hydrogen singlet in its nmr spectrum.

(c) A $C_{10}H_{10}$ compound that yields a single product when ozonolyzed and undergoes dehydrogenation to naphthalene.

8.5 Comparison of Reactivities of Various Carbon Frameworks

Although it is dangerous to place unquestioning faith in many of the generalizations of organic chemistry, certain ones, nevertheless, are useful to the person who is encountering this material for the first time. Despite the exceptions, generalizations provide very useful focal points around which to organize one's thoughts. Recognizing the potential pitfalls, we suggest the following:

1. The characteristic reaction of the saturated alkanes is *free radical* substitution; that of the alkenes and alkynes is *addition* (both ionic and free radical); and that of aromatic compounds is *electrophilic substitution.*
2. The relative reactivity of C—H bonds with respect to hydrogen atom abstraction decreases in the following order:

$$C{=}C{-}\overset{|}{\underset{|}{C}}{-}H \simeq Ar\overset{|}{\underset{|}{C}}{-}H > R_3CH > R_2CH_2 > RCH_3 > CH_4$$

3. The resistance to oxidation of various carbon frameworks decreases in the following order: arene > alkane > alkyne > alkene. This must be accepted with reservation, however, for the reaction conditions and the oxidizing reagent both play an important role. For example, toluene is oxidized by potassium permanganate at the methyl group rather than at the benzene ring, but ozone attacks the benzene ring preferentially.

The compounds whose reactions are discussed in this chapter are those composed solely of carbon and hydrogen. We have already seen, however, that hetero atoms and groups can be attached to carbon frameworks (see Sections 3.12 and 4.10), and it is not surprising to find that the chemistry of the original carbon framework may be changed as a result. For example, chloroform ($CHCl_3$) has a hydrogen that is sufficiently acidic to be removed by strong bases; tetrafluoroethylene ($F_2C{=}CF_2$) is inert to electrophilic addition and, instead, reacts readily with nucleophiles; benzoic acid ($C_6H_5CO_2H$) is much less susceptible to electrophilic substitution than benzene; and *p*-hydroxytoluene ($HO—C_6H_4—CH_3$) undergoes oxidation in the aromatic ring rather than the methyl group. Thus, the attached groups (*i.e.*, the functional groups) may alter the properties of the carbon frameworks even to the point of reversing the chemical behavior that is observed in the parent systems. Succeeding chapters deal with the structure of various functional groups and the effects these have on the physical and chemical properties of the systems to which they are attached.

8.6 Study Guide

I. Addition reactions
 A. Addition reactions of alkenes
  1. Electrophilic addition reactions of alkenes
   (a) Addition of halogen acids
    (1) Mechanism of reaction; initiation by attachment of H^+
    (2) Orientation of addition—Markovnikov's rule
   (b) Addition of water—catalyzed by trace of strong mineral acid
   (c) Addition of halogen—initiation by attachment of X^+
  2. Free radical addition of hydrogen bromide to alkenes
   (a) Mechanism of reaction—initiation by attachment of Br·
   (b) Orientation of addition—anti-Markovnikov
   (c) Concept of the "chain reaction"
  3. Addition of hydrogen to alkenes
  4. Addition of carbenes to alkenes
 B. Addition reactions of alkynes
  1. Electrophilic addition of halogens, halogen acids, and water (to give vinyl alcohols that interconvert to carbonyl compounds)
  2. Addition of hydrogen
  3. Addition of carbenes
 C. Addition reactions of arenes—catalytic hydrogenation
II. Substitution reactions
 A. Substitution reactions of alkanes
  1. Radical chlorination and bromination with molecular halogen
   (a) Mechanism of reaction
    (1) Initiation by formation of halogen atoms
    (2) Chain reaction process
   (b) Bond selectivity
    (1) Benzylic (Ar—C—H) and allylic (C=C—C—H) bonds most reactive
    (2) Alkane reactivity: $R_3CH > R_2CH_2 > RCH_3$
  2. Radical bromination with N-bromosuccinimide
 B. Substitution reactions of alkynes
  1. Reaction of 1-alkynes with metals (*e.g.*, Na) to form RC≡C—metal and H_2
  2. Reaction of 1-alkynes with strong bases (*e.g.*, KNH_2) to form RC≡C—metal + NuH (*e.g.*, NH_3)
  3. Reaction of 1-alkynes with $Ag(NH_3)_2^+$ and $Cu(NH_3)_2^+$ to form RC≡C—Ag and RC≡C—Cu (diagnostic tests for 1-alkynes)

C. Substitution reactions of arenes
   1. Electrophilic substitution reactions of arenes
      (a) Halogenation (with X_2)
      (b) Nitration (with $HNO_3 + H_2SO_4$)
      (c) Sulfonation (with H_2SO_4)
      (d) Alkylation (with $RX + AlCl_3$) } Friedel–Crafts reactions
      (e) Acylation (with $RCOCl + AlCl_3$) } Friedel–Crafts reactions
   2. Mechanism of electrophilic substitution reactions of arenes
      (a) Formation of electrophile; *e.g.*, NO_2^+
      (b) Addition of electrophile to aromatic ring
      (c) Expulsion of H^+ from intermediate to reform the aromatic system

III. Elimination reactions—dehydrogenation of six-membered alicyclic compounds to form aromatic compounds

IV. Oxidation reactions
   A. Oxidation reactions of alkanes
      1. Selective in RH compounds only if RH contains a particularly reactive C—H bond (*e.g.*, R_3C—H, $\backslash\!\!/$C=C—C—H, or Ar—C—H)
      2. Selective in Ar—R compounds, Ar being resistant to oxidation and R being converted to $-CO_2H$ to give $ArCO_2H$
   B. Oxidation reactions of alkenes and alkynes
      1. Oxidative cleavage of multiple bond to give RCHO, R_2CO, and/or RCO_2H
      2. Ozone a particularly good reagent for oxidizing alkenes
   C. Oxidation reactions of arenes
      1. Benzene ring resistant to oxidation under conditions that convert a CHRR′ side chain to $-CO_2H$
      2. Benzene ring susceptible to ozonolysis, but much more slowly than alkenes and alkynes

PROBLEMS

1. Suggest reasonable intermediate compounds that might be involved in the cracking of the following starting compounds to elemental carbon and atomic hydrogen:

(a)  [bicyclobutane] $\longrightarrow$ 4 C + 6 H

(b) 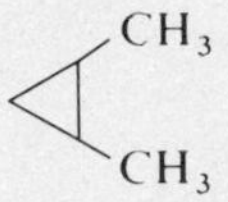 [1,2-dimethylcyclopropane, CH_3, CH_3] $\longrightarrow$ 5 C + 10 H

(c)  [cyclohexane] $\longrightarrow$ 6 C + 12 H

2. Draw the structure of the major organic product from each of the following reactions:

(a) $CH_3CH_2CH_2CH_2CH_3 \; + \; Cl_2 \xrightarrow{\Delta}$

(b) CH_3 + $Br_2 \xrightarrow[h\nu]{}$

(c) CH_3 + $Br_2 \xrightarrow[FeBr_3]{}$

(d) CH_3 CH_3 + O N—Br O $\xrightarrow[\Delta]{}$

(e) $CH{=}CH_2$ + $H_2 \xrightarrow[Pd/C]{}$

(f) —H + $KMnO_4/NaOH$ $\xrightarrow{\text{Under mild conditions}}$ $\xrightarrow{\text{Under strenuous conditions}}$

(g) CH_3 CH_3 + HBr $\longrightarrow$

(h) $(CH_3)_2C{=}CH_2 \; + \; HCl \longrightarrow$

(i) CH_3— —CH_3 + $HNO_3 \xrightarrow[H_2SO_4]{}$

(j) $CH_3\overset{CH_3}{\overset{|}{C}}HCH{=}CH_2 \; + \; HBr \xrightarrow[\text{Benzoyl peroxide}]{}$

(k) $CH_3CH{=}CHCH_2CH_3 \; + \; CH_2N_2 \xrightarrow[h\nu]{}$

(l) $H_2C{=}CHCH_2CH_2CH_2CH{=}C(CH_2CH_3)_2 \; + \; O_3 \xrightarrow[CH_2Cl_2,\ \text{pyridine}]{}$

(m) + $H_2O \xrightarrow[Hg^{++}\ \text{catalyst, } H_2SO_4]{}$

(n) $C_6H_5C{\equiv}CH \; + \; (CH_3)_2N{-}Na \longrightarrow$

(o) + Se $\xrightarrow[\Delta]{}$

3. In the free radical chlorination of *n*-butane at 250°, what are the relative concentrations of $CH_3CH_2CH_2\dot{C}H_2$ and $CH_3CH_2\dot{C}HCH_3$ that would be formed based on bond energies (see Table 2.3), assuming ΔS to be the same in both cases? If the rates of reaction of R· and Cl_2 are the same for both radical species, what would be the relative yields of products? Why is there a discrepancy between the calculated and the observed? Why is the agreement between calculated and observed closer in the case of the bromination reaction?

4. Using the bond energy values in Table 2.3, calculate the energetics of the fluorination of propane to *n*-propyl fluoride.

5. In the following examples, a pair of test tubes, each containing one or the other of the two isomeric compounds, is presented to you. What simple chemical test could be performed on each of the members of the pair to permit a clear structural differentiation? What spectral determination could be performed on each of the members of the pair to permit a clear structural differentiation?

(a) $(CH_3)_2CHC{\equiv}CH$ and $CH_3CH_2C{\equiv}CCH_3$

(b)* 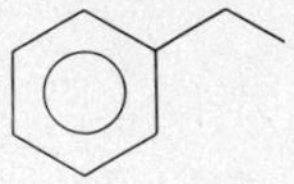 and

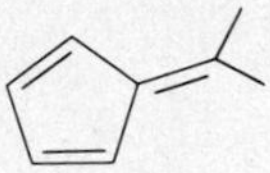

(c) 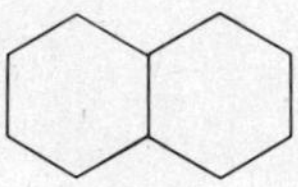 and

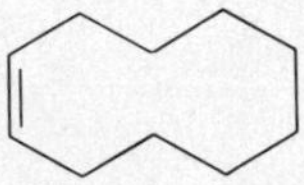

(d) 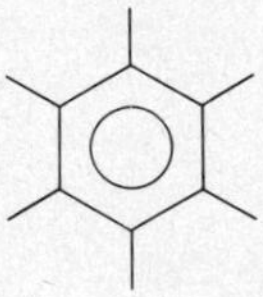 and

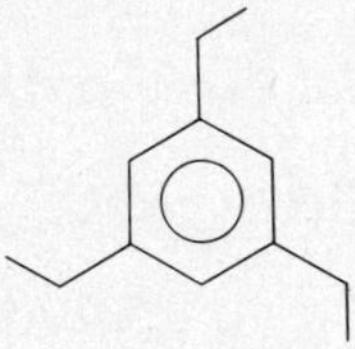

(e) 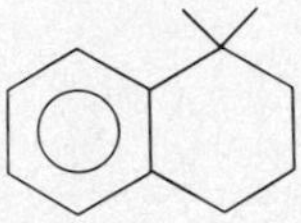 and

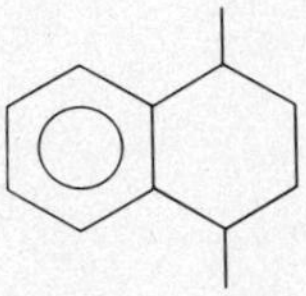

(f) 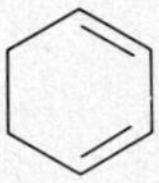 and

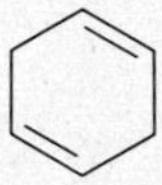

6. Starting with elemental carbon and any inorganic reagents design syntheses for the following compounds:

(a) $HC{\equiv}CH$ (b) $H_2C{=}CH_2$ (c) CH_3CHO

(d) CH_3CH_3 (e) CH_3CH_2Br (f) HCHO

(g) $\overset{O}{CH_2CH_2}$ (epoxide) (h) $HOCH_2CH_2OH$ (i) $Br_2CHCHBr_2$

(j) $ClCH{=}CHCl$ (k) CH_3CH_2OH

7. What bromination products can be formed in the reaction of *cis*-2-heptene with N-bromosuccinimide? What minor products might be formed as the result of chain-terminating reactions?

---

* As an abbreviation device, lines are often drawn to represent alkyl groups, a single line for a methyl (as in methylcyclohexane 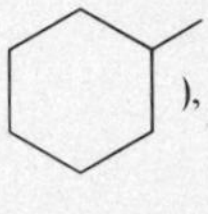), a singly angled line for ethane (as in ethylbenzene 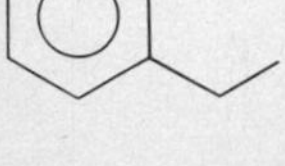), etc.

8. Outline, in as much detail as possible, the probable mechanisms for the following reactions:

(a) $(CH_3)_2C{=}CH_2 + HI \longrightarrow (CH_3)_3C{-}I$

(b) 1-methylcyclopentene $-CH_3$ + HBr $\xrightarrow{\text{Peroxides}}$ 2-bromo-1-methylcyclopentane (Br, $-CH_3$)

(c) $C_6H_5{-}CH(CH_3)_2 + Cl_2 \xrightarrow{h\nu} C_6H_5{-}C(Cl)(CH_3)_2 + HCl$

(d) naphthalene + $Br_2 \xrightarrow{FeBr_3}$ 1-bromonaphthalene + HBr

9. For each of the following reactions, none of which has yet been discussed in the text, indicate whether you would class it as an addition, substitution, elimination, or oxidation process:

(a) $(CH_3)_2C{=}O + H_2 \xrightarrow{Pd/C} (CH_3)_2CHOH$

(b) naphthalene + $CH_3COCl \xrightarrow{AlCl_3}$ 2-naphthyl$-C(=O)-CH_3$

(c) $(CH_3)_3C{-}Br + C_2H_5ONa \longrightarrow (CH_3)_2C{=}CH_2 + C_2H_5OH + NaBr$

(d) cyclohexene $\xrightarrow{Na_2Cr_2O_7/H_2SO_4}$ $HO_2C(CH_2)_4CO_2H$ (CO_2H, CO_2H)

(e) cyclohexanone (=O) + HCN $\longrightarrow$ 1-hydroxycyclohexanecarbonitrile (OH, $-CN$)

(f) $(CH_3)_3CH + HNO_3 \xrightarrow{\text{Heat}} (CH_3)_3C{-}NO_2 + H_2O$

(g) $C_2H_5Br + NaCN \longrightarrow C_2H_5CN + NaBr$

(h) benzene + $3\ Cl_2 \xrightarrow{h\nu}$ $C_6H_6Cl_6$ (cyclohexane ring with H and Cl on each carbon)

COMMENTS ON SYNTHESIS PROBLEMS

From this chapter on to the end of the book, considerable emphasis is placed on synthesis problems, for they provide a particularly good and rigorous test of one's knowledge of organic chemistry. Although there are no completely reliable rules for how to solve a synthesis problem, the most generally applicable advice is to first consider the product and work backward from it to the starting material. In all but the very simplest cases, it is difficult or impossible to immediately perceive the entire outline of a synthesis, but it is not so difficult to guess at what the last step in the process might be. Working backward, then, a step at a time, a trail is revealed which, when turned around, leads from starting materials to product.

A more detailed discussion of this approach to the solution of synthesis problems is presented in Section 4 of Chapter 18. Although the full appreciation of the material in Chapter 18 depends on a knowledge of the chemistry in the previous 17 chapters, readers who have just finished Chapter 8 might derive useful information from a quick perusal of its contents.

Chapter 9

Halides: aliphatic and aromatic

What is the end of study? Let me know?
Why, that to know, which else we should not know.
Things hid and barr'd, you mean, from common sense?
Ay, that is study's god-like recompense.

JOHN MILTON
Love's Labour's Lost

The discussion of the chemistry of functional groups commences with the hetero atoms that fall in the right-hand column of the periodic table, *viz.*, the halogens. We consider first the general structural features of compounds containing halogens, and follow this by discussing nomenclature, physical properties, chemical reactions, and certain significant examples of the class.

9.1 Structure and Nomenclature of Halides

The electron configuration of fluorine (Fig. 2.5) is described as $1s^2 2s^2 2p_x^2 2p_y^2 2p_z^1$, indicating that only one orbital, *i.e.*, $2p_z$, is incompletely filled and available for bonding. Fluorine, then, is predicted to be monovalent. In combination with carbon it can form only a single σ bond, interpreted as arising from overlap of a hybrid orbital on carbon and a p orbital on fluorine, as pictured in Fig. 9.1. Similarly, the other halogens have an incompletely filled p orbital in the outer electron shell and also have the capacity only for single bonding to carbon. The length and strength of the C-halogen bonds are dependent, to some extent, on whether the carbon is sp^3-, sp^2-, or sp-hybridized. For example, the C—Cl bond of methyl chloride has a length of 1.78 Å and a bond dissociation energy of 81 kcal/mole, whereas the C—Cl bond of chlorobenzene has a length of 1.69 Å and a bond dissociation energy of 88 kcal/mole. The shorter and stronger bond in aryl chlorides arises partly from the difference in hybridization of the orbitals on carbon (*i.e.*, sp^3 *vs.* sp^2; see p. 110) and partly from resonance interaction between the nonbonded electrons on chlorine and the π system of the aromatic ring, as Fig. 9.1 depicts.

There are few, if any, limitations to the kind of carbon framework to which halogens can be attached. The halogen can be attached to sp^3-hybridized

Bonding in Alkyl Halides

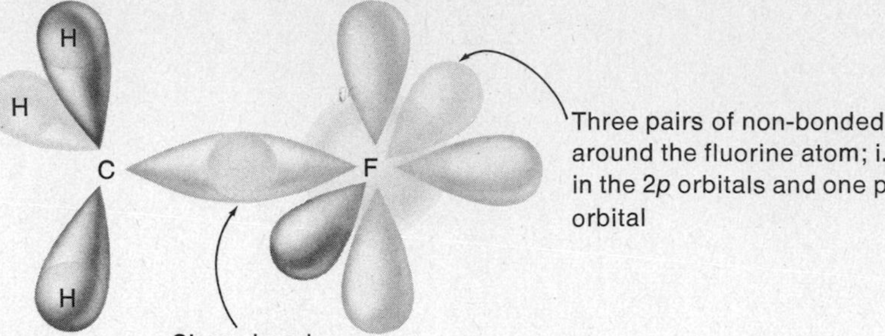

Orbital representation of methyl fluoride

Bonding in Aryl Halides

Resonance structures

Orbital representation

Fig. 9.1. Bonding in alkyl and aryl halides.

carbon carrying three hydrogens (*e.g.*, CH_3Br), two hydrogens (*e.g.*, CH_3CH_2Br, designated as a *primary alkyl halide*), one hydrogen [*e.g.*, $(CH_3)_2CHBr$, designated as a *secondary alkyl halide*], or no hydrogens [*e.g.*, $(CH_3)_3CBr$, designated as a *tertiary alkyl halide*]. Cyclic alkane frameworks can also bear halogens as, for instance, —Cl (cyclopropyl chloride) and —I (cyclohexyl iodide), designated as *cycloalkyl halides*. Halogens attached to sp^2-hybridized carbons are represented by compounds such as $H_2C{=}CHCl$ (vinyl chloride or chloroethene) and —F (phenyl fluoride or fluorobenzene). Attachment to *sp*-hybridized carbon is possible, but compounds such as $HC{\equiv}CCl$ are rarely used in laboratory experiments or industrial operations because of their explosive nature.

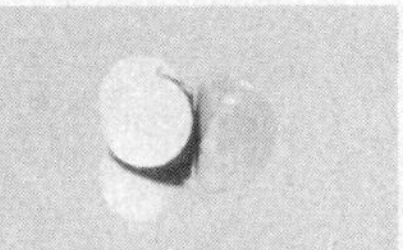

Methyl fluoride

Chlorobenzene

Most of the types of compounds arising from the attachment of two or more halogen atoms to carbon frameworks, such as $RCHX_2$, RCX_3, and $RXC{=}CXR$, correspond to stable, isolable species. Among the few polyhalogen compounds that are *not* stable are those having iodine atoms on adjacent sp^3-hybridized carbons. Although chlorine and bromine add effectively to alkenes to give 1,2-dichloro and 1,2-dibromo compounds, respectively, the equilibrium for the addition of iodine is not favorable; in fact, at room temperature 1,2-diiodo compounds decompose rapidly to the corresponding alkene and molecular iodine:

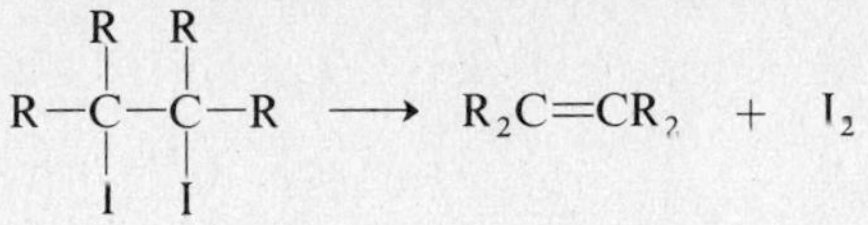

$$R{-}\underset{\large I}{\overset{\large R}{C}}{-}\underset{\large I}{\overset{\large R}{C}}{-}R \longrightarrow R_2C{=}CR_2 + I_2$$

According to one nomenclature system, halogen compounds are named as alkyl, alkenyl, alkynyl, or aryl derivatives of the corresponding halogen acid. Thus, C_2H_5Br is ethyl bromide, $H_2C{=}CHCH_2I$ is allyl iodide, and $C_6H_5{-}Cl$ is phenyl chloride. For the less complex members of the class, this system is quite satisfactory and is frequently employed. For more complex structures, however, we resort to the IUPAC system of nomenclature, using the halogen as a substituent group and affixing the names fluoro, chloro, bromo, or iodo. The compounds just mentioned are named in the IUPAC system as bromoethane, 3-iodopropene, and chlorobenzene, respectively. If two or more substituents (alike or different) are attached to the parent compound, the numbering of the chain starts from the terminus that is *closer to a substituent*. For example,

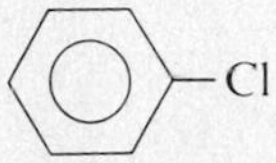

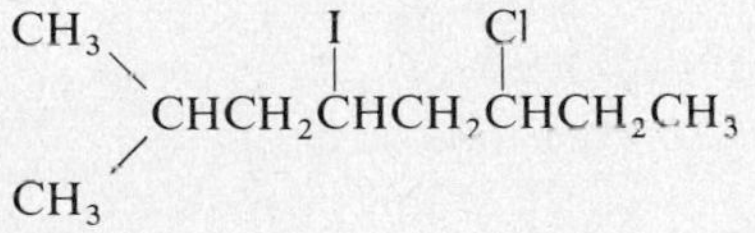

$$(CH_3)_2CHCH_2\underset{}{\overset{I}{C}}HCH_2\overset{Cl}{C}HCH_2CH_3$$

is named 6-chloro-4-iodo-2-methyloctane, the substituents being arranged in alphabetical order. If two substituents are equally placed with respect to the termini, the numbering is determined by the group that takes alphabetical precedence. Thus,

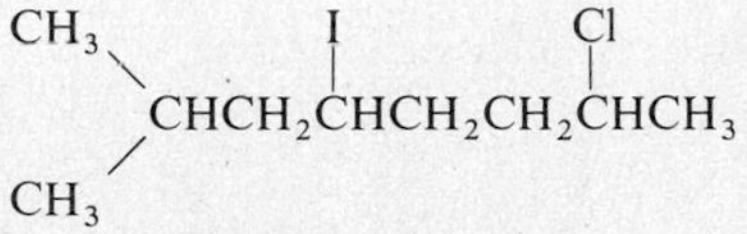

$$(CH_3)_2CHCH_2\overset{I}{C}HCH_2CH_2\overset{Cl}{C}HCH_3$$

is named 2-chloro-5-iodo-7-methyloctane.

The trihalomethanes (CHX_3) are often referred to as haloforms; thus, $CHCl_3$ is chloroform, CHI_3 is iodoform, etc.

PROBLEM 9-1

Draw and name all the compounds with the molecular formula C_4H_9Cl.

9.2 *Physical Properties of Halides*

The C—X bond is polar because of the greater electronegativity of halogen relative to carbon. The dipole moments for the halomethanes, for example, are as follows: CH_3F (1.808 D), CH_3Cl (1.86 D), CH_3Br (1.78 D), CH_3I (1.64 D). Despite their polarity and the fact that these compounds have three pairs of nonbonded electrons in the valence shell of the halogen atom, their boiling points are approximately the same as those of alkanes of corresponding molecular weight and carbon framework. For example, *n*-propyl chloride (mol wt 78) boils at 46.6°, *n*-pentane (mol wt 72) boils at 36°, and *n*-hexane (mol wt 88) boils at 68°. Similarly, the water solubility of halogen compounds is not significantly increased over that of the hydrocarbons of corresponding molecular weight, the increased polarity notwithstanding. This probably results from the size of the halogen atoms, which prevents as effective an intermolecular dipole–dipole interaction as is possible with smaller, less sterically encumbered molecules, such as water. Organic halides are soluble in most organic solvents, and some of the lower members are themselves useful solvents. Perhaps the most characteristic physical attribute of the alkyl halides is their density; the iodoalkanes, bromoalkanes, and polychloroalkanes are all heavier than water, as illustrated by the following densities: CH_2Cl_2 (1.336), C_2H_5Br (1.440), CH_3I (2.279), CI_4 (4.32) (CI_4 is one of the densest known compounds of carbon).

9.3 Spectral Properties of Halides

The stretching frequency for a carbon–halogen bond in the infrared decreases as the atomic weight of the halogen increases; C—F-containing compounds absorb at 1400–1000 cm^{-1}, C—Cl compounds at 800–600 cm^{-1}, C—Br compounds at 600–500 cm^{-1}, and C—I compounds near 500 cm^{-1}. The most commonly used type of infrared spectrometers cover the range from 4000 to 650 cm^{-1}, permitting the detection of only C—F and C—Cl bonds. For example, the spectrum of 1-bromo-3-chloropropane obtained on a conventional instrument shows the C—Cl but not the C—Br stretching band. Instruments capable of measuring frequencies below 650 cm^{-1} are available, however, and their use is becoming increasingly widespread, allowing the detection of C—Br and C—I stretching bands as well.

Since the halogens are electronegative elements relative to carbon, they cause downfield shifts in the nmr of protons that are proximate to them (see Table 6.3).

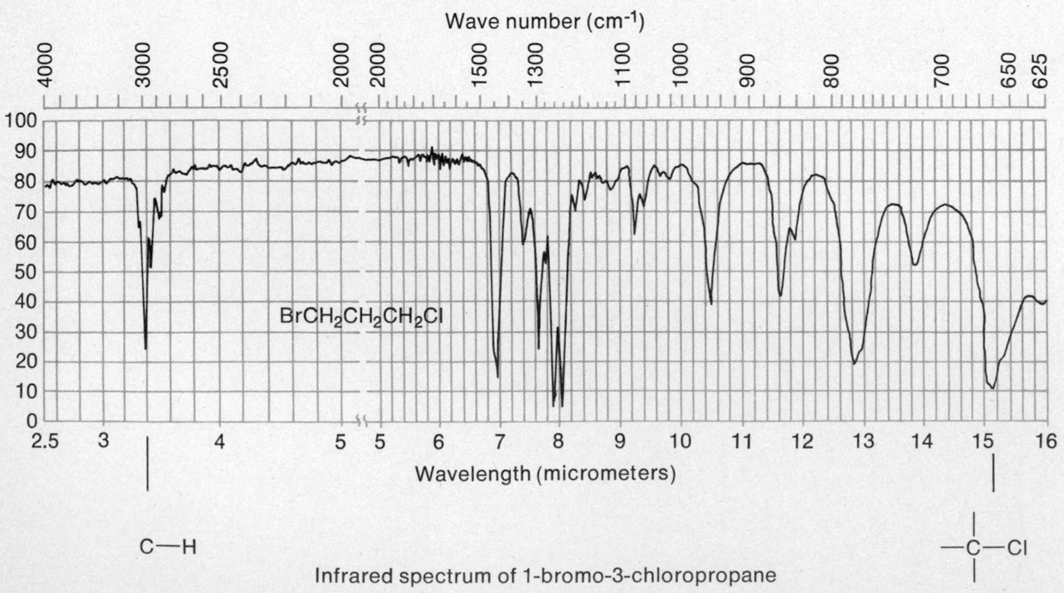

Infrared spectrum of 1-bromo-3-chloropropane

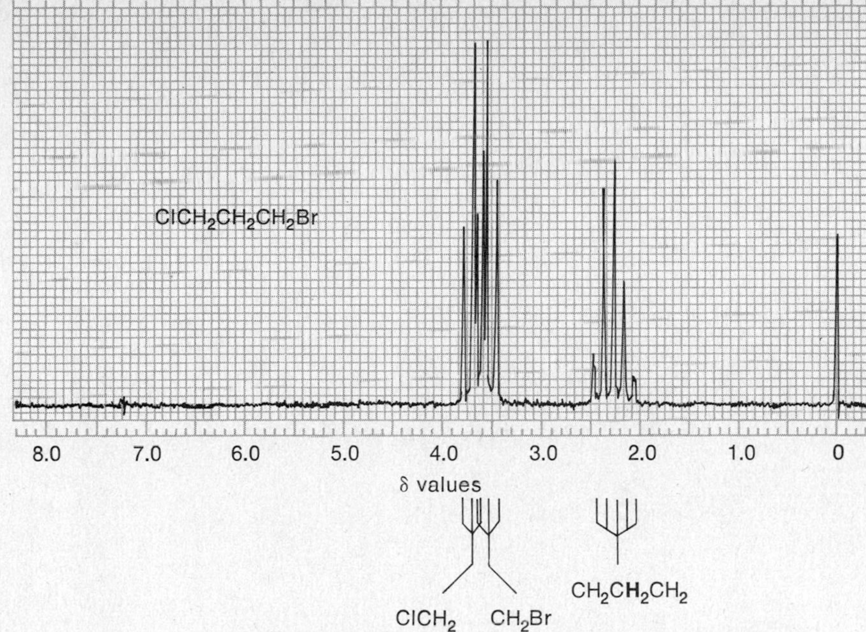

Nmr spectrum of 1-bromo-3-chloropropane

The magnitude of the shift increases from iodine to fluorine. The nmr spectrum of 1-bromo-3-chloropropane, for example, shows *two sets* of triplets (overlapping) for the CH_2X groupings, one triplet centered at δ 3.55 for the CH_2Br protons and the other at δ 3.70 for the CH_2Cl protons; in addition, a quintet pattern for the middle methylene group appears at higher field.*

The halogens all have odd atomic numbers, and consequently possess nuclear magnetic characteristics. The nuclear spins of chlorine, bromine, and iodine are equal to or greater than 1, and these nuclei do not interact very strongly with hydrogens; however, fluorine, with a spin of $\frac{1}{2}$, rather closely resembles hydrogen in its nuclear magnetic properties. For example, the nmr spectrum of CF_3CH_2OH is very similar to that of CH_3CH_2OH with respect to the CH_2 resonance. The only difference is that the separation between the lines of the quartet is characteristic of the F—C—C—H splitting constant (*i.e.*, $J = 9$ Hz) rather than the H—C—C—H splitting constant (*i.e.*, $J = 7$ Hz). Fluorine, although identical with hydrogen in having a spin of $\frac{1}{2}$, comes into resonance at a different combination of magnetic field strength and radio frequency, accounting for the absence of a triplet pattern in the hydrogen resonance spectrum. If the instrument is adjusted to detect fluorine resonances rather than hydrogen resonances, however, the spectrum of 2,2,2-trifluoroethanol then appears as a triplet, the CF_3 group being split by the pair of equivalent adjacent hydrogens.

---

* Since $ClCH_2$ and $BrCH_2$ are not equivalent groups, a triplet of triplets pattern might have been anticipated for the C-2 CH_2 group. The coupling constant between the $—CH_2—$ and $ClCH_2$ hydrogens, however, is almost the same as the coupling constant between the $—CH_2—$ and $BrCH_2$ hydrogens. As a result, the four hydrogens at C-1 and C-3 act as an equivalent set and, by the $(n + 1)$ rule, split the C-2 hydrogens into a five-line pattern.

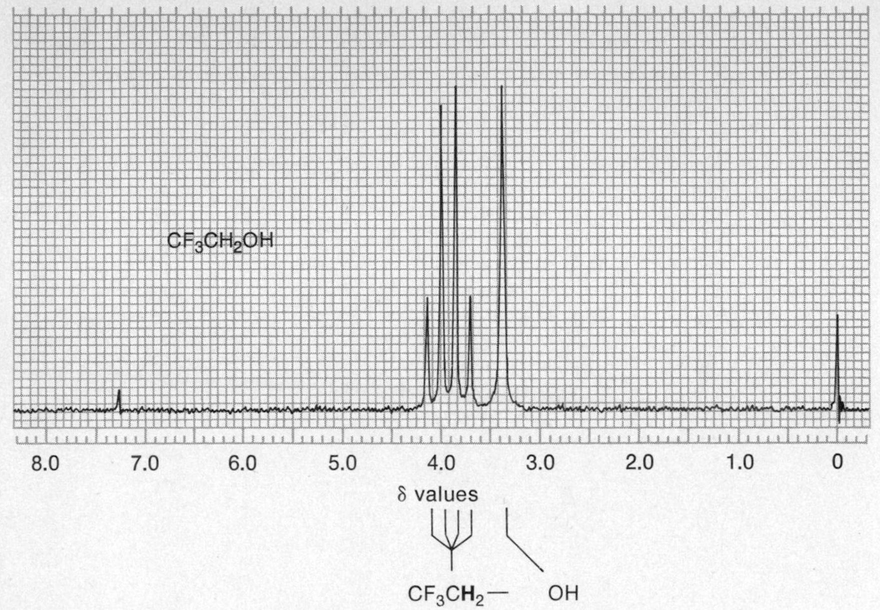

Nmr spectrum of 2,2,2-trifluoroethanol

PROBLEM 9-2

One of the compounds in Problem 9-1 on p. 000 can exist in nonsuperimposable mirror image forms. Draw the nmr spectrum that would be expected for this compound.

9.4 *Synthesis of Halides*

Three methods for the synthesis of halides are discussed in previous chapters, including (a) the action of chlorine or bromine (or N-bromosuccinimide) on alkanes in the presence of heat or light to yield alkyl chlorides and bromides (see Section 8.2a) (b) the action of halogens on aromatic compounds in the presence of Lewis acids to yield aryl chlorides, bromides, and iodides (see Fig. 8.13), and (c) the addition of halogen acids to alkenes to yield the corresponding alkyl halides (see Section 8.1). Illustrative of both methods (a) and (b) is the chlorination of toluene; if the reaction is carried out in the presence of light (free radical pathway), substitution of hydrogen atom by chlorine atom takes place at the methyl carbon, but if the reaction is carried out in the presence of ferric chloride (ionic pathway), substitution of proton by chlorine cation (Cl^+) occurs at phenyl carbon (Fig. 9.2). Illustrative of method (c) is the reaction of 1-methylcyclohexene with hydrogen bromide, the direction of the addition being dependent on the presence (see Fig. 8.14) or absence of peroxides (Fig. 9.2). A fourth, and often used method [method (d)] for preparing alkyl halides involves the action on alcohols of various reagents, such as halogen acids, the halides of phosphorus (PCl_3, PCl_5, and PBr_3), and the thionyl halides ($SOCl_2$ and $SOBr_2$) (Fig. 9.2). Chapter 10 discusses the mechanism of the reaction between alcohols and halogen acids.

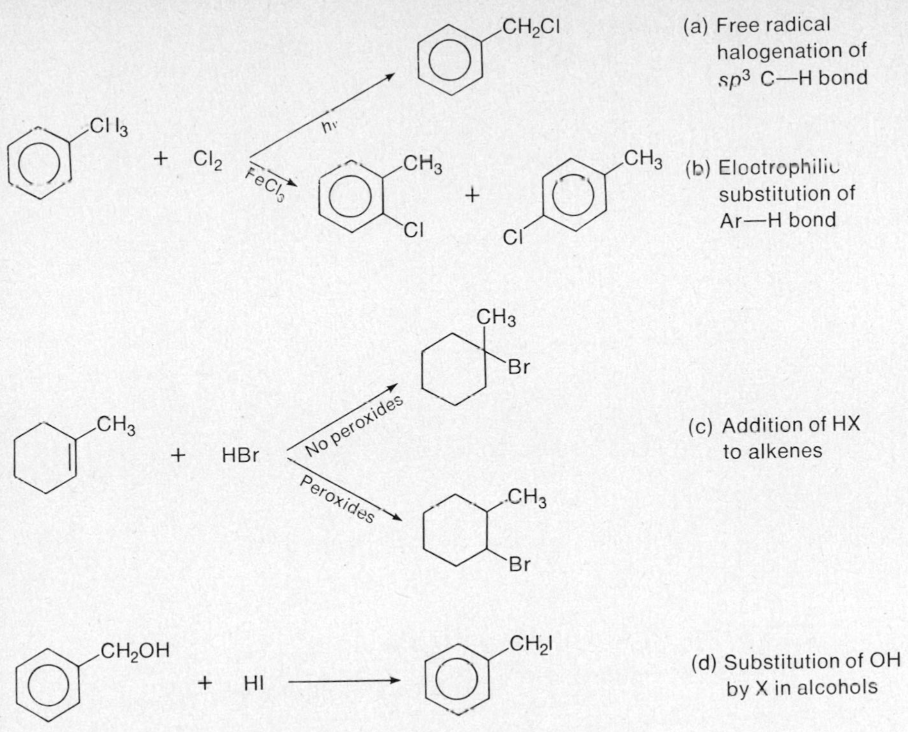

(a) Free radical halogenation of sp^3 C—H bond

(b) Electrophilic substitution of Ar—H bond

(c) Addition of HX to alkenes

(d) Substitution of OH by X in alcohols

Fig. 9.2. Methods for the synthesis of halides.

9.5 Substitution Reactions of Halides: Methods for Introducing Functional Groups and for Forming Carbon–Carbon Bonds

Alkyl halides are exceedingly useful for the synthesis of other compounds, largely because of the ease with which the halogen can be replaced by various atoms, groups, or other carbon frameworks. In the majority of reactions, the displacing entity is a nucleophile (*i.e.*, a Lewis base), and frequently it is an anion; the halogen departs from the molecule as an anion. Reactions in which a nucleophile displaces halogen are called *nucleophilic substitution reactions* (or nucleophilic displacements). A variety of compounds containing hetero atoms are capable of acting as nucleophiles toward alkyl halides; for example, H_2O, H_2S, ROH, and RSH all have this capacity. Even more reactive as nucleophiles, however, are the corresponding anions HO^-, HS^-, RO^-, and RS^-, which can be produced from the parent compounds by the action of an alkali metal or strong base (Fig. 9.3).

Amines are considerably more nucleophilic than water or alcohols and are effective displacing agents in the neutral form (*i.e.*, RNH_2 rather than RNH^-). Primary amines are formed by the action of ammonia on alkyl halides, but secondary amines, tertiary amines, and quaternary ammonium salts are also formed as the result of further action of the alkyl halide. This is an experimental drawback of the synthesis, for the tendency for an amine to undergo successive alkylation often makes it difficult to stop the reaction at the desired point (Fig. 9.4). A way for minimizing this difficulty is to use an excess of one of the reactants. For example, if dimethylamine is the desired product from monomethylamine,

$$CH_3CH_2CH_2CH_2Br + OH^- \longrightarrow CH_3CH_2CH_2CH_2OH + Br^-$$

(From NaOH)

$$CH_3I + (CH_3)_3CO^- \longrightarrow CH_3OC(CH_3)_3 + I^-$$

(From K + $(CH_3)_3COH$)

$$C_6H_5{-}CH_2Cl + HS^- \longrightarrow C_6H_5{-}CH_2SH + Cl^-$$

(From NaOH + H_2S)

$$(CH_3)_2CHCH_2Br + CH_3CH_2S^- \longrightarrow (CH_3)_2CHCH_2SCH_2CH_3 + Br^-$$

(From NaOH + CH_3CH_2SH)

Fig. 9.3. Oxygen and sulfur nucleophiles as displacing agents in nucleophilic substitution reactions with alkyl halides.

an excess of monomethylamine is used to increase the likelihood that it rather than the dimethylamine will react with the methyl iodide.

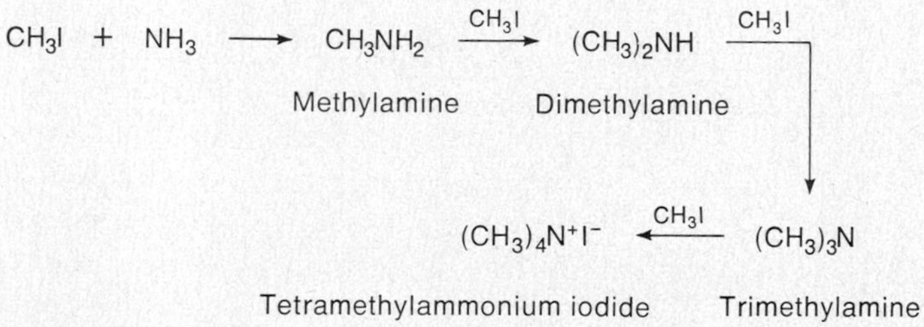

$$CH_3I + NH_3 \longrightarrow \underset{\text{Methylamine}}{CH_3NH_2} \xrightarrow{CH_3I} \underset{\text{Dimethylamine}}{(CH_3)_2NH} \xrightarrow{CH_3I} \underset{\text{Trimethylamine}}{(CH_3)_3N} \xrightarrow{CH_3I} \underset{\text{Tetramethylammonium iodide}}{(CH_3)_4N^+I^-}$$

Fig. 9.4. Nitrogen nucleophiles as displacing agents for alkyl halides.

PROBLEM 9-3

What would you predict to be the product from the reaction of methylamine (CH_3NH_2) and 1,4-dibromobutane?

Methods for forming C—C bonds are important in organic chemistry, for they provide the means by which carbon frameworks are constructed. Particularly useful among these are nucleophilic substitutions in which alkyl halides react with cyanide and acetylenide anions to yield nitriles and substituted acetylenes, respectively (Fig. 9.5). These reactions are significant because the new C—C bonds introduce C≡N and C≡C groups, which are susceptible to facile structural alteration, affording a variety of types of derived compounds.

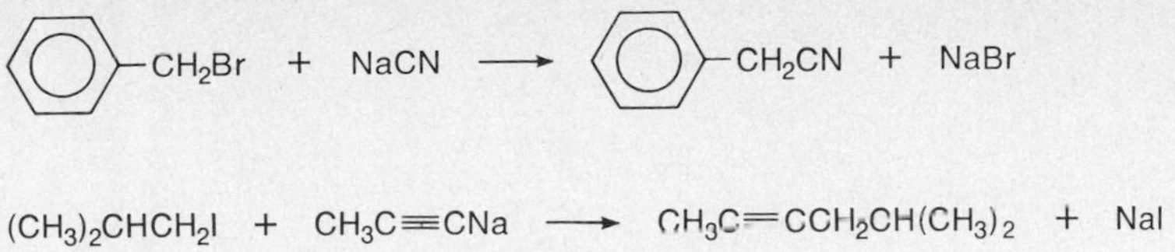

$$C_6H_5{-}CH_2Br \quad + \quad NaCN \longrightarrow C_6H_5{-}CH_2CN \quad + \quad NaBr$$

$$(CH_3)_2CHCH_2I \quad + \quad CH_3C{\equiv}CNa \longrightarrow CH_3C{\equiv}CCH_2CH(CH_3)_2 \quad + \quad NaI$$

Fig. 9.5. Carbon nucleophiles as displacing agents for alkyl halides.

Although, for the sake of brevity, it has been convenient to make broad generalizations in the survey of reactions presented thus far, it should be realized that almost all organic reactions suffer from certain limitations. Not all halogen-containing compounds, for example, undergo substitution reactions with nucleophilic reagents; some undergo elimination reactions instead, and others simply fail to react at all. Thus, halides that are highly branched at the carbon bearing the halogen atom (designated as the α carbon) tend to give alkenes (via the elimination of HX), and halides that are highly branched at the carbon adjacent to the carbon bearing the halogen (designated as the β carbon) along with aromatic and vinyl halides react so sluggishly with nucleophiles that these reactions have no synthetic utility. For example, *t*-butyl bromide reacts with

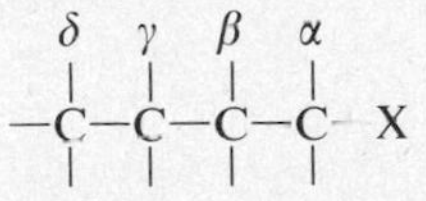

$$\overset{\delta}{-\text{C}}-\overset{\gamma}{\text{C}}-\overset{\beta}{\text{C}}-\overset{\alpha}{\text{C}}-\text{X}$$

Greek letter designation of carbons in an alkyl chain

sodium ethoxide to yield 2-methylpropene as the major product (see Section 9.7), and chlorobenzene, vinyl chloride, and neopentyl bromide all fail to react with sodium cyanide. The tendency for tertiary halides to give alkenes rather than substitution products and the tendency for aryl halides, vinyl halides, and neopentyl-type halides to fail to react constitute *important restrictions on the nucleophilic substitution reaction* and should be kept clearly in mind.

The mechanism of the nucleophilic substitution reaction is one of the most thoroughly explored facets of organic chemistry. Two major reaction pathways have been discerned on the basis of (a) measurements of the rates at which various nucleophilic processes take place (*i.e.*, the kinetics of the reactions) and the way in which the rates respond to changes in concentrations of the reactants, and (b) the stereochemical changes that take place in various nucleophilic substitution processes. Let us look just briefly into these mechanisms, which are dealt with in greater detail in Chapter 21.

The two major pathways for nucleophilic substitution reactions can be characterized as "one-step" and "two-step" processes, and they are illustrated by the reaction coordinate diagrams shown in Fig. 9.6. The one-step pathway involves no intermediates and only one activated complex, whereas the two-step process involves a discrete intermediate and two activated complexes. When we study the kinetics of reactions of the one-step variety we find that the rate is responsive to the concentrations of *both the RX compound and the incoming nucleophile Y*. This situation is referred to as a *second-order reaction* and is indicative of a *bimolecular process*. When we study the kinetics of reactions of the

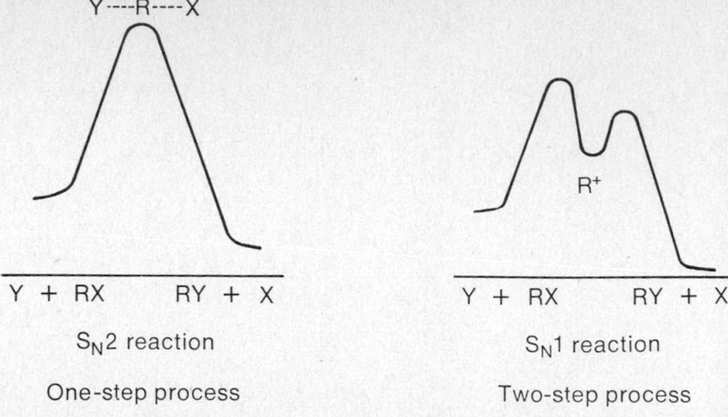

Fig. 9.6. Reaction coordinates for nucleophilic substitution reactions.

two-step variety, on the other hand, we find that the rate is responsive *only to the concentration of the RX compound* and not to that of the incoming nucleophile, Y. This situation is referred to as a *first-order reaction* and is indicative of a *unimolecular process.* On the basis of these kinetic differences, the two types of nucleophilic substitution reactions have acquired the symbols S_N2 (*S*ubstitution, *N*ucleophilic, bimolecular) for the one-step process and S_N1 (*S*ubstitution, *N*ucleophilic, unimolecular) for the two-step process.

Insight into what happens along the one-step and two-step pathways has been gained from studies of the stereochemical changes that take place when optically active compounds are used in which the chiral center is at the site of substitution—*i.e.*, compounds containing three different R groups in addition to the leaving group, X, attached to the chiral center (RR′R″CX). It turns out that reactions classed as S_N2 retain the optical activity of the starting compound but have the arrangement of the groups around the chiral carbon in the product opposite (*i.e.*, enantiomeric) to that of the starting compound (referred to as inversion of absolute configuration). Reactions classed as S_N1, on the other hand, lose the optical activity of the starting compound and yield optically inactive, racemic products.

To explain both the inversion of configuration and the bimolecular nature of the one step, S_N2 reactions, chemists have postulated that the incoming nucleophile, Y, attacks at the backside of the carbon to which the leaving group, X, is attached, giving an activated complex in which the new C—Y bond is partly formed and the old C—X bond is partly broken. When the bond making and bond breaking processes proceed to completion, a product is obtained in which the new group is attached to the face opposite that of the displaced group, resulting in inversion of configuration (Fig. 9.7). The participation of both the RX compound and the nucleophile, Y, in the activated complex is commensurate with a bimolecular process.

To explain both the loss of optical activity and the unimolecular nature of the two step, S_N1 reactions, chemists have postulated that the R—X bond is *completely broken before the R—Y bond is formed*, leading to an intermediate

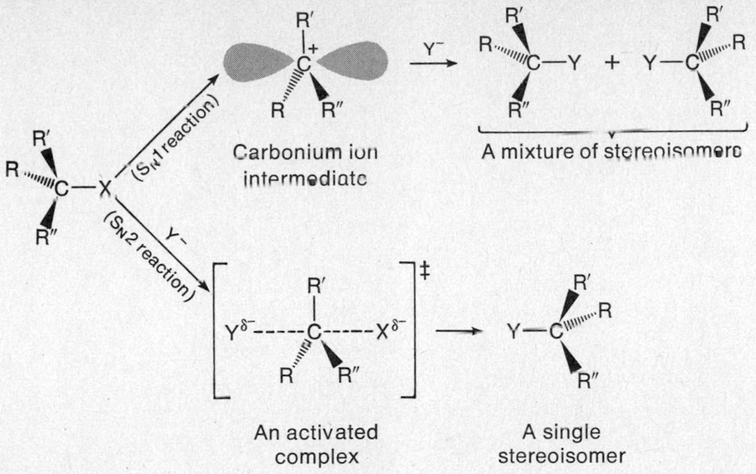

Fig. 9.7. Stereochemical pathway of S_N1 and S_N2 displacement reactions.

carbonium ion, which then reacts with the nucleophile, Y, in a second (and faster) step to form the product. Since carbonium ions are sp^2-hybridized, the tetrahedral character of the reactive site in the starting compound is lost and optical activity disappears. The incoming Y group can approach either "face" of the carbonium ion, yielding precisely equal amounts of the two enantiomeric forms of the product and, hence, a racemate. The lack of participation of the Y group until *after* the intermediate carbonium ion is formed in the slower of the two steps (*i.e.*, the rate-determining step) is commensurate with a unimolecular process.

PROBLEM 9-4

When the enantiomer of 1-deuterio-1-iodoethane shown below is treated with sodium cyanide, an S_N2 reaction occurs. Draw the structure of the product.

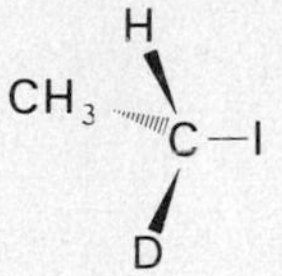

Whether the S_N1 or S_N2 pathway is involved in any particular instance depends on a variety of factors, including structure of reactant, solvent, and reaction conditions. In general, organic halides capable of forming reasonably stable carbonium ions on loss of the halide ion react via the S_N1 process (*e.g.*, tertiary alkyl, allylic, and benzylic halides), whereas primary and secondary alkyl halides react via the S_N2 process. Vinyl and aryl halides undergo substitution reactions very sluggishly and by mechanisms other than nucleophilic displacement. This difference in reactivity provides a useful method for differentiating between types of organic halides. Thus, those capable of reacting via the S_N1 process form precipitates of silver halide when treated with a solution of silver nitrate in aqueous alcohol, those capable of reacting via the S_N2 process form precipitates of sodium halide when treated with a solution of sodium iodide in acetone, and those incapable of reacting by either the S_N1 or S_N2 process fail to form precipitates under either condition. Alkyl fluorides also fail to yield precipitates under either of these conditions.

$$RX + Ag^+ \longrightarrow AgX\text{ (precipitate)} + [R^+] \longrightarrow \text{Products from carbonium ion}$$

$$RX + NaI \longrightarrow NaX\text{ (precipitate)} + RI$$

The proscription against using aryl halides in nucleophilic substitution reactions (p. 235) has already been cited. It is surprising to learn, therefore, that when chlorobenzene is treated with sodium amide in liquid ammonia, a substitution product is formed. To all outward appearances, a straightforward nucleophilic substitution has occurred, but a careful study of this and related reactions shows that quite a different mechanism with several steps and intermediates is actually involved. The first of the intermediates is generated by the base-induced removal of a hydrogen adjacent to the halogen atom to yield a carbanion. The carbanion then expels the halogen as a halide ion, leaving a neutral carbon framework containing a "triple" bond in an aromatic ring (Fig. 9.8). This species is called "benzyne" and represents still another of the

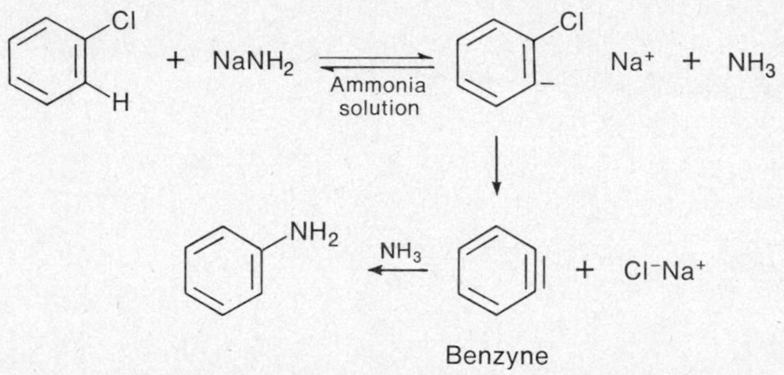

Cl, H + $NaNH_2$ ⇌ (Ammonia solution) Cl, − Na^+ + NH_3

↓

NH_2 ← (NH_3) Benzyne + Cl^-Na^+

Fig. 9.8. Displacement reaction involving a benzyne intermediate.

reactive intermediates of organic chemistry. Benzyne is extremely reactive and seeks to stabilize itself by adding reagents across the "triple" bond. In this particular example, it is ammonia that adds to furnish the stable product aniline. The overall process, then, is a three-step abstraction-elimination-addition sequence and not a bimolecular S_N2 process.

9.6 Organometallic Reagents from Halides: Methods for Forming Carbon–Carbon Bonds

Shortly before the turn of the century, a French chemist named Philippe Barbier noticed to his surprise that a reaction takes place when methyl iodide, acetone, and magnesium are mixed. More interested in other projects that were under way in his laboratory at the time, Professor Barbier turned the investigation of this phenomenon over to a young student named Victor Grignard. As the result of very careful and well-organized experimentation, Grignard quickly dissected the reaction into its constituent parts. He discovered that when methyl iodide and magnesium alone are mixed, a new compound is formed, which then reacts with added acetone to yield the product that was originally isolated when all three of the ingredients were mixed at the outset. In the course of this work, he also discovered that diethyl ether is a particularly good solvent for inducing the initial reaction between the alkyl halide and the magnesium. After receiving a Ph.D. degree in 1901 for his accomplishment, Grignard continued investigations of the reaction, and over the course of the next decade was so successful in demonstrating its wide-ranging scope and utility for organic syntheses that in 1912 he was awarded the Nobel Prize. Chemists continue to honor him by referring to reactions of this type as "Grignard reactions" and to the compound that results from the action of magnesium on a halogen compound as a "Grignard reagent."

Many other chemists in addition to Grignard have been immortalized by having a reaction named after them. Although organic chemists are often chided for according this honor too easily and frequently, it is a convenient shorthand way for designating a particular reaction. It is also a constant reminder that chemistry is a *human* endeavor. "Name reactions" are a characteristic facet of the language of organic chemistry. The student should learn a few of the name reactions at this stage of instruction and more of them during the detailed study of reactions and syntheses at a later stage in the course.

The structural restrictions on the use of halides in S_N1 and S_N2 reactions do not apply to Grignard reactions. Almost all the primary, secondary, and tertiary alkyl chlorides, bromides, and iodides, as well as vinyl and aryl bromides and iodides, react with magnesium to form the organomagnesium compounds (Grignard reagents). These have proved to be among the most versatile and widely used types of organic reagents, and they are encountered frequently throughout the remainder of the course. Despite their extensive use and despite considerable study of their chemistry, the composition of an ethereal solution of a Grignard reagent is still not completely understood. For convenience, the structure of the Grignard reagent is usually written as RMgX, although a solution actually contains several entities including R_2Mg, MgX_2, and $(RMgX)_n$. Diethyl ether is a Lewis base and magnesium-containing compounds are Lewis acids, so Grignard reagents exist in ethereal solution in a solvated form, represented as

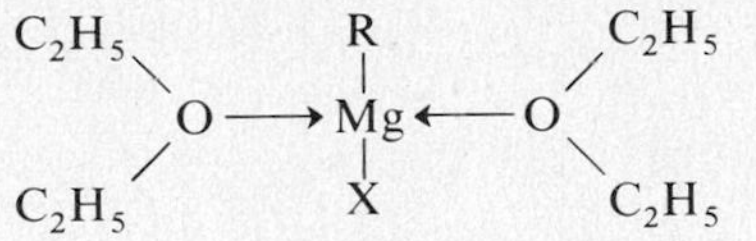

$$(C_2H_5)_2O \longrightarrow \underset{X}{\overset{R}{Mg}} \longleftarrow O(C_2H_5)_2$$

The bond between the carbon and the metal in Grignard reagents is polar, the carbon being the negative end and the metal the positive end of the dipole. Thus, Grignard reagents are sources of carbon nucleophiles, and S_N2 reactions can be carried out with Grignard reagents and primary alkyl halides, allyl halides, and benzyl halides (Fig. 9.9). Also, Chapter 10 discusses Grignard reagents as nucleophiles in addition reactions of carbonyl compounds.

$$CH_3Br + Mg \longrightarrow CH_3MgBr \xrightarrow{C_6H_5CH_2Br} C_6H_5CH_2CH_3 + MgBr_2$$

Fig. 9.9. The Grignard reagent as a nucleophile in an S_N2 displacement reaction.

The formation of Grignard reagents, in common with other organic reactions, suffers certain limitations. Thus, the presence in the halide of groups reactive toward Grignard reagents precludes reaction with magnesium. Since Grignard reagents are reactive toward compounds containing hydrogen attached to a hetero atom (often referred to as "active hydrogen") and toward compounds containing carbonyl groups, it is generally not possible to prepare Grignard reagents from compounds like $BrCH_2CH_2OH$ (*i.e.*, active hydrogen) or $BrCH_2CH_2CHO$ (*i.e.*, carbonyl group). The exceedingly rapid reaction of methylmagnesium iodide with compounds such as H_2O, ROH, RCO_2H, and RNH_2, yielding methane gas as one of the products, is a frequently used test for the presence of "active hydrogen".

Other metals, such as lithium and sodium, can also react with halides to form Grignard-like reagents (*i.e.*, $RX + Li \rightarrow RLi + LiX$), and compounds of this general type, containing a metal and an organic moiety, are referred to as *organometallic reagents.*

PROBLEM 9-5

The C—Mg bond in a Grignard reagent is appreciably polarized. Taking cognizance of this fact, draw three resonance structures for methylmagnesium iodide, two of which are charge-separated resonance structures. Knowing that the Grignard reagent is effective in S_N2 reactions, which charge-separated resonance structure do you think contributes more significantly to the resonance hybrid?

9.7 Elimination Reactions of Halides

If *t*-butyl bromide is treated with potassium hydroxide in anticipation of converting it to *t*-butyl alcohol, the intention will be only partly realized. In addition to *t*-butyl alcohol, a product resulting from the elimination of hydrogen bromide is also formed (Fig. 9.10); isobutene (2-methylpropene), in fact, is the *major*

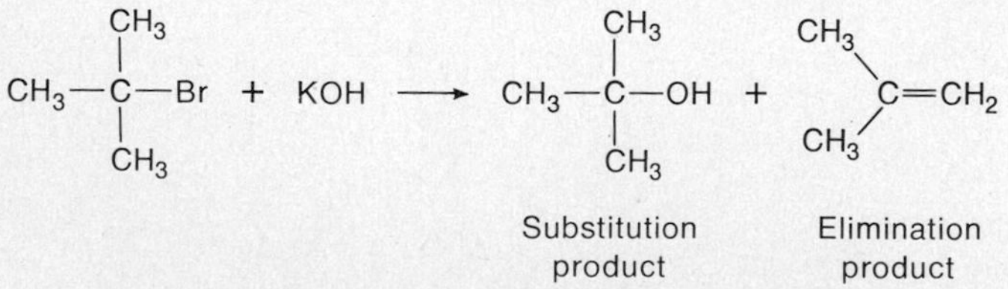

$$(CH_3)_3C{-}Br + KOH \longrightarrow \underset{\text{Substitution product}}{(CH_3)_3C{-}OH} + \underset{\text{Elimination product}}{(CH_3)_2C{=}CH_2}$$

Fig. 9.10. Substitution and elimination products from *t*-butyl bromide.

component of the reaction mixture. Elimination reactions often accompany S_N1 and S_N2 reactions and may be viewed either as disagreeable side reactions or as the desired process, depending on the experimenter's aspirations. One of the important methods for synthesizing alkenes and alkynes, in fact, employs the elimination reaction, so it often behooves the chemist to seek ways for facilitating this pathway. Through the proper choice of alkyl halide, nucleophilic displacing agent, and reaction conditions, either substitution or elimination can often be achieved. With respect to the choice of alkyl halide, it has already been indicated that the tendency to undergo elimination increases from primary to secondary to tertiary halide. With respect to the choice of nucleophilic displacing agent, the tendency to form the elimination product increases as the basicity and size of the reagent increases. Thus, more isobutene is formed when *t*-butyl bromide reacts with potassium hydroxide than with sodium cyanide. The effects of reaction conditions are less easily categorized; understanding them requires an appreciation of certain subtleties of mechanism discussed in Chapter 21.

Because of their close relationship with nucleophilic substitution reactions as well as their importance in synthesis, elimination reactions have been well studied mechanistically. Their mechanisms have been categorized in a fashion analogous to those of substitution processes. Thus, the E1 reaction is a unimolecular process that, like its S_N1 counterpart, involves a two-step sequence in which an intermediate carbonium ion is formed. The E2 reaction is a bimolecular process that, like its S_N2 counterpart, involves a single step without formation of an intermediate. The E2 reaction can be interpreted as involving the simultaneous abstraction of a proton by the nucleophilic reagent and expulsion of the neighboring halide ion, as illustrated in Fig. 9.11.

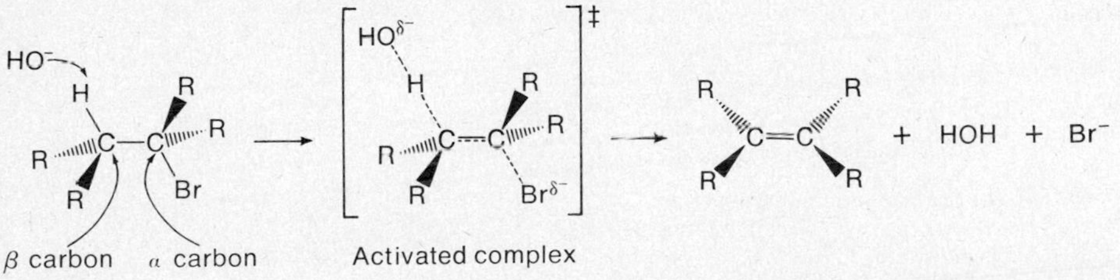

β carbon α carbon Activated complex + HOH + Br^-

Fig. 9.11. Mechanism of the E2 reaction.

Alkyl halides containing hydrogens on more than one β carbon atom have the possibility of forming two or more isomeric alkenes. For example, 2-bromobutane reacts with potassium *t*-butoxide to form a mixture of *cis*- and *trans*-2-butene (*i.e.*, removal of the hydrogens from the C-3 position) and 1-butene (*i.e.*, by removal of the hydrogen from the C-1 position) (Fig. 9.12). The two isomeric

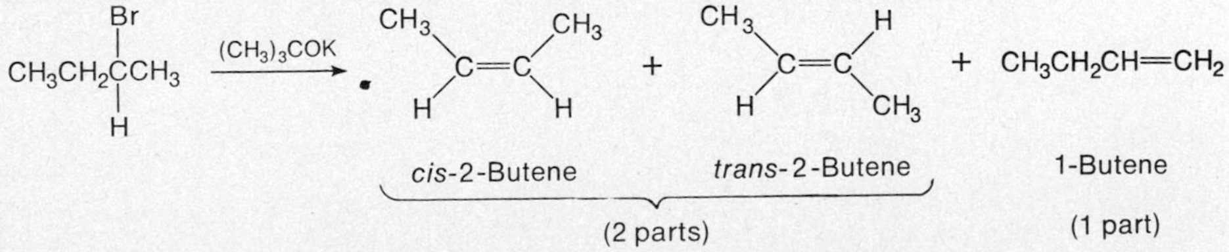

$CH_3CH_2CBrHCH_3 \xrightarrow{(CH_3)_3COK}$ *cis*-2-Butene + *trans*-2-Butene (2 parts) + $CH_3CH_2CH{=}CH_2$ 1-Butene (1 part)

Fig. 9.12. The dehydrobromination of 2-bromobutane.

butenes are not formed in equal amounts, 2-butene predominating by a factor of approximately 2:1. These compounds differ with respect to the number of alkyl groups attached to the double bond, 2-butene having two methyl groups and 1-butene having a single ethyl group. Many years ago a Russian chemist named Saytzev realized that in some elimination reactions the more highly substituted alkenes are formed in larger amounts. This generally holds true for the dehydration of alcohols and the dehydrohalogenation of alkyl halides to alkenes, so these reactions are said to follow the Saytzev rule (Fig. 9.12).

PROBLEM 9-6

Compounds A, B, and C all have the molecular formula $C_7H_{15}Br$. Compound A reacts with sodium hydroxide to give a $C_7H_{16}O$ alcohol, compound B reacts with sodium hydroxide to give a single C_7H_{14} alkene, and compound C reacts with sodium hydroxide to give a mixture containing a $C_7H_{16}O$ alcohol and two C_7H_{14} alkenes. Suggest possible structures for compounds A, B, and C.

9.8 Halides of Particular Utility and Interest

One of the important uses of certain low molecular weight halogen compounds involves the preparation of polymers, which are high molecular weight compounds composed of a large number of small, repeating units. For example, when vinyl chloride is treated with certain catalysts, a long-chain, chlorine-substituted hydrocarbon (*i.e.*, a macromolecule) called polyvinyl chloride is produced:

$$n\,H_2C{=}CHCl \longrightarrow XCH_2(\underset{\displaystyle Cl}{\underset{|}{C}}HCH_2)_{n-1}\underset{\displaystyle Cl}{\underset{|}{C}}HY$$

Polyvinyl chloride

(The X and Y groups at the termini come from other compounds that are present during the polymerization, such as the compounds that initiate the polymerization and the solvent molecules.)

Pure polyvinyl chloride, with molecular weights as high as 1.5 million, is a hard, tough material useful for making panels, pipes (in place of metal pipes), and various molded objects. When mixed with certain low molecular weight compounds called plasticizers, polyvinyl chloride becomes softer and in this form finds use in the manufacture of fabrics resembling leather. In similar fashion, tetrafluoroethylene polymerizes to a completely fluorinated, high molecular weight polymer called Teflon, which is used commercially in a variety of ways, one of which is the coating of frying pans to prevent food from sticking to the surface:

$$n\,F_2C{=}CF_2 \longrightarrow XCF_2(CF_2CF_2)_{n-1}CF_2Y$$

Teflon

An enormous number of materials, both animate and inanimate, that we encounter in our daily lives are polymers. Almost all of these, whether of biological or synthetic origin, can be classified as addition polymers or condensation polymers. Polyvinyl chloride and Teflon are examples of addition polymers, one unit of monomer simply adding to another; Nylon (see Sec. 16.5) is an example of a condensation polymer, one monomer unit joining with another with concomitant extrusion of a water molecule. Together, these two types of polymers comprise one of the biggest and most important areas of organic chemistry.

The first two members of the alkyl halides have considerable commercial utility. Methyl chloride, prepared by the free radical chlorination of methane, is used in the production of silicones and as a solvent for various polymerization processes. Ethyl chloride, produced either by the free radical chlorination of ethane or by the addition of hydrogen chloride to ethylene, is largely consumed in the production of tetraethyl lead, used as a gasoline additive.

Polyhalogenated compounds are used in a variety of commercial applications. Chloroform (the trichloro derivative of methane) and carbon tetrachloride (the tetrachloro derivative of methane) are among the more familiar examples in this category. Chloroform has had some use as an anesthetic, and carbon tetrachloride has been widely employed as a solvent and dry-cleaning agent. Both compounds are hazardous, however, for they can cause liver damage when inhaled or ingested into the body. Still another hazard with chloroform is its light-induced oxidation to the highly toxic gas phosgene—*i.e.*,

$$CHCl_3 + \tfrac{1}{2}O_2 \longrightarrow O{=}CCl_2 + HCl$$

Dichlorodifluoromethane, Cl_2CF_2, on the other hand, is nontoxic, nonflammable, and noncorrosive, and it is useful as a refrigerant liquid (bp −28°), known in the commercial trade as Freon-12. An interesting compound containing three different halogens is $F_3CCHBrCl$, known as Halothane and now widely used instead of ether as an anesthetic.

The toxicity of halogen compounds has been employed on a large scale in the form of insecticides. Undoubtedly the most famous is 1,1,1-trichloro-2,2-di-(4-chlorophenyl)-ethane (DDT)

$$(4\text{-}ClC_6H_4)_2CH{-}CCl_3$$

DDT

DDT

This compound, although remarkably effective as an insecticide and, therefore, responsible for vastly improving the world's agricultural production and reducing the incidence of malaria in many countries,* has the disadvantage of not being readily degraded to nontoxic materials. It lingers in the ground, in the water, on leaves, and on fruits and vegetables. Significant amounts of it have, as a result, been ingested by fish and animals the world around, and the "background level" of DDT in the ecology of the planet is no longer negligible. What the long-term

* For example, the incidence of malaria in Ceylon was reduced from 2.8 million cases in 1946 to fewer than 200 cases in 1961 as the result of the use of DDT. When the DDT spraying program was abandoned in the mid 1960's, however, the incidence of malaria again climbed to 2.5 million cases in 1969.

effects of this will be are not yet clear, but since DDT shares with many other halogen-containing compounds the capacity to cause liver damage, it is possible that the effects may be significant. For that reason, DDT is no longer looked on with favor by the general public. Efforts are being made to find other insecticides that are as effective as DDT but which lack its long-term hazards to health and ecological balance.

9.9 *Study Guide*

I. Structure
   A. σ Bond between carbon framework and halogen
   B. Delocalization in ArX and C=C—X (where X = halogen atom)
II. Nomenclature
   A. Halide as principal name; *i.e.*, alkyl halide, aryl halide
   B. Halogen as substituent—IUPAC nomenclature; *i.e.*, haloalkane
   C. α, β, γ, . . . designations in $—CH_2CH_2CH_2X$
III. Physical properties
   A. Bond polarity—dipole moment
   B. Boiling point—comparison with alkanes
   C. Water solubility—low
   D. Density—high
IV. Spectral properties
   A. Ultraviolet/visible—nothing noteworthy
   B. Infrared—C—X stretching frequency in fingerprint region (toward low frequency end)
   C. Nuclear magnetic resonance—only fluorine of interest; similar to hydrogen with respect to coupling (halogen causes downfield shift of adjacent hydrogens)
V. Methods of synthesis
   A. Methods based on inorganic reagent used
      1. Methods using halogens
         (a) $X_2 + R_3CH \rightarrow R_3C—X$ (free radical halogenation of alkanes)
         (b) $X_2 + ArH \rightarrow ArX$ (electrophilic substitution of arenes)
      2. Methods using halogen acids
         (a) HX + C=C → H—C—C—X (electrophilic addition to alkenes)
         (b) HX + ROH → RX (nucleophilic substitution with alcohols)
   B. Methods based on organic compound used
      1. From alkanes: $R_3CH + X_2 \rightarrow R_3CX$
      2. From alkenes: C=C + HX → H—C—C—X
      3. From arenes: $ArH + X_2 \rightarrow ArX$
      4. From alcohols: ROH + HX → RX
VI. Reactions
   A. Substitution reactions
      1. Mechanism
         (a) S_N1 process
         (b) S_N2 process
         (c) Kinetics of processes

(d) Stereochemistry of processes
(e) Substitution versus elimination

2. Reagents capable of effecting displacement
   (a) Oxygen nucleophiles
      (1) $RX + HO^- \rightarrow ROH$ (synthesis of alcohols)
      (2) $RX + R'O \rightarrow ROR'$ (synthesis of ethers)
   (b) Sulfur nucleophiles
      (1) $RX + HS^- \rightarrow RSH$ (synthesis of thiols)
      (2) $RX + R'S^- \rightarrow RSR'$ (synthesis of sulfides)
   (c) Nitrogen nucleophiles
      (1) $RX + R'NH_2 \rightarrow RNHR'$ (synthesis of amines)
      (2) $RX + R'_3N \rightarrow R\overset{+}{N}R'_3X^-$ (synthesis of quaternary ammonium salts)
   (d) Carbon nucleophiles
      (1) $RX + {}^-CN \rightarrow RCN$ (synthesis of nitriles)
      (2) $RX + {}^-C{\equiv}CR' \rightarrow RC{\equiv}CR'$ (synthesis of alkynes)
3. Synthetic utility
   (a) Preparation of other functionalized compounds
   (b) Limitations (*e.g.*, with tertiary and neopentyl halides)
   (c) Nonreactivity of vinyl and aryl halides

B. Organometallic reagents
   1. Formation of Grignard reagents
   2. Formation of alkali-metal derivatives
   3. Reactions of organometallics
      (a) With "active hydrogen" compounds (*e.g.*, H_2O, ROH, RSH, RCO_2H, R_2NH, etc.)
      (b) With carbonyl compounds (synthesis of alcohols)

C. Elimination reactions
   1. Mechanism
      (a) E1 process
      (b) E2 process
      (c) Saytzev rule
      (d) Elimination versus substitution
   2. Synthetic utility; preparation of alkenes

PROBLEMS

1. Name each of the following compounds (a) as a derivative of the appropriate halogen acid and (b) by the IUPAC system:

(a) CH_3CH_2F (b) $(CH_3)_2CHBr$ (c) $(CH_3)_2CHCH_2I$

(d) $(CH_3)_3CCH_2Cl$ (e) $CH_3CH_2\underset{\displaystyle Cl}{\underset{|}{C}}HCH_3$ (f) Cl

(g) Br (h) F

2. Draw the structures corresponding to each of the following names.

(a) Isooctyl bromide
(b) *t*-Pentyl chloride
(c) Cycloheptyl iodide
(d) Vinyl chloride
(e) Allyl bromide
(f) 2-Chloropentane
(g) 2,3-Dichloro-2-methylbutane
(h) *m*-Dibromobenzene
(i) *p*-Iodotoluene
(j) 1-Fluoro-4,6,7-trimethylnaphthalene

3. Name each of the following structures:

(a) $CH_3(CH_2)_4CH_2Cl$

(b) $(CH_3)_3CCH_2CH_2Br$

(c) 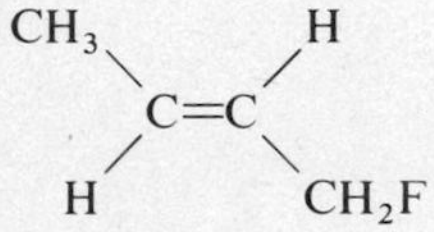 CH_3 H C=C H CH_2F

(d) HCI_3

(e) $(CH_3)_2CC{\equiv}CH$ (with Cl on the central C)

(f)  Cl H H Cl

(g) 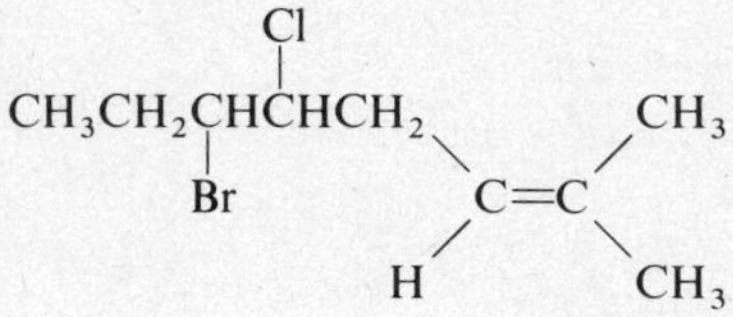 $CH_3CH_2CHCHCH_2$ (Cl, Br) C=C CH_3 H CH_3

(h) Br

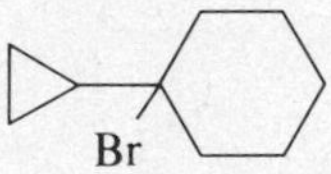

(i) 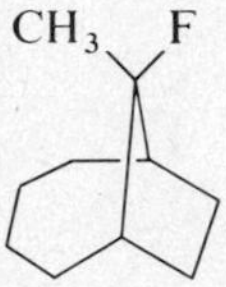 CH_3 F

(j) 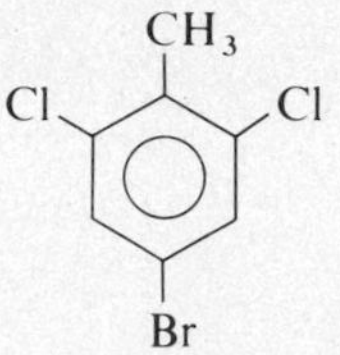 CH_3 Cl Cl Br

(k) 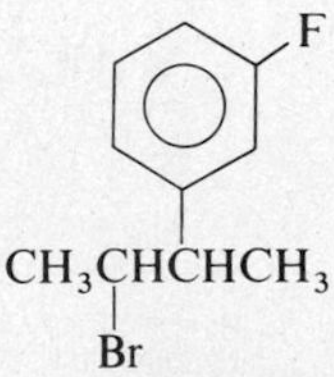 F $CH_3CHCHCH_3$ Br

(l) 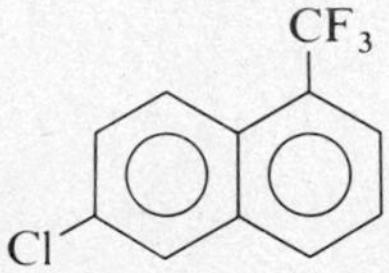 CF_3 Cl

4. Draw the structures corresponding to each of the following names:

(a) 5-Chloro-1-cyclopentyl-3,4-dimethylhexane
(b) 1-Chloro-4-(4-chlorocyclohexyl)benzene
(c) Dichlorodivinylmethane
(d) (2*Z*,4*E*)-6-Bromo-6-vinyl-2,4-nonadien-7-yne
(e) *cis*-3,6-Dichlorocyclohexene
(f) (*Z*)-4-Bromo-3-bromoethyl-3-penten-1-yne
(g) 2-(3-Chloro-1-methyl-1-propenyl)-naphthalene
(h) 9,10-Dibromo-9,10-dihydrophenanthrene
(i) 2-(3,5-Dichlorophenyl)-1-fluoroethane

5. Draw the structure of the product that is produced by the action of the following reagents on 1-methylcyclopentene, CH_3

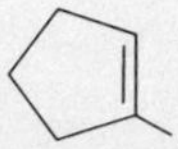

(a) H_2 in the presence of a palladium catalyst
(b) HCl
(c) HBr in the presence of peroxides
(d) Br_2
(e) H_2O in the presence of a mineral acid catalyst
(f) O_3 in the presence of CH_2Cl_2/pyridine

6. The following conversions from the compounds shown on the left to the compounds shown on the right can be effected in one step. Indicate what reagents are necessary to achieve this.

(a) $CH_3CH{=}CH_2 \longrightarrow (CH_3)_2CHBr$

(b) 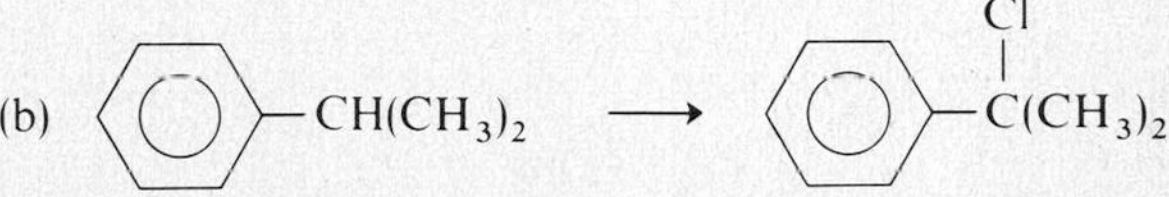

(c) —OH ⟶ —I

(d) Br ⟶

(e) $H_2C{=}CHCH_2Cl \longrightarrow H_2C{=}CHCH_2CN$

(f) $C_2H_5Br \longrightarrow (C_2H_5)_4\overset{+}{N}Br^-$

(g) $(CH_3)_2CHCH_2Cl \longrightarrow (CH_3)_2CHCH_2SH$

(h) —Cl ⟶ —NH_2

7. The following conversions from the compounds shown on the left to the compounds shown on the right require more than one step. Indicate what sequences of reactions are necessary to achieve this, show the reagents, and draw the structures of the intermediate compounds that are produced.

(a) $(CH_3)_2CHCH_2Cl \longrightarrow (CH_3)_3CCl$

(b) ⟶ —N(H)CH_3

(c) —$CH(CH_3)_2$ ⟶ —C(=CH_2)CH_3

(d) $H_2C{=}CHCH_3 \longrightarrow H_2C{=}CHCH_2CH_3$

(e) $CH_3CH_2\underset{\displaystyle H}{\underset{|}{C}}(CH_3)_2 \longrightarrow CH_3CH_2\underset{\displaystyle D}{\underset{|}{C}}(CH_3)_2$

8. Starting with elemental carbon as the only source of carbon and using any necessary inorganic reagents, synthesize each of the following compounds. (*Note*: An intermediate prepared in one scheme can be used in the preparation of a later compound without duplicating its preparation.)

(a) CH_3CH_2Br

(b) $(CH_3CH_2)_2S$

(c) $H_2C{=}CHCH_2CH_3$

(d) $CH_3\underset{\displaystyle O}{\underset{\|}{C}}CH_2CH_3$

(e) $CH_3CH_2CH_2CH_2CN$

(f) $CH_3CH_2\underset{\displaystyle H}{\underset{|}{N}}CH_2CH_2CH_2CH_3$

9. In the following examples you are presented with two test tubes, each containing one or the other of a pair of isomeric compounds. Outline a simple chemical test you could perform to distinguish between the compounds, and describe what the result of the test would be.

(a) o-bromotoluene (Br, CH_3 on benzene ring) and benzyl bromide ($C_6H_5CH_2Br$)

(b) $(CH_3)_3CCHClC_6H_5$ and $(CH_3)_2CHCH_2CHClC_6H_5$

(c) $(CH_3)_3CBr$ and $(CH_3)_2CHCH_2Br$

(d) iodocyclohexane and $H_2C{=}CHCH_2CHICH_2CH_3$

(e) $CH_3CH{=}CHCl$ and $H_2C{=}CHCH_2Cl$

10. Draw the structure of the major product(s) in each of the following reactions. If no reaction is anticipated, indicate this by the designation "NR".

(a) 2-naphthyl-CH_2OH + HBr $\longrightarrow$

(b) $CH_3OCH{=}CH_2$ + HBr $\longrightarrow$

(c) $(CH_3)_3C$–C_6H_4–$CH(CH_3)_2$ (para) + Cl_2 $\xrightarrow[h\nu]{}$

(d) o-$C_6H_4(CH_2Br)_2$ + 2 $NaSCH_2CH_3$ $\longrightarrow$

(e) $CH_3(CH_2)_4CH_2Br$ + $(CH_3)_3COK$ $\longrightarrow$

(f) o-($CH{=}CHBr$)C_6H_4(CH_2Cl) + KCN $\longrightarrow$

(g) CH_3–C_6H_4–Br (para) + $NaNH_2$ $\xrightarrow[NH_3]{}$

(h) 1-iodo-1-methylcyclohexane (CH_3, I) + KOH $\longrightarrow$

11. Draw the structures for each of the compounds in the following reaction schemes:

(a) $C_6H_5CH_2CH_3 \xrightarrow[h\nu]{Br_2} A \xrightarrow{(CH_3)_3COK} B \xrightarrow{Br_2} C \xrightarrow{(CH_3)_3COK} D$

(b) $HC{\equiv}CH \xrightarrow[NH_3]{NaNH_2} E \xrightarrow{CH_3I} F \xrightarrow[NH_3]{NaNH_2} G$

$G \xrightarrow[(2)\ H_2O]{(1)\ (CH_3)_2C{=}O} H \xrightarrow{SOCl_2} I$

(c) $E \xrightarrow{(CH_3)_2CHI} J \xrightarrow[Pd/BaSO_4]{1\text{ mole }H_2} K \xrightarrow[Peroxide]{HBr} L$

$L \xrightarrow[Ether]{Mg} M \xrightarrow{H_2O} N$

12. Draw structures for compounds A and B described in the following problems.

(a) Treatment of 2-bromo-3-methyl-3-phenylbutane with potassium *t*-butoxide (E2 conditions) produces compound A. Compound A rapidly reacts with bromine to give a dibromo compound. The nmr spectrum of compound A has multiplets at δ 7.3, 5.8, and 5.1 and a singlet at δ 1.1 with relative intensities of 5:1:2:6.

(b) Reaction of 2-bromo-3-methyl-3-phenylbutane under E1 conditions yields compound B as the major product. Compound B, isomeric with compound A [see (a)], has an nmr spectrum containing a multiplet at δ 7.5 and three singlets at δ 1.65, 1.58, and 1.50; The relative intensities are 5:3:3:3.

13. A hydrocarbon (C_8H_{10}) reacts with chlorine in the presence of aluminum chloride to produce a single product A with molecular formula C_8H_9Cl. Treatment of the hydrocarbon with one molar equivalent of bromine in the presence of light produces a single product B (C_8H_8BrCl). The nmr spectrum of B displays a multiplet at δ 7.3 and singlets at 4.4 and 2.35 with an intensity ratio of 3:2:3. What are the structures of the original hydrocarbons, A and B?

14. A C_6H_{14} hydrocarbon reacts with bromine in the presence of light to produce a $C_6H_{13}Br$ compound in high yield. When the bromine-containing compound is treated with potassium *t*-butoxide, two isomeric alkenes designated as Y and Z are produced. Alkene Y displays a band in the infrared region at 895 cm^{-1}, and alkene Z has a band at 826 cm^{-1}. On the basis of this information, suggest possible structures for Y and Z. What additional information might you want to be able to establish definitively the structures of Y and Z as well as the C_6H_{14} hydrocarbon?

COMMENTS ON CHARACTERIZATION (IDENTIFICATION, DIFFERENTIATION) PROBLEMS

The organic chemist working in the laboratory frequently faces the problem of establishing the identity of organic compounds. It may be a compound he has isolated from a natural source; it may be a compound in the product of a competitor; it may simply be the compound in a bottle from which the label has disappeared. Though the magnitude of the effort can vary tremendously, from simple tests requiring a few minutes to perform to extensive tests requiring days, weeks, or months to perform, the goal is the same in all cases; *viz.*, to establish the identity of the material in an unequivocal fashion. This exercise is often called "characterization" or "identification" or "differentiation."

In approaching a characterization (identification, differentiation) problem, the modern chemist will probably first measure the spectral characteristics of the unknown compound. Sometimes, however, spectra are not particularly informative and sometimes they cannot even be obtained. Recognizing this, many of the problems in this and the following chapters are designed to require only chemical methods. In such instances the two kinds of tests that should generally be applied first are (a) the solubility characteristics of the compound; *i.e.*, is it soluble in dilute aqueous acid, sodium hydroxide, and/or sodium bicarbonate? and (b) the elemental composition of the compound; *i.e.*, does it contain halogen, nitrogen, or sulfur?

The qualitative determination of the elemental composition is carried out by mixing a small sample of the compound with a piece of metallic sodium in a test tube, heating the test tube rather strongly in a Bunsen flame (causing the sodium to react with the organic compound and converting the halogen, nitrogen, and sulfur to salts of sodium), adding the contents of the test tube to water, and carrying out an inorganic qualitative analysis for halide, cyanide, and sulfide.

On the basis of the solubility characteristics and the elemental composition, the compound can be placed in one or another of the chemical categories, at which point more specific tests (*e.g.*, decoloration of a solution of bromine in carbon tetrachloride as a test for the presence of unsaturation) can then be applied to narrow down the possibilities and establish the identity.

Chapter **10**

Hydroxy compounds: alcohols and phenols

There is held to be no surer test of civilization than the increase per head of the consumption of alcohol and tobacco. Yet alcohol and tobacco are recognizably poisons, so that their consumption has only to be carried far enough to destroy civilization altogether.

HAVELOCK ELLIS
The Dance of Life (1923)

J. STRAUSS
Prince Orlofsky, in *Die Fledermaus*

The term "alcohol" is a generic name pertaining to any compound in which a carbon framework is attached to a hydroxyl group (OH group). Generally, however, its use has been restricted to those compounds in which the hydroxyl group is attached to an sp^3-hybridized carbon atom, the term "phenol" being used for those compounds in which the hydroxyl group is attached to an aromatic framework. Although there are certain similarities between alcohols and phenols, the differences are sufficiently great that it is reasonable to place these two types of compounds in separate classes.

Alcohols and phenols can be considered as derivatives of water in which one of the hydrogens has been replaced by a carbon framework; many of the properties of these compounds do, indeed, resemble those of water. The compounds also, of course, have characteristics deriving from the carbon framework portion of the molecules, and the extent of this contribution to the physical properties of the compound increases as the molecular weight of the compound increases.

The electron configuration of oxygen is described as $1s^2 2s^2 2p_x^2 2p_y^1 2p_z^1$, indicating that a pair of $2p$ orbitals are incompletely filled and available for bonding. Oxygen then, is predicted to be divalent. When oxygen is combined via single bonds to two other entities, such as hydrogen in the water molecule, it is not clear whether the atomic description or a hybridized description of the electron configuration should be invoked. The atomic description of the electron configuration requires an H—O—H bond angle of 90°, because p orbitals are mutually perpendicular (orthogonal) to one another. The sp^3-hybridized description, on the other hand, requires an angle of 109°28′. The measured bond angle in the water molecule is 104.5°, and since this is closer to the angle required by the hybridized oxygen atom than that required by the unhybridized oxygen atom, an arbitrary choice in favor of the hybridized oxygen atom is made for the purpose of this discussion.

Attachment of one hydrogen atom to an oxygen atom produces a hydroxyl group (OH), which can be attached to any of the carbon frameworks. When the hydroxyl group is attached to an sp^3-hybridized carbon, "ordinary" alcohols are produced, an orbital representation of which is shown in Fig. 10.1. Attachment

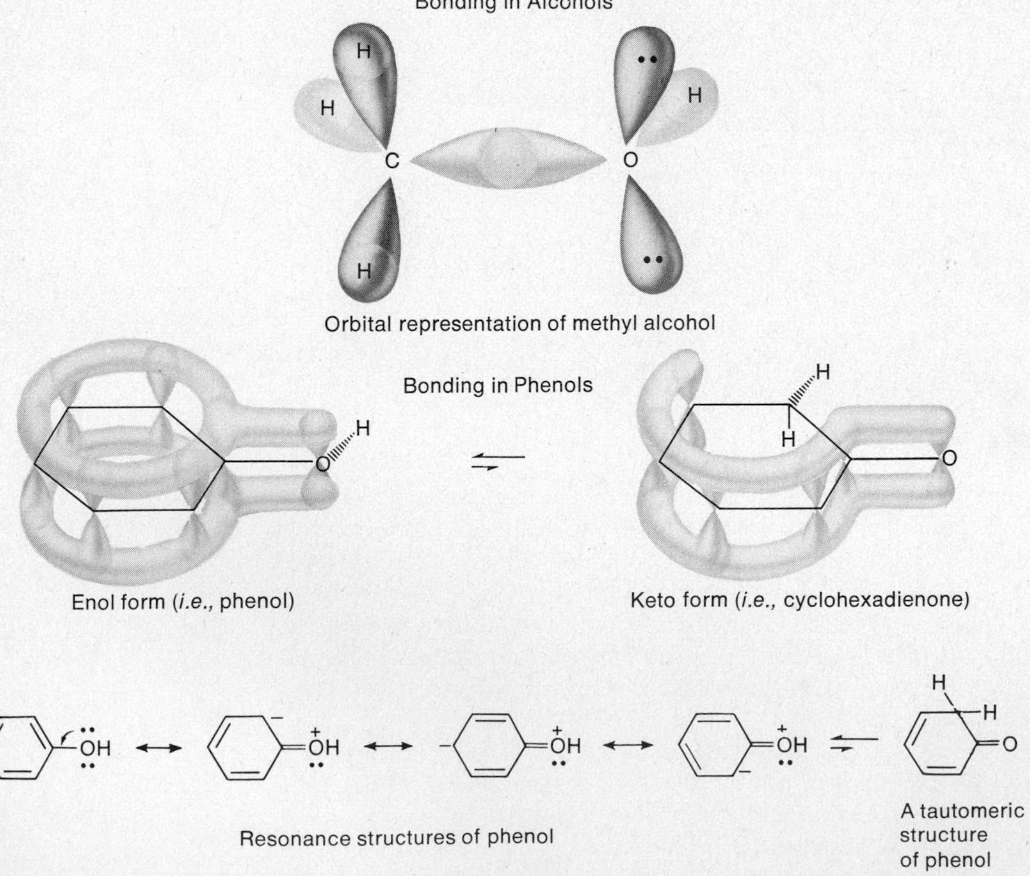

Fig. 10.1. Bonding in alcohols and phenols.

of a hydroxyl group to an sp^2- or sp-hybridized carbon atom, however, produces compounds (alkenols and alkynols, respectively) that in most cases are less stable than the isomeric aldehydes, ketones, or ketenes.

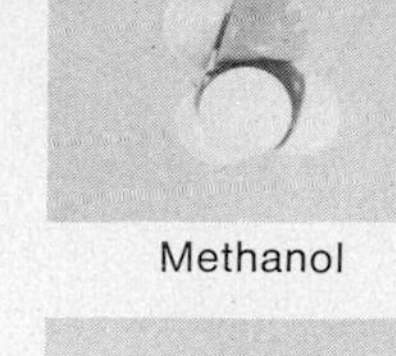

Methanol

Phenol

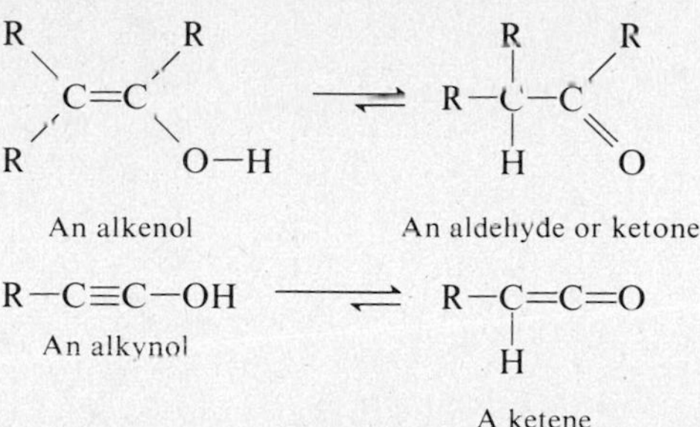

Alkenols (and alkynols), usually simply called "enols," are said to be tautomers of the isomeric carbonyl compounds, and the equilibrium between the species is said to be a *tautomeric equilibrium*; the phenomenon is referred to as *tautomerism*. The reason that the carbonyl compounds are more stable than the alkenol or alkynol is apparent from bond energy calculations. Considering only the bonds that change in the process, $R_2C{=}C(R)(O{-}H)$ has one C=C bond (146 kcal/mole), one C—O bond (85 kcal/mole), and one O—H bond (111 kcal/mole); and $R_2CHC(R){=}O$ has one C—C bond (83 kcal/mole), one C—H bond (99 kcal/mole), and one C=O bond (179 kcal/mole). The difference between the sums of these values is 19 kcal/mole in favor of the carbonyl compound, corresponding to an equilibrium constant for the reaction of approximately 10^8.

A special kind of enol is formed when the hydroxyl group is attached to an aromatic ring. Phenyl alcohol (usually referred to as phenol—*i.e.*, the enol with a phenyl ring), it turns out, is much more stable in the enol form than in the tautomeric ketone form. This, also, can be explained in terms of bond energies. The bond energy advantage of 19 kcal/mole possessed by the ketone (specifically cyclohexadienone in this case) is more than offset by the advantage of about 36 kcal/mole that the phenol derives from the resonance stabilization (*i.e.*, aromaticity) of a completely delocalized π system (Fig. 10.1). Thus, the tautomeric equilibrium in this case strongly favors the phenol, and only a very small amount of the ketone is present in the equilibrium mixture. Since the length and strength of the C—O bond is a function of the hybridization of the carbon, just as in the case of C-halogen bonds (see Section 9.1), the C—O bonds in phenols are shorter and stronger than those in ordinary alcohols.

The structure of several representative examples of alcohols and phenols are shown in Fig. 10.2.

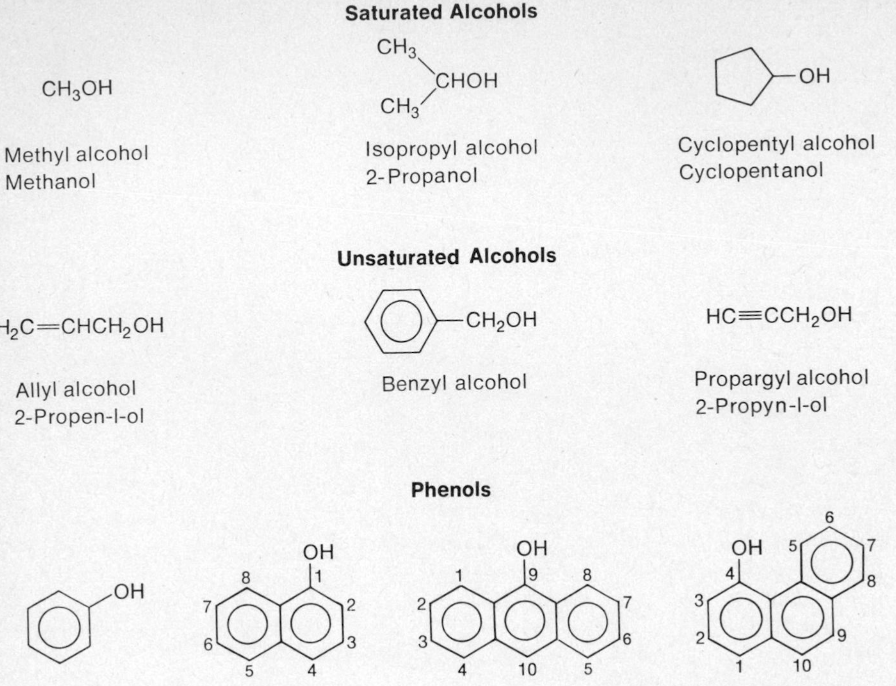

Fig. 10.2. Structures of representative alcohols and phenols.

10.2 Nomenclature of Alcohols and Phenols

Several systems for naming alcohols are available. One of these, called the "alcohol system," simply involves adding the word "alcohol" after the appropriate alkyl designation of the hydrocarbon portion. Thus, CH_3OH is methyl alcohol, CH_3CH_2OH is ethyl alcohol (an example of a *primary alcohol*), $(CH_3)_2CHOH$ is isopropyl alcohol (an example of a secondary alcohol), and $(CH_3)_3COH$ is tertiary-butyl alcohol (an example of a tertiary alcohol). A second method, designated as the "carbinol system," names the parent member, CH_3OH, as "carbinol" and considers more complex alcohols as derivatives of carbinol. Thus, $(CH_3)_3COH$ is trimethylcarbinol, $(C_6H_5)_2CHOH$ is diphenylcarbinol, $CH_3CH_2\underset{\displaystyle OH}{\underset{|}{C}}HCH_2C_6H_5$ is benzylethylcarbinol (note the alphabetical ordering of substituents), etc. Not all alcohols are easily named by the "alcohol" and "carbinol" systems, particularly those having complex frameworks. The most generally applicable nomenclature is the IUPAC system, which is applied according to the following rules:

1. The longest continuous chain *containing the hydroxyl group is selected* as the basis for designating the parent compound, and the actual compound is then named as a derivative of the parent compound. The parent compound, if saturated, is named according to the alkane nomenclature, but the suffix "ane" is replaced by "anol." For example, CH_3OH is methanol,

$CH_3CH_2CH_2CH_2OH$ is butanol, (cyclohexyl)–OH is cyclohexanol, etc.

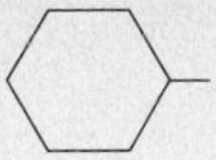

2. The position of the OH group in the parent compound is designated by counting from that end of the chain which assigns the OH group the lower number. Thus, $CH_3CH(OH)CH_2CH_2CH_3$ is 2-pentanol rather than 4-pentanol.
3. The numbering of the parent compound having been established by the position of the OH group, the substituents are specified by names, assigned numbers, and ordered by the same rules that apply to alkanes (see p. 63). For example, the compound at the left side of Fig. 10.3 is named 7,7-dimethyl-4-(2-phenylethyl)-3-octanol.

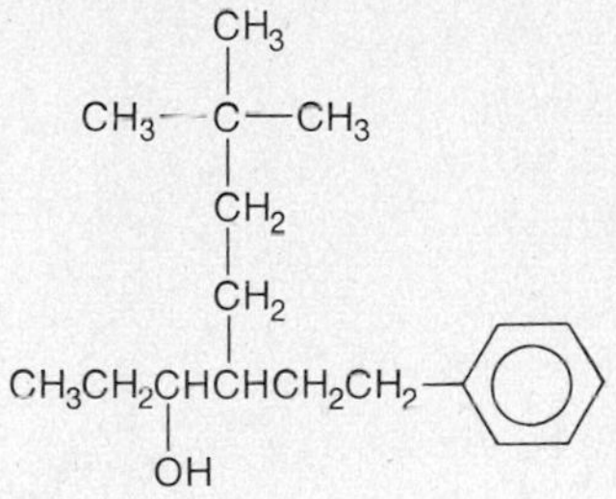

7,7-Dimethyl-4-(2-phenylethyl)-3-octanol

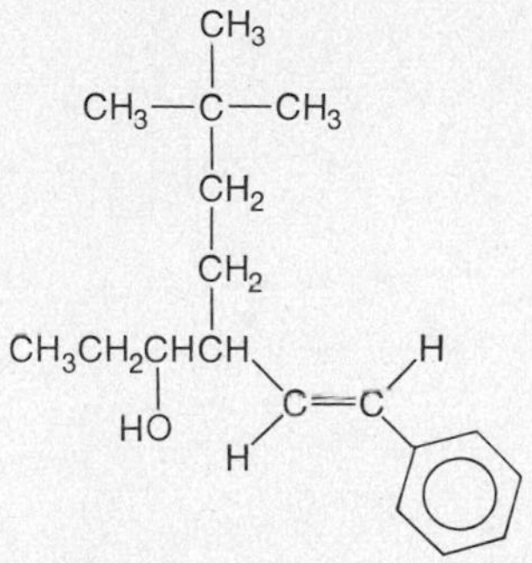

trans-4-(3,3-Dimethylbutyl)-6-phenyl-5-hexen 3-ol

Fig. 10.3. Nomenclature of alcohols with complex frameworks.

For more complex structures containing two or more hydroxyl groups or other functions as well as hydroxyl groups, additional IUPAC rules have been formulated:

1. When only one functional group is present in the molecule, the parent compound is chosen so as to contain that group. When more than one function of the same kind is present, the parent compound is chosen to contain as many of these as possible. For example,

   $CH_3CH(OH)CH(OH)CH(CH_2CH_2CH(CH_3)_2)CH_2OH$ is named 3-(3-methylbutyl)pentane-1,2,4-triol.

   [equally correct is the designation of the position numbers of the hydroxyl functions before the alkane designation; *e.g.*, 3-(3-methylbutyl)-1,2,4-pentanetriol].
2. When two or more functions different in kind are present, the parent compound is chosen to contain the one of greatest preference (designated

as the "principal function"), and the parent chain is numbered to give the principal function the lowest number. For example

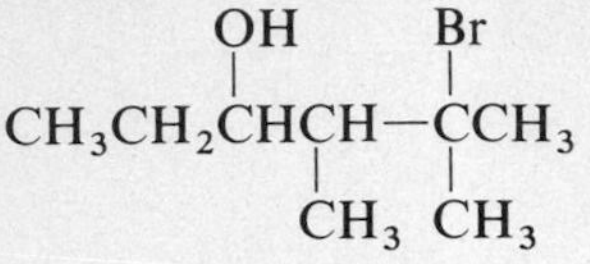

$$\underset{}{CH_3CH_2\overset{OH}{\overset{|}{C}}H\underset{\underset{CH_3}{|}}{C}H-\overset{Br}{\overset{|}{\underset{\underset{CH_3}{|}}{C}}}CH_3}$$ is named 5-bromo-4,5-dimethyl-3-hexanol.

3. The order of *decreasing preference* for functions is as follows: acids (carboxylic, sulfonic, etc.), acid halides, amides, aldehydes, nitriles, ketones, alcohols, phenols, thiols, amines, ethers, thioethers (sulfides), sulfoxides, and sulfones (for a more detailed list, see Chapter 18).
4. Whenever possible, double and triple bonds should be included in the parent compound, even if this does not correspond to the longest carbon chain. The numbering of the system is determined by the principal function, as illustrated by the alcohols in Fig. 10.3.

The parent member of the system in which a hydroxyl group is attached to a benzene ring is called phenol, and various substituted relatives can be named as *derivatives of phenol.* Alternatively, the OH group can be specified as a *substituent*

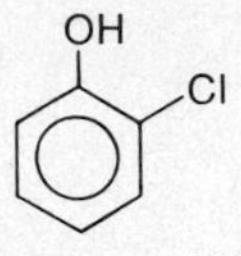

o-Chlorophenol
2-Chlorophenol
2-Chloro-1-hydroxybenzene

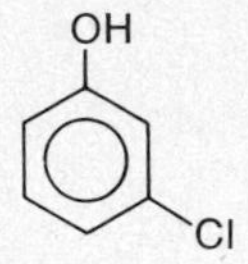

m-Chlorophenol
3-Chlorophenol
3-Chloro-1-hydroxybenzene

p-Chlorophenol
4-Chlorophenol
4-Chloro-1-hydroxybenzene

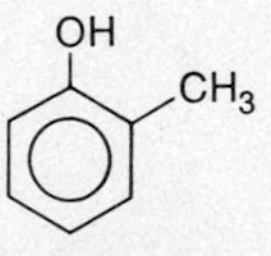

o-Hydroxytoluene
1-Hydroxy-2-methylbenzene
o-Methylphenol
2-Methylphenol
o-Cresol

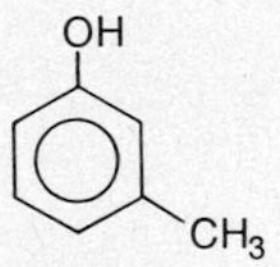

m-Hydroxytoluene
1-Hydroxy-3-methylbenzene
m-Methylphenol
3-Methylphenol
m-Cresol

p-Hydroxytoluene
1-Hydroxy-4-methylbenzene
p-Methylphenol
4-Methylphenol
p-Cresol

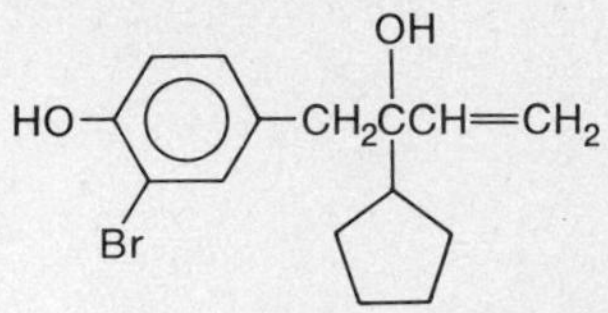

1-(3-Bromo-4-hydroxyphenyl)-2-cyclopentyl-3-butene-2-ol

Fig. 10.4. Nomenclature of phenols, including systematic and common names.

and designated as "hydroxy." When a phenolic moiety is included in a compound that is named by the IUPAC system, the hydroxyl group usually is specified as a substituent. These several possibilities are illustrated by the compounds in Fig. 10.4.

As we have already seen, the names of organic compounds sometimes appear as a single word and sometimes as two or more words. The single-word designation is used when the compound is specified as a derivative of a *particular* compound; the multiple-word designation is used when the compound is specified as a member of a *class* of compounds. Thus, $(CH_3)_2CHCH(OH)CH(CH_3)_2$ is named 2,4-dimethyl-3-pentanol (one word—*i.e.*, a derivative of the particular compound "pentanol") or diisopropylmethyl alcohol (two words—*i.e.*, as a specific member of the class of compounds called "alcohols").

PROBLEM 10-1

Draw and name all the $C_4H_{10}O$ alcohols. Draw and name all the stable C_4H_6O alcohols. Draw and name all of the $C_8H_{10}O$ phenols.

10.3 Physical Properties of Alcohols and Phenols

The comparison of the boiling points of hydroxyl-containing compounds, such as water and alcohols, with nonpolar compounds of comparable molecular weight shows dramatic differences. For example, methanol (mol wt 32) has a boiling point of 65°, whereas ethane (mol wt 30) has a boiling point of −88°—a difference of 154°! This extraordinary disparity in boiling points between methanol and ethane is most readily explained on the basis of increased polarity and a phenomenon known as "hydrogen bonding."

Oxygen is considerably more electronegative than either carbon or hydrogen, and it permanently polarizes an alcohol molecule in the fashion shown in Fig. 10.5. Intermolecular dipole–dipole interaction is, therefore, to be expected. It takes place primarily through the H—O portion of the bond rather than through the O—R portion of the bond, partly for steric reasons (hydrogen is a small atom) and partly because hydrogen has no interior sheath of electrons through which

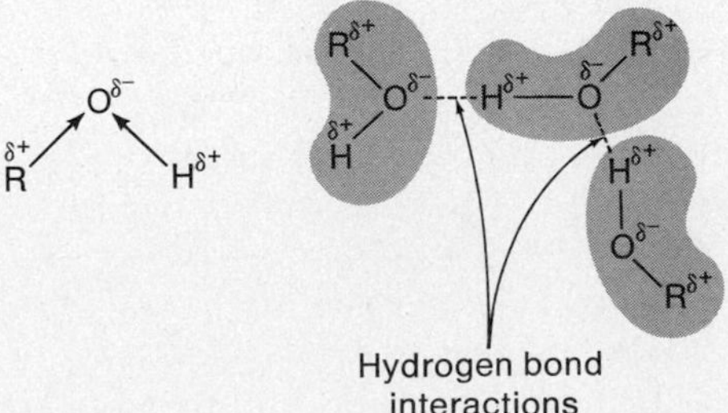

Fig. 10.5. Hydrogen bonding interactions in alcohols. The δ^+ and δ^- symbols indicate fractional positive and negative charges, respectively; the dashed line indicates a hydrogen bond.

the positive charge on a nucleus must operate. Thus, alcohol molecules are held together as the result of an O—H--O interaction, in which a hydrogen is sandwiched between a pair of oxygen atoms. This is referred to as a hydrogen bond. Although it is less than 10% as strong as most covalent bonds, it is still sufficiently strong to increase significantly the intermolecular attraction, which is reflected in the rather high boiling points of alcohols.

Perhaps the simplest (although incomplete) interpretation of the hydrogen-bond phenomenon views it as an acid-base interaction. If a strong acid such as HCl is brought into contact with the moderately strong base NH_3, a proton is transferred more or less completely from Cl^- to NH_3 to produce the salt $NH_4^+Cl^-$. On the other hand, if an extremely weak acid such as CH_4 is brought into contact with NH_3, there is no proton transfer whatsoever. If, now, a compound of acidity somewhere between the very strong acid HCl and the very weak acid CH_4 is brought into contact with NH_3, a *partial transfer* of a proton takes place. This is the situation that exists when, for instance, an alcohol and ammonia interact. The proton is partially, not completely, transferred from the oxygen of the alcohol to the nitrogen of NH_3; the hydrogen remains partially bonded to the oxygen and becomes partially bonded to the nitrogen. In this condition, a hydrogen bond is said to exist between the alcohol and the ammonia. The requirements, then, for an effective hydrogen bond are a good Lewis base and a proton acid of moderate strength. Typical examples of hydrogen bonds are O—H--O (about 6 kcal/mole), ≡C—H--O (2–3 kcal/mole), O—H--N (4–7 kcal/mole), N—H--O (2–3 kcal/mole), and N—H--N (about 6 kcal/mole); the numbers in parentheses correspond to the approximate strength of the particular hydrogen bond.

The intermolecular attraction involved in hydrogen bond formation arises entirely from the OH portion of an alcohol. The hydrocarbon portion contributes to intermolecular attraction only via van der Waals forces, with the result that the *differences* in boiling points between adjacent members of a homologous series of alcohols are about the same as between those of a homologous series of hydrocarbons. The boiling points of methanol and methane, for example, are strikingly different, but the increments per CH_2 group for their homologs are similar.

Phenols have the same capabilities for hydrogen bond formation as alcohols. Thus, phenol (mol wt 94) boils at 182°, whereas toluene (mol wt 92) boils at 111°. The 71° difference between these boiling points is almost the same as the 69° difference between the boiling points of $C_6H_5CH_2OH$ (benzyl alcohol, bp 205°) and $C_6H_5CH_2CH_3$ (ethylbenzene, bp 136°). The lower basicity of the oxygen in phenol (reducing its ability to act as a hydrogen bond acceptor) is offset by the greater acidity of its hydrogen (increasing its ability to act as a hydrogen bond donor), making hydrogen bonding in phenols and alcohols comparable.

The high dipole moment and the capacity for intermolecular hydrogen bonding displayed by alcohols and phenols also affect their melting points. The lowest molecular weight straight-chain hydrocarbon that is solid at room temperature is the 20-carbon compound eicosane (mol wt 282, mp 36°), whereas the 14-carbon alcohol *n*-tetradecyl alcohol (mol wt 214) melts at 38°. The much lighter but more highly branched 4-carbon alcohol *t*-butyl alcohol (mol wt 74) melts at 25°. Even more striking differences are seen in the phenols and the aryl hydrocarbons of corresponding molecular weight where, for instance, phenol melts at 41° and toluene melts at −95°.

In an earlier discussion of solubility phenomena (see Section 5.2d), we pointed out that a pair of liquids is immiscible if the intermolecular forces between like molecules are greater than the intermolecular forces between unlike molecules. Hexane, for example, is virtually insoluble in water, because the van der Waals force between hexane molecules is greater than that between hexane and water molecules. If, however, one of the hydrogens of hexane is replaced by a hydroxyl group, the resulting molecule (hexanol) is considerably more water soluble than hexane. Hydrogen bond formation between the hydroxyl group of the hexanol and the water molecule provides sufficient attraction to offset the van der Waals attraction between the hydrocarbon portion of hexanol molecules; as a result, hexanol is "pulled" into solution. As the size of the hydrocarbon portion decreases, the van der Waals interaction diminishes and the effectiveness of the hydrogen bond formation increases; water solubility increases accordingly. Thus, methanol, ethanol, 1-propanol, and 2-propanol are all *infinitely miscible* with water, 1-butanol dissolves to the extent of 7.9 g/100 ml water, 1-pentanol to the extent of 2.3 g/100 ml of water, and 1-hexanol to the extent of 0.6 g/100 ml of water. By arbitrary definition, compounds that are less soluble than 3 g/100 ml of water are said to be insoluble. Interestingly, phenol, similar in molecular weight to hexanol, is very much more water soluble, dissolving to the extent of 9.3 g/100 ml of water.

PROBLEM 10-2

Suggest a laboratory method for separating the following pairs of compounds: (a) C_2H_5OH and $C_{12}H_{25}OH$; (b) cyclohexyl chloride and cyclohexanol

---

10.4 Spectral Properties of Alcohols and Phenols

Alcohols are almost transparent in the ultraviolet spectrum. Phenols, however, have moderately strong absorptions, arising from the benzene ring, and the corresponding phenoxides absorb somewhat more strongly. For example, phenol has a band at 210 nm (ε 6200), and its ion, $C_6H_5O^-$, has bands at 235 nm (ε 9400) and 287 nm (ε 2600).

The highest frequency absorption that is generally observed in the infrared spectrum of organic compounds arises from the oxygen–hydrogen stretching vibration; the free, non-hydrogen bonded OH group appears at approximately 3600 cm^{-1}. To the extent that the hydroxyl group is hydrogen bonded, however, the absorption moves to lower frequency and may go as low as 2500 cm^{-1} in compounds in which the hydrogen bonding is intramolecular. In general, however, O—H absorption occurs at higher frequency than most C—H absorptions, and a band above 3300 cm^{-1} is usually a good indication that an O—H (or an N—H) moiety is present in the molecule. In addition to the O—H stretching absorption, a C—O stretching absorption is also characteristic of hydroxyl-containing compounds. In phenols, this occurs near 1200 cm^{-1} and in alcohols

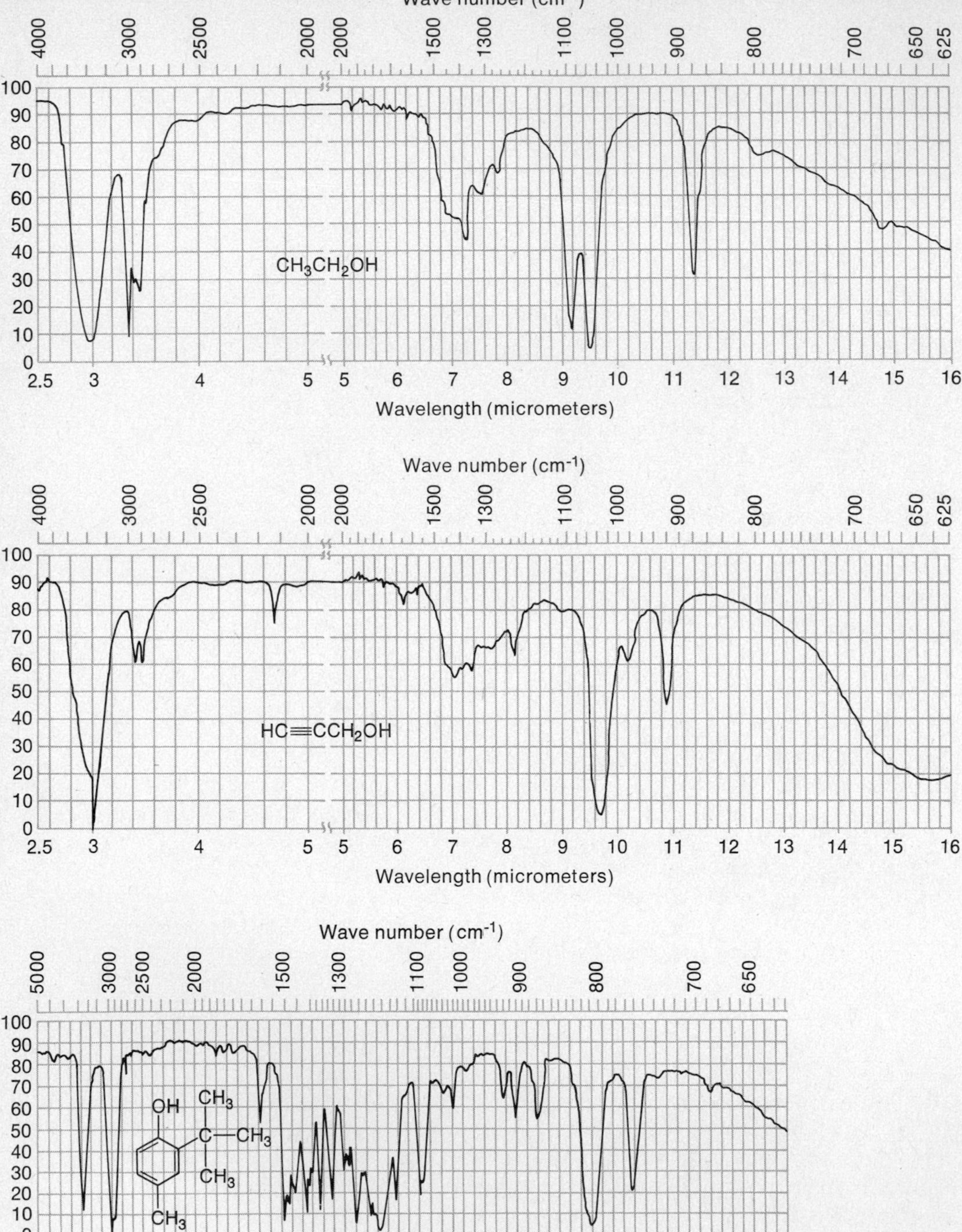

Fig. 10.6. Infrared spectra of ethanol, propynol, and 2-*t*-butyl-4-methylphenol.

at 1100 ± 50 cm^{-1} (Fig. 10.6). A number of other absorptions often occur in this region of the spectrum, though, so the appearance of a band in the 1200–1000 cm^{-1} region is not, in itself, a sure indication that a C—O bond is present in the molecule.

The electronegativity of oxygen causes protons that are proximate to it to be shifted downfield in the nmr spectrum. The nmr spectrum of ethanol, for example, shows a resonance for O—**H** at about δ 5.4 (dependent on concentration and temperature), a resonance for the methylene hydrogens at δ 3.70, and a resonance for the methyl group at δ 1.72, indicating that as the distance between the oxygen and the proton increases (*i.e.*, O**H**, OC**H**$_2$, OCH$_2$C**H**$_3$), the downfield shift diminishes.

Spin–spin coupling between the protons on the carbon and the oxygen should be expected for the system H—C—O—H. In carefully purified samples, this is, in fact, observable (see Fig. 10.7); highly purified ethanol, for example, shows a

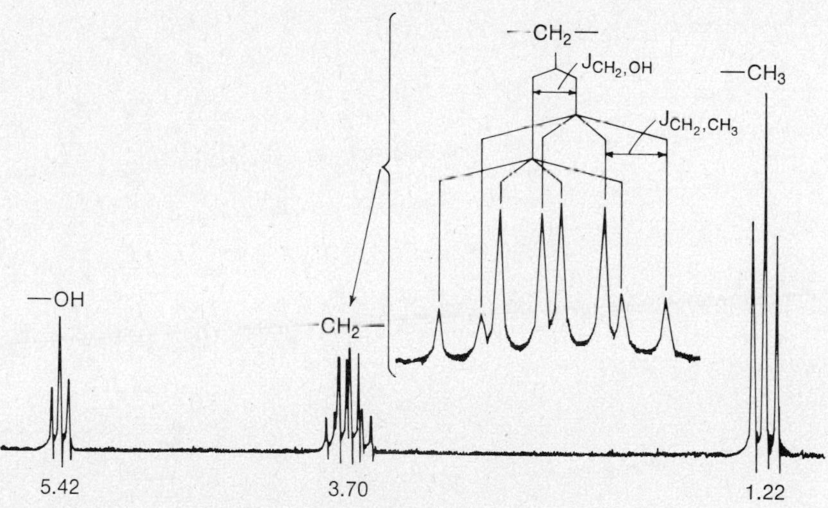

Fig. 10.7. 100-MHz nmr spectrum of highly purified ethanol.

triplet OH resonance arising from the interaction between the hydrogens on the hydroxyl and methylene groups. Reciprocally, the methylene hydrogens couple not only with the methyl hydrogens but also with the hydroxyl hydrogen and show a doublet of quartets (Fig. 10.7). In ordinary, less well-purified alcohol, however, the hydroxyl resonance is a singlet, and the methylene resonance is a quartet. The change is caused by an exchange process of the protons attached to the oxygen atoms. Rather than remaining on any particular oxygen atom during the time that is required for an nmr transition to occur (about 10^{-2} sec), the protons "attached" to oxygen move from one molecule to another. As a result, the hydroxyl resonance is seen as an *average* of the several positions that the perambulatory protons occupy during the time of an nmr transition, causing a disappearance of the coupling between the hydroxyl hydrogen and other hydrogens in the molecule (Fig. 10.8). The rate of proton exchange can be reduced by using certain solvents, such as dimethyl sulfoxide (see Fig. 17.7), which allows the more complex splitting patterns to be observed.

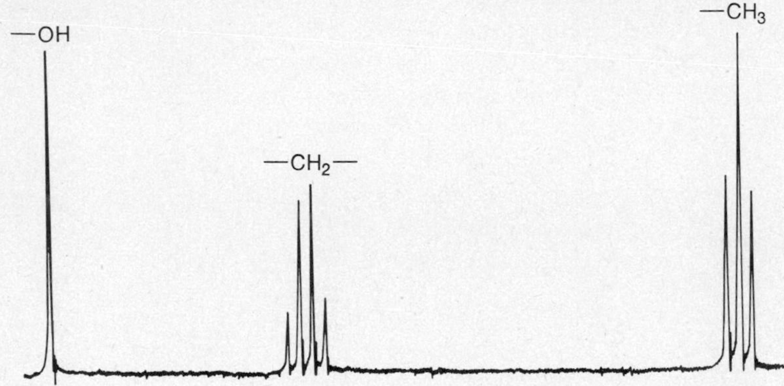

Fig. 10.8. 100-MHz nmr spectrum of ethanol containing a trace of acid.

PROBLEM 10-3

Compounds A and B both have the molecular formula $C_3H_4Cl_2O$. Both show a strong band in the 3500 cm^{-1} region of the infrared, and both show three resonance envelopes in the nmr with one of these being near δ 5. In compound A, the resonance near δ 5 is a singlet, and in compound B it is a triplet. Suggest reasonable structures for compounds A and B.

10.5 *Synthesis of Alcohols and Phenols*

The preparation of alcohols via the acid-catalyzed addition of water to alkenes (see Section 8.1) and via the nucleophilic substitution of halogen in alkyl halides by hydroxyl (see Section 9.5) has already been discussed. Another method of great synthetic utility involves the addition of reagents across the C=O bond of aldehydes, ketones, and esters. The reactions take place in a stepwise fashion, a nucleophile (Nu$^-$) first becoming attached to the carbonyl carbon, followed by attachment of an electrophile (E$^+$) to the carbonyl oxygen. With aldehydes and ketones, the reaction terminates after the addition of one equivalent of

$$\mathrm{RR'C{=}O} + \mathrm{E{:}Nu} \longrightarrow \mathrm{RR'C(Nu)(O{-}E)} \xrightarrow{H_2O} \mathrm{RR'C(Nu)(O{-}H)}$$

Aldehydes (R′ = H)
Ketones (R′ = alkyl or aryl)

reagent. With esters, however, the reaction frequently involves the addition of more than one equivalent of the reagent, because the initially formed product decomposes to an aldehyde or ketone, which undergoes further addition:

$$\text{R(RO)C=O (Esters)} + \text{E–Nu} \longrightarrow \text{R(Nu)C(RO)(O–E)} \longrightarrow \text{R–C(=O)–Nu} + \text{RO–E}$$

$$\text{R–C(=O)–Nu} \xrightarrow{\text{E–Nu}} \text{R(Nu)}_2\text{C–O–E} \xrightarrow{H_3O^+} \text{R(Nu)}_2\text{C–O–H}$$

Among the most often used E–Nu reagents are Grignard reagents (RMgX) and metal hydride reagents [of which lithium aluminum hydride ($LiAlH_4$) is a typical example]; the nucleophilic portions of these reagents are R and H, respectively, these becoming attached to the carbonyl carbon in the course of the reaction:

$$\text{RR'C=O} + \textbf{R}\text{:MgX} \longrightarrow \text{RR'}\textbf{R}\text{C–OMgX} \xrightarrow{H_3O^+} \text{RR'}\textbf{R}\text{C–OH}$$

$$\text{RR'C=O} + Li^+[\text{H–AlH}_2\text{–H}]^- \longrightarrow [\text{RR'HC–O–Al}]^- Li^+ \xrightarrow{H_3O^+} \text{RR'HC–OH}$$

The metal salts of the alcohols (*e.g.*, $-\overset{|}{\underset{|}{C}}-$OMgX from Grignard reagents, $\left[-\overset{|}{\underset{|}{C}}-O\right]_4 Al^- Li^+$ from lithium aluminum hydride), are converted to the corresponding alcohols simply by treatment with aqueous acid in the normal work-up procedure in the laboratory.

By starting with the appropriate aldehyde, ketone, or ester, we can synthesize various primary, secondary, and tertiary alcohols by these procedures, as summarized in Fig. 10.9 and illustrated by specific examples in Fig. 10.10.

We should reiterate an admonition in the use of Grignard reagents cited earlier—viz., Grignard reagents react rapidly with "active hydrogen," more rapidly, in fact, than they add to carbonyl groups. If, *e.g.*, 4-hydroxycyclohexanone is treated with methylmagnesium iodide, the first equivalent of Grignard reagent is consumed by the hydroxyl group to form methane and the MgI salt of 4-hydroxycyclohexanone (Fig. 10.11). If the magnesium salt of the alcohol is sufficiently soluble in the solvent that is used for the reaction (diethyl ether in most cases), a second equivalent of methylmagnesium iodide can then react at the carbonyl group to form a dimagnesium salt, which, upon treatment with water, yields 1-methyl-1,4-cyclohexanediol. Frequently, however, the initially formed magnesium salts are insufficiently soluble to react further, precluding any carbonyl addition reaction.

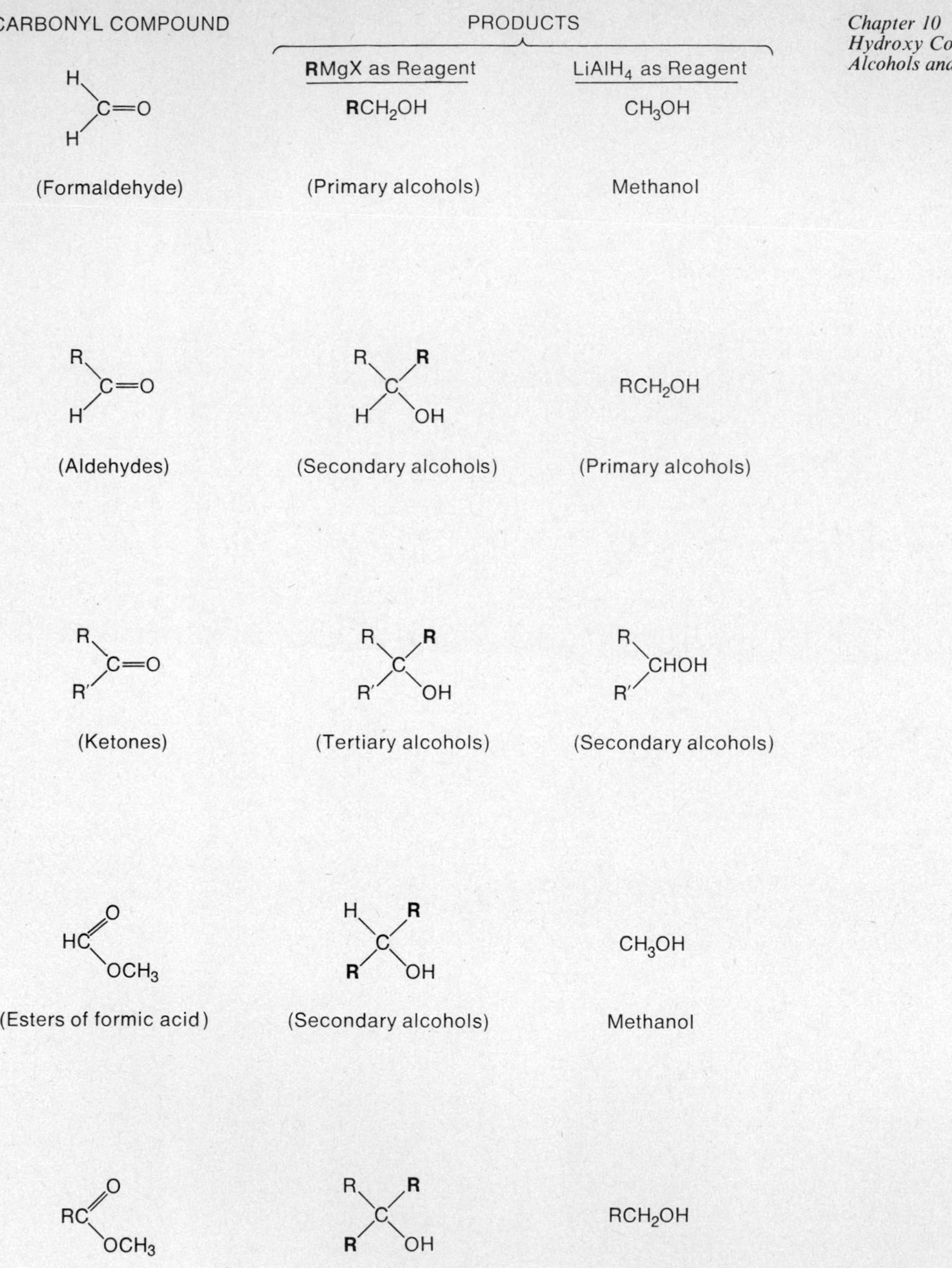

Fig. 10.9. Synthesis of primary, secondary, and tertiary alcohols by the reaction of Grignard reagents and lithium aluminum hydride on aldehydes, ketones, and esters.

Method 1: Addition of H_2O to alkenes

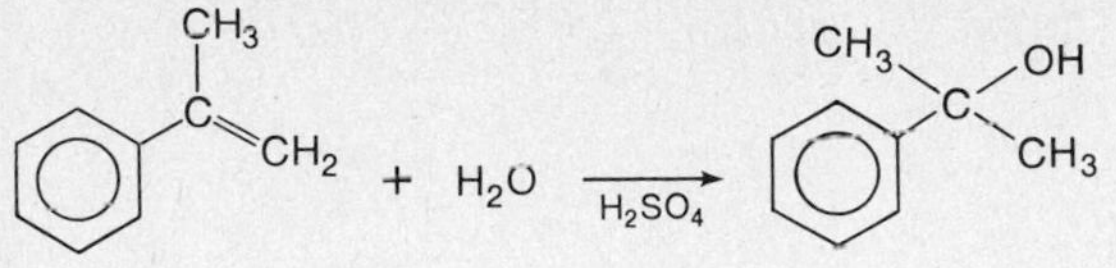

$C_6H_5C(CH_3){=}CH_2 + H_2O \xrightarrow{H_2SO_4} C_6H_5C(CH_3)_2OH$

Method 2: Displacement of halogen by hydroxyl

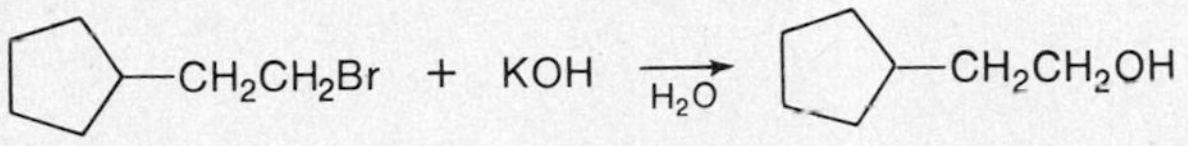

$C_5H_9{-}CH_2CH_2Br + KOH \xrightarrow{H_2O} C_5H_9{-}CH_2CH_2OH$

Method 3: Reduction of carbonyl compounds

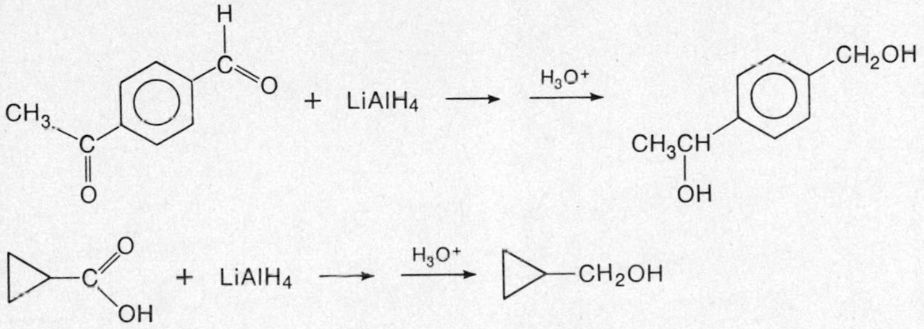

$+ \; LiAlH_4 \longrightarrow \xrightarrow{H_3O^+}$ CH_2OH, $CH_3CH(OH)$

$C_3H_5{-}C(=O)OH + LiAlH_4 \longrightarrow \xrightarrow{H_3O^+} C_3H_5{-}CH_2OH$

Method 4: Addition of organometallic compounds to carbonyl compounds

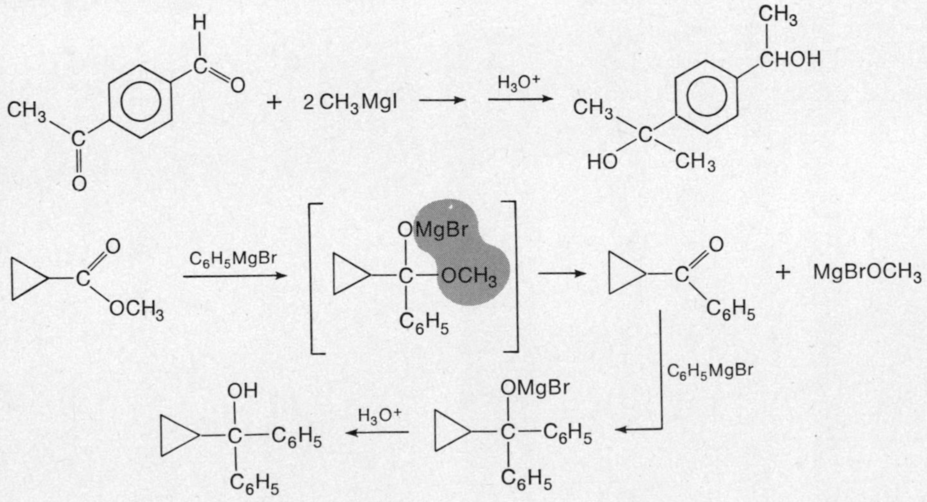

$+ \; 2\,CH_3MgI \longrightarrow \xrightarrow{H_3O^+}$ CH_3CHOH, $(CH_3)_2C(OH)$

$C_3H_5{-}C(=O)OCH_3 \xrightarrow{C_6H_5MgBr} [C_3H_5{-}C(OMgBr)(OCH_3)C_6H_5] \longrightarrow C_3H_5{-}C(=O)C_6H_5 + MgBrOCH_3$

$\xrightarrow{C_6H_5MgBr} C_3H_5{-}C(OMgBr)(C_6H_5)_2 \xrightarrow{H_3O^+} C_3H_5{-}C(OH)(C_6H_5)_2$

Fig. 10.10. Methods for the synthesis of alcohols.

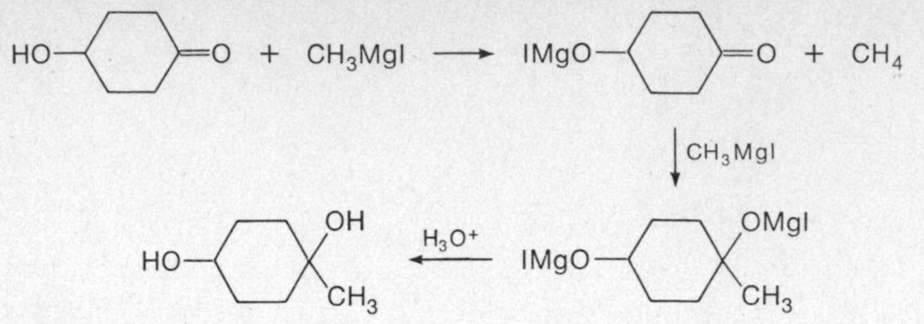

$HO-C_6H_{10}=O + CH_3MgI \longrightarrow IMgO-C_6H_{10}=O + CH_4$

$\downarrow CH_3MgI$

$HO-C_6H_{10}(OH)(CH_3) \xleftarrow{H_3O^+} IMgO-C_6H_{10}(OMgI)(CH_3)$

Fig. 10.11. Preferential reaction of Grignard reagents with "active hydrogen."

PROBLEM 10-4

An experiment carried out by Barbier at the turn of the century is described on p. 238. Indicate what reactions actually took place in this experiment.

PROBLEM 10-5

What sequence of reactions would you use to effect the following conversions:

(a) $H_2C{=}CH_2 \longrightarrow CH_3CH_2CH_2OH$

(b) $C_6H_5CH_3 \longrightarrow C_6H_5C(CH_3)_2OH$

The synthesis methods that suffice for alcohols do not work in the preparation of phenols. The most generally applicable procedure for obtaining a phenol involves the circuitous sequence of reactions illustrated in Fig. 10.12, in which benzene is nitrated (electrophilic substitution—see Fig. 8.13), the nitrobenzene is reduced to the corresponding amine, and the amine is converted to the *diazonium salt* by the action of nitrous acid. Diazonium salts are versatile intermediates, for the $-N_2^+X^-$ moiety can be replaced by a variety of functional groups, including hydroxyl. Gentle heating of an aqueous solution of a diazonium salt is usually sufficient to effect the substitution by hydroxyl.

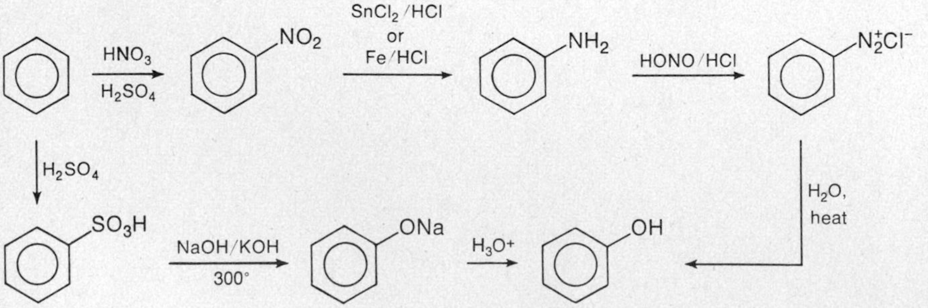

$C_6H_6 \xrightarrow[H_2SO_4]{HNO_3} C_6H_5NO_2 \xrightarrow{SnCl_2/HCl \text{ or } Fe/HCl} C_6H_5NH_2 \xrightarrow{HONO/HCl} C_6H_5N_2^+Cl^- \xrightarrow{H_2O,\ heat} C_6H_5OH$

$C_6H_6 \xrightarrow{H_2SO_4} C_6H_5SO_3H \xrightarrow[300^\circ]{NaOH/KOH} C_6H_5ONa \xrightarrow{H_3O^+} C_6H_5OH$

Fig. 10.12. Methods for the synthesis of phenols.

Another synthesis of phenols involves the second sequence of reactions shown in Fig. 10.12, in which the first step is sulfonation of benzene (electrophilic substitution—see Fig. 8.13) followed by treatment of the sulfonic acid with

NaOH—KOH at elevated temperature. The strenuous conditions required for this method often preclude its use if base-sensitive substituents are present in the molecule.

10.6 Reactions of Alcohols and Phenols Involving the O—H Bond

Alcohols, like water, contain an oxygen-bound hydrogen that is sufficiently acidic to react with alkali metals or with strong bases to form alkoxide salts. Quantitative measurements of acidity show that methanol is comparable to water in acid strength with a pK_a of 15.5, that ethanol has a pK_a of 17, and that *t*-butyl alcohol has a pK_a of 19 (see Table 7.1). In contrast, phenol has a pK_a of only 10, which means that it is about 10 million times more acidic than ethanol. This striking difference is attributed to the relative stabilities of the anions from alcohols and phenols. Whereas the negative charge of the anion derived from ethanol is constrained to remain *localized* on the oxygen atom, the negative charge of the anion derived from phenol can be delocalized into the aromatic ring (Fig. 10.13), providing resonance stabilization.

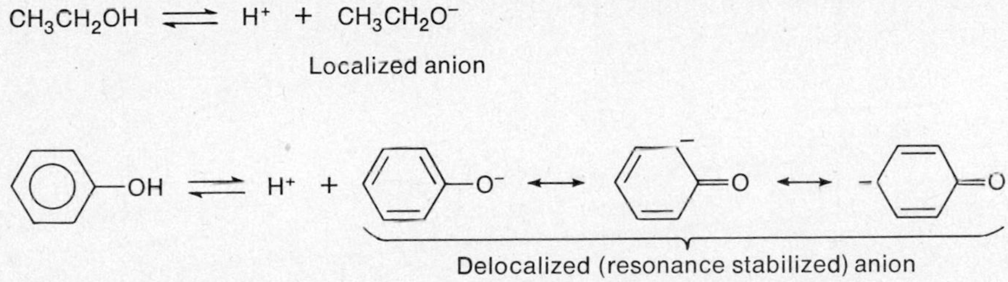

$$CH_3CH_2OH \rightleftharpoons H^+ + CH_3CH_2O^-$$

Localized anion

$$C_6H_5OH \rightleftharpoons H^+ + C_6H_5O^- \longleftrightarrow \longleftrightarrow$$

Delocalized (resonance stabilized) anion

Fig. 10.13. Localized and delocalized (resonance-stabilized) anions.

Alcohols and phenols react alike with (a) alkali metals to form metal alkoxides and hydrogen (*e.g.*, $CH_3OH + Li \rightarrow CH_3OLi + \frac{1}{2}H_2$), (b) Grignard reagents to form magnesium alkoxides and the hydrocarbon corresponding to the Grignard reagent (*e.g.*, $C_2H_5OH + CH_3MgI \rightarrow C_2H_5OMgI + CH_4$), and (c) lithium aluminum hydride to form lithium aluminum alkoxides and hydrogen (*e.g.*, $4C_6H_5OH + LiAlH_4 \rightarrow LiAl(OC_6H_5)_4 + 4H_2$). Only phenols, however, are acidic enough to react with *aqueous* bases ($C_6H_5OH + NaOH \rightarrow C_6H_5O^-Na^+ + H_2O$).

Because alcohols are not sufficiently acidic to react with aqueous bases, they are no more soluble in an aqueous solution of sodium hydroxide than they are in pure water. Phenols, on the other hand, react with aqueous base to form the corresponding phenoxides. In a solution of pH 10, for example, phenol is 50% in the phenolic form (C_6H_5OH) and 50% in the phenoxide form ($C_6H_5O^-$). As the pH of the solution increases (*i.e.*, solution becomes more basic), the ratio of phenoxide to phenol increases; at pH 13 (about 0.1 *N* NaOH), the ratio is about 1000, corresponding to approximately 99.9% in the phenoxide form. Since phenoxide is a much stronger base than phenol itself, it is a better acceptor moiety in hydrogen bond formation. It is the hydrogen bond interaction between solute and water that is responsible for the water solubility of phenols and alcohols, so the increased efficiency of phenoxide in this respect increases its solubility; thus, phenol can be brought into aqueous solution by conversion to its corresponding phenoxide ion through the action of hydroxide ion.

The ability of phenols to react with aqueous base and to dissolve in water is an experimentally useful phenomenon; not only does it afford a very simple way for distinguishing between an alcohol (insoluble in aqueous base) and a phenol (soluble in aqueous base), it also provides an effective way for isolating and purifying phenols. A mixture of phenol and cyclohexanol, for instance, can be separated cleanly by dissolving the mixture in ether and extracting the ether solution with dilute aqueous sodium hydroxide. The cyclohexanol, which is much more soluble in ether than in water, remains in the ether layer, and the phenol goes into the water layer as sodium phenoxide. The phenol can be recovered by acidifying the aqueous layer, thereby converting the sodium phenoxide back to phenol, and extracting the aqueous layer with ether. The phenol, more soluble in ether than in water, passes into the ether layer and can be recovered by evaporation of the ether.

An important reaction of alcohols and phenols involves the formation of esters of organic acids (*i.e.*, carboxylic acids) and inorganic acids (*i.e.*, mineral acids). Methanol, for example, reacts with acetic acid to form methyl acetate and with sulfuric acid to form methyl sulfate and dimethyl sulfate (Fig. 10.14). The mechanisms of these reactions are discussed in Chapter 29.

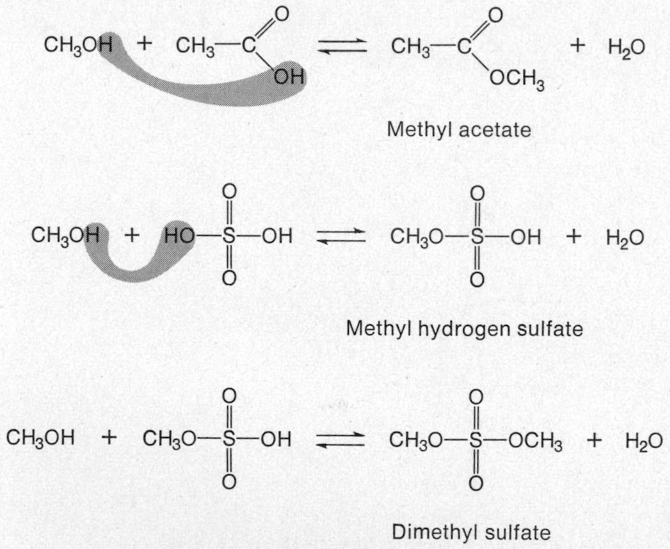

$$CH_3OH + CH_3{-}C(=O){-}OH \rightleftharpoons CH_3{-}C(=O){-}OCH_3 + H_2O$$

Methyl acetate

$$CH_3OH + HO{-}S(=O)_2{-}OH \rightleftharpoons CH_3O{-}S(=O)_2{-}OH + H_2O$$

Methyl hydrogen sulfate

$$CH_3OH + CH_3O{-}S(=O)_2{-}OH \rightleftharpoons CH_3O{-}S(=O)_2{-}OCH_3 + H_2O$$

Dimethyl sulfate

Fig. 10.14. Esterification of alcohols.

10.7 Reactions of Alcohols Involving the R—OH Bond

In an earlier discussion on the effects of the strength of the bonds between functional groups attached to sp^2- and sp^3-hybridized carbon atoms we noted that the difference in bond strengths resulted in substantial changes in chemical reactivity (*cf.* p. 235). Similar differences in reactivity also occur between alcohols and phenols. Only alcohols undergo facile replacement of the hydroxyl function; the following reactions, therefore, are restricted to alcohols and are not observed with phenols.

Halogen acids react with alcohols, as noted in one of the methods of preparation of alkyl halides (see Section 9.4). It is postulated that the reaction is initiated by the addition of a proton from the halogen acid (halogen acids are strong acids) to the hydroxyl group (a Lewis base) to give an "oxonium salt" (*e.g.*, $R\overset{+}{O}H_2$). The $-\overset{+}{O}H_2$ moiety is a far better "leaving group" than —OH, and even a nucleophile as weak as halide anion is capable of displacing it from the carbon to which it is attached. With certain alcohols (especially tertiary alcohols, benzyl alcohols, and allyl alcohols), even water is sufficiently nucleophilic to displace the $-\overset{+}{O}H_2$ moiety, forming a "solvated" carbonium ion intermediate. Both the initially formed oxonium ion and the subsequently formed solvated carbonium ion have the potential of reacting with a halide ion to yield the product. Reaction via the oxonium ion (an S_N2 pathway) is more likely with primary alcohols, whereas reaction via the solvated carbonium ion (S_N1 pathway) is more likely with tertiary alcohols and alcohols capable of producing resonance-stabilized carbonium ions, such as benzyl alcohol and allyl alcohol; secondary alcohols fall between these extremes and may react along either or both pathways (Fig. 10.15).

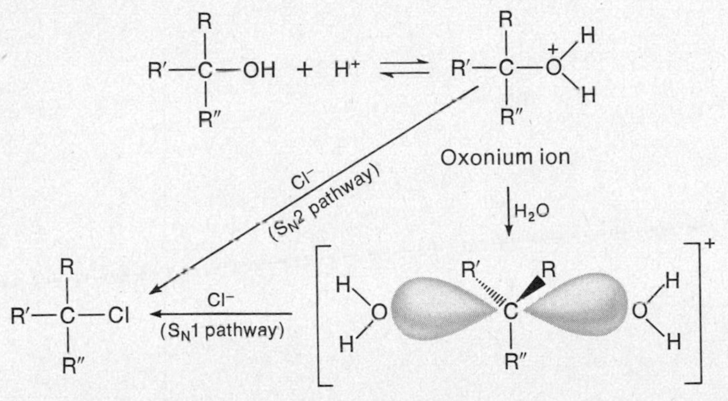

Fig. 10.15. Nucleophilic substitution reactions of alcohols.

The reaction of halogen acids with alcohols to form halides is not altogether general, and lest the student be lulled into a false security, we mention some of the pitfalls that can be encountered. One of the several avenues of stabilization open to a carbonium ion is the intramolecular acquisition of a pair of electrons *without* the attached group. Most of the examples where this is observed are those in which the attached group is hydrogen, resulting in the elimination of a proton and the formation of an alkene. Thus, with alcohols, just as with alkyl halides (see Section 9.7), substitution reactions are frequently accompanied by elimination reactions. By the appropriate choice of conditions, the elimination pathway can be made to predominate, and the dehydration of alcohols provides one of the standard methods for preparing alkenes (Fig. 10.16). Sulfuric acid, for example, is especially effective in inducing the dehydration of alcohols.

The intramolecular acquisition of electrons by a carbonium ion may also include the group with which the electrons are associated, leading to a rearrangement of the carbon framework of the molecule. This is particularly likely to occur if the alcohol from which the carbonium ion is derived is highly branched at the β-carbon (for nomenclature, see p. 235). For example, neopentyl alcohol reacts with hydrogen bromide to yield 2-bromo-2-methylbutane as the major product, the result of methyl group migration (Fig. 10.16). This can be interpreted

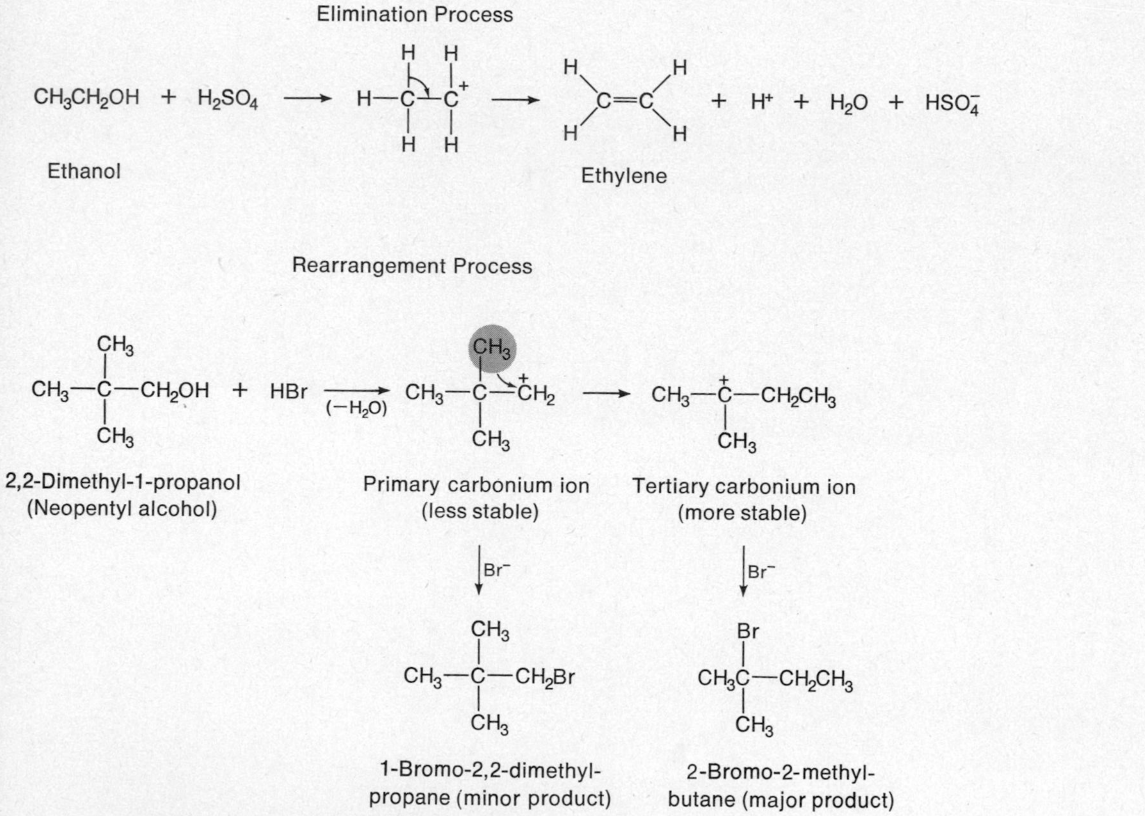

Elimination Process

$$CH_3CH_2OH + H_2SO_4 \longrightarrow H{-}CH_2{-}\overset{+}{C}H_2 \longrightarrow H_2C{=}CH_2 + H^+ + H_2O + HSO_4^-$$

Ethanol — Ethylene

Rearrangement Process

$$CH_3{-}C(CH_3)_2{-}CH_2OH + HBr \xrightarrow[(-H_2O)]{} CH_3{-}C(CH_3)_2{-}\overset{+}{C}H_2 \longrightarrow CH_3{-}\overset{+}{C}(CH_3){-}CH_2CH_3$$

2,2-Dimethyl-1-propanol (Neopentyl alcohol) — Primary carbonium ion (less stable) — Tertiary carbonium ion (more stable)

$$\downarrow Br^- \qquad\qquad \downarrow Br^-$$

$$CH_3{-}C(CH_3)_2{-}CH_2Br \qquad\qquad CH_3C(Br)(CH_3){-}CH_2CH_3$$

1-Bromo-2,2-dimethylpropane (minor product) — 2-Bromo-2-methylbutane (major product)

Fig. 10.16. Elimination and rearrangement reactions accompanying nucleophilic substitution reactions of alcohols.

on the basis of relative carbonium ion stabilities. The initially formed primary carbonium ion can be transformed to the more stable tertiary carbonium ion by migration of an adjacent methyl group and its bonding electrons. The *intra*molecular migration process occurs more rapidly than the *inter*molecular step in which a bromine anion is acquired by the carbonium ion. Skeletal alterations of this type are often referred to as *Wagner–Meerwein rearrangements*, in recognition of the German chemists who discovered them and studied them in detail.

PROBLEM 10-6

When 3,3-dimethyl-1-butene (draw its structure) is treated with hydrogen chloride, the major product is 2-chloro-2,3-dimethylbutane (draw its structure) rather than 2-chloro-3,3-dimethylbutane (draw its structure). How might this be explained?

The relative reactivities of primary, secondary, and tertiary alcohols in reactions involving cleavage of the RO—H bond and cleavage of the R—OH bond follow opposite trends. Because the acidity of the O—H group diminishes from primary to secondary to tertiary alcohols, the rates of reactions involving RO—H cleavage fall in that order. For example, methanol ($pK_a = 15.5$) reacts almost explosively with metallic potassium to form potassium methoxide, whereas *t*-butyl alcohol ($pK_a = 19$) must be refluxed for many hours to convert it to potassium *t*-butoxide. Conversely, nucleophilic substitution reactions involving R—OH cleavage occur much more readily with tertiary alcohols than with primary alcohols. Benzyl alcohols and allyl alcohols also react with great facility in nucleophilic displacement reactions because of the stability of benzyl and allyl cations (Fig. 10.17).

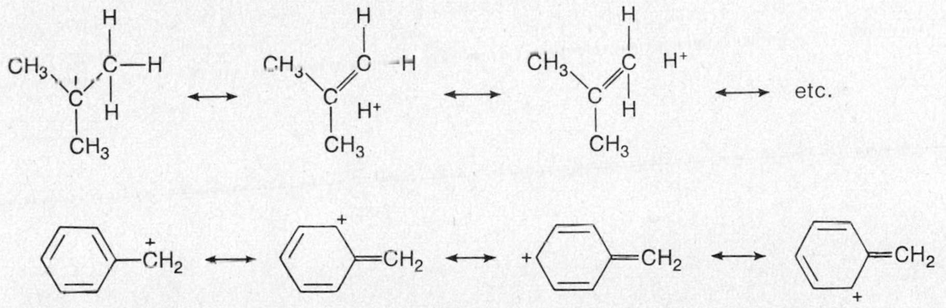

Fig. 10.17. Resonance stabilization of the *t*-butyl cation (by hyperconjugation) and the benzyl cation (by interaction with the aromatic ring).

10.8 Oxidation Reactions of Alcohols and Phenols

Alcohols represent the first step above the alkanes in the oxidation sequence of carbon compounds. In the transformation of methane to methanol, the carbon is considered to have undergone oxidation in the obvious sense of having acquired an oxygen in place of a hydrogen. In actual practice, this conversion would probably be effected by a stepwise sequence of reactions involving the free radical halogenation of methane to form a methyl halide, followed by replacement of the halogen by a hydroxyl group by treatment with aqueous sodium hydroxide; *e.g.*, $CH_4 \rightarrow CH_3Cl \rightarrow CH_3OH$. In which step, we might ask, has the oxidation really taken place? According to the literal interpretation of the term "oxidation," the answer would be that it has occurred at the second step, since only at this point does oxygen enter the system. More accurately, in terms of electron densities, however, oxidation takes place at the halogenation step. If oxidation is defined as the loss of electrons (partial or complete) and reduction as the acquisition of electrons (partial or complete), the carbon in CH_3Cl is more oxidized than the carbon in CH_4. This is due to the greater electron-withdrawing power of the chlorine atom compared with the hydrogen atom.

Primary, secondary, and tertiary alcohols differ in their behavior toward oxidizing agents, providing a means of distinguishing among them. Primary alcohols yield aldehydes upon controlled oxidation; however, with stronger oxidizing agents, aldehydes are readily converted to carboxylic acids ($RCH_2OH \rightarrow RCHO \rightarrow RCO_2H$). Secondary alcohols yield ketones that are generally stable against further oxidation ($R_2CHOH \rightarrow R_2C{=}O$). Tertiary alcohols cannot undergo oxidation to aldehydes or ketones of the same carbon content ($R_3COH \rightarrow$ No reaction) unless skeletal rearrangement occurs under the conditions of the reaction. Thus, by identifying the oxidation product from an alcohol as (a) an aldehyde or carboxylic acid (from primary alcohols), (b) a ketone (from secondary alcohols), or (c) recovered starting material (from tertiary alcohols), we discern the structure of the alcohol (Fig. 10.18).

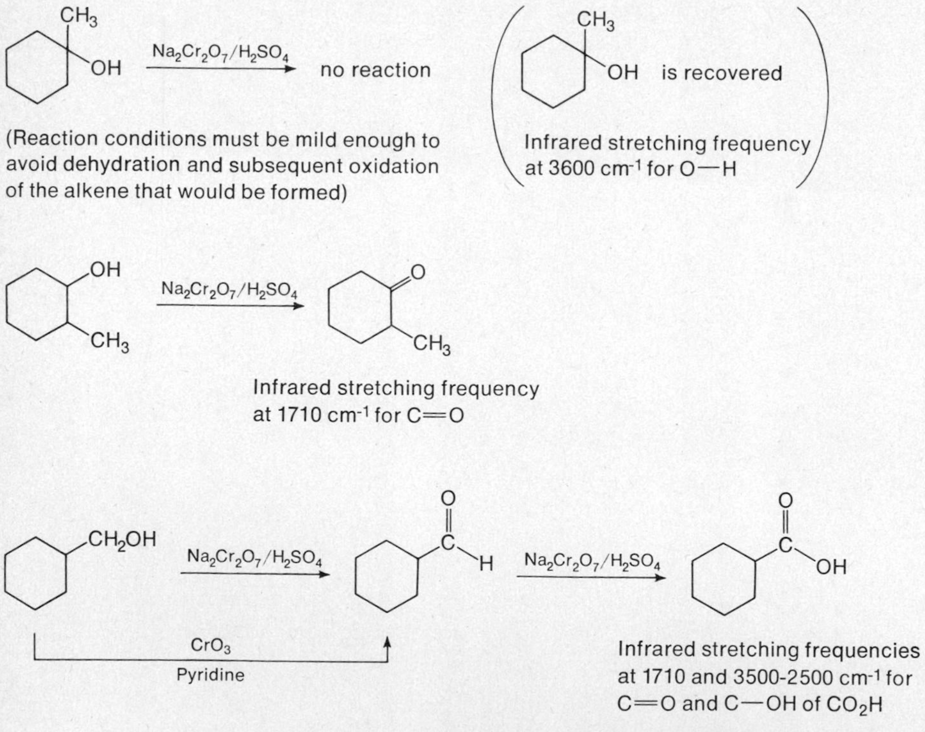

CH_3, OH — $Na_2Cr_2O_7/H_2SO_4$ → no reaction

(CH_3, OH is recovered

Infrared stretching frequency at 3600 cm^{-1} for O—H)

(Reaction conditions must be mild enough to avoid dehydration and subsequent oxidation of the alkene that would be formed)

OH, CH_3 — $Na_2Cr_2O_7/H_2SO_4$ → O, CH_3

Infrared stretching frequency at 1710 cm^{-1} for C=O

CH_2OH — $Na_2Cr_2O_7/H_2SO_4$ → C(=O)H — $Na_2Cr_2O_7/H_2SO_4$ → C(=O)OH

CrO_3 / Pyridine

Infrared stretching frequencies at 1710 and 3500-2500 cm^{-1} for C=O and C—OH of CO_2H

Fig. 10.18. Oxidation as a means of differentiation between primary, secondary, and tertiary alcohols.

The mixture of sodium dichromate and sulfuric acid, usually referred to as "chromic acid" (Fig. 10.18), is a rather strenuous oxidizing reagent that takes primary alcohols all the way to the carboxylic acids, the intermediate aldehydes undergoing oxidation more rapidly than the primary alcohols. Through the use of more selective oxidizing reagents, the oxidation of a primary alcohol can be interrupted at the aldehyde stage. For example, chromium trioxide in pyridine solution is useful in this respect.

PROBLEM 10-7

Remembering the electrophilic addition possibilities of alkenes (see Fig. 8.12) and the reactions of halides and alcohols, indicate how the following reaction sequence could be carried out using ethylene as the only available source of carbon compounds:

$$H_2C{=}CH_2 \longrightarrow CH_3CHO \longrightarrow CH_3CH_2CH(OH)CH_3 \longrightarrow CH_3CH_2C(OH)(CH_3)CH_2CH_3$$

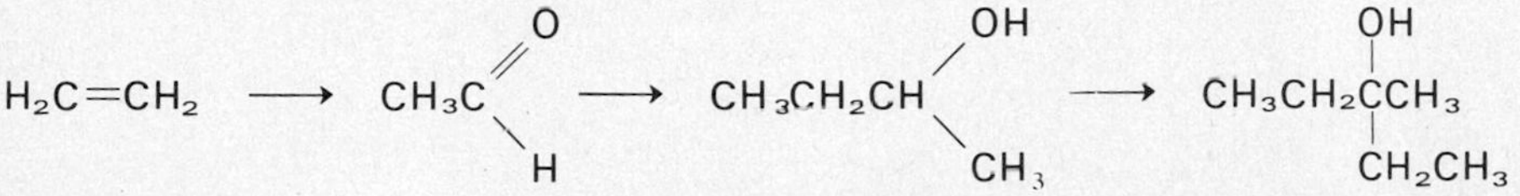

PROBLEM 10-8

Taking cognizance of the free radical substitution possibilities of alkanes (see Fig. 8.3) and the reactions of halides and alcohols, indicate how the following reaction sequence could be carried out using toluene as the only source of carbon compounds:

$$C_6H_5CH_3 \longrightarrow C_6H_5CH_2OH \longrightarrow C_6H_5CH(OH)CH_2C_6H_5 \longrightarrow C_6H_5CHBrCHBrC_6H_5$$

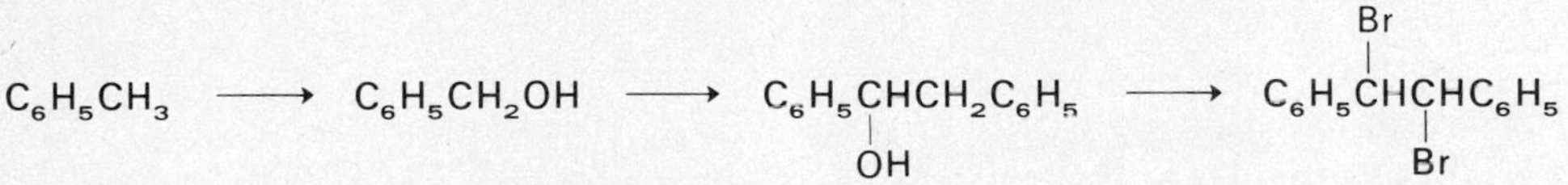

In a sense, phenols correspond to tertiary alcohols and might be expected to be resistant to oxidation. In actual fact, however, phenol is easily oxidized without disruption of its carbon skeleton to a compound known as *p*-benzoquinone (Fig. 10.19).

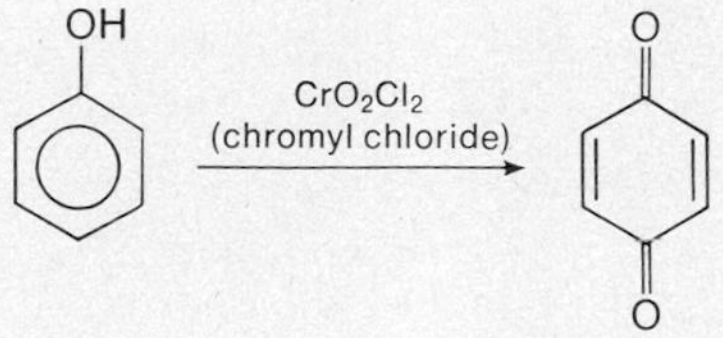

CrO_2Cl_2 (chromyl chloride)

1,4-Benzoquinone

Fig. 10.19. Oxidation of phenol to 1,4-benzoquinone.

10.9 Electrophilic Substitution Reactions of Phenols

Phenols resemble alcohols with respect to reactions involving the O—H bond but differ from alcohols with respect to reactions involving the R—OH (*i.e.*, Ar—OH) bond as well as the carbon framework itself. The hydroxyl function of a phenol interacts to a significant extent with its carbon framework, as expressed by the resonance structures shown in Fig. 10.20. As a result of the attachment of a hydroxyl function to a benzene ring, the ring acquires a greater overall electron density than in benzene itself, and this extra electron density is concentrated at the positions *ortho* and *para* to the hydroxyl group. It is at these

$Br_2 + Br_2 \rightleftharpoons Br^+ + Br_3^-$

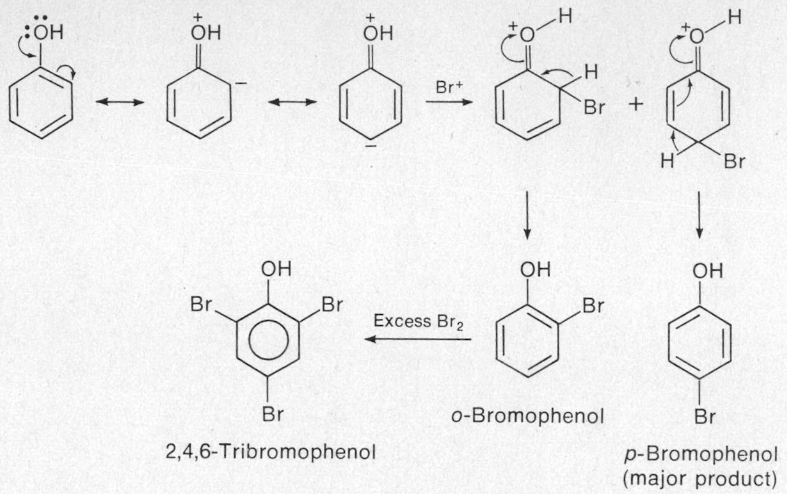

Fig. 10.20. Bromination of phenol.

positions, therefore, that electrophilic substitution reactions occur, for the incoming electrophile seeks the positions of highest electron density. Electrophilic substitution, it will be recalled (see p. 213), is initiated by the formation of an electrophilic reagent, carried forward by the interaction of the electrophile with the aromatic ring, and concluded by the expulsion of a proton to regenerate the aromatic system. Thus, treatment of phenol with a solution containing 1 equivalent of bromine in carbon tetrachloride yields a mixture from which 80 to 84% of *p*-bromophenol can be isolated as a low-melting solid, the remainder being almost entirely the *o*-bromophenol. If an excess of bromine is used, the product is 2,4,6-tribromophenol. In similar fashion, various other electrophiles react with phenol to give mixtures of *o*- and *p*-substituted phenols. (Fig. 10.21). Because the reactions of phenol give *o*- and *p*-substituted phenols, the hydroxyl group is said to be *ortho,para* directing; because the electrophilic substitution reactions of phenol proceed more rapidly than those of benzene, due to the increased electron density in the aromatic ring of phenol, the hydroxyl group is said to be an "activating" function.

PROBLEM 10-9

Starting with benzene, suggest a method for preparing (a) *p*-nitrophenol, and (b) 1-hydroxy-2,4,6-trimethylbenzene.

10.10 *Alcohols and Phenols Containing Two or More Hydroxyl Groups*

Alcohols and phenols containing two or more hydroxyl groups are said to by "polyhydric" (a contraction of polyhydroxylic). Polyhydric alcohols (also called "polyols") are specified as diols, triols, etc., in accord with the number of

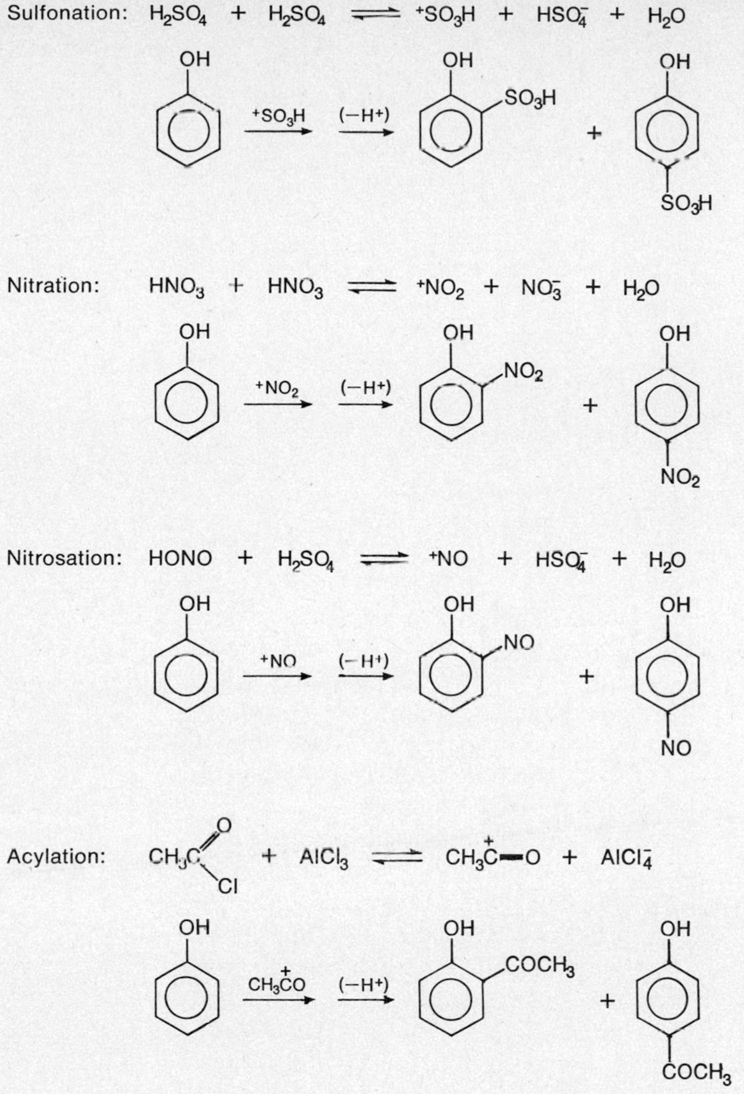

Sulfonation: $$H_2SO_4 + H_2SO_4 \rightleftharpoons {}^{+}SO_3H + HSO_4^- + H_2O$$

OH $\xrightarrow{^{+}SO_3H}$ $\xrightarrow{(-H^+)}$ OH, SO_3H + OH, SO_3H

Nitration: $$HNO_3 + HNO_3 \rightleftharpoons {}^{+}NO_2 + NO_3^- + H_2O$$

OH $\xrightarrow{^{+}NO_2}$ $\xrightarrow{(-H^+)}$ OH, NO_2 + OH, NO_2

Nitrosation: $$HONO + H_2SO_4 \rightleftharpoons {}^{+}NO + HSO_4^- + H_2O$$

OH $\xrightarrow{^{+}NO}$ $\xrightarrow{(-H^+)}$ OH, NO + OH, NO

Acylation: $$CH_3C(=O)Cl + AlCl_3 \rightleftharpoons CH_3\overset{+}{C}{=}O + AlCl_4^-$$

OH $\xrightarrow{CH_3\overset{+}{C}O}$ $\xrightarrow{(-H^+)}$ OH, $COCH_3$ + OH, $COCH_3$

Fig. 10.21. Electrophilic substitution reactions of phenol.

hydroxyl groups that they contain. Diols have acquired still another name and are called "glycols," the name having been derived in a circuitous fashion from glycerol ($HOCH_2CHOHCH_2OH$), which has a sweet taste (Greek, *glykys*, "sweet"). Compounds in which two hydroxyl groups are attached to the same carbon (*i.e.*, *gem*-diols) are usually unstable, the molecules losing the elements of water to form the corresponding carbonyl compound ($R_2C(OH)_2 \rightarrow R_2C{=}O + H_2O$). Virtually all other modes of attachment of two or more OH groups to carbon skeletons, however, lead to stable compounds, several typical examples of which are shown in Fig. 10.22.

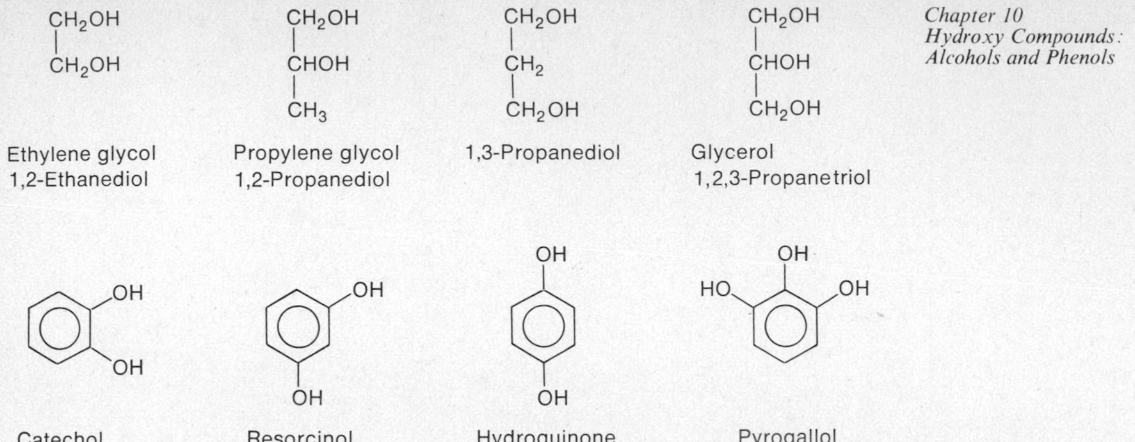

Fig. 10.22. Structures of polyhydric alcohols and phenols.

10.11 *Alcohols and Phenols of Particular Utility and Interest*

Methanol, the parent member of the alcohols, is among the leaders with respect to commercial production by the chemical industry. In 1973, for instance, 7 billion pounds of this material were manufactured in the United States alone, primarily by the high-pressure, high-temperature hydrogenation of carbon monoxide in the presence of a zinc oxide-chromium oxide catalyst. Some of the commercially produced methanol is used as an antifreeze, but the major portion is converted either to formaldehyde or to methyl chloride.

Ethanol, often called grain alcohol, is commercially produced by fermentation processes and also by the hydration (*i.e.*, addition of water) of ethylene. "Ordinary alcohol" is 92 to 95% ethanol, the remainder being water. Pure alcohol (called "absolute alcohol") cannot be prepared by fractional distillation of 95% alcohol, for a constant boiling mixture of ethanol and water (called an "azeotrope") is obtained. Absolute alcohol, therefore, must be prepared by other means. One of the simplest methods, used on a commercial scale, involves the addition of benzene to the "wet" alcohol, followed by careful fractional distillation of the *three-component* mixture. In the vapor phase, a ternary azeotrope (*i.e.*, an azeotrope comprising three substances—in this instance, ethanol, water, and benzene) forms, which has a lower boiling point than any of the individual components of the mixture and which, therefore, is removed by fractional distillation. The binary mixture that is left after all the water has been removed via the ternary azeotrope is then submitted to further fractionation to separate the absolute ethanol from the benzene.

Many naturally occurring products possess hydroxyl functions, including sugars, certain steroids, and certain terpenes that are discussed in later chapters. An interesting example of a naturally occurring terpene alcohol is the sex attractant of the boll weevil. The female boll weevil produces 2-(3,3-dimethylcyclohexylidene)ethanol, which acts in exceedingly low concentration [*i.e.*, microgram (10^{-6} g) amounts] as a sex attractant for the male of the species.

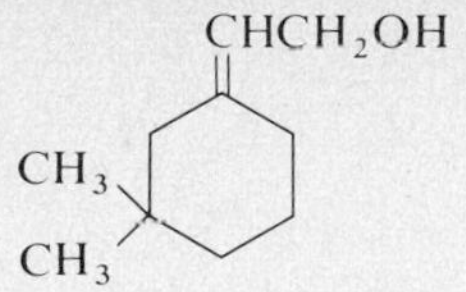

CHCH$_2$OH

CH$_3$

CH$_3$

2-(3,3-Dimethylcyclohexylidene)ethanol

Attempts are being made to take advantage of this phenomenon in the case of the boll weevil and many other insects as a means for insect control—*i.e.*, by luring the insect to a "synthetic" source of the attractant, where the insect is then captured and killed by means that have little or no ecological disadvantage, in contrast to the DDT-type of insect control.

Ethylene glycol, the simplest of the polyols, is miscible with water in all proportions and is extensively used as an antifreeze. It is commercially produced from ethylene by conversion to ethylene oxide either by direct oxidation or by treatment with hypochlorous acid followed by base; hydrolysis of the ethylene oxide then yields the product (Fig. 10.23). Ethylene glycol has the disadvantage

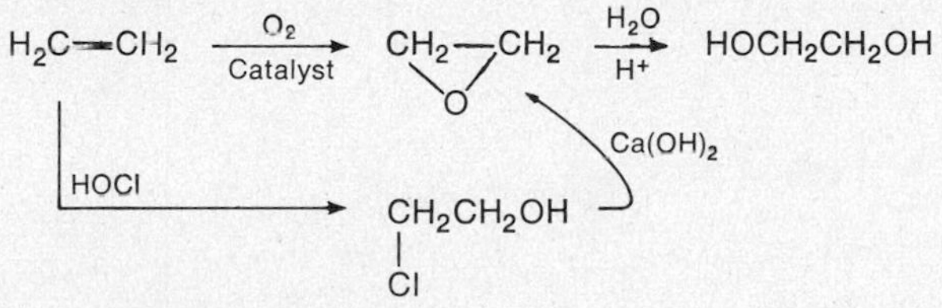

$H_2C{=}CH_2$ $\xrightarrow[\text{Catalyst}]{O_2}$ $CH_2{-}CH_2$ (O) $\xrightarrow[H^+]{H_2O}$ $HOCH_2CH_2OH$

HOCl: $H_2C{=}CH_2 \rightarrow$ CH_2CH_2OH (Cl) $\xrightarrow{Ca(OH)_2}$ ethylene oxide

Fig. 10.23. Commercial synthesis of ethylene glycol.

of being quite toxic when ingested. Propylene glycol, on the other hand, is nontoxic to human beings, although it is poisonous to certain lower forms of life.

Glycerol occurs naturally as the triester of long-chain carboxylic acids called fatty acids. The triesters, known as glycerides or fats (see Section 34.1), yield glycerol and soap (*i.e.*, the sodium salts of the fatty acids) when hydrolyzed with aqueous sodium hydroxide; the process represents one of the oldest of practical organic reactions and one that continues to be used on a commercial scale. In addition to its natural source, glycerol is also produced commercially by a reaction sequence involving high-temperature chlorination of propene, conversion of the 3-chloropropene to the corresponding epoxide (called epichlorohydrin), and hydrolysis of the cyclic oxide (Fig. 10.24). Some of the glycerol so produced is used in the cosmetic, pharmaceutical, and food industries, but the major fraction is converted to the trinitrate ester (known as nitroglycerin) by reaction with nitric acid.

Polyhydroxy compounds that are commercially and biologically important occur in abundance in nature. Among the most significant of these are the sugars (carbohydrates, saccharides), whose chemistry is discussed in detail in Chapter 32.

Phenol has a large and varied commercial use in the synthesis of various derived compounds, including antioxidants, dyes, phenolic resins, and certain pharmaceutical products, such as salicylic acid. A number of commercial processes have been developed for the production of phenol. One of the most

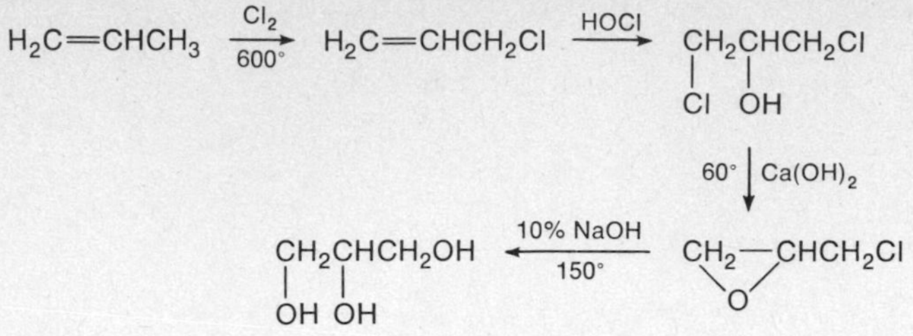

$$H_2C{=}CHCH_3 \xrightarrow[600^\circ]{Cl_2} H_2C{=}CHCH_2Cl \xrightarrow{HOCl} CH_2(Cl)CH(OH)CH_2Cl$$

$$\xrightarrow[60^\circ]{Ca(OH)_2} CH_2{-}CHCH_2Cl \text{ (epoxide, O bridging)} \xrightarrow[150^\circ]{10\%\ NaOH} CH_2(OH)CH(OH)CH_2OH$$

Fig. 10.24. Commercial synthesis of glycerol.

recent of these involves the action of oxygen at 130° on an aqueous alkaline emulsion containing cumene and metal catalysts, which yields cumene hydroperoxide. Treatment of the hydroperoxide with dilute sulfuric acid induces a molecular rearrangement, leading to phenol and acetone (Fig. 10.25).

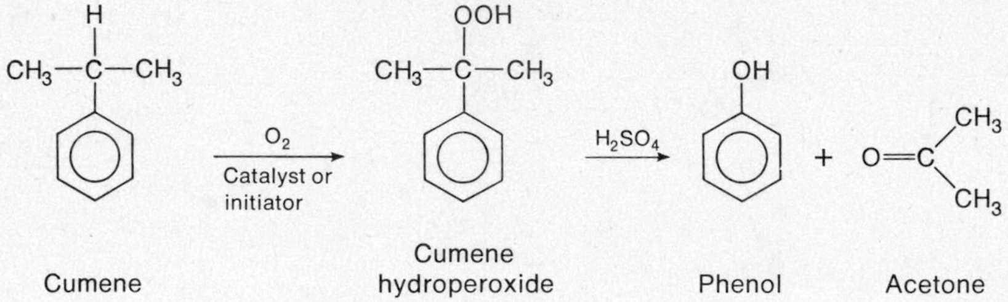

$$C_6H_5C(H)(CH_3)_2 \xrightarrow[\text{Catalyst or initiator}]{O_2} C_6H_5C(OOH)(CH_3)_2 \xrightarrow{H_2SO_4} C_6H_5OH + O{=}C(CH_3)_2$$

Cumene — Cumene hydroperoxide — Phenol — Acetone

Fig. 10.25. Commercial production of phenol.

10.12 *Study Guide*

I. Structure of alcohols and phenols
   A. Bond between carbon framework and oxygen
   B. Tautomerism
      1. Vinyl alcohols less stable than isomeric carbonyl compounds
      2. Phenols more stable than isomeric cyclohexadienones
      3. Bond strength and resonance energy considerations

II. Nomenclature of alcohols and phenols
   A. "Alcohol system"
   B. "Carbinol system"—based on carbinol (*i.e.*, methanol) as parent compound
   C. IUPAC system
      1. Rules for simple alkanols
      2. Rules for polyols
      3. Order of functional group preferences
   D. Phenols

III. Physical properties of alcohols and phenols
   A. Boiling points
      1. Hydrogen bonding in alcohols
      2. Hydrogen bonding in phenols
   B. Melting points
   C. Solubility—lower molecular weight members soluble in water

IV. Spectral properties of alcohols and phenols

Section 10.12 Study Guide

A. Ultraviolet/visible—alcohols generally transparent above 200 nm; phenols have weak-to-moderately strong absorption bands above 200 nm

B. Infrared

1. O—H stretching near $3600\ cm^{-1}$
2. C—O stretching near $1100\ cm^{-1}$

C. Nuclear magnetic resonance

1. Oxygen is electronegative—therefore, downfield shifts of H attached to C—OH
2. Intermolecular exchange of hydrogen between hydroxyl groups

V. Synthesis of alcohols and phenols

A. Methods for alcohols (organized on basis of type of reaction involved)

1. Hydration of alkenes $\left(\gt C=C \lt \; + \; H_2O \xrightarrow[H^+]{} H-\overset{|}{\underset{|}{C}}-\overset{|}{\underset{|}{C}}-OH \right)$
2. Nucleophilic displacement of alkyl halides ($RX + OH^- \rightarrow ROH$)
3. Nucleophilic addition to carbonyl compounds
   (a) Reduction with $LiAlH_4$
      (1) RCO_2H and $RCO_2R' + LiAlH_4 \rightarrow RCH_2OH$ (synthesis of primary alcohols)
      (2) $RCHO + LiAlH_4 \rightarrow RCH_2OH$ (synthesis of primary alcohols)
      (3) $R_2C=O + LiAlH_4 \rightarrow R_2CHOH$ (synthesis of secondary alcohols)
   (b) Reaction with Grignard reagents (RMgX)
      (1) $RCHO + RMgX \rightarrow R_2CHOH$ (synthesis of secondary alcohols)
      (2) $R_2C=O + RMgX \rightarrow R_3COH$ (synthesis of tertiary alcohols)
      (3) $RCO_2R' + 2RMgX \rightarrow R_3COH$ (synthesis of tertiary alcohols—two of R groups from RMgX)
      (4) Exceptions
         $H_2C=O + RMgX \rightarrow RCH_2OH$ (synthesis of primary alcohols)
         $HCO_2R + 2RMgX \rightarrow R_2CHOH$ (synthesis of secondary alcohols—both R groups from RMgX)

B. Methods of synthesis of alcohols (organized on basis of type of product formed)

1. Methods for synthesis of primary alcohols
   (a) Nucleophilic displacement of alkyl halides ($RCH_2X + OH^- \rightarrow RCH_2OH$)
   (b) Hydration of alkenes—only for CH_3CH_2OH from $H_2C=CH_2$
   (c) Reduction of RCHO and RCO_2H with $LiAlH_4$
   (d) Reaction of $H_2C=O$ with RMgX
2. Methods for synthesis of secondary alcohols
   (a) Hydration of alkenes ($RCH=CH_2 + H_2O \rightarrow RCHOHCH_3$)

(b) Nucleophilic displacement of alkyl halides ($R_2CHX + OH^- \rightarrow R_2CHOH$, accompanied by some alkene formation)
(c) Reduction of $R_2C{=}O$ with $LiAlH_4$
(d) Reaction of RCHO and HCO_2R' with RMgX

3. Methods for synthesis of tertiary alcohols
   (a) Hydration of alkenes

$$\left(R_2C{=}CH_2 + H_2O \longrightarrow R_2C\begin{matrix} OH \\ CH_3 \end{matrix} \right)$$

   (b) Nucleophilic displacement of alkyl halides—product is mostly the alkene and not the alcohol
   (c) Reaction of $R_2C{=}O$ and RCO_2R' with RMgX

C. Methods of synthesis of phenols
   1. Nitration-reduction-diazotization sequence
   2. Sulfonation-sodium hydroxide fusion sequence

VI. Reactions of alcohols and phenols

A. Reactions involving the O—H bond
   1. With alkali metals
   2. With Grignard reagents
   3. With $LiAlH_4$
   4. With aqueous base—conversion of phenols to phenoxides
   5. With carboxylic acids and mineral acids to form esters
   6. Order of reactivity: $RCH_2OH > R_2CHOH > R_3COH$

B. Reactions involving the R—OH bond
   1. Ar—OH bond stronger than R—OH bond—therefore, less reactive
   2. Nucleophilic displacement reactions → alkyl halides
      (a) Phenols do not undergo nucleophilic displacement
      (b) Nucleophilic displacement in ROH requires acid catalysis
   3. Dehydration reaction (elimination process) → alkenes
   4. Rearrangement reactions—carbonium ion intermediates
   5. Order of reactivity
      (a) $R_3COH > R_2CHOH > RCH_2OH$
      (b) $ArCH_2OH$ and $R_2C{=}CHCH_2OH$ especially reactive

C. Oxidation-reduction reactions
   1. Primary alcohols → aldehydes and carboxylic acids
   2. Secondary alcohols → ketones
   3. Tertiary alcohols → no reaction
   4. Phenols → quinones

D. Electrophilic substitution reactions of phenols
   1. Mechanism of reaction
   2. *Ortho,para* orientation—OH an "activating" group
   3. Examples of reactions
      (a) Sulfonation
      (b) Nitration

(c) Nitrosation
(d) Alkylation
(e) Acylation

VII. Polyhydric alcohols, polyhydric phenols, polynuclear phenols

PROBLEMS

1. Name each of the following compounds, using the "alcohol system," the "carbinol system," and the IUPAC system

(a) $(CH_3)_2CHOH$ (b) $(CH_3)_2CHCH_2CH_2OH$ (c) $H_2C{=}CHCH_2OH$

(d) $C_6H_5{-}CH_2OH$ (e) $CH_3CH_2CH(OH)CH_3$ (f) $(CH_3)_2C(OH)CH_2CH_3$

2. Name each of the following compounds, using the "carbinol system" and the IUPAC system

(a) cyclobutyl$-CH(OH)CH_3$ (b) $C_6H_5{-}CH(OH)CH_3$ (c) (cyclopropyl)$_3$COH

(d) $(CH_3)_2CHCH_2CH(OH)CH_3$ (e) $(C_6H_5)_2CHOH$

3. Name each of the following structures according to the IUPAC rules:

(a) $(CH_3)_2CHCH_2CH(CH_3)CH_2OH$

(b) m-Cl$C_6H_4{-}CH_2{-}C(OH)(CH_3){-}CH_2CH_3$

(c) $(CH_3)_2C(OH)CH_2CH_2CHClC(CH_3)_3$

(d) $HC{\equiv}C{-}CH(C_6H_5){-}CH{=}CH{-}CH_2OH$ (H and H on C=C)

(e) $HO{-}C_6H_4{-}C(OH)(CH_2CH{=}CH_2)_2$

(f) cyclopentane ring bearing *H*, OH and H, C_2H_5 (wedge/hash stereochemistry)

(g) bicyclic structure bearing H and OH

(h) $HOCH_2CH_2C(CH_3)(CH_2CH_2OH)CH_2CH_2OH$

4. Name each of the following structures as a substituted phenol or naphthol

(a) 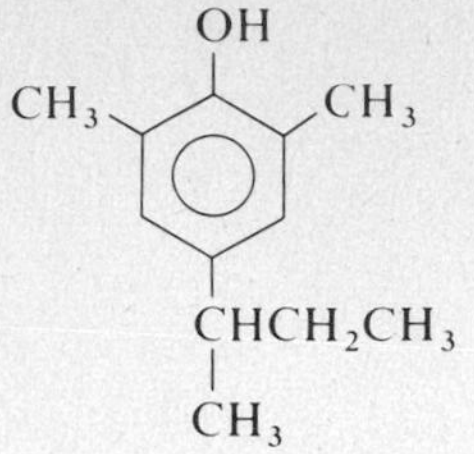 OH; CH_3; CH_3; $CHCH_2CH_3$; CH_3

(b) 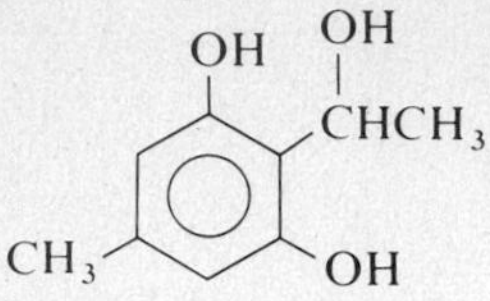 OH; OH; $CHCH_3$; CH_3; OH

(c) 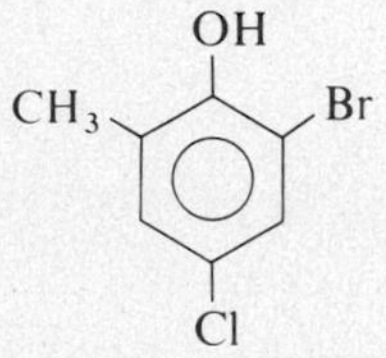 OH; CH_3; Br; Cl

(d) 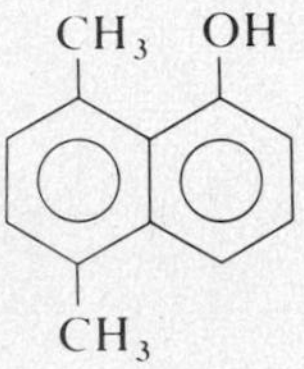 CH_3; OH; CH_3

5. Draw the structures corresponding to the following names

(a) Isohexyl alcohol
(b) Neopentyl alcohol
(c) Cyclopentyl alcohol
(d) Diisopropylcarbinol
(e) 2-Pentanol
(f) 2-Phenylethanol
(g) Cyclohexylcarbinol
(h) Triethylcarbinol
(i) 2-Methyl-3-pentanol
(j) Cyclobutanol
(k) 3-Phenyl-2,3-dimethyl-2-butanol

6. Draw the structure corresponding to the following names:

(a) *cis*-2-Buten-1-ol
(b) (*E*)-4-Bromo-2,3-dimethyl-2-penten-1-ol
(c) *trans*-3-Phenyl-4-hexen-2,3-diol
(d) 4-(3-Hydroxypropyl)-1-naphthol
(e) 2-Trichloromethyl-1,4-dihydroxybenzene
(f) *endo*-2-Hydroxy-2-methylbicyclo[2.2.1]hept-5-ene
(g) 2-(*p*-Tolyl)-3-butyn-2-ol
(h) 9-Hydroxymethyl-9,10-dihydroanthracene
(i) *cis*-3,3-Dimethyl-1,2-cyclobutanediol
(j) 3,4-Dihydroxyphenylethane-1,2-diol

7. For each of the following structures, draw all of the tautomeric structures:

(a) 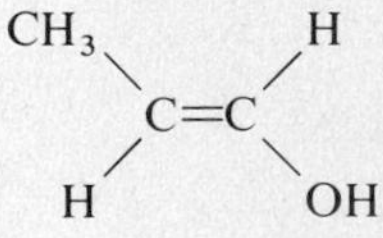 CH_3, H, C=C, H, OH

(b) OH

(c) O, O

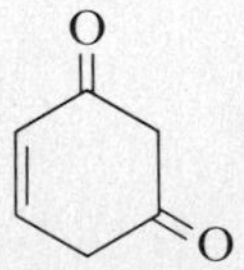

(d) 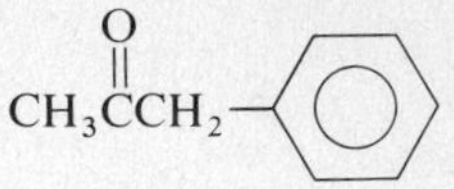 $CH_3\overset{O}{\overset{\|}{C}}CH_2$—

(e) 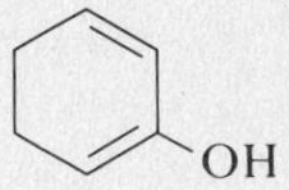 OH

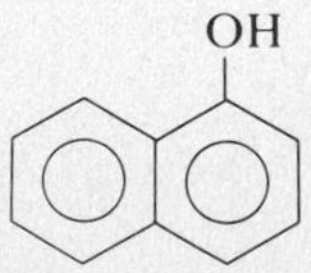

8. Draw the important resonance-contributing structures for each of the following:

(a) 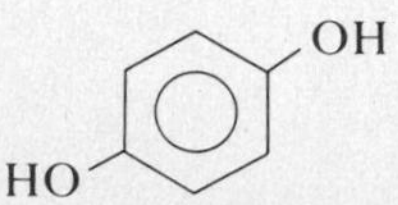 OH; HO

(b) 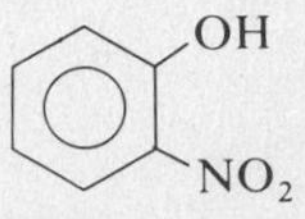 OH; NO_2

(c) 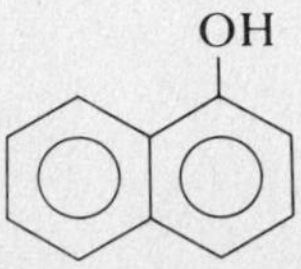 OH

(d) 

9. The following conversions from the compound shown on the left to the compound shown on the right can be effected in one step. Indicate what reagents are necessary to achieve this.

(a) cyclopentanone (=O) ⟶ cyclopentanol (H, OH)

(b) $(CH_3)_2C{=}O \longrightarrow (CH_3)_3COH$

(c) $C_6H_5CH_2Br \longrightarrow C_6H_5CH_2OH$

(d) cyclooctene ⟶ cyclooctanol (OH)

(e) $CH_3C(=O)OCH_3 \longrightarrow CH_3C(OH)(CH_2CH_3)_2$

(f) $C_6H_5C(=O)OCH_2CH_3 \longrightarrow C_6H_5CH_2OH$

(g) $CH_3-C_6H_4-SO_3H \longrightarrow CH_3-C_6H_4-OH$

(h) cyclobutanol (OH) ⟶ cyclobutene

(i) $C_6H_5CH(OH)CH_3 \longrightarrow C_6H_5C(=O)CH_3$

(j) p-xylene (CH_3, CH_3) ⟶ nitro-p-xylene (CH_3, NO_2, CH_3)

10. The following conversions from the compounds shown on the left to the compounds shown on the right require more than one step. Indicate what sequences of reactions are necessary to achieve this, show the reagents, and draw the structures of the intermediate compounds that are produced.

(a) $HC{\equiv}CCH_3 \longrightarrow (CH_3)_2CHOH$

(b) $(CH_3)_2C{=}CH_2 \longrightarrow (CH_3)_2CHCH_2OH$

(c) p-xylene (CH_3, CH_3) ⟶ 2,5-dimethylphenol (CH_3, OH, CH_3)

(d) $C_6H_5CO_2H \longrightarrow C_6H_5CH_2Br$

(e) $(C_6H_5)_2C{=}O \longrightarrow (C_6H_5)_2C{=}CHC_6H_5$

11. Draw the structure of the major product from each of the following reactions:

(a) 1-methylcyclopentene CH_3 + H_2O $\xrightarrow[H_2SO_4]{}$

(b) $(CH_3)_2CHCH_2CH_2CH_2I$ + KOH $\longrightarrow$

(c) cycloheptanone (=O) + $LiAlH_4$ $\longrightarrow$ $\xrightarrow{H_3O^+}$

(d) $(CH_3)_2C{=}O$ + C_6H_5–MgBr $\longrightarrow$ $\xrightarrow{H_3O^+}$

(e) 2-naphthalene–SO_3H $\xrightarrow[\text{Heat}]{NaOH-KOH}$

(f) CH_3–C_6H_4–CO_2H + $LiAlH_4$ $\longrightarrow$ $\xrightarrow{H_3O^+}$

(g) $C_6H_5CO_2CH_3$ + Cl–C_6H_4–MgBr $\longrightarrow$ $\xrightarrow{H_3O^+}$

(h) HO–C_6H_4–CO_2H + CH_3MgI $\longrightarrow$ $\xrightarrow{H_3O^+}$

(i) $C_6H_5C(CH_3)_2$ (with Br on the C) + KOH $\longrightarrow$

(j) 1-naphthol (OH) $\xrightarrow{Na_2Cr_2O_7/H_2SO_4}$

12. The following compounds undergo rearrangement of the carbon framework when treated with aqueous mineral acids to give, depending on the conditions, an alcohol, an alkene, or a mixture thereof. Indicate the probable structure of the rearranged carbon framework in each case.

(a) $CH_3-C(CH_3)_2-CH(OH)CH_3$ $\xrightarrow[H^+]{}$

(b) cyclohexyl–CH_2OH $\xrightarrow[H^+]{}$

(c) 2,2-dimethylcycloheptanol (CH_3, CH_3, OH) $\xrightarrow[H^+]{}$

(d) $C_6H_5C(C_6H_5)_2CH_2OH$ $\xrightarrow[H^+]{}$

Chapter 10 Problems

13. Using chemical and/or spectral methods (as indicated in each case), show how the following pairs of isomeric compounds can be differentiated.

(a) $HC{\equiv}CCH_2CH_2OH$ and $CH_3C{\equiv}CCH_2OH$ — Chemical methods only

(b) $H_2C{=}CHCH(OH)CH_3$ and cyclobutanol (OH on cyclobutane ring) — Chemical methods only

(c) benzyl alcohol ($C_6H_5CH_2OH$) and 2-methylphenol (ring with OH and CH_3) — Chemical methods only

(d) $(CH_3)_2CHOH$ and $CH_3CH_2CH_2OH$ — Chemical and spectral methods

(e) $C_6H_5C(CH_3)_2OH$ and ring bearing CH_3, CH_2OH, CH_3 — Chemical and spectral methods

14. Starting with alkanes, alkenes, alkynes, and/or alkyl halides containing no more than two carbon atoms, devise syntheses for the following compounds:

(a) CH_3CH_2OH (b) $(CH_3)_2CHOH$ (c) $(CH_3)_3COH$

(d) $CH_3CH(OH)CH_2CH_3$ (e) $(CH_3)_2C(OH)CH_2CH_3$

15. Starting with cyclohexanone, and using any other organic and/or inorganic compounds that you think are necessary, devise syntheses for the following compounds:

(a) cyclohexanol (OH) (b) 1-methylcyclohexanol (OH, CH_3) (c) chlorocyclohexane (Cl)

(d) cyclohexene (e) 1-bromo-1-ethylcyclohexane (CH_2CH_3, Br) (f) methoxycyclohexane (OCH_3)

16. Define each of the following terms as briefly and concisely as possible. Do not use a specific example in defining the term, but cite a typical example to illustrate the definition.

(a) Tautomer (b) Secondary alcohol (c) Hydrogen bond
(d) Polyol (e) Azeotrope (f) Absolute ethanol

17. On the basis of the analytical, physical, and spectral information given, suggest a structure that is commensurate with *all* the data for each of the following compounds:

(a) Compound A has a molecular weight of 60 and a bp of 82°. It has a strong ir absorption band at 3600 cm^{-1}, and its nmr spectrum shows a singlet at δ 4.75, a septet at δ 3.9, and a doublet at δ 1.1, with area ratios of 1:1:6.

(b) Compound B is a liquid boiling at 220° and possessing the molecular formula $C_8H_{10}O$. Its ir spectrum contains very strong absorption bands at 3400 and 1050 cm^{-1}, and moderately strong bands at 1600, 1495, and 1450 cm^{-1}. Its nmr spectrum includes a singlet at δ 7.1, a singlet at δ 4.1, a triplet at δ 3.6, and a triplet at δ 2.65, with area ratios of 5:1:2:2.

(c) Compound C is a liquid with a molecular weight of 93 $\pm$ 2 and a boiling point of 161°; a qualitative elemental analysis shows it to contain chlorine. Its ir and nmr spectra are as shown.

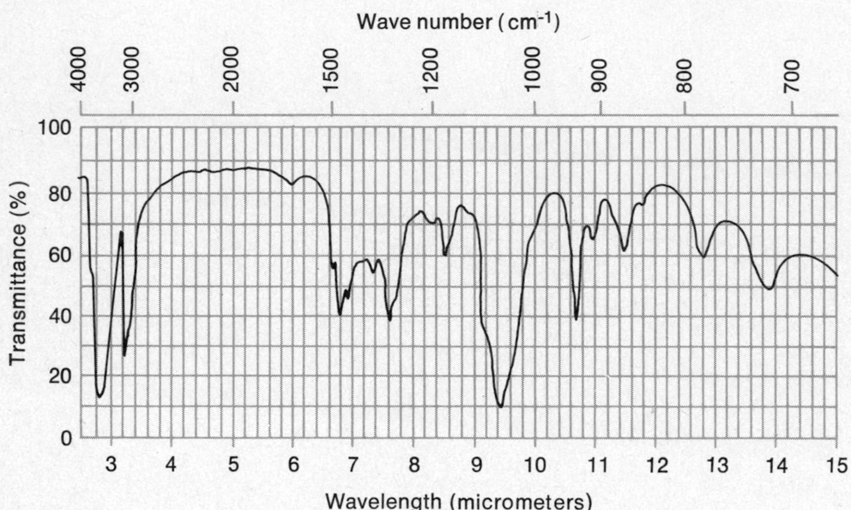

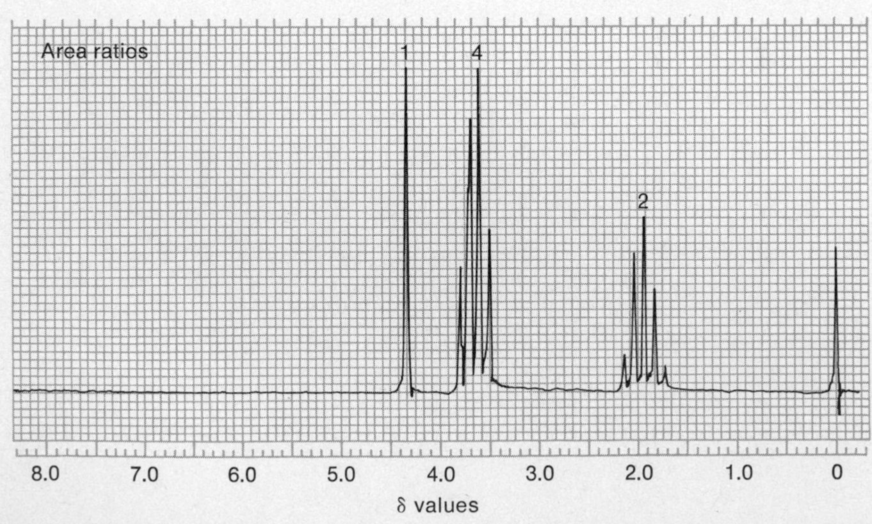

(d) The reaction between 3-methylphenol and isopropyl bromide in the presence of aluminum chloride yields a compound possessing the following uv, ir, and nmr spectra.

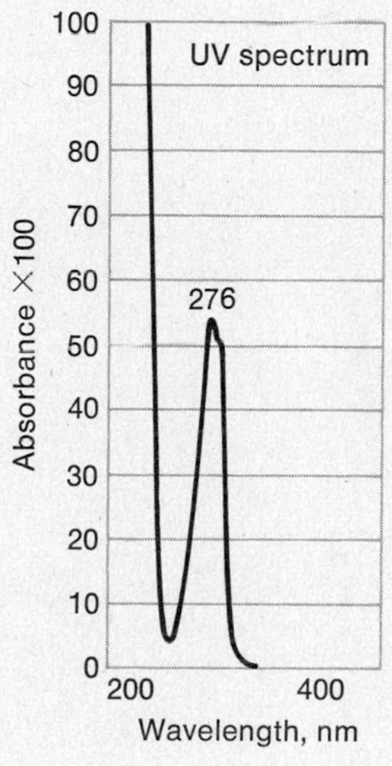

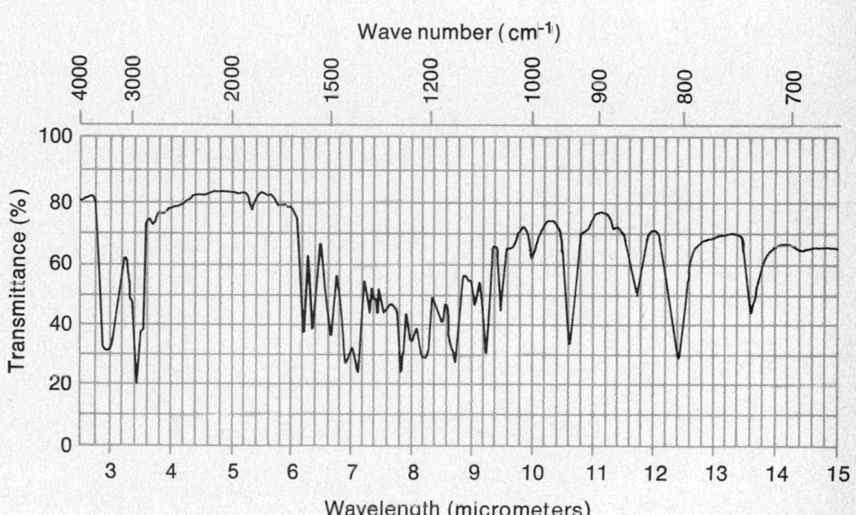

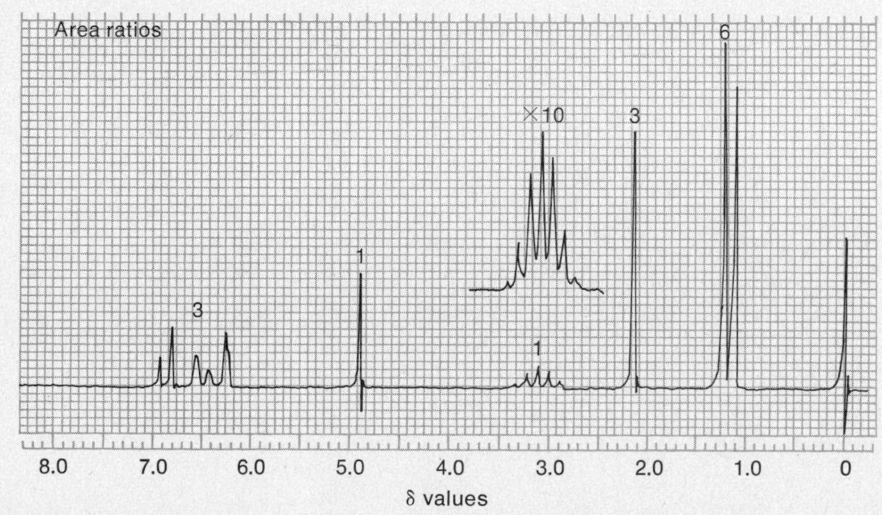

18. Many interesting organic compounds have been obtained from various molds and fungi. For example, from a 150 day old culture of the Bacidiomycete *Polyporus anthracophilus*, a wood rotting fungus found in Australia, a number of acetylenic compounds have been isolated by a group of English chemists. Among these compounds is one called Matricarianol, which is a $C_{10}H_{10}O$ solid melting at 105–106° and possessing the following spectral characteristics:

uv spectrum: a series of intense bands between 217 and 312 nm.
ir spectrum: particularly strong bands at 3600, 955, and 945 cm^{-1}.
nmr spectrum: a 3-proton doublet at δ1.7, a 2-proton doublet at δ3.7, and other resonances for the remaining 5 hydrogens.

When Matricarianol is treated with hydrogen in the presence of a palladium catalyst, 6 mole equivalents of hydrogen are absorbed, and a $C_{10}H_{22}$ alcohol is formed. When Matricarianol is carefully oxidized a $C_{10}H_8O_2$ carboxylic acid is formed. Piecing together these spectral and chemical bits of information, propose a complete structure for Matricarianol.

Chapter **11**

Ethers

The era of the surgeon with a bottle in each hand ended at the Massachusetts General Hospital on October 16, 1846 when the Boston dentist, William Thomas Green Morton, administered ether to a patient who then underwent an operation without feeling pain.

L. J. LUDOVICI
The Discovery of Anaesthesia

Ethers can be considered as derivatives of water in which both of the hydrogens have been replaced by carbon frameworks. Virtually all types of carbon frameworks, including alkyl, alkenyl, alkynyl, and aryl, can be joined to oxygen atoms, structural restrictions being minimal in this case. Unlike water and the alcohols, ethers lack an oxygen–hydrogen bond, and for this reason differ significantly from their hydroxylic counterparts with respect to both physical and chemical properties. Certain similarities persist, nevertheless, as will become apparent in the following discussion.

11.1 Structure and Nomenclature of Ethers

Several typical examples of ethers, selected from the very wide range of possibilities, are shown in Fig. 11.1. One of the schemes for naming these compounds designates the carbon frameworks by their appropriate alkyl, alkenyl, alkynyl, or aryl specifications preceding the word "ether." Thus, $CH_3OCH_2CH_3$ is ethyl methyl ether (alphabetical ordering of groups) and C_6H_5–$OCH(CH_3)_2$ is isopropyl phenyl ether. If the two attached groups are identical, the prefix "di" is usually affixed, although some chemists consider this an unnecessary redundancy. Thus $CH_3CH_2OCH_2CH_3$ is diethyl ether or, simply, ethyl ether, and C_6H_5–O–C_6H_5 is diphenyl ether or phenyl ether, etc. For more complex structures, the IUPAC nomenclature is essential. Under the rules of this system, ethers are named as

$CH_3CH_2OCH(CH_3)_2$

Ethyl isopropyl ether
2-Ethoxypropane
2-Methyl-3-oxapentane

$H_2C{=}CHOCH{=}CH_2$

Divinyl ether
3-Oxapenta-1,4-diene

$C_6H_5CH_2O{-}C_6H_{11}$

Benzyl cyclohexyl ether
Benzyloxycyclohexane

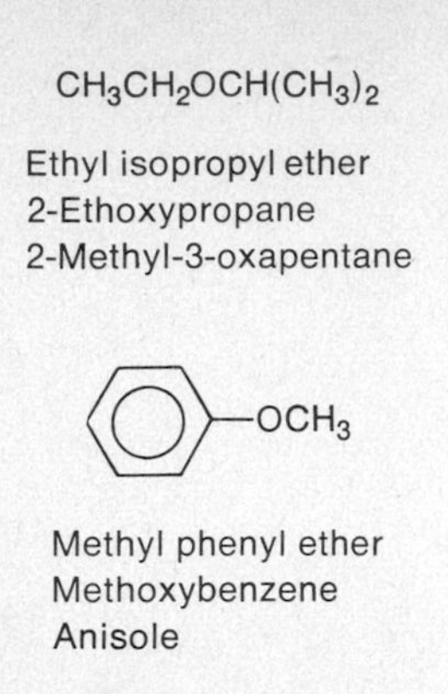

Methyl phenyl ether
Methoxybenzene
Anisole

Diphenyl ether

$CH_3CH_2OC{\equiv}CH$

Ethyl ethynyl ether
Ethoxyethyne
3-Oxa-1-pentyne

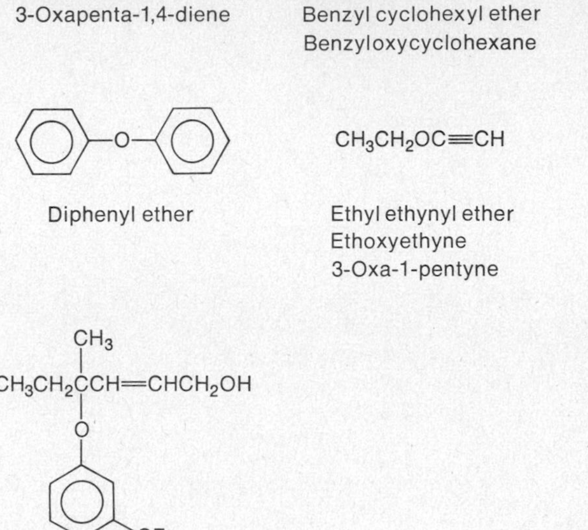

4-Methyl-4-(*m*-trifluoromethylphenoxy)-2-hexen-1-ol

Diethyl ether

Diphenyl ether

Fig. 11.1. Structures and names of ethers.

functional groups; *e.g.*, CH_3O- is a methoxy group, $(CH_3)_2CHO-$ is an isopropoxy group, C_6H_5O- is a phenoxy group, $(CH_3)_2CHCH_2CH_2O-$ is an isopentoxy or a 3-methylbutoxy group, $C_6H_5CH_2O-$ is a benzyloxy group, etc. The ether function ($-O-$) may also be indicated by the incorporation of the prefix "oxa" before the name of the carbon framework along with the appropriate position number, implying a replacement of a methylene group ($-CH_2-$) by an oxygen atom; this method of nomenclature is generally reserved for compounds that cannot be conveniently named by the other systems as, for example

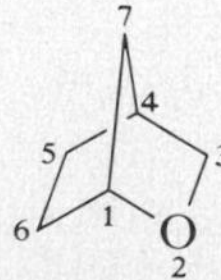

which is designated as 2-oxabicyclo[2.2.1]heptane.

Since ethers, unlike alcohols, contain a *pair* of carbon frameworks attached to the oxygen, cyclic ethers are possible. We call compounds containing one or more hetero atoms and one or more carbons arranged in a cyclic array *heterocycles*, to distinguish them from carbocycles, which contain only carbon atoms in a ring. Numerous examples of heterocyclic ethers are known, including the three-, four-, five-, and six-membered compounds illustrated in Fig. 11.2. In numbering the ring positions of cyclic ethers, we designate oxygen as the 1 position (unless the "oxa" nomenclature for a bicyclic compound is used).

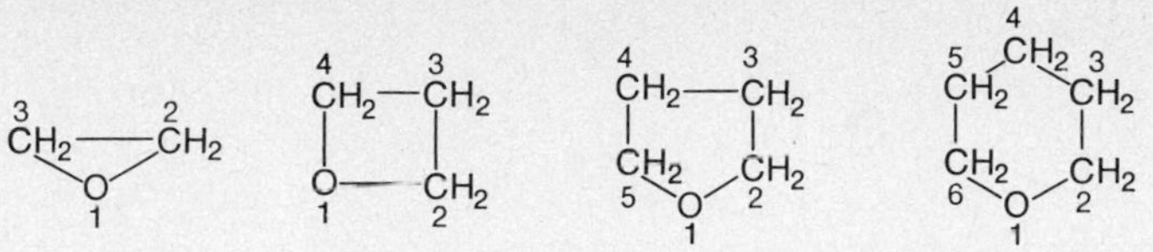

Ethylene oxide	Trimethylene oxide	Tetramethylene oxide	Pentamethylene oxide
Oxirane	Oxetane	Tetrahydrofuran	Tetrahydropyran

Fig. 11.2. Structures of cyclic ethers.

PROBLEM 11-1

Draw and name all the ethers having the molecular formula C_3H_4O.

Heterocyclic rings can, of course, contain unsaturated bonds, and a particularly interesting example in the oxygen series is the compound called furan, in which a five-membered ring contains a pair of double bonds (Fig. 11.3). The possibility of orbital overlap between a nonbonded orbital of oxygen and the π system to which the oxygen is attached has already been discussed in the case of phenols (see Fig. 10.1). Furan is particularly well adapted for such overlap, because it is a

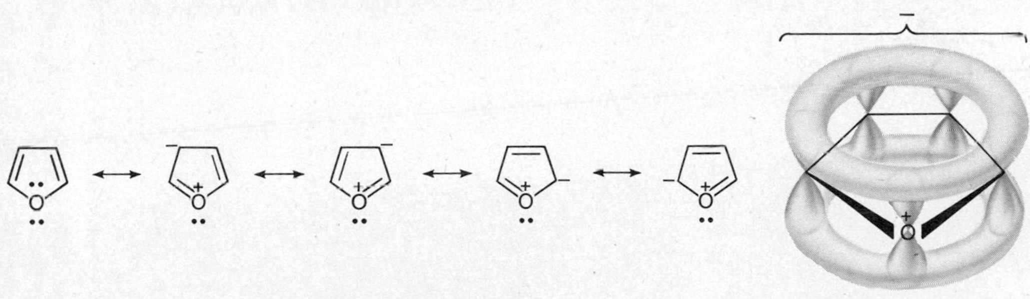

Fig. 11.3. Structure of furan.

planar molecule; thus, the interaction of a *p* orbital on oxygen and the π orbitals of the C=C bonds produces a completely delocalized π system (Fig. 11.3). As a result of the delocalization, the C—O bonds of furan are particularly strong; the system is more stable than cyclopentadiene (where complete delocalization around the entire ring is not possible), and the compound bears some resemblance to benzene in its physical and chemical properties. It is clear, however, that delocalization of the electrons from oxygen into the π system *reduces the electron density on oxygen, increases the electron density on the carbons,* and creates *charge separation.* For this reason, the delocalization is less extensive than in benzene, and furan possesses only 20 kcal/mole of resonance energy in contrast to the 36 kcal/mole possessed by benzene.

11.2 Physical Properties of Ethers

Although ethers have a dipole moment as the result of the electron-withdrawing characteristics of the oxygen atom, they boil considerably lower than the corresponding alcohols. Thus, CH_3OCH_3 (mol wt 46) boils at $-24°$, whereas CH_3CH_2OH (mol wt 46) boils at 78°. This happens because alcohols can hydrogen bond intermolecularly, possessing both a protic hydrogen and a basic oxygen, whereas ethers cannot hydrogen bond with themselves, lacking a protic hydrogen. The inability of ethers to act as hydrogen bond donors also results in a lower water solubility than alcohols. Ethers do possess an oxygen atom that can act as a Lewis base, however, so they can act as hydrogen bond acceptors with proton donors such as water. As a result, the low molecular weight ethers are moderately water soluble (diethyl ether dissolves to the extent of 7 g in 100 ml of water), and are quite soluble in strong acids, such as sulfuric acid. The experimentalist using diethyl ether as an extraction solvent in the laboratory should remember that when mixing 100 ml of diethyl ether with 100 ml of water and shaking the mixture to effect an extraction, about 7 to 8 ml of the ether dissolves in the water; when the mixture separates into layers, the lower water layer contains considerable ether and the upper ether layer contains some water. To insure complete extraction of material from the water layer into the ether layer, therefore, he must either carry out multiple, sequential extractions (subsequently combining the ether extracts) or reduce the solubility of ether in the aqueous phase by adding copious amounts of an inorganic salt such as sodium chloride.

11.3 Spectral Properties of Ethers

The presence of an ether linkage in a molecule is less easily determined by spectral means than is the presence of a hydroxyl function. The only characteristic ir absorption bands occur in the 1150–1060 cm^{-1} region for dialkyl ethers and in the 1270–1230 cm^{-1} region for alkyl aryl ethers and for diaryl ethers. The effect of the ether oxygen on the chemical shifts in the nmr of hydrogens at the α and β positions is almost the same as that of the hydroxyl function.

PROBLEM 11-2

A C_4H_6O compound shows only C—H stretching deformation bands in the 2900-cm^{-1} region of the infrared and has an nmr spectrum consisting of a downfield triplet pattern and an upfield doublet pattern. Suggest a structure for the compound.

11.4 Synthesis of Ethers

One of the commercial methods for the synthesis of diethyl ether involves the action of sulfuric acid on ethanol. The products of the reaction of sulfuric acid with ethanol are very sensitive to the reaction conditions, providing an excellent illustration of the importance of experimental conditions in determining the outcome of a chemical process. If the reaction is conducted at room temperature, the sulfate esters of ethanol are formed [$CH_3CH_2OSO_3H$ is ethyl hydrogen sulfate and $(CH_3CH_2O)_2SO_2$ is diethyl sulfate]; if the reaction is carried out at 140° in the presence of excess ethanol, the product is diethyl ether; if the reaction is carried out at 180°, the product is ethylene (Fig. 11.4). Higher molecular weight

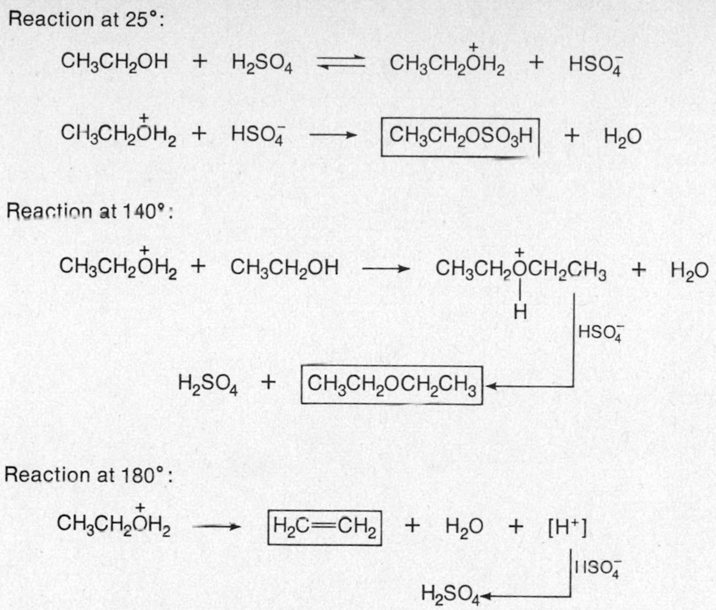

Reaction at 25°:

$$CH_3CH_2OH + H_2SO_4 \rightleftharpoons CH_3CH_2\overset{+}{O}H_2 + HSO_4^-$$

$$CH_3CH_2\overset{+}{O}H_2 + HSO_4^- \longrightarrow \boxed{CH_3CH_2OSO_3H} + H_2O$$

Reaction at 140°:

$$CH_3CH_2\overset{+}{O}H_2 + CH_3CH_2OH \longrightarrow CH_3CH_2\overset{+}{O}(H)CH_2CH_3 + H_2O$$

$$\xrightarrow{HSO_4^-} H_2SO_4 + \boxed{CH_3CH_2OCH_2CH_3}$$

Reaction at 180°:

$$CH_3CH_2\overset{+}{O}H_2 \longrightarrow \boxed{H_2C{=}CH_2} + H_2O + [H^+]$$

$$[H^+] \xrightarrow{HSO_4^-} H_2SO_4$$

Fig. 11.4. Reactions of ethanol and sulfuric acid.

ethers can be prepared in a similar fashion, but the synthesis is limited to the preparation of symmetrical ethers.

The preparation of unsymmetrical ethers is most readily accomplished by the action of alkoxides or aryloxides on alkyl halides. The reaction is a typical example of a nucleophilic substitution process (see Section 9.5), and it is known as the *Williamson ether synthesis.* In carrying out Williamson syntheses, judicious combinations of alkoxide and alkyl halide should be made. For example, benzyl *t*-butyl ether might be synthesized by the reaction of potassium benzyloxide and *t*-butyl bromide or by the reaction of potassium *t*-butoxide and benzyl bromide. A moment's reflection on the restrictions of the S_N2 reaction (see Section 9.5) should persuade the reader that the second of these two alternatives is clearly preferable (Fig. 11.5); primary halides (*e.g.*, benzyl bromide) undergo

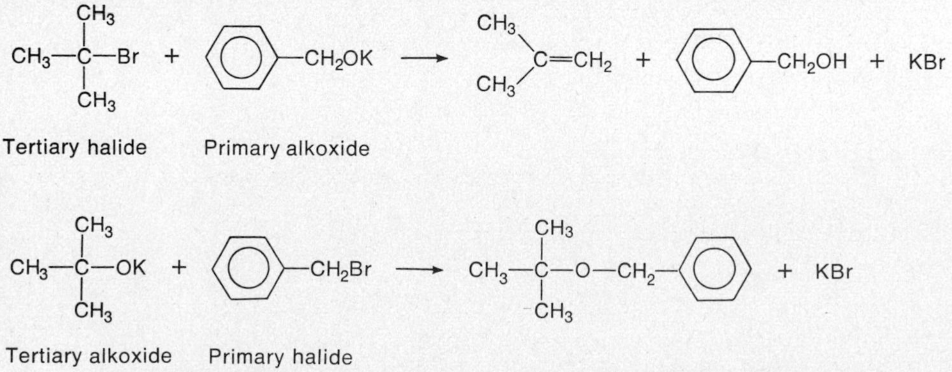

$$(CH_3)_3C{-}Br + C_6H_5{-}CH_2OK \longrightarrow (CH_3)_2C{=}CH_2 + C_6H_5{-}CH_2OH + KBr$$

Tertiary halide   Primary alkoxide

$$(CH_3)_3C{-}OK + C_6H_5{-}CH_2Br \longrightarrow (CH_3)_3C{-}O{-}CH_2{-}C_6H_5 + KBr$$

Tertiary alkoxide   Primary halide

Fig. 11.5. Comparison of procedures for the preparation of benzyl *t*-butyl ether by the Williamson ether synthesis.

S_N2 reactions without complication, whereas tertiary halides, such as *t*-butyl bromide, usually undergo elimination rather than substitution.

When the Williamson synthesis is employed for the preparation of alkyl aryl ethers, it is necessary to use the aryloxide and the alkyl halide rather than the aryl halide and the alkoxide; recall the prescription, cited earlier (see Section 9.5), against using aryl halides in S_N2 reactions. For example, a standard procedure for making anisole (methyl phenyl ether) involves mixing phenol with aqueous sodium hydroxide and adding either methyl iodide or, more conveniently and cheaply, dimethyl sulfate (Fig. 11.6).*

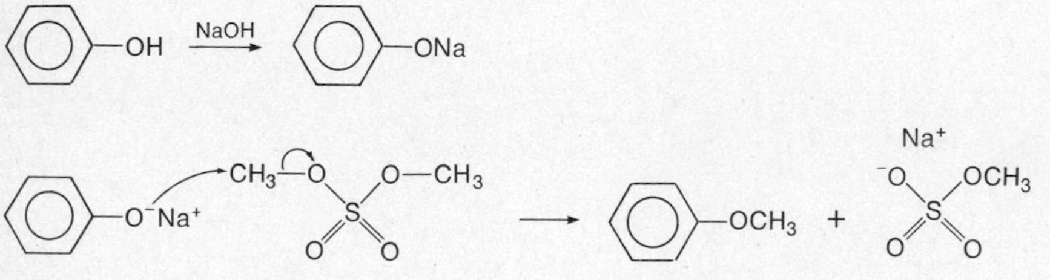

$$C_6H_5{-}OH \xrightarrow{NaOH} C_6H_5{-}ONa$$

$$C_6H_5{-}O^-Na^+ + CH_3{-}O{-}SO_2{-}O{-}CH_3 \longrightarrow C_6H_5{-}OCH_3 + Na^+\ {}^-O{-}SO_2{-}OCH_3$$

Anisole

Fig. 11.6. Preparation of anisole from sodium phenoxide and dimethyl sulfate.

PROBLEM 11-3

Taking cognizance of the reactions of alkynes (see Section 8.1) and the methods for preparing alcohols, halides, and ethers, indicate how the following reaction sequence could be carried out:

$$HC{\equiv}CH \longrightarrow CH_3CHO \longrightarrow CH_3CH_2OH \longrightarrow CH_3CH_2Br \longrightarrow CH_3CH_2OCH_2CH_3$$

PROBLEM 11-4

Remembering the free radical and electrophilic addition reactions of alkenes, suggest a method for converting 2-methylpropene to *t*-butyl isobutyl ether.

The three membered heterocyclic ethers, the oxiranes, are conveniently prepared in good yield by the action of peracids on alkenes (Fig. 11.7). Peracids $\left(RC(=O){-}O{-}OH\right)$ are related to carboxylic acids $\left(RC(=O){-}OH\right)$ in the same way that water (H—OH) is related to hydrogen peroxide (H—OOH); like

---

* Dimethyl sulfate is the dimethyl ester of sulfuric acid; methyl iodide can be considered to be the ester of hydriodic acid. Dimethyl sulfate and methyl iodide react in a comparable fashion; the iodide anion is the leaving group in CH_3I, and the methyl sulfate anion (and, ultimately, the sulfate anion) is the leaving group in $(CH_3O)_2SO_2$ [also written $(CH_3)_2SO_4$]. For all practical purposes, then, dimethyl sulfate reacts in S_N2 reactions in the same way as alkyl halides (Fig. 11.6).

hydrogen peroxide, the peracids are effective oxidizing agents, readily giving up an oxygen atom to various organic substances such as alkenes.

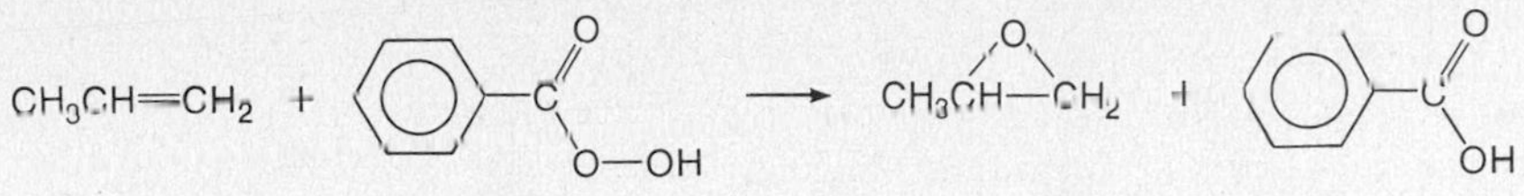

$CH_3CH{=}CH_2$ + Perbenzoic acid → 2-Methyloxirane + Benzoic acid

Fig. 11.7. Preparation of 2-methyloxirane *via* perbenzoic acid.

11.5 Reactions of Ethers

As mentioned in Section 11.2, ethers are Lewis bases and can be protonated with strong acids, such as sulfuric acid and the halogen acids. The R—O bond in protonated ethers, $R{-}\overset{+}{O}(H){-}R$, is much weaker than in the parent ether; *i.e.*, R—O—R is a good leaving group (*cf.* reactions of alcohols in Section 10.7), and even weak nucleophiles such as HSO_4^- can effect a displacement. Particularly useful as a reagent for cleaving alkyl ethers is hydriodic acid, since it is strong enough to protonate the ether effectively and iodide ion is an especially good nucleophile (Fig. 11.8). In the cleavage of a dialkyl ether, such as ethyl methyl ether, two sets of initial cleavage products are possible: methanol and ethyl iodide, or ethanol and methyl iodide. Because of the greater ease of nucleophilic attack at the methyl carbon atom, the second set of products is formed in preponderance. The alcohols produced by either of the initial cleavage steps, however, can themselves react with hydriodic acid to form the corresponding iodides, and the products that are actually obtained are methyl iodide and ethyl iodide. Phenols, on the other hand, do not react in nucleophilic substitution

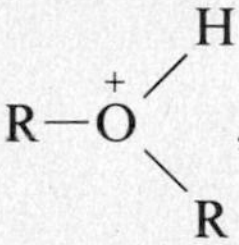

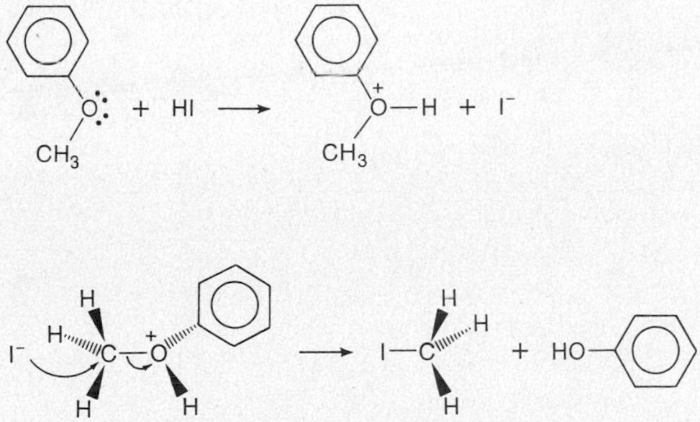

$C_6H_5\ddot{O}CH_3$ + HI → $C_6H_5\overset{+}{O}(CH_3){-}H$ + I^-

I^- + $H_3C{-}\overset{+}{O}(H)C_6H_5$ → $I{-}CH_3$ + $HO{-}C_6H_5$

Fig. 11.8. Hydrogen iodide cleavage of anisole.

reactions (see Section 10.7), so the products isolated from the reaction of methyl phenyl ether (anisole) with hydriodic acid are phenol and methyl iodide. The ether cleavage reaction provides a method for structure elucidation, for the structure of the phenols and/or alkyl iodides produced can often furnish insight into the structure of the parent ether.

To effect a nucleophilic displacement on the C—O bonds of ethers, we must, in most instances, enhance the leaving ability of the —OR group by protonating the oxygen, as indicated by the reactions just discussed. Ethylene oxide (an oxirane), however, is an exception to this requirement, for even in the unprotonated form it reacts with nucleophiles to undergo S_N2 reactions. The ease of this reaction is ascribed to the bond angle strain of the three-membered ring (estimated to be about 27 kcal/mole),* a strain that is relieved in the course of the ring-opening displacement reaction; the bond angle strain provides a "driving force" for the reaction. Among the many nucleophiles with which ethylene oxide reacts are Grignard reagents and lithium aluminum hydride (Fig. 11.9).

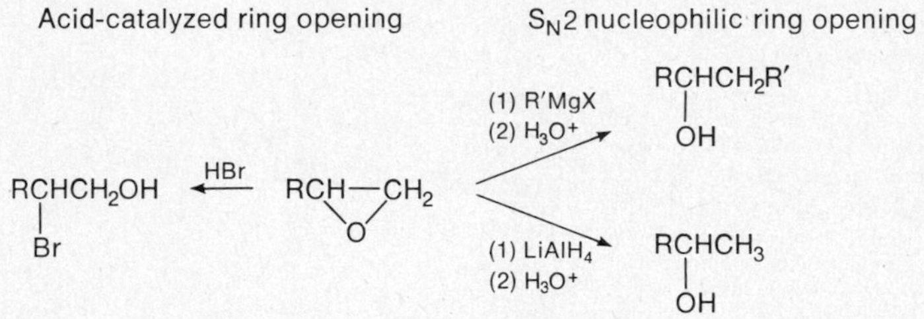

Acid-catalyzed ring opening | S_N2 nucleophilic ring opening

$RCH(Br)CH_2OH \xleftarrow{HBr} RCH{-}CH_2$ (oxirane, ring O)

$RCH{-}CH_2$ (oxirane) $\xrightarrow{(1)\ R'MgX;\ (2)\ H_3O^+} RCH(OH)CH_2R'$

$RCH{-}CH_2$ (oxirane) $\xrightarrow{(1)\ LiAlH_4;\ (2)\ H_3O^+} RCH(OH)CH_3$

Fig. 11.9. Acid-catalyzed and S_N2 nucleophilic ring opening reactions of oxiranes.

Because of the strain energy of oxiranes, they react with acidic reagents even more rapidly than acyclic ethers do, producing *β*-substituted alcohols. The direction of opening of an oxirane in the S_N2 and acid-catalyzed processes differs. The less highly substituted carbon (sterically more accessible) is the site of the attack in the S_N2 process, whereas the more highly substituted carbon (more stable carbonium ion) is the site of attack in the acid-catalyzed process (Fig. 11.9).

Alkyl aryl ethers behave in a fashion similar to phenols with respect to electrophilic substitution reactions. The alkoxy group increases the electron density of the aryl ring through delocalization of the nonbonded electrons on oxygen into the π electron system, and facile reaction with electrophiles occurs at the *o* and *p* positions (Fig. 11.10). Thus, alkoxy functions, like the hydroxyl function, are classified as *o*-, *p*-directing and activating in electrophilic substitution reactions.

* Ring strain diminishes with increasing ring size, although oxetane possesses almost as much strain as ethylene oxide, similar to the situation between cyclopropane and cyclobutane (see p. 79). The strain energies of tetrahydrofuran and tetrahydropyran are estimated to be about 6 and 1.5 kcal/mole.

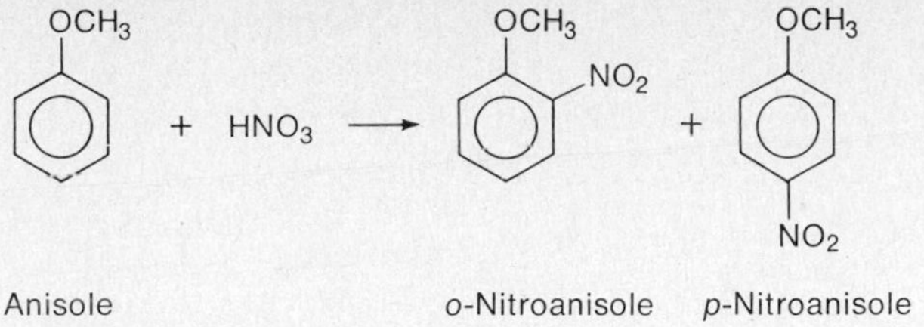

Anisole + HNO_3 → o-Nitroanisole + p-Nitroanisole

Fig. 11.10. Nitration of anisole by electrophilic substitution.

PROBLEM 11-5

Compounds A and B have the molecular formula $C_8H_{10}O$. When compound A is treated with boiling hydriodic acid, a C_6H_6O compound and ethyl iodide are formed. When compound B is similarly treated, a C_7H_7I compound and methyl iodide are formed. Suggest structures for compounds A and B.

Ethers in which there is a hydrogen attached to the α-carbon are susceptible to air oxidation (often called autoxidation). For example, diethyl ether that has been exposed to air for relatively long periods of time (*i.e.*, weeks to months) contains 1-ethoxyethylhydroperoxide as a result of this reaction;

$$CH_3CH_2OCH_2CH_3 + O_2 \longrightarrow CH_3CH_2OCH(OOH)CH_3$$

1-Ethoxyethylhydroperoxide

When ether contaminated with peroxide is distilled, the lower boiling ether comes over first, leaving the higher boiling peroxide in the distillation flask. If the distillation process is carried too far, an explosive concentration of peroxide is reached and a detonation occurs. To avoid this, ether suspected of containing peroxides should be treated, before distillation, with reagents such as sodium iodide, ferrous sulfate, sodium sulfite, or lithium aluminum hydride that reduce the hydroperoxides to nonexplosive compounds.

11.6 Ethers of Particular Utility and Interest

Diethyl ether is famous in the laboratory as the most frequently employed solvent in the preparation of Grignard reagents and as the most often used extracting solvent. To the world at large, it is probably most famous for its anesthetic properties. Discovered and rediscovered as an anesthetic in the 1840's, ether has been of inestimable service to man in alleviating pain during the relatively short durations required in surgical operations. It does so by depressing the activity of the central nervous system, probably by altering the permeability of the cells through the formation of association complexes with compounds at the surfaces of the cells. Other materials have been discovered to possess similar activity: included among the anesthetics that have been used at one time or another are nitrous oxide, ethylene, divinyl ether, cyclopropane, and chloroform. All these have now been displaced by the haloalkane compound known as Halothane or Fluothane (see p. 243), which is currently the most widely used anesthetic.

A series of commercially useful solvents are prepared by the reaction of alcohols with ethylene oxide, which yields ethers of ethylene glycol and polyethylene glycol. The reaction of equivalent amounts of methanol and ethylene oxide, for example, yields

$$CH_3OCH_2CH_2OH$$

(the monomethyl ether of ethylene glycol known as methyl cellosolve). Further reaction of another equivalent of ethylene oxide with ethylene glycol monomethyl ether yields

$$CH_3OCH_2CH_2OCH_2CH_2OH$$

(the monomethyl ether of diethylene glycol known as methyl carbitol), and continuation of the process yields monomethyl ethers of triethylene glycol, tetraethylene glycol, etc. Conversion of the terminal hydroxyl group of these compounds to an ether group yields materials known as "glymes":

$CH_3OCH_2CH_2OCH_3$ is "monoglyme,"

$CH_3OCH_2CH_2OCH_2CH_2OCH_3$ is "diglyme," etc.

The process whereby ethylene oxide is added to nucleophiles such as alcohols to introduce OCH_2CH_2— units is called "oxyethylation." Phenols, for instance, can be "oxyethylated" to give products that can act as nonionic detergents because of their ability to dissolve in both aqueous and nonaqueous phases. Some oxyethylation products are shown in Fig. 11.11.

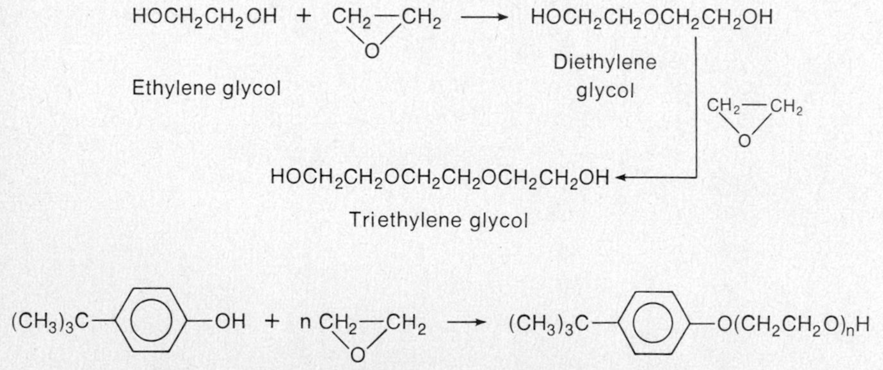

$$HOCH_2CH_2OH + CH_2\text{—}CH_2 \text{ (O)} \longrightarrow HOCH_2CH_2OCH_2CH_2OH$$

Ethylene glycol — Diethylene glycol

$$HOCH_2CH_2OCH_2CH_2OH \xrightarrow{CH_2\text{—}CH_2 \text{ (O)}} HOCH_2CH_2OCH_2CH_2OCH_2CH_2OH$$

Triethylene glycol

$$(CH_3)_3C\text{—}C_6H_4\text{—}OH + n\, CH_2\text{—}CH_2 \text{ (O)} \longrightarrow (CH_3)_3C\text{—}C_6H_4\text{—}O(CH_2CH_2O)_nH$$

Fig. 11.11. Oxyethylation of alcohols and phenols.

A number of alkyl aryl ethers have been isolated from natural sources. Illustrative examples, shown in Fig. 11.12, are anethole (the major component of anise oil), guaiacol (found in the distillate of the resin from American tropical trees of the genus *Guaiacum* and used in medicine and varnishes), and eugenol (in the oil of cloves).

One of the most important heterocyclic ethers is tetrahydrofuran. A commercial method for its preparation starts with furfural, a material that can be produced in enormous quantity by the acid-catalyzed dehydration of the pentose

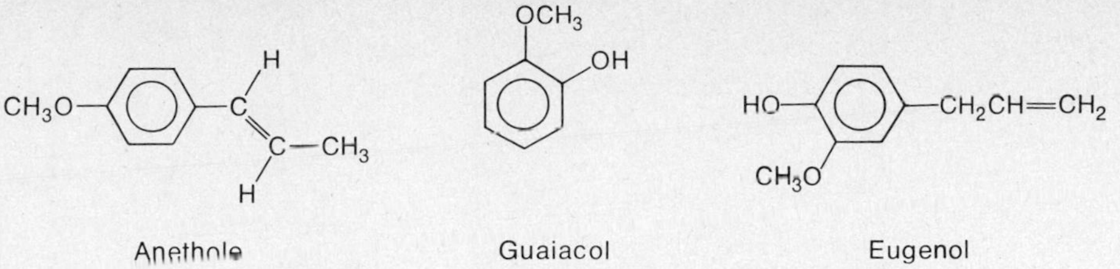

Anethole Guaiacol Eugenol

Fig. 11.12. Naturally occurring aryl alkyl ethers.

sugars in oat hulls and corn cobs. Treatment of furfural with a special catalyst at 400° removes the aldehyde group and yields furan, which, in turn, can be reduced with hydrogen and a metal catalyst to tetrahydrofuran (Fig. 11.13). Tetrahydrofuran has some use as a commercial solvent but is more extensively consumed as an intermediate in the production of one of the constituents of Nylon (see p. 414).

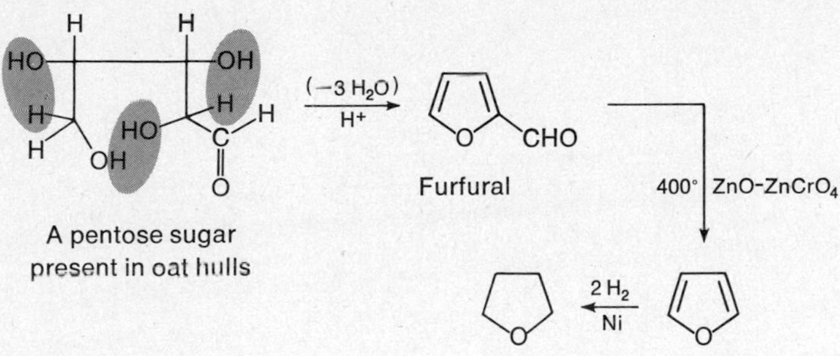

Fig. 11.13. Synthesis of tetrahydrofuran from oat hulls.

11.7 Study Guide

I. Structure
  A. All carbon frameworks possible
  B. Cyclic ethers
II. Nomenclature
  A. "Ether" system
  B. IUPAC system
  C. "Oxa" system
  D. Cyclic ethers
III. Physical properties
  A. Boiling points—lower than corresponding alcohols
  B. Water solubility—less than corresponding alcohols
IV. Spectral properties
  A. Ultraviolet/visible—transparent above 200 nm
  B. Infrared—C—O stretching band near 1100 cm^{-1}
  C. Nuclear magnetic resonance—comparable to alcohols
V. Synthesis
  A. Intermolecular dehydration of alcohols
  B. Nucleophilic displacement of alkyl halides (Williamson synthesis)
    1. General reaction: $RONa + R'X \rightarrow ROR' + NaX$

2. Limitations
   (a) Secondary halides (R_2CHX) give appreciable amounts of alkene
   (b) Tertiary halides (R_3CX) give almost entirely alkene as product
3. Aryl halides (ArX) and vinyl halides ($C{=}CX$) do not react.

VI. Reactions
   A. Acid-catalyzed cleavage (with hydriodic acid)
   B. Nucleophilic displacements with ethylene oxide
   C. Air oxidation (formation of hydroperoxides)
   D. Electrophilic ring substitution of aryl ethers

PROBLEMS

1. Using any nomenclature system that is appropriate, name each of the following structures:

(a) $(CH_3)_2CHOCH_3$

(b) 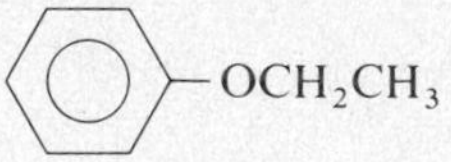 $-OCH_2CH_3$

(c) $(CH_3)_3CO\underset{\displaystyle CH_3}{\underset{|}{C}}HCH_2CH_3$

(d) $C_6H_5CH_2OCH_2CH{=}CH_2$

(e) 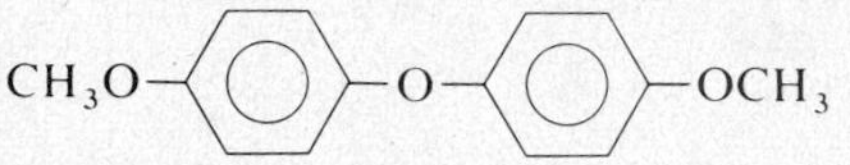 CH_3O- $-O-$ $-OCH_3$

(f) $HC{\equiv}C\underset{\displaystyle OCH_3}{\underset{|}{C}}HCH_2\overset{\displaystyle CH_3}{\overset{|}{C}}HCH{=}CH_2$

(g) $(CH_3)_2CH\underset{\displaystyle CH_3}{\underset{|}{C}}HCH_2OCH_2CH_3$

(h) CH_3 O

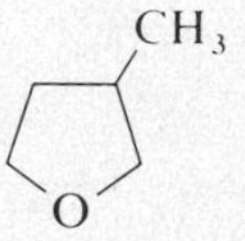

(i) 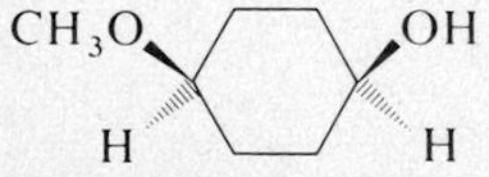 CH_3O OH H H

(j) 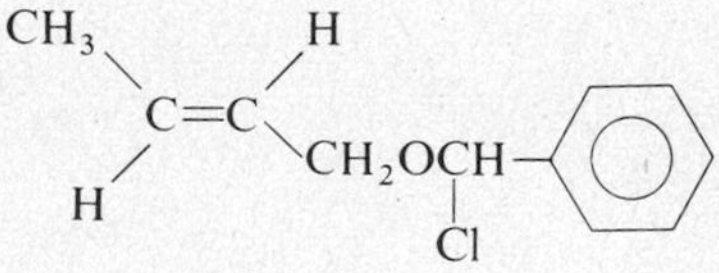 CH_3 H $C{=}C$ H $CH_2O\underset{\displaystyle Cl}{\underset{|}{C}}H-$

(k) $H_2C{=}CHOC{\equiv}CC(CH_3)_3$

(l) $CH_3OCH_2CH_2OCH_2CH_3$

(m) 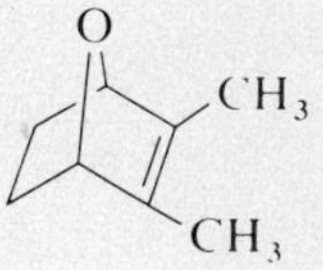 O CH_3 CH_3

2. Draw the structures corresponding to the following names:

(a) Isopropyl neopentyl ether
(b) Cyclobutyl isobutyl ether
(c) 1-Chloroethyl ethyl ether
(d) 2,2-Dimethoxypropane
(e) *p*-Methoxyphenol
(f) 1-Chloro-3-methoxy-2-propanol
(g) *trans*-2-Hydroxy-1-methoxyspiro[2.3]hexane
(h) 2-(*p*-Chlorophenoxy)-ethyl bromide
(i) *trans*-2-Butenyl 2,4,6-tribromophenyl ether
(j) *cis*-8-Oxabicyclo[4.3.0]nonane
(k) 3,3-Dimethyloxetane
(l) *trans*-2-Methyl-3-phenyloxirane

3. The following conversions from the compounds shown on the left to the compounds shown on the right can be effected in one step. Indicate what reagents are necessary to achieve this.

(a) —OH $\longrightarrow$ —OCH_2CH_3

(b) —OH $\longrightarrow$ —OCH_2CH_2OH

(c) —$OCH(CH_3)_2$ $\longrightarrow$ —OH + $(CH_3)_2CHI$

(d) O $\longrightarrow$ $ICH_2CH_2CH_2CH_2I$

(e) $CH_3CH{-}CH_2$ (O) $\longrightarrow$ $CH_3CH_2CH_2OH$

4. The following conversions from the compounds shown on the left to the compounds shown on the right require more than one step. Indicate what sequences of reactions are necessary to achieve this, show the reagents, and draw the structures of the intermediate compounds that are produced.

(a) $CH_3CH{=}CH_2$ $\longrightarrow$ $(CH_3)_2CHOCH(CH_3)_2$

(b) —CH_3 $\longrightarrow$ —CH_2OCH_2—

(c) CH_3CH_2OH and CH_3OH $\longrightarrow$ $CH_3CH_2OCH_3$

(d) $H_2C{=}CHCH_2Br$ $\longrightarrow$ $H_2C{=}CHCH_2OCH_2CH{=}CH_2$

(e) $HC{\equiv}CCH_3$ $\longrightarrow$ $HC{\equiv}CCH_2OCH(CH_3)_2$

5. Draw the structure of the major product formed in each of the following reactions:

(a) $(CH_3)_2C{=}CH_2$ + CH_3OH $\xrightarrow{H^+}$

(b) —ONa + —CH_2CH_2Br $\longrightarrow$

(c) OH, —OH, OH + $(CH_3)_2SO_4$ $\xrightarrow{NaOH}$

(d) OCH_3, —CH_2OCH_3 + HI (excess) $\longrightarrow$

(e) $H_2C{=}CHLi$ + O $\longrightarrow$ $\xrightarrow{H_3O^+}$

(f) (2-methyltetrahydrofuran, ring O with $-CH_3$) + HI (1 equivalent) $\longrightarrow$

(g) $CH_3(C_6H_5)C\overset{O}{—}CH_2$ (epoxide) + $H_2C{=}CHCH_2MgBr \longrightarrow \xrightarrow{H_3O^+}$

6. Starting with alkanes, alkenes, alkynes, alkyl halides, or alcohols containing no more than 3 carbon atoms, benzene, and/or phenol, along with any necessary inorganic reagents, synthesize the following compounds:

(a) $(CH_3)_2CHOCH_3$

(b) 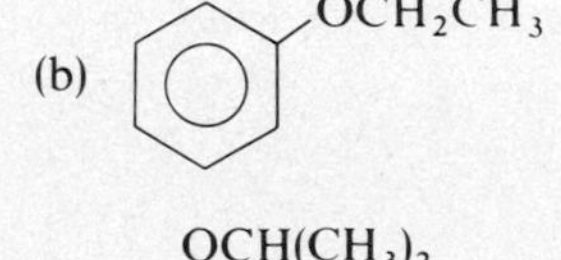

(c) Br (para-substituted benzene ring) OCH_3

(d) $OCH(CH_3)_2$ (para-substituted benzene ring) NO_2

(e) $(CH_3CH_2CH_2CH_2)_2O$

7. Starting with any alkane, alkene, alkyne, alkyl halide, or alkanol containing four or fewer carbon atoms, benzene, phenol, and any aryl halide and any inorganic reagent, outline reasonable syntheses for each of the following:

(a) $(CH_3)_2CHOCH(CH_3)_2$

(b) $C_6H_5CH_2OCHCH_2CH_3$ with CH_3 on the CH

(c) $CH_3CH_2OCH_2CH_2OCH_3$

(d) $CH_3CH_2CH_2COCH_3$ with CH_2CH_3 and CH_3 on the central C

(e) $(CH_3)_2CH-C_6H_4-OCH_2CH{=}CHCH_3$ (para-substituted benzene ring)

8. Arrange the following compounds in the order of increasing boiling point:

(a) $CH_3OCH_2CH_2OCH_3$ (b) $ClCH_2CH_2Cl$ (c) $ClCH_2CH_2OH$

(d) $HOCH_2CH_2OH$ (e) $CH_3OCH_2CH_2OH$

9. Draw structures for the compounds designated by letters in each of the following reaction schemes.

(a) $CH_3CH_2OH + Na \longrightarrow \frac{1}{2}H_2 + \mathbf{A} \xrightarrow{(CH_3)_2CHBr} \mathbf{B}$

$\mathbf{B} \xrightarrow{Br_2} \mathbf{C} + HBr$

$\mathbf{D}$ (formula $C_5H_{10}O$) $\xleftarrow{(CH_3)_3COK} \mathbf{C} + HBr$

(b) 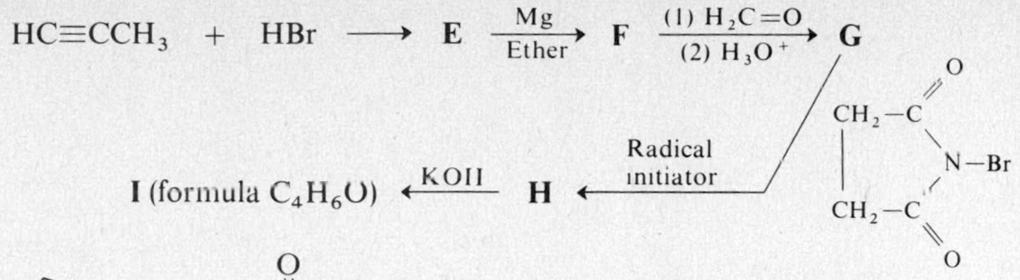 $HC{\equiv}CCH_3$ + HBr ⟶ **E** $\xrightarrow[\text{Ether}]{\text{Mg}}$ **F** $\xrightarrow[(2)\ H_3O^+]{(1)\ H_2C{=}O}$ **G**

G $\xrightarrow[\text{initiator}]{\text{Radical}}$ (N-bromosuccinimide) **H** $\xrightarrow{\text{KOH}}$ **I** (formula C_4H_6O)

(c) cyclopentene + $CF_3\overset{O}{\overset{\|}{C}}{-}OOH$ ⟶ **J** $\xrightarrow{CH_3ONa}$ **K** $\xrightarrow{C_6H_5CH_2Br}$ **L**

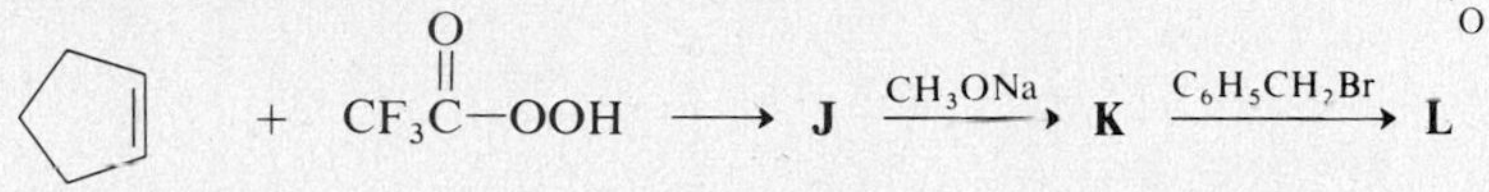

10. Using chemical and/or spectral methods (as indicated in each case), show how the following pairs of isomeric compounds can be differentiated.

(a) 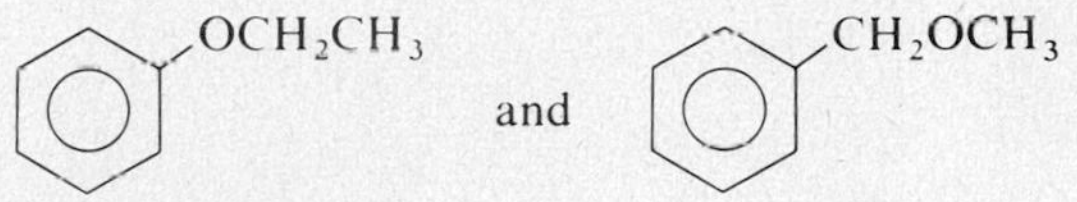 $C_6H_5OCH_2CH_3$ and $C_6H_5CH_2OCH_3$ — Chemical methods only

(b) $(CH_3)_2CHOCH(CH_3)_2$ and $(CH_3CH_2CH_2)_2O$ — Spectral methods only

(c) $H_2C{=}CHCH_2OCH_3$ and $CH_3CH{=}CHCH_2OH$ — Chemical and spectral methods

(d) 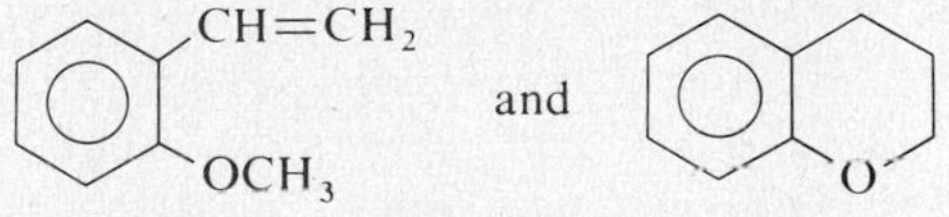 o-$CH{=}CH_2$, OCH_3 benzene and chroman (benzo-fused ring with O) — Chemical and spectral methods

(e) 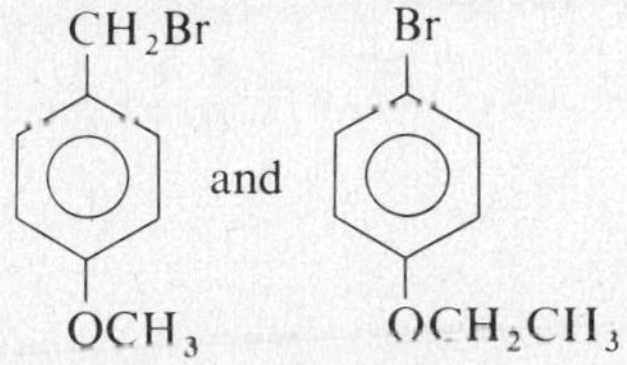 p-CH_2Br, OCH_3 benzene and p-Br, OCH_2CH_3 benzene — Chemical and spectral methods

11. In the following examples, you are presented with four test tubes, each containing one of the four compounds of the isomeric set. What chemical tests would you perform to make a structural assignment to each of the compounds in the set?

(a) 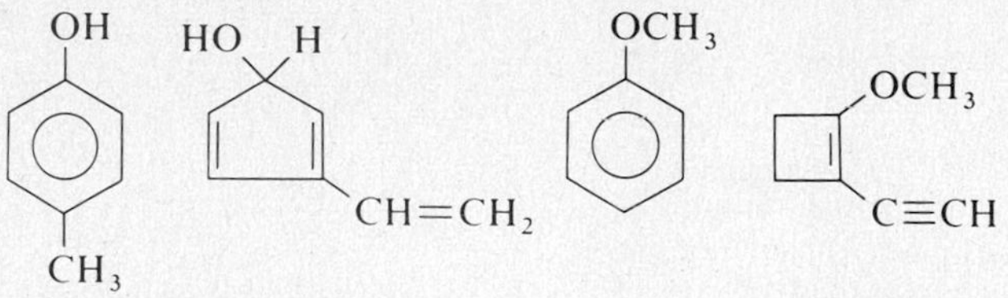 p-$CH_3C_6H_4OH$; cyclopentadiene ring bearing HO, H and $CH{=}CH_2$; $C_6H_5OCH_3$; cyclobutene bearing OCH_3 and $C{\equiv}CH$

(b) $CH_3OCH_2CH_2OCH_3$ $\qquad$ $HOCH_2CH_2CH_2CH_2OH$

$HOCH_2CH_2CH_2OCH_3$ $\qquad$ $HO\underset{CH_3}{\underset{|}{C}}H{-}\underset{CH_3}{\underset{|}{C}}HOH$

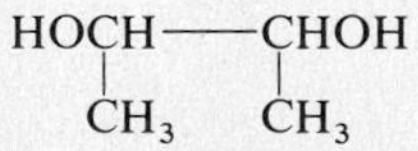

(c) $ClCH{=}CHCH_2CH_2OH$ $\qquad$ tetrahydrofuran ring (CH_2–CH_2–CH_2–$CHCl$–O)

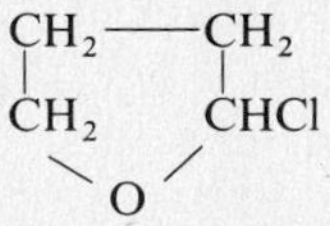

$ClCH_2CH{=}CHCH_2OH$ $\qquad$ cyclobutane ring ClCH–CH_2–CHOH–CH_2

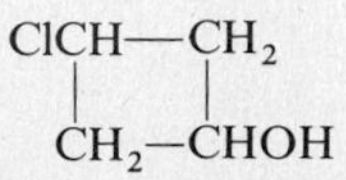

12. A pair of isomeric $C_9H_{12}O$ compounds possess the nmr spectra shown below. Suggest structures compatible with these data.

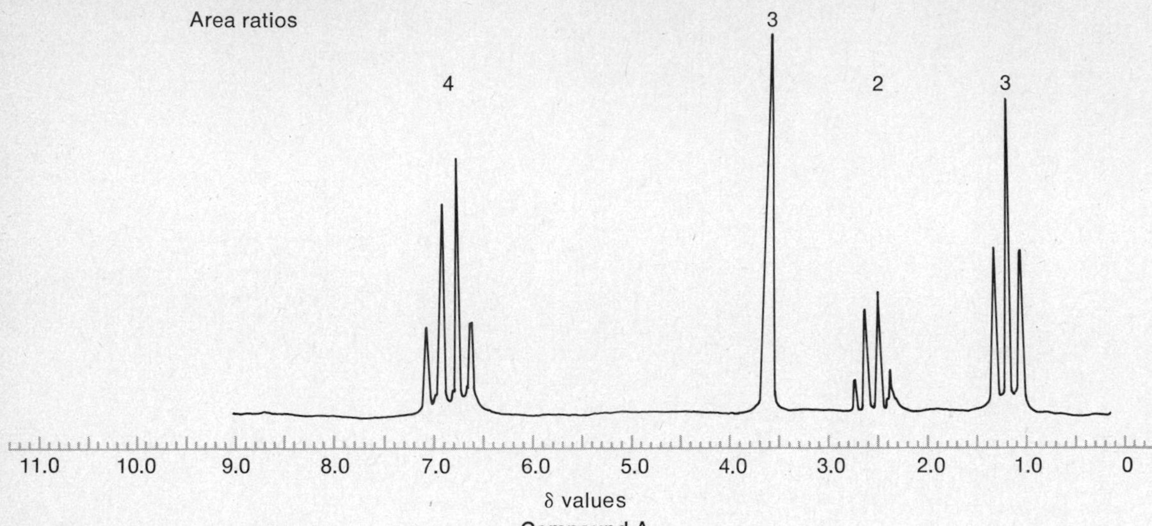

Compound A

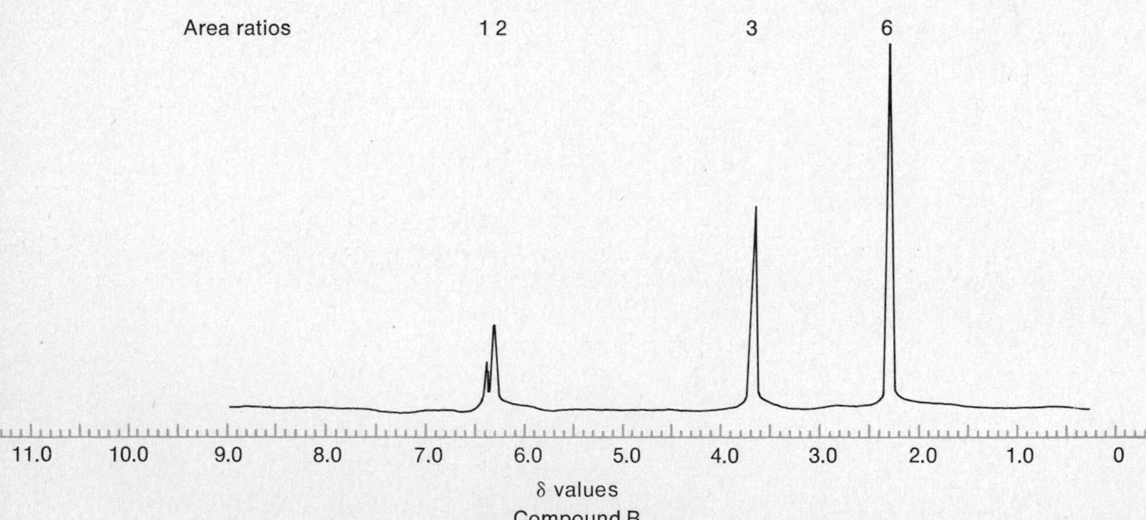

Compound B

Chapter 12

Amines

"Non olet?" (Translation: "Does it not betray itself by its odor?")

CICERO
Orator

Progressing toward the left along the top row of the periodic chart, we next consider the structure and chemistry of compounds resulting from the attachment of nitrogen to carbon frameworks by single bonds. The compounds of this class are called *amines.* Just as the alcohols, phenols, and ethers are considered as derivatives of water, the amines are considered as derivatives of ammonia, in which one, two, or all three of the hydrogens are replaced by carbon frameworks.

12.1 Structure and Nomenclature of Amines

The electron configuration of nitrogen (Fig. 2.5) is described as $1s^2 2s^2 2p_x^1 2p_y^1 2p_z^1$, indicating that three $2p$ orbitals are incompletely filled and available for bonding. Nitrogen, then, is predicted to be trivalent, with bond angles of 90°. When nitrogen is single bonded to hydrogen and/or carbon frameworks, however, the bond angles are close to 109° [*i.e.*, 107° for NH_3 and 108° for $(CH_3)_3N$], suggesting that nitrogen is sp^3 hybridized in these cases.

Attachment of two hydrogen atoms to a nitrogen produces an amino group ($-NH_2$) that can be attached to any of the carbon frameworks. When the amino group is attached to an sp^3-hybridized carbon, a primary amine (RNH_2) is produced, an orbital representation of which is shown in Fig. 12.1. If, now, another carbon framework is affixed to the nitrogen in place of one of the hydrogens, a secondary amine (R_2NH) is produced, and if still another carbon framework is affixed, a tertiary amine (R_3N) results. The prefixes "primary," "secondary," and "tertiary" specify the degree of branching *at the nitrogen atom* and are used in the same way that we have used them to specify the degree of branching at carbon (see p. 61). The different contexts in which these profixes are used should be kept in mind. When applied, for example, to the alcohol series, the prefix denotes the degree of branching *at the carbon* attached to the hydroxyl group; but, when used to specify the *class* to which an amine belongs, it denotes

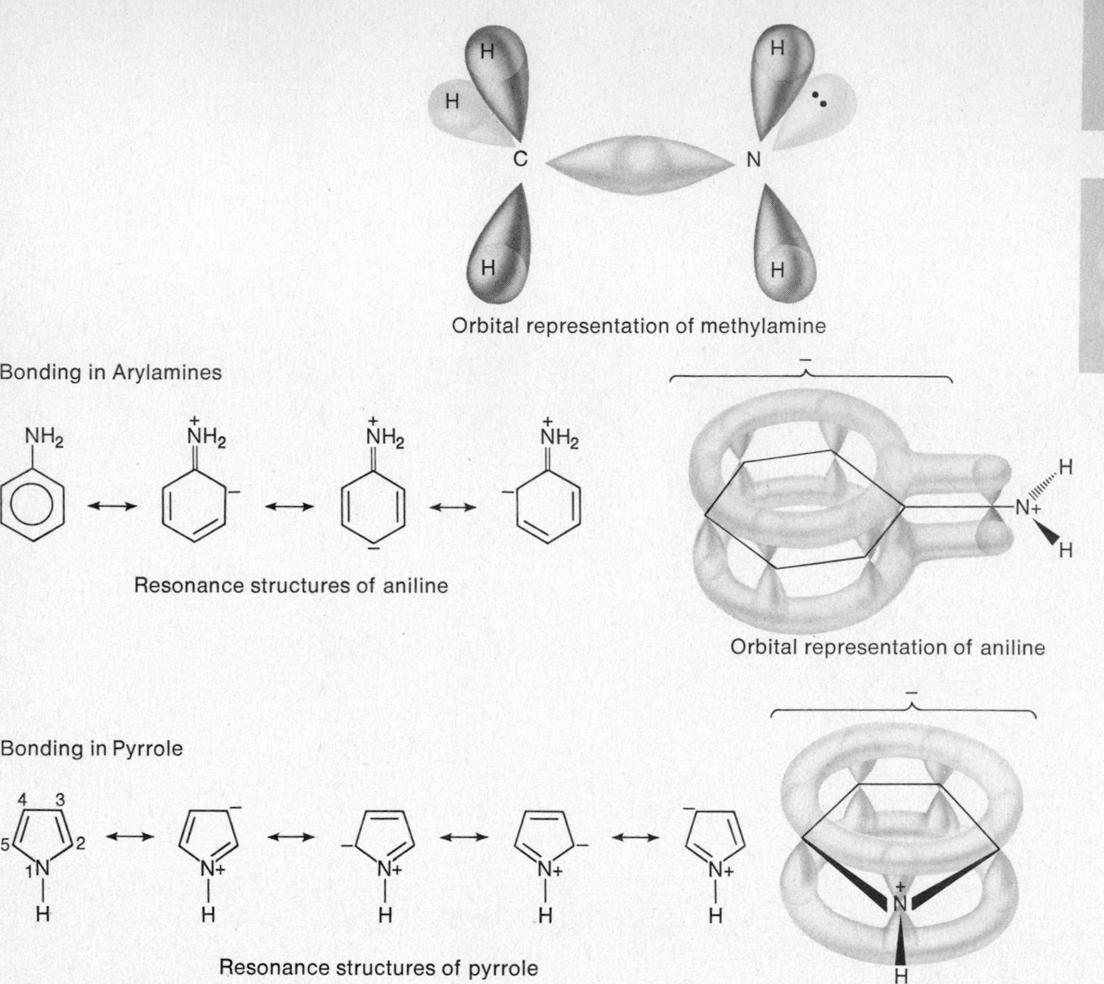

Fig. 12.1. Bonding in alkylamines, arylamines, and pyrrole.

the degree of branching *at the nitrogen.* Thus, $(CH_3)_3COH$ is *t*-butyl alcohol, but $(CH_3)_3CNH_2$ although named *t*-butylamine, belongs to the class of primary amines. Several illustrative examples of amines, including some in which the nitrogen is incorporated into a ring, are shown in Fig. 12.2.

In the primary, secondary, and tertiary amines, the nitrogen atom possesses a pair of nonbonded electrons, and it is possible for these to engage in bonding. For example, if trimethylamine reacts with methyl iodide, the quaternary compound tetramethylammonium iodide is formed. Like its inorganic analog, ammonium chloride, it is salt-like in character. The fourth bond to the nitrogen in the quaternary compounds arises from the contribution of the two nonbonded electrons, resulting in a positive charge on the nitrogen, which is counterbalanced

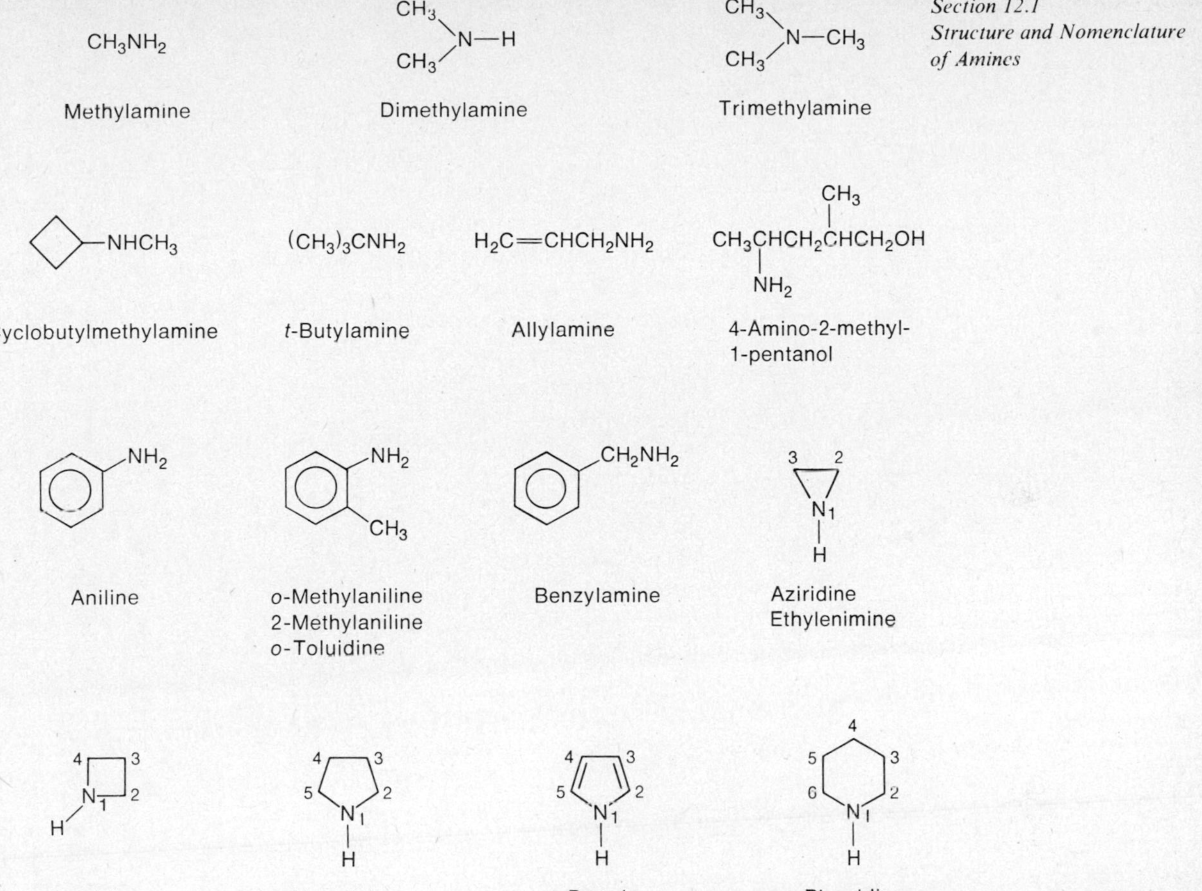

Fig. 12.2. Structures and names of some amines.

by the negative charge on the halide ion. Bonds of this type, in which one of the atoms contributes a greater number of electrons than the other, are said to be *coordinate covalent.* Once the interaction between CH_3I and $(CH_3)_3N$ has taken place to form $(CH_3)_4N^+I^-$, however, all four C—N bonds are equivalent; no one of them can be singled out and stated to be coordinate-covalent, but all four have 25% coordinate-covalent character.

Tetramethylammonium iodide

As in the case of the hydroxyl group, attachment of an amino group to an sp^2- or sp-hybridized carbon produces compounds that in most cases are less stable than the isomeric tautomeric compounds containing a C=N bond:

$$\underset{\textbf{Amino form}}{>C{=}\overset{|}{C}{-}NH_2} \rightleftharpoons \underset{\textbf{Imino form}}{-\overset{|}{\underset{H}{C}}{-}\overset{|}{C}{=}N{-}H}$$

Also in similarity with their hydroxyl counterparts, amino functions attached to aromatic rings correspond to compounds that are stable in the amino form. This is ascribed to the resonance stabilization (aromaticity) provided by the aromatic ring, as illustrated by the orbital and resonance structure representations in Fig. 12.1. In a rather similar vein, the five-membered cyclic unsaturated amine known as pyrrole is stable in the amino form, also as the result of resonance stabilization, as depicted in Fig. 12.1. Like its oxygen counterpart, furan, pyrrole is more stable than its classical, localized bond structure would suggest, and it possesses about 22 kcal/mole of resonance energy. It shows aromatic properties, as indicated by its nmr characteristics (see Section 12.2) and by its tendency to undergo electrophilic substitution reactions.

The alkylamines are named in the IUPAC system by designating the alkyl group in the appropriate fashion and then affixing the suffix "amine." Thus, CH_3NH_2 is methylamine, $(CH_3)_2NH$ is dimethylamine, $CH_3NHCH_2CH_3$ is ethylmethylamine, etc. Alternatively, the suffix "amine" is added to the name of the carbon framework, with omission of the terminal "e" of the name of the carbon framework where it occurs. Thus, CH_3NH_2 is methanamine, $CH_3CH_2\underset{\underset{CH_3}{|}}{C}HNH_2$ is 2-butanamine, etc. When the $R_2N—$ group is attached to a phenyl ring, the compound can be named as a derivative of aniline, $C_6H_5NH_2$; *i.e.*, $C_6H_5N(CH_3)_2$ is N,N-dimethylaniline, the prefixes "N" indicating that the methyl groups are attached to the nitrogen instead of elsewhere in the molecule. Certain other common names have also been associated with some of the amines, as Fig. 12.2 indicates. In more complex cases, the amino functions appear as substituents rather than as part of the parent compound; *e.g.*, $(CH_3)_2N—$ is designated as a dimethylamino function. Alternatively, the amino function (—NH—) can be indicated by the incorporation of the prefix "aza" before the name of the carbon framework along with the appropriate position number, implying replacement of a methylene group ($—CH_2—$) by an NH group; this method of nomenclature is generally reserved for compounds that cannot be conveniently named by the other systems as, for example,

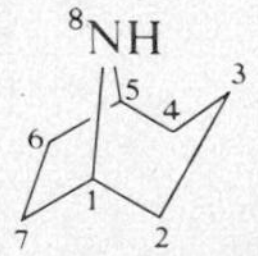

which is 8-azabicyclo[3.2.1]octane.

PROBLEM 12-1

Draw and name all the acyclic C_4H_9N amines. Draw and name all the $C_8H_{11}N$ anilines.

12.2 Physical Properties of Amines

Methylamine (mol wt 31) boils at −7°, a value between that of the less polar ethane (mol wt 30), which boils at −88°, and the more polar methanol (mol wt 32), which boils at 64°. Clearly, methylamine is less strongly intermolecularly associated than methanol, almost certainly as the result of the weaker hydrogen bonding in the amine. The ability of primary and secondary amines to form hydrogen bonds, although weaker than those in alcohols, is indicated by the contrast between their boiling points and those of the isomeric tertiary amines. For example, *n*-propylamine ($CH_3CH_2CH_2NH_2$) boils at 49°, ethylmethylamine ($CH_3NHCH_2CH_3$) boils at 35°, and trimethylamine [$(CH_3)_3N$] boils at 3°. Tertiary amines (lacking N—H) bear the same relationship to primary and secondary amines (containing N—H) that ethers (lacking O—H) bear to alcohols (containing O—H); the parallel is reflected not only in boiling point comparisons but in water solubility comparisons as well. Thus, just as ethers are less water soluble than their isomeric alcohols, tertiary amines are less water soluble than their isomeric primary and secondary amines. All the lower molecular weight amines, however, are water soluble, for even in tertiary amines the nitrogen can still function as a hydrogen bond acceptor and engage in hydrogen bonding with water. The hydrogen bond between a tertiary amine and water, in fact, is stronger than that between an ether and water, because the nonbonded electrons on nitrogen are less tightly held than those on oxygen. Conversely, hydrogens attached to nitrogen are less acidic than those attached to oxygen. Apparently, the greater basicity of the nitrogen atoms in amines does not quite offset the lower acidity of the N—H bond in primary and secondary amines, for the strength of hydrogen bonding in primary and secondary amines is less than that of the analogous alcohols.

Odor is a property possessed by many organic compounds, and it can be used in a variety of ways. Certain compounds and mixtures of compounds with odors pleasing to most human noses are used as perfumes. Other compounds with less pleasing odors, not only to human noses but to those of other species as well, are used by certain animals as a defense mechanism. Many amines, particularly the lower molecular weight ones, belong in the second category and possess odors that range from somewhat unpleasant to absolutely obnoxious. Odor can sometimes be used to advantage by the organic chemist, for the nose is a remarkably sensitive detection device. On many occasions, the amine-like structure of a compound has been inferred from the characteristic fish-like odor of the compound itself or of some of its degradation products.

12.3 Spectral Properties of Amines

The infrared spectra of amines resemble those of alcohols and ethers. Primary amines have two N—H stretching deformation bands in the 3500–3300 cm^{-1} region, secondary amines have one, and tertiary amines, of course, have none. Primary, secondary, and tertiary amines all have C—N stretching deformation bands in the 1350–1000 cm^{-1} region (see examples in Fig. 12.3). If a compound is known to contain nitrogen, therefore, the presence of a band in the vicinity of 3400 cm^{-1} is suggestive of the presence of an N—H function. The absorptions in the 1350–1000 cm^{-1} region, however, must be interpreted with discretion, because many other types of stretching and bending deformations also give rise to absorptions in this region.

The nmr characteristics of amines also resemble those of alcohols and ethers. Nitrogen is electronegative relative to carbon but less so than oxygen, so hydrogens

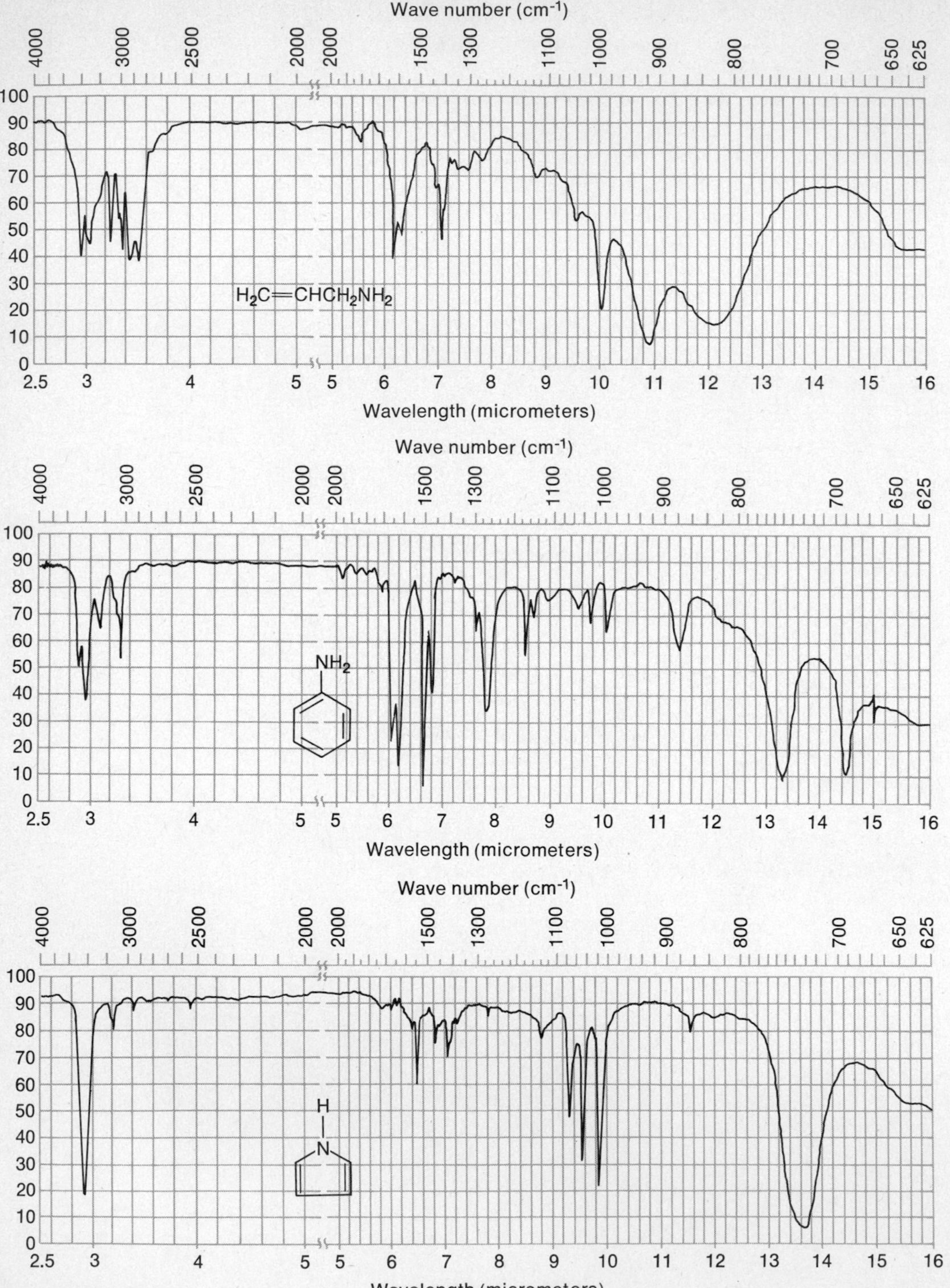

Fig. 12.3. Infrared spectra of allylamine, aniline, and pyrrole.

proximate to nitrogen are shifted downfield to a lesser extent in amines than in alcohols and ethers. For example, the protons of the $—CH_2OH$ group in allyl alcohol resonate at δ 4.13, whereas those of the $—CH_2NH_2$ group of allylamine resonate at δ 3.30. The position of resonance of the N—H varies widely, depending on the acidity of the hydrogen and also on the concentration of the solution on which the nmr measurement is made. In alkylamines, the N—H resonance occurs in the δ 1.5–2.0 region. In anilines, it is shifted downfield to about δ 3.5, and in pyrrole it is shifted even further downfield to approximately δ 8, where it appears as a very broad, almost undetectable band (Fig. 12.4). The positions of the C—H resonances in pyrrole are shifted considerably downfield

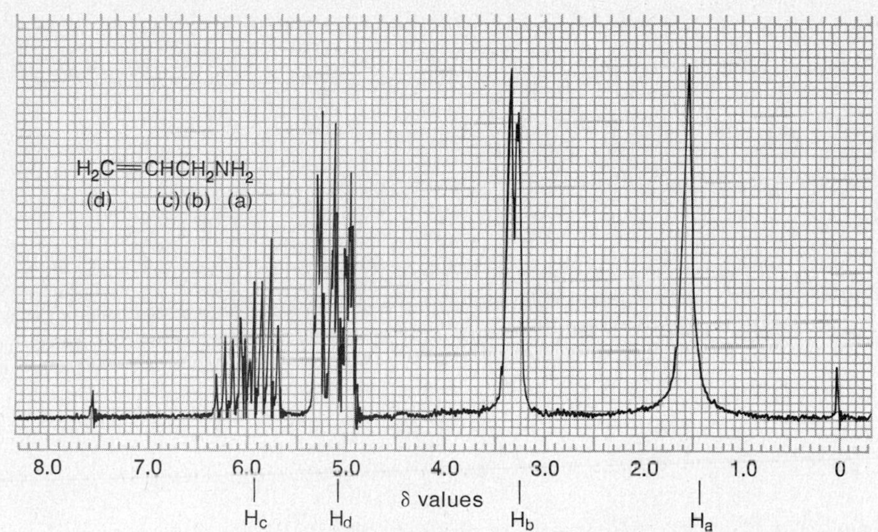

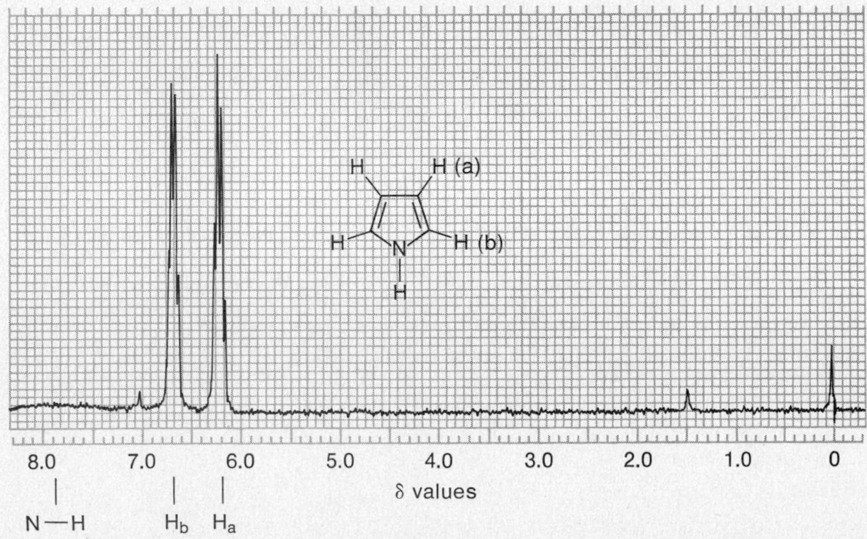

Fig. 12.4. Nuclear magnetic resonance spectra of allylamine and pyrrole.

from ordinary =C—H resonances and are very close to those of Ar—H, commensurate with the aromatic properties ascribed to this ring.

Although the alkylamines are essentially transparent in the accessible ultraviolet, aromatic amines such as aniline and heterocyclic amines such as pyrrole have moderately strong absorptions. Aniline itself absorbs at 230 nm with an absorptivity of 8600, N,N-dimethylaniline absorbs at 251 nm with an absorptivity of 12,900, and pyrrole absorbs at 210 nm with an absorptivity of 5100. Thus, a strong absorption band in the 210–250 nm region cannot categorically be taken as evidence for the presence of a 1,3-diene system, for it may arise from an aniline moiety.

PROBLEM 12-2

A $C_9H_{13}N$ compound yields benzoic acid when subjected to oxidation with sodium dichromate and sulfuric acid, suggesting several possible structures. Indicate how ir and nmr spectra might be used to differentiate between these possibilities.

12.4 Synthesis of Amines

The two principal methods for making amines involve (a) nucleophilic substitution reactions using ammonia or amines as the nucleophiles, and (b) reduction of more highly oxidized nitrogen compounds. The action of ammonia and amines on alkyl halides results in the introduction of one or more alkyl groups onto the nitrogen atom, as illustrated by the reaction of isobutylamine and methyl iodide to form isobutylmethylamine (Fig. 12.5). Since it is sometimes difficult to avoid

Via Nucleophilic Substitution

$$(CH_3)_2CHCH_2NH_2 + CH_3I \longrightarrow (CH_3)_2CHCH_2NHCH_3 + (CH_3)_2CHCH_2N(CH_3)_2 + (CH_3)_2CHCH_2\overset{+}{N}(CH_3)_3\ I^-$$

Via Reduction

(a) Indirect (two-step) preparation from alkyl halides (reduction of —CN)

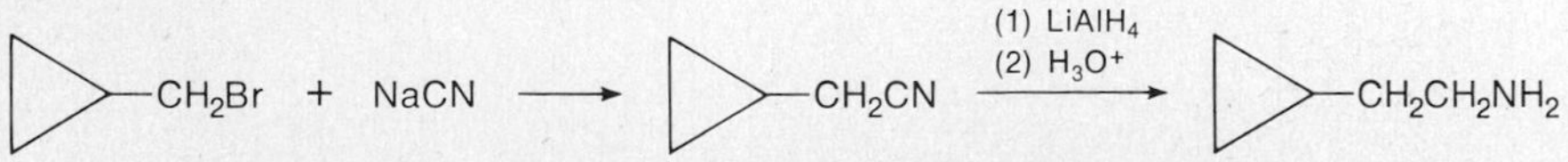

$$\triangleright\!-CH_2Br + NaCN \longrightarrow \triangleright\!-CH_2CN \xrightarrow[(2)\ H_3O^+]{(1)\ LiAlH_4} \triangleright\!-CH_2CH_2NH_2$$

(b) Indirect (two-step) preparation from aromatic compounds (reduction of $—NO_2$)

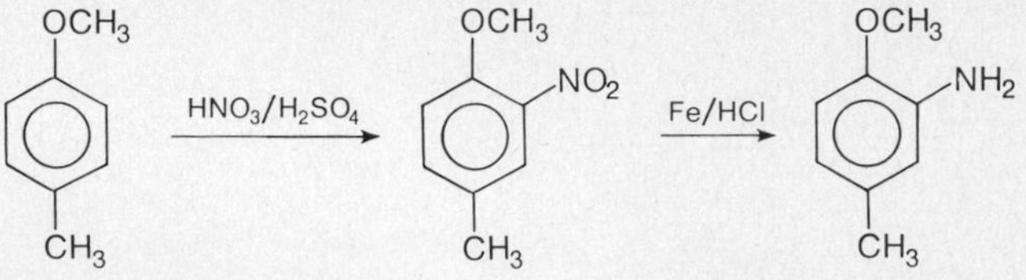

$$p\text{-}CH_3OC_6H_4CH_3 \xrightarrow{HNO_3/H_2SO_4} \text{4-}CH_3\text{-2-}NO_2C_6H_3OCH_3 \xrightarrow{Fe/HCl} \text{4-}CH_3\text{-2-}NH_2C_6H_3OCH_3$$

Fig. 12.5. Synthesis of amines by nucleophilic substitution and reduction methods.

the introduction of more than one alkyl group by this procedure, however, it is more generally applicable to the synthesis of tertiary amines and quaternary ammonium salts (*e.g.*, see p. 307). Methods better adapted to the synthesis of primary amines involve the reduction of nitriles, obtained by the action of sodium cyanide on alkyl halides, or by the reduction of nitro compounds, obtained in the aromatic series by the nitration of arenes (Fig. 12.5).

PROBLEM 12-3

Remembering the free radical substitution reactions of alkanes, indicate a reaction sequence for converting toluene to 2-phenylethylamine.

PROBLEM 12-4

Starting with benzene, devise a method for synthesizing *p*-aminophenol.

12.5 Reactions of Amines

The chemistry of amines is largely concerned with the nonbonded electrons on the nitrogen atom. As a consequence of these electrons, amines are Lewis bases (nucleophiles) and are reactive toward electrophiles, including charged electrophiles such as H^+ and NO^+ as well as covalently bonded entities such as alkyl halides and carbonyl compounds. Let us first consider a practical laboratory application of the basicity of amines by discussing a method for separating a mixture of benzylamine ($C_6H_5CH_2NH_2$), benzyl alcohol ($C_6H_5CH_2OH$), and *p*-cresol ($CH_3{-}C_6H_4{-}OH$). The mixture can be separated easily and cleanly into its components by dissolving it in ether and extracting first with aqueous sodium hydroxide to remove the *p*-cresol and then with aqueous hydrochloric acid to remove the benzylamine. Remaining in the ether is the neutral benzyl alcohol. Acidification of the sodium hydroxide solution followed by ether extraction yields an ether solution of the *p*-cresol, and basification of the hydrochloric acid solution followed by ether extraction yields an ether solution of benzylamine. Evaporation of the ether from these three solutions then yields the pure components (Fig. 12.6). The success of this separation depends on the

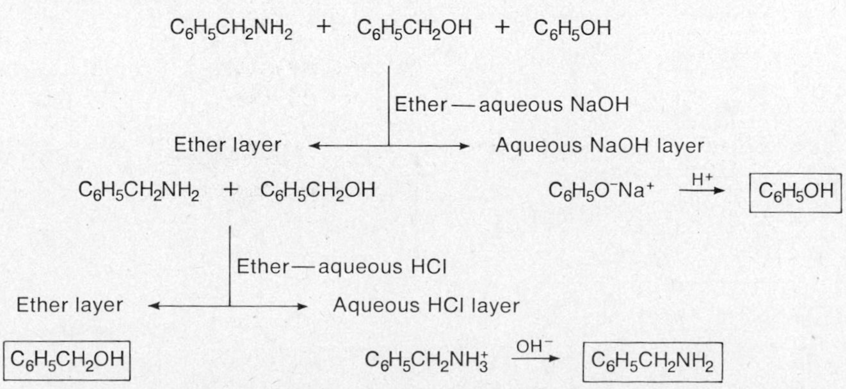

Fig. 12.6. Separation of benzylamine, benzyl alcohol, and *p*-cresol.

acidity of *p*-cresol (see p. 267) and the basicity of benzylamine that allows it to be protonated

$$(C_6H_5CH_2NH_2 \quad + \quad H^+ \rightleftharpoons C_6H_5CH_2\overset{+}{N}H_3)$$

and brought into aqueous solution as a result of the enhanced hydrogen bonding ability of the protonated species;

$$C_6H_5CH_2\overset{+}{N}H_2\text{—}H\cdots OH_2$$

Most amines share this property with benzylamine, although there are rather striking differences in basicities of various amines, as the following data indicate.

One of the ways for expressing the basicity of an amine focuses on its reaction with water*

$$R_3N + H_2O \rightleftharpoons R_3NH^+ + OH^-$$

$$K_b = \frac{[R_3NH^+][OH^-]}{[R_3N][H_2O]} \quad \text{and} \quad pK_b = -\log K_b$$

Measured in this way, alkylamines have pK_b values of 3 to 5; arylamines have pK_b values of 9 to 10; and pyrrole has a pK_b of 13.60 (Table 12.1). Another, more generally useful, way for expressing the basicity of amines is often employed, however. It focuses on the dissociation reaction of the protonated amine (see Table 7.1):†

$$R_3NH^+ + B \rightleftharpoons R_3N + BH^+$$

$$K_{BH^+} = \frac{[R_3N][BH^+]}{[R_3NH^+][B]} \quad \text{and} \quad pK_{BH^+} = -\log K_{BH^+}$$

Specified in this fashion, alkylamines have pK_{BH^+} values‡ of 9 to 11, arylamines have pK_{BH^+} values of 4 to 5, and pyrrole has a pK_{BH^+} of 0.40 (Table 12.1).

The explanation for the considerably reduced basicity of aniline and the drastically reduced basicity of pyrrole as compared with alkylamines is implicit in the structural formulations shown in Fig. 12.2. To the extent that the non-bonded electrons on the nitrogen are delocalized into the π systems in aniline and pyrrole, the electrons are less available for interaction with protons (or other electrophiles).

In addition to the proton, other electrophiles can also react with amines. Important among these is the nitrosonium ion (NO^+), generated by the action of mineral acid on nitrous acid (*i.e.*, $HCl + HONO \rightleftharpoons H_2O + Cl^- + NO^+$). When an amine is treated with an aqueous acidic solution of nitrous acid, a

---

* As explained earlier (see Section 7.1), it is customary to include the water concentration as part of the equilibrium constant in aqueous solution, so $K_b = K'[H_2O]$ and $K_{BH^+} = K''[H_2O]$.

† In aqueous solution, "B" is H_2O.

‡ The pK values for the dissociation of protonated amines are frequently expressed as pK_a values, implying pK_{BH^+}. To avoid confusion, however, we shall use the pK_{BH^+} designation in this text when it is applicable.

Table 12.1 pK_b AND pK_{BH^+} VALUES OF AMINES

Compound	pK_b	pK_{BH^+}
NH_3	4.75	9.25
CH_3NH_2	3.37	10.63
$(CH_3)_2NH$	3.22	10.78
$(CH_3)_3N$	4.20	9.80
$C_6H_5NH_2$	9.37	4.63
 pyrrole (N–H)	13.60	0.40

reaction occurs to produce molecular nitrogen in the case of a primary amine, a colored N-nitroso compound in the case of a secondary amine, and an unstable salt that reverts to the starting compound in the case of a tertiary amine (Fig. 12.7). This, then, provides a simple method of characterization; primary amines produce a gas (bubbles in the reaction mixture), secondary amines produce a color, and tertiary amines revert to the starting materials.

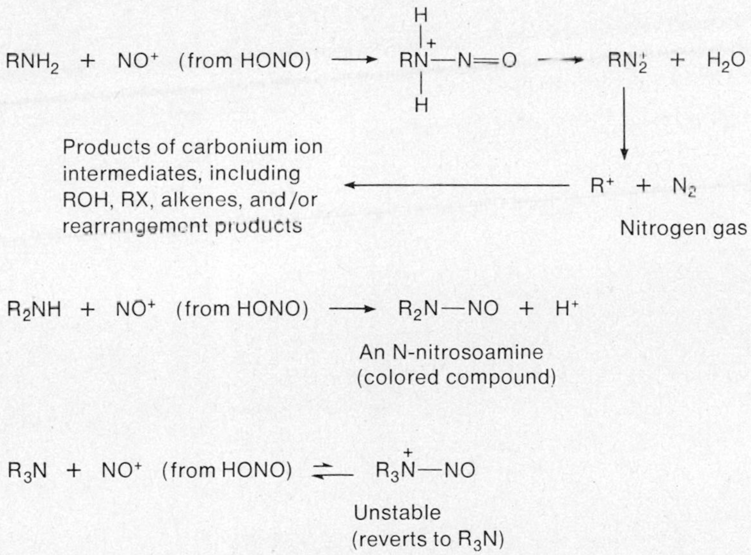

$$RNH_2 + NO^+ \text{ (from HONO)} \longrightarrow RN^+H_2{-}N{=}O \longrightarrow RN_2^+ + H_2O$$

$$RN_2^+ \longrightarrow R^+ + N_2 \text{ (Nitrogen gas)} \longrightarrow$$ Products of carbonium ion intermediates, including ROH, RX, alkenes, and/or rearrangement products

$$R_2NH + NO^+ \text{ (from HONO)} \longrightarrow R_2N{-}NO + H^+$$

An N-nitrosoamine (colored compound)

$$R_3N + NO^+ \text{ (from HONO)} \rightleftharpoons R_3\overset{+}{N}{-}NO$$

Unstable (reverts to R_3N)

Fig. 12.7. The reaction of primary, secondary, and tertiary amines with nitrous acid.

Amines can engage in nucleophilic substitution reactions with alkyl halides to give secondary and tertiary amines and quaternary ammonium salts (see Section 12.4), and in nucleophilic addition reactions with carbonyl compounds to give imines and amides, as discussed in a later chapter (see Section 15.4). Quaternary ammonium compounds have proved useful as intermediates in the preparation of alkenes, for they undergo an elimination reaction in the presence

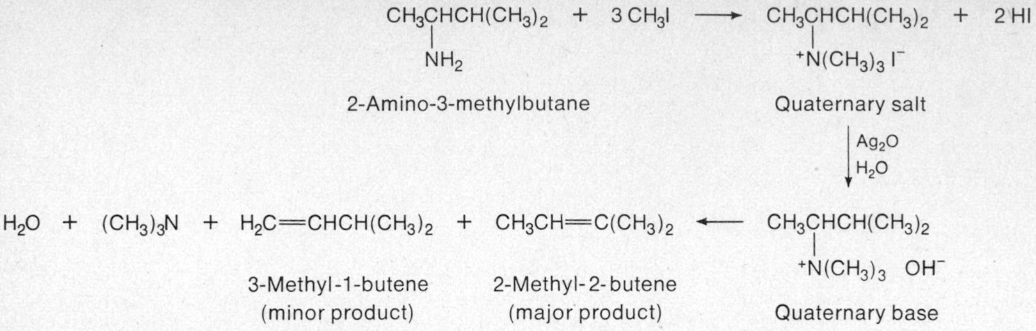

$$CH_3CHCH(CH_3)_2 \text{ (with } NH_2 \text{ on C-2)} + 3\,CH_3I \longrightarrow CH_3CHCH(CH_3)_2 \text{ (with } {}^+N(CH_3)_3\,I^- \text{ on C-2)} + 2\,HI$$

2-Amino-3-methylbutane — Quaternary salt

$\downarrow Ag_2O,\ H_2O$

$$H_2O + (CH_3)_3N + H_2C{=}CHCH(CH_3)_2 + CH_3CH{=}C(CH_3)_2 \longleftarrow CH_3CHCH(CH_3)_2 \text{ (with } {}^+N(CH_3)_3\ OH^- \text{ on C-2)}$$

3-Methyl-1-butene (minor product) — 2-Methyl-2-butene (major product) — Quaternary base

Fig. 12.8. The alkylation of amines and the Hofmann elimination reaction of quaternary ammonium hydroxides.

of various bases. For example, when 2-amino-3-methylbutane is treated with an excess of methyl iodide, trimethyl(3-methyl-2-butyl)ammonium iodide is formed (Fig. 12.8). This material is a water-soluble, high-melting solid that possesses typical salt-like properties. Treatment of the quaternary salt with silver oxide replaces the iodide ion by a hydroxide ion and yields the corresponding quaternary ammonium hydroxide, which decomposes to trimethylamine and a mixture of alkenes on heating. The deamination reaction, classified as an E2 process, takes place because of the enhanced leaving ability of the nitrogen function that arises from the positive charge that it carries. It turns out that the alkene formed in larger amount is 3-methyl-1-butene. This corresponds to the less highly substituted isomer, opposite to the prediction of the Saytzev rule (see Fig. 9.12); the reaction is an example of the operation of the *Hofmann rule*. Thus, quaternary ammonium compounds undergo elimination reactions to yield the less highly substituted alkenes (*Hofmann pathway*), whereas alkyl halides and alcohols undergo elimination reactions to yield the more highly substituted alkenes (*Saytzev pathway*).

PROBLEM 12-5

Compounds A and B both have the molecular formula $C_6H_{15}N$. When treated with methyl iodide to form the quaternary salt and then decomposed by the action of strong base, compound A yields a $C_5H_{13}N$ compound and C_2H_4. Similar treatment of compound B yields a C_3H_9N compound and C_4H_8. Suggest structures for compounds A and B on the basis of these data.

PROBLEM 12-6

A $C_5H_{11}N$ compound is soluble in dilute aqueous hydrochloric acid and shows no stretching deformation bands above 3000 cm^{-1} in the infrared. When it is converted to the quaternary salt and subjected to treatment with strong base, a $C_6H_{13}N$ compound is formed, which yields formaldehyde as one of the products of ozonolysis. Suggest a structure that is compatible with these data. If more than one structure is possible, suggest how a differentiation might be made by means of nmr spectra.

Much of the important chemistry of the alcohols involves cleavages of the O—H and R—O bonds (see Sections 10.6–7). The chemistry of the amines, however, only seldom involves cleavages of N—H or R—N bonds. Amines are far less acidic than alcohols, having dissociation constants in the vicinity of 10^{-33} to 10^{-36} (*i.e.*, pK_a values of 33–36), and they undergo replacement of the hydrogen only with very strong bases (*e.g.*,

$$R_2NH \ + \ CH_3MgI \longrightarrow R_2NMgI \ + \ CH_4$$

i.e., primary and secondary amines possess active hydrogen). The R—N bond of amines is even less susceptible to displacement by nucleophiles than the C—O bond of alcohols, and only in quaternary ammonium compounds do we find examples of reactions in which this bond is cleaved, as illustrated by the elimination process shown in Fig. 12.8.

With alkylamines, just as with the alcohols, the chemistry usually involves only the functional group and not the carbon framework. With aromatic amines (anilines), as with phenols, however, the carbon framework can be the site of reaction. Thus, when N,N-dimethylaniline is treated with bromine, an electrophilic substitution occurs to yield 2,4,6-tribromo-N,N-dimethylaniline (Fig. 12.9).

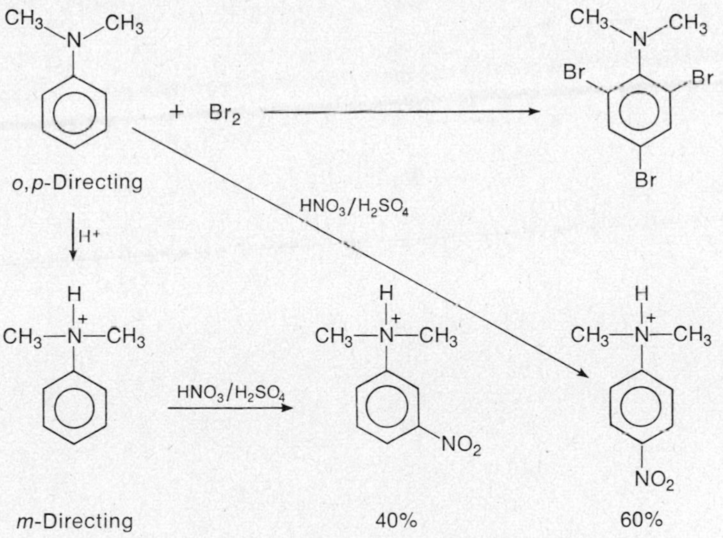

CH3 N CH3 + Br2 → CH3 N CH3 Br Br Br

o,p-Directing

HNO3/H2SO4

H+

H CH3—N+—CH3 —HNO3/H2SO4→ H CH3—N+—CH3 NO2 H CH3—N+—CH3 NO2

m-Directing 40% 60%

Fig. 12.9. Electrophilic substitution reactions of N,N-dimethyllaniline.

The reaction proceeds far more rapidly than the bromination of benzene itself, which, coupled with the fact that the bromines enter the aniline molecule at the *ortho* and *para* positions, indicates that amino functions, like hydroxyl and alkoxyl functions, are *ortho*, *para* directing and activating. This, of course, is what would be predicted on the basis of the resonance structures of aniline (see Fig. 12.1), which include charge-separated forms in which the negative charge resides at the *ortho* and *para* positions. Caution with respect to certain other electrophilic aromatic substitution reactions of aniline is necessary, however. Nitration, for example, is carried out in a strongly acidic medium, under which condition the aniline molecule is converted to the anilinium cation. Whereas the amino group is *ortho, para* directing and activating, the positively charged ammonium group is *meta* directing and deactivating. Consequently, nitration of dimethylaniline gives a significant amount of the *meta*-nitro product (Fig. 12.9).

PROBLEM 12-7

Starting with benzene, devise a method for synthesizing *p*-bromo-N,N-dimethylaniline.

12.6 Amines of Particular Utility and Interest

Methylamine, dimethylamine, and trimethylamine all have commercial applications. Methylamine is used primarily for the production of an insecticide called Sevin; dimethylamine is used for a variety of derived products, including dimethylformamide (see Section 15.4); and trimethylamine is used in the synthesis of an animal feed constituent known as choline chloride, formed by reaction with 2-chloroethanol;

$$(CH_3)_3N + ClCH_2CH_2OH \longrightarrow [(CH_3)_3\overset{+}{N}CH_2CH_2OH]\ Cl^-$$

Among the simple aromatic amines, aniline is by far the most commercially important. Its early use in the production of the first dyestuff, mauve, has already been mentioned (see p. 14), and aniline continues today to be used as a starting material for the synthesis of commercial dyes, such as malachite green. In addition, it is the starting material for many other types of products, including accelerators for the vulcanization of rubber, antioxidants, pharmaceuticals, and photographic developers.

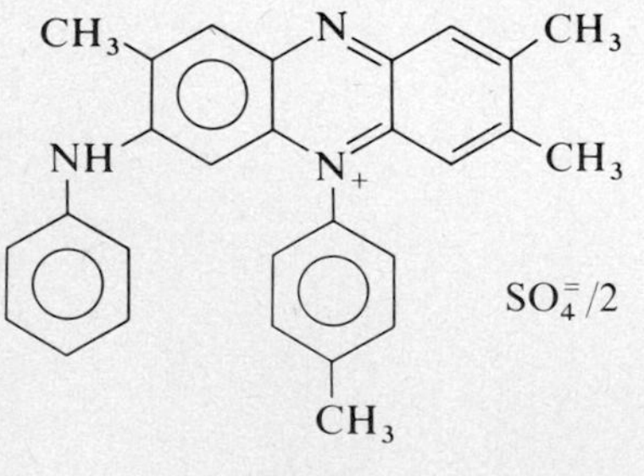

Mauve

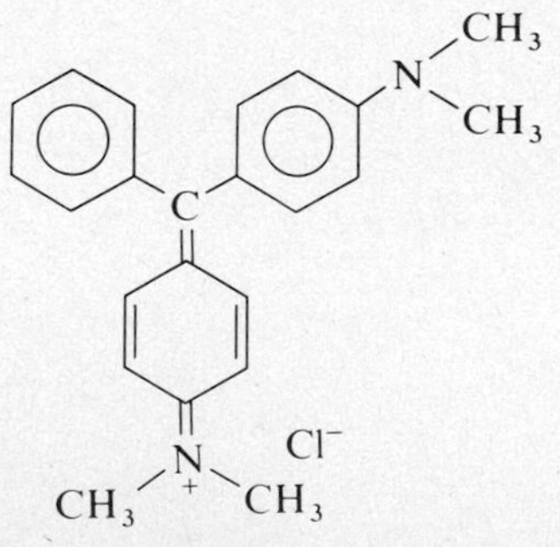

Malachite green

Dye chemistry, like polymer chemistry, is accorded less space in this text than is warranted by its commercial importance. Arising in the middle of the nineteenth century with Perkin's discovery of mauve, the dye industry has grown to become one of the largest organic chemical enterprises in the world. Until World War I it was a German monopoly, but the exigencies of war forced other nations, particularly the United States, to develop its own dye technology. Today, the production of dyestuffs is a worldwide endeavor, and enormous quantities of all conceivable colors are being manufactured for all the wide variety of natural and synthetic fibers and fabrics that are currently available.

Many of the useful amino compounds are polyfunctional, containing two or more amino groups or other functional groups in addition to an amino group. Ethylenediamine (1,2-diaminoethane), produced by the action of ammonia vapor at 150° on 1,2-dichloroethane, is converted by the action of chloroacetic acid (see Chapter 14) to a compound called ethylenediaminetetraacetic acid and known in the trade as Versene. Versene has the remarkable ability to capture certain metal cations by folding itself around them in a clawlike fashion, as illustrated in Fig. 12.10. Compounds with this capability are known as chelating

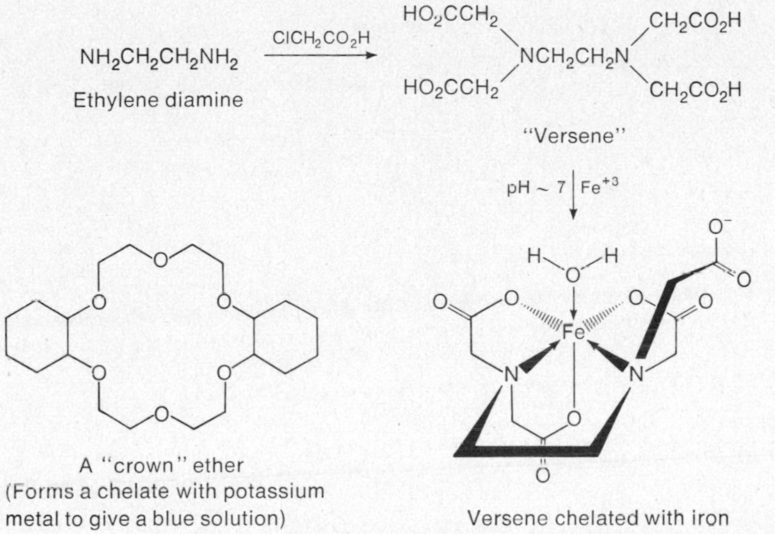

$$NH_2CH_2CH_2NH_2 \xrightarrow{ClCH_2CO_2H} (HO_2CCH_2)_2NCH_2CH_2N(CH_2CO_2H)_2$$

Ethylene diamine

"Versene"

pH ~ 7, Fe^{+3}

A "crown" ether
(Forms a chelate with potassium metal to give a blue solution)

Versene chelated with iron

Versene

Fig. 12.10. Chelate compounds with metals and ions.

agents (Greek, *chele*, "claw"), and the ones containing amino functions are often particularly effective because of the strong Lewis base properties of the nitrogen atom. Certain oxygen compounds can also act in a similar capacity, and a series of cyclic polyethers called "crown ethers" has recently become available, which can sequester certain metals and their cations. The specificity of capture depends on the size of the metal or metal cation and on the size of the hole in the middle of the crown ether.

Low molecular weight hydroxyamines are commercially produced by the action of ammonia on ethylene oxide; one equivalent of ethylene oxide yields ethanolamine, two equivalents yield diethanolamine, and three equivalents yield triethanolamine.

$$NH_3 \xrightarrow{\text{ethylene oxide}} NH_2CH_2CH_2OH \xrightarrow{\text{ethylene oxide}} NH(CH_2CH_2OH)_2 \xrightarrow{\text{ethylene oxide}} N(CH_2CH_2OH)_3$$

Choline, a biologically important hydroxyamine, has already been discussed (see p. 318). Esterification (see Section 15.3) of the hydroxyl group of choline produces a compound known as acetylcholine;

$$(CH_3)_3\overset{+}{N}CH_2CH_2OC(=O)CH_3$$

This compound is important in the physiology of many living organisms, for it is involved in the transfer of nerve impulses across the junctures (synapses) between neighboring nerve cells. The details of the mechanism of this physiological process are fairly well understood. Release of one or more molecules of acetylcholine at one side of the synapse allows transmission of the nerve impulse to the other side of the synapse. To avoid continuous stimulation (continuous transmission) of the impulse, however, enzymes (known as acetylcholine esterases) that possess the ability to catalyze the hydrolysis of acetylcholine to the physiologically inactive choline are present in the region surrounding the synapse. If acetylcholine is not present at the synapse, or if the esterase enzymes are not in the vicinity of the synapse, the transmission of nerve impulse malfunctions. Among the chemicals that can disrupt the transmission are the curare alkaloids (which displace acetylcholine at the nerve junctions) and compounds such as isopropyl methylfluorophosphonate

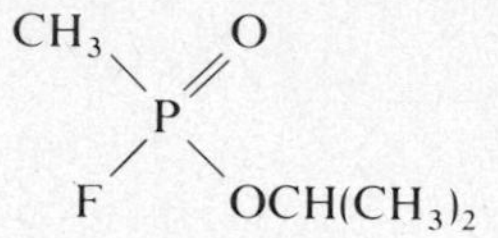

$$CH_3P(=O)(F)OCH(CH_3)_2$$

(which interfere with the enzyme action). Both of these types of compounds are active in extremely small doses and, therefore, are exceedingly dangerous. Other chemicals have similar but less drastic effects and can be used medicinally. For instance,

$$[(C_2H_5)_3\overset{+}{N}(CH_2)_6\overset{+}{N}(CH_3)_3]\ 2\,Br^-$$

(commercially known as "Hexamethonium") has been used to modify the function of the nerve ganglia and reduce high blood pressure and nervous irritability.

If an amine rather than ammonia is treated with ethylene oxide, the corresponding N-substituted ethanolamine is produced. Thus, dimethylamine and ethylene oxide yield N,N-dimethylethanolamine, a compound from which the antihistamine known as "Benadryl" can be obtained by treatment with bromodiphenylmethane:

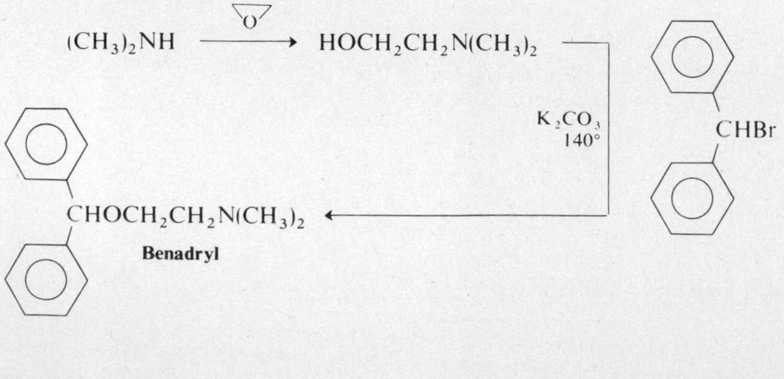

$$(CH_3)_2NH \xrightarrow{\text{ethylene oxide}} HOCH_2CH_2N(CH_3)_2 \xrightarrow[140^\circ]{K_2CO_3,\ (C_6H_5)_2CHBr} (C_6H_5)_2CHOCH_2CH_2N(CH_3)_2$$

Benadryl

A number of compounds possessing a β-phenylethylamine structure have the ability to induce activity in the sympathetic nervous system—*i.e.*, to dilate the pupils of the eye, to strengthen the heartbeat, and to increase blood pressure. Among this group of compounds, known as sympathomimetic amines, are Amphetamine (Benzadrine), Ephedrine, Epinephrine (Adrenalin), Norepinephrine (Arterenol), and Phenylephrine (Neo-Synephrine). Still another compound based on the same fundamental structure is Chloromycetin (Chloramphenicol), a substance which is isolated from a soil bacterium and which has useful antibiotic properties. It was the first antibiotic to be commercially synthesized on a large scale and was used rather widely for a number of years. Its toxic side effects, however, limit its utility, and today it is administered only when other antibiotics are not effective.

$C_6H_5-CH_2CH(CH_3)NH_2$

Amphetamine

$C_6H_5-CH(OH)CH(CH_3)NHCH_3$

Ephedrine

$(HO)_2C_6H_3-CH(OH)CH_2NHCH_3$

Epinephrine

$(HO)_2C_6H_3-CH(OH)CH_2NH_2$

Norepinephrine

$HO-C_6H_4-CH(OH)CH_2NHCH_3$

Phenylephrine

$O_2N-C_6H_4-CH(OH)CH(CH_2OH)NHC(=O)CHCl_2$

Chloromycetin

12.7 *Study Guide*

I. Structure
   A. Bonding between carbon framework and nitrogen
   B. Tautomerism in systems containing the structure C=C—NHR
      1. Vinyl amines less stable than isomeric imines
      2. Aromatic amines (anilines) more stable than cyclohexadienyl imines
   C. Cyclic amines

II. Nomenclature
   A. "Amine" as the parent name; *e.g.*, methylamine
   B. Amino groups as substituents; *e.g.*, dimethylamino-
   C. "Aza" system
   D. Primary, secondary, and tertiary amine designations

III. Physical properties
   A. Boiling points—lower than corresponding alcohols
   B. Water solubility—lower than corresponding alcohols
   C. Odor—lower molecular weight members odoriferous

IV. Spectral properties
   A. Ultraviolet/visible—generally transparent above 200 nm
   B. Infrared
      1. N—H stretching near 3500 cm^{-1}: 2 bands for $-NH_2$, one band for $>NH$
      2. C—N stretching near 1200 cm^{-1}

C. Nuclear magnetic resonance
   1. Nitrogen is electronegative—therefore, downfield shifts
   2. N—H resonance position related to acidity of N—H bond

V. Synthesis
   A. Nucleophilic displacement of alkyl halides
      1. Primary amines: $RX + NH_3 \rightarrow RNH_2$
      2. Secondary amines: $RX + RNH_2 \rightarrow R_2NH$
      3. Tertiary amines: $RX + R_2NH \rightarrow R_3N$
      4. Quaternary ammonium salts: $RX + R_3N \rightarrow R_4N^+ X^-$
   B. Reduction
      1. Of nitriles ($RCN \rightarrow RCH_2NH_2$)
         (alkyl nitriles prepared by $RX + NaCN \rightarrow RCN$)
      2. Of nitro compounds ($RNO_2 \rightarrow RNH_2$)
         (aromatic nitro compounds prepared by $ArH + HNO_3 \rightarrow ArNO_2$)

VI. Reactions
   A. Reactions involving N—H bond
      1. Very weak acids
      2. React with RMgX ($R_2N\text{—}H + RMgX \rightarrow R_2N\text{—}MgX + RH$)
   B. Reactions involving nonbonded electrons on nitrogen
      1. Protonation
         (a) Basicity of amines; comparison of alkylamines and aromatic amines
         (b) pK_{BH^+} (*i.e.*, pK_a of R_3NH^+) measurements
         (c) Solubility in aqueous acid
      2. Nucleophilic displacement on alkyl halides—see V.A
   C. Reactions involving R—N bond
      1. $R\text{—}NR_2$ and $Ar\text{—}NR_2$ do not undergo nucleophilic displacement reactions.
      2. $R\text{—}\overset{+}{N}R_3$ can undergo elimination reactions
   D. Electrophilic substitution reactions of anilines
      1. Amino function (R_2N—) is *ortho,para* directing and activating
      2. Protonated amino function ($R_2\overset{+}{N}H$—) group is *meta* directing and deactivating

PROBLEMS

1. Name each of the following compounds

(a) $CH_3CH_2NH_2$ (b) $(CH_3)_2CHNH_2$ (c) $CH_3CH_2CH_2NHCH_3$

(d) $CH_3CH_2CH(NH_2)CH_2CH_3$ (e) $(CH_3)_2NCH_2CH_3$ (f) $(CH_3)_3CNH_2$

(g) $C_6H_5NH_2$ (benzene ring bearing NH_2) (h) $C_6H_{11}N(CH_3)_2$ (cyclohexane ring bearing $N(CH_3)_2$) (i) $C_6H_5NHCH_2CH_3$ (benzene ring bearing $NHCH_2CH_3$)

2. Draw the structures corresponding to the following names:

(a) Propylamine
(b) *sec*-Butylamine
(c) 2-Methyl-2-butanamine
(d) Ethylisopropylamine
(e) Cyclobutylamine
(f) Di-*n*-propylamine
(g) 4-Ethylaniline
(h) 2-Amino-2-methylbutane
(i) Diphenylamine
(j) Dimethylaminocyclohexane

3. Name each of the following compounds:

(a) $H_2C{=}CHCH_2NH_2$

(b) $HC{\equiv}CCHCH_2CH_3$ (with NH_2 on the CH)

(c) $CH_3CHCH_2CHCH_2OH$ (with OCH_3 on C-2 and NH_2 on C-4)

(d) $(H_2C{=}CH)_2NCH_3$

(e) $(CH_3O{-}C_6H_4{-})_3N$ (para-substituted)

(f) benzene ring with NHCH$_3$ and CH$_3$ (meta)

(g) piperidine ring (N–H) with two CH$_3$ on C-3

(h) pyrrole ring with N–CH$_3$

(i) $H_2C{=}CHCHCH_2CH_2{-}C_6H_4{-}F$ (with $CH_2N(CH_3)_2$ on the CH; para-F)

(j) aziridine ring N$-CH_2CHCH_2CH{=}C(CH_3)_2$ (with NH_2 on the CH)

(k) $CH_3CH_2\overset{+}{N}H_3Br^-$

(l) bicyclic amine (N at bridgehead)

(m) azetidine ring with N–CH$_3$; C-3 bearing CH$_3$ and phenyl

(n) $(CH_3)_2CHCH_2{-}\overset{+}{N}(CH_2CH_3)(H){-}CH_3$ Cl^-

4. Draw the structures corresponding to the following names:

(a) Allylethylisopropylamine
(b) Di-(*m*-trifluoromethylphenyl)amine
(c) N-(2,4-Dimethylphenyl)-2-methylcyclopentylimine
(d) Benzyltrimethylammonium chloride
(e) 2-Dimethylaminonaphthalene
(f) N-(2,5-Dimethylhexyl)-N-methylaniline
(g) (2E,6E)-4,5-Diamino-4,5-dimethyl-2,6-octadiene
(h) N,3,3-Trimethylazetidine
(i) *trans*-2,6-Diethylpiperidine

(j) Cyclohexylcyclopropylneohexylamine
(k) 8-Benzyl-8-azabicyclo[4.3.0]non-3-ene
(l) 7-Cyclopropylaminobicyclo[4.1.0]heptane

5. Draw resonance structures for each of the following compounds:

(a) 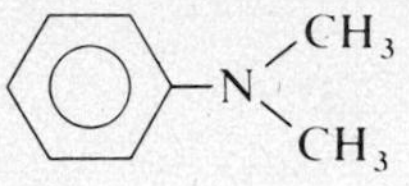 $C_6H_5-N(CH_3)_2$

(b) 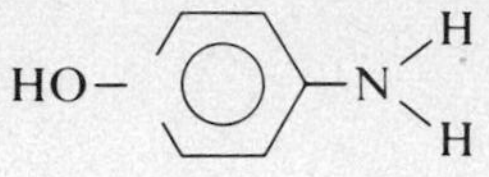 $HO-C_6H_4-NH_2$

(c) $(H_2C{=}CH)_2NH$

(d) 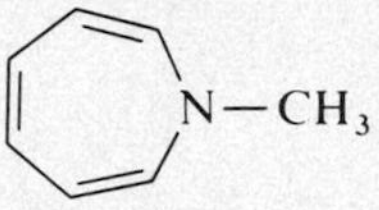 $N-CH_3$ (N-methylazepine ring)

6. Draw the structure of the major product in each of the following reactions:

(a) 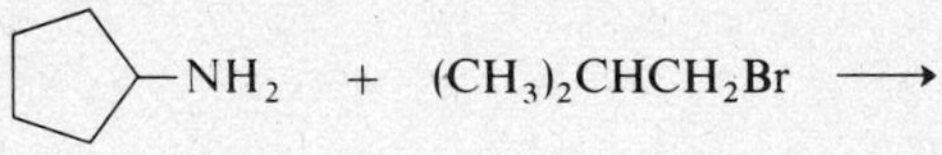 $C_5H_9-NH_2 + (CH_3)_2CHCH_2Br \longrightarrow$

(b) $C_6H_4(CN)(CH_3) + LiAlH_4 \longrightarrow \xrightarrow{H_3O^+}$

(c) $C_6H_5-C(CH_3)_2-Br + (CH_3)_2NH \longrightarrow$

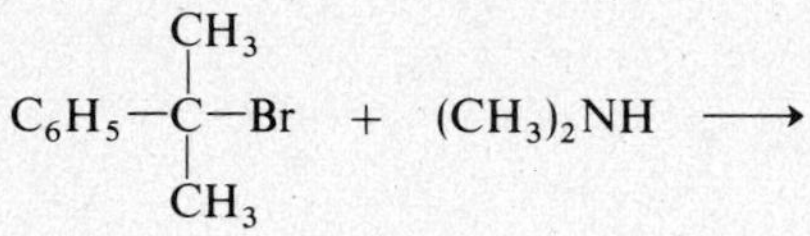

(d) 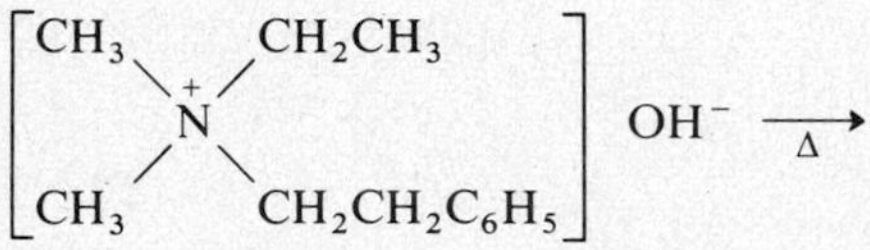 $\left[(CH_3)_2\overset{+}{N}(CH_2CH_3)(CH_2CH_2C_6H_5) \right] OH^- \xrightarrow[\Delta]{}$

(e) 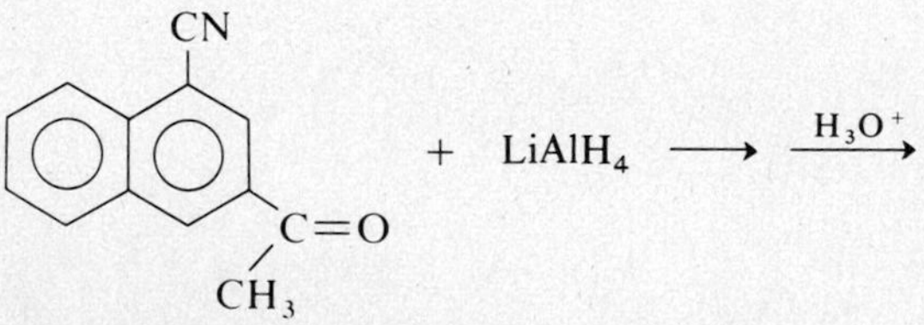 naphthalene with CN and $C(=O)CH_3$ substituents $+ LiAlH_4 \longrightarrow \xrightarrow{H_3O^+}$

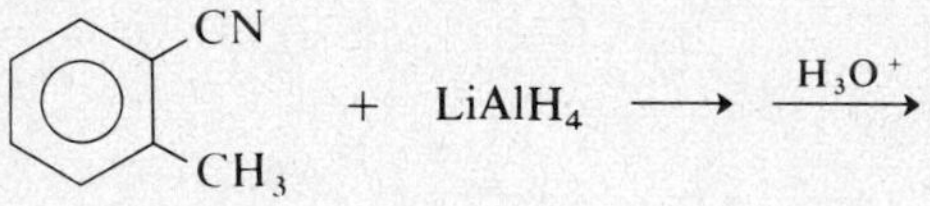

(f) 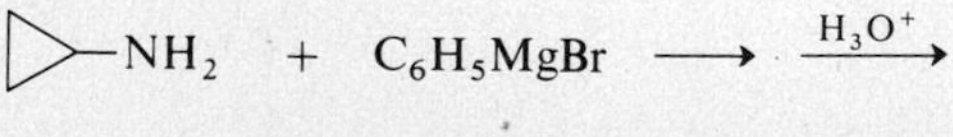 $C_3H_5-NH_2 + C_6H_5MgBr \longrightarrow \xrightarrow{H_3O^+}$

7. Indicate the reactions by means of which the compound on the left-hand side can be transformed to the compound on the right-hand side

(a) 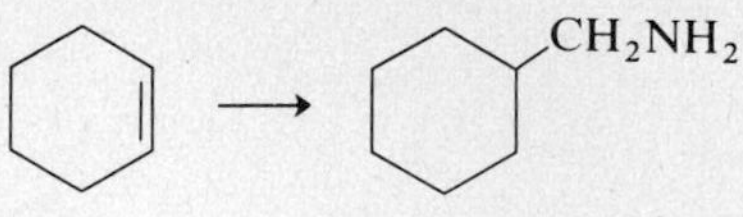 cyclohexene $\longrightarrow$ $C_6H_{11}CH_2NH_2$

(b) 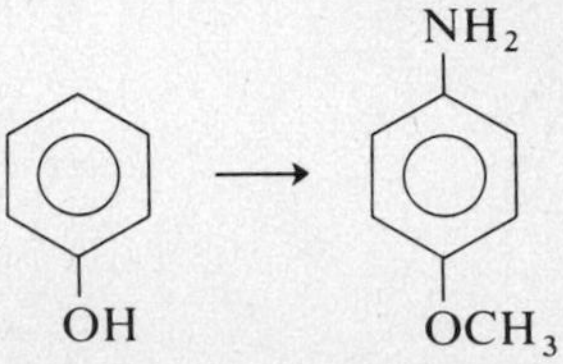 $C_6H_5OH \longrightarrow$ $NH_2-C_6H_4-OCH_3$

(c) [benzene] ⟶ [N-methyl-N-benzylaniline: $C_6H_5N(CH_3)CH_2C_6H_5$]

(d) $CH_3CH_2OH \longrightarrow CH_3CH_2CH_2N(CH_2CH_3)_2$

(e) $(CH_3)_2C{=}CH_2 \longrightarrow (CH_3)_2CHCH_2NHCH_3$

8. Starting with propene as the only source of carbon compounds, outline a synthesis for each of the following compounds:

(a) $H_2C{=}CHCH_2NH_2$ (b) $(CH_3)_2CHNH_2$ (c) $CH_3CH_2CH_2NH_2$

(d) $(CH_3)_2CHCH_2NH_2$ (e) $[(H_2C{=}CHCH_2)_4\overset{+}{N}]Br^-$

9. Starting with any alkane, alkene, alkyne, alkyl halide, or alkanol containing four or fewer carbon atoms, benzene, toluene, *o*-, *m*-, and *p*-xylene, and any inorganic reagent, outline reasonable syntheses for each of the following compounds:

(a) $H_2C{=}CHCH_2NH_2$ (b) $(CH_3)_2CHCH_2CH_2NH_2$

(c) [benzene ring with CH_3, $NHCH_2CH_2CH_3$, and CH_3 substituents] (d) $CH_3\underset{OH}{CH}{-}\underset{NH_2}{CH}CH_3$

(e) $Br{-}C_6H_4{-}N(CH_3)_2$ (f) [azetidine ring]$N{-}CH_3$

(g) $C_6H_5{-}CH_2NH_2$ (h) $H_2NCH_2CH_2CH_2CH_2CH_2NH_2$

10. Using spectral and/or chemical data, show how you could distinguish between the compounds in each of the following isomeric pairs:

(a) [pyrrole ring, N–CH_3] and [dihydropyridine ring, N–H] (b) [bicyclic ring with bridgehead N] and [bicyclic ring with C=C, N–H]

(c) $C_6H_5{-}CH_2N(CH_3)_2$ and $CH_3{-}C_6H_4{-}N(CH_3)_2$

(d) [pyrrolidine ring, N–CH_2CH_3] and [piperidine ring, N–CH_3] (e) [azepane ring, N–H] and [cyclohexane ring with CH_3 and NH_2]

11. A pair of isomeric $C_{11}H_{17}N$ compounds possess the nmr spectra shown below. Suggest structures for these two compounds that are in accord with these data.

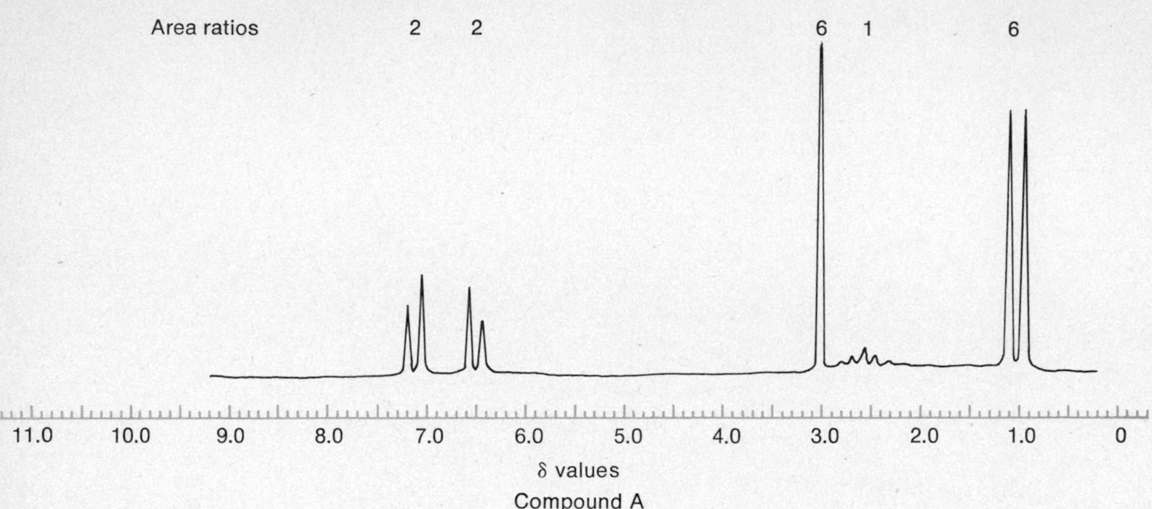

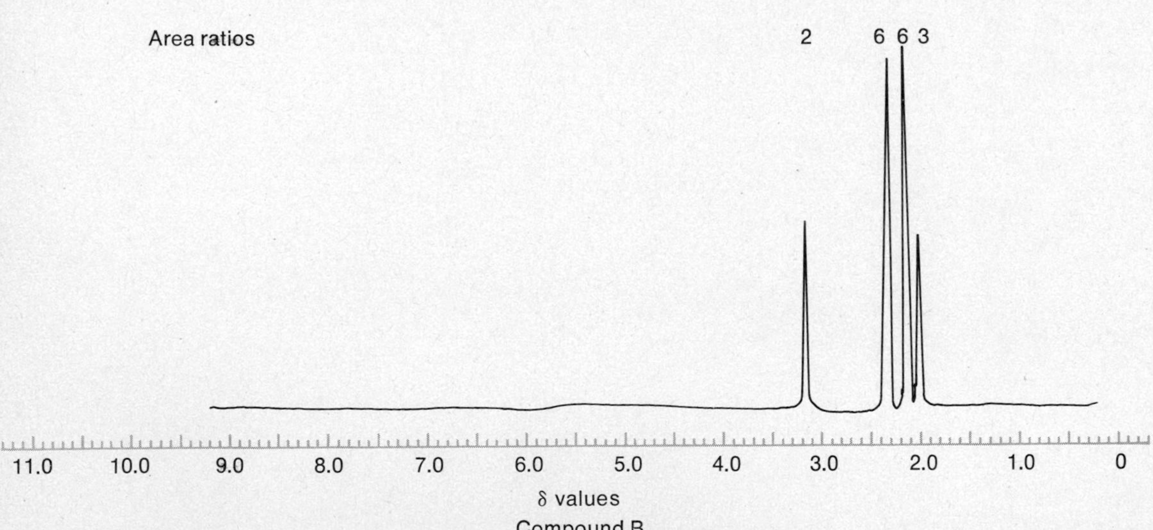

12. Draw the structures of each of the unknown compounds in the following reaction schemes:

(a) $CH_3CH_2CH{=}CH_2 \xrightarrow{C_6H_5CO_3H} \mathbf{A} \xrightarrow{(CH_3)_2CHNH_2} \mathbf{B}$

$\mathbf{B} \xrightarrow{SOCl_2} \mathbf{C} \xrightarrow{NaOH} \mathbf{D}$ (Formula $C_7H_{15}N$; does not possess a band in the infrared near 3500 cm^{-1})

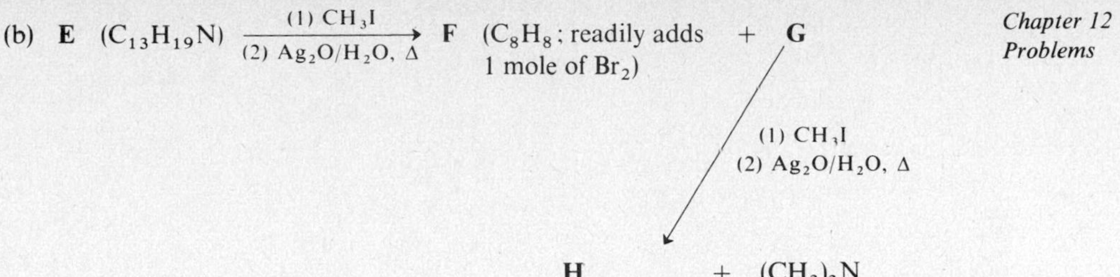

(C_4H_6; does not absorb in the accessible region of the ultraviolet)

13. Provide a reasonable explanation for the trend in base strengths of the following compounds (expressed in pK_{BH^+} values):

NH_2 (cyclohexyl)	CH_3O / NH_2	NH_2	F / NH_2	O_2N / NH_2
10.64	5.29	4.62	4.52	2.96

14. Define the following terms as briefly and concisely as possible. Do not use a specific example in defining the term, but cite a typical example to illustrate the definition

(a) Amine-imine tautomerism (b) Heterocyclic amine (c) pK_{BH^+} of an amine
(d) Hofmann rule (e) Chelate compound

15. One of the products of the breakdown of protein-containing material (*i.e.*, putrefaction) is a malodorous compound called putrescine. The molecular formula of putrescine is $C_4H_{12}N_2$, and when the compound is treated with nitrous acid it releases 2 mole-equivalents of nitrogen gas. Given this information, write all the possible structures for putrescine. For each of these structures, indicate how many sets of equivalent hydrogens there are. By means of nmr, it is possible to differentiate among these several possibilities on the basis of the number of resonance envelopes observed, each envelope corresponding to a set of equivalent hydrogens. It turns out that putrescine has the fewest sets of equivalent hydrogens (*i.e.*, the greatest number of equivalent hydrogens). What is its structure?

Chapter **13**

Aldehydes and ketones

Carbonyl compounds are divided
Into classes through which we'll be guided
  Beginning at first
  By our being well-versed
In the aldones, two sub-classes elided

CDG, 1975

The ability of the carbonyl group $\rangle C{=}O$ to combine not only with a large variety of carbon frameworks but also with hetero atoms and hetero atom groups results in the proliferation of carbonyl compounds throughout the organic chemical world. As we shall see in later chapters, the carbonyl group is present in many substances of biological and commercial importance, and carbonyl compounds provide the essential ingredients for a large number of organic syntheses. The chemistry of carbonyl compounds, therefore, occupies a central place in the study of organic chemistry.

Carbonyl compounds, represented by the general structure $\begin{matrix} G \\ & \rangle C{=}O \\ G' \end{matrix}$ are conveniently classified on the basis of the type of groups (*i.e.*, G and G′) that are attached to the carbonyl function; for the purpose of the present discussion, the attached groups are classified as follows:

1. Type A groups: Hydrogen or organic frameworks in which the carbon attached to the C=O is sp^3 hybridized. This includes H— and alkyl groups, such as CH_3—, CH_3CH_2—, $(CH_3)_3C$—, and substituted alkyl groups, such as $C_6H_5CH_2$—, Cl_3C—, $HOCH_2CH_2$—, etc.
2. Type B and **B′** groups: Organic frameworks to which the carbon attached to the C=O is sp^2 or sp hybridized. If the attached group is C=C— or C≡C—, it is designated as a type **B** group; if it is an aryl ring it is classed as a type **B′** group.

3. Type C groups: Groups other than carbon or hydrogen. This includes —halogen, —OH, —NH_2, etc.
4. Type D groups: Groups in which G and G′ are ligands on the identical atom. This includes oxygen itself and nitrogen-containing and carbon-containing moieties double bonded to C=O (*e.g.*, as in ketene, $H_2C{=}C{-}O$).

Various combinations of the attached groups G and G′ lead to all the known types of carbonyl compounds. The present chapter is concerned only with those compounds in which the G groups are hydrogen (type A group), alkyl functions (type A group), and alkenyl, alkynyl, or aryl functions (type B and B′ groups), these combinations representing aldehydes (*i.e.*, type AA and AB aldehydes) and ketones (*i.e.*, type AA, AB, and BB ketones). The following two chapters, then, deal with compounds in which type C and type D groups are attached to the carbonyl function—*viz.*, carboxylic acids and carboxylic acid derivatives.

PROBLEM 13-1

Classify each of the following compounds as type AA, AB, BB, etc.

$CH_3(H)C{=}O$ $\quad$ $H_2C{=}CH(CH_3)C{=}O$ $\quad$ $C_2H_5O(Cl)C{=}O$ $\quad$ $C_6H_5COCH_3$ $\quad$ $C_6H_5{-}CH{=}C{=}O$ $\quad$ $CH_3C(=O)OC_2H_5$

13.1 Structure and Nomenclature of Aldehydes and Ketones

The description of the double bond between carbon and oxygen in the carbonyl group can make use of either of two states of hybridization for the oxygen. In one case, the oxygen is considered to be unhybridized, forming a σ-bond by axial overlap between a $2p$ orbital on oxygen and a sp^2-hybridized orbital on carbon, along with a π bond by lateral overlap of $2p$ orbitals on the two atoms; the remaining two pairs of nonbonded electrons on oxygen, in this case, reside in $2s$ and $2p$ orbitals (Fig. 13.1). In the other case, the oxygen is considered to be sp^2 hybridized, forming a double bond by axial overlap of sp^2-hydridized orbitals on each of the atoms along with lateral overlap of $2p$ orbitals on each of the atoms; the remaining two pairs of nonbonded electrons on oxygen, in this case, both reside in sp^2-hybridized orbitals (Fig. 13.1). Spectroscopic evidence indicates that the first description is probably the preferred one.

The carbonyl group can be attached to almost all carbon frameworks to give stable compounds. If two hydrogens or one hydrogen and one carbon framework are attached to the carbonyl group, the compound is called an *aldehyde*; if two carbon frameworks are attached to the carbonyl group, the compound is called a *ketone*.

A number of aldehydes and ketones have acquired common names. The first five members of the aldehyde series are known as formaldehyde, acetaldehyde, propionaldehyde, butyraldehyde, and valeraldehyde; the parent member of the

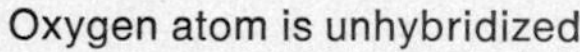

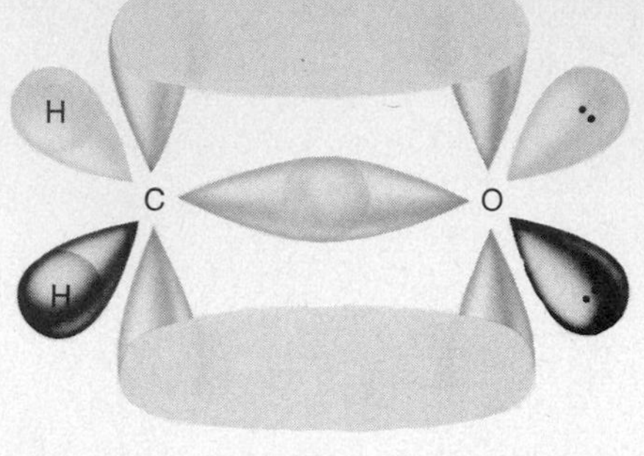

Fig. 13.1. Orbital representations of formaldehyde in which the oxygen atom is unhybridized and sp^2-hybridized.

ketones is known as acetone. A nomenclature system, useful in certain instances, employs acetaldehyde $CH_3C\genfrac{}{}{0pt}{}{\diagup O}{\diagdown H}$ * and acetone, $CH_3\overset{\overset{\displaystyle O}{\|}}{C}CH_3$,* for the parent names and specifies the nature and position of substituents on these frameworks. Thus $C_6H_5CH_2C\genfrac{}{}{0pt}{}{\diagup O}{\diagdown H}$ is called phenylacetaldehyde, and $HOCH_2\overset{\overset{\displaystyle O}{\|}}{C}CH_2OH$ is called α,α'-dihydroxyacetone. An extension, used primarily in the aldehydes, employs the appropriate higher aldehyde as the parent and specifies the identity and position of the substituent by a Greek letter prefix; for example, $Cl\overset{\beta}{C}H_2\overset{\alpha}{C}H_2C\genfrac{}{}{0pt}{}{\diagup O}{\diagdown H}$ is β-chloropropionaldehyde.

Still another system for naming ketones specifies the C=O portion as "ketone" and identifies the attached carbon frameworks by the appropriate alkyl, alkenyl, alkynyl, or aryl designation. Thus, $CH_3\overset{\overset{\displaystyle O}{\|}}{C}CH_2CH_3$ is ethyl methyl ketone, $C_6H_5-CH_2\overset{\overset{\displaystyle O}{\|}}{C}-CH_2-C_6H_5$ is dibenzyl ketone, and

* In linear representation, acetaldehyde is CH_3CHO, acetone is CH_3COCH_3, etc. The reader should learn to recognize the carbonyl function when it is presented in this fashion, for it will be frequently encountered.

$CH_3\overset{O}{\overset{\|}{C}}CH_2CH{=}CH_2$ is allyl methyl ketone. The attached groups are sequenced either by size or by the alphabet; we shall use the alphabetical ordering.

The most generally applicable nomenclature for aldehydes and ketones is the IUPAC system, which is applied according to the following rules:

1. The longest continuous chain *containing the carbonyl function* is selected as the basis for designating the parent compound, and the actual compound is then named as a derivative of the parent compound. The parent compound, if saturated, is named according to the alkane nomenclature, but with the final "e" replaced by "al" in the case of aldehydes or "one" in the case of ketones.
2. Aldehyde groups *must be terminal*, so when they are treated as the principal function they are always carbon atom number 1 (and the number designation, therefore, is omitted in the naming). Keto groups, however, are nonterminal, so when they are treated as the principal function their position must be specified by a number, the lower of the two possible numbers taking precedence. The number can appear either before the parent name or immediately before the suffix "one."
3. Substituents are then specified by name and number in the usual way. It is not always possible to include the carbonyl groups in the parent compound, however, and in this case they must be specified as substituents. The =O (attached to carbon) function is designated as a "keto" or "oxo" group. For example, $\mathrm{OHC{-}CH_2CH_2\overset{O}{\overset{\|}{C}}CH_2{-}CHO}$ (drawn as $\mathrm{H(O{=})CCH_2CH_2C(=O)CH_2C({=}O)H}$) is named 3-ketohexandial or 3-oxohexandial. The $\mathrm{H{-}C{=}O}$ function is designated in either of two ways:

   (a) "Formyl" (or "methanoyl") designation: The aldehyde specification appears as a prefix. For example, $\mathrm{H(O{=})CCH_2CH_2\underset{CHO}{\underset{|}{C}}HCH_2C({=}O)H}$ is named 3-formylhexandial (or 3-methanoylhexandial).

   (b) "Carboxaldehyde" designation: The aldehyde specification appears as a suffix, this nomenclature being generally used when the aldehyde function is directly attached to a cycloalkane ring. For example, $\square{-}CHO$ (cyclobutane ring–CHO) is named cyclobutanecarboxaldehyde.

When one carbon framework is attached to a carbonyl group, *i.e.*, $\mathrm{G{-}C{=}O}$ (G–C(=O)–), the function is referred to as an *acyl group* (if G is alkanyl, alkenyl, or alkynyl) or an *aroyl group* (if G is aryl). For example, $\mathrm{H{-}C{=}O}$ (H–C(=O)–) is the formyl group,

CH_3—C=O is the acetyl group, and C_6H_5—C=O is the benzoyl group, the names being derived from those of the corresponding carboxylic acids (see Section 14.1) by replacing the suffix "ic" or "oic" by "yl" or "oyl". Acyl and aroyl designations can be used as substituent names, as illustrated by the following examples.

$C_6H_{11}COCH_3$	$C_6H_5COCH_3$	$C_6H_5COCH_2CH_3$	$C_6H_5COC_6H_5$
Cyclohexyl methyl ketone **Acetylcyclohexane**	**Methyl phenyl ketone** **Acetylbenzene** **Benzoylmethane** **Acetophenone**	**Ethyl phenyl ketone** **Propionylbenzene** **Benzoylethane** **Propiophenone**	**Diphenyl ketone** **Benzoylbenzene** **Benzophenone**

Substituted aromatic aldehydes and ketones are frequently named as derivatives of the parent compound. Thus, Br—C_6H_4—CHO is named *p*-bromobenzaldehyde (or 4-bromobenzaldehyde), and [naphthalene ring with CH_3 at 1, CHO at 2, CH_3 at 4] is named 1,4-dimethyl-2-naphthaldehyde. In the case of aromatic ketones, the name of the parent member is constructed by combining the suffix "phenone" with the appropriate acyl or aroyl prefix, replacing the "yl" of the substituent name by "o"; for example, acetylbenzene, propionylbenzene, and benzoylbenzene are called acetophenone, propiophenone, and benzophenone, respectively. Thus, [benzene ring with F at meta position]—C(=O)—CH_3 is named *m*-fluoroacetophenone (or 3-fluoroacetophenone).

A number of aldehydes and ketones are illustrated in Figs. 13.2 and 13.3, where their common and systematic names are given and their classification as type AA, AB, BB, etc., carbonyl compounds is indicated.

The bonding between carbon and oxygen in the carbonyl group is pictured as the combination of a σ bond and a π bond (see Fig. 13.1). The C=O bond differs quantitatively from the C=C bond, however, in being inherently polar and stronger. The polarity arises from the greater electronegativity of oxygen compared with that of carbon, resulting in a dipole in which oxygen is the negative and carbon the positive end. Because of the significant dipole moment,

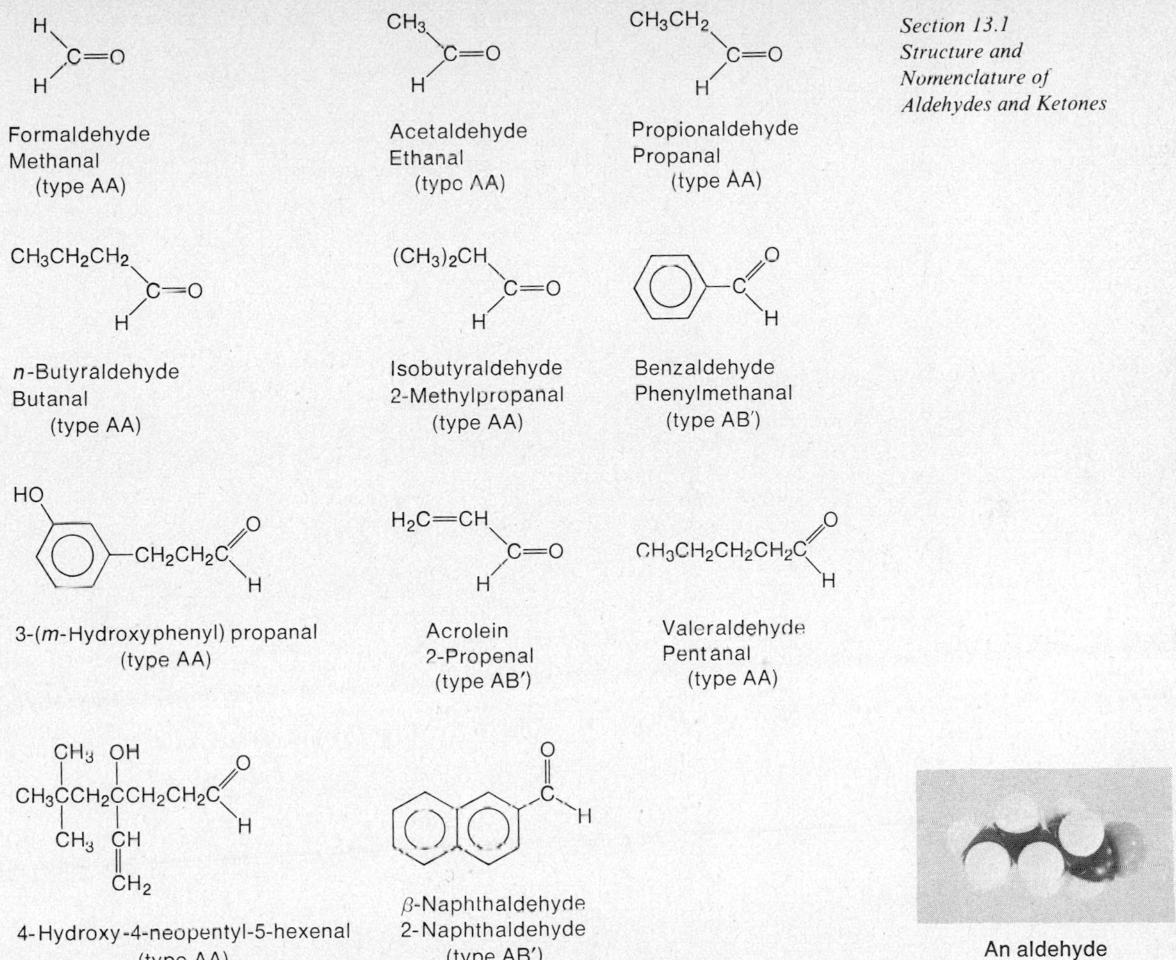

Fig. 13.2. Structures and names of aldehydes.

a single structure for a carbonyl compound is an incomplete representation. Acetone, for example, should be shown as a resonance hybrid of two resonance structures (Fig. 13.4) to account properly for its properties. In carbonyl compounds carrying only type A groups, the positive charge is localized on the carbon, but if a type B or B′ group is present, the charge is delocalized into the neighboring π system. A complete structural representation for benzaldehyde, for example, includes not only the usual resonance structures for the benzene ring and the simple charge-separated resonance structure for the carbonyl group but other charge-separated resonance structures as well (Fig. 13.4). As we shall see in the following discussion, the extent to which the attached groups interact with the carbonyl function is reflected in the varying physical and chemical properties of the various types of aldehydes and ketones.

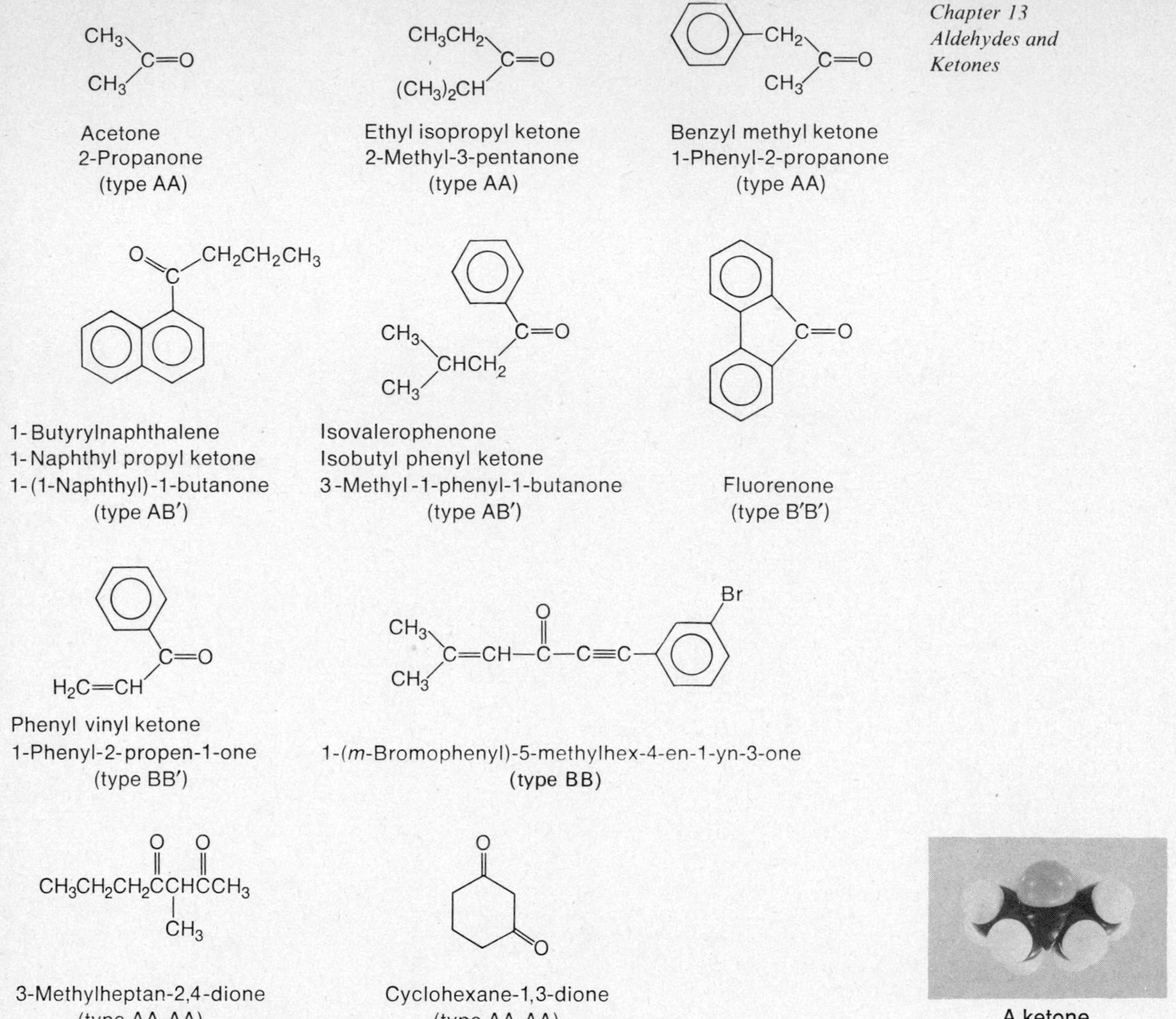

Fig. 13.3. Structures and names of ketones.

PROBLEM 13-2

Draw and name all the C_4H_8O aldehydes and ketones. Draw and name all the $C_9H_{10}O$ compounds that contain a benzene ring and are classified as type AB′ compounds.

13.2 Physical Properties of Aldehydes and Ketones

Aldehydes and ketones, like ethers, cannot be hydrogen bond donors and are expected to boil lower than the corresponding alcohols. They are, however, more polar than the ethers; acetone, for example has a dipole moment of 2.9 D, whereas dimethyl ether has a dipole moment of 1.3 D. As a consequence, alde-

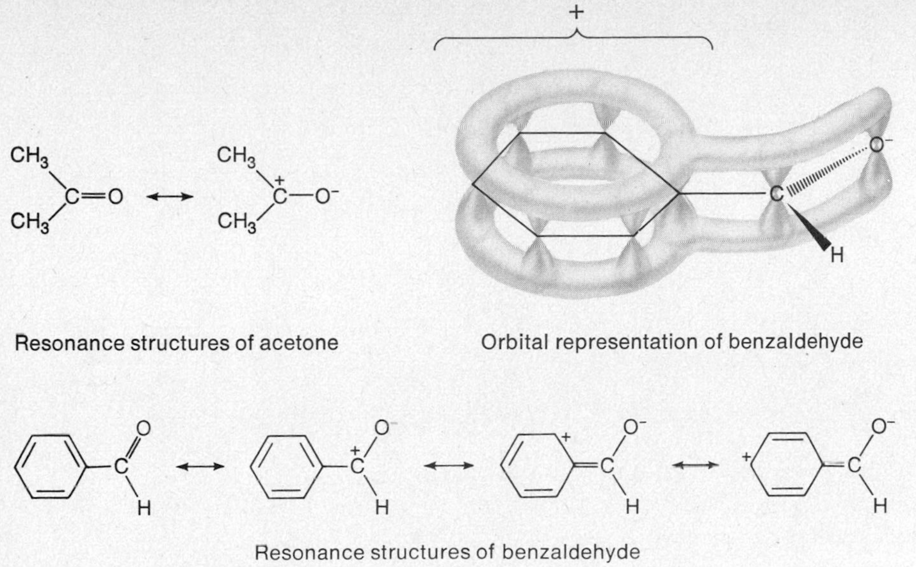

Fig. 13.4. Resonance structures of acetone and benzaldehyde.

hydes and ketones boil higher than the corresponding ethers; *e.g.*, acetone (mol wt 58) boils at 56°, ethyl methyl ether (mol wt 60) boils at 8°, and *n*-propanol (mol wt 60) boils at 97°. Also like ethers, aldehydes and ketones are Lewis bases capable of acting as acceptors in hydrogen bond formation. Therefore, they are moderately soluble in water, ketones being somewhat more soluble than aldehydes; *e.g.*, $CH_3CH_2COCH_3$ (2-butanone) has a solubility of 26 g/100 ml of water, $CH_3CH_2CH_2CHO$ (*n*-butanal) a solubility of 7 g/100 ml water, and $CH_3CH_2OCH_2CH_3$ (diethyl ether) a solubility of 7 g/100 ml water.

13.3 *Spectral Properties of Aldehydes and Ketones*

The most characteristic spectral feature of type AA carbonyl compounds is an intense absorption band in the 1740–1705-cm^{-1} region of the infrared arising from the C=O stretching deformation (Table 13.1). In general, aldehydes absorb at a slightly higher frequency (near 1730 cm^{-1}) than ketones (near 1715 cm^{-1}), although structural features can alter this considerably, as indicated by the effect that ring size has on the stretching frequency of cycloalkanones (Table 13.1). In addition to the carbonyl stretching deformation band, aldehydes also show a band near 2750 cm^{-1}, a region where few other absorption bands appear. Type AB and BB aldehydes and ketones show carbonyl stretching frequencies about 25–40 cm^{-1} lower in frequency than their corresponding saturated counterparts, providing a means for distinguishing between type AA carbonyl compounds on the one hand and type AB or BB carbonyl compounds on the other (Table 13.1).

The effect of the carbonyl group on nmr resonances is a downfield shift of the proximate hydrogens. If the hydrogen is directly attached to the carbonyl group, as it is in aldehydes, the chemical shift is about δ 10, and the appearance of a resonance at this downfield position provides a definitive means for differentiating aldehydes from ketones. If the hydrogen is removed by one carbon,

Table 13.1 CARBONYL STRETCHING DEFORMATION FREQUENCIES OF ALDEHYDES AND KETONES

Compound Type	C=O Stretching Frequency, cm^{-1}	Compound Type	C=O Stretching Frequency, cm^{-1}
RCHO	1740–1720	RCH=CH(R)C=O	1685–1665
R_2CO and cyclohexanones	1725–1705		
Cyclopropanones	1810	(RCH=CH)$_2$C=O	1670–1660
Cyclobutanones	1775		
Cyclopentanones	1740	Ar(R)C=O	1700–1680
RCH=CHCHO	1705–1680	$Ar_2C=O$	1670–1660
RCH=CHCH=CHCHO	1680–1660		

as it is in the α position of ketones or aldehydes, the chemical shift is only about δ 2 (see Table 6.3).

Type AA aldehydes and ketones possess only a weak absorption band near 280 nm arising from an $n \rightarrow \pi^*$ transition; the $\pi \rightarrow \pi^*$ transition falls well below 200 nm. The attachment of a type **B** group to a carbonyl function produces a conjugated system, and for the reasons outlined earlier (see Section 7.1) the position of the $\pi \rightarrow \pi^*$ transition moves to longer wavelengths. The most striking spectral difference between type AA and type AB or BB aldehydes and ketones, therefore, appears in the ultraviolet region. For example, ketone $CH_3CH_2CH_2COCH_3$ does not absorb strongly between 200 and 270 nm, but ketone $CH_3CH{=}CHCOCH_3$ absorbs strongly at 227 nm with an extinction coefficient of 6500.

PROBLEM 13-3

A C_9H_9ClO compound is optically active, shows a triplet pattern at δ 10.1 in the nmr, and can be oxidized to benzoic acid. Suggest a structure for this compound.

PROBLEM 13-4

Compounds A and B both have the molecular formula C_5H_8O, they are both inherently achiral, and neither one shows any absorption in the region of the nmr characteristic of vinyl hydrogens. Compound A has an infrared absorption at 1740 cm^{-1}, and compound B at 1775 cm^{-1}. Suggest reasonable structures for compounds A and B that are compatible with these data.

Aldehydes and ketones stand midway in the oxidation sequence of carbon compounds; alcohols are their neighbors on the more reduced side and carboxylic acids on the more oxidized side. Many of the synthesis methods, therefore, start with either alcohols or carboxylic acids. For the oxidation of alcohols, see Fig. 10.18 and some further examples in Fig. 13.7. Illustrative of a reduction method is a sequence called the *Rosenmund reduction* (Fig. 13.7), which involves the

13.4 *Synthesis of Aldehydes and Ketones*

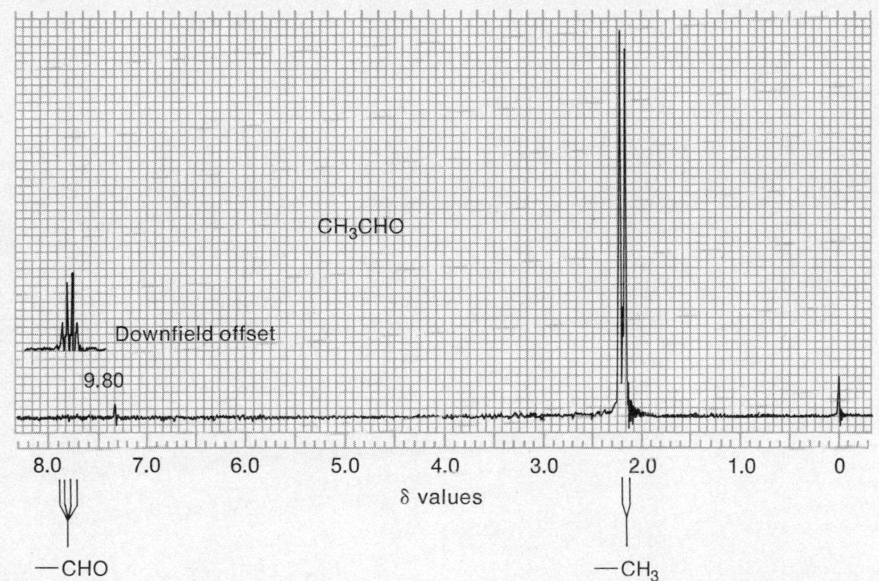

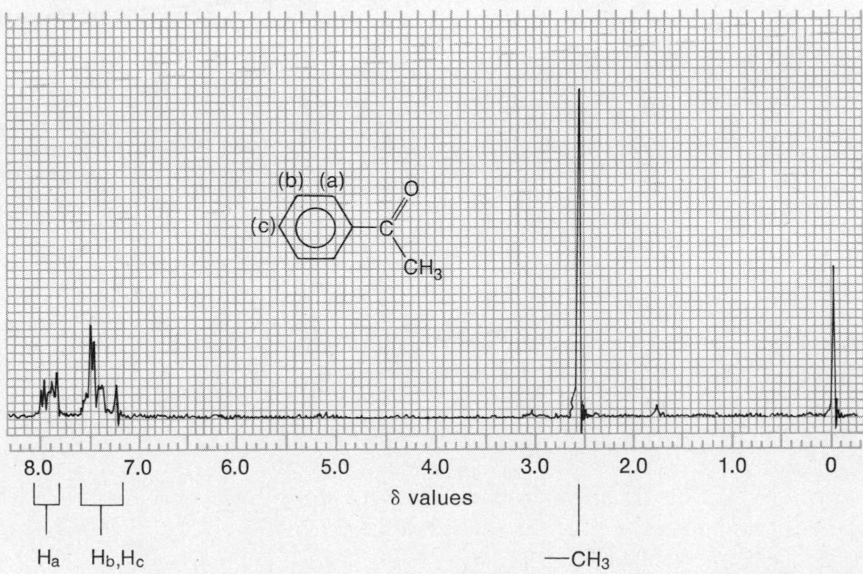

Fig. 13.5. Nuclear magnetic resonance spectra of acetaldehyde and acetophenone.

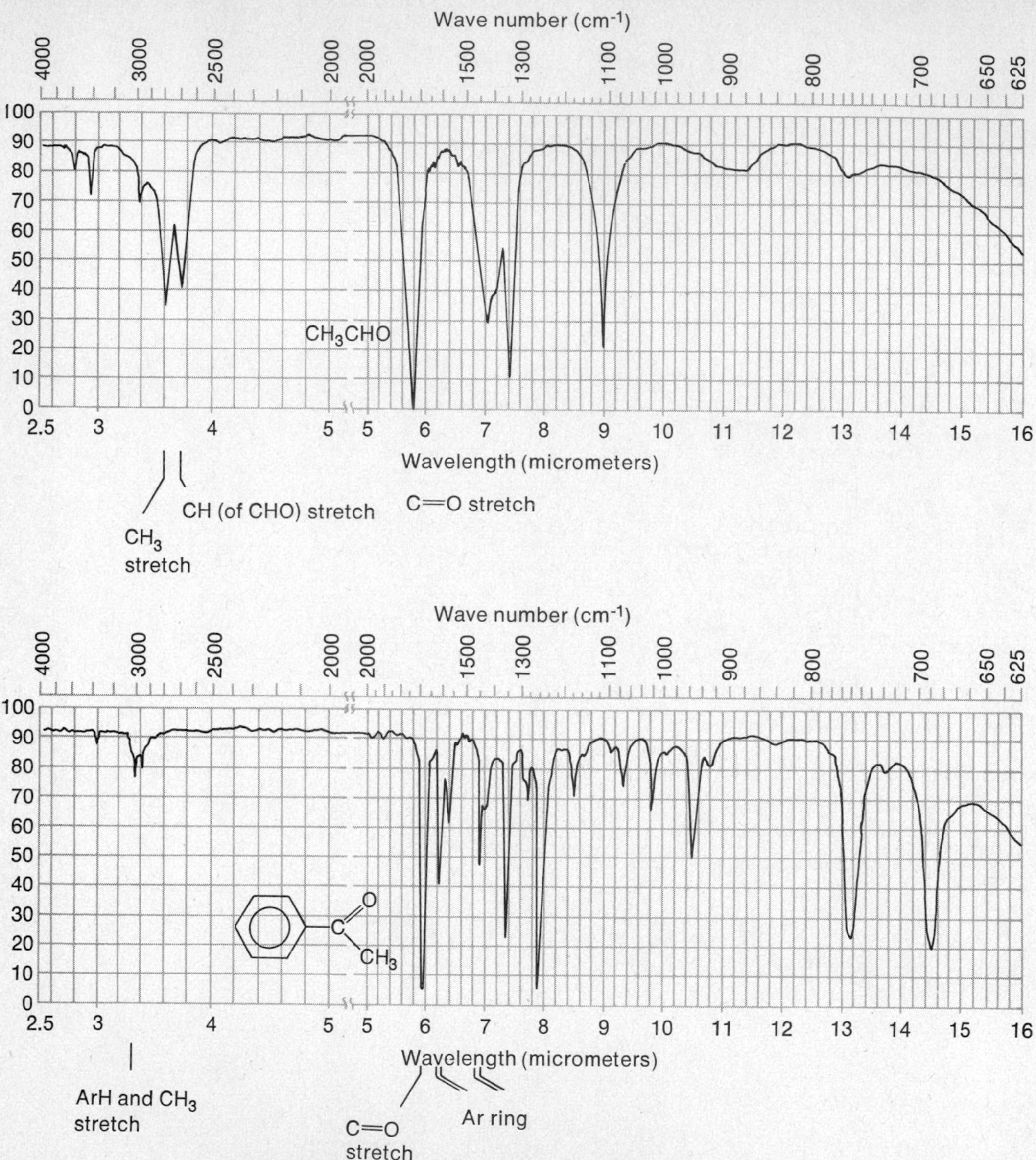

Fig. 13.6. Infrared spectra of acetaldehyde and acetophenone.

conversion of carboxylic acids to the acid chlorides (see Section 15.1c) followed by replacement of the halogen by hydrogen, using molecular hydrogen, a special catalyst consisting of palladium metal deposited on barium sulfate, and some sulfur compound (*e.g.*, tetramethylthiourea) to moderate the catalyst and prevent the reduction from proceeding all the way to the alcohol.

Oxidative cleavage of carbon–carbon double bonds by ozone or other strong oxidizing agents (see Section 8.4) is another process that yields aldehydes and

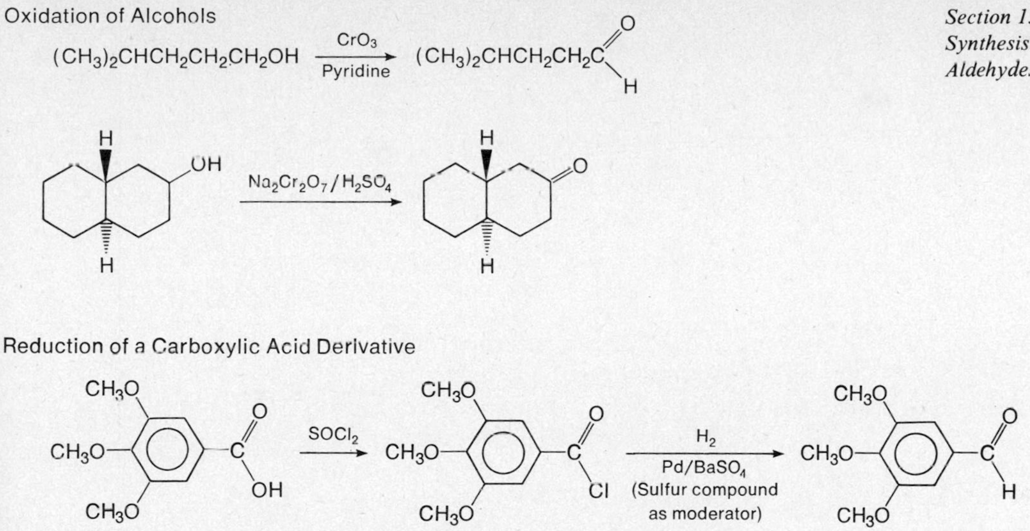

Oxidation of Alcohols

$$(CH_3)_2CHCH_2CH_2CH_2OH \xrightarrow[\text{Pyridine}]{CrO_3} (CH_3)_2CHCH_2CH_2CHO$$

$$\xrightarrow{Na_2Cr_2O_7/H_2SO_4}$$

Reduction of a Carboxylic Acid Derivative

$$\xrightarrow{SOCl_2} \qquad \xrightarrow[\text{(Sulfur compound as moderator)}]{H_2,\ Pd/BaSO_4}$$

Fig. 13.7. Methods and preparation of aldehydes and ketones.

ketones. If the alkene is acyclic, two carbonyl compounds are produced, but if the double bond is part of a ring, a single dicarbonyl compound results (Fig. 13.8).

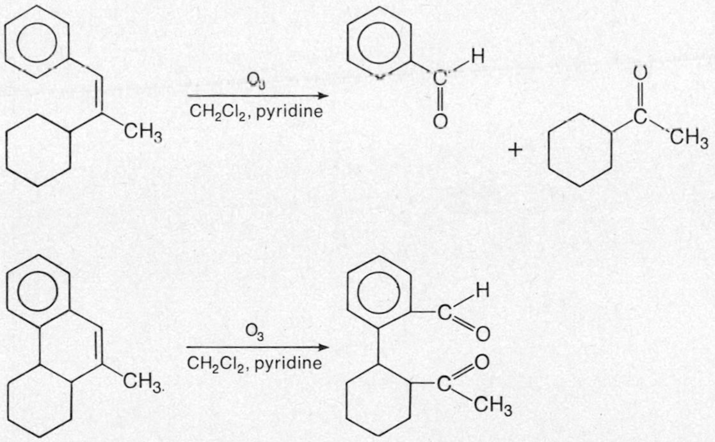

$$\xrightarrow[CH_2Cl_2,\ \text{pyridine}]{O_3}$$

$$\xrightarrow[CH_2Cl_2,\ \text{pyridine}]{O_3}$$

Fig. 13.8. Synthesis of aldehydes and ketones by oxidation of alkenes.

Aromatic aldehydes and ketones can be synthesized from aromatic hydrocarbons by the Friedel–Crafts acylation reaction (see Fig. 8.24). The reagents involved in this reaction are an aromatic hydrocarbon, a carboxylic acid chloride, and a Lewis acid catalyst such as aluminum chloride. The catalyst and the acid chloride form a complex that yields the electrophile which attacks the aromatic ring:

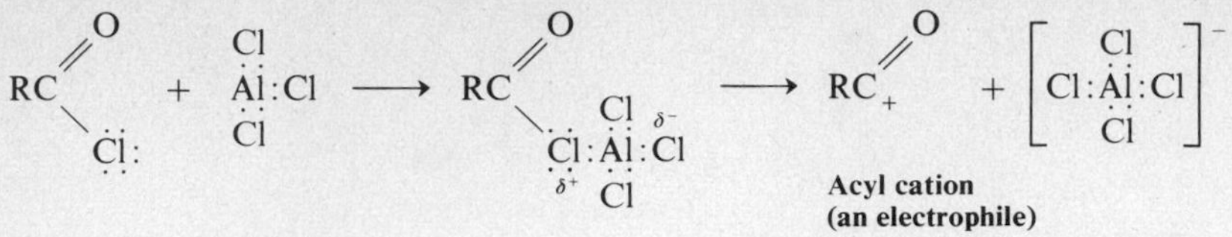

$$\mathrm{RC(=O)\ddot{C}l:} + \mathrm{AlCl_3} \longrightarrow \mathrm{RC(=O)}\overset{\delta+}{\mathrm{Cl}}:\overset{\delta-}{\mathrm{AlCl_3}} \longrightarrow \mathrm{RC^{+}(=O)} + [\mathrm{Cl:AlCl_3}]^{-}$$

Acyl cation
(an electrophile)

The acid chloride that is required if the product is to be an aldehyde is formyl chloride, HC(=O)Cl an unstable compound that must be prepared *in situ* (see footnote, p. 207) from carbon monoxide and hydrogen chloride. Thus, when toluene, CO, HCl, and $AlCl_3$ are mixed the product is *p*-methylbenzaldehyde (*p*-tolualdehyde) (Fig. 13.9). Other acid chlorides react in a comparable fashion to yield aromatic ketones, such as benzophenone and acetophenone.

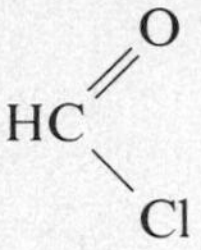

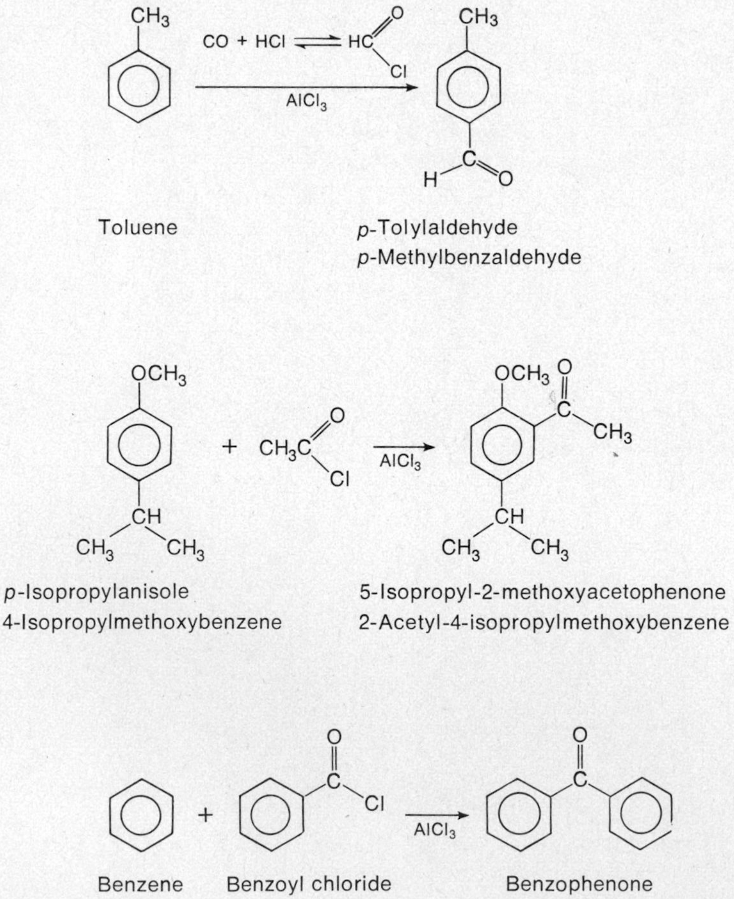

CO + HCl ⇌ HC(=O)Cl; $AlCl_3$

Toluene → *p*-Tolylaldehyde
p-Methylbenzaldehyde

p-Isopropylanisole
4-Isopropylmethoxybenzene
+ $CH_3C(=O)Cl$ ($AlCl_3$) →
5-Isopropyl-2-methoxyacetophenone
2-Acetyl-4-isopropylmethoxybenzene

Benzene + Benzoyl chloride ($AlCl_3$) → Benzophenone

Fig. 13.9. Synthesis of aromatic aldehydes and ketones by Fridel-Crafts acylation reactions.

PROBLEM 13-5

Suggest methods for carrying out the following conversions: (a) toluene to benzaldehyde; (b) propyne to acetone; and (c) benzene to *p*-hydroxybenzaldehyde.

13.5 Nucleophilic Addition Reactions of Aldehydes and Ketones

As a result of the polarization of the carbonyl group, $\rangle C{=}O \leftrightarrow \rangle \overset{+}{C}{-}\overset{-}{O}$, aldehydes and ketones have a marked tendency to add nucleophilic species (Lewis bases) to the carbonyl carbon, followed by the addition of an electrophilic species (Lewis acids) to the carbonyl oxygen; the reactions are classed as *1,2-nucleophilic additions*:

$$\rangle C{=}O + Nu^- \rightleftharpoons \rangle C(O^-)(Nu) \overset{E^+}{\rightleftharpoons} \rangle C(O{-}E)(Nu)$$

The position of the overall equilibrium is dependent on the nucleophile; the stronger is the nucleophile, the farther the reaction proceeds to completion. Thus, certain carbon nucleophiles add so effectively that, for all practical purposes, the reactions are irreversible, whereas halogen nucleophiles add so ineffectively that, for all practical purposes, the reactions don't proceed at all.

1,2-NUCLEOPHILIC ADDITION REACTIONS INVOLVING CARBON NUCLEOPHILES. A variety of compounds can act as carbon nucleophiles in addition reactions with carbonyl functions to form new carbon–carbon bonds. Included among these are hydrogen cyanide, acetylides (*e.g.*, $HC{\equiv}CNa$), Grignard reagents (*e.g.*, C_6H_5MgBr), and various other organometallic reagents (*e.g.*, CH_3Li). Illustrative examples involving these reagents are shown in Fig. 13.10. Reactions of this type provide useful intermediates in synthesis sequences, and the reader who takes time out at this point to consider the ramifications that are possible with various combinations of aldehydes, ketones, and carbon nucleophiles is ready to play the game of "paper synthesis."

1.2-NUCLEOPHILIC ADDITION REACTIONS INVOLVING NITROGEN NUCLEOPHILES. Nitrogen nucleophiles add to carbonyl groups much less effectively than carbon nucleophiles, and the equilibrium constants are generally unfavorable to the formation of α-hydroxyamines. When primary amines add to the carbonyl group, however, the α-hydroxyamines that are formed can lose water to form imines, the equilibrium constant for the dehydration step often being sufficiently large that the overall equilibrium constant is favorable for product formation. Particularly useful amines in this respect are phenylhydrazine, 2,4-dinitrophenylhydrazine, semicarbazide, and hydroxylamine (for structures, see Fig. 13.12). Phenylhydrazine, for example, reacts with acetone to form the phenylhydrazone of acetone by the sequence of steps shown in Fig. 13.11. The first step is an addition of the terminal nitrogen of the hydrazine moiety, acting as a nucleophile,

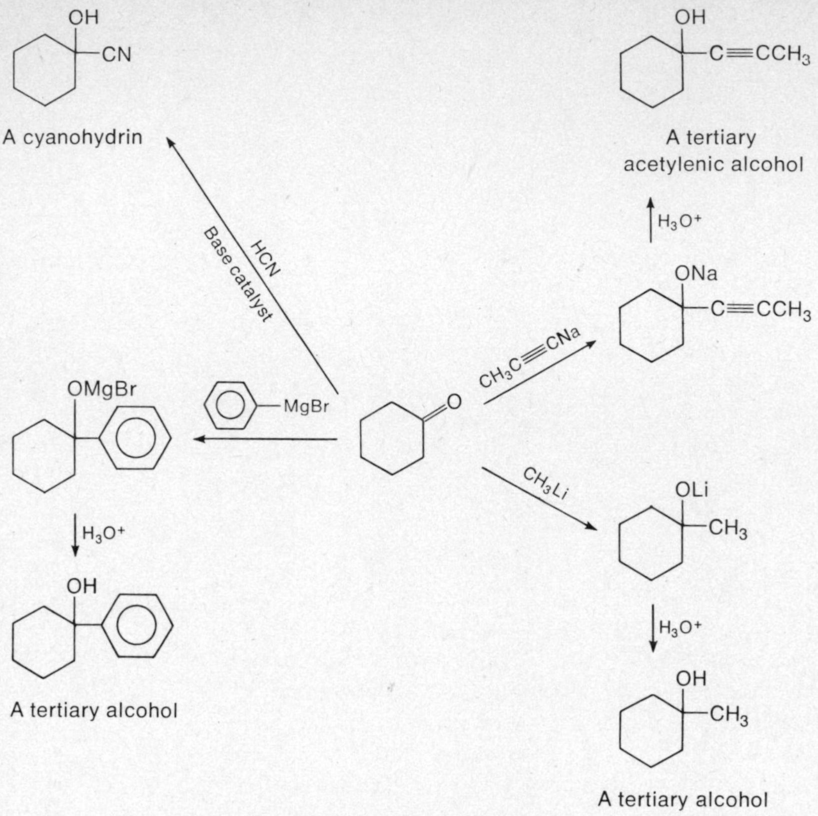
OH
CN
A cyanohydrin
HCN
Base catalyst
OH
C≡CCH3
A tertiary
acetylenic alcohol
H3O+
ONa
C≡CCH3
CH3C≡CNa
OMgBr
MgBr
O
H3O+
OH
A tertiary alcohol
CH3Li
OLi
CH3
H3O+
OH
CH3
A tertiary alcohol

Fig. 13.10. Reaction of cyclohexanone with carbon nucleophiles.

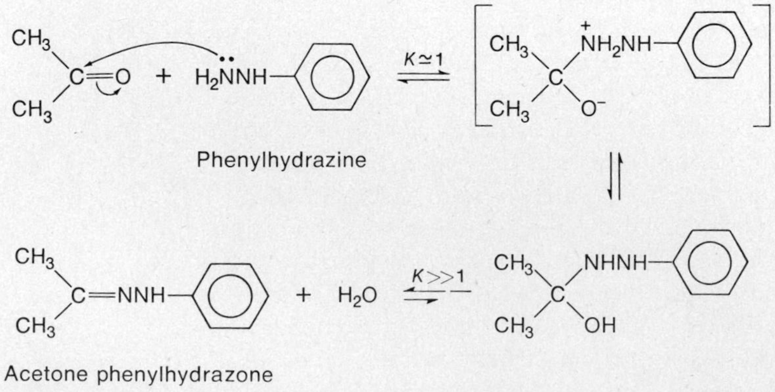
CH3
C=O
CH3
+
H2NNH
Phenylhydrazine
K≃1
CH3
NH2NH
C
CH3
O−
CH3
C=NNH
CH3
Acetone phenylhydrazone
+
H2O
K>>1
CH3
NHNH
C
CH3
OH

Fig. 13.11. Reaction of acetone with phenylhydrazine to form a phenylhydrazone.

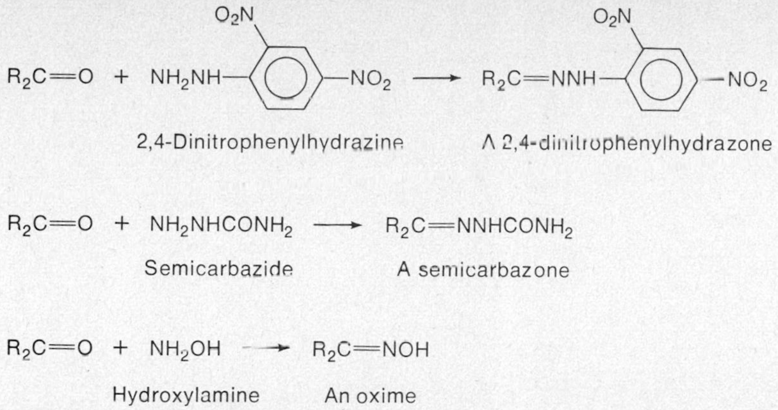

$$R_2C{=}O + NH_2NH{-}C_6H_3(NO_2)_2 \longrightarrow R_2C{=}NNH{-}C_6H_3(NO_2)_2$$

2,4-Dinitrophenylhydrazine — A 2,4-dinitrophenylhydrazone

$$R_2C{=}O + NH_2NHCONH_2 \longrightarrow R_2C{=}NNHCONH_2$$

Semicarbazide — A semicarbazone

$$R_2C{=}O + NH_2OH \longrightarrow R_2C{=}NOH$$

Hydroxylamine — An oxime

Fig. 13.12. 2,4-Dinitrophenylhydrazone, semicarbazone, and oxime derivatives of aldehydes and ketones.

to the carbonyl carbon, acting as the electrophilic center, to give an α-hydroxyamine. The equilibrium constant for this step is about 1. The second step is an intramolecular loss of a molecule of water resulting in the formation of a double bond between carbon and nitrogen, the equilibrium constant for this step being very much greater than 1. The overall reaction, therefore, is "pulled" essentially to completion, permitting the product to be isolated in good yield as a bright yellow, crystalline, high-melting, readily crystallized solid. These characteristics make "derivatives" of this sort valuable, for they are pure compounds that have retained the carbon framework of the original aldehyde or ketone, possess melting points which can be compared with those of known compounds, and can be converted back to the original aldehyde or ketone by hydrolysis. In similar fashion, 2,4-dinitrophenylhydrazine, semicarbazide, and hydroxylamine react with aldehydes and ketones to give 2,4-dinitrophenylhydrazones, semicarbazones, and oximes, respectively (Fig. 13.12).

1,2-NUCLEOPHILIC ADDITION REACTIONS INVOLVING OXYGEN NUCLEOPHILES. Oxygen nucleophiles add to carbonyl groups much less effectively than carbon nucleophiles and nitrogen nucleophiles. In fact, the equilibrium constants for the formation of hydrates (*i.e.*, by the addition of water) and hemiacetals or hemiketals (*i.e.*, by the addition of alcohols) from aldehydes and ketones are generally so unfavorable that the reactions have no preparative value. Certain very strongly polarized aldehydes and ketones, however, do form stable hydrates and hemiacetals or hemiketals. Trichloroacetaldehyde (chloral), for example, possesses a carbonyl carbon that is particularly electron-poor as the result of electron withdrawal not only by the oxygen atom but also by the trichloromethyl group. Consequently, it is reactive enough to form a moderately stable bond to the oxygen atom of water (to form chloral hydrate) or alcohols (to form hemiacetals) (Fig. 13.13).

If the reaction between an aldehyde or ketone and alcohol is carried out in the presence of a trace of mineral acid, the reaction does not stop at the hemiacetal

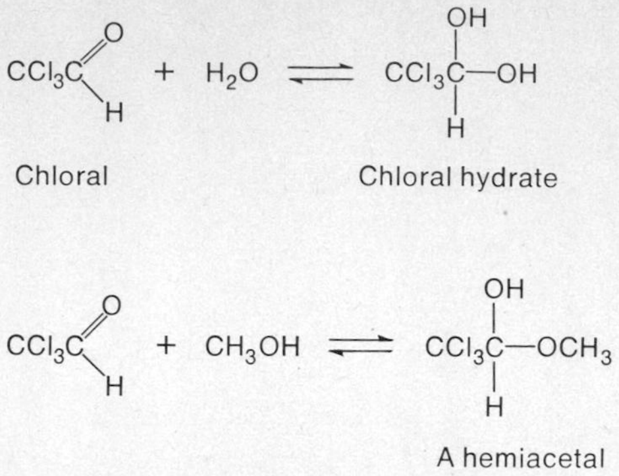

$$CCl_3CHO + H_2O \rightleftharpoons CCl_3CH(OH)_2$$

Chloral — Chloral hydrate

$$CCl_3CHO + CH_3OH \rightleftharpoons CCl_3CH(OH)OCH_3$$

A hemiacetal

Fig. 13.13. Reaction of chloral with oxygen nucleophiles to form addition products.

or hemiketal stage but proceeds further with the involvement of a second equivalent of alcohol to form the acetal or ketal (Fig. 13.14). The reaction is interpreted in terms of an intermediate carbonium ion ($R\overset{+}{C}HOCH_3$), resulting from the acid-catalyzed removal of water from the initially formed hemiactal or hemiketal, which then reacts with another molecule of alcohol in a fashion comparable to

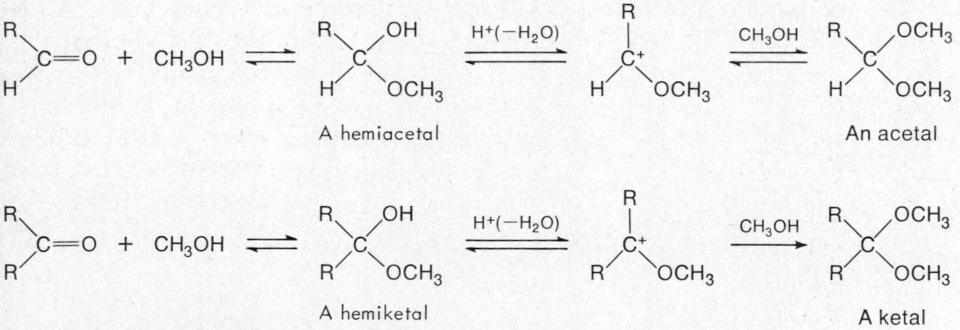

$$RHC{=}O + CH_3OH \rightleftharpoons RHC(OH)(OCH_3) \xrightleftharpoons{H^+(-H_2O)} RH\overset{+}{C}(OCH_3) \xrightleftharpoons{CH_3OH} RHC(OCH_3)_2$$

A hemiacetal — An acetal

$$R_2C{=}O + CH_3OH \rightleftharpoons R_2C(OH)(OCH_3) \xrightleftharpoons{H^+(-H_2O)} R_2\overset{+}{C}(OCH_3) \xrightarrow{CH_3OH} R_2C(OCH_3)_2$$

A hemiketal — A ketal

Fig. 13.14. Reaction of aldehydes and ketones with alcohols to form acetals.

the acid-catalyzed conversion of alcohols to ethers (see Fig. 11.4). The acid-catalyzed conversion of the hemiacetal or hemiketal to the acetal or ketal serves the same purpose as the dehydration step in the formation of imines from aldehydes and ketones, introducing a step with a sufficiently favorable equilibrium constant to offset the very unfavorable equilibrium constant of the initial addition step. Although acetals and ketals can be readily synthesized, isolated, and purified, they are easily converted back to the parent aldehydes or ketones by treatment with dilute aqueous acid. For this reason, synthesis sequences often involve acetal or ketal formation as a means for "protecting" aldehyde or ketone functions, allowing reagents that might react with a free carbonyl group to be used to effect reactions elsewhere in the molecule and then allowing the free carbonyl group to be regenerated by hydrolysis.

PROBLEM 13-6

If you wanted to convert $C_6H_5CH(OH)CHO$ to $C_6H_5C(=O)CHO$, how might you do it, realizing that the aldehyde function is very readily oxidized?

1,4-NUCLEOPHILIC ADDITION REACTIONS (CONJUGATE ADDITIONS).* α,β-Unsaturated aldehydes and ketones,

$$\rangle C{=}\overset{|}{C}{-}C\overset{\textstyle O}{\underset{\textstyle R}{\lessgtr}}$$

(*i.e.*, type AB and type BB carbonyl compounds) can react with nucleophiles not only via a 1,2 pathway (*direct addition*), as has been discussed, but also via a 1,4 pathway (*conjugate addition*), in which the nucleophile becomes affixed to the β carbon. This is a consequence of charge delocalization in the conjugated system, the positive charge residing partly on the carbonyl carbon and partly on the β carbon. For example, 4-methyl-3-penten-2-one (mesityl oxide) reacts with carbon nucleophiles such as hydrogen cyanide and with hetero atom nucleophiles such as dimethylamine to form 4-substituted-4-methyl-2-pentanones (Fig. 13.15). Although in the overall sense, the reaction is an addition to the α and β positions of the α,β-unsaturated carbonyl compound, there is good evidence to indicate that the initially formed product is, indeed, a vinyl alcohol resulting from a 1,4 addition. The vinyl alcohol, however, rapidly undergoes a tautomeric shift to the isomeric, and much more stable, carbonyl compound. The equilibrium constants for conjugate additions are generally more favorable to product formation than those of direct carbonyl additions. Thus, even weak nucleophiles, such as amines and alcohols, which do not give stable products via direct carbonyl addition with type AA carbonyl compounds, frequently yield stable 1,4-addition products with type AB, BB, and BC type carbonyl compounds. The greater stability of the 1,4-addition products, as estimated from bond energy values (see Table 2.3), is largely the result of the retention of the very strong C=O bond in the products, in contrast to the 1,2-addition products in which the C=O bond is no longer present.

13.6 Aldol Condensation Reactions of Aldehydes and Ketones

A hydrogen attached to a carbon adjacent to a carbonyl group (*i.e.*, an α-hydrogen) of an aldehyde or ketone can be abstracted by a sufficiently strong base. Although the pK_a for the C—H bonds in simple hydrocarbons is 40 or greater, that for the

* When the terms "1,2 addition" and "1,4 addition" are used, the numbering system implied is one in which the numbering starts on one terminus of the unsaturated system. Thus, in the system C=C—C=O, numbering starts on the oxygen; *i.e.*, $\overset{4}{C}{=}\overset{3}{C}{-}\overset{2}{C}{=}\overset{1}{O}$. It has no relation to IUPAC numbering and is used simply as a convenience.

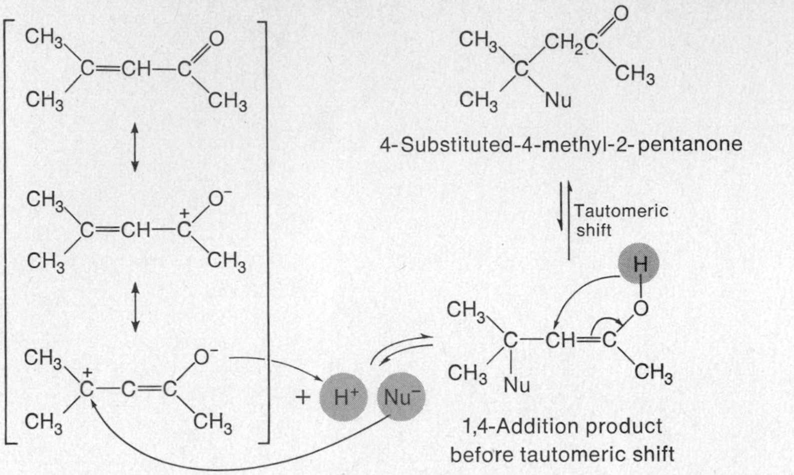

Fig. 13.15. Reaction of an α,β-unsaturated ketone with nucleophiles in a 1,4-addition (conjugate addition) reaction.

α C—H bond in acetaldehyde, for example, is about 19. This enormous increase in acidity is attributed to the ability of the charge in the anion derived from acetaldehyde (*i.e.*, $[CH_2CHO]^-$) to be delocalized so that it resides on both carbon and oxygen (Fig. 13.16). Anions of this sort are often referred to as *enolate anions* or *enolates.* Carbonyl compounds from which resonance-stabilized anions can be derived are said to possess 'active methylene" groups. These anions are capable of acting as carbon nucleophiles in the same fashion as the nucleophiles that have been discussed in the previous sections, adding either by 1,2 or 1,4 pathways to aldehydes and ketones. Thus, certain aldehydes and ketones, such as acetaldehyde and acetone, respectively, can furnish the nucleophile as well as the carbonyl group to which the nucleophile can add; *i.e.*, they can undergo self-condensation reactions. Treatment of acetaldehyde with aqueous sodium hydroxide, for example, promotes the sequence of reactions shown in Fig. 13.17, to give a product called acetaldol. Reactions of this general type, therefore, are usually referred to as *aldol condensations.*

* With stronger nucleophiles such as carbanions from RMgX and hydride ions from $LiAlH_4$, the proton does not accompany the nucleophile but is furnished in a separate step; *e.g.*,

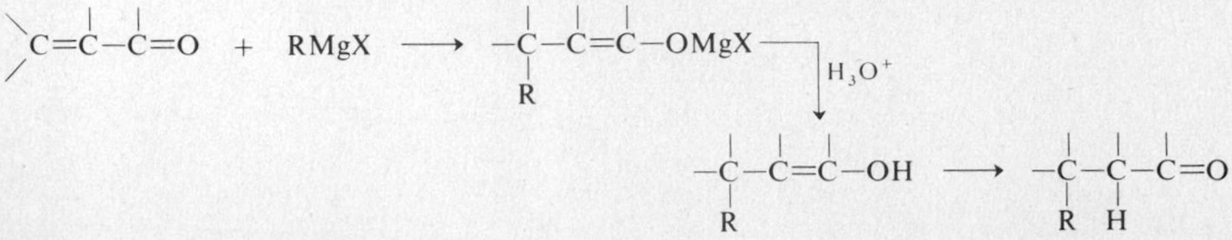

$$\mathrm{>C{=}C{-}C{=}O + RMgX \longrightarrow -C(R){-}C{=}C{-}OMgX \xrightarrow{H_3O^+} -C(R){-}C{=}C{-}OH \longrightarrow -C(R){-}C(H){-}C{=}O}$$

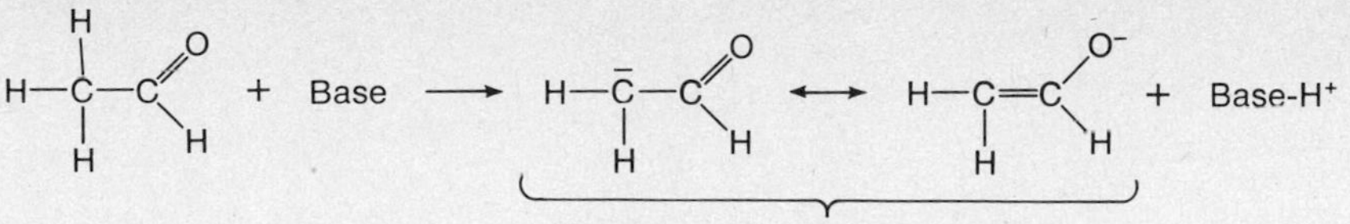

$$H{-}\underset{H}{\overset{H}{C}}{-}C(=O)H \;+\; \text{Base} \longrightarrow \underbrace{H{-}\underset{H}{\bar{C}}{-}C(=O)H \longleftrightarrow H{-}\underset{H}{C}{=}C(O^-)H}_{\text{Resonance structures of the anion}} \;+\; \text{Base-H}^+$$

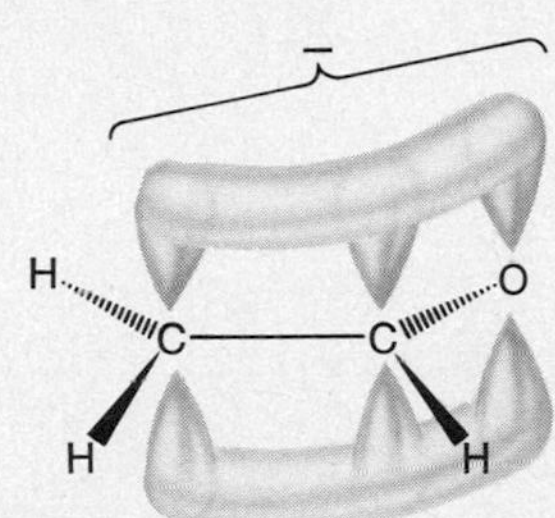

Orbital representation of the anion

Fig. 13.16. The resonance-stabilized enolate anion of acetaldehyde.

Removal of the α hydrogens of an aldehyde or ketone yields a resonance-stabilized anion; anions resulting from the removal of hydrogens at the β, γ, δ, and other carbons of the molecule, however, cannot be stabilized by charge delocalization. Thus, if propionaldehyde is mixed with aqueous sodium hydroxide, the reaction product is $CH_3CH_2\underset{OH}{CH}CH(CH_3)(CHO)$ and *not* $CH_3CH_2\underset{OH}{CH}CH_2CH_2CHO$.

Acetone reacts in an entirely comparable fashion; in the presence of a base, it yields a resonance-stabilized (delocalized) anion, which can condense with

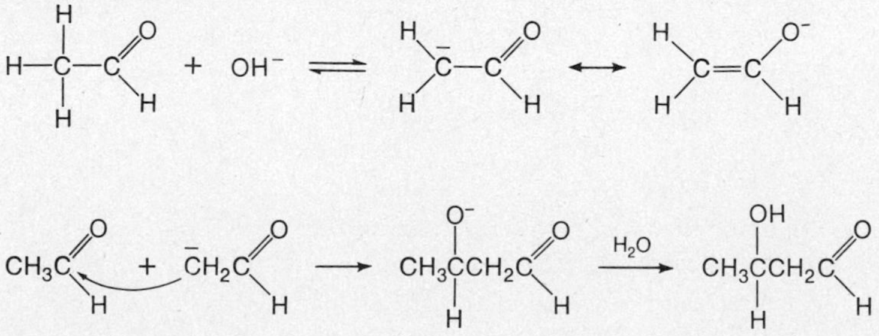

$$H{-}\underset{H}{\overset{H}{C}}{-}C(=O)H \;+\; OH^- \rightleftharpoons H_2\bar{C}{-}C(=O)H \longleftrightarrow H_2C{=}C(O^-)H$$

$$CH_3C(=O)H \;+\; \bar{C}H_2C(=O)H \longrightarrow CH_3\underset{H}{\overset{O^-}{C}}CH_2C(=O)H \xrightarrow{H_2O} CH_3\underset{H}{\overset{OH}{C}}CH_2C(=O)H$$

Acetaldol
(3-Hydroxybutanal)

Fig. 13.17. The self-condensation reaction of acetaldehyde; the aldol condensation.

another molecule of acetone to yield a product called diacetonc alcohol (4-hydroxy-4-methyl-2-pentanone):

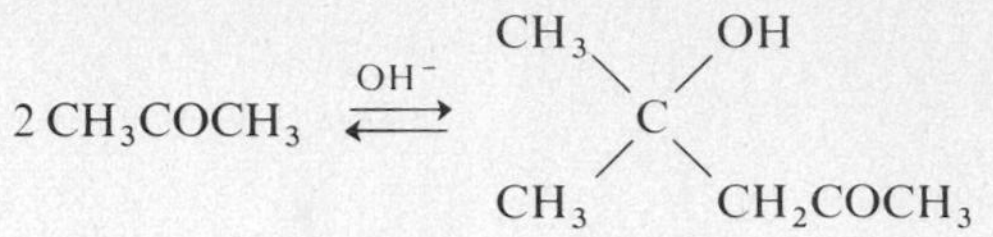

$$2\,CH_3COCH_3 \overset{OH^-}{\rightleftharpoons} (CH_3)_2C(OH)CH_2COCH_3$$

The equilibrium constant for the overall process in this instance, however, is rather unfavorable to product formation (about 2% conversion), and special methods have to be used to drive the reaction to completion.

The products from the aldol condensations of aldehydes and ketones contain a hydroxyl group in a position β to the carbonyl group, and compounds of this type undergo extremely facile intramolecular dehydration in the presence of a trace of mineral acid. Thus, acetaldol can be dehydrated to $CH_3CH{=}CHCHO$ (crotonaldehyde), and $(CH_3)_2C(OH)CH_2C(=O)CH_3$ can be dehydrated to $(CH_3)_2C{=}CHC(=O)CH_3$ (mesityl oxide), providing a method for the synthesis of α,β-unsaturated aldehydes and ketones.

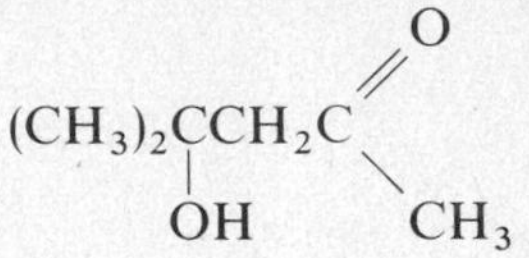

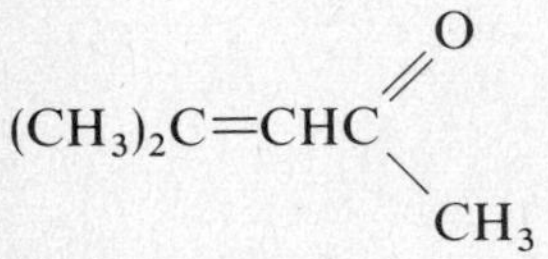

PROBLEM 13-7

What is the aldol product from phenylacetaldehyde? What compound does it yield on dehydration? Would you expect the dehydration to take place easily or with difficulty? Explain.

13.7 Oxidation and Reduction Reactions of Aldehydes and Ketones

Aldehydes and ketones are susceptible to reduction to the corresponding primary and secondary alcohols by a variety of reagents. Metal hydrides, such as lithium aluminum hydride ($LiAlH_4$) and sodium borohydride ($NaBH_4$), and molecular hydrogen in the presence of certain catalysts (*e.g.*, platinum or palladium) are particularly effective (Fig. 13.18). Most selective among these is sodium borohydride, which has no effect on C=C bonds, C≡C bonds, CO_2H (carboxyl) groups, or CO_2R (ester) functions.

Perhaps the most characteristic difference between aldehydes and ketones is their susceptibility to oxidation, aldehydes being much more readily converted to carboxylic acids than are ketones. Among the reagents that are quite specific for the oxidation of $RCHO \rightarrow RCO_2H$ are cupric ion (as sodium cupritartrate, known as *Fehling's solution* and prepared from copper sulfate, sodium hydroxide, and tartaric acid) and silver ion (as ammoniacal silver hydroxide, known as *Tollens' reagent* and prepared from silver hydroxide and ammonium hydroxide). The cupric ion of Fehling's solution is reduced to red cuprous oxide (Cu_2O), and

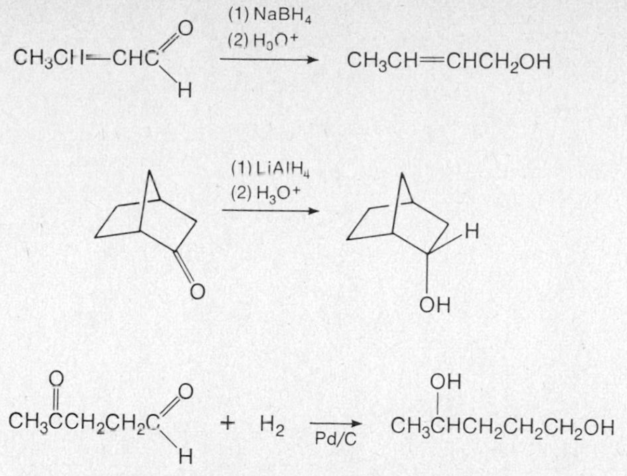

$$CH_3CH{=}CHCHO \xrightarrow[(2)\ H_3O^+]{(1)\ NaBH_4} CH_3CH{=}CHCH_2OH$$

$$\text{norcamphor} \xrightarrow[(2)\ H_3O^+]{(1)\ LiAlH_4} \text{norborneol (H, OH)}$$

$$CH_3COCH_2CH_2CHO + H_2 \xrightarrow{Pd/C} CH_3CH(OH)CH_2CH_2CH_2OH$$

Fig. 13.18. Reduction of aldehydes and ketones.

the silver ion of Tollen's reagent is reduced to metallic silver, which appears as a black colloidal suspension or as a mirror on the sides of the flask. Both tests provide telltale evidence for the presence of an aldehyde. Thus, a simple way for distinguishing between aldehydes and ketones is to treat the samples with either Fehling's or Tollens' reagent and note whether a precipitate forms; *i.e.*,

$$CH_3CH_2CHO + Cu^{+2}\ (\text{or } Ag^+) \longrightarrow CH_3CH_2CO_2H + \underline{Cu_2O}\ (\text{or } \underline{Ag^0})$$

$$CH_3COCH_3 + Cu^{+2}\ (\text{or } Ag^+) \longrightarrow \text{No reaction}$$

These reagents are quite selective in their action and do not oxidize other unsaturated functions, such as carbon–carbon double and triple bonds. For example, crotonaldehyde ($CH_3CH{=}CHCHO$) is oxidized to crotonic acid ($CH_3CH{=}CHCO_2H$) by Fehling's or Tollens' reagent, whereas the stronger reagents, such as potassium permanganate, yield acetic acid:

$$CH_3CH{=}CHCHO \xrightarrow{Ag^+ \text{ or } Cu^{+2}} CH_3CH{=}CHCO_2H$$

$$CH_3CH{=}CHCHO \xrightarrow{KMnO_4/NaOH} 2\,CH_3CO_2H + 2\,CO_2$$

Hydroxyl functions, unless they are α to a carbonyl group, are similarly unaffected by Fehling's or Tollens' reagent:

$$CH_3CH(OH)CH_2CH_2CHO \xrightarrow{Ag^+ \text{ or } Cu^{+2}} CH_3CH(OH)CH_2CH_2CO_2H$$

Air is an effective oxidant for many aldehydes (autoxidation), and special precautions must be taken if aldehydes are to be stored for any length of time. The reaction is another example of a free radical chain process that is initiated by trace amounts of free radicals (Fig. 13.19).

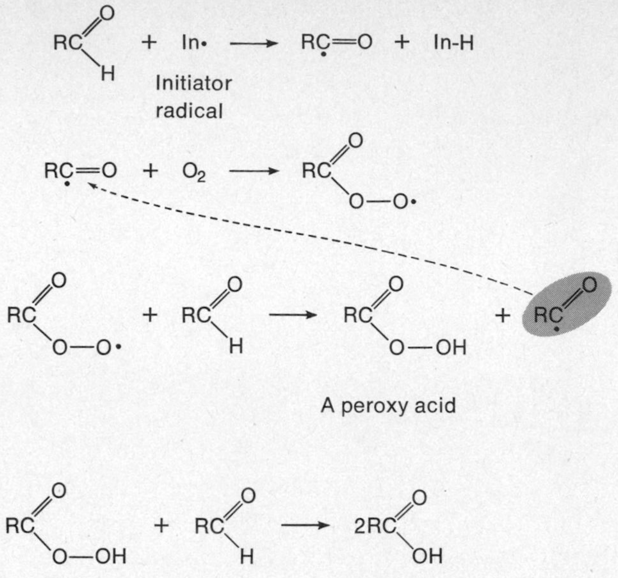

$$RCHO + In\cdot \longrightarrow R\dot{C}{=}O + In\text{-}H$$

Initiator radical

$$R\dot{C}{=}O + O_2 \longrightarrow RC(=O)O{-}O\cdot$$

$$RC(=O)O{-}O\cdot + RCHO \longrightarrow RC(=O)O{-}OH + R\dot{C}{=}O$$

A peroxy acid

$$RC(=O)O{-}OH + RCHO \longrightarrow 2RC(=O)OH$$

Fig. 13.19. Air oxidation of aldehydes.

13.8 *Electrophilic Substitution Reactions of Aromatic Aldehydes and Ketones*

The electrophilic substitution of benzene (see p. 213) is initiated by the interaction of an electrophile with the π system of benzene to form an intermediate cation which then expels a proton to regenerate the aromatic ring. When alkyl, OH, OR, or NR_2 groups are attached to the benzene ring, the reaction proceeds *faster* than with benzene itself, and the incoming electrophile becomes attached to the *o*- and *p*-positions; alkyl, OH, OR, and NR_2 groups are said to be activating and *o,p*-directing. When a carbonyl group is attached to the benzene ring, the opposite condition obtains; *i.e.*, the electrophilic substitution proceeds *more slowly* than with benzene itself, and the incoming electrophile becomes attached to the *meta*-position. Thus, carbonyl functions are said to be deactivating and *meta*-directing. For example, when acetophenone is treated with bromine in the presence of $FeBr_3$, *meta*-bromoacetophenone is slowly produced (Fig. 13.20).

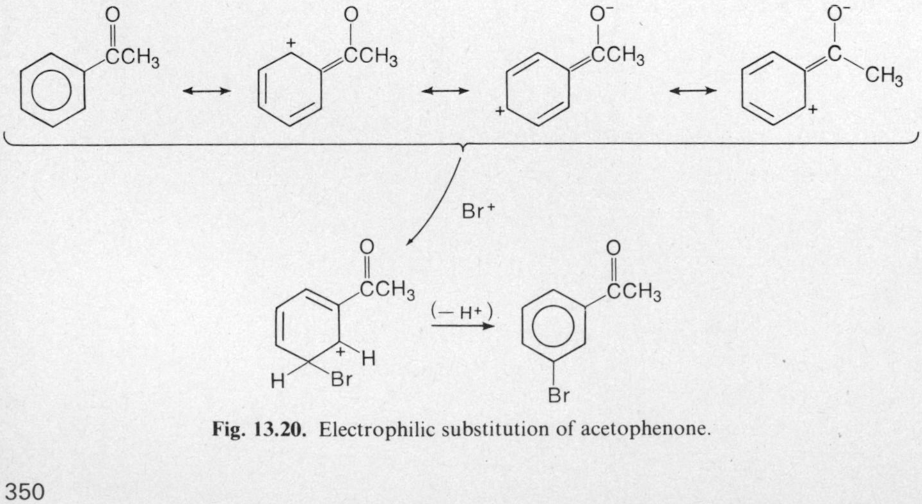

$$C_6H_5CCH_3(=O) \longleftrightarrow \longleftrightarrow \longleftrightarrow \quad \xrightarrow{Br^+} \quad \xrightarrow{(-H^+)} \quad m\text{-}BrC_6H_4CCH_3(=O)$$

Fig. 13.20. Electrophilic substitution of acetophenone.

The lower rate and the *meta*-orientation can be interpreted in terms of resonance interaction whereby electron density is drained away from the benzene ring toward the carbonyl oxygen. This results in less electron density on the benzene ring (hence a slower reaction with the electrophile), with the electron deficiency being greater at the *ortho*- and *para*-positions than at the *meta*-position (hence, selective reaction at the *meta* position).

13.9 Compounds Containing Two or More Aldehyde or Ketone Functions

If two carbonyl functions are separated by two or more methylene groups, they behave in an essentially independent fashion; no *qualitatively* new characteristics are imparted to the compound. If the carbonyl functions are *directly attached* to each other or if only *one methylene group separates them*, however, certain special features are present. For example, α-dicarbonyl compounds* are much more susceptible to oxidative cleavage than ordinary aldehydes and ketones. Thus, 2,3-butanedione is converted to acetic acid by the action of either potassium permanganate or periodic acid:

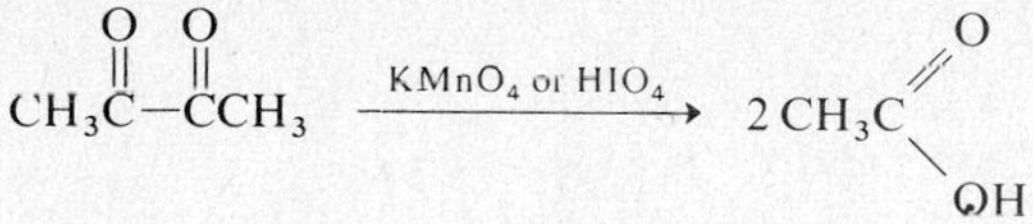

$$CH_3\overset{\overset{O}{\|}}{C}-\overset{\overset{O}{\|}}{C}CH_3 \xrightarrow{KMnO_4 \text{ or } HIO_4} 2\,CH_3C\begin{matrix}\diagup\!\!\!\!\diagup O \\ \diagdown OH\end{matrix}$$

The characteristic feature of β-dicarbonyl compounds $\left(R\overset{\overset{O}{\|}}{C}-\underset{|}{\overset{|}{C}}-\overset{\overset{O}{\|}}{C}R\right)$ containing a hydrogen attached to the carbon between the carbonyl groups is the ability to exist in stable enol forms. In monoaldehydes and ketones, the equilibrium constant for the reaction

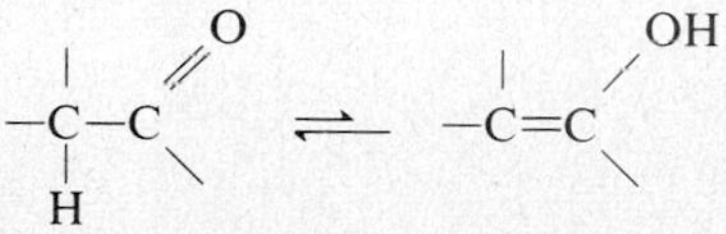

$$-\underset{\underset{H}{|}}{\overset{|}{C}}-C\begin{matrix}\diagup\!\!\!\!\diagup O \\ \diagdown\end{matrix} \rightleftharpoons -\overset{|}{C}=C\begin{matrix}\diagup OH \\ \diagdown\end{matrix}$$

is very small, the equilibrium being almost completely to the carbonyl side. The attachment of a second carbonyl group to the α carbon, however, greatly increases the equilibrium constant. For example, 2,4-pentanedione $(CH_3\overset{\overset{O}{\|}}{C}CH_2\overset{\overset{O}{\|}}{C}CH_3)$ is 15% in the enol form in aqueous solution and 92% in the enol form in hexane solution. An important factor accounting for the increased stability of the enol form of β-diketones (or β-ketoaldehydes) as contrasted with monoketones or aldehydes is the conjugated system (*i.e.*, $O=\overset{|}{C}-\overset{|}{C}=\overset{|}{C}-OH$) that the enol contains. Whereas a bond energy calculation (see Section 10.1) for mono-

---

* Compounds possessing the structure $-\overset{\overset{O}{\|}}{C}-\overset{\overset{O}{\|}}{C}-$ are called *α-dicarbonyl compounds*; those having the structure $-\overset{\overset{O}{\|}}{C}-C-\overset{\overset{O}{\|}}{C}-$ are called *β-dicarbonyl compounds*, etc.

aldehydes and monoketones indicates an advantage of approximately 11 kcal/mole for the carbonyl form, the bond energy advantage of the β-dicarbonyl form is offset to a considerable extent by the delocalization of the π system in the enol form of the β-dicarbonyl compound. Another factor favoring the enol form is the intramolecular hydrogen bonding that can occur to produce a chelate structure [see p. 319 (Fig. 13.21)]. Variations in the amount of intramolecular *vs.* intermolecular hydrogen bonding are probably responsible for the differences that are observed for the extent of enolization of β-dicarbonyl compounds in nonpolar, nonhydrogen-bonding solvents such as hexane, on the one hand and polar, hydrogen-bonding solvents such as water, on the other.

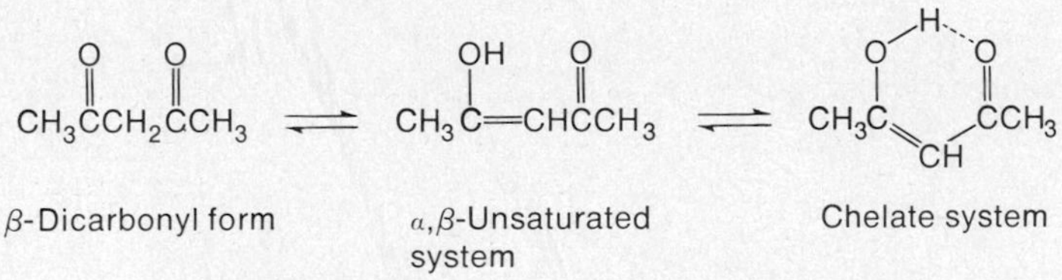

$$CH_3\overset{O}{\overset{\|}{C}}CH_2\overset{O}{\overset{\|}{C}}CH_3 \rightleftharpoons CH_3\overset{OH}{\overset{|}{C}}{=}CH\overset{O}{\overset{\|}{C}}CH_3 \rightleftharpoons CH_3C(-O\cdots H\cdots O=)CCH_3 \text{ (ring via } {=}CH{-})$$

β-Dicarbonyl form α,β-Unsaturated system Chelate system

Fig. 13.21. Enolization of 2,4-pentanedione.

β-Dicarbonyl systems are more acidic than simple carbonyl compounds, because the derived anions are strongly resonance stabilized. Whereas the pK_a value for the dissociation of acetone,

$$CH_3COCH_3 \rightleftharpoons [CH_3COCH_2]^- + H^+,$$

is about 20 (see Table 7.1), that for the dissociation of 2,4-pentanedione,

$$CH_3COCH_2COCH_3 \rightleftharpoons [CH_3COCHCOCH_3]^- + H^+,$$

is 8.9; 2,4-pentanedione is an even stronger acid than phenol. In the presence of a strong base, such as sodium ethoxide, therefore, 2,4-pentanedione is essentially completely converted to the resonance-stabilized enolate anion shown in Fig.

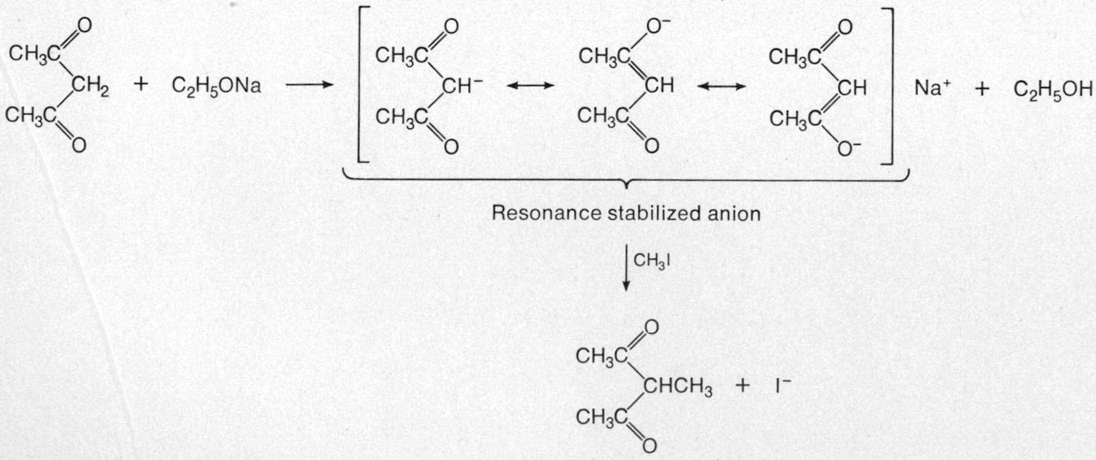

$$(CH_3C{=}O)_2CH_2 + C_2H_5ONa \longrightarrow \left[(CH_3C{=}O)_2CH^- \longleftrightarrow CH_3C(O^-){=}CH{-}C(=O)CH_3 \longleftrightarrow CH_3C(=O){-}CH{=}C(O^-)CH_3\right] Na^+ + C_2H_5OH$$

Resonance stabilized anion

$$\downarrow CH_3I$$

$$(CH_3C{=}O)_2CHCH_3 + I^-$$

Fig. 13.22. Alkylation of 2,4-pentanedione enolate.

13.22. This anion can be used as an effective nucleophile, and its reaction with alkyl halides is a typical nucleophilic substitution process.

PROBLEM 13-8

Cyclohexane-1,3-dione reacts with methyl iodide in the presence of sodium ethoxide. What is the structure of the product? When *t*-butyl iodide is used instead of methyl iodide, the cyclohexane-1,3-dione is recovered largely unchanged, and the major product is 2-methylpropene. Explain.

13.10 *Aldehydes and Ketones of Particular Interest and Utility*

Formaldehyde, the simplest of the aldehydes, is considerably more reactive than its higher homologs. In aqueous solution, for example, it exists almost entirely in the hydrated form, *i.e.*,

$$H_2C{=}O + H_2O \rightleftharpoons H_2C(OH)_2$$

Acetaldehyde, on the other hand, is only about 50% in the hydrated form in aqueous solution, and the higher homologs are even less so. The special reactivity of formaldehyde is probably due to the lack of steric hindrance toward nucleophilic attack.

Formaldehyde has a marked tendency to react with itself to form linear polymers (designated as paraformaldehyde) or a cyclic trimer (designated as trioxane); *i.e.*,

$$\underset{\textbf{Paraformaldehyde}}{HO(CH_2O)_xH} \xleftarrow{H_2O} H_2C{=}O \xrightarrow[H^+,\,\Delta]{} \underset{\textbf{Trioxane}}{(CH_2O)_3}$$

Formaldehyde, produced commercially by the oxidation of methanol and by the oxidation of hydrocarbon mixtures, is used for a variety of items of commerce, including the polymer Delrin [a linear polymer containing a large number of HCHO units and protected at the termini by esterification (see Section 15.3)] and a compound called hexamethylenetetramine, produced by reaction with ammonia; *i.e.*,

$$HCHO \longrightarrow HO(CH_2O)_nH \xrightarrow{(CH_3C)_2O} \underset{\textbf{Delrin}}{CH_3\overset{O}{\overset{\|}{C}}O(CH_2O)_n\overset{O}{\overset{\|}{C}}CH_3}$$

$$6\,HCHO + 4\,NH_3 \longrightarrow \underset{\textbf{Hexamethylenetetramine}}{(CH_2)_6N_4}$$

Hexamethylenetetramine has some medicinal use as a urinary antiseptic but finds wider use as the starting material for the preparation of an explosive known as Cyclonite or RDX.

Acetaldehyde is produced commercially by the silver-catalyzed air oxidation of ethanol,

$$CH_3CH_2OH \quad + \quad O_2 \xrightarrow[{[Ag]}]{} CH_3CHO \quad + \quad H_2O.$$

by the cupric chloride-palladium chloride-catalyzed air oxidation of ethylene,

$$H_2C{=}CH_2 \quad + \quad O_2 \xrightarrow[CuCl_2-PdCl_2]{} CH_3CHO.$$

and by the acid-catalyzed hydration of acetylene,

$$HC{\equiv}CH \quad + \quad H_2O \xrightarrow[H^+]{} [CH_2{=}CHOH] \longrightarrow CH_3CHO$$

The main use of acetaldehyde is as an intermediate in the synthesis of other materials, including acetic acid, acetic anhydride, ethyl acetate, *n*-butanol, acrylonitrile, and acrylic esters.

One of the interesting reactions that acetaldehyde undergoes is a "mixed" (also called a "crossed") aldol condensation with formaldehyde to yield tetrakis-(hydroxymethyl)-methane* (known as pentaerythritol). In general, mixed aldol condensations are unsatisfactory, because several products are formed (*i.e.*, two self-condensation products and two mixed condensation products). Formaldehyde, however, has no α hydrogens and so cannot self-condense in an aldol type reaction. Because it has a more reactive carbonyl group than acetaldehyde, the reaction proceeds exclusively via the anion from acetaldehyde and the carbonyl group of formaldehyde to yield tris-(hydroxymethyl)acetaldehyde.* At this point, another type of reaction, known as a mixed Cannizzaro reaction, takes place. It is an oxidation-reduction process, in which formaldehyde is oxidized to formic acid and tris-(hydroxymethyl)acetaldehyde is reduced to pentaerythritol

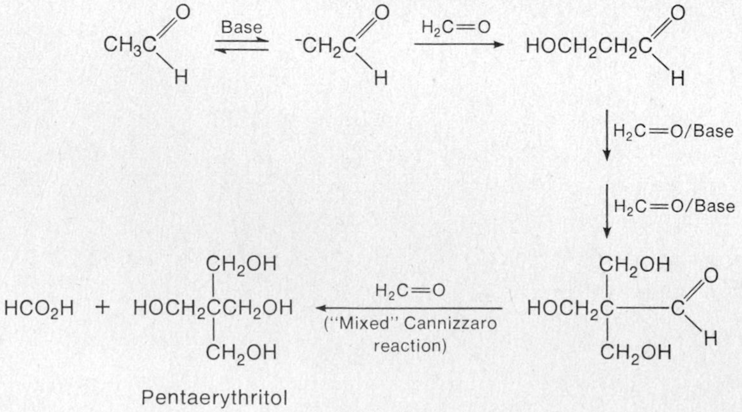

$$CH_3CHO \underset{}{\overset{\text{Base}}{\rightleftharpoons}} {}^-CH_2CHO \xrightarrow{H_2C{=}O} HOCH_2CH_2CHO$$

$$\downarrow H_2C{=}O/\text{Base}$$

$$\downarrow H_2C{=}O/\text{Base}$$

$$HCO_2H \quad + \quad \underset{\text{Pentaerythritol}}{C(CH_2OH)_4} \xleftarrow[\text{("Mixed" Cannizzaro reaction)}]{H_2C{=}O} (HOCH_2)_3C{-}CHO$$

Fig. 13.23. Synthesis of pentaerythritol.

* When two or more identical compound-functions (*e.g.*, hydroxymethyl) are present in a compound, the multiplying designations are *bis-*, *tris-*, *tetrakis-*, etc. rather than *di-*, *tri-*, *tetra-*, etc.

(Fig. 13.23). Large amounts of pentaerythritol are used in the production of types of plastics known as alkyd resins and polyester resins.

Chloral (Cl_3CCHO), the trichloro derivative of acetaldehyde, is prepared from ethanol and chlorine and is one of the starting materials in the manufacture of DDT (see Section 9.8). Chloral hydrate (see p. 344) is a quick-acting soporific, allegedly the essential ingredient of a "Mickey Fin" but also used medicinally as a sleep inducer.

Acetone is the most commercially important of the aliphatic ketones. It is produced by dehydrogenation of isopropyl alcohol,

$$(CH_3)_2CHOH \longrightarrow (CH_3)_2C{=}O + H_2,$$

and as a coproduct in the synthesis of phenol by the oxidation of cumene (see Fig. 10.25). In addition to its use as a commercial solvent, it provides the starting material for a variety of derived products, including $CH_3COCH_2CH(CH_3)_2$ (isobutyl methyl ketone) and $H_2C{=}C(CH_3)CO_2CH_3$ (methyl methacrylate).

The parent member of the aromatic aldehydes is benzaldehyde, a compound that occurs in nature as the glycoside amygdalin (see Fig. 13.24). An intriguing

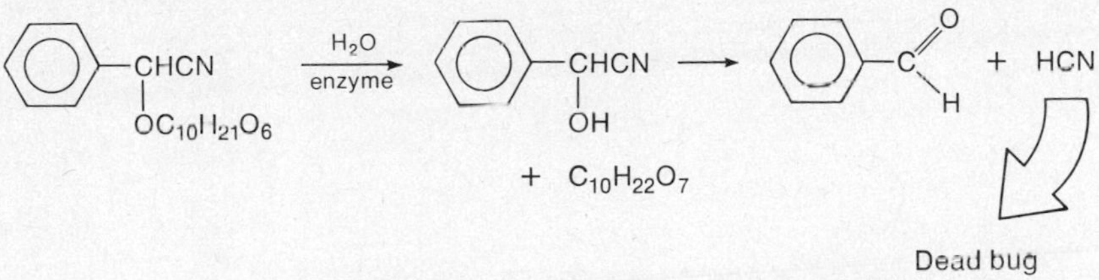

$$C_6H_5CH(CN)OC_{10}H_{21}O_6 \xrightarrow[\text{enzyme}]{H_2O} C_6H_5CH(CN)OH + C_{10}H_{22}O_7 \longrightarrow C_6H_5CHO + HCN$$

Dead bug

Fig. 13.24. Mechanism of chemical defense of the millipede.

use of a glycoside resembling amygdalin is observed in the creature known as the millipede. When threatened by an enemy at close range, the species *Apheloria corrugata* injects into a sac at the bottom of its body an enzyme that catalyzes the hydrolysis of the glycoside to benzaldehyde cyanohydrin, which then immediately decomposes to benzaldehyde and hydrogen cyanide (Fig. 13.24). Pointing its sac at the enemy, the millipede releases the hydrogen cyanide, which envelops the enemy in lethal fumes, allowing the millipede to escape to fight another day. Chemical warfare at the insect level long predates man's discovery of this method of destruction.

Vanillin, the most widely used of synthetic flavoring compounds, is a derivative of benzaldehyde. One of its major commercial sources today is the lignin sulfonates from the waste sulfite liquors of the paper industry; air oxidation of an alkaline mixture of this material oxidizes the coniferyl alcohol to vanillin; *i.e.*,

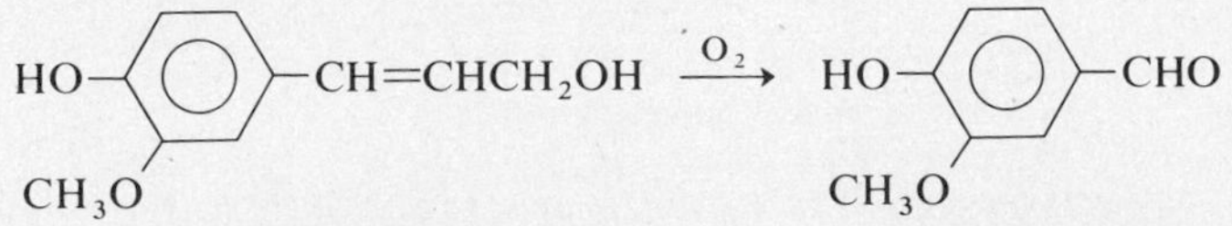

$$HO(CH_3O)C_6H_3{-}CH{=}CHCH_2OH \xrightarrow{O_2} HO(CH_3O)C_6H_3{-}CHO$$

Coniferyl alcohol → **Vanillin**

The simplest member of the aryl ketones is acetophenone, which has been commercially produced by the Friedel–Crafts reaction of benzene and acetyl chloride (see Fig. 13.8). Today, however, its principal commercial source is ethylbenzene, which is converted to acetophenone by air oxidation in the presence of manganese diacetate as a catalyst. This process is a good example of the special reactivity of bonds attached to a benzylic carbon atom (see Section 8.4a).

$$C_6H_5CH_2CH_3 + O_2 \xrightarrow[130^\circ,\ 50\ \text{psi}]{Mn(OAc)_2} C_6H_5COCH_3$$

Acetophenone

Styrene, a versatile industrial product that is used in the preparation of a variety of types of plastics, is produced from acetophenone by reduction to 1-phenylethanol followed by dehydration; *i.e.*,

$$C_6H_5COCH_3 \xrightarrow{[H]^*} C_6H_5CH(OH)CH_3 \xrightarrow[(-H_2O)]{} C_6H_5CH{=}CH_2$$

Acetophenone **1-Phenylethanol** **Styrene**

A cyclic ketone of considerable commercial importance is cyclohexanone. It is prepared either by air oxidation of cyclohexane, using special catalysts, or by the oxidation of cyclohexanol, which, in turn, is prepared by the reduction of phenol.

$$\text{Phenol} \xrightarrow{[H]^*} \text{Cyclohexanol} \xrightarrow{[O]^*} \text{Cyclohexanone} \xleftarrow{O_2} \text{Cyclohexane}$$

The main use of cyclohexanone is in the preparation of caprolactam, which is one of the ingredients in the production of the fiber Nylon-6, a compound closely related to Nylon-6,6. (See p. 414.)

Among the many naturally occurring ketones are several large ring compounds, including 3-methylcyclopentadecanone (known as muscone), isolated from the scent glands of the musk deer, and cycloheptadec-9-enone (known as civetone), isolated from the scent glands of the civet cat. These materials, along with cyclopentadecanone itself, find use in perfumery.

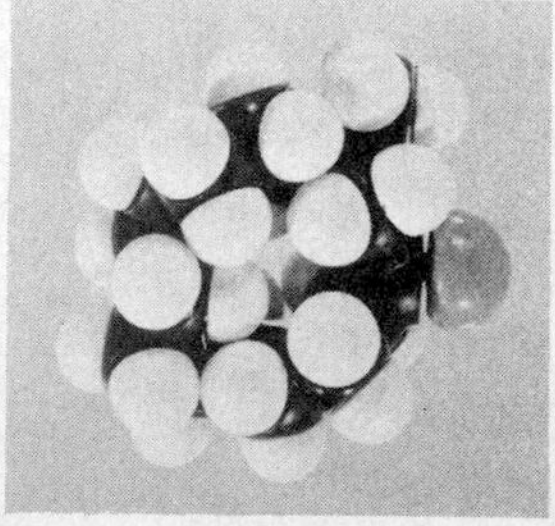

Muscone

* To indicate a reduction or oxidation process, a bracketed hydrogen (*i.e.*, [H]) or oxygen (*i.e.*, [O]) is often used. The bracketed letters imply any reducing or oxidizing agent that is appropriate in the particular case. The reduction of phenol to cyclohexanol, for example, can be effected with H_2 in the presence of a platinum catalyst; the oxidation of cyclohexanol to cyclohexanone can be effected with $Na_2Cr_2O_7/H_2SO_4$.

Muscone

Civetone

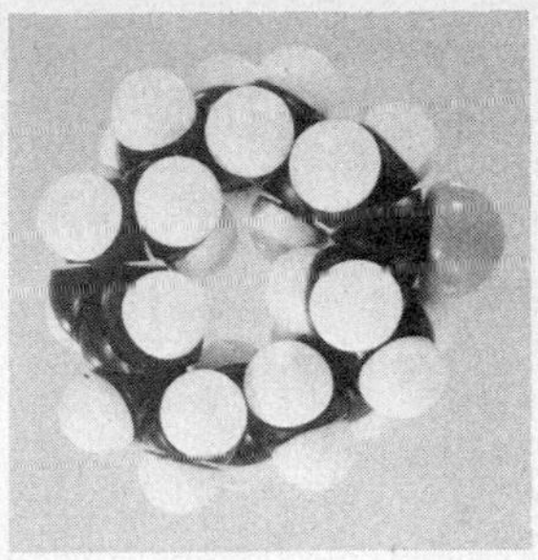
Civetone

A heterocyclic aldehyde that has previously been mentioned for its commercial importance is furfural (see Fig. 11.12). Furfural is used on a large scale as an industrial solvent, as a component in phenol-formaldehyde-type plastics (*i.e.*, it replaces formaldehyde), and as an intermediate in the synthesis of other useful organic compounds, including tetrahydrofuran.

Some ketones of type BB possess unusual structures and have properties that are different from those of most aldehydes and ketones. One of these is 2,4,6-cycloheptatrienone (tropone), a compound having an unusually low carbonyl stretching frequency in the infrared (1638 cm^{-1}) and a great reluctance to form carbonyl derivatives. These properties are commensurate with a charge-separated structure in which the positive charge is delocalized over the entire cycloheptatriene ring (Fig. 13.25). As we discussed previously (see Fig. 4.17), a positively charged cycloheptatriene system (*i.e.*, a cycloheptatrienylium cation) conforms to the Hückel rule and is aromatic. Thus, to attain this structure, the carbon–oxygen bond becomes considerably more polarized than in simple aldehydes and ketones, resulting in a high dipole moment for the molecule and a reduction in the double bond character of the carbonyl function and, concomitantly, a reduction in its reactivity toward nucleophiles.

Tropone

Fig. 13.25. Structure of 2,4,6-cycloheptatriene (tropone).

Another group of compounds classed as type BB are the quinones (see Fig. 10.19). 1,4-Benzoquinone (2,5-cyclohexadiene-1,4-dione) is prepared commercially by the manganese dioxide oxidation of aniline;

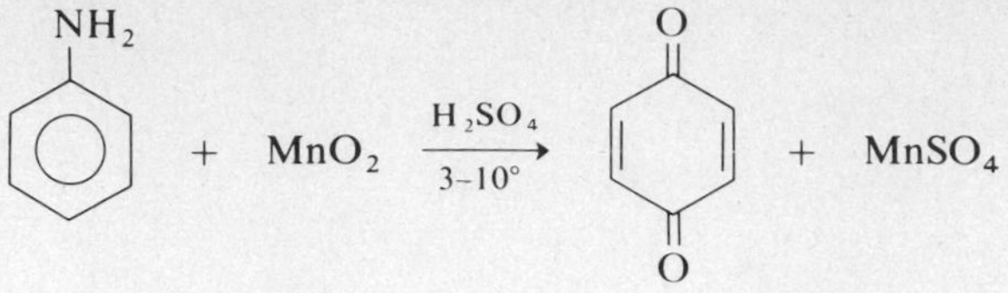

$$C_6H_5NH_2 + MnO_2 \xrightarrow[3\text{–}10°]{H_2SO_4} \text{1,4-benzoquinone} + MnSO_4$$

Quinones of many other aromatic frameworks are also known, including the following typical examples:

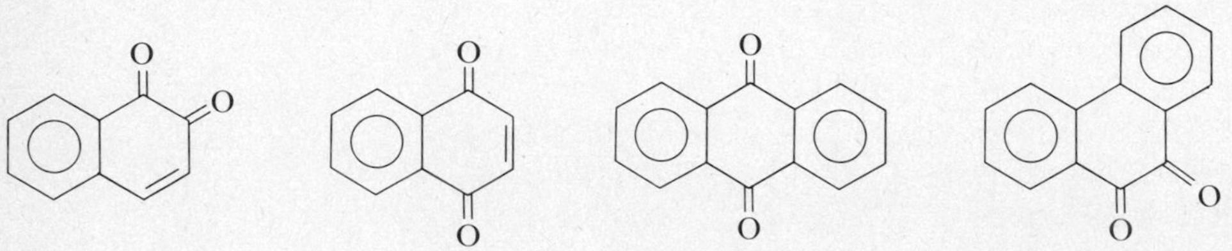

1,2-Naphthoquinone 1,4-Naphthoquinone 9,10-Anthraquinone 9,10-Phenanthraquinone

Although a comparison of the structures of quinones with that of tropone might suggest a similarity in behavior, the quinones are not aromatic but react like α,β-unsaturated carbonyl compounds. For example, 1,4-benzoquinone reacts with hydrogen chloride to form 2-chloro-1,4-dihydroxybenzene (chlorohydroquinone), the reaction involving a 1,4 addition (*i.e.*, hydrogen going to oxygen and chlorine to carbon) followed by a tautomeric shift to produce the aromatic system;

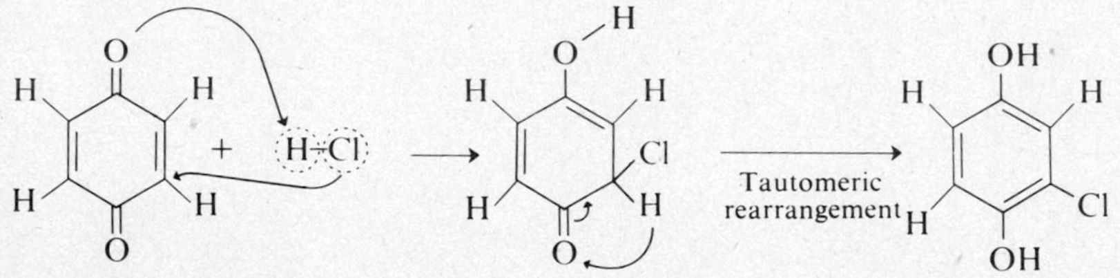

Tautomeric rearrangement

13.11 Study Guide

I. Classification of carbonyl compounds
II. Structure of aldehydes and ketones
  A. Bonding of C=O bond; resonance structures
  B. Tautomerism—see vinyl alcohols
  C. Cycloalkanones
  D. Conjugated aldehydes and ketones (α,β-unsaturated)
  E. Aromatic aldehydes and ketones
III. Nomenclature of aldehydes and ketones
  A. Common (trivial) names
  B. "Parent member-based" systems
    1. "Acetaldehyde" system for aldehydes
    2. "Acetone" system for ketones
    3. α, β, γ, etc., designation of position
  C. "Ketone" system
  D. IUPAC system (including "carboxaldehyde" system)

E. "Phenone" system for ArCOR compounds
F. Carbonyl-containing functional groups
   1. RCO — acyl groups
   2. ArCO — aroyl groups

IV. Physical properties of aldehydes and ketones
   A. Dipole moments—fairly large
   B. Boiling points—higher than corresponding ethers; lower than alcohols
   C. Water solubility—moderately soluble

V. Spectral properties of aldehydes and ketones
   A. Ultraviolet/visible
      1. RCHO and R_2CO absorb weakly near 280 nm
      2. ArCHO, ArCOR, Ar_2CO, and α,β-unsaturated aldehydes and ketones absorb strongly in the 200–270 nm region
   B. Infrared
      1. C=O stretch for RCHO near 1730 cm^{-1}
      2. C=O stretch for R_2CO near 1715 cm^{-1}
      3. C=O stretch of cycloalkanones a function of ring size and conjugation
   C. Nuclear magnetic resonance
      1. Resonance near δ 10 for RCHO and ArCHO
      2. Carbonyl function is electronegative—α hydrogens experience downfield shift.

VI. Synthesis of aldehydes and ketones
   A. Oxidation methods
      1. Oxidation of alcohols
         (a) $RCH_2OH \rightarrow RCHO$ (aldehyde synthesis)
         (b) $R_2CHOH \rightarrow R_2C{=}O$ (ketone synthesis)
      2. Ozonolysis of alkenes
         (a) $RCH{=}CHR \rightarrow RCHO$ (aldehyde synthesis)
         (b) $R_2C{=}CR_2 \rightarrow R_2C{=}O$ (ketone synthesis)
         (c) $R_2C{=}CHR \rightarrow R_2C{=}O + RCHO$ (mixture of products)
   B. Reduction methods—aldehydes only ($RCO_2H \rightarrow RCOCl \rightarrow RCHO$)
   C. Electrophilic substitution reactions—for aromatic aldehydes and ketones only
      1. $ArH + CO + HCl + AlCl_3 \rightarrow ArCHO$ (aldehyde synthesis)
      2. $ArH + RCOCl + AlCl_3 \rightarrow ArCOR$ (ketone synthesis)

VII. Reactions of aldehydes and ketones
   A. Nucleophilic addition reactions
      1. Water as the nucleophile → hydrates
      2. Alcohol as the nucleophile → hemiacetals, hemiketals, acetals, ketals
      3. Nitrogen compounds as the nucleophile
         (a) Phenylhydrazine → phenylhydrazones
         (b) 2,4-Dinitrophenylhydrazine → 2,4-dinitrophenylhydrazones
         (c) Semicarbazide → semicarbazones
         (d) Hydroxylamine → oximes

4. Carbon compounds as the nucleophile
   (a) HCN → cyanohydrins
   (b) Grignard reagents and other organometallic reagents → alcohols (see Chapter 10)
5. Mechanism of nucleophilic addition
   (a) Direct addition (1,2 addition)
   (b) Conjugate addition (1,4 addition)

B. Aldol condensation reactions
1. Carbanion formation from aldehydes and ketones
2. Self-condensation reaction: carbanion + aldehyde or ketone → aldol or ketol product

C. Oxidation-reduction reactions
1. Reduction
   (a) RCHO → RCH_2OH (synthesis of primary alcohols—see Chapter 10)
   (b) $R_2C{=}O$ → R_2CHOH (synthesis of secondary alcohols—see Chapter 10)
2. Oxidation
   (a) RCHO → RCO_2H
   (b) $R_2C{=}O$ → no reaction (unless carbon skeleton rearranges and/or C—C bonds undergo cleavage)

VIII. Polycarbonyl compounds
A. α-Dicarbonyl compounds—susceptible to oxidative cleavage
B. β-Dicarbonyl compounds
1. Enolization of $-\overset{O}{\overset{\|}{C}}CHR\overset{O}{\overset{\|}{C}}-$ compounds
2. Acidity of β-dicarbonyl compounds
3. Chelate structures for enol form

PROBLEMS

1. Using the "acetaldehyde" or "ketone" system and the IUPAC system of nomenclature, name each of the following compounds:

(a) CH_3CHO (b) $(CH_3)_2CHCHO$ (c) $CH_3\overset{O}{\overset{\|}{C}}CH_3$

(d) $CH_3\overset{O}{\overset{\|}{C}}CH_2CH_3$ (e) $C_6H_5-CH_2CHO$ (f) $C_6H_5-\overset{O}{\overset{\|}{C}}CH_2CH_3$

(g) $CH_3-C_6H_4-CHO$ (h) $C_6H_5-\overset{O}{\overset{\|}{C}}CH_3$ (i) $(CH_3)_2CH\overset{O}{\overset{\|}{C}}CH_3$

2. Draw the structure corresponding to each of the following names:

(a) Formaldehyde
(b) Isohexanal
(c) Isobutyl methyl ketone
(d) Cyclohexyl ethyl ketone
(e) Trimethylacetaldehyde
(f) Diphenyl ketone
(g) 4-Isopropylbenzaldehyde
(h) 3-Methyl-2-hexanone

3. Using any appropriate system of nomenclature, name each of the following compounds:

(a) 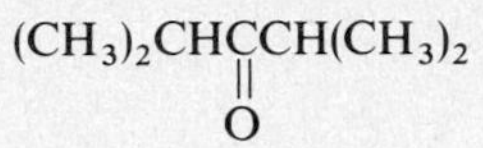 $(CH_3)_2CHC(=O)CH(CH_3)_2$

(b) 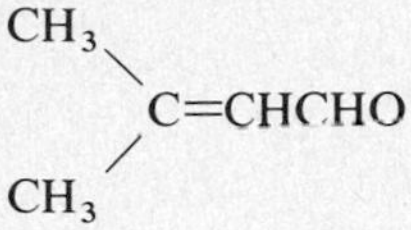 $(CH_3)_2C{=}CHCHO$

(c) 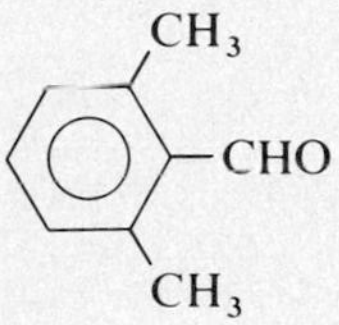 [benzene ring with CHO, flanked by two CH_3 groups]

(d) 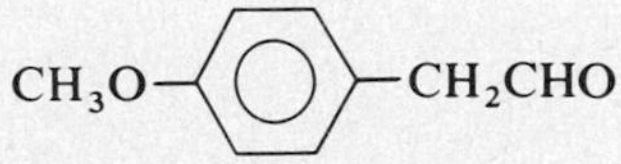 $CH_3O-C_6H_4-CH_2CHO$

(e) 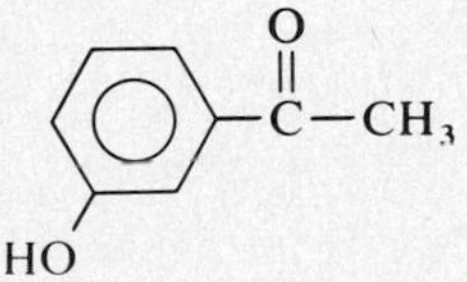 [benzene ring with $-C(=O)-CH_3$ and HO]

(f) 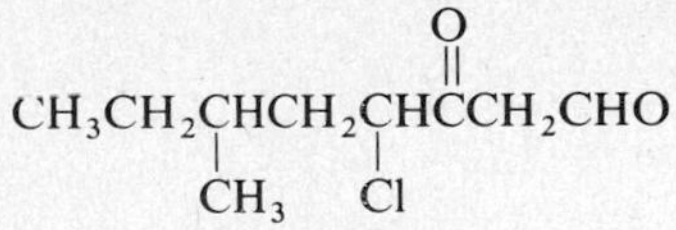 $CH_3CH_2CH(CH_3)CH_2CH(Cl)C(=O)CH_2CHO$

(g) 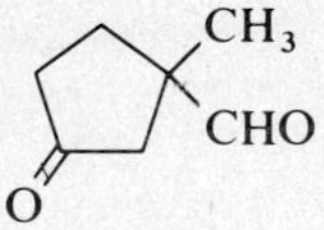 [cyclopentanone ring bearing CH_3 and CHO]

(h) 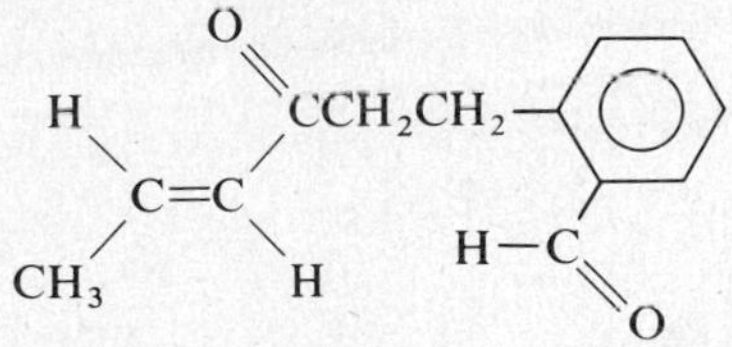 [H/CH_3 C=C H / $C(=O)CH_2CH_2-$ benzene ring with H–C=O]

(i) 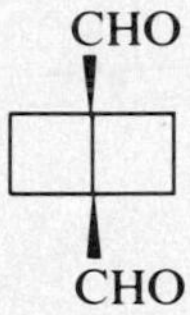 [cyclobutane ring with two CHO groups]

(j) 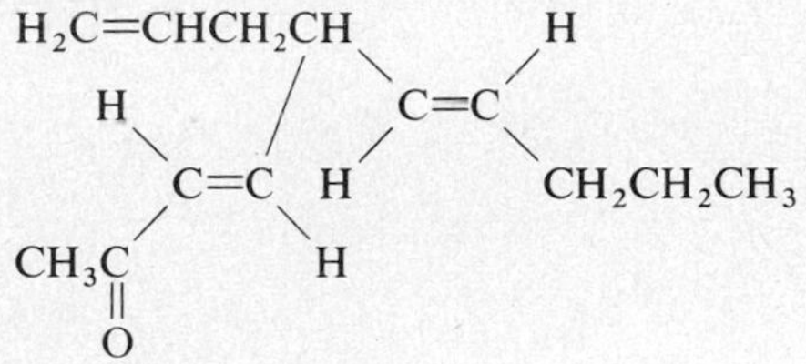 [$H_2C{=}CHCH_2CH$, H, C=C, H, $CH_2CH_2CH_3$; H, C=C, H, $CH_3C(=O)$, H]

(k) 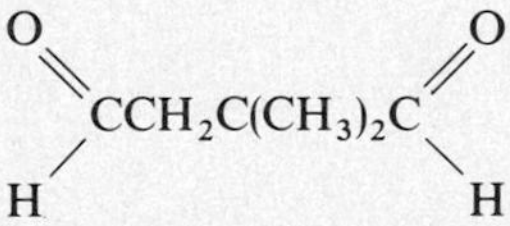 $OHCCH_2C(CH_3)_2CHO$

(l) 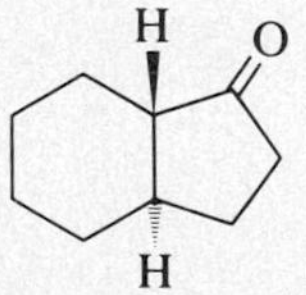 [bicyclic ketone with H, H, O]

(m) 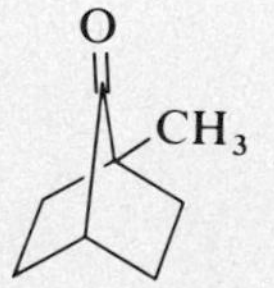 [bicyclic ketone with O and CH_3]

(n)  $CH_3C(=O)-C_6H_4-C(=O)CH_3$

4. Draw the structure for each of the following compounds:

(a) Ethyl vinyl ketone
(b) 5-Hexenal
(c) 3-Methyl-4-oxo-2-pentenal
(d) β-Phenylpropionaldehyde
(e) Isobutyrophenone
(f) 4-Chloroheptan-2,5-dione

(g) Cyclohept-3-enecarboxaldehyde
(h) 2-(4-Acetylphenyl)-6-heptyn-3-one
(i) Bicyclo[8.1.0]undecane-11-carboxyaldehyde
(j) *cis*-7-Methyl-1,2,6-octatriene-4,5-dione
(k) 4-Methyl-1-naphthaldehyde
(l) 4-Oxocyclohexanecarboxaldehyde
(m) Triphenylacetaldehyde
(n) 5-Phenyl-*trans*-3-penten-2-one

5. Draw the structure of the product from each of the following reactions:

(a) cyclopentane-C(H)(OH) + $Na_2Cr_2O_7$ + H_2SO_4 $\longrightarrow$

(b) $(CH_3)_2C{=}CH(C_6H_5)$ + O_3 $\longrightarrow$

(c) benzene + CH_3CH_2COCl $\xrightarrow[AlCl_3]{}$

(d) $H_2C{=}O$ + $(CH_3)_2CHMgBr$ $\longrightarrow$ $\xrightarrow{H_3O^+}$

(e) $HC(=O)CH_2CH_2CH_2C(=O)H$ + $NaBH_4$ $\longrightarrow$ $\xrightarrow{H_3O^+}$

(f) CH_3CH_2CHO + NaOH $\longrightarrow$

6. Draw the structure of the major product in each of the following reactions:

(a) 2-naphthyl-C(=O)Cl + H_2 $\xrightarrow[\text{(sulfur moderator)}]{Pd/BaSO_4}$

(b) bicyclic alkene (octalin) + O_3 $\xrightarrow[\text{Pyridine, } CH_2Cl_2]{}$

(c) p-xylene (CH_3, CH_3) + CO + HCl $\xrightarrow[AlCl_3]{}$

(d) furan-2-CHO (O, CHO) + NH_2NH–C_6H_5 $\longrightarrow$

(e) $C_6H_5COCH_3$ (acetophenone) $+ \ C_6H_5MgBr \longrightarrow \xrightarrow{H_3O^+}$

(f) o-$C_6H_4(CH_2CHO)_2$ $+ \ NaOH \longrightarrow$

(g) $C_6H_5CH{=}CHCHO + NaBH_4 \longrightarrow \xrightarrow{H_3O^+}$

(h) 1,3-cyclopentanedione $+ \ (CH_3)_3CCH_2CH_2CH_2Br \xrightarrow{C_2H_5ONa}$

(i) $C_6H_5CHO + C_6H_5\overset{O}{\overset{\|}{C}}CH_2\overset{O}{\overset{\|}{C}}C_6H_5 \xrightarrow{NaOH}$

7. The following conversions from the compound shown on the left to the compound shown on the right can be affected in one step. Indicate what reagents are necessary to achieve this.

(a) $CH_3COCH_3 \longrightarrow (CH_3)_3COH$

(b) $C_6H_5COCl \longrightarrow C_6H_5\overset{O}{\overset{\|}{C}}C_6H_5$

(c) 1-methylcyclopentene $\longrightarrow CH_3\overset{O}{\overset{\|}{C}}CH_2CH_2CH_2CHO$

(d) cyclobutanone $\longrightarrow$ 1-hydroxycyclobutanecarbonitrile (OH, CN)

(e) $C_6H_5COCH_3 \longrightarrow C_6H_5\underset{CH_3}{\underset{|}{C}}{=}NNHCONH_2$

8. For each of the following interconversions, indicate the steps that are involved in going from the starting material to the product:

(a) $CH_3CHO \longrightarrow CH_3CH{=}CHCH_2OH$

(b) benzene $\longrightarrow C_6H_5\underset{CH_3}{\underset{|}{\overset{OH}{\overset{|}{C}}}}{-}CH_3$

(c) cyclohexene $\longrightarrow$ cyclohexanone

(d) $C_6H_5-C\equiv CH \longrightarrow C_6H_5-C(OH)(CH_3)-CN$

(e) 3-methylindene $\longrightarrow$ o-$C_6H_4(CH(CH_3)OH)(CH_2CH_2OH)$

(f) $CH_3O-C_6H_5 \longrightarrow CH_3O-C_6H_4-CH{=}NNH-C_6H_3(NO_2)-NO_2$

(g) $C_6H_6 \longrightarrow C_6H_5-CH(OCH_3)_2$

(h) $(CH_3)_2C{=}CH_2 \longrightarrow (CH_3)_2CHCHO$

(i) $C_6H_5OH \longrightarrow$ 1-ethynylcyclohexanol ($C_6H_{10}(OH)-C\equiv CH$)

9. Draw all the tautomeric structures possible for each of the following compounds:

(a) $CH_3COCH_2CH_3$

(b) $HO-C_6H_4-OH$ (meta)

(c) $CH_3CH{=}CHCCH_3$ (with C=O on the C shown as $\overset{O}{\|}$C)

(d) $HO-C_6H_4-C(=O)-CH_3$ (para)

(e) 2-(hydroxymethylene)cyclopentane-1,3-dione ($=C(H)OH$)

10. Draw the important resonance structures for each of the following compounds:

(a) $C_6H_5-CH{=}CH\overset{O}{\overset{\|}{C}}CH_3$

(b) $^-O-C_6H_4-\overset{O}{\overset{\|}{C}}-CH_3$

(c) $(HO)(CH_3)C{=}C(H)-\overset{O}{\overset{\|}{C}}-CH_3$

(d) 2-acetylcyclopentane-1,3-dione anion ($=C(-\overset{O}{\overset{\|}{C}}-CH_3)$)

11. Arrange the following compounds in the expected order of increasing boiling point:

(a) $H_2C{=}CHOCH{=}CH_2$

(b) $HOCH_2C{\equiv}CCH_3$

(c) $CH_3CH{-}CHC(=O)H$

(d) $H_2C{-}CHC(=O)CH_3$

(e) $(CH_3)_3CCH_3$

(f) $CH_3(CH_2)_3CH_3$

12. Starting with 1-butene as the only source of carbon compounds, show how the following compounds can be synthesized.

(a) CH_3CH_2CHO

(b) $CH_3CH_2\overset{O}{\overset{\|}{C}}CH_3$

(c) $CH_3CH_2CH_2OH$

(d) $CH_3CH_2CH_2CH_2OH$

(e) $CH_3CH_2\overset{OH}{\overset{|}{C}}HCH_2CH_2CH_3$

13. Starting with any alkane, alkene, alkyne, alkyl halide, alkanol, alkanal, or alkanone containing four or fewer carbon atoms, benzene, toluene, *o*-, *m*-, and *p*-xylene and any necessary inorganic reagents, outline a reasonable synthesis for each of the following compounds:

(a) $CH_3CH_2C(OH)(CH_3)CH_3$

(b) $CH_3CH(OH){-}C_6H_5$

(c) $C_6H_5{-}CH_2C(OH)(CH_3)CH_3$

(d) $CH_3{-}C_6H_4{-}C(OH)(CH_3)CH_2CH(CH_3)_2$ (para)

(e) $CH_3CH{=}CHCH_2\overset{OH}{\overset{|}{C}}HCH_3$

(f) benzene ring bearing CH_3, CH_2OH (adjacent), and CH_3 (para to the first CH_3)

(g) $CH_3{-}C_6H_4{-}CH{=}CH\overset{O}{\overset{\|}{C}}CH_3$ (para)

(h) benzene ring bearing $COCH_3$ and Br (meta)

14. A pair of isomeric $C_9H_{10}O$ compounds possess the nmr spectra shown below. Suggest a structure for each of these compounds that is in accord with these data.

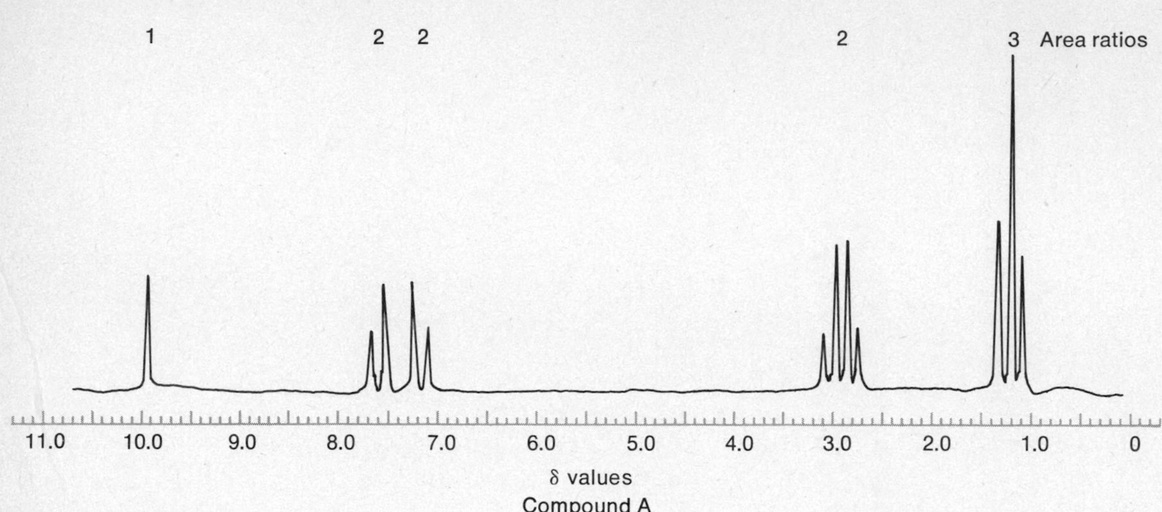

Compound A

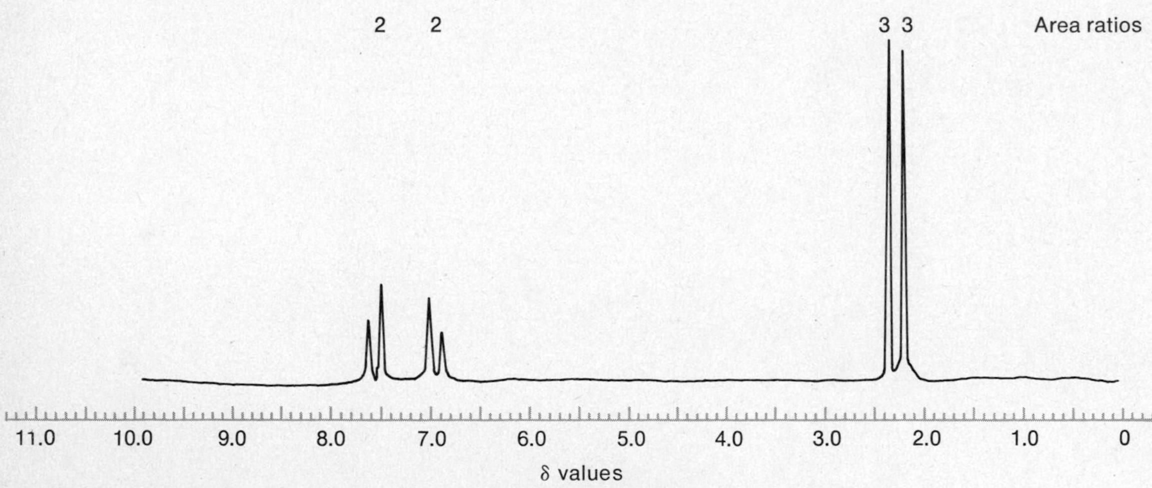

Compound B

15. In each of the following cases, a compound is being investigated which is known to be one of the four isomeric structures that is drawn. Using chemical and/or spectral methods (as instructed), indicate how a definitive structural assignment could be made in each instance.

(a) $HC{\equiv}CCH_2CH_2CH_2CHO$ $\quad$ $CH_3C{\equiv}CCH_2\overset{\overset{\large O}{\|}}{C}CH_3$ $\quad$ Chemical methods only.

$CH_3C{\equiv}CCH_2CH_2CHO$ $\quad$ [bicyclo[2.2.0]hexan-2-one drawn as two fused squares with C=O]

(b) Spectral methods only.

(c) Chemical and/or spectral methods.

16. Arrange the following compounds in the order of increasing equilibrium constant for the formation of the cyanohydrin from the carbonyl compound: *i.e.*,

$$>C{=}O + HCN \rightleftharpoons >C(OH)(CN) \qquad K = \frac{[>C(OH)(CN)]}{[>C{=}O][HCN]}$$

Explain the reasons for the order you choose.

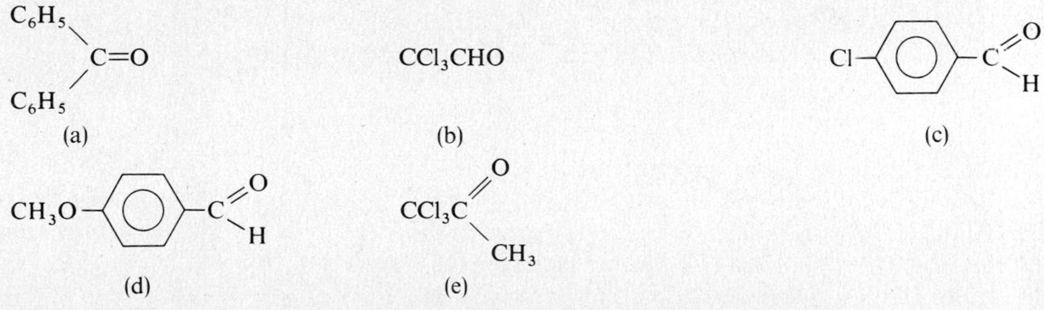

17. Tropone (see Fig. 13.25) is an aromatic compound and possesses about 30 kcal/mole of resonance energy. What other cyclic, unsaturated ketones might be expected to have aromatic properties comparable to those of tropone? Why?

18. Define the following terms as briefly and concisely as possible. Do not use a specific example in defining the term, but cite a typical example to illustrate the definition

(a) Rosenmund reduction (b) Acyl cation (c) Conjugate addition
(d) Enolate (e) Resonance stabilized anion

Chapter 14

Carboxylic acids

Everye white will have its blacke
and everye sweete its soure.

SIR CARLINE
15th-Century Ballad

Attachment of a carbon framework (R) and a hydroxyl group (OH) to a carbonyl function produces the type AC and BC carbonyl compounds that are known as *carboxylic acids*,

$$\begin{array}{l} R \\ \quad \diagdown \\ \quad\; C{=}O \\ \quad \diagup \\ HO \end{array}$$

Although aldehydes, ketones, and carboxylic acids are all members of the carbonyl family and share certain physical and chemical attributes, some of the quantitative differences between these types of compounds are sufficiently large that it is convenient to discuss carboxylic acids and their derivatives as a class by themselves.

14.1 Structure and Nomenclature of Carboxylic Acids

An adequate representation of the carboxyl group requires several resonance structures, including not only the charge-separated structure involving the polarization of the C═O bond but also the charge-separated structure involving the delocalization of the nonbonded electrons on the hydroxyl oxygen (Fig. 14.1). One of the consequences of this delocalization is the weakening of the O—H bond, increasing its tendency to dissociate to an oxygen anion and a proton. It is the considerable extent to which this dissociation takes place, in fact, that accounts for the acidity of the carboxylic acids (Fig. 14.2). In aqueous solution, carboxylic acids are sufficiently dissociated into RCO_2^- anions and H^+ cations to give solutions that taste sour, change blue litmus paper to red, and react with certain metals (*e.g.*, zinc) to release hydrogen and form the corresponding metal salt of the carboxylic acid. The greatly enhanced dissociation of carboxylic acids

Fig. 14.1. Resonance structures and orbital representations for carboxylic acids.

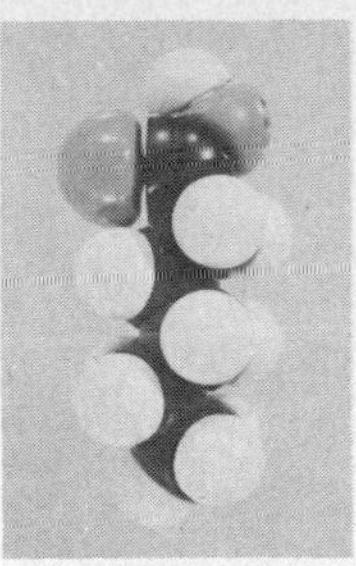
Hexanoic acid

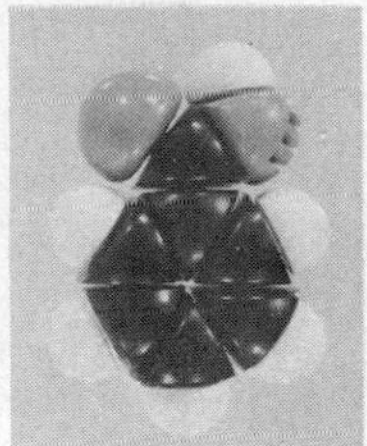
Benzoic acid

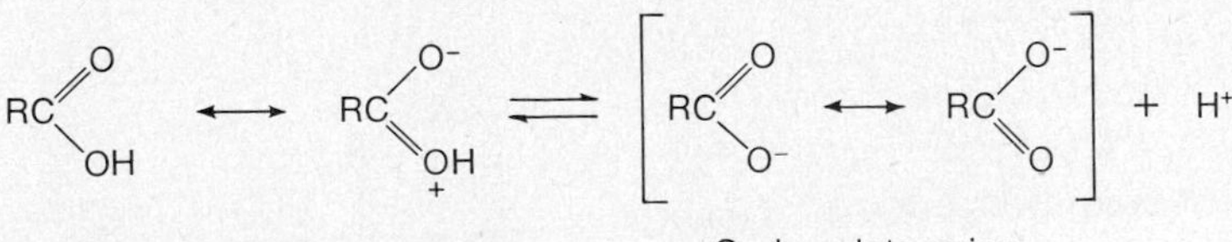

Fig. 14.2. Dissociation of carboxylic acids and resonance structures of the carboxylate anion.

(pK_a values of 3.5–5) as compared with the corresponding alcohols (pK_a values of 16–19) must be ascribed both to the labilizing effect of the carbonyl group on the attached O—H bond and to its stabilizing effect on the RCO_2^- anion (the carboxylate ion). Whereas the anion from the dissociation of an alcohol,

$$RCH_2OH \rightleftharpoons RCH_2O^- + H^+$$

is a charge-localized species, the negative charge being constrained to remain on a single oxygen atom, that from the dissociation of a carboxylic acid is a charge-delocalized species; the negative charge is equally distributed between two oxygen atoms as illustrated by the orbital representation in Fig. 14.3. The increased stability resulting from the delocalization is comparable to that present in an aromatic system. The carboxylate anion, like the benzene ring, has major contributing resonance structures that are precisely equal in energy, a circumstance particularly conducive to effective stabilization. The resonance energy of a carboxylate anion is estimated to be about 15 kcal/mole.

Carboxylic acids, some of which are called fatty acids because they were first isolated by the hydrolysis of naturally occurring fats, can be named by a variety of nomenclature systems. For all the lower molecular weight straight-chain acids, as well as for some of the higher molecular weight members, common names are frequently used (Table 14.1).

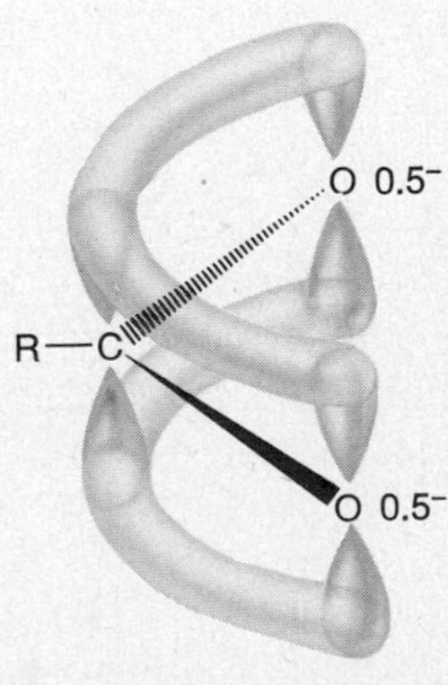

Fig. 14.3. Orbital representation of the carboxylate anion.

Table 14.1 COMMON NAMES OF STRAIGHT-CHAIN CARBOXYLIC ACIDS

No. of Carbon Atoms	Common name	Natural source
1	Formic	Ants (Latin, *formica*)
2	Acetic	Vinegar (Latin, *acetum*)
3	Propionic	Milk, butter, cheese; (Gr. *protos*, "first" + *pion*, "fat")
4	Butyric	Butter (Latin, *butyrum*)
5	Valeric	Valerian root (Latin, *valere*, "to be strong")
6	Caproic	Goats (Latin, *caper*, "goat") and other animal fats
7	Enanthic	Floral waxes
8	Caprylic	Coconut oil, palm oil, and animal fats (including goat)
9	Pelargonic	Essential oil of *Pelargonium roseum*, fusel oil
10	Capric	Coconut oil, palm oil, and animal fats (including goat)
12	Lauric	Seed fats of Lauraceae (laurel kernal oil), coconut oil
14	Myristic	Nutmeg (*Myristica fragrans*)
16	Palmitic	Palm oil and other vegetable oils
18	Stearic	Animal fats; (Gr. , *stear*, "suet")
20	Arachidic	Peanuts (*Arachis hypogea*)

If a methyl branch occurs at the penultimate position, the prefix "iso" can be used, but branching at any other position requires the designation of the identity of the substituent and its point of attachment, specified by a Greek letter. Thus, $(CH_3)_2CHCO_2H$ is isobutyric acid, $CH_3CH_2\underset{\displaystyle CH_3}{\underset{|}{C}}HCO_2H$ is α-methylbutyric acid, $ClCH_2CH_2CO_2H$ is β-chloropropionic acid, etc. Similarly, double methyl branching at the penultimate position allows the prefix "neo" to be used; *e.g.*, $(CH_3)_3CCH_2CO_2H$ is named neocaproic acid.

Another nomenclature that is sometimes useful is called the "acetic acid system"; it specifies compounds as derivatives of acetic acid. For example, the compound $CH_3-\overset{\displaystyle CH_3}{\underset{\displaystyle CH_3}{\overset{|}{\underset{|}{C}}}}-CO_2H$ is trimethylacetic acid.

Inevitably, carboxylic acids occur that cannot be named by these conventions, and recourse must be made to the more generally applicable IUPAC system, which involves the following rules:

1. The longest continuous chain *containing the carboxyl group* (CO_2H) is selected as the basis for designating the parent compound, and the actual compound is then named as a derivative of the parent compound. The parent compound is specified by dropping the final "e" from the alkane designation and adding the suffix "oic." For instance, CH_3CO_2H is named ethanoic acid.
2. The carbon of the CO_2H group is assigned number 1, and the positions and identities of attached groups are specified accordingly. For example, $CH_3CH_2\underset{\displaystyle Cl}{\underset{|}{C}}HCO_2H$ is 2-chlorobutanoic acid. The carboxyl function takes

precedence over other carbonyl functions and, therefore, is designated as the principal function in polycarbonyl compounds. Thus, the compound $CH_3CH(CHO)C(=O)CO_2H$ is 3-formyl-2-oxobutanoic acid.

3. If double or triple bonds are present, the parent compound is chosen to include the carboxyl group and the maximum number of double and/or triple bonds. The position of the multiple bond is indicated by the lower numbered of the carbons between which the multiple bond occurs. The final "e" from the alkene or alkyne designation is dropped, and the suffix "oic" is added. Thus, $H_2C{=}C(CH_2CH_3)CO_2H$ is named 2-ethyl-2-propenoic acid, and $HC{\equiv}C{-}CH(Cl)CO_2H$ is 2-chloro-3-butynoic acid. If other functions in addition to the carbonyl group are present, the inclusion of unsaturation in the main chain takes precedence over the substituents in determining the specification and numbering of the main chain. For example,

$$CH_3CH_2C(=O)CH(CH{=}CH_2)CO_2H$$

is 2-(1-oxopropyl)-3-butenoic acid.

4. If two CO_2H groups are present, both can usually be included in the parent compound and the suffix "dioic acid" is applied. Thus,

$$HO_2C(CH_2)_6CO_2H$$

is octanedioic acid.

5. Aromatic acids are named by attaching the suffix "oic" to an appropriate prefix derived from the name of the aromatic hydrocarbon.

Thus, $C_6H_5{-}CO_2H$ is benzoic acid,

$C_{10}H_7{-}CO_2H$ (naphthalene ring numbered 1–8, CO_2H at position 2) is 2-naphthoic acid,

$C_{14}H_9{-}CO_2H$ (phenanthrene ring numbered 1–10, CO_2H at position 2) is 2-phenanthroic acid, and

$Br{-}C_6H_4{-}CO_2H$ is 4-bromobenzoic acid (or, alternatively, *p*-bromobenzoic acid).

6. The CO_2H group is designated as a *substituent group* in either of two ways:
   (a) "Carboxy" designation: If it is not possible to include the carboxyl group as part of the parent compound, it can be designated as a "carboxy" function. For example,

$$HO_2C(CH_2)_2\underset{\displaystyle CO_2H}{\underset{|}{C}}H(CH_2)_2CO_2H$$

   is 4-carboxyheptanedioic acid. The carboxyl group stands near the top of the list of functional group priority (see Table 18.5), however, so its designation as a substituent is less frequently necessary than with most of the other functional groups.
   (b) "Carboxylic acid" designation: Cycloalkanes containing directly attached carboxyl groups are named as cycloalkanecarboxylic acids, following the convention that has been described for the analogous aldehydes (see p. 331). For example,

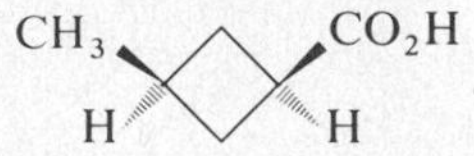

   CH_3 CO_2H
   H H

   is *cis*-3-methylcyclobutanecarboxylic acid. Note that with the carboxylic acid designation, the numbering commences on the carbon to which the carboxyl group is attached, not on the carboxyl group itself.

PROBLEM 14-1

Draw and name all the $C_5H_{10}O_2$ carboxylic acids. Draw and name all the $C_9H_{10}O_2$ benzoic acids. Draw and name all the $C_4H_6O_3$ carboxylic acids.

14.2 Physical Properties of Carboxylic Acids

Carboxylic acids contain an acidic hydrogen and are capable of acting as donors in hydrogen bond formation. The oxygen atoms of a carboxylic group are Lewis bases, so they can also act as hydrogen bond acceptors. As a result, carboxylic acids form quite stable dimers (Fig. 14.4), and those from the lower molecular weight acids, such as acetic acid, are stable enough to exist even in the vapor state. Carboxylic acids have higher boiling points than most other compounds of comparable molecular weight that lack the ability to form stable dimers. For example, acetic acid (mol wt 60) boils at 118°; 1-propanol (mol wt 60), which is less acidic and, therefore, less effective as a hydrogen bond donor, boils at 97°; and ethyl methyl ether (mol wt 60), which has no acidic hydrogens, boils at 8°.

O---H—O
R—C C—R
O—H---O

Fig. 14.4. Hydrogen bond dimer of carboxylic acids.

The water solubility of carboxylic acids is greater than that of the corresponding alcohols. This has been ascribed to partial dissociation of the acid to a carboxylate ion and a proton, both of these species being more readily solvated by water than the undissociated carboxylic acid. The water solubility might also be ascribed to the additional oxygen atom that is present in a carboxylic acid and to the consequent increase in its ability to act as an acceptor species in hydrogen bond formation (Fig. 14.5).

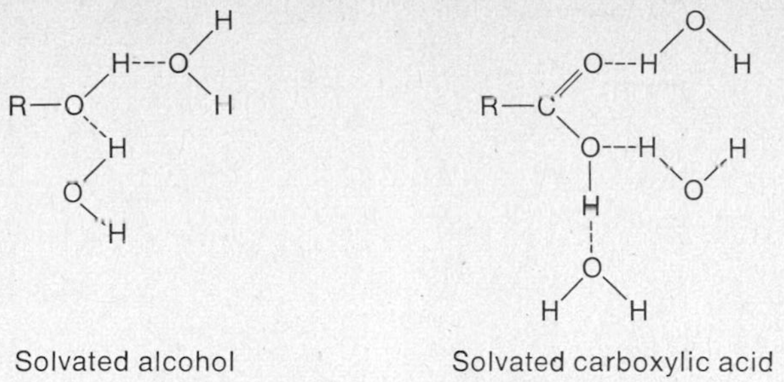

Solvated alcohol

Solvated carboxylic acid

Fig. 14.5. Solvation of alcohols and carboxylic acids.

The lower molecular weight carboxylic acids, in addition to giving aqueous solutions that taste sour, also excite another of man's senses; they have an acrid odor, affecting the olfactory nerves. Butyric, caproic, and capric acids are particularly disagreeable in this respect. The higher molecular weight acids are much less volatile and, consequently, are odorless.

14.3 *Spectral Properties of Carboxylic Acids*

Carboxylic acids show an O—H stretching deformation band in the 3500–2500 cm^{-1} region, which is generally very broad because of extensive intermolecular hydrogen bonding. The carbonyl stretching deformation band appears near 1710 cm^{-1} for saturated acids (type AC) and near 1680 cm^{-1} for α,β-unsaturated and aryl carboxylic acids (type BC and B′C) (see Table 14.2 and Fig. 14.6).

Table 14.2 INFRARED STRETCHING DEFORMATION BANDS OF CARBOXYLIC ACIDS

Compound Type	Absorption Position
$R-C(=O)OH$	3500–2500 cm^{-1} (O—H stretching deformation) Approx. 1710 cm^{-1} (C=O stretching deformation)
$R_2C=CRC(=O)OH$	3500–2500 cm^{-1} (O—H) Approx. 1680 cm^{-1} (C=O)
$Ar-C(=O)OH$	3500–2500 cm^{-1} (O—H) Approx. 1690 cm^{-1} (C=O)

We can also distinguish between a type AC and a type BC carboxylic acid by means of ultraviolet spectral measurements. Saturated carboxylic acids, such as butyric acid ($CH_3CH_2CH_2CO_2H$), are essentially transparent in the accessible ultraviolet, whereas α,β-unsaturated acids, such as crotonic acid ($CH_3CH{=}CHCO_2H$), possess absorption maxima near 205 nm ($\varepsilon \sim 14{,}000$).

The nmr spectra of carboxylic acids show resonances at very low field (δ 10–15) for the hydrogen of the carboxyl group (see Fig. 14.6) and resonances at about δ 2 for the hydrogens attached to the α-carbon.

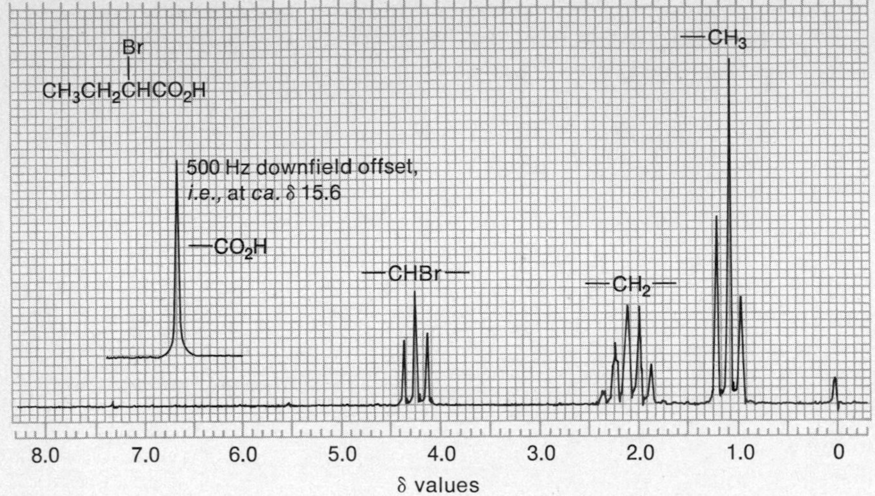

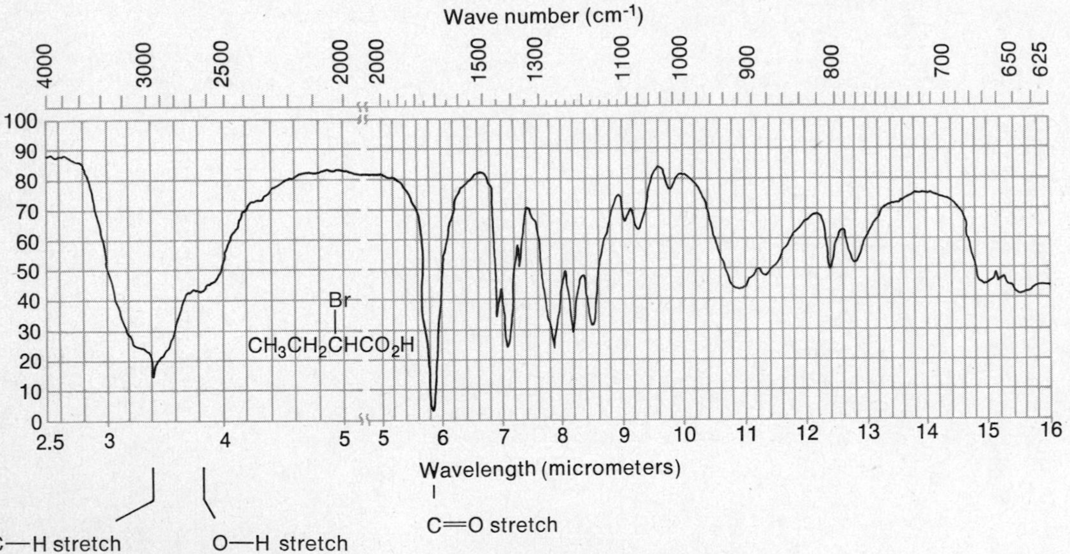

Fig. 14.6. Nuclear magnetic resonance and infrared spectra of 2-bromobutyric acid.

PROBLEM 14-2

Compounds A, B, and C are carboxylic acids all having the molecular formula $C_7H_{10}O_2$. Compounds A and B both have five vinyl hydrogens (as indicated by the nmr spectra), and both show an infrared band near 900 cm^{-1}. Compound C has only four vinyl hydrogens and lacks the infrared band at 900 cm^{-1}. Compound A is transparent in the ultraviolet region above 200 nm, compound B absorbs strongly near 200 nm, and compound C absorbs strongly at a wavelength well above 200 nm. Suggest structures that are compatible with these data.

14.4 *Synthesis of Carboxylic Acids*

Since acids represent the end of the oxidation line, short of going all the way to carbon dioxide, they are generally prepared by starting with less highly oxidized compounds. Numerous examples of such procedures have been given in previous sections (see Figs. 8.19–8.22 and 10.18 and Section 13.7), in which alkenes, alkynes, alcohols, aldehydes, ketones, and alkylbenzenes are oxidized to carboxylic acids. In further illustration of this general preparative method, the representative examples in Fig. 14.7 are cited.

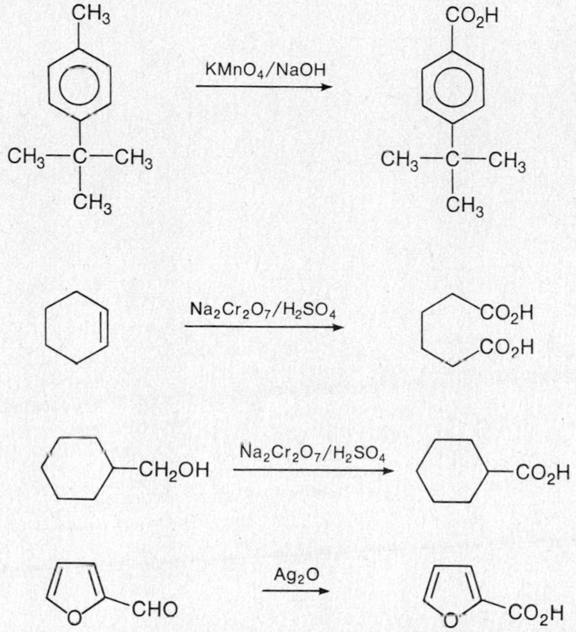

Fig. 14.7. Preparation of carboxylic acids by oxidation.

Alkyl halides are useful intermediates for the preparation of carboxylic acids. In one method, the alkyl halide is converted to an organometallic compound, usually the Grignard reagent, which is then treated with carbon dioxide to yield the salt of the carboxylic acid. The action of mineral acid on the salt produces the carboxylic acid itself (Fig. 14.8). In an alternative method, the alkyl halide is treated with sodium cyanide (see Fig. 9.5) to yield the nitrile that is then hydrolyzed in aqueous mineral acid to the corresponding carboxylic acid. Of these two procedures, the first is less restricted in its application; the cyanide displacement reaction is an S_N2 process, which fails with alkyl halides that are highly branched at the α or β positions and which also fails with vinyl and aryl halides (see p. 235). The organometallic route, however, is more general in its application, allowing, for example, bromobenzene to be converted to benzoic acid (Fig. 14.8).

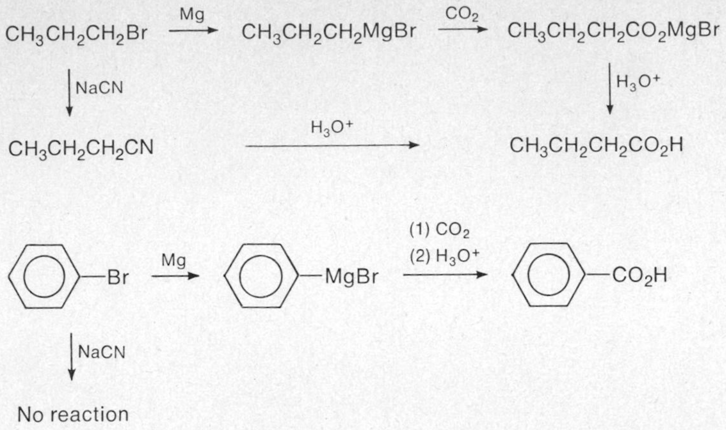

$$CH_3CH_2CH_2Br \xrightarrow{Mg} CH_3CH_2CH_2MgBr \xrightarrow{CO_2} CH_3CH_2CH_2CO_2MgBr$$

$$CH_3CH_2CH_2Br \xrightarrow{NaCN} CH_3CH_2CH_2CN \xrightarrow{H_3O^+} CH_3CH_2CH_2CO_2H$$

$$CH_3CH_2CH_2CO_2MgBr \xrightarrow{H_3O^+} CH_3CH_2CH_2CO_2H$$

$$C_6H_5\text{—Br} \xrightarrow{Mg} C_6H_5\text{—MgBr} \xrightarrow[(2)\ H_3O^+]{(1)\ CO_2} C_6H_5\text{—CO}_2H$$

$$C_6H_5\text{—Br} \xrightarrow{NaCN} \text{No reaction}$$

Fig. 14.8. Synthesis of aliphatic and aromatic carboxylic acids from halogen compounds.

PROBLEM 14-3

Suggest methods for effecting the conversion of (a) 2-methylpropene to isovaleric acid, and (b) benzene to *p*-methylbenzoic acid (*p*-toluic acid).

14.5 Reactions of Carboxylic Acids

The most striking chemical feature of carboxylic acids is their ability to dissociate in aqueous solution to carboxylate anions and protons (Fig. 14.2). The equilibrium constant for this dissociation is expressed as

$$K_a = \frac{[RCO_2^-][H^+]}{[RCO_2H]}$$

for which K_a can be given as a pK_a value (see Section 7.1). In an aqueous solution containing equal concentrations of the free acid, RCO_2H, and its anion, RCO_2^-, the hydrogen ion concentration of the solution will be equal to the K_a of the acid (*i.e.*, pH $= pK_a$). Thus, by measuring the pH of a solution containing equal concentrations of, for instance, benzoic acid and sodium benzoate (assumed to be completely dissociated in aqueous solution), we can ascertain the pK_a of benzoic acid. By this and other methods, the pK_a values of a host of carboxylic acids have been measured and found to be generally in the vicinity of 3.5–5 (see Table 7.1). Formic acid, for instance, has a pK_a of 3.77, acetic acid a pK_a of 4.76, and benzoic acid a pK_a of 4.20. Carboxylic acids are thus considerably weaker than mineral acids (pK_a values 2 to -10) but considerably stronger acids than phenols (pK_a values approximately 10) and very much stronger acids than water and alcohols (pK_a values about 16–19).

In solutions of pH above 6, carboxylic acids are almost completely dissociated to the carboxylate ions, which, being more strongly hydrogen bonded to the solvent than the free carboxylic acid, are more water soluble. Carboxylic acids, therefore, can be dissolved even in very weak aqueous base. Extraction of an ether solution of benzoic acid with aqueous sodium bicarbonate, for instance, results in almost complete passage of the benzoic acid from the ether phase (present as $C_6H_5CO_2H$) into the aqueous phase (present as $C_6H_5CO_2^-$). A very effective means for separating organic acids from nonacidic materials is thereby afforded by the simple device of base extraction. By the proper adjustment of the basicity of the extracting agent, materials of *different* aciditics may also be separated. Thus, an ether solution of phenol, benzoic acid, and benzyl alcohol can be cleanly separated by extracting first with aqueous sodium bicarbonate and then with aqueous sodium hydroxide. The bicarbonate solution has a pH of about 8.4; at this basicity, benzoic acid (pK_a of 4.20) is more than 99.9% in the benzoate form, but phenol (pK_a of 10) is only about 1% in the phenoxide form. The benzoate anion passes into the aqueous phase, therefore, leaving the phenol in the ether phase. If a solution of 0.1 *N* sodium hydroxide (pH of about 13) is now used to extract the ether, the phenol passes into the aqueous phase as the phenoxide ion (see p. 267). At this point, then, the experimenter has three solutions: one containing sodium benzoate in water, another containing sodium phenoxide in water, and a third containing benzyl alcohol in ether. The benzoic acid and phenol can be recovered from the respective aqueous solutions by acidification (which converts the anions to the corresponding parent compounds) followed by ether extraction. Evaporation of the ether from each of the three solutions containing benzoic acid, phenol, and benzyl alcohol, respectively, yields the separated compounds in essentially pure form.

We can also use the solubility of carboxylic acids in sodium bicarbonate and phenols in sodium hydroxide solutions to quickly characterize these classes of compounds. We should be careful, however, for many of the higher molecular weight acids and phenols form anions that are insoluble in water; in these compounds, the water-soluble portion of the molecule (*i.e.*, the carboxylate and phenoxide anionic groups) are unable to carry the large water-insoluble portion of the molecule into the aqueous phase. We can sometimes overcome the problem by using an aqueous-alcoholic solution of base rather than a completely aqueous solution, although some acids and phenols are insoluble even in this mixture.

Carboxylic acids, standing at the end of the oxidation sequence, can be reduced successively to aldehydes, alcohols, and alkanes. Reduction to aldehydes is usually effected indirectly by way of the acid chloride (see Fig. 13.6). Reduction to the alcohol, however, can be carried out by direct reduction of the carboxylic acid with lithium aluminum hydride as the reducing agent. Reduction to alkanes is usually carried out in two steps by way of the aldehyde or alcohol. The hydrogen of the carboxyl group is an "active hydrogen," so the carboxylic acid and lithium aluminum hydride first form a lithium aluminate salt of the acid (Fig. 14.9). Subsequently, a nucleophilic addition of hydride (*i.e.*, H^-) from lithium aluminum hydride to the carbonyl function of $RCO_2AlH_3^-$ takes place (see the hydride reduction of aldehydes and ketones in Section 13.7) to form the lithium aluminum salt of a 1,1-diol (Fig. 14.9). Under the conditions of the reaction, however, this salt is unstable, decomposing to an aldehyde, which then undergoes further reduction to yield the lithium aluminate salt of the alcohol. Hydrolysis in the workup procedure converts this to the alcohol itself.

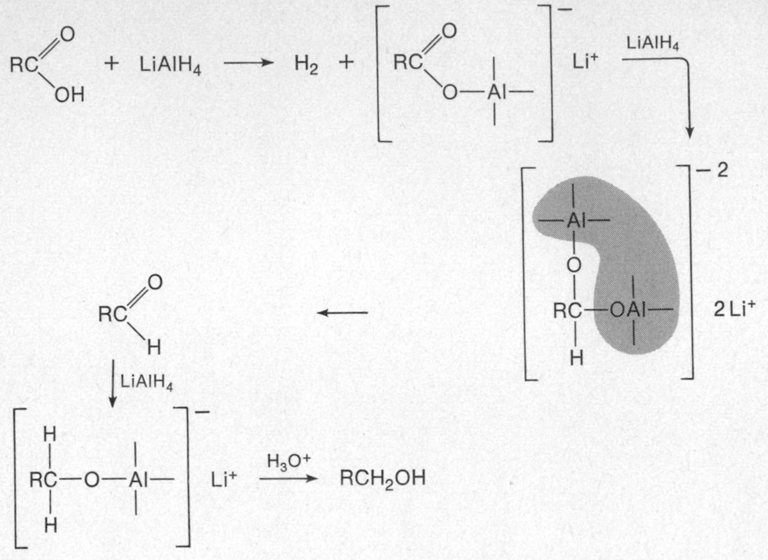

$$RCOOH + LiAlH_4 \longrightarrow H_2 + [RC(=O)-O-Al]^- Li^+ \xrightarrow{LiAlH_4} [RCH(-OAl)-O-Al]^{-2}\ 2Li^+ \longrightarrow RCHO \xrightarrow{LiAlH_4} [RCH_2-O-Al]^- Li^+ \xrightarrow{H_3O^+} RCH_2OH$$

Fig. 14.9. Reduction of carboxylic acids to alcohols with lithium aluminum hydride.

Aromatic carboxylic acids undergo electrophilic substitution reactions to form *m*-substituted products. For example, benzoic acid yields *m*-bromobenzoic acid when treated with bromine in the presence of ferric bromide catalyst (Fig. 14.10). It does not, however, react with alkyl halides and chlorides under Friedel–Crafts conditions; the carboxyl group deactivates the ring toward electrophilic attack (see p. 214), and R^+ and RCO^+ electrophiles are not sufficiently reactive to overcome this disadvantage.

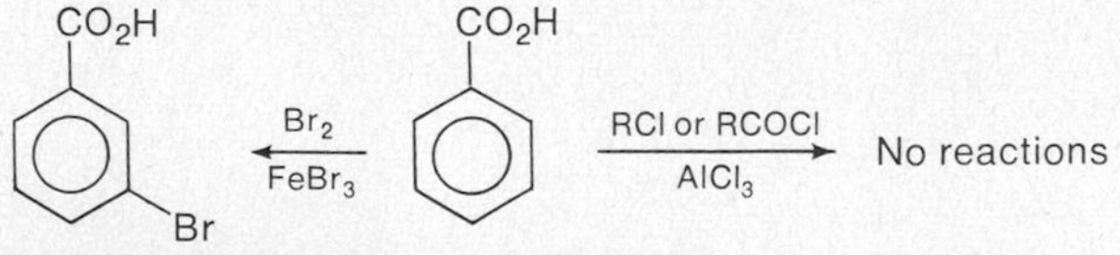

$$m\text{-}BrC_6H_4CO_2H \xleftarrow[FeBr_3]{Br_2} C_6H_5CO_2H \xrightarrow[AlCl_3]{RCl \text{ or } RCOCl} \text{No reactions}$$

Fig. 14.10. Electrophilic substitution of benzoic acid.

Carboxylic acids undergo a variety of reactions involving conversion to carboxylic acid derivatives, which are discussed in Chapter 15.

14.6 Polycarboxylic Acids

The lowest molecular weight dicarboxylic acid has the common name "oxalic acid" and the IUPAC name "ethanedioic acid." It is a high-melting solid (mp 189°) that is less soluble in water (9 g/100 ml) than we would anticipate from its molecular weight. It is easily oxidized, and sodium oxalate has been used as a primary reference for the standardization of oxidizing agents such as potassium permanganate. When heated rapidly to above 150°, it decomposes with the formation of CO_2, CO, and H_2O (Fig. 14.11).

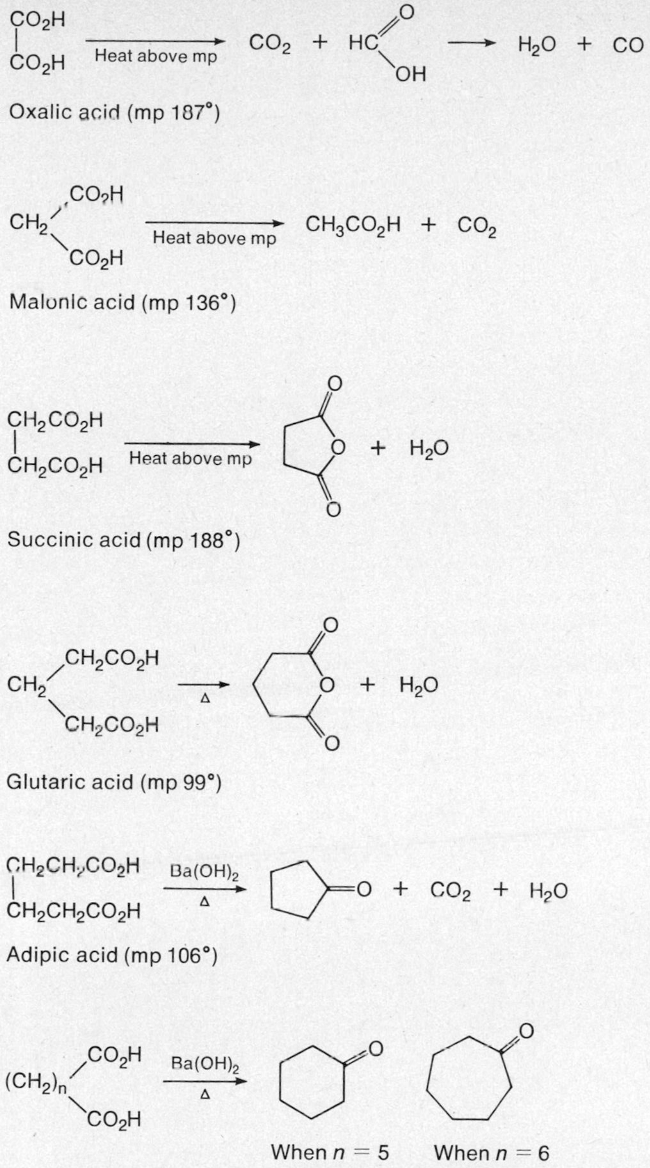

$$\underset{\text{Oxalic acid (mp 187°)}}{\begin{matrix}CO_2H\\|\\CO_2H\end{matrix}} \xrightarrow[\text{Heat above mp}]{} CO_2 + HC(=O)OH \longrightarrow H_2O + CO$$

$$\underset{\text{Malonic acid (mp 136°)}}{CH_2(CO_2H)_2} \xrightarrow[\text{Heat above mp}]{} CH_3CO_2H + CO_2$$

$$\underset{\text{Succinic acid (mp 188°)}}{\begin{matrix}CH_2CO_2H\\|\\CH_2CO_2H\end{matrix}} \xrightarrow[\text{Heat above mp}]{} \text{succinic anhydride} + H_2O$$

$$\underset{\text{Glutaric acid (mp 99°)}}{CH_2(CH_2CO_2H)_2} \xrightarrow[\Delta]{} \text{glutaric anhydride} + H_2O$$

$$\underset{\text{Adipic acid (mp 106°)}}{\begin{matrix}CH_2CH_2CO_2H\\|\\CH_2CH_2CO_2H\end{matrix}} \xrightarrow[\Delta]{Ba(OH)_2} \text{cyclopentanone} + CO_2 + H_2O$$

$$(CH_2)_n(CO_2H)_2 \xrightarrow[\Delta]{Ba(OH)_2} \underset{\text{When } n = 5}{\text{cyclohexanone}} \quad \underset{\text{When } n = 6}{\text{cycloheptanone}}$$

Fig. 14.11. Thermal reactions of alkanedioic acids.

The next higher homolog, malonic acid (propanedioic acid) melts lower (mp 136°) than oxalic acid and is considerably more water soluble (74 g/100 ml). When heated above its melting point, it decomposes to acetic acid and carbon dioxide (Fig. 14.11), a reaction that is characteristic of all molecules containing a carbonyl group β to the carboxyl group *i.e.*,

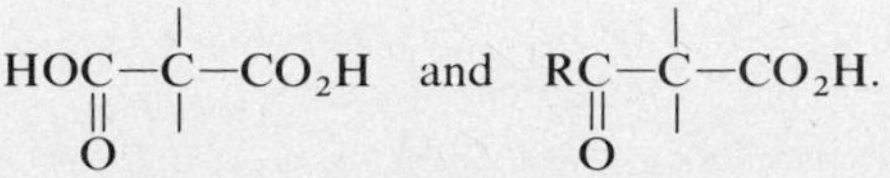

$$HO\underset{\underset{O}{\|}}{C}-\overset{|}{\underset{|}{C}}-CO_2H \quad \text{and} \quad R\underset{\underset{O}{\|}}{C}-\overset{|}{\underset{|}{C}}-CO_2H.$$

Succinic acid (butanedioic acid) and glutaric acid (pentanedioic acid) are the first members of the series of dicarboxylic acids capable of forming stable cyclic anhydrides (Fig. 14.11). Heating induces the expulsion of water and the formation of the anhydride, a process that is hastened by the addition of reagents that can react with water and prevent the reverse reaction; thionyl chloride, for instance, serves this purpose admirably ($SOCl_2 + H_2O \rightarrow SO_2 + 2HCl$).

Adipic acid (hexanedioic acid), like glutaric and succinic acids, can form a cyclic anhydride. Another intramolecular reaction is also available to adipic acid, however. When it is heated in the presence of barium hydroxide, both carbon dioxide and water are expelled, and cyclopentanone is produced. In similar fashion, pimelic acid (heptanedioic acid) and suberic acid (octanedioic acid) yield cyclohexanone and cycloheptanone, respectively, when strongly heated in the presence of barium hydroxide (Fig. 14.11). With dicarboxylic acids having eight or more carbon atoms in the chain, the yield of cyclization products falls almost to zero, and polymeric anhydrides are formed instead. We can often deduce the type of framework present in a dicarboxylic acid of unknown structure by noting whether decarboxylation (with formation of an acid), dehydration (with formation of an anhydride), decarboxylation and dehydration (with formation of a ketone), or polymeric anhydride formation takes place.

Three benzenedicarboxylic acids can exist. The *ortho*-isomer (1,2-benzenedicarboxylic acid) has the common name *phthalic* acid, the *meta*-isomer (1,3-benzenedicarboxylic acid) the common name *isophthalic acid*, and the *para*-isomer (1,4-benzenedicarboxylic acid) the common name *terephthalic acid.* Only the first of these—*viz.*, phthalic acid—can form a cyclic anhydride, so we easily differentiate it from the other two by this means (Fig. 14.12).

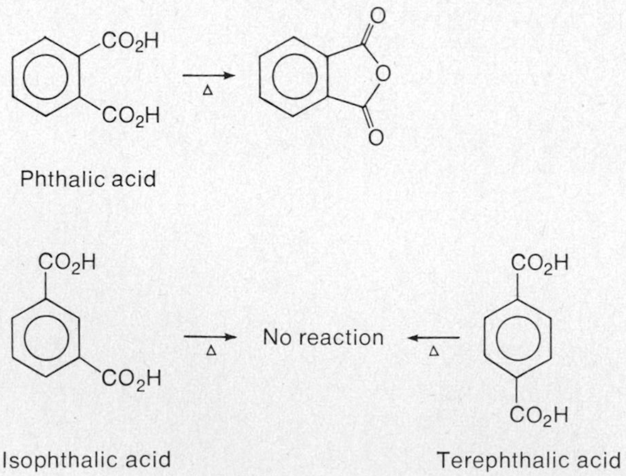

CO_2H
CO_2H
Δ
O
O
O
Phthalic acid

CO_2H
CO_2H
Δ
No reaction
Δ
CO_2H
CO_2H
Isophthalic acid
Terephthalic acid

Fig. 14.12. Benzenedicarboxylic acids.

Important clues to the type of framework present in compounds of unknown structure are often provided by conversion of the compound to a benzene polycarboxylic acid (by side chain oxidation) followed by anhydride formation. For example, we can choose between the alternative structures (a) and (b) in Fig. 14.13 for the naturally occurring product curvularin, a mold metabolite, by

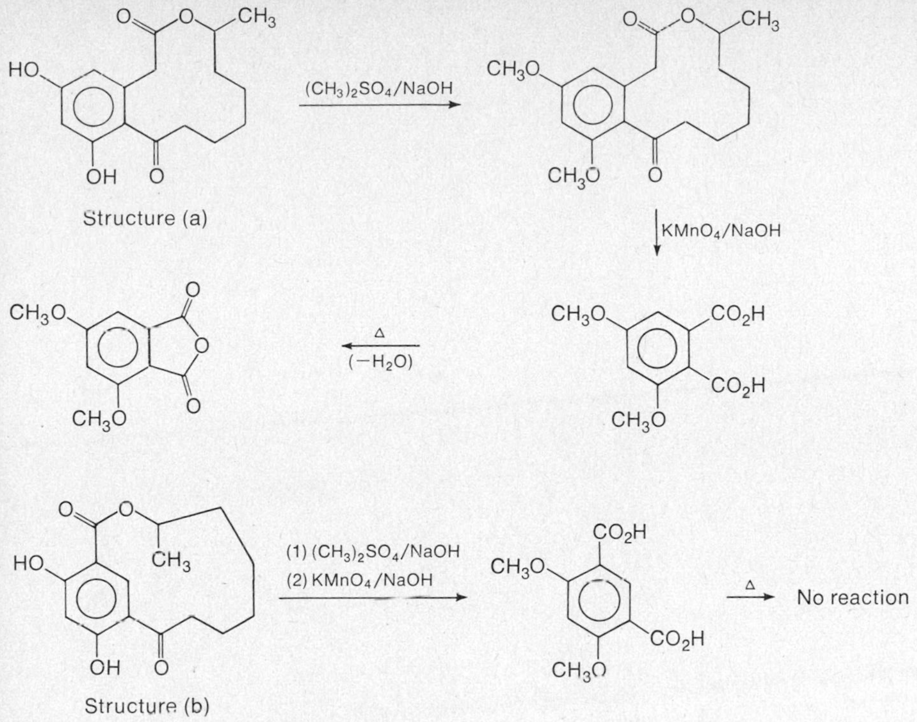

Fig. 14.13. Structure proof of curvularin.

converting the compound to the diether [with dimethyl sulfate and aqueous sodium hydroxide (see Fig. 11.6)] and then oxidizing the diether with potassium permanganate. The product from structure (a) forms a cyclic anhydride, whereas that from structure (b) does not. The naturally occurring material yields a dicarboxylic acid that forms a cyclic anhydride, thereby indicating that curvularin possesses structure (a).

PROBLEM 14-4

A $C_{10}H_{10}O_3$ compound is soluble in dilute aqueous sodium bicarbonate, contains a chiral carbon, and can be oxidized to a $C_8H_6O_4$ dicarboxylic acid that forms a cyclic anhydride upon heating. Suggest a structure that is compatible with these data. If more than one structure is possible, indicate how nmr spectral data could be used to distinguish among them.

14.7 Carboxylic Acids of Particular Interest and Utility

Formic acid, the lowest molecular weight member of the carboxylic acid series, was discovered in the seventeenth century in the red ant (*Formica rufa*), from which it derives its name. It is fairly widespread in the plant kingdom and has been shown to be present, for instance, in the stinging nettle, in pine needles, and in certain fruits. It is manufactured in quantity by the reaction of carbon monoxide and aqueous sodium hydroxide under 200–250 psi pressure at 160–200°. The initial product from this reaction is sodium formate from which formic acid can be obtained by acidification with sulfuric acid. The major commercial application is in the textile industry, where it is used in dyeing and finishing processes.

Acetic acid, the next homolog, is commercially the most important of the carboxylic acids. In the dilute solution known as vinegar acetic acid has been known for many centuries, probably before recorded history (see p. 3). It is one of the first organic compounds to be isolated in pure form. Because many microorganisms have the ability to convert various organic compounds to acetic acid, the compound occurs widely in nature. Sour milk and sour wine, for instance, are but two of the better-known examples of the results of this biological activity called fermentation. One of the several industrial processes for acetic acid production, for instance, employs the oxidation of ethanol by microbial action. Several other industrial syntheses of acetic acid are also used, including the air oxidation of acetaldehyde (prepared, in turn, by hydration of acetylene—see Fig. 8.7), the air oxidation of ethanol, the air oxidation of *n*-butane,

$$CH_3CH_2CH_2CH_3 + O_2 \longrightarrow CH_3CH_2COCH_3 \longrightarrow 2\,CH_3CO_2H,$$

and the "carbonylation" of methanol in the presence of a cobalt catalyst,

$$CO + CH_3OH \longrightarrow CH_3CO_2H \text{ at 750 psi and 210°.}$$

Most of the commercially produced acetic acid is converted either to acetic anhydride or vinyl acetate.

Benzoic acid is one of the very important compounds in the history of organic chemistry as a result of the efforts of Liebig and Wöhler to establish its composition in 1832. It is commercially prepared by the oxidation of toluene, by the hydrolysis of trichloromethylbenzene,

$$C_6H_5CCl_3 + 2\,H_2O \longrightarrow C_6H_5CO_2H + 3\,HCl,$$

and by the partial decarboxylation of phthalic acid. Its major commercial use is in the form of its sodium salt, an effective food preservative.

Known as salicylic acid, *o*-hydroxybenzoic acid is produced commercially by the action of carbon dioxide on sodium phenoxide (the *Kolbe reaction*):

$$C_6H_5ONa \xrightarrow{CO_2} o\text{-}HOC_6H_4CO_2Na \xrightarrow{H_3O^+} o\text{-}HOC_6H_4CO_2H$$

Salicylic acid

Esterification (see Section 15.3) of the carboxyl group of salicylic acid with methanol yields the compound known as oil of wintergreen, used medicinally in liniments. Esterification of the hydroxyl group with acetic acid yields the compound known as aspirin (acetylsalicylic acid). Aspirin is probably the most widely used drug in the world. Americans alone swallow 42 tons of the compound every day, seeking psychic advantage of the ability of aspirin to increase the threshold of pain while not interfering with the level of mental acuity.

Aspirin

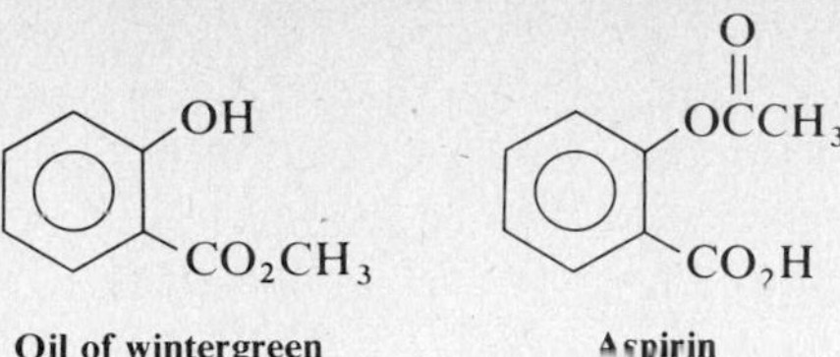

Oil of wintergreen **Aspirin**

14.8 *Study Guide*

I. Structure of carboxylic acids
   A. Bonding in RCO_2H
   B. Bonding in RCO_2^-; resonance stabilization
II. Nomenclature of carboxylic acids
   A. Common names for carboxylic acids
   B. "Acetic acid" system for carboxylic acids
   C. IUPAC system (including "carboxylic acid" system)
III. Physical properties of carboxylic acids
   A. Boiling points—higher than corresponding alcohols
   B. Water solubility—greater than corresponding alcohols; give acidic solutions
   C. Soluble in aqueous bicarbonate
IV. Spectral properties of carboxylic acids
   A. Ultraviolet/visible—only α,β-unsaturated acids absorb above 200 nm
   B. Infrared
      1. O—H stretching at 3500–2500 cm^{-1}
      2. C=O stretching at 1710–1680 cm^{-1}
   C. Nuclear magnetic resonance
      1. O—H resonance near δ 10–15
      2. Protons proximate to CO_2H shifted downfield
V. Synthesis of carboxylic acids
   A. By oxidation
      1. $ArR \rightarrow RCO_2H$ (oxidation of aryl-substituted hydrocarbons)
      2. $RCH_2OH \rightarrow RCO_2H$ (oxidation of primary alcohols)
      3. $RCHO \rightarrow RCO_2H$ (oxidation of aldehydes)
      4. $RCH{=}CHR \rightarrow RCO_2H$ (oxidation of alkenes)
      5. $RC{\equiv}CR \rightarrow RCO_2H$ (oxidation of alkynes)
   B. From halogen compounds
      1. Carbonation of Grignard reagents
         $RX\,(\text{or } ArX) \rightarrow RMgX\,(\text{or } ArMgX) \rightarrow RCO_2MgX\,(\text{or } ArCO_2MgX)$
      2. Cyanation of alkyl halides followed by hydrolysis
         $RX + CN^- \rightarrow RCN \rightarrow RCO_2H$
VI. Reactions of carboxylic acids
   A. Dissociation: $RCO_2H \rightarrow RCO_2^- + H^+$ (solubility in aqueous base)
   B. Reduction
      1. $RCO_2H \rightarrow RCHO$ (indirect methods required)
      2. $RCO_2H \rightarrow RCH_2OH$ (directly with $LiAlH_4$)
      3. $RCO_2H \rightarrow RCH_3$ (indirect methods required)

VII. Polycarboxylic Acids
   A. Oxalic acid—undergoes facile oxidation
   B. Malonic acid—loses CO_2 on heating to produce monocarboxylic acid
   C. Succinic and glutaric acids—form cyclic anhydrides on heating
   D. Adipic and pimelic acids—form cyclic ketones and CO_2 on heating with barium hydroxide

PROBLEMS

1. Using any appropriate nomenclature, name the following compounds:
   (a) HCO_2H (b) $(CH_3)_2CHCO_2H$ (c) $(CH_3)_3CCO_2H$
   (d) $CH_3CH_2CH(CH_3)CO_2H$ (e) $C_6H_5CO_2H$ (f) $CH_3CH(C_3H_5)CH(CH_3)CO_2H$ (cyclopropyl on C-3)

2. Draw the structure of the compound corresponding to each of the following names:
   (a) Isovaleric acid (b) Caproic acid (c) Triphenylacetic acid
   (d) 4,4-Dimethylpentanoic acid (e) 4-Methylbenzoic acid (f) 3-Phenylpropionic acid

3. Using any appropriate system of nomenclature, name the following carboxylic acids:
   (a) $CH_3CH_2CH(Cl)CH_2CH_2CO_2H$ (b) $(CH_3)_2C{=}CHCO_2H$
   (c) $HO_2CCH{=}C(CH_2CO_2H)_2$ (d) $HO{-}C_6H_4{-}CO_2H$ (para)
   (e) naphthalene with CO_2H at C-1 and CH_3 at C-6
   (f) $CH_3CH(CH_3)CH_2CH(CH_3)CH(CH_2CH(CH_3)_2)CH_2CH(CO_2H)CH_3$
   (g) cyclobutane with CO_2H, H, CO_2H, H (stereo wedges)
   (h) cyclopentadienyl$-CO_2H$
   (i) $H_2C{=}CHC{\equiv}C{-}CH{=}CH{-}CO_2H$ (H and CO_2H trans)
   (j) decalin with CO_2H and OH at ring-fusion carbons

4. Draw the structure for each of the following carboxylic acids:
(a) Ethylmethylphenylacetic acid
(b) α-Methylpelargonic acid
(c) γ-Phenylcaproic acid
(d) β-Phenylglutaric acid
(e) Naphthalene-1,4-dicarboxylic acid
(f) 5-Chloromethyl-3,7-dimethyloctanoic acid
(g) 1,1-Cycloheptanedicarboxylic acid
(h) 4-(4-Hydroxy-1-cyclohexenyl)-5-hexenoic acid
(i) Bicyclo[1.1.0]butane-2,2,4,4-tetracarboxylic acid
(j) (2*Z*,5*E*)-7-Methyl-2,5,8-nonatrienoic acid

5. Whereas a pair of arrows is used to represent the equilibrium between two forms of a carboxylic acid, *i.e.*,

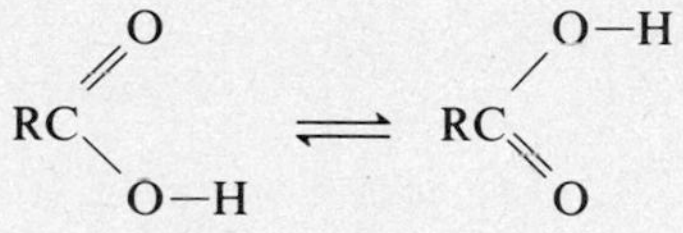

$$RC(=O)-O-H \rightleftharpoons RC(-O-H)=O$$

the double-headed arrow is used to depict the situation in the corresponding anion, *i.e.*,

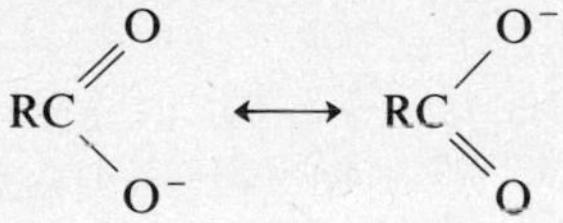

$$RC(=O)-O^- \longleftrightarrow RC(-O^-)=O$$

Explain the reasons for this.

6. For each of the following reactions, draw the structure of the major product:

(a) 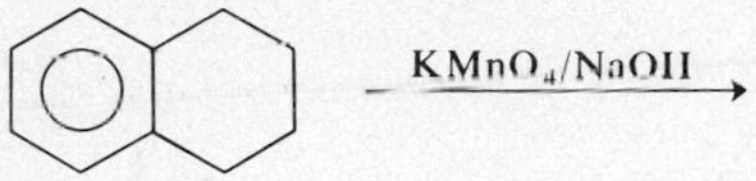 [tetralin] $\xrightarrow{KMnO_4/NaOH}$

(b) 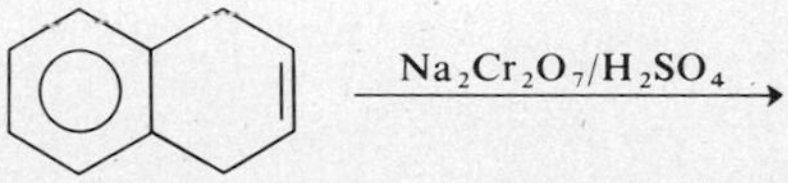 [1,2-dihydronaphthalene] $\xrightarrow{Na_2Cr_2O_7/H_2SO_4}$

(c) 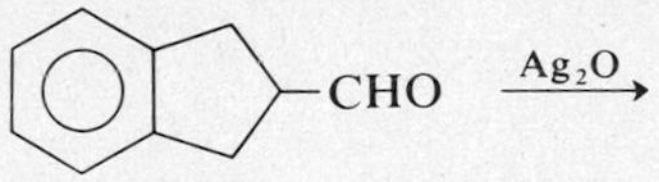 [indane]–CHO $\xrightarrow{Ag_2O}$

(d) $NCCH_2CH_2CN + H_2O \xrightarrow{H^+ \text{ or } OH^-}$

(e)  [cycloheptyl]–MgBr + $CO_2 \longrightarrow \xrightarrow{H_3O^+}$

(f) $CH_3\overset{O}{\overset{\|}{C}}CH_2CH_2CO_2H + LiAlH_4 \longrightarrow \xrightarrow{H_3O^+}$

(g) 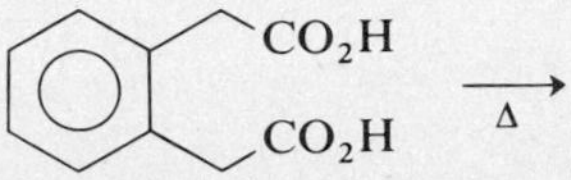 [benzene ring with two CH_2–CO_2H / CO_2H chains] $\xrightarrow{\Delta}$

(h) 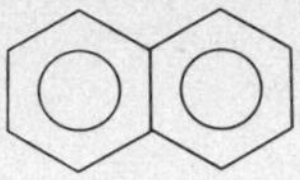 [naphthalene] + CH_3COCl $\xrightarrow{AlCl_3}$

7. The following conversions from the compounds shown on the left to the compounds shown on the right require more than one step. Indicate what sequences of reactions are necessary to achieve this, show the reagents, and draw the structures of the intermediate compounds that are produced.

(a) $H_2C{=}CHCH_3 \longrightarrow H_2C{=}CHCH_2CO_2H$

(b) 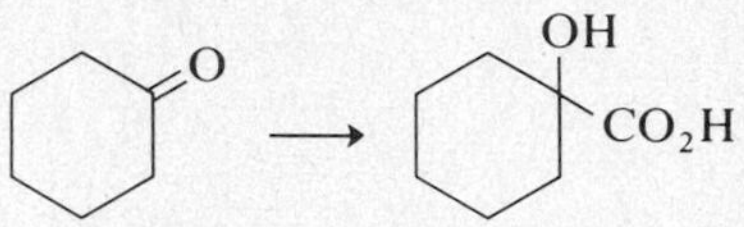 [cyclohexanone] ⟶ [1-hydroxycyclohexanecarboxylic acid: OH, CO_2H]

(c) 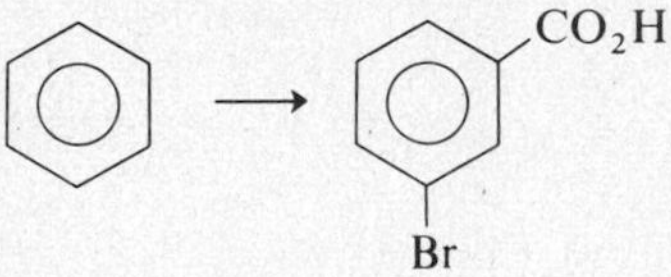 [benzene] ⟶ [m-bromobenzoic acid: CO_2H, Br]

(d) 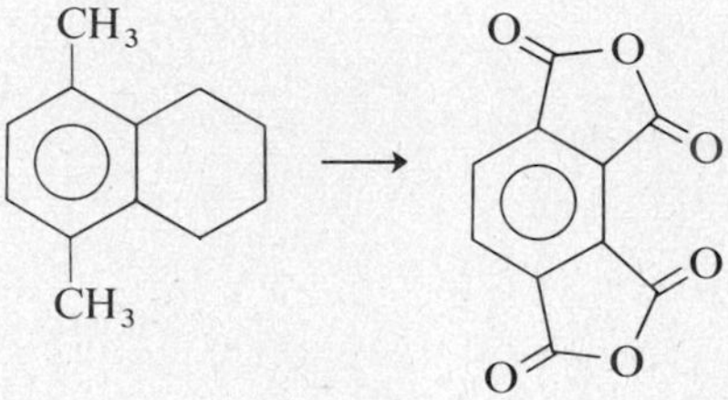 [5,8-dimethyltetralin: CH_3, CH_3] ⟶ [benzene-1,2,3,4-tetracarboxylic dianhydride]

(e) $(CH_3)_2C{=}CH_2 \longrightarrow (CH_3)_3CCO_2H$

8. Available starting materials for the following syntheses include compounds of the general structure RH, RCH_3, RCH_2OH, RCHO, and RCO_2H where R can be H, CH_3, CH_3CH_2, and C_6H_5. Starting with organic compounds chosen from this group, and using any necessary inorganic reagents, design a synthesis for each of the following compounds:

(a) $CH_3CH_2CH_2CO_2H$

(b) $(CH_3)_2C(OH)CO_2H$

(c)  [p-benzenedicarboxylic acid: CO_2H, CO_2H]

(d) 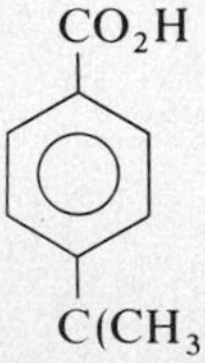 [p-tert-butylbenzoic acid: CO_2H, $C(CH_3)_3$]

(e) $CH_3CH{=}CHCO_2H$

9. Using simple chemical tests, describe how the following pairs of isomeric compounds can be differentiated.

(a) 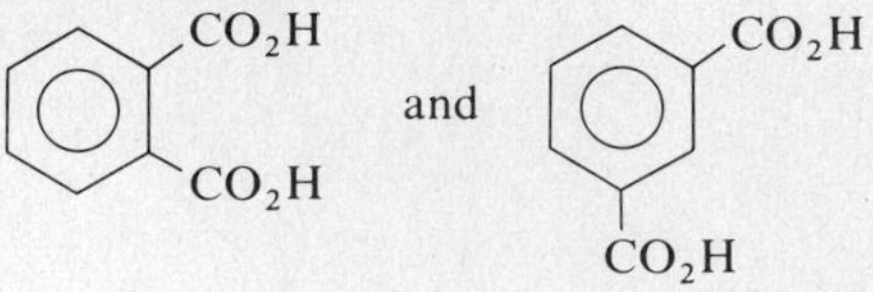 [o-phthalic acid: CO_2H, CO_2H] and [isophthalic acid: CO_2H, CO_2H]

(b) $(CH_3)_2NCH_2CO_2H$ and $H_2N(CH_2)_3CO_2H$

(c) 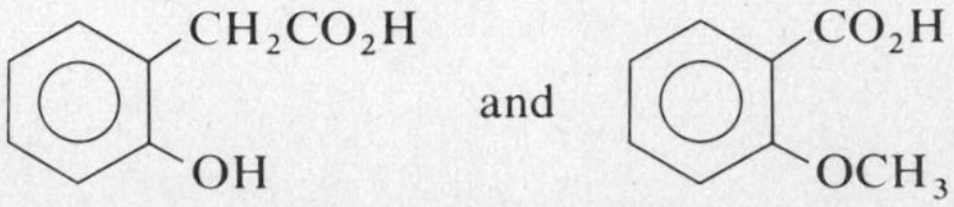 [o-hydroxyphenylacetic acid: CH_2CO_2H, OH] and [o-methoxybenzoic acid: CO_2H, OCH_3]

(d) $(CH_3)_2CHCH{=}CHCO_2H$ and cyclopentane$-CO_2H$

(e) cyclohexane-1,1-dicarboxylic acid (CO_2H, CO_2H) and cyclohexane-1,2-dicarboxylic acid (CO_2H, CO_2H)

10. Four unlabeled bottles contain pure samples of the following isomeric compounds:

(a) p-$CH_3C_6H_4CO_2H$ (b) p-$HOC_6H_4C(=O)CH_3$ (c) $C_6H_3(OH)_2CH{=}CH_2$ (d) $C_6H_4(OCH_2CH_2O)$

On the basis of chemical tests alone, indicate how the identity of the material in each bottle can be established.

11. Each of five bottles contains a pure sample of one of the following isomeric $C_7H_{10}O_2$ carboxylic acids:

$(CH_3)_2C{=}CHCH{=}CHCO_2H$ **(a)** $H_2C{=}C{=}CHCH_2CH_2CH_2CO_2H$ **(b)** $HC{\equiv}CCH_2CH_2CH_2CH_2CO_2H$ **(c)**

(d) bicyclic cyclopropane (H, H, H, CO_2H) **(e)** cyclobutane (H, CO_2H, $CH{=}CH_2$, H)

Using chemical tests and no more than one spectral determination (applied at whatever stage in the analysis you think most effective), indicate how the identity of the material in each of these bottles can be positively identified.

12. A pair of isomeric $C_5H_8O_4$ carboxylic acids possess the nmr spectra shown below. Suggest a structure for each of these compounds that is commensurate with these data.

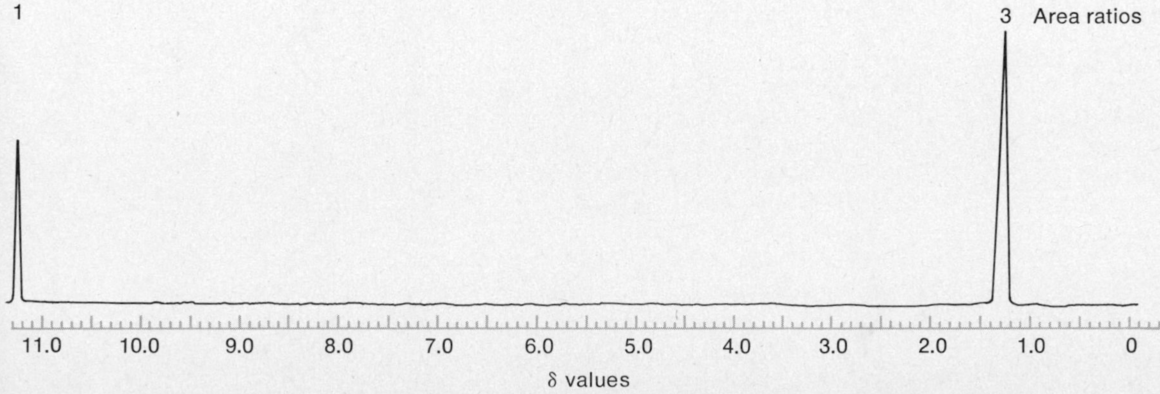

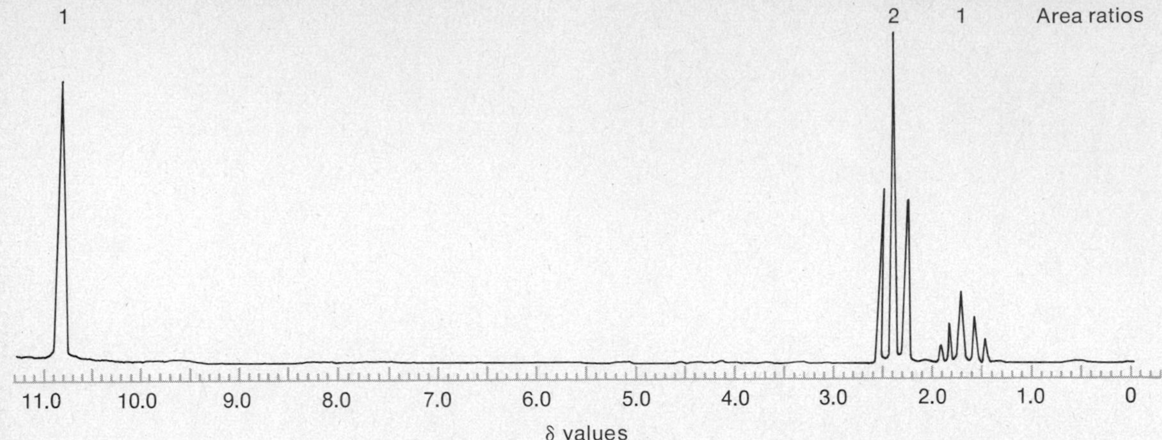

13. A compound known as dictyopterene has been isolated from the brown algae *Dictyopteris* and has been shown to be 6-butyl-1,4-cycloheptadiene (see structure).

$CH_3CH_2CH_2CH_2$ H

Dictyopterene

There is evidence, however, that this material is the result of an isomerization occurring during the isolation procedure, and that the compound actually present in the algae is an isomer. A compound that has been shown to undergo isomerization to dictyopterene when heated has the molecular formula $C_{11}H_{18}$ and undergoes ozonolysis to yield formaldehyde, pentanal, and a $C_5H_6O_2$ compound. The $C_5H_6O_2$ compound reacts with silver oxide to yield a $C_5H_6O_4$ compound, which loses water on heating to form a $C_5H_4O_3$ compound containing two sets of equivalent hydrogens and no vinyl hydrogens (as indicated by nmr). On the basis of this information, suggest a structure for the $C_{11}H_{18}$ compound that can be isomerized to dictyopterene. [*Note*: Don't let the structure of dictyopterene guide your thinking too strongly, for the pattern of the isomerization (*i.e.*, the mechanism) is not self-evident to anyone unfamiliar with reactions of this sort.]

14. Carry out the following calculations:

(a) Calculate the concentration of carboxylate ion when 1 mole of isobutyric acid ($pK_a = 4.86$) is added to 1 liter of water at pH 3, 4, 5, 6, and 7.

(b) Calculate the pH of the solution that is obtained when 0.5 mole of *o*-nitrobenzoic acid ($pK_a = 2.17$) and 0.5 mole of *p*-nitrobenzoic acid ($pK_a = 3.44$) are added to water and the total volume is adjusted to 1 liter.

(c) Calculate the relative amounts of anion in equilibrium with the corresponding conjugate acid for a solution prepared from 1 liter of water buffered at pH 7 and 1 mole each of phenol ($pK_a = 10$), benzoic acid ($pK_a = 4.20$), and *o*-nitrophenol ($pK_a = 7.23$).

Chapter 15

Carboxylic acid derivatives

And so do his sisters and his cousins and his aunts.

GILBERT AND SULLIVAN
H.M.S. Pinafore

Attachment of a carbon framework (R) and a hetero atom or group (Y) other than a hydroxyl group to a carbonyl function produces compounds known as *carboxylic acid derivatives*, $\overset{R}{\underset{Y}{>}}C{=}O$. When Y is halogen (X) the compounds are *acid halides*, $\overset{R}{\underset{X}{>}}C{=}O$ (X usually being chlorine). When Y is alkylcarboxy, $RC(=O)O-$, or arylcarboxy $ArC(=O)O-$, the compounds are *anhydrides*, $RC(=O)-O-C(=O)R(Ar)$. When Y is alkoxy (RO–) or aryloxy (ArO–), the compounds are *esters*, $\overset{R}{\underset{(ArO)RO}{>}}C{=}O$. When Y is amino ($R_2N$–), the compounds are *amides*, $\overset{R}{\underset{R_2N}{>}}C{=}O$. Acid halides, anhydrides, esters, and amides are at the same oxidation state as carboxylic acids and are converted to carboxylic acids by hydrolysis. *Ketenes* ($R_2C{=}C{=}O$), which are at the same oxidation level as carboxylic acids and are converted to acids on hydrolysis, are also classed as carboxylic acid derivatives.

15.1 *Acid Halides*

15.1a. STRUCTURE AND NOMENCLATURE OF ACID HALIDES. Acid halides can be considered as acyl, *i.e.*, $RC(=O)-$, or aroyl, *i.e.*, $ArC(=O)-$, derivatives of the hydrogen halides in the same sense that carboxylic acids can be considered as the acyl and aroyl derivatives of water. Common names have been given to the lower molecular weight acid halides, and IUPAC names can be assigned to all members of the family (Table 15.1). In cases where it is necessary to designate the $>C=O$ (with X attached) group as the parent function, the compound is called a *alkanecarbonyl halide*. When it is treated as a substituent, it is specified as a *halocarbonyl* group. The following discussion of acid halides deals mainly with the acid chlorides, for organic chemists use them to the virtual exclusion of acid fluorides, bromides, and iodides.

Table 15.1 NOMENCLATURE OF ACID HALIDES

Compound	Common Name	IUPAC Name
CH_3COCl	Acetyl chloride	Ethanoyl chloride
CH_3CH_2COCl	Propionyl chloride	Propanoyl chloride
$CH_3CH_2CH_2COCl$	*n*-Butyryl chloride	Butanoyl chloride
$(CH_3)_2CHCOCl$	Isobutyryl chloride	2-Methylpropanoyl chloride
C_6H_5-COCl (benzene ring)	Benzoyl chloride	Benzoyl chloride
$C_6H_{11}-COCl$ (cyclohexane ring)		Cyclohexanecarbonyl chloride
$(CH_3)_3CCOCl$	Trimethylacetyl chloride	2,2-Dimethylpropanoyl chloride
$HO_2C(CH_2)_2CH(COCl)(CH_2)_2CO_2H$		4-Chlorocarbonylheptandioic acid

15.1b. PHYSICAL AND SPECTRAL PROPERTIES OF ACID HALIDES. Lacking an acidic hydrogen, acid halides are less strongly intermolecularly associated than the corresponding acids, and are, therefore, lower boiling and less water soluble. Whereas acetic acid boils at 118°, acetyl chloride boils at 51° despite its higher molecular weight. Acetic acid is completely miscible with water, but acetyl chloride is only moderately soluble. As the molecular weight increases, the water solubility and the rate of hydrolysis of acid chlorides diminish.

Acid halides are distinguished from the corresponding acids by a carbonyl stretching band at a significantly higher frequency in the infrared. Whereas type AC carboxylic acids absorb at 1725–1700 cm^{-1}, the corresponding acid chlorides absorb at 1815–1770 cm^{-1} (see Table 15.2 on p. 394).

15.1c. SYNTHESIS OF ACID HALIDES. Acid halides are almost invariably synthesized from the corresponding carboxylic acids by the action of any one of several inorganic reagents, including phosphorus trichloride (PCl_3), phosphorus pentachloride (PCl_5), and thionyl chloride ($SOCl_2$) (Fig. 15.1).

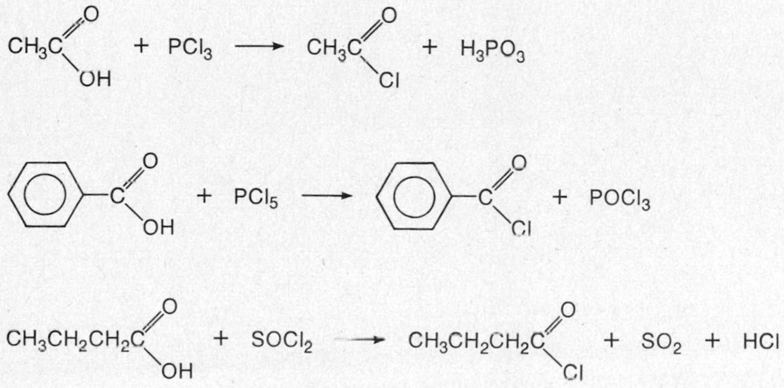

$$CH_3C(=O)OH + PCl_3 \longrightarrow CH_3C(=O)Cl + H_3PO_3$$

$$C_6H_5C(=O)OH + PCl_5 \longrightarrow C_6H_5C(=O)Cl + POCl_3$$

$$CH_3CH_2CH_2C(=O)OH + SOCl_2 \longrightarrow CH_3CH_2CH_2C(=O)Cl + SO_2 + HCl$$

Fig. 15.1. Synthesis of acid chlorides from carboxylic acids.

15.1d. REACTIONS OF ACID HALIDES. We might predict that the electrophilic character of the carbonyl carbon in acid halides should be greater than in aldehydes and ketones because of the electron-withdrawing effect (*i.e.*, inductive effect) of the halogen atom. This does, in fact, prove to be the case, although to some extent this effect is counterbalanced by the electron-releasing effect of the nonbonded electrons of the halogen [*e.g.*, resonance structure Fig. 15.2(c)];

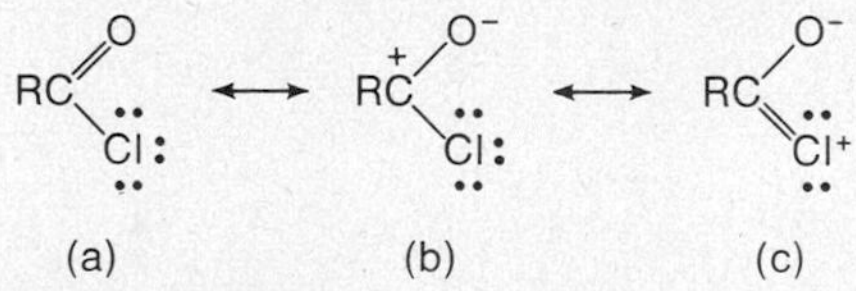

$$RC(=O)\ddot{C}l: \longleftrightarrow R\overset{+}{C}(O^-)\ddot{C}l: \longleftrightarrow RC(O^-)=Cl^+$$

(a) (b) (c)

Fig. 15.2. Resonance structures of acid chlorides.

acid halides are exceedingly reactive toward nucleophilic reagents. The products of reaction, however, are different from those from aldehydes and ketones (see Section 13.5), for the initial step, producing an addition product, is succeeded by a second step, in which the halogen is eliminated and the carbonyl group is regenerated. The overall reaction is a *substitution* process, which proceeds via nucleophilic addition followed by elimination. The hydrolysis of acetyl chloride, for instance, can be depicted in this fashion (Fig. 15.3).

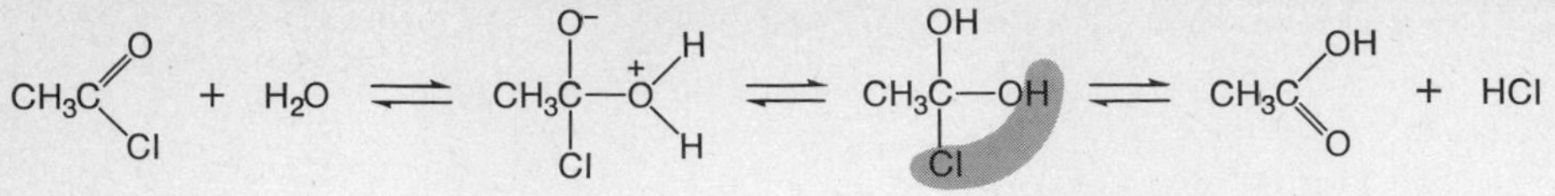

$$CH_3C(=O)Cl + H_2O \rightleftharpoons CH_3C(O^-)(Cl)\text{—}\overset{+}{O}H_2 \rightleftharpoons CH_3C(OH)(Cl)\text{—}OH \rightleftharpoons CH_3C(=O)OH + HCl$$

Fig. 15.3. Hydrolysis of acetyl chloride.

A variety of other nucleophiles also react with acid chlorides via addition-elimination reactions to give various carboxylic acid derivatives. Alcohols and phenols react to give esters (esterification reaction); the sodium salts of carboxylic acids react to yield anhydrides; ammonia, primary, and secondary amines react to produce amides (Fig. 15.4).

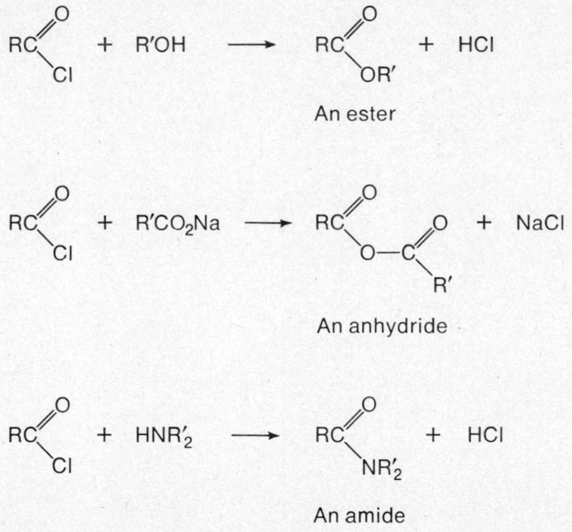

$$RC(=O)Cl + R'OH \longrightarrow RC(=O)OR' + HCl$$

An ester

$$RC(=O)Cl + R'CO_2Na \longrightarrow RC(=O)\text{—}O\text{—}C(=O)R' + NaCl$$

An anhydride

$$RC(=O)Cl + HNR'_2 \longrightarrow RC(=O)NR'_2 + HCl$$

An amide

Fig. 15.4. Synthesis of carboxylic acid derivatives from acid chlorides.

The synthesis of aromatic ketones by the Friedel–Crafts acylation reaction,

$$ArH + RCOCl \xrightarrow[AlCl_3]{} ArCOR$$

is an important application of acid chlorides, which we have already discussed. The mechanism of this reaction is outlined in Fig. 8.14, and examples of the reaction are presented in Fig. 13.9.

Strong reducing agents, such as lithium aluminum hydride, convert acid chlorides to the corresponding alcohols—*e.g.*, $RCOCl \xrightarrow{LiAlH_4} RCH_2OH$. With certain other reducing agents, however, the reduction can be interrupted at the aldehyde stage, as illustrated by the previously discussed (see Fig. 13.7) Rosenmund reduction,

$$RCOCl + H_2 \longrightarrow RCHO + HCl$$

PROBLEM 15-1

Starting with benzene and acetylene as the only available carbon compounds, devise a synthesis of acetophenone that involves the use of an acid chloride in one of the steps.

15.2 *Anhydrides*

Carboxylic anhydrides can be prepared by the reaction of a carboxylic acid or its salt with an acid chloride, or by the reaction of carboxylic acids with ketenes:

$$RCO_2H \quad + \quad R'COCl \longrightarrow R\overset{O}{\overset{\|}{C}}-O-\overset{O}{\overset{\|}{C}}R' \quad + \quad HCl$$

$$RCO_2^- \quad + \quad R'COCl \longrightarrow R\overset{O}{\overset{\|}{C}}-O-\overset{O}{\overset{\|}{C}}R' \quad + \quad Cl^-$$

$$RCO_2H \quad + \quad H_2C{=}C{=}O \longrightarrow R\overset{O}{\overset{\|}{C}}-O-\overset{O}{\overset{\|}{C}}CH_3$$

Anhydrides are named by adding the word "anhydride" to the name of the acid or acids from which they are derived. For example, the action of CH_3CO_2Na (sodium acetate) on CH_3COCl (acetyl chloride) yields $(CH_3CO)_2O$, which is called acetic anhydride; the action of $CH_3CH_2CO_2H$ (propionic acid) on $CH_2{=}C{=}O$ (ketene) yields

$$\begin{array}{l} CH_3CH_2C({=}O) \\ \quad \diagdown \\ \qquad O \\ \quad \diagup \\ CH_3C({=}O) \end{array}$$

which is named acetic propionic anhydride (or ethanoic propanoic anhydride in the IUPAC system).

Anhydrides, like acid halides, lack acidic hydrogens and boil considerably lower than carboxylic acids of corresponding molecular weight. Acetic anhydride (mol wt 102), for instance, boils at 140°, whereas valeric acid (mol wt 102) boils at 187°. Water solubility, also, is lower for anhydrides than for the comparable acids, dissolution in water being accompanied by hydrolysis to the corresponding carboxylic acid.

Anhydrides are distinguished from carboxylic acids as well as from acid halides by their infrared characteristics; a pair of carbonyl stretching bands is present, one at 1850–1800 cm^{-1} and the other at 1790–1740 cm^{-1}, whereas acids and acid halides possess only a single carbonyl stretching absorption (Table 15.2).

The reactions of anhydrides resemble those of acid chlorides: they react with water to form acids, with alcohols to form esters (esterification), with amines to form amides, with aromatic compounds in the presence of aluminum chloride

Table 15.2. SPECTRAL CHARACTERISTICS OF CARBOXYLIC ACID DERIVATIVES

Type of Compound	NMR δ values	IR cm^{-1}	UV nm (ε)
Acid chlorides: RC(=O)Cl	a	Near 1800 (R = alkyl) Near 1765 (R = aryl)	Approx. 235 (10–100)
Anhydrides: RC(=O)–O–C(=O)R	a	Near 1785 and 1725 (two bands) (R = alkyl) Near 1785 and 1725 (two bands) (R = aryl)	
Esters: RC(=O)OR	a	Near 1735 (R = alkyl) Near 1715 (R = aryl)	Approx. 205 (10–100) (R = alkyl) 208 (14,000) (R = $H_2C{=}CH-$) 230 (16,000) (R = C_6H_5)
Amides: RC(=O)NH_2	a 5–6.5	3500 and 3400 (two N—H stretching bands) 1650 (C=O stretch)	205–215 (10–100)
RC(=O)N(H)R′	6–8.2	Approx. 3400 (one N—H stretching band) 1680–1630 (C=O stretch)	"
RC(=O)N(H)Ar	7.8–9.4	Approx. 3400 (one N—H stretching band) 1680–1630 (C=O stretch)	"
Ketenes: $R_2C{=}C{=}O$		2150	

a H—C(—)—C(=O)—Y resonances at about δ 2.2–2.5.

to form ketones (Friedel–Crafts acylation), and with lithium aluminum hydride to form alcohols (Fig. 15.5). The essential difference between acid chlorides and anhydrides is a quantitative one. Acid chlorides (*i.e.*, the acyl derivatives of the strong mineral acid, hydrochloric acid) are more reactive than acid anhydrides (*i.e.*, the acyl derivatives of the weaker organic acids, carboxylic acids).

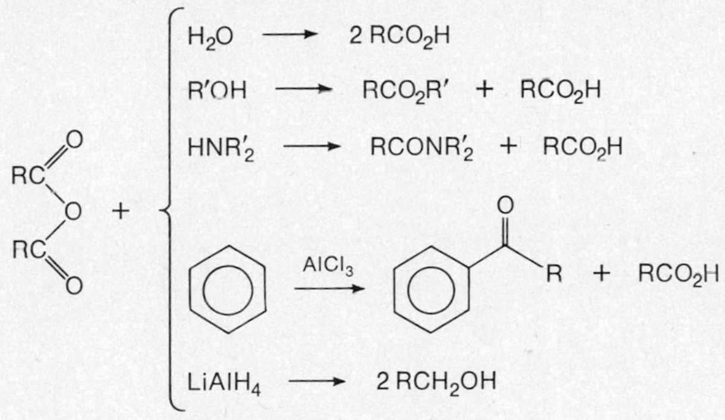

$$(RCO)_2O + \begin{cases} H_2O \longrightarrow 2\,RCO_2H \\ R'OH \longrightarrow RCO_2R' + RCO_2H \\ HNR'_2 \longrightarrow RCONR'_2 + RCO_2H \\ C_6H_6 \xrightarrow{AlCl_3} C_6H_5COR + RCO_2H \\ LiAlH_4 \longrightarrow 2\,RCH_2OH \end{cases}$$

Fig. 15.5. Reactions of anhydrides.

PROBLEM 15-2 How could the synthesis in Problem 15-1 on p. 393 be modified to involve an anhydride rather than an acid chloride?

15.3 Esters

15.3a. STRUCTURE AND NOMENCLATURE OF ESTERS. The names of esters must specify both the carboxylic acid portion and the alcohol portion of the molecule. Using common name designations, $CH_3C(=O)OCH_3$, for instance, is named methyl acetate, the "methyl" indicating that the alcohol portion is derived from methanol, the "acet" indicating that the carboxylic acid portion is derived from acetic acid, and the "ate" indicating that the compound is an ester. In similar fashion,

$CH_3CH_2C(=O)OCH(CH_3)_2$ is named isopropyl propionate,

$(CH_3)_2\text{-}CHCH_2C(=O)OCH_2C_6H_5$ is benzyl isovalerate, etc. The IUPAC nomenclature is employed in the same way; the systematic IUPAC names are substituted for the common names. Thus, methyl acetate is methyl ethanoate in the IUPAC system, benzyl isovalerate is benzyl 3-methylbutanoate,

$$CH_3CH_2CH(CH_3)CH_2C(=O)OCH_2CH(CH_3)CH_2CH_3$$

is 2-methylbutyl 3-methylpentanoate, etc. If the "carboxylic acid" nomenclature is necessary to specify the acid portion, the suffix "ic" gives way to "ate." Thus,

$$\text{3,3-}F_2\text{-cyclopentyl}-C(=O)OCH_2C(CH_3)_3$$

is named neopentyl 3,3-difluorocyclopentanecarboxylate. In the event that an ester function must be specified as a functional group, it is designated as an "alkoxycarbonyl" group; *e.g.*, $-C(=O)OCH_2CH_3$ is an ethoxycarbonyl group, $-C(=O)OCH_2C_6H_5$ is a benzyloxycarbonyl group, etc.

PROBLEM 15-3

Draw and name all the $C_4H_8O_2$ esters.

15.3b. PHYSICAL AND SPECTRAL PROPERTIES OF ESTERS. Esters usually are considerably more volatile and less water soluble than the corresponding acids. Ethyl acetate (mol wt 88), *e.g.*, boils at 77° and is only partially miscible with water, whereas butyric acid (mol wt 88) boils at 163° and is completely miscible with water. Esters are characterized in the infrared spectrum by a carbonyl stretching band at 1750–1735 cm^{-1}; as with other carbonyl compounds, the frequency of this band is diminished by about 30 cm^{-1} if unsaturation is present at the α,β positions (*i.e.*, type BC esters). Type AC esters are almost transparent in the ultraviolet, but type BC esters absorb strongly in the vicinity of 220 nm (see Table 15.2 on p. 394).

Esters often have a fragrant, pleasant, fruitlike odor. The chemist should not be seduced into sloppy laboratory technique by the olfactory allure of esters, however, for social disaster may await him. A particularly striking example is ethyl cyclohexanecarboxylate, which possesses a rather pleasant aroma. The ester is sufficiently volatile, though, to be wafted about by currents of air, which can carry it onto the skin of the unwary experimenter. There, in contact with the moisture of the epidermis, it undergoes hydrolysis to cyclohexanecarboxylic acid, which is a compound of exceptionally repugnant odor and which, because of its relative nonvolatility, is difficult to dissipate by evaporation. Individuals unfortunate enough to have become polluted with ethyl cyclohexanecarboxylate, or other esters of similar ilk, may find themselves *personae non gratae* for some scores of hours.

15.3c. SYNTHSIS OF ESTERS. Acid chlorides and anhydrides react with alcohols and phenols to produce esters, the process being referred to as esterification. The equilibrium constants for these reactions are much greater than

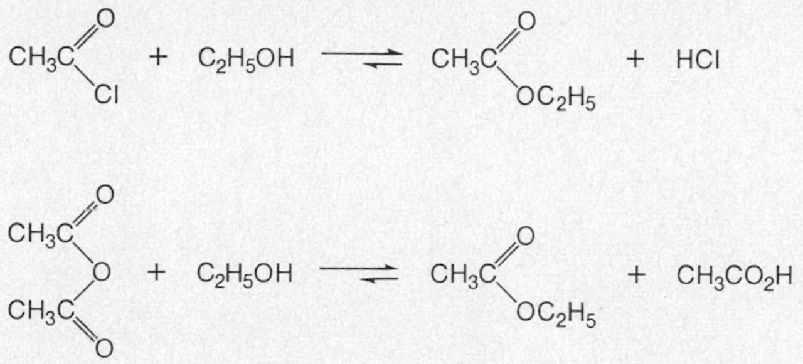

$$CH_3C(=O)Cl + C_2H_5OH \rightleftharpoons CH_3C(=O)OC_2H_5 + HCl$$

$$(CH_3CO)_2O + C_2H_5OH \rightleftharpoons CH_3C(=O)OC_2H_5 + CH_3CO_2H$$

Fig. 15.6. Formation of esters from acid chlorides and anhydrides.

1, and the reactions are essentially irreversible (Fig. 15.6). Another procedure for esterification involves the action of an alcohol on the carboxylic acid itself:

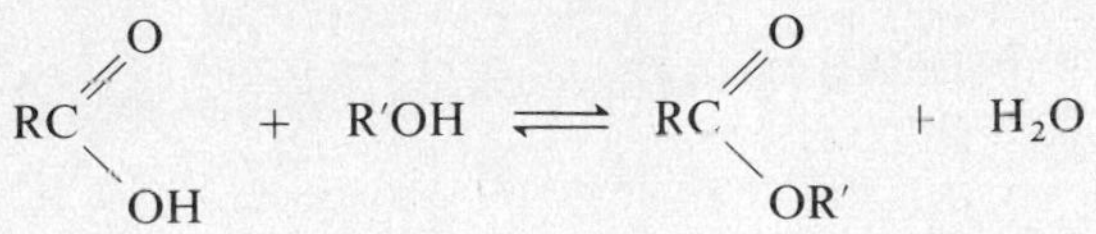

$$RC(=O)OH + R'OH \rightleftharpoons RC(=O)OR' + H_2O$$

In this case, however, the equilibrium constant is not particularly favorable to product formation, for water and alcohols are comparable to each other in nucleophilic strength. Procedures must be employed, therefore, to shift the equilibrium in favor of ester formation if satisfactory yields are to be obtained. We can achieve this, taking cognizance of the law of mass action, either by increasing the concentration of one of the compounds on the left side of the equation or by decreasing the concentration of one of the compounds on the right side of the equation. Thus, one method involves using a large excess of either the acid or the alcohol, whichever is more readily available. Another method involves the removal of water by "azeotropic distillation." We can prepare ethyl esters, for example, in almost quantitative yield by carrying out the reaction in benzene solution (containing a trace of sulfuric acid catalyst to increase the rate at which equilibrium is established) in an apparatus equipped with a device for allowing the distillate from the refluxing solution to collect in a separate container before returning to the reaction mixture. Benzene, ethanol, and water form a ternary azeotrope (see p. 276), which is homogeneous in the vapor phase but which separates into two liquid phases on condensation. The lighter of these is a mixture of alcohol and benzene, and the heavier one is a mixture of alcohol and water. If the heavier phase is discarded and only the lighter phase returned to the reaction mixture, water is thereby removed and the concentration of ester in the equilibrium mixture increases. The apparatus shown in Fig. 15.7 permits the preparation of ethyl esters in very high yields by this technique.

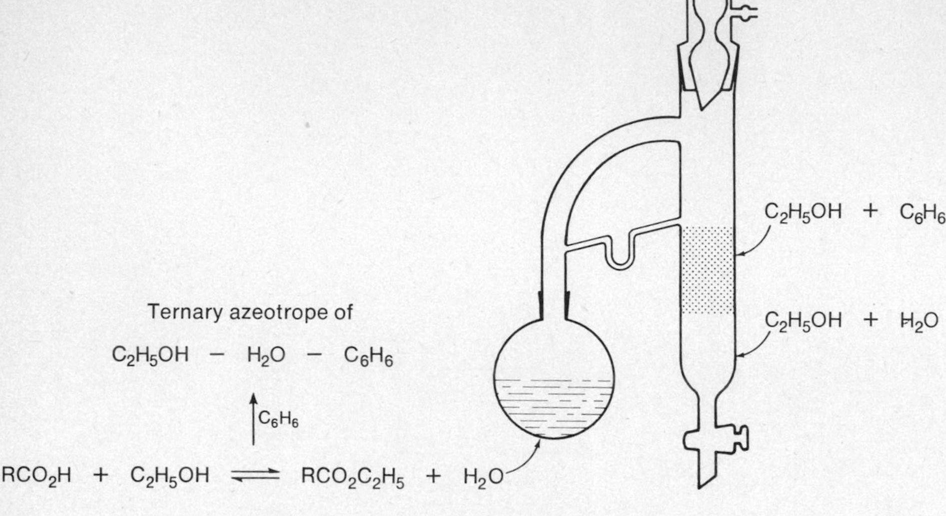

Fig. 15.7. Azeotropic synthesis of ethyl esters.

15.3d. REACTIONS OF ESTERS. In common with all carboxylic acid derivatives, esters can be hydrolyzed to the corresponding carboxylic acids, although the equilibrium constants for ester hydrolysis ($RCO_2R' + H_2O \rightleftharpoons RCO_2H + R'OH$) are much smaller than those for acid halide or anhydride hydrolysis. The effect of the less favorable equilibrium constant can be minimized in the same ways that have been discussed for the esterification reaction, *i.e.*, by using an excess of one of the reactants or by removing one of the products. The former method involves using a large excess of water, carrying out the acid-catalyzed hydrolysis with a dilute solution of a mineral acid (*e.g.*, HCl or H_2SO_4) (Fig. 15.8). The latter

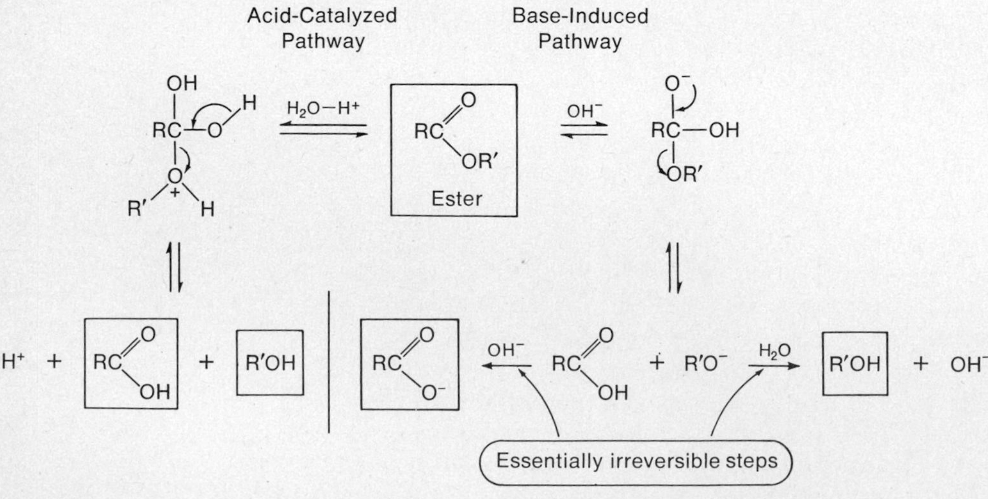

Fig. 15.8. Acid-catalyzed and base-induced hydrolysis of esters.

method involves "removing" the carboxylic acid in the form of its carboxylate ion by carrying out the hydrolysis with aqueous base (*e.g.*, sodium hydroxide solution) (Fig. 15.8). The initially formed carboxylic acid is transformed to the carboxylate ion in the basic solution; even in moderately weak base, the equilibrium constant for this step is very large (see Section 14.5), so the carboxylic acid is converted almost completely to its carboxylate ion. This displaces the hydrolysis equilibrium in the direction of product formation, and the reaction proceeds virtually to completion. Clearly, a full mole-equivalent of base is required for every ester group that is hydrolyzed, so the reaction is classed as a base-induced rather than a base-catalyzed process. The base-induced hydrolysis of esters is commonly referred to as *saponification* (Latin, *sapo*, "soap"), because it is a type of reaction used in soap making (see Sections 10.11 and 15.8).

Nucleophiles stronger than alcohols react effectively with esters, and the equilibrium constants for the reactions are favorable to product formation. Of particular note in this respect are ammonia, primary amines, and secondary amines, all of which react with esters to yield amides (Fig. 15.9).

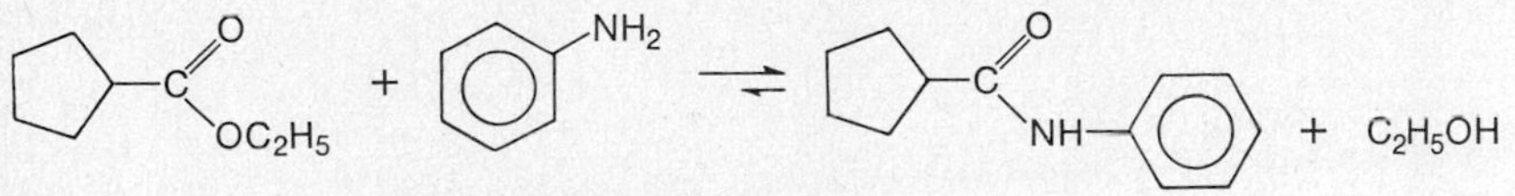

$$C_5H_9C(=O)OC_2H_5 + C_6H_5NH_2 \rightleftharpoons C_5H_9C(=O)NH{-}C_6H_5 + C_2H_5OH$$

Fig. 15.9. Synthesis of amides from esters.

Hydride ion is also an effective nucleophile in addition-elimination reactions of esters, yielding alcohols as the final products when lithium aluminum hydride is used as the source of hydride; the initially formed aldehydes are not isolable under these conditions and are rapidly reduced to the alcohols (Fig. 15.10).

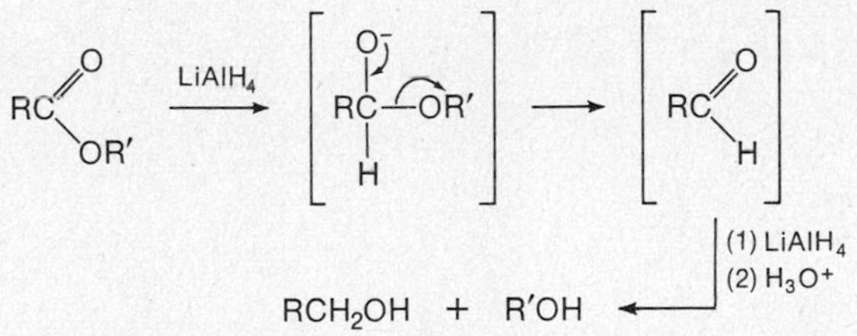

$$RC(=O)OR' \xrightarrow{LiAlH_4} \left[RC(O^-)(H){-}OR' \right] \longrightarrow \left[RC(=O)H \right] \xrightarrow{(1)\ LiAlH_4;\ (2)\ H_3O^+} RCH_2OH + R'OH$$

Fig. 15.10. Lithium aluminum hydride reduction of esters.

Type AC esters containing one or more α hydrogens undergo self-condensation reactions via a pathway reminiscent of the aldol condensation of aldehydes and ketones (see Section 13.6). Thus, ethyl acetate reacts with sodium ethoxide to form a resonance-stabilized anion ($[CH_2CO_2C_2H_5]^-$), which engages in an addition-elimination sequence with the carbonyl group of another molecule of ethyl acetate to form ethyl 3-oxobutanoate (acetoacetic ester). The overall process is known as a *Claisen condensation* (Fig. 15.11) and is useful for synthesizing β-keto esters.

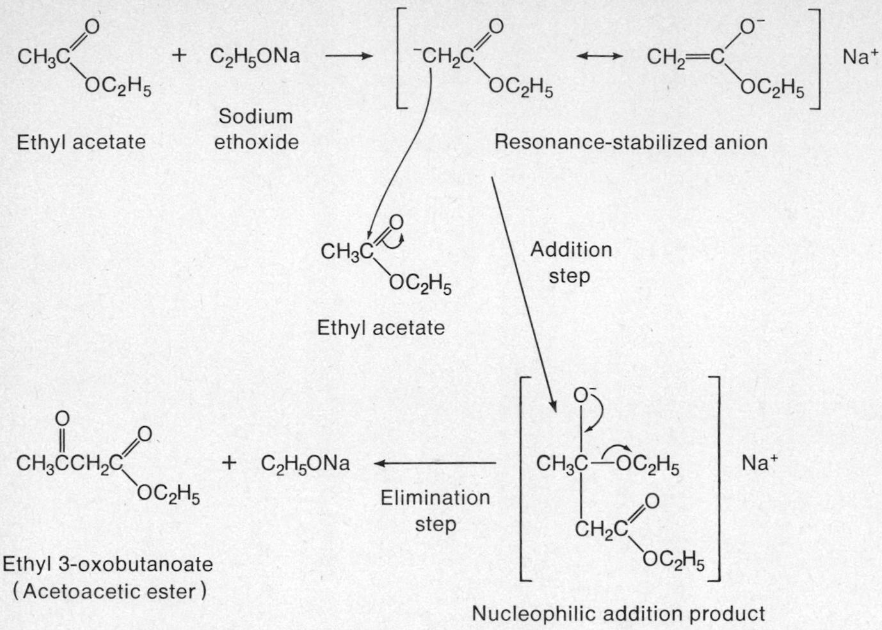

$$CH_3C(=O)OC_2H_5 + C_2H_5ONa \longrightarrow \left[{}^-CH_2C(=O)OC_2H_5 \longleftrightarrow CH_2{=}C(O^-)OC_2H_5 \right] Na^+$$

Ethyl acetate; Sodium ethoxide; Resonance-stabilized anion

$CH_3C(=O)OC_2H_5$ Ethyl acetate — Addition step →

$$\left[CH_3C(O^-)(OC_2H_5)CH_2C(=O)OC_2H_5 \right] Na^+$$

Nucleophilic addition product

Elimination step →

$$CH_3C(=O)CH_2C(=O)OC_2H_5 + C_2H_5ONa$$

Ethyl 3-oxobutanoate (Acetoacetic ester)

Fig. 15.11. Self-condensation reaction of ethyl acetate (Claisen condensation).

Type AC esters react by 1,2-addition or by addition-elimination, but type BC esters (*i.e.*, α,β-unsaturated esters) show a decided preference for 1,4-addition analogous to α,β-unsaturated aldehydes and ketones (see Fig. 13.15). Ethyl crotonate, for example, forms β-substituted esters in reactions with hetero atom nucleophiles, such as alcohols and amines, as well as with carbon nucleophiles such as hydrogen cyanide (Fig. 15.12).

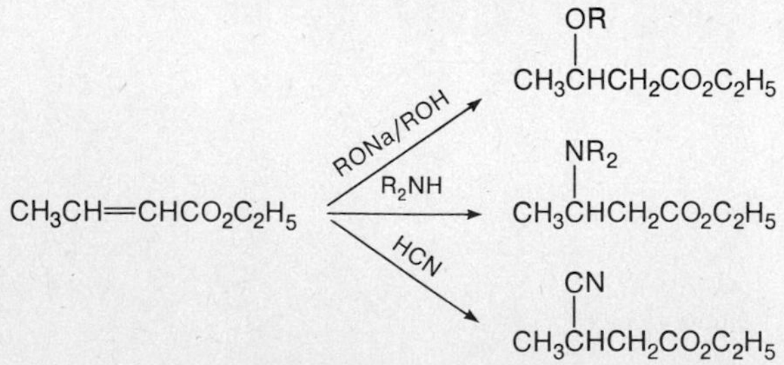

$$CH_3CH{=}CHCO_2C_2H_5 \xrightarrow{RONa/ROH} CH_3CH(OR)CH_2CO_2C_2H_5$$

$$CH_3CH{=}CHCO_2C_2H_5 \xrightarrow{R_2NH} CH_3CH(NR_2)CH_2CO_2C_2H_5$$

$$CH_3CH{=}CHCO_2C_2H_5 \xrightarrow{HCN} CH_3CH(CN)CH_2CO_2C_2H_5$$

Fig. 15.12. 1,4-Addition reactions (conjugate additions) of ethyl crotonate with oxygen, nitrogen, and carbon nucleophiles.

PROBLEM 15-4

What is the structure of the product obtained from the action of sodium ethoxide on ethyl phenylacetate?

PROBLEM 15-5

Indicate the reagents and reaction conditions that could be used to effect each of the steps in the following sequence:

$$HC{\equiv}CH \longrightarrow CH_3CHO \longrightarrow CH_3CH{=}CHCHO \longrightarrow CH_3CH{=}CHCO_2H$$

$$\downarrow$$

$$\underset{\displaystyle CH_3CHCH_2CO_2H}{\overset{CO_2H}{|}} \longleftarrow \underset{\displaystyle CH_3CHCH_2CO_2C_2H_5}{\overset{CN}{|}} \longleftarrow CH_3CH{=}CHCO_2C_2H_5$$

15.4 *Amides*

15.4a. Structure and Nomenclature of Amides. Amides can be considered as the nitrogen analogs of carboxylic acids; just as acetic acid $CH_3C(=O)OH$ is the acetyl derivative of water, acetamide $CH_3C(=O)NH_2$ is the acetyl derivative of ammonia. Amides of the general structure $RC(=O)N(R')R'$ are sometimes designated as primary (when both R′ groups are H), secondary (when one R′ group is H and the other is alkyl or aryl), or tertiary (when neither R′ group is H). It is preferable, however, to designate them as amides, N-mono-substituted amides, and N,N-disubstituted amides, respectively.

Amides are named in relation to the carboxylic acid of corresponding structure by replacing the designation "ic acid" or "oic acid" by "amide." If we use common names, $HC(=O)NH_2$ is formamide, $CH_3C(=O)NH_2$ is acetamide, and $C_6H_5CH_2CH_2C(=O)NH_2$ is β-phenylpropionamide. In the IUPAC nomenclature, these compounds are named methanamide, ethanamide, and 3-phenylpropanamide, respectively (note that the final "e" on the alkane name is dropped and replaced by "amide"). If alkyl or aryl groups are attached to the nitrogen, they are specified by the appropriate name and prefixed with the "N-" designation. Thus, $CH_3C(=O)NHCH_3$ is N-methylacetamide (or N-methylethanamide),

$(CH_3)_2CHC(=O)N(CH_3)_2$ is N,N-dimethylisobutyramide (or N,N-dimethyl-2-methylpropanamide), and C_4H_7–$C(=O)N(CH_2CH_3)CH(CH_3)C_6H_5$ (cyclobutyl) is N-ethyl-N-(1-phenylethyl)-cyclobutanecarboxamide (note that in the "carboxylic acid" nomenclature, the "amide" suffix replaces the "ylic acid" portion of the name).

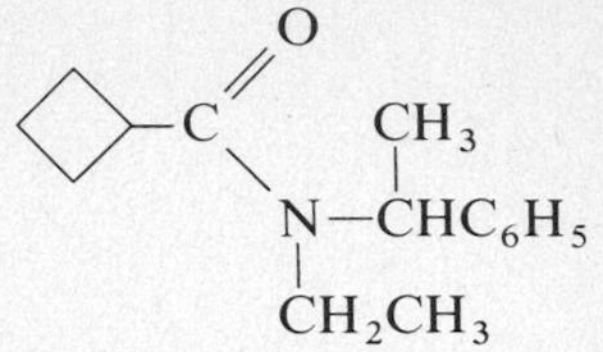

If the amide group is named as a substituent it acquires the designation "carboxamido." For example,

$$HO_2CCH_2CH_2\underset{\displaystyle CH_2CON(CH_3)_2}{\underset{|}{C}}H\overset{\displaystyle O}{\overset{||}{C}}CH_2CH_2CO_2H$$

is named 4-oxo-5-(N,N-dimethylcarboxamidomethyl)octanedioic acid.

PROBLEM 15-6

Draw and name all the $C_9H_{11}NO$ amides containing a benzoyl moiety.

15.4b. Physical and spectral properties of amides. Amides are more water soluble than the corresponding amines, alcohols, and esters, and they have surprisingly high melting and boiling points. Even formamide, the parent member of the series, melts at 2° and boils (with decomposition) at 200°. Acetamide, its next higher homolog, melts at 82° and boils at 221°, and benzamide, the parent member of the aromatic series, melts at 130° and boils at 290°. For the amides just cited, two factors are responsible for the great water solubility and high melting points, *viz.*, the ability to act both as donor and acceptor in hydrogen bond formation and the unusually great dipolar character.

That the dipolar factor is particularly important is illustrated by the tertiary amides, which, although incapable of acting as hydrogen bond donors, are still higher boiling than comparable carboxylic acids or carboxylic acid derivatives. Thus, $CH_3CON(CH_3)_2$ boils at 175°, lower than CH_3CONH_2 (bp 221°) but slightly higher than butyric acid (bp 164°) and much higher than ethyl acetate (bp 77°), which are compounds of comparable molecular weight.

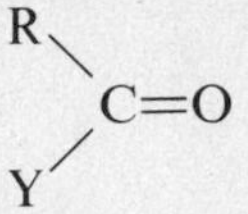

In compounds of the general structure $\underset{Y}{\overset{R}{>}}C{=}O$ where Y is a hetero atom function (*i.e.*, a type C group), two competing effects are operative as a result of the hetero atom Y. One is an electron withdrawal arising from the greater electronegativity of Y as compared with carbon or hydrogen, and the other is an electron release arising from the resonance interaction of the nonbonded electrons on Y with the π system of the carbonyl group. The balance between these effects is dependent on the particular hetero atom and is manifested in certain physical and chemical properties of the systems. The length, strength, and dipolar character of the carbon–oxygen bond, for example, are reflected in the infrared stretching frequencies for the carbonyl group, and it is interesting to compare the various R(Y)C=O compounds in this respect. Acid chlorides absorb at about 1800 cm^{-1}, esters at about 1735 cm^{-1}, carboxylic acids at about 1715 cm^{-1}, amides at about 1650 cm^{-1}, and carboxylate anions at about 1575 cm^{-1} (see Table 15.2 on p. 394). These differences can be explained on the basis of the assumptions that (a) the greater the electron-withdrawing power of the hetero atom the higher is the carbonyl stretching deformation frequency, and (b) the greater the resonance interaction of the hetero atom the lower is the carbonyl stretching deformation frequency.

In the case of the strongly electronegative and weakly nucleophilic chlorine atom, the electron-withdrawing effect outweighs the resonance effect (see Fig. 15.2), and the carbonyl stretching deformation frequency is higher than that of an aldehyde or ketone. In the case of the much less strongly electronegative but much more nucleophilic amine moiety, on the other hand, the resonance effect outweighs the electron-withdrawing effect, and the carbonyl stretching frequency is lower than that of an aldehyde or ketone. With OH (carboxylic acids) and OR (esters), the two effects are about equal, whereas for the O^- group (carboxylate anions), the full formal negative charge greatly enhances the resonance interaction, and the carbonyl stretching frequency is reduced even beyond that of the amides. Amides, then, must be represented by several resonance structures (Fig. 15.13), in which structure (c) contributes very significantly to the resonance hybrid. One of the consequences of this is a large permanent dipole moment (*e.g.*, μ = 3.84 D for formamide, compared with μ = 1.92 D for ethyl formate), which, in turn, enhances intermolecular attractions and increases melting and boiling points.

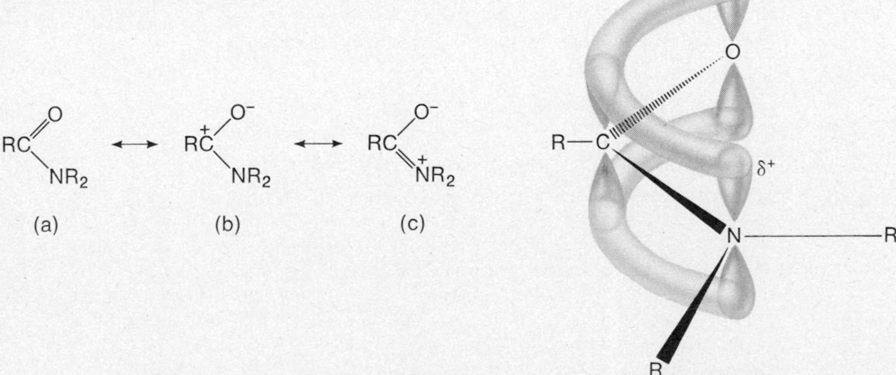

Fig. 15.13. Resonance structures and orbital representation of amides.

Another consequence of the resonance interaction between the amide nitrogen and the carbonyl group is the presence of a sufficiently large barrier to rotation around the C—N bond to permit the detection of conformational isomers by nmr spectroscopy. Thus, when the nmr spectrum of N,N-dimethylbenzamide is measured at −26°, a pair of three-proton singlet resonances (Fig. 15.14) is

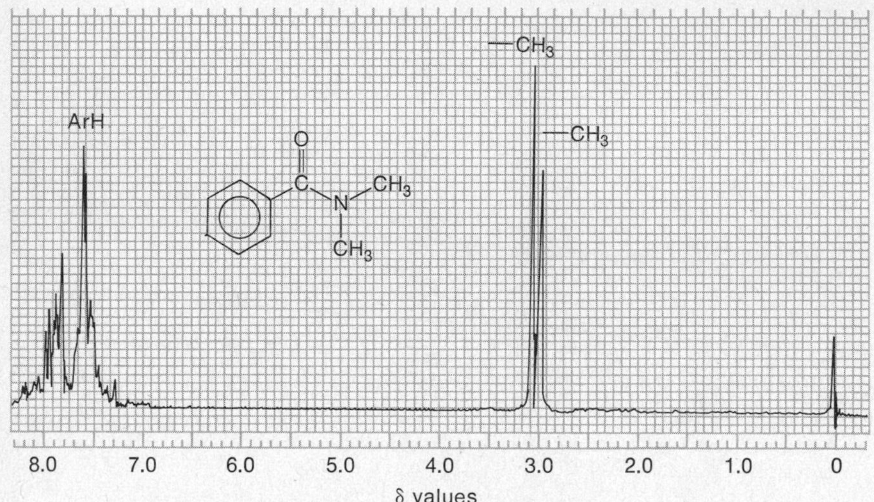

Fig. 15.14. Nuclear magnetic resonance spectrum of N,N-dimethylbenzamide (at − 26°).

observed, because the two N—CH_3 groups are in different environments relative to the rest of the molecule. At room temperature, only a single six-hydrogen resonance is observed, however, for at this higher temperature the rate of rotation around the C—N bond is fast on the nmr time scale and the N—CH_3 groups are in an "averaged" environment* as a result.

PROBLEM 15-7

Compounds A, B, and C have the molecular formula C_4H_9NO. All three can be hydrolyzed to amines and carboxylic acids, compound A giving a $C_4H_8O_2$ acid, compound B a $C_3H_6O_2$ acid, and compound C a $C_2H_4O_2$ acid. Suggest structures for these compounds. If more than one structure is possible in any case, indicate how nmr spectral data could be used to make the differentiation.

15.4c. SYNTHESIS OF AMIDES. Amides are prepared, as indicated previously (see Figs. 15.4, 15.5, and 15.9), by the action of ammonia, primary amines, and

* The approximate rate of rotation around the bond at the temperature at which the two resonances just coalesce into a single resonance can be calculated by the formula, rate (in rotations/sec) = $\pi(\Delta\delta/\sqrt{2})$, where $\Delta\delta$ is the chemical shift difference (in Hertz) between the separate resonances below the "coalescence temperature." For example, in the case of N,N-dimethylbenzamide, the $\Delta\delta$ at −26° is 9 Hz, so at this temperature the rate of rotation is less than 20 rotations per second and at room temperature it is greater than 20 rotations per second.

secondary amines on acid chlorides, anhydrides, and esters. All these reactions are classified as nucleophilic addition-elimination processes, in which the amine acts as the nucleophile (Fig. 15.15). The equilibrium favors amide formation, for R_2NH is a much stronger nucleophile than Cl^-, RCO_2H, or ROH.

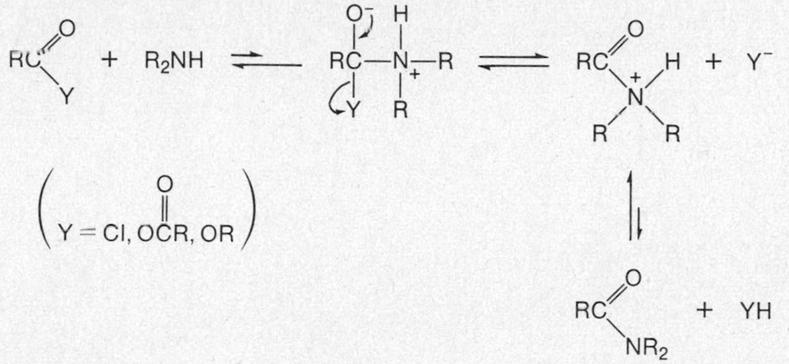

$$RC(=O)Y + R_2NH \rightleftharpoons RC(O^-)(Y)-N^+(H)(R)-R \rightleftharpoons RC(=O)-N^+(H)R_2 + Y^- \rightleftharpoons RC(=O)NR_2 + YH$$

$$\left(Y = Cl,\ O\overset{O}{\overset{\|}{C}}R,\ OR\right)$$

Fig. 15.15. Reaction of carboxylic acid derivatives with amines to form amides.

15.4d. Reactions of amides. The chemical consequence of the significant contribution of the resonance structure Fig. 15.13(c) to the resonance hybrid of an amide is a reduction in the "carbonyl character" of the molecule. Amides, for example, are much less readily hydrolyzed to the corresponding carboxylic acid than esters, acid chlorides, or anhydrides. The equilibrium constant for the reaction

$$RCONR_2 + H_2O \rightleftharpoons RCO_2H + R_2NH$$

is less than 1. In acidic or basic solution, however, the hydrolysis of amides proceeds essentially to completion, because under these conditions one of the products of reaction is removed, *i.e.*, by formation of the ammonium ion ($R_2NH_2^+$) in acidic solution and the carboxylate anion (RCO_2^-) in basic solution (Fig. 15.16).

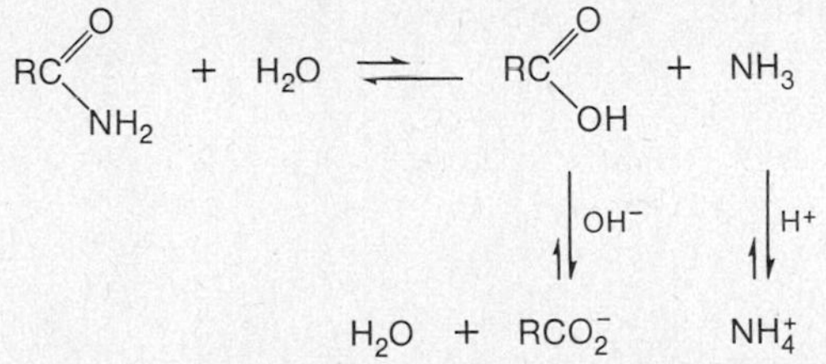

$$RC(=O)NH_2 + H_2O \rightleftharpoons RC(=O)OH + NH_3$$

$$RC(=O)OH \xrightarrow{OH^-} H_2O + RCO_2^- \qquad NH_3 \xrightarrow{H^+} NH_4^+$$

Fig. 15.16. Base-induced and acid-induced hydrolysis of amides.

Thus, just as in the saponification of esters, the reaction can be forced to completion despite an unfavorable equilibrium in the intermediate step. Remember, as a general principle, that if a reaction step with an unfavorable equilibrium constant is *succeeded* by a step in which the equilibrium constant is sufficiently favorable to offset the prior step(s), the reaction will proceed in the forward direction and be synthetically useful.

The substitution of one of the hydrogens of water by an acyl group increases the acidity of the other hydrogen by a factor of more than 10^{10}. The same result occurs when an acyl group replaces one of the hydrogens of ammonia, amides being very much more acidic than ammonia and amines. Whereas ammonia has a pK_a of 33, acetamide has a pK_a of 15 and is comparable in acidity to water and methanol.

The substitution of two of the hydrogens of ammonia or a primary amine further enhances the acidity of the remaining hydrogen. Compounds of the general structure RC(=O)—N(H)—C(=O)R are called *imides*, the best known examples of which are the cyclic compounds succinimide, (CH₂C(=O))₂NH and its aromatic counterpart phthalimide, $C_6H_4(CO)_2NH$ which have pK_a values of 9.6 and 7.4, respectively; they are acidic enough to react with moderately strong aqueous base to form the anions.

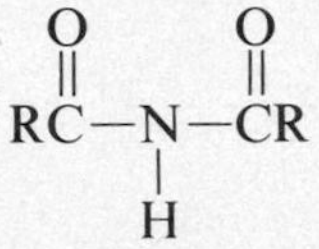

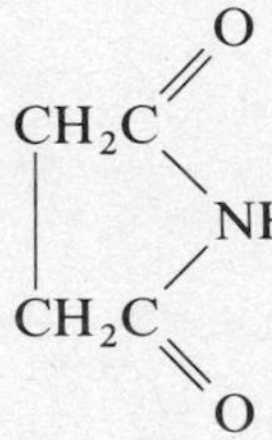

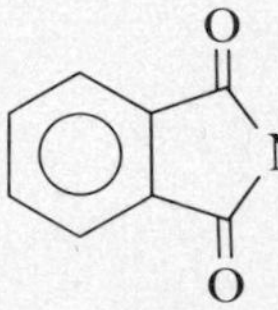

The nonbonded electrons on the nitrogen atom are responsible for the Lewis base character of nitrogen compounds (see p. 313), and to the extent that these electrons are delocalized from the nitrogen in amides, the basicity of amides is reduced. Thus, whereas alkylamines (pK_{BH^+} is 10–11) and aromatic amines (pK_{BH^+} is about 5) are sufficiently basic to react with aqueous mineral acids to form salts, amides (pK_{BH^+} is about -1) are not protonated to a significant extent by aqueous acids and, consequently, do not dissolve in them. Chemical separations of amines from amides, on the one hand, and of carboxylic acids from amides, on the other, are easily achieved by appropriate extractions with ether and aqueous acid and aqueous base, respectively.

PROBLEM 15-8

Devise a method for separating a mixture containing phthalimide, benzoic acid, aniline, and ethyl benzoate.

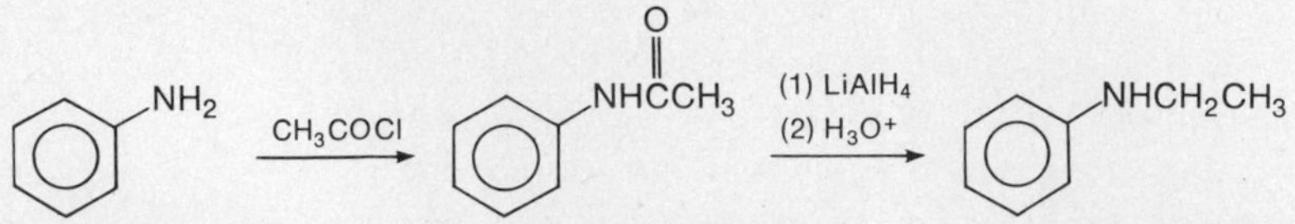

$C_6H_5NH_2 \xrightarrow{CH_3COCl} C_6H_5NHCOCH_3 \xrightarrow[(2)\ H_3O^+]{(1)\ LiAlH_4} C_6H_5NHCH_2CH_3$

Fig. 15.17. Synthesis of N-substituted amines by lithium aluminum hydride reduction.

Amides are reduced with lithium aluminum hydride to the corresponding amines rather than the alcohols, as is the case in the lithium aluminum hydride reduction of carboxylic acids, acid halides, anhydrides, and esters:

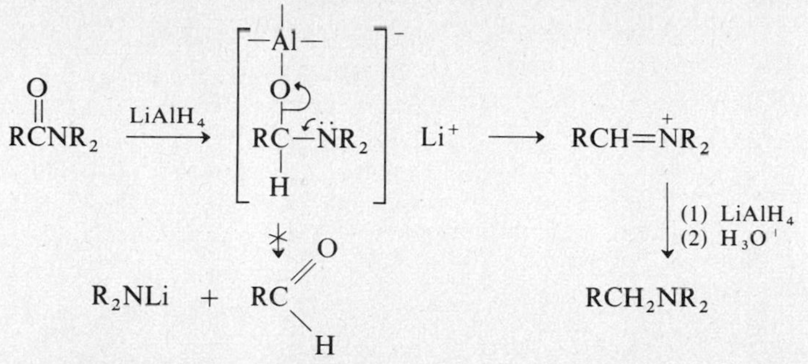

$$\text{RC(=O)NR}_2 \xrightarrow{\text{LiAlH}_4} \left[\text{RC(H)(O–Al}^{-}\text{)–}\ddot{\text{N}}\text{R}_2\right]\text{Li}^+ \longrightarrow \text{RCH}{=}\overset{+}{\text{N}}\text{R}_2 \xrightarrow[(2)\ \text{H}_3\text{O}^+]{(1)\ \text{LiAlH}_4} \text{RCH}_2\text{NR}_2$$

$$\not\rightarrow \quad \text{R}_2\text{NLi} + \text{RCHO}$$

The reduction of an amide to an amine provides an especially useful method for introducing a single alkyl group onto a nitrogen, obviating the difficulty of overalkylation when the nucleophilic substitution method involving alkyl halides is used (see Fig. 9.4). For example, N-ethylaniline, uncontaminated with N,N-diethylaniline, can be prepared from aniline by acetylation to form N-phenylacetamide (acetanilide), followed by reduction of the amide (Fig. 15.17).

15.5 Ketenes

Ketenes are anhydro-carboxylic acids (meaning a carboxylic acid minus a molecule of water) and are transformed to carboxylic acids upon treatment with water. The parent member of the series can be prepared by two methods, *viz.*, the high-temperature dehydration of acetic acid and the pyrolysis of acetone. Ketene itself, as well as its higher homologs, can also conveniently be synthesized by dehydrohalogenation of acid chlorides, using a weakly nucleophilic organic base such as triethylamine (Fig. 15.18).

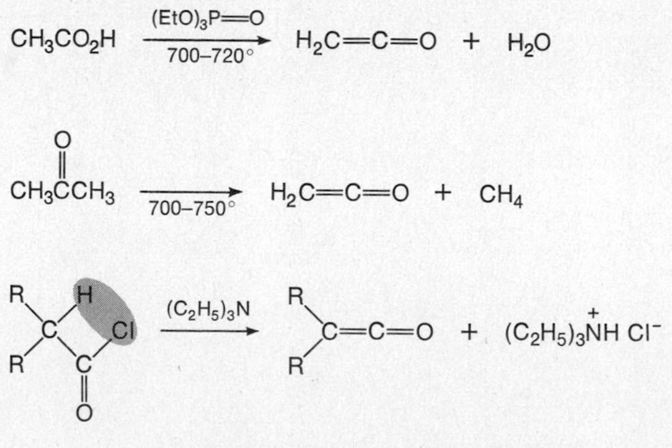

$$\text{CH}_3\text{CO}_2\text{H} \xrightarrow[700\text{–}720^\circ]{(\text{EtO})_3\text{P}{=}\text{O}} \text{H}_2\text{C}{=}\text{C}{=}\text{O} + \text{H}_2\text{O}$$

$$\text{CH}_3\text{C(=O)CH}_3 \xrightarrow[700\text{–}750^\circ]{} \text{H}_2\text{C}{=}\text{C}{=}\text{O} + \text{CH}_4$$

$$\text{R}_2\text{C(H)–C(=O)Cl} \xrightarrow{(\text{C}_2\text{H}_5)_3\text{N}} \text{R}_2\text{C}{=}\text{C}{=}\text{O} + (\text{C}_2\text{H}_5)_3\overset{+}{\text{N}}\text{H Cl}^-$$

Fig. 15.18. Preparation of ketenes.

Ketene is a poisonous gas boiling at -41°; higher molecular weight ketenes are liquids or even solids. Ketenes are characterized by an unusually high frequency carbonyl stretching deformation band in the infrared at about 2150

cm^{-1} (see Table 15.2 on p. 394). However, remember that the carbonyl carbon of a ketene is *sp* hybridized, whereas it is sp^2 hybridized in all the other classes of carbonyl compounds; this difference is also reflected in a shorter and stronger carbon–oxygen bond in ketenes.

Ketenes react readily with a variety of nucleophiles of the general structure HY to form carboxylic acids (if HY is HOH), acid halides (if HY is HX), anhydrides (if HY is HO_2CR—see p. 393), esters (if HY is HOR), and amides (if HY is HNR_2) (Fig. 15.19). Because of ketene's high reactivity, the reactions take place at low temperature, providing one of the mildest procedures for the acetylation of an alcohol, phenol, or amine.

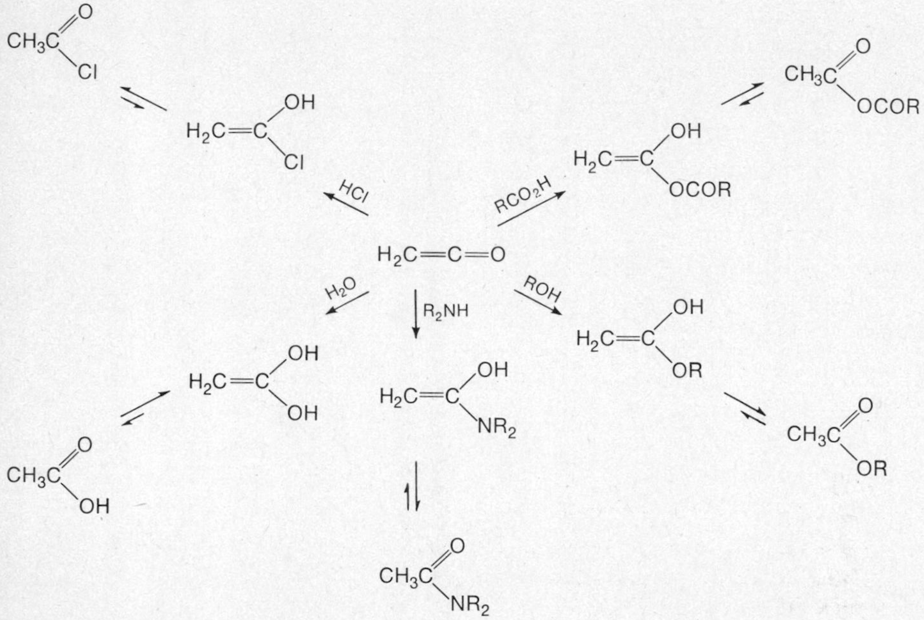

$$H_2C{=}C{=}O \xrightarrow{HCl} H_2C{=}C(OH)Cl \rightleftharpoons CH_3C(=O)Cl$$

$$H_2C{=}C{=}O \xrightarrow{RCO_2H} H_2C{=}C(OH)OCOR \rightleftharpoons CH_3C(=O)OCOR$$

$$H_2C{=}C{=}O \xrightarrow{H_2O} H_2C{=}C(OH)OH \rightleftharpoons CH_3C(=O)OH$$

$$H_2C{=}C{=}O \xrightarrow{ROH} H_2C{=}C(OH)OR \rightleftharpoons CH_3C(=O)OR$$

$$H_2C{=}C{=}O \xrightarrow{R_2NH} H_2C{=}C(OH)NR_2 \rightleftharpoons CH_3C(=O)NR_2$$

Fig. 15.19. Addition reactions of ketene.

15.6 *Derivatives of Carbonic Acid*

Compounds containing two hetero atom groups (*i.e.*, type C groups) attached to a carbonyl function include the following:

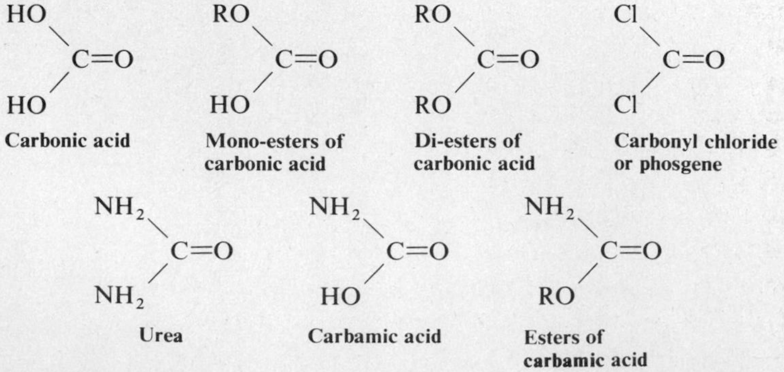

$(HO)_2C{=}O$	Carbonic acid
$(RO)(HO)C{=}O$	Mono-esters of carbonic acid
$(RO)_2C{=}O$	Di-esters of carbonic acid
$Cl_2C{=}O$	Carbonyl chloride or phosgene
$(NH_2)_2C{=}O$	Urea
$(NH_2)(HO)C{=}O$	Carbamic acid
$(NH_2)(RO)C{=}O$	Esters of carbamic acid

These compounds vary considerably with respect to stability. Carbonic acid and carbamic acid, for example, are unstable at room temperature, decomposing to carbon dioxide and water in the case of carbonic acid or carbon dioxide and ammonia in the case of carbamic acid. Carbonic acid esters and carbamic acid esters, on the other hand, are stable under ordinary conditions and decompose only when subjected to hydrolysis. Carbamic acid esters (also called urethanes) are most conveniently prepared by the action of alcohols on isocyanic acid (HNCO), which gives the parent member of the series, or on organic isocyanates (RNCO), which give the higher molecular weight members of the series (Fig. 15.20).

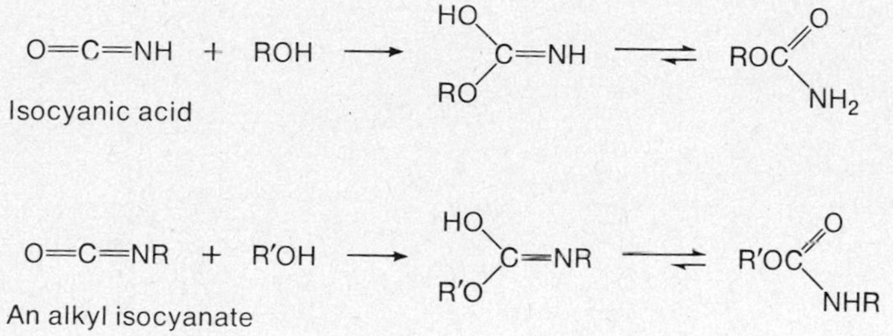

O=C=NH + ROH ⟶ (HO)(RO)C=NH ⇌ ROC(=O)NH$_2$

Isocyanic acid

O=C=NR + R'OH ⟶ (HO)(R'O)C=NR ⇌ R'OC(=O)NHR

An alkyl isocyanate

Fig. 15.20. Synthesis of urethanes.

PROBLEM 15-9

A compound having the molecular formula $C_7H_{14}O$ reacts with sodium to liberate hydrogen but is insoluble in aqueous acid or base. Its nmr spectrum shows a one-hydrogen triplet in the vinyl region, and ozonolysis of the compound yields a C_3H_6O ketone and a $C_4H_8O_2$ compound, which is readily oxidized to a $C_4H_8O_3$ compound. The $C_4H_8O_2$ compound loses water upon heating and forms a C_4H_6O compound, which has an infrared absorption band near 1750 cm^{-1}. Suggest a structure for the $C_7H_{14}O$ compound.

15.7 Cyclic Carboxylic Acid Derivatives

Ester and amide groups, in contrast to carboxylic acid and acid halide groups, can occur in ring structures. The intramolecular elimination of water from the hydroxyl and carboxylic acid groups in hydroxy acids yields cyclic esters known as *lactones* (Fig. 15.21); the intramolecular elimination of water from the amino and carboxyl groups in amino acids yields cyclic amides known as *lactams* (Fig. 15.22). The size of the lactones and lactams formed in this fashion is restricted to five-membered, six-membered, and seven-membered rings. Attempts to obtain smaller rings by intramolecular dehydration lead to alkene formation or dimerization; attempts to obtain larger rings often lead to polymerization (*i.e.*, intermolecular rather than intramolecular dehydration), although special methods for preparing large-ring lactones are known.

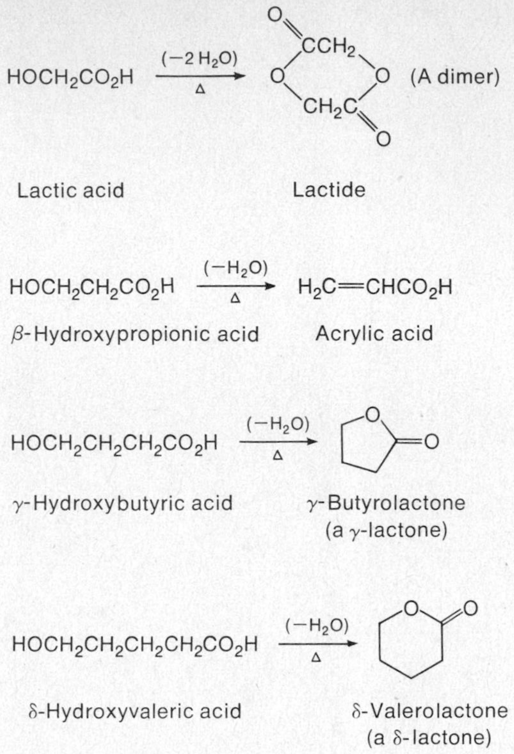

$$HOCH_2CO_2H \xrightarrow[\Delta]{(-2H_2O)} \text{(cyclic dimer: } O{=}C(CH_2)\text{–O–}CH_2C(=O)\text{–O)} \quad \text{(A dimer)}$$

Lactic acid — Lactide

$$HOCH_2CH_2CO_2H \xrightarrow[\Delta]{(-H_2O)} H_2C{=}CHCO_2H$$

β-Hydroxypropionic acid — Acrylic acid

$$HOCH_2CH_2CH_2CO_2H \xrightarrow[\Delta]{(-H_2O)} \text{(five-membered lactone)}$$

γ-Hydroxybutyric acid — γ-Butyrolactone (a γ-lactone)

$$HOCH_2CH_2CH_2CH_2CO_2H \xrightarrow[\Delta]{(-H_2O)} \text{(six-membered lactone)}$$

δ-Hydroxyvaleric acid — δ-Valerolactone (a δ-lactone)

Fig. 15.21. Dehydration of hydroxy acids.

15.8 Carboxylic Acid Derivatives of Particular Interest and Utility

Among the highly reactive carboxylic acid derivatives, acetic anhydride and phthalic anhydride are commercially the most important. Acetic anhydride is used primarily for the synthesis of cellulose acetate and other esters of acetic acid that do not form readily from the free acid and the alcohol. Phthalic anhydride, produced by the vanadium-catalyzed air oxidation of *o*-xylene or naphthalene, is used extensively in the plastics industry in the manufacture of plasticizers as well as certain types of resins.

A variety of natural products contain ester functions. These can be grouped into three classes: (1) the "fruit essences," esters of low and medium molecular weight alcohols combined with low and medium molecular weight carboxylic acids; (2) the "fats" and "oils," esters of glycerol (see Fig. 10.24) with medium and higher molecular weight carboxylic acids; and (3) the "waxes," esters of higher molecular weight alcohols with higher molecular weight carboxylic acids. Examples of group 1 include butyl butyrate, which has a pineapple odor, isoamyl *n*-valerate, which has an apple odor, and isoamyl acetate, which has a banana odor. The odor of the natural products themselves, however, is due to more than a single chemical. The volatile oil of pineapple obtained from winter fruit, for instance, contains ethyl acetate, acetaldehyde, methyl *n*-valerate, methyl isovalerate, methyl isocaproate, and methyl caprylate. "Fats," which are solid esters, and "oils," which are liquid esters, are often referred to as "glycerides"

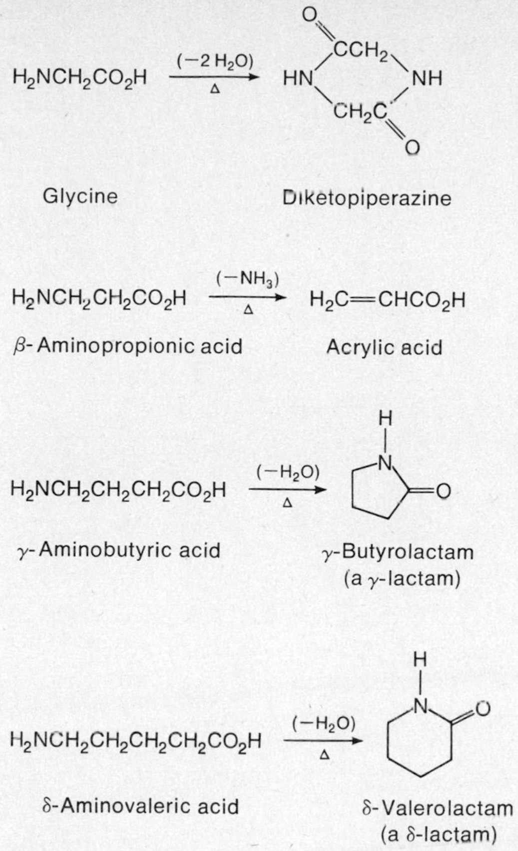

$$H_2NCH_2CO_2H \xrightarrow[\Delta]{(-2\,H_2O)} \text{Diketopiperazine}$$

Glycine — Diketopiperazine

$$H_2NCH_2CH_2CO_2H \xrightarrow[\Delta]{(-NH_3)} H_2C{=}CHCO_2H$$

β-Aminopropionic acid — Acrylic acid

$$H_2NCH_2CH_2CH_2CO_2H \xrightarrow[\Delta]{(-H_2O)} \gamma\text{-Butyrolactam}$$

γ-Aminobutyric acid — γ-Butyrolactam (a γ-lactam)

$$H_2NCH_2CH_2CH_2CH_2CO_2H \xrightarrow[\Delta]{(-H_2O)} \delta\text{-Valerolactam}$$

δ-Aminovaleric acid — δ-Valerolactam (a δ-lactam)

Fig. 15.22. Dehydration (and deamination) of amino acids.

(see Chapter 34). A typical example of a naturally occurring wax is beeswax, the material used by bees in the construction of the honeycomb. It is a solid, melting at 60–82°, consisting primarily of the esters formed from C_{26} and C_{28} acids and C_{30} and C_{32} alcohols.

The naturally occurring esters have a variety of applications. Some of the fruit essences are used in flavoring and perfumery. Many of the fats and oils, including, beef fat, mutton fat, lard, olive oil, palm oil, etc., are important foodstuffs. As mentioned previously (see p. 277), saponification of glycerides yields glycerol and the sodium salt of long-chain fatty acids, the latter being known as "soaps" and used for hundreds of years as cleansing aids. Certain waxes, such as carnauba wax, are used as ingredients in various kinds of polishes. Among the commercially important synthetic esters are ethyl acetate and butyl acetate, used as solvents in the formulation of lacquers. Some of the higher molecular weight synthetic esters are used as plasticizers for resins and plastics and are also used as integral building blocks for the plastics themselves.

A naturally occurring ester of complex structure and important commercial application as pyrethrin, the most potent insecticidal constituent of pyrethrum flowers. Pyrethrin is very effective against household insects and has relatively little toxicity toward mammals.

Pyrethrin

Coumarin is a cyclic ester (*i.e.*, a lactone) that is isolated from the tonka bean and various other plant sources. The laboratory synthesis of coumarin was achieved by W. H. Perkin (see p. 14), who marketed the compound as the first synthetic perfume, calling it "Jockey Club" and "Odor of New-Mown Hay."

Coumarin

Of the simple amides, only N,N-dimethylformamide and N,N-dimethylacetamide have significant commercial importance, being used as solvents in the processing of acrylic fibers. In the laboratory, dimethyl formamide has also been widely employed as a solvent, for its high dielectric constant usually has a facilitating effect on reactions involving charged species or reactions in which it is necessary for charge (or charge separation) to develop in the transition state. Many S_N2 substitution reactions fall into this category and are markedly enhanced in rate when carried out in dimethylformamide solution.

An interesting and biologically important amide derived from pyridine (see Fig. 16.1) is the compound known as nicotinamide, so named because the carboxylic acid to which it is related can be obtained by side-chain oxidation of nicotine (Fig. 15.23), which is an alkaloid found in tobacco. In 1937, it was

Nicotinamide adenine dinucleotide (NAD^+)

Nicotine $\xrightarrow{HNO_3}$ Nicotinic acid; Nicotinamide; NAD^+ (Ribose—ADP)

Fig. 15.23. Nicotinamide and related compounds.

realized that the lack of nicotinic acid or its amide in the human diet results in a disease known as pellagra. It is now known that nicotinamide is part of a coenzyme named nicotinamide adenine dinucleotide (NAD), the building blocks of which are nicotinamide, the sugar ribose (see Section 32.5), and adenosine diphosphate (see Fig. 32.41). The nicotinamide portion is attached to the sugar portion at the nitrogen atom as a quaternary ammonium moiety (Fig. 15.23). The function of this coenzyme, operating in collaboration with certain enzymes, is to catalyze the transfer of hydrogen between various compounds in the living system. As a result of industry's success in selling the public on the importance of nicotinamide in the diet, fortified wheat flour and vitamin preparations provide a rather large market for this material.

A cyclic amide (*i.e.*, a lactam) of great pharmaceutical importance is penicillin, which contains a β-lactam ring.

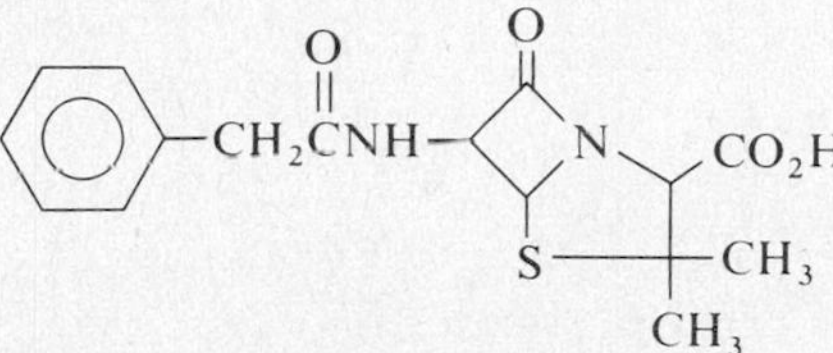

Penicillin-G

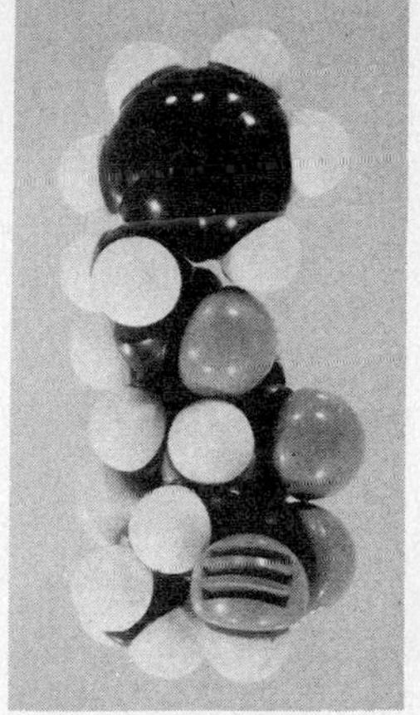

Penicillin

Although β-lactams have been known since 1912, they were considered to be so exotic that little thought was given to this ring system as a biological structural possibility when penicillin was being investigated in the 1940's. As a consequence, the structure proof study followed several blind alleys before X-ray crystallographic data proved beyond doubt that penicillin had to contain a β-lactam ring. With this and other lessons now before us, chemists have come to realize that nature often constructs molecules that seem bizarre by the conservative standards of classical organic chemistry.

Urea (NH_2CONH_2) occupies a special place in organic chemistry because of its historical interest, its biological importance, and its commercial utility. Its synthesis from the inorganic compound ammonium cyanate was achieved by Wöhler in 1828, and we have already discussed the impact of this simple reaction,

$$NH_4OCN \longrightarrow O{=}C(NH_2)_2$$

on the subsequent course of organic chemistry (see Section 1.2). In mammals, urea is the final product from the metabolism of most nitrogen compounds, and it is excreted in moderately large quantities in the urine (from which it was first isolated in 1773). Today it is manufactured on an enormous scale, most of it being used as a fertilizer but smaller amounts going into the manufacture of urea-formaldehyde plastics. A part of it also goes to the manufacture of barbiturates, which are compounds having the ability to act as depressants on the central nervous system, and which are therefore useful as sedatives and soporifics. Barbiturates are made by the condensation of urea with esters of malonic acid, as illustrated in Fig. 15.24 by the specific examples of barbituric acid and phenobarbital.

An interesting property of urea is its ability to form crystalline compounds with molecules that possess a certain geometry. For example, when *n*-decane is added to a saturated solution of urea in methanol, the precipitate that separates can be shown to contain *n*-decane and urea. The *n*-decane can be recovered from the complex simply by dissolving the urea portion in water or the *n*-decane portion in ether. The isomeric 5-methylnonane, on the other hand, does not form a

$$CH_2(CO_2C_2H_5)_2 + (NH_2)_2C{=}O \xrightarrow{C_2H_5ONa} \text{Barbituric acid}$$

Barbituric acid

$$C_6H_5(CH_3CH_2)C(CO_2C_2H_5)_2 + (NH_2)_2C{=}O \xrightarrow{C_2H_5ONa} \text{Phenobarbital}$$

Phenobarbital

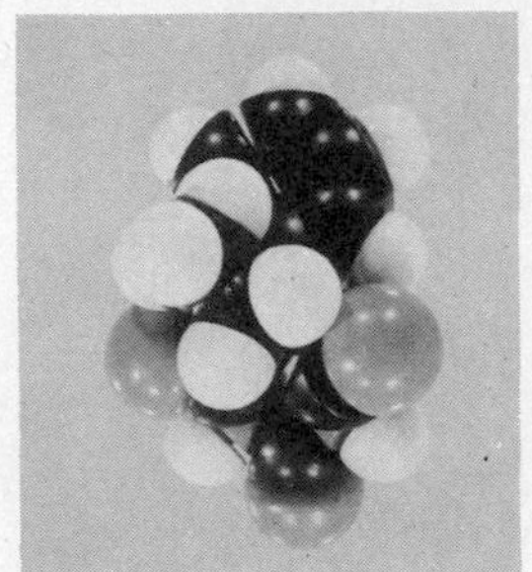

Phenobarbital

Fig. 15.24. Synthesis of barbituric acid and phenobarbital.

precipitate with urea and, in fact, a mixture of *n*-decane and 5-methylnonane can be separated cleanly in this fashion. Straight-chain alkanes, alcohols, ethers, aldehydes, nitriles, etc., containing six or more carbons all form precipitates with urea, whereas shorter analogs or compounds containing branching fail to do so. Studies of these precipitates have shown that they consist of a cylinder composed of urea molecules arranged in a helical fashion and forming a 5-Å hole or channel in which the long-chain "guest" molecule resides. Molecules of the wrong geometry, such as branched chain compounds, do not fit in the channel made by the "host" urea molecules and, as a result, are unacceptable guests. Complexes of this sort are now known for a variety of "host" and "guest" combinations and are usually referred to as *inclusion complexes* or *clathrate compounds.*

In 1935, **Wallace Carothers** of the du Pont Company discovered that high molecular weight compounds can be prepared by heating diamines with dicarboxylic acids, the reaction proceeding with the expulsion of water between the amino and carboxyl groups to form amide bonds. The macromolecule that is formed is classed as a condensation polymer (see p. 242 for polymer classification). Because of the silk-like properties of fibers made from 1,6-hexanediamine (hexamethylenediamine) and hexandioic acid (adipic acid), the polymer was introduced on the market in 1940 and given the name Nylon (more specifically, Nylon-6,6). Although first touted only as a substitute for natural silk, Nylon, along with the many other synthetic fibers that have followed it on the market since 1950, now provide synthetic substitutes for all the natural fibers. Dacron, for example, is a polymer made from dimethyl terephthalate and ethylene glycol that can be extruded in fibers and spun into fabrics resembling cotton and wool. Synthetic fibers and fabrics have become an important commercial enterprise, and a sizeable fraction of the chemical industry is devoted to their manufacture.

$n\,NH_2(CH_2)_6NH_2 \;+\; n\,HO_2C(CH_2)_4CO_2H$

1,6-Hexanediamine **Adipic acid**

$$\longrightarrow H[NH(CH_2)_6NH\overset{O}{\overset{\|}{C}}(CH_2)_4\overset{O}{\overset{\|}{C}}]_nOH + (2n-1)H_2O$$

Nylon-6,6

$n\,CH_3O_2C-C_6H_4-CO_2CH_3 \;+\; n\,HOCH_2CH_2OH$

Dimethyl terephthalate **Ethylene glycol**

$$\longrightarrow H[OCH_2CH_2O\overset{O}{\overset{\|}{C}}-C_6H_4-\overset{O}{\overset{\|}{C}}]_nOH + 2n\,CH_3OH$$

Dacron

15.9 Study Guide

I. Structure and nomenclature of carboxylic acid derivatives
   A. Acid halides
   B. Anhydrides
   C. Esters
   D. Amides
   E. Ketenes
   } Common names; Acetic acid nomenclature; Carboxylic acid nomenclature; IUPAC nomenclature

II. Physical and spectral properties of carboxylic acid derivatives (see Table 15.2)
   A. Acid halides
   B. Anhydrides
   C. Esters
   D. Amides
   E. Ketenes
   } Boiling points; Melting points; Water solubility; Ultraviolet spectra; Infrared spectra; Nmr spectra

III. Synthesis of carboxylic acid derivatives
   A. Acid halides—from RCO_2H and inorganic compounds (PCl_3, PCl_5, $SOCl_2$)
   B. Anhydrides—from acid halides
   C. Esters—from acid halides and anhydrides
   D. Amides—from acid halides, anhydrides, and esters
   E. Ketenes—from acid halides (dehydrohalogenation) and by pyrolysis of $CH_3\overset{O}{\overset{\|}{C}}R$ compounds

IV. Reactions of carboxylic acid derivatives
   A. Acid halides
   B. Anhydrides
   C. Esters
   D. Amides
   E. Ketenes
   } Hydrolysis to carboxylic acids; Reduction by lithium aluminum hydride; Nucleophilic displacement reactions

PROBLEMS

1. Using any appropriate nomenclature, name each of the following compounds:

(a) CH_3CH_2COCl

(b) 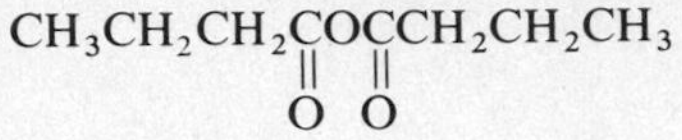 $CH_3CH_2CH_2\overset{}{\underset{O}{\underset{\|}{C}}}O\underset{O}{\underset{\|}{C}}CH_2CH_2CH_3$

(c) $HC(=O)OCH_2CH_3$

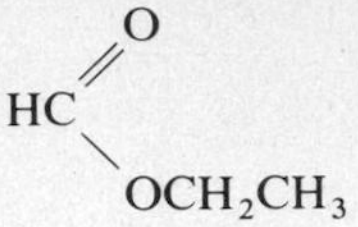

(d) 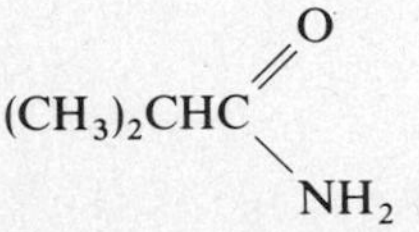 $(CH_3)_2CHC(=O)NH_2$

(e) 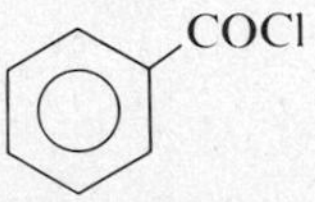 C_6H_5COCl

(f) 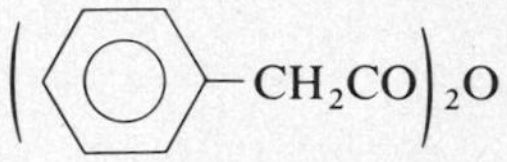 $(C_6H_5CH_2CO)_2O$

(g) 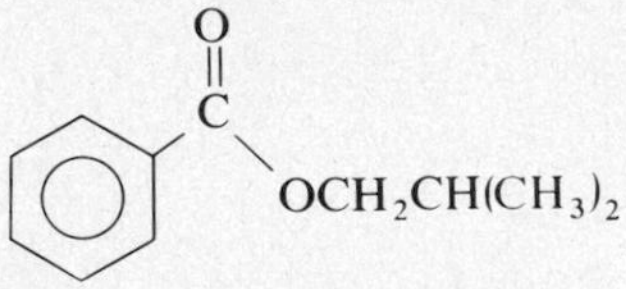 $C_6H_5C(=O)OCH_2CH(CH_3)_2$

(h) 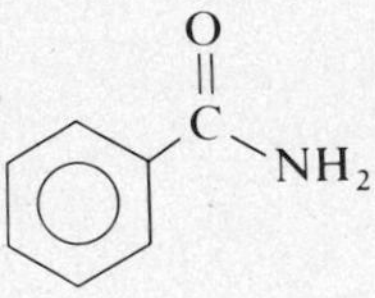 $C_6H_5C(=O)NH_2$

2. Draw the structure corresponding to each of the following names:

(a) Isovaleryl chloride
(b) Benzoic anhydride
(c) Methyl acetate
(d) Isopropyl butyrate
(e) 3-Methylpentanamide
(f) *p*-Bromobenzamide
(g) 2-Phenylbutryrl chloride
(h) 2-Phenylethyl formate

3. Using any appropriate system of nomenclature, name each of the following carboxylic acid derivatives:

(a) $(CH_3)_2CHCOCl$

(b) 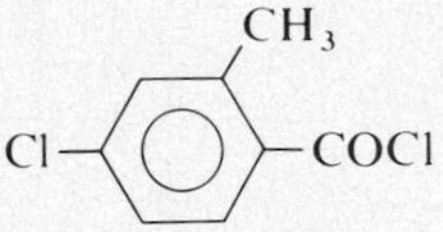 Cl—$C_6H_3(CH_3)$—COCl

(c) $CH_3\underset{CH_3}{\underset{|}{C}}HCH_2CH_2\underset{C_6H_5}{\underset{|}{C}}HCOCl$

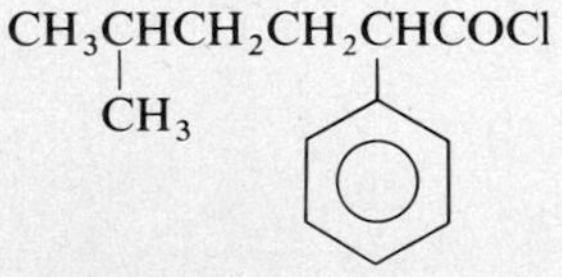

(d) 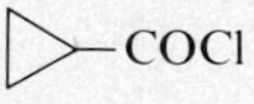 cyclopropyl—COCl

(e) 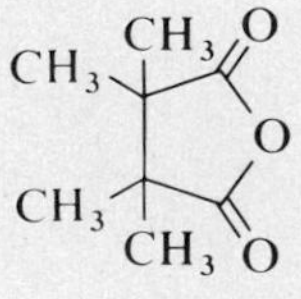 tetramethylsuccinic anhydride ring: CH_3, CH_3, O, O, CH_3, CH_3, O

(f) $HC{\equiv}CCO_2CH_2C_6H_5$

(g) 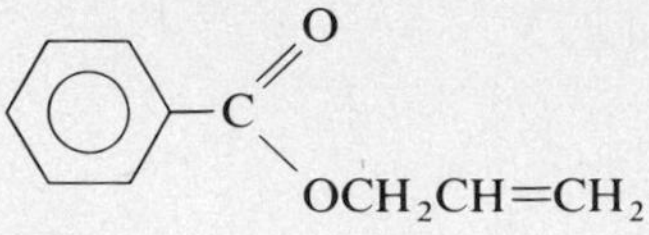 $C_6H_5C(=O)OCH_2CH{=}CH_2$

(h) 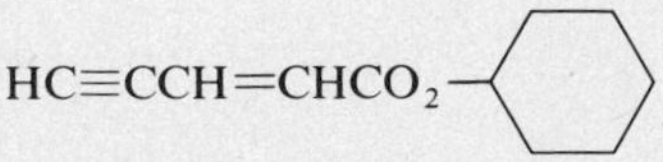 $HC{\equiv}CCH{=}CHCO_2$—cyclohexyl

(i) 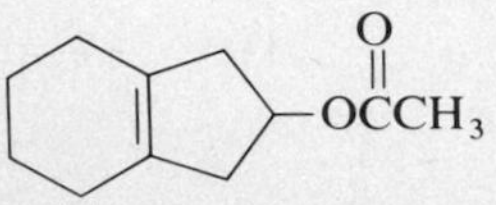 bicyclic ring—$O\overset{O}{\overset{\|}{C}}CH_3$

(j) 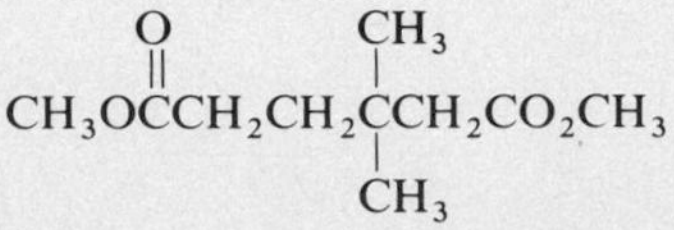 $CH_3O\overset{O}{\overset{\|}{C}}CH_2CH_2\underset{CH_3}{\underset{|}{\overset{CH_3}{\overset{|}{C}}}}CH_2CO_2CH_3$

(k) $ClCH_2CH_2CH_2CONH_2$

(l) $C_6H_{11}C(=O)NHC_2H_5$ (cyclohexyl–C(=O)–NHC_2H_5)

(m) $CH_3C(=O)N(CH_3)CH_2CH(CH_3)CH_2CH_3$

(n) m-$CH_3OC_6H_4C(=O)NH_2$ (C(=O)NH_2, OCH_3)

(o) $(CH_3)_3CCONH-C_6H_5$

(p) CONH$_2$, H

(q) $HO_2CCH_2C{\equiv}C-C(=O)-CHCH_2CO_2H$ with $C(=O)NHCH_3$ on CH

4. Draw the structure for each of the following carboxylic acid derivatives:

(a) 2-Naphthoyl chloride
(b) Stearoyl chloride
(c) Triethylacetyl chloride
(d) Cyclododecanecarbonyl chloride
(e) Benzyl phenylacetate
(f) Cyclobutylmethyl 4-vinyl-2,5-hexadienoate
(g) 1-Bicyclo[3.3.0]octyl isobutyrate
(h) Tetramethyl 1,1,2,2-ethanetetracarboxylate
(i) β-Methylvaleramide
(j) N,N-Diethylcycloheptanecarboxamide
(k) N-(3,3-Diphenylbutyl)-propanamide
(l) *p*-Trifluoromethylbenzamide
(m) Tris-(*p*-methoxyphenyl)-acetamide
(n) *cis*-Bicyclo[2.2.0]hexan-1,4-dicarboxamide
(o) 2-Chloro-6-nitro-N,N-dimethyl-3,4-octadienamide

5. Which of the following compounds would be predicted to have the (a) highest boiling point, (b) lowest boiling point, (c) greatest water solubility, (d) lowest water solubility, (e) largest dipole moment, and (f) smallest dipole moment?

CO_2H, HO_2C, CO_2H

(a)

O, O, O, CO_2H

(b)

O, C, OCH_3, C, O, OCH_3

(c)

CH_3

(d)

6. For each of the following reactions, draw the structure of the major product:

(a) $CH_3O-C_6H_4-CO_2H + SOCl_2 \longrightarrow$

(b) CH_3COCl + HN(piperidine) $\longrightarrow$

(c) $(CH_3CO)_2O + LiAlH_4 \longrightarrow \xrightarrow{H_3O^+}$

(d) $(CH_3CO)_2O + HO-C_6H_4-CH_2OH \longrightarrow$

(e) $C_6H_4(CO_2CH_3)_2$ (meta) $+ LiAlH_4 \longrightarrow \xrightarrow{H_3O^+}$

(f) cyclohexyl-COCl $+ (C_2H_5)_3N \longrightarrow$

(g) cyclohexylidene=C=O $+ (CH_3)_2NH \longrightarrow$

(h) $CH_3-C_6H_5 + CH_3COCl \xrightarrow{AlCl_3}$

(i) $C_6H_5-CON(CH_3)_2 + H_2O \xrightarrow{NaOH}$

(j) phthalimide (NH) $+ CH_3MgBr \longrightarrow$

(k) $C_6H_5-C(=O)-N(CH_2)_3$ (azetidine) $+ LiAlH_4 \longrightarrow \xrightarrow{H_3O^+}$

7. For each of the following interconversions, indicate the steps involved in going from the starting material to the product:

(a) $C_6H_5-CHO \longrightarrow C_6H_5-COCl$

(b) Cyclohexyl–CH_2OH ⟶ Cyclohexyl–$\overset{O}{\overset{\|}{C}}$–$OCH_3$

(c) Cyclopropyl–CH_2Br ⟶ Cyclopropyl–$CH_2CH_2NH_2$

(d) $(CH_3)_3CCH_2CO_2CH_3$ ⟶ $(CH_3)_3CCH_2CH_2Br$

(e) $(CH_3)_2C{=}O$ ⟶ $(CH_3)_2C(OH)CO_2H$

8. A compound is known to have the molecular formula $C_6H_{10}O_3$ and to possess one of the following structures. In the scorebox shown, indicate by a (+) for a positive reaction and a (−) for a negative reaction how you would expect each compound to react with each reagent or test. Are these tests sufficient to establish the structure definitively? If not, add one or more tests to the list.

A: tetrahydropyran-2-yl–CO_2H (ring O adjacent to the carbon bearing CO_2H)

B: $CH_3OCH_2CH{=}CHCO_2CH_3$

C: $CH_3\overset{O}{\overset{\|}{C}}CH_2CH_2CO_2CH_3$

D: $H(O{=})C–C(CH_3)_2–CH_2CO_2H$

E: $HOCH_2CH{=}CHCH_2CO_2CH_3$

F: tetrahydrofuran-2-yl–CO_2CH_3 (ring O adjacent to the carbon bearing CO_2CH_3)

	A	*B*	*C*	*D*	*E*	*F*
Solubility in $NaHCO_3$ soln						
Reaction with $C_6H_5NHNH_2$						
Reaction with dilute HCl						
Reaction with $Na_2Cr_2O_7 + H_2SO_4$						
Reaction with O_3						

9. In each of the following cases, a compound is being investigated which is known to be one of the four isomeric structures that is drawn. Using chemical and/or spectral methods (as instructed), indicate how a definite structural assignment can be made in each instance.

(a) Chemical methods only.

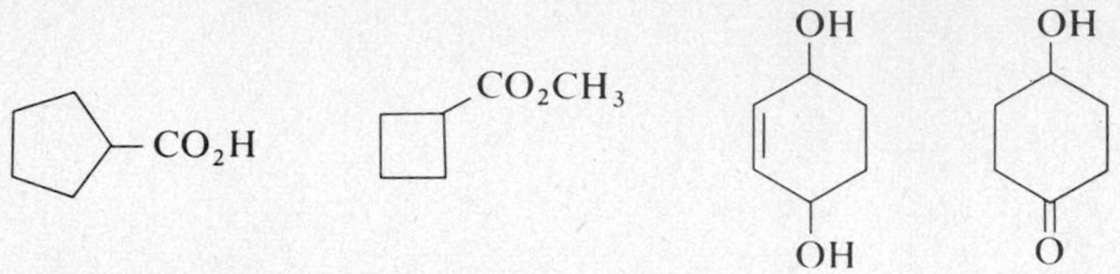

Cyclopentyl–CO_2H cyclobutyl–CO_2CH_3 (OH, OH cyclohexene-1,4-diol)   (OH, O 4-hydroxycyclohexanone)

(b) Spectral methods only.

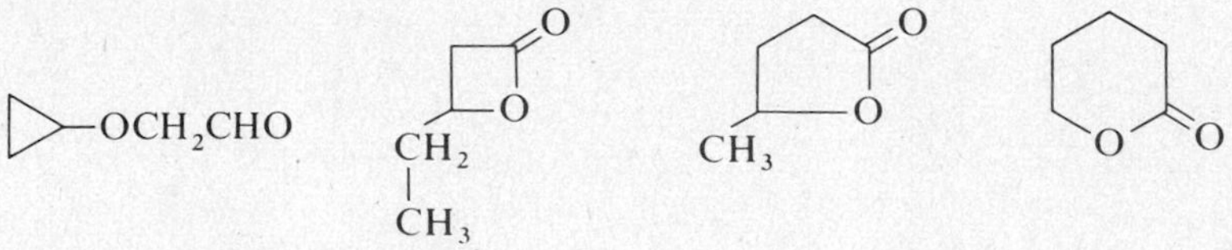

cyclopropyl–OCH_2CHO (β-lactone with CH_2CH_3)   (γ-lactone with CH_3)   (δ-valerolactone)

(c) Chemical and/or spectral methods.

(COCl, CH_3, OH benzene)   (CO_2H, CH_3, Cl benzene)   (CO_2CH_3, Cl benzene)   (CHO, Cl, OCH_3 benzene)

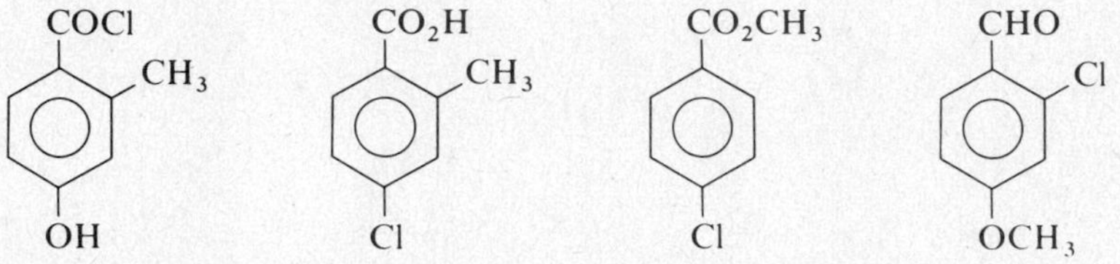

10. Available starting materials include (a) any alkanes, alkenes, alkynes, alkyl halides, alkanols, alkanals, alkanones, or alkanoic acids containing four or fewer carbons, (b) benzene, toluene, *o*-, *m*-, and *p*-xylene, and (c) any inorganic compounds. Starting with materials chosen from this group, outline syntheses for each of the following compounds:

(a) $CH_3CH_2\overset{O}{\overset{\|}{C}}OCH(CH_3)_2$ (b) 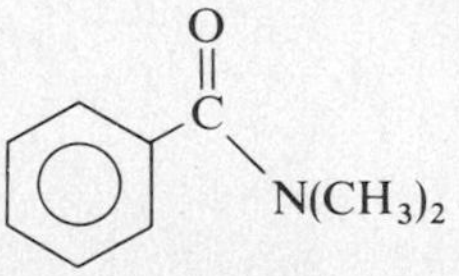 $C_6H_5\overset{O}{\overset{\|}{C}}N(CH_3)_2$ (c) (COCl, COCl *para*-benzene)

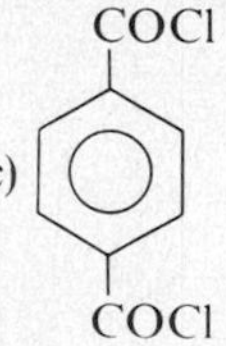

(d) $CH_3CH_2CH_2\underset{CH_3}{\underset{|}{C}}HCOCl$ (e) 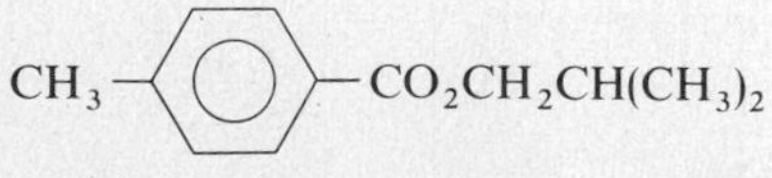 CH_3–C_6H_4–$CO_2CH_2CH(CH_3)_2$

(f) 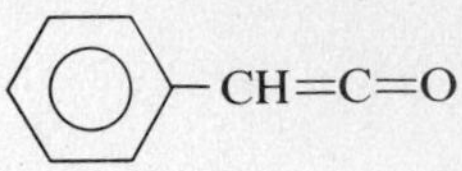 C_6H_5–CH=C=O

11. For each of the following examples, indicate a structure that is compatible with all of the data given:

(a) A $C_6H_{10}O_2$ compound boiling at 145°, which is susceptible to hydrolysis to a carboxylic acid and an alcohol and has the following nmr spectrum:

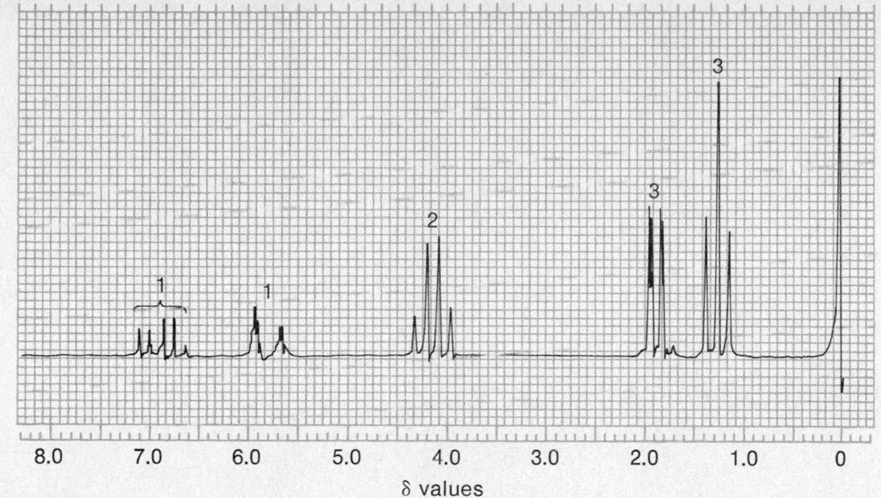

(b) A $C_8H_8O_2$ compound melting at 77°, soluble in aqueous sodium bicarbonate solution and possessing the following nmr spectrum:

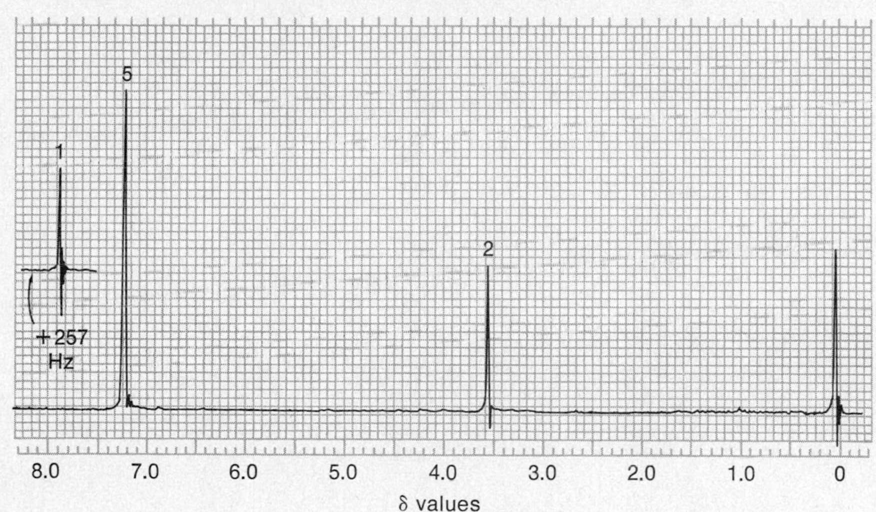

12. Arrange the following compounds with respect to increasing rate of base-induced hydrolysis, and explain the reasons for your ordering:

CO_2CH_3 / $(CH_3)_2CH$ **(a)**

$CO_2C(CH_3)_3$ **(b)**

$COCl$ / $(CH_3)_3C$ **(c)**

O / C=O / $(CH_3)_2CH$ **(d)**

13. Define the following terms as briefly and concisely as possible. Do not use a specific example in defining the term, but cite a typical example to illustrate the definition.

(a) Azeotropic distillation
(b) Claisen condensation
(c) A lactone
(d) A lactam
(e) A clathrate

Chapter **16**

Nitrogen-containing compounds other than amines

Not chaos-like together crush'd and bruis'd,
But, as the world, harmoniously confused:
Where order in variety we see
And where though all things differ, all agree.

ALEXANDER POPE
Windsor Forest

In addition to amines (see Chapter 12), in which nitrogen is single bonded to carbon frameworks, organic compounds containing nitrogen in a variety of other combinations are possible. Included among these are compounds in which nitrogen is double bonded or triple bonded to carbon, double bonded to one or two oxygens, or single and multiple bonded to itself. It is with these types of compounds that the present chapter deals.

16.1 *Compounds Containing Nitrogen Multiple Bonded to Carbon (Imines and Nitriles)*

If nitrogen is hybridized in the sp^2 fashion, it can bond with an sp^2-hybridized carbon atom to form the nitrogen analog of the C=C bond (*i.e.*, C=N) (Fig. 16.1). Compounds containing this unit are called *imines*, and the RN= moiety is called an *imino group.* Imines are named by using the suffix "imine" and a prefix specifying the carbon framework that is double bonded to the nitrogen. For example, $(CH_3)_2C{=}NH$ is 2-propylimine, the "2" indicating the point of attachment of the imino function to the propyl group. If the nitrogen is also single bonded to another group, this is indicated by the prefix "N" and the specification of the group. Thus,

$$\begin{matrix} CH_3CH_2 \diagdown & \\ & C{=}NCH_3 \\ CH_3 \diagup & \end{matrix}$$

is N-methyl-2-butylimine. When used as a substituent, the RN= group is specified as an imino function if R is H, or an N-substituted imino function if R

Bonding in Imines

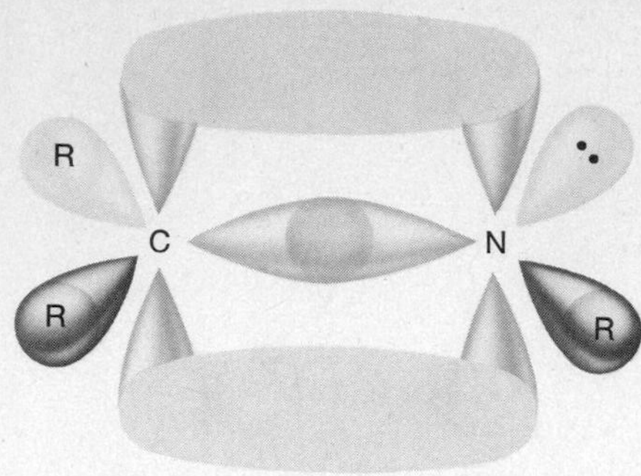

Orbital representation of an imine

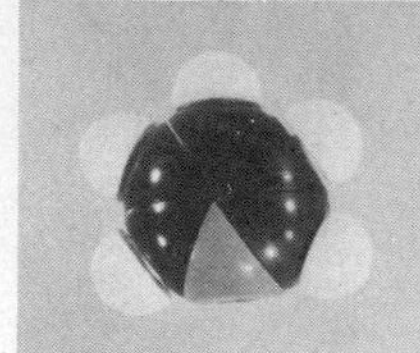

Pyridine

Bonding in pyridine

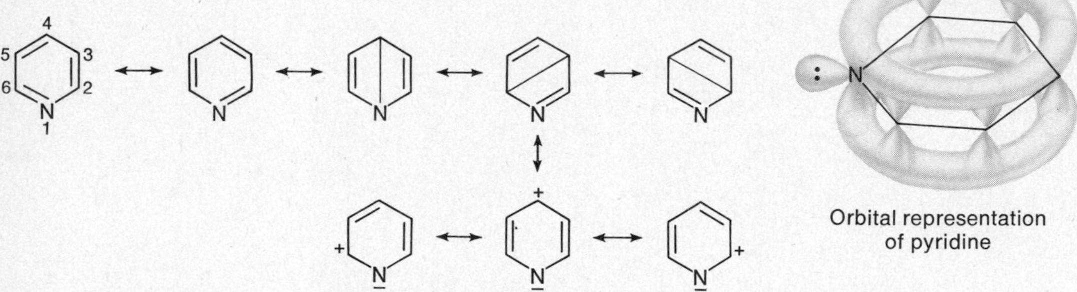

Resonance structures of pyridine

Fig. 16.1 Bonding in imines and pyridine.

is some other group. Imines are most frequently encountered in the form of phenylhydrazones, semicarbazones, and oximes (see Figs. 13.10 and 13.11), which are stable, high-melting crystalline solids. In these instances, the R′ group is a nitrogen or oxygen moiety, resulting in compounds that are fairly stable in neutral or alkaline solution (although in aqueous acid they can be converted back to the original aldehydes or ketones). If the R′ group is an alkyl or aryl group, the compounds are usually much less resistant to hydrolysis and are more difficult to prepare and purify. The C=N bonds of imines have infrared spectral characteristics very similar to those of the C=O bond, showing absorptions in the vicinity of 1670 cm^{-1}.

Pyridine (Fig. 16.1) is a cyclic imine that occurs to the extent of about 0.1% in coal tar. It is the nitrogen analog of benzene and, like benzene, is considerably more stable than a classical bond-localized structure would indicate. Pyridine contains a delocalized π system (resonance stabilization energy 23 kcal/mole), and its physical and chemical properties are in complete accord with those of a true aromatic compound. Thus, pyridine is resistant to oxidation and to addition reactions but undergoes electrophilic substitution reactions, although with extreme difficulty because of the deactivating influence of the strongly electronegative nitrogen atom. Pyridine, with a pK_{BH^+} of 5.20, is considerably less basic than aliphatic amines but comparable in basicity to aniline (pK_{BH^+} is 4.60). Thus, it dissolves in mineral acids and acts as a nucleophile toward alkyl halides, forming quaternary pyridinium salts.

Other types of compounds containing nitrogen double bonded to carbon include the isocyanides (RN=C), the isocyanates (RN=C=O), and the carbodiimides (RN=C=NR). The isocyanides, also known as carbylamines, are formed by heating a primary amine with chloroform and sodium hydroxide:

$$RNH_2 + CHCl_3 + 3\,NaOH \longrightarrow RN{=}C{:} + 3\,NaCl + 3\,H_2O$$

They have very strong odors and, consequently, provide a sensitive test for

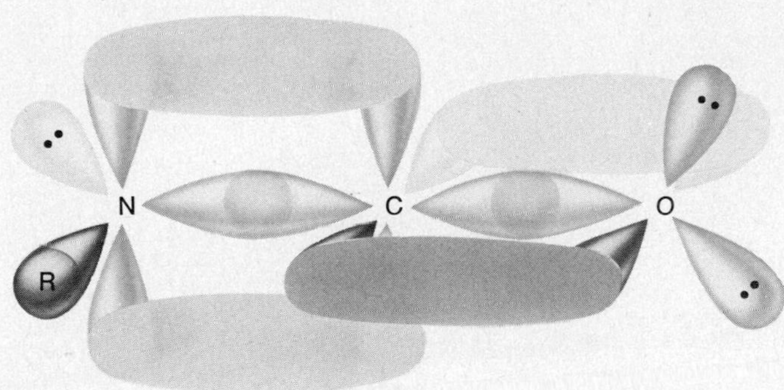

Orbital representation of an isocyanate

detecting the presence of primary amines. Isocyanates can be prepared by the action of phosgene ($COCl_2$) on primary amines:

$$3\,RNH_2 + COCl_2 \longrightarrow RN{=}C{=}O + 2\,RNH_3^+Cl^-$$

Isocyanates resemble ketene in their reactions (see Fig. 15.19); for example, phenyl isocyanate reacts with methanol to form methyl N-phenylcarbamate:

$$C_6H_5N{=}C{=}O + CH_3OH \longrightarrow \left[C_6H_5N{=}C\begin{matrix} OH \\ OCH_3 \end{matrix} \right] \longrightarrow C_6H_5NHC(=O)OCH_3$$

Carbodiimides, also called dialkylmethanediimines, are the products of dehydration of ureas—*e.g.*, by the action of benzenesulfonyl chloride and triethylamine:

$$RNHC(=O)NHR \xrightarrow[(C_2H_5)_3N]{C_6H_5SO_2Cl} RN{=}C{=}NR + H_2O$$

They have found extensive use in syntheses involving intermolecular dehydration, such as the formation of esters from carboxylic acids and alcohols, or the formation of amides from carboxylic acids and amines.

$$RCO_2H + R'OH + C_6H_{11}{-}N{=}C{=}N{-}C_6H_{11} \longrightarrow RCO_2R' + C_6H_{11}{-}NHC(=O)NH{-}C_6H_{11}$$

Bonding between an *sp*-hybridized nitrogen atom and an *sp*-hybridized carbon atom produces the *cyano group*, $-C{\equiv}N$, the nitrogen analog of the $C{\equiv}C$ bond. The parent member of this series is hydrogen cyanide, HCN, but the first member that is ordinarily considered to be an organic compound is CH_3CN. It belongs to the class of compounds known as *nitriles* or *cyanides*. Nitriles are named in relation to the carboxylic acid of corresponding structure by replacing the suffix "ic acid" or "oic acid" with "onitrile."

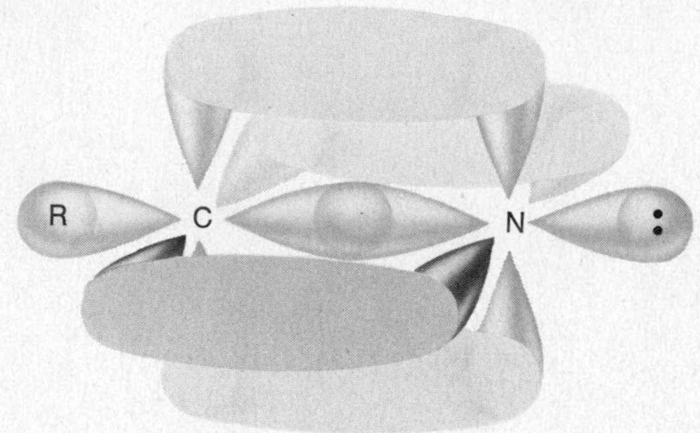

Orbital representation of a nitrile

Thus, CH_3CN is acetonitrile in the "common name" system and is ethanonitrile in the IUPAC system; C_6H_5—CN (benzene ring)—CN is benzonitrile; (cyclohexane ring)—CN is cyclohexanecarbonitrile. If the CN group is named as a substituent, it is designated as "cyano." The compound $N{\equiv}CCH_2CH{=}CHC({=}O)NH_2$, for example, is named 4-cyano-2-butenamide.

PROBLEM 16-1

Draw and name all the nitriles having the molecular formula C_4H_6ClN.

Alkyl nitriles can be prepared by the action of sodium or potassium cyanide on alkyl halides (but remember the restrictions!) in S_N2 displacement reactions (see Fig. 9.5), and alkyl as well as aryl nitriles can be prepared by dehydration of amides or oximes (Fig. 16.2). Arylnitriles can also be prepared by the action of cuprous cyanide on aryldiazonium salts (see Section 16.2 and Figs. 16.2 and 16.6).

Because nitrogen is more electronegative than carbon, carbon–nitrogen bonds are dipolar in character. Methylamine has a dipole moment of 1.3 D, and nitriles (*i.e.*, three carbon–nitrogen bonds) are about three times as dipolar, with acetonitrile having a dipole moment of 4.0 D; as a result, nitriles are unusually high-boiling materials. Acetonitrile (mol wt 41) boils at 82°, considerably higher, for instance, than dimethylamine (mol wt 45), which boils at 7.5°, and comparable

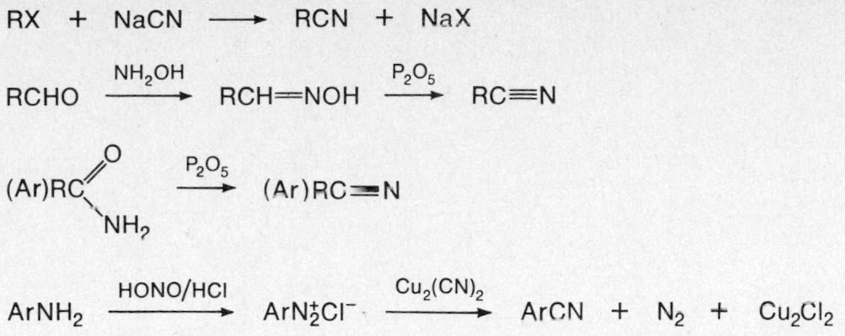

$$RX + NaCN \longrightarrow RCN + NaX$$

$$RCHO \xrightarrow{NH_2OH} RCH{=}NOH \xrightarrow{P_2O_5} RC{\equiv}N$$

$$(Ar)RC(=O)NH_2 \xrightarrow{P_2O_5} (Ar)RC{\equiv}N$$

$$ArNH_2 \xrightarrow{HONO/HCl} ArN_2^+Cl^- \xrightarrow{Cu_2(CN)_2} ArCN + N_2 + Cu_2Cl_2$$

Fig. 16.2. Preparation of alkyl- and arylnitriles.

to formic acid (mol wt 46), which boils at 100°. Nitriles are not, however, particularly water soluble; lacking hydrogen, they cannot act as hydrogen donors, and are such weak bases that they are not particularly effective as hydrogen bond acceptors.

If nitrogen is known to be present in a compound (*i.e.*, from qualitative or quantitative elemental analysis), its presence in the form of a cyano group is easily discerned by a characteristic absorption band in the infrared spectrum at 2260–2240 cm^{-1}. This is close to the region of absorption of the C≡C bond of unsymmetrically substituted acetylenes (2140–2100 cm^{-1}) but sufficiently separated from it that there is seldom any difficulty in establishing the identification (see Table 16.1 on p. 437).

Nitriles contain a nonbonded pair of electrons and might be expected to possess basic properties comparable to those of amines. In contrast to amines, however, they are exceedingly weak bases and show no tendency whatsoever to dissolve in aqueous acid. Whereas the pK_{BH^+} values for typical alkylamines are about 10, those for nitriles are about −10; *i.e.*, 20 orders of magnitude difference. A phenomenon similar to that suggested for the C—H bond characteristics in acetylene may explain this striking difference. If it is assumed that the nitrogen atom is *sp* hybridized in nitriles, the nonbonded pair of electrons resides in an *sp*-hybridized orbital. The greater *s* character of an *sp*-hybridized orbital as compared with an sp^3-hybridized orbital (*i.e.*, as in amines) brings the electrons closer to the nitrogen; they are, consequently, more tightly held, making it more difficult for them to engage in bonding with an electrophile. Stated conversely, the acidity of a protonated nitrile, *i.e.*, $RC{\equiv}\overset{+}{N}H \rightleftharpoons RC{\equiv}N + H^+$ is greater than the acidity of a protonated amine, *i.e.*, $R_3\overset{+}{N}H \rightleftharpoons R_3N + H^+$ by about the same amount that the acidity of RC≡CH is greater than that of R_3CH (*i.e.*, 20 orders of magnitude).

PROBLEM 16-2

Devise a reaction sequence by means of which toluene can be converted to phenylacetamide, using a nitrile as an intermediate.

PROBLEM 16-3

The following reaction sequence involves a nitrile intermediate at one point. Indicate how each of the steps in the sequence could be effected, realizing that more than one step might be involved between the compounds that are pictured.

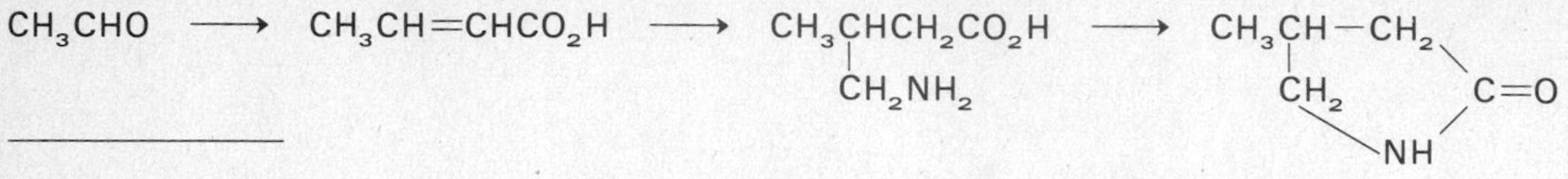

$$CH_3CHO \longrightarrow CH_3CH{=}CHCO_2H \longrightarrow CH_3CH(CH_2NH_2)CH_2CO_2H \longrightarrow \text{(cyclic lactam: } CH_3CH{-}CH_2{-}C{=}O,\ CH_2{-}NH \text{ ring)}$$

The two most generally used reactions of nitriles are their hydrolysis to the corresponding acids and their reduction to the corresponding primary amines. Either aqueous acid or aqueous base can be employed to accomplish the hydrolysis, and lithium aluminum hydride is particularly effective in the reduction (Fig. 16.3). These reactions afford very good methods for introducing carboxyl and amino functions into aliphatic and aromatic systems and should be remembered in designing synthesis sequences.

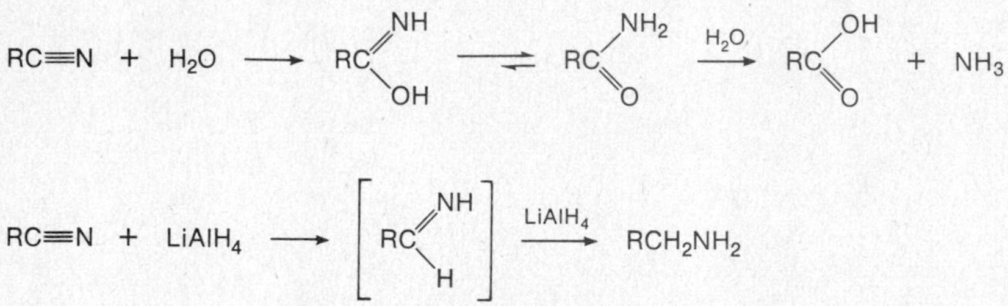

$$RC{\equiv}N + H_2O \longrightarrow RC({=}NH)OH \rightleftharpoons RC({=}O)NH_2 \xrightarrow{H_2O} RC({=}O)OH + NH_3$$

$$RC{\equiv}N + LiAlH_4 \longrightarrow \left[RC({=}NH)H\right] \xrightarrow{LiAlH_4} RCH_2NH_2$$

Fig. 16.3. Conversion of nitriles to acids and amines.

16.2 Compounds Containing Nitrogen–Nitrogen Bonds

Compounds containing nitrogen–nitrogen single, double, and triple bonds are well known. The parent member of the family containing a N—N single bond is called hydrazine or *diazane*; *i.e.*, a *diaza*alk*ane* (H_2NNH_2)—*cf.* p. 429. One, two, three, or four of the hydrogens of hydrazine can be replaced by alkyl, aryl, or acyl and aroyl functions, giving well-known types of compounds (Fig. 16.4). The alkyl- and aryl-substituted hydrazines may be named either as alkylhydrazines or alkyldiazanes.

The parent member of the series containing N=N bonds is called diimide or diazene (HN=NH); it is an unstable compound that decomposes at room temperature to yield molecular nitrogen and hydrazine, NH_2NH_2. It can be generated *in situ* and has found interesting applications as a reducing agent for the conversion of alkenes and alkynes to alkanes. The dialkyl and diaryl derivatives of diimide (*i.e.*, dialkyl and diaryldiazenes), on the other hand, can be isolated and characterized. Azomethane (dimethyldiazene), for example, is a yellow gas that is stable up to 200°, at which temperature it decomposes to methyl radicals and nitrogen,

$$CH_3N{=}NCH_3 \longrightarrow 2\,CH_3 + N_2$$

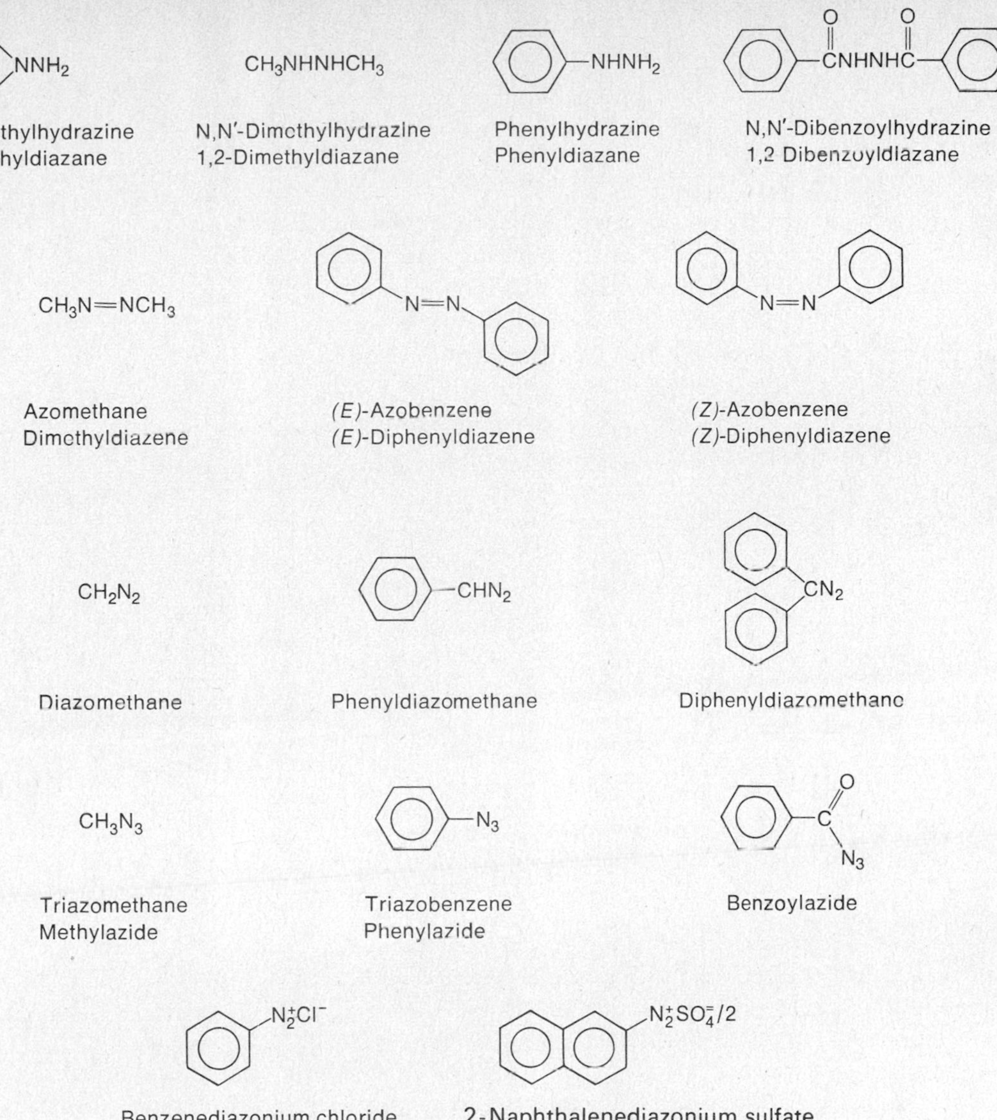

Fig. 16.4. Structures of compounds containing nitrogen-nitrogen bonds.

the methyl radicals, of course, react rapidly to form ethane and other products. Azobenzene (diphenyldiazene, $C_6H_5N{=}NC_6H_5$) is a stable, orange-red solid, which exists as (*Z*) and (*E*) isomers comparable to 1,2-diphenylethylene. The most generally useful method for preparing aromatic azo compounds involves the action of an aryldiazonium salt on an aromatic ring which is activated toward electrophilic substitution by a hydroxyl or amino group (Fig. 16.5). The process, used extensively in the synthesis of dyes, is known as a "coupling reaction" and is a typical example of an electrophilic substitution on an aromatic nucleus (see Figs. 8.13 and 8.14).

$$C_6H_5-N_2^+Cl^- + C_6H_5-OH \xrightarrow[(2)\ H_3O^+]{(1)\ NaOH} C_6H_5-N{=}N-C_6H_4-OH$$

Fig. 16.5. Formation of azo compounds from aryldiazonium salts.

Classes of compounds containing N≡N bonds include the diazoalkanes, the aryldiazonium salts, and the azides. Diazomethane, the simplest member of the diazoalkanes, is a yellow, highly toxic and explosive gas which, despite these undesirable characteristics, is a useful reagent for a variety of reactions, including the preparation of esters from carboxylic acids ($RCO_2H + CH_2N_2 \rightarrow N_2 + RCO_2CH_3$). An adequate representation of diazomethane requires a number of charge-separated resonance contributing structures; *i.e.*,

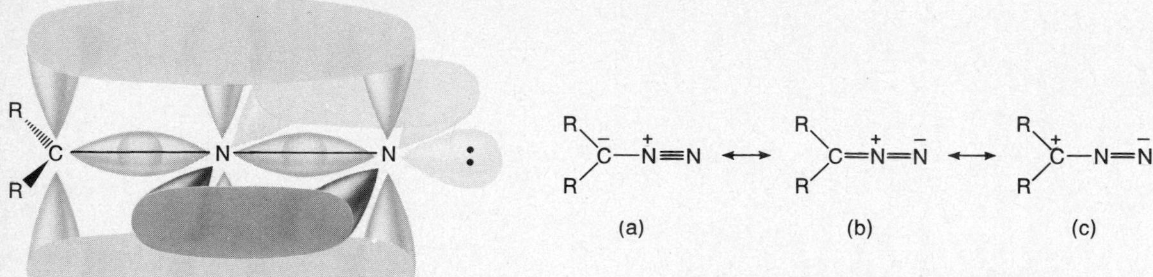

Resonance structures of a diazoalkane

Orbital representation of a diazoalkane

When heated, irradiated with light of the appropriate wavelength, or exposed to certain copper-containing catalysts, diazomethane loses molecular nitrogen and forms carbene,

$$CH_2N_2 \longrightarrow CH_2: + N_2$$

(see Fig. 7.14). Carbenes are exceedingly reactive species, and, as discussed previously, can add to alkenes to form cyclopropanes (see Fig. 8.7).

Analogous in some respects to the diazoalkanes are the triazo compounds (azides, azido compounds), which are derivatives of hydrazoic acid (HN_3). Triazo compounds are similar to diazoalkanes in possessing linear structures that require a number of charge-separated, resonance-contributing structures for their proper representation; *i.e.*,

$$R-\ddot{\overset{-}{N}}-\overset{+}{N}{\equiv}N: \longleftrightarrow R-\underset{\cdot\cdot}{N}{=}\overset{+}{N}{=}\underset{\cdot\cdot}{\overset{-}{N}}: \longleftrightarrow R-\underset{\cdot\cdot}{\overset{+}{N}}-\ddot{N}{=}\underset{\cdot\cdot}{\overset{-}{N}}:$$

In a fashion similar to the diazoalkanes, triazoalkanes decompose when heated or irradiated to yield molecular nitrogen and nitrenes, the nitrogen analogs of carbenes ($RN_3 \rightarrow R{-}N: + N_2$). Nitrenes are comparable in reactivity to carbenes and, for example, add to alkenes to produce aziridines:

$$>C{=}C< \; + \; R\underset{\cdot\cdot}{N}: \longrightarrow \text{(aziridine ring: two C and NR)}$$

Although the diazonium salts have a superficial structural similarity to the diazoalkanes, they are actually quite different in composition and chemistry. They can be considered as quaternary ammonium compounds (Fig. 16.6), in

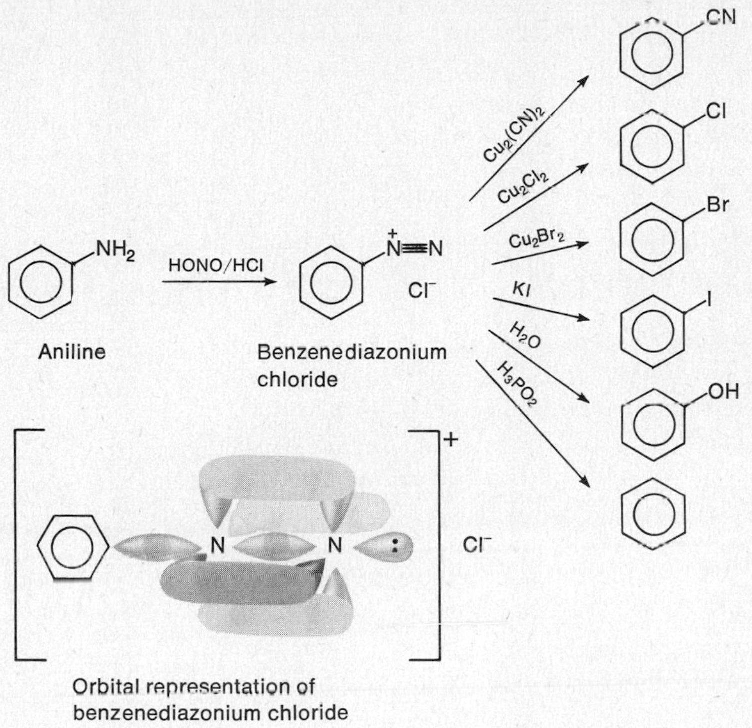

Fig. 16.6. Synthesis of aromatic nitriles and other substituted aromatic compounds via aryldiazonium salts.

which three of the four ligands of the central nitrogen are attached to the adjacent nitrogen and the fourth ligand to a carbon framework. If the attached framework is an alkyl group, the compounds are so unstable that they decompose almost immediately upon formation. If the attached framework is an aromatic group, however, the compounds are sufficiently stable to be able to exist at or below room temperature for a long enough time to permit convenient laboratory manipulation. The usual method of synthesis of the aryldiazonium salts involves the action of nitrous acid and a mineral acid (HCl or H_2SO_4) on an aromatic amine. The resulting aqueous solution of the aryldiazonium salt can be used directly in the subsequent reaction without the necessity of isolating the dry salt (sometimes explosive), and treatment with any of a number of nucleophilic reagents leads to replacement of nitrogen by the nucleophile. Because of the ease with which aromatic amines are prepared (*e.g.*, by nitration followed by reduction) and because of the variety of available nucleophiles (Fig. 16.6), aryldiazonium salts have been enormously important in organic synthesis since their discovery by Peter Griess in 1858.

Nitroso compounds contain one trivalent nitrogen atom double bonded to one oxygen atom and are represented by the general structure RN=O. If the carbon to which the NO group is attached carries a hydrogen, the nitroso compound is thermodynamically unstable relative to the tautomeric oxime, *i.e.*,

$$R_2CHN{=}O \rightleftharpoons R_2C{=}NOH$$

The only stable nitroso compounds in which the R group is an alkyl framework, therefore, carry the NO group in a tertiary position, *i.e.*, $R_3C{-}NO$. More widely occurring are compounds in which the nitroso group is attached to an aromatic framework, such as *p*-nitroso-N,N-dimethylaniline, which is obtained by the action of nitrous acid on N,N-dimethylaniline:

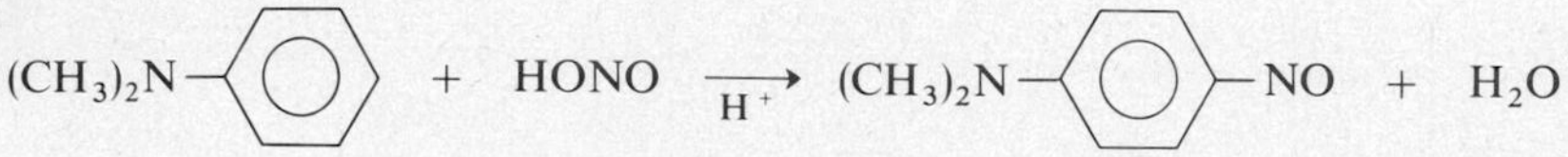

$$(CH_3)_2N{-}C_6H_5 + HONO \xrightarrow[H^+]{} (CH_3)_2N{-}C_6H_4{-}NO + H_2O$$

Nitro compounds contain one nitrogen atom multiple bonded to two oxygen atoms and are represented by the general structure RNO_2. Although our first inclination might be to depict the nitro group as containing two double bonded oxygens, *i.e.*,

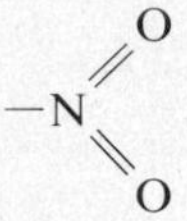

$$-N{\genfrac{}{}{0pt}{}{\diagup\!\!\!\diagup O}{\diagdown\!\!\!\diagdown O}}$$

we must remember that there is abundant evidence to indicate that first-row elements can accommodate no more than eight electrons in their outer shell; the nitrogen atom in the structure shown has 10 electrons in the outer shell. We can circumvent this difficulty by making one of the nitrogen–oxygen linkages a single bond—*i.e.*, a coordinate-covalent bond in which the nitrogen provides both the electrons of the bond. We can, for example, picture the nitro group as arising from the combination of a nitroso group with an atom of oxygen; *i.e.*,

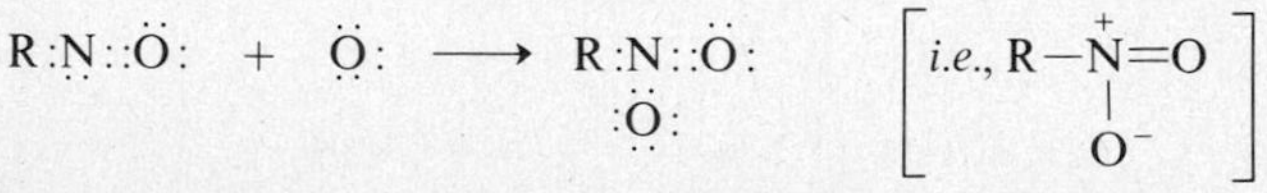

$$R{:}\ddot{N}{::}\ddot{O}{:} + {:}\ddot{O}{:} \longrightarrow R{:}N{::}\ddot{O}{:} \ (\text{with } {:}\ddot{O}{:} \text{ on N}) \qquad \left[\textit{i.e.}, R{-}\overset{+}{N}(-O^-){=}O \right]$$

In furnishing both the electrons for the bond, the nitrogen atom acquires a positive charge and the oxygen atom a negative charge; a charge-separated, highly polar bond is produced. This structure by itself is insufficient, however, because bond length measurements of nitro compounds show that there is only one nitrogen–oxygen distance and that it is somewhere between a N=O and N—O bond. This dilemma is resolved by postulating that the nitro group is a delocalized structure in which the π system extends over the nitrogen atom and both of the oxygen atoms (Fig. 16.7). The negative charge, thus, is shared equally by the two oxygen atoms, depicted by a pair of resonance structures in Fig. 16.7.

Nitro compounds are named by treating the nitro moiety as a functional group and specifying its position on the parent compound; for example, CH_3NO_2 is nitromethane, $(CH_3)_2CHNO_2$ is 2-nitropropane, $C_6H_5{-}NO_2$ is nitro-

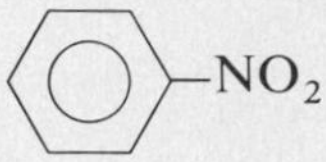

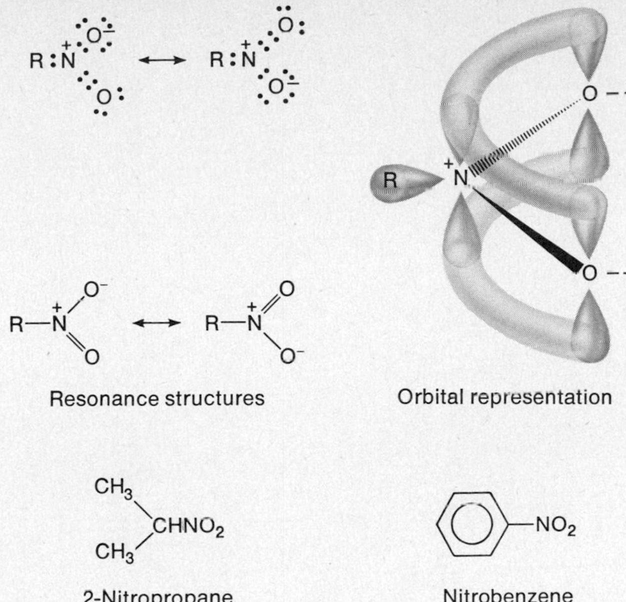

Fig. 16.7. Resonance structure and orbital representation of nitro group.

benzene, and CH_3O—C_6H_4—NO_2 is 4-methoxy-l-nitrobenzene (also known as *p*-nitroanisole).

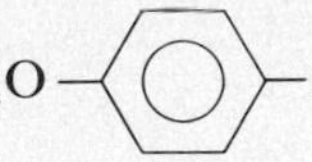

Nitroalkanes can be prepared by the direct nitration of alkanes and by S_N2 displacement reactions of alkyl halides with inorganic nitrites. The nitrite anion, however, can react either as a nitrogen nucleophile, to give a nitro compound, or as an oxygen nucleophile, to give an alkyl nitrite (Fig. 16.8). A species of this sort is called an "ambident anion," meaning that it can react in an equivalent fashion at two or more of its constituent atoms.

The most convenient method for the preparation of aromatic nitro compounds involves the action of a nitric acid–sulfuric acid mixture on aromatic compounds

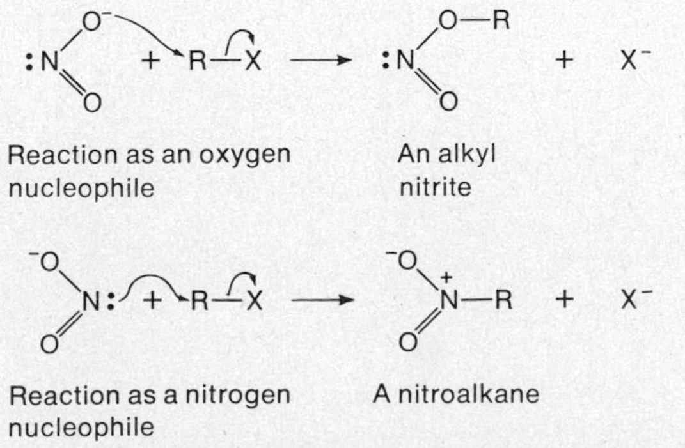

Reaction as an oxygen nucleophile — An alkyl nitrite

Reaction as a nitrogen nucleophile — A nitroalkane

Fig. 16.8. Reactions of nitrite anion with alkyl halides.

(see Figs. 8.13 and 10.21). Although the initially formed mono-nitro compound is susceptible to further nitration, the second nitro group enters the molecule much less easily as a consequence of the electron-attracting characteristics of the nitro group (implied by the resonance structures for nitrobenzene shown in Fig. 16.9). Withdrawal of electrons from the benzene ring diminishes the ability

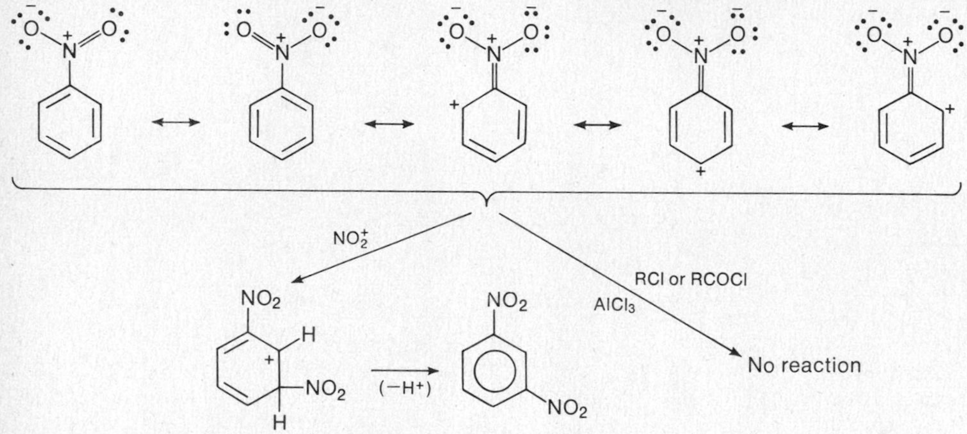

NO_2^+ RCl or RCOCl $AlCl_3$

NO_2 H NO_2 H $\xrightarrow{(-H^+)}$ NO_2 NO_2 No reaction

Fig. 16.9. Electrophilic substitution of nitrobenzene.

of the ring to act as a Lewis base and reduces the rate of the reaction with the second incoming electrophile, NO_2^+. Thus, the nitro group resembles the carbonyl group (see p. 350 and 378) and acts as a deactivating and *m*-directing function in electrophilic substitution reactions. For example, nitrobenzene reacts with nitric acid–sulfuric acid to slowly yield *m*-dinitrobenzene, and *m*-dinitrobenzene requires very high temperatures and long reaction times to produce 1,3,5-trinitrobenzene. Nitrobenzene, like benzoic acid (see p. 378), fails to react with alkyl halides and acid chlorides under Friedel-Crafts conditions.

Nitro compounds, because of the coordinate–covalent character of the N—O bond of the nitro group, have high dipole moments (nitromethane, $\mu = 3.4$ D; nitrobenzene, $\mu = 4.3$ D), and high boiling points (nitromethane bp, 101.5°; nitrobenzene bp 209°). For reasons similar to those cited for nitriles (see Section 16.1), however, nitro compounds are not particularly water soluble. Like nitriles, they cannot act as hydrogen bond donors, and they are only weakly basic. Sufficiently strong acids, however, are capable of interacting with the nitro group, as exemplified by the complex that is formed between aluminum chloride (a very strong Lewis acid) and nitrobenzene:

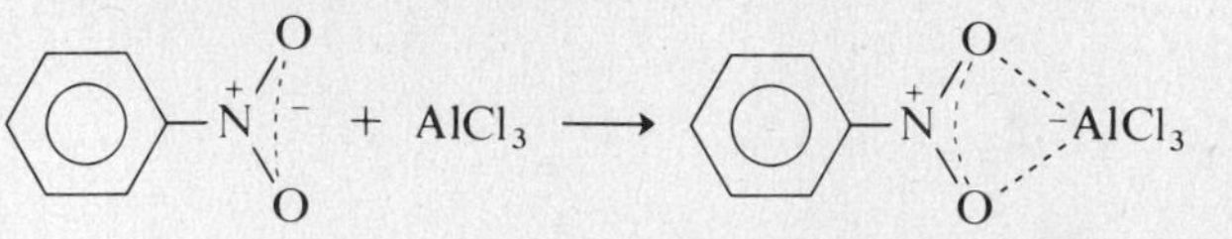

$$C_6H_5NO_2 + AlCl_3 \longrightarrow C_6H_5NO_2 \cdot AlCl_3$$

Nitro groups, like carboxyl groups, are near the end of an oxidation line and can undergo further oxidation only with cleavage of C—N bonds. They are, of course, susceptible to reduction. Reduction all the way to primary amines

($RNO_2 \rightarrow RNH_2$) can be effected with a variety of reagents, including (a) molecular hydrogen and a catalyst such as platinum, (b) metals such as iron or tin and a mineral acid, and (c) sodium hydrosulfite ($Na_2S_2O_4$) and base. The reduction product of an alkyl nitro compound is a primary alkylamine; that from an aromatic nitro compound is an aromatic primary amine. The latter represents a particularly important synthesis intermediate, for aromatic amines can, in turn, be converted into a variety of other types of compounds (see Fig. 16.6). The conversion of anisole to *p*-methoxybenzylamine, for example, can be accomplished by the sequence of reactions outlined in Fig. 16.10.

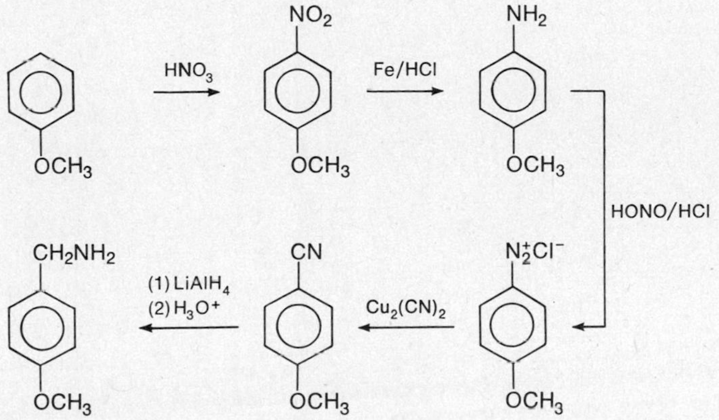

Fig. 16.10. Synthesis of *p*-methoxybenzylamine from anisole.

Nitroalkanes of the general structure RCH_2NO_2 and R_2CHNO_2 have surprisingly acidic α hydrogens, attributed to the electron-withdrawing power of the nitro group coupled with its ability to delocalize the negative charge of the anion that is formed upon ionization; *i.e.*,

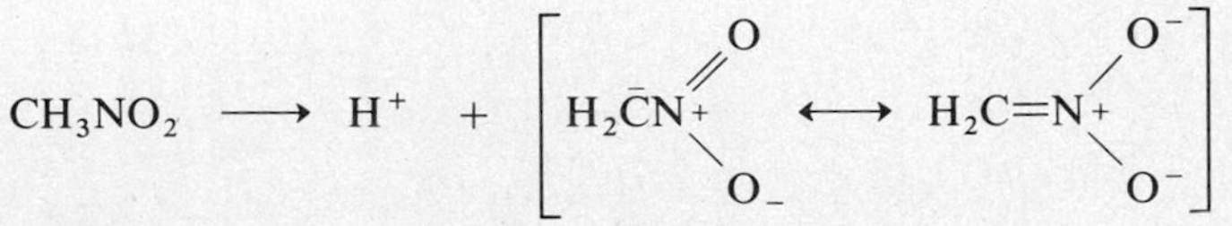

$$CH_3NO_2 \longrightarrow H^+ + \left[H_2\bar{C}N^+\!\begin{matrix}\nearrow O \\ \searrow O^-\end{matrix} \longleftrightarrow H_2C{=}N^+\!\begin{matrix}\nearrow O^- \\ \searrow O^-\end{matrix} \right]$$

Thus, nitromethane belongs to the class of compounds containing "active methylene" groups and adds in aldol-like condensations to the carbonyl groups of aldehydes and ketones to yield β-hydroxynitro compounds. For example, three equivalents of formaldehyde react in the presence of sodium hydroxide with one equivalent of nitromethane, which contains three acidic hydrogens, to yield tris-(hydroxymethyl)-nitromethane (Fig. 16.11); reduction of this compound yields the corresponding amine, tris-(hydroxymethyl)-methylamine, a compound known as "tris" and used extensively as a buffer reagent in biochemical systems.

PROBLEM 16-4

Devise a method for converting benzene to *p*-nitrobenzonitrile, realizing that direct nitration of benzonitrile yields the *meta* isomer.

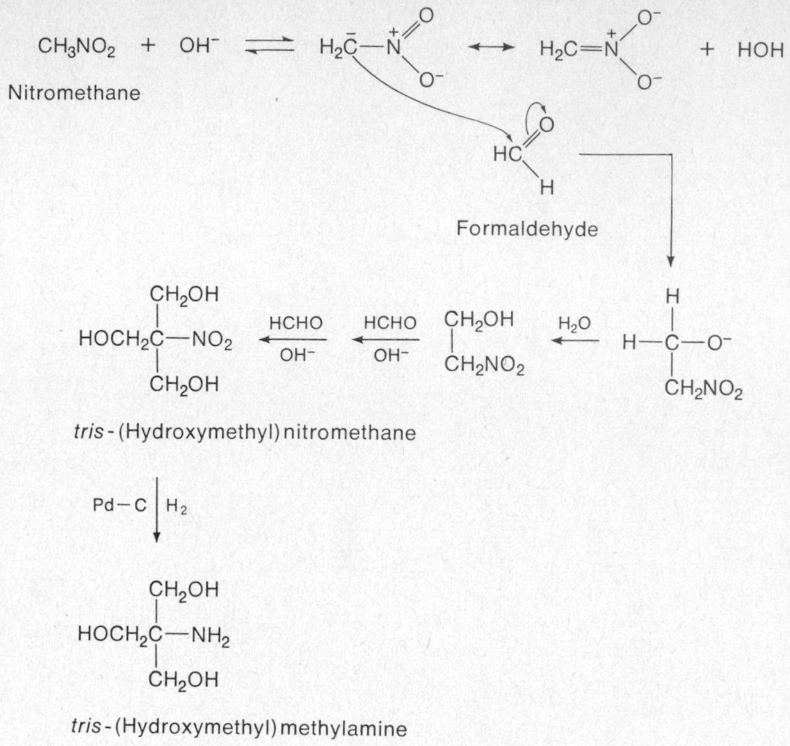

$$CH_3NO_2 + OH^- \rightleftharpoons H_2\bar{C}-\overset{+}{N}(=O)O^- \longleftrightarrow H_2C=\overset{+}{N}(O^-)O^- + HOH$$

Nitromethane

$$HCHO$$

Formaldehyde

$$H_2\bar{C}NO_2 + HCHO \longrightarrow H-\underset{CH_2NO_2}{\overset{H}{C}}-O^- \xrightarrow{H_2O} \underset{CH_2NO_2}{CH_2OH} \xrightarrow[OH^-]{HCHO} \xrightarrow[OH^-]{HCHO} HOCH_2C(CH_2OH)_2-NO_2$$

tris-(Hydroxymethyl)nitromethane

$$\downarrow Pd-C,\ H_2$$

$$HOCH_2C(CH_2OH)_2-NH_2$$

tris-(Hydroxymethyl)methylamine

Fig. 16.11. Aldol-type condensation of nitromethane.

PROBLEM 16-5

Compounds A and B both have the molecular formula C_9H_9N. Compound A has an infrared absorption band at 2230 cm^{-1} and compound B has one at 1640 cm^{-1}. The nmr spectrum of compound A has two resonance envelopes in the region upfield from $\delta 3.0$, one being a two-hydrogen quartet and the other a three-hydrogen triplet. Compound B has all its nmr resonances downfield from $\delta 5.0$. Hydrolysis of compound A yields a C_9 carboxylic acid, and hydrolysis of compound B yields aniline and a C_3 compound. On the basis of this information, suggest structures for compounds A and B. If more than one structure is compatible with all these data, suggest additional chemical and/or spectral methods that might be used to make the differentiation.

16.4 *Spectral Properties of Compounds Containing Multiple Bonded Nitrogen*

The uv/visible, ir, and nmr spectral characteristics of several of the types of nitrogen-containing compounds discussed in this chapter are shown in Table 16.1.

Table 16.1 NMR, IR, AND UV SPECTRAL CHARACTERISTICS OF COMPOUNDS CONTAINING MULTIPLE BONDED NITROGEN

Type of Compound Class Name	Structure	NMR δ Values	IR cm^{-1}	UV nm (ε)
Imines	$R_2C{=}N{-}R$		1670	195 (2000) (R = alkyl)
				225 (R = alkyl)
Nitriles	$R{-}C{\equiv}N$	2–2.5[a]	2260–2240	210 (10,000) $R = H_2C{=}CH{-}$)
	$Ar{-}C{\equiv}N$		2240–2220	224 (13,000)
Diazenes	$R{-}N{=}N{-}R$		1630–1575	Approx. 350 (low)
Diazonium salts	$Ar{-}N_2^+$		2260	
Diazoalkanes	R_2CN_2		2100	350 (30,000)
Azides	$R{-}N_3$		2130	
Nitroso compounds	$R{-}N{=}O$		1550	
Nitro compounds	$R{-}NO_2$	4.3[a]	1570 1500	
	$Ar{-}NO_2$		1570–1500	268 (8000)
Pyridines	N	C-2 8.50 C-3 7.06 C-4 7.46	Similar to benzene	195 (7500) 252 (2750)

[a] For proton in α position to functional group.

16.5 *Multiple Bonded Nitrogen Compounds of Particular Interest and Utility*

Included among the nitriles of considerable commercial importance are HCN (hydrogen cyanide), CH_3CH (acetonitrile), $H_2C{=}CHCN$ (acrylonitrile), and $NC(CH_2)_4CN$ (adiponitrile). Hydrogen cyanide, manufactured by the platinum–rhodium-catalyzed oxidation of a mixture of ammonia and methane,

$$CH_4 + NH_3 + \tfrac{3}{2}O_2 \longrightarrow HCN + 3\,H_2O,$$

is used primarily for the synthesis of other compounds, such as acetone cyanohydrin (see Section 13.5); these in turn, are intermediates in the production of synthetic rubbers, plastics, and fibers. Acrylonitrile, an industrially important commodity because of its use in the production of the synthetic fibers Orlon and Acrylan, is manufactured commercially from ethylene and propylene. Starting with ethylene, epoxidation, treatment of the epoxide with hydrogen cyanide, and dehydration yield acrylonitrile;

$$H_2C{=}CH_2 \longrightarrow H_2C\overset{O}{\frown}CH_2 \xrightarrow{HCN} HOCH_2CH_2CN \xrightarrow{(-H_2O)} H_2C{=}CHCN$$

Starting with propylene, vapor phase ammonation and oxidation yields acrylonitrile;

$$CH_3CH{=}CH_2 \ + \ NH_3 \ + \ O_2 \longrightarrow H_2C{=}CHCN \ + \ H_2O$$

Acetonitrile, produced from acetic acid and ammonia over alumina at 400°,

$$CH_3CO_2H \ + \ NH_3 \xrightarrow[Al_2O_3]{} CH_3CN \ + \ 2\,H_2O,$$

is a useful commercial solvent. Adiponitrile, produced by the reaction of 1,4-dichloro-2-butene with sodium cyanide followed by reduction,

$$ClCH_2CH{=}CHCH_2Cl \ + \ NaCN \longrightarrow NCCH_2CH{=}CHCH_2CN \longrightarrow NC(CH_2)_4CN,$$

is further reduced to 1,6-diaminohexane, which is one of the ingredients in the manufacture of Nylon-6,6.

Alkyl nitro compounds, prepared commercially by direct nitration of alkanes, are used as solvents and also as intermediates in the synthesis of amino alcohols (see Fig. 16.11). Aromatic nitro compounds are well known because of the capacity for certain polynitro substances, such as 1,3,5-trinitrobenzene, to undergo violent detonation. However, 1,3,5-trinitrobenzene is difficult to prepare because of the "deactivating" effect of the nitro groups. Thus, in the nitration of benzene, the second nitro group is more difficult to introduce than the first, and the third much more difficult than the second. To ameliorate the difficulty, toluene rather than benzene is used as the starting material; the methyl group is *o,p* directing and activating, and to some extent counteracts the deactivating effect of the nitro group, allowing three nitro groups to be introduced to produce 2,4,6-trinitrotoluene (TNT). This material has been extensively used as a military explosive, and until the advent of nuclear weapons, "TNT" epitomized the horrors of war and destruction. Now, of course, its once seemingly awesome power fades into insignificance in the face of the hydrogen bomb, the destructive capacities of which are rated in millions of tons of TNT.

The nitration of phenol, when carried out to completion, yields 2,4,6-trinitrophenol. The electron-withdrawing effect of the three nitro groups greatly increases the acidity of the phenolic hydroxyl group, making the compound comparable in acidity ($pK_a = 0.38$) to the mineral acids. For this reason, it goes under the name of picric acid. One of the interesting properties of picric acid and other polynitroarenes, including 1,3,5-trinitrobenzene, is the ability to form molecular complexes with many aromatic compounds. For example, picric

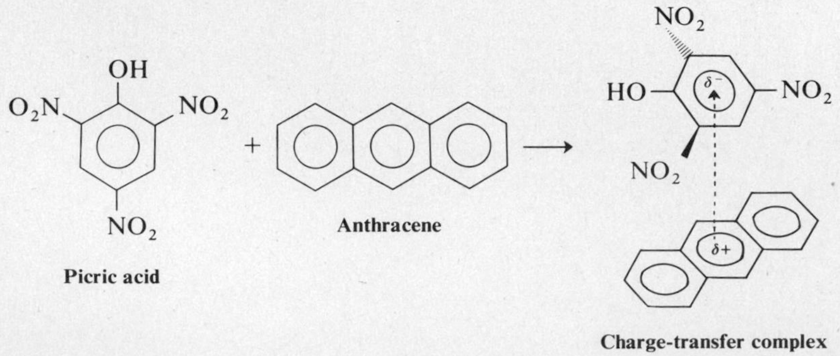

OH, O_2N, NO_2, NO_2

Picric acid

+

Anthracene

$\longrightarrow$

NO_2, HO, NO_2, NO_2, δ^-, $\delta+$

Charge-transfer complex

acid and 1,3,5-trinitrobenzene both form red, crystalline, high-melting compounds with anthracene, in which the bond between the participants is the result of the interaction between the electron-poor ring of the polynitroarene (acting as a Lewis acid) and the electron-rich ring of the anthracene (acting as a Lewis base). Compounds of this type are often referred to as "charge-transfer complexes."

16.6 *Study Guide*

I. Compounds containing nitrogen multiple bonded to carbon
   A. Imines
      1. General structure $R_2C{=}NR'$ (*e.g.*, semicarbazones, phenylhydrazones, etc.)
      2. Pyridine; an aromatic imine—the nitrogen analog of benzene
   B. Nitriles
      1. General structure RCN
      2. Properties: high dipole moments, high bp and mp, low water solubility, extremely low basicity
      3. Synthesis: by S_N2 reactions and dehydration of amides
      4. Reactions: hydrolysis to acids, reduction to amines

II. Compounds containing nitrogen–nitrogen bonds
   A. Hydrazines (diazanes); R_2NNR_2
   B. Azo compounds (diazenes, diimides); $RN{=}NR$
   C. Aliphatic diazo compounds; R_2CN_2
   D. Aromatic diazonium salts; $ArN_2^+X^-$
   E. Azides; RN_3

III. Compounds containing nitrogen multiple bonded to oxygen
   A. Nitroso compounds; $RN{=}O$ (stable in $R{=}NOH$ form if R is primary or secondary) and $ArN{=}O$
   B. Nitro compounds; RNO_2 and $ArNO_2$
      1. Synthesis: RNO_2 via S_N2 reactions, $ArNO_2$ via nitration of ArH
      2. Properties: high dipole moments, high bp and mp, low water solubility, very low basicity
      3. Reactions: reduction to amines, aldol condensations of R_2CHNO_2

PROBLEMS

1. Among the types of nitrogen compounds discussed in this chapter, the four that are most frequently encountered in laboratory work are the pyridines, the nitriles, the aryldiazonium salts, and the nitro compounds. Name the following examples of compounds falling in this category.

(a) 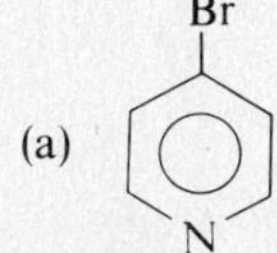

(b) $(CH_3)_2CHCN$

(c) C_6H_4 with $N_2^+Cl^-$ and $CH(CH_3)_2$ (meta)

(d) Naphthalene with NO_2

(e) $N{\equiv}C{-}C_6H_4{-}NO_2$

(f) $(CH_3)_2CHCH(CH_3)CH_2{-}C_6H_4{-}N_2^+Br^-$

2. Assign structures to each of the following pyridines, nitriles, aryldiazonium salts, and nitro compounds:

(a) 4-*sec*-Butylpyridine
(b) 3,3-Dimethylpentanonitrile
(c) 1-Azulenediazonium bromide
(d) 1,4,7-Trinitronaphthalene
(e) 2-Nitro-4-cyanobenzenediazonium sulfate

3. Using any appropriate system of nomenclature, name each of the following compounds:

(a) $CH_3CH_2CH(CH_3)CH_2CH(CH_3)N_2$ — printed as $CH_3CH_2\underset{CH_3}{C}HCH_2\underset{CH_3}{C}N_2$

(b) $CH_3{-}C(CH_3)_2{-}NO_2$

(c) $CH_3O{-}C_6H_4{-}N{=}NH$

(d) $CH_3CH_2(H)C{=}C(H)C({=}O)N_3$

(e) $O{=}N{-}C_6H_4{-}N(CH_3)(C_6H_5)$

(f) $Cl{-}C_6H_4{-}N{=}NCH_2CH_3$

(g) Bicyclic azo compound (N=N bridge)

(h) $O_2N{-}C_6H_4{-}N(CH_3)NH_2$

4. Draw the structure corresponding to each of the following names:

(a) 2-Methyl-3-phenylbutyldiazane
(b) Diazocyclohexane
(c) (*E*)-1-Methyl-2-neopentyldiazene
(d) Benzoyldiazomethane
(e) *p*-Dinitrosobenzene
(f) 2-Nitromethyl-1,4-butanediol
(g) *p*-Nitrophenylazide

5. Draw all the resonance structures that you think are necessary to provide a reasonably good approximation of the true structure of each of the following compounds:

(a) 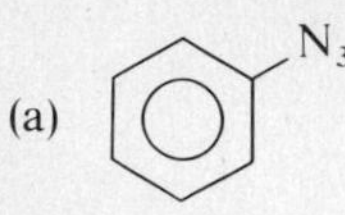

(b) $C_6H_5{-}C({=}O){-}N(CH_3)_2$

(c) 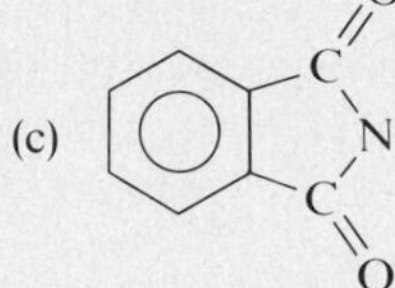

(d) $CH_3O{-}C_6H_4{-}NO_2$

(e) $C_6H_5{-}NO$

6. For each of the following reactions, draw the structure of the major product:

(a) cyclopropyl$-CH_2N{=}C{=}O + CH_3OH \longrightarrow$

(b) 1-naphthyl$-N_2^+Cl^-$ $+ Cu_2(CN)_2 \longrightarrow$

(c) $N{\equiv}C(CH_2)_4C{\equiv}N + H_2 \xrightarrow{Pd/C}$

(d) $(CH_3)_2CHNO_2 + HCHO \xrightarrow{NaOH}$

(e) $CH_3O-C_6H_5 + HONO \xrightarrow{H^+}$

(f) $CH_3C(=O)-C_6H_4-NO_2$ (meta) $\xrightarrow{Fe/HCl}$

(g) $(CH_3)_2C(CO_2C_2H_5)_2 + (H_2N)_2C{=}O \longrightarrow$

7. For each of the following interconversions, indicate the steps involved in going from the starting material to the product:

(a) $C_6H_5CO_2H \longrightarrow C_6H_5CN$

(b) benzene $\longrightarrow$ C_6H_5CN

(c) $C_6H_5OCH_3 \longrightarrow$ 2,4-diaminoanisole (OCH_3, NH_2, NH_2)

(d) 3-cyanopyridine (CN, N) $\longrightarrow$ 3-pyridyl$-CON(CH_3)_2$

(e) $C_6H_5NO_2 \longrightarrow C_6H_5OH$

8. Available starting materials include (a) any alkanes, alkenes, alkynes, alkyl halides, alkanols, alkanals, alkanones, alkanoic acids, or alkylamines containing four or fewer carbons, (b) benzene, toluene, *o*-, *m*-, and *p*-xylene, and (c) any inorganic compounds. Starting with materials chosen from this group, outline a synthesis for each of the following compounds:

(a) $(CH_3)_2CHCH_2CN$

(b) 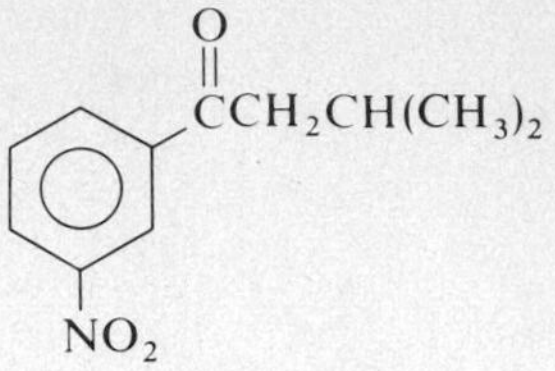 [structure: benzene ring with $\overset{O}{\overset{\|}{C}}CH_2CH(CH_3)_2$ and meta NO_2]

(c) [structure: benzene ring with OCH_3 and para NO]

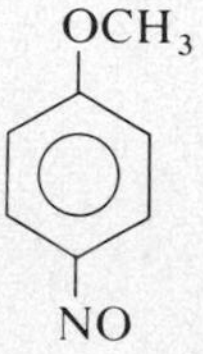

(d) 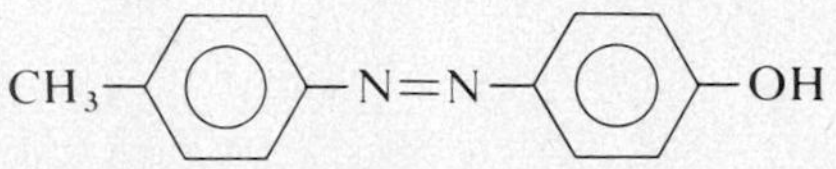 CH_3—C_6H_4—N=N—C_6H_4—OH

(e) 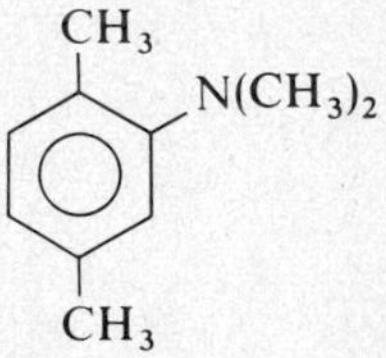 [structure: benzene ring with CH_3, adjacent $N(CH_3)_2$, and CH_3 para to the first CH_3]

9. Using the indicated chemical and/or spectral determination, describe how the following pairs of isomeric compounds can be differentiated:

(a)  [structure: benzene ring with CN] and [structure: pyridine ring with C≡CH para to N] — Chemical methods only

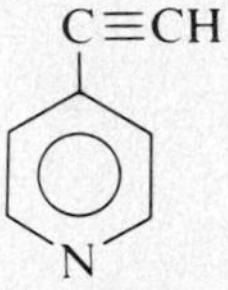

(b)  [structure: benzene ring with para NO_2 groups] and [structure: benzene ring with meta NO_2 groups] — Spectral methods only

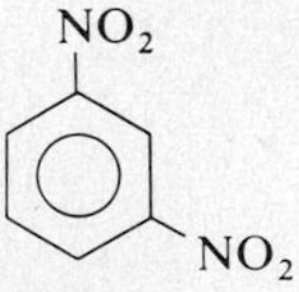

(c) $(CH_3)_2CHCH_2CN$ and $(H_2C{=}CH)_2NCH_3$ — Chemical and spectral methods

(d) 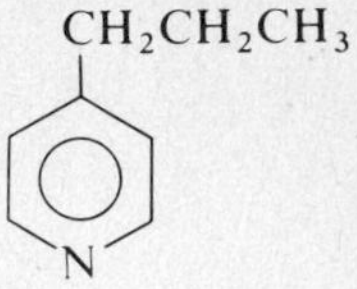 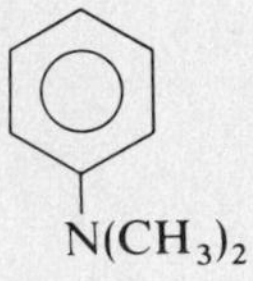 [structure: pyridine ring with $CH_2CH_2CH_3$ para to N] [structure: benzene ring with $N(CH_3)_2$] — Chemical methods only

(e) 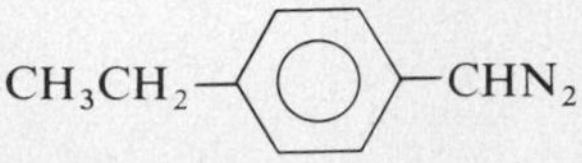 CH_3CH_2—C_6H_4—CHN_2 and CH_3—C_6H_4—$\underset{CH_3}{\underset{|}{C}}N_2$ — Spectral methods only

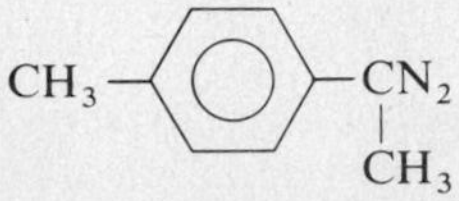

10. In the following examples, you are presented with four test tubes, each containing one of the four compounds of the isomeric set. What chemical and/or spectral determinations would you perform to make a structural assignment to each of the compounds in the set?

(a) $CONH_2$ / OH; CO_2H / NH_2; NO_2 / CH_3; CH_2OH / NO

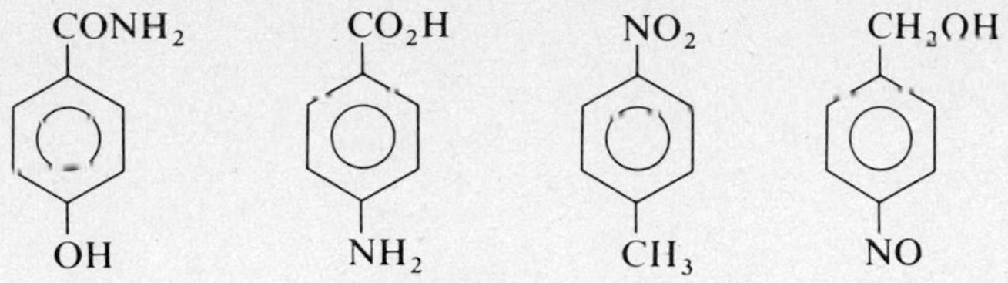

(b) CHO / N / CH_2OH; CH_3 / NO_2; OH / CH_3 / NO; OCH_3 / NO

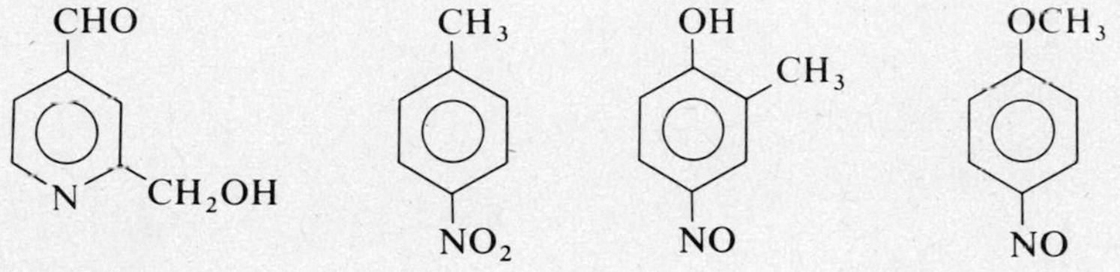

Chapter 17

Sulfur-containing compounds

I counted two and seventy stenches,
All well defined, and several stinks.

COLERIDGE
Koln

Sulfur, appearing just below oxygen in group VI of the periodic chart, resembles oxygen in many respects and is able to take the place of oxygen in a wide variety of organic structures. Sulfur, though, can have valence states higher than two, making possible the existence of certain classes of sulfur-containing compounds that have no oxygen counterparts. The discussion in this chapter centers first on those compounds of sulfur for which there are direct oxygen analogs and then proceeds to compounds containing sulfur in higher valence states.

17.1 Thiols (Mercaptans) and Sulfides (Thioethers)

17.1a. STRUCTURE AND NOMENCLATURE OF THIOLS AND SULFIDES. The electron configuration of atomic sulfur is $1s^2 2s^2 2p^6 3s^2 3p_x^2 3p_y^1 3p_z^1$, similar in the outer valence shell to oxygen except that the valence shell is at level 3 rather than level 2. Because of this difference, sulfur, although behaving qualitatively like oxygen, shows quantitative divergences when it is incorporated into carbon frameworks in its divalent form. This is particularly apparent in the case of alkylthiols (RSH) and arylthiols (ArSH), which are considerably more acidic than their oxygen counterparts. Ethanethiol (CH_3CH_2SH), for example, has a pK_a of 9.5 and dissolves in dilute sodium hydroxide solution; benzenethiol (C_6H_5SH) has a pK_a of 7.8 and dissolves in aqueous sodium carbonate solution. This increased acidity is ascribed to the greater size of $3p$ orbitals as compared with $2p$ orbitals, resulting in less effective (*i.e.*, more diffuse) overlap with the $1s$ orbital of hydrogen. We must argue that this more than offsets the lower electronegativity of sulfur, which would affect the acidity in the opposite direction. Sulfur also differs from oxygen in another respect. It possesses vacant d orbitals that are sufficiently low in energy that they are available for bonding, conferring on sulfur oxidation states greater than two.

The common names for compounds containing the —SH function employ the designation "mercaptan," preceded by the specification of the alkyl or aryl group; for example, CH_3SH is methyl mercaptan, C_6H_5SH is phenyl mercaptan, etc. In the IUPAC system, the suffix "thiol" (the contraction of *thio*alcoh*ol*) is added to the name for the hydrocarbon group; thus, CH_3SH is methanethiol, and C_6H_5SH is benzenethiol. Occasionally, the prefix "thio" is placed before the name of the corresponding alcohol or phenol; for instance, C_6H_5SH is often named thiophenol. When named as a substituent, the —SH function is designated as a "mercapto" group as in the compound,

$$\underset{\displaystyle SH}{HC{\equiv}C\underset{|}{C}HCO_2H}$$

called 2-mercapto-3-butynoic acid.

The attachment of two carbon frameworks to a sulfur atom gives compounds known as *sulfides* or *thioethers* (R—S—R). A particularly interesting example is the cyclic sulfide thiophene (Fig. 17.1), the sulfur analog of furan (see Fig. 11.3).

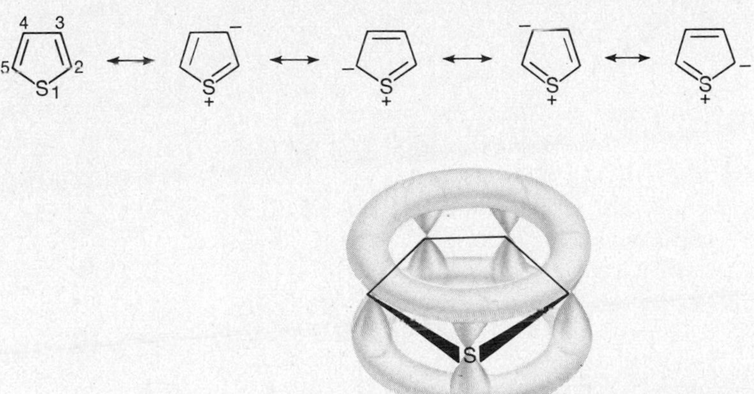

Fig. 17.1. Resonance structures and orbital representation of thiophene.

It occurs in small amounts in the benzene obtained from coal tar, and because of the similarity in physical and chemical properties, thiophene and benzene are rather difficult to separate. Thiophene, like furan, has aromatic properties and requires several resonance structures for its adequate representation.

Sulfides may be named by specifying the alkyl or aryl groups attached to the sulfur; for example, CH_3SCH_3 is dimethyl sulfide, $CH_3SCH_2CH(CH_3)_2$ is isobutyl methyl sulfide, and $C_6H_5SC_6H_5$ is diphenyl sulfide. In more complex compounds, the sulfur and one of its attached groups are designated as a substituent function and specified as an "alkylthio" or "arylthio" group;

$$\underset{\displaystyle CH_3}{CH_3CH_2SCH_2\underset{|}{C}HCH_2CH(CH_3)_2}$$

for example, is named 1-ethylthio-2,4-dimethylpentane. The replacement of —CH_2— by —S— is indicated by the prefix *thia*.

In addition to thiophene, other cyclic sulfides are known; the three-membered sulfide,

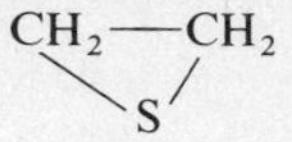

$$\begin{array}{c} CH_2\text{—}CH_2 \\ \diagdown\ \diagup \\ S \end{array}$$

is called thiirane, thiacyclopropane, or ethylene episulfide; the four-membered sulfide, $\begin{array}{l} CH_2\text{—}CH_2 \\ \mid \quad\ \ \mid \\ CH_2\text{—}S \end{array}$ is called thietane, and the five-membered sulfide $\begin{array}{c} CH_2\text{——}CH_2 \\ \mid \qquad\quad \mid \\ CH_2 \qquad CH_2 \\ \diagdown\ \diagup \\ S \end{array}$ (*i.e.*, the tetrahydro derivative of thiophene) is called thiolane, thiacyclopentanone, or tetrahydrothiophene.

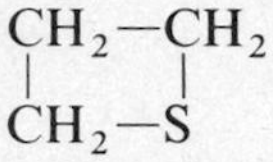

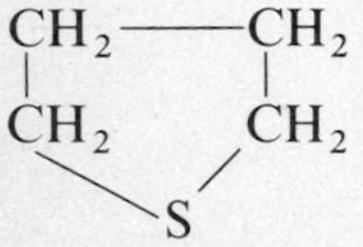

PROBLEM 17-1 Draw and name all possible compounds having the molecular formula $C_4H_{10}S$.

17.1b. PHYSICAL PROPERTIES OF THIOLS AND SULFIDES. As mentioned, thiols and thiophenols are more acidic than the corresponding alcohols and phenols. Despite this, however, they are less effective than alcohols in the formation of hydrogen bonds. Consequently, they boil lower and are less soluble in water than the corresponding alcohols. For example, ethanethiol boils at only 37° and is soluble only to the extent of 1.5 g/100 ml H_2O; ethanol boils at 78° and is completely miscible with water. The reduced basicity, like the increased acidity, is attributed to the greater dispersion of electron density in the outer valence shell of sulfur as compared with oxygen. The lower molecular weight thiols are toxic and possess a notoriously disagreeable odor, their only virtue being that they are rather volatile and tend to dissipate more quickly than certain other noxious materials. In the higher molecular weight compounds (above approximately nine carbon atoms), the odors, curiously, become pleasant.

17.1c. SYNTHESIS AND REACTIONS OF THIOLS AND SULFIDES. Alkanethiols (alkyl mercaptans) can be prepared by S_N2 reactions involving alkyl halides and sodium hydrogen sulfide (NaSH) (see Fig. 9.3) and are readily accessible compounds. Arenethiols (aryl mercaptans), however, require indirect synthesis routes and are generally prepared by reduction of higher valence state compounds. Benzenesulfonyl chloride (see Fig. 17.11), for example, reacts with zinc and sulfuric acid to yield benzenethiol.

$$2\,C_6H_5SO_2Cl \;+\; 6\,Zn \;+\; 5\,H_2SO_4 \longrightarrow 2\,C_6H_5SH \;+\; ZnCl_2 \;+\; 5\,ZnSO_4 \;+\; 4\,H_2O$$

Sulfides can be prepared by the action of sodium sulfide on alkyl halides,

$$2\,RX \;+\; Na_2S \longrightarrow R\text{—}S\text{—}R \;+\; 2\,NaX,$$

and by the action of the salts of thiols on alkyl halides

$$RX + R'SNa \longrightarrow R{-}S{-}R' + NaX$$

(Fig. 17.2), the first procedure being restricted to the preparation of symmetrical sulfides, but the second allowing the preparation of unsymmetrical sulfides.

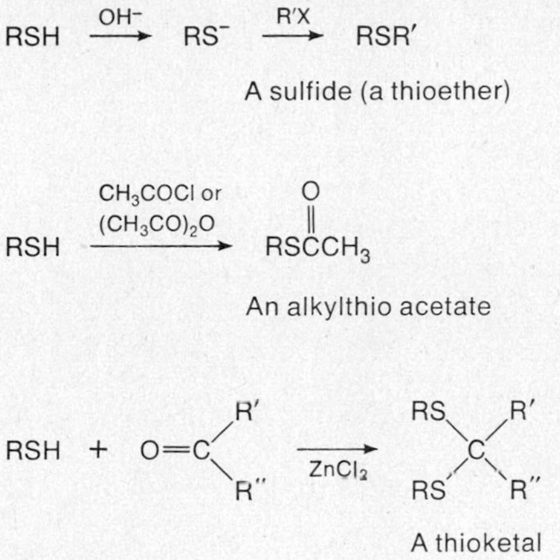

$$RSH \xrightarrow{OH^-} RS^- \xrightarrow{R'X} RSR'$$

A sulfide (a thioether)

$$RSH \xrightarrow[(CH_3CO)_2O]{CH_3COCl \text{ or}} RS\overset{O}{\overset{\|}{C}}CH_3$$

An alkylthio acetate

$$RSH + O{=}C(R')(R'') \xrightarrow[ZnCl_2]{} (RS)_2C(R')(R'')$$

A thioketal

Fig. 17.2. Reactions of thiols (mercaptans).

Thiols resemble alcohols in their reactions. Treatment with base followed by an alkyl halide yields sulfides, as illustrated, affording the sulfur counterpart of the Williamson ether synthesis (see Fig. 11.5); treatment with acetyl chloride or acetic anhydride yields the acetyl derivatives; treatment with aldehydes or ketones in the presence of acid catalysts yields the thioacetals or thioketals (Fig. 17.2).

A reaction of thiols that has no direct counterpart in the alcohols is the oxidation to the disulfide. For example, when ethanethiol is treated with iodine, the products are ethanedisulfide and hydrogen iodide; *i.e.*,

$$2\,C_2H_5SH + I_2 \longrightarrow C_2H_5S{-}SC_2H_5 + 2\,HI$$

The counterpart oxygen structure, diethyl peroxide ($C_2H_5O{-}OC_2H_5$), exists, but it is not synthesized by direct oxidation of ethanol. The ease of formation of the S—S disulfide bond reflects its greater stability as compared with the O—O bond; *i.e.*, 73 kcal/mole for the S—S bond in ethanedisulfide as contrasted with 37 kcal/mole for the O—O bond in diethyl peroxide. Air alone is capable of effecting the oxidation of thiols to disulfides, and care must often be exercised to protect against this.

For the reason already given, the sulfur atom in thiols and thioethers is less basic toward protic acids than the oxygen atom in the analogous alcohols and ethers; divalent sulfur compounds are, nevertheless, *more nucleophilic* than their oxygen counterparts. This is ascribed partly to the greater polarizability of the electrons in the third level as compared with the second level (*i.e.*, the electrons in the third level are farther from the nucleus and, therefore, less firmly held) and partly to the decreased steric hindrance arising from the increased space around the sulfur. Dimethyl sulfide, for example, is sufficiently nucleophilic to react in a fashion analogous to a tertiary amine, and effects an S_N2 displacement on methyl iodide to give trimethylsulfonium iodide (Fig. 17.3).

$$(CH_3)_2\ddot{S}: + CH_3{-}I \longrightarrow (CH_3)_3\overset{+}{S}\ I^-$$

Trimethylsulfonium iodide

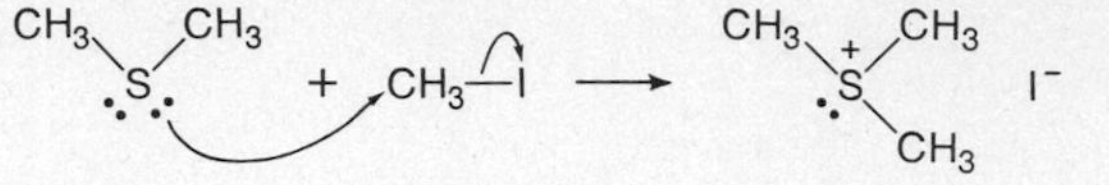

Fig. 17.3. Formation of trimethylsulfonium iodide.

As indicated in Section 17.1, thiophene possesses aromatic properties (see Fig. 17.1), chemically verified by the fact that it undergoes electrophilic substitution reactions. For example, halogenation, nitration, and Friedel–Crafts acylation all proceed in a fashion entirely comparable to benzene (Fig. 17.4).

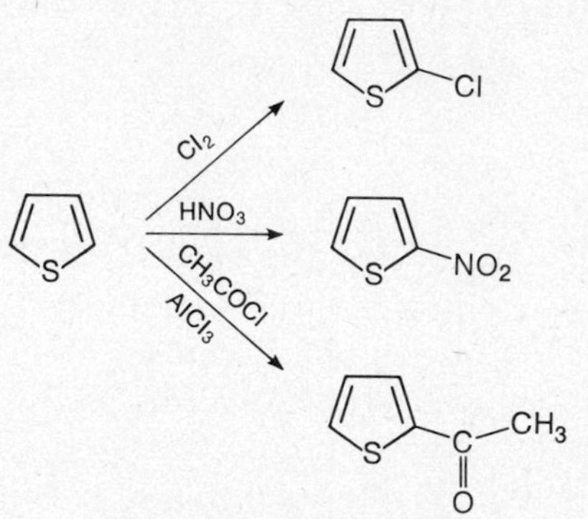

Thiophene $\xrightarrow{Cl_2}$ 2-chlorothiophene (S, Cl)

Thiophene $\xrightarrow{HNO_3}$ 2-nitrothiophene (S, NO_2)

Thiophene $\xrightarrow[AlCl_3]{CH_3COCl}$ 2-acetylthiophene (S, $C(=O)CH_3$)

Fig. 17.4. Electrophilic substitution reactions of thiophene.

A sulfide that has gained notoriety, although for dubious reasons, is bis-(2-chloroethyl) sulfide, better known as mustard "gas" despite the fact that it is a liquid boiling at 217°. One of the methods for its manufacture involves the action of hydrogen sulfide on ethylene oxide to give bis-(2-hydroxyethyl) sulfide, which is then treated with hydrogen chloride (Fig. 17.5).

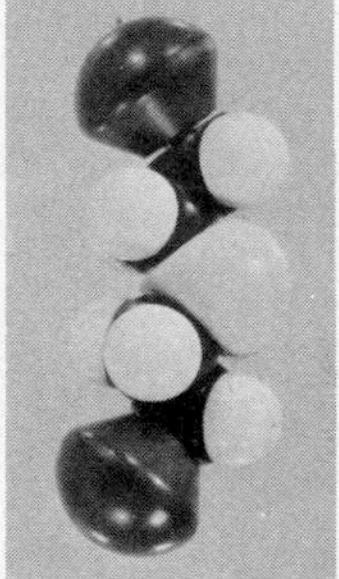

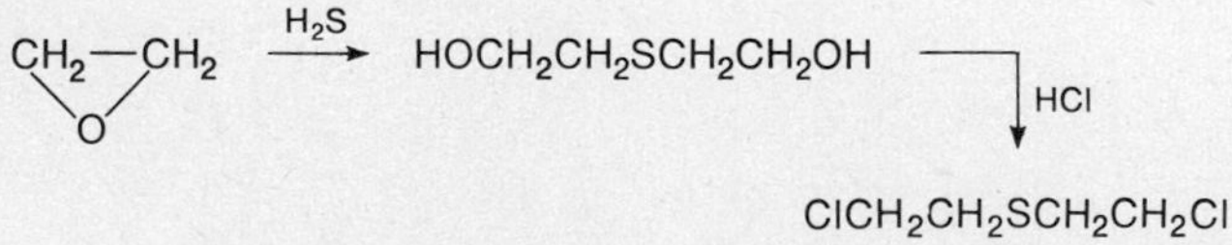

Mustard gas

$$CH_2{-}CH_2\ (\text{ethylene oxide, O}) \xrightarrow{H_2S} HOCH_2CH_2SCH_2CH_2OH \xrightarrow{HCl} ClCH_2CH_2SCH_2CH_2Cl$$

Fig. 17.5. Synthesis of mustard gas.

An interesting reaction that thiols and sulfides both undergo is "desulfurization" under the influence of molecular hydrogen and a catalyst known as Raney nickel, prepared by treating a nickel–aluminum alloy with sodium hydroxide which removes most of the aluminum and leaves the nickel in a finely divided state (Fig. 17.6).

$$C_5H_{11}SC_5H_{11} + 2\,H_2 \xrightarrow[\text{catalyst}]{\text{Raney nickel}} 2\,C_5H_{12} + H_2S$$

Dipentyl sulfide

$$\text{2,5-dimethylthiophene} + 4\,H_2 \xrightarrow[\text{catalyst}]{\text{Raney nickel}} CH_3(CH_2)_4CH_3 + H_2S$$

2,5-Dimethylthiophene

Fig. 17.6. Desulfurization of sulfur compounds with hydrogen and Raney nickel catalyst.

PROBLEM 17-2

Realizing that thiols and alcohols behave in a comparable fashion and remembering that alcohols react with aldehydes and ketones to form acetals (see Fig. 13.14), suggest a method for converting cyclohexanone to cyclohexane.

17.2 Thioaldehydes Thioketones, and Thiocarboxylic Acids and Derivatives

Most thioaldehydes and thioketones are unstable in the monomeric form and tend to trimerize to 1,3,5-trithianes, *e.g.*,

$$3\,R_2C{=}S \longrightarrow$$

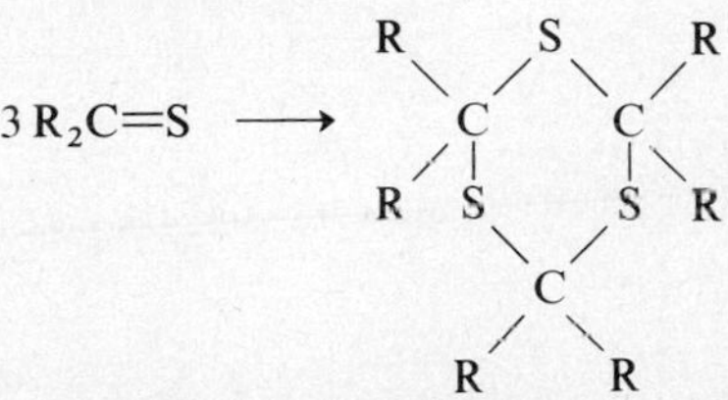

A trithiane (a 1,3,5-trithiacyclohexane)

Dimethyl thioketone (thioacetone), for example, forms a trimer of this type, although di-*n*-propyl thioketone is relatively stable as the monomer, in which form it reacts in much the same fashion as the analogous ketone.

Thiocarboxylic acids, $RC(=O)SH$, thiocarboxylic esters, $RC(=O)SR$ (prepared by the action of acid halides or anhydrides on thiols as illustrated in Fig. 17.2), thioamides, $RC(=S)NH_2$, dithio acids, $RC(=S)SH$, and dithio esters, $RC(=S)SR$, all are known types of compounds. They are unpopular materials, however, because of their objectionable odors. The dithio acids, in particular, are notorious for their stench.

The best known of the sulfur analogs of type CC carbonyl compounds (see p. 328) is thiourea, which can be made by a Wöhler reaction [*i.e.*, $NH_4SCN \rightarrow (NH_2)_2C{=}S$]. Thiourea is a water-soluble solid melting at 180–182° which, like urea, forms inclusion complexes (see Section 15.8). Forming a helix into which a guest molecule can fit, thiourea provides a larger channel than urea and is able, consequently, to accept and more effectively hold long chain molecules that are branched as well as molecules with cyclic structures.

17.3 Compounds Containing Sulfur in Higher Valence States

17.3a. SULFOXIDES AND SULFONES. Thiols and sulfides can be converted to compounds in which one, two, or even three oxygen atoms are bonded to the sulfur, increasing the valence of sulfur to four or six. Dimethyl sulfide, for example, when treated with 1 mole-equivalent of hydrogen peroxide, is converted to dimethyl sulfoxide; with an excess of hydrogen peroxide and at a somewhat higher temperature, a second atom of oxygen is added to produce dimethyl sulfone as illustrated in Fig. 17.7.

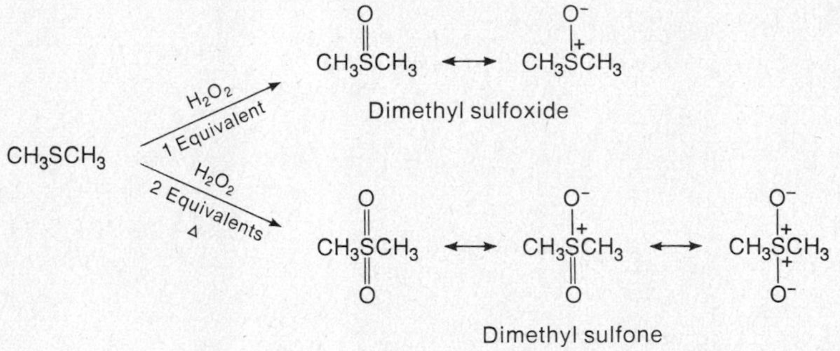

$CH_3SCH_3 \xrightarrow[\text{1 Equivalent}]{H_2O_2} CH_3S(=O)CH_3 \longleftrightarrow CH_3\overset{+}{S}(O^-)CH_3$ (Dimethyl sulfoxide)

$CH_3SCH_3 \xrightarrow[\text{2 Equivalents}, \Delta]{H_2O_2} CH_3S(=O)_2CH_3 \longleftrightarrow CH_3\overset{+}{S}(O^-)(=O)CH_3 \longleftrightarrow CH_3\overset{2+}{S}(O^-)_2CH_3$ (Dimethyl sulfone)

Fig. 17.7. Conversion of sulfides to sulfoxides and sulfones.

PROBLEM 17-3

Starting with ethylene as the only source of carbon compound, devise a synthesis of diethylsulfone.

Sulfoxides and sulfones are named by designating the carbon frameworks attached to the —SO— and $—SO_2—$ functions and following them by the terms sulfoxide or sulfone: $CH_3SOCH_2CH_3$ is ethyl methyl sulfoxide, $CH_3SO_2C_6H_5$ is methyl phenyl sulfone. As substituents, the RSO— and $RSO_2—$ functions are designated as alkylsulfinyl and alkylsulfonyl groups.

The sulfur–oxygen bonds in dimethyl sulfoxide and dimethyl sulfone are coordinate–covalent bonds and are highly polar, requiring a set of resonance structures (see Fig. 17.7) for an adequate representation. Dimethyl sulfoxide, with a dipole moment of 3.9 D, is a good solvent for polar materials, partly because of its high dielectric constant of 45, and partly because the high electron density on its periphery (*i.e.*, on the oxygen atom) allows it to effectively solvate cations (E^+). The anion (Nu^-) that tags along with the solvated cation, however, is not itself significantly solved; to introduce the element of pornography that characterizes many books today, it is referred to as a "naked" anion (Fig. 17.8).

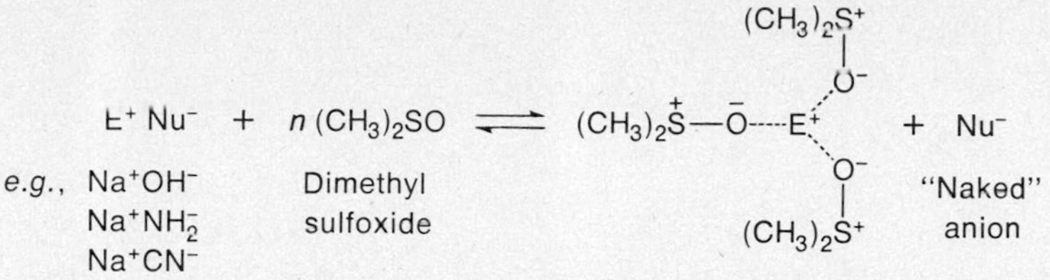

$$E^+Nu^- + n\,(CH_3)_2SO \rightleftharpoons (CH_3)_2\overset{+}{S}-\overset{-}{O}\cdots E^+ \left(\cdots O^- - \overset{+}{S}(CH_3)_2\right)_2 + Nu^-$$

e.g., Na^+OH^-, $Na^+NH_2^-$, Na^+CN^- — Dimethyl sulfoxide — "Naked" anion

Fig. 17.8. Dimethylsulfoxide (DMSO) as a solvent for ionic compounds.

As a result of its lack of solvation, the anion is highly reactive and seeks to bond to whatever ligands are available. Thus, compounds such as sodium hydroxide, sodium amide, and sodium cyanide become extraordinarily strong nucleophiles in dimethyl sulfoxide solution as compared with aqueous or alcoholic solution. Acceleration factors of several orders of magnitude for S_N2 reactions carried out in dimethyl sulfoxide have been reported, and the compound has been extensively used in the laboratory as a solvent. Similarly, the cyclic sulfone

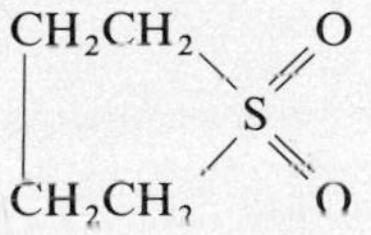

$$\begin{matrix} CH_2CH_2 & & O \\ | & \diagdown S \diagup\!\!\!\!\!\diagup & \\ CH_2CH_2 & & O \end{matrix}$$

known as sulfolane (thiacyclopentane-1,1-dioxide) has a high dielectric constant ($\varepsilon = 40$) and a high dipole moment ($\mu = 4.2$ D), and it also is a good solvent in which to carry out nucleophilic displacement reactions.

17.3b. Sulfenic, sulfinic, and sulfonic acids. A number of oxysulfur acids are known in which the H or R group of thiols, sulfides, sulfoxides, and sulfones, is replaced by a hydroxyl function. Replacement of the sulfur-bonded hydrogen of a thiol by —OH produces a class of compounds called *sulfenic acids* (RSOH). Sulfenic acids are very unstable, undergoing disproportionation to other sulfur-containing compounds. The analogs derived from sulfoxides in which an R is replaced by a hydroxyl function are called *sulfinic acids* and are represented by the general structure

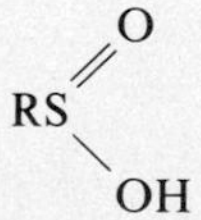

$$RS(=O)OH$$

One of the methods by which they can be prepared involves the action of SO_2 on a Grignard reagent (*cf.* reaction of RMgX with CO_2—Fig. 14.8) followed by acidification of the resulting sulfinic acid salt:

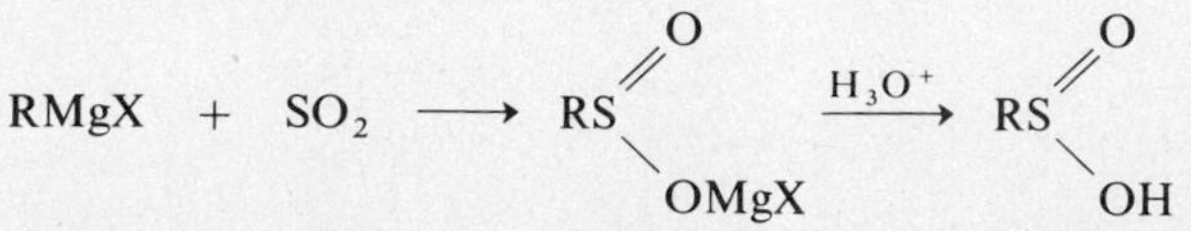

$$RMgX + SO_2 \longrightarrow RS(=O)OMgX \xrightarrow{H_3O^+} RS(=O)OH$$

These compounds, which can be considered as derivatives of sulfurous acid [$O{=}S(OH)_2$], are moderately strong acids that are unstable and decompose on standing. Oxidation of sulfinic acids produces sulfonic acids, the analogs of sulfones in which a hydroxyl rather than an R group is attached to the sulfur. These compounds have the general structure

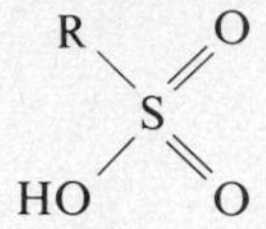

$$\mathrm{R{-}S({=}O)_2{-}OH}$$

and can be considered as derivatives of sulfuric acid

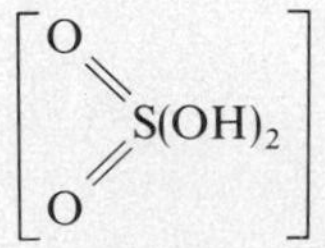

$$\left[\mathrm{O_2S(OH)_2}\right]$$

They are very strong acids and have pK_a values of about −2. Although numerous examples of alkanesulfonic acids are known, the arenesulfonic acids have been studied more extensively; this can probably be ascribed to the facile preparation of arenesulfonic acids by the direct sulfonation of aromatic compounds (see Fig. 8.13).

The oxysulfur acids are named by adding the suffix "sulfenic," "sulfinic," or "sulfonic" to the name of the carbon framework and following it by the term "acid." Thus, CH_3SOH is methanesulfenic acid, $C_6H_5SO_2H$ is benzenesulfinic acid, and $CH_3CH_2CH_2CH(SO_3H)CH_3$ is 2-pentanesulfonic acid. If the sulfonic acid function is named as a substituent, it is designated as a "sulfo" function. For example, $C_6H_4(CO_2H)(SO_3H)$ (ortho) is called 2-sulfobenzoic acid.

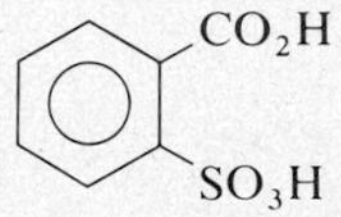

The same comments that pertain to the nitration reaction of benzene and other aromatic compounds also hold true for the sulfonation reaction. Treatment of benzene with fuming sulfuric acid* at room temperature results in an electrophilic substitution reaction in which an $-SO_3H$ group replaces a hydrogen of the aromatic ring (see Fig. 8.13). Considerably more strenuous conditions are necessary, however, to introduce additional sulfonic acid groups, for in similarity to the nitro group, the sulfo group is a *meta*-directing and *deactivating* moiety.

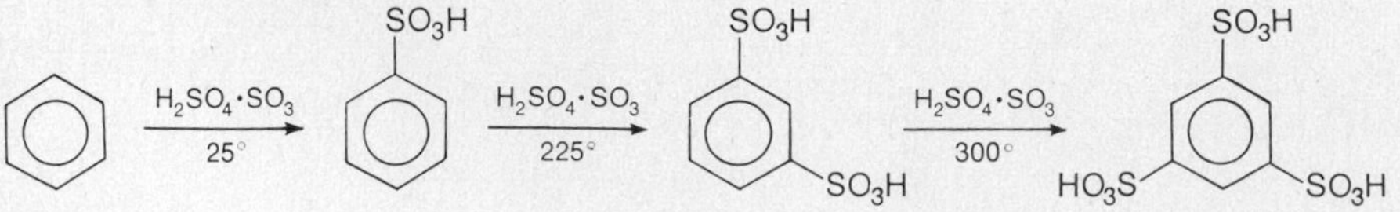

$$C_6H_6 \xrightarrow[25^\circ]{H_2SO_4\cdot SO_3} C_6H_5SO_3H \xrightarrow[225^\circ]{H_2SO_4\cdot SO_3} 1,3\text{-}C_6H_4(SO_3H)_2 \xrightarrow[300^\circ]{H_2SO_4\cdot SO_3} 1,3,5\text{-}C_6H_3(SO_3H)_3$$

Fig. 17.9. Sulfonation of benzene.

---

* Fuming sulfuric acid is concentrated sulfuric acid containing dissolved sulfur trioxide, generally indicated as $H_2SO_4{\cdot}SO_3$.

Temperatures of approximately 225 and 300°, respectively, are required, therefore, to introduce the second and third groups (Fig. 17.9).

Section 17.3
Compounds Containing Sulfur in Higher Valence States

A mixture of the sodium salt of benzenesulfonic acid and solid sodium hydroxide when strongly heated yields sodium phenoxide (often referred to as a "fusion" reaction); acidification of the product then yields phenol. In similar fashion, fusion of sodium benzenesulfonate with sodium cyanide yields benzonitrile (Fig. 17.10).

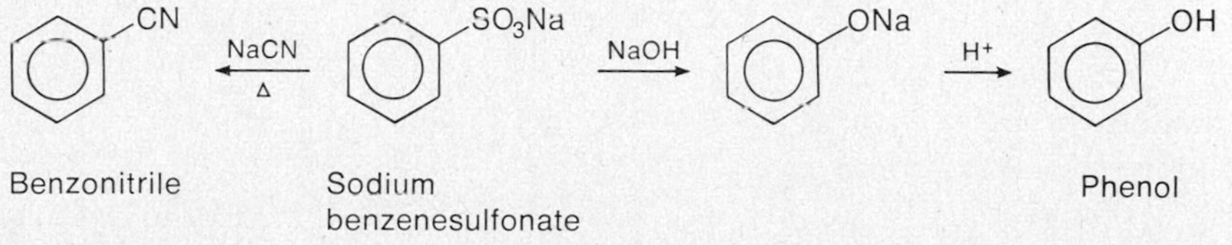

Benzonitrile ← ($NaCN$, Δ) — Sodium benzenesulfonate ($C_6H_5SO_3Na$) → ($NaOH$) C_6H_5ONa → (H^+) Phenol

Fig. 17.10. Preparation of phenol and benzonitrile from sodium benzenesulfonate.

Using methods similar to those that have been described for the preparation of carboxylic acid derivatives (see Chapter 15), we can make sulfonic acid *derivatives.* Thus, the action of PCl_3 or $SOCl_2$ (*cf.* Fig. 15.1) on sulfonic acids yields the corresponding sulfonic acid chlorides called *sulfonyl chlorides*, treatment of acid chlorides with alcohols yields esters called *sulfonates*, and treatment of the acid chlorides or the esters with amines yields amides called *sulfonamides* (Fig.

Benzenesulfonic acid ($C_6H_5SO_3H$) ⇄ (PCl_3 / H_2O) Benzenesulfonyl chloride ($C_6H_5SO_2Cl$) ← ($HOSO_2Cl$) benzene

Benzenesulfonic acid → (150°, H_3O^+) benzene

Benzenesulfonyl chloride → (ROH) An alkyl benzenesulfonate ($C_6H_5SO_2OR$)

Benzenesulfonyl chloride → (R_2NH) A benzenesulfonamide ($C_6H_5SO_2NR_2$)

An alkyl benzenesulfonate → (R_2NH) A benzenesulfonamide

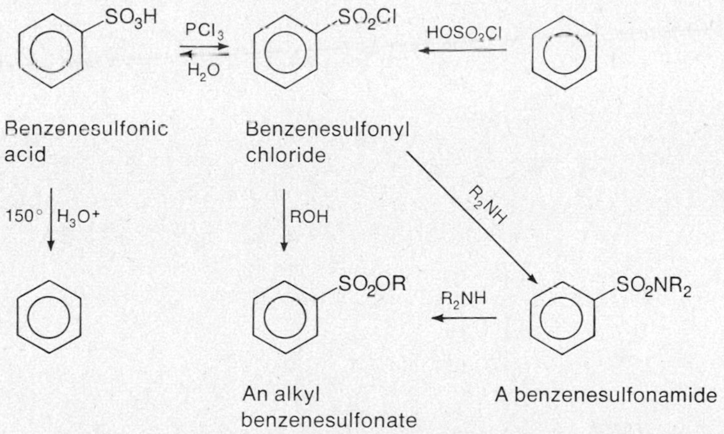

Fig. 17.11. Reactions of benzenesulfonic acid and its derivatives.

17.11). A key compound in the interrelationships shown in Fig. 17.11 is benzenesulfonyl chloride, which can be converted to the free acid, to the ester, or to the amide. In addition to its availability from the free acid, it can also be synthesized directly from benzene by the action of chlorosulfonic acid ($HOSO_2Cl$) in an electrophilic aromatic substitution reaction.

PROBLEM 17-4

On the basis of what you know about the chemistry of acid chlorides such as benzoyl chloride, what would you predict to be the product from the reaction of benzenesulfonyl chloride, toluene, and aluminum chloride?

Most of the electrophilic substitution reactions of benzenoid compounds (see Fig. 8.13) are essentially irreversible; once the incoming substituent is affixed to the aromatic ring it is not easily displaced by hydrogen except by indirect methods. Benzenesulfonic acid, however, is an exception, for it loses the sulfonic acid group and produces benzene when heated with aqueous acid at an elevated temperature (Fig. 17.11).

Benzenesulfonamides carrying one or more hydrogens attached to the nitrogen are appreciably acidic. Because of the strong electron-withdrawing power of the SO_2 group, as well as its ability to accept a pair of nonbonded electrons from the adjacent nitrogen atom into one of its vacant *d* orbitals, facile dissociation to the anion occurs; benzenesulfonamide, for example, dissolves in moderately strong aqueous sodium hydroxide solution. A classical method used to distinguish between primary, secondary, and tertiary amines, called the *Hinsberg test*, is based on this property. Thus, primary amines react with benzenesulfonyl chloride to form a benzenesulfonamide that is soluble in aqueous base; secondary amines react with benzenesulfonyl chloride to form benzenesulfonamides that do not contain an N—H bond and so are insoluble in aqueous base; and tertiary amines do not react with benzenesulfonyl chloride (Fig. 17.12).

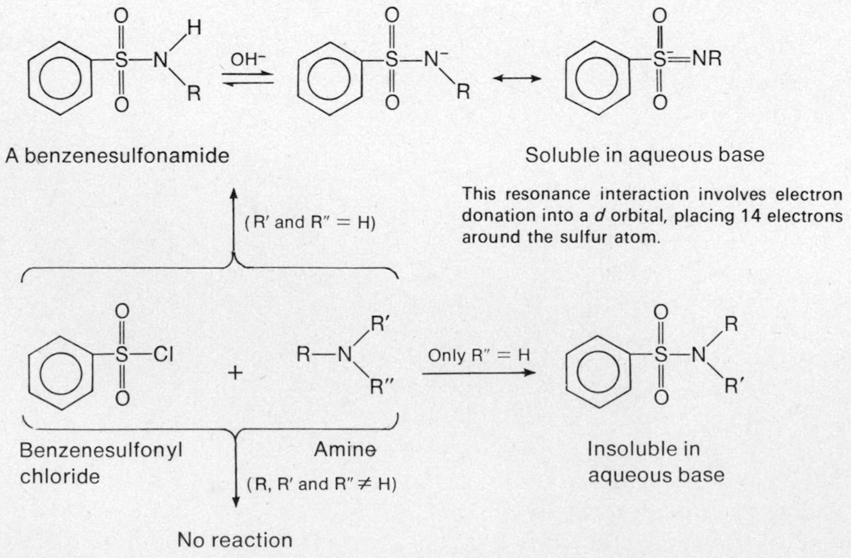

$$C_6H_5\text{—}SO_2\text{—}NHR \underset{}{\overset{OH^-}{\rightleftharpoons}} C_6H_5\text{—}SO_2\text{—}N^-R \longleftrightarrow C_6H_5\text{—}\overset{-}{S}(O)(O){=}NR$$

A benzenesulfonamide

Soluble in aqueous base

This resonance interaction involves electron donation into a *d* orbital, placing 14 electrons around the sulfur atom.

(R′ and R″ = H)

$$C_6H_5\text{—}SO_2\text{—}Cl + R\text{—}NR'R'' \xrightarrow{\text{Only } R'' = H} C_6H_5\text{—}SO_2\text{—}NRR'$$

Benzenesulfonyl chloride

Amine

Insoluble in aqueous base

(R, R′ and R″ ≠ H)

No reaction

Fig. 17.12. Differentiation of primary, secondary, and tertiary amines by means of reaction with benzenesulfonyl chloride (Hinsberg test).

17.4 *Spectral Properties of Sulfur-Containing Compounds*

The ir and nmr spectral properties of several of the types of sulfur-containing compounds discussed in this chapter are shown in Table 17.1. In general, the uv properties are of little use in structural characterizations, for only those compounds in which an oxidized sulfur moiety, *i.e.*,

$$\gt S{=}O \quad \text{or} \quad \gt S({=}O)_2$$

is attached to an unsaturated group show any significant absorption in the accessible region of the uv.

Table 17.1 NMR AND IR SPECTRAL PROPERTIES OF SULFUR-CONTAINING COMPOUNDS

Type of Compound	NMR δ Values	IR cm^{-1}
RSH	1.3–1.7	2600–2550 (S—H stretching deformation)
ArSH	2.5–4.0	2600–2550 (S—H stretching deformation)
RSO_3H and $ArSO_3H$	11–12	1260–1150, 1080–1010 (S=O stretching deformation)
RSO_3R'		1420–1330, 1200–1145 (S=O stretching deformation)
$RSO_2NR'_2$		1370–1330, 1180–1160 (S=O stretching deformation)
RS—C—H	2–2.5	
RSO—C—H	2.5–2.8	1090–1020 (S=O stretching deformation)
RSO_2—C—H	2.8–3.1	1350–1310, 1160–1120 (S=O stretching deformation)

PROBLEM 17-5

Compounds A and B both have the molecular formula $C_6H_{10}O_2S$. Compound A has a pair of singlet resonances in the nmr, and compound B has three resonance envelopes comprising two doublets and a multiplet. The infrared spectrum of compound A shows a strong absorption band near 1715 cm^{-1} and that of compound B a strong absorption band at 1050 cm^{-1}. Compound A is susceptible to oxidation by hydrogen peroxide, and compound B undergoes cleavage when treated with ozone. Suggest structures for compounds A and B.

17.5 *Sulfur-Containing Compounds of Particular Interest and Utility*

Some of the arenesulfonic acids are industrially important because of the water solubility of their sodium salts. Many commercial dyestuffs, for example, contain sulfonic acid groups as a means for rendering the compounds water soluble. A typical illustration of an industrial dye is Acid Yellow I.

$$\text{Acid Yellow I: naphthalene with OH, } O_2N, SO_3Na, NO_2$$

Acid Yellow I

A variety of synthetic detergents also contain a sulfonic acid group for the purpose of conferring water-soluble properties on compounds having a long-chain hydrocarbon portion, the sulfonic acid group thus performing the function of the carboxyl group of an ordinary soap. The sulfonic acid compounds, an early commercial example of which is the sodium salt of 2-(*p*-sulfophenyl)-4,6,8-trimethylnonane, have the advantage over soaps in that their calcium, magnesium, and iron salts are water soluble (*i.e.*, they can be used in "hard" water).

$$CH_3CHCH_2CHCH_2CHCH_2CHCH_3 \quad (CH_3,\ CH_3,\ CH_3,\ C_6H_4SO_3Na)$$

Sodium salt of 2-(*p*-sulfophenyl)-4,6,8-trimethylnonane

Some of them, including the one just named, have the ecological disadvantage, however, of being nonbiodegradable. Microorganisms are incapable of metabolizing them to carbon dioxide and water; consequently, they accumulate in lakes and rivers, where their presence may be detrimental to the ecology. Since 1965, biodegradable synthetic detergents have appeared on the market, and this difficulty has been partly alleviated, although other problems yet remain.

Certain benzenesulfonamides, of which sulfanilamide is the prototype, have antibiotic properties and are valuable medicines. In fact, in 1932 the modern era of chemotherapy was ushered in with the discovery of the therapeutic qualities of sulfanilamide. Although the use of sulfa drugs has diminished as a result of the introduction of other kinds of antibiotics such as penicillin, they remain the recommended therapy for certain afflictions. For example, 2-sulfanilamidopyrimidine is useful in the treatment of meningitis.

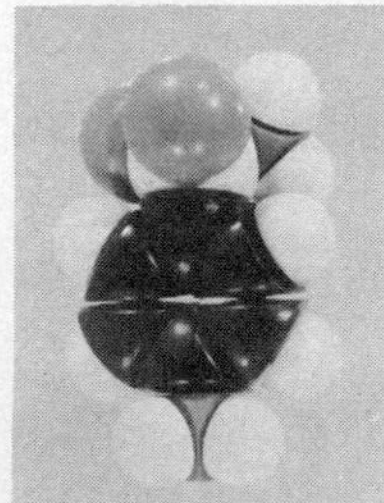

Sulfanilamide

$$H_2N-C_6H_4-SO_2NH_2 \qquad H_2N-C_6H_4-SO_2NH-C_4H_3N_2$$

Sulfanilamide | **2-Sulfanilamidopyrimidine**

17.6 Study Guide

I. Thiols (mercaptans) and sulfides (thioethers)
   A. Structure—sulfur able to expand its valence shell
   B. Nomenclature

C. Physical properties—more acidic, yet lower boiling and less water soluble
D. Synthesis and reactions
   1. Utility of S_N2 reactions
   2. Reactions of mercaptans—substitution, addition, addition-elimination
   3. Reactions of sulfides—formation of sulfonium salts
   4. Reactions of thiophene—electrophilic substitution
   5. "Desulfurization" of mercaptans and sulfides

II. Thioaldehydes, thioketones, and thiocarboxylic acids and derivatives

III. Compounds containing sulfur in higher valence states
   A. Sulfoxides and sulfones
      1. Nomenclature
      2. Synthesis
      3. Properties
   B. Sulfenic, sulfinic, and sulfonic acids
      1. Nomenclature
      2. Synthesis
      3. Reactions
         (a) Fusion reactions; with NaOH and NaCN
         (b) Electrophilic substitution of arylsulfonic acids
         (c) Sulfonic acid derivatives
         (d) Hinsberg method for differentiating primary, secondary, and tertiary amines

IV. Spectral properties of sulfur-containing compounds

PROBLEMS

1. Name each of the following structures, using any appropriate nomenclature:

(a) $(CH_3)_2CHCH_2CH_2SH$

(b) C_6H_5—CH_2SCH_3 (benzene ring bonded to CH_2SCH_3)

(c) F—C_6H_4—SH (F and SH *para* on a benzene ring)

(d) CH_3CH_2(H)C=C(H)$CHCH_2CH_2OH$, with SH on the CH carbon (CH_3CH_2 and H on one alkene carbon; H and $CH(SH)CH_2CH_2OH$ on the other)

(e) $HC{\equiv}CCHCH_2$(H)C=C(H)CO_2H, with SCH_3 on the CH carbon ($HC{\equiv}CCH(SCH_3)CH_2$ and H on one alkene carbon; CO_2H and H on the other)

(f) thiophene ring (S) bearing CO_2H at the 3-position

(g) $CH_3CH_2\overset{O}{\overset{\|}{S}}CHCH_2CH_3$, with CH_3 on the CH carbon

(h) C_6H_5—SO_2H (benzene ring bonded to SO_2H)

(i) 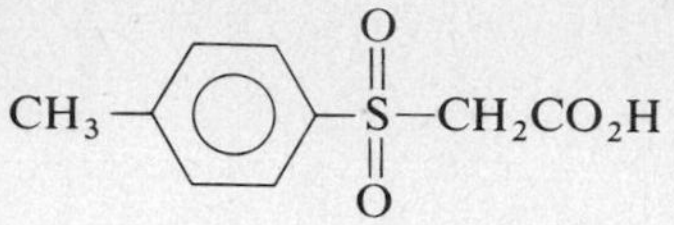 CH_3–C_6H_4–SO_2–CH_2CO_2H

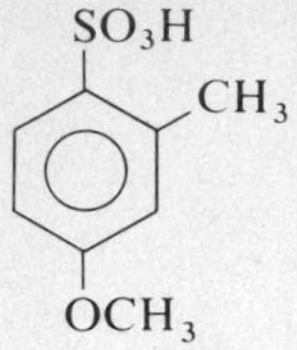

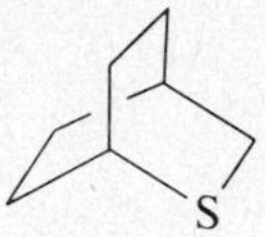

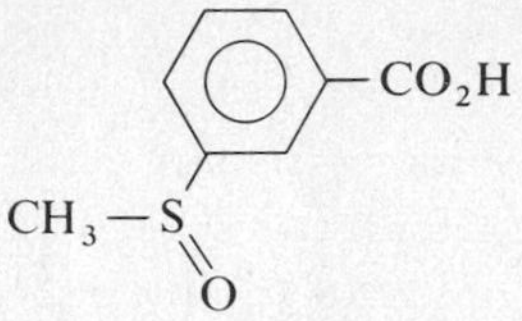

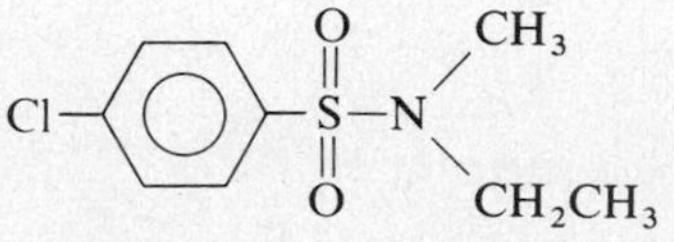

(j) SO_3H, CH_3, OCH_3

(k) S

(l) CO_2H, CH_3–S, O

(m) Cl–C_6H_4–SO_2–N(CH_3)(CH_2CH_3)

(n) $CH_3CH(SH)CH_2CH_2CH(SH)CH_3$

2. Draw the structure corresponding to each of the following names:

(a) 6-Methylheptyl mercaptan
(b) β-Mercaptopropionic acid
(c) 2,4,6-Trimethylbenzenethiol
(d) *trans*-4-Benzyl-4-methylthio-2-pentenal
(e) Diisopropyl sulfide
(f) N,N-Dimethylthiophene-3-carboxamide
(g) Diphenylsulfone
(h) Methyl isopentanesulfonate
(i) 8,8-Bismethylthiobicyclo[3.2.1]octane
(j) Naphthalene-2,7-disulfonic acid
(k) 2-Methylsulfonylcyclohexanone
(l) Benzyl methyl disulfide
(m) Trifluoromethanesulfonyl chloride
(n) 3,3-Dimethylthietane-1,1-dioxide

3. Draw the structure of the major product in each of the following reactions.

(a) $CH_3CH_2Br + C_6H_5CH_2SH \xrightarrow{NaOH}$

(b) 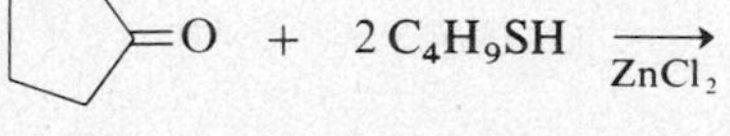 cyclopentanone (=O) $+ 2\,C_4H_9SH \xrightarrow[ZnCl_2]{}$

(c) tetrahydrothiophene (S) $+ CH_3I \longrightarrow$

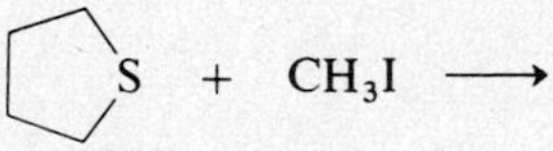

(d) 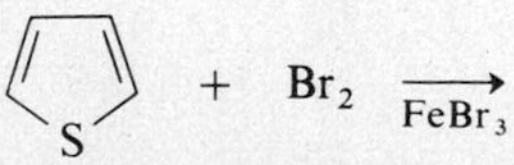 thiophene (S) $+ Br_2 \xrightarrow[FeBr_3]{}$

(e) 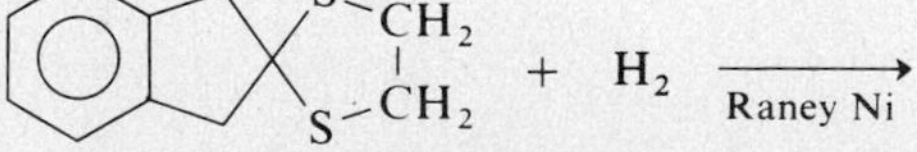 S–CH_2, S–CH_2 $+ H_2 \xrightarrow[\text{Raney Ni}]{}$

(f) 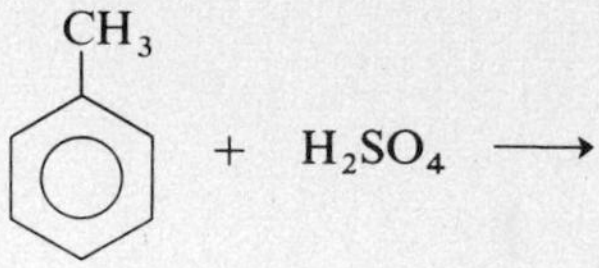 toluene (CH_3) $+ H_2SO_4 \longrightarrow$

(g) $C_6H_5SO_2NHCH_3 + CH_3MgI \longrightarrow$

(h) 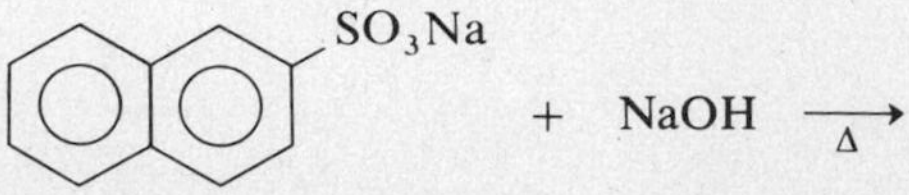 naphthalene–SO_3Na $+ NaOH \xrightarrow[\Delta]{}$

(i) 1,4-dithiane (S ring S) + H_2O_2 (1 equivalent) $\longrightarrow$

(j) CH_3CHCO_2H (with SH on C-2) + I_2 $\longrightarrow$

(k) $CH_3CH_2-C(CH_3)_2-Br$ + C_6H_5-SNa $\longrightarrow$

(l) $C_6H_5-SCH_3$ + HNO_3 $\xrightarrow{H_2SO_4}$

4. In each of the following interconversions, indicate what steps are necessary to convert the material on the left to the material on the right:

(a) cyclohexene $\longrightarrow$ dicyclohexyl sulfide (cyclohexyl–S–cyclohexyl)

(b) $C_6H_5CO_2H \longrightarrow C_6H_5CH_2SCH_3$

(c) benzene $\longrightarrow$ $C_6H_5-SO_2N(CH_3)_2$

(d) $(CH_3)_2C{=}CH_2 \longrightarrow (CH_3)_3CS(=O)OH$

(e) thiophene $\longrightarrow$ 2-thienyl–$\overset{OH}{|}{CHCH_2CH_3}$

5 Starting with any alkane, alkene, alkyne, or any monosubstituted alkane containing four or fewer carbons, benzene, toluene, *o*-, *m*-, or *p*-xylene, and any necessary inorganic reagents, outline reasonable syntheses for each of the following compounds.

(a) $CH_3CH_2SCH_3$

(b) $(CH_3)_2CHCH_2\overset{O}{\overset{\|}{S}}CH_2CH(CH_3)_2$

(c) $(CH_3)_2CHCH(CH_3)SCH_2CH_3$

(d) $C_6H_5-CH_2SCH_2-C_6H_5$

(e) $CH_3CH_2-C_6H_4-SO_2OCH_3$ (para)

(f) $CH_3-C_6H_4-SCH(CH_3)_2$ (para)

(g) $(CH_3CH_2)_2C(SCH_2C_6H_5)_2$

(h) $HS-C_6H_4-SCH_3$ (para)

6. Draw the important resonance structures for each of the following compounds:

(a) $H_2C{=}CHSCH{=}CH_2$ (b) $C_6H_5{-}SH$ (c) 2-nitrothiophene ($-NO_2$)

(d) CH_3SO_3H (e) $CH_3S(O)_2CH_2^-$ (f) $CH_3S(O){-}C_6H_4{-}O^-$

7. (a) Arrange the following compounds in order of increasing dipole moment:

$(CH_3)_2SO$ $(CH_3)_2S$ [tetrahydrothiophene 1,1-dioxide] [1,4-dithiane 1,4-dioxide, O=S ... S=O]

(b) Arrange the following compounds in order of increasing acid strength:

$C_6H_5CO_2H$ C_6H_5OH $C_6H_5\overset{+}{N}H_3$ p-$NO_2C_6H_4CO_2H$ C_6H_5SH $C_6H_5SO_3H$ p-$CH_3OC_6H_4SH$

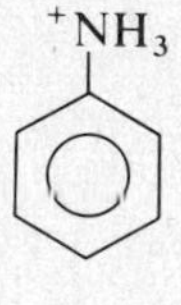

8. In each of the following pairs of compounds, indicate how you would be able to distinguish between the isomeric compounds by chemical *and* spectral means:

(a) p-$CH_3C_6H_4SH$ and $C_6H_5SCH_3$

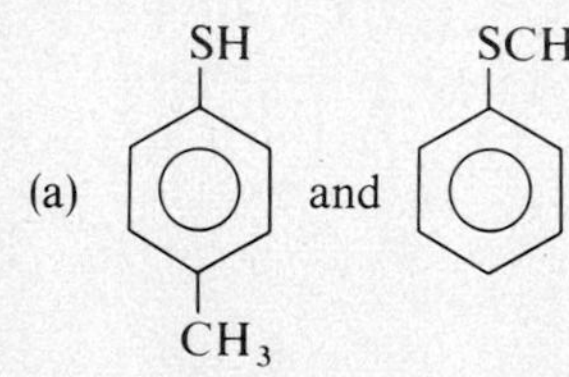

(b) p-$CH_3C_6H_4SO_3H$ and $C_6H_5SO_2OCH_3$

(c) [tetrahydrothiopyran-4-one] and [tetrahydrothiopyran 1-oxide]

(d) $CH_3SCH_2CH_2OH$ and $HSCH_2CH_2OCH_3$

9. Draw structures for each of the unknown compounds in the following reaction schemes:

(a) $(CH_3)_2C{=}CH_2$ + [N-bromosuccinimide, N—Br] $\xrightarrow{h\nu}$ A $\xrightarrow[h\nu]{HBr}$ B $\xrightarrow{Na_2S}$ C $\xrightarrow[CH_3CO_2H]{\text{Excess } H_2O_2}$ D ($C_4H_8SO_2$; possesses a pair of absorption bands at 1345 and 1140 cm^{-1})

(b) $E\ (C_6H_{14}S) \xrightarrow{CH_3I} F \xrightarrow[(CH_3)_3COH]{(CH_3)_3COK} (CH_3)_2S + G + H$

G: (C_5H_{10}; nmr spectrum displays two singlets at δ 1.62 and 1.64, a doublet at 1.56, and a quartet at 5.18, with relative intensities 3:3:3:1)

H: (C_5H_{10}; nmr spectrum displays a doublet at δ 1.2, and multiplets at 1.9, 4.8, and 5.3, with relative intensities 6:1:2:1)

10. Propose a structure that is compatible with the given data.

(a) A compound with the molecular formula C_7H_8OS is soluble in dilute sodium hydroxide and has the following ir and nmr spectra:

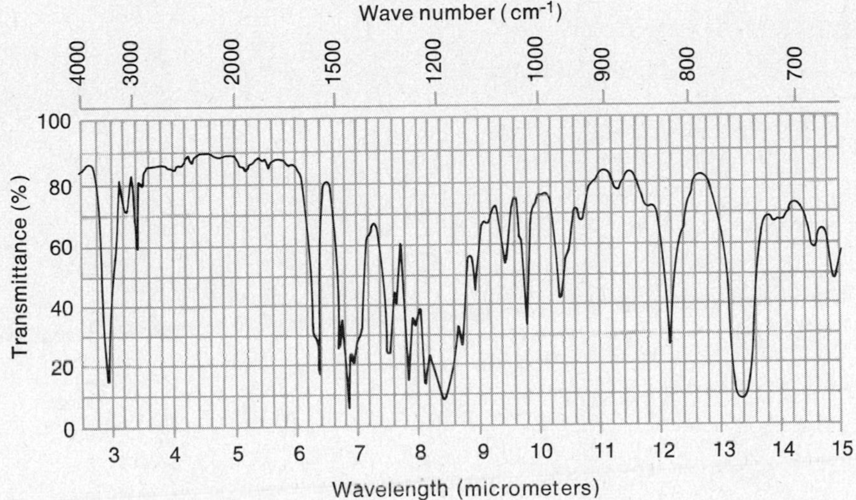

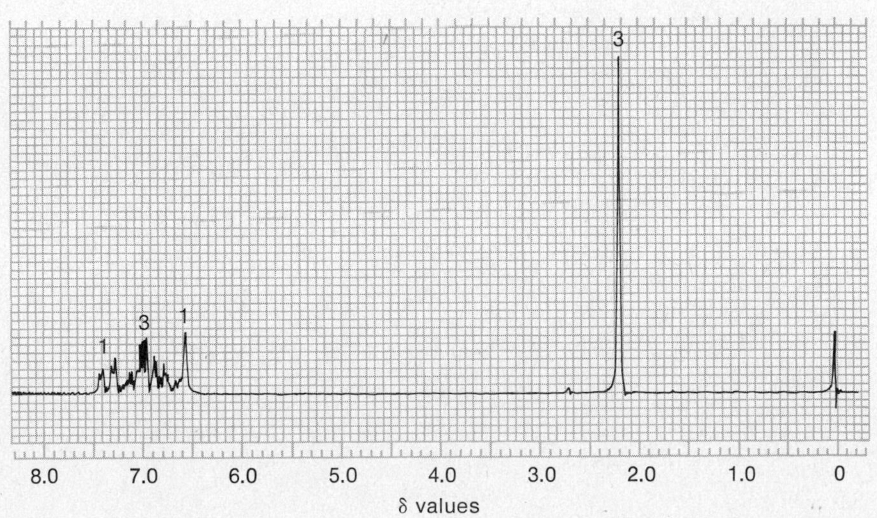

(b) A compound with the molecular formula $C_4H_2Br_2S$ is a stable, high-boiling liquid (bp 210°), which has a single resonance at δ 6.65 in the nmr.

Chapter **18**

Summary of physical properties, nomenclature, structure elucidation, and synthesis

"不以規矩，不成方圓"

Translation: "Without rules there can be no perfection." Literally, Without the compass and the square, squares and circles cannot be formed."

Sayings of Mencius (Chinese)

18.1 *Comparative Data on Physical Properties of Organic Compounds*

We have now discussed all the major classes of organic compounds, and we have cited some of the physical properties for each. It is instructive to compare these various types of compounds with respect to certain of their physical properties; for this purpose we have assembled the data in Table 18.1. In most cases the parent member (*i.e.*, the one with the lowest molecular weight) is listed, along with a member with a molecular weight close to 100 and a member containing an aromatic ring. The classes of compounds are listed in the order that they have been discussed in the text.

Table 18.1 COMPARATIVE PHYSICAL PROPERTIES OF REPRESENTATIVE COMPOUNDS OF VARIOUS CLASSES (CLASSES ARE ARRANGED ACCORDING TO THE ORDER OF DISCUSSION IN CHAPTERS 8–17)

	Mol Wt	MP	BP[a]	Density	Refractive Index	H_2O solubility
Alkanes:						
CH_4	16	−182	−162	0.415 (liq)		Slightly soluble
n-C_7H_{16}	100	−91	98	0.6838	1.3876	Insoluble
Alkenes:						
$H_2C{=}CH_2$	28	−169	−104		1.363	Slightly soluble
$C_2H_5CH{=}CHC_3H_7$	98	−137	96	0.6981	1.4043	Insoluble

Table 18.1—cont.

	Mol Wt	Mp	BP[a]	Density	Refractive Index	H_2O solubility
Alkynes:						
$HC\equiv CH$	26	−81	−84	0.6181 (liq)	1.0005	Slightly soluble
$C_2H_5C\equiv CC_3H_7$	96		105	0.7527	1.4220	Insoluble
Arenes:						
Benzene	78	5.5	80	0.8787	1.5011	Slightly soluble
Naphthalene	128	80	210	1.145	1.4003	Insoluble
Halides:						
CH_3F	34	−142	−78	0.8428		Soluble
$C_5H_{11}Cl$	106	−99	108	0.8828	1.4128	Insoluble
C_6H_5Cl	112	−45	132	1.1064	1.5248	Insoluble
Alcohols, phenols:						
CH_3OH	32	−98	65	0.7914	1.3288	Miscible
n-$C_6H_{13}OH$	102	−47	158	0.8136	1.4178	Slightly soluble
C_6H_5OH	94	43	182	1.0722	1.5509	Soluble
Ethers:						
CH_3OCH_3	46	−138	−25			Soluble
$C_3H_7OC_3H_7$	102	−122	91	0.7360	1.3832	Slightly soluble
$C_6H_5OCH_3$	108		153	0.9954	1.5179	Insoluble
Amines:						
CH_3NH_2	31	−93	−6.3	0.699		Very soluble
$(CH_3)_2NH$	45	−96	7.4	0.6804	1.350	Very soluble
$(CH_3)_3N$	59	−117	3.5	0.6079		Very soluble
$C_6H_{13}NH_2$	101	−19	129	0.763	1.4255	Slightly soluble
$C_6H_5NH_2$	93	−6.2	184	1.0216	1.5863	Soluble
Aldehydes:						
$HCHO$	30	−92	−21	0.815		Soluble
$C_5H_{11}CHO$	100	−56	128			Slightly soluble
C_6H_5CHO	106	−55	178	1.0415	1.5463	Slightly soluble
Ketones:						
CH_3COCH_3	58	−95	56	0.7908	1.3588	Miscible
$C_3H_7COC_2H_5$	100		123	0.8149	1.3990	Slightly soluble
$C_6H_5COCH_3$	120	19	202	1.0281	1.5363	Insoluble
Carboxylic acids:						
HCO_2H	46	8	101	1.220	1.3714	Miscible
$C_4H_9CO_2H$	102	−34	186	0.939	1.4086	Soluble
$C_6H_5CO_2H$	122	122	249	1.2659	1.504	Slightly soluble
Acid chlorides:						
CH_3COCl	78	−112	51	1.1039	1.3898	Reacts
C_3H_7COCl	106	−89	102	1.0277	1.4121	Reacts
C_6H_5COCl	140	−1	197	1.2105	1.5537	Reacts
Anhydrides:						
$(CH_3CO)_2O$	102	−73	136	1.0820	1.3906	Very soluble

Table 18.1—cont.

	Mol Wt	Mp	BP[a]	Density	Refractive Index	H_2O solubility
Esters:						
$CH_3CO_2CH_3$	74	−98	57	0.9723	1.3617	Very soluble
$C_3H_7CO_2CH_3$	102	−95	102	0.8984	1.3870	Slightly soluble
$C_6H_5CO_2CH_3$	136	−12	195	1.0937	1.5205	Insoluble
Amides:						
$HCONH_2$	45	2.5	105^{11}	1.134	1.4453	Miscible
$C_4H_9CONH_2$	101	101		1.023		Soluble
$C_6H_5CONH_2$	121	132	290	1.0792		Slightly soluble
Nitriles:						
CH_3CN	41	−45	80	0.7856	1.3441	Miscible
$C_5H_{11}CN$	97	−45	160	0.8093	1.4115	Insoluble
C_6H_5CN	103	−13	191	1.0102	1.5289	Slightly soluble (hot)
Nitro compounds:						
CH_3NO_2	61	−17	100	1.1371	1.3817	Soluble
$C_4H_9NO_2$	103		153	0.9710	1.4103	Slightly soluble
$C_6H_5NO_2$	123	5.7	211	1.9867	1.5562	Slightly soluble
Imines (oximes):						
$H_2C{=}NOH$	45		84			Soluble
$C_4H_9CH{=}NOH$	101	52				Slightly soluble
$C_6H_5CH{=}NOH$	121	130		1.145		Soluble (hot)
Diazenes:						
$CH_3N{=}NCH_3$	58	−78	1.5			
$C_6H_5N{=}NC_6H_5$	182	68	296	1.203	1.6266	Slightly soluble
Diazo compounds:						
CH_2N_2	42	−145	−23			
Triazo compounds:						
$C_6H_5N_3$	119		70^{11}	1.0880	1.5589	Insoluble
Thiols:						
CH_3SH	48	−123	6	0.8665		Slightly soluble
$C_5H_{11}SH$	104	−75	127	0.8375	1.4459	Insoluble
C_6H_5SH	110	71	169	1.0728	1.5879	Insoluble
Sulfonic acids:						
CH_3SO_3H	96	20	167^{10}	1.4812	1.4317	Very soluble
$C_6H_5SO_3H$	158	65				Soluble
Sulfonyl chlorides:						
$C_6H_5SO_2Cl$	177	14	252	1.3842		Decomposes
Sulfonamides:						
$C_6H_5SO_2NH_2$	157	150				Insoluble

[a] Superscript numbers indicate the reduced pressure (in mm.) at which the boiling point was recorded.

18.2 *Summary of the Rules of Systematic Nomenclature*

We introduced the systematic nomenclature for naming carbon frameworks according to the IUPAC rules in Chapter 3, and discussed the extension of these rules to the naming of compounds containing functional groups in Chapters 9–17. To emphasize the utility of these rules, and to provide a convenient means for reviewing them, we present a brief summary in this section. Included are broad generalizations rather than detailed working rules, however, and the reader should consult the sections in the earlier chapters dealing individually with the various functional groups for the more specific applications. Even more detailed information is to be found in the references cited at the end of this section.

18.2a. General rules for the naming of organic compounds.

Parent Compound. The portion of a molecule that is chosen as the basis for assigning a name to the structure is called the parent compound. In some instances, the parent name coincides with the complete name; *e.g.*, pentane for $CH_3(CH_2)_3CH_3$, pentanol for $CH_3CH_2CH_2CH_2CH_2OH$, etc. In other instances, however, the parent name requires prefix designations to specify the position and identity of substituent functions. The parent compound is chosen to contain the principal function and as much as possible of the "continuous chain" of the carbon framework to which the principal function is attached. If two or more functions of the same kind are present, the parent compound is chosen to contain as many of these as possible. If the compound contains two or more functions of unlike kind, the rule of "precedence of functions" (see Table 18.5) determines which function(s) [designated as the principal function(s)] appears in the parent compound and which function(s) [designated as subordinate function(s)] appear as the substituent(s). If C=C or C≡C bonds are present, as many as possible of these bonds are included in the parent compound even if this does not allow the longest carbon chain to be assigned to the parent compound.

Numbering of Positions of Functional Groups. The parent compound is numbered to give the principal function the lowest possible number. In the absence of a principal function, or in the event that two principal functions are symmetrically situated in the parent compound, the numbering is established by the position of unsaturation in the parent compound, or, in the absence of unsaturation, by the substituents in the parent compound; *e.g.*,

$$(CH_3)_2C{=}CHC{\equiv}CCH(OH)C(CH_3)_3$$

2,2,7-Trimethyl-6-octen-4-yn-3-ol

$$HO_2CCH{=}CHCH_2C(=O)CO_2H$$

5-Oxo-2-hexenedioic acid

Compounds with More Than One Functional Group. In many instances, a compound contains more than one functional group, and we must choose one of the functions as the principal function in order to establish a name for the parent compound. The name of the principal function is incorporated as a suffix and the names of the subordinate functions as prefixes in the name of the parent compound; however, carbon–carbon unsaturation is always incorporated in the parent name if possible, in conjunction with the suffix designation of the principal function: *e.g.*

$$\underset{\text{2-Amino-3-buten-1-ol}}{H_2C{=}CH\overset{NH_2}{\overset{|}{C}}HCH_2OH} \qquad \underset{\text{1,3-Dihydroxy-4-pentyn-2-one}}{HC{\equiv}C\overset{OH}{\overset{|}{C}}H\underset{O}{\underset{\|}{C}}CH_2OH}$$

2-Amino-3-buten-1-ol **1,3-Dihydroxy-4-pentyn-2-one**

The principal function is determined by the rule of "precedence of functions," summarized in Table 18.5.

If a principal function cannot be designated for a molecule, the numbering is determined by the terminus that is closer to a substituent. If two substituents are equally placed with respect to the two termini, the numbering is determined by the substituent that has alphabetical precedence; *e.g.*,

$$CH_3\underset{CH_3}{\underset{|}{C}}H\overset{Cl}{\overset{|}{C}}HCH_2CH_3 \qquad CH_3\underset{CH_3}{\underset{|}{C}}HCH_2\overset{Cl}{\overset{|}{C}}HCH_3$$

3-Chloro-2-methylpentane **2-Chloro-4-methylpentane**

Order of Prefix Designation. The system currently recommended by *Chemical Abstracts* for ordering functional groups alphabetically is used in this text; one of the virtues of this method is that it makes indexing easier.

18.2b. NAMES OF CARBON FRAMEWORKS.

Straight-Chain Hydrocarbons. Table 18.2 gives the names of the straight-chain hydrocarbons designated as parent compounds. For alkenes, the "ane" suffix is changed to "ene," and for alkynes, the "ane" suffix is changed to "yne."

Table 18.2 NAMES OF STRAIGHT-CHAIN HYDROCARBONS[a]

CH_4	Methane	$C_{11}H_{24}$	Undecane	$C_{25}H_{52}$	Pentacosane
C_2H_6	Ethane	$C_{12}H_{26}$	Dodecane	$C_{30}H_{62}$	Triacontane
C_3H_8	Propane	$C_{13}H_{28}$	Tridecane	$C_{40}H_{82}$	Tetracontane
C_4H_{10}	Butane	$C_{14}H_{30}$	Tetradecane	$C_{50}H_{102}$	Pentacontane
C_5H_{12}	Pentane	$C_{15}H_{32}$	Pentadecane	$C_{100}H_{202}$	Hectane
C_6H_{14}	Hexane	$C_{16}H_{34}$	Hexadecane	$C_{200}H_{402}$	Dictane
C_7H_{16}	Heptane	$C_{17}H_{36}$	Heptadecane		
C_8H_{18}	Octane	$C_{18}H_{38}$	Octadecane		
C_9H_{20}	Nonane	$C_{19}H_{40}$	Nonadecane		
$C_{10}H_{22}$	Decane	$C_{20}H_{42}$	Eicosane		

[a] The underlined portions indicate that part of the name which combines with the appropriate suffix and prefix for other names.

Branched-Chain Hydrocarbons. A number of branched-chain structures can also be used as the parent compound; however, the use of these designations is restricted to alkanes containing no more than six carbons according to the IUPAC rules. The presence of a terminal $(CH_3)_2CH-$ group in an otherwise straight-chain hydrocarbon is denoted by the prefix "iso" followed by the name of the alkane corresponding to the *total* number of carbon atoms present. The presence of a $(CH_3)_3C-$ group at the terminus of an otherwise straight-chain hydrocarbon is denoted by the prefix "neo," *e.g.*,

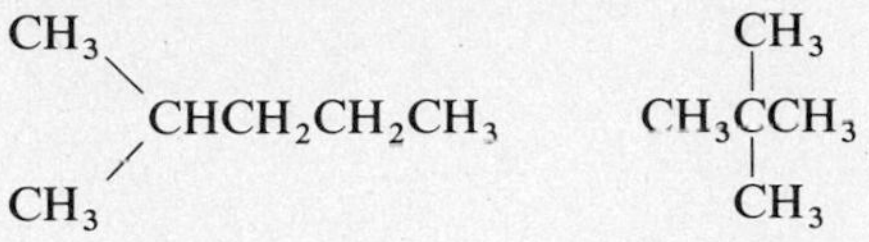

$$(CH_3)_2CHCH_2CH_2CH_3 \qquad CH_3C(CH_3)_2CH_3$$

Isohexane **Neopentane**

Cyclic and Polycyclic Hydrocarbons. Monocyclic hydrocarbons are indicated by the prefix "cyclo" followed by the name of the alkane corresponding to the number of carbon atoms contained *in* the cycle. Polycyclic hydrocarbons can appear as "fused," "bridged," or "spiro" structures. (See Section 3.11 for the methods of designating and numbering these systems); *e.g.*,

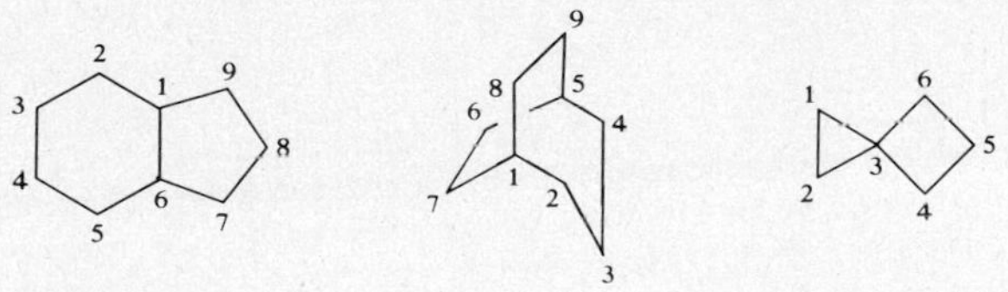

A "fused"-ring bicyclic hydrocarbon

A "bridged"-ring bicyclic hydrocarbon

A "spiro" bicyclic hydrocarbon

Aromatic Hydrocarbons. The names of the four lowest molecular weight benzenoid aromatic hydrocarbons are

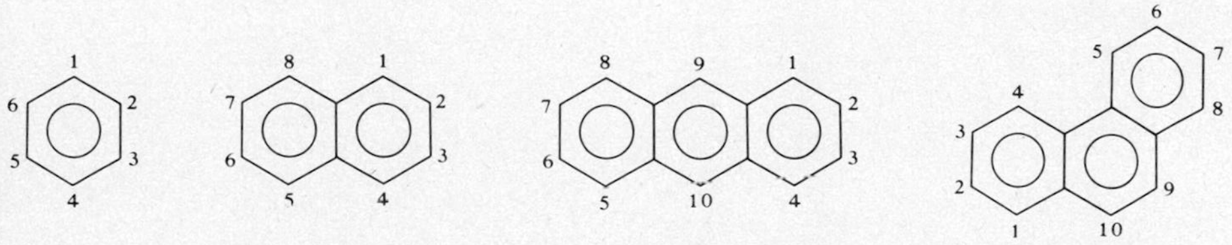

Benzene **Naphthalene** **Anthracene** **Phenanthrene**

18.2c. NAMES OF HETEROCYCLIC FRAMEWORKS. The chemistry of heterocyclic compounds bears sufficient resemblance to that of the compounds discussed in the previous chapters that only cursory treatment has been accorded them. They are exceedingly important compounds, nevertheless, and require certain special rules for their systematic naming. Although these rules are too extensive to outline in detail, we give a few to provide some elementary guidelines for the nomenclature of compounds of these types.

Monocyclic Rings Containing One Heteroatom. In assigning the name of a heterocyclic compound containing oxygen, nitrogen, or sulfur, the identity of the hetero atom is indicated by "oxa," "aza," and "thia," for oxygen, nitrogen, and sulfur, respectively (except when they precede a vowel, in which case the final "a" of the prefix is dropped), and the size of the ring by the appropriate suffix chosen from Table 18.3; *e.g.*,

Oxirane **N-Methylaziridine** **N-Methylazirine** **Thiolane**

Table 18.3 SUFFIX DESIGNATIONS FOR MONOCYCLIC HETEROCYCLIC COMPOUNDS

Number of Atoms in Ring	Saturated Ring		Unsaturated Ring	
	X = N	X = O or S	X = N	X = O or S
3	-iridine	-irane	-irine	-irene
4	-etidine	-etane	-ete	-ete
5	-olidine	-olane	-ole	-ole
6	[a]	-ane	-ine	-in
7	[a]	-epane	-epine	-epin

[a] In these cases, the prefix "perhydro" is affixed to the name of the unsaturated analog (*e.g.* NH is hexahydropyridine, also called piperidine).

Many of the unsaturated five-membered heterocyclic compounds containing a single hetero atom are commonly designated by their trivial, or common, names instead of by the IUPAC names derived by use of the appropriate prefixes and suffixes, as the following examples illustrate:

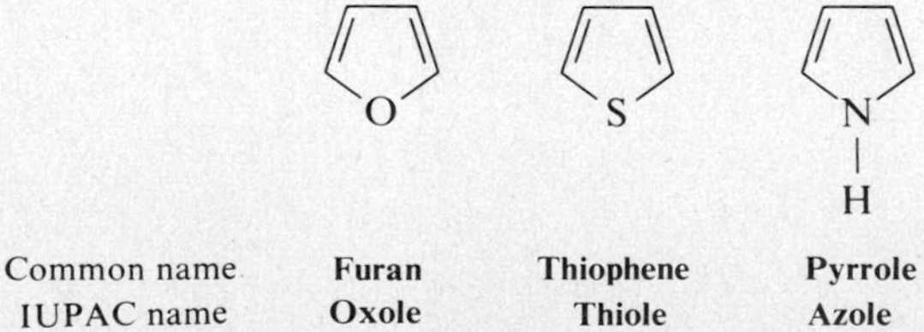

The corresponding saturated analogs are commonly referred to as tetrahydrofuran, tetrahydrothiophene, and pyrrolidine, respectively, instead of by the IUPAC names, oxolane, thiolane, and azolidine.

A system for naming compounds containing hetero atoms (or "foreign elements") that is gaining increasing acceptance employs the "replacement procedure." In this system a compound is named on the basis of its carbon framework analog, and the nature of the replacement of the carbons by hetero atoms in this framework is indicated by the appropriate prefixes and numbers. For example, the acyclic compound $CH_3CH_2SCH_3$ is named 2-thiabutane, the "thia" indicating the presence of sulfur and the "2" indicating its position in the carbon framework. This nomenclature is particularly useful for heterocyclic systems, for it avoids the necessity of remembering the suffixes listed in Table 18.3. Thus, oxirane is oxacyclopropane, thiolane is thiacyclopentane, pyrrole is aza-2,4-cyclopentadiene, etc.

Monocyclic Rings Containing Two or More Heteroatoms. If two or more hetero atoms of like kind are present in the ring, the positions are designated by number, beginning with the 1 position at one of the hetero atoms such that the hetero atoms are assigned the lowest set of numbers, and indicating the number of hetero atoms with the appropriate prefixes "di," "tri," etc.

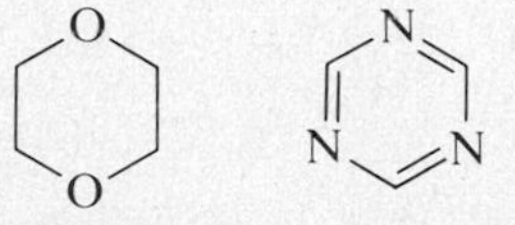

1,4-Dioxane **1,3,5-Triazine**

If two or more hetero atoms of unlike kind are present in the ring, their order of citation is by descending group number of the periodic table and increasing atomic number within each group of the periodic table. The positions are indicated by designating the ranking hetero atom (the first designated) as the 1 position, and numbering around the ring such that the hetero atoms are assigned the lowest set of numbers.

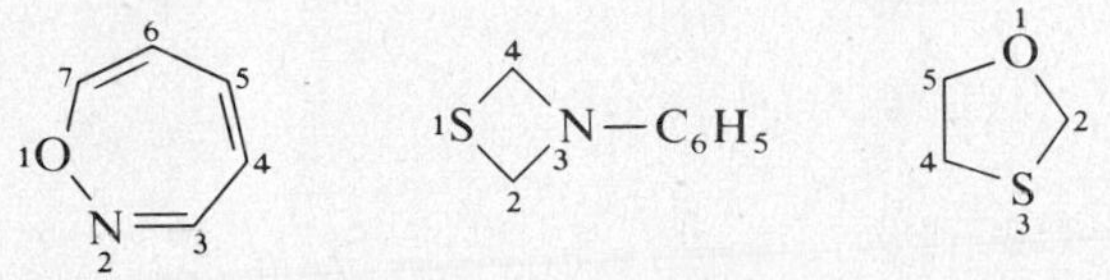

1,2-Oxazepine **N-Phenyl-1,3-thiazetidine** **1,3-Oxathiolane**

In the "replacement procedure" 1,4-dioxane is named 1,4-dioxacyclohexane, 1,3,5-triazine is 1,3,5-triazabenzene, 1,2-oxazepine is 1-oxa-2-aza-2,4,6-cycloheptatriene, N-phenyl-1,3-thiazetidine is 3-phenyl-1-thia-3-azacyclobutane, and 1,3-oxathiolane is 1-oxa-3-thiacyclopentane.

18.2d. FUNCTIONAL GROUPS. The term "functional group" or "function" refers to an atom or group of atoms within a molecule that confers some distinctive chemical and/or physical property on the molecule. In the broadest sense, all the atoms or groups within a molecule can be considered to be functions. The two principal methods for incorporating functional groups into the names of compounds are called the "radicofunctional system" and the "substitutive system." In the radicofunctional system the identity of the functional group is expressed as the final word of a multiword name, and the preceding words designate the remainder of the molecule. Thus, $CH_3CH_2CH_2OH$ is *n*-propyl alcohol, $CH_3COCH_2CH_3$ is ethyl methyl ketone, etc. In the substitutive system each substituent atom or group is specified either as a prefix or a suffix to the name of the parent compound. Thus, $CH_3CH_2CH_2OH$ is 1-propanol, $CH_3COCH_2CH_3$ is 2-butanone, etc.

Hydrocarbon Functions. Functions derived by the removal of hydrogen from an alkane framework are indicated by the name of the alkane with replacement of the suffix "ane" by "yl." If the alkyl group is unbranched, it is not necessary to affix the prefix *n.* If it is branched, we can so indicate either by using the systematic IUPAC nomenclature for the alkyl group or by affixing the appropriate prefix. The IUPAC system restricts the use of prefixes in this respect to alkyl groups having six or fewer carbon atoms and includes the following listed examples. Frequently, however, chemists disregard this rule and apply the iso and neo prefixes to larger alkyl groups as well.

$(CH_3)_2CH-$	**Isopropyl**	$(CH_3)_2CHCH_2CH_2-$	**Isopentyl**
$(CH_3)_2CHCH_2-$	**Isobutyl**	$(CH_3)_3CCH_2-$	**Neopentyl**
$CH_3CH_2(CH_3)CH-$	***s*-Butyl (*sec*-Butyl)**	$CH_3CH_2C(CH_3)_2-$	***t*-Pentyl**
$(CH_3)_3C-$	***t*-Butyl (*tert*-Butyl)**	$(CH_3)_2CHCH_2CH_2CH_2-$	**Isohexyl**

Groups derived by removal of a hydrogen atom from an alkene framework acquire the suffix "enyl," and those from an alkyne the suffix "ynyl," the position of the unsaturation being indicated by the appropriate position numbers:

$$\overset{5}{C}H_3\overset{4}{C}H{=}\overset{3}{C}H\overset{2}{C}H_2\overset{1}{C}H_2- \qquad H\overset{5}{C}{\equiv}\overset{4}{C}\overset{3}{C}H_2\overset{2}{C}H_2\overset{1}{C}H_2- \qquad H_2\overset{5}{C}{=}\overset{4}{C}H\overset{3}{C}H{=}\overset{2}{C}H\overset{1}{C}H_2-$$

3-Pentenyl **4-Pentynyl** **2,4-Pentadienyl**

$H_2C{=}CH-$	$CH_3CH{=}CHCH_2-$	$HC{\equiv}CCH_2-$	$C_6H_5CH_2-$
Vinyl-	***cis*- and *trans*-Crotyl or 2-Buten-1-yl**	**Propargyl-**	**Benzyl-**

Groups derived by the removal of two hydrogens from the same carbon atom of an alkane framework acquire the suffix "ylidene," and those derived by removing three hydrogens from the same carbon acquire the suffix "ylidyne." Thus, $CH_3CH{=}$ is ethylidene, $CH_3CH_2C{\equiv}$ is propylidyne, etc. Groups derived by removing hydrogens from each of the terminal carbons of an alkane are named as follows:

$-CH_2CH_2-$	$-CH_2CH_2CH_2-$	$-CH_2CH_2CH_2CH_2-$	etc.
Ethylene	**Trimethylene**	**Tetramethylene**	

For example, ethylene dibromide is $BrCH_2CH_2Br$ (note that dibromoethylene is a different compound, *i.e.*, $BrCH{=}CHBr$).

Aromatic Functions. Groups derived by removal of a hydrogen from an aromatic ring acquire the suffix "yl"; *e.g.*,

Phenyl **1-Naphthyl (α-Naphthyl)** **2-Naphthyl (β-Naphthyl)**

Hetero Atom Functions. In the substitutive system of nomenclature the majority of functions have both prefix and suffix designations; *e.g.*, the hydroxyl function (—OH) is designated as "hydroxy" when specified by a prefix and "-ol" when specified as a suffix. A few groups, however, are specified only by prefix designations (*i.e.*, "nitro" for "$-NO_2$," "hydro" for "—H," etc.), and a few are specified only by suffix designations (*i.e.*, "ene" for "C=C" and "yne" for "C≡C"). In the radicofunctional system some, but not all, of the hetero atom functions provide the basis for a name. Table 18.4 tabulates these specifications for a variety of hetero atom functional groups.

Table 18.4 SUFFIX, PREFIX, AND RADICOFUNCTIONAL DESIGNATIONS OF FUNCTIONAL GROUPS

Functional Group	Suffix Designation	Prefix Designation	Radicofunctional Name
—F, —Cl, —Br, —I		Fluoro, Chloro, Bromo, Iodo	Fluoride, Chloride, Bromide, Iodide
—OH	-ol	Hydroxy	Alcohol
—O—			Ether
—OR		Alkoxy	
—OAr		Aryloxy	
—OOH		Hydroperoxy	
—C(=O)H	-al or -carbaldehyde	Formyl	
>C=O	-one	Oxo	Ketone
$-CO_2H$	-oic acid or -carboxylic acid	Carboxy	
—C(=O)Halogen	-oyl(-yl) or -carbonyl halide	Halocarbonyl or Haloformyl	
$-C(=O)NH_2$	-amide or -carboxamide	Carboxamido or Carbamoyl	
—C(=O)OR		Alkoxycarbonyl	
$-C(=O)CH_3$		Acetyl or Ethanoyl	
—C(=O)R		Alkanoyl	
$-OC(=O)CH_3$		Acetoxy or Ethanoyloxy	

Table 18.4—cont.

Functional Group	Suffix Designation	Prefix Designation	Radicofunctional Name
$-NH_2$, $-NHR$, $-NR_2$	-amine	Amino	
$-NHNH_2$	-hydrazine or -diazane	Hydrazino or Diazanyl	
$=N_2$		Diazo	
$-N_3$		Azido	Azide
$-CN$	-nitrile or -carbonitrile	Cyano	Cyanide
$-NC$		Isocyano or Carbylamino	Isocyanide
$-OCN$		Cyanato	Cyanate
$-NCO$		Isocyanato or Carbonylamino	Isocyanide
$-NO$		Nitroso	
$-NO_2$		Nitro	
$-NHC(=O)CH_3$		Acetamido	
$-SH$	-thiol	Mercapto	Mercaptan
$-S-$			Sulfide
$-SR$		Alkylthio	
$-SAr$		Arylthio	
$>S=O$	-sulfoxide	Sulfinyl	
$>SO_2$	-sulfone	Sulfonyl	
$-SO_3H$	-sulfonic acid	Sulfo	
$-SO_2Cl$	-sulfonyl chloride	Chlorosulfonyl	
$-SO_2NH_2$	-sulfonamide	Sulfonamido	
$-SO_2R$		Alkylsulfonyl	
$-C(=S)H$	-thial or -carbothial	Thioformyl	
$>C=S$	-thione	Thioxo	Thioketone
$-SCN$		Thiocyanato	Thiocyanate
$-NCS$		Isothiocyanato	Isothiocyanate

Order of Precedence of Functional Groups. Functional groups can be placed in one of five categories, in order of decreasing precedence: (I) "onium" functions, (II) peroxides, (III) groups that *must* be a terminus, (IV) groups that need not be a terminus, and (V) nonprecedence groups. Table 18.5 shows the order of precedence in each of the first four of these classes.

18.2e. NOMENCLATURE REFERENCES. The nomenclature rules presented in the earlier chapters on the individual functions and summarized in this section are sufficient to name most, although not all, the compounds encountered in this text. Where they are insufficient, the complete rules for organic nomenclature as issued by the Commission on the Nomenclature of Organic Chemistry of the

Table 18.5. ORDER OF PRECEDENCE OF FUNCTIONAL GROUPS

I. "Onium" groups: $R_4\overset{+}{N}—$, $R_3\overset{+}{O}—$, $R_3\overset{+}{S}—$, $R_3\overset{+}{C}—$, $R\overset{+}{N_2}—$
II. Peroxides: RO—OR and RO—OH
III. Groups that must be a terminus
  A. Acid groups: $—CO_2H$, $—SO_3H$, $—SO_2H$, —SOH
  B. Acid derivative groups: —COX, $—CONR_2$
  C. Formyl: —CHO
  D. Cyano: —CN
IV. Groups that need not be a terminus
  A. Oxo: $\rangle C{=}O$
  B. Hydroxy: —OH
  C. Mercapto: —SH
  D. Amino: $—NH_2$, —NHR, $—NR_2$
  E. Imino: $\rangle C{=}NH$, $\rangle C{=}NR$
  F. Alkoxy (and aryloxy): —OR (or —OAr)
  G. Alkylthio (or arylthio): —SR (or —SAr)
  H. Sulfinyl: $\rangle S{=}O$
  I. Sulfonyl: $\rangle SO_2$
  J. Hydro: —H
V. Nonprecedence groups: halogens, nitro ($—NO_2$), nitroso (—NO), azo (—N=N—), diazo ($\rangle CN_2$).

International Union of Pure and Applied Chemistry (references 1–4)* should be consulted. Reference 5 contains a compilation of the names of all known carbocyclic and heterocyclic ring systems.

1. *Nomenclature of Organic Chemistry*: *Section A. Hydrocarbons*; *Section B. Fundamental Heterocyclic Systems*, 2nd ed., Butterworth & Co., Ltd., London, 1966. [This also appears in *Journal of the American Chemical Society*, *82*, 5545 (1960).]
2. *Nomenclature of Organic Chemistry*: *Section C. Characteristic Groups Containing Carbon, Hydrogen, Oxygen, Nitrogen, Halogen, Sulfur, Selenium, and/or Tellurium*, Butterworth & Co., Ltd., London, 1966.
3. *IUPAC Tentative Rules for the Nomenclature of Organic Chemistry. Section E. Fundamental Stereochemistry*, *Journal of Organic Chemistry*, *35*, 2849 (1970).
4. *IUPAC-IUB Revised Tentative Rules for Nomenclature of Steroids*, *Journal of Organic Chemistry*, *34*, 1517 (1969).

* Section D of *Nomenclature of Organic Chemistry*, devoted to the nomenclature of organometallic compounds, has not yet been published.

5. *The Ring Index* by A. M. Patterson, L. T. Capell, and D. F. Walker, Chemical Abstracts Service, American Chemical Society, Washington, D.C., 1960.
6. *Nomenclature of Organic Compounds* by J. H. Fletcher, O. C. Dermer, and R. B. Fox, Advances in Chemistry Series No. 126, American Chemical Society, Washington, D.C., 1974.

18.3 *Guidelines for Structure Elucidation*

The structure of organic compounds is the focal point of Part I of this text, and methods for deducing structures are the essential ingredient for all discussions of this facet of organic chemistry. The patterns for structure elucidation vary, depending on the nature of the particular problem to be solved. Despite this, certain general guidelines are usually employed, and the following sequence of steps is usually followed:

1. Qualitative determination of elements present.
2. Quantitative determination of elements present → empirical formula.
3. Molecular weight determination → molecular formula.
4. Determination of physical characteristics, particularly spectral characteristics.
5. Determination of chemical characteristics.
   (a) Tests for functional groups.
   (b) Degradation to compounds of known structure.
6. Unequivocal synthesis.

The molecular formula of a compound can provide a useful clue to the compound's possible structure. Not only does it indicate what elements are present, but it also allows the "degrees of unsaturation" to be calculated by considering the hydrogen-to-carbon ratio (see p. 167). Thus, a C_4H_{10} compound must be a saturated alkane; a C_4H_6 compound can contain one ring and one double bond, two double bonds, or one triple bond. In general, we can calculate the degrees of unsaturation by the formula

$$N = \frac{\sum_i n_i(v_i - 2) + 2}{2}$$

where N is the degrees of unsaturation, n_i is the number of atoms of element i, and v_i is the valence of element i (*i.e.*, carbon has a valence of 4, hydrogen a valence of 1, oxygen a valence of 2, sulfur a valence of 2, nitrogen a valence of 3, *etc.*). Thus, a $C_4H_{10}O$ compound contains no degrees of unsaturation and must be an acyclic alcohol or ether, a C_4H_8S compound contains one degree of unsaturation, a C_4H_7N compound contains two degrees of unsaturation, *etc.*

As an introduction to the techniques of structure analysis, let us first consider a compound possessing the molecular formula C_4H_8O. From this molecular formula we know that the compound contains one degree of unsaturation—*i.e.*,

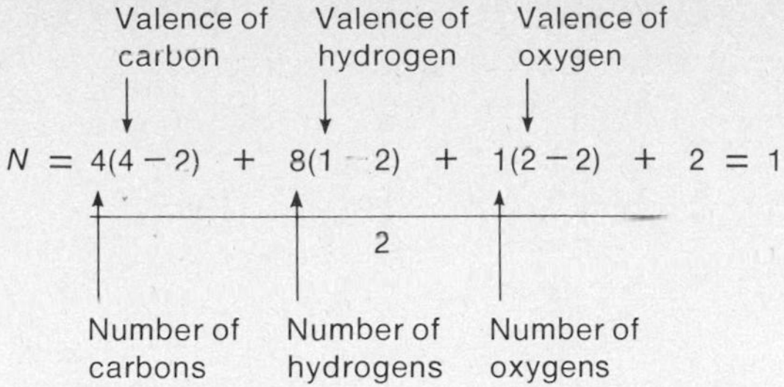

which can be present as a carbocyclic ring, a heterocyclic ring, a C=C bond, or a C=O bond. To decide among these alternatives, we carry out the following test reactions and note the results:

1. The compound reacts with lithium aluminum hydride to produce a different compound, which releases hydrogen gas when treated with metallic sodium.
2. The compound reacts with semicarbazide hydrochloride to form a semicarbazone.
3. The compound *does not* undergo facile oxidation.

Tests 1 and 2 suggest the presence of a carbonyl function of an aldehyde or ketone (RCHO or $R_2C{=}O$); test 3 rules out an aldehyde. Since the C=O group accounts for the one degree of unsaturation that is present in the molecule, the carbon framework must have a C_4H_8 composition; the only structural possibility that accommodates *all* these data is 2-butanone, $CH_3CH_2\overset{\displaystyle O}{\overset{\|}{C}}CH_3$.

As an alternative to chemical tests, we could use spectral measurements in the example just described to provide a definitive assignment of structure. From the ir absorption at 1715 cm^{-1}, we can discern the presence of a carbonyl function, and from the nmr spectrum (consisting of a singlet from the CH_3C(=O)— group, a quartet from the $-CH_2-$ group, and a triplet from the CH_3CH_2- group), we can establish the structure of the carbon framework.

Let us next turn to a slightly more complex example, a $C_{10}H_{12}$ compound exhibiting the following chemical characteristics:

1. A dilute solution of bromine in carbon tetrachloride is decolorized when treated with the compound.
2. Strenuous oxidation of the compound yields benzoic acid.
3. Ozonolysis of the compound yields two materials, one possessing the molecular formula C_2H_4O and the other the molecular formula C_8H_8O.

The reaction with bromine (test 1) and the reaction with ozone (test 3) both indicate the presence of a C=C bond in the molecule, and the formation of benzoic acid on strenuous oxidation (test 2) establishes the presence of a benzene ring. These data, as well as the molecular formula indicating five degrees of unsaturation, are commensurate, then, with a compound containing one C=C bond and one phenyl ring. The formation of a two-carbon compound upon ozonolysis indicates that the C=C bond must be in the center of a four-carbon unit to which is attached a phenyl ring, *viz.*,

$CH_3CH{=}C(C_6H_5)CH_3$ (phenyl ring on the central carbon) or $CH_3CH{=}CHCH_2{-}C_6H_5$

(a) (b)

To choose between alternatives (a) and (b), we must determine the structure of the C_8H_8O ozonolysis product, which we can do by testing the ease of oxidation of the compound. If facile oxidation to an acid takes place, the ozonolysis product must be $C_6H_5CH_2CHO$, and structure (b) must be correct; if oxidation does not take place easily, however (as we shall assert to be the case), the ozonolysis product must be $C_6H_5COCH_3$, and structure (a) must be correct.

As in the previous example, spectral methods provide definitive information concerning the structure of the $C_{10}H_{12}$ compound. Structure (a) shows an nmr spectrum containing a five-hydrogen resonance for the aromatic hydrogens, a three-hydrogen singlet for the $CH_3C{=}$ group, a three-hydrogen doublet for the $CH_3CH{=}$ group, and a one-hydrogen quartet for the $CH_3CH{=}$ hydrogen. Structure (b) would show a somewhat different pattern of resonance bands in the nmr.

A considerably more complex example is represented by a $C_{12}H_{19}NO_2$ compound, which is soluble in dilute hydrochloric acid and gives the following results when tested with a variety of reagents:

1. Upon treatment with metallic sodium, hydrogen is evolved.
2. Oxidation with periodic acid (HIO_4) produces a $C_{10}H_{13}NO$ compound that is soluble in dilute hydrochloric acid and insoluble in sodium bicarbonate solution, and that contains $CH_3CO{-}$ and $(CH_3)_2N{-}$ functions, as indicated by its nmr spectrum, which shows a three-hydrogen singlet at δ 2.3 and a six-hydrogen singlet at δ 3.0.
3. Treatment with nitrous acid does not result in the formation of a new compound.

On the basis of these data, we can piece together an acceptable structure in the following manner. The molecular formula indicates that the original compound contains four degrees of unsaturation, and the presence of an aromatic ring accounts for all these. The liberation of hydrogen upon treatment with sodium indicates the presence of an $-OH$, $-CO_2H$, or $-NH-$ group. The original compound is not soluble in sodium bicarbonate solution, however, thus ruling out the presence of a $-CO_2H$ function. The lack of reaction with nitrous acid similarly eliminates an $-NH_2$ or $-NHR$ group from consideration, leaving $-OH$ as the group possessing the acidic hydrogen. The 10 carbon atoms in the oxidation product from test 2 are most reasonably accounted for as a phenyl ring, a $(CH_3)_2N{-}$ function, and a $CH_3CO{-}$ group; the nmr spectrum of this material corresponds best to a *para*-disubstituted compound; *i.e.*, *p*-dimethylaminoacetophenone. At this point in the analysis, 10 of the 12 carbon atoms in the original compound have been accounted for in the structure:

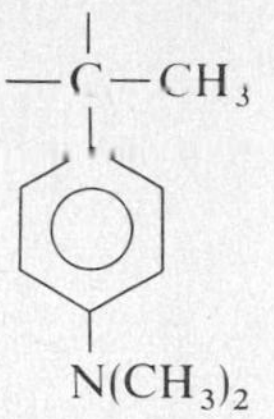

$$-\overset{|}{\underset{|}{C}}-CH_3 \text{ on } p\text{-}C_6H_4\text{-}N(CH_3)_2$$

The facile oxidation of the original compound by periodic acid indicates the presence of a 1,2-diol. The most reasonable structure that accommodates *all* these data, then, is

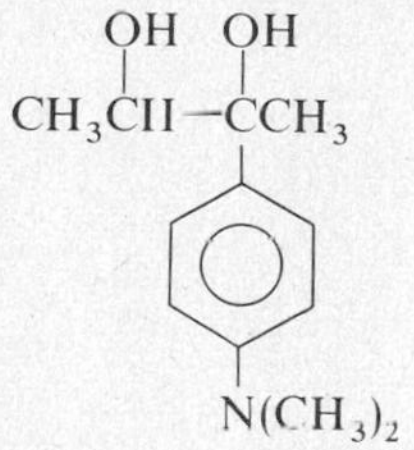

$$CH_3\overset{OH}{\overset{|}{C}}H-\overset{OH}{\overset{|}{C}}CH_3 \text{ with } p\text{-}C_6H_4N(CH_3)_2 \text{ on the second carbon}$$

It has been accepted as axiomatic for many years that the ultimate proof of structure can only be provided by an unequivocal synthesis of the compound from starting materials of known structure. To illustrate the application of this dictum, we outline in the next section syntheses for the three compounds that have just been discussed.

18.4 Elementary Guidelines for Devising Organic Syntheses

Chapters 8–17 have presented a variety of organic reactions for the conversion of one type of compound into another. Although these represent only a small fraction of the thousands of organic reactions known today, they are sufficient in number to allow us to devise some reasonably complex syntheses as demonstrated by the problems at the ends of these chapters. To give hard and fast rules for the design of even such simple syntheses as these presents a problem, however, and for more complex syntheses it is an impossible task. Nevertheless, certain useful guidelines can be developed, and it is the purpose of this discussion to point out some of these by analyzing representative syntheses that can be designed by using the reactions presented in Chapters 8–17.

Although organic synthesis is sometimes practiced as a pure art form, it is directed more frequently to the production of a compound that is sought for a specific purpose. The natural product chemist views a total synthesis of a naturally occurring material as putting the capstone on its proof of structure. The physical organic chemist often requires a synthesis to secure a material ideally suited for testing a theory. The biochemist may require a synthesis to obtain an appropriate enzyme inhibitor. For the industrial chemist, a synthesis provides a new product that can be marketed at a profit. In all cases, although with varying emphasis, the practical considerations uppermost in the mind of the chemist are (a) How much do the starting materials cost? (b) How much time does the process require? (c) What is the overall yield of the desired product? In other words, how expensive is the product in terms of chemical and human resources? The ideal synthesis is one that employs only very cheap starting materials, that produces very high yields in each step of the sequence, and that proceeds very rapidly and requires little time on the part of the chemist. Since almost all syntheses fall short of these goals in practice, the central problem in the design of a synthesis is how to bring practice as close as possible to the ideal.

It has been stated that the ability to synthesize a system—to reconstitute it from its component parts—is the most stringent test of the understanding of and control over that system. Many readers have probably taken a clock apart in their childhood but few, we suspect, have put it back together again. To take a clock apart mindlessly is, of course, not difficult. To take it apart in a fashion that reveals the way in which it works is more challenging, and to put it back together, taking advantage of what was learned in its dissection, is most difficult of all. In similar fashion, to devise a good synthesis is difficult, demanding a thorough knowledge of organic chemistry and testing one's mastery of the tools of one's trade. The basic tools are provided by the material in the previous chapters of this book; we shall now attempt to organize this material in such a fashion that it can be easily reviewed by the reader and then applied with skill to the solution of the problems at the end of this chapter.

18.4a. AVAILABLE STARTING MATERIALS. The first essential question to be answered in the design of a synthesis is, "Where does one start?" It depends, of course, on what chemicals are on hand as "available starting materials," and differs from one laboratory to another, one city to another, one country to another, and one period of history to another. Numerous chemical supply houses operate today in various parts of the world, making available to the chemist several thousand compounds that can be purchased for prices ranging from a few cents per pound to many dollars per gram. In the more sophisticated syntheses discussed in later chapters, we assume that the chemist has access to any of the chemicals that are generally available from chemical supply houses. For the present purpose, though, we take a much more restricted view, limiting the available starting compounds to methane, ethane, propane, ethene, propene, ethyne, and benzene. Let us begin, then, first by stockpiling our own chemical supply house by synthesizing a series of useful starting materials from these compounds, using reactions discussed in previous chapters.

Compounds derived directly from methane, ethane, and propane:

Synthesis of alkyl halides

$$\begin{cases} CH_4 \ + \ Cl_2 \ \text{(or } Br_2) \longrightarrow CH_3Cl \ \text{(or } CH_3Br) \\ CH_3CH_3 \ + \ Cl_2 \ \text{(or } Br_2) \longrightarrow CH_3CH_2Cl \ \text{(or } CH_3CH_2Br) \\ CH_3CH_2CH_3 \ + \ Cl_2 \ \text{(or } Br_2) \end{cases}$$

$$\begin{Bmatrix} CH_3CH_2CH_2Cl \\ + \\ (CH_3)_2CHCl \end{Bmatrix} \text{ or } \begin{Bmatrix} CH_3CH_2CH_2Br \\ + \\ (CH_3)_2CHBr \end{Bmatrix}$$

Mixtures requiring separation by fractional distillation

Compounds derived directly from ethene and propene:

Synthesis of alcohols

$$H_2C{=}CH_2 + H_2O \xrightarrow{H^+} CH_3CH_2OH$$

$$CH_3CH{=}CH_2 + H_2O \xrightarrow{H^+} (CH_3)_2CHOH$$

Synthesis of epoxides

$$H_2C{=}CH_2 \xrightarrow{[O]^*} \overset{O}{CH_2{-}CH_2}$$

$$CH_3CH{=}CH_2 \xrightarrow{[O]^*} CH_3\overset{O}{CH{-}CH_2}$$

Synthesis of alkyl halides

$$H_2C{=}CH_2 + HX \longrightarrow CH_3CH_2X \quad (X = Cl, Br, I)$$

$$CH_3CH{=}CH_2 + HX \longrightarrow (CH_3)_2CHX \quad (X = Cl, Br, I)$$

$$CH_3CH{=}CH_2 + HBr \xrightarrow{Peroxides} CH_3CH_2CH_2Br$$

Synthesis of 1,2-dihaloalkanes

$$H_2C{=}CH_2 + X_2 \longrightarrow XCH_2CH_2X \quad (X = Cl, Br)$$

$$CH_3CH{=}CH_2 + X_2 \longrightarrow CH_3\overset{X}{\overset{|}{C}}HCH_2X \quad (X = Cl, Br)$$

Compounds derived directly from ethyne:

Synthesis of acetaldehyde

$$HC{\equiv}CH + H_2O \xrightarrow{H^+,\ Hg^{++}} CH_3CHO$$

Synthesis of sodium acetylide

$$HC{\equiv}CH + Na \longrightarrow HC{\equiv}CNa$$

Synthesis of dihalides and tetrahalides

$$HC{\equiv}CH + X_2 \longrightarrow \underset{(X = Cl, Br, I)}{XCH{=}CHX} \longrightarrow \underset{(X = Cl, Br)}{CHX_2CHX_2}$$

Compounds derived directly from benzene:

Synthesis of mono-substituted benzenes

$$C_6H_6 \xrightarrow[FeX_3]{X_2} C_6H_5{-}X \quad (X = Cl, Br, I)$$

$$C_6H_6 \xrightarrow{H_2SO_4} C_6H_5{-}SO_3H$$

$$C_6H_6 \xrightarrow[H_2SO_4]{HNO_3} C_6H_5{-}NO_2$$

$$C_6H_6 \xrightarrow[AlCl_3]{RX} C_6H_5{-}R \quad (R = CH_3, C_2H_5, C_3H_7)$$

---

* The conversion of an alkene to an epoxide can be effected by molecular oxygen in the presence of a suitable catalyst, by the action of hydrogen peroxide, or by the action of organic peracids such as $C_6H_5CO_3H$ (obtained by treating $C_6H_5CO_2H$ with H_2O_2).

For the purpose of the synthesis schemes to be discussed, Table 18.6 gives the list of the names of starting materials available from our modest chemical supply house.

Table 18.6 STARTING MATERIALS AVAILABLE FROM METHANE, ETHANE, PROPANE, ETHENE, ETHYNE, PROPENE, AND BENZENE

Saturated hydrocarbons: methane, ethane, and propane
Unsaturated hydrocarbons: ethene, propene, and ethyne
Aromatic hydrocarbons: benzene
Alkyl halides: RX where R = CH_3-, CH_3CH_2-, $CH_3CH_2CH_2-$, and $(CH_3)_2CH-$, and X = Cl, Br, I
Alcohols: ethanol, 2-propanol
Aldehydes: acetaldehyde
Epoxides: oxirane and 2-methyloxirane
Polyhalides: 1,2-dihaloethane, 1,2-dihalopropane, 1,2-dihaloethene, 1,1,2,2-tetrahaloethane
Substituted benzenes: halobenzenes, nitrobenzene, toluene, ethylbenzene, propylbenzene, isopropylbenzene, benzenesulfonic acid (for summary of reactions, see pp. 214, 222)

18.4b. ELABORATION OF CARBON FRAMEWORKS. Although syntheses do not always follow the same general pattern, the usual sequence of events involves the elaboration of the carbon framework of the molecule as the first phase, followed by appropriate functional group alteration. Of prime importance for constructing carbon frameworks, therefore, are reactions in which new carbon–carbon bonds are formed. Table 18.7 lists the methods discussed in the previous chapters for accomplishing this. Later chapters present several additional procedures, but for the moment we shall limit ourselves to those which should already be familiar.

18.4c. MANIPULATION OF FUNCTIONAL GROUPS. A variety of methods for effecting interconversions between various types of functional groups have been discussed in the previous chapters, and many more are introduced in succeeding chapters. In Table 18.8, these methods are classified according to the type of reaction that is involved and include (a) substitution processes, (b) addition processes, and (c) elimination processes. The left-hand portion of the table indicates the type of bonds that are involved. The middle column shows the specific types of reactions in generalized fashion, and the right-hand portion of the table indicates the types of products resulting from the functional group interchange. Thus, Tables 18.7 and 18.8 present capsule summaries of most of the reactions appearing in Chapters 8–17, and we urge the reader to review them carefully as a preparation for their use in the synthesis problems at the end of this chapter.

18.4d. EXAMPLES OF SYNTHESIS PROBLEMS. For the first example of a synthesis problem, let us consider the way in which 2-butanone might be constructed from compounds in the list of available materials (Table 18.6). With the exception of benzene, none of the materials in this list has more than three carbons, so it is obvious that some method of carbon–carbon bond formation will be necessary for the construction of the appropriate four-carbon framework of

Table 18.7 METHODS FOR CREATING NEW CARBON–CARBON BONDS

Starting Material	Reaction	Product
Alkyl halides	$RX + CN^- \longrightarrow RCN$	Nitriles
	$RX \xrightarrow{Mg} RMgX \xrightarrow{CO_2} RCO_2MgX \xrightarrow{H_3O^+} RCO_2H$	Carboxylic acids
Aromatic halides	$ArX \xrightarrow{Mg} ArMgX \xrightarrow{CO_2} ArCO_2MgX \xrightarrow{H_3O^+} ArCO_2H$	Carboxylic acids
Aldehydes and ketones	$RCHO + R'(\text{or Ar})MgX \longrightarrow R(CHOH)R'(Ar)$	Secondary alcohols
	$RR'C{=}O + R''(\text{or Ar})MgX \longrightarrow R'{-}C(R)(R''(Ar)){-}OH$	Tertiary alcohols
	$RR'C{=}O + RC{\equiv}CNa \longrightarrow RR'C(OH)(C{\equiv}CR)$	Secondary (R′=H) and tertiary alcohols
Carboxylic acid derivatives	$RC(=O)Y + R'(\text{or Ar})MgX \longrightarrow R{-}C(R'(Ar))_2{-}OH$	Tertiary alcohols (two R′ groups from R′MgX)
Aromatic compounds	$ArH + RX \xrightarrow{AlCl_3} Ar{-}R$	Alkylbenzenes
	$ArH + RCOCl \xrightarrow{AlCl_3} ArCOR$	Acyl (or aroyl) benzenes
	$ArN_2^+ + CN^- \xrightarrow{CuCN} ArCN$	Aryl nitriles
Alkenes	$>C{=}C< + CH_2N_2 \xrightarrow[h\nu \text{ or } \Delta]{} -C(CH_2)C-$ (cyclopropane ring)	Cyclopropanes
Ethylene oxide	$CH_2{-}CH_2$ (O bridge) $+ RMgX \longrightarrow \xrightarrow{H_3O^+} RCH_2CH_2OH$	Primary alcohols

2-butanone. Now, the pathway to the compound that is to be synthesized is rarely immediately apparent. Almost inevitably, it is necessary to "dissect" the final product in such a fashion that the approach to its synthesis becomes revealed. The first step in devising a synthesis, then, is to decide what the *last* reaction in the sequence might be; *e.g.*, in this case, we must consider from what compound 2-butanone can be directly obtained. On the basis of the information presented so far, the most logical answer is that 2-butanone can be obtained by oxidizing

Table 18.8 METHODS FOR THE INTERCHANGE OF FUNCTIONAL GROUPS

SUBSTITUTION PROCESSES:	PRODUCTS
(1) $-\overset{\vert}{\underset{\vert}{C}}-H \longrightarrow -\overset{\vert}{\underset{\vert}{C}}-X$	
$RH + X_2 \longrightarrow RX + HX$ (X = Cl, Br)	Alkyl halides
$>C=\overset{\vert}{C}-\overset{\vert}{\underset{\vert}{C}}-H$ + N-bromosuccinimide (NBr) $\xrightarrow[\text{initiator}]{\text{Radical}}$ $>C=\overset{\vert}{C}-\overset{\vert}{\underset{\vert}{C}}-Br$ + succinimide (NH)	Allyl halides
$C_6H_5-H + E^+ \longrightarrow C_6H_5-E + H^+$ (E = halogen, NO_2, SO_3H)	Substituted aromatics
(2) $-\overset{\vert}{\underset{\vert}{C}}-\text{Halogen} \longrightarrow -\overset{\vert}{\underset{\vert}{C}}-Y$	
$RX + R'O^- \longrightarrow R-O-R'$ (R′ = H, alkyl, aryl)	Alcohols, ethers
$RX + R'S^- \longrightarrow R-S-R'$ (R′ = H, alkyl, aryl)	Thiols, sulfides
$RX + R'_2NH \longrightarrow R-NR'_2$ (R′ = H, alkyl, aryl)	Amines
(3) $-\overset{\vert}{\underset{\vert}{C}}-OH \longrightarrow -\overset{\vert}{\underset{\vert}{C}}-X$	
$ROH + HX \longrightarrow RX + H_2O$	Alkyl halides
(4) $-C(=O)X \longrightarrow -C(=O)Y$ (X = halogen)	
Hydrolysis:	
$RCOCl + H_2O \longrightarrow RCO_2H + HCl$	Carboxylic acids
$(RCO)_2O + H_2O \longrightarrow RCO_2H$	Carboxylic acids
$RCO_2R' + H_2O \longrightarrow RCO_2H + R'OH$	Carboxylic acids
$RCONR'_2 + H_2O \longrightarrow RCO_2H + R'_2NH$	Carboxylic acids
Alcoholysis:	
$RCOCl + R'OH \longrightarrow RCO_2R' + HCl$	Esters
$(RCO)_2O + R'OH \longrightarrow RCO_2R' + RCO_2H$	Esters
Aminolysis:	
$RCOCl + R'_2NH \longrightarrow RCONR'_2$	Amides
$(RCO)_2O + R'_2NH \longrightarrow RCONR'_2 + RCO_2H$	Amides
$RCO_2R + R'_2NH \longrightarrow RCONR'_2 + ROH$	Amides

Table 18.8—cont.

Hydrogenolysis:

Reaction	Product
$RCOCl + [H] \longrightarrow RCHO + HCl$	Aldehydes

(5) $-C\equiv N \longrightarrow -CO_2H$

Reaction	Product
$RCN + H_2O \longrightarrow RCO_2H$ $ArCN + H_2O \longrightarrow ArCO_2H$	Carboxylic acids

ADDITION PROCESSES:

(1) Addition to $\rangle C=C \langle$

Reaction	Product
$\rangle C=C \langle + H_2 \longrightarrow H-\overset{\vert}{\underset{\vert}{C}}-\overset{\vert}{\underset{\vert}{C}}-H$	Alkanes
$\rangle C=C \langle + HX \longrightarrow H-\overset{\vert}{\underset{\vert}{C}}-\overset{\vert}{\underset{\vert}{C}}-X$ (X = halogen)	Alkyl halides
$\rangle C=C \langle + H_2O \longrightarrow H-\overset{\vert}{\underset{\vert}{C}}-\overset{\vert}{\underset{\vert}{C}}-OH$	Alcohols
$\rangle C=C \langle + X_2 \longrightarrow X-\overset{\vert}{\underset{\vert}{C}}-\overset{\vert}{\underset{\vert}{C}}-X$ (X = Cl, Br)	1,2-Dihalo compounds
$\rangle C=C \langle + [O] \longrightarrow -\overset{\vert}{C}\underset{O}{\diagdown\diagup}\overset{\vert}{C}-$	Oxiranes

(2) Addition to $-C\equiv C-$

Reaction	Product
$-C\equiv C- + H_2 \longrightarrow H-\overset{\vert}{C}=\overset{\vert}{C}-H \longrightarrow H_2\overset{\vert}{C}-\overset{\vert}{C}H_2$	Alkenes, alkanes
$-C\equiv C- + HX \longrightarrow H-\overset{\vert}{C}=\overset{\vert}{C}-X$ (X = halogen)	Vinyl halides
$-C\equiv C- + H_2O \longrightarrow H-\overset{\vert}{\underset{H}{C}}-\underset{\vert}{C}=O$	Aldehydes, ketones
$-C\equiv C- + X_2 \longrightarrow X-\overset{\vert}{C}=\overset{\vert}{C}-X \longrightarrow X_2\overset{\vert}{C}-\overset{\vert}{C}X_2$	Di- and tetrahalides

(3) Addition to $\rangle C=O$

Reaction	Product
$RCHO + LiAlH_4 \longrightarrow RCH_2OH$	Primary alcohols
$R_2C=O + LiAlH_4 \longrightarrow R_2CHOH$	Secondary alcohols
$(R)(R')C=O + HCN \longrightarrow (R)(R')C(OH)(CN)$	Cyanohydrins

(4) Addition to $>C{=}\overset{|}{C}{-}\overset{|}{C}{=}O$

$$>C{=}\overset{|}{C}{-}\overset{|}{C}{=}O \;+\; NuH \longrightarrow -\underset{Nu}{\overset{|}{C}}-\underset{H}{\overset{|}{C}}-\overset{|}{C}{=}O$$ β-Substituted aldehydes, ketones, and esters

Nu = CN^-, R_2N^- (from R_2NH), RO^- (from ROH), X^- (from XH)

(5) Addition by "bond insertion"

$$RX + Mg \longrightarrow RMgX$$
$$ArX + Mg \longrightarrow ArMgX$$ Grignard reagents

ELIMINATION PROCESSES:

(1) $-\underset{H}{\overset{|}{C}}-\underset{Y}{\overset{|}{C}}- \longrightarrow >C{=}C< \;+\; HY$

$$R_2\underset{H}{C}-\underset{X}{CR_2} \xrightarrow{\text{Base}} R_2C{=}CR_2 + HX \quad (X = \text{halogen})$$ Alkenes

$$R_2\underset{H}{C}-C(=O)Cl \xrightarrow{\text{Base}} R_2C{=}C{=}O + HCl$$ Ketenes

$$R_2\underset{H}{C}-\underset{OH}{CR_2} \xrightarrow{\text{Acid}} R_2C{=}CR_2 + H_2O$$ Alkenes

(2) ArR + $KMnO_4$ (or other strong oxidizing agents) ⟶ $ArCO_2H$ — Aromatic acids

(3) $RCH_2OH + [O] \longrightarrow RCHO \longrightarrow RCO_2H$ — Aldehydes, acids

(4) $R_2CHOH + [O] \longrightarrow R_2C{=}O$ — Ketones

(5) Bond cleavage processes

$$>C{=}C< + O_3 \longrightarrow >C{=}O + O{=}C<$$ Aldehydes, ketones

2-butanol. So, a last step in the sequence could be

$$CH_3CH_2CHOHCH_3 + [O] \longrightarrow CH_3CH_2COCH_3$$

The next problem (*i.e.*, the next step in the dissection) is how to synthesize 2-butanol. Reviewing the methods presently available to us, we find that they include (a) treatment of RX with OH^-, (b) addition of H_2O to alkenes, (c) reduction of ketones and aldehydes, and (d) the addition of Grignard reagents to aldehydes, ketones, and carboxylic acid derivatives. Of these four methods, only one, *viz.*, the Grignard reaction, involves the formation of carbon–carbon bonds. Since carbon–carbon bond formation is necessary at some point in the sequence, the possibility of simultaneously introducing the OH group and joining the carbon–carbon bond is an attractive one. Two different combinations of carbonyl compound and Grignard reagent lead to 2-butanol, *viz.*,

$$\text{(a)}\quad CH_3CH_2C(=O)H + CH_3MgX \longrightarrow \xrightarrow{H_3O^+} CH_3CH_2CH(OH)CH_3 \xrightarrow{[O]} CH_3CH_2C(=O)CH_3$$

$$\text{(b)}\quad CH_3CH_2MgX + HC(=O)CH_3 \longrightarrow \xrightarrow{H_3O^+} CH_3CH_2CH(OH)CH_3$$

Either of these reactions is satisfactory and will proceed in good yield. In the context of the chemicals listed in Table 18.6, however, the second alternative is preferred; acetaldehyde is included in the list, and ethylmagnesium bromide is readily prepared from ethyl bromide (which is on the list). Thus, by working backward from the final product, we have arrived at a sequence of reactions that involves the conversion of ethyl bromide to the corresponding Grignard reagent, the interaction of the Grignard reagent with acetaldehyde to form 2-butanol, and the oxidation of 2-butanol to 2-butanone, the desired product.

The synthesis of 2-butanone is a simple example, but it illustrates the basic principle of unraveling a synthesis problem by working backward from the final product to the available starting materials. To provide further emphasis and illustration of this hindsight approach, let us explore the possible synthetic approaches to 2-phenyl-2-butene, the second compound discussed in the previous section on structure elucidation. As before, we must first ask what the last step in the reaction sequence might be; *i.e.*, how can alkenes be prepared? Surveying Tables 18.7 and 18.8, we see that alkenes can be made by partial reduction of alkynes, dehydrohalogenation of alkyl halides, and dehydration of alcohols. The first method must be excluded, however, for it can lead only to alkenes that carry one (or more) hydrogen on *each* of the carbons of the double bond (*i.e.*, $RC{\equiv}CR + H_2 \rightarrow RCH{=}CHR$). The second method can also be excluded, although less categorically so, because alkyl halides are generally made by the addition of HX to alkenes (but, the alkene is the desired product in this case) or by the reaction of HX with alcohols (which, themselves, can be dehydrated to yield alkenes). As a working hypothesis, then, let us choose the dehydration of an alcohol as the best means for introducing the double bond in the present instance. Again, we are faced with the problem of synthesizing an alcohol, and again the possibility of constructing the carbon framework concurrently with the generation of the hydroxyl function should be considered. One of the appropriate combinations of reagents should be immediately apparent in the light of the synthesis of 2-butanone previously discussed—*viz.*, the reaction between

2-butanone and phenylmagnesium bromide:

$$\underset{}{CH_3CH_2\overset{\overset{\displaystyle O}{\|}}{C}CH_3} \; + \; C_6H_5MgBr \longrightarrow \xrightarrow{H_3O^+} CH_3CH_2\underset{\underset{\displaystyle C_6H_5}{|}}{\overset{\overset{\displaystyle OH}{|}}{C}}{-}CH_3 \xrightarrow{(-H_2O)} CH_3CH{=}\underset{\underset{\displaystyle C_6H_5}{|}}{C}{-}CH_3$$

Mixture of (*Z*) and (*E*) isomers

Thus, one synthetic sequence starting with materials in Table 18.6 involves the conversion of ethyl bromide to ethylmagnesium bromide, the interaction of the Grignard reagent with acetaldehyde to yield 2-butanol, the oxidation of 2-butanol to 2-butanone, the interaction of 2-butanone with phenylmagnesium bromide (obtained from bromobenzene and magnesium) to yield 2-phenyl-2-butanol, and dehydration of this alcohol to 2-phenyl-2-butene.

If 2-butanone is on hand, the procedure just described would certainly be the one to choose. If 2-butanone is not available, however, other approaches should also be considered. In addition to the combination of 2-butanone as the carbonyl component and phenylmagnesium bromide as the Grignard reagent, two other combinations lead to the same tertiary alcohol:

$$\textbf{(a)}\; CH_3CH_2MgBr \; + \; O{=}C\begin{matrix} \diagup CH_3 \\ \diagdown C_6H_5 \end{matrix} \longrightarrow \xrightarrow{H_3O^+} CH_3CH_2\underset{\underset{\displaystyle C_6H_5}{|}}{\overset{\overset{\displaystyle OH}{|}}{C}}{-}CH_3$$

$$\textbf{(b)}\; CH_3CH_2C\begin{matrix} \mathbin{/\!\!/} O \\ \diagdown C_6H_5 \end{matrix} \; + \; CH_3MgI \longrightarrow \xrightarrow{H_3O^+} CH_3CH_2\underset{\underset{\displaystyle C_6H_5}{|}}{\overset{\overset{\displaystyle OH}{|}}{C}}{-}CH_3$$

The acetophenone required for alternative (a) can be prepared in several ways, including a Friedel–Crafts acylation involving benzene and acetyl chloride. The acetyl chloride can be obtained by any of the following methods:

$$CH_3Br \longrightarrow CH_3MgBr \xrightarrow{CO_2} CH_3CO_2MgBr \xrightarrow{H_3O^+} CH_3CO_2H \xrightarrow{SOCl_2} CH_3COCl$$

$$CH_3Br \; + \; KCN \longrightarrow CH_3CN \xrightarrow{H_2O} CH_3CO_2H \xrightarrow{SOCl_2} CH_3COCl$$

$$CH_3CHO \; + \; [O] \longrightarrow CH_3CO_2H \xrightarrow{SOCl_2} CH_3COCl$$

Of the three methods, the oxidation of acetaldehyde (an available starting material) involves the fewest steps. A second synthetic sequence, therefore, involves the oxidation of acetaldehyde to acetic acid, the conversion of acetic acid to acetyl chloride, the acylation of benzene with acetyl chloride and aluminum chloride to yield acetophenone, reaction of acetophenone with ethylmagnesium bromide to yield 2-phenyl-2-butanol, and dehydration to 2-phenyl-2-butene. The reader is urged to outline the steps for process (b) and to also consider still other synthesis schemes for 2-phenyl-2-butene.

Finally, let us turn our synthetic expertise to the more difficult problem of putting together the third compound that was discussed in the previous section on structure elucidation, *viz*., 2-(4-dimethylaminophenyl)-2,3-butanediol. Taking cognizance of the reactions employed in the two examples just discussed and working backward, one step at a time, from the final product to available starting materials, we can devise the sequence outlined as follows:

HNO_3 Fe/HCl CH_3Br

NO_2 NH_2

O CH_3
C

CH_3CH_2MgBr CH_3COCl
$AlCl_3$

$N(CH_3)_2$ $N(CH_3)_2$

OH
$CH_3CH_2CCH_3$ $CH_3CH{=}CCH_3$

H_3O^+ $C_6H_5CO_3H$
$(-H_2O)$

$N(CH_3)_2$ $N(CH_3)_2$

OH OH O
$CH_3CH{-}CCH_3$ $CH_3CH{-}CCH_3$

H_2O
NaOH

$N(CH_3)_2$ $N(CH_3)_2$

The basic tools of synthesis presented in Part I are sufficient for most simple compounds and sometimes even for more complex compounds. In many instances, however, methods other than those that have been discussed are necessary to achieve the optimum yield of product. For example, in the synthesis just outlined, the conversion of the alkene to the epoxide with perbenzoic acid might be complicated by simultaneous oxidation at the nitrogen atom; other methods of 1,2-hydroxylation might be preferable. Details such as this are vital to the success of a multistep synthesis where poor yields at any of the steps along the way are devastating with respect to the overall yield of product. Chapters 20–31 provide more of these details necessary in the design of a good synthesis.

NOMENCLATURE PROBLEMS

1. Name each of the following structures according to the IUPAC rules:

(a) 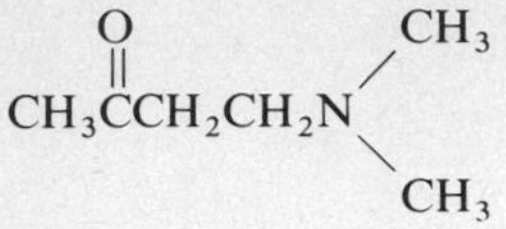 $CH_3\overset{O}{\overset{\|}{C}}CH_2CH_2N(CH_3)_2$

(b) $HO_2C(H)C{=}C(H)C{\equiv}C\overset{O}{\overset{\|}{C}}CH_2CH_2CH(CH_3)_2$

(c) CH_3-N (five-membered ring with S)

(d) pyrazine ring with two CH_3 groups

(e) benzene ring with $(CH_3)_3C$, NO_2, OH, CH_3

(f) bicyclic structure with CH_2OH, CH_3, H, H

(g) dioxolane ring with CH_3, H, H; $C{=}C$ with H, H, CH_3

(h) $C_6H_5CH_2-\overset{+}{N}(CH_3)(CH_3)-CH(CH_3)CH_2CH_3 \quad I^-$

(i) benzene ring with $N(CH_3)_2$, CH_3, CH_2CH_3, $CH_2CH(CH_3)_2$

(j) $(CH_3)_2CHCH(SH)\overset{O}{\overset{\|}{C}}OC_2H_5$

(k) $CH_3C(OH)(CH_3)-C{\equiv}C\overset{O}{\overset{\|}{C}}CH_3$

(l) $CH_3CH_2CH_2CH(CH_3)-\overset{+}{S}(CH_3)-CH_2C_6H_5 \quad I^-$

(m) tetrahydrofuran ring with $-C(=O)H$

(n) cyclohexane ring with $-\overset{O}{\overset{\|}{C}}CH_2SO_2C_2H_5$

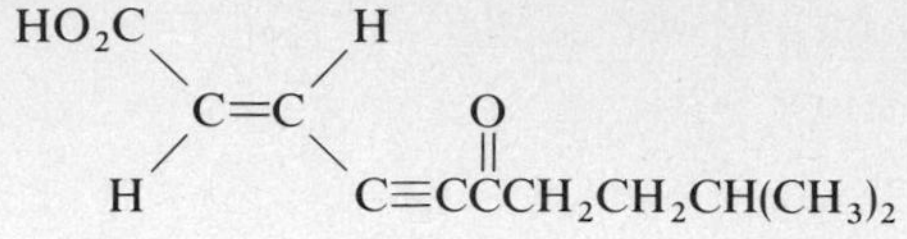

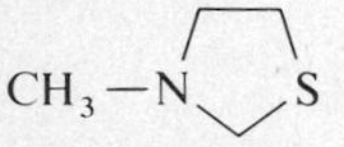

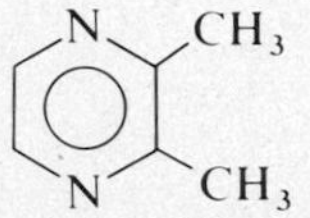

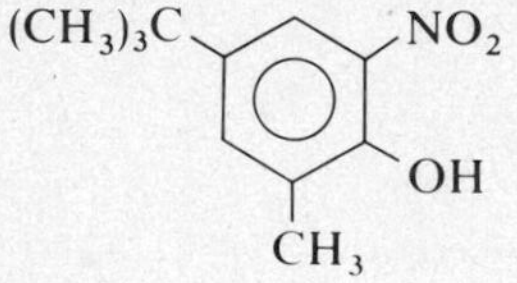

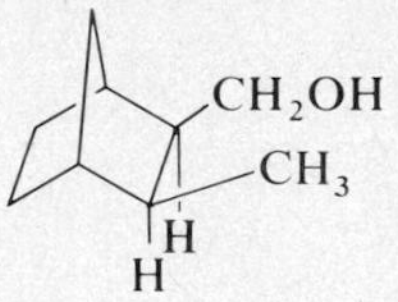

(o) $CH_3CH(I)CH(OH)CH_2NH_2$

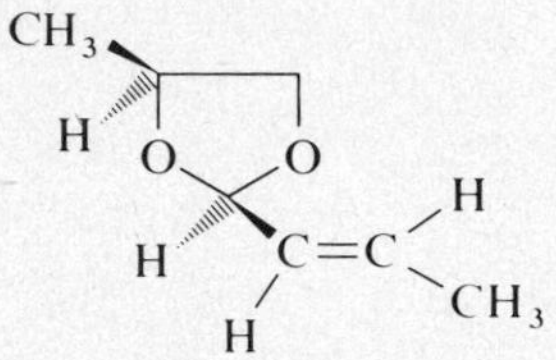

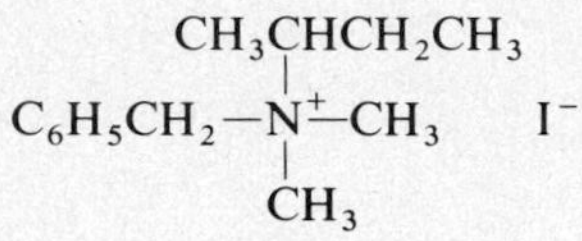

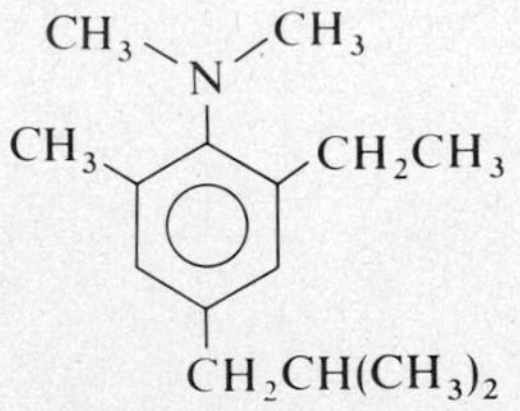

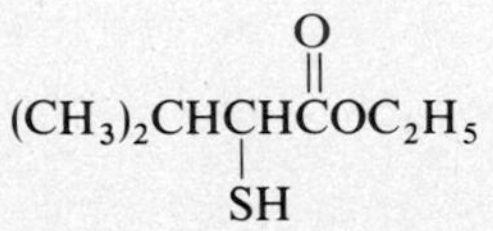

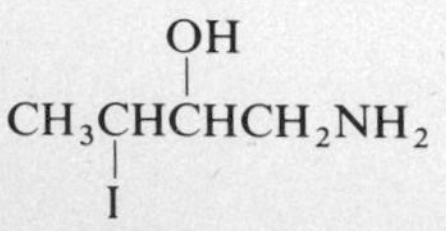

(p) 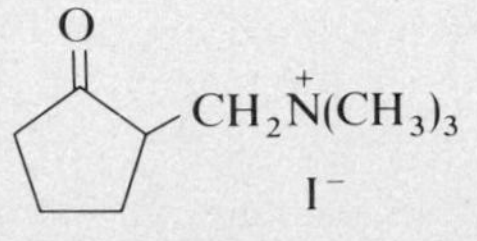 cyclopentanone ring with $CH_2\overset{+}{N}(CH_3)_3 \quad I^-$

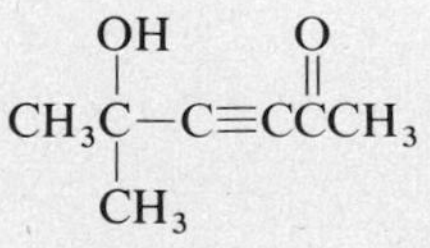

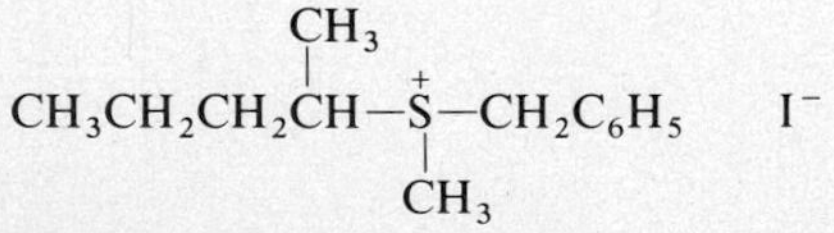

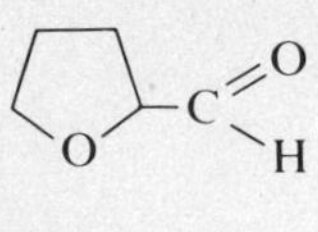

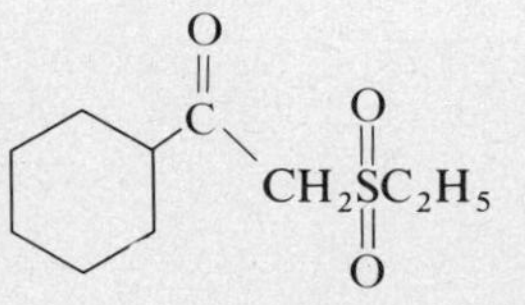

(q) CO_2H N H

(r) $C_6H_5CH_2SCH_2CH_2\underset{\displaystyle N(CH_3)_2}{CH}\overset{\displaystyle O}{\overset{\|}{C}}CH(CH_3)_2$

(s) N Cl H

(t) $(CH_3)_2CH$ C=C H H $\overset{O}{\overset{\|}{C}}CH_2C$ O H

2. Draw structures corresponding to each of the following names:
   (a) 4-(*p*-Isopentylphenyl)-4-oxobutyric acid
   (b) 1,4-Dimethyl-8-nitronaphthalene
   (c) Methyl 5-chloromethyl-2-furoate
   (d) 3-Cyanocyclopentanone
   (e) *exo*-6-Hydroxy-8-methyl-8-azabicyclo[3.2.1]-octan-3-one
   (f) *cis*-3,4-Dimethyl-1,2-dithietane
   (g) 3,4-Diacetoxybenzoic acid
   (h) 4,4-Dimethylthiocyclohexanone
   (i) 1-*exo*-2,4-Tri-*t*-butylbicyclo[2.2.0]hexane
   (j) Triethyloxonium fluoborate
   (k) 4-(N,N-Dimethylcarboxamido)-pentanoic acid
   (l) 4-*t*-Butylmethylenecyclohexane
   (m) 2,5-Divinylfuran
   (n) 2-Ethyl-1,3-oxazole
   (o) 1-Iodo-2-methyl-2-nitropropane
   (p) 2,4-Dibromo-1-methoxy-6-nitronaphthalene
   (q) 2,3-Dichlorobutanoyl azide
   (r) N-(*p*-Methoxyphenyl)trifluoromethanesulfonamide
   (s) N-(2-Hydroxyethyl)-aziridine
   (t) 3,5-Diphenyl-1,2,4-thiadiazole

STRUCTURE PROOF PROBLEMS

SET I

1. A compound with the molecular formula $C_9H_7ClO_2$ (compound A) is insoluble in water but reacts with water to give a compound with the molecular formula $C_9H_8O_3$ (compound B). Compound B is stable, is soluble in aqueous sodium bicarbonate solution, reacts with phenylhydrazine to form a solid derivative, but fails to react with Fehling's solution. Upon strenuous oxidation, compound B yields compound C with molecular formula $C_8H_6O_4$. Compound C loses water upon heating and yields an acid anhydride with the molecular formula $C_8H_4O_3$. On the basis of these data, suggest structures for compounds A, B, and C.

2. Compound A possesses the molecular formula $C_7H_{16}O$, is slightly water soluble, and undergoes smooth dehydration to compound B, which has the molecular formula C_7H_{14}. Ozonolysis of compound C yields two materials, one a compound with molecular formula C_3H_6O, which is stable to further oxidation, and the other a compound of molecular formula C_4H_8O, which is easily oxidized to $C_4H_8O_2$. Suggest suitable structures for compound A and its various degradation products.

3. (a) A pair of isomeric hydrocarbons (compounds A and B) have the molecular formula C_6H_6. When compound A is treated with ozone followed by Tollens' or Fehling's reagent, a tetracarboxylic acid is produced. When this acid is heated, it loses 2 moles of carbon dioxide and yields a cyclic anhydride. When compound B is subjected to similar treatment, a dicarboxylic acid is formed, which upon heating forms a cyclic anhydride without loss of carbon dioxide. Propose reasonable structures for compounds A and B, and outline the sequence of degradation reactions.

(b) A pair of isomeric hydrocarbons (compounds C and D) have the molecular formula C_6H_8. When compound C is treated with ozone followed by Tollens' or Fehling's reagent, a dicarboxylic acid is formed. When this acid is heated it loses carbon dioxide and gives acetic acid. When compound D is subjected to similar treatment, a pair of dicarboxylic acids is formed. One of these decomposes to carbon monoxide, carbon dioxide, and water upon heating, whereas the other yields a cyclic anhydride. Propose reasonable structures for compounds C and D, and outline the sequence of degradation reactions.

4. A deuterium-containing compound was subjected to carbon, hydrogen, and deuterium analysis and found to have the molecular formula $C_{20}H_{13}D_3$. When this material was treated with ozone, two new compounds were formed. One of these had the molecular formula C_7H_5DO and reacted with ammoniacal silver hydroxide to yield a carboxylic acid with the molecular formula $C_7H_6O_2$. [*Note*: Whereas the C—H bond does not undergo exchange of H for D in aqueous solution, the O—H bond undergoes rapid exchange of H for D (or vice versa) in water; consequently, any O—D bonds formed in this reaction would be changed to O—H bonds in the presence of water.] The other compound had the molecular formula $C_{13}H_8D_2O$ and reacted with the Grignard reagent from 4-deuteriobromobenzene to yield a tertiary alcohol in which all the R groups are identical.

5. A compound A of unknown structure has the molecular formula C_8H_{12}. When it was subjected to complete hydrogenation, 2 moles of H_2 were absorbed. The resulting product B, of molecular formula C_8H_{16}, and the original compound A, when heated with palladium, gave *o*-xylene (*i.e.*, 1,2-dimethylbenzene) as the product. Reaction of compound A with ozone, followed by decomposition of the resulting ozonide by one of the standard procedures gave 2 moles of a *single* substance, C, per mole of A. Write structural formulas for A, B, and C.

6. Compound A, of molecular formula $C_6H_{13}Br$, when refluxed with alcoholic KOH gives compound B (molecular formula C_6H_{12}). On oxidation with alkaline $KMnO_4$, compound B is converted to a carboxylic acid (C) and a ketone (D). Compound B cannot exist as *cis, trans* isomers. Addition of HBr to compound B in the presence of peroxide gives compound A once again. Write structural formulas for A, B, C, and D.

7. Compound A has the molecular formula $C_{10}H_{16}N_2O_2$ and does not dissolve in either dilute aqueous base or dilute aqueous acid. Upon hydrolysis, however, it is converted to a $C_{10}H_{16}O_4$ compound (compound B), which readily dissolves in dilute aqueous base. Compound B undergoes ozonolysis to give only one product (compound C), a material with the molecular formula $C_5H_8O_3$, which very readily loses carbon dioxide to yield compound D. Compound D possesses the molecular formula C_4H_8O and is stable to oxidation with Fehling's solution. Suggest suitable structures for compounds A, B, C, and D.

SET II

1. A compound is known to have the molecular formula $C_6H_{10}O_2$ and to possess one of the following structures. Indicate what chemical tests and reactions could be used to decide which of these structures properly represents the compound.

(a) HO–cyclohexanone; (b) HO–cyclopentane–CHO; (c) cyclobutane–CH_2CO_2H

(d) cyclic lactone (ring O, C=O); (e) cyclic ether ketone (ring O, C=O); (f) cyclobutane–$\overset{O}{\overset{\|}{C}}OCH_3$

2. A compound possessing the molecular formula $C_{11}H_9ClO_3$ was shown to possess an aldehyde group, a carboxylic acid group, and a C=C group. Treatment of the compound with an alcoholic solution of silver nitrate resulted in the rapid formation of a precipitate of silver chloride, indicative of the presence of a displaceable chlorine atom. Treatment of the compound with Tollens' reagent resulted in the formation of a dibasic acid, which readily formed a cyclic anhydride upon heating. Treatment of the compound with ozone resulted in the formation of a dialdehyde, which yielded a C_9 tribasic acid upon oxidation with Tollens' reagent. Of the compounds drawn, indicate the one that you think best represents these data. For each of the other compounds, state, as briefly as possible, the reason you think it should be excluded as a suitable structure.

(a) CHO; HO_2C; $CH{=}CHCH_2Cl$

(b) CHO; HO_2C; $CH{=}CHCH_2Cl$

(c) CHO; HO_2C; $C{=}CHCl$ with CH_3

(d) $\overset{O}{\overset{\|}{C}}$–H; CO_2H; $CHCH{=}CH_2$ with Cl

(e) HO_2C; Cl; $\overset{O}{\overset{\|}{C}}$–H; $CH{=}CHCH_3$

3. Through an unfortunate accident, the labels have been removed from the bottles in your medicine chest. Not wanting to replenish your stock with fresh samples from the corner druggist, you instead resort to organic analysis. Within minutes you have succeeded in correctly identifying the contents in each of the five bottles, having remembered that medications on hand included aspirin for headaches, phenobarbital for going to sleep, amphetamine for staying awake, hexylresorcinol for preventing infection from surface wounds, and sodium sulfanilate as a remedy for acute upper respiratory infections.

Without the use of a balance, a thermometer, a pH meter, or a spectrophotometer, indicate what simple chemical test(s) and observations(s) you would use to identify each of these materials correctly.

Aspirin	Phenobarbital	Amphetamine	Hexylresorcinol	Sodium sulfanilate
$OCCH_3$ (C=O), CO_2H	H, O, N, O, C_2H_5, NH, O	CH_2CHCH_3, NH_2	OH, OH, C_6H_{13}	NH_2, SO_3Na

SYNTHESIS PROBLEMS

1. Starting with compounds included in Table 18.6, synthesize the following compounds in as short and efficient a manner as possible. None of these syntheses requires more than two steps if compounds prepared in previous questions are used [*e.g.*, using the product in (e) as starting material for (h)].

(a) $CH_3CH_2CH_2CH_3$

(b) CH_3–C_6H_4–CH_3 (para)

(c) C_6H_{11}–$CH(CH_3)_2$

(d) $CH_3CH{=}CHCH_3$

(e) $C_6H_5CH{=}CH_2$

(f) $CH_3C(OH)CH(CH_3)_2$ (CH bearing OH)

(g) $CH_3CH{=}CHCH_2Br$

(h) $C_6H_5CH_2CH_2Br$

(i) $CH_3CHClCH(CH_3)_2$

(j) $CH_3OCH_2CH_2OH$

(k) $(CH_3)_2C{=}O$

(l) $(CH_3)_2C(OH)$–C_6H_4–CH_3

(m) $CH_3C(O)CH(CH_3)_2$

(n) $C_6H_5CO_2H$

(o) CH_3CO_2H

(p) $C_6H_5C(O)CH_3$

(q) $C_6H_5C(OH)(CH_3)CH_2CH_3$

(r) $(CH_3)_2C(OH)CO_2H$

(s) $CH_3C(O)OCH_2CH_2OCH_3$

(t) C_6H_5COCl

(u) $(CH_3)_2NH$

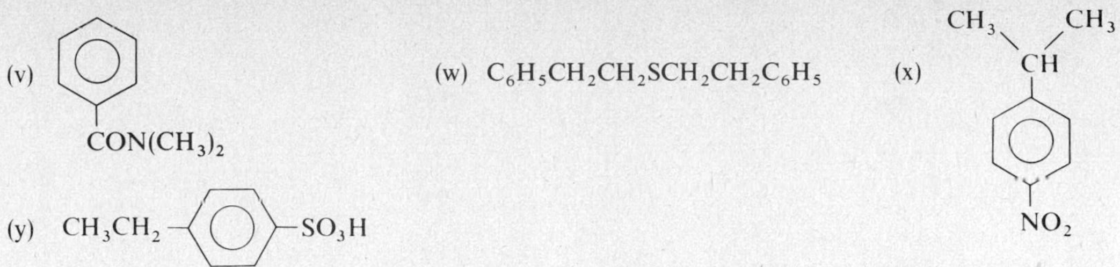

2. The following syntheses are more difficult than those in Problem 1 and require several steps in most instances. The starting materials may include compounds listed in Table 18.6 and compounds prepared in Problem 1.

(a) $(CH_3)_2CBr{-}C(CH_3)_2Br$

(b) $(C_6H_5)_2C(OH){-}C_6H_4{-}CH_3$ (p-tolyl)

(c) $(CH_3)_2CH{-}C_6H_4{-}NHCH_2CH_2OH$ (para)

(d) $CH_3CH_2\overset{O}{\overset{\|}{C}}{-}C_6H_4{-}CH_2CH_2CH_3$ (para)

(e) $CH_3CH_2C(OH)(CH_3)CH(CH_3)_2$

(f) $CH_3CH_2{-}C_6H_4{-}SO_2CH_2CH(CH_3)_2$ (para)

(g) $C_6H_5CH{=}C(C_6H_5){-}CH_2{-}C_6H_5$

(h) $CH_3CH_2CH_2CO_2CH_2{-}C_6H_5$

Chapter 19

Stereochemistry: the architecture of organic molecules

To hold as 'twere the mirror up to Nature.

WILLIAM SHAKESPEARE
Hamlet

With the investigations of Wöhler and Liebig early in the nineteenth century (see p. 10) came the first realization that elemental composition alone does not uniquely define a substance but that alternative arrangements of the same elements within a molecule can give substances with different properties. Out of this awareness was born the concept of isomerism, a concept that has become an integral part of the structural theory of organic compounds. Isomerism, as we have seen, can be divided into two major categories; *viz.*, structural and stereochemical. Although some of the elements of stereochemistry are discussed in Chapters 3 and 4, a more detailed treatment is essential because of the important part that stereochemical factors play in the chemical, physical, and physiological behavior of compounds—from the conformational isomerism of molecules as simple as ethane to the mechanism of the immune response in molecules as complicated as proteins.

19.1 *Double-Bond Diastereomerism*

Double-bond diastereomerism, also referred to as (*Z*),(*E*) isomerism, or, in the older terminology, as *cis,trans* or geometric isomerism, arises as a consequence of the high barrier to rotation around double bonds. Substituents attached to the carbons of a C=C bond, for example, remain essentially fixed in space under ordinary conditions, the energy required to effect rotation around a C=C bond being far greater than that available at room temperature. Alkenes that are unsymmetrically substituted on *both* carbons, therefore, can exist in two stereoisomeric forms, designated as *cis* and *trans* in simple cases or (*Z*) and (*E*) in the systematic terminology (see p. 96) (Fig. 19.1).

cis-2-Pentene
(*Z*)-2-Pentene

trans-2-Pentene
(*E*)-2-Pentene

cis-2-Bromo-2-butene
(*E*)-2-Bromo-2-butene

(1*E*,5*Z*)-Cyclodeca-1,5-diene-1,6-dicarboxylic acid

Fig. 19.1. Stereochemical designation of alkene diastereoisomers.

PROBLEM 19-1

Draw and name all the $C_4H_6Cl_2$ compounds that exist in (*Z,E*) diastereomeric forms.

Double bonds involving elements other than carbon may also allow the existence of diastereomers. For example, an unsymmetrically substituted oxime, such as 2-pentanone oxime, has two isomers called *syn* and *anti* in the older terminology or (*Z*) and (*E*) in the newer nomenclature;

(*Z*)-2-Pentanone oxime **(*E*)-2-Pentanone oxime**

The diastereomeric relationship between the (*Z*) and (*E*) isomers of alkenes, as discussed earlier (see p. 96), confers on them a difference in all their physical and chemical properties. For example, (*E*)-1,2-dichloroethene (*trans* isomer) boils at 47°, melts at −50°, and has zero dipole moment, whereas (*Z*)-1,2-dichloroethene (*cis* isomer) boils at 60°, melts at −80°, and has a dipole moment of 1.85 D. A chemical difference between (*E*) and (*Z*) isomers is illustrated by the 2-butenedioic acids: (*E*)-2-butenedioic acid (fumaric acid) remains unchanged upon heating, whereas (*Z*)-2-butenedioic acid (maleic acid) undergoes facile conversion to the cyclic anhydride (Fig. 19.2). Because of differences in physical and chemical properties, (*Z*) and (*E*) isomers can usually be separated readily by standard laboratory methods such as distillation, crystallization, and chromatography. The assignment of configuration [*i.e.*, the specification of the isomers as (*Z*) or (*E*)], however, is not always easy. In certain symmetrical compounds,

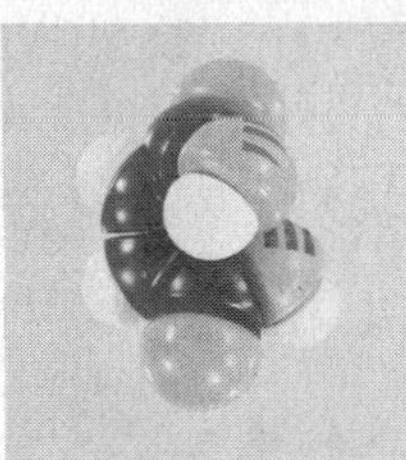

Maleic acid

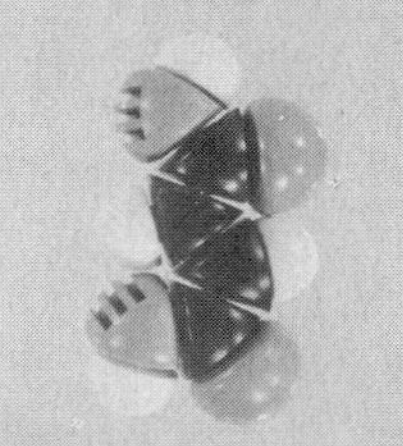

Fumaric acid

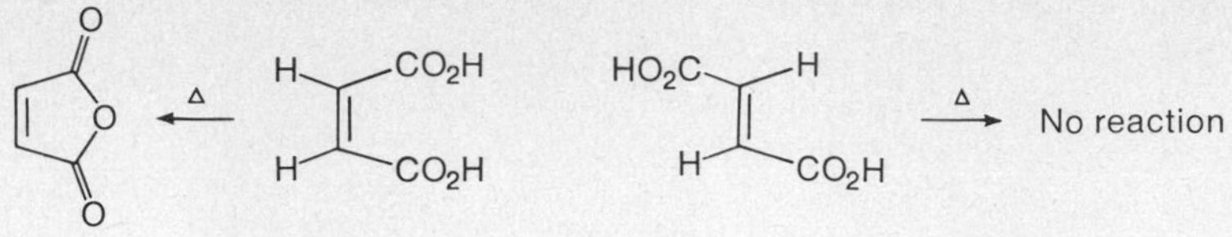

Fig. 19.2. Chemical differences of (*Z*) and (*E*) isomers.

such as the dichloroethylenes, the measurement of the dipole moment may suffice. In general, though, other procedures must be resorted to. If the compound has the structure RCH=CHR′, one of the most useful methods is the measurement of the nmr spectrum; if the geometry is (*Z*), the coupling constant between the two vinyl hydrogens is 8–14 Hz; if the geometry is (*E*), the coupling constant is 11–18 Hz. The range results from the differing effect of attached groups, electronegative groups decreasing the magnitude of the coupling constant. Assignment of configuration on the basis of the coupling constant is most safely applied when both the (*Z*) and (*E*) isomers are on hand. Certain bands in the infrared spectrum (see Table 6.2) are also useful in making (*Z*,*E*) assignments. In compounds where functional groups are present that are capable of intramolecular reaction, the occurrence or nonoccurrence of a chemical reaction can provide the basis for the (*Z*,*E*) assignment, as illustrated in Fig. 19.2.

In contrast to the carbon–carbon single bond, where the barrier to rotation is usually less than 10 kcal/mole, the barrier to rotation for the carbon–carbon double bond* is about 45 kcal/mole. The probability of traversing this barrier under ordinary conditions is almost zero, so (*Z*,*E*) isomers are very stable to interconversion. Nevertheless, it is possible in many instances to effect interconversion of (*Z*) and (*E*) isomers by using appropriate catalysts. Certain free radicals are among the reagents that can act in this capacity. A mechanism by

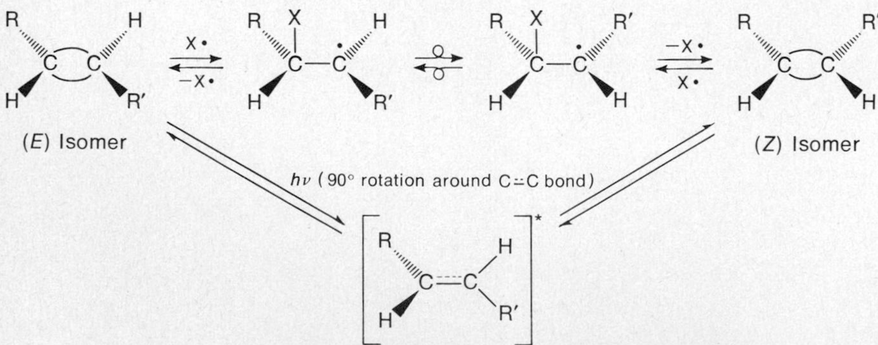

Fig. 19.3. Interconversion of (*Z*) and (*E*) isomers. [The free radical X· is, typically, an iodine atom; the superscript asterisk (*) denotes an electronically excited state.]

---

* To this point in the text, the carbon–carbon double bond has been represented as C=C. In many of the figures in this chapter, however, a "bent-bond" representation, C C or C C, will be employed as a means for indicating more explicitly the plane of the π bond.

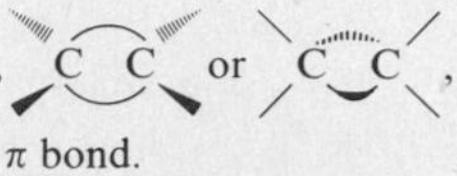

which free radicals may induce isomerization involves the addition of the radical to the double bond, thereby changing the bond to a single bond and lowering its energy barrier to rotation; rotation followed by expulsion of the free radical and reformation of the double bond produces the other isomer (Fig. 19.3). Another method for inducing (*Z,E*) isomerization makes use of the ability of alkenes to absorb light. One result of an electronic transition may be a reduction in the strength of bonding between the carbon atoms and a concomitant reduction in the barrier to rotation. Thus, when the molecule that is in the excited state falls back to the ground state, it gives either the (*E*) or the (*Z*) isomer.

Compounds containing more than one unsymmetrically substituted alkene unit have several stereoisomeric possibilities. For example, ethyl 2,4,hexadienoate possesses two double bonds, both of which are unsymmetrically substituted (Fig. 19.4). Four diastereomers, therefore, can exist, designated as (*E,E*)-, (*E,Z*)-, (*Z,E*)-, and (*Z,Z*)-. The more symmetrical compound 2,4-hexadiene ($CH_3CH{=}CHCH{=}CHCH_3$), however, has only three (*Z,E*) isomers, for the (*E,Z*) and (*Z,E*) isomers are identical.

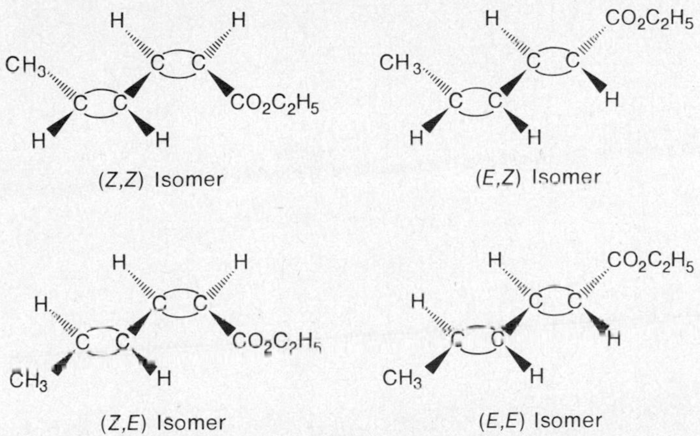

Fig. 19.4. (*Z,E*) isomers of ethyl 2,4-hexadienoate.

Cumulative systems containing an odd number of double bonds also have the potentiality for (*Z,E*) isomerism if both termini are unsymmetrically substituted. For example, 2,3,4-hexatriene and 2,3,4,5,6-octapentaene (Fig. 19.5) both have (*Z*)

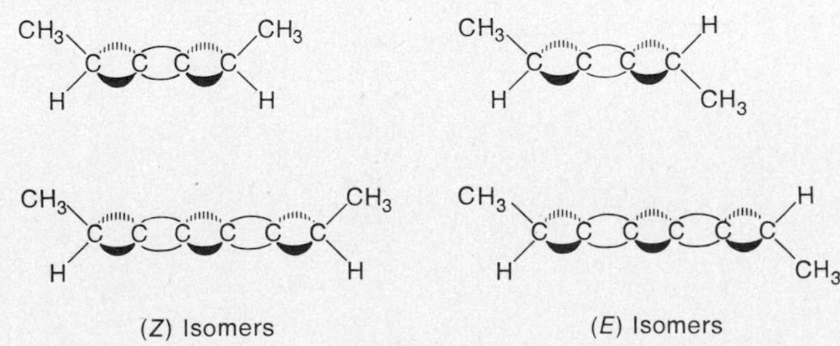

Fig. 19.5. (*Z*) and (*E*) isomers of cumulenes containing an odd number of double bonds.

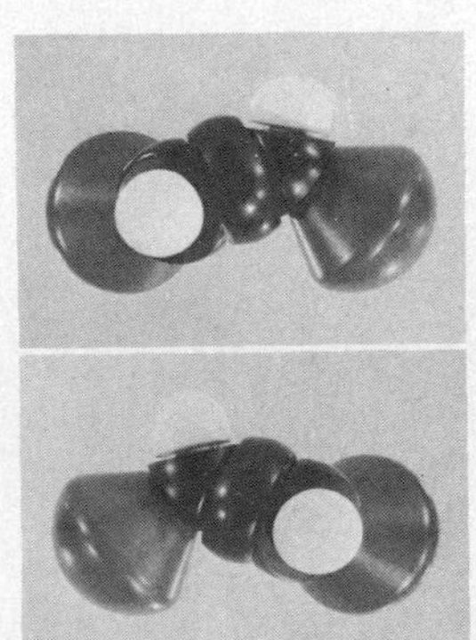

Nonsuperimposable allenes

and (*E*) isomers. Cumulative systems containing an even number of double bonds, on the other hand, give enantiomers if both termini are unsymmetrically substituted; 2,3-pentadiene is a simple example of this circumstance, which is discussed in greater detail in Section 19.8.

19.2 *Symmetry Properties of Molecules; Mirror Image Isomers*

In 550 B.C., the Greek philosopher Simplicius said, "The atoms move in the void and catching each other up jostle together, and some recoil in any direction that may chance, and others become entangled with one another in various degrees according to the symmetry of their shapes and sizes and positions and order, and they remain together and thus the coming into being of composite things is effected." The varieties of symmetry that result from these "entanglements" to give what we now call molecules have been codified, some 25 centuries later, in a rather simple fashion. By means of only a few "symmetry operations" the symmetry elements of organic compounds can be discerned and classified. The required operations are tests for the presence of simple axes of symmetry, planes of symmetry, centers of symmetry, and four-fold alternating axes of symmetry. The results of these tests are classified as follows: (a) a compound that possesses a plane of symmetry, a center of symmetry, or a fourfold alternating axis of symmetry is said to be *symmetric*; (b) a compound that lacks any of these symmetry elements but contains one or more simple axes of symmetry is said to be *dissymmetric*; and (c) a compound that lacks all these symmetry elements is said to be *asymmetric*. The important item to remember for the present discussion is that dissymmetric and asymmetric compounds possess *nonsuperimposable mirror image forms*, whereas symmetric compounds possess *superimposable mirror image forms*. Let us explore the application of these symmetry operations to a variety of types of organic molecules.

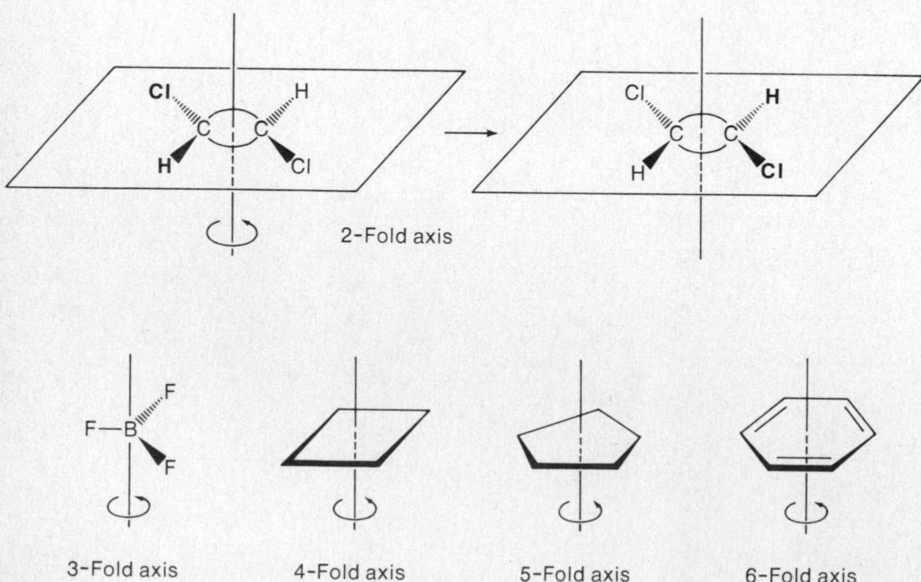

Fig. 19.6. Molecules containing simple axes of symmetry.

A molecule possesses a *simple axis of symmetry* if a line can be passed through it in such a fashion that rotation by a certain number of degrees around the line results in an orientation that is indistinguishable from the original molecule. For example, (*E*)-1,2-dichloroethene possesses a simple axis of rotation that passes through the midpoint of the molecule, perpendicular to the plane described by the atoms of the molecule. Rotation by 180° around the axis results in an orientation (Fig. 19.6) indistinguishable from the original. Boron trifluoride, in similar fashion, has an axis passing through its midpoint, around which a 120° rotation produces an orientation indistinguishable from the original. The difference between the axes of rotation for these two molecules is specified by what is called the multiplicity (*i.e.*, the "foldedness") of the rotation, the multiplicity (designated by n) being defined as equal to $360°/\theta$, where θ is the number of degrees of rotation required for superposition with the original. Thus, the axis of rotation in (*E*)-1,2-dichloroethene is twofold, and that in boron trifluoride is threefold. The planar forms of cyclobutane and cyclopentane have fourfold and fivefold simple axes of rotation, respectively, and benzene has a sixfold simple axis of rotation.

PROBLEM 19-2

What are the simple axes of symmetry in the following molecules?

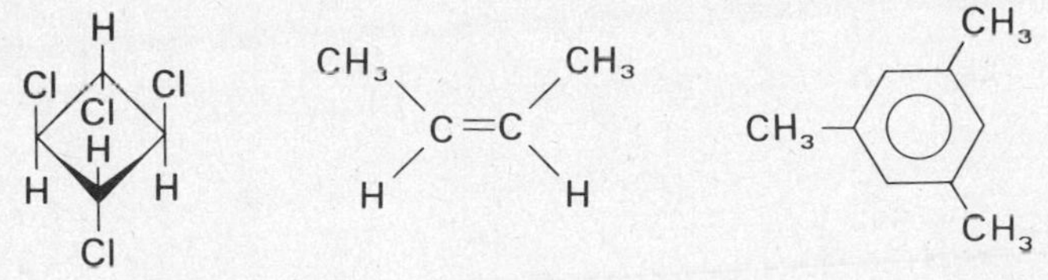

A molecule possesses a *plane of symmetry* if (a) all the atoms of the molecule are in the same plane, or (b) a plane can be passed through the molecule so that it divides the molecule into mirror image halves, one half-reflecting the other. For example, (*E*)-1,2-dichloroethene possesses a plane of symmetry which is the plane of the molecule and includes all the atoms. The planar form of cyclobutane has five symmetry planes, one passing through the four carbons in the plane of the molecule and four passing through the molecule perpendicular to the plane of the ring in the fashion shown in Fig. 19.7. Some molecules have symmetry planes in addition to simple axes of rotation, whereas other molecules have a symmetry plane as their only symmetry element (Fig. 19.7).

PROBLEM 19-3

What are the planes of symmetry in the following molecules?

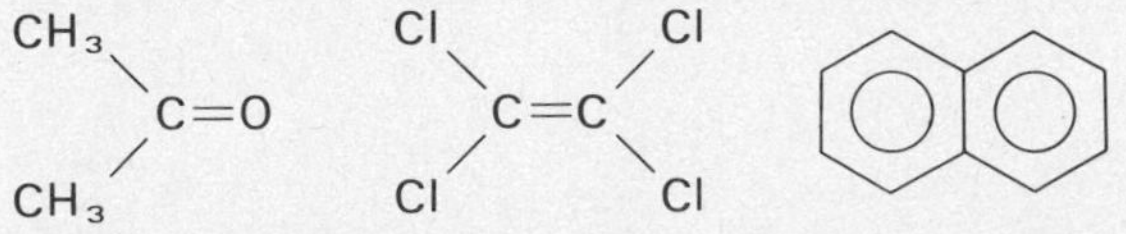

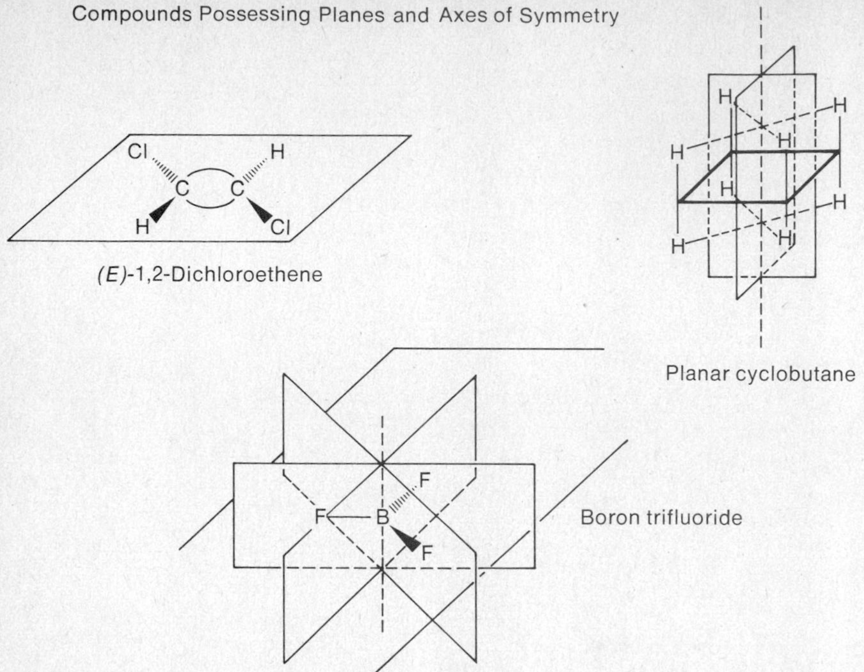

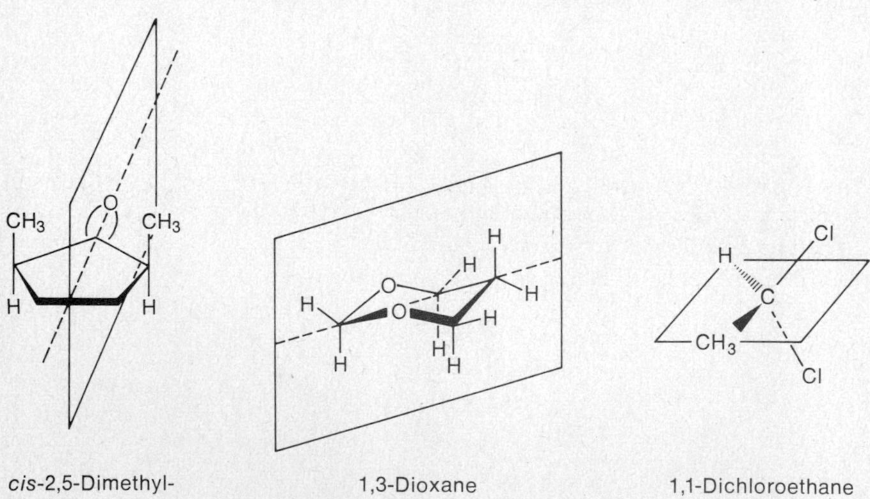

Fig. 19.7. Molecules containing planes of symmetry.

A molecule possesses a *center of symmetry* (point of symmetry) if all straight lines that can be passed through the center of the molecule encounter identical atoms at identical distances from the center. For example, the isomer of 1,3-dichloro-2,4-difluorocyclobutane shown in Fig. 19.8 possesses a center of symmetry as its only symmetry element. A center of symmetry is equivalent to a twofold alternating axis of symmetry. A molecule is said to possess an alternating axis of symmetry if an orientation indistinguishable from the original is obtained when (a) the molecule is rotated θ degrees around an axis passing through the molecule, and (b) the rotated molecule is reflected in a mirror that is perpendicular to the axis of rotation in step (a). For example, when the compound shown in Fig. 19.8 is rotated 180° around an axis passing through the center of the molecule and then reflected in a mirror perpendicular to this axis, a molecule superimposable on the original is obtained. The multiplicity of the alternating axis is determined by the extent of the rotation in step (a) (*i.e.*, $\frac{360}{180} = 2$; thus, a twofold axis).

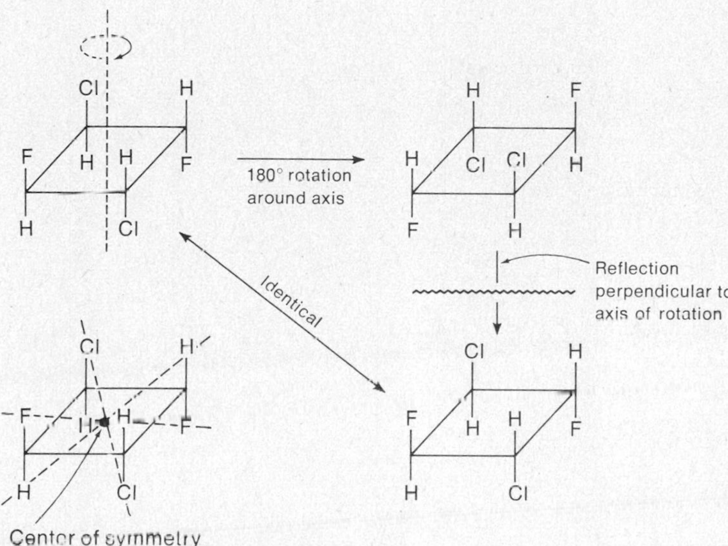

Fig. 19.8. Center of symmetry (two-fold alternating axis of symmetry) in 1,3-dichloro-2,4-difluorocyclobutane.

A fourfold alternating axis of symmetry is present in a molecule if a 90° rotation around an axis followed by a reflection perpendicular to that axis results in an orientation indistinguishable from the original. Relatively few organic compounds are known in which this symmetry element is the only one present. One type of compound in which this circumstance occurs is shown in Fig. 19.9.

Molecules may contain more than one element of symmetry, as we have already seen in the case of some of the compounds in Fig. 19.7, which contain several planes and axes of symmetry. One of the most symmetrical of organic molecules is cubane, which possesses the symmetry of a cube; *i.e.*, six twofold simple axes, four simple threefold axes, three simple fourfold axes, nine planes, and one twofold alternating axis of symmetry (center of symmetry).

Cubane

The necessary and sufficient condition for a molecule to have a nonsuperimposable mirror image is that it contain *no* plane, center, or fourfold alternating axis of symmetry. Several examples of such molecules are shown in Fig. 19.10. This requirement does not preclude the presence of a simple axis of symmetry,

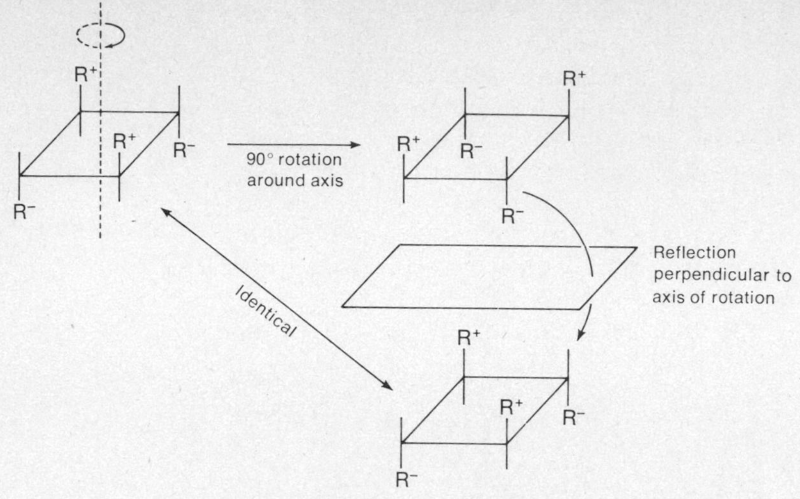

Fig. 19.9. Four-fold alternating axis of symmetry in a 1,2,3,4-tetrasubstituted cyclobutane molecule, in which the substituent groups, R, bear a nonsuperimposable mirror image relationship to one another (*i.e.*, R^+ is the nonsuperimposable mirror image of R^-).

Fig. 19.10. Compounds containing nonsuperimposable mirror images.

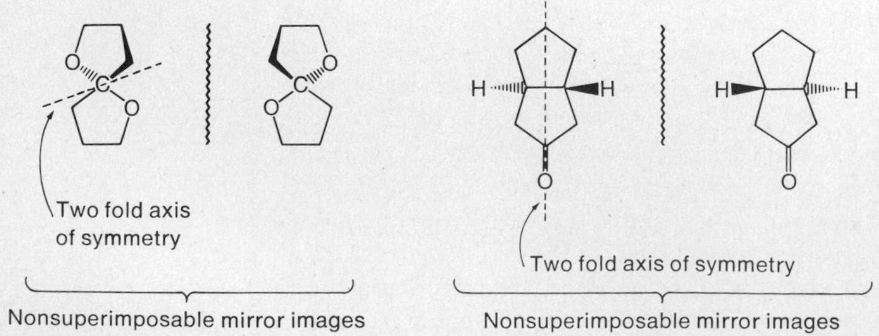

Fig. 19.11. Nonsuperimposable mirror image pairs of compounds containing a simple symmetry axis.

however, and there are numerous examples of compounds having nonsuperimposable mirror images in which this symmetry element is present (Fig. 19.11). To distinguish these compounds from those that are truly asymmetric (*i.e.*, which contain no symmetry element of any kind), we class them as dissymmetric molecules. Although the distinction is important in some stereochemical considerations, in the majority of instances the only concern is whether the mirror images are superimposable or nonsuperimposable. To simplify matters, therefore, we have come to use the term *chirality*. Any molecule that has a nonsuperimposable mirror image is classed as a *chiral molecule*; any molecule that has a superimposable mirror image is classed as an *achiral molecule* (*i.e.*, without chirality). Thus, all the compounds in Figs. 19.10 and 19.11 are chiral molecules, whereas those in Figs. 19.6–19.9 are achiral molecules.

PROBLEM 19-4

Which of the following compounds have nonsuperimposable mirror image forms?

$(CH_3)_2CHOH$ [structure: five-membered ring with O, bearing CH_3 and H]   [structure: cyclobutane bearing H, CH_3 and CH_3, H]

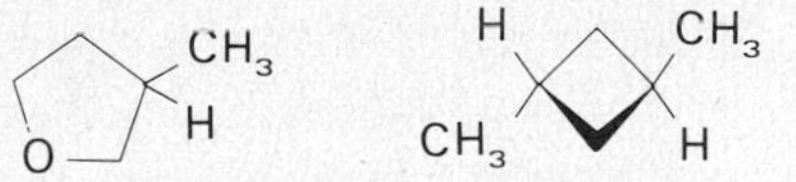

For those compounds that do not have nonsuperimposable mirror image forms, what are the symmetry elements that are present?

19.3 Compounds Containing One Chiral Center

The chirality of a compound can often be associated with one or more specific atoms, although in some instances a larger portion of the molecule, such as an axis or a plane, must be designated as the seat of chirality. Compounds in which the chirality is associated with a specific atom are said to possess a chiral center; if the specific atom is carbon, the carbon is referred to as a *chiral carbon atom* (equivalent to the older term "asymmetric carbon"). Examples of compounds containing a single chiral carbon atom are illustrated in Fig. 19.10, and their occurrence in nonsuperimposable mirror image forms has already been discussed (see Section 3.7). The physical and chemical consequences of the subtle structural difference have not yet been considered, however. Under "ordinary conditions" defined as an achiral environment, in which only achiral reagents are used, no difference whatsoever can be discerned between a pair of nonsuperimposable mirror images. Thus, the mirror image pairs for the compounds in Figs. 19.10 and 19.11 have identical boiling points, melting points, refractive indices, absorption spectra, and chemical reactivity if achiral conditions prevail. *Only in a chiral environment can any difference between a pair of nonsuperimposable mirror image molecules be discerned.*

One of the ways for creating a chiral environment involves the use of polarized light. A beam of light has associated with it an electrical vector and a magnetic vector, and in ordinary light the planes of these vectors are randomly oriented. If the beam is passed through prisms made of calcite crystals (Nicol prisms), it emerges with the electrical and magnetic vectors in separate, mutually perpendicular planes; the light is said to be plane polarized (Fig. 19.12). If, then, the

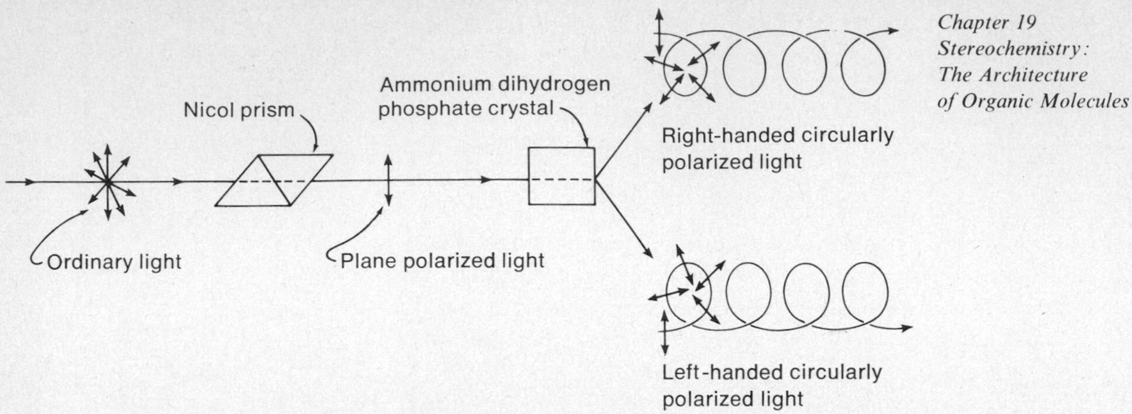

Fig. 19.12. Production of plane-polarized and circularly polarized light.

beam of plane-polarized light is passed through a crystal of ammonium dihydrogen phosphate, across which is imposed a high-voltage potential (1,000–5,000 V), the plane of polarization is given a twist in a direction that depends on the polarity of the potential that is imposed across the crystal (Fig. 19.12). In this fashion, a beam of light can be produced that is *circularly polarized* in either a right-handed or a left-handed helical sense. The two beams bear a nonsuperimposable mirror image relationship to one another and, therefore, are chiral.

As a result of its chirality, a beam of circularly polarized light can "see" a difference between a pair of nonsuperimposable mirror image molecules. For example, an aqueous solution containing one of the nonsuperimposable mirror image pairs of lactic acid (see Fig. 19.10), which we shall designate as the (+) isomer, has a refractive index of 1.10011 when measured with right-hand circularly polarized light. Its mirror image, which we shall designate as the (−) isomer, has a refractive index of 1.10017 under identical conditions. Conversely, when the measurements are made with left-hand circularly polarized light, the (+) isomer has a refractive index of 1.10017 and the (−) isomer has a refractive index of 1.10011. In the presence of circularly polarized light, then, there is a measurable difference between the refractive indices of the two members of a nonsuperimposable mirror image pair. Now, plane-polarized light is simply the superposition of two beams of circularly polarized light, one of right-hand and one of left-hand helicity. Passage of a beam of plane-polarized light through a solution of (+)- or (−)-lactic acid, therefore, results in an *unequal* interaction between the molecule and the right-hand and left-hand components of the beam, which produces a difference in the orientation of the plane of polarization between the incident beam and the emergent beam (Fig. 19.13). The change is in a clockwise direction for the (+) isomer and a counterclockwise direction for the (−) isomer, the *absolute magnitude* of the change being identical for the two isomers.

An instrument for measuring the change in the plane of polarization of a beam of light was invented over a century ago and is known as a polarimeter (Fig. 19.14).

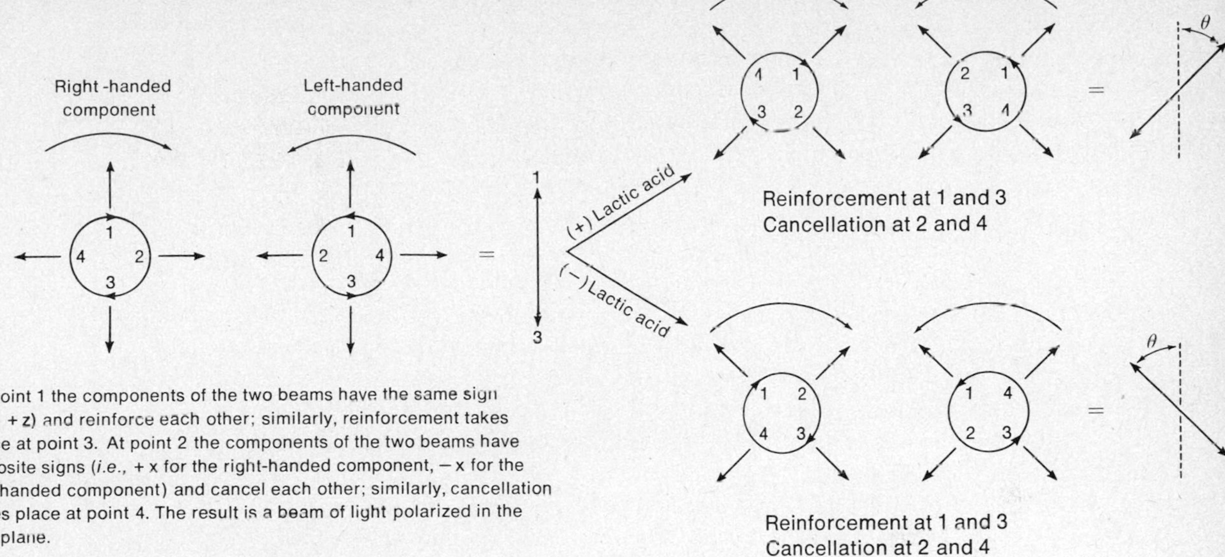

Fig. 19.13. Change in plane of polarization in a beam of plane-polarized light on passage through a solution containing chiral molecules.

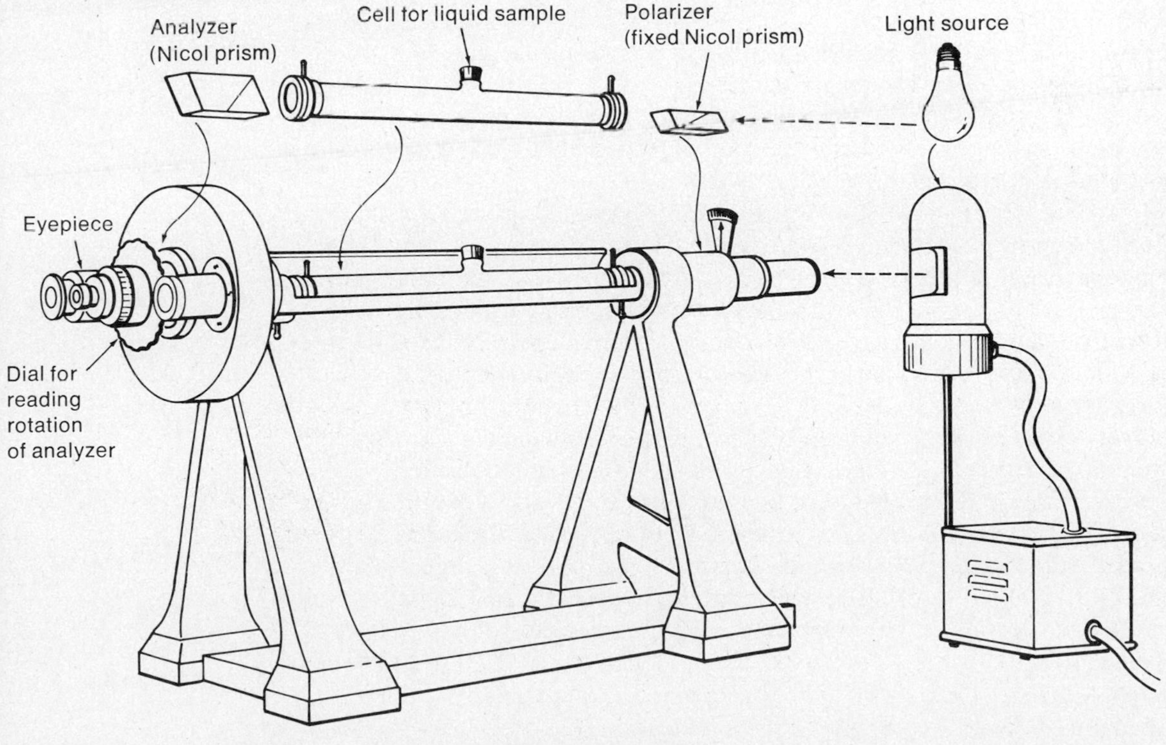

Analyzer (Nicol prism)

Cell for liquid sample

Polarizer (fixed Nicol prism)

Light source

Eyepiece

Dial for reading rotation of analyzer

Fig. 19.14. A polarimeter and its component parts.

In its simplest form, it consists of a pair of Nicol prisms placed on either side of a solution cell. Light passing through the first Nicol prism (the polarizer prism) becomes plane polarized and, if suffering no shift in the direction of its polarization on passage through the sample, enters the second Nicol prism (the analyzer prism) with its original plane of polarization. In this circumstance, if the second Nicol prism is placed at right angles to the first, none of the plane-polarized light impinging on it passes through, and a viewer at this end of the instrument sees no emergent light whatsoever; the prisms are said to be "crossed." If, however, the plane of polarization changes on traversing the solution, the Nicol prisms in the originally "crossed" position no longer cancel out all the light; some emerges and can be detected by the viewer. By rotating the second prism a certain number of degrees, we reach a position again where none of the light emerging from the solution passes through the second prism, and the viewer once more notes darkness. The direction in which the second prism has to be moved specifies the sign of the rotation of the sample, and the number of degrees that it must be moved specifies the magnitude of rotation. The polarimeter (Fig. 19.14), then, provides a quantitative measure of the "optical activity" of the solution, the magnitude of the rotation (*i.e.*, the optical activity) being dependent on the particular compound being measured, on its concentration, and on the length of the absorption cell;

$$[\alpha] = \frac{\alpha}{cl} \qquad \text{Expression 19-1}$$

where $[\alpha]$ is the specific rotation, α is the observed rotation in degrees, c is the concentration of grams per milliliter of solution, and l is the length of the absorption cell in decimeters. The magnitude of rotation is not always linearly dependent on concentration, however, for it is also a function of the solvent, the temperature, and the wavelength of light used in making the measurement. It is advisable, therefore, to include all these data when reporting a rotation; *e.g.*, $[\alpha]_D{}^{25°} = +38.1°$ (EtOH, $c = 0.167$ g/ml),* where the superscript, 25°, defines the temperature and the subscript D indicates that the sodium "D" line (589 nm) was used as the source of monochromatic light. If the null position for the analyzer requires a clockwise change, the compound is said to be dextrorotatory, and the rotation is given a positive sign; if it requires a counterclockwise change, the compound is said to be levorotatory, and the rotation is given a negative sign.

Lactic acid is one of a very large number of compounds that occur naturally in an optically active form. First isolated from sour milk by Sheele in 1780, it is produced in nature by the action of microorganisms on lactose. With *Bacillus bulgaricus*, the lactic acid produced is dextrorotatory and has a specific rotation of $+3.82°$; with *Bacillus aerogenes*, on the other hand, it is levorotatory and has a specific rotation of $-3.82°$. The two nonsuperimposable mirror image isomers of lactic acid are said to have opposite *configurations* and are referred to as

* Although the concentration factor is incorporated into the calculation of $[\alpha]$, it is frequently specified when reporting rotations, because the degree of association of the solute, and hence the value of $[\alpha]$, may be concentration dependent.

"*enantiomers*" or "*optical antipodes*." When lactic acid is synthesized under simple laboratory conditions, however, optically inactive material is produced unless a chiral environment is employed (see Section 19.11). Pyruvic acid (CH_3COCO_2H), for example, undergoes reduction with sodium borohydride to yield lactic acid containing exactly equal amounts of (+) enantiomer and (−) enantiomer (Fig. 19.15). A mixture of equal amounts of enantiomers such as this is called a *racemic modification* or a *racemate*.

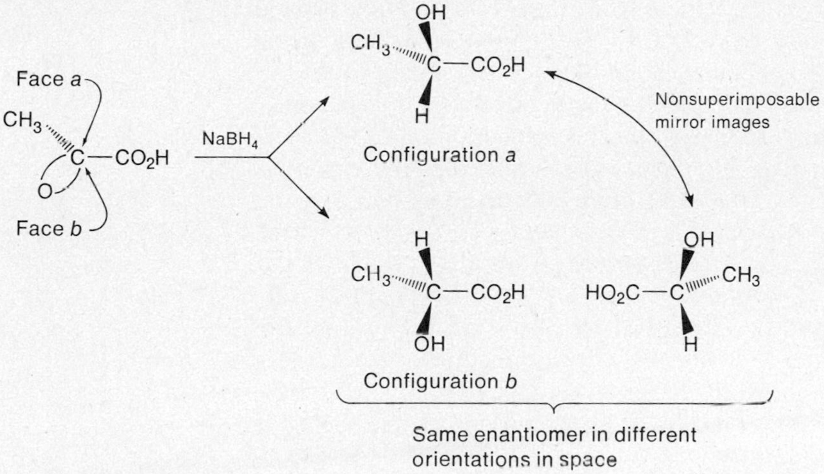

Fig. 19.15. Reduction of pyruvic acid to optically inactive (racemic) lactic acid.

The difference between the synthesis of lactic acid by the microorganism and by the sodium borohydride reduction of pyruvic acid is that the former is carried out under chiral conditions, living systems being strongly chiral, whereas the latter is carried out under achiral conditions. Pyruvic acid is an achiral molecule (it has a plane of symmetry), and the tendency for hydride addition to occur at face *a* of the carbonyl group is precisely the same as that at face *b* under achiral conditions. Equal amounts of the two enantiomers are produced, and the mixture has no net effect on plane-polarized light (rotation ± 0). Only through the use of chiral reagents can the two enantiomers be separated from each other in a process known as "resolution" (see Section 19.11). Thus, if optically inactive starting compounds are used and if achiral reagents and achiral conditions are employed, the products *must* be optically inactive (*i.e.*, racemic), even though individual molecules in the mixture may possess chiral centers, chiral axes, or chiral planes.

19.4 Configuration

A characteristic feature of enantiomers, as has been indicated, is their ability to change the plane of plane-polarized light, one enantiomer causing a right-handed rotation (dextrorotatory) and the other a left-handed rotation (levorotatory). Although the experimentally measurable difference provides one very useful way for designating the members of an enantiomeric pair, it leaves much to be desired as a nomenclature device. The magnitude and sometimes even the sign of the rotation are dependent on the experimental conditions (*i.e.*, temperature, concentration, wavelength of light, solvent) under which the measurement of optical rotation is made; furthermore, the sign of the rotation bears no easily discernable relation to the actual arrangement of the groups at the chiral center. The difference between the dextrorotatory and levorotatory enantiomers arises from the arrangements in space of the atoms and groups attached to the chiral center (or other chiral element—see later discussion), the arrangements defining the *configuration* of the enantiomers. But which configuration belongs to the dextrorotatory enantiomer and which to the levorotatory enantiomer is *not* revealed by measuring the direction of rotation. As an alternative to the (+) and (−) optical rotation designations for specifying the members of an enantiomeric pair, a system has been devised whereby the three-dimensional arrangement of the groups around the chiral portion(s) of a molecule can be denoted explicitly.

To introduce this stereochemical nomenclature system, let us first consider the arrangement of four different groups around a chiral carbon in terms of a simple color analogy. Let us build a structure in which a white ball, yellow ball, red ball, and black ball are tetrahedrally bonded to a center point. There are two ways in which the four balls can be attached, corresponding to a pair of nonsuperimposable mirror image structures. Let us arbitrarily establish a "sequence rule," which states that darker colors take priority over lighter colors; *i.e.*, black > red > yellow > white. Finally, let us formulate a "chirality rule," which states that (a) the object is viewed along an axis that places the ball of lowest priority (*i.e.*, white) away from the viewer, and (b) the direction, clockwise or counterclockwise, is noted that one's eye must travel in order to progress from the ball of highest priority to the one of second priority and then to the one of third priority (*i.e.*, from black to red to yellow). If the direction is clockwise, the object is defined as having the (*R*) (Latin, *rectus*, "right") chirality; if the direction is counterclockwise, it is defined as having the (*S*) (Latin, *sinister*, "left") chirality (Fig. 19.16).

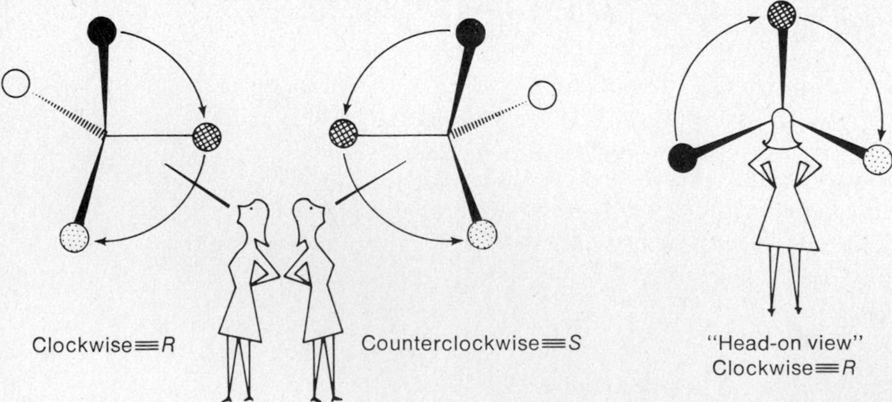

Fig. 19.16. Application of the sequence and chirality rules to an enantiomeric pair of objects; the sequence is

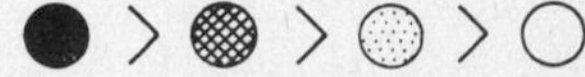

In applying a sequence rule to chemical compounds, we must establish an order of priority of the substituents attached to the chiral carbon. The scheme that has been developed by Cahn, Ingold, and Prelog employs the priority rules that were enumerated on p. 96 in connection with the (*Z*,*E*) nomenclature, *viz.* (a) higher atomic number precedes lower atomic number, and (b) higher atomic mass number precedes lower atomic mass number when the atomic numbers are the same. Thus, the two enantiomers of α-chloroethyl methyl ether (Fig. 19.17)

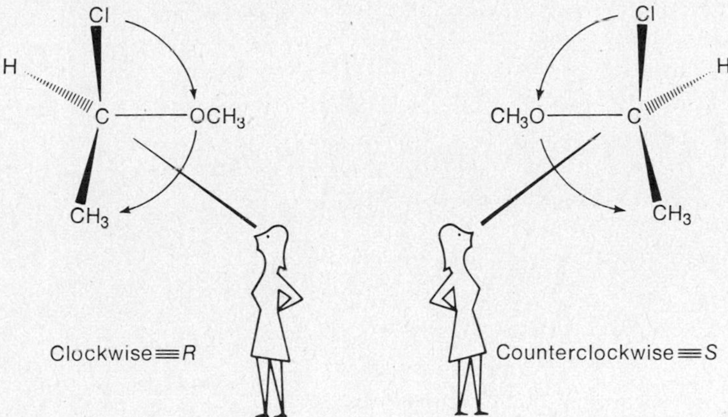

Fig. 19.17. Application of sequence and chirality rule to α-chloroethyl methyl ether.

are designated as (*R*) and (*S*) on the basis of (a) the sequence as defined by the decreasing order of priority of the groups attached to the chiral center; *i.e.*, Cl > O > C > H, and (b) application of the chirality rule, viewing the compound along the C—H axis with the H *away* from the viewer. In similar fashion the enantiomers of *sec*-butyl methyl ether can be specified as (*R*) and (*S*), although here the technique of "outward exploration" (see p. 96) must be used, because

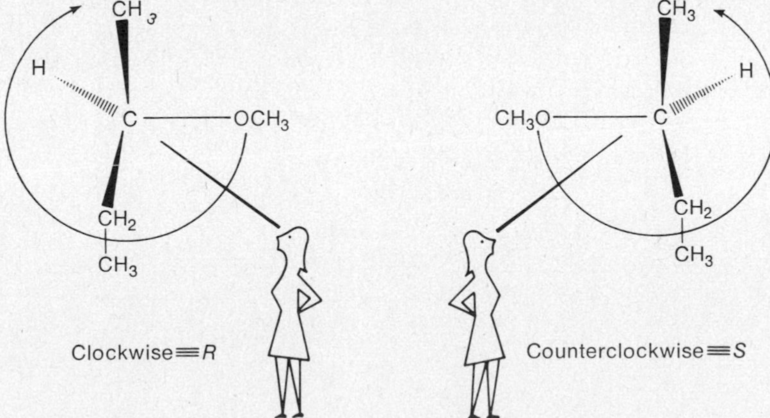

Fig. 19.18. Application of the sequence and chirality rule to *sec*-butyl methyl ether.

two *identical* atoms (*i.e.*, carbon) are attached to the chiral center. Thus, progressing outward from the methyl carbon, the next atom is H; progressing outward from the ethyl carbon, the next atom is C. Carbon takes precedence over hydrogen, so ethyl takes precedence over methyl, and the sequence is established as O > Et > Me > H (Fig. 19.18).

Still another complication arises in glyceraldehyde, where, in comparing the CH=O and CH_2OH groups, we see that not only are the first atoms identical (*i.e.*, carbon) but also the second atoms as well (*i.e.*, hydrogen and oxygen). As a solution, the sequence rules state that a double or triple bonded atom is equivalent to two or three of the same atom single bonded; *e.g.*, >C=O is equivalent to >C(O)(O), C≡N is equivalent to —C(N)(N)—N etc.* Thus, in glyceraldehyde the —CH=O group is equivalent to (H)C(O)(O) and takes precedence over the (H)(OH)C(H) group, which carries only one oxygen attached to the carbon, thereby establishing the sequence as HO > CHO > CH_2OH > H (Fig. 19.19).

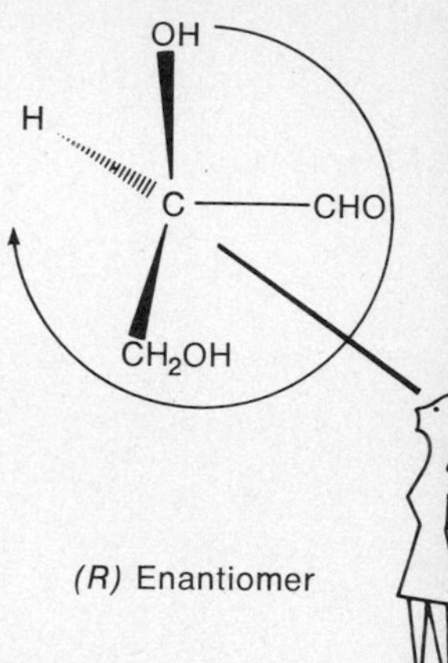

Fig. 19.19 Application of sequence and chirality rule to glyceraldehyde.

A phenyl group is equivalent to

```
H—C—C
  |
 —C—C
  |
H—C—C
```

and, for example, takes precedence over an isopropyl or *t*-butyl group. Thus, in 3-methyl-2-nitro-2-phenylbutane, the sequence of the groups attached to the chiral center is NO_2 > phenyl > $(CH_3)_2CH$ > CH_3 (Fig. 19.20).

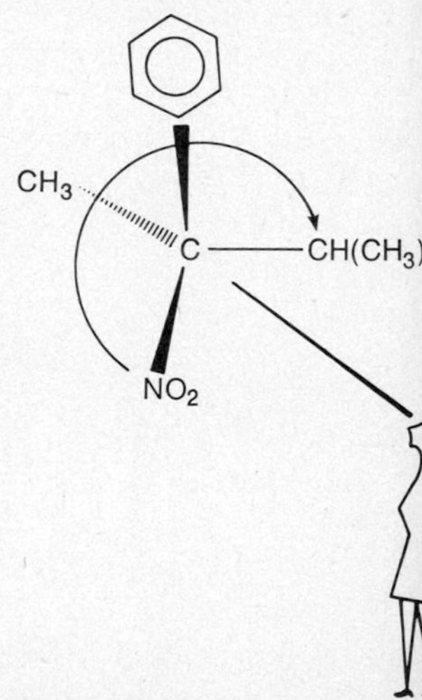

Fig. 19.20. (*R*)-3-Methyl-2-nitro-2-phenylbutane.

Chiral compounds containing isotopic substitutions may require rule (b) to be invoked, whereby the sequence is determined on the basis of mass number differences. For example, a compound that is important in the study of various biological mechanisms is 1-deuterioethanol, in which chirality results from the substitution of one of the C-1 hydrogens by a deuterium. The specification of the chirality for the two enantiomers is shown in Fig. 19.21, based on the precedence sequence O > C > D > H.

---

* The "complementation procedure" for assessing the precedence of multiple-bonded groups should actually be applied to both ends of the bond. In the case of the C=O group, for example, it involves a duplicate representation of the oxygen (*i.e.*, the carbon end of the bond) as well as a duplicate representation of the carbon (*i.e.*, the oxygen end of the bond) to give >C(O—C)(O). Thus, a —CHO group takes precedence over a $—CH(OH)_2$ group by virtue of the complementation of the oxygen by the carbon (*i.e.*, C takes precedence over H).

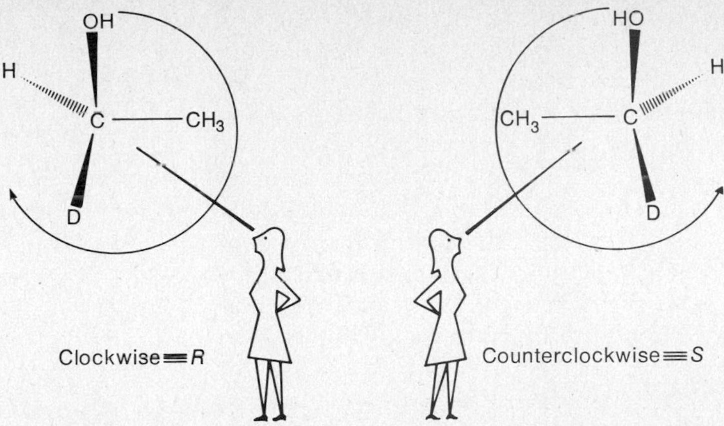

Fig. 19.21. Application of sequence and chirality rule to 1-deuterioethanol.

The sequence rules illustrated in these examples are inadequate to deal with every situation encountered in organic chemistry, and several additional rules must be invoked on occasion. For the present purposes, however, those that have just been discussed are adequate to illustrate how the scheme is applied, and they are capable of accommodating most of the examples that appear in subsequent chapters.

The lack of any easily discernible correlation between the sign of rotation of an optically active compound and its configuration (often called its *absolute configuration*) was recognized by Emil Fischer. Knowing of no experimental way for attacking the problem, Fischer "solved" it by fiat, arbitrarily declaring the dextrorotatory enantiomer of glyceraldehyde to have the absolute configuration represented by the structure shown at the left side of Fig. 19.22. This

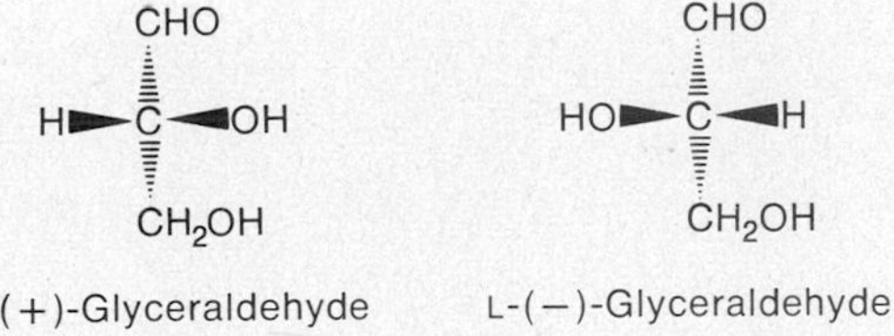

Fig. 19.22. Absolute configurations of glyceraldehyde enantiomers.

enantiomer, which in the (*R*),(*S*) nomenclature is the (*R*) compound, was designated by Fischer as the D compound; its mirror image, which in the (*R*),(*S*) nomenclature is the (*S*) compound, was designated by Fischer as the L compound. The D,L system is the oldest of the stereochemical notations and has been applied to a variety of compounds of the general structure RCHXR′, in which the R—C—R portion constitutes the parent chain in the IUPAC specification. To determine which configuration corresponds to D and which to L, the molecule is oriented in space so that (a) the main chain is vertically oriented with C-1 at the top, and (b) the chiral carbon carries the horizontal groups projecting toward the viewer and the vertical groups projecting away from the viewer, as depicted in Fig. 19.22.

If the molecule, when viewed in this orientation, has the X group (*i.e.*, the OH group in Fig. 19.22) on the right-hand side, it corresponds to the D configuration; If it has the X groups on the left-hand side, it corresponds to the L configuration. Although far less general in its application than the (*R*),(*S*) nomenclature, the D and L designations are extensively used, particularly in the carbohydrate family and amino acid family (see Chapters 32 and 33). In fact, enantiomeric pairs are frequently referred to as D,L pairs (or, in the older literature, as d,l pairs), even when the D,L nomenclature is not truly applicable. Although it is difficult to alter tradition, it would really be preferable if all (+),(−) pairs were simply called *enantiomeric pairs*, obviating the necessity of restricting the D,L pair designation to certain classes of compounds to which it correctly applies and removing the temptation to use it inappropriately.

PROBLEM 19-5

The D,L nomenclature is applicable to 2,3-dihydroxypropanal (glyceraldehyde) but not to 2,3-dihydroxy-2-methylpropanal. Explain. Draw the (*R*) and (*S*) configurations of 2,3-dihydroxy-2-methylpropanal.

D-(+)-Glyceraldehyde provides a reference point for a variety of stereochemical correlations. For example, oxidation of the aldehyde function to a carboxyl function yields levorotatory glyceric acid. Since the conversion from glyceraldehyde involves no change in configuration at the chiral center, however, the product must have the same absolute configuration as D-(+)-glyceraldehyde (Fig. 19.23). D-(−)-Glyceric acid, in turn, can be converted to levorotatory lactic

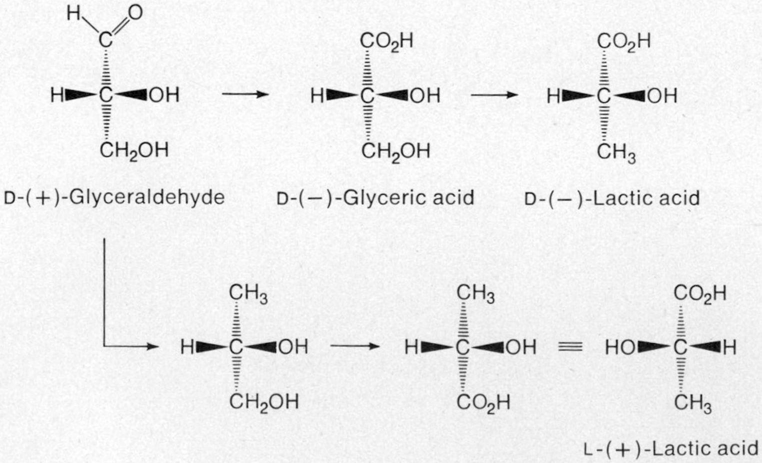

D-(+)-Glyceraldehyde → D-(−)-Glyceric acid → D-(−)-Lactic acid

→ L-(+)-Lactic acid

Fig. 19.23. Stereochemical correlations.

acid in a series of reactions involving the conversion of the CH_2OH group to a CH_3 group. Thus, (−)-lactic acid must also possess the D configuration. Of course, to apply such correlations properly, we must understand the chemistry of the interconversions. It is possible, for example, to convert D-(+)-glyceraldehyde to (+)-lactic acid by (a) conversion of the CHO group to a CH_3 group, and (b) conversion of the CH_2OH group to a CO_2H group, from which the erroneous conclusion might be drawn that (+)-lactic acid has the D configuration. Advantage has been taken of stereochemical correlations of the type illustrated in Fig. 19.23 to determine the absolute configurations of various molecules by relating them to the enantiomers of tartaric acid, the absolute configurations of which are known.

In 1951, Bijvoet and his co-workers accomplished the feat of determining the absolute configurations of the enantiomers of the potassium rubidium salt of tartaric acid by means of X-ray crystallographic techniques. Since the tartaric acids had already been related stereochemically to the glyceraldehydes, the absolute configurations of D- and L-glyceraldehyde were automatically revealed. It turned out that Fischer had made the correct guess 75 years earlier; dextrorotatory glyceraldehyde, designated as D in the D,L notation and (*R*) in the (*R*),(*S*) notation, does, indeed, have the absolute configuration illustrated in Figs. 19.19, 19.22, and 19.23.

19.5 Compounds Containing Two Chiral Centers

Molecules can contain more than one chiral carbon atom, and the complexity of the stereoisomerism increases as the number of chiral centers increases. For example 2,3,4-trihydroxybutanoic acid contains two chiral carbons and can exist in four different stereochemical forms (Fig. 19.24). Forms (a) and (b)

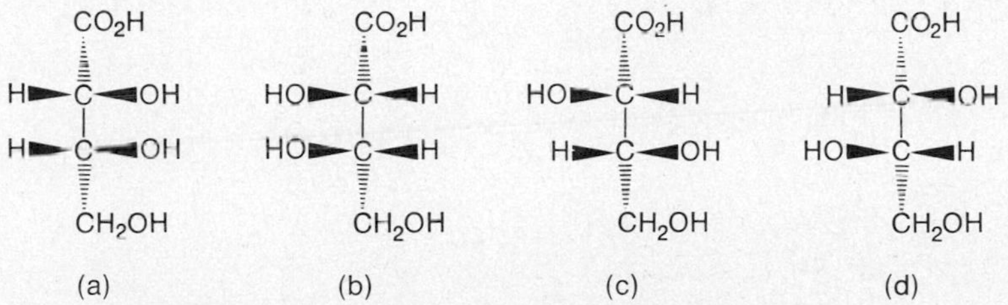

(a)	(b)	(c)	(d)
CO_2H	CO_2H	CO_2H	CO_2H
H–C–OH	HO–C–H	HO–C–H	H–C–OH
H–C–OH	HO–C–H	H–C–OH	HO–C–H
CH_2OH	CH_2OH	CH_2OH	CH_2OH

Fig. 19.24. Stereoisomers of 2,3,4-trihydroxybutanoic acid.

bear a nonsuperimposable mirror image relationship to each other, as do forms (c) and (d); thus, (a) and (b) constitute one enantiomeric pair, and (c) and (d) constitute a second enantiomeric pair. The properties of (a) and (b) are identical in an achiral environment. The relationship of (a) to (c), however, is *not* a mirror image one; a real object can have only one mirror image, and (b), not (c), is the mirror image of (a). Similarly, (d) is not the mirror image of (a) but is the mirror image *only* of (c). The relationship between (a) and (c), (a) and (d), (b) and (c), and (b) and (d) is a diastereomeric one, and these compounds are said to be *diastereomers.* The subtlety of the mirror image relationship no longer obtains, and diastereomers differ from each other in physical and chemical properties under *ordinary, achiral* conditions. As will become apparent in Section 19.11, it is the creation of diastereomeric relationships that provides the basis for most of the methods for resolving optically inactive mixtures of enantiomers (*i.e.*, racemates) into the optically active forms.

To this point in the discussion, pseudo-three-dimensional representations have been drawn to depict the architecture of the compounds under consideration. These provide the clearest and most explicit pictures of what the molecules actually look like. Unfortunately, however, they become increasingly hard to draw as the stereochemical complexity of the molecule increases. To circumvent this problem, Emil Fischer invented a more formalized scheme late in the nineteenth century. It is known as the Fischer projection representation, and is a two-dimensional rendering of the three-dimensional molecule oriented in space so that two of the substituents attached to the chiral center are vertical and the other two are horizontal, the *vertical groups projecting away* from the viewer and the *horizontal groups projecting toward* the viewer. Thus, the Fischer projection of the two enantiomeric forms of lactic acid are those shown in Fig. 19.25.

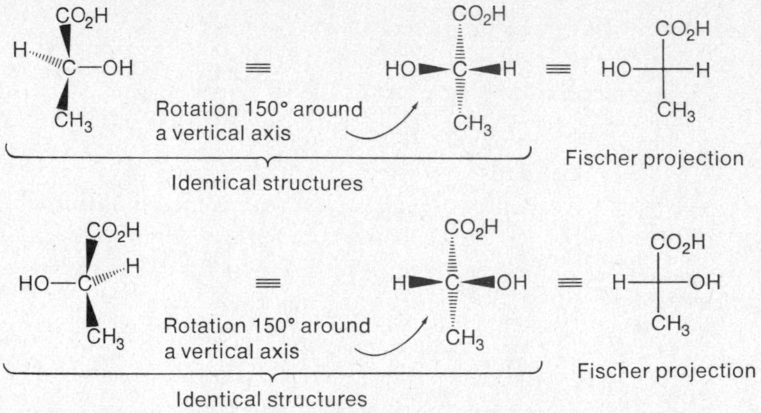

CO_2H / H–C–OH / CH_3 ≡ CO_2H / HO–C–H / CH_3 ≡ CO_2H / HO–┼–H / CH_3

Rotation 150° around a vertical axis

Identical structures

Fischer projection

CO_2H / HO–C–H / CH_3 ≡ CO_2H / H–C–OH / CH_3 ≡ CO_2H / H–┼–OH / CH_3

Rotation 150° around a vertical axis

Identical structures

Fischer projection

Fig. 19.25. Fischer projection representations of lactic acid.

Drawn in similar fashion, the two chiral centers in 2,3,4-trihydroxybutanoic acid are shown in Fischer projection representation in Fig. 19.26, with the molecule oriented so that the two OH groups and two H atoms are horizontal and are projecting toward the viewer and the CH_2OH and CO_2H groups are vertical and are projecting away from the viewer. It is important to remember that the Fischer projection representation implies *only* these spatial orientations; other orientations give misleading and incorrect projection pictures.

PROBLEM 19-6

Draw Fischer representations for the following compounds. For which of these compounds do enantiomers exist? Draw their Fischer projection.

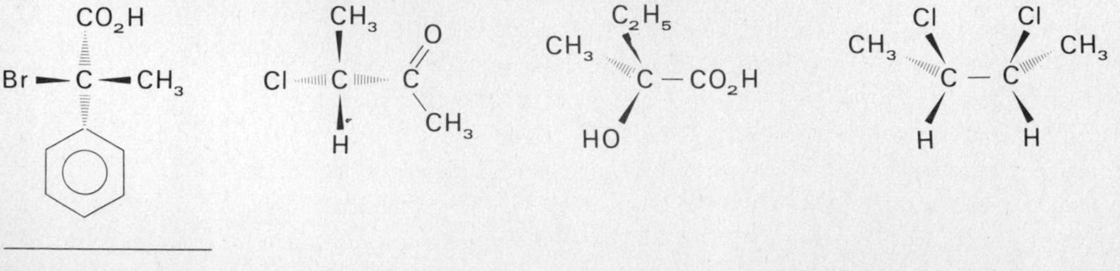

CO_2H / Br–C–CH_3 / C_6H_5 (phenyl)

CH_3 / Cl–C–C(=O)CH_3 / H

C_2H_5 / CH_3–C–CO_2H / HO

Cl, CH_3, H–C–C–Cl, CH_3, H

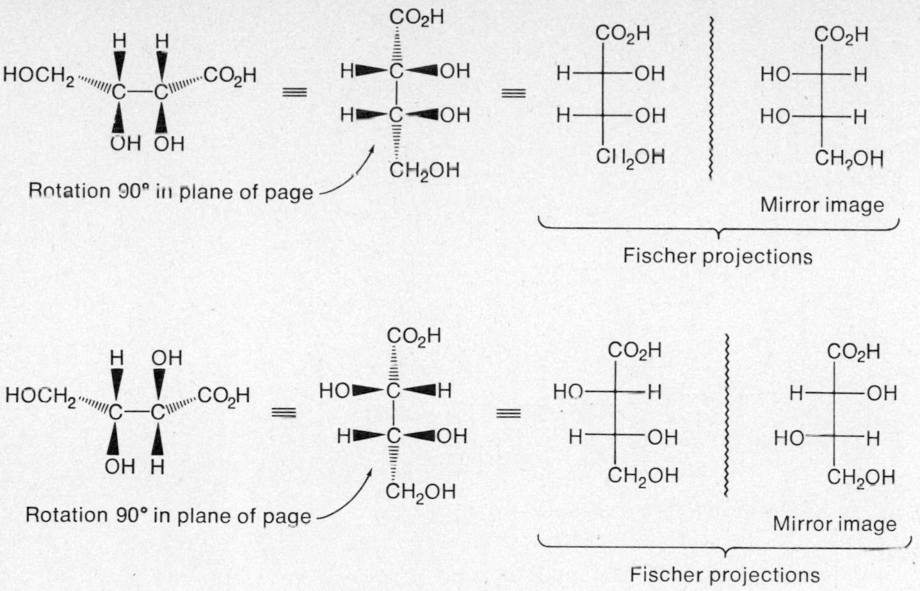

Fig. 19.26. Fischer projection representations of 2,3,4-trihydroxybutanoic acid.

Using plane projection representations and keeping in mind the three-dimensional objects that they depict, let us consider the reactions involved in the synthesis of 2,3,4-trihydroxybutanoic acid from glyceraldehyde; *i.e.*,

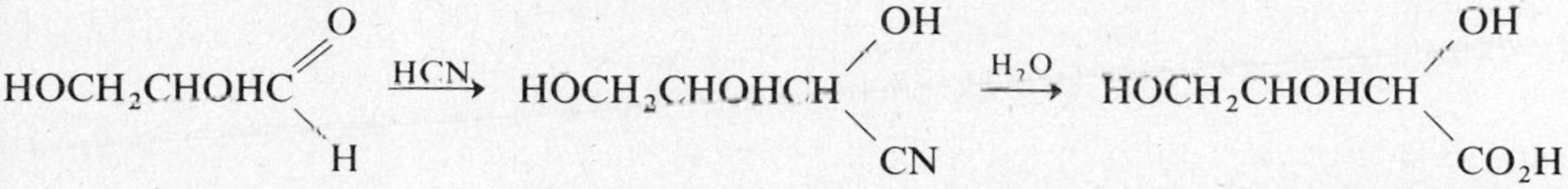

$$HOCH_2CHOHC(=O)H \xrightarrow{HCN} HOCH_2CHOHCH(OH)CN \xrightarrow{H_2O} HOCH_2CHOHCH(OH)CO_2H$$

(Fig. 19.27). If either of the optically active forms of glyceraldehyde is the starting compound, the product is an optically active mixture of two of the four structures shown in Fig. 19.27, *viz*., forms (a) and (c) or (b) and (d), depending on whether (+)- or (−)-glyceraldehyde is used. But (a) is diastereomeric with (c), and (b) is diastereomeric with (d); this means that (a) and (c) [or (b) and (d)] do not necessarily form in equal amounts, and that they can be separated by ordinary chemical or physical means. For example, (+)-glyceraldehyde might yield a mixture containing 60% of (a) ([α] = +53°) and 40% of (c) ([α] = +21°); reciprocally, (−)-glyceraldehyde would then yield a mixture containing 60% of (b) ([α] = −53°) and 40% of (d) ([α] = −21°). In contrast, if optically inactive (±)-glyceraldehyde (racemic material) is used as the starting material, an optically *inactive* mixture is produced, which contains 60% of the enantiomeric pair (a) and (b) and 40% of the enantiomeric pair (c) and (d). The diastereomeric relationship between the enantiomers is still present, nevertheless [*i.e.*, the (a)–(b) pair is diastereomeric with the (c)–(d) pair], and the (a)–(b) pair is separable from the (c)–(d) pair by ordinary chemical and physical methods. Thus, optically active starting materials yield mixtures of optically active diastereomers, and optically inactive starting materials yield comparable mixtures of optically inactive

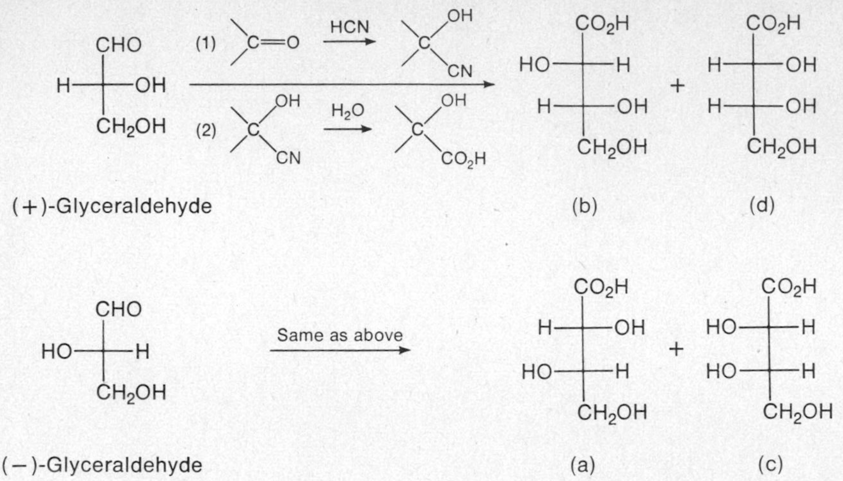

Fig. 19.27. Conversion of glyceraldehyde to 2,3,4-trihydroxybutanoic acid.*

diastereomers (in accord with the proscription that optically active products *cannot* be produced unless the environment is chiral). When either (+)- or (−)-glyceraldehyde is used as the starting material, the chiral environment is provided by the compound itself.

Oxidation of the terminal hydroxymethyl group of 2,3,4-trihydroxybutanoic acid yields 2,3-dihydroxybutanedioic acid (tartaric acid), a compound that still possesses two chiral centers but which are now identical in structure; *i.e.*, the same kinds of groups are attached to C-2 and C-3. Inspection of the pseudo-three-dimensional representations (a)–(d) in Fig. 19.28 reveals that (c) and (d) are enantiomeric but that (a) and (b) are superimposable and, therefore, identical. This is predictable when it is realized that (a) [*identical with* (*b*)] possesses a plane of symmetry that divides the molecule into mirror image halves. Consequently, it cannot exist in nonsuperimposable mirror image forms and cannot, under any circumstance, be resolved into enantiomeric forms. It is *inherently* optically inactive and is referred to as a *meso* form; this is different from an optically inactive mixture of equal amounts of the (+)- and (−)-forms, which is *not* a *meso* form but is a racemic mixture capable of being resolved into its optically active enantiomers. Thus, tartaric acid exists in a dextrorotatory form (mp 170°, $[\alpha] = +12°$), a levorotatory form (mp 170°, $[\alpha] = -12°$), and a *meso* form (mp 146°, $[\alpha] = \pm 0°$).

The term *meso* (Greek, *mesos*, "middle") applies to any inherently archiral member of a stereoisomeric set. Thus, *meso* compounds can be members not only of stereoisomeric sets in which some of the members are chiral, but also of stereoisomeric sets in which none of the members is chiral. For example, the (*Z*) and (*E*) isomers of unsymmetrically substituted alkenes can be classed as

* Although the "cyanohydrin method" for lengthening carbon chains has been used in many instances, its application to glyceraldehyde is reported only in the very old literature without experimental details. The data recorded in this discussion, therefore, are fabricated for the purpose of illustrating enantiomeric and diastereomeric relationships in this series.

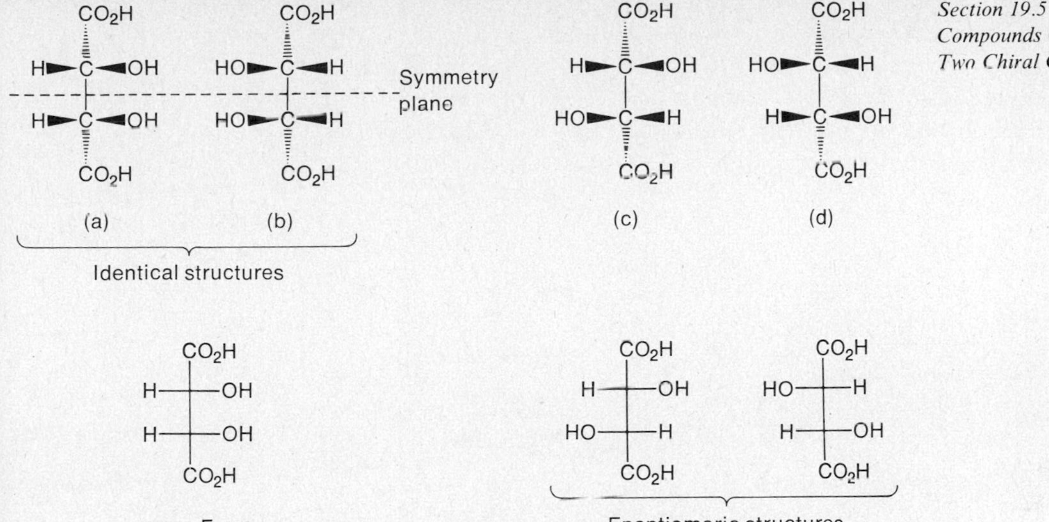

Fig. 19.28. Stereoisomers of tartaric acid.

meso compounds, although the usage of the term in this context is seldom encountered.

By pointing out that only one of the infinite number of conformations of *meso*-tartaric acid possesses a plane of symmetry, the reader might challenge the statement that the *meso* form *must* be optically inactive. A conformation such as Fig. 19.29(a′) contains *no* symmetry element and does, indeed, have the potentiality of being resolved into the enantiomeric forms (a′) and (a″).

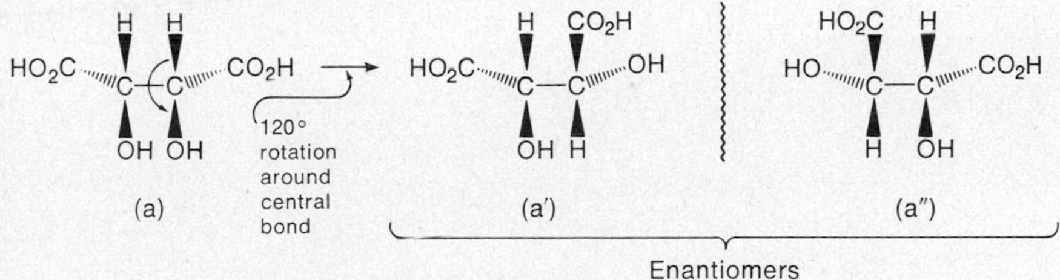

Fig. 19.29. Enantiomeric conformations of *meso*-tartaric acid.

As discussed earlier, however (see p. 67), the rapidity with which these forms interconvert precludes the separation and isolation of just one of the enantiomeric forms. Instead, we are dealing with a rapidly interconverting set of conformers, including one that possesses a plane of symmetry and another a center of symmetry. Over the period of time that is required to effect the resolution and measurement of optical activity, all the molecules in the system pass through these symmetrical conformations many times. As a result, the system contains precisely equal amounts of the enantiomeric forms of all conformations. This fact is implicit, then, in the representation of *meso*-tartaric acid in the particular conformation that possesses a plane of symmetry.

The conversion of D-(+)-glyceraldehyde to a diastereomeric mixture of 2,3,4-trihydroxybutanonitrile followed by hydrolysis and oxidation to the corresponding dicarboxylic acids (Fig. 19.30) yields a mixture of optically inactive *meso*-tartaric acid and levorotatory tartaric acid. It is this stereochemical correlation that permitted the absolute configuration of the glyceraldehyde to be assigned from the X-ray-determined absolute configurations of the tartaric acids.

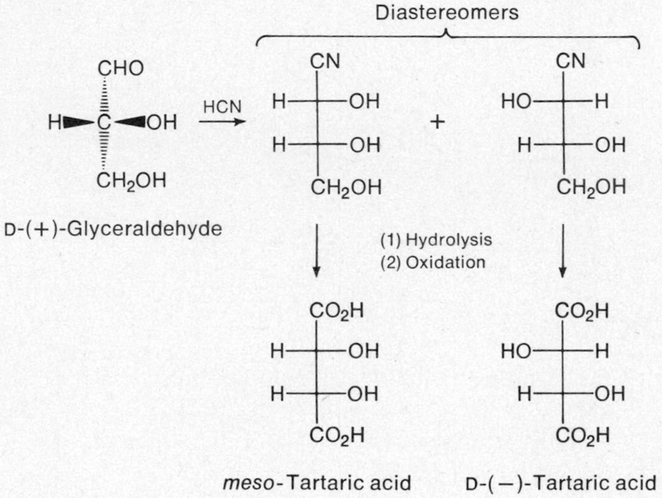

Diastereomers

D-(+)-Glyceraldehyde $\xrightarrow{HCN}$ (2,3,4-trihydroxybutanonitrile diastereomers) $\xrightarrow{\text{(1) Hydrolysis (2) Oxidation}}$ *meso*-Tartaric acid + D-(−)-Tartaric acid

Fig. 19.30. Stereochemical correlation of D-(+)-glyceraldehyde with levoratatory tartaric acid.

PROBLEM 19-7

In the sequence of reactions shown in Fig. 19.30, a mixture of *meso* and levorotatory tartaric acid is obtained from D-(+)-glyceraldehyde. Can this mixture be separated by ordinary laboratory means? Explain why or why not.

In applying the (*R*),(*S*) nomenclature to compounds containing two or more chiral centers, each of the chiral centers is designated as (*R*) or (*S*), following the rules described in Section 19.4. For example, (−)-tartaric acid is named (2*S*, 3*S*)-2,3-dihydroxybutanedioic acid (Fig. 19.31).

Another stereochemical notation that is frequently applied to compounds containing two adjacent chiral centers uses the sugars erythrose and threose as the reference points. Compounds possessing the configurations corresponding to those in D- or L-erythrose are referred to as *erythro* isomers, and those possessing the configurations corresponding to those in D- or L-threose are referred to as *threo* isomers (Fig. 19.32).

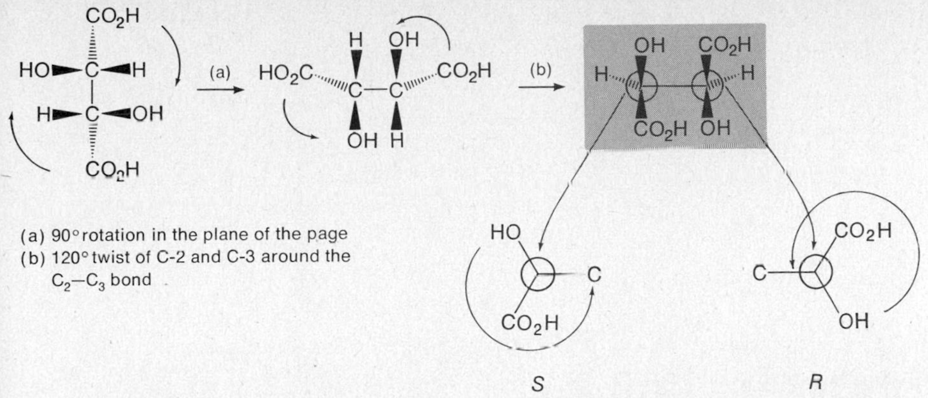

(a) 90° rotation in the plane of the page
(b) 120° twist of C-2 and C-3 around the C_2—C_3 bond

Fig. 19.31. (−)-Tartaric acid specified as (2*S*, 3*S*)-2,3,-dihydroxybutanedioic acid.

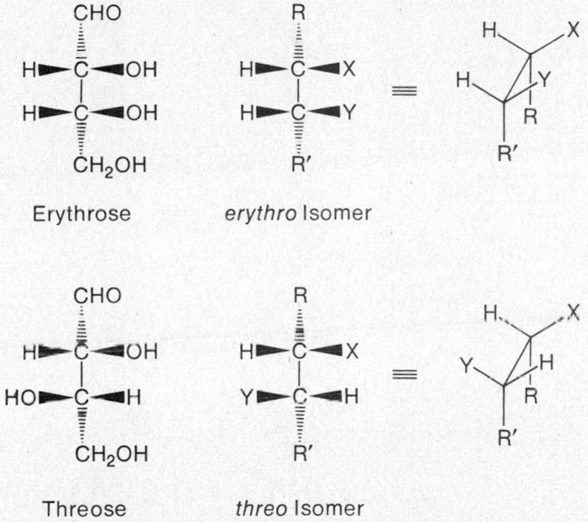

Erythrose *erythro* Isomer

Threose *threo* Isomer

Fig. 19.32. *Erythro* and *threo* stereochemical notations.

19.6 Compounds Containing Three or More Chiral Centers

2,3,4,5-Tetrahydroxypentanoic acid contains three structurally different chiral centers, and eight stereoisomers are possible (Fig. 19.33), including the enantiomeric pairs (a)–(b), (c)–(d), (e)–(f), and (g)–(h). Each member of the stereoisomeric set (a) through (h) is enantiomeric with one other compound in the set and diastereomeric with the other six; the enantiomeric pairs (a)–(b), (c)–(d), (e)–(f), and (g)–(h) are diastereomeric with each other. Thus, (a) can be separated from (c), (b) from (d), (a)–(b) from (c)–(d), etc., by ordinary chemical or physical means, but resolution methods must be used to separate (a) from (b), (c) from (d), etc.

A compound with two structurally different chiral centers has four stereoisomers; a compound with three structurally different chiral centers has eight stereoisomers; a compound with *n* structurally different chiral centers has 2^n

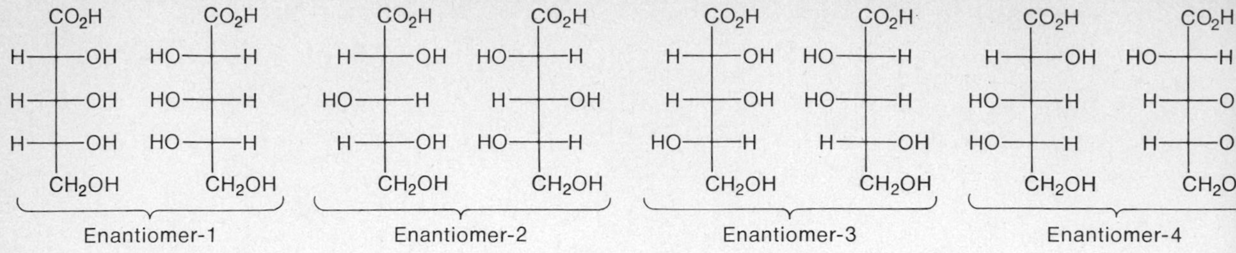

Fig. 19.33. Stereoisomers of 2,3,4,5-tetrahydroxypentanoic acid.

stereoisomers comprising $2^n/2$ enantiomeric pairs. The structure shown in Fig. 19.34, for example, has eight chiral centers and can exist in 256 optically active forms (128 enantiomeric pairs). One of these 256 possibilities corresponds to the naturally occurring compound cholesterol. If some of the chiral centers

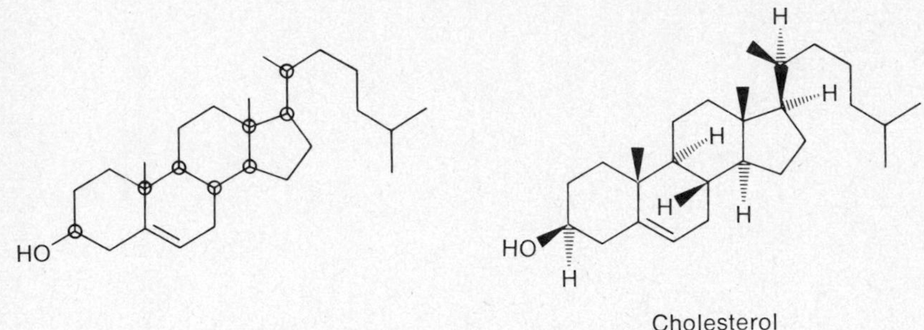

Fig. 19.34. Stereoisomers for a compound having eight chiral centers.

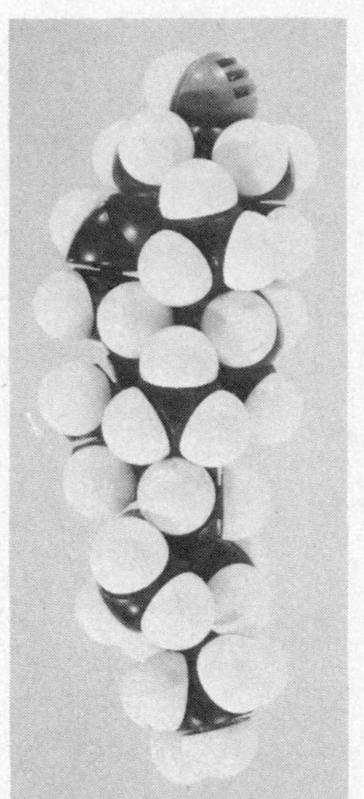

Cholesterol

PROBLEM 19-8

Would you expect naturally occurring cholesterol to be optically active or inactive? Explain.

in the compound are structurally identical, however, the total number of stereoisomers is lower than the 2^n value, as we have already seen in the case of the tartaric acids. Similarly, 2,3,4-trihydroxypentanedioic acid, obtained by oxidation of 2,3,4,5-tetrahydroxypentanoic acid, has only one enantiomeric pair

meso Form-1	*meso* Form-2	Enantiomeric pair	
CO_2H	CO_2H	CO_2H	CO_2H
H—OH	H—OH	H—OH	HO—H
H—OH	HO—H	H—OH	HO—H
H—OH	H—OH	HO—H	H—OH
CO_2H	CO_2H	CO_2H	CO_2H

Fig. 19.35. Stereoisomers of 2,3,4-trihydroxypentanedioic acid.

and two *meso* forms (Fig. 19.35). Each of the *meso* forms has a plane of symmetry that includes the H, C, and OH atoms and groups at C-3 and which divides the remainder of the molecule into mirror image halves. The two *meso* forms are nonidentical and are diastereomeric, because the OH group that falls in the symmetry plane can be attached to the carbon atom in two different configurations.

19.7 Cyclic Systems Containing Chiral Centers

The stereochemistry of cyclic systems closely resembles that of their acyclic counterparts. The number of stereoisomers in the cyclic system is no greater and may, in fact, be lower because of the constraints that the ring places on the architecture of the system. The smallest of the cycloalkanes is represented by the compound 2-hydroxymethyl-1-cyclopropanecarboxylic acid, which has two chiral centers that are differently substituted. Four stereoisomers are possible, consisting of two pairs of enantiomers (Fig. 19.36). This stereochemistry is

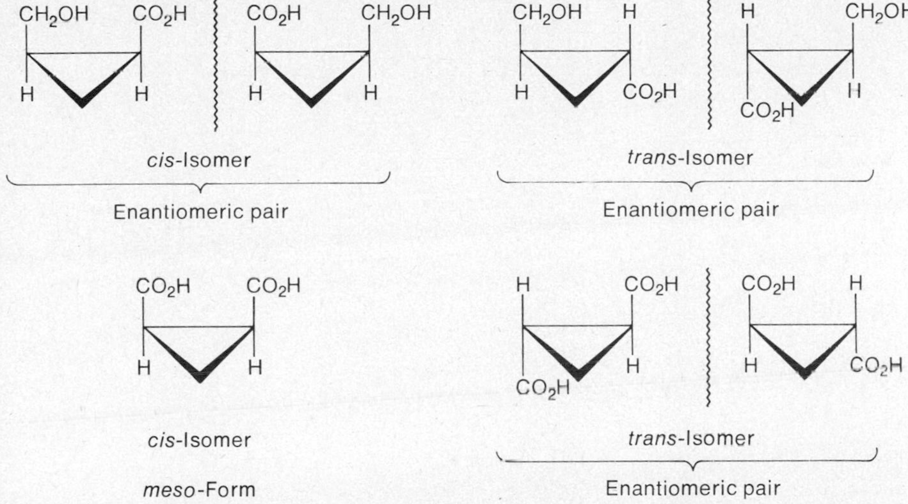

Fig. 19.36. Stereoisomerism in substituted cyclopropanes.

precisely the same as that of its acyclic counterpart 2,3-dimethyl-4-hydroxybutanoic acid. Oxidation of the hydroxymethyl group of the cyclopropyl compound yields cyclopropane-1,2-dicarboxylic acid, a compound containing two identically substituted chiral centers and having, therefore, one pair of enantiomers and one *meso* form. Again, this is identical with the stereochemistry of the acyclic analog, 2,3-dimethylbutanedioic acid.

The prefixes *cis* and *trans*, as discussed previously (see p. 72), are used to designate the orientations of substituents on cyclic structures, *cis* indicating that the substituents are on the same side of the ring and *trans* indicating that they are on opposite sides of the ring. With rigid planar ring systems, such as cyclopropane, the designations "same side" and "opposite side" are clear and unequivocal. With flexible nonplanar rings, such as cyclohexane, the designations remain unequivocal although not always clear at first glance. For example, the *trans* isomer of cyclohexane-1,2-dicarboxylic acid can exist in a variety of rapidly interconverting ring conformations, including ones in which the carboxyl groups are diequatorially aligned or diaxially aligned. In the more stable diequatorial conformer, the carboxyl groups may not appear to be *trans*, although they really are (*e.g.*, note the *trans* relationship between the hydrogens diaxially aligned at these positions). The mirror image form of the *trans* compound is nonsuperimposable on the original, so an enantiomeric pair exists. The *cis* isomer also consists of a rapidly interconverting set of conformers, of which the most stable ones are the equatorial-axial structures shown in Fig. 19.37; the mirror image of

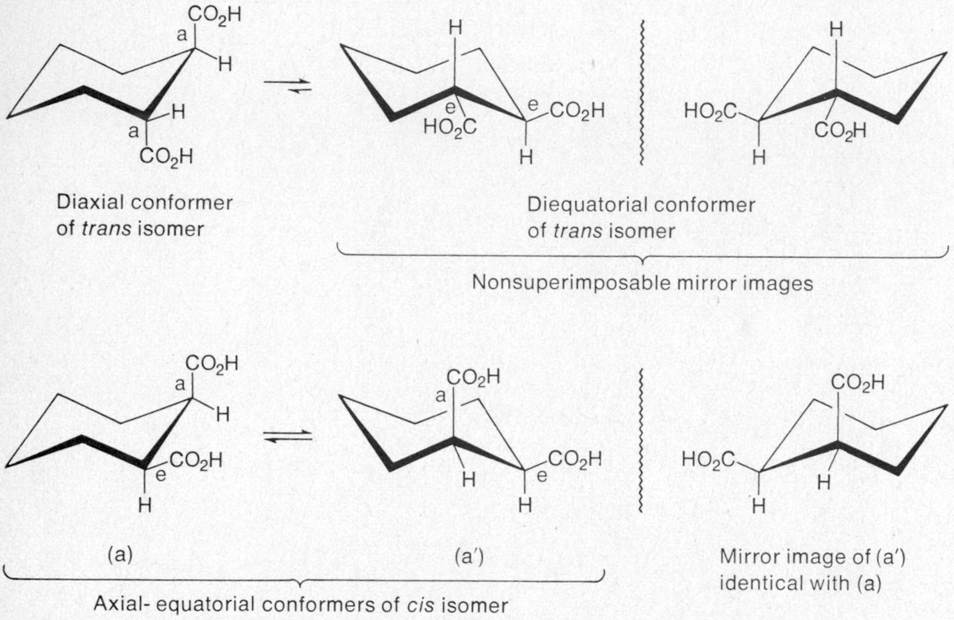

Diaxial conformer of *trans* isomer

Diequatorial conformer of *trans* isomer

Nonsuperimposable mirror images

(a)

(a′)

Mirror image of (a′) identical with (a)

Axial- equatorial conformers of *cis* isomer

Fig. 19.37. Stereoisomerism of cyclohexane-1,2-dicarboxylic acids.

any particular conformer is superimposable on the interconverted conformer [*i.e.*, the mirror image of (a′) is identical with (a) in Fig. 19.37], and it corresponds to a *meso* form. The *meso* character of the *cis* isomer is more readily perceived by considering the boat form, through which the system must pass in interconverting from one chair form to the other. In the boat form of the *cis* isomer, a plane of symmetry divides the molecule into mirror image halves as shown in the diagram:

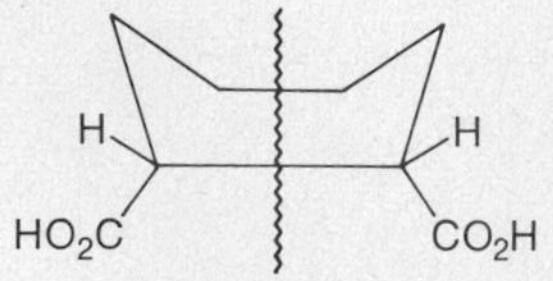

Because of the rapid rate of conformer interconversion, chirality arising from conformational contributions can be neglected, as in the acyclic counterparts. For purposes of delineating the stereochemistry of cyclic systems, therefore, it

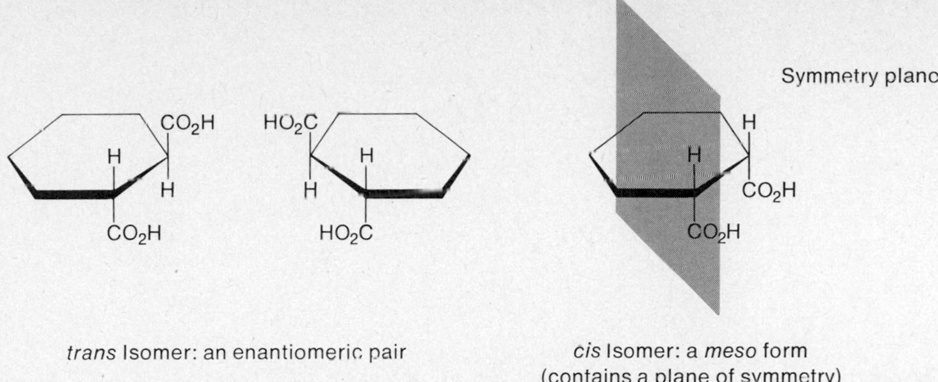

Fig. 19.38. Planar representation of cyclohexane ring.

is often convenient to depict the cyclohexane ring as planar (Fig. 19.38), realizing, of course, that at most moments in time it is not planar.

Conversion of cyclohexane-1,2-dicarboxylic acid to a half-ester produces a compound containing two differently substituted chiral centers, for which four stereoisomers are possible (two enantiomeric pairs). If these groups are moved to the 1,4-positions on the ring, however, a plane of symmetry is present, enantiomerism disappears, and only two *meso* forms (diastereoisomers) remain (Fig. 19.39).

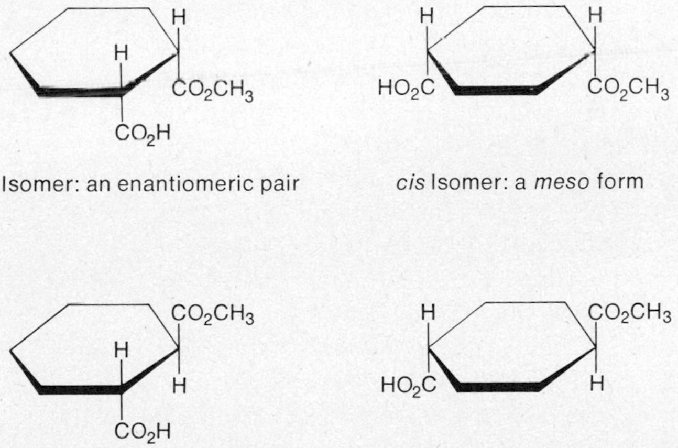

cis Isomer: an enantiomeric pair

cis Isomer: a *meso* form

trans Isomer: an enantiomeric pair

trans Isomer: a *meso* form

Fig. 19.39. Stereoisomers of 1,2- and 1,4-disubstituted cyclohexanes.

The attachment of three or more groups to a cycloalkane ring increases the stereoisomeric possibilities and complicates the nomenclature problem. For example, 3,5-dimethylcyclohexanol possesses a pair of *meso* forms and a pair of enantiomers (Fig. 19.40), and each form can be specified in terms of the (*R*) or

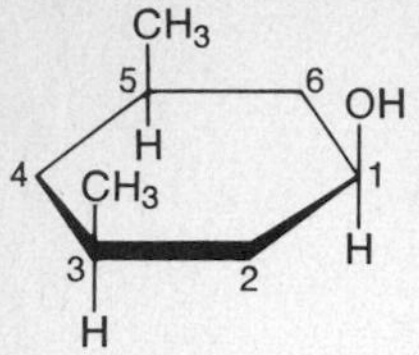

cis-3,*cis*-5-Dimethylcyclohexan-*r*-1-ol
meso form

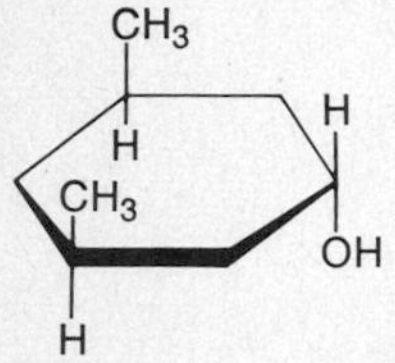

trans-3,*trans*-5-Dimethylcyclohexan-*r*-1-ol
meso form

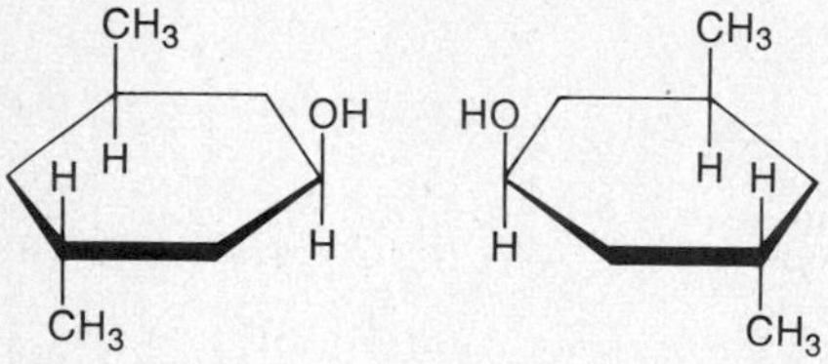

cis-3,*trans*-5-Dimethylcyclohexan-*r*-1-ol
enantiomeric pair

Fig. 19.40. Stereoisomers of 3,5-dimethylcyclohexanol and stereo nomenclature for trisubstituted cyclohexanes.

(*S*) configuration at each of the three centers of chirality.* It is convenient also to have a means of specifying the relation between the substituents in terms analogous to the *cis,trans* prefixes that are used for the disubstituted cycloalkanes. One scheme for doing this retains the *cis,trans* prefixes and specifies a reference group within the molecule (designated by the prefix "r"), chosen by application of the priority sequence rules. For 3,5-dimethylcyclohexanol, the hydroxyl group is the reference group, and the *cis* or *trans* relationship of the methyl groups to the hydroxyl group is specified in the manner indicated in Fig. 19.40.

Cyclic systems containing two or more rings joined in the "fused" or "bridged" fashion may have certain constraints placed on the number of stereoisomers that are possible. In bicyclobutane, for example, the rings can be joined only in the *cis* fashion, because a *trans* junction involves an inordinate amount of strain. Thus, 1,2-dimethylbicyclobutane (Fig. 19.41) can exist in only four stereoisomeric forms despite the fact that it has three chiral centers. A similar constraint is present in the fused-ring systems of bicyclo[2.1.0]pentane and bicyclo[2.2.0]hexane but disappears in bicyclo[3.2.0]heptane and larger systems. For example, 1-methylbicyclo[4.3.0]nonane can exist in all four of the stereoisomeric forms (Fig. 19.42). Small-ring, bridged-ring systems, such

* Carbon-1, containing the OH group, is called a pseudochiral atom in the two meso forms, for its chirality depends on the stereochemical difference of C-3 and C-5 (*i.e.*, C-3 is *S* and C-5 is (*R*)]. To handle the stereochemical nomenclature in this situation, the priority rules state that (*R*) precedes (*S*). Thus, the *cis,cis* compound has an (*S*) configuration at C-1, and the *trans,trans* compound has an (*R*) configuration. In the *cis-trans* isomer C-1 is achiral, because it now contains two identical groups [*i.e.*, C-3 and C-5 are both (*R*)].

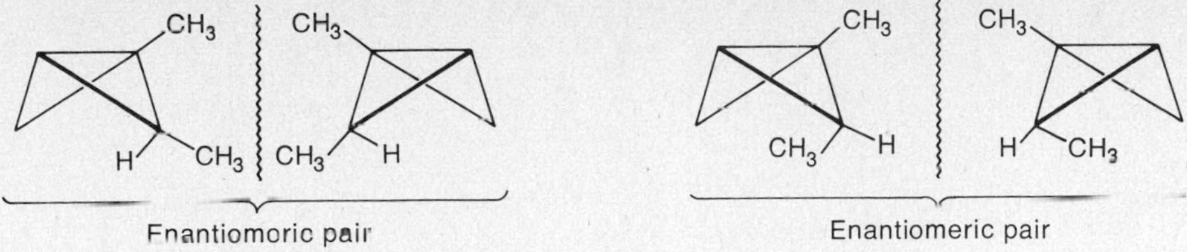

Fig. 19.41. Stereoisomers of 1,2-dimethyl[1.1.0]bicyclobutane.

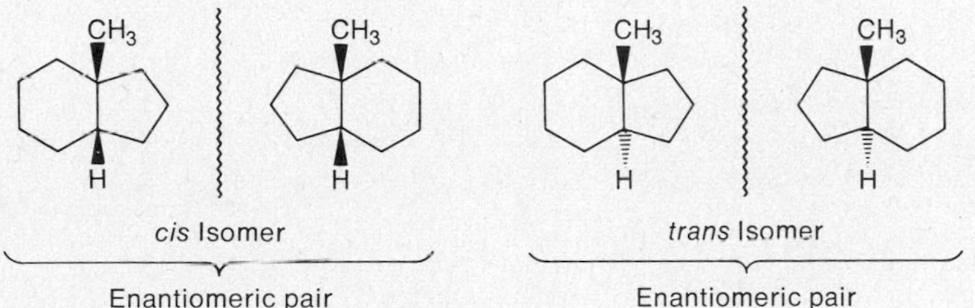

Fig. 19.42. Stereoisomers of 1-methylbicyclo[4.3.0]nonane.

as the highly strained compound bicyclo[1.1.1]pentane as well as the considerably less strained compound bicyclo[2.2.2]octane, are known only in the *cis* fused form. However, as mentioned previously (see p. 84), the more flexible bicyclo-[4.4.1]undecane is known in both the *cis* and *trans* forms (Fig. 19.43).

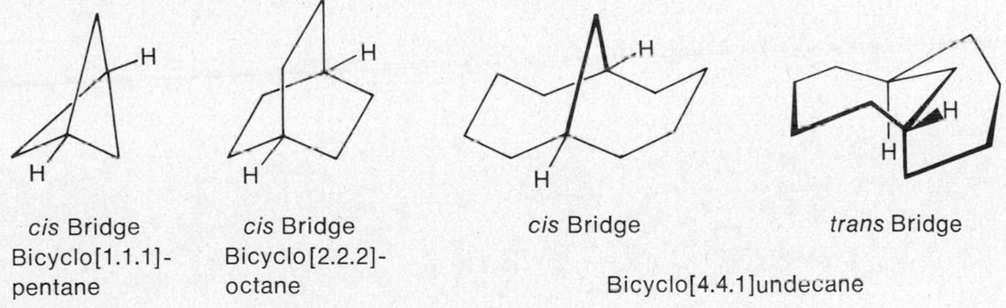

Fig. 19.43. Steric constraints in bridged-ring systems.

19.8 *Compounds Containing a Chiral Axis*

All the chiral compounds discussed in Sections 19.3–19.7 contain one or more chiral carbons and have their chirality specified at one or more *centers.* In some compounds possessing nonsuperimposable mirror images, it is not possible to identify a chiral center, and it is necessary to focus on a larger portion of the molecule. For example, 2,3-pentadiene (dimethylallene) has a pair of nonsuperimposable forms (Fig. 19.44) but does not contain a chiral center. It contains, instead, a *chiral axis* that includes the allene system, C=C=C. Because of the geometry of the 1,2-diene system, the groups attached to the termini of the allene occupy mutually perpendicular planes. If each terminus carries a pair of unlike groups (although the sets may be identical, as in the case of 2,3-pentadiene), the result is a compound capable of existing in enantiomeric forms.

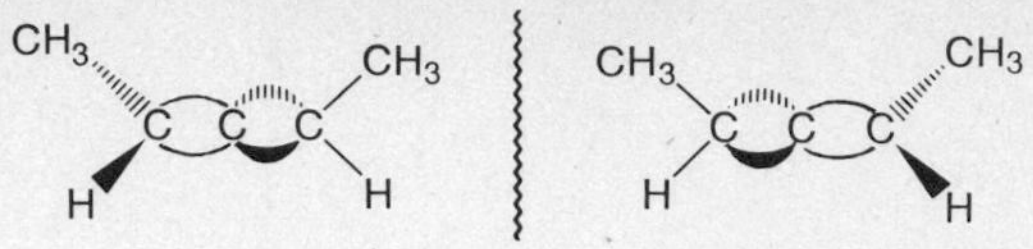

Fig. 19.44. Enantiomeric forms of 2,3-pentadiene.

This possibility was predicted by van't Hoff and LeBel in their original formulation of the theory of the tetrahedral carbon atom, for it follows logically from this concept. It was not until 61 years later, however, that the first allene was actually resolved and obtained in optically active forms. Other frameworks capable of holding unlike pairs of groups in mutually perpendicular planes have stereoisomeric possibilities comparable to those of allenes. For example, 2,6-dimethylspiro[3.3]heptane and 1-ethylidene-3-methylcyclobutane exist in nonsuperimposable mirror image forms and contain chiral axes (Fig. 19.45).

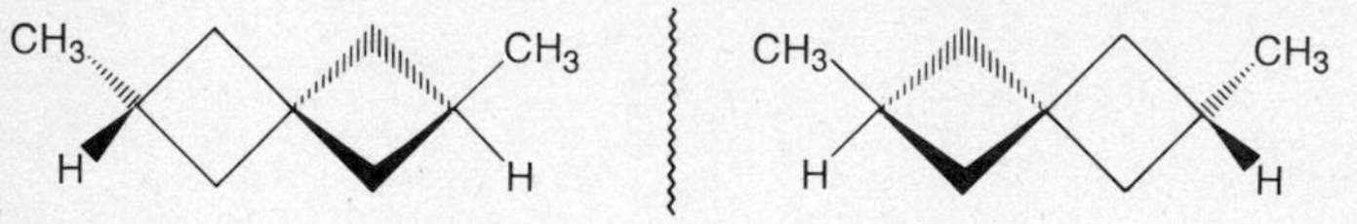

2,6-Dimethylspiro[3,3]heptane

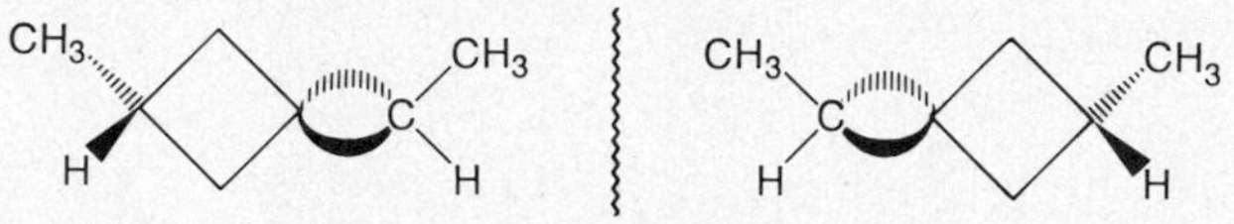

1-Ethylidene-3-methyl cyclobutane

Fig. 19.45. Enantiomers of compounds having a chiral axis.

PROBLEM 19-9

Draw all the possible stereoisomers for 1,3,4-hexatriene, 2,3,4-hexatriene, and 1,3,5-hexatriene.

The reader with a shrewd eye for symmetry may have realized that 2,3-pentadiene and 2,6-dimethylspiro[3.3]heptane both have symmetry axes passing through the center of the system. These axes are not easy to discern, however, for they are located at a 45° angle to the planes described by the terminal groups, as indicated by the "end-on" projection representations in Fig. 19.46.

The existence of enantiomeric pairs in allenes and spirans is most readily perceived when the chiral axis is a rigid framework, as in the compounds depicted in Figs. 19.44 and 19.45. For isomerism of this type to be possible, however, it is *not* necessary that the groups attached to the chiral axis be in planes that are mutually perpendicular; it is sufficient that the system simply be able to prevent a plane of symmetry from being acquired. Thus, flexible systems such as spiro

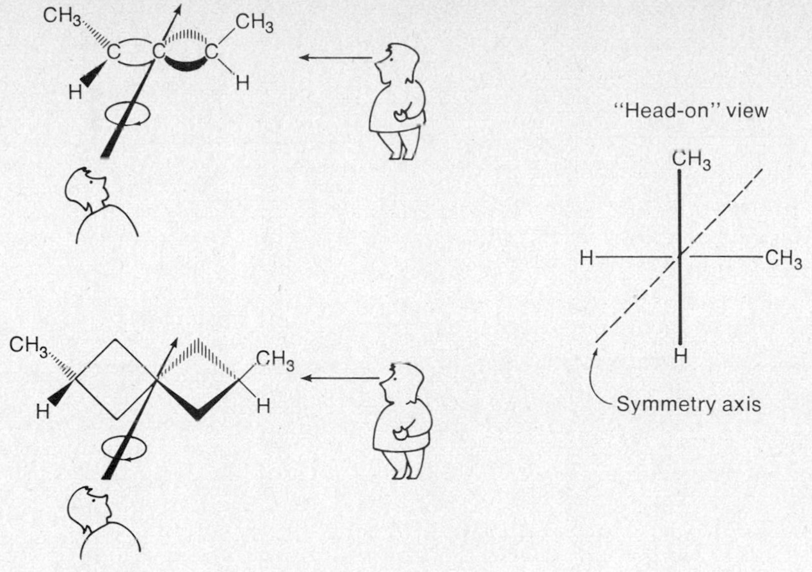

Fig. 19.46. Two-fold rotation axis of symmetry in 2,3-pentadiene and 2,6-dimethylspiro[3,3]heptane.

compounds of the type shown in Fig. 19.47 can have nonsuperimposable mirror image pairs even when n and m are greater than two.

The compounds pictured in Figs. 19.44–19.47 are "inherently chiral," and the possibility of enantiomeric forms was predicted from the tenets of the tetrahedral carbon theory. Forces other than direct bonding forces may also

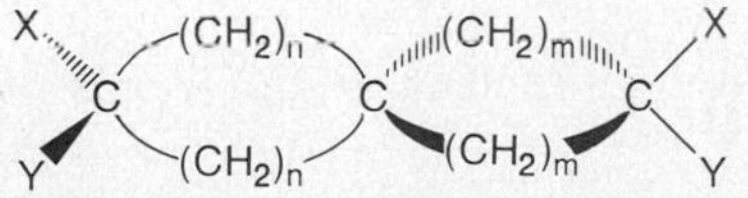

Fig. 19.47. Flexible spiro systems (n and $m > 2$).

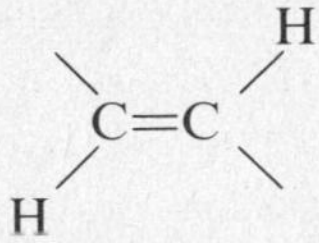

put constraints on molecules and give rise to the possibility of nonsuperimposable mirror images. The best examples of this phenomenon are observed with *trans*-cyclooctene and certain substituted biphenyls, where, because rotation around the bonds in the systems is severely restricted, the molecules are capable of existing in nonsuperimposable mirror image conformations. In *trans*-cyclooctene, interconversion of one enantiomer to the other requires a double rotation process, *viz*., a rotation of the methylene groups at C-5 and C-6 "through the ring" followed by a similar rotation of the H/C=C/H portion of the molecule—a process that is more easily vizualized by use of molecular models. In *trans*-cyclooctene, the energy barrier for this process is sufficiently great that the rotation is slow enough to permit preparation and characterization of optically active *trans*-cyclooctene at room temperature. Larger-ring cycloalkenes are also capable of existing in enantiomeric conformations, but the barriers to rotation are considerably lower (*i.e.*, the "inside holes" of the rings are larger) and the pure enantiomers have only a short lifetime at room temperature.

In biphenyl compounds, the two phenyl rings may be restricted from being coplanar if the steric interactions between groups attached to the 2-, 2′-, 6-, and 6′- positions are sufficiently large. If both rings are unsymmetrically substituted, the compound contains a chiral axis and enantiomers can exist. Whether the rotation around the bond between the phenyl groups will be sufficiently hindered to restrict the freedom of rotation and allow resolution of the enantiomers and measurement of optical activity depends on the sizes of the functions attached to the *ortho* positions. This is something that cannot be predicted without recourse to experiment and which, therefore, was *not* anticipated by van't Hoff and LeBel in 1874. Little did they realize, for example, that 2,2′-dicarboxy-6,6′-dinitrobiphenyl would be resolvable into a pair of enantiomers (Fig. 19.48). Nor could

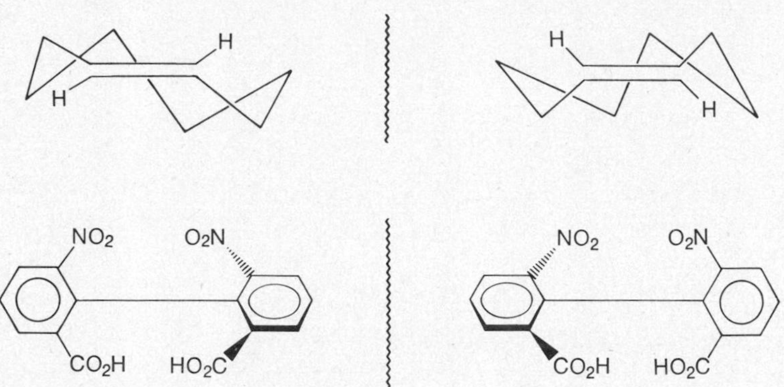

Fig. 19.48. Enantiomers of *trans*-cyclooctene and 2,2′-dicarboxy-6,6′-dinitrobiphenyl.

Hindered biphenyl

it be predicted without recourse to experiment that 2,2′-dinitrobiphenyl *cannot* be resolved at room temperature. The crucial point in these situations is the degree to which the functions at the *ortho* positions interfere with each other and keep the phenyl rings from achieving coplanarity. When nitro and carboxyl functions occupy the *ortho* positions, the interference is sufficient to prevent the rings from easily becoming coplanar, but when nitro and hydrogen functions occupy the *ortho* positions, the interference is reduced to the point where rapid rotation around the bond between the phenyl groups takes place at room temperature. A condition reminiscent of that in *meso*-tartaric acid is then present; *i.e.*, although at any given moment in time chiral conformations may be present, the system can pass through an achiral planar conformation with relative ease, and the mixture is a rapidly equilibrating collection of equal numbers of enantiomeric conformers. Whether an unsymmetrically substituted biphenyl is capable of resolution, therefore, depends on the sizes of the *ortho* groups, and the critical size can be determined only by experiment. It turns out that if the sum of the radii of the interfering groups at the *ortho* positions is greater than about 2.90 Å, resolution into optically stable enantiomers is possible, but resolution is not possible if the sum is less than this value. Typical group radii along with group interaction data are given in Table 19.1.

PROBLEM 19-10

Which of the following biphenyl-type compounds would you predict to be resolvable into enantiomeric forms at room temperature?

CH_3 O_2N NO_2 CH_3 CH_3 CH_3 H F F H CH_3 N CH_3 O_2N NO_2 N

Table 19.1 INTERACTIONS BETWEEN GROUPS IN THE *o,o′* POSITIONS OF BIPHENYL

Group Radii, Å		Interactions	Sum of Radii, Å	Difference from 2.90 Å	Resolvable
H	0.94	H,H	1.88	−1.02	No
CO_2H	1.56	H,NO_2	2.86	−0.04	No
CH_3	1.73	F,CO_2H	2.95	0.05	Yes (but easily racemized)
F	1.39	H,Br	3.05	0.15	Yes
Cl	1.89	NO_2,CO_2H	3.48	0.38	Yes
Br	2.11				
I	2.20				
OH	1.45				
NH_2	1.56				
NO_2	1.92				

The requirements for a biphenyl to be resolvable are (a) sufficiently large groups in the *ortho* positions that free rotation around the central bond is slow (on the laboratory time scale) and (b) unsymmetrical substitution in *both* rings of the biphenyl system. For instance, 2,6-dicarboxy-2′,6′-dinitrobiphenyl satisfies requirement (a) but not requirement (b); although the rotation is sufficiently hindered, the molecule is not chiral, because the rings are not unsymmetrically substituted. This system could be altered to satisfy requirement (b) by adding a methyl group to the 3 and 3′ positions of the rings (Fig. 19.49).

The possibility for chirality in so simple a molecule as *n*-butane was mentioned in Section 3.7 (see Fig. 3.10), and it is clear from the present discussion that the synclinal form of *n*-butane possesses a chiral axis. The rapid interconversion

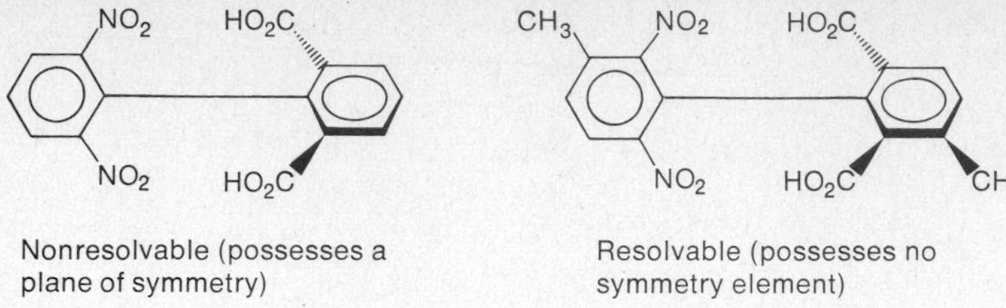

Nonresolvable (possesses a plane of symmetry)

Resolvable (possesses no symmetry element)

Fig. 19.49. Resolvable and nonresolvable biphenyls.

among the chiral synclinal conformations, the achiral synperiplanar conformation (plane of symmetry), and the achiral antiperiplanar conformation (center of symmetry), however, establishes equal numbers of enantiomeric chiral conformations and precludes the isolation of optically stable isomers (Fig. 19.50).

Antiperiplanar conformer
Achiral—*meso* form

Synclinal conformer   Synclinal conformer
Chiral—enantiomeric forms

Synperiplanar conformer
Achiral—*meso* form

Fig. 19.50. Chiral and achiral forms of *n*-butane.

19.9 Compounds Containing a Chiral Plane

A few molecules are known in which neither a center nor an axis of chirality can be specified but which are, nevertheless, capable of existing in nonsuperimposable mirror image structures. Such compounds are said to contain *planes of chirality* and are exemplified by hexahelicene (Fig. 19.51). A molecular model of hexahelicene* shows that the terminal benzene rings cannot occupy the same plane

Hexahelicene

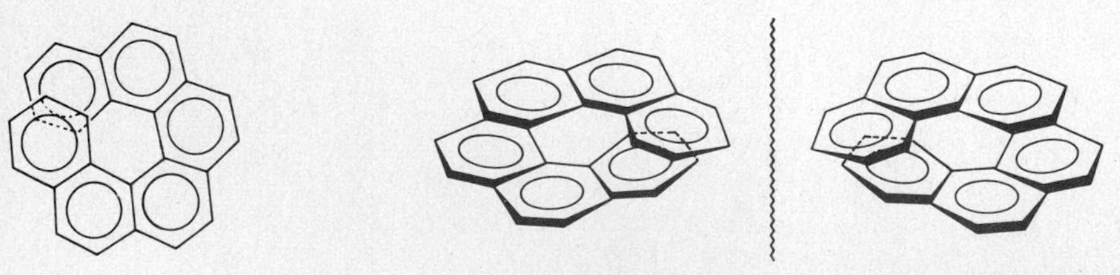

Nonsuperimposable mirror images

Fig. 19.51. Enantiomeric forms of hexahelicene.

* The difference between hexahelicene and coronene (see Fig. 4.21) should be recognized. In coronene *every* benzene ring is attached to *two* other benzene rings, and the molecule is a circle. In hexahelicene, the terminal rings are connected to only one benzene ring, and the molecule is not a circle but a helix.

without interfering with one another. Consequently, the molecule is forced to become nonplanar, and two nonsuperimposable mirror image forms are possible—one with a right-handed helix and the other with a left-handed helix. A plane of chirality is present, in which the middle rings (*i.e.*, 3 and 4) lie in the plane, and the terminal rings (*i.e.*, 1 and 6) fall above and below the plane, respectively. The resolution of hexahelicene has been achieved, and the enantiomers are extraordinarily optically active, possessing specific rotations of about 3,700°.

PROBLEM 19-11

Referring to the formula on p. 506 for calculating specific rotations, what concentration of hexahelicene would be required to give the observed rotation of 45° in a 1-dm tube? If a concentration 10 times greater were used, what would be the observed rotation and the apparent specific rotation? If a concentration 10 times smaller were used, what would be the observed and specific rotations? What does this suggest with respect to the technique of establishing a specific rotation for an optically active compound?

19.10 Stereoisomeric Relationships of Groups in Molecules

Equivalent Atoms, Groups, and Faces. Atoms, groups, or faces* are defined as equivalent if they can be interchanged by rotation around a simple axis to give a structure indistinguishable from the original. In acetic acid, for example, the three hydrogens of the methyl group are equivalent, for a 120° rotation around the C—CO_2H axis produces a compound indistinguishable from the original. Similarly, the methyl groups as well as the two faces of the carbonyl group in acetone are equivalent, as demonstrated by a 180° rotation around the axis of the C=O bond (Fig. 19.52).

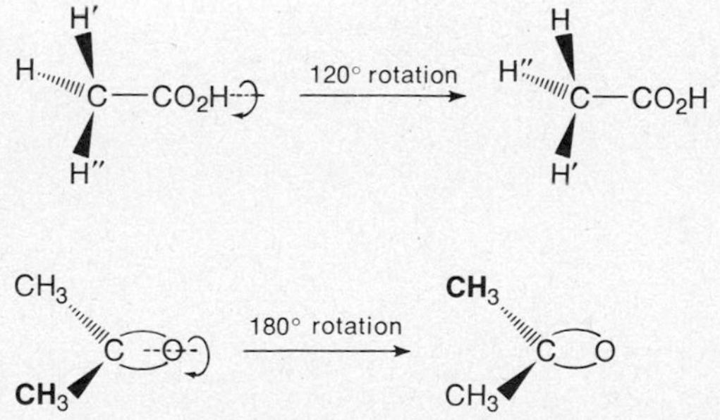

Fig. 19.52. Equivalent relationships in acetic acid and acetone.

---

* The term "faces" refers to "sides" of a multiple bond or a ring. For example, the carbonyl group and the cyclobutane ring oriented as shown can be said to have an upper face and a lower face.

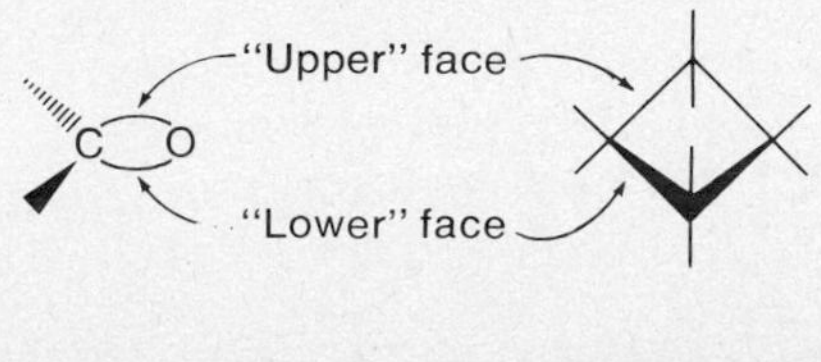

A simple test for equivalency involves the substitution of the atoms, groups, or faces in question by a test group, G; if the resulting structures are indistinguishable, the atoms, groups, or faces are equivalent. For example, substitution of either of the methyl groups of acetone by the test group G gives the identical compound (Fig. 19.53).

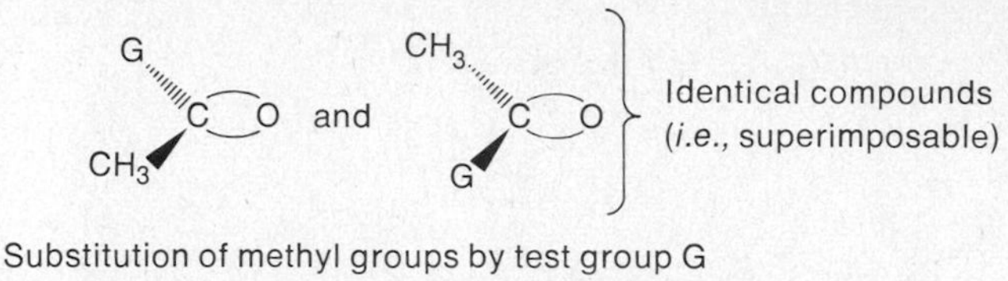

Identical compounds (*i.e.*, superimposable)

Substitution of methyl groups by test group G

Fig. 19.53. Substitution test for equivalent relationships.

PROBLEM 19-12

Which hydrogens in cyclobutanone are equivalent and which are not?

Enantiotopic Atoms, Groups, and Faces. Atoms, groups, and faces are defined as enantiotopic if they can be interchanged only by a rotation-reflection operation (*i.e.*, an alternating axis of symmetry operation) to give a structure indistinguishable from the original. The majority of compounds in which enantiotopic relationships exist have a plane of symmetry. For example, in hydroxyacetic acid (glycolic acid), a plane passing through the hydroxyl group and the OH of the carboxyl function, through the two carbon atoms, and through the carbonyl oxygen divides the molecule into mirror image halves in which the two hydrogens of the hydroxymethyl group reflect each other (Fig. 19.54).

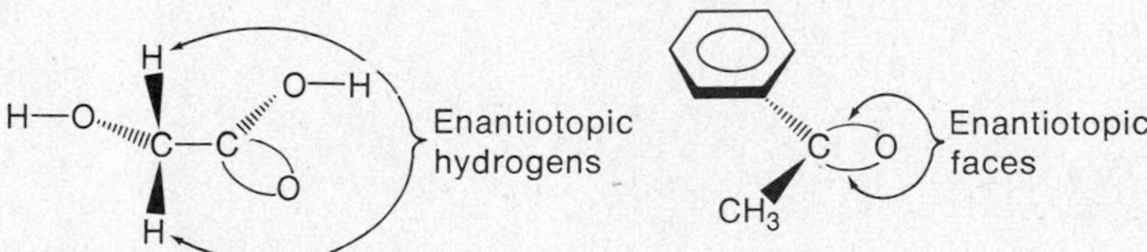

Fig. 19.54. Enantiotopic relationships between atoms in hydroxyacetic acid (lactic acid) and between faces in acetophenone.

Application of the substitution test for enantiotopic relationships can be carried out as described for equivalent relationships. If the structures that result from the substitution by the test group G are enantiomers the groups are enantiotopic, as illustrated for the hydrogens in hydroxyacetic acid and the carbonyl group faces in acetophenone (Fig. 19.55).

PROBLEM 19-13

When cyclobutanone is treated with perbenzoic acid it yields γ-butyrolactone (see Fig. 15.21 for structure). Which hydrogens in γ-butyrolactone are enantiopically related? Which are equivalently related? Which are neither of these?

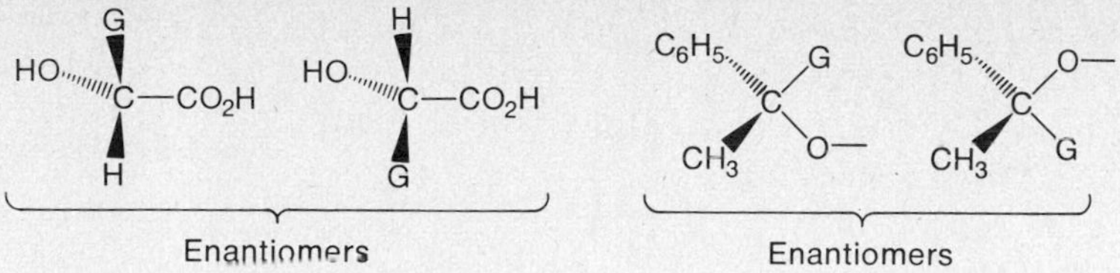

Enantiomers Enantiomers

Fig. 19.55. Substitution test for enantiotopic relationships in hydroxyacetic acid (lactic acid) and in acetophenone.

Diastereotopic Atoms, Groups, and Faces. Atoms, groups, and faces are defined as diastereotopic if they have the same structure but cannot be interchanged by any symmetry operation. In instances where the diastereotopic relationship is not immediately obvious, it can be revealed by the substitution test. If the structures that result from the substitution by the test group G are diastereomers, the relationship between them is diastereotopic—as indicated by the substitution test which produces a pair of diastereomers (Fig. 19.56).

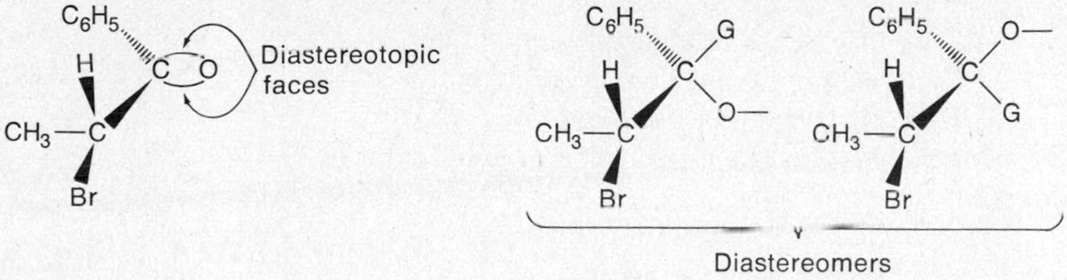

Diastereotopic faces

Diastereomers

Fig. 19.56. Diastereotopic relationships in α-bromopropiophenone.

PROBLEM 19-14

What is the relation between the two faces of the carbonyl group in γ-butyrolactone? What is the relation between the two faces of the carbonyl group in α-methyl-γ-butyrolactone?

Prochirality. If the replacement of one of a pair of enantiotopically or diastereotopically related groups by a different group creates a chiral center, the original center is said to be prochiral. For example, the hydrogens of the methylene group in ethanol are enantiotopic; replacement of one of them by deuterium creates a chiral center and yields (*R*)- or (*S*)-deuterioethanol, depending on which of the hydrogens is replaced (Fig. 19.57). Thus, the methylene group of ethanol is said to be *prochiral.*

PROBLEM 19-15

Which carbon atoms, if any, are prochiral in 1-butanol?

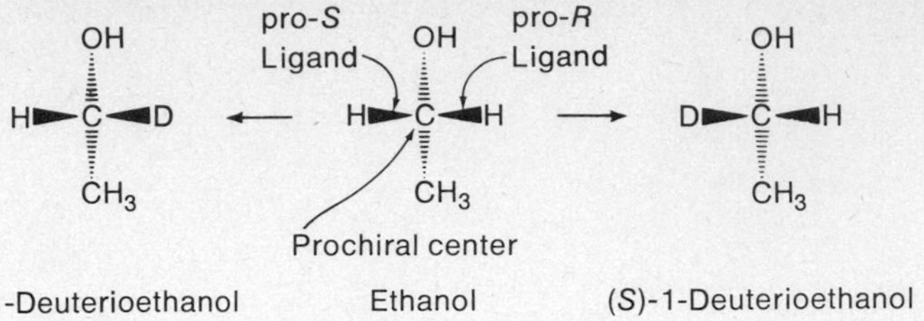

(*R*)-1-Deuterioethanol   Ethanol   (*S*)-1-Deuterioethanol

Fig. 19.57. Prochirality of ethanol; conversion to (*R*)- and (*S*)-1-deutereoethanol.

Chiral Test Groups. If *achiral* test groups (*e.g.*, indicated by "G" in the preceding examples) are employed, equivalently related entities (*i.e.*, atoms, groups, or faces) produce no stereoisomers, enantiotopically related entities produce enantiomers, and diastereotopically related entities produce diastereomers (see previous discussion). If *chiral* test groups are used, this situation remains unchanged with respect to equivalently related and diastereotopically related entities. For enantiotopically related entities, however, chiral test groups lead to diastereomers rather than to enantiomers. For example, attachment of a chiral test group (designated as $G^{(R)}$, the (*R*) specifying that it has (*R*) chirality) to the two enantiotopically related faces of acetophenone produces a pair of diastereomers. Compound (a) in Fig. 19.58 is *not* the mirror image of compound (b); the mirror image of compound (a) is the compound in which *both* chiral centers have the opposite configuration [*i.e.*, compound (c)].

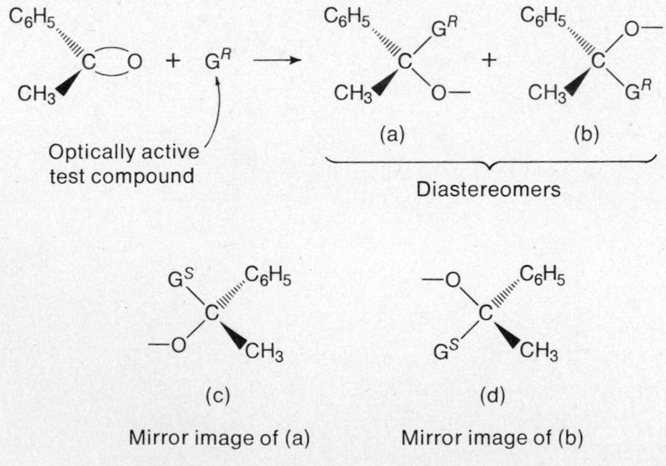

Fig. 19.58. Substitution test with chiral test reagent and enantiotopically related faces.

19.11 *The Resolution of Mixtures of Enantiomers (Racemates) into Optically Active Compounds*

The separation of an optically inactive enantiomeric pair into its constituent optically active enantiomers is called "resolution," and a chiral agent of some sort is required in the process. In the majority of instances, the chiral agent is another compound that is already optically active. For example, when optically active (+)-1-phenylethylamine reacts with optically inactive (±)-lactic acid, a

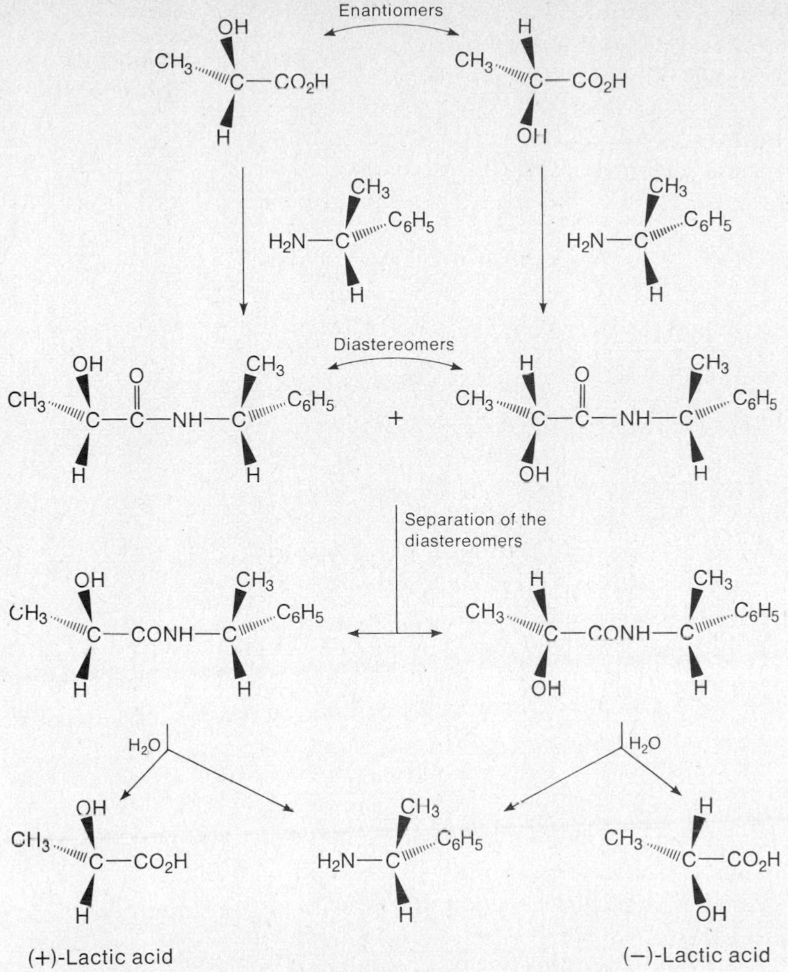

Enantiomers

Diastereomers

Separation of the diastereomers

(+)-Lactic acid

(−)-Lactic acid

Fig. 19.59. Resolution of (±)-lactic acid.

pair of diastereomeric amides is produced (Fig. 19.59). Since diastereomers differ in physical and chemical properties, separation can be effected by conventional means. Thus, fractional crystallization of the mixture yields one diastereomer in flask *A* and the other in flask *B*. Hydrolysis of the amide in flask *A* then produces a mixture of (+)-lactic acid and the (+)-amine, and hydrolysis of the amide in flask *B* produces a mixture of (−)-lactic acid and (+)-amine. Separation of the optically active lactic acids from the amine in each of the flasks can be effected by precipitation of the acid as the barium salt, filtration, and regeneration of the free acid by treatment with sulfuric acid. Through the agency of the chiral reagent (+)-1-phenylethylamine, then, (±)-lactic acid can be resolved into its optically active enantiomers, the resolving agent being recoverable in its original form.

The sequence of events outlined for the resolution of (±)-lactic acid is representative of the manner in which the majority of resolutions are carried out.

It is based on the production of diastereomers and is reminiscent of the interaction of chiral compounds with molecules containing enantiotopically related groups (see Fig. 19.58). The only difference between the two systems is that in the examples discussed in Section 19.10, the relationships were enantiotopic by *internal* comparison, whereas in the case of enantiomers the relationship is enantiotopic by *external* comparison. The method depends on finding a chiral reagent that can attach itself in some fashion to the enantiomeric pair to yield a mixture of diastereomers; *i.e.*,

$$(Z)^{(+)} + (E)^{(+)}(E)^{(-)} \longrightarrow (Z)^{(+)}(E)^{(+)} + (Z)^{(+)}(E)^{(-)}$$

where $Z^{(+)}$ is the optically active resolving agent and $E^{(+)}E^{(-)}$ is the enantiomeric pair to be resolved; (+) and (−) specify the dextrorotatory and levorotatory enantiomers, respectively. Separation by chemical and/or physical methods of the pair of diastereomers is then effected (*i.e.*, $Z^{(+)}E^{(+)}$ in one flask and $Z^{(+)}E^{(-)}$ in another), and the separated diastereomers are decomposed to the original constituents (*i.e.*, $(Z)^{(+)}(E)^{(+)} \rightarrow (Z)^{(+)} + (E)^{(+)}$ in one flask and $(Z)^{(+)}(E)^{(-)} \rightarrow (Z)^{(+)} + (E)^{(-)}$ in the other). The problem faced by the laboratory chemist attempting to effect a resolution is the choice of the proper resolving agent. A good reagent (a) must be able to attach itself to the compound to be resolved; (b) should give a pair of diastereomers that differ sufficiently in properties to be readily separated, preferably by crystallization; and (c) should be easily removed after the diastereomers have been separated. The choice of the resolving agent is determined, of course, by the functional groups present in the compound to be resolved; *i.e.*, the "handles" it possesses. The "handle" in the case of lactic acid is the carboxyl group, which allows attachment to amines (via amide formation or salt formation) or to alcohols (via ester formation). Numerous optically active amines and alcohols are available (many of natural origin), and they provide useful resolving agents for many acids. Conversely, optically active acids can be used to effect the resolution of amines or alcohols. A technique that is often employed for the resolution of alcohols involves the reaction with phthalic anhydride to form the half-ester, followed by resolution via an optically active amine (Fig. 19.60).

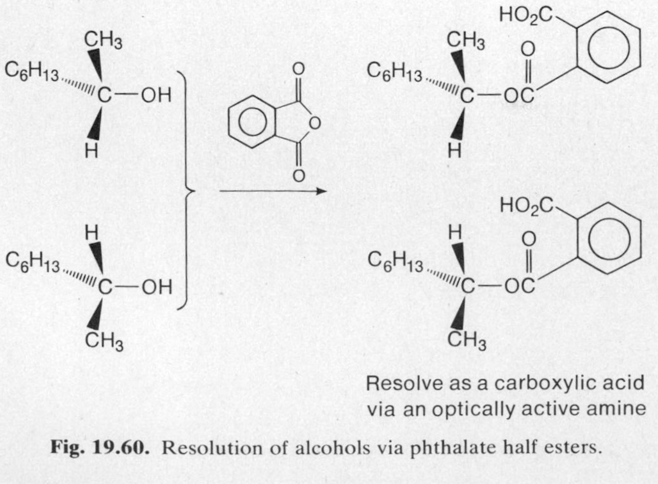

Fig. 19.60. Resolution of alcohols via phthalate half esters.

Various other modes of attachment of the resolving agent to the compound to be resolved are possible, the only requirement being that the "bond" be strong enough to survive the means used for separating the diastereomers. It need not even be a covalent linkage, a good case in point being the means by which hexahelicene has been resolved. A "molecular complex" was formed between an optically active, electron poor π system afforded by a tetranitro compound prepared from fluorenone (see Fig. 13.3 for the structure of fluorenone) and the electron rich π system present in hexahelicene. The pair of diastereomeric molecular complexes was separated by fractional crystallization (Fig. 19.61).

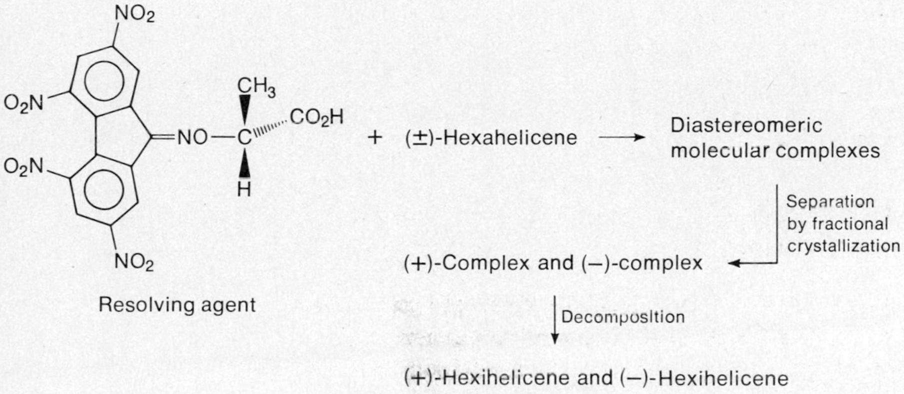

NO_2
O_2N
O_2N
NO_2
CH_3
$=NO-C$ CO_2H
H
Resolving agent

+ (±)-Hexahelicene → Diastereomeric molecular complexes

Separation by fractional crystallization

(+)-Complex and (−)-complex

Decomposition

(+)-Hexihelicene and (−)-Hexihelicene

Fig. 19.61. Resolution of hexahelicene via complex formation.

PROBLEM 19-16

Let us suppose that you have spent the last three years of your life devising and carrying out the synthesis of a natural product, which is a tertiary amine compound. Suggest a method by means of which you might resolve this racemic product into its enantiomers so that a direct comparison with the optically active natural material might be made.

Among the most highly chiral compounds known are the enzymes, which are substances having remarkable abilities for catalyzing the reactions that take place in living systems (see Section 33.6). Sometimes enzymes can be employed as resolving agents, and, where applicable, they lead to unusually clean and effective resolutions. For example, the laboratory preparation of phenylalanine,

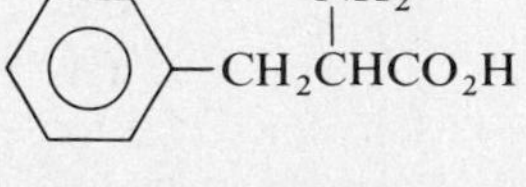

NH_2
$C_6H_5-CH_2CHCO_2H$

yields an optically inactive enantiomeric mixture. If the mixture is converted to the N-acetyl derivative and the enzyme known as "acylase" is then allowed to act on it, only one enantiomer undergoes hydrolysis. The result is a mixture of

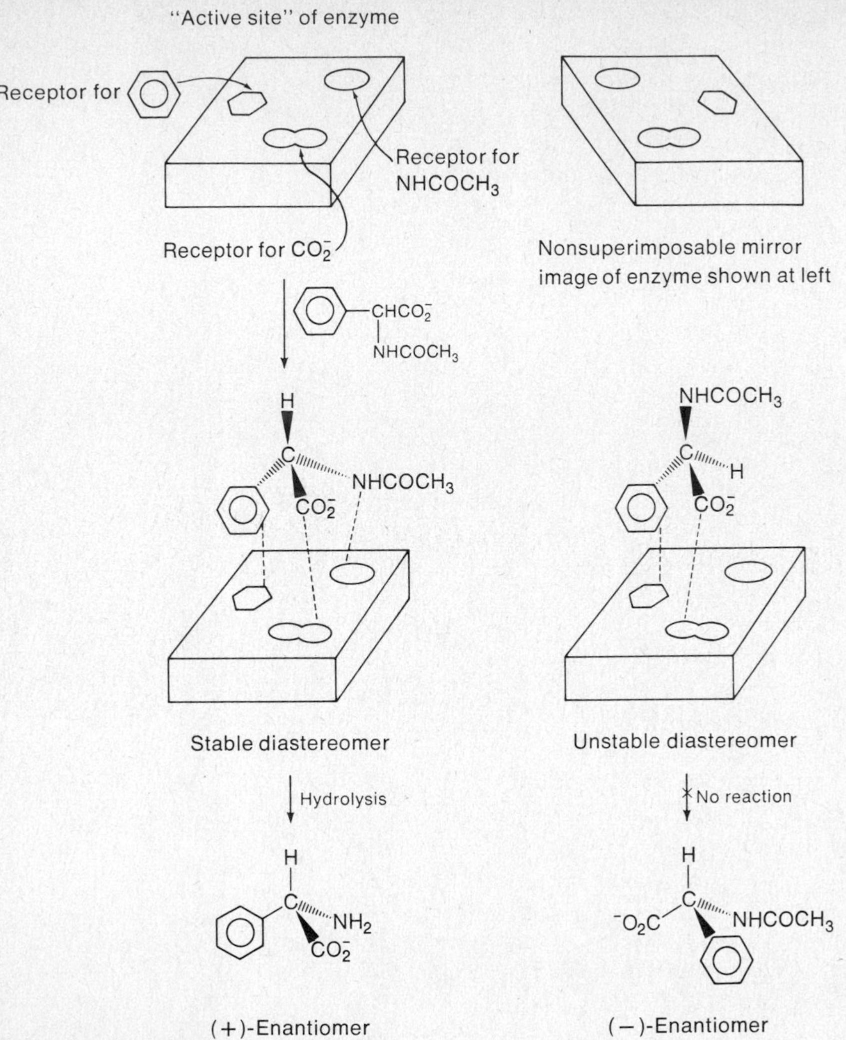

Fig. 19.62. Diastereomeric enzyme-substrate complex formation.

(+)-phenylalanine and (−)-N-acetylphenylalanine, compounds that are structurally different and easily separated (*e.g.*, by extraction of the free amine into aqueous acid). The phenylalanine that is obtained is virtually 100% optically pure.*

The intimate details of the mechanism of enzyme action are only now becoming understood in a few cases, but it is quite certain that a substrate† such as

* Optical purity is defined in terms of the excess of one enantiomer over the other. It is expressed as a percentage of the total and ranges from 0 (for a racemate) to 100 (for a pure enantiomer). A mixture containing 70% of enantiomer (a) and 30% of enantiomer (b), for example, has an optical purity [in terms of enantiomer (a)] of 70–30/100 = 40%.

† The term "substrate" is frequently used, particularly in reactions involving enzymes, to denote the reactants.

(±)-N-acetyl phenylglycine interacts with the optically active, chiral enzyme to form a diastereomeric pair of entities, as schematically illustrated in Fig. 19.62. Considerable evidence has accumulated to suggest that in many instances there is at least a three-point attachment between the substrate and the enzyme and that, as a consequence, a very large difference in the stability between the two diastereomeric complexes can prevail. If complex formation is a necessary antecedent to catalytic action, only the enantiomer that forms a stable complex with the enzyme will undergo chemical alteration; *e.g.*, hydrolysis of the N-acetyl bond in the specific example under discussion.

Curiously, the first resolution ever effected in the laboratory involved none of these procedures. In 1848, Louis Pasteur, fresh from his Ph.D. degree at the École Normale in Paris, was engaged in a crystallographic study of two compounds called tartaric acid and racemic acid, which were alleged to be identical but had different effects on plane-polarized light. In the course of his study, Pasteur noticed that the sodium ammonium salt of racemic acid (optically inactive) was present in two different crystalline forms, bearing a nonsuperimposable mirror image relationship to each other. Laboriously separating the left-handed from the right-handed crystals of racemic acid under a microscope, he discovered that solutions made from the separated crystals were optically active. One solution rotated plane-polarized light to the right in a fashion identical with that of tartaric acid, and the other rotated plane-polarized light to the left. With remarkable insight, Pasteur realized that the difference in geometry between the two crystal forms must reflect a mirror image relationship between the molecules themselves, and as early as 1860 he formulated theories concerning the asymmetric carbon atom. Pasteur's efforts extended even further into the realm of stereochemistry, for in addition to the method of hand picking of crystals as a means of resolution, he also discovered other methods of resolution, including the method of diastereomers (see pp. 534–536) and the biochemical method (see p. 537). Although better known to the general public for his contributions to medicine, Louis Pasteur must certainly be recognized as one of the geniuses of chemistry and the founder of modern stereochemistry.

Louis Pasteur
(1822–1895)

For the resolution of an enantiomeric pair to be possible, it is necessary, as we have stressed, that a chiral reagent be involved. Where, one might ask, is the chiral agent in Pasteur's resolution of racemic tartaric acid by the hand separation of crystals? The answer is that it is man himself. Man is a highly chiral entity, composed of thousands of different compounds, almost all of which are optically active. If we were racemic rather than chiral we would not, presumably, be able to distinguish right-handedness from left-handedness or be able to differentiate between the crystalline forms pictured in Fig. 19.63.

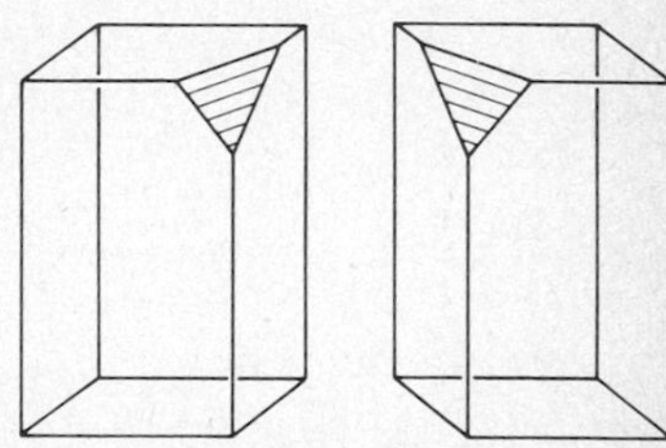

Fig. 19.63. Nonsuperimposable mirror image pair of crystals.

19.12 *Chiral Syntheses*

Chiral syntheses (referred to in the older terminology as asymmetric syntheses) are divided into two types: (a) partial chiral syntheses (also called relative chiral syntheses) and (b) absolute chiral syntheses. Partial chiral syntheses simply involve chemical reactions that take place with compounds that are already optically active; *i.e.*, reactions of optically active compounds containing atoms, groups, or faces that are diastereotopically related and, therefore, produce diastereomeric products. For example, sodium borohydride reduction of the ketoamide obtained by the reaction of (+)-1-phenylethylamine and pyruvic acid yields a mixture of diastereomeric alcohols (*cf*. Fig. 19.59), which can be separated by ordinary chemical methods (Fig. 19.64). Furthermore, the two diastereomers

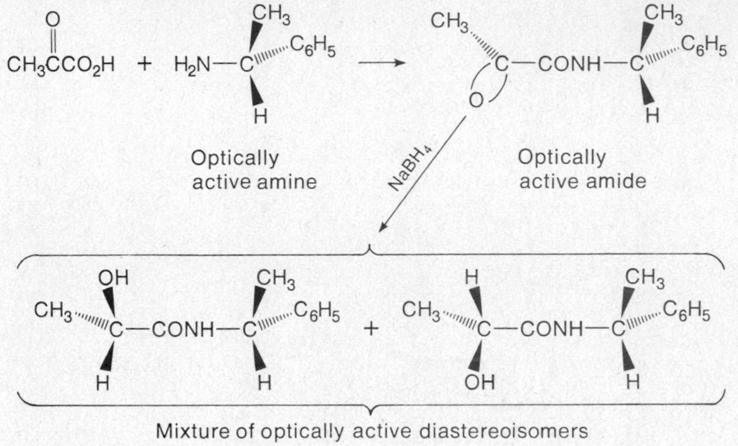

CH_3CCO_2H + $H_2N—C(CH_3)(C_6H_5)H$ ⟶ $CH_3C(O)—CONH—C(CH_3)(C_6H_5)H$

Optically active amine

Optically active amide

$NaBH_4$

$CH_3C(OH)(H)—CONH—C(CH_3)(C_6H_5)H$ + $CH_3C(H)(OH)—CONH—C(CH_3)(C_6H_5)H$

Mixture of optically active diastereoisomers

Fig. 19.64. Partial chiral synthesis.

are not produced in equal amount. The two faces of the carbonyl group are diastereotopically related, and the rate of reaction of reagents (*i.e.*, hydride in this example) is not the same at the two faces. This conversion of the optically active ketoamide to a mixture of optically active diastereomers is an example of a partial chiral synthesis.

Absolute chiral syntheses involve the formation of optically active compounds *without the intervention* of other optically active compounds; *i.e.*, without the aid of the usual resolving agents. One of the most interesting of the few known examples of absolute chiral syntheses involves the light-induced cyclization of 1,2-diarylethylenes to dihydrophenanthrene derivatives (Fig. 19.65) followed by

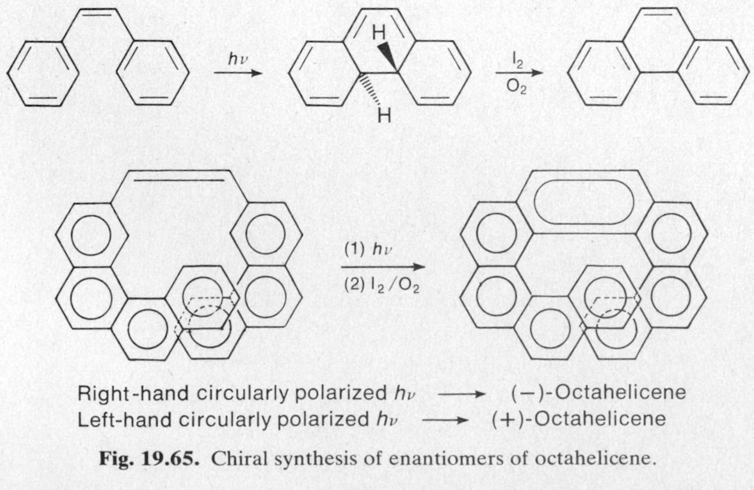

$h\nu$

I_2 / O_2

(1) $h\nu$

(2) I_2/O_2

Right-hand circularly polarized $h\nu$ ⟶ (−)-Octahelicene
Left-hand circularly polarized $h\nu$ ⟶ (+)-Octahelicene

Fig. 19.65. Chiral synthesis of enantiomers of octahelicene.

oxidation (via I_2 and O_2) to the phenanthrene. When the 1,2-diarylethylene shown in Fig. 19.65 is subjected to this sequence of reactions using *circularly polarized light to induce the cyclization*, optically active octahelicene is produced (see chirality of hexahelicene, p. 530); right-hand circularly polarized light yields the (−) enantiomer in 2% optical purity, and left-hand circularly polarized light yields the (+) enantiomer in 2% optical purity. The chiral reagent in this "resolution," of course, is the circularly polarized light, the two forms of which bear a nonsuperimposable mirror image relationship to one another (Fig. 19.12). The circularly polarized light serves the same purpose as an optically active compound in a conventional resolution, "combining" with (+) and (−) enantiomers of the diarylethylene to form a pair of excited states that are diastereomerically related and, therefore, form and decompose at different rates.

19.13 Racemization and Inversion of Configuration

When a chiral compound is synthesized in an achiral environment (*i.e.*, ordinary conditions involving no optically active compounds, no optically active solvents, and no circularly polarized light), the two enantiomers are produced in precisely equal amount, and the compound is optically inactive (a racemic system). It can exist as a simple physical mixture, as a solid solution, or as a complex in which the two enantiomers are held together by secondary forces, such as hydrogen bonds. For example, (+)- and (−)-tartaric acid each melt at 170°, whereas the mixture of equal parts of the (+)- and (−)-enantiomers melts at 206°. The racemic system in this instance is referred to as a *racemic compound*.

The conversion of an optically active enantiomer to an optically inactive mixture containing both enantiomeric forms is referred to as *racemization*. In the majority of instances, racemization is not an easy process, for it requires an achiral intermediate that is usually much higher in energy than the enantiomers. We encounter an exception to this restriction, however, in those compounds in which a carbonyl function is attached to a chiral center that also carries a hydrogen. Optically active 2-methylcyclohexanone, for example, can be readily racemized by the action of a catalytic amount of acid or base. The mechanism of the racemization involves the conversion of the ketone to an enol or enolate anion in which the chirality of the α carbon is destroyed by the double bond; reketonization then yields the (+) and (−) enantiomers in equal amount (Fig. 19.66). This low-energy pathway for inversion of configuration is restricted to a chiral center to which *both a carbonyl group and a hydrogen* are attached.

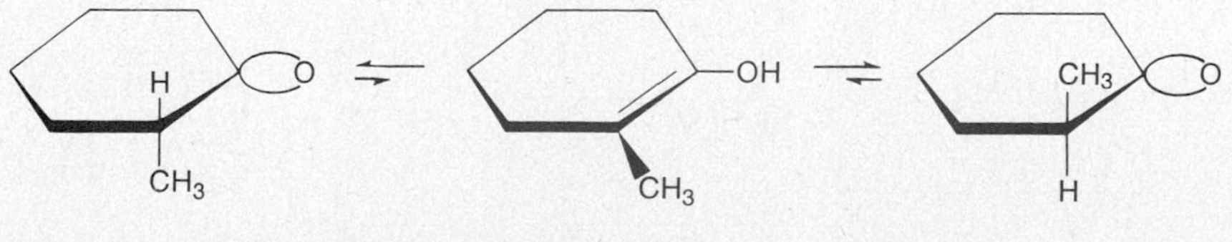

Fig. 19.66. Racemization of 2-methylcyclohexanone.

Inversion of configuration via enolization is not restricted to compounds containing only one chiral carbon. If other chiral carbons are also present in the molecule but are not similarly susceptible to configurational inversion, the enolization process leads not to a racemic mixture but to a mixture of optically active diastereoisomers (if the starting compound is optically active). For example, the action of base on (2*S*,4*R*)-2,4-dimethylcyclohexanone yields a mixture of the (2*S*,4*R*) and (2*R*,4*R*) isomers of 2,4-dimethylcyclohexanone, the result of configurational inversion at C-2 and configurational retention at C-4 (Fig. 19.67). A process in which only one chiral center is changed in a system containing two or more chiral centers is called *epimerization*; diastereomers that are related in having identical configurations at all but one chiral center are called *epimers*. Thus, the two isomers of 2,4-dimethylcyclohexanone shown in Fig. 19.67 are epimers.

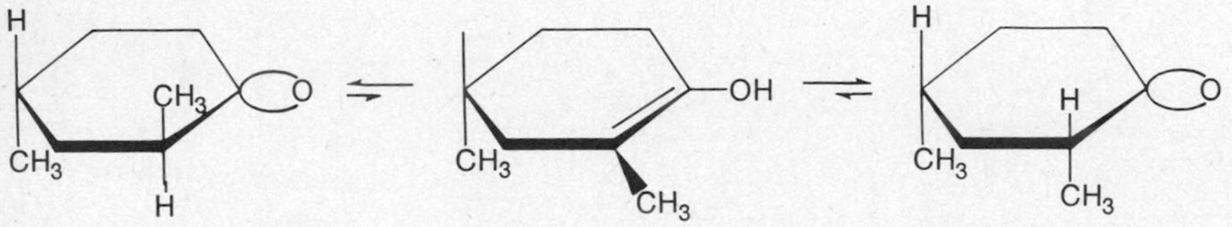

Fig. 19.67. Epimerization of 2,4-dimethylcyclohexanone.

PROBLEM 19-17

An optically active sample of diethyl 2,3,4-trimethylpentanedioate is treated with an ethanolic solution of sodium ethoxide. What stereoisomers are present after this treatment? Are they optically active? Can they be separated by ordinary chemical means? Explain.

19.14 Study Guide

I. Double-bond diastereomerism
   A. Requirements for double-bond diastereomerism
      1. Rigid framework capable of holding groups in two different orientations
      2. Unsymmetrical substitution at both portions of rigid framework
   B. (*Z*),(*E*) Nomenclature system
   C. Interconversion of double-bond diastereomers
   D. Double-bond diastereomerism in polyene systems

II. Symmetry properties of molecules; mirror image isomerism
   A. Symmetry elements
      1. Simple axes of symmetry
      2. Planes of symmetry
      3. Centers of symmetry (twofold alternating axes of symmetry)
      4. Fourfold alternating axes of symmetry
   B. Symmetry classifications
      1. Symmetric molecules—possess a plane, center, or fourfold alternating axis of symmetry
      2. Dissymmetric molecules—contain a simple axis of symmetry but no plane, center, or fourfold alternating axis of symmetry

3. Asymmetric molecules contain no element of symmetry

C. Chiral molecules
   1. Definition—molecules that are not superimposable with their mirror image forms (*i.e.*, asymmetric or dissymmetric molecules)
   2. Elements of chirality
      (a) Center of chirality (synonymous with asymmetric center)
      (b) Axis of chirality
      (c) Plane of chirality

III. Compounds containing one chiral center
   A. Physical and chemical properties of nonsuperimposable mirror images (enantiomers)
      1. Identical properties in an achiral environment
      2. Different properties in a chiral environment
   B. Plane-polarized light
      1. A composite of right-hand and left-hand circularly polarized light
      2. Measurement of optical activity
   C. Formation of chiral molecules from achiral molecules
      1. In an achiral environment, both nonsuperimposable mirror image forms obtained in equal amount
      2. Racemic mixture (racemate)

IV. Configuration
   A. Configuration notation
      1. (*R*),(*S*) notation
         (a) Sequence rule
         (b) Chirality rule
      2. D,L notation
   B. Stereochemical correlations
   C. Absolute configuration
      1. As established by interconversion with a reference compound
      2. As established by X-ray methods

V. Compounds containing two chiral centers
   A. Pictorial representation
      1. Pseudo three-dimensional representations
      2. Fischer projection (plane projection) representations
   B. Diastereomers
      1. Distinction from enantiomers
      2. Chemical properties—different even in an achiral environment
   C. *Meso* forms
   D. *Erythro* and *threo* designations

VI. Compounds containing three or more chiral centers
   A. 2^n Rule applies if all chiral centers are differently substituted
   B. Number of enantiomeric pairs = $2^n/2$

VII. Cyclic systems containing chiral centers
   A. Distinction between *cis,trans* isomerism and optical isomerism in cyclic systems
   B. Stereoisomers in substituted monocyclic carbocyclic compounds

C. Stereoisomers in bicyclic carbocyclic compounds

VIII. Compounds containing a chiral axis
  A. Requirements for existence of enantiomers
    1. Both ends of axis are unsymmetrically substituted.
    2. Groups at end of axis are noncoplanar.
  B. Systems containing chiral axes (if unsymmetrically substituted)
    1. Cumulenes, $RR'(C{=}C)_nCRR'$, where $n = 2, 4, 6, \ldots$
    2. Spirans
    3. Alkylidenecycloalkanes
    4. Biphenyls—noncoplanarity only if *o*-, *o'*- groups large enough
    5. Synclinal conformations of acyclic compounds
    6. *trans*-Cycloalkenes

IX. Compounds containing a chiral plane—hexahelicene

X. Stereoisomeric relationships of groups in molecules
  A. Equivalent atoms, groups, and faces
  B. Enantiotopic atoms, groups, and faces
  C. Diastereotopic atoms, groups, and faces
  D. Prochirality
  E. Chiral test groups

XI. Resolution of enantiomeric mixtures
  A. Conversion to diastereomers and separation by chemical or physical methods
  B. Biochemical method (enzymatic method)
  C. Hand picking of crystals

XII. Chiral synthesis
  A. Partial chiral synthesis
  B. Absolute chiral synthesis

XIII. Racemization and inversion of configuration
  A. Racemization
  B. Epimerization

REFERENCES

ELIEL, E. L., *Elements of Stereochemistry*, John Wiley & Sons, Inc., New York, 1969.

———, *Stereochemistry of Carbon Compounds*, McGraw-Hill Book Company, New York, 1962.

ELIEL, E. L., N. L. ALLINGER, S. J. ANGYAL, and G. A. MORRISON, *Conformational Analysis*, John Wiley & Sons, Inc., New York, 1965.

GRUNDY, J., *Stereochemistry*, Butterworth & Co., Ltd., London, 1964.

MISLOW, K., *Introduction to Stereochemistry*, W. A. Benjamin, Inc., New York, 1965.

MORRISON, J. D. and H. S. MOSHER, *Asymmetric Organic Reactions*, Prentice-Hall, Inc., Englewood Cliffs, 1971.

NATTA, G. and M. FARINA, *Stereochemistry*, Longman Group Ltd., London, 1972.

PROBLEMS

1. Using pseudo-three-dimensional representations, indicate the results of carrying out the indicated operations on the following objects. Which of the following objects have nonsuperimposable mirror images?

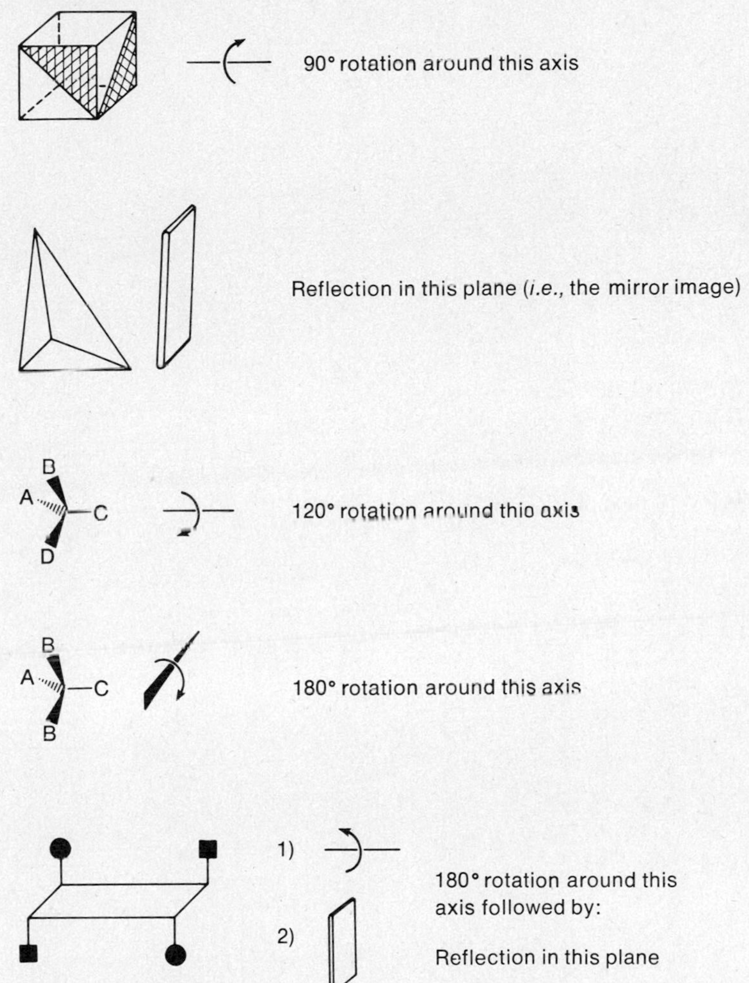

2. For each of the following objects, indicate whether the mirror image is superimposable or nonsuperimposable with the original.
   (a) The Statue of Liberty
   (b) A baseball
   (c) A die (*i.e.*, one member of a pair of dice)
   (d) The Washington Monument
   (e) The letter *S*

3. For each of the following three-dimensional objects, draw the mirror image and indicate whether the two mirror images are superimposable:

(a) 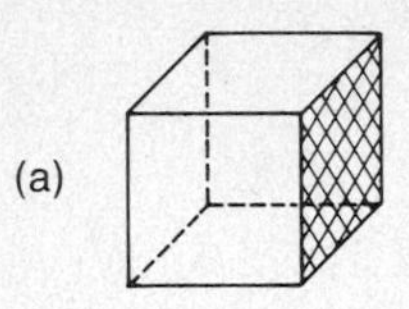

(b) 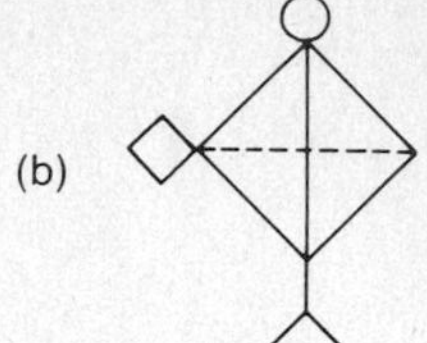

(c) 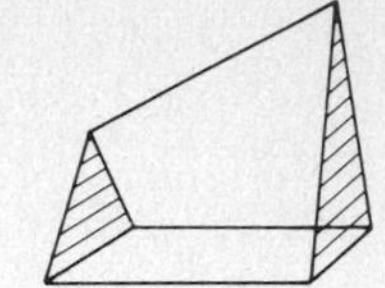

(d) 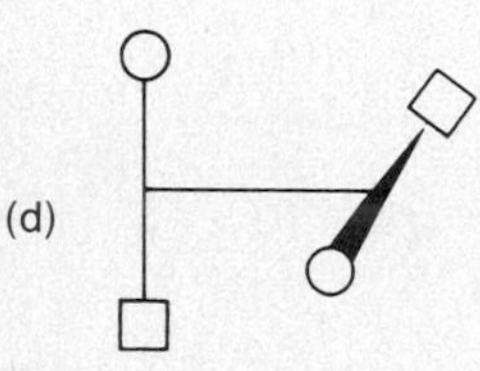

(e) 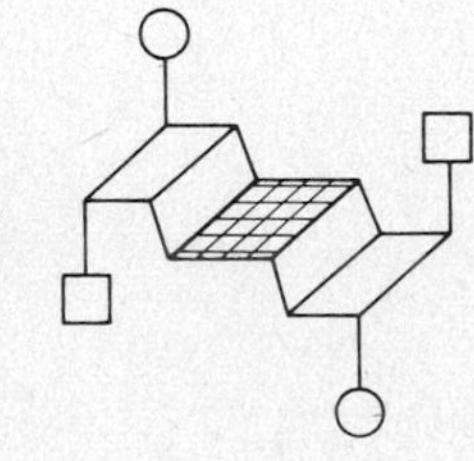

4. Using pseudo-three-dimensional drawings, show all the symmetry elements present in each of the following compounds:

(a) CH_3Br

(b) (cyclobutane ring bearing CH_3 and CH_3)
meso Isomer

(c) $CH_3CH_2CH_2CH_3$
Antiperiplanar conformation

(d) $CH_3CHBrCO_2H$ (Br on the second carbon)

(e) H, H, O, O (bicyclic diketone)

5. Using pseudo-three-dimensional representations, draw and label [*i.e.*, (*Z*),(*E*), member of an enantiomeric pair, or *meso* form] all the stereoisomers for the following compounds:

(a) $CH_3CH_2C(Br)(OCH_3)CH_3$

(b) $CH_3CH{=}CH{-}C_6H_5$

(c) $CH_3CH(Br)CH_2CH(Br)CH_3$

(d) (cyclopentene ring bearing CH_3 and CH_3 on the double-bond carbons)

(e) $CH_3CH{=}CHCH(OH)CH_3$

(f) (cyclohexane ring bearing CH_3, OH, and a phenyl group)

(g) $CH_3CH{=}C{=}CH{-}C_6H_5$

(h) (biphenyl bearing Br, CH_3, NO_2, Br)

(i) $CH_3{-}$(bicyclobutane)${-}CH_3$

(j) 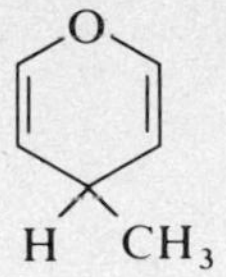 4-methyl-4H-pyran structure: ring with O; H, CH_3

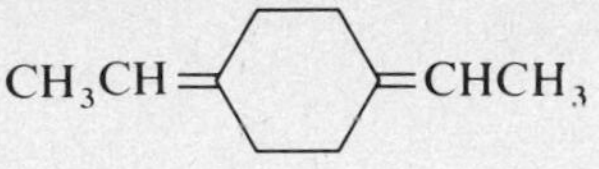

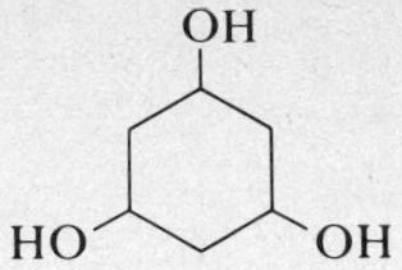

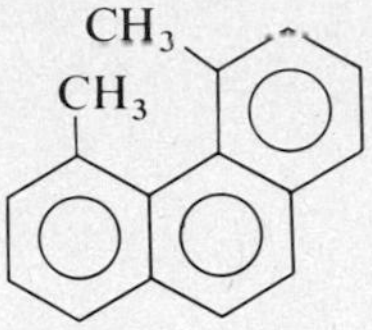

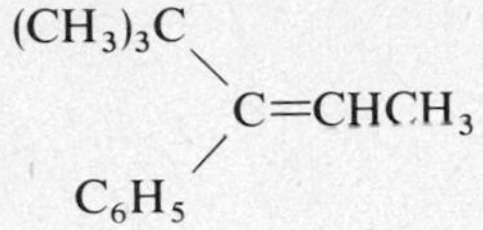

(k) $CH_3CH{=}$(cyclohexane)$=CHCH_3$

(l) HO, OH, OH (cyclohexanetriol)

(m) CH_3, CH_3

(n) $(CH_3)_3C$, $C{=}CHCH_3$, C_6H_5

(o) CH_3, $C{=}NNH$–(ring)–NO_2, CH_2, CH_3, NO_2

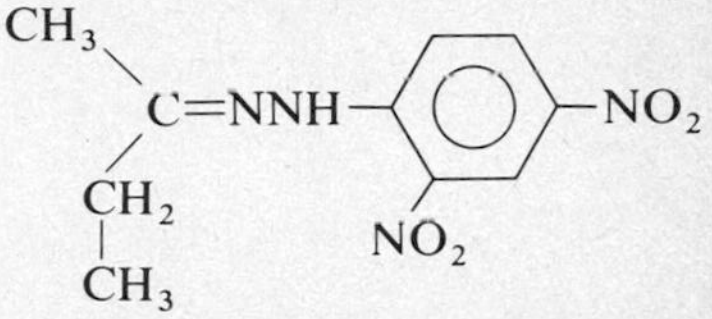

(p) 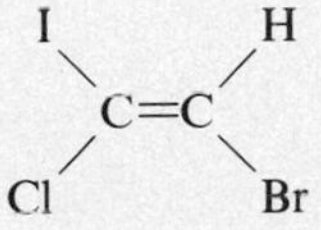 I, H, C=C, Cl, Br

(q) 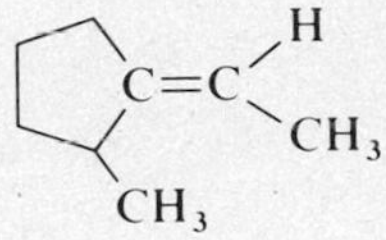 C=C, H, CH_3, CH_3

(r) 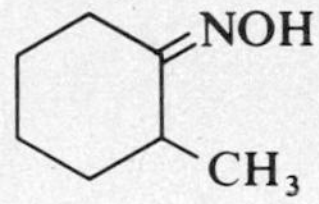 NOH, CH_3

(s) 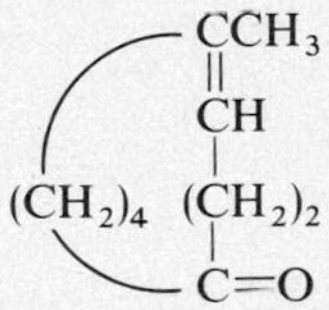 CCH_3, CH, $(CH_2)_4$, $(CH_2)_2$, C=O

(t) 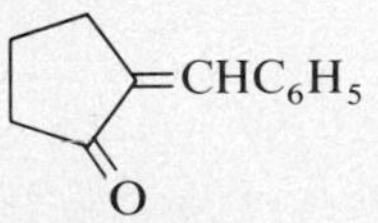 $=CHC_6H_5$, O

6. Name the following structures, indicating the absolute stereochemistry at the chiral center(s):

(a) 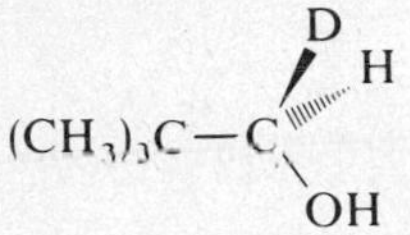 $(CH_3)_3C{-}C$ D, H, OH

(b) 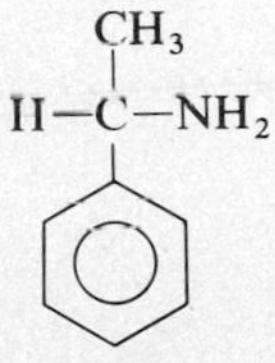 CH_3, $H{-}C{-}NH_2$, (phenyl) (Fischer projection)

(c) 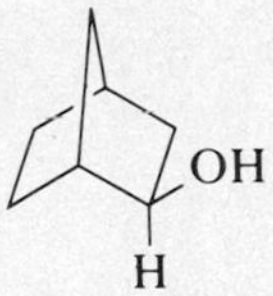 OH, H

(d) 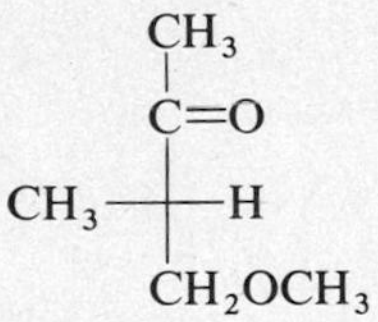 CH_3, C=O, CH_3—H, CH_2OCH_3 (Fischer projection)

(e) 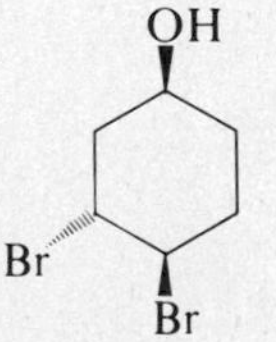 OH, Br, Br

(f) 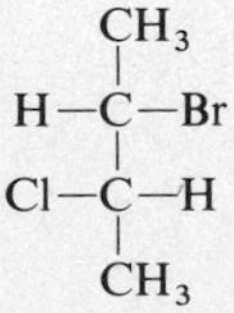 CH_3, H—C—Br, Cl—C—H, CH_3 (Fischer projection)

(g) 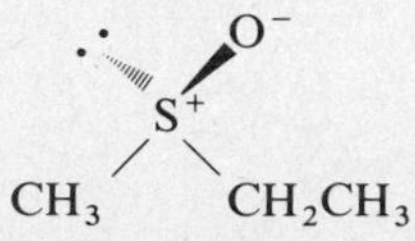 O^-, S^+, CH_3, CH_2CH_3

(h) 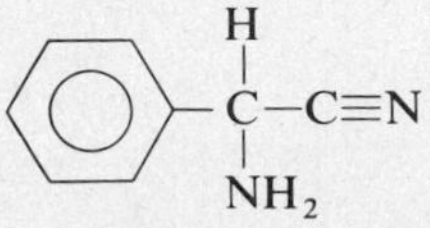 H, (phenyl)—C—C≡N, NH_2 (Fischer projection)

(i) 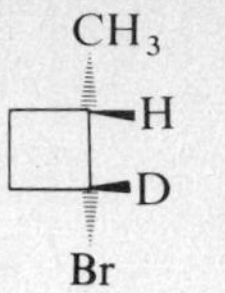 [cyclobutane structure: CH_3, H, D, Br]

(j) 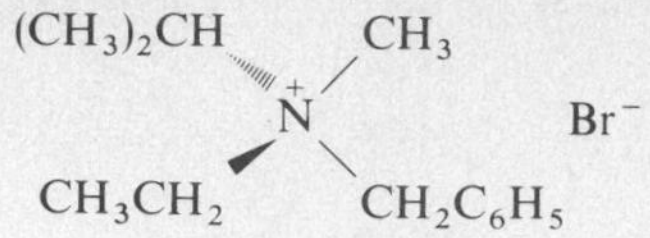 $(CH_3)_2CH$, CH_3, $\overset{+}{N}$, Br^-, CH_3CH_2, $CH_2C_6H_5$

(k)

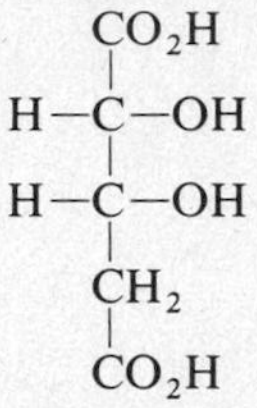

$$\begin{array}{c} CO_2H \\ | \\ H-C-OH \\ | \\ H-C-OH \\ | \\ CH_2 \\ | \\ CO_2H \end{array}$$

(l) 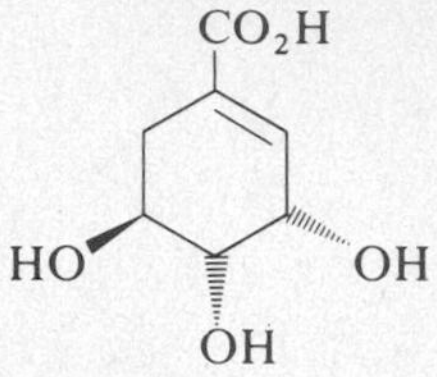 [cyclohexene structure: CO_2H, HO, OH, OH]

7. Draw the structures corresponding to each of the following names.

(a) (*R*)-2-Butanol
(b) (*S*)-Deuteriotritioacetic acid
(c) (*R*)-*exo*-2-Hydroxybicyclo[3.2.0]heptane
(d) (4*S*)-4-Bromo-2-chloro-(*E*)-2-pentene
(e) (*E*)-2-Butylidenecyclopentanone
(f) (2*R*,3*R*,4*S*)-2,3-Dibromo-4-chlorohexane
(g) (*S*)-3-Methylcyclohexanone
(h) (1*R*,3*R*)-3-[(1*S*)-*sec*-Butyl]-chlorocyclohexane
(i) (*S*)-Benzyl ethyl sulfoxide
(j) *trans*-5-Chloro-*r*-1,*cis*-3-cyclohexanedicarboxylic acid
(k) *r*-1-Chlorocyclohexane-*cis*-3,*trans*-5-dicarboxylic acid
(l) *meso*-3,4-Diacetoxyhexane
(m) *erythro*-1-Chloro-1-phenyl-2-propanol
(n) *meso*-1,2,3-Trihydroxycyclopentane

8. An experiment carried out many years ago by the English chemists Kenyon and Phillips involved the following sequence of reactions:

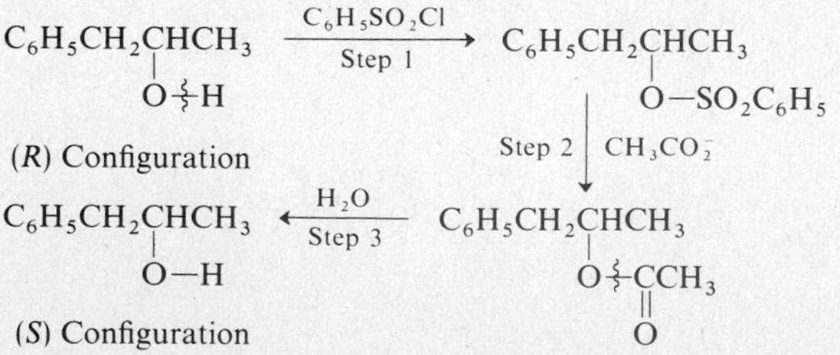

$$\underset{(R)\text{ Configuration}}{C_6H_5CH_2\underset{\underset{O\,\S\,H}{|}}{C}HCH_3} \xrightarrow[\text{Step 1}]{C_6H_5SO_2Cl} C_6H_5CH_2\underset{\underset{O-SO_2C_6H_5}{|}}{C}HCH_3$$

$$\downarrow \text{Step 2} \quad CH_3CO_2^-$$

$$\underset{(S)\text{ Configuration}}{C_6H_5CH_2\underset{\underset{O-H}{|}}{C}HCH_3} \xleftarrow[\text{Step 3}]{H_2O} C_6H_5CH_2\underset{\underset{O\,\S\,\underset{\underset{O}{\|}}{C}CH_3}{|}}{C}HCH_3$$

The bond being cleaved is indicated in steps 1 and 3. What can be inferred about the stereochemistry of the second step? Using pseudo-three-dimensional representations, depict the stereochemical course of the second step.

9. Treatment of 2,5-heptanedione with ethylmagnesium iodide followed by water yields 6-hydroxy-6-methyl-3-octanone, which, upon treatment with methylmagnesium iodide followed by water, yields 3,6-dimethyl-3,6-octanediol. If resolution of enantiomers is carried out after the first step and each of the enantiomers carried separately through the second step, what products will be obtained?

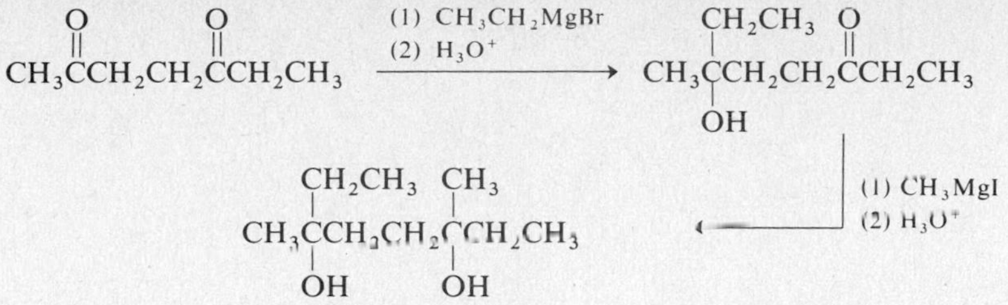

10. One enantiomer of a chiral compound, for which the stylized representation shown will be used, undergoes enzyme-catalyzed hydrolysis. Assuming that the enzyme has sites which are complementary to the groups attached to the chiral compound (the substrate) and that it has an X group which must come close to the carbonyl of the ester function in order to catalyze the hydrolysis, which enantiomer will remain unhydrolyzed after treatment with the enzyme? Using pseudo-three-dimensional representations, draw the structure of the enzyme-substrate complex.

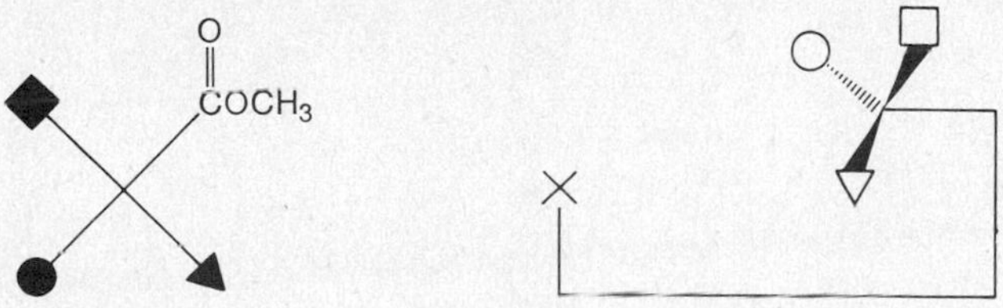

Substrate Enzyme

11. N,N-Dimethyl-2-azido-2-methylpropanamide undergoes light-induced decomposition to an imine. If right-hand circularly polarized light is used to induce the decomposition will the product of 50% decomposition show any optical activity? Explain.

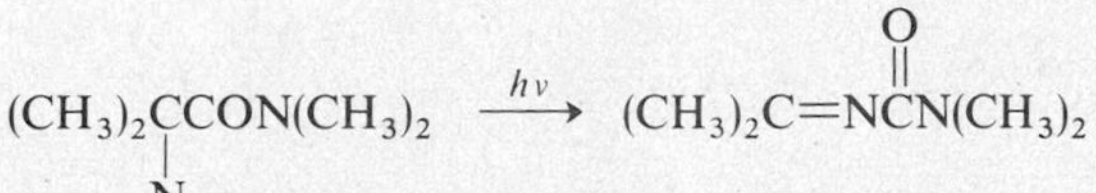

12. The triol designated as compound *A* undergoes ozonolysis and hydrolysis to yield (+)-glyceraldehyde and compound *B*. When compound *B* is oxidized with Tollens' reagent, (+)-tartaric acid is the product.

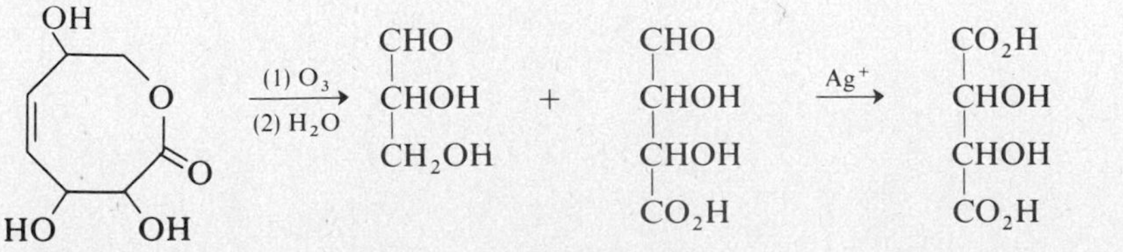

Compound *A* (+)-Glyceraldehyde Compound *B* (+)-Tartaric acid

(a) Draw a pseudo-three-dimensional representation for (+)-glyceraldehyde.
(b) Draw a pseudo-three-dimensional representation for (+)-tartaric acid.
(c) Using the projection representation shown, indicate the absolute configurations for the chiral centers in compound *A*.

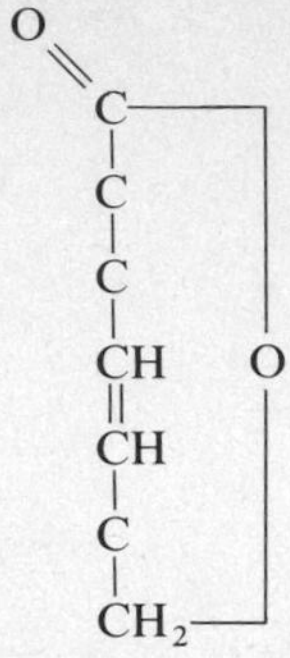

Projection representation 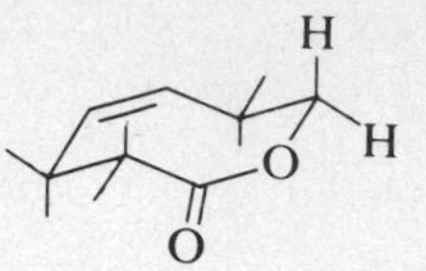   Pseudo-three-dimensional representation

13. In each of the following reactions, the starting material is optically active and possesses the absolute configuration that is shown by the pseudo-three-dimensional representations. Draw the pseudo-three-dimensional representation of the product in each case, and indicate whether it is optically active.

(a) 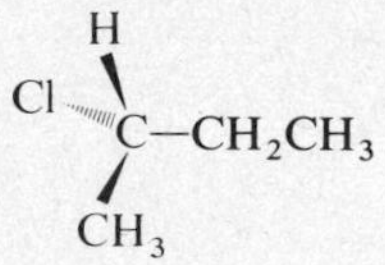 $+$ KOH $\longrightarrow$

(b) 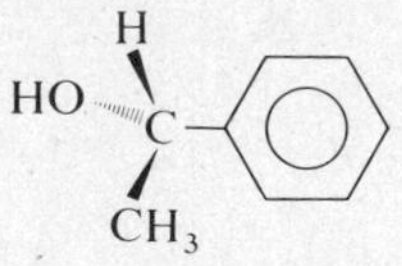 $+$ HCl $\longrightarrow$

(c) 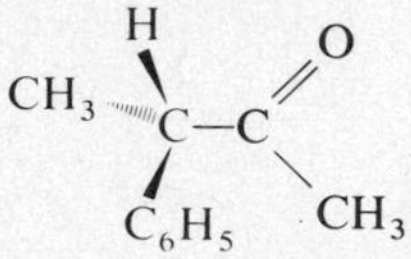 $+$ C_6H_5MgBr $\longrightarrow$

(d) 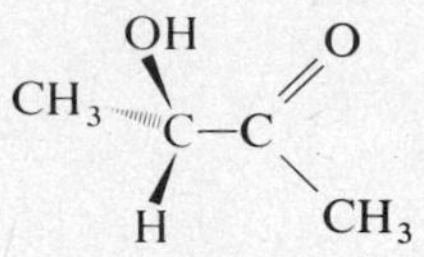 $+$ $LiAlH_4$ $\longrightarrow$

(e) 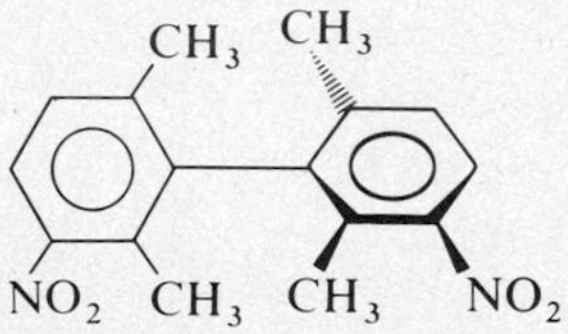 $\xrightarrow[H_2SO_4]{HNO_3}$

(f) 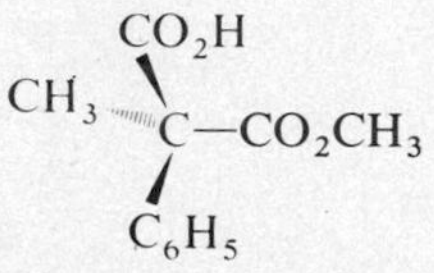 $+$ H_2O $\xrightarrow{NaOH}$

(g) 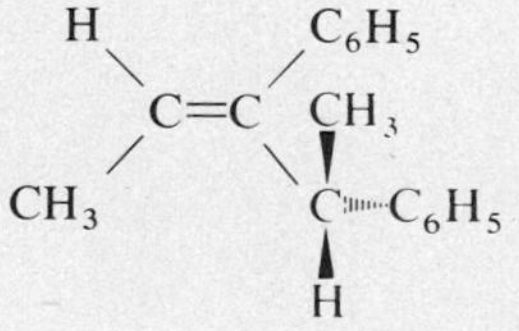 $+$ HCl $\longrightarrow$

(h) 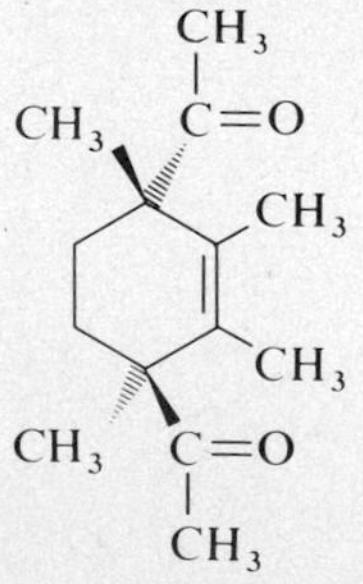 $+$ O_3 $\longrightarrow$

(i) 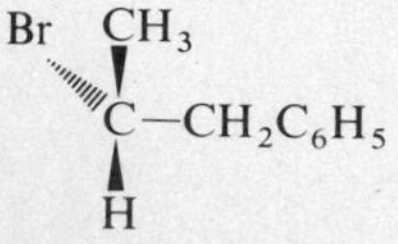 $\xrightarrow{KOH}$

(j) 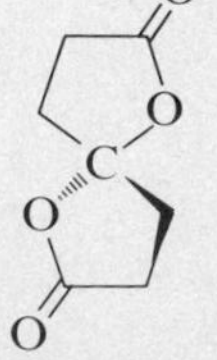 $\xrightarrow[NaOH]{H_2O}$

14. A compound of structure (A) is optically inactive and cannot be resolved. When it is oxidized to compound (B), however, treatment with an optically active base followed by fractional crystallization and hydrolysis yields an optically active form. Further oxidation of this optically active form to compound (C) yields an optically inactive product. Explain these results.

$CH_3SCH_2CO_2H$ Compound (A)

$CH_3S(=O)CH_2CO_2H$ Compound (B)

$CH_3S(=O)_2CH_2CO_2H$ Compound (C)

15. The following reaction sequence involves the introduction of chiral carbon atoms at several points. Using pseudo-three-dimensional representations, indicate the stereochemistry of the intermediates in this reaction sequence. If the resolution of the enantiomers is effected whenever racemic mixtures are produced and if each of the enantiomers is then taken through the subsequent steps, indicate the stereochemical outcome in each case.

Cyclohexanone $\xrightarrow{(1)\ CH_3MgI,\ (2)\ H_3O^+}$ 1-methylcyclohexanol (OH, CH_3) $\xrightarrow{H_2SO_4}$ 1-methylcyclohexene (CH_3) $\xrightarrow{CrO_3}$ 3-methylcyclohex-2-enone (CH_3, O) $\xrightarrow{(1)\ CH_3MgI,\ (2)\ H_3O^+}$ (CH_3; CH_3, OH) $\xrightarrow{H_2,\ Pd/C}$ (CH_3; CH_3, OH) $\xrightarrow{H_2SO_4}$ (CH_3; CH_3) $\xrightarrow{H_2,\ Pd/C}$ (CH_3; CH_3)

PART II

The Mechanism, Scope and Limitations, and Synthetic Utility of Organic Reactions

Part I has dealt with most of the major areas of organic chemistry—in considerable detail in some cases, but in rather sketchy fashion in others. It is the purpose of Part II to elaborate on those areas that are lightly touched on in Part I so that Parts I and II together will provide the reader with a well-rounded treatment of basic organic chemistry.

Part I concentrates on the structure of organic compounds and includes discussions of bonding theory, the composition and constitution of carbon frameworks and functional groups, nomenclature, and stereochemistry. Part II concentrates on the reactions of organic compounds and includes discussions of their mechanisms, their scope and limitations, and their utility in synthesis processes.

The first chapter of Part II enumerates the various ways that chemists have devised for studying the mechanisms and for classifying the various types of organic reactions. The remainder of Part II then surveys the several classes of compounds that have already been discussed in Part I, starting with those in which the functional group is single bonded to a carbon framework (halides, alcohols, phenols, ethers, and amines) and progressing to those containing multiple bonds (alkenes, alkynes, carbonyl compounds, and arenes). Some of the headings for Chapters 21–31 may look familiar, because they are the same as those already encountered in Part I. Indeed, it is hoped that they will seem to be old friends, for the presentation in Part II is based on the premise that the reader has acquired the knowledge from Part I that is the necessary foundation for this more advanced and detailed discussion.

As one progresses to higher and higher levels of understanding in any particular field of study, attention must inevitably be directed to the smaller facets of the field. Thus, in studying the scope and limitations of organic reactions, we are sharpening the focus of our attention and are becoming concerned with finer details. We are interested not only in the general features of a particular type of reaction, as in Part I, but now also with the precise details of applications to specific instances. We are interested in finding out where the reaction is particularly useful and where it fails, where it can be trusted and where it must be distrusted. We are striving to assemble the information that can provide us with the ability to design methods of construction of increasingly complex organic molecules.

The ability to construct things requires mastery over the materials of construction. For the organic chemist, the materials of construction are organic reactions, and the process of construction is known as organic synthesis. Successful organic synthesis—sometimes called a game, sometimes an art, and sometimes a science—requires a mastery of organic reactions. Attention in Part II, therefore, is directed not only to organic reactions *per se* but also to their combination into synthesis sequences. This double focus is reflected in the questions at the ends of the chapters, some dealing with the mechanism, scope, and limitations of reactions and others dealing with the application of these reactions to the synthesis of various compounds. In approaching the organic synthesis questions, we suggest that the reader review the discussion in Chapter 18.

Chapter **20**

Reaction mechanisms

A man is wise with the wisdom of his time only, and ignorant with its ignorance. Observe how the greatest minds yield in some degree to the superstition of their age.

HENRY DAVID THOREAU
Journal, 1853

A reaction mechanism seeks to picture in as minute detail as possible the sequence of changes that take place in the electronic structure of the reactants and the positions of the nuclei when a compound undergoes transformation to a different compound. Although structural theory has produced an accurate description of the positions of nuclei as well as the nature of the electronic states of reactants and products of a reaction, this information does not necessarily afford a clear picture of the structures of intermediates and activated complexes that are involved in going from reactant to product. One of the central problems of the study of reaction mechanisms, therefore, is to devise means for obtaining this information.

Ideally, a reaction mechanism should describe the changes in the positions of all the atoms in all the reactants involved during the course of the reaction. This has been achieved, however, only for the simplest of reactions (*e.g.*, $H\cdot + H_2 \rightarrow H_2 + H\cdot$). In more complex reactions information concerning the atomic motions only at the reaction site has been gained; furthermore, it is quite probable that for molecules the size of most organic reactants a variety of equal energy pathways exist between reactant and activated complex, adding complication to the problem of discerning a mechanism in complete detail. Even the relatively limited ideas that we currently have about how organic reactions occur represent major accomplishments, requiring a battery of sophisticated techniques to gather data from which information about the structures of intermediates and activated complexes can be gained. These techniques include (a) determination of the structure(s) of the product(s), (b) measurement of rates at which reactions occur (*i.e.*, kinetic measurements), (c) studies of the effects on reaction rates resulting from changes in the structure of reactants and/or solvent, (d) use of isotopically substituted compounds, (e) observation of the stereochemical course of the reaction, and (f) detection of suspected intermediates. The detection of intermediates can often be accomplished by adding other reagents ("trapping reagents"), by carrying out the reaction at low temperature, or by employing various spectroscopic techniques. In the first portion of this chapter, we discuss each of these general methods for the investigation of reaction mechanisms and present specific examples to illustrate their utility.

If a reaction mechanism is acceptable it must be commensurate with the results gained from *all* the techniques used to study it; failure of a mechanism to be commensurate with any piece of data is sufficient to rule it out of consideration. Thus, mechanisms are generally evolved by excluding the ones that are untenable on the basis of the various experimental data and by looking at what is left. The choice among the mechanisms that are left may then require the exercise of subjective judgment, and it is at this point that the will and wit of the investigator become a part of science. The wise investigator will adhere to the principle of Occam's razor, which states that complexity should not be multiplied beyond necessity; *i.e.*, the simplest mechanism that accommodates *all* the data should be chosen.

HOW TO STUDY THE MECHANISM OF A REACTION

20.1 Reaction Kinetics as a Means for Studying Reaction Mechanisms

When the sugar, sucrose, is treated with dilute mineral acid it undergoes hydrolysis to two smaller sugars, glucose and fructose:

$$\underset{\text{Sucrose}}{C_{12}H_{22}O_{11}} + H_2O \longrightarrow \underset{\text{Glucose}}{C_6H_{12}O_6} + \underset{\text{Fructose}}{C_6H_{12}O_6}$$

The specific rotation of the original solution changes as the formation of glucose and fructose progresses, and the idea of using the change as a means for studying the course of the reaction was realized in 1850 by a German chemist named Wilhelmy. Following the reaction polarimetrically, Wilhelmy measured the way in which the rate of the reaction decreases with time, and he was able to fit his data to a simple mathematical expression. Thus, he became the first kineticist of the modern era of chemistry and introduced a method for investigating reactions that has become one of the most powerful tools available to chemists.

PROBLEM 20-1

Suggest other methods by means of which the kinetics of this reaction might be studied.

Reaction kinetics deal with the measurement of rates at which chemical processes occur. The time scale is extraordinary, ranging from reactions that occur almost instantaneously to those that are too slow to proceed to any significant extent on a human time scale. To study the kinetics of any of these processes, wherever they fall in this spectrum of rates, one must measure the concentration of the reactants or the products at various times during the course of the reaction—*i.e.*, measure the concentrations as a function of time. The capabilities for such measurement have been greatly expanded in recent years as the result of striking advances in analytical techniques. Thus, by means of the older techniques, such as titrametric, gravimetric, colorimetric, and polarimetric measurements, and the new applications of stopped-flow, fast-flow quenching, and shock-tube techniques, the majority of organic reactions are amenable to kinetic study. Let us, then, see what kinds of information we can derive from kinetic data.

20.1a. CORRELATION OF RATES WITH STRUCTURE. In our discussion of the use of kinetics for studying reaction mechanisms, let us consider the specific examples of a 0.1 *M* solution of sodium ethoxide in ethanol reacting, in turn, with an equal volume of 0.1 *M* solutions of methyl bromide, neopentyl bromide, and *t*-butyl bromide to produce the three corresponding ethers (see Fig. 11.2)*; *i.e.*,

$$CH_3Br + C_2H_5ONa \longrightarrow CH_3OC_2H_5 + NaBr$$

$$(CH_3)_3CCH_2Br + C_2H_5ONa \longrightarrow (CH_3)_3CCH_2OC_2H_5 + NaBr$$

$$(CH_3)_3CBr + C_2H_5ONa \longrightarrow (CH_3)_3COC_2H_5 + \underset{\text{Major product}}{(CH_3)_2C{=}CH_2} + NaBr$$

Let us first consider the reaction of methyl bromide. If an aliquot of this reaction mixture is withdrawn as soon as the solution is prepared, analysis (*i.e.*, by nmr or gc† methods) shows that the concentration of methyl bromide is 0.05 *M*. If similar measurements are made at later times, the concentration of methyl bromide is found to be smaller. A plot of the concentration *vs.* time (Fig. 20.1) shows a rapid decrease in concentration at the beginning of the reaction and a continuous slowing down in the rate of change in concentration as the reaction progresses. After several hours, the concentration of methyl bromide has fallen almost to zero, and for all practical purposes the reaction is complete. Alternatively, we could have measured the concentration of the product, ethyl methyl ether, in which case we would obtain a plot reciprocal to that of the methyl bromide concentration, starting at zero concentration at zero time and increasing to about 0.05 *M* after several hours.

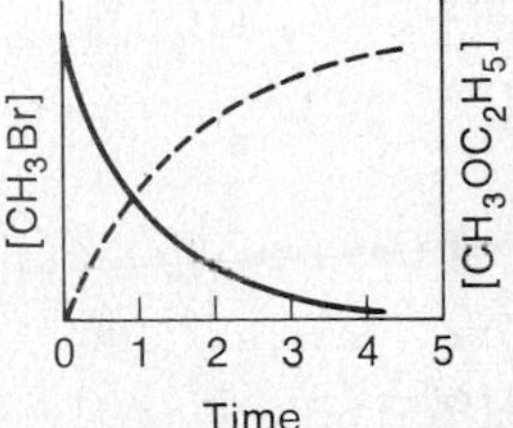

Fig. 20.1. Plot of concentration of methyl bromide (solid line) and ethyl methyl ether (dashed line) *vs.* time.

PROBLEM 20-2

Suggest methods for measuring the kinetics of the reaction of methyl bromide with sodium ethoxide; *i.e.*, methods for measuring the concentration of the reactants and/or the products.

---

* In the reaction of *t*-butyl bromide with sodium ethoxide, 2-methylpropene is also formed in large amount.

† "gc" is the abbreviation for gas chromatography, a method of analysis in which the vaporized mixture is passed through a column packed with a material appropriate for separating the components of the particular mixture.

In performing the same experiment with neopentyl bromide, we discover that over the span of time which is sufficient to convert essentially all the methyl bromide to ethyl methyl ether, practically none of the neopentyl bromide is converted to ethyl neopentyl ether. In fact, if Julius Caesar had prepared the neopentyl bromide solution shortly before his death in 44 B.C., only about half the neopentyl bromide would have been consumed over the intervening centuries to the present time; neopentyl bromide reacts approximately 10 million times less rapidly than methyl bromide.

In the third experiment, involving *t*-butyl bromide, we discover that the rate at which the product, *t*-butyl ethyl ether, is formed is only slightly less than the rate of product formation in the first experiment with methyl bromide. Thus, although the reactions of methyl bromide, neopentyl bromide, and *t*-butyl bromide are identical in that they each involve the conversion of an alkyl bromide to an ether, they differ in the rates at which they occur. Since the conditions for the reactions are identical, the only thing that can reasonably account for the difference in rates is the structure of the carbon frameworks to which the bromine is attached. Measurement of the kinetics of the reactions, then, provides a way for establishing correlations between the structures of the reactants and the way in which the reactants undergo particular reactions. It is out of such correlations that ultimately emerge the hypotheses which chemists call reaction mechanisms. For example, a hypothesis that accommodates the difference in rate between the methyl bromide and neopentyl bromide reactions postulates that an ethoxide anion approaches the "backside" of the C—Br carbon. In methyl bromide, the ethoxide anion encounters three hydrogens along the "backside" pathway; since hydrogen atoms are small, there is little steric hindrance. In neopentyl bromide, on the other hand, the ethoxide anion encounters two hydrogens and the very bulky *t*-butyl group as well. As a consequence of the severe steric hindrance at the "backside" afforded by the *t*-butyl group, the ethoxide anion is deflected from the preferred pathway and is forced to come in along a less efficient pathway (Fig. 20.2), commensurate with the very much slower reaction of neopentyl bromide.

PROBLEM 20-3

Knowing that *t*-butyl bromide gives large amounts of 2-methylpropene and relatively small amounts of *t*-butyl ethyl ether, is it better to measure the kinetics of ether formation by measuring the change in concentration of *t*-butyl bromide or *t*-butyl ethyl ether? Explain.

The mechanism that serves to explain the *difference* in rate between methyl bromide and neopentyl bromide fails to explain the *similarity* in rate between methyl bromide and *t*-butyl bromide. Methyl groups are larger than hydrogen and would be expected to provide considerably more hindrance to the "backside" approach of the ethoxide anion, greatly reducing the rate of reaction of *t*-butyl bromide. Since this is not what is observed, another mechanism must be

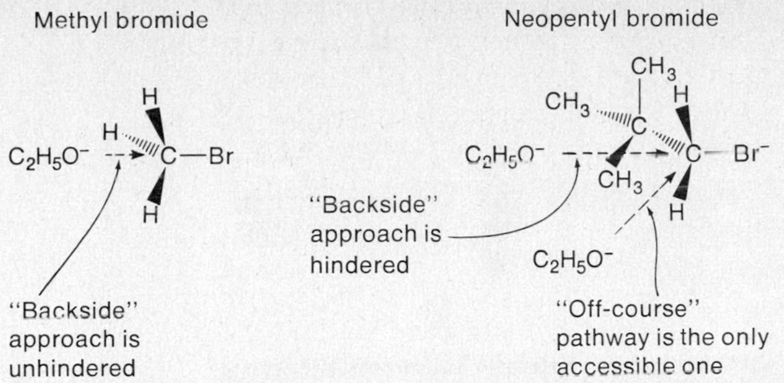

Fig. 20.2. "Backside" pathway for reaction of ethoxide with methyl bromide and neopentyl bromide.

operative in the case of *t*-butyl bromide, and to gain some insight into this we must look at the way in which the rates of these reactions depend on the concentrations of the reactant alkyl halides and sodium ethoxide.

20.1b. Concentration Dependence of Reaction Rates

The Order of a Reaction. The manner in which the rate of the reaction of methyl bromide changes as the reaction progresses is shown in Fig. 20.1. Clearly, as the concentration of methyl bromide decreases, the rate of the reaction diminishes; the rate is directly proportional to the concentration of the reactant. This situation exists for most, although not all, organic reactions and can be expressed as follows:

$$\text{Observed rate} = k\ (\text{concentration factor}) \qquad \text{expression (20–1)}$$

The proportionality factor, k, known as the *specific rate*, is independent of the concentration of reactants or products and is a characteristic constant for the particular reaction under a given set of experimental conditions.

PROBLEM 20-4

Consider two reactions that have the same concentration dependence. The specific reaction rate for one reaction is 3.6×10^{-4} sec^{-1} and that for the other is 7.1×10^2 sec^{-1}. Which is the faster reaction, and how much faster is it?

The "concentration factor" can assume a variety of forms. For example, if we measure the rate of reaction of methyl bromide with sodium ethoxide in two experiments, one in which the starting concentration of methyl bromide is 0.1 M and one in which the concentration is 0.2 M (with the ethoxide concentration 0.1 M in both cases), we discover that the reaction occurs twice as fast in the 0.2 M solution of methyl bromide. The rate of ether formation is directly proportional to the concentration of methyl bromide *raised to the first power*, in which instance the reaction is said to be *first order with respect to methyl bromide*. Now, if we carry out another pair of experiments, this time with starting concentrations of ethoxide ion of 0.1 *M* and 0.2 *M* (keeping the methyl bromide concentration at 0.1 *M* in both cases), we find that the reaction proceeds twice as fast in the 0.2 *M* solution. Thus, the rate is also directly proportional to the concentration of ethoxide ion *raised to the first power*; it is *first order with respect to ethoxide*. The reaction, therefore, is a second-order process—*i.e.*, first order with respect to methyl bromide and also first order with respect to ethoxide. This is expressed in the following manner:

$$\text{Rate of reaction of methyl bromide with sodium ethoxide in ethanol} = k[CH_3Br][C_2H_5O^-]$$

Similar studies with neopentyl bromide, although difficult to carry out because of the extreme sluggishness of the reaction, lead to the same conclusion. Investigation of the *t*-butyl bromide reaction shows that here also the rate of ether formation is dependent on the alkyl bromide concentration raised to the first power; however, in contrast to the methyl and neopentyl bromide reactions, the rate is *independent* of the concentration of the ethoxide ion; *i.e.*, the reaction is no faster in 1.0 *M* sodium ethoxide than in 0.1 *M* sodium ethoxide. The reaction is *zero order* in ethoxide. In its totality, therefore, this reaction is an overall first-order process and is expressed as follows:

$$\text{Rate of reaction of } t\text{-butyl bromide with sodium ethoxide in ethanol} = k[(CH_3)_3CBr]$$

Although first-order and second-order rate expressions of the type illustrated are the ones most commonly encountered in organic chemistry, numerous other possibilities exist. For example, a reaction that shows a dependence of reaction rate on the concentration of a single reactant raised to the second power is also a second-order reaction; a reaction that shows a first-order dependence on each of three reactants is a third-order process; a reaction that shows a first-order dependence on one reactant and a second-order dependence on a second reactant is also a third-order process. Sometimes the rate expression shows a much more complex concentration dependence, as illustrated by the last entry in Fig. 20.3.

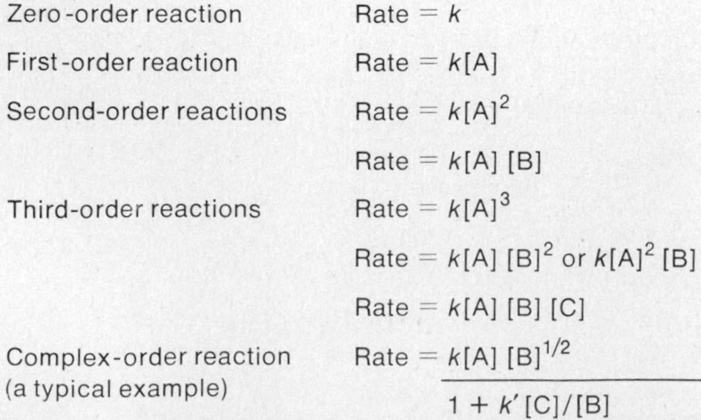

Zero-order reaction	Rate $= k$
First-order reaction	Rate $= k[A]$
Second-order reactions	Rate $= k[A]^2$
	Rate $= k[A][B]$
Third-order reactions	Rate $= k[A]^3$
	Rate $= k[A][B]^2$ or $k[A]^2[B]$
	Rate $= k[A][B][C]$
Complex-order reaction (a typical example)	Rate $= \dfrac{k[A][B]^{1/2}}{1 + k'[C]/[B]}$

Fig. 20.3. Examples of rate expressions.

Reaction Order and Mechanism. The overall order of a reaction as well as the order with respect to each of the reactants can provide valuable information concerning the mechanism of a reaction, the difference between the reactions of methyl bromide and *t*-butyl bromide with ethanol being a case in point. The second-order kinetics of the methyl bromide reaction are commensurate with the backside approach mechanism just discussed (see Fig. 20.1), but the first-order kinetics of the *t*-butyl bromide reaction indicate that the ethoxide is *not* involved in the rate-determining step. A mechanism that is commensurate with this fact is one in which an intermediate carbonium ion is formed (see p. 189) in a slow step, followed by combination of the carbonium ion with ethoxide anion (or ethanol and loss of a proton) in a rapid step to form the product. Thus, in the methyl bromide reaction, the concentration of both the alkyl halide and the ethoxide ion play a part in the rate-determining step (only a single step in this case), whereas in the *t*-butyl bromide reaction the concentration of only the alkyl halide plays a part in the rate-determining step; the ethoxide moiety does not enter the picture until later. A point suggested by these facts which should be kept in mind is that the rate that one measures is the rate of the *slowest* step (the so-called *rate-determining step*). Thus, if a reaction proceeds stepwise, only those processes that occur *before* the slowest step will exert a concentration effect on the rate expression.

PROBLEM 20-5

In the reaction of *t*-butyl bromide leading to a mixture of ether and alkene, will the rate of disappearance of starting material depend on the specific reaction rates of ether and alkene formation? Explain.

Molecularity of a Reaction. The "molecularity" of a reaction specifies the number of participants that are present in the activated complex of the rate-determining step. For example, the activated complex (see p. 186) for the reaction of methyl bromide and ethoxide ion has the structure shown. It contains one

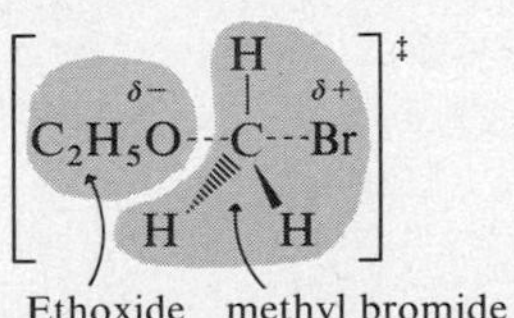

Activated complex

molecule of methyl bromide and one molecule of ethoxide ion (*i.e.*, it is bimolecular), and the reaction is referred to as a bimolecular reaction. The activated complex for the rate-determining step of the reaction of *t*-butyl bromide and ethanol has the structure shown.

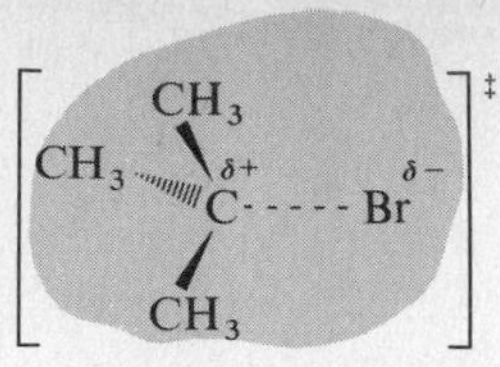

Activated complex

It contains only one molecule (*i.e.*, *t*-butyl bromide, in which the C—Br bond is partially broken and the carbonium ion is partially formed), and the reaction is referred to as a unimolecular reaction.

PROBLEM 20-6

On the assumptions that (a) ether formation from *t*-butyl bromide is first order in *t*-butyl bromide and zero order in sodium ethoxide, (b) alkene formation is first order in both *t*-butyl bromide and sodium ethoxide, and (c) the rates of ether and alkene formation are not the same, write an expression for the rate of disappearance of *t*-butyl bromide.

The direct inference of molecularity from reaction order can sometimes be obscured if one or more of the reactants does not undergo appreciable change in concentration during the course of the reaction. For example, if the reaction of methyl bromide and ethoxide is carried out with 0.01 *M* of methyl bromide in 1 *M* sodium ethoxide, the concentration of the sodium ethoxide changes by only 1% over the course of the reaction; for all practical purposes it remains constant, and the rate expression might appear to show a concentration dependence only on the methyl bromide. The kinetics of the reaction carried out under these conditions are said to be *pseudo-first-order*. Thus, we can make reactions having an actual kinetic order greater than one show pseudo-first-order kinetics simply by employing all but one of the reactants in large excess.

20.1c. ACTIVATION PARAMETERS FROM REACTION RATES. By measuring the rate of a reaction at two (or more) different temperatures, we can calculate the magnitude of two quantities known as the *enthalpy of activation* (symbolized by $\Delta H^{\ddagger}$) and the *entropy of activation* (symbolized by $\Delta S^{\ddagger}$), as discussed on p. 187. The enthalpy of activation, usually quoted in units of kilocalories per mole, is approximately equal to the energy that must be added to the reactants to elevate them to the transition state. The entropy of activation, usually quoted in entropy units (e.u. or calories/degree/mole), is a measure of the ease of passage of the reactants through the transition state to form products when they have sufficient energy to do so. The values for these parameters for the reactions of methyl bromide, neopentyl bromide, and *t*-butyl bromide with ethoxide in ethanol, for example, are shown in Table 20.1. It is beyond the scope of the present discussion to interpret these data in any detail. An organic chemist well schooled in the intricacies of reaction kinetics can advance reasonably cogent arguments to show that these data are commensurate with bimolecular processes for the formation of ether from methyl bromide and neopentyl bromide and with a unimolecular process for the formation of ether from *t*-butyl bromide.

The enthalpy of activation and entropy of activation both contribute to free energy of activation ($\Delta G^\ddagger = \Delta H^\ddagger - T\Delta S^\ddagger$), and it is the magnitude of this parameter that determines the rate at which a reaction proceeds. The less positive is $\Delta G^\ddagger$, the faster is the rate of the reaction ($\Delta G^\ddagger = -RT \ln K^\ddagger$)

Table 20.1 ACTIVATION PARAMETERS FOR THE REACTION OF SODIUM ETHOXIDE IN ETHANOL WITH METHYL BROMIDE, NEOPENTYL BROMIDE, AND *t*-BUTYL BROMIDE

Alkyl Halide	$\Delta H^\ddagger$ (kcal/mole)	$\Delta S^\ddagger$ (eu)	$-T\Delta S^\ddagger$ (kcal/mole)	$\Delta G^\ddagger$ (kcal/mole)
Methyl bromide	19.4	− 8.6	2.6	22.0
Neopentyl bromide	26.5	−19.9	6.0	32.5
t-Butyl bromide[a]	24.7	+ 6.4	−1.9	22.8

[a] Derived from rate data for the unimolecular reaction of *t*-butyl bromide in methanol.

20.1d. SOLVENT EFFECTS ON REACTION RATES. The solvent in which a reaction is carried out may have an effect on the reaction rate, ranging from almost negligible to extremely large. From the magnitude and nature of the effect of a particular solvent, valuable inferences concerning reaction mechanism can often be drawn.

For example, in the reactions of methyl, neopentyl, and *t*-butyl bromide with sodium ethoxide in ethanol, we observe that changes in the dielectric constant of the solvent (see Table 21.5) produce substantial changes in the rates but do not change the kinetic order. In the ethoxide reactions of methyl and neopentyl bromide, a decrease in the dielectric constant of the solvent (for example, by the addition of a nonreacting, low dielectric constant solvent such as benzene or hexane to the ethanol) results in an increase in the rate of reaction. This suggests that in these reactions a reduction of charge density occurs in going from the initial state to the transition state; *i.e.*, the activated complex carries less charge per unit volume than does the ethoxide ion, and thus is less dependent on stabilization by interaction with polar solvent molecules. This is commensurate with a one-step process involving a bimolecular activated complex (see p. 561). In the ethoxide reaction of *t*-butyl bromide, however, a decrease in the dielectric constant of the reaction medium results in a decrease in the rate of ether formation. This suggests that an increase in charge density occurs in going from the reactant to the activated complex, commensurate with a two-step process involving the rate-determining formation of a carbonium ion.

In drawing mechanistic inferences from solvent effects the following generalities are useful:

1. An increase in the dielectric constant of the solvent results in an increase in the rates of those reactions in which the charge density is greater in the activated complex than in the reactant molecule(s).

2. An increase in the dielectric constant of the solvent results in a decrease in the rates of those reactions in which the charge density is lower in the activated complex than in the reactant(s).
3. A change in the dielectric constant of the solvent will have a negligible effect on the rates of those reactions that involve little or no change in the charge density on going from reactant(s) to the activated complex.

20.2 Substituent Effects as a Means for Studying Reaction Mechanisms

Comparison of the chemical behavior of a series of structurally related compounds has been a favorite occupation of chemists for many generations; rules such as the Saytzev rule (see p. 242), the Hofmann rule (see p. 316), and Bredt's rule (see p. 98) all had their genesis in comparisons of this sort. Since the efficacy of comparison is no less powerful today than in times past, one of the components of a complete mechanistic study continues to be an investigation of the behavior of appropriately substituted analogs. The utility of such investigations, however, is dependent on our knowledge of the kinds of steric* and electronic effects that various substituent groups exert in various circumstances.

Steric effects arise from the direct, physical interaction either between groups in the same molecule (intramolecular interactions) or between groups in different molecules as the molecules approach one another (intermolecular interactions). The magnitude of the steric effect is closely related to the sizes of the interacting groups that are forced into close proximity in the activated complex. Generally, a reduction in the extent of steric crowding in going from reactant to activated complex increases the reaction rate, and an increase in the steric crowding decreases the reaction rate. For example, one of the factors contributing to the greater ease of formation of the *t*-butyl cation in comparison with the methyl cation may be the greater relief of strain that is involved in going from the tetrahedral reactant (bond angles = 109°28′) to the trigonal activated complex (bond angles 120°) in the *t*-butyl series, the steric interference between proximate methyl groups being considerably greater than that between proximate hydrogens.

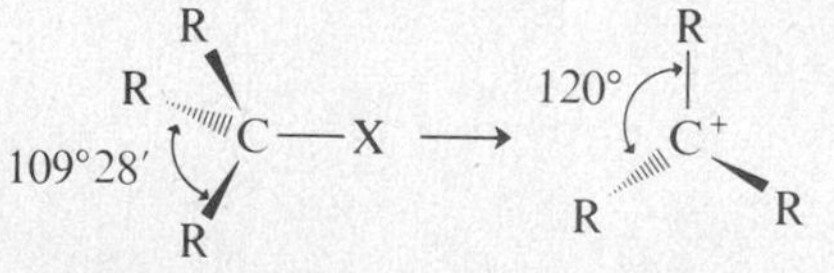

$$R_3C{-}X \longrightarrow R_3C^+$$

(109°28′) (120°)

Tetrahedral reactant **Trigonal intermediate**

PROBLEM 20-7

Arrange the following cations in order of increasing ease of formation:

$(C_2H_5)_3C^+$ $[(CH_3)_3C]_3C^+$ 1-methylcyclohexyl cation (cyclohexane ring with CH_3 and + on the same carbon)

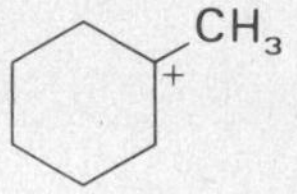

* Steric effects are actually electronic in origin, arising from *nonbonded* electronic interactions. It is convenient, however, to follow the common practice of considering steric and electronic effects as separate phenomena.

The electronic effects of groups fall into two categories: (a) those referred to as *inductive effects*, which operate through σ and π bonds to induce a permanent dipole in the regions of the molecule close to the group, and (b) those referred to as *resonance effects*, which operate mainly through π bonds to induce a transient or permanent dipole in the regions of the molecule that are conjugated with the group.

20.2a. INDUCTIVE EFFECT. The inductive effect arises from electronic influences that operate either *through bonds* in the molecule (both σ- and π-type bonds) or through space.* It causes an unequal sharing of the electrons in bonds. It falls off rather rapidly with distance and is significant only when the substituent is quite close in space to the moiety on which it is exerting its influence. Using the bond transmission pathway to illustrate this point, we can show the electron density in the bonds of an alkane chain containing an electron-withdrawing hetero atom in terms of the electron pairs that are progressively closer to the center of the bond as the bond is increasingly distant from the *Y* group:

$$\mathrm{Y:\overset{|}{\underset{|}{C}}:\overset{|}{\underset{|}{C}}\ :\ \overset{|}{\underset{|}{C}}\ :\ \overset{|}{\underset{|}{C}}-}$$

PROBLEM 20-8

In the compounds shown would you expect the effect of the chlorine on the circled hydrogen to be the same or different if the mode of transmission is through bonds or through space? Explain.

Cl H   Cl H H H $ClCH_2CH_2CH_2CH_2$ H

The extent to which a scientific discipline can be made quantitative by the specific and accurate assignment of numbers to various phenomena is a measure of the level of advancement of that discipline. Organic chemistry has been and continues to be qualitative in many of its aspects. With the advent of kinetic measurements more than a century ago, however, and with the recent advances in analytical techniques, quantitative measurements of the phenomena of organic chemistry have improved remarkably. One of the phenomena that was quantitatively measured very early in the development of what we now refer to as *physical organic chemistry* was the inductive effect. The strengths of carboxylic acids (*i.e.*, their dissociation constants) can be measured easily and accurately, and the application of these measurements to the quantitative assessment of the

* The effect operating through the space between the interacting portions of the molecule is called the *field effect*. The extent to which electronic interactions are transmitted via the inductive effect or field effect is still uncertain. For simplicity, only the term "inductive" effect is used in this discussion.

inductive effect was made many years ago (Table 20.2). In a series of α-substituted acetic acids of the general structure Z—CH_2CO_2H, substituents that are more electron withdrawing than hydrogen (*i.e.*, Z = H) give stronger acids than CH_3CO_2H, and substituents that are less electron withdrawing than hydrogen give weaker acids than CH_3CO_2H; the magnitude of the dissociation constant, therefore, is directly related to the magnitude of the inductive effect. Thus, electronegative hetero atoms (*i.e.*, halogens) and groups that contain electronegative hetero atoms (*e.g.*, hydroxy, alkoxy, and nitro) are classed as electron-withdrawing functions, whereas alkyl groups appear to behave as electron-releasing functions* (Table 20.2).

Table 20.2 DISSOCIATION CONSTANTS, EXPRESSED AS pK_a VALUES, FOR

$$Z\text{—}CH_2CO_2H \rightleftharpoons Z\text{—}CH_2CO_2^- + H^+$$

Acids Stronger Than CH_3CO_2H		Acids Weaker Than CH_3CO_2H	
Z	pK_a	Z	pK_a
$-NO_2$	1.68	$-CH_3$	4.88
$-\overset{+}{N}(CH_3)_3$	1.83	$-C(CH_3)_3$	5.05
—CN	2.46		
—F	2.66		
—Br	2.86		
—Cl	2.86	Reference acid:	
—I	3.12	$pK_{CH_3CO_2H} = 4.76$	
—OH	3.83		

PROBLEM 20-9

Arrange in the order of increasing acidity 2-chloro-, 3-chloro-, and 4-chlorobutanoic acid. Explain the basis for your ordering.

Several other methods have also been used to obtain quantitative assessments of the inductive effects of various atoms and groups. For example, the extent to which a proton is deshielded in the nmr spectrum (*i.e.*, its downfield shift) is closely related to the electron density around the proton which, in turn, is related to the electron-releasing or electron-withdrawing effects of the attached groups (Table 20.3).

* Measurements of acidities in the gas phase have shown that propionic acid is actually a stronger acid than acetic acid. Its apparent lower acidity in solution is, therefore, probably not due to the electronic effects of the methyl group but to the greater steric inhibition of solvation of the carboxylate anion $CH_3CH_2CO_2^-$ in comparison with $CH_3CO_2^-$. Alkyl groups appear to be essentially neutral functions in their electronic demands, acting either as weakly electron-donating or withdrawing groups as the reaction requires.

Table 20.3 CHEMICAL SHIFTS OF THE NMR RESONANCES OF METHYL HYDROGENS IN Z—CH_3 COMPOUNDS

Z	δ	Z	δ
$-NO_2$	4.28	$-N(CH_3)_2$	2.2
—F	4.26	—I	2.16
—OH	3.47	$-COCH_3$	2.1
—Cl	3.05	$-CO_2H$	2.07
—Br	2.68	—CN	2.0
—SH	2.44	$-CH_3$	0.9
$-C_6H_5$	2.3	—H	0.23

PROBLEM 20-10

Is there a perfect correspondence between the magnitude of the inductive effect as measured by pK_a and nmr? If there are discrepancies, are they qualitative or quantitative ones?

20.2b. RESONANCE EFFECT. The resonance effect is a polarization (*i.e.*, static) as well as a polarizability (*i.e.*, dynamic) phenomenon, characteristic of a molecule in its rest state and also in its perturbed state as, for example, during the course of a reaction. Like the inductive effect, it causes an unequal sharing of the electrons by atoms. Unlike the inductive effect, the resonance effect usually operates most effectively through π bonds and falls off much less rapidly with increasing distance between the group exerting the effect and the portion of the molecule in which the effect is felt. A typical example of the resonance effect is provided by a comparison of the acidities of *m*-nitrophenol ($pK_a = 8.40$) and *p*-nitrophenol ($pk_a = 7.14$). Knowing that the nitro group is electron withdrawing via the inductive effect (see Tables 20.2 and 20.3) and knowing that the inductive effect falls off with distance, one might predict that *p*-nitrophenol should be a weaker acid than the *meta* isomer. That the reverse is true is ascribed to the resonance effect of the nitro group in stabilizing the negative charge on the oxygen of the phenoxide ion, as illustrated in Fig. 20.4. Phenol itself is seven orders of magnitude more acidic than an alkanol because of the delocalization of the electron density *away* from the oxygen atom and on to the *o*- and *p*- positions of the phenyl ring (see p. 267).

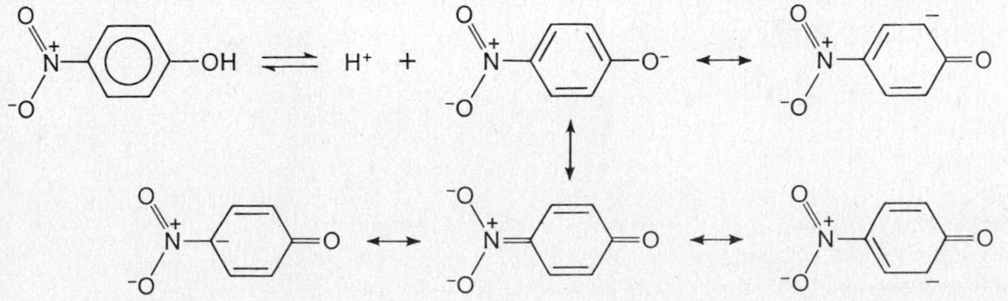

Fig. 20.4. Dissociation of *p*-nitrophenol and resonance structures for *p*-nitrophenoxide ion.

If a nitro group is attached to a position that is *ortho* or *para* to the hydroxyl group, even more extensive delocalization occurs; consequently, the *o*-nitrophenoxide and *p*-nitrophenoxide ions are *more* stable than the phenoxide ion (*i.e.*, *o*- and *p*-nitrophenol are more acidic than phenol; Fig. 20.4). The *m*-nitrophenoxide ion, however, is less stable than the *o*-nitrophenoxide and *p*-nitrophenoxide ions, for the negative charge *cannot* be delocalized into the nitro group but is constrained to remain in the ring (Fig. 20.5).*

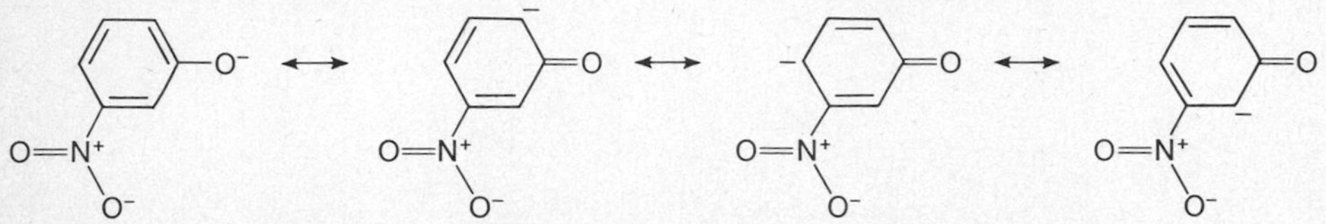

Fig. 20.5. Resonance structures for *m*-nitrophenoxide anion.

PROBLEM 20-11

Explain why *p*-methoxybenzyl cation is more stable than *m*-methoxybenzyl cation.

20.2c. QUANTITATIVE CORRELATIONS OF SUBSTITUENT EFFECTS: THE HAMMETT EQUATION. The quantitative assessment of resonance and inductive effects on chemical reactivity came primarily from the studies of Louis Hammett at Columbia University in the 1940's. In a typical experiment carried out in these studies, Hammett measured (a) the ionization constants of *p*-methoxybenzoic acid, *p*-methylbenzoic acid, benzoic acid, and *p*-nitrobenzoic acid, and (b) the rates of hydrolysis of the methyl esters of these same four *p*-substituted benzoic acids. He observed that when he plotted the log of the ionization constants of the acids on the abscissa and the log of the rates of hydrolysis of the esters on the ordinate he obtained a straight line (Fig. 20.6). By simple algebraic manipulation of the analytic expression for a straight line, Hammett devised a formula that relates the ionization constants and the rate constants for this system; *i.e.*,

$$\log \frac{k_{X-C_6H_4CO_2CH_3}}{k_{H-C_6H_4CO_2CH_3}} = \rho \log \frac{K_{X-C_6H_4CO_2H}}{K_{H-C_6H_4CO_2H}}.$$

He carried out many other experiments of a similar sort and eventually came to the conclusion that a general formula was possible, encompassing many such sets of reactions. The formula, now referred to as the *Hammett correlation* (or *Hammett equation*), is expressed as:

$$\log \frac{k_{X-C_6H_4-Y}}{K_{H-C_6H_4-Y}} = \rho \log \frac{K_{X-C_6H_4CO_2H}}{K_{H-C_6H_4CO_2H}}$$

---

* Detailed studies of the dissociation of the *p*- and *m*-nitrophenols, as well as a variety of other substituted phenols, indicate that the interpretation of the differences in pK_a values is less straightforward than we have suggested. Although resonance effects are certainly operative, they may often be obscured by solvent effects when the pK_a measurements are made in aqueous solution. Thus, the ΔH values for the dissociation of *p*- and *m*-nitrophenol are virtually identical; the difference appears in the ΔS values for the dissociation, reflecting the important role that the solvent plays in this reaction.

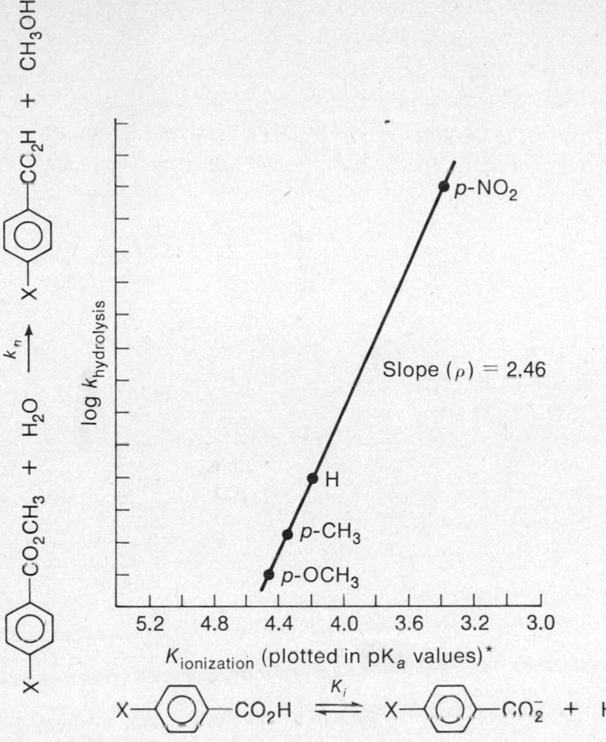

Fig. 20.6.* The Hammett correlation (a Hammett plot) for the hydrolysis of *p*-substituted methyl benzoates.

or, more succinctly, as

$$\left.\begin{array}{l}\log k - \log k_0 = \sigma\rho \\ \text{or} \quad \log k/k_0 = \sigma\rho\end{array}\right\} \begin{array}{l}\text{equivalent}\\ \text{expressions}\end{array} \qquad \text{expression (20-2)}$$

In the Hammett correlation, the k represents either an equilibrium constant or a rate constant of the substituted compound, k_0 represents the equilibrium constant or rate constant for the unsubstituted compound (*i.e.*, the reference compound), and σ and ρ have the meanings described in the following paragraphs.

The term ρ (Greek rho) is called the *reaction constant* and is defined as the slope of the line in the Hammett plot. For example, the data from the hydrolysis of methyl benzoates in 60% aqueous acetone at 0° give a Hammett plot with a slope of +2.46; the ρ value for this reaction, therefore, is +2.46. If the reaction is carried out in a different solvent mixture, or at a different temperature, or with a different ester (*e.g.*, ethyl benzoates), a different ρ value is obtained; thus, the ρ value is a constant that is characteristic for a particular type of reaction under a *specified set of conditions*, some representative examples of which are shown in Table 20.4. Note that the ρ value for the ionization of benzoic acid in water at

* It is convenient to plot the ionization constants on the abscissa as pK_a values. Since $pK_a = \log K_i$, however, it is necessary to plot the values *diminishing from left to right* if the ρ value (*i.e.*, the slope of the line) is to have the proper sign.

25° is +1.000, because the same data are plotted on the abscissa (*i.e.*, $pK_{\text{ionization}}$ is the reference reaction) and the ordinate (*i.e.*, $pK_{\text{ionization}}$ is the "other" reaction as well). The magnitude of ρ, which generally falls between +3 and −3, is a quantitative measure of how sensitive the reaction is to the electronic characteristics of the substituents; its sign indicates whether the reaction is helped or hindered by electron-withdrawing or electron-releasing groups. For example, the +2.46 ρ value for the hydrolysis of methyl benzoates indicates that the reaction is facilitated by electron-withdrawing groups (*i.e.*, positive value) and is quite sensitive to substituent effects (*i.e.*, large value). Conversely, the −1.688 ρ value for the hydrolysis of substituted benzyl chlorides indicates that the reaction is facilitated by electron-releasing groups (*i.e.*, negative value) and is moderately sensitive to substituent effects (*i.e.*, medium value).

Table 20.4 REPRESENTATIVE ρ VALUES FOR THE HAMMETT CORRELATION

Reaction	Solvent	Temperature	ρ Value
$R{-}C_6H_4{-}CO_2H \rightleftharpoons R{-}C_6H_4{-}CO_2^- + H^+$	H_2O	25°	+1.000
$R{-}C_6H_4{-}CO_2H \rightleftharpoons R{-}C_6H_4{-}CO_2^- + H^+$	C_2H_5OH	25°	+ 1.957
$R{-}C_6H_4{-}CO_2CH_3 + NaOH \longrightarrow R{-}C_6H_4{-}CO_2Na$	60% acetone	0°	+2.460
$R{-}C_6H_4{-}CH_2Cl + H_2O \longrightarrow R{-}C_6H_4{-}CH_2OH$	50% acetone	60°	−1.688
$R{-}C_6H_4{-}NH_2 + C_6H_5COCl \longrightarrow R{-}C_6H_4{-}NHCOC_6H_5$	Benzene	25°	−2.781

PROBLEM 20-12

Would you predict the ρ value for the ionization of benzoic acid at 50° to be 1.000? Why or why not?

The *substituent constant* σ (Greek *sigma*) is *independent of the* ρ *constant* and is characteristic of a given substituent. However, σ constants are different when the group is in the *meta* position and the *para* position. Also, the Hammett correlation fails when substituents are attached to the *ortho* position, for steric factors then complicate the issue and lead to nonlinear correlations. The σ constants are simply the difference in pK_a values between benzoic acid and the appropriate substituted benzoic acid in aqueous solution at 25°; *i.e.*,

$$pK_{\text{H}-\text{C}_6\text{H}_4\text{CO}_2\text{H}} - pK_{\text{X}-\text{C}_6\text{H}_4\text{CO}_2\text{H}} = \log \frac{K_{\text{X}-\text{C}_6\text{H}_4\text{CO}_2\text{H}}}{K_{\text{H}-\text{C}_6\text{H}_4\text{CO}_2\text{H}}} = \sigma$$

Thus, the σ constant for the methoxyl group in the *para* position is the difference between 4.20 (pK_a of benzoic acid) and 4.47 (pK_a of *p*-methoxybenzoic acid), which is equal to -0.27. Similarly, the σ constant for the nitro group in the *meta* position is $4.20 - 3.49$ (pK_a of *m*-nitrobenzoic acid), which is equal to $+0.71$. Hydrogen is the reference point for σ constants (its σ value is zero) so groups that have negative σ constants are electron releasing *relative to hydrogen,* and groups with positive σ constants are electron withdrawing *relative to hydrogen.* Extensive tables of σ constants have been compiled, a representative sampling of which is shown in Table 20.5.

PROBLEM 20-13

Would you predict the σ-value for the *p*-methoxyl group to be the same at 25° and 50°? Why or why not?

Table 20.5 SIGMA VALUES (σ) OF *meta* AND *para* SUBSTITUENTS, ARRANGED IN ORDER OF INCREASING VALUE FOR THE SUBSTITUENT IN THE *meta* POSITION[a]

Substituent	Meta σ	Para σ	Substituent	Meta σ	Para σ
$-N(CH_3)_2$	−0.21	−0.83	$-F$	0.34	0.06
$-O^-$	−0.17	−0.32	$-I$	0.35	0.18
$-NH_2$	−0.16	−0.66	$-CO_2H$	0.36	0.27
$-CO_2^-$	−0.10	0.00	$-Cl$	0.37	0.23
$-CH_3$	−0.07	−0.17	$-COCH_3$	0.38	0.50
$-H$	0.00	0.00	$-Br$	0.39	0.23
$-C_6H_5$	0.06	−0.01	$-CN$	0.56	0.66
$-OH$	0.12	−0.37	$-NO_2$	0.71	0.78
$-OCH_3$	0.12	−0.27	$-\overset{+}{N}(CH_3)_3$	0.88	0.82
$-SCH_3$	0.15	0.00			

[a] From C. D. Ritchie and W. F. Sager, *Progress in Physical Organic Chemistry*, **2**, 323 (1964).

20.2d. THE HAMMETT CORRELATION AS A MEANS FOR STUDYING MECHANISMS. Ezra Pound has said that "any general statement is like a check drawn on a bank; its value depends on what is there to meet it." In the case of the general statement embodied in the Hammett correlation, there is a great deal of collateral in the chemical bank to meet it; it is, in fact, one of the most often used and quoted generalizations that the physical organic chemist has at his disposal. It provides him with a way for predicting and assessing the effect of electron withdrawal or electron release on the rate of a particular reaction. To apply the Hammett correlation, one simply has to plot the rate of a few *m*- or *p*-substituted phenyl compounds against the σ values for the *meta* and *para* substituents and note (a) whether a straight line is obtained; if it is, the Hammett correlation obtains; (b) the sign of the slope of the line; if it is positive, the reaction is accelerated by electron-withdrawing groups, and if it is negative, the reaction is accelerated by electron-releasing groups; and (c) the magnitude of the slope; the greater its magnitude, the greater is the sensitivity of the reaction to the electronic effect of substituents.

20.3 Isotopes as a Means for Studying Reaction Mechanisms

Even in some of the earliest mechanism studies, the desirability of being able to "tag" a portion of a molecule in order to study its fate in a reaction was appreciated, and ways to achieve this were sought. For example, methyl groups were substituted for hydrogens, or halogen atoms were substituted for methyl groups as "tags." Clearly, however, a methyl group is not a hydrogen, and a halogen atom is not a methyl group. As "tags," these left much to be desired, for they changed the chemical and physical characteristics of the molecule; their use was simply dictated by the lack of anything better. In 1931 something better was discovered when Harold Urey prepared the first samples of molecular hydrogen and water in which an isotope of hydrogen replaced the ordinary hydrogen atom. Molecular deuterium and deuterium oxide ("heavy water"), thus, were the first members of what has grown to a long list of isotopic compounds. In fact, because of the many isotopic elements that are now available and the many organic synthesis techniques that have been developed, almost any compound that can be synthesized with ordinary elements can also be synthesized with isotopes. As a result, the organic chemist now has at his disposal a variety of "tags" that are almost ideal. Isotopes of a given element are, in general, essentially identical in their chemical behavior. This is true even in the case of the lightest element, despite the fact that deuterium is twice as heavy as hydrogen.

For a typical example of the application of isotopes as "tags" to determine the fate of some portion of a molecule during the course of a reaction, let us consider the nitrous acid-induced conversion of 2-phenylethylamine to 2-phenylethanol:

$$C_6H_5\text{—}CH_2CH_2NH_2 \;+\; HONO \longrightarrow C_6H_5\text{—}CH_2CH_2OH \;+\; N_2 \;+\; H_2O$$

Now, we know that primary amines react with nitrous acid to give diazonium compounds that are very unstable, decomposing to nitrogen and carbonium ions: $RNH_2 + HONO \rightarrow RN_2^+ \rightarrow R^+ + N_2$. We also know that carbonium ions can react with water to yield alcohols: $R^+ + H_2O \rightarrow ROH + H^+$. A reasonable mechanism for the conversion of phenylethylamine to 2-phenylethanol, therefore, postulates the formation of the phenylethyldiazonium compound, its decomposition to phenylethyl cation, and reaction of the cation with water to form the product:

$$C_6H_5CH_2CH_2NH_2 \xrightarrow[H^+]{HONO} C_6H_5CH_2CH_2N_2^+ \xrightarrow{(-N_2)} C_6H_5CH_2CH_2^+ \xrightarrow[(-H^+)]{H_2O} C_6H_5CH_2CH_2OH$$

That something more than this simple interpretation is required was indicated when a sample of phenylethylamine "tagged" with isotopic carbon (^{14}C) at the aminomethylene carbon was used as starting material. If the reaction follows the simple course outlined, the product should be phenylethanol in which *all* of the isotopic carbon is at the carbinol carbon. It was discovered, though, that in only half the molecules was the isotopic carbon at this position; in the other half it was found at the benzyl carbon;

$$C_6H_5-CH_2\overset{*}{C}H_2NH_2 \xrightarrow[H^+]{HONO} \underset{50\%}{C_6H_5-CH_2\overset{*}{C}H_2OH} + \underset{50\%}{C_6H_5-\overset{*}{C}H_2CH_2OH}$$

To explain this new facet of the reaction that was revealed by the isotopic labeling experiment, it was postulated that the intermediate carbonium ion is a "bridged" species, now called a "phenonium" ion, in which the phenyl group is attached to *both* of the methylene carbons in an equivalent fashion. Thus, reaction of water with this carbonium ion takes place at either one of the methylene carbons to an equal extent, yielding the two labeled compounds in equal amount:

equivalent positions

$$C_6H_5-CH_2\overset{*}{C}H_2NH_2 \longrightarrow CH_2-\overset{*}{C}H_2$$

Bridged carbonium ion

Here, again, it should be stressed that the isotope labeling experiment does not *prove* this mechanism, although it does rule out the earlier mechanisms which invoked a simple carbonium ion. Whether a "bridged" species or some other symmetrical entity is involved in this reaction remains a point of controversy.

PROBLEM 20-14

It is conceivable that the position of attachment to the phenyl ring is different in the reactant $C_6H_5CH_2CH_2NH_2$ and the product $C_6H_5CH_2CH_2OH$. How might this possibility be tested experimentally?

We have already stated that isotopes are excellent "tags" because of the close similarity of the chemistry of the members of an isotopic set. There are some quantitative differences, however, of which we can often take advantage. The substitution of one isotope by another, *i.e.*, C—D for C—H, produces a change in the vibrational frequency of the bond and, therefore, in the bond strength. Thus, the rates at which isotopically related bonds undergo reaction may be different. This phenomenon is called the *kinetic isotope effect*, and it is expressed as the ratio of the reaction rates of the substrate containing the lighter-mass isotope and the substrate containing the heavier-mass isotope—*i.e.*, k_H/k_D in the case of hydrogen and deuterium. The kinetic isotope effect is usually greater than unity, although in certain instances it can fall below unity. Its magnitude is a function of the position of the hydrogen (or deuterium) between the atom from which it is departing and the atom to which it is becoming attached, reaching its maximum value in activated complexes in which it is half way between these atoms.

An example of the way in which the kinetic isotope effect can be used as a mechanism probe is provided by the reaction of *t*-butyl bromide with sodium ethoxide in ethanol. In addition to *t*-butyl ethyl ether, which has been our focal point in the previous discussion of this reaction, 2-methylpropene is also formed. If the deuterated analog of *t*-butyl bromide were used in place of the ordinary compound it would turn out that the ratio of 2-methylpropene to *t*-butyl ether would be lower when high concentrations of sodium ethoxide are used. If we compare the rates of formation of products from the proteo and deuterio compounds (*i.e.*, k_H/k_D ratios for alkene formation and ether formation), we would find that the value for ether formation is not very much larger than 1, whereas that for alkene formation is significantly above 1. This difference is ascribed to a difference in mechanism. Whereas the ether is formed via an intermediate carbonium ion (no C—D bond breaking in the rate-determining process), the alkene is formed by attack of ethoxide ion on one of the β hydrogens (or deuteriums) to yield an activated complex of the structure shown, *viz.*, a structure in which a C—H (or C—D) bond is partly broken and an O—H (or O—D) bond is partly formed.

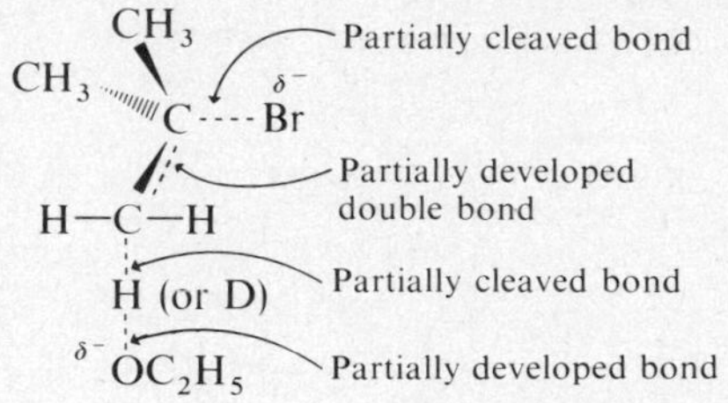

20.4 Stereochemistry as a Means for Studying Reaction Mechanisms

When a chiral center is the site at which a reaction takes place, its stereochemistry can experience any one of several fates: (a) it can be converted to an achiral center; (b) it can be converted to a new chiral center with retention of absolute configuration; (c) it can be converted to a new chiral center with inversion of absolute configuration; or, (d) it can be converted to a new chiral center with both enantiomers being formed in equal amounts (Fig. 20.7). If optically active

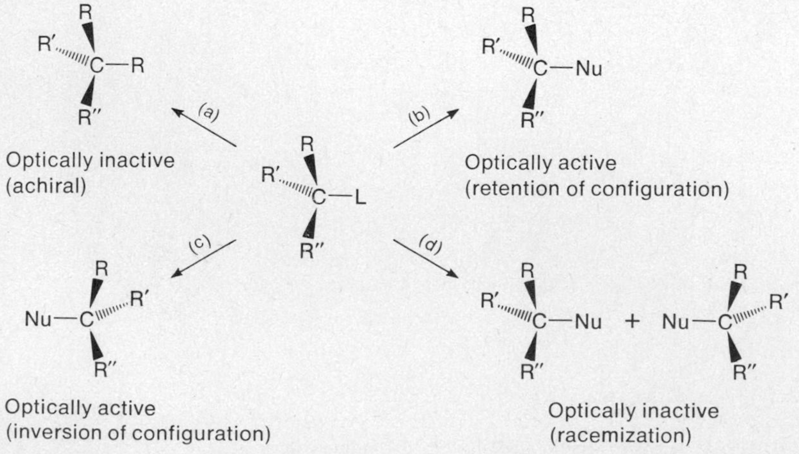

Fig. 20.7. Stereochemical modes of reaction at a chiral center.

starting material is used, fates (a) and (d) will lead to complete loss of optical activity, whereas fates (b) and (c) will result in the formation of optically active product. Thus, the retention or loss of optical activity and retention or inversion of absolute configuration provides insight into reaction mechanisms, for each of these four possibilities requires a different mechanistic interpretation. For this reason, the use of stereochemistry has been one of the most powerful means for studying reaction mechanisms.

To illustrate the application of stereochemistry to the study of reaction mechanisms, let us again return to the reactions of methyl bromide, neopentyl bromide, and *t*-butyl bromide with sodium ethoxide in ethanol. Although none of these molecules is chiral, chiral centers can be introduced by appropriate isotopic substitutions (which do not significantly alter the reaction characteristics of the molecules*) to give the compounds shown in Fig. 20.8. If optically active forms of these compounds were subjected to reaction with sodium ethoxide in ethanol, as previously described, both the methyl bromide and neopentyl

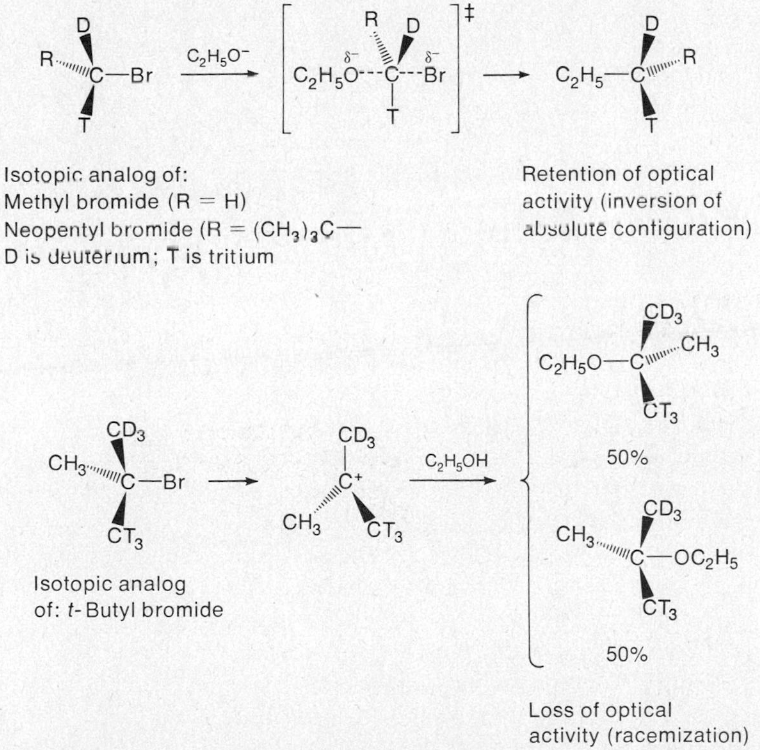

$$\mathrm{RCDT{-}Br} \xrightarrow{C_2H_5O^-} \left[C_2H_5\overset{\delta-}{O}\cdots C(R)(D)(T)\cdots \overset{\delta-}{Br} \right]^{\ddagger} \longrightarrow C_2H_5{-}CDTR$$

Isotopic analog of:
Methyl bromide (R = H)
Neopentyl bromide (R = $(CH_3)_3C$—
D is deuterium; T is tritium

Retention of optical activity (inversion of absolute configuration)

$$CH_3(CD_3)(CT_3)C{-}Br \longrightarrow CH_3(CD_3)(CT_3)C^+ \xrightarrow{C_2H_5OH} \begin{cases} C_2H_5O{-}C(CD_3)(CH_3)(CT_3) & 50\% \\ CH_3(CD_3)(CT_3)C{-}OC_2H_5 & 50\% \end{cases}$$

Isotopic analog of: *t*-Butyl bromide

Loss of optical activity (racemization)

Fig. 20.8. Stereochemical course of the reaction of methyl, neopentyl, and *t*-butyl bromide with sodium ethoxide in ethanol.

---

* Although the isotopically substituted forms of methyl bromide, neopentyl bromide, and *t*-butyl bromide shown in Fig. 20.8 have not actually been prepared, similar compounds have been synthesized and resolved. For example, $CHDTCO_2H$ (*i.e.*, deuteriotritioacetic acid) has been obtained in an optically active form (D is the symbol for deuterium, and T is the symbol for tritium).

bromide would give products which are optically active and in which the absolute configuration is inverted, commensurate with a mechanism that postulates a one-step, concerted process in which the incoming group becomes attached to one face of the chiral center as the leaving group departs from the opposite face (Fig. 20.8). The loss of optical activity in the reaction of *t*-butyl bromide is commensurate with a mechanism that postulates a two-step process, the first step involving the formation of a planar, therefore achiral carbonium ion (Fig. 20.8).

PROBLEM 20-15

Predict the stereochemical outcome of the reaction of optically active C_6H_5CHDCl with (a) $LiAlH_4$ to give $C_6H_5CH_2D$, (b) $LiAlT_4$ to give C_6H_5CHDT, and (c) $LiAlD_4$ to give $C_6H_5CHD_2$.

20.5 The Detection of Reaction Intermediates as a Means for Studying Reaction Mechanisms

Some reactions involve the formation of discrete intermediates during the transformation of reactants to products, whereas others do not (see Section 7.4). When intermediates do occur along the reaction pathway, they are usually very reactive and short lived, either reverting to starting material or proceeding to product with great speed. One of the methods for the detection of reaction intermediates makes use of spectroscopic techniques (*i.e.*, ir, uv, and nmr), for reaction intermediates frequently possess characteristic spectral properties that permit their formation and disappearance to be detected and followed. In many instances, however, the intermediates are present in concentrations that are too low to be measured by spectral techniques.

Another method for the detection of reaction intermediates employs reagents that react with the intermediates to form stable, isolable compounds whose structures reflect the structure of the intermediate. Such compounds are referred to as "trapping reagents." For example, when bicyclo[5.2.1]decan-10-one is irradiated with ultraviolet light, cyclononene and carbon monoxide are produced. We can show that the reaction involves at least two reactive intermediates by carrying out a pair of experiments, one in which piperidine is added to the reaction mixture and another in which a large amount of an alkene such as 2-butene is added to the reaction mixture. In the first instance, the product consists almost entirely of the amide (Fig. 20.9) resulting from the reaction between the piperidine (*i.e.*, the trapping reagent) and the ketene (the first detectable intermediate along the reaction pathway). In the second instance, the product contains some 1,2-dimethylspiro[2.8]undecane resulting from the reaction of the added alkene with the carbene intermediate.

PROBLEM 20-16

Suggest another method for detecting a ketene as a reaction intermediate.

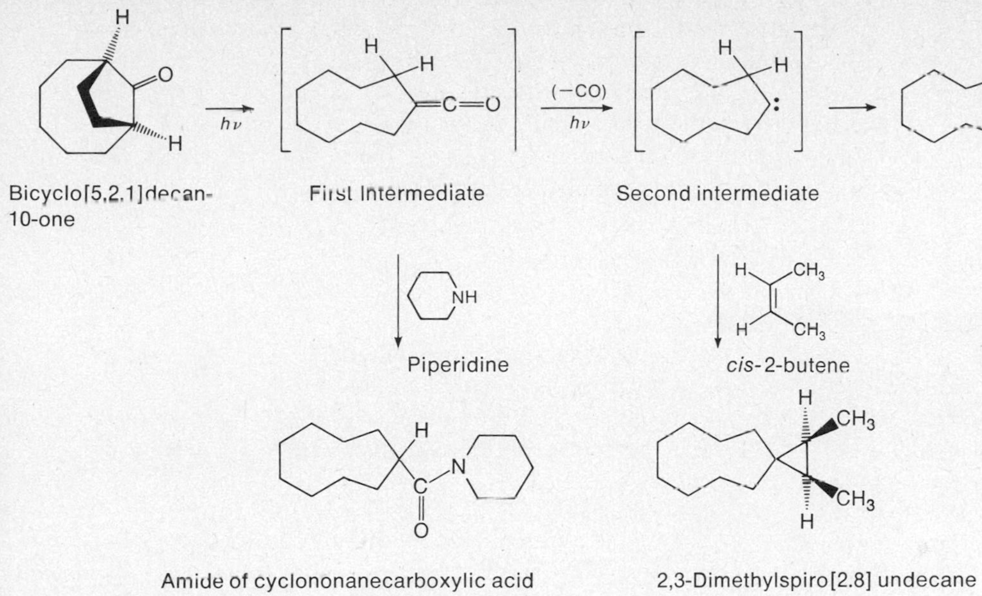

Bicyclo[5.2.1]decan-10-one — $h\nu$ → [First intermediate] — $(-CO)$, $h\nu$ → [Second intermediate] →

Piperidine → Amide of cyclononanecarboxylic acid

cis-2-butene → 2,3-Dimethylspiro[2.8] undecane

Fig. 20.9. Piperidine and 2-butene as trapping reagents in the photolysis reaction of bicyclo[5.2.1]decan-10-one.

The technique of adding a "trapping" reagent can also be applied to the reactions of methyl bromide, neopentyl bromide, and *t*-butyl bromide, which have occupied our attention throughout this chapter. For example, when lithium azide is added to the *t*-butyl bromide reaction mixture, the product consists largely of *t*-butyl azide and the rate of the reaction remains almost unchanged. The lithium azide has trapped the intermediate, which, as has already been suggested by other pieces of evidence, is the *t*-butyl cation. When lithium azide is added to the methyl bromide and the neopentyl bromide reaction mixtures, the products contain large amounts of methyl azide and neopentyl azide, respectively, but the rates of the reactions are substantially *increased*. In these instances, therefore, lithium azide is not trapping an intermediate but is acting in some other capacity. On the basis of the other evidence that we have discussed in connection with this reaction, it is reasonable to postulate that azide competes with ethoxide for the formation of the activated complexes that lead to products.

20.6 Thermodynamic vs. Kinetic Control of Reactions

An important aspect of a reaction, although usually of limited significance with respect to revealing its mechanism, is the question of whether thermodynamic or kinetic control is operative. In a closed system in which all the components of a reaction mixture remain in contact with each other, all reacting systems eventually reach a state of thermodynamic equilibrium. Thus, in a system containing the starting material (compound A) and its two products (compound B and compound C), the amounts of compounds A, B, and C that are present *at equilibrium* are determined by the relative thermodynamic stabilities of these

compounds, *i.e.*,

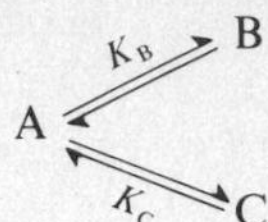

For illustration, let us assume that the equilibrium constant, K_B, for the formation of product B is 100 times greater than the equilibrium constant, K_C, for the formation of product C. Clearly, product B will predominate at equilibrium, accompanied by only about 1% of product C. Under these conditions, B is said to be the *product of thermodynamic control.*

Systems do not reach equilibrium instantaneously, however. Indeed, it sometimes requires an almost infinite time for this to occur. A diamond, for example, is unstable relative to its oxidation products, carbon dioxide and water, but has virtually no tendency to equilibrate with these compounds. The rate at which equilibrium is established is governed by the *free energies of activation.* These vary not only from one system to another but also for the various components of a given system. Thus, the free energies of activation, $\Delta G_B^\ddagger$ and $\Delta G_C^\ddagger$ for the formation of products B and C are not necessarily the same; nor do they necessarily differ from each other by the same increment nor even in the same direction as the free energies of the reaction; *i.e.*, ΔG_B and ΔG_C. Consequently, situations can arise in which the major isolable product is determined not by the relative energies of the reactants and products but by the rate at which the products are formed.

For illustration, let us assume that the rate constants for the forward and reverse reactions of A to B are 1000 and 1 $\sec^{-1}$, respectively. Now, the equilibrium constant for a reaction is equal to the rate constant of the forward reaction divided by the rate constant of the reverse reaction; thus $K_B = 1000/1 = 1000$. Let us next assume that the rate constants for the forward and reverse reactions of A to C are 100,000 and 10,000 $\sec^{-1}$ respectively, leading to an equilibrium constant K_C of $100{,}000/10{,}000 = 10$. The ratio of the equilibrium constants K_B and K_C is $1000/10 = 100$, in accord with the previous assumption that, at equilibrium, product B is present in 100 times greater concentration than product C. But products B and C are not formed at equal rates; initially, product C is formed 100 times more rapidly than product B, which means that if the reaction mixture is worked up well before equilibrium has been reached, C will be the major product. It is said to be the *product of kinetic control.*

$$K_B = \frac{1000}{1} = 1000$$

$$\frac{[B]}{[C]} = \frac{K_B}{K_C} = \frac{1000}{10} = 100$$

$$K_C = \frac{100{,}000}{10{,}000} = 10$$

The reaction coordinate diagram in Fig. 20.10 portrays the free energy relationships that prevail in this system.

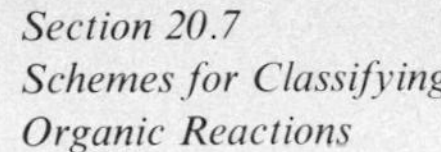

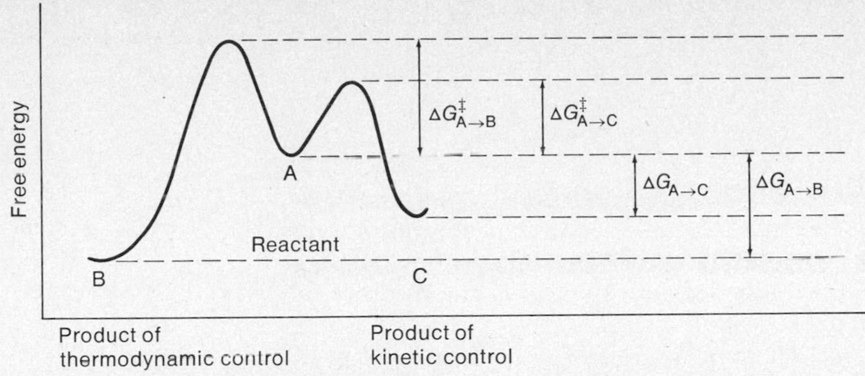

Fig. 20.10. Reaction coordinate diagram for system illustrating products of thermodynamic control and kinetic control.

PROBLEM 20-17

Draw the reaction coordinate for a system in which reactant A goes to products B and C, product C being that of kinetic *and* thermodynamic control. Draw the reaction coordinate for a system in which the equilibrium concentrations of products B and C are the same but in which product B is that of kinetic control.

A classical example of thermodynamic *vs.* kinetic product control involves the treatment of a mixture containing equivalent amounts of cyclohexanone, furfuraldehyde, and semicarbazide. If the reaction mixture is worked up after a few seconds, the product is almost entirely the semicarbazone of cyclohexanone, but if the mixture is allowed to stand for several hours before workup, the product is almost entirely the semicarbazone of furfuraldehyde. Cyclohexanone semicarbazone corresponds to product C in the generalized example that we have just discussed, and furfuraldehyde semicarbazone corresponds to product B.

$NH_2NHCONH_2$ + furfuraldehyde (CHO) ⇌ furfuraldehyde–$CH{=}NNHCONH_2$
Product of thermodynamic control

$NH_2NHCONH_2$ + cyclohexanone (O) ⇌ cyclohexanone=$NNHCONH_2$
Product of kinetic control

20.7 Schemes for Classifying Organic Reactions

20.7a. CLASSIFICATION ON THE BASIS OF TYPE OF REACTION INVOLVED. The three major types of organic reactions are (a) substitution reactions (also called displacement reactions), (b) addition reactions, and (c) elimination reactions. It is usually possible to place any particular reaction into one of these three categories, although sometimes the distinctions are blurred. Also, there are frequently other descriptors that find their way into the designation of reaction type. For example, reactions are often specified as rearrangement reactions, oxidation reactions, reduction reactions, insertion reactions, etc. In almost all instances, however, these can be interpreted in terms of a substitution reaction, an addition reaction, an elimination reaction, or some combination of these.

20.7b. CLASSIFICATION ON THE BASIS OF THE TYPE OF INTERMEDIATE INVOLVED. The reactive intermediates that are most often involved in organic reactions (see Section 7.4) are carbonium ions, free radicals, carbanions, carbenes, and benzynes. Reactions involving these intermediates are frequently referred to as carbonium ion reactions, carbanion reactions, free radical reactions, carbene reactions, and benzyne reactions. However, a carbonium ion reaction might be a substitution reaction, an addition reaction, an elimination reaction, or a rearrangement reaction; and similarly, reactions involving radicals, carbanions, carbenes, and benzynes might be any one of these several reaction types.

An even broader subdivision that is based on the type of intermediate specifies a reaction as *homolytic* (radical intermediates) or *heterolytic* (cation and anion intermediates), the names designating bond cleavages in which the electrons either distribute evenly between the fragments (*i.e.*, homolysis) or unevenly between the fragments (*i.e.*, heterolysis):

$$A:B \longrightarrow A\cdot + B\cdot \quad \text{Homolytic cleavage}$$

$$A:B \longrightarrow A^+ + B:^- \quad \text{Heterolytic cleavage}$$

20.7c. CLASSIFICATION ON THE BASIS OF EXPERIMENTAL PROCEDURE. The two experimental procedures that are most often used to categorize a reaction are *thermal* and *photochemical*. Reactions induced by heat are referred to as thermal reactions, and those induced by irradiation are referred to as photochemical reactions. Realize, however, that all thermal reactions do not belong to the same "type," nor do all photochemical reactions.

20.8 Classification of Reaction Mechanisms

20.8a. SUBSTITUTION REACTIONS.* Substitution reactions involve the replacement of one group (designated as the *leaving* group and symbolized by L in the following reactions) by another group (designated as the *entering group* and symbolized by Nu for nucleophile, X for radical, or E for electrophile in the following reactions). The three most common types of substitution reactions are called nucleophilic substitution, radical substitution, or electrophilic substitution, depending on the nature of the leaving and entering groups.

	Reactant		Entering Group		Product		Leaving Group
Nucleophilic substitution:	R—L	+	Nu:	$\longrightarrow$	R—Nu	+	L:
Radical substitution:	R—L	+	X·	$\longrightarrow$	R—X	+	L·
Electrophilic substitution:	R—L	+	E^+	$\longrightarrow$	R—E	+	L^+

* The word "displace" is frequently used as a synonym for the word "substitute"; consequently, the terms "displacement reactions" and "substitution reactions" are used interchangeably.

In nucleophilic substitution reactions, the leaving and entering groups possess one or more unshared pairs of electrons (Lewis bases); in radical substitution reactions the leaving and entering groups are radical species bearing one unpaired electron; in electrophilic substitution reactions the leaving and entering groups possess a vacant orbital (Lewis acid).

The designation of a substitution reaction as being nucleophilic or electrophilic is rather arbitrary and depends on that part of the reaction system on which one chooses to focus attention. Thus, in the reaction of CH_3Br with C_2H_5ONa to give $CH_3OC_2H_5$, the reaction is (a) a nucleophilic displacement at CH_3 by $C_2H_5O^-$, (b) an electrophilic displacement at Br by Na^+, (c) a nucleophilic displacement at Na by Br^-, and (d) an electrophilic displacement at C_2H_5O by CH_3^+, depending on what parts of the reacting system are being considered. Generally, the carbon framework of one of the reactants is chosen as the reference point, and the nucleophilic or electrophilic character of the portion of that molecule undergoing replacement determines in what class of reaction the process is to be placed.

Illustrative examples of the several types of substitution reactions are shown in Table 20.6.

If a reaction is to be a useful synthesis method it is essential that it give as high a yield of a single product as possible. The degree to which a reaction takes place at one portion of a molecule to the exclusion of other parts of the same molecule is designated as the *situselectivity* (Greek, *situ*, site or place). For example, the reaction of *n*-butyl bromide with sodium ethoxide in ethanol yields *n*-butyl ethyl ether as the major product, reflecting the very high situselectivity of this process. On the other hand, the reaction of *n*-butyl bromide with chlorine under the influence of light gives a mixture of products in which the

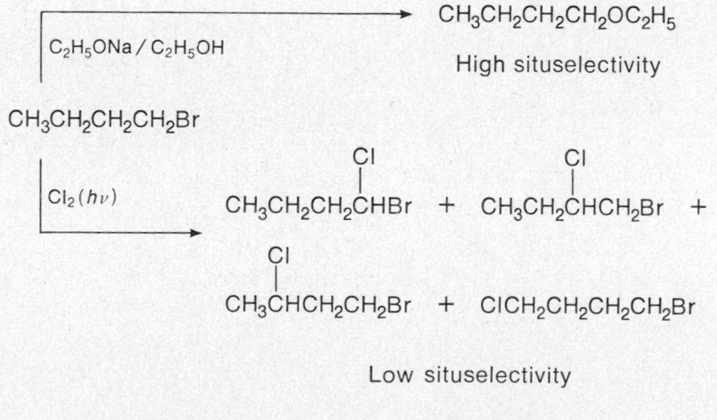

$$CH_3CH_2CH_2CH_2Br \xrightarrow{C_2H_5ONa/C_2H_5OH} CH_3CH_2CH_2CH_2OC_2H_5$$

High situselectivity

$$CH_3CH_2CH_2CH_2Br \xrightarrow{Cl_2(h\nu)} CH_3CH_2CH_2CHClBr + CH_3CH_2CHClCH_2Br + CH_3CHClCH_2CH_2Br + ClCH_2CH_2CH_2CH_2Br$$

Low situselectivity

Fig. 20.11. Situselectivity of organic reactions.

chlorine atom has reacted at various sites in the molecule (Fig. 20.11). Clearly, the latter process is less useful as a preparative method, for a mixture of compounds is produced.

Table 20.6 EXAMPLES OF SUBSTITUTION REACTIONS

Nucleophilic Substitution Reactions:

Nucleophilic substitution at saturated carbon:

$$CH_3{:}Br{:} + {:}C_2H_5O{:}Na \longrightarrow CH_3OC_2H_5 + Na{:}Br{:}$$

Leaving group   Nucleophile

Nucleophilic substitution at unsaturated carbon:

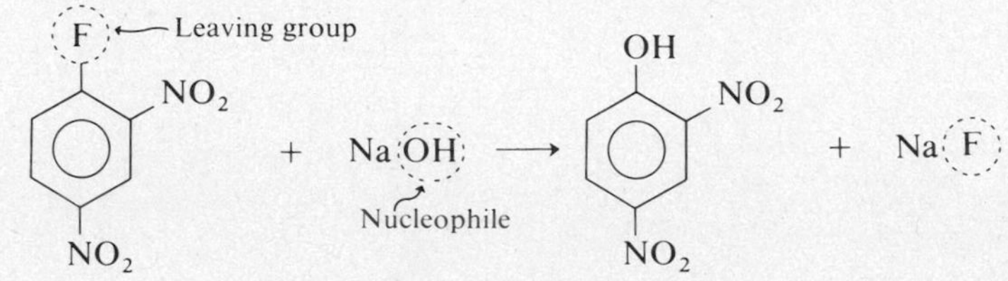

F ← Leaving group; NO_2; NO_2 + Na:OH: ⟶ OH; NO_2; NO_2 + Na:F:

Nucleophile

Nucleophilic substitution at carbonyl carbon:

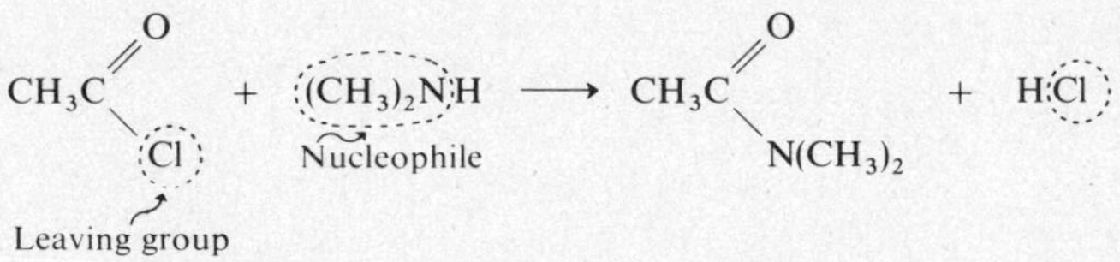

$$CH_3C(=O)Cl + {:}(CH_3)_2N{:}H \longrightarrow CH_3C(=O)N(CH_3)_2 + H{:}Cl{:}$$

Nucleophile

Leaving group

Radical Substitution Reactions:

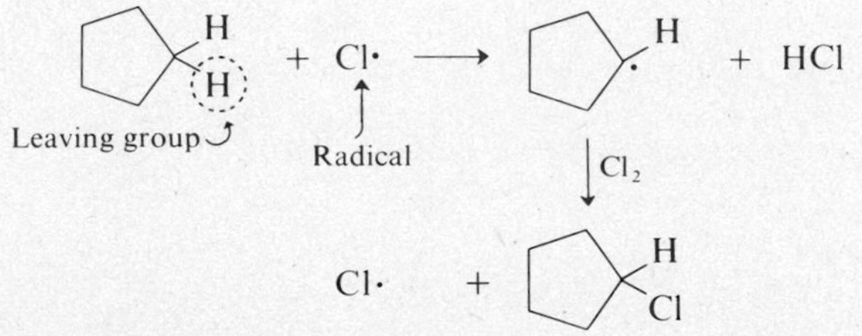

H, H + Cl· ⟶ H, · + HCl

Leaving group   Radical

↓ Cl_2

Cl· + H, Cl

Electrophilic Substitution Reactions:

Electrophilic substitution at saturated carbon:

$$\overset{\delta-}{CH_3}{:}\overset{\delta+}{MgCl}{:} + {:}H^+{:}Cl^- \longrightarrow CH_4 + MgCl_2$$

Leaving group   Electrophile

Electrophilic substitution at unsaturated carbon:

H ← Leaving group; + Cl^+ ⟶ Cl + H^+

Electrophile

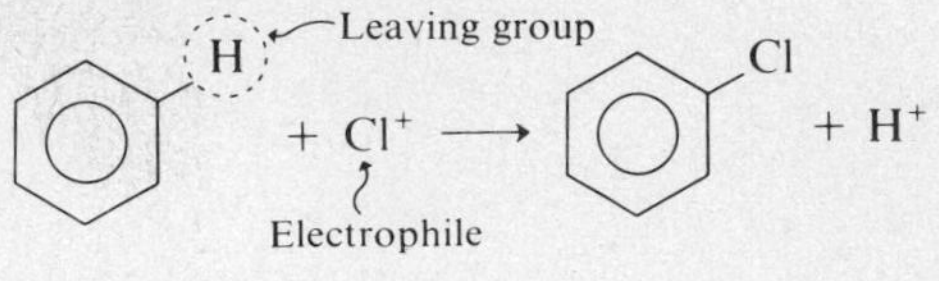

20.8b. ADDITION REACTIONS. Addition reactions are processes in which new bonds are formed at both atoms of a multiple bond, with consequent decrease in the degree of unsaturation of the molecule. They can be classed as (a) electrophilic additions, (b) nucleophilic additions, (c) radical additions, and (d) concerted additions, depending on the nature of the entities that are adding and the sequence in which they add.

Electrophilic addition:

$$\mathrm{{>}C{=}Z + E^+ \longrightarrow \left[{>}\overset{+}{C}{-}Z{-}E\right] \xrightarrow{Nu^-} Nu{-}\overset{|}{\underset{|}{C}}{-}Z{-}E}$$

Electrophile attacks first (cationic intermediate)

Nucleophilic addition:

$$\mathrm{{>}C{=}Z + Nu^- \longrightarrow \left[Nu{-}\overset{|}{\underset{|}{C}}{-}Z^-\right] \xrightarrow{E^+} Nu{-}\overset{|}{\underset{|}{C}}{-}Z{-}E}$$

Nucleophile attacks first (anionic intermediate)

Radical addition:

$$\mathrm{{>}C{=}Z + X\cdot \longrightarrow \left[{>}\dot{C}{-}Z{-}X\right] \xrightarrow{X{-}X} X{-}\overset{|}{\underset{|}{C}}{-}Z{-}X + X\cdot}$$

Radical intermediate

Concerted addition:

$$\mathrm{{>}C{=}Z + A{-}B \longrightarrow \left[\begin{array}{c}{-}\overset{|}{C}{\cdots}Z\\ \vdots\quad\vdots\\ A{\cdots}B\end{array}\right]^{\ddagger} \longrightarrow A{-}\overset{|}{\underset{|}{C}}{-}Z{-}B}$$

Activated Complex

No intermediate

In electrophilic addition, the first step is the attachment of an electrophile, followed in a second step by the attachment of a nucleophile. In nucleophilic addition, the sequence is reversed, and the nucleophile attacks first, followed by the electrophile. Radical addition involves attachment of a radical in the first step followed by a radical abstraction reaction in the second step. Concerted additions proceed with the synchronous formation of both of the new bonds without the intervention of intermediates and without the development of significant charge density or free electron density in the structures involved along the reaction coordinate. The reagents that are added in concerted reactions are often multiple bonded, leading to cyclic products in *cycloaddition* processes.

Illustrative examples of the several types of addition reactions are shown in Table 20.7.

The types of unsaturated groups that are most commonly involved in addition reactions are C=C, C≡C, C=O, C=NR, and C≡N. Although C=C and C≡C bonds are, of themselves, symmetrical, they may be unsymmetrically substituted (*e.g.*, RCH=CHR′, RC≡CR′); C=O, C=NR, and C≡N bonds are inherently unsymmetrical, regardless of the attached groups. In the majority of instances, therefore, two structurally different modes of addition are possible if the reagent being added (referred to as the *addend*) is unsymmetrical. The direction of the addition in such instances is called the *regioselectivity* of the reaction. For example, 2-methylpropene can react with hydrogen chloride to

Table 20.7 EXAMPLES OF ADDITION REACTIONS

Electrophilic addition:

$$(CH_3)_2CH{=}CH_2 \; + \; \underset{\text{Electrophile}}{H^+}Cl^- \longrightarrow (CH_3)_2\overset{\displaystyle Cl}{\overset{|}{C}}HCH_3$$

Nucleophilic addition:

$$(CH_3)_2C{=}O \; + \; \underset{\text{Nucleophile}}{\overset{\delta^-}{CH_3}}\overset{\delta^+}{Mg}I \longrightarrow (CH_3)_2C\begin{matrix} \diagup OMgI \\ \diagdown CH_3 \end{matrix}$$

Radical addition:

$$H_2C{=}CH_2 \quad \underset{\text{Radical}}{Cl\cdot}\cdot Cl \longrightarrow ClCH_2CH_2Cl$$

Concerted addition:

$$CH_2{=}CH{-}CH{=}CH_2 \; + \; CH_2{=}CH{-}\overset{\displaystyle O}{\overset{\|}{C}}{-}CH_3 \longrightarrow \text{cyclo-}(CH{=}CH{-}CH_2{-}CH(\overset{\displaystyle O}{\overset{\|}{C}}CH_3){-}CH_2{-}CH_2)$$

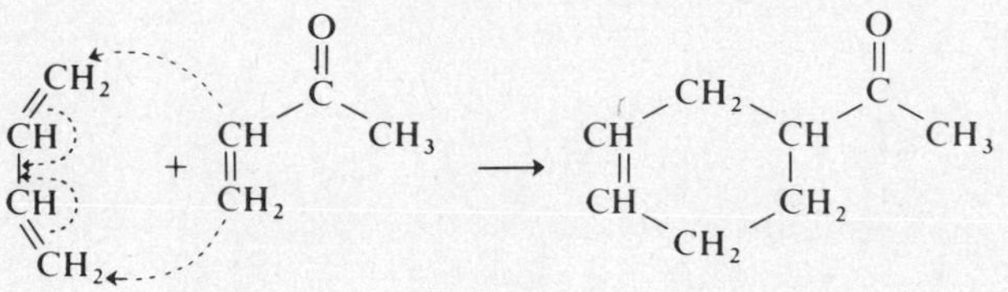

give either *t*-butyl chloride or isobutyl chloride. The ratio of *t*-butyl chloride to isobutyl chloride is a measure of the regioselectivity of this reaction, and the fact that this ratio is considerably greater than unity is formulated as Markovnikov's rule (see p. 204). In a mechanistically similar reaction, hypochlorous acid can add to 2-methylpropene to give 1-chloro-2-methyl-2-propanol or 2-chloro-2-methyl-1-propanol. In this case, the ratio of products is much closer to unity, indicating that the regioselectivity for hypochlorous acid addition is lower than that for hydrogen chloride addition (Fig. 20.12).

In C=C, C≡C, and C=NR compounds, two *stereochemically* different modes of addition are structurally discernible if the bonds are unsymmetrically

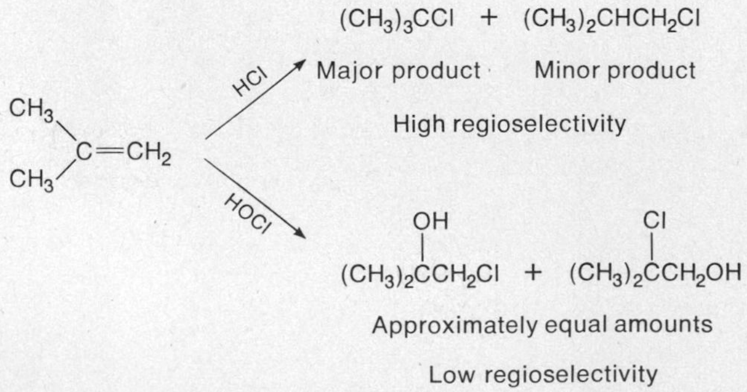

$$(CH_3)_2C{=}CH_2 \xrightarrow{HCl} \underset{\text{Major product}}{(CH_3)_3CCl} + \underset{\text{Minor product}}{(CH_3)_2CHCH_2Cl}$$

High regioselectivity

$$(CH_3)_2C{=}CH_2 \xrightarrow{HOCl} (CH_3)_2\overset{\displaystyle OH}{\overset{|}{C}}CH_2Cl + (CH_3)_2\overset{\displaystyle Cl}{\overset{|}{C}}CH_2OH$$

Approximately equal amounts

Low regioselectivity

Fig. 20.12. Regioselectivity of organic reactions.

substituted. These are designated as *syn* addition if the groups are added on the same face of the multiple bond and *anti* addition if the groups are added at opposite faces of the multiple bond.* The degree to which one stereochemistry predominates is specified as the *stereoselectivity* of the reaction. For example, *cis*-2-butene reacts with chlorine in a highly stereoselective *anti* addition to form (±)-2,3-dichlorobutane and with perbenzoic acid in a highly stereoselective *syn* addition to form *cis*-2,3-dimethyloxirane (*meso* compound) (Fig. 20.13).

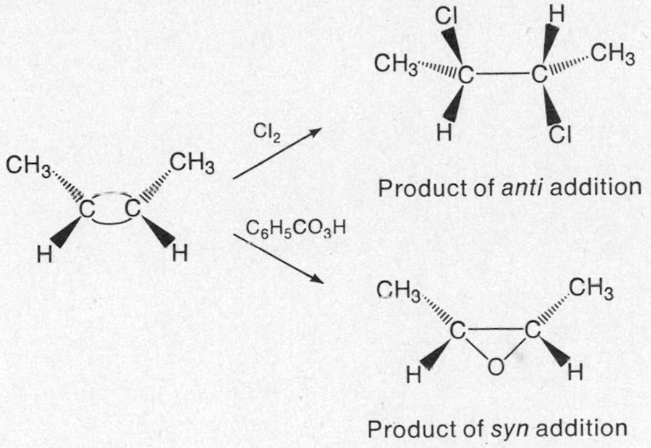

Cl_2 → Product of *anti* addition

$C_6H_5CO_3H$ → Product of *syn* addition

Fig. 20.13. Stereoselectivity of organic reactions.

The term *stereospecificity*† is used in defining another aspect of stereochemistry of some reactions, applicable to systems in which diastereomers yield stereochemically different products. For example, *trans*-2-butene reacts with chlorine and perbenzoic acid in a fashion comparable to the *cis* isomer, but it yields stereoisomers of the products obtained from the *cis* isomer; chlorination yields the *meso* form of 2,3-dichlorobutane, and epoxidation yields *trans*-2,3-dimethyloxirane (an enantiomeric mixture). With respect to the diastereomeric set of *cis*- and *trans*-2-butene, these reactions are said to be stereospecific, one diastereomer yielding one stereoisomer and the other diastereomer yielding a different stereoisomer; with respect to the compounds individually, these reactions are said to be highly stereoselective (Fig. 20.14).

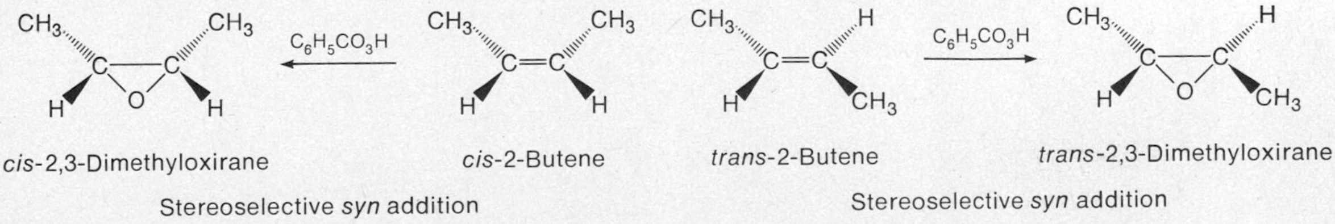

cis-2,3-Dimethyloxirane ← $C_6H_5CO_3H$ ← *cis*-2-Butene; *trans*-2-Butene → $C_6H_5CO_3H$ → *trans*-2,3-Dimethyloxirane

Stereoselective *syn* addition — Stereoselective *syn* addition

Fig. 20.14. Stereospecificity of organic reactions.

---

* To avoid confusion with the specification of the double bond diastereomerism of the starting material and/or product, the terms *syn* and *anti* are preferred over the older terms *cis* and *trans* when the stereochemistry of addition processes is discussed.

† The term "stereospecificity" is sometimes used by chemists in another sense, referring to a reaction in which the stereoselectivity is essentially 100%.

20.8c. ELIMINATION REACTIONS. Elimination reactions are the converse of addition reactions. They involve the *removal* of functions from a molecule, with a consequent increase in the degree of unsaturation of the molecule. One method for classifying these reactions is based on the proximity relationship between the groups that depart from the carbon framework in the elimination process. If the leaving groups are attached to the same atom, the process is classed as an *α elimination*; if they are attached to adjacent atoms, it is a *β elimination*; if they are attached in a 1,3 relationship, it is a γ elimination;

α Elimination:

$$R_2CAB \longrightarrow A + B + R_2C: \quad \textit{Formation of a carbene}$$

β Elimination

$$R_2C(A){-}CR_2(B) \longrightarrow A + B + R_2C{=}CR_2 \quad \textit{Formation of an alkene}$$

γ Elimination:

$$R_2C(A){-}C{-}CR_2(B) \longrightarrow A + B + \text{cyclo-}(R_2C{-}C{-}CR_2) \quad \textit{Formation of a cyclopropane}$$

Table 20.8 EXAMPLES OF ELIMINATION REACTIONS

α-Elimination Reaction:

$$H_2CHCl + (CH_3)_3COK \longrightarrow H_2C: + (CH_3)_3COH + KCl$$

Carbene: A reactive species that cannot be isolated

β-Elimination Reactions:

Ionic reaction:

$$CH_3{-}C(CH_3)(Br){-}CH_2{-}H + C_2H_5ONa \xrightarrow[C_2H_5OH]{} (CH_3)_2C{=}CH_2 + NaBr + C_2H_5OH$$

Thermal reaction:

$$H{-}CH_2CH_2OC(=O)CH_3 \xrightarrow[\text{Heat}]{} H_2C{=}CH_2 + CH_3CO_2H$$

γ-Elimination Reaction:

$$C_6H_5{-}CH(H)CH_2CH_2{-}F + C_2H_5ONa \longrightarrow C_6H_5{-}CH\langle(CH_2)_2 + C_2H_5OH + NaF$$

Elimination reactions are also classified on the basis of their mechanistic pathway as *ionic* or *thermal* (*i.e.*, nonionic). Reactions following an ionic pathway can be further subdivided on the basis of the timing and sequence of the events that transpire during the elimination process, as discussed in Chapter 25.

Illustrative examples of the several types of elimination reactions are shown in Table 20.8.

Considerations of regioselectivity and stereoselectivity come into play in many elimination reactions. With respect to regioselectivity, the Saytzev rule (see p. 242) and the Hofmann rule (see p. 316) have been used as descriptors. The regioselectivity and stereoselectivity of these processes are considered in detail in Chapter 25 on the alkenes.

20.8d. Rearrangement reactions. In considering organic reactions at the lowest level of sophistication, it is generally assumed that the carbon frameworks involved remain structurally invariant throughout the course of the reaction. In the majority of instances, this is a reasonably good assumption, although we have already seen examples of cases where it fails. Thus, the reaction of neopentyl alcohol with hydrogen bromide (see Fig. 10.16) proceeds with skeletal rearrangement and exemplifies the type of alteration that is most often encountered, *viz.*, that involving migration of an atom or group Z (*e.g.*, Z = halogen, RO—, RS—, R_2N—, $-\overset{|}{\underset{|}{C}}-$, or H) from one atom, A, to an adjacent atom, B, *with a pair of bonding electrons.* Such rearrangements are generally categorized as nucleophilic 1,2-shifts;

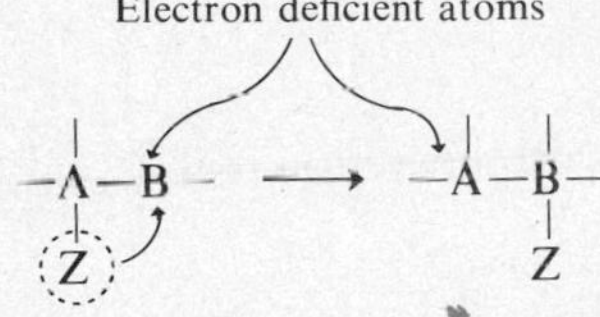

The essential requirement for the reaction is that the atom B have only a sextet of electrons in its outer valence shell. When atom B is carbon or nitrogen, the electron sextet may exist in neutral species (*i.e.*, carbenes and nitrenes) as well as in cationic species (*i.e.*, carbonium and nitrenium ions); when atom B is oxygen, only the cation provides this possibility.

Electron-deficient carbon species:

R_3C^+ (R—C(R)—R, +)
Carbonium ion

$R_2C:$ (R—C(R):)
Carbene

Electron-deficient nitrogen species:

$R_2N:^+$ (R—N(R):, +)
Nitrenium ion

R—N̈:
Nitrene

Electron-deficient oxygen species:

R—Ö:$^+$

Table 20.9 Examples of rearrangement reactions

Ionic Rearrangements (Nucleophilic 1,2-Rearrangements):

Rearrangements involving migration to electron-deficient carbon:

$$(CH_3)_3CCH_2OH + HBr \longrightarrow (CH_3)_2C(Br)CH_2CH_3 + H_2O$$

Rearrangements involving migration to electron-deficient nitrogen:

$$C_6H_5C(=NOH)CH_3 \xrightarrow{H^+} C_6H_5NHC(=O)CH_3$$

Rearrangement involving migration to electron-deficient oxygen:

$$C_6H_5C(=O)CH_3 + C_6H_5CO_3H \longrightarrow C_6H_5OC(=O)CH_3 + C_6H_5CO_2H$$

Thermal Rearrangements (Concerted Reactions):

$$C_6H_5OCH_2CH{=}CH_2 \longrightarrow \text{(6-allyl-2,4-cyclohexadienone, with H at C-6)} \longrightarrow o\text{-}HOC_6H_4CH_2CH{=}CH_2$$

Table 20.9 shows examples of typical nucleophilic 1,2 rearrangement reactions. The mechanisms of these processes are discussed in detail at the appropriate point in later chapters, where the utility of the reactions is considered.

In addition to ionic rearrangements, exemplified by nucleophilic 1,2 shifts, rearrangements induced by heat or light are known, in which the reactant changes to product in a concerted fashion, without the development of charge and without the intervention of intermediates. One example of a reaction of this type is illustrated in Table 20.9. This and other examples are considered in more detail in later chapters.

20.8e. Oxidation-reduction reactions. Many of the reactions of inorganic compounds involve the complete transfer of one or more electrons from one species to another, the species that lose the electrons being said to have been oxidized and the species that gain the electrons being said to have been reduced:

$$A\cdot + B \longrightarrow A^+ + \cdot B^-$$

Only a few organic reactions can be classified as oxidation-reduction processes in this sense. For example, metallic sodium reacts with naphthalene in liquid ammonia to give sodium ion (*i.e.*, sodium is oxidized) and naphthalene radical anion (*i.e.*, naphthalene is reduced).

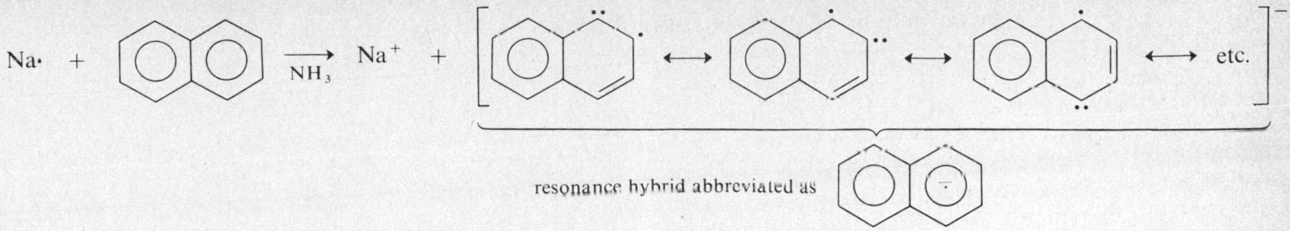

Na· + [naphthalene] $\xrightarrow[NH_3]{}$ Na^+ + [resonance structures of naphthalene radical anion] ⟷ ⟷ ⟷ etc. $]^-$

resonance hybrid abbreviated as

The majority of reactions involving organic compounds do *not* involve the complete transfer of electrons from one molecule (or atom) to another, and therefore require the concept of oxidation and reduction to be interpreted in a somewhat different fashion. One approach is to consider whether the electron density at a particular portion of a molecule has decreased or increased as the result of the reaction. If it has decreased, the molecule has been oxidized; if it has increased, the molecule has been reduced. For example, the reaction of bromine with toluene to yield benzyl bromide can be viewed as an oxidation process from the standpoint of the toluene molecule. Whereas the electrons of the C—H bond of the methyl group of toluene are almost equally shared by the carbon and hydrogen, the electrons of the C—Br bond in the product are displaced toward the bromine; *i.e.*, the electron density on the benzyl carbon is lower in the product than in the starting material, and in this sense the carbon framework of the molecule can be said to have been oxidized:

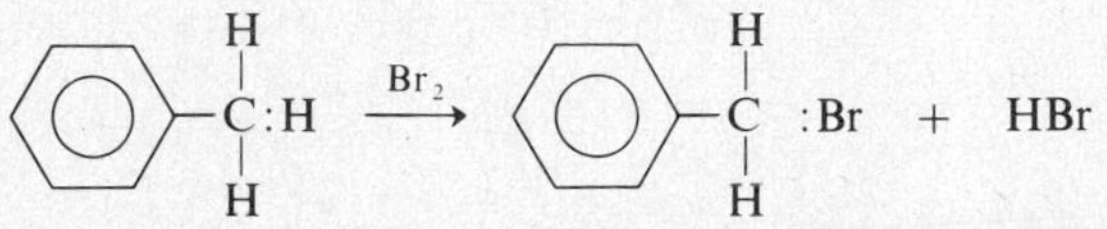

$$C_6H_5\text{—}CH_2\text{:}H \xrightarrow{Br_2} C_6H_5\text{—}CH_2\ \text{:Br} + HBr$$

Another basis for assessing whether a reaction is an oxidation or reduction process is to consider the hydrogen content of the reactant and products. Products of oxidation have less hydrogen than the reactant; products of reduction have more hydrogen than the reactant. In the previous example, benzyl bromide has one less hydrogen than toluene, and by this criterion it can be said to represent an oxidation product.

Neither of these criteria is without flaw, however. The first one is too encompassing in its scope, for almost all organic reactions involve some change in the electron density at the reaction site. The second one is too restricted in its scope and, furthermore, not always true. For example, the dehydration of ethanol to ethylene yields a product containing less hydrogen than the reactant, but the process is not ordinarily considered to be an oxidation. Thus, the guidelines for classifying reactions as oxidations or reductions are not always well defined. We are forced, therefore, to choose from among the reactions that adhere to one or the other of the preceding criteria only the ones that organic chemists have somewhat arbitrarily chosen to identify as oxidations, such as those illustrated in Table 20.10. The mechanism of an oxidation or reduction reaction can usually be put into the context of a substitution, addition, or elimination process, as the examples in subsequent chapters illustrate.

Table 20.10 EXAMPLES OF OXIDATION AND REDUCTION REACTIONS

Oxidation Reactions:

$$C_6H_5CH_3 \xrightarrow{Na_2Cr_2O_7/H_2SO_4} C_6H_5CO_2H$$

$$CH_3CH{=}CHCH_3 + KMnO_4 \longrightarrow CH_3CH(OH){-}CH(OH)CH_3$$

Reduction Reactions:

$$(C_3H_7)_2C{=}O + Zn{-}Hg \xrightarrow{HCl} (C_3H_7)_2CH_2$$

$$C_{10}H_{12} \text{ (tetralin)} + H_2 \xrightarrow[\text{Catalyst}]{} C_{10}H_{18} \text{ (decalin)}$$

20.9 *Study Guide*

I. Definition of a reaction mechanism
II. Techniques for studying reaction mechanisms
  A. Determination of structures of reactants and products
  B. Measurement of rates at which reactions occur (kinetics)
  C. Studies of the effects on reaction rates of changes in structures
  D. Use of isotopically substituted compounds
  E. Observation of stereochemical course of reaction
  F. Detection of suspected intermediates
III. Reaction kinetics as a means for studying reaction mechanisms
  A. Correlation of rates with structure
    1. Different members of same family of compounds react at different rates.
    2. Differences in rates may be interpreted in various ways.
  B. Concentration dependence of reaction rates
    1. The order of a reaction
      (a) First-order reactions; rate $= k[A]$
      (b) Second-order reactions; rate $= k[A][B]$ or rate $= k[A]^2$
      (c) Pseudo-first-order reactions
      (d) Reaction order and mechanism
    2. The molecularity of a reaction—number of participants in the activated complex
      (a) Unimolecular reactions—one entity in activated complex
      (b) Bimolecular reactions—two entities in activated complex
    3. Rate-determining step—slowest step in a reaction sequence
  C. Activation parameters from reaction rates
    1. Obtained by measuring rate as a function of temperature
    2. Enthalpy of activation—energy to elevate system to transition state

3. Entropy of activation—related to the ease of passage of the system through the transition state
4. Thermodynamic relationships; $\Delta G^{\ddagger} = \Delta H^{\ddagger} - T\Delta S^{\ddagger}$

D. Solvent effects on reaction rates—related to whether the activated complex is more polar or less polar than the reactants (*i.e.*, changes in charge density)

IV. Substituent effects as a means for studying reaction mechanisms
- A. Steric effects
- B. Electronic effects
  1. Inductive effect
     (a) A polarization effect.
     (b) Operates through bonds and/or through space.
     (c) Falls off rapidly with distance.
     (d) Quantitative measurement:
        (1) From ionization constants
        (2) From nmr shifts
  2. Resonance effect
     (a) Operates through π bonds in conjugated systems.
     (b) May augment or oppose the inductive effect.
- C. Quantitative correlations of substituent effects: The Hammett equation
  1. Hammett equation: $\log k - \log k_0 = \sigma\rho$
  2. The substituent constant, σ—characteristic of the substituent and independent of the type of reaction
  3. The reaction constant, ρ—characteristic of the type of reaction and independent of the substituent
  4. The Hammett equation as a means for studying mechanisms

V. Isotopes as a means for studying reaction mechanisms
- A. Isotopes as "tags"—isotope labeling
  1. Chemistry of isotopes of a given element essentially identical
  2. Example—reaction of phenylethylamine and nitrous acid
- B. Kinetic isotope effect
  1. Isotopically substituted compounds may react at different rates if rate-determining step involves cleavage of bond to isotopic element
  2. Example—base-induced dehydrobromination of *t*-butyl bromide

VI. Stereochemistry as a means for studying reaction mechanisms
- A. Conversion of a chiral center to an achiral center (loss of optical activity)
- B. Conversion of a chiral center to a new chiral center with retention of absolute configuration (retention of optical activity)
- C. Conversion of a chiral center to a new chiral center with inversion of absolute configuration (retention of optical activity)
- D. Conversion of a chiral center to a new chiral center with both enantiomers being formed in equal amount (loss of optical activity; racemization)

VII. Detection of reaction intermediates as a means for studying reaction mechanisms
   A. Spectroscopic methods
   B. Use of "trapping" reagents

VIII. Thermodynamic and kinetic control of reactions
   A. Thermodynamic product control—major product is the most stable product.
   B. Kinetic product control—major product is the one formed most rapidly.
   C. Example—formation of semicarbazones from mixture of furfuraldehyde and cyclohexanone.

IX. Schemes for classifying organic reactions
   A. Classification on the basis of type of reaction involved
      1. Substitution reactions
      2. Addition reactions
      3. Elimination reactions
   B. Classification on the basis of the type of intermediate involved
      1. Types of intermediates: carbonium ions, free radicals, carbanions, carbenes, and benzynes
      2. Types of bond cleavage: homolytic and heterolytic
   C. Classification on the basis of experimental procedure
      1. Thermal reactions
      2. Photochemical reactions

X. Classification of reaction mechanisms
   A. Substitution reactions
      1. Nucleophilic substitution
      2. Radical substitution
      3. Electrophilic substitution
      4. Situselectivity
   B. Addition reactions
      1. Electrophilic addition
      2. Nucleophilic addition
      3. Radical addition
      4. Concerted addition
      5. Regioselectivity
      6. Stereoselectivity and stereospecificity
   C. Elimination reactions
      1. α-Elimination
      2. β-Elimination
      3. γ-Elimination
   D. Rearrangement reactions: 1,2 nucleophilic rearrangements
   E. Oxidation-reduction reactions

REFERENCES

General References:

ALDER, R. W., R. BAKER, and J. M. BROWN, *Mechanisms in Organic Chemistry*, John Wiley and Sons, Inc., New York, 1971. A 378 page book that is readily comprehensible by the beginning student but which is more extensive and detailed in its coverage than the books by Breslow and by Sykes; includes many references to the original literature.

BRESLOW, R., *Organic Reaction Mechanisms*, 2nd. ed., W. A. Benjamin, Inc., Menlo Park, California, 1969. A 272 page paperback written for the student in elementary organic chemistry seeking a more extended treatment of reaction mechanisms than available in elementary textbooks; includes a few references to the original literature.

FERGUSON, L. N., *The Modern Structural Theory of Organic Chemistry*, Prentice-Hall, Inc., Englewood Cliffs, N.J., 1963. A good treatment of a variety of topics in organic reactions, structure, and mechanisms written for the advanced undergraduate or graduate student.

GOULD, E. S., *Mechanism and Structure in Organic Chemistry*, Henry Holt and Company, New York, 1959. One of the best introductions to the more quantitative and sophisticated aspects of organic reaction mechanisms written for the advanced undergraduate or graduate student.

HAMMETT, L. P., *Physical Organic Chemistry*, 2nd. ed., McGraw-Hill Book Company, New York, 1970. A quantitative approach to reaction mechanisms, written by a scientist who was instrumental in starting the conversion of organic chemistry from an art to a science.

HINE, J., *Physical Organic Chemistry*, 2nd. ed., McGraw-Hill Book Company, New York, 1962. A thoroughly documented and scholarly treatment of physical organic chemistry, written for the graduate student and the professional chemist.

INGOLD, C. K., *Structure and Mechanism in Organic Chemistry*, 2nd. ed., Cornell University Press, Ithaca, 1969. A large (1266 pages) book that contains the definitive treatment of certain aspects of reaction mechanisms, written by one of the founders of physical organic chemistry.

KOSOWER, E. M., *An Introduction to Physical Organic Chemistry*, John Wiley & Sons, Inc., New York, 1968. Presents data and ideas not found in other texts, although selective and somewhat limited in its coverage of topics.

LIBERLES, A., *Introduction to Theoretical Organic Chemistry*, Macmillan Company, New York, 1968. A fairly extensive and sophisticated coverage of physical organic chemistry, including molecular structure, kinetics, spectra, and reaction mechanisms.

MARCH, J., *Advanced Organic Chemistry: Reactions, Mechanisms, and Structure*, McGraw-Hill Book Company, New York, 1968. A large book replete with information and references to the original literature, written for the serious student of organic chemistry.

STEWART, R., *The Investigation of Organic Reactions*, Prentice-Hall, Inc., Englewood Cliffs, N.J., 1966. A 125 page paperback written for the beginning student; a good complement to chapter 20 of the present textbook.

SYKES, P., *The Search for Organic Reaction Pathways*, Longman Group, Ltd., London, 1972. A 247 page paperback written for the student in elementary organic chemistry seeking a more extended treatment of reaction mechanisms than available in elementary textbooks.

WHEELER, O. H., *Physical Organic Chemistry*, Elsevier Publishing Company, New York, 1966. A 172 page paperback written for the beginning student; a good complement to chapter 20 of the present textbook.

WHELAND, G. W., *Advanced Organic Chemistry*, 3rd. ed., John Wiley & Sons, Inc., New York, 1960. Although growing old, one of the best and most precise treatments of a variety of areas of mechanistic organic chemistry; recommended reading for those interested in style as well as chemistry.

WIBERG, K. B., *Physical Organic Chemistry*, John Wiley & Sons, Inc., New York, 1964. The most sophisticated of the references in this list in its treatment of physical organic chemistry.

References Dealing with Equilibria and Kinetics:

FROST, A. A., and R. G. PEARSON, *Kinetics and Mechanism*, 2nd. ed., John Wiley & Sons, Inc., New York, 1961.

LEFFLER, J. E., and E. GRUNWALD, *Rates and Equilibria of Organic Reactions*, John Wiley & Sons, Inc., New York, 1963.

References Dealing with Catalysis:

BENDER, M. L., *Mechanisms of Homogeneous Catalysis from Protons to Proteins*, John Wiley & Sons, Inc., New York, 1971.

JENCKS, W. P., *Catalysis in Chemistry and Enzymology*, McGraw-Hill Book Company, New York, 1969.

References Dealing with Mechanisms of Oxidation-Reduction Reactions:

RINEHART, K. L., *Oxidation and Reduction of Organic Compounds*, in *Foundations of Modern Organic Chemistry Series*, Prentice-Hall, Inc., Englewood Cliffs, N.J., 1973.

STEWART, R., *Oxidation Mechanisms: Applications to Organic Chemistry*, W. A. Benjamin, Inc., Menlo Park, California, 1964.

PROBLEMS

1. In the following pairs of reactions, will reaction A or reaction B proceed more rapidly?
(a) Reaction A ($\Delta H^{\ddagger} = 25$ kcal/mole; $\Delta S^{\ddagger} = 10$ eu; $T = 27°C$)
Reaction B ($\Delta H^{\ddagger} = 35$ kcal/mole; $\Delta S^{\ddagger} = 10$ eu; $T = 27°C$)
(b) Reaction A ($\Delta H^{\ddagger} = 25$ kcal/mole; $\Delta S^{\ddagger} = 10$ eu; $T = 200°C$)
Reaction B ($\Delta H^{\ddagger} = 25$ kcal/mole; $\Delta S^{\ddagger} = -15$ eu; $T = 200°C$)
(c) Reaction A ($\Delta H^{\ddagger} = 25$ kcal/mole; $\Delta S^{\ddagger} = 10$ eu; $T = 27°C$)
Reaction B ($\Delta H^{\ddagger} = 35$ kcal/mole; $\Delta S^{\ddagger} = -15$ eu; $T = 27°C$)
(d) Reaction A ($\Delta H^{\ddagger} = 25$ kcal/mole; $\Delta S^{\ddagger} = -10$ eu; $T = 27°C$)
Reaction B ($\Delta H^{\ddagger} = 35$ kcal/mole; $\Delta S^{\ddagger} = 15$ eu; $T = 27°C$)
[*Note*: Remember that $\Delta G = \Delta H - T\Delta S$ and that 1 eu = 1 cal/deg/mole. Thus, at 300°K, the $T\Delta S$ term is 3000 cal/mole (3 kcal/mole) if ΔS is 10 eu.]

2. Indicate the order of the following reactions:
(a) The rate of conversion of triphenylmethyl chloride to methyl triphenylmethyl ether in methanol solution is dependent only on the first power of the concentration of the alkyl halide.

(b) Under certain conditions, the rate of reaction of triphenylmethyl chloride with methanol in benzene solution to yield methyl triphenylmethyl ether is dependent on the first power of the alkyl halide concentration and the second power of the methanol concentration.

(c) The rate of the reaction of diethylamine with ethylene oxide to yield N-(2-hydroxyethyl)-N,N diethylamine is dependent on the first power of the concentration of both the diethylamine and the ethylene oxide.

3. (a) What is the apparent molecularity of the reaction of triphenylmethyl chloride with methanol in Problem 2(a)? What is it in Problem 2(b)? If it is not the same in the two cases, how can the difference be explained?

(b) The reaction of ethylene oxide with hydrogen bromide has the following rate expression: rate $= k[C_2H_4O][H^+][Br^-]$. What is the molecularity of the reaction? Suggest a possible course of events to explain your answer.

4. Compounds A and B react to yield compound C. Four kinetic experiments are performed with various concentrations of compounds A and B, yielding the following results:

Experiment 1: $[A] = 4\,M$, $[B] = 10^{-2}\,M$ Experiment 2: $[A] = 4\,M$, $[B] = 0.1\,M$

Relative reaction rate for reaction 2/reaction 1 = 100

Experiment 3: $[A] = 10^{-2}\,M$, $[B] = 4\,M$ Experiment 4: $[A] = 0.1\,M$, $[B] = 4\,M$

Relative reaction rate for reaction 4/reaction 3 = 1

Write the rate expression for the reaction $A + B \rightarrow C$.

5. Many enzyme-catalyzed reactions have rate expressions of the following form:

$$\text{Rate} = \frac{k[A][B][\text{enzyme}]}{K + [A]}$$

where k and K are constants and A and B are the reactants.

(a) In some circumstances, the value of K is very much larger than that of [A]. What is the order with respect to A? What is the order with respect to B?

(b) In other circumstances, the value of K is very much less than that of [A]. What is the order with respect to A? What is the order with respect to B?

(c) What is the overall order of the reaction in cases (a) and (b)?

6. For each of the following pairs of compounds draw resonance structures to illustrate the difference in the interaction between the two attached groups when they are oriented *m*- and *p*-, respectively.

(a) $m\text{-}CH_3C(=O)\text{-}C_6H_4\text{-}O^-$ and $p\text{-}CH_3C(=O)\text{-}C_6H_4\text{-}O^-$

(b) $m\text{-}(CH_3)_2N\text{-}C_6H_4\text{-}\bar{N}H$ and $(CH_3)_2N\text{-}C_6H_4\text{-}\bar{N}H$ (para)

(c) [structure: m-carboxybenzoate] CO_2^- and HO_2C–[p-phenylene]–CO_2^- (with HO_2C in meta position)

7. Making use of the Hammett correlation, calculate the quantities asked for in each of the following examples:

(a) Benzoic acid is a very much weaker acid in methanol than in water; in water, the pK_a is 4.20, whereas the pK_a in methanol is 9.40. The ρ value for the ionization of benzoic acids in methanol is +1.537. What is the pK_a of *p*-nitrobenzoic acid in methanol?

(b) The pK_a of benzeneboronic acid [$C_6H_5B(OH)_2$] in 25% ethanol at 25° is 9.70. The ρ value for the ionization is +2.164. What is the pK_a of *m*-fluorobenzeneboronic acid?

(c) The plot of the log K_{BH^+} values for anilinium ions *vs.* σ values is as shown. What is the ρ value for the ionization of anilinium ions?

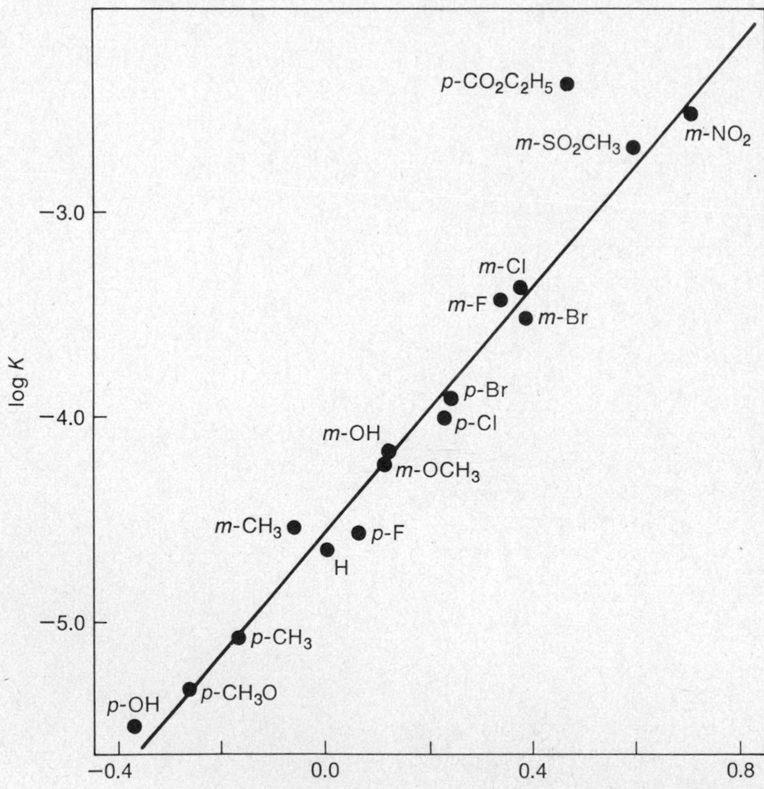

8. A variety of hydrolysis reactions have rate expressions of the following form:

$$\text{Rate} = k_0 + k_H[H_3O^+] + k_{OH}[OH^-]$$

where k_0 is the specific rate for the water-catalyzed reaction (sometimes called the "spontaneous reaction"), k_H is the specific rate for the proton-catalyzed reaction, and k_{OH} is the specific rate for the hydroxide ion-catalyzed reaction.

(a) Draw the pH-rate profile for a reaction in which $k_0 = k_H = k_{OH} = 1$.
(b) Draw the pH-rate profile for a reaction in which $k_0 = 10^{-3}$, $k_H = 1$, $k_{OH} = 1$.
(c) Draw the pH-rate profile for a reaction in which $k_0 = 10^{-5}$, $k_H = 1$, $k_{OH} = 10^2$.
(d) Draw the pH-rate profile for a reaction in which $k_0 = 10^{-3}$, $k_H = 10^4$, $k_{OH} = 10^2$.

9. Predict which member of the following pairs of compounds is more acidic, and indicate the reasoning on which your answer is based:
(a) 2,3-Dichloropropionic acid or 3,3-dichloropropionic acid
(b) Dinitromethane or trinitromethane
(c) Cyclohexylacetic acid or phenylacetic acid
(d) 2-Chloro-3-cyanopropionic acid or 3-chloro-2-cyanopropionic acid
(e) *m*-Carboxyphenyltrimethylammonium bromide or *p*-carboxyphenyltrimethylammonium bromide
(f) *m*-Hydroxyacetophenone or *p*-hydroxyacetophenone
(g) *p*-Methoxybenzamide or *p*-nitrobenzamide

10. The rate of a reaction of the type shown was studied as a function of the substituent in the *para* position, and the following data were collected:

$$S{-}C_6H_4{-}\overset{|}{\underset{|}{C}}{-}X + Y^- \longrightarrow S{-}C_6H_4{-}\overset{|}{\underset{|}{C}}{-}Y + X^-$$

S Substituent	$-N(CH_3)_2$	$-OH$	$-OCH_3$	$-Br$	$-CN$	$-NO_2$
Rate, $\times 10^{-3}$	5.57	2.82	2.40	2.00	2.78	3.05

(a) Construct a Hammett plot from these rate data and the ionization constants of the appropriate benzoic acids.
(b) Can a p value be assigned to this reaction over the whole range of *para* substituents? Can more than one ρ value be assigned to the reaction?

11. The reaction of semicarbazide with benzaldehyde and various substituted benzaldehydes to form semicarbazones (see Fig. 13.11) has been studied in considerable detail. It has been shown that the rate of the reaction is dependent on the pH, proceeding most rapidly at about pH 4 and diminishing in rate above pH 4 as well as below pH 4. When a Hammett plot is constructed for reactions carried out below pH 4, a ρ value of $+1.81$ is obtained, but when a Hammett plot is constructed for reactions carried out above pH 4, a ρ value of -1.74 is obtained.
(a) What does the positive ρ value indicate about the electronic requirements of the reaction when it is carried out below pH 4?
(b) What does the negative ρ value indicate about the electronic requirements of the reaction when it is carried out above pH 4?
(c) Explain how the information gained from the Hammett plots at pH values above and below 4 shows that the reaction must involve at least two steps.
(d) On the assumption that one of the steps is acid catalyzed, would you predict it to be the first or second step in the process? Explain.
(e) On the assumption that one of the steps involves a nucleophilic addition to the aldehyde carbonyl group, would you predict it to be the first or second step? Explain.
(f) On the assumption that one of the steps is acid catalyzed, how can you account for a rate maximum at pH 4? That is, why does the rate not continue to increase below pH 4 as the $[H^+]$ continues to increase?

12. Reactant A yields product B and C with rate constants of 10^{-3} and 10^{-6} sec^{-1}, respectively. The back reactions of products B and C to reactant A proceed at rates of 10^{-6} and 10^{-5} sec^{-1}, respectively. Draw a reaction coordinate diagram for this system, and indicate which product is that of thermodynamic control and which is that of kinetic control.

13. Classify the following reactions as falling in one (or more) of the following categories: (a) substitution reaction, (b) addition reaction, (c) elimination reaction, (d) rearrangement reaction, (e) insertion reaction, or (f) oxidation-reduction reaction.

(a) $(C_6H_5)_3COH + HCl \longrightarrow (C_6H_5)_3CCl + H_2O$

(b) $(C_6H_5)_2C{=}CHBr \xrightarrow{C_2H_5ONa} C_6H_5{-}C{\equiv}C{-}C_6H_5$

(c) $(C_6H_5)_3CCH_2OH + HCl \longrightarrow (C_6H_5)_2C(Cl)CH_2C_6H_5 + H_2O$

(d) $(CH_3)_2C{=}C(CH_3)_2 + Cl_2 \longrightarrow (CH_3)_2C(Cl){-}C(Cl)(CH_3)_2$

(e) decalin $+ O_2 \xrightarrow{\text{Radical initiator}}$ decalin-OOH (OOH at ring-fusion carbon)

(f) $(CH_3)_2C{=}O + NaBH_4 \longrightarrow (CH_3)_2CHOH$

(g) furan + maleic anhydride $\longrightarrow$ Diels–Alder adduct (oxa-bridged bicyclic anhydride, with H, H)

(h) $C_6H_5{-}CH{-}CH{-}C_6H_5$ (epoxide, O bridging) $+ (C_6H_5)_3P \longrightarrow C_6H_5{-}CH{=}CH{-}C_6H_5 + (C_6H_5)_3P{=}O$

14. Which of the reactions in Problem 13 are best classified as ionic and which as non-ionic processes?

15. Explain why the following reaction sequence is generally considered to be a nucleophilic substitution from the standpoint of the CH_3I and an electrophilic substitution from the standpoint of the $CH_2(CO_2CH_3)_2$.

$$CH_2(CO_2CH_3)_2 + CH_3ONa \longrightarrow {}^-CH(CO_2CH_3)_2 + Na^+ + CH_3OH$$

$$^-CH(CO_2CH_3)_2 + CH_3I \longrightarrow CH_3CH(CO_2CH_3)_2 + I^-$$

16. For each of the following reactions, comment on the regioselectivity, situselectivity, stereoselectivity, and stereospecificity of the process:

(a) cyclohexene + CH_2N_2 $\xrightarrow{h\nu}$ 4-methylcyclohexene + 4-methylcyclohexene + 1-methylcyclohexene + cis-bicyclo[4.1.0]heptane (H, H, CH_2)

(b)

$C_6H_5(CH_3)C{=}C(CH_3)C_6H_5$ ((*E*) isomer) + H_2 $\xrightarrow{Pd-C}$ $C_6H_5(CH_3)CHCH(CH_3)C_6H_5$ ((±) isomer) (98% yield)

$C_6H_5(CH_3)C{=}C(CH_3)C_6H_5$ ((*Z*) isomer) + H_2 $\xrightarrow{Pd-C}$ $C_6H_5(CH_3)CHCH(CH_3)C_6H_5$ (*meso* isomer) (99% yield)

(c) 6-methoxy-1-tetralone (CH_3O, O) $\xrightarrow{Zn-Hg/HCl}$ 6-methoxytetralin (CH_3O) + 6-hydroxytetralin (HO)

(d) 3-phenylcyclohex-2-en-1-ol (OH) + HCl ⟶ 1-chloro-1-phenyl-3-chlorocyclohexane (Cl, Cl, H) + 1-chloro-1-phenyl-3-chlorocyclohexane (Cl, H, Cl)

Chapter 21

Halides: alkyl, alkenyl, and aryl halides

> What if one does say the same things,—of course in a little different form each time,—over and over? If he has anything to say worth saying, that is just what he ought to do.
>
> OLIVER WENDELL HOLMES
> *Over the Teacups*

The first class of organic compounds that we shall consider in a more advanced and detailed fashion than in Part I are the halides. These compounds, including the alkyl, alkenyl, and aryl halides, are exceedingly useful synthesis intermediates as, indeed, we have already seen in several examples in previous chapters. It is important, therefore, that we be familiar both with their methods of preparation and with the various ways in which they react; their chemistry constitutes an important ingredient in our quest for proficiency in the art of organic synthesis.

The halides are also interesting for another reason. They provide the compounds on which a variety of reaction mechanism studies have been carried out, furnishing us with information that not only is intellectually satisfying but also gives us insight into how to improve and extend the utility of halides as synthesis reagents.

All three of these aspects of the halides are dealt with in this chapter, starting with methods of preparation, progressing then to a discussion of the mechanisms of nucleophilic substitution reactions, and concluding with an account of the reactions of halides.

METHODS OF PREPARATION OF HALIDES

21.1 Methods of Preparation of Alkyl Halides

21.1a. CONVERSION OF C—H TO C—HALOGEN. Although the preparation of alkyl halides, particularly alkyl chlorides, by the free radical halogenation of hydrocarbons finds some industrial application, its use in the laboratory is generally limited to those compounds which contain a C—H bond that is made especially reactive by an adjacent double bond, triple bond, or aromatic ring. Allylic and benzylic bromination, for example, can often be smoothly effected through the use of N-bromosuccinimide (NBS) (see Fig. 8.11), the reagent acting as a source of molecular bromine that is maintained at a low but steady concentration as the reaction progresses. The reaction does not always proceed in a straightforward manner, however, for the intermediate allyl radicals generally have the capability of reacting at both the 1- and 3-carbon atoms of the resonance-stabilized allylic systems. As Fig. 21.1 illustrates, 1-pentene yields both 3-bromo-1-pentene and 1-bromo-2-pentene when treated with N-bromosuccinimide in the presence of benzoyl peroxide, which acts as a radical initiator.

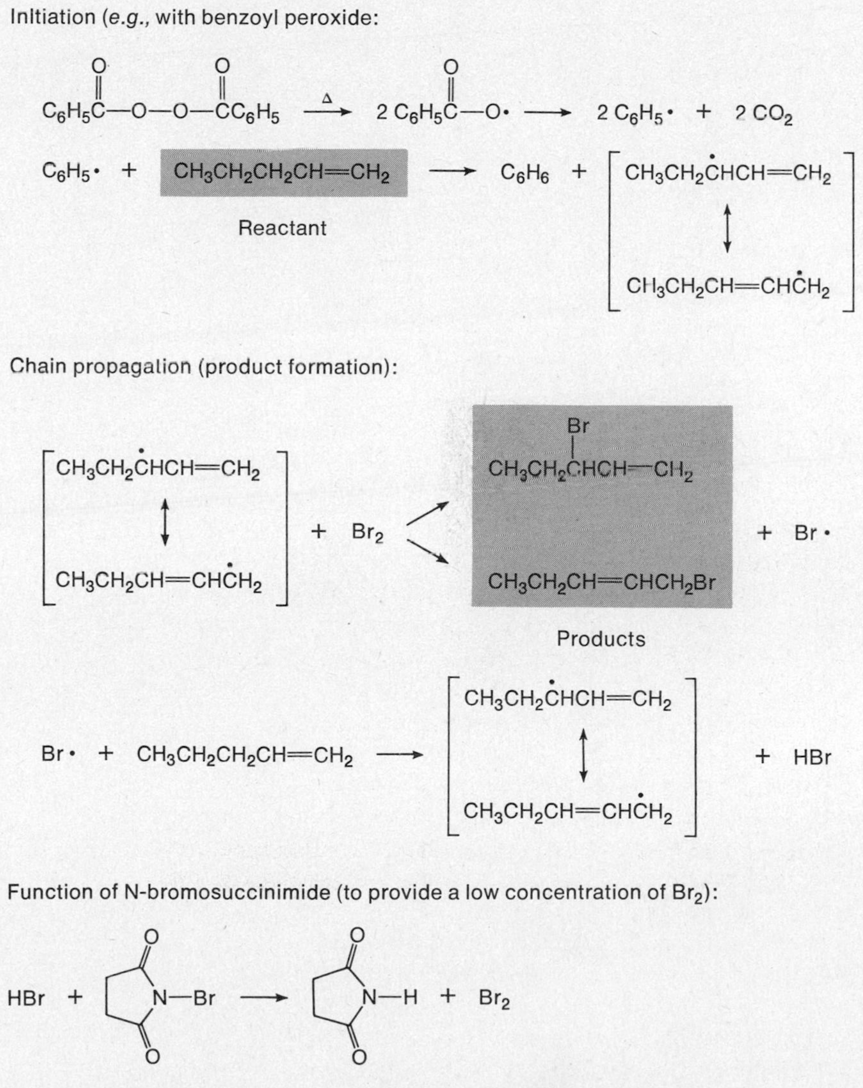

Initiation (*e.g.*, with benzoyl peroxide:

$$C_6H_5\overset{O}{\overset{\|}{C}}-O-O-\overset{O}{\overset{\|}{C}}C_6H_5 \xrightarrow{\Delta} 2\ C_6H_5\overset{O}{\overset{\|}{C}}-O\cdot \longrightarrow 2\ C_6H_5\cdot + 2\ CO_2$$

$$C_6H_5\cdot + \underset{\text{Reactant}}{CH_3CH_2CH_2CH{=}CH_2} \longrightarrow C_6H_6 + \left[\begin{array}{c} CH_3CH_2\dot{C}HCH{=}CH_2 \\ \updownarrow \\ CH_3CH_2CH{=}CH\dot{C}H_2 \end{array}\right]$$

Chain propagation (product formation):

$$\left[\begin{array}{c} CH_3CH_2\dot{C}HCH{=}CH_2 \\ \updownarrow \\ CH_3CH_2CH{=}CH\dot{C}H_2 \end{array}\right] + Br_2 \longrightarrow \underset{\text{Products}}{\begin{array}{c} CH_3CH_2CH(Br)CH{=}CH_2 \\ CH_3CH_2CH{=}CHCH_2Br \end{array}} + Br\cdot$$

$$Br\cdot + CH_3CH_2CH_2CH{=}CH_2 \longrightarrow \left[\begin{array}{c} CH_3CH_2\dot{C}HCH{=}CH_2 \\ \updownarrow \\ CH_3CH_2CH{=}CH\dot{C}H_2 \end{array}\right] + HBr$$

Function of N-bromosuccinimide (to provide a low concentration of Br_2):

$$HBr + \text{(succinimide ring)}N-Br \longrightarrow \text{(succinimide ring)}N-H + Br_2$$

Fig. 21.1. Allylic bromination via N-bromosuccinimide.

The halogenation of hydrocarbons generally suffers from one or another limitation. Radical chlorination is not highly situselective and usually produces mixtures of difficultly separable isomeric chlorides (see Fig. 8.4). Radical fluorination is even worse in this respect, the reaction being so exceedingly exothermic (and, therefore, undiscriminating) that it must be carried out under special conditions involving very dilute solutions at low temperatures. Iodination, on the other hand, is an endothermic process and, therefore, thermodynamically unfavorable. Thus, bromination turns out to be the most useful free radical halogenation procedure and is most conveniently carried out in the laboratory by means of N-bromosuccinimide.

PROBLEM 21-1

The chlorination of optically active $C_6H_5CH(CH_3)CH_2Cl$

yields optically inactive $C_6H_5CCl(CH_3)CH_2Cl$

whereas bromination yields $C_6H_5CBr(CH_3)CH_2Cl$

in which some optical activity is retained. What does this suggest about the geometry of a radical species, and how can these data be explained?

PROBLEM 21-2

Arrange the following reactions in order of increasing situselectivity:

(a) $C_6H_5CH_2CH_2C_6H_5$ + N-bromosuccinimide
(b) $C_6H_5CH{=}CHCH_3$ + N-bromosuccinimide
(c) $C_6H_5(CH_2)_3CH_3 + Cl_2$
(d) $C_6H_5(CH_2)_3CH_3 + F_2$

Explain the reasons for your arrangement.

21.1b. CONVERSION OF C—OH TO C—HALOGEN. The phosphorus chlorides and bromides are useful reagents for converting alcohols to the corresponding chlorides or bromides (Fig. 21.2). By "tagging" the position adjacent to the hydroxyl carbon with deuterium, we can show that the reaction proceeds with inversion of configuration, probably as the result of the formation of an intermediate phosphorus ester (i.e., $ROPBr_4$ when PBr_5 is used), which undergoes a bimolecular reaction (S_N2 reaction) with bromide ion (see Section 21.4 for details of S_N2 reactions). Thionyl chloride ($SOCl_2$), when used in pyridine solution, reacts in a similar fashion and yields alkyl chlorides with inversion of configuration via the attack of chloride ion on an intermediate chlorosulfite ester (ROSOCl) (Fig. 21.2).

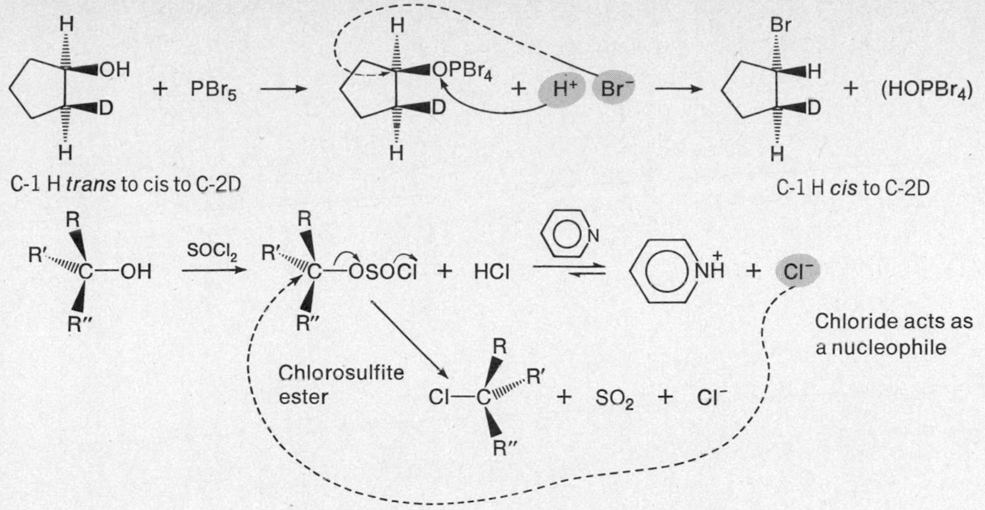

C-1 H *trans* to cis to C-2D

C-1 H *cis* to C-2D

Chloride acts as a nucleophile

Chlorosulfite ester

Fig. 21.2. Conversion of alcohols to alkyl halides with inversion of configuration.

If the reaction between alcohols and thionyl chloride is carried out in benzene rather than pyridine solution, a different stereochemical course is followed and the configuration of the starting material is retained. We explain this by postulating that under these conditions the initially formed chlorosulfite ester dissociates to a "tight ion pair," which subsequently decomposes to the alkyl halide and sulfur dioxide in such a fashion that the incoming chlorine atom enters at the same face of carbon as the departing oxygen atom (Fig. 21.3). Reactions of this type are classified as S_Ni (Substitution, Nucleophilic, internal).

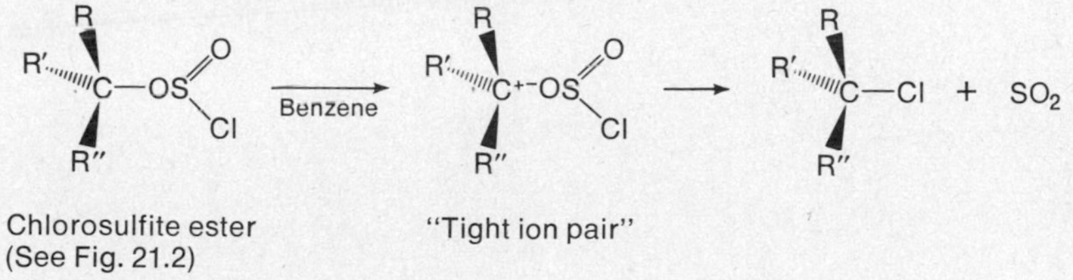

Chlorosulfite ester (See Fig. 21.2)

"Tight ion pair"

Fig. 21.3. Conversion of alcohols to alkyl chlorides with retention of configuration.

A two-step sequence for converting alcohols to alkyl chlorides starts with the formation of the *p*-toluenesulfonate ester of the alcohol* (by treatment with

---

* The *p*-toluenesulfonyloxy group (abbreviated tosyloxy) has the structure

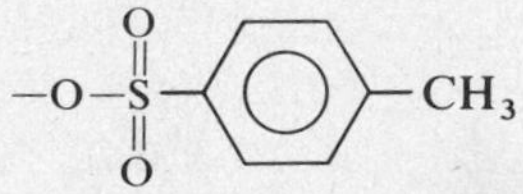

$$-O-\overset{\overset{O}{\|}}{\underset{\underset{O}{\|}}{S}}-C_6H_4-CH_3$$

and behaves in nucleophilic substitution reactions in a manner virtually identical with that of halogens. Thus, alkyl halides and alkyl tosylates can be considered members of the same family of compounds, so far as their reactions are concerned.

p-toluenesulfonyl chloride), followed by reaction with lithium chloride in dimethylsulfoxide (DMSO) solution, which effects an S_N2 reaction (Fig. 21.4).

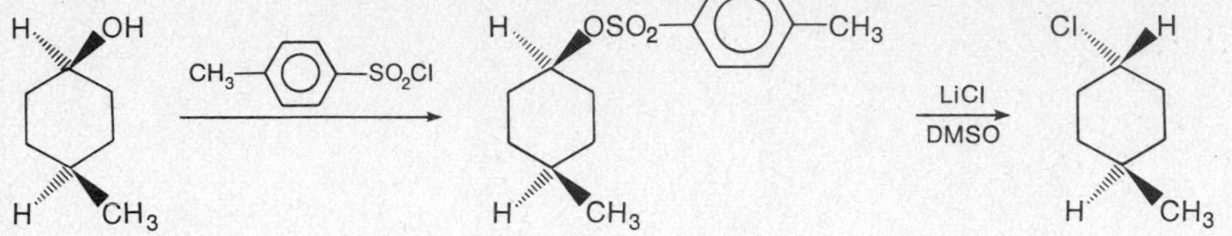

Fig. 21.4. Conversion of alcohols to alkyl chlorides by conversion to the p-toluenesulfonate, followed by treatment with lithium chloride.

PROBLEM 21-3

What, if any, are the mechanistic similarities of the reactions shown in Figs. 21.2 and 21.4?

The oldest method for converting alcohols to alkyl halides employs the direct action of the halogen acids on the alcohols (see Fig. 10.15). Although the majority of alcohols react with halogen acids, the method is limited by low stereoselectivity and a marked tendency to skeletal rearrangement. The probability of nonstereospecific substitution and skeletal rearrangement generally increases from primary to secondary to tertiary alcohols. Even some primary alcohols react with skeletal rearrangement, particularly those carrying doubly or triply substituted carbon atoms adjacent to the hydroxyl carbon (see Fig. 10.16). For example, 1-methylcyclopentylmethanol reacts with hydrogen bromide to give 1-bromo-1-ethylcyclopentane and 1-bromo-1-methylcyclohexane as the major products (Fig. 21.5).

PROBLEM 21-4

The action of mineral acids on alcohols can be used in ring expansion reactions (see Fig. 21.5) and also in ring contraction reactions. Predict the structure of the product resulting from the action of HBr on 2,2-diphenylcycloheptanol. Outline the mechanism of the reaction.

A special application of the conversion of alcohols to alkyl chlorides occurs in the *chloromethylation* reaction, which involves the action of formaldehyde and hydrogen chloride on arenes in the presence of the Lewis acid catalyst, zinc chloride (Fig. 21.6).

21.1c. CONVERSION OF C—C TO C—HALOGEN. The treatment of solutions of carboxylic acids in aqueous sodium hydroxide with water soluble inorganic

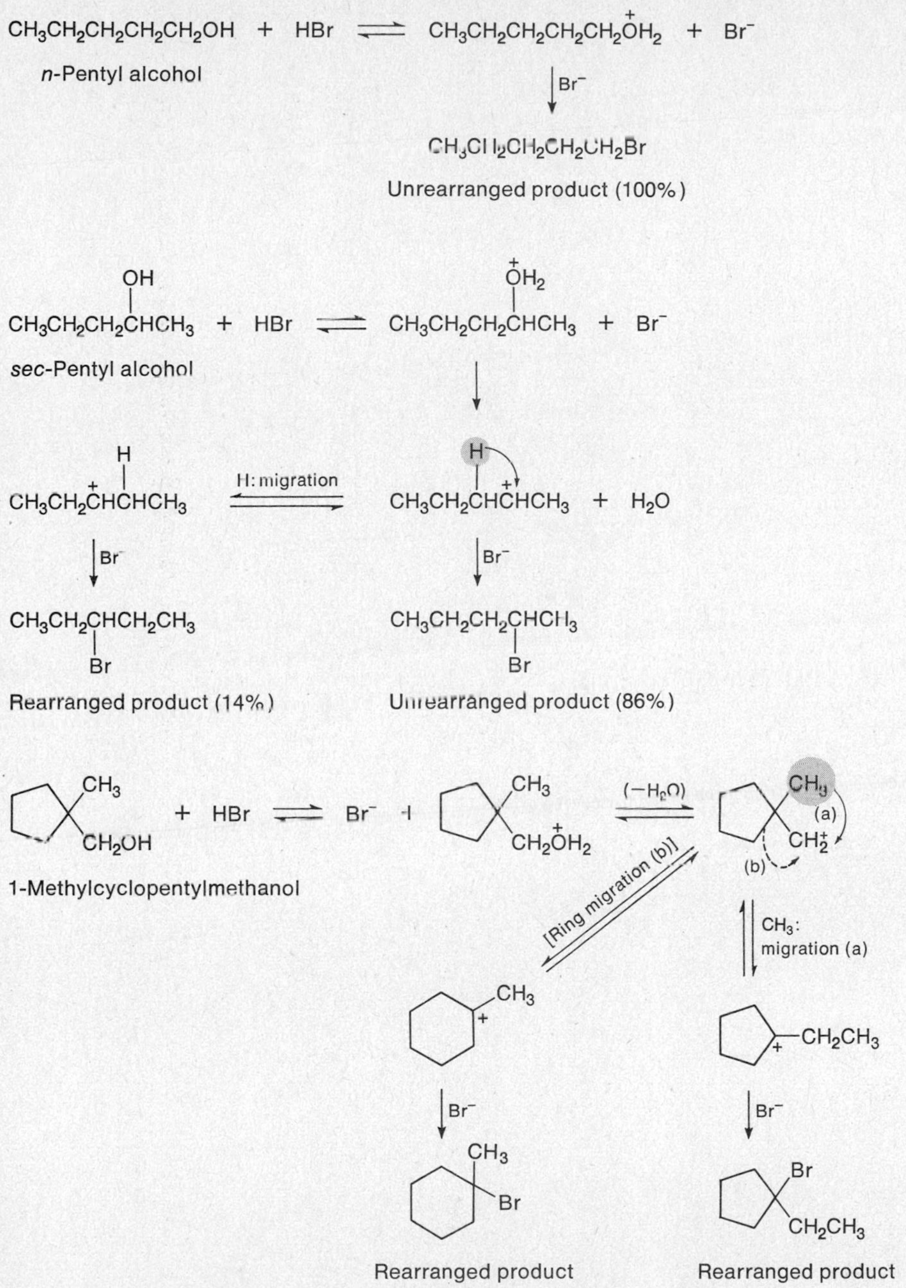

$CH_3CH_2CH_2CH_2CH_2OH$ + HBr ⇌ $CH_3CH_2CH_2CH_2CH_2\overset{+}{O}H_2$ + Br^-

n-Pentyl alcohol

↓ Br^-

$CH_3CH_2CH_2CH_2CH_2Br$

Unrearranged product (100%)

OH

$CH_3CH_2CH_2CHCH_3$ + HBr ⇌ $CH_3CH_2CH_2CH(\overset{+}{O}H_2)CH_3$ + Br^-

sec-Pentyl alcohol

H: migration

$CH_3CH_2\overset{+}{C}HCH_2CH_3$ ⇌ $CH_3CH_2CH_2\overset{+}{C}HCH_3$ + H_2O

↓ Br^- ↓ Br^-

$CH_3CH_2CHBrCH_2CH_3$ $CH_3CH_2CH_2CHBrCH_3$

Rearranged product (14%) Unrearranged product (86%)

1-Methylcyclopentylmethanol + HBr ⇌ Br^- + (1-methylcyclopentyl)$CH_2\overset{+}{O}H_2$ ⇌ $(-H_2O)$ (1-methylcyclopentyl)CH_2^+

[Ring migration (b)] (b) (a)

CH_3: migration (a)

1-methylcyclohexyl cation; 1-ethylcyclopentyl cation

↓ Br^- ↓ Br^-

1-bromo-1-methylcyclohexane; 1-bromo-1-ethylcyclopentane

Rearranged product Rearranged product

Fig. 21.5. Conversion of alcohols to alkyl halides with halogen acids.

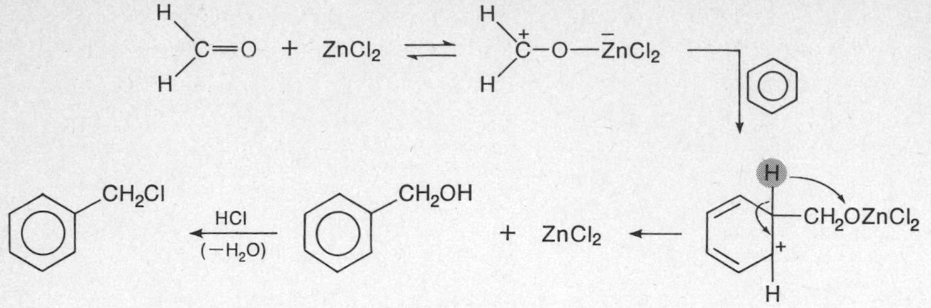

Fig. 21.6. Preparation of benzyl chloride via chloromethylation.

silver, mercury, or thallium salts yields the silver, mercury, or thallium carboxylates which precipitate from solution and can be removed by filtration. When these are dried and then mixed with bromine and heated, an acyl hypobromite $RC(=O)OBr$ is formed which subsequently undergoes homolytic cleavage to an alkyl radical, carbon dioxide, and a bromine atom. Recombination of the alkyl radicals and bromine atoms yields the alkyl bromide, usually as a mixture of stereoisomers (Hunsdiecker reaction) (Fig. 21.7).

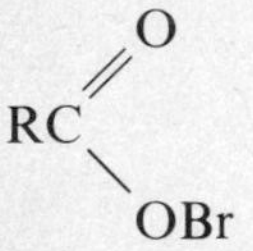

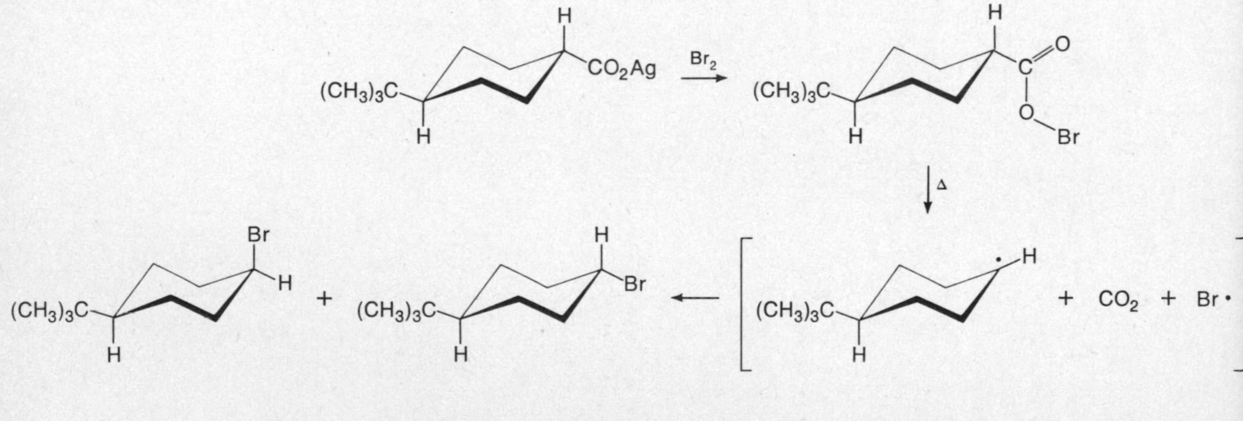

Fig. 21.7. The Hunsdiecker reaction.

21.1d. CONVERSION OF C=C to C—C—X AND X—C—C—X (X = HALOGEN). Alkenes react with halogen acids to form alkyl halides and with halogens to form vicinal (Latin, *vicinalis*, neighboring) dihalides, examples of which are cited in earlier chapters (see Figs. 8.3–8.7, 9.2). The mechanisms and utility of these reactions are discussed in Chapter 26.

21.2 Methods of Preparation of Alkenyl Halides (Vinyl Halides)

Alkenyl halides (vinyl halides) can be prepared by (a) the partial dehydrohalogenation of geminal (Latin, *geminus*, twin) dihalides, prepared by the action of phosphorus pentachloride on ketones, (b) the partial dehydrohalogenation of vicinal dihalides, and (c) the addition of halogen acids to alkynes, to produce vinyl halides, or the addition of halogens to alkynes to produce vinyl dihalides (Fig. 21.8).

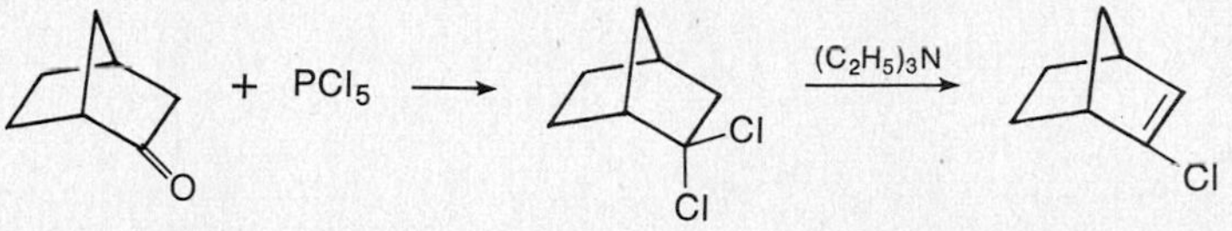

Bicyclo[2.2.1]heptan-2-one $+ PCl_5 \longrightarrow$ (2,2-dichloro) $\xrightarrow{(C_2H_5)_3N}$ 2-Chlorobicyclo[2.2.1]hept-2-ene

$$HC{\equiv}CH + HCl \longrightarrow H_2C{=}CHCl$$

Vinyl chloride

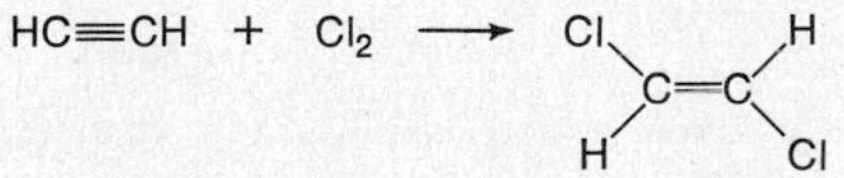

$$HC{\equiv}CH + Cl_2 \longrightarrow ClHC{=}CHCl$$

trans-1,2-Dichloroethene

Fig. 21.8. Methods for the preparation of vinyl halides.

PROBLEM 21-5

Suggest a method for converting 1-butyn-3-one to 1,2-dibromo-3-chlorobuta-1,3-diene.

21.3 Methods of Preparation of Aryl Halides

21.3a. VIA ELECTROPHILIC SUBSTITUTION. All the aryl halides can be prepared by electrophilic substitution reactions with the halogens. The fluorination reaction is so exothermic that it is necessary to carry it out in dilute acetonitrile solution at low temperature (*e.g.*, $-20°$), bubbling the fluorine gas diluted with helium or nitrogen slowly through the solution (Fig. 21.9).

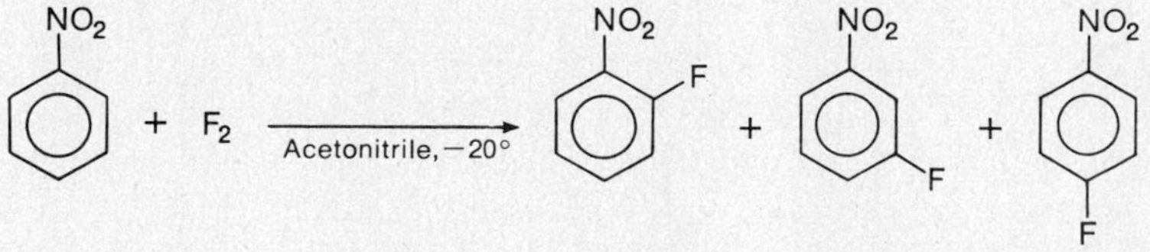

Nitrobenzene $+ F_2 \xrightarrow{\text{Acetonitrile, } -20°}$ o-fluoronitrobenzene (13%) + m-fluoronitrobenzene (78%) + p-fluoronitrobenzene (9%)

Fig. 21.9. The direct fluorination of nitrobenzene.

The equilibrium for the electrophilic iodination reaction ($ArH + I_2 \rightleftharpoons ArI + HI$), on the other hand, is sufficiently unfavorable that it requires displacement in the direction of product formation by the addition of a base such as sodium bicarbonate, an oxidizing agent such as nitric acid or perbenzoic acid, or a precipitating agent such as silver ion (Fig. 21.10).

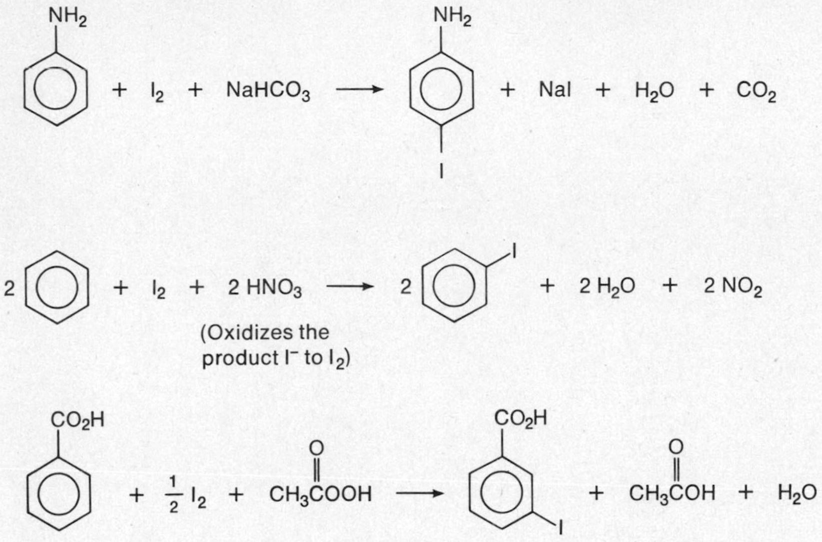

$$C_6H_5NH_2 + I_2 + NaHCO_3 \longrightarrow p\text{-}IC_6H_4NH_2 + NaI + H_2O + CO_2$$

$$2\,C_6H_6 + I_2 + 2\,HNO_3 \longrightarrow 2\,C_6H_5I + 2\,H_2O + 2\,NO_2$$

(Oxidizes the product I^- to I_2)

$$C_6H_5CO_2H + \tfrac{1}{2}I_2 + CH_3\overset{O}{\overset{\|}{C}}OOH \longrightarrow m\text{-}IC_6H_4CO_2H + CH_3\overset{O}{\overset{\|}{C}}OH + H_2O$$

Fig. 21.10. Electrophilic iodination of aromatic compounds.

PROBLEM 21-6

Suggest a method for converting *p*-iodoanisole to anisole.

Between these extremes fall the direct chlorination and bromination reactions, which are the most generally useful for the preparation of aryl halides. When *o,p*-directing, activating substituents are present on the benzene ring (*e.g.*, CH_3O-), chlorine and bromine are sufficiently reactive to effect halogenation by themselves. When *m*-directing, deactivating substituents are present (*e.g.*, $-NO_2$), however, it is necessary to add a Lewis acid, such as $AlCl_3$, $FeCl_3$, or $FeBr_3$, to increase the concentration of halogen cation (*e.g.*, $Cl_2 + AlCl_3 \rightleftharpoons Cl^+ + AlCl_4$) (Fig. 21.11).

21.3b. VIA ARYLDIAZONIUM SALTS. The position on the benzene ring to which the halogen atom becomes attached in the direct halogenation reaction is determined by the directive influence of the groups already present on the ring. If we wish to introduce halogens at other positions on the ring, we must use different methods. The most generally applicable of these involves nucleophilic substitution reactions with aryldiazonium salts. For example, benzenediazonium salts react with potassium iodide to form iodobenzenes, iodine entering

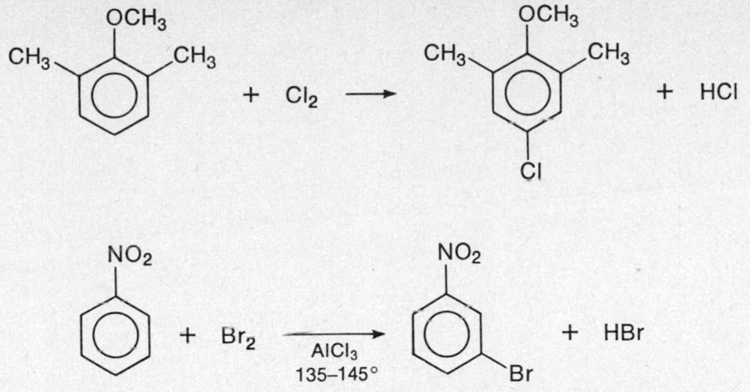

Fig. 21.11. Chlorination and bromination of substituted benzenes.

the ring at the site of the departing N_2^+ group. Similarly, benzenediazonium salts react with cuprous bromide to form bromobenzene and with cuprous chloride to form chlorobenzene in processes known as *Sandmeyer reactions* (see Fig. 16.6). To obtain fluorobenzene, a two-step sequence is required, involving treatment first with fluoroboric acid to produce phenyldiazonium fluoroborate followed by heating, which causes the elimination of molecular nitrogen and boron trifluoride (Fig. 21.12).

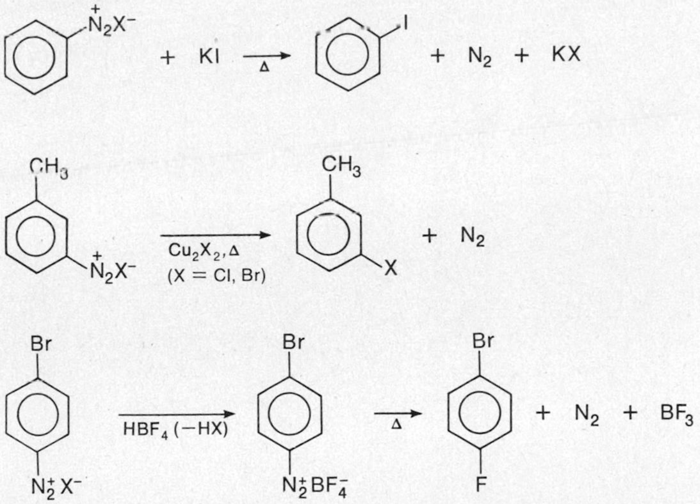

Fig. 21.12. Preparation of aryl halides via aryldiazonium salts.

PROBLEM 21-7

Suggest reaction sequences for converting toluene to *p*-bromobenzoic acid and *m*-bromobenzoic acid.

Table 21.1 METHODS OF PREPARATION OF ALKYL, ALKENYL, AND ARYL HALIDES

Starting Materials	Reagents	Product	Comments
Alkyl halides:			
Alkanes (R—H)	X_2	R—X	F_2 requires special conditions; produces mixtures Cl_2 produces mixtures Br_2 selective for $R_3C{-}H$ I_2 no reaction
Alkenes ($>C{=}C{-}\overset{\mid}{\underset{\mid}{C}}{-}H$)	$(CH_3)_3C{-}O{-}X$ (X=Cl, Br)	$>C{=}C{-}\overset{\mid}{\underset{\mid}{C}}{-}X$ (X=Cl, Br)	
	O N—Br O	$>C{=}C{-}\overset{\mid}{\underset{\mid}{C}}{-}Br$	
Alkenes ($>C{=}C<$)	HX (X = F, Cl, Br, I)	$H{-}\overset{\mid}{\underset{\mid}{C}}{-}\overset{\mid}{\underset{\mid}{C}}{-}X$ (X = F, Cl, Br, I)	Markovnikov regioselectivity, *anti*-Stereoselectivity: ionic reaction anti-Markovnikov with HBr, *anti*-Stereoselectivity: radical reaction
Alcohols (ROH)	HX (X = Cl, Br, I)	R—X (X = Cl, Br, I)	Rearrangements with $R_3C{-}\overset{\mid}{\underset{\mid}{C}}{-}OH$ and $Ar{-}\overset{\mid}{\underset{\mid}{C}}{-}\overset{\mid}{\underset{\mid}{C}}{-}OH$
	PX_3, PX_5 (X = Cl, Br)	R—X (X = Cl, Br)	Inversion of configuration
	SOX_2 (X = Cl, Br)	R—X (X = Cl, Br)	Retention of configuration in benzene Inversion of configuration in pyridine
Carboxylic acids (RCO_2H)	Br_2 on Ag, Hg, or Tl salt	R—Br	Nonstereoselective
Alkenyl halides:			
gem-Dihalides ($-\overset{H}{\underset{\mid}{C}}{-}\overset{X}{\underset{X}{C}}-$)	Base	$>C{=}C<^{X}$	Generally restricted to dichlorides and dibromides
vic-Dihalides ($-\overset{\mid}{\underset{X}{C}}{-}\overset{\mid}{\underset{X}{C}}-$)	Base	$>C{=}C<^{X}$	Generally restricted to dichlorides and dibromides
Aryl halides:			
Aromatic hydrocarbons (ArH)	X_2	Ar—X	F_2 requires special conditions; Cl_2 and Br most useful I_2 requires special conditions Orientation dependent on directive influence of groups on aromatic ring
Aryldiazonium salts (ArN_2^+)	HBF_4 and heat, Cu_2Cl_2, Cu_2Br_2, KI	Ar—F Ar—Cl, Ar—Br Ar—I	Position of halogen determined by position of N_2^+ function

NUCLEOPHILIC SUBSTITUTION REACTIONS

Nucleophilic substitution is the most characteristic reaction of the alkyl halides, and it is instructive to consider the mechanisms by which it occurs. Since nucleophilic substitution is by no means limited to the halides, however, the following discussion includes certain other classes of compounds as well.

21.4 Nucleophilic Substitution Reactions at Saturated Carbon Atoms

21.4a. GENERAL FEATURES. Two mechanisms for nucleophilic substitution, discussed in Chapters 9 (see Sec. 9.5) and 20, are designated as S_N1 and S_N2 processes. The S_N1 process, which is unimolecular and kinetically first order, involves the formation of carbonium ion intermediates and proceeds with loss of stereochemical integrity* at the center of substitution. The S_N2 process, which is bimolecular and kinetically second order, involves no intermediates (only an activated complex) and proceeds with inversion of absolute configuration at the center of substitution. The clear distinction between these is that the S_N2 pathway is a concerted one, meaning that a continuously linked transformation of reactant to product takes place *without* the intervention of intermediates, whereas the S_N1 pathway is a nonconcerted one, meaning that a noncontinuous transformation of reactant to product takes place *with* the formation of intermediates. Unfortunately, the distinction is not always so clean as this categorical classification might suggest, as the following example indicates.

The possibility that in some instances S_N1 reactions are concerted processes is indicated by an experiment involving the ester obtained from neopentyl alcohol and *p*-toluenesolfonyl chloride (called neopentyl tosylate) (Fig. 21.13). This compound undergoes nucleophilic substitution with skeletal rearrangement, a frequent occurrence in S_N1 reactions and a rare one in S_N2 reactions. If we use a neopentyl tosylate containing a deuterium in place of one of the α hydrogens, we are dealing with a compound containing a chiral center and which, therefore, permits the study of the rearrangement reaction with optically active starting material. It turns out that when optically active deuterioneopentyl tosylate reacts with acetic acid—*i.e.*, the incoming nucleophile is an acetate moiety, $CH_3C(=O)O-$ the products are optically active esters of acetic acid in which the absolute configuration at the chiral center of the product is opposite to that of the starting material; inversion of absolute configuration has taken place. This might seem to suggest an S_N2 process, for if true carbonium ion intermediates were involved we would anticipate that optical activity should be lost. On the basis of kinetic measurements and other observations, however, the reaction is

* The term "stereochemical integrity" refers to the extent to which optical purity (see footnote, p. 538) is retained in the product. If complete retention or inversion of configuration occurs in a reaction, the optical purity of reactant and product are identical, and stereochemical integrity has been maintained. If partial retention or partial inversion of configuration occurs in a reaction, the optical purity of the product is lower than that of the reactant, and stereochemical integrity has not been maintained.

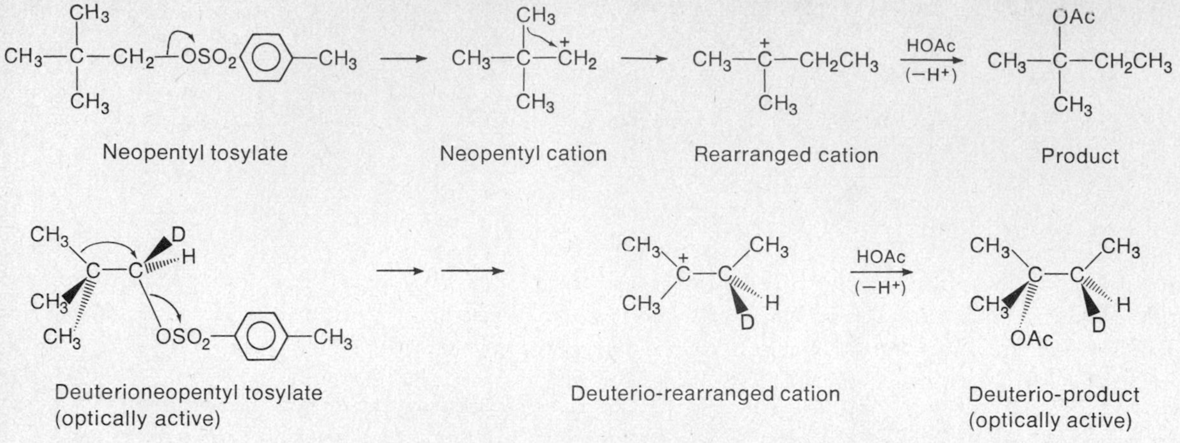

Neopentyl tosylate → Neopentyl cation → Rearranged cation → Product

Deuterioneopentyl tosylate (optically active) → → Deuterio-rearranged cation → Deuterio-product (optically active)

Fig. 21.13. Evidence for concerted methyl migration in the reaction of optically active deuterio-neopentyl tosylate with acetic acid (abbreviated HOAc).

classed as an S_N1 process and is viewed as being initiated by dissociation of the tosyloxy group to give an incipient carbonium ion center to which an adjacent methyl group migrates, via a backside approach, as the tosyloxy group is departing. The reaction is considered to have "carbonium ion character" but to take place in an essentially concerted fashion, thus blurring the clean distinction between S_N1 and S_N2 processes.

PROBLEM 21-8

Treatment of (+)-2-octyl iodide with radioactive iodide anion, $*I^-$, yields (±)-2-octyl iodide containing radioactive iodine. The rate of racemization is twice the rate of incorporation of radioactive iodine. Explain why.

21.4b. EFFECT OF STRUCTURE OF THE CARBON FRAMEWORK, R, ON S_N1 AND S_N2 REACTIONS. Carbon frameworks capable of stabilizing carbonium ions facilitate the S_N1 process, the stabilization generally increasing in the order $CH_3^+ \ll RCH_2^+ < R_2CH^+ \simeq C{=}C{-}C^+ < R_3C^+ < ArC^+$. Stable carbonium ions are planar, and those that cannot acquire a planar configuration form with great difficulty. For example, 1-bromobicyclo[2.2.1]heptane reacts in an S_N1 process extremely slowly, because the bridgehead cation is prevented by excessive strain from becoming planar.

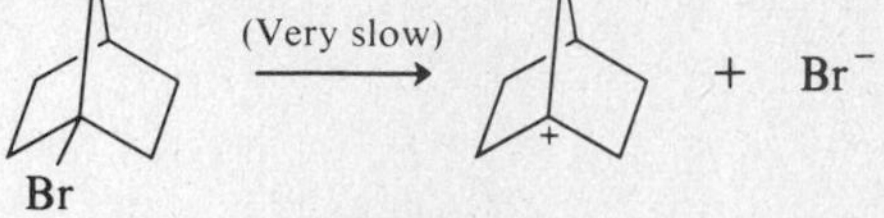

(Very slow) + Br^-

Bridgehead cation
Planar configurations would put great strain on the system

PROBLEM 21-9

Arrange the cations derived by loss of bromide ion from the following alkyl bromides in the order of increasing stability: 1-bromobicyclo[2.2.1]heptane, 1-bromobicyclo[2.2.2]octane, 1-bromobicyclo[3.2.2]nonane, 1-bromobicyclo[4.4.0]decane. Explain the basis for your arrangement.

In the S_N2 process, the entering group approaches the backside of the carbon carrying the leaving group, and the ease with which the substitution takes place is diminished by anything that hinders this approach. Thus, groups attached to the α carbon, the β carbon, or otherwise situated to interfere with the backside approach of the entering group retard S_N2 reactions, the reactivity of alkyl frameworks generally increasing because of steric reasons in the order $R_3C < R_2CH < RCH_2 < CH_3$ (*i.e.*, the opposite of the reactivity sequence for S_N1 reactions). Examples in which remote substituents retard S_N2 reactions are illustrated in Fig. 21.14. In the neopentyl compound, the very bulky *t*-butyl

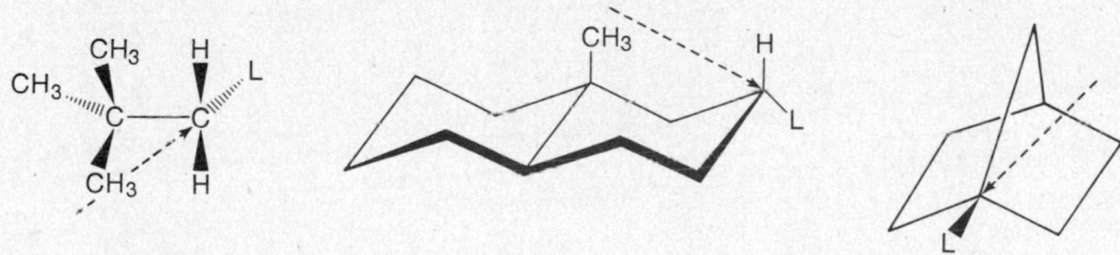

Fig. 21.14. Examples of molecules that react very slowly in S_N2 reactions because of steric hindrance to backside approach by the incoming nucleophile, indicated by the dashed lines.

group interferes with the approach of the nucleophile to the backside of the primary carbon atom (see Fig. 20.2). In the bicyclo[4.4.0]decane, the axial methyl group at the bridgehead position (often referred to as an *angular methyl group*) similarly impedes the approach of the nucleophilic reagent to the backside of the reaction center, whereas in the 1-substituted bicyclo[2.2.1]heptane it is impossible for the nucleophile to approach the backside of C-1 from the "inside" of the molecule.

It is not always easy to predict whether a reaction will follow the S_N1 or the S_N2 pathway. As a general rule, the S_N2 pathway is more likely, because it involves only charge dissipation or charge dispersal rather than full charge development in progressing from reactants to activated complex. The S_N1 pathway, on the other hand, involves charge development, and this requires considerably more energy. Thus, only when the carbon framework is especially capable of forming a stable carbonium ion and/or when severe steric interference to the backside approach of the nucleophile exists will the S_N1 process be favored.

Whether a reaction follows the S_N1 or S_N2 pathway is dependent on the particular carbon framework, as these examples have illustrated; Table 21.2 lists the reactivities of various carbon frameworks.

PROBLEM 21-10

Chloromethyl ether, CH_3OCH_2Cl, reacts with water in an S_N1 process 1600 times more rapidly than its sulfur counterpart CH_3SCH_2Cl. What does this suggest about the relative abilities of oxygen and sulfur to stabilize a cation by electron delocalization? What orbitals on oxygen and sulfur are involved in the electron delocalization?

Table 21.2 RELATIVE REACTIVITIES OF VARIOUS FRAMEWORKS IN S_N1 AND S_N2 REACTIONS, ARRANGED IN ORDER OF INCREASING REACTIVITY

S_N1 Reaction	S_N2 Reaction
 1-Norbornyl–L (bicyclic bridgehead, L at bridgehead)	 1-Norbornyl–L (bicyclic bridgehead, L at bridgehead)
$RCH_2{-}L$ where R is an electron-withdrawing group, such as $R{-}C(=O){-}$	$(CH_3)_3CCH_2{-}L$
	$(CH_3)_3C{-}L$
$CH_3{-}L$	$(CH_3)_2CH{-}L$
$CH_3CH_2{-}L$	$CH_3CH_2{-}L$
$(CH_3)_2CH{-}L$	$CH_3{-}L$
$H_2C{=}CHCH_2{-}L$	$CH_3OCH_2{-}L$
$(CH_3)_3C{-}L$	$H_2C{=}CHCH_2{-}L$
$CH_3OCH_2{-}L$	$C_6H_5CH_2{-}L$
$H_2C{=}CHC(CH_3)_2{-}L$	$RC(=O)CH_2{-}L$
$C_6H_5CH_2{-}L$	
$(C_6H_5)_2CH{-}L$	
$(C_6H_5)_3C{-}L$	

21.4c. REARRANGEMENT OF CARBON FRAMEWORKS IN S_N1 REACTIONS. One of the earliest examples of a molecular rearrangement and certainly now one of the best studied was discovered late in the nineteenth century by the German chemist Wagner, who found that a terpene called camphene hydrochloride undergoes a rearrangement to an isomeric terpene called isobornyl chloride (Fig. 21.15). The details of this seemingly baffling reaction were worked out in the 1920's by another German chemist, Hans Meerwein. In fact, the foundations for our modern concepts of carbonium ion theory had their genesis in the experiments of Meerwein, who interpreted the rearrangement in terms of electron-deficient "intermediates" (Fig. 21.15).* A host of Wagner–Meerwein-type rearrangements involving carbonium ion "intermediates" are now known (see p. 270 for another example).

The greater the tendency to follow the S_N1 pathway the greater is the likelihood of a Wagner–Meerwein rearrangement. Also influencing the probability of rearrangement is the ease with which the group adjacent to the carbonium ion

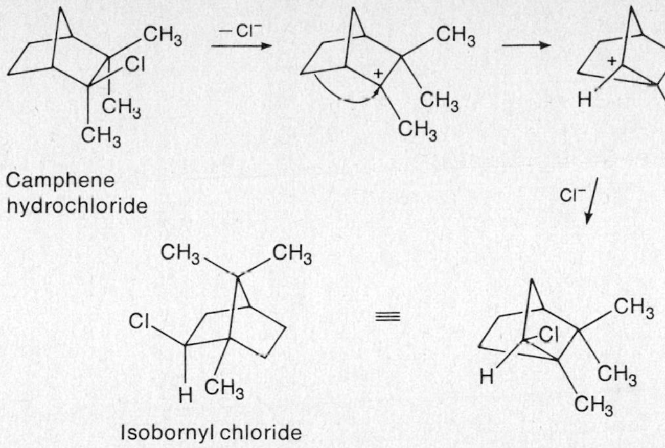

Fig. 21.15. Wagner-Meerwein rearrangement of camphene hydrochloride.

center can migrate *with a pair of electrons*, this being characterized as the "migratory aptitude" of the group. In general, it decreases in the following order:

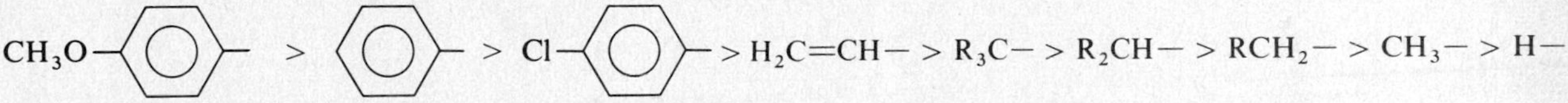

$$\text{CH}_3\text{O}-\text{C}_6\text{H}_4- > \text{C}_6\text{H}_5- > \text{Cl}-\text{C}_6\text{H}_4- > \text{H}_2\text{C}{=}\text{CH}- > \text{R}_3\text{C}- > \text{R}_2\text{CH}- > \text{RCH}_2- > \text{CH}_3- > \text{H}-$$

although numerous exceptions to this sequence are known. Aryl groups are particularly prone to migrate, and the pathway for this process has received considerable study. It has been postulated that a structure exists along the reaction pathway in which the aryl group (*e.g.*, a phenyl group) is bonded to *both* the migration origin and the migration terminus. This structure, which may be a true intermediate or simply an activated complex, is called a *phenonium ion* and is represented by the structures shown in Fig. 21.16 for the 2,2-diphenylethyl cation. Among the pieces of evidence in support of this idea is the radioactive labeling experiment described on p. 573.

* It may seem puzzling that a rearrangement such as that shown in Fig. 21.15 should occur, since it involves the conversion of a tertiary carbonium ion to a secondary carbonium ion. Remember, though, that an unfavorable equilibrium of an intermediate step in a reaction sequence does not necessarily determine the overall equilibrium. In the camphene hydrochloride to isobornyl chloride transformation the equilibrium constant is greater than unity as the result of fewer nonbonded interactions in isobornyl chloride than in camphene hydrochloride (*i.e.*, methyl–methyl eclipsing and methyl–chlorine eclipsing). Depicting this rearrangement in terms of discrete carbonium ion intermediates, its reaction coordinate could be shown as

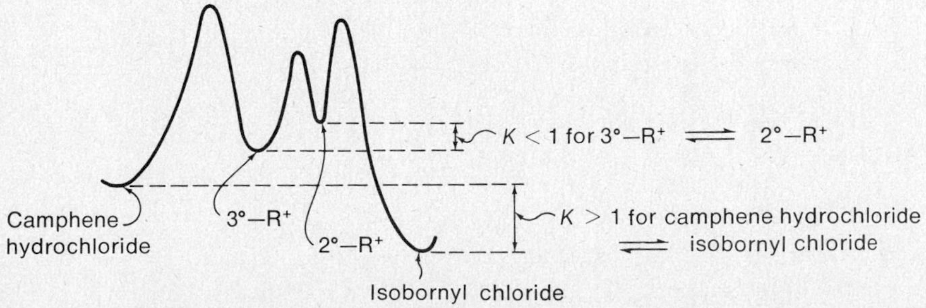

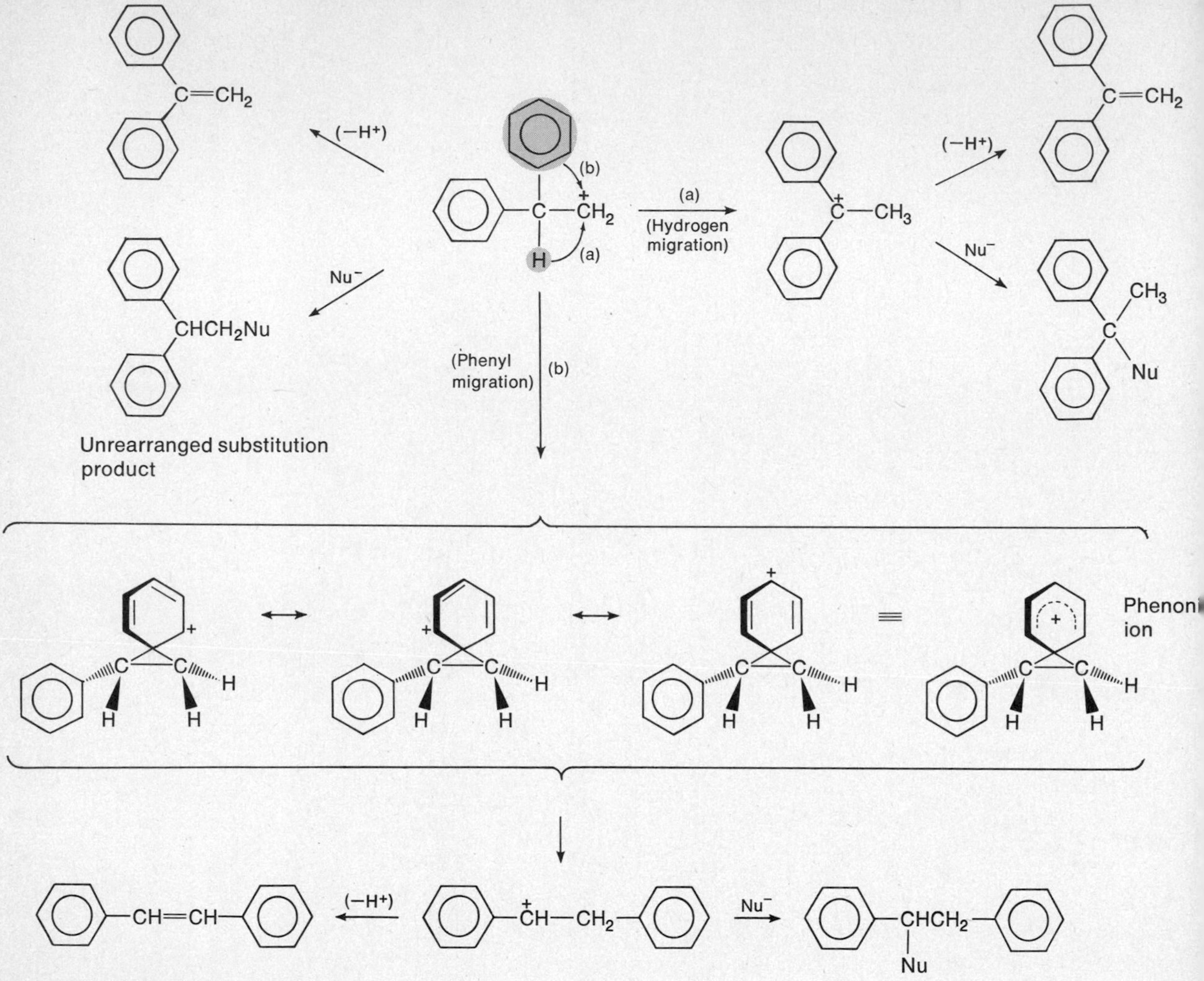

(−H+) (a) (Hydrogen migration) (−H+) Nu⁻ Nu⁻ (b) (Phenyl migration)

Unrearranged substitution product

Phenonium ion

(−H+) Nu⁻

Fig. 21.16. Rearrangement of 2,2-diphenylethyl cation via hydrogen migration or phenyl migration.

PROBLEM 21-11

Predict the structure of the major product from the action of nitrous acid in acetic acid solution on $C_6H_5CDCH_2NH_2$ (with a p-OCH_3-phenyl group on the C bearing D)

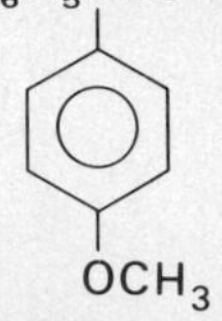

PROBLEM 21-12

Design an isotope labeling experiment by means of which the formation of the compound $(C_6H_5)_2C{=}CH_2$ by the two pathways shown in Fig. 21.16 could be differentiated.

21.4d. EFFECT OF NUCLEOPHILE, NU, ON S_N1 AND S_N2 REACTIONS. The stronger the incoming nucleophile, Nu, the greater is the tendency for the reaction to proceed via the S_N2 pathway; conversely, the weaker the incoming nucleophile the greater is the tendency for the reaction to proceed via the S_N1 pathway. The nucleophilic reactivity (termed *nucleophilicity*) of a reagent is a function of a variety of factors, including its charge, its basicity, its polarizability, and its size. Thus, a negatively charged nucleophile is more reactive than its neutral counterpart, a strong base is more reactive than a weak base, a more polarizable group is a better nucleophile than a less polarizable group, and a small nucleophile is generally more reactive than a larger, more sterically hindered nucleophile.

The reactivity of nucleophiles derived from atoms of the same row of the periodic table approximately parallels their basicity. For example, the nucleophilicity of the anions derived from first-row elements increases in the order $F^- < RO^- < R_2N^- < R_3C^-$; the basicity increases in the same order, basicity being inversely related to the acidity of the conjugate acid (*i.e.*, the acidity order of the conjugate acids is $HF > ROH > R_2NH > R_3CH$). Nucleophilicity, however, is a measure of the ability of an entity to attach itself to carbon (*i.e.*, displacement on carbon; *e.g.*, $CH_3O^- + CH_3I \rightarrow CH_3OCH_3 + I^-$), whereas basicity is a measure of its ability to attach itself to hydrogen (*i.e.*, displacement on hydrogen; *e.g.*, $CH_3O^- + H{-}OH \rightarrow CH_3OH + OH^-$). Consequently, changes in the structure of the displacing agent do not necessarily change its basicity and nucleophilicity in precisely the same fashion. Changes in size of the nucleophile, for example, may have a more pronounced effect on nucleophilicity than on basicity because of the very different steric demands for displacement on carbon as contrasted with displacement on hydrogen. Thus, the nucleophilicity of alkoxides decreases in the order $CH_3O^- > C_2H_5O^- > (CH_3)_2CHO^- > (CH_3)_3CO^-$, although their basicities follow the opposite order. Partly because of their geometry and minimal steric demands,* linear nucleophiles such as CN^- and N_3^- are particularly reactive.

* Comparisons of basicities and nucleophilicities are complicated by solvent effects. Gas phase measurements show that the basicities of halide ions decrease in the order $F^- > Cl^- > Br^- > I^-$, the same order that is shown for the nucleophilicities of these species in aprotic solvents, such as dimethylsulfoxide $[(CH_3)_2SO]$, dimethylformamide $[HCON(CH_3)_2]$, and hexamethylphosphoramide $([(CH_3)_2N]_3PO)$. In protic solvents such as alcohols, however, the nucleophilicities of halide ions decrease in the order $I^- > Br^- > Cl^- > F^-$. In the hydroxylic solvents, extensive hydrogen bonding between the solvent and the halide ion occurs, thus reducing the nucleophilic character of the halide ion. Because the order of the basicity of the halide ions is $F^- > Cl^- > Br^- > I^-$, the fluoride anion will be more extensively hydrogen bonded with the solvent than will the other halide ions: thus, it will be a poorer nucleophile than the other halide ions in hydrogen-bond donor solvents.

PROBLEM 21-13

Arrange the following anions in the order of (a) increasing nucleophilicity and (b) increasing basicity:

Cl^-, $(CH_3)_3CO^-$, $C_3H_2CH_2O^-$, $(CH_3CH_2)_2N^-$, $(C_3H_2)_3C^-$, CH_3^-, and

```
          O
          ||
          C
    HC ⁄    ＼
    ||        N⁻
    HC ＼    ⁄
          C
          ||
          O
```

Differences in polarizability are also responsible for the failure of basicities and nucleophilicities to change in a precisely parallel fashion. Polarizability is a measure of the ease with which the electron cloud surrounding the nucleophile can be distorted: the more polarizable the electron cloud the greater is the nucleophilicity. For example, although CH_3O^- is a stronger base than CH_3S^-, the thiomethoxide ion is a considerably more effective nucleophile. The sulfur atom of the thiomethoxide, even though larger than oxygen, is more polarizable and is more capable of "squeezing" into the necessary volume of space required in the activated complex for the substitution process to occur easily.

Table 21.3 lists a representative group of nucleophiles.

Table 21.3 NUCLEOPHILES ARRANGED IN THE ORDER OF INCREASING REACTIVITY WITH RESPECT TO NUCLEOPHILIC DISPLACEMENT OF BROMIDE IN METHYL BROMIDE

Nucleophile	Relative Reactivity	Nucleophile	Relative Reactivity
H_2O	1.00	N_3^-	1,000
NO_3^-	1.02	$(NH_2)_2C{=}S$	1,250
ROH	*	HO^-	1,600
R_2S	*	RO^-	*
F^-	10	$C_6H_5NH_2$	3,100
$CH_3CO_2^-$	52.5	SCN^-	5,900
HCO_2^-	56.5	I^-	10,200
Cl^-	102	$HS^-(RS^-)$	12,600
$C_6H_5O^-$	316	CN^-	12,600
C_5H_5N (pyridine)	400	$SO_3^=$	12,600
R_3N	*	$S_2O_3^=$	220,000
Br^-	775		

* Values not accurately known.

21.4e. EFFECT OF THE LEAVING GROUP, L, ON S_N1 AND S_N2 REACTIONS. The halogens are good leaving groups, being more or less easily displaced by other nucleophiles. They are not alone in being able to act in this fashion, however, and a variety of other functions can also serve as leaving groups. Our discussion at this point, therefore, expands beyond the halides to include certain other types of compounds as well. The "leaving ability" of groups varies considerably, as might be expected, and these differences give rise to still another factor affecting the reaction pathway. The greater the "leaving ability" of a group the greater is the tendency for the reaction to follow the S_N1 pathway; conversely, the poorer the leaving ability, the greater is the tendency to follow the S_N2 pathway. In general, the leaving ability parallels the acidity of the conjugate acid of the leaving group. Thus, Cl^- is a better leaving group than OH^- (*i.e.*, HCl is a stronger acid than HOH), and $C_6H_5O^-$ and is a better leaving group than CH_3O^- (*i.e.*, C_6H_5OH is a stronger acid than CH_3OH). The leaving ability also correlates with the electronegativity of the group, diminishing from group VII elements (halogens) to group VI elements (oxygen, sulfur) to group V elements (nitrogen) to group IV elements (carbon).

Groups that are positively charged in the starting compound and, therefore, can leave as neutral entities are much better leaving groups than their uncharged counterparts. For example, the loss of R_3N from $R{-}\overset{+}{N}R_3$ occurs far more readily than the loss of R_2N^- from $R{-}NR_2$. Similarly, the loss of H_2O from $R{-}\overset{+}{O}H_2$ occurs with far greater facility than the loss of OH^- from R—OH. Thus, the nucleophilic substitution of an alcohol generally takes place only in solutions that are sufficiently acidic to transfer a proton to the alcohol, thereby converting

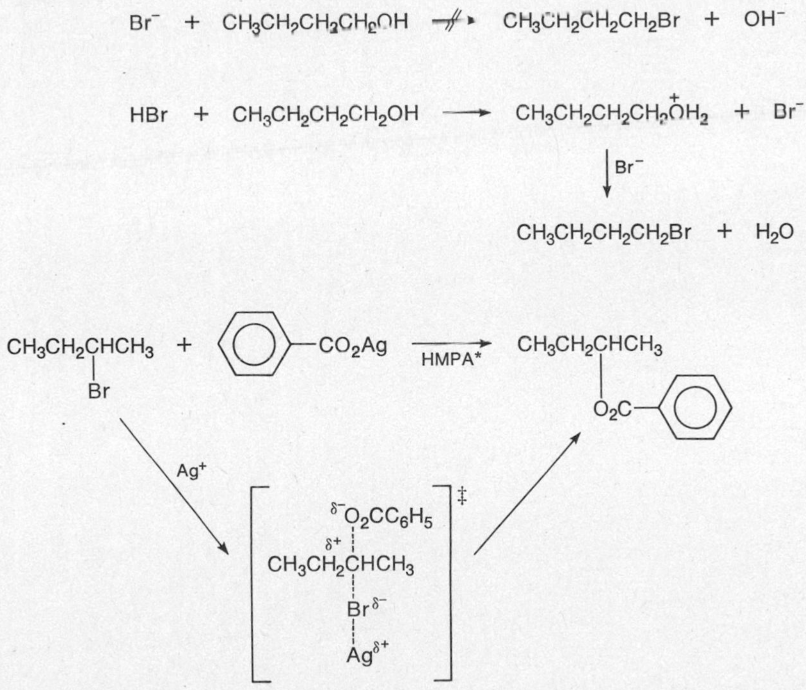

$$Br^- + CH_3CH_2CH_2CH_2OH \not\longrightarrow CH_3CH_2CH_2CH_2Br + OH^-$$

$$HBr + CH_3CH_2CH_2CH_2OH \longrightarrow CH_3CH_2CH_2CH_2\overset{+}{O}H_2 + Br^-$$

$$\xrightarrow{Br^-} CH_3CH_2CH_2CH_2Br + H_2O$$

$$CH_3CH_2CH(Br)CH_3 + C_6H_5{-}CO_2Ag \xrightarrow[\text{HMPA*}]{} CH_3CH_2CH(O_2C{-}C_6H_5)CH_3$$

via $\xrightarrow{Ag^+}$ $\left[\begin{array}{c} {}^{\delta-}O_2CC_6H_5 \\ \vdots \\ CH_3CH_2\overset{\delta+}{C}HCH_3 \\ \vdots \\ Br^{\delta-} \\ \vdots \\ Ag^{\delta+} \end{array}\right]^{\ddagger}$

Fig. 21.17. Examples of acid-catalyzed S_N2 reactions. (HMPA is hexamethylphosphoramide: for structure see footnote on p. 617.

it to the oxonium salt $R\overset{+}{O}H_2$. In similar fashion, alkyl halides are readily converted to esters upon treatment with carboxylate anions if silver ions are also present, the silver ion acting as a Lewis acid that attaches itself to the halogen (acting as a Lewis base) and facilitates its departure (Fig. 21.17).

PROBLEM 21-14

Which member of the following pairs of compounds contains the better nucleophilic leaving group:

(a) $RC(=O)OR'$ or $RC(=O)\overset{+}{O}(H)R'$ (b) RI or RF (c) R_2S or $R-\overset{+}{S}R_2$

A list of functional groups is shown in Table 21.4, where they are arranged in the order of increasing leaving ability.

Table 21.4 LEAVING GROUPS ARRANGED IN THE ORDER OF INCREASING LEAVING ABILITY

Mechanism	Class of Compound		Leaving Group	
S_N2	$R-NR'_2$	Amines	$^-NR'_2$	Dialkylamide anion
	$R-OR'$	Dialkyl ethers	$^-OR'$	Alkoxide
	$R-OH$	Alcohols	^-OH	Hydroxide
	$R-OAr$	Alkyl aryl ethers	^-OAr	Phenoxide
	$R-SR'$	Dialkyl sulfides	$^-SR'$	Thioalkoxide
	$R-OCOR'$	Esters	$^-OCOR'$	Carboxylate anion
	$R-F$	Alkyl fluorides	^-F	Fluoride (rel rate = 0.001)
	$R-Cl$	Alkyl chlorides	^-Cl	Chloride (rel rate = 0.02)
	$R-\overset{+}{N}R'_3$	Alkyl ammonium salts	NR'_3	Trialkylamine
	$R-\overset{+}{S}R'_2$	Alkyl sulfonium salts	SR'_2	Dialkylsulfide (rel rate = 0.5)
S_N1 or S_N2	$R-\overset{+}{O}(R')H$	O-Protonated ethers	$R'OH$	Alcohol
	$R-\overset{+}{O}H_2$	Protonated alcohols	H_2O	Water (rel rate = 1)
	$R-Br$	Alkyl bromides	^-Br	Bromide (rel rate = 1)
	$R-I$	Alkyl iodides	^-I	Iodide (rel rate = 3)
	$R-OSO_2Ar$	Alkyl aryl sulfonates	$^-OSO_2Ar$	Arylsulfonate anion (rel rate = 6)
	$R-OC(=O)CH_2CF_3$	Alkyl 3,3,3-trifluoropropionates	$^-OC(=O)CH_2CF_3$	3,3,3-Trifluoropropionate
	$R-OC(=O)CF_3$	Alkyl trifluoroacetates	$^-OC(=O)CF_3$	Trifluoroacetate
	$R-\overset{+}{O}R'_2$	Trialkyl oxonium ion	R'_2O	Dialkyl ether
	$R-OSO_2F$	Alkyl fluorosulfonates	$^-OSO_2F$	Fluorosulfonate (rel rate = 10^5)
	$R-OSO_2CF_3$	Alkyl trifluoromethanesulfonates	$^-OSO_2CF_3$	Trifluoromethanesulfonate
S_N1	$R-N_2^+$	Alkyl diazonium ion	N_2	Nitrogen

21.4f. EFFECT OF SOLVENT ON S_N1 AND S_N2 REACTIONS. The solvent that is used in a nucleophilic substitution reaction can have a profound effect on the rate of the reaction. In reactions in which product formation is favored by charge development in the transition state (*e.g.*, $R{-}L \rightarrow [R^{\delta+}\cdots L^{\delta-}]^{\ddagger}$), solvents of high dielectric constant (see Table 21.5) favor the process; conversely, in reactions in which reduction or dispersal of charge in the transition state favors product formation (*e.g.*, $Nu^- + R{-}L \rightarrow [Nu^{\delta-}\cdots R\cdots L^{\delta-}]^{\ddagger}$), solvents of low dielectric constant will favor the process. Thus, S_N1 reactions and those S_N2 reactions involving uncharged nucleophiles are greatly accelerated by high dielectric constant solvents, because charge is developed in the transition state; S_N2 reactions involving negatively charged nucleophiles, on the other hand, take place more rapidly in low dielectric constant solvents. The effects of solvent polarity on these and other types of S_N1 and S_N2 reactions are summarized in Table 21.6.

Table 21.5 DIELECTRIC CONSTANTS OF SOME COMMON ORGANIC SOLVENTS

Solvent	Dielectric Constant (ε)	Solvent	Dielectric Constant (ε)
Cyclohexane	2.015	2-Propanol	18.3
Carbon tetrachloride	2.238	1-Propanol	20.1
Benzene	2.38	Acetone	20.7
Diethyl ether	4.34	Ethanol	24.3
Chloroform	4.806	Methanol	32.63
Acetic acid	6.15	Dimethylformamide	37.0
Methyl acetate	6.68	Dimethylsulfoxide	46.3
Pyridine	12.3	Water	78.5

Table 21.6 EFFECT OF SOLVENT POLARITY ON S_N1 AND S_N2 REACTIONS

Type	Reactants	Charge Distribution in Activated Complex	Change in Activated Complex	Effect on Rate of Increase in Solvent Polarity
S_N1	$R{-}L$	$R^{\delta+}$ ---- $L^{\delta-}$	Increased	Large increase
S_N1	$R{-}L^+$	$R^{\delta+}$ ---- $L^{\delta+}$	Dispersed	Small decrease
S_N2	$R{-}L + Nu^-$	$Nu^{\delta-}$ ---- R ---- $L^{\delta-}$	Dispersed	Small decrease
S_N2	$R{-}L + Nu$	$Nu^{\delta+}$ ---- R ---- $L^{\delta-}$	Increased	Large increase
S_N2	$R{-}L^+ + Nu^-$	$Nu^{\delta-}$ ---- R ---- $L^{\delta+}$	Decreased	Large decrease
S_N2	$R{-}L^+ + Nu$	$Nu^{\delta+}$ ---- R ---- $L^{\delta+}$	Dispersed	Small decrease

21.4g. SOLVOLYSIS REACTIONS. Many solvents are nucleophilic and, therefore, can react with compounds containing leaving groups to yield stable products. Such reactions are called *solvolysis reactions*. The hydrolysis, ethanolysis, and acetolysis reactions shown in Fig. 21.18 are typical examples.

21.4h. INTRAMOLECULAR NUCLEOPHILIC SUBSTITUTION (NEIGHBORING GROUP PARTICIPATION). The nucleophilic substitutions discussed in the preceding sections have all involved an *inter*molecular reaction between a nucleophilic

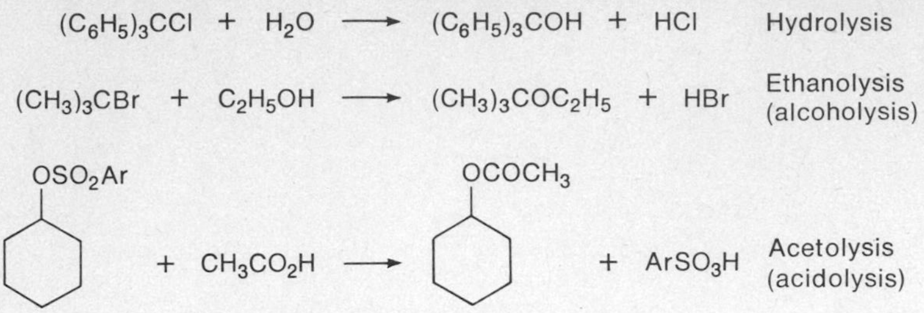

$(C_6H_5)_3CCl + H_2O \longrightarrow (C_6H_5)_3COH + HCl$ Hydrolysis

$(CH_3)_3CBr + C_2H_5OH \longrightarrow (CH_3)_3COC_2H_5 + HBr$ Ethanolysis (alcoholysis)

$C_6H_{11}OSO_2Ar + CH_3CO_2H \longrightarrow C_6H_{11}OCOCH_3 + ArSO_3H$ Acetolysis (acidolysis)

Fig. 21.18. Examples of solvolysis reactions.

reagent and the R—L compound. In suitably constituted molecules, *intra*-molecular counterparts of these reactions are possible, as, for instance, in the reaction between 2-chloro-3-butanol and sodium methoxide to form 2,3-dimethyloxirane. The sodium methoxide abstracts a proton from 2-chloro-3-butanol to give the corresponding alkoxide, which then effects an intramolecular nucleophilic displacement of the neighboring chlorine to yield the oxirane:

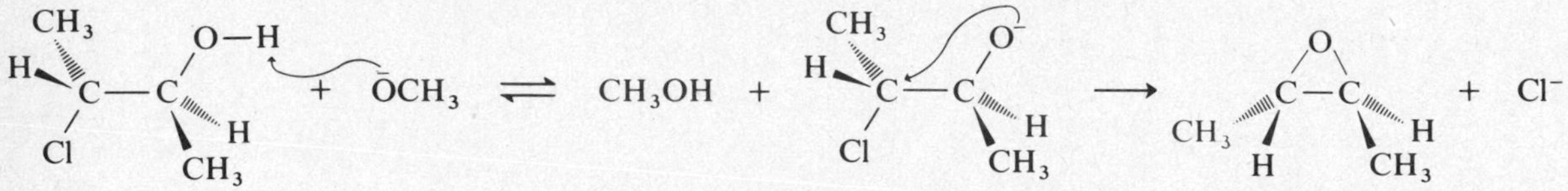

$$CH_3CHCl\text{—}CH(OH)CH_3 + {}^-OCH_3 \rightleftharpoons CH_3OH + CH_3CHCl\text{—}CH(O^-)CH_3 \longrightarrow CH_3CH\text{—}CHCH_3\ (\text{oxirane}) + Cl^-$$

PROBLEM 21-15

Predict the structure of the product that is formed by the action of Ag_2O on $ClCH_2CH_2\overset{+}{N}H_3Cl^-$.

Reactions of this type are exceedingly useful for the preparation of a wide variety of cyclic compounds, and many examples are known in which stable products are formed in intramolecular displacement processes. In many other cases, however, the product of intramolecular displacement is unstable and undergoes subsequent reaction to yield acyclic products. The formation of cyclic intermediates in these instances must usually be inferred from the kinetics and the stereochemistry of the reaction. For example, the *cis* and *trans* isomers of 1-acetoxy-2-tosyloxycyclohexane undergo solvolysis in acetic acid to produce 1,2-diacetoxycyclohexane at markedly different rates and along stereochemically different pathways. Surprisingly, the *cis* and *trans* isomers both yield *trans*-1,2-diacetoxycyclohexane, but the *trans* isomer forms it 1,000 times more rapidly than the *cis* isomer. Thus, in a slow reaction the *cis* isomer undergoes substitution with *inversion* of configuration (*i.e.*, via an S_N2 process), whereas the *trans* isomer undergoes substitution in a rapid reaction with apparent *retention* of configuration. The rate enhancement as well as the stereochemical outcome with the *trans* isomer can be explained in terms of an interaction with the neighboring

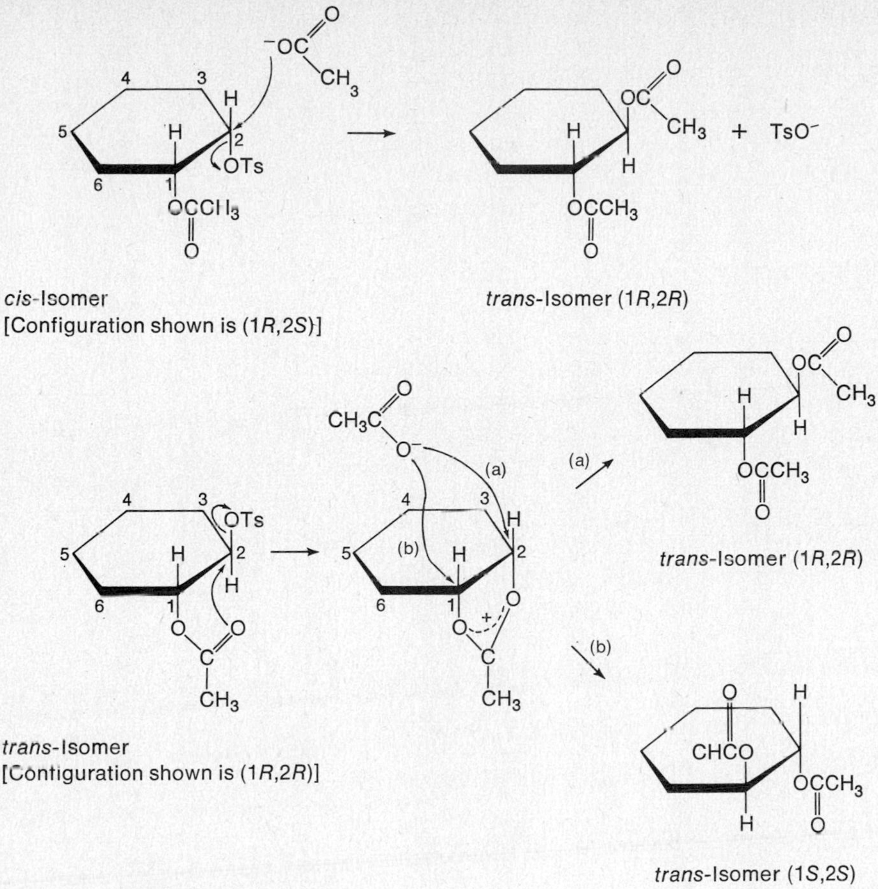

cis-Isomer
[Configuration shown is (1R,2S)]

trans-Isomer (1R,2R)

trans-Isomer (1R,2R)

trans-Isomer
[Configuration shown is (1R,2R)]

trans-Isomer (1S,2S)

Fig. 21.19. Acetoylysis of *cis*- and *trans*-1-acetoxy-2-tosyloxcyclohexane.

acetoxy function that leads to a cyclic intermediate (Fig. 21.19).* This phenomenon is referred to as *neighboring group participation*, and the enhancement in the rate of reaction is termed anchimeric assistance (Greek, *anchi*, adjacent, and *meros*, part).

The extent of anchimeric assistance is dependent on the distance between the neighboring group and the reaction site; *i.e.*, the size of the ring that is formed in the cyclic intermediate (or, in some cases, in the product as well). It has been observed that the ease of ring formation decreases in the following order of ring size: $3 > 5 > 6 > 7 > 4$. The extent of anchimeric assistance also depends on the stereochemical relationships within the molecule, for it is necessary that the neighboring group approach the backside of the reaction center. It is for this reason that only the *trans* isomer in Fig. 21.19 shows acetoxy group participation, for it alone can assume a conformation that allows the acetoxy group to be *anti*

* The cyclohexane ring is represented in Fig. 21.19 as a planar structure, although in actual fact it exists predominantly in nonplanar chair conformations. It is a useful simplifying assumption for the present discussion, however, to draw the cyclohexane ring as a planar structure; to do so does not change the argument.

(*i.e.*, *trans* diaxial) to the tosyloxy group. Now, intramolecular processes are usually much faster than the corresponding intermolecular analogs, and, consequently, the tosyloxy group is more rapidly displaced in the *trans* isomer of 1-acetoxy-2-tosyloxycyclohexane (intramolecular displacement) than in the *cis* isomer (intermolecular displacement). The cyclic compound that results from the intramolecular displacement, however, is not stable; it very rapidly reacts in an intermolecular fashion with available nucleophiles that open the ring and yield substitution products. In the example illustrated in Fig. 21.19, this second step can occur (a) with attack of acetate at C-2, in which case there is overall *retention* of configuration at *both* C-1 and C-2, or (b) with attack of acetate at C-1, in which case there is overall *inversion* of configuration at *both* C-1 and C-2. The products resulting from these alternative pathways are enantiomers and are formed in equal amount, because the cyclic intermediate is a symmetrical entity in this particular case. Thus, if optically active starting material is used, the product is optically inactive (racemic).

In the more general situation involving neighboring group participation, the cyclic intermediate is not symmetrical, nor is the entering group identical with the neighboring group. For example, the reaction of (2*S*,3*S*)-3-bromo-2-pentanol (Fig. 21.20) with hydrogen chloride proceeds via a cyclic "bromonium

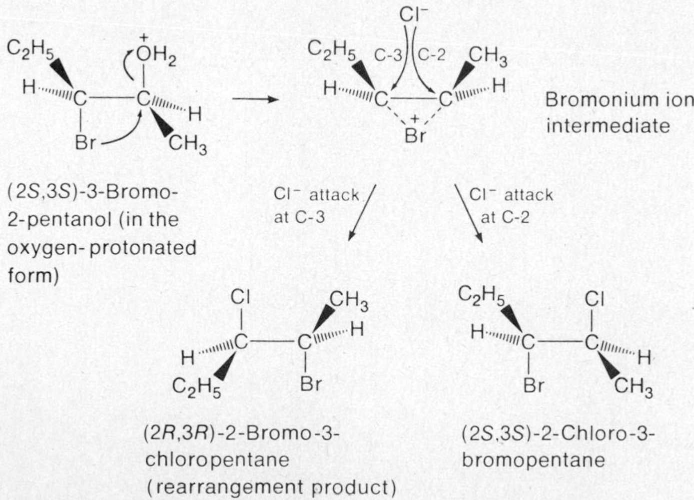

Fig. 21.20. Neighboring group participation involving an unsymmetrical cyclic bromonium ion intermediate.

ion" intermediate, which can react with chloride ion at C-2 to yield (2*S*,3*S*)-2-chloro-3-bromopentane or at C-3 to yield the "rearrangement product" (2*R*,3*R*)-2-bromo-3-chloropentane. These two products have opposite configurations at both C-2 and C-3. If optically active starting material is used, a mixture of optically active diastereomers is produced. If hydrogen bromide rather than hydrogen chloride is employed, however, the two products bear an enantiomeric relationship to one another. If optically active starting material is used in this case, the product is a mixture of unequal amounts of the two enantiomers (*i.e.*, partial racemization). The usual hallmarks of neighboring group participation, then, are (a) rate enhancement and (b) double inversion of configuration, the latter leading to stereochemical results that depend on the symmetry of the cyclic intermediate and the structure of the neighboring group and the entering nucleophile.

PROBLEM 21-16

What would you predict to be the stereochemical outcome of the following reactions:

(a) $CH_3CH(Cl)CH(OH)CH_3$ (2*S*,3*S*-isomer) + HCl

(b) 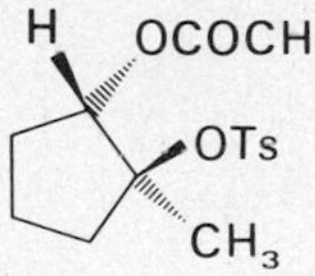 1-methyl-2-acetoxycyclopentyl tosylate (H, $OCOCH_3$; OTs, CH_3) + $CH_3CO_2^-$

(c) 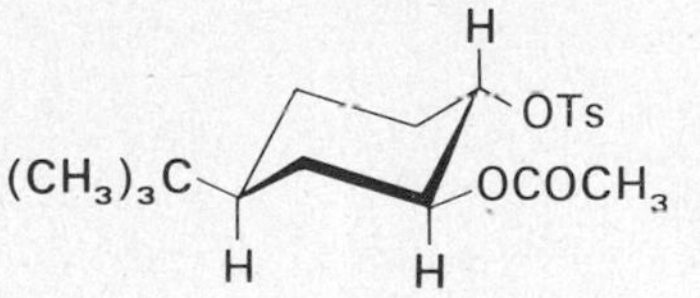 $(CH_3)_3C$-cyclohexane bearing OTs and $OCOCH_3$ (H, H, H) + $CH_3CO_2^-$

Functional groups capable of neighboring group participation include almost all those having a hetero atom. Thus, CO_2^-, CO_2R, COR, CHO, OR, OH, NR_2, SR, SH, I, Br, and Cl (but not F) are known to possess this capability. In addition, carbon–carbon unsaturated groups, particularly aromatic rings (see Figure 21.16), are effective neighboring groups.

21.4i. Neighboring group participation, anchimeric assistance, and nonclassical carbonium ions. Experiments on the solvolysis of the arylsulfonate esters of the *exo* and *endo* isomers of 2-bicyclo[2.2.1]heptanol (Fig. 21.21) have shown that the *exo* isomer reacts several hundred times more rapidly than the *endo* isomer, a situation reminiscent of the *trans* and *cis* isomers of 1-acetoxy-2-tosyloxycyclohexane (Fig. 21.19) and suggestive of a neighboring group effect. But, bicycloheptane contains no π bonds or nonbonded pairs of electrons, the usual requirements for neighboring group participation. It turns out that in certain suitably constituted systems, even the electrons of σ bonds can act as neighboring groups. Thus, in the *exo* isomer, the electrons of the C-1/C-6 bond, situated at the backside of the carbon to which the leaving group is attached, assist the departure of the leaving group and form a structure that is represented as a "bridged" carbonium ion (Fig. 21.21). Evidence for the formation of a symmetrical structure at some point along the reaction pathway between the starting material and the product is provided by stereochemical studies. Starting with optically active material, the product of solvolysis is optically inactive, a fact most readily interpreted in terms of a symmetrical bridged cation, which undergoes either intermolecular attack by acetate at C-2 to form the product in which overall configuration is retained, or attack at C-1 to form the product in

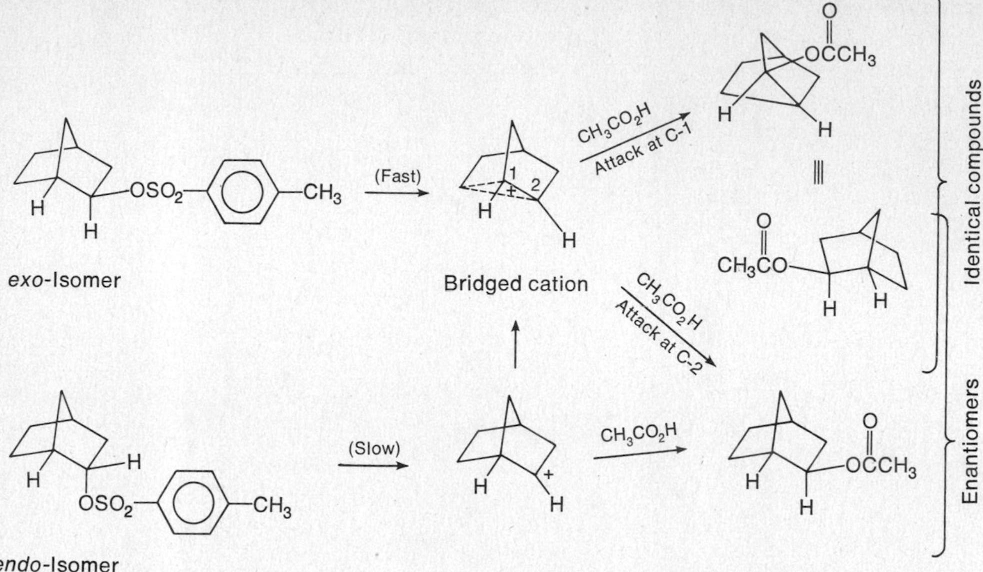

Fig. 21.21. Anchimeric assistance in the solvolysis of *exo*-2-bicyclo[2.2.1]-heptyl *p*-toluenesulfonate, leading to the formation of the nonclassical, bridged cation.

which overall configuration is inverted. Since these two pathways are equally probable and since the two products are enantiomers, optical activity is lost (*i.e.*, racemization occurs). Solvolysis of the *endo* isomer, on the other hand, yields an optically active product.*

The nature of the bridged cation has been the subject of heated debate for a number of years. According to one interpretation, it is considered to be an equilibrium mixture of rapidly interconverting localized carbonium ions, often referred to as "classical carbonium ions." According to another interpretation, it is considered to be a single entity, a delocalized carbonium ion that is often referred to as a "nonclassical carbonium ion"; evidence in favor of the nonclassical ion has been provided by ^{13}C nmr studies and by photoelectron spectrometric measurement.

Equilibrating classical structures

Nonclassical representation

* Although a carbonium ion may be involved in this process, the product is optically active because of the presence of chiral centers at C-1 and C-7. In contrast to the *exo* isomer, a symmetrical intermediate is *not* formed; *i.e.*, it is an example of a partial chiral synthesis (see p. 539).

Nonclassical carbonium ions have been proposed in a number of different systems, but there is little conclusive evidence to either support or refute their reality in the majority of instances. Although the phenomenon of anchimeric assistance is commensurate with the notion of a stable, resonance-delocalized, cyclic intermediate, the possibility that the intermediate might actually be a mixture of rapidly equilibrating carbonium ions cannot be discarded on the basis of most of the presently available experimental evidence.

PROBLEM 21-17

A "nonclassical" carbonium ion has been postulated for the rearrangement of $(CH_3)_3CCH_2^+$ to $(CH_3)_2\overset{+}{C}CH_2CH_3$. Draw its structure.

REACTIONS OF HALIDES

21.5 Substitution Reactions of Halides

21.5a. Conversion of C—Halogen to C—H

Via Lithium Aluminum Hydride. Alkyl chlorides, bromides, and iodides undergo replacement of halogen by hydrogen when treated with lithium aluminum hydride in refluxing tetrahydrofuran solution (Fig. 21.22). The reaction is an

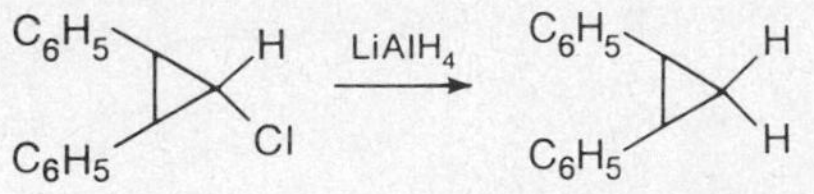

C_6H_5, C_6H_5 / H, Cl cyclopropane $\xrightarrow{LiAlH_4}$ C_6H_5, C_6H_5 / H, H cyclopropane

Fig. 21.22. Replacement of halogen by hydrogen via lithium aluminum hydride.

S_N2 process in which hydride (H^-) is the nucleophile and is subject to the usual restrictions on this type of reaction. The situselectivity is limited, because lithium aluminum hydride is a powerful reducing agent that also reacts with carbonyl groups, nitro groups, cyano groups, etc.

Via Tin Hydrides. Alkyl halides, including fluorides, can be reduced by tin hydride reagents (R_3SnH) (Fig. 21.23). These reactions involve a radical chain

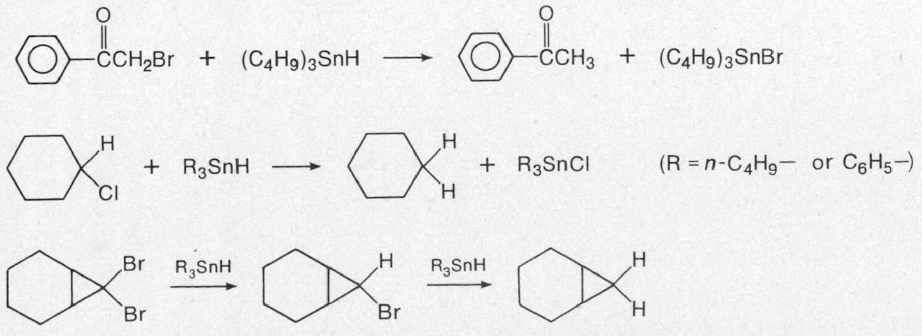

$C_6H_5COCH_2Br + (C_4H_9)_3SnH \longrightarrow C_6H_5COCH_3 + (C_4H_9)_3SnBr$

cyclohexyl(H)(Cl) $+ R_3SnH \longrightarrow$ cyclohexyl(H)(H) $+ R_3SnCl$ $\quad (R = n\text{-}C_4H_9\text{—}$ or $C_6H_5\text{—})$

bicyclic (Br)(Br) $\xrightarrow{R_3SnH}$ bicyclic (H)(Br) $\xrightarrow{R_3SnH}$ bicyclic (H)(H)

Fig. 21.23. Examples of tin hydride reductions of alkyl halides.

mechanism (Fig. 21.24) and are partially situselective in that carbonyl groups are not affected. Double and triple bonds undergo addition of tin hydride reagents, however.

Initiation: $(In)_2 \xrightarrow[\text{or } \Delta]{h\nu} 2\ In\cdot$ In = initiator species

$In\cdot + R{-}X \longrightarrow In{-}X + R\cdot$

Propagation steps:

$R\cdot + R'_3Sn{-}H \longrightarrow R{-}H + R'_3Sn\cdot$

$R'_3Sn\cdot + R{-}X \longrightarrow R'_3Sn{-}X + R\cdot$

Fig. 21.24. Radical chain mechanism for the reduction of an alkyl halide with triphenyl- or tributyltin hydride.

PROBLEM 21-18

The reduction of $(CH_2)_n$ [bicyclic C(H)–C(Cl)(F)–C(H)] with $(C_4H_9)_3SnH$ yields $(CH_2)_n$ [bicyclic C(H)–C(H)(F)–C(H)]

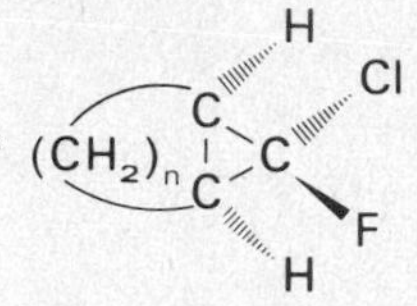

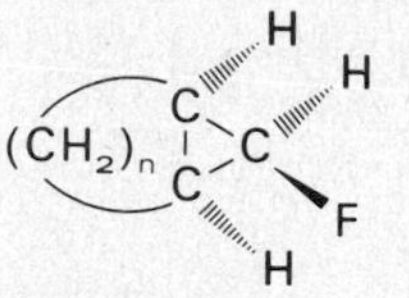

What does this suggest about the geometry of the intermediate radical?

Via Hydrogen. Molecular hydrogen in the presence of platinum or palladium catalysts can effect the hydrogenolysis of C—Halogen bonds. This is particularly effective in the hydrogenolysis of allyl and benzyl halides (Fig. 21.25).

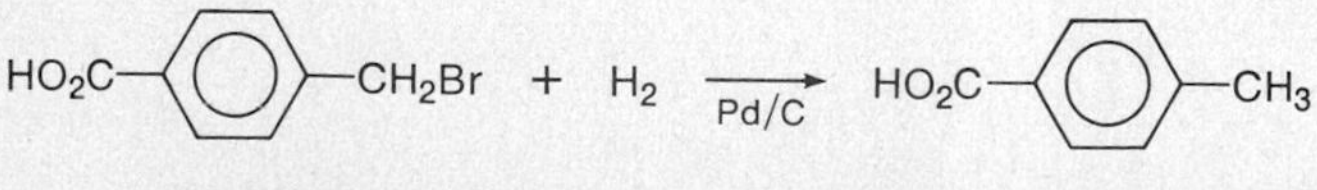

$HO_2C{-}C_6H_4{-}CH_2Br + H_2 \xrightarrow{Pd/C} HO_2C{-}C_6H_4{-}CH_3$

Fig. 21.25. Hydrogenolysis of a benzyl halide.

Via Grignard Reagents. Halogens attached to aromatic rings are not readily replaced in most reduction procedures. A reaction sequence that accomplishes this result involves converting an aryl halide to the corresponding Grignard reagent, followed by treatment with a proton donor such as water. For this procedure to be applicable, it is necessary, of course, that other functions capable of preventing the formation of or reacting with the Grignard reagent be absent (Fig. 21.26).

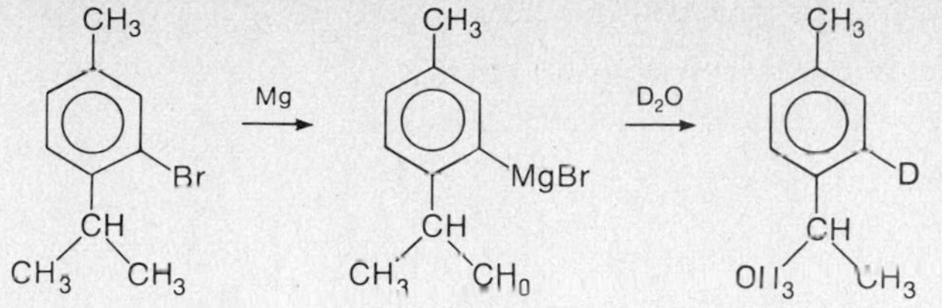

Fig. 21.26. Conversion of an aryl bromide to an arene via the Grignard reagent.

PROBLEM 21-19

Devise a method for sequentially removing the halogen atoms from

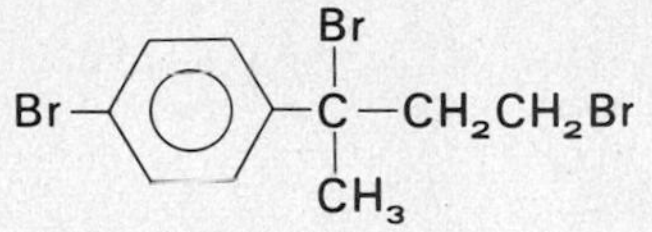

21.5b. Conversion of C—Halogen to C—C

Wurtz and Wurtz–Fittig Reactions. Alkyl bromides and iodides react with metallic sodium to form, in some cases, a hydrocarbon containing twice the number of carbons that are present in the alkyl halide (Fig. 21.27). This is

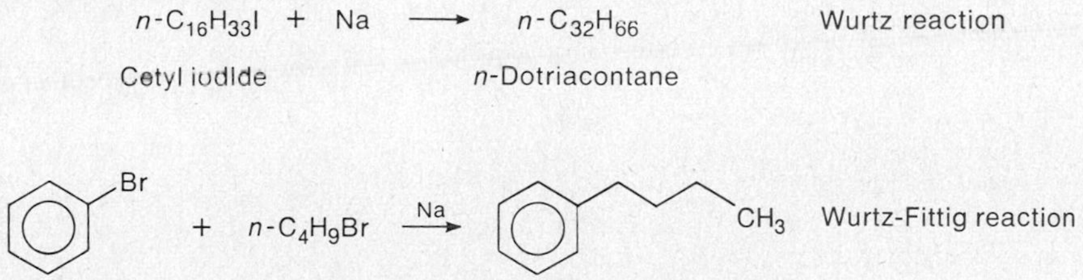

$n\text{-}C_{16}H_{33}I + Na \longrightarrow n\text{-}C_{32}H_{66}$ Wurtz reaction

Cetyl iodide $\quad$ *n*-Dotriacontane

$C_6H_5Br + n\text{-}C_4H_9Br \xrightarrow{Na} C_6H_5CH_2CH_2CH_2CH_3$ Wurtz-Fittig reaction

Fig. 21.27. Wurtz and Wurtz–Fittig reactions.

called the *Wurtz reaction*; it can be interpreted as an S_N2 process and is subject, therefore, to all the usual limitations:

$$RX + Na \longrightarrow RNa + NaX$$

$$RNa + RX \longrightarrow R{-}R + NaX$$

Thus, secondary halides give only poor yields of coupling product, and tertiary halides usually give little or no coupling product, forming the products of elimination instead. The reaction is used primarily for the synthesis of symmetrical alkanes that are unbranched near the midpoint of the molecule.

If two different alkyl halides, RX and R′X, are treated with sodium, a mixture containing RR, R′R′, and RR′ is obtained. The mixture is difficult to separate into its pure components, so the "mixed Wurtz" reaction is not the method of choice for the synthesis of unsymmetrical alkanes. If an alkyl halide is treated with an aryl halide, however, the product mixture contains only R—R and Ar—R, and the latter is often the major product as the result of the easier formation of ArNa from ArX and Na. This type of mixed Wurtz reaction is called a *Wurtz–Fittig reaction* and has some limited utility in the preparation of certain alkylbenzenes. Both the Wurtz and Wurtz–Fittig reactions are precluded if other groups are present that are sensitive to the action of metallic sodium or organosodium compounds; since this includes the majority of functional groups, the scope of the reactions is severely limited.

Coupling Reactions of Organometallic Reagents. The symmetrical coupling of Grignard reagents can be effected by the action of silver bromide, and even Grignard reagents from tertiary halides can be employed (Fig. 21.28). Unsymmetrical coupling can be achieved between Grignard reagents and benzyl and allyl halides (Fig. 21.28).

Symmetrical coupling:

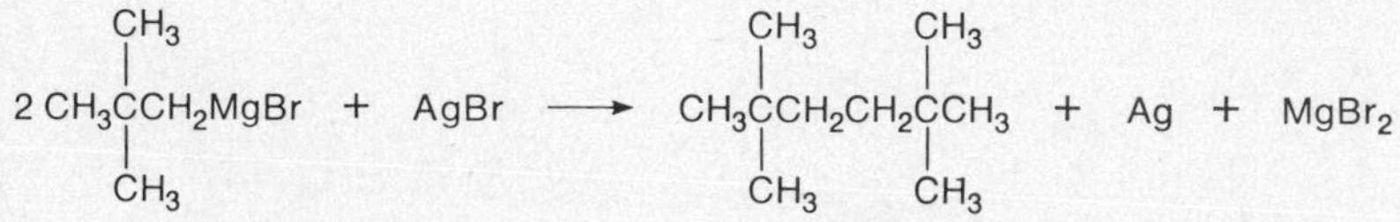

$$2\ CH_3C(CH_3)_2CH_2MgBr + AgBr \longrightarrow CH_3C(CH_3)_2CH_2CH_2C(CH_3)_2CH_3 + Ag + MgBr_2$$

Unsymmetrical coupling:

$$C_6H_5CH_2Cl + (CH_3)_3CMgBr \longrightarrow C_6H_5CH_2C(CH_3)_3 + MgBrCl$$

Fig. 21.28. Coupling reactions of Grignard reagents.

PROBLEM 21-20

Suggest a mechanism for each of the reactions shown in Fig. 21.28.

A useful variant involves the coupling of lithium dialkycopper compounds, which can be prepared by the action of alkyllithium compounds, obtained from primary or secondary halides on cuprous iodide:

$$RX + 2\,Li \longrightarrow RLi + LiX$$

$$4\,RLi + Cu_2I_2 \longrightarrow 2\,R_2CuLi + 2\,LiI$$

These reagents are more reactive in the coupling reaction than the corresponding Grignard reagents and couple not only with alkyl halides (primary, secondary, and tertiary) but also with vinyl and aryl halides. The reactions with alkyl halides proceed with inversion of configuration and are S_N2-type processes; they

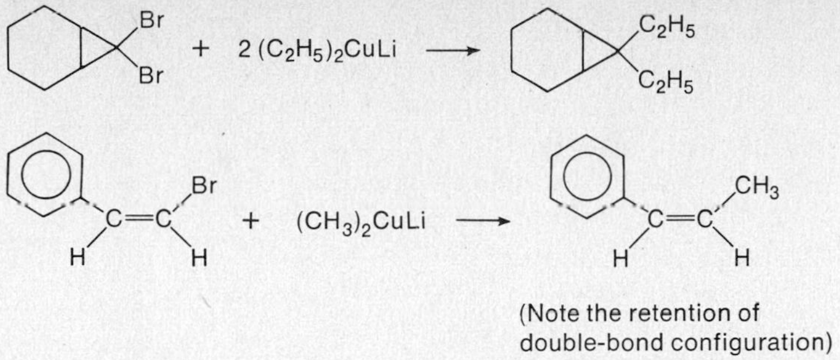

Fig. 21.29. Coupling reactions of lithium dialkylcopper reagents.

appear to be less subject to the usual limitations of S_N2 reactions, however (Fig. 21.29).

Substitution processes involving aryl halides can be effected through the use of copper metal, with aryl iodides being the most useful compounds in this respect. Treatment of aryl iodides with copper yields biphenyls (*Ullman reaction*), probably via arylcopper intermediates;

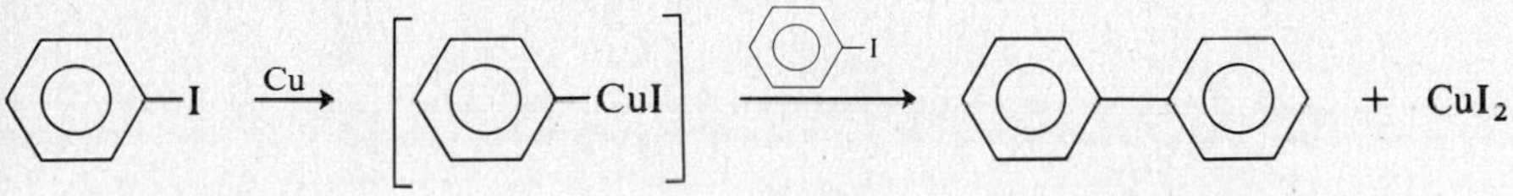

Reaction with Cyanide. Alkyl halides react with metal cyanides, usually NaCN or KCN, to yield nitriles via S_N2 processes (see Table 21.7).

Friedel–Crafts Reactions. A frequently used method for attaching an alkyl group to an aromatic ring is the Friedel–Crafts alkylation reaction (see Fig. 8.13

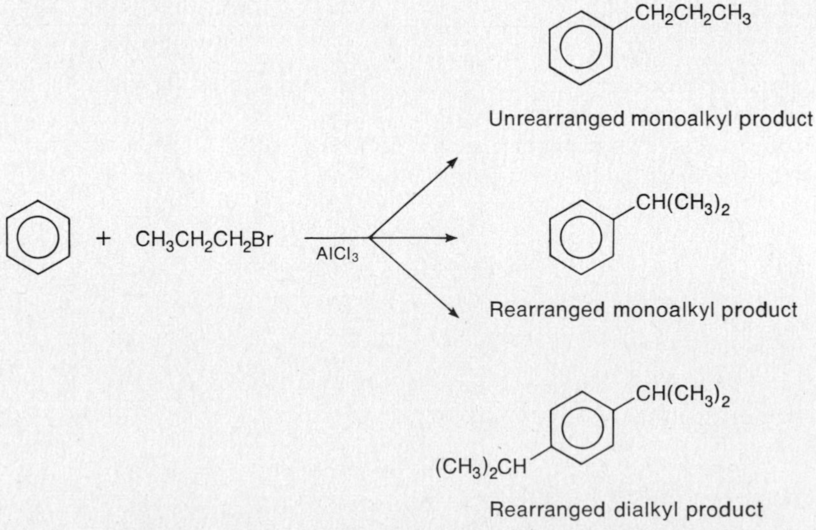

Fig. 21.30. Friedel–Crafts reaction of benzene and *n*-propyl bromide.

and 21.30). The major drawbacks of the reaction are (a) the tendency for the alkyl halide to undergo skeletal rearrangement (*i.e.*, Wagner–Meerwein type rearrangement of carbonium ions), and (b) the tendency for more than one alkyl group to become attached to the aromatic ring (alkyl groups are *o,p*-directing and activating with respect to electrophilic substitution reactions; see Fig. 21.30).

Intramolecular Nucleophilic Substitution Reactions. Intramolecular substitution reactions involving the attack of a carbanion on a C—Halogen center provides a method for synthesizing cycloalkanes. Among the ways for generating carbanion centers are the action of base on compounds containing acidic C—H bonds (see Chapter 30 for further details) and the action of a metal on a C—Halogen bond (Fig. 21.31).

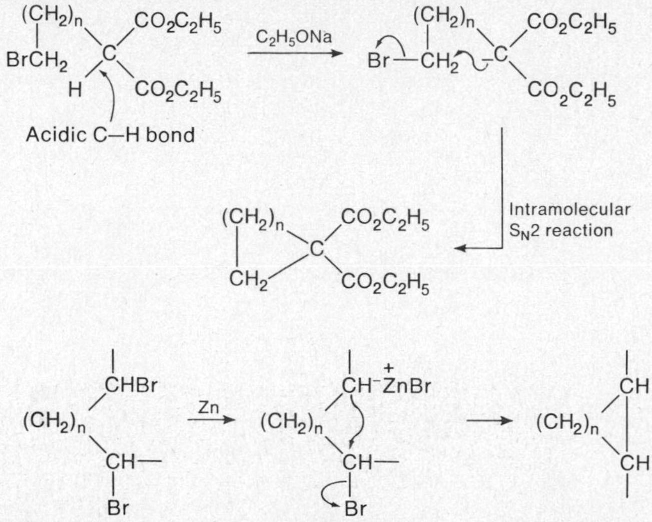

Fig. 21.31. Preparation of cycloalkanes via intramolecular nucleophilic substitution reactions.

21.5c. CONVERSION OF C—HALOGEN TO C—HETERO ATOM. A variety of hetero atom nucleophiles can engage alkyl halides in nucleophilic displacement reactions, leading to alcohols, ethers, esters, thiols, sulfides, and amines (Table 21.7).

PROBLEM 21-21

Explain the relative rates of nucleophilic displacement for the following alkyl halides:

$C_6H_5CH_2X$	$H_2C{=}CHCH_2X$	CH_3X	CH_3CH_2X
120	40	3	1

$(CH_3)_2CHX$	$(CH_3)_3CCH_2X$
0.025	0.00001

Aryl halides and vinyl halides, for all practical purposes, are unreactive in S_N2 reactions. Nevertheless, it is possible to replace the halogen atom of an aryl halide with various other groups under certain conditions. One such condition prevails when strong electron-withdrawing groups are *ortho* and *para* to the halogen atom in the benzene ring. For example, 2,4-dinitrofluorobenzene reacts with sodium hydroxide to yield 2,4 dinitrophenol (Fig. 21.32). This reaction is

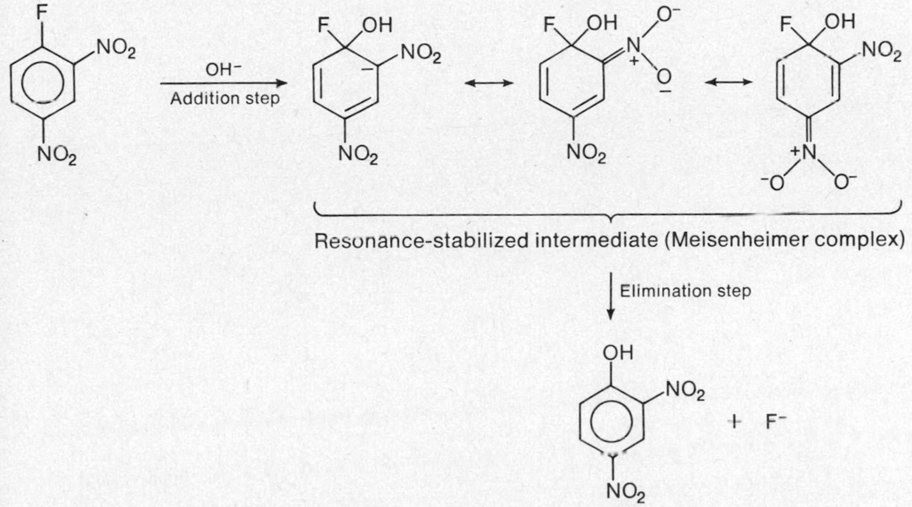

Fig. 21.32. Nucleophilic displacement reaction of an aryl halide via addition-elimination sequence.

thought to take place not by a direct S_N2 displacement but by an addition-elimination sequence, the hydroxide anion adding to yield an intermediate compound (called a *Meisenheimer complex*), which then loses fluoride ion to regenerate the aromatic system. The reason that electron-withdrawing groups facilitate this process is that they provide resonance stabilization for the negative charge in the intermediate.

Another condition that leads to halogen displacement in aryl halides occurs when reactions are carried out in solutions of sodium amide ($NaNH_2$) or other strong bases in liquid ammonia. Chlorobenzene, for example, reacts with $NaNH_2$ in ammonia to yield aniline. By means of various types of experiments, it has been well demonstrated that this reaction proceeds by an elimination-addition sequence (*i.e.*, just the opposite order of events of those depicted in Fig. 21.32). Sodium amide, a very strong base, effects a displacement on the hydrogen atom adjacent to the carbon to which the halogen is attached, yielding a phenyl anion. This then loses chloride ion to produce a neutral species known as benzyne. Benzyne is an extremely reactive entity that rapidly adds reagent to restore the benzenoid system. In this case, it adds ammonia to yield aniline (Fig. 21.33).

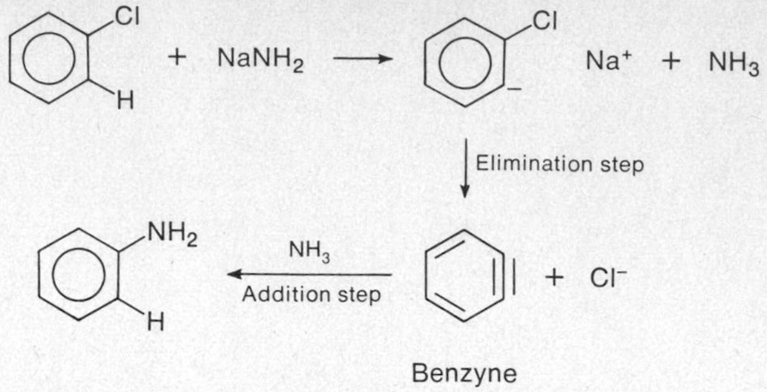

C_6H_5Cl (ortho H) + $NaNH_2$ ⟶ (ortho carbanion, Cl) Na^+ + NH_3

Elimination step ↓

Aniline (NH_2, ortho H) ⟵ (NH_3, Addition step) Benzyne + Cl^-

Benzyne

Fig. 21.33. Nucleophilic displacement reaction of an aryl halide via elimination-addition sequence ("benzyne" mechanism).

PROBLEM 21-22

If chlorobenzene labeled with ^{14}C at the 1 position is used in this reaction, what is the labeling pattern of the aniline that is formed? If chlorobenzene labeled with ^{14}C at the 2 position is used, what is the labeling pattern of the product?

Benzyne is an example of a class of reactive intermediates collectively known as arynes. Their structure remains controversial, for they are so reactive that they do not have a long enough life to be studied easily, even by very rapid spectroscopic techniques. They are frequently represented as having a triple bond in the ring. It is not a triple bond in the usual sense, however, for it is constructed from two adjacent sp^2-hybridized carbons, in which the "third" bond arises from the overlap of a pair of *hybrid* orbitals in the plane of the aromatic ring. Alternatively, benzyne might be represented in terms of a pair of charge-separated resonance structures (Fig. 21.34).

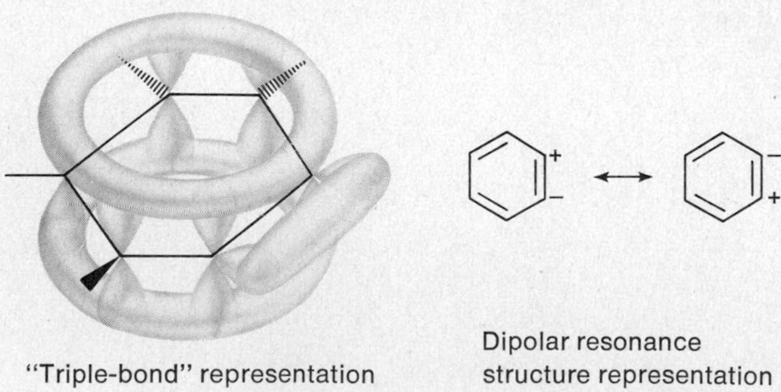

Fig. 21.34. Representations of benzyne.

21.6 β-Elimination Reactions

Halogen compounds undergo a variety of types of elimination reactions; these are discussed in detail in later chapters. The most often encountered elimination reactions occur in alkyl, alkenyl, and aryl halides that carry C—H bonds adjacent to the C—Halogen (*i.e.*, β C—H bonds), allowing the elimination of H—Halogen from adjacent carbons to take place to produce an alkene. Elimination reactions of this type often accompany nucleophilic substitution reactions of alkyl halides and reduce the yield of substitution product.

Table 21.7 REACTIONS OF ALKYL, ALKENYL, AND ARYL HALIDES

Type of Reaction	Starting Materials	Reagents	Products	Comments
Formation of Organometallic Reagents	RX, C=C—X, Ar—X ($X = Cl, Br, I$)	Mg, Li, Na	R—Metal, C=C—Metal, Ar—Metal (RMgX, RLi, RNa, etc.)	Order of halogen reactivity I > Br > Cl ≫ F
Nucleophilic Substitution	RX ($X = Cl, Br, I$)	H^- (*e.g.*, $LiAlH_4$)	R—H	*Via* S_N1 pathway—skeletal rearrangements possible; usually nonstereospecific
		HO^- (and H_2O)	R—OH	
		$R'O^-$ (and R′OH)	R—OR′	
		ArO^- (and ArOH)	R—OAr	
		$R'CO_2^-$	R—OCOR′	
		HS^-	R—SH	
		$R'S^-$ and ArS^-	R—SR′ and R—SAr	*Via* S_N2 pathway—usually no skeletal rearrangement; inversion of configuration
		R'_2S	$R—\overset{+}{S}R'_2$	
		NH_3 (and NH_2^-)	$R—NH_2$	
		$R'NH_2$ (and RNH^-)	R—NHR′	
		R'_2NH (and R'_2N^-)	$R—NR'_2$	
		R'_3N	$R—\overset{+}{N}R'_3$	Competition with β-elimination reaction (increases from primary to secondary to tertiary RX)
		CN^-	R—CN	
		R'^- (from R′MgX or R'_2CuLi)	R—R′	
		Ar^- (from ArMgX or Ar_2CuLi)	R—Ar	
		$R'C{\equiv}C^-$	$R—C{\equiv}CR'$	
		$ZZ'\bar{C}—R'$ ($Z, Z' = —\overset{O}{\overset{\parallel}{C}}R''$, —CN, $—NO_2$, etc.)	$ZZ'C(R)(R')$	
		ArH (with $AlCl_3$)	R—Ar	
Elimination	$H—\overset{\mid}{\underset{\mid}{C}}—\overset{\mid}{\underset{\mid}{C}}—X$	NaOH, NaOR	$>C=C<$	Competes with S_N reactions (increases from primary to secondary to tertiary RX)
	$H—\overset{\mid}{C}=\overset{\mid}{C}—X$	NaOR	$—C{\equiv}C—$	
	o-halobenzene (X and H on adjacent ring carbons)	RLi	benzyne	Nonisolable—reacts by cycloaddition or nucleophilic addition

21.7 *Formation of Organometallic Reagents*

Upon treatment with metals such as magnesium and lithium, alkyl, alkenyl, and aryl halides can be converted to organometallic compounds that are among the most useful reagents which the organic chemist has at his disposal. Numerous examples of their use have already been cited, and many more appear in subsequent sections of this text. Although remarkably versatile and general in their application, there are certain limitations to their preparation and use. For example, organometallic reagents cannot be prepared from halides containing groups that react with organometallic reagents; this includes groups containing acidic hydrogen (*e.g.*, —OH, —SH, $-CO_2H$, >NH) and groups containing unsaturation (*e.g.*, >C=O, —CN, $-NO_2$). Compounds having halogens at adjacent positions, or a halogen adjacent to an alkoxy group, cannot be converted to organometallic reagents, for they undergo an elimination reaction to yield alkenes:

$$\underset{\displaystyle OCH_3}{CH_3CH_2\overset{}{C}HCH_2Br} \xrightarrow{Mg} \underset{\displaystyle OCH_3}{CH_3CH_2CH}\!-\!CH_2\overset{+}{M}gBr \longrightarrow CH_3CH_2CH{=}CH_2 + Mg\begin{matrix}Br\\ OCH_3\end{matrix}$$

21.8 *Study Guide*

I. Methods of preparation of halides
   A. Methods of preparation of alkyl halides
      1. Conversion of C—H to C—Halogen
         (a) Reagents: Br_2 and N-bromosuccinimide for bromination; Cl_2 for chlorination
         (b) Mechanisms—free radical chain reactions
         (c) Situselectivity—usually low in chlorination unless allylic or benzylic bonds are present
      2. Conversion of C—OH to C—Halogen
         (a) Reagents: PCl_3, PCl_5, $SOCl_2$, or HCl for alkyl chlorides; PBr_3, PBr_5 or HBr for alkyl bromides; HI for alkyl iodides.
         (b) Mechanisms—nucleophilic displacements (S_N1, S_N2, and S_Ni)
         (c) Rearrangements likely with HI, HBr, or HCl reagents
         (d) Chloromethylation—synthesis of benzyl chlorides
      3. Conversion of C—C to C—Halogen
         (a) Reagents: Br_2 on a salt of a carboxylic acid
         (b) Mechanism: free radical intermediates
      4. Conversion of C=C to C—C—X and X—C—C—X (X = halogen); addition of halogens or halogen acids.
   B. Methods of preparation of alkenyl halides (vinyl halides)
      1. Dehydrohalogenation of *gem* dihalides
      2. Dehydrohalogenation of *vic* dihalides
   C. Methods of preparation of aryl halides
      1. Via electrophilic substitution—action of halogens on aromatic compounds
      2. Via aryldiazonium salts—action of cuprous halides or halide ions on ArN_2^+

II. Nucleophilic substitution reactions at saturated carbon atoms *Section 21.8 Study Guide*

A. General features

1. S_N1 pathway—unimolecular, kinetically first order; involves carbonium ion intermediates
2. S_N2 pathway—bimolecular, kinetically second order; no intermediates
3. Distinction between S_N1 and S_N2 not absolutely sharp in some cases

B. Effect of structure of carbon framework on S_N1 and S_N2 reactions

1. Stability of carbonium ions in S_N1 reactions
   (a) Degree of substitution: $3° > 2° > 1° > CH_3^+$ (decreasing reactivity)
   (b) Adjacent system: $ArCH_2^+ > R_3C^+ > C{=}C{-}CH_2^+ \simeq R_2CH^+$ (decreasing reactivity)
   (c) Planarity—bridgehead carbonium ions cannot form if planarity cannot be achieved
2. Backside approach in S_N2 reactions
   (a) Effect of alkyl groups at α position: $CH_3 > 1° > 2° > 3°$ (decreasing reactivity)
   (b) Effect of alkyl groups at β position; neopentyl compounds very unreactive
   (c) Effect of more remote groups on backside approach
3. Rearrangement of carbon framework in S_N1 reactions
   (a) Wagner–Meerwin rearrangement
   (b) Migratory aptitudes: $Ar{-} > C{=}C{-} > R_3C{-} > R_2CH{-} > RCH_2{-} > CH_3{-} > H$ (in order of decreasing tendency to migrate)

C. Effect of nucleophile on S_N1 and S_N2 reactions

1. Good nucleophiles more likely to give S_N2 reaction: poor nucleophiles more likely to give S_N1 reaction
2. Negatively charged nucleophiles more effective than their uncharged analogs
3. Highly polarizable nucleophiles better than weakly polarizable ones
4. Nucleophilicity parallels basicity in the same row of the periodic table
5. Nucleophilicity and basicity are not necessarily affected in the same way by changes in structure

D. Effect of leaving group on S_N1 and S_N2 reactions

1. Strongly electronegative groups better leaving groups than more weakly electronegative ones; group VII elements > group VI elements > group V elements > group IV elements
2. Positively charged groups better leaving groups than their uncharged analogs; best leaving group is N_2^+
3. The better the leaving group the greater is the tendency for S_N1 reaction

E. Effect of solvent on S_N1 and S_N2 reactions

1. If charge is developed in transition state, reaction is faster in high dielectric solvents than in low dielectric solvents
2. If charge is diffused or reduced in transition state, reaction is slower in high dielectric solvents than in low dielectric solvents

F. Solvolysis reactions—solvent also a reactant and acts as nucleophile

G. Neighboring group participation and anchimeric assistance

1. Formation of isolable products from neighboring group participation
2. Effect of neighboring groups on rate and stereochemistry of reaction
   (a) Required conformation for neighboring group effect—antiperiplanar
   (b) Rate of reaction increased by neighboring group participation; anchimeric assistance
   (c) Stereochemistry: one inversion of configuration occurs in formation of cyclic intermediate and a second inversion of configuration occurs in transformation of intermediate to product
3. Effective neighboring groups—contain nonbonded electrons or π bonds
4. Nonclassical carbonium ions

III. Reactions of halides

A. Conversion of C—Halogen to C—H

1. Via lithium aluminum hydride—an S_N2 reaction
2. Via tin hydrides—a radical chain reaction
3. Via hydrogen—requires a catalyst
4. Via Grignard reagents—a method for converting Ar—X to Ar—H

B. Conversion of C—Halogen to C—C

1. Wurtz reaction—action of Na on RX
   (a) S_N2 displacement reaction process
   (b) Limited, in most cases, to primary halides
   (c) Wurtz–Fittig reaction—action of Na on mixture of RX and ArX
2. Coupling reactions of organometallic reagents
   (a) Grignard reagents in coupling reactions
   (b) Lithium dialkylcopper reagents (R_2CuLi)
      (1) Couple with alkyl halides (with inversion of configuration)
      (2) Couple with vinyl halides (with retention of configuration)
      (3) Couple with aryl halides
3. Reaction of alkyl halides with CN^-—an S_N2 reaction
4. Friedel–Crafts reactions—aromatic electrophilic substitution with carbonium ions

C. Conversion of C—Halogen to C—Hetero atom

1. S_N2 displacement reactions with alkyl halides—see Table 21.7
2. Nucleophilic displacements with aryl halides
   (a) Aryl halides carrying electron-withdrawing groups in *o* and *p* positions

(b) Benzyne mechanism

D. Elimination reactions—yield alkenes (see Chapter 25)

E. Formation of organometallic reagents
   1. Formed by action of metals (*e.g.*, Mg, Li) on RX and ArX
   2. Limitations—cannot form organometallic reagents if:
      (a) Active hydrogen is present (*e.g.*, —OH, —SH, CO_2H, >NH groups).
      (b) Lithium dialkylcopper reagents (R_2CuLi)
      (c) Halogen or alkoxyl group is adjacent to X group in RX.

REFERENCES

Bartlett, P. D., ed., *Nonclassical Ions*, W. A. Benjamin, Inc., Menlo Park, California, 1965.

Breslow, R., *Organic Reaction Mechanisms*, 2nd ed., W. A. Benjamin, Inc., Menlo Park, California, 1969, Chapters 3 and 5.

Fieser, L. F. and M. Fieser, *Advanced Organic Chemistry*, Reinhold Publishing Corp., New York, 1961, Chapters 10 and 23.

Hine, J., *Physical Organic Chemistry*, 2nd ed., McGraw-Hill Book Company, New York, 1962, Chapters 6, 16, and 17.

Ingold, K. U., and B. P. Roberts, *Free Radical Substitution Reactions*, John Wiley & Sons, Inc., New York, 1971.

Jolles, Z. E., ed., *Bromine and its Compounds*, Academic Press, Inc., New York, 1966.

March, J., *Advanced Organic Chemistry: Reactions, Mechanisms and Structure*, McGraw-Hill Book Company, New York, 1968, Chapters 10 and 11.

Musgrave, W. R. K., in *Rodd's Chemistry of Carbon Compounds*, 2nd ed., Vol. IA, ed. by S. Coffey, Elsevier Publishing Co., New York, 1964, pp. 478–526.

Streitwieser, A., *Solvolytic Displacement Reactions*, McGraw-Hill Book Company, New York, 1962.

PROBLEMS

1. Draw the structures of the products from each of the following reactions, and indicate which product is expected to be formed in the largest amount.

(a) $(C_6H_5)_2C{=}CH{-}C_6H_{11}$ + N-bromosuccinimide ⟶

(b) $C_6H_5CH(CH_3)CH_2CH_3$ + Cl_2 $\xrightarrow{h\nu}$

(c) 1-methylcyclohexyl-CH_2OH + HBr $\longrightarrow$

(d) cyclohexane with CO_2Ag and CH_3 (H, H) + Br_2 $\xrightarrow[\Delta]{}$

(e) bicyclic alcohol (H, OH, H, H) + PBr_5 $\longrightarrow$

2. Suggest a reasonable mechanism for the formation of the major product in each of the reactions in Problem 1.

3. Give a specific example of an S_N2 reaction involving the action of the following types of nucleophiles on an alkyl halide:

(a) An oxygen nucleophile.
(b) A nitrogen nucleophile.
(c) A sulfur nucleophile.
(d) An sp^3-hybridized carbon nucleophile.
(e) An sp^2-hybridized carbon nucleophile.
(f) An sp-hybridized carbon nucleophile.

4. For each of the following nucleophilic substitution reactions predict whether the most likely pathway is S_N1 or S_N2. Give the reasoning on which your prediction is based.

(a) $CH_3I + OH^- \longrightarrow CH_3OH + I^-$

(b) $(CH_3)_3CCl + CH_3OH \longrightarrow (CH_3)_3COCH_3 + HCl$

(c) decalin with CH_3, H, I, H + $CH_3CO_2^-$ $\longrightarrow$ decalin with CH_3, $OCOCH_3$, H, H + I^-

~~~ indicates that the configuration is not known

(d) $C_6H_5CH_2Br + CH_3S^- \longrightarrow C_6H_5CH_2SCH_3 + Br^-$

(e) $C_6H_5C(CH_3)_2Br + H_2O \longrightarrow C_6H_5C(CH_3)_2OH + HBr$
~~~

(f) $C_6H_5CH(CH_3)Br + C_5H_5N \longrightarrow C_6H_5CH(CH_3)\overset{+}{N}C_5H_5 + Br^-$

(g) $(CH_3)_2C{=}CHCH_2Br + C_6H_5NH_2 \longrightarrow (CH_3)_2C{=}CHCH_2\overset{+}{N}H_2{-}C_6H_5 \quad Br^-$

(h) $C_6H_5\overset{O}{\overset{\|}{C}}CH_2I + CH_3OH \longrightarrow C_6H_5\overset{O}{\overset{\|}{C}}CH_2OCH_3 + HI$

(i) 1,1-dimethyl-2-hydroxycyclohexane (CH_3, CH_3, OH) $+ HBr \longrightarrow$ 1-bromo-1,2-dimethylcyclohexane (CH_3, Br, CH_3) $+ H_2O$

(j) 1-methyl-1-($CH_2\overset{+}{N}(CH_3)_3$)cyclopentane $+ N_3^- \longrightarrow$ 1-methyl-1-(CH_2N_3)cyclopentane $+ (CH_3)_3N$

5. For each of the reactions in Problem 4, predict the effect that increasing solvent polarity will have on the rate of the reaction.

6. For each of the following reactions, draw a reaction coordinate that you think is commensurate with the probable reaction mechanism:

(a) $(C_6H_5)_3CCl + H_2O \longrightarrow (C_6H_5)_3COH + HCl$

(b) $C_6H_5CH_2Br + N_3^- \longrightarrow C_6H_5CH_2N_3 + Br^-$

(c) $(CH_3)_3CCH_2OH + HCl \longrightarrow (CH_3)_2\underset{Cl}{\underset{|}{C}}CH_2CH_3 + H_2O$

(d) norbornyl bromide (Br, H) $\xrightarrow[CH_3CO_2^-]{CH_3CO_2H}$ norbornyl acetate ($OCOCH_3$, H) $+ Br^-$

(e) $C_6H_5Br \xrightarrow[NH_3]{NH_2^-} C_6H_5NH_2 + Br^-$

7. For each of the following pairs of nucleophiles indicate which one is the more reactive. Explain the reasons for your choice.

(a) $CH_3{-}C_6H_4{-}O^-$ or $O_2N{-}C_6H_4{-}O^-$ (b) NH_3 or CH_3NH_2

(c) $^-CH(CO_2C_2H_5)_2$ or $^-CH_2CO_2C_2H_5$ (d) H_2O or CH_3OH

(e) $(CH_3)_2CHCH_2OH$ or $(CH_3)_3COH$

(f) CH_3O^- or $C_6H_5O^-$

(g) C_6H_5SH or C_6H_5OH

(h) pyridine (N) or 2-methylpyridine (N, CH_3)

8. For each of the following reactions involving the formation of a new C—C or C—H bond, indicate what you consider to be a plausible mechanism.

(a) 1,1-bis(bromomethyl)cyclohexane (CH_2Br, CH_2Br) $\xrightarrow{Zn}$ spiro[2.5]octane

(b) 2-(bromomethyl)naphthalene (CH_2Br) $\xrightarrow{LiAlH_4}$ 2-methylnaphthalene (CH_3)

(c) 2-bromo-2-methylindane (CH_3, Br) $\xrightarrow{(1)\ Mg,\ (2)\ D_2O}$ 2-deuterio-2-methylindane (CH_3, D)

9. For each of the following pairs of compounds, indicate which would be expected to be more reactive toward nucleophilic displacement by iodide ion. Explain the reasons for your choice.

(a) $C_6H_5CH_2Br$ or $O_2N{-}C_6H_4{-}CH_2Br$

(b) $CH_3CH_2CH_2CH_2Br$ or $(CH_3)_2CHCH_2Br$

(c) $HOCH_2CH_2Br$ or CH_3OCH_2Br

(d) $NC{-}C_6H_4{-}Br$ (para) or $C_6H_4{-}Br$ with NC (meta)

(e) bicyclic bromide (Br, H) or bicyclic bromide (H, Br)

(f) cyclohexane chair with CH_3, H, $(CH_3)_3C$, H, H, Br or cyclohexane chair with CH_3, H, H, $(CH_3)_3C$, H, Br

(g) cyclobutyl–Br or cyclopentyl–Br

(h) 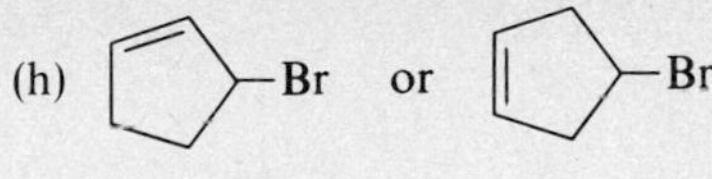

(i) 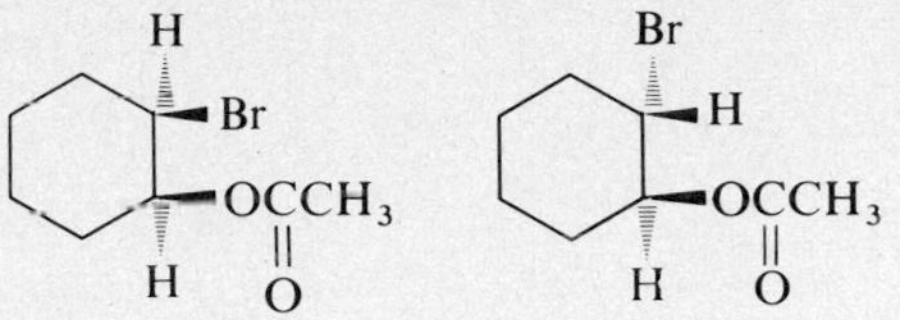 H, Br, $OCCH_3$ (C=O), H   H, Br, $OCCH_3$ (C=O), H

(j) 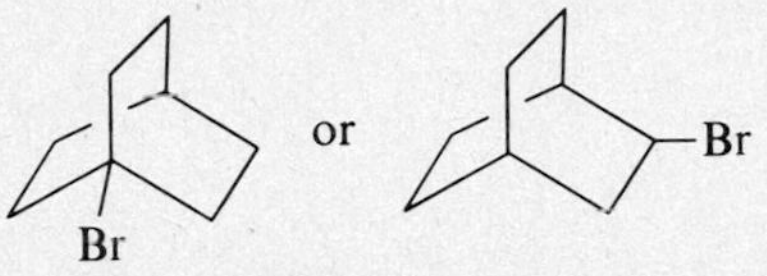 Br   or   Br

10. For each of the following reactions, give two plausible mechanisms and discuss, in as much detail as possible, all the ways that you can conceive for distinguishing between these mechanisms and for supporting what you believe to be the correct mechanism.

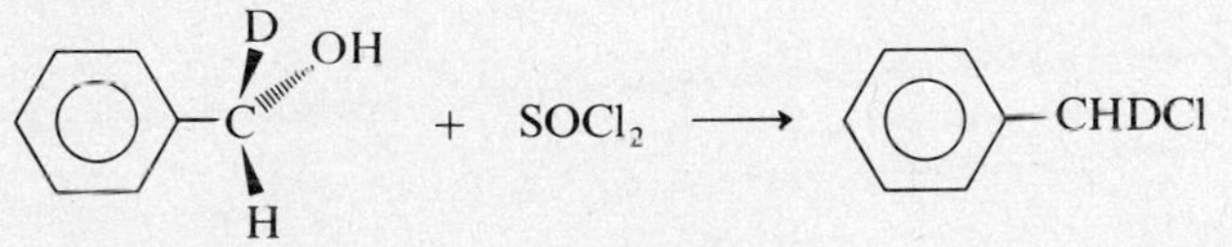

(a) 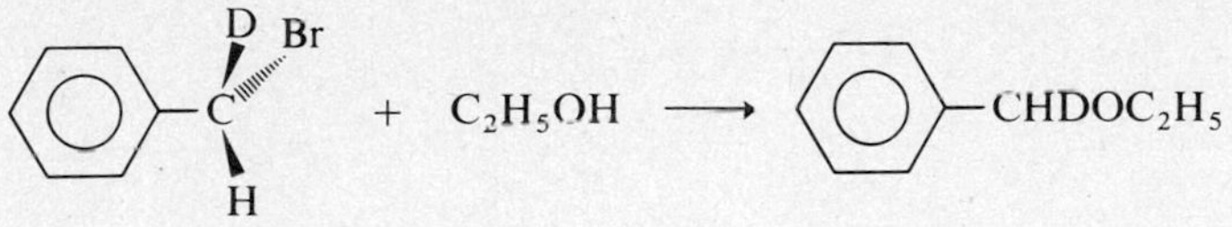 C_6H_5–C(D)(Br)(H) + C_2H_5OH ⟶ C_6H_5–$CHDOC_2H_5$

(b) C_6H_5–C(D)(OH)(H) + $SOCl_2$ ⟶ C_6H_5–CHDCl

(c) C_6H_5Br $\xrightarrow[NH_3]{NaNH_2}$ $C_6H_5NH_2$

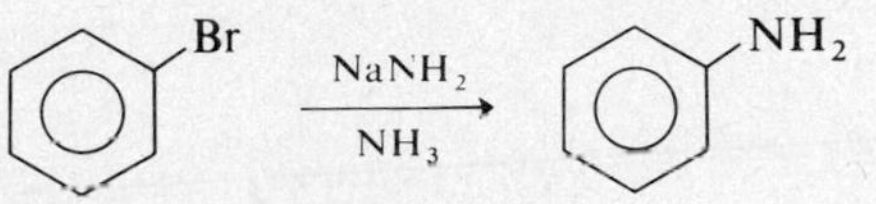

11. The chlorination of propene with *t*-butyl hypochlorite can be initiated by light and is known to involve a chain process:

$$H_2C{=}CHCH_3 \;+\; (CH_3)_3C{-}O{-}Cl \;\longrightarrow\; H_2C{=}CHCH_2Cl \;+\; (CH_3)_3COH$$

(a) Write a mechanism showing the initiation, propagation, and termination steps in this free radical chain reaction.

(b) On the basis of bond energy calculations, using the values given as follows, calculate the ΔH's for the propagation steps. Are the reactions exothermic or endothermic?

Bond energy values:	O—H	111 kcal/mole	Cl—O	52 kcal/mole
	C—H	97 kcal/mole	C—O	85 kcal/mole
	Cl—H	103 kcal/mole	Cl—C	81 kcal/mole

(c) Small amounts of acetone and methyl chloride are isolated as side products in this reaction. Suggest a mechanism for their formation.

12. On the basis of the data given, deduce the structures involved in each of the following examples:

(a) A $C_{10}H_{10}BrCl$ compound is inherently achiral, reacts with alcoholic silver nitrate to deposit silver chloride, and undergoes ozonolysis to yield a C_3H_5ClO compound and a C_7H_5BrO compound. When the latter is treated with Tollen's reagent and the resulting carboxylic acid is converted to the silver salt and then heated with bromine, a dibromobenzene is obtained which forms only one mononitro derivative.

(b) A $C_9H_5Cl_2N$ compound reacts with ammoniacal silver hydroxide to give a precipitate of metallic silver, does not react with alcoholic silver nitrate to form silver chloride, and can be oxidized to pyridine-2-carboxylic acid.

13. Which of the following compounds can form stable Grignard reagents?

(a) Br (on benzene ring fused to O–CH₂–O ring)

(b) benzene ring with CH_2Br and CH_2Cl

(c) Br–cyclohexane–C(OH) with $CH(CH_3)$–CH_3

(d) Br on bicyclic ring with S

(e) $(CH_3)_3C\overset{I}{C}HCH_2CH_3$

(f) Cl on bicyclic ring with =O

14. Indicate the steps by means of which the compound shown at the left can be converted to the compound shown at the right.

(a) cyclohexene ⟶ 1-bromocyclohexene (Br)

(b) benzene ⟶ $C_6H_5C(Cl)(CH_3)CH_3$

(c) benzene ⟶ $C_6H_5CH{=}CHCH_2Br$

(d) 1-methylcyclohexane-CO_2H (CH_3) ⟶ 1-methylcyclohexane-CH_2Br (CH_3)

(e) $(CH_3)_2C{=}CH_2 \longrightarrow (CH_3)_2CHCH_2CH_2CH(CH_3)_2$

(f) $(CH_3)_2CHCH_2OH \longrightarrow (CH_3)_2CHCH_2D$

(g) $C_6H_5C{\equiv}CC_6H_5 \longrightarrow C_6H_5CH{=}C(C_6H_5)(CH_3)$

(h) cyclohexane-CH_2CO_2H ⟶ cyclohexane-CH_2OH

(i) $\mathrm{(CH_3)(H)C(OCOCH_3)-CH_2CH_2SCH_3} \longrightarrow \mathrm{(CH_3)(H)C(Cl)-CH_2CH_3}$

15. Starting with benzene and/or any alkane, alkene, or alkyne containing no more than four carbons, along with any inorganic reagents or special organic reagents (*e.g.*, N-bromosuccinimide), synthesize the following compounds:

(a) $CH_3CH(SH)CH_2CH_3$ (b) $CH_3C{\equiv}CCH_2NH_2$ (c) C_6H_5D

(d) $C_6H_5-C_6H_5$ (e) $CH_3-C(CH_3)_2-OH$

16. Available starting materials include (a) any alkanes, alkenes, alkynes, alkyl halides, alkanols, alkanals, alkanones, alkanoic acids or alkanoic acid derivatives containing four or fewer carbon atoms, (b) benzene and any monosubstituted benzene or naphthalene, (c) special organic reagents, and (d) any inorganic compounds. Starting with materials chosen from this group, outline syntheses for each of the following compounds:

(a) $p\text{-}CH_3O-C_6H_4-CHClCH_3$ (b) 3-bromocyclohexene (Br)

(c) $(CH_3)_2CHCH_2CH_2I$ (d) 2-bromo-4-nitroaniline (NH_2, Br, NO_2)

(e) $p\text{-}F-C_6H_4-CH_3$ (f) $(CH_3)_2CHCH(Br)-C_6H_4-CH(Cl)CH(CH_3)_2$

(g) $C_6H_5-C(Cl){=}C(CH_3)_2$ (h) $BrCH_2CH_2Cl$

17. Chlorine-containing compounds have become widely used as insecticides, herbicides, and fungicides. Probably the best known of these is 2,2-di-(4-chlorophenyl)-1,1,1-trichloroethane (DDT) (see p. 243), which is prepared by the reaction of chlorobenzene with chloral in the presence of sulfuric acid. The major product is the *p,p*-isomer, and the minor product is the *o,p*-isomer:

$$Cl\text{-}C_6H_4 + Cl_3CCHO \xrightarrow{H_2SO_4} (p\text{-}ClC_6H_4)_2CH\text{-}CCl_3 + (o\text{-}ClC_6H_4)(p\text{-}ClC_6H_4)CH\text{-}CCl_3$$

If you were the first one to have carried out this reaction, what methods would you have used to decide which compound is the *p,p*-isomer and which is the *o,p*-isomer.

18. The following organic halides are inexpensive starting materials (*i.e.*, less than 50 ¢/lb): methyl chloride, methyl bromide, ethyl chloride, *n*-butyl chloride, chloroform, carbon tetrachloride, allyl chloride, benzyl chloride, benzal chloride ($C_6H_5CHCl_2$), and *o*-chlorotoluene. Using one or more of these as a starting material, along with any other compounds and reagents that you think are necessary, devise a synthesis for each of the following compounds:

(a) $H_2C{=}CHCH_2CH_2CH{=}CH_2$

(b) $C_6H_5\text{-}C(CH_3)_3$

(c) $C_6H_4(CH_3)(D)$ (*ortho*)

(d) $C_6H_5\text{-}CH_2CH_2\text{-}C_6H_5$

(e) $CH_3CH_2CH_2CH_2CH_2CH{=}CH_2$

19. Explain the following observations:

(a) 3-Bromocyclohexene reacts with NaOH more rapidly than 4-bromocyclohexene, which, in turn, reacts with NaOH very much more rapidly than bromocyclohexane.

(b) (*R*)-2,3-dimethyl-3-pentanol reacts with $SOCl_2$ to form (*R*)-3-chloro-2,3-dimethylpentane.

(c) The ratio of 2-chloro-2-methylpropane to 1-chloro-2-methylpropane in the free radical chlorination of isobutane is 7.0 at 100° and 2.6 at 600°.

(d) Neither free radical fluorination nor free radical iodination are useful laboratory procedures for the preparation of alkyl fluorides or iodides.

(e) The rate of the reaction of 1-bromo-3-phenylpropane with NaCN is faster in dimethylsulfoxide than in ethanol.

(f) *t*-Butyl bromide solvolyzes in aqueous-ethanol in a first-order reaction producing 2-methylpropene and *t*-butyl alcohol. Addition of sodium hydroxide produces only a very slight increase in reaction rate; however, the reaction is still first order (in *t*-butyl bromide), but the ratio of 2-methylpropene to *t*-butyl alcohol is substantially different.

(g) The solvolysis of 3-phenyl-2-pentyl *p*-toluenesulfonate in aqueous-ethanol produces 3-phenyl-2-pentanol of retained configuration at C-2 and C-3 and 2-phenyl-3-pentanol of inverted configuration at each center.

(h) 6-Chloro-2-methyl-3-hexanone reacts with silver ion in 80% aqueous-ethanol to produce 6-hydroxy-2-methyl-3-hexanone 536 times faster than does *n*-butyl chloride to give *n*-butanol.

(i) The following reaction proceeds 35 times less rapidly when R is CH_3 rather than H.

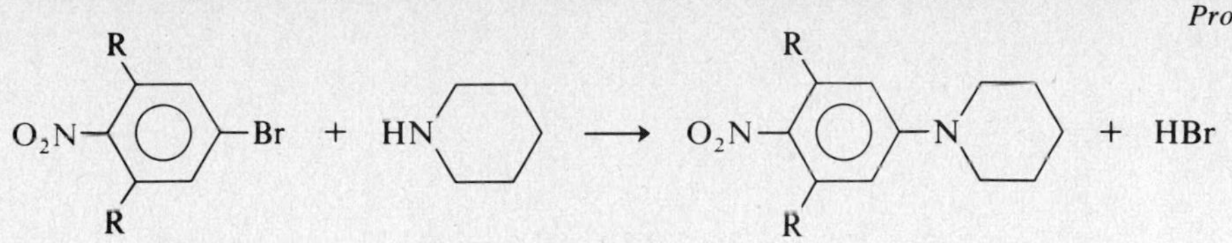

O_2N–(C₆H₂R₂)–Br + HN(piperidine) ⟶ O_2N–(C₆H₂R₂)–N(piperidine) + HBr

(j) 2-Chloromethylfuran reacts with sodium cyanide to yield 2-cyano-5-methylfuran.

(k) The following reaction proceeds with retention of configuration.

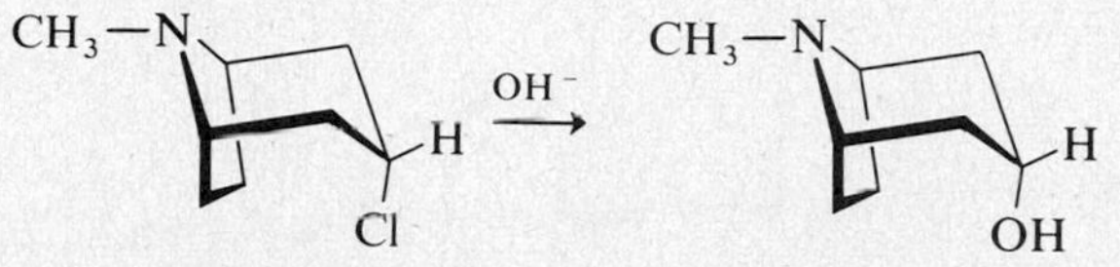

CH_3—N (bicyclic, H, Cl) $\xrightarrow{OH^-}$ CH_3—N (bicyclic, H, OH)

(l) The order of reactivity of solvolysis in 80% ethanol for the following compounds is as noted:

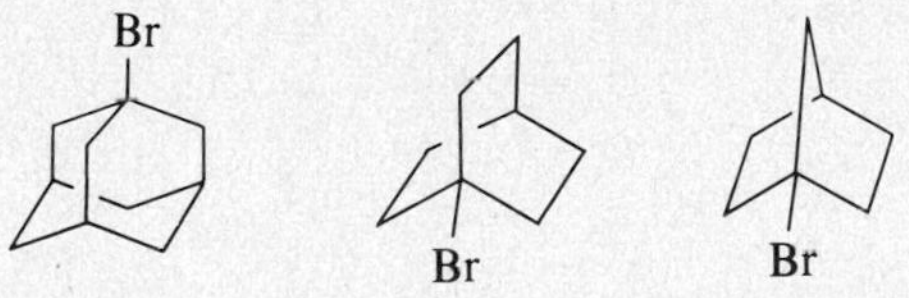

$(CH_3)_3CBr$	1-bromoadamantane (Br)	bicyclo[2.2.2]octyl bromide (Br)	norbornyl bromide (Br)
1.0	10^{-3}	10^{-6}	10^{-13}

Chapter **22**

Hydroxy compounds: alcohols and phenols

After people have repeated a phrase a great number of times they begin to realize it has meaning and may even be true.

H. G. WELLS
The Happy Turning

In this chapter, we continue the discussion of compounds containing functional groups that are single bonded to carbon frameworks. The essential features of the chemical behavior of the alcohols and phenols have been presented in Chapter 10, and now we are ready to consider these two classes of closely related compounds from a more advanced and detailed standpoint. We shall do so by directing our attention to the scope, limitations, and mechanisms of their reactions with the aim of assessing their utility in synthesis reactions. The chapter is divided into two major parts, the first one dealing with methods for preparing alcohols and phenols and the second dealing with their reactions.

METHODS OF PREPARATION OF ALCOHOLS AND PHENOLS

22.1 *Conversion of* C—H *to* C—OH

The direct oxidation of C—H to C—OH is of very limited application, succeeding only if the starting compound contains one C—H bond that is more labile than all the other C—H bonds in the molecule. For example, isopropyl benzene reacts more rapidly at the benzylic C—H bond than elsewhere in the molecule, forming 2-phenyl-2-propanol when treated with potassium permanganate (see Fig. 8.16) and 2-phenyl-2-propylhydroperoxide when treated with air and a catalyst.

22.2 *Conversion of* C—*Halogen to* C—OH

The hydrolysis of alkyl halides via the S_N1 or S_N2 pathway is quite satisfactory for the preparation of primary alcohols. It is somewhat less useful for the preparation of secondary alcohols, however, and generally much less useful for the preparation of tertiary alcohols because of competition from the elimination reaction that forms the alkene (see Chapter 25 for a detailed discussion). Alkene formation can often be minimized by substituting a two-step sequence for the direct procedure, converting the halide first to the corresponding ester of acetic acid by treatment with acetate anion as the nucleophile followed, then, by hydrolysis of the ester. Acetate anion, which is a much weaker base than hydroxide, has a greater tendency to act as a nucleophile toward carbon to induce a substitution reaction than toward hydrogen to induce an elimination reaction (Fig. 22.1). As a counter ion, Ag^+ is often used because of its ability to act as an electrophilic catalyst.

$$CH_3CH_2CH(Br)CH_3 \xrightarrow[CH_3CO_2H]{CH_3CO_2Ag} CH_3CH_2CH(OCOCH_3)CH_3 \xrightarrow[OH^-]{H_2O} CH_3CH_2CH(OH)CH_3$$

$$CH_3CH_2CH(Br)CH_3 \xrightarrow{OH^-,\ H_2O} CH_3CH_2CH(OH)CH_3 + CH_3CH{=}CHCH_3$$

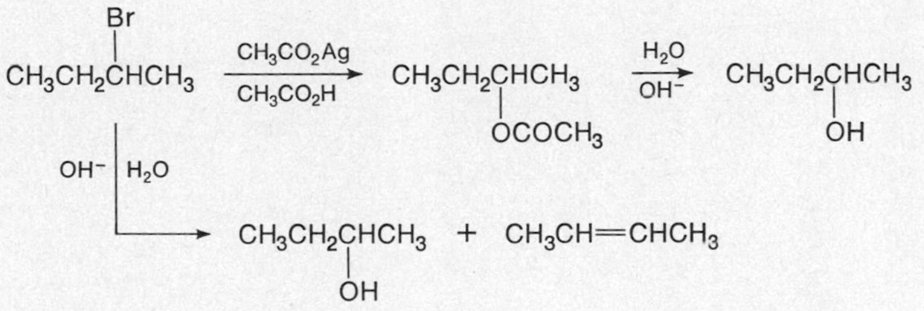

Fig. 22.1. One-step and two-step sequences for converting alkyl halides to alcohols.

PROBLEM 22-1

Explain how Ag^+ acts as an electrophilic catalyst in the reaction of RX with CH_3CO_2Ag.

Another two-step conversion of alkyl halides to alcohols proceeds by formation of the Grignard reagent followed by treatment with oxygen. For example, 1-bromo-2,2,3,3-tetramethylbutane is converted to the corresponding alcohol in reasonably good yield (60–90%) by this procedure, whereas direct hydrolysis with aqueous sodium hydroxide is too slow to be useful because of the β-branching effect (see Section 21.4b). The reaction involves the initial formation of the halomagnesium salt of the hydroperoxide ($RMgX + O_2 \rightarrow ROOMgX$), which

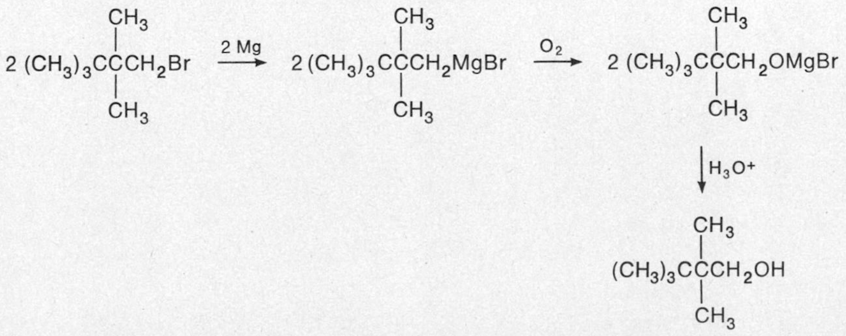

$$2\,(CH_3)_3CC(CH_3)_2CH_2Br \xrightarrow{2\,Mg} 2\,(CH_3)_3CC(CH_3)_2CH_2MgBr \xrightarrow{O_2} 2\,(CH_3)_3CC(CH_3)_2CH_2OMgBr \xrightarrow{H_3O^+} (CH_3)_3CC(CH_3)_2CH_2OH$$

Fig. 22.2. Conversion of alkyl halides to alcohols via Grignard reagents.

then reacts with more Grignard reagent to form the halomagnesium salt of the alcohol (ROOMgX + RMgX → 2 ROMgX) (Fig. 22.2). An alternative procedure that is particularly suitable for the preparation of phenols involves *t*-butyl peroxide as the oxidizing agent.

PROBLEM 22-2

Still another alternative method involves reaction of the Grignard reagent with trimethyl borate, followed by oxidation with hydrogen peroxide:

$$\text{RMgBr} \xrightarrow{(\text{CH}_3\text{O})_3\text{B}} (\text{CH}_3\text{O})_2\text{B—R} \xrightarrow[\text{NaOH}]{\text{H}_2\text{O}_2} \text{ROH}$$

Suggest a mechanism for the first reaction in this sequence.

The hydrolysis of aryl halides is generally unsatisfactory for the preparation of phenols, for nucleophilic substitution at aromatic carbon does not proceed with facility except with phenyl halides that carry strongly electron-withdrawing substituents. As mentioned previously, 2,4-dinitrofluorobenzene reacts with sodium hydroxide to yield 2,4-dinitrophenol (see Fig. 21.32). Under sufficiently strenuous conditions, though, even simple aryl halides such as chlorobenzene react with sodium hydroxide to yield phenols. One of the industrial preparations of phenol, in fact, involves the action of sodium hydroxide on chlorobenzene at 360°:

$$\text{C}_6\text{H}_5\text{—Cl} + \text{NaOH} \xrightarrow[5{,}000\ \text{psi}]{360^\circ} \text{HCl} + \text{C}_6\text{H}_5\text{—ONa} \xrightarrow[\text{H}_2\text{O}]{\text{HCl}} \text{C}_6\text{H}_5\text{—OH} + \text{NaCl}$$

As discussed in Section 21.5c, these reactions are not of the S_N2 type but involve either an addition-elimination sequence or an elimination-addition sequence (benzyne intermediate).

PROBLEM 22-3

Outline the mechanism for the conversion of chlorobenzene to phenol. Would this be a satisfactory method for the conversion of *p*-chlorotoluene to pure *p*-methylphenol? Why or why not?

22.3 *Conversion of C—Nitrogen to C—OH*

In contrast to halogens, most nitrogen moieties are very poor leaving groups. The only nucleophilic displacement reactions involving C—Nitrogen compounds are those of quaternary ammonium compounds (R_4N^+) and diazonium compounds (RN_2^+), *viz.*, compounds in which the nitrogen carries a positive charge. Diazonium compounds, in particular, provide a means for converting amines to alcohols or phenols. The action of nitrous acid on primary aliphatic amines yields alkyldiazonium compounds that are exceedingly unstable, decomposing almost immediately to molecular nitrogen and a carbonium ion. In aqueous solution, one of the pathways of stabilization available to the carbonium ion is acquisition of a water molecule to form an alcohol (Fig. 22.3). Although the reaction is sometimes a good preparative method for alcohols, skeletal rearrangement and alkene formation often reduce its utility (Fig. 22.3).

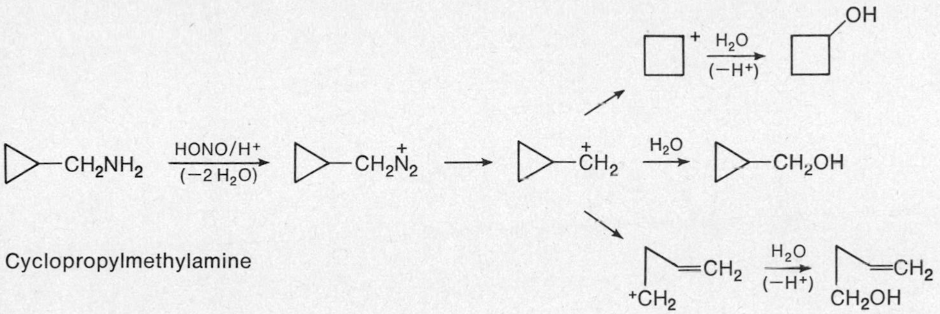

Cyclopropylmethylamine

Fig. 22.3. Conversion of alkylamines to alcohols by treatment with nitrous acid.

PROBLEM 22-4

The cations $\triangleright\!-\overset{+}{C}H_2$ and (cyclobutyl cation), which react to produce identical mixtures containing three- and four-membered ring compounds, have been postulated to exist as a "nonclassical" carbonium ion. Suggest a structure for this species.

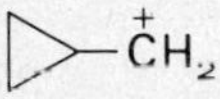

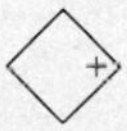

Aryldiazonium compounds are much more stable than their aliphatic counterparts and can be prepared and kept for a limited time in aqueous solution at room temperature. Slowly at room temperature, and more rapidly at elevated temperature, the aryldiazonium salts react with water to form phenols, providing an excellent synthesis method (Fig. 22.4).

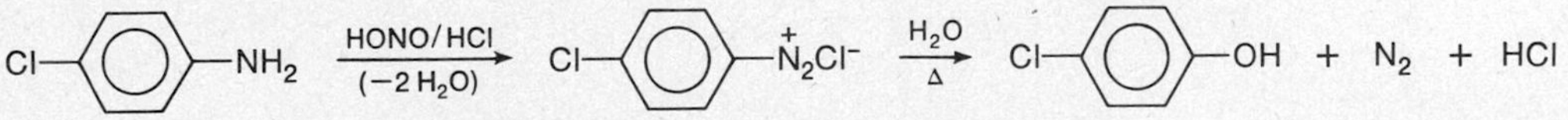

Fig. 22.4. Conversion of arylamines to phenols.

22.4 Conversion of C—Sulfur to C—OH

The only synthetically useful reaction that falls in this category involves the action of molten sodium hydroxide and potassium hydroxide on arylsulfonic acids at elevated temperatures. Like the aryldiazonium method for the synthesis of phenols, it is applicable not only to benzenesulfonic acids (see Fig. 10.12) but to polycyclic aromatic sulfonic acids as well (Fig. 22.5).

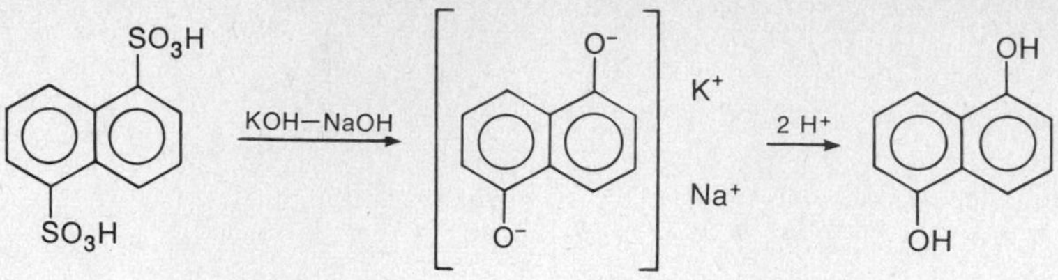

Fig. 22.5. Conversion of arylsulfonic acids to phenols.

22.5 Conversion of C—C to C—OH

Under the influence of perbenzoic acid, aldehydes and ketones are converted to esters in a process known as the *Baeyer–Villiger reaction*. The reaction is an example of a nucleophilic 1,2 rearrangement (see p. 615), in which the migration takes place to an electron-deficient oxygen center. The method is particularly applicable for the preparation of phenols, for alkyl aryl ketones rearrange with migration of the aryl group rather than the alkyl group to yield the corresponding aryl ester from which the phenol can be obtained by hydrolysis (Fig. 22.6).

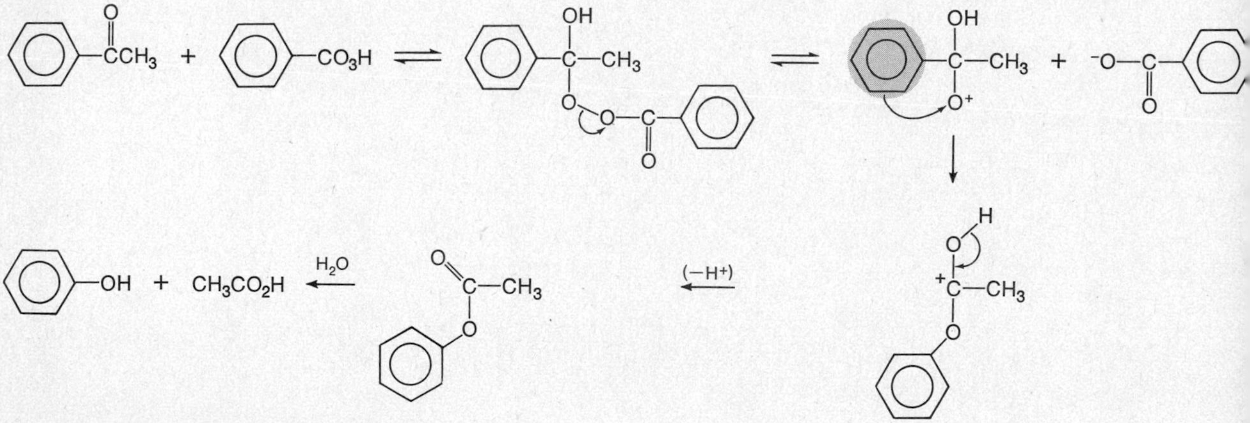

Fig. 22.6. Oxidation of aromatic ketones with perbenzoic acid (Baeyer–Villiger reaction).

PROBLEM 22-5

Oxygen cations are rather unstable and reactive species. What do you think is the major "driving force" for the dissociation of

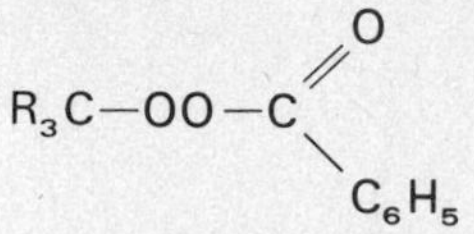

$$R_3C{-}OO{-}C(=O)C_6H_5$$

to an oxygen-containing cation and a carboxylate anion?

22.6 Conversion of $\rangle C{=}C\langle$ **to** $-\overset{|}{\underset{|}{C}}-\overset{|}{\underset{|}{C}}-OH$, $HO-\overset{|}{\underset{|}{C}}-\overset{|}{\underset{|}{C}}-OH$, **and Related Compounds**

22.6a. ACID-CATALYZED ADDITION OF WATER. The addition of the elements of water to the double bond of an alkene provides a useful synthesis of alcohols. The oldest of several procedures simply involves the acid-catalyzed addition of water (see Fig. 8.4). The addition occurs with Markovnikov regioselectivity, so in most instances the hydroxyl group becomes attached to the more highly substituted carbon of the double bond (Fig. 22.7). Because carbonium ions are involved as intermediates, however, skeletal rearrangements often take place, and this possibility must be taken into account (Fig. 22.7).

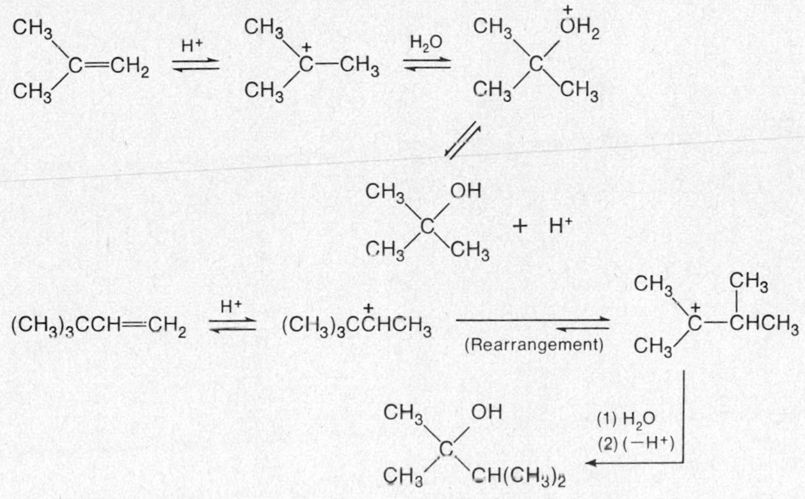

$$(CH_3)_2C{=}CH_2 \underset{}{\overset{H^+}{\rightleftharpoons}} (CH_3)_2\overset{+}{C}{-}CH_3 \overset{H_2O}{\rightleftharpoons} (CH_3)_3C{-}\overset{+}{O}H_2 \rightleftharpoons (CH_3)_3C{-}OH + H^+$$

$$(CH_3)_3CCH{=}CH_2 \overset{H^+}{\rightleftharpoons} (CH_3)_3C\overset{+}{C}HCH_3 \underset{\text{(Rearrangement)}}{\rightleftharpoons} (CH_3)_2\overset{+}{C}{-}CH(CH_3)CH_3 \xrightarrow{(1)\ H_2O,\ (2)\ (-H^+)} (CH_3)_2C(OH){-}CH(CH_3)_2$$

Fig. 22.7. Acid-catalyzed addition of water to alkenes.

22.6b. HYDROXYMERCURATION-REDUCTION PROCEDURE FOR THE ADDITION OF WATER. The complication of skeletal rearrangement can be eliminated by using a process known as hydroxymercuration which involves treatment of an alkene with mercuric acetate in aqueous solution followed by reduction with sodium borohydride (Fig. 22.8). The initial addition of mercuric acetate takes

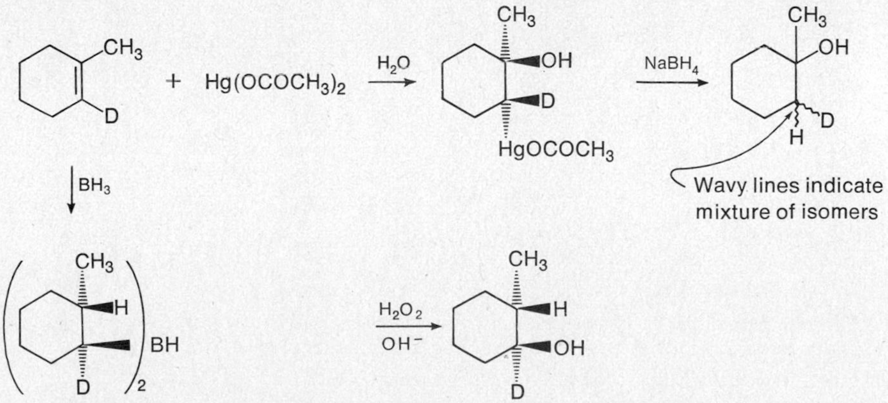

Fig. 22.8. Addition of water to alkenes by hydroxymercuration (overall result is Markovnikov addition) and by hydroboration (overall result is anti-Markovnikov addition).

place with Markovnikov regioselectivity and *anti*-stereoselectivity; the electrophilic portion, $HgOAc^+$, becomes attached to the less highly substituted carbon, and the nucleophile derived from the solvent, OH^-, becomes attached to the more highly substituted carbon. Replacement of the —HgOAc moiety by hydrogen can be effected by sodium borohydride reduction, which, unfortunately, tends to proceed in a nonstereoselective fashion. The overall result of the procedure is to add water to a double bond in Markovnikov fashion.

22.6c. HYDROBORATION-OXIDATION PROCEDURE FOR THE ADDITION OF WATER. The regioselectivity of the addition of water to an alkene can be reversed by using a procedure called hydroboration, which involves the addition of borane (BH_3) to an alkene. Diborane (B_2H_6), obtained by the action of sodium borohydride on boron trifluoride ($3\,NaBH_4 + 4\,BF_3 \rightarrow 3\,NaBF_4 + 2\,B_2H_6$), dissociates to the monomer (BH_3) in tetrahydrofuran solution, the monomer having the capability of adding to a carbon–carbon double bond in such a fashion that a $—BH_2$ fragment becomes affixed to one of the carbons and a hydrogen to the other. Depending on the substituents attached to the double bond, the initially formed alkylborane (RBH_2) may react with a second equivalent of alkene to form a dialkylborane (R_2BH) and with a third equivalent to form a trialkylborane (R_3B); in general, $RCH{=}CH_2$, $R_2C{=}CH_2$, and $RCH{=}CHR$ alkenes form R_3B; $R_2C{=}CHR$ alkenes form R_2BH; and $R_2C{=}CR_2$ alkenes form RBH_2. Treatment of the boranes with alkaline hydrogen peroxide converts them to the corresponding alcohols, the C—B bond undergoing conversion to C—OH. Since the addition of BH_3 to the alkene takes place with Markovnikov regiospecificity (*i.e.*, the electrophilic boron becomes attached to the less highly substituted carbon and the nucleophilic H to the more highly substituted carbon), the overall result is the addition of water to the alkene in anti-Markovnikov fashion (Fig. 22.8). The overall reaction proceeds in a highly stereoselective fashion, the addition of borane occurring in a *syn* fashion and the replacement of the boron moiety by the hydroxyl group occurring with retention of configuration.

PROBLEM 22-6

If borane prepared from $NaBD_4$ and BF_3 is used, what is the product of overall addition of D_2O via the hydroboration process to (*E*)-2-phenyl-2-butene? In how many stereoisomeric forms can the product exist?

22.6d. HYDROBORATION-CARBONYLATION PROCEDURES. If the trialkylboranes obtained from the reaction of diborane and alkenes are treated with carbon monoxide, a variety of types of alcohols can be obtained, depending on the conditions under which the reaction is carried out. If carried out in the presence of lithium borohydride, the product is a primary alcohol. If water is present, a secondary alcohol is produced. In the absence of water or a reducing agent, the intermediate boraepoxide rearranges and polymerizes to $(O{=}BCR_3)_n$, which can

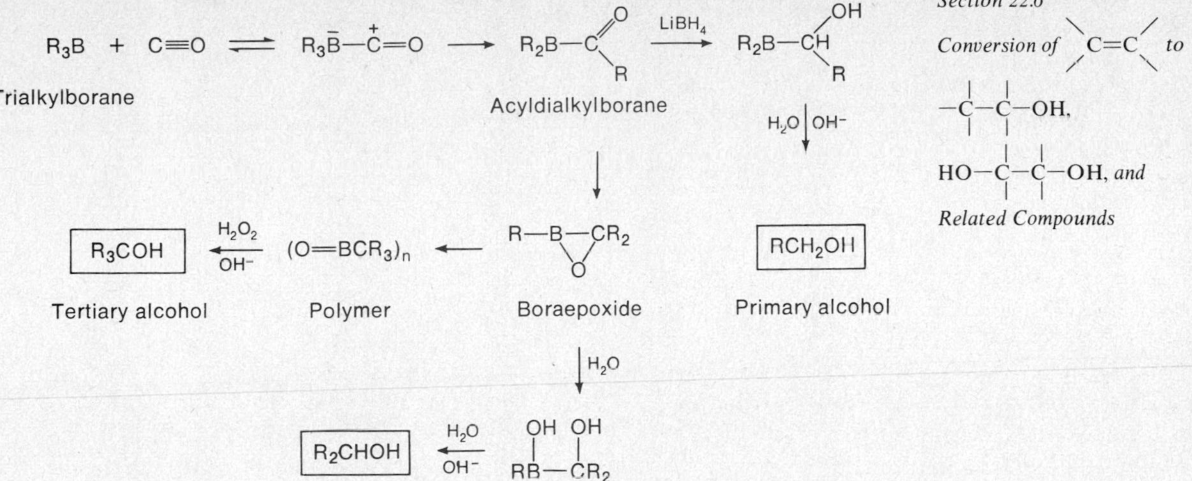

Fig. 22.9. Formation of alcohols via carbonylation of trialkylboranes.

be oxidized with alkaline hydrogen peroxide to a tertiary alcohol. The mechanisms by which these various products are thought to arise are shown in Fig. 22.9. The initially formed charge-separated adduct undergoes a migration of an R group to form an acyldialkylborane. This is postulated to undergo a second alkyl group migration to form a boraepoxide, followed by a third alkyl group migration to form a trialkylmethylboron oxide polymer which yields the tertiary alcohol upon oxidative workup. In the presence of a reducing agent, however, the sequence is interrupted after a single migration at the acyldialkylborane state, and in the presence of water it is interrupted at the boraepoxide stage after two migrations have taken place. These reactions are exceedingly useful for the conversion of alkenes (a) to primary alcohols containing one additional carbon atom or (b) to secondary and tertiary alcohols of more complex structure. In all cases the alkyl groups migrate from boron to carbon with retention of configuration (Fig. 22.10).

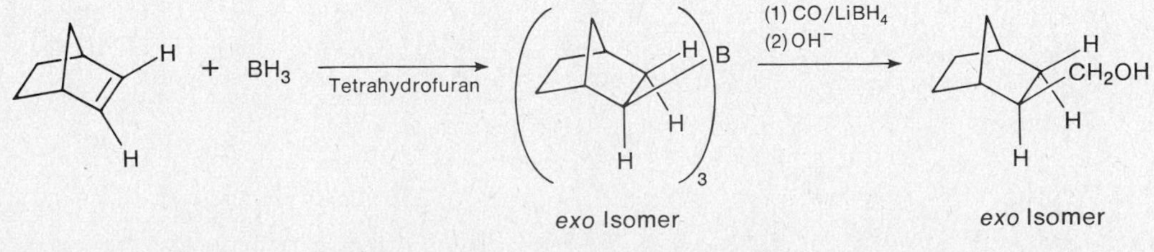

+ BH_3 → Tetrahydrofuran → (1) CO/$LiBH_4$ (2) OH^- →

exo Isomer   *exo* Isomer

Fig. 22.10. Hydroboration-carbonylation of bicyclo[2.2.1]hept-2-ene.

PROBLEM 22-7

The tertiary tricyclic alcohol

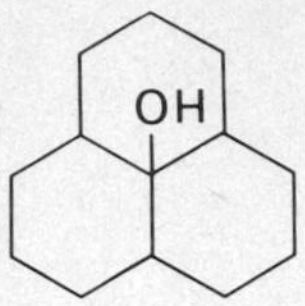

can be obtained from the readily available 1,5,9-cyclododecatriene by the hydroboration-carbonylation procedure. Show the sequence of reactions involved in this synthesis.

22.6e. HYDROFORMYLATION (CONVERSION OF C=C TO C—C—CH_2OH). Still another method for preparing alcohols from alkenes involves the addition of carbon monoxide and hydrogen under the catalytic influence of certain metal carbonyls. The process, known as *hydroformylation* or the *Oxo process*, has considerable commercial importance for the preparation of aldehydes and carboxylic acids as well as alcohols. The reaction is postulated to involve the addition of the metal atom and a hydrogen atom to the carbons of an alkene to give a compound that undergoes intramolecular rearrangement to yield an acyl compound (Fig. 22.11). In the presence of molecular hydrogen, hydrogenolysis then occurs to yield an aldehyde that can be isolated, or that, if an excess of hydrogen is present, undergoes further reduction to the corresponding alcohol.

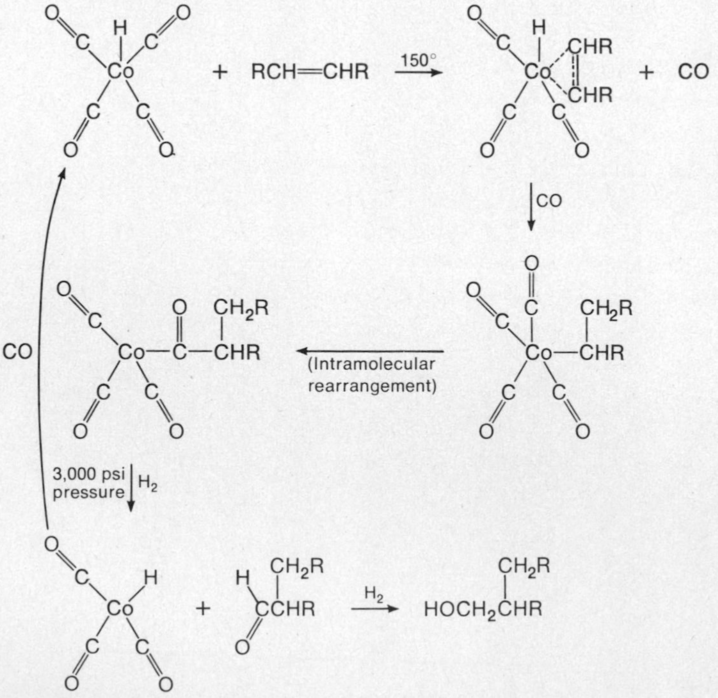

Fig. 22.11. Preparation of alcohols via hydroformylation process.

As a laboratory procedure, the method is severely limited in its application because of the tendency for skeletal rearrangements to occur.

Section 22.6

Conversion of $\rangle C{=}C\langle$ *to* $-\overset{|}{\underset{|}{C}}-\overset{|}{\underset{|}{C}}-OH$, $HO-\overset{|}{\underset{|}{C}}-\overset{|}{\underset{|}{C}}-OH$, *and Related Compounds*

22.6f. VICINAL DIHYDROXYLATION OF ALKENES. Alkenes can be converted to 1,2-diols (*vic*-glycols) in a variety of ways. One that provides the diol directly makes use of potassium permanganate in neutral or mildly alkaline (pH 9–11) solution, the reaction involving the *syn* addition of MnO_4^- to the double bond to give a manganate ester that undergoes hydrolysis to the *cis*-1,2-diol (Fig. 22.12).

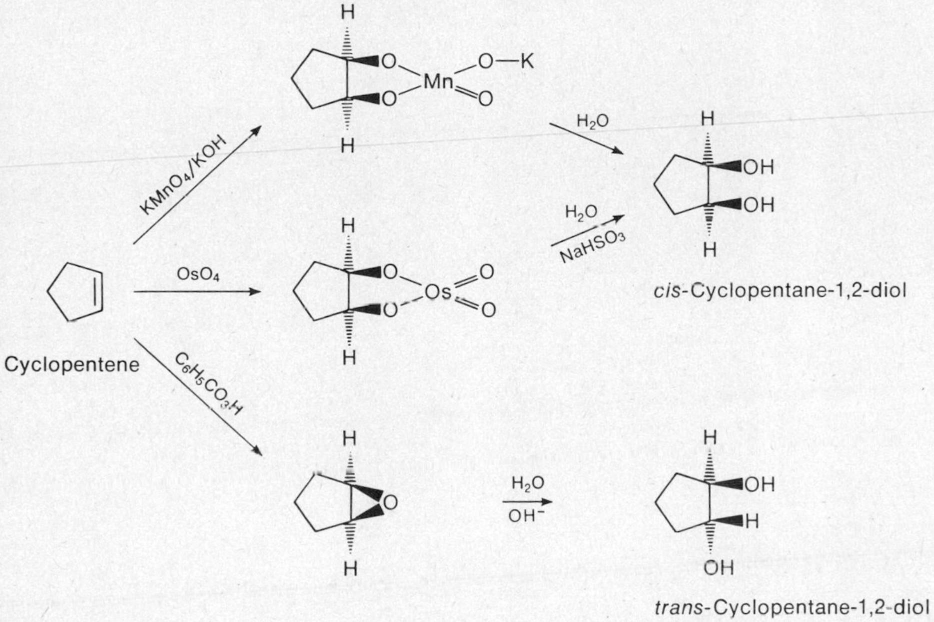

Fig. 22.12. Preparation of 1,2-diols from alkenes.

In similar fashion, osmium tetroxide adds to alkenes in *syn* fashion to yield an isolable cyclic osmium ester, which produces the *cis*-1,2-diol upon hydrolysis (Fig. 22.12). In both of these instances, the stereochemistry of the product is established in the initial reaction in which a cyclic ester is formed; the subsequent hydrolysis does not involve rupture of the C—O bonds. A third procedure, leading to *trans*- rather than *cis*-1,2-diols, involves the action of peracids on alkenes to form oxiranes (1,2-epoxides). When oxiranes are treated with aqueous acid or base, ring opening occurs via a nucleophilic substitution process that involves the rupture of one of the C—O bonds of the oxirane with inversion of configuration at that carbon (Fig. 22.12).

PROBLEM 22-8

If D_2O instead of H_2O is used in the reaction shown in Fig. 22.12, is the stereochemistry of the product changed? Explain.

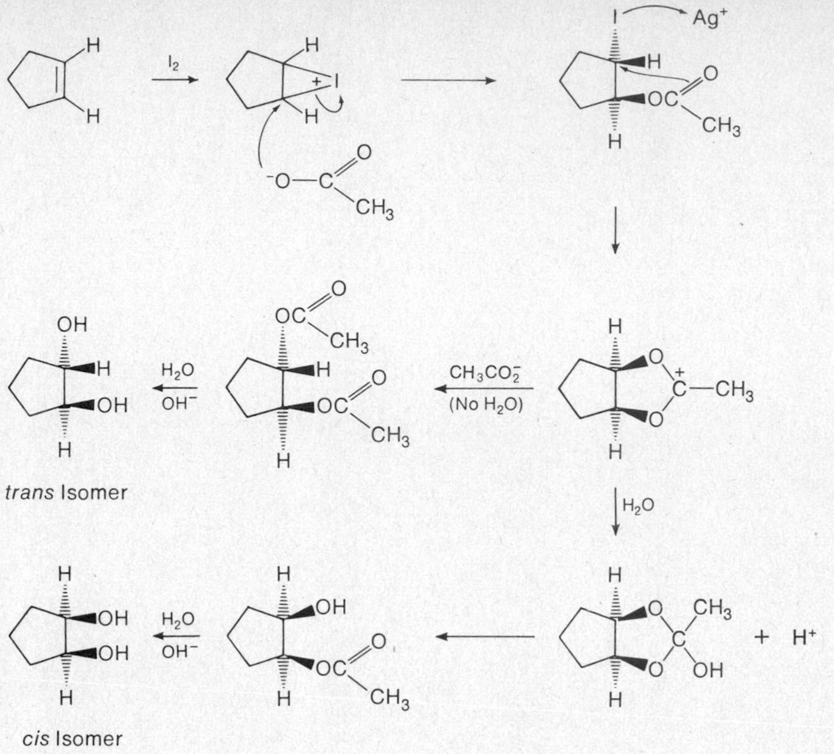

Fig. 22.13. Formation of *cis*- and *trans*-1,2-diols from alkenes with iodine and silver acetate.

A fourth procedure, leading either to the *cis*- or *trans*-diol, involves the action of iodine and silver acetate (or benzoate) on the alkene followed by hydrolysis. If the addition is carried out in the *absence* of water, it follows an *anti* pathway and yields *trans* products; if it is carried out in the *presence* of water, it follows a *syn* pathway and yields *cis* products (Fig. 22.13).

22.6g. Preparation of 1,3-diols from alkenes. A general procedure for the synthesis of 1,3-dihydroxy compounds (1,3-glycols) involves the acid-catalyzed reaction between alkenes and formaldehyde (*Prins reaction*), which in some

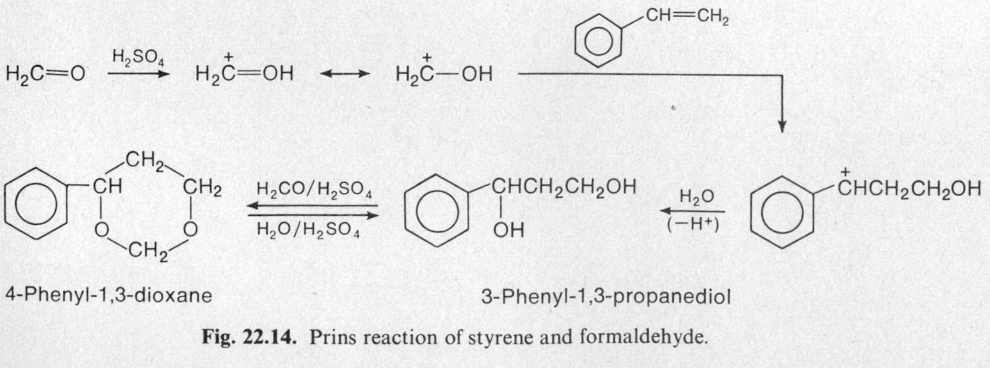

Fig. 22.14. Prins reaction of styrene and formaldehyde.

instances leads directly to the 1,3-diol but in other instances yields a 1,3-dioxane (from further reaction of the 1,3-diol with formaldehyde) from which the diol can be obtained by hydrolysis (Fig. 22.14).

22.7 *Conversion of C=O to C—OH by Nucleophilic Addition*

Carbonyl compounds react with nucleophiles to form addition products, which, in the majority of cases, are alcohols (see Chapter 29 for a detailed discussion). Among the most effective nucleophiles are hydride ion (provided by reagents such as lithium aluminum hydride and sodium borohydride) and carbanions (provided by organometallic reagents such as RMgX and RLi):

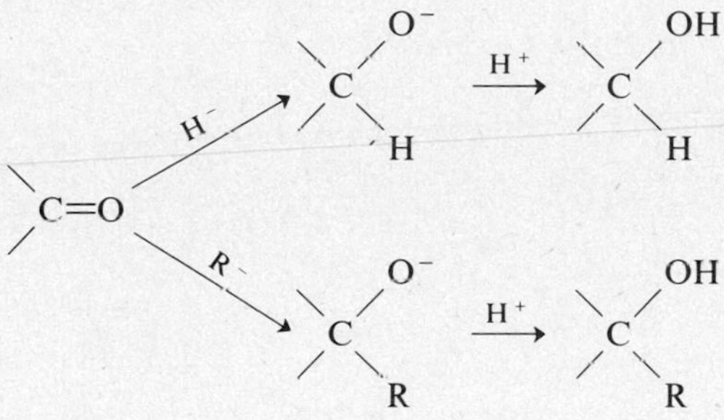

$$>C{=}O \xrightarrow{H^-} >C(O^-)H \xrightarrow{H^+} >C(OH)H$$

$$>C{=}O \xrightarrow{R^-} >C(O^-)R \xrightarrow{H^+} >C(OH)R$$

22.7a. HYDRIDE REDUCTION OF CARBONYL COMPOUNDS. Several methods are available for the hydride reduction of carbonyl compounds to the corresponding alcohols, some limited to aldehydes and ketones but others also applicable to carboxylic acids and derivatives. One of the oldest of these methods makes use of isopropyl alcohol acting on an aldehyde or ketone in the presence of aluminum isopropoxide. Known as the *Meerwein–Pondorff–Verley reduction*, it involves a hydride transfer from the aluminum isopropoxide (yielding acetone) to the aldehyde or ketone (yielding the aluminum salt of the alcohol), as outlined in Fig. 22.15. The equilibrium constant for this process is close to unity. To improve

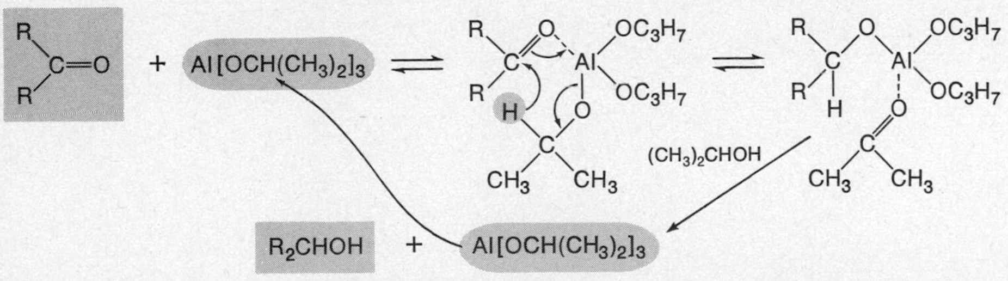

$R_2C{=}O$ + $Al[OCH(CH_3)_2]_3$ ⇌ [cyclic complex: $R_2C{=}O \cdot Al(OC_3H_7)_2$–$OCH(CH_3)_2$] ⇌ [$R_2CH$–$O$–$Al(OC_3H_7)_2 \cdots O{=}C(CH_3)_2$] ; $(CH_3)_2CHOH$ → R_2CHOH + $Al[OCH(CH_3)_2]_3$

Fig. 22.15. Preparation of alcohols by the Meerwein–Pondorff–Verley reduction of aldehydes and ketones.

its utility as a reduction method, therefore, we use an excess of isopropyl alcohol (taking advantage of the law of mass action), and we remove the acetone that is formed as the reaction proceeds by fractional distillation (bp of acetone is 56°; bp of isopropyl alcohol is 82°).

PROBLEM 22-9

The Meerwein–Pondorff–Verley reduction of cyclohexyl methyl ketone with optically active 3-methyl-2-butanol as the hydrogen donor and aluminum 3-methyl-2-butoxide as the catalyst yields optically active 1-cyclohexylethanol. Explain why the product is optically active.

Among the more recently discovered reagents for reductions are various metal hydrides such as lithium aluminum hydride ($LiAlH_4$) and sodium borohydride ($NaBH_4$). For the most part, these reagents have displaced the Meerwein–Pondorff–Verley procedure as the method of choice for the reduction of aldehydes and ketones to alcohols. Although both $LiAlH_4$ and $NaBH_4$ are a source of hydride ion, they differ significantly in their reducing power. Sodium borohydride is a relatively mild reducing agent and can be used in aqueous or alcoholic solution. Lithium aluminum hydride, on the other hand, is a much more powerful reducing agent and reacts violently with protic compounds such as water and alcohols. Therefore, it must be used in aprotic solvents such as diethyl ether, tetrahydrofuran, or 1,2-dimethoxyethane. This difference in reactivity is reflected in the fact that sodium borohydride will reduce aldehydes and ketones to alcohols but is without action on carboxylic acid derivatives, whereas lithium aluminum hydride reduces aldehydes, ketones, carboxylic acid derivatives (except amides, which are reduced to amines), and carboxylic acids to alcohols (see Figs. 14.9 and 22.16), as well as numerous other functions, such as nitro, nitroso,

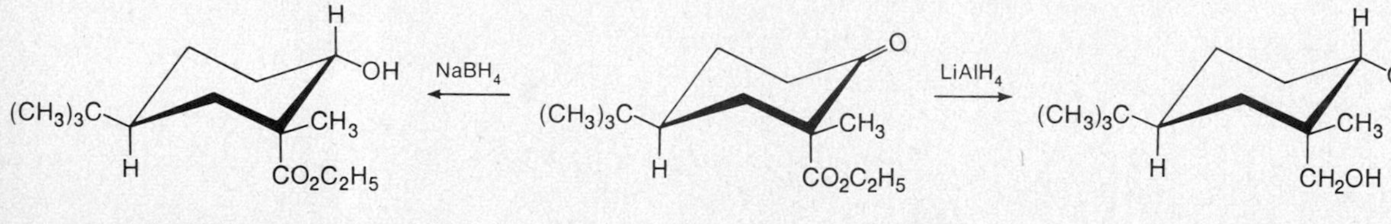

Fig. 22.16. Preparation of alcohols by the metal hydride reduction of carbonyl compounds.

imino, and cyano groups. Sodium borohydride and other metal hydrides do not reduce C=C and C≡C bonds unless they are in conjugation with a carbonyl group (*e.g.*, α,β-unsaturated carbonyl compounds).

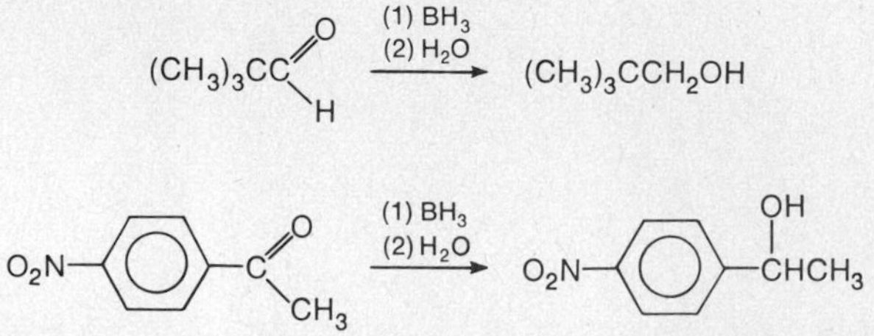

$$(CH_3)_3CCHO \xrightarrow[(2)\ H_2O]{(1)\ BH_3} (CH_3)_3CCH_2OH$$

$$O_2N\text{-}C_6H_4\text{-}COCH_3 \xrightarrow[(2)\ H_2O]{(1)\ BH_3} O_2N\text{-}C_6H_4\text{-}CH(OH)CH_3$$

Fig. 22.17. Preparation of alcohols by diborane reduction of aldehydes and ketones.

Also capable of reducing carbonyl compounds to alcohols are borane, monoalkylboranes, and dialkylboranes (Fig. 22.17), but these reagents, in contrast to the metal hydrides, attack C=C and C≡C bonds. The relative susceptibility of groups to reduction by the boranes decreases in the following order: $RCO_2H > R_2C{=}CR_2 > R_2C{=}O > RC{\equiv}N > RCO_2R'$. These reagents do not reduce acid chlorides (RCOCl), nitro compounds (RNO_2), or halides (R-Halogen).

22.7b. ADDITION OF ORGANOMETALLIC REAGENTS TO CARBONYL COMPOUNDS. Organometallic compounds such as Grignard reagents (RMgX) and organolithium reagents (RLi) are sources of carbon nucleophiles that can add (a) to formaldehyde to give primary alcohols, (b) to aldehydes or esters of formic acid (HCO_2R) to give secondary alcohols, and (c) to ketones and esters (other than formates) to give tertiary alcohols (see Fig. 10.9). Although these reactions usually proceed smoothly and in good yield, side reactions sometimes do occur. Principal among these are reduction and enolization reactions, both of which result in the failure to attach the R moiety of the Grignard reagent to the carbonyl carbon. The reduction reaction is interpreted in terms of a six-membered ring-activated complex, in which a β hydrogen of the organometallic reagent is transferred to the carbonyl carbon:

Mg, CR_2, O, CR_2, CR_2, H → CR_2=CR_2 + Mg–O–CR_2–H $\xrightarrow{H_2O}$ R_2CHOH

The products are the alcohol corresponding to the ketone or aldehyde and the alkene corresponding to the organometallic reagent. The enolization reaction is interpreted in terms of nucleophilic attack on the α hydrogen, a process that competes with the "normal" pathway involving nucleophilic attack on the carbonyl carbon atom:

Pathway (a)

Pathway (b)

Pathway (a)

Pathway (b)

R^-M^+

Nucleophilic addition to C=O (addition pathway)

H_2O

Alcohol

H_2O

Starting compound

+ RH

Basic abstraction of α—H (enolization pathway)

The products, after hydrolysis, are the original carbonyl compound and the hydrocarbon corresponding to the organometallic reagent. As steric hindrance around the carbonyl group increases and as the bulk of the Grignard reagent increases, the incidence of reduction and enolization increases, as the data in Fig. 22.18 indicate.

$$\underbrace{(CH_3)_2CH\overset{O}{\overset{\|}{C}}CH(CH_3)_2 \;+\; RMgX}$$
$$\downarrow$$

	Addition Product, %	Enolization Product, %	Reduction Product, %
CH_3CH_2MgBr	79	1	20
$CH_3CH_2CH_2MgCl$	47	2	51
$(CH_3)_2CHMgCl$	0	28	72

Fig. 22.18. Addition product, enolization product, and reduction product formed from the reaction of Grignard reagents with diisopropyl ketone.

PROBLEM 22-10

Which pair of carbonyl compounds and Grignard reagents from among the following compounds would be most likely to yield both an enolization and a reduction product? Explain.

$$(CH_3)_2CHCH_2\overset{O}{\overset{\|}{C}}CH_2CH(CH_3)_2,$$

$$(CH_3)_3C\overset{O}{\overset{\|}{C}}CH_2CH(CH_3)_2,$$

$$(CH_3)_3C\overset{O}{\overset{\|}{C}}C(CH_3)_3,$$

$$C_6H_5CH_2CH_2MgCl,$$

$$C_2H_5MgCl,$$

$$(CH_3)_2CHCH_2MgCl$$

22.8 *Conversion of C=O to C—OH by Catalytic Hydrogenation*

Most aldehydes, ketones, and carboxylic acid derivatives can be reduced to the corresponding alcohols with molecular hydrogen and the appropriate metal catalyst. The reaction involves the attachment of the carbonyl compound and the hydrogen to the surface of the metal catalyst (chemisorption) in such a fashion that the addition reaction is facilitated.

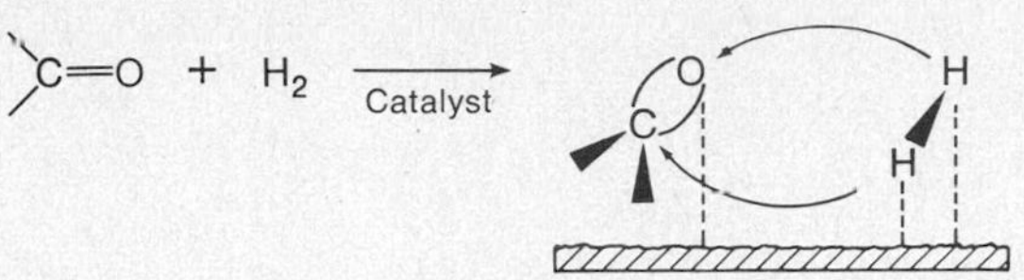

>C=O + H_2 —Catalyst→ [surface complex] → H—C—O—H

Depending on the compounds involved and the conditions of the reaction, the hydrogen atoms may add simultaneously to the carbon and oxygen or in a stepwise fashion (*i.e.*, desorption after one hydrogen atom has been transferred followed by readsorption for the transfer of the second hydrogen atom). The catalysts that are most often used are, in order of descending activity, platinum (as finely divided metal or deposited on carbon), palladium (usually deposited on carbon), and Raney nickel (Fig. 22.19). In general, carbonyl compounds undergo hydrogenation more slowly than alkenes and alkynes.

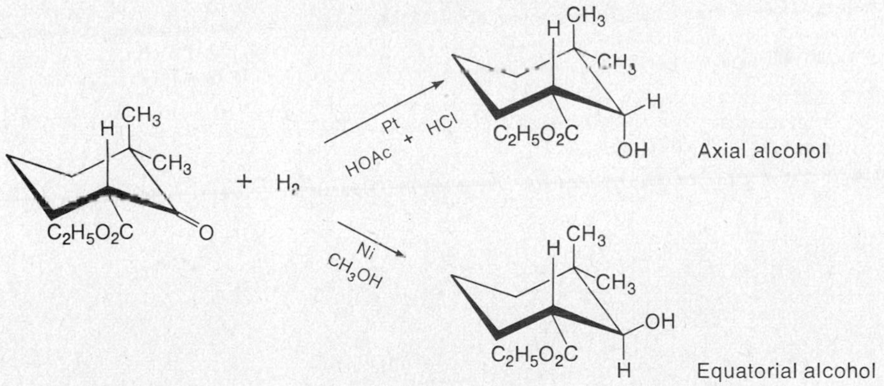

Pt, HOAc + HCl → Axial alcohol

Ni, CH_3OH → Equatorial alcohol

Fig. 22.19. Preparation of alcohols by catalytic reduction of ketones.

22.9 *Conversion of C=O to HO—C—C—OH by Bimolecular Reduction*

Aldehydes and ketones undergo reductive dimerization in the presence of magnesium, magnesium amalgam, or a mixture of magnesium and magnesium iodide to form dimers from which the 1,2-diols are obtained on hydrolysis (Fig. 22.20). The product formed from acetone by this procedure has a "platelike" crystalline structure and was called a *pinacol* (Greek, *pinax*, plate) by the chemists who first obtained it. The name has subsequently been generically applied to all symmetrically substituted 1,2-diols, and reductions of this sort have come to be called *pinacol reductions.* The reaction is thought to involve the transfer of one electron from the magnesium atom to the carbonyl group of the aldehyde or ketone to produce a radical anion. Two radical anions then undergo electron pairing at the carbon atoms to form the magnesium salt of the pinacol.

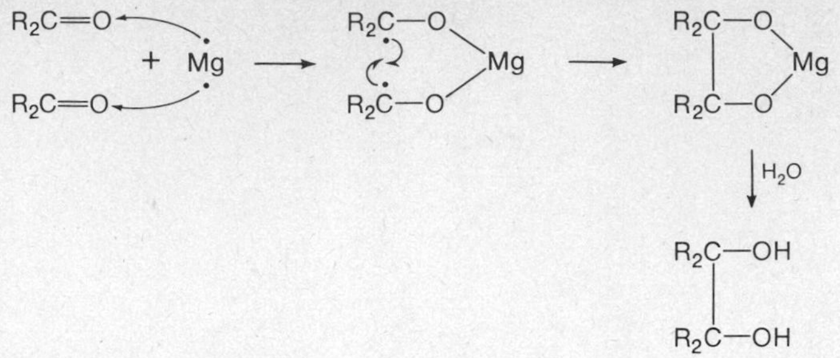

$$\begin{array}{l} R_2C{=}O \\ \quad + \; Mg \\ R_2C{=}O \end{array} \longrightarrow \begin{array}{l} R_2C{-}O \\ \qquad\quad Mg \\ R_2C{-}O \end{array} \longrightarrow \begin{array}{l} R_2C{-}O \\ \;\;| \qquad\; Mg \\ R_2C{-}O \end{array} \xrightarrow{H_2O} \begin{array}{l} R_2C{-}OH \\ \;\;| \\ R_2C{-}OH \end{array}$$

Fig. 22.20. Preparation of 1,2-diols by bimolecular reduction of aldehydes and ketones. The single-headed curved arrows indicate the movement of single electrons (in contrast to double-headed curved arrows, which indicate the movement of pairs of electrons); these are often referred to as "fish hook arrows."

PROBLEM 22-11

What product do you think might be formed by the action of magnesium on 2,6-heptanedione?

Table 22.1 summarizes the methods that are useful for the synthesis of alcohols and phenols.

Table 22.1 METHODS FOR THE PREPARATION OF ALCOHOLS AND PHENOLS

Type of Product	Starting Materials	Reagents	Products	Comments
Monoalcohols	$Ar\overset{R}{\underset{R}{C}}{-}H$	Oxygen	$Ar\overset{R}{\underset{R}{C}}{-}OOH$	
		$KMnO_4$ or $Na_2Cr_2O_7 + H_2SO_4$	$Ar\overset{R}{\underset{R}{C}}{-}OH$	Severely limited in application
	Alkyl halides (R—X)	OH^-	R—OH	Limited to primary RX and som secondary RX; some secondar and most tertiary RX eliminat HX; order of halogen reactivity $I > Br > Cl > F$
		(1) OAc^-; (2) H_2O	R—OH	Reduces extent of competing elim ination
		(1) Mg; (2) O_2; (3) H_2O	R—OH	Minimizes skeletal rearrangemen and allows replacement in frame works that are sluggish in S_N reactions
	Amines (RNH_2)	HONO, H_2O	R—OH	Skeletal rearrangement and alken formation very probable

Table 22.1—*cont.*

Type of Product	Starting Material	Reagents	Products	Comments
	Alkenes ($R_2C{=}CH_2$)	H_3O^+	$R_2C(OH){-}CH_3$	Markovnikov regioselectivity; *anti* stereoselectivity or nonstereoselective; skeletal rearrangements possible
		(1) H_2O, Hg^{++}; (2) $NaBH_4$, OH^-	$R_2C(OH){-}CH_3$	Markovnikov regioselectivity, nonstereoselective; no skeletal rearrangement
		(1) BH_3; (2) H_2O_2, OH^-	R_2CHCH_2OH	anti-Markovnikov regioselectivity; *syn* stereospecificity; no skeletal rearrangement
		$CO + H_2Co_2(CO)_8$ catalyst	$R_2CHCH_2{-}HOCH_2$	Skeletal rearrangement probable
	Aldehydes (RCHO)	$LiAlH_4$, $NaBH_4$, BH_3	RCH_2OH	$NaBH_4$ most selective
		$[(CH_3)_2CHO]_3Al$	RCH_2OH	Generally has no advantage over hydride method
		H_2 + catalyst	RCH_2OH	
		(1) R′MgX; (2) H_3O^+	$RR'CHOH$	R′ can be alkyl, alkenyl, alkynyl, aryl
	Ketones ($RR'C{=}O$)	$LiAlH_4$, $NaBH_4$, BH_3	$RR'CHOH$	$NaBH_4$ most selective
		$[(CH_3)_2CHO]_3Al$	$RR'CHOH$	Generally has no advantage over hydride methods
		H_2 + catalyst	$RR'CHOH$	
		(1) R″MgX; (2) H_3O^+	$RR'R''C{-}OH$	Reduction and/or enolization if severe steric hindrance is present (bulky ketone and/or bulky Grignard reagent)
	Carboxylic acids and carboxylic acid derivatives ($RC(=O)Y$)	$LiAlH_4$	RCH_2OH	$NaBH_4$ ineffective in most cases
	(Y = Halogen, OCOR, OR, OH)	(1) R′MgX; (2) H_3O^+	$RR'R'C{-}OH$	RCO_2H decomposes 1 mole-equivalent of R′MgX (may react further). Two of R groups in tertiary alcohol come from Grignard reagent

Table 22.1—*cont.*

Type of Product	Starting Material	Reagents	Products	Comments
	Trialkylboranes (R_3B) prepared from ($>C=C<$ + BH_3)	CO, $LiBH_4$	RCH_2OH	
		CO, H_2O	R_2CHOH	
		(1) CO; (2) H_2O_2, OH^-	R_3COH	
Diols	Alkenes ($>C=C<$)	(1) OsO_4 or $KMnO_4$ (2) H_2O	—C(OH)—C(OH)—	*cis*-1,2-Diol formed
		(1) RCO_3H; (2) H_2O	—C(OH)—C(OH)—	*trans*-1,2-Diol formed
		I_2—AgOAc/H_2O	HO—C—C—OH	*cis*-1,2-Diol in presence of H_2(*trans*-1,2-diol in absence of H_2O
		$RCHO/H_3O^+$	—C(OH)—C—CH(OH)R	
	Aldehydes and ketones ($R_2C=O$)	Mg, Mg(Hg) Mg(MgI_2)	$R_2C(OH)$—$C(OH)R_2$	Bimolecular reduction
Phenols	Aryl halides (Ar—X)	OH^-	ArOH	Proceeds with facility only wh strongly electron-withdrawing groups are present *o*- and *p*- to hal gen atom; otherwise, extreme strenuous conditions required
	Aryl amines ($ArNH_2$)	HONO, H_2O	ArOH	Generally applicable to ar amines; not particularly affect by the presence of other function groups
	Arylsulfonic acids ($ArSO_3H$)	(1) NaOH—KOH; (2) H_3O^+	ArOH	Requires very strenuous condition generally not applicable if oth functional groups are present
	Aryl alkyl ketones	(1) $C_6H_5CO_3H$; (2) H_2O	ArOH	Baeyer–Villiger reaction

REACTIONS OF ALCOHOLS AND PHENOLS

22.10 *Reactions Involving the* RO—H *Bond of Alcohols or the* ArO—H *Bond of Phenols*

The important reactions involving the RO—H bond include the reaction of phenols with hydroxide ion to form phenoxides (ArO^-) (see Fig. 10.13), the reaction of phenols and alcohols with metals above hydrogen in the emf series to form phenoxides or alkoxides (RO^-) (see p. 267), and the reactions with carboxylic acids (see Fig. 10.14), anhydrides (see Fig. 15.5), acid chlorides (see Fig. 15.4), and ketenes (see Fig. 15.19) to form esters. An example of ester formation that illustrates the differences in reactivity of primary and tertiary alcohols involves the reaction of acetyl chloride with methanol (to form methyl acetate) and *t*-butyl alcohol (to form *t*-butyl acetate), respectively. Whereas methanol reacts in the expected fashion, *t*-butyl alcohol yields *t*-butyl chloride as the major product accompanied by only a trace of the ester. This result is explained by the sequence of reactions shown in Fig. 22.21. Esterification of *t*-butyl alcohol with acetyl

Reaction	R=H	R=CH_3
$R_3COH + CH_3COCl \longrightarrow R_3COC(=O)CH_3 + HCl$	Fast	Slow
$HCl + R_3COH \longrightarrow R_3CCl + H_2O$	Slow	Fast
$H_2O + CH_3COCl \xrightarrow{\text{(Fast)}} CH_3CO_2H + HCl$		
Product	$CH_3OC(=O)CH_3$	$(CH_3)_3CCl$

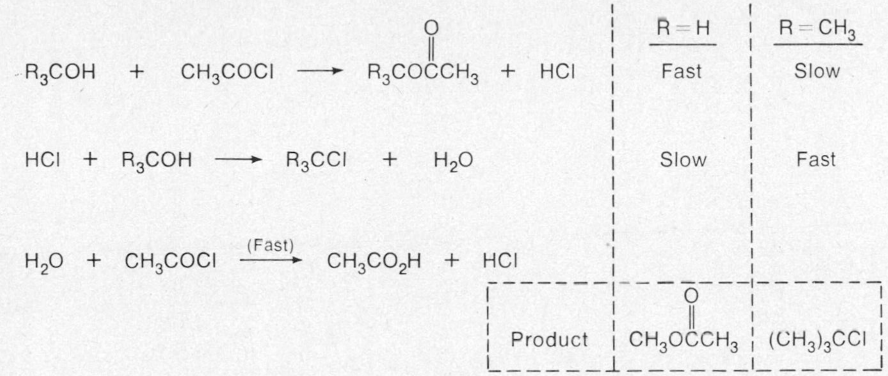

Fig. 22.21. Reaction of acetyl chloride with methanol and *t*-butyl alcohol.

chloride yields *t*-butyl acetate and hydrogen chloride in a reaction that we will characterize as "slow" relative to certain other reactions that subsequently take place. The hydrogen chloride that is produced then engages in a "fast" reaction with *t*-butyl alcohol to yield *t*-butyl chloride and water. And, the water so formed engages in another "fast" reaction with acetyl chloride to form acetic acid and more hydrogen chloride, thus initiating a chain process in which *t*-butyl chloride and acetic acid are the major products. A similar circumstance does not arise with methanol, however, because (a) in O—H bond-breaking reactions (*e.g.*, ester formation), the order of reactivity is primary > secondary > tertiary, whereas (b) in C—OH bond-breaking reactions (*e.g.*, alkyl halide formation), the order of reactivity is tertiary > secondary > primary. Thus, with methanol the reaction in which the ester is formed is fast in comparison with the reaction in which the alkyl halide is formed.

We can avoid the complication that arises in the attempted esterification of *t*-butyl alcohol with acetyl chloride by adding pyridine to the reaction mixture to trap the hydrogen chloride as pyridinium hydrochloride ($C_5H_5\overset{+}{N}H\ Cl^-$). An even better method for obtaining the acetate from *t*-butyl alcohol or other tertiary alcohols is to use ketene as the acetylating agent. Ketene is a very reactive compound, allowing the reaction to be carried out under neutral conditions at room temperature or even lower temperature if necessary;

$$ROH + H_2C{=}C{=}O \longrightarrow ROC(=O)CH_3$$

PROBLEM 22-12

Write a mechanism for the reaction of phenol and ketene to form phenyl acetate.

Alcohols can be converted to sulfonate esters by the action of arylsulfonyl chlorides in the presence of pyridine;

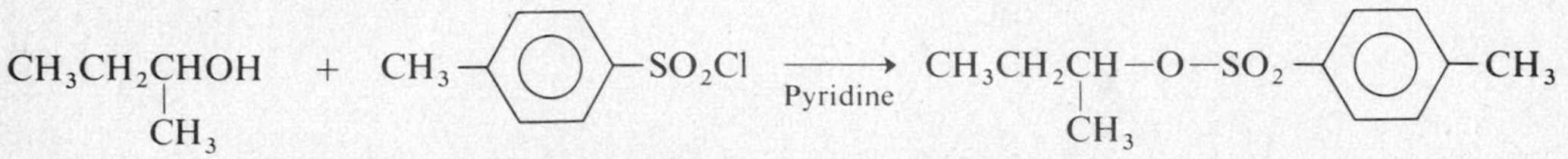

$$CH_3CH_2\underset{\displaystyle CH_3}{\underset{|}{C}}HOH + CH_3-C_6H_4-SO_2Cl \xrightarrow[\text{Pyridine}]{} CH_3CH_2\underset{\displaystyle CH_3}{\underset{|}{C}}H-O-SO_2-C_6H_4-CH_3$$

p-Toluenesulfonyl chloride (Tosyl chloride) **2-Butyl p-toluenesulfonate (2-Butyl tosylate)**

22.11 Reactions Involving the R—OH Bond of Alcohols

22.11a. NUCLEOPHILIC SUBSTITUTION REACTIONS. The hydroxyl function of an alcohol is a very poor leaving group and can be displaced by nucleophiles only if the oxygen carries a positive charge, *e.g.*, by protonation or by association with a Lewis acid. Typical examples of nucleophilic displacements involving alcohols include the reactions with halogen acids to form alkyl halides (see Figs. 10.15 and 21.5) and with aromatic compounds in the presence of Lewis acids such as aluminum chloride (Friedel–Crafts alkylation; see Fig. 8.13) to form alkylbenzenes. An intramolecular example of the latter process is illustrated in Fig. 22.22.

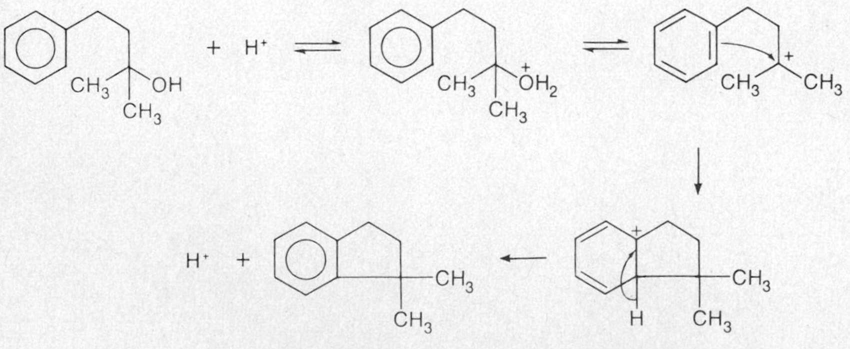

Fig. 22.22. Intramolecular Friedel–Crafts alkylation reaction with an alcohol as the starting material.

The Ar—OH bond is considerably stronger than the R—OH bond, and even protonation or complexation with aluminum chloride is insufficient to promote nucleophilic displacement reactions in phenols.

PROBLEM 22-13

The action of sulfuric acid on 4-phenylbutanol yields tetralin and 1-methylindan. Explain.

Tetralin 1-Methylindan

22.11b. Dehydration reactions. Alcohols can be dehydrated to alkenes by the action of strong acids, usually sulfuric acid or *p*-toluenesulfonic acid, or by conversion to an acetate or xanthate ester followed by pyrolysis of the ester. The acid-catalyzed dehydration is the reverse of the acid-catalyzed addition of water to alkenes (see Fig. 8.4) and involves carbonium ion intermediates. The process, therefore, is prone to yield rearranged products, although this circumstance can sometimes be used to advantage, as illustrated by the synthesis of phenanthrene that is outlined in Fig. 22.23.

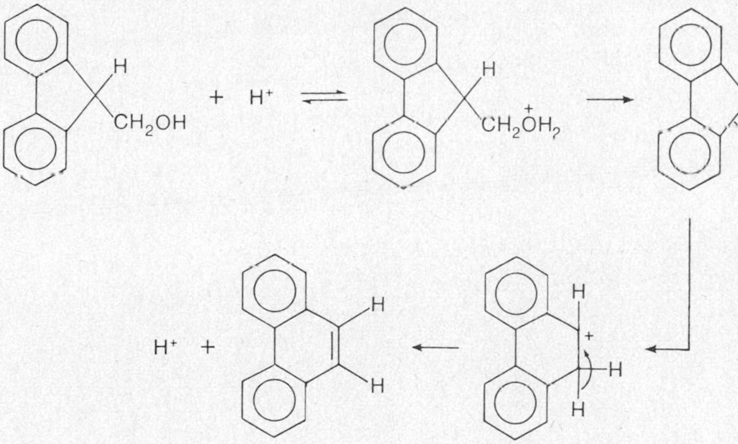

H CH_2OH + H^+ ⇌ H $CH_2\overset{+}{O}H_2$ → H $\overset{+}{C}H_2$

H^+ + H H ← H + H H

Fig. 22.23. Acid-catalyzed dehydration of an alcohol, occurring with skeletal rearrangement.

Dehydration of alcohols *via* pyrolysis of their acetate or xanthate esters involves a thermal elimination reaction. It is thought to take place through six-membered ring-activated complexes of the type illustrated in Fig. 22.24, in which the breaking and making of bonds takes place in a completely concerted fashion without the intervention of intermediates between the reactant and the product. The stereochemical aspects and the regioselectivity of these reactions are discussed in Chapter 25 on the alkenes.

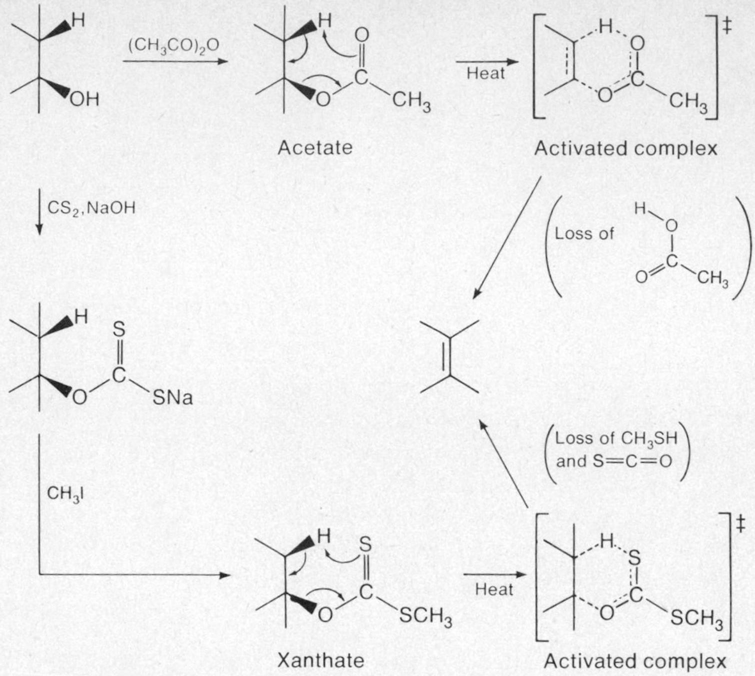

Fig. 22.24. Dehydration of alcohols by pyrolysis of acetate and xanthate esters.

22.11c. OXIDATION REACTIONS. As shown in Fig. 10.18, sodium dichromate mixed with sulfuric acid can oxidize primary alcohols to aldehydes (and, ultimately, to carboxylic acids), and can oxidize secondary alcohols to ketones. Studies of this reaction show that the rate-determining oxidation step involves a water molecule acting as a Lewis base to abstract the hydrogen attached to the hydroxyl carbon with departure of the chromium-containing moiety (Fig. 22.25).

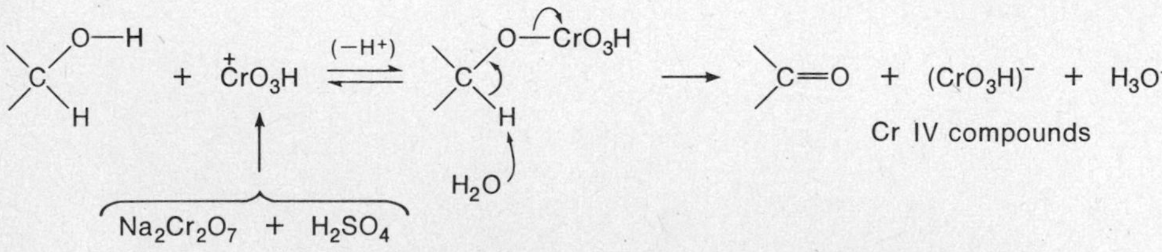

Fig. 22.25. Mechanism of the chromic acid oxidation of alcohols to aldehydes and ketones.

That the hydrogen abstraction is the rate-determining step is indicated by the six-fold reduction in the rate of oxidation that is observed when the deuterio compound (*i.e.*, R_2CDOH) is used in place of the proteo compound (*i.e.*, R_2CHOH), the ratio of k_H/k_D being approximately 6.

Numerous reagents have been tested for the selective oxidation of primary alcohols to aldehydes. Among the more successful of these are (a) a mixture of chromium trioxide and pyridine (known as *Sarrett's reagent*) and potassium ferrate (Fig. 22.26). Both are fairly situselective, having little effect on C=C,

$$C_6H_{13}CH_2OH + CrO_3 \cdot 2\,C_5H_5N \xrightarrow[25^\circ]{CH_2Cl_2} C_6H_{13}CHO$$

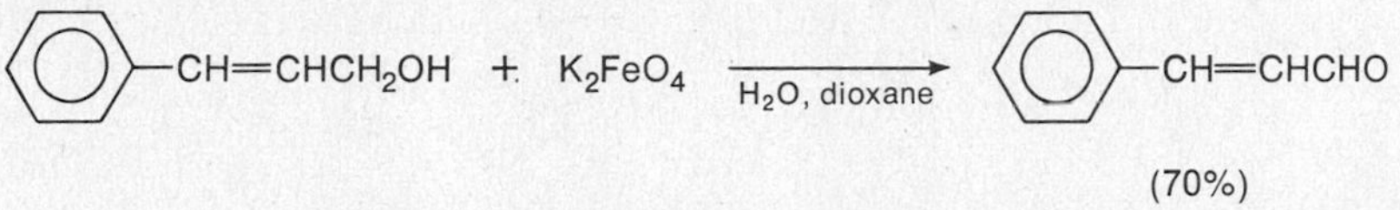

$$C_6H_5\text{—}CH{=}CHCH_2OH + K_2FeO_4 \xrightarrow{H_2O,\ \text{dioxane}} C_6H_5\text{—}CH{=}CHCHO$$

(70%)

Fig. 22.26. Oxidation of primary alcohols to aldehydes.

C≡C, or Ar bonds, although they do oxidize primary and secondary amines, thiols, and sulfides. These reagents also oxidize secondary alcohols to ketones, but it usually is unnecessary to resort to them because oxidation by means of chromate suffices. A number of recipes for chromate oxidations have been used, a typical one consisting of chromium trioxide in aqueous sulfuric acid, which is added to an acetone solution of the secondary alcohol (*Jones' reagent*):

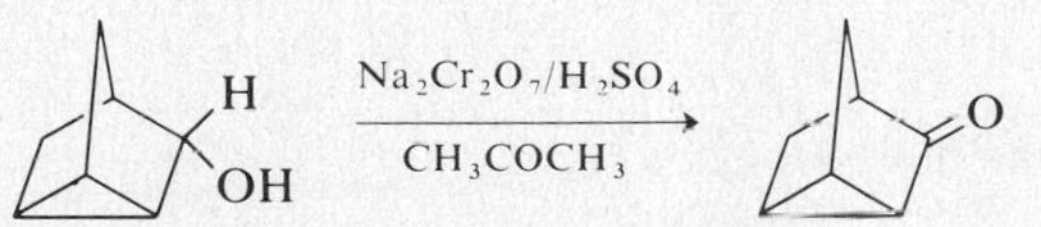

$$\text{(bicyclic secondary alcohol, H, OH)} \xrightarrow[CH_3COCH_3]{Na_2Cr_2O_7/H_2SO_4} \text{(bicyclic ketone, =O)}$$

(In acetone solution)

A mild, situselective reagent capable of oxidizing primary or secondary allyl or benzyl alcohols to the corresponding aldehydes or ketones is manganese dioxide suspended in an inert hydrocarbon solvent.

$$CH_3CH{=}CHCH{=}CH\overset{CH_3}{\overset{|}{C}}{=}CHCH_2OH \xrightarrow[\text{Petroleum ether, 4 hr, }20^\circ]{MnO_2} CH_3CH{=}CHCH{=}CH\overset{CH_3}{\overset{|}{C}}{=}CHC(=O)H \quad 80\%$$

A highly situselective procedure for converting primary alcohols to aldehydes or secondary alcohols to ketones is the treatment of the alcohols with a ketone such as acetone in the presence of aluminum *t*-butoxide. The method, known as the *Oppenauer oxidation*, is mechanistically identical with the Meerwein–Pondorff–Verley reduction (see Fig. 22.15). The difference is simply that our focus of attention is now on the alcohol undergoing oxidation (at the expense of the acetone) rather than on the ketone (*i.e.*, acetone in this case) undergoing reduction. Carried out as an oxidation procedure, the reaction is pushed in the direction of product formation by the use of a large excess of acetone (Fig. 22.27).

PROBLEM 22-14

Why is aluminum *t*-butoxide preferable to aluminum isopropoxide in the Oppenauer oxidation?

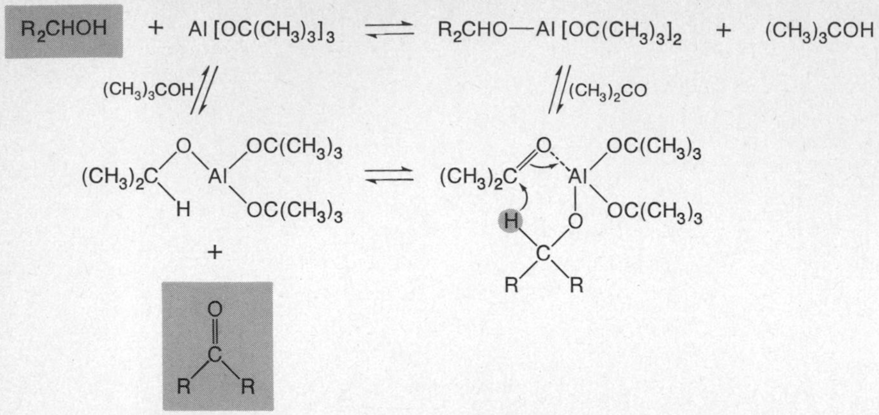

$$R_2CHOH + Al[OC(CH_3)_3]_3 \rightleftharpoons R_2CHO{-}Al[OC(CH_3)_3]_2 + (CH_3)_3COH$$

Fig. 22.27. Oxidation of alcohols by the Oppenauer method.

The Oppenauer method is more successful for the preparation of ketones than aldehydes, for the aldehydes that are formed are often susceptible to aldol condensation (see Section 13.6) under the basic conditions of the reaction. The Oppenauer oxidation is particularly useful when the reactant contains acid-sensitive functions that could not survive the strongly acidic conditions of the acid chromate methods.

22.12 Reactions Involving the Aromatic Ring of Phenols

The hydroxyl group of phenols is *o,p*-directing in electrophilic substitution reactions and is strongly activating. With weak electrophiles, such as aryldiazonium cations, monosubstitution of phenol takes place at the *para* position, but with more reactive electrophiles, such as nitronium and halogen cations, di- and tri- substitution tends to occur (Fig. 22.28). Polysubstitution can be minimized by

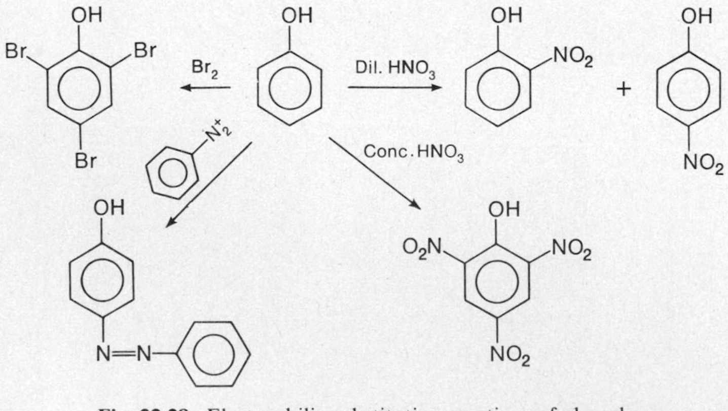

Fig. 22.28. Electrophilic substitution reactions of phenols.

using the acetate ester of the phenol and/or by using milder conditions (*i.e.*, dilute nitric acid at room temperature for nitration).

The stability of the phenoxide ion is interpreted in terms of a charge-delocalized resonance hybrid, in which the negative charge is concentrated not only on the oxygen but at the *ortho* and *para* carbons as well;

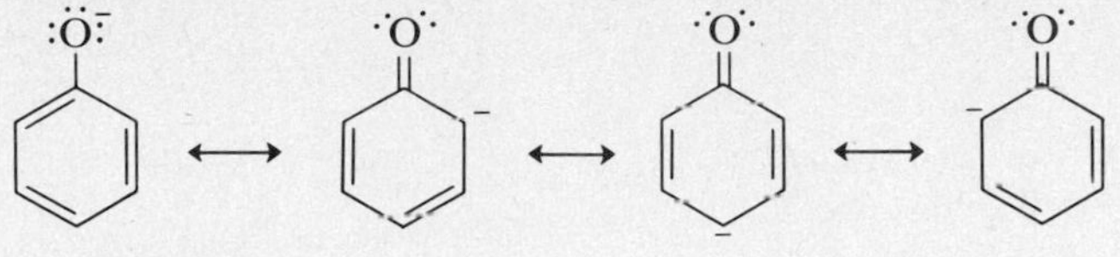

The phenoxide ion, therefore, should have the ability to react both at oxygen and at carbon as a nucleophile (it is an ambident anion), which is precisely what is observed. For example, reactions with alkyl halides may proceed either via the O-alkylation pathway to give alkyl aryl ethers (see Fig. 11.6) or via the C-alkylation pathway to give *o*-substituted phenols, depending on the reaction conditions (Fig. 22.29). In polar aprotic solvents, such as acetone and dimethylsulfoxide,

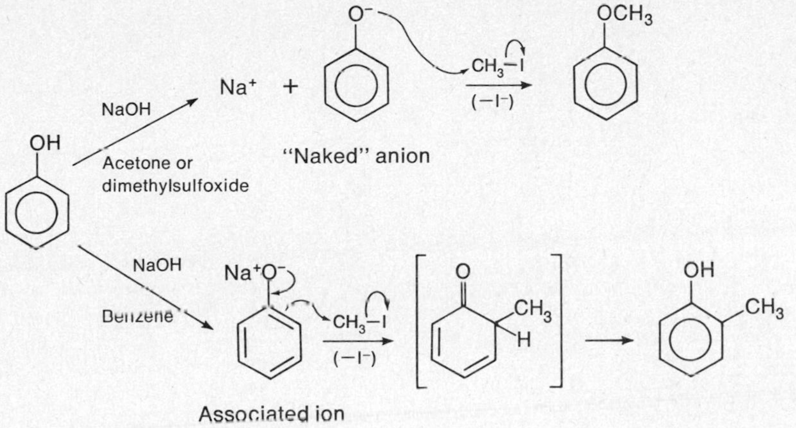

NaOH; Acetone or dimethylsulfoxide; Na^+ +; "Naked" anion; CH_3—I; $(-I^-)$; OCH_3

NaOH; Benzene; Na^+O^-; CH_3—I; $(-I^-)$; Associated ion; CH_3; H; OH; CH_3

Fig. 22.29. Illustration of the O and C alkylation of phenoxide ion.

O-alkylation is favored, whereas in nonpolar solvents, such as benzene, C-alkylation is favored. This difference is ascribed to differences in the degree to which the phenoxide ion is associated with the accompanying cation (*e.g.*, Na^+). In nonpolar solvents, the anion and cation are closely associated, with the result that the anion is less reactive than in polar, aprotic solvents which can better sustain charge separation but which do not solvate the anion (see p. 451). It is then argued that the more reactive the anion, the more likely it is to attach itself to the point of highest electron density (*i.e.*, the oxygen). The reasons for this, however, are still somewhat obscure, and the behavior of ambident anions such as phenoxide is an area of current and active study.

Several other synthetically useful reactions of phenols also involve phenoxide ion as a carbon nucleophile, including reactions with chloroform (Fig. 22.30), aldehydes (Fig. 22.31), and carbon dioxide (Fig. 22.32). When a solution of phenol in chloroform is treated with aqueous sodium hydroxide, *o*-hydroxybenzaldehydes are produced (*Reimer–Tiemann reaction*) by the pathway outlined in Fig. 22.30 in which the dichloromethyl group is converted to a formyl group by nucleophilic displacement of one of the chlorine atoms by a hydroxyl group

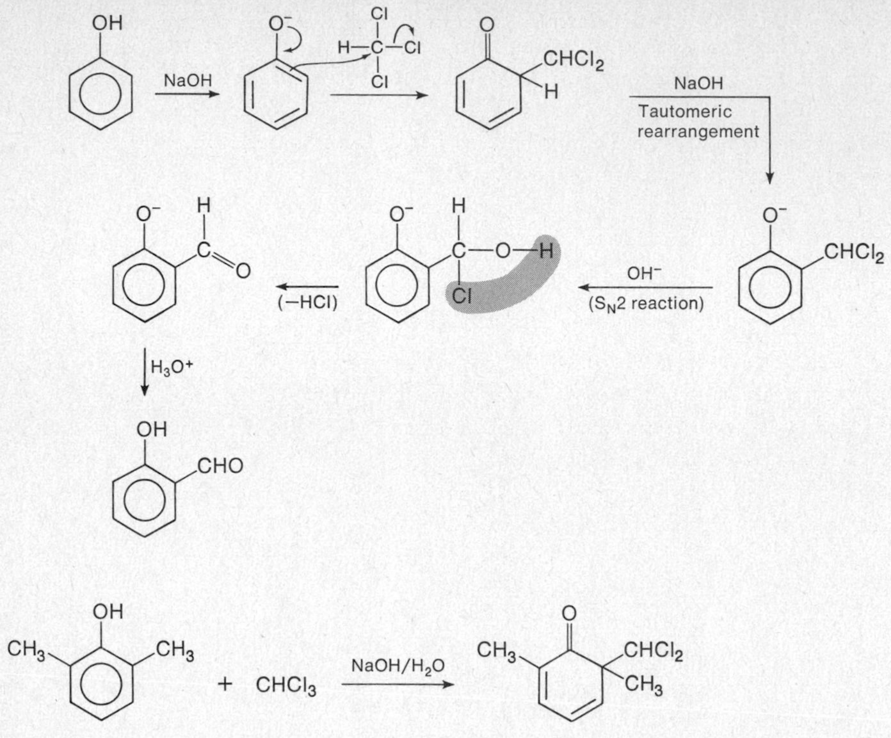

OH NaOH O⁻ H—C—Cl Cl Cl CHCl2 H NaOH Tautomeric rearrangement

O⁻ H C=O (−HCl) O⁻ H C—O—H Cl OH⁻ (S_N2 reaction) O⁻ CHCl2

H_3O^+

OH CHO

OH CH_3 CH_3 + $CHCl_3$ $NaOH/H_2O$ O CH_3 $CHCl_2$ CH_3

Fig. 22.30. Formation of *o*-hydroxybenzaldehyde via the Reimer–Tiemann reaction.

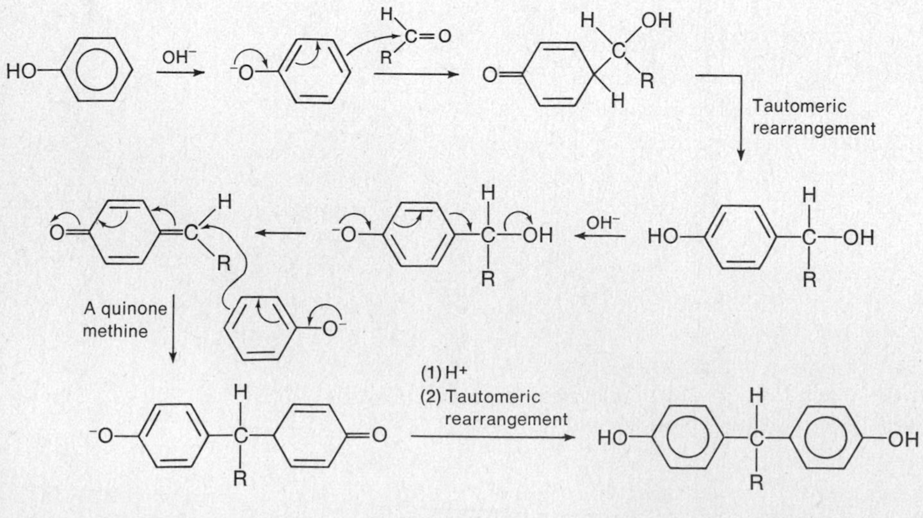

HO OH⁻ ⁻O H C=O R O= H OH C R H Tautomeric rearrangement

O= H C R ⁻O H C OH R OH⁻ HO H C—OH R

A quinone methine O⁻

(1) H^+
(2) Tautomeric rearrangement

⁻O H C R =O HO H C R OH

Fig. 22.31. Formation of a diphenylmethane in the base-catalyzed reaction of phenol with an aldehyde.

followed by loss of hydrogen chloride from the chloroalcohol. The *o*-substituted product is preferred to the extent that even if both *ortho* positions of the phenol are blocked, the product is a cyclohexadienone that is incapable of tautomerizing to a phenol.

Many phenols react with aldehydes in the presence of base via a nucleophilic addition process to form *p*-hydroxybenzyl alcohols:

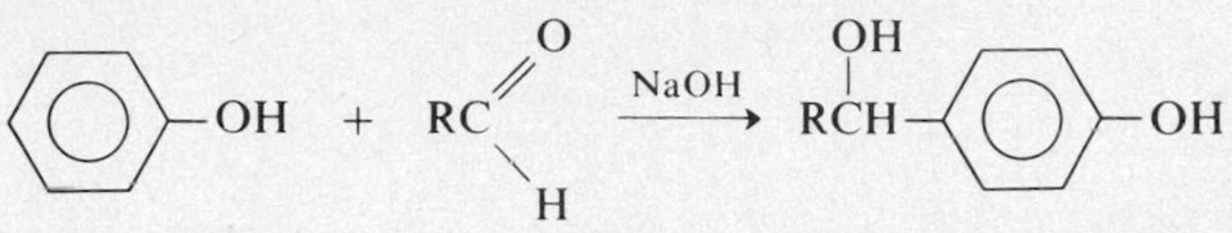

$$C_6H_5\text{-}OH + RCHO \xrightarrow{NaOH} RCH(OH)\text{-}C_6H_4\text{-}OH$$

Under vigorous conditions or with particularly reactive aldehydes such as formaldehyde, the reaction usually progresses further to form diphenylmethanes (Fig. 22.31). This can be interpreted as the result of a base-induced elimination of water from the initially formed hydroxymethyl compound to form a quinonemethine intermediate, which subsequently undergoes a 1,6-conjugate addition of phenoxide to yield the product. In the case of phenol and formaldehyde, this process can be repeated many times over, with the result that a high molecular weight, highly cross-linked polymer known as Bakelite is formed.

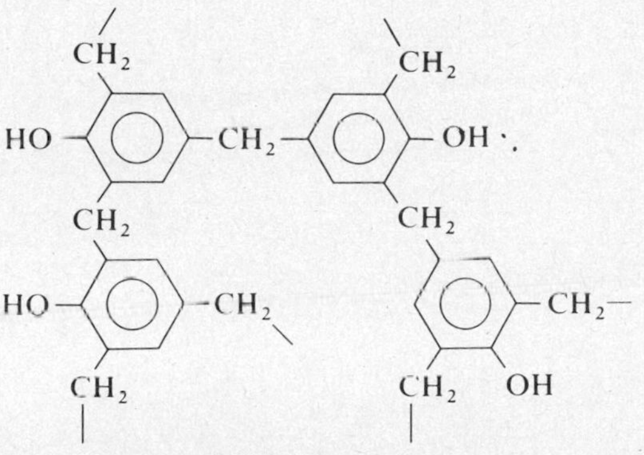

CH₂ CH₂ HO—⟨ ⟩—CH₂—⟨ ⟩—OH CH₂ CH₂ HO—⟨ ⟩—CH₂ ⟨ ⟩—CH₂— CH₂ CH₂ OH

Phenol-formaldehyde polymer (Bakelite)

Thus, the process that we first met in skeleton form when we discussed a case history of an organic reaction in Section 1.4 now acquires flesh; we can understand it not only in structural detail, something that escaped its discoverer Adolph von Baeyer, but also in mechanistic detail, something that escaped its exploiter Leo Baekeland.

PROBLEM 22-15

p-Methylphenol and formaldehyde react in the presence of sodium hydroxide to form a cyclic tetramer. Suggest a structure for the product.

Phenoxide nucleophiles are also capable of adding to carbon dioxide. At room temperature, the phenoxide acts as an oxygen nucleophile to give the very

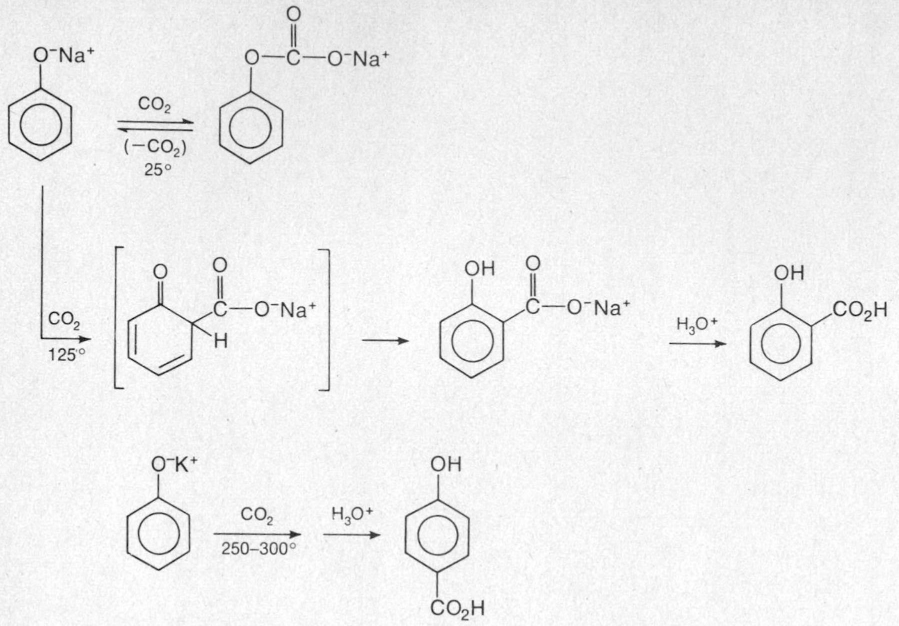

Fig. 22.32. Reaction of phenoxide with carbon dioxide.

unstable sodium salt of phenylcarbonate; at higher temperatures, however, the phenoxide acts as a carbon nucleophile, yielding the *ortho*-carboxylation product at 125° and the *para*-carboxylation product at 250–300° (Fig. 22.32).

Phenol itself is sufficiently nucleophilic to react with carbonyl compounds in the presence of acids strong enough to convert the carbonyl compounds to the oxozonium salts. For example, phenol reacts with acetone to form a compound

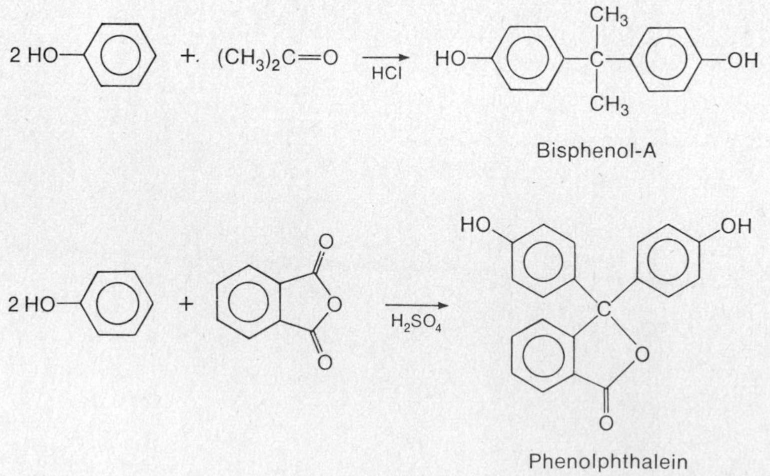

Bisphenol-A

Phenolphthalein

Fig. 22.33. Acid-catalyzed condensations of phenol with carbonyl compound.

known as bisphenol-A and with phthalic anhydride to form a compound known as phenolphthalein (Fig. 22.33). Phenolphthalein is noteworthy for its ability to act as an acid-base indicator, existing as a colorless, neutral form below pH 10 and a red, dianionic form above pH 10. At very high pH it is converted to a colorless, trianionic form:

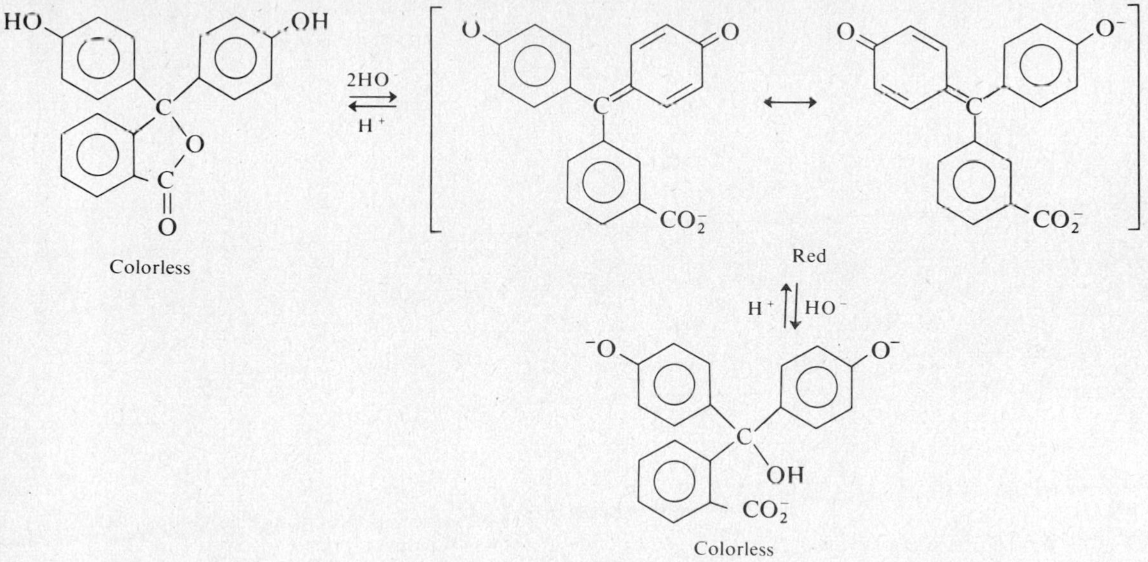

Colorless

Red

Colorless

22.13 Reactions of 1,2-Diols

22.13a. PINACOL REARRANGEMENT. A particularly well-studied carbonium ion rearrangement involves 1,2-dihydroxy compounds referred to as pinacols (for preparation, see Section 22.9). The simplest example is the acid-catalyzed conversion of pinacol itself to *t*-butyl methyl ketone (Fig. 22.34). Although conveniently depicted in terms of carbonium ion intermediates, there is considerable evidence that this reaction is actually a concerted process. The structural outcome of a pinacol rearrangement is determined by three considerations: (a) the

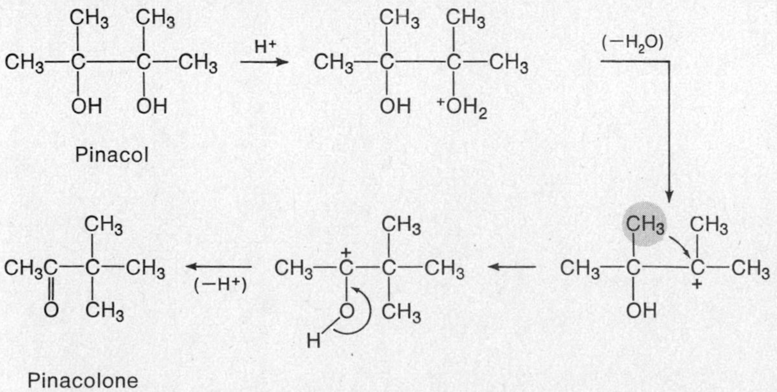

Pinacol

Pinacolone

Fig. 22.34. Pinacol rearrangement of 2,3-dimethyl-2,3-butanediol (pinacol) to *t*-butyl methyl ketone.

site of carbonium ion formation; (b) the migratory aptitude of the neighboring groups (hydrogen, alkyl, aryl); and (c) the stereochemical requirements of the reaction.

The carbonium ion forms at the site at which the cation is better stabilized. For example, 2,2-diphenyl-1,2-ethanediol undergoes acid-catalyzed rearrangement to diphenylacetaldehyde via a resonance-stabilized 1,1-diphenyl-2-hydroxyethyl cation rather than the less stable 2,2-diphenyl-2-hydroxyethyl cation (Fig. 22.35).

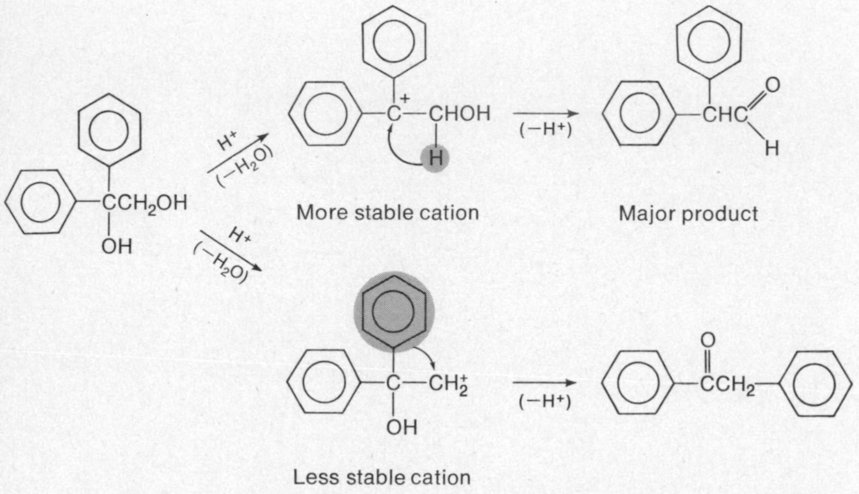

More stable cation

Major product

Less stable cation

Fig. 22.35. Pinacol rearrangement of 2,2-diphenyl-1,2-ethanediol.

The migratory aptitudes of aryl groups generally are greater than those of alkyl groups. Thus, the diol obtained by reductive dimerization of acetophenone rearranges with the migration of the phenyl group rather than the methyl group to yield 3,3-diphenyl-2-butanone as the major product;

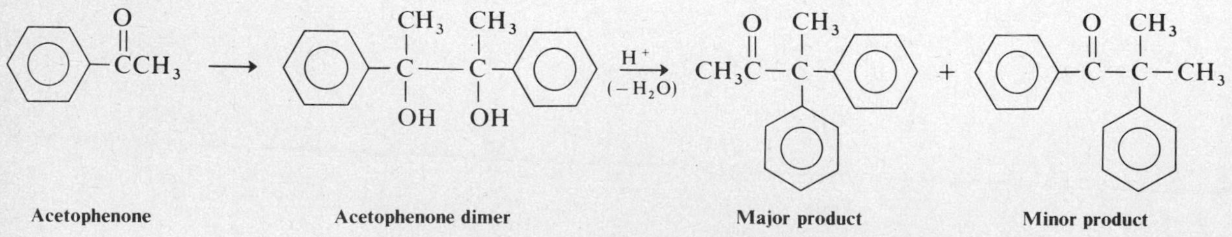

Acetophenone **Acetophenone dimer** **Major product** **Minor product**

The migratory aptitude of hydrogen in the pinacol rearrangement is more difficult to assess, for in some cases hydrogen migrates less rapidly and in other cases more rapidly than alkyl groups.

Migratory aptitudes, arising from the electronic differences of groups, are superseded, however, by the requirement that the migrating and leaving groups be *anti* to each other (*i.e.*, antiperiplanar conformation). This is illustrated in the

anti relationship between R and $^+OH_2$

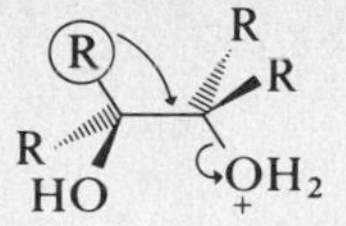

reaction of nitrous acid with a pair of epimers of 2-amino-4-*t*-butylcyclohexanol, the amino function in this case serving as a leaving group in place of a second hydroxyl group (Fig. 22.36). In the epimer in which the hydroxyl and amino

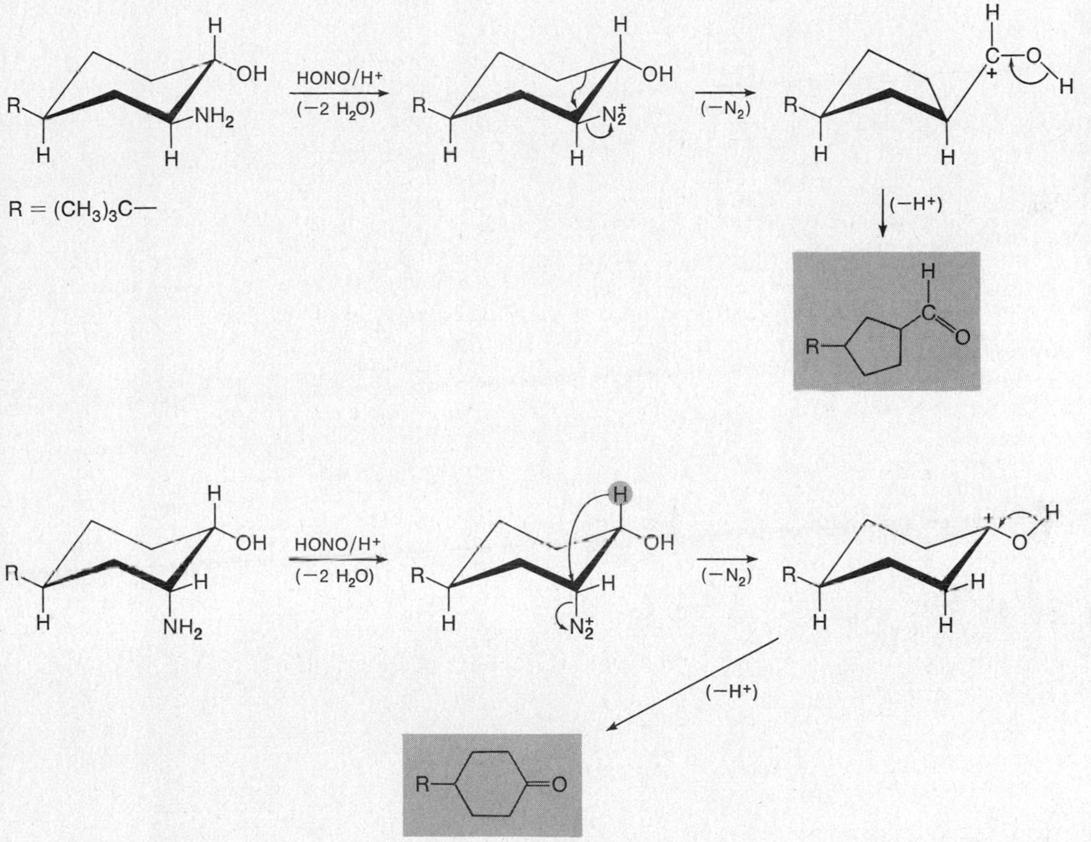

R = $(CH_3)_3C$—

Fig. 22.36. The effect of stereochemical features on the course of a pinacol-type rearrangement.

functions are both equatorial, the group that is *anti* to the amino function is a C—C bond of the ring; migration of a ring carbon takes place as a result, and a ring-contracted cyclopentanecarboxaldehyde is produced. In the epimer in which the hydroxyl function is equatorial and the amino function is axial, the group that is *anti* to the amino function is a C—H bond; migration of a hydrogen takes place as a result, and a cyclohexanone is produced.

PROBLEM 22-16

The stereoisomer of 2-amino-4-*t*-butylcyclohexanol in which the three substituents can be equatorially aligned can also exist in a conformation in which they are all axially aligned. What would be the expected product from the action of nitrous acid on this conformer? Why is this product not likely to form in high yield?

Pinacol rearrangements of 1,2-diols and pinacol-like rearrangements of 1-amino-2-hydroxy compounds are frequently applied in ring expansion and ring contraction reactions, where they find considerable synthetic application.

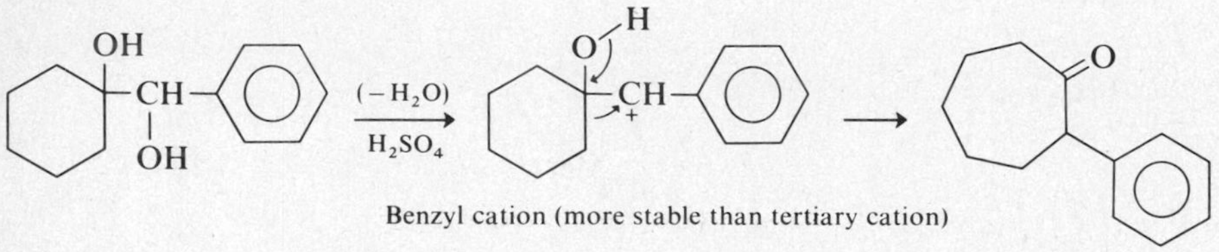

$(-H_2O)$, H_2SO_4

Benzyl cation (more stable than tertiary cation)

22.13b. OXIDATION REACTIONS. Vicinal glycols (1,2-diols) are susceptible to oxidative cleavage. Periodic acid and lead tetraacetate are particularly useful reagents for effecting the cleavage, because aldehydes, if they are formed as primary oxidation products, do not undergo further oxidation to the carboxylic acids with these reagents (Fig. 22.37).

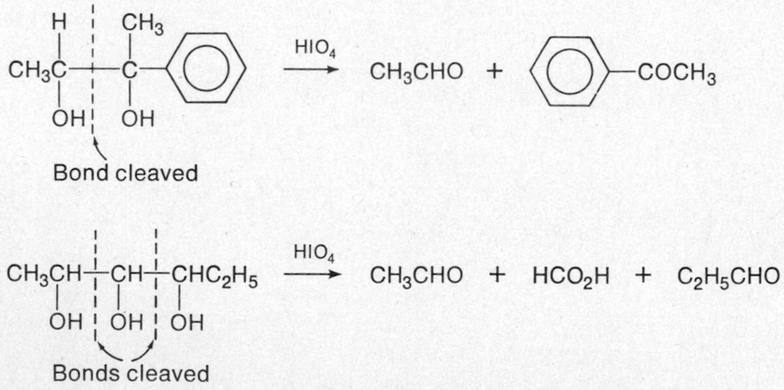

$CH_3C(H)(OH)$–$C(CH_3)(OH)C_6H_5$ $\xrightarrow{HIO_4}$ CH_3CHO + $C_6H_5COCH_3$

Bond cleaved

$CH_3CH(OH)$–$CH(OH)$–$CH(OH)C_2H_5$ $\xrightarrow{HIO_4}$ CH_3CHO + HCO_2H + C_2H_5CHO

Bonds cleaved

Fig. 22.37. Oxidative cleavage of 1,2-diols and 1,2,3-triols with periodic acid.

The periodate and lead tetraacetate oxidative cleavage reactions both are thought to involve cyclic intermediates, which break down with scission of the C—C bond and concomitant transformation of the hydroxyl groups to carbonyl groups (Fig. 22.38).

The choice between periodic acid and lead tetraacetate as the oxidant is usually made on the basis of the solubility characteristics of the material being oxidized; for water-soluble compounds, periodic acid is the better reagent, whereas for compounds that are soluble in organic solvents lead tetraacetate is the reagent of choice. Because most of the members of the carbohydrate family

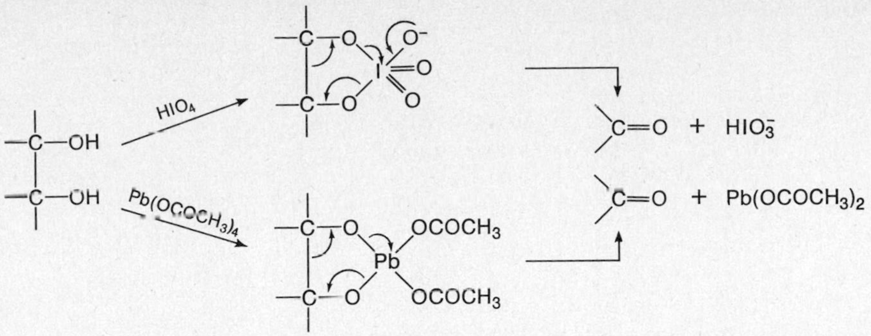

Fig. 22.38. Mechanism of the oxidative cleavage of 1,2-diols with periodic acid and lead tetraacetate.

are water soluble, periodic acid is extensively used in the structure elucidation of these compounds (see Fig. 32.18). For example, methyl glucoside (a monomethyl ether of the sugar glucose) consumes 2 mole-equivalents of periodic acid to yield formic acid and a dialdehyde as products (Fig. 22.39).

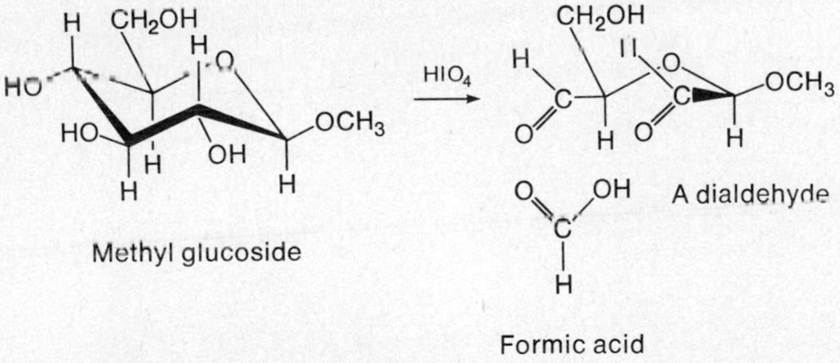

Fig. 22.39. Oxidative cleavage of methyl glucoside with periodic acid.

A number of other 1,2 difunctional compounds undergo oxidative cleavage when treated with periodic acid or lead tetraacetate, including compounds in which $-NH_2$, $-OH$, or $>C{=}O$ groups are present in any of the possible combinations; *i.e.*, α-aminoketones, 1-amino-2-hydroxy compounds, α-hydroxyaldehydes, α-diketones, *etc.*,

$$R-\underset{R}{\overset{OH}{C}}-\overset{O}{\overset{\|}{C}}-R \xrightarrow{HIO_4} R_2C{=}O + HO-\overset{O}{\overset{\|}{C}}-R$$

Table 22.2 summarizes the reactions of alcohols and phenols.

Table 22.2 REACTIONS OF ALCOHOLS AND PHENOLS

Type of Reaction	Starting Materials	Reagents	Products	Comments
Reactions of RO—H Bond:	ROH	Alkali metals (*e.g.*, Na, K, Li)	RO-Metal (*e.g.*, RONa, ROK, ROLi)	Hydrolyzed back to ROH
		R′COCl and $(R'CO)_2O$	$R'CO_2R$	Order of reactivity: primary > secondary > tertiary
		$R'CH{=}C{=}O$	$R'CH_2CO_2R$	
		$ArSO_2Cl$	$ROSO_2Ar$	
		$R'R''C{=}O$	$R'R''C(OH)(OR)$	Products usually unstable except in special cases
	RO^-	R′X	ROR′	See Table 21.7
Reactions of R—OH Bond:	ROH	HX, PX_3, PX_5, SOX_2	RX	Order of reactivity: tertiary > secondary > primary Restricted to X = Cl, Br, and I
		$R'R''C{=}O$ + acid catalyst	$R'R''C(OR)_2$	Products stable to base but easily converted to starting materials by aqueous acid
	$-\overset{\mid}{\underset{H}{C}}-\overset{\mid}{\underset{\mid}{C}}-OH$	Acid catalysts	$>C{=}C<$	Highly susceptible to molecular rearrangement
	RCH_2OH	$Na_2Cr_2O_7 + H_2SO_4$, $CrO_3{\cdot}2C_5H_5N$, or Oppenauer oxidation	RCO_2H RCHO	Via RCHO
	R_2CHOH	$Na_2Cr_2O_7 + H_2SO_4$, $CrO_3{\cdot}2C_5H_5N$, or Oppenauer oxidation	$R_2C{=}O$	
	$>C{=}\underset{\mid}{C}-\overset{H}{\underset{\mid}{C}}-OH$	K_2FeO_4 or MnO_2	$>C{=}\underset{\mid}{C}-\underset{\mid}{C}{=}O$	Specific for allyl and benzyl alcohols
	$Ar\overset{H}{\underset{\mid}{C}}-OH$	K_2FeO_4 or MnO_2	$Ar-\underset{\mid}{C}{=}O$	
Reactions of 1,2-Diols	$R-\overset{OH}{\underset{R}{C}}-\overset{OH}{\underset{R}{C}}-R$	Acid catalysts	$R_3CC(=O)R$	Migratory aptitudes are stereochemically dependent
		HIO_4 or $Pb(OAc)_4$	$R_2C{=}O + O{=}CR_2$	Specific for 1,2-diols
		$R'R''C{=}O$	$R'R''C$ bonded through two O atoms to R_2C-CR_2 (cyclic ketal)	Products stable to base but easily converted to starting materials by aqueous acid

Table 22.2—*cont.*

Type of Reaction	Starting Materials	Reagents	Products	Comments
Reactions of Phenols.	 phenol (OH)	Electrophiles (e.g., X^+, NO_2^+, SO_3H^+, R^+, RCO^+, ArN_2^+)	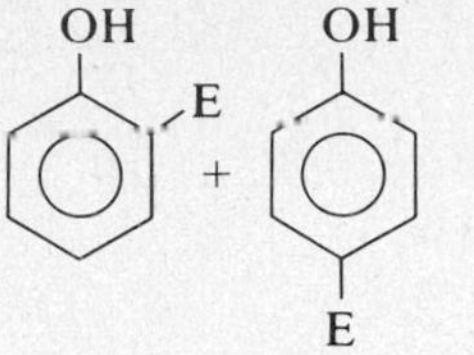 o-E-phenol + p-E-phenol	Entering electrophilic groups go *ortho* and *para* to hydroxyl group
	 phenoxide (O^-)	(1) RX; (2) H^+	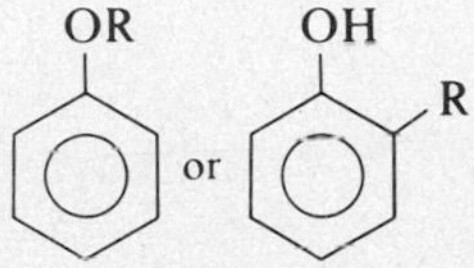 OR-benzene or o-R-phenol	O-alkylation or C-alkylation dependent on R and reaction conditions
		(1) $CHCl_3$, NaOH; (2) H^+	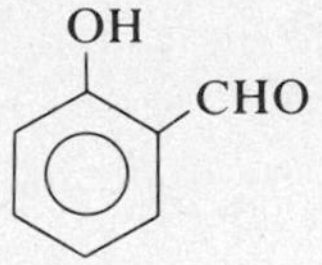 o-CHO-phenol	Reimer–Tiemann reaction
		R_2CO	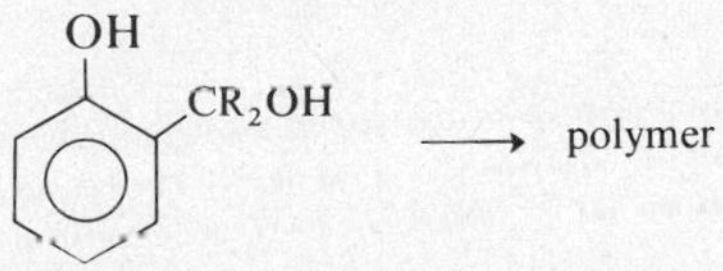 o-CR_2OH-phenol ⟶ polymer	Difficult to terminate at monomer stage; polymer forms readily when R=H
		(1) CO_2; (2) H^+	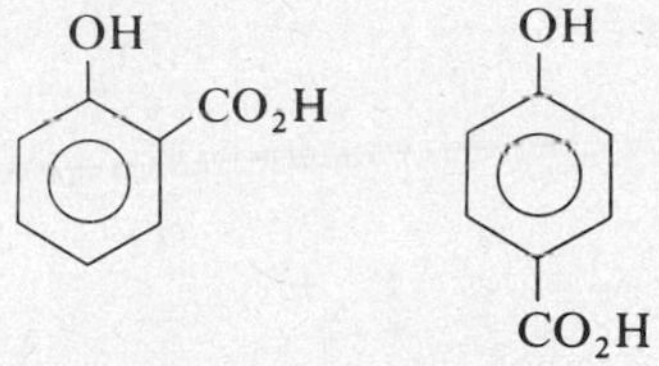 o-CO_2H-phenol, p-CO_2H-phenol	*o*-Product formed at lower temperature; *p*-product formed at higher temperature

22.14 *Study Guide*

I. Methods of preparation of alcohols and phenols
   A. Conversion of C—H to C—OH—very limited in application
   B. Conversion of C—Halogen to C—OH
      1. Hydrolysis of alkyl halides—S_N1 and S_N2 pathways
      2. Two-step sequence (RX → ROAc → ROH) to avoid E1 and E2 reactions
      3. Oxidation of Grignard reagents
      4. Hydrolysis of aryl halides
         (a) With aryl halides substituted in *o*- and *p*-positions with strongly electron-withdrawing groups
         (b) Via benzyne intermediates
   C. Conversion of C—Nitrogen to C—OH
      1. Action of nitrous acid on alkylamines—yields alcohols and other products; generally not a good method for the preparation of alcohols

2. Conversion of arylamines to aryldiazonium compounds followed by hydrolysis—yields phenols; a good preparative method

D. Conversion of C—Sulfur to C—OH
1. Action of NaOH—KOH on arylsulfonic acids
2. Limited to compounds containing groups that can withstand the strenuous conditions of the reaction

E. Conversion of C—C to C—OH; action of peracids on aldehydes and ketones

F. Conversion of C=C to C—C—OH, HO—C—C—OH, and related compounds
1. Acid-catalyzed addition of water; Markovnikov regioselectivity
2. Hydroxymercuration-reduction method; Markovnikov regioselectivity, no rearrangement, non-stereoselective
3. Hydroboration-oxidation method; anti-Markovnikov regioselectivity, *syn* addition
4. Hydroboration-carbonylation procedures—adds one carbon to the original framework
5. Hydroformylation—not particularly useful as a laboratory method
6. Vicinal dihydroxylation of alkenes
   (a) Potassium permanganate—yields *cis* glycols via cyclic manganate ester
   (b) Osmium tetraoxide—yields *cis* glycols via cyclic osmate ester
   (c) Perbenzoic acid—yields *trans* glycols via oxirane
   (d) Iodine and silver acetate—yields *cis* or *trans* glycols
7. Acid-catalyzed addition of formaldehyde to alkenes

G. Conversion of C=O to C—OH by nucleophilic addition
1. Hydride reduction of carbonyl compounds
   (a) Meerwein–Pondorff–Verley reduction—hydride transfer from isopropyl alcohol as source of hydride in the presence of aluminum isopropoxide
   (b) Metal hydrides—lithium aluminum hydride, sodium borohydride
   (c) Diborane
2. Addition of organometallic reagents to carbonyl compounds
   (a) Products are alcohols (primary, secondary, or tertiary depending on the carbonyl compound)
   (b) Side reactions
      (1) Reduction—yields alcohol corresponding to carbonyl compound and the alkene corresponding to the Grignard reagent
      (2) Enolization—yields starting carbonyl compound and the alkane corresponding to the Grignard reagent

H. Conversion of C=O to C—OH by catalytic reduction

I. Conversion of C=O to HO—C—C—OH by bimolecular reduction; pinacol reduction

II. Reactions of alcohols and phenols

A. Reactions involving the RO—H bond or ArO—H bond

1. Reactions with metals and strong bases to give RO^- or ArO^-
2. Order of reactivity of alkanols: $CH_3OH > RCH_2OH >$ $> R_2CHOH > R_3COH$

B. Reactions involving the R—OH bond
1. —OH a poor leaving group; protonation to $-\overset{+}{O}H_2$ required for nucleophilic displacement to occur
2. Dehydration of alcohols
   (a) Acid-catalyzed—rearrangements likely to occur
   (b) Pyrolysis of acetate or xanthate esters—rearrangements do not occur
3. Oxidation of alcohols
   (a) Mechanism— C—H bond cleavage the rate-determining step
   (b) Reagents for oxidation of RCH_2OH to RCHO (without formation of RCO_2H)
   (c) Selective oxidation of allylic alcohols

C. Reactions involving the aromatic ring of phenols
1. Hydroxyl group is *o,p*-directing and activating in electrophilic substitution reactions
2. Reactions of phenoxide ion
   (a) C-alkylation *vs.* O-alkylation
   (b) Reaction with chloroform
   (c) Reaction with aldehydes and ketones (*e.g.*, formaldehyde to give Bakelite)
   (d) Reaction with carbon dioxide

D. Pinacol rearrangements of 1,2-diols
1. Controlling factors
   (a) Site of carbonium ion formation
   (b) Migratory aptitude of neighboring groups
   (c) Stereochemical requirements in constrained molecules (antiperiplanar conformation required)
2. Migratory aptitudes: aryl > alkyl > hydrogen
3. Pinacol-like rearrangements of 1-amino-2-hydroxy compounds

E. Oxidation reactions
1. Reagents: $Pb(OCOCH_3)_4$ for organic solvent-soluble compounds; HIO_4 for water-soluble compounds
2. Products: aldehydes and ketones

REFERENCES

COFFEY, S., ed., *Rodd's Chemistry of Carbon Compounds*, 2nd ed., Vols. IB, ID, IF, Elsevier Publishing Co., New York, 1965–67.

FIESER, L. F., and M. FIESER, *Advanced Organic Chemistry*, Reinhold Publishing Corp., New York, 1961, Chapter 22.

MONICK, J. A., *Alcohols—Their Chemistry, Properties, and Manufacture*, Reinhold Publishing Corp., New York, 1968.

PATAI, S., ed., *The Chemistry of the Ether Linkage*, Interscience Publishers, New York, 1967.

———, *The Chemistry of the Hydroxyl Group*, Parts 1 and 2, John Wiley & Sons, Inc., New York, 1971.

PROBLEMS

1. Draw the structure of the major product from each of the following reactions:

(a) $HCO_2C_2H_5 + C_6H_5MgBr \longrightarrow$

(b) 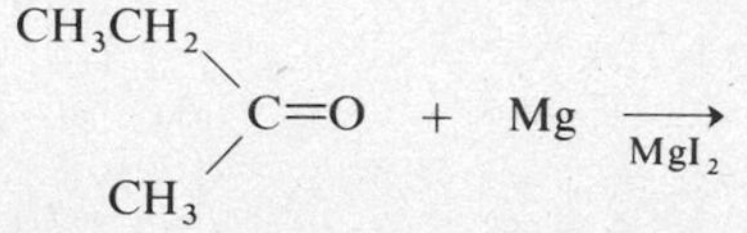 CH_3CH_2, CH_3 C=O + Mg $\xrightarrow[MgI_2]{}$

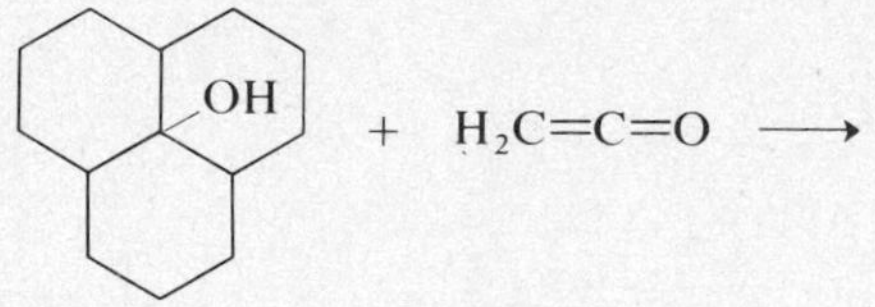

(c) OH + $H_2C{=}C{=}O \longrightarrow$

(d) 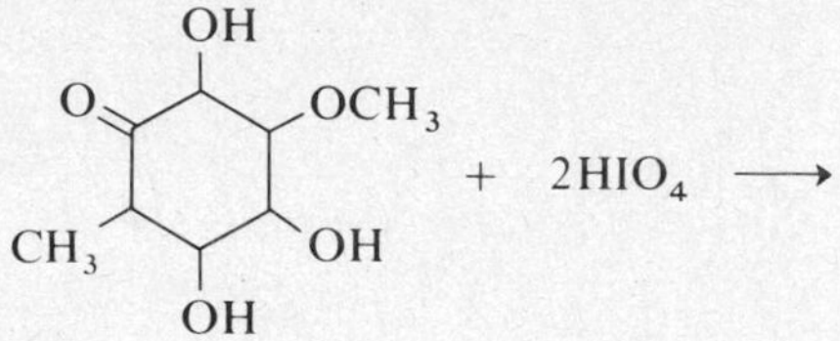 OH, O, OCH_3, CH_3, OH, OH + $2HIO_4 \longrightarrow$

(e) 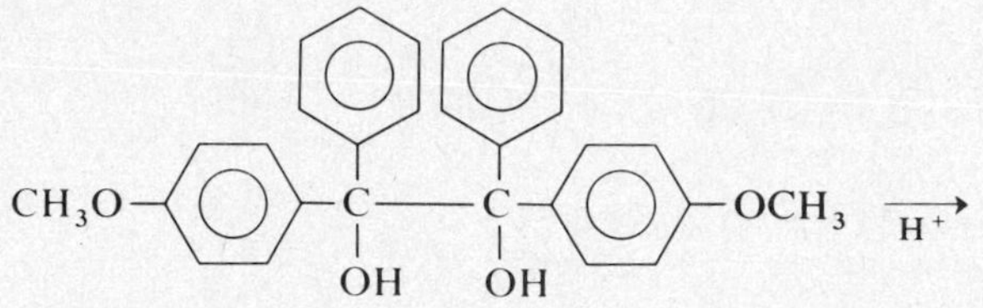 CH_3O—C—C—OCH_3 (OH, OH) $\xrightarrow[H^+]{}$

(f) 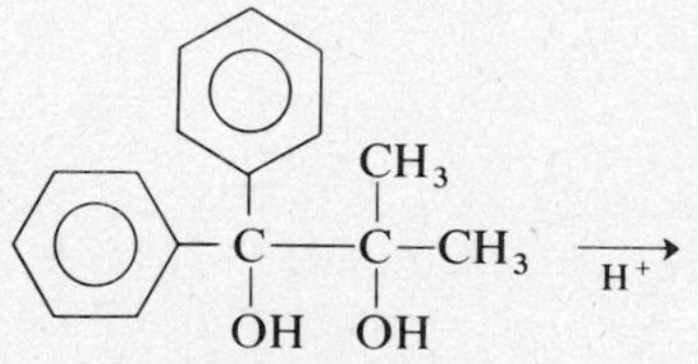 CH_3, C—C—CH_3 (OH, OH) $\xrightarrow[H^+]{}$

(g) 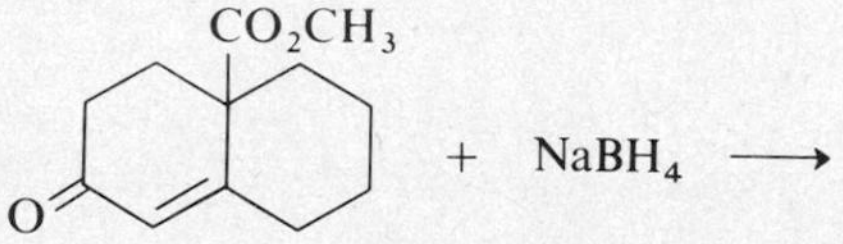 CO_2CH_3, O + $NaBH_4 \longrightarrow$

(h) D, CH_3 $\xrightarrow{OsO_4}$ $\xrightarrow[NaHSO_3]{H_2O}$

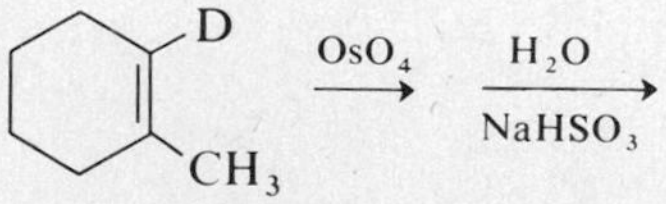

(i) 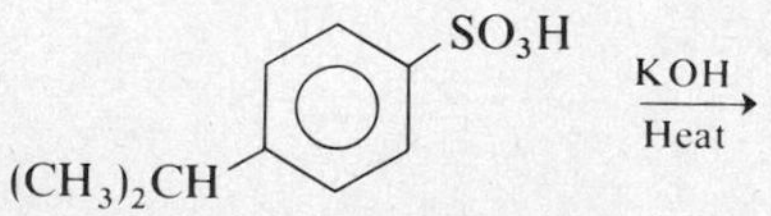 SO_3H, $(CH_3)_2CH$ $\xrightarrow[Heat]{KOH}$

(j) 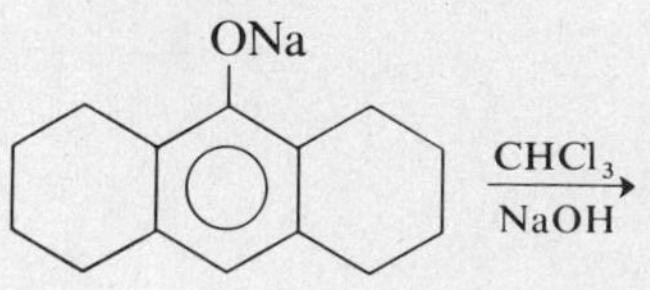 ONa $\xrightarrow[NaOH]{CHCl_3}$

2. Suggest a reasonable mechanism for each of the reactions in Problem 1.

3. Give a specific example of an S_N2 reaction involving the action of the following types of nucleophiles on propylene oxide (2-methyloxirane):
   (a) An oxygen nucleophile
   (b) A nitrogen nucleophile
   (c) A sulfur nucleophile
   (d) A hydrogen nucleophile
   (e) An sp^3-hybridized carbon nucleophile

4. On the basis of the data given, deduce the structures involved in each of the following examples:

(a) A $C_8H_{14}O_2$ compound is optically active, insoluble in aqueous acid or base, and can be dehydrated to a $C_8H_{12}O$ compound that shows no strong absorption above 200 nm in the ultraviolet. Treatment with sodium borohydride yields a mixture of $C_8H_{16}O_2$ compounds, one of which is optically active and one of which is optically inactive.

(b) A compound known as linalool is isolated from various plant sources including the flowers of the Philippine ylang–ylang tree. It is optically active, has the molecular formula $C_{10}H_{18}O$, and undergoes ozonolysis to yield formaldehyde, acetone, and a $C_6H_{10}O_3$ compound. Treatment of the $C_6H_{10}O_3$ compound with sodium borohydride yields a $C_6H_{14}O_3$ compound susceptible to periodic acid cleavage, yielding formaldehyde and a $C_5H_{10}O_2$ compound. When the latter is treated with bromine in the presence of sodium hydroxide it undergoes the haloform reaction (characteristic of $RCOCH_3$ compounds) and yields tribromomethane (bromoform) and a carboxylic acid. Oxidation of the carboxylic acid with sodium dichromate and sulfuric acid produces succinic acid.

(c) When a methylene chloride extract of 850,000 virgin female moths was processed by means of gas chromatography, 1.5 mg of a $C_{18}H_{32}O_2$ compound was isolated, which proved to be a powerful sex attractant for the male of the species. Known as "propylure," it is optically inactive, insoluble in aqueous acid and aqueous base, and undergoes hydrolysis to acetic acid and a C_{16} alcohol. Ozonolysis yields a mixture containing 4-heptanone, butanedial, and a $C_7H_{12}O_3$ compound, which can be hydrolyzed to acetic acid and a $C_5H_{10}O_2$ compound. Oxidation of the latter yields glutaric acid. What is the structure of propylure?

5. In each of the following cases, a compound is being investigated that is known to be one of the three isomeric structures depicted. Using simple chemical methods, indicate how a definitive structural assignment can be made in each instance.

(a) 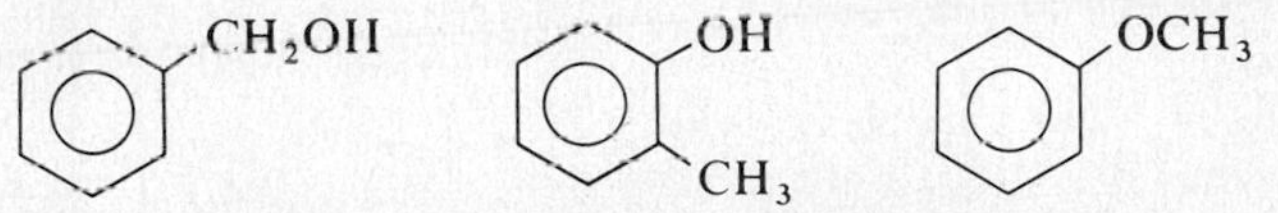 CH_2OH OH, CH_3 OCH_3

(b) 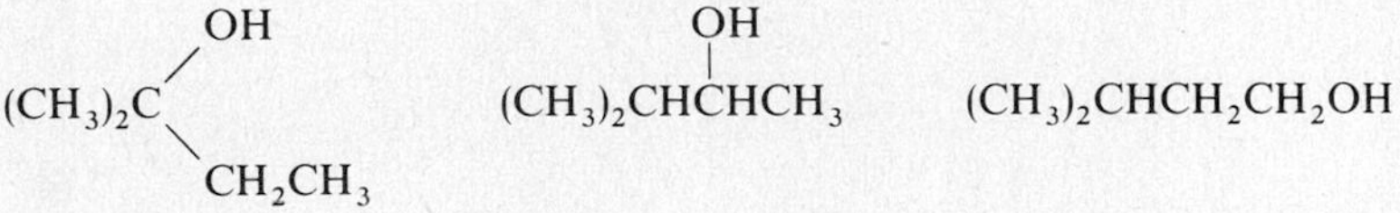 $(CH_3)_2C(OH)CH_2CH_3$ $(CH_3)_2CHCH(OH)CH_3$ $(CH_3)_2CHCH_2CH_2OH$

(c) 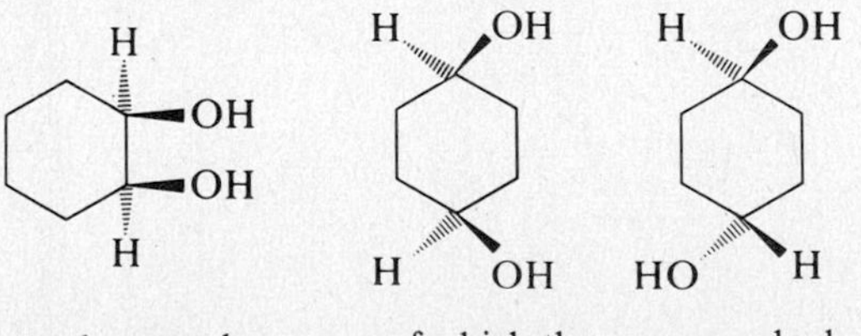 H, OH, OH, H   H, OH, H, OH   H, OH, HO, H

6. Indicate the steps by means of which the compounds shown at the left can be converted to the compounds shown at the right.

(a) 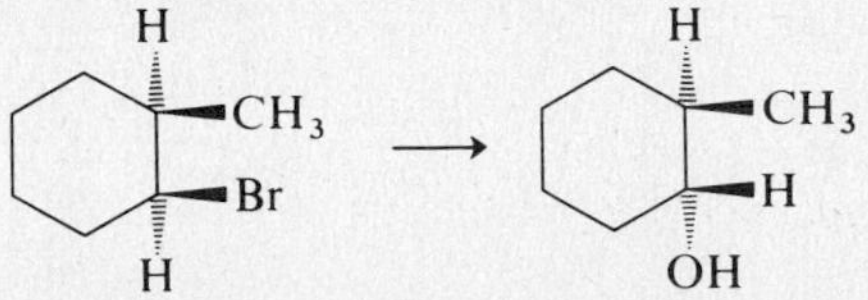 H, CH_3, Br, H $\longrightarrow$ H, CH_3, H, OH

(b) cyclohexene $\longrightarrow$ $(\text{cyclohexyl})_3COH$

(c) cyclopentene $\longrightarrow$ cis-1,2-cyclopentanediol (H, OH; OH, H)

(d) $(CH_3)_2C{=}CH_2 \longrightarrow (CH_3)_2C(OH)CH_2CH(OH)C_6H_5$

(e) quinoline-2-CO_2H $\longrightarrow$ quinoline-2-CHO

7. Outline the reactions by means of which methylenecyclopentane, cyclopentane$=CH_2$, can be converted to the following compounds:

(a) 1-methylcyclopentanol (OH, CH_3) (b) cyclopentyl$-CH_2OH$ (c) cyclopentyl$-CH_2CH_2OH$

8. For each of the following compounds, indicate what carbonyl compound and what Grignard reagent could be used for its preparation:

(a) $CH_3C(OH)(CH_2CH_3)CH_2CH_2CH_3$ (Three different combinations of carbonyl compound and Grignard reagent)

(b) $(CH_3)_2CHOH$ (Two different combinations of carbonyl compound and Grignard reagent)

(c) cyclobutane bearing $C(CH_3)_2OH$ and $C(CH_3)_2OH$ on the same carbon

(d) cyclopropyl$-CH_2CH(OH)-C_6H_4-CH(OH)CH_2-$cyclopropyl

9. Available starting materials include (a) any alkanes, alkenes, alkynes, alkyl halides, alkanols, alkanals, alkanones, alkanoic acids, or alkanoic acid derivatives containing four or fewer carbon atoms, (b) benzene and any monosubstituted benzenes, naphthalene, phenanthrene, pyridine, furfural, (c) special organic reagents, and (d) any inorganic compounds. Starting with materials chosen from this group, outline syntheses for each of the following compounds:

(a) $(CH_3)_2CHCH_2CH_2OH$

(b) $CH_3CH_2C(OH)(C_6H_5)CH_3$

(c) $CH_3CH{=}CHCH_2C(OH)(CH_3)CH_3$

(d) $(C_6H_5)_2CHOH$

(e) cyclohexyl–CH_2OH

(f) cycloheptane ring bearing H, CH_3, H, OH

(g) 1-(phenylethynyl)cyclohexanol: cyclohexane ring with OH and $-C{\equiv}C-C_6H_5$

(h) $(CH_3)_2C(OH)CH_2CH_2CH_2OH$

(i) $HO-C(C_6H_5)(CH_2CH{=}CH_2)-CH_2CH{=}CH_2$

(j) $HO-C(C_6H_5)_2$(1-naphthyl)

10. The following alcohols are cheap starting materials, costing less than fifty cents a pound: methanol, ethanol, *n*-propyl alcohol, isopropyl alcohol, *n*-butyl alcohol, *sec*-butyl alcohol, *t*-butyl alcohol, *n*-amyl alcohol, *n*-hexyl alcohol, *n*-octyl alcohol, *n*-decyl alcohol, myristyl alcohol, cetyl alcohol, stearyl alcohol, ethylene glycol, propylene glycol, glycerine, pentaerythritol, cyclohexanol, tetrahydrofurfuryl alcohol, allyl alcohol, benzyl alcohol. Using one of more of these as starting materials, along with any other compounds that you think are necessary, devise a synthesis for each of the following compounds:

(a) $CH_3(CH_2)_{16}CH_2OCH_2CH_2OH$

(b) $CH_3CH_2CH_2CH_2CH$—CH_2—CH_2 (cyclopropane ring)

(c) $C_6H_5-\overset{O}{\overset{\|}{C}}CH_2CH_2CH_2CH_3$

(d) cyclohexane ring with OH and $-CH_2CH{=}CH_2$

(e) $CH_3CH_2C(Br)(CH_3)CH_2-C_6H_5$

11. Using Tables 39.1 to 39.5 as guidelines to readily available starting materials, devise syntheses for the following compounds. In addition to readily available organic compounds, any inorganic materials and special organic reagents may also be used.

(a) $CH_3(CH_2)_{15}CH{=}CH_2$

(b) $CH_3(CH_2)_{15}CH(OH)CH_3$

(c) $CH_3(CH_2)_{13}CH(OH)CH_2CH_2OH$

(d) $HO(CH_2)_{10}OH$

(e) $CH_3C(OH)(C_6H_5){-}CH_2CH_2{-}C(OH)(C_6H_5)CH_3$

(f) $(CH_3)_3C{-}C_6H_4{-}OH$

(g) $CH_3{-}C(OH)(C_5H_4N){-}CH_3$

(h) cyclooctane-1,2,5,6-tetraol (HO, HO, OH, OH)

(As a mixture of stereoisomers)

Chapter 23

Ethers

A fact is like a sack which won't stand up when it is empty. In order that it may stand up, one has to put into it the reason and sentiment which have caused it to exist.

LUIGI PIRANDELLO
Six Characters in Search of an Author, 1921

A number of interesting natural products as well as certain materials of commercial importance are ethers, so compounds containing this functional group are sometimes the terminal point of a synthesis. Since ethers lack an acidic hydrogen and are much less reactive than alcohols and phenols, however, they are not so frequently used as synthesis intermediates. One important exception to this generality are the oxiranes (epoxides), which are much more reactive than most ethers and which, consequently, are very useful synthesis intermediates. As in Chapter 22, this chapter also is divided into two major parts, the first dealing with methods for preparing ethers and the second dealing with their reactions.

METHODS OF PREPARATION OF ETHERS

23.1 Dehydration of Alcohols

Symmetrical dialkyl ethers can be prepared by the acid-catalyzed intermolecular dehydration of some alcohols (see Fig. 11.4). If an alcohol is mixed with an alkene which can yield a more stable cation than that derived from the alcohol, an unsymmetrical ether can form, as illustrated by the acid-catalyzed reaction between 2-methylpropene and ethanol (Fig. 23.1).

PROBLEM 23-1

Predict the structure of the product that would result from the action of sulfuric acid on $H_2C{=}CHCH_2CH_2CH_2CH_2OH$.

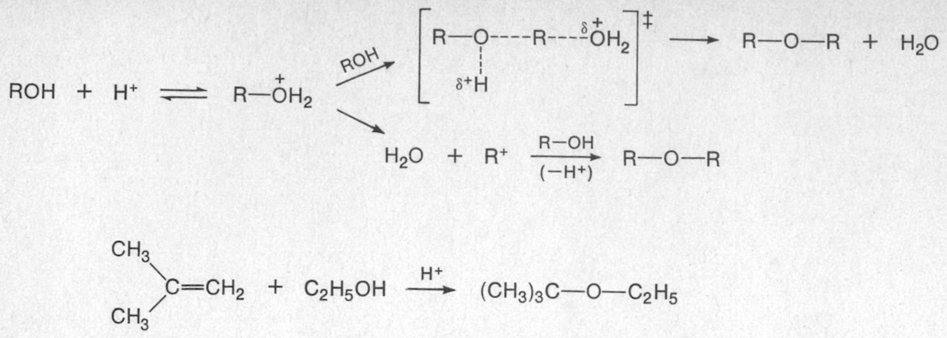

$$ROH + H^+ \rightleftharpoons R\text{—}\overset{+}{O}H_2 \xrightarrow{ROH} \left[R\text{—}\underset{\delta^+ H}{O}\text{---}R\text{---}\overset{\delta+}{O}H_2 \right]^{\ddagger} \longrightarrow R\text{—}O\text{—}R + H_2O$$

$$R\text{—}\overset{+}{O}H_2 \longrightarrow H_2O + R^+ \xrightarrow[(-H^+)]{R\text{—}OH} R\text{—}O\text{—}R$$

$$(CH_3)_2C{=}CH_2 + C_2H_5OH \xrightarrow{H^+} (CH_3)_3C\text{—}O\text{—}C_2H_5$$

Fig. 23.1. Preparation of symmetrical and unsymmetrical dialkyl ethers via acid-catalyzed processes.

23.2 Via Nucleophilic Displacement Reactions

23.2a. Nucleophilic displacement by alkoxides and phenoxides. Unsymmetrical, as well as symmetrical, ethers can be prepared by the action of alkoxides and aryloxides on alkyl halides (see Fig. 11.5) or alkyl sulfonates (see Fig. 11.6). For alkoxides, the solvents of choice are those that are aprotic but highly polar (*e.g.*, dimethylsulfoxide). For phenoxides, prepared by the action of sodium hydroxide on phenols, water or alcohol solutions are generally the most satisfactory.

PROBLEM 23-2

Would you expect the acid-catalyzed reaction of methanol with

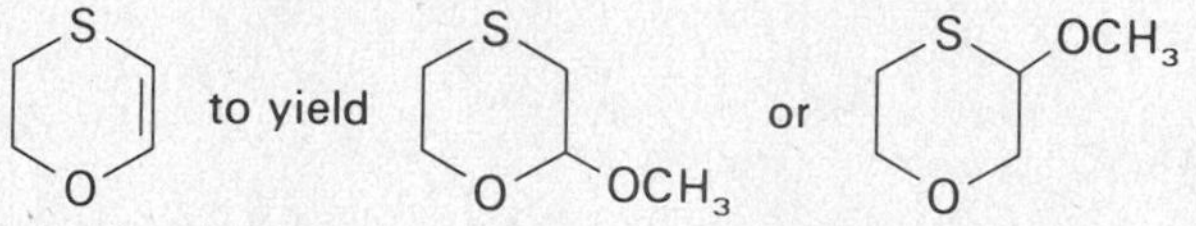

[2,3-dihydro-1,4-oxathiin] to yield [2-methoxy-1,4-oxathiane (OCH_3 on carbon next to O)] or [3-methoxy-1,4-oxathiane (OCH_3 on carbon next to S)]

Useful data for making this prediction can be found in Problem 21–10 on p. 614.

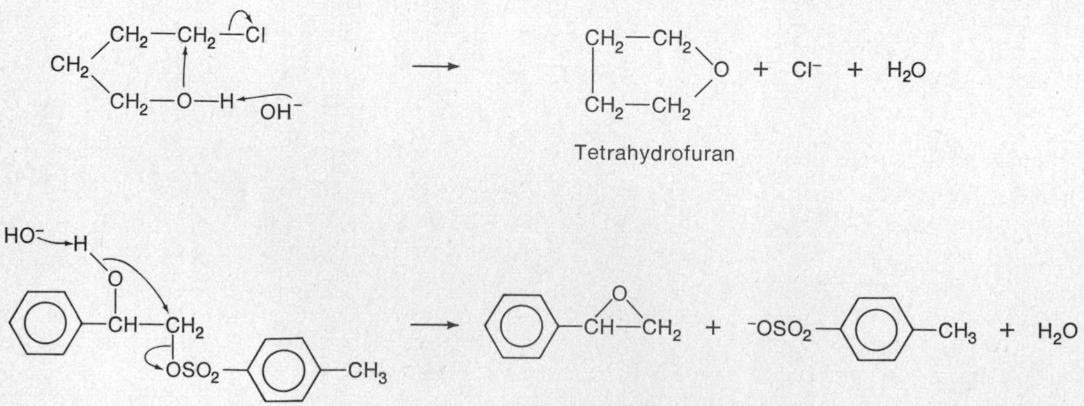

$$HO\text{—}CH_2CH_2CH_2CH_2\text{—}Cl \xrightarrow{OH^-} \text{(tetrahydrofuran)} + Cl^- + H_2O$$

Tetrahydrofuran

$$C_6H_5\text{—}CH(OH)\text{—}CH_2\text{—}OSO_2\text{—}C_6H_4\text{—}CH_3 \xrightarrow{HO^-} C_6H_5\text{—}\underset{O}{CH\text{—}CH_2} + {}^-OSO_2\text{—}C_6H_4\text{—}CH_3 + H_2O$$

2-Phenyloxirane

Fig. 23.2. Preparation of cyclic ethers by intramolecular nucleophilic displacements.

If the hydroxyl function and the halide or sulfonate function are in the same carbon framework, intramolecular displacement is possible as illustrated by the formation of the oxirane and tetrahydrofuran ring systems shown in Fig. 23.2. This method is generally restricted to the preparation of cyclic ethers having no more than seven atoms in the ring.

Although aryl halides are usually unreactive in nucleophilic displacement reactions, *o*- and *p*-nitrosubstituted aryl halides react with alkoxides and phenoxides to form ethers (Fig. 23.3). In the presence of cuprous oxide at elevated temperature, even unactivated aryl halides may react with phenols to yield diaryl ethers, as illustrated by the examples in Fig. 23.3.

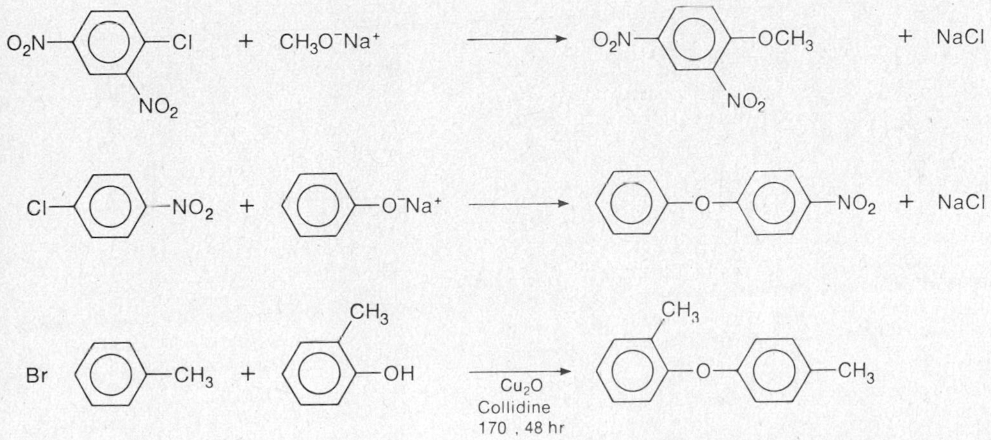

$$O_2N\text{-}C_6H_3(NO_2)\text{-}Cl + CH_3O^-Na^+ \longrightarrow O_2N\text{-}C_6H_3(NO_2)\text{-}OCH_3 + NaCl$$

$$Cl\text{-}C_6H_4\text{-}NO_2 + C_6H_5\text{-}O^-Na^+ \longrightarrow C_6H_5\text{-}O\text{-}C_6H_4\text{-}NO_2 + NaCl$$

$$Br\text{-}C_6H_4\text{-}CH_3 + (CH_3)C_6H_4\text{-}OH \xrightarrow[\text{170, 48 hr}]{Cu_2O\text{, Collidine}} (CH_3)C_6H_4\text{-}O\text{-}C_6H_4\text{-}CH_3$$

Fig. 23.3. Preparation of aromatic ethers from aryl halides.

23.2b. NUCLEOPHILIC DISPLACEMENT BY ALCOHOLS AND PHENOLS. Diazoalkanes* are sufficiently basic to abstract protons from many protic compounds (*i.e.*, compounds containing acidic hydrogens), the rate of protonation increasing as the acidity of the proton donor increases. The protonated diazoalkane, an alkyldiazonium ion, is an exceedingly unstable entity, losing molecular nitrogen to form a carbonium ion which then becomes affixed to whatever nucleophile is available. Thus, the overall reaction is a replacement of the nitrogen of the diazoalkane by the hydrogen and accompanying nucleophilic portion of the protic

* Diazoalkanes can be prepared in a number of ways. A typical procedure, illustrated by the preparation of diazomethane, involves the conversion of an N-alkylamide to an N-alkyl-N-nitrosoamide by the action of nitrous acid followed by base-induced decomposition of the nitrosoamide:

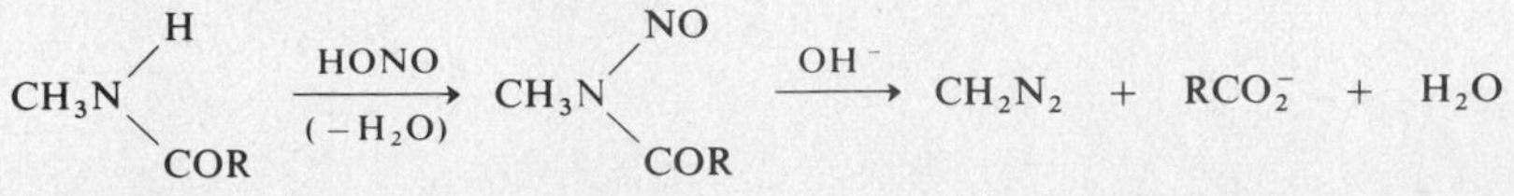

$$CH_3N(H)(COR) \xrightarrow[(-H_2O)]{HONO} CH_3N(NO)(COR) \xrightarrow{OH^-} CH_2N_2 + RCO_2^- + H_2O$$

compound, as illustrated for diazomethane:

$$NuH + \overset{-}{C}H_2-\overset{+}{N}\equiv N \rightleftharpoons Nu^- + CH_3-N_2^+ \longrightarrow CH_3^+ + N_2$$

$$CH_3^+ + Nu^- \longrightarrow CH_3Nu$$

or

$$CH_3N_2^+ + Nu^- \longrightarrow CH_3NU + N_2$$

$$CH_3^+ + NuH \longrightarrow CH_3Nu + H^+$$

Although alcohols are not acidic enough to protonate diazomethane sufficiently rapidly to make the reaction a useful one, many phenols are readily converted to aryl methyl ethers in this fashion:

$$\text{1,3-}C_6H_4(OH)_2 + 2\,CH_2N_2 \longrightarrow \text{1,3-}C_6H_4(OCH_3)_2 + 2\,N_2$$

PROBLEM 23-3

What is the structure of the product resulting from the action of 1 mole-equivalent, 2 mole-equivalents, and 3 mole-equivalents of diazomethane on 2,3-dihydroxy-6-nitrobenzoic acid?

By using fluoboric acid as the donor of a proton to diazomethane, even alcohols can be etherified, as illustrated by the conversion of cyclohexanol to cyclohexyl methyl ether:

$$CH_2N_2 + HBF_4 \longrightarrow BF_4^- + CH_3N_2^+ \xrightarrow[(-N_2)]{C_6H_{11}OH} C_6H_{11}-\overset{+}{O}(H)-CH_3$$

$$C_6H_{11}-\overset{+}{O}(H)-CH_3 \xrightarrow{BF_4^-} HBF_4 + C_6H_{11}-OCH_3$$

23.2c. Nucleophilic displacements on aryldiazonium salts. Alcohols are among the nucleophiles with which aryldiazonium salts will react, leading to aryl ethers:

$$C_6H_5N_2^+ + ROH \longrightarrow C_6H_5OR + N_2 + H^+$$

23.3 Via Addition of Oxygen to Alkenes

Alkenes react with aliphatic peracids (*e.g.*, peracetic acid, CH_3CO_3H) and with aromatic peracids (*e.g.*, perbenzoic acid, $C_6H_5CO_3H$) to form oxiranes (epoxides):

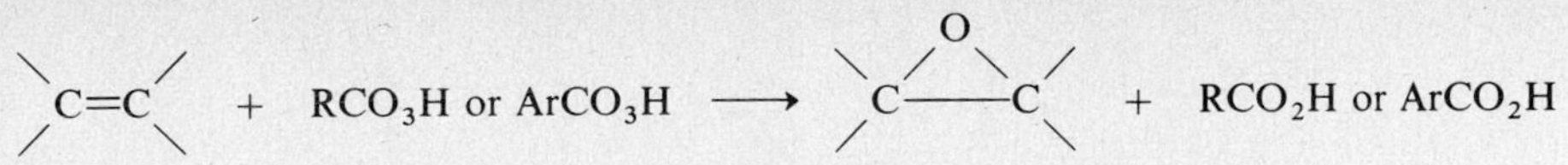

$$\rangle C{=}C\langle + RCO_3H \text{ or } ArCO_3H \longrightarrow \rangle C\overset{O}{\diagdown\!\diagup}C\langle + RCO_2H \text{ or } ArCO_2H$$

The utility of this reaction in the preparation of 1,2-diols is illustrated in Fig. 22.12, and the mechanism of the reaction is depicted in Fig. 26.14

Table 23.1 summarizes the methods that are useful for the synthesis of ethers.

Table 23.1 METHODS FOR THE PREPARATION OF ETHERS

Starting Materials	Reagents	Products	Comments
Alcohols (ROH)	H^+	R—O—R	Only for preparation of symmetrical ethers
	$CH_2N_2 + HBF_4$	$R{-}O{-}CH_3$	
Alcohols and alkenes	H^+	R—O—R′	Applicable if alkene forms a more stable carbonium ion than the one derived from ROH
Alkenes ($\rangle C{=}C\langle$)	RCO_3H or $ArCO_3H$	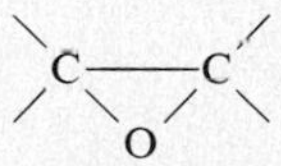 $\rangle C\underset{O}{\diagdown\!\diagup}C\langle$	
Phenols (ArOH)	R_2CN_2	$Ar{-}O{-}CHR_2$	Diazomethane, CH_2N_2, is most often used
Aryldiazonium compounds (ArN_2^+)	ROH	Ar—O—R	
Alkyl halides (RX)	R′ONa ArONa	R—O—R′ R—O—Ar	Limited to primary RX and some secondary RX; some secondary and most tertiary RX eliminate HX
Hydroxy-halo compounds $HO{-}C{-}(C)_n{-}C{-}X$	Base	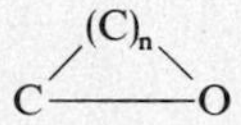 $(C)_n$ ring: C—O	Restricted to compounds with n no greater than 5

REACTIONS OF ETHERS

23.4 Cleavage Reactions

Most ethers are relatively inert under ordinary reaction conditions and, as a consequence, are often used as solvents in carrying out reactions. Ethers are by no means completely inert, however, and are susceptible to cleavage by protic acids (Fig. 23.4, see also Fig. 11.7), by Lewis acids, and by organometallic compounds (Fig. 23.4). An interesting example of the effect of Lewis acids on ethers

---

PROBLEM 23-4

What would be the expected products from the hydriodic acid cleavage of isopropyl neopentyl ether?

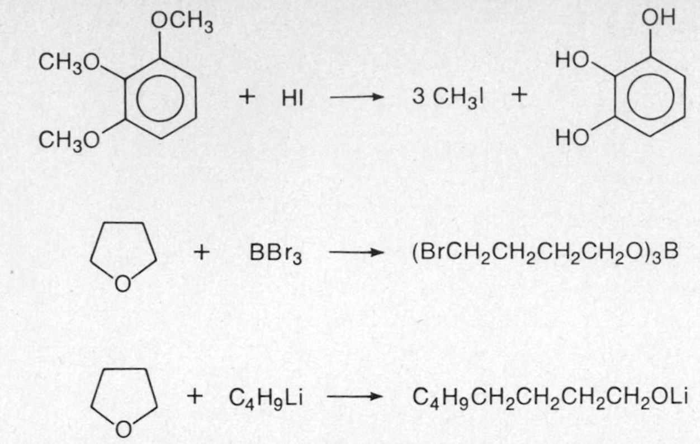

$$\text{1,2,3-(CH}_3\text{O)}_3\text{C}_6\text{H}_3 + \text{HI} \longrightarrow 3\,\text{CH}_3\text{I} + \text{1,2,3-(HO)}_3\text{C}_6\text{H}_3$$

$$\text{THF} + \text{BBr}_3 \longrightarrow (\text{BrCH}_2\text{CH}_2\text{CH}_2\text{CH}_2\text{O})_3\text{B}$$

$$\text{THF} + \text{C}_4\text{H}_9\text{Li} \longrightarrow \text{C}_4\text{H}_9\text{CH}_2\text{CH}_2\text{CH}_2\text{CH}_2\text{OLi}$$

Fig. 23.4. Cleavage of ethers by protic acids, Lewis acids, and organometallics.

occurs in the preparation of ketones by the action of organocadmium compounds (R_2Cd) on acid chlorides. The organocadmium compound is prepared by the action of cadmium chloride on an ethereal solution of a Grignard reagent ($2\,RMgX + CdCl_2 \rightarrow R_2Cd + 2\,MgXCl$). Before the acid chloride is added, however, all the ether must be removed and replaced with benzene to avoid a magnesium halide-catalyzed reaction between the ether and the acid chloride:

$$C_2H_5OC_2H_5 \;+\; RCOCl \xrightarrow[MgX_2]{} RCO_2C_2H_5 \;+\; C_2H_5Cl$$

PROBLEM 23-5

Suggest a mechanism for the reaction of diethyl ether and an acid chloride in the presence of $MgCl_2$.

Vinyl ethers (enol ethers) constitute a special class of ethers that are unusually susceptible to hydrolysis by aqueous acid. The ease with which these react with water is ascribed to the enhanced tendency for the double bond to undergo protonation. This, in turn, is the result of the electron-releasing effect of the attached oxygen and the resonance stabilization that this provides to the cation:

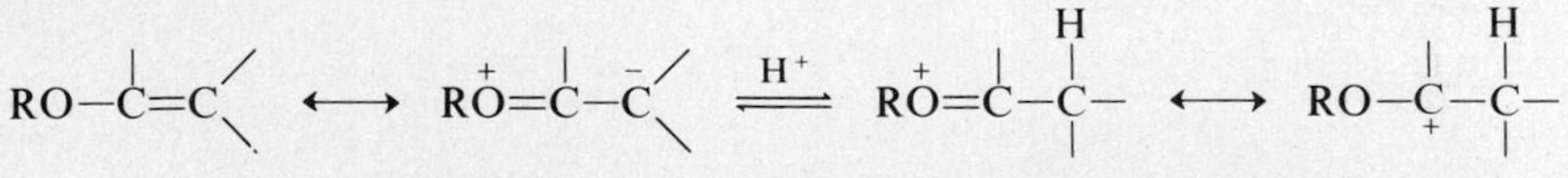

$$\underset{\text{Vinyl ether (enol ether)}}{RO{-}C{=}C \longleftrightarrow R\overset{+}{O}{=}C{-}\overset{-}{C}} \overset{H^+}{\rightleftharpoons} \underset{\text{Protonated vinyl ether}}{R\overset{+}{O}{=}C{-}C(H) \longleftrightarrow RO{-}\overset{+}{C}{-}C(H)}$$

The protonated vinyl ether adds a molecule of water to form a hemiacetal (or hemiketal), which is an unstable compound that decomposes to the aldehyde or ketone:

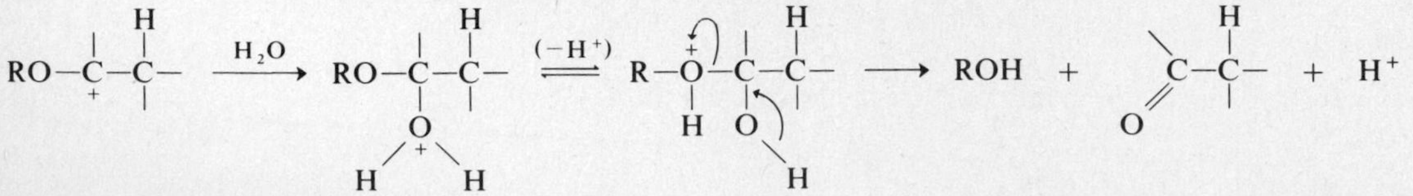

$$RO{-}\overset{+}{C}{-}C(H){-} \xrightarrow{H_2O} RO{-}C(\overset{+}{O}H_2){-}C(H){-} \overset{(-H^+)}{\rightleftharpoons} R{-}\overset{+}{O}(H){-}C(OH){-}C(H){-} \longrightarrow ROH + O{=}C{-}C(H){-} + H^+$$

For example, 1-methoxy-1,4-cyclohexadiene, obtained by the action of metallic sodium in ammonia solution on methoxybenzene (anisole), is readily converted to 3-cyclohexenone by the action of dilute aqueous acid:

$$\text{C}_6\text{H}_5\text{OCH}_3 \xrightarrow[\text{NH}_3]{\text{Na}} \text{1-methoxy-1,4-cyclohexadiene} \xrightarrow{\text{H}_3\text{O}^+} \text{3-cyclohexenone}$$

A vinyl ether

PROBLEM 23-6

Under more strenuous treatment with acid or base, 3-cyclohexen-1-one rearranges to 2-cyclohexen-1-one. Why?

23.5 *Nucleophilic Substitution Reactions*

Nucleophilic displacement of the C—O bond is no more facile in ethers than in alcohols, and it is usually necessary to protonate or complex the oxygen to induce reaction (see Fig. 23.4 and Section 22.11a). The exceptions to this general rule are the oxirane compounds in which the excessive strain of the small ring is sufficient to make them susceptible to nucleophilic displacement even in the unprotonated form. Oxiranes react at the most sterically accessible carbon atom with a number of nucleophiles in typical S_N2 processes,

$$R_2C\overset{O}{\frown}CH_2 + \text{Nu—H} \longrightarrow R_2\overset{OH}{\overset{|}{C}}\text{—CH}_2\text{—Nu.}$$

These reactions proceed with inversion of configuration at the carbon to which the nucleophile becomes attached and with retention of configuration at the neighboring carbon, as the following example illustrates:

H, OH^-, H, O → OH, H, H, O^-

Inversion of configuration

Retention of configuration

Their utility in synthesis processes is illustrated by the reactions shown in Fig. 23.5 which includes the preparation of alcohols, hydroxy ethers, hydroxythioethers, hydroxyamines, and hydroxyesters.

PROBLEM 23-7

Suggest a sequence of reactions for converting benzyl bromide to 3-phenylpropyl bromide.

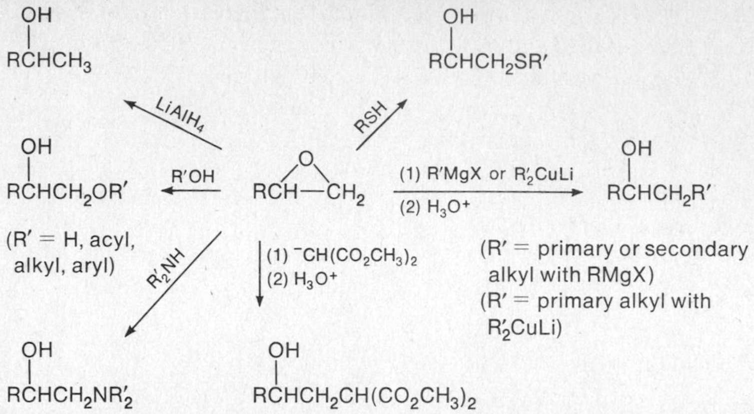

Fig. 23.5. Nucleophilic substitution reactions of oxiranes.

Even unstrained ethers undergo nucleophilic displacement if they are protonated (see Fig. 11.8), so oxiranes are especially reactive in the presence of protic or Lewis acids. Reactions under these conditions follow the same stereochemical course as those carried out on the unprotonated oxirane, proceeding with inversion of configuration at the carbon to which the nucleophile becomes attached. They are best viewed as concerted processes in which there is sufficient carbonium ion character along the reaction pathway to direct the nucleophile to the carbon that is better able to sustain a positive charge. Thus, in contrast to the ring-opening reactions with the neutral oxiranes, those with the protonated oxiranes lead to the attachment of the nucleophile at the more highly substituted carbon atom:

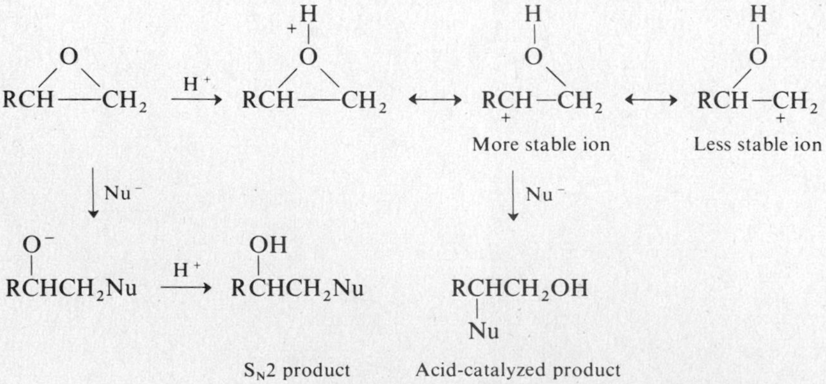

23.6 Oxidation Reactions

Ethers are susceptible to air oxidation which converts them to hydroperoxides (see p. 297). Since hydroperoxides are very explosive, they represent a laboratory hazard. Ethers that have been exposed to the atmosphere for any length of time should be treated with a good reducing agent (*e.g.*, aqueous potassium iodide, lithium aluminum hydride, or ferrous sulfate) to destroy the peroxides. Ethers that have not been treated in this fashion should never be distilled!

23.7 Deoxygenation Reactions of Oxiranes

The deoxygenation of ethers is limited to the oxiranes, which yield alkenes when treated with triphenylphosphine or triethylphosphite. Phosphorus, a group V element falling just below nitrogen in the periodic table, resembles nitrogen in its nucleophilic characteristics. The deoxygenation reactions, thus, are the result of a nucleophilic attack by the phosphorus reagents at the oxirane carbon, which proceeds with inversion of configuration and leads to a charge-separated intermediate (often referred to as a *betaine*). This then decomposes via a four-membered cyclic transition state which requires a 180° rotation around the C—C bond to establish the appropriate conformation. The stereochemical result of this sequence of events is to put the R groups attached to the oxirane carbons in a different stereochemical relationship in the oxirane and alkene. Thus, if they are *cis* in the oxirane they become *trans* in the alkene. Since the formation of oxiranes from alkenes proceeds in a *syn* fashion, this provides a method for inverting the configuration of the groups attached to a C=C bond, as illustrated by the example in Fig. 23.6.

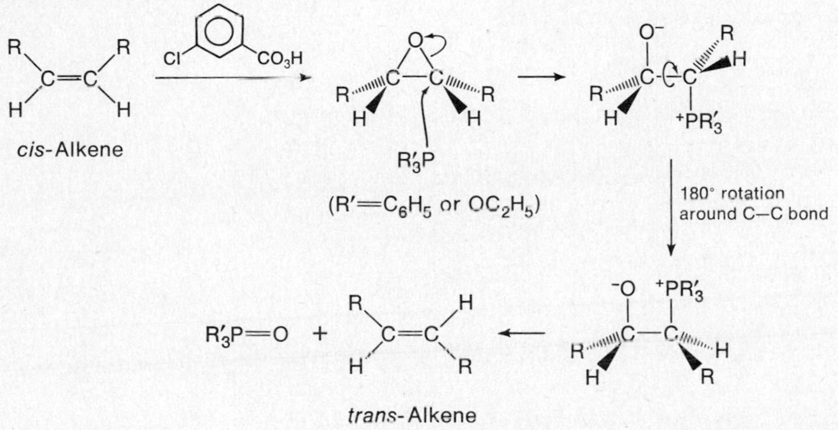

cis-Alkene

$(R'=C_6H_5$ or $OC_2H_5)$

180° rotation around C—C bond

$R'_3P=O$ + *trans*-Alkene

Fig. 23.6. Interconversion of double-bond diastereomers via expoxidation-deoxygenation sequence.

PROBLEM 23-8

What would you predict to be the outcome of the application of the reaction sequence shown in Fig. 23.6 to *cis*-cyclohexene?

23.8 Rearrangement Reactions

23.8a. Ionic rearrangement reactions. Instead of being cleaved with very strong bases such as organolithium compounds (see Fig. 23.4), some ethers rearrange under these conditions (*Wittig rearrangement*). For example, treatment of dibenzyl ether with phenyllithium yields 1,2-diphenylethanol as the result of a reaction that is probably best rationalized in terms of radical and radical anion intermediates (Fig. 23.7).

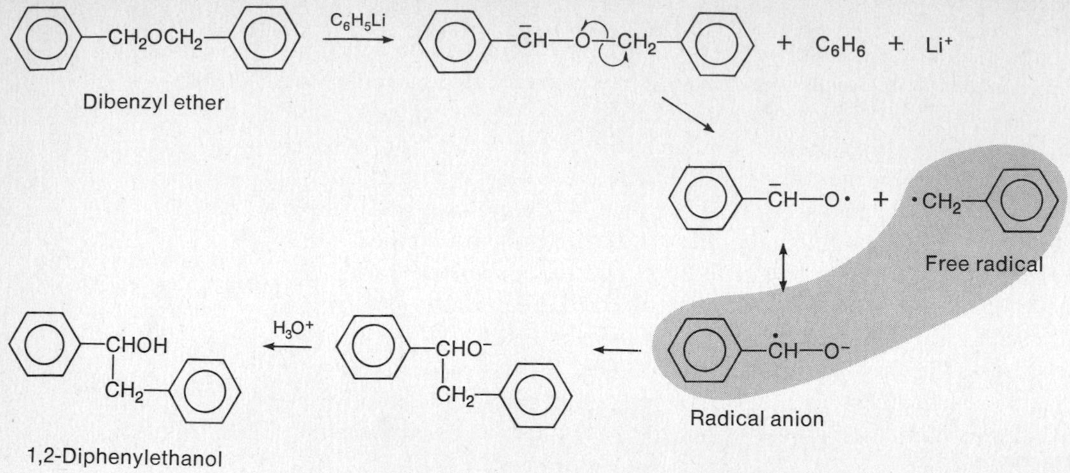

Dibenzyl ether

1,2-Diphenylethanol

Radical anion

Free radical

Fig. 23.7. Wittig rearrangement of dibenzyl ether.

Treatment of oxiranes with Lewis acids in aprotic solvents often induces isomerization to aldehydes or ketones:

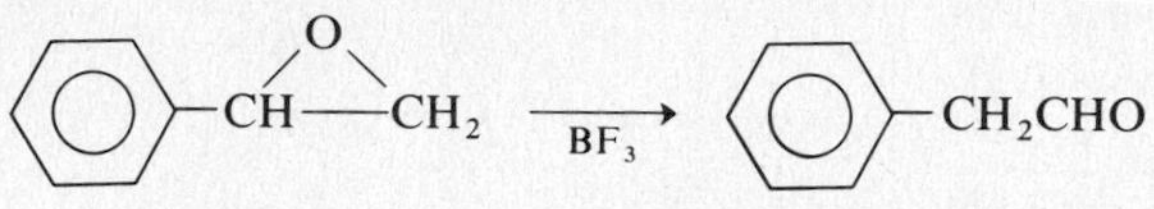

$$C_6H_5\text{-}CH(O)CH_2 \xrightarrow{BF_3} C_6H_5\text{-}CH_2CHO$$

In the presence of strong bases, oxiranes usually undergo nucleophilic substitution as described in Section 23.4, although in some cases rearrangement to allylic alcohols takes place. For example, the action of lithium diethylamide, a very strong base but a poor nucleophile, on *trans*-1-ethyl-2-isopropyloxirane yields *trans*-5-methyl-2-hexen-4-ol in a highly stereoselective process:

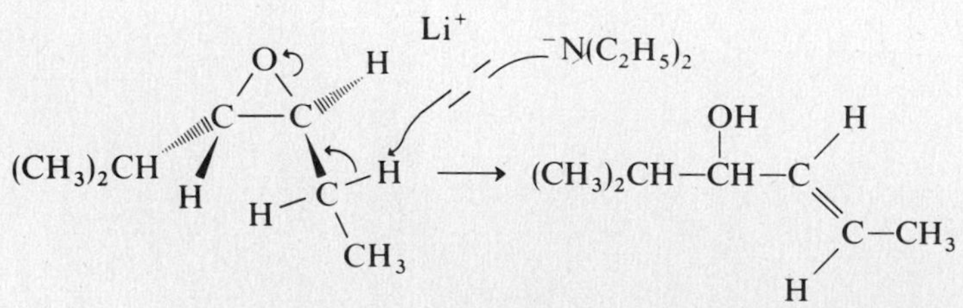

$$(CH_3)_2CH\text{-oxirane-}CH_2CH_3 \xrightarrow{Li^+\ {}^-N(C_2H_5)_2} (CH_3)_2CH{-}CH(OH){-}CH{=}CH{-}CH_3$$

23.8b. Thermal rearrangement reactions. Allyl aryl ethers constitute a special class of aromatic ethers which rearrange, when heated, to *o*- and *p*-substituted phenols (*Claisen rearrangement*). The reaction, which proceeds via a concerted process involving a six-membered cyclic transition state, yields *o*-substituted product if at least one *ortho* position is available, or *p*-substituted products if both *ortho* positions are blocked by substituents (Fig. 23.8).

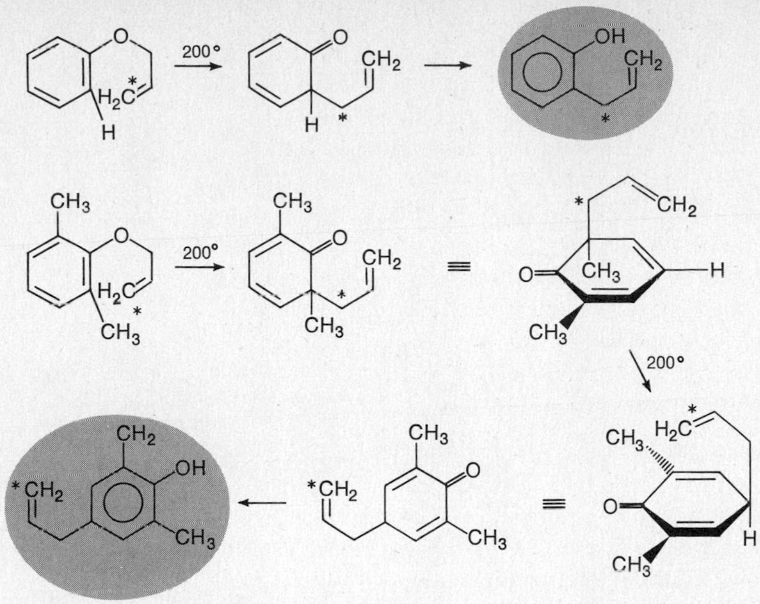

Fig. 23.8. Claisen rearrangement of allyl phenyl ether and allyl 2,6-dimethylphenyl ether labeled with ^{14}C (indicated by asterisks).

Table 23.2 summarizes the reactions of ethers.

Table 23.2 REACTIONS OF ETHERS

Type of Reaction	Starting Materials	Reagents	Products	Comments
Reactions of R—OR Bond	R—O—R′	HI	RI + R′I	Strenuous conditions; skeletal rearrangements frequent
	Ar—O—R	HI	ArOH + RI	
	>C=C(—)—OR (vinyl ethers)	H_2O, H^+	—CH(—)—C=O + ROH	Very facile reaction
Reaction of oxiranes	H—C(H)—O—C(H)—R (oxirane)	Nu—H	Nu—C(H)(H)—C(R)(H)—OH	See Table 21.3 for list of nucleophiles; Nu attaches to less hindered carbon
		Nu, H^+	HO—C(H)(H)—C(R)(H)—Nu	Nu attaches to more highly substituted carbon
		BF_3	—C(—)—C(—)=O	
		R_3P	>C=C<	Method for effecting *cis-trans* interconversion of double bonds

I. Methods of preparation of ethers
   A. Dehydration of alcohols
   B. Via nucleophilic displacement reactions
      1. Via nucleophilic displacements by alkoxides and phenoxides
         (a) Intermolecular reaction—preparation of unsymmetrical ethers
         (b) Intramolecular reaction—preparation of cyclic ethers (oxiranes, tetrahydrofurans, etc.)
         (c) Reaction with aryl halides
      2. Nucleophilic displacements on aryldiazonium salts
   C. Alcohols and phenols with diazomethane; alcohols require HBF_4
II. Reactions of ethers
   A. Cleavage reactions
      1. —OR a poor leaving group; protonation to $-\overset{+}{O}(R)H$ required for nucleophilic displacement to occur
      2. Vinyl ethers unusually susceptible to acid-catalyzed hydrolysis
   B. Nucleophilic substitution reactions—reactions of oxiranes
      1. In absence of acid—reaction follows S_N2 pathway occurring at the less hindered carbon with inversion of configuration
      2. In presence of acid—reaction follows S_N1 pathway occurring at more highly substituted carbon with inversion of configuration
   C. Oxidation reactions—formation of peroxides
   D. Deoxygenation reactions of oxiranes—with R_3P
   E. Rearrangement reactions
      1. Ionic rearrangements—similar to pinacol rearrangements
      2. Thermal rearrangement of allyl phenyl ethers

REFERENCE

PATAI, S., ed., *The Chemistry of the Ether Linkage*, John Wiley & Sons, Inc., New York, 1967.

PROBLEMS

1. Draw the structure of the major product(s) from each of the following reactions:

(a) $ClCH_2CO_2Na + CH_3ONa \longrightarrow$

(b) $CH_3C(OCH_3){=}CHCO_2CH_3 + H_3O^+ \longrightarrow$

(c) 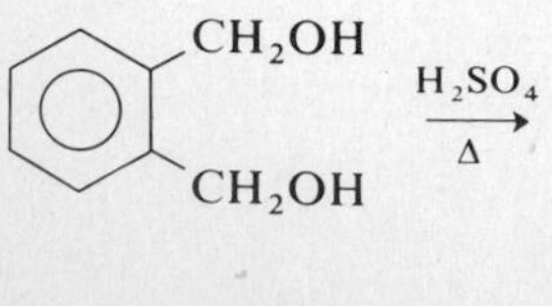 1,2-benzenedimethanol ($C_6H_4(CH_2OH)_2$) $\xrightarrow[\Delta]{H_2SO_4}$

(d) 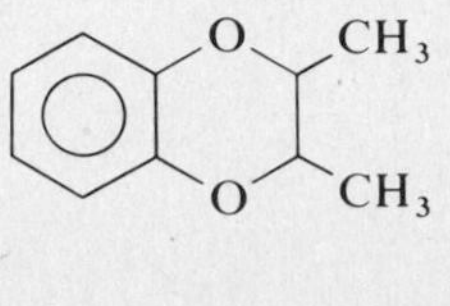 2,3-dimethyl-1,4-benzodioxane (benzene ring fused to O–CH(CH_3)–CH(CH_3)–O) $+ HI \longrightarrow$

(e) $CH_3CH_2CH_2OH + (C_2H_5O)_2SO_2 \xrightarrow{NaOH}$

(f) 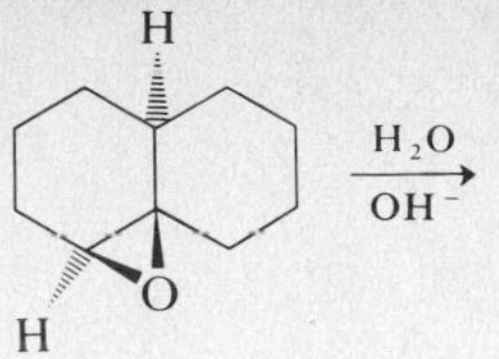 [bicyclic epoxide, H, H, O] $\xrightarrow[OH^-]{H_2O}$

(g) [cyclohexane: H, OSO_2Ar; H, CH_2CH_2OH] $\xrightarrow{NaOH}$

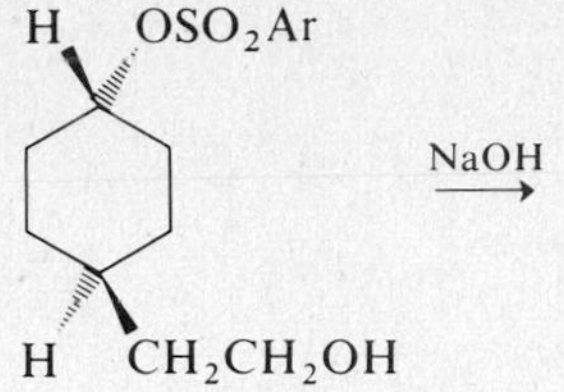

(h) [2-naphthol, OH] $+ C_6H_5CHN_2 \longrightarrow$

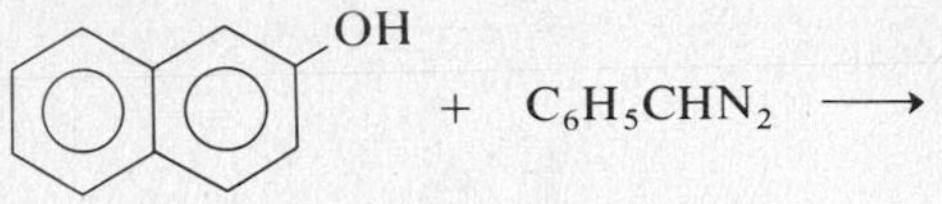

(i) 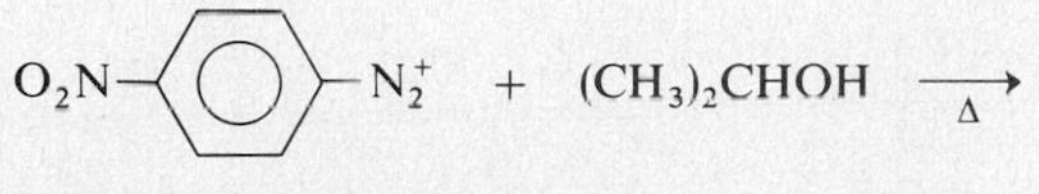 $O_2N-C_6H_4-N_2^+ + (CH_3)_2CHOH \xrightarrow{\Delta}$

(j) 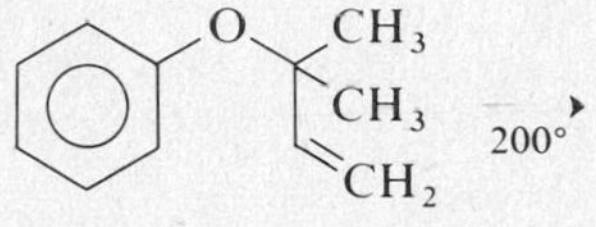 [$C_6H_5O-C(CH_3)_2-CH=CH_2$] $\xrightarrow{200°}$

2. Suggest a reasonable mechanism for each of the reactions in Problem 1.

3. In each of the following cases a compound is being investigated that is known to be one of the three isomeric structures shown. Using simple chemical methods indicate how a structural assignment can be made in each instance.

(a) 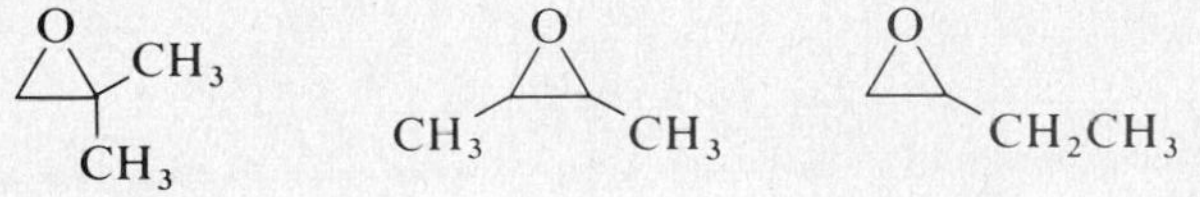 [epoxide with CH_3, CH_3] [epoxide with CH_3, CH_3] [epoxide with CH_2CH_3]

(b) 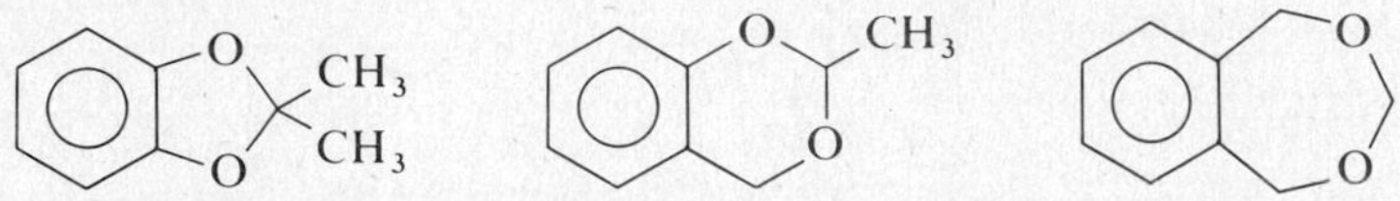 [benzodioxole with CH_3, CH_3] [benzodioxane with O, O, CH_3] [benzodioxepine with O, O]

(c) 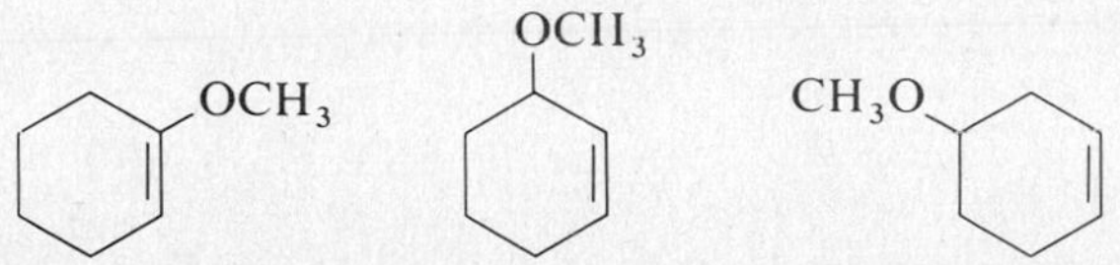 [cyclohexene with OCH_3] [cyclohexene with OCH_3] [cyclohexene with CH_3O]

4. Indicate the steps by means of which the compounds shown at the left can be converted to the compounds shown at the right.

(a) 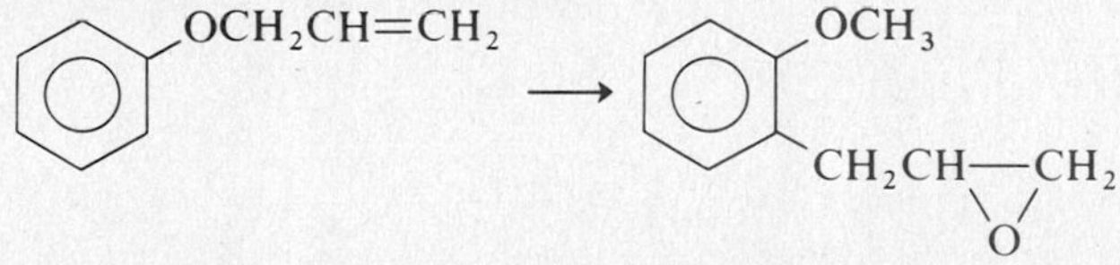 $C_6H_5OCH_2CH{=}CH_2 \longrightarrow$ [benzene with OCH_3 and $CH_2CH\!-\!CH_2$ epoxide (O)]

(b) [$CH_3O-C_6H_5$] $\longrightarrow$ [$CH_3O-C_6H_4-OCH_2CH_3$]

(c) [cyclohexene] $\longrightarrow$ [cyclohexane with H, OH, H, $N(CH_3)_2$]

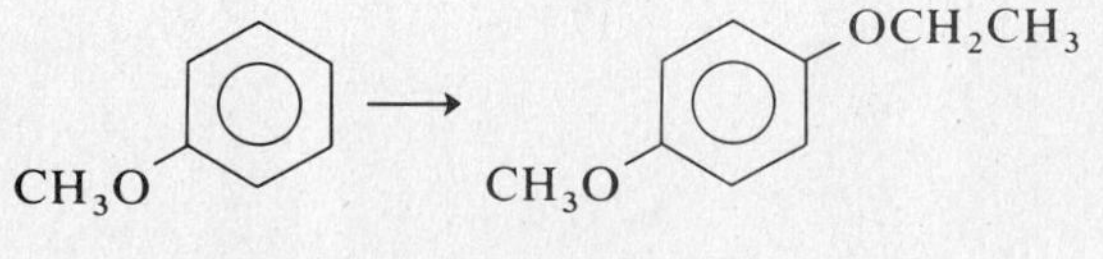

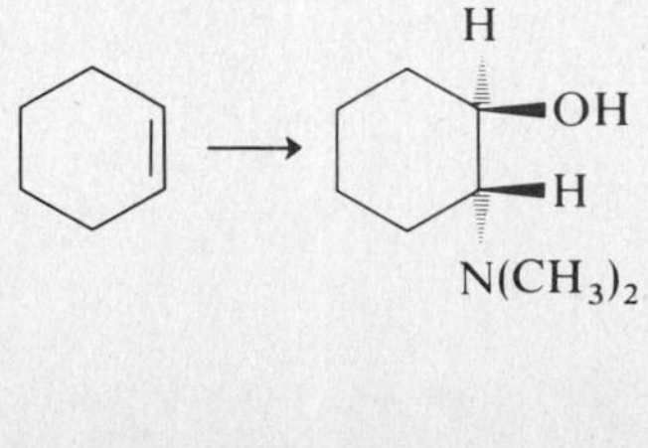

(d) 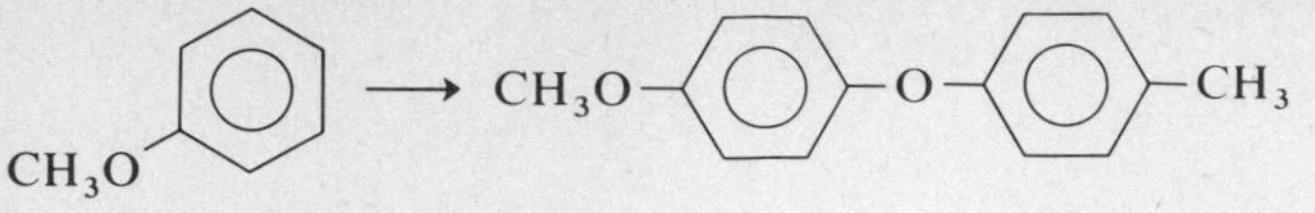 CH_3O–C_6H_5 ⟶ CH_3O–C_6H_4–O–C_6H_4–CH_3

(e) 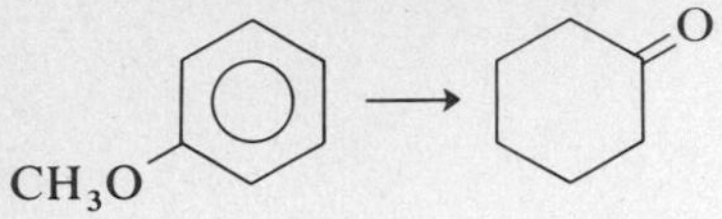 CH_3O–C_6H_5 ⟶ cyclohexanone

5. Available starting materials include (a) any alkanes, alkenes, alkynes, alkyl halides, alkanols, alkanals, alkanones, alkanoic acids, or alkanoic acid derivatives containing four or fewer carbon atoms, (b) benzene and any monosubstituted benzenes, (c) special organic reagents, and (d) any inorganic compounds. Starting with materials chosen from this group, outline syntheses for each of the following compounds:

(a) 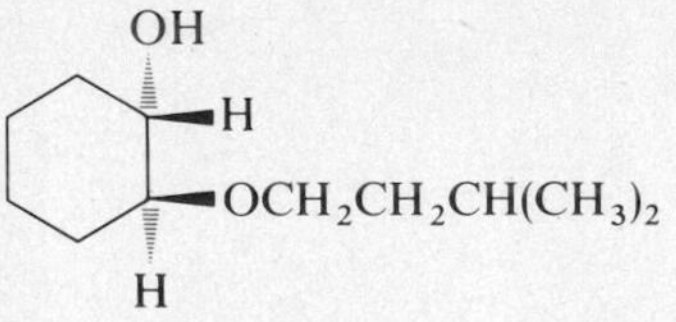 cyclohexane bearing OH, H and $OCH_2CH_2CH(CH_3)_2$, H

(b) 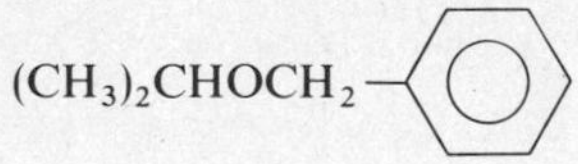 $(CH_3)_2CHOCH_2$–C_6H_5

(c) C_6H_5–$CH(OCH_3)CH_3$

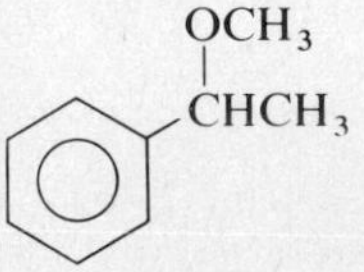

(d) $(CH_3)_2CHCH_2OCH_2CH_2CH_2CH_3$

6. Suggest a plausible explanation for the following observations:

(a) Hexyl benzenesulfonate reacts with sodium methoxide to give hexyl methyl ether in over 90% yield with only a trace of 1-hexene being formed. With potassium *t*-butoxide, however, the yield of *t*-butyl hexyl ether is only 69%, and 20% of 1-hexene is formed.

(b) The sodium salt of 2-naphthol reacts with methyl bromide to give a mixture of methyl 2-naphthyl ether and 1-methyl-2-naphthol.

(c) The compound pictured is a powerful methylating reagent, converting alcohols and phenols to the methyl ethers with great facility.

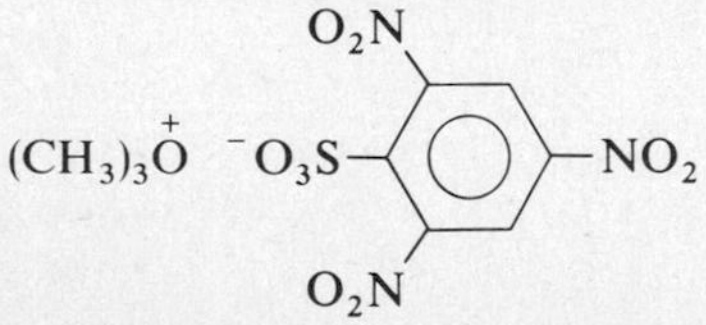

$(CH_3)_3\overset{+}{O}\ \ {}^{-}O_3S$–$C_6H_2(NO_2)_3$ (2,4,6-trinitrobenzenesulfonate)

(d) Thermal rearrangement of allyl 1-cyclohexenyl ether yields 2-allylcyclohexanone.

(e) *o*-Dinitrobenzene reacts with sodium methoxide to yield *o*-nitroanisole and sodium nitrite.

Chapter **24**

Amines

What drugs, what charms, what conjuration and what mighty magic.

WILLIAM SHAKESPEARE
Othello

A wide variety of amines are known, many of natural origin and many more as the result of laboratory synthesis. A surprising number have physiological properties, ranging all the way from deadly poisonous to beneficiently tranquilizing. Because of these interesting attributes, amines are often the end products of synthesis schemes, so methods for their preparation assume a special interest. They are also frequently used as synthesis intermediates, which means that a knowledge of their reactions as well as their preparation is necessary information for the well-stocked arsenal of synthesis methods. This chapter is divided into two major parts, the first one dealing with methods for preparing amines and the second dealing with their reactions.

METHODS OF PREPARATION OF AMINES

24.1 Conversion of C—Halogen to C—NR$_2$

24.1a. PREPARATION OF ALKYL AMINES. Ammonia and amines are rather strong nucleophiles (see Table 21.3) that are capable of effecting S_N2 reactions with alkyl halides and sulfonates (see Fig. 12.5). The extent of the alkylation is frequently difficult to control, with the result that mixtures of products are usually formed. For example, the action of ammonia on benzyl bromide yields not only benzylamine but also dibenzylamine, tribenzylamine and even the quaternary ammonium salt tetrabenzylammonium bromide;

$$C_6H_5CH_2Br + NH_3 \longrightarrow C_6H_5CH_2NH_2 + (C_6H_5CH_2)_2NH + (C_6H_5CH_2)_3N + (C_6H_5CH_2)_4N^+\ Br^-$$

Alkylation beyond the formation of benzylamine can be minimized by using an excess of ammonia, and it can be completely avoided by using potassium phthalimide rather than ammonia as the nucleophile. Phthalimide is sufficiently acidic to lose a proton to hydroxide ion to form the resonance-stabilized phthalimide anion:

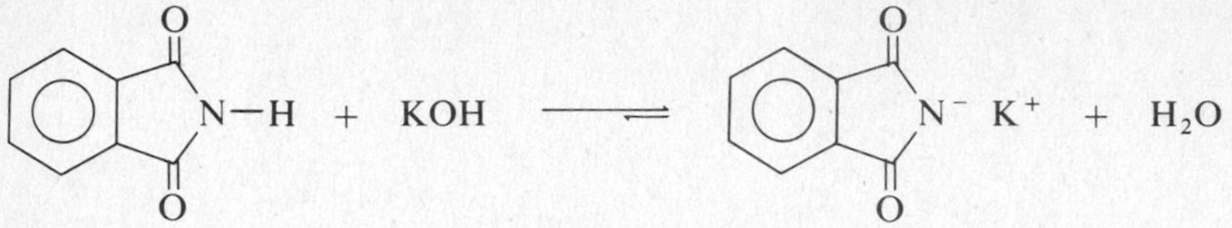

N—H + KOH ⇌ N⁻ K⁺ + H_2O

PROBLEM 24-1

Draw the resonance structures for the phthalimide anion that account for its stability as an anion.

The phthalimide anion, acting as a nucleophile, can engage in an S_N2 reaction with alkyl halides to yield N-substituted phthalimides from which the primary amines may subsequently be obtained by cleavage of the phthalimide with water or, preferably, hydrazine. The process is called the *Gabriel synthesis*:

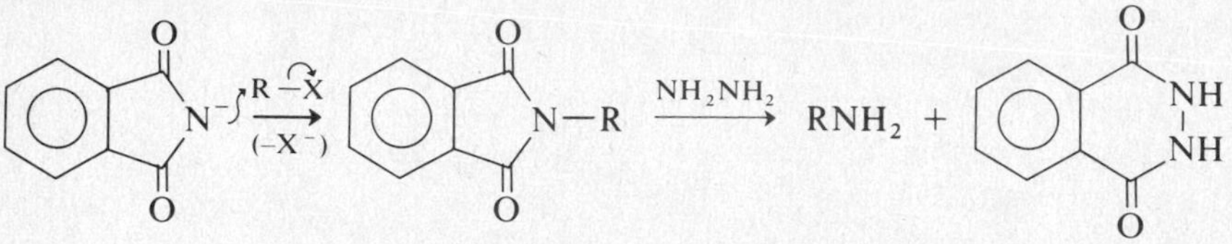

N⁻ —R—X, (−X⁻)→ N—R —NH_2NH_2→ RNH_2 + (NH, NH)

24.1b. Preparation of aryl amines. Although most aryl halides do not react with ammonia under conditions which convert alkyl halides to amines, aryl halides carrying electron-withdrawing groups *ortho* and *para* to the halogen atom form anilines with facility (*cf.* Section 21.5c). In the presence of sodium amide in liquid ammonia, even unactivated aryl halides react via benzyne intermediates to yield anilines (see Fig. 21.33). This reaction is not situselective, however, as illustrated by the action of sodium amide in liquid ammonia or *o*-bromoanisole to yield *m*-aminoanisole (*m*-anisidine) and on *p*-bromoanisole to yield a mixture of *m*-amino- and *p*-aminoanisole (*m*- and *p*-anisidine) (Fig. 24.1). It is this lack of situselectivity that first singled out these reactions as being something other than straightforward S_N2 displacements. It is this lack of situselectivity that provides one of the elements of proof for the benzyne mechanism.

PROBLEM 24-2

Suggest a method for deciding whether the rate-determining step in the reaction of bromobenzene with sodium amide in ammonia is the removal of the proton to form the benzene anion. Suggest a method for determining to what extent the first step of the sequence is a reversible reaction.

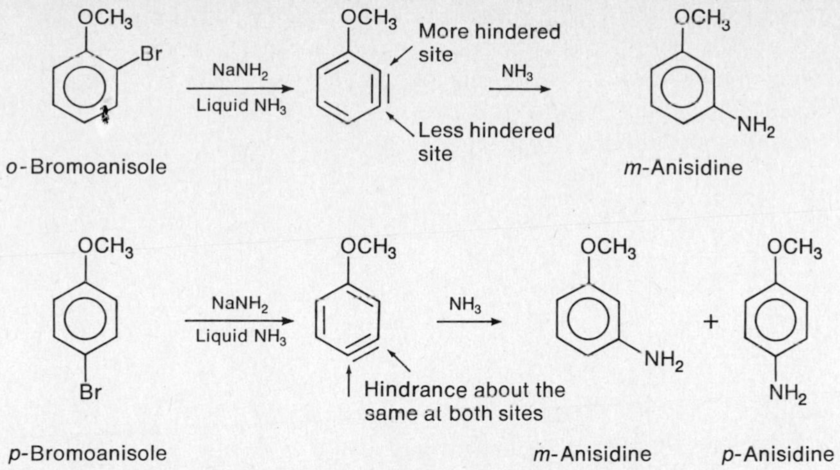

o-Bromoanisole → ($NaNH_2$, Liquid NH_3) → [benzyne; More hindered site / Less hindered site] → (NH_3) → m-Anisidine

p-Bromoanisole → ($NaNH_2$, Liquid NH_3) → [benzyne; Hindrance about the same at both sites] → (NH_3) → m-Anisidine + p-Anisidine

Fig. 24.1. Preparation of anilines from aryl halides by nucleophilic displacement reactions involving benzyne intermediates.

24.2 Conversion of C—C to C—NR_2

24.2a. Hofmann degradation of amides. The action of a mixture of aqueous sodium hydroxide and bromine on an amide of the general structure $RCONH_2$ usually results in the loss of carbon dioxide and the formation of an amine:

$$RC(=O)NH_2 + Br_2 + 2\,NaOH \longrightarrow RNH_2 + CO_2 + 2\,NaBr + H_2O$$

The reaction is called the *Hofmann degradation* and can be rationalized as shown in Fig. 24.2. In the first step, the amide, acting as a nitrogen nucleophile, effects a displacement on bromine to form the N-bromoamide. This then loses the elements of hydrogen bromide (an α-elimination reaction) to produce a nitrene (*i.e.*, the nitrogen analog of a carbene—see p. 587). Rearrangement of the nitrene

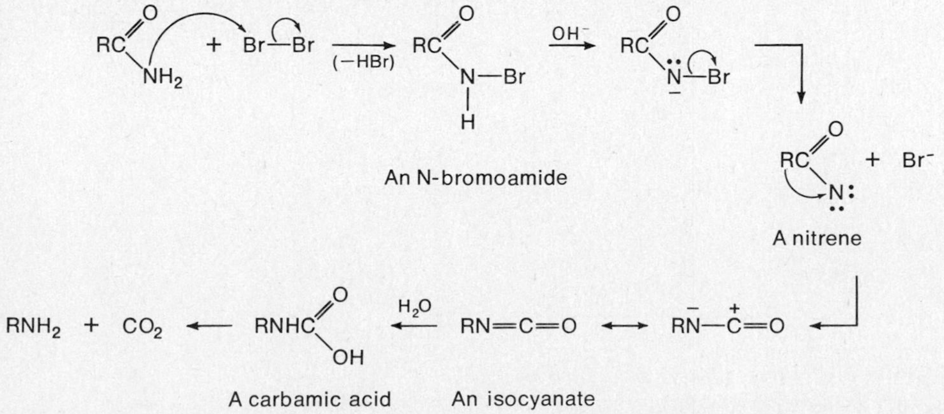

$$RC(=O)NH_2 + Br—Br \xrightarrow{(-HBr)} RC(=O)N(H)—Br \xrightarrow{OH^-} RC(=O)\ddot{N}^-—Br \longrightarrow RC(=O)\ddot{N}: + Br^-$$

An N-bromoamide — A nitrene

$$\longrightarrow R\overset{-}{N}—\overset{+}{C}=O \longleftrightarrow RN=C=O \xrightarrow{H_2O} RNHC(=O)OH \longrightarrow RNH_2 + CO_2$$

An isocyanate — A carbamic acid

Fig. 24.2. Hofmann degradation of amides to amines.

yields the isocyanate, which reacts with water to form a carbamic acid. Carbamic acids are very unstable, however, rapidly losing carbon dioxide and leaving the corresponding amine. The Hofmann degradation is quite general in its application, yielding amines from amides in which the R group is alkyl (primary, secondary, or tertiary) or aryl. It proceeds with retention of configuration of the migrating group; *e.g.*,

$$\text{(norbornyl)}-CONH_2 + Br_2 \xrightarrow[H_2O]{NaOH} \text{(norbornyl)}-NH_2$$

24.2b. Reactions of azides. Acyl and aroyl azides rearrange by thermal or photochemical processes to yield isocyanates via nitrene intermediates (Fig. 24.3).

$$RC(=O)\overset{-}{N}-\overset{+}{N}\equiv N \xrightarrow{h\nu \text{ or } \Delta} N_2 + RC(=O)\ddot{N}: \longrightarrow RN{=}C{=}O$$

$$RN{=}C{=}O \xrightarrow{H_2O} RNHCO_2H \longrightarrow RNH_2 + CO_2$$

Fig. 24.3. Rearrangement of acid azides.

If the reaction is carried out in the presence of water, the isocyanate behaves as described for the Hofmann degradation (see Fig. 24.1), forming a carbamic acid that loses carbon dioxide and yields the amine. Any one of several methods can be used for generating the azide. The action of hydrazoic acid (HN_3) on a carboxylic acid, the action of sodium azide (NaN_3) on an acid chloride, or the action of hydrazine on an ester followed by treatment of the resulting hydrazide ($RCONHNH_2$) with nitrous acid, all produce the acyl or aryl azide ($RCON_3$ or $ArCON_3$);

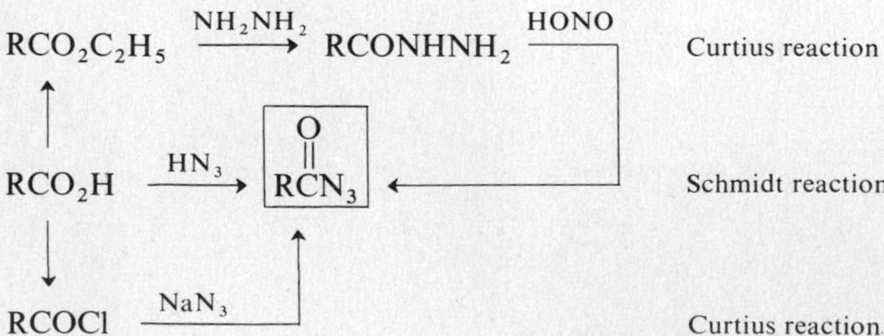

24.2c. Beckmann rearrangement. The oximes of aldehydes and ketones are susceptible to acid-catalyzed rearrangement to amides (*Beckmann rearrangement*), which, upon hydrolysis, yield the constituent amine and carboxylic acid:

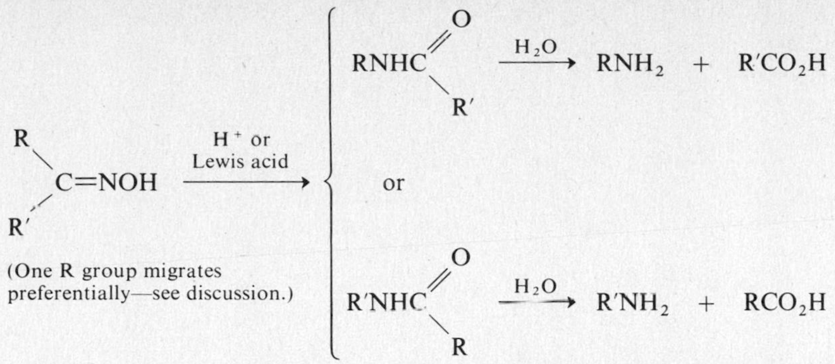

$$\underset{\text{(One R group migrates preferentially—see discussion.)}}{\overset{R}{\underset{R'}{\diagdown}}C{=}NOH} \xrightarrow[\text{Lewis acid}]{H^+ \text{ or}} \left\{ \begin{array}{l} RNHC(=O)R' \xrightarrow{H_2O} RNH_2 + R'CO_2H \\ \text{or} \\ R'NHC(=O)R \xrightarrow{H_2O} R'NH_2 + RCO_2H \end{array} \right.$$

The acid catalysts generally employed for the Beckmann rearrangement are polyphosphoric acid and phosphorus pentachloride. The course of the reaction can be delineated in terms of a series of intermediate cations (Fig. 24.4), the

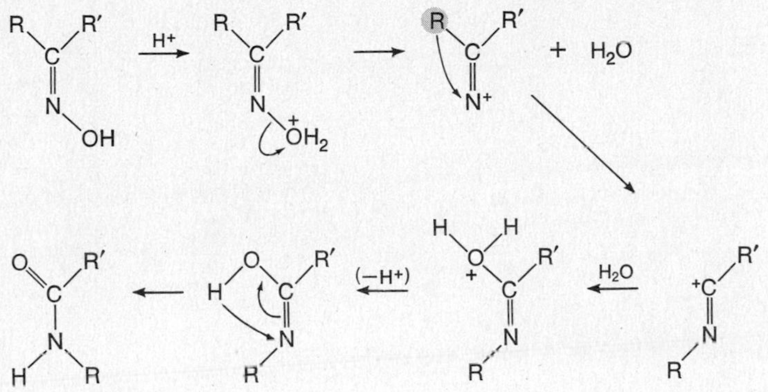

Fig. 24.4. Beckmann rearrangement of oximes.

reaction being initiated by the formation of an electron-deficient nitrogen center (a nitrenium cation), to which an adjacent alkyl or aryl group migrates with the formation of a carbonium ion. Addition of water to the carbonium ion, loss of a proton, and tautomeric rearrangement then yields the amide. Although for the bookkeeping aspects of the reaction it is convenient to depict discrete intermediates, in actual fact the various cations probably do not develop fully. In support of this idea is the strict stereochemical requirement to which the reaction conforms: only that R group that is *anti* to the hydroxyl group on the nitrogen atom undergoes migration. Thus, the amine that is obtained upon hydrolysis of the amide formed in the Beckmann rearrangement is determined by the stereochemistry of the starting oxime, the group that is *anti* to the hydroxyl group being the one that ends up in the amine. For example, the *E* and *Z* isomers of the oxime of bicyclo[4.3.0]nonan-7-one rearrange to give different products (Fig. 24.5). With oximes from alkyl aryl ketones, the aryl group is usually *anti* to the hydroxyl group, with the result that the products of rearrangement and

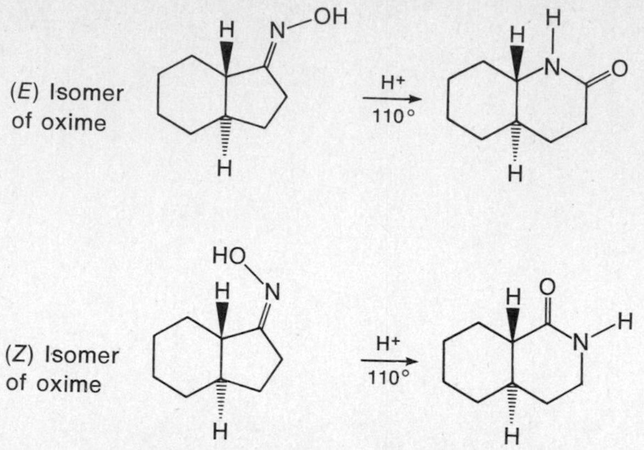

(*E*) Isomer of oxime $\xrightarrow[110^\circ]{H^+}$

(*Z*) Isomer of oxime $\xrightarrow[110^\circ]{H^+}$

Fig. 24.5. Rearrangement of isomeric oximes.

hydrolysis are an alkanecarboxylic acid and an aniline:

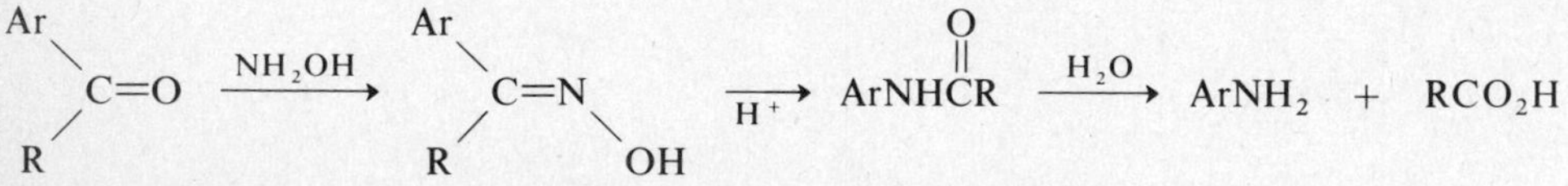

$$\underset{R}{\overset{Ar}{>}}C{=}O \xrightarrow{NH_2OH} \underset{R}{\overset{Ar}{>}}C{=}N\diagdown OH \xrightarrow{H^+} ArNH\overset{O}{\overset{\|}{C}}R \xrightarrow{H_2O} ArNH_2 + RCO_2H$$

Like the Hofmann degradation, the rearrangement of azides and the Beckmann rearrangement proceed with retention of configuration of the migrating group.

PROBLEM 24-3

Depict the rearrangement of benzophenone oxime to N-phenylbenzamide in terms of phenonium ion intermediates.

24.3 *Conversion of* C=C *to* —C—C—N

The acid-catalyzed addition of amines to alkenes is precluded, because amines are so basic that they become completely protonated in the presence of strong acids, thereby losing their nucleophilic properties. Nitriles, on the other hand, are much less basic than amines and add to alkenes in the presence of mineral acid catalysts despite their greatly reduced nucleophilicity (*Ritter reaction*). The reaction can be interpreted in terms of protonation of the alkene to form a carbonium ion followed by attachment to the nitrogen atom of the nitrile to produce a new carbonium ion; subsequent attack by a water molecule followed by loss of a proton and a tautomeric shift produces the amide, from which the amine can be obtained by hydrolysis. The reaction is particularly useful for preparing primary amines, in which the amino group is attached to a tertiary alkyl framework (Fig. 24.6). Compounds of this type cannot be obtained by substitution reactions with *t*-alkyl halides and ammonia, because elimination of hydrogen halide occurs as the major reaction.

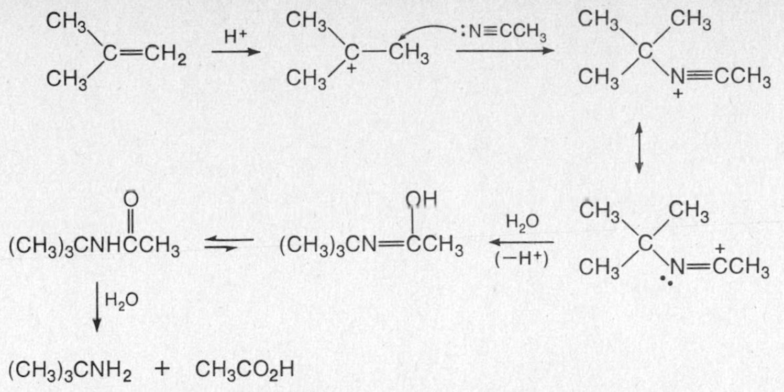

$$(CH_3)_2C{=}CH_2 \xrightarrow{H^+} (CH_3)_2\overset{+}{C}{-}CH_3 \xrightarrow{:N{\equiv}CCH_3} (CH_3)_3C{-}\overset{+}{N}{\equiv}CCH_3 \longleftrightarrow (CH_3)_3C{-}\ddot{N}{=}\overset{+}{C}CH_3$$

$$(CH_3)_3C{-}\ddot{N}{=}\overset{+}{C}CH_3 \xrightarrow[(-H^+)]{H_2O} (CH_3)_3CN{=}\overset{OH}{\overset{|}{C}}CH_3 \rightleftharpoons (CH_3)_3CNH\overset{O}{\overset{\|}{C}}CH_3 \xrightarrow{H_2O} (CH_3)_3CNH_2 + CH_3CO_2H$$

Fig. 24.6. Preparation of primary amines via the Ritter reaction.

24.4 *Conversion of* C=O *to* —C—N

Ammonia, primary amines, and secondary amines add to aldehydes and ketones to give 1-amino-1-hydroxy compounds, which are the nitrogen analogs of hydrates, hemiacetals, and hemiketals:

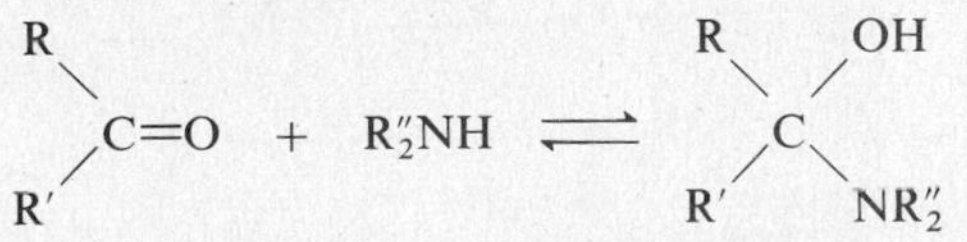

$$RR'C{=}O + R''_2NH \rightleftharpoons RR'C(OH)NR''_2$$

The equilibrium constants for these reactions usually are not large enough to allow the products to be isolated. With certain nucleophiles, however, the adducts react further, undergoing replacement of the hydroxyl group by the nucleophile to form isolable amines. This displacement process can be interpreted as the result of the formation of an iminium intermediate ($R_2C{=}\overset{+}{N}H_2$) followed by attack of the nucleophile at the strongly electrophilic iminium carbon atom:

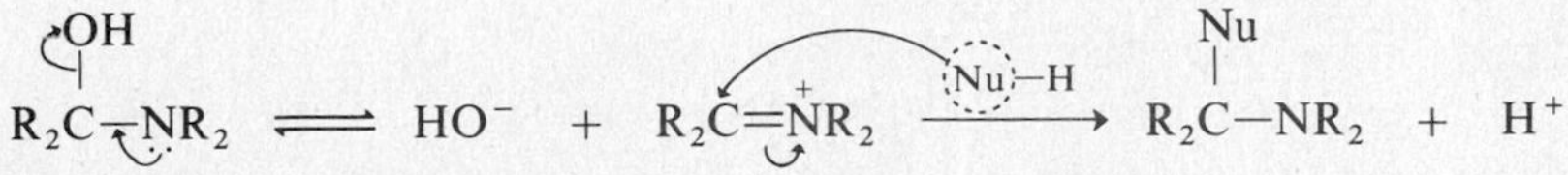

$$R_2C(OH){-}\ddot{N}R_2 \rightleftharpoons HO^- + R_2C{=}\overset{+}{N}R_2 \xrightarrow{Nu{-}H} R_2C(Nu){-}NR_2 + H^+$$

PROBLEM 24-4

Why would you expect the iminium ion to be much more susceptible to nucleophilic attack than the imine itself?

A typical example of this type of reaction is observed when cyclohexanone is treated with ammonia and formic acid to yield cyclohexylamine. The nucleophile in this case is hydride, provided by the formate anion produced from formic acid (*Leuckart reaction*) (Fig. 24.7). As an alternative to formate anion for the reducing agent, molecular hydrogen in the presence of a platinum catalyst can be used for the replacement of the hydroxyl group (Fig. 24.8).

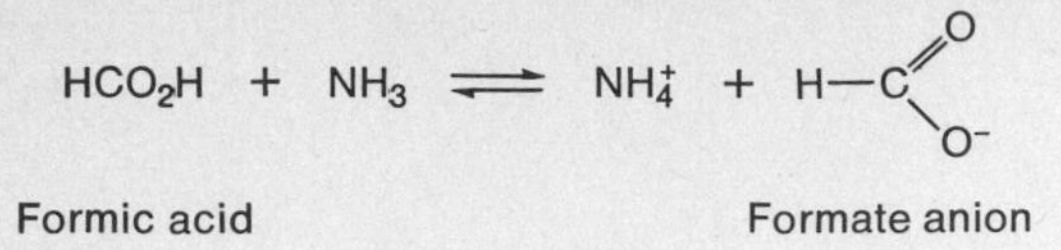
HCO2H + NH3 ⇌ NH4+ + H—C(=O)O−
Formic acid
Formate anion

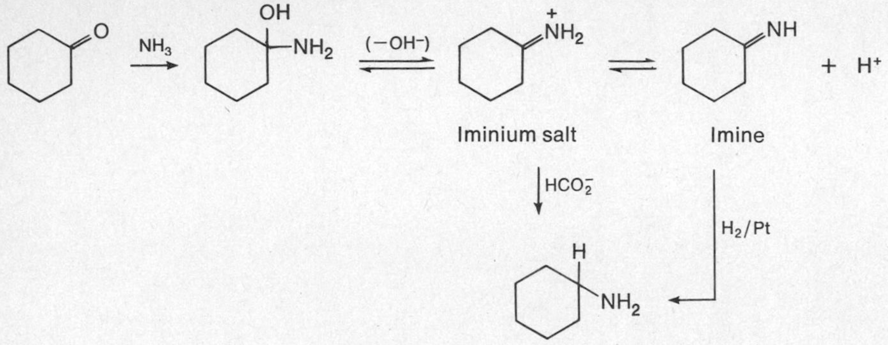
NH3
OH
NH2
(−OH−)
NH2
Iminium salt
NH
Imine
+ H+
HCO2−
H2/Pt
H
NH2

Fig. 24.7. Preparation of cyclohexyl amine via reductive amination.

Formaldehyde is frequently employed as the carbonyl component in reductive amination reactions. The process, which is known as the *Eschweiler–Clarke* reaction, leads to N-methyl products. Thus, a mixture of cyclohexylamine, formaldehyde, and formic acid yields cyclohexyldimethylamine (Fig. 24.8). The reactions shown in Figs. 24.7 and 24.8 can be combined into a "one-flask" process, in which cyclohexanone, formaldehyde, ammonia, and formic acid are mixed and heated to yield cyclohexyldimethylamine directly.

Certain carbon nucleophiles also have the ability to add to imines (or iminium compounds). In particular, carbon nucleophiles derived from "active methylene" compounds serve this purpose. For example, if cyclohexanone is mixed with formaldehyde and dimethylamine, an aminoalkylation reaction (*Mannich reaction*) takes place; formaldehyde and dimethylamine react to form first the aminoalcohol and then the iminium compound, and the enol form of cyclohexanone then effects a nucleophilic displacement to yield 2-(N,N-dimethylaminomethyl)-cyclohexanone (Fig. 24.9).

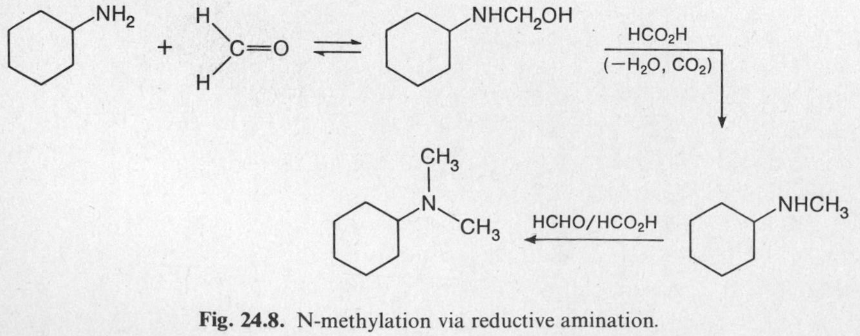
NH2
H
C=O
H
NHCH2OH
HCO2H
(−H2O, CO2)
CH3
N
CH3
HCHO/HCO2H
NHCH3

Fig. 24.8. N-methylation via reductive amination.

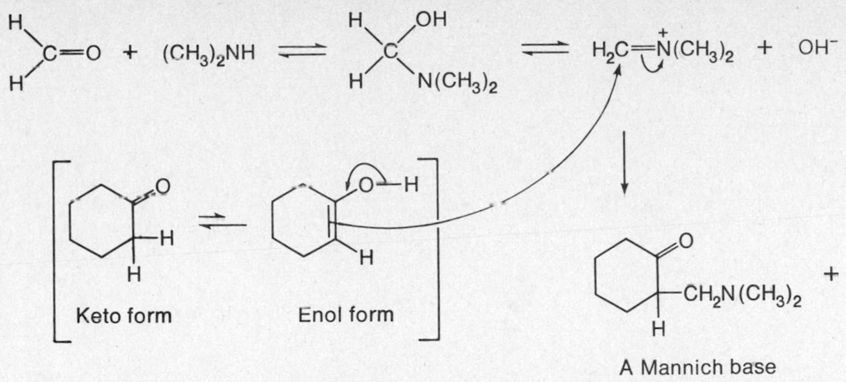

$$H_2C{=}O + (CH_3)_2NH \rightleftharpoons H_2C(OH)N(CH_3)_2 \rightleftharpoons H_2C{=}\overset{+}{N}(CH_3)_2 + OH^-$$

Keto form ⇌ Enol form → A Mannich base + H^+

Fig. 24.9. Aminoalkylation of cyclohexanone (Mannich reaction).

PROBLEM 24-5

Why is dimethylamine preferable to methylamine in the Mannich reaction? What would you predict to be the products from the reaction of acetone, methylamine, and formaldehyde?

24.5 Reduction of Oxidized Nitrogen Compounds

Most compounds in which the nitrogen is present in a more highly oxidized state than that represented by amines can be reduced to amines. Thus, nitro compounds, azo compounds, hydrazo compounds, hydrazones, oximes, etc., are all reduced with reagents such as lithium aluminum hydride, molecular hydrogen and a platinum catalyst, or iron and hydrochloric acid to the corresponding amines. This method is particularly important for the preparation of aromatic amines, for the requisite nitro compounds can be obtained by direct nitration of the aromatic hydrocarbon (Fig. 24.10).

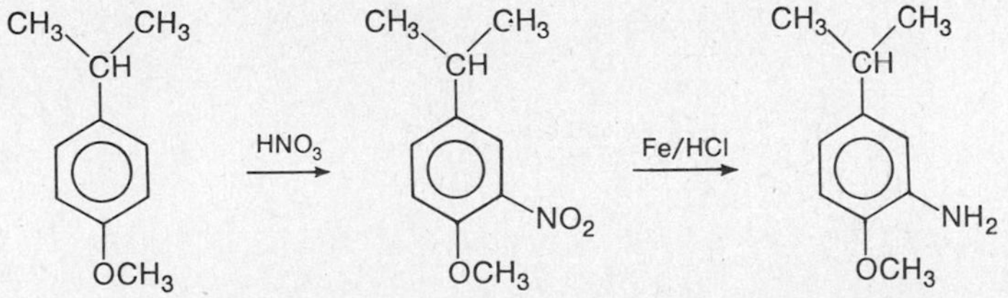

$$\xrightarrow{HNO_3} \qquad \xrightarrow{Fe/HCl}$$

Fig. 24.10. Synthesis of aromatic amines by nitration and reduction.

Because of the synthetic importance of the reduction of aromatic nitro compounds, the reaction has been studied in considerable detail. It has been found that a number of different products in addition to aniline can be obtained in the reduction of nitrobenzene, including those illustrated in Fig. 24.11.

Also of synthetic utility is the selective reduction of one of the nitro groups in a polynitro compound, as illustrated by the examples in Fig. 24.12. Unfortunately,

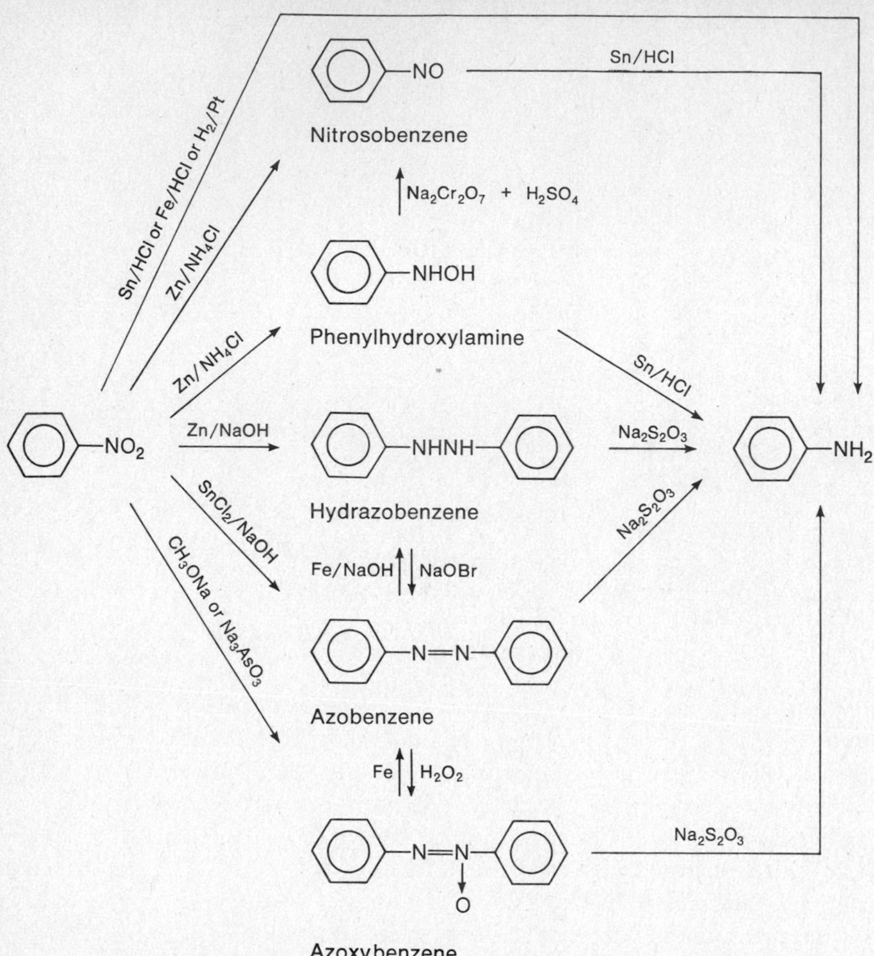

Fig. 24.11. Intermediate oxidation-state nitrogen-containing compounds derivable by reduction of nitroaromatics.

it is usually not possible to predict in advance which nitro group will be selectively reduced in any particular system, and it is necessary to determine this by empirical observation.

Also included in the group of reducible nitrogen moieties is the cyano function; nitriles react with lithium aluminum hydride, or with hydrogen and a platinum catalyst, to yield the corresponding primary amines. A useful sequence for amine preparations involves the introduction of a cyano group into a carbon framework by nucleophilic substitution followed by reduction:

$$RX + CN^- \longrightarrow RCN \longrightarrow RCH_2NH_2$$

$$ArN_2^+ + Cu_2(CN)_2 \longrightarrow ArCN \longrightarrow ArCH_2NH_2$$

Amides of carboxylic acids are reduced with lithium aluminum hydride or diborane to amines (see Fig. 15.17). This provides a convenient method for

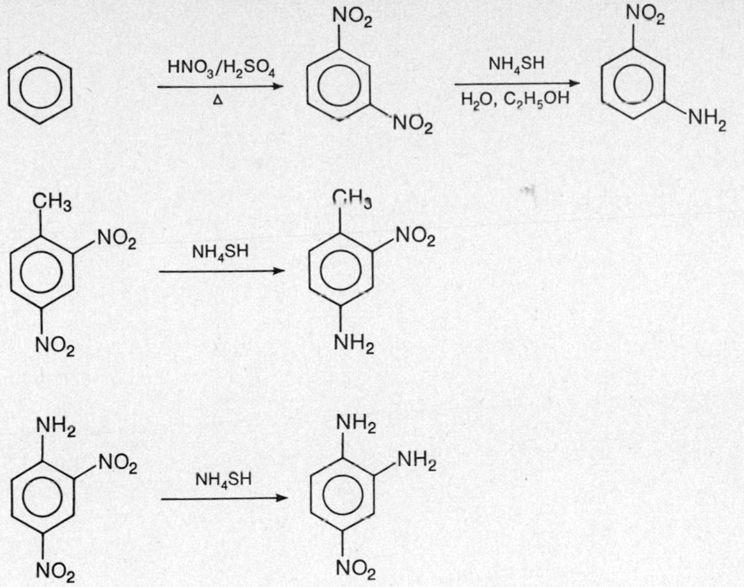

$$C_6H_6 \xrightarrow[\Delta]{HNO_3/H_2SO_4} C_6H_4(NO_2)_2 \xrightarrow[H_2O,\ C_2H_5OH]{NH_4SH} NO_2C_6H_4NH_2$$

$$CH_3C_6H_3(NO_2)_2 \xrightarrow{NH_4SH} CH_3C_6H_3(NO_2)(NH_2)$$

$$NH_2C_6H_3(NO_2)_2 \xrightarrow{NH_4SH} (NH_2)_2C_6H_3NO_2$$

Fig. 24.12. Selective reduction of dinitrobenzenes.

introducing one alkyl group at a time without the complication of polyalkylation:

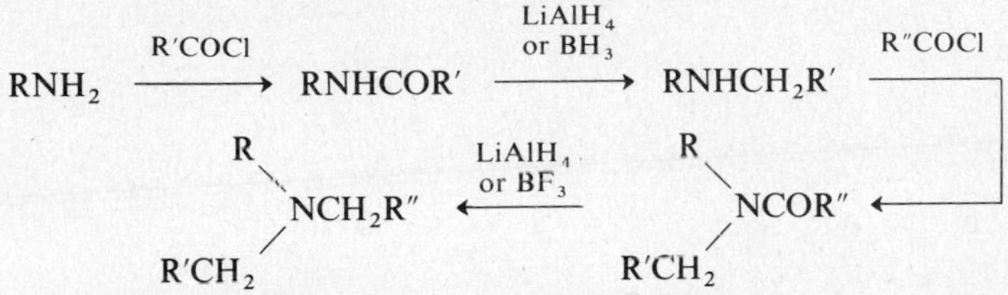

$$RNH_2 \xrightarrow{R'COCl} RNHCOR' \xrightarrow[\text{or } BH_3]{LiAlH_4} RNHCH_2R' \xrightarrow{R''COCl} \underset{R'CH_2}{\overset{R}{>}}NCOR'' \xrightarrow[\text{or } BF_3]{LiAlH_4} \underset{R'CH_2}{\overset{R}{>}}NCH_2R''$$

Table 24.1 summarizes the methods that are useful for the synthesis of amines.

Table 24.1 METHODS OF PREPARATION OF AMINES

Starting Materials	Reagents	Products	Comments
Alkyl halides (R-Halogen)	NH_3 $R'NH_2$ R'_2NH	RNH_2 $RNHR'$ RNR'_2	Polyalkylation likely Reduced yields with secondary alkyl halides; poor yields or no yield with tertiary alkyl halides
	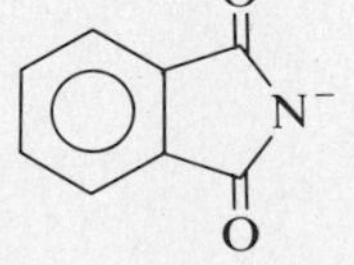 Phthalimide anion (N^-)	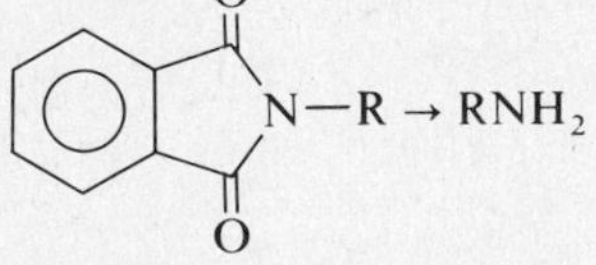 N-alkylphthalimide ($N{-}R$) $\rightarrow RNH_2$	S_N2 process; hydrolysis or hydrazinolysis yields RNH_2
Aryl halides (Ar-Halogen)	$NaNH_2$, liquid NH_3	$ArNH_2$	Benzyne intermediates; reaction is not situselective

Table 24.1—*cont.*

Starting Materials	Reagents	Products	Comments
Carboxylic acid derivatives $\left(RC\begin{smallmatrix}\nearrow O\\ \searrow Y\end{smallmatrix}\right)$[a]			
Amides ($RCONR'_2$)	$LiAlH_4$	$RCH_2NR'_2$	An alternative to direct alkylation—avoids problem of polyalkylation
Amides ($RCONH_2$)	$Br_2 + NaOH$	RNH_2	Hofmann degradation; retention of configuration of migrating group
$RCON_3$ prepared from:	Heat or light (H_2O)	RNH_2	
$RCO_2H + HN_3$			Schmidt degradation; retention of configuration of migrating group
$RCOCl + NaN_3$		RNH_2	Curtius reaction; retention of configuration of migrating group
$RCONHNH_2 + HONO$		RNH_2	
Oximes $\left(\begin{smallmatrix}R\\ R'\end{smallmatrix}\!\!>C{=}NOH\right)$[a]	Protic acids (*e.g.*, polyphosphoric acid), PCl_5, or $POCl_3$	$RNHCOR' \rightarrow RNH_2$ or $R'HNCOR \rightarrow R'NH_2$	Beckmann rearrangement; the R group *anti* to the OH group migrates to nitrogen. The amine is obtained by hydrolysis of the amide.
	Reducing agents (e.g., $LiAlH_4$; $H_2 + Pt$)	$\begin{smallmatrix}R\\ R'\end{smallmatrix}\!\!>CHNH_2$	
Alkenes $\left(>C{=}C<\right)$	RCN + strong acid	$-\overset{\mid}{\underset{\mid}{C}}-\overset{\mid}{\underset{\mid}{CH}}$ (RCONH on C) $\longrightarrow -\overset{\mid}{\underset{\mid}{C}}-\overset{\mid}{\underset{\mid}{CH}}$ (H_2N on C)	Ritter reaction; the amine is obtained by hydrolysis of the amide.
Aldehydes and ketones[a] $\begin{smallmatrix}R\\ R'\end{smallmatrix}\!\!>C{=}O$	$R''_2NH + HCO_2H$	$\begin{smallmatrix}R & & NR''_2\\ & C & \\ R' & & H\end{smallmatrix}$	Reductive amination
Active methylene compounds, $\begin{smallmatrix}Z\\ Z'\end{smallmatrix}\!\!>CH_2$ (where Z,Z' = $>C{=}O$, $-CN$, $-NO_2$, etc.)	$HCHO + R_2NH$	$\begin{smallmatrix}Z\\ Z'\end{smallmatrix}\!\!>CHCH_2NR_2$	Mannich reaction; aminomethylation
Nitro compounds (RNO_2)[a]	Reducing agents (*e.g.*, Fe + HCl; $H_2 + Pt$)	RNH_2	
Nitriles (RCN)[a]	Reducing agents (*e.g.*, $LiAlH_4$; $H_2 + Pt$)	RCH_2NH_2	Product sometimes contaminated by formation of secondary amine—*i.e.*, $(RCH_2)_2NH$

[a] These methods are generally applicable to systems in which R is alkyl or aryl.

REACTIONS OF AMINES

24.6 *Reactions Involving the N—H Bond*

Amines are very much weaker acids than alcohols, having pK_a values in the region of 30–35. They do, though, react with alkali metals and with very strong organic bases such as sodium hydride, Grignard reagents, and organolithium compounds to form the corresponding salts:

$$2\,R_2NH + 2\,K \longrightarrow 2\,R_2N{-}K + H_2$$

$$R_2NH + NaH \longrightarrow R_2N{-}Na + H_2$$

$$R_2NH + C_4H_9Li \longrightarrow R_2N{-}Li + C_4H_{10}$$

Sodium amide ($NaNH_2$) and alkali dialkylamides [*e.g.*, $(CH_3)_2NLi$] are widely used as catalysts in organic reactions, for although they are strongly basic they are not very nucleophilic.

PROBLEM 24-6

Calculate the equilibrium constant for the reaction

$$(CH_3)_2N^- + CH_3OH \rightleftharpoons (CH_3)_2NH + CH_3O^-$$

24.7 *Reactions Involving Amines as Nucleophiles*

24.7a. NUCLEOPHILIC DISPLACEMENT AT SATURATED CARBON ATOMS. The use of ammonia and amines as nucleophilic agents in S_N2 reactions with halides is illustrated by numerous reactions in this and previous chapters (see Fig. 12.5, Table 21.7). A typical example of the application of the N-alkylation process in a synthesis sequence is seen in Fig. 24.13.

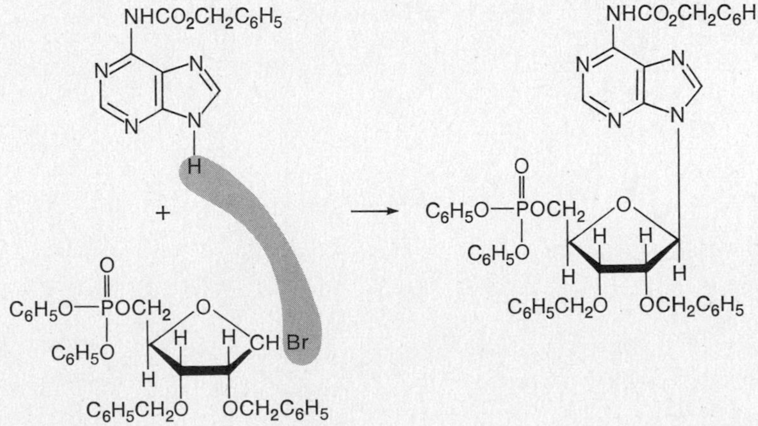

NHCO₂CH₂C₆H₅ ... (figure)

Fig. 24.13. An N-alkylation reaction as one of the steps in a synthesis sequence leading to a naturally occurring nucleotide.

24.7b. Nucleophilic displacement at carbonyl carbon atoms. As indicated previously, ammonia, primary amines, and secondary amines add to aldehydes and ketones to give amino alcohols in processes for which the equilibrium constants are not favorable to product formation. The conversion of the amino alcohols to isolable compounds can be promoted by (a) nucleophilic attack by hydride ion, leading to the products of reductive amination (see Section 24.4), (b) nucleophilic attack by carbon nucleophiles, leading to the products of aminoalkylation (see Section 24.4), (c) dehydration involving the adjacent N—H bond leading to imines, or (d) dehydration involving the adjacent C—H bond leading to enamines (Fig. 24.14). The dehydration reactions are

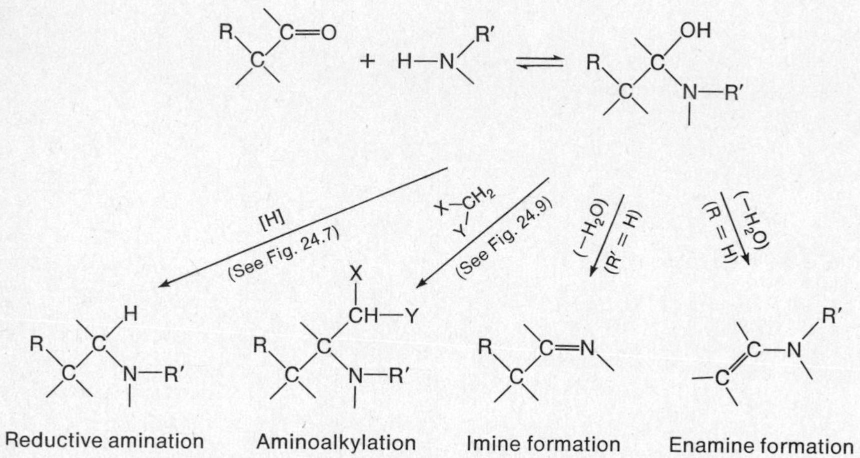

[H]
(See Fig. 24.7)
X—CH2
Y
(See Fig. 24.9)
(−H2O)
(R′ = H)
(−H2O)
(R = H)

Reductive amination Aminoalkylation Imine formation Enamine formation

Fig. 24.14. Conversion of amino alcohols to stable, isolable products.

readily reversible and are usually carried out in such a fashion that the water is removed by azeotropic distillation with benzene as it forms (see Fig. 15.7), thereby driving the reaction to completion. For example, aniline reacts with cyclopentanone to form an imine; dimethylamine reacts with cyclopentanone to form an enamine:

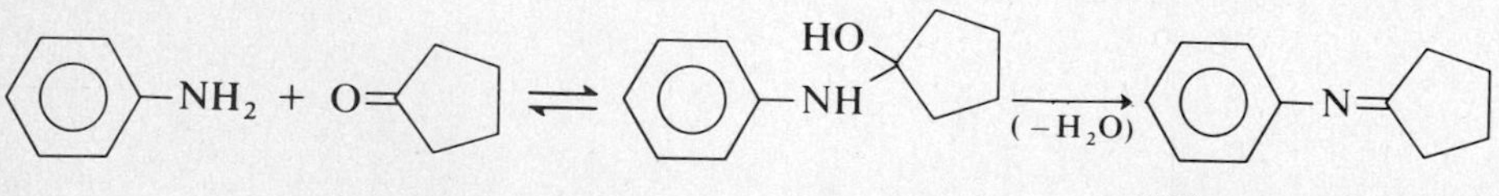

$(-H_2O)$

N-Phenylcyclopentylimine

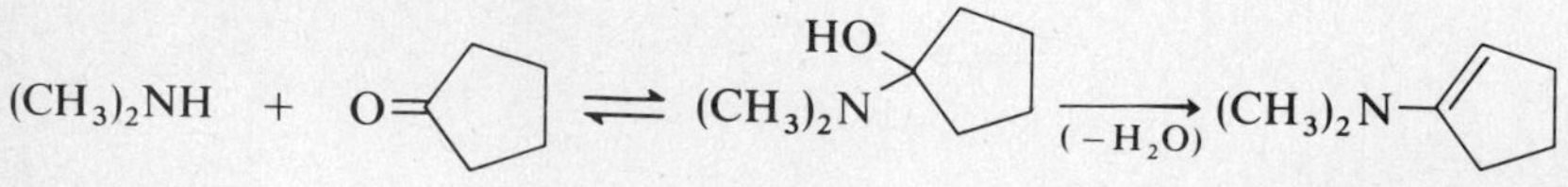

$(CH_3)_2NH + O{=}C_5H_8 \rightleftharpoons (CH_3)_2N(HO)C_5H_8 \xrightarrow{(-H_2O)} (CH_3)_2N{-}C_5H_7$

N,N-Dimethylaminocyclopentene
(An enamine)

Ammonia, primary amines, and secondary amines act as nucleophiles toward carboxylic acid derivatives, engaging in addition-elimination reactions to yield amides (see Figs. 15.4 and 15.5). Tertiary amines also react in a comparable fashion, but the products formed are acyl (or aroyl) ammonium salts that in most instances are very reactive and are not readily isolated:

$$\text{C}_5\text{H}_5\text{N} + CH_3C(=O)Cl \rightleftharpoons CH_3C(O^-)(Cl)-\overset{+}{N}C_5H_5 \longrightarrow CH_3C(=O)-\overset{+}{N}C_5H_5 \; Cl^-$$

Pyridine

1-Acetylpyridinium chloride

1-Acetylpyridinium chloride is very reactive toward nucleophiles, such as H_2O, ROH, etc.

PROBLEM 24-7

Why is $RC(=O)-\overset{+}{N}C_5H_5$ (1-acylpyridinium) more reactive toward nucleophiles than $RC(=O)-NC_5H_{10}$ (1-acylpiperidine)?

24.7c. NUCLEOPHILIC DISPLACEMENT AT HALOGEN ATOM. Alkylamines react with hypochlorous acid or *t*-butyl hypochlorite in the presence of base to form N-chloroamines, compounds that have a so-called "positive halogen" and are strong oxidizing agents:

$$R_2NH + HOCl \xrightarrow{OH^-} R_2N-Cl + H_2O$$

$$RNH_2 + HOCl \xrightarrow{OH} RNHCl \longrightarrow RNCl_2$$

24.7d. NUCLEOPHILIC DISPLACEMENT AT NITROGEN ATOM. Nitrous acid in the presence of mineral acids generates nitroso cations, NO^+, which attack the nitrogen atom of amines to form N-nitroso compounds. With primary amines, the initially formed N-nitroso compounds react further, losing the elements of water to yield diazonium compounds:

$$HONO + H_2SO_4 \rightleftharpoons HSO_4^- + H_2\overset{+}{O}-NO \longrightarrow H_2O + NO^+$$

Nitroso cation

$$RNH_2 + NO^+ \rightleftharpoons RNH_2^+-NO \rightleftharpoons RNH-N=O + H^+$$

$$[R\overset{+}{N}\equiv N: \longleftrightarrow R\ddot{N}=\overset{+}{N}:] + H_2O \longleftarrow \left[RNH-\overset{+}{N}-OH \longleftrightarrow RNH-N=\overset{+}{O}H\right]$$

Diazonium compound

The diazonium compounds formed from primary alkylamines (RNH_2) are not isolable, for they decompose almost immediately to yield nitrogen and carbonium ions. The fate of the carbonium ion depends on its structure and on the reaction conditions; possible pathways include reaction with nucleophiles to form substitution products, expulsion of proton to form alkenes, or rearrangement followed by substitution or elimination. Examples illustrating these events are given in the chapter on alcohols (see Figs. 22.3 and 22.35). Still another example is shown in Fig. 24.15, which illustrates the use of the reaction in ring expansion processes, and which also emphasizes the fact that mixtures of products are usually obtained.

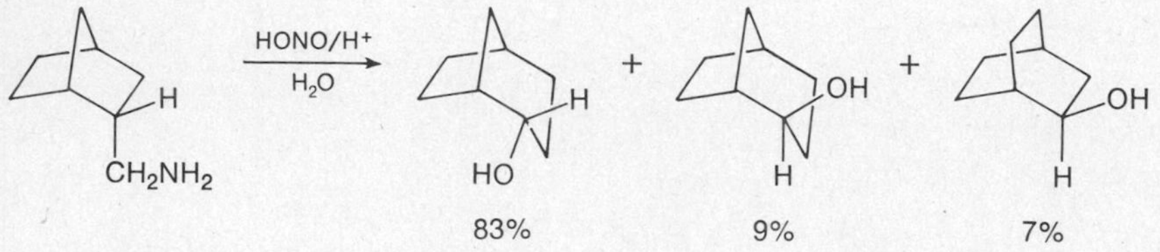

CH_2NH_2 —HONO/H^+, H_2O→ HO (83%) + OH (9%) + OH (7%)

Fig. 24.15. Reaction of *endo*-2-aminomethylbicyclo[2.2.1]heptane with nitrous acid.

PROBLEM 24-8

Outline the mechanism of the reaction shown in Fig. 24.15.

With secondary amines, the initially formed N-nitroso compounds are stable, yellow substances that are not basic enough to dissolve in dilute aqueous acid. Tertiary amines react with nitrous acid to form N-nitrosonium salts. Attempts to isolate these, however, usually cause either decomposition back to the original amine or to the N-nitroso compound corresponding to that of a secondary amine, one of the R groups originally attached to the nitrogen ending up as an aldehyde or ketone. The different behavior of primary, secondary, and tertiary amines toward nitrous acid provides a quick and convenient test for determining the type of amine with which one is dealing (see Fig. 12.7).

The diazonium compounds obtained from primary aromatic amines ($ArNH_2$), although considerably more stable than those from aliphatic amines, are very reactive and rarely are isolated. Instead, they are prepared in aqueous solution and used directly, *in situ*, in nucleophilic substitution reactions, providing a variety of possibilities for synthesis (see Fig. 16.6).

In addition to being very susceptible to nucleophilic displacement, aryldiazonium compounds can also act as electrophiles in electrophilic aromatic substitution reactions. They are notably weak in this respect, however, and react only with strongly activated aromatic compounds such as phenols and anilines. For example, the *bis*-diazonium salt obtained from 4,4′-diaminobiphenyl (benzidine) effects an electrophilic substitution reaction on salicylic acid to yield the bright yellow dye known as Chrysamine G:

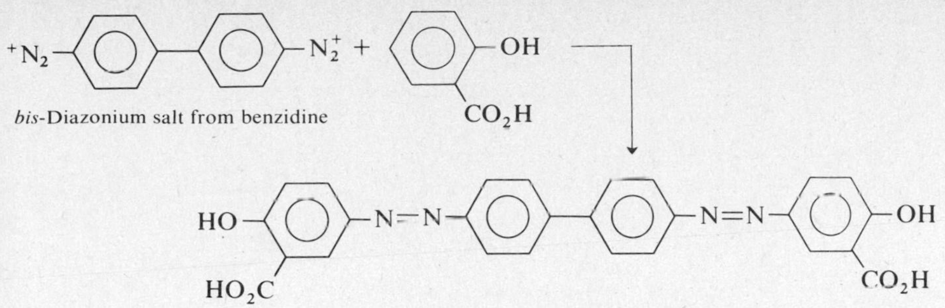

$^{+}N_2$–(biphenyl)–N_2^{+} + m-hydroxybenzoic acid (OH, CO_2H) →

bis-Diazonium salt from benzidine

HO–(ring, HO_2C)–N=N–(biphenyl)–N=N–(ring, CO_2H)–OH

Chrysamine G (a yellow dye)

N-Alkylanilines react with nitrous acid to form N-nitroso compounds, and N,N-dialkylanilines react in the *para* position of the aromatic ring to yield *p*-nitrosoanilines (Fig. 24.16).

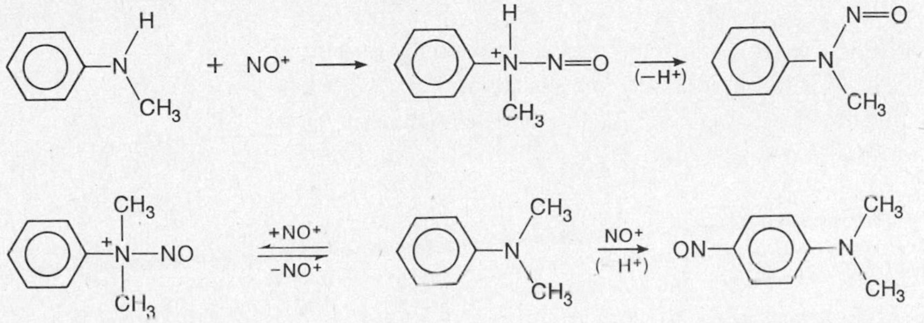

Fig. 24.16. Reaction of N-alkyl and N,N-dialkylanilines with nitrous acid.

24.8 Oxidation Reactions

Except in a few special cases the oxidation of amines has relatively little preparative value. Amines in which the nitrogen is attached to a primary or secondary position of an alkyl framework (*i.e.*, $RCH_2-N\langle$ and $R_2CH-N\langle$) react with aqueous potassium permanganate in neutral solution to produce imines, which subsequently hydrolyze to the corresponding aldehydes or ketones. Amines in which an $-NH_2$ group is attached to a tertiary position of an alkyl framework (*i.e.*, R_3C-NH_2) react under these same conditions to produce the corresponding nitro compounds (Fig. 24.17). Probably the most useful oxidative synthesis method using amines as starting materials is the conversion of N,N,N-trisubstituted amines to amine oxides with peroxy acids (RCO_3H) (Fig. 24.17). Mono- and disubstituted amines also react with peroxy acids, but complex mixtures of products are usually obtained.

24.9 Reactions Involving the R—NR_2 Bond

24.9a. Nucleophilic Displacement reactions. The carbon–nitrogen bond is even more resistant to nucleophilic displacement than is the carbon–oxygen bond. A positive charge on the nitrogen, obtained by protonation, quaternization, or conversion to the amine oxide, greatly enhances the leaving ability of the nitrogen moiety. Even so, there are only a few synthetically useful nucleophilic

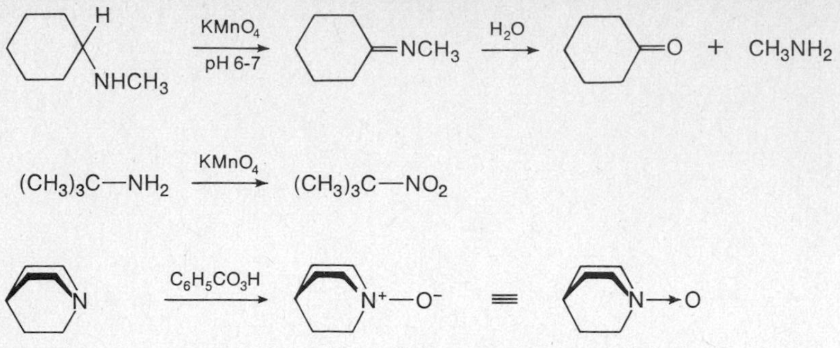

$$\text{C}_6\text{H}_{10}(\text{H})(\text{NHCH}_3) \xrightarrow[\text{pH 6-7}]{\text{KMnO}_4} \text{C}_6\text{H}_{10}{=}\text{NCH}_3 \xrightarrow{\text{H}_2\text{O}} \text{C}_6\text{H}_{10}{=}\text{O} + \text{CH}_3\text{NH}_2$$

$$(\text{CH}_3)_3\text{C—NH}_2 \xrightarrow{\text{KMnO}_4} (\text{CH}_3)_3\text{C—NO}_2$$

$$\text{quinuclidine (N)} \xrightarrow{\text{C}_6\text{H}_5\text{CO}_3\text{H}} \text{N}^+\text{—O}^- \equiv \text{N}\rightarrow\text{O}$$

An amine oxide

Fig. 24.17. Oxidation reactions of amines.

displacement reactions involving amines. One of these that has been employed in the elucidation of structure of amines provides a way for selectively removing alkyl groups from the nitrogen atom. Treatment of a tertiary amine with cyanogen bromide (BrC≡N) produces an alkyl bromide and an N-cyano compound, probably as the result of an intermediate quaternary salt that is attacked by bromide anion at the least hindered alkyl group in an S_N2 process (*von Braun reaction*);

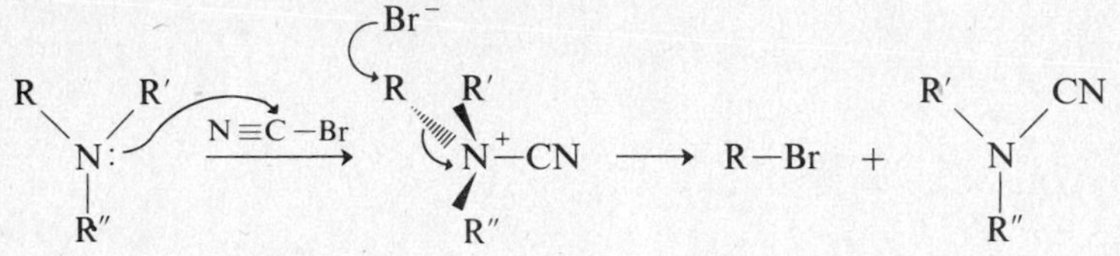

$$\text{RR}'\text{R}''\text{N:} \xrightarrow{\text{N}\equiv\text{C}-\text{Br}} [\text{RR}'\text{R}''\overset{+}{\text{N}}\text{—CN}]\ \text{Br}^- \longrightarrow \text{R—Br} + \text{R}'\text{R}''\text{N—CN}$$

PROBLEM 24-9

Why is the $-\overset{|}{\underset{|}{\text{N}}}{}^{+}-\text{O}^-$ moiety a better leaving group than $-\text{N}\langle$?

24.9b. Elimination reactions. Although quaternary ammonium groups are not particularly susceptible to displacement by external nucleophiles, internal displacement occurs rather readily to produce alkenes and tertiary amines (*Hofmann elimination*—see Fig. 12.8) (Fig. 24.18). Amine oxides undergo facile pyrolytic decomposition (*Cope elimination*) when heated to about 150° (Fig. 24.18), a temperature that is considerably lower than that necessary to convert acetic acid esters to alkenes (see Fig. 22.24). Amine oxide pyrolysis, therefore, is the recommended procedure in instances where thermally unstable alkenes are the desired product.

24.9c. Benzidine rearrangement. 4,4′-Diaminobiphenyl (benzidine) is formed when 1,2-diphenyldiazane (hydrazobenzene—see Fig. 24.11) is treated

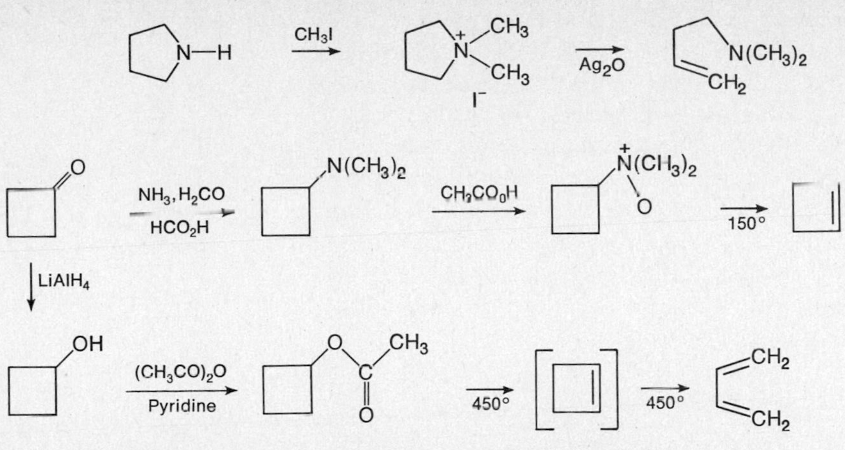

Fig. 24.18. Elimination reactions of quaternary ammonium compounds, tertiary amine oxides, and acetic acid esters (acetates).

with strong acid. The mechanism of this rearrangement has been the subject of considerable study over a period of many years. The kinetics of the reaction show a second-order dependence on hydrogen ion concentration, suggesting that the rate-determining step involves a diprotonated 1,2-diphenyldiazane. It is postulated that the dication then undergoes homolytic cleavage at the N—N bond to form a pair of radical cations (Fig. 24.19), which reorient themselves with

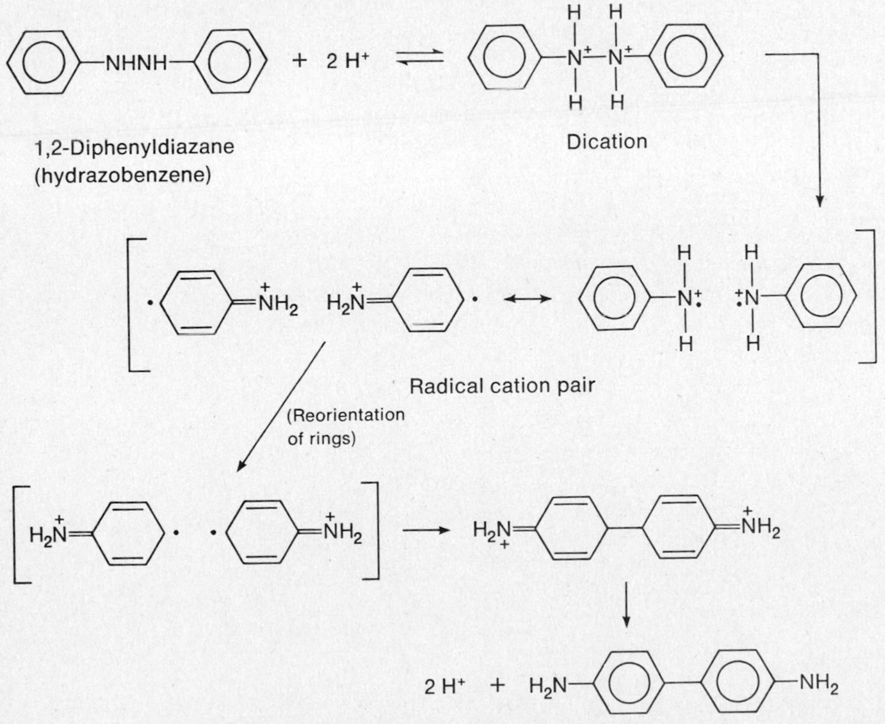

1,2-Diphenyldiazane (hydrazobenzene)

Dication

Radical cation pair

(Reorientation of rings)

Fig. 24.19. Mechanism of the benzidine rearrangement.

respect to each other and recombine, joining at their *para* positions to form a biphenyl. One of the several interesting facets of this reaction concerns the nature of this reorientation step. Do the radical cations stay close to each other during this process or do they depart from each other? That is, is the process intramolecular or intermolecules? This question has been answered by carrying out benzidine rearrangements on mixtures of hydrazobenzenes and analyzing the product to determine whether any "crossed product" is present. For example, when a mixture of hydrazobenzene and 2,2′-dimethylhydrazobenzene rearranges, only two products are formed—benzidine and a dimethylbenzidine (Fig. 24.20). None of the crossed product, a monomethylbenzidine, is formed, indicating that the rearrangement process is strictly *intramolecular*.

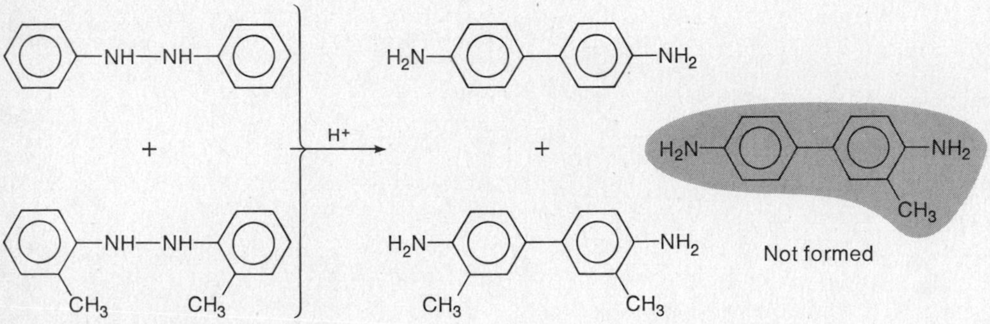

NH—NH
+
NH—NH
CH_3 CH_3
H^+
H_2N NH_2
+
H_2N NH_2
CH_3 CH_3
H_2N NH_2
CH_3
Not formed

Fig. 24.20. Illustration of a cross-over experiment in the benzidine rearrangement.

PROBLEM 24-10

Suggest another method for carrying out the "crossed product" experiment.

Blocking the *para* position of the hydrazobenzenes does not prevent the benzidine rearrangement from taking place; if one *para* position is blocked, the product is called a parasemidine; if both *para* positions are blocked, the product is called an orthosemidine (Fig. 24.21). Table 24.2 summarizes the reactions of amines.

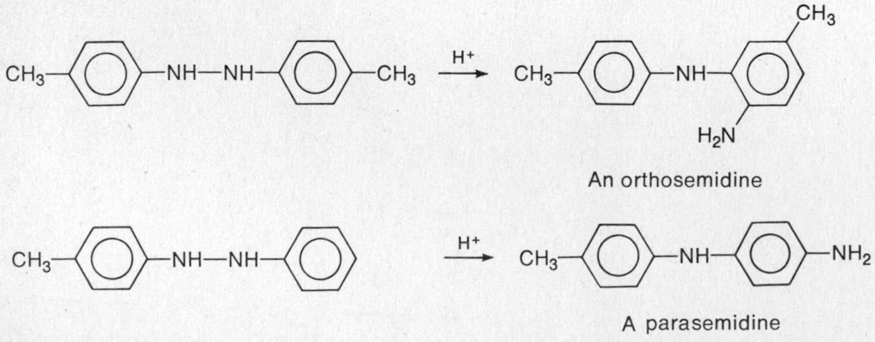

CH_3—NH—NH—CH_3 $\xrightarrow{H^+}$ CH_3—NH— (CH_3, H_2N)

An orthosemidine

CH_3—NH—NH— $\xrightarrow{H^+}$ CH_3—NH—NH_2

A parasemidine

Fig. 24.21. Illustration of the orthosemidine and parasemidine rearrangements.

Table 24.2 REACTIONS OF AMINES

Type of Reactions	Starting Materials	Reagents	Products	Comments
Reaction of $R_2N{-}H$ bond:	NH_3, RNH_2, and R_2NH (where R = alkyl or aryl)	Na, K, Li	R_2N Na, $R_2N{-}K$, $R_2N{-}Li$,	Products are very strong bases, poor nucleophiles
		$R'MgX$, $R'Li$	$R_2N{-}MgX$, $R_2N{-}Li$ (and $R'H$)	
$-\overset{\vert}{\underset{\vert}{N}}:$ as nucleophile in displacement reactions	NH_3, RNH_2, and R_2NH (where R = alkyl or aryl)	$R'X$	$R_2N{-}R'$	Amine alkylation
		$R'_2C{=}O$ + reducing agent (*e.g.*, HCO_2H)	$R_2N{-}CHR'_2$	Reductive amination
		$H{-}\overset{\vert}{\underset{\vert}{C}}{-}C(=O)R$	$>C{=}C(R){-}N<$	Enamine formation
	R_2NH	$ZZ'CH_2$ + HCHO ($Z, Z' = -C(=O)R$, $-CN$, $-NO_2$, etc. + HCHO)	R_2NCH_2CHZZ'	Mannich reaction
	RNH_2	$R'_2C{=}O$	$RN{=}CR'_2$	Imine formation
	NH_3, RNH_2, and R_2NH (where R = alkyl or aryl)	$R'C(=O)Y$ (where Y = Cl, OCOR, OR)	$R_2NC(=O)R'$	Amide formation
	R_3N	$R'C(=O)Y$	$R_3\overset{+}{N}{-}C(=O)R'$	Products usually very unstable
	RNH_2	HONO	$R^+ + N_2$	Products from carbonium ion reactions obtained
	$ArNH_2$	HONO	ArN_2^+	Products very reactive—form Ar—Nu compounds
	R_2NH	HONO	$R_2N{-}NO$	Products neutral
	R_3N	HONO	$R_3\overset{+}{N}{-}NO$	Products non-isolable and revert to starting material

Table 24.2—*cont.*

Type of Reactions	Starting Materials	Reagents	Products	Comments
	R_2NH	Cl_2 or HOCl	$R_2N{-}Cl$	Products strong oxidizing agents
	R_3CNH_2	$KMnO_4$	$R_3C{-}NO_2$	
	R_3N	RCO_3H	$R_3\overset{+}{N}{-}\overset{-}{O}$	
Reaction of N—R bond	$R_3\overset{+}{N}{-}R'$	Nu^-	$Nu{-}R' + R_3N$	Very limited application
	$H{-}\overset{\vert}{\underset{\vert}{C}}{-}\overset{\vert}{\underset{\vert}{C}}{-}\overset{+}{N}R_3$	Base	$\rangle C{=}C\langle + R_3N$	Hofmann elimination
	$H{-}\overset{\vert}{\underset{\vert}{C}}{-}\overset{\vert}{\underset{\vert}{C}}{-}\overset{+}{N}R_2(O^-)$	Heat	$\rangle C{=}C\langle + R_2NOH$	Cope elimination

24.10 *Study Guide*

I. Methods of preparation of amines
- A. Conversion of C—Halogen to C—NR_2
  1. Preparation of alkyl amines
     - (a) Problem of polyalkylation with RX and NH_3, RNH_2, or R_2NH
     - (b) Phthalimide method for synthesis of primary amines
  2. Preparation of aryl amines
     - (a) Nucleophilic displacement with aryl halides substituted in *o* and *p*-positions with electron-withdrawing groups
     - (b) Nucleophilic displacement via benzyne intermediates
- B. Conversion of C—C to C—NR_2
  1. Action of sodium hydroxide and bromine on amides (Hofmann degradation)
     - (a) Mechanism—involves nitrene intermediates
     - (b) Stereochemistry—proceeds with retention of configuration at the migrating group
  2. Thermal or photochemical rearrangement of azides
  3. Acid-catalyzed rearrangement of oximes (Beckmann rearrangement)
     - (a) Mechanism—1,2-nucleophilic rearrangement to electron-deficient nitrogen
     - (b) Stereochemistry—group that migrates is *anti* to the OH group attached to nitrogen of the oxime; group migrates with retention of configuration
- C. Conversion of C=C to C—C—N; acid-catalyzed addition of RCN to alkenes (Ritter reaction)
- D. Conversion of C=O to C—N
  1. Formation of nitrogen analogs of hemiacetals and hemiketals

2. Reductive amination—method for N-alkylation
   (a) Formate ion as the reducing agent (source of hydride ion)
   (b) Molecular hydrogen and a catalyst as the reducing agent
3. Aminoalkylation—synthesis of Mannich bases

E. Reduction of oxidized nitrogen compounds
1. Reduction of nitro compounds—particularly useful for preparation of $ArNH_2$
2. Reduction of nitriles—useful for preparation of RCH_2NH_2

II. Reactions of amines

A. Reactions involving the N—H Bond; formation of strong bases (*e.g.*, R_2N-Na)

B. Reactions involving amines as nucleophiles
1. Nucleophilic displacement at saturated carbon atoms—see Table 21.7
2. Nucleophilic displacement at carbonyl carbon atoms
   (a) Addition reactions—formation of imines, enamines, aminoalkylation products, reductive amination products
   (b) Addition-elimination reaction—formation of amides (from NH_3, primary and secondary amines) or acylammonium salts (from tertiary amines)
3. Nucleophilic displacement at halogen atom—formation of N-halo compounds
4. Nucleophilic displacement at nitrogen atom
   (a) Formation of N-nitroso compounds
   (b) Product formation from alkyl amines and nitrous acid
   (c) Product formation from primary aryl amines and nitrous acid—formation of aryldiazonium salts

C. Oxidation reactions—generally not very useful

D. Reactions involving the $R-NR_2$ bond
1. Nucleophilic displacement reactions—require formation of quaternary salt
2. Elimination reactions—require formation of quaternary salt or amine oxide

E. Benzidine rearrangement

REFERENCES

FIESER, L. F., and M. FIESER, *Advanced Organic Chemistry*, Reinhold Publishing Corp., New York, 1961, Chapters 14 and 21.

GINSBURG, D., *Concerning Amines*, Pergamon Press, Inc., New York, 1967.

PATAI, S., ed., *The Chemistry of the Amino Group*, John Wiley & Sons, Inc., New York, 1968.

SIDGWICK, N. V., *The Organic Chemistry of Nitrogen*, 3rd ed., revised and rewritten by I. T. MILLAR and H. D. SPRINGALL, Clarendon Press, Oxford, England 1966.

SMITH, P. A. S., *The Chemistry of Open-Chain Organic Nitrogen Compounds*, Vols. 1 and 2, W. A. Benjamin, Inc., New York, 1965.

PROBLEMS

1. Draw the structure of the major product(s) from each of the following reactions:

(a) 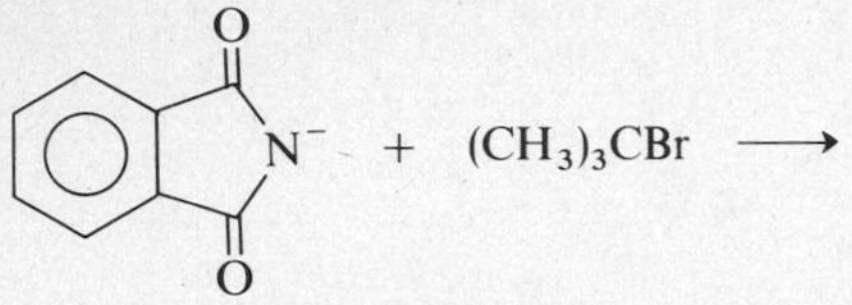 [phthalimide anion, N^-] + $(CH_3)_3CBr$ ⟶

(b) 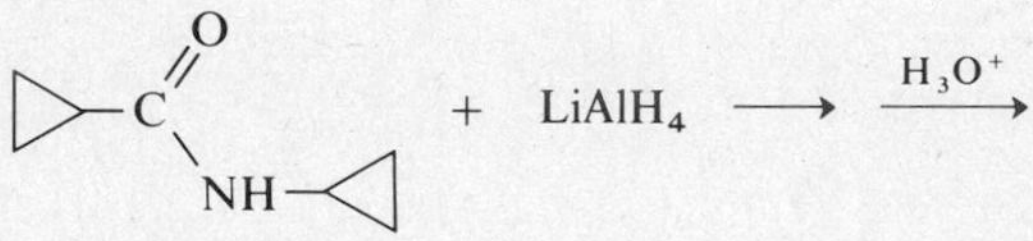 [cyclopropyl–C(=O)–NH–cyclopropyl] + $LiAlH_4$ ⟶ $\xrightarrow{H_3O^+}$

(c) [trans-decalin with axial H, $CONH_2$] + Br_2 $\xrightarrow[H_2O]{NaOH}$

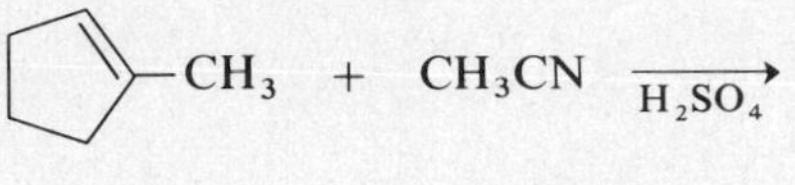

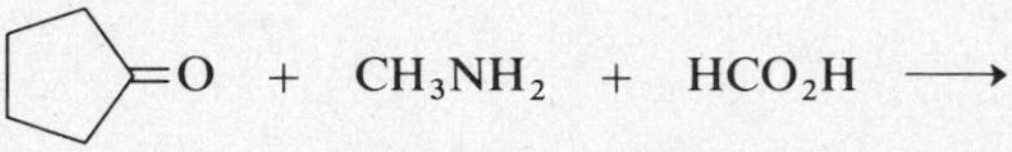

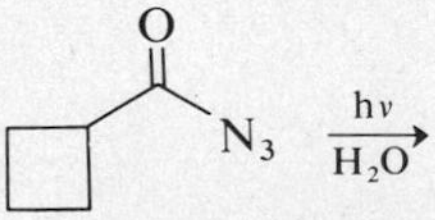

(d) [1-methylcyclopentene, $-CH_3$] + CH_3CN $\xrightarrow[H_2SO_4]{}$

(e) [cyclopentanone, =O] + CH_3NH_2 + HCO_2H ⟶

(f) [C_6H_5–C(=O)–CH_3] + HCHO + HN[pyrrolidine] ⟶

(g) [cyclobutyl–C(=O)–N_3] $\xrightarrow[H_2O]{h\nu}$

(h) [piperidine, NH] + $Br(CH_2)_4Br$ ⟶

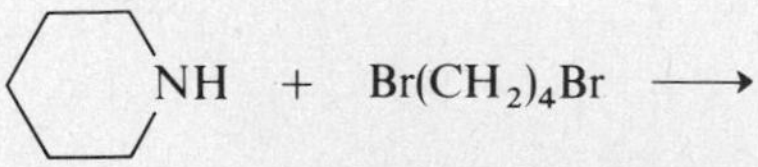

(i) $(CH_3)_3CNH_2$ + HONO $\xrightarrow[H_2O]{H^+}$

(j) 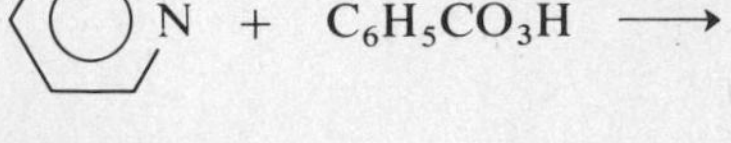 [pyridine, N] + $C_6H_5CO_3H$ ⟶

2. For each of the reactions in Problem 1 write a reasonable mechanism.

3. In each of the following cases a compound is being investigated that is known to be one of the three isomeric structures depicted. Using simple chemical methods indicate how a definitive structural assignment can be made in each instance.

(a) 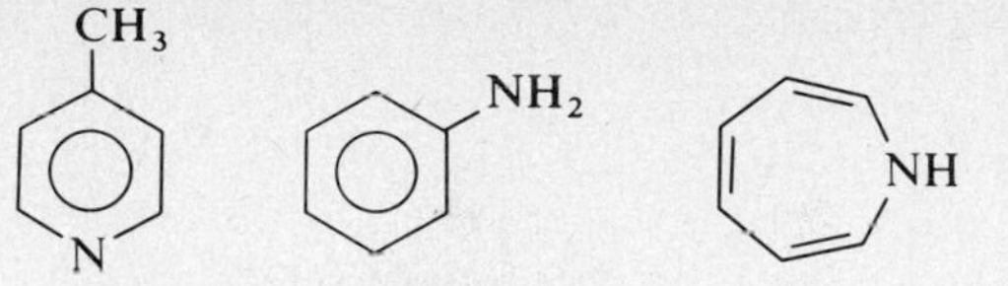 [structures: 4-methylpyridine; aniline ($C_6H_5NH_2$); azepine (NH)]

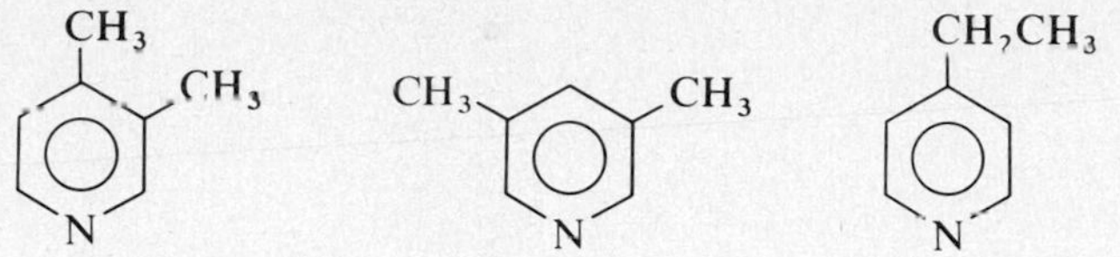

(b) [structures: 3,4-dimethylpyridine; 3,5-dimethylpyridine; 4-ethylpyridine (CH_2CH_3)]

(c) [structures: N-methylaziridine ($N-CH_3$); 2-methylaziridine (NH, CH_3)] $H_2C{=}CHCH_2NH_2$

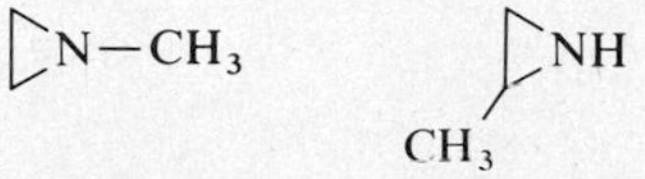

4. Each of the following problems concerns some aspect of the elucidation of structure of naturally occurring amines:

(a) An alkaloid known as piperine occurs in the fruit kernel of the pepper plant *Piper nigrum L.* It has the molecular formula $C_{17}H_{19}NO_3$, is optically inactive, is insoluble in aqueous base or acid (pK_{BH^+} 2.1), and undergoes hydrolysis to yield piperidine and a $C_{12}H_{10}O_4$ carboxylic acid. Ozonolysis of the carboxylic acid yields ethanedial, glyoxylic acid ($OCHCO_2H$), and a $C_8H_6O_3$ compound that forms a semicarbazone and gives a positive test with Fehling's solution. Treatment of the $C_8H_6O_3$ compound with boiling hydriodic acid yields methylene diiodide and a compound that can be shown by its infrared spectrum to be a 1,2,4-trisubstituted benzene. On the basis of these data suggest a structure for piperine. To provide conclusive proof for the structure you have proposed, devise a synthesis of piperine, using vanillin, piperidine, and any of the materials listed in Problems 7 and 8 as starting compounds.

(b) When ornithine labeled with ^{14}C at the carbon α to the carboxyl group [*i.e.*, $H_2N(CH_2)_3\overset{*}{C}H(NH_2)CO_2H$] is fed to tobacco plants, nicotine containing ^{14}C at the carbons designated in the structure,

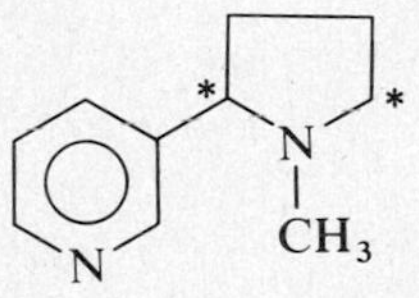

[structure: nicotine, with * at C-2 and C-5 of the pyrrolidine ring; N–CH_3] (* indicates ^{14}C)

is produced. Devise a method whereby this labeling pattern can be ascertained by chemical degradations.

(c) When the alkaloid cocaine is boiled with water, it yields methanol and a compound called benzoylecgonine. When benzoylecgonine is hydrolyzed with dilute hydrochloric acid, it yields benzoic acid and ecgonine. Ecgonine is an optically active material with the molecular formula $C_9H_{15}NO_3$. It undergoes very facile dehydration to anhydroecgonine ($C_9H_{13}NO_2$) and can be oxidized, with loss of carbon dioxide, to tropinone ($C_8H_{13}NO$). Strenuous oxidation of tropinone yields N-methylsuccinimide. Both anhydroecgonine and tropinone can be converted by a series of elimination and reduction reactions to cycloheptane. On the basis of these data deduce the structure of ecgonine, and indicate a possible structure for cocaine.

(d) The alkaloid pinidine is isolated from *Pinus sabiniana*. If you did not know that it has the structure shown,

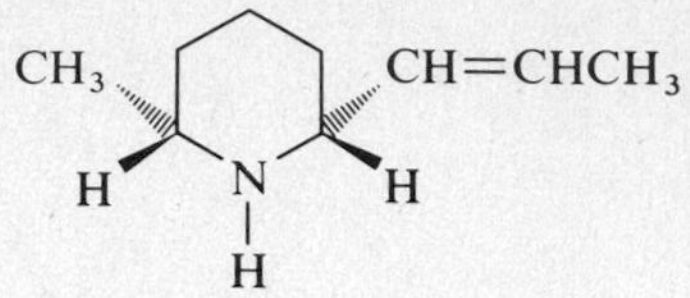

$$\text{CH}_3 \ldots \text{(2,6-disubstituted piperidine, N–H)} \ldots \text{CH=CHCH}_3$$

what methods would you use to establish this structure? Suggest a means for establishing that the configuration between the substituents at the 2 and 6 positions is *cis*. Draw a pseudo-three-dimensional representation of the most stable conformation of the naturally occurring form of pinidine.

(e) The cactus alkaloid macromerine is optically active and has the molecular formula $C_{12}H_{19}NO_3$. On the basis of its nmr spectrum, we can determine that the molecule contains two CH_3O- groups, three Ar—H hydrogens, two $N-CH_3$ groups, and a Ar—C(R)(O—)—H moiety. On the basis of its infrared spectrum we can determine that the molecule contains a 1,2,4-trisubstituted benzene ring and an OH group. On the basis of these data suggest a structure for macromerine.

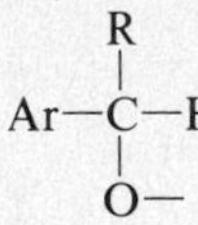

5. For each of the following examples indicate the steps that can be used to convert the compound shown at the left to the compound shown at the right.

(a) $(CH_3)_3CCH_2CH_2Br \longrightarrow (CH_3)_3CCH_2CH_2NHCH_3$

(b) $(C_6H_5)_2CHCO_2C_2H_5 \longrightarrow (C_6H_5)_2CHNH_2$

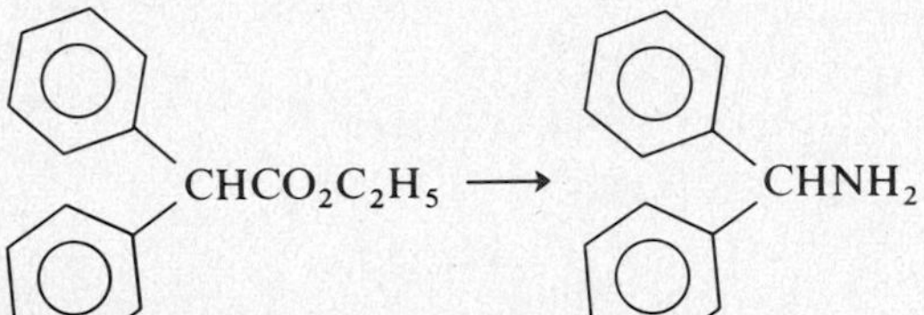

(c) cyclobutanone (=O) $\longrightarrow$ cyclobutyl$-N(CH_3)_2$

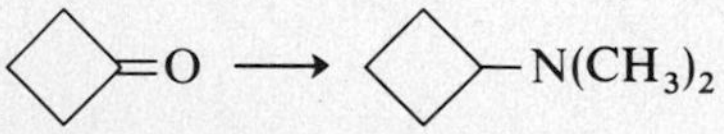

(d) benzene $\longrightarrow$ m-nitro-N-methylaniline ($NHCH_3$, NO_2)

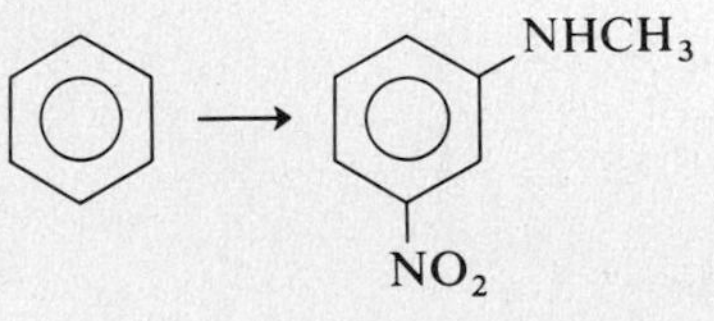

(e) cyclohexylamine (NH_2) $\longrightarrow$ cyclohexene

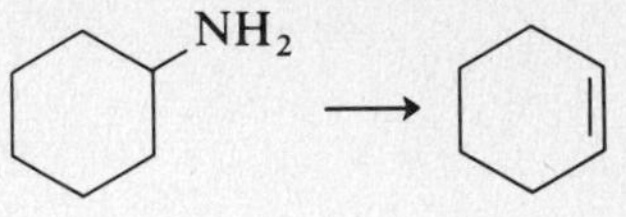

6. Explain the following observations:

(a) The reaction of cyclobutylmethylamine ($C_4H_7CH_2NH_2$) with nitrous acid yields a mixture containing cyclobutylmethanol, methylenecyclobutane, 1-methylcyclobutene, cyclopentene, and cyclopentanol.

(b) The nmr spectrum of 2-aminoethanol consists of a pair of two-hydrogen triplets and a three-hydrogen envelope, the position and shape of which is strongly temperature and concentration dependent.

(c) Treatment of phenylacetamide with bromine in aqueous sodium hydroxide yields benzylamine (Hofmann rearrangement). Treatment of phenylacetamide with bromine in a methanolic solution containing sodium methoxide, however, yields $C_6H_5CH_2NHCO_2CH_3$

(d) *o*-Deuteriofluorobenzene reacts with sodium amide in ammonia to yield aniline more slowly than does *o*-deuteriobromobenzene. When the reaction of the fluoro compound is stopped at 50% conversion to aniline the starting material is found to be largely devoid of deuterium. When the reaction of the bromo compound is stopped at 50% conversion to aniline the starting material still contains a large amount of deuterium.

7. Available starting materials include (a) any alkanes, alkenes, alkynes, alkyl halides, alkanols, alkanals, alkanones, or alkanoic acids containing four or fewer carbon atoms, (b) benzene and any monosubstituted benzenes, naphthalene, anthracene, phenanthrene, pyridine, and furfural, (c) special organic reagents, and (d) any inorganic compounds. Starting with materials chosen from this group, outline syntheses for each of the following compounds:

(a) $(CH_3)_2CHCH_2CH_2NH_2$

(b) C_6H_5–$CH_2N(CH_2CH_2CH_3)(CH_3)$

(c) trans-2-(N-piperidyl)cyclohexanol (structure: piperidine N attached to cyclohexane carbon bearing H; adjacent carbon bearing HO and H)

(d) CH_3CH_2NH–C_6H_4–C_6H_4–$NHCH_2CH_3$

(e) 2-furyl–$\overset{OH}{|}{CHCH_2NH_2}$

(f) 2-($CH_2N(C_2H_5)_2$)cyclohexanone

8. The following amines are inexpensive starting materials, costing less than 50 ¢/lb: dimethylamine, trimethylamine, *t*-butylamine, diethylamine, diisopropylamine, cyclohexylamine, ethylenediamine, aniline, N,N-dimethylaniline, diphenylamine, *o*-toluidine, *p*-toluidine, *o*-nitroaniline, *p*-nitroaniline, 1-naphthylamine, pyridine, 2-methylpyridine, 3-methylpyridine, 4-methylpyridine, and quinoline. Using only these compounds as starting materials, along with any inorganic reagents that might be necessary, devise a synthesis for each of the following compounds:

(a) 1-naphthyl–N=N–C_6H_4–$N(CH_3)_2$

(b) C_6H_5–CH_2NH_2

(c) N⟨pyridine⟩–$\overset{O}{\overset{\|}{C}}NHCH_2CH_2NH\overset{O}{\overset{\|}{C}}$–⟨pyridine⟩N

(d) 2-fluorobenzoic acid (CO_2H, F on adjacent ring positions)

(e) O_2N–C_6H_4–$CH_2\overset{+}{N}(CH_3)_2$ Br^-

9. Using Tables 39.1 to 39.5 as guidelines to readily available starting materials, devise syntheses for the following compounds. In addition to readily available organic compounds, you may also use any inorganic materials and special organic reagents.

(a) C_6H_5–$NHCH(C_2H_5)_2$

(b) $(CH_3)_2N$–C_6H_4–$N(CH_3)_2$

(c) 1-methylcyclohexylamine (ring carbon bearing CH_3 and NH_2)

(d) CH_3–C_6H_4–$N{=}N$–C_6H_4–CH_3

(e) (CH_3, NO_2 on one ring)–$N{=}N$–C_6H_4–OH

Chapter **25**

Alkenes: methods of preparation

> Fortunately science, like the nature to which it belongs is neither limited by time nor by space. It belongs to the world, and is of no country and no age. The more we know, the more we feel our ignorance; the more we feel our ignorance the more we feel how much remains unknown; and in philosophy the sentiment of the Macedonian hero can never apply—there are always new fields to conquer.
>
> SIR HUMPHRY DAVY, 1825

Alkenes are reactive compounds and, consequently, are frequently encountered in synthesis sequences. This chapter discusses a number of methods for preparing compounds containing carbon–carbon double bonds, paying particular attention to methods employing the β-elimination pathway. β-Elimination reactions have been rather extensively investigated, and considerable insight into their mechanisms has been gained. The chapter begins with a discussion of the details of these mechanisms, followed by a consideration of the scope, limitations, and synthetic utility of β-elimination reactions and other methods for preparing alkenes.

25.1 The β-Elimination Reaction

25.1a. GENERAL FEATURES OF β-ELIMINATION REACTIONS

Ionic Elimination Reactions. β-Elimination reactions following an ionic pathway can be classified on the basis of the order in which the electrophilic and nucleophilic leaving groups depart. In reactions classed as E1 (Elimination, unimolecular), the nucleophilic group departs first, leaving a carbonium ion which then loses an electrophilic group, usually a proton, to form the product. Thus, E1 elimination is the counterpart of S_N1 substitution, and whether a reaction follows the S_N1 or E1 pathway depends on the way in which the carbonium ion intermediate reacts *after* it has been formed. Thus, although the overall kinetics of the reaction are determined by the rate of formation of the carbonium ion intermediate, the product mixture is determined by the rates at which the carbonium ion intermediate (a) reacts with a nucleophile to form S_N1 product, (b) loses a proton to form E1 product, or (c) rearranges to a new carbon skeleton and then reacts along an S_N1 or E1 pathway.

Any nucleophilic species present including the original leaving group

$$R_2CHCH_2Nu \xrightarrow[k_1]{\text{(Slow step)}} R_2CH\overset{+}{C}H_2 \xrightarrow[k_2]{Nu'^-} R_2CHCH_2Nu' \quad S_N1$$

$$R_2CH\overset{+}{C}H_2 \xrightarrow[k_3]{(-H^+)} R_2C{=}CH_2 \quad \text{E1}$$

$$R_2CH\overset{+}{C}H_2 \xrightarrow{k_4} R\overset{+}{C}H{-}CH_2R \quad \text{Rearrangement}$$

k_2, k_3, and $k_4 > k_1$

(S_N1/El/rearrangement $= k_2/k_3/k_4$)

PROBLEM 25-1

If a certain carbonium ion reacts via the S_N1, E1, solvolysis, and rearrangement pathways with specific reaction rate constants of 10^2 sec^{-1}, 10^3 sec^{-1}, 5×10^3 sec^{-1}, and 10^4 sec^{-1}, respectively, what are the yields of products formed from each of these pathways?

In the competition between the S_N1 and E1 pathways, the S_N1 pathway is favored by weakly basic, strongly nucleophilic reagents, whereas the E1 pathway is favored by strongly basic and weakly nucleophilic reagents and also by structural features that increase the acidity of the β-hydrogen. Thus, 1,1,2,2-tetraphenylethanol is easily converted to tetraphenylethene (E1 product) by treatment with aqueous hydrobromic acid, whereas similar treatment of 2-methyl-1,1-diphenyl-1-propanol yields comparatively more alkyl bromide via the S_N1 pathway:

$$(C_6H_5)_2CH{-}C(OH)(C_6H_5)_2 \xrightarrow[(-H_2O)]{HBr} (C_6H_5)_2CH{-}\overset{+}{C}(C_6H_5)_2 \xrightarrow{(-H^+)} (C_6H_5)_2C{=}C(C_6H_5)_2$$

E1 product

$$(CH_3)_2CHC(OH)(C_6H_5)_2 \xrightarrow[(-H_2O)]{HBr} (CH_3)_2CH\overset{+}{C}(C_6H_5)_2 \xrightarrow{Br^-} (CH_3)_2CHCBr(C_6H_5)_2$$

S_N1 product

In reactions classed as $E1_{cb}$ (*E*limination, unimolecular with respect to the conjugate base of the starting compound), the electrophilic group departs first, leaving a carbanion which then loses the nucleophilic group to form the product. Reactions following this pathway are rare, one of the few examples being the base-induced dehydrofluorination of 2,2-dichloro-1,1,1-trifluoroethane:

$$F_3C{-}CCl_2H + C_2H_5O^- \underset{}{\overset{\text{Fast step}}{\rightleftharpoons}} F_3C{-}CCl_2^- \xrightarrow{\text{Slow step}} F_2C{=}CCl_2 + F^-$$

Conjugate base of starting material

PROBLEM 25-2

If the reaction of F_3CCHCl_2 with C_2H_5OD is terminated at 50% completion, does the recovered starting material contain deuterium? Explain why or why not.

In reactions classed as E2 (elimination, bimolecular), the electrophilic and nucleophilic groups depart in an essentially simultaneous fashion without the formation of a discrete intermediate. E2 elimination is the counterpart of S_N2 substitution; it is the result of the attack by the nucleophile at the β-hydrogen rather than the α-carbon. The difference between the unimolecular systems (S_N1, E1) and the bimolecular systems (S_N2, E2) is depicted by the reaction coordinate diagrams shown in Fig. 25.1. A typical example of an E2 reaction is the

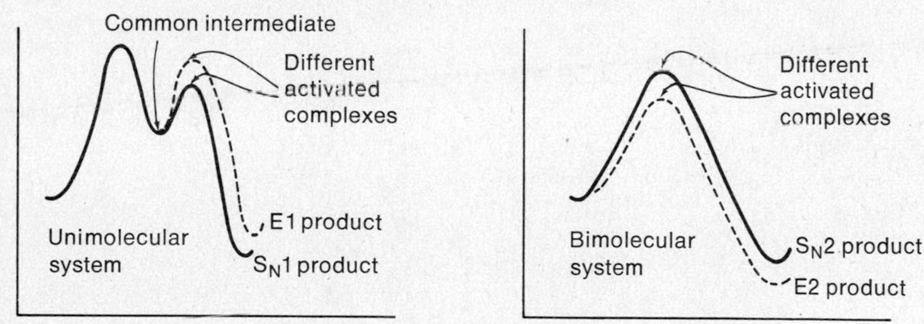

Fig. 25.1. Reaction coordinates for S_N1, E1, S_N2, and E2 reactions.

conversion of *t*-butyl bromide to 2-methylpropene under the influence of sodium ethoxide in ethanol (see p. 574):

$$(CH_3)_2C(Br){-}CH_2{-}H + C_2H_5O^- \longrightarrow \left[(CH_3)_2\overset{\delta-}{C}\cdots Br,\ CH_2\cdots H \cdots \overset{\delta-}{O}C_2H_5 \right]^{\ddagger} \longrightarrow CH_3{-}C(CH_3){=}CH_2 + C_2H_5OH + Br^-$$

S_N2 pathway (attack at C); E2 pathway (attack at H)

E2 activated complex

The competition between the S_N2 and E2 pathways is governed by (a) the amount of steric hindrance at the center bearing the nucleophilic leaving group, (b) the size of the attacking nucleophile, (c) the polarizability of the attacking nucleophile, and (d) the basicity of the attacking nucleophile. Thus, the E2 reaction is favored if there is considerable steric hindrance at the α-carbon and if the attacking nucleophile is large and highly basic. Conversely, the S_N2 reaction is favored if there is minimal steric hindrance at the α-carbon and if the attacking nucleophile is small and highly polarizable. For example, ethyl bromide (unhindered α-carbon) reacts with sodium methylthiolate (a sterically small, highly polarizable nucleophile) to yield ethyl methyl sulfide as the major product, but *t*-butyl bromide (hindered α-carbon) reacts with potassium *t*-butoxide (a sterically large, not highly polarizable base) to yield 2-methylpropene as the major product:

$$CH_3CH_2Br + CH_3SNa \longrightarrow CH_3CH_2SCH_3 + NaBr$$

$$(CH_3)_3CBr + (CH_3)_3COK \longrightarrow (CH_3)_2C{=}CH_2 + (CH_3)_3COH + KBr$$

Thermally Induced Elimination Reactions. A number of elimination reactions are known that do not require ionic reagents or conditions and that do not proceed via ionic intermediates. Such reactions are initiated by heat alone and are referred to as thermal elimination reactions. The two major types of thermal elimination reactions are (a) those in which the leaving groups are *not* bonded to each other in the starting material, and (b) those in which the leaving groups *are* bonded to each other in the starting material and also in the product. The first type is called a *pyrolytic elimination*, and is illustrated by the pyrolysis of ethyl acetate to yield ethylene and acetic acid (see Fig. 22.24):

$$\text{H–CH}_2\text{–CH}_2\text{–O–C(=O)–CH}_3 \xrightarrow{475^\circ} CH_2{=}CH_2 + HO{-}C({=}O){-}CH_3$$

PROBLEM 25-3

What is the mechanism of the pyrolytic elimination of ethyl acetate to ethylene and acetic acid?

The second type is called a *cycloelimination* (or *cycloreversion*), and is illustrated by the thermal elimination of carbon dioxide from 5-oxabicyclo[2.2.1]hept-2-en-6-one:

$$\text{5-oxabicyclo[2.2.1]hept-2-en-6-one} \xrightarrow{275^\circ} \text{cyclopentadiene} + CO_2$$

PROBLEM 25-4

Using the curved-arrow notation draw a plausible electron flow sequence for the decarboxylation reaction shown. What would you predict to be the product from

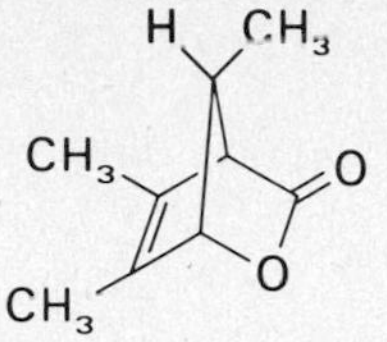

H CH3
CH3 O
O
CH3

25.1b. STEREOCHEMISTRY OF β-ELIMINATION REACTIONS

Stereochemistry of Ionic Elimination Reactions. Two stereochemical features must be considered in most β-elimination reactions, *viz.*, the relationship between the leaving groups in the starting material (*i.e.*, *syn* or *anti*) and the relationship between the groups remaining in the alkene [*i.e.*, (*Z*) or (*E*)]. In E1 reactions the stereochemical relationship between the leaving groups may be of little importance, because the intermediate carbonium ion that is formed has a planar configuration:

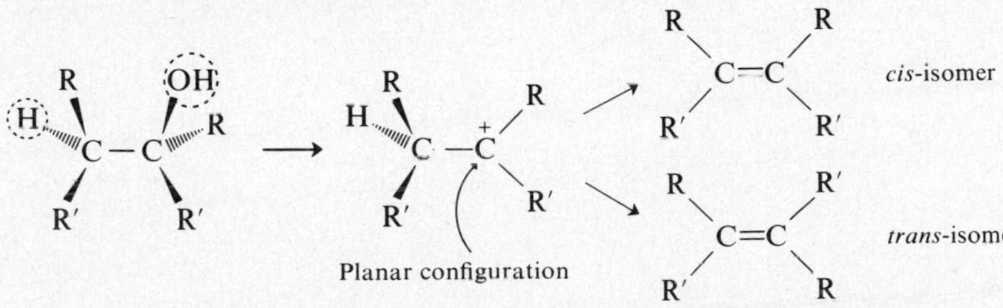

R R
C=C *cis*-isomer
R' R'

R OH R H R
H C—C R → C—C⁺
R' R' R' R'

R R'
C=C *trans*-isomer
R' R

Planar configuration

In some cases, however, a *syn*- or *anti*-elimination pathway is followed in E1 reactions, the particular situation that obtains depending on the structure of the starting material (i.e., its ability to form a stable carbonium ion), the polarity and basicity of the solvent, and the nature of the leaving group.

In E2 reactions, the stereochemical relationship between the leaving groups assumes great importance. Usually, E2 reactions proceed most readily when the leaving groups are in an antiperiplanar relationship, and the process is referred to as *anti*-elimination. If two or more antiperiplanar conformations are possible, the one with the smallest amount of nonbonded interference will be preferred, this generally leading to the product in which the two largest groups are *trans*. For example, 2-iodobutane yields a mixture of 1- and 2-butene, the 2-butene consisting of 78% of the *trans*-isomer and 22% of the *cis*-isomer:

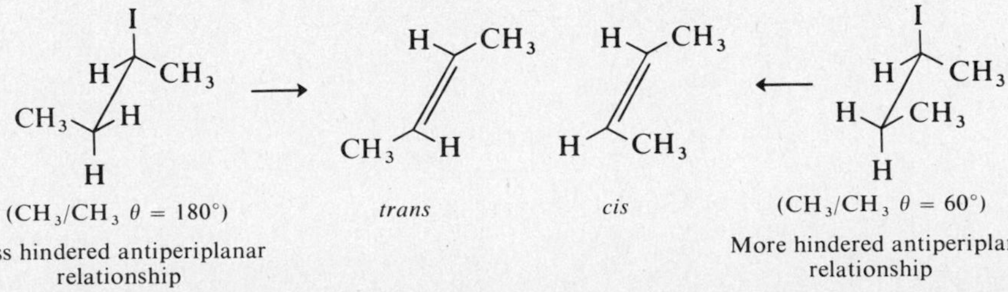

I / H CH3 / CH3 H / H	→	H CH3 / CH3 H	H CH3 / H CH3	←	I / H CH3 / H CH3 / H
(CH_3/CH_3 $\theta = 180°$) Less hindered antiperiplanar relationship		*trans*	*cis*		(CH_3/CH_3 $\theta = 60°$) More hindered antiperiplanar relationship

Although antiperiplanar groups are generally the ones that are eliminated in E2 reactions, in certain systems it is the synperiplanar groups that depart with greater facility, this process being referred to as *syn*-elimination. For example, 2-*exo*-bromo-3-*exo*-deuteriobicyclo[2.2.1]heptane undergoes base induced elimination with the adjacent *exo* deuterium rather than the adjacent *endo* hydrogen, leading to the product of *syn*-elimination:

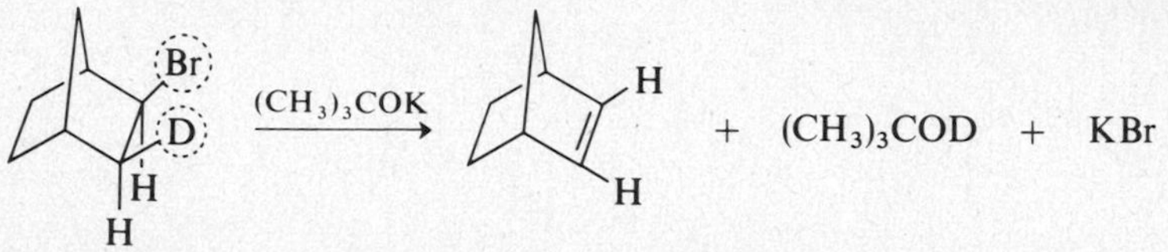

$\xrightarrow{(CH_3)_3COK}$ + $(CH_3)_3COD$ + KBr

Synperiplanar relationship

Whether the E2 reaction proceeds via an antiperiplanar conformation, as is usually the case, or a synperiplanar conformation, the stereochemical requirement that the leaving groups have a certain relationship to each other usually overrides other factors that can control the pathway of an elimination reaction. This is illustrated in the next section dealing with situselectivity.

PROBLEM 25-5

Are the *exo*-Br and *endo*-H at positions 2 and 3 of 2-*exo*-bromobicyclo[2.2.1]-heptane antiperiplanar? If not, how can they become so?

Stereochemistry of Nonionic Elimination Reactions. Pyrolytic elimination reactions of acetates, xanthates, and amine oxides occur via cyclic transition states which require a synperiplanar (*cis*) arrangement of the leaving groups; *i.e.*,

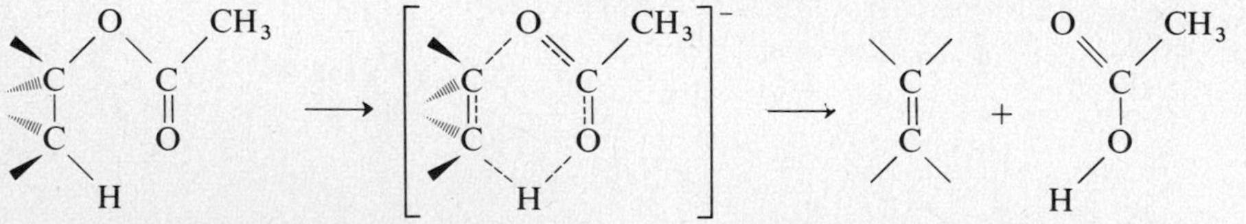

If two or more synperiplanar relationships are possible, the one with the smallest amount of nonbonded interference is preferred, generally leading to the product in which the largest groups are *trans*. For example, pyrolysis of 2-butyl acetate yields a mixture of 1-butene, *cis*-2-butene, and *trans*-2-butene, with the *trans*/*cis* ratio being about 2:

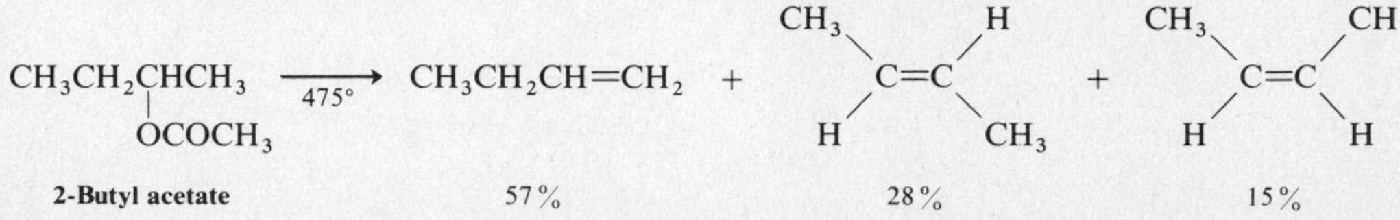

$CH_3CH_2CH(OCOCH_3)CH_3 \xrightarrow{475°} CH_3CH_2CH{=}CH_2 + (trans)\ CH_3CH{=}CHCH_3 + (cis)\ CH_3CH{=}CHCH_3$

2-Butyl acetate 57% 28% 15%

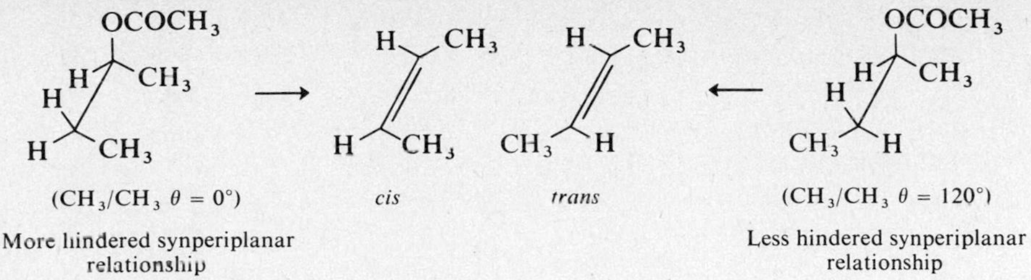

$(CH_3/CH_3\ \theta = 0°)$ → *cis* *trans* ← $(CH_3/CH_3\ \theta = 120°)$

More hindered synperiplanar relationship

Less hindered synperiplanar relationship

As with the E2 reaction, the stereochemical requirement that the leaving groups have a *syn* relationship to each other usually overrides other factors that can control the pathway of the elimination reaction.

25.1c. Regioselectivity* of β-Elimination Reactions

Saytzev Regioselectivity. Ionic elimination reactions involving E1 or E2 reactions of alcohols, alkyl halides, and alkyl sulfonates usually follow the Saytzev pathway to yield the more highly substituted alkenes. For example, the action of sodium ethoxide on 2-bromobutane yields 81% of the more highly substituted product, 2-butene, and only 19% of the less highly substituted product, 1-butene. The extent to which the Saytzev pathway is followed in E2 reactions, however, is quite dependent on the structure of the carbon framework, the nature of the leaving group, and the base that induces the elimination. Thus, as steric hindrance at the β hydrogen increases and as the size of the base increases, the Saytzev pathway becomes less and less favorable. With the highly hindered halide 2-bromo-2,4,4-trimethylpentane and the highly hindered base potassium

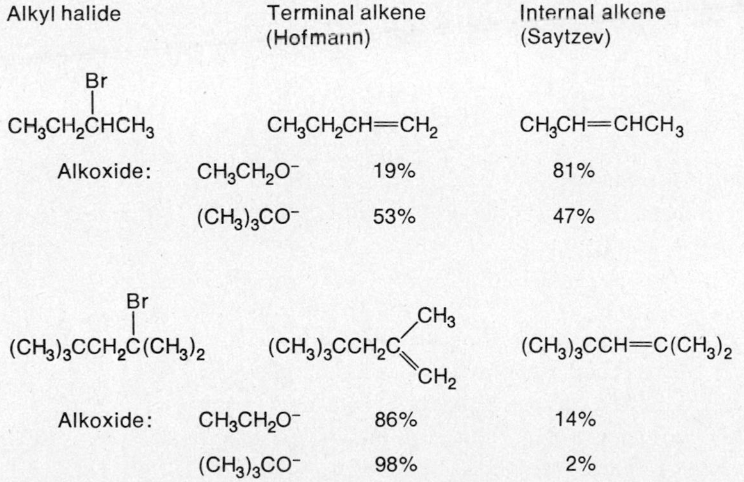

Alkyl halide		Terminal alkene (Hofmann)	Internal alkene (Saytzev)
$CH_3CH_2CH(Br)CH_3$		$CH_3CH_2CH{=}CH_2$	$CH_3CH{=}CHCH_3$
Alkoxide:	$CH_3CH_2O^-$	19%	81%
	$(CH_3)_3CO^-$	53%	47%
$(CH_3)_3CCH_2C(Br)(CH_3)_2$		$(CH_3)_3CCH_2C(CH_3){=}CH_2$	$(CH_3)_3CCH{=}C(CH_3)_2$
Alkoxide:	$CH_3CH_2O^-$	86%	14%
	$(CH_3)_3CO^-$	98%	2%

Fig. 25.2. Effect of steric hindrance on regioselectivity of dehydrogenation.

---

* The direction of elimination might equally well be specified as the "situselectivity" of the reaction. However, since elimination processes are the reverse of addition processes, to which the term "regioselectivity" is applied, "regioselectivity" is the term that is generally used when discussing elimination reactions.

t-butoxide, for example, only 2% of 2,4,4-trimethyl-2-pentene is formed, accompanied by 98% of 2,4,4-trimethyl-1-pentene (Fig. 25.2).

Hofmann Regioselectivity. Ionic elimination reactions involving E2 reactions of quaternary ammonium salts and sulfonium salts usually follow the Hofmann pathway to yield the less highly substituted alkenes (Fig. 25.3). Also, as we have

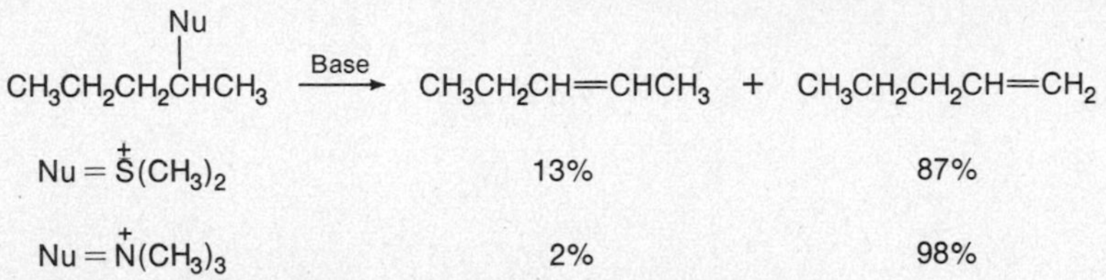

$$CH_3CH_2CH_2\underset{}{\overset{Nu}{\overset{|}{C}}}HCH_3 \xrightarrow{\text{Base}} CH_3CH_2CH{=}CHCH_3 + CH_3CH_2CH_2CH{=}CH_2$$

$Nu = \overset{+}{S}(CH_3)_2$	13%	87%
$Nu = \overset{+}{N}(CH_3)_3$	2%	98%

Fig. 25.3. Regioselectivity in the base-induced elimination reactions of ammonium and sulfonium salts.

seen from the data in Fig. 25.2, the Hofmann pathway may predominate in the case of alkyl halides (or sulfonates) when certain steric factors are present in the carbon framework of the compound undergoing elimination and when the base is large. The reason that the Hofmann pathway is followed in preference to the Saytzev pathway as the size of the attacking base increases is fairly obvious. The Hofmann product is the result of the abstraction of the *least* hindered hydrogen, whereas the Saytzev product is the result of the abstraction of the *most* hindered hydrogen. Thus, the larger the base the more difficult it is for it to get close enough to a hydrogen in a crowded environment to effect a displacement:

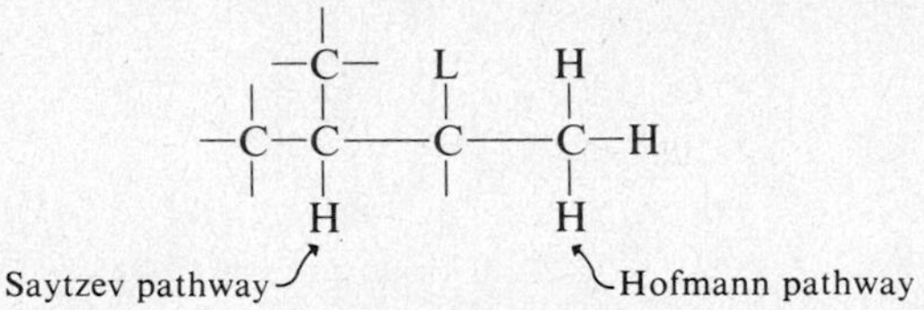

```
         |
       —C—   L    H
     |  |    |    |
    —C—C————C————C—H
     |  |    |    |
        H         H
Saytzev pathway↗   ↖Hofmann pathway
```

The reason that the Hofmann pathway becomes more important as the leaving ability of the nucleophile diminishes is more subtle. One explanation focuses on the stabilities of carbanions in solution,* these decreasing in the order $CH_3^- > RCH_2^- > R_2CH^- > R_3C^-$. It is argued that the poorer the leaving ability of the departing group, the greater must be the electron density that is developed at the β-carbon to force the leaving group out of the molecule—*i.e.*, the greater must be the extent to which the attacking base has removed the hydrogen from the β-carbon. Thus, the β-carbon that can best stabilize a negative charge should react preferentially; as the sequence noted indicates, RCH_2^- is better than R_2CH^- or R_3C^- in this respect. The hydrogen at a primary position, therefore, reacts in preference to hydrogens at secondary or tertiary positions, and the least highly substituted alkene is formed.

* The order that is manifested in solution arises from the way in which the solvent interacts with the carbanion. The intrinsic stabilities of carbanions in the gas phase are probably in the reverse order.

PROBLEM 25-6

The products obtained by the pyrolysis of the N-oxide of *trans*-1-dimethylamino-2-methylcyclopentane and by the action of base on the quaternary salt derived from this compound have different structures. What are they? Are they commensurate with the Hofmann and Saytzev pathways, with the stereochemical requirements, or both?

Regioselectivity vs. Stereochemistry. The stereochemical requirement that the departing electrophilic and nucleophilic groups be *anti* in most E2 reactions and *syn* in pyrolytic processes generally overrides the tendency a particular system might otherwise have to follow the Saytzev or Hofmann pathway. For example, although *cis*-1-bromo-2-methylcyclopentane yields 1-methylcyclopentene (Saytzev product) as the major product, its *trans*-isomer yields 3-methylcyclopentene (Hofmann product) as the major product; the stereochemical requirement that the departing groups be *anti* to each other is the controlling factor (Fig. 25.4). Conversely, *cis*-1-acetoxy-2-methylcyclopentane yields the Hofmann product, and the *trans*-isomer yields the Saytzev product because of the stereochemical requirement that the departing groups be *syn* to each other (Fig. 25.4).

25.2 Alkenes from the Acid-Catalyzed Dehydration of Alcohols

In the presence of strong acids, such as sulfuric, phosphoric, and *p*-toluenesulfonic acids, many alcohols lose water to form alkenes. The mechanism of the reaction involves the formation of an intermediate oxonium ion, which reacts either by an E1 or an E2 pathway. In either case, the Saytzev product is usually formed, although rarely as the exclusive structural isomer. The stereoselectivity is low, and mixtures of *cis*- and *trans*-isomers are usually produced. The procedure finds its most effective use, therefore, in situations where neither situselectivity nor stereoselectivity play a part, such as, for instance, in the dehydration of symmetrical alcohols like 4-methylcyclohexanol:

$$CH_3\text{-}C_6H_{10}\text{-}OH \xrightarrow[H^+ \Delta]{} CH_3\text{-}C_6H_9$$

Still another drawback to the acid-catalyzed dehydration of alcohols is that the reaction is unusually prone to skeletal rearrangement. For example, in the course of the structure proof of one of the naturally occurring steroids, an acid-catalyzed dehydration was carried out on the tertiary alcohol at C-17, as shown in Fig. 25.5. Instead of the anticipated compound in which the double bond is between C-16 and C-17, a rearranged alkene was obtained in which the methyl group originally present at a bridgehead position had migrated to the C-17 position. Thus, inferences concerning the structure of the alcohol made on

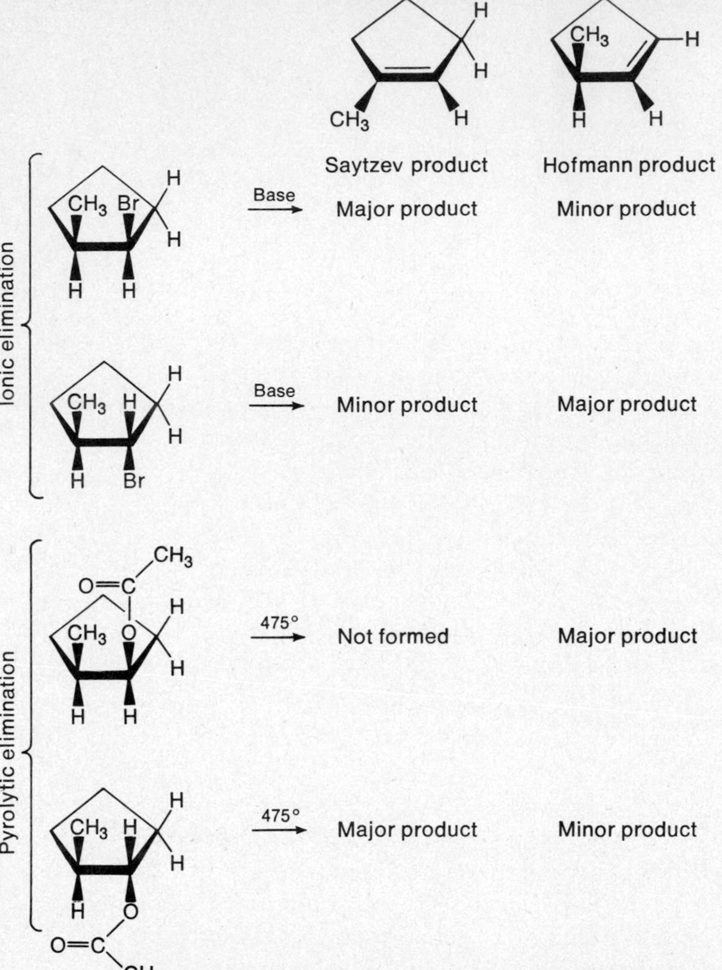

Fig. 25.4. Stereochemical determination of situselectivity in β-elimination reactions.

the basis of the structure of the alkene were incorrect, pointing out a hazard that must always be remembered when reactions involving carbonium ion intermediates are undertaken.

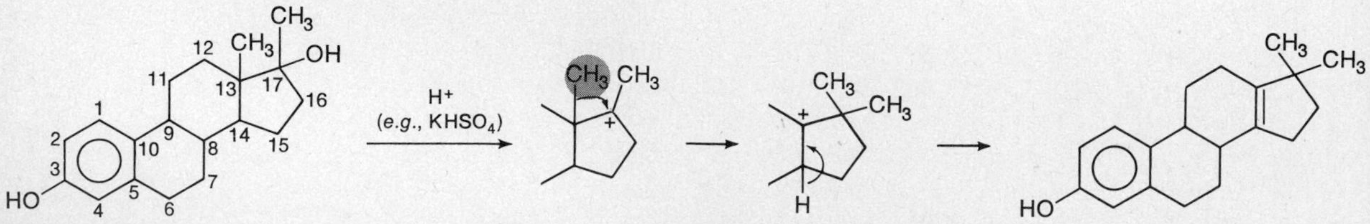

Fig. 25.5. Rearrangement in the acid-catalyzed dehydration of an alcohol.

PROBLEM 25-7

Would the compounds at the left- and right-hand sides of Fig. 25.5 be expected to yield the same or different products on dehydrogenation? (*Note*: The five-membered ring remains intact and does not lose hydrogen.)

25.3 Alkenes from Base-Induced Elimination Reactions

As discussed in Section 25.1, compounds of the general structure H—C—C—Nu, where Nu is a halogen, an arene- or alkanesulfonate, an ammonium, or a sulfonium group, undergo base-induced β-elimination to form alkenes. An interesting example of the application of this method for the introduction of C=C bonds is provided by the synthesis of cyclooctatetraene (see p. 117), a compound that has been of considerable importance in the testing of theories of aromaticity (Fig. 25.6).

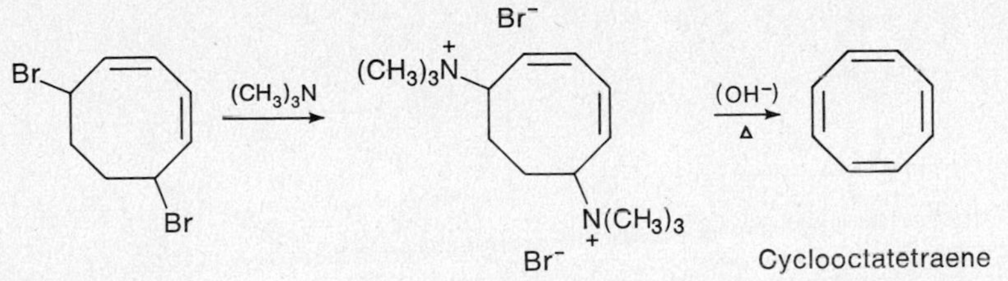

Br, Br (cyclooctene) $\xrightarrow{(CH_3)_3N}$ $(CH_3)_3\overset{+}{N}$, Br^-, $\overset{+}{N}(CH_3)_3$, Br^- $\xrightarrow[\Delta]{(OH^-)}$ Cyclooctatetraene

Fig. 25.6. Synthesis of cyclooctatetraene by base-induced elimination.

25.4 Alkenes from Metal-Induced Elimination Reactions

Another method for developing a negative charge at a carbon β to a nucleophilic leaving group and thereby inducing an elimination reaction involves the action of metals such as zinc on 1,2-dihalides and β-haloethers:

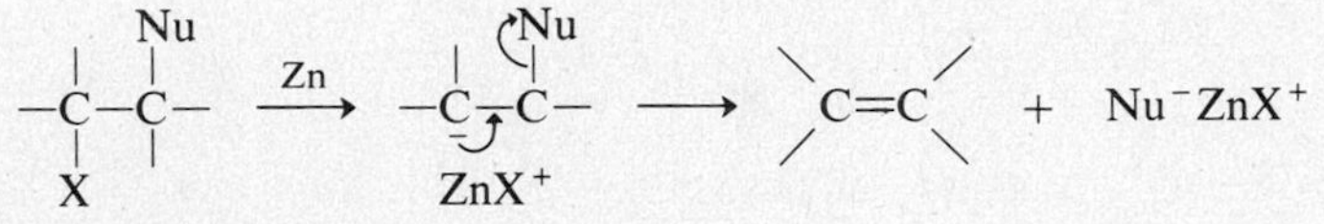

$$-\overset{Nu}{\overset{|}{C}}-\underset{X}{\underset{|}{C}}- \xrightarrow{Zn} -\overset{Nu}{\overset{|}{C}}-\underset{ZnX^+}{\underset{|}{C}}- \longrightarrow \;\rangle C{=}C\langle \; + \; Nu^-ZnX^+$$

Nu = Halogen or —OR

PROBLEM 25-8

Benzynes can be produced by the action of metals on 1,2-dihaloarenes. Indicate the mechanism of this reaction, using the action of magnesium on 1-bromo-2-fluorobenzene as the specific example.

The sequence, outlined in Fig. 25.7, starts with the preparation of an α-chloro ether by the reaction of an aldehyde with an alcohol (to form a hemiacetal) followed by an S_N1 displacement with hydrogen chloride. Reaction of the α-chloro ether with bromine produces an α,β-dibromo ether via an intermediate

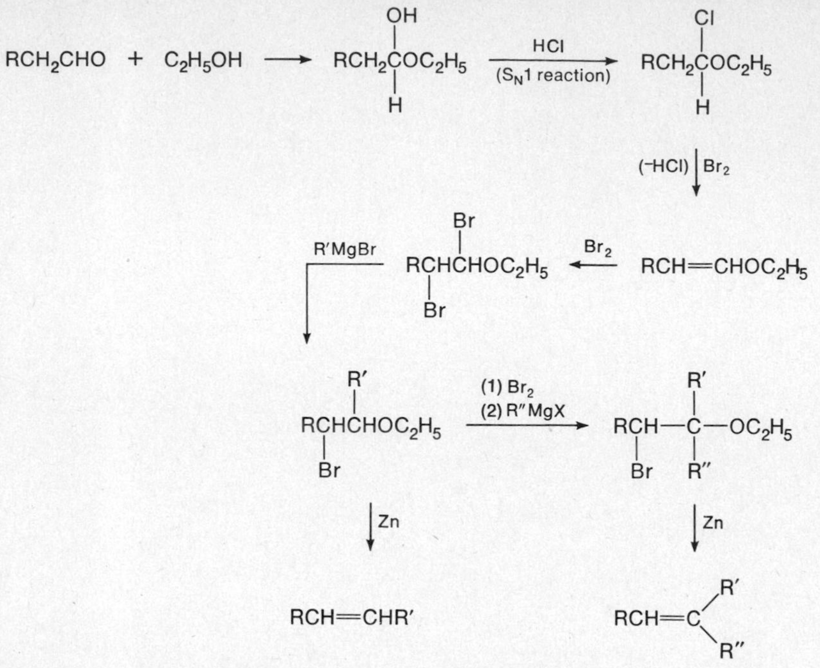

$$RCH_2CHO + C_2H_5OH \longrightarrow RCH_2CH(OH)OC_2H_5 \xrightarrow[(S_N1\ \text{reaction})]{HCl} RCH_2CH(Cl)OC_2H_5$$

$$RCH_2CH(Cl)OC_2H_5 \xrightarrow[(-HCl)]{Br_2} RCH{=}CHOC_2H_5 \xrightarrow{Br_2} RCH(Br)CH(Br)OC_2H_5$$

$$RCH(Br)CH(Br)OC_2H_5 \xrightarrow{R'MgBr} RCH(Br)CH(R')OC_2H_5 \xrightarrow[(2)\ R''MgX]{(1)\ Br_2} RCH(Br){-}C(R')(R''){-}OC_2H_5$$

$$RCH(Br)CH(R')OC_2H_5 \xrightarrow{Zn} RCH{=}CHR'$$

$$RCH(Br){-}C(R')(R''){-}OC_2H_5 \xrightarrow{Zn} RCH{=}C(R')R''$$

Fig. 25.7. The Boord alkene synthesis.

vinyl ether (formed by abstraction of a β-H by Br· followed by loss of Cl·) to which bromine adds. Treatment with a Grignard reagent results in the displacement of the more reactive α-bromine by the R group of RMgX to give a β-bromo ether, the reaction probably proceeding via the resonance stabilized carbonium ion formed by the loss of the α-bromine (facilitated by the magnesium of RMgX acting as a Lewis acid). Zinc-induced elimination at this point yields a disubstituted alkene, or repetition of the process followed by zinc-induced elimination yields a trisubstituted alkene (Fig. 25.7). The method is not suited to the preparation of tetrasubstituted alkenes, however.

Another use of metal-induced dehalogenation is the "protection" of double bonds. For example, bromine can be added to a double bond to keep the bond from being altered during a synthesis sequence, and the double bond subsequently can be regenerated by removal of the bromine via treatment with zinc or magnesium.

PROBLEM 25-9

Suggest a specific example where this sequence of reactions might be useful.

25.5 *Alkenes from Pyrolytic Elimination Reactions*

As Section 25.1 indicates, the pyrolysis of acetates, xanthates, and amine oxides provides a very useful pathway for the synthesis of alkenes from alcohols and amines. These procedures possess the particular virtues of rarely involving skeletal rearrangement and, despite the somewhat elevated temperatures required to effect them, of allowing other sensitive groups within the molecule to survive unscathed. A notable example of the lack of rearrangement is observed in the pyrolysis of 3,6-*bis*-(acetoxymethyl)-cyclohexene to 3,6-dimethylenecyclohexene (Fig. 25.8), an isomer of the far more thermodynamically stable *p*-xylene to which it rapidly isomerizes in the presence of a trace of acid or base.

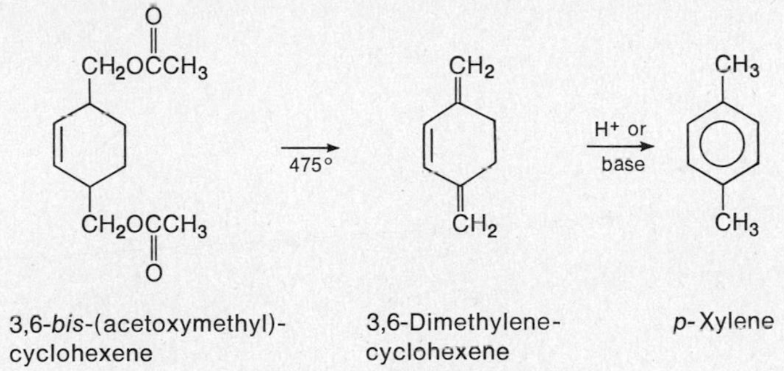

3,6-*bis*-(acetoxymethyl)-cyclohexene $\xrightarrow{475°}$ 3,6-Dimethylene-cyclohexene $\xrightarrow[\text{base}]{H^+ \text{ or}}$ *p*-Xylene

Fig. 25.8. Preparation of alkenes by acetate pyrolysis.

25.6 Alkenes by Oxidative Elimination Reactions

1,2-Dicarboxylic acids undergo oxidation in the presence of lead tetraacetate with the loss of both carboxyl groups and the formation of a double bond. The reaction can be rationalized as involving the formation of a triacetoxylead derivative of the dicarboxylic acid followed by decomposition of this intermediate to the alkene and carbon dioxide. Two plausible pathways for the decomposition are shown:

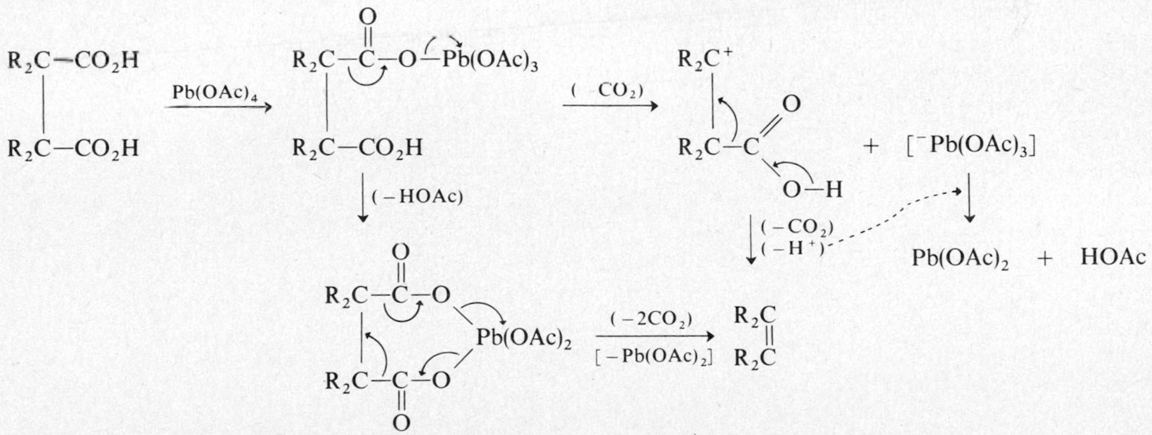

$R_2C(CO_2H)\text{—}C(CO_2H)R_2 \xrightarrow{Pb(OAc)_4} R_2C(\text{—}C(=O)\text{—}O\text{—}Pb(OAc)_3)\text{—}R_2C\text{—}CO_2H \xrightarrow{(-CO_2)} R_2C^+\text{—}R_2C\text{—}C(=O)OH + [^-Pb(OAc)_3] \rightarrow Pb(OAc)_2 + HOAc$

$R_2C^+\text{—}R_2C\text{—}CO_2H \xrightarrow[(-H^+)]{(-CO_2)} R_2C{=}CR_2$

$R_2C(\text{—}C(=O)\text{—}O\text{—}Pb(OAc)_3)\text{—}R_2C\text{—}CO_2H \xrightarrow{(-HOAc)}$ cyclic $R_2C(\text{—}C(=O)O)\text{—}R_2C(\text{—}C(=O)O)Pb(OAc)_2 \xrightarrow[[-Pb(OAc)_2]]{(-2CO_2)} R_2C{=}CR_2$

The method is quite mild and is useful for preparing highly strained compounds, as illustrated by the examples in Fig. 25.9.

25.7 Alkenes by Reduction of Alkynes

Reduction of disubstituted acetylenes affords an excellent method for preparing 1,2-disubstituted alkenes, for the reduction can be carried out stereoselectively to produce either the *cis* or the *trans* alkene, depending on the method used.

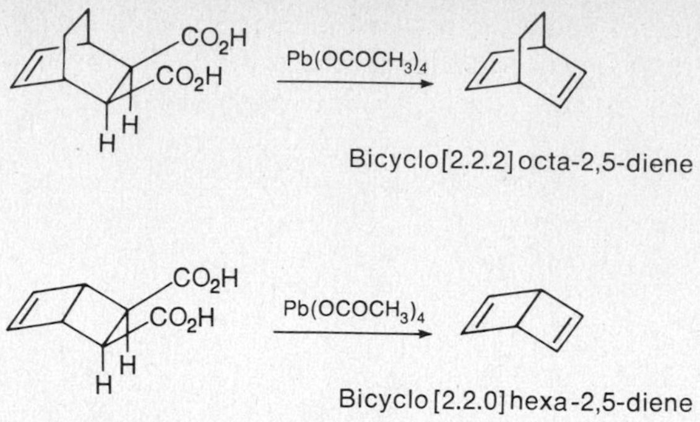

Pb(OCOCH$_3$)$_4$

Bicyclo[2.2.2]octa-2,5-diene

Pb(OCOCH$_3$)$_4$

Bicyclo[2.2.0]hexa-2,5-diene

Fig. 25.9. Preparation of alkenes via lead tetraacetate oxidation of 1,2-dicarboxylic acids.

Molecular hydrogen in the presence of a specific catalyst* reduces disubstituted acetylenes to *cis*-disubstituted alkenes:

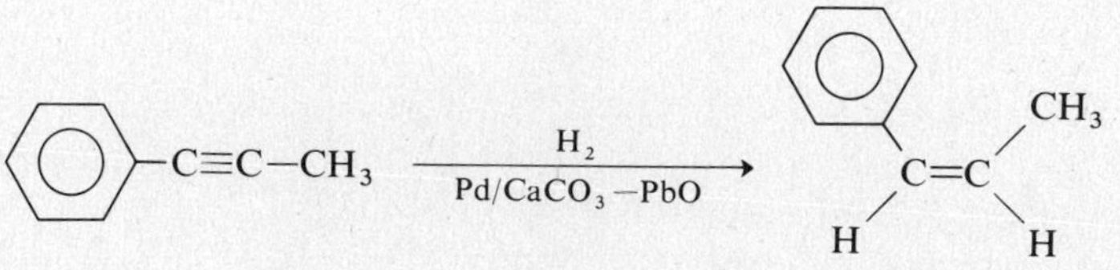

$$C_6H_5-C{\equiv}C-CH_3 \xrightarrow[Pd/CaCO_3-PbO]{H_2} \underset{H}{\overset{C_6H_5}{}}C{=}C\underset{H}{\overset{CH_3}{}}$$

cis-Alkenes are also produced from disubstituted acetylenes by the hydroboration-protonolysis sequence. If we use a bulky dialkylborane such as dicyclohexylborane, prepared from 2 moles of cyclohexene and 1 mole of BH_3, *syn* addition takes place to yield the alkenyldicyclohexylborane, which can then be converted to the alkene by treatment with an organic acid (*e.g.*, propionic acid) under reflux conditions:

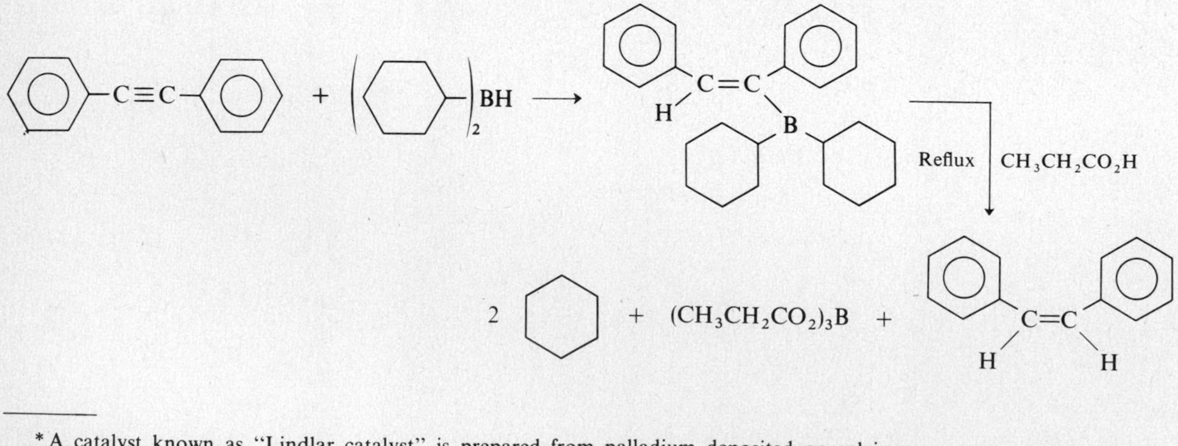

$$C_6H_5-C{\equiv}C-C_6H_5 + (C_6H_{11})_2BH \longrightarrow \underset{H}{\overset{C_6H_5}{}}C{=}C\underset{B(C_6H_{11})_2}{\overset{C_6H_5}{}} \xrightarrow[\text{Reflux}]{CH_3CH_2CO_2H} 2\ C_6H_{12} + (CH_3CH_2CO_2)_3B + \underset{H}{\overset{C_6H_5}{}}C{=}C\underset{H}{\overset{C_6H_5}{}}$$

* A catalyst known as "Lindlar catalyst" is prepared from palladium deposited on calcium carbonate followed by treatment with lead salts which moderate the activity of the catalyst; it is symbolized as Pd/CaCO$_3$—PbO.

trans-Alkenes can be obtained from disubstituted acetylenes by using sodium in liquid ammonia as the reducing agent. An interesting example of the application of this method is the conversion of cyclooctyne to *trans*-cyclooctene:

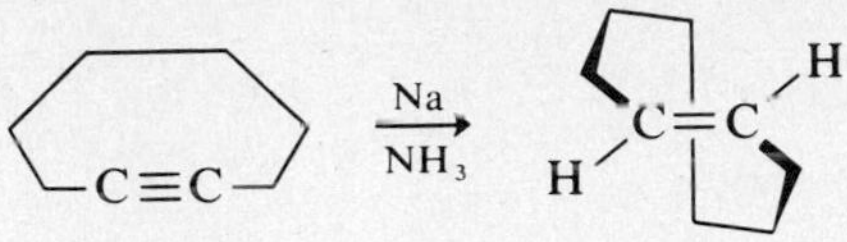

$$\text{Cyclooctyne} \xrightarrow[NH_3]{Na} \textit{trans}\text{-Cyclooctene}$$

Cyclooctyne **trans-Cyclooctene**

25.8 Alkenes by Reduction of Arenes

Although the catalytic hydrogenation of arenes can rarely be stopped short of complete reduction to the cycloalkanes, sodium or lithium metal dissolved in ammonia converts benzenes to cyclohexadienes. For example, treatment of benzene with sodium in liquid ammonia containing a weak proton donor such as methanol or ammonium chloride produces 1,4-cyclohexadiene in a process

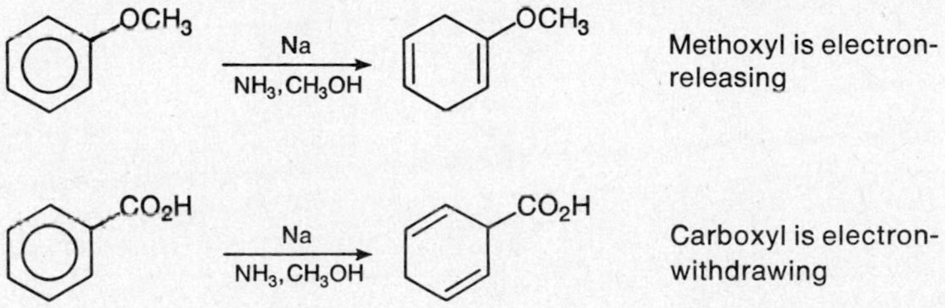

$$C_6H_5OCH_3 \xrightarrow[NH_3,\,CH_3OH]{Na} \text{1-methoxy-1,4-cyclohexadiene}$$

Methoxyl is electron-releasing

$$C_6H_5CO_2H \xrightarrow[NH_3,\,CH_3OH]{Na} \text{2,5-cyclohexadiene-1-carboxylic acid}$$

Carboxyl is electron-withdrawing

Fig. 25.10. Birch reduction of substituted benzenes.

known as the *Birch reduction* (Fig. 25.10). Substituted benzenes give the 1-substituted 1,4-cyclohexadiene if the substituent is electron releasing, or the 3-substituted 1,4-cyclohexadiene if the substituent is electron withdrawing. More extensive reduction of the benzene ring can be achieved with lithium metal in ethylenediamine solution in the absence of a proton donor:

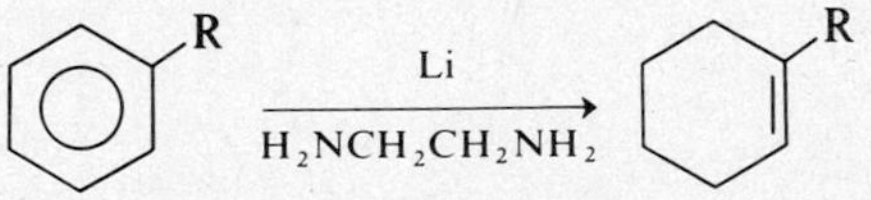

$$C_6H_5R \xrightarrow[H_2NCH_2CH_2NH_2]{Li} \text{1-R-cyclohexene}$$

25.9 Alkenes from Aldehydes and Ketones

25.9a. VIA REDUCTION OR ADDITION OF ORGANOMETALLIC REAGENTS. Aldehydes and ketones can be converted to alcohols either by reduction or by the nucleophilic addition of organometallic compounds such as Grignard reagents. Conversion of the alcohol to the corresponding alkene can then be effected by acid-catalyzed dehydration, by conversion to the acetate or xanthate followed by pyrolysis, or by conversion to the alkyl halide followed by base-induced dehydrohalogenation:

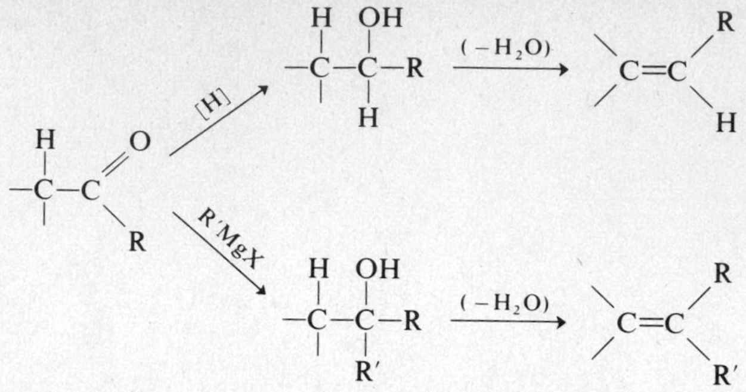

$$\begin{array}{c} H \\ | \\ -C-C\!\!\overset{O}{\diagup\!\!\diagup} \\ | \quad \diagdown R \end{array} \xrightarrow{[H]} -\overset{H}{\underset{|}{\overset{|}{C}}}-\overset{OH}{\underset{H}{\overset{|}{C}}}-R \xrightarrow{(-H_2O)} \;\diagdown C{=}C \diagup^{R}_{H}$$

$$\xrightarrow{R'MgX} -\overset{H}{\underset{|}{\overset{|}{C}}}-\overset{OH}{\underset{R'}{\overset{|}{C}}}-R \xrightarrow{(-H_2O)} \;\diagdown C{=}C \diagup^{R}_{R'}$$

25.9b. VIA PHOSPHORUS YLIDES. Aldehydes and ketones are readily converted to alkenes by reaction with alkylidenetriphenylphosphoranes (phosphorus ylides) in a process known as the *Wittig reaction.* Triphenylphosphine is similar to tertiary amines in its ability to react in an S_N2 process with alkyl halides, giving alkyltriphenylphosphonium salts. When the salts are treated with very strong bases, such as butyllithium or phenyllithium, the moderately acidic hydrogen attached to the carbon adjacent to the phosphonium group is removed, and a species known as a phosphorane (also called a phosphorus ylide or a *Wittig reagent*) is produced:

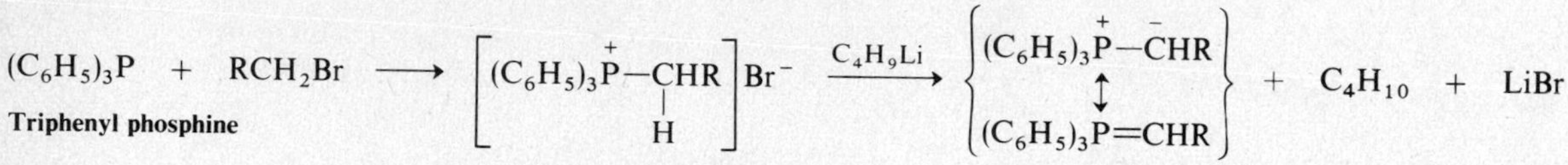

$$\underset{\textbf{Triphenyl phosphine}}{(C_6H_5)_3P} + RCH_2Br \longrightarrow \underset{\textbf{An alkyltriphenylphosponium salt}}{\left[(C_6H_5)_3\overset{+}{P}-\underset{H}{\underset{|}{C}}HR\right]Br^-} \xrightarrow{C_4H_9Li} \underset{\textbf{A phosphorane}}{\left\{\begin{array}{c}(C_6H_5)_3\overset{+}{P}-\overset{-}{C}HR \\ \updownarrow \\ (C_6H_5)_3P{=}CHR\end{array}\right\}} + C_4H_{10} + LiBr$$

$$\begin{array}{c} R \\ \diagdown \\ C{=}O \\ \diagup \\ R' \end{array} + R''\overset{-}{C}H-\overset{+}{P}(C_6H_5)_3$$

$$\rightleftharpoons \quad \begin{array}{c} O^- \; {}^+P(C_6H_5)_3 \\ | \qquad | \\ R,R'\;C-C\;H,R'' \end{array} \qquad\qquad \rightleftharpoons \quad \begin{array}{c} O^- \; {}^+P(C_6H_5)_3 \\ | \qquad | \\ R,R'\;C-C\;R'',H \end{array}$$

$$\downarrow \qquad\qquad\qquad\qquad \downarrow$$

$$\begin{array}{c} R \qquad H \\ C{=}C \\ R' \qquad R'' \end{array} \qquad\qquad \begin{array}{c} R \qquad R'' \\ C{=}C \\ R' \qquad H \end{array}$$

Major product when R′ is smaller than R — Major product when R′ is larger than R

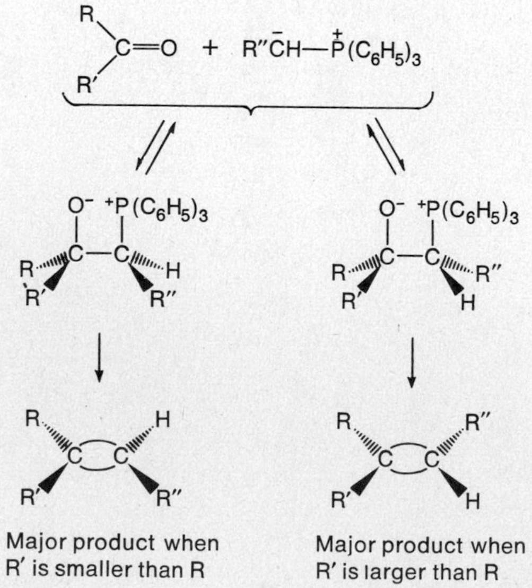

Fig. 25.11. Mechanism of the reaction of a phosphorane (Wittig reagent) with an aldehyde or ketone.

Phosphoranes have the interesting property of reacting with aldehydes and ketones in such a fashion that the alkylidene group of the phosphorane trades places with the carbonyl oxygen of the aldehyde or ketone. The mechanism is thought to be that depicted in Fig. 25.11.

PROBLEM 25-10

Would you predict triphenylphosphine, $(C_6H_5)_3P$, to be a stronger or weaker nucleophile than triphenylamine? What are the reasons for your conclusion?

The Wittig reaction is quite versatile, for not only can primary and secondary alkyl halides be used, but various hetero atom moieties can also be included in the groups of the phosphorane. For example, phosphoranes containing ester groups are readily prepared from α-haloesters:

$$(C_6H_5)_3P + RCH(Br)CO_2C_2H_5 \longrightarrow [(C_6H_5)_3\overset{+}{P}CH(R)CO_2C_2H_5]Br^- \xrightarrow{C_6H_5Li} (C_6H_5)_3\overset{+}{P}-\overset{-}{C}(R)CO_2C_2H_5$$

$$(C_6H_5)_3\overset{+}{P}-\overset{-}{C}(R)CO_2C_2H_5 \xrightarrow{R'CHO} R'CH=C(R)CO_2C_2H_5 + (C_6H_5)_3P=O$$

Some groups are precluded, however, for protic functions such as hydroxyl and carboxyl transfer their proton to the phosphorane and convert it back to the phosphonium salt.

One of the principal advantages of the Wittig method is that the position of the double bond is specifically established by the position of the carbonyl group in the aldehyde or ketone; the process is highly situselective, and molecular rearrangements do not take place. Unfortunately, however, it is not very stereoselective; mixtures of *cis*- and *trans*-isomers are usually formed, although the alkenes having the largest groups *trans* to each other generally predominate.

PROBLEM 25-11

How could the following compounds be prepared via a Wittig reaction:

$C_6H_5CH=CHCH_3$ $\qquad$ (cyclopentylidene)$=CHCH_2CH_2CH=$(cyclopentylidene)

25.10 Alkenes from Vinyl Halides

The halogen of a vinyl halide can be replaced by either hydrogen or an alkyl group to yield the corresponding alkene. Replacement by hydrogen can be effected in a stereoselective fashion by treating the vinyl halide with lithium metal at a low temperature, followed by hydrolysis of the intermediate organolithium compound:

$$C_6H_5(Cl)C{=}C(H)CH_3 + Li \xrightarrow[-35°]{} C_6H_5(Li)C{=}C(H)CH_3 \xrightarrow{H_3O^+} C_6H_5(H)C{=}C(H)CH_3$$

Replacement by an alkyl group can be effected by treating the vinyl halide with lithium dialkylcopper reagents in a stereoselective process that yields alkenes in which the stereochemistry of the starting material is retained (see p. 630):

$$C_6H_5(H)C{=}C(H)Br + (CH_3)_2CuLi \longrightarrow C_6H_5(H)C{=}C(H)CH_3$$

Since vinyl halides are readily prepared by the stereoselective addition of halogen acids to alkynes, this two-step sequence provides a very useful synthesis of alkenes of specified (Z)- or (E)-configuration.

25.11 Preparation of Dienes

The same reactions that give rise to monoalkenes can also be used in the preparation of nonconjugated polyenes. Of particular utility are the pyrolytic elimination reactions, for the possibility of rearrangement to conjugated dienes is thereby minimized (see Fig. 25.8). Similarly, the Wittig reaction with dicarbonyl compounds yields dienes in which the sites of the double bonds are uniquely determined by the positions of the carbonyl groups.

Conjugated dienes, also, can be prepared by methods described in the previous sections, applying them either twice, in tandem, or to compounds already containing a double bond. For example, pyrolysis of 1,4-diacetates or dehydrohalogenation via base-induced elimination of 1,4-dibromides both lead to conjugated diene systems:

$$\text{cis-1,2-bis}(CH_2OAc)\text{cyclohexane} \xrightarrow{\Delta} \text{1,2-bis}(=CH_2)\text{cyclohexane}$$

$$BrCH_2CH_2CH_2CH_2Br \xrightarrow{C_2H_5ONa} H_2C{=}CHCH{=}CH_2$$

PROBLEM 25-12

Suggest a method for synthesizing the diacetate of *cis*-1,2-bis(hydroxymethyl)-cyclohexane starting with *o*-xylene.

Especially useful for the preparation of conjugated dienes are reactions involving elimination of functional groups at allylic positions. Allylic alcohols, obtained by direct oxidation, by hydrolysis of allylic halides, or by the addition of Grignard reagents to the carbonyl group of α,β-unsaturated aldehydes or ketones, can be dehydrated either by acid catalysis or by pyrolysis of the acetate or xanthate ester. The pyrolysis procedures are generally recommended, for they avoid the skeletal and/or double bond migrations that are prone to occur in the presence of acid:

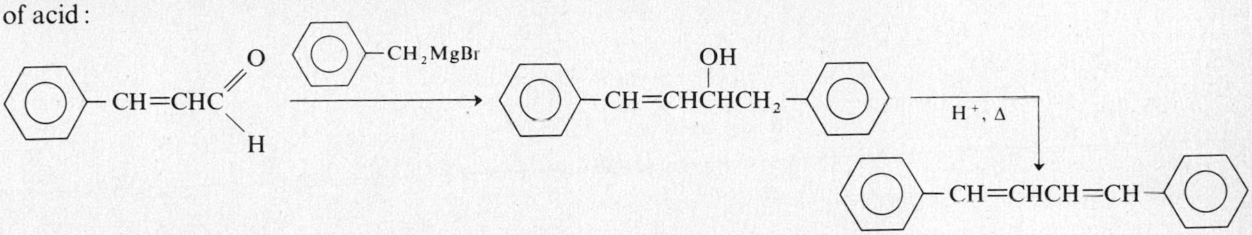

$$C_6H_5CH{=}CHCHO \xrightarrow{C_6H_5CH_2MgBr} C_6H_5CH{=}CHCH(OH)CH_2C_6H_5 \xrightarrow{H^+,\ \Delta} C_6H_5CH{=}CHCH{=}CHC_6H_5$$

Allylic halides are readily prepared by direct halogenation or by means of reagents such as N-bromosuccinimide, and provide convenient precursors to conjugated dienes:

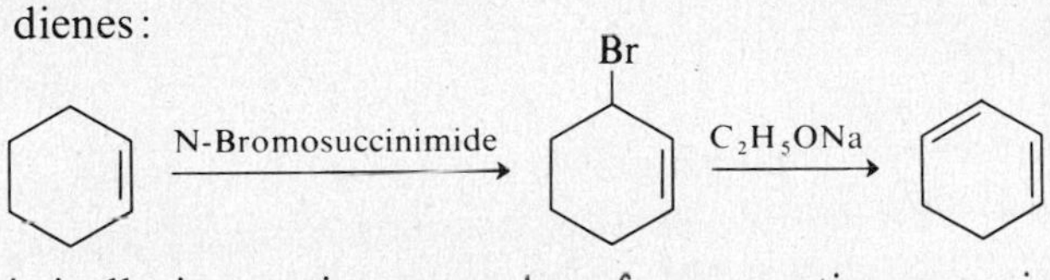

$$\text{cyclohexene} \xrightarrow{\text{N-Bromosuccinimide}} \text{3-bromocyclohexene} \xrightarrow{C_2H_5ONa} \text{1,3-cyclohexadiene}$$

A mechanistically interesting procedure for generating a conjugated diene system involves the action of a metal such as zinc or magnesium on a compound containing a 1,4-dihalo-2-butene system;

$$BrCH_2CH{=}CHCH_2Br \xrightarrow{Zn} ZnBr^+\ {}^-CH_2{-}CH{=}CH{-}CH_2{-}Br \longrightarrow H_2C{=}CHCH{=}CH_2 + ZnBr_2$$

Coupling reactions of alkynes afford still another entry into the conjugated diene system. Acetylene itself, as well as monosubstituted acetylenes, undergo symmetrical coupling reactions in the presence of certain oxidizing agents such as $K_3Fe(CN)_6$ or air in the presence of cuprous salts to give conjugated diynes which can be partially reduced to *cis,cis*-conjugated dienes with molecular hydrogen and the Lindlar catalyst or by the hydroboration-protonolysis sequence (Fig. 25.12).

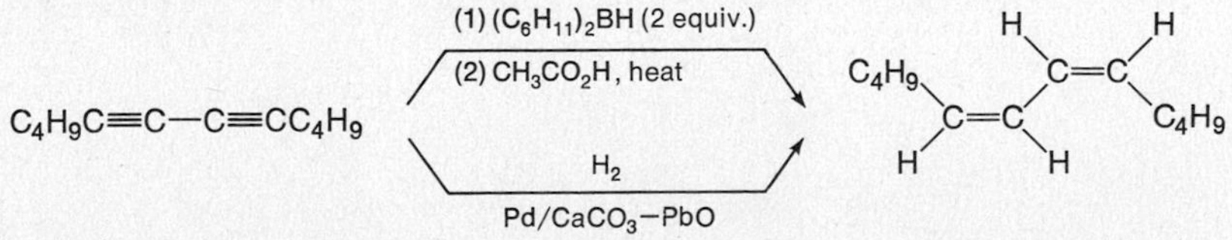

$$C_4H_9C{\equiv}C{-}C{\equiv}CC_4H_9 \xrightarrow[\text{or } H_2,\ Pd/CaCO_3{-}PbO]{(1)\ (C_6H_{11})_2BH\ (2\ \text{equiv.});\ (2)\ CH_3CO_2H,\ \text{heat}} (Z,Z)\text{-}C_4H_9CH{=}CH{-}CH{=}CHC_4H_9$$

Fig. 25.12. Reduction of conjugated diynes to *cis,cis*-dienes.

PROBLEM 25-13

The reaction shown in Fig. 25.12 produces the (Z,Z) isomer. How many stereoisomers are possible?

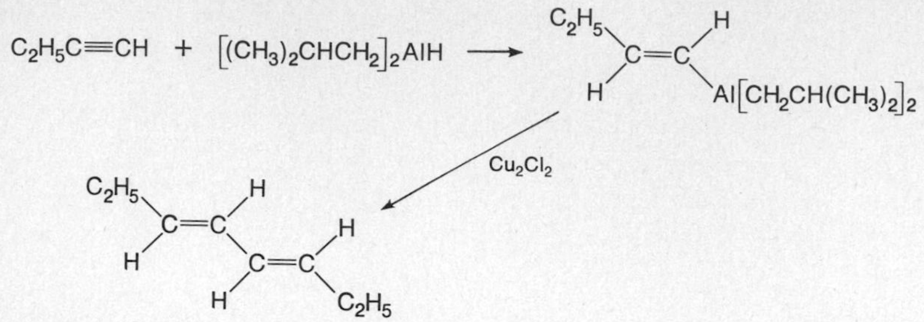

$$C_2H_5C{\equiv}CH + [(CH_3)_2CHCH_2]_2AlH \longrightarrow \underset{H}{\overset{C_2H_5}{}}C{=}C\underset{Al[CH_2CH(CH_3)_2]_2}{\overset{H}{}}$$

$$\xrightarrow{Cu_2Cl_2} \underset{H}{\overset{C_2H_5}{}}C{=}C\overset{H}{}\text{—}\underset{H}{}C{=}C\underset{C_2H_5}{\overset{H}{}}$$

Fig. 25.13. Preparation of *trans,trans*-dienes from monosubstituted alkynes.

Trans,trans-dienes can be obtained by reductive coupling of alkynes. In the presence of dialkylaluminum hydrides monosubstituted acetylenes add the hydride reagent in a *syn* fashion to give intermediates that undergo coupling on treatment with cuprous chloride (Fig. 25.13). When disubstituted alkynes are used, the initially formed adduct reacts further with another molecule of alkyne to form a diene framework from which the diene itself can be obtained by hydrolysis (Fig. 25.14).

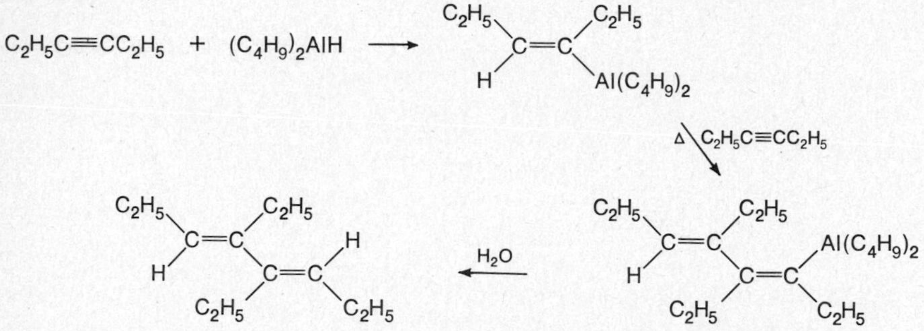

$$C_2H_5C{\equiv}CC_2H_5 + (C_4H_9)_2AlH \longrightarrow \underset{H}{\overset{C_2H_5}{}}C{=}C\underset{Al(C_4H_9)_2}{\overset{C_2H_5}{}}$$

$$\xrightarrow[\Delta]{C_2H_5C{\equiv}CC_2H_5} \underset{H}{\overset{C_2H_5}{}}C{=}C\overset{C_2H_5}{}\text{—}\underset{C_2H_5}{}C{=}C\underset{C_2H_5}{\overset{Al(C_4H_9)_2}{}}$$

$$\xrightarrow{H_2O} \underset{H}{\overset{C_2H_5}{}}C{=}C\overset{C_2H_5}{}\text{—}\underset{C_2H_5}{}C{=}C\underset{C_2H_5}{\overset{H}{}}$$

Fig. 25.14. Preparation of *trans,trans*-dienes from disubstituted acetylenes.

The Wittig reaction (see Section 25.9) is applicable to α,β-unsaturated aldehydes and ketones, thus providing an excellent and quite general method for the preparation of conjugated dienes. A good example of its utility is seen in the synthesis of vitamin A (Fig. 25.15).

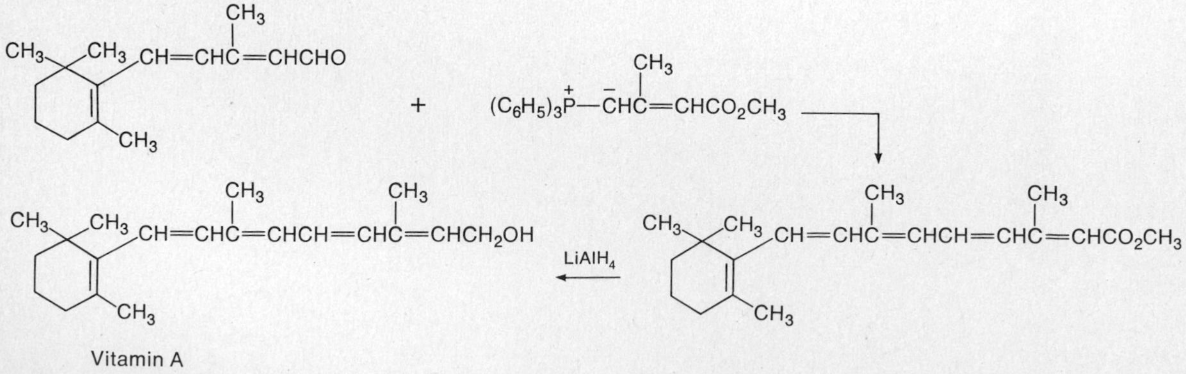

$$\text{(2,6,6-trimethylcyclohexenyl)}\text{—}CH{=}CH\overset{CH_3}{\overset{|}{C}}{=}CHCHO + (C_6H_5)_3\overset{+}{P}\text{—}\overset{-}{C}H\overset{CH_3}{\overset{|}{C}}{=}CHCO_2CH_3 \longrightarrow$$

$$\text{(2,6,6-trimethylcyclohexenyl)}\text{—}CH{=}CH\overset{CH_3}{\overset{|}{C}}{=}CHCH{=}CH\overset{CH_3}{\overset{|}{C}}{=}CHCO_2CH_3$$

$$\xrightarrow{LiAlH_4} \text{(2,6,6-trimethylcyclohexenyl)}\text{—}CH{=}CH\overset{CH_3}{\overset{|}{C}}{=}CHCH{=}CH\overset{CH_3}{\overset{|}{C}}{=}CHCH_2OH$$

Vitamin A

Fig. 25.15. Preparation of vitamin A by means of the Wittig reaction.

25.12 *Preparation of Allenes*

1,2-Dienes constitute a special category of compounds called allenes. Allenes are thermodynamically unstable relative to the corresponding 1,3-dienes, and, consequently, rearrange under the influence of strong bases. A method for preparing allenes that avoids this isomerization is the zinc-induced dehydrohalogenation of 2,3-dihaloalkenes. Starting with an allyl halide, halogen addition yields a 1,2,3-trihalide, which can be converted first to a dihaloalkene by base-induced elimination and then to the allene by treatment with zinc (Fig. 25.16).

Alkynes containing a halogen on the carbon that is directly attached to the triple bond can be converted to allenes by the action of strong nucleophiles such as hydride ion (from lithium aluminium hydride) or carbanions (from Grignard reagents) in S_N2'-type reactions (Fig. 25.17). If potassium *t*-butoxide is used as the nucleophile, an elimination rather than a substitution process takes place, and carbenes of allenes are produced. Thus, treatment of a solution containing a 3,3-dialkyl-3-haloalkyne (often called a propargyl halide) and an alkene with potassium *t*-butoxide yields an alkenylidenecyclopropane, resulting from the addition of the carbene to the alkene (Fig. 25.18). Carbene addition reactions are discussed in more detail in Section 26.3b.

A particularly interesting and useful allene preparation is afforded by the base-induced rearrangement of *gem*-dihalocyclopropanes. For example, the action of methyllithium on 9,9-dibromobicyclo[6.1.0]nonane [prepared by the action of bromoform ($CHBr_3$) and potassium *t*-butoxide on cyclooctene] yields 1,2-cyclononadiene. The reaction involves an α elimination to produce the cyclopropylcarbene which then isomerizes to the allene (Fig. 25.19).

Table 25.1 summarizes the methods that are useful for the synthesis of compounds containing C=C bonds.

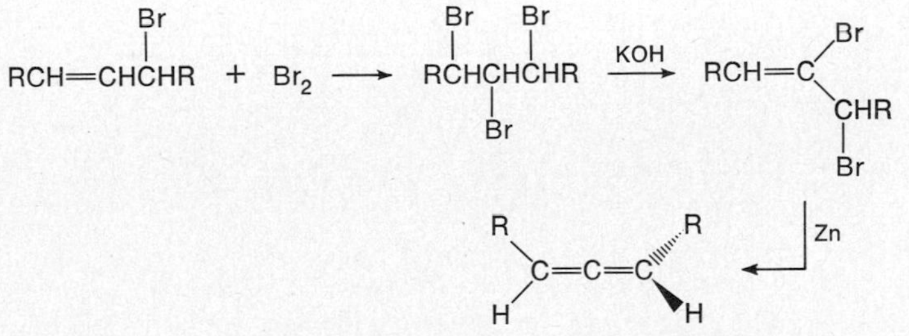

$$RCH{=}CHCH(Br)R + Br_2 \longrightarrow RCH(Br)CH(Br)CH(Br)R \xrightarrow{KOH} RCH{=}C(Br)CH(Br)R \xrightarrow{Zn} (R)(H)C{=}C{=}C(R)(H)$$

Fig. 25.16. Preparation of an allene via a 1,2,3-tribromo compound.

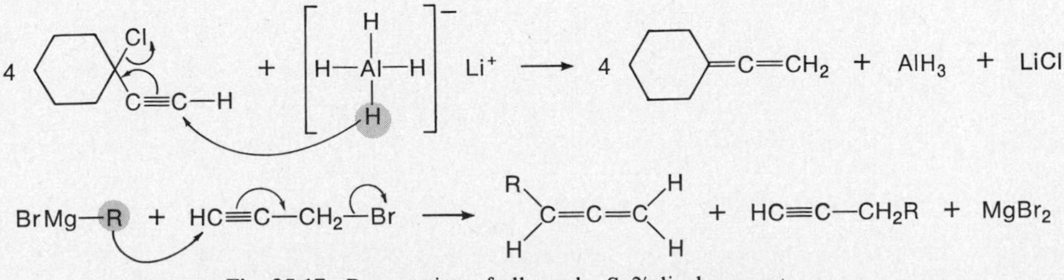

$$4\ \text{(1-chloro-1-ethynylcyclohexane)} + [AlH_4]^- Li^+ \longrightarrow 4\ \text{(cyclohexylidene)}{=}C{=}CH_2 + AlH_3 + LiCl$$

$$BrMg{-}R + HC{\equiv}C{-}CH_2{-}Br \longrightarrow (R)(H)C{=}C{=}C(H)(H) + HC{\equiv}C{-}CH_2R + MgBr_2$$

Fig. 25.17. Preparation of allenes by S_N2' displacement.

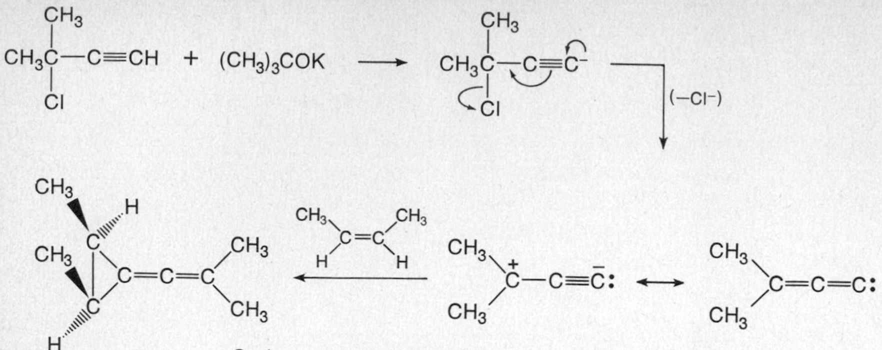

Carbene addition reaction

Fig. 25.18. Preparation of alkenylidenecyclopropanes from 3-haloalkynes and alkenes via a carbene addition reaction.

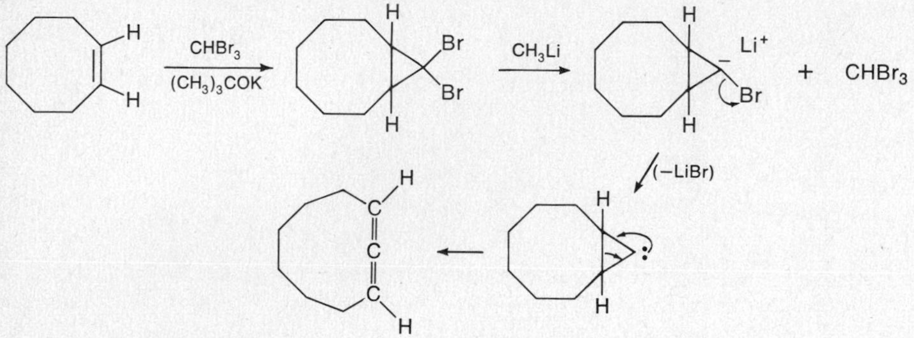

Fig. 25.19. Preparation of allenes from 1,1-dihalocyclopropanes via a carbene rearrangement reaction.

Table 25.1 METHODS FOR THE INTRODUCTION OF C=C BONDS

Classification	Starting Materials	Reagents and Conditions	Products	Comments
Acid-catalyzed dehydration of alcohols	$H{-}CR_2{-}CR_2{-}OH$	Strong acids (*e.g.*, H_2SO_4)	$R_2C{=}CR_2$	Saytzev products, low stereoselectivity, rearrangements possible
Base-induced elimination	$R{-}C(H)(R){-}C(R)(Nu){-}R$	Strong bases [*e.g.*, $(CH_3)_3COK$]	$R_2C{=}CR_2$	Saytzev product when Nu = —Cl, —Br, —I, or $—OSO_2R$; Hofmann products when Nu = $—\overset{+}{N}R_3$ and $—\overset{+}{S}R_2$; Antiperiplanar conformation preferred; synperiplanar conformation reactive if antiperiplanar conformation unattainable
Metal-induced elimination	$R{-}C(E)(R){-}C(R)(Nu){-}R$ E = halogen Nu = halogen or alkoxyl	Metals (*e.g.*, Zn or Mg)	$R_2C{=}CR_2$	Antiperiplanar conformation required Situspecificity determined by positions of E and Nu

Table 25.1—*cont.*

Classification	Starting Materials	Reagents and Conditions	Products	Comments
Pyrolytic elimination	$R_2C(H)-C(Y)R_2$; Y = $-OCOCH_3$, $-OCSSCH_3$, or $-\overset{+}{N}(O^-)(CH_3)_2$	Heat	$R_2C=CR_2$	*Syn* conformation required; Saytzev products
Oxidative elimination	$R_2C(CO_2H)-C(CO_2H)R_2$	$Pb(OCOCH_3)_4$	$R_2C=CR_2$	
Alkyne reduction	$RC{\equiv}CR'$	$H_2/Pd-CaCO_3-PbO$ or R_2BH and protonolysis	$RHC=CHR'$ (R and R′ cis)	
		Na/NH_3	$RHC=CHR'$ (R and R′ trans)	
Arene reduction	C_6H_5Y	$Na/NH_3/ROH$ or NH_4Cl	1-Y-1,4-cyclohexadiene or	Major isomer when Y is electron releasing
			3-Y-1,4-cyclohexadiene	Major isomer when Y is electron withdrawing
	C_6H_5R	Li, $H_2NCH_2CH_2NH_2$	1-R-cyclohexene	Major isomer when R is either electron withdrawing or donating
Wittig reaction	$RR'C=O$	$(C_6H_5)_3\overset{+}{P}-\overset{-}{C}R''R'''$	$RR'C=CR''R'''$	R groups can contain hetero atoms but cannot contain acidic hydrogens; yields product with two largest groups *trans*

25.13 *Study Guide*

I. The β-elimination reaction
  A. General features of β-elimination reactions
    1. Ionic elimination reactions
      (a) E1 reaction
        (1) The elimination counterpart of the S_N1 reaction
        (2) A two-step process involving carbonium ion intermediates

(3) E1 is favored over S_N1 pathway by strongly basic and weakly nucleophilic reagents, and by structural features that increase the stability of the intermediate carbonium ion.

(b) $E1_{cb}$ reaction—very few known examples

(c) E2 reaction

(1) The elimination counterpart of the S_N2 reaction

(2) A one-step process involving no intermediate

(3) E2 favored over S_N2 pathway by steric hindrance at the α-carbon and by large and strongly basic nucleophiles

2. Thermally induced elimination reactions

(a) Leaving groups not bonded—pyrolysis of acetate and xanthate esters and N-oxides

(b) Leaving groups bonded—cycloelimination reactions

3. Stereochemistry of β-elimination reactions

(a) Ionic reactions—usually require the antiperiplanar conformation between leaving groups (*i.e.*, *anti* elimination)

(b) Thermal reactions—usually require the synperiplanar conformation between leaving groups (*i.e.*, *syn* elimination)

(c) Preferred activated complex minimizes steric interactions but must maintain the antiperiplanar or synperiplanar conformation

4. Regioselectivity of β-elimination reactions

(a) Saytzev pathway—yields most highly substituted alkene with good leaving groups

(b) Hofmann pathway—yields least highly substituted alkene with poor leaving groups

(c) Pathway determined by leaving group, by size of base, and by steric hindrance at β positions

(d) Regioselectivity *vs.* stereochemistry—stereochemistry is the determining factor

II. Scope and limitations of methods of preparation of alkenes

A. Alkenes from acid-catalyzed dehydration of alcohols—rearrangements likely

B. Alkenes from base-induced elimination reactions—see I.A.1. (c)

C. Alkenes from metal-induced elimination reactions

1. Action of metals (*e.g.*, zinc) on 1,2-dihalides or 1-halo-2-alkoxy compounds

2. Boord alkene synthesis

D. Alkenes from pyrolytic elimination reactions—see I.A.2

E. Alkenes by oxidative elimination reactions—lead tetraacetate on dicarboxylic acids

F. Alkenes by reduction of alkynes—allows formation of *cis* or *trans* isomer

G. Alkenes by reduction of arenes—Birch reduction

H. Alkenes from aldehydes and ketones

1. Via reduction or addition of organometallic reagents followed by dehydration

2. Via phosphorus ylides (phosphoranes)—Wittig reaction
   (a) Reagents: RCHO or $R_2C{=}O$ and $(C_6H_5)_3P{=}CHR$
   (b) R in phosphorus ylide can be alkyl, aryl, or a functional group (if functional group does not contain an acidic hydrogen)

I. Alkenes from vinyl halides—coupling with R_2CuLi reagent

J. Preparation of dienes
   1. Via methods described above, used in tandem
   2. Coupling reactions of alkynes

K. Preparation of allenes
   1. Dehalogenation of 2,3-dihalo-1-alkenes
   2. Addition of carbenes of allenes to alkenes
   3. Dehydrohalogenation of 1,1-dihalocyclopropanes

REFERENCES

FIESER, L. F., and M. FIESER, *Advanced Organic Chemistry*, Reinhold Publishing Corp., New York, 1961, Chapter 5.

PATAI, S., ed., *The Chemistry of Alkenes*, Vol. 1, John Wiley & Sons, Inc., New York, 1964.

PATAI, S., and J. ZABICKY, *The Chemistry of the Alkenes*, Vol. 2, John Wiley & Sons, Inc., New York, 1970.

PROBLEMS

1. Draw the structure of the major product from each of the following reactions:

(a) $(C_6H_5)_2C(OH)CH_2CH_3 \xrightarrow{H^+}$

(b) CH_3–C(H)(H)–C(H)(OTs)–C_6H_5 (stereo) $\xrightarrow{C_2H_5ONa}$

(c) 1-phenyl-2-acetoxycyclohexane (H, H; $OCOCH_3$) $\xrightarrow{\Delta}$

(d) 1-bromo-1-($CHOC_2H_5$ / CH_3)cyclohexane $\xrightarrow{Zn}$

(e) cyclohexyl–$C{\equiv}CCH_3$ $\xrightarrow[NH_3]{Na}$

(f) cyclopentanone $+ \ (C_6H_5)_3\overset{+}{P}{-}\overset{-}{C}HCH_2CH_3$

(g) cyclooctyne + $(CH_3)_2AlH \longrightarrow$

(h) $(CH_3)_2CHCH(CH_3)-C(CH_3)(C_6H_5)-\overset{+}{N}(CH_3)_2-O^- \xrightarrow{\Delta}$

(i) 2-methoxynaphthalene (OCH_3) $\xrightarrow[NH_3]{Na}$

2. Outline, in as much detail as possible, the probable mechanisms by which the following reactions occur:

(a) $(CH_3)_3CCH(OH)CH_3 \xrightarrow{H^+} (CH_3)_2C{=}C(CH_3)_2$

(b) $H-C(CH_3)(OC(=S)SCH_3)-C(H)(H)-C_6H_5 \longrightarrow$ $C_6H_5(H)C{=}C(H)CH_3$

(c) $BrCH_2CH_2CH_2Br + Zn \longrightarrow$ cyclo-$(CH_2)_3$

(d) $(CH_3)_2C{=}O + (C_6H_5)_3P{=}CHCH_2OCH_3 \longrightarrow (CH_3)_2C{=}CHCH_2OCH_3$

3. The following questions are concerned with the comparative properties of various compounds with respect to their ability to undergo elimination reactions or induce elimination reactions.

(a) Rank the following arylalkyl bromides in the order of increasing ease of E1 elimination:

$CH_3CHBr-C_6H_5$	$CH_3CHBr-C_6H_4-NO_2$	$CH_3CHBr-C_6H_4-OCH_3$	$CH_3CHBr-C_6H_4-CH_3$
(A)	(B)	(C)	(D)

(b) Rank the following compounds with respect to increasing ability to effect E2 elimination reactions:

$(CH_3)_3CO^-Li^+$ (A) $\quad (CH_3)_2N^-Li^+$ (B) $\quad (CH_3)_3CCO_2^-Li^+$ (C)

$(CH_3)_3C^-Li^+$ (D) $\quad (C_6H_5)_3C^-Li^+$ (E)

(c) Rank the following compounds in increasing ease of E1 elimination:

$(CH_3)_2CHF$ (A) $(CH_3)_2CHOSO_2CF_3$ (B) $(CH_3)_2CHN_2^+$ (C) $(CH_3)_2CHBr$ (D)

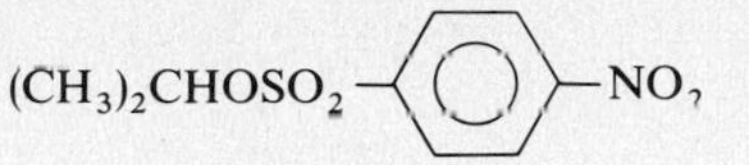

$(CH_3)_2CHOSO_2-C_6H_4-NO_2$ (E) $(CH_3)_2CHOCOCH_3$ (F)

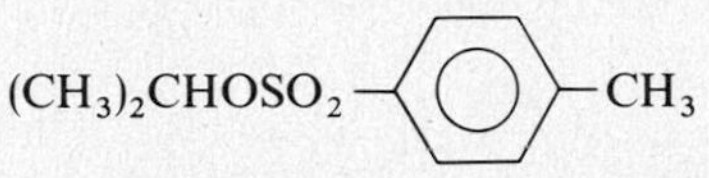

$(CH_3)_2CHOSO_2-C_6H_4-CH_3$ (G)

4. For each of the following reactions, give two plausible mechanisms and discuss in as much detail as possible, all the ways you can think of for distinguishing between these mechanisms and for supporting what you believe to be the correct mechanism:

(a) $C_2H_5(CH_3)(H)C-C(Br)(CH_3)(H)$ + Base ⟶ $(C_2H_5)(CH_3)C=CHCH_3$

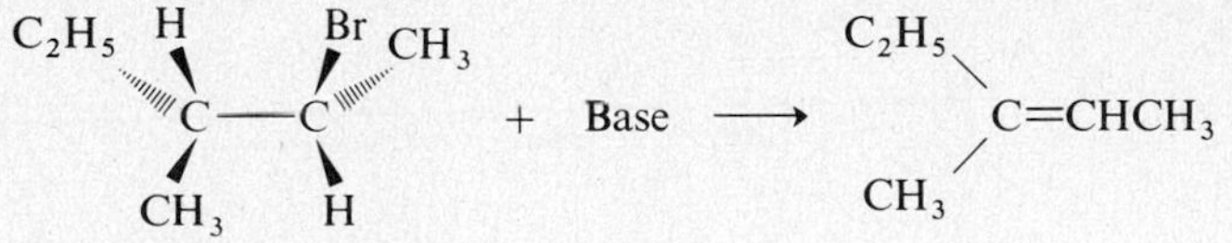

(b) 2-methylcyclohexyl $OC(=S)SCH_3$ (CH₃, H, H as drawn) $\xrightarrow{\Delta}$ 3-methylcyclohexene (CH_3, H) + CH_3SH + $O=C=S$

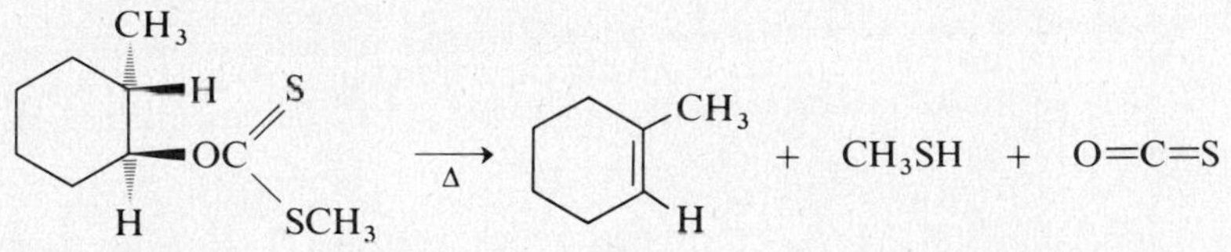

(Chugaev elimination reaction)

5. Treatment of *cis*-2-phenylcyclopentanecarboxamide with bromine in aqueous sodium hydroxide yields *cis*-2-phenylcyclopentylamine:

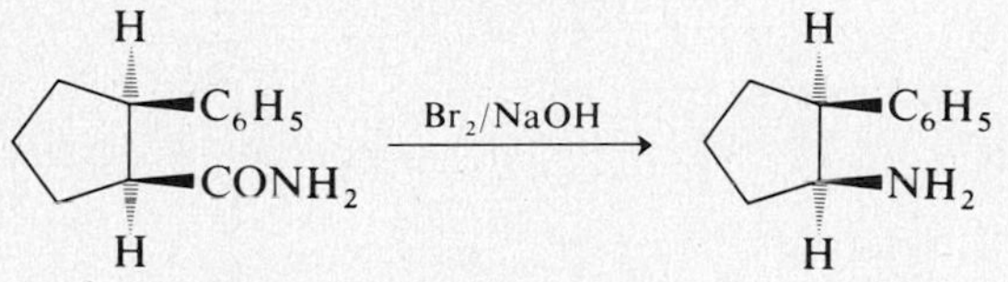

cis-2-(C_6H_5)cyclopentane-$CONH_2$ $\xrightarrow{Br_2/NaOH}$ cis-2-(C_6H_5)cyclopentyl-NH_2

(a) Outline a reasonable mechanism for this reaction.
(b) When the amine is converted to the N,N-dimethylamine oxide and heated, an alkene is obtained. What is its structure and what is the mechanism of its formation?
(c) When the amine is converted to the trimethylammonium salt and treated with base, an alkene is obtained. What is its structure and what is the mechanism of its formation?
(d) When the amine is treated with nitrous acid an alkene is obtained. What is its structure and what is the mechanism of its formation?
(e) If an optically active sample of the amine is used, which, if any, of the alkenes formed by the processes indicated in (b), (c), and (d) will be optically active?

6. The reaction of 2-butanone with $(C_6H_5)_3\overset{+}{P}-\overset{-}{C}HCO_2CH_3$ occurs via a reversible addition of the Wittig reagent to the carbonyl group of 2-butanone to produce a dipolar intermediate that undergoes a concerted *syn* elimination of triphenylphosphine oxide $[(C_6H_5)_3\overset{+}{P}-\overset{-}{O}]$ with production of a substituted alkene. What is the structure of the major product formed in this reaction? Explain.

7. Outline procedures for effecting the following conversions:

(a) $C_6H_5-CH=CH_2 \longrightarrow C_6H_5-C\equiv C-H$

(b) $(CH_3)_2CH(H)C=C(H)CH_3$ (trans) $\longrightarrow$ $(CH_3)_2CH(H)C=C(CH_3)H$ (cis)

(c) CH_3-cyclohexanone ($=O$) $\longrightarrow$ CH_3-cyclohexene$-CH_3$

(d) cycloheptene $\longrightarrow$ cycloheptadiene

(e) 2-methylphenol (CH_3, OH) $\longrightarrow$ methylcyclohexene (CH_3)

(f) $(CH_3)_2C=C(H)CH_3 \longrightarrow (CH_3)_2CHCH=CH_2$

8. Indicate the most logical precursor for each of the following compounds, and show the reaction that is involved in going from precursor to product:

(a) 1,2-bis(methylene)cyclohexane (CH_2, CH_2)

(b) $(C_2H_5)_2C=CHCH_2D$

(c) hexahydronaphthalene carboxylic acid (CO_2H)

(d) [bicyclic alkene structure]

9. Outline procedures for the preparation of each of the five isomeric hexenes using regio- and stereoselective reactions starting from any C_1 or C_2 compounds.

10. Given the following starting materials, indicate the procedure by which they can be converted to the products shown. More than one step may be necessary.

(a) $CH_3C\equiv CH \longrightarrow CH_3CH=CHCH=CH_2$

(b) $C_6H_5-CH_2CHO \longrightarrow C_6H_5-CH=CHCH$

(c) cycloheptanone ($=O$) $\longrightarrow$ cycloheptylidene$=CHCH_2CH(CH_3)_2$

(d) 1-tetralone (O) $\longrightarrow$ 1,2-dihydronaphthalene

(e) [cyclohexane ring with $=CH_2$] ⟶ [cyclohexane ring with CH_2OH] (f) [benzene ring with OCH_3] ⟶ [cyclohexenone, O]

11. For the following syntheses, the available starting materials include (a) any alkane, alkene, alkyne, alkyl halide, alkanol, alkanal, alkanone, alkanoic acid or alkanoic acid derivative containing four or fewer carbon atoms, (b) benzene, toluene, *o*-, *m*-, and *p*-xylene, naphthalene, and (c) any inorganic compounds and special organic reagents such as $(C_6H_5)_3P$, N-bromosuccinimide, etc. Starting with materials chosen from this group, outline syntheses for each of the following compounds:

(a) [naphthalene ring with $CH{=}CHCH_3$]

(b) $CH_3C{\equiv}CCH{=}CHCH_2CH_3$

(c) $(CH_3)_2C{=}CHCH_2CH_3$

(d) $CH_3CH_2CH{=}CHCH{=}CHCH_2CH_3$

(e) [1,4-diphenylbutadiene: C_6H_5–CH=CH–CH=CH–C_6H_5, with H, H labels on each C=C]

(f) $H_2C{=}CHCH_2CH_2CH{=}CH_2$

(g) $H_2C{=}C(CH_3){-}C(CH_3){=}CH_2$

(h) $CH_3CH{=}C{=}CHCH_3$

12. Using Tables 39.1 to 39.5 as guidelines to readily available starting materials, devise syntheses for the following compounds. In addition to readily available organic compounds, any inorganic materials and special organic reagents may also be used.

(a) $(CH_3CH_2CH_2)(CH_3)C{=}CH(CH_2)_8CH_3$

(b) 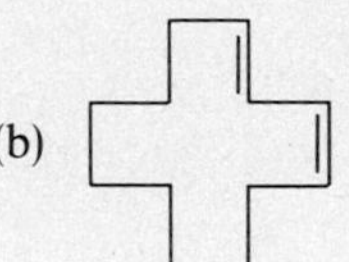

(c) [cyclopentene ring bearing a benzene ring]

(d) [naphthalene ring bearing –CH=CH–C_6H_5, with H labels on each C]

(e) [cyclohexadiene ring with CO_2H]

Chapter **26**

Alkenes: reactions

New parts must then be added: Twill follow next if thou percase would'st vary still its shapes, that by like logic each arrangement still requires its increment of other parts.

LUCRETIUS
Rerum Novarum

With the knowledge in hand of how to introduce carbon–carbon double bonds into organic molecules, let us now turn our attention to the various ways in which double bonds can react. Among the most important and useful of the reactions of alkenes are those in which various entities are added to the double bond. This chapter therefore devotes considerable attention to addition processes, and deals with their mechanisms in more detail than in earlier chapters, where some of these reactions are first introduced. As in previous instances, a knowledge of the mechanisms of a reaction provides insight into its scope, limitations, and utility in synthesis procedures.

26.1 *Reactions in Which the Double Bond Is Retained*

Alkenes undergo *cis,trans* isomerization, positional isomerization, and skeletal rearrangement under certain conditions. The *cis,trans* isomerization can be either chemically initiated by reagents such as iodine, or photochemically initiated by light of the proper wavelength. In each case, an equilibrium mixture of the two isomers is established, although the equilibrium constant usually is different for the chemically and photochemically induced isomerizations; in the former, the *trans* isomer generally predominates, and in the latter the *cis* isomer generally predominates (Fig. 26.1).

Treatment of alkenes with protic acids or Lewis acids can set in motion all three types of isomerization mentioned, as illustrated by the example in Fig. 26.2.

PROBLEM 26-1

Arrange the alkenes in Fig. 26.2 in the order of decreasing thermodynamic stability.

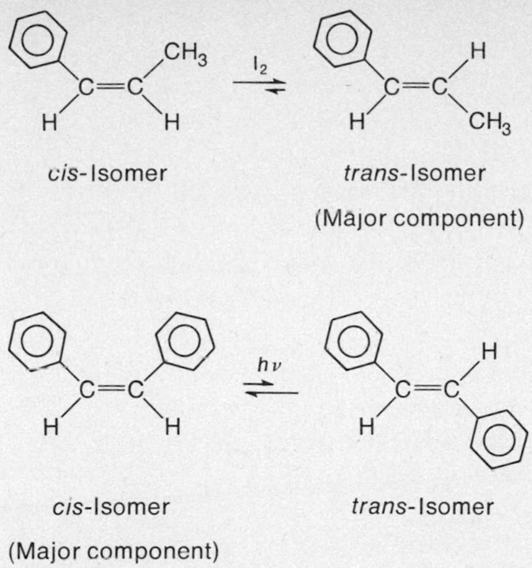
CH3
I2
H
C=C
H
CH3
cis-Isomer
trans-Isomer
(Major component)
hν
cis-Isomer
(Major component)
trans-Isomer

Fig. 26.1. *Cis,trans* isomerization of alkenes.

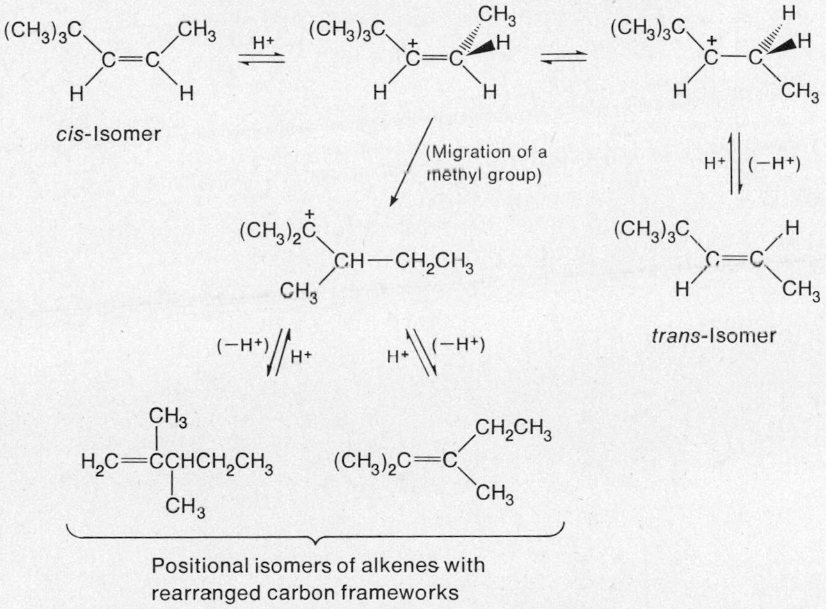
(CH3)3C
CH3
H+
cis-Isomer
(Migration of a
methyl group)
H+
(−H+)
(CH3)2C
CH—CH2CH3
trans-Isomer
H2C=CCHCH2CH3
(CH3)2C=C
CH2CH3
Positional isomers of alkenes with
rearranged carbon frameworks

Fig. 26.2. Acid-catalyzed isomerization of 4,4-dimethyl-2-pentene.

PROBLEM 26-2

Nonconjugated dienes can be converted to conjugated dienes by the action of strong base. What is the driving force for this reaction? Suggest a mechanism for the conversion of 1,4-cyclooctadiene to 1,3-cyclooctadiene by treatment with potassium *t*-butoxide in dimethylsulfoxide for 1 hr at 70°.

26.2a. General features of electrophilic addition reactions. Electrophilic addition is the characteristic reaction of alkenes and alkynes. It is initiated by attachment of an electrophile (Lewis acid) to the alkene or alkyne (Lewis base) to give a carbonium ion which then reacts with a nucleophile. The timing of these two steps varies considerably from one reaction to another. In some instances, the second step is enough slower than the first that a carbonium ion is a true intermediate; in other instances, the second step follows so close on the heels of the first that the overall process takes place in a concerted fashion, *i.e.*, the electrophilic and nucleophilic portions of the reagent become bonded to the alkene simultaneously.

The direction of addition, the regioselectivity, of an unsymmetrical addend (E—Nu) to an unsymmetrically substituted alkene or alkyne usually results in the attachment of the nucleophile to the more highly alkyl-substituted carbon atom of the multiple bond.

$$R_2C{=}CH_2 + E{-}Nu \longrightarrow R{-}\underset{Nu}{\overset{R}{C}}{-}CH_2E$$

The original statement of this fact, known as Markovnikov's rule (see p. 204), correctly predicts the direction of addition for alkyl-substituted alkenes. It is not satisfactory for alkenes containing hetero atoms, however, and a more generally applicable rule states that *the electrophilic addition to a double bond results in the attachment of the nucleophile to the atom more capable of bearing*

$$(CH_3)_2C{=}CHOC_2H_5 + \underset{(E{-}Nu)}{H{-}OC_2H_5} \longrightarrow (CH_3)_2CH\underset{OC_2H_5}{CH}OC_2H_5 \quad \text{(not } (CH_3)_2\underset{OC_2H_5}{C}CH_2OC_2H_5)$$

$$CF_3CH{=}CH_2 + \underset{(E{-}Nu)}{H{-}Cl} \longrightarrow CF_3CH_2CH_2Cl \quad \text{(not } CF_3\underset{Cl}{CH}CH_3)$$

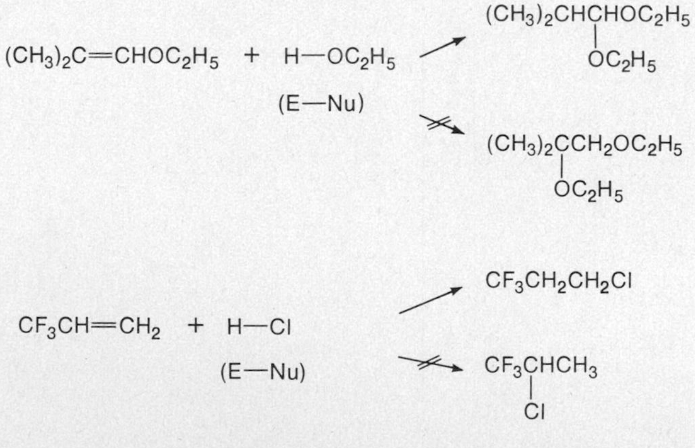

Fig. 26.3. Examples of electrophilic addition reactions that do not obey the Markovnikov rule in its original formulation but do obey it when defined in terms of the mechanism of the reaction.

a positive charge. Figure 26.3 shows examples of electrophilic addition reactions that do not follow Markovnikov's rule as originally formulated but do follow the more general rule.

PROBLEM 26-3

Why is $(CH_3)_2CH\overset{+}{C}HOC_2H_5$ more stable than $(CH_3)_2\overset{+}{C}CH_2OC_2H_5$? Why is $CF_3CH_2\overset{+}{C}H_2$ more stable than $CF_3\overset{+}{C}HCH_3$?

26.2b. Bimolecular electrophilic addition reactions occurring via cationic intermediates. Addition reactions that are kinetically second order (*i.e.*, first order in E—Nu and first order in alkene) are designated as Ad_E2 and are subdivided into reactions involving carbonium ion intermediates and reactions involving cyclic onium ion intermediates:

$$H_2C{=}CR_2 + E^+ \nearrow H\text{—}\underset{E}{\overset{H}{C}}\text{—}\overset{+}{C}R_2 \xrightarrow{Nu^-} H\text{—}\underset{E}{\overset{H}{C}}\text{—}\underset{R}{\overset{R}{C}}\text{—}Nu$$

Carbonium ion intermediate

$$H_2C{=}CR_2 + E^+ \searrow H\text{—}\overset{H}{C}\text{——}\overset{R}{C}\text{—}R \;(\text{bridged by } \overset{+}{E}) \xrightarrow{Nu^-} H\text{—}\underset{E}{\overset{H}{C}}\text{—}\underset{R}{\overset{Nu}{C}}\text{—}R$$

Cyclic onium ion intermediate

The carbonium ion pathway is favored in systems in which the carbon framework is especially capable of delocalizing a positive charge. For example, 1,2-dimethylcyclohexene, which forms an intermediate tertiary cation, and 1-phenylpropene, which forms an intermediate benzyl cation, follow the Ad_E2 pathway via carbonium ion intermediates (Fig. 26.4). The cyclic onium pathway, on the other hand, is favored in systems less able to sustain a free carbonium ion and is particularly likely to be followed when the electrophile is derived from second-, or higher, row elements in the periodic chart. For example, chlorine, bromine, and iodine add to most alkenes by Ad_E2 processes involving cyclic chloronium, bromonium, and iodonium intermediates, respectively (Fig. 26.5). Other electrophiles that add by a presumably similar pattern include the hypohalous acids (HOCl and HOBr), iodine isocyanate (INCO), iodine azide (IN_3), alkyl- and aryl-sulfenyl chlorides (RSCl), and mercuric salts [*e.g.*, $Hg(OCOCH_3)_2$] in the presence of water or alcohols. Addition reactions involving onium ion intermediates occur exclusively via the *anti* pathway.

Although the regioselectivity of the Ad_E2 reactions involving carbonium ions is the same as those involving cyclic onium ions (*i.e.*, Markovnikov addition), the stereoselectivity may sometimes be different. For reactions that involve true carbonium ions, a loss of stereochemical integrity at the carbon bearing the positive charge is expected. In the hydration of 1,2-dimethylcyclohexene, for instance, the *cis* and *trans* isomers are obtained in identical amount (Fig. 26.4). In the addition of deuterium chloride (DCl) to 1-phenylpropene, on the other

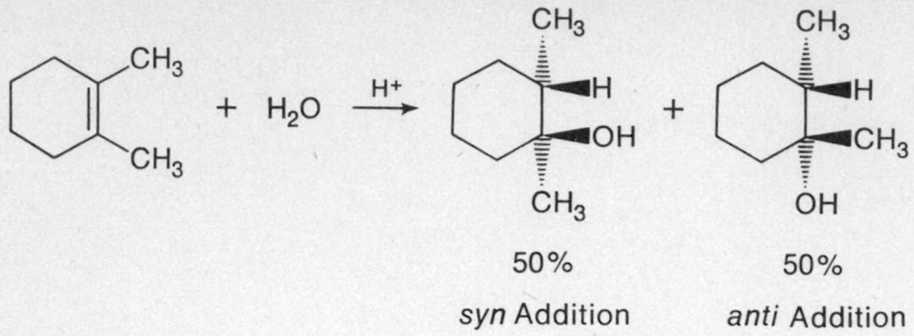

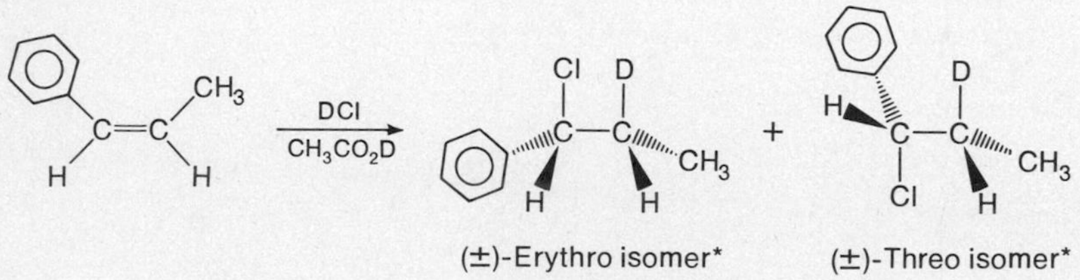

Fig. 26.4. Ad_E2 reactions involving carbonium ion intermediates.

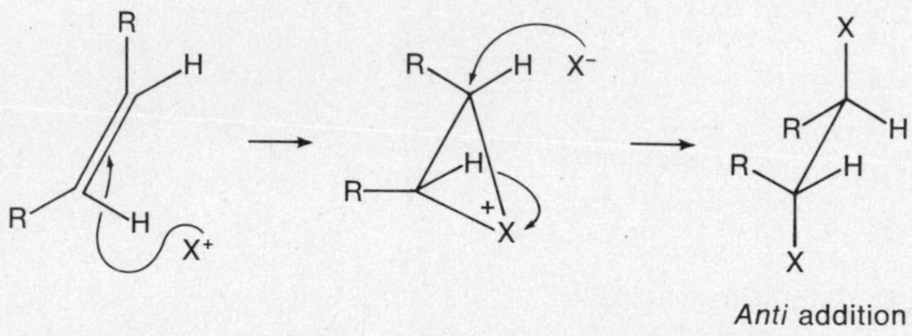

Fig. 26.5. Ad_E2 reactions involving cyclic halonium ion.

hand, the *erythro* isomer is formed in larger amount than the *threo* isomer, in apparent contradiction to the notion of a free carbonium ion. One suggestion advanced to explain this outcome postulates that the initially formed carbonium ion is a "tight-ion pair," in which the nucleophile is so closely associated with the carbonium ion that it attaches itself to the cationic center before the bond can rotate, thus resulting in *syn* addition (Fig. 26.6). In still other examples of reactions alleged to involve carbonium ion intermediates, *anti* addition occurs preferentially, further clouding the true nature of the intermediate.

The intermediate cations formed in Ad_E2 reactions, particularly carbonium ions, can form stable products along several pathways. The straightforward addition pathway involves the addition of a nucleophile to the cationic center. Alternatively, however, the cation can (a) undergo skeletal rearrangement to an

* See p. 518 for a discussion of the *erythro* and *threo* designations. If the reaction is carried out in a chiral environment, equal amounts of enantiomeric forms are obtained; *i.e.*, equal amounts of the (1*R*,2*R*) and (1*S*,2*S*) forms corresponding to the *erythro* form and equal amounts of the (1*R*,2*S*) and (1*S*,2*R*) forms corresponding to the *threo* form. The *erythro* and *threo* forms are not necessarily produced in equal amount, however.

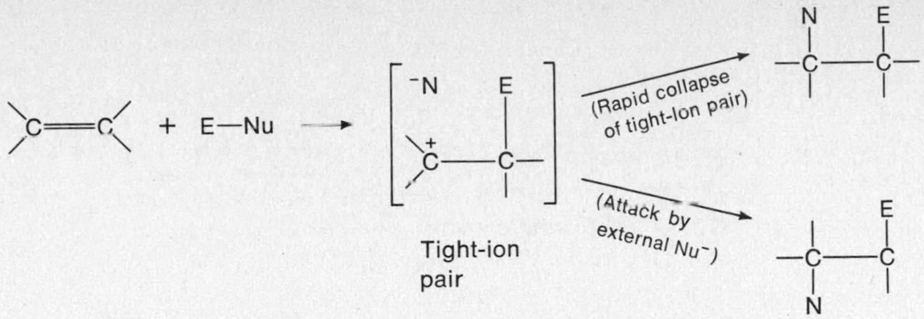

Fig. 26.6. *Syn* addition via "tight-ion pair" carbonium ion intermediates.

isomeric cation, (b) react with nucleophiles other than the one accompanying the electrophile (including the solvent if it is a nucleophile), and (c) expel a proton to form an alkene rather than an addition product (Fig. 26.7). Solvent participa-

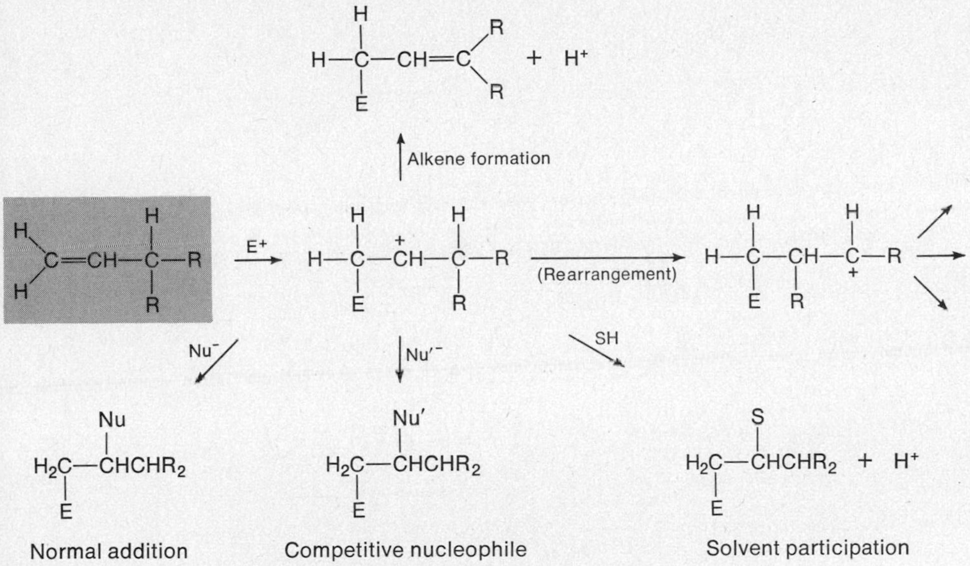

Fig. 26.7. Reaction pathways of carbonium ions.

tion, for example, is noted when the addition of hydrogen chloride to 1-phenylpropene is carried out in acetic acid solution. In addition to the major product, 1-phenyl-1-propyl chloride, 10–15% of the product is 1-phenylpropyl acetate:

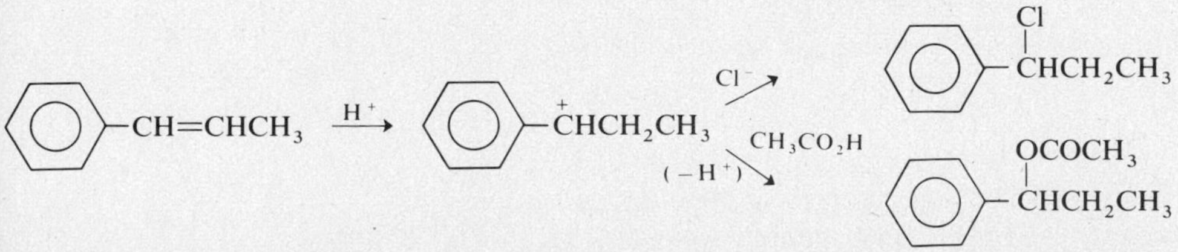

$$C_6H_5CH{=}CHCH_3 \xrightarrow{H^+} C_6H_5\overset{+}{C}HCH_2CH_3 \xrightarrow{Cl^-} C_6H_5CH(Cl)CH_2CH_3$$

$$C_6H_5\overset{+}{C}HCH_2CH_3 \xrightarrow[(-H^+)]{CH_3CO_2H} C_6H_5CH(OCOCH_3)CH_2CH_3$$

The extent of solvent participation depends on the nucleophilicity of the solvent and can be minimized by using solvents such as nitromethane, which is aprotic and nonnucleophilic.

26.2c. TERMOLECULAR ELECTROPHILIC ADDITION REACTIONS. Under certain conditions, the kinetics of electrophilic addition reactions show a greater than first-order dependence on the concentration of the addition reagent, E—Nu. Thus, reactions characterized as Ad_E3, Ad_E4, etc., are known, with the higher-order reactions showing an increasing tendency to occur as the polarity of the solvent decreases. The third-order process has been depicted as involving an activated complex in which one molecule of the addend acts as a nucleophile and another molecule of the addend acts as an electrophile, both molecules associating themselves with the double bond as shown in Fig. 26.8. The addition

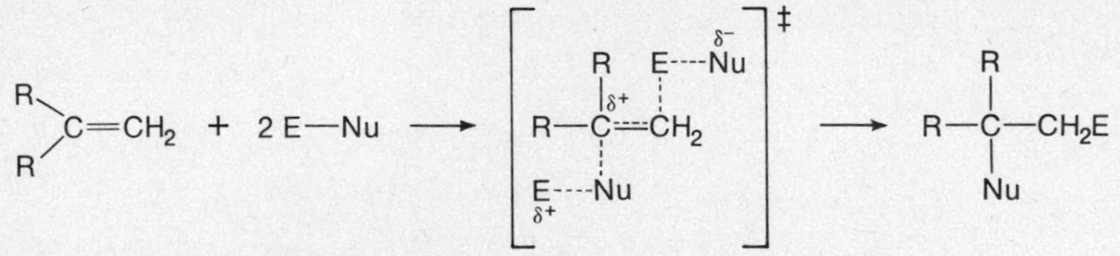

$R_2C{=}CH_2 + 2\,E{-}Nu \longrightarrow [\text{activated complex}]^{\ddagger} \longrightarrow R{-}C(R)(Nu){-}CH_2E$

Fig. 26.8. Mechanisms of Ad_E3 reactions.

of halogen acids to alkenes from which relatively unstable carbonium ions are formed tends to follow the Ad_E3 pathway as, for example, the addition of deuterium bromide to *cis*- and *trans*-2-butene (Fig. 26.9).

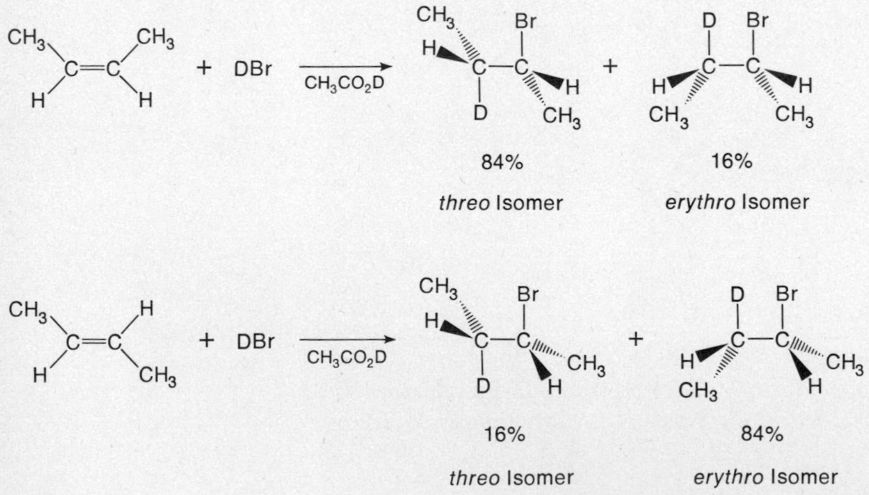

cis-2-butene + DBr $\xrightarrow{CH_3CO_2D}$ 84% *threo* Isomer + 16% *erythro* Isomer

trans-2-butene + DBr $\xrightarrow{CH_3CO_2D}$ 16% *threo* Isomer + 84% *erythro* Isomer

Fig. 26.9. Addition of deuterium bromide to *cis*- and *trans*-2-butene.

26.2d. STEREOCHEMISTRY OF ELECTROPHILIC ADDITION. Most electrophilic additions to alkenes take place, as the preceding examples illustrate, in the *anti* fashion. For example, the addition of bromine to cyclopentene yields *trans*-1,2-dibromocyclopentane via a cyclic bromonium ion intermediate (Fig. 26.10).

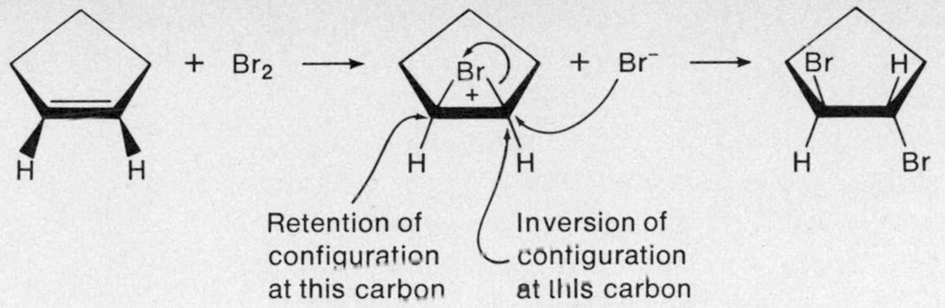

Fig. 26.10. Electrophilic addition of bromine to cyclopentene.

PROBLEM 26-4

What would be the stereochemical outcome of the addition of bromine to *cis*- and *trans*-cyclooctene?

If the cyclopentene moiety is incorporated into the rigid bicyclo[2.2.1]-hept-2-ene, however, the *syn* pathway is lower in energy than the *anti* pathway. For example, the addition of deuterium chloride to this compound yields 2-*exo*-chloro-3-*exo*-deuteriobicyclo[2.2.1]heptane (Fig. 26.11). The preference

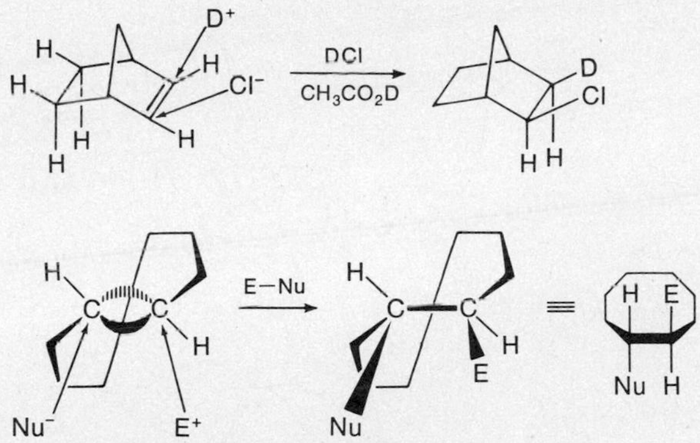

Fig. 26.11. Examples of *syn* addition reactions.

for *syn* addition in this case is ascribed to the excessive strain that is involved in the *anti* addition; when the nucleophile approaches the intermediate carbonium ion an activated complex is formed in which the C—E and C—Nu bonds bear an antiperiplanar relationship, and the bicycloheptane ring system cannot easily accommodate this requirement. Electrophilic *syn* addition is also observed in *trans*-cyclooctene, which yields *trans*-1,2-disubstituted cyclooctanes. The preference for *syn* addition in this case is ascribed to the inaccessibility of the back face of the double bond, which is shielded by the methylene groups of the eight membered ring (Fig. 26.11). To appreciate the stereochemistry of the *syn* addition to *trans*-cyclooctene, it is advisable to use molecular models.

Addition of Protic Acids. All the halogen acids add to alkenes, in most cases via Ad_E2 or Ad_E3 processes, with the additions of hydrogen bromide and hydrogen chloride generally being the most useful for synthesis processes (see Figs. 26.4 and 26.9). Sulfuric acid adds to alkenes to form alkylsulfate esters, which, in the presence of water, undergo hydrolysis to the corresponding alcohols, providing a method for the acid-catalyzed addition of water to alkenes:

$$CH_3CH{=}CH_2 + H_2SO_4 \longrightarrow \underset{\displaystyle OSO_3H}{CH_3CHCH_3} \xrightarrow{H_2O} \underset{\displaystyle OH}{CH_3CHCH_3}$$

PROBLEM 26-5

The addition of carboxylic acids to alkenes provides a method for preparing esters. Outline the mechanism of the reaction of acetic acid with 2-methylpropene in the presence of a trace of sulfuric acid to form *t*-butyl acetate.

Addition of Halogens. All the halogens add to alkenes to give 1,2-dihalides. The addition of fluorine is so exothermic that the reaction is difficult to control, however, and the equilibrium constant for the addition of iodine is generally unfavorable with respect to adduct formation. Thus, only chlorine and bromine additions are generally useful synthesis reactions. These proceed via halonium ion intermediates to yield the products of *anti* addition:

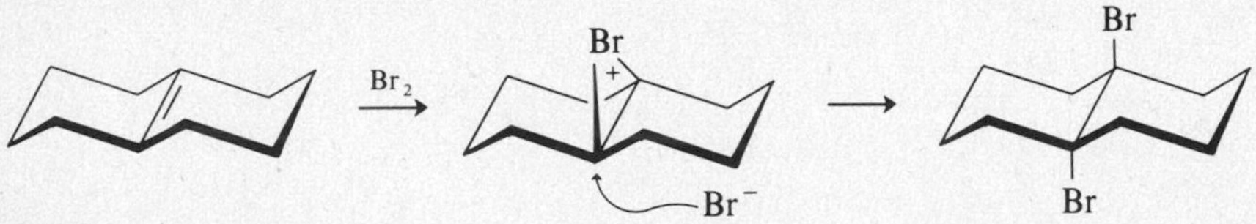

Instead of molecular bromine, a reagent known as pyridinium hydrobromide perbromide,

is often used in bromine addition reactions; it is a stable, crystalline solid that reacts in the same fashion as bromine but is easier to handle. As discussed in Section 31.4, free radical addition (and substitution) may compete with ionic addition of halogen. To minimize this competition, it is advisable to carry out halogenation reactions in the absence of light and free radical initiators.

PROBLEM 26-6

1,3-Butadiene reacts with bromine in both a 1,2- and 1,4-addition fashion. Draw the structures of these two products. The 1,2-addition product is the major one at room temperature, and the 1,4-addition product is the major one at 100°. Which is the product of thermodynamic control and which is the product of kinetic control?

Addition of Oxyhalogen Acids. Hypochlorous and hypobromous acids add to alkenes to form β-halo alcohols (halohydrins) via halonium ion intermediates:

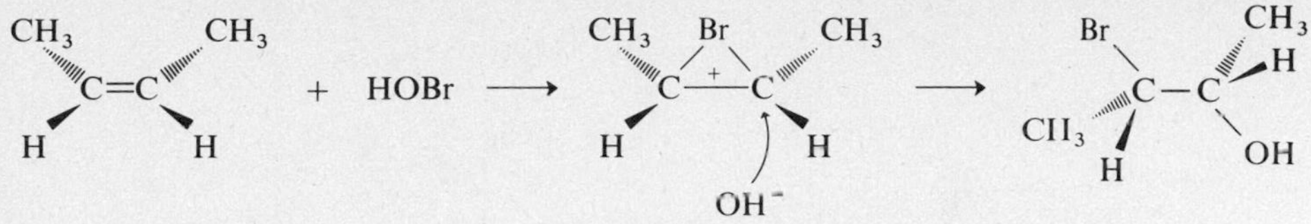

$$CH_3(H)C{=}C(H)CH_3 + HOBr \longrightarrow CH_3(H)C\overset{Br^{+}}{\frown}C(H)CH_3 \;(+\;OH^-) \longrightarrow Br(CH_3)(H)C{-}C(CH_3)(H)OH$$

Addition of Mercuric Salts (Solvomercuration). Alkenes react with mercuric salts (*e.g.*, mercuric acetate, nitrate, or perchlorate) in solvents that are protic as well as nucleophilic (*e.g.*, water and alcohols) to give addition products in which the solvent moiety is attached to the carbon most capable of bearing a positive charge. The reaction is interpreted as involving an intermediate mercurinium ion (Fig. 26.12). The C—Hg bond of the products of *solvomercuration* can be re-

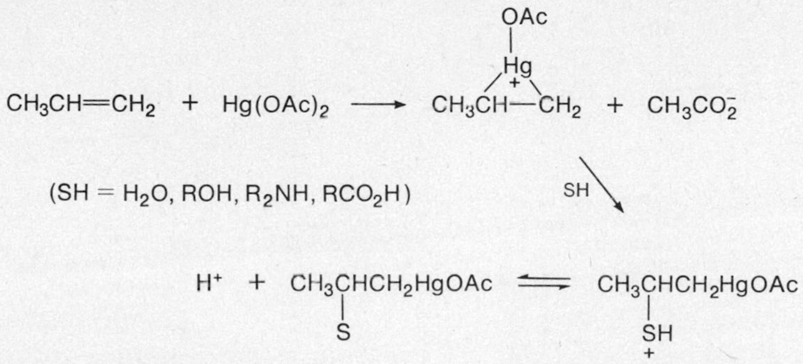

$$CH_3CH{=}CH_2 + Hg(OAc)_2 \longrightarrow CH_3CH\overset{\overset{OAc}{|}}{\underset{+}{Hg}}CH_2 + CH_3CO_2^-$$

$(SH = H_2O, ROH, R_2NH, RCO_2H)$

$$\xrightarrow{SH} \quad CH_3\underset{\underset{+}{SH}}{\underset{|}{C}}HCH_2HgOAc \rightleftharpoons H^+ + CH_3\underset{S}{\underset{|}{C}}HCH_2HgOAc$$

Fig. 26.12. Electrophilic addition of mercury salts and solvent moieties to alkenes (solvomercuration reaction).

placed by a C—H bond by treatment with sodium borohydride, providing a method for the Markovnikov addition of the solvent to the double bond (see Fig. 22.8).

Addition of Carbonium Ions. See polymerization of alkenes (p. 793).

PROBLEM 26-7

Arenes can be added to alkenes in the presence of protic or Lewis acid catalysts. Outline the mechanisms of the sulfuric acid-catalyzed conversion of 2-methyl-5-phenyl-2-pentene to 1,1-dimethyl-1,2,3,4-tetrahydronaphthalene.

Addition of Formaldehyde. Alkenes react with formaldehyde in the presence of mineral acids to produce 1,3-diols and/or the cyclic acetals resulting from the further interaction between the 1,3-diol and formaldehyde (see Fig. 22.14). The mechanism of the reaction is outlined in Fig. 26.13.

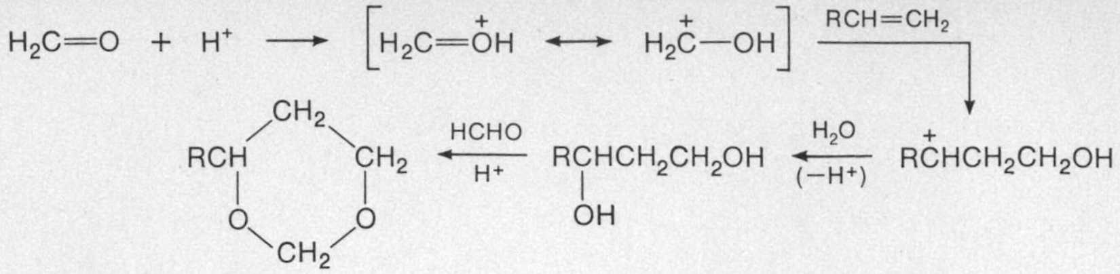

$$H_2C{=}O + H^+ \longrightarrow \left[H_2C{=}\overset{+}{O}H \longleftrightarrow H_2\overset{+}{C}{-}OH\right] \xrightarrow{RCH=CH_2} R\overset{+}{C}HCH_2CH_2OH \xrightarrow[(-H^+)]{H_2O} RCH(OH)CH_2CH_2OH \xrightarrow[H^+]{HCHO} \text{1,3-dioxane (RCH, CH}_2\text{, CH}_2\text{, O, CH}_2\text{, O ring)}$$

Fig. 26.13. Mechanism of the acid-catalyzed addition of formaldehyde to alkenes (Prins reaction).

Addition of Oxygen. Alkenes react with peroxy acids such as perbenzoic acid to yield oxiranes (epoxides; see Fig. 22.12). The reaction, which proceeds in a completely stereoselective *syn* fashion, is an electrophilic addition in which the electrophile is the formal equivalent of an electrophilic oxygen atom (Fig. 26.14).

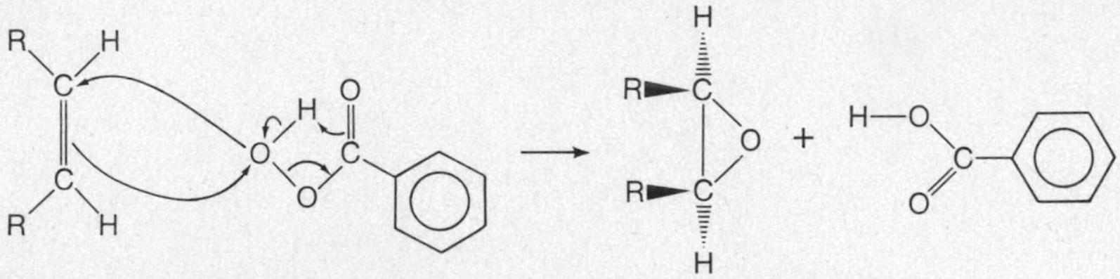

Fig. 26.14. Epoxidation of alkenes with perbenzoic acid.

26.3 Concerted Electrophilic Addition to Alkenes

26.3a. ADDITION OF BORANE (HYDROBORATION). The addition of borane (BH_3) to alkenes to give organoboranes takes place with Markovnikov regioselectivity, the boron atom acting as the electrophile and the hydrogen as the nucleophile. The addition is completely stereoselective and proceeds in the *syn* fashion. The process, referred to as *hydroboration*, is considered to be an example of a concerted electrophilic addition. The alkylboranes are unusually versatile reagents for organic synthesis, and examples of their application appear in a variety of places throughout this text.

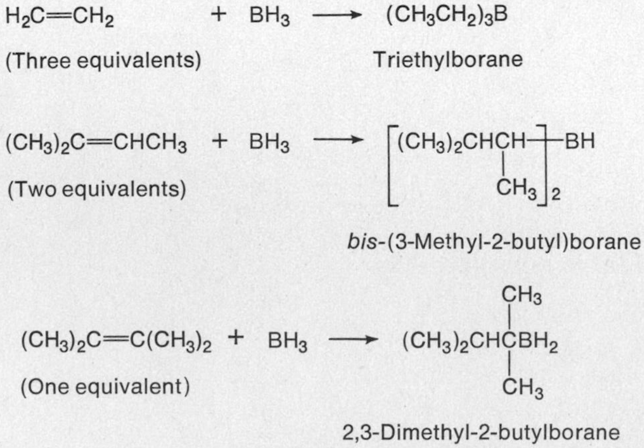

$$H_2C{=}CH_2 \text{ (Three equivalents)} + BH_3 \longrightarrow (CH_3CH_2)_3B \text{ Triethylborane}$$

$$(CH_3)_2C{=}CHCH_3 \text{ (Two equivalents)} + BH_3 \longrightarrow [(CH_3)_2CHCH(CH_3)]_2BH$$

bis-(3-Methyl-2-butyl)borane

$$(CH_3)_2C{=}C(CH_3)_2 \text{ (One equivalent)} + BH_3 \longrightarrow (CH_3)_2CHC(CH_3)_2BH_2$$

2,3-Dimethyl-2-butylborane

Fig. 26.15. Hydroboration of alkenes.

The reaction of an alkene with borane yields a trialkylborane, a dialkylborane, or a monoalkylborane, depending on the steric hindrance around the double bond of the alkene (Fig. 26.15). The boron–carbon bonds that are produced are susceptible to substitution by a variety of functions, including other carbon frameworks (*i.e.*, coupling reactions), amino groups, hydroxyl groups, hydrogen, and halogens (Fig. 26.16; see also Fig. 22.9).

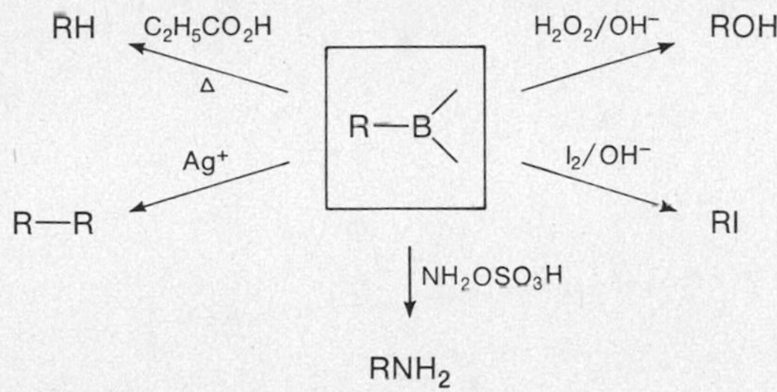

Fig. 26.16. Reactions of organoboranes involving replacement of the boron by carbon frameworks, hydrogen, hetero atoms, and hetero atom groups.

26.3b. Addition of Carbenes

General Features of Reaction. Carbenes (see Fig. 7.14) are electrophilic species, possessing only six electrons in the valence shell and having the capacity for adding to alkenes. The regioselectivity of the addition cannot be determined, however, because the two new bonds that are formed in this process are identical (*i.e.*, C—C bonds). Stereoselectivity, on the other hand, does enter the picture, and the addition can occur either with a high degree or a low degree of stereoselectivity, depending on the spin state (the multiplicity) of the carbene. In general, the stereoselectivity is greater when the carbene is in the singlet state (*i.e.*, nonbonded electrons spin-paired and in the same orbital) than when it is in the triplet state (*i.e.*, nonbonded electrons unpaired and in different orbitals). For example, direct irradiation of diazomethane (CH_2N_2) produces singlet methylene ($H_2C\uparrow\downarrow$) which adds stereoselectively to *cis*-2-butene to give *cis*-1,2-dimethylcyclopropane as the major product; photosensitized* irradiation of diazomethane, on the other hand, produces triplet methylene ($H_2C\uparrow\uparrow$) which adds nonstereoselectively to *cis*-2-butene to give a mixture of *cis*- and *trans*-1,2-dimethylcyclopropane. The singlet addition is considered to be a one-step process without the intervention of an intermediate (*i.e.*, a concerted electrophilic addition), whereas the triplet addition is considered to take place stepwise with the formation of a biradical intermediate in which rotation around the C—C bond occurs at a rate that is competitive with the rate of C—C bond formation (Fig. 26.17); the greater is the rate of rotation relative to the rate of C—C bond formation, the lower is the stereoselectivity. The ground state for methylene is the triplet spin state, so anything that diminishes the rate of reaction with the

* Photosensitization refers to a process in which the compound undergoing reaction is not directly excited by light but receives its energy by transfer from another species which acts as the primary light absorber, that species being known as a photosensitizer.

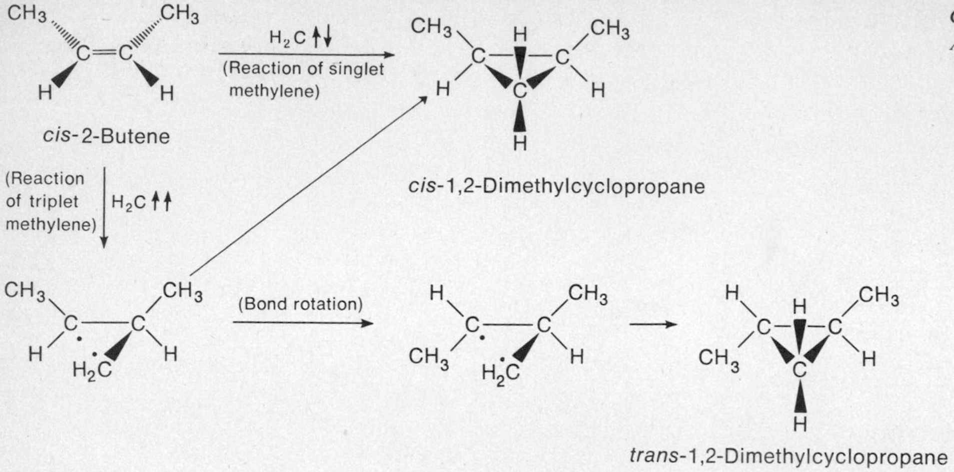

Fig. 26.17. Reaction of singlet and triplet methylene (carbene) with *cis*-2-butene.

alkene prolongs the lifetime of the methylene and increases its chances of interconverting to the triplet species.

PROBLEM 26-8

Are the additions of singlet methylene to *cis*- and *trans*-2-butene to give *cis*- and *trans*-1,2-dimethylcyclopropane, respectively, examples of highly stereoselective reactions, stereospecific reactions, or both?

Generation of Carbenes. Carbenes can be produced by the action of strong base on polyhalogen compounds, such as chloroform. The C—H bond in chloroform is moderately acidic as a result of the cumulative electron-withdrawing effect of three chlorine atoms. Consequently, chloroform reacts with hydroxide ion to form trichloromethyl anion, which subsequently loses chloride anion and leaves the neutral entity dichlorocarbene. This sequence of events, involving removal of a proton to form a carbanion followed by slow expulsion of an anion, puts the reaction in the category of α-elimination reactions of the El_{cb} variety (see p. 735). Evidence for this pathway has been obtained from trapping experiments, in which substances are added to the reaction mixture that can react with the intermediate carbanion before the chloride expulsion can take place. For example, in a reaction carried out in D_2O—NaOD, chloroform isolated from a reaction mixture that has been stopped short of complete conversion to dichlorocarbene is found to be the deuterated species, $CDCl_3$. In a similar vein, a reaction of chloroform carried out in the presence of acetone yields 2-trichloromethyl-2-propanol, resulting from the nucleophilic addition of trichloromethyl anion to acetone (Fig. 26.18).

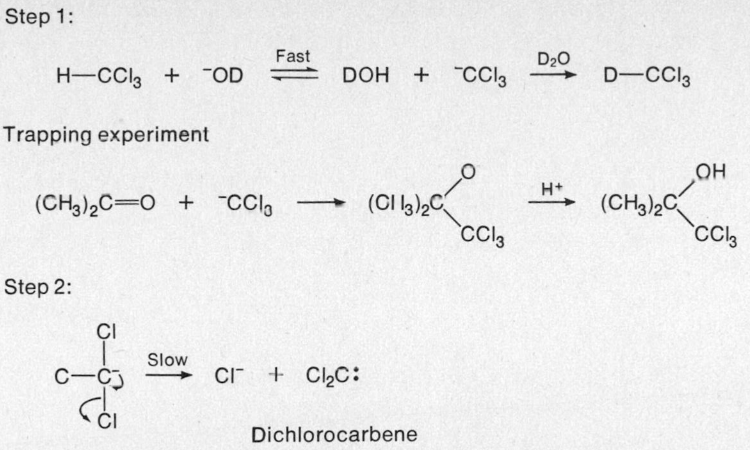

Step 1:

$$H{-}CCl_3 + {}^-OD \underset{}{\overset{\text{Fast}}{\rightleftharpoons}} DOH + {}^-CCl_3 \xrightarrow{D_2O} D{-}CCl_3$$

Trapping experiment

$$(CH_3)_2C{=}O + {}^-CCl_3 \longrightarrow (CH_3)_2C(O^-)CCl_3 \xrightarrow{H^+} (CH_3)_2C(OH)CCl_3$$

Step 2:

$$Cl{-}\overset{Cl}{\underset{Cl}{C^-}} \xrightarrow{\text{Slow}} Cl^- + Cl_2C\!:$$

Dichlorocarbene

Fig. 26.18. α-Elimination reaction of chloroform.

Several other methods are known for forming carbenes via α-elimination reactions. For example, the action of base on trichloroacetic acid yields the carboxylate anion, which loses carbon dioxide, leaving trichloromethyl anion. As in the previous instance, the carbanion expels chloride ion to form dichlorocarbene:

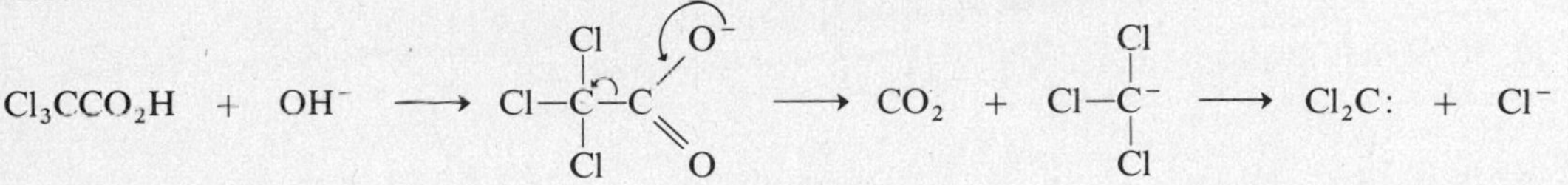

$$Cl_3CCO_2H + OH^- \longrightarrow Cl{-}\overset{Cl}{\underset{Cl}{C}}{-}C(=O)O^- \longrightarrow CO_2 + Cl{-}\overset{Cl}{\underset{Cl}{C^-}} \longrightarrow Cl_2C\!: + Cl^-$$

Removal of two halogen atoms from the same carbon provides still another method for preparing carbenes. The action of methyllithium on carbon tetrabromide leads to dibromocarbene, and the action of zinc-copper couple on methylene diiodide leads to methylene (Fig. 26.19). In these instances, however, there is

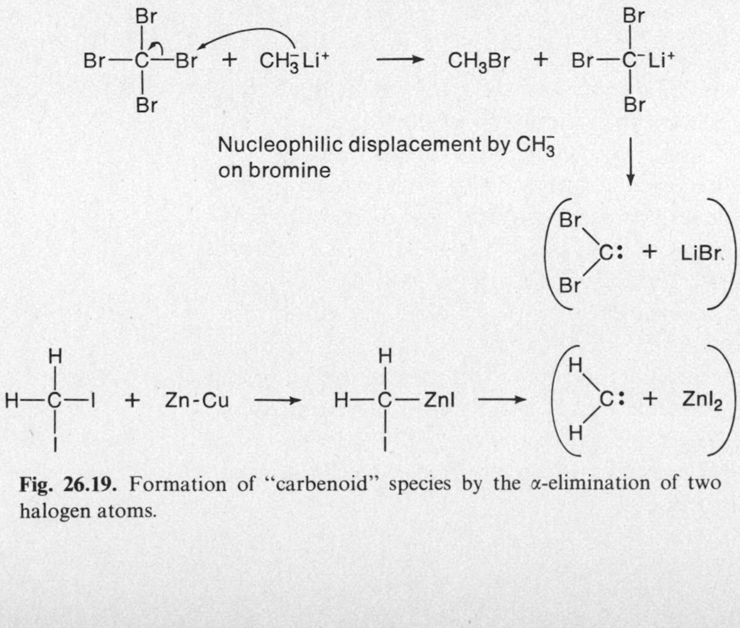

$$Br{-}\overset{Br}{\underset{Br}{C}}{-}Br + CH_3^- Li^+ \longrightarrow CH_3Br + Br{-}\overset{Br}{\underset{Br}{C^-}}Li^+ \longrightarrow (Br_2C\!: + LiBr)$$

Nucleophilic displacement by CH_3^- on bromine

$$H{-}\overset{H}{\underset{I}{C}}{-}I + Zn\text{-}Cu \longrightarrow H{-}\overset{H}{\underset{I}{C}}{-}ZnI \longrightarrow (H_2C\!: + ZnI_2)$$

Fig. 26.19. Formation of "carbenoid" species by the α-elimination of two halogen atoms.

considerable doubt that "free" carbenes are present, and these species are better referred to as "carbenoid," implying that they are in more or less close combination with the metal-containing species that is present in the mixture.

PROBLEM 26-9

Draw a representation of the transition state for the reaction of methyllithium and carbon tetrabromide to form methyl bromide, tribromomethyl anion, and lithium cation.

Still another procedure for generating carbenes involves the action of heat or light of the appropriate wavelength on alkyldiazo compounds and ketenes (Fig. 26.20).

$$CH_2N_2 \xrightarrow{\Delta \text{ or } h\nu} :CH_2 + N_2$$

Diazomethane

$$H_2C{=}C{=}O \xrightarrow{h\nu} :CH_2 + CO$$

Ketene

Fig. 26.20. Generation of carbene (methylene) by heat- or light-induced decomposition of diazomethane and ketene.

PROBLEM 26-10

Draw a pair of resonance structures for CH_2N_2 and for CH_2CO that shows a formal similarity between these two compounds and provides a rationale for their facile decomposition to N_2 and CO, respectively.

Scope and Limitations of Reactions. The generation of carbenes by any of the methods just discussed is generally limited to compounds in which there are no C—H bonds immediately adjacent to the carbene center, for carbenes of the structure $R_2CH\ddot{C}H$ rearrange to alkenes, $R_2C{=}CH_2$, more rapidly than they do anything else. Thus, the most generally used carbenes are dihalocarbenes (from CHX_3 or CX_4), methylene (from CH_2N_2, $H_2C{=}C{=}O$, or CH_2I_2), arylcarbenes (*e.g.*, from $C_6H_5CHN_2$), and carboalkoxycarbenes (*e.g.*, from $N_2CHCO_2CH_3$). Carbene addition reactions are carried out by generating the carbene in the presence of the alkene, the addition occurring with the carbene that is produced *in situ*. They provide an easy and convenient way for introducing a cyclopropane ring at the site of a double bond. For example, the naturally occurring compound thujopsene has been synthesized by the addition of methylene, generated from methylene diiodide and zinc-copper couple, to the double bond of the allylic alcohol shown in Fig. 26.21.

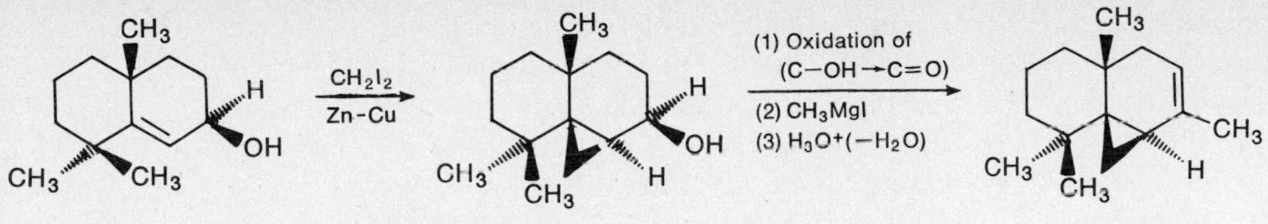

$$CH_2I_2 \xrightarrow{\text{Zn-Cu}} \quad \xrightarrow[\text{(2) } CH_3MgI \text{ (3) } H_3O^+(-H_2O)]{\text{(1) Oxidation of (C-OH} \rightarrow \text{C=O)}}$$

Thujopsene

Fig. 26.21. Synthesis of thujopsene via a carbene addition reaction.

PROBLEM 26-11

The addition of methylene generated from CH_2I_2 and Zn to the double bond of the alkene shown in Fig. 26.21 is a highly stereoselective reaction, occurring *cis* to the hydroxyl group. Suggest a possible reason for this.

26.4 *Radical Addition to Alkenes*

26.4a. GENERAL FEATURES OF RADICAL ADDITION REACTIONS. Radical addition reactions generally occur via chain processes similar to those that have been described for radical substitution reactions. Thus, initiation, self-propagation, and termination steps are involved, as illustrated for the peroxide-induced addition of hydrogen bromide to 1-butene (Fig. 26.22). The regioselectivity in this reaction is determined by the relative stabilities of the two radicals that can form in the first of the propagation steps. Assuming the bromine atom to be the chain-carrying radical, it can form either a 1-bromobutyl radical or a 2-bromobutyl radical when it adds to the double bond. The 1-bromobutyl radical is the

Initiation (initiator radical symbolized by In ·):

$$HBr + In\cdot \longrightarrow H{-}In + Br\cdot$$

Self-propagation:

$$CH_3CH_2CH{=}CH_2 + Br\cdot \longrightarrow CH_3CH_2\dot{C}HCH_2Br \text{ or } CH_3CH_2CH(Br)\dot{C}H_2$$

1-Bromobutyl radical (more stable) — 2-Bromobutyl radical (less stable)

$$CH_3CH_2\dot{C}HCH_2Br + HBr \longrightarrow CH_3CH_2CH_2CH_2Br + Br\cdot$$

Termination:

$$2\,CH_3CH_2\dot{C}HCH_2Br \longrightarrow CH_3CH_2CH(CH_2Br){-}CH(CH_2Br)CH_2CH_3$$

$$2\,Br\cdot \longrightarrow Br_2$$

$$CH_3CH_2\dot{C}HCH_2Br + Br\cdot \longrightarrow CH_3CH_2CH(Br)CH_2Br$$

Fig. 26.22. Radical addition of hydrogen bromide to 1-butene.

more stable species, because the radical center is more highly substituted. The stabilities of radicals follow the same order as the stabilities of cations; *i.e.*, tertiary > secondary > primary (see p. 191). Thus, the peroxide-induced reaction proceeds via the 1-bromobutyl radical, with the result that the major product is 1-bromobutane. This is opposite to the regioselectivity of the addition of hydrogen bromide to 1-butene via the ionic pathway (Markovnikov addition) and is said to be anti-Markovnikov.

26.4b. THERMODYNAMIC CONSIDERATIONS IN RADICAL CHAIN ADDITION REACTIONS. Hydrogen bromide is the only hydrogen halide that adds to alkenes in a radical chain process. If the self-propagating phase of a reaction is to occur, it is necessary not only that the overall process be exothermic but also that each step be exothermic, or no more than very slightly endothermic. Calculations of heats of reactions from bond energies for the two steps in the radical addition

Table 26.1 VALUES FOR THE HEATS OF REACTION FOR THE SELF-PROPAGATING STEPS IN THE RADICAL ADDITION OF HYDROGEN HALIDES TO ALKENES

Hydrogen Halide	ΔH Values		
	X· + C=C → X—C—C·	X—C—C· + HX → X—C—C—H + X·	Overall
HF	−53	+36	−17
HCl	−18	+4	−14
HBr	−5	−12	−17
HI	+12	−28	−16

process (Table 26.1) show that only for the addition of HBr are *both* steps exothermic. In addition to hydrogen bromide, a variety of other reagents can add to alkenes and alkynes via radical chain processes, including those shown in Fig. 26.23.

PROBLEM 26-12
Draw reaction coordinates for the reaction sequence

$$\text{X}\cdot \quad + \quad \text{C=C} \longrightarrow \text{X—C—C}\cdot \xrightarrow{\text{HX}} \text{X—C—C—H} \quad + \quad \text{X}\cdot$$

for each of the four halogen atoms, using the data in Table 26.1 as a guideline.

26.4c. ELECTROPHILIC ADDITION VERSUS RADICAL ADDITION REACTIONS. A number of reagents such as HBr, Br_2, and Cl_2 are capable of adding to alkenes and alkynes by either an ionic or a radical pathway. To study the competition between these two pathways, chemists have investigated the reaction of chlorine with 2-methylpropene. The products of this reaction are 1,2-dichloro-2-methylpropane and 3-chloro-2-methylpropene (Fig. 26.24). The formation of the latter

$$RCH{=}CH_2 + A\cdot \longrightarrow R\dot{C}HCH_2A \text{ (more stable) or } RCH(A)CH_2\cdot \text{ (less stable)}$$

$$R\dot{C}HCH_2A + A{-}B \longrightarrow RCH(B)CH_2A + A\cdot$$

Reagent AB	Product
Br—H	RCH_2CH_2Br
X—X (X = Cl or Br)	$RCHXCH_2X$
X_3C—H (X = Cl or Br)	$RCH_2CH_2CX_3$
X_3C—X (X = Cl and/or Br)	$RCH(X)CH_2CX_3$
YS—H (Y = H or R)	RCH_2CH_2SY
RSn—H	$RCH_2CH_2SnR_3$

Fig. 26.23. Compounds capable of undergoing radical addition to alkenes.

compound might suggest a radical pathway, for allylic halogenation via radical intermediates is well known. To prove or disprove a radical pathway, starting material labeled with radioactive carbon (^{14}C) was employed; the products were found to contain the label exclusively at C-1 of the dichloride and C-3 of the allylic chloride. If the allylic chloride had been formed by a radical abstraction pathway, the intermediate radical would be the resonance-stabilized species (A) in Fig. 26.24, and attack at either a labeled or unlabeled carbon would be equally probable. The formation of product with label exclusively at C-3,

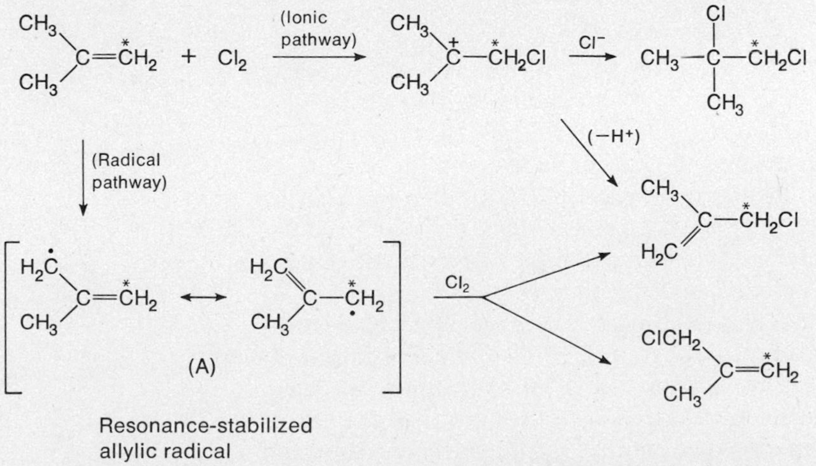

(Ionic pathway): $(CH_3)_2C{=}\overset{*}{C}H_2 + Cl_2 \longrightarrow (CH_3)_2\overset{+}{C}{-}\overset{*}{C}H_2Cl \xrightarrow{Cl^-} CH_3{-}C(Cl)(CH_3){-}\overset{*}{C}H_2Cl$

$(CH_3)_2\overset{+}{C}{-}\overset{*}{C}H_2Cl \xrightarrow{(-H^+)} CH_2{=}C(CH_3){-}\overset{*}{C}H_2Cl$

(Radical pathway): $(CH_3)_2C{=}\overset{*}{C}H_2 \longrightarrow [\,H_2\dot{C}(CH_3)C{=}\overset{*}{C}H_2 \longleftrightarrow H_2C{=}C(CH_3){-}\overset{*}{C}H_2\cdot\,]$ (A)

Resonance-stabilized allylic radical

(A) $\xrightarrow{Cl_2}$ $H_2C{=}C(CH_3){-}\overset{*}{C}H_2Cl$ + $ClCH_2C(CH_3){=}\overset{*}{C}H_2$

Fig. 26.24. Reaction of chlorine with 2-methylpropene (* indicates ^{14}C).

therefore, is taken as evidence *against* the radical pathway and in support of the ionic pathway for the formation of both products.

Subsequent to the investigation with 2-methylpropene, a study of the reaction of chlorine with cyclohexene showed that the composition of the product mixture in this case depends on the conditions of the reaction. When carried out in a nitrogen atmosphere, the three products shown in Fig. 26.25 are obtained, whereas

Reaction Conditions	(A)	(B)	(C)
N_2 atmosphere, 25°	1.95 parts	1.00 parts	0.60 parts
O_2 atmosphere, 25°	2.00 parts	1.00 parts	

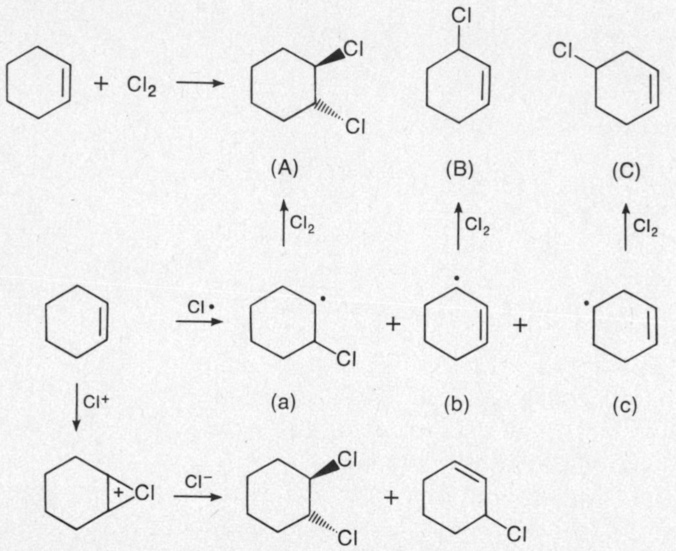

Fig. 26.25. Reaction of chlorine with cyclohexene via radical and ionic pathways.

when carried out in an oxygen atmosphere, only the first two of these are present in the reaction mixture. This difference has been interpreted in terms of a competition between an ionic pathway and a radical pathway, the ionic pathway operating in the presence of oxygen and the radical pathway in the absence of oxygen. Oxygen, as previously mentioned, is an effective chain-terminating species; its effect in this case is to prevent the radical chain reaction from developing, thereby allowing the ionic reaction to take precedence. In the absence of oxygen, the chain reaction is able to proceed, and once it is in the self-propagating phase it quickly outdistances the much slower ionic reaction. The radical pathway involves the formation of a chlorine atom followed by its addition to cyclohexene to produce radical (a) or its abstraction of hydrogen to produce radicals (b) and (c). Each of these intermediate radicals then reacts with chlorine in a chain process to form the three products of the reaction mixture. The ionic pathway, on the other hand, is initiated *only* by addition of a chlorine cation to the double bond to give a chloronium ion intermediate, which can react with chloride ion to give the dichloride, or lose an allylic proton to give the allylic chloride. The balance between the radical and ionic pathways, thus, is delicate, and which pathway is followed depends strongly on the structure of the alkene and on the reaction conditions. As the degree of alkylation at the double bond increases, the ionic pathway gains ascendency. Thus, monosubstituted alkenes react almost exclusively via a radical pathway; disubstituted alkenes can react via either pathway, depending on the reaction conditions; tri- and tetrasubstituted alkenes react only via the ionic pathway, regardless of the reaction conditions.

26.5 Cycloaddition Reactions of Alkenes

26.5a. General features of cycloaddition reactions. The term "cycloaddition" is generally reserved for those processes in which two or more multiply unsaturated groups combine with each other in such a way as to generate a cyclic structure containing fewer multiple bonds than the reactants:

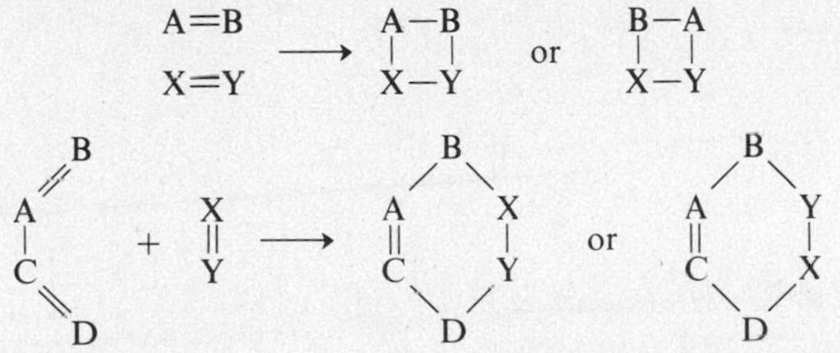

$$\begin{array}{c} A{=}B \\ \\ X{=}Y \end{array} \longrightarrow \begin{array}{c} A{-}B \\ |\quad| \\ X{-}Y \end{array} \quad \text{or} \quad \begin{array}{c} B{-}A \\ |\quad| \\ X{-}Y \end{array}$$

$$\begin{array}{c} \quad B \\ A\!\!/\!\!/ \\ | \\ C\!\!\backslash\!\!\backslash \\ \quad D \end{array} + \begin{array}{c} X \\ \| \\ Y \end{array} \longrightarrow \begin{array}{ccc} & B & \\ A & & X \\ \| & & | \\ C & & Y \\ & D & \end{array} \quad \text{or} \quad \begin{array}{ccc} & B & \\ A & & Y \\ \| & & | \\ C & & X \\ & D & \end{array}$$

Although cycloadditions have been known for many years, the Diels–Alder reaction being the best known example, only recently have the mechanisms of these processes been subjected to careful scrutiny. The recent burgeoning of activity in this respect can be ascribed, at least in part, to observations made by R. B. Woodward and coworkers during investigations of the synthesis of vitamin B_{12}. From the stereochemical outcome of certain photochemical and thermal cyclization reactions carried out in the course of this work, it was realized that existing theories were inadequate to explain the results. In collaboration with the theoretician R. Hoffmann, a theory was subsequently formulated by means of which the mechanistic and stereochemical course of these and other cycloaddition reactions can be not only explained but predicted. The principles of this theory have come to be known as the Woodward–Hoffmann "Conservation of Orbital Symmetry" rules (or, usually, simply as the Woodward–Hoffmann rules). Although only a few years old, they have already assumed great importance, and understanding them has become a necessary part of modern chemical theory. An introductory discussion of these rules is given in Chapter 38.

26.5b. Diels–Alder reactions. The prototype of the concerted cycloaddition process is the reaction between 1,3-dienes and alkenes (the "dienophile") to form cyclohexenes. Discovered early in the twentieth century, this cycloaddition reaction was very extensively studied some years later by the German chemists Diels and Alder, whose names it has come to bear. Thousands of examples of *Diels–Alder reactions* have been published; it is one of the most useful and versatile of methods for organic synthesis, in recognition of which Diels and Alder were awarded a Nobel Prize in 1950.

Diels–Alder reactions frequently give high yields of one particular product, even in instances where several structural and/or diastereoisomers might be formed. For example, the reaction of 1,3-hexadiene and *cis*-2-butenonitrile has the potential for yielding no fewer than eight different compounds:

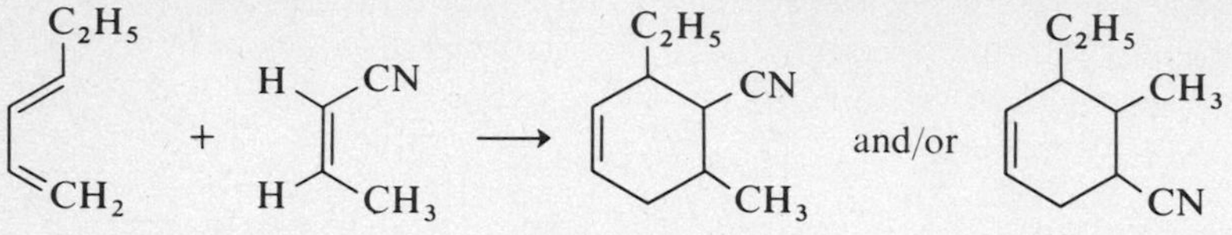

C_2H_5, CH_2 + H, CN, H, CH_3 ⟶ C_2H_5, CN, CH_3 and/or C_2H_5, CH_3, CN

4 Diastereomers **4 Diastereomers**

In actual fact, it yields ($\pm$)-*r*-4-cyano-*cis*-3-ethyl-*cis*-5-methylcyclohexene as the major product,

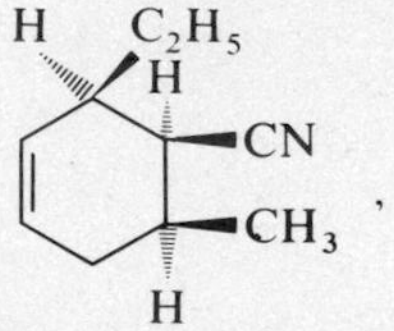

H C_2H_5 H CN CH_3 H ,

which is rationalized in the following fashion. The regioselectivity of the Diels–Alder reaction can be interpreted in terms of the diradical character of the activated complex, even though the reaction is viewed as concerted. Depicting the process as a stepwise sequence simply for the purpose of discussion,* we see that the diradical species leading to the 4-cyano-3-ethyl-5-methylcyclohexene is more stable (*i.e.*, a secondary radical) than the diradical species leading to the 5-cyano-3-ethyl-4-methylcyclohexene (*i.e.*, a primary radical):

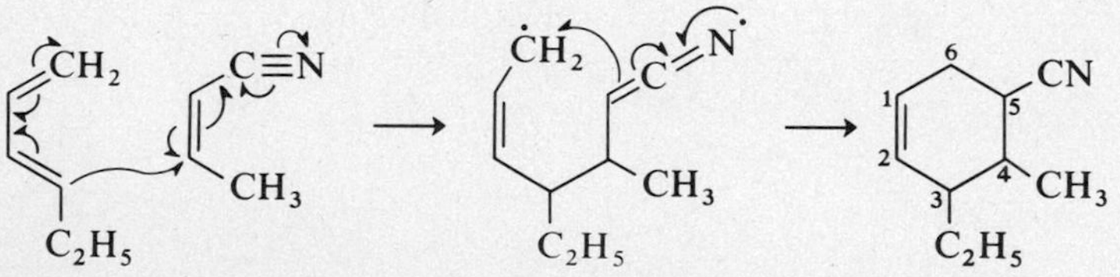

CH_2, C_2H_5, C≡N, CH_3 ⟶ $\dot{C}H_2$, C=N·, CH_3, C_2H_5 ⟶ CN, CH_3, C_2H_5

Less stable   Not formed

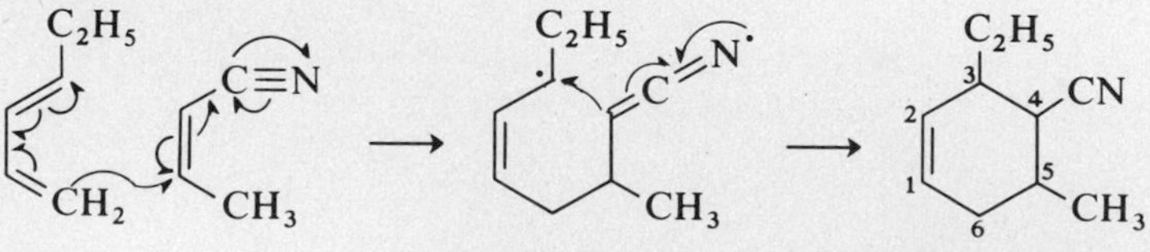

C_2H_5, CH_2, C≡N, CH_3 ⟶ C_2H_5, C=N·, CH_3 ⟶ C_2H_5, CN, CH_3

More stable   Major product

The stereochemistry of the dienophile is retained during the reaction, leading only to those compounds in which the methyl and cyano groups are *cis* to each

* See caption of Fig. 22.20 for explanation of "fish hook" arrows.

other. Thus, the possibilities are reduced to

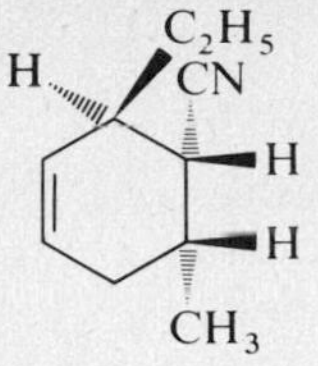

and/or

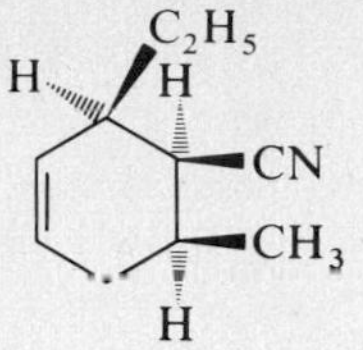

When, as in the preceding example, the dienophile component carries an activating group containing multiple bonds (*e.g.*, cyano), the "principle of maximum overlap of multiple bonds" in the activated complex has an important controlling influence on the stereochemical outcome. According to this principle, the diene and dienophile become associated in the fashion that permits the maximum interaction between the π bonds of the reactants; *i.e.*, orientation (B) in preference to orientation (A):

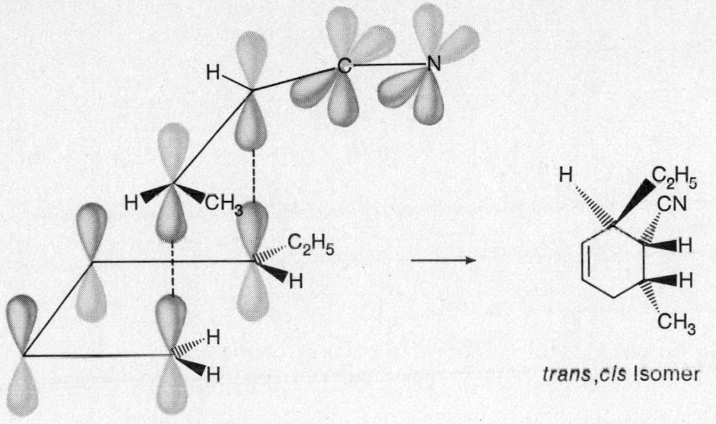

trans,cis Isomer

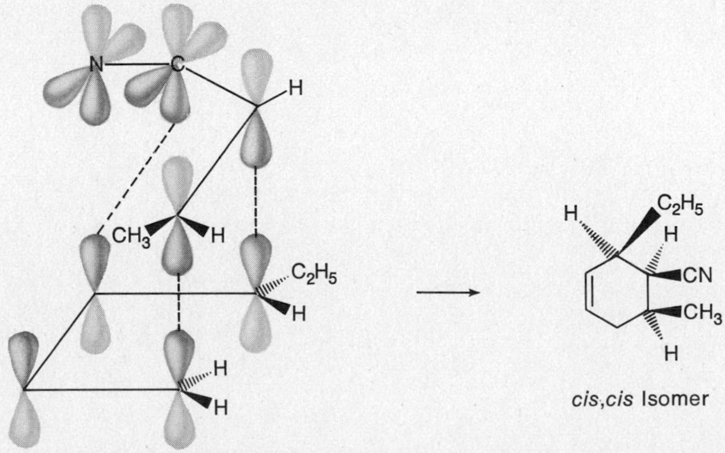

Orientation A

cis,cis Isomer

Orientation B

Of the eight possible isomers, therefore, the *r*-4-cyano-*cis*-3-ethyl-*cis*-5-methylcyclohexene is the major product from the reaction of 1,3-hexadiene and *cis*-2-butenonitrile.

Maximum overlap of multiple bonds in the activated complex leads to the preferential formation of *endo* products when cyclic dienes such as cyclopentadiene are used. If the dienophile does not carry a group that has a multiple bond capable of resonance interaction with the π bond, however, the preference for the *endo* isomer is lost (Fig. 26.26).

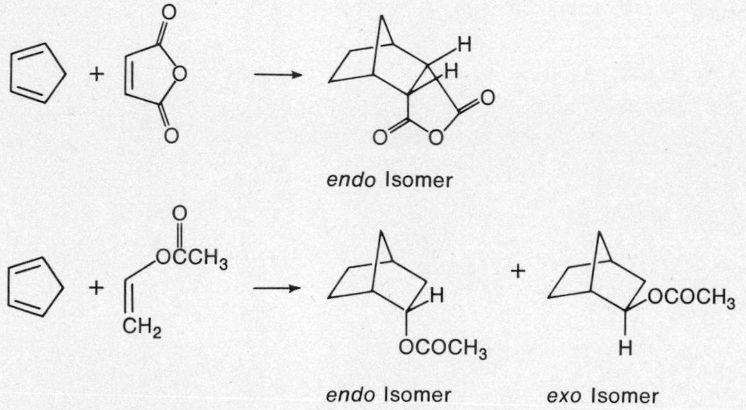

endo Isomer

$OCCH_3$ (O) CH_2 $OCOCH_3$ $OCOCH_3$

endo Isomer *exo* Isomer

Fig. 26.26. Reaction of cyclopentadiene with maleic anhydride and with vinyl acetate.

Although ethylene itself has the potentiality of acting as a dienophile, only alkenes bearing electron-withdrawing groups undergo the Diels–Alder reaction with facility. Included among the best dienophiles are compounds of the general structure X—C=C—Z, where X and/or Z are

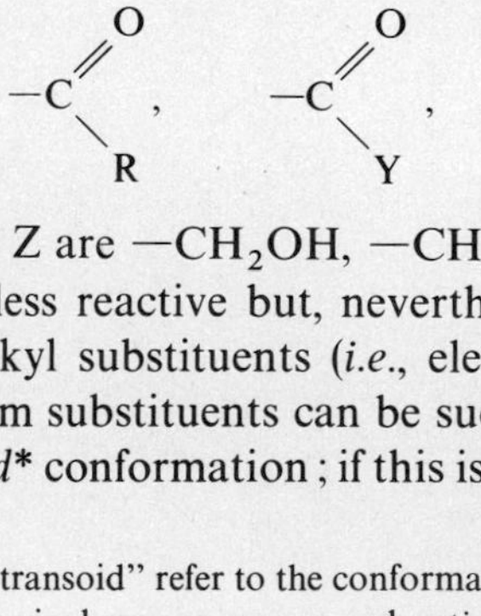

$$-C(=O)R, \quad -C(=O)Y, \quad -C{\equiv}N, \text{ or } -NO_2$$

When X and/or Z are $-CH_2OH$, $-CH_2X$, —C=C, and —Ar, the dienophiles are somewhat less reactive but, nevertheless, useful. The most reactive dienes carry simple alkyl substituents (*i.e.*, electron-releasing effect), although dienes with hetero atom substituents can be successfully used. The diene must be able to attain a *cisoid** conformation; if this is not possible, the diene will not undergo

* "cisoid" and "transoid" refer to the conformations around the C-2/C-3 bond of a 1,3-butadiene system, in which the vinyl groups are *syn* and *anti*, respectively:

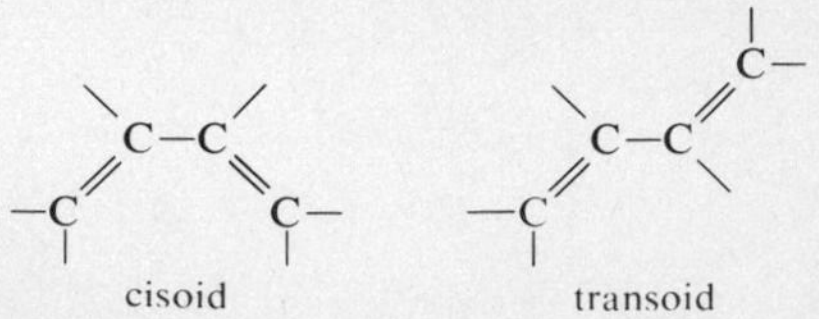

cisoid transoid

a Diels–Alder reaction—e.g.,

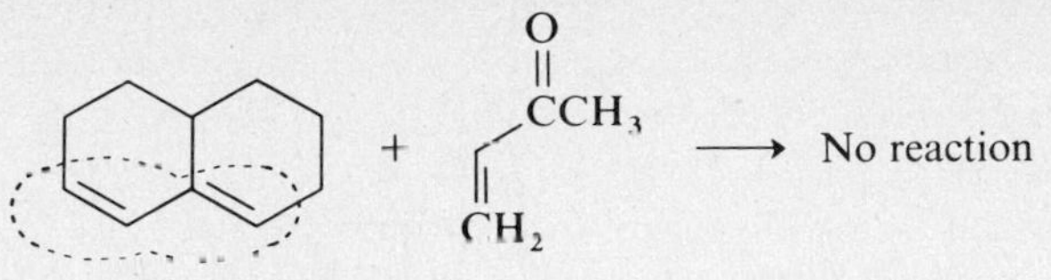

No reaction

Transoid 1,3-butadiene system

An interesting example of the application of Diels–Alder reactions is seen in the synthesis of the naturally occurring product cantharidin (Fig. 26.27). Starting

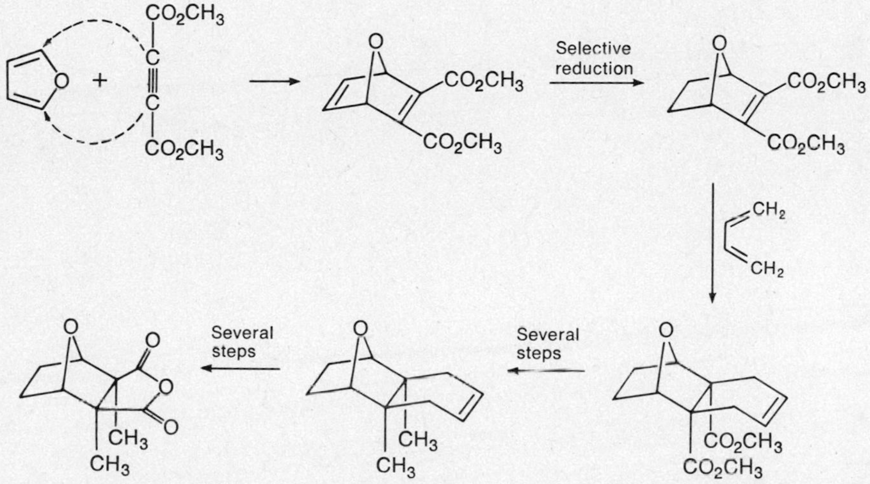

Selective reduction

Several steps

Several steps

Fig. 26.27. Diels–Alder reactions in the synthesis of cantharidin.

with the Diels–Alder reaction between furan and dimethyl butynedioate, the resulting adduct is selectively reduced at the less hindered double bond. This compound is then employed as the dienophile in a second Diels–Alder reaction, this time with butadiene, to yield a tricyclic compound. By a series of reactions which we need not detail, the carbomethoxy groups of this tricyclic compound can be converted to methyl groups, the cyclohexene ring opened, and two carbons removed to yield a compound identical in structure with the naturally occurring cantharidin, a material obtained from dried beetles (*Cantharis vesicatoria*), possessing vesicant activity and, according to some, aphrodisiac activity.

The reader may wonder why this lengthy sequence of reactions is required to synthesize a compound which, it would seem, might be made in two simple steps from furan and dimethylmaleic anhydride followed by reduction:

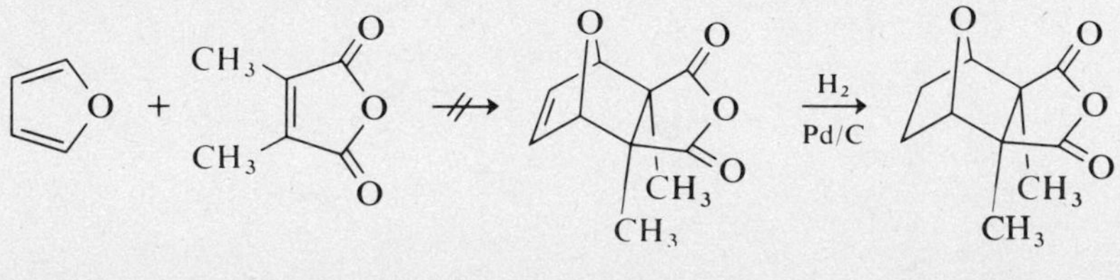

$\xrightarrow[\text{Pd/C}]{H_2}$

Despite its immense versatility, however, even the Diels–Alder reaction has its failures, and this is one of them. Probably because of the hindrance around the double bond in the dienophile, the equilibrium constant does not favor adduct formation, and this simple route to cantharidin is, therefore, precluded.

26.5c. 1,3-DIPOLAR ADDITION REACTIONS. Certain functional groups that contain atoms bearing formal positive and negative charges undergo cyloaddition reactions with alkenes in what are referred to as 1,3-dipolar addition reactions. Several examples of these are illustrated in Fig. 26.28.

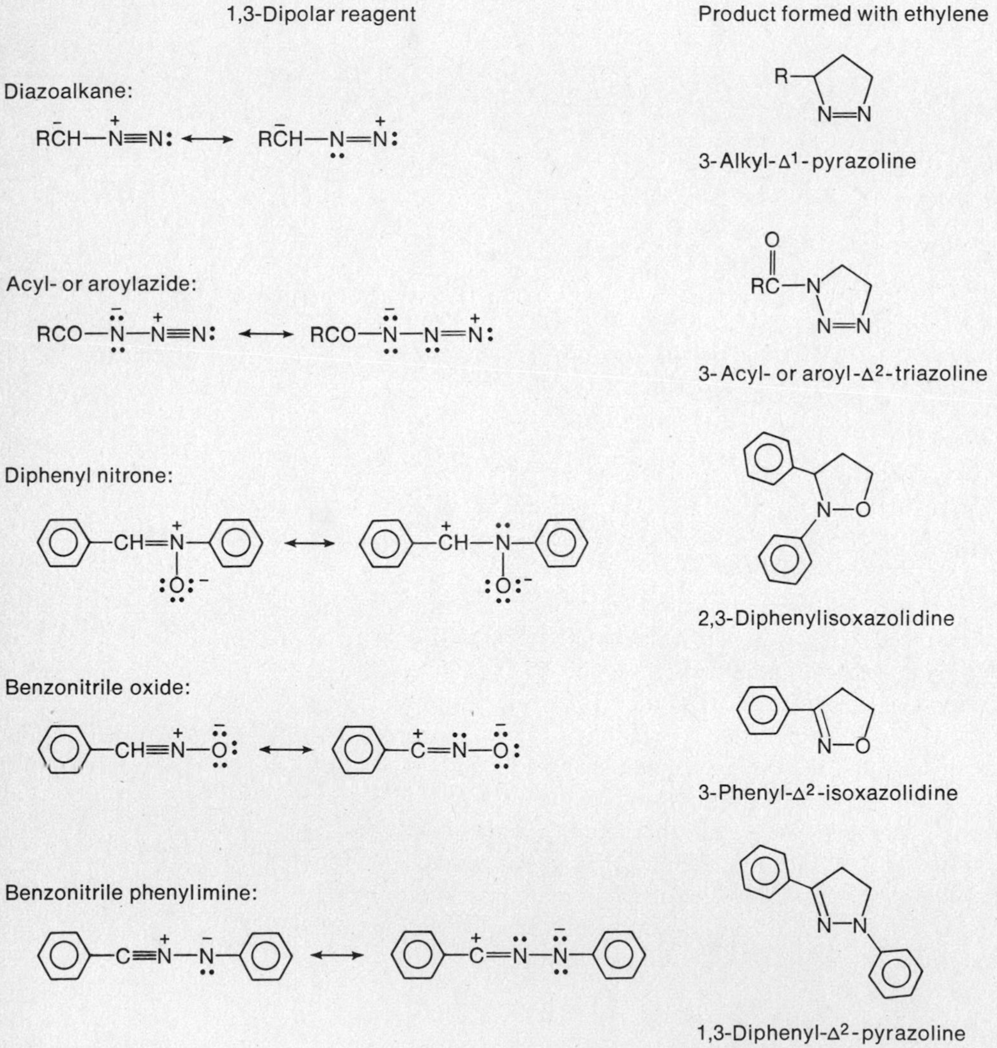

1,3-Dipolar reagent	Product formed with ethylene
Diazoalkane: $R\bar{C}H{-}\overset{+}{N}{\equiv}N{:} \longleftrightarrow R\bar{C}H{-}\ddot{N}{=}\overset{+}{N}{:}$	3-Alkyl-Δ^1-pyrazoline
Acyl- or aroylazide: $RCO{-}\overset{-}{\ddot{N}}{-}\overset{+}{N}{\equiv}N{:} \longleftrightarrow RCO{-}\overset{-}{\ddot{N}}{-}\ddot{N}{=}\overset{+}{N}{:}$	3-Acyl- or aroyl-Δ^2-triazoline
Diphenyl nitrone: $C_6H_5{-}CH{=}\overset{+}{N}(O^-){-}C_6H_5 \longleftrightarrow C_6H_5{-}\overset{+}{C}H{-}\ddot{N}(O^-){-}C_6H_5$	2,3-Diphenylisoxazolidine
Benzonitrile oxide: $C_6H_5{-}C{\equiv}\overset{+}{N}{-}\overset{-}{\ddot{O}}{:} \longleftrightarrow C_6H_5{-}\overset{+}{C}{=}\ddot{N}{-}\overset{-}{\ddot{O}}{:}$	3-Phenyl-Δ^2-isoxazolidine
Benzonitrile phenylimine: $C_6H_5{-}C{\equiv}\overset{+}{N}{-}\overset{-}{\ddot{N}}{-}C_6H_5 \longleftrightarrow C_6H_5{-}\overset{+}{C}{=}\ddot{N}{-}\overset{-}{\ddot{N}}{-}C_6H_5$	1,3-Diphenyl-Δ^2-pyrazoline

Fig. 26.28. Products from the addition of 1,3-dipolar reagents to ethylene. The Δ^n in the names of these compounds indicates the presence of a double bond, the superscript number specifying the lower numbered position of the bond.

26.6a. OXIDATION WITH RETENTION OF THE CARBON–CARBON BOND. The conversion of alkenes to oxiranes and *cis*- and *trans*-1,2-diols is discussed in Section 22.6f (see p. 657).

An industrially important oxidation reaction of an alkene is the conversion of ethylene to acetaldehyde by the *Wacker process*. This is brought about by the action of aqueous hydrogen peroxide, palladium chloride, and cupric chloride on ethylene:

$$H_2C{=}CH_2 + PdCl_2 + H_2O \longrightarrow CH_3CHO + Pd + 2\,HCl$$

$$2\,Pd + 2\,CuCl_2 \longrightarrow 2\,PdCl_2 + 2\,CuCl$$

$$2\,CuCl + H_2O_2 + 2\,HCl \longrightarrow 2\,CuCl_2 + 2\,H_2O$$

Net reaction $H_2C{=}CH_2 + H_2O_2 \xrightarrow[CuCl_2]{PdCl_2} CH_3CHO$

Although the mechanism of the reaction is not completely understood, the following possibility has been suggested:*

$$(H_2C{=}CH_2)\text{-}Pd(OH)Cl_2 \longrightarrow H{-}\underset{CH_3}{\overset{OH}{C}}{-}PdCl \longrightarrow H(C{=}O)CH_3 + Pd + HCl$$

26.6b. OXIDATION WITH CLEAVAGE OF THE CARBON–CARBON BOND. Potassium permanganate in strongly basic solution and sodium dichromate in sulfuric acid solution are strenuous oxidizing agents that have the capacity for cleaving C=C bonds to carboxylic acids or ketones. The permanganate reaction proceeds in a stepwise fashion, yielding first the 1,2-diol, which, if neutral conditions and low temperature are employed, can frequently be isolated (see Fig. 22.12). The 1,2-diol is susceptible to further oxidation, however, with scission of the C—C bond:

$$R_2C{=}CHR \xrightarrow[pH \simeq 7]{KMnO_4} R{-}\underset{OH}{\overset{R}{C}}{-}\underset{OH}{\overset{R}{CH}} \xrightarrow[Basic]{KMnO_4} R_2C{=}O + [O{=}C(R)H] \rightarrow RCO_2H$$

The permanganate oxidation may fail if the alkene is tetrasubstituted, in which case the more strenuous dichromate method must be resorted to. Both of these oxidizing agents lack situselectivity, attacking not only C=C bonds but many other functional groups in molecules as well; this must be taken into account when they are used as oxidizing agents for alkenes.

---

* The designation $\overset{>C}{\underset{>C}{\|}}\dashrightarrow Pd$ indicates an entity, called a π-complex, in which bonding between carbon and palladium occurs as the result of electron donation from the π-bond of the alkene to an empty orbital on the palladium.

A much more selective oxidizing agent is ozone, which cleaves alkenes to aldehydes and/or ketones but shows little reactivity toward most other functional groups. Skeletal rearrangements occur infrequently during ozonolysis, and the method has been extensively used, therefore, as a means for inferring the position of unsaturation and the structure of the alkene from the structures of the carbonyl compounds produced. The reaction is carried out simply by bubbling a stream of oxygen containing a few percent of ozone* through a solution of the alkene, usually at quite low temperature (*i.e.*, $-70°$).

The mechanism of the ozonolysis reaction has challenged chemists for many years. Considerable effort has been expended in attempting to unravel its intricacies, and the problem is made no easier by the reputed explosive nature of the initially formed products. Despite these difficulties, considerable progress has been made, and what appears to be a satisfactory mechanism has evolved. It postulates that the initial step is an electrophilic addition of ozone (a dipolar molecule) to the alkene (acting as a Lewis base), followed by a cyclization of the charge-separated "intermediate" to a cyclic compound containing three contiguous oxygen atoms. Fragmentation of this cyclic compound to a carbonyl compound and a peroxide is followed by a recombination of these fragments to produce an isomeric five-membered ring compound containing three oxygen atoms, one of which is inserted between a pair of carbon atoms. This compound, known as an *ozonide*, is the end product of the action of ozone (Fig. 26.29).

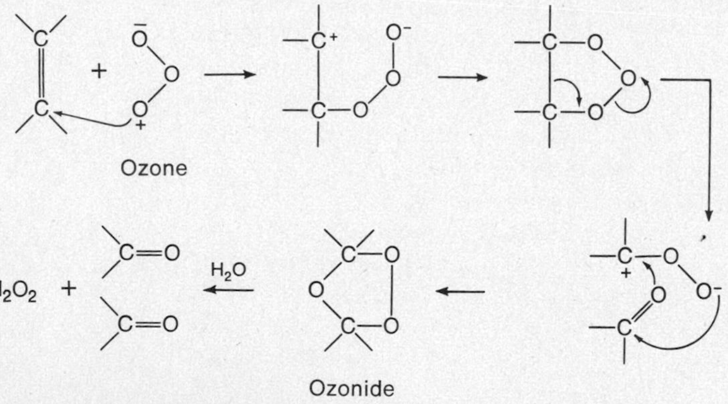

Ozone

H_2O_2 + ... $\xleftarrow{H_2O}$ ... Ozonide

Fig. 26.29. Mechanism of the ozonolysis of alkenes.

PROBLEM 26-13

Is the addition of ozone to an alkene an example of a 1,3-dipolar addition reaction?

The ozonides are rarely isolated; in the solvent-free state at room temperature many tend to be explosive, and for this reason they are decomposed before

* Passage of oxygen (or air) through a tube across which a high voltage electrical discharge is occurring results in the conversion of some of the oxygen to ozone.

product isolation is attempted. Decomposition can be effected with water, although in this case the hydrogen peroxide that is produced may add to the aldehydes or ketones to give hydroxyhydroperoxides, or may oxidize aldehydes to carboxylic acids:

$$\underset{\textbf{Ozonide}}{\text{RHC}(\text{O–O})(\text{O})\text{CHR}} \xrightarrow{H_2O} \text{RCHO} + \text{RCH(OH)–OOH} \longrightarrow \text{RCO}_2\text{H} + \text{H}_2\text{O}$$

Several alternatives to hydrolytic decomposition are available that prevent the formation of hydrogen peroxide. For example, treatment of the ozonide with pyridine removes the "extra" oxygen, which becomes attached to the nitrogen atom to form pyridine-N-oxide:

$$\underset{\textbf{Ozonide}}{\text{RHC}(\text{O–O})(\text{O})\text{CHR}} + \underset{\textbf{Pyridine}}{\text{C}_5\text{H}_5\text{N}} \longrightarrow 2\,\text{RCHO} + \underset{\textbf{Pyridine N-oxide}}{\text{C}_5\text{H}_5\text{N}^+\text{–O}^-}$$

Pyridine diluted with methylene chloride can, in fact, be conveniently used as a solvent for ozonolyses. At $-70°$ the ozonide forms and remains intact in this solvent mixture, but when the mixture is then warmed to room temperature, the ozonide reacts with the pyridine to produce the aldehyde or ketone. Another reagent that serves this same purpose is triethyl phosphite, $(C_2H_5O)_3P$, which removes the "extra" oxygen by conversion to triethyl phosphate, $(C_2H_5O)_3PO$.

Decomposition of the ozonide can also be carried out under conditions that either oxidize or reduce the initially formed products. For example, treatment with zinc and hydrochloric acid or with molecular hydrogen in the presence of a catalyst reduces the aldehydes and ketones to the corresponding alcohols. Conversely, treatment of the ozonide mixture with hydrogen peroxide in formic acid results in the oxidation of the initially formed aldehydes to the corresponding carboxylic acids:

$$\text{R'CH}_2\text{OH} + \text{R}_2\text{CHOH} \xleftarrow[\text{H}_2/\text{Pd–C}]{\text{Zn/HCl or}} \text{R'HC}(\text{O–O})(\text{O})\text{CR}_2 \xrightarrow[\text{HCO}_2\text{H}]{\text{H}_2\text{O}_2} \text{R'CO}_2\text{H} + \text{O=CR}_2$$

Complete cleavage of the C=C bond can also be achieved in a stepwise fashion by first converting the alkene to the 1,2-diol (see Fig. 22.12), followed by treatment with periodic acid or lead tetraacetate (see Section 22.13b):

$$\text{>C=C<} \longrightarrow \text{–C(OH)–C(OH)–} \xrightarrow{\text{HIO}_4 \text{ or Pb(OAc)}_4} \text{>C=O} + \text{O=C<}$$

PROBLEM 26-14

What would you predict to be the product of ozonolysis of 1-methoxycyclohexene (*i.e.*, an enol ether of cyclohexanone)?

26.6c. CATALYTIC REDUCTION OF ALKENES. Hydrogen can be added to C=C bonds by a number of procedures. The one that is most generally used involves the direct addition of molecular hydrogen with the aid of catalysts, such as platinum, palladium, rhodium, or nickel, in a process known as catalytic hydrogenation. If the transfer of hydrogen to both termini of the double bond takes place in one step, the stereochemistry of the addition is *syn*, and this is the course of events in many catalytic hydrogenations. Sometimes, however, the molecule undergoing reduction is released from the catalyst (desorption) after only one hydrogen atom has been added, resulting in a stepwise reaction that may lead to *anti* as well as *syn* addition. Alkenes are generally much more easily reduced than arenes, allowing the selective reduction of compounds containing both types of functions (Fig. 26.30).

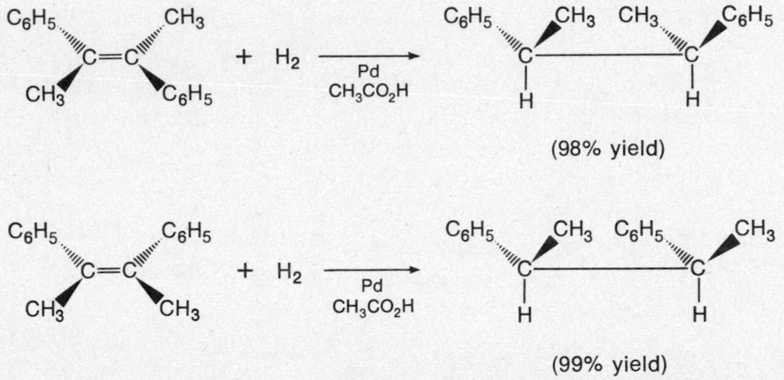

C_6H_5, CH_3, CH_3, C_6H_5 (C=C) + H_2 —(Pd, CH_3CO_2H)→ C_6H_5, CH_3, H (C)—(C) CH_3, C_6H_5, H

(98% yield)

C_6H_5, C_6H_5, CH_3, CH_3 (C=C) + H_2 —(Pd, CH_3CO_2H)→ C_6H_5, CH_3, H (C)—(C) C_6H_5, CH_3, H

(99% yield)

Fig. 26.30. Stereospecific *syn* addition of hydrogen to (*E*)- and *Z*-isomers of 2,3-diphenyl-2-butene.

The rate at which catalytic hydrogenation takes place depends on (a) the steric bulk of the groups attached to the double bond, (b) the electronic nature of the groups attached to the double bond, (c) the particular catalyst, and (d) the reaction conditions. As the extent of substitution at the double bond increases and as the substituents get large in size, the effectiveness of adsorption of the alkene onto the catalyst diminishes; thus, the reactivity decreases in the order $RCH{=}CH_2 > RCH{=}CHR > R_2C{=}CH_2 > R_2C{=}CHR > R_2C{=}CR_2$ and in the order $R = CH_3{-} > C_2H_5{-} > (CH_3)_2CH{-} > (CH_3)_3C{-}$. Effective adsorption onto the catalyst is the result of the Lewis base characteristics of the double bond, so electron-withdrawing groups in the alkene deactivate it toward catalytic hydrogenation. The order of catalyst effectiveness falls approximately in the order $Pt > Pd > Rh \simeq Ru > Ni$. The activity of the catalyst, however, is a function not only of the particular metal involved but also the manner in which the catalyst is prepared, the presence or absence of a "support" (*e.g.*, carbon, barium sulfate, calcium carbonate, etc.), the solvent, the temperature, and the pressure at which the hydrogenation is carried out.

The situselectivity of catalytic hydrogenation is rather limited. In addition to alkenes, most other functional groups containing multiple bonds undergo catalytic reduction, including alkynes, arenes, aldehydes, ketones, carboxylic acid derivatives, nitro compounds, and nitriles. Carboxylic acids, however, are usually unaffected. The relative rates of reduction of various functions are sometimes sufficiently different that selective reduction can be effected. Alkynes can often be selectively reduced to alkenes; alkenes, alkynes, and hetero atom unsaturated groups are all reduced considerably more rapidly than arenes. For example, 1-phenyl-1-propyne is reduced to 1-phenyl-1-propene in the presence of palladium on barium sulfate catalyst and 1 mole-equivalent of hydrogen (Fig. 26.31). If another mole-equivalent of hydrogen is introduced, the alkene is

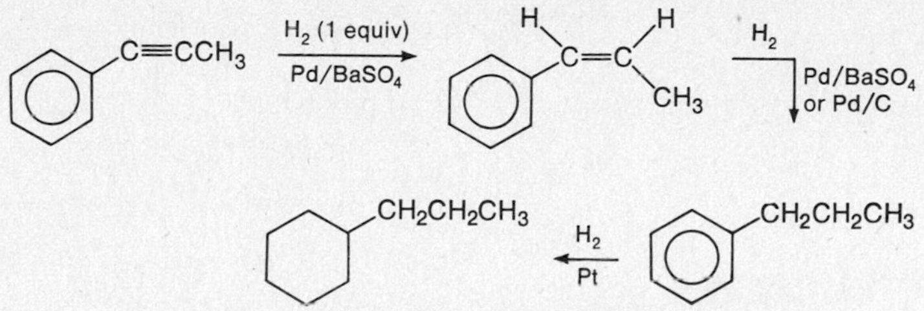

$C_6H_5C{\equiv}CCH_3 \xrightarrow[Pd/BaSO_4]{H_2\ (1\ equiv)} C_6H_5CH{=}CHCH_3$ (H, H cis) $\xrightarrow[Pd/BaSO_4\ or\ Pd/C]{H_2} C_6H_5CH_2CH_2CH_3 \xrightarrow[Pt]{H_2} C_6H_{11}CH_2CH_2CH_3$

Fig. 26.31. Selective reduction of 3-phenyl-2-propyne.

reduced to *n*-propylbenzene. More strenuous conditions are required to saturate the benzene ring to form *n*-propylcyclohexane, this requiring the more active platinum catalyst and a few atmospheres pressure of hydrogen or, alternatively, a nickel catalyst at elevated temperature (100–200°) and high pressure (1000–3000 psi).

Hydrogenations involving metals such as Pt, Pd, Rh, Ru, and Ni take place at the interface between the insoluble catalyst and the solution containing the material to be reduced, and are called *heterogeneous* reactions. *Homogeneous* hydrogenation can be effected by means of soluble catalysts such as RhCl-$[P(C_6H_5)_3]_3$ and $IrCl(CO)[P(C_6H_5)_3]_3$, which are complexes of ruthenium and

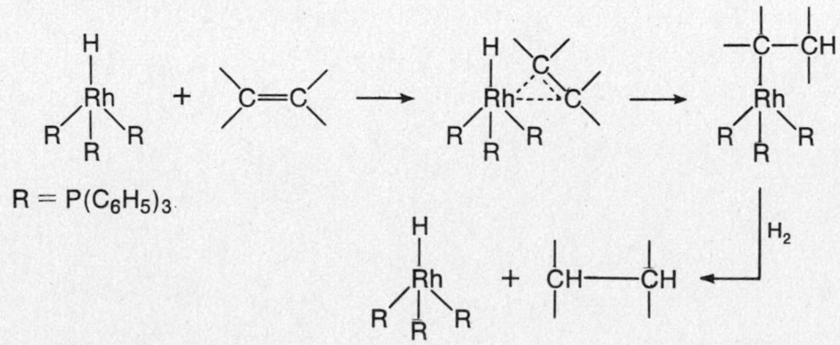

$HRhR_3 + \text{>C=C<} \longrightarrow$ (π-complex of H–RhR_3 with C=C) $\longrightarrow$ R_3Rh–C–CH $\xrightarrow{H_2} HRhR_3 + \text{CH–CH}$

$R = P(C_6H_5)_3$

Fig. 26.32. "Hydrogenation" of an alkene using a homogeneous hydrido transition metal complex.

iridium capable of being converted to a hydrido form in the presence of molecular hydrogen. The alkene forms a complex with the hydrido form, adds molecular hydrogen, and then leaves the complex to make room for another molecule of alkene (Fig. 26.32). Coenzyme B_{12}, a cobalt-containing compound (see Fig. 36.2), serves living systems as a homogeneous catalyst for various reduction processes.

26.6d. NONCATALYTIC REDUCTION OF ALKENES. Diazene (diimide) is a very reactive compound that is conveniently prepared *in situ* by the cupric ion-catalyzed reaction of diazane (hydrazine) with hydrogen peroxide, or by the hydrolysis of potassium azodicarboxylate:

$$H_2NNH_2 + H_2O_2 \xrightarrow[Cu^{++}]{} HN{=}NH + 2\,H_2O$$

Diazane **Diazene (diimide)**

$$KO_2CN{=}NCO_2K \xrightarrow[H_2O]{\text{neutralization}} HO_2CN{=}NCO_2H \longrightarrow HN{=}NH + 2\,KOH + 2\,CO_2$$

Potassium azodicarboxylate **Diazene**

Diazene has the ability to reduce most alkenes by the stereoselective *syn* transfer of its hydrogens to the alkene:

$$HN{=}NH + \mathrm{RR'C{=}CRR'} \longrightarrow \left[\begin{array}{c} N{\equiv}N \\ H \quad\quad H \\ C{=}C \\ R\;R' \quad R\;R' \end{array}\right]^{\ddagger} \longrightarrow N_2 + \mathrm{RR'HC{-}CHRR'}$$

The ease of reduction of the alkene decreases as the extent of substitution at the double bond increases. Carboxyl groups are unaffected by diazene, although aldehydes and ketones are reduced to alcohols:

$$\mathrm{HO_2CCH{=}CHCO_2H} + DN{=}ND \longrightarrow \mathrm{HO_2C(H)(D)C{-}C(D)(H)CO_2H} + N_2$$

Maleic acid **Dideuteriodiazene** **Product of *syn* addition**

Another noncatalytic method for adding hydrogen to alkenes is provided by the hydroboration sequence. The addition of borane followed by protonolysis of the alkylborane that is formed results in the *syn* addition of hydrogen:

$$\mathrm{(C_6H_5)(CH_3)C{=}C(CH_3)(C_6H_5)} + BH_3 \longrightarrow \mathrm{(C_6H_5)(CH_3)(H)C{-}C(B{<})(CH_3)(C_6H_5)} \xrightarrow[\Delta]{RCO_2H} \mathrm{(C_6H_5)(CH_3)(H)C{-}C(H)(CH_3)(C_6H_5)}$$

The situselectivity of the reaction is limited, however, for borane rapidly reduces most carbonyl functions.

26.7 Reactions at the Allylic Position of Alkenes

A C—H bond adjacent to a C=C bond (*i.e.*, an allylic bond) is weaker and more labile than an ordinary C—H bond at an sp^3-hybridized carbon. We have already encountered examples of reactions that are selective for this position in brominations with N-bromosuccinimide (see Fig. 21.1). In like manner, oxidation reactions that are selective for the allylic position can also be carried out.

PROBLEM 26-15

Would the conversion of cyclohexene to cyclohexyl bromide by means of N-bromosuccinimide followed by treatment of the resulting product with hydrogen in the presence of palladium on charcoal be likely to proceed in good yield? Explain why or why not.

For example, alkenes react with oxygen in the presence of light of the appropriate wavelength to yield allylic hydroperoxides. The reaction might be interpreted as involving (a) excitation of oxygen from its ground state (a triplet, ↑Ö:Ö↑) to an electronically excited state (a singlet, ↑Ö:Ö↓), (b) abstraction of an allylic hydrogen to generate an allylic radical, and (c) attachment of the peroxy and allylic radicals to yield the product. Since the peroxy function ends up at what was originally one of the termini of the double bond and the double bond migrates in the process, however, it is probable that the reaction takes place in a concerted fashion without the intervention of dissociated intermediates:

H H + ·Ö:Ö· (Singlet oxygen) ⟶ [H H ·Ö:Ö·] ⟶ H H O O

26.8 Polymerization of Alkenes

26.8a. CLASSIFICATION OF POLYMERS. The time at which organic polymers first appeared on earth is uncertain. Undoubtedly they were among the constituents of the primeval broth out of which life emerged, for living organisms from the smallest virus to man himself depend on polymers to preserve the structural integrity of their cells and organs and to perform the miracles of synthesis that transpire in these cells and organs. The most important among the biological polymers are the polypeptides, the polysaccharides, and the polynucleotides. All these are classed as *condensation polymers*, resulting from the elimination of a small molecule (*e.g.*, water) between each of the monomeric units comprising the polymer;

H—X—OH + H—X—OH + H—X—OH + H—X—OH ⟶ H—X—X—X—X—OH + 3 H_2O

A condensation homopolymer

H—X—OH + H—Y—OH + H—X—OH + H—Y—OH ⟶ H—X—Y—X—Y—OH + 3 H_2O

A condensation heteropolymer (a copolymer)

One of the major developments in organic chemistry in the present century has been the discovery of methods for the laboratory synthesis of useful polymers. Many of these, as Chapter 29 discusses in more detail, are condensation polymers bearing some resemblance to their biological counterparts. Others, however, are *addition polymers*, formed by the addition of one monomer to another *without* the extrusion of a small molecule such as water:

$$\mathrm{X + X + X + X \longrightarrow X{-}X{-}X{-}X}$$

An addition homopolymer

$$\mathrm{X + Y + X + Y \longrightarrow X{-}Y{-}X{-}Y}$$

An addition heteropolymer (a copolymer)

It is with these that we are concerned in this chapter, for the majority of the synthetic addition polymers are made from alkenes.

26.8b. INITIATION OF ALKENE POLYMERIZATION. Alkene polymerization is induced by the addition of an initiator species (In) to the double bond of the alkene, setting in motion a train of events that can be depicted as follows:

$$\mathrm{In + {>}C{=}C{<} \longrightarrow In{-}C{-}C\cdot \xrightarrow{{>}C=C{<}} In{-}C{-}C{-}C{-}C\cdot \xrightarrow{n\ {>}C=C{<}} In{-}C{-}C{-}(C{-}C)_n{-}C{-}C \xrightarrow{\text{Termination step}} Polymer}$$

Cation, radical, or anion species (In)

Cation, radical, or anion at terminal carbon

If the initiator is a cation (*e.g.*, H^+), the initially formed adduct is a carbonium ion, and the polymerization is characterized as cationic. If the initiator is a radical (*e.g.*, RO·), the initially formed adduct is a carbon radical, and the polymerization is characterized as free radical. If the initiator is an anion (*e.g.*, H_3C^-), the initially formed adduct is a carbanion, and the polymerization is characterized as anionic. Alkene polymerizations following all three of these pathways are known, the particular pathway being determined partly by the initiator and partly by the substituents attached to the double bond. The manner in which the growing polymer chain is terminated is also dependent on the initiator and the alkene that are employed, as the several following examples illustrate.

26.8c. CATIONIC POLYMERIZATION OF ALKENES. Although ethylene itself does not undergo facile cationic polymerization, some alkyl-substituted ethylenes do. For example, 2-methylpropene (isobutylene) can be polymerized to "polyisobutylene" by the addition of catalytic amounts of protic acids such as sulfuric and hydrofluoric acid, or Lewis acids such as boron trifluoride-water complex ($BF_3 \cdot H_2O$) (Fig. 26.33). Under certain conditions (*e.g.*, 60% sulfuric acid), the reaction terminates after the first addition to form a mixture of isomeric dimers ("diisobutylenes"), the termination resulting from the extrusion of a proton

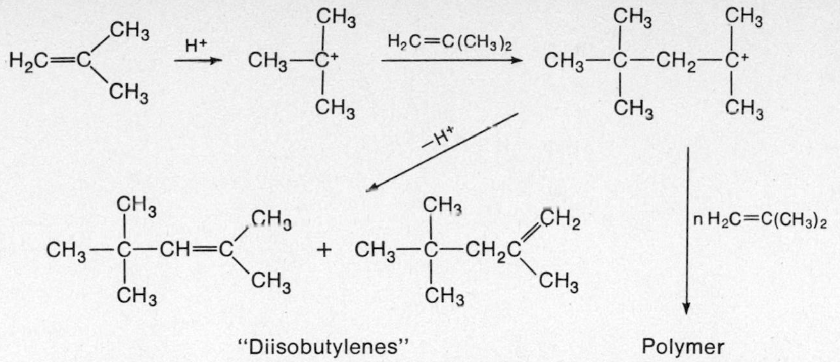

$$H_2C{=}C(CH_3)_2 \xrightarrow{H^+} CH_3{-}\overset{+}{C}(CH_3)_2 \xrightarrow{H_2C=C(CH_3)_2} CH_3{-}C(CH_3)_2{-}CH_2{-}\overset{+}{C}(CH_3)_2$$

$$\xrightarrow{-H^+} CH_3{-}C(CH_3)_2{-}CH{=}C(CH_3)_2 + CH_3{-}C(CH_3)_2{-}CH_2C(=CH_2)CH_3$$

"Diisobutylenes"

$$CH_3{-}C(CH_3)_2{-}CH_2{-}\overset{+}{C}(CH_3)_2 \xrightarrow{n\,H_2C=C(CH_3)_2} \text{Polymer}$$

Fig. 26.33. Cationic dimerization and polymerization of 2-methylpropene (isobutylene).

from the dimer cation. Similarly, the polymeric cation can terminate by the loss of a proton to yield a double bond at or near the end of the polymeric chain.

Propene (propylene) is also susceptible to cationic polymerization, yielding a polymer in which every second carbon atom of the chain is a chiral center;

$$H_2C{=}CHCH_3 \longrightarrow -\overset{*}{C}H(CH_3)-CH_2-\overset{*}{C}H(CH_3)-CH_2-\overset{*}{C}H(CH_3)-CH_2-$$

* indicates a chiral carbon atom

When the polymerization is initiated by a protic or Lewis acid, it occurs in such a fashion that the stereochemical relationship between the chiral centers is completely random; the product is characterized as an *atactic polymer*. Under certain conditions, however, the polymerization occurs in a more stereoregular fashion and gives polymers that are characterized as *syndiotactic* or *isotactic*, depending on the manner of the stereoregularity (Fig. 26.34).

The condition that is conducive to the formation of isotactic polypropylene involves the use of catalysts prepared from aluminum alkyls (R_3Al) and titanium tetrachloride. Generally known as *Ziegler–Natta catalysts*, they were introduced to the chemical world in the 1950's by Karl Ziegler of Germany and Guilio Natta of Italy, who shared the 1963 Nobel Prize for their contributions to polymer chemistry. The exact mechanism by which these catalysts induce stereoregular polymerization is not yet known, although it appears certain that a coordination complex involving the catalyst, the monomer, and the growing polymer plays a crucial role. These reactions are best characterized as coordination polymerizations, for they are not truly cationic, radical, or anionic processes. The Ziegler–Natta catalyst is effective for polymerizing almost all alkenes of the general structure $RR'C{=}CH_2$, yielding materials that are generally higher in melting point and physically stronger than their atactic counterparts (see Table 26.2).

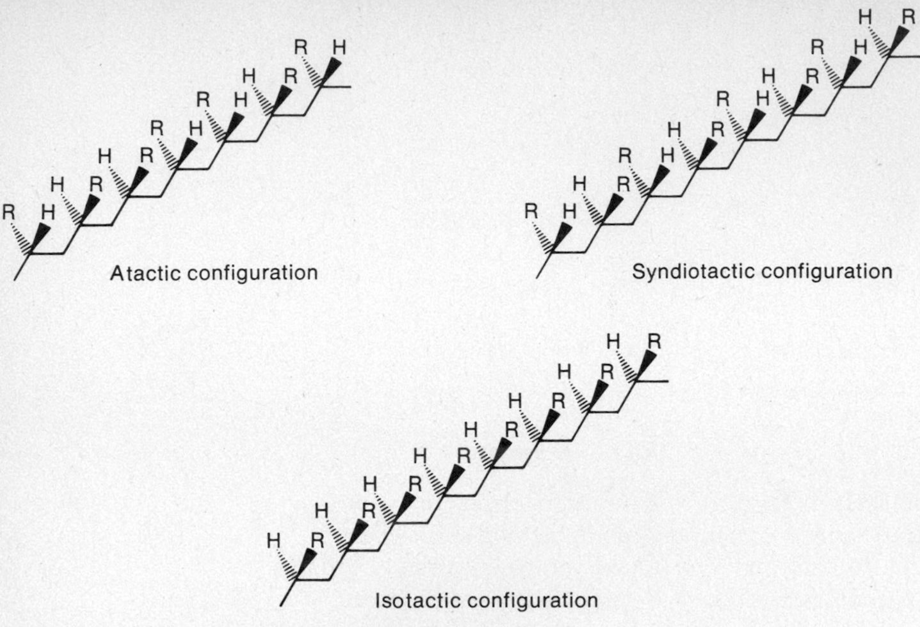

Fig. 26.34. Configurations of polymers containing chiral centers.

Table 26.2 THE STRUCTURE, TRADE NAME, AND USES OF MATERIALS OBTAINED BY CATIONIC AND ZIEGLER–NATTA POLYMERIZATIONS OF ALKENES

Monomer	Catalyst	Structure	Trade Name	Uses
$H_2C{=}CH_2$	Ziegler–Natta	Crystalline	Polyethylene	Films, pipes, etc.
$H_2C{=}CH_2$	Ziegler–Natta	Amorphous	Nordel	Rubber substitute
$CH_3CH{=}CH_2$	Ziegler–Natta	Isotactic	Polypropylene	Fibers, molded objects
$(CH_3)_2C{=}CH_2$	Cationic	Amorphous	Butyl rubber	Tires, inner tubes, etc.
$H_2C{=}C(CH_3){-}CH{=}CH_2$	Ziegler–Natta	*cis*-1,4: Amorphous	Ameripol (natural rubber)	Rubber articles
$C_6H_5{-}CH{=}CH_2$	Ziegler–Natta	Isotactic	Polystyrene	Packing materials

26.8d. FREE RADICAL POLYMERIZATION OF ALKENES. Ethylene, styrene, vinyl chloride, tetrafluoroethylene, and a number of other alkenes undergo radical-induced polymerization when treated with catalytic amounts of compounds that can generate free radicals. For example, addition of a small amount of di-*t*-butyl peroxide to styrene at 100–130° initiates the conversion of styrene to atactic polystyrene (Fig. 26.35). Termination of the growing polymer chain can occur

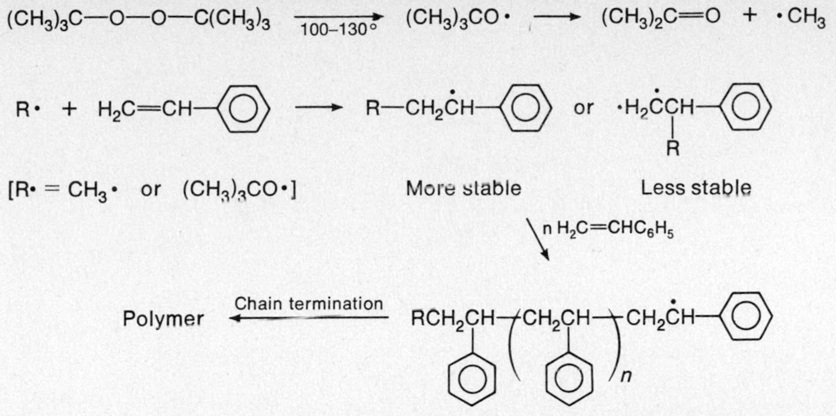

$$(CH_3)_3C{-}O{-}O{-}C(CH_3)_3 \xrightarrow[100\text{–}130°]{} (CH_3)_3CO\cdot \longrightarrow (CH_3)_2C{=}O + \cdot CH_3$$

$$R\cdot + H_2C{=}CH{-}C_6H_5 \longrightarrow \underset{\text{More stable}}{R{-}CH_2\dot{C}H{-}C_6H_5} \text{ or } \underset{\text{Less stable}}{\cdot H_2C\underset{R}{C}H{-}C_6H_5}$$

$[R\cdot = CH_3\cdot$ or $(CH_3)_3CO\cdot]$

$\searrow n\,H_2C{=}CHC_6H_5$

$$\text{Polymer} \xleftarrow{\text{Chain termination}} RCH_2CH(C_6H_5){-}(CH_2CH(C_6H_5))_n{-}CH_2\dot{C}H{-}C_6H_5$$

Fig. 26.35. Radical-induced polymerization of styrene.

by (a) combination of two radical chains, *i.e.*,

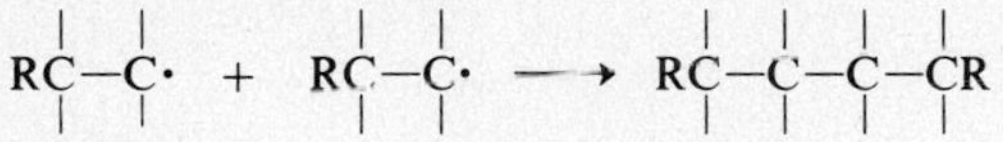

$$R\overset{|}{\underset{|}{C}}{-}\overset{|}{\underset{|}{C}}\cdot + R\overset{|}{\underset{|}{C}}{-}\overset{|}{\underset{|}{C}}\cdot \longrightarrow R\overset{|}{\underset{|}{C}}{-}\overset{|}{\underset{|}{C}}{-}\overset{|}{\underset{|}{C}}{-}\overset{|}{\underset{|}{C}}R$$

(b) disproportionation, *i.e.*,

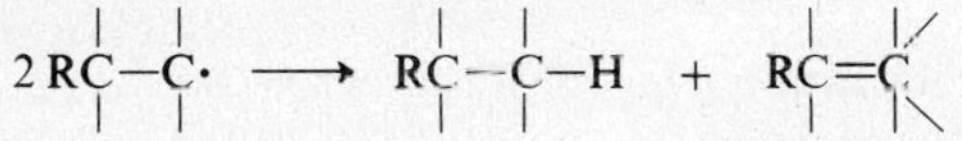

$$2\,R\overset{|}{\underset{|}{C}}{-}\overset{|}{\underset{|}{C}}\cdot \longrightarrow R\overset{|}{\underset{|}{C}}{-}\overset{|}{\underset{|}{C}}{-}H + R\overset{|}{\underset{|}{C}}{=}C\lessgtr$$

or (c) chain transfer, whereby a chain-transfer agent gives up a hydrogen to a growing polymer radical, itself becoming a radical capable of initiating polymerization:

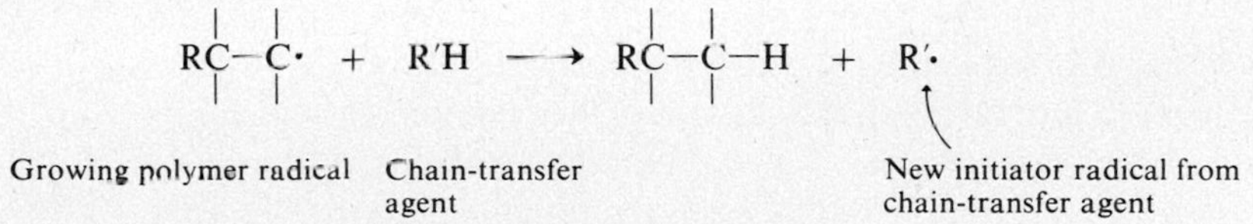

$$\underset{\text{Growing polymer radical}}{R\overset{|}{\underset{|}{C}}{-}\overset{|}{\underset{|}{C}}\cdot} + \underset{\text{Chain-transfer agent}}{R'H} \longrightarrow R\overset{|}{\underset{|}{C}}{-}\overset{|}{\underset{|}{C}}{-}H + \underset{\text{New initiator radical from chain-transfer agent}}{R'\cdot}$$

Copolymerization can frequently be carried out effectively under free radical conditions. For example, when styrene is mixed with maleic anhydride in the presence of an initiator, a copolymer containing styrene and maleic anhydride moieties in alternation is obtained:

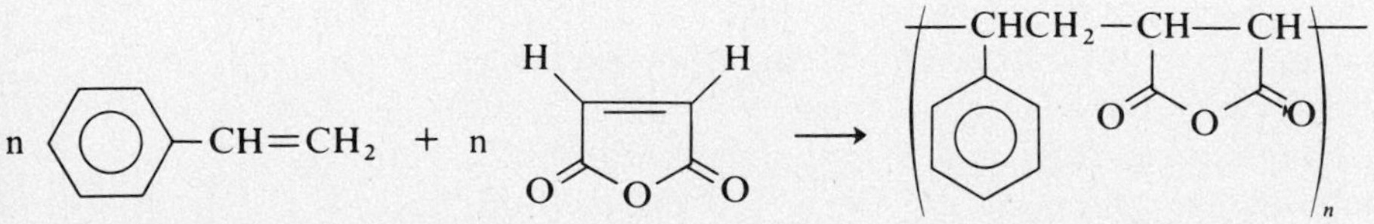

$$n\,C_6H_5{-}CH{=}CH_2 + n\,\text{(maleic anhydride)} \longrightarrow {-}\!\left(CH(C_6H_5)CH_2{-}CH{-}CH\ (\text{anhydride ring})\right)_n\!{-}$$

Copolymerization can also be used as a means for introducing cross linking into the product. Thus, when styrene is polymerized in the presence of 1,4-divinylbenzene, the product consists of a meshlike polymer, in which the polystyrene chains are connected at various points by $-CH-C_6H_4-CH-$ moieties:

$$C_6H_5-CH=CH_2 + CH_2=CH-C_6H_4-CH=CH_2 \longrightarrow$$

$$-CH_2CH(C_6H_5)-CH_2-CH(C_6H_4-)-CH_2-CH(C_6H_5)-CH_2-CH(C_6H_5)-$$

$$-CH(C_6H_4-)-CH_2-CH(C_6H_5)-CH_2-CH(-C_6H_4-)-CH_2-CH(C_6H_5)-CH_2-CH(C_6H_5)-$$

$$-CH_2-CH(-C_6H_4-)-CH_2-CH(C_6H_5)-CH_2-$$

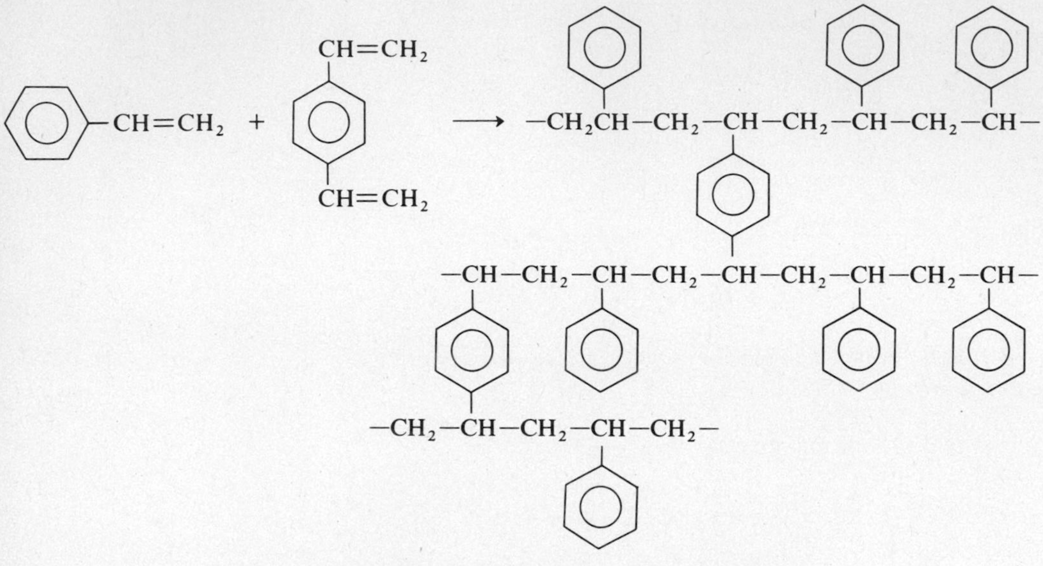

Cross-linked polystyrene

Conjugated dienes undergo facile polymerization, yielding polymers that contain double bonds separated by two carbon atoms:

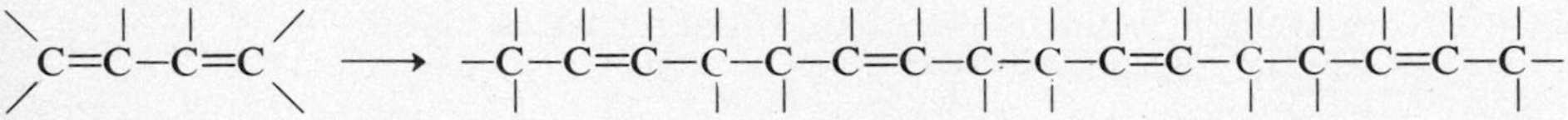

$$\gt C=\overset{|}{C}-\overset{|}{C}=C\lt \longrightarrow -\overset{|}{\underset{|}{C}}-\overset{|}{C}=\overset{|}{C}-\overset{|}{\underset{|}{C}}-\overset{|}{\underset{|}{C}}-\overset{|}{C}=\overset{|}{C}-\overset{|}{\underset{|}{C}}-\overset{|}{\underset{|}{C}}-\overset{|}{C}=\overset{|}{C}-\overset{|}{\underset{|}{C}}-\overset{|}{\underset{|}{C}}-\overset{|}{C}=\overset{|}{C}-\overset{|}{\underset{|}{C}}-$$

The double bonds in the polymer can have either the *cis* or *trans* configuration, the degree of stereoregularity being a function of the conditions of the reaction and the structure of the diene monomer. For example, 2-methylbutadiene (isoprene) yields a product containing mostly *trans* double bonds when the polymerization is radical induced but an essentially all-*cis* polymer when the Ziegler–Natta catalyst is used. The all-*cis* compound closely resembles natural rubber in structure as well as properties, providing man with a synthetic source of a natural polymer that has assumed an important place in his technology.

The products obtained from free radical-induced polymerizations of various alkenes are listed in Table 26.3, along with the chemical and physical properties of these materials. These materials furnish the basic ingredient for ion-exchange resins, because functional groups can be introduced into them by the appropriate chemical reactions. Thus, sulfonation yields a polymer containing sulfonic acid groups that can act as cation-exchange centers (*i.e.*, H^+ for E^+); chloromethylation, $ArH + HCHO + HCl \xrightarrow[ZnCl_2]{} ArCH_2Cl$, followed by treatment with trimethylamine yields a polymer containing $(CH_3)_3\overset{+}{N}HC_2-$ groups which can act as anion-exchange centers (*i.e.*, OH^- for Nu^-).

Table 26.3 THE STRUCTURE, TRADE NAME, AND USES OF MATERIALS OBTAINED BY FREE RADICAL POLYMERIZATION OF ALKENES

Monomer	Structure	Trade Name	Uses
$H_2C{=}CH_2$	Semicrystalline	Polyethylene	Film, containers
$H_2C{=}CHCl$	Atactic, semicrystalline, and crystalline	Polyvinylchloride, Saran	Film, piping, fibers, etc.
$F_2C{=}CFCl$	Atactic, semicrystalline	Kel-F	Valves, gaskets
$F_2C{=}CF_2$	Crystalline	Teflon	Nonreactive surface coatings, valves, gaskets
$H_2C{=}C(Cl)CH{=}CH_2$	Amorphous	Neoprene	Rubber substitute

26.8e. ANIONIC POLYMERIZATION OF ALKENES. For anionic polymerization to take place, it is necessary not only to use very strong bases, such as sodium metal, sodium amide, or an alkyl sodium, but also to employ alkenes bearing functions capable of stabilizing a negative charge. Thus, ethylene and alkyl-substituted ethylenes do not undergo anionic polymerization, but ethylenes carrying electron-withdrawing groups such as —CN and halogen do. For example, acrylonitrile undergoes polymerization under the influence of sodium amide to yield a polycyano compound, the intermediate anions at each step in the process being stabilized by resonance interaction with the adjacent cyano group:

$$H_2C{=}CHCN \xrightarrow{NaNH_2} H_2NCH_2\bar{C}HCN \xrightarrow{H_2C{=}CHCN} H_2NCH_2CH(CN)CH_2\bar{C}HCN \longrightarrow \text{Polymer}$$

Anions, in contrast to radicals and cations, have little or no tendency to disproportionate and cannot undergo dimerization. Consequently, the growing polymer chains in anionic polymerizations do not terminate so readily as those in cationic and radical polymerizations. In fact, in a nonprotic solvent they continue to grow until all the monomer is consumed, acquiring a terminal group only when treated with an electrophile (usually a proton donor). Polymers of this variety are said to be "living polymers," for they have the capability of growing to greater lengths simply by the addition of more monomer without the necessity of more initiator.

26.8f. OLIGOMERIZATION OF ALKENES. Polymers containing only a small number of monomeric units are frequently referred to as oligomers—*i.e.*, dimers, trimers, tetramers, etc. Among the interesting alkene oligomers are the cyclic dimers and trimers of 1,3-butadiene, which can be obtained by the action of a diallylnickel catalyst. Diallylnickel is prepared from allylmagnesium bromide and nickel bromide:

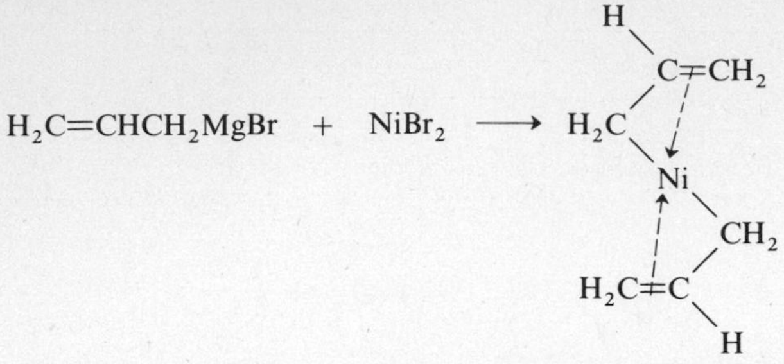

$$H_2C{=}CHCH_2MgBr + NiBr_2 \longrightarrow (\eta^3\text{-}CH_2CHCH_2)_2Ni$$

Diallylnickel (a π-allyl complex)

Butadiene displaces the allyl groups in diallylnickel and undergoes dimerization or trimerization, depending on the reaction conditions, and the diallylnickel is then regenerated:

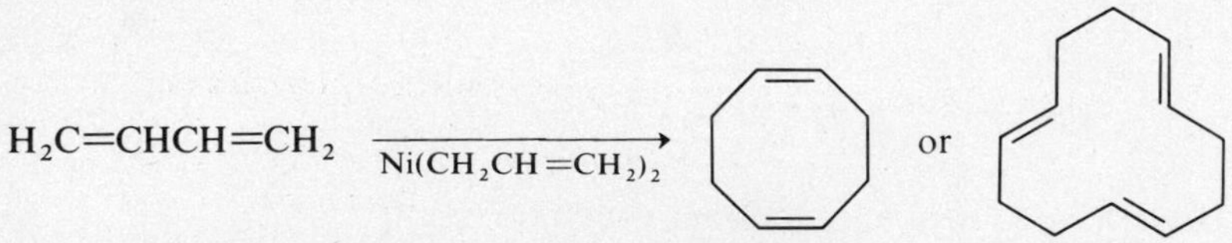

$$H_2C{=}CHCH{=}CH_2 \xrightarrow[Ni(CH_2CH{=}CH_2)_2]{} \text{1,5-Cyclooctadiene} \quad \text{or} \quad \text{1,5,9-Cyclododecatriene}$$

1,5-Cyclooctadiene **1,5,9-Cyclododecatriene**

Simple alkenes can be dimerized in a comparable fashion. For example, ethylene yields 1-butene, and propene yields a mixture of isomeric hexenes when treated with diallylnickel.

Table 26.4 summarizes the reactions of alkenes.

Table 26.4 REACTIONS OF ALKENES

Type of Reaction	Starting Material	Reagents	Products	Comments
Isomerization	$(R)(R')C{=}C(R')(CH_2R)$	Heat, light, or acids	$(R)(R')C{=}C(CH_2R)(R')$	(*E,Z*) isomerization
		Acids	$(R)(R')CHC(R'){=}CHR$	Positional isomerization
Electrophilic addition	$>C{=}C<$	HX (protic acids)	$H{-}\overset{\vert}{\underset{\vert}{C}}{-}\overset{\vert}{\underset{\vert}{C}}{-}X$	Markovnikov regioselectivity; *anti* stereoselectivity
		X_2 (halogens)	$X{-}\overset{\vert}{\underset{\vert}{C}}{-}\overset{\vert}{\underset{\vert}{C}}{-}X$	Restricted to Cl_2 and Br_2, in most cases
		$Hg(OAc)_2$, solvent-H	$\text{Solvent}{-}\overset{\vert}{\underset{\vert}{C}}{-}\overset{\vert}{\underset{\vert}{C}}{-}HgOAc$	C—Hg bond can be converted to C—H bond with $NaBH_4$
		R^+	${-}\overset{+}{\underset{\vert}{C}}{-}\overset{\vert}{\underset{\vert}{C}}{-}R$	Dimerization, trimerization, . . . , polymerization of alkenes

Table 26.4—*cont.*

Type of Reaction	Starting Material	Reagents	Products	Comments
Electrophilic addition	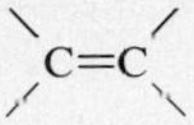 $>C=C<$	HCHO, H^+	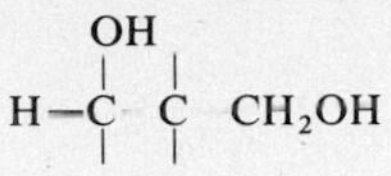 $H-C(OH)-C-CH_2OH$	Prins reaction
		RCO_3H or $ArCO_3H$	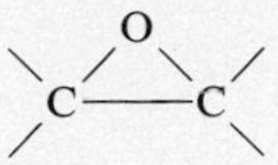 $>C-C<$ (epoxide, bridged by O)	Epoxidation reaction; *syn* addition.
		BH_3	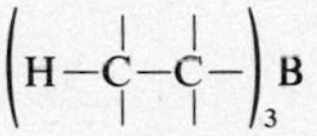 $(H-C-C-)_3B$	Hydroboration reaction; *syn* addition; C—B bond can be converted to various other bonds with appropriate reagents. Overall addition is anti-Markovnikov
		R_2C: (carbenes generated *in situ*)	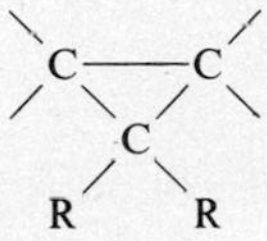 $>C-C<$ (cyclopropane, bridged by CR_2)	Restricted to carbenes that cannot rearrange to alkenes
Radical addition	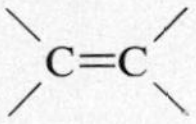 $>C=C<$	HBr	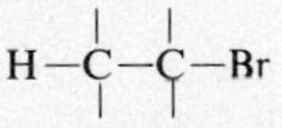 $H-C-C-Br$	anti-Markovnikov regioselectivity
		A—B (*e.g.*, Cl_2, Br_2, $CHCl_3$, $CHBr_3$, CCl_4, CBr_4, H_2S, RSH, R_3SnH)	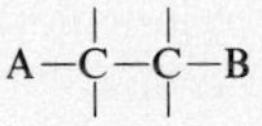 $A-C-C-B$	
Cycloaddition	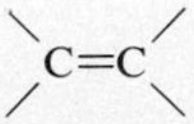 $>C=C<$ (Dienophile)	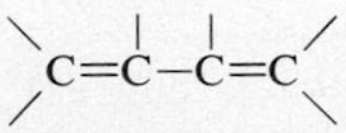 $>C=C-C=C<$ (Diene)	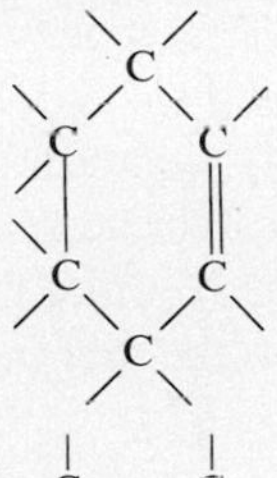 cyclohexene ring (six C, one C=C)	Diels–Alder reaction; for effective reaction, alkene must carry an electron-withdrawing substituent (*e.g.*, C=C—$COCH_3$, C=C—CN)
	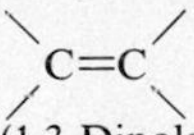 $>C=C<$ (1,3-Dipolarophile)	$\bar{A}-B=\overset{+}{D}$ (1,3-Dipolar reagent)	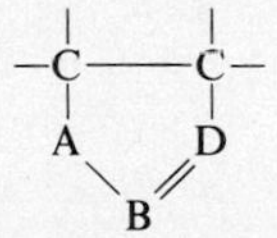 five-membered ring $-C-C-$, A, B, D (B=D)	1,3-Dipolar addition reactions
Oxidation	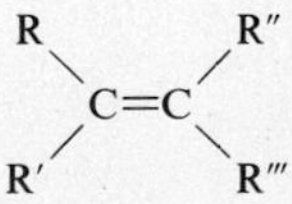 $RR'C=CR''R'''$	OsO_4 or $KMnO_4$ (neutral or weakly basic)	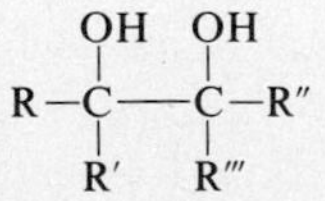 $R-C(OH)(R')-C(OH)(R''')-R''$	See Table 22.1
		$KMnO_4$ (basic), $Na_2Cr_2O_7 + H_2SO_4$, O_3.	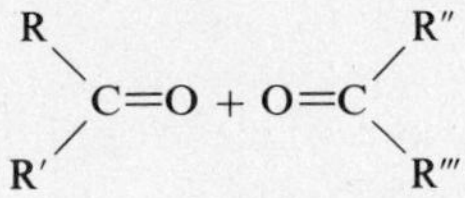 $RR'C=O + O=CR''R'''$	Aldehydes are oxidized to carboxylic acids Ozone is the most selective of the oxidizing agents for complete C=C cleavage

Table 26.4—*cont.*

Type of Reaction	Starting Material	Reagents	Products	Comments
Oxidation	$R_2CH{-}\overset{\vert}{C}{=}C\langle$	O_2, $h\nu$ (sensitizer)	$R_2C{=}C{-}\overset{OOH}{\overset{\vert}{C}}{-}$	Allylic oxidation—occurs with double-bond migration
Catalytic reduction	$\rangle C{=}C\langle$	H_2 + catalyst	$H{-}\overset{\vert}{\underset{\vert}{C}}{-}\overset{\vert}{\underset{\vert}{C}}{-}H$	*syn* Addition in many cases. Catalysts include Pd, Pt, Rh, and Ni
Noncatalytic reduction	$\rangle C{=}C\langle$	HN=NH (diazene generated *in situ*)	$H{-}\overset{\vert}{\underset{\vert}{C}}{-}\overset{\vert}{\underset{\vert}{C}}{-}H$	*syn* Addition
Polymerization	$\rangle C{=}C\langle$	Cationic, anionic, or free radical catalysts	$\cdots{-}\overset{\vert}{\underset{\vert}{C}}{-}\overset{\vert}{\underset{\vert}{C}}{-}\overset{\vert}{\underset{\vert}{C}}{-}\overset{\vert}{\underset{\vert}{C}}{-}\overset{\vert}{\underset{\vert}{C}}{-}\overset{\vert}{\underset{\vert}{C}}{-}\cdots$	See Section 26.8 for details

26.9 *Study Guide*

- I. Reactions in which the double bond is retained
  - A. *cis,trans* Isomerization
  - B. Double-bond migration
- II. Electrophilic additions to alkenes
  - A. General features of electrophilic addition reactions
    1. Regioselectivity—Markovnikov's rule
    2. Bimolecular electrophilic additions via cationic intermediates
       - (a) Ad_E2 reactions involving carbonium ion intermediates
         - (1) Favored by carbon frameworks that can stabilize a carbonium ion (*e.g.*, benzyl)
         - (2) Favored by small electrophiles (*e.g.*, H^+)
         - (3) Stereoselectivity—usually low
       - (b) Ad_E2 reactions involving cyclic onium intermediates
         - (1) Favored by electrophiles from elements in the second, or higher, row of the periodic chart (*e.g.*, Cl^+, Br^+, I^+, $CH_3CO_2Hg^+$, RS^+)
         - (2) Stereoselectivity—*anti* addition
       - (c) Reaction pathways of carbonium ions from alkenes
         - (1) Addition of nucleophile to form addition product
         - (2) Loss of a proton to form an alkene
         - (3) Rearrangement to an isomeric carbonium ion
    3. Termolecular electrophilic addition reactions—Ad_E3 reactions
       - (a) Favored by carbon frameworks that cannot stabilize a carbonium ion
       - (b) Favored with first row electrophiles (*i.e.* H^+)
  - B. Scope and limitations of electrophilic addition reactions
    1. Addition of protic acids
    2. Addition of halogens—most useful with Cl_2 and Br_2
    3. Addition of oxyhalogen acids
    4. Addition of mercuric salts (solvomercuration)

5. Addition of formaldehyde—yields 1,3-glycols
6. Addition of oxygen—yields oxiranes

C. Concerted electrophilic addition to alkenes

1. Addition of borane (hydroboration)
   (a) Regioselectivity—acts as $H_2\overset{+}{B}-\overset{-}{H}$ and adds in Markovnikov fashion
   (b) Stoichiometry of reaction of BH_3 with alkene depends on number of substituents attached to the double bond
   (c) Reactions of alkylboranes
      (1) Conversion to alcohols by oxidative hydrolysis
      (2) Conversion to alkyl iodide by treatment with iodine and base
      (3) Conversion to amines by treatment with NH_2OSO_3H
      (4) Conversion to dimeric hydrocarbons by treatment with Ag^+
2. Addition of carbenes
   (a) Stereoselectivity—depends on spin state of carbene
   (b) Generation of carbenes
      (1) Action of base on haloforms (CHX_3)
      (2) Action of base on trichloroacetic acid
      (3) Action of alkyllithium on tetrahalomethanes
      (4) Action of zinc-copper couple on methylene iodide
      (5) Photolysis of diazomethane or ketene
   (c) Most often-used carbenes: $:CH_2$, $:CX_2$, $:CHAr$, $:CHCO_2C_2H_5$

III. Radical additions to alkenes

A. General features of radical addition reactions
   1. Radical chain reaction
   2. Regioselectivity—anti-Markovnikov

B. Thermodynamic considerations in radical chain addition reactions

C. Electrophilic addition *vs.* radical addition reactions
   1. Radical addition reactions require a free radical initiator
   2. As the degree of alkylation at the double bond increases, the ionic pathway becomes more likely

IV. Cycloaddition reactions of alkenes

A. Woodward–Hoffmann rules

B. Diels–Alder reaction
   1. Reagents
      (a) Dienes—must be able to assume the cisoid conformation
      (b) Dienophiles—general structure is $X-C=C-Z$ where X and/or Z can be $-COR$, $-COY$, $-CN$, $-NO_2$ for greatest reactivity
   2. Regioselectivity—interpreted in terms of stability of diradical intermediates (although reaction is considered to be a concerted process)
   3. Stereochemistry
      (a) Configuration of dienophile is retained in the product

(b) Principle of maximum overlap of multiple bonds influences stereochemical outcome in certain cases

C. 1,3-Dipolar addition reactions

V. Oxidation and reduction reactions of alkenes
  A. Oxidation with retention of C—C bond—see Section 22.6f.
  B. Oxidation with complete cleavage of C=C bond
    1. Potassium permanganate and chromic acid—strenuous oxidants
    2. Ozone—highly situselective
      (a) Mechanism of ozonolysis—dipolar addition, rearrangement, formation of ozonide
      (b) Decomposition of ozonide
        (1) Reductive decomposition if aldehydes are desired products
        (2) Oxidative decomposition if acids are desired products
  C. Catalytic reduction of alkenes
    1. Catalysts: Pt, Pd, Rh, Ru, Ni
    2. Situselectivity—limited, because many other groups are also susceptible to catalytic reduction; approximate reactivity sequence is alkynes > alkenes > arenes
    3. Stereochemistry—frequently, but not always, addition is *syn*
    4. Homogeneous catalysts
  D. Noncatalytic reduction of alkenes
    1. Reagent: diazene (diimide), HN=NH
    2. Regioselectivity—most other groups not reduced by this reagent
    3. Stereochemistry—*syn* addition

VI. Reactions at the allylic positions of alkenes

VII. Polymerization of alkenes
  A. Classification of polymers
    1. Condensation polymers
    2. Addition polymers—characteristic of alkenes
  B. Initiation of polymerization: cationic, radical, anionic, Ziegler–Natta
  C. Cationic polymerization of alkenes
    1. Stereochemistry of polymers: atactic, syndiotactic, isotactic
    2. Stereoregular polymers; Ziegler–Natta coordination catalysts
  D. Free radical polymerization of alkenes
    1. Chain termination processes; radical combination, disproportionation, chain-transfer
    2. Copolymerization
    3. Diene polymerization; stereochemistry of product
  E. Anionic polymerization of alkenes

REFERENCES

AUGUSTINE, R. L., *Catalytic Hydrogenation*, Marcel Dekker, Inc., New York, 1965.

BILLMEYER, F. W., *Textbook of Polymer Science*, 2nd ed., John Wiley & Sons, Inc., New York, 1971.

FIESER, L. F., and M. FIESER, *Advanced Organic Chemistry*, Reinhold Publishing Corp., New York, 1961, Chapter 12.

GARWOOD, R. F., in *Rodd's Chemistry of Carbon Compounds*, 2nd ed., Vol. IA, ed. by S. Coffey, Elsevier Publishing Co., New York, 1964, pp. 398–447.

PATAI, S., ed., *The Chemistry of Alkenes*, Vol. 1, Interscience Publishers, New York, 1964.

PATAI, S., and J. ZABICKY, *The Chemistry of the Alkenes*, Vol. 2, John Wiley & Sons, Inc., New York, 1970.

RAVVE, A., *Organic Chemistry of Macromolecules*, Marcel Dekker, Inc., New York, 1967.

RYLANDER, P. N., *Catalytic Hydrogenation over Platinum Metals*, Academic Press, Inc., New York, 1967.

PROBLEMS

1. Draw the structure of the major product from each of the following reactions:

(a) 1,2-dimethylcyclohexene (CH_3, CH_3) + Br_2 $\longrightarrow$

(b) $\left(\text{C}_6\text{H}_{11}\text{–CH}_2\text{–}\right)_3B \xrightarrow{H_2O_2/OH^-}$

(c) cyclohexane with OH and CH_2HgOAc + $NaBH_4$ $\xrightarrow[NaOH]{}$

(d) C_6H_5–$CH{=}CH_2$ + $H_2C{=}O$ $\xrightarrow[H^+]{H_2O}$

(e) bicyclic alkene with CH_3 + $C_6H_5CO_3H$ $\longrightarrow$

(f) tricyclic alkene + O_3 $\xrightarrow[CH_2Cl_2,\ pyridine]{}$

(g) $CH_3\overset{\overset{\displaystyle O}{\|}}{C}CH_2CH_2CH{=}CH_2 + H_2 \xrightarrow[Pt]{}$

(h) p-benzoquinone (O, O) + cyclopentene $\longrightarrow$

(i) [bicyclo[2.2.1]heptene with D on each alkene carbon] + NCO_2K ‖ NCO_2K $\xrightarrow{HCl}$

(j) [thiophene] + H_2 $\xrightarrow[\text{Raney Ni}]{}$

(k) [C_6H_5(H)C=C(H)CH_3] + IN_3 ⟶

(l) CH_3(H)C=C(H)CH_3 + C_6H_5–SCl ⟶

(m) $HO–CH_2–CH_2–CH_2–CH_2–CH=CH_2$ $\xrightarrow{H^+}$

(n) $O_2N–C_6H_4–CH=CH–C_6H_4–OCH_3$ + HBr ⟶

2. Give a specific example for each of the following:
(a) An Ad_E2 reaction (b) A chloronium ion intermediate (c) A *syn* addition
(d) An *anti* addition (e) An *anti*-Markovnikov addition (f) A Diels–Alder reaction
(g) A 1,3-dipolar addition reaction

3. Give a concise definition for each of the following terms:
(a) Regioselectivity (b) Electrophilic addition reaction (c) Concerted addition reaction
(d) A cycloaddition reaction (e) "Tight-ion pair"

4. For each of the following reactions, indicate in as much detail as possible what you think to be the probable mechanism for the reaction:

(a) [$\Delta^{9,10}$-octalin] + HCl ⟶ [trans-9-chlorodecalin, H and Cl]

(b) [cyclopentene] + Cl_2 $\xrightarrow{h\nu}$ [1,2-dichlorocyclopentane] + [3-chlorocyclopentene]

(c) $C_6H_5–C(=O)Cl$ + [cyclopentene] $\xrightarrow{AlCl_3}$ [2-chlorocyclopentyl phenyl ketone]

(d) 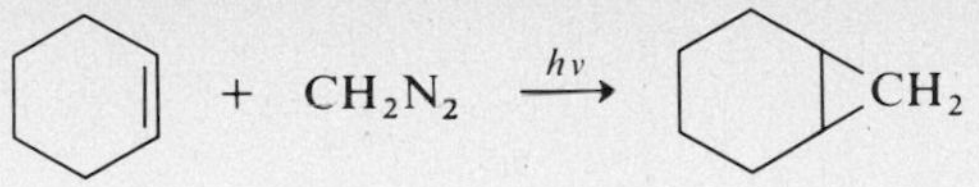 + CH_2N_2 $\xrightarrow{h\nu}$ CH_2

(e) 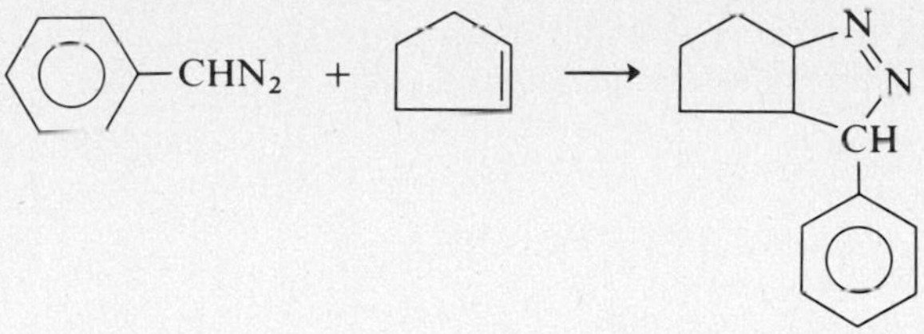 $-CHN_2$ + $\longrightarrow$ N N CH

(f) 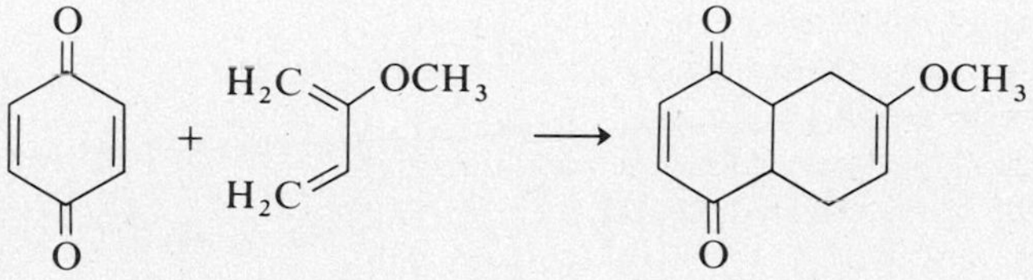 O O + H_2C OCH_3 H_2C $\longrightarrow$ O O OCH_3

5. The double bond of an alkene provides a "handle" that almost invariably facilitates the proof of structure of the alkene.

(a) Each of four test tubes contains a pure sample that is known to be one of the following four isomeric alkenes:

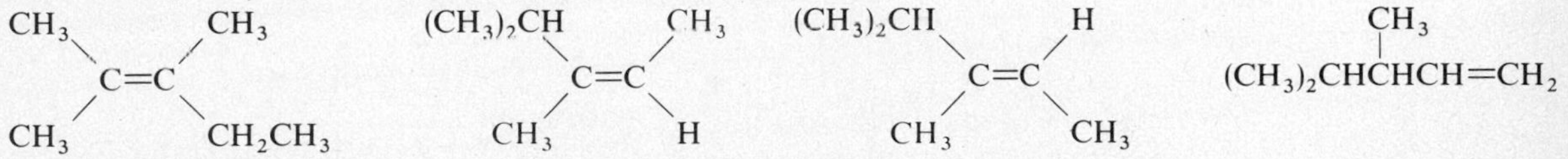

$(CH_3)(CH_3)C{=}C(CH_3)(CH_2CH_3)$ $(CH_3)_2CH(CH_3)C{=}C(CH_3)(H)$ $(CH_3)_2CH(CH_3)C{=}C(H)(CH_3)$ $(CH_3)_2CHCH(CH_3)CH{=}CH_2$

Using chemical and, if necessary, spectral methods, indicate how the contents of each of the four test tubes can be established.

(b) The terpene limonene has the structure shown

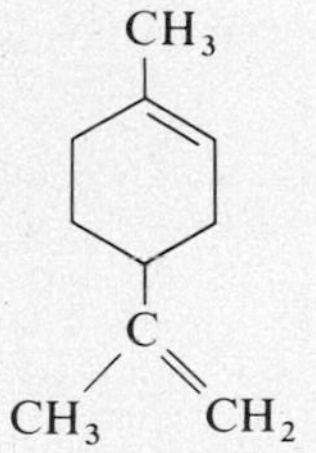

CH_3 C CH_3 CH_2

Imagine that you have just isolated this material from lemon oil and, not realizing that it is a known compound, embark on a structure proof. Outline the procedure that you would use to reach this goal. Compounds obtained by conversion or degradation of limonene can be considered as known substances if they are among the group listed as available starting materials in Tables 39.1–39.5 or can be synthesized from these compounds.

(c) The terpene ocimene has the molecular formula $C_{10}H_{16}$, is optically inactive, and undergoes ozonolysis to yield two equivalents of formaldehyde, one equivalent of a three-carbon aldehyde, and one equivalent of a five-carbon aldehyde. Oxidation of the latter two aldehydes yields pyruvic acid (CH_3COCO_2H) and levulinic acid ($CH_3COCH_2CH_2CO_2H$), respectively. Reduction of ocimene yields a $C_{10}H_{22}$

hydrocarbon that can exist in two enantiomeric forms. What is the structure of ocimene?

(d) A C_6H_8 compound reacts with maleic anhydride to yield an adduct for which *exo* and *endo* isomers are possible. Catalytic reduction of the adduct results in the absorption of one equivalent of hydrogen to yield a material in which the distinction between *exo* and *endo* isomers has disappeared. What is the structure of the C_6H_8 compound?

6. Explain the following observations:
   (a) The rate of the acid-catalyzed addition of water to acrolein ($H_2C{=}CHCHO$) is much greater than to 2-butene.
   (b) The bromination of styrene in methanol solution yields not only 1,2-dibromo-1-phenylethane but 2-bromo-1-methoxy-1-phenylethane as well.
   (c) Methyl vinyl ketone reacts more rapidly with cyclopentadiene than with 1,3-butadiene in the Diels–Alder reaction.
   (d) 1,2-Dimethylcyclopentene reacts with HCl to give the product of *anti* addition, but 1-phenylpropene reacts with DBr to give the product of *syn* addition.
   (e) Bromine reacts about 165 times faster with $C_6H_5CH{=}CHC_6H_5$ than with $C_6H_5CH{=}CHBr$.
   (f) The rate of addition of IBr to olefins is about 1,000 times slower than the rate of addition of BrCl.
   (g) Propene, $H_2C{=}CHCH_3$, undergoes acid-catalyzed hydration to $(CH_3)_2CHOH$ but is inert to aqueous sodium hydroxide; benzenesulfonylethene, $H_2C{=}CHSO_2C_6H_5$, on the other hand, reacts with sodium hydroxide to yield $HOCH_2CH_2SO_2C_6H_5$.
   (h) 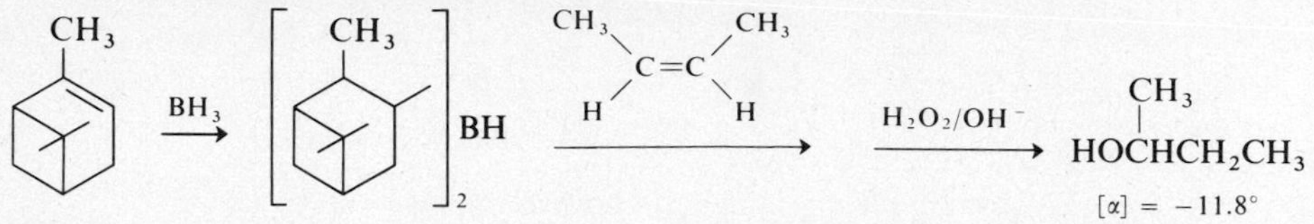 [α-pinene, CH_3] $\xrightarrow{BH_3}$ [CH_3 bicyclic]$_2$BH; + CH_3(H)C=C(H)CH_3 ⟶ $\xrightarrow{H_2O_2/OH^-}$ $HO\overset{CH_3}{C}HCH_2CH_3$ $[\alpha] = -11.8°$
   (i) The enzyme-catalyzed addition of water to deuterium-labeled fumaric acid, HO_2C(D)C=C(H)CO_2H, produces malic acid, $HO_2C\overset{D}{C}H\overset{OH}{C}HCO_2H$, in which the coupling constant of the α,α′ hydrogens is 7.1 Hz, indicating an antiperiplanar relationship. Is the addition *syn* or *anti*? Would you predict the acid-catalyzed addition to proceed with the same stereochemistry? Why or why not?

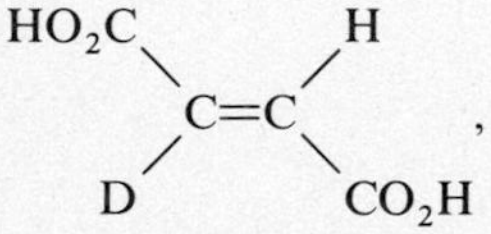

7. Indicate the steps by means of which the compound shown at the left can be converted to the compound shown at the right.
   (a) $(CH_3)_2CHCH{=}CH_2 \longrightarrow (CH_3)_2CHCH_2CH_2OH$
   (b) 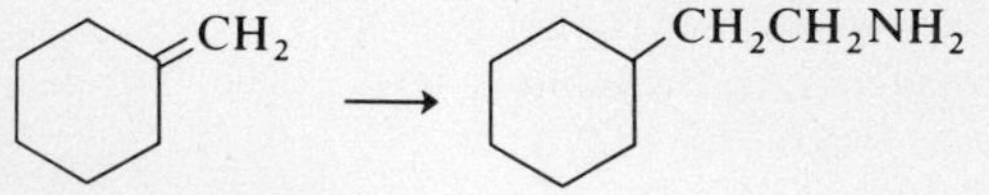 cyclohexane=CH_2 ⟶ cyclohexane–$CH_2CH_2NH_2$
   (c) 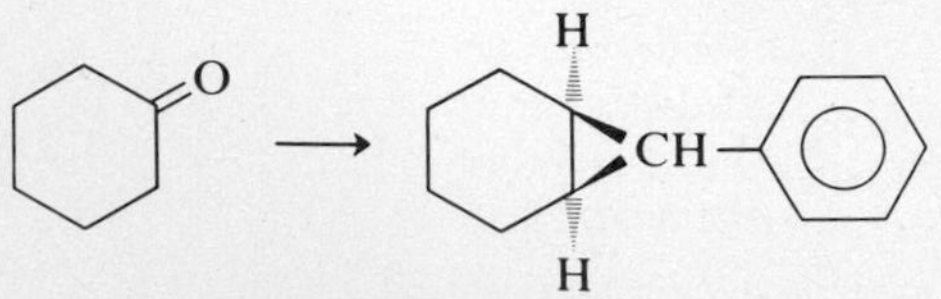 cyclohexanone (=O) ⟶ bicyclic (H, H) CH–C₆H₅ (phenyl)

(d) 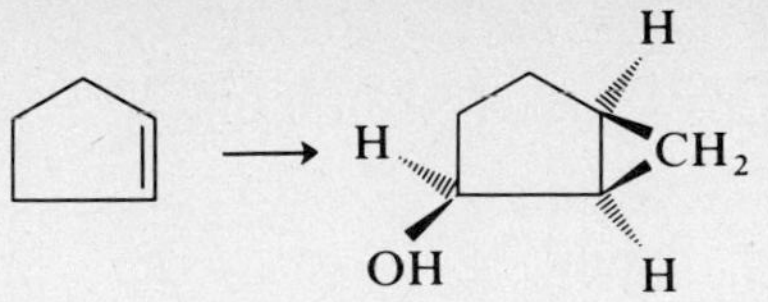 [cyclopentene] ⟶ [bicyclic cyclopropane product with H, OH, CH₂, H, H]

(e) 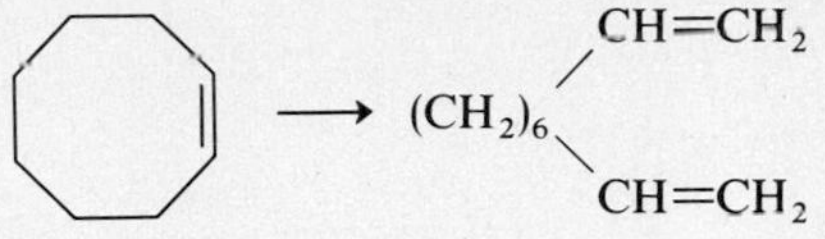 [cyclooctene] ⟶ $(CH_2)_6(CH{=}CH_2)_2$

8. Using Tables 39.1 to 39.5 as guidelines to readily available starting materials, devise syntheses for the following compounds. In addition to readily available organic compounds, any inorganic materials and special organic reagents may be used.

(a) $CH_3(CH_2)_7CHDCH_2D$

(b) $(CH_3)_2CHCHBrCH_2CBr_3$

(c) [cyclooctane-fused 1,3-dioxolane with H, H] $CH(CH_2)_5CH_3$

(d) [cyclohexane with Br and] $CHBrCH_3$

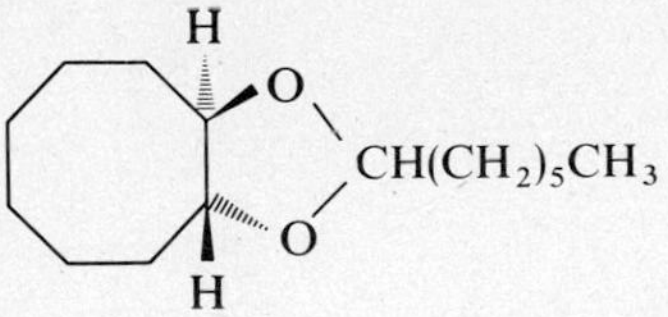

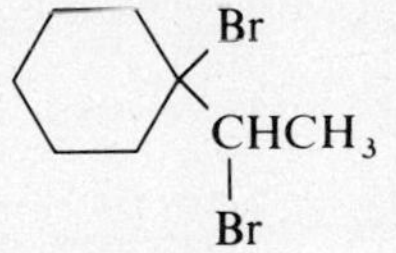

Chapter 27

Alkynes

Look for the stars, you'll say that there are none;
Look for a second time, and, one by one,
You mark them twinkling out with silvery light
And wonder how they could elude the sight!

William Wordsworth
Calm is the Fragrant Air

Alkynes, like alkenes, are reactive compounds, and, for that reason, are very useful synthesis intermediates. Their methods of preparation and their reactions constitute important segments in the rules of the game of organic synthesis.

PREPARATION OF ALKYNES

27.1 *Preparation of Alkynes from Compounds Not Containing C≡C Bonds*

Relatively few methods are available for introducing a C≡C bond into an existing carbon framework. In the majority of cases, as discussed in the next section, complex acetylenes are generally elaborated from simpler compounds in which the C≡C bond is already present.

Acetylene, the parent member of the alkyne family, is prepared commercially by the action of water on calcium carbide, which, in turn, is obtained from the reaction of elemental carbon with calcium oxide (see Fig. 8.12).

$$\mathrm{CaO} \quad + \quad 3\,\mathrm{C} \xrightarrow[2000^\circ]{} \mathrm{CO} \quad + \quad \mathrm{CaC_2} \xrightarrow{\mathrm{H_2O}} \mathrm{HC{\equiv}CH} \quad + \quad \mathrm{Ca(OH)_2}$$

The principal methods for introducing C≡C bonds into preformed carbon skeletons make use of elimination reactions. For example, 1,2-dihalides obtained by the addition of halogen to alkenes can often be doubly dehydrohalogenated by strong base to yield the alkynes (Fig. 27.1). In similar fashion, 1,1-dihalides, which can be prepared by the action of phosphorus halides on ketones, undergo double dehydrohalogenation on treatment with strong base (Fig. 27.2). This

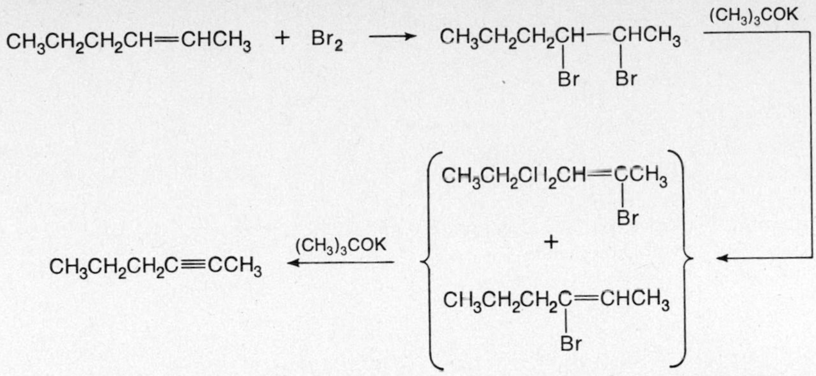

$$CH_3CH_2CH_2CH{=}CHCH_3 + Br_2 \longrightarrow CH_3CH_2CH_2CH(Br){-}CH(Br)CH_3 \xrightarrow{(CH_3)_3COK} \left\{ \begin{array}{c} CH_3CH_2CH_2CH{=}C(Br)CH_3 \\ + \\ CH_3CH_2CH_2C(Br){=}CHCH_3 \end{array} \right\} \xrightarrow{(CH_3)_3COK} CH_3CH_2CH_2C{\equiv}CCH_3$$

Fig. 27.1. Preparation of 2-hexyne from 2-hexene by bromination-dehydrobromination.

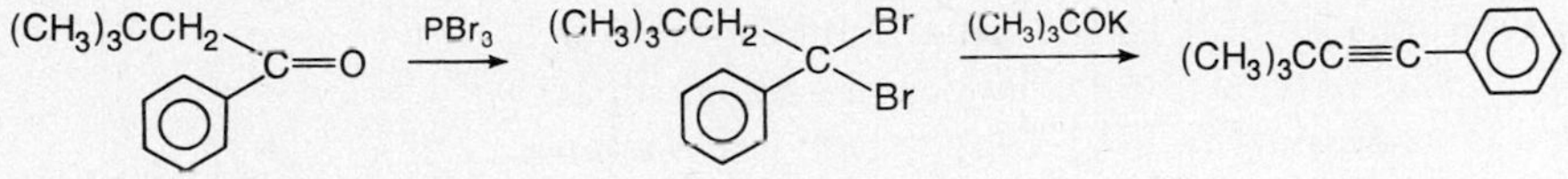

$$(CH_3)_3CCH_2(C_6H_5)C{=}O \xrightarrow{PBr_3} (CH_3)_3CCH_2(C_6H_5)CBr_2 \xrightarrow{(CH_3)_3COK} (CH_3)_3CC{\equiv}C{-}C_6H_5$$

Fig. 27.2. Preparation of 3,3-dimethyl-1-phenyl-1-butyne from neopentyl phenyl ketone.

reaction tends to be less regioselective than the dehydrohalogenation of 1,2-dihalides, giving a mixture of isomers if the 1,1-dihalide is flanked on both sides by methylene groups.

27.2 Preparation of Alkynes from Compounds Already Containing C≡C Bonds

Acetylene and monosubstituted acetylenes react with various metals or with strong bases such as sodium hydride (NaH) and sodium amide ($NaNH_2$) to form the metal derivatives of the acetylene:

$$2\,RC{\equiv}C{-}H + 2\,Na \longrightarrow RC{\equiv}C{-}Na + H_2$$

$$RC{\equiv}C{-}H + NaNH_2 \longrightarrow RC{\equiv}C{-}Na + NH_3$$

$$RC{\equiv}C{-}H + R'MgX \longrightarrow RC{\equiv}C{-}MgX + R'H$$

PROBLEM 27-1

Write two resonance structures for the acetylide anion.

Although the acetylene metallics are less nucleophilic than those organometallic compounds in which the metal is associated with an aliphatic carbon (*i.e.*, RMgX) or an aromatic carbon (*i.e.*, ArMgX), they are still rather reactive nucleophiles and can engage in substitution and addition reactions. By means of these reactions, acetylenic moieties can be incorporated into a wide variety of carbon frameworks, two representative examples of which are shown in Fig. 27.3.

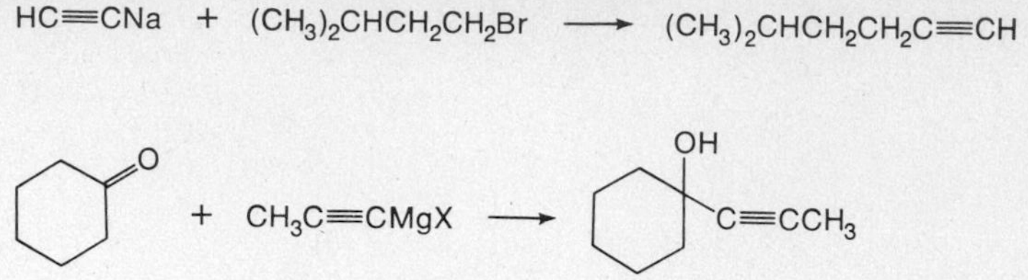

$$HC{\equiv}CNa + (CH_3)_2CHCH_2CH_2Br \longrightarrow (CH_3)_2CHCH_2CH_2C{\equiv}CH$$

$$\text{cyclohexanone} + CH_3C{\equiv}CMgX \longrightarrow \text{1-(propynyl)cyclohexanol (OH, } C{\equiv}CCH_3)$$

Fig. 27.3. Preparation of substituted alkynes via S_N2 displacement and nucleophilic additions with metal derivatives of acetylenes.

PROBLEM 27-2

What would you expect to be the major product from the reaction of $CH_3C{\equiv}CNa$ and *t*-butyl bromide?

The alkylation of metal acetylides, subject to the usual limitations of S_N2 reactions, is restricted to primary and secondary alkyl halides. To attach an aryl group directly to a C≡C bond it is necessary to use a copper acetylide prepared by the action of cuprous ammonium hydroxide on an acetylene (see Section 8.2C), and an aryl iodide. The mechanism of this reaction is not yet well understood.

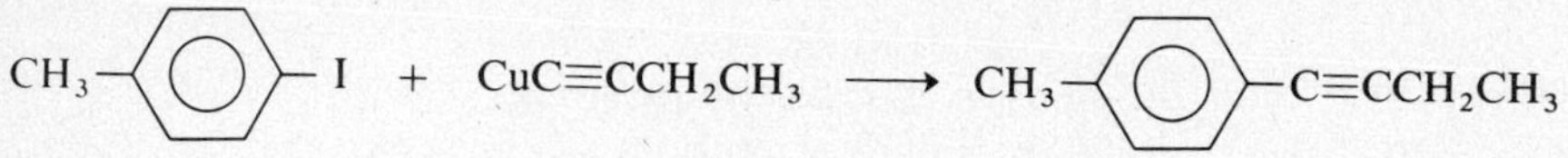

$$CH_3{-}C_6H_4{-}I + CuC{\equiv}CCH_2CH_3 \longrightarrow CH_3{-}C_6H_4{-}C{\equiv}CCH_2CH_3$$

REACTIONS OF ALKYNES

27.3 *Reactions of the Acetylenic* C—H *Bond* (≡C—H)

As mentioned in Section 27.2, acetylene and monosubstituted acetylenes can be converted to derivatives containing metals above hydrogen in the emf series (*e.g.*, Na, K, Li) as well as by metals below hydrogen in the emf series (*e.g.*, Cu, Ag). The first group of acetylides are thermally stable but are converted back to the parent acetylene in the presence of protic compounds such as water and alcohols. The second group of acetylides, on the other hand, are thermally unstable (explosive in the dry state) but are unreactive toward protic compounds; in fact, they can be prepared in aqueous solution from which they precipitate as insoluble substances. The utility of the first group of acetylides in synthesis reactions derives from the nucleophilic character of the acetylenic portion and the ability to engage in S_N2 displacements and nucleophilic additions (see Section 27.2). The utility of the second group of acetylides in synthesis is exemplified by their use in preparing arylalkynes (see Section 27.2) and their use in coupling reactions. Treatment of copper acetylides with potassium ferricyanide, for example, results in the attachment of a pair of acetylenic moieties to produce a conjugated diyne:

$$RC{\equiv}CH \xrightarrow{Cu^+} RC{\equiv}C{-}Cu \xrightarrow{K_3Fe(CN)_6} RC{\equiv}C{-}C{\equiv}CR$$

A more convenient coupling procedure involves treating the acetylene itself with a mixture of a cuprous compound, an amine such as pyridine, and air.

Oxidative coupling takes place, probably via the copper acetylide. This reaction has produced a number of interesting polyunsaturated compounds, including the 18-carbon cyclic compounds shown in Fig. 27.4. The initially formed

$$2\ C_4H_9C{\equiv}CH + \tfrac{1}{2}O_2 \xrightarrow[NH_3]{Cu_2Cl_2} C_4H_9C{\equiv}C{-}C{\equiv}CC_4H_9 + H_2O$$

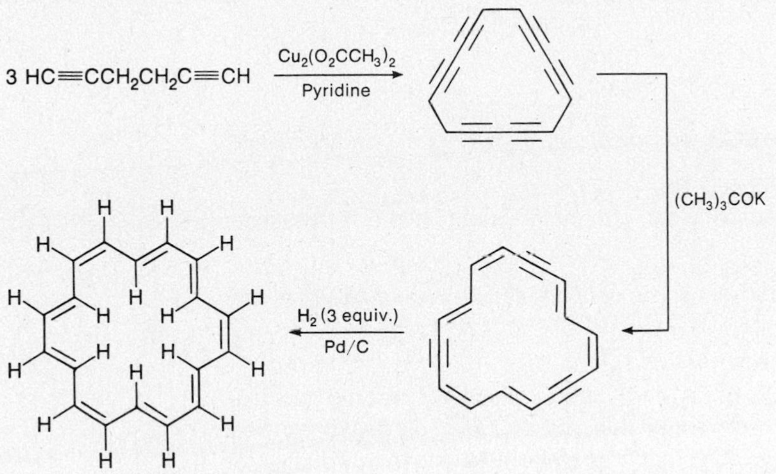

$$3\ HC{\equiv}CCH_2CH_2C{\equiv}CH \xrightarrow[Pyridine]{Cu_2(O_2CCH_3)_2}$$

$(CH_3)_3COK$

H_2 (3 equiv.)
Pd/C

[18]-Annulene

Fig. 27.4. Oxidative coupling of terminal alkynes and the synthesis of [18]-annulene.

octadeca-1,3,7,9,13,15-hexayne can be isomerized to a completely conjugated enyne, which, in turn, can be selectively reduced at the triple bonds to yield octadeca-1,3,5,7,9,11,13,15,17-nonaene. This compound is called an [18]-annulene and is particularly interesting because it is cyclic, planar, and contains 18 electrons in the π system, thereby conforming to the Hückel $(4n + 2)$ rule. As these features predict, it does, indeed, possess certain properties characteristic of an aromatic compound.

PROBLEM 27-3

[18]-Annulene shows two nmr resonances, one at δ −1.8 for the hydrogens on the inside of the ring and one at δ 8.9 for the hydrogens on the outside of the ring. Is this commensurate with an aromatic structure?

Reductive coupling of alkynes by the action of dialkylaluminum hydrides is discussed in Section 25.11 as a means for synthesizing conjugated dienes (see Figs. 25.13 and 25.14).

27.4a. THREE-CARBON TAUTOMERISM. In the presence of base under rather strenuous conditions, alkynes containing methylene groups adjacent to the $C \equiv C$ bond undergo three-carbon tautomeric shifts,* first to the allene and then to the conjugated diene.

$$CH_3CH_2CH_2C \equiv C{-}H \xrightarrow[\text{Base}]{} CH_3CH_2CH{=}C{=}CH_2 \xrightarrow[\text{Base}]{} CH_3CH{=}CH{-}CH{=}CH_2$$

Under the reaction conditions of alkylation and nucleophilic addition, however, rearrangements of this sort usually do not take place.

PROBLEM 27-4

If optically active $CH_3CH_2\underset{\displaystyle CH_3}{\underset{|}{C}}HC \equiv CCH_3$ is isomerized to the corresponding allene, $CH_3CH_2\underset{\displaystyle CH_3}{\underset{|}{C}}{=}C{=}CHCH_3$, can an optically active material be obtained? If the allene, in turn, is isomerized to a conjugated diene, can optically active material be obtained?

27.4b. ELECTROPHILIC ADDITION. Alkynes, like alkenes, undergo electrophilic addition with *anti* stereoselectivity and, in the case of unsymmetrical addends, with Markovnikov regioselectivity, as illustrated by the additions of hydrogen chloride and bromine (Fig. 27.5).

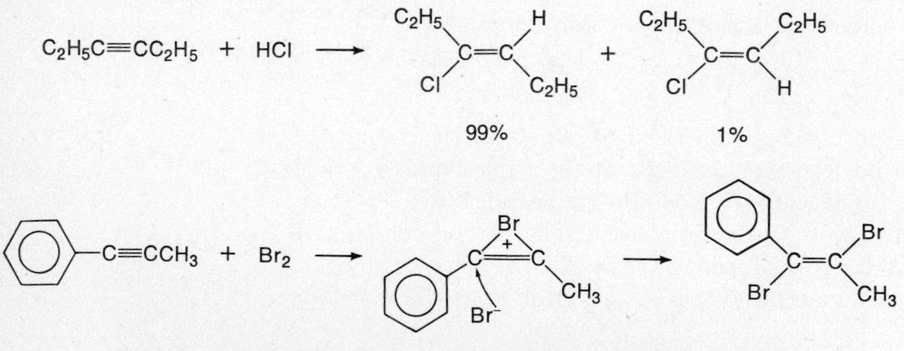

$$C_2H_5C \equiv CC_2H_5 + HCl \longrightarrow \underset{99\%}{(C_2H_5)(Cl)C{=}C(H)(C_2H_5)} + \underset{1\%}{(C_2H_5)(Cl)C{=}C(C_2H_5)(H)}$$

$$C_6H_5{-}C \equiv CCH_3 + Br_2 \longrightarrow [\text{bromonium ion}] + Br^- \longrightarrow (C_6H_5)(Br)C{=}C(Br)(CH_3)$$

Fig. 27.5. Electrophilic addition of halogen acids and halogens to alkynes.

When water is added to a $C \equiv C$ bond, the initially formed enol rearranges to the more stable aldehyde or ketone, affording methods for the synthesis of these compounds. The regioselectivity of the reaction is determined by the method

---

* Tautomeric systems can be described by the general equation

$$H{-}A{-}B{=}C \rightleftharpoons A{=}B{-}C{-}H$$

In the case of keto-enol tautomerism, atoms A and B are carbon and atom C is oxygen. When atoms A, B, and C are all carbon, the system is referred to as a "three-carbon tautomeric system." Numerous other tautomeric systems are also possible, depending on the identity of the atoms A, B, and C.

chosen for adding the elements of water. In the presence of aqueous mineral acid and mercuric salts (added as catalysts), the addition follows Markovnikov regioselectivity, and the carbonyl oxygen ends up on the more highly substituted carbon of the alkyne. By the hydroboration-oxidation sequence, on the other hand, the reaction follows overall anti-Markovnikov regioselectivity, and the carbonyl oxygen ends up on the less highly substituted carbon of the alkyne:

$$RC{\equiv}CH \xrightarrow{H_3O^+,\ Hg^{++}} RC(OH){=}CH_2 \longrightarrow RC(=O)CH_3$$

$$RC{\equiv}CH \xrightarrow{BH_3} (RCH{=}CH)_3B \xrightarrow{H_2O_2/OH^-} RCH{=}CHOH \rightleftharpoons RCH_2CHO$$

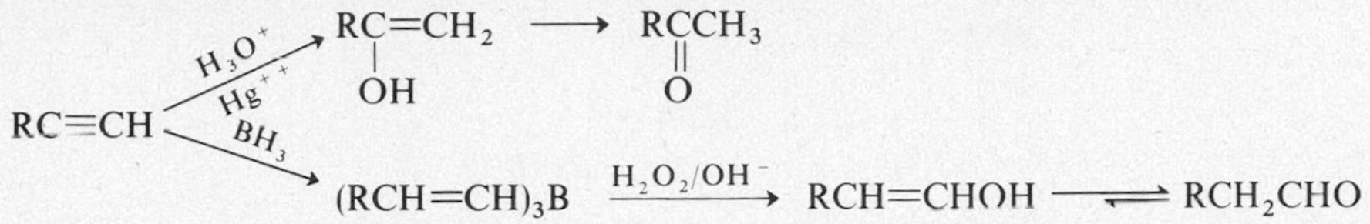

PROBLEM 27-5

Suggest a mechanism for the sulfuric acid catalyzed rearrangement of 1,1-diphenyl-2-propyn-1-ol to 3,3-diphenyl-2-propenal.

The function of mercuric salts in the direct hydration of alkynes probably is to induce a solvomercuration reaction (see Fig. 26.12) to yield the mercury derivative of the ketone (or in the case of acetylene itself, the aldehyde) which then undergoes protonolysis in the acidic solution (Fig. 27.6).

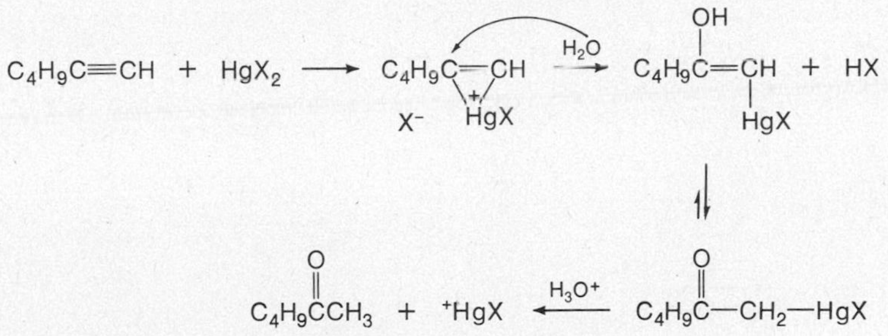

$$C_4H_9C{\equiv}CH + HgX_2 \longrightarrow C_4H_9C{=}CH\ (\text{bridged } \overset{+}{Hg}X)\ X^- \xrightarrow{H_2O} C_4H_9C(OH){=}CH(HgX) + HX$$

$$C_4H_9C(OH){=}CH(HgX) \rightleftharpoons C_4H_9C(=O){-}CH_2{-}HgX$$

$$C_4H_9C(=O){-}CH_2{-}HgX \xrightarrow{H_3O^+} C_4H_9C(=O)CH_3 + {}^+HgX$$

Fig. 27.6. Direct hydration of alkynes with mercury salt catalysis.

In the hydroboration-oxidation sequence for effecting hydration, a bulky dialkylborane such as dicyclohexylborane is used to prevent more than one borane from adding to the C≡C bond (Fig. 27.7). Hydration of an internal C≡C bond by either procedure usually results in the formation of a mixture of ketones.

Perhaps surprisingly, the C≡C bond is a weaker Lewis base than the C=C bond and, therefore, is somewhat less susceptible to electrophilic attack, as illustrated by the selective reaction of perbenzoic acid at the double bonds of deca-2,6-diene-4,6-diyne:

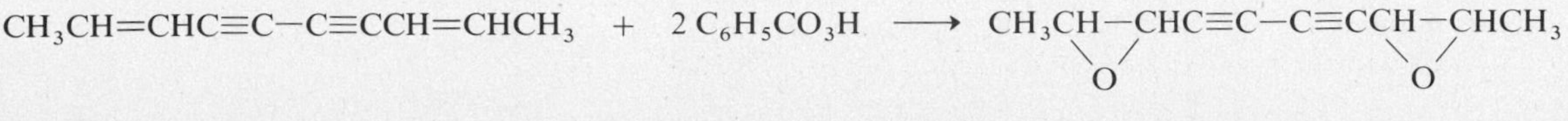

$$CH_3CH{=}CHC{\equiv}C{-}C{\equiv}CCH{=}CHCH_3 + 2\,C_6H_5CO_3H \longrightarrow CH_3\underset{\diagdown O \diagup}{CH{-}CH}C{\equiv}C{-}C{\equiv}C\underset{\diagdown O \diagup}{CH{-}CH}CH_3$$

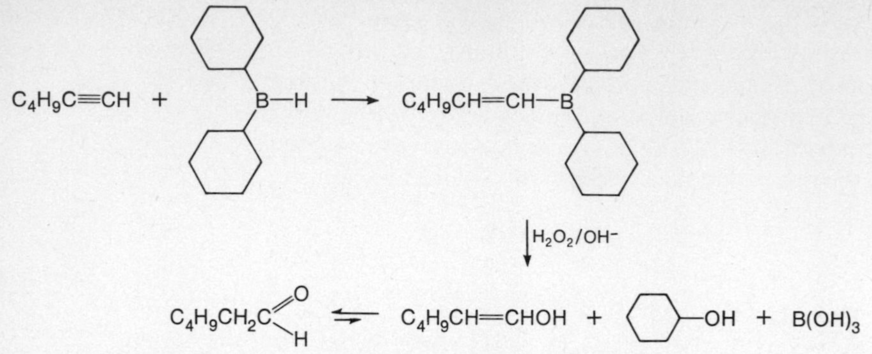

$C_4H_9C{\equiv}CH$ + (cyclohexyl)$_2$B—H → $C_4H_9CH{=}CH{-}B$(cyclohexyl)$_2$

↓ H_2O_2/OH^-

$C_4H_9CH_2CHO$ ⇌ $C_4H_9CH{=}CHOH$ + cyclohexyl—OH + $B(OH)_3$

Fig. 27.7. Hydration of alkynes via hydroboration-oxidation sequence.

27.4c. RADICAL ADDITION. The same types of reagents that add to alkenes via radical processes also add to alkynes. For example, phenylmethylthiol or hydrogen bromide in the presence of peroxide initiators both add in a predominantly *syn* fashion to give the products of anti-Markovnikov regioselectivity (Fig. 27.8).

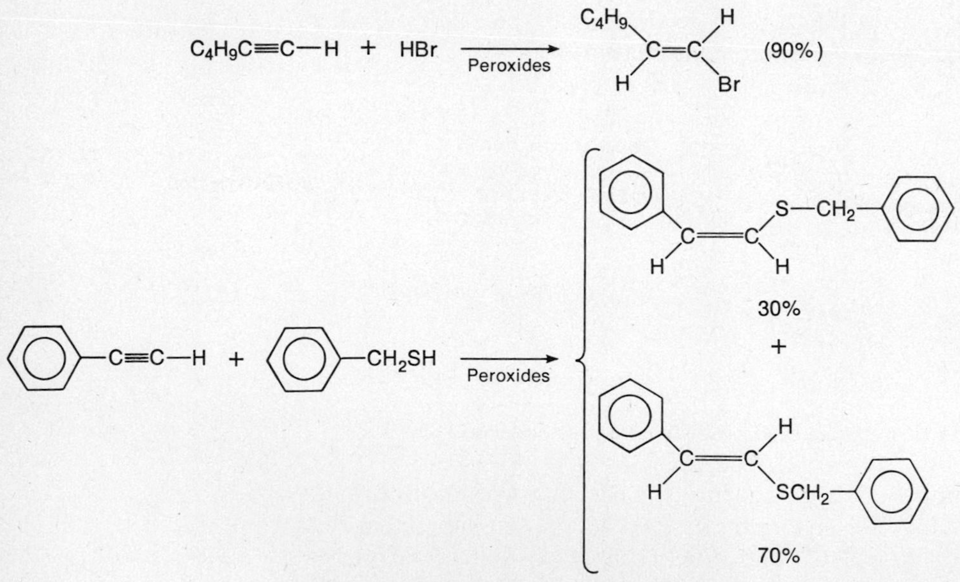

$C_4H_9C{\equiv}C{-}H$ + HBr —Peroxides→ (C_4H_9)(H)C=C(H)(Br) (90%)

$C_6H_5C{\equiv}C{-}H$ + $C_6H_5CH_2SH$ —Peroxides→ (C_6H_5)(H)C=C(H)($SCH_2C_6H_5$) 30% + (C_6H_5)(H)C=C(H)($SCH_2C_6H_5$) 70%

Fig. 27.8. Radical addition reactions of alkynes.

PROBLEM 27-6

Outline the mechanism of the benzoyl peroxide-induced addition of phenylmethylthiol to phenylethene.

27.4d. NUCLEOPHILIC ADDITION. As a consequence of the greater stability of a vinyl carbanion as compared with an alkyl carbanion, C≡C bonds are more susceptible to nucleophilic addition than C=C bonds. A particularly interesting example of this is the reaction between pyrrolidone and acetylene in the presence of potassium hydroxide to yield N-vinylpyrrolidone:

$$\text{pyrrolidone (N–H)} \underset{}{\overset{OH^-}{\rightleftharpoons}} \text{N}^- \text{anion} \xrightarrow{HC\equiv CH} \text{N–CH=}\bar{\text{C}}\text{H} \xrightarrow{H_2O} \text{N–CH=CH}_2 + OH^-$$

Polymerization of N-vinylpyrrolidone in a manner analogous to the polymerization of styrene yields polyvinylpyrrolidone (PVP). a material that finds use in medicine as a blood plasma substitute.

27.4e. OXIDATION OF ALKYNES. Although alkynes are somewhat less susceptible to oxidation than alkenes, as illustrated by the example on p. 815, they are, nevertheless, readily cleaved to the corresponding carboxylic acids by strongly basic potassium permanganate, by sodium dichromate and sulfuric acid, and by ozone. Although we might anticipate that the products of intermediate oxidation states could be isolated from the ozonolysis of alkynes, such is not the case, for the intermediates are more rapidly oxidized than the starting alkyne:

$$RC\equiv CR' \xrightarrow{KMnO_4/NaOH,\ Na_2Cr_2O_7/H_2SO_4,\ \text{or}\ O_3} RCO_2H + R'CO_2H$$

27.5 Cycloaddition Reactions of Alkynes

As already illustrated in the cantharidin synthesis (see Fig. 26.27), alkynes carrying electron-withdrawing groups can function as dienophiles in the Diels Alder reaction to yield cyclohexadienes:

$$CH_2{=}CH{-}CH{=}CH_2 + HC\equiv C\overset{O}{\overset{\|}{C}}CH_3 \longrightarrow \text{1-acetyl-1,4-cyclohexadiene}$$

Under the influence of various metal catalysts, acetylene and substituted acetylenes cyclotrimerize and cyclotetramerize. Cobalt octacarbonyl catalyst induces cyclotrimerization; acetylene yields benzene, monosubstituted acetylenes yield mixtures of trialkylbenzenes, and disubstituted acetylenes yield hexasubstituted benzenes. The reactions do not follow a concerted pathway but proceed via formation of a cobalt-alkyne complex in which alkyne units become bonded to each other, one at a time, in a sequential fashion, until the cycle is formed. Oxidative decomposition of the complex (*e.g.*, with bromine as the oxidant) releases the aromatic compound and completes the process (Fig. 27.9). The formation of 1,2,4-tri-*t*-butylbenzene by this process is particularly interesting because of the strain that exists in this molecule as a result of the proximity of the very bulky *t*-butyl groups (estimated to be approximately 17 kcal/mole of strain energy).

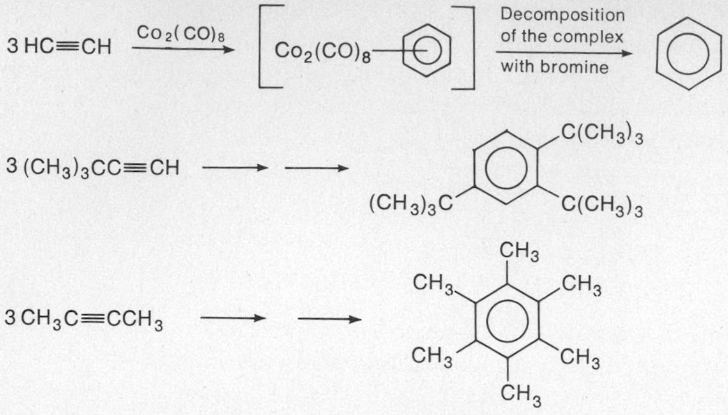

$$3\,HC{\equiv}CH \xrightarrow{Co_2(CO)_8} [Co_2(CO)_8\text{—}C_6H_6] \xrightarrow[\text{with bromine}]{\text{Decomposition of the complex}} C_6H_6$$

$$3\,(CH_3)_3CC{\equiv}CH \longrightarrow \longrightarrow \text{1,2,4-tri-}tert\text{-butylbenzene}$$

$$3\,CH_3C{\equiv}CCH_3 \longrightarrow \longrightarrow \text{hexamethylbenzene}$$

Fig. 27.9. Cyclotrimerization of alkynes.

PROBLEM 27-7

What would you predict to be the product of trimerization of cyclooctyne?

Cyclotetramerization of acetylene to cyclooctatetraene can be achieved by using nickel cyanide as the catalyst:

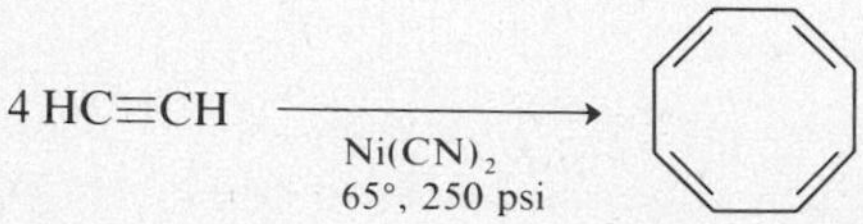

$$4\,HC{\equiv}CH \xrightarrow[65°,\ 250\ \text{psi}]{Ni(CN)_2} \text{cyclooctatetraene}$$

Cyclooctatetraene has an interesting history. It was first synthesized in 1911 by Richard Willstätter, who used a scarce, naturally occurring alkaloid called pseudopelletierine as the starting material. The final product was obtained only in small yield and was characterized simply by showing that it absorbed 4 mole-equivalents of hydrogen and yielded cyclooctane. Then, for the ensuing 35 years, chemists indulged in the game of speculating whether Willstätter had or had not succeeded in synthesizing cyclooctatetraene; it was thought by many that he had actually obtained something else. In 1948, however, all doubt was dispelled when Arthur Cope and coworkers at the Massachusetts Institute of Technology repeated the Willstätter synthesis. Using the more refined techniques of synthesis that had been developed over the intervening years, they showed that Willstätter was correct in every detail and that he had, indeed, synthesized cyclooctatetraene. At about the same time that the Cope synthesis was being completed, it was revealed that Reppe and his coworkers in Germany had developed the nickel cyanide process for cyclotetramerizing acetylene; thus, what had hitherto been one of the most esoteric of compounds, available only in very small quantity and with extraordinary labor, suddenly became available in large quantity with little effort.

Table 27.1 summarizes the methods of preparation and reactions of alkynes.

Table 27.1 PREPARATION AND REACTIONS OF ALKYNES

Type of Reaction	Starting Material	Reagents	Products	Comments
Hydration	CaC_2	H_2O	HC≡CH	Commercial method for production of acetylene
Dehydrohalogenation	X—C(H)—C(H)—X or —C(H)(H)—C(X)—X	$(CH_3)_3COK$ or other very strong bases	—C≡C—	
Alkylation	RC≡C—metal (metal = Na, K, Li, MgX)	R′X	RC≡CR′	S_N2 reaction—limited to primary and secondary alkyl halides
	RC≡CCu	ArI	ArC≡CR	
Anion formation	RC≡CH	Alkali metals or strong bases	RC≡C—metal (metal = Na, K, Li, MgX)	
		$Cu(NH_3)_2^+$ or $Ag(NH_3)_2^+$	RC≡CCu or RC≡CAg	Useful as a diagnostic test for presence of ≡C—H bond
Coupling	RC≡CH	$K_3Fe(CN)_6$ or $Cu_2(OAc)_2$, O_2, and pyridine	RC≡C—C≡CR	
Isomerization	—C(H)—C(H)—C≡C—	Basc	—C(H)—C=C=C(H)< and >C=C(H)—C=C(H)<	
Electrophilic addition	—C≡C—	HX	(X)C=C(H)	Markovnikov regioselectivity; *anti* stereoselectivity
	RC≡CH	H_2O, Hg^{++}	$RCOCH_3$	Markovnikov regioselectivity
	RC≡CH	R'_2BH	$RCH{=}CHBR'_2 \rightarrow RCH_2CHO$	Overall anti-Markovnikov addition of H_2O.
Radical addition	RC≡CH	H—Y (Y = Br, SR)	RCH=CHY	*syn* Stereoselectivity
Nucleophilic addition	HC≡CH	(1) Nu^-; (2) H^+	$H_2C{=}CH{-}Nu$	*anti* Stereoselectivity
Oxidation	RC≡CR′	$KMnO_4$ + NaOH, $Na_2Cr_2O_7$ + H_2SO_4, or O_3	$RCO_2H + R'CO_2H$	C≡C oxidizes less readily than C=C

Table 27.1—*cont.*

Type of Reaction	Starting Material	Reagents	Products	Comments
Cycloaddition	$-C\equiv C-$	$>C=C(-)-C(-)=C<$	cyclohexa-1,4-diene ring (C=C / C=C with two $>C<$)	Diels–Alder reaction; for effective reaction the alkyne must have an electron-withdrawing substituent (*e.g.*, $-C\equiv C-COCH_3$, $-C\equiv C-CN$).
Trimerization	$RC\equiv CR$	(1) $Co_2(CO)_8$; (2) oxidative decomposition of complex	C_6R_6 (hexa-R-benzene)	
Tetramerization	$HC\equiv CH$	$Ni(CN)_2$	cyclooctatetraene	

27.6 Study Guide

I. Methods of preparation of alkynes
  A. Preparation from compounds not containing C≡C bonds (dehydrohalogenation)
  B. Preparation from compounds already containing C≡C bonds
    1. Via S_N2 displacement on alkyl halides
    2. Via nucleophilic addition to carbonyl compounds

II. Reactions of alkynes
  A. Reaction of the ≡C—H bond
    1. Replacement of hydrogen by metals above hydrogen in emf series
    2. Replacement of hydrogen by metals below hydrogen in emf series
    3. Coupling reactions
  B. Reactions of the C≡C bond
    1. Three-carbon tautomerism: bond migration to afford allenes and dienes
    2. Electrophilic addition reactions
      (a) To yield vinyl compounds
      (b) To yield carbonyl compounds (addition of water)
        (1) Markovnikov addition via direct hydration
        (2) Anti-Markovnikov addition via hydroboration-oxidation
    3. Radical addition reactions
    4. Nucleophilic addition reactions
    5. Oxidation (chain cleavage)
  C. Cycloaddition reactions of alkynes
    1. Diels–Alder reaction (to yield cyclohexadienes)
    2. Trimerization reaction (to yield benzenes)
    3. Tetramerization reaction (to yield cyclooctatetraene)

REFERENCES

FIESER, L. F., and M. FIESER, *Advanced Organic Chemistry*, Reinhold Publishing Corp., New York, 1961, Chapter 6.

GARWOOD, R. F., in *Rodd's Chemistry of Carbon Compounds*, 2nd ed., Vol. IA, ed. by S. COFFEY, Elsevier Publishing Co., New York, 1964, pp. 447–77.

HARDIE, D. W. F., *Acetylene: Manufacture and Uses*, Oxford University Press, New York, 1965.

ROYALS, E. E., *Advanced Organic Chemistry*, Prentice-Hall, Inc., Englewood Cliffs, N.J., 1954, Chapter 6.

RUTLEDGE, T. F., *Acetylenes and Allenes: Addition, Cyclization, and Polymerization Reactions*, Reinhold Publishing Corp., New York, 1969.

———, *Acetylenic Compounds: Preparation and Substitution Reactions*, Reinhold Publishing Corp., New York, 1969.

VIEHE, H. G., ed., *Chemistry of Acetylenes*, Marcel Dekker, Inc., New York, 1969.

PROBLEMS

1. Draw the structure of the major product from each of the following reactions:

(a) $CH_3C(Br){=}CHCO_2C_2H_5 + (CH_3)_3COK \longrightarrow$

(b) $C_6H_5{-}CH_2CBr_2CH_2CH_3 + (CH_3)_3COK \longrightarrow$

(c) $CH_3C{\equiv}CNa + (CH_3)_3CI \longrightarrow$

(d) Cyclohexanone $+ C_6H_5{-}CH_2C{\equiv}CMgI \longrightarrow$

(e) $C_6H_5{-}C{\equiv}C{-}C_6H_4{-}OCH_3 + H_2O \xrightarrow[Hg^{++},\ H^+]{}$

(f) $C_6H_5{-}C{\equiv}CH + HBr \xrightarrow[\text{Peroxides}]{}$

(g) 1,3-Cyclohexadiene $+ HC{\equiv}CCO_2C_2H_5 \longrightarrow$

2. Outline, in as much detail as possible, the probable mechanism by which the following reactions occur.

(a) $C_6H_5(H)C{=}C(Br)CH_3 + RO^- \longrightarrow C_6H_5{-}C{\equiv}CCH_3$

(b) $HC{\equiv}CLi + (CH_3)_2C{=}O \longrightarrow (CH_3)_2C(OLi)C{\equiv}CH$

(c) $CH_3C{\equiv}CCH_3 + HBr \longrightarrow (CH_3)(H)C{=}C(Br)(CH_3)$

(d) $HC{\equiv}CH + H_2O \xrightarrow[Hg^{++},\ H^+]{} CH_3CHO$

(e) $C_6H_5{-}C{\equiv}CH + C_2H_5OH \xrightarrow[NaOH]{} C_6H_5{-}CH{=}CHOC_2H_5$

3. Specify the structures of the products in the following reaction schemes:

(a) $C_6H_5{-}C{\equiv}CH + NaH \xrightarrow{Ether} A \xrightarrow{(CH_3)_2CHCHO} B \xrightarrow[Pyridine]{(CH_3CO)_2O} C \xrightarrow{475^\circ} D$

(b) Cyclooctanone $+ LiAlH_4 \longrightarrow \xrightarrow{H_3O^+} E \xrightarrow[ether]{NaH} F \xrightarrow{(1)\ CS_2;\ (2)\ CH_3I} G \xrightarrow[\Delta]{(loss\ of\ O{=}C{=}S\ and\ CH_3SH)} H \xrightarrow{Br_2} I \xrightarrow[(CH_3)_3COH]{(CH_3)_3COK} J$

(c) $CH_3CH_2C{\equiv}CCH_2CH_3 \xrightarrow{HBr} K \xrightarrow{(CH_3)_2CuLi} L \xrightarrow{(1)\ BD_3;\ (2)\ H_2O_2/OH^-} M$

4. The triple bond of an alkyne, like the double bond of an alkene, provides a "handle" that facilitates the proof of structure of the alkyne.

(a) Each of four test tubes contains a pure sample that is known to be one of the following four isomeric alkynes:

o-$C_6H_4(CH_2C{\equiv}CH)(CH_2C{\equiv}CH)$ $\quad$ o-$C_6H_4(CH_2C{\equiv}CH)(C{\equiv}CCH_3)$ $\quad$ o-$C_6H_4(C{\equiv}CCH_3)(C{\equiv}CCH_3)$ $\quad$ o-$C_6H_4(C{\equiv}CH)(CH(CH_3)C{\equiv}CH)$

Using chemical and, if necessary, spectral methods, indicate how the contents of each of the four test tubes can be established.

(b) A compound isolated from a *Basidiomycetes* fungus possesses the molecular formula $C_{13}H_8$. It is optically inactive and absorbs 6 equivalents of hydrogen when treated with hydrogen and a palladium on charcoal catalyst to yield a $C_{13}H_{20}$ compound. The original compound, as well as the reduced compound, yield benzoic acid when oxidized with potassium permanganate. Suggest a structure compatible with these data.

(c) The polyacetylene, $CH_3C{\equiv}C{-}C{\equiv}C{-}C{\equiv}C{-}C{\equiv}C{-}C{\equiv}CCH{=}CH_2$, has been isolated from certain plants of the *Compositae* and *Umbelliferae* genera. Outline the procedure that you would use to establish the structure of this substance. Compounds obtained by conversion or degradation of the polyacetylene can be considered as known substances if they are among the group listed as available starting materials in Problem 7 or can be synthesized from these compounds.

5. Using the compounds shown as the only source of carbon, indicate the reaction sequence by means of which they can be converted to the indicated product:

(a) $C_6H_5{-}COCH_3 \longrightarrow C_6H_5{-}C{\equiv}C{-}C{\equiv}C{-}C_6H_5$

(b) $C_6H_5{-}CH{=}CH{-}C_6H_5 \longrightarrow C_6H_5{-}C{\equiv}C{-}C_6H_5$

(c) $C \longrightarrow CH_3CHO$

(d) $CH_3C{\equiv}CH \longrightarrow (CH_3)_2C(OH){-}CH{=}CHCH_3$

(e) $(CH_3)_2CHC{\equiv}CH \longrightarrow (CH_3)_2CHCH_2CHO$

6. Outline reasonable reaction schemes for the synthesis of each of the following, starting from any C_1 or C_2 compound, benzene, and necessary inorganic reagents:

(a) $C_6H_5{-}C{\equiv}CH$

(b) $CH_3{-}C(CH_3)_2{-}C{\equiv}CCH_2CH_3$

(c) $HC{\equiv}CCH_2CH_2CH_2CH_2CH_2C{\equiv}CH$

(d) $CH_3CH_2C{\equiv}C{-}C{\equiv}CCH_2CH_3$

(e) $CH_3(H)C{=}C(H){-}C{\equiv}C{-}CH(CH_3)_2$ (trans)

7. Available starting materials include (a) any alkanes, alkenes, alkynes, alkyl halides, alkanols, alkanals, alkanones, alkanoic acids or alkanoic acid derivatives containing four or fewer carbon atoms, (b) benzene, toluene, *o*-, *m*-, and *p*-xylene, and (c) any inorganic compounds and special reagents such as $(C_6H_5)_3P$, N-bromosuccinimide, etc. Starting with materials chosen from this group, outline syntheses for each of the following compounds:

(a) Benzene ring bearing CH_3 (top), $C{\equiv}CCH_3$ (adjacent), and CH_3 (para to the first CH_3)

(b) $CH_3CH_2CH_2CH_2(H)C{=}C(H)CH_2CH_2CH_3$ (trans: $CH_3CH_2CH_2CH_2$ and H on one carbon, H and $CH_2CH_2CH_3$ on the other)

(c) $C_6H_5C(OH)(CH_3)C{\equiv}CCH_2C_6H_5$

(d) $CH_3CH_2CH_2CH{=}CHBr$

(e) Benzene ring bearing two $C{\equiv}CCH_2Br$ groups (meta)

Chapter **28**

Carbonyl compounds: methods of synthesis

When we mean to build we first survey the plot.

WILLIAM SHAKESPEARE
Henry IV

As Chapter 13 indicates, the carbonyl group is the most frequently encountered function in organic compounds; if its chemical behavior is comprehended in detail, a significant amount of organic chemistry has been mastered. The carbonyl group confers a qualitative similarity on the physical and chemical properties of all the members comprising this family of compounds. Sometimes the quantitative differences between various members of the family are sufficiently great, however, that a division into distinct subdivisions is a justifiable convenience. Traditionally, aldehydes and ketones have been considered in one category, and carboxylic acids and carboxylic acid derivatives in another. To some extent, this division is employed in the present discussion. But, to emphasize the similarities rather than the differences between carbonyl compounds, the various types of carbonyl compounds are discussed as a single class to as great an extent as possible. In the present chapter, dealing with the methods for synthesizing carbonyl compounds, the discussion is subdivided into methods for the synthesis of aldehydes and ketones, carboxylic acids, and carboxylic acid derivatives. For the synthesis of aldehydes, ketones, and carboxylic acids, the same types of methods are generally employed, involving (a) replacement of C-hetero atoms by C—H (*i.e.*, reduction methods), (b) replacement of C—H by C—O (*i.e.*, oxidation methods), (c) oxidative cleavage of carbon–carbon bonds, (d) formation of carbon–carbon bonds, (e) water as a reactant, and (f) molecular rearrangement. Although the details of the application of these methods differ from one class to another, the fact that the same general procedures encompass all the methods for synthesizing these several types of carbonyl compounds emphasizes the similarity of the aldehydes, ketones, carboxylic acids, and carboxylic acid derivatives.

SYNTHESIS OF ALDEHYDES AND KETONES

An unusually large number of methods for synthesizing aldehydes and ketones have been devised. This is partly a reflection of their synthetic importance, which is due to their utility as starting materials in the preparation of other compounds, and partly a reflection of the fact that a variety of oxidative as well as reductive methods can be employed because aldehydes and ketones stand at the midpoint in the oxidation sequence of organic compounds. Only those methods of synthesis that have rather general applicability are discussed in this section. It should be realized, however, that numerous other methods also exist, which in particular instances may offer certain advantages over the better-known procedures. References containing information on these procedures are listed at the end of this chapter.

28.1 *Methods Involving Reduction of Carboxylic Acid Derivatives*

28.1a. SYNTHESIS OF ALDEHYDES. A variety of methods are available for converting carboxylic acid derivatives to aldehydes. One of the most frequently employed is the *Rosenmund reduction* (see Fig. 13.7), a reaction in which an acid chloride is reduced by molecular hydrogen in the presence of a catalyst. To avoid further reduction of the initially formed aldehyde, we use a palladium catalyst whose potency has been properly adjusted by "poisoning" with some sulfur compound (*e.g.*, tetramethylthiourea) is used. The major restriction to the Rosenmund reduction is the presence in the acid chloride of other functions that undergo catalytic reduction. The catalyst, however, is quite mild in its action, and ester groups, for example, are not affected (Fig. 28.1).

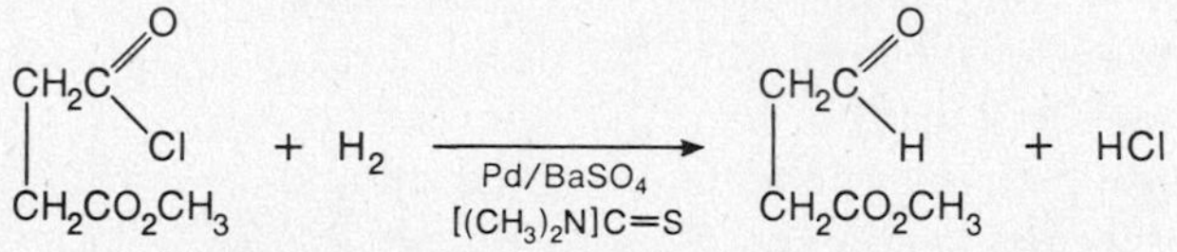

$$CH_3O_2CCH_2CH_2C(=O)Cl + H_2 \xrightarrow[[(CH_3)_2N]C=S]{Pd/BaSO_4} CH_3O_2CCH_2CH_2C(=O)H + HCl$$

Fig. 28.1. Rosenmund reduction of an acid chloride to an aldehyde.

With the advent of hydride reagents, a particularly useful reduction method for converting carboxylic acid derivatives to aldehydes appeared on the scene. Lithium aluminum trialkoxyhydrides, prepared from lithium aluminum hydride and three equivalents of an alcohol, for example, have the capacity for reducing acid chlorides, N,N-dimethylamides, or nitriles to the corresponding aldehyde without further reduction to the alcohol (Fig. 28.2).

PROBLEM 28-1

Why is $LiAl[OC(CH_3)_3]_3H$ better than $LiAl(OC_2H_5)_3H$ for the reduction of RCOCl to RCHO?

$$LiAlH_4 + 3\,ROH \longrightarrow LiAl(OR)_3H + 3\,H_2$$

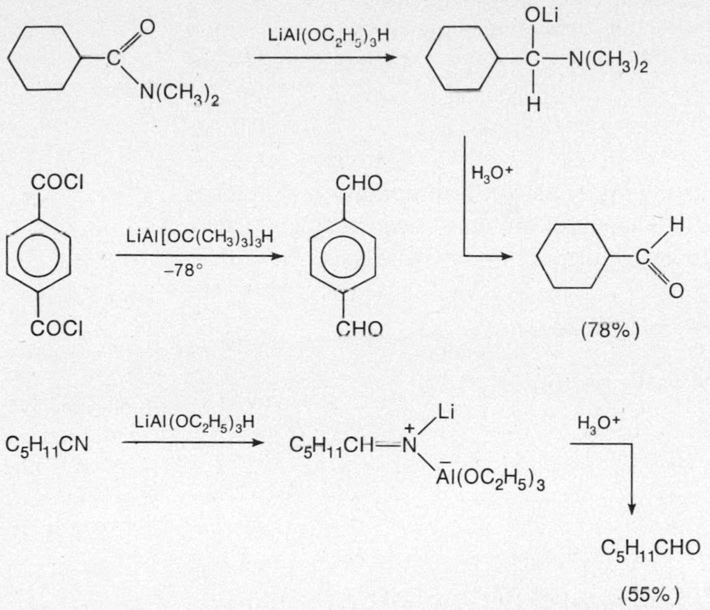

Fig. 28.2. Reduction of carboxylic acid derivatives via lithium trialkoxy hydride.

28.1b. SYNTHESIS OF KETONES. Substitution of the hetero atom moiety of a carboxylic acid derivative by a carbon framework can be accomplished by the use of organometallic reagents, which provide the nucleophilic carbon. Thus, acid chlorides react with Grignard reagents to form ketones. Under ordinary circumstances, however, the initially formed ketones react with a second equivalent of Grignard reagent to form alcohols. To prevent this, a number of means have been devised. For example, if acid chlorides are treated with Grignard reagents at low temperature (about −70°) in the presence of two equivalents of anhydrous ferric chloride, ketones can be obtained in moderate yield (Fig. 28.3). Alternatively, the Grignard reagent can be converted to the corresponding organocadmium reagent, which has sufficiently reduced reactivity to react with acid chlorides but *not* with ketones (Fig. 28.3).

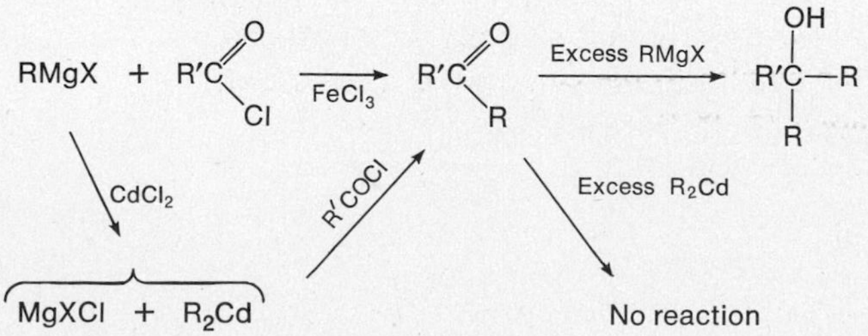

Fig. 28.3. Preparation of ketones from acid chlorides and organometallic reagents.

PROBLEM 28-2

The "inverse" method of Grignard addition involves adding the Grignard reagent dropwise to the acid chloride. Why is this procedure more likely to give good yields of ketone than the "normal" method in which the acid chloride is added to the Grignard reagent?

Superior to acid chlorides as starting materials for the preparation of ketones are nitriles. Nitriles react with one equivalent of Grignard reagent to give imine salts, which have little tendency to react further; upon hydrolysis, the imines produce the ketones (Fig. 28.4).

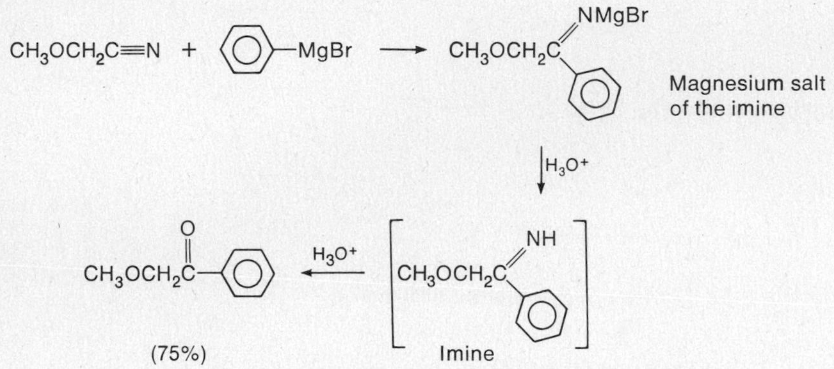

$CH_3OCH_2C{\equiv}N$ + C_6H_5–MgBr ⟶ $CH_3OCH_2C(=NMgBr)C_6H_5$ (Magnesium salt of the imine)

↓ H_3O^+

[$CH_3OCH_2C(=NH)C_6H_5$] (Imine) $\xrightarrow{H_3O^+}$ $CH_3OCH_2C(=O)C_6H_5$ (75%)

Fig. 28.4. Preparation of ketones from nitriles and Grignard reagents.

Even carboxylic acids themselves can be used with organometallic reagents to prepare ketones. With Grignard reagents, carboxylic acids react to form the halomagnesium salt of the acid, and the salt usually precipitates from solution and precludes further reaction. With organolithium reagents, however, the initially produced lithium carboxylate is sufficiently soluble to react with another equivalent of the organolithium reagent to yield a dilithium salt of a *gem*-diol. The dilithium salt resists further reaction and yields the corresponding ketone upon hydrolysis (Fig. 28.5).

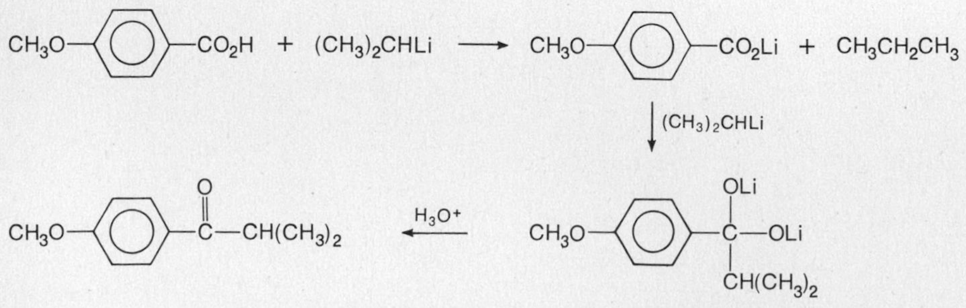

CH_3O–C_6H_4–CO_2H + $(CH_3)_2CHLi$ ⟶ CH_3O–C_6H_4–CO_2Li + $CH_3CH_2CH_3$

↓ $(CH_3)_2CHLi$

CH_3O–C_6H_4–$C(OLi)_2CH(CH_3)_2$ $\xrightarrow{H_3O^+}$ CH_3O–C_6H_4–$C(=O)$–$CH(CH_3)_2$

Fig. 28.5. Reaction of a carboxylic acid with an alkyllithium.

28.2a. OXIDATION OF ACTIVATED METHYL AND METHYLENE GROUPS. Methyl and methylene groups attached to unsaturated functions (*e.g.*, C=C—C—H, Ar—C—H, O=C—C—H) contain C—H bonds that are particularly susceptible to oxidation. Thus, in the presence of appropriate reagents, methyl groups and methylene groups attached to double bonds, benzene rings, or carbonyl functions can be selectively converted to an aldehyde or ketone function, respectively (Fig. 28.6).

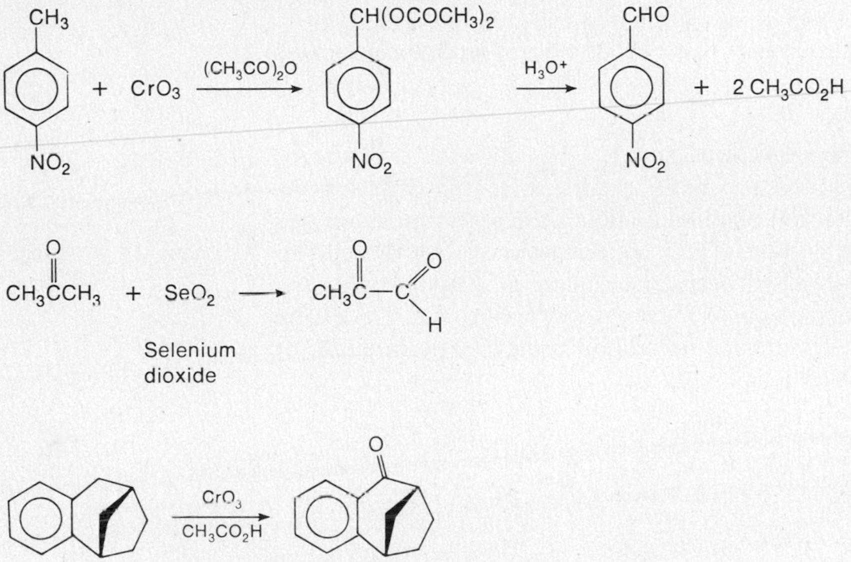

CH_3 NO_2 + CrO_3 $\xrightarrow{(CH_3CO)_2O}$ $CH(OCOCH_3)_2$ NO_2 $\xrightarrow{H_3O^+}$ CHO NO_2 + 2 CH_3CO_2H

CH_3COCH_3 + SeO_2 ⟶ CH_3COCHO

Selenium dioxide

$\xrightarrow[CH_3CO_2H]{CrO_3}$

Fig. 28.6. Oxidation of activated methyl and methylene groups.

28.2b. OXIDATION OF PRIMARY AND SECONDARY ALCOHOLS. The oxidation of primary alcohols to aldehydes and of secondary alcohols to ketones is discussed in Section 22.11c, where the use of chromate reagents, potassium ferrate, and the Oppenauer method are mentioned for the general case and manganese dioxide

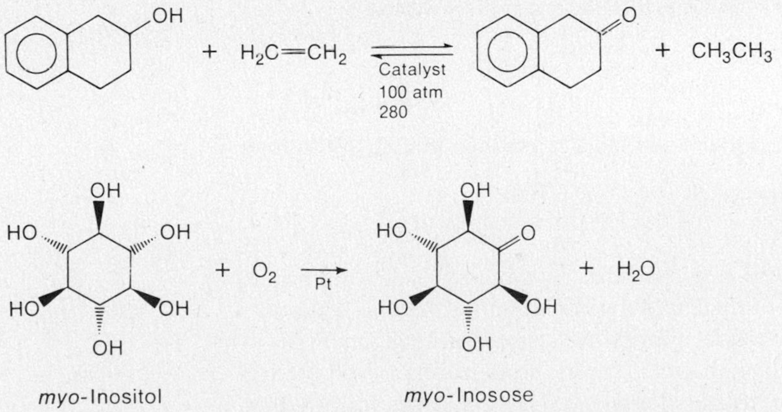

OH + $H_2C{=}CH_2$ ⇌ (Catalyst, 100 atm, 280) O + CH_3CH_3

myo-Inositol + O_2 $\xrightarrow{Pt}$ *myo*-Inosose + H_2O

Fig. 28.7. Oxidation of secondary alcohols to ketones by hydrogen acceptors.

for the special case of allyl and benzyl alcohols. In addition to these, many other oxidizing agents have been used in various special instances, such as ethylene (to form ethane) or oxygen (to form water) as hydrogen acceptors in hydrogen transfer processes (Fig. 28.7).

PROBLEM 28-3

Draw a pseudo-three-dimensional representation of the most stable conformation of inositol. Is the C—H bond involved in the oxidation *meso*-inositol (see Fig. 28.7) equatorial or axial? How many equivalent C—H bonds are present in the molecule?

28.2c. Oxidation of RCH_2—halogen and R_2CH—halogen compounds. Dimethyl sulfoxide, frequently used as a polar, aprotic solvent in organic reactions, is a reactive molecule that can oxidize certain alkyl halides and sulfonates to aldehydes in a situselective fashion. The reactions proceed with those compounds that contain one or more α hydrogens and that are capable of undergoing S_N2 reactions with dimethyl sulfoxide to form a sulfoxonium salt. Under the action of a base such as triethylamine, the sulfoxonium salt then decomposes to dimethyl sulfide and the aldehyde (Fig. 28.8).

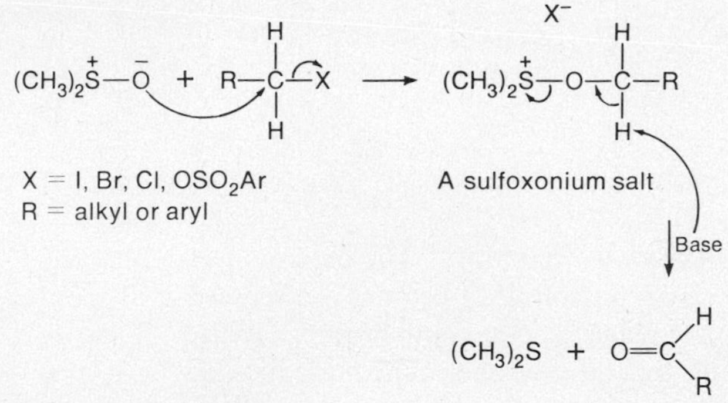

$$(CH_3)_2\overset{+}{S}—\overset{-}{O} + R—CH_2—X \longrightarrow (CH_3)_2\overset{+}{S}—O—CH_2—R \;\; X^-$$

X = I, Br, Cl, OSO_2Ar
R = alkyl or aryl

A sulfoxonium salt

$$\xrightarrow{\text{Base}} (CH_3)_2S + O{=}C(H)R$$

Fig. 28.8. Dimethyl sulfoxide oxidation of RCH_2X compounds to aldehydes.

PROBLEM 28-4

Why is the C—H bond in the sulfoxonium salt more labile than that in the starting compound RCH_2X?

Although primary alcohols are not oxidized by dimethyl sulfoxide alone, the addition of dicyclohexylcarbodiimide and pyridinium phosphate to a mixture of the alcohol and dimethyl sulfoxide affords the corresponding aldehyde in high yield (Fig. 28.9). The reaction probably involves the addition of the dimethyl sulfoxide, acting as a nucleophile, to the dicyclohexylcarbodiimide, followed by interaction of the resulting compound with the alcohol to form a sulfoxonium

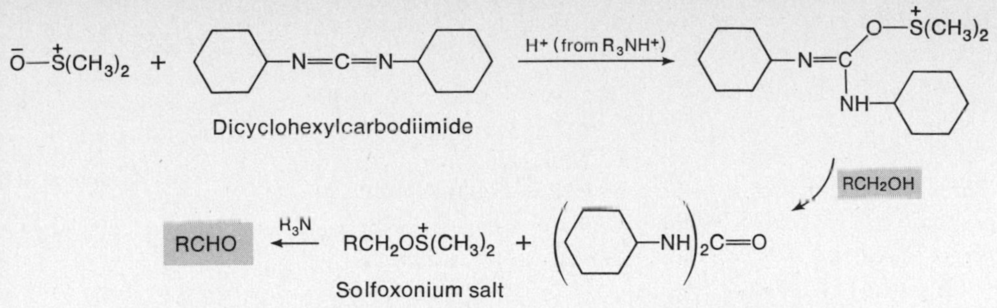

Dicyclohexylcarbodiimide

Solfoxonium salt

Fig. 28.9. Dimethyl sulfoxide oxidation of primary alcohols to aldehydes.

salt. In a fashion comparable to that shown in Fig. 28.8, the sulfoxonium salt breaks down to dimethyl sulfide and the aldehyde. This procedure has also been applied to the synthesis of ketones, as illustrated by the conversion of the alkaloid yohimbine to yohimbinone:

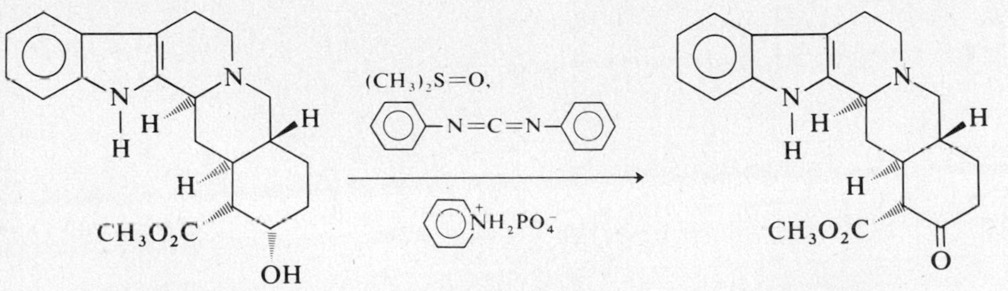

$(CH_3)_2S{=}O$, $C_6H_5N{=}C{=}NC_6H_5$, $C_5H_5\overset{+}{N}H_2PO_4^-$

Yohimbine **Yohimbinone**

28.2d. OXIDATION OF ALKYL NITRO COMPOUNDS (RCH_2NO_2 AND R_2CHNO_2). Nitroalkanes having an α-hydrogen react with base to form salts that can be converted to aldehydes or ketones by treatment with strong mineral acid (Nef reaction). Alternatively, if neither R group is hydrogen, potassium permanganate can be used to produce the corresponding ketones:

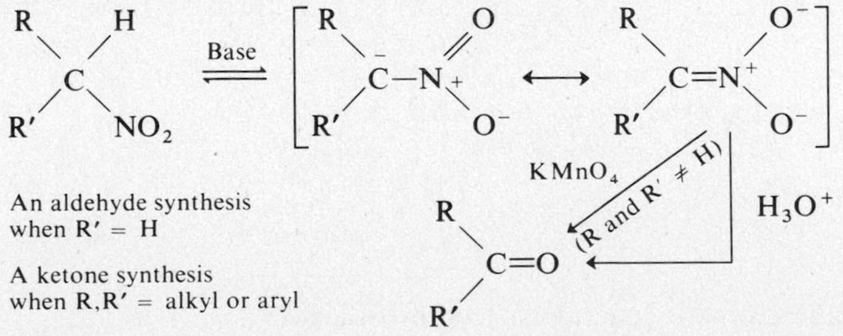

Base

$KMnO_4$ (R and R′ ≠ H)

H_3O^+

An aldehyde synthesis when R′ = H

A ketone synthesis when R,R′ = alkyl or aryl

Since nitromethane undergoes facile, base-induced reaction with aldehydes and ketones to form nitro alcohols, as discussed in Chapter 29, we can take advantage of this reaction, in tandem with the hydrolysis reaction just described, to extend the chain length of compounds of the sugar series. For example, the five-carbon sugar D-arabinose can be converted to a mixture of the six-carbon sugars D-glucose and D-mannose (Fig. 28.10).

PROBLEM 28-5

Would you expect D-glucose and D-mannose to be produced in equal amounts by the reaction sequence shown in Fig. 28.10? Explain.

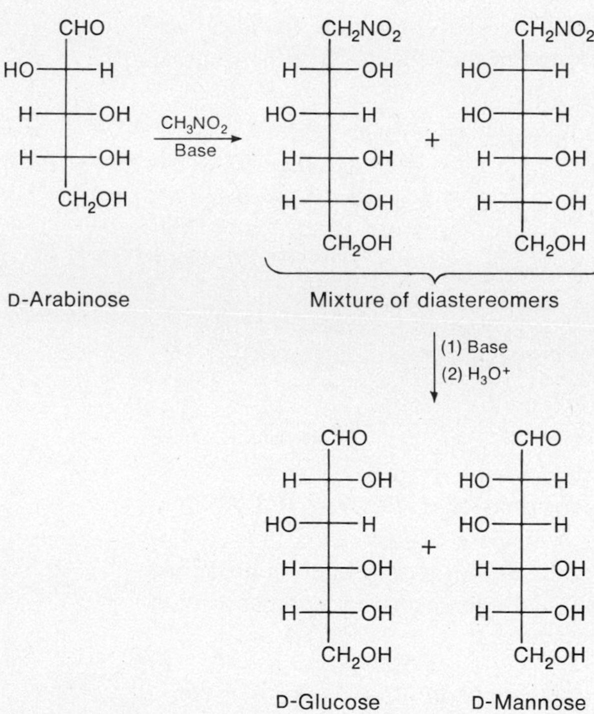

Fig. 28.10. Chain-lengthening method for sugars involving nitromethane condensation followed by conversion of the nitromethyl group to an aldehyde function.

28.2e. OXIDATION OF AMINES. Amines of the general structure R_2CHNR_2 are oxidized to ketones in warm, buffered solutions of potassium permanganate in aqueous *t*-butyl alcohol. When primary or secondary amines are used, imines are formed as intermediates; when tertiary amines are used, enamines are formed as intermediates. Both imines and enamines yield ketones on hydrolysis (Fig. 28.11).

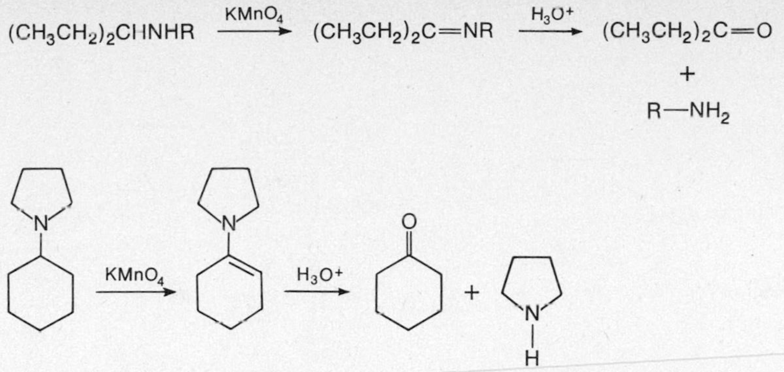

$$(CH_3CH_2)_2CHNHR \xrightarrow{KMnO_4} (CH_3CH_2)_2C{=}NR \xrightarrow{H_3O^+} (CH_3CH_2)_2C{=}O + R{-}NH_2$$

KMnO$_4$ → H$_3$O$^+$ → + N–H

Fig. 28.11. Formation of ketones by oxidation of alkylamines.

28.3 *Methods Involving Oxidation with Cleavage of Carbon–Carbon Bonds*

28.3a. OZONOLYSIS. Depending on the substitution at the C=C bond, the products of ozonolysis of alkenes are formaldehyde, aldehydes, and/or ketones (see pp. 788–9). Unless the alkene is symmetrically substituted, a mixture of carbonyl products is obtained, reducing the utility as a synthetic method. This difficulty is minimized, of course, if the products are easily separated, as is the case with terminal alkenes, which yield formaldehyde as one of the products (Fig. 28.12).

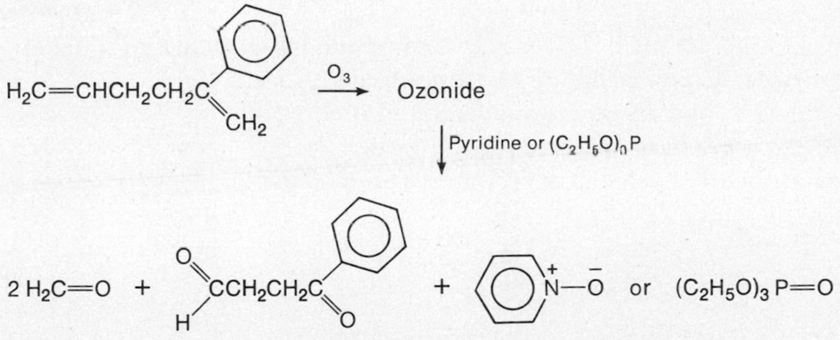

$H_2C{=}CHCH_2CH_2C(C_6H_5){=}CH_2$ $\xrightarrow{O_3}$ Ozonide

Pyridine or $(C_2H_5O)_3P$

$2\,H_2C{=}O$ + $HC(=O)CH_2CH_2C(=O)C_6H_5$ + $C_5H_5\overset{+}{N}{-}\overset{-}{O}$ or $(C_2H_5O)_3P{=}O$

Fig. 28.12. Ozonolysis of unsymmetrically substituted alkenes.

28.3b. GLYCOL CLEAVAGE. Depending on the extent of substitution of *vic*-diols, the products of cleavage with periodic acid or lead tetraacetate (see Section 22.13b) are aldehydes and/or ketones. As with ozonolysis, the synthetic utility is compromised in many instances because unsymmetrically substituted *vic*-diols yield mixtures of products. An interesting variation on the glycol cleavage that compresses the hydroxylation and cleavage reactions into a one-flask process starts with an alkene and involves the use of sodium metaperiodate (a source of periodic acid) and a small amount of osmium tetroxide. The osmium tetroxide hydroxylates the alkene (see Fig. 22.12), and the periodic acid not only cleaves the *vic*-diol but also oxidizes the reduced osmium-containing compound back to osmium tetroxide, thereby allowing osmium tetroxide to be used in catalytic amounts (Fig. 28.13).

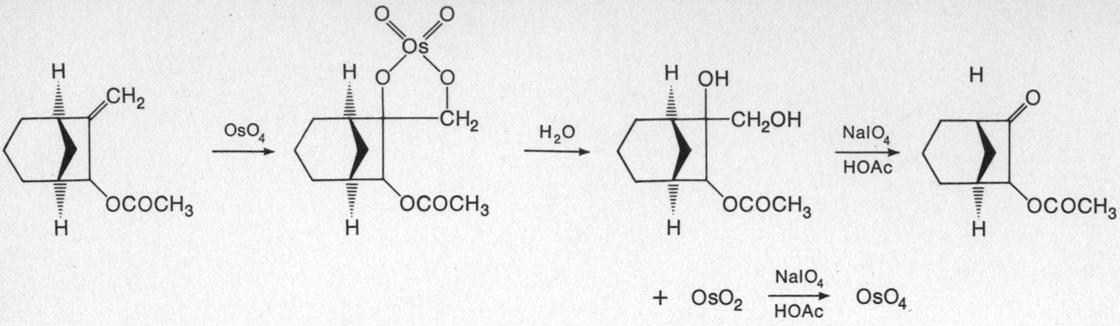

Fig. 28.13. Hydroxylation-glycol cleavage sequence, converting alkenes to aldehydes and ketones.

PROBLEM 28-6

What would be the expected product from the OsO_4—$NaIO_4$ oxidation of 1,2-dimethylenecyclohexane?

28.4 *Methods Involving the Formation of Carbon–Carbon Bonds*

28.4a. Acylation of Aromatic Compounds

Aldehyde Synthesis. Formyl chloride, HCOCl, is not a stable compound and cannot be used as such in the Friedel–Crafts acylation to produce aldehydes. If carbon monoxide, hydrogen chloride, and an aromatic compound are mixed with aluminum chloride, however, a reaction takes place as though formyl chloride were involved, and an aromatic aldehyde is produced (*Gattermann–Koch reaction*):

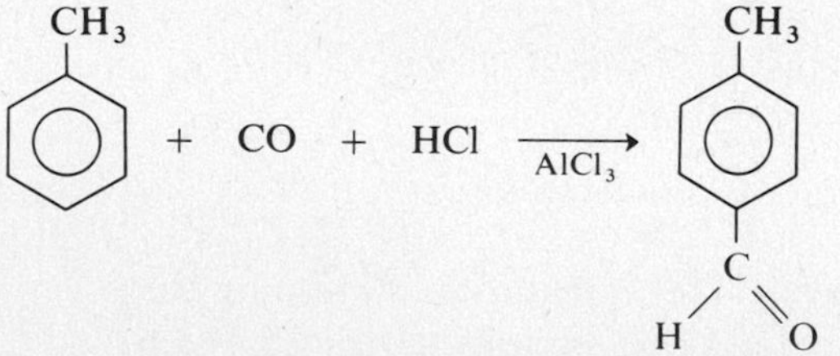

$$C_6H_5CH_3 + CO + HCl \xrightarrow{AlCl_3} p\text{-}CH_3C_6H_4CHO$$

PROBLEM 28-7

Formyl fluoride, HCOF, is a stable gas. Is there any reason that it could not be used in the Friedel–Crafts formylation reaction?

If the aromatic compound carries a strongly electron-releasing group, such as HO—, RO—, and R_2N—, the formylation by this procedure usually leads not to the aldehyde but to a triarylmethane derivative as the result of the reaction of two additional molecules of the aromatic compound with the initially formed aldehyde:

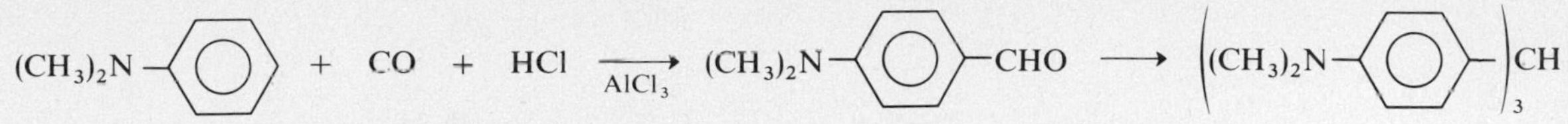

$$(CH_3)_2N-C_6H_4 + CO + HCl \xrightarrow[AlCl_3]{} (CH_3)_2N-C_6H_4-CHO \longrightarrow \left((CH_3)_2N-C_6H_4-\right)_3CH$$

This difficulty can be circumvented by using as the formylating agent (a) a mixture of hydrogen cyanide, hydrogen chloride, and zinc chloride catalyst (*Gattermann reaction*), or (b) formyl transfer compounds, such as N,N-dimethylformamide or N-methylformanilide in the presence of phosphorus oxychloride (*Vilsmeier reaction*) (Fig. 28.14). An interesting example of the Vilsmeier formylation,

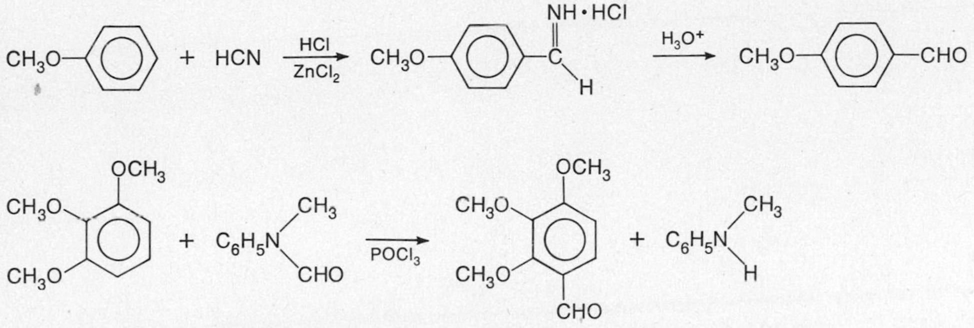

$$CH_3O-C_6H_5 + HCN \xrightarrow[ZnCl_2]{HCl} CH_3O-C_6H_4-C(=NH\cdot HCl)H \xrightarrow{H_3O^+} CH_3O-C_6H_4-CHO$$

$$(CH_3O)_3C_6H_3 + C_6H_5N(CH_3)CHO \xrightarrow{POCl_3} (CH_3O)_3C_6H_2-CHO + C_6H_5N(CH_3)H$$

Fig. 28.14. Introduction of formyl groups into strongly activated benzene rings.

illustrating its susceptibility to the activating influence of the aromatic ring, is seen in the monoformylation of 1-(3,4,5-trimethoxyphenyl)-3-(4-methoxyphenyl)-propane:

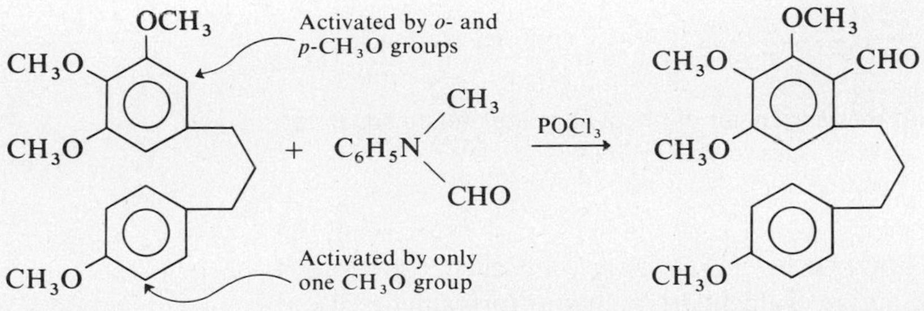

$$(CH_3O)_3C_6H_2-CH_2CH_2CH_2-C_6H_4-OCH_3 + C_6H_5N(CH_3)CHO \xrightarrow{POCl_3} (CH_3O)_3C_6H(CHO)-CH_2CH_2CH_2-C_6H_4-OCH_3$$

Activated by *o*- and *p*-CH_3O groups

Activated by only one CH_3O group

Ketone Synthesis. The Friedel–Crafts acylation reaction is illustrated in Figs. 8.13 and 13.9. Closely related to the Friedel–Crafts acylation reaction is a process known as the *Fries rearrangement*, whereby phenolic esters rearrange under the influence of Lewis acid catalysts such as aluminum chloride to *o*- and *p*-acylphenols. The mechanism of the reaction is not entirely certain; it appears that the rearrangement of the *o*-acylphenol is an intramolecular process, whereas that to the *p*-acylphenol is an intermolecular process in which transient acyl cations are formed (Fig. 28.15).

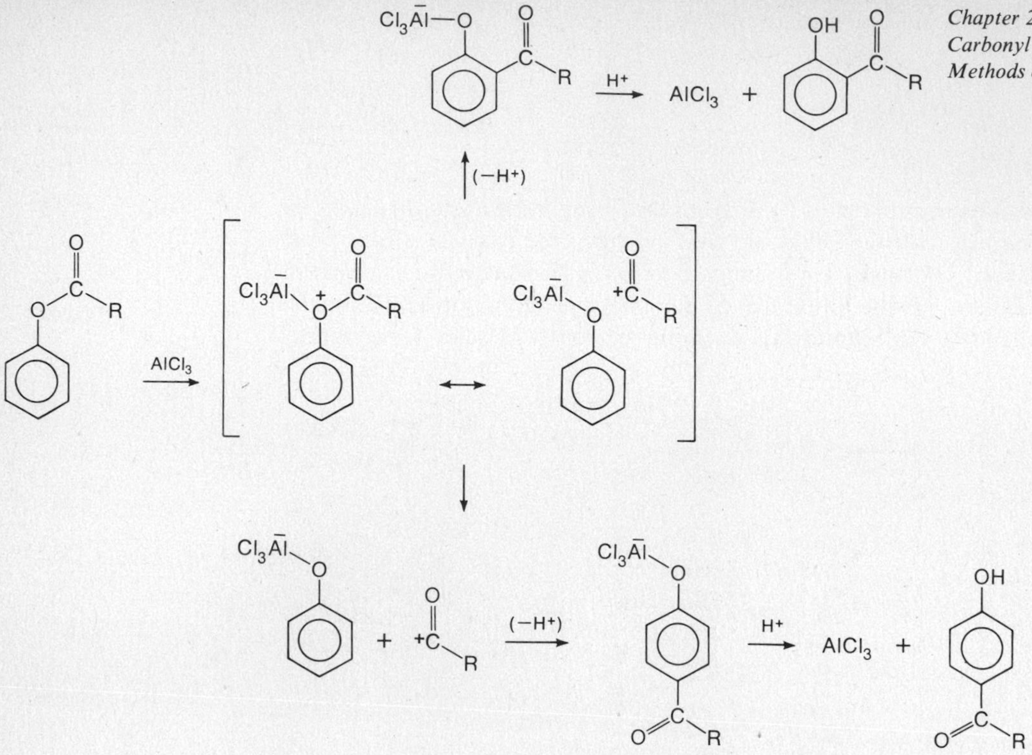

Fig. 28.15. The mechanism of the Fries rearrangement.

PROBLEM 28-8

Suggest an experiment that could test whether the *ortho*- and *para*-rearrangements are intramolecular or intermolecular.

PROBLEM 28-9

The isomeric products of the Fries rearrangement of phenyl acetate can be separated by fractional distillation. Which isomer boils lower? Why?

28.4b. ALKYLATION OF 1,3-DITHIANES. 1,3-Dithianes are cyclic thioacetals that can be obtained by the interaction of aldehydes (including formaldehyde) and propane-1,3-dithiol:

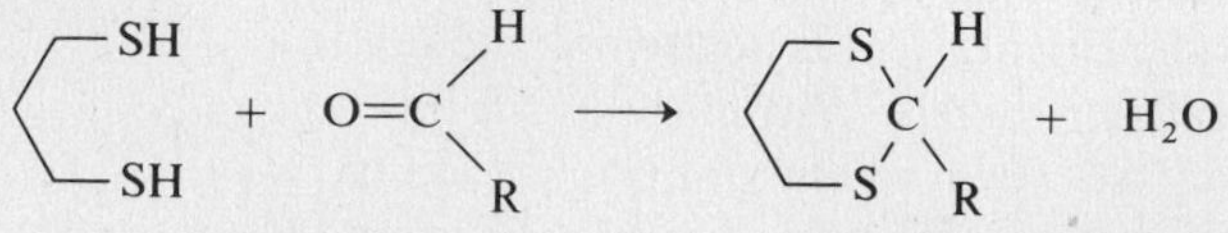

$$HS(CH_2)_3SH + O{=}C(H)R \longrightarrow \text{1,3-dithiane (2-H, 2-R)} + H_2O$$

These compounds are sufficiently acidic to react with strong bases such as butyllithium to form lithium derivatives, which, when treated with alkyl halides, undergo alkylation at what was originally the carbonyl carbon of the aldehyde.

Subsequent hydrolysis of the substituted dithianes, facilitated by mercuric chloride, then yields the aldehydes or ketones (Fig. 28.16).

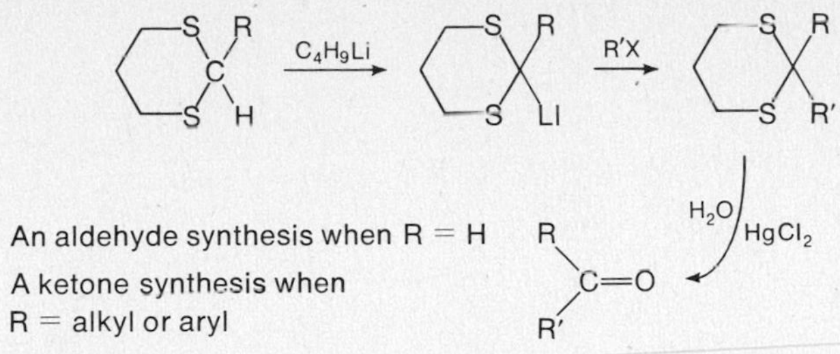

C_4H_9Li R′X H_2O $HgCl_2$

An aldehyde synthesis when R = H

A ketone synthesis when R = alkyl or aryl

Fig. 28.16. Preparation of aldehydes and ketones via alkylation and hydrolysis of 1,3-dithianes.

PROBLEM 28-10

Why is the C—H bond at the 2 position of a 1,3-dithiane more acidic than the C—H bond in methane?

PROBLEM 28-11

What purpose does the Hg^{++} serve in the hydrolysis of 1,3-dithianes?

28.4c. ALKYLATION OF ACETOACETIC ESTER. An important method, limited to the synthesis of ketones, involves the action of alkyl halides on the sodium salt of acetoacetic ester (ethyl 3-ketobutanoate) followed by hydrolysis under controlled conditions. The sodium salt of acetoacetic ester, obtained by the action of a strong base (*e.g.*, sodium ethoxide) on the ester, provides a carbon nucleophile that can engage in S_N2 reactions with alkyl halides, resulting in the introduction of an alkyl group onto the α-carbon of acetoacetic ester. This process can be repeated to introduce a second alkyl group, taking cognizance in both cases of the limitations of S_N2-type reactions. The alkylated acetoacetic esters are then hydrolyzed either in dilute acid solution or in dilute basic solution followed by acidification to yield the corresponding β-keto acids. Although these can be isolated at low temperatures, they are unstable at room temperature or slightly above and decompose to ketones and carbon dioxide (Fig. 28.17). In its simplest application, then, the acetoacetic ester method provides a way for synthesizing compounds of the general structures RCH_2COCH_3 and $RR'CHCOCH_3$. Its versatility can be extended if other β-keto esters are used as starting materials.

PROBLEM 28-12

In the preparation of 3,4-dimethyl-2-pentanone via the acetoacetic ester method, it is better to introduce the isopropyl group before the methyl group. Why?

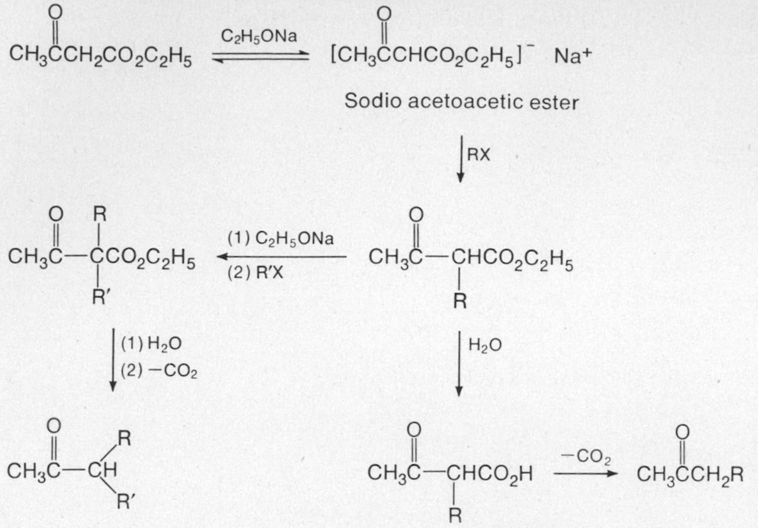

$$CH_3\overset{O}{\overset{\|}{C}}CH_2CO_2C_2H_5 \xrightleftharpoons{C_2H_5ONa} [CH_3\overset{O}{\overset{\|}{C}}CHCO_2C_2H_5]^- \ Na^+$$

Sodio acetoacetic ester

$$\downarrow RX$$

$$CH_3\overset{O}{\overset{\|}{C}}-\underset{R'}{\overset{R}{C}}CO_2C_2H_5 \xleftarrow[(2)\ R'X]{(1)\ C_2H_5ONa} CH_3\overset{O}{\overset{\|}{C}}-\underset{R}{C}HCO_2C_2H_5$$

$$\downarrow (1)\ H_2O,\ (2)\ -CO_2 \qquad\qquad \downarrow H_2O$$

$$CH_3\overset{O}{\overset{\|}{C}}-CH(R)(R') \qquad CH_3\overset{O}{\overset{\|}{C}}-\underset{R}{C}HCO_2H \xrightarrow{-CO_2} CH_3\overset{O}{\overset{\|}{C}}CH_2R$$

Fig. 28.17. Acetoacetic ester method for the preparation of ketones.

28.4d. FORMYLATION OF GRIGNARD REAGENTS. A useful method, limited to the synthesis of aldehydes, involves the action of Grignard reagents on ethyl orthoformate. The reaction takes place by displacement of one of the ethoxy groups to yield an acetal, from which the aldehyde can then be obtained by acid-catalyzed hydrolysis (Fig. 28.18).

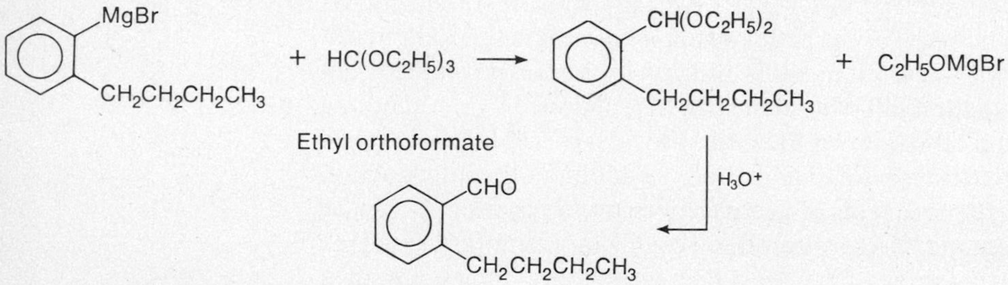

$$o\text{-}(CH_2CH_2CH_2CH_3)C_6H_4MgBr + HC(OC_2H_5)_3 \longrightarrow o\text{-}(CH_2CH_2CH_2CH_3)C_6H_4CH(OC_2H_5)_2 + C_2H_5OMgBr$$

Ethyl orthoformate

$$\xrightarrow{H_3O^+} o\text{-}(CH_2CH_2CH_2CH_3)C_6H_4CHO$$

Fig. 28.18. Synthesis of aldehydes via Grignard reagents and ethyl orthoformate followed by hydrolysis.

PROBLEM 28-13

The reaction of RMgX with $HC(OC_2H_5)_3$ is an example of a nucleophilic displacement reaction involving OC_2H_5 as the leaving group. The reaction proceeds in the case of the ortho ester, but not in the case of simple ethers? Write mechanisms for the reactions and explain the difference in reactivity.

28.5 Methods Involving the Addition of Water to Alkynes

28.5a. SYNTHESIS OF ALDEHYDES. To synthesize aldehydes by the hydration of alkynes, it is necessary to add the elements of water to the triple bond in anti-Markovnikov fashion. This is accomplished via the hydroboration-hydrogen peroxide oxidation procedure (see Fig. 27.7), leading to a vinyl alcohol which rearranges to the corresponding aldehyde (Fig. 28.19).

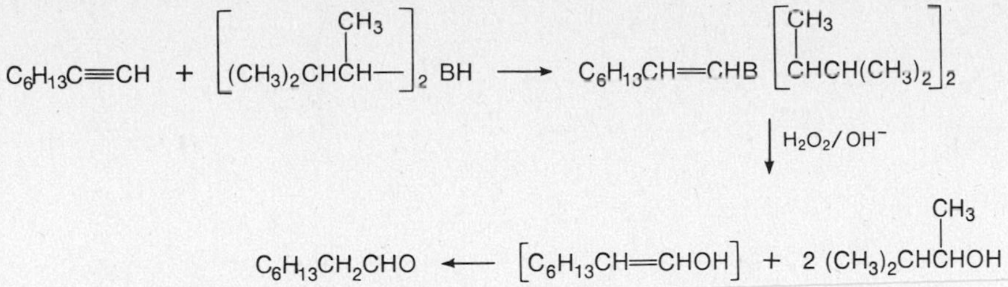

$$C_6H_{13}C{\equiv}CH + \left[(CH_3)_2CH\overset{CH_3}{\overset{|}{C}}H{-}\right]_2BH \longrightarrow C_6H_{13}CH{=}CHB\left[\overset{CH_3}{\overset{|}{C}}HCH(CH_3)_2\right]_2$$

$$\downarrow H_2O_2/OH^-$$

$$C_6H_{13}CH_2CHO \longleftarrow \left[C_6H_{13}CH{=}CHOH\right] + 2\ (CH_3)_2CH\overset{CH_3}{\overset{|}{C}}HOH$$

Fig. 28.19. Synthesis of aldehydes by the hydration of alkynes via the hydroboration process-oxidation sequence.

28.5b. Synthesis of ketones. Monosubstituted alkynes undergo mercuric ion-catalyzed addition of water to yield methyl ketones, but disubstituted alkynes often yield mixtures of ketones unless the two substituents are identical (Fig. 28.20).

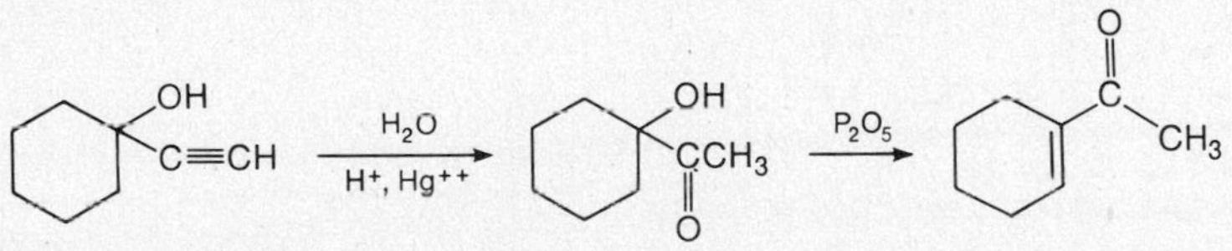

$$\text{1-ethynylcyclohexanol (OH, }C{\equiv}CH) \xrightarrow[H^+,\ Hg^{++}]{H_2O} \text{1-hydroxycyclohexyl }\overset{O}{\overset{\|}{C}}CH_3 \xrightarrow{P_2O_5} \text{1-cyclohexenyl }\overset{O}{\overset{\|}{C}}CH_3$$

(For method of preparation see Fig. 27.3). 84% yield

Fig. 28.20. Preparation of ketones via hydration of alkynes.

28.6a. Rearrangement of pinacols. The pinacol rearrangement (see Section 22.13a) yields ketones in most cases, although diols of the general structure $R_2C(OH)CH_2OH$ may yield aldehydes upon acid-catalyzed rearrangement:

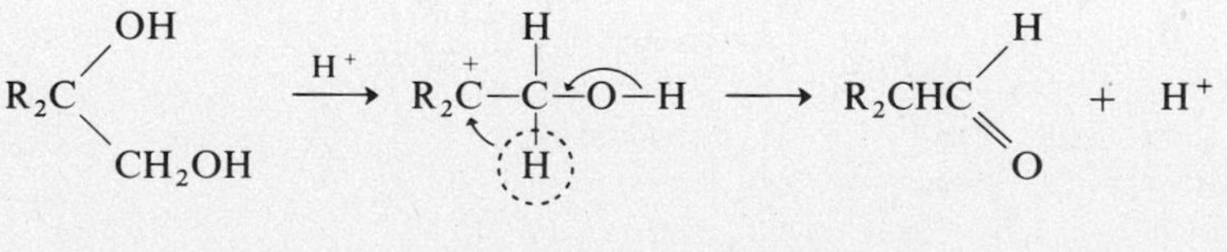

$$R_2C(OH)CH_2OH \xrightarrow{H^+} R_2\overset{+}{C}{-}\overset{H}{\overset{|}{C}}{-}O{-}H \longrightarrow R_2CHC(H){=}O + H^+$$

28.6 Methods Involving Molecular Rearrangement

28.6b. Rearrangement of oxiranes. Closely related to the pinacol rearrangement is the rearrangement of oxiranes, which are converted to aldehydes or ketones under the influence of Lewis acid catalysts or lithium dialkylamides. For example, when 1,2-diphenyloxirane (stilbene epoxide) is treated with boron trifluoride, it rearranges to diphenylacetaldehyde; when it is treated with lithium diethylamide it is converted to 1,2-diphenylethanone (deoxybenzoin) (Fig. 28.21).

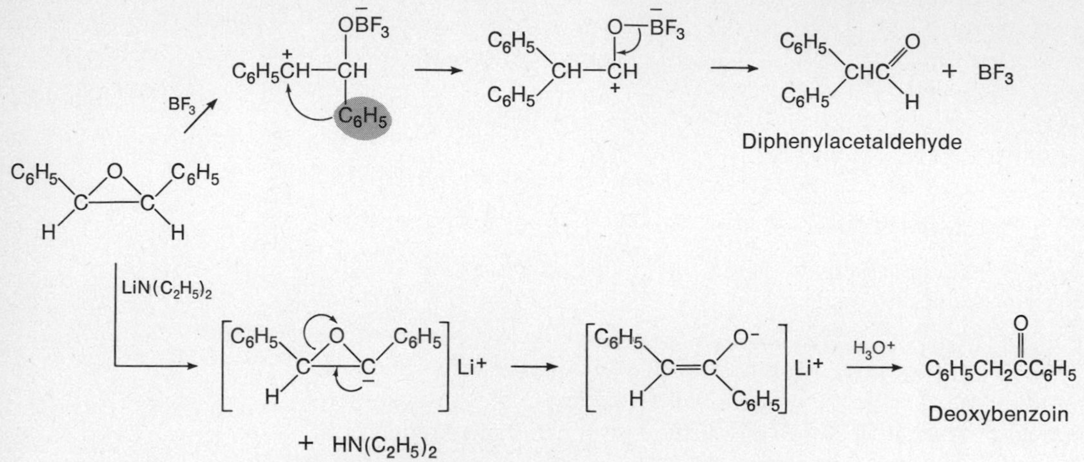

Fig. 28.21. Preparation of aldehydes and ketones by rearrangement of oxiranes.

A particularly useful application of the oxirane rearrangement, most often applied as an aldehyde synthesis makes use of α,β-epoxyesters known as glycidic esters. Glycidic esters, obtained by the action of ethyl α-chloroacetate on aldehydes and ketones in the presence of sodium amide, rearrange with loss of carbon dioxide when treated with aqueous acid (Fig. 28.22).

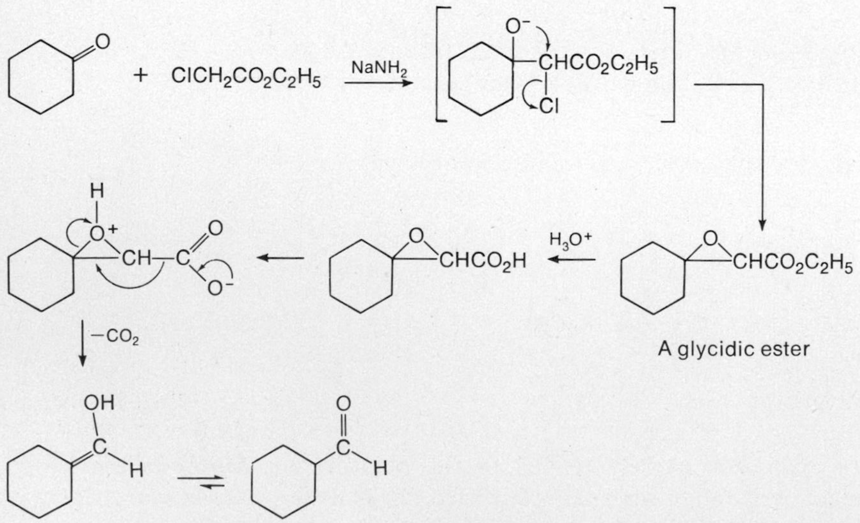

Fig. 28.22. Glycidic ester method for the synthesis of aldehydes.

28.7 Methods Involving Oxidation of Alkanes, Primary Alcohols, and Aldehydes

Primary alcohols and aldehydes are smoothly converted to carboxylic acids by strong oxidizing reagents, such as chromic acid (sodium dichromate and sulfuric acid), potassium permanganate, and nitric acid. More selective reagents for the conversion of aldehydes to carboxylic acids include silver ion and cuprous ion. Methyl groups are less susceptible to oxidation than aldehyde groups, but when attached to benzene rings they can be transformed to the corresponding carboxylic acid group. Larger alkyl groups attached to benzene rings also undergo facile oxidation (except groups that are triply branched at the benzyl carbon—*e.g.*, *t*-butyl), although in these cases the oxidation yields products containing fewer carbon atoms than the starting material (Fig. 28.23).

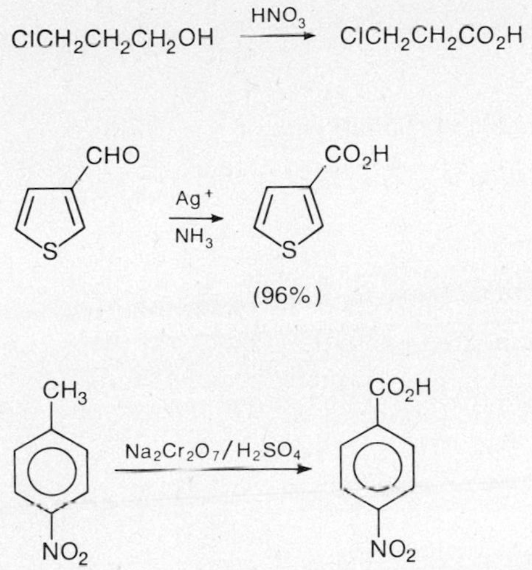

$$ClCH_2CH_2CH_2OH \xrightarrow{HNO_3} ClCH_2CH_2CO_2H$$

3-Thiophenecarboxaldehyde (CHO) $\xrightarrow[NH_3]{Ag^+}$ 3-thiophenecarboxylic acid (CO_2H) (96%)

p-Nitrotoluene (CH_3, NO_2) $\xrightarrow{Na_2Cr_2O_7/H_2SO_4}$ p-nitrobenzoic acid (CO_2H, NO_2)

Fig. 28.23. Preparation of carboxylic acids by oxidation of CH_3, CH_2OH, and CHO functions.

PROBLEM 28-14

What would you predict to be the isolable product from the action of sodium dichromate and sulfuric acid on 3-hydroxy-5-hexenal?

28.8 Methods Involving Cleavage of Carbon–Carbon Bonds

28.8a. Oxidation of Alkenes and Alkynes. Alkynes and alkenes containing at least one vinyl hydrogen yield carboxylic acids upon treatment with strong oxidizing agents such as sodium dichromate/sulfuric acid or strongly basic potassium permanganate. Advantage is taken of this reaction in one of the steps in the *Barbier–Wieland degradation*, a reaction sequence in which a carboxylic acid is converted to its next lower homolog (Fig. 28.24).

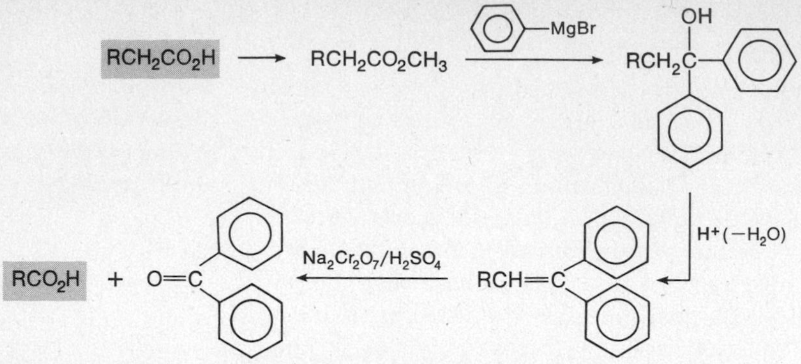

$$RCH_2CO_2H \longrightarrow RCH_2CO_2CH_3 \xrightarrow{C_6H_5MgBr} RCH_2C(OH)(C_6H_5)_2$$

$$\xrightarrow{H^+(-H_2O)} RCH{=}C(C_6H_5)_2 \xrightarrow{Na_2Cr_2O_7/H_2SO_4} RCO_2H + O{=}C(C_6H_5)_2$$

Fig. 28.24. Barbier–Wieland degradation for converting a carboxylic acid to its lower homolog.

PROBLEM 28-15

What is the product of the Barbier–Wieland degradation starting with

$$C_6H_5CH_2CH(CH_3)CO_2H?$$

28.8b. OXIDATION OF SECONDARY ALCOHOLS AND KETONES. The oxidation of secondary alcohols and ketones to carboxylic acids is of little general utility, although in a few special cases it finds application. For example, cyclohexanol and cyclohexanone both yield adipic acid upon oxidation:

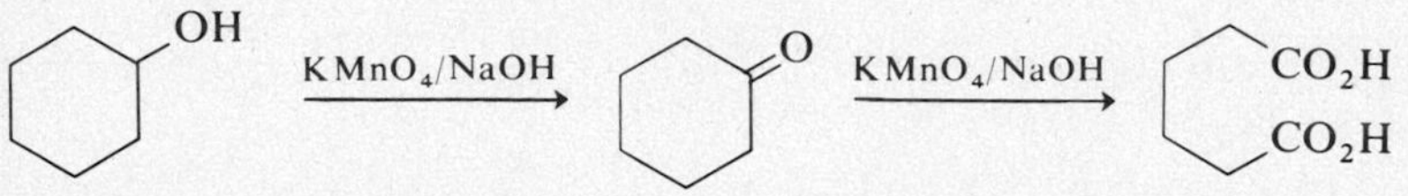

$$\text{cyclohexanol} \xrightarrow{KMnO_4/NaOH} \text{cyclohexanone} \xrightarrow{KMnO_4/NaOH} HO_2C(CH_2)_4CO_2H$$

Methyl ketones constitute a special class of compounds that can frequently be oxidized to carboxylic acids by selective loss of the methyl group in a process known as the *haloform reaction.* Treatment of methyl ketones, $RCOCH_3$, with halogen and aqueous sodium hydroxide yields the corresponding 1,1,1-trihaloketones. In the presence of strong base, these cleave to yield haloforms (CHX_3) and the sodium salt of the carboxylic acid, from which the free acid is obtained upon neutralization (Fig. 28.25). The mechanism of the cleavage reaction is discussed in Section 29.11.

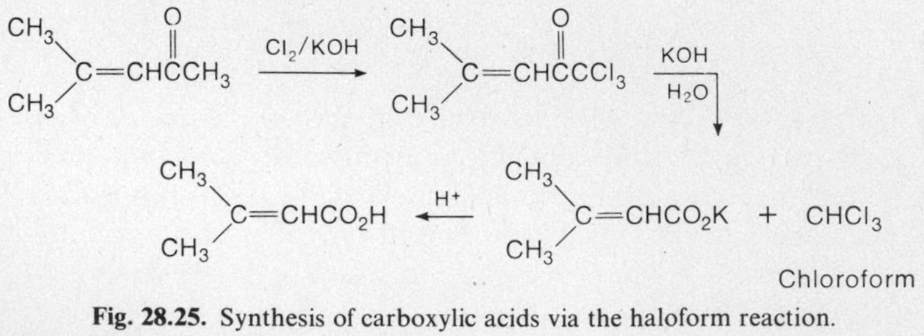

$$(CH_3)_2C{=}CHCOCH_3 \xrightarrow{Cl_2/KOH} (CH_3)_2C{=}CHCOCCl_3 \xrightarrow[H_2O]{KOH} (CH_3)_2C{=}CHCO_2K + \underset{\text{Chloroform}}{CHCl_3}$$

$$(CH_3)_2C{=}CHCO_2K \xrightarrow{H^+} (CH_3)_2C{=}CHCO_2H$$

Fig. 28.25. Synthesis of carboxylic acids via the haloform reaction.

28.9a. ALKYLATION OF ACTIVE METHYLENE COMPOUNDS. The alkylation of acetoacetic ester, and other β-keto esters, and the subsequent cleavage of the alkylated acetoacetic ester to ketones by the action of dilute acid or base (referred to as the "ketone" mode of cleavage) is discussed in Section 28.4c. If the cleavage is effected with concentrated aqueous base, however, it takes a different course and yields a substituted acetic acid (referred to as the "acid" mode of cleavage), thus providing a method for the synthesis of mono- and disubstituted acetic acids (Fig. 28.26).

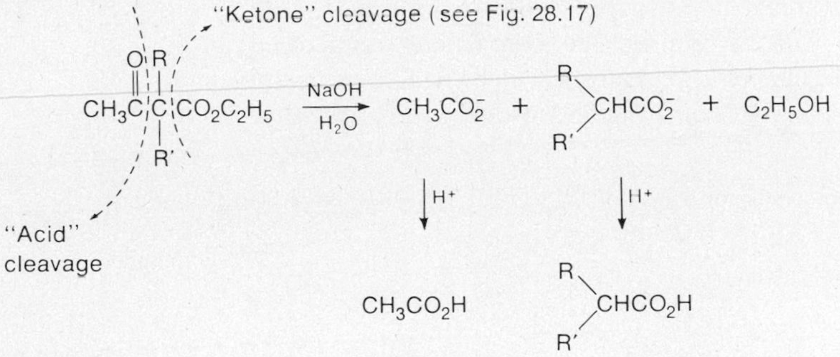

"Ketone" cleavage (see Fig. 28.17)

$$CH_3C(=O)C(R)(R')CO_2C_2H_5 \xrightarrow[H_2O]{NaOH} CH_3CO_2^- + RR'CHCO_2^- + C_2H_5OH$$

"Acid" cleavage

$$CH_3CO_2^- \xrightarrow{H^+} CH_3CO_2H \qquad RR'CHCO_2^- \xrightarrow{H^+} RR'CHCO_2H$$

Fig. 28.26. Acetoacetic ester method for the synthesis of substituted acetic acids.

Closely related to the acetoacetic ester sequence is the malonic ester synthesis, in which diethyl malonate (malonic ester) is alkylated by treatment with base and an alkyl halide followed by hydrolysis to yield α-alkylmalonic acids. When these are heated, they lose carbon dioxide to produce alkanoic acids of the general structure RCH_2CO_2H. If a second alkylation is carried out before hydrolysis and decarboxylation, alkanoic acids of the general structure $RR'CHCO_2H$ are obtained (Fig. 28.27).

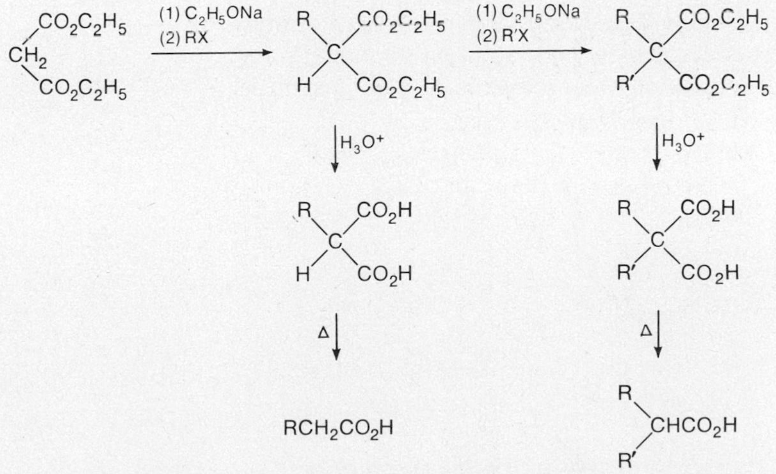

$$CH_2(CO_2C_2H_5)_2 \xrightarrow[(2)\ RX]{(1)\ C_2H_5ONa} RHC(CO_2C_2H_5)_2 \xrightarrow[(2)\ R'X]{(1)\ C_2H_5ONa} RR'C(CO_2C_2H_5)_2$$

$$RHC(CO_2C_2H_5)_2 \xrightarrow{H_3O^+} RHC(CO_2H)_2 \xrightarrow{\Delta} RCH_2CO_2H$$

$$RR'C(CO_2C_2H_5)_2 \xrightarrow{H_3O^+} RR'C(CO_2H)_2 \xrightarrow{\Delta} RR'CHCO_2H$$

Fig. 28.27. Malonic ester synthesis of substituted acetic acids.

PROBLEM 28-16

What are the products of the action of diethyl sodiomalonate on 1-methyl-1-bromocyclohexane? Can diethyl phenylmalonate be prepared by the phenylation of diethyl sodiomalonate?

28.9b. CARBOXYLATION OF ORGANOMETALLIC REAGENTS. Organometallic compounds (including Grignard reagents, organolithium compounds, and organosodium compounds) react with carbon dioxide (conveniently generated from Dry Ice) to form the salts of carboxylic acids from which the free acids are obtained upon neutralization, providing a versatile and widely used method of synthesis (Fig. 28.28).

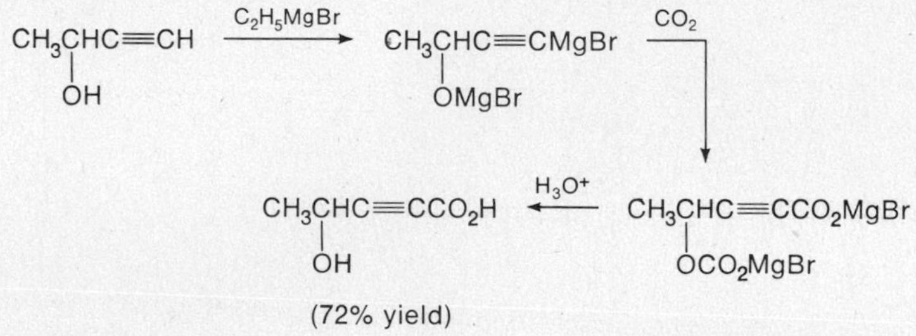

$$CH_3CH(OH)C{\equiv}CH \xrightarrow{C_2H_5MgBr} CH_3CH(OMgBr)C{\equiv}CMgBr \xrightarrow{CO_2} CH_3CH(OCO_2MgBr)C{\equiv}CCO_2MgBr \xrightarrow{H_3O^+} CH_3CH(OH)C{\equiv}CCO_2H$$

(72% yield)

Fig. 28.28. Synthesis of carboxylic acids via carboxylation of Grignard reagents.

28.9c. NITRILE FORMATION FOLLOWED BY HYDROLYSIS. The nitrile group can be introduced into molecules in a variety of ways, the most usual ones involving nucleophilic addition of hydrogen cyanide to ketones or aldehydes (to form cyanohydrins) and the nucleophilic substitution on alkyl halides with metal cyanides (to form alkyl nitriles). Acid- or base-induced hydrolysis of the nitriles yields the corresponding carboxylic acids (Fig. 28.29). If the hydrogen cyanide addition to aldehydes or ketones is carried out in a solution containing ammonium hydroxide and ammonium chloride, the product is an α-amino nitrile,

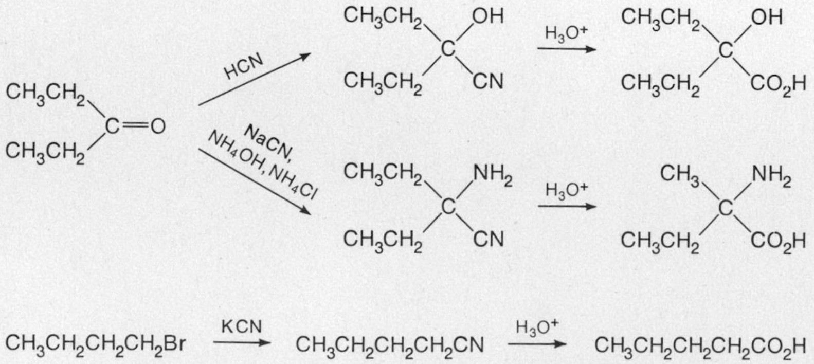

$$(CH_3CH_2)_2C{=}O \xrightarrow{HCN} (CH_3CH_2)_2C(OH)CN \xrightarrow{H_3O^+} (CH_3CH_2)_2C(OH)CO_2H$$

$$(CH_3CH_2)_2C{=}O \xrightarrow{NaCN,\ NH_4OH,\ NH_4Cl} (CH_3CH_2)_2C(NH_2)CN \xrightarrow{H_3O^+} (CH_3)(CH_3CH_2)C(NH_2)CO_2H$$

$$CH_3CH_2CH_2CH_2Br \xrightarrow{KCN} CH_3CH_2CH_2CH_2CN \xrightarrow{H_3O^+} CH_3CH_2CH_2CH_2CO_2H$$

Fig. 28.29. Synthesis of carboxylic acids via hydrolysis of nitriles.

hydrolysis of which yields the corresponding α-amino acid. This sequence of reactions is referred to as the *Strecker synthesis* and has found considerable use in the preparation of naturally occurring α-amino acids.

PROBLEM 28-17

Application of the Strecker synthesis to phenylacetaldehyde yields phenylalanine, one of the naturally occurring amino acids. How many stereoisomers are possible for this compound? Does the Strecker synthesis produce optically active material? If not, how might optically active D- and L-phenylalanine be obtained?

28.10 Methods Involving Hydrolysis

In addition to the hydrolysis of nitriles, the reaction of water with carboxylic acid derivatives, such as acid chlorides, acid anhydrides, esters, amides, and ketenes, also yields carboxylic acids. Still another hydrolysis procedure involves the action of water on 1,1,1-trihalo compounds:

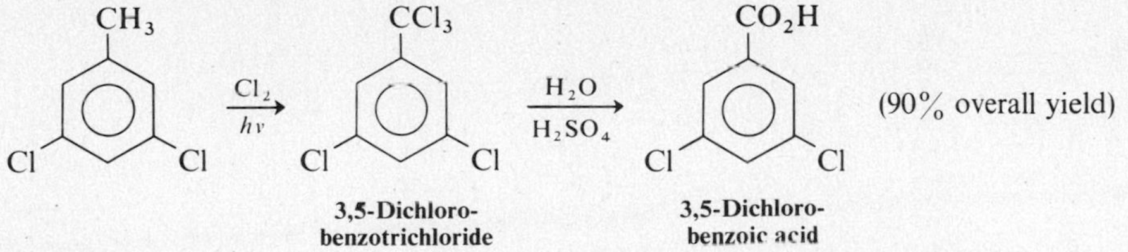

$$3,5\text{-}Cl_2C_6H_3CH_3 \xrightarrow[h\nu]{Cl_2} 3,5\text{-}Cl_2C_6H_3CCl_3 \xrightarrow[H_2SO_4]{H_2O} 3,5\text{-}Cl_2C_6H_3CO_2H \quad \text{(90\% overall yield)}$$

3,5-Dichloro-benzotrichloride; 3,5-Dichloro-benzoic acid

28.11 Methods Involving Molecular Rearrangement

28.11a. Rearrangement of diazoketones. A useful method for converting a carboxylic acid to its next higher homolog, known as the *Arndt–Eistert sequence*, involves a rearrangement of a diazoketone (*Wolff rearrangement*) as a key step. The sequence involves the conversion of the acid to its acid chloride, treatment of the acid chloride with diazomethane to form the diazoketone, and rearrangement

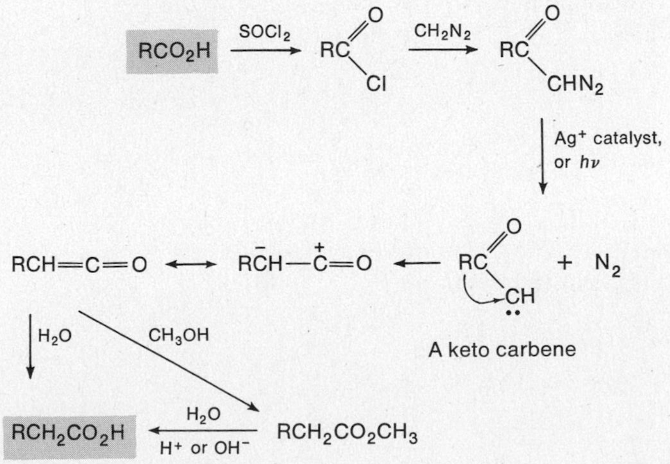

$$RCO_2H \xrightarrow{SOCl_2} RC(=O)Cl \xrightarrow{CH_2N_2} RC(=O)CHN_2 \xrightarrow{Ag^+ \text{ catalyst, or } h\nu} RC(=O)\ddot{C}H + N_2$$

A keto carbene

$$RC(=O)\ddot{C}H \longrightarrow R\bar{C}H{-}\overset{+}{C}{=}O \longleftrightarrow RCH{=}C{=}O$$

$$RCH{=}C{=}O \xrightarrow{H_2O} RCH_2CO_2H$$

$$RCH{=}C{=}O \xrightarrow{CH_3OH} RCH_2CO_2CH_3 \xrightarrow[H^+ \text{ or } OH^-]{H_2O} RCH_2CO_2H$$

Fig. 28.30. The Arndt–Eistert sequence for converting a carboxylic acid to its higher homolog.

of the diazoketone to the homologous acid (in aqueous solution) or to the ester (in alcoholic solution) which can be hydrolyzed to the acid (Fig. 28.30). The second of these alternatives is often preferable because the overall yields are higher. The Wolff rearrangement is an example of a nucleophilic 1,2-rearrangement process involving migration to an electron-deficient carbon. It is thought to proceed with loss of nitrogen (promoted by heat, light, or catalysts) to yield a ketocarbene which rearranges to a ketene. Although the ketene can be isolated if the reaction is carried out in an aprotic medium, in the presence of protic compounds such as water and alcohols it reacts immediately to form acids or esters (Fig. 28.30).

28.11b. Rearrangement of oximes (Beckmann rearrangement). The rearrangement of oximes to amides followed by hydrolysis of the amides is cited among the methods for preparing amines (see Section 24.2c). The other product of hydrolysis is a carboxylic acid, and the method is occasionally used as a synthesis of acids:

$$C_6H_5(R)C{=}N{-}OH \xrightarrow{PCl_5} C_6H_5{-}NH\overset{O}{\overset{\|}{C}}R \xrightarrow[H^+ \text{ or } OH^-]{H_2O} C_6H_5{-}NH_2 + RCO_2H$$

PROBLEM 28-18

Outline the mechanism of the Beckmann rearrangement.

28.11c. Baeyer–Villiger rearrangement. The peracid oxidation of ketones to esters and the subsequent hydrolysis of the esters is discussed as a method for preparing phenols (see Fig. 22.6). As with the Beckmann rearrangement, the carboxylic acid obtained from this hydrolysis is sometimes the objective of the synthesis:

$$C_6H_5(R)C{=}O \xrightarrow{C_6H_5CO_3H} C_6H_5{-}O\overset{O}{\overset{\|}{C}}R \xrightarrow[H^+ \text{ or } OH^-]{H_2O} C_6H_5{-}OH + RCO_2H$$

28.11d. Benzilic acid rearrangement. The action of strong aqueous base on benzil induces a molecular rearrangement to benzilic acid:

$$C_6H_5{-}\overset{O}{\overset{\|}{C}}{-}\overset{O}{\overset{\|}{C}}{-}C_6H_5 \xrightarrow{NaOH} (C_6H_5)_2C(OH)CO_2^-Na^+ \xrightarrow{H_3O^+} (C_6H_5)_2C(OH)CO_2H$$

Benzil **Benzilic acid**

Other 1,2-diketones react in a comparable fashion, and the reactions are referred to as *benzilic acid rearrangements.* The mechanism is thought to involve nucleophilic addition of hydroxide followed by a 1,2 shift of an R group concomitant with the formation of a carboxylic acid group (Fig. 28.31).

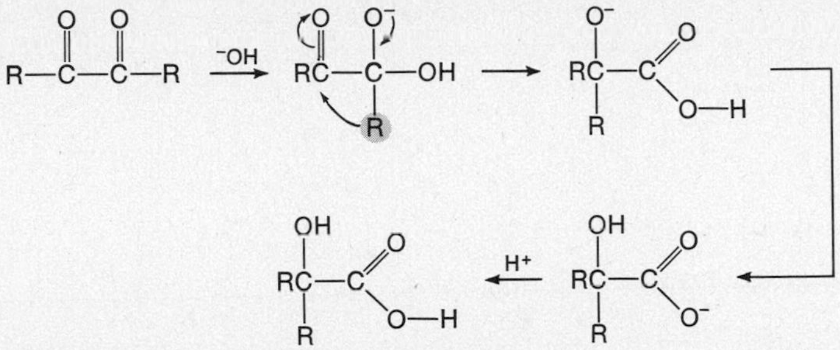

Fig. 28.31. Synthesis of carboxylic acids via the benzilic acid rearrangement.

28.11e. FAVORSKII REACTION. α-Haloketones often undergo skeletal rearrangement in the presence of strong bases to yield carboxylic acids or carboxylic acid derivatives. For example, 2-chlorocyclohexanone reacts with aqueous sodium hydroxide to yield cyclopentanecarboxylic acid. Through the use of 2-chlorocyclohexanone labeled with ^{14}C at carbons 1 and 2, it has been shown that a symmetrical intermediate must be involved to account for the labeling pattern in the product, the generally accepted intermediate being a cyclopropanone (Fig. 28.32). The Favorskii reaction has found special use in the preparation of strained, small-ring compounds, for 2-halocyclopentanones form cyclobutanecarboxylic acids, and 2-halocyclobutanones form cyclopropanecarboxylic acids (Fig. 28.33).

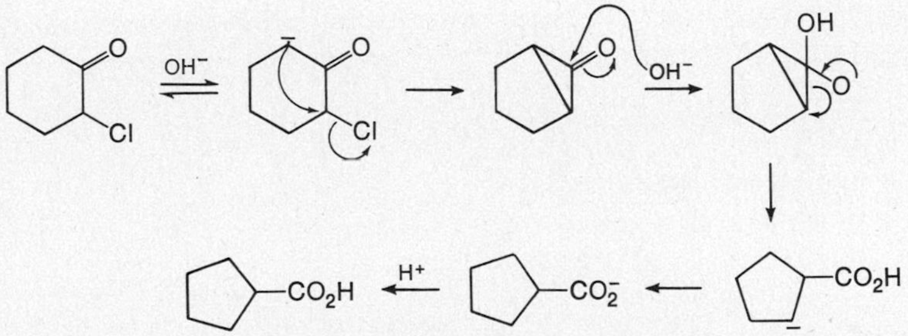

Fig. 28.32. Mechanism of the Favorskii reaction.

PROBLEM 28-19

If 2-chlorocyclohexanone is doubly labeled with ^{14}C at C-1 and C-2, what is the labeling pattern of the Favorskii product if a symmetrical intermediate is involved?

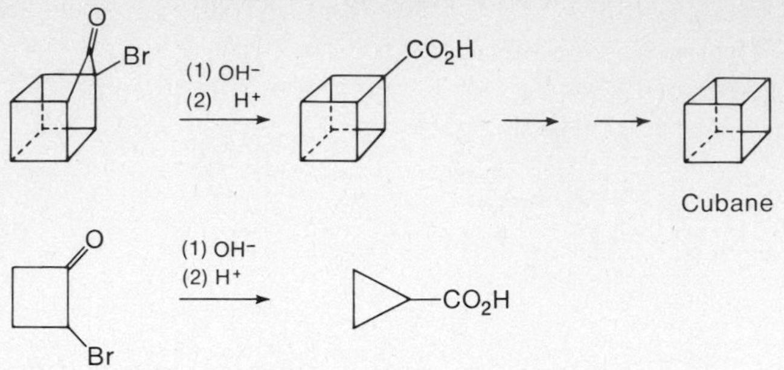

(1) OH^-
(2) H^+

CO_2H

Cubane

(1) OH^-
(2) H^+

CO_2H

Br

2-Bromocyclobutanone   Cyclopropanecarboxylic acid

Fig. 28.33. Synthesis of carboxylic acids via the Favorskii reaction.

SYNTHESIS OF CARBOXYLIC ACID DERIVATIVES

Compounds of the general structure $RC(=O)Y$, where Y is halogen (acid halides), —OCOR (anhydrides), —OR (esters), or $—NR_2$ (amides), are called carboxylic acid derivatives. All the methods for their synthesis that are discussed in this section involve as starting materials carbonyl compounds that are at the oxidation level of a carboxylic acid. Since ketenes are at this same oxidation level, they are also included among this group of carboxylic acid derivatives.

28.12 *Synthesis of Acid Chlorides (Acyl and Aroyl Chlorides)*

Although acyl and aroyl fluorides, chlorides, bromides, and iodides are all known species, only the acyl and aroyl chlorides (usually referred to as acid chlorides) have any general utility as synthesis intermediates. The three reagents that are most often used for their preparation from carboxylic acids are phosphorus

$$CH_3CH_2CO_2H + PCl_3 \longrightarrow CH_3CH_2COCl + H_3PO_3$$

bp 180°   bp 80°   bp 200° (dec)

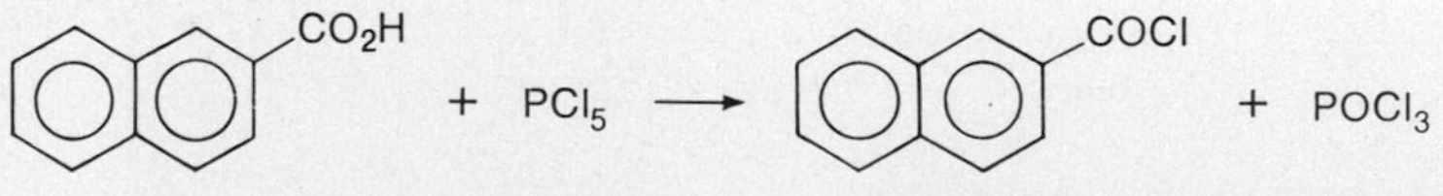

CO_2H + PCl_5 ⟶ $COCl$ + $POCl_3$

subl 162°   bp 304°   bp 107°

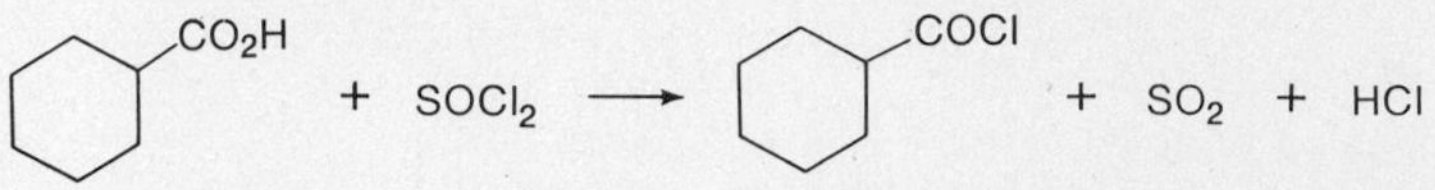

CO_2H + $SOCl_2$ ⟶ $COCl$ + SO_2 + HCl

bp 77°   bp 184°

Fig. 28.34. Synthesis of carboxylic acid chlorides.

trichloride (PCl_3), phosphorus pentachloride (PCl_5), and thionyl chloride ($SOCl_2$), the choice between these reagents often being determined by the ease of separation of the products in the reaction mixture. The phosphorus halides are more reactive than thionyl chloride, but thionyl chloride has the advantage that the by-products of reaction are sulfur dioxide and hydrogen chloride, both of which escape as gases during the course of the reaction. The products from the phosphorus halides are phosphorous acid (bp about 200°) and phosphorus oxychloride (bp 107°), which may codistill with the acid chloride and contaminate it (Fig. 28.34).

28.13 Synthesis of Anhydrides

Although cyclic anhydrides such as succinic anhydride and phthalic anhydride can be prepared from the dicarboxylic acids simply by heating, a more generally applicable synthesis of anhydrides involves the interaction of an acid chloride with a carboxylate ion, which is generated by the action of an inorganic base (*e.g.*, potassium hydroxide) or an organic base (*e.g.*, pyridine) on the carboxylic acid (Fig. 28.35). When pyridine is used, the pyridine probably reacts with the acid chloride to form an acylpyridinium chloride which is the actual acylating agent.

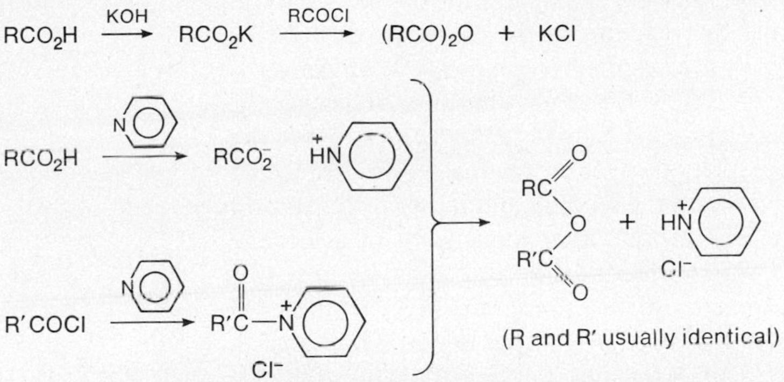

$$RCO_2H \xrightarrow{KOH} RCO_2K \xrightarrow{RCOCl} (RCO)_2O + KCl$$

$$RCO_2H \xrightarrow{\text{pyridine}} RCO_2^- \; HN^+C_5H_5$$

$$R'COCl \xrightarrow{\text{pyridine}} R'C(=O)\text{—}N^+C_5H_5 \; Cl^-$$

$$\longrightarrow RC(=O)\text{—}O\text{—}C(=O)R' + HN^+C_5H_5 \; Cl^-$$

(R and R′ usually identical)

Fig. 28.35. Synthesis of anhydrides from carboxylic acids and acid chlorides.

Ketenes are even more effective acylating agents than acid chlorides and react with the free acids in an essentially irreversible process to yield the anhydride. With either the acid chloride or the ketene method, it is possible to prepare unsymmetrical anhydrides. Usually, however, these are unstable relative to the corresponding symmetrical anhydrides, and redistribute upon distillation to a mixture of the symmetrical anhydrides (Fig. 28.36). Through the use of certain

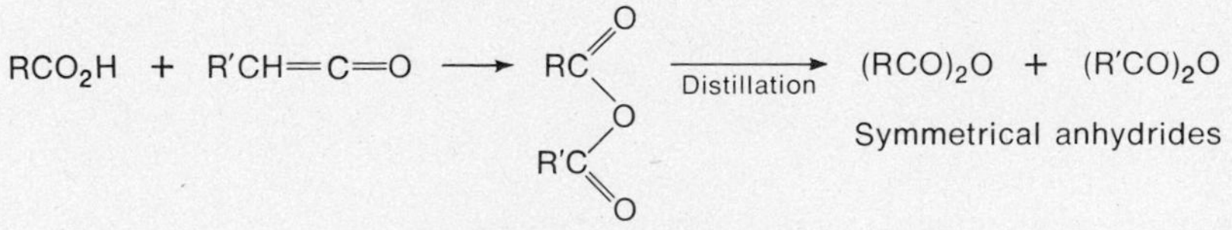

$$RCO_2H + R'CH{=}C{=}O \longrightarrow \underset{\text{Unsymmetrical anhydride}}{RC(=O)\text{—}O\text{—}C(=O)R'} \xrightarrow{\text{Distillation}} \underset{\text{Symmetrical anhydrides}}{(RCO)_2O + (R'CO)_2O}$$

Fig. 28.36. Synthesis of anhydrides via ketenes.

"dehydrating" agents, anhydrides can be prepared using only the acids as the starting materials. For example, dicyclohexylcarbodiimide and trifluoroacetic anhydride both have this capacity, the water that is formed in the reaction being incorporated into these reagents to give dicyclohexylurea and trifluoroacetic acid, respectively (Fig. 28.37).

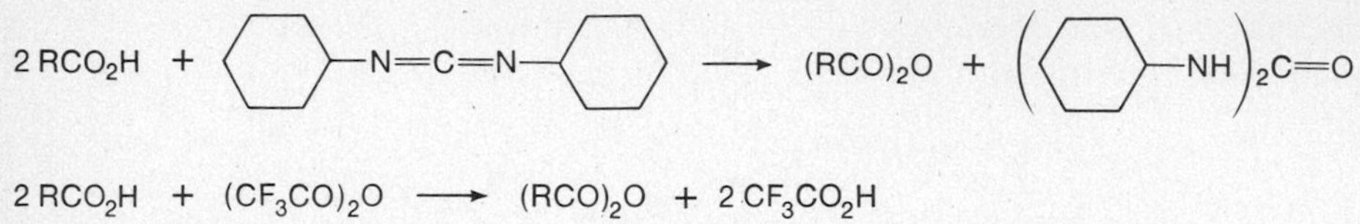

$$2\,RCO_2H + C_6H_{11}{-}N{=}C{=}N{-}C_6H_{11} \longrightarrow (RCO)_2O + (C_6H_{11}{-}NH)_2C{=}O$$

$$2\,RCO_2H + (CF_3CO)_2O \longrightarrow (RCO)_2O + 2\,CF_3CO_2H$$

Fig. 28.37. Synthesis of anhydrides via intermolecular dehydration of carboxylic acids.

28.14 Synthesis of Esters

28.14a. REACTIONS OF CARBOXYLIC ACIDS WITH ALCOHOLS. Alcohols react with carboxylic acids in the presence of catalytic amounts of mineral acid to form esters (*Fischer esterification*). The equilibrium constant for the reaction is usually rather close to unity, however, so the yields of ester are strongly dependent on the structure of the reactants. Tertiary alcohols, for example, furnish esters only in poor yield by this method. To drive the reaction in the direction of product formation, either an excess of carboxylic acid or alcohol can be used (whichever is more readily available) to take advantage of the law of mass action. Alternatively, if water, which is one of the products of reaction, can be removed as it forms, the reaction can be driven virtually to completion. This can be readily accomplished in the preparation of ethyl esters by carrying the reaction out in benzene solution in an apparatus (see Fig. 15.7) that allows the vapor (consisting of a ternary azeotrope of benzene, ethanol, and water) to condense and separate into two layers before being returned to the reaction flask. The lower layer of the condensate consists of a mixture of ethanol and water, and the upper layer consists of a mixture of ethanol and benzene. The lower layer can be withdrawn, thereby removing the water from the reaction, and the upper layer is allowed to flow back into the reaction mixture.

28.14b. REACTION OF CARBOXYLIC ACID DERIVATIVES WITH ALCOHOLS. The reactions of alcohols with acid halides and anhydrides are mechanistically similar to their reactions with carboxylic acids. With acid halides and anhydrides, however, the equilibrium constant for ester formation is very much larger than 1, and the yields are generally very high. Similarly, alcohols react with ketenes to give esters in reactions that, for all practical purposes, are irreversible (Fig. 28.38).

PROBLEM 28-20

What would you predict to be the product from the reaction of thionyl chloride and *cis*-4-hydroxycyclohexanecarboxylic acid?

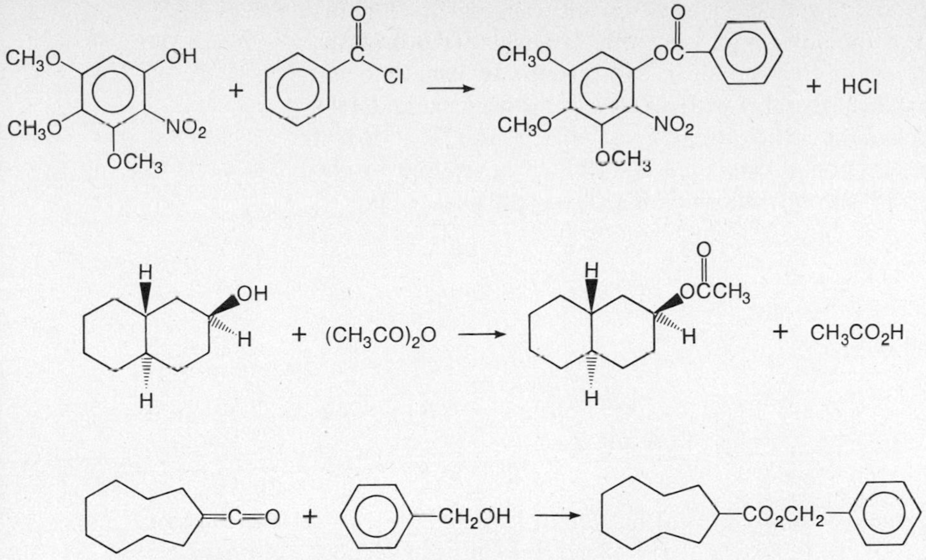

$$+ \; HCl$$

$$+ \; (CH_3CO)_2O \longrightarrow \; + \; CH_3CO_2H$$

$$-C{=}O \; + \; C_6H_5CH_2OH \longrightarrow \; -CO_2CH_2C_6H_5$$

Fig. 28.38. Synthesis of esters via alcohols and carboxylic acid derivatives.

28.14c. REACTION OF CARBOXYLIC ACIDS WITH ALKYL HALIDES, DIAZOALKANES, AND ALKENES. Carboxylate ions are moderately strong nucleophiles and can effect displacements on compounds that have good leaving groups. Thus, sodium carboxylates, silver carboxylates, trialkylammonium carboxylates, *etc.*, react with most primary and many secondary alkyl halides in S_N2 reactions to yield the corresponding esters:

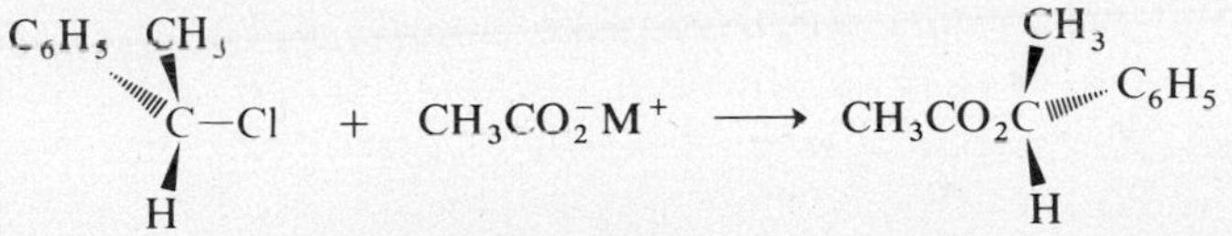

$$(C_6H_5)(CH_3)(H)C{-}Cl \; + \; CH_3CO_2^-M^+ \longrightarrow CH_3CO_2C(CH_3)(C_6H_5)(H)$$

A particularly good leaving group is the diazonium moiety, obtained in the case of aliphatic compounds by the addition of a proton to a diazoalkane (see p. 694). Carboxylic acids are sufficiently acidic to readily effect this protonation, as a result of which they react very smoothly and rapidly with diazoalkanes such as diazomethane to yield esters:

$$RCO_2H \; + \; CH_2N_2 \longrightarrow RCO_2^- \; + \; \overset{+}{C}H_3N_2 \xrightarrow{\text{(Very fast)}} \overset{+}{C}H_3 \; + \; N_2$$

$$\overset{+}{C}H_3 \xrightarrow[S_N1 \text{ pathway}]{RCO_2^-} RCO_2CH_3$$

$$RCO_2^- \; + \; \overset{+}{C}H_3N_2 \underset{S_N2 \text{ pathway}}{\longrightarrow} RCO_2CH_3 \; + \; N_2$$

This method of esterification is particularly useful when the R group is large; *i.e.*, when there is steric hindrance to the carbonyl addition that is required in the methods described in Sections 28.14a and b, and/or when the acid is sensitive to acid- or base-induced rearrangement, the reaction proceeding in an essentially neutral environment at room temperature or below.

Whether esterification via diazoalkanes involves the carboxylate ion (as illustrated) or the free acid as the nucleophile, or both, and whether it involves $CH_3N_2^+$ or $^+CH_3$, or both, as the electrophile, is not certain. In a mechanistically similar fashion, carboxylic acids react with alkenes in the presence of Lewis acids to form esters via attack of the carboxylic acid on the carbonium ion formed from the alkene. For example, *t*-butyl esters are conveniently prepared by the action of 2-methylpropene and boron trifluoride on a carboxylic acid:

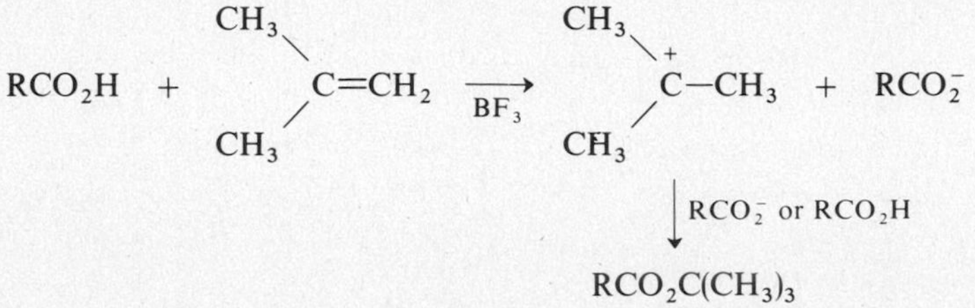

$$RCO_2H + (CH_3)_2C{=}CH_2 \xrightarrow{BF_3} (CH_3)_2\overset{+}{C}{-}CH_3 + RCO_2^-$$

$$\xrightarrow{RCO_2^- \text{ or } RCO_2H} RCO_2C(CH_3)_3$$

28.15 *Synthesis of Amides*

Amides can sometimes be prepared simply by heating a mixture of an amine and a carboxylic acid, the reaction being particularly effective in intramolecular examples:

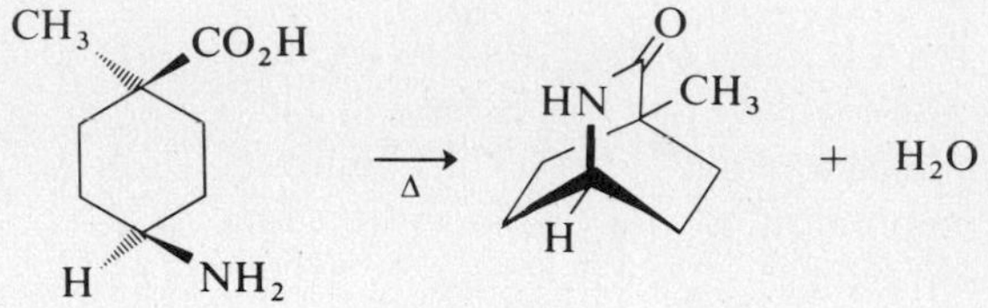

$$\text{cis-4-amino-1-methylcyclohexanecarboxylic acid} \xrightarrow{\Delta} \text{bicyclic lactam} + H_2O$$

More general in application, however, is the reaction of ketenes, acid chlorides, anhydrides, or esters with ammonia, primary amines, or secondary amines. The equilibrium constants for the reactions decrease in the order $R_2C{=}C{=}O >$ $RCOCl > (RCO)_2O > RCO_2R$, but even in the case of esters it is much larger than unity (Fig. 28.39).

$$R_2C{=}C{=}O + R'_2NH \longrightarrow R_2CHCONR'_2$$

$$RCOCl + 2R'_2NH \longrightarrow RCONR'_2 + H_2\overset{+}{N}R'_2\ Cl^-$$

$$(RCO)_2O + 2R'_2NH \longrightarrow RCONR'_2 + H_2\overset{+}{N}R'_2\ {}^-O_2CR$$

$$RCO_2R'' + R'_2NH \longrightarrow RCONR'_2 + R''OH$$

Fig. 28.39. Synthesis of amides.

PROBLEM 28-21

What product is obtained if the Arndt–Eistert sequence, starting with benzoyldiazomethane, is carried out in piperidine solution? Explain.

Primary amides are readily obtained by partial hydrolysis of nitriles, the reagent of choice for this purpose being aqueous sodium hydroxide containing 6 to 12% hydrogen peroxide:

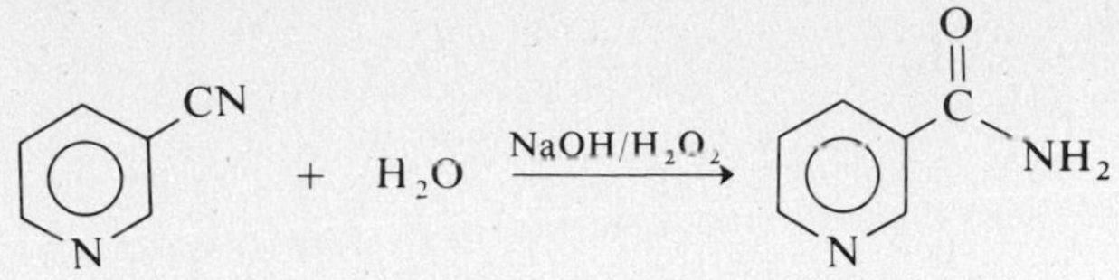

$$\text{3-cyanopyridine (CN)} + H_2O \xrightarrow{NaOH/H_2O_2} \text{3-pyridyl-}C(=O)NH_2$$

28.16 Synthesis of Ketenes

Ketenes are at the same oxidation level as carboxylic acids and, like acid chlorides, anhydrides, esters, and amides, they yield carboxylic acids when treated with water. Ketene itself can be obtained by pyrolysis of almost any compound that contains an acetyl group. Generally, both commercially and in the research laboratory, the acetyl compound employed is acetone, which decomposes to ketene at 700–750° (Fig. 28.40). Larger ketenes are best prepared by the action of trialkylamines on acid chlorides, the yields in many cases being as high as 90% (Fig. 28.40).

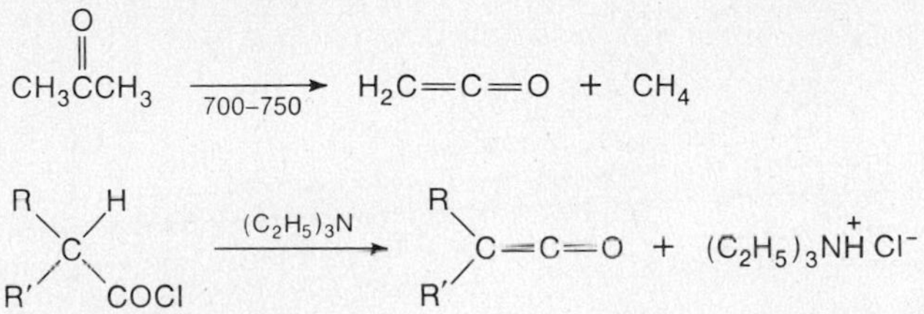

$$CH_3\overset{O}{\overset{\|}{C}}CH_3 \xrightarrow[700\text{–}750]{} H_2C{=}C{=}O + CH_4$$

$$RR'CH{-}COCl \xrightarrow{(C_2H_5)_3N} RR'C{=}C{=}O + (C_2H_5)_3\overset{+}{N}H\ Cl^-$$

Fig. 28.40. Synthesis of ketenes.

The products from the reactions illustrated in Fig. 28.40 are often the ketene dimers rather than the ketene monomers. Ketene itself, for example, exists at room temperature as the β-lactone, $\begin{array}{l}CH_2{-}C{=}O\\ \;|\quad\;\;|\\ H_2C{=}C{-}O\end{array}$, which can be "cracked" to the ketene monomer by heating at 500—600°. Disubstituted ketenes dimerize to tetrasubstituted cyclobutan-1,3-diones:

$$R_2C{=}C{=}O \longrightarrow \begin{array}{l}R_2C{-}C{=}O\\ \;\;\;|\quad\;\;|\\ O{=}C{-}CR_2\end{array}$$

Monosubstituted ketenes dimerize in both these fashions.

SYNTHESIS OF DICARBONYL COMPOUNDS

PROBLEM 28-22

How many stereoisomers are possible for the dimer of ethylmethylketene? Can any of them be optically active? Why or why not?

28.17a. OXIDATION OF METHYL AND METHYLENE GROUPS ADJACENT TO CARBONYL FUNCTIONS. As illustrated in Fig. 28.6, compounds containing methyl or methylene groups adjacent to a carbonyl function can be oxidized to α-dicarbonyl compounds with selenium dioxide:

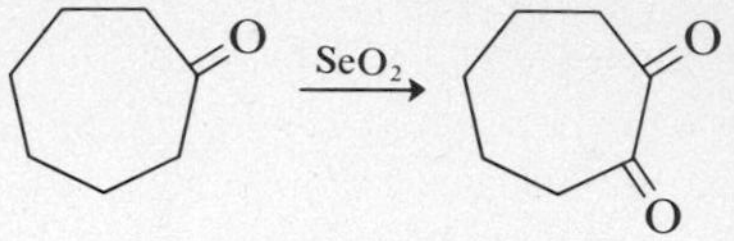

$$\text{cycloheptanone} \xrightarrow{SeO_2} \text{cycloheptane-1,2-dione}$$

28.17b. OXIDATION OF α-HYDROXYKETONES. α-Hydroxyketones can be prepared by the reductive dimerization of esters. The reaction, called the *acyloin condensation*, is carried out in the presence of metallic sodium and is thought to occur on the surface of the metal where one-electron transfers take place to produce radical anions. Combination of the radical anions yields an intermediate that gains another pair of electrons and then expels two alkoxide anions to form an α-diketone. Under the reducing conditions of the reaction, however, the diketone adds two more electrons to yield, after treatment of the mixture with water, an α-hydroxyketone (Fig. 28.41). Oxidation back to the α-diketone can be effected with mild oxidants, such as silver ion and copper ion [*e.g.*, $Cu(OCOCH_3)_2$].

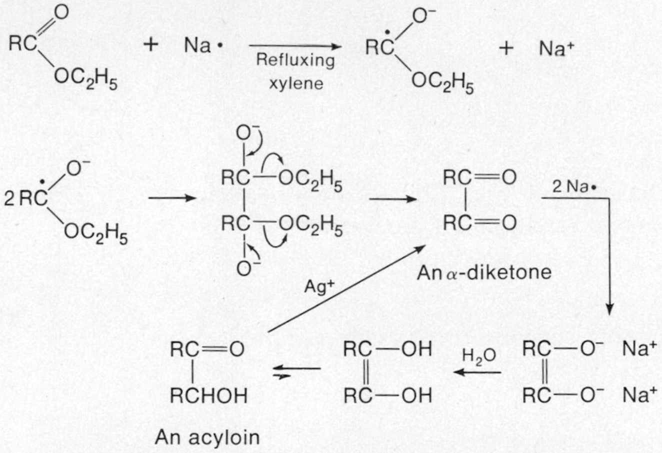

$$RC(=O)OC_2H_5 + Na\cdot \xrightarrow[\text{xylene}]{\text{Refluxing}} R\dot{C}(O^-)OC_2H_5 + Na^+$$

$$2\,R\dot{C}(O^-)OC_2H_5 \longrightarrow RC(O^-)(OC_2H_5){-}C(O^-)(OC_2H_5)R \longrightarrow RC(=O){-}C(=O)R \text{ (An α-diketone)}$$

$$RC(=O)C(=O)R \xrightarrow{2\,Na\cdot} RC(O^-\,Na^+){=}C(O^-\,Na^+)R \xrightarrow{H_2O} RC(OH){=}C(OH)R \rightleftharpoons RC(=O){-}CHOHR \text{ (An acyloin)}$$

$$\text{An acyloin} \xrightarrow{Ag^+} \text{An α-diketone}$$

Fig. 28.41. Mechanism of the acyloin condensation.

The most interesting application of the acyloin condensation is the synthesis of cyclic α-hydroxyketones of ring sizes of eight carbons and greater. For reasons that are not entirely clear, these larger ring carbocyclic compounds are obtained in surprisingly good yield by this method. For example, 2-hydroxycyclodecanone (sebacoin) is prepared in 66% yield by the action of metallic sodium on a dilute, xylene solution of diethyl sebacate:

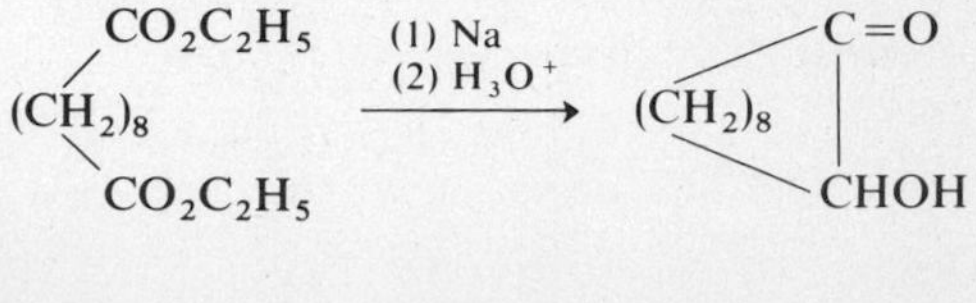

$$(CH_2)_8(CO_2C_2H_5)_2 \xrightarrow[(2)\ H_3O^+]{(1)\ Na} (CH_2)_8\langle C{=}O,\ CHOH\rangle \text{ (ring)}$$

28.18a. SELF-CONDENSATION OF ESTERS (CLAISEN AND DIECKMANN CONDENSATIONS). The prototype of the simple ester condensation (*the Claisen condensation*) is the conversion of ethyl acetate to ethyl acetoacetate (acetoacetic ester) by the action of sodium ethoxide (see Fig. 15.11). A particularly useful application of the Claisen condensation involves the preparation of cyclic keto esters by the action of strong bases on esters of dicarboxylic acids, the reaction in such cases being referred to as a *Dieckmann condensation*. For example, treatment of dimethyl adipate with sodium methoxide yields methyl 2-ketocyclopentanecarboxylate (Fig. 28.42). Hydrolysis produces 2-ketocyclopentanecarboxylic acid, a β-keto

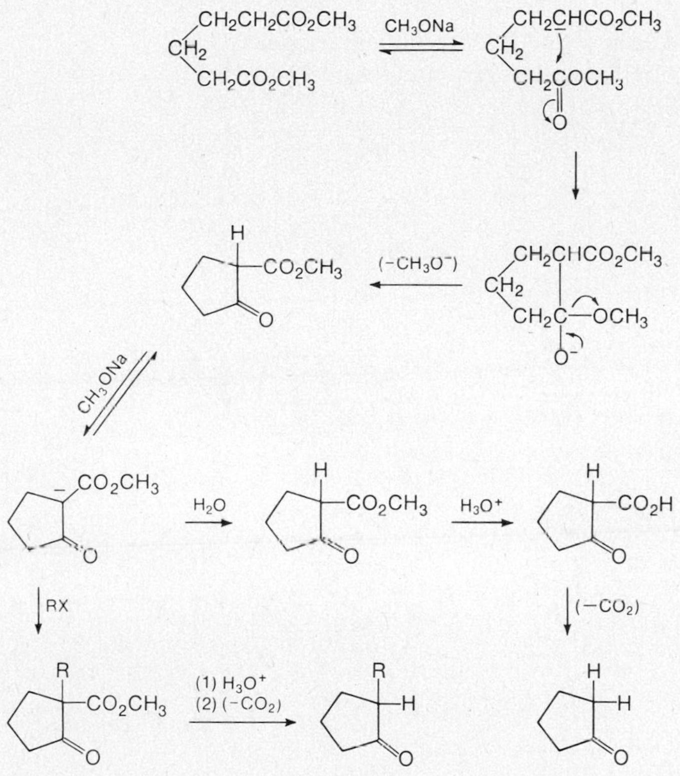

Fig. 28.42. Dieckmann condensation for the preparation of cyclic ketones.

acid, which loses carbon dioxide to form cyclopentanone. Alternatively, the initially formed product can be treated with an alkyl halide prior to hydrolysis, and a 2-alkylcyclopentanone can be obtained (*i.e.*, an acetoacetic ester-type ketone synthesis).

Reactions involving active methylene compounds, such as the acetoacetic ester alkylation and the Claisen and Dieckmann condensations, are initiated by the action of a base. For reactions carried out in aqueous solution, hydroxide ion is the strongest base available. Generally, however, Claisen and Dieckmann condensations are carried out in nonaqueous solutions, and a variety of bases can be used under these conditions. Among the useful bases are, in the order of increasing basicity: sodium methoxide, sodium ethoxide, potassium *t*-butoxide, sodium amide, sodium hydride, and potassium triphenylmethide [$(C_6H_5)_3CK$]. In some instances, any of these bases will suffice, and there is little reason to choose one over the other. In many instances, however, certain bases offer special advantages. In the reaction illustrated in Fig. 28.42, for example, if sodium ethoxide rather than sodium methoxide is used, some "ester interchange" takes place, and the product consists of a mixture of the methyl and ethyl esters of 2-ketocyclopentanecarboxylic acid; sodium methoxide, therefore, is the preferable of these two bases. Alternatively, potassium *t*-butoxide might be used, for the *t*-butoxide ion is sufficiently nonnucleophilic that little or no ester interchange occurs.

The Claisen condensation of ethyl acetate and the Dieckmann condensation of diethyl adipate both yield anions of β-keto esters that are considerably more resonance stabilized than anions of the starting esters [$K_3 > K_1$ in Fig. 28.43];

Step 1:

$$RR'CHCO_2CH_3 + \text{Base} \overset{K_1}{\rightleftharpoons} RR'\bar{C}CO_2CH_3 + \text{Base-H}^+$$

Step 2:

$$RR'\bar{C}CO_2CH_3 + R''\overset{O}{\overset{\|}{C}}OCH_3 \overset{K_2}{\rightleftharpoons} RR'C(CO_2CH_3)(\overset{O}{\overset{\|}{C}}R'') + CH_3O^-$$

Step 3: (if R = H)

$$RR'C(CO_2CH_3)(\overset{O}{\overset{\|}{C}}R'') + \text{Base} \overset{K_3}{\rightleftharpoons} R'\bar{C}(CO_2CH_3)(\overset{O}{\overset{\|}{C}}R'') + \text{Base-H}^+$$

Fig. 28.43. Equilibria in the Claisen condensation.

Consequently, they are more acidic, favoring the formation of product (Fig. 28.43). Methyl isobutyrate, however, yields a product containing no hydrogens on the carbon between the carbonyl functions which, therefore, is no more acidic than the starting ester, *i.e.*, $K_3 \simeq K_1$ in Fig. 28.43. To minimize this disadvantage, we use bases that increase the equilibrium constant for step 1, offsetting the less favorable equilibrium constant in step 3. For example, either sodium hydride or sodium amide yields a sodium enolate of the starting ester, with concomitant formation of hydrogen and ammonia, respectively. The hydrogen or ammonia escapes from the reaction mixture and, consequently step 1 is pushed to completion. Alternatively, bases, such as potassium triphenylmethide, that are very much stronger than the enolate anions can be used to increase the equilibrium constant of step 1 and promote good yields of product.

The Dieckmann condensation is limited in its application to the preparation of five-, six-, and seven-membered rings. In rings with less than five members, bond strain hinders the reaction; in rings with more than seven members, non-bonded interference between hydrogens (*trans*-annular interaction) as well as entropy factors (*i.e.*, distance between reacting centers) come into play. A modification of the Dieckmann condensation has been developed, however, that does permit the synthesis of larger ring compounds in reasonably good yield. Called the *Thorpe–Ziegler condensation*, it involves the action of lithium N-methylanilide (chosen because of its high basicity and its ether solubility) on a very dilute solution of an α,ω-dinitrile in ether. The initially formed α-cyanoimines undergo hydrolysis to yield ammonia, carbon dioxide, and the corresponding ketone (Fig. 28.44).

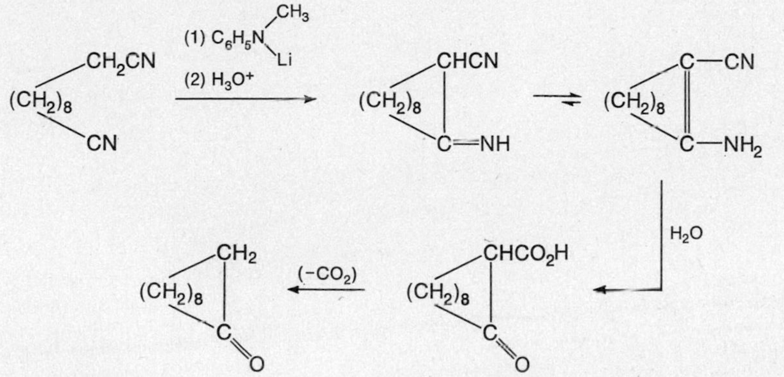

Fig. 28.44. Rhorpe–Ziegler method for preparing cyclic ketones.

28.18b. Mixed ester condensations. "Mixed" ester condensations involving esters reacting with aldehydes or ketones can sometimes be effected in good yield to produce β-keto aldehydes and β-diketones, depending on the ester that is used (Fig. 28.45).

Table 28.1 summarizes the methods of synthesis of carbonyl containing compounds.

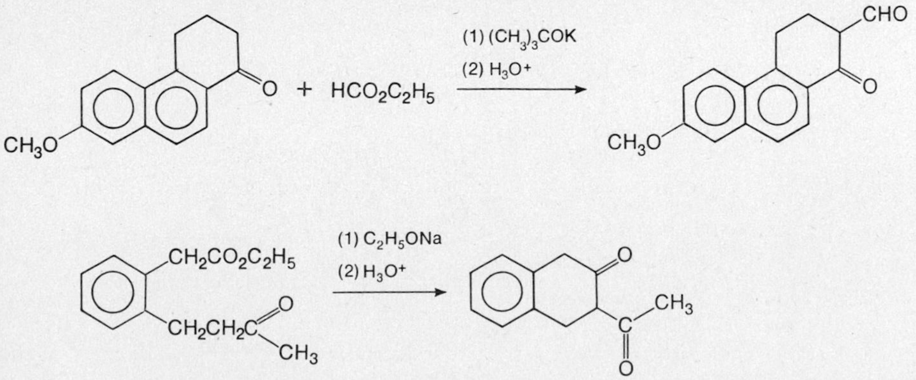

Fig. 28.45. Preparation of β-ketoaldehydes and β-diketones via "mixed" ester condensations.

Table 28.1 METHODS OF PREPARATION OF CARBONYL COMPOUNDS

Type of Product	Starting Material	Reagents	Products	Comments
Aldehydes	RCOCl	Pd—$BaSO_4$ moderated with $[(CH_3)_2N]_2S$	RCHO	Sometimes difficult to moderate properly to avoid RCH_2OH formation
	RCOCl	(1) $LiAl[OC(CH_3)_3]_3H$; (2) H_3O^+	RCHO	
	$RCONR'_2$	(1) $LiAl[OC(CH_3)_3]_3H$; (2) H_3O^+	RCHO	
	RCN	(1) $LiAl[OC(CH_3)_3]_3H$; (2) H_3O^+	RCHO	
	$ArCH_3$	(1) CrO_3—HOAc; (2) H_2O	ArCHO	
	ArH	CO, HCl, $AlCl_3$	ArCHO	See Table 31.1 for more details
	RCH_2OH	CrO_3/pyridine, or K_2FeO_4	RCHO	See Table 22.2 for more details
	RCH_2X	(1) $(CH_3)_2S{=}O$; (2) base	RCHO	
	RCH_2NO_2	(1) Base; (2) H_3O^+	RCHO	Useful in sugar series for chain elongation
	RCH=CHR′	O_3	RCHO + R′CHO	Useful diagnostic as well as preparative method
	RCH(OH)—CH(OH)R′	HIO_4 or $Pb(OAc)_4$	RCHO + R′CHO	Specific for 1,2-diols
	$ZZ'CH_2$	(1) $HCO_2C_2H_5$ + base; (2) H_3O^+	ZZ′CHCHO	
	(active methylene compound)	SeO_2	ZZ′C=O	
	RMgBr	(1) $HC(OC_2H_5)_3$; (2) H_3O^+	RCHO	
	RX	(1) 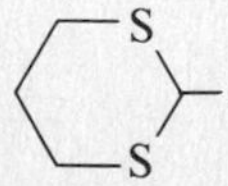 2-lithio-1,3-dithiane (1,3-dithian-2-yl—Li); (2) H_2O, $HgCl_2$	RCHO	
	RC≡CH	(1) R_2BH; (2) H_2O_2/OH^-	RCH_2CHO	anti-Markovnikov addition of water
	$R_2C(OH)$—CH_2OH	Acid catalysts	R_2CHCHO	Pinacol rearrangement
	$R_2C{=}O$	(1) $ClCH_2CO_2C_2H_5$ + $NaNH_2$; (2) H_3O^+	R_2CHCHO	Glycidic ester synthesis
Ketones	$ArOCH_3$	(1) Na/NH_3, ROH; (2) H_3O^+	Cyclohexenones	Birch reduction
	R_2CHOH	$Na_2Cr_2O_7/H_2SO_4$	$R_2C{=}O$	See Table 31.1 for more details
	R_2CHX	(1) $(CH_3)_2S{=}O$; (2) base	$R_2C{=}O$	

Table 28.1—*cont.*

Type of Product	Starting Material	Reagents	Products	Comments
	R_2CHNR_2	$KMnO_4$	$R_2C{=}O$	Restricted in application
	R_2CHNO_2	(1) Base; (2) H_3O^+	$R_2C{=}O$	Nef reaction
	$R_2C{=}CR'_2$	O_3	$R_2C{=}O + R'_2C{=}O$	Useful diagnostic as well as preparative method
	$R_2C(OH){-}CR'_2(OH)$	HIO_4 or $Pb(OAc)_4$	$R_2C{=}O + R'_2C{=}O$	Specific for 1,2-diols
	ArH	$RCOCl + AlCl_3$	ArCOR	
	$CH_3COCH_2CO_2C_2H_5$	(1) $RX + C_2H_5ONa$; (2) $H_3O^+(-CO_2)$	CH_3COCH_2R or CH_3COCHR_2	Acetoacetic ester synthesis—usual restrictions of S_N2 reactions
	$RCH_2CO_2C_2H_5$	(1) Base; (2) H_3O^+	$RCH_2C(=O)CH(R)CO_2C_2H_5$	Claisen condensation
	$CH_2CO_2C_2H_5$ / $(CH_2)_nCO_2C_2H_5$ (linked)	(1) Base; (2) H_3O^+	cyclic $(CH_2)_n$ ring containing $CHCO_2C_2H_5$ and $C{=}O$	Dieckmann condensation
		(1) Na; (2) H_3O^+	cyclic $(CH_2)_n$ ring containing CH_2CHOH and $C{=}O$	Acyloin condensation
	RCOCl	(1) R'_2Cd; (2) H_3O^+	RCOR′	Little tendency to form tertiary alcohol
	RCN	(1) R′MgBr; (2) H_3O^+	RCOR′	
	RCO_2H	(1) R′Li; (2) H_3O^+	RCOR′	
	RX	(1) 2-R′-2-Li-1,3-dithiane; (2) H_2O, $HgCl_2$	RCOR′	
	$RC{\equiv}CH$	H_2OHgSO_4, H_2SO_4	$RCOCH_3$	Markovnikov addition
	$R_2C(OH){-}CR_2(OH)$	Acid catalysts	$R_3CC(=O)R$	Pinacol rearrangement
	$R_2C{-}CR_2$ (oxirane, O bridging)	Acid catalysts	$R_3CC(=O)R$	Oxirane rearrangement
Carboxylic acids	RCH_2OH	$KMnO_4/NaOH$ or $Na_2Cr_2O_7/H_2SO_4$	RCO_2H	
	RCHO	$KMnO_4/NaOH$, Cu^+, or Ag^+	RCO_2H	
	$RCH{=}CHR'$	$KMnO_4/NaOH$ or $Na_2Cr_2O_7/H_2SO_4$	$RCO_2H + R'CO_2H$	
	$RCOCH_3$	Halogen/NaOH	RCO_2H	Haloform reaction

Table 28.1—*cont.*

Type of Product	Starting Material	Reagents	Products	Comments
	$CH_3COC(R)(R')CO_2C_2H_5$	(1) Concentrated base; (2) H_3O^+	$RR'CHCO_2H$	"Acid" cleavage of acetoacetic esters
	$CH_2(CO_2C_2H_5)_2$	(1) $RX + C_2H_5ONa$; (2) H_3O^+; (3) heat	RCH_2CO_2H or $RR'CHCO_2H$	Malonic ester syntheses; restrictions of S_N2 reactions
	RX	(1) Mg; (2) CO_2; (3) H_3O^+	RCO_2H	No restrictions on structure of R
	RX	(1) CN^-; (2) H_3O^+ or OH^-	RCO_2H	Restrictions of S_N2 reactions
	$RCOCHN_2$	(1) Δ, hν, or $Ag^+/R'OH$; (2) H_2O	RCH_2CO_2H	
	$R_2C{=}NOH$	(1) Polyphosphoric acid or PCl_5; (2) H_2O	RCO_2H ($+RNH_2$)	Beckmann rearrangement
	$R_2C{=}O$	(1) $C_6H_5CO_3H$ (2) H_2O	RCO_2H	Baeyer–Villiger rearrangement
	$RC(=O)-C(=O)R$	(1) NaOH; (2) H_3O^+	$R_2C(OH)CO_2H$	Benzilic acid rearrangement
	$-\overset{\vert}{\underset{\vert}{C}}-X$ with adjacent $C{=}O$	(1) NaOH; (2) H_3O^+	$-\overset{\vert}{\underset{\vert}{C}}-CO_2H$	Favorskii rearrangement
Acid chlorides	RCO_2H	PCl_3, PCl_5, or $SOCl_2$	RCOCl	
Anhydrides	RCO_2H	R'COCl + pyridine	$RC(=O)OC(=O)R'$	Unsymmetrical anhydrides disproportionate to symmetrical anhydrides on distillation
	RCO_2H	$R'CH{=}C{=}O$	$RC(=O)OC(=O)CH_2R'$	
Esters	RCO_2H	$R'OH + H^+$	RCO_2R'	Equilibrium constants near unity
	RCOCl or $(RCO)_2O$	R'OH	RCO_2R'	Equilibrium constants very large
	$RCH{=}C{=}O$	R'OH	RCH_2CO_2R'	Equilibrium constants very large
Amides	RCOCl	NHR'_2	$RCONR'_2$	Equilibrium constants large
	$(RCO)_2O$	NHR'_2	$RCONR'_2$	Equilibrium constants large
	RCO_2R''	NHR'_2	$RCONR'_2$	Equilibrium constants moderately large
	$RCH{=}C{=}O$	NHR'_2	$RCH_2CONR'_2$	Equilibrium constant very large
Ketenes	$R_2CHCOCl$	$(C_2H_5)_3N$	$R_2C{=}C{=}O$	Product often dimerizes

Table 28.1—*cont.*

Type of Product	Starting Material	Reagents	Products	Comments
α-Dicarbonyl compounds	$RC(=O)CH_2R'$	SeO_2	$RC(=O)-C(=O)R'$	α-Ketoaldehydes if $R' = H$ α-Diketones if $R' =$ alkyl or aryl
	$RC(=O)CH(OH)R'$	Ag^+	$RC(=O)-C(=O)R'$	α-Hydroxyketones prepare via reductive dimerization of esters (acyloin condensation)
β-Dicarbonyl compounds	$RCH_2CO_2C_2H_5$	C_2H_5ONa	$RCH_2C(=O)CH(R)CO_2C_2H_5$	Claisen condensation
	$(CH_2)_n$ bearing $CO_2C_2H_5$ and $CH_2CO_2C_2H_5$	C_2H_5ONa	$(CH_2)_n$ ring closed by $C{=}O$ and $CHCO_2C_2H_5$	Dieckman condensation; Works best when n = 3, 4, or 5; not applicable to larger rings
	$(CH_2)_n$ bearing CN and CH_2CN	(1) $C_6H_5N(CH_3)Li$; (2) H_2O	$(CH_2)_n$ ring closed by $C{=}O$ and CH_2	Thorpe–Ziegler method; Works best when n = 3, 4, or 5, but also applicable to larger rings
	$RCH_2-C(=O)R$	$R'CO_2C_2H_5/C_2H_5ONa$	$RCH(C(=O)R')-C(=O)R$	Formylation when $R' = H$ Acylation when $R' =$ alkyl

28.19 *Study Guide*

I. Synthesis of aldehydes and ketones
  A. Methods involving reduction of carboxylic acid derivatives
    1. Synthesis of aldehydes
      (a) Compounds reduced: RCOCl. $RCONR_2$. or RCN
      (b) Reducing agents: H_2 and a catalyst, or lithium aluminum trialkoxyhydrides
    2. Synthesis of ketones
      (a) Reaction of acid chlorides or nitriles with Grignard reagents
      (b) Reaction of acid chlorides with organocadmium reagents
      (c) Reaction of carboxylic acids with organolithium reagents
  B. Methods involving oxidation without cleavage of carbon–carbon bonds
    1. Oxidation of activated methyl and methylene groups
    2. Oxidation of primary and secondary alcohols
    3. Oxidation of alkyl halides with dimethyl sulfoxide
    4. Oxidation of primary alcohols with dimethyl sulfoxide, dicyclohexylcarbodiimide, and pyridinium phosphate

5. Oxidation of alkyl nitro compounds
   (a) Salt of nitro compound converted to aldehyde or ketone either by hydrolysis or by permanganate oxidation
   (b) Application to the extension of the carbon chain in sugars
6. Oxidation of amines—limited in scope

C. Methods involving oxidation with cleavage of carbon–carbon bonds
   1. Ozonolysis of alkenes
   2. Glycol cleavage—with lead tetraacetate or periodic acid

D. Methods involving the formation of carbon–carbon bonds
   1. Acylation of aromatic compounds
      (a) Aldehyde synthesis
         (1) Reagents: ArH, CO, HCl, $AlCl_3$ (Gattermann reaction)
         (2) Other formylating agents: N,N-dimethylformamide or N-methylformanilide with $POCl_3$ (Vilsmeier reaction)
      (b) Ketone synthesis
         (1) Friedel–Crafts acylation
         (2) Fries rearrangement
   2. Alkylation of 1,3-dithianes
   3. Alkylation of acetoacetic ester
      (a) Alkylation step—an S_N2 reaction
      (b) Hydrolysis step—yields a β-keto acid which decarboxylates
   4. Formylation of Grignard reagents

E. Methods involving the addition of water to alkynes
   1. Hydration of acetylene yields acetaldehyde
   2. Hydration of monosubstituted acetylenes
      (a) Acid-catalyzed hydrolysis yields methyl ketones
      (b) Hydroboration-hydrogen peroxide oxidation procedure yields aldehydes

F. Methods involving molecular rearrangement
   1. Rearrangement of pinacols
   2. Rearrangement of oxiranes; glycidic ester method for aldehydes

II. Synthesis of carboxylic acids

A. Methods involving oxidation of alkanes, primary alcohols, and aldehydes

B. Methods involving cleavage of carbon–carbon bonds
   1. Oxidation of alkenes and alkynes
   2. Oxidation of secondary alcohols and ketones

C. Methods involving the formation of carbon–carbon bonds
   1. Alkylation of acetoacetic ester followed by cleavage with strong base
   2. Alkylation of malonic ester followed by hydrolysis and decarboxylation
   3. Carboxylation of organometallic reagents
   4. Formation and hydrolysis of cyano compounds
      (a) S_N2 reaction of alkyl halides and cyanide to form alkyl nitriles
      (b) Addition of hydrogen cyanide to aldehydes and ketones to form cyanohydrins

(c) Strecker synthesis of α-amino acids

D. Methods involving hydrolysis—hydrolysis of benzotrihalides

E. Methods involving molecular rearrangement

1. Rearrangement of diazoketones (Arndt–Eistert sequence)
2. Rearrangement of oximes (Beckmann rearrangement)
3. Peracid oxidation of ketones (Baeyer–Villiger rearrangement)
4. Benzilic acid rearrangement of α-diketones
5. Favorskii reaction of α-chloroketones

III. Synthesis of carboxylic acid derivatives

A. Synthesis of acid chlorides; action of PCl_3, PCl_5, or $SOCl_2$ on carboxylic acid

B. Synthesis of anhydrides

1. Action of carboxylate ion on an acid chloride
2. Action of ketene on a carboxylic acid—yields a mixed anhydride
3. Action of dicyclohexylcarbodiimide on a carboxylic acid

C. Synthesis of esters

1. Reaction of carboxylic acids with alcohols—equilibrium constant near unity
2. Reactions of acid chlorides and anhydrides with alcohols—equilibrium constant very favorable to ester formation
3. Reaction of carboxylate ions with alkyl halides
4. Reactions of carboxylic acids with diazoalkanes—very mild method
5. Reaction of carboxylic acids with alkenes—electrophilic addition

D. Synthesis of amides

1. Reaction of ammonia, primary amines, or secondary amines with acid chlorides, anhydrides, esters, or ketenes
2. Partial hydrolysis of nitriles

IV. Synthesis of ketenes

A. Cracking of compounds containing a $-COCH_3$ group

B. Action of trialkylamines on acid chlorides

V. Synthesis of dicarbonyl compounds

A. Synthesis of α-dicarbonyl compounds

1. Oxidation of methyl and methylene groups adjacent to carbonyl functions
2. Oxidation of α-hydroxyketones
   (a) α-Hydroxyketones (acyloins) prepared by reductive dimerization of esters
   (b) Intramolecular acyloin condensation—allows synthesis of large carbocyclic rings

B. Synthesis of β-dicarbonyl compounds

1. Self-condensation of esters (or nitriles)
   (a) Intermolecular reaction—Claisen condensation
   (b) Intramolecular reaction
      (1) Dieckmann condensation—generally limited to the preparation of five-, six-, and seven-membered rings

(2) Thorpe–Ziegler condensation—can be used for the preparation of rings larger than seven membered
(c) Equilibrium considerations in Claisen condensations
2. Mixed ester condensations

REFERENCES

BUEHLER, C. A., and D. E. PEARSON, *Survey of Organic Syntheses*, John Wiley & Sons, Inc., New York, 1970, Chapters 10–18.

HOUBEN-WEYL, *Methoden der Organischen Chemie*, 4th ed., Vol. VII, *Part 1: Aldehydes*, ed. by E. Müller, Georg Thieme Verlag, Stuttgart, 1954.

PROBLEMS

1. Cite specific examples in which the following reagents are employed to useful synthetic advantage:

(a) "Poisoned" palladium on barium sulfate catalyst (b) Lithium aluminum tri-*t*-butoxy hydride (c) Selenium dioxide
(d) Cupric acetate (e) Dimethyl sulfoxide (f) Periodic acid

2. Cite specific examples for each of the following types of reactions:

(a) Rosenmund reduction (b) Nef reaction (c) Glycidic ester synthesis
(d) Birch reduction (e) Oppenauer oxidation (f) Claisen condensation
(g) Dieckmann condensation (h) Acyloin condensation (i) Barbier–Wieland degradation
(j) Arndt–Eistert sequence (k) Beckmann rearrangement (l) Baeyer–Villiger rearrangement
(m) Benzilic acid rearrangement (n) Favorskii reaction (o) Fries rearrangement

3. Compare and contrast the acetoacetic ester method and the malonic ester method with respect to the preparation of substituted acetic acids. Comment on the scope and limitations of each of these methods.

4. Compare and contrast the acetoacetic ester method and the 1,3-dithiane method with respect to the preparation of ketones. Comment on the scope and limitations of each of these methods.

5. Complete the following reactions:

(a) 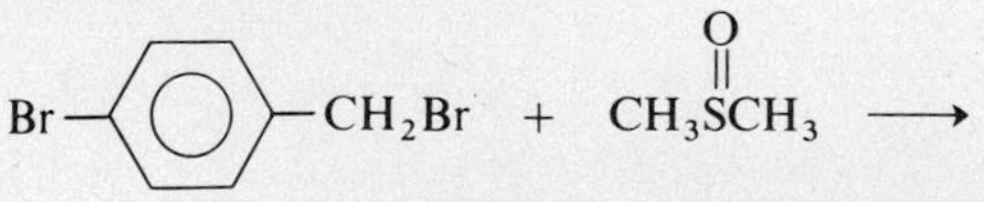 Br–C_6H_4–CH_2Br + $CH_3\overset{O}{\overset{\|}{S}}CH_3$ $\longrightarrow$

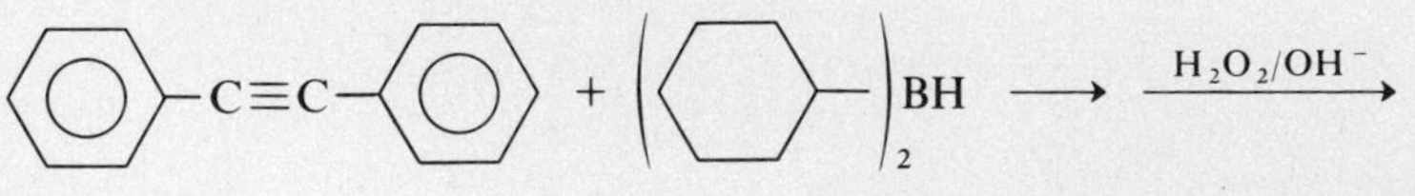

(b) C_6H_5–C≡C–C_6H_5 + $(C_6H_{11})_2BH$ $\longrightarrow$ $\xrightarrow{H_2O_2/OH^-}$

(c) [bicyclic ketone with angular CH_3, O and H] + CF_3CO_3H $\longrightarrow$

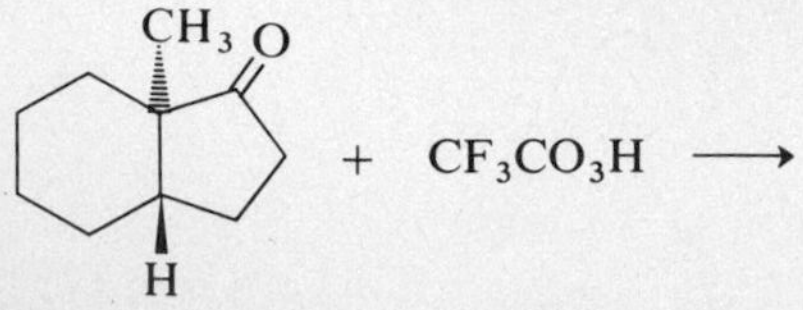

(d) $(CH_3)_3C\overset{\overset{\large O}{\|}}{C}CH_3$ + $3\,Br_2$ + NaOH ⟶

(e) $C_6H_5(H)C{=}C{=}O$ + $C_6H_5C(CH_3)_2OH$ ⟶

(f) $CH_3CH_2\underset{CH_3}{CH}\overset{O}{\overset{\|}{C}}\underset{CH_3}{CH}CO_2CH_3$ + Concentrated NaOH ⟶

(g) $CH_3O_2C\underset{CH_3}{CH}(CH_2)_5CO_2CH_3$ + Na $\xrightarrow{\text{Xylene, }140^\circ}$

(h) 1-hydroxycyclopentyl–$\overset{O}{\overset{\|}{C}}-CH_3$ + HIO_4 ⟶

(i) cyclopropyl–$\overset{O}{\overset{\|}{C}}-OH$ + $(CH_3)_2CHCH_2Li$ ⟶

(j) $C_6H_5(H)C\!-\!O\!-\!C(H)CH_3$ (epoxide) + BF_3 ⟶

(k) cyclic α-diazo ketone ($=O$, $=N_2$) $\xrightarrow[CH_3OH]{h\nu}$

(l) phenanthrenequinone ($=O$, $=O$) + Concentrated KOH ⟶

6. Propose a reasonable mechanism for each of the reactions in Problem 5.

7. Outline, to the extent possible on the basis of the information that is available to you, the scope and limitations of the following types of reactions:
(a) The ozonolysis reaction as a method for the preparation of aldehydes
(b) The pinacol rearrangement as a method for the preparation of aldehydes
(c) The Birch reduction as a method for the preparation of ketones
(d) The Dieckmann condensation as a method for the preparation of ketones
(e) The Beckmann rearrangement as a method for the preparation of carboxylic acids

8. Given the following starting materials, what is the method of choice for converting them to the indicated products?

(a) o-$C_6H_4(CO_2H)(CO_2CH_3)$ ⟶ o-$C_6H_4(CHO)(CO_2CH_3)$

(b) $C_6H_5COCH_2C_6H_5$ ⟶ $C_6H_5COCOC_6H_5$

(c) p-$O_2NC_6H_4CH_2Br$ ⟶ p-$O_2NC_6H_4CHO$

(d) $(CH_3)_2CHCH_2Cl$ ⟶ $(CH_3)_2CHCH_2COCH_3$

(e) $(CH_3)_2C{=}O$ ⟶ $(CH_3)_3CCOCH_3$

(f) 6,6-dimethylcyclohexenyl-$CH_2CH_2C(=O)CH_3$ ⟶ 6,6-dimethylcyclohexenyl-$CH_2CH_2CH(CH_3)CHO$

(g) $(CH_3)_2C(CH_2CH_2Br)_2$ ⟶ 4,4-dimethylcyclohexanone

(h) 1-cyclohexenyl-$COCH_3$ ⟶ 1-cyclohexenyl-CO_2H

(i) trans-2-phenylcyclopentane-CO_2H ⟶ trans-2-phenylcyclopentane-CH_2CO_2H

(j) $(CH_3)_2CHCO_2H$ ⟶ 2,2,4,4-tetramethylcyclobutane-1,3-dione

(k) 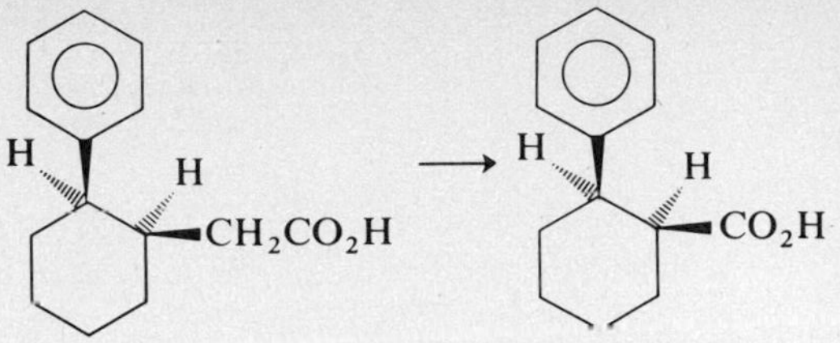 [structure: 2-phenylcyclohexyl-CH_2CO_2H] → [structure: 2-phenylcyclohexyl-CO_2H]

9. Starting with any compound listed in Tables 39.1 to 39.5, along with any necessary inorganic reagents and special organic reagents, devise syntheses for each of the following aldehydes:

(a) 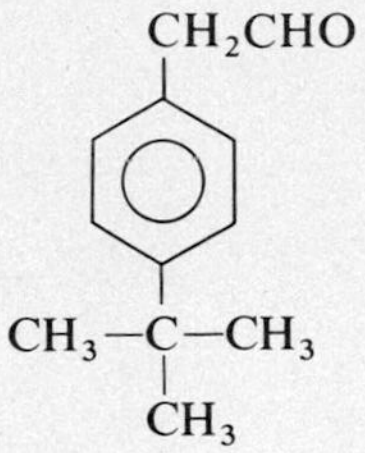 CH_2CHO on a benzene ring, para to $CH_3-C(CH_3)-CH_3$

(b) $CH_3(CH_2)_{12}CHO$

(c) 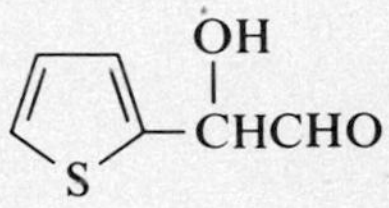 2-thienyl-$\overset{OH}{CH}CHO$

(d) 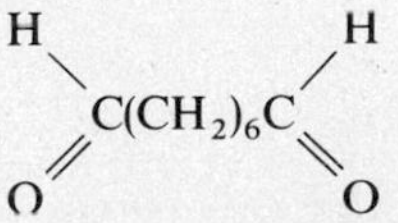 $H(O=)C(CH_2)_6C(=O)H$

(e) 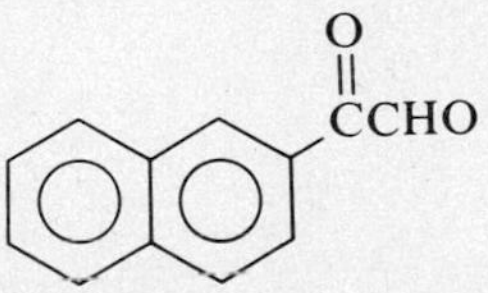 2-naphthyl-$\overset{O}{\overset{\|}{C}}CHO$

10. Starting with any compound listed in Tables 39.1 to 39.5, along with any inorganic reagents and special organic reagents, devise syntheses for each of the following ketones:

(a) 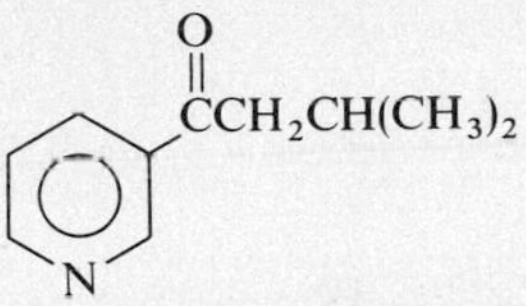 3-pyridyl-$\overset{O}{\overset{\|}{C}}CH_2CH(CH_3)_2$

(b) 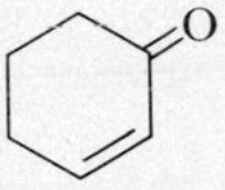 [structure: cyclohexenone]

(c) [structure: cyclic 1,2-dione]

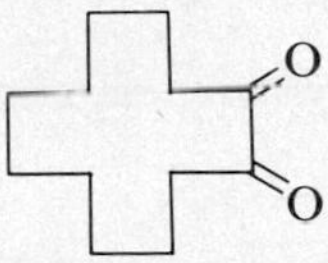

(d) 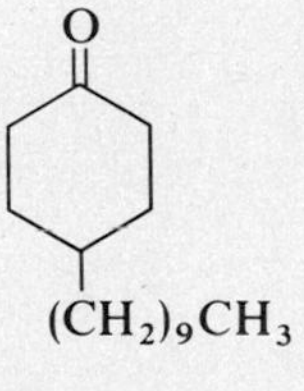 [structure: 4-$(CH_2)_9CH_3$-cyclohexanone]

(e) 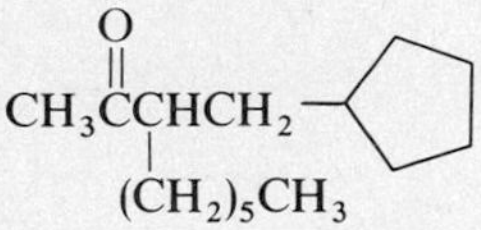 $CH_3\overset{O}{\overset{\|}{C}}CHCH_2$-cyclopentyl, with $(CH_2)_5CH_3$ on CH

11. Starting with any compound listed in Tables 39.1 to 39.5, along with any necessary inorganic reagents and special organic reagents, devise syntheses for each of the following carboxylic acids:

(a) $CH_3(CH_2)_{13}CO_2H$

(b) 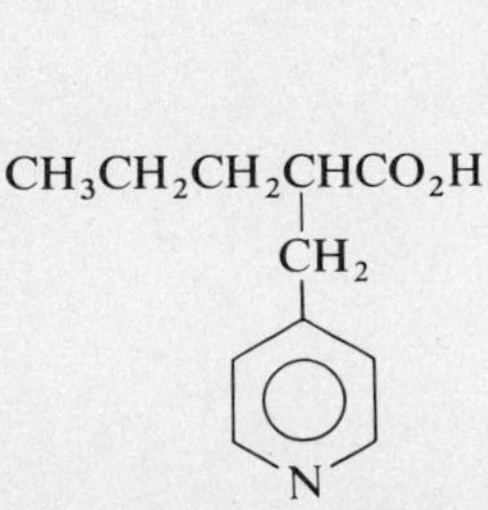 $CH_3CH_2CH_2CHCO_2H$, with CH_2-(4-pyridyl) on CH

(c) 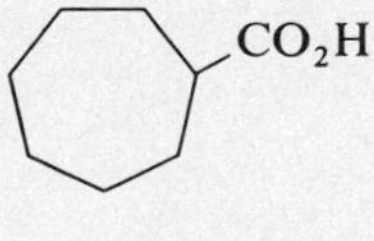 cycloheptyl-CO_2H

(d) p-methoxybenzoic acid (CO_2H / benzene ring / OCH_3)

(e) 2-pyrrolyl–$\underset{\displaystyle NH_2}{CHCO_2H}$

12. Starting with (a) any monofunctional aliphatic compounds containing four or fewer carbon atoms, (b) any monofunctional benzenoid compound, and (c) any necessary inorganic reagents and special organic reagents, devise a reasonable synthesis scheme for each of the following compounds:

(a) $CH_3\overset{O}{\overset{\|}{C}}$–$C_6H_4$–$\overset{O}{\overset{\|}{C}}CH_3$ (para)

(b) cyclobutyl–$\overset{O}{\overset{\|}{C}}CH_2CH_3$

(c) C_6H_5–$\overset{O}{\overset{\|}{C}}$–$\underset{CH_3}{CH}CH_2CH_3$

(d) $(CH_3)_2CHCH_2\overset{O}{\overset{\|}{C}}CH_2CH(CH_3)_2$

(e) $(CH_3)_3CC(=O)H$

(f) $(CH_3)_2CHO$–$\overset{O}{\overset{\|}{C}}$–$CH_2CH(CH_3)_2$

(g) C_6H_5–$CH_2\underset{CH_2CH_2CH_3}{CH}CO_2H$

(h) CH_3O–C_6H_4–$NH\overset{O}{\overset{\|}{C}}CH_3$ (para)

13. Outline plausible mechanisms for the following reactions:

(a) 1-acetyl-5-hydroxycyclohexene ($COCH_3$, OH) $\xrightarrow{KOH}$ 3-acetylcyclohexanone ($COCH_3$, O)

(b) cyclohexyl–COCl $\xrightarrow{(C_2H_5)_3N}$ dispiro cyclobutane-1,3-dione (two cyclohexane rings, two C=O)

(c) $RCH_2NO_2 \xrightarrow[H_2SO_4]{H_2O} RCO_2H + NH_2OH$

(d) $(CH_3)_2C(Br)(COBr) \xrightarrow{Zn} (CH_3)_2C{=}C{=}O$

(e) 1-methylcyclohexene (CH_3) $\xrightarrow{BH_3}\xrightarrow{Na_2Cr_2O_7/H_2SO_4}$ 2-methylcyclohexanone (CH_3, O)

Chapter **29**

Carbonyl compounds: Reactions at the carbonyl group

Let us grasp the situation
  Solve the complicated plot
Quiet calm deliberation
  Disentangles every knot.

GILBERT AND SULLIVAN
The Gondoliers, Act II

The reactions of carbonyl compounds can be broadly divided into two groups; those which involve the carbonyl group directly and those which involve the position adjacent to the carbonyl group. The present chapter deals with reactions involving the carbonyl group directly and includes (a) those in which a nucleophile adds in a 1,2 fashion (direct addition), (b) those in which a nucleophile adds in a 1,4 fashion (conjugate addition), (c) those in which an addition-elimination sequence takes place, and (d) those in which some other type of reaction is involved.

As a basis for organizing the discussion of the reactions of nucleophiles with carbonyl compounds, the nucleophiles are divided into three classes: (a) hetero atom nucleophiles, in which the nucleophilic center is a hetero atom (usually halogen, oxygen, sulfur, or nitrogen), (b) carbon atom nucleophiles, in which the nucleophilic center is a carbon atom, and (c) hydride as a nucleophile.

GENERAL CHARACTERISTICS OF 1,2-NUCLEOPHILIC ADDITION REACTIONS

1,2-Nucleophilic addition (direct nucleophilic addition) involves the attachment of a nucleophile to the carbonyl carbon atom and an electrophile to the oxygen atom, the sequence of attachments usually being in that order:

$$\rangle C{=}O + Nu^- \longrightarrow \rangle C\langle^{Nu}_{O^-} \xrightarrow{E^+} \rangle C\langle^{Nu}_{O-E}$$

Direct addition is the characteristic reaction of type AA carbonyl compounds but is frequently observed in various types of AB and BB type carbonyl compounds as well.

PROBLEM 29-1

Give specific examples of type AA, AB, and BB carbonyl compounds.

29.1 *Catalysis of 1,2-Nucleophilic Addition Reactions*

1,2-Nucleophilic addition reactions are susceptible both to base and acid catalysis. In the case of base catalysis, the function of the base is to generate a more reactive nucleophile from its less nucleophilic conjugate acid. For example, the addition of hydrogen cyanide to aldehydes and ketones requires at least a trace of base for the reaction to proceed with any facility. The nucleophilicity of HCN is negligible, whereas the nucleophilicity of CN^- is appreciable, and a base increases the concentration of the latter (Fig. 29.1). Acid catalysis involves the

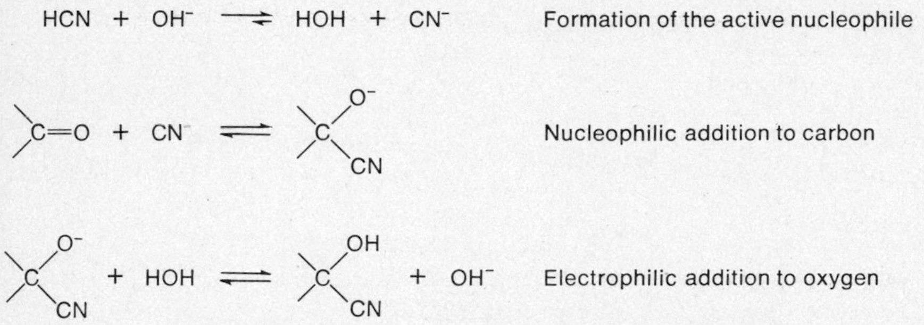

$$HCN + OH^- \rightleftharpoons HOH + CN^- \qquad \text{Formation of the active nucleophile}$$

$$>C{=}O + CN^- \rightleftharpoons >C(O^-)(CN) \qquad \text{Nucleophilic addition to carbon}$$

$$>C(O^-)(CN) + HOH \rightleftharpoons >C(OH)(CN) + OH^- \qquad \text{Electrophilic addition to oxygen}$$

Fig. 29.1. Base-catalyzed addition of HCN to aldehydes and ketones.

coordination of a Lewis acid or a proton with the oxygen atom of the carbonyl group *prior* to the attachment of the nucleophile to the carbon atom. The effect of the coordination is to increase the electrophilic character of the carbon atom and make it more reactive toward nucleophiles. For example, the addition of water to acetaldehyde is catalyzed by mineral acids (Fig. 29.2). The proton becomes affixed to the oxygen atom of acetaldehyde, and the cation that is produced

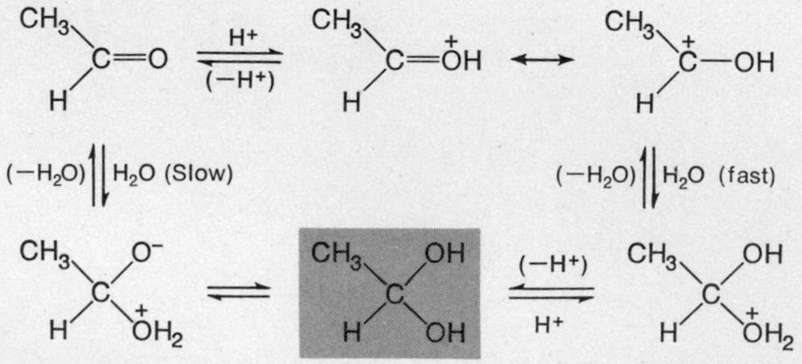

$$CH_3(H)C{=}O \underset{(-H^+)}{\overset{H^+}{\rightleftharpoons}} CH_3(H)C{=}\overset{+}{O}H \longleftrightarrow CH_3(H)\overset{+}{C}{-}OH$$

$$\text{(from } CH_3(H)C{=}O\text{): } \underset{(-H_2O)}{\overset{H_2O\ (\text{Slow})}{\rightleftharpoons}} \qquad \text{(from } CH_3(H)\overset{+}{C}{-}OH\text{): } \underset{(-H_2O)}{\overset{H_2O\ (\text{fast})}{\rightleftharpoons}}$$

$$CH_3(H)C(O^-)(\overset{+}{O}H_2) \rightleftharpoons \boxed{CH_3(H)C(OH)(OH)} \underset{H^+}{\overset{(-H^+)}{\rightleftharpoons}} CH_3(H)C(OH)(\overset{+}{O}H_2)$$

Fig. 29.2. Acid-catalyzed hydration of acetaldehyde.

is very much more reactive toward the weakly nucleophilic water molecule than is acetaldehyde itself. In the absence of acid, the rate at which water adds to the carbonyl group is rather slow; in the presence of acid it becomes so fast that the rate-determining step in the overall process is the hydrogen transfer from the mineral acid to the acetaldehyde. This is quite exceptional, for hydrogen transfer processes are so exceedingly fast that in almost all instances some other step in the sequence is slower and, hence, rate determining.

PROBLEM 29-2

The rate of the uncatalyzed addition of water to the C=O bond is much greater than to the C=C bond, but the equilibrium constant for the latter process is greater than the former. Draw a reaction coordinate reflecting these data.

29.2 Equilibria in 1,2-Nucleophilic Addition Reactions

29.2a. RANGE OF EQUILIBRIUM CONSTANTS FOR 1,2-NUCLEOPHILIC ADDITION REACTIONS. Equilibrium constants for the addition of nucleophiles to C=O bonds vary by many orders of magnitude. In some instances, they are so small that for all practical purposes the reactions don't take place. In other instances, they are so large that the reactions are essentially irreversible. Between these extremes are those reactions, such as cyanohydrin formation, which have equilibrium constants that differ from unity by only a few orders of magnitude. Although *all* nucleophilic additions are actually reversible, it is often convenient to designate only those with equilibrium constants of about 10^4 or smaller as reversible, while specifying those with very high equilibrium constants as irreversible. Included among the "reversible" nucleophilic reactions are additions of (a) weak nucleophiles, such as the halogen acids, water, alcohols, and amines, and (b) moderately strong nucleophiles, such as CN^- and HSO_3^-. In general, only the moderately strong nucleophiles add with sufficient facility to give isolable products.

29.2b. CORRELATION OF STRUCTURE WITH EQUILIBRIUM CONSTANTS FOR 1,2-NUCLEOPHILIC ADDITION REACTIONS. Because the equilibrium constants for the addition of hydrogen cyanide to aldehydes and ketones frequently fall in the experimentally measurable range (*e.g.*, between about 10^3 and 10^{-3}), cyanohydrin formation has been extensively used as a means for obtaining structure-reactivity correlations for 1,2-nucleophilic addition reactions. On the basis of these studies, the following generalizations have been adduced for the equilibrium constants for the system:

$$RR'C{=}O + H{-}Nu \rightleftharpoons RR'C(OH)(Nu)$$

Steric Factors. As the steric bulk of R, R′, and/or Nu increases, the equilibrium constant decreases. The R—C—R′ bond angle in the starting compound is approximately 120°, whereas that in the product is about 109°, resulting in increased steric interference between the R, R′, and Nu groups in the product. For example, the equilibrium constant for the addition of HCN to $CH_3COCH_2CH_3$ is 38, whereas that for $(CH_3)_3CCOC(CH_3)_3$ is too small to measure (Fig. 29.3).

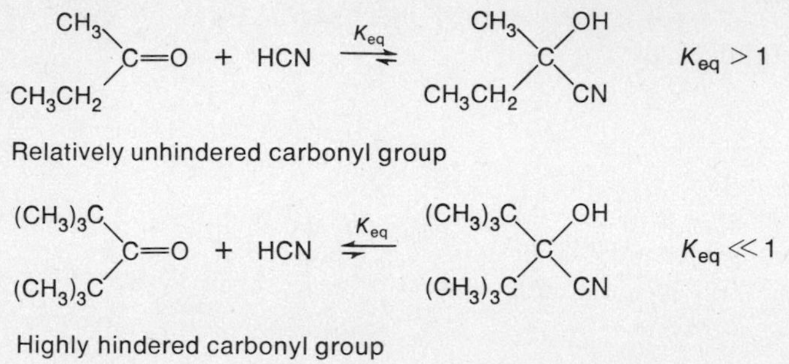

$$CH_3(CH_3CH_2)C{=}O + HCN \xrightleftharpoons{K_{eq}} CH_3(CH_3CH_2)C(OH)CN \qquad K_{eq} > 1$$

Relatively unhindered carbonyl group

$$[(CH_3)_3C]_2C{=}O + HCN \xrightleftharpoons{K_{eq}} [(CH_3)_3C]_2C(OH)CN \qquad K_{eq} \ll 1$$

Highly hindered carbonyl group

Fig. 29.3. Addition of HCN to hindered and unhindered ketones.

Electronegativity Factors. As the electronegativity of the R and R′ groups increases, the equilibrium constant increases. This is the result of electron withdrawal by R and R′ which increases the electrophilic character of the carbon atom, enhancing its susceptibility to nucleophilic attack and increasing the rate of the forward reaction relative to the reverse reaction. For example, the equilibrium constant for the addition of HCN to benzaldehyde is 210, whereas that for *m*-bromobenzaldehyde is 530 as the result of the inductive electron withdrawal by the bromine atom (Fig. 29.4).

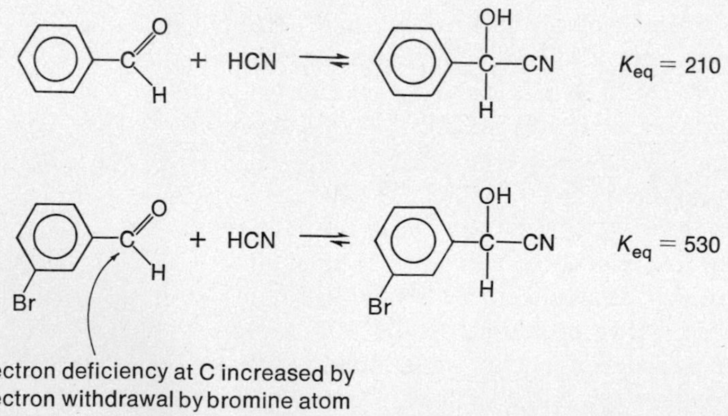

$$C_6H_5CHO + HCN \rightleftharpoons C_6H_5CH(OH)CN \qquad K_{eq} = 210$$

$$m\text{-}BrC_6H_4CHO + HCN \rightleftharpoons m\text{-}BrC_6H_4CH(OH)CN \qquad K_{eq} = 530$$

Electron deficiency at C increased by electron withdrawal by bromine atom

Fig. 29.4. Addition of HCN to aldehydes with and without electron-withdrawing substituents.

Resonance Interaction. As the ability of the R and R′ groups to engage the carbonyl group in resonance interaction increases, the equilibrium constant decreases. Thus, groups containing π bonds (*i.e.*, aryl groups) and groups containing nonbonded electrons (*i.e.*, —halogen, —OR, $—NR_2$, etc.) reduce the reactivity of the carbonyl group toward nucleophilic addition. For example, the equilibrium constant for the addition of HCN to acetaldehyde is too large to measure (*i.e.*, probably greater than 10^4), whereas that for benzaldehyde is only 210; the equilibrium constant for the addition of HCN to $CH_3CH_2COCH_3$ is 38, whereas that for $C_6H_5COCH_3$ is only 0.77 (Fig. 29.5).

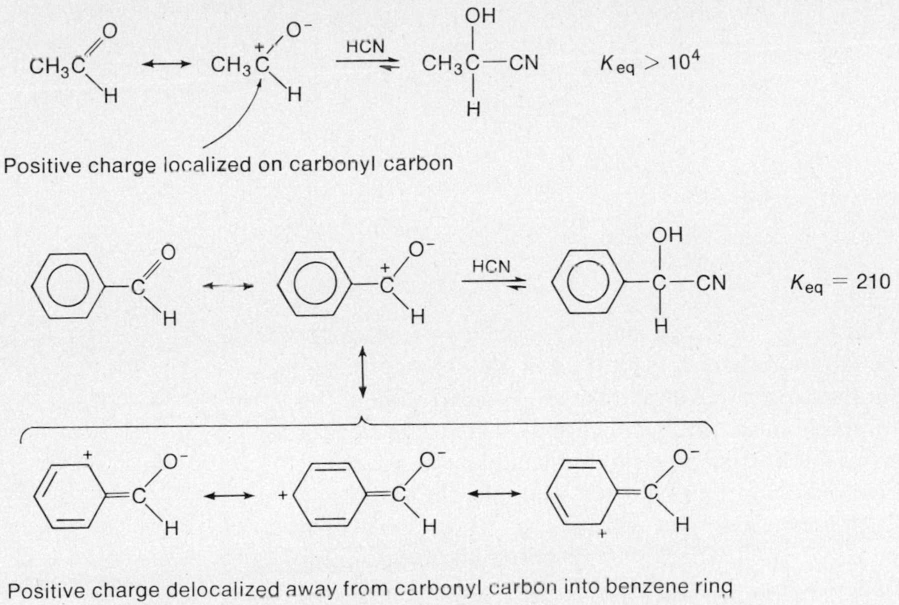

$$CH_3C(=O)H \longleftrightarrow CH_3\overset{+}{C}(O^-)H \xrightarrow{HCN} CH_3C(OH)(H)\text{—}CN \qquad K_{eq} > 10^4$$

Positive charge localized on carbonyl carbon

$$C_6H_5C(=O)H \longleftrightarrow C_6H_5\overset{+}{C}(O^-)H \xrightarrow{HCN} C_6H_5C(OH)(H)\text{—}CN \qquad K_{eq} = 210$$

Positive charge delocalized away from carbonyl carbon into benzene ring

Fig. 29.5. Addition of HCN to charge-localized and charge-delocalized aldehydes.

PROBLEM 29-3

On the basis of the cyanohydrin equilibrium data shown on this page, how would you expect the K_{eq} for the cyanohydrin formation of

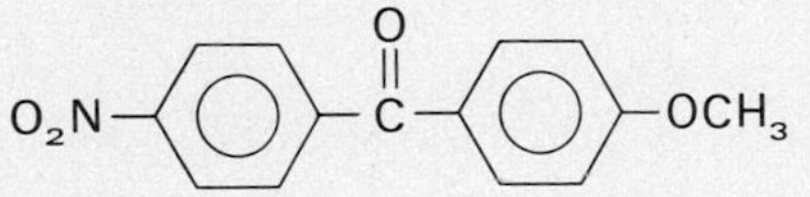

$$O_2N\text{—}C_6H_4\text{—}C(=O)\text{—}C_6H_4\text{—}OCH_3$$

to compare with that of benzophenone?

Nucleophilicity Factors. As the nucleophilicity of Nu decreases, the equilibrium constant decreases. The same trends that have been discussed for nucleophilic substitution reactions (see Table 21.3 for nucleophilicity values) operate in this case also. Thus, nucleophilicity increases from halogen to oxygen to nitrogen to carbon; negatively charged species are stronger nucleophiles than their neutral counterparts (*i.e.*, their conjugate acids); highly polarizable nucleophiles are more effective than less polarizable ones. For example, the equilibrium constant for the addition of water to acetaldehyde is approximately 1, whereas that for the addition of HCN to acetaldehyde is too large to measure (Fig. 29.6).

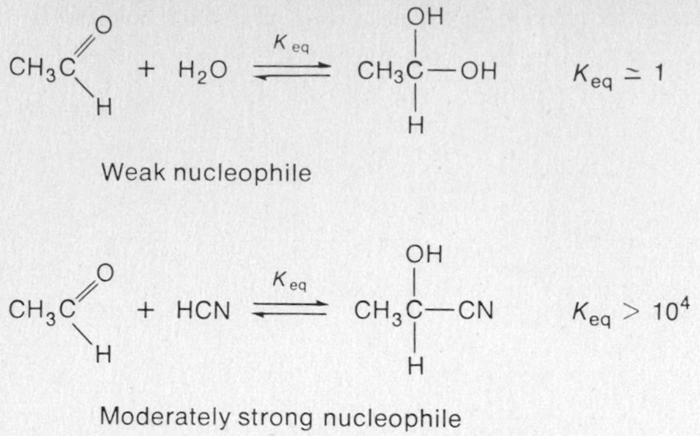

$$CH_3C(=O)H + H_2O \overset{K_{eq}}{\rightleftharpoons} CH_3C(OH)(H)—OH \qquad K_{eq} \simeq 1$$

Weak nucleophile

$$CH_3C(=O)H + HCN \overset{K_{eq}}{\rightleftharpoons} CH_3C(OH)(H)—CN \qquad K_{eq} > 10^4$$

Moderately strong nucleophile

Fig. 29.6. Addition of a weak nucleophile and a moderately strong nucleophile to acetaldehyde.

Bond Angle Factors. As the R—C—R′ bond angle changes, two factors are generally operative, *viz.*, (a) angle strain, which is reduced in going from the sp^2 hybridization of the carbon in the carbonyl group to the sp^3 hybridization of the carbon in the adduct, and (b) nonbonded strain, which may increase in certain cyclic compounds in going from the starting material to the adduct as the result of an increased number of eclipsed conformations. For example, the very large bond angle strain in cyclopropanone (preferred angle is 120°; the actual angle is 60°, so the bond angle strain is 120 − 60, which is 60°) is partially relieved upon adduct formation (preferred angle is 109°; actual angle is 60°, so the bond angle strain is 109 − 60°, which is 49°). Although there are more eclipsed conformations in the adduct than in the starting ketone, the relief of angle strain more than offsets the increase in nonbonded strain. Cyclopropanone is extraordinarily reactive in nucleophilic addition, and even so weak a nucleophile as water adds effectively. Cyclohexanone also adds nucleophiles readily, the equilibrium constant for its reaction with HCN to form a cyanohydrin being about 10,000. In this case, the change in bond angle strain is rather modest (bond angle strain in the ketone is 11° and in the adduct is 0°), and the increase in nonbonded strain is zero (all the conformations in the adduct being of the staggered variety). The equilibrium constants for the addition of HCN to cyclobutanone and cyclopentanone are somewhat lower than those for cyclopropanone and cyclohexanone; in these compounds, the nonbonded strain factors approximately counterbalance the angle strain factors (Fig. 29.7).

PROBLEM 29-4

Would you expect the K_{eq} for cyanohydrin formation from 2-methoxycyclohexanone to be greater or smaller than that from cyclohexanone? Explain the considerations on which your answer is based.

Although the factors just noted can exert dramatic reversals in reactivity, the order of reactivity that is generally observed for 1,2-nucleophilic additions

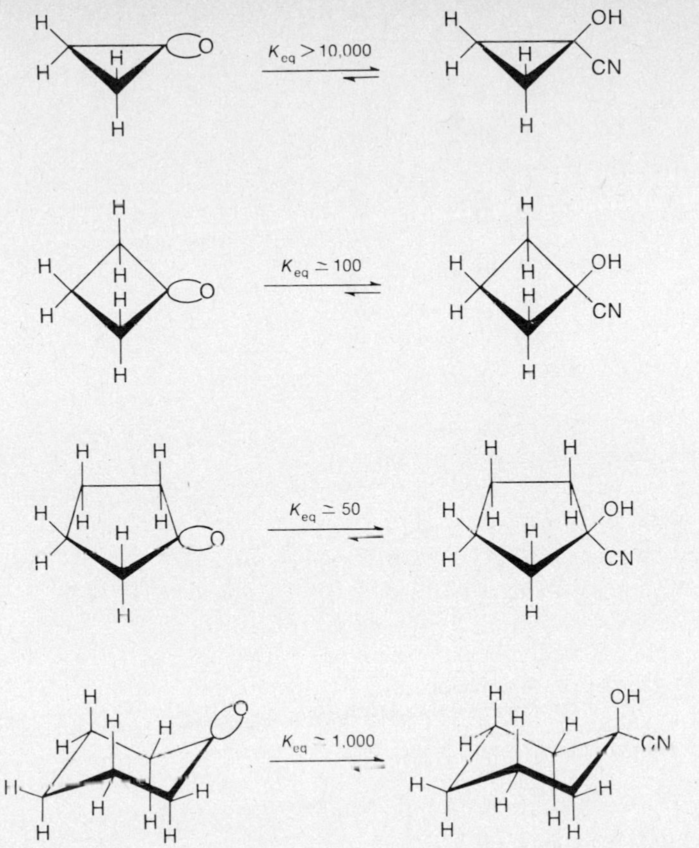

Fig. 29.7. Equilibrium constants for the addition of hydrogen cyanide to cyclic ketones

to aldehydes and ketones is cyclopropanones > aldehydes > cyclohexanones > cyclobutanones > cyclopentanones > dialkyl ketones > alkyl aryl ketones > diaryl ketones.

SCOPE AND LIMITATIONS OF 1,2-NUCLEOPHILIC ADDITION REACTIONS

29.3
Addition of Hetero Atom Nucleophiles

29.3a. ADDITION TO HIGHLY REACTIVE ALDEHYDES AND KETONES. Most hetero atom nucleophiles add to aldehydes and ketones ineffectively, establishing equilibria in which the equilibrium constants are well below unity. Only those carbonyl compounds in which strong electron-withdrawing groups are close to the carbonyl function yield isolable addition products (Fig. 29.8). An exception

---

PROBLEM 29-5

Why does CCl_3CHO add water more effectively than CH_3CHO?

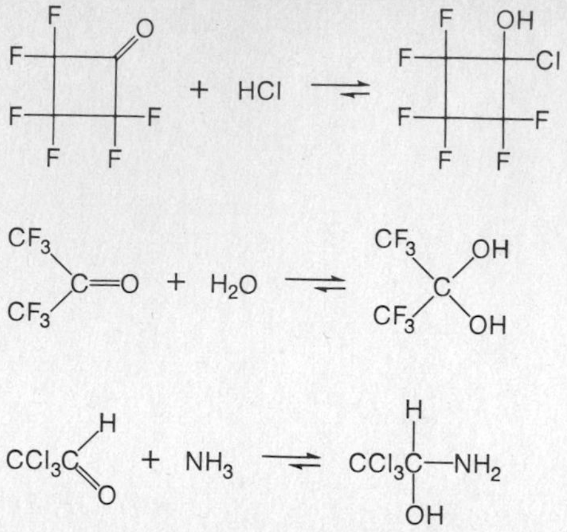

$$C_4F_6O + HCl \rightleftharpoons C_4F_6(OH)Cl$$

$$(CF_3)_2C{=}O + H_2O \rightleftharpoons (CF_3)_2C(OH)_2$$

$$CCl_3CHO + NH_3 \rightleftharpoons CCl_3CH(OH)NH_2$$

Fig. 29.8. Addition of hetero-atom nucleophiles to strongly activated aldehydes and ketones.

to this general pattern of hetero atom nucleophilic addition is sodium bisulfite, in which the sulfur is a sufficiently strong nucleophile to add to most aldehydes as well as to some of the more reactive ketones (*e.g.*, cyclohexanone) (Fig. 29.8). The bisulfite addition products are ionic in character and, therefore, rather insoluble in organic solvents, providing a means for separating aldehydes, and some ketones, from other types of compounds. Regeneration of the original aldehyde is readily accomplished by treatment with either aqueous base or acid (Fig. 29.9).

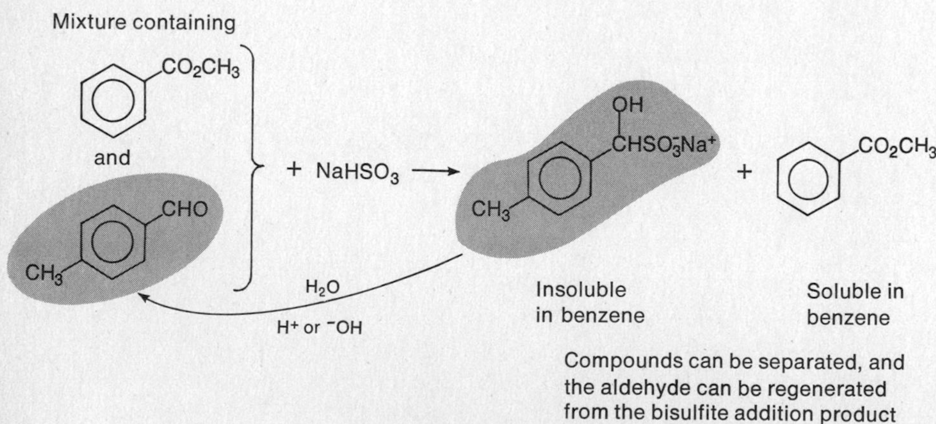

Fig. 29.9. Formation of bisulfite addition products from aldehydes.

29.3b. ADDITION-DEHYDRATION REACTIONS. If the initially formed adduct can lose the elements of water, the unfavorable equilibrium constant of the first step can often be offset by a very favorable equilibrium constant in the dehydration step. With alcohols as nucleophiles, the dehydration can be achieved *inter*-molecularly in an acid-catalyzed process leading to acetals or ketals (Fig. 29.10).

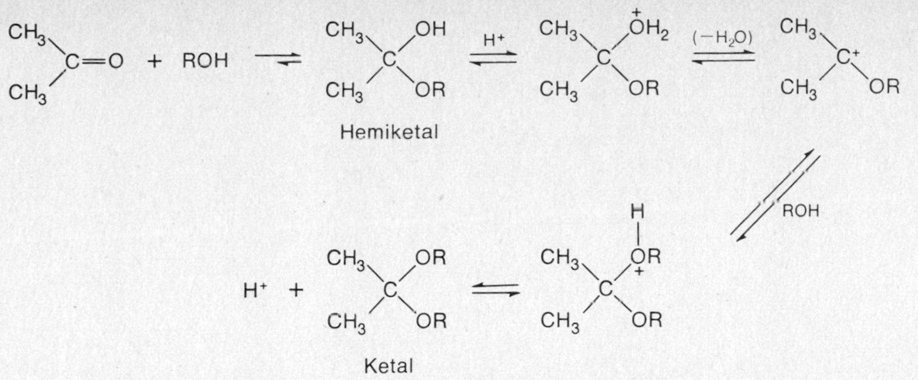

Fig. 29.10. Mechanism of ketal formation.

Thiols (mercaptans) react with aldehydes and ketones in a fashion comparable to alcohols, the initial addition step in this case being more favorable than with alcohols because of the considerably greater nucleophilicity of sulfur. The resulting thioacetals or thioketals can be hydrogenolyzed with molecular hydrogen in the presence of Raney nickel catalyst, oxidized, or converted back to the original aldehyde or ketone (Fig. 29.11). Because acetals and ketals, as well as

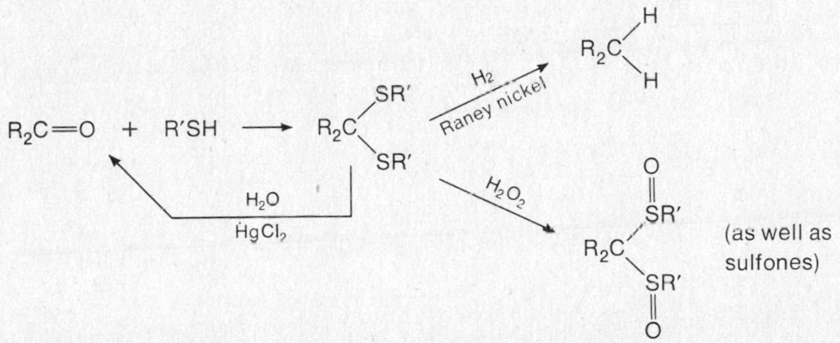

Fig. 29.11. Formation and reactions of thiocetals and thioketals.

thioacetals and thioketals, are stable under basic conditions and are readily converted back to the original aldehydes or ketones (with aqueous acid in the case of acetals or ketals, or aqueous mercuric chloride in the case of thioacetals or thioketals), they are useful intermediates for sequences in which it is necessary to "protect" a carbonyl function while operations elsewhere in the molecule are performed (Fig. 29.12).

The adducts from primary amines and aldehydes or ketones can lose water intramolecularly to form imines. The overall addition-dehydration sequence is particularly favorable with amines such as hydrazine, phenylhydrazine, hydroxylamine, and semicarbazide (see Figs. 13.11 and 13.12), which give derivatives that are especially useful for characterizing aldehydes and ketones (Fig. 29.13).

The adducts from secondary amines and aldehydes or ketones can also lose water intramolecularly if an α hydrogen is present; the products are enamines (Fig. 29.14) (see Fig. 24.14).

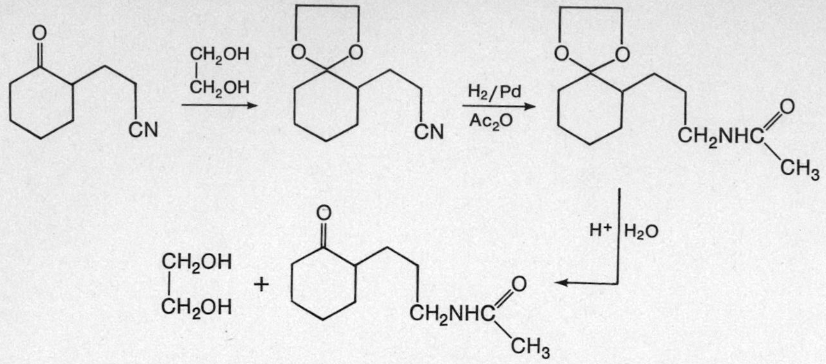

CH_2OH / CH_2OH; CN; H_2/Pd, Ac_2O; $CH_2NHC(=O)CH_3$; H^+ H_2O; CH_2OH / CH_2OH +

Fig. 29.12. Use of a ketal intermediate to "protect" a carbonyl group.

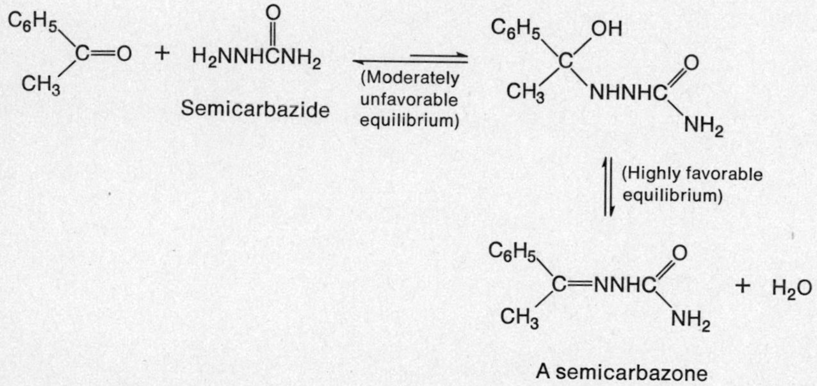

$C_6H_5(CH_3)C{=}O$ + $H_2NNHC(=O)NH_2$ (Semicarbazide) ⇌ (Moderately unfavorable equilibrium) $C_6H_5(CH_3)C(OH)NHNHC(=O)NH_2$

(Highly favorable equilibrium)

$C_6H_5(CH_3)C{=}NNHC(=O)NH_2$ + H_2O

A semicarbazone

Fig. 29.13. Formation of semicarbazone via addition-elimination sequence.

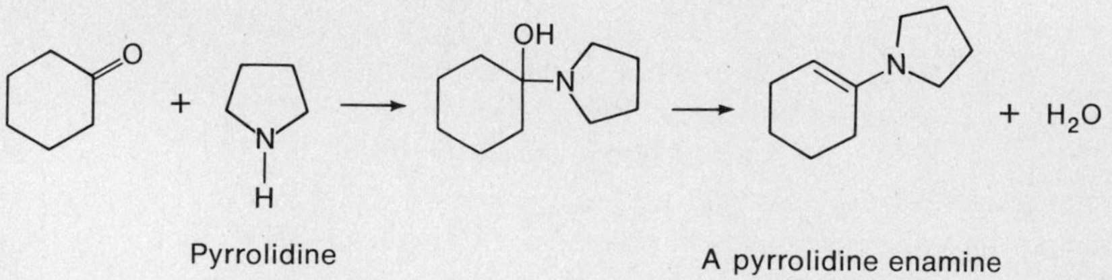

+ → → + H_2O

Pyrrolidine — A pyrrolidine enamine

Fig. 29.14. Formation of the enamine of cyclohexanone.

Ammonia itself reacts with aldehydes and ketones to give imines of the general structure $R_2C{=}NH$, which may undergo further reaction, as exemplified by the reactions with formaldehyde and benzaldehyde:

$$6\,HCHO + 4\,NH_3 \longrightarrow \text{(hexamethylenetetramine, } (CH_2)_6N_4)$$

Hexamethylenetetramine

$$3\,C_6H_5{-}CHO + 2\,NH_3 \longrightarrow C_6H_5{-}CH(N{=}CH{-}C_6H_5)_2$$

PROBLEM 29-6

Suggest a mechanism for the formation of $C_6H_5CH(N{=}CHC_6H_5)_2$ from C_6H_5CHO and NH_3.

Whereas the equilibrium constants are usually a little less than unity for the addition of hetero atom nucleophiles to aldehydes and ketones, and a great deal less than unity for the addition to carboxylic acids and their derivatives, ketenes add nucleophiles in reactions for which the equilibrium constants are very much greater than unity. This can be ascribed (a) to the very high degree of strain in the "two-membered" ring ketone structure (*i.e.*, a cycloethanone) and (b) to the tautomeric rearrangement of the initially formed vinyl alcohol to the much more stable carbonyl compound. All hetero atom nucleophiles, therefore, react with ketenes to give carboxylic acid derivatives—*i.e.*, acid halides from HX, anhydrides from RCO_2H, esters from ROH, and amides from RNH_2:

$$R_2C{=}C{=}O + H{-}Nu \rightleftharpoons R_2C{=}C(Nu){-}OH \rightleftharpoons R_2CH{-}C({=}O){-}Nu$$

PROBLEM 29-7

If you wanted to try to measure the rate of the addition of a nucleophile to a ketene, what ketene and what nucleophile might be reasonable choices to bring the rate of reaction within a measurable range? What physical method might be used to follow the reaction?

29.3c. Disproportionation reactions. Aldehydes that do not contain α hydrogens (*e.g.*, benzaldehyde) react with concentrated sodium hydroxide solutions to yield mixtures of the sodium salt of the carboxylic acid and the corresponding alcohol. The disproportionation, an oxidation-reduction process known as the *Cannizzaro reaction*, involves the addition of hydroxide ion to the carbonyl group followed by the transfer of a hydride ion from the addition product to the carbonyl group of another molecule of aldehyde (Fig. 29.15). Although most aromatic aldehydes undergo the Cannizzaro reaction, *o*- and *p*-hydroxybenzaldehydes fail to do so.

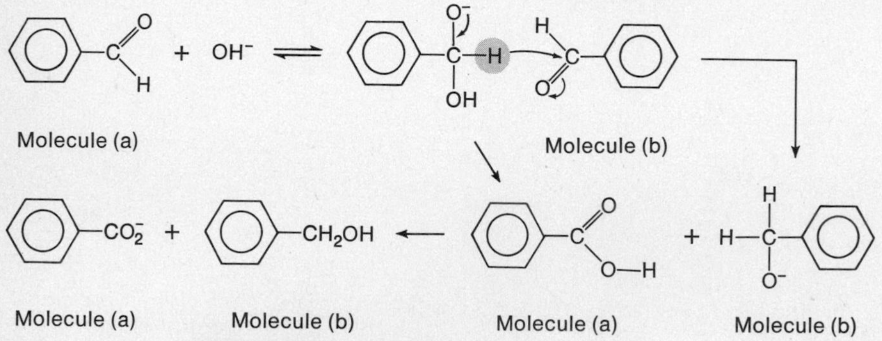

Molecule (a) Molecule (b) Molecule (a) Molecule (b) Molecule (a) Molecule (b)

Fig. 29.15. The Cannizzaro reaction of benzaldehyde.

Formaldehyde, like benzaldehyde, lacks α hydrogens. It has the interesting property of being able to participate in "crossed Cannizzaro" reactions, in which it serves as a reducing reagent toward a different aldehyde. For example, $(HOCH_2)_3CCHO$ reacts with formaldehyde in strongly basic solution to yield pentaerythritol, $(HOCH_2)_3CCH_2OH$ (see Fig. 13.23), and benzaldehyde reacts under similar conditions to form benzyl alcohol:

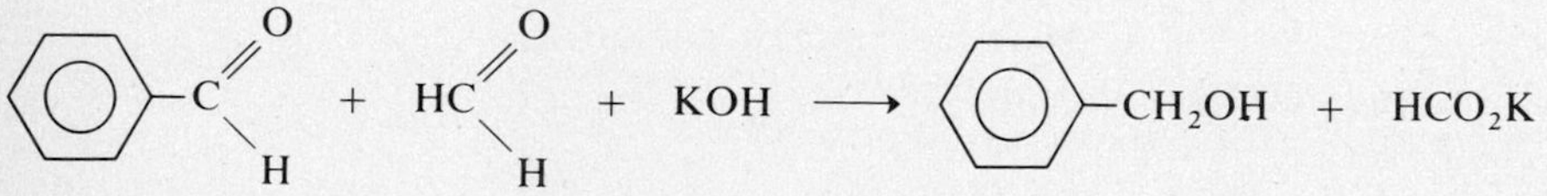

$$C_6H_5CHO + HCHO + KOH \longrightarrow C_6H_5CH_2OH + HCO_2K$$

29.4 Addition of Carbon Nucleophiles

29.4a. CYANIDE AS A CARBON NUCLEOPHILE. The base-catalyzed addition of hydrogen cyanide to aldehydes and ketones yields cyanohydrins, examples of which are discussed in Section 29.2b. Cyanide is only a moderately strong nucleophile, and the equilibrium constants for carbonyl addition are not always favorable to product formation. Most aldehydes, many alkanones, and many cycloalkanones, however, do form stable cyanohydrins that are useful as synthesis intermediates, for they can be converted to a variety of derived products (Fig. 29.16).

29.4b. STRONGLY POLARIZED COMPOUNDS AS SOURCES OF CARBON NUCLEOPHILES. Diazoalkanes, phosphorus ylides and sulfur ylides are compounds in which, to some extent, the carbon atom is the negative end of a dipole, as indicated by the resonance structures shown in Fig. 29.17. As nucleophiles they add to

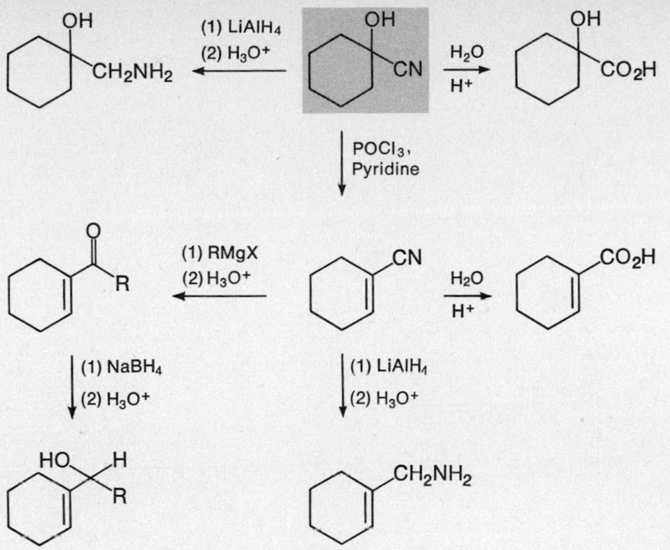

Fig. 29.16. The synthetic utility of cyanohydrins.

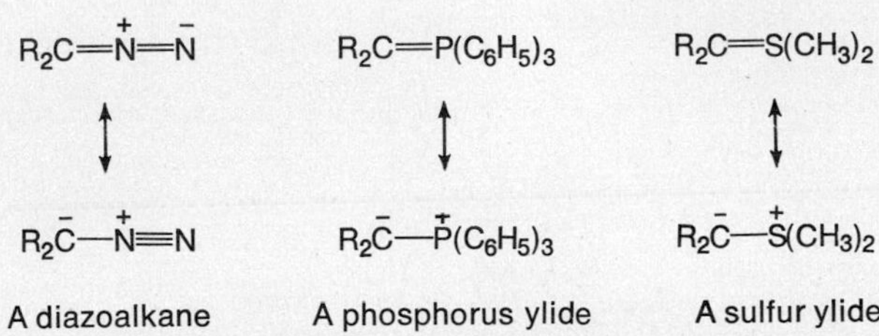

Fig. 29.17. Resonance structures for diazoalkanes, phosphorus ylides, and sulfur ylides which indicate their carbanionic character.

the carbonyl carbon of aldehydes and ketones via direct nucleophilic addition to give intermediates that lose the nitrogen, phosphorus, or sulfur moieties to produce aldehydes, ketones, oxiranes, or alkenes (see Wittig reaction—Section 25.9b) (Fig. 29.18).

PROBLEM 29-8

The reaction of diazomethane with 2-chlorocyclohexanone gives more epoxide than the reaction of diazomethane with cyclohexanone. Suggest a reason for this. Of the two isomeric chlorocycloheptanones possible, which would be predicted to be formed in greater amount? Why?

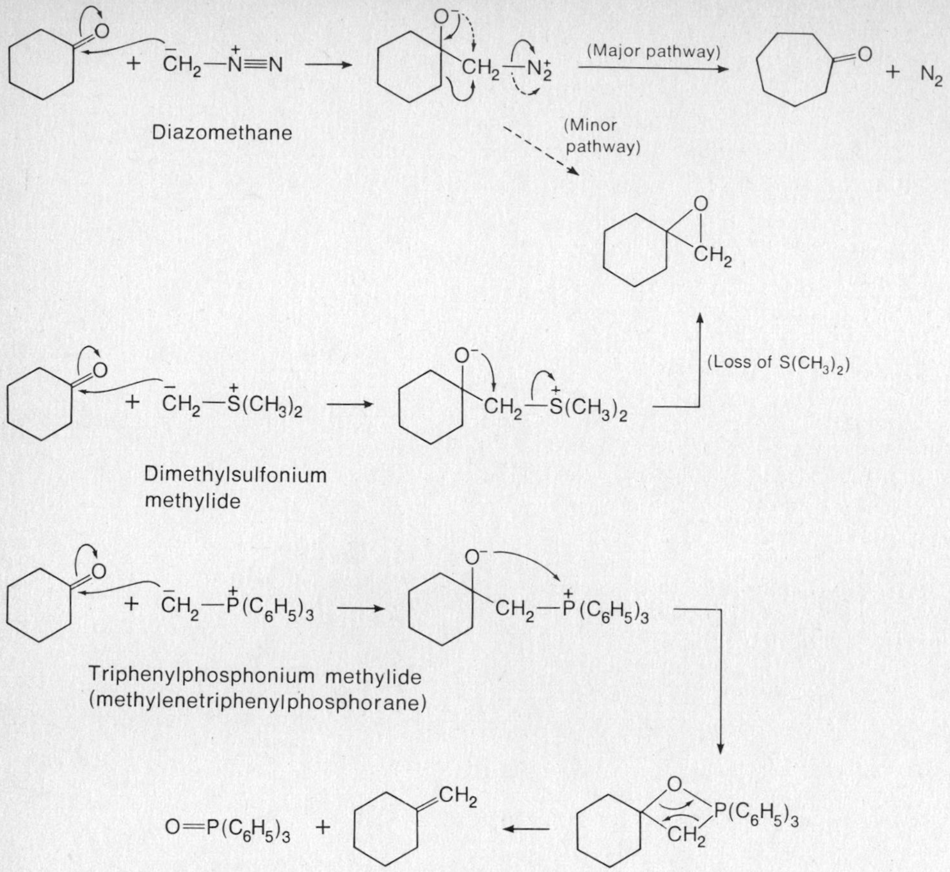

Diazomethane

(Major pathway)

(Minor pathway)

(Loss of $S(CH_3)_2$)

Dimethylsulfonium methylide

Triphenylphosphonium methylide (methylenetriphenylphosphorane)

$O{=}P(C_6H_5)_3$ +

Fig. 29.18. Reactions of nucleophiles from strongly polarized compounds with cyclohexanone.

29.4c. ORGANOMETALLIC COMPOUNDS AS SOURCES OF CARBON NUCLEOPHILES. Among the best-known and most widely used of the carbon atom nucleophiles are those derived from organometallic compounds, particularly organomagnesium reagents (Grignard reagents) and organolithium reagents. The reactions of these compounds with aldehydes and ketones as a means for preparing alcohols are shown in Fig. 13.10 and Table 22.1 and are discussed in Section 22.7b. The organomagnesium and organolithium compounds are, in most instances, entirely comparable in reactivity. In situations where both 1,2- and 1,4-addition (see Section 29.6) are possible, however, the organolithium reagents tend to give a higher proportion of 1,2-addition product (Fig. 29.19).

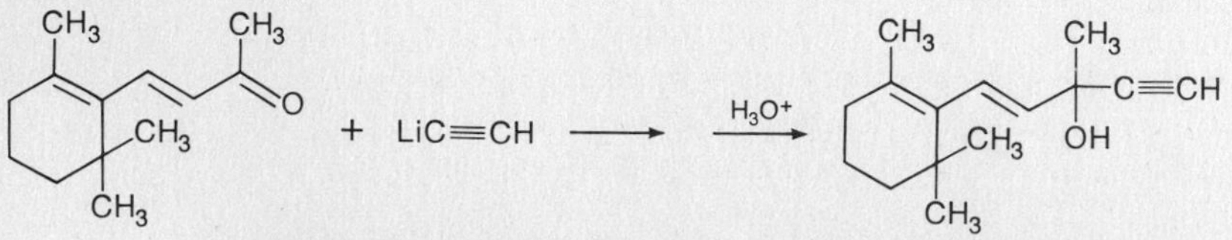

+ $LiC{\equiv}CH$ ⟶ $\xrightarrow{H_3O^+}$

Fig. 29.19. Direct nucleophilic addition of lithium acetylide.

29.4d. ACTIVE METHYLENE COMPOUNDS AS SOURCES OF CARBON NUCLEOPHILES. A very important source of carbon nucleophiles are active methylene compounds, of which acetoacetic ester and malonic ester are typical examples. Since the formation of these nucleophiles involves reactions that occur at the position adjacent to the carbonyl function, we defer discussion of their preparation and reactions to Chapter 30.

29.4e. BENZOIN CONDENSATION. Aldehydes that lack α hydrogens undergo a reaction in the presence of cyanide ion that produces "dimeric" α-hydroxyketones. The process is called the *benzoin condensation*, and it is thought to involve (a) the direct addition of hydrogen cyanide to the carbonyl group of the aldehyde, (b) loss of the α hydrogen of the cyanohydrin formed in step (a), (c) addition of the carbanion formed in step (b) to the carbonyl group of another molecule of the aldehyde, (d) transfer of a proton from one oxygen atom to the other oxygen atom of the condensation product, and (e) expulsion of hydrogen cyanide (Fig. 29.20). Similar reactions occur in biological systems, vitamin B_2 (thiamine pyrophosphate) serving the catalytic role in a fashion comparable to cyanide.

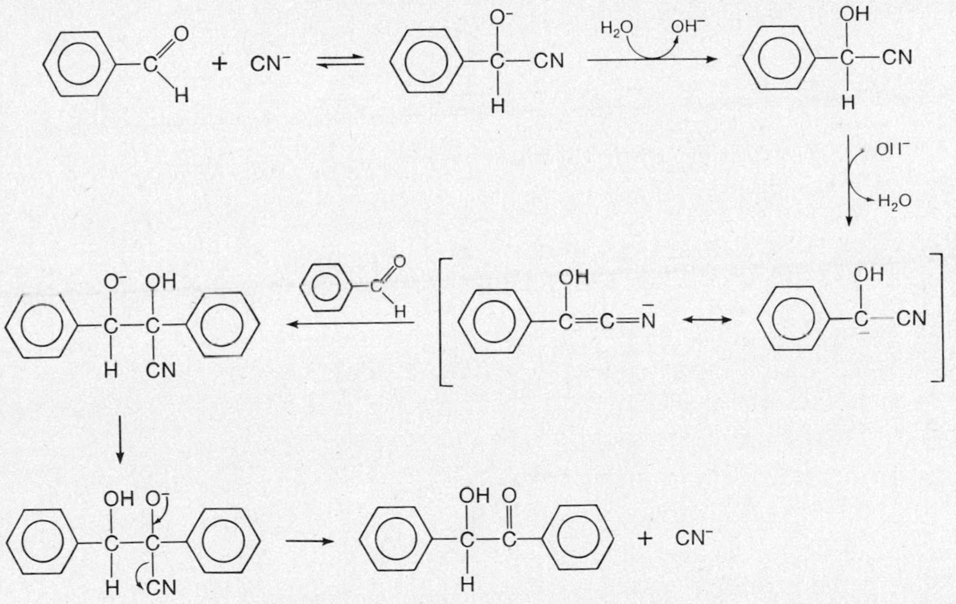

Fig. 29.20. The benzoin condensation of benzaldehyde.

29.5 Addition of Hydride

Hydride is a strong nucleophile with the ability to add to carbonyl functions. The most usual sources of hydride are certain complex metal hydrides, the nucleophilic reactivity of the hydride depending on the particular metal hydride. Lithium aluminum hydride, for example, is a considerably more powerful reducing agent than sodium borohydride; it reduces almost all carbonyl compounds, including carboxylic acids, whereas sodium borohydride reduces aldehydes,

Table 29.1 FUNCTIONAL GROUP SELECTIVITY WITH METAL HYDRIDE REDUCING AGENTS (+ MEANS THAT REDUCTIONS TAKE PLACE, − MEANS THAT NO REACTION OCCURS)

Functional Group	$LiAlH_4$	$LiBH_4$	$NaBH_4$
RCHO	+	+	+
$R_2C{=}O$	+	+	+
RCOCl	+	+	+
RCO_2R'	+	+	−
RCO_2H	+	±	−
RCN	+	−	−
RNO_2	+	−	−
$ArNO_2$	−	±	−
R-Halogen	+	±	−

ketones, and acid chlorides but does not reduce carboxylic acids (Table 29.1). By the appropriate choice of metal hydride, selective reductions can often be carried out (Fig. 29.21).

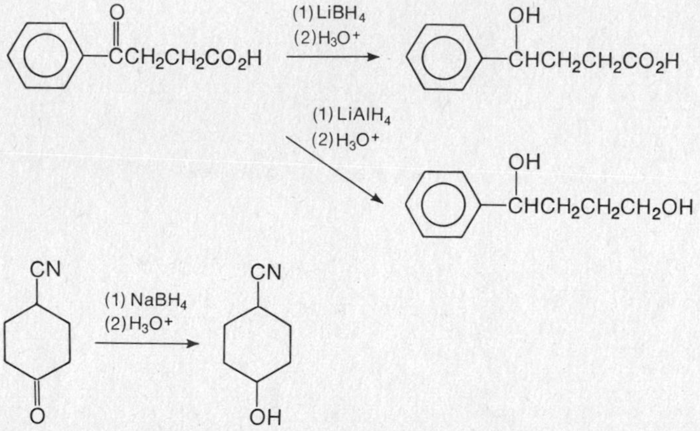

$$C_6H_5COCH_2CH_2CO_2H \xrightarrow[(2)\,H_3O^+]{(1)\,LiBH_4} C_6H_5CH(OH)CH_2CH_2CO_2H$$

$$C_6H_5COCH_2CH_2CO_2H \xrightarrow[(2)\,H_3O^+]{(1)\,LiAlH_4} C_6H_5CH(OH)CH_2CH_2CH_2OH$$

$$\text{4-oxocyclohexanecarbonitrile} \xrightarrow[(2)\,H_3O^+]{(1)\,NaBH_4} \text{4-hydroxycyclohexanecarbonitrile}$$

Fig. 29.21. Metal hydride reductions.

1,4-NUCLEOPHILIC ADDITION REACTIONS

29.6 *General Characteristics of 1,4-Nucleophilic Addition Reactions*

1,4-Nucleophilic addition (conjugate nucleophilic addition) involves the attachment of a nucleophile to the β-carbon of an α,β-unsaturated carbonyl compound and attachment of an electrophile to the oxygen atom. It occurs as a consequence of charge delocalization whereby the electron deficiency resides not only at the carbonyl carbon but also at the β-carbon. It is the characteristic reaction of type AB and BB carbonyl compounds:

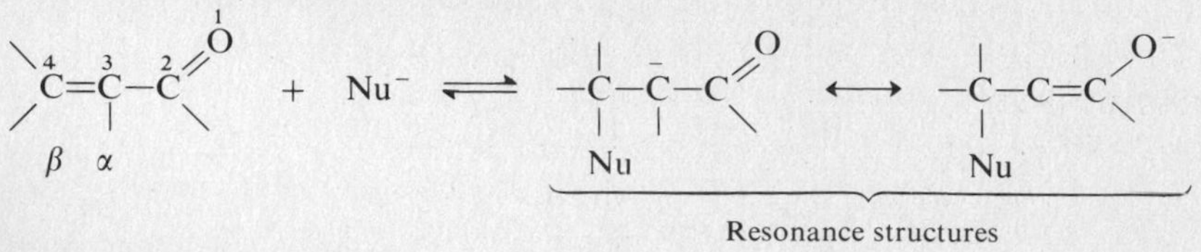

$$\underset{\beta\ \ \alpha}{\overset{4}{C}{=}\overset{3}{C}{-}\overset{2}{C}{=}\overset{1}{O}} + Nu^- \rightleftharpoons \underbrace{-\underset{Nu}{C}{-}\overset{-}{C}{-}C{=}O \longleftrightarrow -\underset{Nu}{C}{-}C{=}C{-}O^-}_{\text{Resonance structures}}$$

The α,β-unsaturated ketone 3-butene-2-one (methyl vinyl ketone), for example, reacts with hydrogen cyanide to yield the product of 1,2 addition at low temperature (the kinetically controlled product), but the product of 1,4 addition at room temperature (the thermodynamically controlled product). (Fig. 29.22). The product resulting from conjugate addition appears to be a 1,2

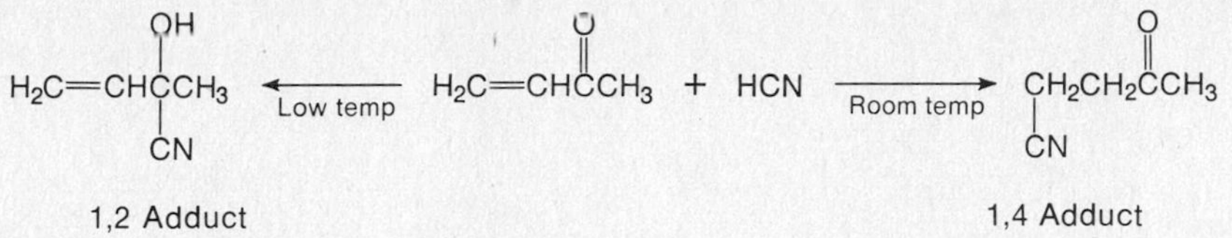

$$\underset{\text{1,2 Adduct}}{H_2C{=}CHC(OH)(CN)CH_3} \xleftarrow{\text{Low temp}} H_2C{=}CHC(=O)CH_3 + HCN \xrightarrow{\text{Room temp}} \underset{\text{1,4 Adduct}}{CH_2(CN)CH_2C(=O)CH_3}$$

Fig. 29.22. 1,2- and 1,4-Nucleophilic addition to methyl vinyl ketone.

addition to the carbon–carbon double bond but actually is the result of a 1,4 addition followed by a tautomeric rearrangement:

$$\overset{4}{C}{=}\overset{3}{C}{-}\overset{2}{C}{=}\overset{1}{O} \underset{}{\overset{CN^-}{\rightleftharpoons}} {-}C(CN){-}C{=}C{-}O^- \overset{H^+}{\rightleftharpoons} {-}C(CN){-}C{=}C{-}OH \rightleftharpoons {-}C(CN){-}C(H){-}C{=}O$$

Addition of nucleophile to atom 4 — Addition of electrophile to atom 1 — Tautomeric rearrangement

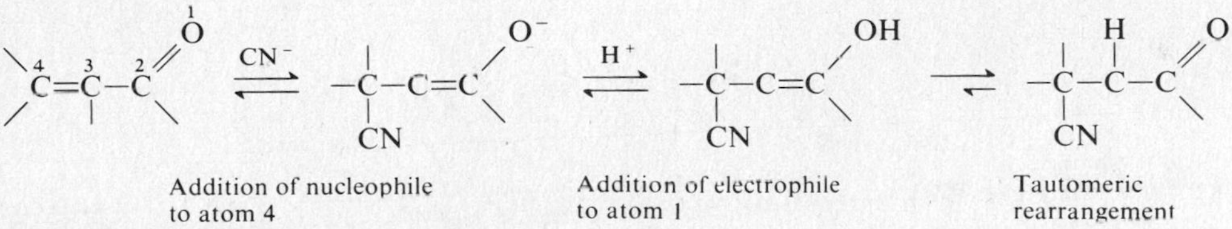

Because of the large equilibrium constants for tautomeric rearrangements, conjugate addition of proton-containing nucleophiles such as hydrogen cyanide is generally favored over direct addition. Furthermore, a carbonyl function such as an ester group that does not react via direct addition with nucleophiles like cyanide, undergoes facile 1,4 addition if present in an α,β-unsaturated compound (Fig. 29.23).

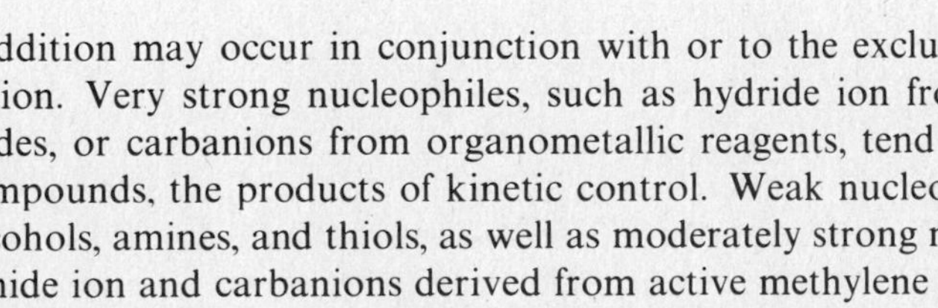

$$CH_3CH_2CH_2CO_2C_2H_5 + HCN \longrightarrow \text{No reaction}$$

$$CH_3CH{=}CHCO_2C_2H_5 + HCN \longrightarrow CH_3CH(CN)CH_2CO_2C_2H_5$$

Fig. 29.23. 1,4-Nucleophilic addition to an α,β-unsaturated ester.

Direct addition may occur in conjunction with or to the exclusion of conjugate addition. Very strong nucleophiles, such as hydride ion from complex metal hydrides, or carbanions from organometallic reagents, tend to give 1,2-addition compounds, the products of kinetic control. Weak nucleophiles, such as water, alcohols, amines, and thiols, as well as moderately strong nucleophiles, such as cyanide ion and carbanions derived from active methylene compounds, are much more likely to give 1,4-addition compounds, the products of thermodynamic control.

29.7 Conjugate Addition Reactions Involving Hetero Atom Nucleophiles

As mentioned, the equilibrium constants for conjugate addition are usually considerably larger than for direct addition, and even weak nucleophiles may add in a conjugate fashion. For example, even so weak a nucleophile as chloride reacts via conjugate addition with 1,4-benzoquinone to yield 2-chloro-1,4-dihydroxybenzene (see p. 358). Similarly, ethanol shows little tendency toward direct addition to ester carbonyl groups but adds in a conjugate fashion to ethyl acrylate to yield ethyl 2-ethoxypropionate:

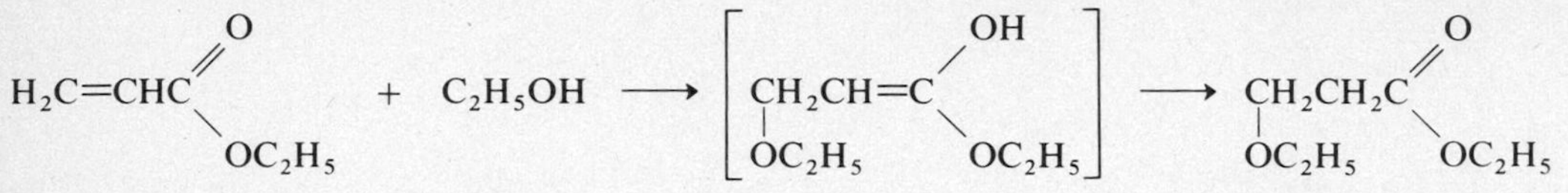

$$\underset{\text{Ethyl acrylate}}{H_2C{=}CHC({=}O)OC_2H_5} + C_2H_5OH \longrightarrow \left[\underset{OC_2H_5}{CH_2}CH{=}C(OH)OC_2H_5\right] \longrightarrow \underset{\text{Ethyl 2-ethoxypropionate}}{\underset{OC_2H_5}{CH_2}CH_2C({=}O)OC_2H_5}$$

PROBLEM 29-9

Phenylhydrazine reacts with ethyl crotonate to yield

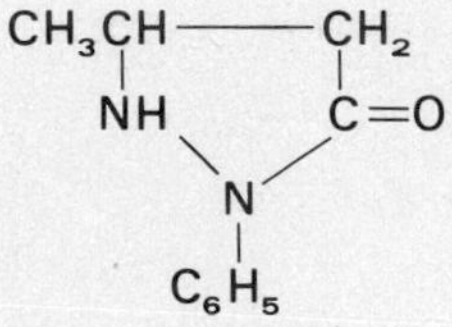

$$\begin{array}{ccc} CH_3CH & \text{—} & CH_2 \\ | & & | \\ NH & & C{=}O \\ & \diagdown \quad \diagup & \\ & N & \\ & | & \\ & C_6H_5 & \end{array}$$

Write a mechanism for the reaction.

29.8 Conjugate Addition Reactions Involving Carbon Frameworks

Carbon atom nucleophiles, including strongly polarized compounds such as diazomethane, lithium dialkylcuprates, cyanide, or active methylene compounds such as diethyl malonate, all are very reactive with respect to conjugate addition reactions (Fig. 29.24).

Another interesting and useful source of carbon frameworks for conjugate addition reactions are the trialkylboranes, derived from the interaction of borane and alkenes (see Section 26.3a). These compounds undergo rapid 1,4-addition to α,β-unsaturated aldehydes and ketones. The reaction appears to be mechanistically different from those illustrated in Fig. 29.24, however, and is thought to involve a radical chain process. The overall sequence, starting with the alkene, provides a convenient method for attaching a carbon chain to one terminus of a double bond. The method has the disadvantage, though, of using only one of the three alkyl groups attached to the boron. Also, it may not be highly regioselective when unsymmetrical alkenes are used in the preparation of the trialkylborane (Fig. 29.25).

29.9 Methods for the Preparation of α,β-Unsaturated Carbonyl Compounds

Having discussed the conjugate addition reactions of various α,β-unsaturated carbonyl compounds, it is appropriate for us now to consider some of the ways in which these useful compounds can be synthesized.

Diazoalkane:

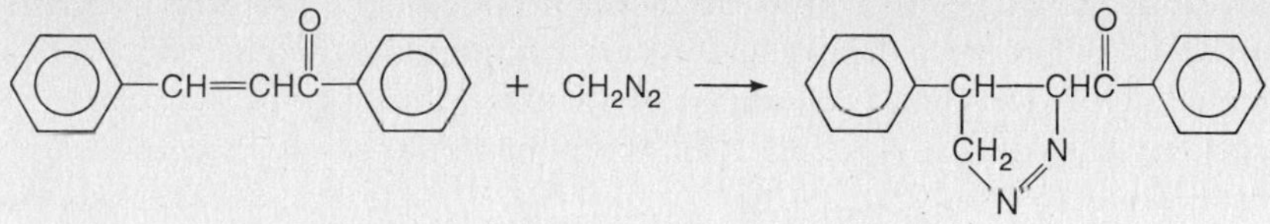
CH=CHC
O
+ CH2N2
CH—CHC
CH2
N
N

Lithium dialkylcuprate:

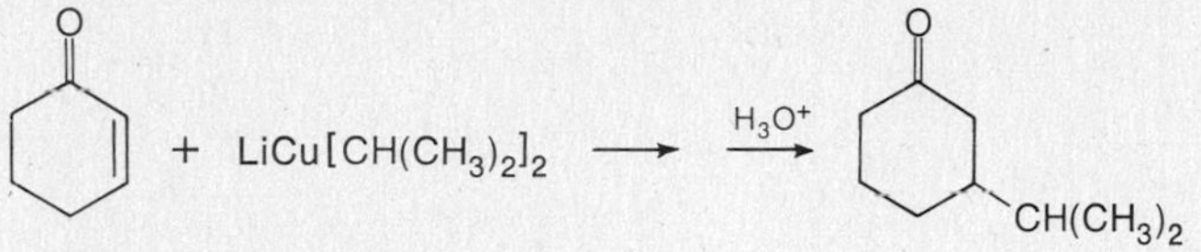
O
+ LiCu[CH(CH3)2]2
H3O+
CH(CH3)2

Cyanide:

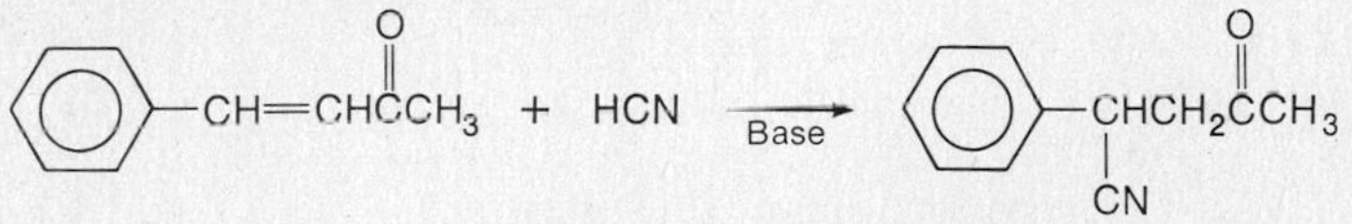
CH=CHCCH3
+ HCN
Base
CHCH2CCH3
CN

Active methylene compounds (Michael addition reaction):

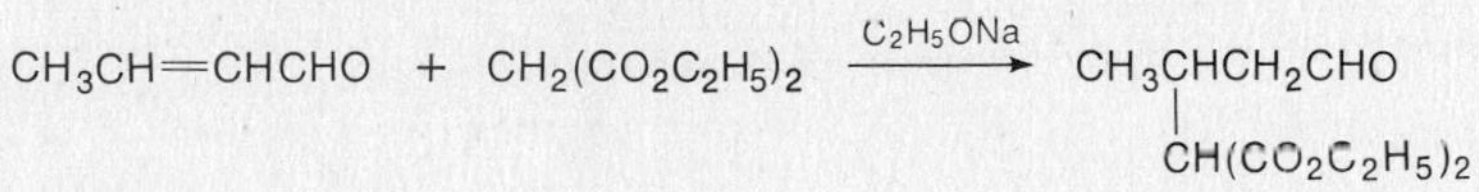
CH3CH=CHCHO + CH2(CO2C2H5)2
C2H5ONa
CH3CHCH2CHO
CH(CO2C2H5)2

Fig. 29.24. Conjugate addition of carbon atom nucleophiles to α,β-unsaturated compounds.

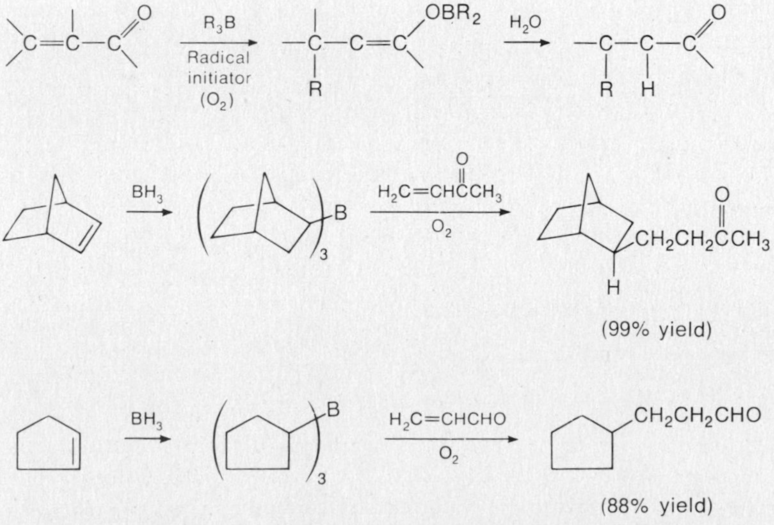
R3B
Radical initiator (O2)
OBR2
H2O
R
R H
BH3
H2C=CHCCH3
O2
CH2CH2CCH3
H
(99% yield)
H2C=CHCHO
CH2CH2CHO
(88% yield)

Fig. 29.25. Conjugate addition of trialkylboron compounds.

29.9a. VIA THE MANNICH REACTION. The Mannich reaction, involving the action of formaldehyde and an amine on a ketone (see Fig. 24.9), yields a β-amino ketone which can be converted to the corresponding quaternary ammonium compound by treatment with methyl iodide. Under the influence of a strong base, compounds of this type undergo a β-elimination reaction with loss of the tertiary amine moiety to form α,β-unsaturated ketones (Fig. 29.26).

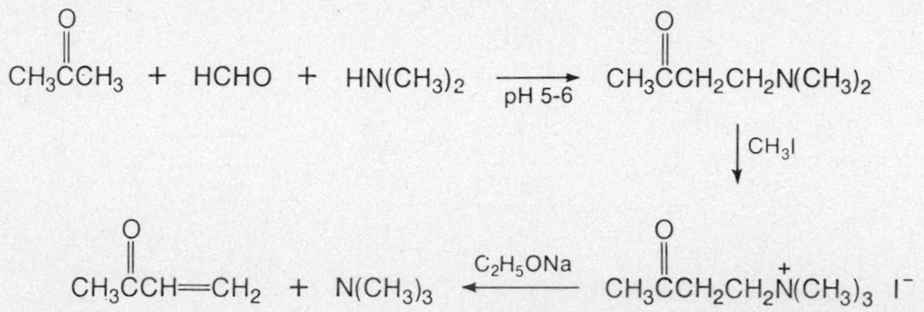

$$CH_3\overset{O}{\overset{\|}{C}}CH_3 + HCHO + HN(CH_3)_2 \xrightarrow[\text{pH 5-6}]{} CH_3\overset{O}{\overset{\|}{C}}CH_2CH_2N(CH_3)_2$$

$$\downarrow CH_3I$$

$$CH_3\overset{O}{\overset{\|}{C}}CH{=}CH_2 + N(CH_3)_3 \xleftarrow{C_2H_5ONa} CH_3\overset{O}{\overset{\|}{C}}CH_2CH_2\overset{+}{N}(CH_3)_3\ I^-$$

Fig. 29.26. Preparation of methyl vinyl ketone by the Mannich reaction followed by deamination.

29.9b. VIA THE REFORMATSKY REACTION. The reaction of an aldehyde or ketone with ethyl α-bromoacetate in the presence of zinc yields, after hydrolysis of the initially formed product, a β-hydroxy ester in a process known as the *Reformatsky reaction.* The reaction can be rationalized in terms of the formation of an organozinc compound in which the anionic portion is a resonance-stabilized anion. The organozinc reagent reacts with aldehydes and ketones but it is not reactive enough to add to ester functions. The anion that is formed in this case is the same kind that is produced by the action of base on active methylene compounds, and it simply represents an alternative method for obtaining it. The initially formed β-hydroxy esters from the Reformatsky reaction readily lose water upon heating or treatment with acid catalysts, yielding the α,β-unsaturated esters (mixed, sometimes, with the β,γ-unsaturated esters) (Fig. 29.27).

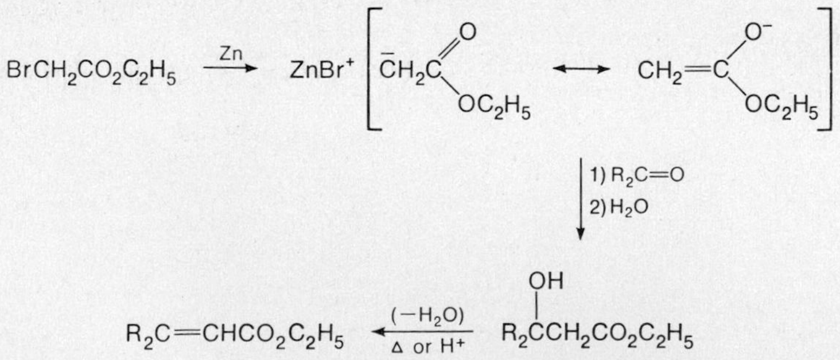

$$BrCH_2CO_2C_2H_5 \xrightarrow{Zn} ZnBr^+ \left[\bar{C}H_2C(=O)OC_2H_5 \longleftrightarrow CH_2{=}C(O^-)OC_2H_5 \right]$$

$$\downarrow \text{1) } R_2C{=}O \quad \text{2) } H_2O$$

$$R_2C{=}CHCO_2C_2H_5 \xleftarrow[\Delta \text{ or } H^+]{(-H_2O)} R_2\overset{OH}{\overset{|}{C}}CH_2CO_2C_2H_5$$

Fig. 29.27. Preparation of α,β-unsaturated esters via the Reformatsky reaction.

PROBLEM 29-10

Explain why ethyl β-bromopropionate does not undergo the Reformatsky reaction.

29.9c. VIA THE WITTIG REACTION. An alternative procedure to the Reformatsky reaction that leads directly to the α,β-unsaturated ester without the necessity of a dehydration step involves a Wittig reagent (see Section 25.9b) prepared from an α-halo ester (Fig. 29.28).

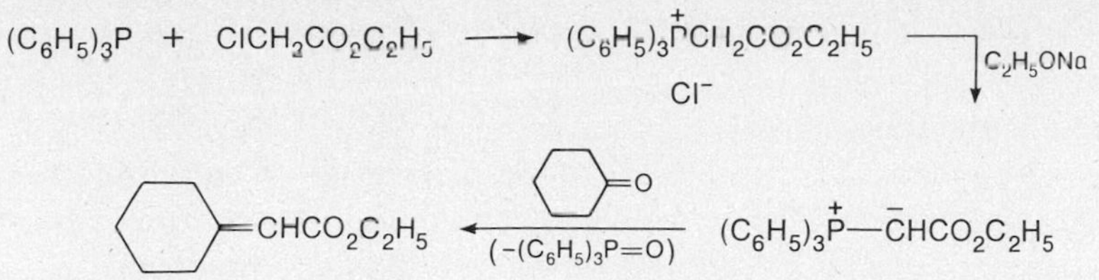

$$(C_6H_5)_3P + ClCH_2CO_2C_2H_5 \longrightarrow (C_6H_5)_3\overset{+}{P}CH_2CO_2C_2H_5\ Cl^- \xrightarrow{C_2H_5ONa} (C_6H_5)_3\overset{+}{P}\text{—}\overset{-}{C}HCO_2C_2H_5$$

$$(C_6H_5)_3\overset{+}{P}\text{—}\overset{-}{C}HCO_2C_2H_5 \xrightarrow[(-(C_6H_5)_3P{=}O)]{\text{cyclohexanone}} \text{cyclohexylidene}{=}CHCO_2C_2H_5$$

Fig. 29.28. Preparation of α,β-unsaturated esters via the Wittig reaction.

PROBLEM 29-11

The formation of the Wittig reagent from $ClCH_2CO_2C_2H_5$ requires a less strong base (*i.e.*, C_2H_5ONa) than the formation of the Wittig reagent from CH_3Cl (*i.e.*, C_4H_9Li). Explain why this is so.

29.9d. VIA α-HALOGENATION–DEHYDROHALOGENATION. Introduction of a halogen atom onto the position adjacent to a carbonyl function followed by base-induced dehydrohalogenation provides a fairly general method for synthesizing α,β-unsaturated carbonyl compounds (Fig. 29.29). Among the bases

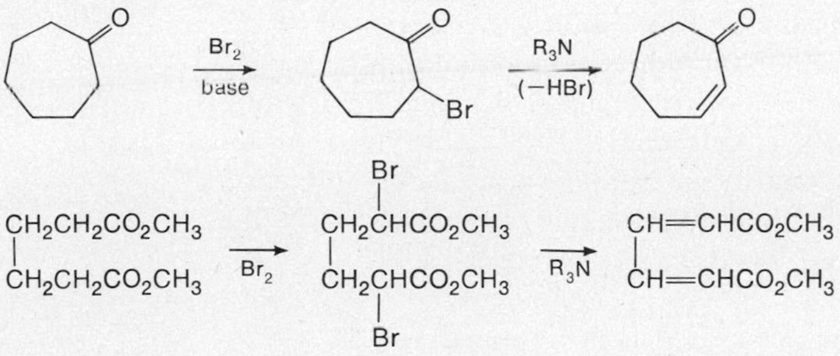

$$\text{cycloheptanone} \xrightarrow[\text{base}]{Br_2} \text{2-bromocycloheptanone} \xrightarrow[(-HBr)]{R_3N} \text{2-cycloheptenone}$$

$$\begin{matrix} CH_2CH_2CO_2CH_3 \\ | \\ CH_2CH_2CO_2CH_3 \end{matrix} \xrightarrow{Br_2} \begin{matrix} Br \\ | \\ CH_2CHCO_2CH_3 \\ | \\ CH_2CHCO_2CH_3 \\ | \\ Br \end{matrix} \xrightarrow{R_3N} \begin{matrix} CH{=}CHCO_2CH_3 \\ | \\ CH{=}CHCO_2CH_3 \end{matrix}$$

Fig. 29.29. Preparation of α,β-unsaturated carbonyl compounds via the elimination of HX.

that are effective for this purpose are tertiary amines, such as 2,4,6-trimethylpyridine (collidine), 1,4-diazabicyclo[2.2.2]octane (DABCO), and 1,5-diazabicyclo[4.3.0]non-5-ene (DBN):

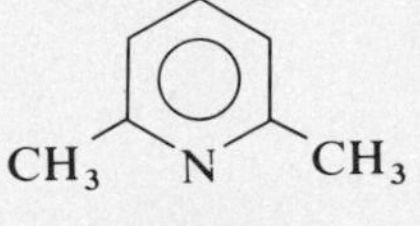

γ-Collidine 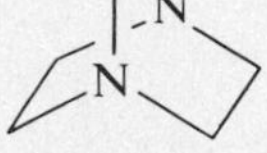 **DABCO** 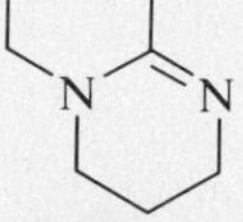 **DBN**

29.9e. VIA ALDOL-TYPE CONDENSATIONS. Aldol-type condensations followed by dehydration provide a variety of approaches to the synthesis of α,β-unsaturated carbonyl compounds. These are discussed in Chapter 30.

ADDITION-ELIMINATION REACTIONS

29.10 General Characteristics of Addition-Elimination Reactions

The addition-elimination sequence is the characteristic reaction of nucleophiles with type AC and BC carbonyl compounds. Carboxylic acids and carboxylic acid derivatives undergo reactions with nucleophiles in which a group initially attached to the carbonyl function is replaced by another group. Although it is possible that these reactions might proceed by direct displacement (*e.g.*, an S_N2 process), it is thought that in most instances they proceed in a stepwise fashion. In the first step, the nucleophile adds to the carbonyl group to form a so-called "tetrahedral intermediate" (*i.e.*, an sp^3-hybridized carbon) and in the second step the leaving group departs to regenerate the carbonyl group (*i.e.*, a trigonal or sp^2-hybridized carbon):

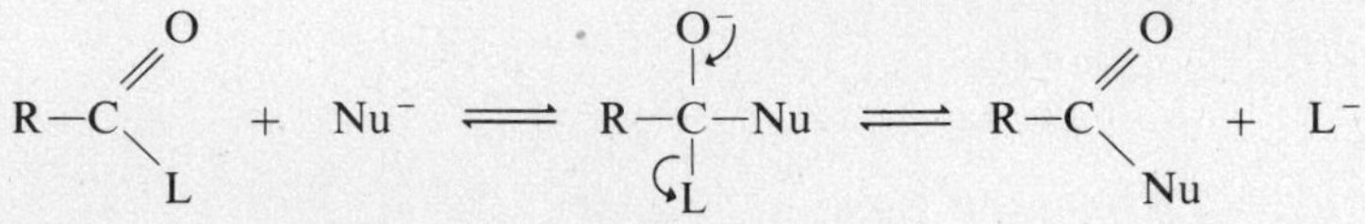

$$\mathrm{R{-}C(=O){-}L} + \mathrm{Nu^-} \rightleftharpoons \mathrm{R{-}C(O^-)(L){-}Nu} \rightleftharpoons \mathrm{R{-}C(=O){-}Nu} + \mathrm{L^-}$$

Tetrahedral intermediate

29.10a. STRUCTURE-REACTIVITY CORRELATIONS. The overall equilibrium in the addition-elimination process is determined by the individual equilibria for the addition and elimination steps, and these, in turn, are governed by the characteristics of the carbonyl function, the entering nucleophile, and the leaving group. With respect to the carbonyl function, the manner in which the L group interacts with it is an important factor in determining the equilibrium constant for the addition step. Electron withdrawal by the L group via inductive interaction activates the carbonyl group to nucleophilic addition, but electron release by the L group via resonance interaction deactivates the carbonyl group (Fig. 29.30).

Activation to nucleophilic addition via inductive electron withdrawal:

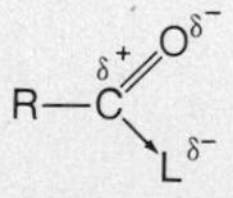

$$\mathrm{R{-}C^{\delta+}(=O^{\delta-}){-}L^{\delta-}}$$

Deactivation to nucleophilic addition via resonance interaction:

$$\mathrm{R{-}C(=O){-}\ddot{L}} \longleftrightarrow \mathrm{R\overset{+}{C}(O^-){-}\ddot{L}} \longleftrightarrow \mathrm{RC(O^-){=}L^+}$$

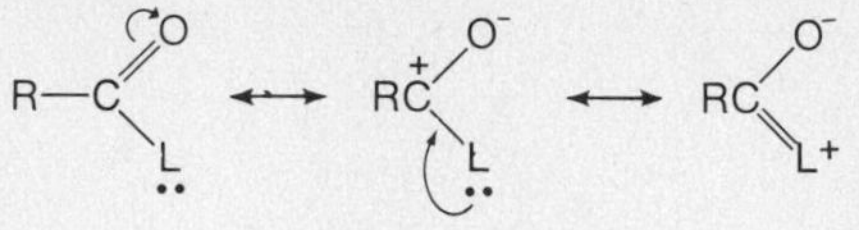

Fig. 29.30. Interaction between L group and C=O bond in $\mathrm{RC(=O)L}$ compounds.

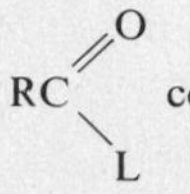

Most of the functions that are capable of acting as leaving groups, L, in the elimination step interact with the carbonyl group in both these fashions. The admixture of these two effects varies from one group to another in the following fashion. When L is chlorine (*i.e.*, acid chlorides), the electron-withdrawing effect outweighs the resonance effect, and the group is strongly activated with respect to nucleophilic addition. When L is alkoxy (*i.e.*, esters), the two effects are approximately equal, and esters show about the same ability as ketones to add nucleophiles. When L is amino (*i.e.*, amides), the resonance effect outweighs the inductive effect, and the group is deactivated with respect to nucleophilic addition. Thus, the equilibrium constant for the addition step decreases in the order: $RCOCl > RCO_2R > RCONR_2$. With respect to the elimination step, the order of reactivity follows the same pattern that is observed in nucleophilic substitution reactions. Chloride, the most stable anion of the group (*i.e.*, its conjugate acid is the strongest acid), is the most easily eliminated; alkoxide, intermediate in stability, is less readily eliminated; amide ion, least stable of this group, is eliminated with the greatest difficulty. The equilibrium constant for the elimination step, therefore, decreases in the following order: adduct from $RCOCl >$ adduct from $RCO_2R >$ adduct from $RCONR_2$. Combining the equilibrium constants for the addition and elimination steps (Table 29.2), we see clearly that the overall equilibrium constant decreases in this same order: $RCOCl > (RCO)_2O > RCO_2R > RCONR_2$.

PROBLEM 29-12

Explain why, despite the low value for $K_{overall}$ for the addition-elimination reaction of amides with nucleophiles, the hydrolysis of amides goes essentially to completion in either acidic or basic medium.

Table 29.2 EFFECT OF THE LEAVING GROUP (L) ON THE ADDITION-ELIMINATION REACTION OF RC(=O)L COMPOUNDS

L in RC(=O)L	Inductive Effect of L	Resonance Effect of L	Stability of L^-	K_{addn}	K_{elim}	$K_{overall}$
—Cl and —OC(=O)R	Large	Small	Large	Large	Large	Very large
—OR	Medium	Medium	Medium	Medium	Medium	Medium
$—NR_2$	Small	Large	Small	Small	Small	Very small

29.10b. PRODUCT DETERMINATION. The type of product that is obtained from the addition-elimination reaction is determined by both the leaving group, L, and the entering nucleophile, Nu. With weak to moderately strong nucleophiles, the reaction proceeds only as far as the initially formed substitution product, $RC(=O)Nu$, but with stronger nucleophiles, the reaction may proceed further. Organometallic reagents, for example, are so reactive that a second addition reaction occurs, and the salts of tertiary alcohols are formed (Fig. 29.30). Similarly, lithium aluminum hydride reduces carboxylic acids, acid chlorides, anhydrides, and esters to primary alcohols. Amides, however, react in a different fashion as a result of the loss of the oxygen moiety rather than the nitrogen moiety from the initially formed addition product, leading to amines (Fig. 29.31).

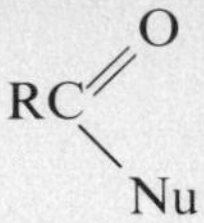

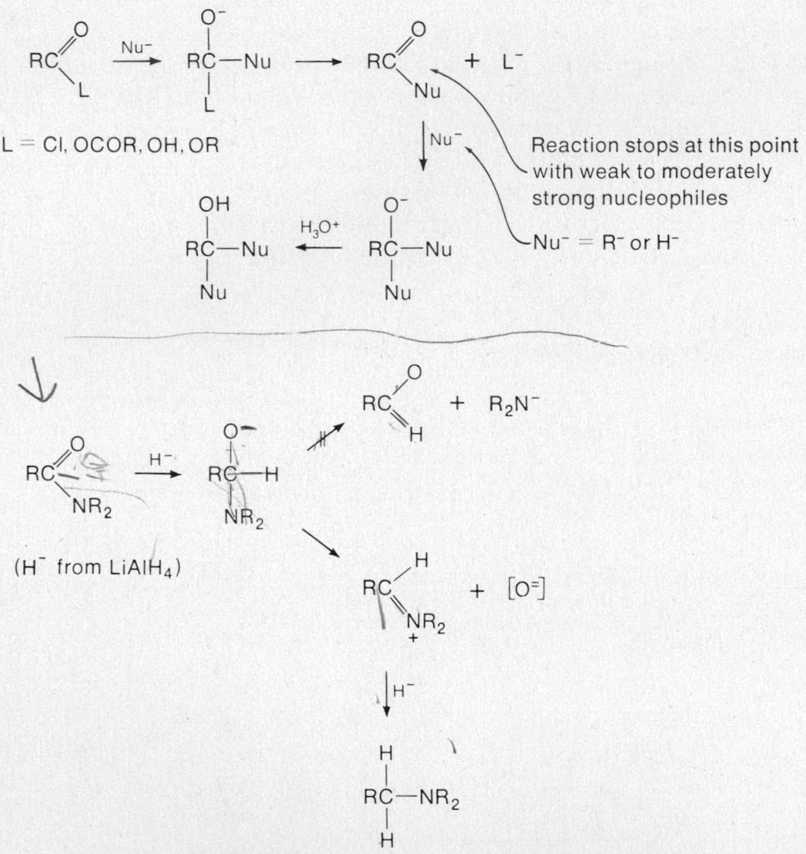

RC(=O)L —Nu⁻→ RC(O⁻)(Nu)L → RC(=O)Nu + L⁻

L = Cl, OCOR, OH, OR

Reaction stops at this point with weak to moderately strong nucleophiles

RC(=O)Nu —Nu⁻→ RC(O⁻)(Nu)Nu —H_3O^+→ RC(OH)(Nu)Nu

Nu⁻ = R⁻ or H⁻

RC(=O)NR₂ —H⁻→ RC(O⁻)(H)NR₂ ⇸ RC(=O)H + R_2N^-

RC(O⁻)(H)NR₂ → RC(H)=$\overset{+}{N}R_2$ + [O⁼] —H⁻→ RC(H)(H)NR₂

(H^- from $LiAlH_4$)

Fig. 29.31. Reaction of carboxylic acid derivatives with nucleophiles.

29.11 *Addition-Elimination Reactions Involving Hetero Atom Nucleophiles*

Numerous examples of addition-elimination reactions involving hetero atom nucleophiles are discussed in previous portions of the text, including the hydrolysis of carboxylic acid derivatives to acids, the aminolysis of acid chlorides, anhydrides, and esters to amides, and the alcoholysis of acid chlorides and anhydrides to esters. Closely related to these "olysis" reactions are "interchange" reactions, in which an exchange of one L group for another takes place at the carbonyl function. For example, esters in the presence of alcohols establish an equilibrium mixture that contains the original ester as well as the ester of the added alcohol (Fig. 29.32). Similarly, interchanges between carboxylic acids and anhydrides and between acids and esters are well known (Fig. 29.32).

$$RCO_2R' + R''OH \rightleftharpoons RCO_2R'' + R'OH$$

$$RCO_2H + (R'CO)_2O \rightleftharpoons (RCO)_2O + R'CO_2H$$

$$RCO_2R' + R''CO_2H \rightleftharpoons RCO_2H + R''CO_2R'$$

$$RCOCl + R'CO_2H \rightleftharpoons RCO_2H + R'COCl$$

Fig. 29.32. "Interchange" reactions of carboxylic acids and carboxylic acid derivatives.

Although the addition-elimination reaction is characteristic of type AC and BC carbonyl functions, some type AA carbonyl functions also undergo this reaction. This involves the cleavage of a carbon–carbon bond (rather than a carbon-hetero atom bond) and occurs only when the expelled carbon moiety is especially capable of sustaining a negative charge. One of the most important examples is the base-induced cleavage of β-dicarbonyl compounds. For example, acetoacetic ester reacts with concentrated, aqueous sodium hydroxide to yield acetic acid and ethanol (see Fig. 28.26), an outcome that can be rationalized as the result of the addition of hydroxide ion to the β-carbonyl group followed by the elimination of the resonance-stabilized anion of ethyl acetate or acetic acid (Fig. 29.33).

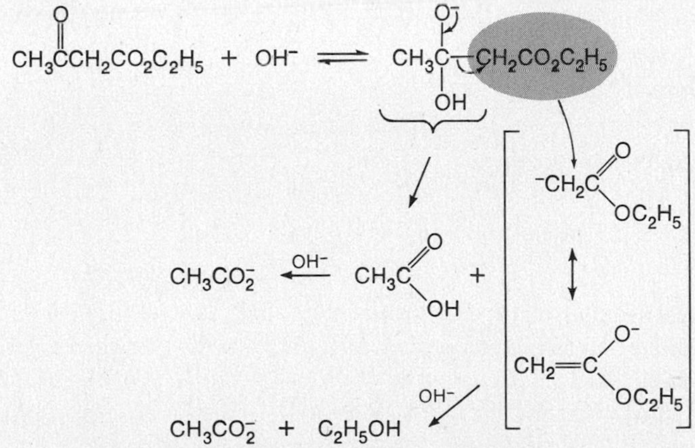

$$CH_3\overset{O}{\overset{\|}{C}}CH_2CO_2C_2H_5 + OH^- \rightleftharpoons CH_3\overset{O^-}{\underset{OH}{C}}-CH_2CO_2C_2H_5$$

$$\longrightarrow CH_3C(=O)OH + \left[{}^-CH_2C(=O)OC_2H_5 \longleftrightarrow CH_2{=}C(O^-)OC_2H_5 \right]$$

$$CH_3C(=O)OH \xrightarrow{OH^-} CH_3CO_2^-$$

$$\left[\ldots\right] \xrightarrow{OH^-} CH_3CO_2^- + C_2H_5OH$$

Fig. 29.33. Addition-elimination reaction of acetoacetic ester.

Another important example of an addition-elimination reaction involving a type AA carbonyl function is the haloform reaction, which takes place with 1,1,1-trihalomethyl ketones in the presence of strong base (see Fig. 28.25). The expelled moiety in this case is the trihalomethyl anion, which is stabilized as the

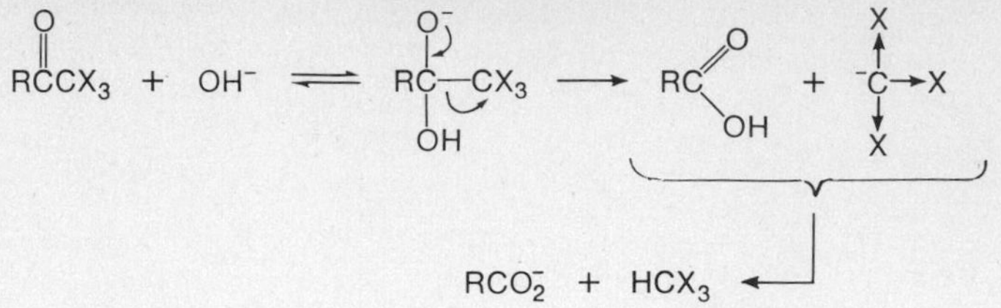

$$RC(=O)CX_3 + OH^- \rightleftharpoons RC(O^-)(OH)CX_3 \longrightarrow RC(=O)OH + {}^-CX_3 \longrightarrow RCO_2^- + HCX_3$$

Fig. 29.34. Addition-elimination reaction of 1,1,1-trihalomethyl ketones (haloform reaction).

result of the electron-withdrawing character of the attached halogen atoms (Fig. 29.34). The trihaloketones that are required for this reaction are generally obtained by the action of alkaline solutions of halogens on methyl ketones or the corresponding secondary alcohols:

$$RCHOHCH_3 \xrightarrow[OH^-]{X_2} RCOCH_3 \xrightarrow[OH^-]{X_2} RCOCX_3 \xrightarrow{OH^-} RCO_2^- + HCX_3$$

The products of the overall reaction are the carboxylic acid (see Fig. 28.25 for a synthesis application) and the haloform. If iodine is employed as the halogen, iodoform is produced; its characteristic color and odor can be used as a diagnostic test for the presence of a $RCHOHCH_3$ or $RCOCH_3$ group in a compound.

PROBLEM 29-13

What would you predict to be the product from the action of bromine and sodium hydroxide on $CH_3COCH_2COC_6H_5$. Explain the basis for your answer.

PROBLEM 29-14

A haloform reaction is obtained from $CH_3COCH_2CO_2C_2H_5$, but the haloform does not arise from the CH_3CO moiety of the molecule. Explain.

29.12 *Addition-Elimination Reactions Involving Carbon Atom Nucleophiles*

Carbon atom nucleophiles, including strongly polarized compounds such as diazomethane, organometallic compounds such as Grignard reagents, and active methylene compounds such as diethyl malonate, all are capable of entering into addition-elimination reactions with various carboxylic acid derivatives (Fig. 29.35). The products with diazomethane are diazoketones, useful as synthesis intermediates in the Arndt–Eistert sequence (see Fig. 28.30). The products from the Grignard reaction are tertiary alcohols (see Table 22.1), the initially formed ketone reacting with another equivalent of Grignard reagent via direct addition. The products from the reactions with active methylene compounds are β-dicarbonyl compounds, and these reactions (referred to as Claisen condensations) are discussed in greater detail in Chapter 30.

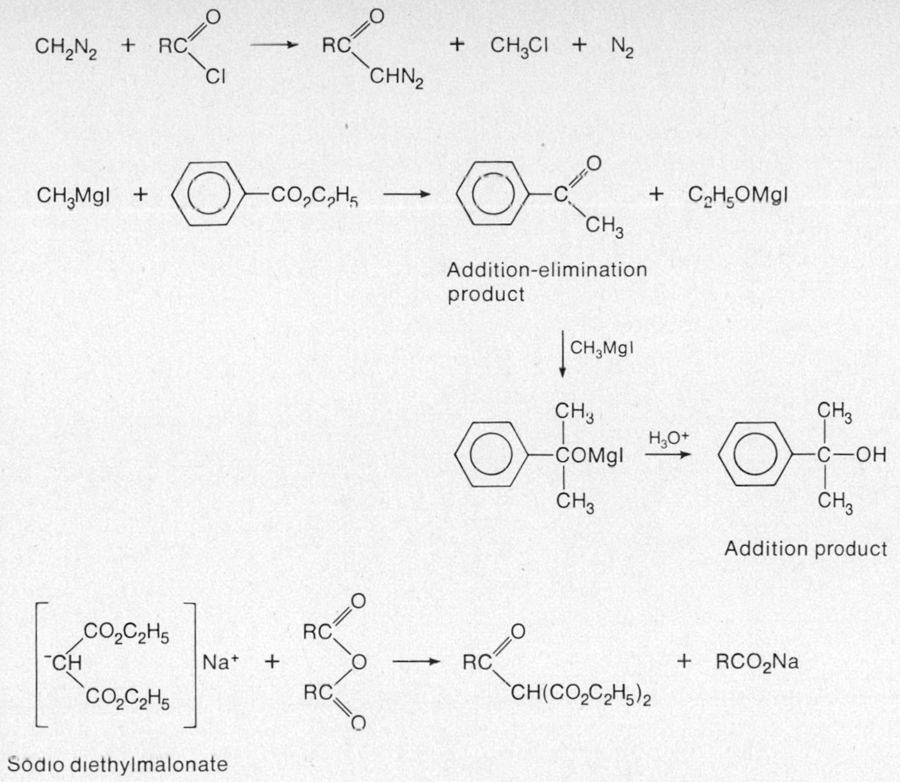

$$CH_2N_2 + RC(=O)Cl \longrightarrow RC(=O)CHN_2 + CH_3Cl + N_2$$

$$CH_3MgI + C_6H_5{-}CO_2C_2H_5 \longrightarrow C_6H_5{-}C(=O)CH_3 + C_2H_5OMgI$$

Addition-elimination product

$$\xrightarrow{CH_3MgI} C_6H_5{-}C(CH_3)_2OMgI \xrightarrow{H_3O^+} C_6H_5{-}C(CH_3)_2{-}OH$$

Addition product

$$\left[{}^{-}CH(CO_2C_2H_5)_2\right]Na^+ + (RCO)_2O \longrightarrow RC(=O)CH(CO_2C_2H_5)_2 + RCO_2Na$$

Sodio diethylmalonate

Fig. 29.35. Addition-elimination reactions with carbon atom nucleophiles.

PROBLEM 29-15

Knowing that diazoketones react with halogen acids (*i.e.*, $RCOCHN_2 + HX \longrightarrow RCOCH_2X + N_2$), explain the necessity for using two equivalents of diazomethane in the preparation of diazoketones.

29.13 Addition-Elimination Reactions Involving Hydride as the Nucleophile

Lithium aluminum hydride reacts with carboxylic acids and carboxylic acid derivatives to give products that are the result of an initial addition-elimination reaction followed by a direct nucleophile addition of hydride (see Fig. 29.31). With all these compounds except amides, the products are primary alcohols; with amides, the products are amines (see Fig. 29.31). Less reactive metal hydrides are more selective in their action; sodium borohydride, for example, reduces acid chlorides to alcohols but has no effect on esters, amides, or carboxylic acids (see Table 29.1). The reaction of metal hydrides with carboxylic acid derivatives can be terminated at the addition-elimination stage (*i.e.*, to give aldehydes) if the reactivity of the hydride is properly modified (*e.g.*, by conversion of lithium aluminum hydride to lithium aluminum trialkoxyhydrides, as illustrated in Fig. 28.2).

OTHER REACTIONS INVOLVING THE CARBONYL GROUP

29.14 Oxidation Reactions Occurring at the Carbonyl Carbon Atom

The facile oxidation of aldehydes as a means for preparing carboxylic acids is illustrated in Fig. 28.23. The Baeyer–Villiger reaction as a means for preparing carboxylic acids from either aldehydes or ketones is discussed in Section 28.11c. This reaction proceeds best with peracids derived from very strong carboxylic acids, trifluoroperacetic acid being particularly good in this respect. The nucleophilic 1,2 rearrangement that takes place in the course of the reaction occurs with retention of configuration of the migrating carbon atom and is sensitive to migratory aptitudes which decrease in the order aryl > vinyl > *t*-alkyl > *sec*-alkyl > *n*-alkyl > CH_3 > H (Fig. 29.36).

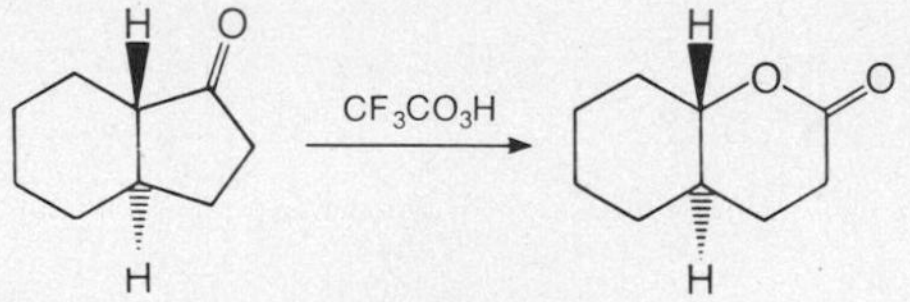

CF_3CO_3H

Fig. 29.36. Baeyer–Villiger oxidation of bicyclo[4.3.0] non-7-one.

PROBLEM 29-16

Outline the mechanism of the reaction shown in Fig. 29.36.

29.15 Carbonyl Replacement Reactions

The carbonyl group in aldehydes and ketones can be converted to a methyl or methylene group by several general methods, including (a) reduction with zinc amalgam and hydrochloric acid (*Clemmensen reduction*), (b) conversion to the hydrazone followed by treatment with base at an elevated temperature (*Wolff–Kishner reduction*), (c) treatment with hydrazine and potassium hydroxide in diethylene glycol at 200° (Huang–Minlon modification of the Wolff–Kishner reduction), and (d) conversion to the thioacetal or thioketal followed by Raney nickel-catalyzed hydrogenolysis (Fig. 29.37). Each method has its particular virtues and shortcomings. The Clemmensen reduction is quite specific for the reduction of aldehydes and ketones, although acid-labile groups (*e.g.*, esters) may be affected by the acidic conditions of the reaction. The Wolff–Kishner method is more situselective than the Clemmensen reduction but has the disadvantage of being difficult to carry out on a large scale and incompatible with the presence of base sensitive groups. The Huang–Minlon modification of the Wolff–Kishner method circumvents the first difficulty by using diethylene glycol as the solvent, a compound that is sufficiently high boiling to provide the necessary temperature and polar enough to dissolve potassium hydroxide, the basic catalyst. The mechanism of the Wolff–Kishner (and Huang–Minlon) reaction is depicted in Fig. 29.38. The sequence involving the thioacetal or thioketal is the most gentle of the carbonyl replacement methods, although it does require acid catalysis (*e.g.*,

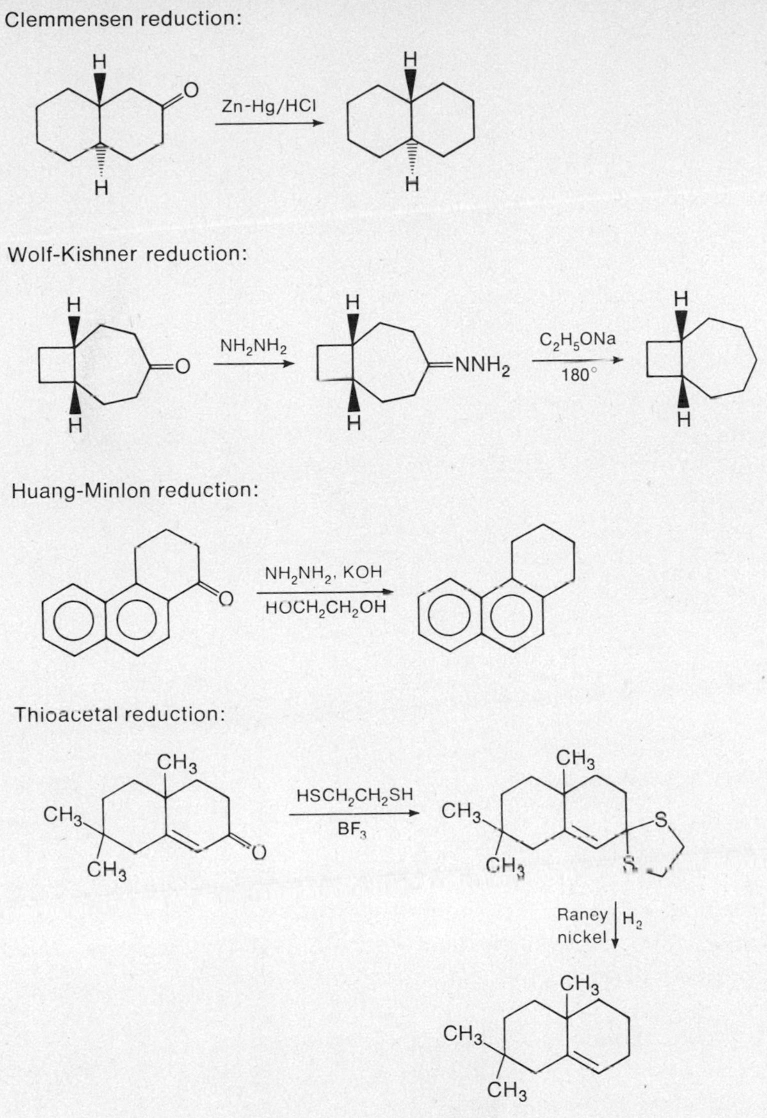

Clemmensen reduction:

Zn-Hg/HCl

Wolf-Kishner reduction:

NH_2NH_2 =NNH$_2$ C_2H_5ONa 180°

Huang-Minlon reduction:

NH_2NH_2, KOH $HOCH_2CH_2OH$

Thioacetal reduction:

$HSCH_2CH_2SH$ BF_3

Raney nickel H_2

Fig. 29.37. Methods for converting $R_2C{=}O$ to R_2CH_2.

boron trifluoride) in the formation of the thioacetal or thioketal. As the example in Fig. 29.37 illustrates, a double bond can be preserved in the molecule undergoing reduction.

Other methods for converting the carbonyl group of an aldehyde or ketone to a methyl or methylene group include reduction to the corresponding alcohol followed by replacement of the hydroxyl group (a) indirectly by conversion to the halide or tosylate followed by direct replacement by hydrogen, or by dehydration followed by hydrogenation, or (b) directly via hydrogenolysis, when applicable. For example, benzophenone can be reduced to diphenylmethane by the action of molecular hydrogen in the presence of a platinum catalyst. The

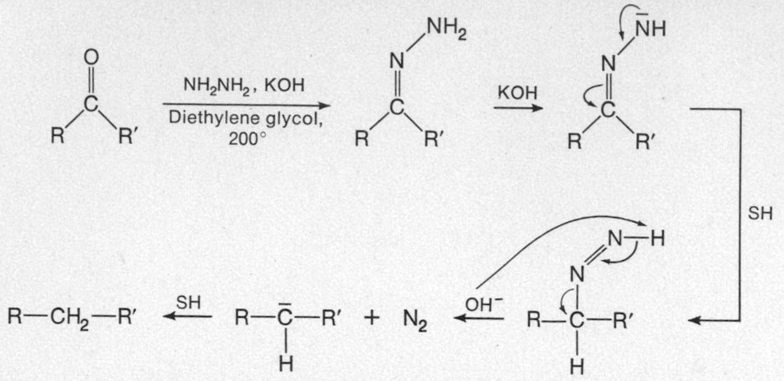

Fig. 29.38. Mechanism of the Wolff–Kishner reduction.

initially formed diphenylmethanol is a benzyl alcohol, and, consequently, undergoes hydrogenolysis to the alkane:

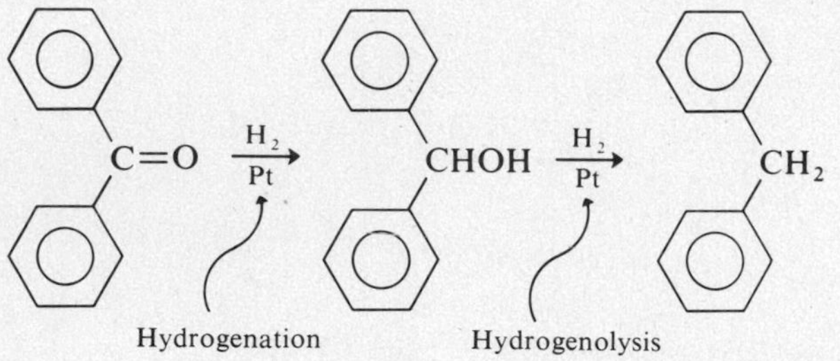

$$(C_6H_5)_2C{=}O \xrightarrow[Pt]{H_2} (C_6H_5)_2CHOH \xrightarrow[Pt]{H_2} (C_6H_5)_2CH_2$$

Hydrogenation Hydrogenolysis

29.16 Polymerization Reactions of Carbonyl Compounds

As mentioned previously (see Section 26.8a), polymers can be broadly categorized as addition polymers or condensation polymers. The majority of polymers prepared from carbonyl compounds are of the condensation type, although a few involving formaldehyde as a component belong to the addition category.

29.16a. POLYMERS INVOLVING FORMALDEHYDE AS A COMPONENT. In aqueous solution, formaldehyde undergoes addition polymerization to form a material called paraformaldehyde, with water acting as the initiator:

$$(n + 2)\ HCHO + H_2O \longrightarrow HOCH_2(OCH_2)_nOCH_2OH$$

Although paraformaldehyde can be isolated as a solid, it readily reverts to formaldehyde upon heating, and consequently provides a convenient source of this material. This facile decomposition can be prevented if the terminal hydroxyl groups are "capped" by conversion to acetoxy groups by the action of acetic anhydride. The resulting product, $CH_3CO_2CH_2(OCH_2)_nOCH_2OCOCH_3$, is a very tough, high-melting plastic marketed by the duPont company under the trade name of Delrin.

Formaldehyde is also a component in a number of other commercially important polymers, the oldest of which are the phenol-formaldehyde resins patented in 1909 by L. H. Baekeland and known as Bakelites (see p. 17). Bakelites are produced by the reaction of phenol and formaldehyde, phenol reacting at both of its *ortho* positions as well as its *para* position to yield highly cross-linked polymers, noted for their hardness (see p. 798).

Certain nitrogen-containing compounds react with formaldehyde in a comparable fashion, yielding cross-linked polymers possessing a variety of useful properties. For example, urea yields urea-formaldehyde resins, which, when mixed with a filler (*e.g.*, cellulose), can be molded under pressure at elevated temperature into the desired shape; 2,4,6-triamino-1,3,5-triazine (melamine) with formaldehyde yields melamine-formaldehyde resins, which are cast into sheets, used as surface coatings, or molded into dishware (Melmac):

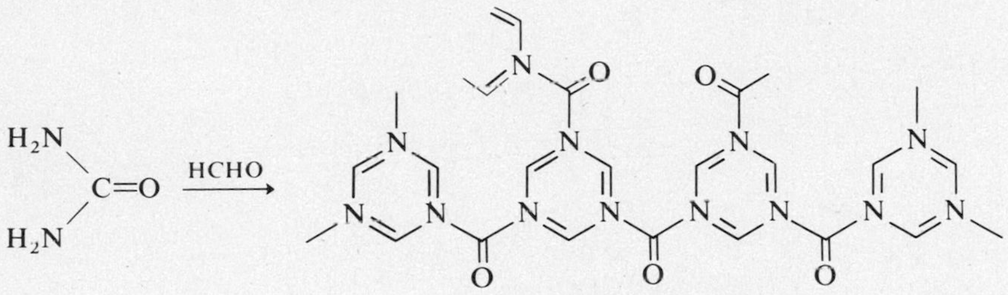

$$(H_2N)_2C{=}O \xrightarrow{HCHO} \text{Urea-formaldehyde polymer}$$

Urea **Urea-formaldehyde polymer**

$$\text{Melamine} \xrightarrow{HCHO} \text{Melamine-formaldehyde polymer}$$

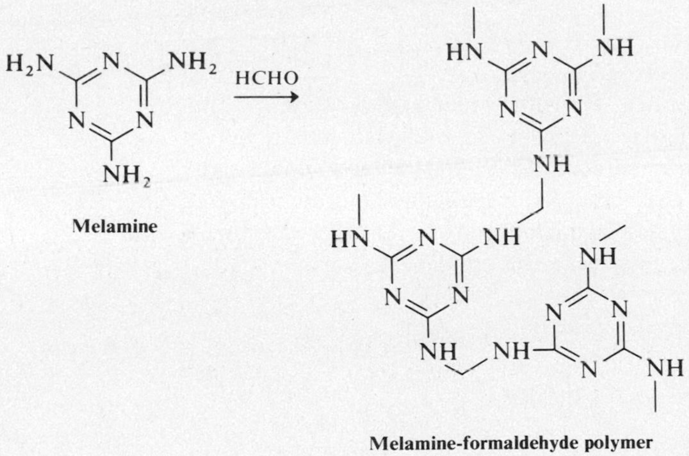

Melamine **Melamine-formaldehyde polymer**

29.16b. Polymers involving acetone as a component. Acetone shows no tendency to form addition polymers comparable to those derived from formaldehyde. It is a component of some commerically important polymers, nevertheless, because of its reaction with phenol to form bisphenol-A (see Fig. 22.33). Bisphenol-A can be converted to an "epoxy monomer" by the action of 2-chloromethyloxirane (epichlorohydrin), and polymerization of the monomer can be induced by trifunctional amines such as diethylenetriamine to yield epoxy resins, useful in low molecular weight form as adhesives and in higher molecular weight form as surface coatings:

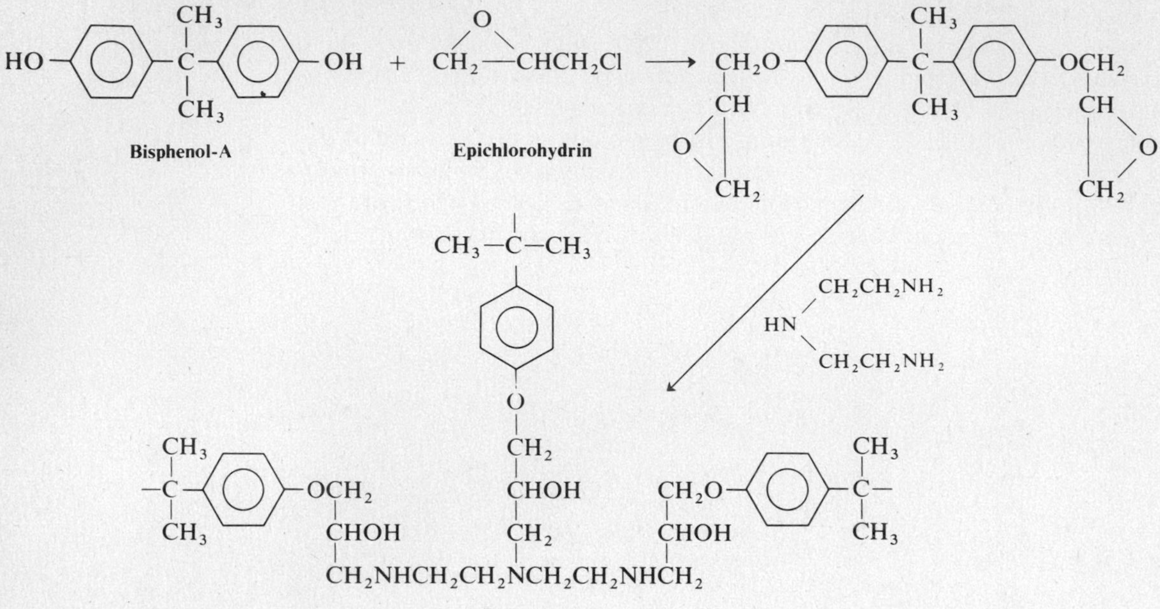

$$HO-C_6H_4-C(CH_3)_2-C_6H_4-OH \;+\; \underset{\text{O}}{CH_2-CH}CH_2Cl \;\longrightarrow\; \underset{\text{O}}{CH_2-CH}CH_2O-C_6H_4-C(CH_3)_2-C_6H_4-OCH_2\underset{\text{O}}{CH-CH_2}$$

Bisphenol-A **Epichlorohydrin**

$$\xrightarrow{HN(CH_2CH_2NH_2)_2}$$

$$-C(CH_3)_2-C_6H_4-OCH_2CHOH-CH_2NHCH_2CH_2N(CH_2CHOH-CH_2-O-C_6H_4-C(CH_3)_2-)CH_2CH_2NHCH_2-CHOH-CH_2O-C_6H_4-C(CH_3)_2-$$

Epoxy resin

29.16c. Polyesters and polyamides. Polyfunctional acids, esters, or anhydrides react with polyfunctional alcohols to form high molecular weight polyesters, some of which have commercial utility. Particularly noteworthy among these is the compound marketed under the trade name Dacron, prepared from dimethyl terephthalate and ethylene glycol (see p. 415). Another example is the compound prepared from maleic anhydride and ethylene glycol:

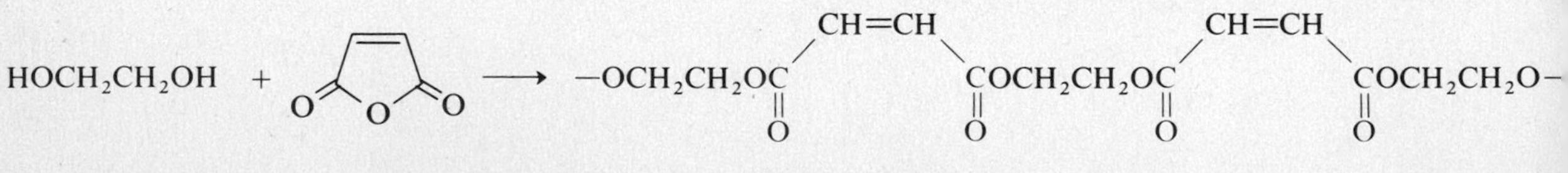

$$HOCH_2CH_2OH \;+\; \text{maleic anhydride} \;\longrightarrow\; -OCH_2CH_2O\overset{O}{\overset{\|}{C}}-CH{=}CH-\overset{O}{\overset{\|}{C}}OCH_2CH_2O\overset{O}{\overset{\|}{C}}-CH{=}CH-\overset{O}{\overset{\|}{C}}OCH_2CH_2O-$$

This linear polymer can be cross-linked by heating it with styrene, yielding a material that when used to impregnate fiberglass gives a remarkably strong and tough plastic useful for making boat hulls, automobile bodies, and other items requiring great strength and resiliency.

In a comparable fashion, polyfunctional acids, esters, or anhydrides react with polyamines to form high molecular weight polyamides, the Nylons being the best-known examples (see p. 415).

Polymers having elastic properties, known as polyurethane elastomers, contain both amide- and ester-type linkages and are prepared by the action of diisocyanates on low molecular weight polymers of the polyester type:

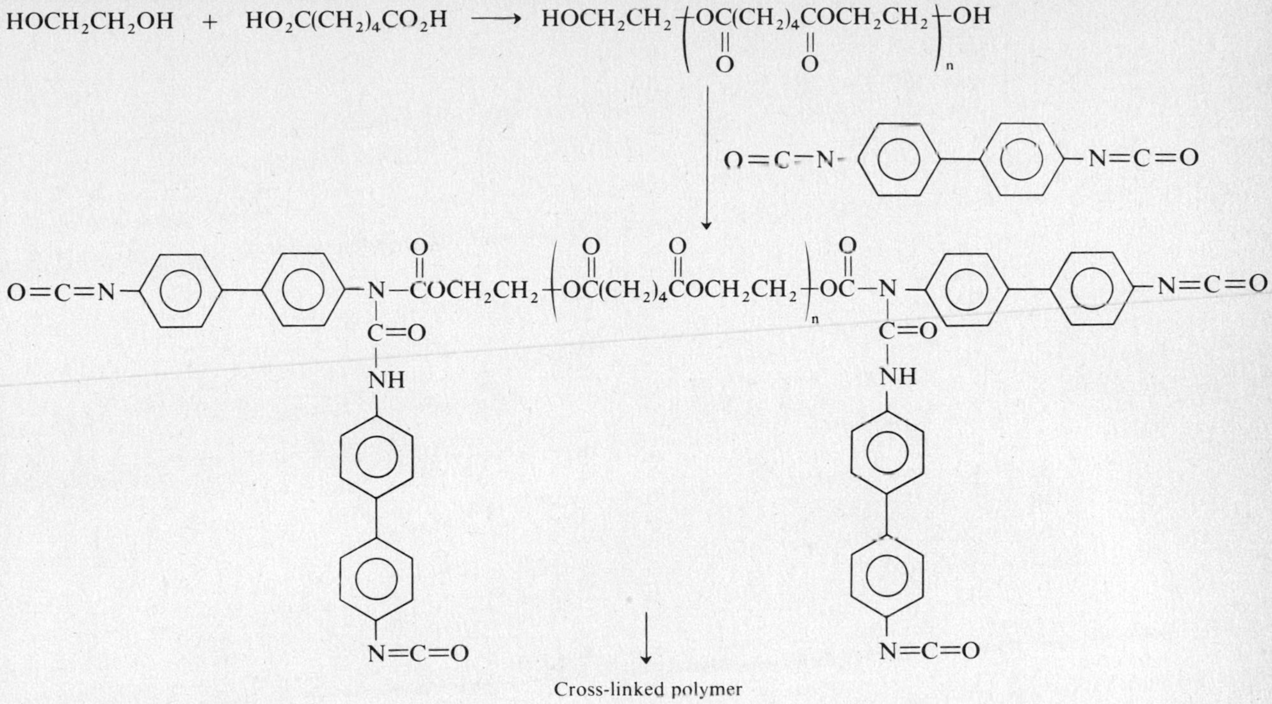

$$HOCH_2CH_2OH + HO_2C(CH_2)_4CO_2H \longrightarrow HOCH_2CH_2\left(OC(=O)(CH_2)_4C(=O)OCH_2CH_2\right)_n OH$$

$$\downarrow \; O{=}C{=}N{-}C_6H_4{-}C_6H_4{-}N{=}C{=}O$$

$$O{=}C{=}N{-}C_6H_4{-}C_6H_4{-}N(C(=O)NH{-}C_6H_4{-}C_6H_4{-}N{=}C{=}O){-}C(=O)OCH_2CH_2\left(OC(=O)(CH_2)_4C(=O)OCH_2CH_2\right)_n OC(=O){-}N(C(=O)NH{-}C_6H_4{-}C_6H_4{-}N{=}C{=}O){-}C_6H_4{-}C_6H_4{-}N{=}C{=}O$$

$$\downarrow$$

Cross-linked polymer

Table 29.3 summarizes the reactions of carbonyl compounds occurring at the C—O group.

Table 29.3 REACTIONS OF CARBONYL COMPOUNDS AT THE CARBONYL GROUP

Type of Reaction	Starting Materials	Reagents	Products	Comments
Direct addition of hetero-atom nucleophiles	$R_2C{=}O$	H—Nu (Nu = halogen, —OH, —OR, —SH, —SR, —NH_2, —NR_2)	$R_2C(OH)(Nu)$	Useful synthesis reaction only when $R_2C{=}O$ is unusually reactive
		ROH or RSH (acid catalyst)	$R_2C(OR)(OR)$ or $R_2C(SR)(SR)$	Useful for the protection of $R_2C{=}O$; $R_2C{=}O$ can be regenerated by hydrolysis with H_3O^+
		H_2NNHY (Y = —H, —OH, —Ar, —$CONH_2$)	$R_2C{=}NNHY$	Useful derivatives for characterizing carbonyl compounds
	$-\overset{\mid}{C}H-C(=O)R$	R_2NH	$-\overset{\mid}{C}{=}C(R)NR_2$ (enamine)	

Table 29.3—*cont.*

Type of Reaction	Starting Materials	Reagents	Products	Comments
	ArCHO	Base	$ArCH_2OH + ArCO_2H$	Cannizzaro reaction
Direct addition of carbon-atom nucleophiles	$R_2C{=}O$	CH_2N_2	$RCH_2(R)C{=}O$	Homologation reaction
		$\overset{-}{C}H_2\overset{+}{S}(CH_3)_2$	$R_2C(-O-)CH_2$ (epoxide)	
		$R\overset{-}{C}H-\overset{+}{P}(C_6H_5)_3$	$R_2C{=}CHR$	Wittig reaction
		R′MgX	$R_2C(OH)R'$	See Table 22.1 for more details
		HCN (base)	$R_2C(OH)CN$	
		Active methylene carbanions (see Table 30.1)		
Direct addition of hydride	$R_2C{=}O$	$LiAlH_4$, $NaBH_4$	R_2CHOH	
Conjugate addition reactions	$>C{=}C(-)-C({=}O)-$	H—Nu (Nu = halogen, —OH, —OR, —SH, —SR, $-NH_2$, $-NR_2$)	$-C(Nu)-CHC({=}O)-$	Even weak nucleophiles give stable products
		CH_2N_2	$-C(CH_2-N{=}N-)-C(N)C({=}O)-$ (pyrazoline)	
		(1) R_2CuLi; (2) H_2O or R_3B + radical initiator	$-C(R)-CHC({=}O)-$	Uses only one R group of R_3B
		HCN	$-C(CN)-CHC({=}O)-$	
		$ZZ'CH_2$ (active methylene compounds)	$-C(CHZZ')-CHC({=}O)-$	Michael addition reactions

Table 29.3—*cont.*

Type of Reaction	Starting Materials	Reagents	Products	Comments
Addition-elimination reactions	RC(=O)Cl	H—Nu (Nu = —OH, —OR, —OCOR, —NH_2, —NR_2)	RC(=O)Nu	"Olysis" reactions of acid chlorides
	$(RCO)_2O$	H—Nu (Nu = —OH, —OR, —NH_2, —NR_2)	RC(=O)Nu	"Olysis" reactions of anhydride
	RC(=O)OR′	H—Nu (Nu = —OH, —OR′, —NH_2, —NR'_2)	RC(=O)Nu	"Olysis" reactions of esters
	RC(=O)NR'_2	H_2O	RC(=O)OH	Hydrolysis of amides
	RC(=O)Y	ZZ′CH_2 (active methylene compounds)	RC(=O)CHZZ′	Claisen condensation (when Y = —OR)
	$RCOCH_3$	Halogen + NaOH (H_2O)	RCO_2H + CHX_3	Haloform reaction
	RCOCl	CH_2N_2	$RCOCHN_2$	
	RC(=O)Y	R′MgX	RC(OH)(R′)—R′	See Table 22.1 for more details
	RC(=O)Y	$LiAlH_4$	RCH_2OH (except when Y = —NR_2 in which case product is RCH_2NR_2)	
Other reactions	$R_2C{=}O$	CF_3CO_3H	ROC(=O)R	Baeyer–Villiger oxidation (migratory aptitudes apply)
		Zn—Hg/HCl	R_2CH_2	Clemmensen reduction
		(1) NH_2NH_2; (2) C_2H_5ONa	R_2CH_2	Wolff–Kishner reduction
		NH_2NH_2/KOH (diethylene glycol solvent)	R_2CH_2	Huang–Minlon modification of Wolff–Kishner reduction
		(1) conversion to $R_2C(SR)_2$; (2) H_2, Raney nickel	R_2CH_2	

I. General characteristics of 1,2-nucleophilic addition reactions
  A. Catalysis of 1,2-nucleophilic addition reactions
    1. Base catalysis—generation of a more reactive nucleophile
    2. Acid catalysis—generation of a more reactive carbonyl group
  B. Equilibria in 1,2-nucleophilic addition reactions
    1. Range of equilibrium constants for 1,2-nucleophilic addition reactions
      (a) Reversible reactions K_{eq} less than 10^4
      (b) Essentially irreversible reactions: K_{eq} greater than 10^4
    2. Correlation of structure with equilibrium constants
      (a) Steric factors—increasing steric hindrance lowers K_{eq}
      (b) Electronegativity factors—increasing electron-withdrawal increases K_{eq}
      (c) Resonance interaction—increasing resonance interaction lowers K_{eq}
      (d) Nucleophilicity factors—increasing nucleophilicity increases K_{eq}
      (e) Bond angle factors
    3. Reactivity sequence: cyclopropanones > aldehydes > cyclohexanones > cyclobutanones > cyclopentanones > dialkyl ketones > alkyl aryl ketones > diaryl ketones

II. Scope and limitations of 1,2-nucleophilic addition reactions
  A. Addition of hetero atom nucleophiles
    1. Addition to highly reactive aldehydes and ketones
      (a) Most hetero atom nucleophiles not effective in general; sodium bisulfite an exception—gives isolable products with aldehydes and some ketones
      (b) Aldehydes and ketones with strong electron-withdrawing groups add hetero atom nucleophiles
    2. Addition-dehydration reactions
      (a) Intermolecular dehydration—formation of acetals and ketals
      (b) Intramolecular dehydration
        (1) Formation of imines (*e.g.*, semicarbazones, phenylhydrazones, etc.)
        (2) Formation of enamines
        (3) Reaction of aldehydes with ammonia
    3. Disproportionation reactions; Cannizzaro reaction
  B. Addition of carbon nucleophiles
    1. Cyanide ion as a carbon nucleophile; cyanohydrin formation
    2. Strongly polarized compounds as sources of carbon nucleophiles—diazoalkanes, phosphorus ylides, sulfur ylides.
    3. Organometallic compounds as sources of carbon nucleophiles—see previous chapters
    4. Active methylene compounds as sources of carbon nucleophiles—see Chapter 30.
  C. Addition of hydride ion—affords possibilities for selective reduction

III. 1,4-Nucleophilic addition reactions
   A. General characteristics of 1,4-nucleophilic addition reactions
      1. Reaction pathway—nucleophilic attachment at β-carbon, electrophilic attachment at oxygen, tautomeric rearrangement to carbonyl compound when the electrophile is a proton
      2. Overall equilibrium constant generally more favorable to 1,4 addition than to 1,2 addition
   B. Conjugate addition reactions involving hetero atom nucleophiles
   C. Conjugate addition reactions involving carbon frameworks
      1. Carbon atom nucleophiles—diazoalkanes, lithium dialkylcuprates, hydrogen cyanide, active methylene compounds
      2. Trialkylboranes as sources of carbon frameworks for 1,4 addition
   D. Methods of preparation of α,β-unsaturated carbonyl compounds
      1. Mannich reaction followed by quaternization and elimination
      2. Reformatsky reaction followed by dehydration
      3. Wittig reaction using $(C_6H_5)_3P{=}CHCOR$ reagents
      4. α-Halogenation followed by dehydrohalogenation
      5. Aldol-type condensations followed by dehydration—see Chapter 30

IV. Addition-elimination reactions
   A. General characteristics of addition-elimination reactions
      1. Structure-reactivity correlations
         (a) Inductive electron-withdrawal increases reactivity of carbonyl group
         (b) Resonance interaction decreases reactivity of carbonyl group
         (c) Reactivity sequence: $RCOCl > (RCO)_2O > RCO_2R > RCONR_2$
      2. Product determination
         (a) Weak to moderately strong nucleophiles (*i.e.*, hetero atom nucleophiles) give carbonyl products
         (b) Strong nucleophiles (*i.e.*, hydride and organometallic-derived nucleophiles) give alcohols, the result of further reaction with the initially formed carbonyl product (except in the case of hydride and amides which yield amines)
   B. Addition-elimination reactions involving hetero atom nucleophiles
      1. "Olysis" reactions—hydrolysis, aminolysis, etc.
      2. "Interchange" reactions
      3. Addition-elimination reaction of acetoacetic ester
      4. Addition-elimination reaction of 1,1,1-trihalomethyl ketones (haloform reaction)
   C. Addition-elimination reactions involving carbon atom nucleophiles
      1. Diazoalkanes
      2. Organometallic reagents
      3. Active methylene compounds (*e.g.*, diethyl sodiomalonate)
   D. Addition-elimination reactions involving hydride as the nucleophile

V. Other reactions involving the carbonyl group
   A. Oxidation reactions occurring at the carbonyl carbon atom

   1. Oxidation of aldehydes and ketones to carboxylic acids
   2. Oxidation of aldehydes and ketones to esters with peracids (Baeyer–Villiger oxidation)
B. Carbonyl replacement reactions
   1. Clemmensen reduction
   2. Wolff–Kishner reduction
   3. Huang–Minlon modification of Wolff–Kishner reduction
   4. Conversion to thioacetal or thioketal followed by hydrogenolysis
C. Polymerization reactions of carbonyl compounds
   1. Polymers involving formaldehyde as a component
      (a) Self-condensation polymers (Delrin)
      (b) Phenol-formaldehyde polymers (Bakelite)
      (c) Urea-formaldehyde polymers
      (d) Melamine-formaldehyde polymers
   2. Polymers involving acetone as a component; bisphenol-A
   3. Polyesters and polyamides
      (a) Dimethyl terephthalate-ethylene glycol polymer (Dacron)
      (b) Maleic anhydride-ethylene glycol polymer
      (c) Polyol-diisocyanate polymers (polyurethane elastomers)

REFERENCES

COFFEY, S., ed., *Rodd's Chemistry of Carbon Compounds*, 2nd ed., Vol. IC, ID, Elsevier Publishing Co., New York, 1965.

FIESER, L. F., and M. FIESER, *Advanced Organic Chemistry*, Reinhold Publishing Corp., New York, 1961, Chapters 11, 12, 24, and 25.

GUTSCHE, C. D., *The Chemistry of Carbonyl Compounds*, in *Foundations of Modern Organic Chemistry Series*, Prentice-Hall, Inc., Englewood Cliffs, N.J., 1967.

HOUSE, H. O., *Modern Synthetic Reactions*, 2nd ed., W. A. Benjamin, Inc., Menlo Park, Calif., 1972.

PATAI, S., ed., *The Chemistry of the Carbonyl Group*, Vol. 1, John Wiley & Sons, Inc., New York, 1966.

PATAI, S., and J. ZABICKY, eds., *The Chemistry of the Carbonyl Group*, Vol. 2, John Wiley & Sons, Inc., New York, 1970.

PATAI, S., ed., *The Chemistry of Carboxylic Acids and Esters*, John Wiley & Sons, Inc., New York, 1969.

ROYALS, E. E., *Advanced Organic Chemistry*, Part II: "The Chemistry of Carbonyl Compounds", Prentice-Hall, Inc., Englewood Cliffs, N. J., 1954.

PROBLEMS

1. Draw the structure of the major product from each of the following reactions:

(a) [bicyclic diol structure] $\xrightarrow[\text{Pyridine}]{CrO_3}$

(b) $C_6H_5CH_2Br + (CH_3)_2\overset{+}{S}-\overset{-}{O} \longrightarrow$

(c) $CH_3O-C_6H_4-COCl + H_2 \xrightarrow[\text{[(CH_3)_2N]_2C=S}]{Pd/BaSO_4}$

(d) 2-lithio-1,3-dithiane + $C_6H_5CH_2Br \longrightarrow$

(e) $C_6H_5COCH_3 \xrightarrow{SeO_2}$

(f) cyclopentanone hydrazone ($=NNH_2$) + KOH $\xrightarrow{\Delta}$

(g) $CH_3CH{=}CHCOC_6H_5 + H_2S \longrightarrow$

(h) 1-indanone $\xrightarrow{C_6H_5CO_3H}$

(i) cyclopentanone ($=O$) + $BrCH_2CO_2C_2H_5 \xrightarrow{Zn}$

(j) furfural (2-furyl CHO) $\xrightarrow{KCN}$

(k) 2-chlorocyclohexanone $\xrightarrow{NaOH}$

(l) $CH_3CH(OH)C(=O)CH_2CH(OH)CH_2OH + HIO_4 \longrightarrow$

(m) [bicyclic enone structure with H atoms at ring fusion] $\xrightarrow{\text{Zn-Hg/HCl}}$

(n) $(CH_2)_8(CO_2C_2H_5)_2$ + Na $\xrightarrow[\text{xylene}]{\text{Refluxing}}$

(o) $C_6H_5CH{=}CHCOCH_3$ + $CH_3CH_2CH_2NO_2$ $\xrightarrow{C_2H_5ONa}$

2. Suggest a reasonable mechanism for each of the reactions in Problem 1 except parts (c), (e), (l), (m), and (n).

3. Indicate the steps by means of which the compound shown at the left in each of the following examples can be converted to the compound shown at the right:

(a) $C_6H_5C{\equiv}CH \longrightarrow C_6H_5CH_2CHO$

(b) $C_6H_5CH_2Cl \longrightarrow C_6H_5CH_2COCH_2C_6H_5$

(c) $Br(CH_2)_5Br \longrightarrow$ cyclohexanone

(d) cyclopentyl$-CH_2Br \longrightarrow$ cyclopentyl$-CH_2COCH_3$

(e) benzene $\longrightarrow C_6H_5CH(OH)SO_3Na$

4. Indicate what reaction or series of reactions could be used to convert cyclohexanone into each of the following compounds:

(a) cyclohexane

(b) cyclohexane spiro-epoxide (O, CH_2)

(c) cyclopentyl$-CO_2H$

(d) cyclohexyl$-CH_2CO_2C_2H_5$

(e) 2-methylenecyclohexanone ($=O$, $=CH_2$)

(f) seven-membered lactone (ring O, $=O$)

(g) cyclohexyl$-CHO$

(h) cyclohexylidene$=CHCH_2CH_3$

(i) cycloheptanone

5. Indicate what reaction or series of reactions could be used to convert benzaldehyde into each of the following compounds:

(a) $C_6H_5C(=O)CH(OH)C_6H_5$ (C₆H₅–CCH(OH)–C₆H₅)

(b) $C_6H_5CH{=}CHC(=O)C_6H_5$

(c) $C_6H_5CH(OH)CH_2NH_2$

(d) $C_6H_5C(=O)CH_3$

(e) $C_6H_5CH{=}CHCO_2H$

(f) $C_6H_5CH{=}CHCH_2CH(CH_3)_2$

6. For each of the following series of compounds, predict the order of decreasing equilibrium constant for cyanohydrin formation:

(a) $(CH_3)_2C{=}O$ $[(CH_3)_3C]_2C{=}O$ C_6H_5CHO $O_2N{-}C_6H_4{-}CHO$ $CH_3O{-}C_6H_4{-}CHO$ CCl_3CHO

(b) 2,2,6,6-tetramethylcyclohexanone (CH₃, CH₃, O, CH₃, CH₃); 3,3-dimethylcyclohexanone (CH₃, CH₃); bicyclic ketone with H, H (=O); cis-fused bicyclic cyclopentanone with H, H (=O)

(c) 4,4-dimethylcyclohexanone (CH_3, CH_3); $(CH_3)_2N{-}C_6H_4{-}C(=O)CH_3$; 3,3-dimethylcyclobutanone (CH_3, CH_3); CH_3CHO; $(CH_3)_2CHC(=O)CH(CH_3)_2$

7. Arrange the reactions of acetone with the following nucleophiles in the order of the increasing value of the equilibrium constant: CH_3MgI, $(CH_3)_3COH$, CH_3OH, HCN, $H_2NNHCONH_2$.

$$(CH_3)_2C{=}O + E{-}Nu \underset{}{\overset{K_{eq}}{\rightleftharpoons}} (CH_3)_2C(O{-}E)(Nu)$$

8. Explain the following observations:

(a) Treatment of acetic acid labeled with ^{18}O in the carbonyl group with aqueous acid produces acetic acid in which the ^{18}O is equally distributed between the carbonyl and hydroxyl oxygens.

(b) Treatment of acetone with ^{18}O-labeled H_2O in the presence of a mineral acid results in the formation of ^{18}O-labeled acetone.

(c) Hexafluorocyclobutanone reacts with water to form a stable hydrate.

(d) The equilibrium constant for the reaction

$$R_2C{=}O \quad + \quad Na \quad \rightleftharpoons \quad (R_2CO)^{\overline{\cdot}} \quad + \quad Na^+$$

(Radical anion)

is greater when R is $O_2N-C_6H_4-$ than when R is $CH_3O-C_6H_4-$

(e) The action of methylmagnesium bromide on triphenylmethyl benzoate $C_6H_5CO_2C(C_6H_5)_3$ yields 1,1,1-triphenylethane and the magnesium salt of benzoic acid.

(f) 2,4,6-$[CH(CH_3)_2]_3C_6H_2-C{=}O-C_6H_4-OCH_3$ + $(CH_3)_3CMgCl$ ⟶ 2,4,6-$[CH(CH_3)_2]_3C_6H_2-C{=}O-C_6H_4-C(CH_3)_3$

(g) The benzilic acid rearrangement of the ^{14}C-labeled diketone shown yields a product in which essentially all the label is in the α carbon.

$CH_3O-C_6H_4-\overset{O}{\overset{\|}{C^*}}-\overset{O}{\overset{\|}{C}}-C_6H_5$ $\xrightarrow{KOH}$ $CH_3O-C_6H_4-\overset{OH}{\overset{|}{C^*}}(C_6H_5)-CO_2H$

* indicates ^{14}C

(h) 2-Carbethoxy-1,3-indandione ($-CO_2C_2H_5$) + CH_2N_2 (Excess) ⟶ 1-hydroxy-2-($CO_2C_2H_5$)-4-(OCH_3)naphthalene

(i) The action of dimethylsulfoxonium methylide, $(CH_3)_2\overset{+}{S}(=O)=\overset{-}{C}H_2$ on 2-cyclohexenone yields bicyclo[4.1.0]heptan-2-one.

(j) The action of dimethyl malonate on methyl 2,4-hexadienoate in the presence of sodium methoxide yields dimethyl 6-carbomethoxy-5-methyl-3-heptenedioate.

(k) Phenylpyruvaldehyde, $C_6H_5\overset{O}{\overset{\|}{C}}CHO$, rearranges under the influence of aqueous sodium hydroxide to mandelic acid, $C_6H_5\overset{OH}{\overset{|}{C}}HCO_2H$.

(l) Some compounds containing a 3-buten-1-ol moiety undergo thermal rearrangement, yielding an alkene and an aldehyde or ketone as products. For example, 2-(1-octenyl)-cyclopentanol rearranges to 5-tridecenal via what is assumed to be a six-membered cyclic transition state:

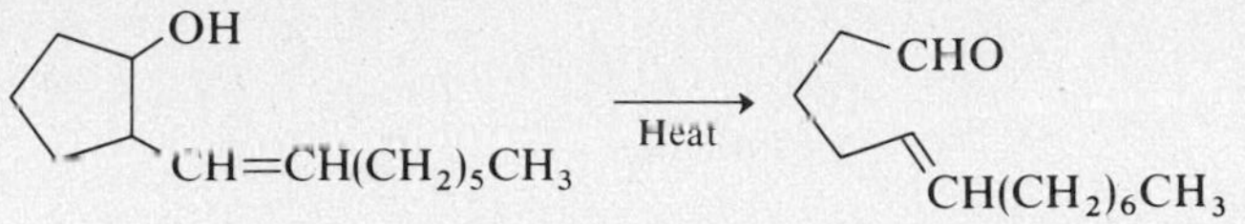

$$\text{(cyclopentane ring with OH and } CH{=}CH(CH_2)_5CH_3\text{)} \xrightarrow{\text{Heat}} \text{CHO} \ldots CH(CH_2)_6CH_3$$

Outline the mechanism of this reaction, and predict the products that would be formed from a similar reaction with ricinoleic acid,

$$CH_3(CH_2)_5\underset{\displaystyle OH}{\underset{|}{C}}HCH_2CH{=}CH(CH_2)_7CO_2H$$

9. A compound called "Cue-lure" shows strong sex attractant properties for the Melon fly *Dacus cucurbitae*. It has the molecular formula $C_{12}H_{14}O_3$, is optically inactive, and is insoluble in aqueous acid or base. Upon prolonged exposure to aqueous acid or base, however, it is converted to a water-soluble $C_2H_4O_2$ compound and an aqueous sodium hydroxide-soluble $C_{10}H_{12}O_2$ compound. The $C_{10}H_{12}O_2$ compound shows a positive haloform test, exchanges five carbon-bound hydrogens for deuterium when treated with $CH_3OD{-}NaOD$, and reacts with dimethyl sulfate to form a $C_{11}H_{14}O_2$ compound, which can be oxidized to an aqueous sodium bicarbonate-soluble $C_8H_8O_3$ compound. The $C_8H_8O_3$ compound shows a pair of doublets in the Ar—H region of the nmr and reacts with bromine to form two isomeric monobromo compounds. On the basis of these data, formulate a structure for "Cue-lure." To provide the finishing touch to your proof of structure, devise a synthesis of "Cue-lure," using as starting materials any compounds that you consider to be readily available.

10. Massiolactone, which is now known to have the structure shown,

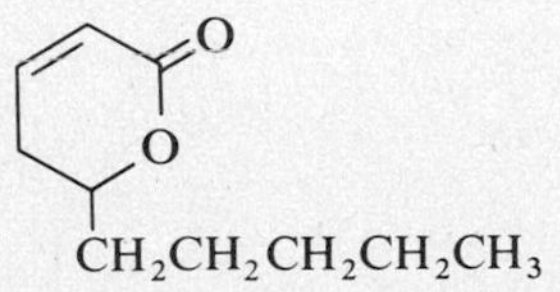

$$\text{(6-membered lactone ring: C=O, ring O, C=C)} - CH_2CH_2CH_2CH_2CH_3$$

has been found in the bark of certain members of the laurel family (*Lauraceae*) and has been used as an ingredient of native medicines in New Guinea. More recently, it has been isolated from several species of Australian ants, where it presumably serves as part of the ant's chemical defense mechanism against predators. Let us suppose that it is your privilege to be the first scientist to have a sufficiently large supply of this material to embark on a structure proof. Outline the procedures that you would use to achieve this goal. Any compounds in Tables 39.1–39.5 can be considered to be available as authentic samples for comparison with degradation products from massiolactone.

11. For the following syntheses, the available starting materials include (a) any alkanes, alkenes, alkynes, alkyl halides, alkanols, alkanals, alkanones, or alkanoic acids containing four or fewer carbon atoms, (b) benzene and any monosubstituted benzenes, (c) special organic reagents, and (d) any inorganic compounds. Starting with materials chosen from this group, outline syntheses for each of the following compounds.

(a) 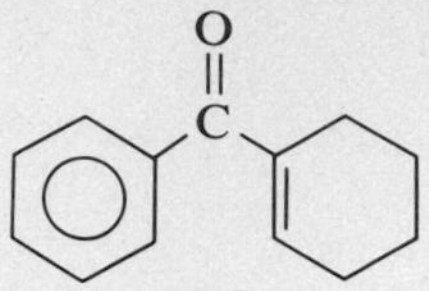 $C_6H_5COC_6H_9$ (phenyl 1-cyclohexenyl ketone)

(b) $(CH_3)_2CHCH{=}CHCH_2CH_3$

(c) 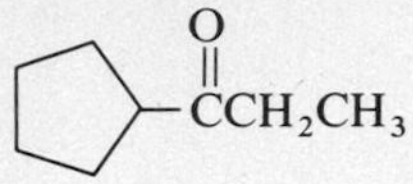 $C_5H_9\text{-}\overset{O}{\overset{\|}{C}}CH_2CH_3$ (cyclopentyl)

(d) 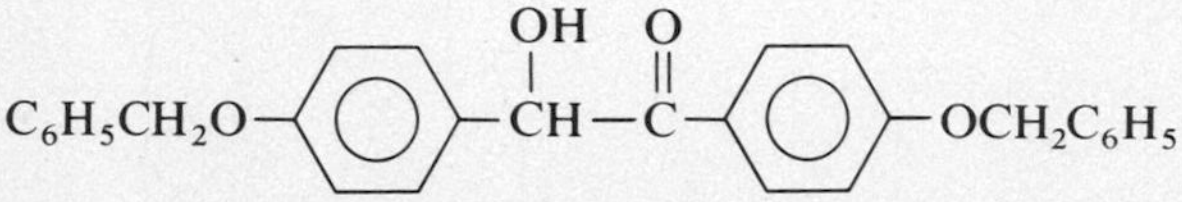 $C_6H_5CH_2O\text{-}C_6H_4\text{-}\overset{OH}{\overset{|}{C}H}\text{-}\overset{O}{\overset{\|}{C}}\text{-}C_6H_4\text{-}OCH_2C_6H_5$

(e) 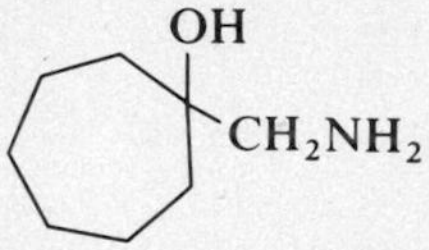 1-hydroxy-1-(CH_2NH_2)cycloheptane (OH, CH_2NH_2)

12. The following carbonyl compounds are inexpensive starting materials, costing less than 50c/lb:

Aldehydes: Formaldehyde, acetaldehyde, *n*-butyraldehyde, isobutyraldehyde, acrolein, benzaldehyde, chloral, glyoxal

Ketones: Acetone, ethyl methyl ketone, cyclohexanone, mesityl oxide, acetophenone

Acids: Formic acid, acetic acid, propionic acid, *n*-butyric acid, isobutyric acid, caproic acid, caprylic acid, capric acid, lauric acid, myristic acid, palmitic acid, stearic acid, acrylic acid, crotonic acid, benzoic acid, *p*-nitrobenzoic acid, oxalic acid, succinic acid, adipic acid, fumaric acid, phthalic acid, chloroacetic acid, trichloroacetic acid, lactic acid

Using only these compounds as starting materials, along with any inorganic reagents that might be necessary, devise a synthesis for each of the following compounds:

(a) 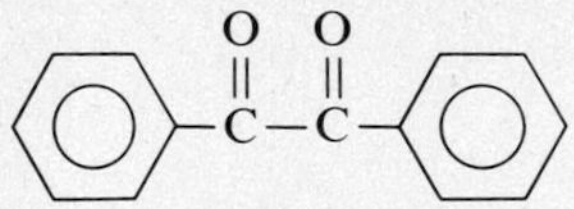 $C_6H_5\text{-}\overset{O}{\overset{\|}{C}}\text{-}\overset{O}{\overset{\|}{C}}\text{-}C_6H_5$

(b) $C_6H_5\text{-}\overset{OH}{\overset{|}{C}}(CH_3)CH_2NH_2$

(c) $HO_2C\text{-}C_6H_4\text{-}NH\overset{OH}{\overset{|}{C}H}CCl_3$ (para)

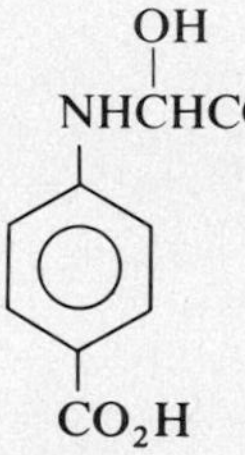

(d) 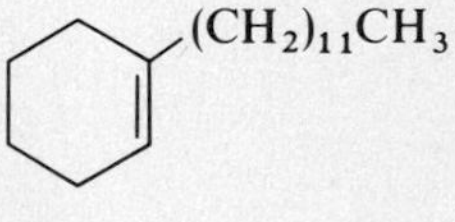 1-$(CH_2)_{11}CH_3$-cyclohexene

(e) 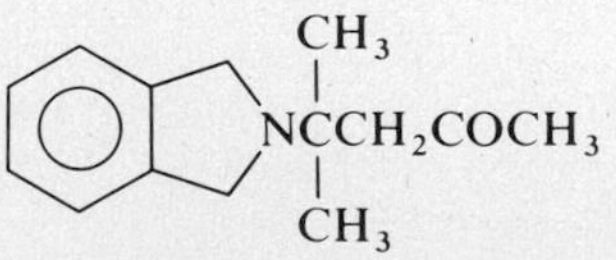 isoindoline-$N\overset{CH_3}{\overset{|}{C}}(CH_3)CH_2COCH_3$

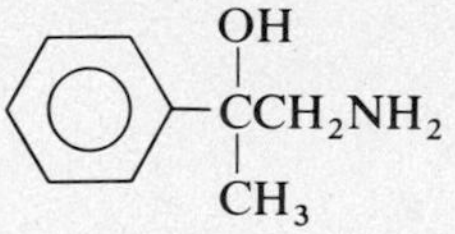

Chapter **30**

Carbonyl compounds: Reactions adjacent to the carbonyl group; other reactions of carbonyl compounds

See plastic Nature working to this end
The single atoms each to other tend
Attract, attracted to, the next in place
Form'd and impell'd its neighbor to embrace

ALEXANDER POPE
Essay on Man

Carbonyl compounds are vulnerable to attack by other compounds not only at the carbonyl group but at the adjacent position as well. This is a consequence of the labilizing effect that the carbonyl group has on the bonds at the α-position, interpretable in terms of the delocalization of electron density away from the α-carbon and onto the carbonyl oxygen. The present chapter deals mainly with reactions in which hydrogen atoms attached to the α-carbon of carbonyl compounds are replaced by (a) hetero atoms or hetero atom functions and (b) carbon atom frameworks. In addition, certain other reactions occurring at the α-carbon are also considered.

REACTIONS ADJACENT TO THE CARBONYL GROUP

30.1 *Active Methylene Compounds as Sources of Carbon Nucleophiles*

The heterolytic cleavage of the C—H bond in compounds of the general structure H—C—Y is induced by the attack of a base on the hydrogen, and it is facilitated by Y groups that are capable of delocalizing the negative charge that develops on the carbon atom. The carbonyl group is one of the functions that serves this purpose.

$$-\overset{|}{\underset{\underset{H}{|}}{C}}-C\overset{\diagup O}{\diagdown} + \text{Base} \rightleftharpoons -\overset{|}{\underset{-}{C}}-C\overset{\diagup O}{\diagdown} \longleftrightarrow -\overset{|}{C}=C\overset{\diagup O^-}{\diagdown} + \text{Base-H}^+$$

An enolate anion, a resonance-stabilized anion

PROBLEM 30-1

Draw resonance structures showing charge delocalization in the following anions:

$[CH_2CN]^-$ $\quad$ $[C_6H_5CHNO_2]^-$ $\quad$ $[CH_3CHSO_2CH_3]^-$

In addition to carbonyl functions, other groups can also serve this same purpose, including >C=NR, —C≡N, $-NO_2$, $-SO_2R$, >C=C<, —C≡C, and —Ar. Compounds containing these groups attached to a carbon atom bearing one or more hydrogen atoms are often referred to as *active methylene compounds.* The degree of lability of the C—H bonds at the α-position in many of these compounds (*i.e.*, their acidities) have been measured (Table 30.1). From these data, it is apparent that the acidity is a function of the particular group attached to the α carbon and the number of such activating groups in the compound.

PROBLEM 30-2

Estimate the magnitude of the equilibrium constant for the reaction

$$CH_3SOCH_3 + NaNH_2 \rightleftharpoons CH_3SOCH_2Na + NH_3$$

How could the formation of CH_3SOCH_2Na be pushed to completion?

Table 30.1 ACIDITIES OF ACTIVE METHYLENE COMPOUNDS (BOLD-FACE **H** IS THE ACIDIC PROTON)

Compound	pK_a
$CH_3\overset{O}{\overset{\|}{S}}CH_3$	33.5
$(C_6H_5)_3CH$	31.5
$CH_3CON(CH_3)_2$	30
$CH_3CO_2C_2H_5$	24
CH_3COCH_3	19
$CH_2(CO_2C_2H_5)_2$	13
$CH_2(CN)_2$	11
$CH_3COCH_2CO_2C_2H_5$	10.7
$CH_2(CN)CO_2C_2H_5$	9
$CH_2(COCH_3)_2$	9

The anions from active methylene compounds are stabilized by delocalization, the negative charge residing not only on the methylene carbon but on other atoms as well. Thus, they are "ambident" anions, and have the potentiality of reacting at more than one site. Generally, however, the isolable addition products are those in which the methylene carbon atom has served as the nucleophilic center. In the case of carbonyl compounds, this mode of reaction can be rationalized in terms of heats of reaction (ΔH). The reaction for the formation of a new C—C bond (*i.e.*, the ambident anion acting as a nucleophile at carbon) is more exothermic than the reaction for the formation of a new C—O bond (*i.e.*, the ambident anion acting as a nucleophile at oxygen) (Fig. 30.1).

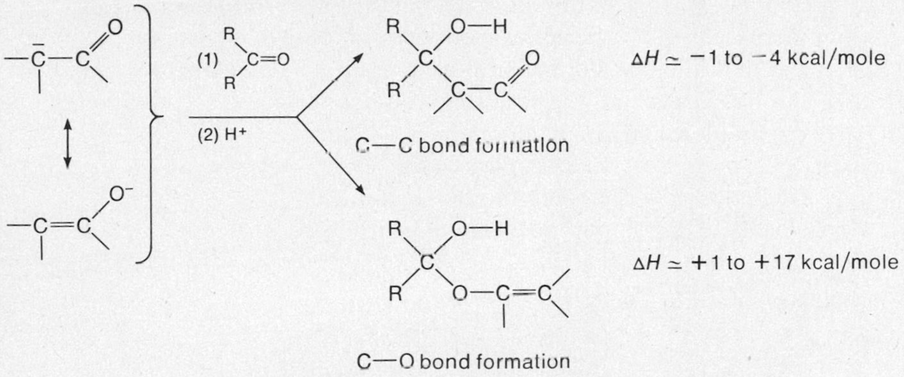

(1) $R_2C{=}O$; (2) H^+

C—C bond formation: $\Delta H \simeq -1$ to -4 kcal/mole

C—O bond formation: $\Delta H \simeq +1$ to $+17$ kcal/mole

Fig. 30.1. Heats of reaction for an active methylene ambident anion reacting via C—C bond and C—O bond-forming pathways.

Under some conditions, however, the oxygen atom of active methylene compounds can be the point of attachment of the incoming group (*i.e.* in a kinetically controlled process). For example, when aldehydes containing α-hydrogens are heated in the presence of acetic anhydride and potassium acetate, an enol acetate forms. A more generally applicable synthesis of enol acetates involves the treatment of active methylene compounds, particularly aldehydes and ketones, with isopropenyl acetate (*i.e.*, the enol acetate of acetone) in the presence of an acid catalyst (Fig. 30.2).

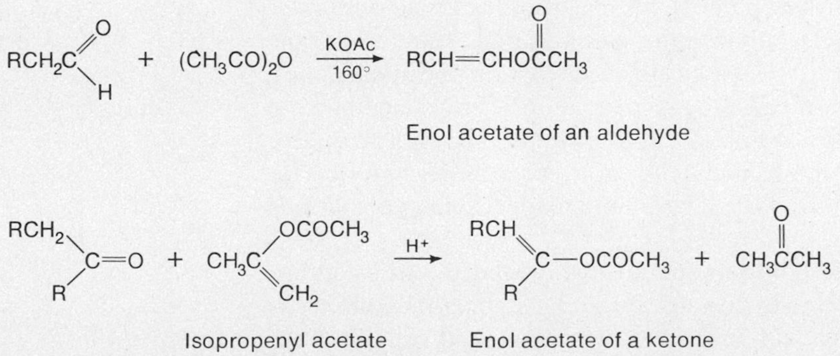

$$RCH_2CHO + (CH_3CO)_2O \xrightarrow[160^\circ]{KOAc} RCH{=}CHOCCH_3 \ (\text{C=O})$$

Enol acetate of an aldehyde

$$RCH_2(R)C{=}O + CH_3C(OCOCH_3){=}CH_2 \xrightarrow{H^+} RCH{=}C(R){-}OCOCH_3 + CH_3CCH_3 \ (\text{C=O})$$

Isopropenyl acetate — Enol acetate of a ketone

Fig. 30.2. Preparation of enol acetates from active methylene compounds.

Because of the tendency for active methylene compounds to dissociate to protons and resonance-stabilized anions, they undergo a variety of reactions with electrophilic reagents. These are discussed in the following sections, starting with reactions in which the electrophiles are hetero atoms. Following this, the reactions with carbon atom electrophiles are discussed, including (a) reactions occurring at sp^3-hybridized carbon atoms, (b) reactions occurring by 1,2-addition at carbonyl carbon atoms, (c) reactions occurring by 1,4-addition in α,β-unsaturated carbonyl systems, and (d) reactions occurring by addition-elimination at carbonyl carbon atoms.

30.2 Replacement of α-Hydrogen by Hetero Atoms

30.2a. HALOGENATION. The base-induced halogenation of active methylene compounds involves a nucleophilic displacement by the active methylene anion on the halogen molecule. More acidic than the starting material as a result of the electron-withdrawing effect of the halogen atom, the resulting compound rapidly reacts further with the halogen and leads to α,α-dihalo and α,α,α-trihalo compounds. The first steps in the haloform reaction, for example (see Fig. 29.33), follows this course, yielding the 1,1,1-trihalomethyl ketone. Another example is provided by cyclohexanone, which reacts with bromine in sodium hydroxide solution to yield 2,2-dibromocyclohexanone rather than 2,6-dibromocyclohexanone (Fig. 30.3).

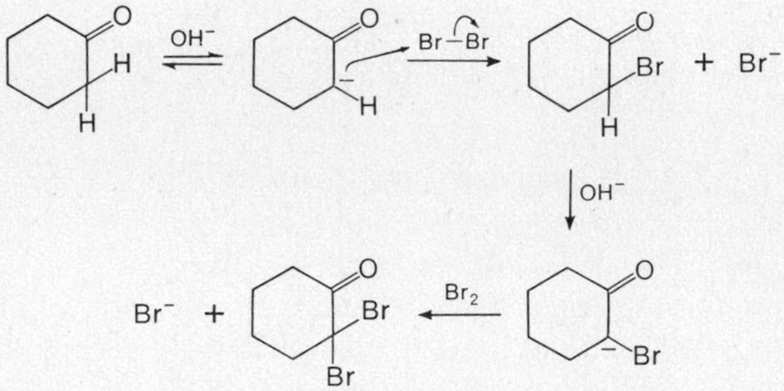

Fig. 30.3. Base-catalyzed bromination of cyclohexanone.

Halogenation of active methylene compounds can also be carried out under acid-catalyzed conditions, the nucleophile in this case being the enol form of the carbonyl compound. In contrast to the base-catalyzed reaction, the introduction of the halogen decreases the rate of subsequent enolization, thereby enabling the monohalo compound to be readily isolated. Furthermore, the enol may also change the position of entry of the second halogen as, for example, with cyclohexanone, which yields 2,6-dibromocyclohexanone under acidic conditions (Fig. 30.4).

The rate of the acid-catalyzed halogenation is directly related to the equilibrium concentration of the enol form. In the case of carboxylic acids, the concentration is vanishingly small, and reaction with halogen and an acid catalyst takes place exceedingly slowly. Addition of a trace of phosphorus or phosphorus halide, however, greatly increases the rate and endows the reaction with preparative value (*Hell–Volhard–Zelinsky reaction*). The function of the phosphorus (which is converted to phosphorus halide by the halogen) or phosphorus halide is to change the carboxylic acid to the carboxylic acid halide. The enol concentration

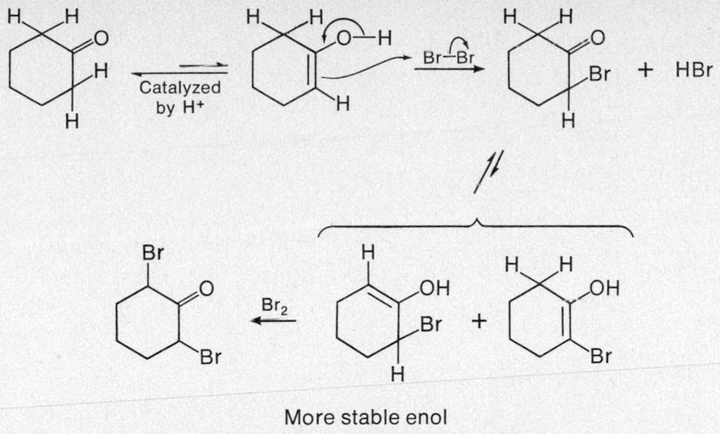

More stable enol

Fig. 30.4. Acid-catalyzed bromination of cyclohexanone.

is greater for the acid halide than for the acid, thereby making the halogenation proceed more rapidly. Subsequent "interchange" between the α-halo acid halide and the carboxylic acid (see Fig. 29.32) yields the α-halo acid and the unsubstituted acid halide, allowing the reaction to proceed with only a catalytic amount of phosphorus halide being required (Fig. 30.5).

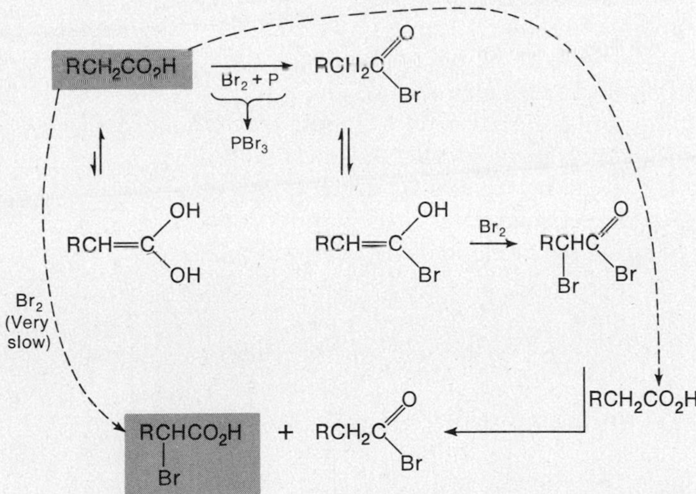

Fig. 30.5. Bromination of carboxylic acids catalyzed by phosphorus (Hell–Volhard–Zelinsky reaction).

PROBLEM 30-3

Why is the concentration of the enol in equilibrium with an acid chloride greater than that for the corresponding acid?

30.2b. OXIDATION. The susceptibility of active methyl and methylene groups to oxidation is illustrated in Fig. 28.6, selenium oxide usually being the reagent of choice. An alternative that is sometimes used involves treatment of the active methylene compound with nitrous acid, the initially formed α-nitroso compound undergoing tautomeric rearrangement to the oxime which can be hydrolyzed to the aldehyde or ketone (or reduced to the amine) (Fig. 30.6).

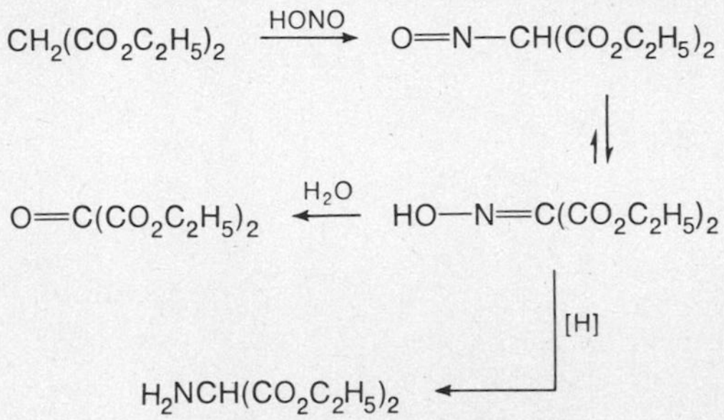

$$CH_2(CO_2C_2H_5)_2 \xrightarrow{HONO} O{=}N{-}CH(CO_2C_2H_5)_2$$

$$\rightleftharpoons$$

$$O{=}C(CO_2C_2H_5)_2 \xleftarrow{H_2O} HO{-}N{=}C(CO_2C_2H_5)_2$$

$$\xrightarrow{[H]} H_2NCH(CO_2C_2H_5)_2$$

Fig. 30.6. Preparation and reactions of diethylnitrosomalonate.

30.3 Active Methylene Nucleophiles in Substitution Reactions with Alkyl Halides (Alkylation Reactions)

30.3a. REACTIONS WITH SINGLY ACTIVATED METHYLENE COMPOUNDS. In the presence of strong bases, the concentration of the enolate anions from singly activated active methylene compounds is usually sufficiently high to allow S_N2 reactions with alkyl halides and sulfonates to take place. For examplc, treatment of acetophenone with methyl iodide in the presence of sodium methoxide yields propiophenone as the initially formed product. Although the product is slightly less acidic than the starting material, the anion of the product is more reactive than the anion of the starting material, and polyalkylated material tends to form (Fig. 30.7). To some extent polyalkylation can be reduced by using a stronger base [*e.g.*, $(C_6H_5)_3CK$], which increases the concentration of the acetophenone anion, resulting in an increased rate of formation of monoalkylated product.

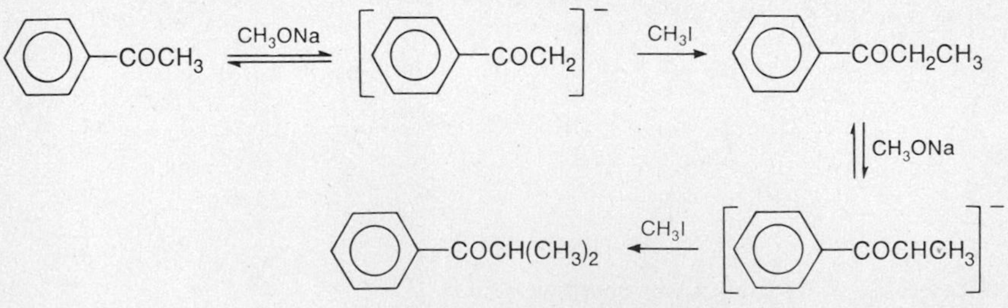

$$C_6H_5COCH_3 \underset{}{\overset{CH_3ONa}{\rightleftharpoons}} [C_6H_5COCH_2]^- \xrightarrow{CH_3I} C_6H_5COCH_2CH_3$$

$$C_6H_5COCH_2CH_3 \overset{CH_3ONa}{\rightleftharpoons} [C_6H_5CO\bar{C}HCH_3]^- \xrightarrow{CH_3I} C_6H_5COCH(CH_3)_2$$

Fig. 30.7. Base-induced methylation of acetophenone.

PROBLEM 30-4

What would you predict to be the product of the reaction of cyclohexanone with 1,3-dibromopropane in the presence of two equivalents of potassium *t*-butoxide?

"Blocking Groups" in Alkylation Reactions. Another problem often encountered in alkylation reactions occurs when more than one active methylene position is present in the reactant, permitting mixtures of products and/or the "wrong" alkylation product to form. For example, *trans*-bicyclo[4.4.0]decan-2-one, which has acidic hydrogens at C-1 and C-3, reacts with methyl iodide in the presence of sodium amide almost exclusively at C-3 (Fig. 30.8). If the C-1 methyl compound is desired it is necessary to employ a sequence of reactions whereby the C-3 position is "blocked" in such a way that alkylation at C-1 can take place, subsequent to which the blocking group at C-3 can be removed. One of the several procedures for doing this is illustrated in Fig. 30.8.

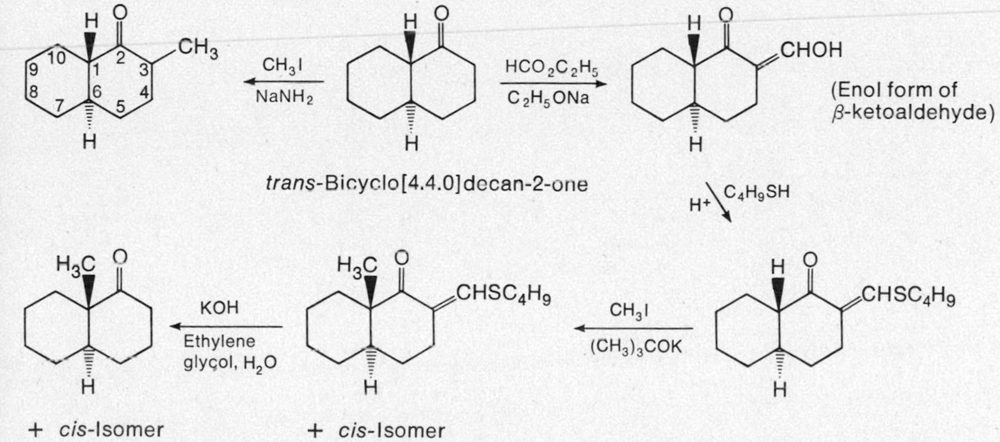

Fig. 30.8. Use of a "blocking" group to direct the course of the alkylation of trans-bicyclo[4.4.0]decan-2-one.

Alkylation via Enamines. The problems arising from polyalkylation and lack of situselectivity are jointly avoided by converting the aldehyde or ketone to the corresponding enamine (see Fig. 29.14) and then subjecting the enamine to alkylation (Fig. 30.9). Hydrolysis of the iminium salt affords a compound in

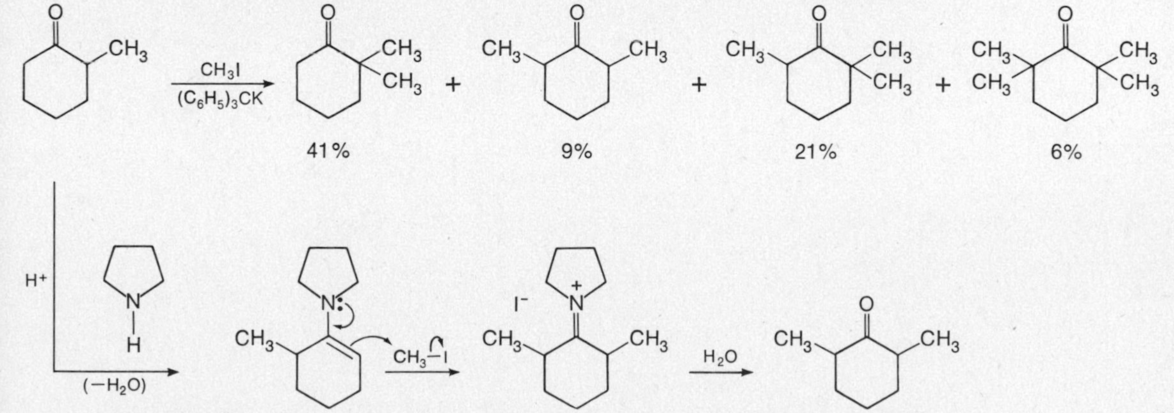

Fig. 30.9. Alkylation of 2-methylcyclohexanone and the enamine of 2-methylcyclohexanone.

which a single alkyl group has been introduced. For example, 2-methylcyclohexanone yields 2,6-dimethylcyclohexanone as the major product, in contrast to the outcome of the direct alkylation (Fig. 30.9). This is attributed to the double bond in the enamine forming at the position *away from* the methyl group rather than *toward* the methyl group, probably to minimize steric interference between the methyl group and the groups attached to the nitrogen.

Reaction with acrylonitrile:

$$\text{CH}_3\text{-(pyrrolidine enamine)} + H_2C{=}CHCN \longrightarrow \xrightarrow{H_2O} \text{CH}_3\text{-cyclohexanone-}CH_2CH_2CN$$

Reaction with 3-buten-2-one:

$$\text{(enamine)} + H_2C{=}CH{-}\overset{O}{\overset{\|}{C}}CH_3 \xrightarrow[\Delta,\ 24\ hr]{\text{Benzene}} \text{(iminium)}{-}CH_2{-}CH{=}C(O^-)CH_3 \longleftrightarrow \text{(iminium)}{-}CH_2\bar{C}H\overset{O}{\overset{\|}{C}}CH_3 \rightleftharpoons \text{(iminium)}{-}CH_2CH_2C(=O)\bar{C}H_2 \longrightarrow \text{(amino ketol)}$$

$$\xrightarrow[\Delta]{NaOAc,\ CH_3CO_2H,\ H_2O} \longrightarrow \text{(octalone)} + \text{pyrrolidine (N–H)}$$

Reaction with acid chlorides:

$$\text{(morpholine enamine)} + RCOCl \longrightarrow \xrightarrow{H_2O} \text{2-acylcyclohexanone (COR)}$$

Fig. 30.10. Conjugate addition and acylation reactions via enamines.

Enamines, in addition to being useful synthesis intermediates in alkylations, are useful in other reactions as well, engaging in conjugate additions and acylations (Fig. 30.10). With methyl vinyl ketone, they react by a combination of conjugate addition and alkylation to give cyclohexenones, affording an alternative to the Robinson annelation process (see Fig. 30.26).

PROBLEM 30-5

Suggest a mechanism for the conversion of

$$\overset{\displaystyle O}{\overset{\|}{RC}}-\overset{\displaystyle CHSC_4H_9}{\overset{\|}{CR}} \quad \text{to} \quad \overset{\displaystyle O}{\overset{\|}{RC}}-CH_2R,\ HCHO \ \text{and}\ C_4H_9SH$$

by the action of aqueous potassium hydroxide.

30.3b. REACTIONS WITH DOUBLY ACTIVATED ACTIVE METHYLENE COMPOUNDS. Compounds containing a methylene group between two activating functions, including β-diketones such as acetylacetone, β-keto esters such as ethyl acetoacetate (acetoacetic ester), and 1,1-diesters such as diethyl malonate (malonic ester), are very much more acidic than their monocarbonyl counterparts, and they readily form anions (see Table 30.1). The use of these compounds in the synthesis of ketones and carboxylic acids is discussed in Sections 28.4c and 28.9a. Conversion of a singly activated active methylene compound to one that is doubly activated can be used as still another way for achieving situselective alkylation. For example, condensation of 2-methylcyclohexanone with diethyl carbonate yields 2-carbethoxy-6-methylcyclohexanone, base-catalyzed alkylation of which proceeds exclusively at the 2 position. Hydrolysis of the β-keto ester followed by decarboxylation yields 2,6-dimethylcyclohexanone (Fig. 30.11). In this case the carbocthoxy group plays the role of an *activating group*.

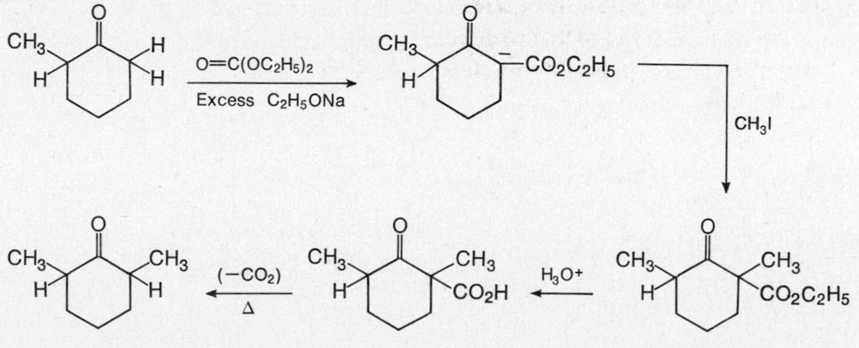

Fig. 30.11. Use of the carboethoxy group as an activating group in the alkylation of a ketone.

If a β-dicarbonyl compound such as 2,4-pentanedione (acetylacetone) is treated with two equivalents of a very strong base, such as sodium amide or sodium hydride, hydrogen is removed not only from the doubly activated methylene group but also from the singly activated methyl group (Fig. 30.12). The anion at the singly activated carbon (*i.e.*, C-1) is considerably more basic and, hence, more reactive, than the anion at the doubly activated carbon (*i.e.*, C-3). Therefore, when the dianion is treated with an alkyl halide, alkylation at the more reactive C-1 position occurs, providing an alternative to the alkylation procedure involving the monoanion, which leads to alkylation at C-3.

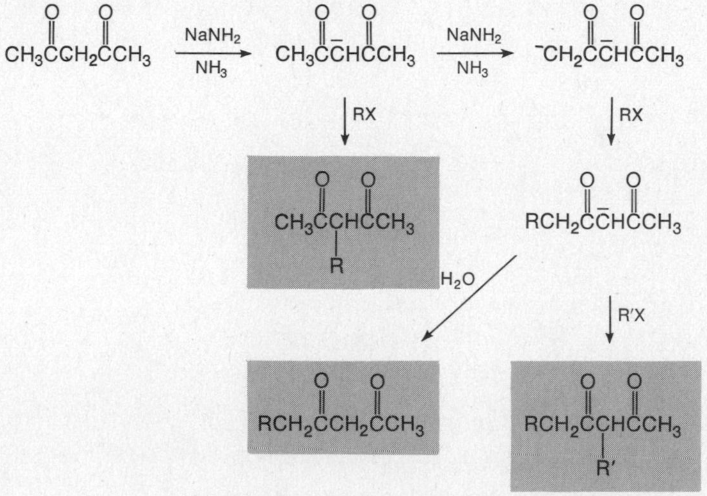

$$CH_3\overset{O}{\overset{\|}{C}}CH_2\overset{O}{\overset{\|}{C}}CH_3 \xrightarrow[NH_3]{NaNH_2} CH_3\overset{O}{\overset{\|}{C}}\bar{C}H\overset{O}{\overset{\|}{C}}CH_3 \xrightarrow[NH_3]{NaNH_2} {}^{-}CH_2\overset{O}{\overset{\|}{C}}\bar{C}H\overset{O}{\overset{\|}{C}}CH_3$$

$$CH_3\overset{O}{\overset{\|}{C}}\bar{C}H\overset{O}{\overset{\|}{C}}CH_3 \xrightarrow{RX} CH_3\overset{O}{\overset{\|}{C}}\underset{R}{\underset{|}{C}H}\overset{O}{\overset{\|}{C}}CH_3$$

$$^{-}CH_2\overset{O}{\overset{\|}{C}}\bar{C}H\overset{O}{\overset{\|}{C}}CH_3 \xrightarrow{RX} RCH_2\overset{O}{\overset{\|}{C}}\bar{C}H\overset{O}{\overset{\|}{C}}CH_3$$

$$RCH_2\overset{O}{\overset{\|}{C}}\bar{C}H\overset{O}{\overset{\|}{C}}CH_3 \xrightarrow{H_2O} RCH_2\overset{O}{\overset{\|}{C}}CH_2\overset{O}{\overset{\|}{C}}CH_3$$

$$RCH_2\overset{O}{\overset{\|}{C}}\bar{C}H\overset{O}{\overset{\|}{C}}CH_3 \xrightarrow{R'X} RCH_2\overset{O}{\overset{\|}{C}}\underset{R'}{\underset{|}{C}H}\overset{O}{\overset{\|}{C}}CH_3$$

Fig. 30.12. Formation and alkylation of monoanion and dianion from acetylacetone.

In a fashion analogous to the formation of the dianion from *β*-dicarbonyl compounds, carboxylic acids can form dianions in the presence of sufficiently strong bases. Treatment of isovaleric acid with sodium diisopropylamide, for example, yields a carboxylate ion that loses an *α*-proton in the presence of the even stronger base *n*-butyllithium to form the dianion. Alkylation of the dianion with *n*-butyl bromide followed by hydrolysis yields 2-isopropylhexanoic acid, providing a very convenient alternative to the malonic ester synthesis of acids (see Fig. 28.27) (Fig. 30.13).

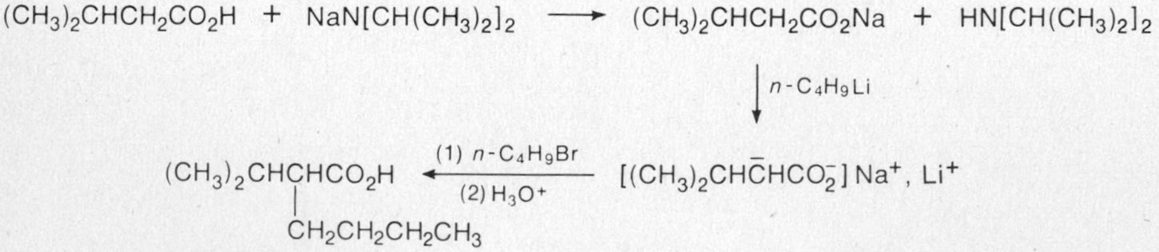

$$(CH_3)_2CHCH_2CO_2H + NaN[CH(CH_3)_2]_2 \longrightarrow (CH_3)_2CHCH_2CO_2Na + HN[CH(CH_3)_2]_2$$

$$(CH_3)_2CHCH_2CO_2Na \xrightarrow{n\text{-}C_4H_9Li} [(CH_3)_2CH\bar{C}HCO_2^-]\,Na^+,\ Li^+$$

$$[(CH_3)_2CH\bar{C}HCO_2^-]\,Na^+,\ Li^+ \xrightarrow[(2)\ H_3O^+]{(1)\ n\text{-}C_4H_9Br} (CH_3)_2CH\underset{CH_2CH_2CH_2CH_3}{\underset{|}{C}H}CO_2H$$

Fig. 30.13. Base-catalyzed alkylation of the dianion of isovaleric acid.

30.3c. Alkylation of *α*,*β*-unsaturated carbonyl compounds. The activating effect of a group, such as a carbonyl function, can be transmitted to nonproximate regions of a molecule by means of vinyl groups. For example, the γ-hydrogen in the system $\overset{\gamma}{\text{H—C}}—\overset{\beta}{\text{C}}=\overset{\alpha}{\text{C}}—\text{C=O}$ is comparable in acidity to the α-hydrogen in the system $\text{H}—\overset{\alpha}{\text{C}}—\text{C=O}$. This phenomenon is referred to as the *principle of vinylogy*. Thus, when bicyclo-[4.4.0]dec-1-en-3-one is treated with base, an ambident anion is formed in which the electron density is concentrated on the oxygen atom and the C-2 and C-10 carbon atoms (Fig. 30.14). Treatment of the anion with methyl iodide results in alkylation almost exclusively at the 2-position, however, this position having the highest electron density. Repetition of the process yields 2,2-dimethylbicyclo[4.4.0]dec-10-en-3-one.

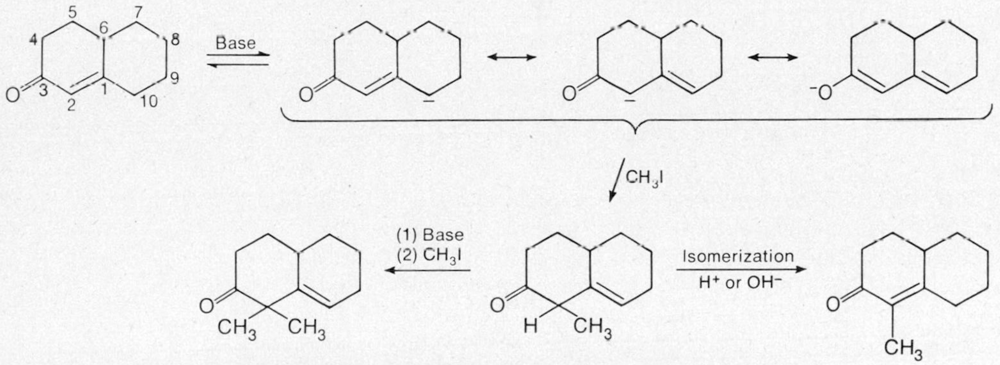

Fig. 30.14. Alkylation of an α,β-unsaturated ketone.

30.4 Active Methylene Nucleophiles in 1,2-Addition Reactions with Carbonyl Groups

30.4a. Aldol Condensations Leading to β-Hydroxyaldehydes and β-Hydroxyketones

Self-Condensation Reactions. The classic example of the self-condensation reaction is that of acetaldehyde (Fig. 30.15), which yields acetaldol (also known as aldol, out of which has arisen the term *aldol condensation* for this type of reaction

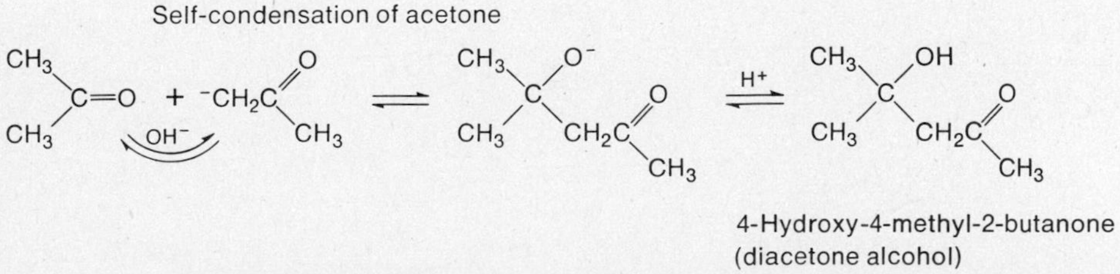

Self-condensation of acetaldehyde

3-Hydroxybutanal (acetaldol)

Self-condensation of acetone

4-Hydroxy-4-methyl-2-butanone (diacetone alcohol)

Fig. 30.15. Self-condensation reactions of acetaldehyde and acetone.

in general). Acetone also undergoes a comparable self-condensation reaction in the presence of base. Unlike the acetaldehyde condensation, however, where the equilibrium constant for product formation is quite large, that for the acetone condensation is only 0.02. Thus, to obtain the condensation product, called diacetone alcohol (Fig. 30.15), it is necessary to carry out the reaction in a special apparatus that keeps the product, once formed, away from the catalyzing base, thereby preventing the product from reverting to the reactant acetone (*i.e.*, re-establishing the equilibrium concentration). The difference in yield of product in the acetaldehyde and acetone self-condensations is due primarily to the greater hindrance around the carbonyl group in acetone. The anions from active methylene compounds are, in general, only moderately effective nucleophiles and, consequently, are quite susceptible to the steric environment of the carbonyl groups with which they react.

PROBLEM 30-6

In the self-condensation reaction of acetaldehyde, the formation of the anion is the rate-determining step. In the self-condensation reaction of acetone, the condensation of the anion with acetone is the rate-determining step. In which of these condensations would you expect to find C—D bonds in the product if the reaction is carried out in D_2O? Explain.

Mixed Condensation Reactions. If a mixture of aldehydes, or a mixture of an aldehyde and a ketone, is subjected to aldol condensation, four products can be formed unless the structures of the reactants are such as to require that one of them serve only as the electrophilic carbonyl component and the other only as the active methylene nucleophilic component. For example, formaldehyde lacks α-hydrogens and can, therefore, serve only as the carbonyl component, as in the crossed Cannizzaro reaction (see p. 880). Thus, in the presence of calcium hydroxide, formaldehyde and isobutyrophenone yield a product in which the active methylene carbon of isobutyrophenone has become attached to the carbonyl carbon of formaldehyde, this type of reaction being designated as a *Tollens' condensation* (Fig. 30.16). In a similar vein, nitromethane, which lacks a

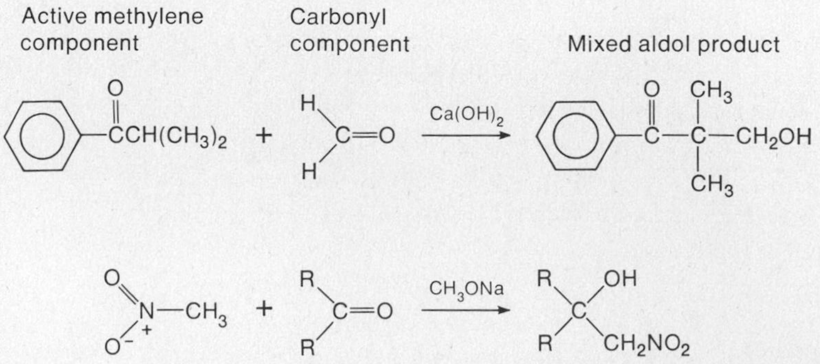

Active methylene component		Carbonyl component		Mixed aldol product
$C_6H_5COCH(CH_3)_2$	+	$H_2C{=}O$	$\xrightarrow{Ca(OH)_2}$	$C_6H_5CO{-}C(CH_3)_2{-}CH_2OH$
$O_2N^+{-}CH_3$ (O, O⁻)	+	$R_2C{=}O$	$\xrightarrow{CH_3ONa}$	$R_2C(OH){-}CH_2NO_2$

Fig. 30.16. Synthetically useful mixed aldol condensations.

carbonyl group and has no tendency to undergo self-condensation, reacts as an active methylene component and adds to aldehydes and ketones to give nitro alcohols (Fig. 30.16; see also Fig. 28.10).

30.4b. Aldol condensations leading to α,β-unsaturated carbonyl compounds

Claisen–Schmidt Condensation. The frequently unfavorable equilibrium constant for the addition step in aldol condensations can often be offset by the removal of the elements of water from the initially formed β-hydroxy compound. This is particularly true if aromatic aldehydes serve as the carbonyl components, for the aromatic ring labilizes the benzylic hydroxyl group and, thereby, promotes dehydration. Furthermore, the resulting double bond is resonance stabilized not only by resonance interaction with the carbonyl group, but with the benzene ring as well, leading to a large equilibrium constant for the dehydration step. For example, the mixed aldol condensation between benzaldehyde and acetaldehyde provides the α,β-unsaturated compound, cinnamaldehyde, in good yield as the result of the large equilibrium constant for the dehydration step, which effectively counterbalances the favorable equilibrium constant for the addition of the enolate anion of acetaldehyde to acetaldehyde (*i.e.*, the self-condensation reaction (Fig. 30.17). Reactions of this type are called *Claisen–Schmidt condensations.*

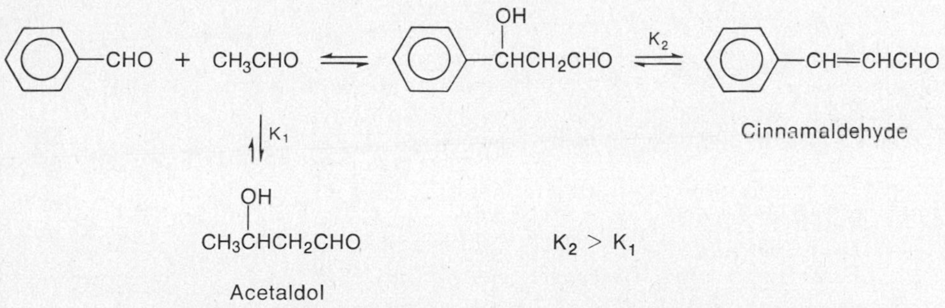

$$C_6H_5CHO + CH_3CHO \rightleftharpoons C_6H_5CH(OH)CH_2CHO \overset{K_2}{\rightleftharpoons} C_6H_5CH{=}CHCHO$$

Cinnamaldehyde

$$2\,CH_3CHO \overset{K_1}{\rightleftharpoons} CH_3CH(OH)CH_2CHO$$

Acetaldol

$$K_2 > K_1$$

Fig. 30.17. Mixed aldol condensation between benzaldehyde and acetaldehyde (Claisen–Schmidt reaction).

Perkin Condensation. A number of reactions similar in type to the Claisen–Schmidt reaction are known and are usually identified by the names of the chemists who either discovered them or exploited their possibilities. The oldest of these reactions is the *Perkin condensation*, discovered in 1868 by W. H. Perkin. It involves a high-temperature (175–180°) interaction between an aromatic aldehyde, an anhydride, and the potassium salt of the carboxylic acid corresponding to the anhydride. The potassium carboxylate serves as the basic catalyst, the anhydride serves as the active methylene component, and the aldehyde serves as the carbonyl component (Fig. 30.18). The synthesis is restricted to aldehydes lacking α hydrogens, for those that contain α hydrogens form enol acetates under the conditions of the reaction (see Fig. 30.2).

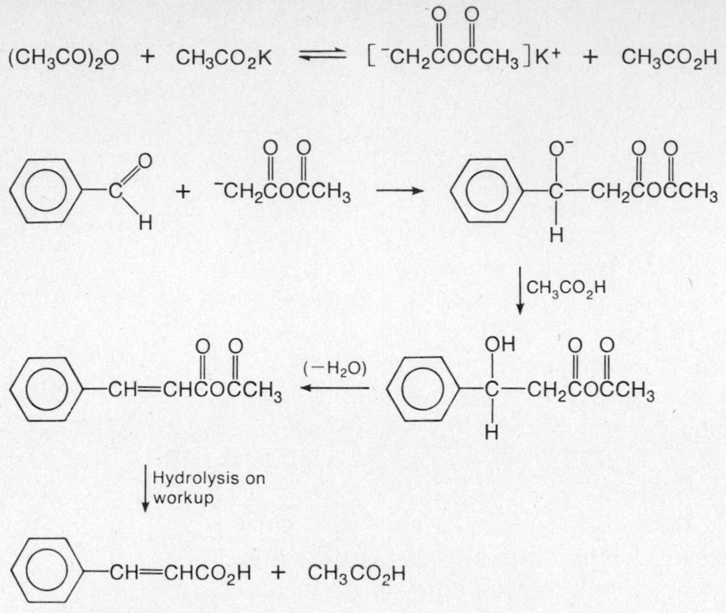

$$(CH_3CO)_2O \;+\; CH_3CO_2K \;\rightleftharpoons\; [\,^{-}CH_2\overset{\overset{O}{\|}}{C}O\overset{\overset{O}{\|}}{C}CH_3\,]K^+ \;+\; CH_3CO_2H$$

$$C_6H_5\text{—}\overset{\overset{O}{/\!/}}{C}\text{—}H \;+\; ^{-}CH_2\overset{\overset{O}{\|}}{C}O\overset{\overset{O}{\|}}{C}CH_3 \;\longrightarrow\; C_6H_5\text{—}\underset{\underset{H}{|}}{\overset{\overset{O^-}{|}}{C}}\text{—}CH_2\overset{\overset{O}{\|}}{C}O\overset{\overset{O}{\|}}{C}CH_3$$

$$\downarrow CH_3CO_2H$$

$$C_6H_5\text{—}CH{=}CH\overset{\overset{O}{\|}}{C}O\overset{\overset{O}{\|}}{C}CH_3 \;\xleftarrow{(-H_2O)}\; C_6H_5\text{—}\underset{\underset{H}{|}}{\overset{\overset{OH}{|}}{C}}\text{—}CH_2\overset{\overset{O}{\|}}{C}O\overset{\overset{O}{\|}}{C}CH_3$$

$$\downarrow \text{Hydrolysis on workup}$$

$$C_6H_5\text{—}CH{=}CHCO_2H \;+\; CH_3CO_2H$$

Cinnamic acid

Fig. 30.18. The Perkin condensation with benzaldehyde.

PROBLEM 30-7

For years there was a controversy as to whether the active methylene component in the Perkin condensation is derived from the sodium acetate or the acetic anhydride. Suggest an experiment by means of which this question might be resolved.

Knoevenagel–Doebner Condensation. An alternative to the Perkin condensation, leading to the same types of products, is the *Knoevenagel–Doebner condensation* involving the interaction of aromatic aldehydes and malonic acid in the presence of organic bases such as pyridine (Fig. 30.19). Malonic acid possesses a much more reactive methylene group than acetic anhydride and allows the reaction to proceed at lower temperatures and in generally higher yields. A more recent modification, introduced by Arthur Cope, employs the even more reactive active methylene compound cyanoacetic ester and uses ammonium acetate as the basic catalyst; the reaction is forced to completion by azeotropic removal of the water as it forms (Fig. 30.19). Both these modifications of the Knoevenagel reaction are more versatile than the Perkin condensation, being applicable to unreactive aryl ketones such as benzophenone as well as aliphatic aldehydes and ketones.

Stobbe–Johnson Condensation. An interesting and synthetically useful aldol-type reaction known as the *Stobbe–Johnson condensation* involves aldehydes or ketones as the carbonyl components and succinic esters as the active methylene

Knoevenagel-Doebner reaction:

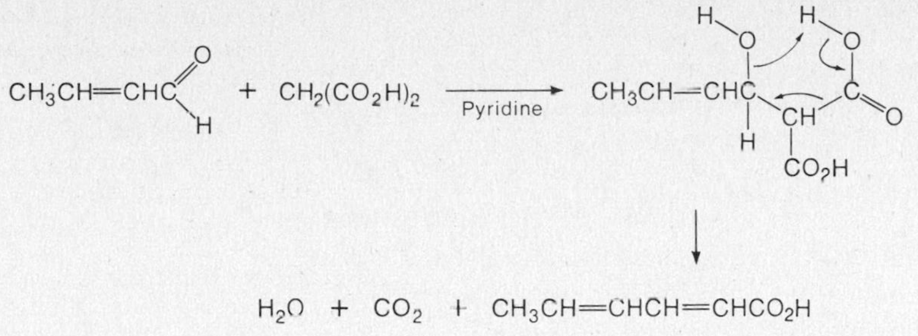

$$CH_3CH{=}CHCHO + CH_2(CO_2H)_2 \xrightarrow{\text{Pyridine}} CH_3CH{=}CHCH(OH)CH(CO_2H)CO_2H \longrightarrow H_2O + CO_2 + CH_3CH{=}CHCH{=}CHCO_2H$$

Knoevenagel-Cope reaction:

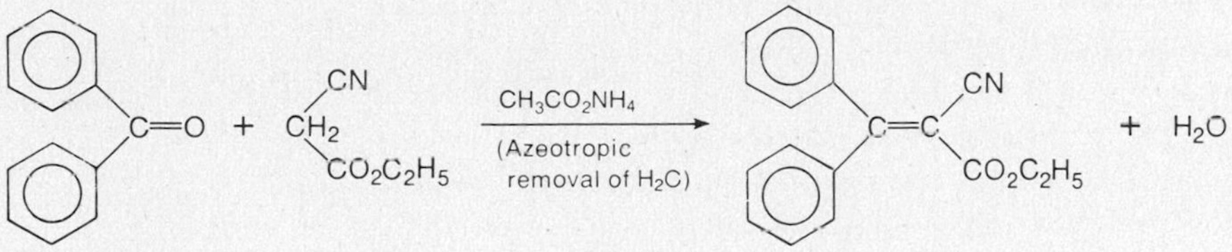

$$(C_6H_5)_2C{=}O + CH_2(CN)CO_2C_2H_5 \xrightarrow[\text{(Azeotropic removal of } H_2O)]{CH_3CO_2NH_4} (C_6H_5)_2C{=}C(CN)CO_2C_2H_5 + H_2O$$

Fig. 30.19. Knoevenagel–Doebner and Knoevenagel–Cope modifications of the Perkin-type condensation.

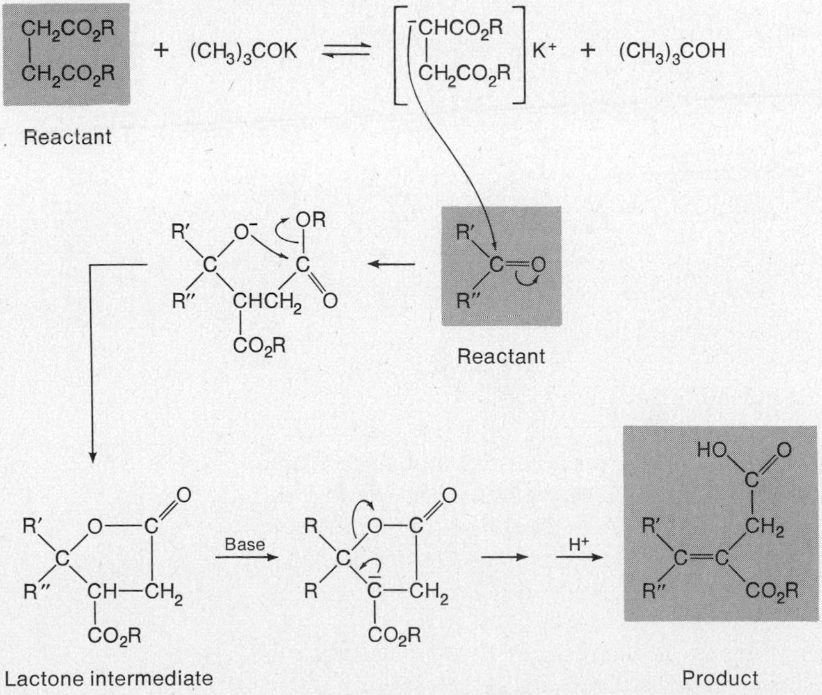

$$\underset{\text{Reactant}}{CH_2CO_2R\text{–}CH_2CO_2R} + (CH_3)_3COK \rightleftharpoons [{}^-CHCO_2R\text{–}CH_2CO_2R]K^+ + (CH_3)_3COH$$

$$\underset{\text{Reactant}}{R'R''C{=}O} \longrightarrow R'R''C(O^-)CHCH_2C(=O)OR \text{ (with } CO_2R)$$

$$\underset{\text{Lactone intermediate}}{\text{lactone}} \xrightarrow{\text{Base}} \longrightarrow \xrightarrow{H^+} \underset{\text{Product}}{R'R''C{=}C(CO_2R)CH_2CO_2H}$$

Fig. 30.20. Stobbe–Johnson condensation of aldehydes and ketones.

components, di-*t*-butyl succinate being the preferred choice (to minimize the self-condensation reaction of the ester). The reaction has been shown to involve a lactone as an intermediate, the lactone undergoing ring opening during the course of the reaction to yield an unsaturated mono acid-mono ester (called a "half" ester) as the product (Fig. 30.20). The process provides a means for introducing a *three-carbon chain* at the site of an aldehyde or ketone carbonyl group.

30.4c. "DIRECTED" ALDOL CONDENSATIONS. A method has been devised that circumvents the problem of the formation of product mixtures from mixed aldol condensations as well as the problem of the unreactivity of ketones toward carbanions derived from monoactivated active methylene compounds. Called a "directed" aldol condensation, the procedure involves conversion of the aldehyde or ketone that is to serve as the active methylene component to the corresponding imine by treatment with a primary amine (Fig. 30.21). Under the influence of a strong base, such as lithium diethylamide, the imine is converted to its anion, comparable to an enolate anion, which is then treated with the aldehyde or ketone that is to serve as the carbonyl component, leading to formation of a β-hydroxylimine. Upon treatment with aqueous acid, the imine hydrolyzes to the corresponding aldehyde or ketone and undergoes dehydration to yield the α,β-unsaturated compound.

30.4d. ACID-CATALYZED ALDOL CONDENSATIONS. In the presence of acids or bases in catalytic amounts, active methylene compounds are an equilibrium mixture of their carbonyl and enol forms:

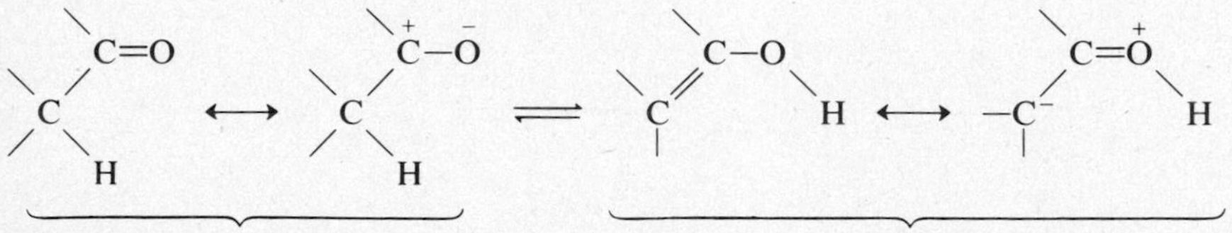

$$\underbrace{\mathrm{H{-}C{-}C{=}O} \longleftrightarrow \mathrm{H{-}C{-}\overset{+}{C}{-}\overset{-}{O}}}_{\text{Carbonyl form}} \rightleftharpoons \underbrace{\mathrm{C{=}C{-}O{-}H} \longleftrightarrow \mathrm{{-}\overset{-}{C}{-}C{=}\overset{+}{O}{-}H}}_{\text{Enol form}}$$

The enol forms in many cases are sufficiently nucleophilic at the α carbon to add to a carbonyl group of another molecule in which the carbonyl group is activated by protonation:

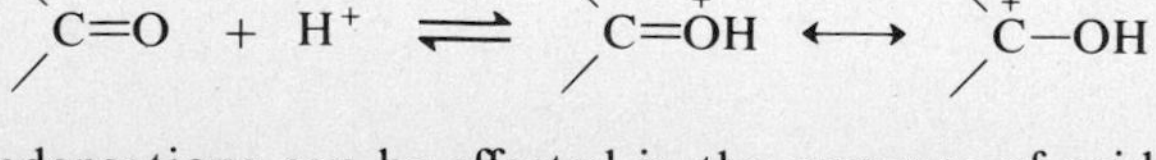

$$\mathrm{{>}C{=}O} + \mathrm{H^+} \rightleftharpoons \mathrm{{>}C{=}\overset{+}{O}H} \longleftrightarrow \mathrm{{>}\overset{+}{C}{-}OH}$$

Thus, aldol condensations can be effected in the presence of acids; the initially formed β-hydroxy compounds undergoing acid-catalyzed dehydration to yield α,β-unsaturated carbonyl compounds, driving the reaction to completion (Fig. 30.22). In compounds that can enolize in more than one way, the enol that is most highly substituted is the one that usually leads to the major product, as illustrated by the example of ethyl methyl ketone in Fig. 30.22.

Base-catalyzed aldol condensations do not necessarily show the same situselectivity as their acid-catalyzed counterparts. For example, the base-catalyzed condensation of benzaldehyde with 2-butanone occurs via the carbanion derived

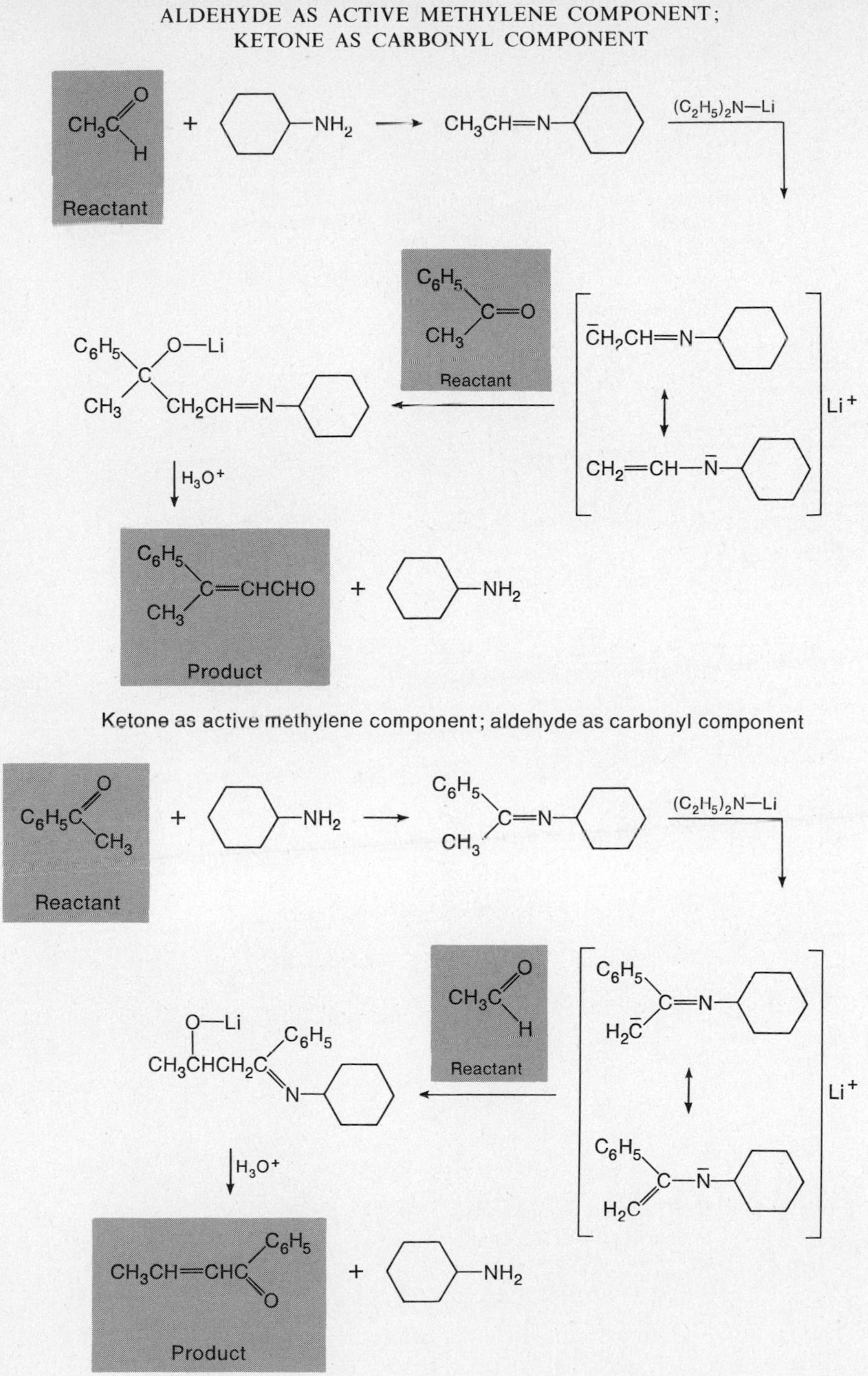

ALDEHYDE AS ACTIVE METHYLENE COMPONENT; KETONE AS CARBONYL COMPONENT

Reactant: CH_3CHO + cyclohexyl$-NH_2$ ⟶ $CH_3CH{=}N-$cyclohexyl $\xrightarrow{(C_2H_5)_2N-Li}$ [$\bar{C}H_2CH{=}N-$cyclohexyl ⟷ $CH_2{=}CH-\bar{N}-$cyclohexyl] Li^+

$\xrightarrow{\text{Reactant: } C_6H_5(CH_3)C{=}O}$ $C_6H_5C(CH_3)(O-Li)CH_2CH{=}N-$cyclohexyl

$\xrightarrow{H_3O^+}$ Product: $C_6H_5(CH_3)C{=}CHCHO$ + cyclohexyl$-NH_2$

Ketone as active methylene component; aldehyde as carbonyl component

Reactant: $C_6H_5COCH_3$ + cyclohexyl$-NH_2$ ⟶ $C_6H_5(CH_3)C{=}N-$cyclohexyl $\xrightarrow{(C_2H_5)_2N-Li}$ [$C_6H_5(H_2\bar{C})C{=}N-$cyclohexyl ⟷ $C_6H_5(H_2C{=})C-\bar{N}-$cyclohexyl] Li^+

$\xrightarrow{\text{Reactant: } CH_3CHO}$ $CH_3CH(O-Li)CH_2C(C_6H_5){=}N-$cyclohexyl

$\xrightarrow{H_3O^+}$ Product: $CH_3CH{=}CHC(O)C_6H_5$ + cyclohexyl$-NH_2$

Fig. 30.21. Mixed aldol condensations via "directed aldol" condensation.

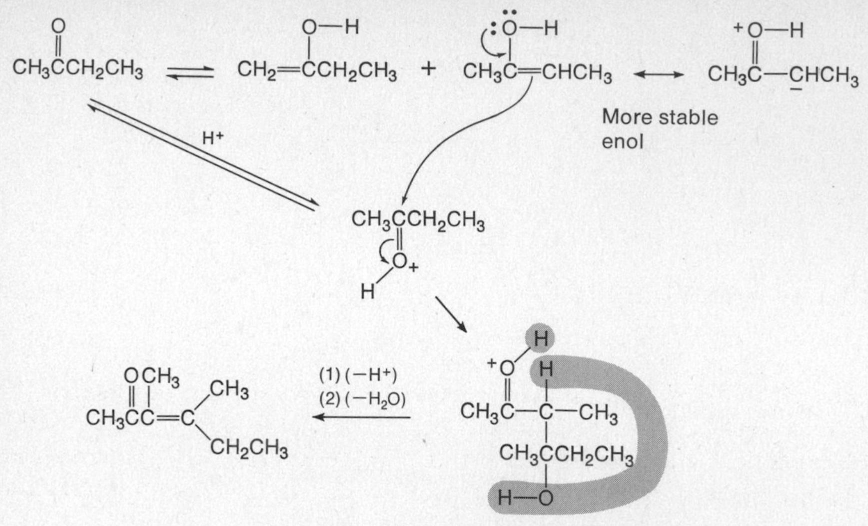

Fig. 30.22. Acid-catalyzed aldol condensation of ethyl methyl ketone.

from the methyl rather than the methylene group, yielding

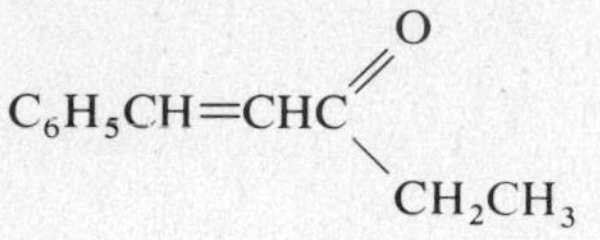

$$C_6H_5CH{=}CHC(=O)CH_2CH_3$$

as the product.

30.4e. INTRAMOLECULAR ALDOL CONDENSATION. Dicarbonyl compounds in which the carbonyl groups are in a 1,4 or a 1,5 relationship to each other, undergo intramolecular aldol condensations to yield cyclic products (Fig. 30.23). When there is the possibility of forming rings of different sizes, five- and six-membered rings are usually preferred over smaller or larger rings.

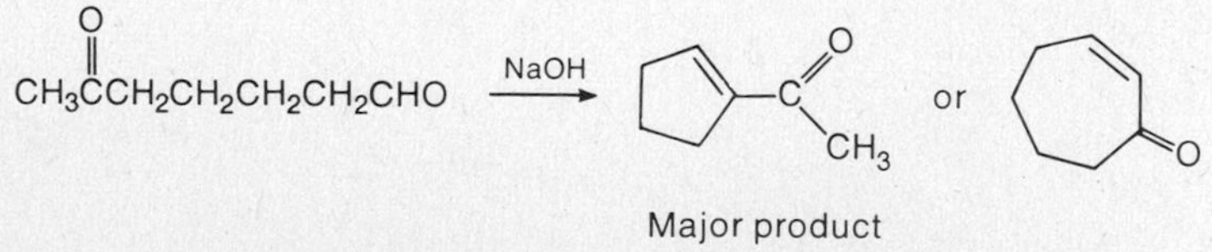

Fig. 30.23. Intramolecular aldol condensation.

30.4f. GLYCIDIC ESTER CONDENSATION. The glycidic ester condensation, often called the *Darzens glycidic ester condensation*, represents another example of a means for driving an aldol-type reaction to completion. The process involves the addition of a carbanion derived from an α-haloester to an aldehyde or ketone, the oxygen anion thus formed then effecting an intramolecular displacement of the halogen to yield an α,β-epoxyester called a glycidic ester (Fig. 30.24). An important synthesis application of glycidic esters involves the preparation of aldehydes, as illustrated in Fig. 28.22.

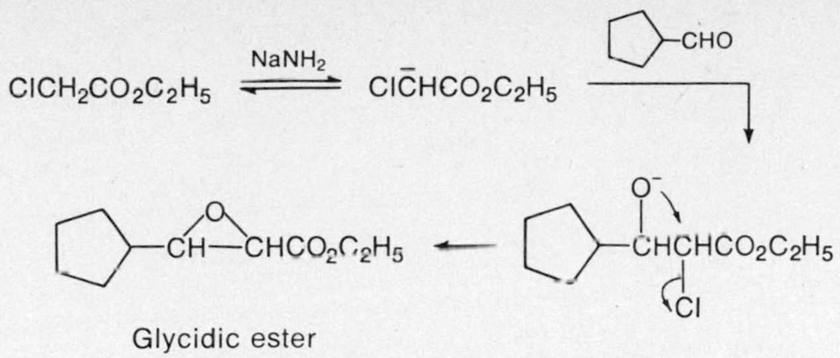
NaNH2
ClCH2CO2C2H5
ClCHCO2C2H5
CHO
O⁻
CHCHCO2C2H5
Cl
O
CH—CHCO2C2H5
Glycidic ester

Fig. 30.24. Darzens glycidic ester synthesis.

30.5 Active Methylene Nucleophiles in 1,4-Addition Reactions with α,β-Unsaturated Carbonyl Compounds

The 1,4-addition reactions of α,β-unsaturated carbonyl compounds with carbon atom nucleophiles is discussed in Section 29.8, where the base-induced reaction of crotonaldehyde and diethyl malonate (known as a *Michael addition reaction*) is depicted (see Fig. 29.24). An important procedure that makes use of the Michael reaction is the *Robinson annelation process*, whereby a six-membered ring can be added on to a cycloalkanone in a one-flask reaction. For example, treatment of cyclohexanone with 3-keto-1-butyltrimethylammonium iodide (*i.e.*, the quaternary salt obtained from the Mannich base prepared from acetone, formaldehyde, and dimethylamine followed by treatment with methyl iodide) in the presence of sodium ethoxide yields bicyclo[4.4.0]dec-1-en-3-one (Fig. 30.25). The reaction involves (a) the *in situ* conversion of the Mannich base to methyl vinyl ketone, (b) the conjugate addition of the anion from cyclohexanone to methyl vinyl ketone, (c) the intramolecular aldol condensation of the resulting 1,5-diketone, and (d) dehydration of the β-hydroxyketone to the α,β-unsaturated ketone.

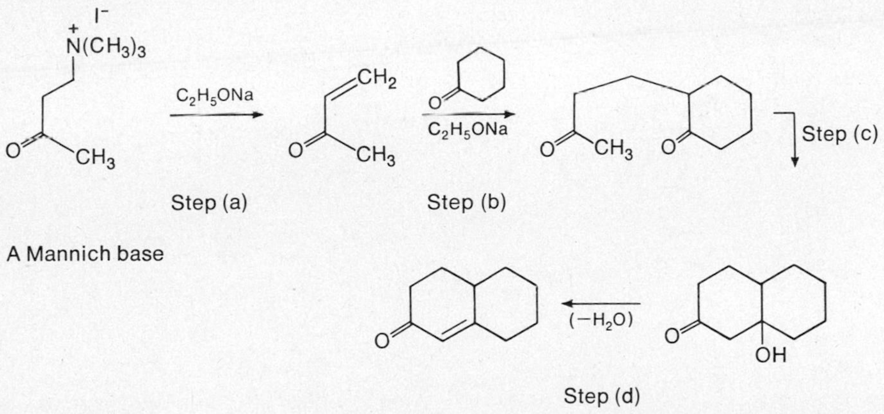
I⁻
N(CH3)3
O
CH3
C2H5ONa
Step (a)
A Mannich base
CH2
O
CH3
C2H5ONa
Step (b)
CH3
O
Step (c)
(−H2O)
OH
Step (d)

Fig. 30.25. Robinson annelation process (Robinson–Mannich base synthesis).

PROBLEM 30-8

What product would you predict to result from the action of $(CH_3)_3\overset{+}{N}CH_2CH_2COCH_3$ on 2-carbomethoxy-6-methylcyclohexanone in the presence of sodium methoxide?

Another interesting process developed by Robinson, which also combines aldol-type reactions and Michael reactions, involves the interaction of 1,4-butanedial (succindialdehyde), diethyl β-ketoglutarate, and methylamine. In a dilute, aqueous solution at pH 7 and a temperature of about 35° (*i.e.*, "physiological conditions") these materials react to form a bicyclic compound containing a tropinone ring (Fig. 30.26).

Shaded portion corresponds to the tropinone ring

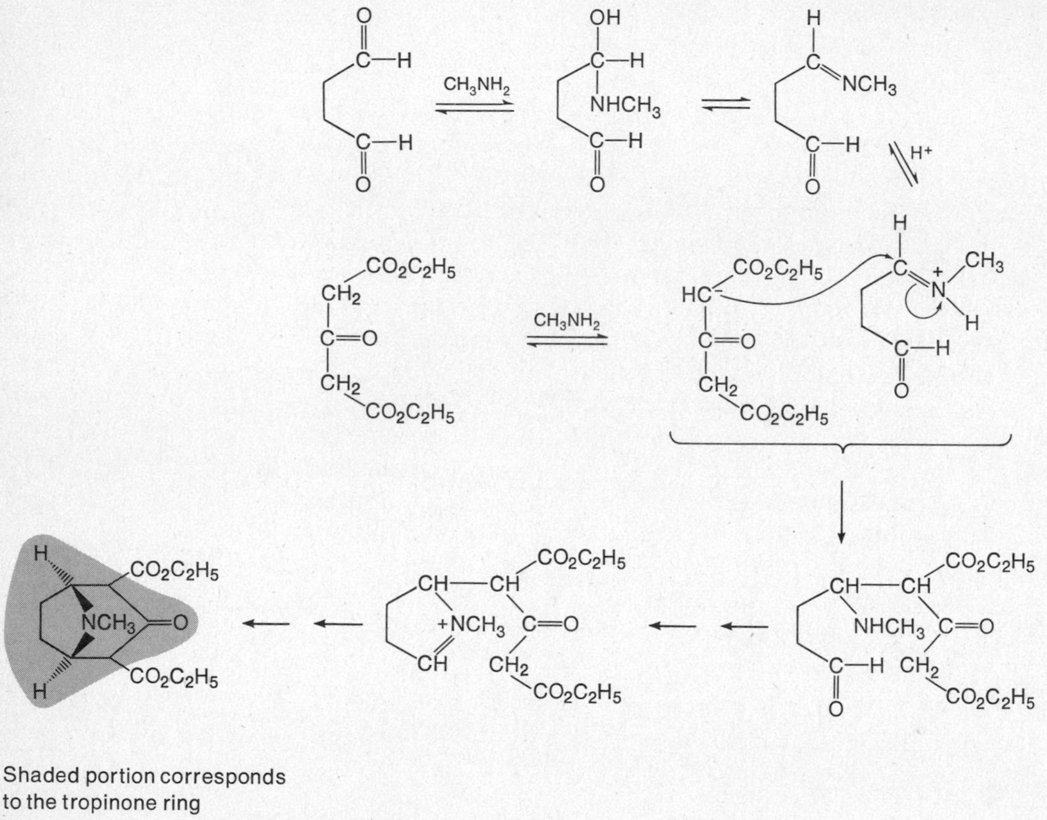

Fig. 30.26. Robinson's synthesis of a tropinone compound.

30.6 *Active Methylene Nucleophiles in Addition-Elimination Reactions*

Type AC and BC carbonyl compounds react with active methylene nucleophiles derived from carbonyl compounds to give β-dicarbonyl compounds, the process being known as the *Claisen condensation.* The simplest example of this process is the self-condensation reaction of ethyl acetate, which yields ethyl acetoacetate as shown in Fig. 15.11. Other examples of the Claisen condensation that have already been discussed include the reaction of the anion derived from diethyl malonate with anhydrides to yield β-keto esters (see Fig. 29.35) and the base-induced cyclization of diesters to yield cyclic β-keto esters, this process being known as the Dieckmann condensation (see Section 28.18a).

PROBLEM 30-9

Di-*t*-butyl succinate is used in the Stobbe–Johnson condensation to minimize the self-condensation of the ester. What would you predict to be the structure of the self-condensation product of dimethyl succinate?

The Claisen condensation, although resembling the aldol condensation, differs with respect to the amount of base that is required. Whereas a catalytic amount of base is sufficient for the aldol condensation (the product being comparable in acidity to the starting material), a full equivalent is required in the Claisen condensation. This is because the product is a doubly activated active methylene compound, and, consequently, a much stronger acid than the starting compound which reacts with one equivalent of base to form the anion (see Fig. 28.43).

Constraints similar to those that apply to the aldol condensation also apply to the Claisen condensation; "mixed" Claisen condensations yield mixtures of products unless one of the reagents can act only as the carbonyl component or as the active methylene component. For example, diethyl carbonate and diethyl oxalate are esters lacking α-hydrogens and, therefore, capable only of acting as the carbonyl component; good yields of a single Claisen product can be obtained

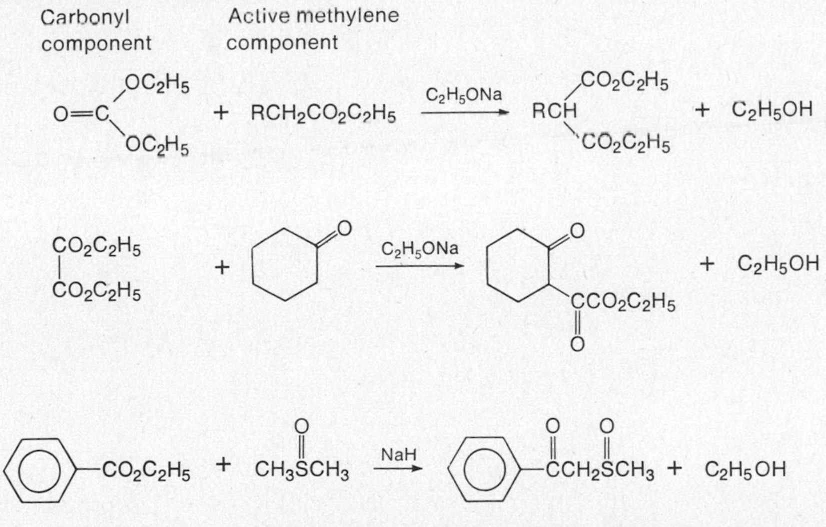

Carbonyl component | Active methylene component

$$O{=}C(OC_2H_5)_2 + RCH_2CO_2C_2H_5 \xrightarrow{C_2H_5ONa} RCH(CO_2C_2H_5)_2 + C_2H_5OH$$

$$\underset{CO_2C_2H_5}{\overset{CO_2C_2H_5}{|}} + \text{cyclohexanone} \xrightarrow{C_2H_5ONa} \text{2-(}\underset{\|}{C}CO_2C_2H_5\text{)cyclohexanone} + C_2H_5OH$$

$$C_6H_5{-}CO_2C_2H_5 + CH_3\overset{O}{\overset{\|}{S}}CH_3 \xrightarrow{NaH} C_6H_5{-}\overset{O}{\overset{\|}{C}}CH_2\overset{O}{\overset{\|}{S}}CH_3 + C_2H_5OH$$

Fig. 30.27. "Mixed" Claisen-type condensations.

from condensations of these esters with active methylene compounds (Fig. 30.27). Conversely, the anion from dimethylsulfoxide has little tendency to undergo self-condensation; therefore, it serves only as the active methylene component in reactions with carboxylic acid derivatives (Fig. 30.27).

PROBLEM 30-10

The active methylene compound $C_6H_5COCH_2SOCH_3$ can be alkylated in the usual fashion. Suggest a method for converting it to

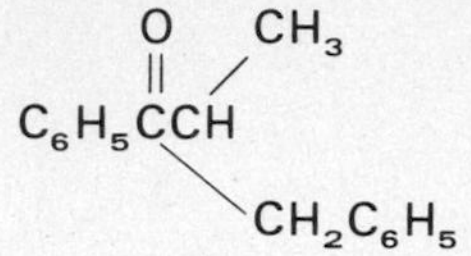

$$C_6H_5\overset{O}{\overset{\|}{C}}CH(CH_3)(CH_2C_6H_5)$$

PROBLEM 30-11

Would you expect

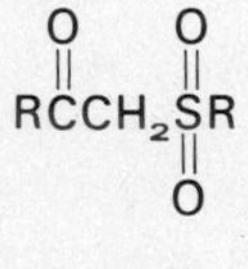

$$R\overset{O}{\overset{\|}{C}}CH_2\overset{O}{\overset{\|}{\underset{O}{\underset{\|}{S}}}}R$$

to undergo base-induced cleavage to RCO_2^- and CH_3SO_2R, or to $RCOCH_3$ and RSO_3^-? Explain.

As illustrated by the Dieckmann condensation (see Fig. 28.42), intramolecular Claisen condensation reactions are possible. Another example is the base-induced cyclization of δ-keto esters to cyclic β-diketones (Fig. 30.28).

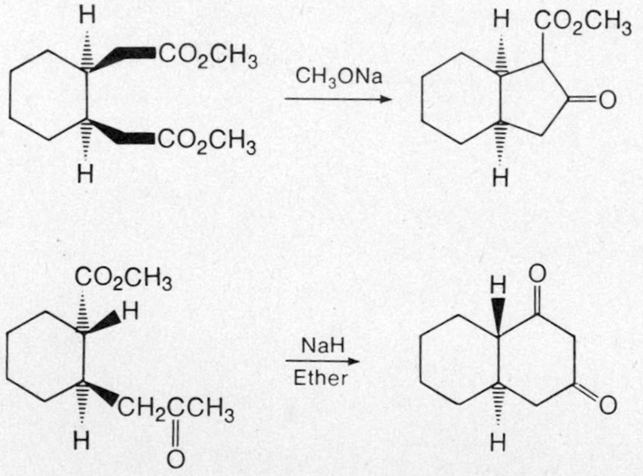

Fig. 30.28. Intramolecular Claisen-type condensations.

OTHER REACTIONS OF CARBONYL COMPOUNDS

In addition to the reactions enumerated in the previous sections, which apply with considerable generality to all types of carbonyl compounds, a number of other reactions are known that are more specific to particular types of carbonyl compounds.

30.7 Decarbonylation Reactions

α-Dicarbonyl compounds sometimes lose carbon monoxide (decarbonylation) when treated with strong acids such as sulfuric acid or when heated under certain conditions. For example, the α-keto esters obtained from the Claisen condensations with diethyl oxalate (see Fig. 30.27) lose carbon monoxide when they are mixed with powdered soft glass and this mixture then heated. The overall sequence thus provides a method of synthesis of β-keto esters (Fig. 30.29).

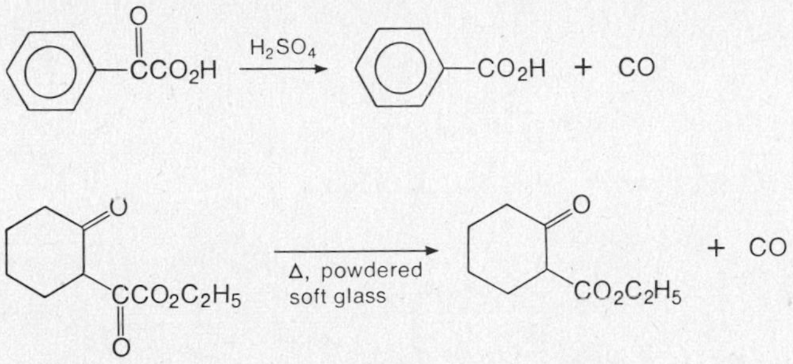

$$C_6H_5COCO_2H \xrightarrow{H_2SO_4} C_6H_5CO_2H + CO$$

$$\text{2-oxocyclohexyl}-COCO_2C_2H_5 \xrightarrow[\text{soft glass}]{\Delta,\ \text{powdered}} \text{2-oxocyclohexyl}-CO_2C_2H_5 + CO$$

Fig. 30.29. Decarbonylation of 1,2-dicarbonyl compounds.

30.8 Decarboxylation Reactions

Certain types of carboxylic acids lose carbon dioxide (decarboxylation) with considerable ease. Notable among these are β-keto acids (*i.e.*, acetoacetic acid), 1,1-dicarboxylic acids (*i.e.*, malonic acid), and α,β-epoxy acids (*i.e.*, glycidic acid), which lose carbon dioxide upon heating (room temperature or slightly above is sufficient for β-keto acids; about 150° is necessary for malonic acids). All of these decarboxylations can be rationalized in terms of an intramolecular proton transfer and simultaneous expulsion of carbon dioxide via a cyclic six-membered activated complex (Fig. 30.30).

In a reaction that bears some resemblance to the Dieckmann cyclization, dibasic acids in which the carboxyl groups are separated by four or five atoms undergo decarboxylation when their calcium or thorium salts are strongly heated. The products of the reaction, in addition to carbon dioxide, are the five- or six-membered cycloalkanones (see Fig. 14.11).

Still another method for inducing decarboxylation involves treatment with lead tetraacetate, examples of which are the conversions of 1,2-dicarboxylic acids to alkenes, as Fig. 25.9 illustrates. The action of lead tetraacetate on a monoacid (in a solution of acetic acid) causes the replacement of the carboxyl group by an acetoxyl group, leading to esters of acetic acid (Fig. 30.31). The treatment of carboxylic acids with lead tetraacetate in the presence of lithium chloride results in the formation of alkyl chlorides (compare with the Hunsdiecker method involving the action of bromine on the silver or thallium salts of carboxylic acids—see Fig. 21.7).

PROBLEM 30-12

Suggest a mechanism for the conversion of RCO_2H to RCl by the action of lead tetraacetate in the presence of lithium chloride.

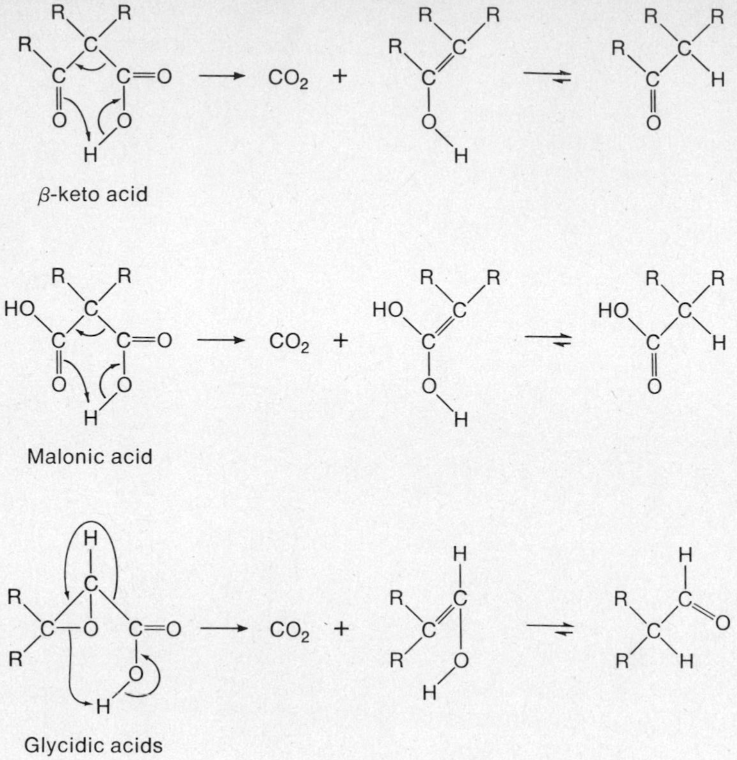

β-keto acid

Malonic acid

Glycidic acids

Fig. 30.30. Decarboxylation of β-keto, malonic, and glycidic acids.

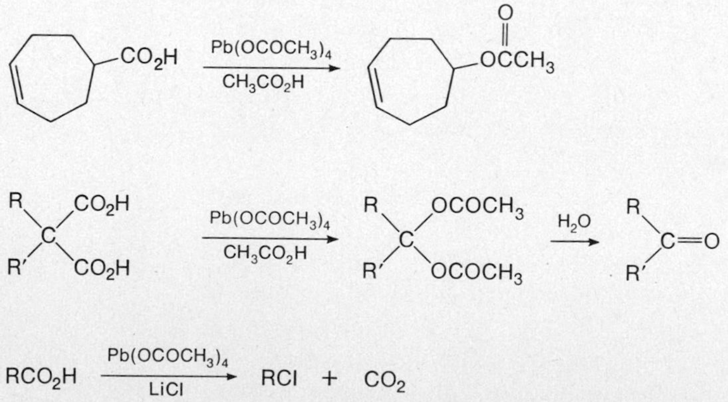

$$RCO_2H \xrightarrow[LiCl]{Pb(OCOCH_3)_4} RCl + CO_2$$

Fig. 30.31. Decarboxylation of carboxylic acids with lead tetraacetate.

30.9 Reductive Dimerization Reactions

The reductive dimerization reaction involving the carbonyl group is illustrated by the pinacol reduction of aldehydes and ketones (see Section 22.9). A reductive dimerization involving the α-carbon can be effected by treating the enolate anions of active methylene compounds with iodine, as illustrated by the dimerization of β-keto esters (Fig. 30.32).

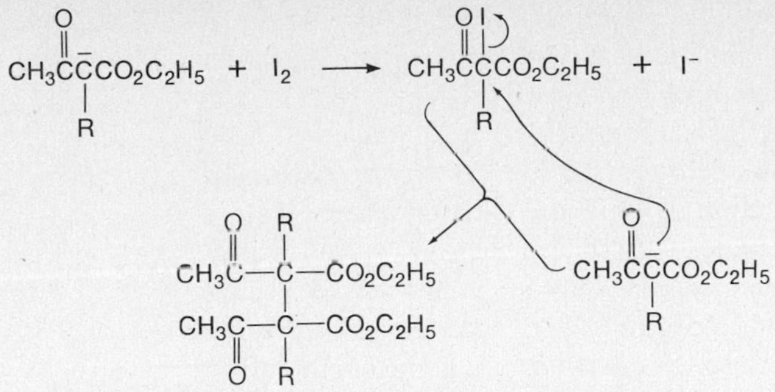

$$CH_3\overset{O}{\overset{\|}{C}}\overset{-}{\underset{R}{C}}CO_2C_2H_5 + I_2 \longrightarrow CH_3\overset{O}{\overset{\|}{C}}\underset{R}{\overset{I}{C}}CO_2C_2H_5 + I^-$$

$$CH_3\overset{O}{\overset{\|}{C}}\overset{-}{\underset{R}{C}}CO_2C_2H_5 \longrightarrow \begin{array}{l} CH_3\overset{O}{\overset{\|}{C}}-\overset{R}{\overset{|}{C}}-CO_2C_2H_5 \\ CH_3\underset{O}{\underset{\|}{C}}-\underset{R}{\underset{|}{C}}-CO_2C_2H_5 \end{array}$$

Fig. 30.32. Reductive dimerization of acetoacetic esters.

PROBLEM 30-13

What would you predict to be the structure of the compound obtained by the action of sodium ethoxide on the product of the reaction shown in Fig. 30.33?

30.10 *Cycloaddition Reactions*

Aldehydes and some ketones react with (a) ketenes to give addition products (four-membered ring compounds), (b) 1,3-dipolar reagents to give 1,3-cycloaddition products (five-membered ring compounds), and (c) dienes to give 1,4-cycloaddition products (six-membered ring compounds) (Fig. 30.33).

1,2-Cycloaddition

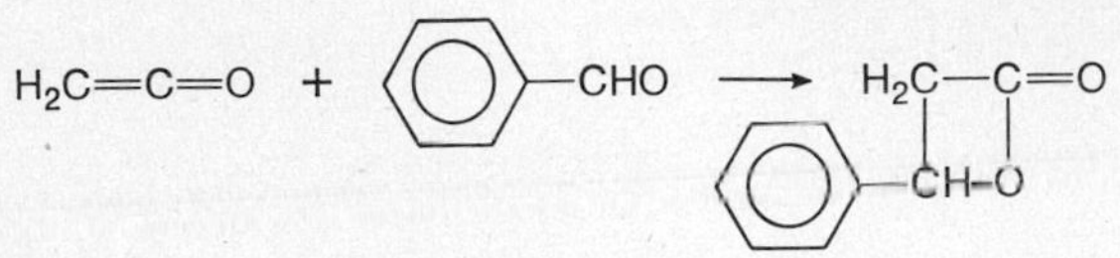

$$H_2C{=}C{=}O + C_6H_5CHO \longrightarrow \begin{array}{l} H_2C-C{=}O \\ \;|\quad\;\;| \\ C_6H_5-CH-O \end{array}$$

1,3-Cycloaddition

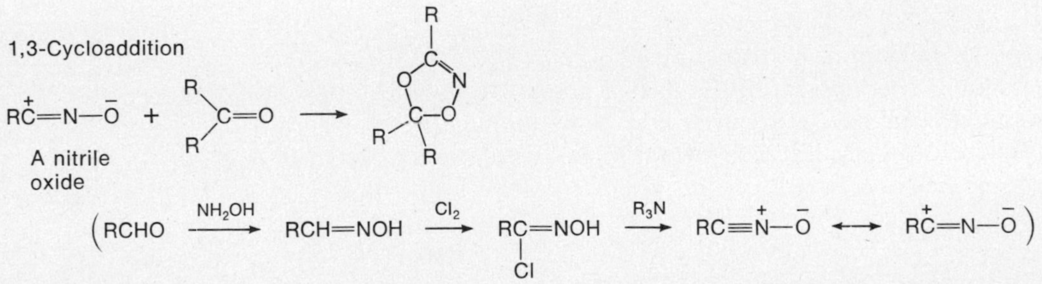

$$R\overset{+}{C}{=}N-\overset{-}{O} + R_2C{=}O \longrightarrow \text{(five-membered ring: } R{-}C{=}N{-}O{-}CR_2{-}O\text{)}$$

A nitrile oxide

$$\left(RCHO \xrightarrow{NH_2OH} RCH{=}NOH \xrightarrow{Cl_2} R\underset{Cl}{\underset{|}{C}}{=}NOH \xrightarrow{R_3N} RC{\equiv}\overset{+}{N}-\overset{-}{O} \longleftrightarrow R\overset{+}{C}{=}N-\overset{-}{O}\right)$$

1,4-Cycloaddition

$$CH_3CH{=}CH-C(CH_3){=}CH_2 + H_2C{=}O \longrightarrow \text{2-methyl-4-methyl-3,6-dihydro-2H-pyran (}CH_3\text{ ring substituents)}$$

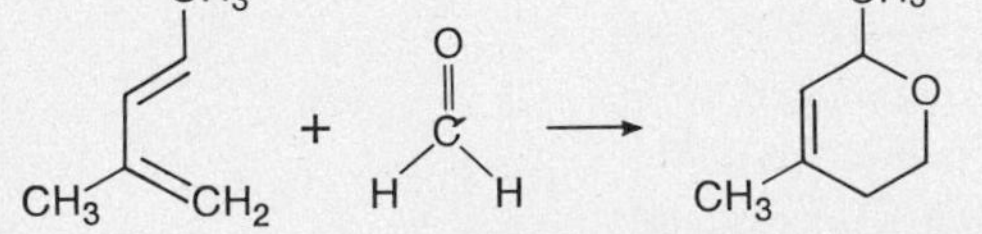

Fig. 30.33. Cycloaddition reactions of carbonyl compounds.

30.11 Rearrangement Reactions

Carbonyl compounds undergo a variety of rearrangement reactions. Many of these are discussed elsewhere in the text, including the Beckmann rearrangement of the oximes of aldehydes and ketones (see Sections 24.2c and 28.11b), the Baeyer–Villiger oxidation of aldehydes and ketones (see Fig. 22.6 and Section 28.11c), the Wolff rearrangement of diazoketones (see Fig. 28.30), the Favorskii rearrangement of α-haloketones (see Section 28.11e), and the Curtius, Hofmann, and Schmidt reactions of carboxylic acid derivatives (see Section 24.2b). Tautomeric rearrangements also appear frequently, generally in cases where enols are involved as either immediate precursors to products (*e.g.*, in conjugate addition reactions) or reactive intermediates that initiate a reaction (*e.g.*, in the bromination of ketones). Aromatic enols (phenols) represent another facet of enol stability, being much more stable than the corresponding ketones. This difference in stability provides the driving force for certain reactions, such as the acid-catalyzed dienone-phenol rearrangement. An example of this reaction is observed when 6-methylbicyclo[4.4.0]deca-1,4-dien-3-one is treated with sulfuric acid, the product being a phenol (Fig. 30.34). The rearrangement can be explained in

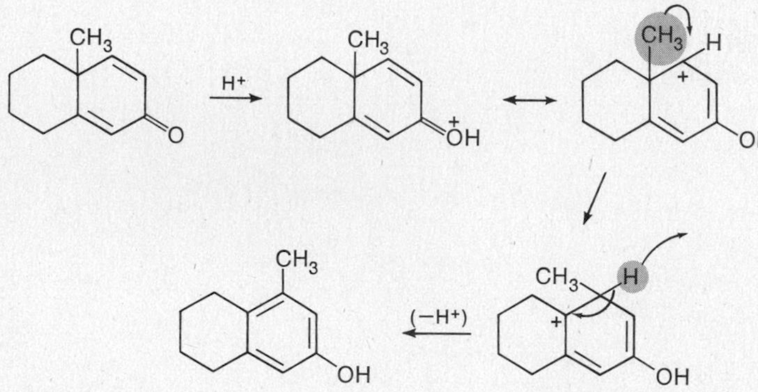

Fig. 30.34. Dienone-phenol rearrangement of 6-methylbicyclo[4.4.0]deca-1,4-dien-3-one.

terms of a cationic intermediate, in which the positive charge resides at a position β to the carbonyl group and adjacent to the methyl group. Migration of the methyl group to this electron-deficient center (i.e., a 1,2-nucleophilic rearrangement) forms an isomeric cation that can generate the more stable phenolic system by expelling a proton. Rearrangements of this sort are general for 4,4-disubstituted cyclohexa-2,5-dienones.

30.12 Photochemical Reactions

The carbonyl group absorbs strongly in the 175 nm region of the ultraviolet ($\pi \rightarrow \pi^*$ transition) and weakly in the 280 nm region ($n \rightarrow \pi^*$ transition). The more interesting of these absorptions is the one near 280 nm, because the photochemistry of the electronically excited state resulting from this absorption has been the most extensively studied. Ketones, in particular, have been well investigated in this regard, and to give a brief synopsis of the somewhat involved photochemistry of carbonyl compounds, we single out the ketones for special consideration.

A ketone in the electronically excited state resulting from an $n \rightarrow \pi^*$ transition almost always reacts by one, or more, of the five pathways illustrated in Fig. 30.35. Three of these pathways (*i.e.*, I, II, III) are intramolecular in character and are independent of external reagents, whereas the other two (*i.e.*, IV and V) require an external hydrogen donor or alkene to proceed. Which of the pathways is followed depends on a number of factors, including the structure of the ketone, the way in which the reaction is carried out (*i.e.*, in the vapor phase or in solution),

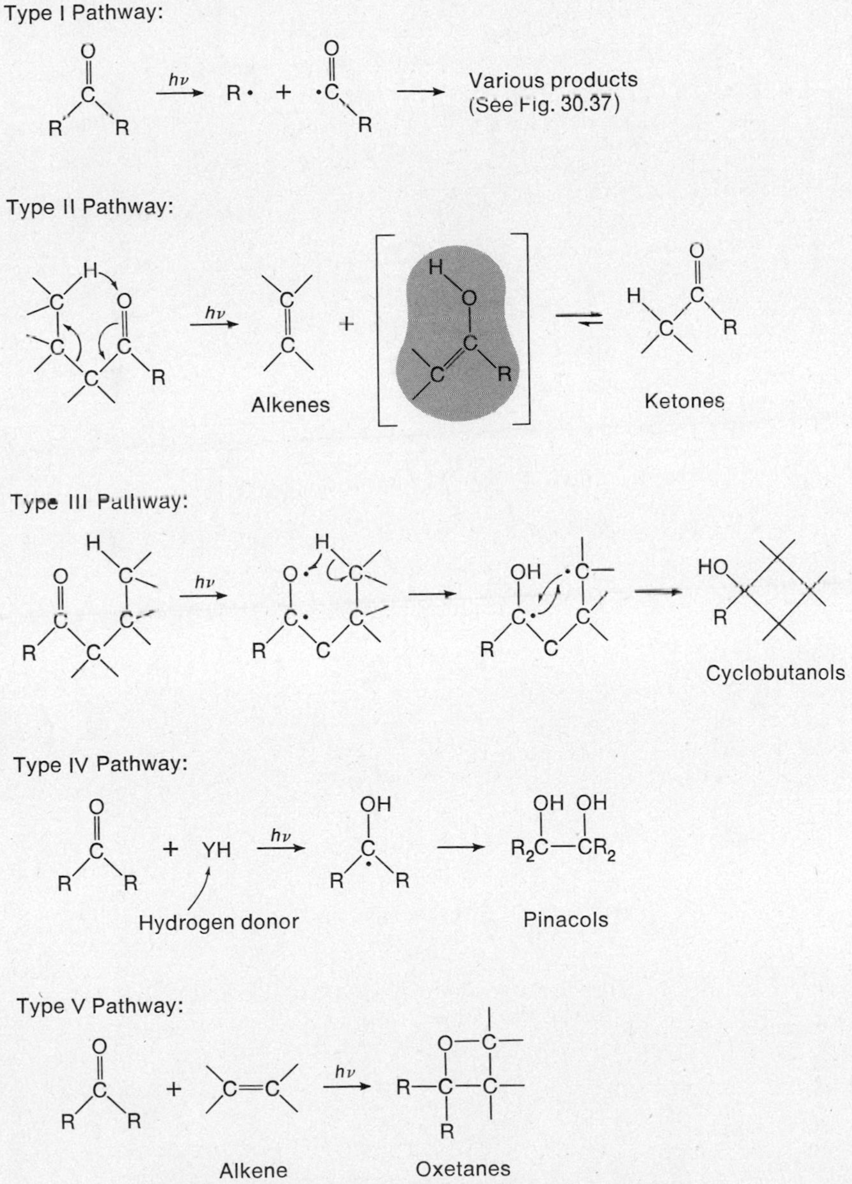

Type I Pathway:

Various products (See Fig. 30.37)

Type II Pathway:

Alkenes

Ketones

Type III Pathway:

Cyclobutanols

Type IV Pathway:

Hydrogen donor

Pinacols

Type V Pathway:

Alkene

Oxetanes

Fig. 30.35. The five major photochemical reactions of ketones.

and the presence or absence of other reagents (*e.g.*, hydrogen donors or alkenes). Pathways II–V are usually straightforward, leading to aldehydes (or ketones) and alkenes in the case of pathway II, cyclobutanols in the case of pathway III, pinacols in the case of pathway IV, and oxetanes in the case of pathway V. Pathway I, on the other hand, can lead to a variety of products along any one, or more, of the three pathways shown in Fig. 30.36.

The several specific examples shown in Fig. 30.37 illustrate the interesting variety of possibilities in the photochemical behavior of ketones.

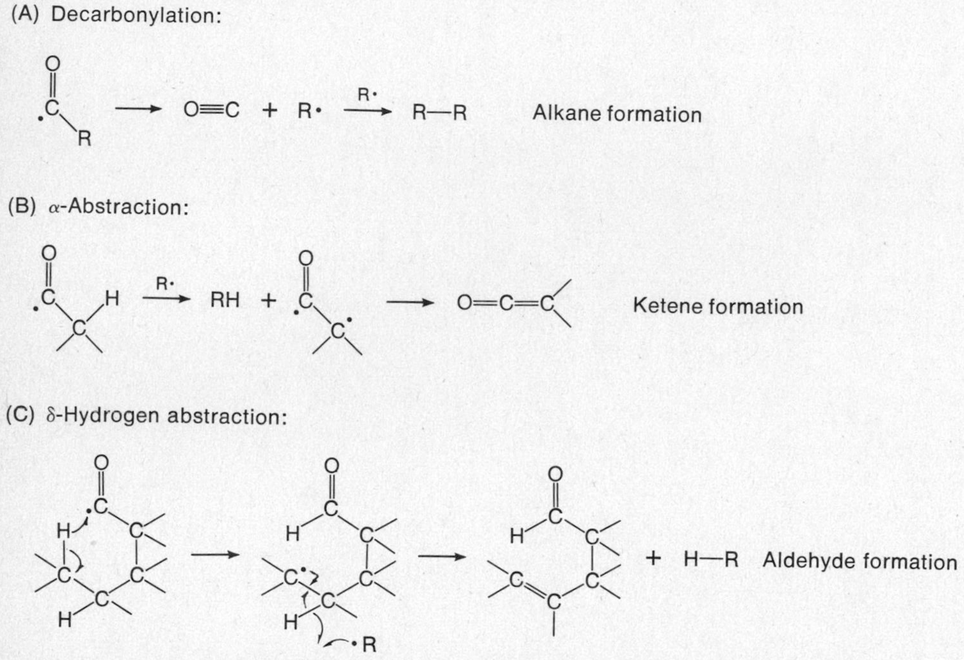

(A) Decarbonylation:

O≡C + R· —R·→ R—R Alkane formation

(B) α-Abstraction:

—R·→ RH + ... → O=C=C Ketene formation

(C) δ-Hydrogen abstraction:

→ → + H—R Aldehyde formation

Fig. 30.36. Reactions of the acyl radicals formed in type I photolysis of ketones (see Fig. 30.35).

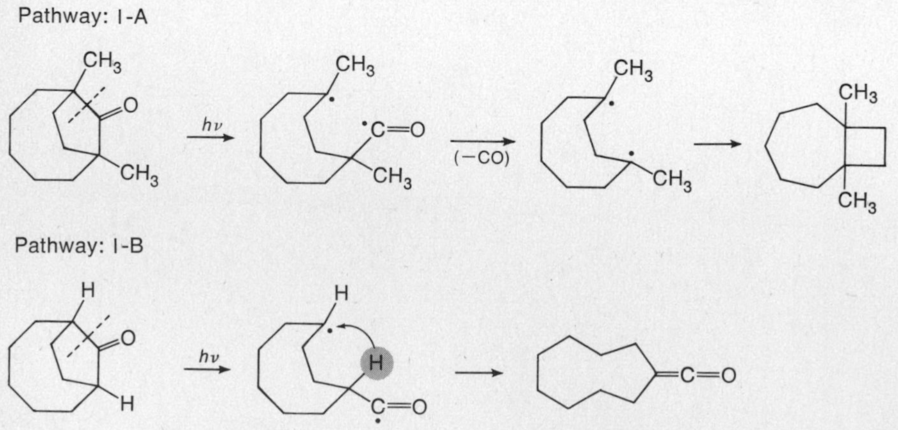

Pathway: I-A

CH_3 ... CH_3 —$h\nu$→ ·C=O ... (−CO) → ... CH_3 → CH_3 CH_3

Pathway: I-B

H ... H —$h\nu$→ H ... C=O → =C=O

Fig. 30.37. Photochemical reactions of ketones.

Pathway: I-C

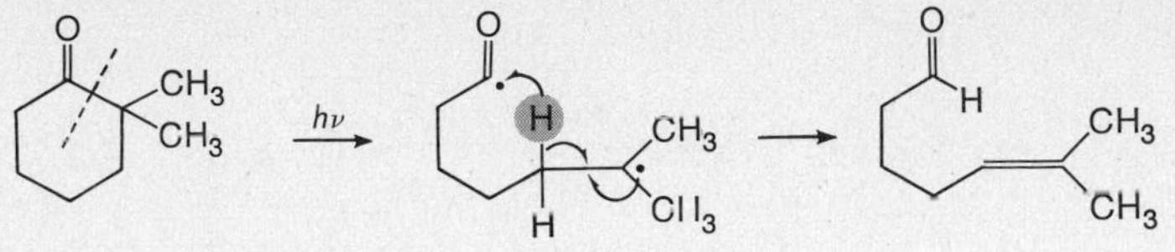
hν

Pathway: II

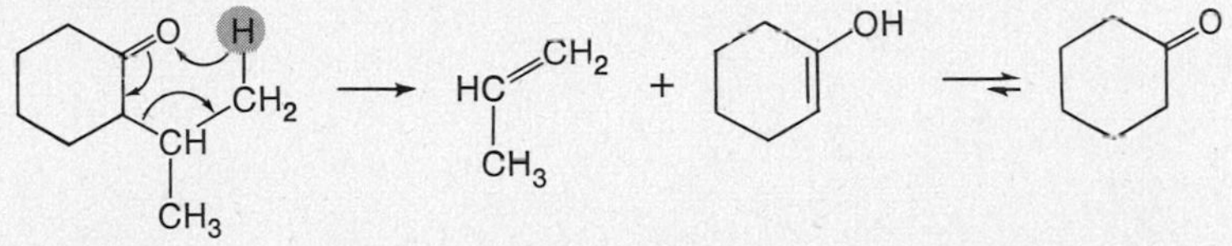

Pathway: III

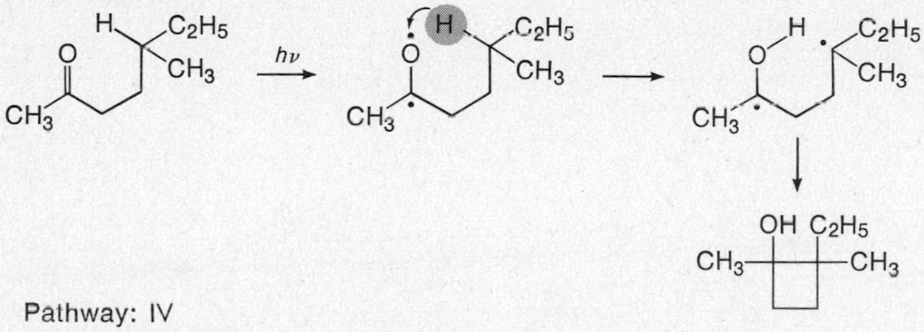
hν

Pathway: IV

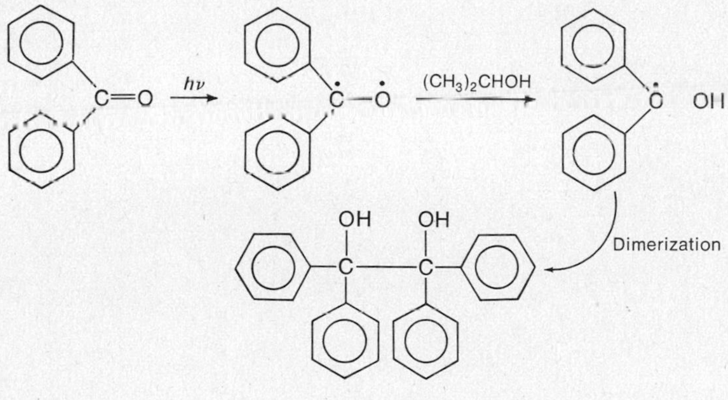
hν
(CH3)2CHOH
Dimerization

Pathway: V

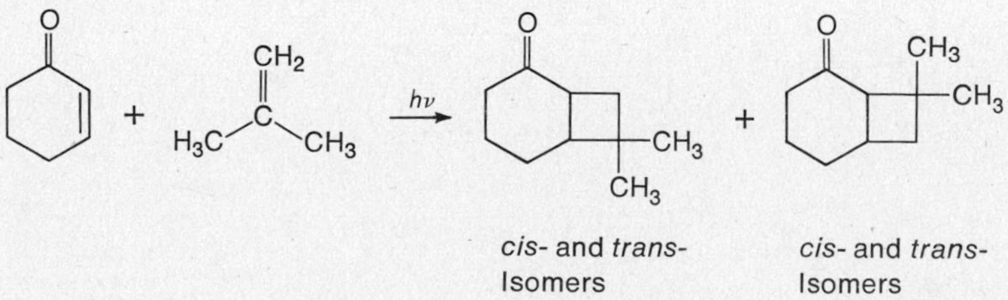
hν
cis- and *trans*-
Isomers
cis- and *trans*-
Isomers

Fig. 30.37. (Continued.)

Table 30.2 summarizes the reactions of carbonyl compounds occurring at the α-position, as well as miscellaneous reactions.

Table 30.2 Reactions of carbonyl compounds adjacent to the carbonyl group and other reactions

Type of Reaction	Starting Materials	Reagents	Products	Comments
α-Halogenation	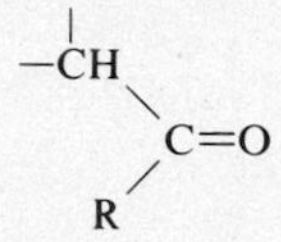 $-CH-C(=O)R$	X_2 (acid- or base-catalyzed)	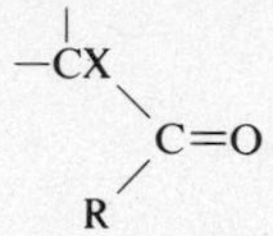 $-CX-C(=O)R$	
α-Nitrosation	$-CH_2-C(=O)R$	HONO	$-CH(N{=}O)-C(=O)R \rightleftharpoons -C(=NOH)-C(=O)R$	
α-Alkylation	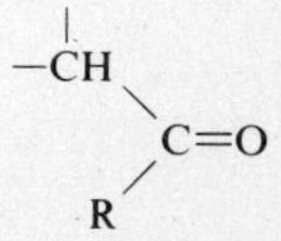 $-CH-C(=O)R$	(1) Base: (2) R′X	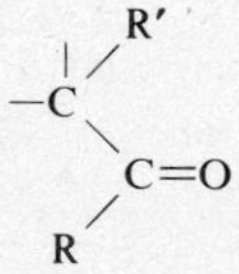 $-C(R')-C(=O)R$	S_N2 reaction—subject to usual limitations
1,2-Addition to carbonyl groups				
	$R_2C{=}O$	ZCH_2Z' (active methylene compounds)	$R_2C(OH)-CH(Z)Z'$	Aldol-type condensation
	ArCHO	CH_3CHO + base	ArCH=CHCHO	Claisen–Schmidt condensation
		$(RCH_2CO)_2O + RCH_2CO_2K$	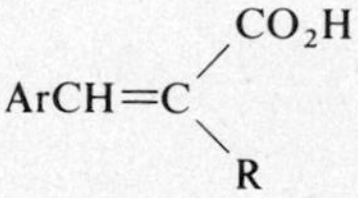 $ArCH{=}C(CO_2H)R$	Perkin condensation
		$CH_2(CO_2H)_2$ or 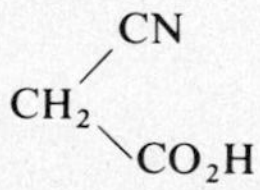 $CH_2(CN)CO_2H$	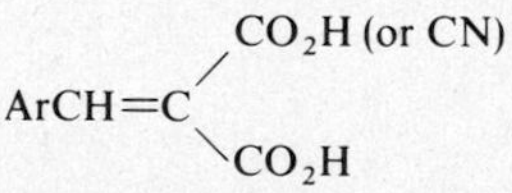 $ArCH{=}C(CO_2H\text{ (or CN)})CO_2H$	Knoevenagel and Cope modifications of Perkin condensation
	$R_2C{=}O$	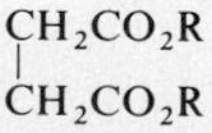 $CH_2CO_2R-CH_2CO_2R$ + $(CH_3)_3COK$	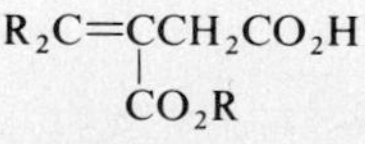 $R_2C{=}C(CO_2R)CH_2CO_2H$	Stobbe–Johnson condensation
		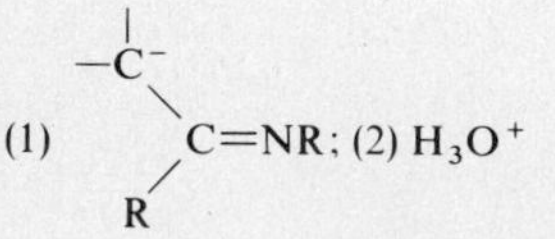 (1) $-C^{-}-C(=NR)R$; (2) H_3O^+	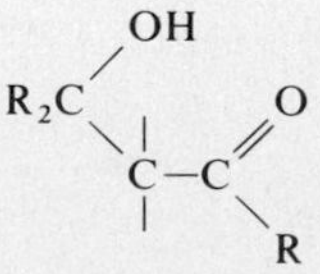 $R_2C(OH)-C-C(=O)R$	"Directed" aldol condensations
		$ClCH_2CO_2C_2H_5$, $NaNH_2$	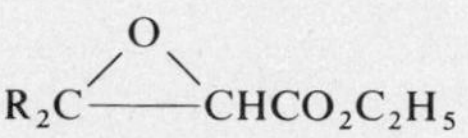 $R_2C\overset{O}{—}CHCO_2C_2H_5$ (epoxide)	Glycidic ester condensation
	ArCHO	KCN, $HOCH_2CH_2OH$, Δ	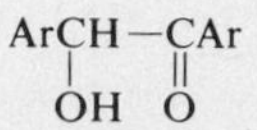 $ArCH(OH)-C(=O)Ar$	Benzoin condensation

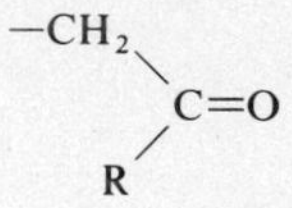

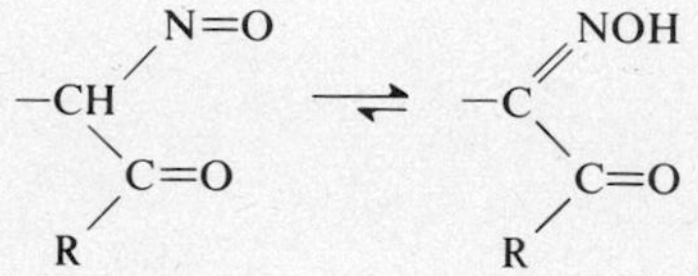

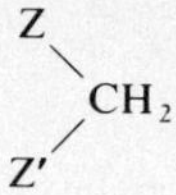

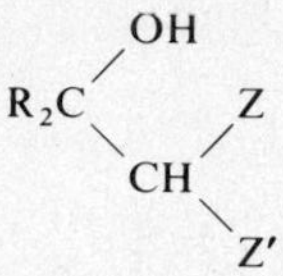

Table 30.2—*cont.*

Type of Reaction	Starting Materials	Reagents	Products	Comments
Addition-elimination of carbonyl groups	$RC(=O)Y$	ZCH_2Z', base (active methylene compounds)	$RC(=O)CH(Z)Z'$	Claisen condensation
	$(CH_2)_n$ bridging $C(=O)OR$ and $CH_2C(=O)OR$	RONa	cyclic $(CH_2)_n$–$C(=O)$–$CHC(=O)OR$	Dieckmann condensation
Other reactions	$RC(=O)-C(=O)R$	Strong acid or heat	$RC(=O)R$ + CO	Decarbonylation
	$RC(=O)CH_2CO_2H$	Heat	$RC(=O)CH_3$	Decarboxylation
	RCO_2H	$Pb(OCOCH_3)_4$ in acetic acid solution	$ROCOCH_3$	
	Reductive dimerization—see Fig. 30.33			
	Cycloadditions—see Fig. 30.34			
	Photochemical reactions—see Figs. 30.36–30.38			

30.13 *Study Guide*

I. Active methylene compounds as sources of carbon nucleophiles
   A. Formation of resonance-stabilized anions (ambident anions)
   B. Definition of active methylene compounds
   C. Acidities of active methylene compounds
   D. Reaction of ambident anions as carbon nucleophiles and oxygen nucleophiles
II. Replacement of α hydrogen by hetero atoms
   A. Halogenation
      1. Base catalyzed
      2. Acid catalyzed
      3. Hell–Volhard–Zelinzky halogenation of carboxylic acids
   B. Oxidation
      1. Direct oxidation with selenium dioxide
      2. Oxidation via nitrosation and hydrolysis of resulting oxime
III. Active methylene nucleophiles in substitution reactions with alkyl halides
   A. Alkylation of singly activated active methylene compounds
      1. Problems often encountered
         (a) Introduction of more than one alkyl group; polyalkylation
         (b) Alkylation at "wrong" α-carbon atom
      2. Use of blocking groups to direct the alkylation
      3. Use of activating groups to direct the alkylation
      4. Use of enamines in alkylation reactions

- B. Alkylation of doubly activated active methylene compounds
  1. Alkylation of monoanions—occurs at carbon between the activating groups
  2. Alkylation of dianions—occurs at position adjacent to only one of the activating groups
- C. Alkylation of α,β-unsaturated carbonyl compounds at the α-position; principle of vinylogy

IV. Active methylene nucleophiles in 1,2-addition reactions with carbonyl groups
- A. Aldol condensations leading to β-hydroxyaldehydes and β-hydroxyketones
  1. Aldol condensation of acetaldehyde and acetone
  2. Mixed aldol condensation involving formaldehyde or nitromethane
- B. Aldol condensations leading to α,β-unsaturated carbonyl compounds
  1. Claisen–Schmidt condensation
  2. Perkin condensation
  3. Knoevenagel–Doebner and Knoevenagel–Cope condensations
  4. Stobbe–Johnson condensation
  5. "Directed" aldol condensations
  6. Acid-catalyzed aldol condensations
  7. Intramolecular aldol condensations

V. Active methylene nucleophiles in 1,4-addition reactions with α,β-unsaturated carbonyl compounds
- A. Robinson annelation process
- B. Robinson tropinone synthesis

VI. Active methylene nucleophiles in addition-elimination reactions
- A. Intermolecular reaction—Claisen condensation
- B. Intramolecular reaction—Dieckmann condensation
- C. Mixed Claisen condensations with diethyl carbonate or diethyl oxalate

VII. Other reactions of carbonyl compounds
- A. Glycidic ester condensation
- B. Benzoin condensation
- C. Decarbonylation reactions
- D. Decarboxylation reactions
- E. Reductive dimerization reactions
- F. Cycloaddition reactions; 1,2-, 1,3-, and 1,4-cycloadditions
- G. Rearrangements reactions; dienone-phenol rearrangement
- H. Photochemical reactions

REFERENCES

COFFEY, S., ed., *Rodd's Chemistry of Carbon Compounds*, 2nd ed., Vol. IC, ID, Elsevier Publishing Co., New York, 1965.

GUTSCHE, C. D., *The Chemistry of Carbonyl Compounds*, in *Foundations of Modern Organic Chemistry Series*, Prentice-Hall, Inc., Englewood Cliffs, N.J., 1967.

HOUSE, H. O., *Modern Synthetic Reactions*, 2nd ed., W. A. Benjamin, Inc., Menlo Park, Calif., 1972.

PATAI, S., ed., *The Chemistry of the Carbonyl Group*, Vol. 1, John Wiley & Sons, Inc., New York, 1966.

PATAI, S., ed., *The Chemistry of Carboxylic Acids and Esters*, John Wiley & Sons, Inc., New York, 1969.

PATAI, S., and J. ZABICKY, eds., *The Chemistry of the Carbonyl Group*, Vol. 2, John Wiley & Sons, Inc., New York, 1970.

ROYALS, E. E., *Advanced Organic Chemistry*, Part II: "The Chemistry of Carbonyl Compounds," Prentice-Hall, Inc., Englewood Cliffs, N.J., 1954.

PROBLEMS

1. Draw the structure of the major product from each of the following reactions:

(a) o-$C_6H_4(CH_2CO_2C_2H_5)_2$ $\xrightarrow{C_2H_5ONa}$

(b) 5,5-dimethylcyclohexane-1,3-dione (CH_3, CH_3) $\xrightarrow{Cl_2,\ NaOH}$ $\xrightarrow{H_3O^+}$

(c) C_6H_5—$CH_2CO_2C_2H_5$ $\xrightarrow{C_2H_5ONa}$

(d) $CH_3\overset{O}{\overset{\|}{C}}CH_2CO_2C_2H_5 + (CH_3)_2CHCH_2Br \xrightarrow{C_2H_5ONa}$

(e) $CH_3\overset{O}{\overset{\|}{C}}CH_2\overset{O}{\overset{\|}{C}}CH_3 + Br(CH_2)_4Br \xrightarrow{2\ C_2H_5ONa}$

(f) $CH_3\overset{O}{\overset{\|}{C}}CH_2CH_2CH_2\overset{O}{\overset{\|}{C}}CH_3 \xrightarrow{C_2H_5ONa}$

(g) $CH_3CH_2CHO \xrightarrow{NaOH}$

(h) C_6H_5—$\overset{O}{\overset{\|}{C}}$—$CH_3$ + C_6H_5—CHO $\xrightarrow{NaOH}$

(i) O_2N—C_6H_4—CHO + $(CH_3CO)_2O$ $\xrightarrow{CH_3CO_2Na}$

(j) 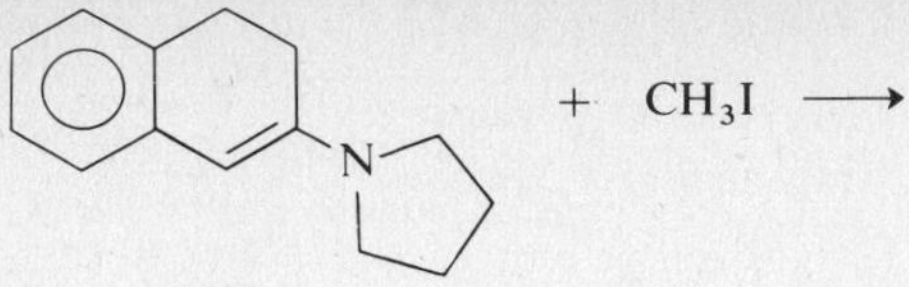 [enamine of 3,4-dihydronaphthalen-2-yl with pyrrolidine] + CH_3I ⟶

(k) 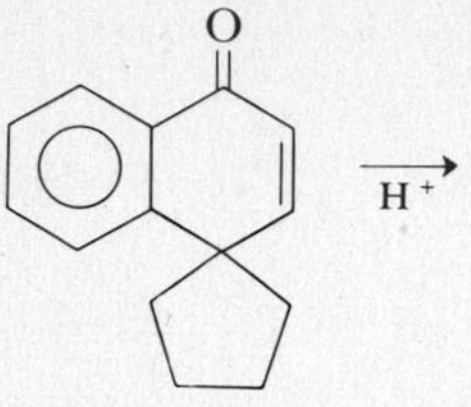 [spiro(cyclopentane-1,4'-naphthalen-1'(4'H)-one)] $\xrightarrow{H^+}$

(l) 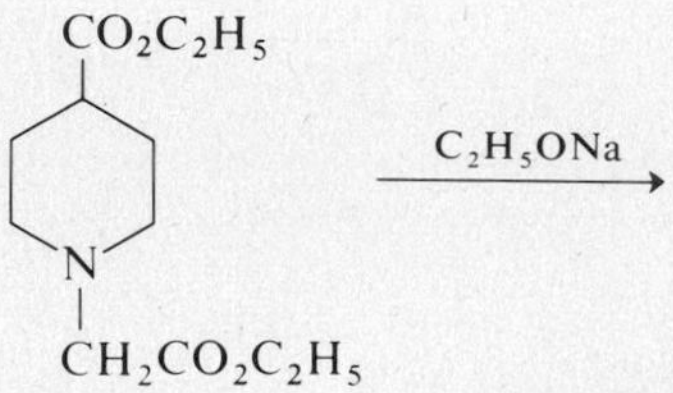 [ethyl 1-(ethoxycarbonylmethyl)piperidine-4-carboxylate: ring bearing $CO_2C_2H_5$, N bearing $CH_2CO_2C_2H_5$] $\xrightarrow{C_2H_5ONa}$

(m) 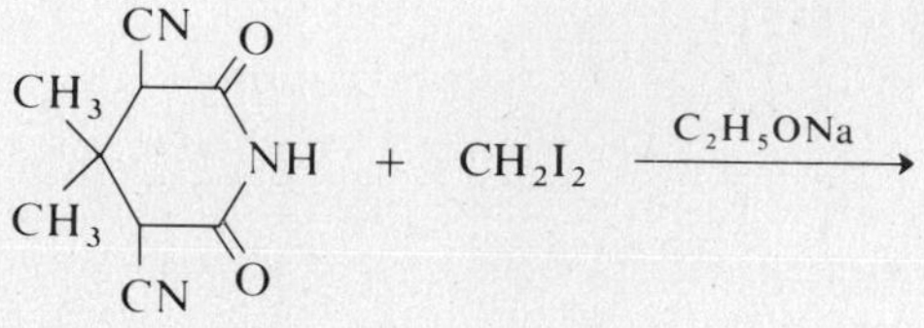 [4,4-dimethyl-3,5-dicyano-2,6-piperidinedione: CN, CH_3, CH_3, CN, O, NH, O] + CH_2I_2 $\xrightarrow{C_2H_5ONa}$

2. Propose a reasonable mechanism for each of the following reactions:

(a) 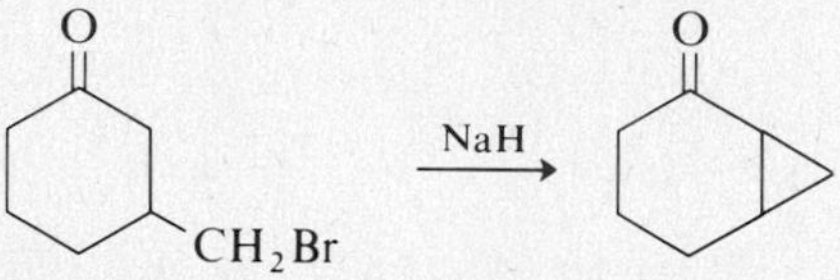 [3-(bromomethyl)cyclohexanone, CH_2Br] $\xrightarrow{NaH}$ [bicyclo[4.1.0]heptan-2-one]

(b) 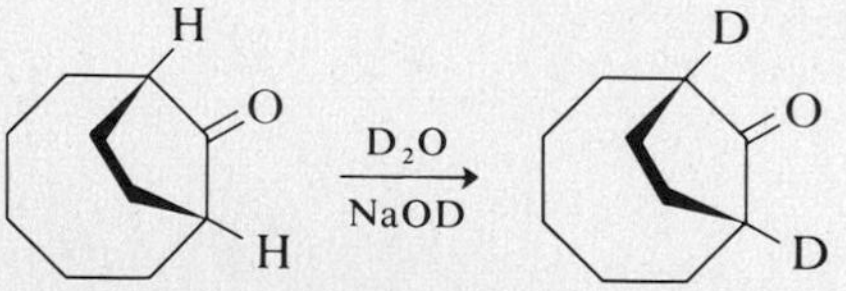 [bicyclic ketone with bridgehead H, H, and O] $\xrightarrow[NaOD]{D_2O}$ [same with bridgehead D, D]

(c) $C_6H_5CH_2C(=O)CH_3$ + Br_2 $\xrightarrow{NaOH}$ $C_6H_5CBr_2-C(=O)CH_3$

(d) [cyclohex-2-enone] $\xrightarrow{CH_3I/NaNH_2}$ [6,6-dimethylcyclohex-3-enone, CH_3 CH_3]

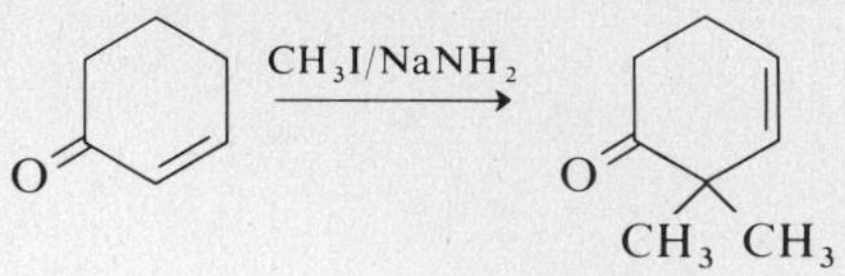

(e) 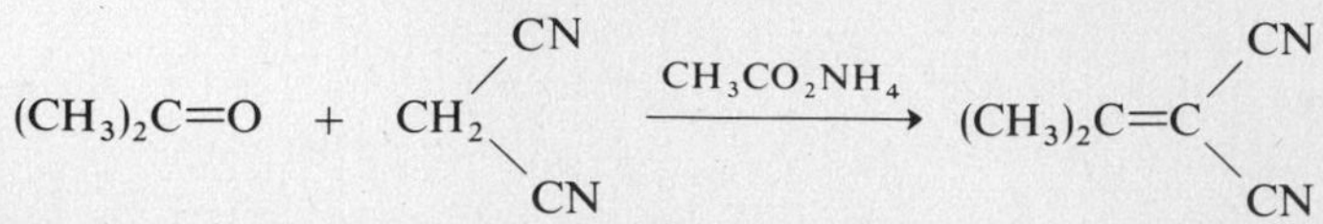 $(CH_3)_2C{=}O$ + $CH_2(CN)_2$ $\xrightarrow{CH_3CO_2NH_4}$ $(CH_3)_2C{=}C(CN)_2$

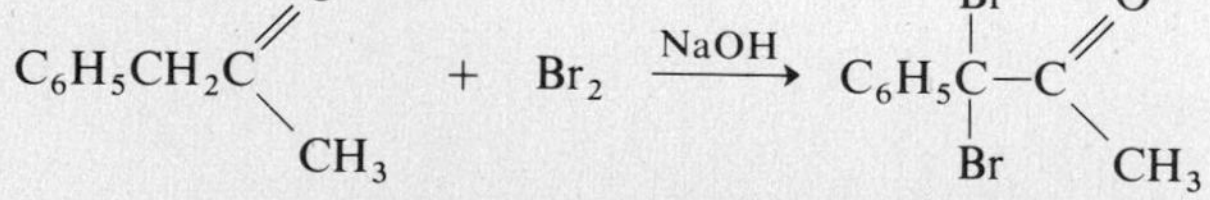

(f) 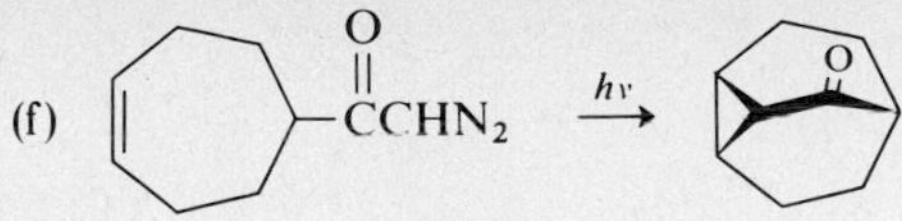

3. Explain the following observations:

(a) Compound A decarboxylates readily at room temperature, whereas compound B is stable even at temperatures well above room temperature.

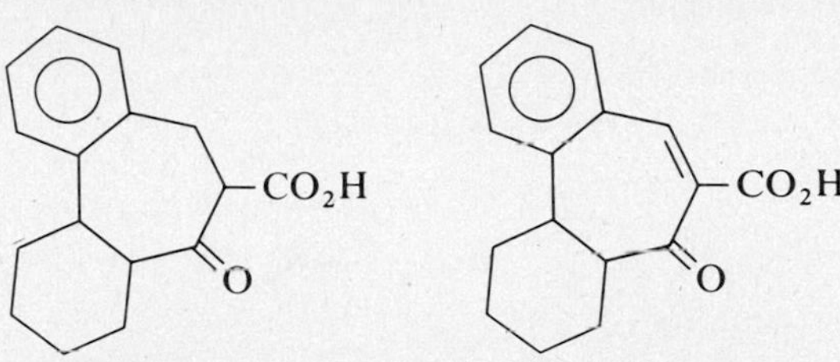

Compound A   Compound B

(b) Bicyclo[2.2.2]octan-2,6-dione undergoes almost no exchange of deuterium for hydrogen when treated with CH_3OD—NaOD over the period of time that is sufficient to effect the complete exchange of one deuterium for one hydrogen in 2-ethylcyclohexane-1,3-dione.

(c) Ketene reacts with hydrogen cyanide to form 1-cyanovinyl acetate,

$$CH_3CO_2C(=CH_2)CN$$

(d) The action of sodium hydride on the methylsulfonate ester of 6-(2-hydroxyethyl)-bicyclo[2.2.2]octan-2-one yields twistanone,

=O

(e) The action of triethylamine on 2,8-dibromocyclooctanone yields bicyclo[5.1.0]oct-1(7)-en-8-one, =O

(f) The action of sodium hydroxide on $C_6H_5COCH_2COCH_3$ yields acetophenone and sodium acetate rather than acetone and sodium benzoate.

(g) The action of aqueous sodium hydroxide on 2-isopropylidene-5-methylcyclohexanone (pulegone) yields 3-methylcyclohexanone and acetone:

CH$_3$ … O, C, CH$_3$ CH$_3$ —Aqueous NaOH→ CH$_3$ … O + $(CH_3)_2C{=}O$

(h) The application of the Perkin condensation to salicylaldehyde yields coumarin,

$$\text{o-HOC}_6\text{H}_4\text{CHO} \xrightarrow[\text{CH}_3\text{CO}_2\text{K}]{(\text{CH}_3\text{CO})_2\text{O}} \text{coumarin}$$

(i) The action of methylmagnesium bromide on 2,4,6-trimethylacetophenone yields, after hydrolysis, methane and the starting ketone.

4. Indicate the steps by means of which the compounds shown at the left in each of the following examples can be converted to the compounds shown at the right:

(a) $CH_3COCH_2CO_2C_2H_5 \longrightarrow CH_3COCH(CH_2CH_3)CH_2CH_2CH_3$

(b) $CH_3COCH_2COCH_3 \longrightarrow C_6H_5COCH(COCH_3)_2$

(c) $CH_3CH_2CO_2H \longrightarrow CH_3CH_2COCH(CH_3)CO_2C_2H_5$

(d) cyclohexanone $\longrightarrow$ 2-(CHOCCH$_3$)cyclohexanone [2-acetylcyclohexanone drawn as =CHOC(=O)CH$_3$]

(e) 1-decalone $\longrightarrow$ 8a-methyl-2-methyl-1-decalone (CH_3 at ring junction, CH_3 at C-2)

(f) $CH_3COCH_2CO_2C_2H_5 \longrightarrow CH_3\overset{O}{\overset{\|}{C}}CH(CH_3)-CH(CH_3)\overset{O}{\overset{\|}{C}}CH_3$

(g) 2-naphthaldehyde (CHO) $\longrightarrow$ 2-naphthyl-$CH_2CH(CH_3)CO_2H$

(h) $(C_6H_5)_2C{=}O \longrightarrow (C_6H_5)_2C(CH_2NH_2)(CO_2C_2H_5)$

(i) $C_6H_5COCH_3 \longrightarrow C_6H_5CH(CH_3)CH(CO_2H)CH_2CO_2H$

(j) 2-furyl-$COCH_3 \longrightarrow$ 2-furyl-$COCH{=}CHCH_2CH_3$

(k) $(CH_3)_2C{=}CHCOCH_3 \longrightarrow$ 5,5-dimethylcyclohexane-1,3-dione (ring with CH_3, CH_3 and two C=O)

5. Indicate what reactions or series of reactions can be used to convert cyclohexanone to each of the following compounds:

(a) 2-propylcyclohexanone ($CH_2CH_2CH_3$)

(b) octalone (bicyclic α,β-unsaturated ketone)

(c) cyclohexanone with $CO_2C_2H_5$ and CH_3 on the α carbon

(d) cyclohexane-1,2-diol bearing $CH_2CH_2C_6H_5$, OH, OH, H

(e) 2,6-bis(CHC_6H_5)cyclohexanone

(f) cyclohex-2-enone

(g) 2-methylcyclopentanone (CH_3)

6. For each of the following systems, draw the structures of the products, and indicate whether the equilibrium constant is greater or less than 1:

(a) $CH_3COCH_3 \; + \; CH_3ONa \rightleftharpoons$

(b) cyclopentane-1,3-dione $+ \; (C_6H_5)_3CNa \rightleftharpoons$

(c) $[CH_3COCHCOCH_3]Na \; + \; CH_3COCH(CH_3)COCH_3 \rightleftharpoons$

(d) $[CH(CO_2C_2H_5)_2]Na \; + \; CH_2(CN)_2 \rightleftharpoons$

(e) $C_4H_9Li \; + \; CH_3S(=O)CH_3 \rightleftharpoons$

7. Suggest a chemical method by means of which the following pairs of compounds might be differentiated:

(a) 2,2-dimethylcyclohexanone (CH_3, CH_3) and 2,6-dimethylcyclohexanone (CH_3, CH_3)

(b) $CH_3COCH(CH_2CH_3)CO_2C_2H_5$ and $CH_3COC(CH_3)_2CO_2C_2H_5$

(c) $CH_3COCH_2CH_2CN$ and $CH_3CH_2COCH_2CN$

(d) $C_6H_5COCH_2CH_3$ and $C_6H_5CH_2COCH_3$

(e) 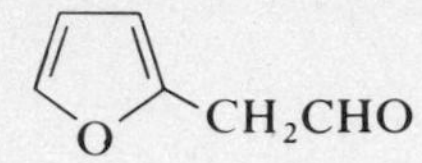 CH_2CHO (furan-2-yl) and CH_3, CHO (furan)

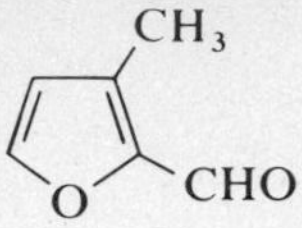

For each pair of compounds, indicate how their nmr spectra could provide a means for differentiation.

8. For the following syntheses, the available starting materials include (a) any alkanes, alkenes, alkynes, alkyl halides, alkanols, alkanals, alkanones, alkanoic acids, or alkanoic acid derivatives containing four or fewer carbon atoms, (b) benzene and any monosubstituted benzenes, (c) special organic reagents, and (d) any inorganic compounds. Starting with materials chosen from this group, outline syntheses for each of the following compounds:

(a) 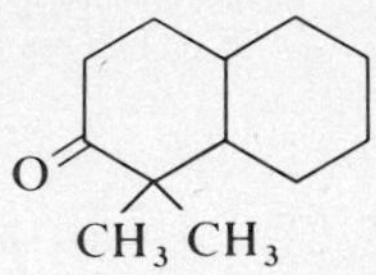 O, CH_3 CH_3 (bicyclic ketone)

(b) 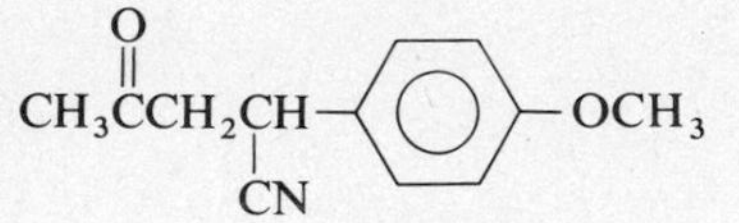 $CH_3\overset{O}{\overset{\|}{C}}CH_2\underset{CN}{\underset{|}{C}}H-C_6H_4-OCH_3$

(c) $CH_2CH_2CH_3$, $CO_2C_2H_5$, O (cyclopentanone)

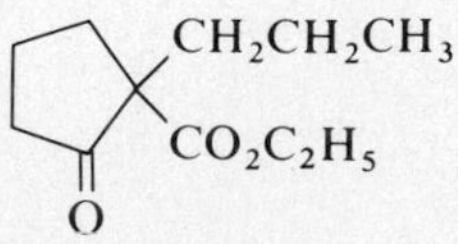

(d) 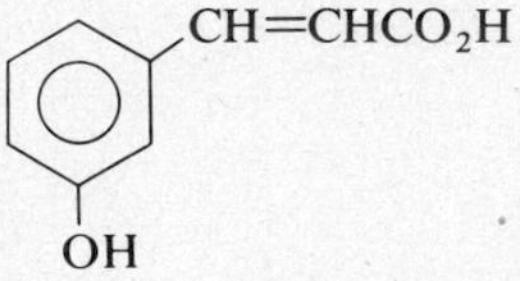 $CH{=}CHCO_2H$, OH (benzene ring)

(e) 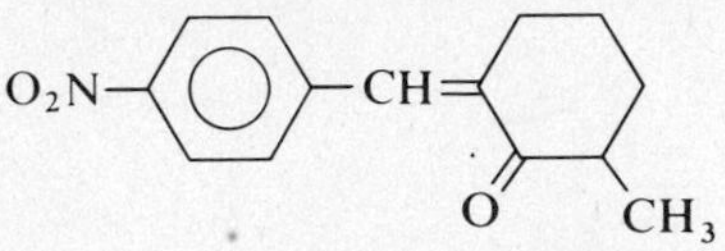 $O_2N-C_6H_4-CH{=}$ (cyclohexanone), O, CH_3

9. The following carbonyl compounds are inexpensive starting materials, costing less than 50¢/lb:

Aldehydes: Formaldehyde, acetaldehyde, *n*-butyraldehyde, isobutyraldehyde, acrolein, benzaldehyde, chloral, glyoxal

Ketones: Acetone, ethyl methyl ketones, cyclohexanone, 4-methyl-3-buten-2-one (mesityl oxide), acetophenone

Acids: Formic acid, acetic acid, propionic acid, *n*-butyric acid, isobutyric acid, caproic acid, caprylic acid, capric acid, lauric acid, myristic acid, palmitic acid, stearic acid, acrylic acid, crotonic acid, benzoic acid, *p*-nitrobenzoic acid, oxalic acid, succinic acid, adipic acid, fumaric acid, phthalic acid, chloroacetic acid, trichloroacetic acid, lactic acid.

Using only these compounds as starting materials, along with any inorganic reagents that might be necessary, devise a synthesis for each of the following compounds:

(a) 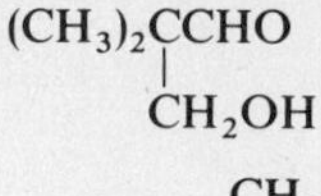 $(CH_3)_2\underset{CH_2OH}{\underset{|}{C}}CHO$

(b) $C_6H_5-CH{=}\underset{CH_2CH_3}{\underset{|}{C}}CHO$

(c) O, $\overset{}{\underset{O}{\underset{\|}{C}}}CO_2C_2H_5$ (cyclohexanone)

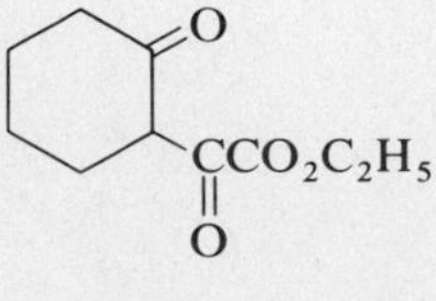

(d) 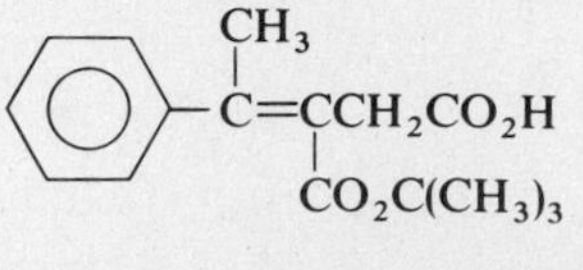 $C_6H_5-\overset{CH_3}{\overset{|}{C}}{=}\underset{CO_2C(CH_3)_3}{\underset{|}{C}}CH_2CO_2H$

(e) 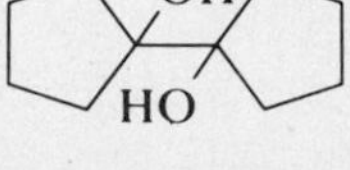 OH, HO (1,1'-bicyclopentyl-1,1'-diol)

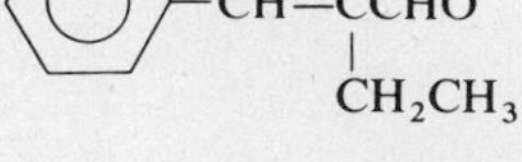

10. An optically active $C_{12}H_{20}$ hydrocarbon undergoes dehydrogenation to 1-methylnaphthalene and reacts with ozone to yield a $C_{12}H_{20}O_2$ diketone. Treatment of the diketone with sodium hydroxide produces a $C_{12}H_{18}O$ compound, which undergoes a haloform reaction to yield iodoform and a $C_{11}H_{16}O_2$ carboxylic acid. In the presence of hydrogen and a palladium on carbon catalyst, the carboxylic acid absorbs one equivalent of hydrogen to give a compound that, when subjected to a Barbier–Wieland degradation, yields a $C_{10}H_{16}O$ compound. When this material is treated with D_2O-NaOD, three equivalents of deuterium are incorporated. Write structures for these compounds that are compatible with all the preceding data.

Chapter **31**

Benzenoid aromatic compounds

There's hardly a thing that man can name
Of use or beauty in life's small game
But you can extract from alembic or jar
From the physical basis of black coal tar
Oil and ointment, wax and wine
And the lovely colors called aniline
You can make anything from a salve to a star
If you only know how, from black coal tar.

Punch Magazine, 1884

When bituminous coal is heated to 1000–1300° in the absence of air, carbonization takes place along with the concomitant formation of a gas (coal gas) and a viscous, black material (coal tar), which accounts for about 3% of the original weight of the coal. Careful fractional distillation of the coal tar reveals that it is a mixture of a rather large number of substances, the great majority of which fall in the class of compounds called *aromatic*. Some of these are monocyclic benzenoid aromatics, some are polycyclic benzenoid aromatics, others contain nitrogen and are heterocyclic aromatics, and still others contain oxygen and are phenols. Representative examples of the well over 100 fully characterized components of coal tar are shown in Fig. 31.1.

Coal tar was originally considered to be a disagreeable by-product in the production of coal gas, used for years as a source of both heat and illumination, and still used today as a source of heat. With the emergence of a chemical industry in the middle of the nineteenth century, and with its emphasis on the production of synthetic dyestuffs, which, for the most part, were based on aromatic compounds, the demand for the components of coal tar shot upward. With this emerging industry, so dependent on the constituents of coal tar, the study of the chemistry of aromatic compounds began in earnest, and has continued unabated to the present day. It constitutes one of the major areas of interest in organic chemistry. This century-long investigation has not only led to a vastly increased knowledge of the myriad ways in which syntheses built around aromatic materials can be carried out, but has also provided great insight into the nature of chemical

I. Benzenoid Aromatics:

Monocyclic:

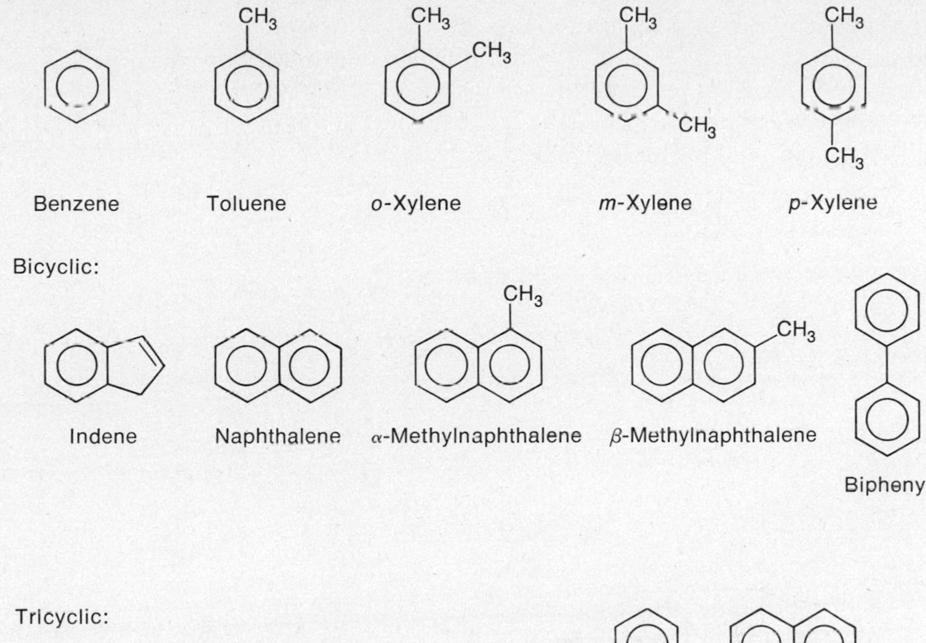

Tricyclic:

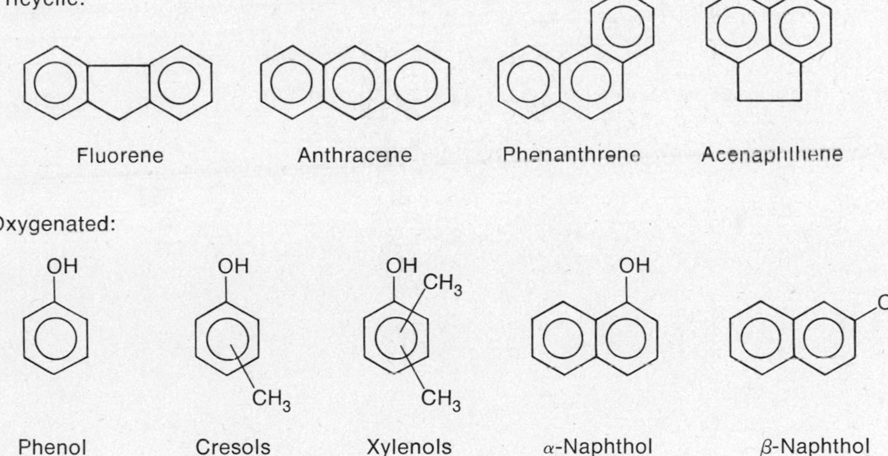

II. Heterocyclic Aromatics:

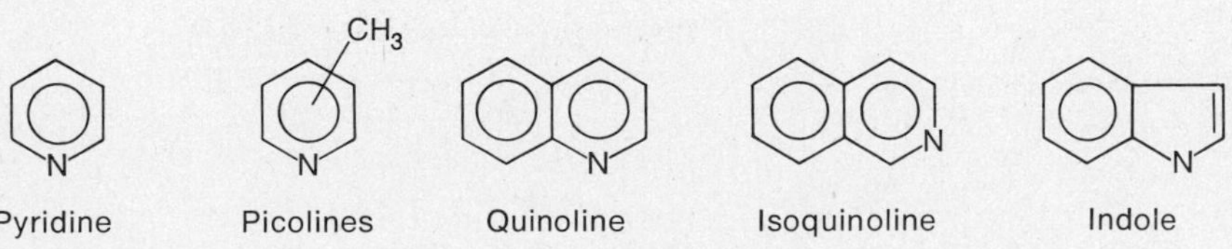

Fig. 31.1. Representative examples of aromatic constituents of coal tar.

bonding (see Chapter 4). Our attention in this chapter is directed to those aromatic compounds that contain one or more benzene rings and that, therefore, are referred to as *benzenoid aromatics*. We shall first consider methods for preparing benzenoid aromatic compounds and then discuss their reactions in terms of electrophilic substitutions, nucleophilic substitutions, and oxidation and reduction processes. Readers interested in pursuing the topic of aromaticity in greater detail are referred to Chapter 37 in Part IV of the text, which deals with nonbenzenoid aromatic compounds and includes both carbocyclic and heterocyclic examples.

METHODS OF PREPARATION OF BENZENOID AROMATIC COMPOUNDS

31.1 Preparation of Aromatic Compounds from Noncyclic or Alicyclic Precursors

31.1a. CYCLOTRIMERIZATION OF ALKYNES. See Fig. 27.9.

PROBLEM 31-1

Starting with acetylene, outline a preparation of 1,2,4-triisopropylbenzene.

31.1b. DEHYDROGENATION OF ALICYCLIC COMPOUNDS

General Features of Dehydrogenation Processes. Cyclohexanes, cyclohexenes, and cyclohexadienes can be dehydrogenated to benzenes by either catalytic or chemical means:

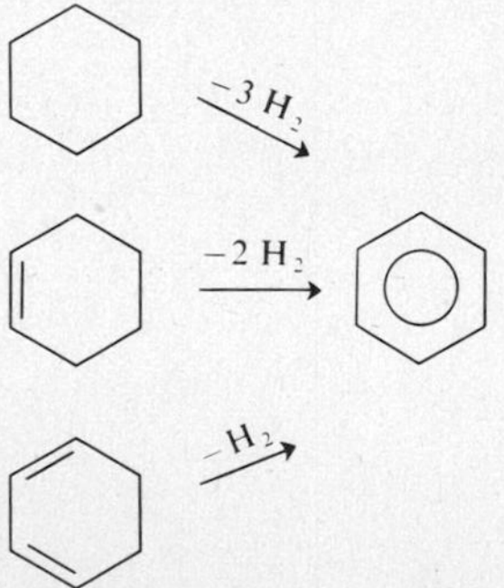

Catalytic dehydrogenations are carried out at elevated temperatures, generally between 175 and 350°. The required temperature, which is a function of the ease with which hydrogen is lost from the molecule, increases as the degree of unsaturation of the alicyclic compound decreases. Thus, cyclohexane requires a higher temperature than cyclohexene which, in turn, requires a higher temperature than cyclohexadiene. The presence of a quaternary carbon in the alicyclic system further increases the difficulty of dehydrogenation, 1,1-dimethylcyclohexane requiring an even higher temperature than cyclohexane to undergo loss of methane and dehydrogenation to toluene:

$$\text{1,1-dimethylcyclohexane} \longrightarrow \text{toluene} + CH_4$$

The metals usually used for catalytic dehydrogenation are platinum and palladium, the same metals that are also effective as hydrogenation catalysts. Their dual utility arises from the fact that a temperature-dependent equilibrium exists between a cycloalkane, its aromatic analog, and hydrogen:

$$\text{cyclohexane} \underset{}{\overset{\text{Pt or Pd}}{\rightleftharpoons}} \text{benzene} + 3\,H_2$$

At room temperature and in the presence of a hydrogen atmosphere, the equilibrium is displaced to the left to favor the alicyclic compound; at elevated temperatures, on the other hand, and under conditions whereby the hydrogen is removed from the system (*e.g.*, by a stream of inert gas passing through the mixture to sweep out the hydrogen), the equilibrium is displaced to the right to favor the aromatic compound.

Dehydrogenations by chemical methods generally employ sulfur, selenium, or benzoquinones as the hydrogen acceptors. Sulfur and selenium require elevated temperatures usually in the range of 250–350°, and are capable of effecting the dehydrogenation of almost all six-membered carbocyclic compounds, including those containing quaternary carbons:

$$\text{(1,1,3-trimethyl-5-acetylcyclohexane)} + \text{S (or Se)} \xrightarrow{\Delta} \text{(3,5-dimethylethylbenzene)} + H_2S\ (\text{or } H_2Se) + CH_4$$

The catalytic methods using platinum or palladium and the chemical methods using sulfur or selenium all require high temperatures, and the majority of functional groups do not survive this treatment unscathed. Alcohols are almost invariably dehydrated, ketones and aldehydes are reduced to the corresponding alkyl groups, and carboxyl groups are often lost as carbon dioxide. Much milder reagents, but restricted in their use to easily dehydrogenated compounds, are certain benzoquinones. 1,4-Benzoquinones carrying electron-withdrawing substituents, such as chloro or cyano groups, are more stable in the hydroquinone form than in the quinone form and, consequently, have a considerable propensity to add hydrogen and serve as dehydrogenating agents:

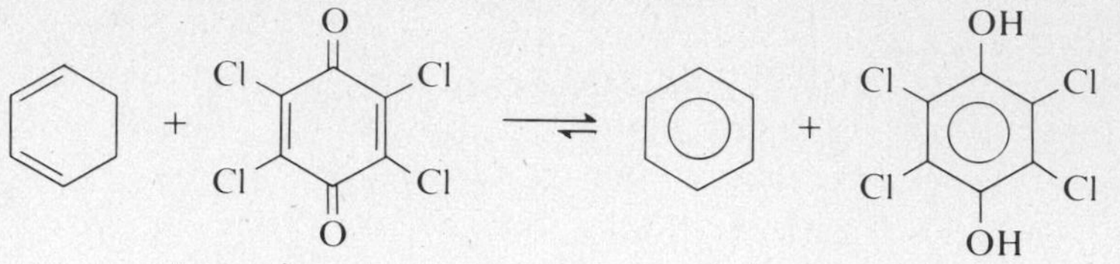

2,3,5,6-Tetrachloro-1,4-benzoquinone

2,3,5,6-Tetrachloro-1,4-hydroquinone

Diels–Alder-Dehydrogenation Sequence for Benzenes. Since six-membered carbocyclic rings can be dehydrogenated to aromatic rings, synthesis sequences that produce alicyclic rings of this size also have the potentiality of affording benzenoid aromatic compounds. For example, the Diels–Alder reaction can be effectively used in the preparation of a variety of benzenoid aromatic compounds (Fig. 31.2).

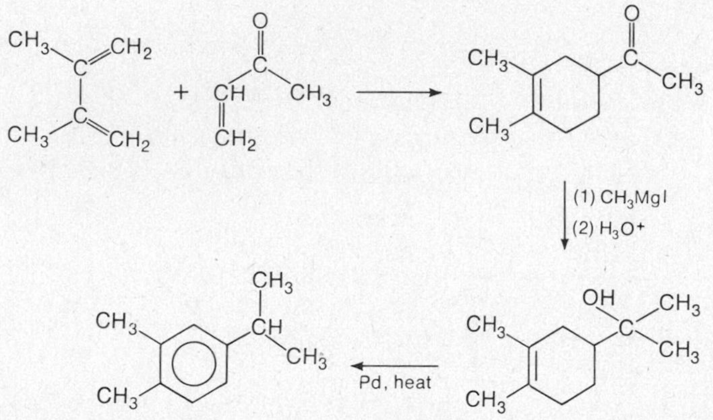

(1) CH_3MgI
(2) H_3O^+

Pd, heat

Fig. 31.2. Diels–Alder dehydrogenation sequence for the synthesis of benzene compounds.

PROBLEM 31-2

The Diels–Alder product from 2,3-dimethyl-1,3-butadiene and 1-buten-3-one (see Fig. 31.2) can be hydrogenated by the *syn* addition of hydrogen. How many products would be expected from this reaction? If the product mixture is equilibrated by treatment with base, what is the stereochemical configuration of the major product? Draw a pseudo-three-dimensional representation of its conformation.

Succinoylation-Dehydrogenation Sequence for Naphthalenes. The most versatile method for the synthesis of naphthalenes involves a Friedel–Crafts succinoylation sequence, in which (a) a benzene compound is treated with succinic anhydride in the presence of a Lewis acid catalyst (usually $AlCl_3$) to yield a 4-phenyl-4-oxobutanoic acid, (b) the 4-phenyl-4-oxobutanoic acid is reduced (*e.g.*, via a

Clemmensen reduction) to a 4-phenylbutanoic acid, (c) the 4-phenylbutanoic acid is cyclized to a 1,2,3,4-tetrahydronaphthalen-1-one (an α-tetralone) by an intramolecular Friedel–Crafts reaction, and (d) the α-tetralone is either directly dehydrogenated to a naphthalene or first treated with a Grignard reagent (which introduces a substituent at the site of the carbonyl function) and then dehydrogenated (Fig. 31.3). The method allows considerable flexibility; substituted benzenes can be used as starting materials unless they contain electron-withdrawing groups that prevent the Friedel–Crafts reaction from taking place, and various R groups can be introduced via the Grignard reaction with the α-tetralone.

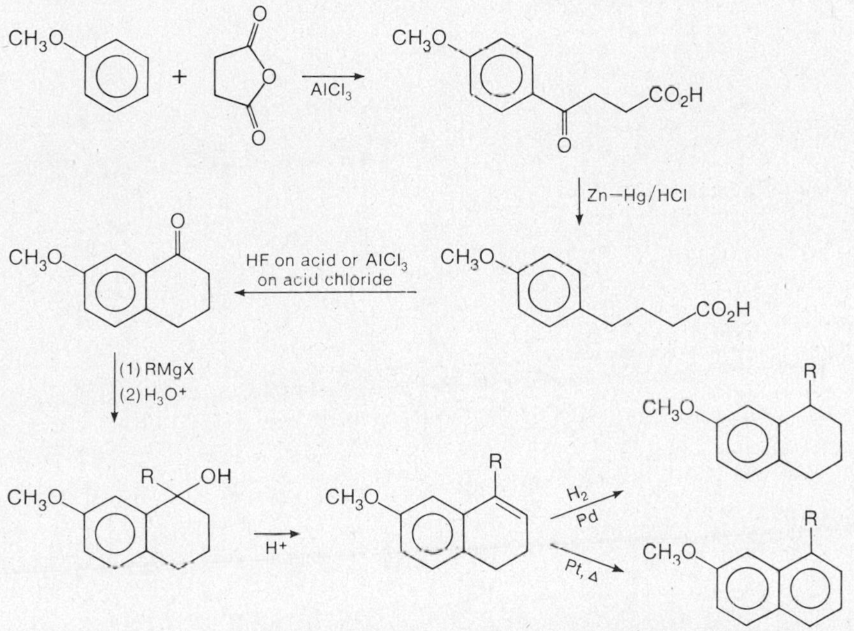

Fig. 31.3. Preparation of naphthalenes via succinoylation sequence applied to benzenes.

Aroylation–Aromatization Sequence for Anthracenes. *o*-Benzoyltoluene (obtained via a Friedel–Crafts reaction between benzene and *o*-methylbenzoyl chloride) and *o*-benzoylbenzoic acid (obtained via a Friedel–Crafts reaction between benzene and phthalic anhydride) both yield anthracene when heated in the presence of zinc dust (Fig. 31.4). The first of these reactions, which can also be induced by heat alone (*Elbs reaction*), can be viewed simply as a dehydration; the second one, which can also be induced by heating in the presence of phosphorus and hydriodic acid, can be viewed as a reduction-dehydration sequence. If substituted benzenes are employed as starting materials, substituted anthracenes can be prepared by these procedures, although the substituents that can be incorporated are limited to ones that can withstand the high temperatures required.

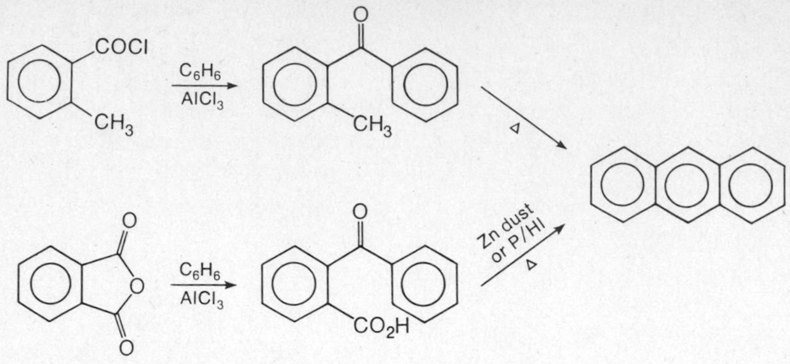

Fig. 31.4. Synthesis of anthracene via aroylation-aromatization sequence.

PROBLEM 31-3

Outline the mechanism of the reaction of benzene, phthalic anhydride, and aluminum chloride to give 9,10-anthraquinone.

Chain Introduction-Cyclization-Dehydrogenation Sequence for Phenanthrene. A number of synthesis sequences for phenanthrenes have been devised in the course of structure elucidations of various naturally occurring compounds that yield substituted phenanthrenes on dehydrogenation. One of these involves the introduction of a butanoic acid side chain into the 1-position of naphthalene by the sequence of reactions shown in Fig. 31.5. Cyclization followed by dehydrogenation, either before or after treatment with a Grignard reagent to introduce a substituent into the 1-position, then yields the phenanthrene.

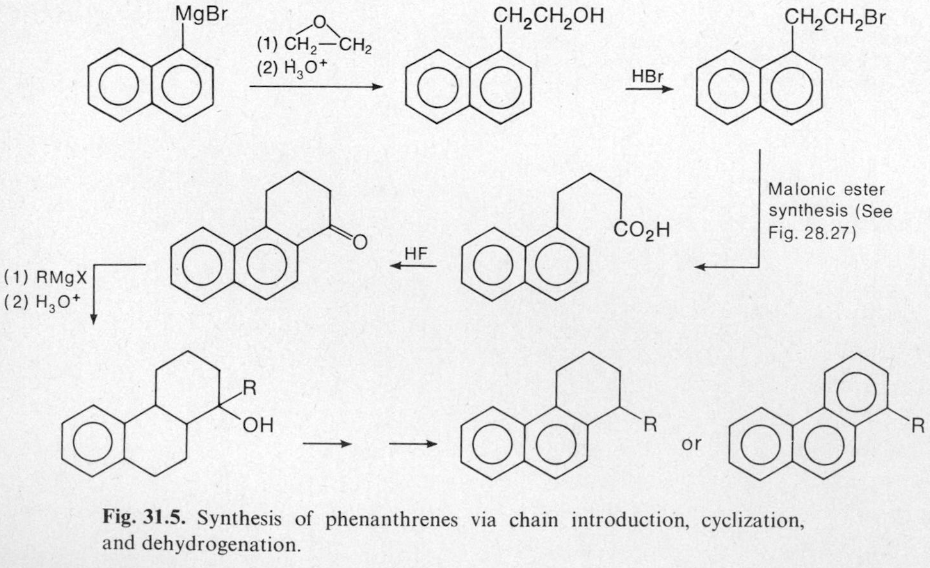

Fig. 31.5. Synthesis of phenanthrenes via chain introduction, cyclization, and dehydrogenation.

PROBLEM 31-4

If $CH_3CH\overset{O}{—}CH_2$ (epoxide) were used in the reaction sequence shown in Fig. 31.5, what would be the structure of the final product? What complications in the sequence might arise in this case?

31.2 *Preparation of Aromatic Compounds from Aromatic Precursors*

The large-scale production of coal tar makes available to the chemist an inexpensive and plentiful supply of a variety of aromatic compounds that can be used as starting materials for the preparation of more complex aromatic compounds. The two principal methods for introducing substituents onto the aromatic ring make use of electrophilic substitution reactions and nucleophilic substitution reactions (with aryldiazonium compounds). Examples are found in earlier chapters, and the mechanisms of the processes are discussed later in this chapter. These methods are summarized in Table 31.1.

Table 31.1 METHODS FOR THE SYNTHESIS OF AROMATIC COMPOUNDS

Classification	Starting Material	Reagents and Conditions	Products	Comments
Benzene and alkylbenzenes	RC≡CR	Cobalt catalyst	Alkylbenzenes	Allows preparation of highly substituted benzenes (see Section 27.5)
	Alicyclic compounds (e.g., cyclohexane)	Pt, Pd, S, or Se	Benzenes	Requires high temperatures—many functional groups altered or lost during dehydrogenation
		Quinones (*e.g.*, chloranil)	Benzenes	Mild method—works only on very easily dehydrogenated compounds
	ArH	RX + $AlCl_3$ (Friedel–Crafts alkylation)	Ar—R	Suffers from rearrangement of R and formation of polyalkylation products
Chloromethylbenzenes	ArH	HCHO, HCl, $ZnCl_2$ (chloromethylation)	$ArCH_2Cl$	
Aromatic ketones	ArH	RCOCl + $AlCl_3$	ArC(=O)R	Friedel–Crafts acylation; clean reaction; no rearrangement of R and no polyacylation products; succinolylation reaction important See Table 28.1 for general methods of ketone synthesis
Aromatic aldehydes	ArH	CO, HCl, $AlCl_3$	ArCHO	(Gatterman–Koch reaction)
		CO, HCN, $ZnCl_2$	ArCHO	(Gatterman reaction)
		$C_6H_5N(CH_3)CHO$ + $POCl_3$	ArCHO	(Vilsmeier reaction); requires strongly activated aromatic nucleus
	$ArCH_3$	$Ce(NH_4)_2(NO_3)_6$	ArCHO	Selective if more than one methyl group in Ar compound See Table 28.1 for general methods of aldehyde synthesis

Table 31.1—*cont.*

Classification	Starting Material	Reagents and Conditions	Products	Comments
Aromatic acids	ArX	(1) Mg; (2) CO_2; (3) H_3O^+	$ArCO_2H$	
	$ArNH_2$	(1) HONO; (2) CuCN; (3) H_2O (H^+ or OH^-)	$ArCO_2H$	
	ArR	$KMnO_4$/NaOH, $Na_2Cr_2O_7$/H_2SO_4, or HNO_3	$ArCO_2H$	Most side chains oxidized under these conditions except t-alkyl groups
Aromatic halides	ArH	Halogen	ArX	Cl_2 and Br_2 work best
	ArN_2^+	KI	ArI	
		Cu_2X_2	ArX	Particularly for ArBr and ArCl
		(1) HBF_4: (2) heat	ArF	
Phenols	$ArSO_3H$	NaOH—KOH, Δ	ArOH	High-temperature reaction
	ArN_2^+	H_2O, Δ	ArOH	
	ArCOR	(1) $C_6H_5CO_3H$; (2) H_2O	ArOH	
Aryl ethers	ArOH	RX + NaOH	ArOR	
		R_2SO_4 + NaOH	ArOR	
	ArOH	R_2CN_2	$ArOCHR_2$	
	ArN_2^+	ROH	ArOR	
Aromatic nitro compounds	ArH	HNO_3/H_2SO_4	$ArNO_2$	
Aromatic amines	$ArNO_2$	Fe/HCl, H_2/Pt, Sn/HCl, etc.	$ArNH_2$	
	$ArCON_3$	(1) heat or $h\nu$; (2) H_2O	$ArNH_2$	Schmidt and Curtius rearrangements
	$ArCONH_2$	Br_2/NaOH	$ArNH_2$	Hofmann degradation
	ArCOR	(1) NH_2OH; (2) H^+, Lewis acid; (3) H_2O	$ArNH_2$	Beckmann rearrangement
Aromatic sulfonic acids	ArH	H_2SO_4	$ArSO_3H$	
		(1) $ClSO_3H$; (2) H_2O	$ArSO_3H$	

ELECTROPHILIC AROMATIC SUBSTITUTION

31.3 General Features of Electrophilic Aromatic Substitution Reactions

The characteristic reaction of aromatic compounds is electrophilic substitution. The mechanism of these reactions, discussed briefly in Section 8.2d (see Fig. 8.13), is outlined in greater detail in Fig. 31.6, where it is shown to involve (a) formation of the electrophile, E^+, (b) association of the electrophile (a Lewis acid) with the aromatic ring (a Lewis base) to form a structure in which the electrophile is loosely associated with the aromatic ring and not attached to any specific position (designated as a *π complex*), (c) rearrangement of the π complex to a structure in which the electrophile becomes attached to a specific carbon of the aromatic system (designated as a *σ complex* or *benzenonium ion*), and (d) regeneration of the aromatic system by expulsion of a proton. Which of these four steps is rate determining varies from one system to another, although in most instances either the formation of the electrophile or the formation of the benzenonium ion is the slow step in the overall sequence, the formation of the π complex and the expulsion of the proton generally being rapid processes.

Step 1: Formation of the electrophilic reagent:

By spontaneous dissociation: $E—Nu \rightleftharpoons E^+ + Nu^-$

By acid-catalyzed dissociation: $E—Nu + A \rightleftharpoons E—\overset{+}{N}u—\overset{-}{A} \rightleftharpoons E^+ + Nu—A^-$

Lewis acid

Step 2 π-complex Step 3

(−H+)

Step 4 Benzenonium ion (σ-complex)

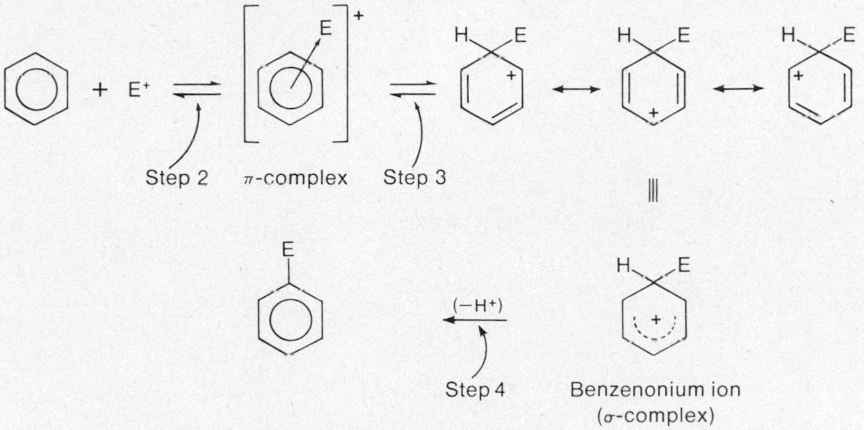

Fig. 31.6. General mechanism for aromatic electrophilic substitution.

A substituent on a benzene ring can affect not only the rate at which an electrophilic substitution reaction occurs (for the general features of the mechanism see Section 8.2d) but also the position at which the incoming group enters (for examples, see pp. 274 and 434). For example, the methoxyl group in anisole strongly accelerates the nitration reaction and directs the incoming nitro group to the *ortho* and *para* positions, whereas the carboxyl group in benzoic acid strongly decelerates the nitration reaction and directs the incoming nitro group to the *meta*-position. The methoxyl group is said to be *activating* and *ortho,para-directing*; the carboxyl group is said to be *deactivating* and *meta-directing*. The course of events in each of these instances can be rationalized on the basis of the ability of the substituents to aid or hinder the stabilization of the intermediate benzenonium ion.

31.3a. NITRATION OF ANISOLE. Let us first consider the nitration of anisole in more detail. Attachment of the electrophile, the nitronium cation (NO_2^+) to the *ortho*, *meta*, and *para* positions of the benzene ring produces the three benzenonium ions shown in Fig. 31.7. In the *ortho* and *para* benzenonium ions, the positive charge is stabilized by resonance interaction with the nonbonded electrons of the ether oxygen; in the *meta* benzenonium ion, however, this mode of stabilization is not possible. As a consequence of the greater stability of the *ortho*- and *para*-benzenonium ions, the *o*- and *p*-substituted products are preferentially formed. Because of the resonance interaction between the phenyl ring and the oxygen atom of the methoxyl group (see Fig. 10.1), the entire ring in

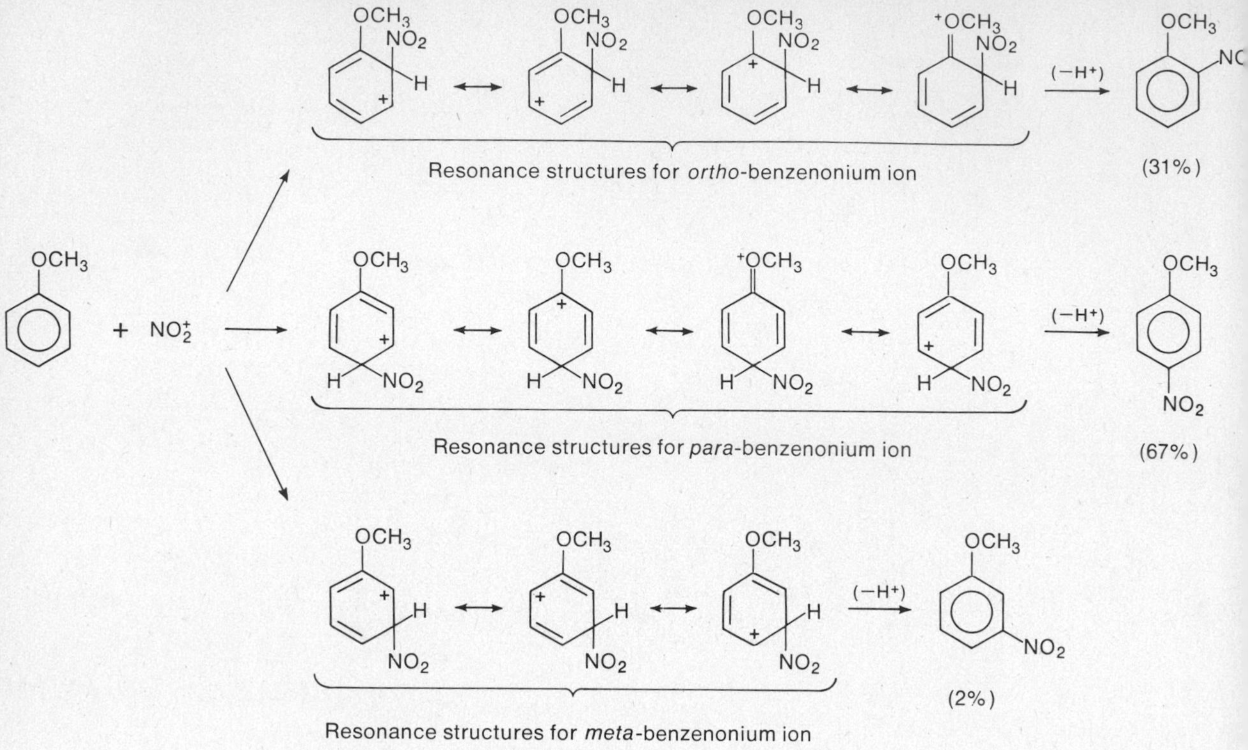

Fig. 31.7. Benzenonium ions formed from nitronium ion and anisole.

anisole possesses higher electron density than benzene itself (*i.e.*, it is a stronger Lewis base). Thus, even the *meta* position in anisole has a higher electron density than any of the carbon atoms of benzene and, consequently has an inherently greater reactivity than benzene (Fig. 31.7).

31.3b. NITRATION OF BENZOIC ACID. Next, let us discuss the nitration of benzoic acid in the same fashion by considering the three benzenonium ions that result from the attachment of the nitronium cation (NO_2^+) to the *ortho*, *meta*, and *para* positions of benzoic acid (Fig. 31.8). In the *ortho* and *para* ions, one of the resonance structures bears the positive charge *on* the carbon attached to the carboxyl group. Since the carboxyl group is electron withdrawing, this is equivalent to placing two positive charges next to each other and represents a particularly unfavorable situation. In the *meta* ion, on the other hand, none of the resonance structures has this feature, so in a relative sense the resonance hybrid for the *meta* ion is more stable than the resonance hybrids for the *ortho* and *para* ions. As a result, the *meta*-nitro compound is formed preferentially. Thus, although the entire ring in benzoic acid is *less* electron-rich than in benzene,

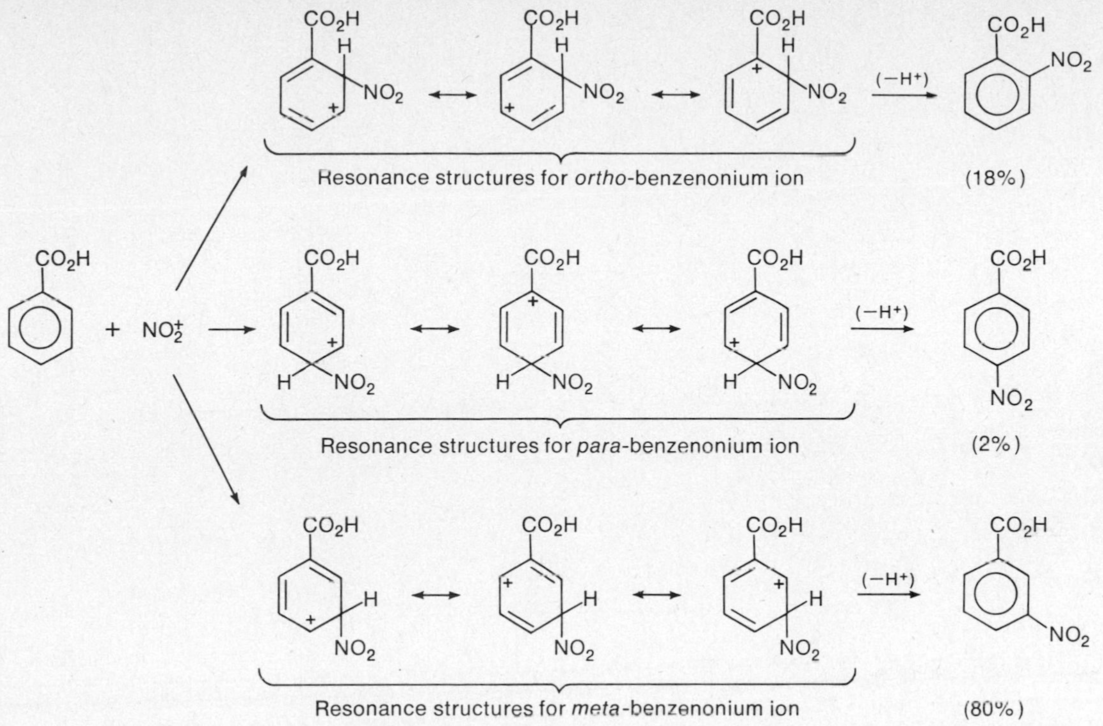

Resonance structures for *ortho*-benzenonium ion (18%)

Resonance structures for *para*-benzenonium ion (2%)

Resonance structures for *meta*-benzenonium ion (80%)

Fig. 31.8. Benzenonium ions formed from nitronium ion and benzoic acid.

the *meta* position is electron-rich *relative* to the *ortho* and *para* positions. Thus, it is the preferred site of attack by the nitronium ion.

PROBLEM 31-5

Predict the product of nitration of (a) 1,3-dimethoxybenzene, (b) 1,3-dinitrobenzene, and (c) 1-methoxy-3-nitrobenzene.

31.3c. Classification of orientation effects. Groups that are electron releasing via the resonance pathway are generally activating and *ortho,para*-directing in electrophilic aromatic substitution reactions. Groups that are electron-withdrawing via the resonance and/or inductive processes are generally deactivating and *meta*-directing. A third class of substituents exists, which are deactivating, yet *ortho,para*-directing. Included in this group are chloro-, bromo-, and iodo-substituents. The inductive electron-withdrawing effect of these substituents outweighs their ability to donate electron density by resonance; consequently, they are deactivating groups. They are capable of stabilizing the positive charge imposed on the system when an electrophile is attached to the *ortho* or *para* positions but not the *meta* position;

31.3d. QUANTITATIVE COMPARISONS OF EFFECT OF GROUPS ON REACTIVITY. The effect of substituents on the reactivity of aromatic compounds toward electrophilic substitution varies enormously, as indicated by the relative rate data for the nitration of substituted benzenes (Table 31.2). The activating-deactivating and the directing properties of various groups with respect to aromatic electrophilic substitution reactions are given in Table 31.3.

Table 31.2 RELATIVE REACTIVITIES OF NITRATION AT THE POSITION *para* TO THE SUBSTITUENT GROUP

$$\text{Y–C}_6\text{H}_5 \xrightarrow{HNO_3} \text{Y–C}_6\text{H}_4\text{–NO}_2$$

Substituent (Y)	Relative rate	Substituent (Y)	Relative rate
$-N(CH_3)_2$	$\sim 2 \times 10^{11}$	$-I$	0.18
$-OCH_3$	$\sim 2 \times 10^{5}$	$-F$	0.15
$-CH_3$	24.5	$-Cl$	0.033
$-C(CH_3)_3$	15.5	$-Br$	0.030
$-CH_2CO_2C_2H_5$	3.8	$-NO_2$	6×10^{-8}
$-H$ (benzene)	1.0	$-\overset{+}{N}(CH_3)_3$	1.2×10^{-8}

Table 31.3 ACTIVATING-DEACTIVATING PROPERTIES AND DIRECTIVE PROPERTIES OF GROUPS WITH RESPECT TO AROMATIC ELECTROPHILIC SUBSTITUTION REACTIONS

Activating Groups in Order of Decreasing Degree of Activation (All Are *ortho,para*-Directing)	Deactivating Groups in Order of Increasing Deactivation (All Are *meta*-Directing, Except the Halogens)
$-O^-$	$(-H)$
$-NR_2$	$-CH_2Cl$
$-NHR$	$-I$
$-NH_2$	$-Cl$
$-OH$	$-Br$
$-NHCOR$	$-CHCl_2$
$-OR$	$-CCl_3$
$-OCOR$	$-CONR_2$
$-SR$	$-CO_2R$
$-CH_3$	$-CO_2H$
$-C_2H_5$	$-COR$
$-CH(CH_3)_2$	$-CHO$
$-C(CH_3)_3$	$-SO_3H$
—aryl	$-CN$
$-CO_2^-$	$-CF_3$
$-CH_2CO_2R$	$-NO_2$
—F (activating in the *para* position only)	$-\overset{+}{N}R_3$
$(-H)$	

31.3. *ortho,para* RATIO. Electrophilic substitution reactions involving aromatic compounds bearing *ortho,para*-directing functions produce mixtures of *ortho*- and *para*-disubstituted products in varying ratios. For example nitration of toluene yields a mixture in which the *ortho* isomer predominates by a ratio of 1.57/1.0, whereas nitration of *t*-butylbenzene yields a mixture in which the *para* isomer predominates by a ratio of 4.6/1.0.

R-C6H5 —HNO_3/H_2SO_4→ *o*-NO_2 product + *p*-NO_2 product

Major product when R = CH_3 (ortho)   Major product when R = $C(CH_3)_3$ (para)

The ratio of *ortho* to *para* disubstitution depends on (a) the reactivity of the electrophilic reagent, (b) the steric bulk of the electrophilic reagent, and (c) the steric bulk of the substituent on the benzene ring. In general, the more reactive the electrophilic reagent (see Table 31.4), the lower is the selectivity for the *ortho* and/or the *para* position; the greater the bulk of the electrophilic reagent and the greater the bulk of the substituent attached to the benzene ring, the greater is the amount of the *para* isomer formed. This is ascribed to the greater steric hindrance encountered during formation of *ortho* product compared with *para* product (see Table 31.5).

Table 31.4 ELECTROPHILIC REAGENTS FOR AROMATIC ELECTROPHILIC SUBSTITUTION REACTIONS, ARRANGED APPROXIMATELY IN THE ORDER OF DECREASING REACTIVITY

Reagent		Electrophilic Species	Reaction Type
$NO_2^+ ClO_4^-$	(nitronium perchlorate)	NO_2^+	Nitration
$NO_2^+ BF_4^-$	(nitronium fluoborate)	NO_2^+	Nitration
$ClSO_2OH$	(chlorosulfonic acid)	$ClSO_2^+$	Chlorosulfonation
$H_2SO_4{\cdot}SO_3$	(fuming sulfuric acid)	SO_3	Sulfonation
$HNO_3 + H_2SO_4$	(nitric + sulfuric acid)	NO_2^+	Nitration
HNO_3	(nitric acid)	NO_2^+	Nitration
$RCOCl + AlCl_3$	(acid chloride)	RCO^+	Acylation
$X_2 + AlX_3$	(X = halogen except F)	X^+	Halogenation
$X_2 + Fe$	(X = halogen except F)	X^+	Halogenation
X_2	(X = halogen except F)	X^+	Halogenation
HOX	(hypohalous acid, X = Cl, Br)	X^+	Halogenation
$RX + AlCl_3$	(alkyl halide)	R^+	Alkylation

Table 31.5 *ortho,para* PRODUCT RATIO IN THE NITRATION OF ALKYLBENZENES

$$\left(\text{R–C}_6\text{H}_5 \xrightarrow{HNO_3} \text{R–C}_6\text{H}_4\text{–NO}_2 \right)$$

Alkyl Substituent R	*Ortho, Para* Product Ratio
Methyl	1.57
Ethyl	0.93
Isopropyl	0.48
t-Butyl	0.22

31.4 Scope and Limitations of Electrophilic Substitution Reactions

31.4a. ELECTROPHILIC SUBSTITUTION REACTIONS OF BENZENES. The significant features of electrophilic substitution reactions of benzenes are outlined in previous chapters in which the methods of preparation of aryl halides, phenols, aryl ethers, anilines, aromatic aldehydes, ketones, carboxylic acids, and carboxylic acid derivatives are discussed. The principal aspects that the experimenter must keep in mind when employing these reactions are (a) the orienting influence of groups already present on the ring, and (b) the deactivating effect exerted by *meta*-directing groups, precluding certain electrophilic substitution processes such as Friedel–Crafts reactions.

31.4b. ELECTROPHILIC SUBSTITUTION REACTIONS IN NAPHTHALENE. Naphthalene differs from benzene in having carbon–carbon bonds of different lengths, the bond between C-1 and C-2 being shorter than that between C-2 and C-3. This is interpreted to mean that resonance structure **I**, in which both rings are benzenoid,

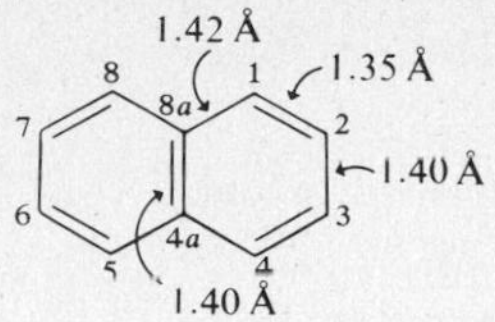

contributes more significantly to the resonance hybrid than structures **II** and **III**, in which only one ring is benzenoid and the other ring is quinoid.

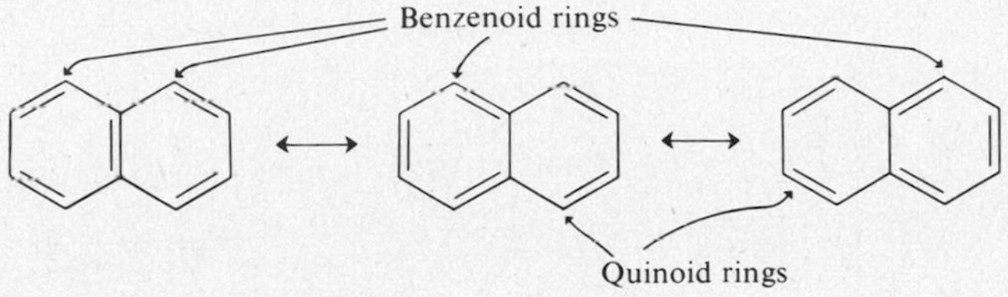

I II III

The resonance energy of naphthalene is about 61 kcal/mole, a value that is less than twice the amount of a single benzene ring (36 kcal/mole). As a result, naphthalene is somewhat less aromatic than benzene.

Naphthalene undergoes electrophilic substitution with the same array of reagents that react with benzene, and it reacts more rapidly as a result of its lower degree of aromaticity. Whereas the attachment of an electrophile to a benzene ring results in the disruption of an aromatic ring having 36 kcal/mole of resonance energy, the attachment of an electrophile to naphthalene results in the disruption of an aromatic ring having only 61 − 36 = 25 kcal/mole of resonance energy (*i.e.*, one of the benzene rings remains intact in the carbonium ion produced by electrophilic attack on naphthalene).

The reaction of an electrophile with naphthalene usually occurs more rapidly at the 1-position than at the 2-position, as indicated by the bromination and nitration reactions leading to 1-bromonaphthalene and 1-nitronaphthalene, respectively (Fig. 31.9). The preference for the 1-position can be explained in terms of the resonance structures for the carbonium ions formed when an electrophile becomes

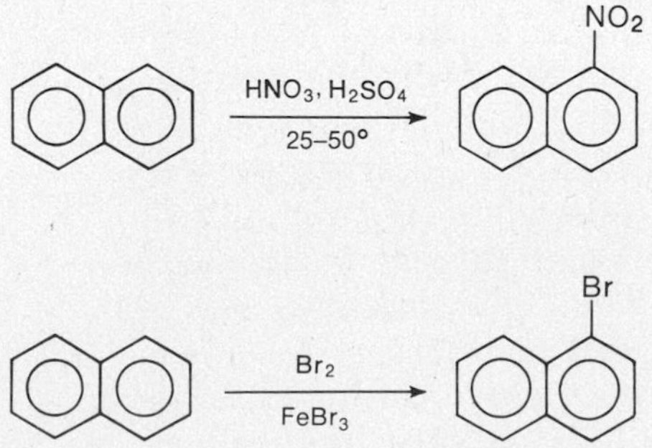

Fig. 31.9. Examples of electrophilic substitution with naphthalene.

attached at the 1- or 2-position (Fig. 31.10). Attachment at the 1-position produces a carbonium ion for which two of the resonance structures have a benzenoid ring, whereas attachment at the 2-position produces a carbonium ion for which only one of the resonance structures has a benzenoid ring; all the other structures in this second case are quinoid.

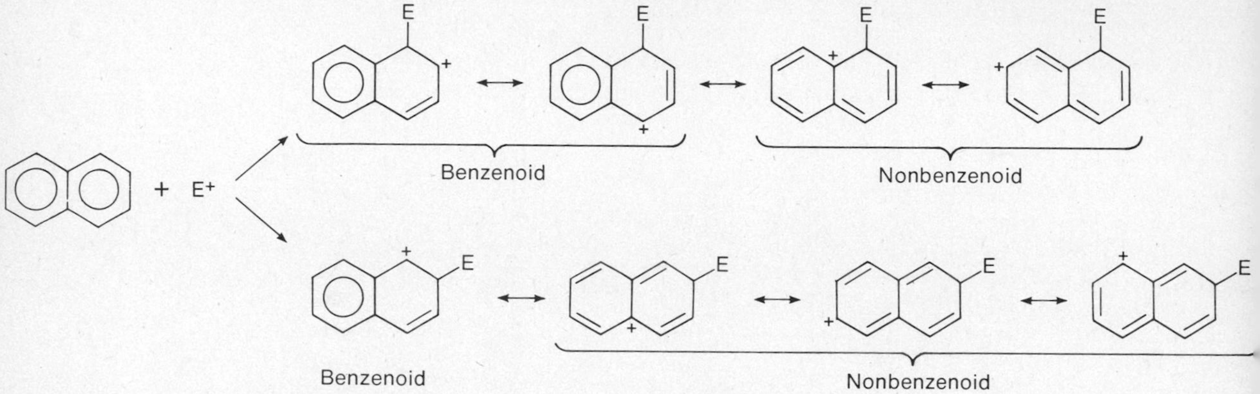

Benzenoid

Nonbenzenoid

Benzenoid

Nonbenzenoid

Fig. 31.10. Carbonium ions from electrophilic attack on naphthalene.

The 1-position of naphthalene is 570 times more reactive than a position of benzene, and the 2-position of naphthalene is 50 times more reactive than a position of benzene in the nitration reaction. Thus, although electrophilic substitution tends to take place preferentially at the 1-position, the 2-position is by no means unreactive. In some cases, in fact, electrophilic substitution at the 2-position can be made to be the dominant pathway. This arises from the circumstance that although reaction at the 1-position is faster, the equilibrium constant for this pathway may be lower than that for the 2-position because of the steric interference arising from the hydrogen at the 8-position, often referred to as the *peri*-position. Thus, under conditions where the equilibrium constant for the formation of the 1-substituted naphthalene is not particularly large, and where an equilibrium is established between the starting materials and *all* the possible products, the more stable 2-substituted naphthalene will be present in larger amount (*i.e.*, thermodynamic product control *vs.* kinetic product control (see Section 20.6)), A good example is seen in the sulfonation of naphthalene. which, at low temperatures, yields 1-naphthalenesulfonic acid, the product of kinetic control, but at higher temperatures yields 2-naphthalenesulfonic acid, the product of thermodynamic control (Fig. 31.11). The Friedel–Crafts acetylation is subject to similar control, the low temperature reaction in carbon disulfide yielding 1-acetylnaphthalene and the higher temperature reaction in nitrobenzene yielding 2-acetylnaphthalene (Fig. 31.11). In this case, both the temperature and the solvent control the situselectivity. Carbon disulfide has little tendency to solvate the acetyl cation, whereas nitrobenzene is a more polar molecule and shows a considerable tendency to do so. The solvated acetyl cation is much larger than its unsolvated counterpart and, consequently much

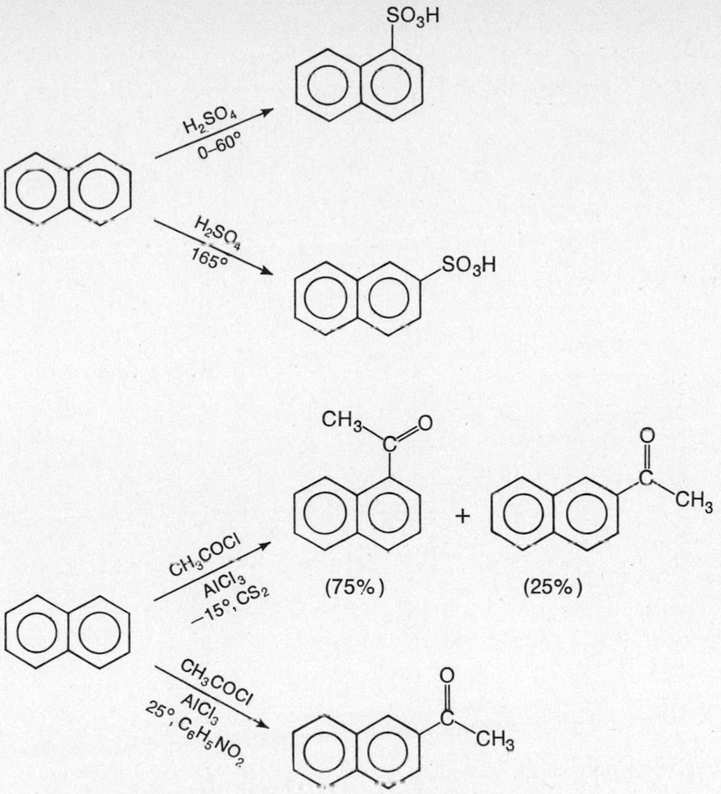

Fig. 31.11. Electrophilic substitution at the 2-position of naphthalene.

more susceptible to the steric hindrance arising from the hydrogen at the 8-position. Substitution at the 2-position takes place as a result;

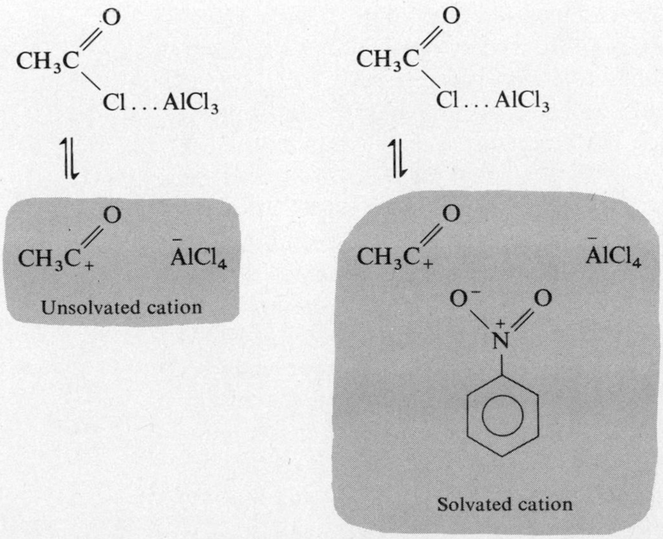

31.4c. Electrophilic substitution reactions of anthracene. The bond lengths in anthracene are commensurate with a resonance hybrid in which structures **I** and **II**, containing two benzenoid rings and one quinoid ring, contribute

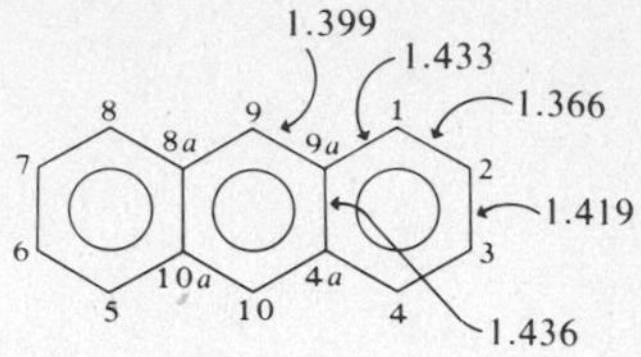

more heavily than structures **III** and **IV**, containing one benzenoid ring and two quinoid rings:

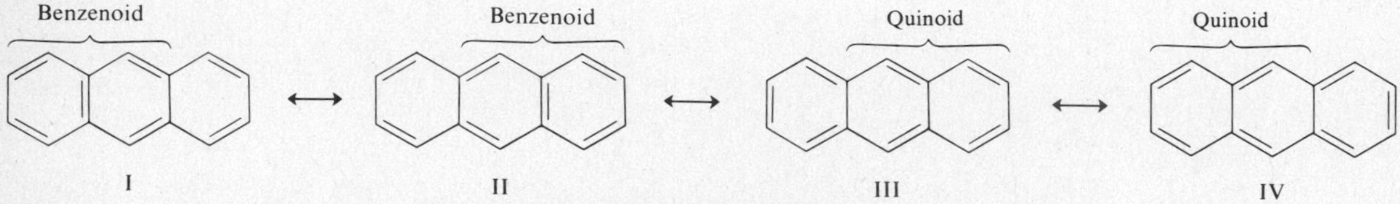

The resonance energy of anthracene is 84 kcal/mole. This averages to 26 kcal/mole per ring, which is substantially lower than that of benzene and somewhat lower than that of naphthalene.

Anthracene contains three sets of equivalent positions (*i.e.*, 1, 4, 5, 8; 2, 3, 6, 7; and 9, 10), and the point of most rapid electrophilic attachment can be predicted by considering the three carbonium ions that can be formed. Attachment of

---

PROBLEM 31-6

What are the symmetry elements of the anthracene molecule?

---

the electrophile at the 1- and 2-positions generates carbonium ions in which naphthalenoid systems are retained in the more heavily contributing resonance structures. Attachment of the electrophile at the 9-position generates a carbonium ion in which two benzenoid systems are retained in the more heavily contributing resonance structure (Fig. 31.12). Since the resonance energy of two benzene rings exceeds that of a naphthalene ring, it is predicted that substitution at the 9-position should take place most rapidly. As in previous examples, however, steric factors as well as electronic factors must be considered. A substituent at the 9-position is flanked on either side by C—H bonds that are parallel to it; a substituent at the 1-position, on the other hand, has only one adjacent bond parallel to it, and a substituent at the 2-position has none. As a result, there is competition between kinetic product control and thermodynamic product control, and the product that is isolated depends on the reactivity of the electrophile and the reaction conditions. Essentially irreversible electrophilic substitution reactions, such as bromination, yield the 9-substituted products, whereas more reversible electrophilic substitution reactions, such as sulfonation, yield the 1- and/or 2-substituted products (Fig. 31.13).

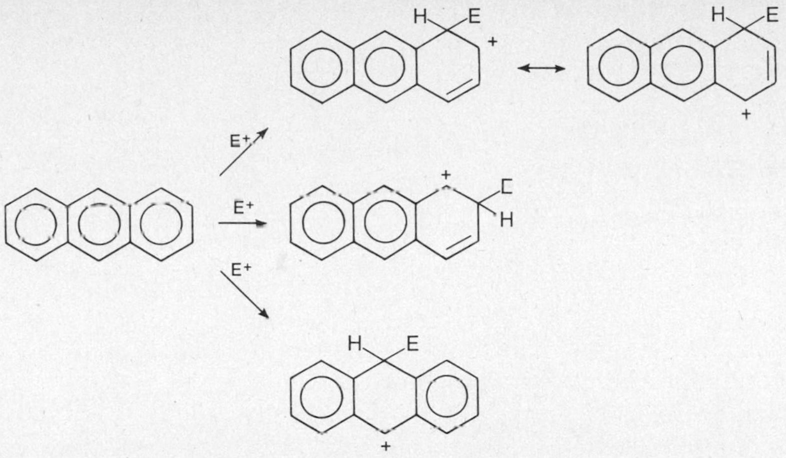

Fig. 31.12. Carbonium ions resulting from the electrophilic attack on anthracene.

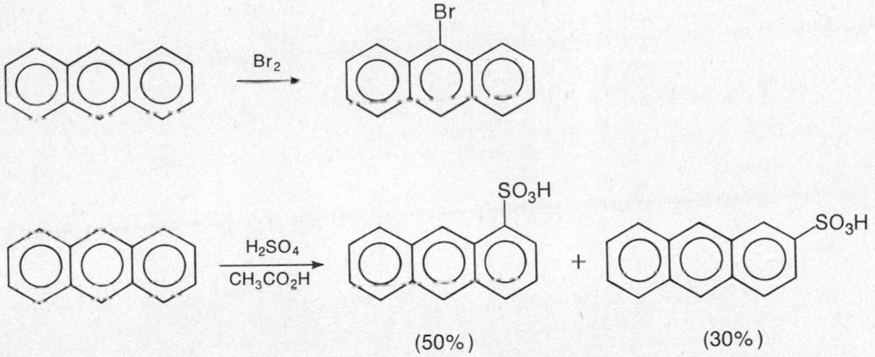

Fig. 31.13. Electrophilic substitution reactions of anthracene.

The picture is actually less tidy than this rationalization suggests, however, for if the bromination is carried out at low temperature and in the absence of base or acid, it is possible to isolate 9,10-dibromo-9,10-dihydroanthracene, the product of *addition* of bromine to the 9,10-positions. This compound undergoes a very facile loss of hydrogen bromide to yield 9-bromoanthracene, suggesting that the substitution product is really not the result of an aromatic electrophilic substitution in the usual sense but, rather, the result of an addition-elimination process. The formation of the addition product *vs.* substitution product can be rationalized on the basis of very simple bond energy considerations (see Table 2.3):

Addition:

$$\ce{>C=C<} + \ce{Br-Br} \rightleftharpoons \ce{-C(Br)-C(Br)-} \qquad \Delta H = -27 \text{ kcal/mole}$$

Substitution:

$$\ce{-C-H} + \ce{Br-Br} \rightleftharpoons \ce{-C-Br} + \ce{HBr} \qquad \Delta H = -10 \text{ kcal/mole}$$

In the absence of any counteracting influences, the addition process is thermodynamically favored over the substitution process by 17 kcal/mole. Addition to an aromatic ring, however, alters the resonance energy of the system. In the case of benzene it reduces it from 36 kcal/mole to 4–5 kcal/mole (*i.e.*, the resonance energy of a 1,3-butadiene); thus, the 27 kcal/mole bond energy advantage in the addition process is more than offset by the loss of resonance energy in the benzene system, so the addition reaction with benzene becomes endothermic and thermodynamically less favorable than the substitution reaction in which the resonance energy is retained. In the case of anthracene, the change in resonance energy involved in the formation of the addition product is equal to the resonance energy difference between that of anthracene (84 kcal/mole) and that of two benzenes ($36 \times 2 = 72$ kcal/mole); it amounts to only 12 kcal/mole. Thus, the bond energy advantage for the addition process in this case is offset by only 12 kcal/mole, so the addition reaction remains exothermic by about 15 kcal/mole. That the addition and substitution processes are about equally balanced in the anthracene system is indicated by the fact that both types of products can be isolated:

[Scheme: anthracene (H at 9 and 10) $\xrightarrow{Br^+}$ cation (H, Br at 9; + at 10 with H); $\overset{Br^-}{\rightleftharpoons}$ 9,10-dibromo-9,10-dihydroanthracene (H, Br; H, Br); $\overset{(-H^+)}{\rightleftharpoons}$ 9-bromoanthracene (Br; H)]

31.4d. Electrophilic substitution reactions of phenanthrene. Phenanthrene reacts in a fashion strikingly similar to that of anthracene, undergoing electrophilic substitution preferentially at the 9-position but also at the 1, 2, 3, and 4-positions. Under carefully controlled conditions, it adds bromine to yield 9,10-dibromo-9,10-dihydrophenanthrene, which readily loses hydrogen bromide to form 9-bromophenanthrene (Fig. 31.14).

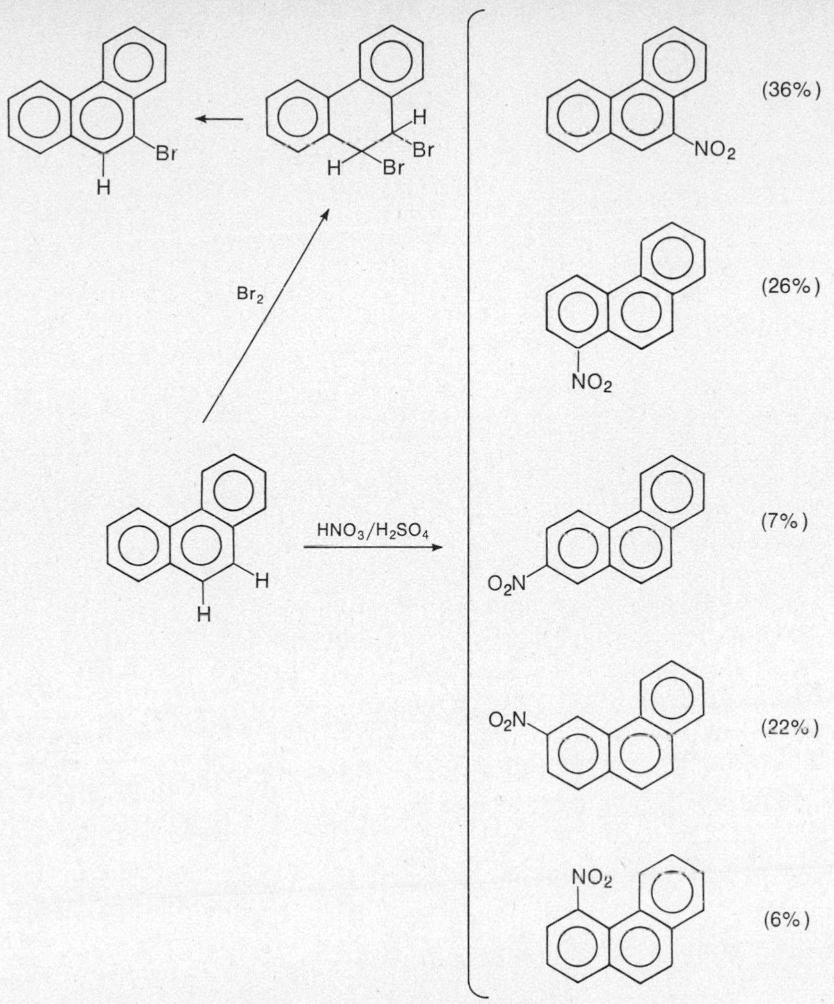

Fig. 31.14. Reactions of phenanthrene.

NUCLEOPHILIC AROMATIC SUBSTITUTION

31.5 Nucleophilic Substitution Reactions with Aryldiazonium Compounds

Molecular nitrogen is such a good leaving group that aryldiazonium compounds are very susceptible to nucleophilic substitution, the nitrogen being replaceable by halogen, oxygen, nitrogen, sulfur, and carbon functions (see Fig. 16.6 and Table 31.1). The method is completely general in its application, useful not only for preparing substituted benzenes but substituted polycyclic aromatic compounds as well (Fig. 31.15).

31.6 Nucleophilic Substitution Reactions of Naphthalenes

As discussed previously, most leaving groups are not sufficiently labile when attached to benzene nuclei to allow simple nucleophilic displacements to take place (see Section 21.5c). The presence of electron-withdrawing groups *ortho*

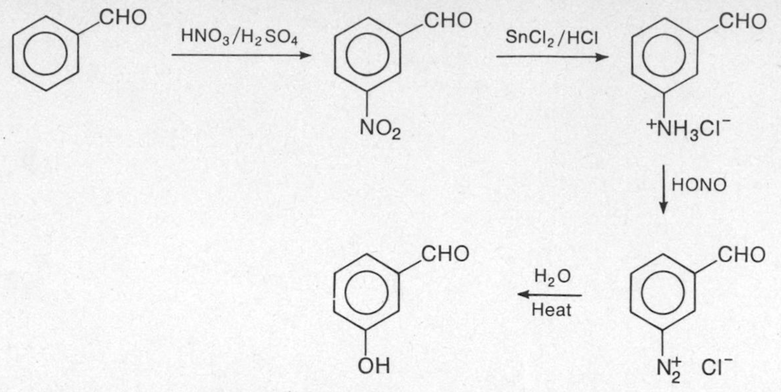

Fig. 31.15. Nucleophilic substitution reaction of an aryldiazonium salt in the preparation of *m*-hydroxybenzoic acid.

and *para* to certain leaving groups such as halogen, however, makes possible an addition-elimination sequence that leads to substitution products (see Fig. 21.32). In the naphthalene series, because of the lower degree of aromaticity of the ring system, reactions following this pathway take place even without the aid of electron-withdrawing groups. For example, the 1- and 2-naphthalenesulfonic acids can be converted to the corresponding naphthols or naphthylamines by treatment under mild conditions with sodium hydroxide and ammonium salts, respectively (Fig. 31.16). This reaction is particularly useful for the preparation of 2-naphthylamine (*Bucherer reaction*), which cannot be obtained via nitration and reduction, because only the 1-nitro compound is formed by the direct nitration of naphthalene.

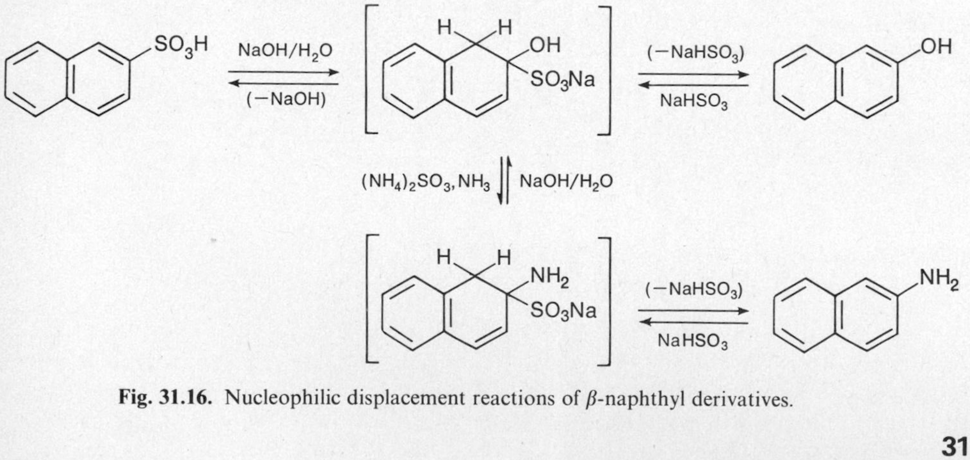

Fig. 31.16. Nucleophilic displacement reactions of β-naphthyl derivatives.

31.7 Nucleophilic Substitution Reactions Involving Benzyne Intermediates

The action of sodium amide, or other very strong bases, on aryl halides causes dehydrohalogenation and the formation of benzynes (see Figs. 21.33, 24.1). Benzynes are exceedingly reactive species and add any nucleophile that happens to be in their environment, including the bases used to generate them. For example, treatment of fluorobenzene with phenyllithium yields *o*-diphenyllithium, the result of the addition of phenyllithium to the benzyne that is formed by the base-induced loss of hydrogen fluoride. If the reaction mixture containing the *o*-diphenyllithium is treated with acid, biphenyl is obtained; if it is first treated with carbon dioxide, carbonation takes place to yield 2-phenylbenzoic acid upon acidification (Fig. 31.17).

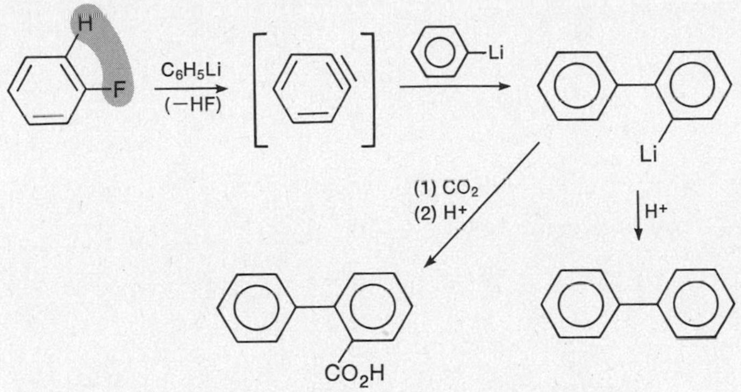

C_6H_5Li (−HF); (1) CO_2 (2) H^+; H^+

Fig. 31.17. Reaction of phenyllithium with fluorobenzene.

As an alternative to the strong base-induced route to benzyne, a procedure has been developed that allows its production under essentially neutral conditions. Treatment of *o*-aminobenzoic acid (anthranilic acid) with nitrous acid yields *o*-carboxybenzenediazonium cation, which, when treated with base, yields the corresponding zwitterion. By expelling two very stable fragments—carbon dioxide and nitrogen—this zwitterion forms benzyne, and the driving force to do so is great enough that the dry salt often undergoes violent decomposition (from which behavior it has acquired the nickname "*ortho*-di-boom, boom!"). To avoid this hazard, the salt is used as a slurry in a nonpolar solvent such as benzene rather than in the dry state. Heating the slurry to 50° is sufficient to initiate the decomposition:

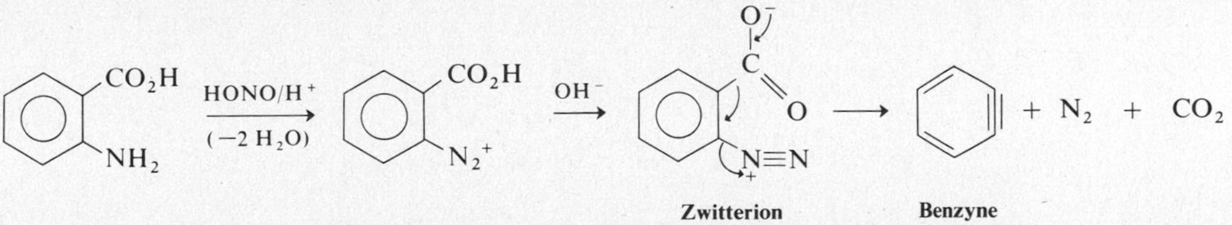

CO_2H, NH_2 —HONO/H^+ (−2 H_2O)→ CO_2H, N_2^+ —OH^-→ O^-–C=O, $N{\equiv}N^+$ → benzyne + N_2 + CO_2

Zwitterion **Benzyne**

If benzyne is generated in this fashion in the presence of nucleophilic reagents, nucleophilic addition occurs as soon as the benzyne forms, and products such as those shown in Fig. 31.18 are formed.

If benzyne is not provided with a nucleophile with which to react, it will react with itself to form biphenylene (Fig. 31.19). In comparable fashion, it undergoes cycloaddition reactions with other compounds containing carbon–carbon unsaturation, including alkenes to form benzocyclobutenes, benzene to form benzocyclooctatetraene, and with anthracene to form triptycene (Fig. 31.19).

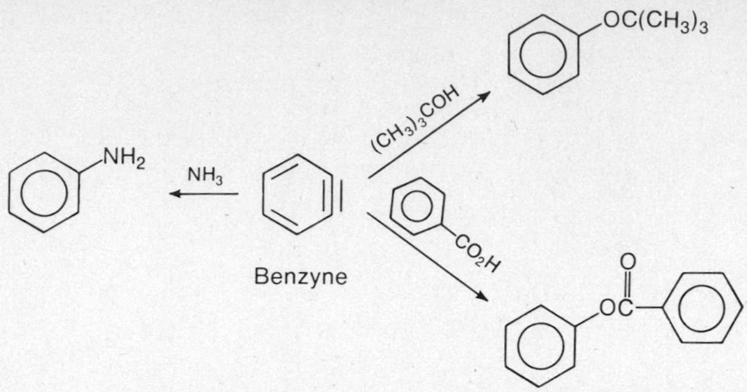

Benzyne

$(CH_3)_3COH$

$OC(CH_3)_3$

NH_2

NH_3

CO_2H

O

OC

Fig. 31.18. Reaction of benzyne with nucleophiles.

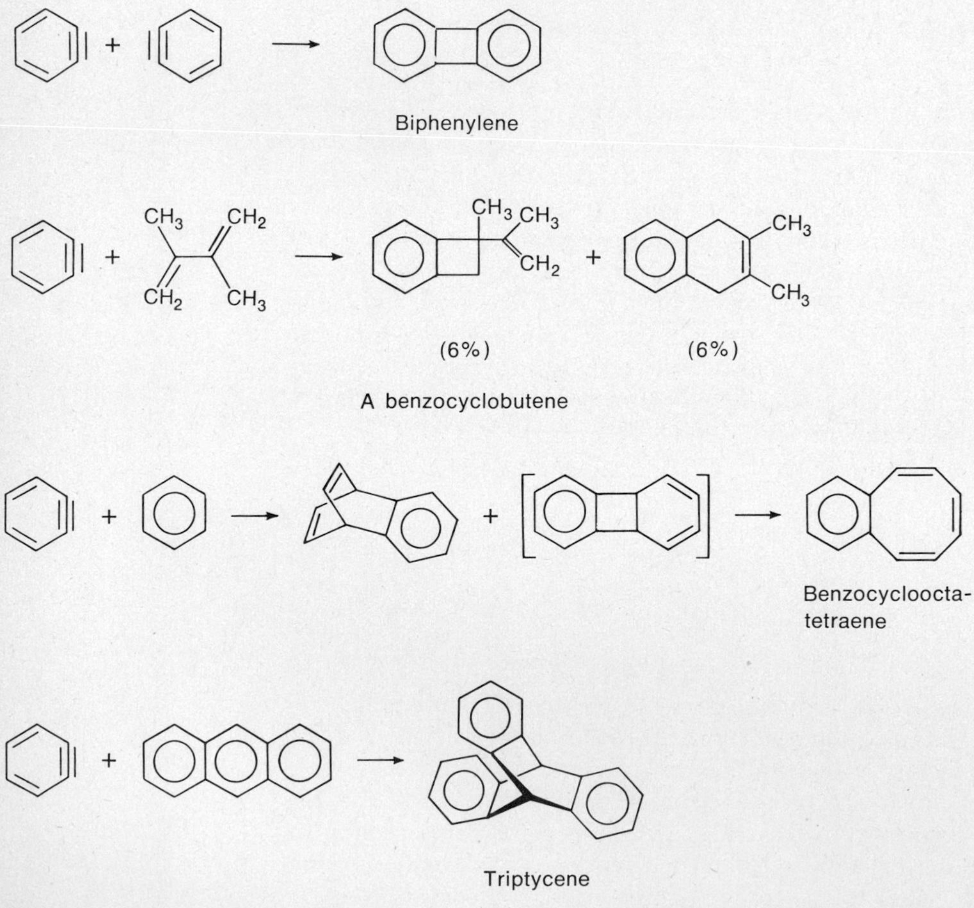

Biphenylene

CH_3 CH_2

CH_2 CH_3

CH_3 CH_3

CH_2

CH_3

CH_3

(6%)

(6%)

A benzocyclobutene

Benzocycloocta-
tetraene

Triptycene

Fig. 31.19. Cycloaddition reactions of benzyne.

PROBLEM 31-7

What would you predict to be the structure of the product obtained from the reaction of benzyne and phenanthrene?

PROBLEM 31-8

Explain the formation of 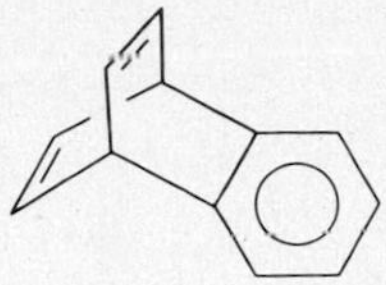 from benzyne and benzene.

OXIDATION AND REDUCTION REACTIONS OF AROMATIC RINGS

31.8 Reduction of Aromatic Rings

Partial reduction of benzenes can be effected via the Birch reduction (see Section 25.8) to yield 1,4-dihydrobenzenes. The method can also be applied to polycyclic aromatic compounds. For example, 1-naphthol undergoes reduction in the presence of lithium metal dissolved in liquid ammonia to provide a quantitative yield of 5,8-dihydro-1-naphthol:

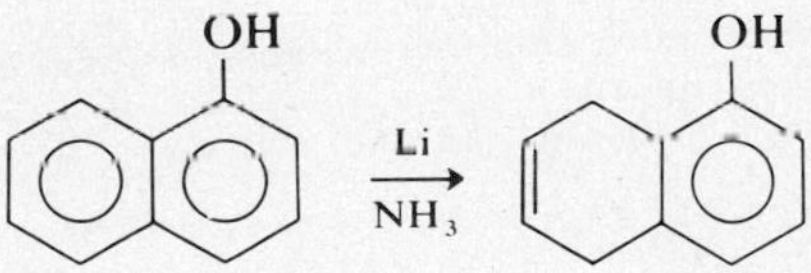

Aromatic hydrocarbons are susceptible to catalytic hydrogenation, the ease and selectivity of hydrogen addition increasing as the number of aromatic rings increases. Whereas it is difficult to stop the catalytic hydrogenation of benzene short of complete reduction to the corresponding cyclohexanes, naphthalene can be selectively hydrogenated to the tetrahydro compound, tetralin; anthracene and phenanthrene are very easily reduced to their 9,10-dihydro compounds (Fig. 31.20).

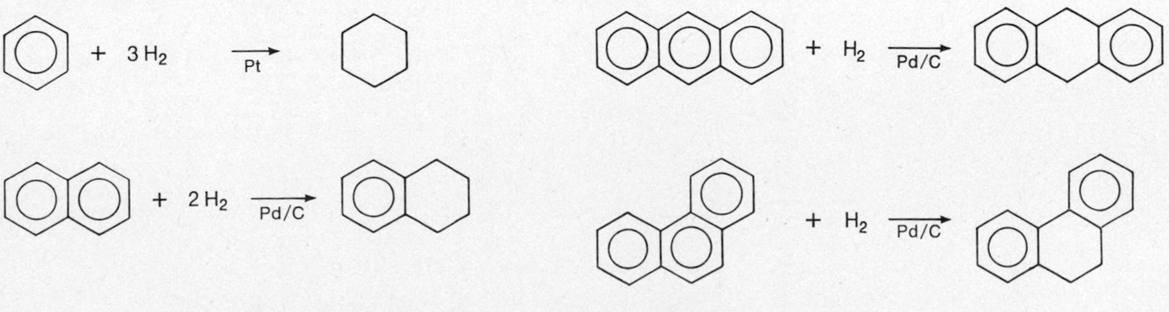

Fig. 31.20. Catalytic hydrogenation of aromatic hydrocarbons.

31.9a. OXIDATION OF AROMATICS. The benzene ring is very resistant to oxidation, as indicated by the fact that most alkylbenzenes undergo oxidation to the corresponding benzoic acids (see Fig. 28.23). The only types of side chains that are oxidized less readily than the benzene ring itself are those that are completely substituted at the benzylic carbon atom (*i.e.*, do not contain a benzylic C—H bond). For example, strenuous oxidation of benzotrifluoride with sodium dichromate and sulfuric acid yields trifluoroacetic acid:

$$C_6H_5CF_3 \xrightarrow{Na_2Cr_2O_7/H_2SO_4} CF_3CO_2H + 6\,CO_2 + 2\,H_2O$$

Although benzene rings are resistant to ozone under the conditions that ordinarily suffice for the ozonolysis of alkenes, prolonged exposure to ozone results in cleavage of benzene rings. This has had occasional application as a diagnostic procedure such as, for example, in the elucidation of the stereochemistry of the 2-phenylbicyclo[2.2.1]heptanes; the stereochemistry of the bicyclo[2,2,1]-heptane-2-carboxylic acids is known, so ozonolysis of the 2-phenyl compounds allows a correlation between the two series to be made:

$$\text{2-phenylbicyclo[2.2.1]heptane (H)} \xrightarrow[(2)\ H_2O_2/HOAc]{(1)\ O_3} \text{bicyclo[2.2.1]heptane-2-}CO_2H\text{ (H)}$$

Electron-releasing groups attached to a benzene ring activate it toward oxidation, and, under the influence of strenuous oxidizing conditions, phenols and anilines are usually oxidized to intractable mixtures of substances. Under mild conditions, or with more selective reagents, products can be isolated in which the carbon framework of the benzene ring has been retained but the aromatic character has been lost. Mild oxidation of phenol with acid dichromate or of aniline with manganese dioxide and sulfuric acid (see p. 358), for example, results in the formation of 1,4-benzoquinone (*p*-benzoquinone):

$$C_6H_5OH \xrightarrow{Na_2Cr_2O_7/H_2SO_4} O{=}C_6H_4{=}O \xleftarrow{MnO_2/H_2SO_4} C_6H_5NH_2$$

Even more susceptible to oxidation to quinones are benzenes that are disubstituted in the *ortho* or *para* positions with hydroxyl or amino groups. Thus, *p*-aminophenol is very readily converted to 1,4-benzoquinone, and 1,2-dihydroxybenzene (catechol) can be oxidized to 1,2-benzoquinone (*o*-benzoquinone):

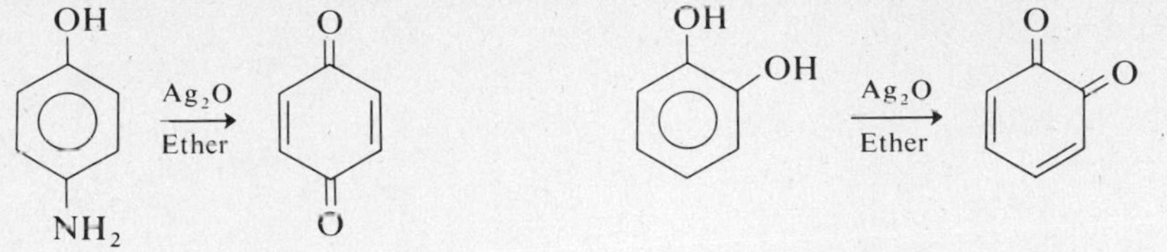

31.9b. OXIDATION OF NAPHTHALENES. Naphthalene, being less aromatic than benzene, is more readily oxidized. Treatment with strong oxidizing agents, such as basic permanganate or acid dichromate, converts naphthalene first to 1,4-naphthoquinone and then to phthalic acid or phthalic anhydride, depending on the reaction conditions:

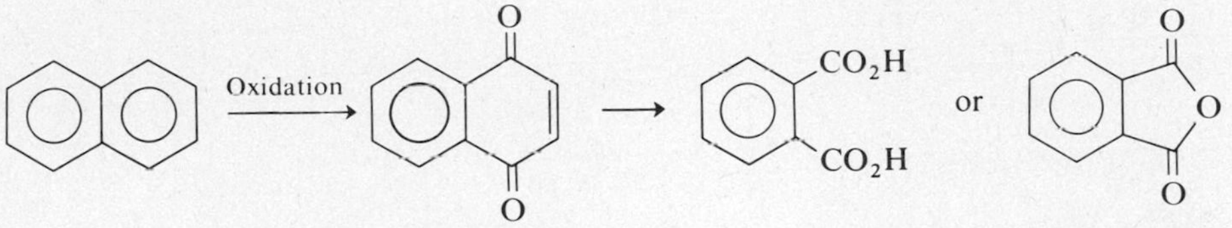

In a commercial process, oxygen in the presence of vanadium pentoxide catalyst is used to convert naphthalene to phthalic anhydride, naphthoquinone being a minor but isolable product in this reaction.

The ease with which the oxidation of a substituted aromatic ring takes place varies with the substituent; electron-releasing groups activate the ring toward oxidation, and electron-withdrawing groups deactivate it. An interesting example of this is seen in the naphthalene series where 1-nitronaphthalene undergoes oxidation to 3-nitrophthalic acid, but 1-aminonaphthalene undergoes oxidation to phthalic acid:

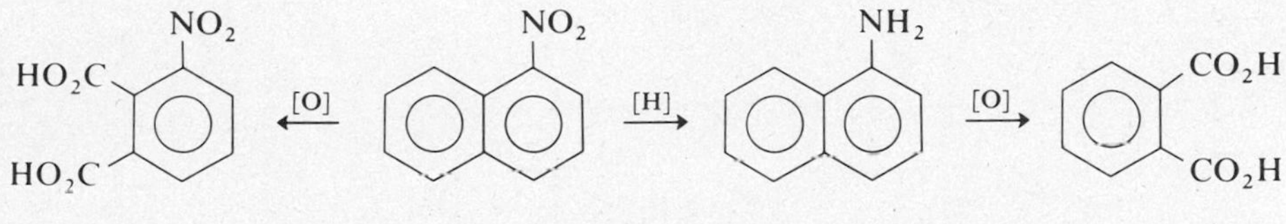

In contrast to alkylbenzenes, alkylnaphthalenes are oxidized by acid dichromate to quinones rather than to naphthalenecarboxylic acids. Thus, 2-methylnaphthalene affords 2-methyl-1,4-naphthoquinone in good yield if the reaction conditions are kept mild enough to prevent the further oxidation to phthalic acid:

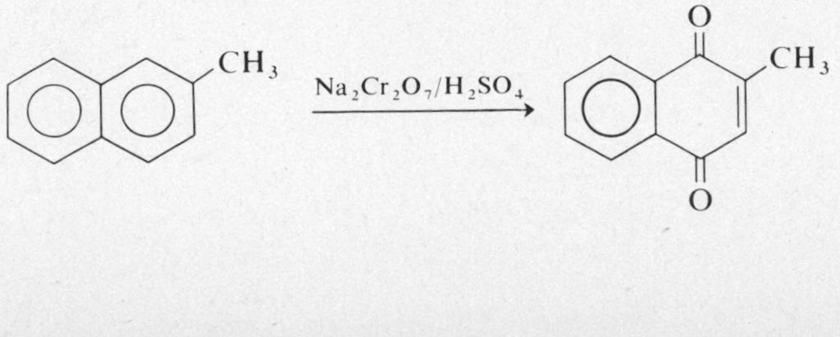

31.9c. OXIDATION OF ANTHRACENE AND PHENANTHRENE. Anthracene and phenanthrene are both converted to their 9,10-quinones by the action of chromic acid:

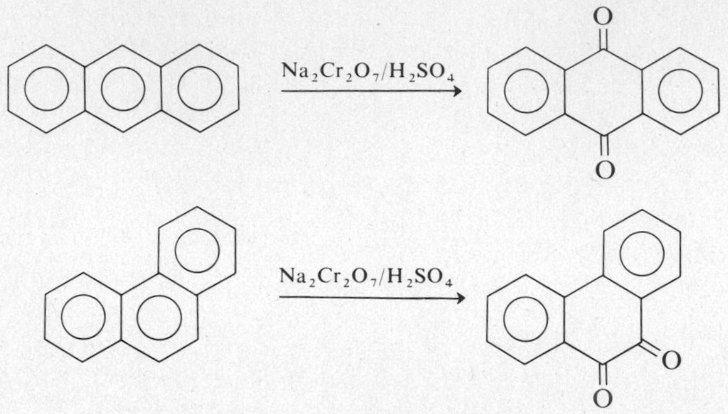

$$\text{Anthracene} \xrightarrow{Na_2Cr_2O_7/H_2SO_4} \text{9,10-anthraquinone}$$

$$\text{Phenanthrene} \xrightarrow{Na_2Cr_2O_7/H_2SO_4} \text{9,10-phenanthrenequinone}$$

PHOTOCHEMICAL REARRANGEMENT OF BENZENE RINGS

31.10 Photochemical Rearrangement of Benzene Rings

When irradiated with ultraviolet light, benzenoid aromatics undergo bond rearrangements that are referred to as valence isomerizations or valence tautomerizations. For example, benzene itself isomerizes to benzvalene, Dewar benzene, and fulvene (Fig. 31.21). The trisubstituted benzene 1,2,4-tri-*t*-butylbenzene, rearranges to a Dewar benzene which, in turn, reacts further to form a valence tautomeric structure called a prismane (Fig. 31.22). Benzvalenes, Dewar

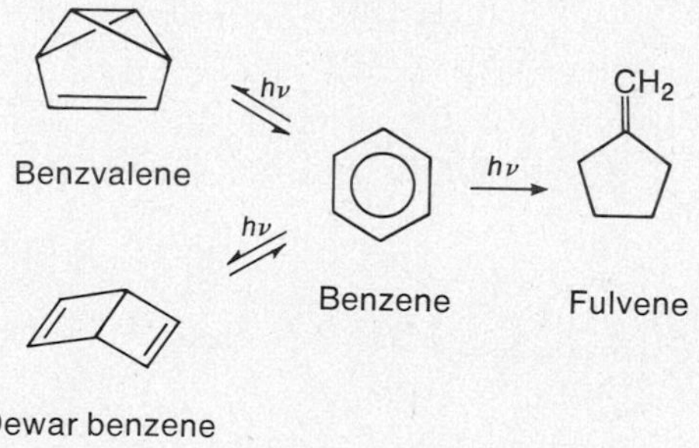

Fig. 31.21. Photoisomerization of benzene.

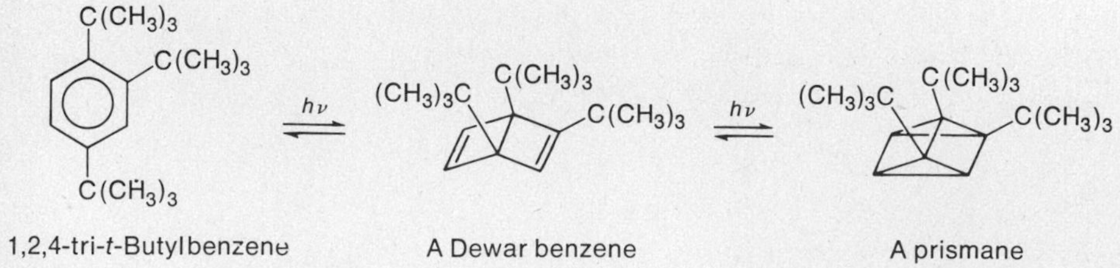

Fig. 31.22. Photoisomerization of 1,2,4-tri-*t*-butylbenzene to the valence tautomeric Dewar benzene and prismane structures.

benzenes, and prismanes all are thermodynamically unstable with respect to benzene and, in the absence of irradiation, slowly revert to the more stable species. In reverting, however, the bonds that are formed to produce the valence isomers are not necessarily the same ones that are broken in the isomerization back to the benzene. For example, the photolysis of *o*-di-*t*-butylbenzene yields a mixture of *m*- and *p*-di-*t*-butylbenzene completely devoid of the *ortho* isomer, a result interpretable in terms of prismane valence isomers that revert to the *meta* and *para* compounds rather than to the much more sterically hindered *ortho* compound (Fig. 31.23).

Section 31.11 Study Guide

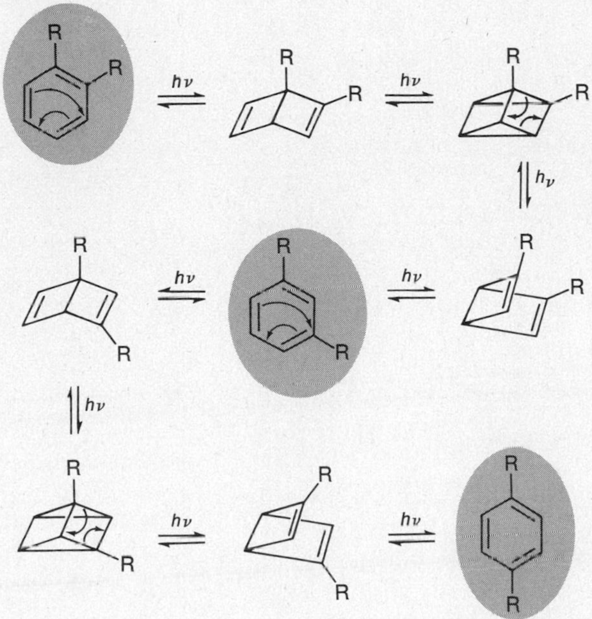

Fig. 31.23. Photoisomerization of *o*-disubstituted benzenes to *m*- and *p*-disubstituted benzenes via prismane intermediates.

31.11 Study Guide

I. Methods of preparation of benzenoid aromatic compounds
   A. Preparation of aromatic compounds from noncyclic or alicyclic precursors
      1. Trimerization of acetylenes—see Fig. 27.9
      2. Dehydrogenation of alicyclic compounds
         (a) Catalytic methods
            (1) Catalysts—palladium and platinum
            (2) Temperatures—125–300°
            (3) Equilibrium between hydrogenated form (*i.e.*, cyclohexane) and dehydrogenated form (*i.e.*, benzene)—low temperature favors hydrogenated form; high temperature favors dehydrogenated form

(b) Chemical methods
   (1) Sulfur and selenium (forming H_2S and H_2Se) at 250–350°
   (2) Quinones (*i.e.*, tetrachlorobenzoquinone)—works only if compound is easily dehydrogenated (*i.e.*, cyclohexadiene)
(c) Ease of dehydrogenation: cyclohexadienes > cyclohexenes > cyclohexanes > structures that contain quaternary carbon atoms

3. Preparation of cyclohexanes, cyclohexenes, and cyclohexadienes via Diels–Alder reactions followed by dehydrogenation
4. Preparation of naphthalenes by succinoylation-dehydrogenation sequence
5. Preparation of anthracenes by aroylation-aromatization sequence
6. Preparation of phenanthrenes by chain introduction-cyclization-dehydrogenation sequence

B. Preparation of aromatic compounds from aromatic precursors—see Table 31.1 for a list of reactions

II. Electrophilic aromatic substitution reactions

A. General features of electrophilic aromatic substitution reactions

1. Reaction sequence
   (a) Formation of electrophilic species—usually a cation such as Br^+, NO_2^+, etc.
   (b) Formation of π-complex
   (c) Formation of σ-complex
   (d) Loss of proton to regenerate the aromatic ring
2. Nitration of anisole
   (a) Methoxyl group is *o,p*-directing—electron density greater at *ortho* and *para* positions than at the *meta* position, and the methoxyl group stabilizes the *o*- and *p*-benzenonium ions
   (b) Methoxyl group is activating—*all* the positions of the benzene ring possess more electron density than those in benzene itself
3. Nitration of benzoic acid
   (a) Carboxyl group is *m*-directing—electron density is greater at *m* position than at *o* and *p* positions; the carboxyl group cannot stabilize the benzenonium ion intermediates
   (b) Carboxyl group is deactivating—*all* the positions of the benzene ring possess less electron density than those in benzene itself
4. Classification of orientation effects
   (a) *ortho,para*-directing and activating (*e.g.*, $CH_3O—$)
   (b) *ortho,para*-directing and deactivating (*e.g.*, —Cl)
   (c) *meta*-directing and deactivating (*e.g.*, $—NO_2$)
   (d) *meta*-directing and activating—no known examples
5. *ortho,para* Ratio
   (a) Reactivity of electrophile—greater reactivity leads to lower situselectivity
   (b) Steric bulk of electrophile—the larger the electrophile the greater the amount of *para* product

(c) Steric bulk of substituent—the larger the substituent the greater the amount of *para* product

B. Scope and limitations of electrophilic substitution reactions
  1. Electrophilic substitution reactions of benzenes—see II.A
  2. Electrophilic substitution reactions of naphthalene
    (a) Resonance structures of naphthalene
    (b) Substitution at 1-position—product of kinetic control
    (c) Substitution at 2-position—product of thermodynamic control
  3. Electrophilic substitution reactions of anthracene
    (a) Resonance structures of anthracene
    (b) Substitution at 9-position—product of kinetic control
    (c) Substitution at 1- and 2-positions—products of thermodynamic control
    (d) Addition-elimination sequence as a possible pathway for electrophilic substitution of anthracene
  4. Electrophilic substitution reactions of phenanthrene; similar to those of anthracene

III. Nucleophilic aromatic substitution reactions
  A. Nucleophilic substitution reactions with aryldiazonium compounds
  B. Nucleophilic substitution reactions of naphthalenes
  C. Nucleophilic substitution reactions involving benzyne intermediates
    1. Formation of benzynes
      (a) Dehydrohalogenation of aryl halides
      (b) Loss of CO_2 and N_2 from *o*-carboxybenzenediazonium cation
    2. Reactions of benzynes
      (a) Addition of nucleophiles
      (b) Cycloaddition reactions

IV. Oxidation and reduction reactions of aromatic rings
  A. Partial reduction with alkali metals in ammonia solution (Birch reduction)
  B. Partial or complete reduction with hydrogen and a catalyst
  C. Oxidation of benzenes—oxidation of ring occurs only if attached group contains no benzyl hydrogen atoms
  D. Phenols and anilines produce quinones
  E. Oxidation of naphthalenes, anthracenes, and phenanthrenes—rings are more susceptible to oxidation than the benzene ring

V. Photochemical rearrangement of benzene rings
  A. Formation of valence isomers
  B. Isomerization of substituted benzenes

REFERENCES

CLAR, E., *Polycyclic Hydrocarbons*, Vols. 1 and 2, Academic Press, Inc., New York, 1964.

COFFEY, S., ed., *Rodd's Chemistry of Carbon Compounds*, 2nd ed., Vol. III, Part A: *Aromatic Compounds. General Introduction*, Elsevier Publishing Co., New York, 1971.

De La Mare, P. B. D., and J. H. Ridd, *Aromatic Substitution: Nitration and Halogenation*, Academic Press, Inc., New York, 1959.

Fieser, L. F., and M. Fieser, *Advanced Organic Chemistry*, Reinhold Publishing Corp., New York, 1961, Chapter 18.

Garratt, P. J., *Aromaticity*, McGraw-Hill Book Company, New York, 1971.

Hoggett, J. G., R. B. Moodie, J. R. Penton, and K. Schofield, *Nitration and Aromatic Reactivity*, Cambridge University Press, New York, 1971.

Norman, R. O. C., and R. Taylor, *Electrophilic Substitution in Benzenoid Compounds*, Elsevier Publishing Co., New York, 1965.

Stock, L. M., *Aromatic Substitution Reactions*, in *Foundations of Modern Organic Chemistry Series*, Prentice-Hall, Inc., Englewood Cliffs, N.J., 1968.

Tomlinson, M., *An Introduction to the Chemistry of Benzenoid Compounds*, Pergamon Press, New York, 1971.

PROBLEMS

1. For each of the following monosubstituted benzenes, draw the structure of the major product formed in the reaction shown:

(a) $CH_3O-C_6H_5 + (CH_3)_2C{=}CH_2 \xrightarrow{AlCl_3}$

(b) $(CH_3)_3\overset{+}{N}-C_6H_5 + HNO_3 \xrightarrow{H_2SO_4}$

(c) $C_6H_5-NH\overset{O}{\overset{\|}{C}}-C_6H_5 + HNO_3 \xrightarrow{H_2SO_4}$

(d) $C_6H_5-CO_2CH_3 \xrightarrow[H_2SO_4]{HNO_3}$

(e) $C_6H_5-Cl + H_2SO_4{\cdot}SO_3 \longrightarrow$

(f) $C_6H_5-SO_3H + Br_2 \xrightarrow{FeBr_3}$

(g) $C_6H_5-O-CH_2-C(=CH_2)CH_3 \xrightarrow{\Delta}$

2. For each of the following reactions, which involve polysubstituted benzenes as reactants and/or products, draw the structure of the major organic product.

(a) 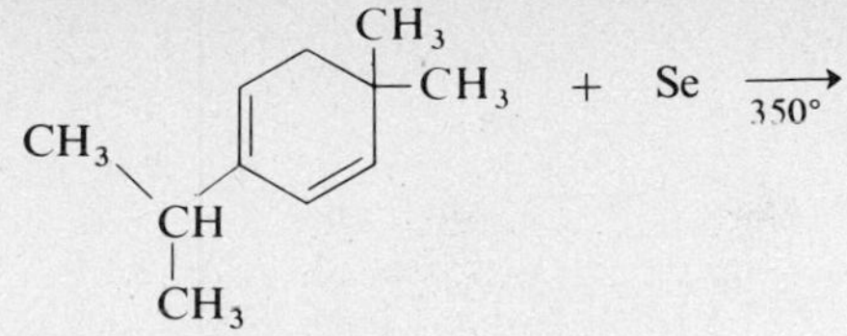 [structure: 1-isopropyl-4,4-dimethylcyclohexa-1,3-diene bearing CH_3, CH_3, and CH_3–CH–CH_3] + Se $\xrightarrow{350°}$

(b) $C_2H_5C{\equiv}CH$ + $Co_2(CO)_8$ $\longrightarrow$ $\xrightarrow{Br_2\ \text{oxidation}}$

(c) 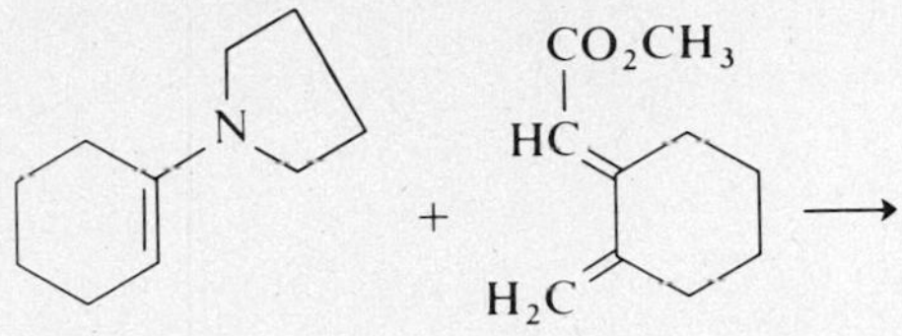 [1-(pyrrolidin-1-yl)cyclohexene, N] + [cyclohexane bearing $=CHCO_2CH_3$ (HC, CO_2CH_3) and $=CH_2$ (H_2C)] $\longrightarrow$

(d) CH_3O–C_6H_4–$C(=N{-}OH)CH(CH_3)_2$ $\xrightarrow{PCl_5}$

(e) [HO-naphthalene bearing NO_2] $\xrightarrow[\text{Excess oxidizing agent}]{Na_2Cr_2O_7/H_2SO_4}$

(f) CH_3O–C_6H_3(Br)(CH_2CO_2H) + $LiAlH_4$ $\longrightarrow$ $\xrightarrow{H_3O^+}$

(g) [benzene bearing CNH_2 (C=O) and CH_3] + Br_2 + NaOH $\longrightarrow$

(h) [benzene bearing Br, O_2N, NO_2, CH_3] + KCN $\xrightarrow[C_2H_5OH]{}$

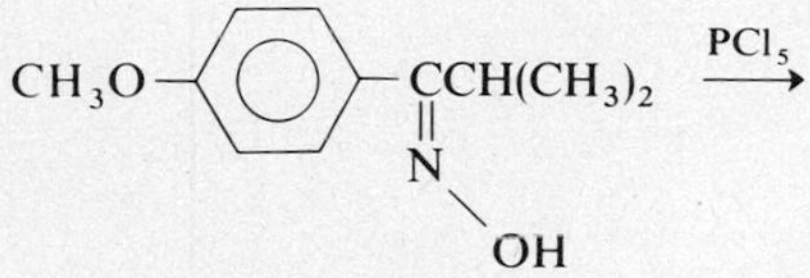

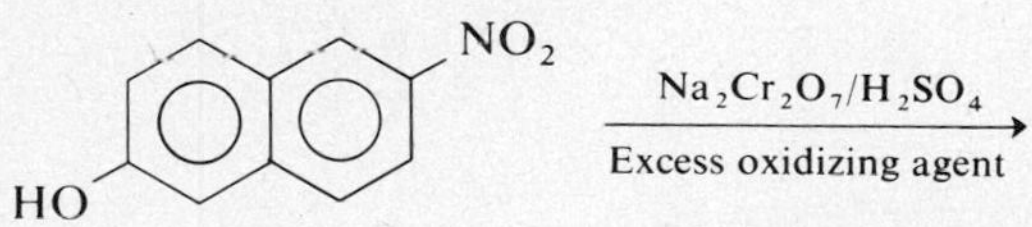

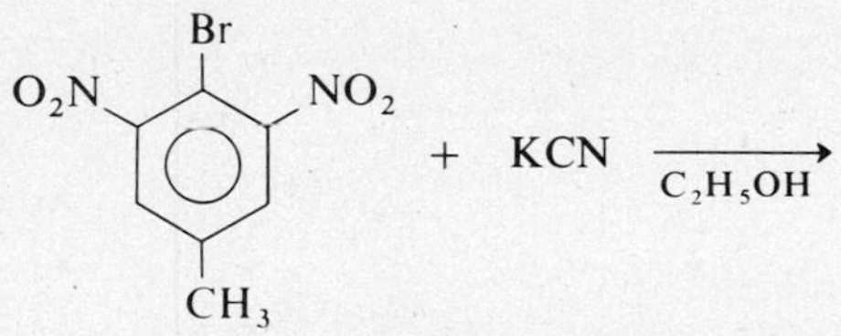

(i) 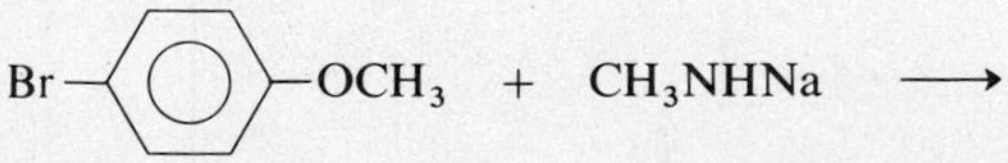 Br–C_6H_4–OCH_3 + CH_3NHNa $\longrightarrow$

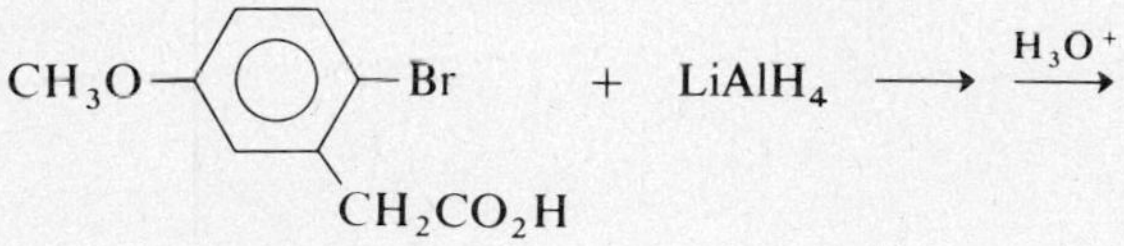

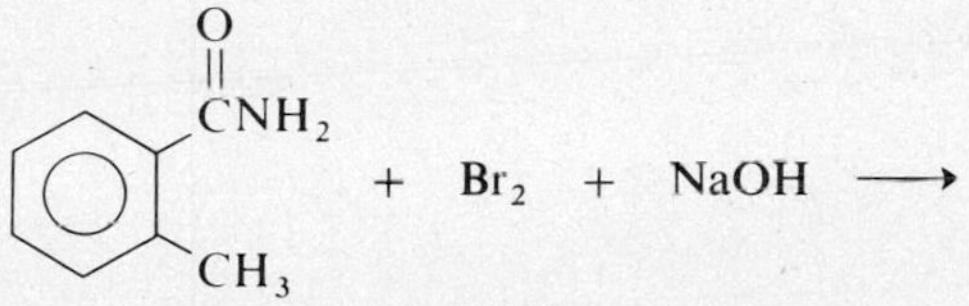

(j) 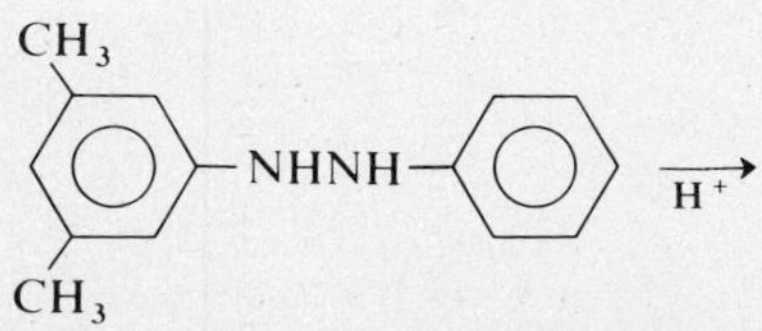 [3,5-dimethylphenyl, CH_3, CH_3]–NHNH–C_6H_5 $\xrightarrow{H^+}$

(k) 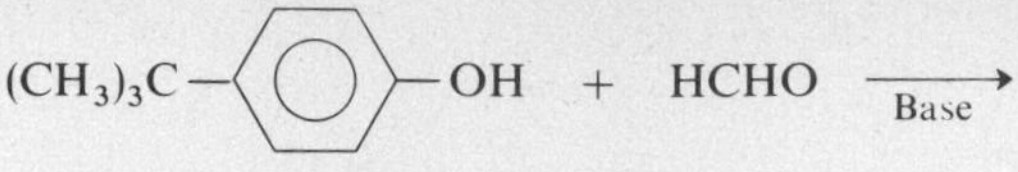 $(CH_3)_3C-C_6H_4-OH$ + HCHO $\xrightarrow[\text{Base}]{}$

3. For each of the following polycyclic aromatic compounds, draw the structure of the major product formed in the reaction shown:

(a) 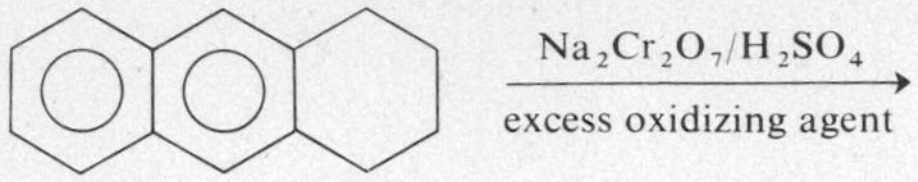 $\xrightarrow[\text{excess oxidizing agent}]{Na_2Cr_2O_7/H_2SO_4}$

(b) 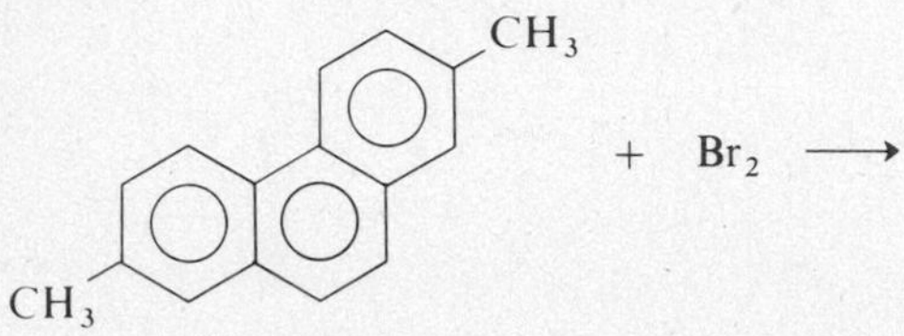 CH_3 CH_3 + Br_2 $\longrightarrow$

(c) 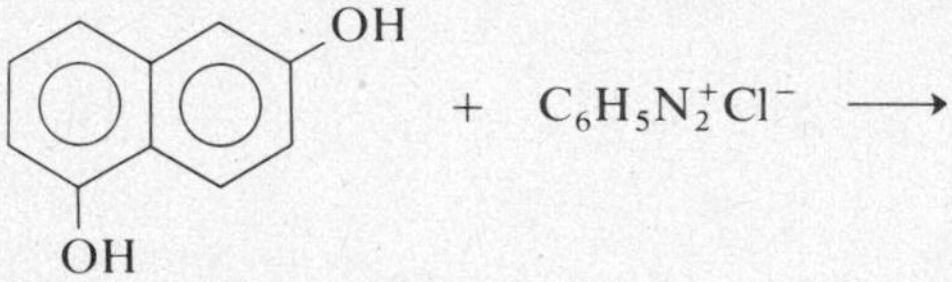 OH OH + $C_6H_5N_2^+Cl^-$ $\longrightarrow$

(d) 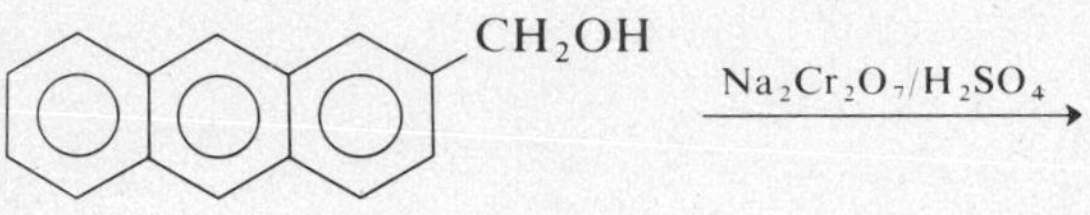 CH_2OH $\xrightarrow{Na_2Cr_2O_7/H_2SO_4}$

(e) 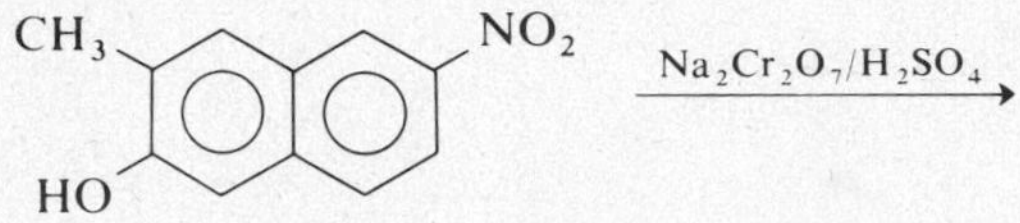 CH_3 HO NO_2 $\xrightarrow{Na_2Cr_2O_7/H_2SO_4}$

(f) 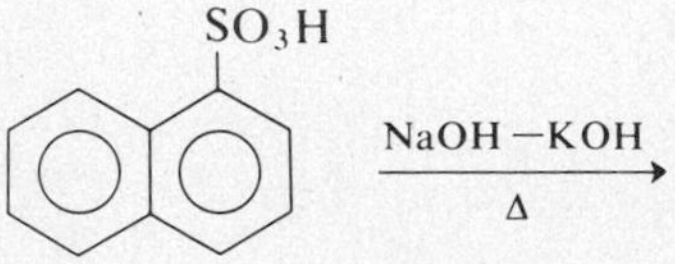 SO_3H $\xrightarrow[\Delta]{NaOH-KOH}$

4. Wilhelm Koerner, a chemist who lived in the late nineteenth and early twentieth centuries, devised a method for deducing the structures of "orientation isomers" of benzene that has come to be known as the *Koerner absolute method.* It consists of counting the number of isomers obtained when an additional substituent is introduced into a benzene ring already containing two or more substituents. For example, *o*-xylene yields 1,2-dimethyl-3-nitro- and 1,2-dimethyl-4-nitrobenzene (*i.e.*, two isomers), *m*-xylene yields 1,3-dimethyl-2-nitro, 1,3-dimethyl-4-nitro-, and 1,3-dimethyl-5-nitrobenzene (*i.e.*, three isomers), and *p*-xylene yields 1,4-dimethyl-2-nitrobenzene (*i.e.*, only one isomer). Thus, if we work backward from the result, the isomer yielding only one nitro compound must be the *para* isomer, the one yielding two nitro compounds must be the *ortho* isomer, and the one yielding three nitro compounds must be the *meta* isomer.

(a) Can the Koerner absolute method be used to differentiate the *ortho*, *meta*, and *para* isomers of the bromotoluenes? Why or why not?

(b) Three isomers of tribromobenzene have been synthesized, one melting at 44°, one at 87°, and one at 119°. Nitration of these three isomers yields three compounds from the isomer that melts at 44°, two compounds from the isomer that melts at 87°, and one compound from the isomer that melts at 119°. On the basis of this information, assign the orientation to the three isomers.

(c) Can the Koerner absolute method be used to establish the structures of the tetramethylbenzenes? Why or why not? If it cannot be used, suggest a procedure that might allow the structural assignments to be made.

(d) The three isomers of bromonitrobenzene yield the following compounds upon nitration: isomer A yields compounds a, b, c, and d, isomer B yields compounds a, c, e, and f; isomer C yields compounds b and e. When *o*-dinitrobenzene is brominated, compounds a and e are obtained. What structural assignments can be made to isomers A, B, and C on the basis of these data?

5. Outline, in as much detail as possible, the probable mechanisms by which the following reactions occur:

(a) $C_6H_6 + HCHO + HCl \xrightarrow{ZnCl_2} C_6H_5\text{—}CH_2Cl$

(b) $C_6H_5\text{—}N_2^+ + KI \longrightarrow C_6H_5\text{—}I + N_2 + K^+$

(c) $C_6H_5\text{—}CH_2CH_3 + Br_2 \xrightarrow{h\nu} C_6H_5\text{—}CHBrCH_3$ (CHCH$_3$ with Br)

$C_6H_5\text{—}CH_2CH_3 + Br_2 \xrightarrow{FeBr_3} Br\text{—}C_6H_4\text{—}CH_2CH_3$

(d) $O_2N\text{—}C_6H_2(NO_2)_2\text{—}Cl + OH^- \longrightarrow O_2N\text{—}C_6H_2(NO_2)_2\text{—}OH + Cl^-$

(e) 2,6-dimethylphenyl allyl ether (CH$_3$, O—CH$_2$*—CH=CH$_2$, CH$_3$) $\xrightarrow{\Delta}$ 4-allyl-2,6-dimethylphenol (CH$_3$, OH, CH$_3$, *CH$_2$—CH=CH$_2$ as H$_2$C=)

* indicates ^{14}C

6. The Friedel–Crafts alkylation of benzene with benzyl chloride and aluminum chloride to form diphenylmethane is postulated to involve the two intermediates shown in the following sequence:

$$C_6H_5\text{—}CH_2Cl + AlCl_3 \longrightarrow C_6H_5\text{—}\overset{\delta^+}{CH_2}Cl\cdots\overset{\delta^-}{AlCl_3}$$

Intermediate 1

$C_6H_5-CH_2Cl\cdots AlCl_3 + C_6H_6 \longrightarrow$ [arenium ion: H, $CH_2-C_6H_5$, H, +] $+ AlCl_3$

Intermediate 2

The formation of intermediate 2 is the rate-determining step in the overall process, and the reaction is first order in $C_6H_5CH_2Cl$, $AlCl_3$, and benzene.

(a) On the assumption that the product is more stable than the reactant, draw a reaction coordinate diagram for the process.

(b) If ΔG for the reaction is -6 kcal/mole, is the equilibrium constant for the reaction greater or smaller than unity? What is the approximate value of the equilibrium constant at 50°?

(c) Using benzyl chloride as the reference point, is the $\Delta G^{\ddagger}$ for the formation of intermediate 1 greater or less than that for the formation of intermediate 2?

(d) According to the reaction coordinate diagram that you have drawn for the reaction, would you anticipate a large or a small kinetic isotope effect if C_6D_6 were used instead of C_6H_6? Explain.

(e) Write a rate expression for the process.

(f) Suggest a structure for the activated complex that occurs between intermediate 1 and intermediate 2.

7. Given the starting compound shown at the left in each of the following examples, outline a procedure for converting it to the compound shown at the right. More than one step will be necessary in most, if not all, cases.

(a) $H_2C{=}CHC(=CH_2)CH_3 \longrightarrow$ benzene ring bearing CH_3, CH_2OH, CH_2OH

(b) benzene $\longrightarrow$ $C_6H_5CH_2CH_2CH_2CH(CH_3)_2$

(c) benzene $\longrightarrow$ benzene ring bearing NO_2 and CN (meta)

(d) $CH_3-C_6H_4-NO_2 \longrightarrow CH_3-C_6H_4-F$

(e) $C_6H_5C_2H_5 \longrightarrow$ benzene ring bearing C_2H_5 and Br (ortho) (as the only isomer formed)

(f) $C_6H_5CH_3 \longrightarrow$ benzene ring bearing CO_2H and CO_2H (meta)

(g) NH_2 (benzene) ⟶ $N(CH_3)_2$ … CHO

(h) CH_3 … $C(CH_3)_3$ ⟶ NH_2 … $C(CH_3)_3$

(i) CH_3 ⟶ CH_3 … CH_3

(j) ⟶ $\overset{O}{\overset{\|}{O C}}CH_2CH_3$

(k) ⟶ O … O

(l) ⟶ OCH_2CH_2OH

(m) ⟶ $CH_2N(CH_3)_2$

8. In each of the following cases, a compound is being investigated that is known to be one of the isomeric structures shown. Using chemical and/or spectral methods (as instructed), indicate how a definitive structural assignment can be made in each instance.

(a) Chemical methods only:

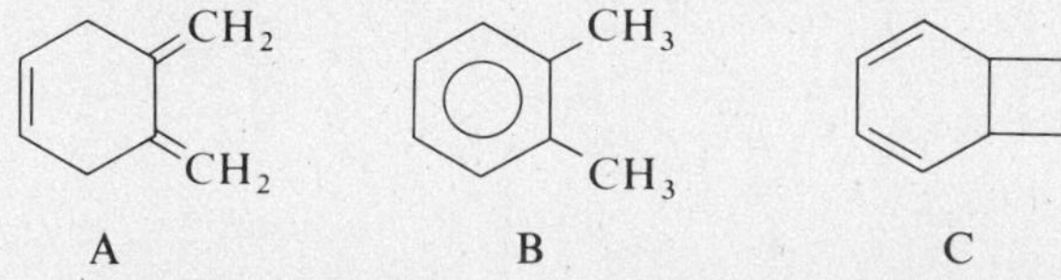

(b) Spectral methods only:

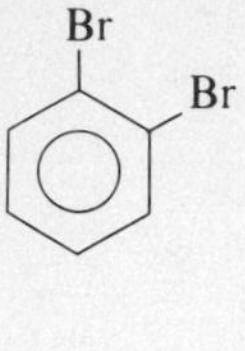

D 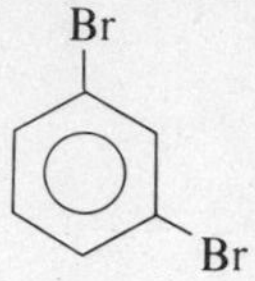 E  F

(c) Chemical and/or spectral methods:

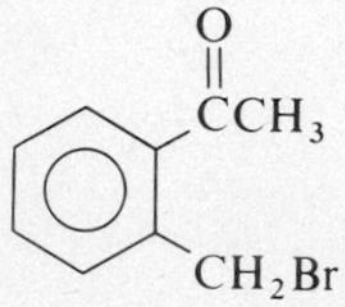

G 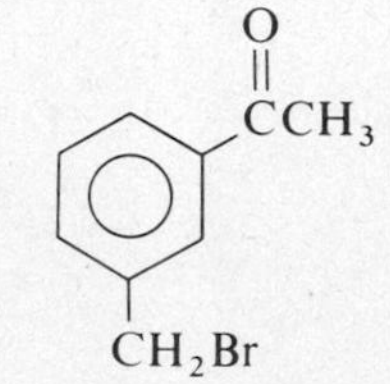 H 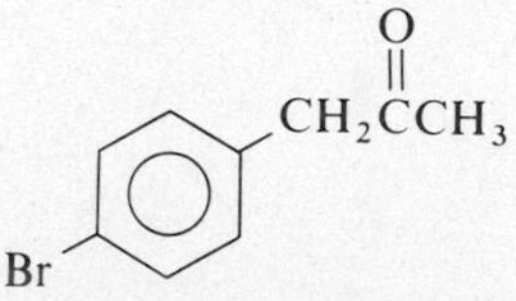 I

9. The compounds in each of the following examples are substances that possess some commercial and/or biological interest:

(a) A $C_{11}H_{13}N_3O_6$ compound was prepared in 1891 by a chemist named Baur. It is called "Baur musk" because it has an odor resembling that of natural musk, and it is used in perfuming cheap soaps. When subjected to oxidation with sodium dichromate and sulfuric acid, Baur musk yields a $C_{11}H_{11}N_3O_8$ carboxylic acid. Reduction of Baur musk yields a $C_{11}N_{19}N_3$ triamine, which fails to react with 2,3-butanedione (*Hint*: *o*-Diaminobenzene reacts with this reagent to form a bicyclic heterocyclic compound.) Oxidation of the triamine with sodium dichromate and sulfuric acid yields trimethylacetic acid as one of the products. On the basis of these data, suggest a structure for Baur musk.

(b) A $C_{18}H_{20}O_2$ compound known as diethylstilbesterol was used for a number of years as a feed additive in the diet of chickens, causing them to become fatter and, presumably, tastier. It has also been used in humans in certain kinds of estrogen therapy and has recently come into prominence as the ingredient in a "morning-after pill," although its possible carcinogenicity discourages its use in this respect. It is soluble in aqueous sodium hydroxide but not in aqueous sodium bicarbonate. Ozonolysis yields a $C_6H_{10}O_2$ compound in almost quantitative amount. The ozonolysis product is soluble in aqueous sodium hydroxide, yields a 2,4-dinitrophenylhydrazone, and fails to react with Tollens' reagent. When stilbesterol or its ozonolysis product is treated with dimethyl sulfate and the resulting compound is then subjected to strenuous oxidation, *p*-methoxybenzoic acid is obtained. What is the structure of diethylstilbesterol?

(c) Eugenol, isolated from the oil of cloves, affords a starting material for a commercial synthesis of vanillin (see p. 355). Imagine that you are the first chemist to isolate eugenol and want to determine its structure, which we now know to be that shown:

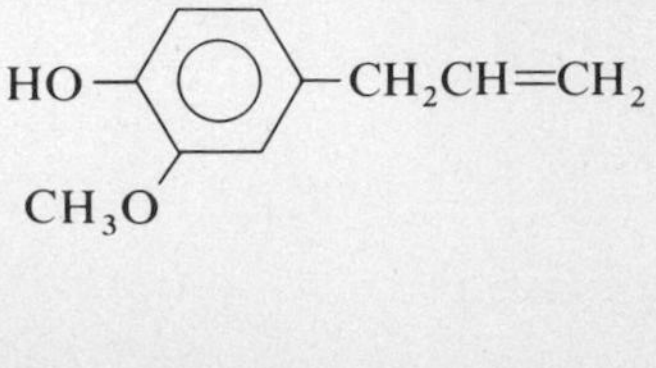

What procedures would you use if you did not have access to spectrophotometers? What procedures would you use if you did have access to spectrophotometers?

10. In each of the following examples, polycyclic aromatic and/or hydroaromatic compounds are involved:

(a) A $C_{13}H_{11}NO_4$ compound is soluble in aqueous sodium bicarbonate but insoluble in hydrochloric acid. When treated with iron and hydrochloric acid, it undergoes conversion to a $C_{13}H_{13}NO_2$ compound that is soluble in hydrochloric acid and loses water upon heating to form a lactam. Oxidation of the $C_{13}H_{13}NO_2$ compound with sodium dichromate and sulfuric acid yields a $C_8H_5NO_6$ compound that loses water upon heating to form a cyclic anhydride. A similar oxidation of the original $C_{13}H_{11}NO_4$ compound yields a $C_{10}H_6O_8$ compound, which loses two equivalents of water upon heating to form a dianhydride. On the basis of these data, suggest structures for the various compounds involved in this sequence of reactions.

(b) A $C_{15}H_{16}O$ compound can be oxidized to phthalic acid and can be dehydrated to 1-methylphenanthrene. When it is ozonolyzed, it yields formaldehyde and a $C_{14}H_{14}O_2$ compound that rapidly exchanges two hydrogen atoms for two deuterium atoms when treated with CH_3OD containing some CH_3ONa. Suggest a structure for the $C_{15}H_{16}O$ compound.

(c) A $C_{19}H_{18}O_2$ compound reacts with sodium to liberate hydrogen, with 2,4-dinitrophenylhydrazine to form a 2,4-dinitrophenylhydrazone, and with Tollens' reagent to form a $C_{19}H_{16}O_3$ carboxylic acid. Its infrared spectrum, however, shows a weaker than usual absorption suggesting the presence of a HO—C—C—C—C=O moiety. Ozonolysis yields ethyl methyl ketone and a $C_{15}H_{10}O_2$ compound that produces 1-methylanthracene on dehydrogenation. Oxidation of the parent compound with sodium dichromate and sulfuric acid yields a mixture of phthalic acid and benzene-1,2,3-tricarboxylic acid. On the basis of this information suggest structures for these compounds.

11. Available starting materials include (a) any alkanes, alkenes, alkynes, alkyl halides, alkanols, alkanals, alkanones, or alkanoic acids containing four or fewer carbon atoms, (b) benzene, toluene, *o*-, *m*-, and *p*-xylene, biphenyl, naphthalene, anthracene, and phenanthrene, (c) special organic reagents, and (d) any inorganic compounds. Starting with materials chosen from this group, outline syntheses for each of the following compounds:

(a) 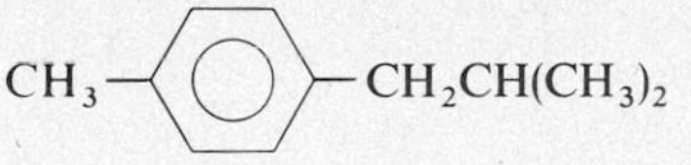 CH_3–(benzene ring)–$CH_2CH(CH_3)_2$

(b) 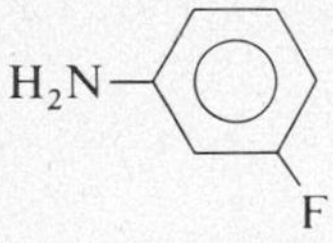 H_2N–(benzene ring, F in meta position)

(c) CH_3CH_2–(benzene ring)–$CH{=}CHCH_3$

(d) 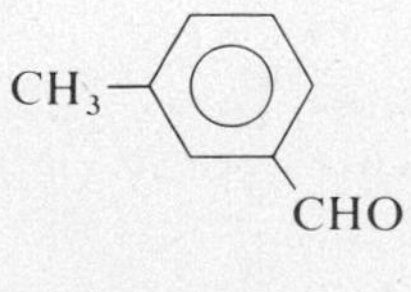 CH_3–(benzene ring, CHO in meta position)

(e) 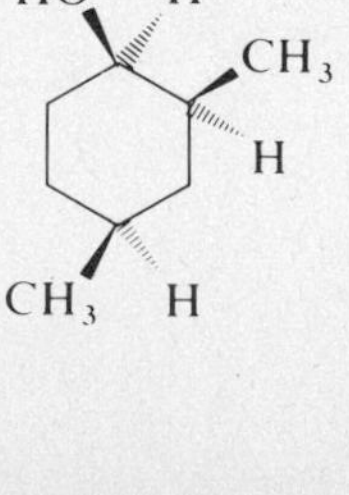 cyclohexane ring bearing HO, H; CH_3, H; CH_3, H

(f) CH_3O–(benzene ring)–NHNH–(benzene ring)–OCH_3

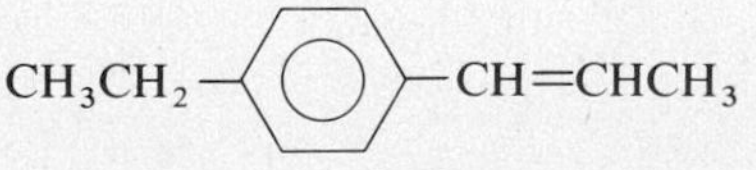

(g) $C_6H_4(CN)CH(CH_3)_2$ (ortho)

(h) 2,6-dibromo-4-chlorophenol: OH, Br, Br, Cl

(i) $C_6H_4(Br)OCH_3$ (meta)

(j) $CH_3\overset{O}{\overset{\|}{C}}-C_6H_4-\overset{CH_3}{\overset{|}{C}}HCH_2CH_3$

(k) CH_3O, CH, CH_3, CH_3

(l) O

(m) NH_2, $NHCOCH_3$

12. Each of the compounds shown possesses some commercial and/or biological interest. Using the same starting materials listed in Problem 11, devise syntheses for these compounds.

(a) CH_3, O_2N, NO_2, $C(CH_3)_3$, NO_2

Baur musk

(b) HO, CH_2CH_3, C=C, CH_3CH_2, OH

Diethylstilbesterol

(c) HO—C_6H_3—CHO, CH_3O

Vanillin

(d) CO_2H, $O\overset{}{C}CH_3$ (C=O)

Aspirin

(e) O, NH, S, O, O

Saccharin

13. Explain the following observations:

(a) Hexadeuteriobenzene undergoes nitration at essentially the same rate as ordinary benzene.

(b) 1,2,4,5-Tetramethylbenzene (durene) reacts with 2 moles of acetyl chloride and aluminum chloride to give a *p*-diacetyl compound, whereas benzene reacts with only 1 mole to give the monoacetyl compound, acetophenone.

(c) N,N-Dimethylaminotoluene undergoes nitration in mildly acidic solution to give 1-dimethylamino-4-methyl-2-nitrobenzene and in strongly acidic solution to give 1-dimethylamino-4-methyl-3-nitrobenzene.

(d) 1,2,4,5-Tetraisopropylbenzene reacts with nitric acid in sulfuric acid to give 1,2,4-triisopropyl-5-nitrobenzene.

(e) 2-Phenylpropene undergoes acid-catalyzed dimerization to the compound shown:

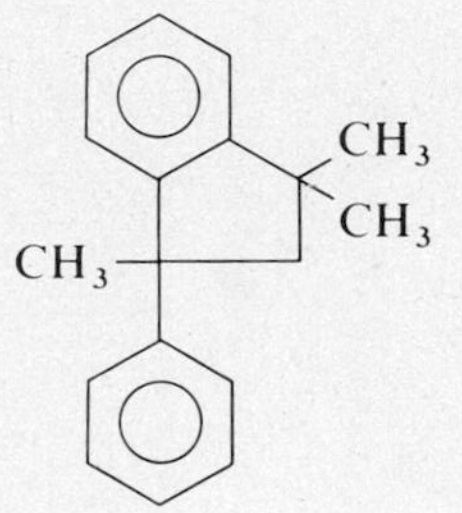

PART III

Natural Products

Part III is concerned with the structures and properties of naturally occurring compounds. Organic chemistry, by its original definition and early history, is the study of life-produced compounds. Despite this, however, very little of the material in Parts I and II has focused on the life-related aspects of organic chemistry, the reason being that the fundamental principles of organic chemistry are more simply and more clearly illustrated by chemical systems that are less complex than those that have been devised by Nature. Naturally occurring compounds tend to be structurally complex; to have used them to illustrate the phenomena discussed in Parts I and II would have tended to obscure rather than clarify the points being made. Now, with the basic principles well in hand and with an understanding of how to apply them to simple systems, we can turn our attention to the marvelously intricate structures that have evolved from the primeval broth (see p. 9) and that play a key role in life on the planet Earth.

Chapter **32**

Carbohydrates: life's connection with the sun

Awake! for morning in the bowl of night
Has flung the stone that puts the stars to flight.
And, lo the Hunter of the East
Has caught the sultan's turret in a noose of light.

OMAR KHAYYAM
The Rubaiyat

32.1 *The Carbon Pathway in Photosynthesis*

Omar Khayyam's Hunter of the East, our sun, maintains all life on Earth—from the simplest virus to the most complex man. All the energy required for life-sustaining processes comes from the sun, directly or indirectly, and an important link in the connection between sun and man is provided by a class of organic compounds called *carbohydrates.* Evolving out of the primeval broth after eons of stewing were complex systems in plants and microorganisms capable of "fixing" carbon dioxide from the atmosphere by incorporating it into higher molecular weight compounds, which have the general formula $C_n(H_2O)_m$ and which, in a literal although chemically misleading sense, are hydrates of carbon—hence, carbohydrates. The overall process by which carbon dioxide fixation occurs, known as photosynthesis, is exceedingly complex, and many of its details are not yet completely understood. One of its most studied aspects is the "carbon pathway of photosynthesis," involving a cyclic set of chemical reactions in which a carbon dioxide molecule is condensed with a five-carbon compound, ribulose-1,5-diphosphate, to form two molecules of phosphoglycerate (Fig. 32.1). The 3-phosphoglycerate (phosphoglyceric acid exists at physiological pH in the carboxylate and phosphate anionic form) is reduced to glyceraldehyde-3-phosphate, which isomerizes, in part, to dihydroxyacetone phosphate. An aldol condensation (see Section 13.6 and Chapter 30) between glyceraldehyde-3-phosphate and dihydroxyacetone phosphate then occurs to yield fructose-1,6-diphosphate. At this point, the system has several alternatives; some of the fructose-1,6-diphosphate is converted to sucrose, some is converted to glucose, and some is converted back to ribulose-1,5-diphosphate and remains in the carbon cycle of photosynthesis. Every time one cycle is completed, one molecule of carbon dioxide is "fixed"; thus, to produce one molecule of the six-carbon compound glucose, six cycles are required.

All the compounds in the carbon pathway of photosynthesis are more highly reduced than carbon dioxide and, therefore, are richer in energy. Thus, the photosynthesis reaction,

$$6\,CO_2 \;+\; 6\,H_2O \;\longrightarrow\; C_6H_{12}O_6 \;+\; 6\,O_2$$

is endothermic and is *not* a spontaneous reaction. It requires energy to drive it, and it is the sun that provides this energy. Employing a complex mechanism that involves, at one point, the light-absorbing compound known as chlorophyll, the living plant uses the energy from the sun to "split" water; the products are molecular oxygen and, in a formal sense, hydride anions,

$$H_2O \;+\; 2\,e^- \;\longrightarrow\; \tfrac{1}{2}O_2 \;+\; 2\,H^-$$

the latter being incorporated into complex organic compounds capable of carrying out various reduction processes in the living system. In this fashion, the electromagnetic energy of the sunlight is converted into the energy of the chemical bond.

The ability to carry on photosynthesis is limited to plants and certain bacteria. Animals, including man, are incapable of directly utilizing the sun's energy and must, instead, acquire life-sustaining energy by ingesting compounds that can be oxidized (metabolized) in exothermic reactions. Animals, by biological inheritance, are obligatory predators that require plants for their survival. An ecological balance between plants and animals has, of course, been achieved; plants convert carbon dioxide and water to molecular oxygen and to compounds of carbon, which animals metabolize back to carbon dioxide and water. The cycle is not 100% efficient, however, and life processes on Earth are kept going because of the constant input of radiant energy from the sun. Without it, life could never have emerged, and without it life could not be sustained.

32.2 The Structure of the Aldehyde Form of Glucose

Most of the compounds in the carbon pathway of photosynthesis (Fig. 32.1) belong to the carbohydrate family. One of the most important of these from the historical as well as the biochemical standpoint is glucose, which will be discussed in detail to illustrate the structural and chemical aspects of the carbohydrates in general. Glucose occurs in two crystalline forms, designated as α and β. The α form melts at 146° and has a specific rotation of +112°, and the β form melts at 150° and has a specific rotation of +18.7°. Both forms have the molecular formula $C_6H_{12}O_6$. Since they are readily interconvertible in solution, they *appear* to react in an entirely equivalent fashion. For the moment, therefore, we will not distinguish between the two forms; the α and β prefixes are omitted in the first portion of the following discussion.

A molecular formula of $C_6H_{12}O_6$ indicates the presence of one degree of unsaturation (see p. 474), which can be present as a carbocyclic ring, a heterocyclic ring, a C=C bond, or a C=O bond. The carbon skeleton of glucose can be

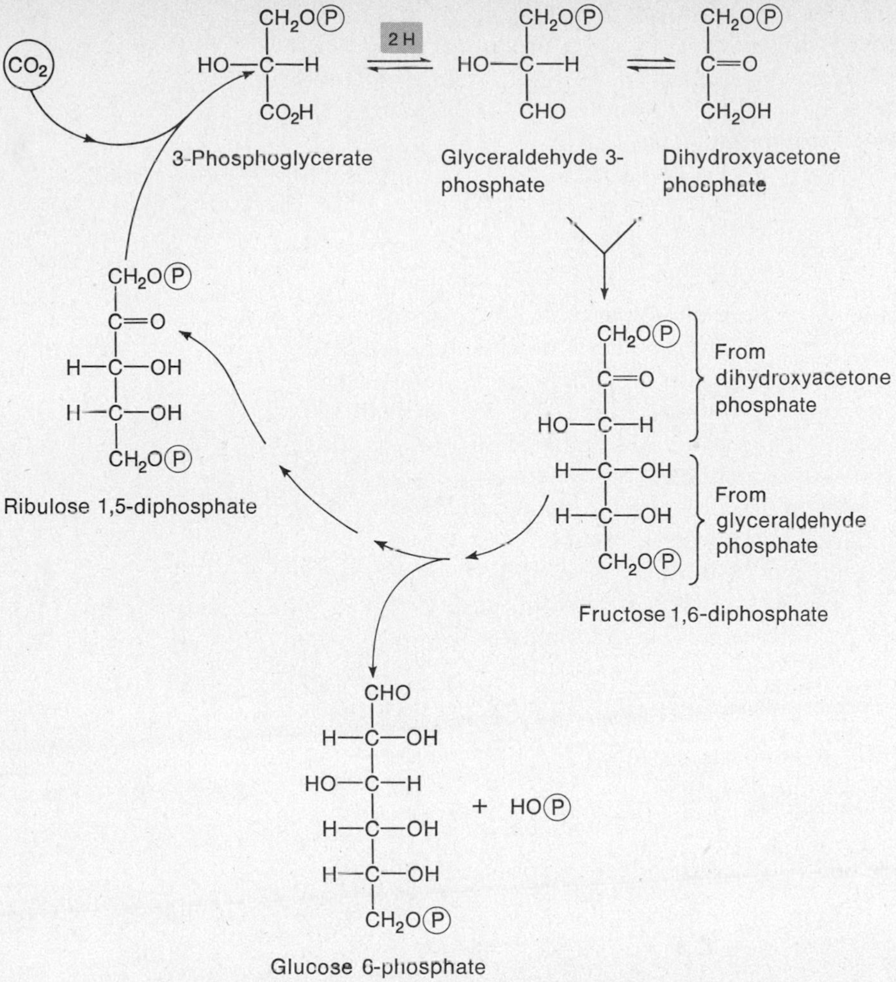

Fig. 32.1. The carbon pathway of photosynthesis Ⓟ stands for $-\overset{\displaystyle O}{\overset{\|}{\underset{\displaystyle OH}{\underset{|}{P}}}}-OH$.

Under physiological conditions, the phosphoric acid group is ionized, existing in a ratio of approximately 1 : 1 of RPO_3H^- and $RPO_3^=$.

inferred from the formation of *n*-hexane when glucose is treated with the strong reducing mixture consisting of hydrogen iodide and phosphorus. The presence of a carbonyl group is indicated by the formation of a mono-oxime and a mono-cyanohydrin when glucose is treated with hydroxylamine and hydrogen cyanide, respectively. That the carbonyl group is present as an aldehyde function is indicated by the oxidation of glucose (with bromine water) to a carboxylic acid possessing the molecular formula $C_6H_{12}O_7$. Reduction of glucose (*e.g.*, with sodium amalgam or sodium borohydride) yields a $C_6H_{14}O_6$ compound (containing no degrees of unsaturation), which reacts with acetic anhydride to form a hexa-acetate, indicative of a hexahydroxy compound. Although a few compounds are known in which two OH groups are attached to the same carbon, they are usually unstable and undergo dehydration to the corresponding carbonyl compound. It is a reasonable inference, therefore, that the $C_6H_{14}O_6$ compound is 1,2,3,4,5,6-hexahydroxyhexane (known as glucitol) and that glucose is the corresponding 2,3,4,5,6-pentahydroxyhexanal, *i.e.*,

$$\begin{array}{c} HC{=}O \\ | \\ (CHOH)_4 \\ | \\ CH_2OH \end{array}$$

By means of the chemistry just adumbrated, the basic structure of glucose was elucidated about 100 years ago. The more subtle details of its structure, including its stereochemistry and the existence of α and β forms, remained the problem of the late nineteenth and early twentieth centuries. The task of determining the stereochemical details of glucose was initiated in 1888 by Emil Fischer, and the result of this effort represents, even by modern standards, one of the most elegant contributions in the literature of organic chemistry. It was partly for these achievements that Fischer received the Nobel Prize in 1902.

The compound 2,3,4,5,6-pentahydroxyhexanal contains four chiral carbon atoms (at C-2, C-3, C-4, and C-5) and has the possibility of existing in $2^4 = 16$ stereoisomers, represented by the Fischer projection formulas in Fig. 32.2. This

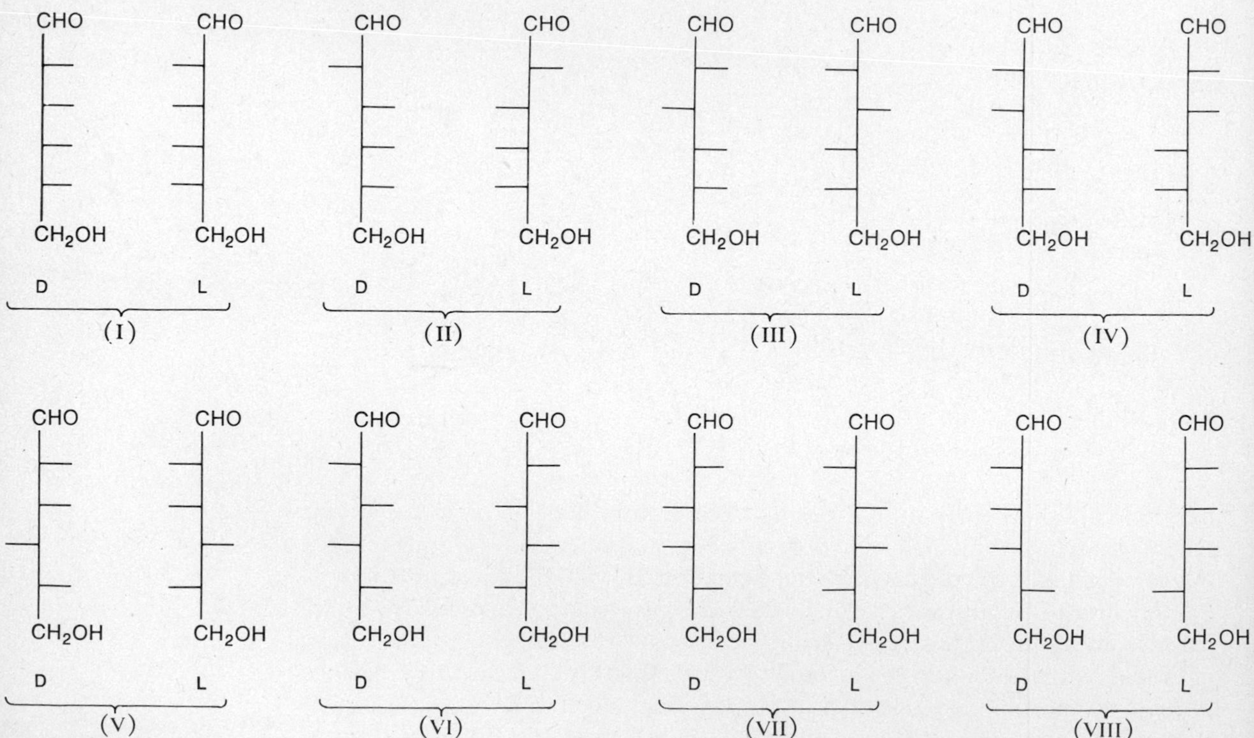

Fig. 32.2. Fischer projection representations of 2,3,4,5-6-pentahydroxyhexanal. (Horizontal dashes correspond to OH groups.)

set contains eight *structurally different* sugars, each of which exists as a pair of enantiomers. It was Fischer's arbitrary choice (see Section 19.4) to designate those sugars having the *same configuration as* (+)-*glyceraldehyde* at their highest numbered chiral carbon atom as belonging to the D series, and those similarly related to (−)-glyceraldehyde as belonging to the L series. Naturally occurring glucose is stereochemically related to (+)-glyceraldehyde; therefore, it is designated as D-glucose and is represented in the Fischer projection with the hydroxyl group at C-5 projecting to the right.

D-(+)-Glucose can be prepared from the five-carbon sugar D-(−)-arabinose by a chain-lengthening reaction (*Kiliani–Fischer synthesis*), which also yields the isomeric D-(+)-mannose. The formation of D-glucose and D-mannose from D-arabinose indicates that (1) the configurations at C-2, C-3, and C-4 of arabinose are *identical* with the configurations at C-3, C-4, and C-5 of both glucose and mannose, and (2) the configuration at C-2 of glucose is *opposite* to that of mannose; *i.e.*, the compounds are epimeric at C-2 (Fig. 32.3). D-Arabinose, D-glucose, and

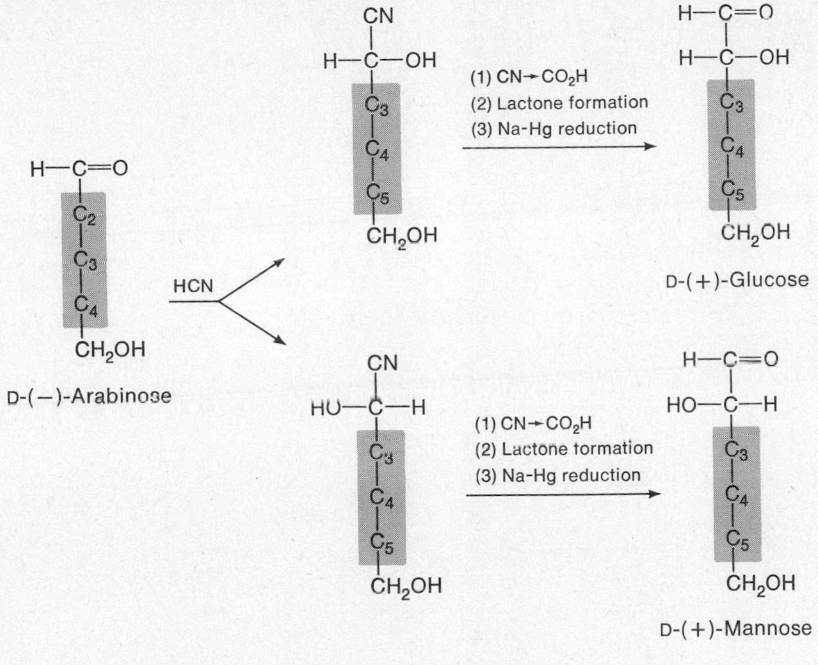

Fig. 32.3. Conversion of D-arabinose to D-glucose and D-mannose.

D-mannose all yield optically active dicarboxylic acids when both of the terminal carbons are oxidized. From this we can deduce that (1) the 2,3,4-trihydroxyglutaric acid from arabinose must have structure (c) or (d) in Fig. 32.4, for structures (a) and (b) both have planes of symmetry and correspond to optically inactive *meso* forms, and (2) the 2,3,4,5-tetrahydroxyadipic acids from glucose and mannose must have structures (g) and (h) in Fig. 32.4 by relation to arabinose. Structure (e) is excluded because it is a *meso* form, and structure (f) is excluded because glucose and mannose are epimers at C-2 and have identical configurations

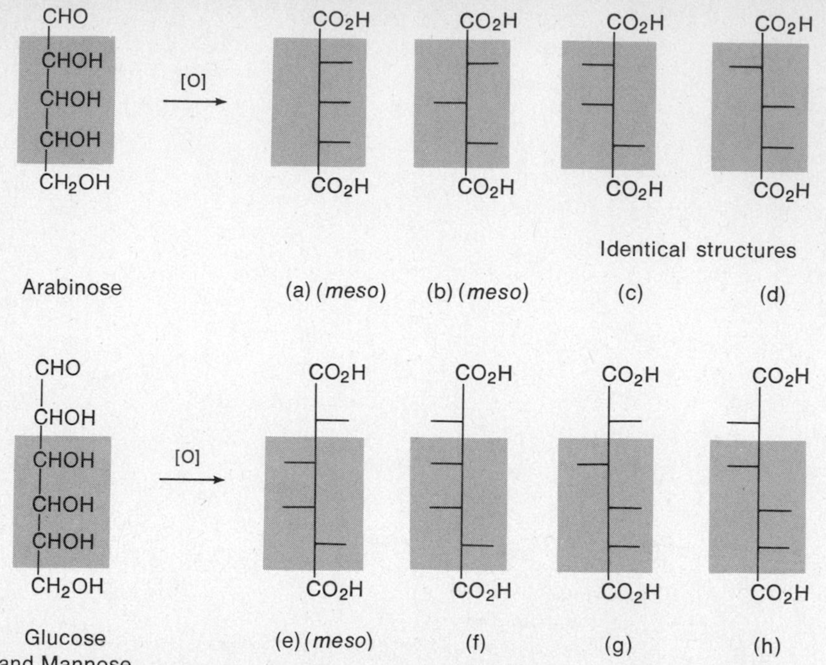

Fig. 32.4. Dicarboxylic acids from arabinose, glucose, and mannose.

at C-3, C-4, and C-5. Glucose and mannose, therefore, must be represented by structures corresponding to the dicarboxylic acids (g) and (h) (*viz.*, D-III and D-IV in Fig. 32.2), and arabinose must be represented by the structure corresponding to the dicarboxylic acid (d) in Fig. 32.4.*

CHO CHO

CH_2OH CH_2OH

D-III D-IV

The choice between D-III and D-IV for the proper representation for D-glucose was made by Fischer on the basis of an interesting relationship that exists between the dicarboxylic acids that correspond to the D- and L-series of the six-carbon aldehyde sugars. Because of the particular way in which the chiral centers are

* The dicarboxylic acids (c) and (d) are indistinguishable, because the two terminal groups are identical. In arabinose, however, the two termini of the molecule are not identical, so a compound with either of the configurations shown at the right would give the dicarboxylic acid (c) [which is identical with (d)].

CHO → (C) (D) ← CHO

CH_2OH Same dicarboxylic acid CH_2OH

Since the nonidentity of the termini is preserved when arabinose is converted to glucose and mannose (*i.e.*, the sixth carbon added at the CHO terminus, not the CH_2OH terminus), both these configurational sets must, *a priori*, be taken into account as possibilities for glucose and mannose.

arranged in D-III and L-V, it turns out that they both yield the *same* optically active dicarboxylic acid (Fig. 32.5). The dicarboxylic acid derived from D-IV,

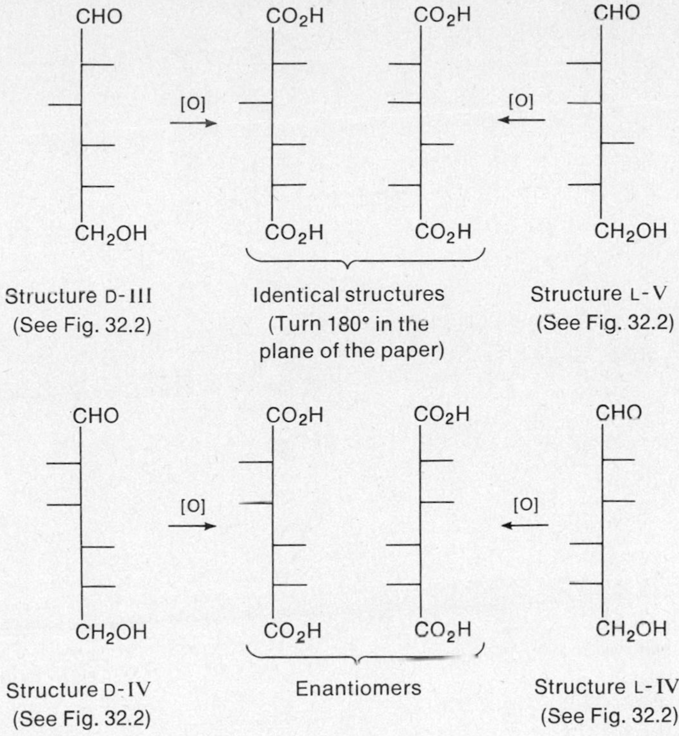

Fig. 32.5. Stereochemical relationship between dicarboxylic acids from D-glucose, D-mannose, L-gulose, and L-mannose.

however, is *not* related in this fashion to any of the dicarboxylic acids from the other fifteen compounds shown in Fig. 32.2. Thus, Fischer reasoned, if one of the compounds of the glucose-mannose pair can be oxidized to a dicarboxylic acid identical with that obtained from one of the other compounds in the series shown in Fig. 32.2, it *must* have the configuration corresponding to D-III.* Verification of his reasoning was subsequently provided when identical optically active dicarboxylic acids were obtained from D-glucose and the sugar known as L-gulose, thereby establishing D-glucose as having the structure D-III and D-mannose, by difference, as having structure D-IV. Even with the advantage of more than three-quarters of a century of chemical and stereochemical reasoning to aid us, Fischer's argument remains a bit elusive in its subtlety and striking in its elegance. In 1891 it was an accomplishment of genius.

---

* Other pairs in the series of Fig. 32.2 also give identical dicarboxylic acids. Thus, D-I and L-I yield the same optically *inactive* dicarboxylic acid, as do D-VII and L-VII. And, of course, L-III and D-V yield the identical dicarboxylic acid; *i.e.*, the enantiomer of the one obtained from D-III and L-V.

32.3 *The Structure of the Cyclic Forms of Glucose*

It was stated at the beginning of Section 32.2 that crystalline glucose exists in α and β forms, which possess different physical properties (*e.g.*, different melting points, specific rotations, etc.). The question arises as to how these differences can be accommodated to a structure for glucose, for there is nothing in the aldehyde structure (compound III of Fig. 32.2) that would appear to supply the answer. One clue is provided by the ir and nmr spectra of glucose; surprisingly, there is no indication of the presence of an aldehyde group, for the expected carbonyl stretching band at about 1720 cm^{-1} in the ir and the expected nmr resonance at δ 9–10 are both missing. Thus, although glucose, in forming an oxime and a cyanohydrin, *behaves like an aldehyde*, it shows no spectral evidence of an aldehyde group. Another clue comes from a consideration of one of the general reactions of aldehydes, *viz.*, their tendency to add alcohols to form hemiacetals (see Figs. 13.14 and 29.10). In the case of glucose, the two functions for this reaction are located *within the same molecule*, and an intramolecular reaction can take place. The result is the formation of a cyclic hemiacetal, which represents the most stable structure for the glucose molecule (Fig. 32.6).

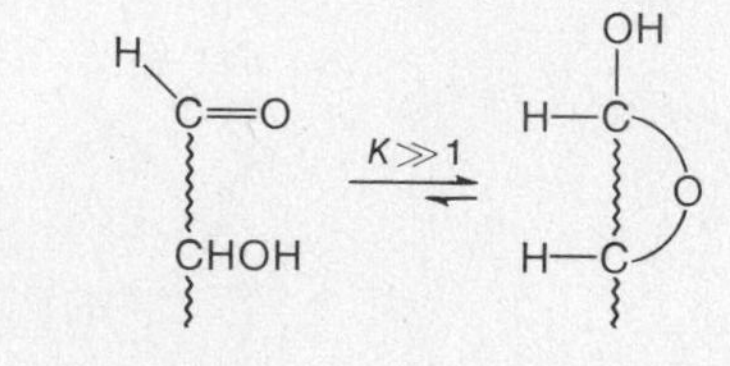

Aldehyde form ($K \gg 1$) Cyclic hemiacetal form

Fig. 32.6. Open-chain and hemiacetal structures.

The cyclic hemiacetal form of glucose is in equilibrium with the aldehyde form, and although the equilibrium favors the hemiacetal (K_{eq} for hemiacetal formation is greater than 200), sufficiently strong nucleophiles can react with the aldehyde form to yield "carbonyl products." Thus, both hydroxylamine and cyanide ion are much more effective nucleophiles than the "internal" hydroxyl groups of the glucose molecule and are able to shift the equilibrium in favor of the oxime or cyanohydrin, respectively (Fig. 32.7).

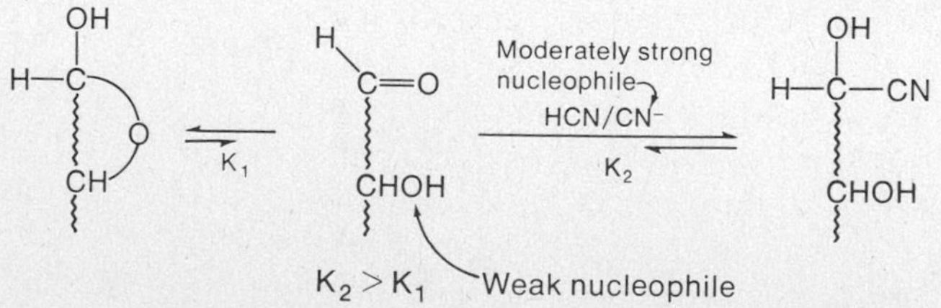

Moderately strong nucleophile HCN/CN^- (K_2); Weak nucleophile (K_1); $K_2 > K_1$

Fig. 32.7. Formation of carbonyl derivatives from glucose.

The aldehyde form of glucose contains five hydroxyl groups, and the next question is which of these is most likely to add to the aldehyde function—*i.e.*, how large is the heterocyclic ring in the hemiacetal structure?

We answer this question by converting the four hydroxyl groups (not including the hemiacetal hydroxyl) to ether functions and then subjecting the tetramethoxy compound to nitric acid oxidation, the vulnerable points of the molecule being the terminal aldehyde group and the carbon bearing the unmethylated hydroxyl group (*i.e.*, the one involved in the hemiacetal formation). Under these conditions, the aldehyde group is converted to a carboxyl group *without* C—C bond cleavage, whereas the CHOH group is converted to a carboxyl group *with* C—C bond cleavage. By determining the length of the carbon chain in the derived dicarboxylic acid, we can determine the position of the $\rangle$CHOH group, from which we can infer the size of the ring in the hemiacetal (Fig. 32.8). In this manner, it

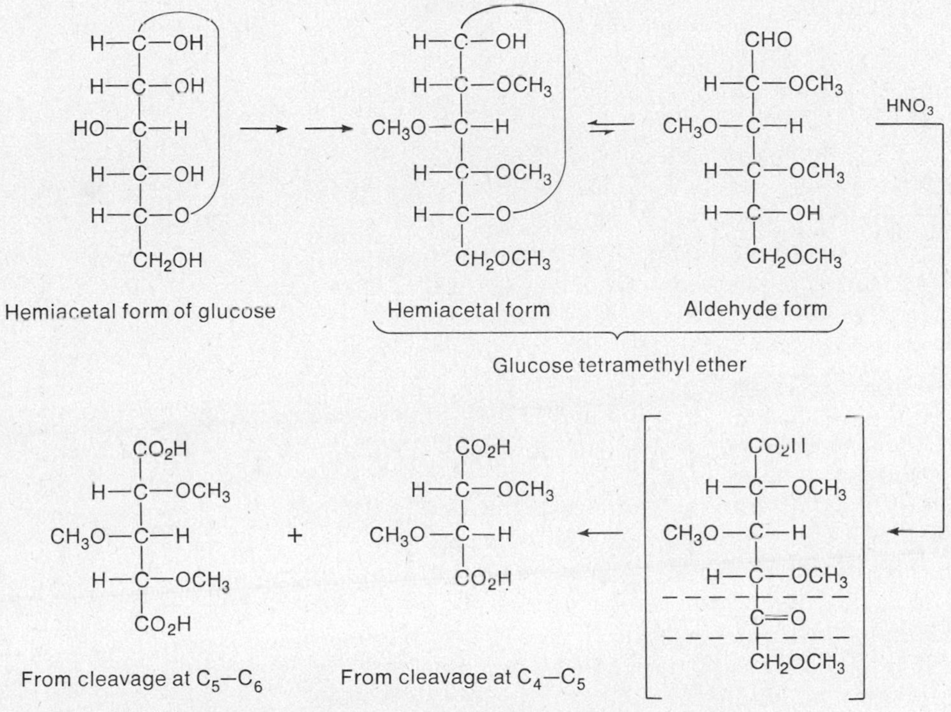

Hemiacetal form of glucose: H—C—OH / H—C—OH / HO—C—H / H—C—OH / H—C—O (ring) / CH_2OH → → Hemiacetal form: H—C—OH / H—C—OCH_3 / CH_3O—C—H / H—C—OCH_3 / H—C—O (ring) / CH_2OCH_3 ⇌ Aldehyde form: CHO / H—C—OCH_3 / CH_3O—C—H / H—C—OCH_3 / H—C—OH / CH_2OCH_3

Hemiacetal form ⏟ Aldehyde form: Glucose tetramethyl ether

HNO_3 → [CO_2H / H—C—OCH_3 / CH_3O—C—H / H—C—OCH_3 / - - - C=O - - - / CH_2OCH_3]

→ CO_2H / H—C—OCH_3 / CH_3O—C—H / H—C—OCH_3 / CO_2H (From cleavage at C_5—C_6) + CO_2H / H—C—OCH_3 / CH_3O—C—H / CO_2H (From cleavage at C_4—C_5)

Fig. 32.8. Determination of ring size of hemiacetal form of glucose.

was established that the most stable forms of glucose are characterized by the presence of a cyclic six-membered hemiacetal—*i.e.*, the hydroxyl group at C-5 is the one that preferentially interacts with the aldehyde carbonyl:

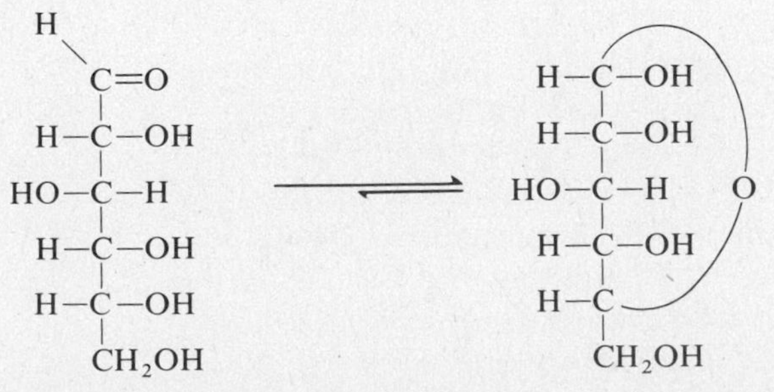

H—C=O / H—C—OH / HO—C—H / H—C—OH / H—C—OH / CH_2OH ⇌ H—C—OH / H—C—OH / HO—C—H / H—C—OH / H—C (ring O) / CH_2OH

If we use a pseudo-three-dimensional *planar ring* representation (a *Haworth projection*) instead of the Fischer projection for the hemiacetal form of glucose, an explanation for the existence of the α- and β-forms becomes immediately apparent (Fig. 32.9). The "new" hydroxyl group in the cyclic hemiacetal form

Fig. 32.9. Haworth projection representations of the α- and β-forms of D-glucose.

can be either below or above the plane of the ring. The two forms are epimers, having identical configurations at C-2, C-3, C-4, and C-5, and opposite configurations only at C-1; they correspond to the α- and β-forms, respectively. In the sugar series, this special case of epimer formation at C-1 is called *anomer* formation; the α- and β-forms of glucose are anomers of each other, and C-1 is designated as the *anomeric carbon.* Remember that epimers are diastereomers and, therefore, differ in chemical and physical properties, as the α- and β-forms of glucose indeed do.

When the α-form of D-glucose is dissolved in water, a specific rotation of +112° is measured. This value diminishes with time, however, eventually reaching a constant value of +52.7°. When the β-form of D-glucose is dissolved in water, a specific rotation of +18.7° is measured, but this increases with time, eventually reaching the constant value of +52.7°. In solution, the two forms interconvert and reach an equilibrium in which about 36% of the α-form and 64% of the β-form are present. This phenomenon, involving a change in rotation from the originally observed value to the equilibrium value, is called *mutarotation.* In addition to the α- and β-forms, which are the major contributors to the equilibrium mixture, other forms are also present as the result of cyclic hemiacetal formation involving hydroxyl groups at other carbons. For example, the anomers of the five-membered cyclic hemiacetal are present in small but finite amounts (Fig. 32.10).

The name *glucose* specifies the particular diastereomer of 2,3,4,5,6-pentahydroxyhexanal that has the relative configurations at C-2, C-3, C-4, and C-5 shown by enantiomeric structures III in Fig. 32.2; the prefix D specifies that it is configurationally related at C-5 to D-(+)-glyceraldehyde. An additional convention, however, is required to denote the fact of hemiacetal formation, and this is provided by adding the suffix "pyranose" (from the six-membered oxygen heterocycle *pyran*) for the six-membered hemiacetal and "furanose" (from the five-membered oxygen heterocycle *furan*) for the five-membered hemiacetal. The α- and β-designations then define the fifth chiral center at C-1. Thus, the cyclic structures in Fig. 32.10 acquire the names α-D-glucopyranose, β-D-glucopyranose, α-D-glucofuranose, and β-D-glucofuranose.

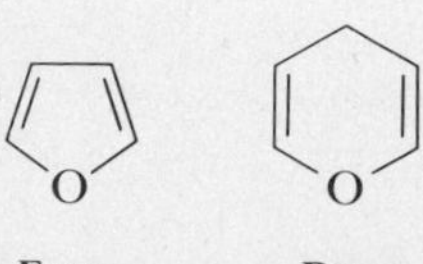

Furan **Pyran**

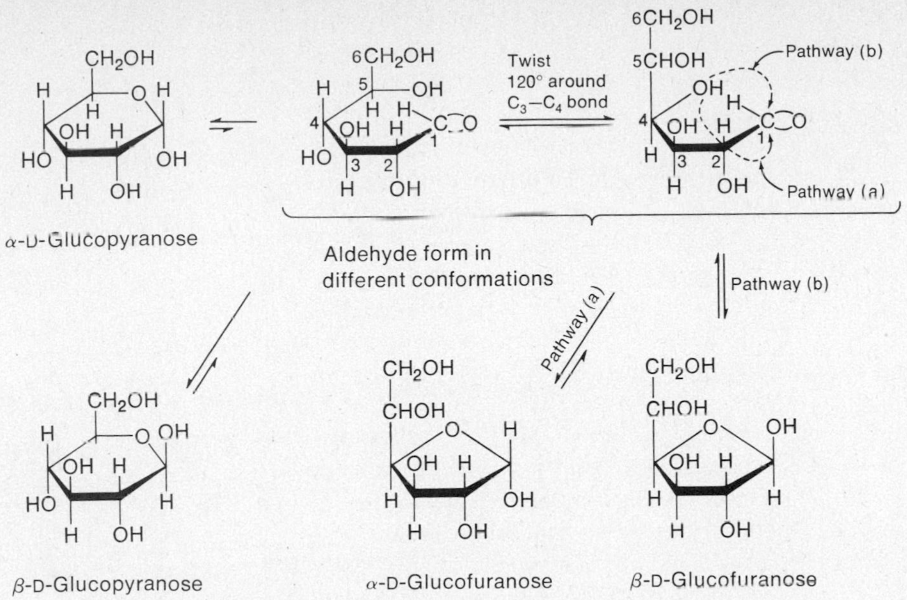

α-D-Glucopyranose

Aldehyde form in different conformations

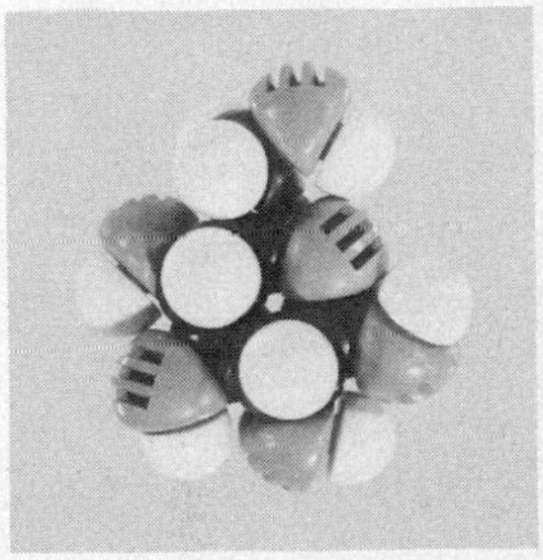

β-D-Glucopyranose α-D-Glucofuranose β-D-Glucofuranose

Fig. 32.10. Haworth projections of five- and six-membered hemiacetal and open-chain forms of D-(+)-glucose in aqueous solution.

β-D-Glucopyranose

The Haworth projections, although useful, do not accurately depict the actual three-dimensional architecture of glucose, particularly in the case of the pyranose forms. Nonaromatic-membered rings do *not* exist in a planar form (see Section 3.9) but in *nonplanar conformations*, of which the chair form is generally the most stable. Thus, real molecules of α-D-glucopyranose and β-D-glucopyranose actually have the conformations depicted in Fig. 32.11.

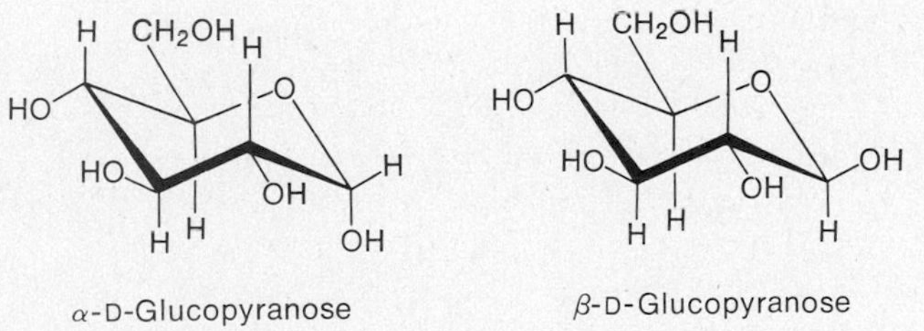

α-D-Glucopyranose β-D-Glucopyranose

Fig. 32.11. Conformational representations of α- and β-D-glucopyranose.

32.4 The Structures of Hexoses

The term *carbohydrate* applies generally to polyhydroxyaldehydes or ketones in which the ratio of hydrogen to oxygen is approximately 2 to 1. They are also referred to as *sugars*, as *saccharides* (Latin, *saccharum*, sugar), or as *glycoses*. In the glycose terminology, the specific sugar can be denoted by replacing the generic prefix "glyc" with another prefix while retaining the suffix "ose"—*i.e.*, *gluc*ose, *arabin*ose, etc. Glucose, containing six carbons in the carbon framework, is classed as a hexose; arabinose, containing five carbons in the carbon framework, is classed as a pentose: etc. Both glucose and arabinose are sugars in which an aldehyde function is present, so they can be classed as aldoses (*i.e.*, *aldehyde* + *ose* for sugar) and designated as aldohexoses and aldopentoses, respectively. Other types of sugars contain ketone rather than aldehyde functions and are classed as ketoses—*e.g.*, ketohexoses, ketopentoses, etc.

Glucose is but one of the eight possible enantiomeric pairs of 2,3,4,5,6-pentahydroxyhexanal (Fig. 32.2). All the members of the other seven enantiomeric pairs are diastereomeric with D- and L-glucose and possess different physical and chemical properties. Each one is a *structurally different* compound, and this difference is specified by the assignment of a particular name. Thus, in Fig. 32.2 compound (I) is allose, compound (II) is altrose, compound (III) is glucose, compound (IV) is mannose, compound (V) is gulose, compound (VI) is idose, compound (VII) is galactose, and compound (VIII) is talose. In each instance, a pair of enantiomers exists, and this is specified in the usual fashion by the prefix (+) or (−), or in the absolute configurational nomenclature by D or L. Remember, though, that the (+) and (−) designations derive from experimental measurements and bear no direct relationship to the D and L specifications.

Allose, altrose, mannose, gulose, idose, galactose, and talose share with glucose the ability to form cyclic hemiacetals, and all these sugars exist predominantly in the hemiacetal form both in solution and in the crystalline state. The *pyranose form* for the D configuration of each of these aldohexoses is illustrated in Fig. 32.12, where the β-anomer is the one in which the hydroxyl group at C-1 is

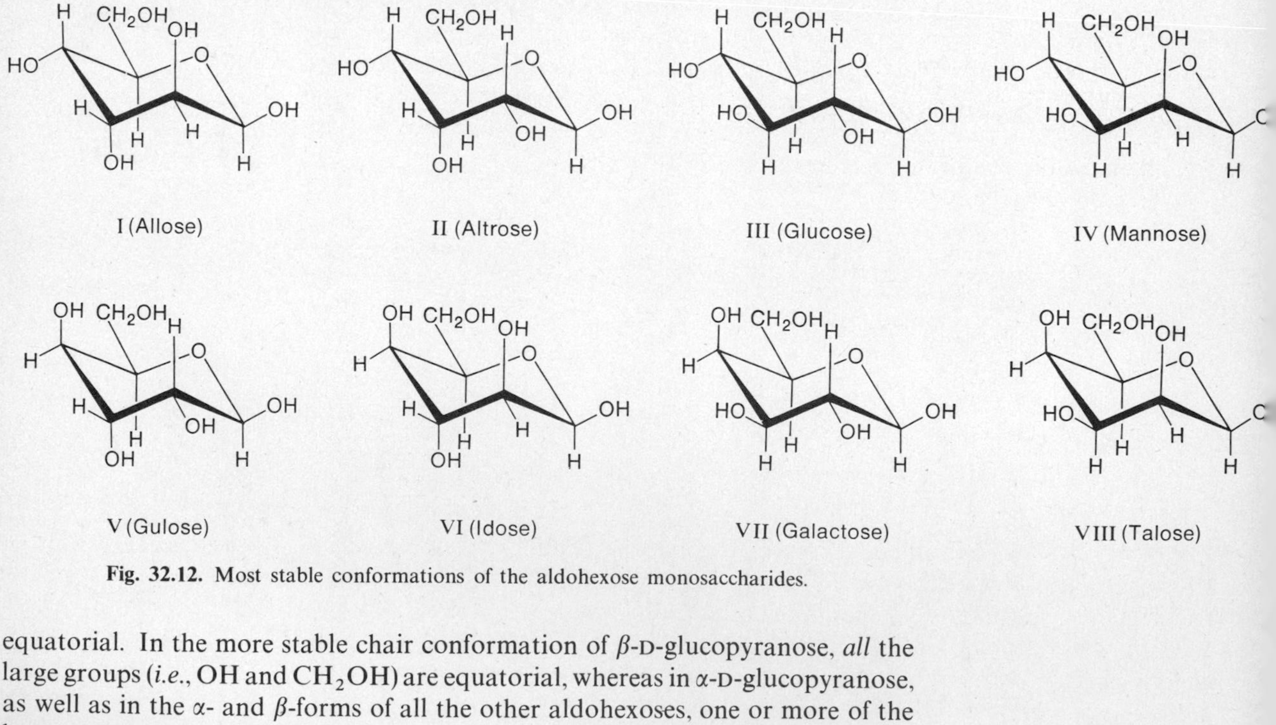

I (Allose) | II (Altrose) | III (Glucose) | IV (Mannose)

V (Gulose) | VI (Idose) | VII (Galactose) | VIII (Talose)

Fig. 32.12. Most stable conformations of the aldohexose monosaccharides.

equatorial. In the more stable chair conformation of β-D-glucopyranose, *all* the large groups (*i.e.*, OH and CH_2OH) are equatorial, whereas in α-D-glucopyranose, as well as in the α- and β-forms of all the other aldohexoses, one or more of the large groups is forced into the axial alignment. This may account for the fact that glucose is the most prevalent of the naturally occurring sugars.

The terminal carbon atoms of a pentahydroxyhexane are but two of the six possible sites for a carbonyl function. If the carbonyl group is located at C-2 or C-3 (note that C-4 ≡ C-3 and C-5 ≡ C-2) rather than at C-1, a class of sugars designated as *ketohexoses* results. Of these, the best known is the 2-ketohexose D-(−)-fructose, which has the interesting distinction of having the sweetest taste of all the sugars. Fructose crystallizes in a single form, which possesses a pyranose

$$\begin{array}{c} CH_2OH \\ | \\ C{=}O \\ | \\ HO{-}C{-}H \\ | \\ H{-}C{-}OH \\ | \\ H{-}C{-}OH \\ | \\ CH_2OH \end{array}$$

D-(−)-Fructose

structure, although in solution it forms an equilibrium mixture containing the α- and β-anomers of the pyranose form, the α- and β-anomers of the furanose form, and the open chain keto form (Fig. 32.13).

β-D-Fructopyranose conformers

α-D-Fructopyranose conformers

β-D-Fructofuranose

α-D-Fructofuranose

β-D-Fructofuranose

Fig. 32.13. Five- and six-membered hemiketal forms of D-(−)-fructose.

32.5 *The Structures of Other Monosaccharides*

The term "sugar" or "saccharide" refers to any member of the carbohydrate series regardless of its complexity. Saccharide units specify the simplest aldose or ketose moieties present in a carbohydrate molecule. Thus, glucose, fructose, and other hexoses contain a single six-carbon framework and are classed as *monosaccharides.* Some sugars, however, contain two or more saccharide units joined through ether linkages (acetal linkages), and these compounds are designated as *disaccharides, trisaccharides, tetrasaccharides,* etc., on up to high molecular weight compounds known as *polysaccharides.*

Monosaccharides containing fewer than six carbons and more than six carbons occur naturally. The simplest compound that is ordinarily considered to be a member of the saccharide family is glyceraldehyde, and it provides the stereochemical reference point for establishing the configurational relationships in the aldose family of saccharides. From glyceraldehyde a pair of four-carbon monosaccharides, known as erythrose and threose, can be obtained; from each

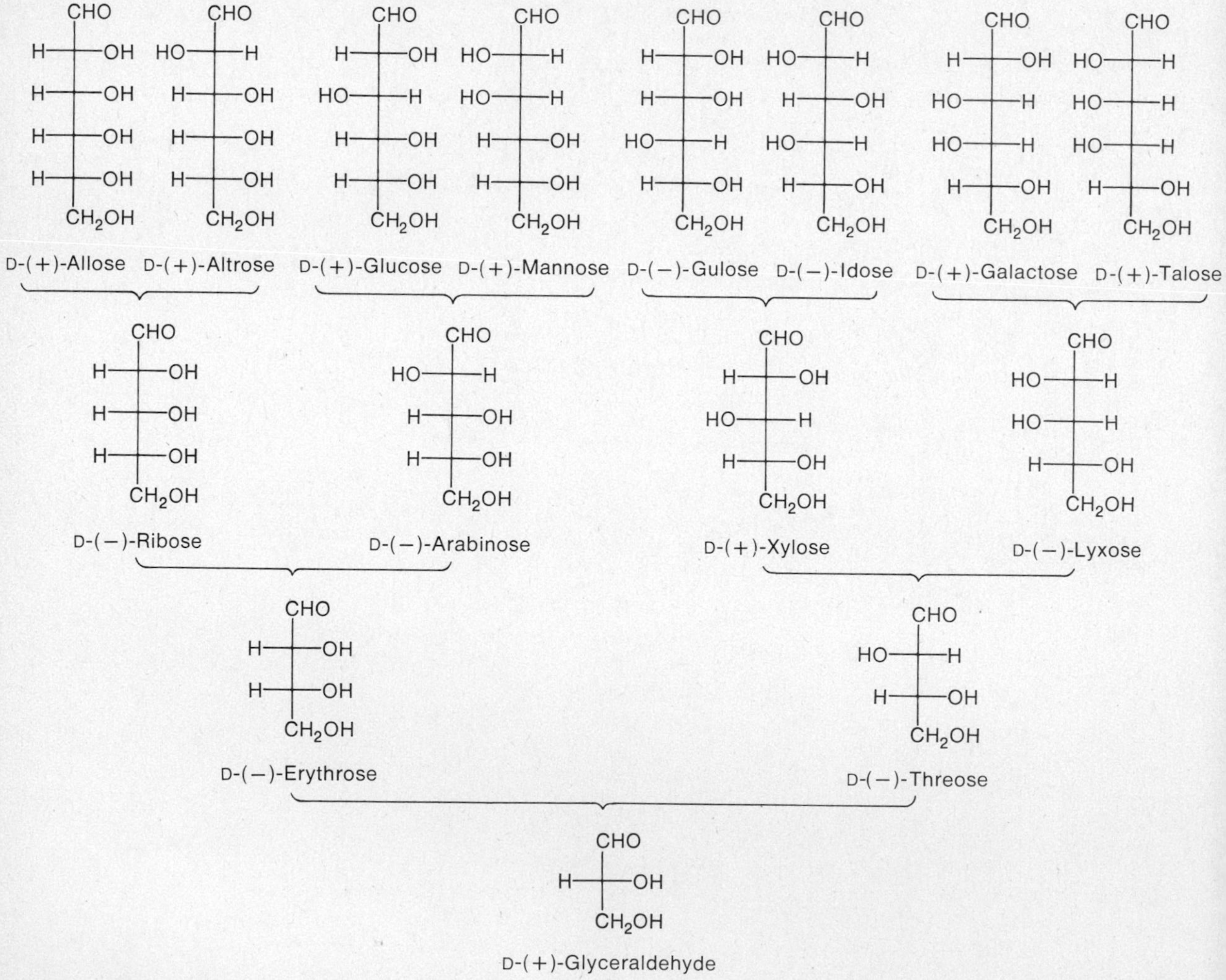

D-(+)-Allose D-(+)-Altrose D-(+)-Glucose D-(+)-Mannose D-(−)-Gulose D-(−)-Idose D-(+)-Galactose D-(+)-Talose

D-(−)-Ribose D-(−)-Arabinose D-(+)-Xylose D-(−)-Lyxose

D-(−)-Erythrose D-(−)-Threose

D-(+)-Glyceraldehyde

Fig. 32.14. Stereochemical relationships in the aldose monosaccharides.

of these, in turn, a pair of five-carbon saccharides can be generated (*i.e.*, ribose and arabinose from erythrose, and xylose and lyxose from threose); and from each of the aldopentoses a pair of aldohexoses can be formed (Fig. 32.14). The sequence need not, of course, stop at the aldohexoses, but can, theoretically, be carried on indefinitely to produce aldoses containing seven carbons, eight carbons, nine carbons, etc. Only a few of these higher monosaccharides are known, however.

Ketoses contain one less chiral carbon than the corresponding aldoses, so the number of stereoisomers is reduced by a factor of 2. Thus, the parent triose, dihydroxyacetone, has no stereoisomers; the only tetrose is erythrulose; the two pentoses are ribulose and xylulose, and the four hexoses are psicose, fructose, sorbose, and tagatose (Fig. 32.15).

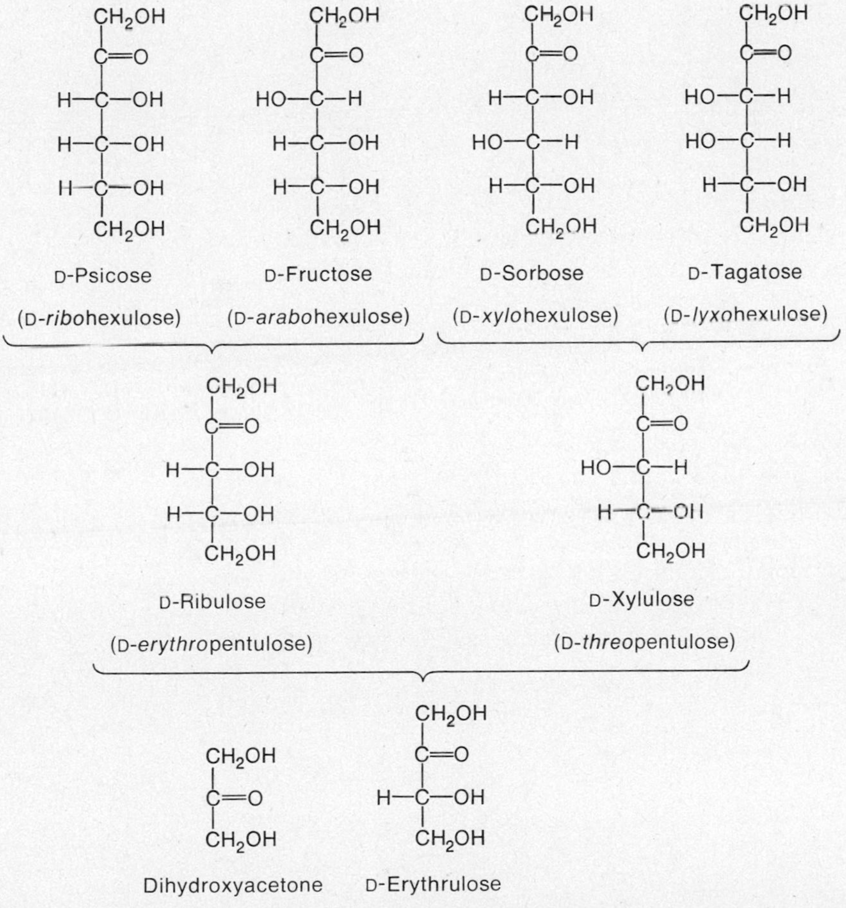

D-Psicose (D-*ribo*hexulose): CH_2OH, C=O, H—C—OH, H—C—OH, H—C—OH, CH_2OH

D-Fructose (D-*arabo*hexulose): CH_2OH, C=O, HO—C—H, H—C—OH, H—C—OH, CH_2OH

D-Sorbose (D-*xylo*hexulose): CH_2OH, C=O, H—C—OH, HO—C—H, H—C—OH, CH_2OH

D-Tagatose (D-*lyxo*hexulose): CH_2OH, C=O, HO—C—H, HO—C—H, H—C—OH, CH_2OH

D-Ribulose (D-*erythro*pentulose): CH_2OH, C=O, H—C—OH, H—C—OH, CH_2OH

D-Xylulose (D-*threo*pentulose): CH_2OH, C=O, HO—C—H, H—C—OH, CH_2OH

Dihydroxyacetone: CH_2OH, C=O, CH_2OH

D-Erythrulose: CH_2OH, C=O, H—C—OH, CH_2OH

Fig. 32.15. Stereochemical relationships in the ketose monosaccharides.

Several naturally occurring monosaccharides have hydrogen-to-oxygen ratios somewhat greater than 2:1 and are known as deoxy sugars.* Examples of

* The prefix "deoxy" indicates that the compound contains one less oxygen atom than the reference or parent compound. For example, $C_6H_5CH_2COC_6H_5$ is often called "deoxybenzoin," benzoin being $C_6H_5CHOHCOC_6H_5$.

$C_6H_{12}O_5$ sugars are L-(+)-rhamnose, which is the 6-deoxy analog of L-mannose, and L-(−)-fucose, which is the 6-deoxy analog of L-galactose. An especially important deoxy sugar is the five-carbon compound 2-deoxy-D-ribose (Fig. 32.16), which is one of the building blocks of certain nucleosides, nucleotides, and nucleic acids (*e.g.*, deoxyribonucleic acid, DNA).

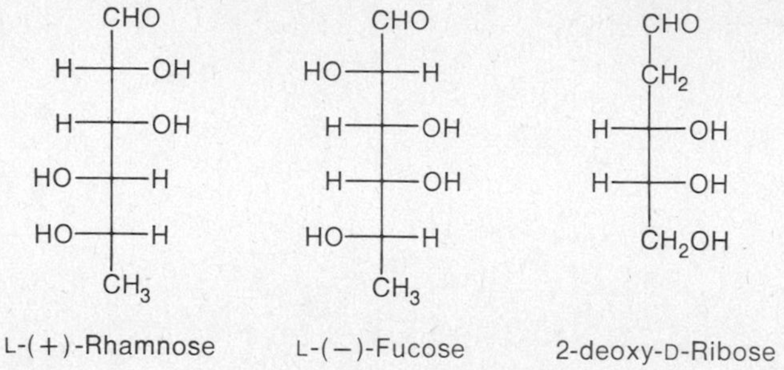

L-(+)-Rhamnose   L-(−)-Fucose   2-deoxy-D-Ribose

Fig. 32.16. Deoxy sugars.

In depicting the structures of the monosaccharides it is often useful to show them in the open-chain aldehyde or keto form. However, both in the crystalline state and in solution, all these compounds exist primarily in cyclic hemiacetal and hemiketal forms, rather than the open chain forms.

32.6 *The Chemistry of the Monosaccharides*

32.6a. Oxidation Reactions

Without Cleavage of the Carbon Chain. The difference between simple aldehydes and ketones with respect to ease of oxidation (see Section 13.7) carries over to the oxidation of the carbonyl group of aldoses and ketoses; aldoses undergo facile oxidation to the corresponding carboxylic acids, but ketoses are more resistant to oxidation. For example, bromine water oxidizes aldoses to the corresponding carboxylic acids, known as *glyconic acids*, but has no effect on ketoses (Fig. 32.17). With more strenuous oxidizing agents such as nitric acid

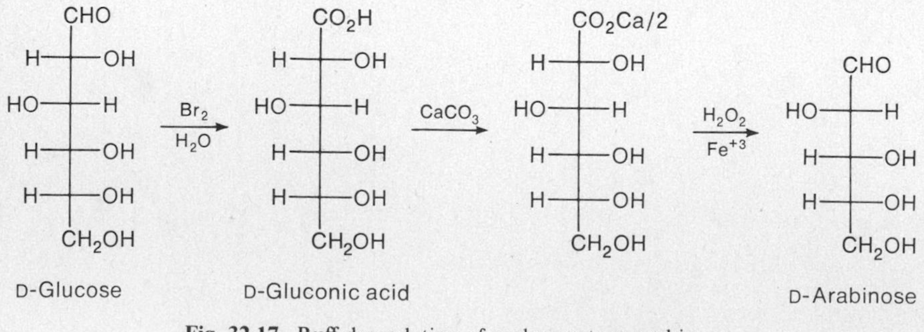

D-Glucose → (Br_2, H_2O) → D-Gluconic acid → ($CaCO_3$) → $CO_2Ca/2$ salt → (H_2O_2, Fe^{+3}) → D-Arabinose

Fig. 32.17. Ruff degradation of D-glucose to D-arabinose.

both the aldehyde group and the terminal hydroxymethyl group are oxidized to carboxyl groups, yielding dibasic acids known as *glycaric acids* (see Figs. 32.4 and 32.5). Aldoses are also oxidized by a variety of metal ions, including copper and silver ions. The most often used reagents for this purpose are Fehling's solution and Tollens' reagent, the qualitative indication of oxidation by these reagents being the formation of red cuprous oxide with Fehling's solution and metallic silver with Tollens' reagent (see p. 348).

With Cleavage of the Carbon Chain. The bromine water oxidation provides a method for degrading an aldose by one carbon unit at a time, for the carboxylic acid produced from the aldose can be converted to the calcium salt and treated with a mixture of hydrogen peroxide and a ferric salt to yield the lower homologous aldose. This sequence of reactions is called the *Ruff degradation*, in recognition of the chemist who discovered it (Fig. 32.17).

A chain-cleaving reaction that has been especially useful in the modern era of saccharide structure determination is the periodic acid oxidation of *vic*-glycols (see Section 22.13b), one example of which is illustrated in Fig. 22.39 (see p. 681). Another example is the oxidation of methyl β-D-fructofuranoside, which consumes 1 mole-equivalent of periodic acid to yield a dialdehyde (Fig. 32.18).

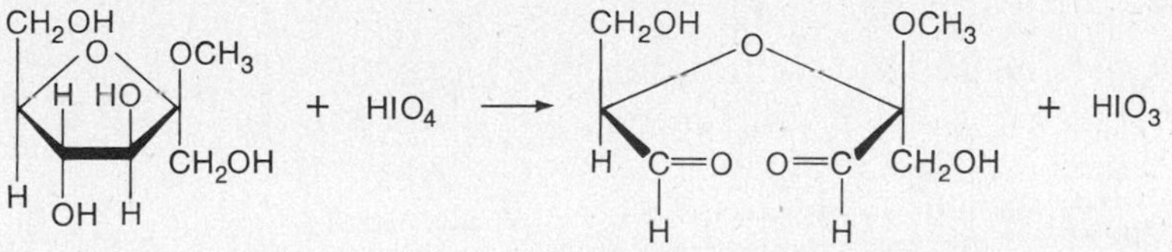

Methyl β-D-fructofuranoside + HIO_4 ⟶ dialdehyde + HIO_3

Methyl β-D-fructofuranoside

Fig 32.18. Periodic acid oxidation of methyl β-D-fructofuranoside.

32.6b. REDUCTION REACTIONS. Reduction to alcohols, a characteristic reaction of simple aldehydes and ketones, can be carried out on the carbonyl groups of aldoses and ketoses. For example, chemical or catalytic reduction of D-glucose and D-mannose yields D-glucitol and D-mannitol, respectively, both of which are naturally occurring compounds (Fig. 32.19).

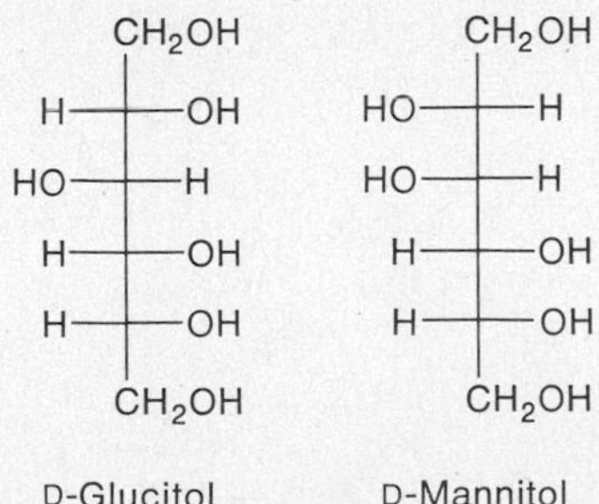

Fig. 32.19. Reduction products of aldohexoses.

32.6c. REACTIONS AT THE CARBONYL GROUP

Oxidation and Reduction of the Carbonyl Group. See preceding discussion.

Addition of Oxygen Nucleophiles. The tendency for intramolecular nucleophilic addition of hydroxyl functions to carbonyl functions in sugar molecules to form cyclic hemiacetals or hemiketals has been discussed (see p. 1004). In the presence of alcohols and protic or Lewis acids, the cyclic hemiacetals and hemiketals can be converted to mixed acetals and ketals known as *glycosides*, in which one of the oxygen functions comes from the sugar molecule and the other from the alcohol molecule (Fig. 32.20). Glycosides can be converted back to the

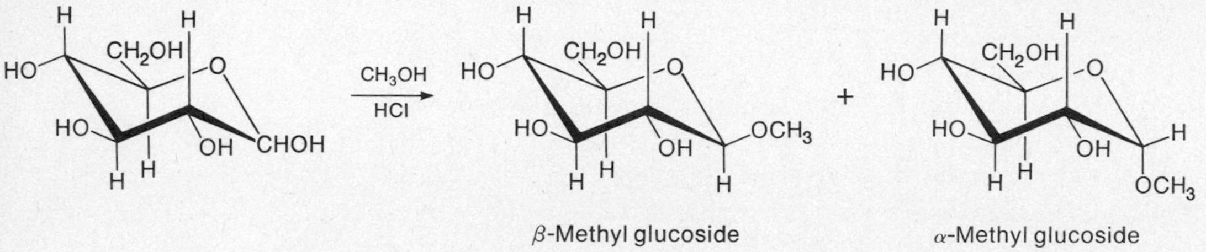

β-Methyl glucoside α-Methyl glucoside

Fig. 32.20. Formation of methyl glucoside.

constituent alcohol and hemiacetal or hemiketal by acid-catalyzed hydrolysis, but in neutral or basic medium they are stable. Thus, β-methylglucoside is not susceptible to oxidation by Fehling's or Tollens' solution, because it is *not* in equilibrium with a free aldehyde form. Whereas glucose itself is a *reducing sugar*, reducing the copper or silver ion of Fehling's or Tollens' solution, glucosides are *nonreducing sugars.*

Addition of Nitrogen Nucleophiles. Aldoses and ketoses react with carbonyl reagents such as hydroxylamine (to form oximes) and phenylhydrazine (to form phenylhydrazones). Phenylhydrazine, if used in excess, converts the initially formed phenylhydrazone to a *bis*-phenylhydrazone called an *osazone*, in which a second phenylhydrazine moiety becomes attached at a position adjacent to the first. Thus, although mannose and glucose yield different monophenylhydrazones, they form identical *bis*-phenylhydrazones, Chirality at the 2 position is destroyed in the process (Fig. 32.21), and this circumstance has been used to advantage in deducing the stereochemical relationships among the monosaccharides.

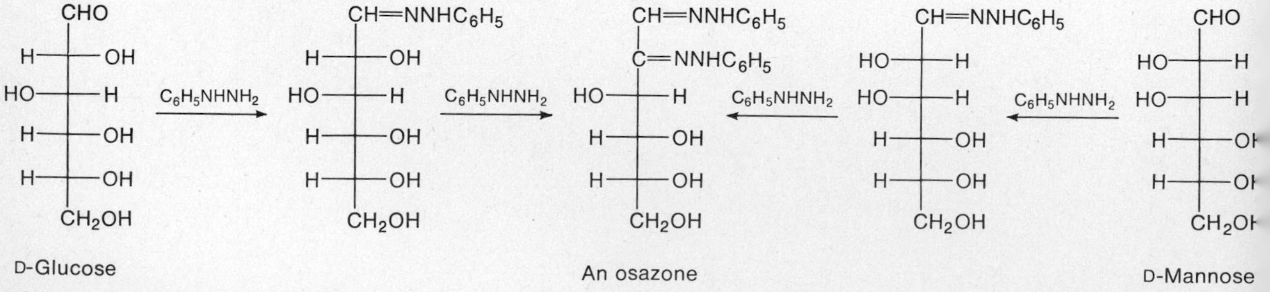

D-Glucose An osazone D-Mannose

Fig. 32.21. Monophenylhydrazones and *bis*-phenylhydrazones (osazones) from D-glucose and D-mannose.

Addition of Carbon Nucleophiles. Hydrogen cyanide adds to the carbonyl group of aldoses and ketoses to yield cyanohydrins (see Section 29.2b), and certain active methylene compounds behave in a comparable fashion. Particularly useful is the condensation reaction with nitromethane to form nitro compounds, in which the nitro group can be converted to an aldehyde function (see Fig. 28.10 for a specific example). The hydrogen cyanide and nitromethane condensations both provide methods for increasing the length of the carbon chain of sugar molecules by one carbon at a time.

32.6d. Reactions at the Hydroxyl Group

Cyclic Hemiacetal and Hemiketal Formation. See preceding discussion.

Ester Formation. In the presence of acetic anhydride and zinc chloride catalyst, all five of the hydroxyl groups of glucopyranose undergo esterification to form a glucose pentaacetate (Fig. 32.22). A mixture of anomers is obtained, but by appropriate adjustment of the reaction conditions either the α- or β-anomer can be made to predominate.

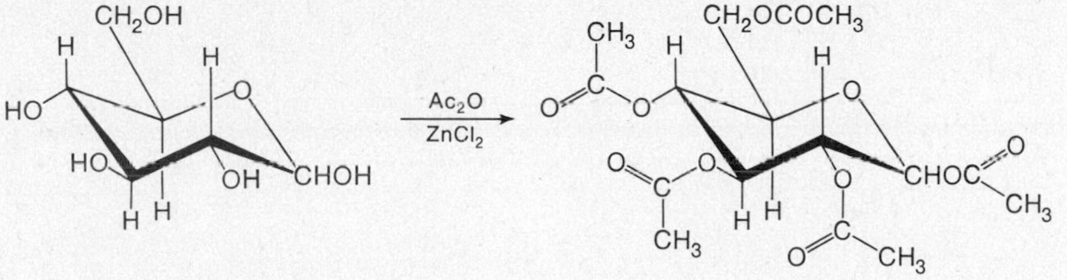

Mixture of α- and β-anomers $\xrightarrow[ZnCl_2]{Ac_2O}$ Mixture of α- and β-anomers

Fig. 32.22. Conversion of glucose to glucose tetraacetate.

Ether Formation. The anomeric hydroxyl function can be converted to an ether by treatment with alcohols in the presence of acid, as previously described (see Fig. 32.20). The remaining hydroxyl groups (*i.e.*, the other four hydroxyl groups of methyl glucoside) can then be converted to ether functions either by treatment with methyl iodide in the presence of silver oxide or with dimethyl sulfate in the presence of sodium hydroxide. In contrast to the anomeric ether group, ether groups at positions 2, 3, 4, and 6 are not susceptible to mild acid-catalyzed hydrolysis; it is possible, therefore, to prepare a tetramethyl ether of glucose by the reaction sequence shown in Fig. 32.23.

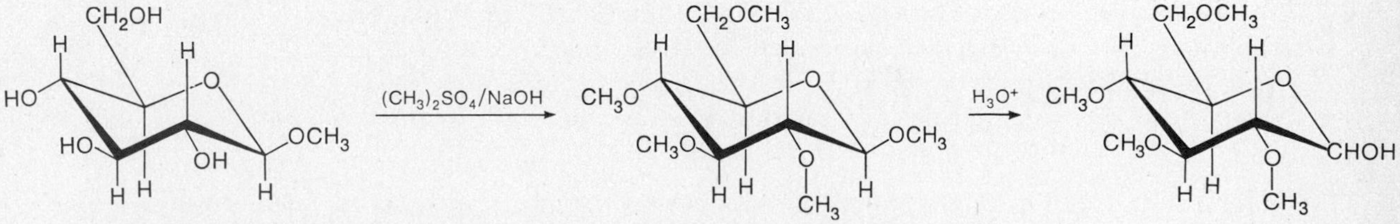

$\xrightarrow{(CH_3)_2SO_4/NaOH}$ $\xrightarrow{H_3O^+}$

Fig. 32.23. Methylation of methyl glucoside to form 2,3,4,6-tetramethylglucose.

Cyclic Acetal and Ketal Formation. The hydroxyl functions of saccharides can react with *external* carbonyl groups (*i.e.*, intermolecular rather than intramolecular reaction) to form acetals and ketals. For example, D-galactose reacts with acetone in the presence of an acid catalyst to form the *bis*-ketal, 1,2,5,6-di-O-isopropylidene-α-D-galactose (Fig. 32.24), the product arising from the less stable galactofuranose isomer that is in equilibrium with the more stable galactopyranose isomer.

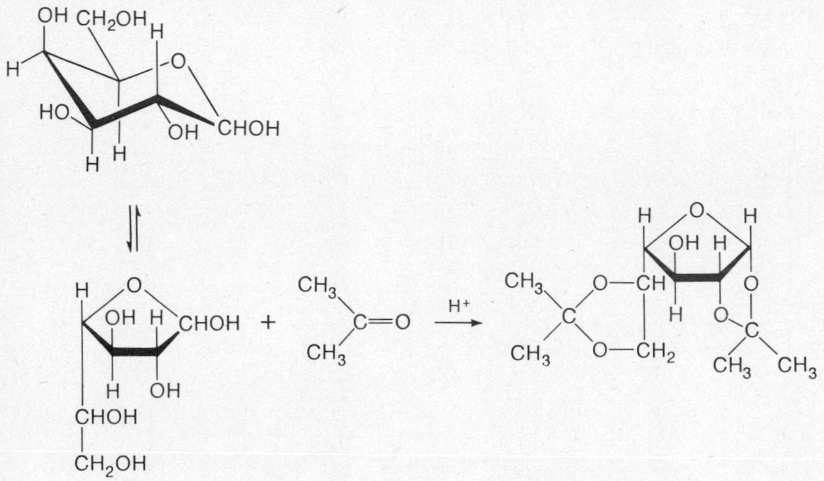

Fig. 32.24. Ketal formation between galactose and acetone.

Dehydration. Aldoses and ketoses are moderately stable in dilute acid at room temperature, but dehydration takes place under more strenuous conditions. Pentoses, for example, when heated with 12% hydrochloric acid produce a distillate containing an almost quantitative yield of furfural (Fig. 32.25). Hexoses, in a comparable fashion, yield 5-hydroxymethylfurfural as the initially formed product, but under the conditions of the reaction this decomposes to levulinic and formic acids (Fig. 32.26).

32.6e. REARRANGEMENT REACTIONS. Aldoses and ketoses are very susceptible to base-catalyzed tautomeric rearrangements (often referred to as *Lobry de Bruyn-Alberda van Eckenstein transformations*). Glucose, for example, undergoes tautomerization when treated with aqueous base to give a mixture containing mannose and fructose along with some of the original glucose (Fig. 32.25). Several other types of reactions, including skeletal rearrangements, fragmentations, and oxidations, also take place when sugars are treated with base; the study of these reactions has been one of the most perplexing areas of investigation in the field of carbohydrates, for exceedingly complex mixtures must be dealt with.

32.7 The Structure and Chemistry of the Disaccharides

Alkyl glycosides (see Fig. 32.21) are compounds containing an acetal or ketal linkage between a sugar alcohol and a simple alcohol: *e.g.*, glucose and methanol. If the second alcohol is another sugar rather than a simple alkanol, the compound

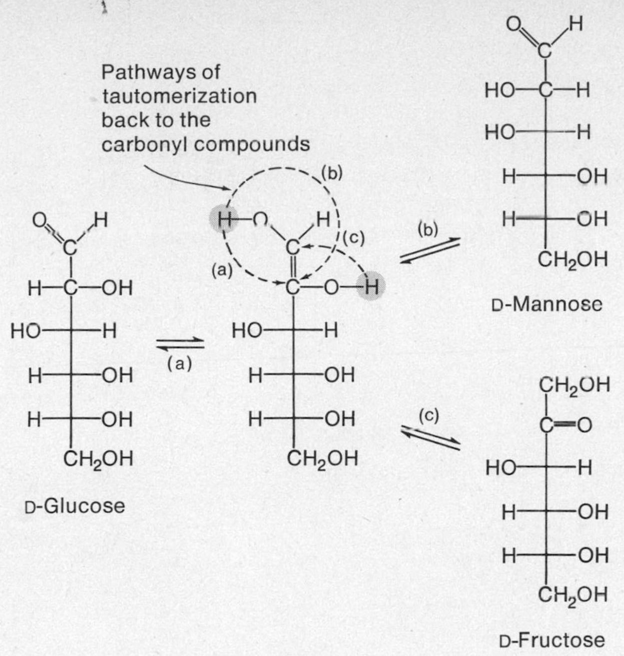

Pathways of tautomerization back to the carbonyl compounds

D-Glucose

D-Mannose

D-Fructose

Fig. 32.25. Base-catalyzed isomerization of glucose (tautomerization).

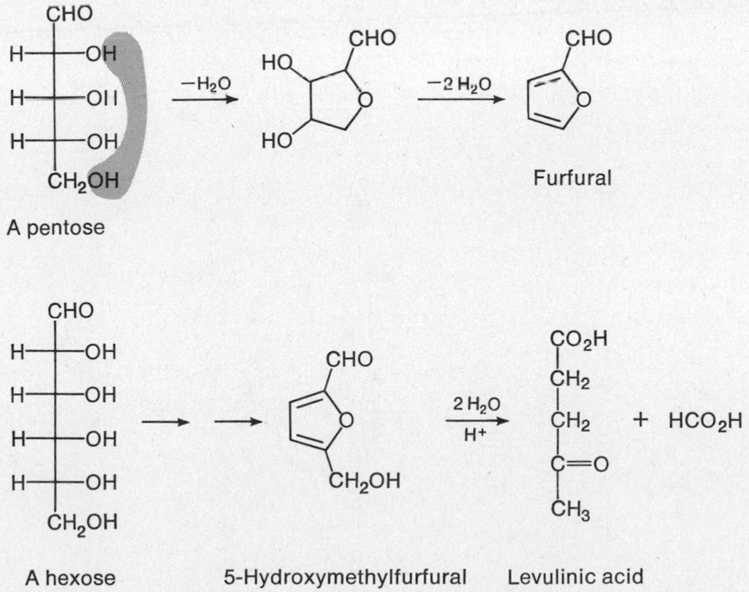

A pentose

Furfural

A hexose

5-Hydroxymethylfurfural

Levulinic acid

$+ \; HCO_2H$

Fig. 32.26. Acid-catalyzed dehydration of aldoses.

is a *disaccharide*. For example, a molecule of glucose can be joined at the C-1 position to any one of several of the oxygen atoms of another molecule of glucose to give a disaccharide containing two glucose units. If the attachment is between the hydroxyl in an α arrangement at C-1 of the glucose units and the oxygen at C-4 of a second glucose unit the compound is known as (+)-maltose. In the systematic nomenclature, it is specified as 4-O-(α-D-glucopyranosyl)-D-glucopyranose to indicate that it contains a D-glucopyranose ring substituted at the C-1 hydroxyl in the α configuration by another D-glucopyranose ring through its oxygen at C-4 (Fig. 32.27). The glucose unit in the left-hand portion of the

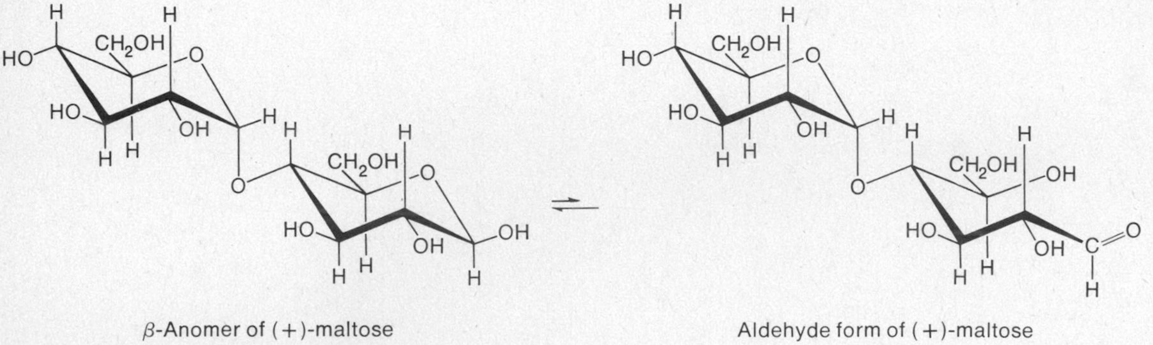

β-Anomer of (+)-maltose ⇌ Aldehyde form of (+)-maltose

Fig. 32.27. Structure of hemiacetal and open-chain forms of (+)-maltose.

structure shown in Fig. 32.27 is in glycoside form (*i.e.*, a full acetal) and, consequently, does *not* exist in equilibrium with an open chain form in solution. The glucose unit in the right-hand portion, on the other hand, retains the hemiacetal structure at C-1 and is capable of interconverting among the open chain and anomeric forms (Fig. 32.27). As a result, the disaccharide maltose exhibits the same characteristics that are typical of a monosaccharide, reacting readily with phenylhydrazine to form phenylhydrazones and osazones, mutarotating in solution, and undergoing oxidation by Fehling's or Tollens' reagent (*i.e.*, it is a reducing sugar).

A simple change in the mode of attachment of the right-hand glucose unit of maltose from the α-configuration to the β-configuration, results in a different disaccharide (Fig. 32.28). The common name for this compound is (+)-cellobiose, and the systematic name is 4-O-(β-D-glucopyranosyl)-D-glucopyranose [identical with the systematic name for (+)-maltose except for the specification of the

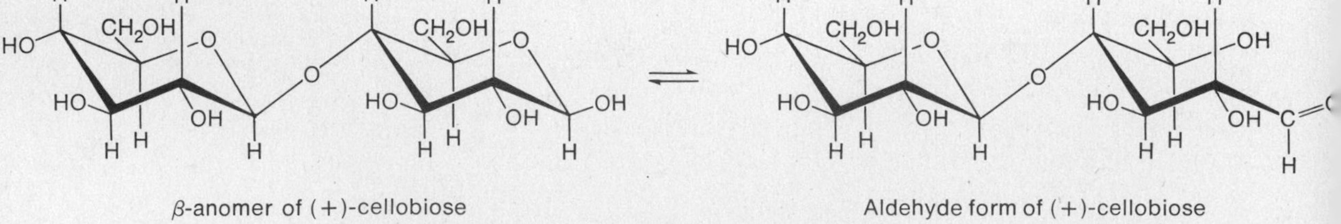

β-anomer of (+)-cellobiose ⇌ Aldehyde form of (+)-cellobiose

Fig. 32.28. Structure of hemiacetal and open-chain forms of (+)-cellobiose.

anomeric configuration as β rather than α)]. Cellobiose, like maltose, exists in solution as an equilibrium mixture of anomeric and open-chain forms as a result of the hemiacetal structure of the right-hand glucose unit; like maltose, it is classed as a reducing sugar (Fig. 32.28).

Attachment of the C-1 hydroxyl group of one glucose unit to hydroxyl groups at C-1, C-2, C-3, or C-6 of a second glucose unit is also possible. A particularly interesting example is the disaccharide trehalose, in which the second glucose unit is attached through its C-1 hydroxyl group (Fig. 32.29). This

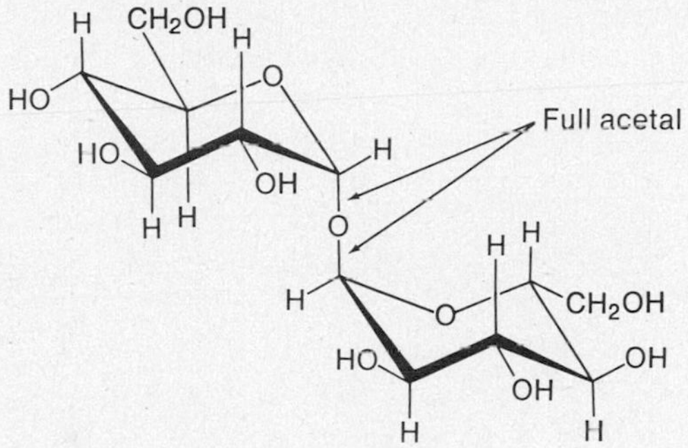

Fig. 32.29. Structure of (+)-trehalose.

compound, specified in the systematic nomenclature as α-D-glucopyranosyl-α-D-glucopyranoside, differs from maltose and cellobiose in that *neither* of its anomeric carbons is capable of equilibrating in solution with the open-chain form; *i.e.*, both are in the full acetal form. Consequently, trehalose does not form a phenylhydrazone or osazone, does not mutarotate, and does not undergo oxidation with Fehling's and Tollens' reagents (*i.e.*, it is a *nonreducing* sugar).

The determination of the structures of disaccharides such as maltose, cellobiose, and trehalose presents some special problems. The identity of the constituent monosaccharide units can be easily established by hydrolysis, glucose being the only product in all three of these examples. The failure of trehalose to be oxidized by Fehling's or Tollens' reagents indicates that the linkage between the two glucose units in this disaccharide must be between the C-1 carbon of *both* molecules. The reducing character of maltose and cellobiose, on the other hand, indicates that in these disaccharides the linkage is not at C-1 and must be at C-2, C-3, C-4, C-5, or C-6. The choice among these several possibilities can be made by converting all the hydroxyl groups in the disaccharide to methoxyl groups (see Fig. 32.23) and then hydrolyzing the disaccharide to monosaccharides and noting at which carbon atom the free hydroxyl group is located—*i.e.*, the C-4 position in the case of the methylated glucose units from maltose and cellobiose. The specification of the configuration at the C-1 position of the glycosidic unit (*i.e.*, the left-hand unit in the structures illustrated) can be made by means of enzyme-catalyzed hydrolysis, for it is known that the enzyme emulsin is specific in its action for β-glycosidic linkages, whereas the enzyme maltase is specific for α-glycosidic linkages. Originally, of course, chemical means had to be used to establish the α and β configurations, but once these were known the maltase and emulsin specificities could then be determined and the enzymes subsequently employed to make stereochemical assignments.

The most familiar member of the disaccharides is (+)-sucrose. Hydrolysis of sucrose yields one molecule of glucose and one molecule of fructose, thereby specifying the monosaccharide units of the disaccharide. It is a nonreducing sugar, indicating that the linkage between the monosaccharide units must be between the anomeric carbons, *i.e.*, C-1 of glucose and C-2 of fructose. Establishing the sizes of the rings and the configuration of the anomeric carbons proved to be very difficult problems which were ultimately solved only by a combination of chemical degradations, X-ray analysis, and finally in 1953 by the total synthesis of the disaccharide. From these combined data, it was concluded that the glucose ring is six membered (*i.e.*, pyranoside), the fructose ring is five membered (*i.e.*, furanoside), the linkage to C-1 in glucose is α, and the linkage to C-2 in fructose is β (Fig. 32.30).

Sucrose

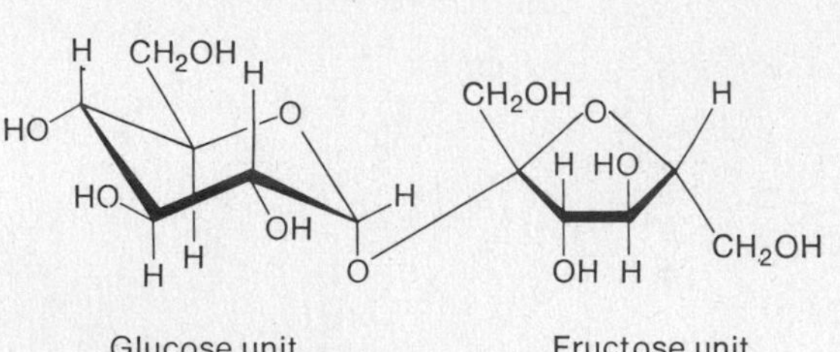

Fig. 32.30. Structure of the disaccharide (+)-sucrose.

The disaccharides just discussed are all obtained from natural sources. (+)-Maltose is produced by the hydrolysis of starch catalyzed by the enzyme diastase (isolated from malt). Cellobiose is obtained by hydrolysis of the high molecular weight polymer cellulose. Trehalose occurs as the free disaccharide and is widely distributed in bacteria, yeasts, fungi, lichens, algae, and some insects; an excellent source is the resurrection plant, a pteridophyte that grows abundantly in the Southwest. Lactose, also known as milk sugar, is present in the milk of almost all mammals and is also found in some fruits, such as that of the sapodilla tree, from which chicle is obtained.

Sucrose, the most abundant and important of the disaccharides, occurs in all parts of photosynthetic plants but is generally not found in animals. The principal sources are sugar cane, sugar beets, and sugar maples; as far back as 300 A.D., it had already been isolated in crystalline form. Of the many pure organic compounds on the market today, sucrose is produced and sold in the largest quantity.

32.8 *The Structure and Chemistry of the Polysaccharides*

Monosaccharides are polyhydroxy compounds, and their joining together via ether linkages need not be limited to the formation of disaccharides. Several naturally occurring trisaccharides are known, such as raffinose, which contains galactose, glucose, and fructose units (Fig. 32.31). The adding on of saccharide units can be continued further to give tetrasaccharides, pentasaccharides, and on up to very high molecular weight compounds known as polysaccharides.

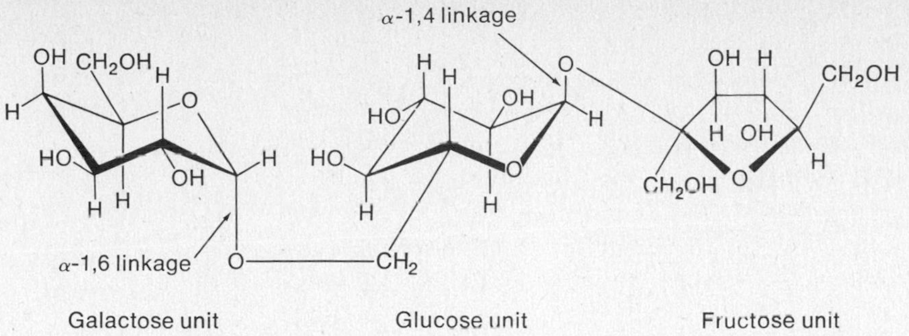

Fig. 32.31. Structure of the trisaccharide (+)-raffinose.

With the variety of naturally occurring monosaccharides available, the number of different polysaccharides that might be formed is virtually unlimited, and living systems might well have used this device for storing the information that is necessary to transmit genetic characteristics from one generation to the next. Instead, as we shall see in Chapter 35, Nature chose another kind of polymer for this task, and most of the polysaccharides that occur naturally are surprisingly simple with respect to the combinations of monosaccharide units that comprise the polymer chain. Several, in fact, are constructed from a single type of monosaccharide; for example, both starch and cellulose are made up entirely from glucose units.

Although starch and cellulose contain only glucose units, they differ significantly in their molecular architecture, for there are many ways in which a collection of glucose molecules can be fitted together into a polymeric structure. Let us look first into the structural details of starch, which, it is now realized, is not a single molecule but a mixture of two different types of molecules. One of these, present to the extent of about 20–30% in a typical starch such as that obtained from corn or potatoes, is a water-soluble substance called *amylose*; the other, present to the extent of 70–80% is a water-insoluble fraction called *amylopectin.* Some insight into the manner in which the glucose units are joined in these molecules is obtained from enzyme-catalyzed hydrolysis, a method that is frequently more selective than simple acid-catalyzed hydrolysis (which yields only glucose). An enzyme called amylase acting on amylose affords a 70% yield of the disaccharide (+)-maltose, the structure of which has already been discussed (see Fig. 32.27). From this result, we infer that amylose consists of a chain of glucose units joined in the α-1,4 manner. Although the formation of maltose does not preclude the possibility that the maltose units are joined to each other via β-linkages (*i.e.*, that amylose contains alternating α-1,4, and β-1,4 linkages), the failure to observe any cellobiose (see Fig. 32.28) in the enzyme-catalyzed hydrolysis argues against this. Why, though, does the amylase-catalyzed hydrolysis give only 70% yield rather than the quantitative yield that is characteristic of most enzyme-catalyzed reactions? The answer to this is suggested by an experiment in which a quantitative production of maltose is obtained when amylase *plus* a glucosidase are used as catalysts; from this it is inferred that an occasional glucosyl unit occurs as a "branch" on the main chain, preventing amylose-catalyzed hydrolysis at this point until the glucose side chain is cleaved by the action of the glucosidase.

Additional confirmation for the structure of amylose is obtained by methylation of the free hydroxyl groups (*e.g.*, see Fig. 32.23), followed by hydrolysis of the acetal linkages to yield 2,3,6-tri-O-methyl-D-glucose as the major product. A small amount of 2,3,4,6-tetra-O-methyl-D-glucose is also obtained, however, arising from the end group of amylose, in which there is a free C-4 hydroxyl group (*i.e.*, the left-hand glucosyl unit in the amylose molecule pictured in Fig. 32.32). From the ratio of the yield of the tri-O-methyl and tetra-O-methyl

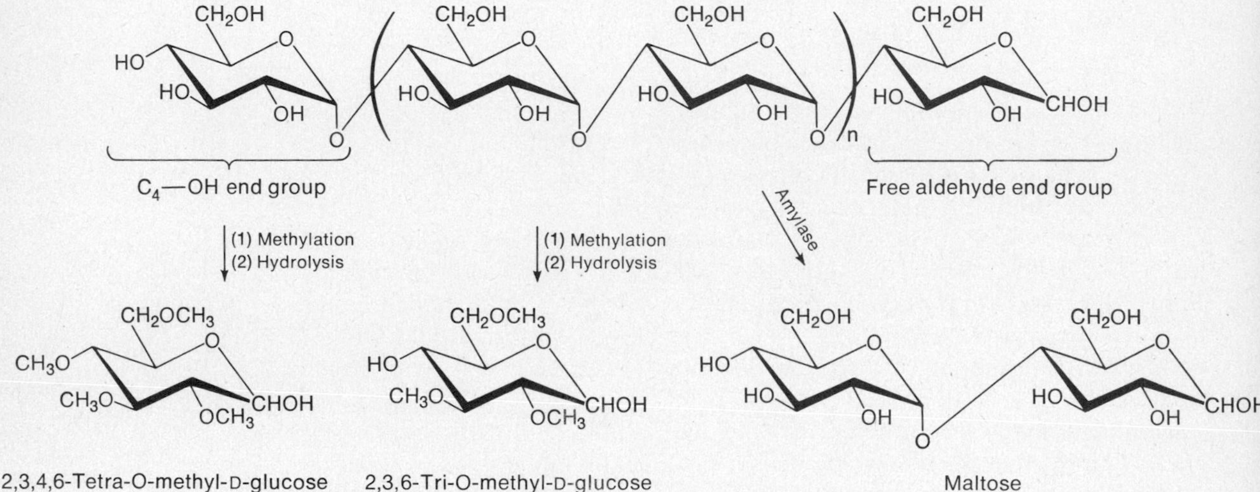

Fig. 32.32. Structure of amylose.

compounds, the number of glucosyl units per molecule can be calculated. For example, if the yield of the tri-O-methyl compound is 96% and that of the tetra-O-methyl compound is 0.6%, there are 96/0.6 = 160 glucosyl units, corresponding to a molecular weight of about 27,000. Other methods of molecular weight determination confirm those obtained by the end-group analysis and give values ranging from 20,000 to as high as 2 million for amylose from various natural sources.

On the basis of these various data, amylose is pictured as an essentially linear molecule to which single glucose units are attached at a few points along the main chain. The linear polymer does not exist in a "stretched-out" conformation, however, but in a coiled arrangement, in which each turn of the coil contains about six glucose units (Fig. 32.33).

The action of amylase on amylopectin, the other constituent of starch, also yields (+)-maltose, indicating a structure similar to that pictured in Fig. 32.33 for amylose. However, the yield of maltose is only 50% and is not increased if the glucosidase is added to the hydrolysis mixture. Although end group analysis via methylation, hydrolysis, and comparison of the yields of the trimethyl and tetramethyl ethers indicates only about 20 glucose units per chain, molecular weight determinations by other methods indicate values ranging from 1 to 6

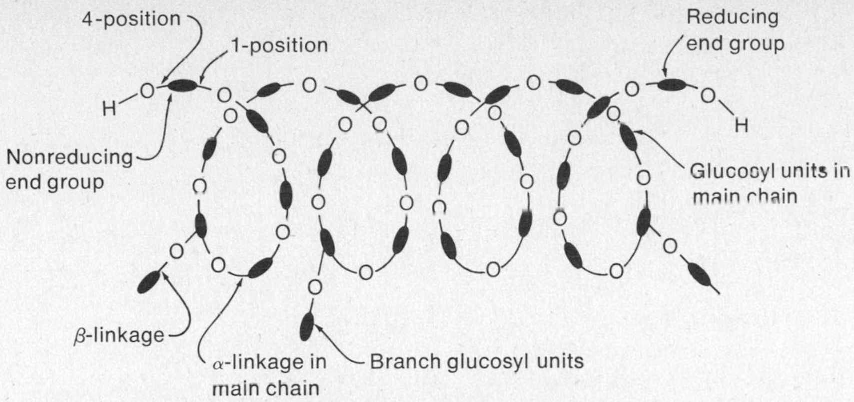

Fig. 32.33. Structure of amylose.

million. These results, as well as the observation that 2,3-di-O-methyl-D-glucose is also obtained in large amounts from the methylation experiments, can be accommodated by a branched structure for amylopectin in which each of the branches contains about 20 glucose units joined in the α-1,4 fashion, the branches joined to one another by α-1,6 linkages (Fig. 32.34).

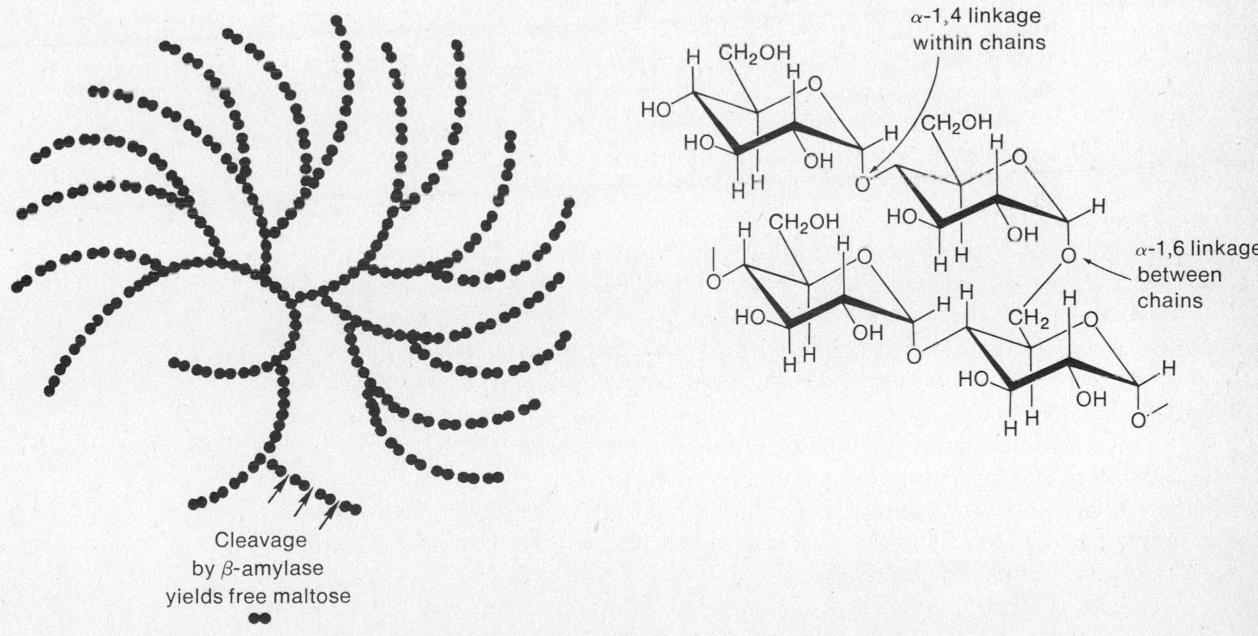

Fig. 32.34. Structure of amylopectin.

Cellulose, like amylose, is a straight-chain polymer composed of glucose units attached in a 1,4-fashion; it differs from amylose, however, in having the β-configuration rather than the α-configuration at the anomeric carbon atom.

β-linkage

Cellulose unit

X-Ray measurements indicate that the glucose units of cellulose are twisted with respect to each other and that the long chains (molecular weights from 1 to 2 million) are twined around one another in a ropelike fashion.

Twisted cellulose chains

Both starch and cellulose are naturally occurring substances, starch providing the plant with a means for storing carbohydrate material for use when needed in metabolism and cellulose providing the plant with a structural material for the construction of cell walls. A few analogs exist in the animal kingdom; a polysaccharide known as *glycogen*, similar to the amylose portion of starch, is the storage material in animals; a polysaccharide known as *tunicin* or animal cellulose has been isolated from marine animals known as tunicates. For commercial production, corn is the main source of starch, with potatoes furnishing a lesser amount; the combined total yearly production is about 3 billion pounds. Starch is used primarily as a food, both as the polymer and in the form of its hydrolysis product corn syrup or crystalline glucose although large quantities also are used as stiffening agents and adhesives. Cotton is composed of 98% cellulose, and wood is composed of approximately 50% cellulose, the other 50% consisting of hemicelluloses, lignin, mineral salts, sugars, fat, resin, and protein in an exceedingly complex structure. Thus, cellulose has been one of Nature's most widely used organic chemicals in the housing and clothing of mankind throughout the ages, and paper (derived from wood) has provided man with a means for communicating his thoughts and wrapping his packages for many centuries. Altered forms of cellulose have been introduced into the market place, the most common of these being *rayon* and *Cellophane*, which are prepared by rendering cellulose water soluble through treatment with carbon disulfide and sodium hydroxide, which forms a sodium xanthate derivative of cellulose containing the moieties

$$-OC(=S)SNa$$

The cellulose is then reprecipitated by acidifying the alkaline solution as it is forced through a hole (to form the fiber *rayon*) or through a slit (to form the film, *Cellophane*).

Some of the reactions of starch and cellulose are interesting as well as useful. For the laboratory chemist, one of the best known of these is the interaction that takes place between iodine and starch to give the intensely blue starch-iodine complex. It has been demonstrated that this "reaction" involves the insertion of

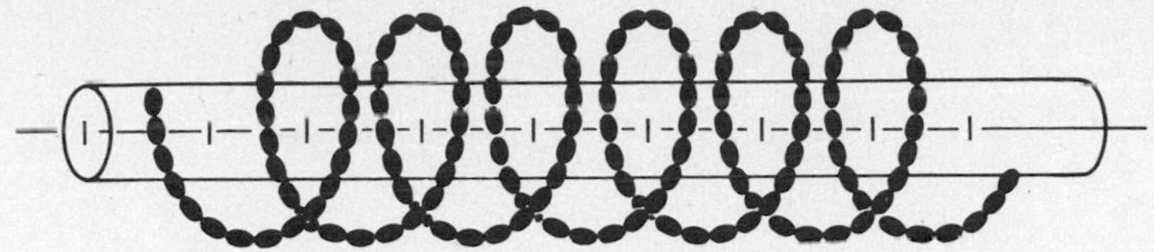

Starch-iodine complex

iodine molecules into the channel provided by the coiled structure present in the water-soluble amylose portion of starch, the dimensions of the channel being just right for the iodine to be accommodated and held by van der Waals attractive forces. A similar phenomenon has been discussed earlier with respect to the inclusion complexes formed between urea and long-chain molecules (see p. 413). Starch can be converted to organic and inorganic esters, but none of these has found widespread commercial use. Ester derivatives of cellulose, on the other hand, have very great utility; cellulose acetate (rayon) is widely used as a fiber for cloth, and cellulose nitrate (commonly called nitrocellulose) in various forms is a powerful explosive (*e.g.*, guncotton and smokeless powder).

Starch, glycogen, and cellulose are made up of only one type of monosaccharide unit and can be classed as homopolysaccharides. Heteropolysaccharides also exist in Nature, and a particularly interesting and complex type of heteropolysaccharide is present in microorganisms and is responsible for their immunological specificity. Bacteria that are foreign to the host animal that harbors them cause an immune response which is remarkably specific for the particular strain of bacteria. Scientists have discovered that this specificity of response is associated with that part of the bacterial cell wall called the *antigen*, which has been shown to be composed of polysaccharides, lipids, and proteins. The polysaccharide portion contains several different pentoses and hexoses attached as appendages to a long-chain polysaccharide (*cf.* the amylose structure, in which glucose units are the appendages).

32.9 Structure of Natural Products Related to Saccharides

A number of naturally occurring compounds correspond to reduced or oxidized aldohexoses. Examples of reduced compounds include sorbitol and mannitol (see Fig. 32.19), and a particularly important example of an oxidized material is ascorbic acid (vitamin C). Ascorbic acid, which is the lactone of 2-deoxy-2-oxo-L-gulonic acid, is related to glucose in the fashion indicated by the synthesis outlined in Fig. 32.35.

Another group of substances closely related to the ordinary saccharides are analogs that contain a nitrogen instead of an oxygen at certain positions. For example, the amino sugars D-glucosamine and D-galactosamine occur widely as constituents of certain polysaccharides (such chitin, which is the structural substance for the shells of crustacea, insects, and fungi), brain gangliosides, and glycoproteins (Fig. 32.36).

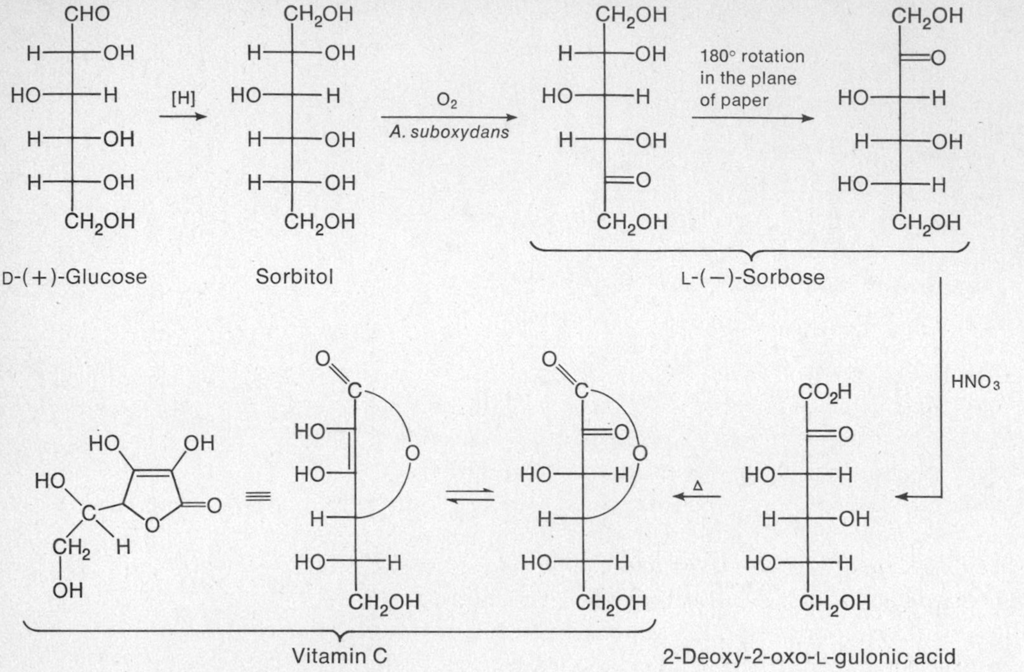

Fig. 32.35. Synthesis of vitamin C from glucose.

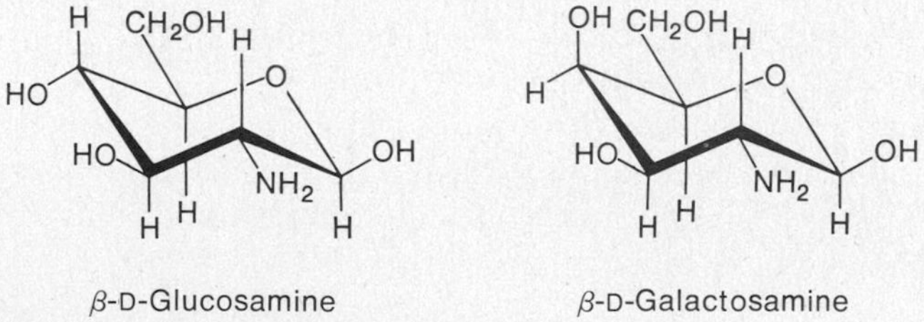

Fig. 32.36. Amino sugars.

An interesting molecule, which contains an amino sugar as well as two other saccharide-related moieties, is the antibiotic streptomycin (Fig. 32.37). The carbocyclic unit in streptomycin is an example of still another type of compound that bears a close structural relationship to the sugars. A number of polyhydroxycyclohexanes or cyclitols are known, the hexahydroxy compounds or inositols being the most interesting (Fig. 32.38). Typical examples include *myo*-inositol (*meso*-inositol), probably present in all living cells as a unit in combination with other moieties, and *scyllo*-inositol (isolated from acorns, coconut palm leaves, and dogwood). The left-hand unit in streptomycin (see Fig. 32.37) is a diamino derivative of *scyllo*-inositol.

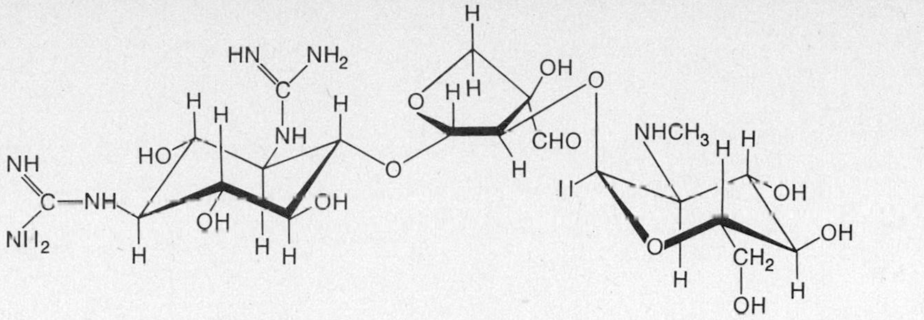

Carbocyclic analog of a diamino sugar

Dialdehyde analog of a 5-deoxyhexose

N-methyl-L-glucosamine

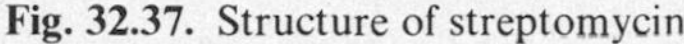

Fig. 32.37. Structure of streptomycin.

Streptomycin

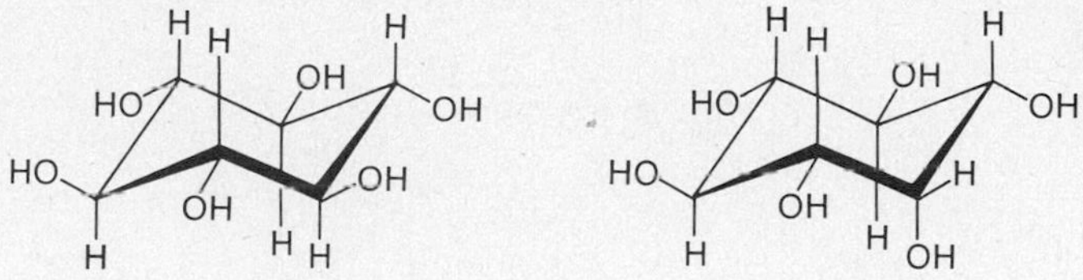

scyllo-Inositol

myo-Inositol

Fig. 32.38. Structure of two inositols.

32.10 *Structure of Other Saccharide-Containing Natural Products*

Many natural products contain saccharide units attached to nonsaccharide moieties, the latter referred to as the *aglycone* part of the molecule. A variety of plant pigments, such as the red pigment of the rose, for example, occur as glycosides. The same 3,5-diglucoside of the anthocyanin that is responsible for the red of the rose is also responsible for the blue of the cornflower, the color changing from red to blue when ferric or aluminum ions are also present (Fig. 32.39). Many other compounds also occur in nature as glycosides; one of these has been mentioned previously in connection with the insect defense mechanism whereby hydrogen cyanide is generated from benzaldehyde cyanohydrin, which, in turn,

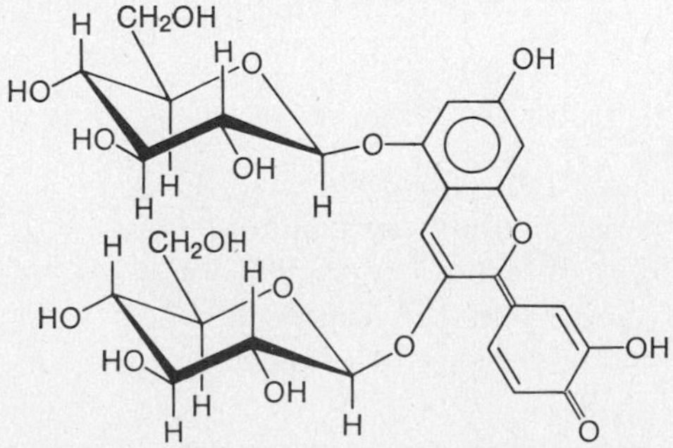

Fig. 32.39. Structure of an anthocyanin glucoside.

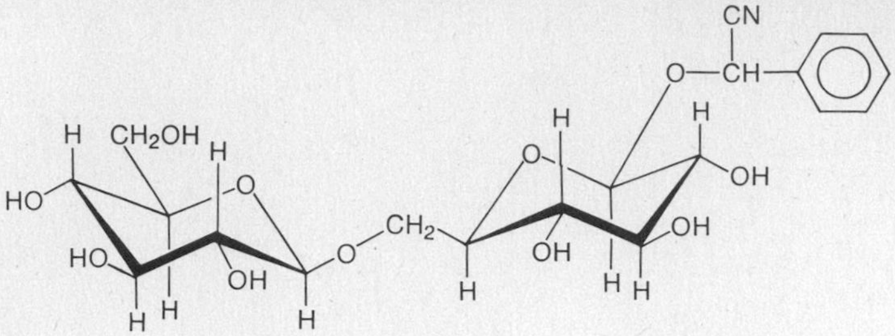

Fig. 32.40. Structure of amygdalin.

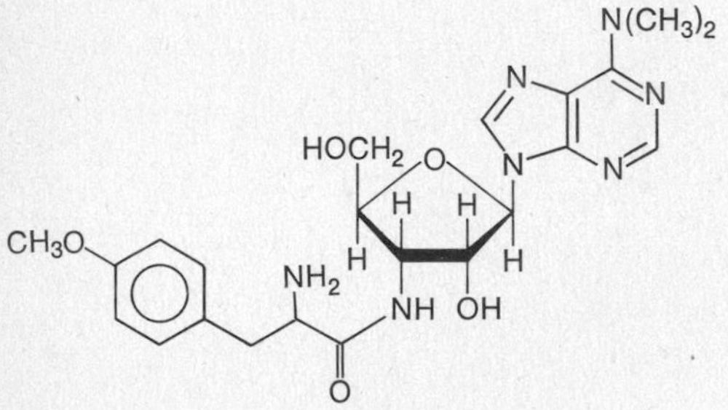

Puromycin

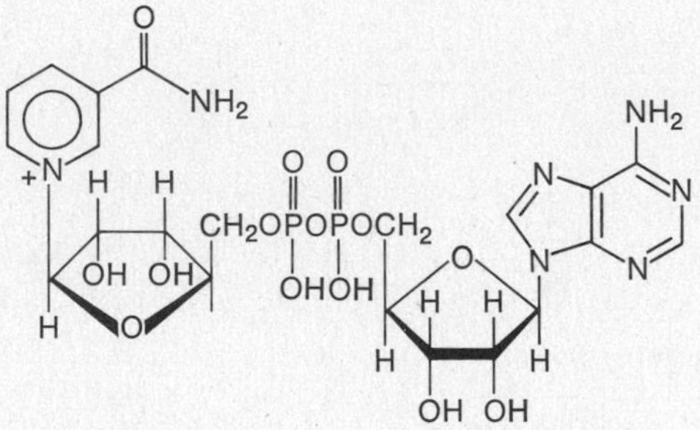

Nicotinamide adenine dinucleotide (NAD^+)

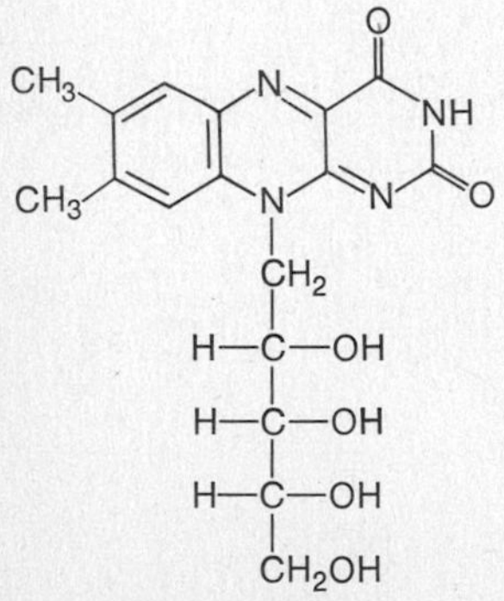

Riboflavin

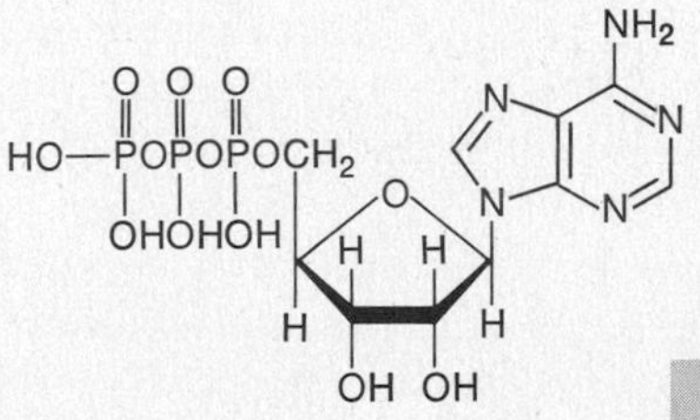

Adenosine triphosphate (ATP)

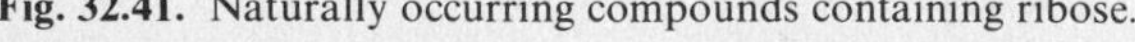

Fig. 32.41. Naturally occurring compounds containing ribose.

Adenosine triphosphate

is obtained by enzymatic hydrolysis of the corresponding glycoside (see Fig. 13.24). A similar glycoside occurs in the seed of the bitter almond and is known as amygdalin. On hydrolysis, it yields 1 mole of benzaldehyde, 1 mole of hydrogen cyanide, and 2 moles of glucose, the latter comprising the disaccharide gentiobiose (Fig. 32.40).

Two exceedingly important naturally occurring sugars are D-ribose (see Fig. 32.14) and 2-deoxy-D-ribose (see Fig. 32.16) which are found in N-glycosidic combination with various organic bases in antibiotics such as puromycin (which contains the amino analog of ribose), in coenzymes such as nicotinamide adenine dinucleotide (NAD) and riboflavin, and in nucleotides such as adenosine monophosphate (AMP) (Fig. 32.41), the latter being the monomeric unit of polymers known as nucleic acids (see Chapter 35).

REFERENCES

ASPINALL, G. O., *Polysaccharides*, Pergamon Press, New York, 1970.

BARKER, R., *Organic Chemistry of Biological Compounds*, Prentice-Hall, Inc., Englewood Cliffs, N.J., 1971, Chapter 5.

BASSHAM, J. A., The Path of Carbon in Photosynthesis, *Bio-Organic Chemistry*: *Readings from Scientific American*, W. H. Freeman and Co., Publishers, San Francisco, 1968, Article 24.

DAVIDSON, E. A., *Carbohydrate Chemistry*, Holt, Rinehart & Winston, Inc., New York, 1967.

FLORKIN, M., and E. H. STOTZ, eds., *Comprehensive Biochemistry*, Vol. 5, *Carbohydrates*, Elsevier Publishing Co., New York, 1963.

FREUDENBERG, K., Emil Fischer and His Contribution to Carbohydrate Chemistry, in M. L. Wolfrom, and R. S. Tipson, *Advances in Carbohydrate Chemistry*, Vol. 21, Academic Press, Inc., New York, 1966, p. 1–38.

GUTHRIE, R. D., and J. HONEYMAN, *Introduction to the Chemistry of Carbohydrates*, 3rd ed., Oxford University Press, New York, 1968.

LEHNINGER, A. L., Energy Transformation in the Cell, *Bio-Organic Chemistry*: *Readings from Scientific American*, W. H. Freeman and Co., Publishers, San Francisco, 1968.

MCILROY, R. J., *Introduction to Carbohydrate Chemistry*, Butterworth & Co., Ltd., London, 1967.

PIGMAN, W., *The Carbohydrates*, Academic Press, Inc., New York, 1957.

STODDART, J. F., *Stereochemistry of Carbohydrates*, John Wiley & Sons, Inc., New York, 1971.

PROBLEMS

1. A $C_5H_{10}O_4$ compound reacts with hydroxylamine to form an oxime and with sodium borohydride to form a $C_5H_{12}O_4$ compound which is optically active and which yields a tetraacetate upon treatment with acetic anhydride. In the presence of methanol and a trace of mineral acid, the $C_5H_{10}O_4$ compound yields a $C_6H_{12}O_4$ compound which undergoes conversion with periodic acid to a $C_6H_{10}O_4$ compound. When this is subjected to acid-catalyzed hydrolysis, glyoxal (CH_2OHCHO) and D-lactaldehyde ($CH_3CHOHCHO$) are produced in equal amount. From these data deduce a logical structure for the $C_5H_{10}O_4$ compound. Is the structure proof definitive or is there more than one structural possibility?

2. A $C_7H_{14}O_7$ compound reacts with phenylhydrazine to yield an osazone that is identical with the osazone obtained from either of the epimers that result from the Kiliani–Fischer synthesis (see p. 1001) with D-allose as the starting material. Oxidation of the $C_7H_{14}O_7$ compound with nitric acid yields an optically inactive $C_7H_{12}O_9$ dicarboxylic acid. From these data deduce a logical structure for the $C_7H_{14}O_7$ compound. Is the structure proof definitive or is there more than one possibility?

3. Which sugars in Fig. 32.2 yield identical osazones?

4. Under the influence of the enzyme aldolase, glyceraldehyde phosphate and dihydroxyacetone phosphate undergo an aldol condensation to yield fructose diphosphate as the only product (see Fig. 32.1). Under ordinary laboratory conditions would this be true? If not, what other products would you expect to form?

5. Draw the structures of the products of the following reactions:

(a) CHO / HO—H / H—OH / CH_2OH + CH_3OH $\xrightarrow{H^+}$

(b) CH_3O [pyranoside ring: H, CH_2OH, H, –O, OCH_3, H, H, OH, OH H] + HIO_4 $\longrightarrow$

(c) CHO / H—OH / HO—H / H—OH / CH_2OH + HNO_3 $\longrightarrow$

(d) CHO / H—OH / HO—H / H—OH / CH_2OH + Br_2 + H_2O $\longrightarrow$

(e) Amygdalin + $(CH_3CO)_2O$

(f) $HOCH_2$ [furanose ring: O, $OCH_2C_6H_5$, H, HO, H, CH_2OH, OH, H] + $(CH_3)_2SO_4$ $\xrightarrow{NaOH}$

(g) Galactose + C_6H_5CHO $\longrightarrow$

(h) Gulose $\xrightarrow{NaOH}$ Equilibrium mixture of tautomers

6. Define each of the following terms and give an appropriately illustrative example:

(a) Mutarotation (b) Intramolecular hemiacetal (c) Nonreducing sugar
(d) Anomer (e) Glycoside

7. Draw the Haworth projection and the pseudo-three-dimensional conformational representations for each of the following compounds:

(a) α-L-glucopyranose
(b) β-D-methylmannoside
(c) 4-O-(α-D-glucopyranosyl)-β-D-mannopyranose

8. Galactose containing ^{14}C at the C-1 position was converted by a series of reactions to a monomethyl ether. Treatment of the ether with an excess of periodic acid yielded formic acid in which the radioactivity per mole of formic acid is 25% as great as the radioactivity per mole of the original galactose. Draw a structure for the methyl ether.

9. For each of the following reaction sequences draw the structures of the compounds indicated by letters.

(a) D-Glucose $\xrightarrow[HCl]{CH_3OH}$ A $\xrightarrow{HIO_4}$ B $\xrightarrow[NaOH/H_2O]{CH_3NO_2}$ C $\xrightarrow[Pt]{H_2}$ D $\xrightarrow{H_3O^+}$ E

(b) D-Altrose $\xrightarrow[H_2O]{Br_2}$ A $\xrightarrow{CaCO_3}$ B $\xrightarrow{H_2O_2}$ C $\xrightarrow{Ag^+}$ D

(c) L-Gulose $\xrightarrow[NaOH]{(CH_3)_2SO_4}$ A $\xrightarrow{H_3O^+}$ B $\xrightarrow{HCN}$ C $\xrightarrow[Pt]{H_2}$ D

(d) D-Arabinose $\xrightarrow{H_2SO_4}$ A $\xrightarrow{Ag^+}$ B $\xrightarrow{CH_2N_2}$ C $\xrightarrow{LiAlD_4}$ D

10. Kojibiose is a reducing disaccharide that is formed when the mold *Aspergillis oryzae* acts on starch. Maltase-catalyzed hydrolysis of kojibiose yields D-glucose as the only product; maltase does not catalyze the hydrolysis to the monosaccharide. Oxidation of kojibiose with bromine water yields kojibionic acid which, when treated with dimethyl sulfate and sodium hydroxide followed by acid-catalyzed hydrolysis, yields a mixture of 2,3,6-trimethylglucose and 2,3,4,6-tetramethylglucose. On the basis of these data draw the structure of kojibiose in the β-anomeric form.

11. Salicin (see p. 17), a glycoside obtained from the bark of the willow tree, yields D-glucose and saligenin (*o*-hydroxybenzyl alcohol) when treated with water in the presence of emulsin. Treatment with dimethyl sulfate and sodium hydroxide produces a pentamethylsalicin which yields, on acid hydrolysis, 2,3,4,6-tetramethylglucose and O-methoxymethylphenol. On the basis of these data draw a structure for salicin.

12. The ancient Egyptians and Romans used an extract of madder root to dye their cotton fabrics red. The chromophoric moiety responsible for the red color is alizarin (1,2-dihydroxy-9,10-anthraquinone), present in the plant as a glycoside known as ruberythric acid. Ruberythric acid does not react with Fehling's or Tollens' solutions and, upon careful hydrolysis, yields alizarin and the disaccharide primeverose. Treatment of primeverose with dimethyl sulfate and sodium hydroxide followed by acid-catalyzed hydrolysis yields a mixture of equal amounts of 2,3,4-trimethyl-D-xylose, 2,3,4-trimethyl-D-glucose, and 2-hydroxyl-1-methoxy-9,10-anthraquinone. On the basis of these data draw a structure for ruberythric acid. Is this a definitive structure or are there other possibilities?

13. Vanillin can be obtained by curing vanilla beans, the curing process involving an enzyme-catalyzed hydrolysis of the β-D-glucoside of vanilla. Describe the techniques that you would employ to prove the structure of this glycoside.

Chapter **33**

Peptides: life's all-purpose polymers

"Can you do addition?" the White Queen asked. "What's one and one and one and one and one and one and one and one and one and one?" "I don't know," said Alice, "I lost count."

LEWIS CARROLL
Through the Looking Glass

The anaerobic (without oxygen) metabolism of glucose yields pyruvic acid, which the living cell can reduce to lactic acid, decarboxylate to acetaldehyde and carbon dioxide, shunt into other metabolic cycles, or convert by amination to 2-aminopropionic acid—*i.e.*,

$$CH_3\overset{\overset{\displaystyle O}{\|}}{C}CO_2H \longrightarrow CH_3\overset{\overset{\displaystyle NH_2}{|}}{C}HCO_2H$$

2-Aminopropionic acid, also known as *alanine*, is one of the approximately 20 commonly occurring α-amino acids that are the building blocks for biological polymers known as *proteins*. The primordial amino acids and proteins that appeared on Earth, however, must have had a more direct genesis and quite possibly arose from the action of sunlight on the constituents of the primeval atmosphere, which, presumably, included nitrogen, hydrogen, carbon monoxide, carbon dioxide, methane, and ammonia. Recent experiments carried out under controlled laboratory conditions have provided strong support for this hypothesis, and it is generally agreed that the primeval broth did, indeed, contain a variety of amino acids which, in turn, condensed to form proteinlike compounds (Fig. 33.1). Formed originally from these simple building blocks and subsequently, as Life emerged, from more complicated precursors, the proteins are found today in every living cell and virus and are truly ubiquitous materials. The word protein comes from the Greek *proteios* meaning "of first importance," and is, therefore, most aptly chosen.

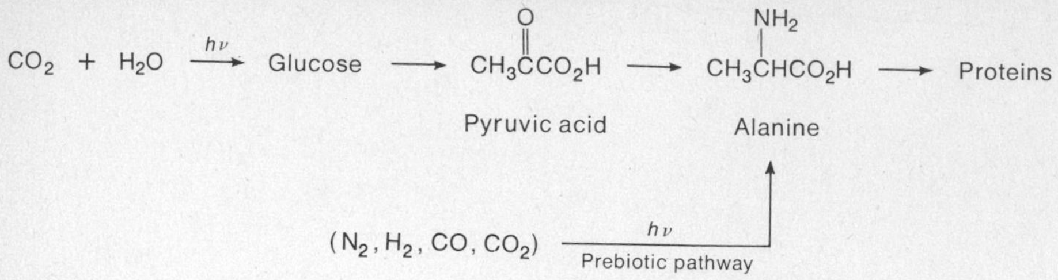

$$CO_2 + H_2O \xrightarrow{h\nu} \text{Glucose} \longrightarrow \underset{\text{Pyruvic acid}}{CH_3\overset{O}{\overset{\|}{C}}CO_2H} \longrightarrow \underset{\text{Alanine}}{CH_3\overset{NH_2}{\overset{|}{C}}HCO_2H} \longrightarrow \text{Proteins}$$

$$(N_2, H_2, CO, CO_2) \xrightarrow[\text{Prebiotic pathway}]{h\nu} \text{Alanine}$$

Fig. 33.1. Prebiotic synthesis of amino acids and proteins.

33.1 *Classification of the Proteins*

Proteins are high molecular weight polymers composed of amino acid units. They have been classified by distinctions made on the basis of their gross architecture, their composition, or their physiological function. In shape, some proteins, such as silk fibroin and keratin, are long and threadlike and are called *fibrous*; other proteins, such as egg albumin, casein, and the enzymes, are more spherical and are called *globular*. In classifying proteins according to composition, we use two major categories: *simple proteins*, which yield only α-amino acids on hydrolysis, and *conjugated proteins*, which yield other compounds in addition to α-amino acids on hydrolysis. Included among the simple proteins are the *albumins* (soluble in water and dilute salt solutions—*e.g.* egg albumin), the *globulins* (insoluble in water but soluble in dilute salt solutions—*e.g.*, blood fibrinogen), the *histones* (soluble in water but insoluble in dilute salt solutions—*e.g.*, globin), and the *scleroproteins* (insoluble in water, in dilute salt solutions, and in most other solvents—*e.g.*, keratin or hair and skin). Included among the conjugated proteins are the *phosphoproteins* (phosphoric acid-protein bond—*e.g.*, casein in milk), the *glycoproteins* (carbohydrate-protein bond—*e.g.*, mucin in saliva), the *chromoproteins* (pigmented moiety-protein bond—*e.g.*, hemoglobin in blood), the *nucleoproteins* (nucleic acid-protein bond—*e.g.*, proteins occurring in the nuclei of plant and animal cells), and the *lipoproteins* (lipid-protein bond—*e.g.*, proteins in the myelin sheath of nerves). The third classification of proteins relies on differences in function in the living system. The *structural proteins* are those found in skin, cartilage, bone, etc.; the *contractile proteins* are present in skeletal muscle; the *enzymes*, *hormones*, *antibodies*, and *blood proteins* are present in other parts of the organism and perform still other specialized functions.

33.2 *The α-Amino Acids*

All proteins, regardless of the class to which they belong, yield a mixture of α-amino acids upon hydrolysis. The amino acids are the building blocks of the proteins, and their chemistry must be understood before the chemistry of the proteins can be dealt with.

All of the commonly occurring α-amino acids except proline and hydroxyproline contain a primary amino group at the α-carbon; all except the achiral molecule glycine have the same absolute configuration at the chiral α-carbon,*

* D-Amino acids do occur in nature, although only rarely, generally in mold metabolites.

designated as L in the D,L nomenclature*. The R groups attached at the α-carbon atom, (Table 33.1) that are responsible for the virtually limitless variation in

$$\begin{array}{c} CO_2^- \\ | \\ H_3\overset{+}{N}-C-H \\ | \\ R \end{array}$$

General formula of
L-amino acids

protein structure, can be divided into four types: nonpolar, polar but neutral, acidic, and basic. Nonpolar R groups are hydrophobic and reduce the water solubility of the amino acids containing them, this influence increasing as the size of the R group increases; thus, isoleucine is considerably less water soluble than glycine and alanine. On the other hand, neutral but polar groups tend to increase water solubility through hydrogen bond formation, and acidic and basic groups are even more effective in this respect because of their ability to sustain a charge at pH 7, thereby greatly enhancing hydrogen bonding capabilities. The effects of the R groups are carried over the protein molecules, and certain portions of a protein become hydrophobic, other portions become moderately hydrophilic, and still other portions become highly hydrophilic.

Table 33.1 STRUCTURES OF NATURALLY OCCURRING α-AMINO ACIDS, $R\underset{}{C}HCO_2^-$ (with $^+NH_3$ on the α-carbon)

Name	3-Letter Abbreviation	1-Letter Abbreviation	Structure of R Group
Class I: Nonpolar R Groups.			
Glycine	Gly	G	H—
Alanine	Ala	A	CH_3—
Valine	Val	V	$(CH_3)_2CH$—
Leucine	Leu	L	$(CH_3)_2CHCH_2$—
Isoleucine	Ile	I	$CH_3CH_2CH(CH_3)$—
Phenylalanine	Phe	F	C_6H_5—CH_2—
Proline	Pro	P	[pyrrolidine ring: NH, H, CO_2H][a]

* In the (R,S) nomenclature, the majority of the naturally occurring amino acids having the L configuration are designated as S at the α-carbon. As a consequence of the way in which the sequence rule works, however, some are designated as R; L-cysteine, for example, has the R configuration at the α-carbon.

Table 33.1—*cont.*

Name	3-Letter Abbreviation	1-Letter Abbreviation	Structure of R Group
Class II: Polar but Neutral R Groups (at pH = 7):			
Serine	Ser	S	$HOCH_2-$
Threonine	Thr	T	$CH_3CH(OH)-$
Cysteine	Cys	C	$HSCH_2-$
Methionine	Met	M	$CH_3SCH_2CH_2-$
Tryptophan	Try	W	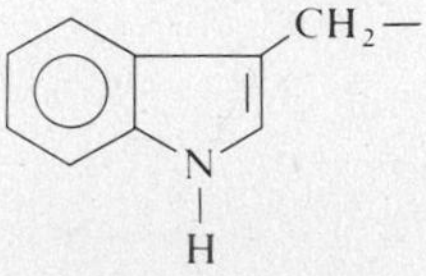 (indol-3-yl)CH_2- (ring N–H)
Tyrosine	Tyr	Y	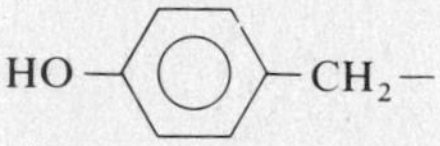 $HO-C_6H_4-CH_2-$
Hydroxyproline	Hyp		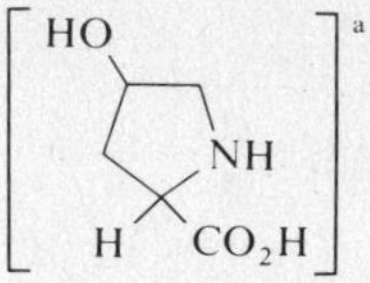 [4-hydroxypyrrolidine ring: HO, NH, H, CO_2H][a]
Glutamine	Gln	Q	$H_2NC(=O)CH_2CH_2-$
Class III: Acid R Groups:			
Aspartic acid	Asp	D	HO_2CCH_2-
Glutamic acid	Glu	E	$HO_2CCH_2CH_2-$
Class IV: Basic R Groups:			
Lysine	Lys	K	$H_2NCH_2CH_2CH_2CH_2-$
Arginine	Arg	R	$H_2NC(=NH)NHCH_2CH_2CH_2-$
Histidine	His	H	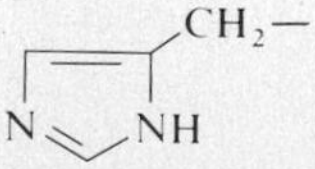 (imidazol-4-yl)CH_2- (N, NH)

[a] The amino acid is pictured in its entirety, because the α-amino group is incorporated into a ring structure.

The absolute configuration of the naturally occurring (+)-alanine was originally established by relating its enantiomer, (−)-alanine, to the same optically active form of lactic acid that is obtained from D-(+)-glyceraldehyde (Fig. 33.2). The interrelationship is predicated on the assumption that the hydroxide and ammonia displacement reactions of (−)-bromopropionic acid both proceed with *inversion* of configuration (*i.e.*, S_N2 reactions). That this assumption

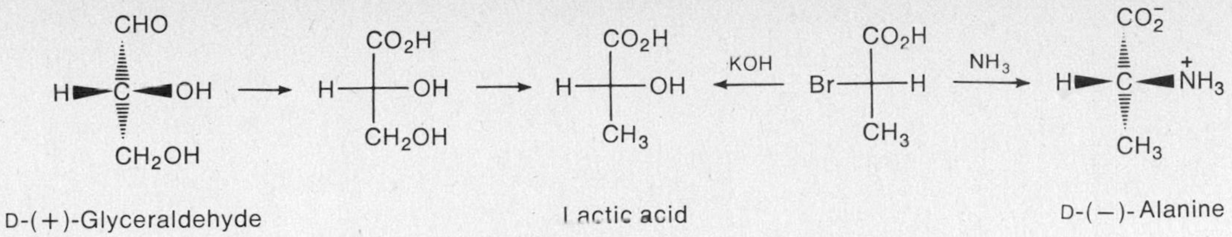

D-(+)-Glyceraldehyde → Lactic acid → D-(−)-Alanine

Fig. 33.2. Stereochemical relationship between glyceraldehyde and alanine.

is correct and that the natural form of alanine must, therefore, have the L-configuration has been established more recently by X-ray crystallographic techniques (see Section 19.4). By reactions not involving the making or breaking of bonds at the chiral α-carbon atom, (+)-alanine has been interrelated with the other commonly occurring amino acids, and all have been shown to have the same absolute configuration.

Amino acids exist primarily in an anionic, cationic, or zwitterionic form, depending on the pH, but to only a vanishingly small extent as the uncharged species. At very low pH (below 1), both the carboxyl group and the amino group are almost completely protonated, and the predominant species is the cation. As the pH is increased, the carboxyl group ($pK_a \simeq 2$) loses a proton more readily than the ammonium group ($pK_a \simeq 9.5$), and in the region between pH 3 and 8 the predominant species is the zwitterion. At very high pH (above 11), neither the carboxyl group nor the amino group is protonated to any extent, and the predominant species is the anion (Fig. 33.3). The pH at which the concentration of

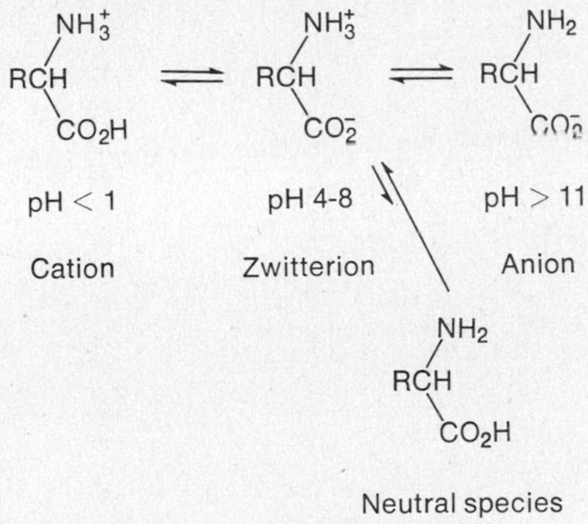

$RCH(NH_3^+)CO_2H$ (pH < 1, Cation) $\rightleftharpoons$ $RCH(NH_3^+)CO_2^-$ (pH 4-8, Zwitterion) $\rightleftharpoons$ $RCH(NH_2)CO_2^-$ (pH > 11, Anion)

$RCH(NH_2)CO_2H$ (Neutral species)

Fig. 33.3. Ionic forms of α-amino acids.

the zwitterion is at a maximum is called the *isoelectric point*, and in the case of amino acids having uncharged R groups it is simply the average of the pK_a values of the carboxyl ammonium functions. Alanine, for example, has pK_a of 2.35 for the carboxyl group and 9.69 for the ammonium group and an isoelectric point of

$$\frac{2.35 + 9.69}{2} = 6.02$$

as indicated in the titration curve illustrated in Fig. 33.4.

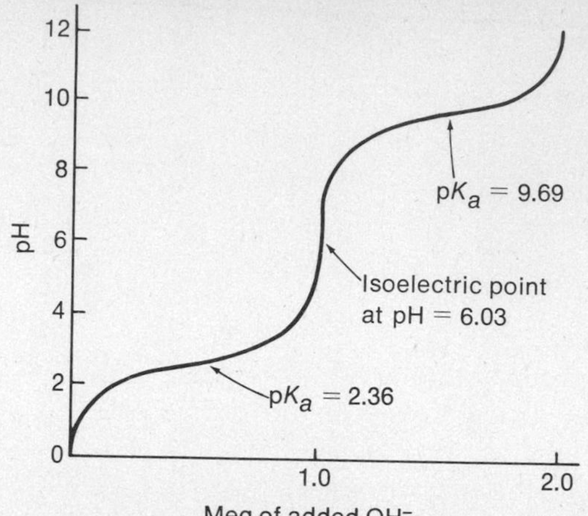

Fig. 33.4. Titration curve of alanine.

33.3 *Small Polypeptides (Oligopeptides)*

Polyfunctional molecules frequently possess the capacity for forming dimers, trimers, tetramers, and, ultimately, polymers. Monosaccharides, as we have seen, split out the elements of water between two or more molecules to form polyethers (acetals) containing saccharide units. In similar fashion, amino acids can lose the elements of water between two or more molecules to form polyamides called *polypeptides.* For example, the amide from a pair of alanine molecules is the dipeptide alanylalanine; the dipeptide from alanine and serine is alanylserine or serylalanine, depending on which amino acid furnishes the carboxyl function

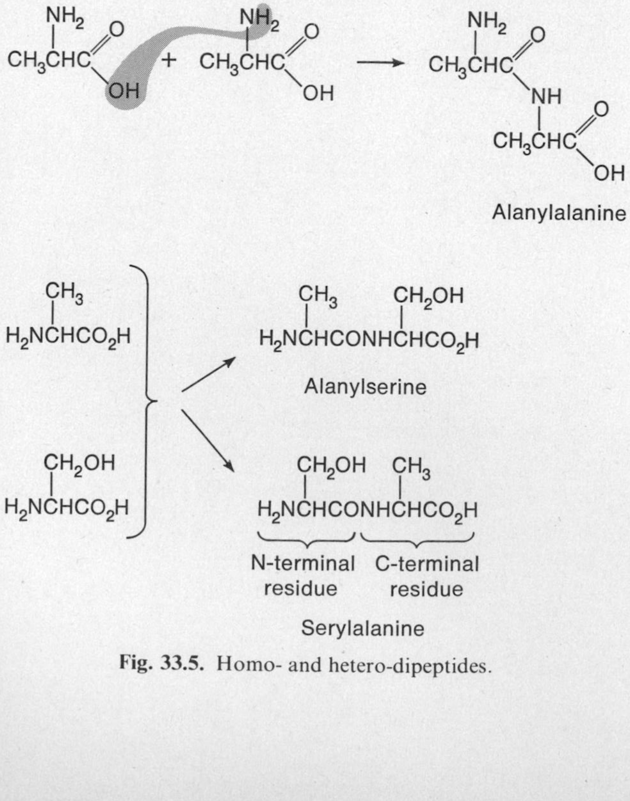

$$CH_3CH(NH_2)C(=O)OH + CH_3CH(NH_2)C(=O)OH \longrightarrow CH_3CH(NH_2)C(=O)NHCH(CH_3)C(=O)OH$$

Alanylalanine

$$H_2NCH(CH_3)CO_2H + H_2NCH(CH_2OH)CO_2H \longrightarrow \begin{cases} H_2NCH(CH_3)CONHCH(CH_2OH)CO_2H & \text{Alanylserine} \\ H_2NCH(CH_2OH)CONHCH(CH_3)CO_2H & \text{Serylalanine} \end{cases}$$

(In serylalanine: $H_2NCH(CH_2OH)CO$ — N-terminal residue; $NHCH(CH_3)CO_2H$ — C-terminal residue)

Fig. 33.5. Homo- and hetero-dipeptides.

and which the amino function to form the peptide bond (Fig. 33.5). In alanylserine, alanine is designated as the amino-terminal residue, usually abbreviated N-terminal, and serine as the carboxyl-terminal residue, usually abbreviated C-terminal; in serylalanine, serine is the N-terminal residue and alanine is the C-terminal residue. By convention, the amino acid residues in a peptide are usually drawn with the N-terminal one at the left and the C-terminal one at the right. Thus, alanylserine is abbreviated as Ala-Ser, and serylalanine is abbreviated as Ser-Ala.

Dipeptides are, themselves, difunctional compounds and can be further joined in peptide linkage at either the N-terminus or C-terminus to give tripeptides, tetrapeptides, etc., on up to very long chain polypeptides. As we have seen in the case of dipeptides, there are alternative ways in which the amino acid residues can occur in the peptide chain, this number rapidly escalating as the number of amino acid units increases. Thus, a tripeptide comprising three different amino acid moieties can have 6 different structures, a tetrapeptide 24 structures, a hexapeptide 720 structures, an octapeptide 40,320 structures, a dodecapeptide 4.79×10^8 structures, etc. The determination of the exact sequence of amino acid units in polypeptides poses a formidable problem, therefore, and not until special analytical procedures were developed was it solved. Now, by the use of the techniques of (a) complete hydrolysis and amino acid identification, (b) molecular weight determination, (c) end-group analysis, and (d) partial hydrolysis to smaller peptides, the sequences in a host of polypeptides have been determined. To discuss the application of these techniques, let us consider how the sequence of the octapeptide, Ala-Pro-Phe-Try-Lys-Ser-Val-Leu, might be established.

First we subject the octapeptide to sufficiently strong hydrolyzing conditions (*e.g.*, 6N HCl at 110° for 48 hr) to break all the amide bonds and release the eight amino acid residues as the free amino acids (Fig. 33.6). Then, using various

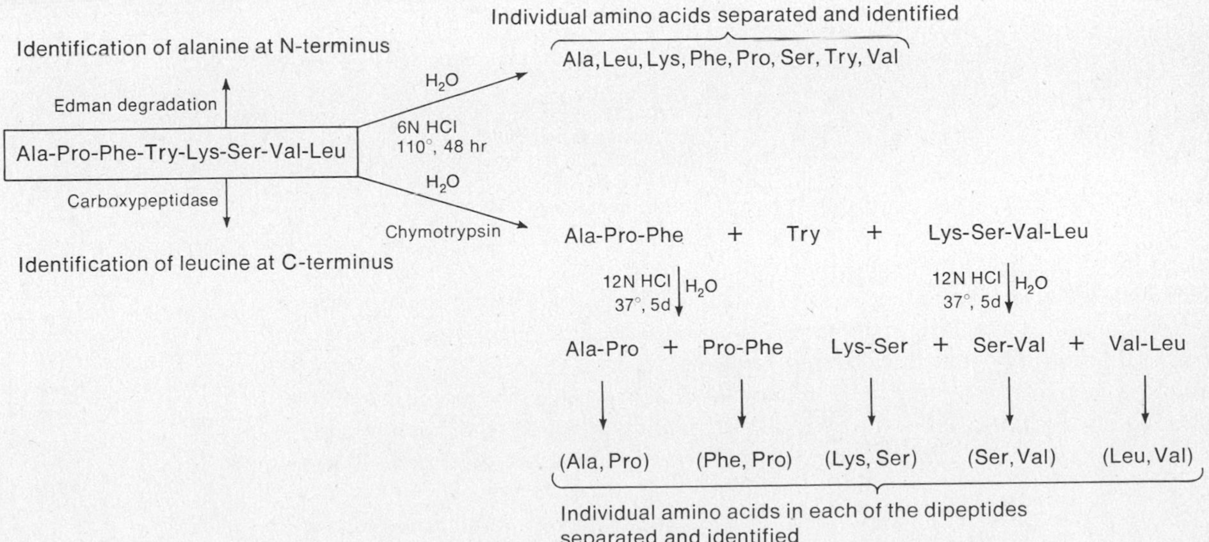

Fig. 33.6. Determination of the structure of an octapeptide.

chromatographic techniques, the eight amino acids are separated, identified, and their relative amounts measured. This analysis reveals that the eight amino acids comprising the peptide (arranged alphabetically) are Ala, Leu, Lys, Phe, Pro, Ser, Try, and Val and that they are present in the ratio 1:1:1:1:1:1:1:1. Such procedures have now become so highly refined that they can be automated, and "amino acid analyzers" are commercially available.

Next a determination of the molecular weight of the polypeptide gives a value of approximately 950 Daltons (exact value is 948). From the molecular weight and the amino acid composition of the polypeptide we know that we are dealing with an octapeptide (and not a hexadecapeptide or longer chain polypeptide). We must now proceed to the more difficult task of determining how the eight amino acid residues are arranged in the octapeptide.

One way to determine amino acid sequence is to selectively remove the end group residue (at either the N- or C-terminus), identify the residue that has been removed, then proceed to the next end group, and so on until all of the residues have been identified. To some extent this is, indeed, possible, for procedures for selectively removing end groups are known. Particularly effective are those for removing the N-terminus residues. The most general of these methods is the *Edman degradation* (Fig. 33.7) which involves (a) conversion of the polypeptide

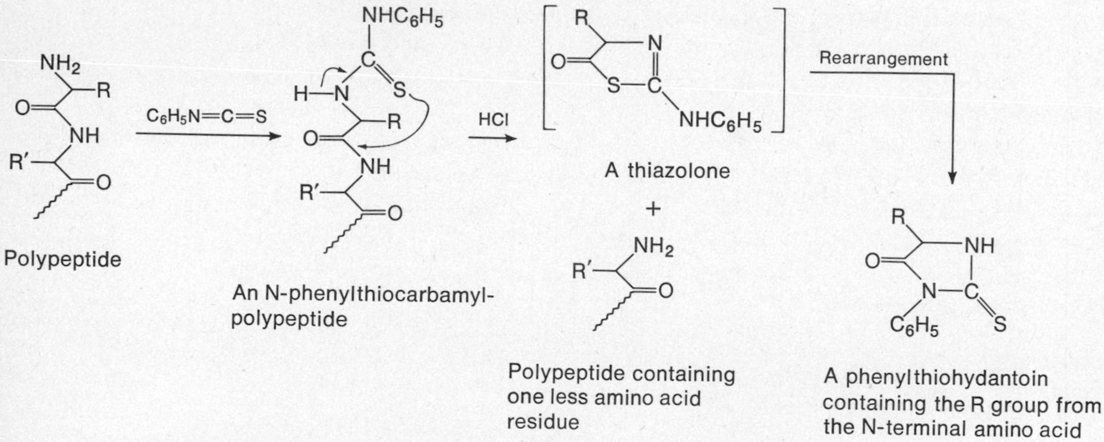

Polypeptide
$C_6H_5N{=}C{=}S$
An N-phenylthiocarbamyl-polypeptide
HCl
A thiazolone
+
Polypeptide containing one less amino acid residue
Rearrangement
A phenylthiohydantoin containing the R group from the N-terminal amino acid

Fig. 33.7. End-group analysis by means of the Edman degradation.

to an N-phenylthiocarbamyl-polypeptide, (b) treatment of this derivative with hydrogen chloride, which converts it to a thiazolone and a peptide containing one less amino acid unit than the original, (c) isolation of the phenylthiohydantoin that is formed by the acid-catalyzed rearrangement of the thiazolone, and (d) identification of the amino acid moiety of the phenylthiohydantoin, thereby identifying the end group of the original polypeptide. Applied to the octapeptide under discussion, the Edman degradation shows the presence of alanine at the N-terminus.

End-group analysis for the C-terminal residues is also possible, the method of choice involving an enzyme that specifically catalyzes the hydrolysis of the amide bond at the C-terminus. Thus, the octapeptide under discussion reacts with water in the presence of carboxypeptidase to yield leucine as the first-released amino acid, indicating that leucine is the residue at the C-terminus. End-group methods have the potentiality of repetitive application, thereby permitting the identification of each amino acid residue in turn. In practice, however, experimental difficulties are usually encountered when this is attempted, and other procedures are usually resorted to in establishing the remainder of the sequence.

The most effective supplement to end-group analysis is the partial hydrolysis of the original polypeptide to smaller peptides, accomplished both by acid-catalyzed and enzyme-catalyzed means. The enzyme chymotrypsin, for example, possesses the ability to specifically catalyze the hydrolysis of the amide bond of an amino acid residue that contains an aromatic ring (*i.e.*, phenylalanine, tryptophan, and tyrosine). Thus the octapeptide is hydrolyzed in the presence of chymotrypsin at the Phe-Try bond and at the Try-Lys bond, yielding the tripeptide Ala-Pro-Phe, the tetrapeptide Lys-Ser-Val-Leu and the free amino acid tryptophan. These three compounds can be separated, the free tyrosine can be easily identified, and tripeptide and tetrapeptide can then be separately subjected to mild acid-catalyzed hydrolysis (*e.g.*, 12N HCl at 37° for 6 days) to yield mixtures of dipeptides. The dipeptides can be separated by chromatographic means, and the composition of each one can be determined by hydrolysis to and identification of the constituent amino acids. In this fashion it can be established that the tripeptide yields the dipeptides (Ala,Pro) and (Pro,Phe) and that the tetrapeptide yields the dipeptides (Lys,Ser), (Ser,Val), and (Val,Leu). Remember, though, that we know the amino acid composition but *not* the sequence in the dipeptides; for this reason we list the amino acids of the dipeptides alphabetically and separate them by a comma rather than a dash. Now, by matching the peptides against one another it is possible to establish two alternative sequences for the tripeptide and for the tetrapeptide;

Tripeptide:

(Ala,Pro)
(Pro,Phe)
(a) Ala-Pro-Phe

(Phe,Pro)
(Pro,Ala)
(b) Phe-Pro-Ala

Tetrapeptide:

(Lys,Ser)
(Ser,Val)
(Val,Leu)
(a) Lys-Ser-Val-Leu

(Leu,Val)
(Val,Ser)
(Ser,Lys)
(b) Leu-Val-Ser-Lys

Choices between these alternatives can now be made on the basis of the end-group information. Since we know that the N-terminal residue of the octapeptide is alanine, the sequence of the tripeptide must be (a) rather than (b). Similarly, since we know that the C-terminal residue of the octapeptide is leucine, the sequence of the tetrapeptide must be (a) rather than (b). Coupling this information with the chymotrypsin-catalyzed hydrolysis, which liberates free tryptophan in addition to the tripeptide and tetrapeptide, we are left with only one possible way for combining the eight amino residues; the structure, therefore, is unequivocally established.

By similar procedures, the structures of a variety of small polypeptides have been determined, including the tripeptide glutathione (a coenzyme and a biological reducing agent), the decapeptide angiotensin-I (which causes an increase in the blood pressure of animals), the nonapeptide bradykinin (which causes a decrease in the blood pressure of animals), and bovine oxytocin (which causes contraction of the uterus) (Fig. 33.8).

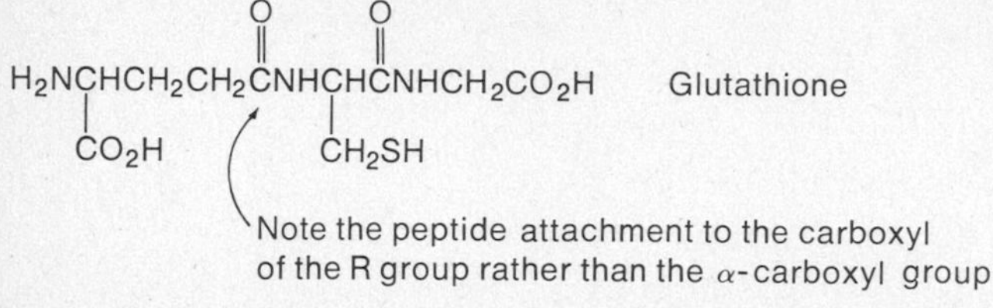

Asp-Arg-Val-Tyr-Ile-His-Pro-Phe-His-Leu

Angiotensin-I

Arg-Pro-Pro-Gly-Phe-Ser-Pro-Phe-Arg

Bradykinin

S——————S

Cys-Tyr-Ile-Glu-Arg-Cys-Pro-Leu-Gly-NH_2 (Terminal group is —C(=O)NH_2 rather than CO_2H)

NH_2 (on Arg)

Bovine oxytocin

Angiotensin-I

Fig. 33.8. Structures of small polypeptides.

33.4 *Large Polypeptides: The Primary Structure of Proteins*

The complications of sequence determination multiply as the length of the chain increases, and the analytical strategy becomes progressively more complex. The general approach for determining protein structure, however, remains essentially that described in Sec. 33.3 for an octapeptide, yielding what is referred to as its *primary structure*; *viz.*, its amino acid sequence. The work most responsible for laying the foundation for the structure determination of polypeptides is that of the English biochemist F. Sanger in the late 1940's and early 1950's. Sanger first developed a method for end-group analysis and then proceeded to apply it to the polypeptide insulin which, he quickly discovered, actually contains two peptide chains attached through disulfide bonds. After separating the chains by reduction of the S—S bonds, each chain was separately analyzed by the methods described above, leading in 1951 to a structure for the shorter chain, in 1953 to a structure of the longer chain, and in 1955 to a complete structure of insulin (Fig. 33.9). For this remarkable achievement, Sanger was awarded the Nobel Prize in 1958. His work provided the impetus for investigations of even more complex polypeptides and has led to complete structures for

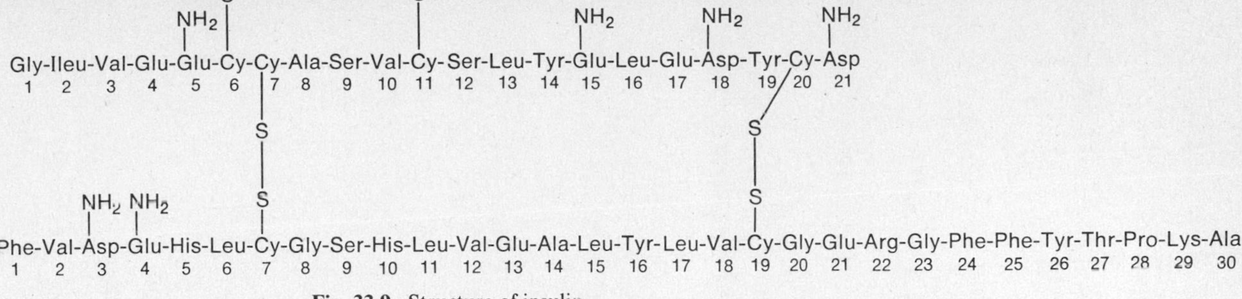

Fig. 33.9. Structure of insulin.

a variety of proteins such as the enzyme ribonuclease (Fig. 33.10) which has 124 residues, the enzyme chymotrypsin which has 241 residues, and the immunoprotein *gamma*-globulin which consists of four chains, two containing 446 residues and another two containing 214 residues.

Lys-Glu-Thr-Ala-Ala-Ala-Lys-Phe-Glu-Arg- 10
Gln-His-Met-Asp-Ser-Ser-Thr-Ser-Ala-Ala- 20
Ser-Ser-Ser-Asn-Tyr-Cys-Asn-Gln-Met-Met- 30
Lys-Ser-Arg-Asn-Leu-Thr-Lys-Asp-Arg-Cys- 40
Lys-Pro-Val-Asn-Thr-Phe-Val-His-Glu-Ser- 50
Leu-Ala-Asp-Val-Gln-Ala-Val-Cys-Ser-Gln- 60
Lys-Asn-Val-Ala-Cys-Lys-Asn-Gly-Gln-Thr- 70
Asn-Cys-Tyr-Gln-Ser-Tyr-Ser-Thr-Met-Ser- 80
Ile-Thr-Asp-Cys-Arg-Glu-Thr-Gly-Ser-Ser- 90
Lys-Tyr-Pro-Asn-Cys-Ala-Tyr-Lys-Thr-Thr- 100
Gln-Ala-Asn-Lys-His-Ile-Ile-Val-Ala-Cys- 110
Glu-Gly-Asn-Pro-Tyr-Val-Pro-Val-His-Phe- 120
Asp-Ala-Ser-Val 124

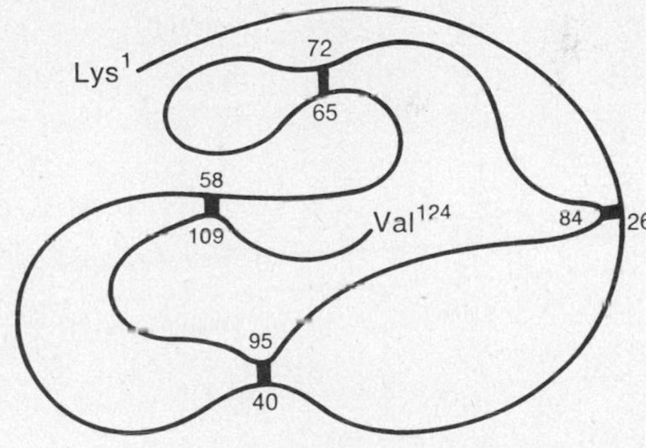

Primary sequence of amino acid residues

Tertiary structure

Fig. 33.10. Structure of ribonuclease.

33.5 Secondary, Tertiary, and Quaternary Structure of Proteins

The amino acid sequence is an essential but still insufficient ingredient in the complete characterization of a polypeptide. Polypeptides, like simple molecules such as butane and cyclohexane, have conformational preferences, and the molecular architecture that a polypeptide assumes is referred to as its *secondary structure* and *tertiary structure*. One of the important factors in determining the conformation of a peptide chain is the peptide link itself. Although the amide moiety is ordinarily written to show a single bond between the carbonyl carbon and the nitrogen, there is actually a considerable amount of double bond character to the C—N bond because of the resonance interaction of the nonbonding electrons on nitrogen with the C=O group (see Fig. 15.13, p. 403). The conformation is fixed in such a fashion that the carbonyl oxygen, the carbonyl carbon, the α-carbon, the nitrogen, and the two atoms directly attached to the nitrogen are all coplanar. Rotation around the (a) and (b) bonds, however, permits groups attached to the α-carbon atom in a peptide chain to occupy positions out of this plane (Fig. 33.11).

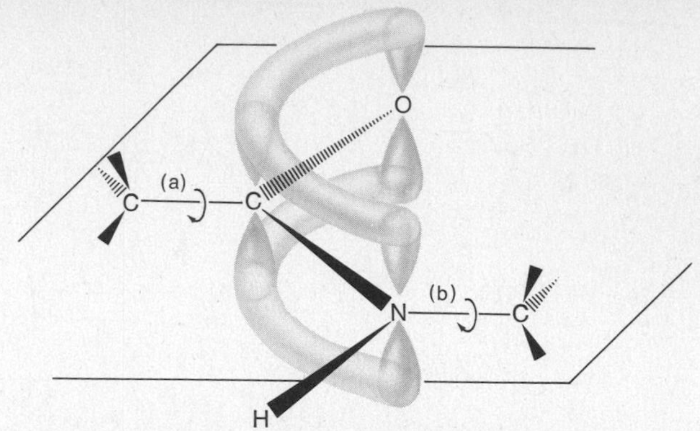

Fig. 33.11. Geometry of the peptide bond.

A second important factor that determines the conformation is the hydrogen bonding that occurs between the carbonyl oxygen atom and the nitrogen atom—*i.e.*, $>C{=}O\cdots H{-}N<$. For example, intermolecular hydrogen bonding might be anticipated between a group of peptide chains that are in extended-chain conformations (zig-zag conformation). That this does, indeed, take place has been shown by X-ray investigations of protein fibers, such as silk fibroin, which have a "pleated sheet" structure (Fig. 33.12). Intramolecular hydrogen bonding

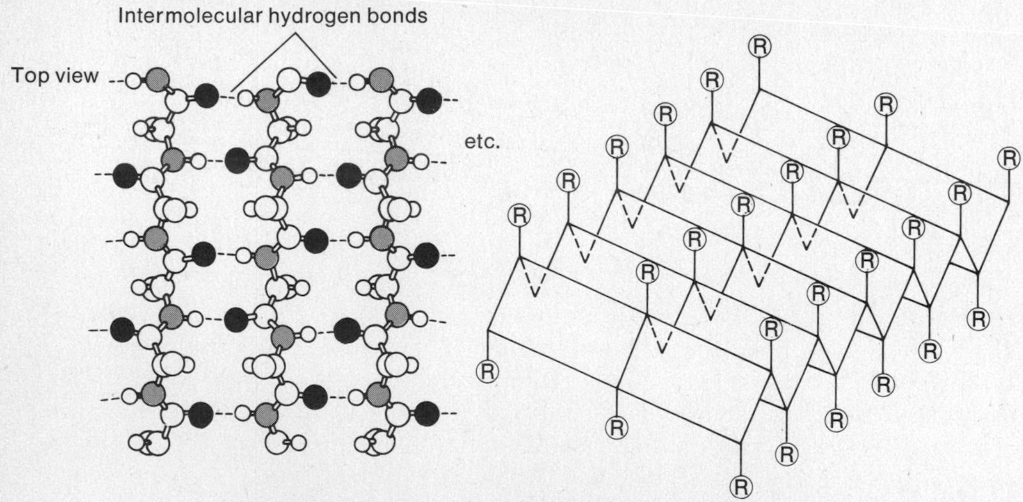

Fig. 33.12. Pleated sheet conformation of polypeptide chains.

might be anticipated if the peptide chain is sufficiently flexible to allow close contact between the carbonyl and amino groups within the molecule. One of the milestones in the history of structural chemistry occurred in the early 1950's, when Linus Pauling and R. B. Corey interpreted the X-ray data from the protein fiber α-keratin in terms of a coiled structure held in a helical conformation by intramolecular hydrogen bonds. Led to this conclusion because it provides the simplest interpretation for the 5.0- to 5.5-Å repeat unit in the X-ray crystallographic picture of α-keratin, they postulated what they termed an α-helix possessing approximately 3.6 amino acid residues per turn of the helix. In this particular conformation every C=O and N—H group in a peptide chain is intramolecularly hydrogen bonded, the bond forming between the C=O and N—H groups in amino acids that are four residues apart in the chain (Fig. 33.13).

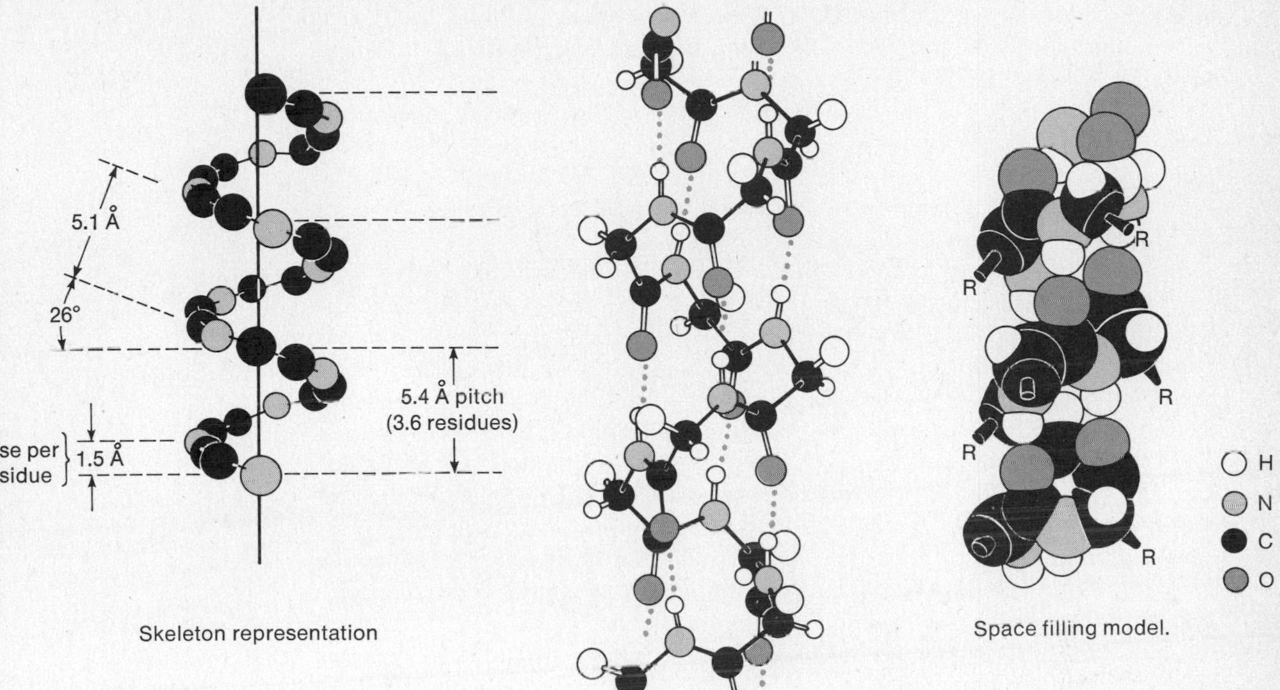

Fig. 33.13. α-Helix conformation of a polypeptide chain.

Although a single hydrogen bond is relatively weak (C=O···H—N bond estimated at 2–5 kcal/mole), the presence of several hydrogen bonds within the same molecule can provide significant stabilization. As a result, the α-helix is a particularly stable conformation for a polypeptide, and it can form spontaneously as the molecule is assembled. In the α-helix structure the peptide chain itself furnishes the core of the helix, and the R groups attached to the α-carbon atoms project outward from this central core like the steps of a spiral staircase.

Certain structural features interfere with the formation of the α helix. Peptides constructed from a mixture of D- and L-amino acids, peptides containing bulky R groups, peptides containing groups that carry charges of like sign at pH 7, and peptides containing proline or hydroxyproline, either do not form α-helices at all or possess interruptions in the α-helix at the sites of the offending amino acid residues. Steric interactions are responsible for each of these phenomena. A chain of mixed D and L configurations throws the R groups into interference with each other; very bulky R groups or proline residues tend to restrict the rotation around the (a) and (b) bonds (see Fig. 33.11) and sometimes, but not in all cases, interfere with helix formation; charged groups of like sign repel each other and distort the conformation that is necessary for α-helix formation. Experiments with synthetic polypeptides have pointed up some of these phenomena in striking fashion. For example, polyalanine forms an α-helix spontaneously, but neither polylysine nor polyglutamic acid forms an α-helix at pH 7; instead, each exists in a random coil conformation as the result of repulsion between the charged groups of like sign. However, α-helices are readily formed when polylysine is placed in solution at pH 11, where it exists almost exclusively in the uncharged amino form, or when polyglutamic acid is placed in a solution at pH 2, where it exists almost exclusively in the uncharged carboxylic acid form. Polyproline is precluded from forming an α-helix, but it forms a larger helix in which there are more amino residues per turn than in the α-helix.

Even the primary structures arising from covalent bonding, combined with the secondary structures arising from hydrogen bonding, may be insufficient to completely characterize a polypeptide. This is particularly true for globular proteins in which the primary chain, whether completely or incompletely in the α-helix form, is wound back on itself in various convoluted fashions to give what is referred to as the molecule's *tertiary structure*. For example, X-ray investigation by J. C. Kendrew and M. F. Perutz (Nobel Prize 1963) of crystalline myoglobin* showed it to have the conformation illustrated in Fig. 33.14. A detailed inspection of this structure revealed that myoglobin contains eight α-helix portions separated from one another by bends, or nonhelical portions, of the peptide chain resulting from the presence of proline residues, isoleucine residues, and residues of like charge, such as lysine or glutamic acid.

The forces responsible for maintaining the tertiary structure in proteins include covalent bonding via S—S bonds as well as noncovalent bonding arising from the entropic advantage that is gained when a molecule turns its hydrophobic portions (nonpolar R groups) toward the inside away from the aqueous environment and its hydrophilic portions (polar R groups) toward the outside where they can form hydrogen bonds with water. Thus, in myoglobin, because of the flexibility provided by the nonhelical portions, a structure can be attained in which almost all the nonpolar groups are located on the interior of the sphere and all the polar groups are located on the outside. The molecule is so compact that there is room for only about four molecules of water in its interior—*i.e.*, an essentially nonaqueous environment.

The stability of the secondary and tertiary structures is, of course, a function of the particular polypeptide. In some molecules, such as ribonuclease, covalent S—S bonds help to preserve the highly convoluted structure. In other molecules, such as myoglobin, only hydrogen bonding and hydrophobic and hydrophilic forces are operative. Consequently the secondary and tertiary structures of myoglobin are more easily broken than that of ribonuclease. The loss of tertiary

---

* 153 amino acid residues; mol wt, 16,700; a protein containing an iron-porphyrin heme group (see p. 1099) and functioning in the living muscle cell as part of the oxygen transport system.

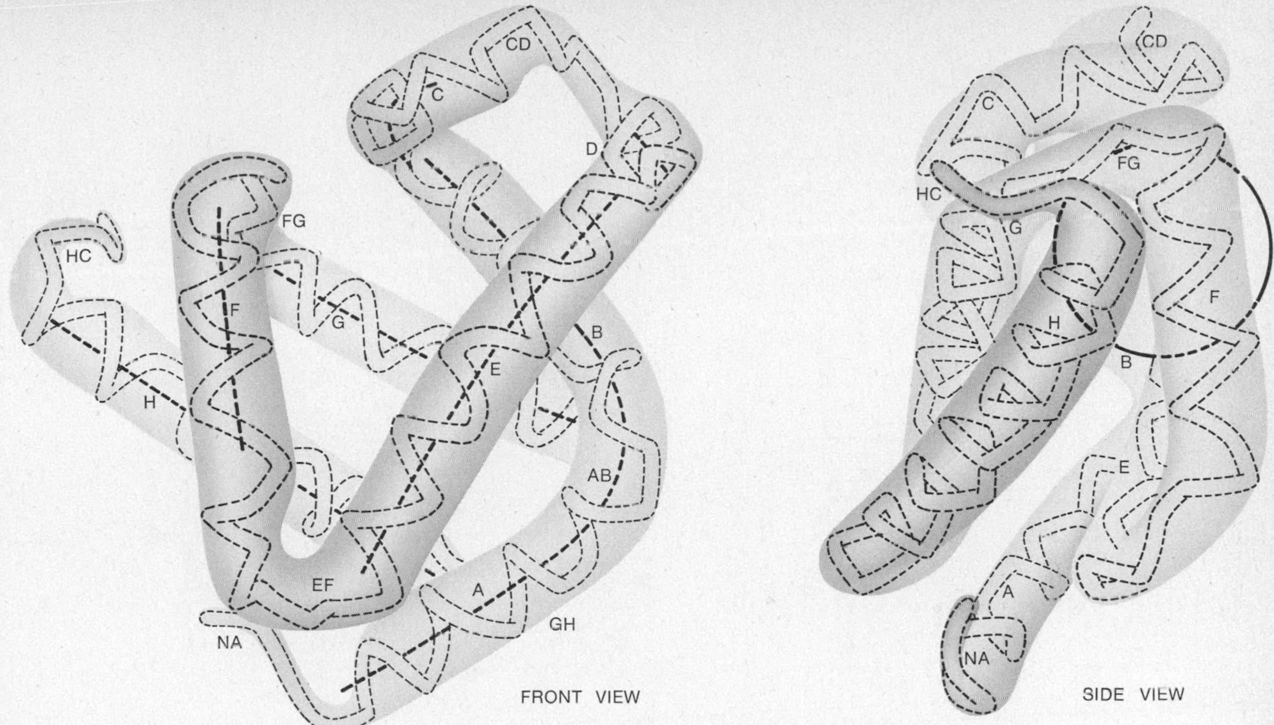

Fig. 33.14. Tertiary structure of myoglobin. The letters A through H specify the α-helical portions of the chain (all right-handed helices) which range in length from 7 residues in helices C and D to 26 in helix H; a total of 121 of the 153 amino residues are in helical regions.

and secondary structure is referred to as *denaturation.* Heat alone is usually sufficient to cause denaturation, and at 40 to 50° the structure of most proteins changes as the chain unfolds. Upon cooling, the original conformation is sometimes restored, particularly if the cooling is carried out very gradually so as to give the molecule ample opportunity to test out the various conformations and find the one of maximum stability.

Although the primary, secondary, and tertiary structures are sufficient to describe adequately many proteins, still another structural detail is required for others. Some proteins are aggregates of individual smaller protein units, held together in a definite array by weak forces. This is referred to as the *quaternary structure* of the protein. Hemoglobin, for instance, consists of four heme proteins (two containing 141 amino acid residues and two containing 146 amino acid residues) combined in a form approximating a tetrahedron (Fig. 33.15).

33.6 *Enzymes*

The living system uses proteins for a variety of purposes, one of which is to catalyze the many reactions required to sustain life. Virtually all the processes occurring in any biological unit are catalyzed by a group of compounds called *enzymes.* All known enzymes are proteins; all enzymes whose tertiary structure

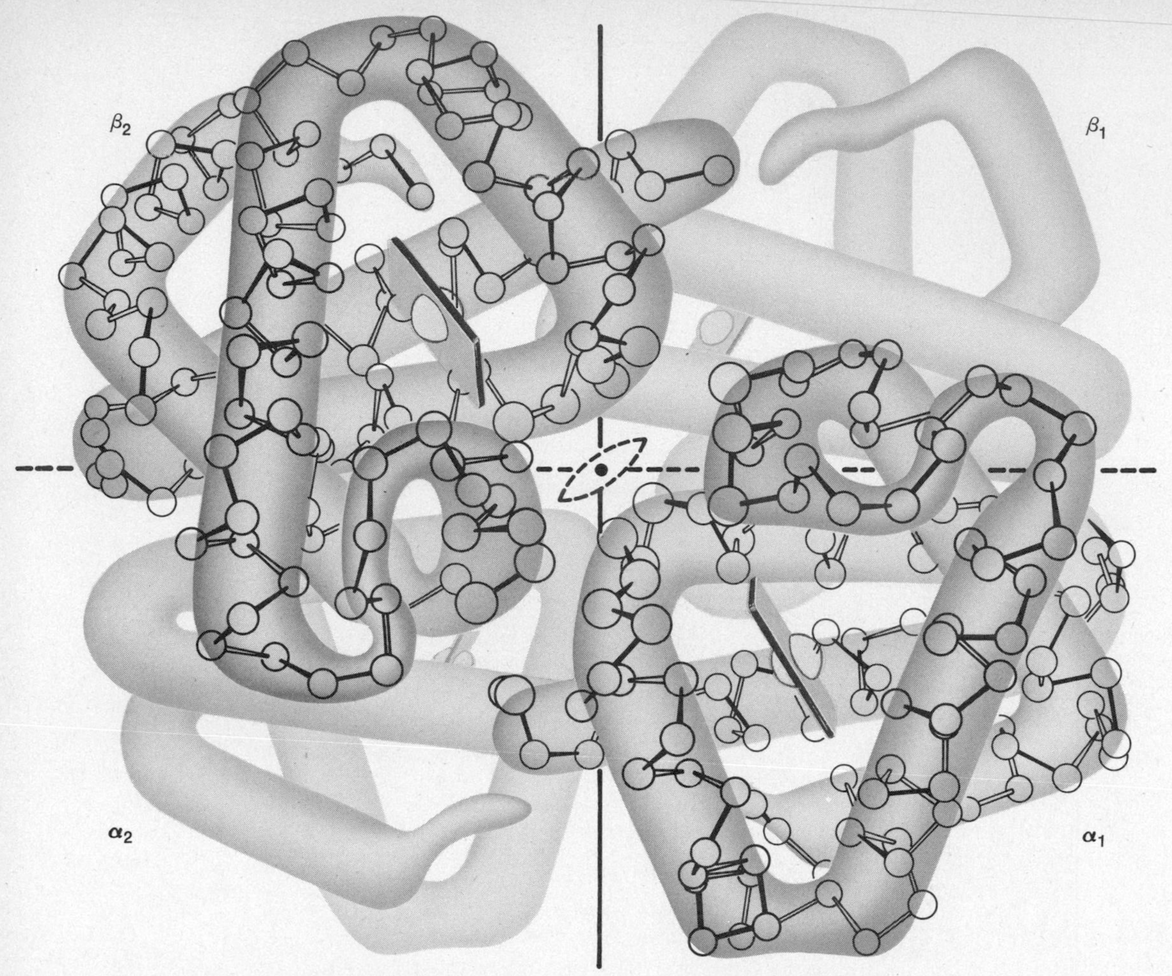

Fig. 33.15. Structure of hemoglobin; α_1, α_2, β_1, β_2 designate the four protein units comprising the hemoglobin molecule.

has been determined are globular in shape; and all enzymes possess, to a greater or lesser degree, the capacity to (a) increase the rate of a reaction, (b) exert chemical selectivity—*i.e.*, choose a particular species from a mixture, (c) exert stereochemical selectivity—*i.e.*, discriminate between enantiomers, and (d) effect catalysis under generally mild conditions—*i.e.*, near pH 7 and a temperature near 37°.* The question of how the enzymes perform these feats of catalysis, which to the organic chemist in the laboratory seem almost miraculous, continues to be one of the most fascinating problems in the field of organobiochemistry.

---

* Some enzymes are capable of operating under considerably more strenuous conditions. Pepsin, for example, can tolerate solutions as acidic as pH 1–2, and the enzymes in certain thermophilic organisms can tolerate temperatures up to 95°C.

A particularly well-studied enzyme is chymotrypsin which, as we have already seen (see Sec. 33.1), is a molecule that specifically catalyzes the hydrolysis of peptide bonds that are adjacent to amino acid residues that carry an aromatic group. Chymotrypsin consists of 241 amino acid residues in a single chain that is folded back on itself into a globular array containing a deep crevice at one region on its surface (Fig. 33.16).

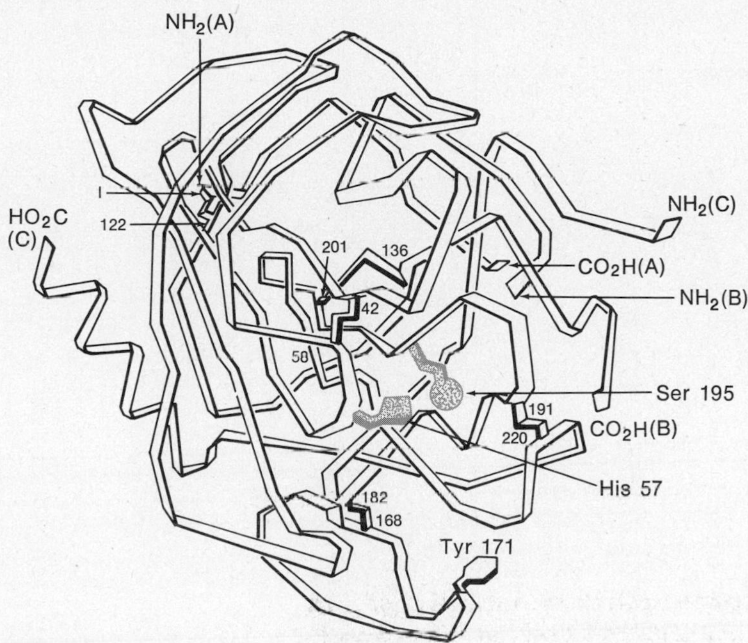

Fig. 33.16. Three-dimensional model of α-chymotrypsin from X-ray analysis.

Although chymotrypsin is less chemically selective than many enzymes and acts on a variety of substrates, including simple amides and esters, its maximum activity is manifested toward peptides containing aromatic amino acid residues, and it shows an almost complete discrimination between D- and L-enantiomers. Its ability to choose between enantiomers is not unusual but is shared by virtually all enzymes and is associated with the contour of the region of the enzyme in which the substrate molecule becomes attached. This area is referred to as the *active site* and is located near the bottom of the crevice in the enzyme. The active site possesses an architecture complementary to that of the substrate, and it binds the substrate through the action of van der Waals forces, electrostatic forces, hydrogen bond formation, or covalent bond formation (or various combinations of these). The active site is chiral, so its combination with the (*R*) and (*S*) enantiomers of a substrate, whatever the mode of binding, corresponds to the formation of a pair of diastereomers (see Section 19.11); all the properties, including the strength of the enzyme-substrate binding, will be different for the two diastereomers. In many cases, in fact, the strength of the enzyme-substrate bind-

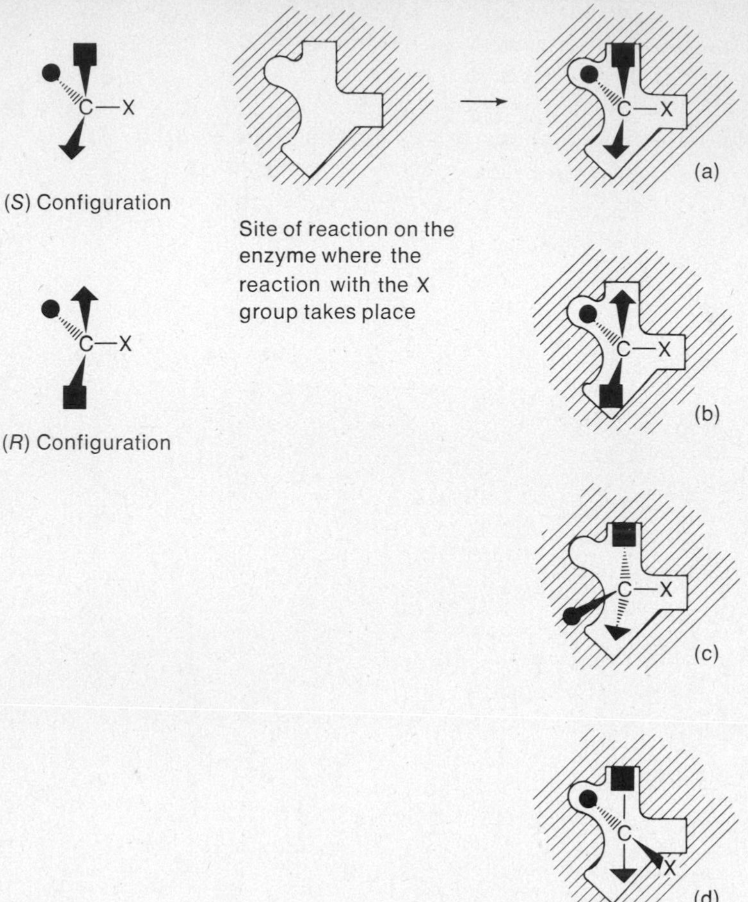

Fig. 33.17. Schematic representation of diastereomeric binding between enzyme and (*R*) and (*S*) substrates: (a) enzyme ~ (S)-substrate complex; (b), (c), and (d) enzyme ~ (*R*)-substrate complex.

ing is so different that the "unnatural" enantiomer is either not bound at all or is bound in an orientation that cannot lead to subsequent reaction (Fig. 33.17). Thus, the (*S*)-configuration of the substrate in Fig. 33.17 chirality sequence arbitrarily chosen as

fits properly into the cavities provided by the enzyme in a fashion that places the X group at that part of the enzyme where chemical reaction can take place.

The (*R*)-configuration, on the other hand, does not fit well. If the X group is to be in the proper position, at least one of the other groups cannot fit properly; conversely, if the other three groups fit into their respective niches, the X group is out of place. It is the ability of the enzymes to respond with great sensitivity to these subtle differences in complementariness that accounts for their extraordinary ability to discriminate between enantiomers. The proper contouring of the surface at the active site is achieved only through the molecular complexity of the tertiary protein structure. Chymotrypsin, for example, loses its enzymatic activity when it is subjected to any condition that causes it to lose its secondary and/or tertiary structure (*i.e.*, to denature).

The chemistry that transpires at the active site of chymotrypsin, once the substrate molecule is in position, has been studied in great detail. It is known that a serine unit at position-195 and a histidine unit at position-57 are close to each other in space in the tertiary structure (see Fig. 33.16), and that they are both involved in the actual catalysis and hydrolysis (i.e., the serine in covalent bonding and the histidine in the transfer of protons between substrate and enzyme). The steps, shown in simplified form in Fig. 33.18, involve (a) the transfer of an "acyl moiety" from the substrate to the enzyme to produce an amino compound and an "acyl enzyme," and (b) hydrolysis of the acyl enzyme to regenerate the original enzyme and liberate the carboxyl compound.

Chymotrypsin shows its optimum activity at pH 8, where it is about 10^9 times more effective as a catalyst than protons or hydroxyl ions. Below pH 8, the activity falls, because the imidazole nitrogen of histidine-57 becomes increasingly protonated, the free base being necessary for catalytic activity. Above pH 8, the activity also falls, because the N-terminal nitrogen becomes decreasingly protonated, the quaternary form being necessary for catalytic activity. The fall in activity above pH 8 is explained in terms of an electrostatic interaction between a carboxylate ion of a C-terminal group or an aspartic or glutamic unit and the ammonium group of the N-terminal residue; at or below pH 8, the N-terminal group is protonated and ion pairing is effective; but above pH 8, the N-terminal group is increasingly in the free base form, and ion pairing diminishes. Apparently, the conformation that is maintained in the ion paired structure is conducive to enzyme activity, whereas the conformation of the non-ion-paired form is not. This is a dramatic example of the critical sensitivity of enzyme activity to the finely detailed structure of the enzyme.

Chymotrypsin is but one of several thousand enzymes that exist in nature. Some are smaller, such as lysozyme (from egg white), which has a molecular weight of 13,930; some are very much larger, such as glutamic dehydrogenase (from bovine liver), which has a molecular weight of 1 million. Some enzymes act on large substrates—*e.g.*, ribonuclease, which catalyzes the hydrolysis of nucleic acids; some act on small substrates—*e.g.*, catalase, which increases the rate at which hydrogen peroxide decomposes to water and oxygen. All enzymes, regardless of their size or substrate specificity, interact in some fashion with the substrate. In some instances it is the R groups of the amino acid units of the enzyme that provide the basis for this interaction; in other instances the enzyme makes use of a second substrate molecule called a *coenzyme*, the interaction in this case taking place, at least in part, between the substrate and the coenzyme

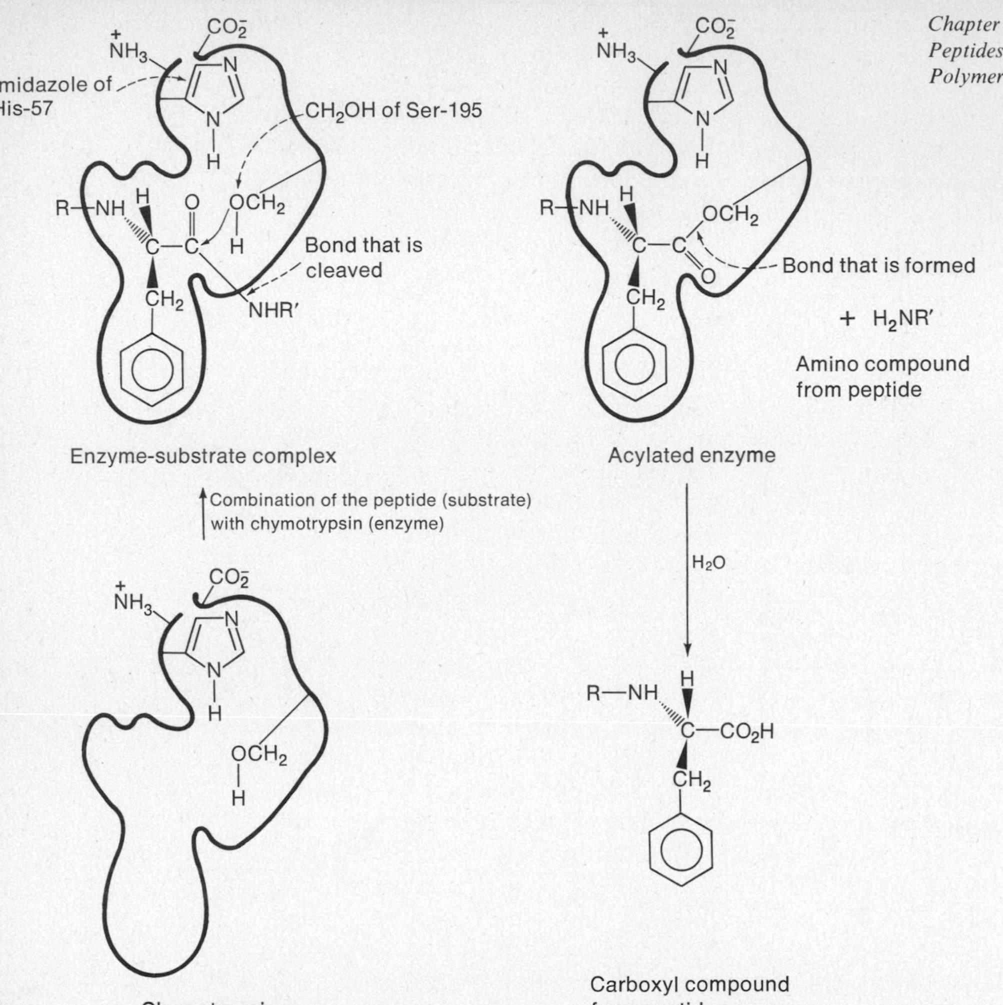

Fig. 33.18. Mechanism of chymotrypsin-catalyzed hydrolysis of a peptide.

molecule, both of which are bound to the enzyme. Nicotinamide adenine dinucleotide (abbreviated NAD^+ in its oxidized form, NADH in its reduced form—see Fig. 32.41), for example, is a typical coenzyme that is involved in a variety of enzyme-catalyzed oxidation-reduction processes; the conversion between 3-phosphoglycerate and glyceraldehyde-3-phosphate in the carbon cycle of photosynthesis (see Fig. 32.1) and the reduction of acetaldehyde to ethanol are both illustrative of processes that require the NAD^+—NADH coenzyme system. The stereochemistry of the acetaldehyde-ethanol interconversion has been studied in some detail. It has been shown that when the reduction is carried out with NADH containing a deuterium atom at the C-4 position of the dihydropyridine ring, the deuterioethanol that is produced is optically active (see Fig. 19.57), illustrating the stereospecificity that is characteristic of most enzyme-catalyzed reactions (Fig. 33.19).

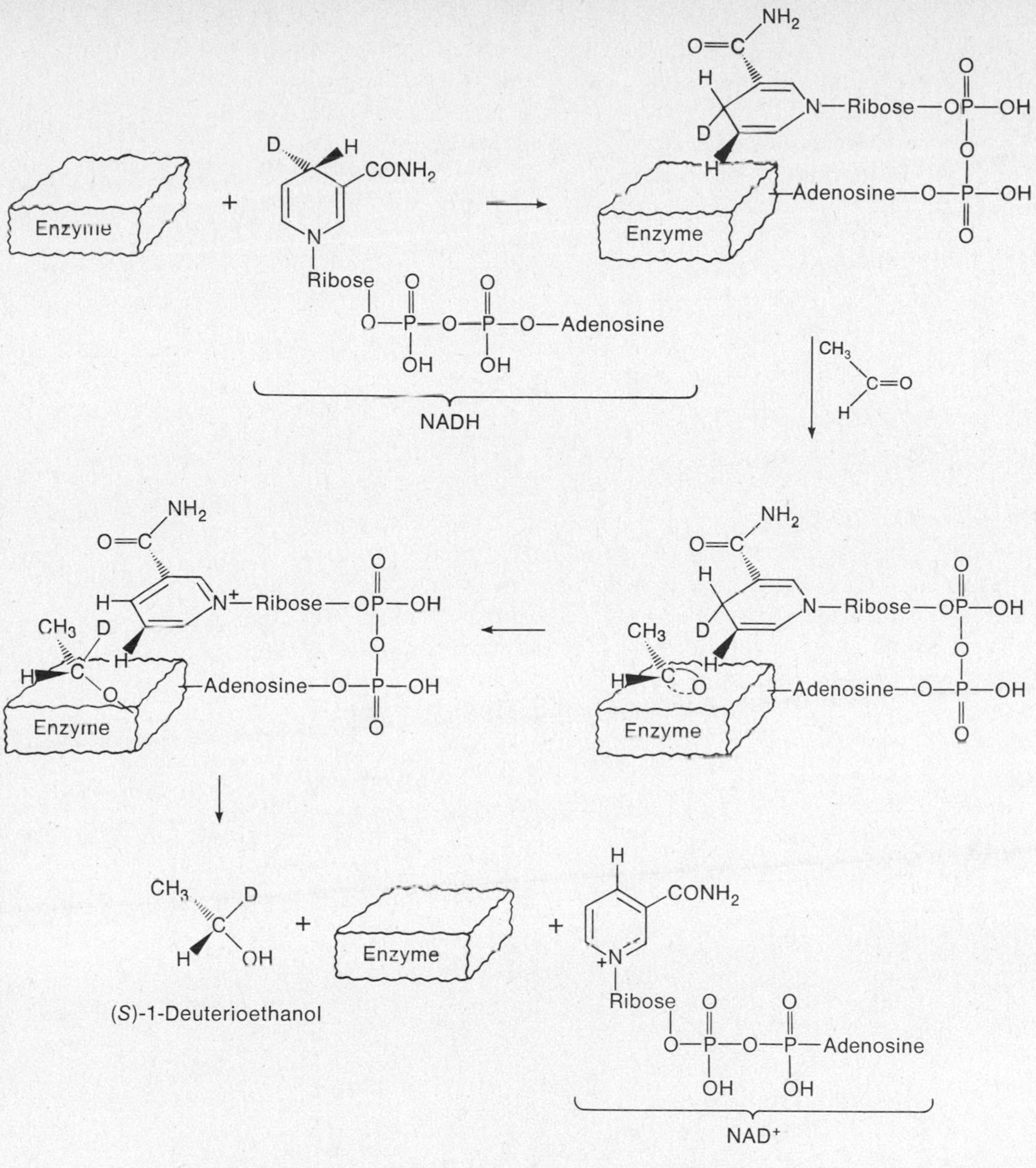

Fig. 33.19. Stereospecific enzyme-catalyzed reduction of acetaldehyde.

33.7 *Amino Acids, Peptides, and Proteins in Combination with Other Compounds*

Many proteins occur in combination with other compounds. For example, hemoglobin, a compound that is involved in the assimilation of oxygen into certain respiring organisms, is a porphyrin-protein complex in which four iron-containing porphyrin molecules are combined with a protein molecule (see Fig. 36.1). Porphyrin-protein complexes are also involved in biological oxidation processes whereby the chemical bond energy of ingested food is made available to the organism.

Lipoproteins occur in various places in the living organism such as the blood plasma and cell membranes. They consist of proteins and lipids (see Chapter 34) combined in a fashion that is not yet completely understood. One of the proposed structures for cell membranes is a lipoprotein with the protein furnishing the "outer coat" and the polar lipids the "inner coat" (Fig. 33.20) of the membrane, the bonding between the protein and the lipid arising from van der Waals attractions, hydrogen bonding, and electrostatic interactions.

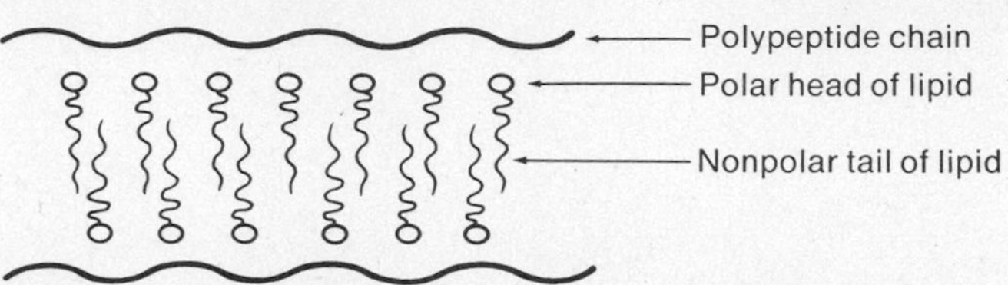

Fig. 33.20. Two-unit membrane model.

Proteins occur in combination with saccharide units in a variety of naturally occurring compounds called glycoproteins. A particularly interesting and important example is the cell wall material of bacteria. The bacterial cell is enclosed in a sacklike structure that essentially is one large molecule composed of polysaccharide chains (containing the repeating disaccharide unit consisting of N-acetyl-D-glucosamine and N-acetylmuramic acid) joined to a polypeptide in cross-linked polymer combination (Fig. 33.21). The resulting structure is netlike,

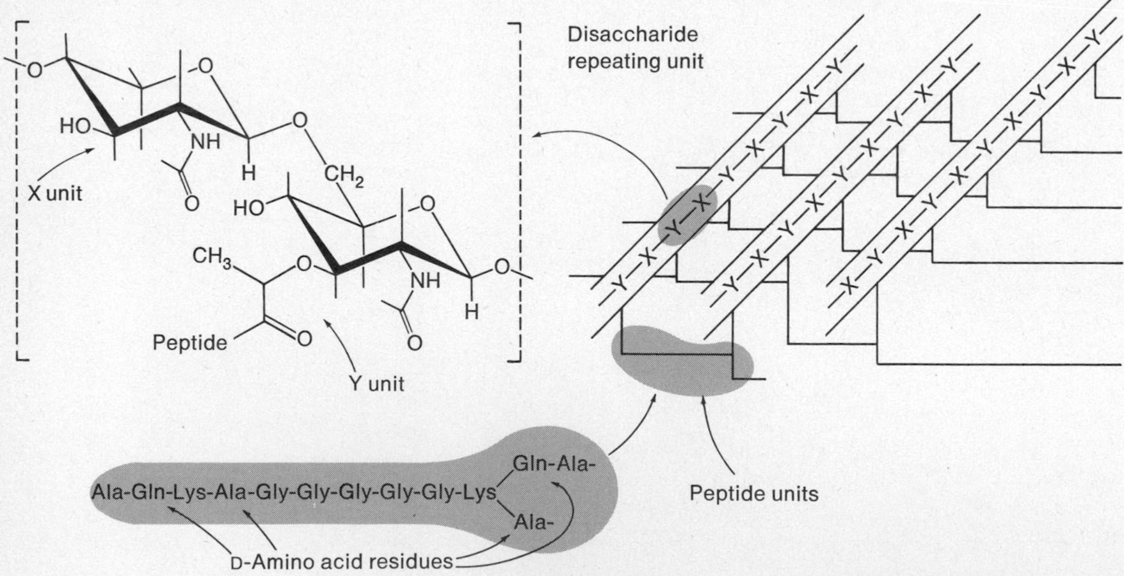

Fig. 33.21. Structure of bacterial cell wall.

sufficiently strong to protect the interior of the cell, yet sufficiently porous to permit substances to pass in and out of the cell. The biosynthesis of the bacterial cell wall is a complicated process in which the disaccharide unit is first assembled and then attached to a short peptide chain, which, in the third stage of the process, becomes cross linked with other units of similar structure to form the final product. Penicillin (sec p. 413), one of the most useful antibiotics currently available to man, has the ability to interfere with the formation of the cross linking and, thereby, to prevent the completion of the construction of the bacterial cell wall.

REFERENCES

AWAPARA, J., *Introduction to Biological Chemistry*, Prentice-Hall, Inc., Englewood Cliffs, N.J., 1968.

BARKER, R., *Organic Chemistry of Biological Compounds*, Prentice-Hall, Inc., Englewood Cliffs, N.J., 1971, Chapter 4.

Bio-Organic Chemistry: Readings from Scientific American, W. H. Freeman and Co., Publishers, San Francisco, 1968.
DOTY, P., Article 2. Proteins.
STEIN, W. H., and S. MOORE, Article 3. The Chemical Structure of Proteins.
THOMPSON, E. O. P., Article 4. The Insulin Molecule.
PERUTZ, M. F., Article 5. The Hemoglobin Molecule.
ZUCKERKANDL, E., Article 6. The Evolution of Hemoglobin.
PHILLIPS, D. C., Article 7. The Three-Dimensional Structure of an Enzyme Molecule.

DICKERSON, R. E., and I. GEIS, *The Structure and Action of Proteins*, Harper & Row Publishers, New York, 1969.

ELMORE, D. T., *Peptides and Proteins*, Cambridge University Press, London, 1969.

FLORKIN, M., and E. H. STOTZ, eds., *Comprehensive Biochemistry*, Vol. 7, *Proteins*, Elsevier Publishing Co., New York, 1963.

KOPPLE, K. D., *Peptides and Amino Acids*, W. A. Benjamin, Inc., New York, 1966.

LANDE, S., Conformation of Peptides: Speculations Based on Molecular Models, *Journal of Chemical Education*, **45**, 587 (1968).

LAW, H. D., *The Organic Chemistry of Peptides*, John Wiley & Sons, Inc., New York, 1970.

LEHNINGER, A. L., *Biochemistry*, Worth Publishers, Inc., New York, 1970, Chapters 3–9.

PROBLEMS

1. Draw the structures of the following compounds:
(a) Glycylleucine (b) Prolylthreonine (c) Aspartylaspartyltryptophan
(d) Ethyllysylargininate (e) Alanylglutamine

2. Indicate the chemical means that you would use to distinguish between the members of the following pairs of compounds:

(a) $CH_3CH(\overset{+}{N}H_3)CO_2^-$ and $CH_3CH(NHCOCH_3)CO_2H$

(b) 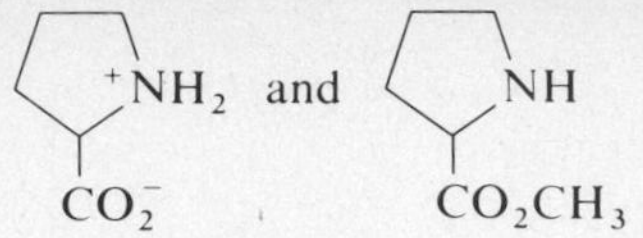 [pyrrolidine ring with $^{+}NH_2$ and CO_2^-] and [pyrrolidine ring with NH and CO_2CH_3]

(c) $C_6H_5CH_2\overset{+NH_3}{\overset{|}{C}}HCO_2^-$ and $CH_3CH_2\underset{CH_3}{\underset{|}{C}}H\overset{+NH_3}{\overset{|}{C}}HCO_2^-$

(d) $HSCH_2\overset{+NH_3}{\overset{|}{C}}HCO_2^-$ and $CH_3SCH_2CH_2\overset{+NH_3}{\overset{|}{C}}HCO_2^-$

(e) $HO_2CCH_2\overset{+NH_3}{\overset{|}{C}}HCO_2^-$ and $H_2N(CH_2)_4\overset{+NH_3}{\overset{|}{C}}HCO_2^-$

3. For each of the following compounds, draw the structure of the species that exists at the designated pH:
(a) Valine at pH 8 (b) Serine at pH 1 (c) Tryptophan at pH 12
(d) Lysine at pH 10 (e) Glutamic acid at pH 3

4. Define each of the following terms and, if possible, give an illustrative example:
(a) An enzyme (b) Tertiary structures (c) Denaturation
(d) α-Helix (e) Lipoprotein

5. Construct a framework model of dihydroxyacetone phosphate, twist it into the conformation shown, and orient it in space with the phosphate group pointed downward.

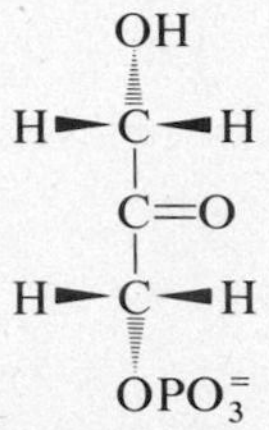

```
      OH
      |
  H—C—H
      |
     C=O
      |
  H—C—H
      |
    OPO3=
```

With the molecule anchored between your knees, convince yourself that the two hydrogens at C_3 are not related to your body in exactly the same way; *i.e.*, one of them is close to your right hand and the other is close to your left hand. What is the relationship between the hydrogens (see p. 533)? Aldolase has the ability to distinguish between these hydrogens. Thus, when dihydroxyacetone phosphate is dissolved in D_2O in the presence of aldolase, C_3-deuteriodihydroxyacetone phosphate is obtained. Is it optically active? Triosephosphate isomerase has the same property as aldolase except that it has the opposite stereospecificity with respect to the hydrogens at C_3. What is the stereochemical outcome when dihydroxyacetone phosphate is dissolved in D_2O in the presence of triosephosphate isomerase?

6. A small peptide that we shall call peptide-A is shown by a study of its hydrolysis products to have the following amino acid composition: (Ala, 2 Gly, Lys, 3 Phe, Ser, Val). End-group analyses indicate that a phenylalanine residue is at the C-terminal end and a glycine residue at the N-terminal end. When subjected to chymotrypsin-catalyzed hydrolysis, two peptides are formed, designated as peptide-A_1 and peptide-A_2. On complete hydrolysis, peptide-A_1 yields 2 glycine, phenylalanine, and valine; peptide-A_2 yields alanine, lysine, phenylalanine, and serine. Partial hydrolysis of peptide-A_1 yields

(Gly, Val) and (Phe, Val); partial hydrolysis of peptide-A_2 yields (Lys, Ser), (Ala, Lys), and (Phe, Ser). What is the structure of peptide-A?

7. Gramicidin-S, a mold metabolite possessing antibiotic properties, has a molecular weight of about 1300. Complete hydrolysis indicates that it contains leucine, ornithine, phenylalanine, proline, and valine. Ornithine is one of the less commonly occurring amino acids and has the structure

$$H_3\overset{+}{N}CH_2CH_2CH_2\underset{}{\overset{NH_2}{\overset{|}{C}}}HCO_2^-$$

End-group analyses show that there is no C-terminal end group, and the only N-terminal group that reacts is the amino group of ornithine, leading to the conclusion that Gramicidin-S is a cyclic peptide. Partial hydrolysis of Gramicidin-S yields the following peptides: (Leu, Phe), (Phe, Pro), (Leu, Orn), (Orn, Val), (Phe, Pro, Val), (Orn, Pro, Val), (Leu, Orn, Val). What is the structure of Gramicidin-S?

8. Draw the structures of the products that are formed in each of the following reactions

(a) 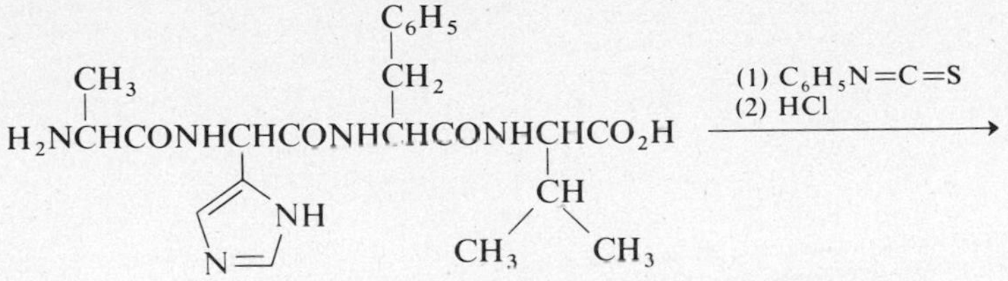 $H_2N\overset{CH_3}{\overset{|}{C}}HCONH\underset{\text{(imidazol-4-yl)}CH_2}{CH}CONH\overset{C_6H_5CH_2}{\overset{|}{C}}HCONH\underset{CH(CH_3)_2}{\underset{|}{C}}HCO_2H \xrightarrow[(2)\ HCl]{(1)\ C_6H_5N{=}C{=}S}$

(b) Tetrapeptide shown in part (a) $\xrightarrow[\text{chymotrypsin}]{H_2O}$

(c) 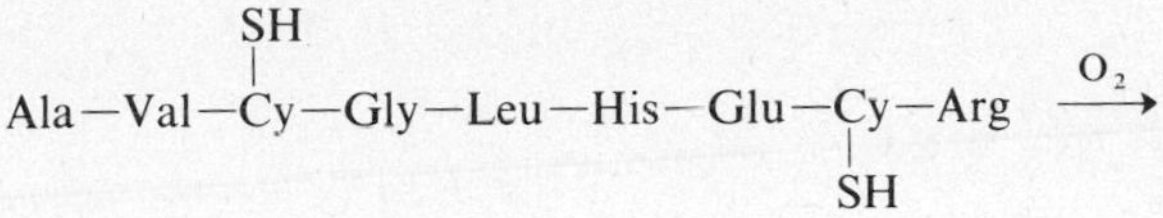 Ala—Val—$\overset{SH}{\overset{|}{\text{Cy}}}$—Gly—Leu—His—Glu—$\underset{SH}{\underset{|}{\text{Cy}}}$—Arg $\xrightarrow{O_2}$

(d) Polylysine (α-helix) $\xrightarrow{\text{adjust to pH 2}}$

(e) $NAD^+ + CH_3CD_2OH \xrightarrow[\text{enzyme}]{}$

Chapter **34**

Acetogenins: lipids, terpenes, steroids, and other compounds

A man must take the fat with the lean.

CHARLES DICKENS
David Copperfield

Carbohydrates and peptides contain hetero atoms, principally oxygen and nitrogen, in relatively high proportion with respect to the carbon atoms; the C/O ratio in sucrose, for example, is $\frac{12}{11}$ and the C/(N + O) ratio in the nonapeptide bradykinin is $\frac{50}{26}$. Consequently, carbohydrates and peptides possess well-developed polar characteristics and are often rather soluble in water and other polar solvents. Conversely, other important types of naturally occurring compounds contain carbon in much higher proportion relative to the hetero atoms present and, therefore, are much less soluble in water and much more soluble in nonpolar solvents. Included among these are fats, waxes, oils, terpenes, and steroidal compounds. All these compounds share the property of having a high carbon content and a low water solubility, although in structural details they differ from one another significantly. The term *lipid* has generally been employed to embrace only the fats, waxes, and oils, but its jurisdiction has sometimes been increased to include all the aforementioned types of compounds. In terms of biochemical etiology, the wider membership is justified, for all these compounds are biosynthesized from two-carbon acetate units (Fig. 34.1), and can be designated, therefore, as *acetogenins* (*i.e.*, having their origin in acetate units). The acetate unit is "carried" in the biosynthetic pathway by coenzyme-A, which is a complex molecule containing a terminal SH group; when an acetate function is attached to the sulfur atom of the coenzyme, *i.e.*,

$$RSH + CH_3CO_2H \longrightarrow RSC(=O)CH_3 + H_2O$$

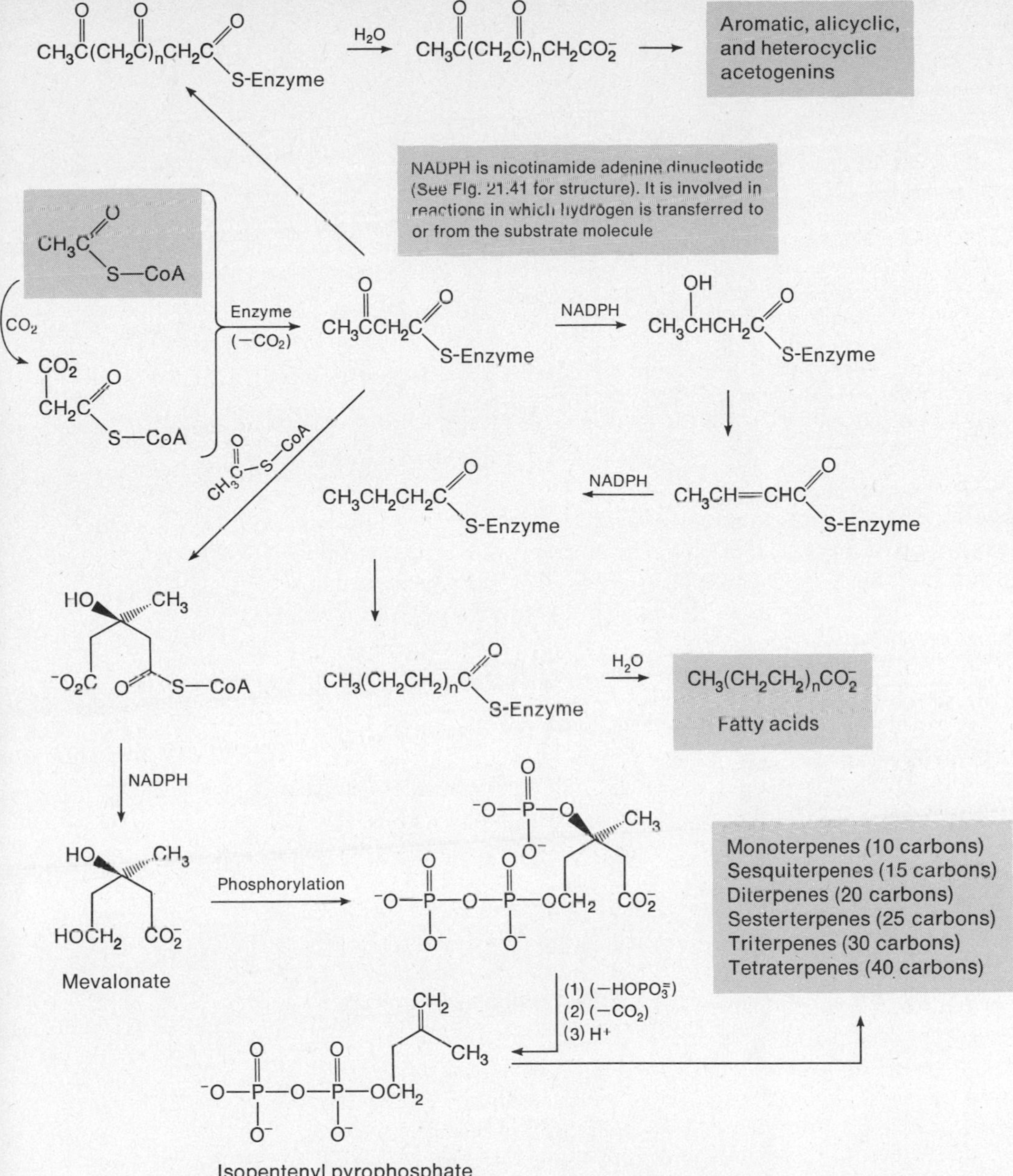

Fig. 34.1. Biosynthesis of acetogenins from acetylcoenzyme-A.

the compound known as acetyl coenzyme-A is obtained (abbreviated CoA—$SCOCH_3$ or Acetyl CoA) (see Fig. 34.1).

34.1 *Fatty Acids, Fats Oils and Waxes*

Fatty acids are carboxylic acids (see Chapter 14) that contain long hydrocarbon chains, saturated or unsaturated. Most of the commonly occurring fatty acids

Table 34.1 NATURALLY OCCURRING FATTY ACIDS

Common Name	Systematic Name	Structure	Carbon Atoms	mp
Lauric acid	*n*-Dodecanoic acid	$CH_3(CH_2)_{10}CO_2H$	12	44.2
Myristic acid	*n*-Tetradecanoic acid	$CH_3(CH_2)_{12}CO_2H$	14	53.9
Palmitic acid	*n*-Hexadecanoic acid	$CH_3(CH_2)_{14}CO_2H$	16	63.1
Stearic acid	*n*-Octadecanoic acid	$CH_3(CH_2)_{16}CO_2H$	18	69.6
Arachidic acid	*n*-Eicosanoic acid	$CH_3(CH_2)_{18}CO_2H$	20	76.5
Lignoceric acid	*n*-Tetracosanoic acid	$CH_3(CH_2)_{22}CO_2H$	24	86.0
Oleic acid	9-Octadecenoic acid	$CH_3(CH_2)_7CH{=}CH(CH_2)_7CO_2H$	18	13.4
Linoleic acid	9,12-Octadecadienoic acid	$CH_3(CH_2)_4CH{=}CHCH_2CH{=}CH(CH_2)_7CO_2H$	18	−5
Linolenic acid	9,12,15-Octadecatrienoic acid	$CH_3CH_2CH{=}CHCH_2CH{=}CHCH_2CH{=}CH(CH_2)_7CO_2H$	18	−11
Arachidonic acid	5,8,11,14-Eicosatetraenoic acid	$CH_3(CH_2)_4CH{=}CHCH_2CH{=}CHCH_2CH{=}CHCH_2CH{=}CH(CH_2)_3CO_2H$	20	−49.5

have an unbranched chain and an even number of carbon atoms, reflecting their biogenesis from two-carbon units (Table 34.1). The configuration at the double bonds in unsaturated fatty acids is usually (*Z*), (*cis*), and, as a result, the extended conformation of these compounds shows bends at the points of unsaturation (Fig. 34.2).

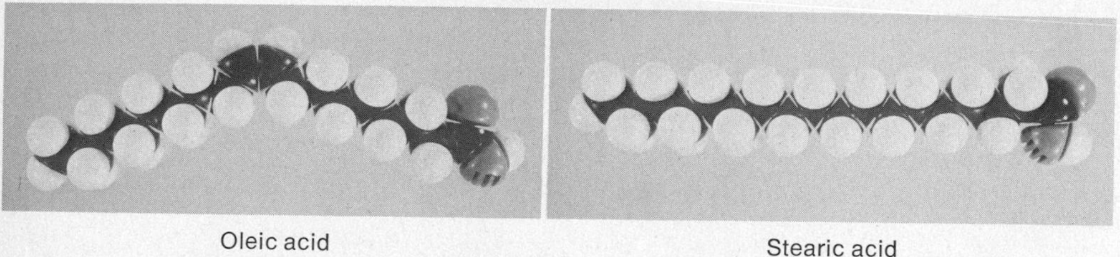

Oleic acid　　　　Stearic acid

Fig. 34.2. Extended conformations of fatty acids.

The fatty acids are examples of compounds that possess a highly hydrophobic portion (the hydrocarbon chain) and a highly hydrophilic portion (the carboxyl group). Such compounds are called *amphipathic* and have the interesting property of forming *micelles* when placed in an aqueous medium. A fatty acid micelle is an aggregate, approximately spherical in shape, composed of a large number of molecules oriented in such a fashion that the hydrocarbon portion points inward and the polar carboxyl groups are arrayed on the surface of the sphere (Fig. 34.3). It is a stable structure; the hydrophobic portions are shielded from the water and can interact with each other via van der Waals forces, whereas, at the same time, the hydrophilic portions are exposed to the water and can be solvated via hydrogen bonding. Conversion of a fatty acid to its sodium or potassium salt increases the polarity of one end of the molecule and increases the ease of micelle formation. Sodium stearate, for example "dissolves" in water to give what might be assumed to be a true solution but that, in fact, (except at

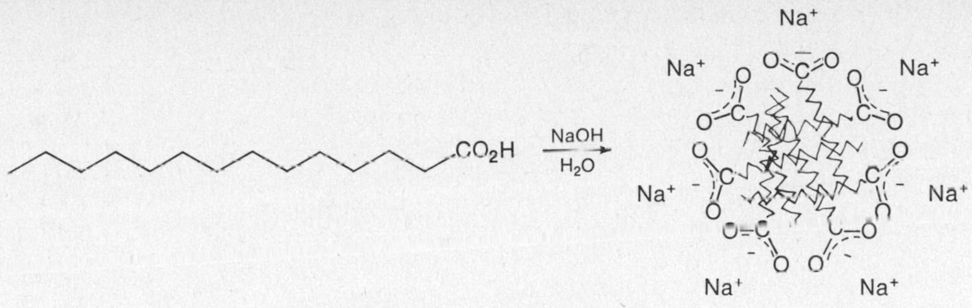

Fig. 34.3. Micelle formation.

very high dilution) is an emulsion in which aggregates of stearate ions are associated with sodium cations at the surface of the micelles. Repulsion between the like-charged micelles prevents them from coalescing and thereby maintains the emulsion. Sodium stearate is a typical soap, and its ability to give micellar "solutions" capable of emulsifying the fat and grease of dirt (bringing the dirt into the interior of the micelle) is responsible for its utility as one of man's oldest cleansing agents.

Fatty acid moieties occur widely in nature as esters of glycerol, and such compounds are called fats, oils, acyl glycerols, or glycerides. Fats are the major energy storage compounds in the animal organism, a function that starch serves in plants. Since glycerol contains three hydroxyl groups, it can form monoglycerides, diglycerides, or triglycerides (more accurately referred to as mono-, di-, or triacylglycerols). All these exist in nature, although the majority of fats and oils are triglycerides. Tristearin (*i.e.*, tristeroylglycerol) is a typical saturated glyceride; triolein (*i.e.*, trioleoylglycerol) is a typical unsaturated glyceride (Fig. 34.4).

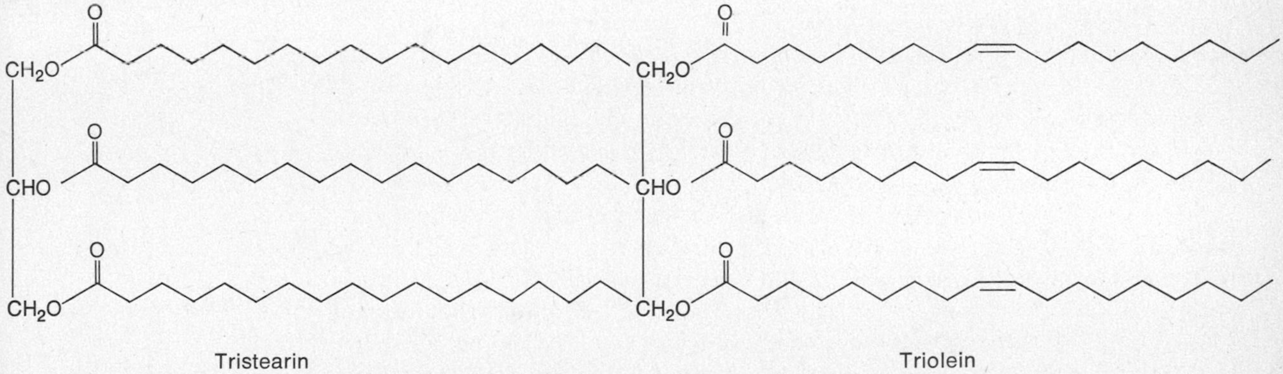

Fig. 34.4. Triglycerides.

The melting points of triglycerides are a function of the fatty acid residues comprising them. Tristearin, for example, is a solid at room temperature and, therefore, is called a fat, whereas triolein is a liquid at room temperature and, therefore, is called an oil. Naturally occurring fats and oils are usually not homoglycerides but are mixed glycerides in which two or even three different fatty acid residues are attached to the same glycerol molecule. Butter, usually classified as a fat, is composed of mixed glycerides containing about 60–70% saturated and 30–40% unsaturated fatty acid residues; cottonseed oil, usually classified as a vegetable oil, is composed of mixed glycerides containing only about 25% saturated and approximately 75% unsaturated fatty acid residues. When fats are hydrolyzed with aqueous sodium hydroxide, the sodium salts of the fatty acids comprising the glycerides are obtained:

$$\begin{array}{l} CH_2OC(=O)R \\ | \\ CHOC(=O)R' \\ | \\ CH_2OC(=O)R'' \end{array} + NaOH \longrightarrow \begin{array}{l} CH_2OH \\ | \\ CHOH \\ | \\ CH_2OH \end{array} \begin{array}{l} + RCO_2Na \\ + R'CO_2Na \\ + R''CO_2Na \end{array}$$

The products are called soaps, and the process, therefore, is referred to as *saponification*, a term that has been adopted by chemists and used now in a much broader sense to denote any base-induced ester hydrolysis. Treatment of cottonseed oil, corn oil, soybean oil, or other unsaturated vegetable oils with molecular hydrogen in the presence of a nickel catalyst at elevated temperature and pressure reduces the double bonds in the unsaturated fatty acid residue and increases the melting point of the material, thereby converting the oil into a fat. Hydrogenated vegetable oils are commercial products familiar to housewives under such brand names as Spry and Crisco. With the possible implication of saturated fats in heart ailments and arterial dysfunction, critical attention is currently being paid by nutritionists to the dietary hazard of hydrogenated oils and fats.

In addition to furnishing 25–50% of the caloric intake of the average man or woman and providing the starting materials for soaps, glycerides have also been put to use in other ways. Linseed oil and tung oil, for example, have a high content of multiply unsaturated glycerides which undergo oxidative polymerization upon exposure to air. This is called a "drying action" and results in the formation of a hard, tough material that has been used for centuries as paint.

Naturally occurring esters of long-chain fatty acids and long-chain monohydric alcohols are known as waxes. Spermaceti wax from the head of the sperm whale is primarily cetyl palmitate, $C_{15}H_{31}CO_2C_{16}H_{33}$ (mp 42–47°); beeswax is a mixture of esters of C_{26} and C_{28} acids and C_{30} and C_{32} alcohols (mp 60–82°); and carnauba wax is a mixture of esters from C_{24} and C_{28} acids with C_{32} and C_{34} alcohols as well as ω-hydroxy acids, $HO(CH_2)_{17-29}CO_2H$ and α,ω-diols, $HO(CH_2)_{22-28}OH$. Lanolin, used as a base for cosmetic preparations and also classed as a wax, is a complex mixture of esters, alcohols, free fatty acids, and hydrocarbons.

34.2 *Phosphoglycerides, Sphingolipids, and Glycolipids*

Phosphoglycerides are compounds containing glycerol esterified with two fatty acid residues and one phosphoric acid residue; the parent member of the group is phosphatidic acid. Other members of the series contain an additional alcohol moiety attached to the phosphoric acid residue and thus are phosphoric acid diesters (Fig. 34.5). The phosphoglycerides resemble the fatty acids in being amphipathic compounds with the capacity for readily forming micelles in an aqueous medium.

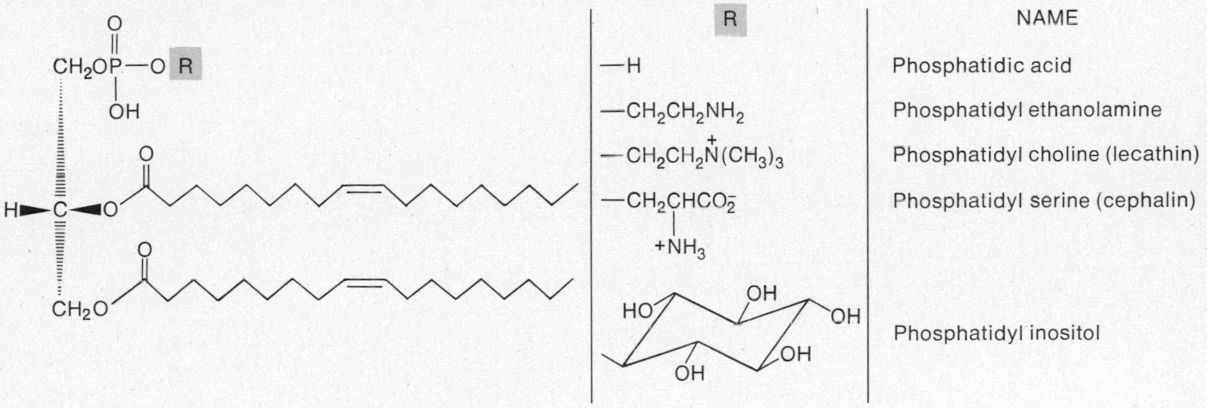

R	NAME
$-H$	Phosphatidic acid
$-CH_2CH_2NH_2$	Phosphatidyl ethanolamine
$-CH_2CH_2\overset{+}{N}(CH_3)_3$	Phosphatidyl choline (lecathin)
$-CH_2CH(^{+}NH_3)CO_2^-$	Phosphatidyl serine (cephalin)
(inositol ring: HO, OH, OH, OH, OH)	Phosphatidyl inositol

Fig. 34.5. Phosphoglycerides.

The glycolipids are compounds of glycerol esterified at C-2 and C-3 with fatty acid residues and joined in glycosidic linkage at C-1, as exemplified by a monogalactosyl diacylglycerol found in certain plants and microorganisms (Fig. 34.6).

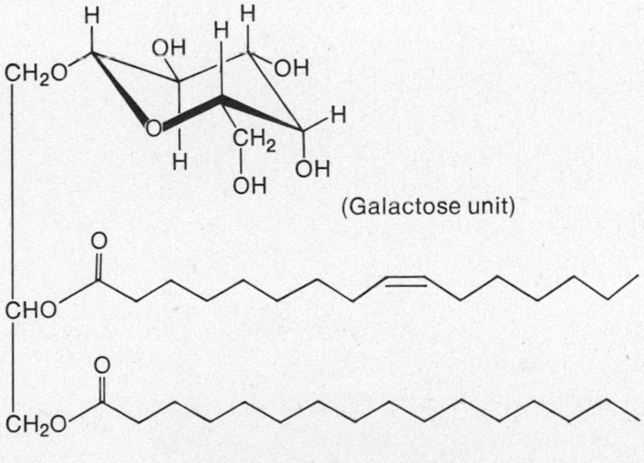

(Galactose unit)

A monogalactosyl diacylglycerol

Fig. 34.6. A glycolipid.

Still another group of compounds, whose architecture is related to the three-carbon framework of glycerol and which resemble phosphoglycerides and glycolipids at C-1, are the sphingolipids, which are found in the membranes of plants and animals and are particularly abundant in brain and nerve tissue. They differ from the glycolipids in having nitrogen instead of oxygen attached at the C-2 position (*i.e.*, an amide linkage) and in having a free hydroxyl group and a carbon–carbon attachment to a long-chain hydrocarbon at the C-3 position (Fig. 34.7).

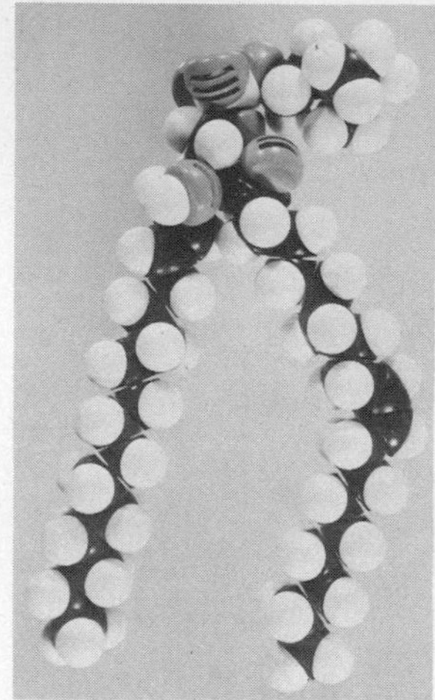

Sphingomyelin

$CH_2OP(=O)(OH)OCH_2CH_2\overset{+}{N}(CH_3)_3$

CHNH–C(=O)–(chain)

HO—CH—(chain)

Sphingomyelin

A cerebroside

Fig. 34.7. Sphingolipids.

34.3 *Terpenes*

Not all acetogenins are polymers of two-carbon units in so obvious a way as the fats and waxes. Terpenes, for example, are more readily perceived to be polymers of a five-carbon moiety referred to as an *isoprene unit*; the great majority of terpenes can be "mentally" dissected into these units (Fig. 34.8). The isoprene units of terpenes, nevertheless, have their genesis in acetylcoenzyme-A, which is converted into the six-carbon compound mevalonic acid. Phosphorylation of mevalonic acid (see Fig. 34.1 for structure) followed by loss of phosphoric acid and carbon dioxide moieties then yields isopentenyl pyrophosphate, which is the biological source of isoprene units and which, under the influence of various enzymes, condenses to form 10-carbon compounds (monoterpenes), 15-carbon

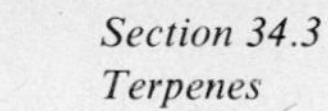

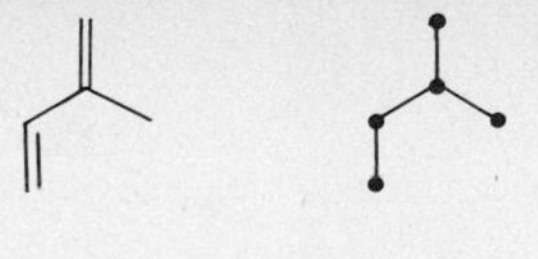

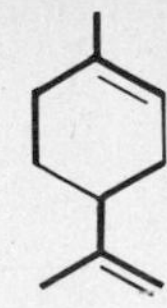

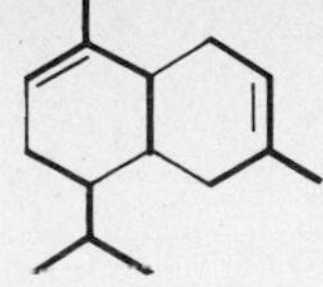

Isoprene 2-Methyl 1,3-butadiene

An isoprene unit

Limonene (a monoterpene from lemon oil)

β-Cadinene (sesquiterpene from the oil of cade)

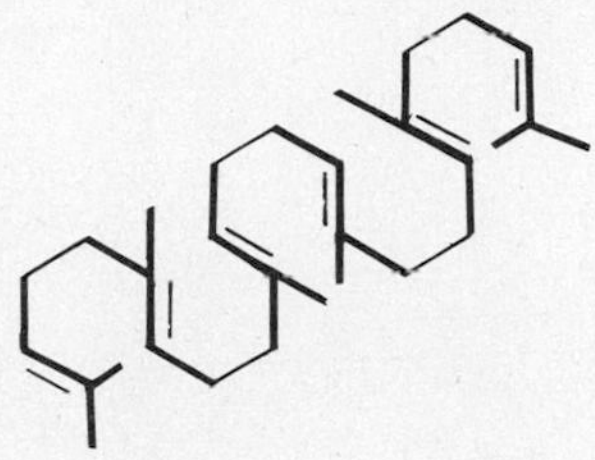

Cembrene (a diterpene from pine oil)

Squalene (a triterpene from shark liver oil)

β-Carotene (a tetraterpene from carrots)

Fig. 34.8. Dissection of terpenes into isoprene units.

compounds (sesquiterpenes, 20-carbon compounds (diterpenes), 25-carbon compounds (sesterterpenes), 30-carbon compounds (triterpenes), and 40-carbon compounds (tetraterpenes) (see Fig. 34.1).

The compounds shown in Fig. 34.8 represent but a few of the many ways in which isoprene units can be fitted together to give acyclic structures such as squalene or cyclic structures such as limonene, cadinene, and cembrene. Although for illustrative purposes the examples chosen for Fig. 34.8 are terpene *hydrocarbons* and contain only carbon and hydrogen, the majority of terpenes contain one or more oxygen functions, generally present as hydroxyl or carbonyl groups (Fig. 34.9).

The monoterpenes are primarily of plant origin, and a very large number have been isolated and characterized. They are generally found in the "essential oils," many of which (*e.g.*, menthol and camphor) are pleasantly fragrant and have been used in perfumes and medicines.

A particularly interesting terpene derivative is the acyclic molecule farnesol pyrophosphate, which is produced *in vivo* by the condensation of two isopentenyl pyrophosphate units (see Fig. 34.1) with a dimethylallyl pyrophosphate unit (isomeric with isopentenyl pyrophosphate). It is known to be the *biological*

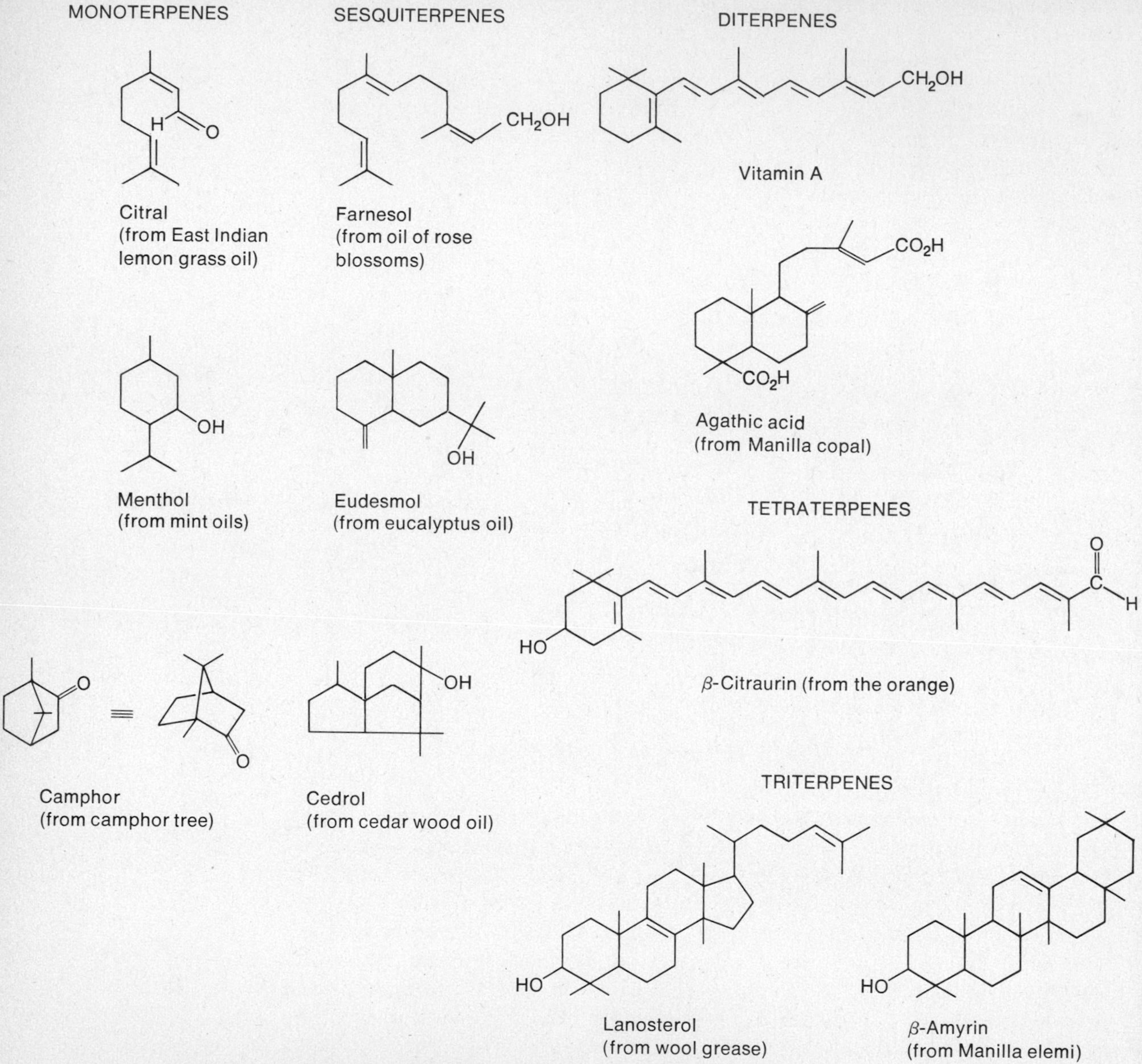

Fig. 34.9. Oxygenated terpenes.

precursor of some of the sesquiterpenes as well as the triterpene squalene. Among the many sesquiterpenes formed from farnesyl pyrophosphate is the bicyclic compound β-vetivone, which is isolated from vetiver oil. The proof of structure of β-vetivone provides a good example of the trials and tribulations that often beset the chemist investigating terpene structures. When β-vetivone is dehydrogenated by heating it with selenium or sulfur at an elevated temperature, 1,5-dimethyl-7-isopropylazulene is formed, and on the basis of this result the terpene was assigned structure (a) in Fig. 34.10. A total synthesis of structure (a), however, yields a product different from β-vetivone, and investigation along other lines reveals that β-vetivone actually has structure (b) in Fig. 34.10. What has happened during dehydrogenation is that a skeletal rearrangement has taken place, obscuring the true structure of the original hydrocarbon. Although dehydrogenation products have provided invaluable evidence in the elucidation of the structures of many natural products, the results of such experiments must be used with discretion, as this example of β-vetivone illustrates.

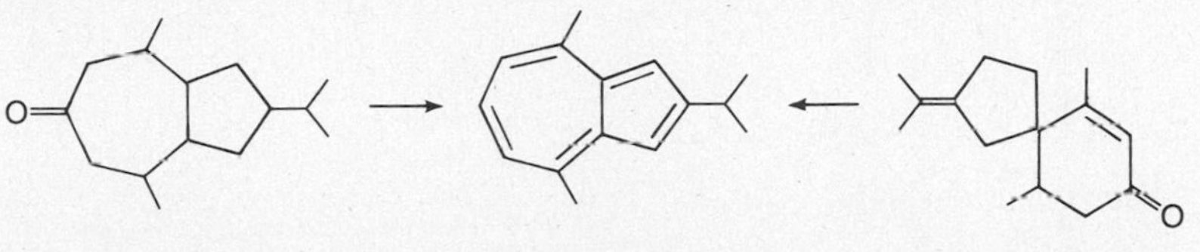

Structure (a) An azulene Structure (b)

Fig. 34.10. Dehydrogenation of β-vetivone [structure (b)] and an isomer of β-vetivone [structure (a)].

One of the most important of the diterpenes is vitamin A, a fat-soluble substance that is an essential ingredient in the physiology of vision in animals. In its more stable form, vitamin A exists as the all-*trans* isomer. *In vivo* oxidation to the corresponding aldehyde followed by enzyme-catalyzed isomerization of the double bond at C-11 from *trans* to *cis* yields neoretinene-b, which combines with a protein called scotopsin to form the visual pigment rhodopsin. When rhodopsin, located in the retinal rods and cones of the eye, is exposed to light, a *cis*-to-*trans* isomerization of neoretinene-b occurs which triggers a response that is transmitted to the brain and perceived as vision (Fig. 34.11). The details of this transformation were worked out by George Wald and his colleagues, for which he received a Nobel Prize in 1967. The precise mechanism by means of which this slight geometric change is transformed into a nerve impulse is not yet known, but the isomerization phenomenon is, nevertheless, an excellent example of the extraordinary sensitivity of the living organism to small changes in the architecture of the molecules comprising it.

Abietic acid is a typical polycyclic diterpene (Fig. 34.12). It occurs as the chief component in rosin and has considerable commercial importance in the sizing of paper, in the manufacture of soap, as a constituent in varnish, and in numerous other applications. The key to unraveling the basic structure of abietic acid is the sulfur-induced dehydrogenation to 1-methyl-7-isopropylphenanthrene (retene); the carboxyl group and the angular methyl group are lost in the process.

The acyclic triterpene squalene, isolated as a major constituent of shark liver oil, is probably present in all organisms. Formed by a head-to-head dimerization of farnesol pyrophosphate, squalene is the biological precursor of the tetracyclic triterpene lanosterol, which, in turn, is the precursor of the steroid cholesterol. The cyclization of squalene is probably initiated by the action of an electrophile (*e.g.*, H^+) on the epoxide, and in the course of the biosynthetic sequence a migration of two methyl groups takes place (Fig. 34.13).

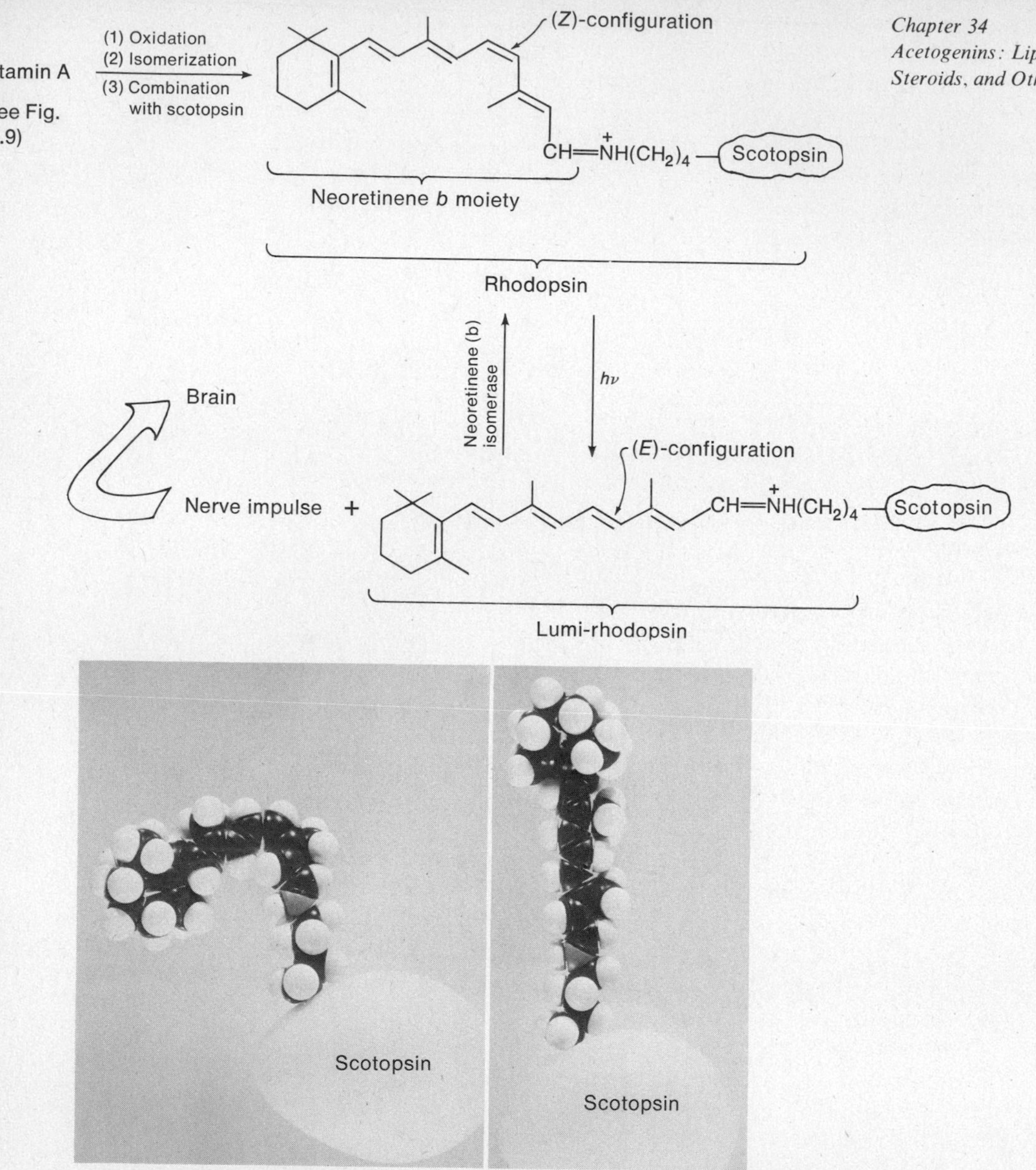

Fig. 34.11. Visual cycle in retinal rods and cones of the eye.

Probably the best-known member of the tetraterpene family is β-carotene, the yellow-red pigment isolated from carrots (hence, the generic name *carotenoids* for compounds related to carotene). In the animal system, β-carotene is converted to vitamin A, and the salutory effects imputed to carrots in curing night blindness are readily understood on the basis of the connection of vitamin A to vision (see Fig. 34.11). In the plant system, β-carotene is one of the important pigments in

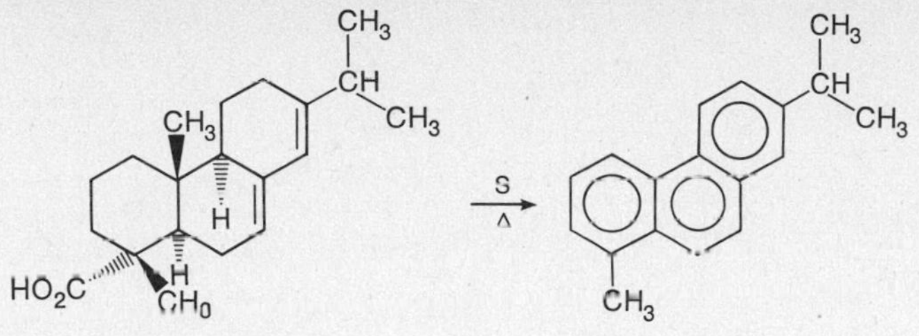

Fig. 34.12. Structure of abietic acid.

photosynthesis, where, in conjunction with chlorophyll, it serves to harness light energy and make it available to the photosynthesizing organism.

In addition to the great variety of well-defined compounds belonging to the terpene family (containing 10, 15, 20, 30, and 40 carbons), higher molecular weight polymers of isoprene are also known in nature. Two of these, rubber and gutta percha, are of particular interest and utility. Upon destructive distillation, both yield isoprene, both contain one double bond per five carbon units in the polymer, and both are catalytically reduced to the same saturated high molecular

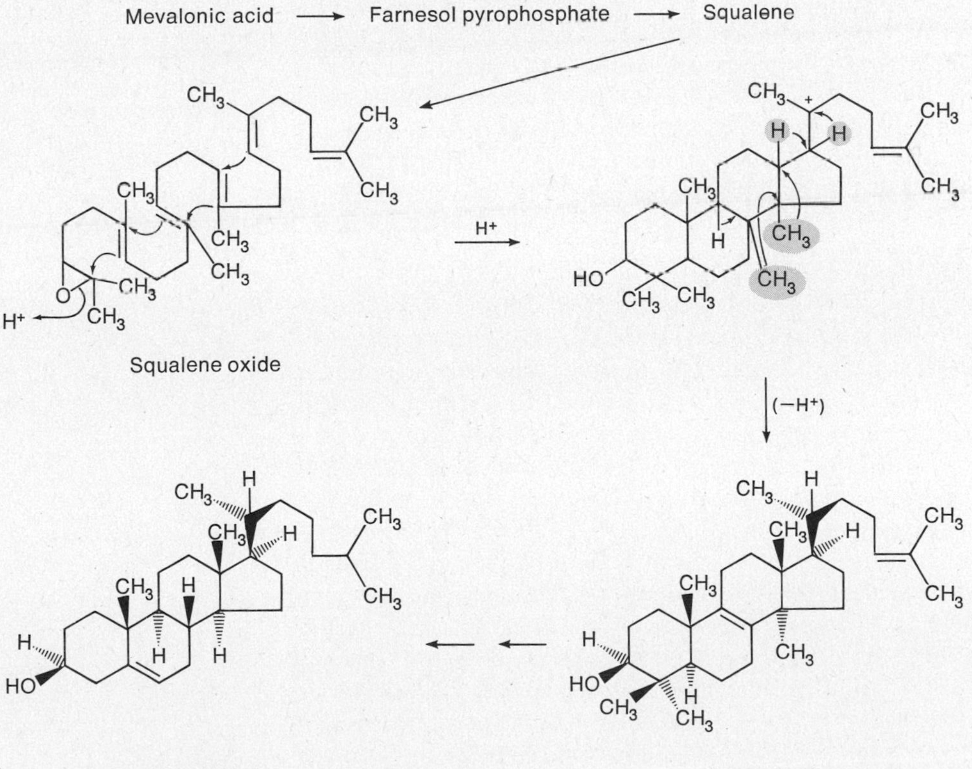

Fig. 34.13. Biosynthesis of lanosterol and cholesterol.

weight compound. On the basis of these data, it is believed that rubber and gutta percha are simply (*E*),(*Z*) isomers of each other, rubber having the (*Z*) configurations along the chain and gutta percha having the (*E*) configurations (Fig. 34.14).

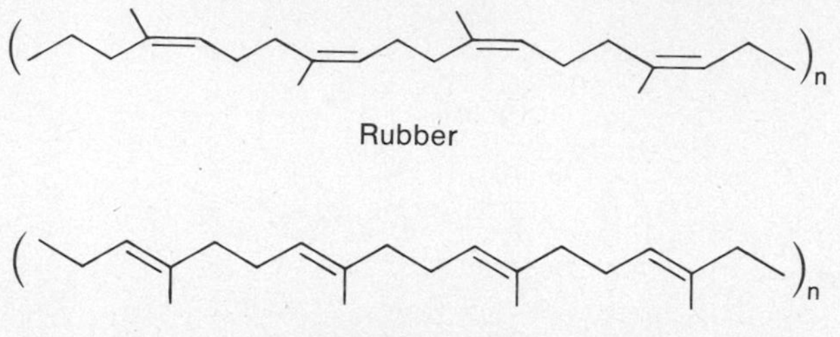

Fig. 34.14. Structure of rubber and gutta percha.

Rubber occurs rather widely in the plant kingdom as a milky suspension known as *latex*. The major source for commerce is the rubber tree, *Hevea brasiliensis*, which accounts for 98% of the world production. Although rubber has been known for several centuries, its commercial use soared with the advent of the automobile and its requisite for rubber tires. Complete dependence on natural rubber has diminished with the development of a variety of synthetic rubber products during the last 30 years. Gutta percha is also formed in the latex cells of a variety of plants. Its prinicpal commercial use is in certain speciality applications, such as covers for golf balls. The most obvious physical difference between rubber and gutta percha is in the elasticity and resiliance of rubber as contrasted with the harder, less resilient gutta percha.

34.4 Steroids

Cholesterol, one of the most ubiquitous compounds in nature, is formed *in vivo* from the tetracyclic triterpene lanosterol (Fig. 34.13), the carbon frameworks of the two compounds differing by three methyl groups. Although derived from a member of the terpene family, cholesterol is *not* classed as a terpene because it contains 27 rather than 30 carbon atoms and cannot be mentally dissected into isoprene units. Instead, it belongs to a class of compounds called *steroids*, all of which are characterized by the presence of a perhydrocyclopentenophenanthrene ring system. On dehydrogenation, steroids such as cholesterol yield methylcyclopentenophenanthrene (Fig. 34.15). Almost all the naturally occurring steroids have methyl groups at the bridgehead positions at C-10 and C-13; almost all have several chiral carbons, the differences between the various members of the steroid family arising from differences in substituents at various positions on the ring. Cholesterol contains eight chiral carbons and can exist in 256 stereoisomeric forms—*i.e.*, 128 racemates (see Fig. 19.34). The only one that occurs naturally, however, has the structure shown in Fig. 34.16, in which the ring junctions are *trans* and the ring system is essentially flat. Reduction of the double bond between C-5 and C-6 introduces another chiral carbon, and the hydrogen attached at

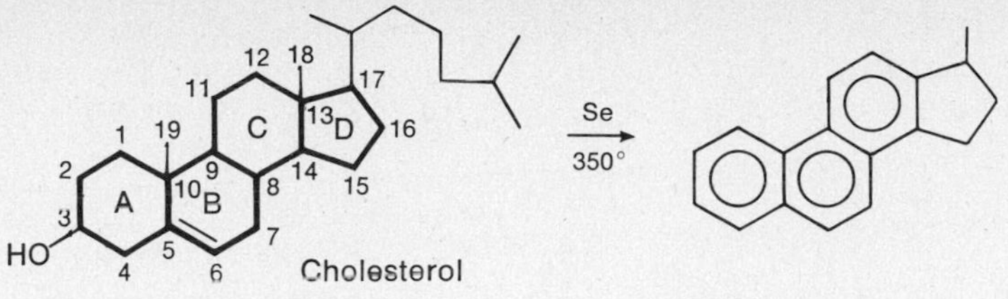

Cholesterol

(Outlined portion is the perhydrocyclopentanophenanthrene ring system)

Se 350°

17-Methyl-15,16-dihydro-17H-cyclopenta[a] phenanthrene*

Fig. 34.15. Structure of steroids.

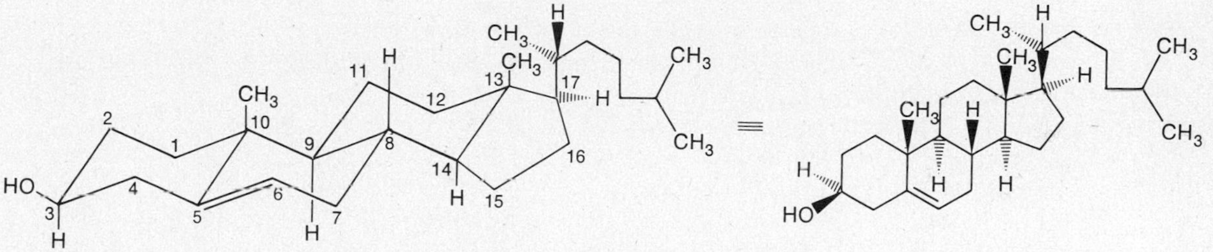

Fig. 34.16. Conformation of cholesterol.

C-5 can enter either from the top of the molecule, to give a *cis* configuration between the A/B rings, or from the bottom of the molecule, to give a *trans* configuration between the A/B rings (Fig. 34.17).

Many of the steroids have physiological properties, and classification into subgroups, therefore, is based partly on biological function and partly on chemical structure. Cholesterol belongs to the subgroup known as sterols and is the biological precursor of some of the other steroids. One of its physiological functions is to provide a storehouse of perhydrocyclopentenophenanthrene rings to which various other groups can be introduced via biosynthesis. Other physiological functions of the abundant supply of cholesterol present in the human body are not yet understood. Under certain circumstances, an excess of cholesterol can, in fact, be deleterious to the body, for it can be the cause of gallstones and, perhaps, arteriosclerosis. The common biogenesis of cholesterol and fatty acids from acetate units makes plausible the connection between fatty acid intake in the diet, blood cholesterol levels, and heart dysfunction (see p. 1062).

* The "17H" indicates that the parent compound contains a CH_2 group at C-17; the "[a]" indicates that the cyclopentene ring is attached to the phenanthrene ring at the side of the phenanthrene ring designated as the "a" side; *i.e.*,

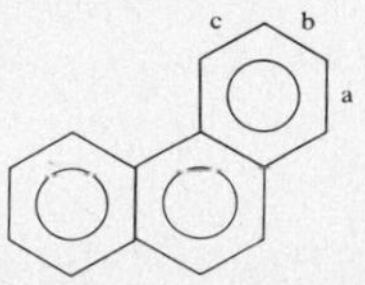

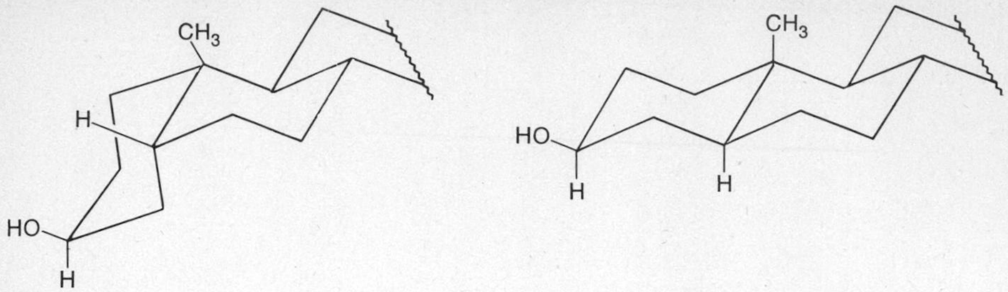

Coprostanol

(A/B Rings *cis*)

Cholestanol

(A/B Rings *trans*)

Fig. 34.17. Configuration and conformation of coprostanol and cholestanol.

Another interesting sterol is ergosterol, first isolated from ergot but more readily obtained from yeast. When ergosterol is irradiated (*e.g.*, sunlight is very effective), it undergoes ring B cleavage and yields precalciferol, which, when heated, isomerizes to calciferol [also known as vitamin D_2 (Fig. 34.18)].

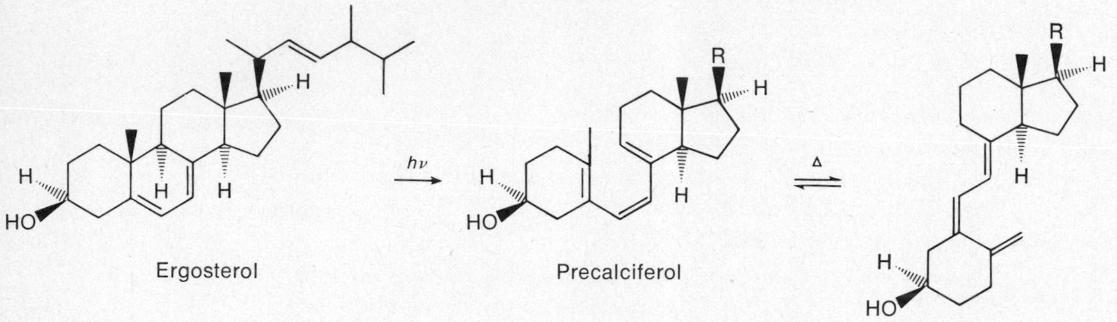

Ergosterol

Precalciferol

Calciferol
(Vitamin D_2)

Fig. 34.18. Conversion of ergosterol to Vitamin D_2.

The bile acids are steroids that function as emulsifying agents for fats and, consequently, aid in the hydrolysis and absorption of fats from the intestinal tract. Cholic acid, a typical steroidal acid occurring in the bile as an amide of glycine or taurine ($NH_2CH_2CH_2SO_3Na$), has been shown by means of ^{14}C-labeling experiments to be derived *in vivo* from cholesterol.

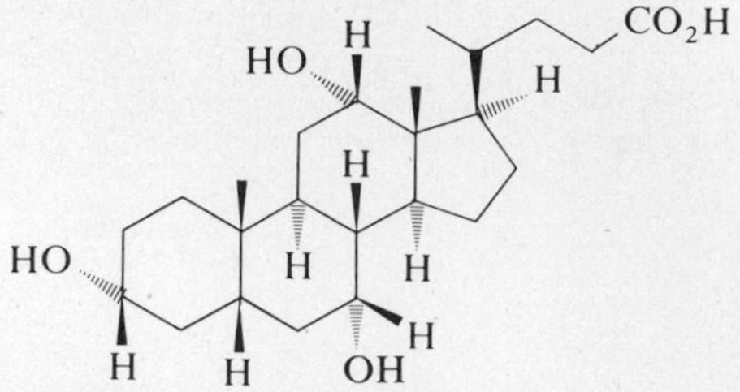

Cholic acid

An important group of steroids consists of the sex hormones and includes female sex hormones such as estradiol, male sex hormones such as testosterone, and pregnancy hormones such as progesterone. Administration of these compounds to humans, even in very small doses, can bring about dramatic changes in physiology, the female hormones inducing certain female body characteristics when given to males and the male hormones inducing certain male body characteristics when given to females. Perhaps foremost on the current scene are certain steroid analogs, such as the acetylenic keto alcohol known as "Norethynodrel," which stops ovulation in nonpregnant women, thereby preventing pregnancy from occurring. These compounds, colloquially known as "The Pill," are being widely used as birth control devices and have had a significant effect on the mores and life style of certain segments of civilized society in the last third of the twentieth century (Fig. 34.19).

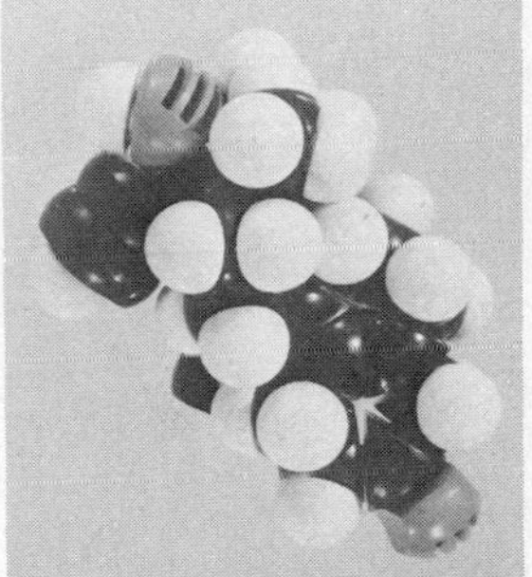

Norethynodrel

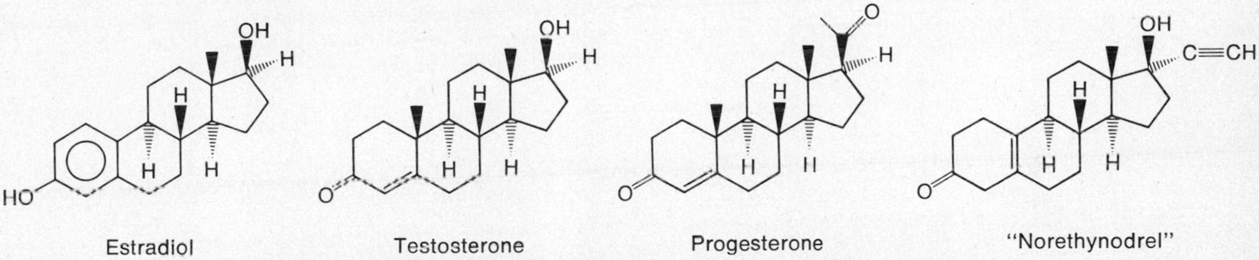

Estradiol   Testosterone   Progesterone   "Norethynodrel"

Fig. 34.19. Structures of steroid sex hormones.

Another important group of steroids occurs in the adrenal cortex and includes such compounds as corticosterone, cortisone, and aldosterone (Fig. 34.20). These compounds are essential for the proper functioning of the animal organism, and their absence can lead to malfunction and death. Therefore, some of them

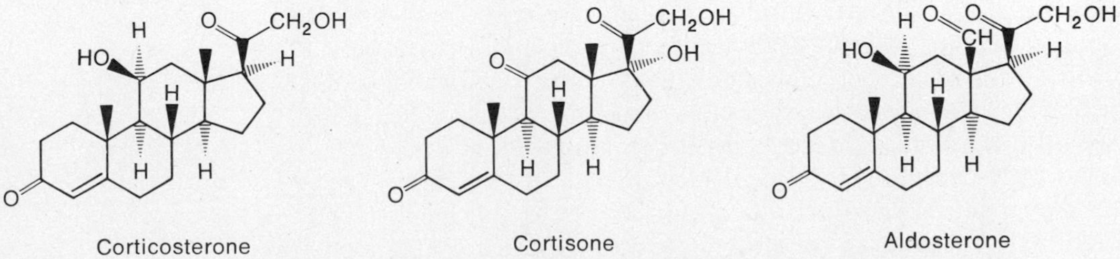

Corticosterone   Cortisone   Aldosterone

Fig. 34.20. Structures of steroids of the adrenal cortex.

have found use as therapeutic agents in the treatment of such maladies as rheumatoid arthritis.

Still other types of steroids occur naturally. Diosgenin, for example, is a steroid sapogenin, so-named because it forms a colloidal aqueous solution that lathers when shaken; digitoxigenin, occurring naturally as a glycoside, has a powerful action on heart muscles and is an example of a cardiac glycoside; bufotalin is the steroid portion of a more complex molecule known as a toad poison. Isolated from the venom secreted in the parotid glands of toads, the physiological effect of bufotalin resembles that of the cardiac glycosides (Fig. 34.21).

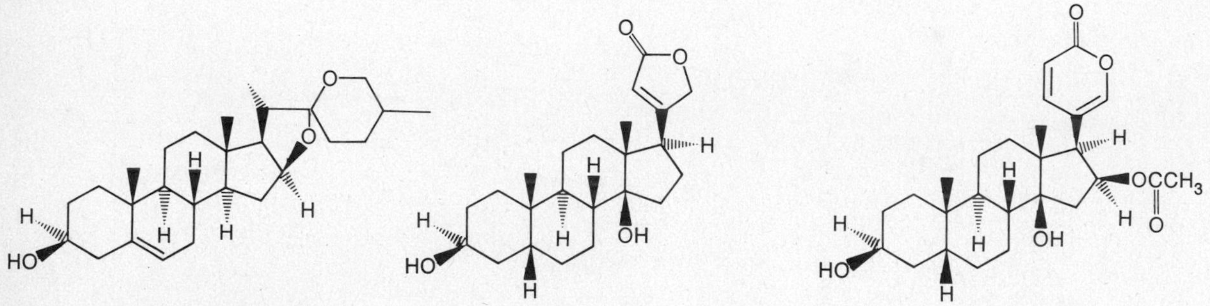

Diosgenin

Digitoxigenin

Bufotalin

Fig. 34.21. Structures of steroid sapogenins, cardiac glycosides, and toad poisons.

34.5 *Prostaglandins*

One of the most recent arrivals on the chemical, biochemical, and medical scenes is the class of compounds called *prostaglandins*. Actually, the existence of these compounds was inferred as early as the 1930's, when studies of human semen and animal seminal tissues indicated the presence of physiogically active substances. On the assumption that they had come from the prostate gland, the name prostaglandins was given to them. It turns out, however, that the prostate gland is not the source but that the seminal vesicles along with many other tissues and fluids in the body are the home of the prostaglandins. Because they are present in such minute amounts, even in the seminal fluid where their concentration is at its highest, their isolation and characterization was a problem that was not solved until the late 1950's and early 1960's. Through the use of new techniques of isolation and the application of new methods of analysis, including paper chromatography, gas chromatography, mass spectrometry, and X-ray crystallography, the structures of more than a dozen prostaglandins have been established. All these are 20-carbon compounds that contain a five-membered carbocyclic ring to which is attached an oxygen function at the 9 position (using a numbering based on arachidonic acid), a seven-carbon fatty acid moiety at the 8 position, and an eight-carbon hydroxyalkyl moiety at the 12 position. The members of the group differ from each other with respect to the number of double bonds and hydroxyl groups that they contain, as indicated by the representative examples shown in Fig. 34.22.

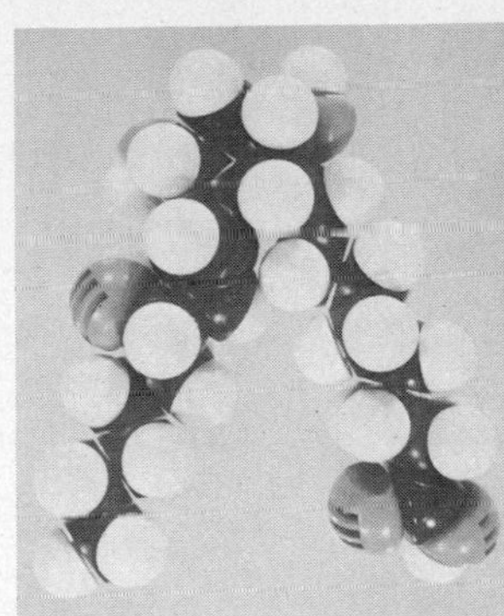
Prostaglandin PGE_1-alpha

PGE_2 PGA_2

$PGF_1\alpha$ 19-Hydroxy-PGB_1

Fig. 34.22. Structures of prostaglandins and their short-hand designations.

It has been shown that the biogenetic precursor of the prostaglandins are 20-carbon unsaturated fatty acids such as arachidonic acid, which undergo *in vivo* cyclization and oxidation:

Arachidonic acid $\xrightarrow{[O]}$

Thus, the necessity for unsaturated fatty acids to be in the normal diet arises from the necessity for prostaglandins to be synthesized, the prostaglandins playing a critical part in a host of physiological phenomena. The full range of involvement of the prostaglandins has yet to be discovered, but already they hold great promise in medicine in the treatment of asthma, in the regulation of blood pressure and the treatment of hypertension, in the regulation of metabolism, and in the control of nervous disorders. For the biochemist and physiologist they also hold great interest, because they are probably intimately involved in cell-to-cell communication, in the transmission of signals across the synapses in the nervous system, and with the functioning of cell membranes in general.

34.6 *Aromatic Acetogenins*

One of the pathways for the biosynthesis of aromatic rings makes use of acetyl-coenzyme-A as a source of two-carbon units and involves a sequence in which the carbonyl oxygens of the acetyl groups are retained during the elaboration of the chain, in contrast to the fatty acid sequence in which the carbonyl oxygens are replaced by hydrogen (see Fig. 34.1). The joining of four acetyl groups in linear combination produces a tetracarbonyl compound in which the possibility of intramolecular aldol condensation exists. If the aldol condensation involves the carbonyl at C-1 and the methylene group at C-6, the naturally occurring phloroacetophenone is formed; alternatively, if the aldol condensation involves the

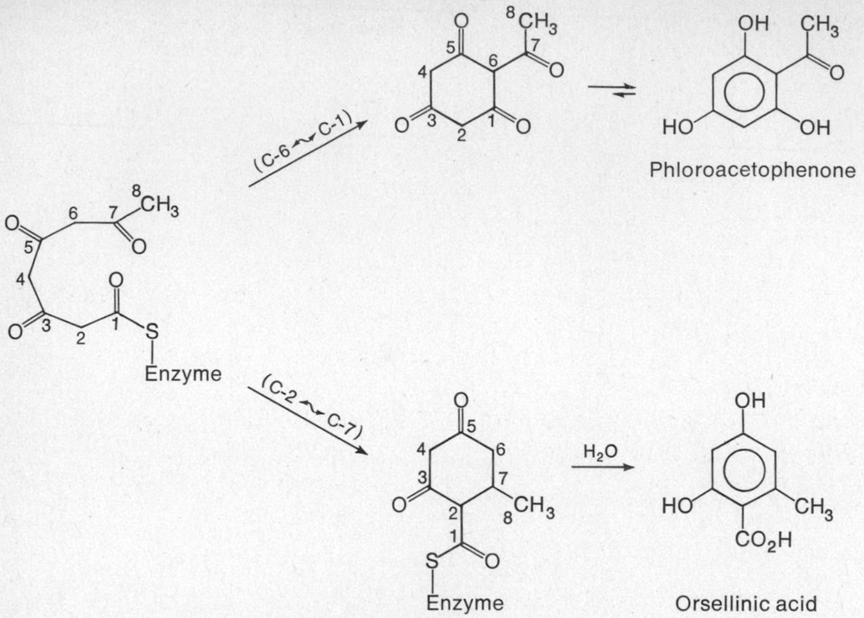

Phloroacetophenone

Orsellinic acid

Fig. 34.23. Aromatic biosynthesis via acetate pathway.

carbonyl at C-7 and the methylene at C-2, the naturally occurring orsellinic acid is formed (Fig. 34.23).

An interesting biosynthesis occurs in the flavones, where one of the aromatic rings arises from acetate units (acetate pathway) and the other from glucose in what is known as the shikimic acid pathway. Attachment of a cinnamic acid

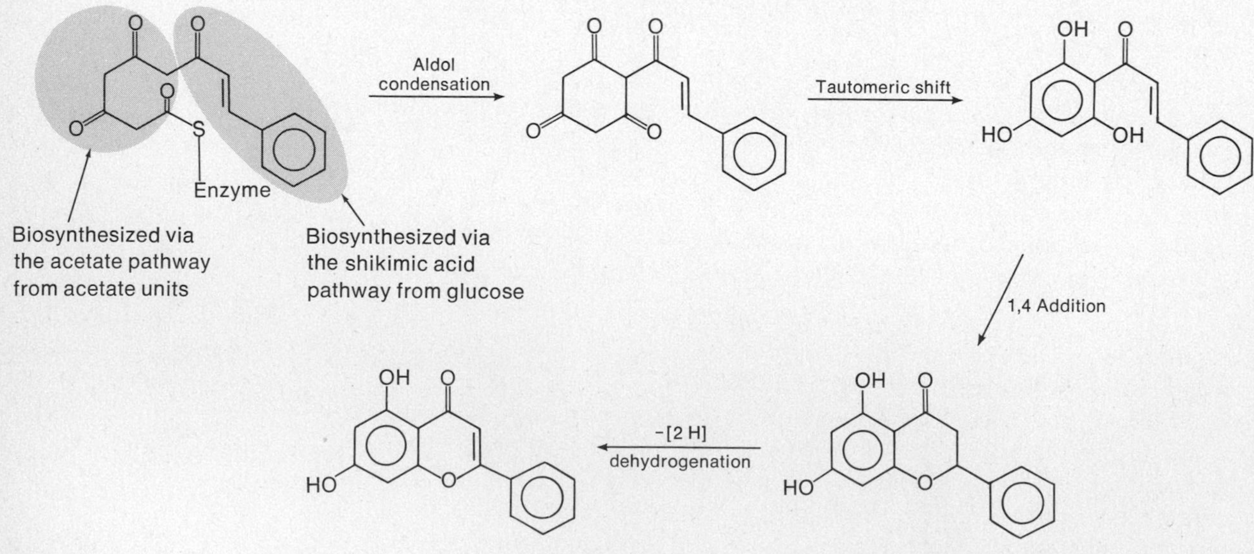

Fig. 34.24. Biosynthesis of flavones.

moiety, formed from glucose, to a triacetyl moiety formed from acetate produces the tetracarbonyl compound shown in Fig. 34.24. Intramolecular aldol condensation followed by intramolecular conjugate addition across the α,β-unsaturated carbonyl system and removal of a pair of hydrogen atoms then yields 5,7-dihydroxyflavone (Fig. 34.24). Closely related to this compound are the anthocyanidin pigments, which are examples of naturally occurring glucosides (see Fig. 32.39).

34.7 Other Acetogenins

In addition to the several types of acetogenins discussed in the previous sections of this chapter, other compounds are known to have their genesis from acetate units. Included in this group are a variety of long-chain, highly unsaturated *mold metabolites* and a variety of large ring ethers called *macrolides*. One of the most interesting mold metabolites is mycomycin, a compound in which optical activity arises from the axial chirality of the allenic structure. A typical example of a macrolide is erythromycin, an antibiotic compound isolated from fungi (Fig. 34.25).

$$HC{\equiv}C{-}C{\equiv}C{-}CH{=}C{=}CH{-}CH{=}CH{-}CH{=}CH{-}CH_2CO_2H$$

Mycomycin (a mold metabolite)

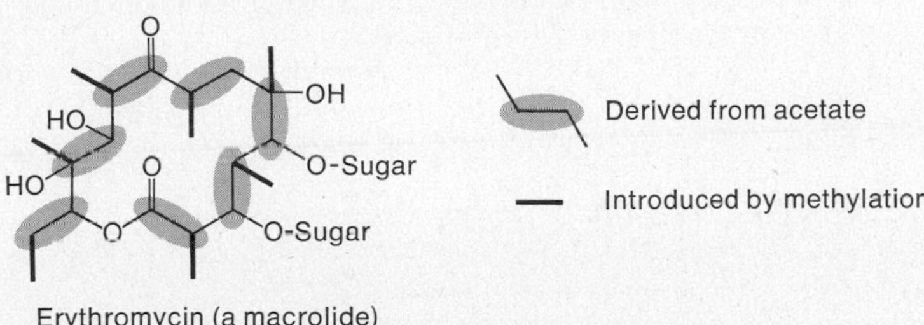

Erythromycin (a macrolide)

Fig. 34.25. Mold metabolite and macrolide acetogenins.

REFERENCES

BARKER, R., *Organic Chemistry of Biological Compounds*, Prentice-Hall, Inc., Englewood Cliffs, N.J., 1971, Chapter 7.

Bio-Organic Chemistry: *Readings from Scientific American*, W. H. Freeman and Co., Publishers, San Francisco, 1968.

- CLEVENGER, S., Article 15. Flower Pigments.
- FIESER, L. F., Article 17. Steroids.
- GREEN, D. E., Article 26. The Metabolism of Fats.
- HUBBARD, R., and A. KROPF, Article 28. Molecular Isomers in Vision.

Florkin, M., and E. H. Stotz, eds., *Comprehensive Biochemistry*, Vol. 6, *Lipids and Amino Acids and Related Compounds*, Elsevier Publishing Co., New York, 1965.

Hanson, J. R., *Introduction to Steroid Chemistry*, Pergamon Press, New York, 1968.

———, *Tetracyclic Diterpenes*, Pergamon Press, New York, 1968.

Lehninger, A. L., *Biochemistry*, Worth Publishers, Inc., New York, 1970, Chapter 10.

Ourisson, G., P. Crabbé, and O. R. Rodig, *Tetracyclic Triterpenes*, Holden-Day, Inc., San Francisco, 1964.

Taylor, W. I., and A. R. Battersby, eds., *Cyclopentanoid Terpene Derivatives*, Marcel Dekker, Inc., New York, 1969.

Templeton, W., *An Introduction to the Chemistry of the Terpenoids and Steroids*, Butterworth & Co., Ltd., London, 1969.

Yates, P., *Structure Determination*, W. A. Benjamin, Inc., New York, 1967.

PROBLEMS

1. For each of the following acetogenins indicate how many stereoisomers (double-bond diastereomers and mirror image isomers) can exist:
(a) Sphingomyelin (b) Mevalonic acid (c) Limonene
(d) Camphor (e) Abietic acid (f) Linolenic acid
(g) Cholic acid (h) Mycomycin

2. Show how the carbon skeleton for each of the following terpenes can be dissected into isoprene units:
(a) Citral (b) Camphor (c) β-Amyrin
(d) Cedrol (e) Abietic acid

3. Available starting materials include any (a) alkanes, alkenes, alkynes, aryl halides, alkanols, alkanones, alkanoic acids, alkanoic acid derivatives, or alkylamines containing four or fewer carbon atoms, (b) benzene, toluene, *o*-, *m*-, and *p*-xylene, naphthalene, and (c) any inorganic compounds. Starting with materials chosen from this group, devise syntheses for the following acetogenins:
(a) Isoprene (b) Menthol (c) Limonene

4. An optically active triglyceride yields one equivalent of glycerol, one equivalent of palmitic acid, and two equivalents of stearic acid on hydrolysis. What is its structure?

5. Show how β-amyrin can be formed by cyclization of squalene.

6. Deduce the structure of each of the following terpenes from the data given:
(a) β-Cyclocitral has a molecular formula $C_{10}H_{16}O$, absorbs at 235 nm ($\varepsilon = 12{,}500$), reacts with Tollens' reagent to form a $C_{10}H_{16}O_2$ carboxylic acid, and can be reduced to a $C_{10}H_{18}$ compound. Dehydrogenation of the carboxylic acid yields *m*-xylene, methane, and carbon dioxide; dehydrogenation of the $C_{10}H_{18}$ compound yields 1,2,3-trimethylbenzene.
(b) β-Phellandrene has the molecular formula $C_{10}H_{16}$ and yields *p*-isopropyltoluene on dehydrogenation. Ozonolysis gives formaldehyde and a $C_9H_{14}O_3$ compound which reacts with periodate to give a $C_8H_{14}O_3$ carboxylic acid.

(c) β-Selinene has the molecular formula $C_{15}H_{24}$, and yields 1-methyl-7-isopropylnaphthalene on dehydrogenation. Ozonolysis gives two equivalents of formaldehyde and a $C_{14}H_{20}O_2$ compound that produces iodoform and a $C_{12}H_{18}O_3$ carboxylic acid on treatment with sodium hydroxide and iodine.

7. Indicate what simple chemical tests could be used to distinguish between the members of the following sets of acetogenins.
   (a) Stearic acid, oleic acid, tristearin, phosphatidic acid, phosphatidyl inositol
   (b) Squalene, farnesol, citral, camphor
   (c) Cholesterol, cholic acid, estradiol, testosterone, progesterone

Chapter **35**

Nucleic acids: the blueprints for life

If, many centuries in the future, men do manage to create life, such artificial life might be—probably would be—nothing but a poor parody or caricature of natural life, of the life that has given us all the world of animals and man. But, now let us get down to problems that are not so hazy.

JEAN ROSTAND
Life, the Great Adventure (1956)

The three types of biopolymers essential for life are the carbohydrates (see Chapter 32), the proteins (see Chapter 33), and the nucleic acids. Through the intermediacy of the carbohydrates the energy of the sun is harnessed for use in the maintenance of life; by means of proteins the reactions involved in the maintenance of life are effectively catalyzed; through the agency of the nucleic acids the patterns of life, its blueprints, are handed from one generation to the next. This chapter deals with nucleic acids, which are the biopolymers that nature has devised for achieving the remarkable feat of genetic transmission.

Nucleic acids are polymers of monomeric units called nucleotides. The nucleotides from nucleic acids can, in turn, be broken down into three smaller units, each of which belongs to a particular class of compounds, *viz.*, a five-carbon sugar, ribose or 2-deoxyribose, an organic base of the purine or pyrimidine type, and phosphoric acid. Before considering the nucleic acids, therefore, we must deal with the nucleotides and their constituent parts.

35.1 Pyrimidines and Purines

A number of heterocyclic compounds have been mentioned in previous chapters (see Figs. 11.2, 11.3, 11.13, 12.1, 12.2, and 15.22–15.24), representing only a few of the several thousand known heterocyclic ring systems. From the biological standpoint, among the most important members of this very large group are the pyrimidines (which contain two nitrogen atoms in a single ring) and the

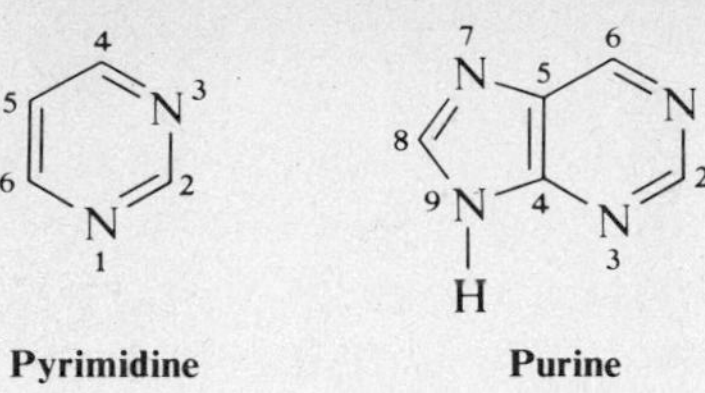

purines (which contain four nitrogens in two fused rings). Attachment of hydroxyl and amino functions (and, in one instance, a methyl group) to these rings gives the organic base units that, together with the ribose or deoxyribose and phosphoric acid, constitute the majority of the nucleotides. There are only five pyrimidine and purine bases that commonly occur in the nucleotides, *viz.*, the pyrimidines uracil, cytosine, and thymine, and the purines adenine and guanine, illustrated in Fig. 35.1.

Uracil

Adenine

Cytosine

Guanine

Thymine

Fig. 35.1. Tautomeric forms of pyrimidine and purine bases of nucleic acids.

Uracil, cytosine, thymine, adenine, and guanine can all exist in tautomeric forms, and the form that predominates depends on the pH of the solution. For the tautomeric pairs illustrated in Fig. 35.1, the form shown on the left-hand side is the one that predominates at the pH of physiological systems (about 7 ± 2 pH units).

The oxygenated pyrimidines uracil and thymine and the oxygenated purine guanine have amidelike tautomers that possess moderately acidic N—H bonds. Uracil and guanine, for example, lose a proton at pH 9–10 and are comparable in acidity to phenol ($pK_a = 10$). The amino groups in cytosine and adenine are only weakly basic ($pK_{BH^+} \simeq 3$–4) and, as a consequence, are not protonated at pH 7. Therefore, the functions required for intermolecular hydrogen bonding are present in these molecules, *i.e.*, moderately acidic proton donors and moderately basic proton acceptors, and this is of crucial importance in determining the conformation of the polymers into which these units are incorporated.

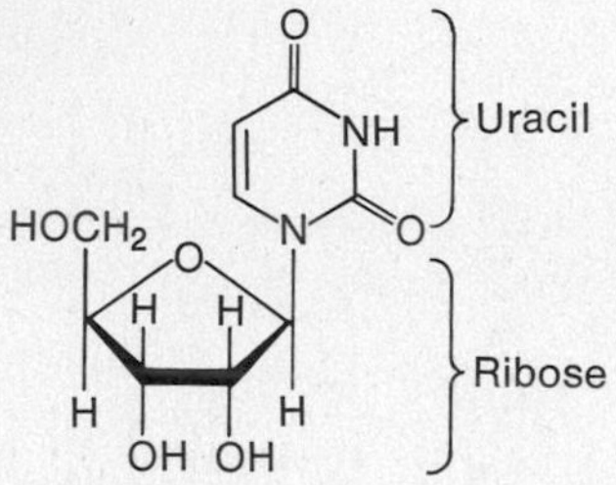

Uridine
(1-β-D-Ribofuranosyluracil)

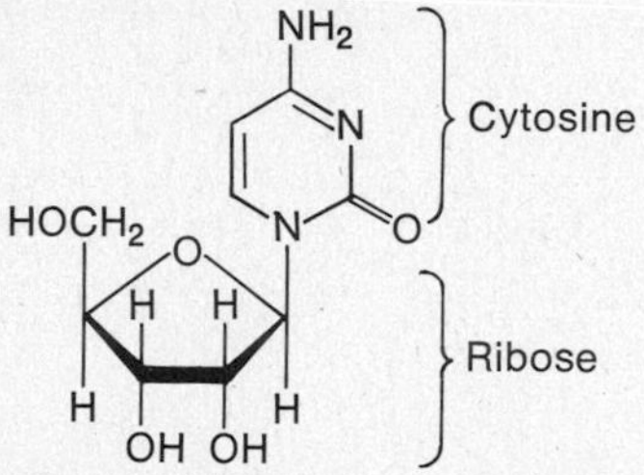

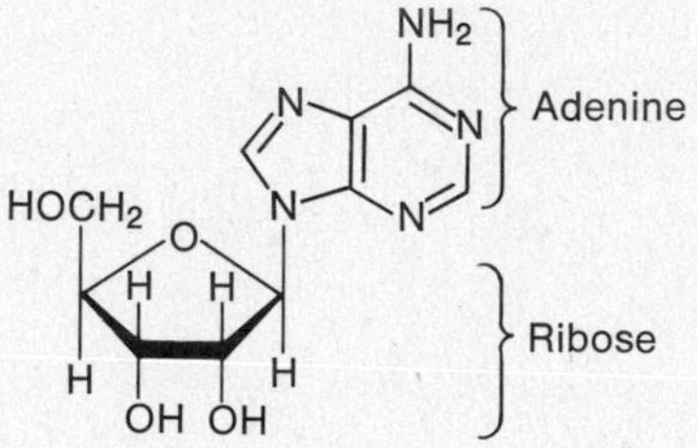

Adenosine
(9-β-D-Ribofuranosyladenine)

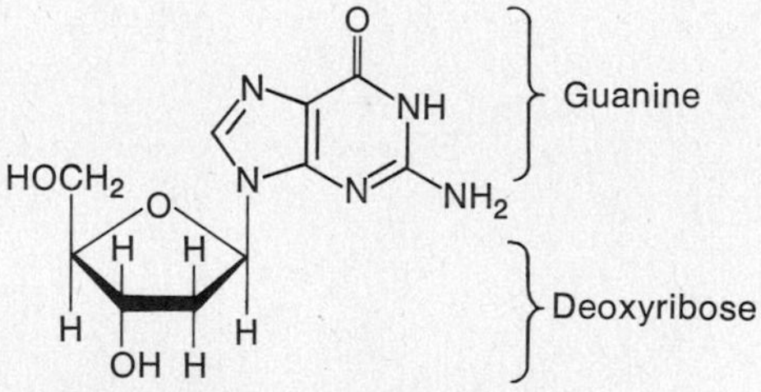

Cytidine
(1-β-D-Ribofuranosylcytosine)

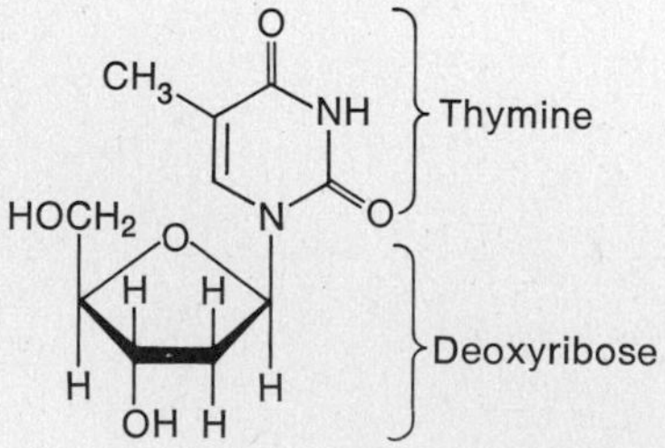

Guanosine
(9-β-D-2-Deoxyribofuranosylguanine)

2-Deoxythymidine
(1-β-D-2-Deoxyribofuranosylthymine)

Fig. 35.2. Structures of ribonucleosides and deoxyribonucleosides.

The formation of a bond between the C-1 position of the five-carbon sugars ribose or 2-deoxyribose and the N-1 position of a pyrimidine or the N-9 position of a purine gives an N-glycoside (a ribofuranosyl-pyrimidine or -purine or a deoxyribofuranosyl-pyrimidine or -purine) known as a *nucleoside.* For example, the attachment of adenine to ribose in this fashion gives the 9-(β-D-ribofuranosyl)-adenine known as adenosine; the attachment of thymine to 2-deoxyribose gives 2-deoxyribofuranosylthymine, known as 2-deoxythymidine (Fig. 35.2). When the nucleosides in turn, are converted to esters by the attachment of a phosphoric acid residue to one of the hydroxyl functions of ribose, *nucleotides* are formed. Phosphorylation of adenosine at the C-5 hydroxyl group of ribose, for example, gives the nucleotide called 5′-adenylic acid*; phosphorylation of thymidine at the C-3 hydroxyl group of 2-deoxyribose gives the nucleotide called 3′-thymidilic acid (Fig. 35.3). Alternatively, these compounds can be named adenosine

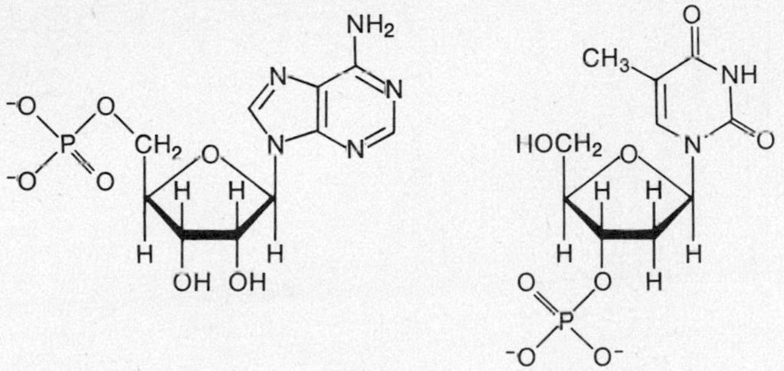

5′-Adenylic acid
[Adenosine 5′-monophosphate (AMP)]

Thymidylic acid
(Thymidine 3′-monophosphate

Fig. 35.3. Structures of ribonucleotides and deoxyribonucleotides.

5′-monophosphate (the 5′ is often omitted, and the abbreviation AMP is frequently used) and thymidine 3′-monophosphate (abbreviated as TMP), respectively. The names of the various nucleosides and nucleotides derivable from the five commonly occurring purines and pyrimidines are shown in Table 35.1.

Attachment of a second phosphoric acid residue to adenosine monophosphate produces adenosine diphosphate (ADP), and the attachment of a third phosphoric acid residue produces the biologically important compound, adenosine triphosphate (ATP) (see Fig. 32.41). It is also possible for cyclic phosphate esters to form. Adenosine-3′,5′-phosphoric acid (cyclic adenylic acid; cyclic AMP), for example, is widely distributed in living systems and is implicated as a mediator for a variety of physiological functions, including the synthesis of proteins, the synthesis of steroids, and the metabolism of carbohydrates and fats.

* The position of the attachment to the sugar unit is indicated by a primed number.

Table 35.1 NAMES OF PURINE AND PYRIMIDINE BASES AND THE NUCLEOSIDES AND NUCLEOTIDES DERIVED FROM THEM BY THE ADDITION OF RIBOSE AND PHOSPHORIC ACID UNITS

Base	Nucleoside[a]	Nucleotide[a]		
Uracil	Uridine	Uridylic acid	Uridine monophosphate	UMP
Cytosine	Cytidine	Cytidylic acid	Cytidine monophosphate	CMP
Thymine	Thymidine	Thymidylic acid	Thymidine monophosphate	TMP
Adenine	Adenosine	Adenylic acid	Adenosine monophosphate	AMP
Guanine	Guanosine	Guanylic acid	Guanosine monophosphate	GMP

[a] Substitution of deoxyribose for ribose yields the corresponding deoxynucleosides and deoxynucleotides, indicated by the prefix *deoxy*. Thymidine, however, is the name of the nucleoside formed from thymine and deoxyribose, so the prefix does not apply in this case.

Adenosine-3′,5′-phosphoric acid
(cyclic AMP)

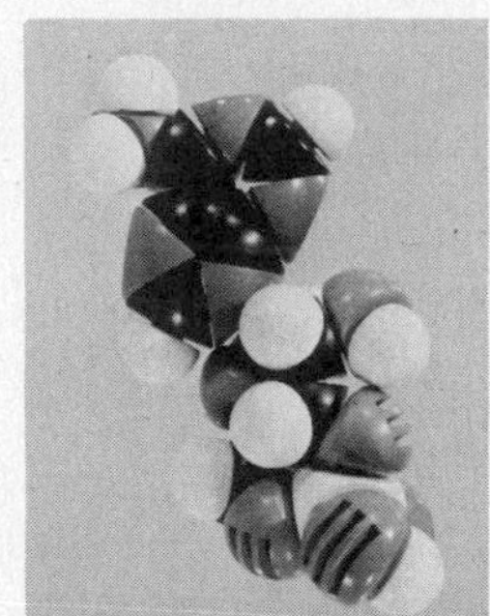

Cyclic AMP

Nucleotides and deoxynucleotides are the monomeric units of the nucleic acids and the deoxynucleic acids, often designated more specifically as the ribonucleic acids (abbreviated RNA) and the deoxyribonucleic acids, (abbreviated DNA), to indicate the presence of ribose and deoxyribose, respectively. The four nucleotides that are most commonly encountered in RNA are adenylic acid (AMP), guanylic acid (GMP), cytidylic acid (CMP), and uridylic acid (UMP). Other variations of these four nucleotides may sometimes be present, including methylated amino functions (*e.g.*, 1-methyladenosine), methylated oxygen functions (*e.g.*, 2′-O-methylguanosine), dihydro rings (*e.g.*, 4,5-dihydrouridine), and attachment to ribose via carbon–carbon rather than carbon–nitrogen bonds (*e.g.*, pseudouridine, in which attachment is at the C-5 position of the pyrimidine ring). More than 30 "unusual" bases have been isolated and identified. Deoxyribonucleic acids (DNA) are constructed mainly from the four deoxynucleotides 2′-deoxyadenylic acid (dAMP), 2′-deoxyguanylic acid (dGMP), 2′-deoxycytidylic acid (dCMP), and thymidylic acid (TMP), with a lower incidence of "unusual" bases than in RNA molecules.

35.3 *Polynucleotides (Nucleic Acids)*

In a fashion similar to that involved in building polysaccharides from monosaccharides, and peptides from amino acids, nucleotides and deoxynucleotides can be polymerized by removing the elements of water between the phosphoric acid portion of one molecule and the hydroxyl group at the C-3 position of a ribose or deoxyribose portion of another molecule. A dinucleotide formed from adenylic acid and cytidylic acid, for example, has the following structure:

Adenylic acid

Cytidylic acid

(Abbreviated representation)

If this process is repeated many times, polynucleotides of the general structure shown in Fig. 35.4 are produced.

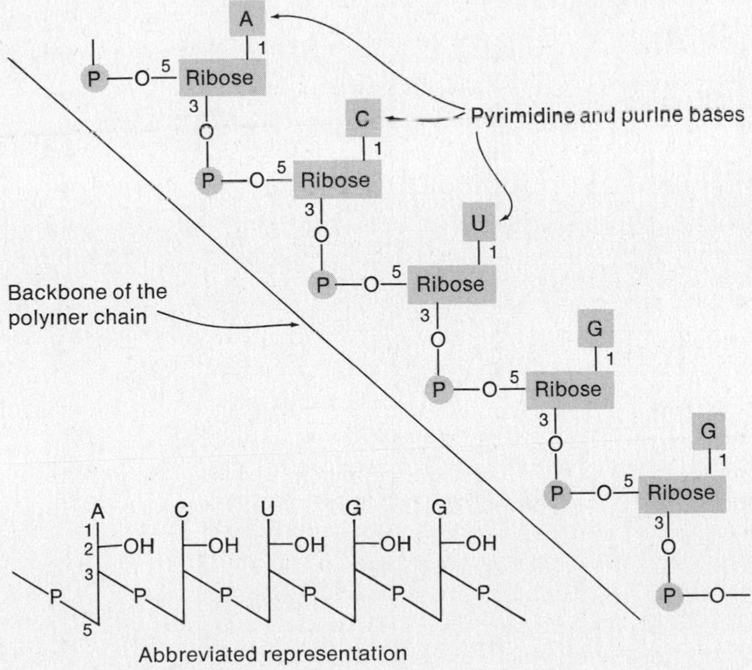

Fig. 35.4. Structure of polyribonucleotides (nucleic acids).

Polynucleotides resemble polypeptides in possessing a backbone portion; *i.e.*, the peptide chain of the polypeptide is comparable to the sugar-phosphate chain in the polynucleotides. Attached to the backbone of the polynucleotides are various pyrimidine and purine bases; *i.e.*, the R groups at the α-carbon of the polypeptides are comparable to the pyrimidine and purine bases in the polynucleotides. As a result of the many possible combinations of amino acids, a great variety of polypeptides are possible; similarly, as a result of the many possible combinations of nucleotide sequences, a great variety of polynucleotides can exist. Like polypeptides, the polynucleotides adopt conformations that are dependent on the identity and the sequence of the substituent groups attached to the backbone chain of the molecule. The most highly publicized of these is the double helix conformation that deoxyribonucleic acid assumes. On the basis of X-ray studies, analysis of the base composition of DNA, and molecular model building, Watson and Crick in 1953 postulated a structure for DNA which consists of a pair of deoxyribonucleic acid strands wound around one another in a right-handed helical fashion and held together by intermolecular hydrogen bonds between the pyrimidine and purine bases. Recognizing an architectural complementariness between certain pairs of bases with respect to hydrogen bonding [adenine, *e.g.*, pairing most effectively with thymine, and guanine pairing most effectively with cytosine (Fig. 35.5)], they postulated that the pyrimidine and

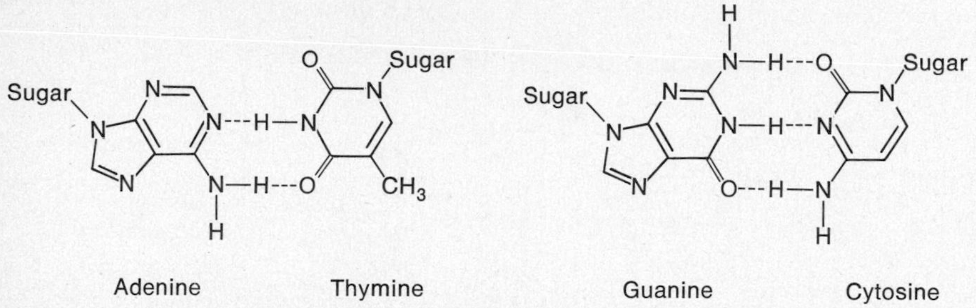

Fig. 35.5. Hydrogen bonding between pyrimidine and purine bases.

purines are on the inside of the double helix and are perpendicular to the axis of the helix, oriented to take maximum advantage of interchain hydrogen bond interactions (Fig. 35.6). Subsequent work has verified the correctness of this picture, and for their pioneering insights Watson and Crick were awarded the Nobel Prize in 1962.

The Watson and Crick hypothesis provided the details for the overall architecture of double-stranded DNA, but it indicated nothing about the sequence of pyrimidine and purine bases in the chain or anything about the size of the DNA molecule. To both of these questions only partial answers are presently available. Some of the lowest molecular weight DNA molecules are found in bacterial viruses. One well-studied example is the DNA of the ϕX174 virus of *E. coli* bacteria, a DNA which is a cyclic molecule and which, in the double-stranded form, has a molecular weight of 3,400,000 and contains about 10,000 nucleotide units. The *E. coli* bacteria on which this virus preys has a much larger DNA with

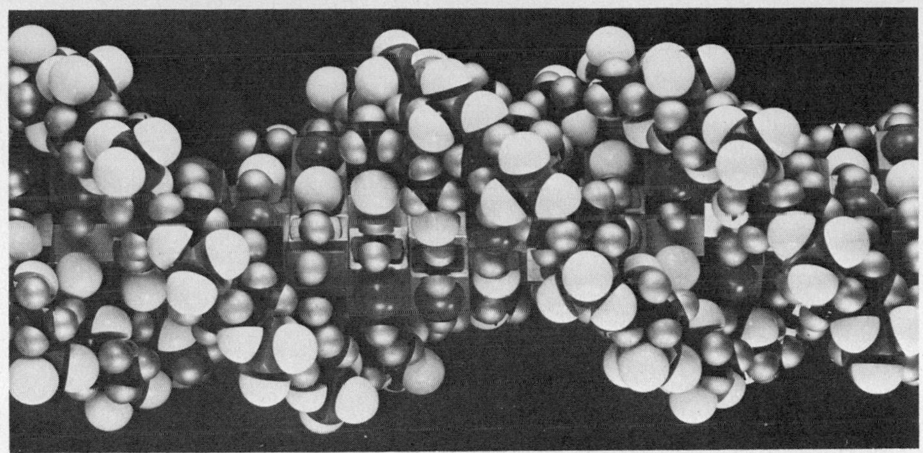

Fig. 35.6. Structure of the double helix of DNA.

a molecular weight of approximately 2,800,000,000, corresponding to about 8,400,000 nucleotide units. The molecules are so enormous that they are exceedingly susceptible to cleavage by mechanical shearing, and simple stirring or passage of their solutions through a pipette is often sufficient to break them into smaller fragments. For this reason, accurate assessments of the size of the DNA in higher organisms are not yet possible, although it has been calculated that the DNA of a mammalian cell must contain about 11 *billion* nucleotide units.

The gigantic chain lengths of DNA molecules pose obvious problems with respect to establishing nucleotide sequences. The kinds of approaches that have proved successful in polypeptide sequencing are, for the most part, inapplicable to the polynucleotides. Of course, complete hydrolysis of DNA can be effected to yield the constituent sugar, phosphate, and nitrogen bases; in fact, it was the observation of 1:1 ratios of adenine to thymine, and of guanine to cytosine, in hydrolysis products that provided an early clue to the complementariness between these pairs of bases. The techniques of *partial* and *selective* hydrolysis by means of certain chemicals or through the agency of enzymes, however, have been less successful in the sequencing of the nucleic acids than the proteins. The determination of the detailed sequence of a large DNA molecule, thus, looms as an organic structure problem of stupendous dimensions, further complicated by the fact that pure material for investigation is available only in very small amounts and, at least in higher organisms, mixed with many other DNA molecules of different yet similar composition.

Deoxyribonucleic acid represents the largest and most complex of the known polynucleotides. Certain smaller polynucleotides are also of great biological significance. Among these are the "messenger ribonucleic acids (mRNA)," so-named because of their imputed capacity for relaying information from the DNA molecule to the site in the cell where protein synthesis takes place. It is very difficult to isolate any specific member of the mRNA family, however, because each specific member is present in very low concentration in the cell, often has a short life, and is admixed with other mRNA molecules of similar

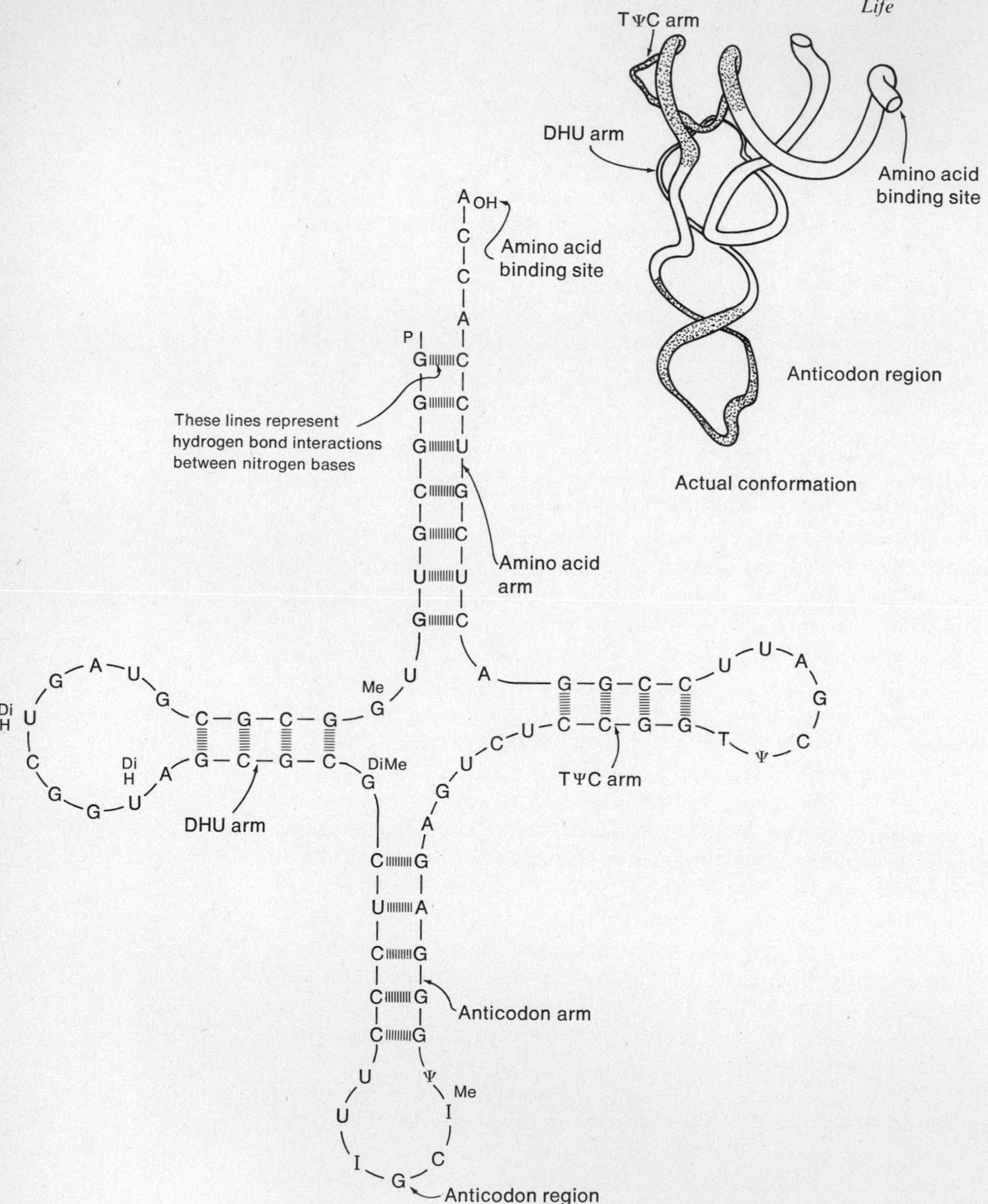

Fig. 35.7. Structure and conformation of yeast alanine tRNA.

structure. Nevertheless, a few structures have been isolated, and their base sequences have been shown by indirect methods of sequencing to be complementary to those of the DNA molecule with which they are associated, a fact that is discussed in more detail in the next section. Another group of even smaller polynucleotides are called "transfer ribonucleic acids" (tRNA), so-named because their function is to transfer amino acids to the site at which messenger RNA is transcribed, leading to the condensation of the amino acids to yield specific proteins. The molecules of tRNA are sufficiently small that complete sequences have been determined and the conformations of the molecules established. The alanine tRNA from yeast, for example, was shown by R. W. Holley (who shared the 1968 Nobel Prize) to consist of a chain of 77 nucleotide units arranged in the sequence shown in Fig. 35.7. One of the consequences of this particular sequence, which includes several "unusual" nucleotides in addition to the four common ones, is the folding of the molecule into a twisted conformation that is often represented as a clover leaf, although its actual shape is less regular than the stylized picture in Fig. 35.7 might suggest.

35.4 ***Biosynthesis of Proteins***

In the present discussion the synthesis of proteins is assumed to start with the amino acids, although these, in turn, have *in vivo* biosynthetic precursors. The first step in the biosynthetic conversion of amino acids to proteins involves a reaction between adenosine triphosphate (ATP) and an amino acid, in which the amino acid becomes attached to the AMP portion of ATP, the other product being pyrophosphate (Fig. 35.8). The aminoacyl adenylate

$$\mathrm{R\overset{\overset{\displaystyle ^{+}NH_3}{|}}{C}H\underset{\underset{\displaystyle O}{\|}}{C}{-}AMP^{=}}$$

then reacts with a transfer RNA molecule, transferring the aminoacyl moiety from

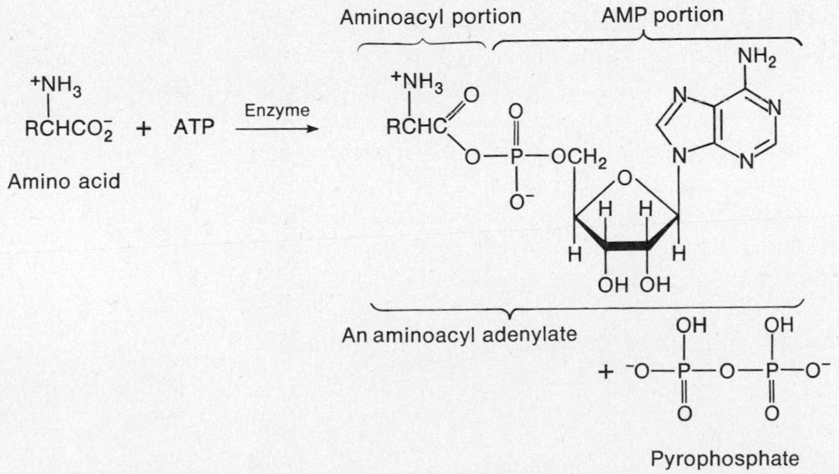

$$\mathrm{R\overset{\overset{\displaystyle ^{+}NH_3}{|}}{C}HCO_2^- + ATP \xrightarrow{\text{Enzyme}} \text{An aminoacyl adenylate} + {}^{-}O{-}\overset{\overset{\displaystyle OH}{|}}{\underset{\underset{\displaystyle O}{\|}}{P}}{-}O{-}\overset{\overset{\displaystyle OH}{|}}{\underset{\underset{\displaystyle O}{\|}}{P}}{-}O^-}$$

Amino acid; Aminoacyl portion; AMP portion; An aminoacyl adenylate; Pyrophosphate

Fig. 35.8. Protein synthesis: formation of the aminoacyl adenylate.

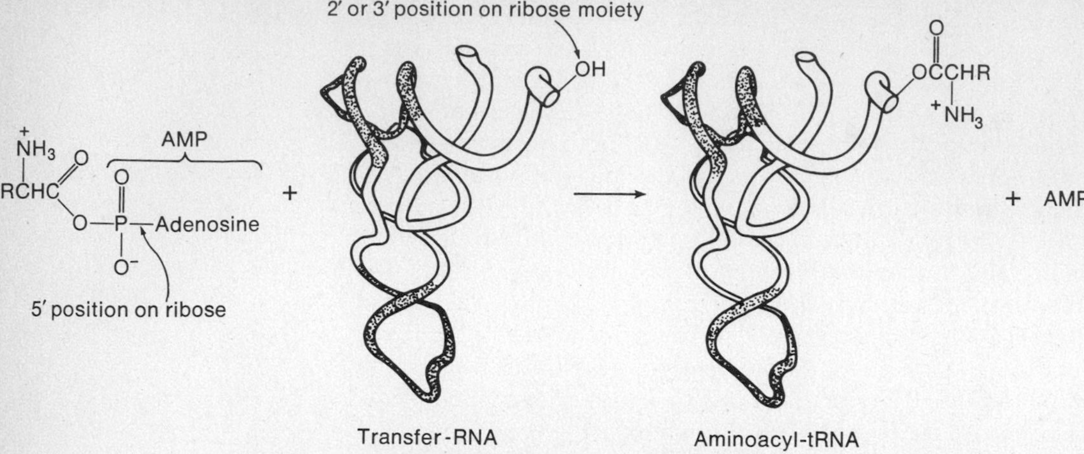

Fig. 35.9. Protein synthesis: formation of aminoacyl-tRNA.

the 5′ position of the sugar moiety of the adenylate to the 2′ or 3′ position of the sugar moiety of one of the nucleotide units in the tRNA (Fig. 35.9). Each amino acid has one or more species of tRNA that is specific for that particular amino acid, as well as an enzyme that is specific for that particular transfer process. The aminoacyl-tRNA compound is next transported to a structure within the cell known as the ribosome, to which it becomes attached in a very specific fashion. Ribosomes are high molecular weight structures (about 4,000,000 for a bacterial ribosome), of which approximately 60% are RNA units and 40% are protein units. They possess the capacity for binding in a specific fashion (a) the mRNA template, which contains the blueprint for the amino acid sequence of the polypeptide to be constructed, (b) the peptide chain that is being formed, and (c) the aminoacyl-tRNA molecules that bring the aminoacyl functions to the synthesis site.

The specificity of interaction between the aminoacyl-tRNA and the messenger RNA, bound to the ribosome, resides at least in part in the complementariness among a sequence of three adjacent nucleotide residues in each of these polynucleotides. Thus, on one portion of the tRNA, designated as the *anticodon region*, three nucleotide residues are present whose pyrimidine or purine bases are complementary to those of three nucleotide residues at a certain site on the mRNA. In this fashion, the mRNA serves as a template that attracts specific tRNA molecules to specific sites on its surface. Since each different tRNA carries a different amino acid, the net result is to establish a sequence of amino acid units which, when amide bonds are formed between them, forms a polypeptide whose sequence has been determined by the sequence at a specific site in the mRNA template.

The actual peptide synthesis is initiated by the association of the aminoacyl-tRNA, specified in Fig. 35.10 as

$$\mathrm{R^1CH}\begin{cases}{}^{+}\mathrm{NH_3}\\ \mathrm{CO_2{-}tRNA_1}\end{cases}$$

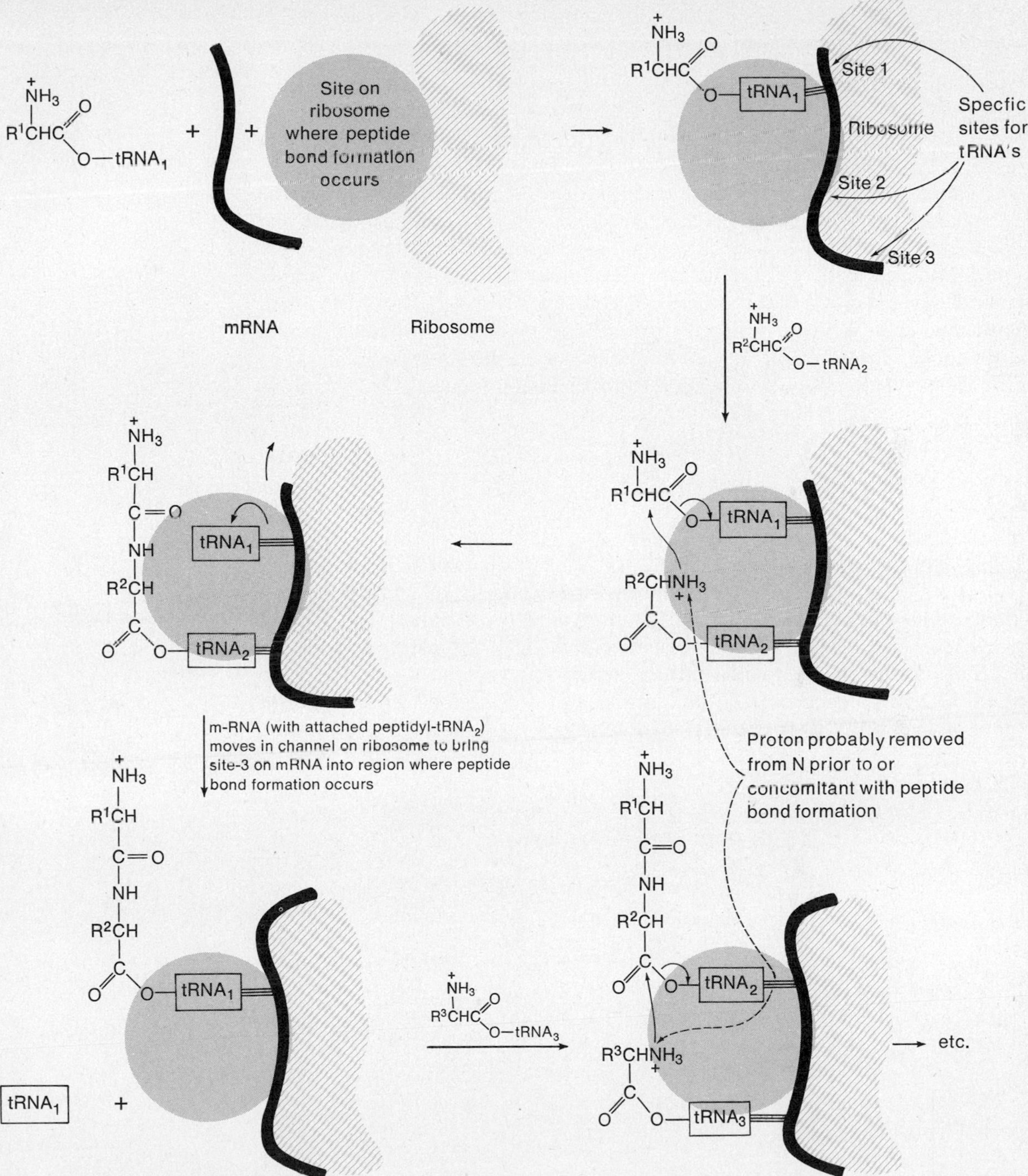

Fig. 35.10. Schematic representation of polypeptide synthesis from amino-acyl tRNA's, mRNA, and ribosome.

with a specific ribosome in such a fashion that site 1 is in the region of the ribosome where catalysis of peptide bond formation occurs. A second aminoacyl-tRNA, specified in Fig. 35.10 as

$$R^2CH(^+NH_3)CO_2{-}tRNA_2$$

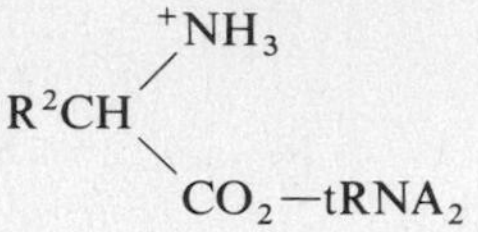

then becomes attached to another specific site on the mRNA (*i.e.*, site 2 in the diagram), and a nucleophilic displacement, involving the amino group of one residue and the carbonyl group of another, takes place between the aminoacyl residues attached to $tRNA_1$ and $tRNA_2$, leading to the formation of an amide bond and the separation of the aminoacyl group from $tRNA_1$. The $tRNA_1$ then departs, and the mRNA to which the peptidyl-$tRNA_2$ remains attached moves in a channel on the ribosome so as to bring site 3 of the mRNA into the region of the ribosome where catalysis of peptide bond formation occurs. A third aminoacyl-tRNA, specified in Fig. 35.10 as

$$R^3CH(^+NH_3)CO_2{-}tRNA_3$$

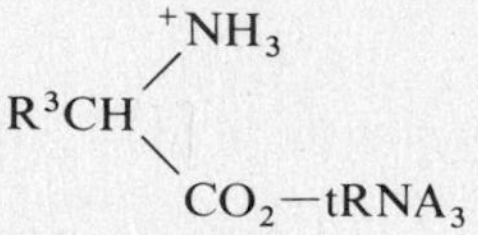

becomes attached to site 3, and the process is repeated in the fashion just described to add another amino acid residue to the growing peptide chain. This continues as many times as there are amino acid residues in the polypeptide that is being constructed. At a certain point, information built into the instructions in the mRNA brings the process to a halt; the completed polypeptide dissociates from the mRNA and the ribosome, to be transported to the region of the organism where it is to perform its particular functions (Fig. 35.11).

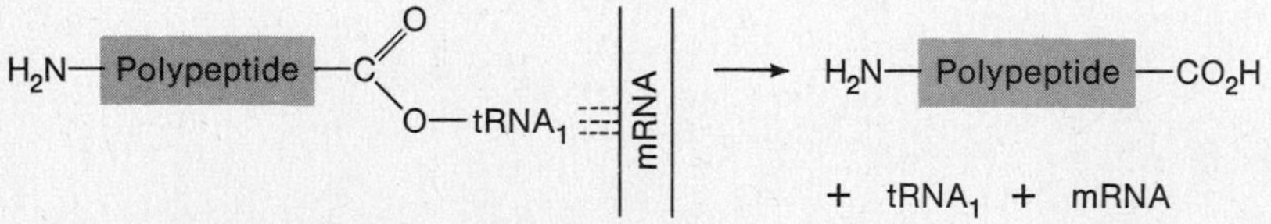

$$H_2N{-}\text{Polypeptide}{-}C(=O){-}O{-}tRNA_1 \cdots \text{mRNA} \longrightarrow H_2N{-}\text{Polypeptide}{-}CO_2H + tRNA_1 + \text{mRNA}$$

Fig. 35.11. Protein synthesis: removal of protein from mRNA and tRNA.

Despite the complexity of the process, the synthesis of a protein takes place with remarkable speed in the living organism. The time required for the synthesis of a polypeptide containing about 150 amino acid residues has been measured in *E. coli* bacteria and found to be about 15–20 sec. The laboratory synthesis of a polypeptide of comparable size has been accomplished, but it required several months for completion—*i.e.*, a rate difference of about 10^5 in favor of the *in vivo* synthesis.

35.5 The Genetic Code and DNA Replication

The ability of the messenger RNA to recognize a particular transfer RNA and affix it to a specific site is attributed to a complementary relationship among a sequence of three adjacent nucleotides on each of these RNA molecules. The necessity for the number three arises from a simple consideration of the ways in which four different nucleotides (*e.g.*, those of which DNA is composed, including adenine, guanine, uracil, and cytosine) can be used to encode information. If only one nucleotide is used, a total of four differentiations can be made; *i.e.*, one for each nucleotide. If pairs of nucleotides are used, 16 different combinations are possible, and 16 differentiations can be made. This is not quite sufficient for the 20 or so different amino acids that comprise a protein, however. But, if a sequence of three nucleotides is used, 64 combinations are possible, more than enough to provide a separate combination for each amino acid. Triplet sequences known as *codons* are used by the messenger RNA to select the appropriate transfer RNA containing a specific amino acid residue and affix it to that site where a particular amino acid is to be incorporated into the growing peptide chain. As the result of investigations by Nirenberg, Lederer, Ochoa, Khorana, and others, the sequences used for the various amino acids are known in almost complete detail. Inspection of Fig. 35.12 shows that there is considerable redundancy in the code and that only a few amino acids have only one codon

	U		C		A		G	
U	UUU	Phe	UCU	Ser	UAU	Tyr	UGU	Cys
	UUC	Phe	UCC	Ser	UAC	Tyr	UGC	Cys
	UUA	Leu	UCA	Ser	UAA	Ochre	UGA	(Umber)
	UUG	Leu	UCG	Ser	UAG	Amber	UGG	Trp
C	CUU	Leu	CCU	Pro	CAU	His	CGU	Arg
	CUC	Leu	CCC	Pro	CAC	His	CGC	Arg
	CUA	Leu	CCA	Pro	CAA	Glu	CGA	Arg
	CUG	Leu	CCG	Pro	CAG	Gln	CGG	Arg
A	AUU	Ile	ACU	Thr	AAU	Asn	AGU	Ser
	AUC	Ile	ACC	Thr	AAC	Asn	AGC	Ser
	AUA	Ile	ACA	Thr	AAA	Lys	AGA	Arg
	AUG	Met	ACG	Thr	AAG	Lys	AGG	Arg
G	GUU	Val	GCU	Ala	GAU	Asp	GGU	Gly
	GUC	Val	GCC	Ala	GAC	Asp	GGC	Gly
	GUA	Val	GCA	Ala	GAA	Glu	GGA	Gly
	GUG	Val	GCG	Ala	GAG	Glu	GGG	Gly

Fig. 35.12. Codons for amino acids.

(*e.g.*, methionine, glutamic acid, tryptophan); several have two codons (*e.g.*, phenylalanine, tyrosine, histidine, etc.), and some have as many as four (*e.g.*, valine, proline, alanine, etc.) or even six codons (*e.g.*, leucine, serine). The differences between the redundant codons generally occur in the third nucleotide of the sequence, the first two being the same in most, but not all, instances. A mRNA molecule with the nucleotide sequence shown in Fig. 35.13 would, for example, provide a template for constructing the polypeptide bradykinin (see Fig. 33.8).

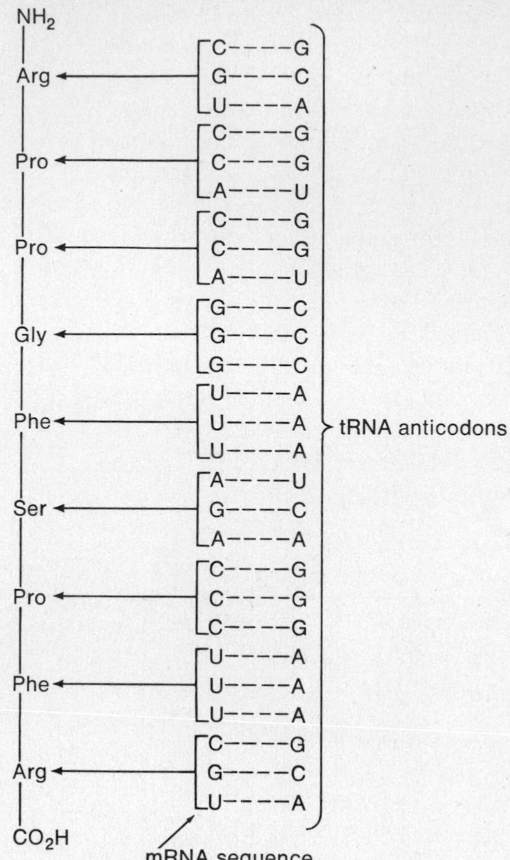

Fig. 35.13. m-RNA template for the nonapeptide bradykinin.

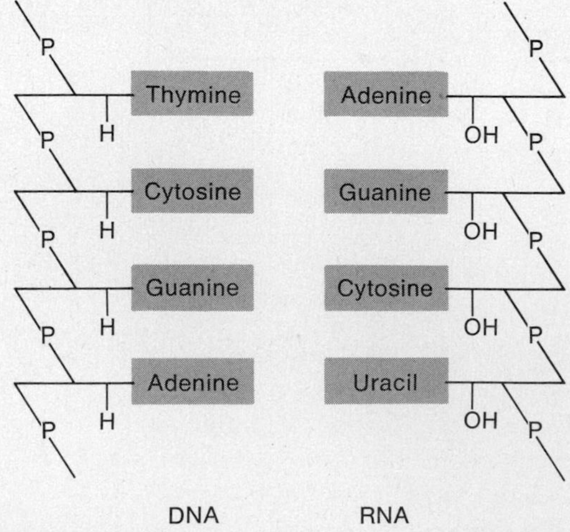

Fig. 35.14. Complementariness between DNA and mRNA.

Complementariness between nucleotides is also used in the biosynthesis of messenger RNA, in which DNA serves as the template; *i.e.*, adenine, guanine, cytosine, and uracil residues in the RNA are complementary to thymine, cytosine, guanine, and adenine residues, respectively, in the DNA (Fig. 35.14). A mixture of ribonucleotides and DNA reacts, in the presence of the enzymes RNA polymerase and DNA ligase, to form a polynucleotide that has a nucleotide sequence complementary to one of the strands of the DNA and identical with that of the other strand, except that uracil replaces thymine. The messenger RNA that is synthesized reproduces a segment of the DNA molecule and uses the sequence of this segment to direct the synthesis of a polypeptide. Thus, the blueprints for building enzymes and other proteins are stored in the nucleotide sequence of the DNA.

It is also through the agency of DNA that the blueprints can be handed down from one generation to another, for DNA has the capacity for self-replication. A mixture of deoxyribonucleotides and DNA reacts, in the presence of one or more DNA polymerases, to form new DNA molecules that are identical with the original. This was demonstrated in 1967 by Kornberg (who shared the 1959 Nobel Prize), who was able to show that the cyclic DNA from the bacterial virus ϕX174 (see p. 1086) acts as a template for the synthesis of new material that is identical in its biological and immunological activity with the original; presumably, therefore, the synthesized material has the same nucleotide sequence as the original. The scheme that has been suggested for DNA synthesis shows the double helix coming apart to form a pair of single strands; each of these then acts as the template for the construction of a complementary strand of DNA which, with the original template, forms a new double helix (Fig. 35.15).

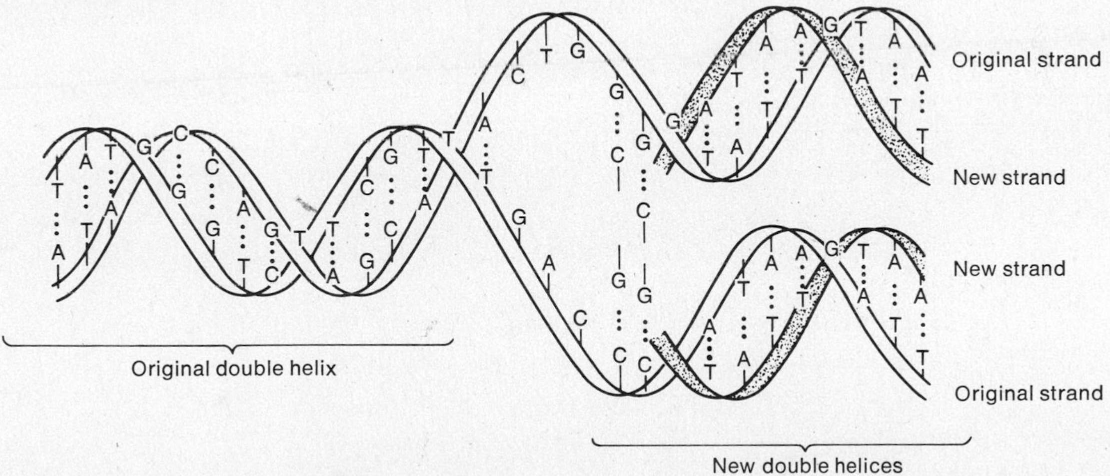

Fig. 35.15. DNA replication.

Enzymes are involved in virtually all biological reactions, including the reactions that synthesize the enzymes themselves. This circularity or feedback is one of the special characteristics of biological systems, and the emergence of organized polynucleotides and polypeptides from the random polymers present in the primeval broth quite probably was the result of complementary interactions among these polymers. The nature of this complementariness, however, is not yet understood. The question of how inanimate chemicals progressed from random to ordered polymers remains unanswered; it is, in fact, the mystery of life. From experiments simulating the conditions on Earth before the emergence of life, we can make reasonable surmises as to the composition of the primeval broth (see p. 9). From fossil records, from comparative biochemical studies, and from an enormous amount of experimentation on living systems, we have discovered many of the details of how life evolved from the simple cell to the complex organism called Man. As yet, we have little insight into the transformation of non-life into life.

This chapter has presented a glimpse into the marvelously complex processes that take place in the living cell when it uses the information stored in a DNA molecule to direct the actual synthesis of the catalysts that make life possible. Although studies of these phenomena have captured the attention of many scientists over the last 30 years, and although numerous Nobel Prizes have been awarded for discoveries in these areas, much remains to be done. Further understanding of the mysteries and complexities of the living organism will depend on an increased comprehension of the precise chemical details of nature.

REFERENCES

Barker, R., *Organic Chemistry of Biological Compounds*, Prentice-Hall, Inc., Englewood Cliffs, N.J., 1971, Chapter 8.

Bio-Organic Chemistry: Readings from Scientific American, W. H. Freeman and Co., Publishers, San Francisco, 1968.
Crick, F. H. C., Article 8. The Structure of the Hereditary Material.
Holley, R. W., Article 9. The Nucleotide Sequence of a Nucleic Acid.
Nirenberg, M. W., Article 29. The Genetic Code.
Fraenkel-Conrat, H., Article 30. The Genetic Code of a Virus.

Florkin, M., and E. H. Stotz, eds., *Comprehensive Biochemistry*, Vol. 8, Elsevier Publishing Co., New York, 1963.

Harbers, E., G. F. Domagk, and W. Muller, *Introduction to Nucleic Acids*, Reinhold Publishing Corp., New York, 1968.

Lehninger, A. L., *Biochemistry*, Worth, New York, 1970, Chapters 12, 28–34.

McGilvery, R. W., *Biochemistry, A Functional Approach*, W. B. Saunders Co., Philadelphia, 1970.

Rhinesmith, H. S., and L. A. Cioffi, *Macromolecules of Living Systems*, Reinhold Publishing Corp., New York, 1968.

PROBLEMS

1. Draw the structure of each of the following compounds:
   (a) 2-Methylpyrimidine-5-carboxylic acid
   (b) 9-Benzyl-8-methylmercaptopurine

(c) 9-β-D-glucosyladenine
(d) 1-α-D-galactosyluracil
(e) 3′-Adenylic acid
(f) Uridine-2′,3′-phosphoric acid
(g) A trinucleotide with the sequence adenosine-cytidine-guanosine
(h) Phenylalanyladenylate

2. Draw the structure of the product that you would predict from each of the following reactions:

(a) Lysine + ATP $\xrightarrow{\text{Enzyme}}$

(b) (imidazolyl)$CH_2CH(^+NH_3)C(=O)$–$AMP^=$ + tRNA $\xrightarrow{\text{Enzyme}}$

(c) ATP + H_2O $\xrightarrow{Ba(OH)_2}$

(d) Cytosine + HONO $\longrightarrow$

(e) 5-Adenylic acid (in H_2O solution) + $AgNO_3$ $\longrightarrow$

3. Define the following terms:
(a) Nucleoside
(b) Nucleotide
(c) Cyclic deoxynucleotide
(d) Transfer-RNA
(e) Messenger-RNA

4. Draw the anticodon sequence of the transfer-RNA and the codon sequence of the mRNA that are required for the synthesis of bovine oxytocin.

5. Draw the structure of the peptide that would be formed by ribosomes responding to each of the codes shown, assuming that the peptide building starts with the left-hand triplet:
(a) UCUUUUAAAGAUCUUGGU
(b) AGCUUUAAGGACCUGGGG
(c) AGAGGUAGGGGGCGAGGAAGGGG

Chapter **36**

Porphyrins, alkaloids, vitamins, and coenzymes

The Lord has created medicines out of the earth; and he that is wise will not abhor them.

Apocrypha, Ecclesiasticus

In addition to the carbohydrates, peptides, acetogenins, and nucleotides, numerous other classes of naturally occurring compounds also exist. The concluding chapter of Part III deals briefly with three of these, each of which constitutes an interesting and important segment of organic chemistry.

36.1 Porphyrins

Pyrrole (see Figs. 12.1 and 12.2), the building block of the porphyrins, is a constituent of coal tar and is present in the oil obtained by the distillation of the material present in animal bone. Because of the delocalization of the nonbonded electrons from the nitrogen into the unsaturated ring system, pyrrole ($pK_{BH^+} =$ 0.4) is a far weaker base than an aliphatic amine such as dimethylamine ($pK_{BH^+} =$ 10.6). Conversely, pyrrole ($pK_a = 9.9$) is a much stronger protic acid than dimethylamine ($pK_a = 33$), and is comparable in acidity to phenol. The skeletal structure of the porphyrins is the 16-membered prophin ring containing four pyrrole units attached at the α-position by methine (*i.e.*, H—C$\genfrac{}{}{0pt}{}{/\!\!/}{\backslash}$) carbons. The porphin ring is planar and contains a cyclic π system, indicated by the heavily shaded double bonds in the structure of hemin in Fig. 36.1, in which there are 18 π electrons, thus conforming to the $4n + 2$ rule (*i.e.*, $n = 4$). The system, therefore, possesses aromatic character, as indicated by a resonance energy that is greater than 400 kcal/mole. Undoubtedly, this remarkably high resonance energy was a key factor in the emergence of porphyrin compounds from the primeval broth, allowing them to assume their present important role as catalysts for the processes whereby living cells assimilate energy, either directly from the sun or indirectly by means of ingested foodstuff.

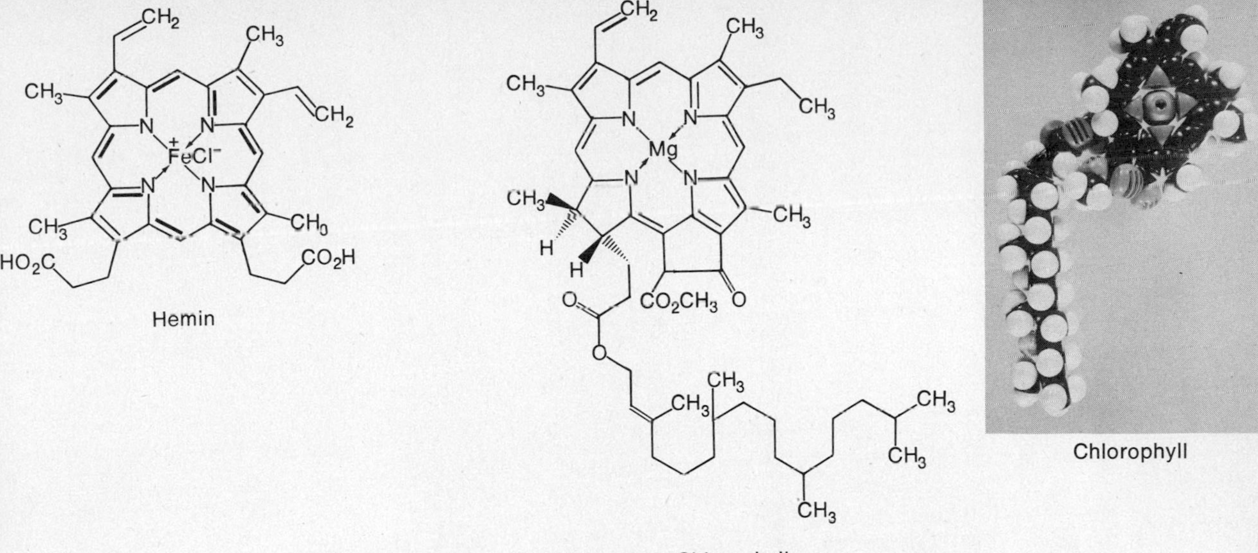

Fig. 36.1. Structures of porphyrins.

Important members of the porphyrin family are hemin, which is the colored pigment of hemoglobin, and chlorophyll, which is the colored component of green leaves (Fig. 36.1). The doughnut-shaped porphin ring can hold various metals in the hole of the doughnut through the combined forces of covalent and coordinate-covalent bonding. Thus, hemin holds an iron atom and chlorophyll holds a magnesium atom, the metals playing an integral part in the reactions of these compounds.

A class of compounds closely related to the porphyrins but differing in having two of the pyrrole rings *directly joined* are the corrins. Coenzyme B_{12} is the best known example, and its structure represents one of the most complex of the nonpolymeric naturally occurring products that have thus far been isolated (Fig. 36.2).

36.2 Alkaloids

Alkaloids are naturally occurring compounds, generally of plant origin, containing one or more nitrogen atoms (usually in a heterocyclic ring) that confer basic properties on the molecules. This general definition is inexact, however, for certain nitrogen compounds that are not very basic, or that do not contain the nitrogen in a ring, are also called alkaloids. Over 2,500 alkaloids have been isolated and their structures elucidated, and the list will undoubtedly grow as more plants are carefully analyzed for their nitrogen-containing constituents. Many alkaloids induce physiological responses when administered to man and animals, which has made them particularly fascinating to man for thousands of years. Some of the oldest drugs and medicines known to man are alkaloids, and the search for new and useful ones continues today.

Alkaloids can be grouped into a variety of families, depending on the type of heterocyclic ring that is present in the molecule. Among the simplest are those

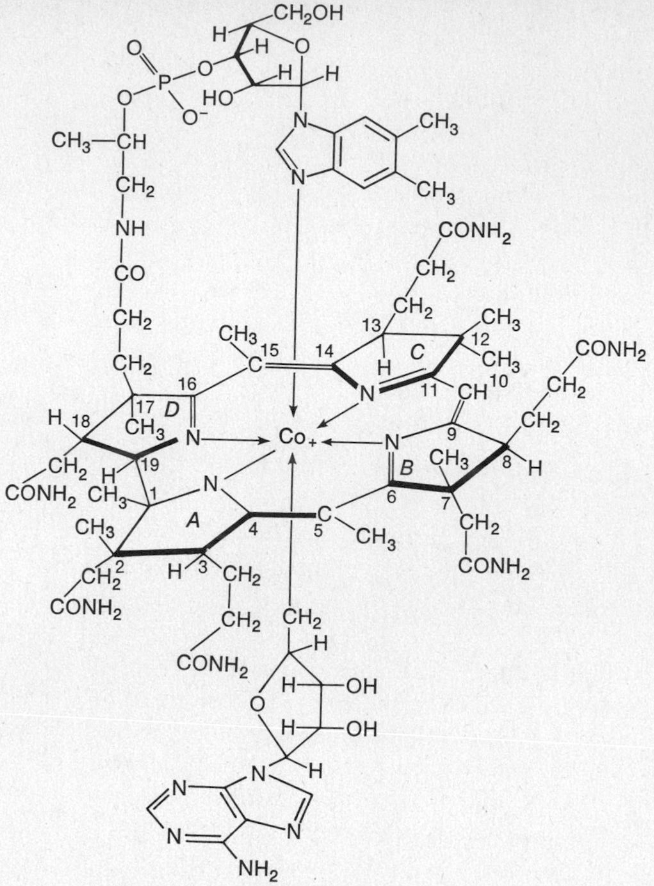

Fig. 36.2. Structure of vitamin B_{12}.

containing a single pyrrolidine (*i.e.*, tetrahydropyrrole) or piperidine (*i.e.*, hexahydropyridine) ring. For example, the compound known as hygrine (Fig. 36.3) is isolated from the leaves of the Peruvian coca shrub, and a compound known as coniine (Fig. 36.3) is isolated from the poison hemlock. Both compounds are optically active liquids that are strongly basic ($pK_{BH^+} \simeq 11$). Coniine is poisonous to humans, causing a gradual paralysis that ends in convulsions and death. Its

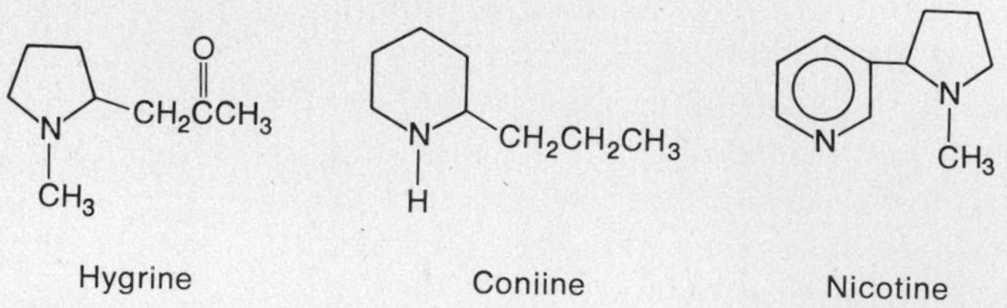

Fig. 36.3. Structures of alkaloids containing five-membered and six-membered heterocyclic rings.

most notorious application in this regard occurred over two millenia ago when Socrates was executed by his countrymen in 400 B.C.

One of the best known of the simple alkaloids, a compound containing both a five-membered and a six-membered heterocyclic ring, is nicotine (Fig. 36.3). Nicotine is one of the principal alkaloids of the tobacco leaf, constituting as much as 1.5% or more of the weight of the leaf. It is a highly toxic substance which in small doses causes respiratory stimulation and in large doses can cause respiratory paralysis and death. Because of the apparent stimulatory action of small doses, millions of persons have adopted the practice of inhaling the smoke from the burning tobacco leaf via cigarette, cigar, or pipe. Because of the lethal action of larger doses, nicotine finds some use as an insecticide.

The majority of alkaloids have polycyclic ring systems, among the simplest of which is the 8-azabicyclo[3.2.1]octane system, the tropane ring system. A well known example of a tropane alkaloid is cocaine (Fig. 36.4), an optically

8-Azabicyclo[3.2.1]-
octane (tropane)

Cocaine

Fig. 36.4. Structure of tropane and cocaine, a tropane alkaloid.

active crystalline solid isolated from the leaves of the coca plant. For centuries the leaves of the coca plant have been chewed for the physical and mental stimulation that ingestion of their alkaloids provides, and more recently inhalation of the smoke of the burning leaves has been employed as the means for intake of the alkaloids into the body. Sherlock Holmes was an inveterate user of cocaine, and the original Coca-Cola contained it as one of its ingredients. Cocaine is an addictive drug and when taken in excessive amounts may lead to mental and physical deterioration.

Among the several alkaloids containing a quinoline ring, quinine (Fig. 36.5) is certainly the most famous. Quinine, one of the alkaloids obtained from the

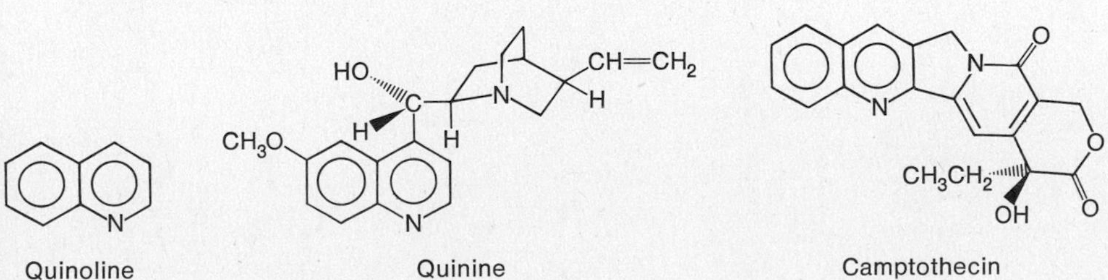

Fig. 36.5. Structure of quinoline and the quinoline alkaloids quinine and camptothecin.

Quinine

bark of the cinchona tree, has been known since 1600 as a remedy for the ravages of malaria, and its analogs quinidine, cinchonine, and cinchonidine also have antimalarial activity. Many other alkaloids also contain a quinoline ring. A recently isolated and characterized compound in this class is camptothecin (Fig. 36.5), a pentacyclic alkaloid found in the Chinese tree *Camptotheca acuminata Nyssacae* and alleged to possess antileukemic and antitumor activity in animals.

The isoquinoline ring system is the basic heterocycle in a number of alkaloids, such as papaverine (Fig. 36.6), isolated as a weakly basic ($pK_{BH^+} = 6.2$) solid from the unripe seed capsule of the opium poppy. It finds some medical application in the treatment of spasms of the stomach or intestines.

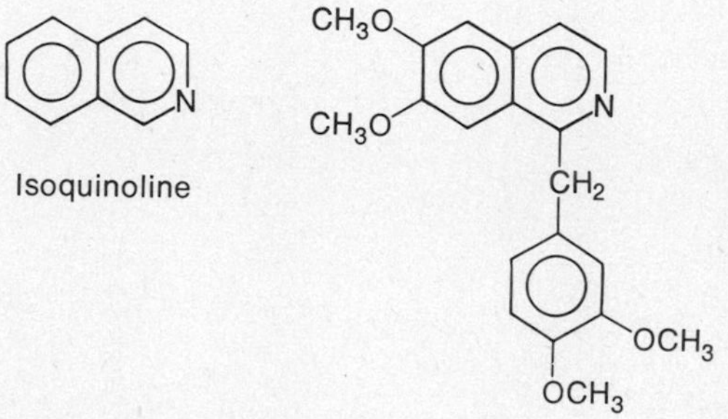

Isoquinoline

Papaverine

Fig. 36.6. Structure of isoquinoline and papaverine, an isoquinoline alkaloid.

A variety of alkaloids are based on the indole nucleus, and one of the best known is lysergic acid (Fig. 36.7). Lysergic acid is a member of the family of ergot alkaloids, isolated from the fungus *Claviceps purpurea*, which infects grass crops, particularly rye. Ingestion of infested rye leads to a disease called ergotism (known in earlier years as "St. Anthony's fire"), which causes severe physical and mental disabilities. In small, regulated amounts, however, lysergic acid is used in the treatment of migraine headaches and for the induction of uterine contraction in childbirth. The diethylamide of lysergic acid, (Fig. 36.7),

Indole

Lysergic acid (R = CO_2H)
Lysergic acid diethylamide, LSD
(R = $CON(CH_2CH_3)_2$)

Strychnine

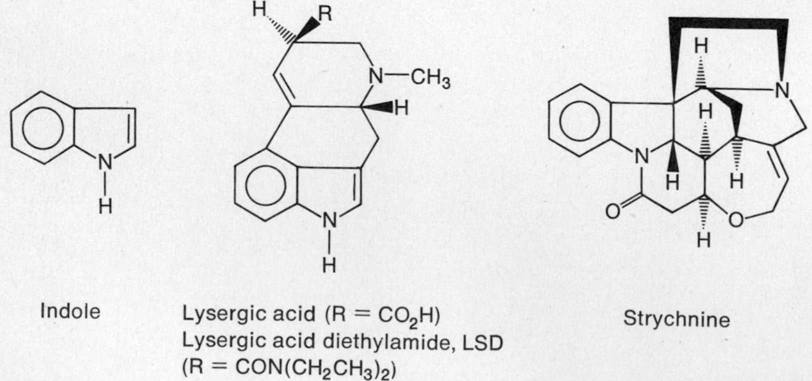

Fig. 36.7. Structure of indole and the indole alkaloids lysergic acid and strychnine.

the notorious LSD, is the most potent psychotomimetic substance known. It has varied effects on humans, causing color hallucinations and psychotic states resembling those of schizophrenia. An indole alkaloid containing an even more convoluted ring system than lysergic acid is strychnine (Fig. 36.7), isolated from the seeds of *Strychnos nux-vomica.* It is extremely poisonous and, for that reason, has been used for arrow poisons by native tribes, for exterminating political enemies in medieval times, and for exterminating vermin today. Its structure was not completely elucidated until 1948, shortly after which a remarkable total synthesis was achieved by R. B. Woodward.

One of the most studied of the alkaloids is morphine, used medicinally for its hypnotic (sleep-inducing) and analgesic (pain relieving) properties. Morphine (from Morpheus, the god of dreams in Greek mythology) was isolated in pure form as long ago as 1804 from the opium poppy *Papaver somniferum* L., and today enormous quantities of opium are processed for the legal as well as the illegal production of morphine. Unfortunately, prolonged intake of morphine leads to addiction, so its continued use as a hypnotic and analgesic is sharply limited. Because of its rather complex ring system (Fig. 36.8), the elucidation of the structure of morphine posed a difficult problem that was not solved until more than 100 years after its original isolation in pure form.

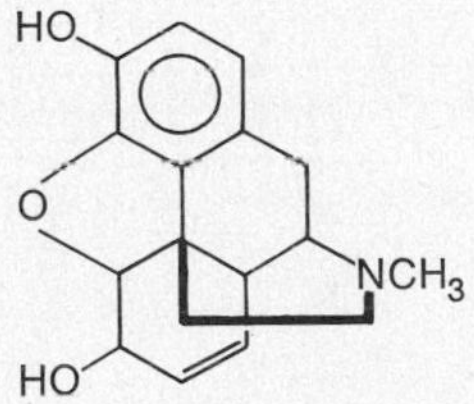

Fig. 36.8. Structure of morphine.

Morphine

In addition to the large number of natural products containing heterocyclic nitrogen, certain other compounds possessing alkaloid-like properties are known, in which nonbasic amide nitrogen is present. A typical example is colchicine, isolated from the autumn crocus, *Colchicum autumnale L.* (Fig. 36.9). Colchicine interferes with the process of mitosis (*i.e.*, cell division) in plants, leading to a

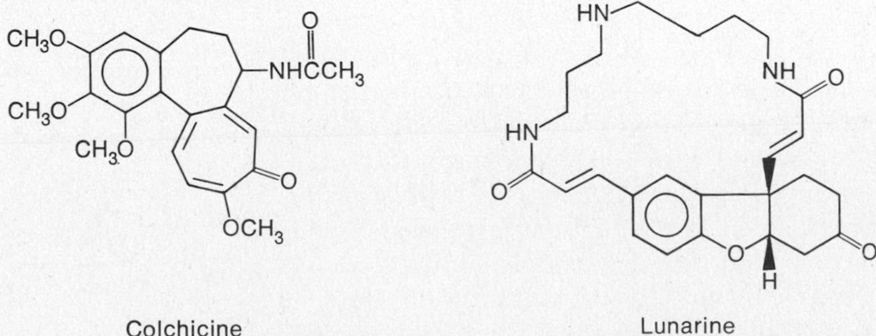

Colchicine

Lunarine

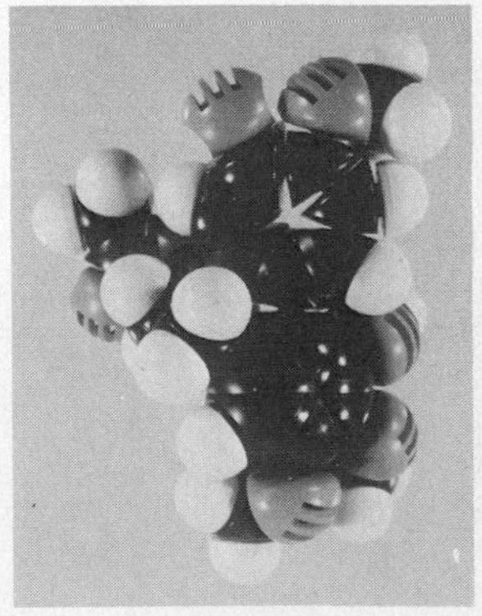

Colchicine

Fig. 36.9. Structures of amide-containing alkaloids.

condition known as polyploidy, in which an increased number of chromosomes (double, quadruple, etc.) are present in the cell. This, in turn, may give rise to altered varieties of plants, conferring some economic interest on colchicine as a means for producing new and useful agricultural variants. Another example of an amide-containing alkaloid is lunarine (Fig. 36.9), isolated from the yew tree, which is poisonous to man and to animals.

Also sometimes included among the alkaloids are certain compounds containing the purine ring system (see Section 35.1). Best known among these are theophylline (in tea), caffeine (in coffee and tea), and theobromine (in cocoa bean) (Fig. 36.10). All are stimulants to the central nervous system and, for that reason, have been enjoyed for many years as beverages.

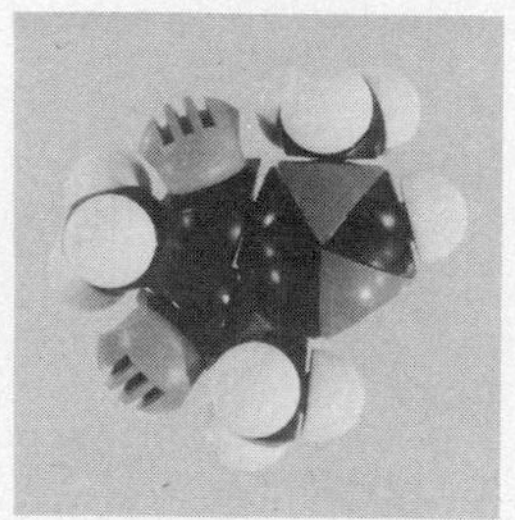

Caffeine

Theophylline   Caffeine   Theobromine

Fig. 36.10. The purine alkaloids theophylline, caffeine, and theobromine.

36.3 *Vitamins and Coenzymes*

In addition to the amino acids, which were probably among the first types of organic molecules to be formed in the primeval broth, other types of molecules must also have been produced. Some of these possessed the ability to catalyze various types of organic reactions, and the ones that were victorious in the "survival of the fittest" as evolution operated at the chemical level in the prebiotic era are known to us today as coenzymes. In conjunction with enzymes, which probably evolved at a later date, coenzymes catalyze a host of reactions that take place in today's living systems. It should be pointed out, though, that while all coenzymes require the association of an enzyme, many enzymes are able to act as catalysts in their own right. Chymotrypsin (see p. 1049), for example, is one of these. Among the enzymes that are not sufficient unto themselves but that require a coenzyme are alcohol dehydrogenase, employing nicotine adenine dinucleotide as the coenzyme (see Fig. 33.19) and the various porphyrin-enzyme complexes involved in metabolism. Numerous other compounds in addition to the porphyrins serve as coenzymes, including the compounds called vitamins.

Nutrition experiments carried out late in the nineteenth century led to the realization that animals require more than carbohydrates, proteins, lipids, and inorganic salts for healthy survival. The necessity for certain specific foods in the diet to avoid such disorders as beriberi became known, and the so-called "accessory factors" present in these foods came eventually to be called *vitamins*. It is now understood that some, or perhaps all, of the vitamins serve as coenzymes for various enzyme-catalyzed reactions and that they correspond to those coenzymes which the animal is incapable of synthesizing within itself. Different animals have different capabilities in this respect; thus, what is a required "accessory factor" for one species may not be so for another. The designation "vitamin" generally refers to those factors that cannot be synthesized by the *human* body and are, therefore, necessary components of *man's diet*.

Vitamin A belongs to the triterpene family and is involved with the visual function (see Fig. 34.11). Vitamin D is indirectly derivable from the triterpenes

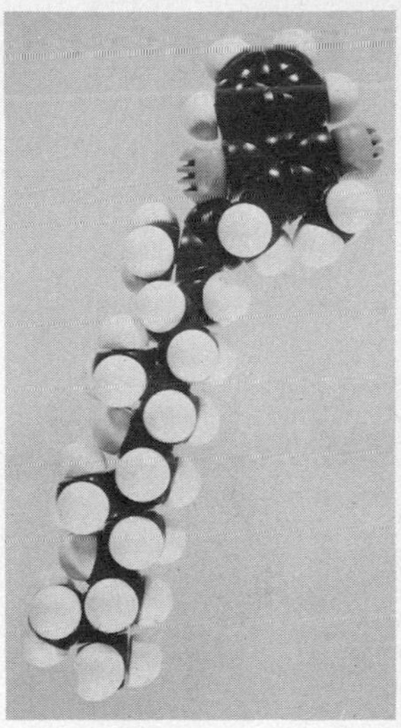

Vitamin E

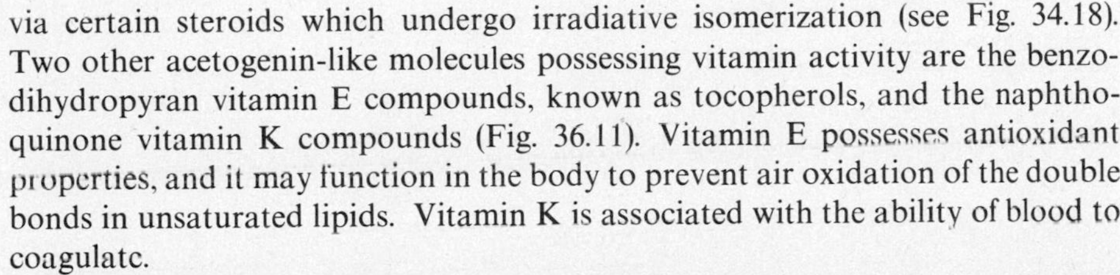

Fig. 36.11. Structures of vitamin E and vitamin K.

via certain steroids which undergo irradiative isomerization (see Fig. 34.18). Two other acetogenin-like molecules possessing vitamin activity are the benzodihydropyran vitamin E compounds, known as tocopherols, and the naphthoquinone vitamin K compounds (Fig. 36.11). Vitamin E possesses antioxidant properties, and it may function in the body to prevent air oxidation of the double bonds in unsaturated lipids. Vitamin K is associated with the ability of blood to coagulate.

The B vitamins include several compounds, designated as B_1, B_2, B_5, B_6, and B_{12}. Vitamin B_1, better known as thiamine (Fig. 36.12), occurs as the pyrophosphate and is involved as a coenzyme in the decarboxylation of pyruvic acid. Its absence in the diet brings on the symptoms called beriberi. Vitamin B_2, known as riboflavin (see Fig. 32.41), is the prosthetic group (*i.e.*, the attached

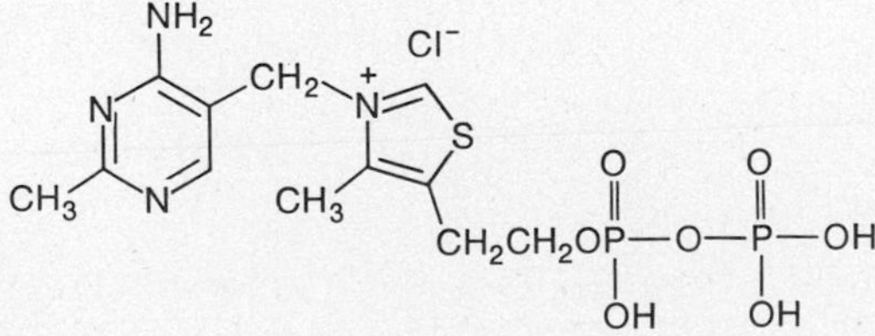

Fig. 36.12. Structure of vitamin B_1 (thiamine) in the form of its pyrophosphate.

group) of the flavoproteins that are involved in the metabolism of α-amino acids and the carbohydrates. Occurring as the monophosphate, it is designated as flavin mononucleotide (FMN in the oxidized form, $FMNH_2$ in the reduced form) (Fig. 36.13).

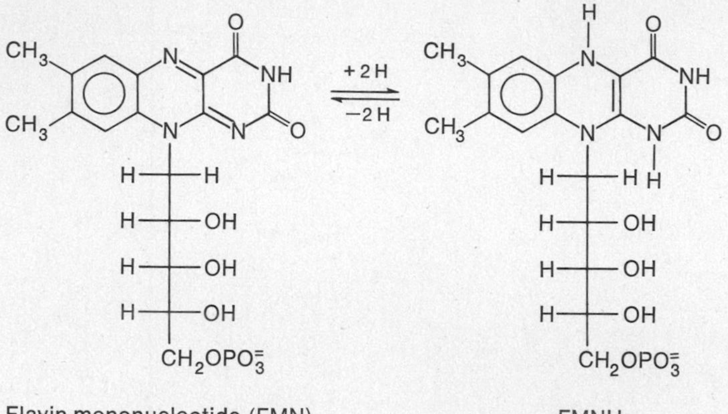

Flavin mononucleotide (FMN) $\underset{-2H}{\overset{+2H}{\rightleftharpoons}}$ $FMNH_2$

Fig. 36.13. Structure of vitamin B_2.

Pantothenic acid, sometimes designated as vitamin B_5, is one of the building blocks of coenzyme-A (Fig. 36.14).

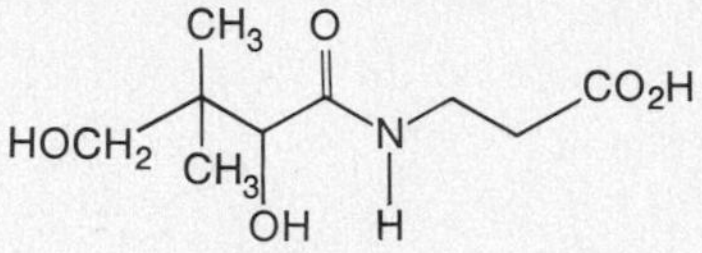

Pantothenic acid

Fig. 36.14. Structure of vitamin B_5.

Vitamin B_6, known as pyridoxal phosphate (Fig. 36.15), occurs as the phosphate and is a coenzyme in a variety of reactions, including the conversion of α-keto acids to α-amino acids, where it acts as a carrier of the amino function.

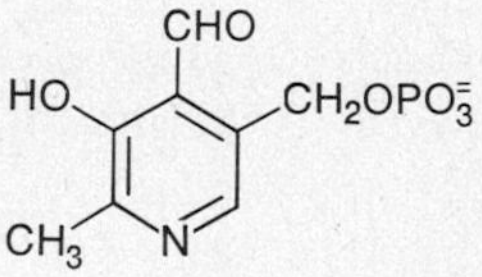

Pyridoxal

Fig. 36.15. Structure of vitamin B_6 in the form of its phosphate.

Vitamin B_{12} in the form of its 5-deoxyadenosyl derivative (see Fig. 36.2) has coenzyme activity and is involved in certain biological rearrangements and methylation reactions. Deficiency of vitamin B_{12} in the diet is a cause of pernicious anemia.

Still another compound that is classed as a member of the vitamin B family is biotin (Fig. 36.16). In association with the proper enzyme, it is involved in the catalysis of carboxylation reactions, such as the conversion of pyruvic acid to oxaloacetic acid,

$$CH_3COCO_2H \rightarrow HO_2CCH_2COCO_2H$$

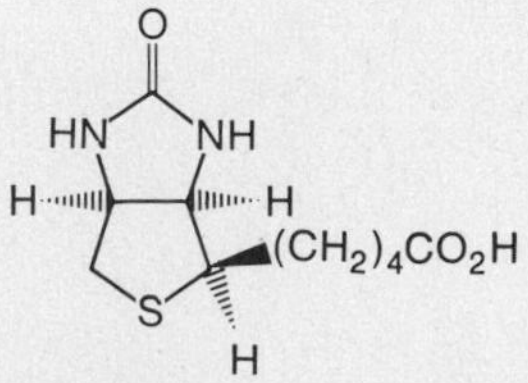

Fig. 36.16. Structure of biotin.

Vitamin C, the only carbohydrate member of the vitamins, has been mentioned earlier (see Fig. 32.35). Folic acid (Fig. 36.17), broadly distributed in plants, is a member of the *pterin* class of chemical compounds. In the form of its tetrahydro

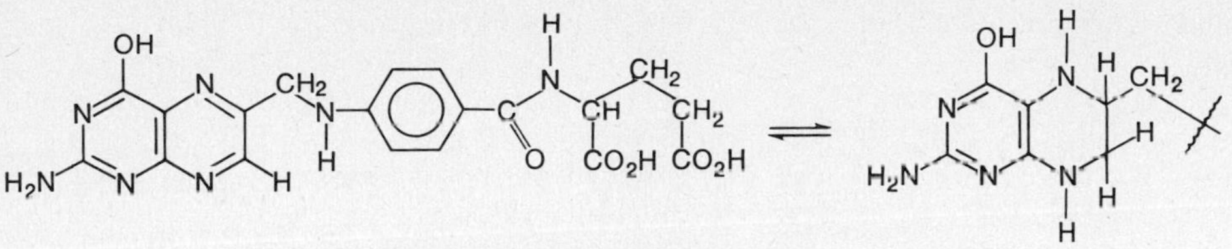

Fig. 36.17. Structures of folic acids.

derivative it is involved in the biosynthesis of pyrimidine and purine rings, where it facilitates the transfer of a one-carbon unit.

Other chemical species in addition to the vitamins have coenzyme capacity and functions. Included among these are various metal ions, porphyrin molecules (see Section 36.1), adenosine triphosphate (see Fig. 32.41), nicotinamide adenine dinucleotide (see Fig. 32.41), and coenzyme-A.

REFERENCES

BARKER, R., *Organic Chemistry of Biological Compounds*, Prentice-Hall, Inc., Englewood Cliffs, N.J., 1971, Chapter 9.

Bio-Organic Chemistry: Readings from Scientific American, W. H. Freeman and Co., Publishers, San Francisco, 1968.
ROBINSON, T., Article 18. Alkaloids.
GATES, M., Article 19. Analgesic Drugs.
BARRON, F., M. E. JARVIK, and S. BUNNEL, Article 20. Hallucinogenic Drugs.

PELLETIER, S. W., ed., *Chemistry of the Alkaloids*, Van Nostrand Reinhold Company, New York, 1970.

SWAN, G. A., *An Introduction to the Alkaloids*, Blackwell Scientific Publications, Oxford, 1967.

VERNON, L. P., and G. R. SEELY, eds., *The Chlorophylls*, Academic Press, Inc., New York, 1966.

WAGNER, A. F., and K. FOLKERS, *Vitamins and Coenzymes*, John Wiley & Sons, Inc., New York, 1964.

PROBLEMS

1. Draw the structure for each of the following compounds:
(a) The semicarbazone of hygrine
(b) The hydrobromide salt of coniine
(c) The benzyl ester analog of cocaine
(d) A dihydroquinine
(e) The enantiomer of camptothecin
(f) 4-Hydroxyindole
(g) Lysergic acid di-*n*-propylamide
(h) Deoxymorphine
(i) 9-Phenyltheophylline
(j) A tetradeuterio vitamin K_1
(k) The isobutyl ester of biotin
(l) Tetradeuteriofolic acid

2. Draw the structures of the products from each of the following reactions:
   (a) The action of two equivalents of ozone on hemin
   (b) The hydrolysis of chlorophyll
   (c) The action of potassium permanganate on nicotine
   (d) The hydrolysis of cocaine
   (e) The Oppenauer oxidation of quinine
   (f) The action of methyl iodide on papaverine
   (g) The action of excess methyl iodide on coniine, followed by treatment with base
   (h) The action of manganese dioxide on morphine
   (i) The action of hydriodic acid on colchicine
   (j) The action of potassium permanganate on vitamin K_1
   (k) The action of hydroxylamine on pyridoxal
   (l) The hydrolysis of pantothenic acid

3. A hospital located on the lowlands near the Mississippi River was inundated by a flood, one result of which was to remove the labels from all the bottles in the storeroom containing various biological compounds. A list of the 10 compounds in one of the boxes in the storeroom remained intact, however, so it was simply necessary to decide which of the ten compounds was in each of the containers. The ten compounds were adenylic acid, biotin, cholesterol, cortisone, glucose, lysergic acid, morphine, nicotine, pyridoxal, and vitamin E. Using qualitative chemical tests (*e.g.*, solubilities, qualitative elemental analyses, etc), how would you decide which compound was in which bottle?

PART IV

Special Topics

Organic chemistry is a big field. In the almost two centuries of its existence as a clearly delineated area of science, a tremendous number of facts and theories have accumulated, and the rate of accumulation is increasing. To acquire a knowledge of these facts and theories and a facility in dealing with them, it is necessary, as we have seen, to take things one step at a time, progressing from the simple to the more complex. In Parts I–III of the text, we have gone from simple structures to very complex ones, from a small collection of synthetically useful reactions to a much larger collection, from easy syntheses to considerably more difficult ones, and from sketchy ideas of reaction mechanisms to more sophisticated treatments. This does not mean, however, that we have furrowed the last row. In almost any field, particularly one that is as fertile as organic chemistry, there is always unploughed ground—always more to learn, Hardly any aspect of organic chemistry has escaped the concentrated attention of the researcher, and the detail in which certain areas are now understood is quite astonishing.

Part IV concerns three of these areas, concentrating its attention on nonbenzenoid aromatic compounds, orbital symmetry considerations, and organic synthesis. Chapter 37 on nonbenzenoid aromatic compounds is an extension of Chapter 31, which deals with benzenoid aromatic compounds and Chapter 4, which deals with aromaticity. Chapter 38 on orbital symmetry considerations provides a rationale for some of the elimination and cycloaddition reactions discussed in Chapters 25 and 26. Chapter 39 on organic synthesis elevates the sophistication of the approach to this important topic, extending the discussion initiated in Chapter 18 and emphasized throughout Part II. Part IV is intended to serve as a bridge between the elementary treatment of topics in Parts I–III and the more detailed treatments that are found in advanced monographs and in the original literature.

Chapter **37**

Nonbenzenoid aromatic compounds

The aromatic sextet has proved a useful hall-mark for the great majority of cases, but it is only one of several similar electron groups which confer a special kind of stability on the molecule.*

SIR ROBERT ROBINSON

Aromaticity is a variously defined term that generally refers to compounds having (a) a high carbon-to-hydrogen ratio, (b) carbon–carbon bond lengths shorter than 1.54 Å but longer than 1.33 Å, (c) considerable resistance to oxidation and a strong tendency to undergo electrophilic substitution, (d) certain uv, ir, and nmr spectral characteristics, and (e) considerable resonance energy, as measured by heats of combustion or heats of hydrogenation (see Section 4.8). The prototype aromatic compounds are benzene and its many polycyclic relatives such as naphthalene, anthracene, and phenanthrene. There are a number of compounds that do not contain benzene rings but that possess most or all of the properties just listed, however, and by these criteria have aromatic character. Such compounds are classed as *nonbenzenoid aromatics* and can be subdivided into (a) those containing carbocyclic rings and (b) those containing heterocyclic rings.

CARBOCYCLIC NONBENZENOID AROMATIC COMPOUNDS

According to the Hückel rule, a planar, cyclic array of atoms corresponds to an aromatic system if it contains $4n + 2$ electrons in its π system. Benzene, naphthalene, and anthracene all conform to this requirement in possessing 6, 10, and

* Taken from the Preface to *Nonbenzenoid Aromatic Compounds*, ed. by D. Ginsberg, Interscience Publishers, New York, 1959.

14 electrons, respectively, in their π systems; they represent particularly stable examples of Hückel aromatic systems because they are neutral entities. The Hückel rule does not require that the species be neutral, however, but only that it contain the right number of electrons. Thus, cyclopropenyl cation, cyclopentadienyl anion, cycloheptatrienyl cation, cyclooctatetraenyl dication, and cyclooctatetraenyl dianion are all cyclic π systems containing $4n + 2$ electrons that are expected to be aromatic. All these systems have been prepared, and they do, indeed, show at least some of the characteristics associated with aromaticity (see Figs. 4.16 and 4.17). The two systems from this group that have been most thoroughly studied and whose chemistry is interesting from the standpoint not only of theory but also of synthetic utility are the cyclopentadienyl anions and the cycloheptatrienyl cations. The following discussion is primarily concerned with members of these two classes of compounds in which the negative and positive charges, respectively, are internally neutralized, special stability thereby being conferred on the molecules. In addition, the aromaticity of an interesting class of macrocyclic polyenes is also considered.

37.1 *Ferrocene*

An experiment was performed in 1951 in which cyclopentadiene was treated with ethylmagnesium bromide and ferric chloride. It was known at that time that cyclopentadiene is sufficiently acidic to decompose ordinary Grignard reagents to form the corresponding cyclopentadienylmagnesium halide. It was not anticipated, however, that ferric chloride would also enter into the reaction, and it was with surprise that the experimenters isolated a crystalline, orange compound that had a molecular formula corresponding to $C_{10}H_{10}Fe$:

$$C_5H_6 + C_2H_5MgBr \longrightarrow C_5H_5MgBr + C_2H_6$$

$$2\,FeCl_3 + 2\,C_2H_5MgBr \longrightarrow 2\,FeCl_2 + C_4H_{10} + 2\,MgBrCl$$

$$2\,C_5H_5MgBr + FeCl_2 \longrightarrow Fe(C_5H_5)_2 + MgBr_2 + MgCl_2.$$

The material was named ferrocene and was found to possess interesting properties. It is stable up to 400°, soluble in the usual range of organic solvents, and sufficiently volatile to steam distill. Chemically, it behaves like an aromatic compound; it

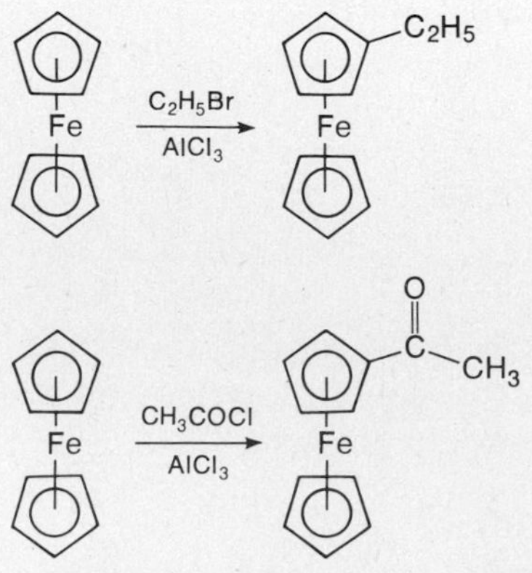

$$\text{Ferrocene} \xrightarrow[\text{AlCl}_3]{C_2H_5Br} \text{ethylferrocene } (C_2H_5 \text{ on one ring})$$

$$\text{Ferrocene} \xrightarrow[\text{AlCl}_3]{CH_3COCl} \text{acetylferrocene } (\text{C(=O)}{-}CH_3 \text{ on one ring})$$

Fig. 37.1. Electrophilic substitution reactions of ferrocene.

is resistant to alkali and acid (in the absence of oxygen) and undergoes electrophilic substitution reactions, as noted in more detail in the following paragraphs. The reconciliation of these properties with a chemical structure was achieved in 1952 by several workers who suggested that ferrocene is a "sandwich compound" in which two cyclopentadienyl rings comprise the bread of the sandwich and the iron atom the filling. This postulate was subsequently verified by X-ray studies, which show ferrocene to be a centrosymmetrical compound in which all the C—Fe distances are identical.

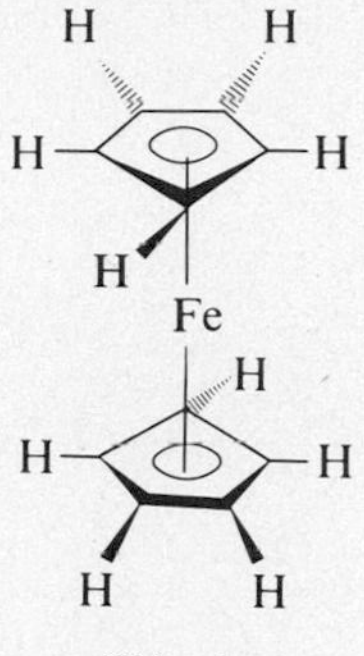

(Side view)

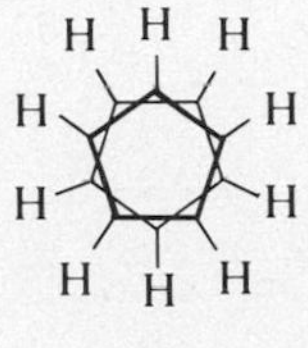

(Top view)

Its aromatic properties can be viewed in terms of a cyclopentadienyl anion whose negative charge is counterbalanced by the positive charge of the iron atom, *i.e.*, a donation of electron density from the π-electron system of the cyclopentadienyl ring to the vacant d orbitals of the iron atom. The bond has axial symmetry (similar to a σ bond), and the cyclopentadiene rings rotate more or less freely with respect to each other. Certain conformations are preferred, however, the staggered ones being of lower energy. This means that for a 1,1′-disubstituted ferrocene, three staggered conformers are possible. In analogy with a 1,2-disubstituted ethane, the rotation is sufficiently rapid at room temperature that these are in mobile equilibrium and cannot be isolated as discrete entities:

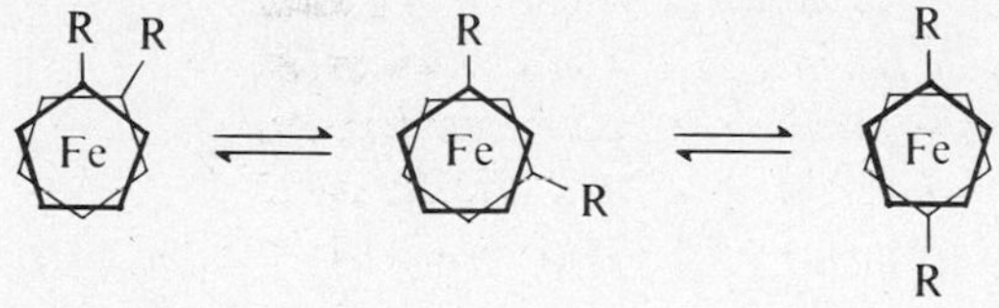

Ferrocene is claimed to be even more aromatic than benzene, because it has an even greater tendency to undergo electrophilic substitution rather than addition. It is more restricted in the range of electrophilic reagents that can be used, though, for those that can act as oxidizing agents (*e.g.*, bromine, nitric acid, etc.) convert ferrocene to the ferricinium ion which decomposes to ferric ion and cyclopentadienyl derivatives. Of particular utility are the Friedel–Crafts alkylations and acylations, by means of which various other substituents can be introduced on the ferrocene ring (Fig. 37.1). Formylation (*e.g.*, with N-methylformanilide—see Fig. 28.14), sulfonation, and mercuration all proceed under rather mild conditions. Electrophilic substitution of a monosubstituted ferrocene can yield three structural isomers, two in which the second substituent is in the same ring as the original substituent and one in which it is in the other ring (Fig. 37.2). Although the ferrocene ring is sensitive to oxidizing agents, reducing agents

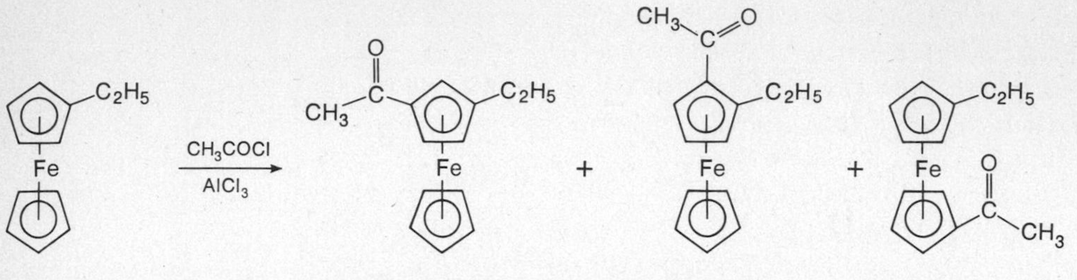

Fig. 37.2. Acylation of ethylferrocene.

are without effect under mild conditions and can be used to alter substituent groups. For instance, acetylferrocene can be reduced to the corresponding alcohol with sodium borohydride (Fig. 37.3).

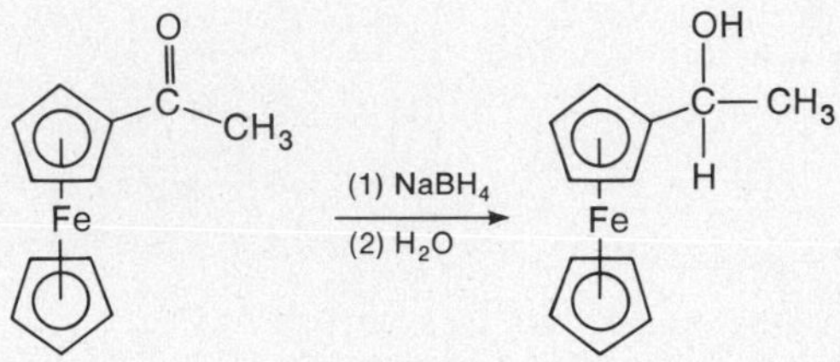

Fig. 37.3. Reduction of acetylferrocene with sodium borohydride.

The discovery of ferrocene opened a new area of chemistry that has come to be called metallocene chemistry, and many examples of metallocenes are now known—*i.e.*, compounds in which a metal atom is sandwiched between various types of carbon moieties, such as cyclobutadienyl rings, cyclopentadienyl rings, benzene rings, carbon monoxide groups, etc.:

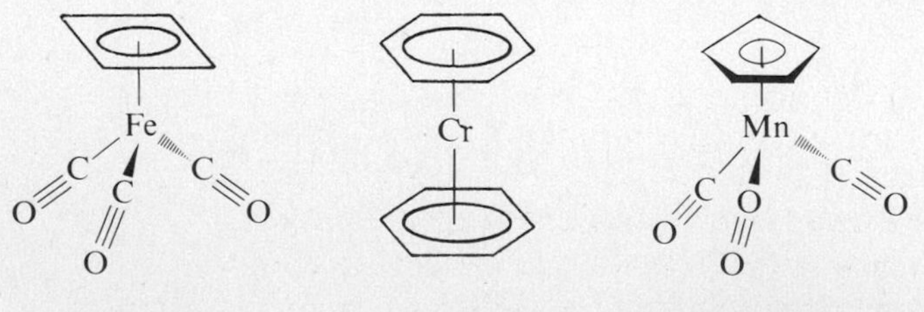

Cyclobutadiene-iron-tricarbonyl **Dibenzene chromium** **Cyclopentadienyl-manganesetricarbonyl**

37.2 *Tropones*

Pasteur once remarked that "chance favors the prepared mind." The accuracy of this judgment is well illustrated by the story of cycloheptatrienylium bromide. Before the turn of the century, a German chemist named Merling prepared cycloheptatriene by degradation of an alkaloid known as tropine. He discovered that tropilidene, which was the name given to cycloheptatriene because of its source in this experiment, reacted with bromine to give a dibromide that slowly lost hydrogen bromide to produce a mixture of benzyl bromide and an isomeric yellow, crystalline solid that was insoluble in ether. This unusual solubility behavior failed to arouse his curiosity, however, and as a consequence, the first example of a nonbenzenoid aromatic compound of the cycloheptatrienylium cation system (tropylium cation system) went unappreciated until 1954 when William Doering exhumed and reinterpreted this experiment. With a mind "prepared" by the concepts of aromaticity that had developed over the intervening decades, Doering realized that the ether insolubility of tropylium bromide is commensurate with a saltlike compound which would result from the dissociation to a bromide ion and a tropylium cation, the latter having the special stability of a $4n + 2$ electron Hückel system.

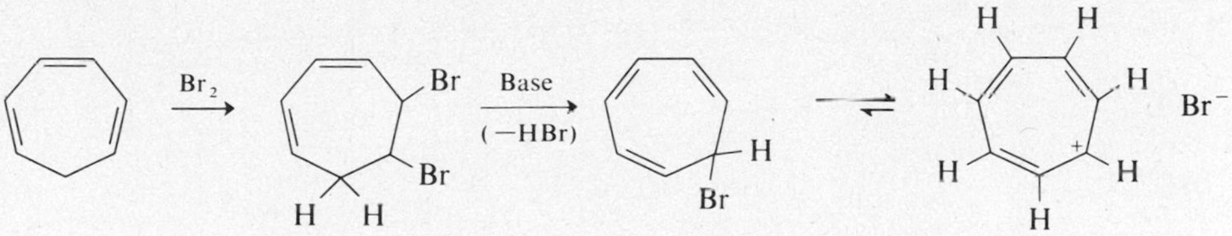

Even before Doering's experiments, the existence of tropylium-type compounds was coming to the attention of chemists from other directions. The structure of the natural product colchicine (see Fig. 36.9) had puzzled chemists for many years, particularly "ring C" of the compound, and in 1945 it was suggested by M. J. S. Dewar that ring C has a cycloheptatrienone structure. He pointed out that in such a structure, the cycloheptatriene ring can assume aromatic character to the extent that the C=O bond is polarized:

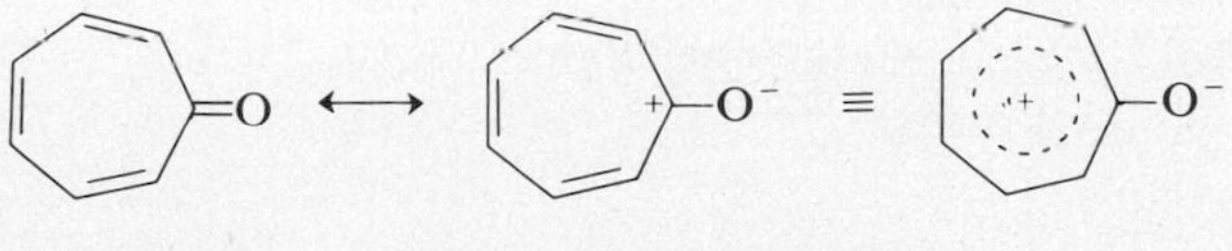

The Dewar formulation of the colchicine structure has been verified through total synthesis, and his suggestion that cycloheptatrienones (tropones) should have aromatic properties has been substantiated by many subsequent studies.

The charge-separated character of tropone agrees with its high dipole moment of 4.3 D, in contrast with simple ketones which have dipole moments of about 2.7 D. Bond length measurements show some variation from one bond to another (Fig. 37.4), but the C—C single bonds are, nevertheless, considerably shorter than ordinary C—C bonds, and the compound is estimated to possess about 25 kcal/mole of resonance energy. The unusual degree of polarization of the carbonyl group is also indicated by the carbonyl stretching frequency in the infrared, which occurs at 1638 cm^{-1}, about 70 cm^{-1} lower than the carbonyl stretching frequency of an ordinary ketone such as cycloheptanone (about 1705 cm^{-1}).

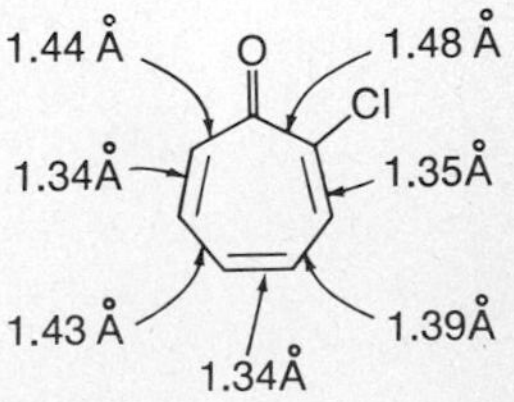

Fig. 37.4. Bond lengths in 2-chlorotropone.

37.2a. METHODS OF PREPARATION OF TROPONES. The synthesis of tropones has presented a challenge to organic chemists, a challenge that has been met by the invention of several routes for elaborating these compounds. One that has a certain amount of flexibility is illustrated in Fig. 37.5. Starting with 3-methoxy-

Fig. 37.5. Synthetic scheme for the preparation of tropone.

benzoic acid, a Birch reduction (see Fig. 25.10) yields a dihydrobenzoic acid which is reduced to the alcohol with lithium aluminum hydride. When the alcohol is converted to its tosylate, a compound is formed with the interesting capability of undergoing an intramolecular nucleophilic substitution reaction, the methoxyl oxygen acting as the nucleophile and the tosylate as the leaving group. The resulting cation then rearranges, with loss of a proton, to form methoxycycloheptatriene. This compound is an enol ether and, consequently, is susceptible to facile hydrolysis to the corresponding ketone, cycloheptadienone. Removal of a pair of hydrogens from cycloheptadienone by bromine oxidation then yields tropone.

37.2b. REACTIONS OF TROPONES. Tropones undergo electrophilic substitution, nucleophilic substitution, nucleophilic addition, and cycloaddition, depending on the particular troponoid compound that is used and the reagents with which it is treated. Although the tendency for the oxygen of tropones to form strong complexes with Lewis acids curtails the application of Friedel–Crafts reactions (because of the increased positive charge on the ring), other electrophilic reactions, such as halogenation, nitration, sulfonation, and azo coupling proceed rather smoothly (Fig. 37.6). In tropone itself, the preferred points of electrophilic attack are the 2- and 4-positions. In 2-hydroxytropone or 2-alkoxytropone, the preferred points of attack are at positions adjacent to the substituent and two carbons removed from the substituent, positions analogous to the *o*- and *p*-positions of phenol. This preference can be easily rationalized in terms of charge-separated resonance structures, in which negative charge is concentrated

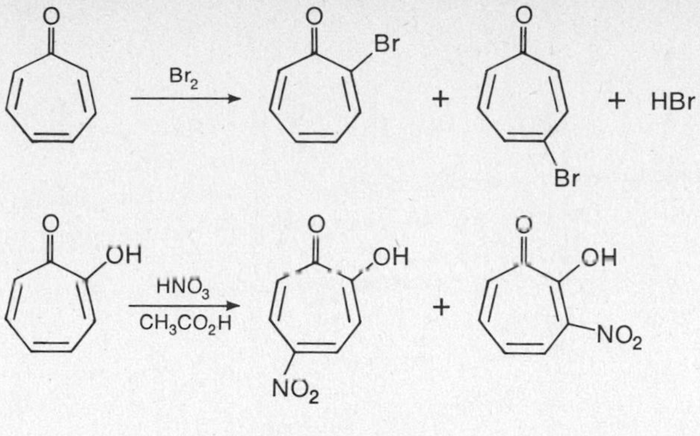

Br_2 ; + HBr

HNO_3 / CH_3CO_2H

Major product

Fig. 37.6. Electrophilic substitution of troponoid compounds.

at these positions:

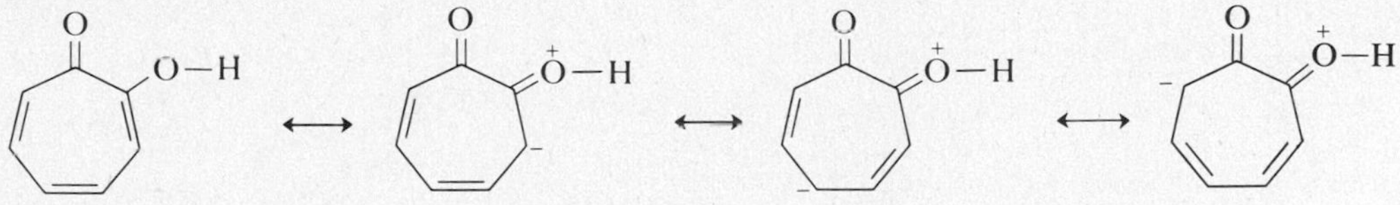

That these resonance structures do, indeed, contribute significantly to the resonance hybrid is indicated by the acidity of 2-hydroxytropone (α-tropolone), which is over 1,000 times greater than that of phenol. The isomeric 3-hydroxytropone (β-tropolone) and 4-hydroxytropone (γ-tropolone) are even more acidic, and all three of these compounds can be considered as vinylogs of a carboxylic acid in which one, two, or three vinyl groups are interposed between the carbonyl and hydroxyl groups (Fig. 37.7).

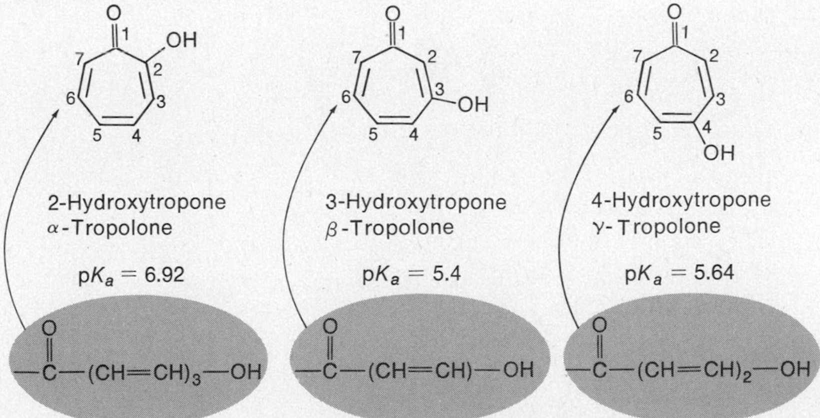

Fig. 37.7. The tropolones and the comparable vinylogous carboxylic acids.

Since the tropone ring bears a partial positive charge, it is reasonable that nucleophilic displacements should take place with some facility. For example,

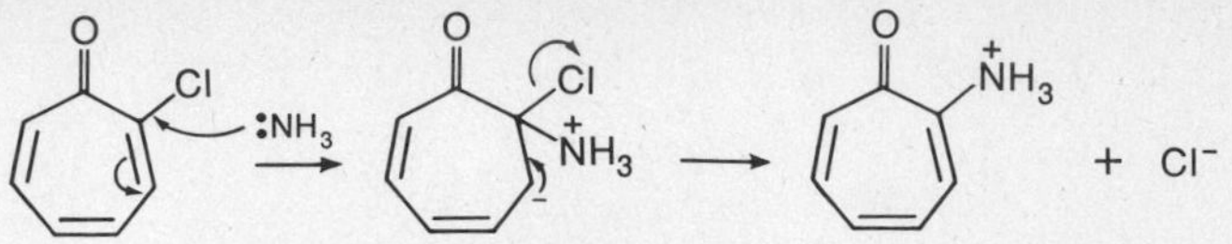

Fig. 37.8. Nucleophilic substitution of 2-chlorotropone.

2-chlorotropone is converted to 2-aminotropone by the action of ammonia (Fig. 37.8). The action of hydroxide, a stronger base, induces a ring contraction reaction in which a benzoic acid is formed (Fig. 37.9).

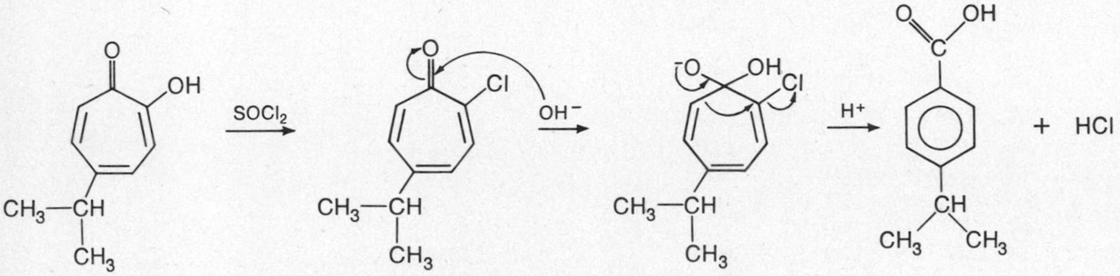

Fig. 37.9. Rearrangement of 2-halotropone on treatment with hydroxide.

Reacting as an unsaturated carbonyl compound, tropone undergoes conjugate addition reactions with Grignard reagents in which the R group of the Grignard reagent becomes attached to the position adjacent to the carbonyl group of the tropone (Fig. 37.10) in what can be considered as a 1,8-conjugate addition.

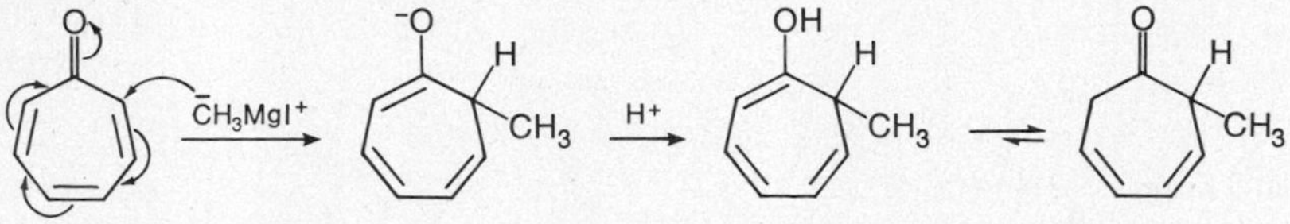

Fig. 37.10. Nucleophilic addition to tropone; 1,8-conjugate addition of methylmagnesium iodine.

Tropones are capable of acting as the diene component in Diels–Alder reactions, as illustrated by the reaction of tropone with maleic anhydride (Fig. 37.11).

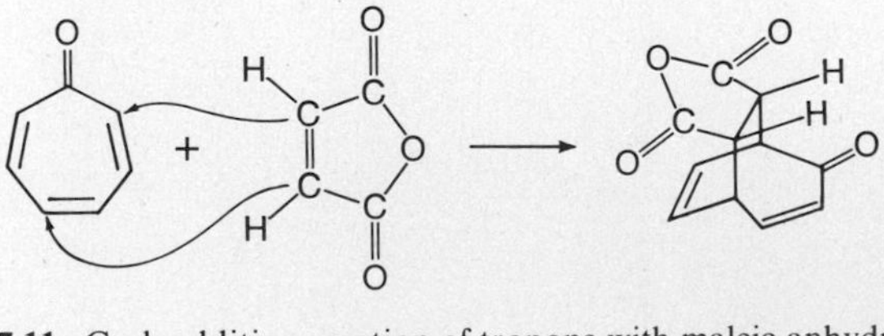

Fig. 37.11. Cycloaddition reaction of tropone with maleic anhydride.

37.3 *Azulene*

A number of naturally occurring substances produce blue hydrocarbons upon dehydrogenation, the carbon skeletons of which are isomeric with naphthalene and consist of a five-membered ring fused to a seven-membered ring. Bond length measurements show that the C—C bond distances are all close to 1.404 Å in the five-membered ring and 1.386 Å in the seven-membered ring. The molecule has a dipole moment of 1.08 D and a resonance energy of 40 kcal/mole, a value considerably lower than that for its isomer, naphthalene, but still appreciable. These data are consistent with a structural formulation for azulene in which the five-membered ring is viewed as a cyclopentadienyl anion and the seven-membered ring as a cycloheptatrienyl cation, each of these rings thereby acquiring aromatic character.

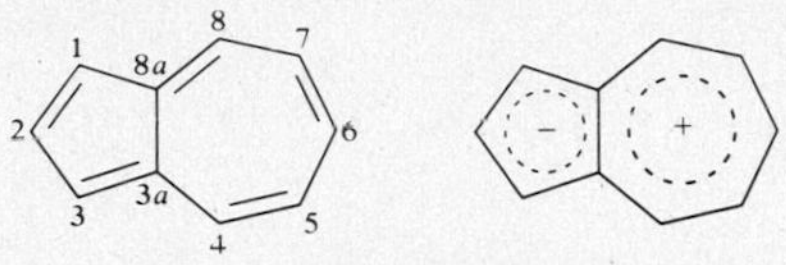

Azulene

Laboratory syntheses of azulene usually involve the elaboration of a bicyclic system containing a five-membered ring fused to a seven-membered ring followed by dehydrogenation to the azulene as the final step in the sequence. An alternative approach, useful primarily for the synthesis of azulene itself, involves the dehydrogenation of cyclodecane under carefully controlled conditions. Despite its considerably greater thermodynamic stability, naphthalene is formed only as a minor side product in this process:

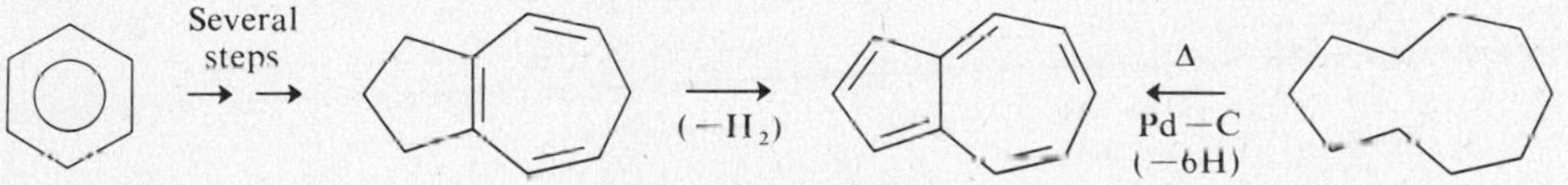

Several steps → → ; $(-H_2)$; Δ, Pd—C $(-6H)$

Azulene is very reactive toward electrophiles and undergoes substitution primarily at the 1 and 3 positions, consistent with the notion that the five-membered ring is electron-rich relative to the seven-membered ring. In fact, azulene is so reactive that it engages in the diazonium coupling reaction which is usually restricted to phenols and amines, as well as a double acylation reaction with acetyl chloride and aluminum chloride (Fig. 37.12).

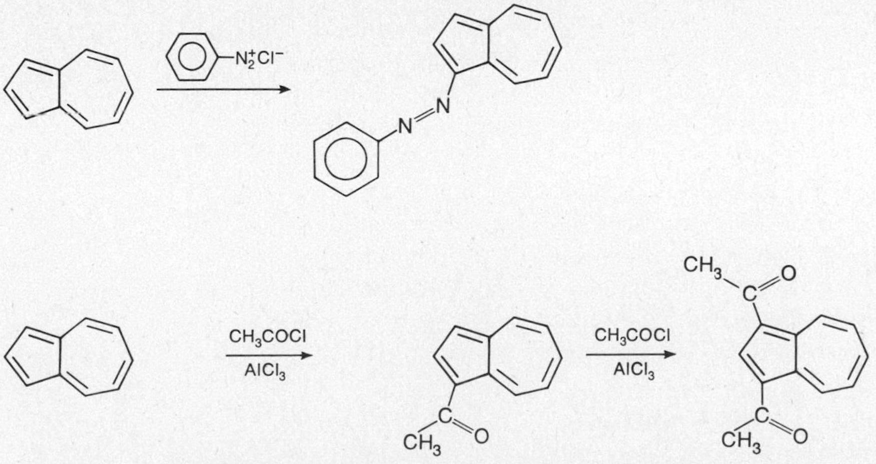

$C_6H_5N_2^+Cl^-$; N=N

CH_3COCl, $AlCl_3$; CH_3COCl, $AlCl_3$; CH_3—C=O

Fig. 37.12. Electrophilic substitution reactions of azulene.

Virtually all aromatic compounds are soluble in concentrated sulfuric acid as a result of their ability to act as Lewis bases and accept protons from the acid. Azulene is unusual in exhibiting this feature in acids much weaker than concentrated sulfuric, dissolving in 60% sulfuric acid and in aqueous hydrochloric acid. This is best explained in terms of the stability of the cycloheptatrienylium cation system, which in azulene itself is only partially developed, but which in the protonated species becomes fully developed:

37.4 Annulenes

Large-ring, conjugated polyenes are called annulenes, and the size of the ring is designated by a number placed in brackets as a prefix. The interest in these compounds derives from the possibility that some of them might be planar and, therefore, aromatic if they contain $4n + 2$ electrons. The [10]-, [14]-, and [18]-annulenes, for example, have received considerable attention from chemists for this reason. The *cis,cis,trans,cis,trans* isomer of the [10]-annulene has, so far, defied synthesis, for the interference between the inward-directed hydrogens attached to the *trans* double bonds is severe. In the [14]-, [16]-, and [18]-annulenes, however, the "hole" in the doughnut* is larger, and the transannular interference between the inward-directed hydrogens is lower. These

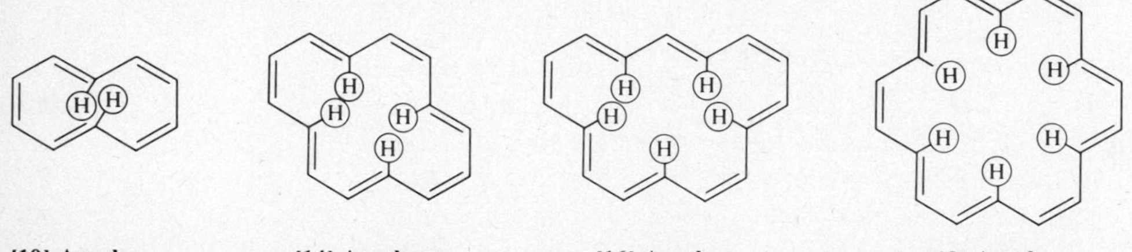

[10]-Annulene **[14]-Annulene** **[16]-Annulene** **[18]-Annulene**

larger annulenes have been synthesized (see Fig. 27.4 for the preparation of the [18]-annulene), and some of these have been shown to have certain properties that can be called aromatic. The [18]-annulene, for example, has been shown to be a planar molecule in which all the carbon–carbon bond distances are identical, in contrast to the [16]-annulene which contains $4n$ π electrons and in which different bond distances are observed for the C—C and C=C bonds. The [18]-annulene, therefore, requires a pair of resonance structures for its representation, and the benzene-like character of the molecule is indicated by the position of the nmr resonance of the hydrogens on the outside of the ring, these appearing at δ 8.9. Heat of combustion measurements indicate a resonance energy of 120 kcal/mole, a value that also is commensurate with the notion of aromaticity.

* Or bagel, depending on one's frame of reference. The annulenes have, in fact, been dubbed "bagelenes," in deference to the chemists who have played a key role in their synthesis and whose laboratory is located in Israel.

Just as the Hückel rule does not preclude aromaticity in rings that carry a charge, neither does it preclude aromaticity in rings that contain a hetero atom. The catalog of organic compounds, in fact, is replete with heterocyclic substances possessing aromatic character. The requirements for aromaticity are the same as in the carbocyclic systems—*i.e.*, a planar, cyclic array of atoms in which $4n + 2$ electrons are present in the π system. Pyridine, for example, is a typical heterocyclic aromatic compound in which the π system comprises five sp^2-hybridized carbons and one nitrogen, the nonbonded electrons of the hetero atom *not* being involved in the π system (Type I). An additional possibility arises in certain heterocyclic systems, however, that is precluded in neutral carbocyclic systems; *viz.*, the contribution of a *pair of electrons* from the hetero atom of the ring to the π system, as exemplified by furan, thiophene, and pyrrole (Type II):

Type I Compounds

Pyridine

Type II Compounds

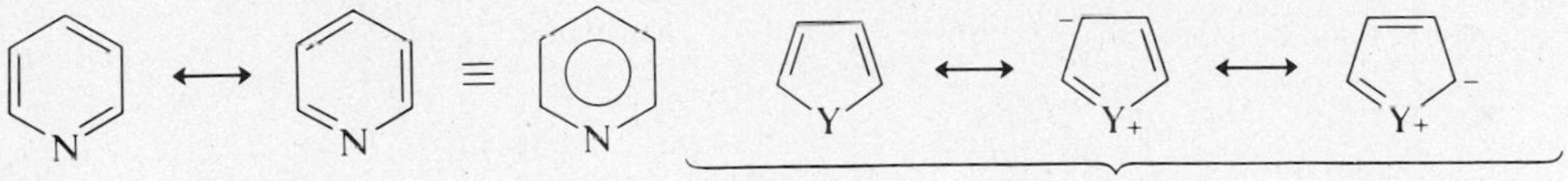

Furan (Y = O) **Thiophene (Y = S)** **Pyrrole (Y = NH)**

In Type II heterocyclics, the nonbonded electrons of the hetero atom *are* involved *in the π system*, and the charge-separated resonance structures contribute very significantly to the resonance hybrid.

Monocyclic compounds containing two, three, and even four hetero atoms and polycyclic compounds containing an even greater number of hetero atoms bring the total number of different types of heterocyclic systems to an almost staggering number. To discuss only a handful of compounds from this list may appear to relegate arbitrarily to an inferior position the many compounds that must go unmentioned. The reactions of pyridine and quinoline as examples of Type I heterocyclic compounds and of furan, thiophene, and pyrrole as examples of Type II heterocyclic compounds, however, are fairly representative of those of the majority of other heterocyclic compounds.

37.5 Pyridines

Pyridine is "isoconjugate" with benzene, the $\gt$C—H unit being replaced by $\gt$N:. This substitution causes only minor changes in structure; the carbon–carbon bond lengths, the spectral characteristics, and the resonance energy of pyridine are very similar to benzene. The nitrogen, however, is electronegative relative to carbon, which induces a dipole moment of 2.26 D in pyridine and alters its chemistry significantly.

37.5a. METHODS OF PREPARATION OF PYRIDINE AND RELATED COMPOUNDS. Pyridine occurs in relatively large quantities in coal tar, which provides one of its commercial sources. It is also produced synthetically, one method making use of acetylene, ammonia, and methoxymethanol (*i.e.*, the hemiacetal of formaldehyde and methanol), and another method involving the action of ammonia on tetrahydrofurfuryl alcohol:

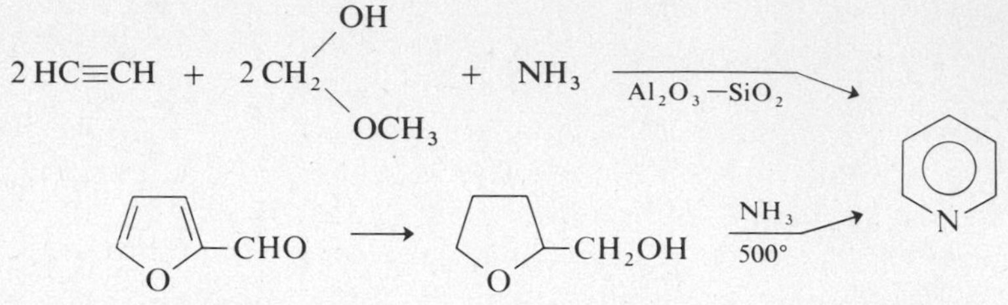

$$2\,HC{\equiv}CH + 2\,CH_2(OH)(OCH_3) + NH_3 \xrightarrow{Al_2O_3-SiO_2}$$

$$\text{Furfural (furan-CHO)} \longrightarrow \text{tetrahydrofuran-}CH_2OH \xrightarrow[500^\circ]{NH_3} \text{Pyridine}$$

Furfural **Pyridine**

A useful laboratory method for the construction of certain substituted pyridines from precursors that do not contain an already formed pyridine ring is known as the *Hantzch synthesis*. It follows the sequence outlined in Fig. 37.13, in which

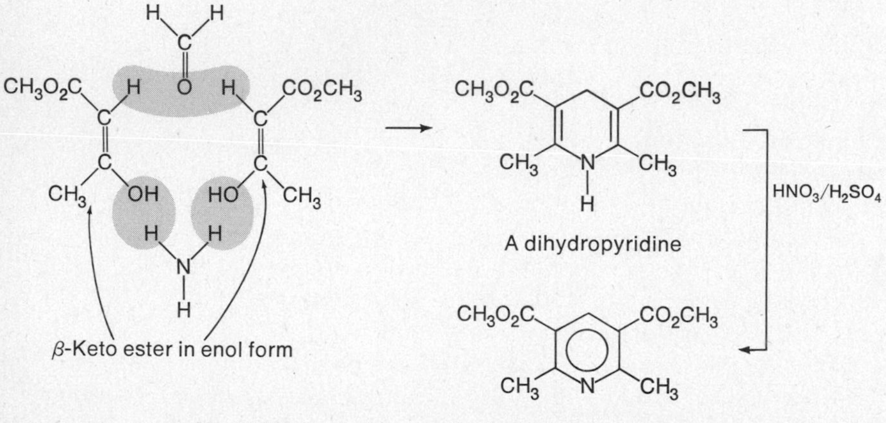

CH_3O_2C CO_2CH_3 CH_3 CH_3 OH HO H H N H

β-Keto ester in enol form

A dihydropyridine

HNO_3/H_2SO_4

Fig. 37.13. Hantzch synthesis of substituted pyridines.

a β-keto ester undergoes aldol-type condensation with formaldehyde and ammonia to yield a dihydropyridine, which can be oxidized (dehydrogenated) by treatment with nitric acid.

For quinoline, which also occurs in coal tar, a method for commercial production known as the *Skraup synthesis* was worked out many years ago. The ingredients are aniline, glycerol, sulfuric acid, and an oxidizing agent such as nitrobenzene. The reaction sequence in this "one-flask" synthesis involves the dehydration of glycerol to acrolein, followed by a conjugate nucleophilic addition of the aniline to the acrolein and an acid-catalyzed cyclization to a dihydroquinoline. Oxidation then produces quinoline (Fig. 37.14).

37.5b. REACTIONS OF PYRIDINES AT THE NITROGEN ATOM. Pyridines, although considerably weaker bases than alkylamines, are sufficiently strong to dissolve

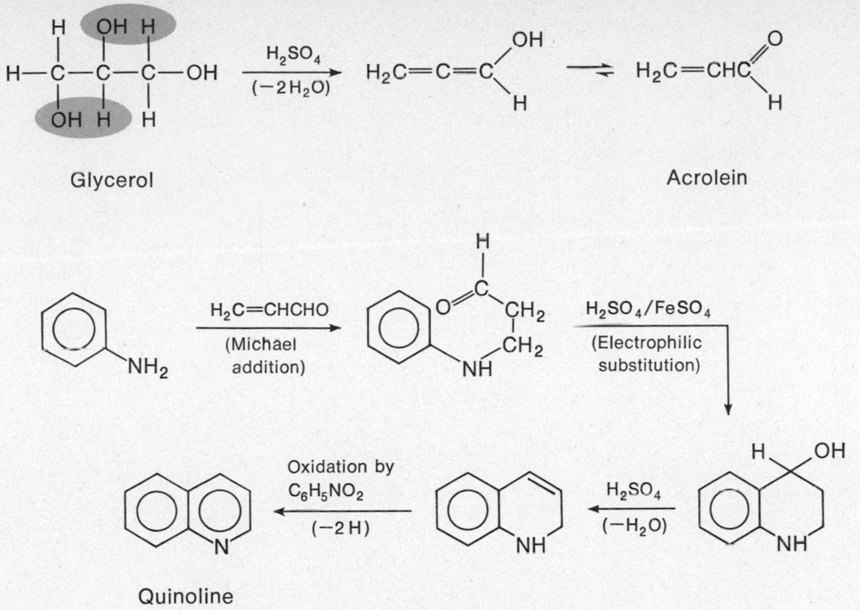

Glycerol $\xrightarrow[(-2H_2O)]{H_2SO_4}$ $H_2C{=}C{=}CH(OH)$ ⇌ $H_2C{=}CHCHO$ Acrolein

Aniline $\xrightarrow[\text{(Michael addition)}]{H_2C=CHCHO}$ → $\xrightarrow[\text{(Electrophilic substitution)}]{H_2SO_4/FeSO_4}$ → $\xrightarrow[(-H_2O)]{H_2SO_4}$ → $\xrightarrow[(-2H)]{\text{Oxidation by } C_6H_5NO_2}$ Quinoline

Fig. 37.14. The Skraup synthesis of quinoline.

in aqueous mineral acids with the formation of pyridinium salts. Pyridines are also sufficiently nucleophilic to effect S_N2 displacements on alkyl halides to form N-alkylpyridinium salts (Fig. 37.15). The pyridinium ring in these compounds has properties somewhat different from those of the parent pyridine, as examples in the next section illustrate.

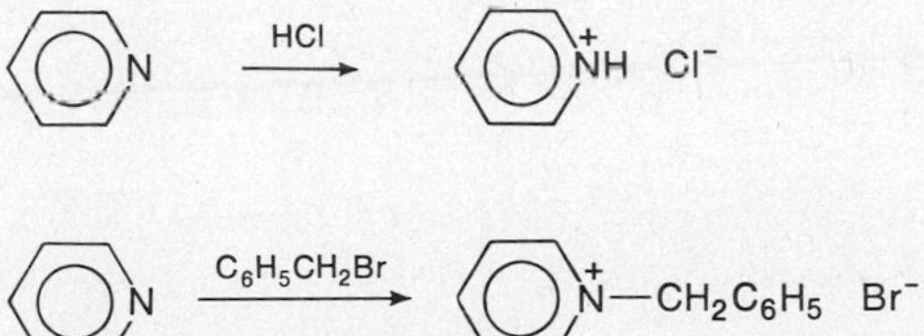

Pyridine $\xrightarrow{HCl}$ pyridinium $\overset{+}{N}H\ Cl^-$

Pyridine $\xrightarrow{C_6H_5CH_2Br}$ $\overset{+}{N}{-}CH_2C_6H_5\ Br^-$

Fig. 37.15. Reactions at the nitrogen atom of pyridine.

Pyridines carrying a formal positive charge on the ring can be generated by conversion to the N-oxides by oxidation with hydrogen peroxide or peracids (Fig. 37.16). As is indicated in the next section, N-oxides can react as nucleophiles at positions -2 and -4 of the pyridine ring, in accord with the resonance structures shown in Fig. 37.16.

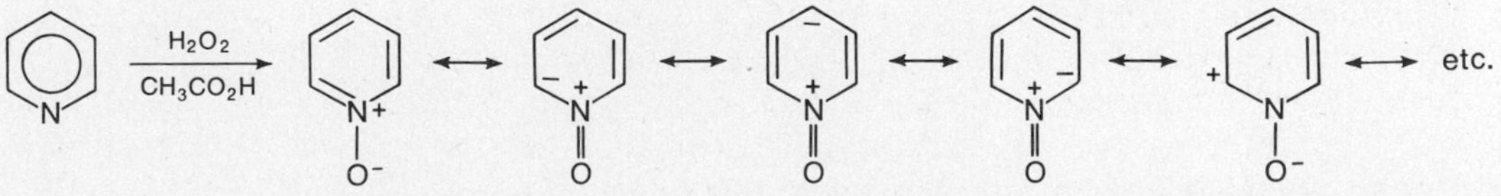

Pyridine $\xrightarrow[CH_3CO_2H]{H_2O_2}$ $N^+{-}O^-$ ⟷ ⟷ ⟷ ⟷ ⟷ etc.

Fig. 37.16. Synthesis and structure of pyridine-N-oxide.

37.5c. Reactions of Pyridines at the Ring Carbon Atoms

Electrophilic Substitution. The nitrogen atom of the pyridine ring is strongly electron withdrawing, deactivating the ring toward electrophilic substitution by enormous factors; it is estimated that pyridine is 10^{12}–10^{18} times less reactive than benzene. Nevertheless, electrophilic substitution can be effected with strong electrophiles, bromination and nitration occurring under strenuous conditions (Fig. 37.17). Less reactive reagents, such as alkyl halides or acyl halides with

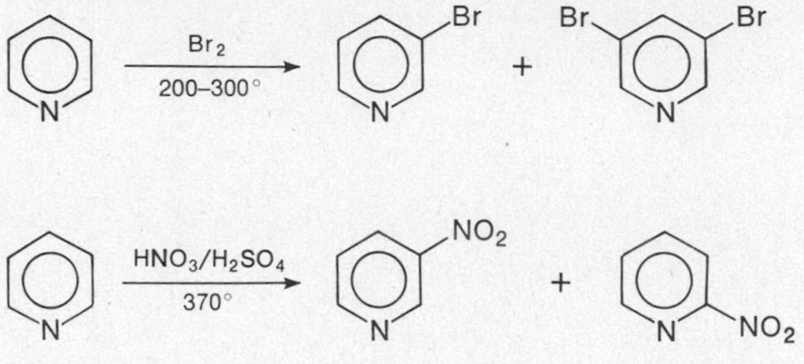

$$\text{pyridine} \xrightarrow[200\text{–}300^\circ]{Br_2} \text{3-bromopyridine} + \text{3,5-dibromopyridine}$$

$$\text{pyridine} \xrightarrow[370^\circ]{HNO_3/H_2SO_4} \text{3-nitropyridine} + \text{2-nitropyridine}$$

Fig. 37.17. Electrophilic substitution reactions of pyridine.

aluminum chloride, have no effect whatsoever on pyridines. The greater reactivity of benzene compared with pyridine is illustrated by quinoline, which undergoes nitration only in the benzene ring to give a mixture of 5-nitro- and 8-nitroquinoline;

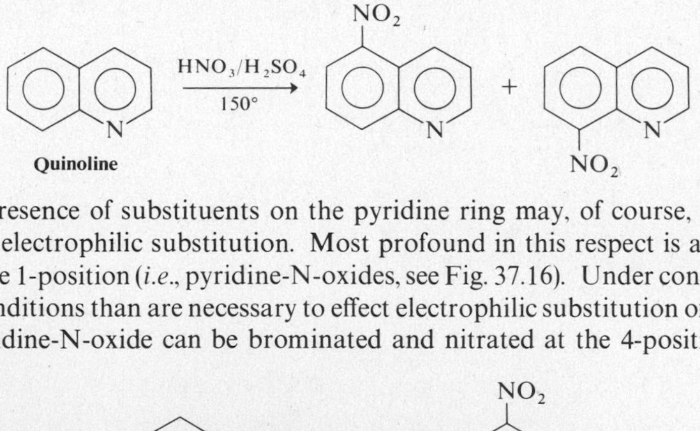

$$\underset{\textbf{Quinoline}}{\text{quinoline}} \xrightarrow[150^\circ]{HNO_3/H_2SO_4} \text{5-nitroquinoline} + \text{8-nitroquinoline}$$

The presence of substituents on the pyridine ring may, of course, affect the course of electrophilic substitution. Most profound in this respect is an oxygen atom at the 1-position (*i.e.*, pyridine-N-oxides, see Fig. 37.16). Under considerably milder conditions than are necessary to effect electrophilic substitution of pyridine itself, pyridine-N-oxide can be brominated and nitrated at the 4-position:

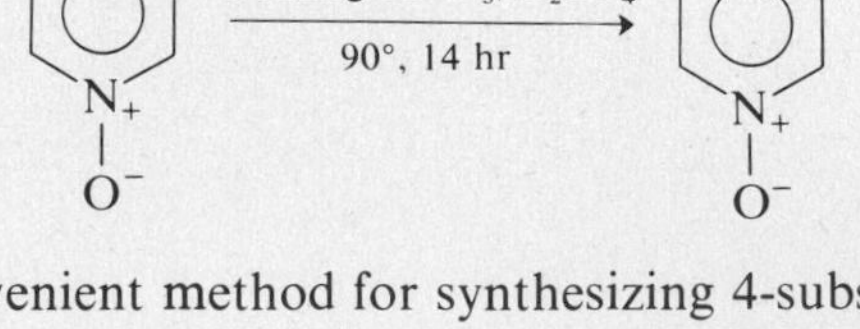

$$\text{pyridine-N-oxide } (N^+\text{–}O^-) \xrightarrow[90^\circ,\ 14\ \text{hr}]{\text{Fuming } HNO_3/H_2SO_4} \text{4-nitropyridine-N-oxide } (NO_2;\ N^+\text{–}O^-)$$

This provides a convenient method for synthesizing 4-substituted pyridines, for the oxygen can be subsequently removed in a variety of ways, such as by treatment with phosphorus trichloride, or hydrogen in the presence of a palladium catalyst.

A hydroxyl group at the 2-, 3-, or 4-position of pyridine also increases the ease with which electrophilic substitution occurs, the incoming electrophile taking up a position adjacent to the hydroxyl function (Fig. 37.18). The 2- and 4-hydroxypyridines are in tautomeric equilibrium with the keto forms, designated as α-pyridone and γ-pyridone, and in solution the equilibrium actually favors the pyridones.

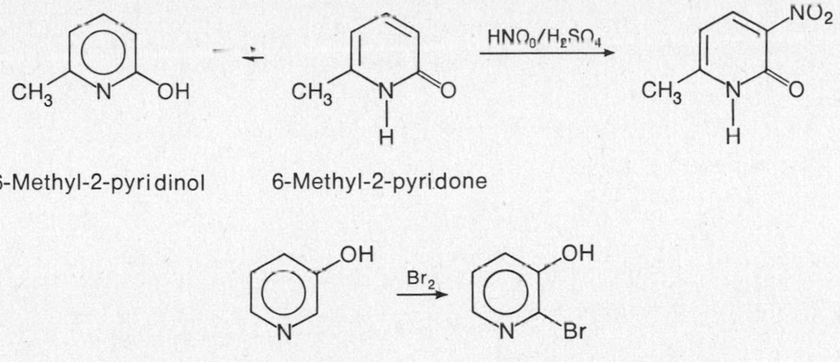

Fig. 37.18. Tautomeric forms and reactions of hydroxypyridines.

Nucleophilic Substitution. The electropositive character of the pyridine ring makes it more susceptible to nucleophilic attack than the benzene ring. For example, 2-halo- and 4-halopyridines undergo facile reactions in which the halogen is replaced by cyanide in an addition-elimination sequence. The 3-halopyridines react more sluggishly, and in the presence of strong bases they yield substitution products via an elimination-addition process involving *pyridyne* intermediates (Fig. 37.19).

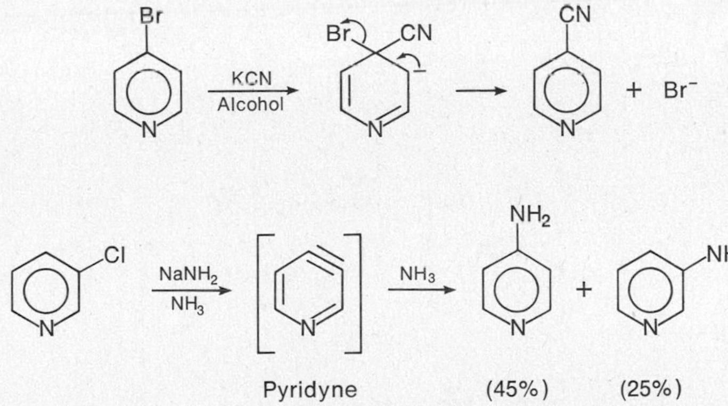

Fig. 37.19. Nucleophilic substitution reactions of halopyridines.

Nucleophilic Addition. Alkyllithium and aryllithium compounds react with pyridine via nucleophilic addition to form 1-lithio-2-substituted dihydropyridines, which undergo rearomatization on heating, thus providing a convenient synthesis of 2-alkyl- and 2-arylpyridines (Fig. 37.20).

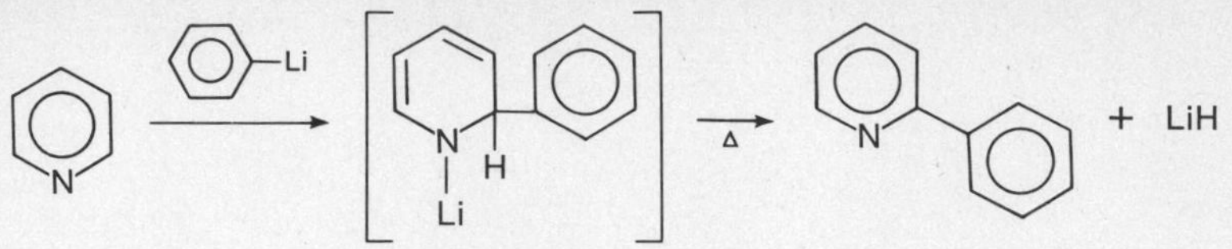

Fig. 37.20. Reaction of pyridine with phenyllithium.

Oxidation and Reduction. Pyridine is quite resistant to reduction but can be converted to piperidine, the hexahydro derivative, with sodium and alcohol or with hydrogen and a catalyst:

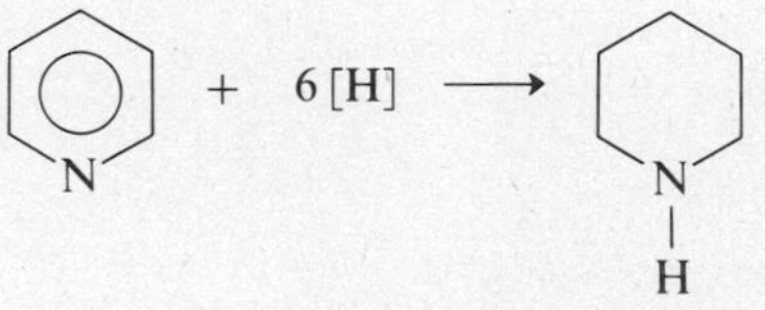

Pyridinium salts are more easily reduced than the related pyridine, as illustrated by the sodium borohydride reduction of 4-carbomethoxy-N-methylpyridinium iodide to the 1,2,5,6-tetrahydro analog (Fig. 37.21). Conversion to the 1,4-dihydro

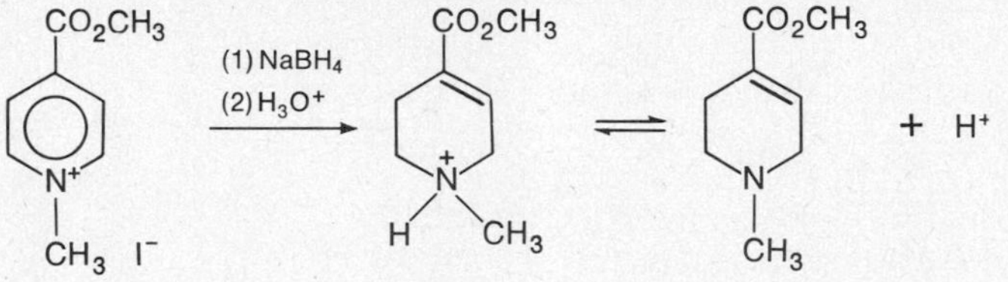

Fig. 37.21. Reduction of a methylpyridinium iodide with sodium borohydride.

derivatives is also possible, as illustrated by several important biological reactions in which nicotinamide adenine dinucleotide (NAD^+; see Fig. 32.41) acts as a hydrogen acceptor in conjunction with certain enzymes and forms the corresponding reduced dihydronicotinamide adenine dinucleotide (NADH).

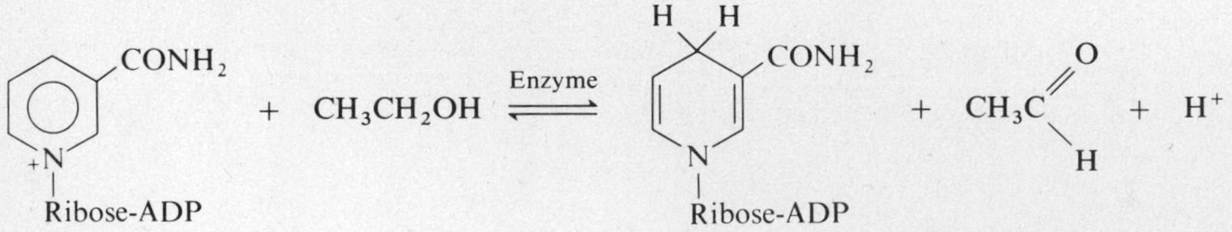

The resistance of the pyridine ring to oxidation is well illustrated by the oxidation of 4-phenylpyridine, which yields 4-carboxypyridine as the sole product, no benzoic acid being formed;

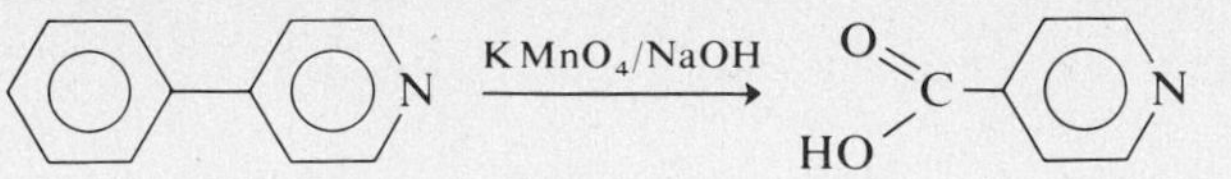

$$C_6H_5\text{-}C_5H_4N \xrightarrow{KMnO_4/NaOH} HO_2C\text{-}C_5H_4N$$

Photochemical Reactions. Valence isomerizations of the type observed in benzene (see Section 31.10) take place when pyridine is irradiated with ultraviolet light. If the reaction is carried out in aqueous solution, the "Dewar pyridine" reacts with water to form 2-amino-3-cyclobutenecarboxaldehyde. Alternatively, the Dewar pyridine can be reduced with sodium borohydride to dihydro-Dewar pyridine (Fig. 37.22).

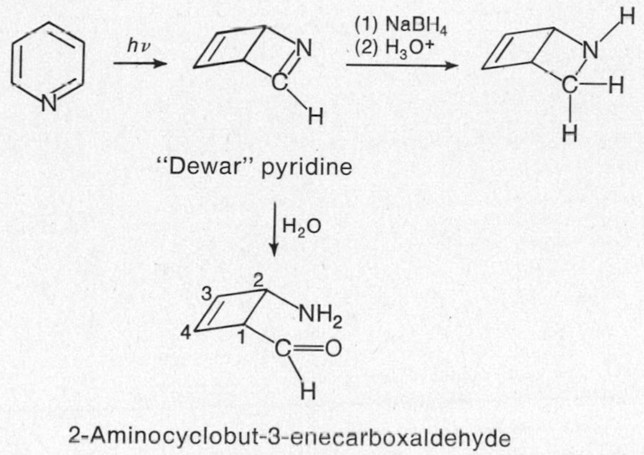

hν; (1) $NaBH_4$ (2) H_3O^+; "Dewar" pyridine; H_2O; NH_2; C=O; H; 2-Aminocyclobut-3-enecarboxaldehyde

Fig. 37.22. Photo-induced valence isomerization of pyridine.

37.5d. Reactions of pyridines at side chain carbon atoms. Alkyl or aryl groups attached to the pyridine ring are converted to carboxyl groups by strenuous oxidation, as illustrated with 4-phenylpyridine. Another example is the oxidation of nicotine to 3-pyridinecarboxylic acid [named nicotinic acid because of its progeniter (see Fig. 15.23)].

The α C—H bond of a 2- or 4-alkylpyridine is quite acidic because of the ability of a pyridine ring to delocalize a negative charge (Fig. 37.23). The anions formed by removal of the α hydrogen possess the ability to engage in S_N2 substitution reactions and in carbonyl addition reactions. In the presence of a base and an aldehyde, for example, 2-methylpyridine yields an alcohol that easily loses water to form a 2-ethenylpyridine (Fig. 37.23).

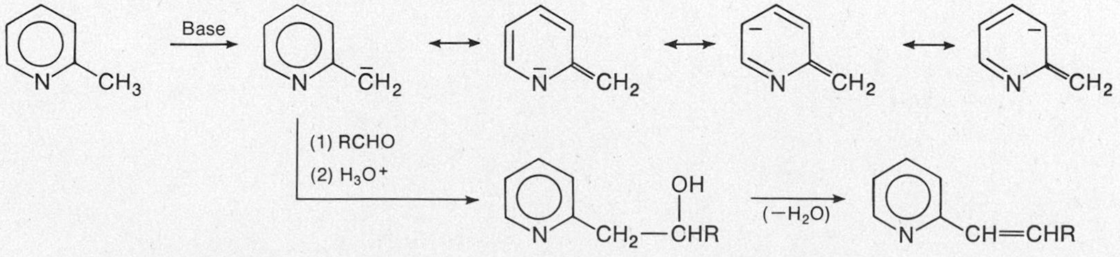

Base; CH_3; $\bar{C}H_2$; CH_2; (1) RCHO (2) H_3O^+; OH; CH_2—CHR; $(-H_2O)$; CH=CHR

Fig. 37.23. Formation of an anion from an alkylpyridine and condensation of the anion with an aldehyde.

37.6 *Furan, Thiophene, and Pyrrole*

Furan, thiophene, and pyrrole are structurally and chemically very similar and possess aromatic character. Their resonance energies are only 11–16 kcal/mole, however, which means that they are much less aromatic than benzene and, in certain respects, much more reactive.

37.6a. METHODS OF PREPARATION OF FURANS, THIOPHENES, AND PYRROLES. Furan is produced commercially by the decarbonylation of furfural, which is derived from the acid-catalyzed hydrolysis and cyclodehydration of C_5 polysaccharides (see Fig. 32.26):

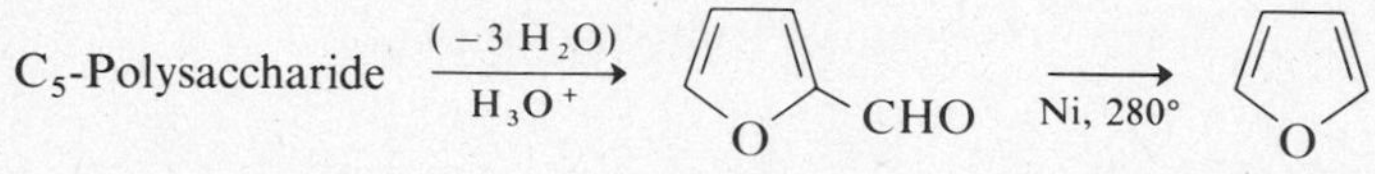

$$C_5\text{-Polysaccharide} \xrightarrow[H_3O^+]{(-3\,H_2O)} \text{Furfural (2-furaldehyde, CHO)} \xrightarrow{\text{Ni, }280°} \text{Furan}$$

Furfural **Furan**

The commercial production of thiophene involves the high-temperature reaction of sulfur with butane or of sulfur dioxide with butenes:

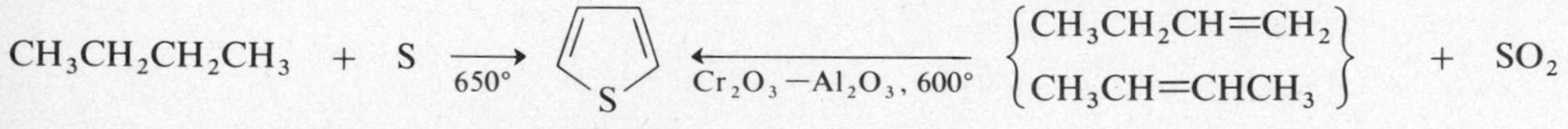

$$CH_3CH_2CH_2CH_3 + S \xrightarrow{650°} \text{Thiophene} \xleftarrow{Cr_2O_3-Al_2O_3,\ 600°} \begin{Bmatrix} CH_3CH_2CH{=}CH_2 \\ CH_3CH{=}CHCH_3 \end{Bmatrix} + SO_2$$

Thiophene

Furan affords the starting material for the commercial production of pyrrole, undergoing conversion to this material upon high-temperature treatment with ammonia in the presence of an aluminum oxide catalyst:

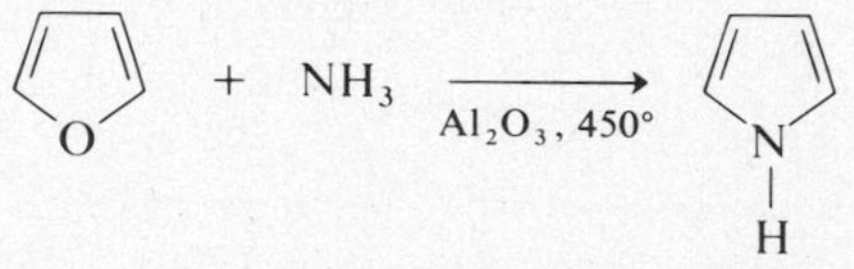

$$\text{Furan} + NH_3 \xrightarrow{Al_2O_3,\ 450°} \text{Pyrrole (N–H)}$$

Pyrrole

A very useful laboratory method for making substituted furans, thiophenes, and pyrroles is known as the *Paal–Knorr synthesis* and employs 1,4-diketones as the starting material. In the presence of sulfuric acid, the product is a furan; in the presence of phosphorus pentasulfide, the product is a thiophene; and in the presence of ammonia or a primary amine, the product is a pyrrole (Fig. 37.24).

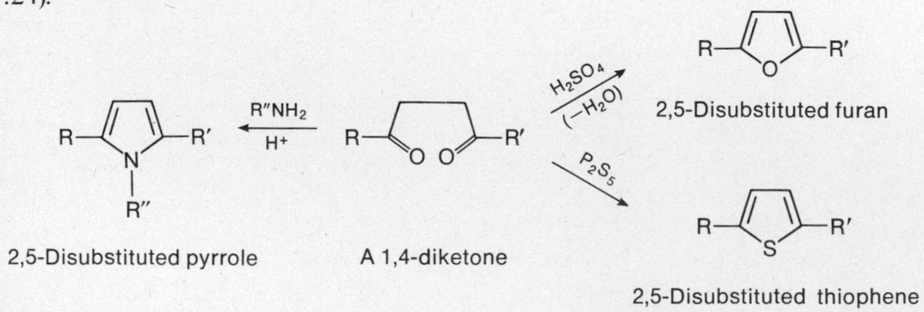

Fig. 37.24. The synthesis of substituted furans, pyrroles, and thiophenes from 1,4-diketones.

Benzofurans and benzopyrroles are compounds in which a benzene ring and a furan or pyrrole ring share a pair of adjacent carbon atoms:

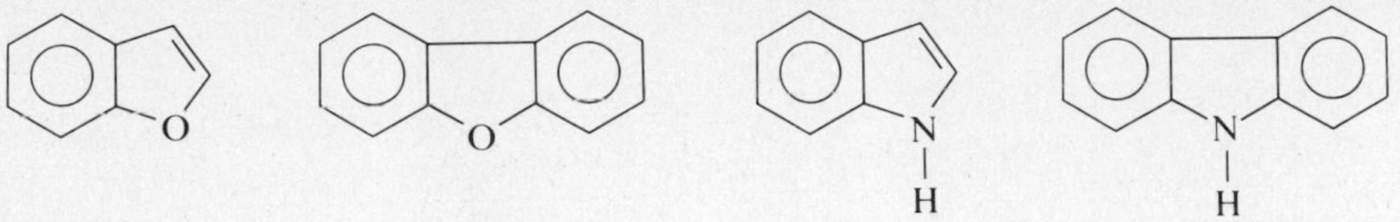

Benzofuran **Dibenzofuran** **Benzopyrrole (Indole)** **Dibenzopyrrole (Carbazole)**

All these ring systems are found in natural products, the indole ring system occurring the most widely (*e.g.*, lysergic acid—see Fig. 36.7). Various methods have been devised for the synthesis of indoles, one of the simplest being the Fischer procedure involving the action of a strong mineral acid on an appropriately substituted phenylhydrazone (Fig. 37.25).

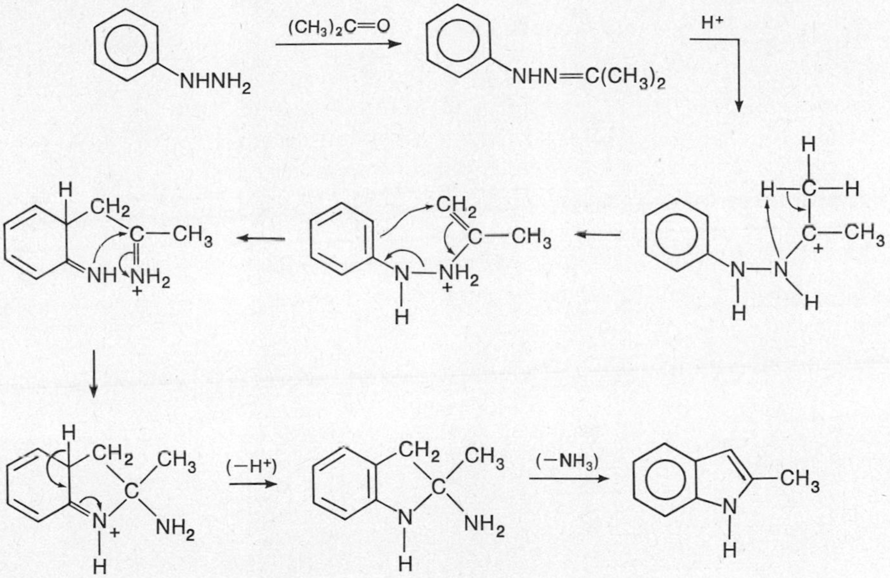

$NHNH_2$ $\xrightarrow{(CH_3)_2C{=}O}$ $NHN{=}C(CH_3)_2$ $\xrightarrow{H^+}$

Fig. 37.25. Fischer indole synthesis via phenylhydrazones.

37.6b. Reactions of Furans, Thiophenes, and Pyrroles

Electrophilic Substitution. Furan, thiophene, and pyrrole are all very susceptible to electrophilic substitution. Furan is the most reactive of the three, and its reactions may get out of hand unless they are carried out under special conditions at low temperatures and high dilutions. Pyrrole is comparable in reactivity to aromatic amines and phenols, and thiophene is the least reactive of the three, although still 300 times more so than benzene. The preferred point of attachment of the electrophile in these compounds is the 2-position, but if the 2- and 5-positions are blocked, substitution at the 3-position will occur (Fig. 37.26).

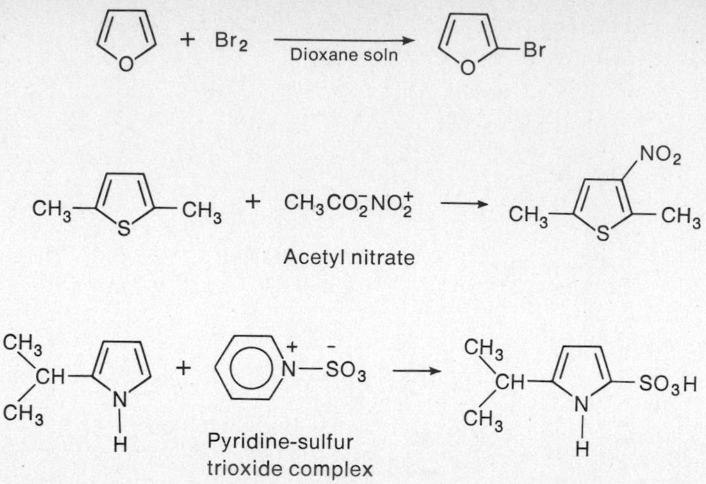

Furan + Br_2 $\xrightarrow{\text{Dioxane soln}}$ 2-bromofuran

2,5-dimethylthiophene + $CH_3CO_2^-NO_2^+$ (Acetyl nitrate) ⟶ 2,5-dimethyl-3-nitrothiophene

2-isopropylpyrrole + pyridine–SO_3 (Pyridine-sulfur trioxide complex) ⟶ 5-isopropylpyrrole-2-SO_3H

Fig. 37.26. Electrophilic substitution reactions of furans, thiophenes, and pyrroles.

Under strongly acidic conditions, these compounds tend to polymerize, and milder methods than those usually employed for nitration and sulfonation of benzene are necessary. Nitration, for example, is best effected with acetyl nitrate (nitronium acetate), and sulfonation with a sulfur trioxide-pyridine complex (Fig. 37.26).

Acyl groups can be introduced into the 2-position, or the 3-position if the 2- and 5-positions are blocked, of furans, thiophenes, and pyrroles by means of the Friedel–Crafts reaction, and the formyl group by means of the Reimer–Tieman (see Fig. 22.30), Gattermann (see p. 835), and Vilsmeier reactions (see Fig. 28.14) (Fig. 37.27).

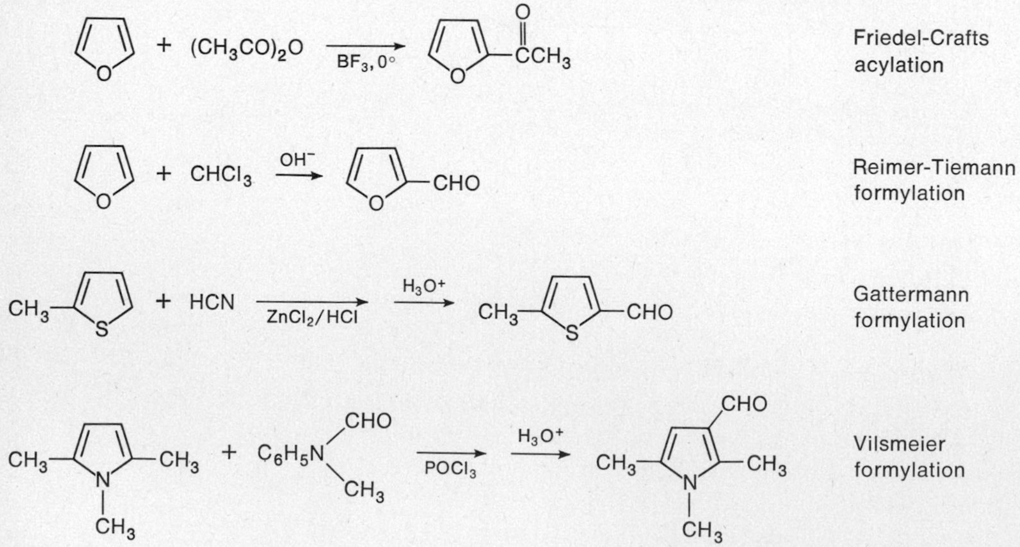

Furan + $(CH_3CO)_2O$ $\xrightarrow{BF_3,\ 0°}$ 2-furyl–$\overset{O}{\|}{C}CH_3$ — Friedel-Crafts acylation

Furan + $CHCl_3$ $\xrightarrow{OH^-}$ 2-furyl–CHO — Reimer-Tiemann formylation

2-methylthiophene + HCN $\xrightarrow{ZnCl_2/HCl}$ $\xrightarrow{H_3O^+}$ 5-methylthiophene-2-CHO — Gattermann formylation

1,2,5-trimethylpyrrole + $C_6H_5N(CHO)CH_3$ $\xrightarrow{POCl_3}$ $\xrightarrow{H_3O^+}$ 1,2,5-trimethylpyrrole-3-CHO — Vilsmeier formylation

Fig. 37.27. Acylation and formylation reactions of furans, thiophenes, and pyrroles.

Reduction Reactions. Furan, thiophene, and pyrrole are easily reduced to the corresponding tetrahydro derivatives by catalytic hydrogenation. Partial reduction of thiophene and pyrrole can be effected by chemical means, such as with sodium or sodium amalgam in alcohol or with zinc and acetic acid:

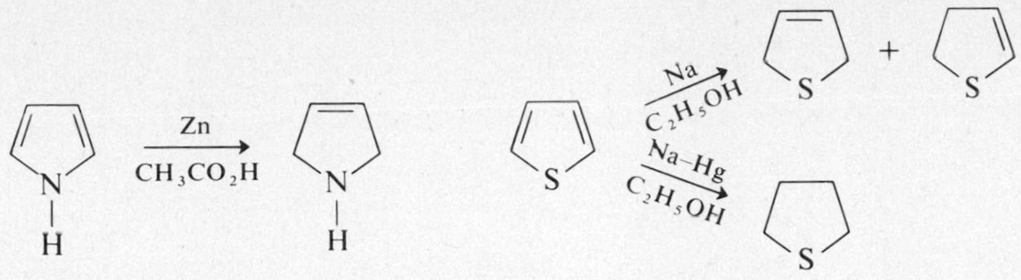
Zn
CH_3CO_2H
N
H
N
H
S
Na
C_2H_5OH
Na–Hg
C_2H_5OH
S
+
S
S

Cycloaddition Reactions. Furan is a particularly reactive diene component in the Diels–Alder reaction (Fig. 37.28; also see Fig. 26.27). Pyrrole is considerably less so, and thiophene shows little or no tendency to engage in Diels–Alder reactions.

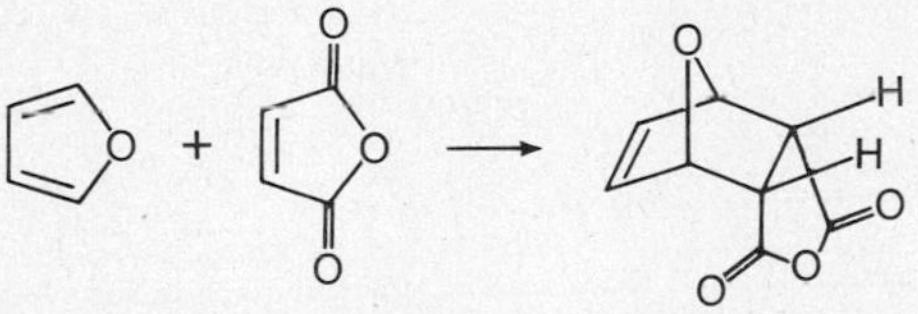

Fig. 37.28. Diels-Alder reaction of furan and maleic anhydride.

Photochemical Reactions. Virtually all heterocyclic compounds containing five-membered rings, regardless of the number and identity of hetero atoms that are present, undergo alteration upon irradiation. For example, 2-methylfuran yields 3-methylfuran and methylcyclopropene along with carbon monoxide under the action of ultraviolet light, presumably via a cyclopropenecarboxaldehyde intermediate (Fig. 37.29).

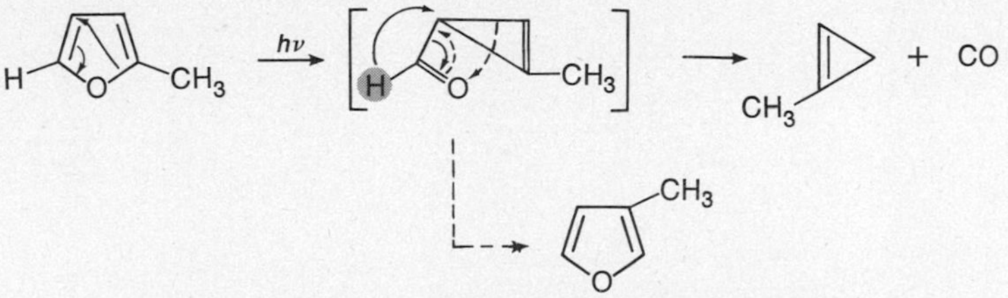
H
CH3
hν
H
CH3
+ CO
CH3
CH3

Fig. 37.29. Photoinduced rearrangement of 2-methylfuran.

REFERENCES

Nonbenzenoid Aromatic Compounds:

GINSBURG, D., ed., *Non-Benzenoid Aromatic Compounds*, Interscience Publishers, New York, 1959.

LLOYD, D., *Alicyclic Compounds*, Elsevier Publishing Co. Inc., New York, 1963, Chapters 9–13.

SNYDER, J. P., ed., *Non-Benzenoid Aromatics*, Vols. 1 and 2, Academic Press, Inc., New York, 1969, 1971.

Heterocyclic Compounds:

ACHESON, R. M., *An Introduction to the Chemistry of Heterocyclic Compounds*, 2nd ed., John Wiley & Sons, Inc., New York, 1967.

ALBERT, A., *Heterocyclic Chemistry*, 2nd ed., Athlone Press, London, 1968.

FITTON, A. O., and R. K. SMALLEY, *Practical Heterocyclic Chemistry*, Academic Press, Inc., New York, 1968.

PAQUETTE, L. A., *Principles of Modern Heterocyclic Chemistry*, W. A. Benjamin, Inc., New York, 1968.

TAYLOR, E. C., *Heterocyclic Chemistry*, in *Foundations of Modern Organic Chemistry Series*, Prentice-Hall, Inc., Englewood Cliffs, N.J., 1968.

PROBLEMS

1. Draw the structure of the major product from each of the following reactions:

(a) [quinoline] + H_2O_2 $\xrightarrow[CH_3CO_2H]{}$

(b) [furan] + [benzene]–COCl $\xrightarrow[AlCl_3]{}$

(c) [tropone]=O + [benzene]–MgBr $\longrightarrow$ $\xrightarrow{H_3O^+}$

(d) [ferrocene] Fe + [succinic anhydride] O $\xrightarrow[AlCl_3]{}$

(e) [tropolone] OH, =O + Br_2 $\longrightarrow$

(f) 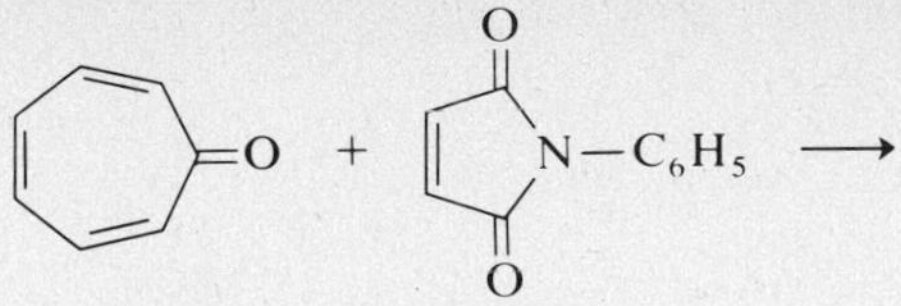 [cycloheptatrienone] + [N-phenylmaleimide] $N-C_6H_5$ ⟶

(g) 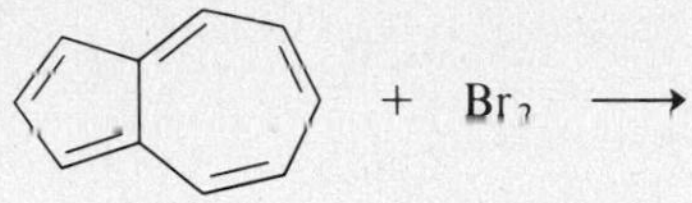 [azulene] + Br_2 ⟶

(h) 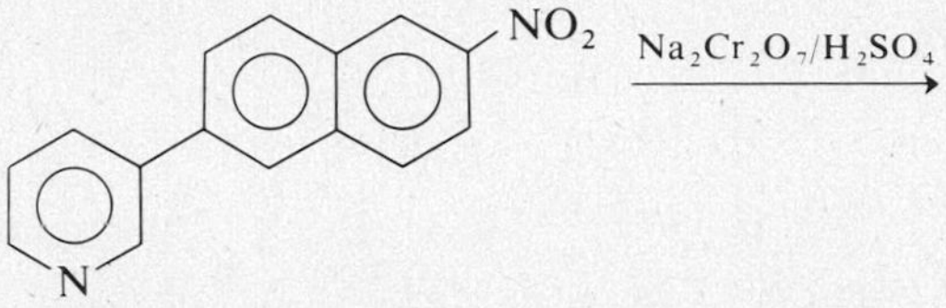 [3-(6-nitro-2-naphthyl)pyridine] $\xrightarrow{Na_2Cr_2O_7/H_2SO_4}$

(i) 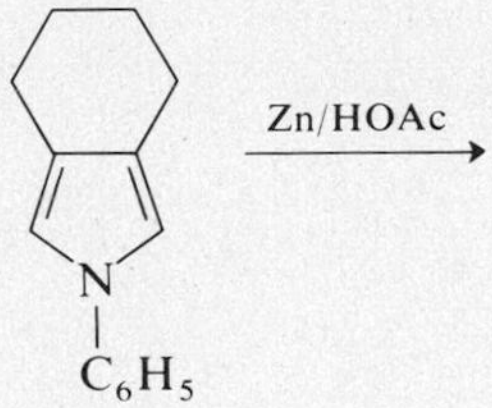 [N-phenyl-4,5,6,7-tetrahydroisoindole] $\xrightarrow{Zn/HOAc}$

2. Outline, in as much detail as possible, the probable mechanisms by which the following reactions occur:

(a) 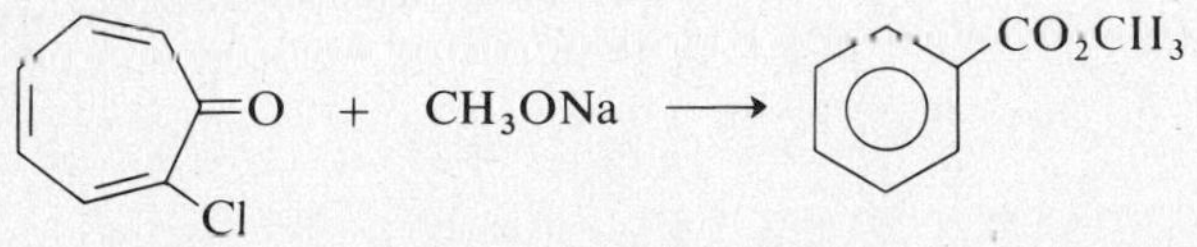 [2-chlorotropone] + CH_3ONa ⟶ [methyl benzoate] CO_2CH_3

(b) 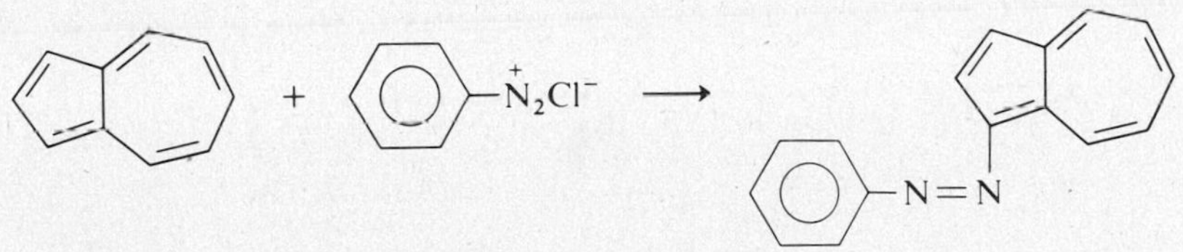 [azulene] + $C_6H_5\overset{+}{N}_2Cl^-$ ⟶ [1-(phenylazo)azulene] $-N{=}N-$

(c) 2 [pyridine] + 2 $(CH_3)_2CHBr$ ⟶ [pyridinium] $\overset{+}{N}-CH(CH_3)_2$ Br^- + $CH_3CH{=}CH_2$ + [pyridinium] $\overset{+}{N}H$ Br^-

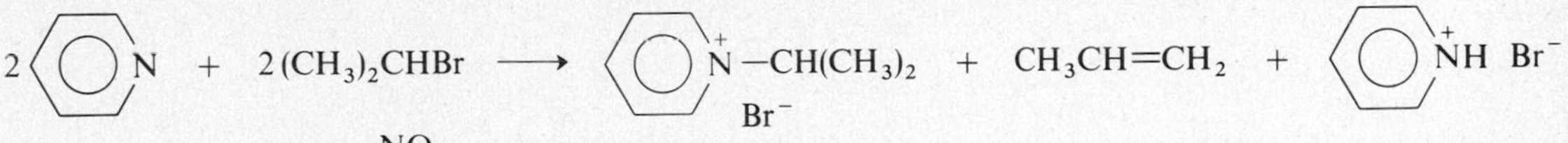

(d) 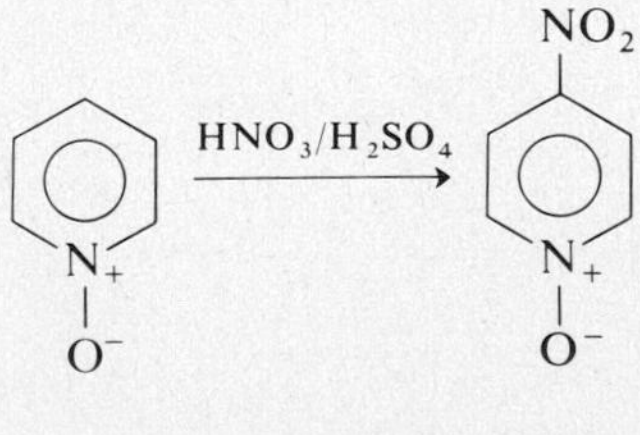 [pyridine N-oxide] $\overset{+}{N}-O^-$ $\xrightarrow{HNO_3/H_2SO_4}$ [4-nitropyridine N-oxide] NO_2, $\overset{+}{N}-O^-$

(e) 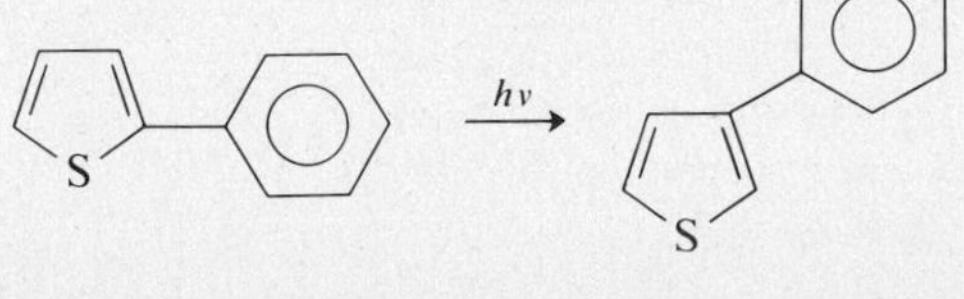 [2-phenylthiophene] $\xrightarrow{h\nu}$ [3-phenylthiophene]

3. In each of the following cases, a compound is being investigated that is known to be one of the four isomeric structures shown. Using simple chemical methods, indicate how a definitive structural assignment can be made in each instance.

(a) (i) 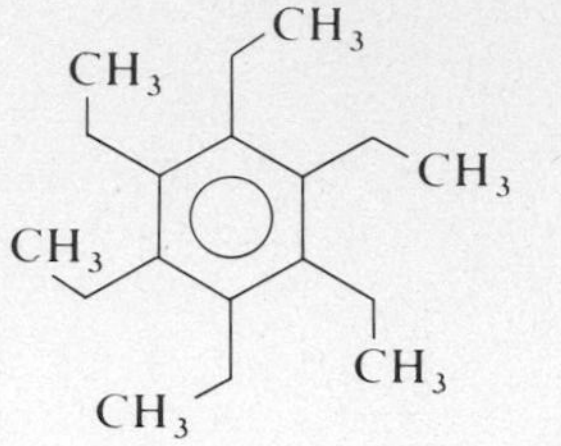 (ii) 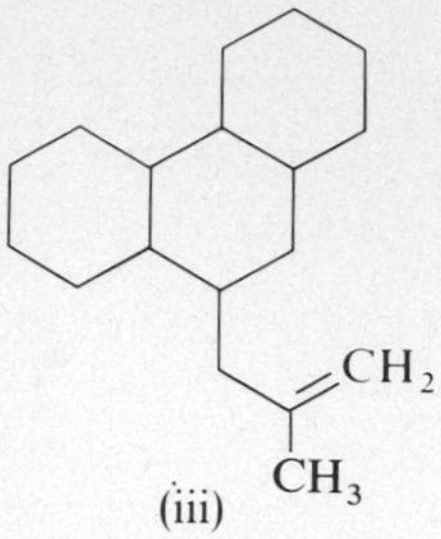 (iii) 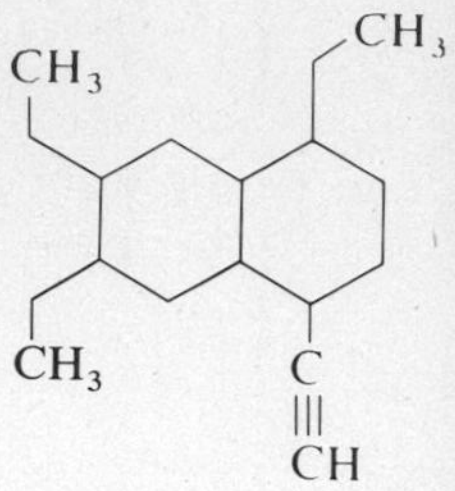 (iv)

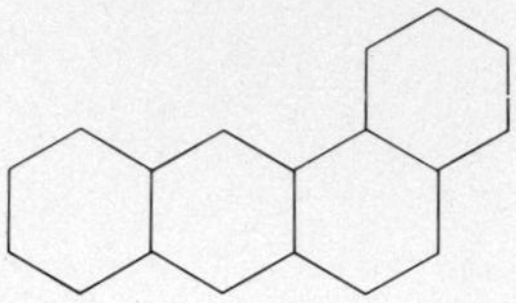

(b) (i) 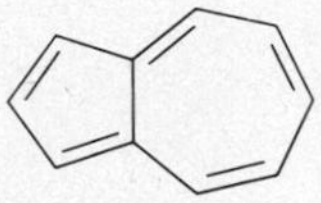 (ii) 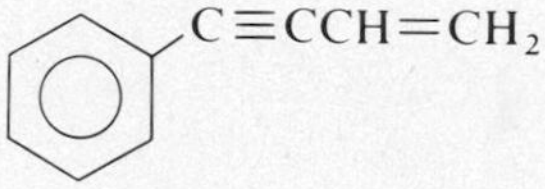 (iii) 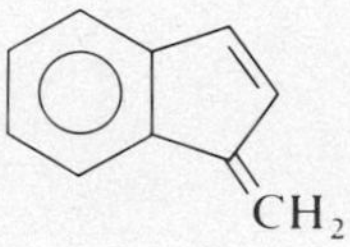 (iv)

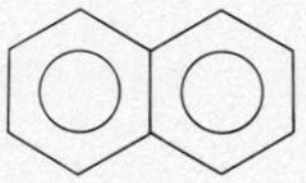

(c) (i) 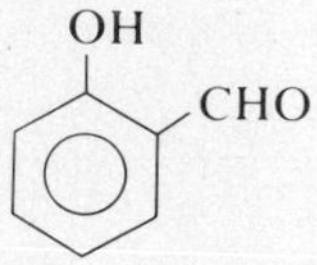 (ii) 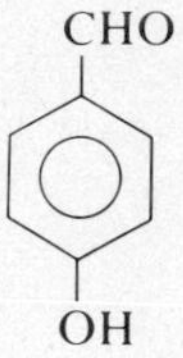 (iii) 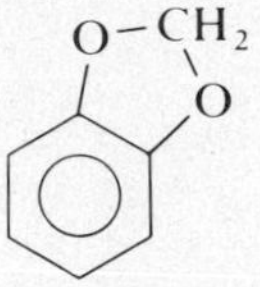 (iv)

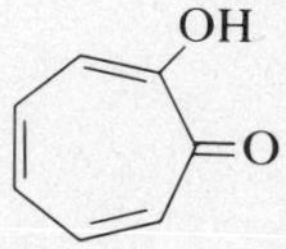

(d) (i) 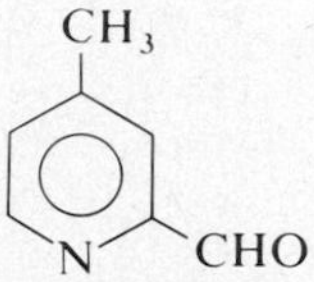 (ii) 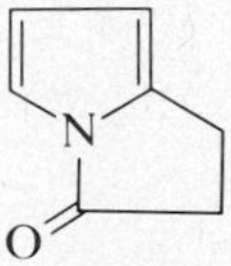 (iii) 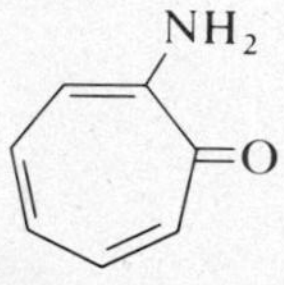 (iv)

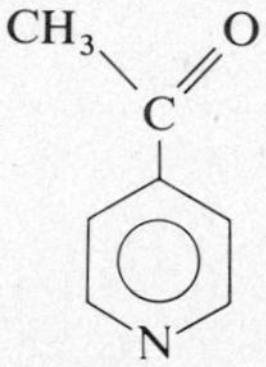

4. On the basis of the data given, deduce the structures that are involved in each of the following examples:

(a) An optically active $C_{14}H_{11}NO_3$ compound is insoluble in aqueous acid and base and undergoes hydrolysis to yield oxalic acid and a $C_{12}H_{13}NO$ compound. When the $C_{12}H_{13}NO$ compound is treated with hydrogen in the presence of a palladium on charcoal catalyst it suffers hydrogenolysis and yields a $C_{12}H_{13}N$ compound that is identical with the material obtained when the phenylhydrazone of cyclohexanone is treated with strong mineral acid.

(b) The alkaloid gentianine is isolated from certain species of *Gentianacae.* It is optically inactive, has the molecular formula $C_{10}H_9NO_2$, is soluble in aqueous acid but insoluble (except on long contact) in aqueous base. Ozonolysis of gentianine yields formaldehyde and a C_9 aldehyde. Vigorous oxidation of gentianine yields a pyridinetricarboxylic acid which is shown by nmr to have equivalent hydrogens at the 2- and 6-positions of the pyridine ring. Hydrolysis of gentianine yields a $C_{10}H_{11}NO_3$ compound that is soluble in aqueous sodium bicarbonate. Oxidation of this compound yields a $C_9H_7NO_6$ tricarboxylic acid which is shown by nmr to have equivalent hydrogens at the 2- and 6-positions of the pyridine ring.

(c) A $C_{18}H_{16}O$ compound is insoluble in aqueous acid or base but is slowly converted by hot, aqueous acid to a $C_{18}H_{18}O_2$ compound that shows a strong absorption at 250 nm in the ultraviolet (characteristic of a ArC=O moiety) and that forms a *bis*-2,4-dinitrophenylhydrazone. When the $C_{18}H_{18}O_2$ compound is treated with concentrated sulfuric acid the original $C_{18}H_{16}O$ compound is formed.

5. Given the following starting materials, outline the procedure by which conversion to the indicated product can be effected. More than one step will be necessary in most instances.

(a) 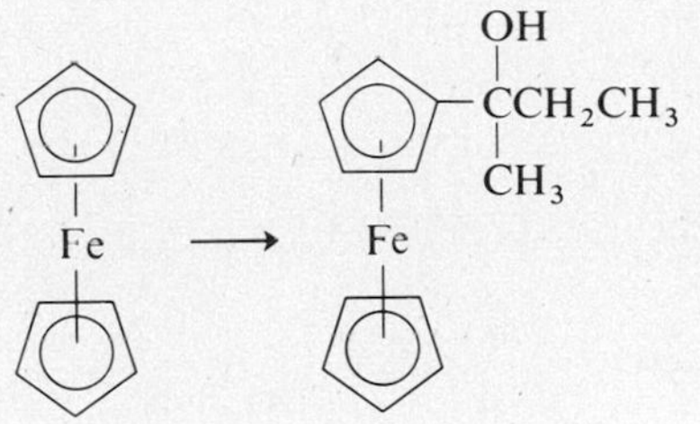 Fe ⟶ Fe (OH, CCH_2CH_3, CH_3)

(b) 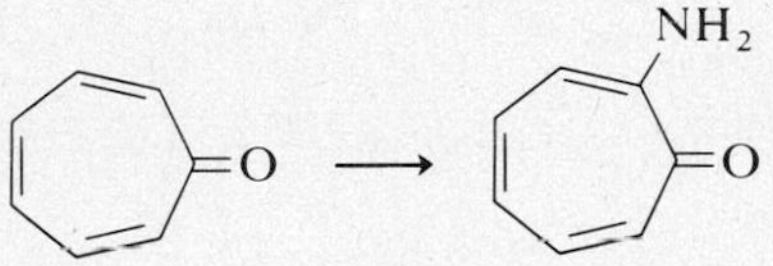 =O ⟶ NH_2 =O

(c) 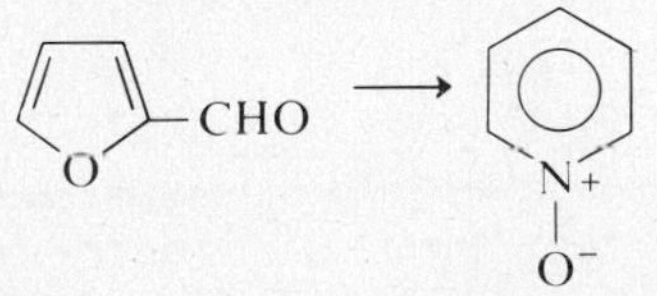 CHO O ⟶ N+ O^-

(d) 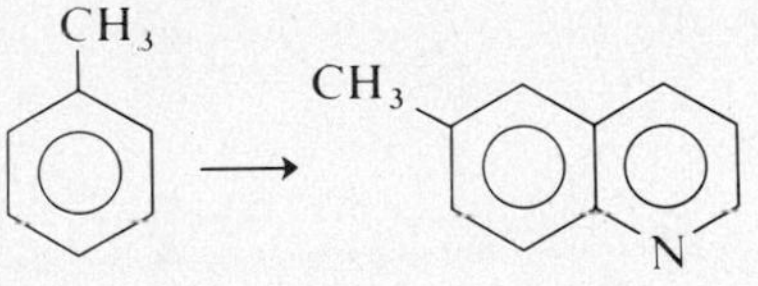 CH_3 ⟶ CH_3 N

(e) 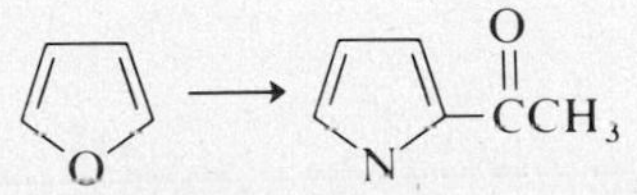 O ⟶ O ‖ CCH_3 N

6. Available starting materials include (a) any alkanes, alkenes, alkynes, alkyl halides, alkanols, alkanals, alkanones, alkanoic acids, or alkanoic acid derivatives containing four or fewer carbon atoms, (b) benzene, toluene, *o*-, *m*-, and *p*-xylene, biphenyl, naphthalene, anthracene, phenanthrene, phenol, aniline, benzoic acid, pyridine, quinoline, furan, furfural, thiophene, and pyrrole, (c) special organic reagents, and (d) any inorganic reagents. Starting with materials chosen from this group, outline syntheses for each of the following compounds:

(a) 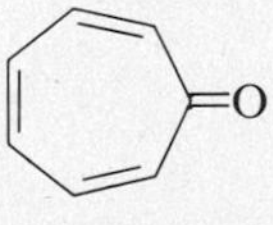 =O

(b) 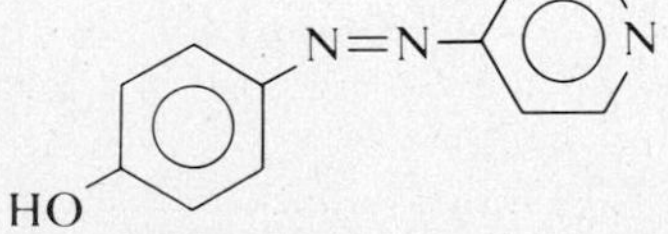 HO N=N N

(c) 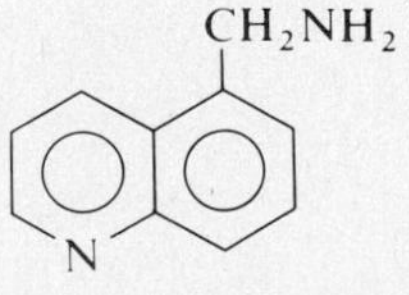 CH_2NH_2 N

(d) 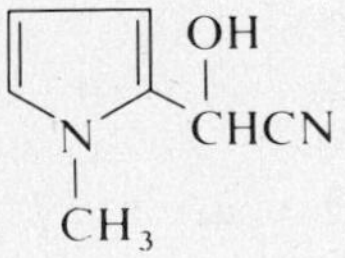 N CH_3 OH CHCN

Chapter **38**

Application of orbital symmetry: Considerations to organic reactions

It always happens that when a man siezes upon a neglected and important idea, people inflamed with the same notion crop up all around.

MARK TWAIN
Life on the Mississippi

One of the most recent and significant advances in organic chemistry is the recognition that many reactions can be interpreted in terms of the symmetry properties of the orbitals that undergo alteration in the transformation of reactant to product. The application of orbital symmetry considerations to organic reactions is a particularly powerful tool for understanding certain cycloaddition reactions (see Section 26.5) and thermal rearrangement reactions. It is with these that the present chapter deals.

CYCLOADDITION REACTIONS

38.1 General Features of Cycloaddition Reactions

Cycloaddition reactions, as explained in Section 26.5, are processes in which two or more multiply unsaturated groups combine to generate a cyclic structure containing fewer multiple bonds than the reactants. The Diels–Alder reaction, for example, is representative of such processes. Particularly interesting to chemists investigating cycloaddition reactions is whether the new bonds are formed simultaneously, *i.e.*, a concerted reaction, or nonsimultaneously, *i.e.*, a nonconcerted reaction. One of the hallmarks of a concerted reaction is its stereoselectivity, one stereoisomer usually being formed to the virtual exclusion of the other stereoisomers. A nonconcerted reaction, on the other hand, often yields mixtures of stereoisomers. For example, if the cycloaddition reaction between dimethyl maleate and 1,3-butadiene proceeds by a concerted pathway, it might be expected to give only the *cis* product, whereas if it proceeds by a nonconcerted pathway a mixture of *cis* and *trans* products are expected (Fig. 38.1). Since only the *cis* product is obtained in the actual experiment, a concerted

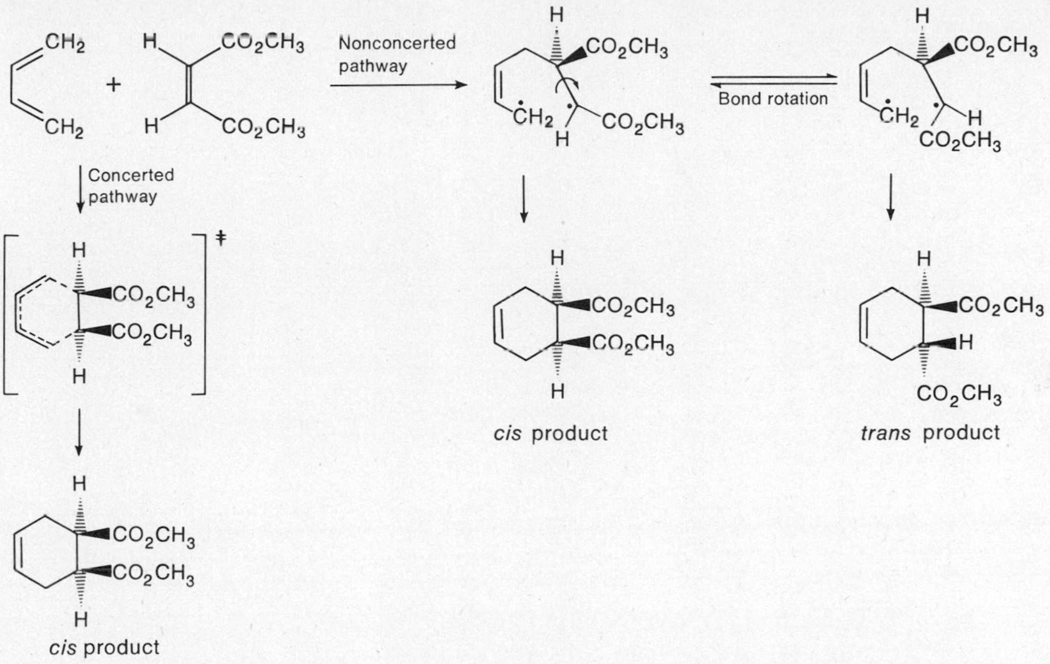

Fig. 38.1. Concerted and nonconcerted reaction pathways for the reaction of butadiene with dimethyl maleate.

pathway is inferred. This conclusion must be accepted with some reservation, however, for if the formation of the second bond in the two-step, nonconcerted process is faster than the rotation around the C—C bond of the intermediate, only a single isomer will be formed. The problem of devising a means for distinguishing between concerted and nonconcerted processes provides a stiff challenge to the experimental chemist.

38.2 Bonding and Symmetry Properties

The first steps in applying orbital symmetry rules are to construct orbital descriptions of the reactants and products and to determine the symmetry properties of the orbitals that are common both to reactants and products. To illustrate the application of these rules, let us consider the cycloaddition reaction of two molecules of ethylene to form cyclobutane. The orientation of the two molecules of ethylene that is most appropriate for this reaction is one in which the π bonds of the two molecules approach each other in a common plane, *e.g.*, the plane of the page (the X plane in Fig. 38.2). Overlap develops between the p orbitals of the two π bonds of the two ethylenes; as the reaction progresses from

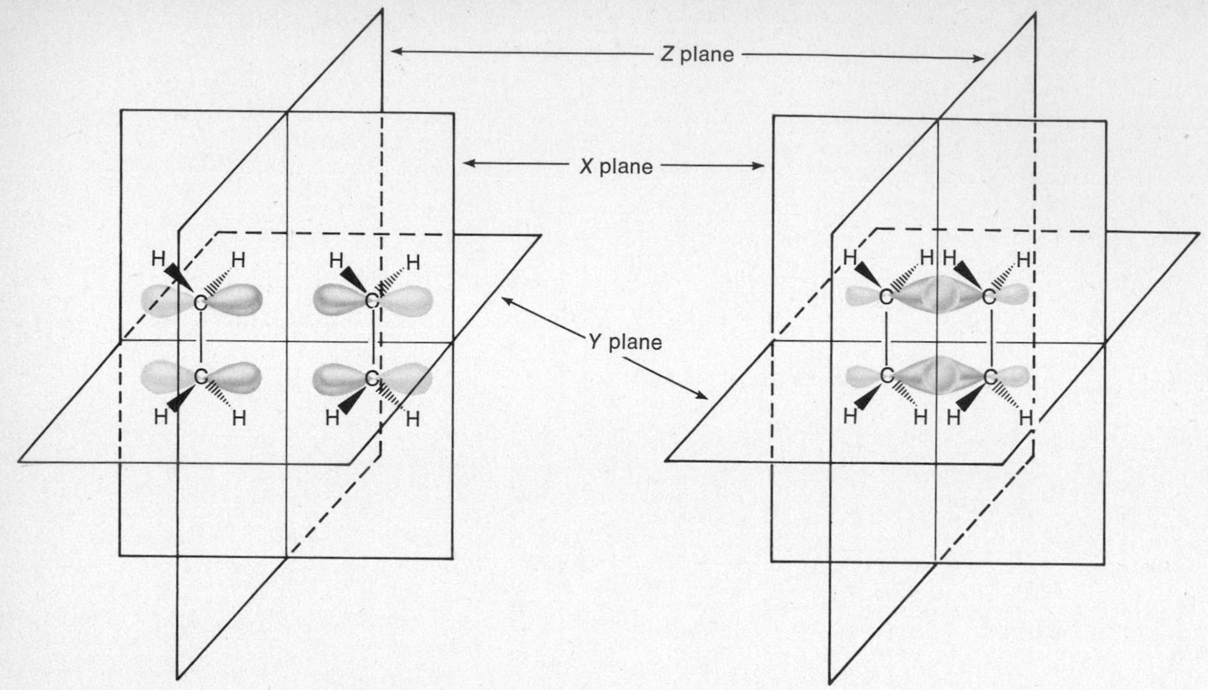

Fig. 38.2. Symmetry elements for the system involving two molecules of ethylene and one molecule of cyclobutane.

reactant to product, the sp^2 hybridization of the carbons of ethylene changes to the sp^3 hybridization of the carbons of cyclobutane (Fig. 38.2). Inspection of the reactants and product, as depicted in Fig. 38.2, shows that there are three symmetry planes with respect to the π orbitals of the ethylene molecules and the σ orbitals of the cyclobutane molecule. It is differences in the symmetries of the orbitals with respect to these planes which interest us, some of the orbital arrangements being symmetric and some being antisymmetric.

Since each ethylene molecule possesses a π and a π^* orbital (see p. 37), four orbital combinations are possible; *i.e.*, $(\pi_1 + \pi_2)$, $(\pi_1 - \pi_2)$, $(\pi_1^* + \pi_2^*)$, and $(\pi_1^* - \pi_2^*)$, given in the order of ascending energy in Fig. 38.3 (the subscripts 1 and 2 are used to designate the orbitals of the individual ethylene molecules or σ bonds of cyclobutane). Now let us consider the orbitals in the product, cyclobutane. The cycloaddition reaction between two molecules of ethylene results in the formation of two new bonds in the product. As with the π bonds of the two molecules of ethylene, four combinations of the σ and σ^* orbitals of the two new *sigma* bonds are possible; *i.e.*, $(\sigma_1 + \sigma_2)$, $(\sigma_1 - \sigma_2)$, $(\sigma_1^* + \sigma_2^*)$, and $(\sigma_1^* - \sigma_2^*)$, given in ascending order of energy in Fig. 38.3.

Having constructed a set of appropriate orbitals in the reactants and products (Fig. 38.3), we are now in a position to assign the symmetry of these orbital combinations. Inspection of the orbitals of the reactants and product (Fig. 38.2) shows that there are three symmetry planes common both to reactants and product, designated as X, Y, and Z. Since all the orbitals lie in the X plane, no

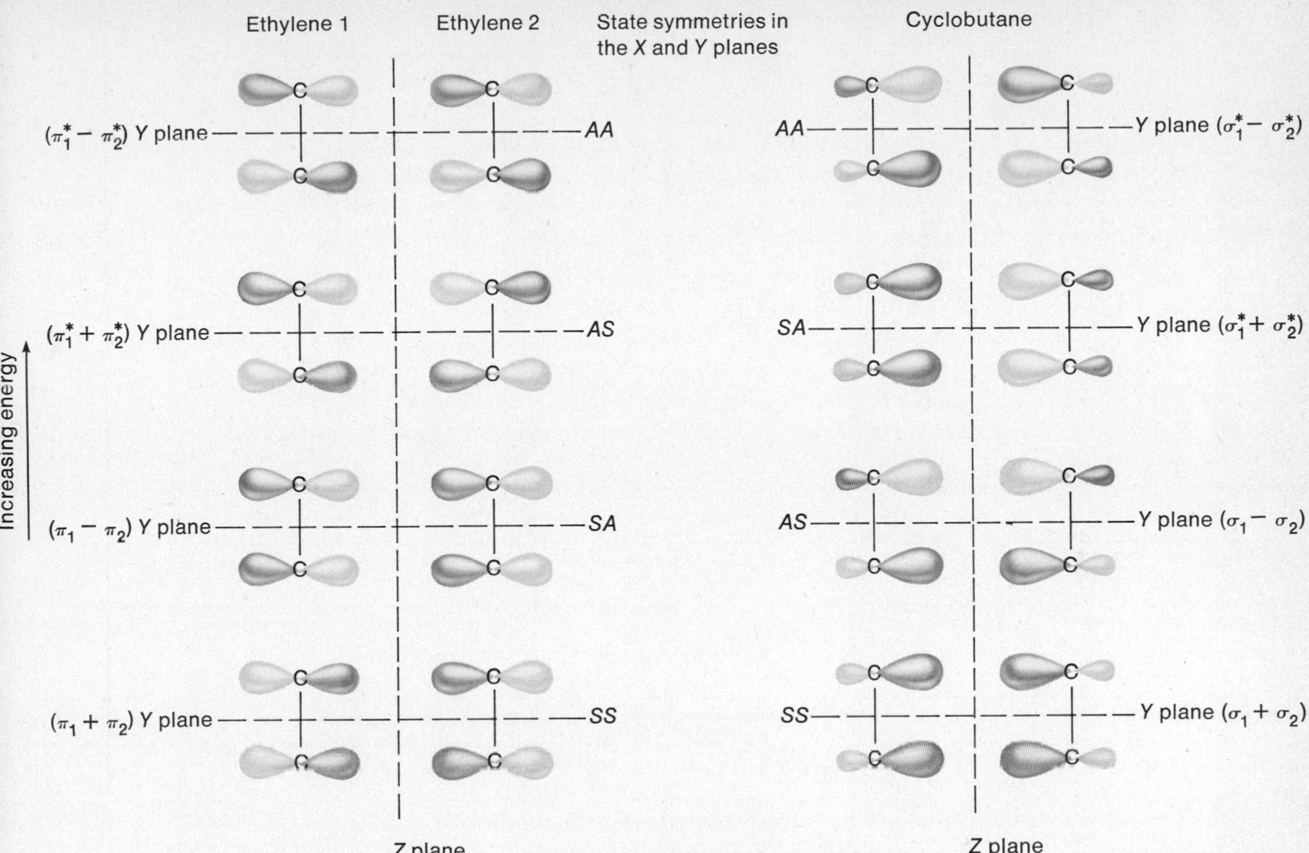

Fig. 38.3. Orbital combinations and symmetries in two reacting ethylenes and the product cyclobutane. The symmetry planes Y and Z are defined in Fig. 29.29.

symmetry *differences* with respect to this plane exist. Therefore, only the symmetries with respect to the Y and Z planes are considered.

The symmetry properties of each of the orbital combinations in the reactants and product can now be assigned on the basis of the reflection properties of the orbitals in the Y and Z planes. For example, reflection in the Y plane of the π orbital of ethylene 1 within the $(\pi_1 + \pi_2)$ set gives the orbitals of ethylene 2, so this reflection is symmetric and is designated by S. Reflection of these same orbitals in the Z plane similarly is symmetric (S). The total state symmetry for the $(\pi_1 + \pi_2)$ set, therefore, is SS. Turning now to the $(\pi_1 - \pi_2)$ set, reflection in the Y plane is symmetric (S). Reflection in the Z plane, however, does not produce an identical mirror image but one in which the signs of all lobes are changed, so the symmetry of this combination with respect to the Z plane is antisymmetric and is designated by A. Thus, the total state symmetry for the $(\pi_1 - \pi_2)$ set is SA (note that SA indicates symmetric with respect to the Y plane and antisymmetric with respect to the Z plane; AS designates a different symmetry—*viz.*, antisymmetric with respect to the Y plane and symmetric with respect to the Z plane).

The other state symmetries of the reactants and product are assigned in the same fashion, the results of which are tabulated in Fig. 38.4. The relationship between the state symmetries of the reactants and product are conveniently summarized in orbital symmetry correlation diagrams, as shown in Fig. 38.5. In the cycloaddition reaction of two molecules of ethylene to produce cyclobutane, four electrons are involved [*i.e.*, two electrons (designated by vertical arrows in Fig. 38.4) from each of the two π bonds corresponding to two electrons in each of the two σ bonds of the product], the four electrons occupying the lowest energy levels in the reactants [*i.e.*, $(\pi_1 + \pi_2)$ and $(\pi_1 - \pi_2)$] and product [*i.e.*, $(\sigma_1 + \sigma_2)$ and $(\sigma_1 - \sigma_2)$]. The total state symmetry of the occupied $(\pi_1 + \pi_2)$ level "correlates" with the total state symmetry of the occupied $(\sigma_1 + \sigma_2)$ level of the product; "to correlate" meaning to possess the same total state symmetry, *i.e.*, *SS* in this case. The $(\pi_1 - \pi_2)$ level of the reactants, with symmetry *SA*, does *not* correlate with the $(\sigma_1 - \sigma_2)$ occupied level of the product, cyclobutane, which has the symmetry *AS*, as indicated in Fig. 38.4.

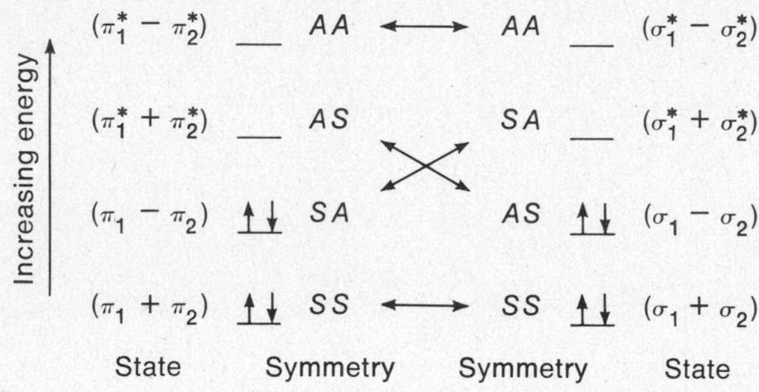

Fig. 38.4. Correlation diagram for the reaction 2 ethylene → cyclobutane.

38.3 The Woodward–Hoffmann Rules

The Woodward–Hoffmann rules state that for a reaction following a concerted pathway, *all* the *occupied* orbitals (*i.e.*, orbitals containing one or two electrons) in the reactant must correlate with occupied orbitals of the *same* symmetry in the product. Clearly, this is not the case in the approach of two molecules of ethylene in a single plane to form cyclobutane (see Fig. 38.2), for the occupied orbitals of the reactant molecules [*i.e.*, $(\pi_1 + \pi_2)$ and $(\pi_1 - \pi_2)$] have symmetries *SS* and *SA*, whereas the occupied orbitals of the product [*i.e.*, $(\sigma_1 + \sigma_2)$ and $(\sigma_1 - \sigma_2)$] have symmetries *SS* and *AS*. The reaction, therefore, is said to be "forbidden," meaning that the low energy, concerted pathway is *unavailable* to it. It does not mean, however, that the reaction cannot take place at all, but simply that it must follow a different pathway involving formation of an intermediate that presumably is a biradical. The reaction may also occur by another pathway with different symmetry and, therefore, a different correlation pattern for orbitals in the reactant and product.

The situation just described can be changed by changing the electron occupancy of the reactants and product. If one of the electrons in the $(\pi_1 - \pi_2)$ and $(\sigma_1 - \sigma_2)$ level (see Fig. 38.4) is elevated to the $(\pi_1^* + \pi_2^*)$ and $(\sigma_1^* + \sigma_2^*)$ level (see Fig. 38.5), the cycloaddition reaction via a concerted pathway becomes allowed. The electronic configurations of ethylene and cyclobutane shown in Fig. 38.5

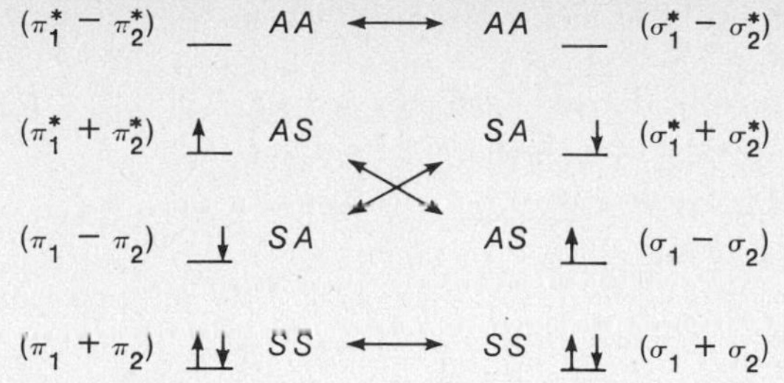

$(\pi_1^* - \pi_2^*)$	__	AA ⟷ AA	__	$(\sigma_1^* - \sigma_2^*)$
$(\pi_1^* + \pi_2^*)$	↑	AS ╳ SA	↓	$(\sigma_1^* + \sigma_2^*)$
$(\pi_1 - \pi_2)$	↓	SA ╳ AS	↑	$(\sigma_1 - \sigma_2)$
$(\pi_1 + \pi_2)$	↑↓	SS ⟷ SS	↑↓	$(\sigma_1 + \sigma_2)$

Fig. 38.5. Correlation diagram for the cycloaddition of electronically excited ethylenes to produce excited cyclobutane.

represent electronically excited states that can be attained by the absorption of electromagnetic radiation in the ultraviolet region (see Section 6.1). Reactions occurring via electronically excited states formed by the absorption of radiant energy are termed *photochemical reactions*; thus, the addition of two molecules of ethylene to each other to form a molecule of cyclobutane is a photochemically allowed process; similarly, the reverse reaction—*viz.*, the cycloreversion* of cyclobutane to ethylene—is a photochemically allowed process. Experimental confirmation of these predictions is provided by the photochemically induced cyclodimerizations of *cis*- and *trans*-2-butene, both of which react when irradiated to give mixtures of isomeric tetramethylcyclobutanes (Fig. 38.6). The reactions of alkenes to form cyclobutanes, therefore, are said to be "thermally forbidden" but "photochemically allowed" processes.

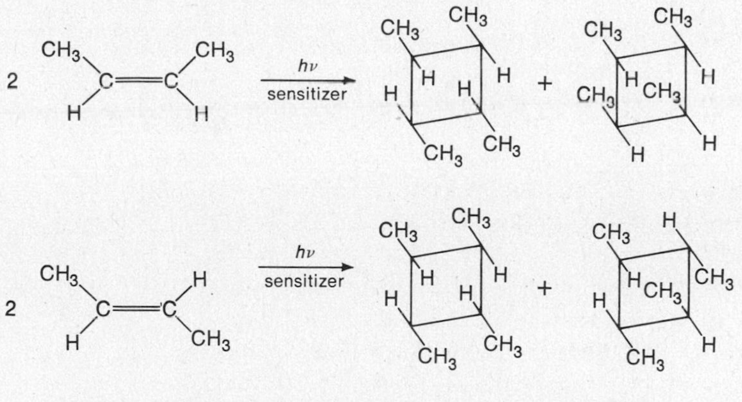

Fig. 38.6. The photo-induced dimerization of *cis*- and *trans*-2-butene.

Let us now analyze in a similar fashion the cycloaddition of a conjugated diene and an alkene to form a cyclohexene. If 1,3-butadiene and ethylene are oriented as shown in Fig. 38.7, the only symmetry element that is common to both the reactants and the product is the Z plane. Assessing the state symmetries

* The conversion of a cyclic compound (*e.g.*, cyclobutane) to its noncyclic precursors (*e.g.*, ethylene) is termed a cycloreversion reaction. The correlation diagrams for the cycloaddition processes (*e.g.*, Figs. 38.4 and 38.5) also apply to the cycloreversion processes.

of the orbitals of the butadiene system and the orbital combinations of the product, we can make the assignments shown in Fig. 38.8. For the reactants, the first three energy levels are doubly filled; similarly, the first three levels of the product are doubly filled. Inspection of Fig. 38.8 shows that *all* the occupied orbitals of the reactants correlate with occupied levels of the product of same symmetry; the reaction, therefore, is thermally allowed.

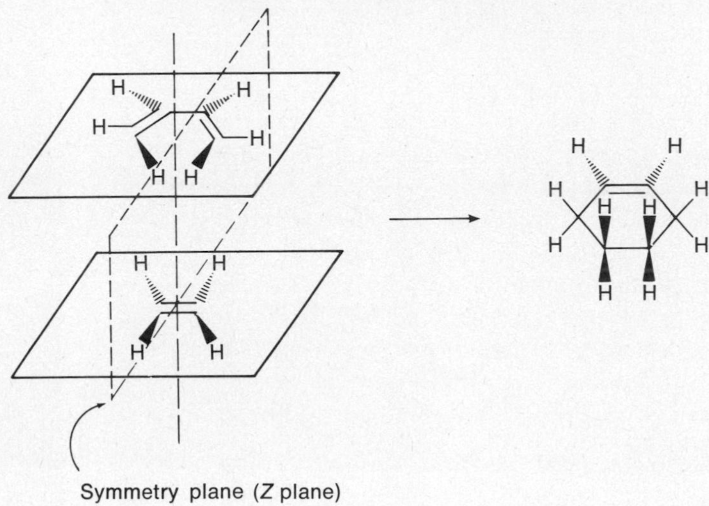

Fig. 38.7. Orientation of 1,3-butadiene and ethylene in the cycloaddition reaction to form cyclohexene.

The cycloaddition reaction between 1,3-butadiene and ethylene in an electronically excited state is not allowed to occur via a concerted pathway, however (Fig. 38.9); the doubly occupied π orbital with S symmetry cannot correlate with a doubly occupied orbital of the same symmetry in the product, and the doubly occupied $(\sigma_1 - \sigma_2)$ orbital of the product possessing A symmetry cannot correlate with a doubly occupied orbital of the reactant with A symmetry. Thus, the concerted reaction of 1,3-butadiene with ethylene is a thermally allowed, but photochemically forbidden, process; *i.e.*, just the reverse trend that is predicted for the cycloaddition of two alkenes to form a cyclobutane.

Application of the approach illustrated in these two examples to other reaction systems has led to the formulation of a general statement of the Woodward–Hoffmann rules for cycloaddition and cycloreversion reactions: *viz.*, *reactions involving* $4n + 2$ *electrons are thermally allowed, photochemically forbidden processes, and reactions involving* $4n$ *electrons are thermally forbidden, photochemically allowed processes.* For example, the reaction of 1,3-butadiene with ethylene can occur either via a $4n$ process to yield vinylcyclobutane, or via a $(4n + 2)$ process to yield cyclohexene; the latter occurs almost exclusively, for it is the

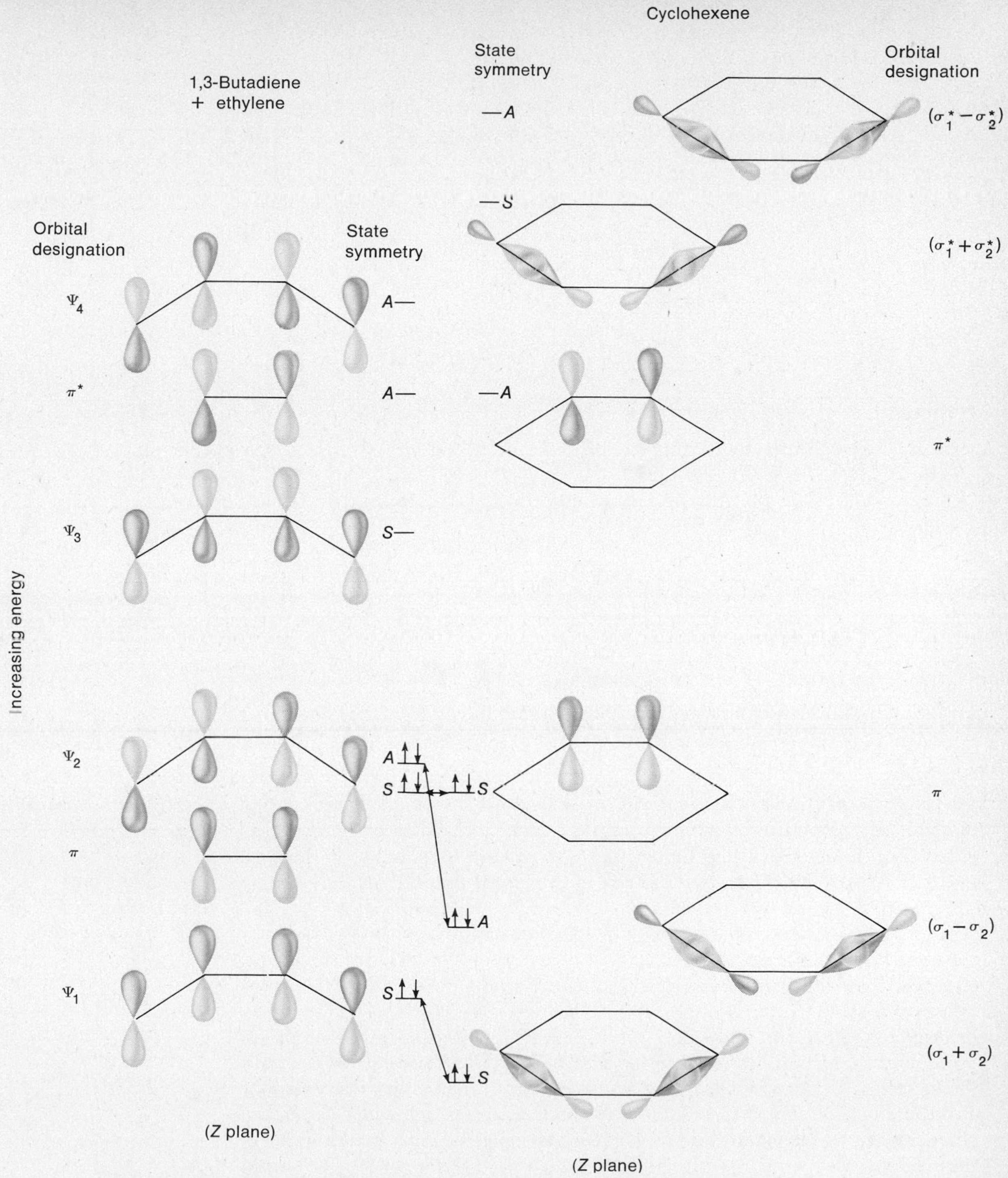

Cyclohexene

State symmetry

Orbital designation

1,3-Butadiene + ethylene

—A $(\sigma_1^* - \sigma_2^*)$

—S $(\sigma_1^* + \sigma_2^*)$

Orbital designation

State symmetry

Ψ_4 A—

π^* A— —A π^*

Ψ_3 S—

Increasing energy

Ψ_2 A

S S π

π

A $(\sigma_1 - \sigma_2)$

Ψ_1 S

S $(\sigma_1 + \sigma_2)$

(Z plane)

(Z plane)

Fig. 38.8. Orbital combinations, symmetries in the Z plane, and correlation diagram for the cycloaddition reaction of 1,3-butadiene with ethylene.

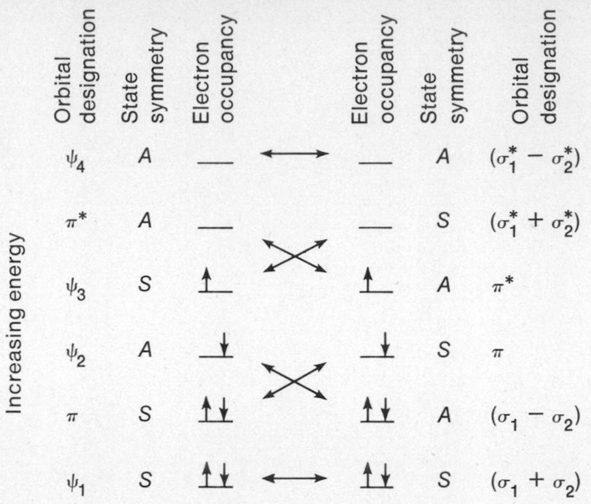

Increasing energy ↑

Orbital designation	State symmetry	Electron occupancy		Electron occupancy	State symmetry	Orbital designation
ψ_4	*A*	—	⟷	—	*A*	$(\sigma_1^* - \sigma_2^*)$
π^*	*A*	—	⤫	—	*S*	$(\sigma_1^* + \sigma_2^*)$
ψ_3	*S*	↑		↑	*A*	π^*
ψ_2	*A*	↓	⤫	↓	*S*	π
π	*S*	↑↓		↑↓	*A*	$(\sigma_1 - \sigma_2)$
ψ_1	*S*	↑↓	⟷	↑↓	*S*	$(\sigma_1 + \sigma_2)$

Fig. 38.9. Excited state correlation diagram for the cycloaddition of 1,3-butadiene and ethylene.

thermally allowed reaction:

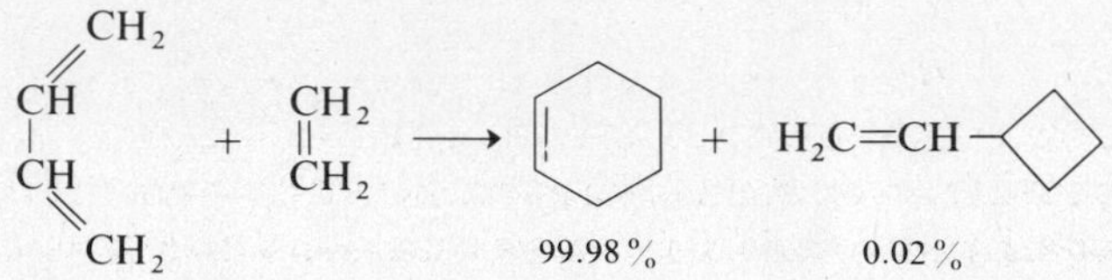

$$CH_2{=}CH{-}CH{=}CH_2 + CH_2{=}CH_2 \longrightarrow \text{cyclohexene } (99.98\%) + H_2C{=}CH{-}\text{cyclobutyl } (0.02\%)$$

The Woodward–Hoffmann rules say nothing about the number of orbitals that are involved, only the number of electrons.

38.4 *The Hückel–Möbius Approach to the Analysis of Cycloaddition Reactions*

The analysis of reaction pathways by means of orbital symmetry correlation diagrams is complicated in many instances by a lack of sufficient symmetry in either the reactant(s) or the product(s) to allow the analysis to be made. To meet this problem, other approaches have been developed that do not rely on the assignment of orbital symmetries. One of these is the "Hückel–Möbius" method. Hückel and Möbius systems are both defined as cyclic arrays of orbitals. They differ, however, with respect to the number of inversions of sign in the wave function describing the array, a change in sign corresponding to an orbital node and being designated by the symbol "$-\beta$." Hückel systems, already familiar to us, are characterized as having either no inversions of sign, *i.e.*, no nodes, or an even number of inversions of sign, *i.e.*, an even number of nodes. For example, the molecular orbitals (MO's) of benzene (see Fig. 4.12) have zero, two, four, or six inversions of sign (Fig. 38.10). Möbius systems, on the other hand, are characterized as having an odd number of inversions of sign. For example, if the *p* orbitals on a cyclic six-carbon framework are twisted in the fashion shown in Fig. 38.11, there are only odd numbers of inversions of sign in the cyclic array—*i.e.*, one, three, or five.

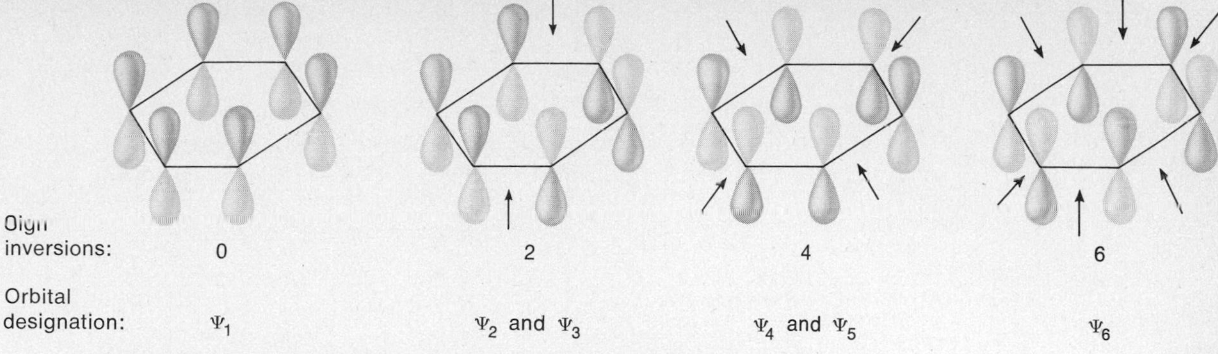

Fig. 38.10. Hückel system of orbitals (the shaded lobes correspond to one sign (*e.g.*, +) and the unshaded lobes to the opposite sign (*e.g.*, −).

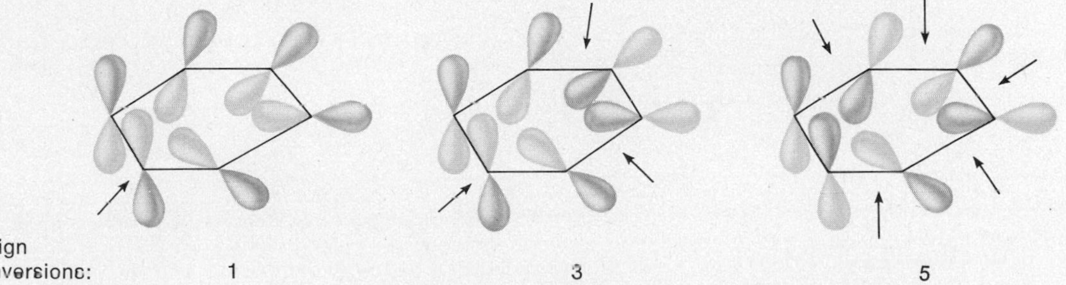

Fig. 38.11. Möbius system of orbitals.

As we have seen in Section 4.7, Hückel systems are particularly stable, *i.e.*, aromatic, when they contain $4n + 2$ electrons, but are not stabilized, *i.e.*, nonaromatic, when they contain $4n$ electrons. With Möbius systems, just the reverse is true. Möbius systems are stabilized (aromatic) when they contain $4n$ electrons and are nonaromatic when they contain $4n + 2$ electrons (Table 38.1). Benzene,

Table 38.1 Characteristics of Hückel and Möbius systems

System	Sign Inversions	Aromatic	Nonaromatic
Hückel	0 or an even number	$4n + 2$ electrons	$4n$ electrons
Möbius	Odd number	$4n$ electrons	$4n + 2$ electrons

for example, represents an aromatic Hückel system and cyclobutadiene a nonaromatic Hückel system. "Möbius benzene," on the other hand, represents a nonaromatic Möbius system and "Möbius cyclobutadiene" an aromatic system (Fig. 38.12).

Although molecules containing Möbius systems are as yet not known, it is useful to apply the Hückel–Möbius concept to bonding in transition states to analyze the allowedness of concerted reaction pathways. In the Hückel–Möbius

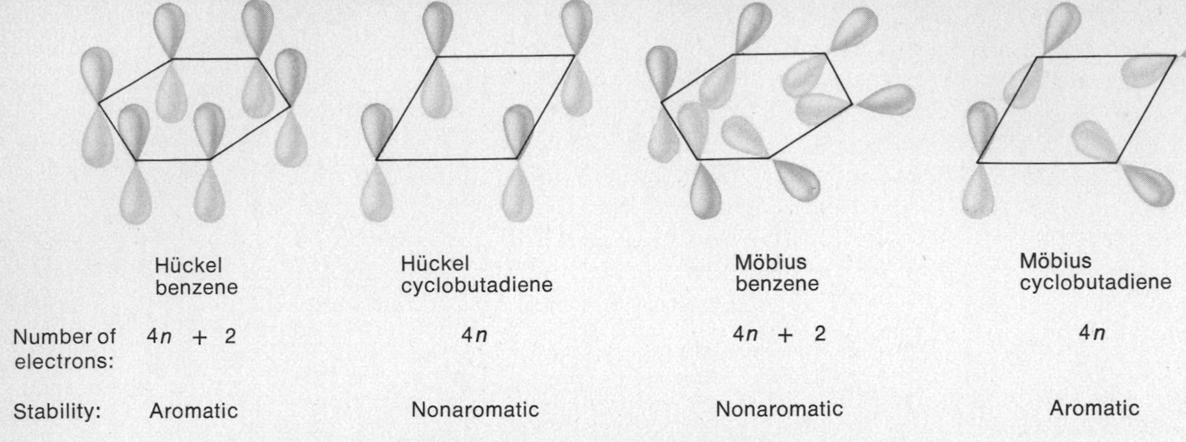

Fig. 38.12. Hückel and Möbius benzene and cyclobutadiene.

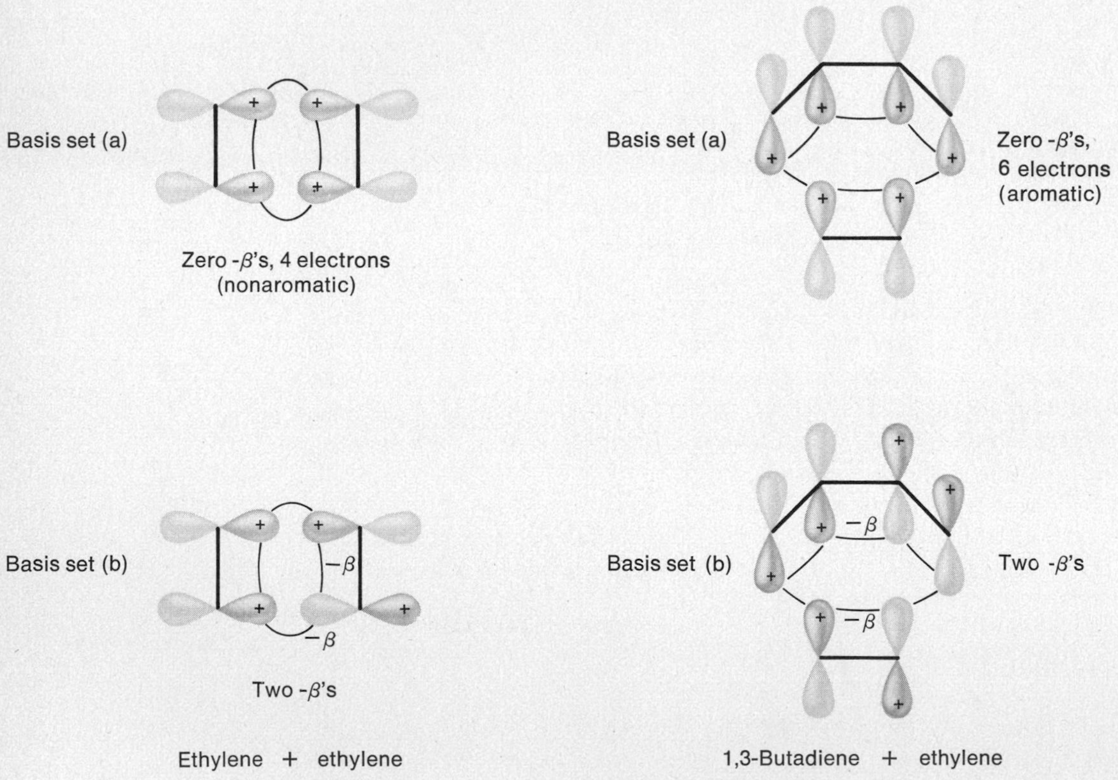

Basis set (a)

Zero -β's, 4 electrons (nonaromatic)

Basis set (a)

Zero -β's, 6 electrons (aromatic)

Basis set (b)

$-\beta$

$-\beta$

Two -β's

Basis set (b)

$-\beta$

$-\beta$

Two -β's

Ethylene + ethylene

1,3-Butadiene + ethylene

Fig. 38.13. Illustration of the application of the Hückel–Möbius concept using arbitrarily chosen basis sets for the cycloaddition of ethylene with ethylene, and 1,3-butadiene with ethylene. (The lines drawn between the lobes of the orbitals indicate where bonding interactions occur in the transition state.)

approach, the allowedness of a concerted reaction is determined (a) by the number of inversions of sign between adjacent atomic orbitals in the basis set* of the wave function describing the transition state, and (b) by the number of electrons involved. Using the Hückel–Möbius approach, let us again consider the reaction of two molecules of ethylene to form cyclobutane. In each of the arbitrarily chosen basis sets for the molecular orbitals of the transition state, there is zero or an even number of inversions of sign, *i.e.*, zero $-\beta$'s or an even number of $-\beta$'s; thus, the transition state is a Hückel system (Fig. 38.13). The system contains four electrons and, therefore, is nonaromatic and is not especially stabilized. If we apply the same kind of analysis to the reaction of 1,3-butadiene with ethylene, the transition state is again a Hückel system. In this case, the transition state contains six electrons and, therefore, is aromatic, corresponding to an especially favorable process.

Concerted reactions are, in most cases, considerably easier to analyze by the Hückel–Möbius approach than by the Woodward–Hoffmann rules. In the Hückel–Möbius approach, it is not necessary to construct all the proper orbitals of the reactant(s) and product(s) (any arbitrary basis set can be used to determine the number of inversions of sign), and it is not necessary to assign orbital and state symmetries.

38.5 Stereochemical Considerations in Cycloaddition Reactions

Additions that take place at the same face of an unsaturated function, *i.e.*, *syn* addition, are characterized as *suprafacial* in the Woodward–Hoffmann terminology, and those that take place at opposite faces of the unsaturated function, *i.e.*, *anti* addition, are characterized as *antarafacial* (indicated by the subscripts *s* and *a*, respectively) (Fig. 38.14).

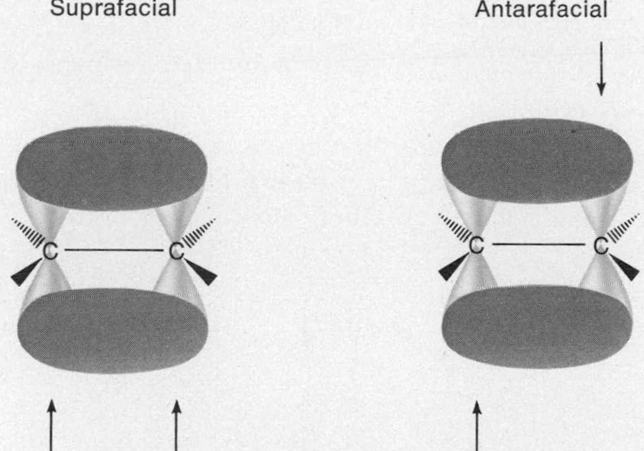

Fig. 38.14. Suprafacial and antarafacial modes of addition to a C=C bond.

* A basis set is the array of atomic orbitals contributing to a given molecular orbital. If two adjacent atomic orbitals differ in the sign of their corresponding upper and lower lobes, there is a sign difference between the corresponding terms in this part of the mathematical expression for the basis set.

The two reactions we have discussed are both classified as suprafacial–suprafacial additions. The ethylene cycloaddition is designated as a $({}_{\pi}2_s + {}_{\pi}2_s)$ process, and the butadiene-ethylene cycloaddition is designated as a $({}_{\pi}4_s + {}_{\pi}2_s)$ process.*

If a pair of ethylene molecules are brought together in a fashion such that a Möbius activated complex forms, the addition becomes a suprafacial-antarafacial process, categorized as $({}_{\pi}2_s + {}_{\pi}2_a)$; according to the Hückel–Möbius analysis, it should be allowed (*i.e.*, a Möbius system containing four electrons). For the addition to be allowed, the ethylene molecules must approach each other in a perpendicular orientation in the fashion illustrated in Fig. 38.15, and completion

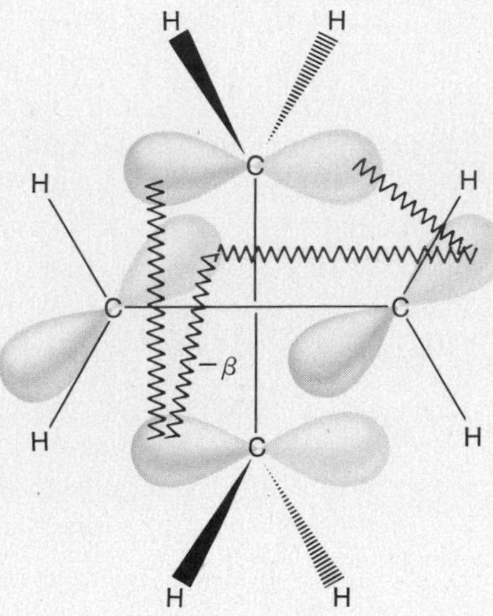

Fig. 38.15. Perpendicular approach of two molecules of ethylene resulting in an aromatic Möbius transition state.

of this pathway requires considerable distortion of the carbon framework. Thus, although the process is allowed (concerted), it is a difficult one and has not yet been observed in simple alkenes. The highly strained triene shown in Fig. 38.16 does, however, undergo a facile thermal reaction in which suprafacial-antarafacial cycloaddition takes place.

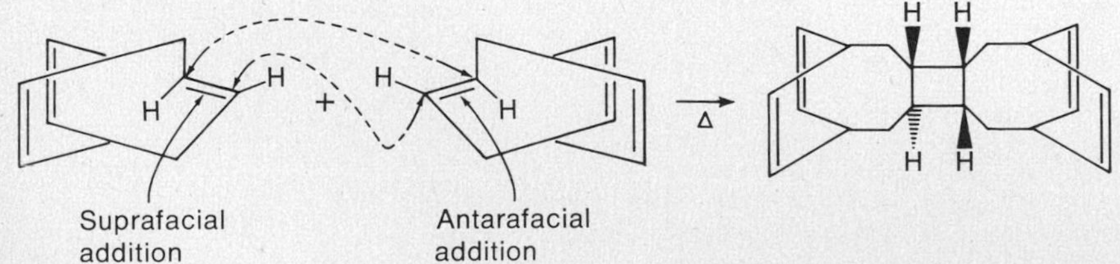

Fig. 38.16. Suprafacial–antarafacial cycloaddition process.

* In these designations, the π indicates that a π-electron system is involved; the numbers indicate the number of π electrons in each reactant, and the subscript after the number (s or a) indicates the stereochemistry of the cycloaddition for each reactant. Cycloaddition reactions are also known in which a σ bond is involved, in which case the first symbol in the designation is σ.

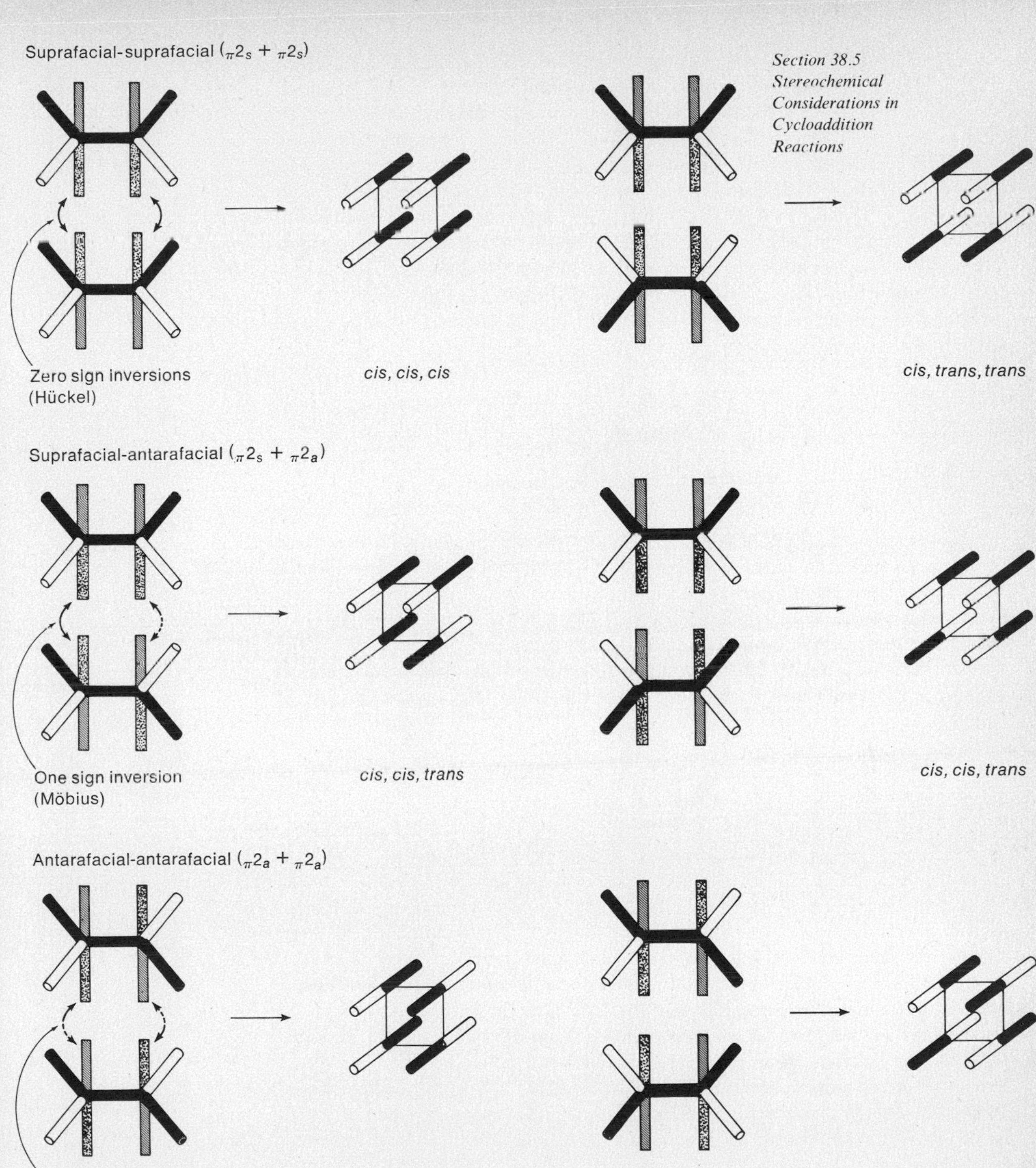

Fig. 38.17. Suprafacial–suprafacial, suprafacial–antarafacial, and antarafacial–antarafacial cycloaddition o' *cis*-2-butene.

The third possibility, an antarafacial–antarafacial cycloaddition [categorized as $({}_{\pi}2_a + {}_{\pi}2_a)$], involves an activated complex that is a Hückel array and, therefore, a forbidden process. Furthermore, the molecular distortion required for it to occur is extremely severe. The stereochemical outcome of this and the other two modes of cycloaddition are shown in Fig. 38.17 for *cis*-2-butene. The required orientation of the double bonds for the $({}_{\pi}2_s + {}_{\pi}2_s)$ process is easy enough to visualize, but that for the $({}_{\pi}2_s + {}_{\pi}2_a)$ and $({}_{\pi}2_a + {}_{\pi}2_a)$ processes are more difficult. It is recommended that a set of molecular framework models be used as an aid in studying the stereochemistry of these reactions. From the molecular framework model set, construct two *cis*-2-butene molecules in which the lobes of each of the *p* orbitals comprising the double bond are shown in two different colors; *i.e.*, as shown in the adjacent diagram.

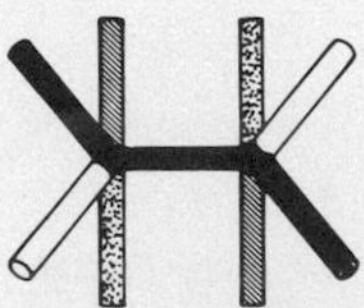

This corresponds to the π bond in the bonding condition. Now, attach these two molecules to each other in a suprafacial–suprafacial fashion. This can be done in two ways, leading to two different stereochemical outcomes—*i.e.*, the all-*cis* isomer and the *cis, trans,trans* isomer (Fig. 38.17). Next, attach the two molecules to each other in a suprafacial–antarafacial fashion and observe the stereochemical outcome. The easiest way to determine the stereochemical outcome in this case is first to twist one of the molecules 180° around the central bond, *i.e.*, as shown in the adjacent diagram,

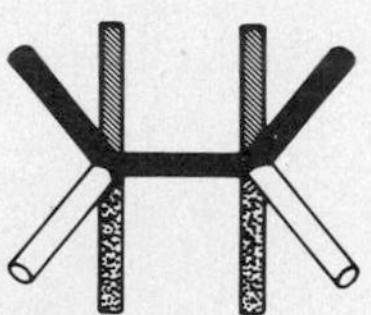

and then let the systems come together in the plane of the π system.* In this case, a single product, the *cis,cis,trans*-tetramethylcyclobutane, forms (Fig. 38.17). Finally, let the two molecules attach themselves to each other in an antarafacial–antarafacial fashion, this time with *both* molecules in the 180° twisted form. In this case, as in the first, two different stereochemical outcomes are possible, leading to the *cis,trans,trans* isomer and the *trans,cis,trans* isomer (Fig. 38.17). As mentioned previously, examples of the photoinduced $({}_{\pi}2_s + {}_{\pi}2_s)$ cycloaddition (see Fig. 38.6) and the thermally induced $({}_{\pi}2_s + {}_{\pi}2_a)$ cycloaddition (see Fig. 38.16) are known, but examples of a $({}_{\pi}2_a + {}_{\pi}2_a)$ cycloaddition have yet to be discovered.

---

* Realize that this is a simplifying formalism and that the entity obtained by twisting by 180° does not correspond to the π bond in the antibonding state but simply to the way in which the bond must twist as the reaction progresses along the antarafacial pathway.

Thermal rearrangements can be classified as (a) *electrocyclic* if the reactant and product contain a *different number* of single and multiple bonds, or (b) *sigmatropic* if the reactant and product contain the *same number* of single and multiple bonds. These reactions, particularly the electrocyclic reactions, are closely related to cycloaddition reactions and have also been recently interpreted in terms of orbital symmetry considerations:

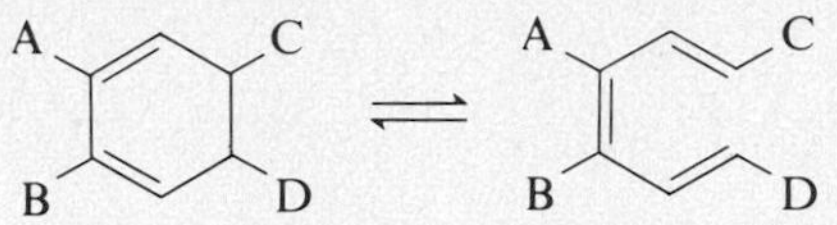

An electrocyclic reaction

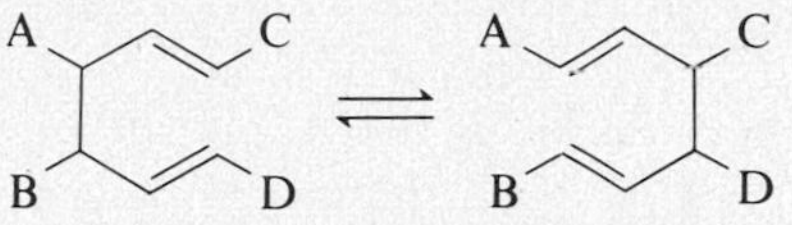

A sigmatropic reaction

38.6 Electrocyclic Thermal Rearrangements

38.6a. ORBITAL SYMMETRY ANALYSIS OF THE CYCLOBUTENE-BUTADIENE INTERCONVERSION. One of the simplest examples of an electrocyclic thermal rearrangement is the interconversion of cyclobutene and 1,3-butadiene (see Chapter 7), and it is particularly interesting to investigate the stereochemistry of this interconversion. Although there is no experimental way for doing this with cyclobutene itself, we can do it with an appropriately "tagged" molecule, such as, for instance, *cis*-3,4-dideuteriocyclobutene. The question we can ask is, "Which of the four possible twist modes are involved when cyclobutene opens to 1,3-butadiene, or when 1,3-butadiene closes to cyclobutene (Fig. 38.18)?"

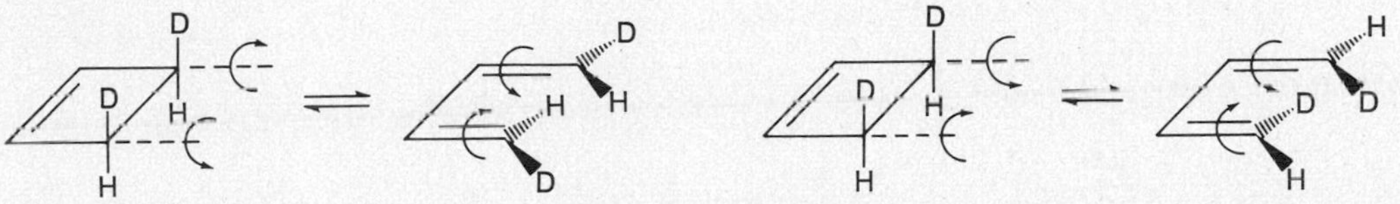

Disrotatory motion

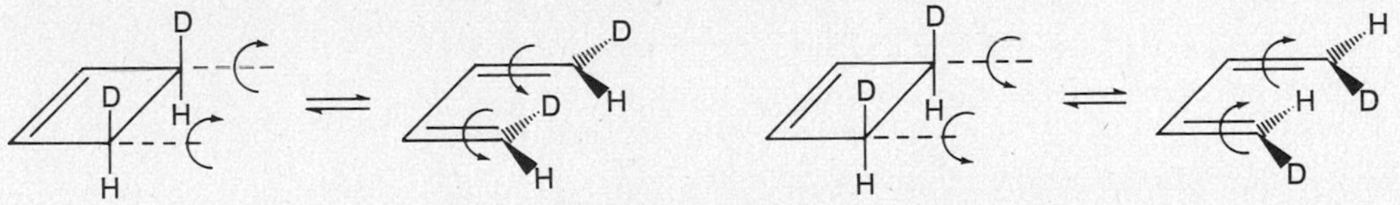

Conrotatory motion

Fig. 38.18. Disrotatory and conrotatory motions in the ring of cyclobutene to 1,3-butadiene (and the ring closing of 1,3-butadiene to cyclobutene).

The means for answering this question have been provided by the Woodward–Hoffmann rules, which are applied to this situation in the same way that they are applied to the cycloaddition reactions (see Section 38.3). As in the case of the cycloaddition reactions, we must first determine what symmetry elements can be used to assign the symmetry properties of the orbitals involved in the reaction. In this case, however, we cannot use the symmetry elements present in the reactant and product to distinguish between the disrotatory and conrotatory modes of reaction, because both would yield the same result. Instead, we must determine the symmetry elements of the orbitals as they change along the reaction coordinate when the transformation from reactant to product occurs:

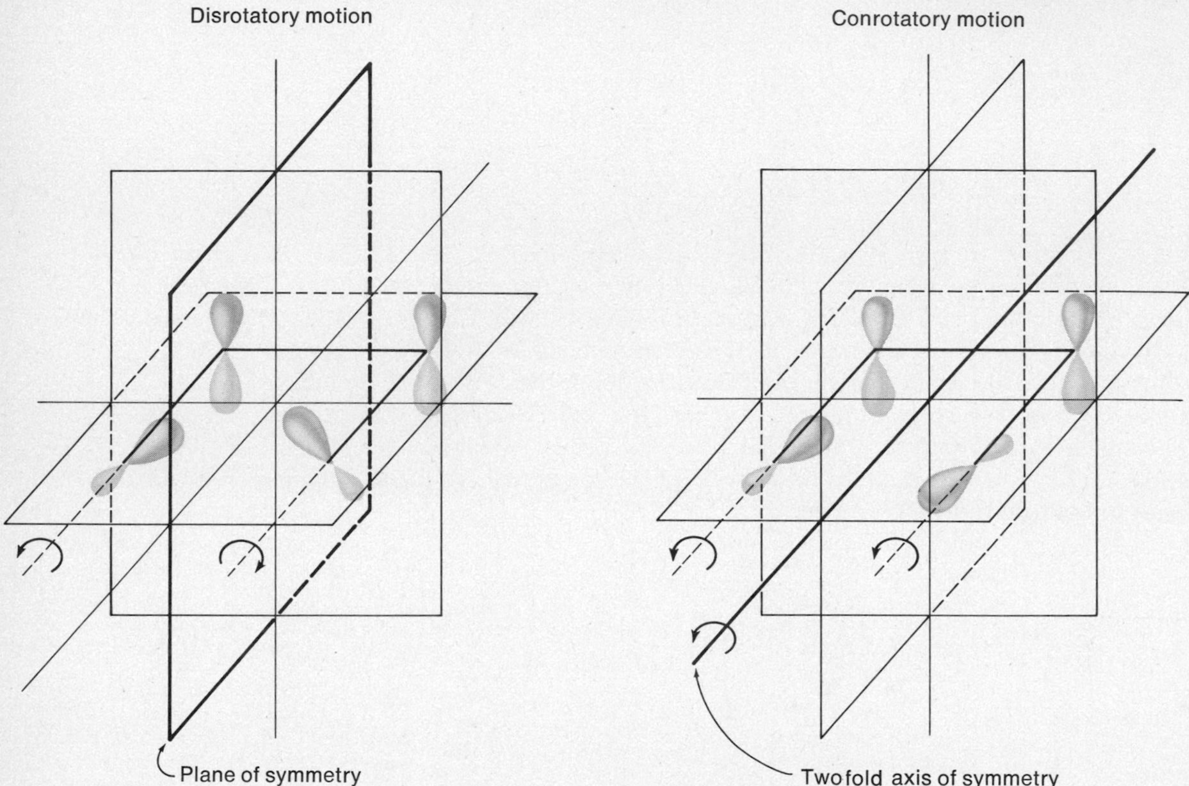

It is now obvious that the assemblages of orbitals along the two reaction modes possess different symmetry elements; in the disrotatory motion, a plane of symmetry exists, whereas in the conrotatory motion a twofold axis of symmetry (involving rotation of 180° around the symmetry axis) is present.

Let us consider first the disrotatory processes. The bonds that undergo alteration in the transformation between cyclobutene and 1,3-butadiene are (a) the π bond between C-1 and C-2 in cyclobutene, (b) the σ bond between C-3 and C-4 in cyclobutene, and (c) the π bonds of butadiene. The symmetries of these bonds along the reaction coordinate can be assigned with respect to the plane of symmetry, and these are tabulated in Fig. 38.19. Using these symmetry assignments, we can construct the correlation diagram shown in Fig. 38.20. For the electronic ground state of cyclobutene, the occupied orbitals are σ and π, and for 1,3-butadiene they are Ψ_1 and Ψ_2. The occupied σ orbital of cyclobutene

Fig. 38.19. Symmetry properties of reacting orbitals in cyclobutene and 1,3-butadiene for the disrotatory motion.

correlates with the occupied Ψ_1 orbital of butadiene, and the occupied π orbital of cyclobutene correlates with the *unoccupied* Ψ_3 orbital of butadiene (Fig. 38.20). According to the Woodward–Hoffmann rules, the reaction is concerted only if all the occupied orbitals of the reactant and product correlate with each other. It must be concluded on the basis of this analysis, therefore, that the disrotatory motion is a forbidden process in this particular system.

Let us next make an analysis for the conrotatory mode of reaction, which contains a twofold axis of rotation as its symmetry element along the reaction coordinate. The symmetry properties of the orbitals about the twofold axis

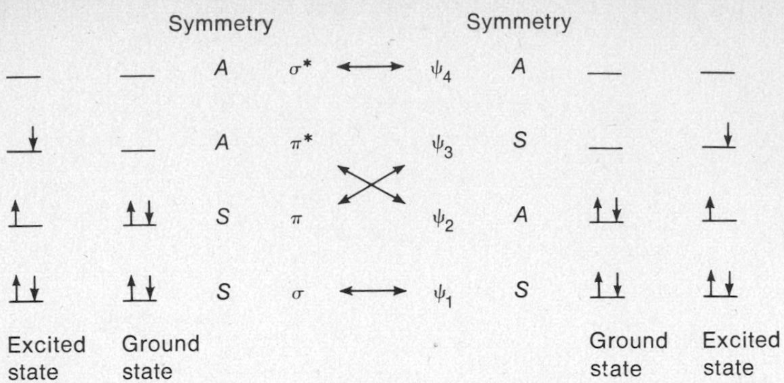

Fig. 38.20. Correlation diagram for disrotatory ring opening of cyclobutene.

are tabulated in Fig. 38.21.* The correlation diagram for the conrotatory transformation (Fig. 38.22) shows that the occupied σ orbital of cyclobutene correlates with the Ψ_2 occupied orbital of butadiene, and that the occupied π orbital of cyclobutene correlates with the occupied Ψ_1 orbital of butadiene. In this case, then, *all* the occupied orbitals of the reactant and product *do* correlate, and the reaction is classed as an allowed process.

38.6b. EXAMPLES OF CYCLOBUTENE-BUTADIENE INTERCONVERSION. Experimental verification of the prediction that the cyclobutene-1,3-butadiene thermal interconversion should take place via a conrotatory motion is provided by numerous examples. For instance, *cis*-3,4-dimethylcyclobutene undergoes thermal rearrangement at 300° to yield a product that contains 99.995% of the *cis,trans* isomer, the result of conrotatory ring opening, and only 0.005% *trans,trans* isomer, the result of disrotatory ring opening (Fig. 38.23). For all practical purposes, therefore, the reaction is completely stereoselective. Another interesting example is the slowness with which bicyclo[2.2.0]hexa-2,5-diene (Dewar benzene) undergoes thermal rearrangement to benzene. Despite the fact that benzene is very much more stable than Dewar benzene (by about 65 kcal/mole), the transformation occurs slowly, because the allowed conrotatory ring opening leads to a benzene containing a *trans* double bond (Fig. 38.23). This corresponds to such a highly strained system that this pathway is precluded, despite its allowedness on the basis of orbital symmetry considerations. The alternative pathway is the nonconcerted, higher energy one involving a biradical intermediate. Here, then, is an excellent example of a reaction that is strongly favored by the thermodynamics of the process but takes place slowly because of the high energy of the activated complex(es) that must be traversed in going from reactant to product—*i.e.*, a thermodynamically favored but kinetically repressed reaction.

38.6c. CYCLOBUTENE-BUTADIENE INTERCONVERSION IN THE ELECTRONICALLY EXCITED STATE. The prediction that the disrotatory ring opening of cyclobutene

---

* Rotation by 180° around an axis passing through an orbital assembly is designated as symmetric (symbolized by *S*) when the rotated assembly is *indistinguishable* from the original and is antisymmetric (symbolized by *A*) when the rotated assembly is *distinguishable* from the original.

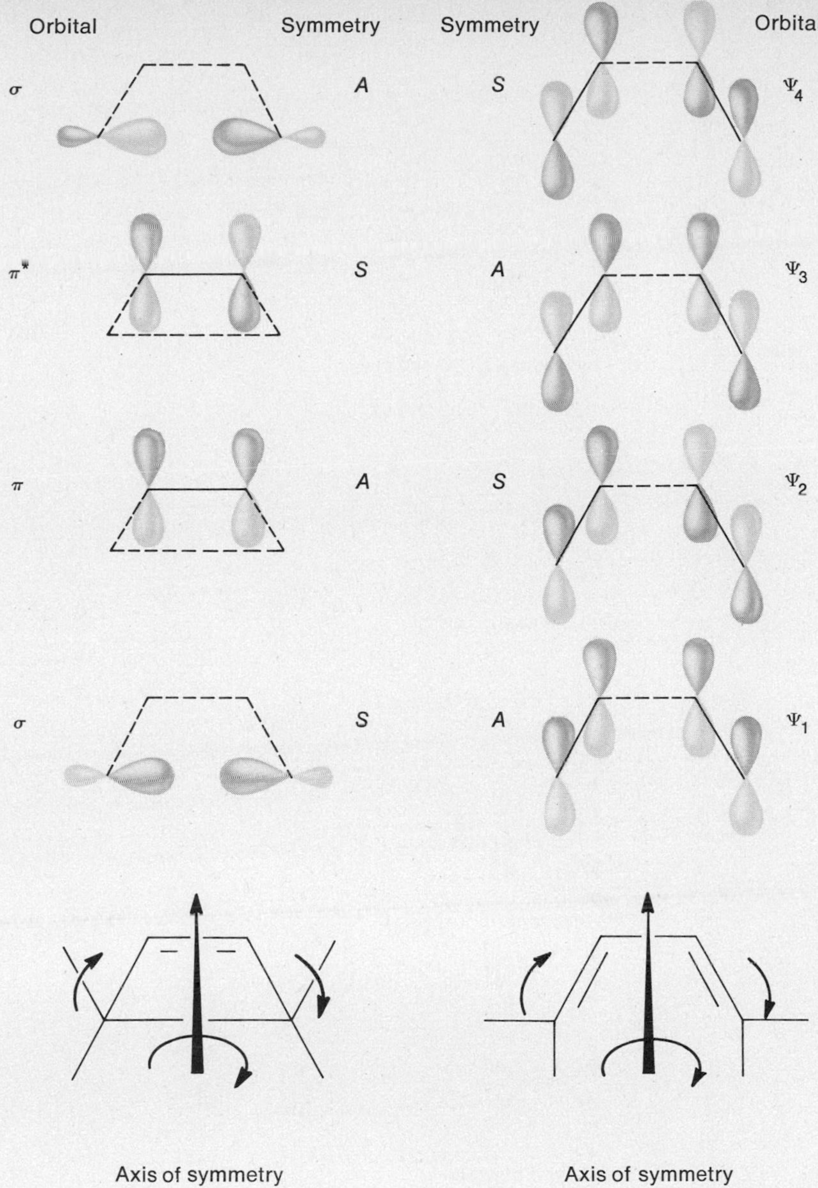

Fig. 38.21. Symmetry properties of reacting orbitals in cyclobutene and 1,3-butadiene for conrotatory motion.

is forbidden and the conrotatory ring opening is allowed pertains to the electronic ground state of the molecule. If an electron is elevated from the highest energy bonding level to the lowest energy antibonding level by absorption of light of the appropriate wavelength, the situation is reversed. Inspection of correlation tables (see Figs. 38.20 and 38.22) shows that for the disrotatory process all the occupied orbitals correlate, *i.e.*, σ, π, and π^* in cyclobutene and Ψ_1, Ψ_2, and Ψ_3

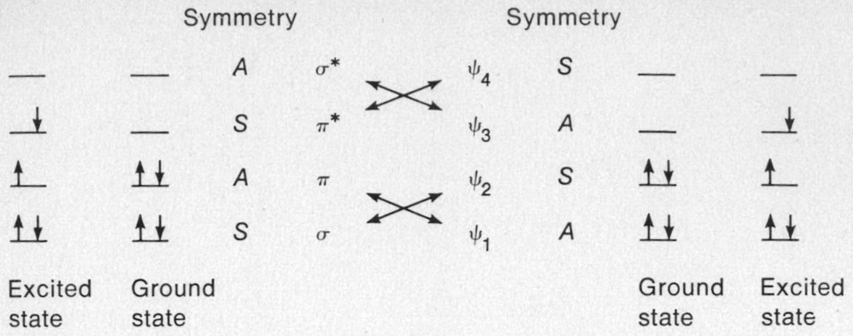

Excited state	Ground state	Symmetry			Symmetry	Ground state	Excited state
—	—	*A*	σ^*	ψ_4	*S*	—	—
↓	—	*S*	π^*	ψ_3	*A*	—	↓
↑	↑↓	*A*	π	ψ_2	*S*	↑↓	↑
↑↓	↑↓	*S*	σ	ψ_1	*A*	↑↓	↑↓

Fig. 38.22. Correlation diagram for conrotatory ring opening of cyclobutene.

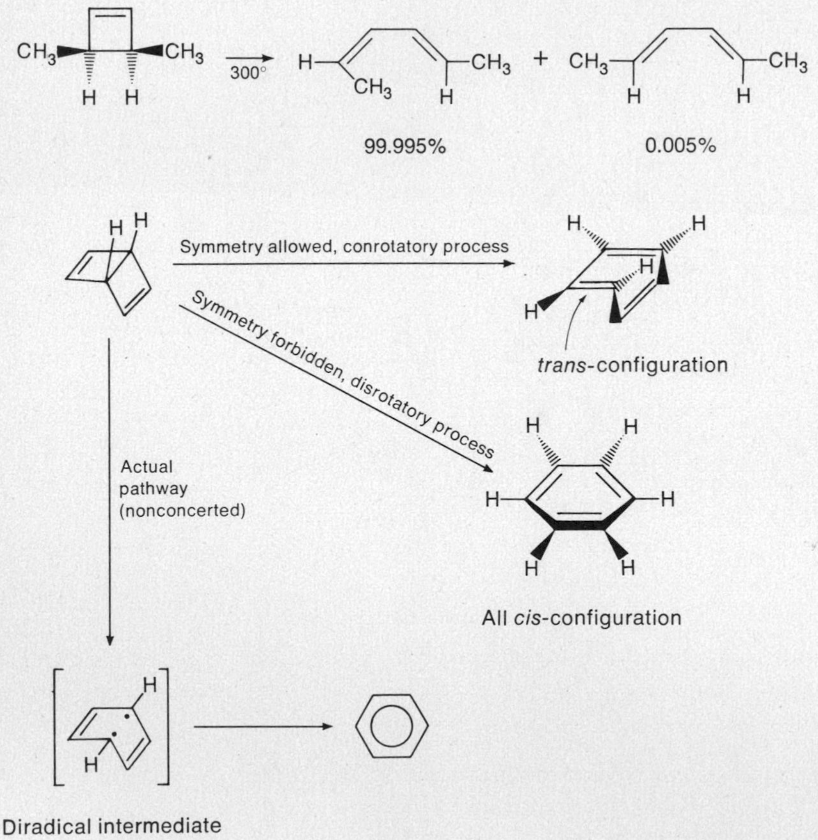

CH_3 CH_3 H H → 300° H— CH_3 —CH_3 H + CH_3— H —CH_3 H

99.995% 0.005%

H H

Symmetry allowed, conrotatory process

H H H H

trans-configuration

Symmetry forbidden, disrotatory process

H H H— —H H H

All *cis*-configuration

Actual pathway (nonconcerted)

H H

Diradical intermediate

Fig. 38.23. Experimental verification of the Woodward–Hoffmann rules in the thermal transformation of bicyclo[2.2.0]hexa-2,5-diene ("Dewar benzene") to benzene.

in butadiene; whereas the occupied orbitals for the conrotatory opening do not all correlate, *i.e.*, π^* does not correlate with Ψ_3.

The Woodward–Hoffmann rules, illustrated in this application to the cyclobutene ⇌ 1,3-butadiene system, can be extended to other systems as well (*i.e.*, 1,3-cyclohexadiene ⇌ 1,3,5-hexatriene, etc.). A summary of the predictions for these various systems is given in Table 38.2, where the number of electrons in the systems is specified in terms of $4n$ and $4n + 2$, n being equal to 0, 1, 2, ...

Table 38.2 WOODWARD–HOFFMANN RULES FOR ELECTROCYCLIC REACTIONS

Electrons in π System of Polyene	Rotatory Motion for Allowed Reaction	
	Ground State	Electronic Excited State
$4n$	Conrotatory	Disrotatory
$4n + 2$	Disrotatory	Conrotatory

38.6d. THE HÜCKEL–MÖBIUS APPROACH TO THE ANALYSIS OF THE CYCLOBUTENE-BUTADIENE INTERCONVERSION. The Hückel–Möbius method of analysis (see Section 38.4) can also be applied to electrocyclic reactions by considering the cyclic activated complexes that fall on the reaction pathway between reactant and product. An activated complex that corresponds to a Hückel system containing four electrons is less favored than the one that corresponds to a Möbius system containing four electrons; an activated complex that corresponds to a Hückel system containing six electrons is more favored than a Möbius system containing six electrons, etc. (see Table 38.1). Thus, the disrotatory ring opening of cyclobutene gives a Hückel system containing four electrons (nonaromatic), whereas the conrotatory ring opening gives a Möbius system containing four electrons (aromatic) (Fig. 38.24). Conversely, the disrotatory ring opening of 1,3-cyclohexadiene gives a Hückel system containing six electrons (aromatic), whereas the conrotatory ring opening gives a Möbius system containing six electrons (nonaromatic) (Fig. 38.24). Again, the Hückel–Möbius approach is much simpler to apply than is the Woodward–Hoffmann approach.

38.7 Sigmatropic Thermal Reactions

38.7a. CLASSIFICATION OF SIGMATROPIC REACTIONS. In sigmatropic rearrangements, the reactant and the product contain the same number of single and double bonds; *i.e.*, the same number of degrees of unsaturation. Three typical examples are the "Cope rearrangement," the "Claisen rearrangement," and the hydrogen transfer reaction illustrated in Fig. 38.25. Sigmatropic reactions can be subdivided on the basis of the numbers of atoms between the σ bond being broken ("old" bond) and the σ bond being made ("new" bond). For example, the Claisen rearrangement of allyl phenyl ether has one oxygen atom and two aromatic ring carbons between the old and new bond on one side, and the three carbons between the old and new bond on the other side.

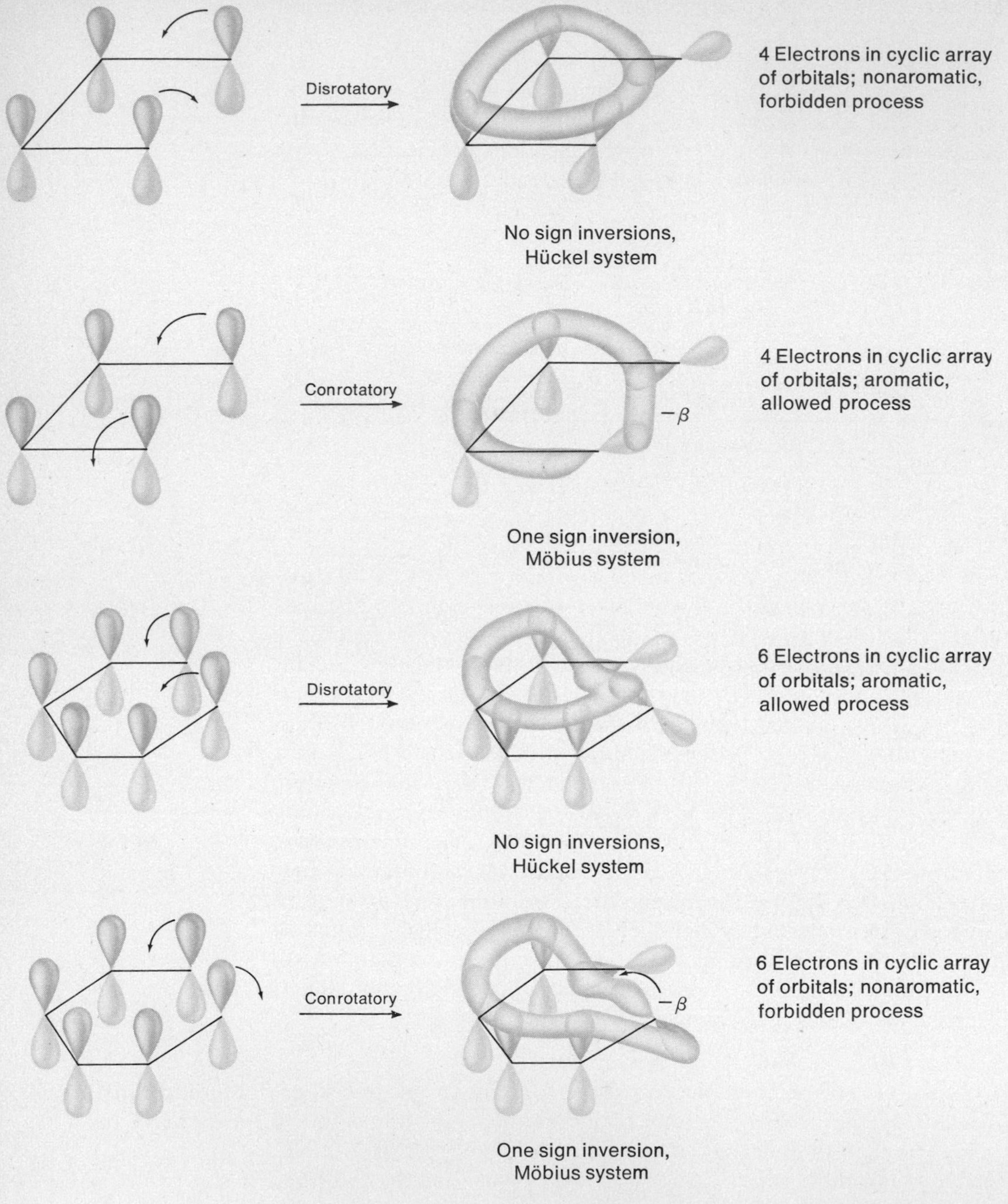

Fig. 38.24. Hückel–Möbius method applied to electrocyclic reactions of cyclobutene and 1,3-cyclohexadiene.

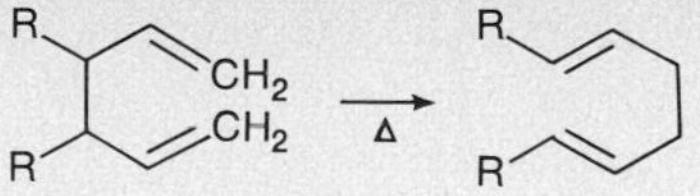

Cope reaction: a [3,3] sigmatropic rearrangement.

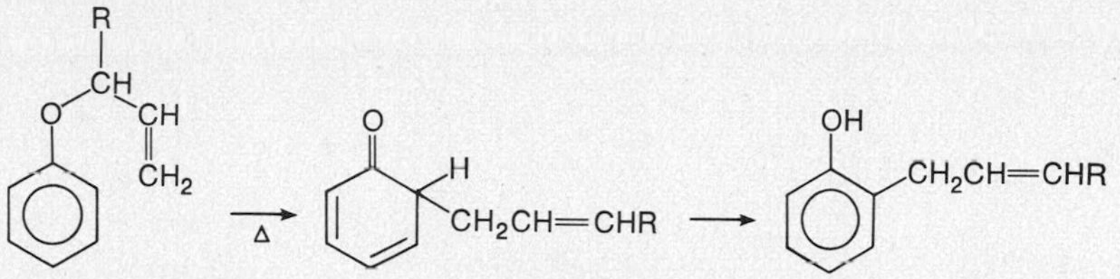

Claisen rearrangement: a [3,3] sigmatropic rearrangement.

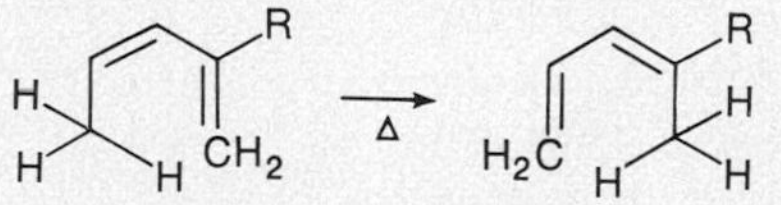

[1,5] Sigmatropic hydrogen shift.

Fig. 38.25. Examples of sigmatropic rearrangements.

"Old" bond

"New" bond

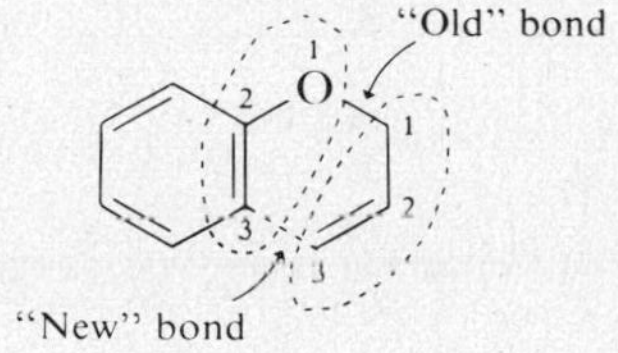

It is classed as a [3,3]sigmatropic shift. Similarly, the sigmatropic shift in 1,3-pentadiene involves five carbons between the old and new bond on one side, and a single hydrogen between the old and new bond on the other side.

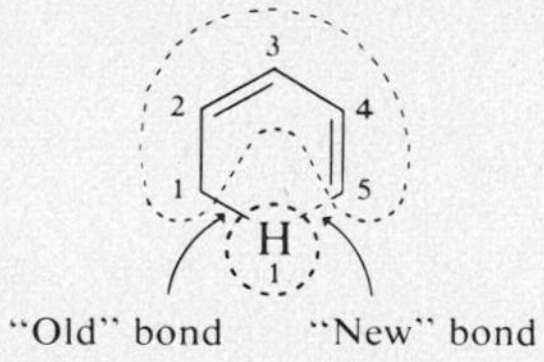

"Old" bond "New" bond

It is classed as a [1,5]sigmatropic shift.

38.7b. SIGMATROPIC REARRANGEMENTS INVOLVING HYDROGEN MIGRATION. Like cycloaddition and electrocyclic reactions, sigmatropic reactions are also amenable to analysis by the Woodward–Hoffmann rules as well as by the Hückel–Möbius method. The latter is somewhat easier in its application to the cases to be discussed and is the approach that is employed here. Let us first consider the

[1,5]sigmatropic shift in 1,3-pentadiene. Although it may not be apparent at first thought, there are two stereochemical pathways by which a hydrogen can migrate from one terminus to the other—*viz.*, across the top face of the pentadiene system, the suprafacial route, or from the top side at one end to the bottom side at the other, the antarafacial route (Fig. 38.26).

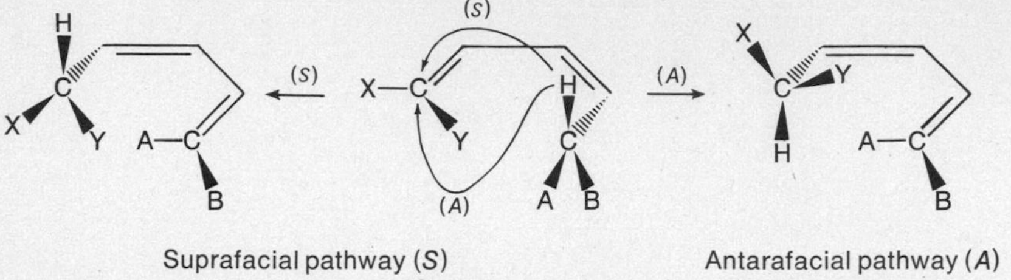

Fig. 38.26. Suprafacial and antarafacial pathways for [1,5]sigmatropic shift.

For each of these pathways, let us consider the cyclic array of orbitals in the activated complex that is formed in going from reactant to product and classify it as a Hückel or a Möbius array. As the representations in Fig. 38.27 reveal, the suprafacial pathway gives a Hückel array and the antarafacial pathway a Möbius array. Since the system contains six electrons in the cyclic array, the former is the allowed (aromatic) pathway, and the reaction is predicted to proceed via the

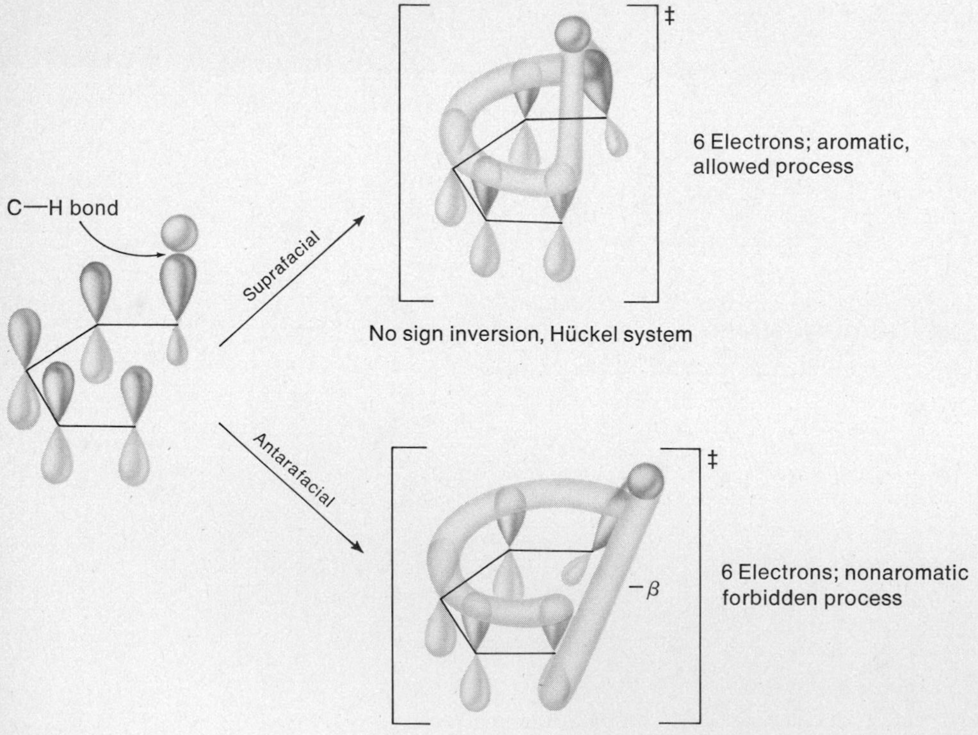

Fig. 38.27. Hückel–Möbius method for the analysis of the [1,5]sigmatropic shift in 1,3-pentadiene.

suprafacial route. The validity of this prediction has been elegantly demonstrated by the reaction in which optically active (6*R*)-6-deuterio-6-methyl-2,4(*E*,*Z*)-octadiene (Fig. 38.28) undergoes thermal rearrangement to optically active

H
C_2H_5 C D—C CH_3 CH_3 ⇌ (Δ) C_2H_5—C H C CH_3 D CH_3

Chiral carbon   Chiral carbon

(6*S*)-2-deuterio-6-methyl-(*E*,*Z*)-2,4-octadiene   (7*S*)-7-deuterio-3-methyl-(*E*,*Z*)-3,5-octadiene

Fig. 38.28. Experimental demonstration of suprafacial hydrogen transfer in a [1,5]sigmatropic rearrangement.

(7*S*)-7-deuterio-3-methyl-3,5(*E*,*Z*)-octadiene, in which the absolute configuration at the new chiral center is in accord with a suprafacial transfer of the hydrogen.

Applying the Hückel–Möbius analysis to a [1,3]sigmatropic hydrogen migration (Fig. 38.29), we easily conclude that the allowed pathway is antarafacial.

Suprafacial process

Hückel system; 4 electrons, aromatic

forbidden process

Antarafacial process

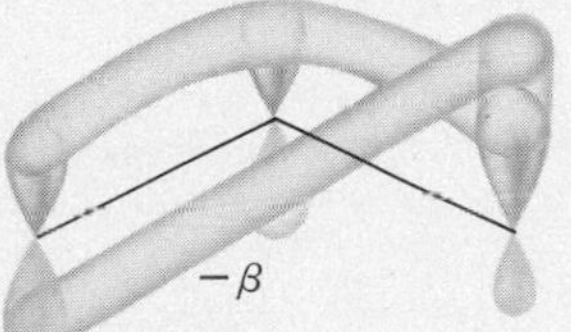

Möbius system; 4 electrons, nonaromatic

allowed process

Fig. 38.29. Hückel–Möbius method for the analysis of the [1,3]sigmatropic hydrogen shift in an allyl system.

Clearly, though, for this pathway to be followed it is necessary to traverse a severely strained activated complex. Thus, although the reaction is allowed by symmetry, it proceeds very slowly because of bond and angle strain in the transition state. An experimental demonstration of this situation is afforded by the reluctance shown by methylenecyclohexadiene to rearrange to the very much more stable toluene:

Suprafacial pathway—symmetry forbidden but sterically feasible

H, H, C, H, H ⇸ C_6H_5—CH_3

Antarafacial pathway—symmetry allowed but sterically forbidden

et us next discuss a [1,3]sigmatropic rearrangement in which a carbon rather n a hydrogen is the migrating entity. Now, in addition to having to consider ie stereochemistry of the migration pathway, we must also consider the stereochemistry of the migrating group; *i.e.*, whether it becomes attached to its new position with retention or inversion of configuration. If retention of configuration is considered to be a suprafacial process with respect to the carbon–carbon bond being broken and if inversion of configuration is considered to be an antarafacial process, four reaction pathways are possible: *i.e.*, (*s*, *s*), (*s*, *a*), (*a*, *s*), and (*a*, *a*). Applying the Hückel–Möbius analysis to this situation (Fig. 38.30), we

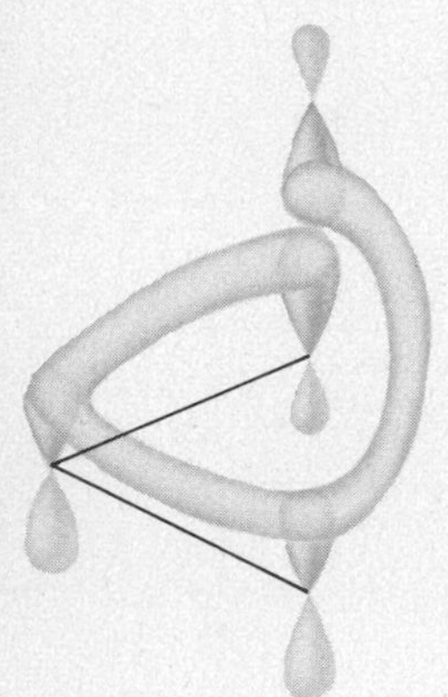

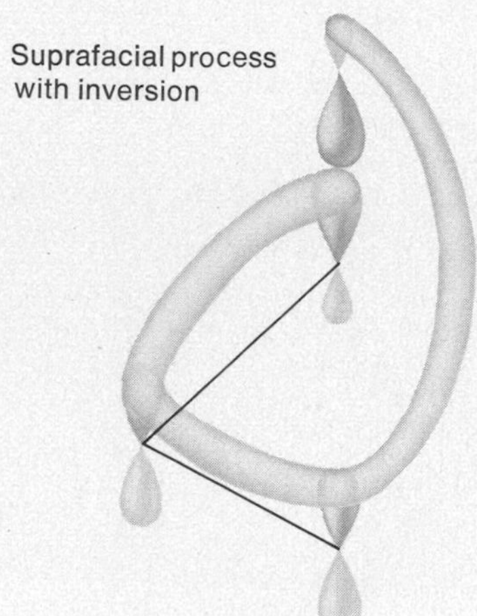

Fig. 38.30. Hückel–Möbius method for the analysis of the [1,3]sigmatropic carbon shift in an allyl system.

can conclude quickly that the (*s*, *s*) pathway involves a Hückel activated complex containing 4 electrons (nonaromatic), and that the (*s*, *a*) pathway involves a Möbius activated complex containing 4 electrons (aromatic). Thus, the reaction should proceed along the suprafacial pathway but with inversion of configuration of the migrating group. This has actually been experimentally demonstrated by thermal rearrangement of the bicyclo-[3.2.0]hept-2-ene derivative shown in Fig. 38.31 to the isomeric bicyclo[2.2.1]hept-2-ene.

Fig. 38.31. Suprafacial–antarafacial pathway of the [1,3]sigmatropic carbon shift in an allyl system.

38.8 *Fluxional Molecules*

The sigmatropic rearrangements illustrated by the examples that have been discussed all require more or less strenuous heating, as high as 300° or more in most cases. Recently, a number of interesting compounds have been prepared that undergo rapid sigmatropic rearrangements even at room temperature. The most celebrated example in this category is a tricyclic bridged ring triene that has acquired the trivial name of "bullvalene" (Fig. 38.32). Bullvalene

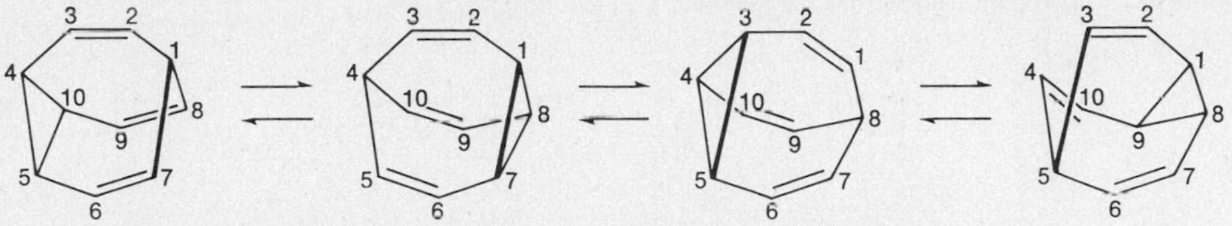

Fig. 38.32. Bullvalene, a fluxional molecule for which a large number of degenerate structures can be written.

cannot be represented by a single structure at room temperature, for its nmr spectrum indicates that it is a mixture of many species, all of which are in rapid equilibrium with each other. The room temperature nmr spectrum of bullvalene displays a single sharp resonance line, indicating that *all* the hydrogens of the molecule are identical *on a time-averaged basis.* This demands that a very rapid interconversion between bullvalene molecules occurs, as Fig. 38.32 illustrates, in which the carbon atoms have been numbered to make it possible to follow their positions through a few of the approximately 10^6 possible isomeric bullvalene structures. Compounds such as bullvalene that undergo very rapid sigmatropic rearrangement are called "fluxional" molecules, several more examples of which are illustrated in Fig. 38.33.

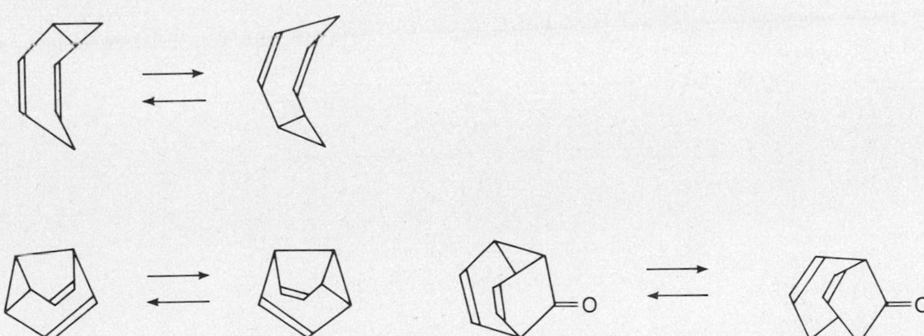

Fig. 38.33. Structures of fluxional molecules.

REFERENCES

LEHR, R. E., and A. P. MARCHAND, *Orbital Symmetry: A Problem-Solving Approach*, Academic Press, Inc., New York, 1972.

WOODWARD, R. B., and R. HOFFMANN, *The Conservation of Orbital Symmetry*, Academic Press, Inc., New York, 1971.

ERMANN, H., The Möbius–Hückel Concept in Organic Chemistry. Application to anic Molecules and Reactions, *Accounts of Chemical Research*, 4, 272 (1971).

PROBLEMS

1. Predict the structures of the products in the following electrocyclic rearrangements:

(a) 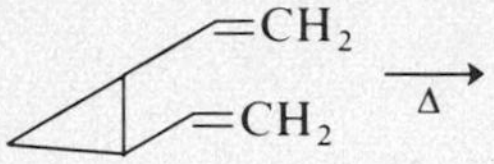 $=CH_2$, $=CH_2$ $\xrightarrow{\Delta}$

(b) 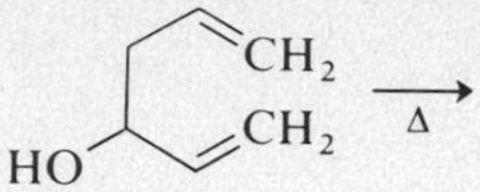 HO, CH_2, CH_2 $\xrightarrow{\Delta}$

(c) 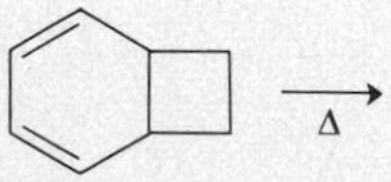 $\xrightarrow{\Delta}$

(d) 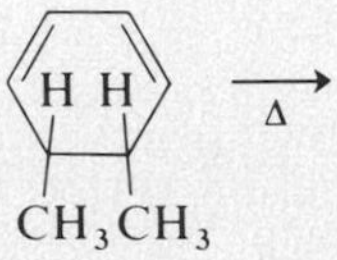 H H, CH_3 CH_3 $\xrightarrow{\Delta}$

(e) 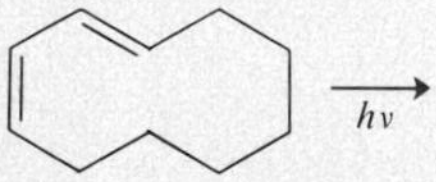 $\xrightarrow{h\nu}$

(f) 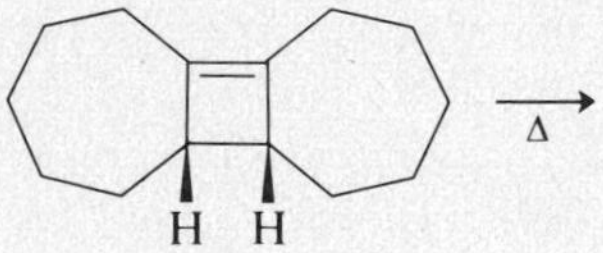 H H $\xrightarrow{\Delta}$

(g) 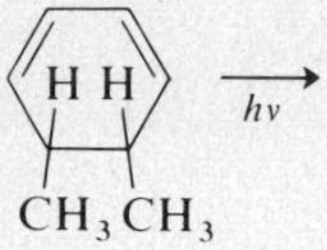 H H, CH_3 CH_3 $\xrightarrow{h\nu}$

(h) CH_3, H, H, CH_3 $\xrightarrow{\Delta}$

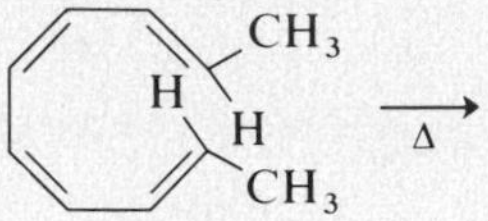

2. Using the Hückel–Möbius concept draw a basis set for the transition state molecular orbitals for the thermally allowed rearrangement processes for parts (a)–(d) in Problem 1.

3. Predict the structures of the products in the following sigmatropic rearrangements:

(a) 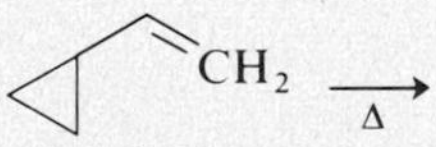 CH_2 $\xrightarrow{\Delta}$

(b) 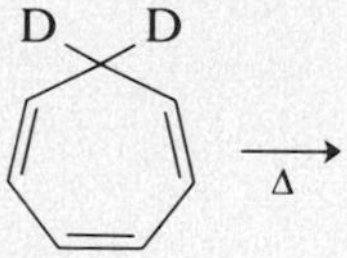 D D $\xrightarrow{\Delta}$

(c) 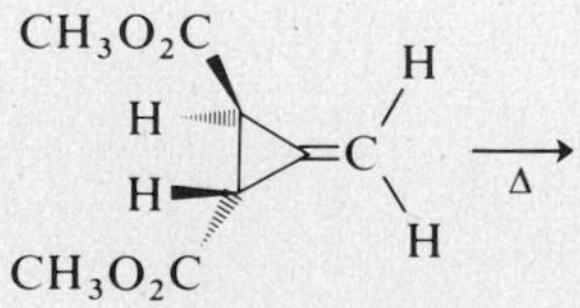 CH_3O_2C, H, H, CH_3O_2C, =C, H, H $\xrightarrow{\Delta}$

(d) H D, H $\xrightarrow{\Delta}$

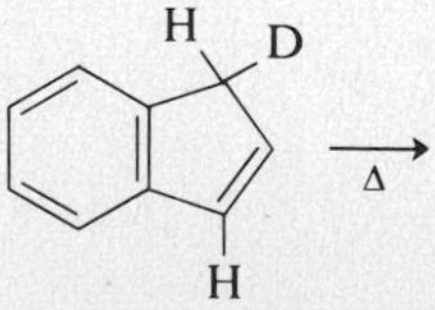

(e) 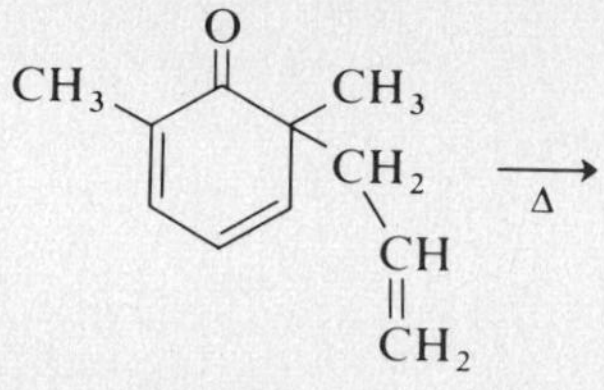 O, CH_3, CH_3, CH_2, CH, CH_2 $\xrightarrow{\Delta}$

4. The reaction of maleic anhydride with cycloheptatriene affords the product shown. Provide a reasonable explanation for the formation of this product.

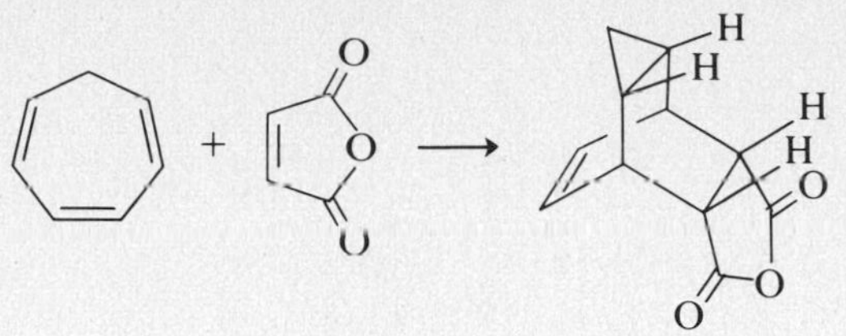

5. When compound A is heated at 154°, it is transformed to compounds C and D via a cyclic tetraene (B). On heating, C is transformed into D. Propose a structure for B, and describe the processes involved in the reactions A ⟶ B, B ⟶ C, and C ⟶ D.

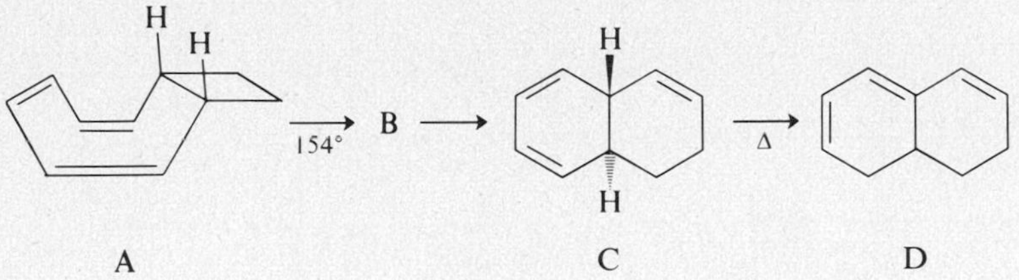

6. 1,2-Nucleophilic rearrangements to electron deficient centers, as in the Wagner–Meerwein rearrangement of carbonium ions, occur with retention of configuration of the migrating group. In contrast, carbanions do not undergo 1,2-migrations. Rationalize these observations on the basis of the Hückel–Möbius concept. (Note: Both 1,2-nucleophilic and 1,2-electrophilic rearrangements are [1,2]sigmatropic processes.)

7. On heating, 1,4-cyclohexadiene undergoes loss of hydrogen via a concerted process to produce benzene. Although 1,3-cyclohexadiene similarly looses hydrogen on heating, the reaction does not proceed via a concerted process. Interpret both of these observations in terms of the Hückel–Möbius concept.

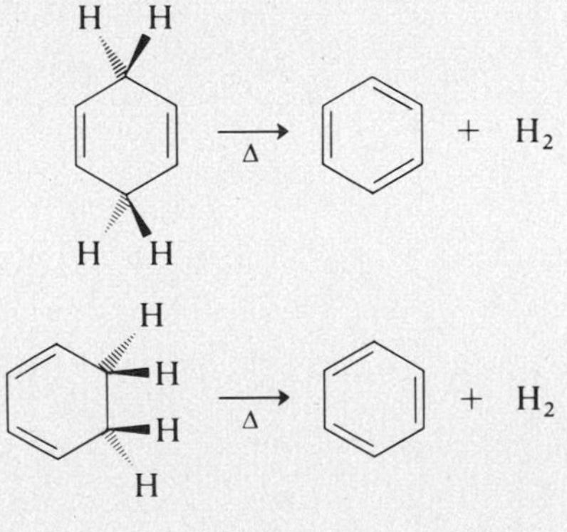

8. The following $C_{12}H_{12}$ compounds have the potential of fragmenting to two molecules of benzene on heating. Which of the compounds are capable of doing so via a concerted process?

9. Bicyclo[2.2.2]octatriene reacts with dimethyl acetylenedicarboxylate to produce the indicated product. Analyze this cycloaddition reaction in terms of the Hückel–Möbius concept. Using the terminology of Woodward and Hoffmann, classify this cycloaddition reaction.

$CH_3O_2CC{\equiv}CCO_2CH_3$

CH_3O_2C C=C CO_2CH_3

10. The following reaction occurs photochemically:

CN, CN, H, CH_2 —$h\nu$→ NC CN, H_2C

Assuming retention of stereochemistry of the migrating group, predict the stereochemistry of the product that would be formed from compound I on irradiation.

CN, CN, H, H, C, D

I

Chapter 39

Organic synthesis

The synthesis of substances occurring in Nature, perhaps in greater measure than activities in any other area of organic chemistry, provides a measure of the condition and powers of the science.

R. B. Woodward
in *Perspectives in Organic Chemistry* (1956)

A deeply ingrained trait of man is his urge to build things, to fashion new objects from the materials available in his environment. The record of civilization, in fact, is the story of man's ever increasing capacity for shaping and refashioning his world by means of the tools that he has invented. Certainly, one of the most interesting of these tools is the organic reaction, and one of the most powerful devices for reshaping our present world comes from the application of this tool to organic synthesis. Even toward the end of the nineteenth century, the impact of organic synthesis on society was being felt as the result of a burgeoning synthetic dye industry and an embryonic drug industry. Today, the products of organic syntheses have become so integral a part of our economy and our daily lives that our awareness of their presence has passed into our subconscious. But, organic syntheses *are* with us on a tremendous scale; and they promise to remain with us so long as our society retains its technological base. Indeed, the impact of organic synthesis will almost surely increase as our knowledge of materials increases and as our insight into biological mechanisms becomes more advanced, expanding our ability to construct new fabrics, new building materials, new agricultural chemicals, and mind-altering and body-altering compounds. The enormous societal and economic consequences inherent in this potentiality represent a pivotal point in the history of civilization.

Organic synthesis is carried out not only on the massive scale of the industrial plant but also on the more modest scale of the laboratory bench. Thousands of chemists throughout the world are devoting their lives to the laboratory study of organic synthesis. Some are employed by industry to explore reactions that will eventually become large-scale industrial processes; others are employed by industry to seek ways to improve reactions that already are being used in the production of salable items. Still others, usually in university or government laboratories, are less immediately involved with the commercial aspects of organic synthesis but, for a variety of reasons, are looking into old ways for making new compounds, new ways for making old compounds, or new ways for making new compounds (see comments on p. 477). Whether working for fortune, fame, or fundamental truths, however, the chemical constraints are the same: one must be concerned with (a) the availability of starting materials, (b) the availability of organic reactions, and (c) the necessity of matching available starting materials with available organic reactions in the most effective manner. This is the crux of the problem of organic synthesis.

39.1 Available Starting Materials

In the beginning, the starting materials in the "primeval broth" consisted of the very simplest molecules, such as methane, water, ammonia, carbon dioxide, hydrogen, and nitrogen. As syntheses occurred under prebiotic conditions, additional compounds accumulated and provided the starting materials for other syntheses. Eventually, the "primeval broth" must have been a veritable storehouse of chemical entities, including a variety of quite complex structures. In similar fashion, the range of starting materials available to the laboratory chemist has grown from very humble beginnings to a list today of over 20,000 compounds that can be obtained from various chemical supply houses at prices ranging from a few cents a pound to many dollars a milligram. To become acquainted with this range of available starting materials, the reader is urged to obtain a catalog from one or more of the suppliers listed:

Aldrich Chemical Company, Inc.
2371 North 30th St.
Milwaukee, Wisconsin 53210

Distillation Products Industries
Eastman Organic Chemicals Department
Rochester 3, New York

Matheson Coleman and Bell
Post Office Box 85
East Rutherford, New Jersey

A perusal of these catalogs will not only instruct the reader in the available starting materials but also in the facts of practical nomenclature. Unfortunately, a given compound frequently parades under a variety of names, some trivial and some systematic. Disagreeable though this may be, it is the situation that exists in the real world of chemistry. It is necessary, therefore, to accept this reality and to be able to find a compound in the catalog regardless of how it is named.

If a compound cannot be purchased, it obviously is not a generally available starting material. Even if it is listed in the catalog of one of the chemical supply houses, however, it is not necessarily an appropriate starting material; in certain situations, particularly in industrial operations, economics enters the picture. In the majority of commercial syntheses, the starting materials must be compounds that can be purchased cheaply (usually ranging from a few cents a pound to less than a dollar a pound), and this severely limits the range of possibilities; it places a special constraint—a financial constraint—on the availability of starting materials. Although laboratory syntheses usually are much less restricted with respect to the cost of starting materials, a limit may exist here also. For example, in a particular situation it might be specified that only those compounds costing less than \$10/100 g be considered as available starting materials (or \$5/100 g or \$15/100 g, *etc*). Because such constraints may exist and may vary because of price fluctuations, it is impossible to give any hard and fast guidelines as to what compounds are and are not available starting materials; the only sure guides are the catalogs of the chemical supply houses, such as those listed on p. 1168. Nevertheless, the following generalities are worth keeping in mind:

1. Small molecules are generally more readily available than large molecules. Although there are some exceptions, straight-chain alkanes, alkenes, alkyl bromides, alkyl chlorides, amines, alcohols, aldehydes, ketones, and carboxylic acids containing six or fewer carbons are inexpensive compounds (Table 39.1).

Table 39.1 READILY AVAILABLE ALKANES, ALKENES, ALKYNES, ALKYL HALIDES, ALCOHOLS, AMINES, ALDEHYDES, KETONES, AND CARBOXYLIC ACIDS

Class	Structure of Carbon Framework	Number of Carbon Atoms in Compound (n)
Alkanes	Straight chain $H(CH_2)_nH$	1–10, 12, 14, 16, 18, 20
	Branched chain	4–8
	Cyclic $(CH_2)_n$	3, 5, 6, 8
Alkenes	Straight chain $H(CH_2)_{n-2}CH{=}CH_2$	2–10
	Branched chain and/or nonterminal C=C	4, 5
	Cyclic $(CH_2)_{n-2}$ CH=CH	6, 8, 12
Alkynes	Straight chain $H(CH_2)_{n-2}C{\equiv}CH$	2
Alkyl halides	Straight chain $H(CH_2)_nX$	1–4 when X = Cl
		1–8, 10 when X = Br
		1, 2 when X = I
	Branched chain	3, 4 when X = Cl
		3–5 when X = Br
		3 when X = I
	Cyclic $(CH_2)_{n-1}CHX$	5, 6 when X = Br
Alcohols	Straight chain $H(CH_2)_nOH$	1–12, 14, 16, 18
	Branched chain	3–5, 7, 8
	Cyclic $(CH_2)_{n-1}CHOH$	5–7, 12
Amines	Straight chain $H(CH_2)_nNH_2$	1–5, 8, 12
	Branched chain	3–5
	Cyclic $(CH_2)_{n-1}$ $CHNH_2$	6
	Heterocyclic $(CH_2)_n$ NH	4, 5

Table 39.1—*cont.*

Class	Structure of Carbon Framework	Number of Carbon Atoms in Compound (n)
Aldehydes	Straight chain $H(CH_2)_{n-1}CHO$	1–8, 10
	Branched chain	4–6, 8
Ketones	Straight chain $H(CH_2)_{n-1}\overset{O}{\overset{\|}{C}}(CH_2)_mH$	3–8 for (n + m)
	Branched chain	5–7 for (n + m)
	Cyclic $(CH_2)_{n-1}$ C=O	5, 6, 12
Carboxylic acids	Straight chain $H(CH_2)_{n-1}CO_2H$	1–8, 10, 12, 14, 16, 18
	Branched chain	4–6
	Cyclic $(CH_2)_{n-2}$ $CHCO_2H$	6

2. Straight-chain molecules are generally more readily available than branched molecules. Although molecules containing six or fewer carbon atoms and possessing the *iso* structure are often relatively inexpensive (Table 39.1), molecules with other modes of branching are less likely to be available.
3. If the carboxylic acid is commercially available most of the corresponding carboxylic acid derivatives are available, although at a somewhat greater or even very much greater cost.
4. A limited number of monofunctional alicyclic compounds are available (Table 39.1), mostly those containing 5-membered, 6-membered, 8-membered, and 12-membered rings.
5. A wide variety of polyfunctional compounds in the aliphatic and alicyclic series are readily available (Tables 39.2 and 39.3).

Table 39.2 Readily available dienes, dihaloalkanes, diols, diamines, dialdehydes, diketones, and dicarboxylic acids

Class	Structure of Carbon Framework	Number of Carbon Atoms in Compound (n)
Dienes	$H_2C{=}CHCH{=}CH{-}R$ (R = H and CH_3)	
	Cyclopentadiene	
	1,5-Cyclooctadiene	
	trans,trans,cis-1,5,9-Cyclododecatriene	
Dihaloalkanes	$Cl(CH_2)_nCl$ and $Br(CH_2)_nBr$	1–6
Diols	$HO(CH_2)_nOH$	2–4, 6
Diamines	$H_2N(CH_2)_nNH_2$	2–4, 6
Dialdehydes	$(CH_2)_{n-1}(CHO)_2$	2, 5
Diketones	$CH_3CO(CH_2)_{n-4}COCH_3$	4, 5, 6
Dicarboxylic acids	$(CH_2)_{n-2}(CO_2H)_2$	2–6, 8, 10

Table 39.3 READILY AVAILABLE DIFUNCTIONAL COMPOUNDS WITH MIXED FUNCTIONS

Class	Structure
Alkenes containing hetero atoms	$H_2C{=}CH{-}R$, where R = $-CH{=}CH_2$, $-CH_2Cl$, $-CH_2OH$, $-CH_2NH_2$, $-CHO$, $-COCH_3$, $CH_2CH_2COCH_3$, $-CO_2H$, $-CN$, $-CH_2CN$ $CH_3CH{=}CH{-}R$, where R = $-CH{=}CH_2$, $-CH_2Cl$, $-CH_2OH$, $CHO{-}CO_2H$
Alkynes containing hetero atoms	$HC{\equiv}C{-}R$, where R = $-CH{=}CH_2$, $-CH_2Cl$, $-CH_2OH$
Dioxygen compounds	CH_3COCH_2OH (acetol), CH_3COCHO (pyruvaldehyde), CH_3COCO_2H (pyruvic acid), $CH_3COCH_2CH_2CO_2H$ (levulinic acid), HO_2CCHO (glyoxylic acid),
Nitrogen–oxygen compounds	$NH_2CH_2CH_2OH$, $NH(CH_2CH_2OH)_2$, $N(CH_2CH_2OH)_3$, $H_2N(CH_2)_nCO_2H$, where n = 1–3, 5, 10
Halogen–oxygen compounds	$ClCH_2CH_2OH$, F_3CCO_2H, $ClCH_2CO_2H$, Cl_2CHCO_2H, Cl_3CCO_2H

6. A wide variety of monosubstituted benzenes are available (Table 39.4), as well as quite a number of di- and trisubstituted benzenes (see catalogs listed on p. 1168). Naphthalene and a limited number of substituted naphthalenes are available (Table 39.4), as well as the parent hydrocarbons for some of the higher polycyclic aromatics, including anthracene and phenanthrene.

Table 39.4 READILY AVAILABLE MONOSUBSTITUTED AROMATIC COMPOUNDS

Monosubstituted benzenes:

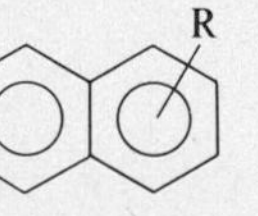

R = $-CH_3$, $-CH_2CH_3$, $-CH_2CH_2CH_3$, $-CH(CH_3)_2$, $-CH_2CH(CH_3)_2$, $-C(CH_3)_3$, $-CH{=}CH_2$, $-F$, $-Cl$, $-Br$, $-I$, $-NH_2$, $-NHCH_3$, $-N(CH_3)_2$, $-NO_2$, $-OH$, $-OCH_3$, $-SH$, $-CH_2OH$, $-CH_2Cl$, $-CH_2Br$, $-CHO$, $-COCH_3$, $-COCH_2CH_3$, $-COCH_2CH_2CH_3$, $-COCH(CH_3)_2$, $-CO_2H$, $-CH_2CO_2H$, $-CH_2CH_2CO_2H$, $-CH{=}CHCO_2H$, $-CN$, $-COCl$, $-CONH_2$, $-SO_3H$

Monosubstituted naphthalenes:

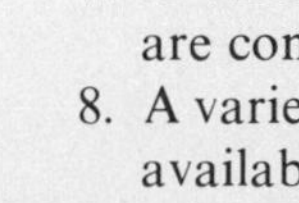

1-Substituted naphthalenes: R = $-CH_3$, $-Br$, $-Cl$, $-NH_2$, $-NO_2$, $-OH$, $-CHO$, $-CO_2H$, $-CH_2CO_2H$, $-CN$

2-Substituted naphthalenes: R = $-NH_2$, $-OH$, $-OCH_3$

7. A variety of heterocyclic molecules containing 5- and 6-membered rings are commercially available (Table 39.5).
8. A variety of special reagents for certain types of reactions are commercially available, although they are often rather expensive.

Table 39.5 READILY AVAILABLE MONOSUBSTITUTED HETEROCYCLIC COMPOUNDS

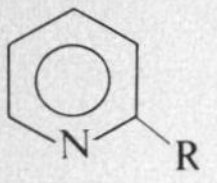 (pyridine ring, 2-R)	2-Pyridyl derivatives:	R = $-CH_3$, $-CH_2CH_3$, $-CH{=}CH_2$, $-CH_2C_6H_5$, $-Cl$, $-Br$, $-NH_2$, $-OH$, $-CHO$, $-CO_2H$, $-CN$, $-CH_2CH_2CH_2OH$
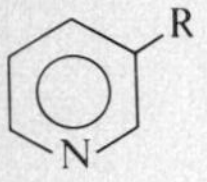 (pyridine ring, 3-R)	3-Pyridyl derivatives:	R = $-CH_3$, $-NH_2$, $-OH$, $-CHO$, $-CO_2H$, $-CONH_2$, $-CN$
 (pyridine ring, 4-R)	4-Pyridyl derivatives:	R = $-CH_3$, $-CH_2CH_3$, $-CH{=}CH_2$, $-CH_2C_6H_5$, $-NH_2$, $-OH$, $-CHO$, $-CO_2H$, $-CN$, $-CH_2CH_2CH_2OH$
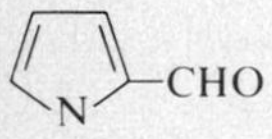 (pyrrole ring)–CHO	2-Pyrrolaldehyde	
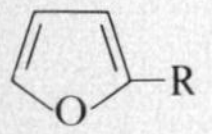 (furan ring)–R	2-Furyl derivatives:	R = $-CHO$, $-CO_2H$, $-CH_2OH$, $-CH_2NH_2$
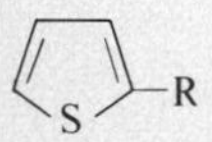 (thiophene ring)–R	2-Thienyl derivatives:	R = $-CHO$, $-CO_2H$, $-Br$, $-Cl$

The list of available starting materials changes with time. Generally, it increases, although there are numerous instances of compounds being on the market for a period of time but subsequently disappearing. Obviously, it is the availability of materials at the moment that determines the course of the synthesis. For example, in the late 1950's, 1-(3′,4′,5′-trimethoxyphenyl)-3-(4′-methoxy-

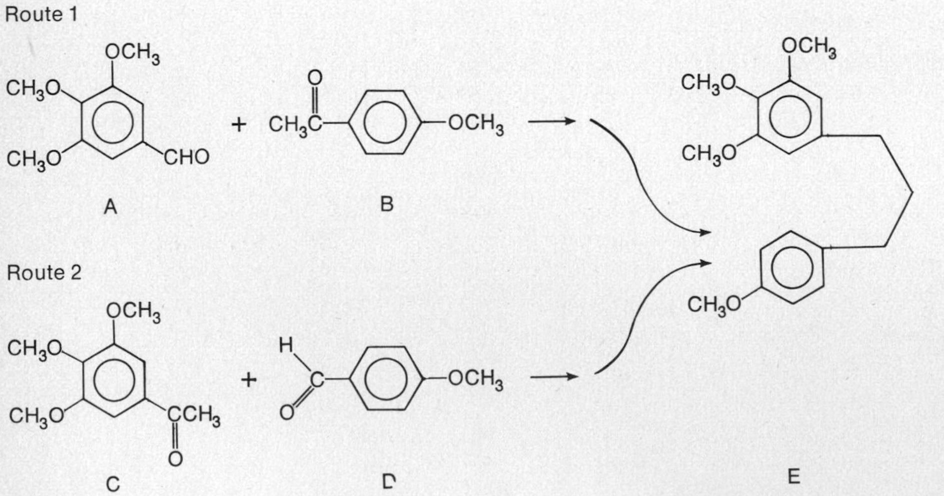

Route 1

A: 3,4,5-$(CH_3O)_3C_6H_2CHO$ + B: $CH_3C(=O)C_6H_4OCH_3$ ⟶

Route 2

C: 3,4,5-$(CH_3O)_3C_6H_2C(=O)CH_3$ + D: $HC(=O)C_6H_4OCH_3$ ⟶

E: 1-(3′,4′,5′-Trimethoxyphenyl)-3-(4′-methoxyphenyl)-propane

Fig. 39.1. Alternative synthesis routes.

phenyl)-propane (compound **E**) in Fig. 39.1 was required as an intermediate in a synthesis sequence, the next step of which was the formylation reaction illustrated on p. 835. A general scheme envisaged for the synthesis involves an aldol-like condensation followed by reduction:

$$ArCHO + CH_3\overset{O}{\overset{\|}{C}}Ar' \longrightarrow ArCH{=}CH\overset{O}{\overset{\|}{C}}Ar' \xrightarrow[Pt]{H_2} ArCH_2CH_2CH_2Ar'$$

Because neither the double bond nor the carbonyl group of the aldol product is retained in the final product, two combinations of ArCHO and $Ar'COCH_3$ are possible (Fig. 39.1). When the synthesis was first carried out in 1957, the commercially available materials were compounds **B** and **D**. Also available was 3,4,5-trihydroxybenzoic acid, and since its conversion to the corresponding trimethoxy compound with dimethyl sulfate (see Fig. 11.6) followed by Rosenmund reduction of the acid chloride (see Fig. 39.1) employs chemistry that had been well established at that time, route 1 was chosen for the synthesis of compound **E**:

$$3,4,5\text{-}(HO)_3C_6H_2CO_2H \xrightarrow[NaOH]{(CH_3)_2SO_4} 3,4,5\text{-}(CH_3O)_3C_6H_2CO_2H \xrightarrow{SOCl_2} 3,4,5\text{-}(CH_3O)_3C_6H_2COCl \xrightarrow[\text{Poisoned Pd/BaSO}_4\text{ catalyst}]{H_2} A$$

Gallic acid

Ten years later, when an additional supply of compound **E** was required for another project, compound **C** had become commercially available. Clearly, the better synthesis now was route 2, for neither starting material requires laboratory preparation in this case.

39.2 Available Organic Reactions

The chemical operations involved in an organic synthesis can be broadly divided into two categories: (a) those involving the building up of the carbon skeleton, and (b) those involving the introduction of the functional groups. Although the two types of operations are interdependent, it is convenient to make a distinction as a means for classifying synthesis reactions. Partial summaries have been made for the reactions discussed in Part I (see Tables 18.7–8) and summaries of the reactions of various functional groups have been presented throughout Part II. A good exercise is the construction of summarizing tables, comparable to Table 18.6, for the reactions that have been discussed in Part II. This is a very effective way for studying reactions and for fixing them in mind in such a way that they become useful tools for devising organic synthesis. As an aid in this undertaking, the following general outline is suggested.

I. Methods for creating and altering carbon–carbon bonds
  A. Substitution reactions
    1. Nucleophilic substitution (S_N1, S_N2) with carbon nucleophiles
    2. Electrophilic substitution with carbon electrophiles (Friedel–Crafts alkylation and acylation)
    3. Carbonyl addition–elimination
      a. Organometallic reagents
      b. Active methylene compounds (Claisen reaction)
  B. Addition reactions
    1. Alkene addition
      a. Electrophilic addition
      b. Carbene addition
      c. Cycloaddition
    2. Carbonyl addition
      a. Direct addition (1,2 addition)
        (1) Strongly polarized carbon compounds
          (a) Wittig reaction
          (b) Diazoalkane reaction
          (c) Sulfur ylide reaction
        (2) Organometallic reagents
        (3) Hydrogen cyanide
        (4) Active methylene compounds (aldol-type reaction)
      b. Conjugate addition
        (1) Strongly polarized carbon compounds
        (2) Organometallic reagents
        (3) Active methylene compounds (Michael reaction)
  C. Elimination reactions (C—C → C=C → C≡C)
  D. Rearrangement reactions

II. Methods for interchange of functional groups
  A. Substitution reactions
    1. Nucleophilic substitution (S_N1, S_N2) with hetero atom nucleophiles
    2. Electrophilic substitution (aromatic substitution with hetero atom electrophiles)
    3. Free radical substitution with hetero atom radicals
    4. Carbonyl addition-elimination reactions—"olysis" reactions
  B. Addition reactions
    1. Electrophilic addition to alkenes and alkynes
    2. Nucleophilic addition to carbonyl functions
    3. Reduction reactions
  C. Elimination reactions
    1. C—C → C=C → C≡C (also listed in category I)
    2. Oxidation processes
  D. Carbon–carbon cleavage reactions
    1. Oxidative cleavage of multiple bonds
    2. Decarboxylation reactions

39.3 Matching Available Starting Materials with Available Organic Reactions

The proper matching of available starting materials with available organic reactions is the central problem of organic synthesis. Its effective solution requires a thorough familiarity with the sources of supply of chemical compounds and a solid working knowledge of organic reactions. The first of these items is relatively easily mastered, although to see all the synthesis possibilities in a given starting material requires a lively imagination. It is the second of these items that is by far the more demanding; in fact, it is rarely, if ever, completely mastered. There are so many organic reactions and so many ways of combining them in any multistep synthesis sequence that the possible combinations are virtually without limit. Designing and carrying through an extended organic synthesis is much like playing a game of chess. As the game progresses beyond the first few moves, which often, although not by requirement, follow a common pattern, it takes on a character that is strongly dependent on the player; it acquires a distinctly personal flavor, which may be helter-skelter in the case of the neophyte who lacks a thorough knowledge of the game or highly organized and well thought out many moves in advance by the expert who knows all the moves and appreciates their long-range consequences. The finesse with which an organic synthesis is designed and executed is completely dependent on the degree to which its architect has mastered organic reactions, the "moves" of the game. A thorough knowledge of the scope, limitations, and mechanisms of organic reactions is the essential ingredient, the *sine qua non*, the "open sesame" to organic synthesis.

The fact that organic synthesis, like chess, is an open-ended endeavor which defies complete mastery should deter no one from trying his hand at it. Like chess, it is an enjoyable game at all levels of accomplishment, becoming more so as one's ability to play it improves. In the remainder of this section, we give a few pointers as to how this improvement might be achieved.

39.3a. RETROSYNTHESIS: WORKING A SYNTHESIS BACKWARD. The necessity of working a synthesis backward is discussed in Chapter 18 (see p. 481). Only in this way can the connection between available starting materials and final product, via available organic reactions, be made. To amplify the examples given in Chapter 18, let us consider the means by which 6-methylbicyclo[4.4.0]decan-3-ol (compound 1) can be constructed.

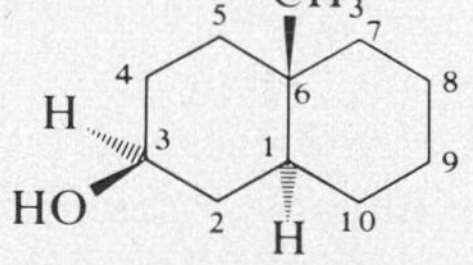

Compound 1
6-Methylbicyclo[4.4.0]decan-3-ol

Let us first concentrate our attention on the hydroxyl function and consider how it might be introduced. One possibility [method (a)] is first to introduce a halogen at C-3 and then to carry out a nucleophilic displacement reaction (Fig. 39.2). Selective introduction of halogen at the 3 position, however, is not feasible, for there is nothing to activate C-3 over other positions in the molecule.

Another possibility [method (b)] is to effect an intramolecular addition of a carbanion to an aldehyde function. The appropriate precursor compound, however, poses a difficult synthesis problem. A third possibility [method (c)] involves the reduction of the corresponding ketone, and this appears to be the best choice in the present instance.

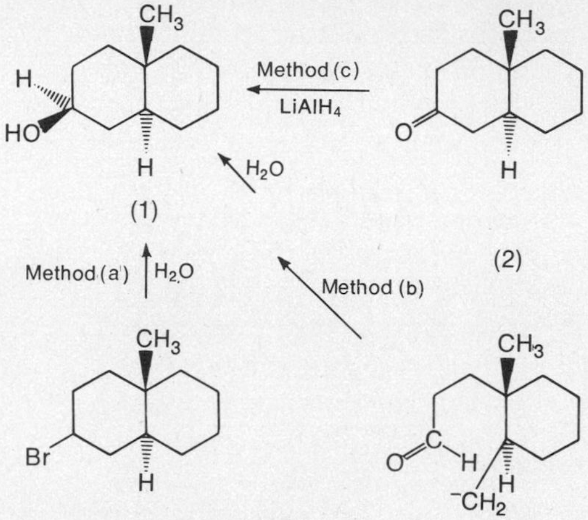

Fig. 39.2. Methods for introduction of hydroxyl functions at C-3.

For the next step in the retrosynthesis, we must consider the means for constructing the alicyclic framework of the molecule; how can the bicyclo[4.4.0]-decane ring system be put together to give the ketone **2**? Among the procedures that should come to mind for constructing six-membered rings are (a) the Dieckmann cyclization, (b) the reduction of aromatic rings (*e.g.*, Birch reduction), (c) the Diels–Alder reaction, and (d) the Robinson annelation procedure (see Fig. 30.25), as sketched in Fig. 39.3. Although the Dieckmann cyclization of **3** represents an entirely feasible approach, the preparation of the requisite diester poses a rather difficult synthesis problem. The difficulty with the Birch reduction of **4** is that it permits no easy way for introducing the methyl group at C-6. The complications with the Diels–Alder reaction between **5** and **6** are (a) the condensation can take place to give the 3-methoxy compound (as shown in Fig. 39.3) and/or the 4-methoxy compound, and (b) additional steps are required to remove the keto group at C-10 (using the numbering system applied to 1) which is necessary to make **6** a sufficiently reactive dienophile. A possible difficulty with the Robinson annelation procedure employing **7** and **8** is that the condensation might take place at the 6 position rather than at the 2 position of 2-methylcyclohexanone. It turns out, however, that the reaction occurs almost exclusively at the 2 position to produce 6-methylbicyclo[4.4.0]dec-1(2)-ene-3-one (number **9** in Fig. 39.3). Thus, the Robinson annelation procedure appears to be the method of choice among the four shown in Fig. 39.3. At this stage, then, we have worked the synthesis backward to the point where available starting materials are at hand (**7** + **8** → **9**; **9** → **2**; **2** → **1**). Acetone, formaldehyde, and dimethyl amine for the synthesis of the Mannich base **7** are directly available, and 2-methylcyclohexa-

Dieckmann cyclization:

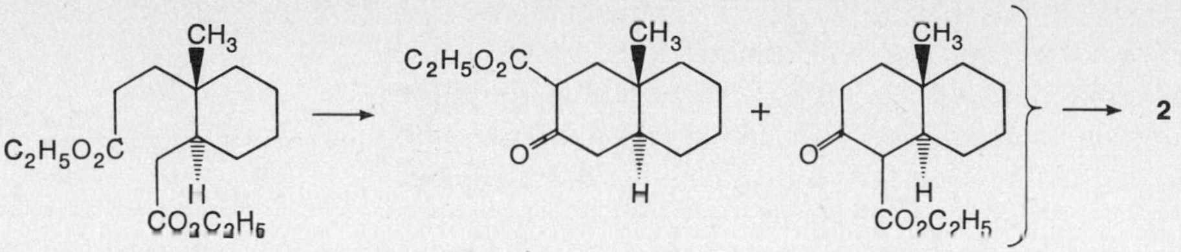

Birch reduction:

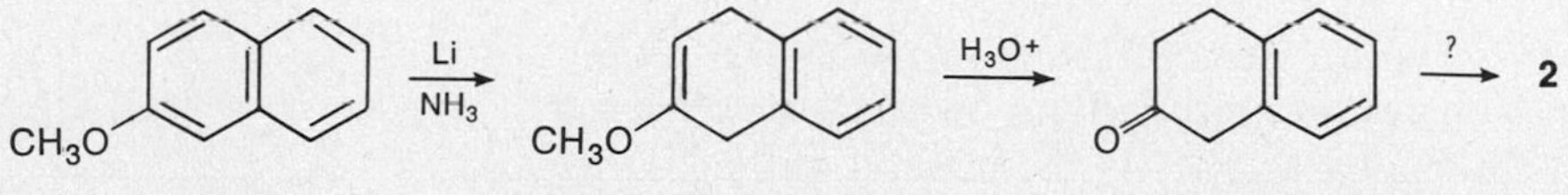

Diels-Alder reaction:

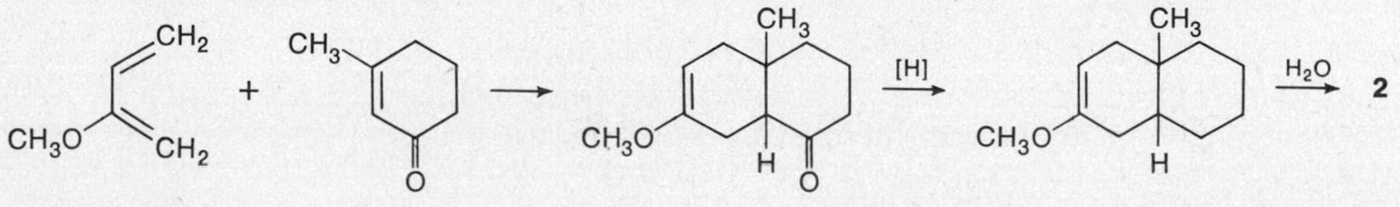

Robinson-Mannich base synthesis:

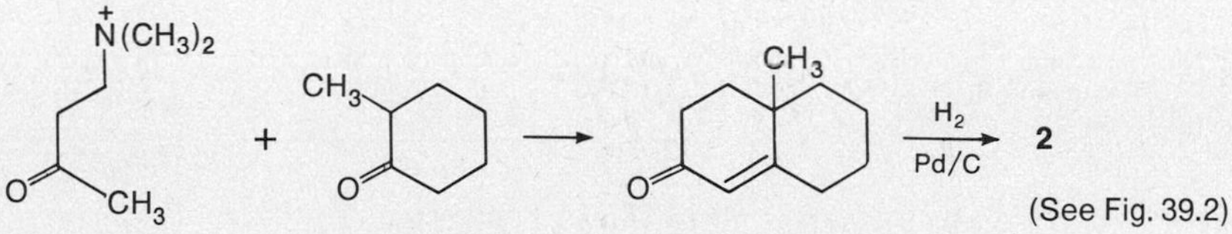

(See Fig. 39.2)

Fig. 39.3. Methods for constructing 6-methylbicyclo[4.4.0]decan-3-one.

none (8) can be made by reduction of 2-methylphenol followed by oxidation:

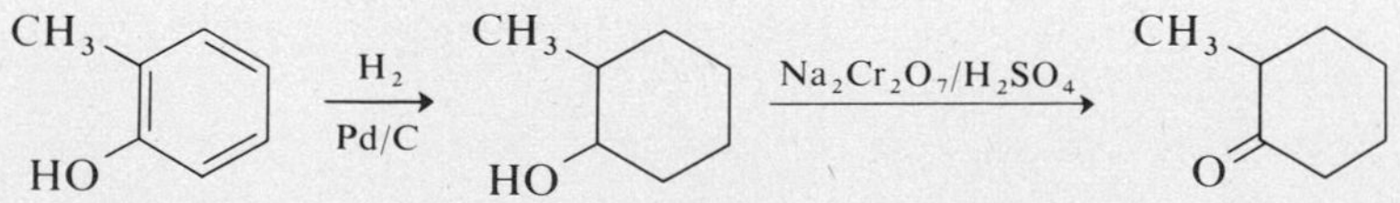

In working the synthesis of compound **1** backward, we have, of course, presented the problem in the simplest, most straightforward fashion. In real life situations, the analysis of a synthesis pathway will not always be so clearly and smoothly revealed as that for compound **1** appears to have been. Frequently, one will be led up blind alleys from which retreat will eventually be necessary. This is part of the game, though, and as skill at playing the game improves, the number of blind alleys that will be followed will diminish. In outlining the synthesis of **1** we have also simplified the discussion by saying nothing about the stereochemistry of the synthesis. Stereochemistry is frequently a matter that must be dealt with, though, and it may even turn out to be a greater challenge than the construction of the framework of the molecule itself. Compound **1** contains three chiral centers and can, therefore, exist in four racemic forms. Although a pair of diastereomers can be produced in the conversion of the unsaturated ketone **9** to the saturated ketone **2**, this reaction takes place in a highly stereoselective fashion to give predominantly the *trans* isomer. Starting with *trans* compound **2**, we can produce another pair of diastereoisomers when the keto group is reduced to a hydroxyl group. In contrast, this reaction is not highly stereoselective, and a mixture of epimers is produced that requires separation by physical means (*e.g.*, column chromatography). By the method outlined, then, two of the four possible diastereomers of 6-methylbicyclo[4.4.0]decan-3-ol are easily prepared. If the other two diastereomers are desired, a different approach to the synthesis would have to be developed because of the high stereoselectivity that favors the *trans* isomer in the reduction of compound **9** to compound **2**.

39.3b. Introduction of "handles": clues to retrosynthesis. Let us now suppose that our synthesis objective is the hydrocarbon 2,2,6-trimethylbicyclo[4.4.0]decane (compound **10**). One of the difficulties in this case arises

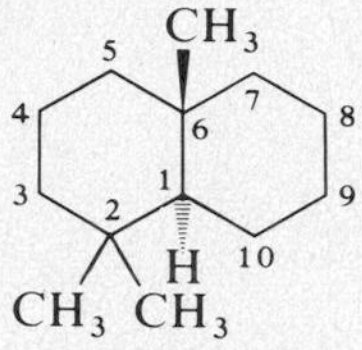

Compound 10
2,2,6-Trimethylbicyclo[4.4.0]decane

from the lack of a functional group on which to focus our attention in devising a retrosynthesis. A technique that is often helpful in such a situation is to introduce, in one's mind, a functional group, into the molecule—a "handle"—in the hope that this will suggest an appropriate retrosynthetic pathway. In the case of compound **10**, for example, we might introduce a keto function at the various positions around the ring system—*e.g.*,

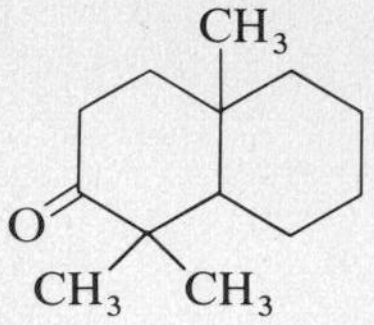

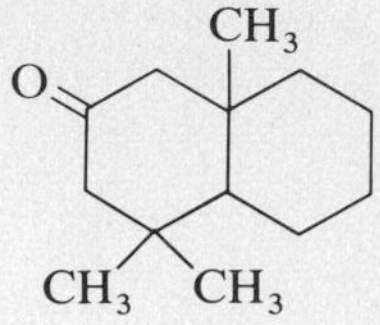

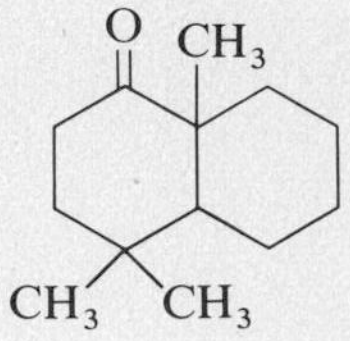

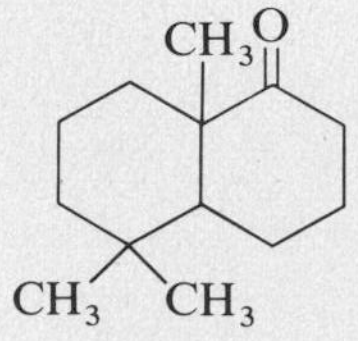

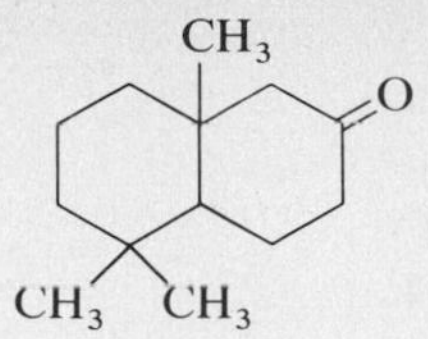

15 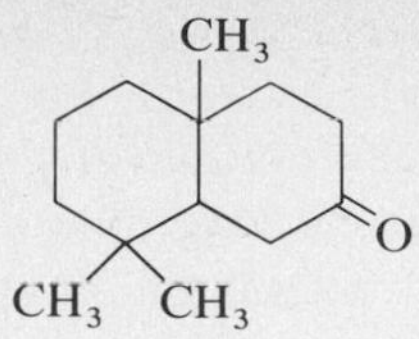 16 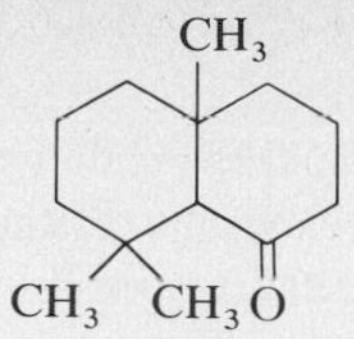 17

Synthesis of any of the compounds **11–17** represents a synthesis of **10**, for removal of the carbonyl function (*i.e.*, by Clemmensen reduction, Wolff–Kishner reduction, Huang–Minlon reduction, or thioketal sequence—see Fig. 29.37) converts all of them to **10**. Among this group of ketones, compound **11** is particularly interesting because it has a carbonyl group adjacent to the methyl groups at C-2. The thought that should be triggered by this fact is the possibility of introducing these methyl groups by an α-alkylation reaction (see Section 30.3).

The mind that is really well-stocked with the tools of organic synthesis will then go beyond this general concept to the more specific recollection of the alkylation reaction pictured in Fig. 30.14, in which bicyclo[4.4.0]dec-1(2)-ene-3-one yields a 2,2-dimethyl compound. This, in fact, is the key step in the synthesis of **10**; starting with ketone **9**, prepared as discussed, 2,2,6-trimethylbicyclo[4.4.0]dec-1(10)-en-3-one (**18**) is produced by the action of methyl iodide and potassium *t*-butoxide. Reduction of the double bond of **18** then yields ketone **11**, from which the desired hydrocarbon **10** can be obtained:

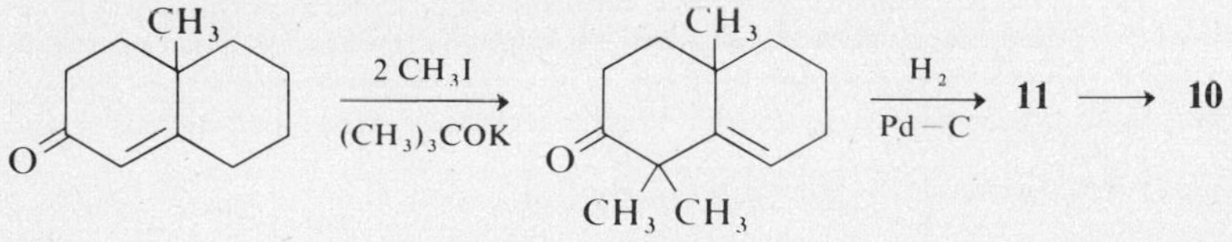

9 $\xrightarrow[(CH_3)_3COK]{2\ CH_3I}$ **18** $\xrightarrow[Pd-C]{H_2}$ **11** ⟶ **10**

Although the synthesis of **10** from ketone **11** probably represents the best synthesis route, it is a useful exercise to consider the other ketones, **12–17**, as possible precursors to **10**. The reader is encouraged to ponder these alternative possibilities to see whether other reasonable syntheses can be devised for compound **10**.

39.3c. Situselectivity: taking advantage of differential reactivity. The concept of situselectivity is implicit in most useful organic reactions—*viz.*, those that occur at specific sites in a molecule and leave other sites unaffected. In some instances, the situselectivity of a process takes on an even more demanding aspect, such as in the synthesis of 7-hydroxy-6-methylbicyclo[4.4.0]decan-3-one **(19)**.

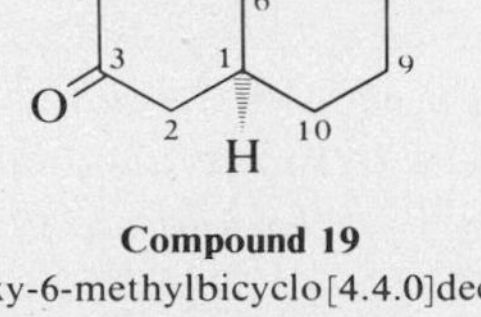

Compound 19
7-Hydroxy-6-methylbicyclo[4.4.0]decan-3-one

Recognizing the close similarity of compound **19** to compound **1**, we realize that for its synthesis we must consider modifications of the method we have discussed for the preparation of **1**. Thus, we conclude that we must use 2-methylcyclohexane-1,3-dione in the condensation reaction with the Mannich base **7** to obtain a compound containing an oxygen function at C-7 (*i.e.*, compound 20 in Fig. 39.4, which is the analog of compound **9** in Fig. 39.3). At this point, we can take

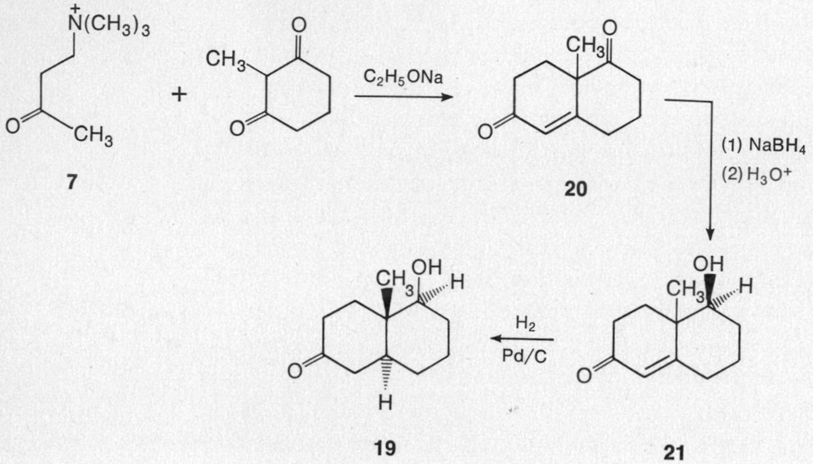

Fig. 39.4. Synthesis of 7-hydroxy-6-methylbicyclo[4.4.0]decan-3-one (**19**).

advantage of situselectivity which results from the greater reactivity of the nonconjugated keto group at C-7 compared with the conjugated keto group at C-3. Thus, reduction with sodium borohydride yields the C-7 hydroxy compound **21**, and catalytic hydrogenation can then be used to remove the double bond and yield the desired compound **19**.

39.3d. THE USE OF THE PROTECTING GROUP (BLOCKING GROUP). A protecting group is a function that can be introduced into a molecule to keep some portion from undergoing alteration while another portion is engaging in reaction; in a subsequent step, the protecting group is then removed, regenerating the original function. A number of examples of the use of protecting groups have already been given in previous sections. The use of a group to "block" the 2 position of bicyclo[4.4.0]decan-2-one and force the methylation reaction to involve the angular C-10 position is portrayed in Fig. 30.8. The use of a ketal function to keep a keto group from becoming involved in a reaction is seen in the reductive acetylation shown in Fig. 29.12. Another example of the use of the ketal function is provided by the synthesis of 3-hydroxy-2,2,6-trimethylbicyclo[4.4.0]decan-7-one (**22**), outlined in Fig. 39.5.

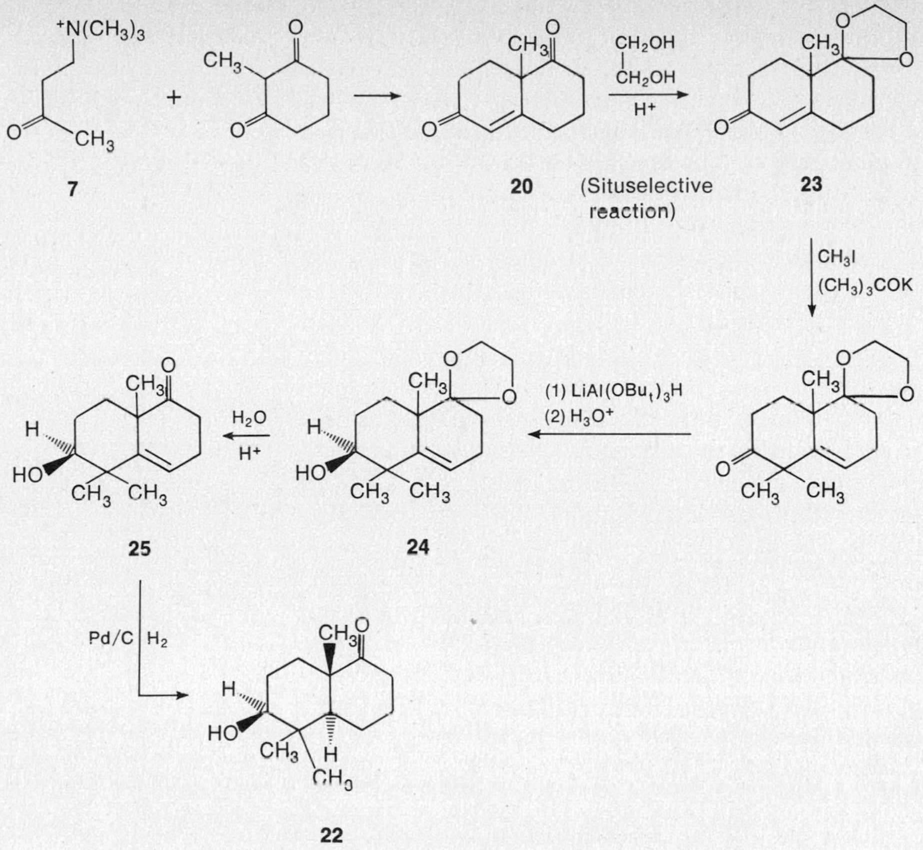

$^{+}N(CH_3)_3$ CH_3 CH_3 CH_2OH CH_2OH H^+ CH_3 CH_3

7 20 (Situselective reaction) 23

CH_3I $(CH_3)_3COK$

(1) $LiAl(OBu_t)_3H$ (2) H_3O^+

H_2O H^+

25 24

Pd/C H_2

22

Fig. 39.5. Synthesis of 3-hydroxy-2,2,6-trimethylbicyclo[4.4.0]decan-7-one (**22**).

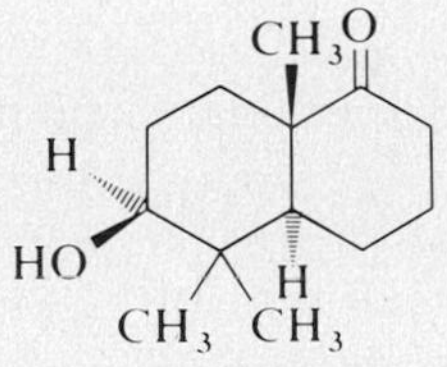

Compound 22
3-Hydroxy-2,2,6-trimethylbicyclo-[4.4.0]decan-7-one

From the discussion in Section 39.3c, we know that the keto group at C-7 is more reactive than the one at C-3 and that ketal formation should, therefore, proceed more rapidly at C-7 than at C-3. Thus, when the diketone **20** is treated with ethylene glycol and a trace of acid catalyst the monoketal **23** is formed. Subsequent methylation and reduction with a mixed metal hydride yields the C-3 alcohol (**24**), from which the C-7 keto group can be regenerated by acid-catalyzed hydrolysis of the ketal; the keto alcohol (**25**) that is produced can then be reduced to the desired compound **22**.

39.4 *Illustrative Examples of the Art of Organic Synthesis*

Despite the great advances that have been made in our understanding of the mechanisms, scope, and limitations of organic reactions, preparative organic chemistry remains, to a considerable extent, an inexact science. Attempts are being made to program information about organic reactions into computers in such a fashion that the computers will be able to assemble this information and provide outlines for suitable synthesis pathways to given compounds. At the moment, though, the ability of computers to do this is extremely limited and organic chemists must still rely on their own wits and ingenuity in the construction of synthesis pathways. Organic synthesis remains an art rather than an exact science in that any particular reaction is usually a little different from its closely related neighbor and has to be treated in an individual manner. Although this is seen by some as an unfortunate aspect of organic chemistry, it does, nevertheless, introduce a quixotic element that protects organic synthesis against the inroads of automation and preserves it as a refuge for the artist-chemist. To give some glimpse of the level of sophistication that chemists have achieved in the art of organic synthesis during the last decade, we shall discuss the following illustrative examples.

39.4a. SYNTHESIS OF LONGIFOLENE, AN ACETOGENIN. Longifolene is a sesquiterpene that can be obtained from the pine tree *Pinus longifolia.* The structure of this unusually convoluted molecule was determined in 1953 by X-ray crystallography, and its total synthesis was accomplished a decade later by E. J. Corey and his coworkers at Harvard University.*

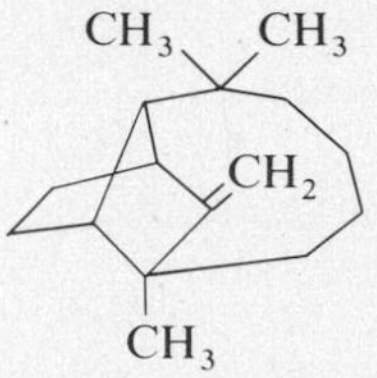

CH_3 CH_3 CH_2 CH_3

Longifolene

For the starting material in this synthesis, Corey and coworkers used compound **23**, prepared as described in the previous section (see Fig. 39.4). This was treated with a Wittig reagent prepared from ethyl bromide (see p. 749), which converted **23** to the diene **26**; the diene was hydroxylated to the glycol **27** with osmium tetroxide (see p. 657; a situselective reaction, the *exo* double bond reacting in preference to the *endo* double bond); and the diol was subjected to pinacol rearrangement to produce the ring-enlarged compound **28**. Acid-catalyzed hydrolysis then removed the ketal grouping to yield the diketone **29**. At this point, one of the key reactions of the synthesis was effected—*viz.*, one involving a base catalyzed reaction in which the position adjacent to the keto group in the six-membered ring acts as an active methylene group and adds in a 1,4-conjugate fashion to the α,β-unsaturated carbonyl moiety in the seven-membered ring—*i.e.*, an intramolecular Michael reaction (see p. 887) to give **30**. As a result of this

* E. J. Corey, M. Ohno, R. B. Mitra, and P. A. Vatakencherry, *Journal of the American Chemical Society*, *86*, 478 (1964).

reaction, the third ring in longifolene is formed, and at this juncture in the synthesis the construction of the cyclic carbon skeleton is complete. All that remains is to add a methyl group and a methylene group and remove the two keto functions. The methyl group was introduced by alkylation with methyl iodide in the presence of the very strong base sodium triphenylmethyl, one of the keto groups was removed by conversion to the cyclic thioketal (another situselective reaction) followed by hydrogenolysis to yield **31**, and the methylene group was introduced (with concomitant disappearance of the other keto group) by the addition of methyllithium followed by dehydration of the resulting alcohol (Fig. 39.6).

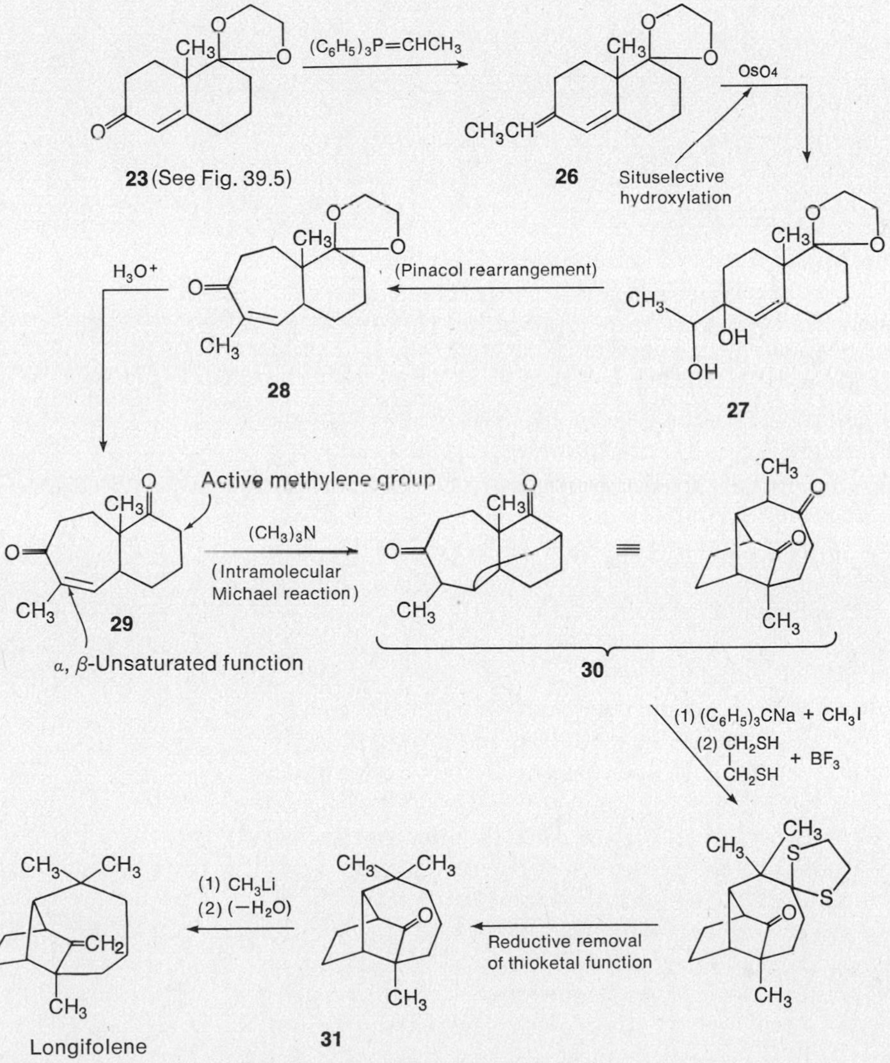

23 (See Fig. 39.5) — $(C_6H_5)_3P{=}CHCH_3$ → **26** — OsO_4 (Situselective hydroxylation) → **27** — (Pinacol rearrangement) → **28** — H_3O^+ → **29** (Active methylene group; α, β-Unsaturated function) — $(CH_3)_3N$ (Intramolecular Michael reaction) → **30** — (1) $(C_6H_5)_3CNa + CH_3I$ (2) CH_2SH–CH_2SH + BF_3 → thioketal — Reductive removal of thioketal function → **31** — (1) CH_3Li (2) $(-H_2O)$ → Longifolene

Fig. 39.6. Synthesis of longifolene.

39.4b. Synthesis of camptothecin, an alkaloid. Let us depart now from the bicyclo[4.4.0]decane system, which has played so prominent a part in the discussion to this point, and consider a compound with a quite different sort of ring structure. Camptothecin is a pentacyclic alkaloid that has been isolated from a tree native to China, *Camptotheca acuminata* nyssaceae. Camptothecin has aroused considerable interest in pharmaceutical circles because in biological testing it has shown antileukemic and antitumor activity. The synthesis of the compound, therefore, takes on special interest inasmuch as synthesis schemes can usually be modified in such a way that analogs of the compound can also be prepared, analogs that may have properties similar to, yet different, from those of the parent compound.

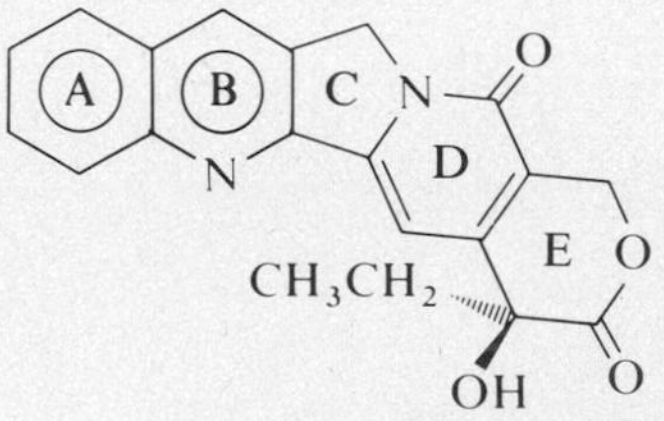

Camptothecin

The synthesis of camptothecin has been accomplished by several research groups, including one under the direction of G. S. Stork of Columbia University.* The synthesis that Stork and his coworkers have developed, uses many of the reactions that have been described in this text and is also of interest in that a new ring annelation procedure was developed in the course of the work.

The total synthesis of camptothecin, outlined in Fig. 39.7, begins with the condensation of *o*-aminobenzaldehyde with 1-ethoxycarbonyl-2-ethoxycarbonylmethyl-3-pyrrolidone (**32**), which yields the quinoline derivative **33**, the reaction involving either an imine aldehyde intermediate or an amino ketone intermediate. Ring D is then introduced by converting the N-ethoxycarbonyl function to a ethoxycarbonylacetyl function

$$-\overset{\overset{\displaystyle O}{\|}}{C}CH_2CO_2C_2H_5$$

to give compound **34** followed by a Dieckmann condensation to give compound **35**. Removal of the ethoxycarbonyl function by hydrolysis and decarboxylation forms **36**, and reduction and acetylation and base-catalyzed deacetylation yields **37**.

At this point in the synthesis rings A, B, C, and D have been constructed, and it remains only to add ring E. The reagent for doing this is ethyl 1-ethoxycarbonyl-1-propyl carbonate (**38**), prepared by the reaction of ethyl α-hydroxybutyrate with ethyl chloroformate. When the anion of **38** is generated by the action of lithium diisopropylamide in tetrahydrofuran at $-70°$ and allowed to react with **37**, a Michael addition takes place, forming a C—C bond between the active methylene carbon of **38** and the β carbon of **37**. The initially formed

* G. S. Stork and A. G. Schultz, *Journal of the American Chemical Society*, *93*, 4074 (1971).

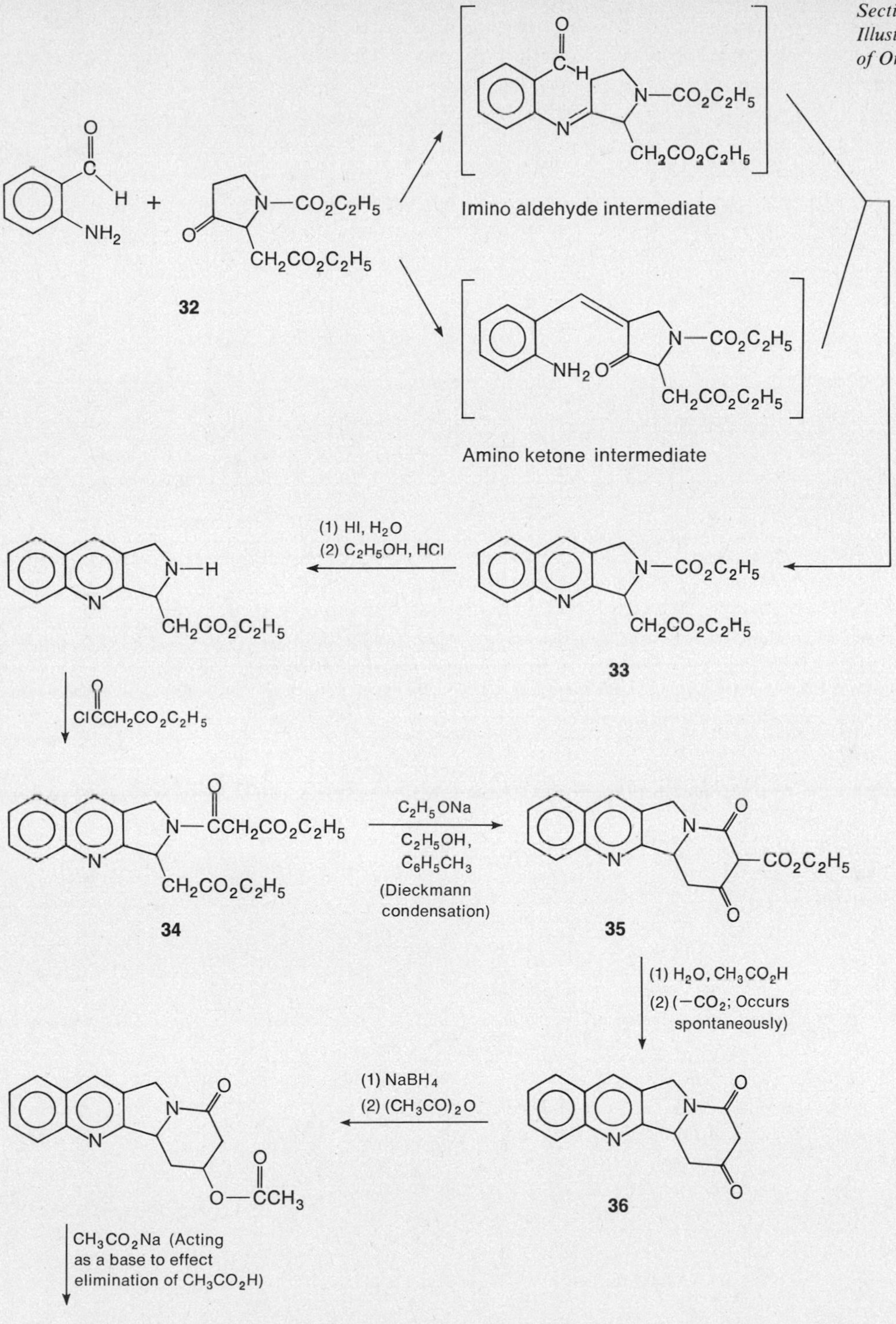

H + $N{-}CO_2C_2H_5$, $CH_2CO_2C_2H_5$, NH_2

32

Imino aldehyde intermediate

Amino ketone intermediate

(1) HI, H_2O
(2) C_2H_5OH, HCl

33

$ClCCH_2CO_2C_2H_5$

34

C_2H_5ONa
C_2H_5OH, $C_6H_5CH_3$
(Dieckmann condensation)

35

(1) H_2O, CH_3CO_2H
(2) ($-CO_2$; Occurs spontaneously)

36

(1) $NaBH_4$
(2) $(CH_3CO)_2O$

CH_3CO_2Na (Acting as a base to effect elimination of CH_3CO_2H)

Fig. 39.7. The synthesis of camptothecin.

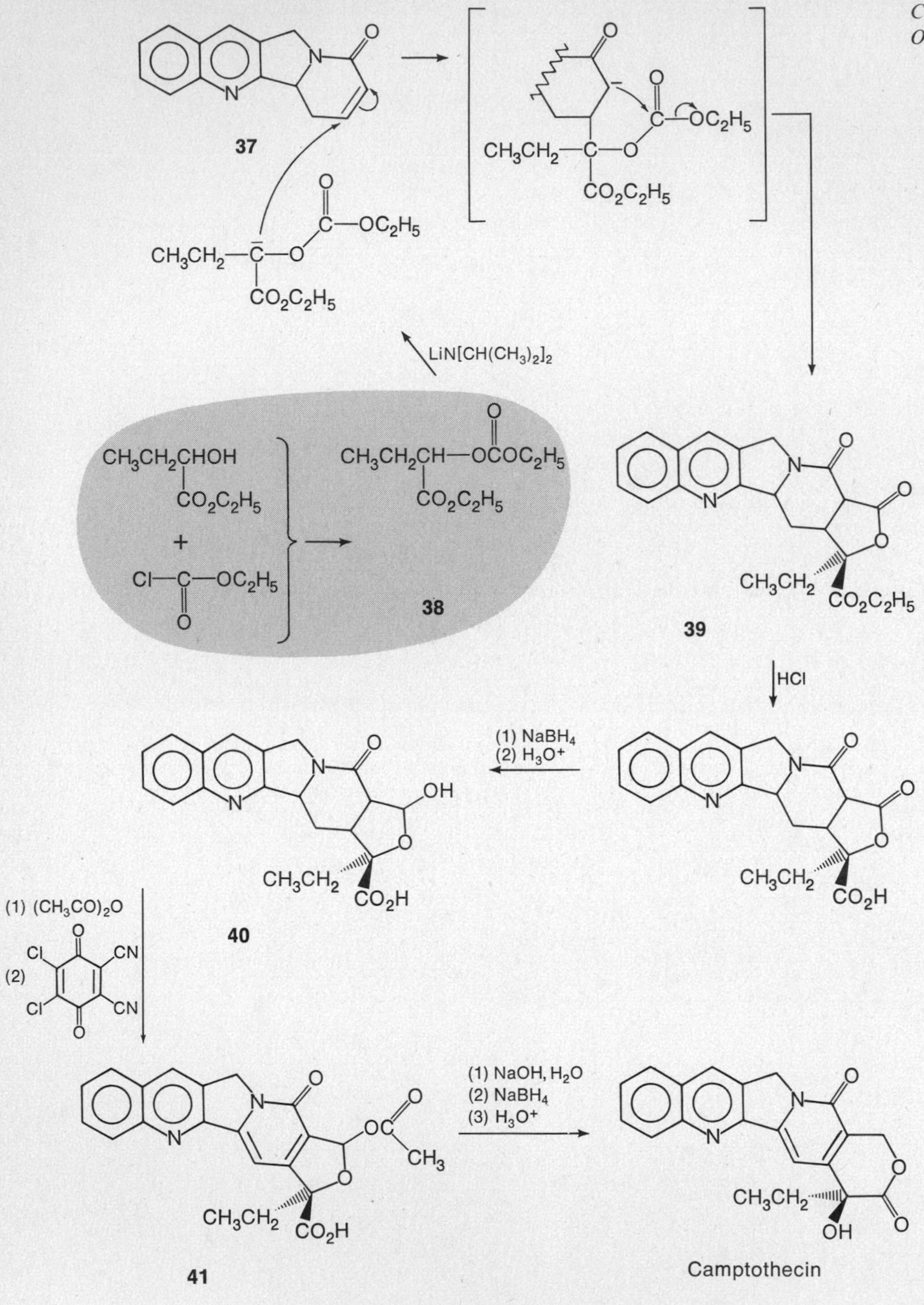

37

$CH_3CH_2-\bar{C}(CO_2C_2H_5)-O-C(=O)-OC_2H_5$

$CH_3CH_2-C(CO_2C_2H_5)-O$

$C-OC_2H_5$

$LiN[CH(CH_3)_2]_2$

CH_3CH_2CHOH ($CO_2C_2H_5$)

+

$Cl-C(=O)-OC_2H_5$

$CH_3CH_2CH(CO_2C_2H_5)-OCOC_2H_5$ (C=O)

38

CH_3CH_2 $CO_2C_2H_5$

39

HCl

CH_3CH_2 CO_2H

(1) $NaBH_4$
(2) H_3O^+

OH

CH_3CH_2 CO_2H

40

(1) $(CH_3CO)_2O$

(2) Cl, Cl, CN, CN (quinone)

CH_3

CH_3CH_2 CO_2H

41

(1) NaOH, H_2O
(2) $NaBH_4$
(3) H_3O^+

CH_3CH_2 OH

Camptothecin

Fig. 39.7. (Continued)

anion from this interaction then effects an intramolecular addition-elimination reaction on the ethoxycarbonyl group to produce the lactone **39** that is isolated from the reaction mixture. This procedure represents a novel method for attaching a five-membered heterocyclic ring, and it promises to have utility beyond this particular application.

With all the carbons for camptothecin now introduced, the final steps involve hydrolysis and sodium borohydride reduction of **39** to the cyclic hemiketal **40**, acetylation and dehydrogenation to **41** with a quinone as the hydrogen acceptor, and hydrolysis and reduction leading to the formation of the six-membered lactone ring of camptothecin.

Camptothecin contains one chiral carbon and, therefore, has the possibility of existing in optically active, nonsuperimposable mirror image forms. The synthesis that has been outlined leads to a racemate, and to obtain the optically active material identical with the naturally occurring (+)-camptothecin, resolution of compound **39** or any compound thereafter in the sequence is necessary.

The syntheses of longifolene (an example of an acetogenin) and camptothecin (an example of an alkaloid) represent but two of the very large number of syntheses of acetogenins and alkaloids that have been carried out by organic chemists over the last several decades. An enormous number of naturally occurring compounds of these types have been isolated and their structures elucidated, and they have provided an irresistible lure for chemists bent on demonstrating their synthetic prowess. As a result of the very extensive effort that has been devoted to the syntheses of such compounds, a host of clever and often very useful methods have been devised for constructing carbon frameworks and for altering the substituents attached to these frameworks. To gain some appreciation of the magnitude of this effort, the reader should consult the references by Ireland, Anand *et al.*, and Danishefsky and Danishefsky, listed at the end of the chapter.

39.4c. SYNTHESIS OF D,L-GLUCOSE, A CARBOHYDRATE. Although more organic chemists have devoted their attention to the synthesis of acetogenins and alkaloids than to other types of natural products, studies directed to the synthesis of carbohydrates, peptides, and nucleotides have proceeded apace, and significant accomplishments have been made. In fact, in what is one of the most spectacular synthesis feats to date, the polypeptide ribonuclease containing 124 amino acid residues has been elaborated *in toto*. As the first example of a synthesis of a compound belonging to the carbohydrate-peptide-nucleotide class of natural products, let us look into the preparation of D,L-glucose, which has recently been achieved by R. K. Brown and his coworkers at the University of Edmonton,* as outlined in Fig. 39.8. The starting material for the synthesis is acrolein, which undergoes a Diels–Alder-type reaction to form acrolein dimer (**42**). Sodium borohydride reduction of **42** yields the corresponding alcohol **43**, which, upon treatment with aqueous acid, undergoes an intramolecular addition of the alcohol to the double bond to produce **44**. Upon bromination, **44** yields a monobromide (**45**), in which bromine enters at a position adjacent to the ketal carbon

* F. Sweet and R. K. Brown, *Canadian Journal of Chemistry*, *46*, 2289 (1968); *48*, 1791 (1970).

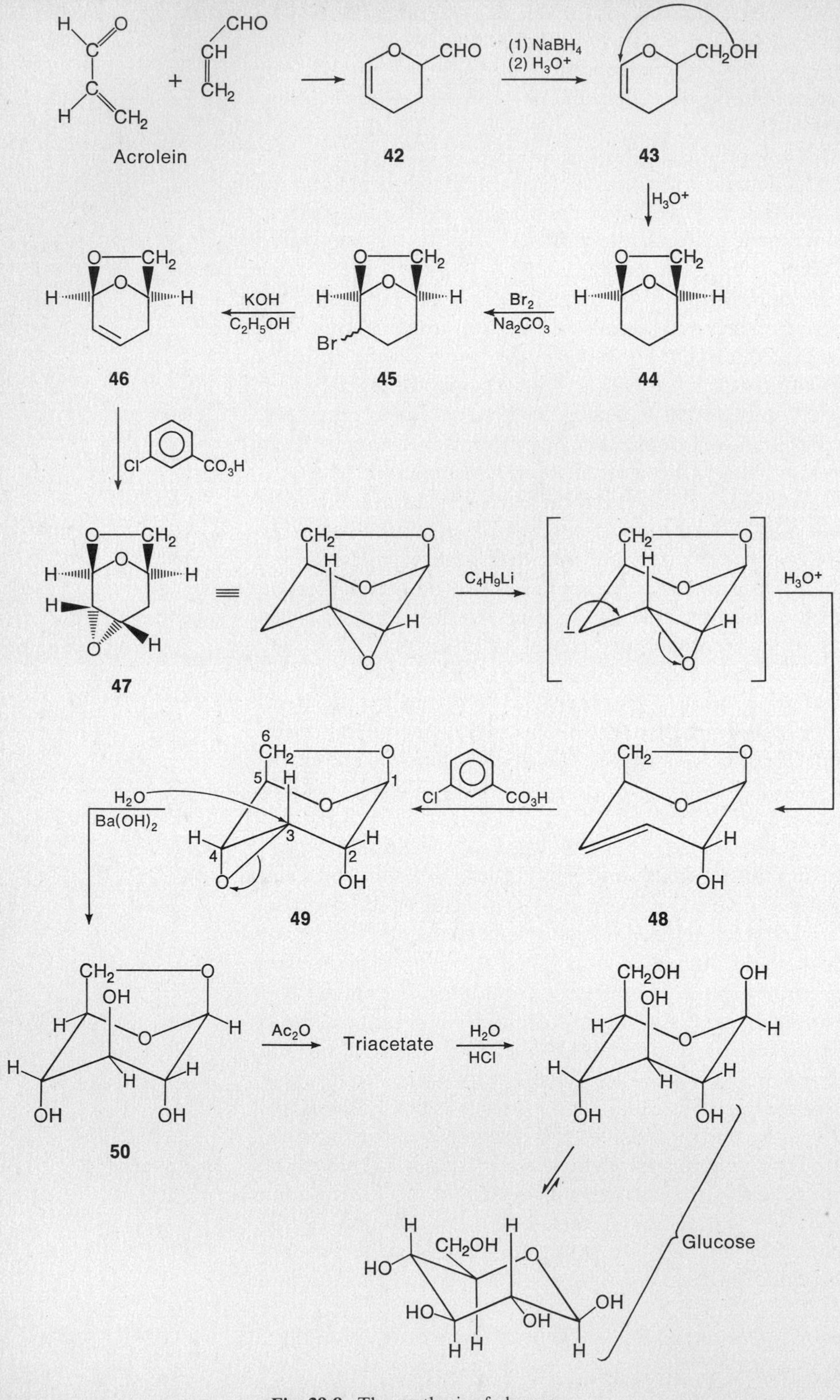

Acrolein + $CH_2{=}CH{-}CHO$ → **42** $\xrightarrow{(1)\ NaBH_4,\ (2)\ H_3O^+}$ **43** $\xrightarrow{H_3O^+}$ **44** $\xrightarrow{Br_2,\ Na_2CO_3}$ **45** $\xrightarrow{KOH,\ C_2H_5OH}$ **46** $\xrightarrow{m\text{-}ClC_6H_4CO_3H}$ **47** $\xrightarrow{C_4H_9Li}$ [intermediate] $\xrightarrow{H_3O^+}$ **48** $\xrightarrow{m\text{-}ClC_6H_4CO_3H}$ **49** $\xrightarrow{H_2O,\ Ba(OH)_2}$ **50** $\xrightarrow{Ac_2O}$ Triacetate $\xrightarrow{H_2O,\ HCl}$ Glucose

Fig. 39.8. The synthesis of glucose.

(as through the system were reacting as an aldehyde—see Boord synthesis of alkenes on p. 744). Base-induced dehydrohalogenation of **45** yields the alkene 46, and epoxidation of **46** with *m*-chloroperbenzoic acid yields the corresponding epoxide **47**. At this point, a base-induced elimination using *n*-butyllithium is carried out, forming **48**. Another epoxidation, again with *m*-chloroperbenzoic acid, yields the epoxide **49**, which undergoes hydrolytic ring opening in the presence of aqueous barium hydroxide to produce the triol **50**. All that is now required is to open the ether linkage between positions C-1 and C-6 (designating the positions by the carbohydrate numbering system), and this is achieved by converting **50** to its triacetate and hydrolyzing the triacetate with 1 N hydrochloric acid. The product that is obtained is D,L-glucose. A particularly interesting aspect of this synthesis is its high degree of stereoselectivity. The transformation of **47** to **48** proceeds in a stereoselective fashion to place the hydroxyl group at C-2 in an axial position, and the epoxidation of **48** yields a *cis*-epoxide (**49**), which undergoes hydrolysis to form a product (**50**), in which the hydroxyl groups at C-3 and C-4 are both in the axial alignment. Then, upon cleavage of the ether linkage between C-1 and C-6, the cyclohexane ring interconverts to the more stable conformer, which places the hydroxyl groups at C-2, C-3, and C-4, as well as the hydroxymethyl group attached to C-5 in the more stable equatorial alignments corresponding to the configuration of glucose.

39.4d. SYNTHESIS OF BRADYKININ, A PEPTIDE. Bradykinin is a small peptide containing nine amino acid residues and possessing hormonal activity that causes a lowering of blood pressure in animals. Its synthesis has been achieved by R. B. Merrifield* at Rockefeller University, using a technique known as the solid-phase method. Before discussing the details of the bradykinin synthesis, however, let us first consider some of the problems inherent in any polypeptide synthesis.

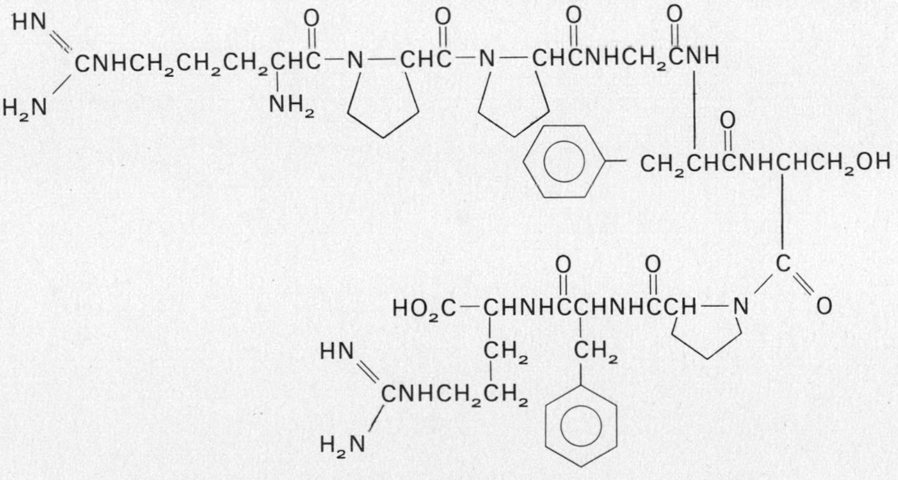

L-Arg-L-Pro-L-Pro-Gly-L-Phe-L-Ser-L-Pro-L-Phe-L-Arg

Bradykinin

* R. B. Merrifield, *Biochemistry*, 3, 1385 (1964).

First, amino acids are bifunctional molecules capable of undergoing facile reaction at the carboxyl function as well as the amino function. Thus, if we wish to form a peptide link between two *different* amino acids, we immediately face the problem associated with any mixed condensation—*viz.*, the formation of mixtures of products:

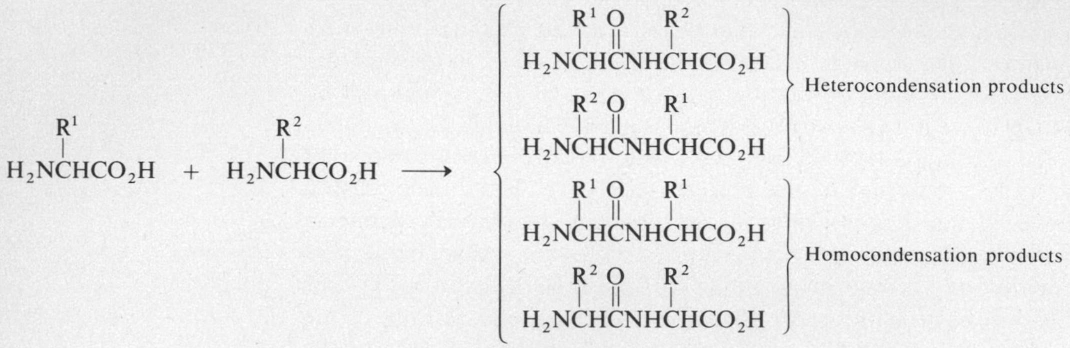

$$H_2NCH(R^1)CO_2H + H_2NCH(R^2)CO_2H \longrightarrow \begin{cases} \left.\begin{array}{l} H_2NCH(R^1)C(=O)NHCH(R^2)CO_2H \\ H_2NCH(R^2)C(=O)NHCH(R^1)CO_2H \end{array}\right\} \text{Heterocondensation products} \\ \left.\begin{array}{l} H_2NCH(R^1)C(=O)NHCH(R^1)CO_2H \\ H_2NCH(R^2)C(=O)NHCH(R^2)CO_2H \end{array}\right\} \text{Homocondensation products} \end{cases}$$

To circumvent this problem, we must "protect" or "block" the amino group of one of the participants and the carboxyl group of the other. Amino groups are frequently protected by means of the *t*-butoxycarbonyl (*t*-BOC) function, prepared by the action of *t*-butyl azidoformate [$(CH_3)_3COCON_3$] or *t*-butyl chloroformate [$(CH_3)_3COCOCl$] on the amino acid. Carboxyl groups can be protected by conversion to the benzyl or *t*-butyl ester. All of these are good protecting groups because they are easily removed under mild acid-catalyzed hydrolysis (Fig. 39.9).

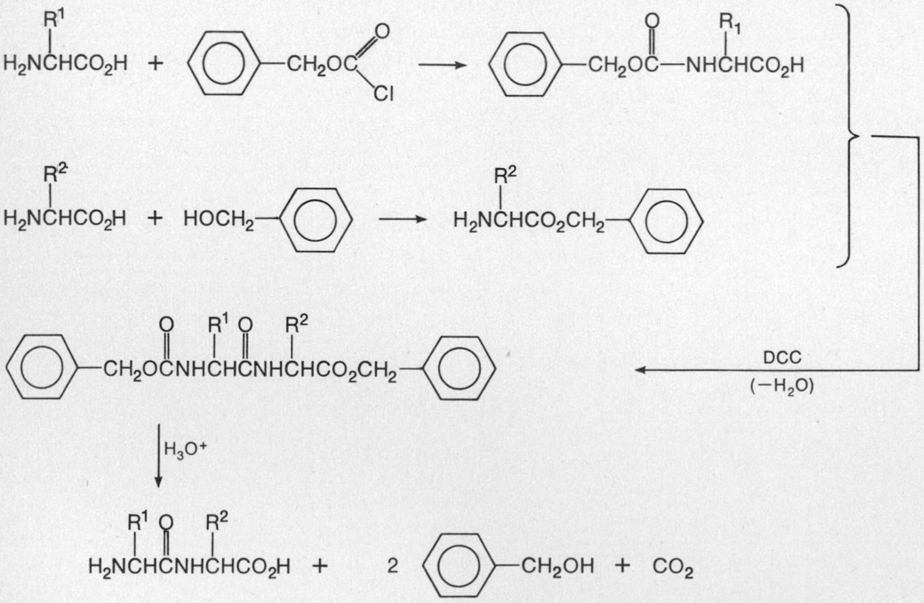

$$H_2NCH(R^1)CO_2H + C_6H_5CH_2OC(=O)Cl \longrightarrow C_6H_5CH_2OC(=O){-}NHCH(R_1)CO_2H$$

$$H_2NCH(R^2)CO_2H + HOCH_2{-}C_6H_5 \longrightarrow H_2NCH(R^2)CO_2CH_2{-}C_6H_5$$

$$\xrightarrow[(-H_2O)]{DCC} C_6H_5{-}CH_2OC(=O)NHCH(R^1)C(=O)NHCH(R^2)CO_2CH_2{-}C_6H_5$$

$$\xrightarrow{H_3O^+} H_2NCH(R^1)C(=O)NHCH(R^2)CO_2H + 2\ C_6H_5{-}CH_2OH + CO_2$$

Fig. 39.9. Use of "protecting" groups in the synthesis of a dipeptide.

A second problem attending not only peptide syntheses but any multistep synthesis is the spectre of diminishing yield. Very few reactions go to 100% completion. For example, if each of the eight peptide-forming reactions necessary for the construction of bradykinin goes in a yield of 80%, the overall yield of product is only 17%; if we do a bit better with each step and achieve 90% yields, the overall yield is 43%; if we do less well with each step and achieve only 70% yields, the overall yield is 5.8%. The devastating effects of poor yields on the outcome of reaction sequences involving 10, 20, 30, or more individual steps is immediately obvious. It is incumbent on the chemist, therefore, to strive for the absolute maximum yield at each stage of a multistep process, for failure to do so will result in vanishingly small amounts of product as the sole reward for the effort expended.

A third and particularly vexing problem in peptide syntheses is the necessity for retaining the configuration at the α carbon atom of each of the amino acid residues. For example, starting with a pair of L-amino acids and using the synthesis scheme outlined in Fig. 39.9, we see that a pair of diastereomeric dipeptides will be produced if partial inversion of configuration occurs at one of the amino acid residues; four diastereomeric dipeptides will be produced if partial inversion of configuration occurs at both of the amino acid residues:

$$\underbrace{\mathrm{L}^1 + \mathrm{L}^2}_{\text{Starting amino acids (L-configuration)}} \longrightarrow \underbrace{\mathrm{L}^1\text{—}\mathrm{L}^2 + \mathrm{D}^1\text{—}\mathrm{L}^2 + \mathrm{L}^1\text{—}\mathrm{D}^2 + \mathrm{D}^1\text{—}\mathrm{D}^2}_{\text{Diastereomeric products}}$$

Although diastereomers are separable, in theory, by ordinary physical means, this is often a difficult matter in practice, particularly with peptides because of the very small chemical and physical differences between diastereomers. To minimize the incidence of configurational inversion (racemization), we must use the mildest possible methods for forming the peptide bond and for removing the protecting groups. The major challenge in peptide synthesis, in fact, has been the development of suitable protecting groups that can be introduced and removed with the minimum of influence on the chiral centers of the peptide and suitable "activating groups" to permit the facile joining of the peptide bond.

A particularly useful reagent for joining the peptide bond is dicyclohexylcarbodiimide (DCC), postulated to "activate" the carboxyl group via the formation of an adduct:

$$RCO_2H + C_6H_{11}\text{—N=C=N—}C_6H_{11} \longrightarrow RC(=O)OC(=N\text{—}C_6H_{11})\text{—NH—}C_6H_{11}$$

DCC

The adduct can react with another molecule of carboxylic acid to yield an anhydride that can condense with an amine to form an amide. Alternatively, the adduct can react directly with the amine to form the amide:

$$\mathrm{RC(=O)\!-\!O\!-\!C(=N\!-\!C_6H_{11})\!-\!NH\!-\!C_6H_{11}} \xrightarrow{RCO_2H} \mathrm{RC(=O)\!-\!O\!-\!C(=O)R} + \mathrm{C_6H_{11}\!-\!NHC(=O)NH\!-\!C_6H_{11}}$$

$$\mathrm{RC(=O)\!-\!O\!-\!C(=N\!-\!C_6H_{11})\!-\!NH\!-\!C_6H_{11}} \xrightarrow{RNH_2} \mathrm{RCONHR} + \mathrm{C_6H_{11}\!-\!NHC(=O)NH\!-\!C_6H_{11}}$$

$$\mathrm{RC(=O)\!-\!O\!-\!C(=O)R} \xrightarrow{RNH_2} \mathrm{RCONHR}$$

(H$^+$; Nu$^-$)

With this insight into the trials and tribulations of peptide synthesis, let us turn our attention to the synthesis of the nonapeptide bradykinin by means of the solid-phase technique developed by R. B. Merrifield. The "solid phase" consists of a chloromethylated copolymer of styrene and divinylbenzene, designated in Fig. 39.9 as

$$\mathrm{ClCH_2\!-\!C_6H_4\!-\!(P)}$$

to which is affixed, in the opening step of the solid-phase synthesis, an N-protected amino acid (designated as *t*-BOC amino acid in Fig. 39.10). The protecting group is then removed from the amino function, and the

$$\text{amino acid}\mathrm{-CH_2\!-\!C_6H_4\!-\!(P)}$$

entity becomes the reactant for the next step involving another N-protected amino acid. Subsequent to the formation of the peptide bond between the second amino acid and the amino acid directly attached to the resin, the N-protecting group is removed; this process is continued in similar fashion until the requisite number of amino acid residues have been introduced. After each step of the process, the peptide-polymer is thoroughly washed to remove unreacted *t*-BOC amino acid, dicyclohexylcarbodiimide (the reagent used to form the peptide bond), and dicyclohexylurea (the product from dicyclohexylcarbodiimide). This, in fact, is the prime virtue of the solid-phase method; the soluble materials of lower molecular weight are easily separated from the "solid phase," to which the growing polymer is attached. Finally, when all the amino acid residues have been added, the peptide is severed from the polymer by reaction with hydrogen bromide in trifluoroacetic acid (Fig. 39.10).

Starting with the nitro analog of arginine, bradykinin was constructed by the solid-phase technique by sequential addition of proline, proline, glycine, phenylalanine, serine, proline, phenylalanine, and arginine (as the nitro analog; *i.e.* NO_2 instead of NH_2) residues (Fig. 39.11). After removal of the nonapeptide from the polymer, the nitro group of the arginine analog was reduced to the corresponding amino group, yielding a bradykinin molecule identical in all respects with the naturally occurring material. The overall yield of the synthesis

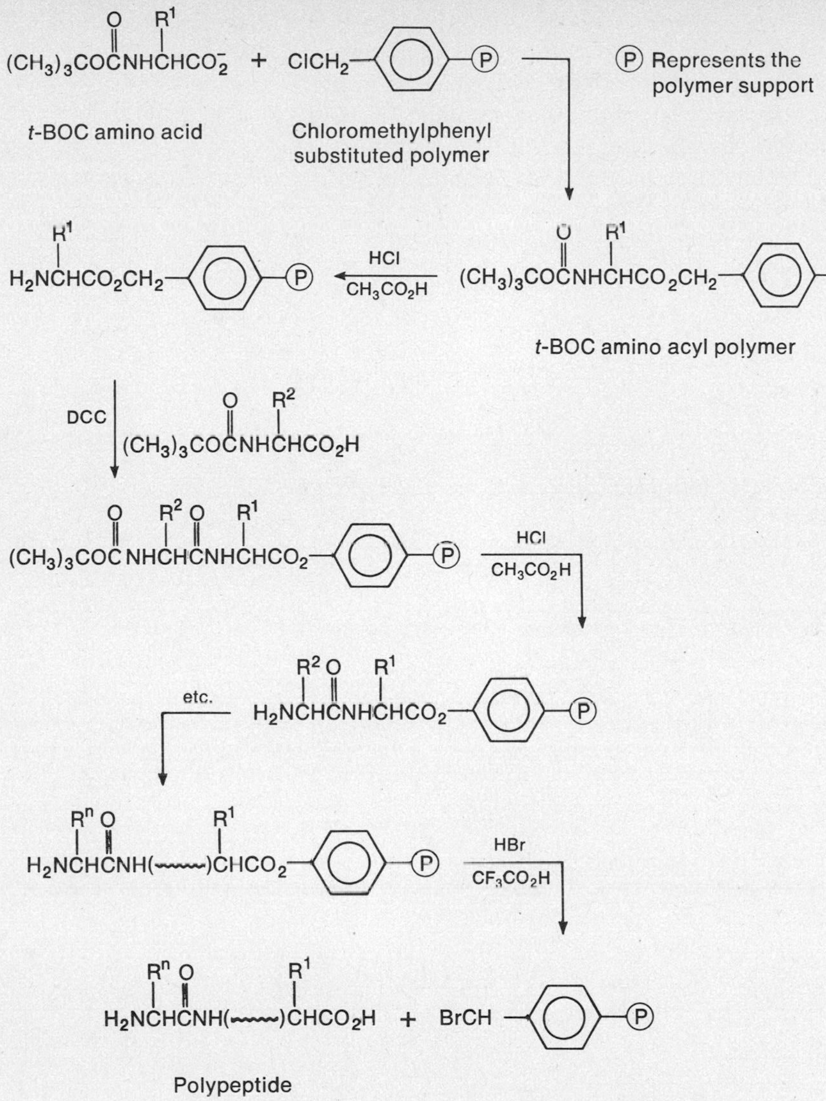

$$(CH_3)_3COC(=O)NHCH(R^1)CO_2^- + ClCH_2-C_6H_4-(P) \longrightarrow (CH_3)_3COC(=O)NHCH(R^1)CO_2CH_2-C_6H_4-(P)$$

t-BOC amino acid; Chloromethylphenyl substituted polymer; (P) Represents the polymer support; t-BOC amino acyl polymer

$$(CH_3)_3COC(=O)NHCH(R^1)CO_2CH_2-C_6H_4-(P) \xrightarrow[CH_3CO_2H]{HCl} H_2NCH(R^1)CO_2CH_2-C_6H_4-(P)$$

$$\xrightarrow[DCC]{(CH_3)_3COC(=O)NHCH(R^2)CO_2H} (CH_3)_3COC(=O)NHCH(R^2)C(=O)NHCH(R^1)CO_2-C_6H_4-(P)$$

$$\xrightarrow[CH_3CO_2H]{HCl} H_2NCH(R^2)C(=O)NHCH(R^1)CO_2-C_6H_4-(P) \xrightarrow{etc.} H_2NCH(R^n)C(=O)NH(\sim\sim)CH(R^1)CO_2-C_6H_4-(P)$$

$$\xrightarrow[CF_3CO_2H]{HBr} H_2NCH(R^n)C(=O)NH(\sim\sim)CH(R^1)CO_2H + BrCH-C_6H_4-(P)$$

Polypeptide

Fig. 39.10. Synthesis of polypeptides via solid-phase technique.

is a remarkable 68%, and the final product shows 100% biological activity, indicating that little, if any, racemization occurs during the course of the synthesis.

Elegant as the synthesis of bradykinin is, it pales by comparison with the synthesis of the enzyme ribonuclease, which contains no less than 124 amino acid residues. This prodigious effort, involving almost 12,000 separate chemical steps, was also achieved by R. B. Merrifield and his group. It was almost completely automated, using a machine (Fig. 39.12) that allows the many steps to be carried out with very little attention on the part of the operator. It should be pointed out, however, that the ribonuclease synthesis differs from the others that have been discussed in that a pure product was not obtained. As the result of less than 100% yields at each step of the way, a mixture of peptides ranging from 124 amino acid residues downward was obtained, and only partial purification has been achieved. The indication that ribonuclease was included in this mixture was obtained by measuring the enzymatic activity, the product showing 17% of the activity of pure ribonuclease. In all the other syntheses described in this section, on the other hand, completely pure product was obtained, demonstrably identical with the natural material by direct chemical and spectral comparisons.

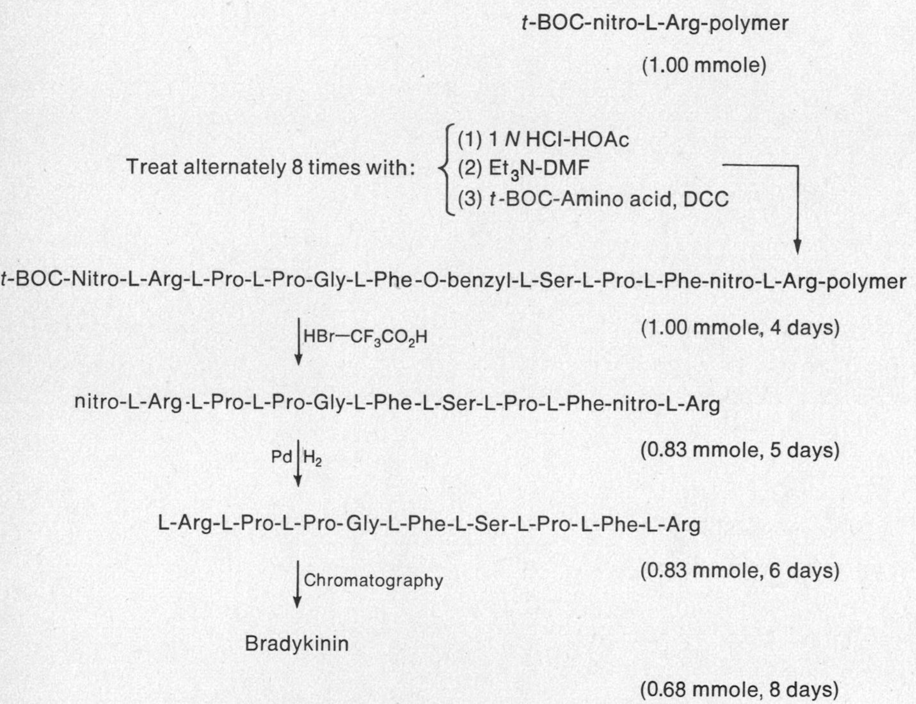

Fig. 39.11. The solid-phase synthesis of bradykinin. Overall yield and time from *t*-BOC-nitro-L-arginyl-polymer to each step is shown in parentheses.

39.4e. SYNTHESIS OF ADENOSINE TRIPHOSPHATE, A NUCLEOTIDE. Although the total synthesis of adenosine triphosphate (ATP) can no longer be considered to be of recent vintage, it is fitting, nevertheless, to conclude this the last chapter of the book with the compound that has been termed the "universal energy carrier" of living systems. Let us take as our synthesis goal not ATP itself but an ATP molecule labeled with radioactive phosphorus (^{32}P) on the phosphate group directly attached to the ribose portion of the molecule (α-phosphorus atom).

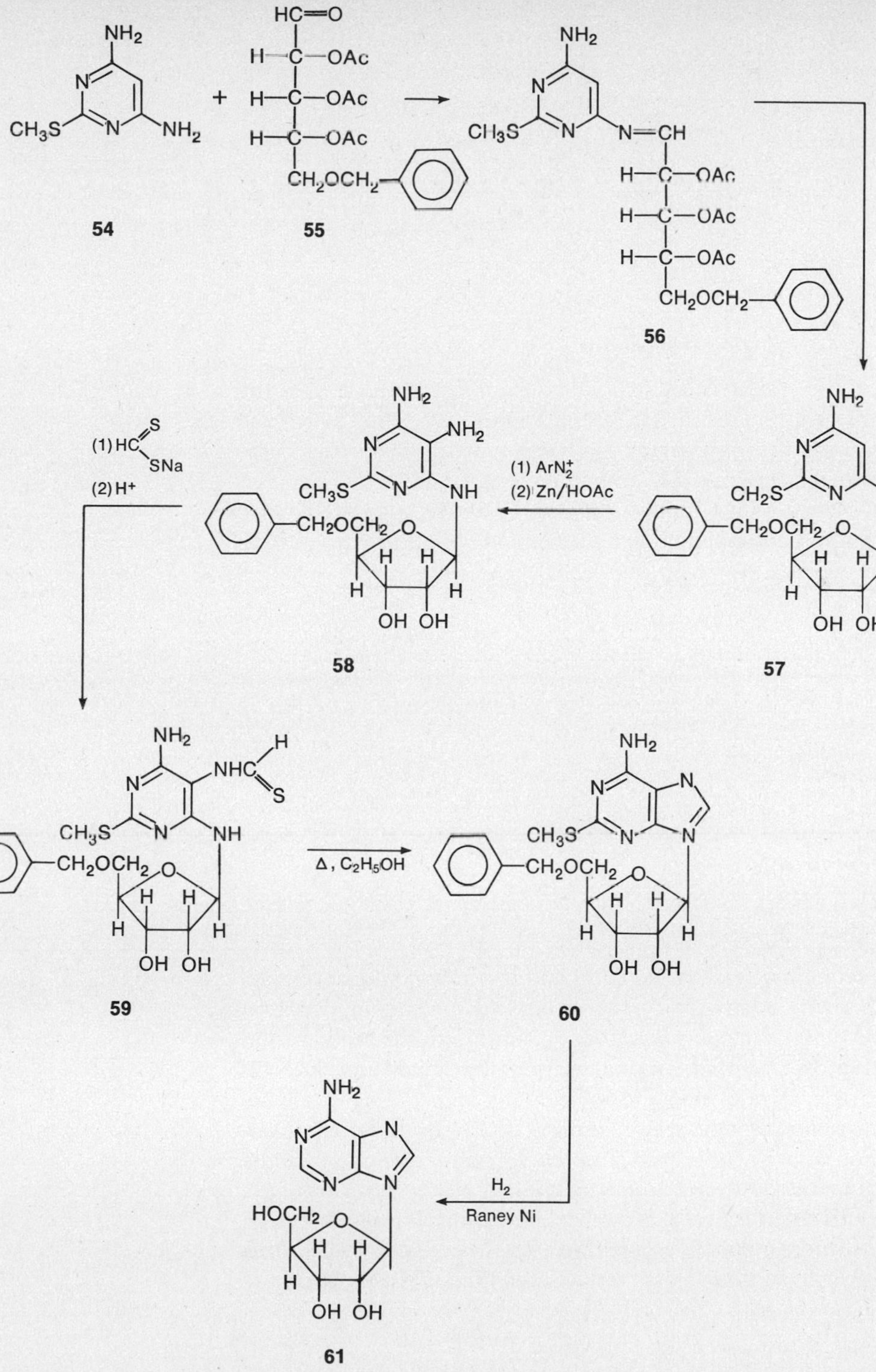

54 + 55 ⟶ 56 ⟶ 57

57 —(1) ArN_2^+ (2) Zn/HOAc⟶ 58

58 —(1) HC(=S)SNa (2) H^+⟶ 59

59 —Δ, C_2H_5OH⟶ 60

60 —H_2, Raney Ni⟶ 61

Fig. 39.12. The synthesis of adenosine (**61**).

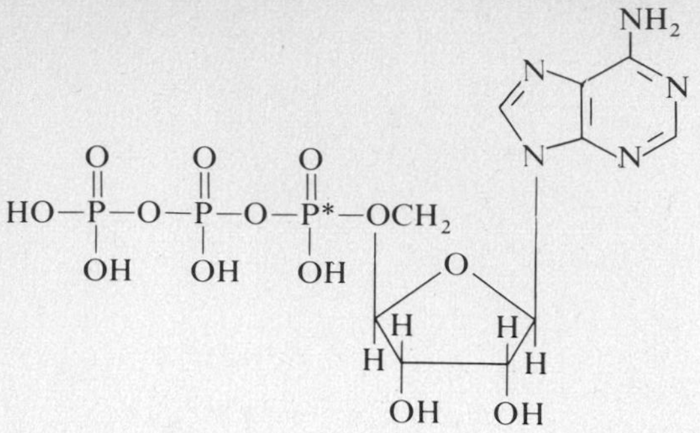

Adenosine triphosphate

For the first portion of the synthesis, we shall go back almost to the turn of the century and outline the way in which the pyrimidine moiety, from which the purine moiety was eventually to be made, was constructed. From the base-catalyzed condensation of thiourea (**51**) and malononitrile (**52**), W. Traube* obtained 4,6-diamino-2-mercaptopyrimidine (**53**), which can be converted to the corresponding 2-methylthio compound (**54**) by treatment with dimethyl sulfate:

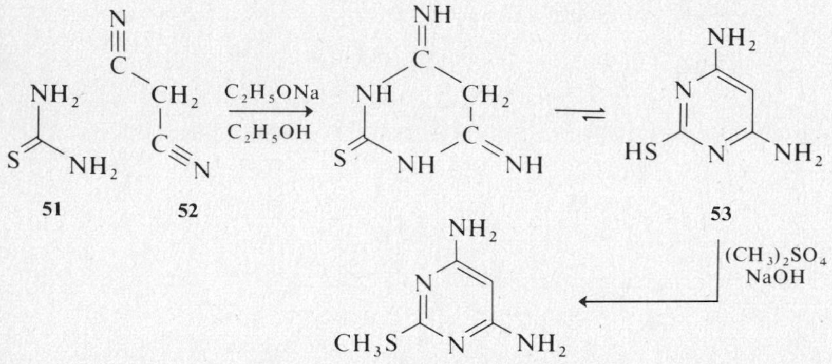

$$\underset{\textbf{51}}{H_2N-C(=S)-NH_2} + \underset{\textbf{52}}{N{\equiv}C-CH_2-C{\equiv}N} \xrightarrow[C_2H_5OH]{C_2H_5ONa} \text{(cyclic intermediate)} \rightleftharpoons \underset{\textbf{53}}{\text{4,6-diamino-2-mercaptopyrimidine (HS, }NH_2, NH_2)} \xrightarrow[NaOH]{(CH_3)_2SO_4} \underset{\textbf{54}}{\text{2-methylthio-4,6-diaminopyrimidine (}CH_3S, NH_2, NH_2)}$$

Using **54** as the starting material, A. R. Todd and his coworkers† developed a procedure for synthesizing adenosine by converting the pyrimidine ring to a purine ring and, concurrently, introducing a ribose residue onto the N-9 position of the purine ring, as Fig. 39.13 outlines. Condensation between **54** and the 5-O-benzyl-2,3,4-triacetate derivative of ribose (**55**) yields the imine (*Schiff base*) **56**, which, upon hydrolytic removal of the acetate groups, cyclizes to the ribofuranoside (**57**). Treatment of **57** with an aryldiazonium salt and reduction of the resulting diazo compound introduces an amino function at the C-5 position of the pyrimidine ring to give **58**. By reaction of **58** with sodium dithioformate followed by acidification, the N-thioformamide compound (**59**) is produced. When this is heated in ethanol, cyclization takes place, with elimination of H_2S, to form the

* W. Traube, *Justus Liebig's Annalen der Chemie*, *331*, 64 (1904).

† G. W. Kenner, C. W. Taylor, and A. R. Todd, *Journal of the Chemical Society* (*London*), 1620 (1949).

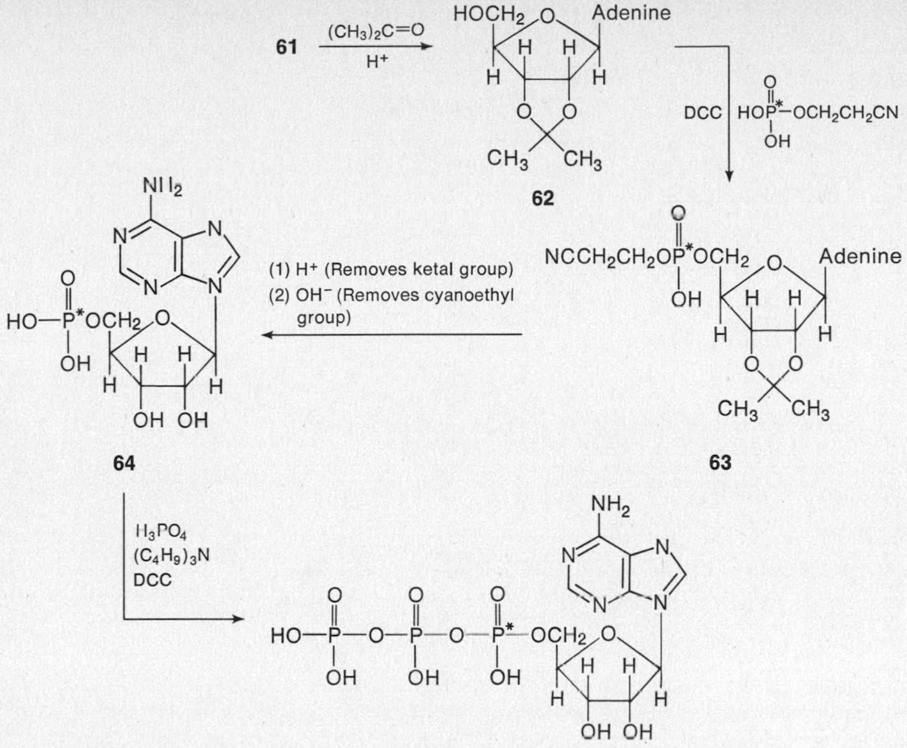

Fig. 39.13. The synthesis of adenosine triphosphate labeled at α-P with ^{32}P.

purine compound (**60**). Finally, the methylthio group at C-2 and the benzyl group on the ribose portion are simultaneously removed by hydrogenolysis of **60** with Raney nickel as catalyst, yielding adenosine (**61**).

The introduction of the phosphate groups onto adenosine to convert it to adenosine triphosphate is carried out in two steps. For the first step, the phosphorylating agent is β-cyanoethyl phosphate, prepared by treatment of phosphoric acid with β-hydroxypropionitrile (in the presence of dicyclohexyl carbodiimide) followed by partial hydrolysis:

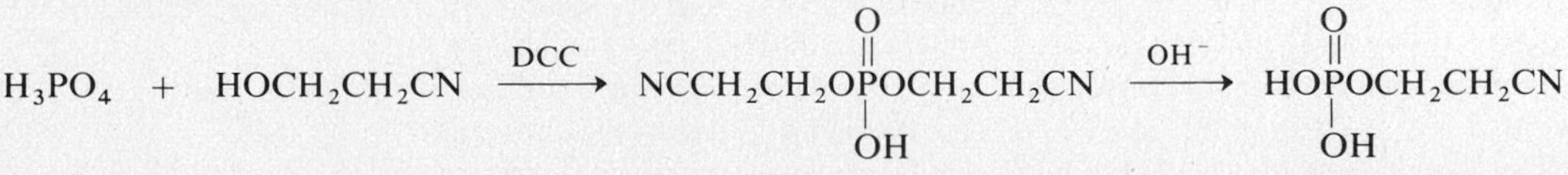

$$H_3PO_4 + HOCH_2CH_2CN \xrightarrow{DCC} NCCH_2CH_2OP(=O)(OH)OCH_2CH_2CN \xrightarrow{OH^-} HOP(=O)(OH)OCH_2CH_2CN$$

β-Cyanoethyl phosphate

When adenosine, protected at the C-2 and C-3 hydroxyl groups via ketal formation (**62**) is treated with β-cyanoethyl phosphate in the presence of dicyclohexylcarbodiimide (DCC), esterification occurs to produce compound **63** from which the ketal group can be removed by treatment with aqueous acid and the cyanoethyl group removed by treatment with base (*i.e.*, a reverse Michael reaction) to leave adenosine monophosphate (**64**) (Fig. 39.14). By using phosphoric acid labeled with ^{32}P, radioactively labeled adenosine monophosphate can be obtained. The remaining two phosphate residues of ATP are then very easily introduced simply by treating adenosine monophosphate (AMP) with phosphoric acid in the presence of tributylamine and dicyclohexylcarbodiimide.

REFERENCES

ANAND, N., J. S. BINDRA, and S. RANGANATHAN, *Art in Organic Synthesis*, Holden-Day, Inc., San Francisco, 1970.

APSIMON, J., ed., *The Total Synthesis of Natural Products*, John Wiley & Sons, Inc., New York, 1973.

BUEHLER, C. A., and D. E. PEARSON, *Survey of Organic Synthesis*, John Wiley & Sons, Inc., New York, 1970.

DANISHEFSKY, S. E., and S. DANISHEFSKY, *Progress in Total Synthesis*, Vol. 1, Meredith Corp., New York, 1971.

FIESER, L. F., and M. FIESER, *Reagents for Organic Synthesis*, Vol. 1, John Wiley & Sons, Inc., New York, 1967.

FIESER, M., and L. F. FIESER, *Reagents for Organic Synthesis*, Vols. 2 and 3, John Wiley & Sons, Inc., New York, 1972.

IRELAND, R. E., *Organic Synthesis*, in *Foundations of Modern Organic Chemistry Series*, Prentice-Hall, Inc., Englewood Cliffs, N.J., 1969.

MONSON, R. S., *Advanced Organic Synthesis*, Academic Press, Inc., New York, 1971.

PARHAM, W. E., *Syntheses and Reactions in Organic Chemistry*, John Wiley & Sons, Inc., 1970.

WOODWARD, R. B., in *Perspectives in Organic Chemistry*, ed. by A. R. Todd, Interscience Publishers Inc., New York, 1956.

PROBLEMS

1. A publication entitled *Organic Syntheses* appears yearly and includes descriptions of synthetic methods that have been carefully checked and are known to be reliable. The structures of some of the compounds whose syntheses are described in the *Fourth Collective Volume of Organic Syntheses* (containing 10 yearly volumes) are shown below, and the pages on which they appear in this volume are given. Before referring to *Organic Syntheses*, work out your own method for the synthesis for each of these compounds and then check to see how closely they correspond to the published ones. In the case of those compounds in groups A and B, your method will probably be similar to the ones outlined in *Organic Syntheses*, whereas this is much less likely for the compounds in group C.

   Group A: Straightforward syntheses involving a single step conversion from readily available starting materials—see Tables 39.1–39.5.

(a) $C_4H_9C{\equiv}CH$
p. 117 in Org. Syn. IV

(b) $C_{16}H_{33}CH(CO_2C_2H_5)_2$
p. 141 in Org. Syn. IV

(c) Br—⟨C₆H₄⟩—⟨C₆H₄⟩—Br
p. 256 in Org. Syn. IV

(d) 9-acetylanthracene ($COCH_3$)

p. 8 in Org. Syn. IV

(e) cyclobutane with $CO_2C_2H_5$, $CO_2C_2H_5$

p. 288 in Org. Syn. IV

(f) CH_3O, OCH_3 benzene ring $-CH{=}CHCO_2H$

p. 327 in Org. Syn. IV

Group B: Straightforward syntheses involving more than one step, starting with readily available materials—see Tables 39.1–39.5.

(a) O, CH_3 (cyclohexenone)

p. 162 in Org. Syn. IV

(b) cycloheptanone $=O$

p. 221 in Org. Syn. IV

(c) CO_2H, Br, Br, Br (benzene ring)

p. 947 in Org. Syn. IV

(d) $C_6H_5C{\equiv}CH$

p. 763 in Org. Syn. IV

(e) $C_6H_5CH{=}CHCH{=}CH_2$

p. 771 in Org. Syn. IV

(f) $(CH_3)_2C{=}C{=}O$

p. 348 in Org. Syn. IV

Group C: Syntheses involving less obvious pathways, starting with readily available compounds—see Tables 39.1–39.5.

(a) $CH_3\overset{O}{\overset{\|}{C}}(CH_2)_4CO_2H$

p. 19 in Org. Syn. IV

(b) benzene ring with CO_2H and $CH_2CH_2CO_2H$

p. 136 in Org. Syn. IV

(c) cyclopropyl$-\overset{O}{\overset{\|}{C}}-$cyclopropyl

p. 278 in Org. Syn. IV

(d) $ICH_2CH_2CH_2CH_2I$

p. 321 in Org. Syn. IV

(e) $C_6H_5-C{\equiv}C-C_6H_5$

p. 377 in Org. Syn. IV

(f) CH_3, CH_3, CH_3, CH_3, CH_3, CH_3 (hexamethylbenzene)

p. 520 in Org. Syn. IV

(g) O (2-phenylcycloheptanone)

p. 780 in Org. Syn. IV

(h) $O_2N-C_6H_4-CH_2CH_2-C_6H_4-NO_2$

p. 367 in Org. Syn. IV

2. Starting with readily available materials (see Tables 39.1–39.5), devise reasonable syntheses for the following compounds:

(a)

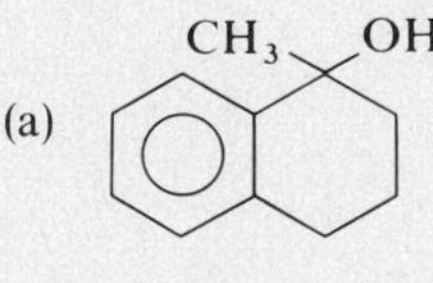

(b) $CH_3O-C_6H_4-NH\overset{O}{\overset{\|}{C}}-C_6H_4-OCH_3$

(c) O, O (cycloheptane-1,2-dione)

(d) $=O$ (2-indanone)

(e) $CH_3(CH_2)_7CHO$

(f) 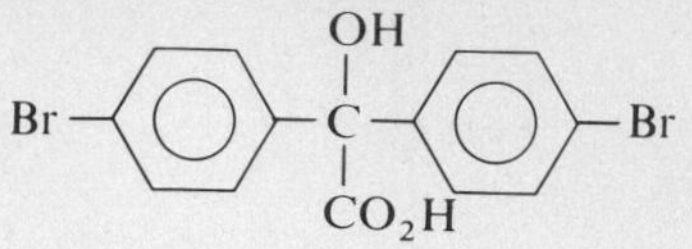 Br–C_6H_4–C(OH)(CO_2H)–C_6H_4–Br

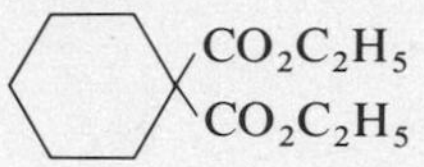

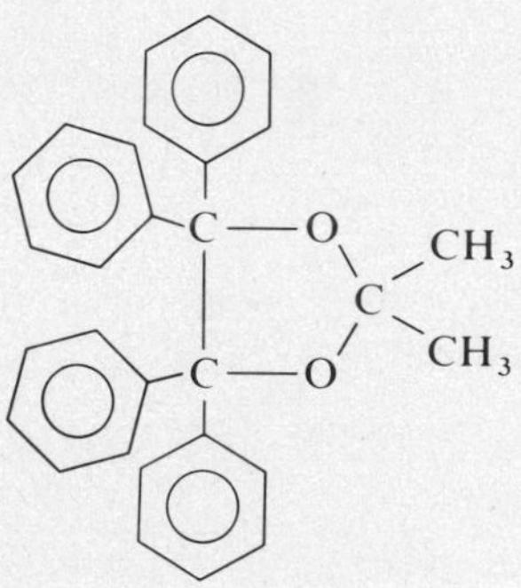

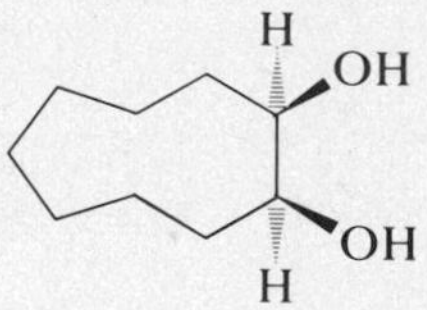

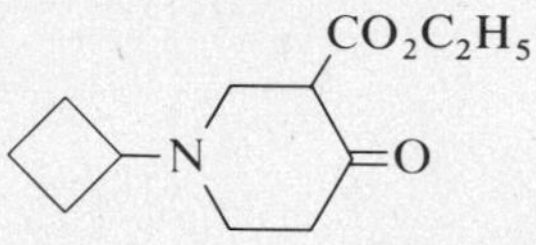

(g) 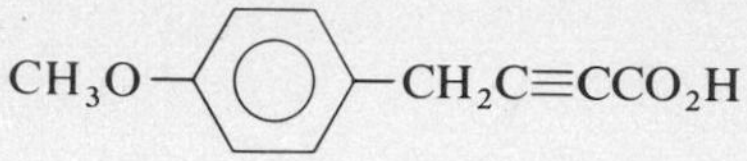 CH_3O–C_6H_4–$CH_2C{\equiv}CCO_2H$

(h) 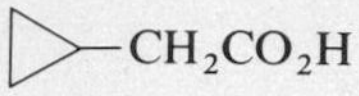 cyclopropyl–CH_2CO_2H

(i) 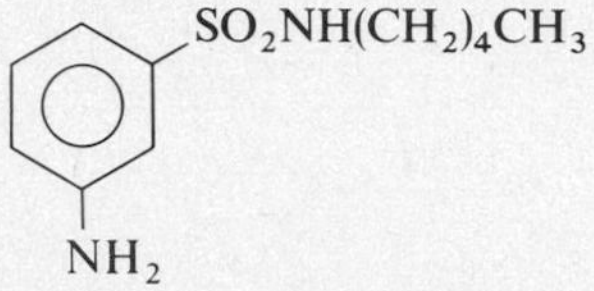 m-H_2N–C_6H_4–$SO_2NH(CH_2)_4CH_3$

(j) cyclohexane-1,1-($CO_2C_2H_5$)$_2$

(k) $(C_6H_5)_2$C—O, C(CH_3)$_2$, $(C_6H_5)_2$C—O

(l) H, OH, OH, H

(m) cyclobutyl–N (piperidine ring), $CO_2C_2H_5$, =O

(n) CH_3, HO, OH, OH

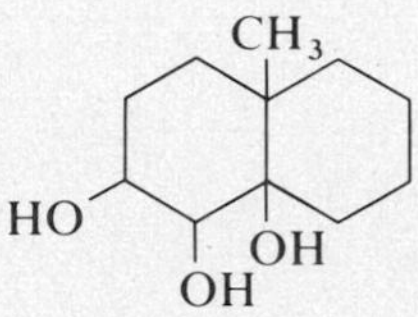

Any stereoisomer

(o) 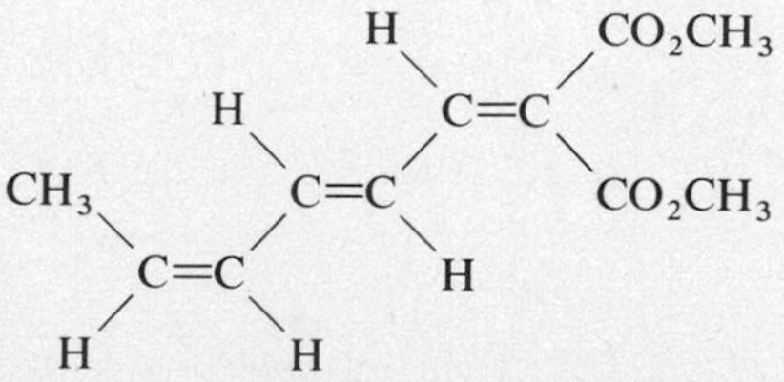 CH_3, C=C, H, H, C=C, H, H, C=C, H, CO_2CH_3, CO_2CH_3, H

3. As part of a "pure" research program that you are carrying out in a university laboratory, it has become of interest to you to measure the rate of nucleophilic displacement of a number of benzyl chlorides. Among the compounds that you want for this purpose are 2-chloro-2-phenylpropane (I) and its *p*-nitro (II) and *p*-dimethylamino (III) analogs, the nitro group being a strong electron-withdrawing function and the dimethylamino group in the *p*-position being a strong electron-releasing function. Also of interest is compound IV, in which the N-acetyl group reduces the electron-releasing capacity of the amino function, and compound V, in which the greatly increased length of the alkyl chain changes the solubility characteristics. Using readily available compounds as starting materials (see Tables 39.1–39.5), devise a synthesis for each of these compounds.

$$\text{I: } C_6H_5\text{–}C(CH_3)_2Cl \quad \text{II: } p\text{-}NO_2C_6H_4\text{–}C(CH_3)_2Cl \quad \text{III: } p\text{-}(CH_3)_2NC_6H_4\text{–}C(CH_3)_2Cl$$

$$\text{IV: } p\text{-}CH_3(COCH_3)NC_6H_4\text{–}C(CH_3)_2Cl \quad \text{V: } p\text{-}(CH_3)_2NC_6H_4\text{–}C(CH_3)(Cl)(CH_2)_{16}CH_3$$

The results of this work are eventually published in a chemical journal and are read by an industrial research chemist, who concludes, on the basis of some of his own work, that compounds combining the structural features of IV and V might prove to be effective herbicides. Therefore, he sets up a program to synthesize and test a variety of such compounds, of which series VI, VII, and VIII are representative examples.

$$\text{VI: } p\text{-}CH_3CH_2(COCH_3)NC_6H_4\text{–}C(CH_3)(Cl)(CH_2)_nCH_3$$
$$\text{VII: } p\text{-}CH_3CH_2(COCH_2Cl)NC_6H_4\text{–}C(CH_3)(Cl)(CH_2)_nCH_3$$
$$\text{VIII: } p\text{-}CH_3CH_2(COCH_2Cl)NC_6H_4\text{–}CH(Cl)(CH_2)_nCH_3$$

VI (n = 10, 11, 12, 14, 16) VII (n = 10, 11, 12, 14, 16) VIII (n = 10, 11, 12, 14, 16)

Using readily available compounds as starting materials (see Tables 39.1–39.5), devise a synthesis for each of these compounds.

The result of the testing of these compounds shows that the most active herbicides are VII (n = 10 and 11) and VIII (n = 10 and 11). It is decided, therefore, to prepare sufficient quantities of these four compounds so that large-scale field tests can be carried out. For this, it is necessary to restrict the starting materials to reasonably inexpensive compounds. Starting with compounds that you think meet this criterion, indicate how these syntheses might be achieved.

Although compound VII (n = 11) proves to be the most active of the four in the field tests, the other three are only slightly less active. It is decided, therefore, to put compound VIII (n = 10) into commercial production. Why was this choice made? Starting only with compounds that you think are very inexpensive, devise a synthesis of compound VIII (n = 10) that might be used for the commercial production of this herbicide.

4. Equilenin, first isolated from the urine of pregnant mares, is the stereochemically least complicated steroid sex hormone, possessing only two chiral centers. Its synthesis was first accomplished by Bachmann, Cole, and Wilds* by the sequence of interconversions

---

* W. E. Bachmann, W. Cole, and A. L. Wilds, *Journal of the American Chemical Society*, *62*, 824 (1940).

depicted in the following flow sheet. Indicate the reactions that you think might have been used for achieving each of these interconversions.*

NH_2 HO_3S I CH_3O CH_2OH CH_3O CO_2H CH_3O $CO_2C_2H_5$ O O CH_3 $CO_2C_2H_5$ O CH_3 $CO_2C_2H_5$ $CO_2C_2H_5$ OH CH_3 CO_2CH_3 CO_2H H CH_3 CO_2CH_3 CO_2CH_3 H CH_3 O H CH_3O CH_3 O H HO

Equilenin

5. Caryophyllene is a terpene hydrocarbon isolated from the oil of cloves. It is structurally interesting because it possesses a nine-membered ring containing a *trans* double bond, the nine-membered ring fused in a *trans* fashion to a four-membered ring. Its complete synthesis has been achieved by E. J. Corey and coworkers† by the sequence of reactions outlined. For those steps in which reagents are not included over the arrow, indicate how the transformation might be effected.

---

* Some of the structures in this and other flow sheets have a wavy line drawn through them. This is to indicate that the remainder of the structure, which isn't shown, remains intact during the reaction being depicted.

† E. J. Corey, R. B. Mitra, and H. Uda, *Journal of the American Chemical Society*, *86*, 485 (1964).

Caryophyllene

6. Anthramycin, a mold metabolite isolated from *Streptomyces refuineus*, is reported to have *in vivo* antibiotic and antitumor activity. The synthesis of this compound has been reported by Leimgruber and coworkers* who prepared it by the following sequence of reactions. For those steps in which reagents are not shown over the arrows, indicate how the transformation might be effected.

---

* W. Leimgruber, A. D. Batcho, and R. C. Czajkowski, *Journal of the American Chemical Society*, *90*, 5641 (1968).

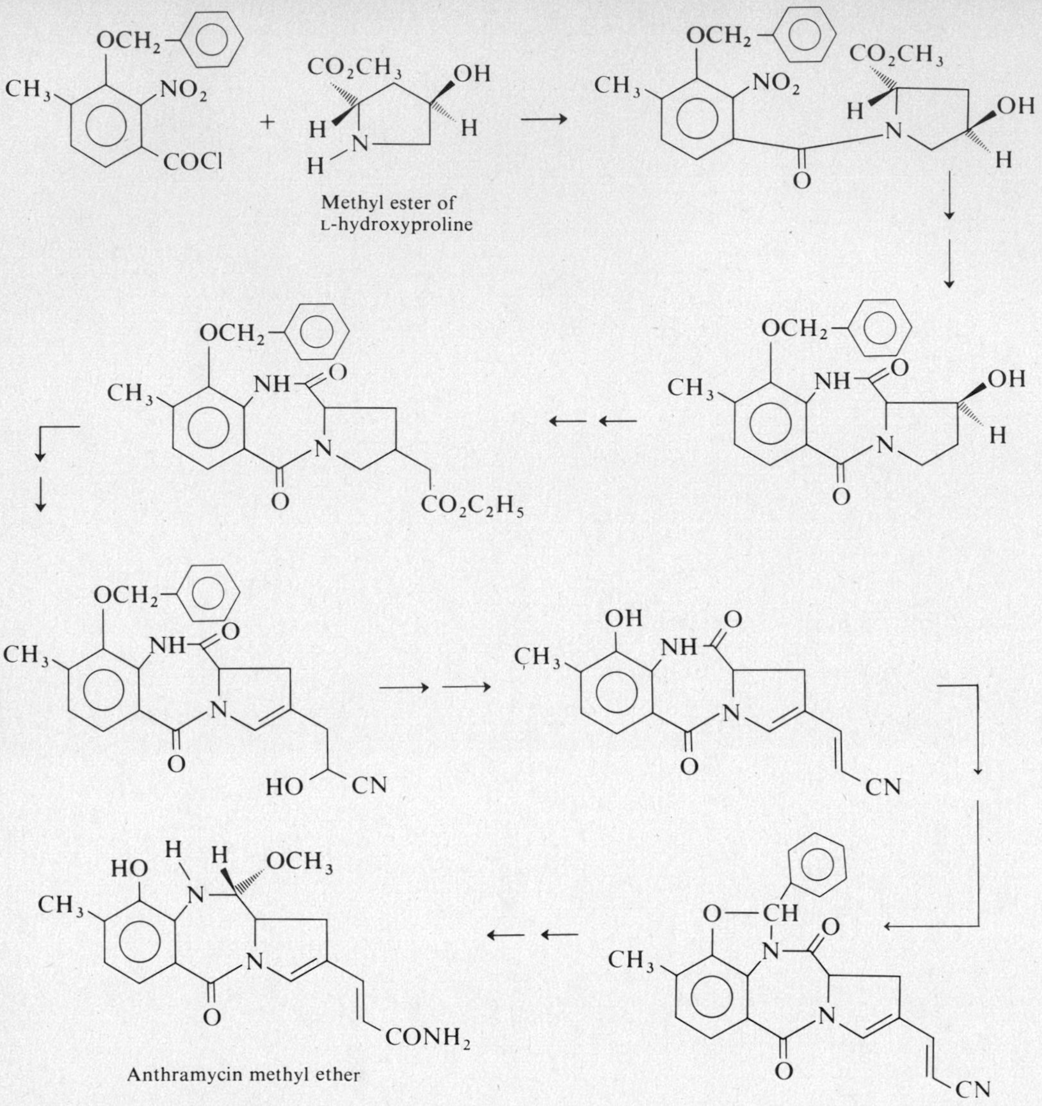

Methyl ester of L-hydroxyproline

Anthramycin methyl ether

7. Cyclopenin is a mold metabolite isolated from *Penicillium cyclopium* and *Penicillium viridicatin.* Its total synthesis has been achieved by Rapoport and coworkers,* by the sequence of reactions illustrated. Indicate how each of the interconversions in this sequence might be effected.

---

* H. Smith, P. Wegfahrt, and H. Rapoport, *Journal of the American Chemical Society, 90,* 1668 (1968).

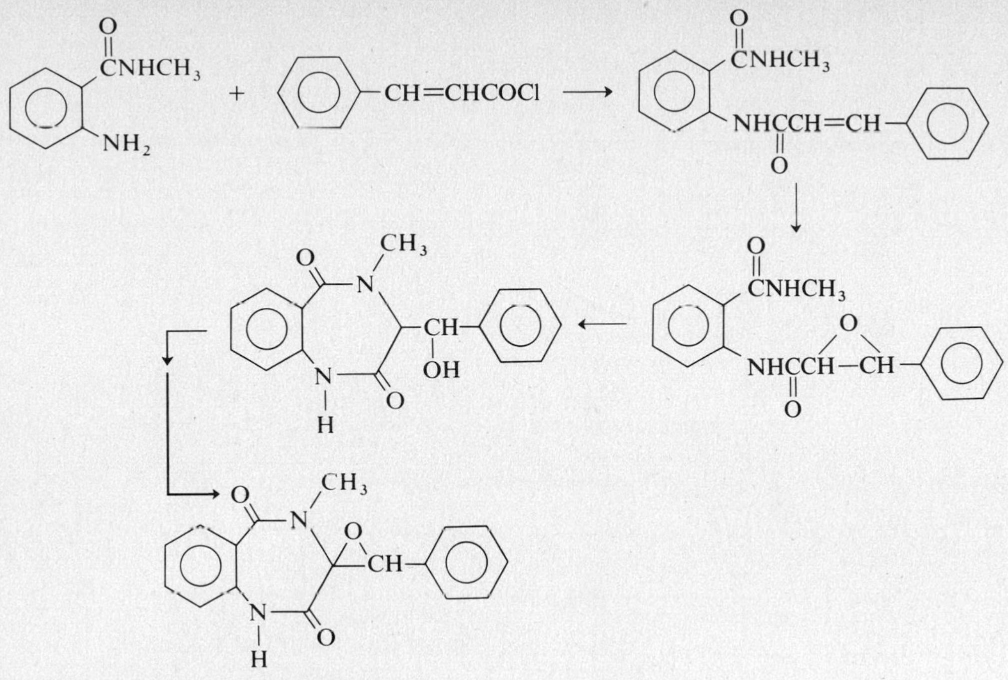

Cyclopenin

Epilogue

There's life all over this universe, but the only life in the solar system is on earth, and in the whole universe we are the only men.

GEORGE WALD
Speech to students of the
Massachusetts Institute of Technology,
March 4, 1969

As George Wald said, in the entire universe we're the only men—
Evolutionary marvels; fifteen billion years this denizen
Required in synthesis; fifteen billion years during which
Elements combined to form, on earth, primeval broth rich
In carbon compounds, out of which there then emerged
A living cell; bent on living and growing it surged
Ahead for eons, and in three billion years evolved intelligent man.
Or is he? Homo sapiens seems to be doing all it can
To stop the game, enlisting science to provide the means
For suicide—to blow three billion years to smitherenes.
But science can as well be put to good and useful ends
And this thought all others in the book just writ transcends.
We ask you, we exhort you with all the force and urgency that we can
To put your talents scientific to proper use to man.

Index

pK_a Values for the reaction $AH + B \rightleftharpoons BH^+ + A^-$ (acid dissociation constants)

Reaction	pK_a
$FSO_3H + B \rightleftharpoons BH^+ + FSO_3^-$	−12
$HI + B \rightleftharpoons BH^+ + I^-$	−10
$HClO_4 + B \rightleftharpoons BH^+ + ClO_4^-$	−10
$H_2SO_4 + B \rightleftharpoons BH^+ + HSO_4^-$	− 9
$(H)(R)C{=}\overset{+}{O}H + B \rightleftharpoons BH^+ + (H)(R)C{=}O$	− 8
$HCl + B \rightleftharpoons BH^+ + Cl^-$	− 7
$R_2C{=}\overset{+}{O}H + B \rightleftharpoons BH^+ + R_2C{=}O$	− 7
$R_2\overset{+}{O}{-}H + B \rightleftharpoons BH^+ + R_2O$	− 3.5
$H_2\overset{+}{O}{-}H + B \rightleftharpoons BH^+ + H_2O$	− 1.7
$HNO_3 + B \rightleftharpoons BH^+ + NO_3^-$	− 1.3
$C_6H_5SO_3H + B \rightleftharpoons BH^+ + C_6H_5SO_3^-$	− 0.6
$CF_3CO_2H + B \rightleftharpoons BH^+ + CF_3CO_2^-$	0
$CCl_3CO_2H + B \rightleftharpoons BH^+ + CCl_3CO_2^-$	0.9
$Cl_2CHCO_2H + B \rightleftharpoons BH^+ + Cl_2CHCO_2^-$	1.3
$H_3PO_4 + B \rightleftharpoons BH^+ + H_2PO_4^-$	2.1
$ClCH_2CO_2H + B \rightleftharpoons BH^+ + ClCH_2CO_2^-$	2.8
$HF + B \rightleftharpoons BH^+ + F^-$	3.2
$HCO_2H + B \rightleftharpoons BH^+ + HCO_2^-$	3.7
$C_6H_5CO_2H + B \rightleftharpoons BH^+ + C_6H_5CO_2^-$	4.2
$C_6H_5\overset{+}{N}H_3 + B \rightleftharpoons BH^+ + C_6H_5NH_2$	4.6
$CH_3CO_2H + B \rightleftharpoons BH^+ + CH_3CO_2^-$	4.8
$C_5H_5\overset{+}{N}H + B \rightleftharpoons BH^+ + C_5H_5N$ (pyridinium / pyridine)	5.2
$H_2CO_3 + B \rightleftharpoons BH^+ + HCO_3^-$	6.5
$H_2S + B \rightleftharpoons BH^+ + HS^-$	7
$O_2N{-}C_6H_4{-}OH + B \rightleftharpoons BH^+ + O_2N{-}C_6H_4{-}O^-$	7.2
$C_6H_4(CO)_2NH + B \rightleftharpoons BH^+ + C_6H_4(CO)_2\bar{N}$ (phthalimide)	7.4
$C_6H_5SH + B \rightleftharpoons BH^+ + C_6H_5S^-$	7.8
$C_6H_5CO_3H + B \rightleftharpoons BH^+ + C_6H_5CO_3^-$	8.2
$CH_3COCH_2COCH_3 + B \rightleftharpoons BH^+ + CH_3CO\bar{C}HCOCH_3$	9.0
$HCN + B \rightleftharpoons BH^+ + CN^-$	9.1
$C_6H_5OH + B \rightleftharpoons BH^+ + C_6H_5O^-$	10.0
$HCO_3^- + B \rightleftharpoons BH^+ + CO_3^{=}$	10.2
$CH_3NH_3^+ + B \rightleftharpoons BH^+ + CH_3NH_2$	10.4
$CH_3CH_2SH + B \rightleftharpoons BH^+ + CH_3CH_2S^-$	10.5
$CH_3NO_2 + B \rightleftharpoons BH^+ + \bar{C}H_2NO_2$	11.0
$CH_3COCH_2CO_2C_2H_5 + B \rightleftharpoons BH^+ + CH_3CO\bar{C}HCO_2C_2H_5$	11.0
$CCl_3CH_2OH + B \rightleftharpoons BH^+ + CCl_3CH_2O^-$	12.2
$CH_2(CO_2C_2H_5)_2 + B \rightleftharpoons BH^+ + {}^-CH(CO_2C_2H_5)_2$	13.0
$C_5H_6 + B \rightleftharpoons BH^+ + C_5H_5^-$ (cyclopentadiene / cyclopentadienide)	14.0
$CH_3CONH_2 + B \rightleftharpoons BH^+ + CH_3CONH^-$	15.0
$CH_3OH + B \rightleftharpoons BH^+ + CH_3O^-$	15.5
$H_2O + B \rightleftharpoons BH^+ + HO^-$	15.7
$CH_3CH_2OH + B \rightleftharpoons BH^+ + CH_3CH_2O^-$	17
$(CH_3)_3COH + B \rightleftharpoons BH^+ + (CH_3)_3CO^-$	19
$CH_3COCH_3 + B \rightleftharpoons BH^+ + CH_3COCH_2^-$	20
$C_6H_5C{\equiv}CH + B \rightleftharpoons BH^+ + C_6H_5C{\equiv}C^-$	21
$(C_6H_5)_2NH + B \rightleftharpoons BH^+ + (C_6H_5)_2N^-$	23
$HC{\equiv}CH + B \rightleftharpoons BH^+ + HC{\equiv}C^-$	26
$C_6H_5NH_2 + B \rightleftharpoons BH^+ + C_6H_5NH^-$	27
$(C_6H_5)_3CH + B \rightleftharpoons BH^+ + (C_6H_5)_3C^-$	31.5
$(C_6H_5)_2CH_2 + B \rightleftharpoons BH^+ + (C_6H_5)_2CH^-$	33
$CH_3CH_2NH_2 + B \rightleftharpoons BH^+ + CH_3CH_2NH^-$	33
$C_6H_5CH_3 + B \rightleftharpoons BH^+ + C_6H_5CH_2^-$	35
$NH_3 + B \rightleftharpoons BH^+ + NH_2^-$	36
$H_2C{=}CH_2 + B \rightleftharpoons BH^+ + H_2C{=}CH^-$	36
$CH_4 + B \rightleftharpoons BH^+ + CH_3^-$	40
$CH_3CH_3 + B \rightleftharpoons BH^+ + CH_3CH_2^-$	42
$C_6H_6 + B \rightleftharpoons BH^+ + C_6H_5^-$	42
$C_6H_{12} + B \rightleftharpoons BH^+ + C_6H_{11}^-$ (cyclohexane / cyclohexyl anion)	45